INDEX OF TABLES

- For users of the three-volume edition, pages 1-668 (Chapters 1-21) are in Volume 1; pages 669-1230 (Chapters 22-39) are in Volume 2; and pages 1231-1460 (Chapters 40-46) are in Volume 3.
- Pages 1231-1460 (Chapters 40-46) are not in the Standard Version of this textbook.

SEARS AND ZEMANSKY'S

UNIVERSITY PHYSICS

WITH MODERN PHYSICS

TENTH EDITION

Hugh D. Young

Carnegie-Mellon University

Roger A. Freedman

University of California, Santa Barbara

CONTRIBUTING AUTHORS

T. R. Sandin

North Carolina A&T State University

A. Lewis Ford

Texas A&M University

ADDISON-WESLEY

An imprint of Addison Wesley Longman

San Francisco • Reading, Massachusetts • New York • Harlow, England
Don Mills, Ontario • Sydney • Mexico City • Madrid • Amsterdam

This book is in the Addison-Wesley Series in Physics

Acquisitions Editor: Sami Iwata

Production Editors: Lisa Weber, Jean Lake

Marketing Manager: Jennifer Schmidt

Director of Marketing: Stacy Treco

Art Coordinators: Lisa Weber, Jean Lake

Composition and Film Buyer: Lillian Hom

Manufacturing Supervisors: Holly Bisso, Virginia Pierce

Publishing Assistants: Anthony J. Asaro, Nancy Gee

Copyeditor: Luana Richards

Proofreader: Martha Ghent

Permissions Editor: Reynold Rieger

Text Designer: Bruce Kortebein

Chapter Opener Designer: Mark Ong, Side by Side Studios

Cover Designer: Yvo Riezebos

Illustrators: James A. Bryant, George V. Kelvin, Gary Torisi, Darwen and Vally Hennings, Karl Miyajima

Photo Editor: Beaura Katherine Ringrose

Composition and Film: Typo-Graphics, Inc.

Cover Printer: Lehigh Press

Printer and Binder: Von Hoffman Press

Library of Congress Cataloging-in-Publication Data

Young, Hugh D.
  University physics / Hugh D. Young, Roger A. Freedman; contributing authors, T. R. Sandin, A. Lewis Ford. — 10th ed., with modern physics.
     p. cm. — (Addison-Wesley series in physics)
  Includes index.
  ISBN 0-201-60336-5
    1. Physics. I. Freedman, Roger A. II. Title. III. Series.

QC21.2.Y67   2000
530—dc21   99-054401

ISBN 0-201-60336-5
 2 3 4 5 6 7 8 9 10-VH-03 02 01 00

PREFACE

This book is the product of a half century of innovation in physics education. When the first edition of *University Physics* by Francis W. Sears and Mark W. Zemansky appeared in 1949, it was revolutionary in its emphasis on the fundamental principles of physics and how to apply them. This Commemorative Tenth Edition continues to emphasize principles and applications as it provides today's students with a broad, rigorous, yet accessible introduction to calculus-based physics. The success of *University Physics* with generations of students and educators in all parts of the world is a testament to the merits of this approach.

Two key objectives guided the writing of this text: helping students develop physical intuition, and helping them build strong problem-solving skills. Also reflected throughout are the results of two decades of research in physics education on the conceptual pitfalls that commonly plague beginning physics students. These pitfalls include the notions that force is required for motion, that electric current is "used up" as it goes around a circuit, and that the product of a body's mass and its acceleration is itself a force. A key focus of this edition is to discuss not only the correct way to analyze a situation or solve a problem, but also the reason why the wrong way (which may have occurred to the student first) is indeed wrong.

The prose style of the book continues to be relaxed and conversational, without being colloquial or excessively familiar. We see the student as our partner in learning, not as an audience to be lectured to from atop a platform. This style makes it much easier for us to convey to the student our own excitement and enthusiasm for the beauty, intellectual challenge, and fundamental unity of physics.

In preparing the Tenth Edition, we have relied heavily on the comments of a great many faculty and students on how best to help them meet the challenges of physics education. Based on these comments, we have designed the following features of this edition.

A GUIDE FOR THE STUDENT

Many physics students experience difficulty simply because they don't know how to make the best use of their textbook. A section entitled "How to Succeed in Physics by Really Trying," which follows this preface, serves as a "user's manual" to all the features of this book. This section, written by Professor Mark Hollabaugh (Normandale Community College), also gives a number of helpful study hints. We strongly encourage *every* student to read this section!

CHAPTER ORGANIZATION

The *Introduction* to each chapter gives specific examples of the chapter's content and connects it with what has come before. At the end of each chapter is a *Summary* of the most important principles introduced in the chapter, along with the associated *Key Equations*. The summary also includes a list of *Key Terms* that the student should have learned to use, with references to the page on which each term is first introduced.

CONTENTS

Some of the most significant content features of this edition include:

- In Chapter 2, motion diagrams help students to distinguish between position, velocity, and acceleration in one-dimensional motion (see pp. 40–41).
- Chapter 12 has been updated with new data on the supermassive black hole at the center of our Milky Way galaxy.
- We discuss the microscopic interpretation of entropy in Chapter 18.
- A qualitative introduction to the ideas behind Gauss's law is given in Chapter 23.
- Chapter 28 on magnetic fields and forces explains the attraction and repulsion of magnets and magnetic materials.
- The discussion of electromagnetic induction in Chapter 30 and of inductance in Chapter 31 has been rewritten to make these essential but challenging concepts more accessible to students.
- Every chapter now includes a selection of photographs that illustrate how physical principles manifest themselves in the natural world and in our technological society.

QUESTIONS AND PROBLEMS

At the end of each chapter is a collection of *Discussion Questions,* intended to probe and extend the student's conceptual understanding, followed by an extensive set of problems. The problems have been revised and their number increased, including many new problems drawn from astrophysics, biology, and aerodynamics. Many problems have a conceptual part in which students must discuss and explain their results. The problems are grouped into *Exercises,* which are single-concept problems keyed to specific sections of the text; *Problems,* usually requiring two or more nontrivial steps; and *Challenge Problems,* intended to challenge the strongest students. Many new questions, exercises, and problems, especially for Chapters 38, 41, 42, 43, 44, and 45, were suggested by Professor A. Lewis Ford (Texas A&M University) and Professor Tom Sandin (North Carolina A&T State University).

PROBLEM-SOLVING STRATEGIES

Problem-Solving Strategy sections, an extremely popular feature of the book, have been retained and strengthened. They have proved to be a very substantial help, especially to the many earnest but bewildered students who "understood the material but couldn't do the problems." (See, for example, pp. 110, 121, and 171.)

EXAMPLES

Each *Problem-Solving Strategy* section is followed immediately by one or more worked-out examples that illustrate the strategy. Several of these are purely qualitative, such as Examples 6–6 (Comparing kinetic energies, p. 173), 8–1 (Momentum vs. kinetic energy, p. 230), and 18–7 (Isentropic processes, p. 576). Many examples are drawn from real-life situations relevant to the student's own experience. Units and correct significant figures in examples are always carried through all stages of numerical calculations.

Example solutions always begin with a statement of the general principles to be used and, when necessary, a discussion of the reason for choosing them. We emphasize modeling in physics, showing the student how to begin with a seemingly complex situation, make simplifying assumptions, apply the appropriate physical principles, and evaluate the final result. Does it make sense? Is it what you expected? How can you check it?

"CAUTION" PARAGRAPHS

In the text of each chapter we have labeled certain paragraphs with the word **CAUTION.** These paragraphs alert the student to common misconceptions or to points of potential confusion. (See, for example, pp. 102, 140, and 167.) We think of them as being similar to the flagged paragraphs in the user's manual for a power drill or a VCR, describing potential sources of trouble when using the equipment.

ACTIVPHYSICS LINKS

An important and unique supplement to *University Physics* is the set of *ActivPhysics 1* and *ActivPhysics 2* CD-ROMs and workbooks, developed by Professors Alan Van Heuvelen and Paul D'Alessandris and published by Addison Wesley Longman. By combining carefully designed interactive simulations with proven pedagogy, *ActivPhysics* helps students become adept at solving problems about dynamic physical phenomena. Icons throughout the text of *University Physics* indicate which of the 200-plus exercises in *ActivPhysics* correspond to specific topics in this book. Exercises 1.1 through 10.10 appear in *ActivPhysics 1,* while Exercises 11.1 through 20.4 are in *ActivPhysics 2.* For more information about the *ActivPhysics* CD-ROMs (compatible with both Macintosh and Windows) and workbooks, see below under "Supplements."

CASE STUDIES

We have included 10 optional sections called *Case Studies,* each building on the material of its chapter. Some (Neutrinos, Black Holes, Photons) emphasize connections between classical and modern physics. Others (Automotive Power, Energy Resources, Power Distribution Systems) have an engineering flavor; still others (Baseball Trajectories, Electric Potential Maps) emphasize computer simulations and include computer exercises for the student. All case studies have corresponding end-of-chapter problems.

NOTATION AND UNITS

Students often have a hard time keeping track of which quantities are vectors and which are not. In this edition, we use boldface italic symbols with an arrow on top for vector quantities, such as $\vec{v}$, $\vec{a}$, and $\vec{F}$; unit vectors have a caret on top, such as $\hat{\imath}$. Boldface $+$, $-$, $\times$, and $=$ signs are used in vector equations to emphasize the distinction between these operations and operations with ordinary numbers.

In this edition SI units are used exclusively. English unit conversions are included where appropriate. The joule is used as the standard unit of energy of all forms, including heat.

FLEXIBILITY

The book is adaptable to a wide variety of course outlines. There is plenty of material for an intensive three-semester or five-quarter course. Most instructors will find that there is too much material for a one-year course, but it is easy to tailor the book to a variety of one-year course plans by omitting certain chapters or sections. For example, any or all of the chapters on relativity, fluid mechanics, acoustics, electromagnetic waves, optical instruments, and several other topics can be omitted without loss of continuity. Some sections that are unusually challenging or somewhat out of the mainstream have been identified with an asterisk preceding the section title; these, too, may be omitted. In any case, no one should feel constrained to work straight through the entire book. We encourage instructors to select the chapters that fit their needs, omitting material that is not appropriate for the objectives of a particular course.

STANDARD, EXTENDED, AND SPLIT VERSIONS

This edition is available in three versions. The Standard version (ISBN 0–201–60322–5) includes 39 chapters, ending with the special theory of relativity. The Extended version (ISBN 0–201–60336–5) adds seven chapters on modern physics, including the physics of atoms, molecules, condensed matter, nuclei, and elementary particles. The Split version includes all 46 chapters in three softbound volumes: Volume 1, Chapters 1–21 (ISBN 0–201–60329–2); Volume 2, Chapters 22–39 (ISBN 0–201–60335–7); and Volume 3, Chapters 40–46 (ISBN 0–201–65663–9).

SUPPLEMENTS

For the Student: The *Online Course Companion Web site* (http://www.awlonline.com/young) offers problem solving tips, interactive quizzes, key concepts for each chapter of *University Physics,* a glossary, tips for success in physics, web links to applications of physical concepts, and much more.

The *ActivPhysics* CD-ROMs and workbooks, developed by Professors Alan Van Heuvelen and Paul D'Alessandris, use interactive simulations and multiple representations to help students become better physics problem-solvers. The CD-ROMs are compatible with both Macintosh and Windows. *ActivPhysics 1* (ISBN 0–201–69482–4) covers the topics of Chapters 1–21, and *ActivPhysics 2* (ISBN 0–201–36111–6) covers the material found in Chapters 22–46. As mentioned above, icons in the text of *University Physics* show the connections between topics in the book and exercises in *ActivPhysics.*

The *Study Guide,* prepared by Professors James R. Gaines and William F. Palmer, reinforces the text's emphasis on problem-solving strategies and student misconceptions. The *Study Guide for Volume 1* (ISBN 0–201–61835–4) covers Chapters 1–21, and the *Study Guide for Volumes 2 and 3* (ISBN 0–201–61834–6) covers Chapters 22–46.

The *Student Solutions Manual,* prepared by Professor A. Lewis Ford, includes completely worked-out solutions for about two-thirds of the odd-numbered problems in *University Physics.* (Answers to all odd-numbered problems are found in this book following the Appendices.) The *Student Solutions Manual for Volume 1* (ISBN 0–201–64394–4) covers Chapters 1–21, and the *Student Solutions Manual for Volumes 2 and 3* (ISBN 0–201–64395–2) covers Chapters 22–46.

For the Instructor: The *Instructor's Solutions Manual,* prepared by Professor Mark Hollabaugh and Dr. Thomas D. Gutierrez, contains worked-out solutions to all exercises, problems, and challenge problems. The *Instructor's Solutions Manual for Volume 1* covers Chapters 1–21 (ISBN 0–201–61836–2) and the *Instructor's Solution Manual for Volumes 2 and 3* covers Chapters 22–46 (ISBN 0–201–61837–0). It is also available as a cross-platform CD-ROM (ISBN 0–201–65679–5). With the CD-ROM, you can read, edit, and print any solutions you choose, as well as post them on your secure, password-protected class web site.

The *Instructor's Guide for an Active Learning Classroom* (ISBN 0–201–65676–0) offers quick strategies for tailoring your course to include active learning techniques. This supplement is ideal for instructors who want to integrate these techniques into their course, but do not have time to create a new teaching plan.

The *Online Course Companion Web site* (http://www.awlonline.com/young) makes it easy to put your course syllabus and assignments on the web and password-protect your course information. Through the site, students can submit assignments to you or your teaching assistants.

The *Instructor's Presentation CD-ROM* contains the full-color line art figures from the text. Images may be exported into other programs, such as PowerPoint.

The *Overhead Transparencies* (ISBN 0–201–61833–8) include 200 four-color figures from the text. These are on acetate for use on an overhead projector.

The *Test Item File* (ISBN 0–201–60344–6), written by Dr. Elliot Farber and Professor Michael Browne, includes multiple-choice and short-answer problems. The accompanying TestGen software (ISBN 0–201–65662–0), compatible with both Macintosh and Windows, makes it easy to edit these test items, assemble them into an exam, and generate an answer key.

ACKNOWLEDGMENTS

In this Commemorative Tenth Edition, we would like to thank the hundreds of reviewers and colleagues who have contributed valuable comments and suggestions over the life of this textbook.

Edward Adelson (Ohio State University), Ralph Alexander (University of Missouri at Rolla), J. G. Anderson, R. S. Anderson, Alex Azima (Lansing Community College), Dilip Balamore (Nassau Community College), Harold Bale (University of North Dakota), Arun Bansil (Northeastern University), John Barach (Vanderbilt University), J. D. Barnett, H. H. Barschall, Albert Bartlett (University of Colorado), Paul Baum (CUNY, Queens College), B. Bederson, Lev I. Berger (San Diego State University), Robert Boeke (William Rainey Harper College), S. Borowitz, A. C. Braden, James Brooks (Boston University), Nicholas E. Brown (California Polytechnic State University, San Luis Obispo), Tony Buffa (California Polytechnic State University, San Luis Obispo), A. Capecelatro, Michael Cardamone (Pennsylvania State University), Duane Carmony (Purdue University), P. Catranides, Roger Clapp (University of South Florida), William M. Cloud (Eastern Illinois University), Leonard Cohen (Drexel University), W. R. Coker (University of Texas, Austin), Malcolm D. Cole (University of Missouri at Rolla), H. Conrad, David Cook (Lawrence University), Gayl Cook (University of Colorado), Hans Courant (University of Minnesota), Bruce A. Craver (University of Dayton), Larry Curtis (University of Toledo), Jai Dahiya (Southeast Missouri State University), Steve Detweiler (University of Florida), George Dixon (Oklahoma State University), Donald S. Duncan, Boyd Edwards (West Virginia University), Robert Eisenstein (Carnegie-Mellon University), William Faissler (Northeastern University), William Fasnacht (U.S. Naval Academy), Paul Feldker (St. Louis Community College), L. H. Fisher, Neil Fletcher (Florida State University), Robert Folk, Peter Fong (Emory University), A. Lewis Ford (Texas A&M University), D. Frantszog, James R. Gaines (Ohio State University), Solomon Gartenhaus (Purdue University), Ron Gautreau (New Jersey Institute of Technology), J. David Gavenda (University of Texas, Austin), Dennis Gay (University of North Florida), James Gerhart (University of Washington), N. S. Gingrich, J. L. Glathart, S. Goodwin, Walter S. Gray (University of Michigan), Howard Grotch (Pennsylvania State University), John Gruber (San Jose State University), Graham D. Gutsche (U.S. Naval Academy), Michael J. Harrison (Michigan State University), Harold Hart (Western Illinois University), Howard Hayden (University of Connecticut), Carl Helrich (Goshen College), Laurent Hodges (Iowa State University), C. D. Hodgman, Michael Hones (Villanova University), Keith Honey (West Virginia Institute of Technology), Gregory Hood (Tidewater Community College), John Hubisz (North Carolina State University), M. Iona, Alvin Jenkins (North Carolina State University), Lorella Jones (University of Illinois), John Karchek (GMI Engineering & Management Institute), Thomas Keil (Worcester Polytechnic Institute), Robert Kraemer (Carnegie-Mellon University), Jean P. Krisch (University of Michigan), Robert A. Kromhout, Robert J. Lee, Alfred Leitner (Rensselaer Polytechnic University), Gerald P. Lietz (De Paul University), Gordon Lind (Utah State University), S. Livingston, Elihu Lubkin (University of Wisconsin, Milwaukee), Robert Luke (Boise State University), Michael Lysak (San Bernardino Valley College), Jeffrey Mallow (Loyola University),

Robert Mania (Kentucky State University), Robert Marchina (University of Memphis), David Markowitz (University of Connecticut), R. J. Maurer, Oren Maxwell (Florida International University), Joseph L. McCauley (University of Houston), T. K. McCubbin, Jr. (Pennsylvania State University), Charles McFarland (University of Missouri at Rolla), Lawrence McIntyre (University of Arizona), Fredric Messing (Carnegie-Mellon University), Thomas Meyer (Texas A&M University), Andre Mirabelli (St. Peter's College, New Jersey), Herbert Muether (S.U.N.Y., Stony Brook), Jack Munsee (California State University, Long Beach), Lorenzo Narducci (Drexel University), Van E. Neie (Purdue University), David A. Nordling (U.S. Naval Academy), L. O. Olsen, Jim Pannell (DeVry Institute of Technology), W. F. Parks (University of Missouri), Jerry Peacher (University of Missouri at Rolla), Arnold Perlmutter (University of Miami), Lennart Peterson (University of Florida), R. J. Peterson (University of Colorado, Boulder), R. Pinkston, Ronald Poling (University of Minnesota), J. G. Potter, C. W. Price (Millersville University), Francis Prosser (University of Kansas), Shelden H. Radin, Michael Rapport (Anne Arundel Community College), R. Resnick, James A. Richards, Jr., John S. Risley (North Carolina State University), Francesc Roig (University of California, Santa Barbara), T. L. Rokoske, Richard Roth (Eastern Michigan University), Carl Rotter (University of West Virginia), S. Clark Rowland (Andrews University), Rajarshi Roy (Georgia Institute of Technology), Russell A. Roy (Santa Fe Community College), Melvin Schwartz (St. John's University), F. A. Scott, L. W. Seagondollar, Stan Shepherd (Pennsylvania State University), Bruce Sherwood (Carnegie-Mellon University), Hugh Siefkin (Greenville College), C. P. Slichter, Charles W. Smith, Malcolm Smith (University of Lowell), Ross Spencer (Brigham Young University), Julien Sprott (University of Wisconsin), Victor Stanionis (Iona College), James Stith (American Institute of Physics), Edward Strother (Florida Institute of Technology), Conley Stutz (Bradley University), Albert Stwertka (U.S. Merchant Marine Academy), Martin Tiersten (CUNY, City College), David Toot (Alfred University), Somdev Tyagi (Drexel University), F. Verbrugge, Helmut Vogel (Carnegie-Mellon University), Thomas Weber (Iowa State University), M. Russell Wehr, Lester V. Whitney, Thomas Wiggins (Pennsylvania State University), George Williams (University of Utah), John Williams (Auburn University), Stanley Williams (Iowa State University), Jack Willis, Suzanne Willis (Northern Illinois University), Robert Wilson (San Bernardino Valley College), L. Wolfenstein, James Wood (Palm Beach Junior College), Lowell Wood (University of Houston), R. E. Worley, D. H. Ziebell (Manatee Community College), George O. Zimmerman (Boston University)

In addition, we both have individual acknowledgments we would like to make.

I want to extend my heartfelt thanks to my colleagues at Carnegie-Mellon, especially Professors Robert Kraemer, Bruce Sherwood, Helmut Vogel, and Brian Quinn, for many stimulating discussions about physics pedagogy and for their support and encouragement during the writing of this new edition. I am equally indebted to the many generations of Carnegie-Mellon students who have helped me learn what good teaching and good writing are, by showing me what works and what doesn't. It is always a joy and a privilege to express my gratitude to my wife Alice and our children Gretchen and Rebecca for their love, support, and emotional sustenance during the writing of this new edition. May all men and women be blessed with love such as theirs. — H. D. Y.

I would like to thank my past and present colleagues at UCSB, including Francesc Roig, Elisabeth Nicol, Al Nash, and Carl Gwinn, for their wholehearted support and for many helpful discussions. I owe a special debt of gratitude to my early teachers Willa Ramsay, Peter Zimmerman, William Little, Alan Schwettman, and Dirk Walecka for showing me what clear and engaging physics teaching is all about, and to Stuart Johnson for inviting me to join this project as a co-author. I am grateful to Nathan Palmer of the Colorado School of Mines for his careful checking of the page proofs for this edition. I want to thank my parents for their continued love and support and for keeping a space open on their bookshelf for this book. Most of all, I want to express my gratitude and love to my wife Caroline, to whom I dedicate my contributions to this book. Hey, Caroline, the new edition's done at last — let's go flying! — R. A. F.

PLEASE TELL US WHAT YOU THINK!

We welcome communications from students and professors, especially concerning errors or deficiencies that you find in this edition. We have devoted a lot of time and effort to writing the best book we know how to write, and we hope it will help you to teach and learn physics. In turn, you can help us by letting us know what still needs to be improved! Please feel free to contact us either by ordinary mail or electronically. Your comments will be greatly appreciated.

September 1999

Hugh D. Young
Department of Physics
Carnegie-Mellon University
Pittsburgh, Pennsylvania 15213
hdy+@andrew.cmu.edu

Roger A. Freedman
Department of Physics
University of California, Santa Barbara
Santa Barbara, California 93106-9530
airboy@physics.ucsb.edu
http://www.physics.ucsb.edu/~airboy/

HOW TO SUCCEED IN PHYSICS BY REALLY TRYING

Mark Hollabaugh, Normandale Community College

Physics encompasses the large and the small, the old and the new. From the atom to galaxies, from electrical circuitry to aerodynamics, physics is very much a part of the world around us. You probably are taking this introductory course in calculus-based physics because it is required for subsequent courses you plan to take in preparation for a career in science or engineering. Your professor wants you to learn physics and to enjoy the experience. He or she is very interested in helping you learn this fascinating subject. That is part of the reason your professor chose this textbook for your course. That is also the reason why Drs. Young and Freedman asked me to write this introductory section. We want you to succeed!

The purpose of this section of *University Physics* is to give you some ideas that will assist your learning. Specific suggestions on how to use the textbook will follow a brief discussion of general study habits and strategies.

PREPARATION FOR THIS COURSE

If you had high school physics, you will probably learn concepts faster than those who have not because you will be familiar with the language of physics. If English is a second language for you, keep a glossary of new terms that you encounter and make sure you understand how they are used in physics. Likewise, if you are farther along in your mathematics courses, you will pick up the mathematical aspects of physics faster. Even if your mathematics is adequate, you may find a book such as Arnold D. Pickar's *Preparing for General Physics: Math Skill Drills and Other Useful Help (Calculus Version)* to be useful. Your professor may actually assign sections of this math review to assist your learning.

LEARNING TO LEARN

Each of us has a different learning style and a preferred means of learning. Understanding your own learning style will help you to focus on aspects of physics that may give you difficulty and to use those components of your course that will help you overcome the difficulty. Obviously you will want to spend more time on those aspects that give you the most trouble. If you learn by hearing, lectures will be very important. If you learn by explaining, then working with other students will be useful to you. If solving problems is difficult for you, spend more time learning how to solve problems. Also, it is important to understand and develop good study habits. Perhaps the most important thing you can do for yourself is to set aside adequate, regularly scheduled, study time in a distraction-free environment.

Answer the following questions for yourself:

• Am I able to use fundamental mathematical concepts from algebra, geometry and trigonometry? (If not, plan a program of review with help from your professor.)
• In similar courses, what activity has given me the most trouble? (Spend more time on this.) What has been the easiest for me? (Do this first; it will help to build your confidence.)
• Do I understand the material better if I read the book before or after the lecture? (You may learn best by skimming the material, going to lecture, and then undertaking an in-depth reading.)
• Do I spend adequate time in studying physics? (A rule of thumb for a class like this is to devote, on the average, 2.5 hours out of class for each hour in class. For a course

meeting 5 hours each week, that means you should spend about 10 to 15 hours per week studying physics.)
- Do I study physics every day? (Spread that 10 to 15 hours out over an entire week!) At what time of the day am I at my best for studying physics? (Pick a specific time of the day and stick to it.)
- Do I work in a quiet place where I can maintain my focus? (Distractions will break your routine and cause you to miss important points.)

WORKING WITH OTHERS

Scientists or engineers seldom work in isolation from one another but rather work cooperatively. You will learn more physics and have more fun doing it if you work with other students. Some professors may formalize the use of cooperative learning or facilitate the formation of study groups. You may wish to form your own informal study group with members of your class who live in your neighborhood or dorm. If you have access to e-mail, use it to keep in touch with one another. Your study group is an excellent resource when reviewing for exams.

LECTURES AND TAKING NOTES

An important component of any college course is the lecture. In physics this is especially important because your professor will frequently do demonstrations of physical principles, run computer simulations, or show video clips. All of these are learning activities that will help you to understand the basic principles of physics. Don't miss lectures, and if for some reason you do, ask a friend or member of your study group to provide you with notes and let you know what happened.

Take your class notes in outline form, and fill in the details later. It can be very difficult to take word for word notes, so just write down key ideas. Your professor may use a diagram from the textbook. Leave a space in your notes and just add the diagram later. After class, edit your notes, filling in any gaps or omissions and noting things you need to study further. Make references to the textbook by page, equation number, or section number.

Make sure you ask questions in class, or see your professor during office hours. Remember the only "dumb" question is the one that is not asked. Your college may also have teaching assistants or peer tutors who are available to help you with difficulties you may have.

EXAMINATIONS

Taking an examination is stressful. But if you feel adequately prepared and are well-rested, your stress will be lessened. Preparing for an exam is a continual process; it begins the moment the last exam is over. You should immediately go over the exam and understand any mistakes you made. If you worked a problem and made substantial errors, try this: Take a piece of paper and divide it down the middle with a line from top to bottom. In one column, write the proper solution to the problem. In the other column, write what you did and why, if you know, and why your solution was incorrect. If you are uncertain why you made your mistake, and how to avoid it again, talk with your professor. Physics continually builds on fundamental ideas and it is important to correct any misunderstandings immediately. Warning: While cramming at the last minute may get you through the *present* exam, you will not adequately retain the concepts for use on the *next* exam.

USING YOUR TEXTBOOK

Now let's take a look at specific features of *University Physics* that will help you understand the concepts of physics. At its heart, physics is not equations and numbers. Physics is a way of looking at the universe and understanding how the universe works and how its various parts relate to each other. And although solving quantitative problems is an important part of physics, it is equally important for you to understand concepts qualitatively. Your textbook will help you in both areas.

First of all, don't be afraid to write in your book. It is more important for you to learn the concepts of physics than to keep your book in pristine condition. Write in the margins, make cross references. Take notes in your notebook as you read. *University Physics* is your primary "reference book" for this course. Refer to it often to help you understand the concepts you hear in lecture. Become familiar with the contents of the appendices and end papers.

CAUTION ▶ Please note that the quantity *m**a*** is *not* a force. All that Eqs. (4–7) and (4–8) say is that the vector *m**a*** is equal in magnitude and direction to the vector sum $\Sigma \vec{F}$ of all the forces acting on the body. It's incorrect to think of acceleration as a force; rather, acceleration is a result of a nonzero net force. It's "common sense" to think that there is a "force of acceleration" that pushes you back into your seat when your car accelerates forward from rest. But *there is no such force;* instead, your inertia causes you to tend to stay at rest relative to the earth, and the car accelerates around you. The "common sense" confusion arises from trying to apply Newton's second law in a frame of reference where it isn't valid, like the non-inertial reference frame of an accelerating car. We will always examine motion relative to *inertial* frames of reference only. ◀

In learning how to use Newton's second law, we will begin in this chapter with examples of straight-line motion. Then in Chapter 5 we will consider more general cases and develop more detailed problem-solving strategies for applying Newton's laws of motion.

CAUTION!

Educational research has found numerous misconceptions or misunderstandings that students frequently have when they study physics. Dr. Freedman has added *Caution!* paragraphs to warn you about these potential pitfalls. Heed them!

WORKED EXAMPLES

Your professor will work example problems in class to illustrate the application of the concepts of physics to real-world problems. You should work through all the examples in the textbook, filling in any missing steps, and making note of things you don't understand. Get help with the concepts that confuse you!

EXAMPLE 6–6

Comparing kinetic energies Two iceboats like the one in Example 5–5 (Section 5–3) hold a race on a frictionless horizontal lake (Fig. 6–9). The two iceboats have masses m and $2m$, respectively. Each iceboat has an identical sail, so the wind exerts the same constant force $\vec{F}$ on each iceboat. The two iceboats start from rest and cross the finish line a distance s away. Which iceboat crosses the finish line with greater kinetic energy?

SOLUTION If you simply use the mathematical definition of kinetic energy, $K = \frac{1}{2}mv^2$ (Eq. (6–5)), the answer to this problem isn't immediately obvious. The iceboat of mass $2m$ has greater mass, so you might guess that the larger iceboat attains a greater

kinetic energy at the finish line. But the smaller iceboat, of mass m, will cross the finish line with a greater speed, and you might guess that *this* iceboat will have the greater kinetic energy. How can we decide?

The correct way to approach this problem is to remember that *the kinetic energy of a particle is just equal to the total work done to accelerate it from rest.* Both iceboats travel the same distance s, and only the horizontal force F in the direction of motion does work on either iceboat. Hence the total work done between the starting line and the finish line is the *same* for each iceboat, $W_{tot} = Fs$. At the finish line, each iceboat has a kinetic energy equal to the work W_{tot} done on it, because each iceboat started from rest. So both iceboats have the *same* kinetic energy at the finish line!

6–9 A race between iceboats.

Problem–Solving Strategy

PROBLEMS USING MECHANICAL ENERGY

1. First decide whether the problem should be solved by energy methods, by using $\Sigma \vec{F} = m\vec{a}$ directly, or by a combination. The energy approach is particularly useful when the problem involves motion with varying forces, motion along a curved path (discussed later in this section), or both. But if the problem involves elapsed time, the energy approach is usually *not* the best choice because this approach doesn't involve time directly.

2. When using the energy approach, first decide what the initial and final states (the positions and velocities) of the system are. Use the subscript 1 for the initial state and the subscript 2 for the final state. It helps to draw sketches showing the initial and final states.

3. Define your coordinate system, particularly the level at which $y = 0$. You will use this to compute gravitational potential energies. Equation (7–2) assumes that the positive direction for y is upward; we suggest that you use this choice consistently.

4. List the initial and final kinetic and potential energies, that is, K_1, K_2, U_1, and U_2. In general, some of these will be known and some will be unknown. Use algebraic symbols for any unknown coordinates or velocities.

5. Identify all nongravitational forces that do work. A free-body diagram is always helpful. Calculate the work W_{other} done by all these forces. If some of the quantities you need are unknown, represent them by algebraic symbols.

6. Relate the kinetic and potential energies and the nongravitational work W_{other} using Eq. (7–7). If there is no nongravitational work, this becomes Eq. (7–4). It's helpful to draw bar graphs showing the initial and final values of K, U, and $E = K + U$. Then solve to find whatever unknown quantity is required.

7. Keep in mind, here and in later sections, that the work done by each force must be represented either in $U_1 - U_2 = -\Delta U$ or as W_{other}, but *never* in both places. The gravitational work is included in ΔU, so do not include it again in W_{other}.

PROBLEM-SOLVING STRATEGIES

One of the features of *University Physics* that first caught my eye as a teacher were the *Problem-Solving Strategy* boxes. This is the advice I would give to a student who came to me for help with a physics problem. Physics teachers approach a problem in a very systematic and logical manner. These boxes will help you as a beginning problem solver to do the same. Study these suggestions in great detail and implement them. In many cases these strategy boxes will tell you *how* to visualize an abstract concept.

CASE STUDIES

Physics relates to the real world, and these *Case Studies* will give you examples of applying physics to real science or engineering problems.

6–6 AUTOMOTIVE POWER

A Case Study in Energy Relations

The power requirements of a gasoline-powered automobile are an important and practical example of the concepts in this chapter. If roads were frictionless and air resistance didn't exist, there would be no need for an automobile to have an engine. All you'd need to go for a drive would be a few strong friends to give you a push to get started and a few other friends at your destination to stop you. (Steering on frictionless roads would be a problem, though.) In the real world, however, a moving car without an engine slows down because of forces that resist its motion. The engine's function is to continuously provide power to overcome this resistance. So to understand how much power is required from a car's engine, we must analyze the forces that act on the car.

Two forces oppose the motion of an automobile: rolling friction and air resistance. We described rolling friction in Section 5–4 in terms of a coefficient of rolling friction μ_r. A typical value of μ_r for properly inflated tires on hard pavement is 0.015. A Porsche 911 Carrera has a mass of 1251 kg and a weight of (1251 kg)(9.80 m/s²) = 12,260 N, and

SUMMARY, REVIEW QUESTIONS AND PROBLEMS

The most important concepts are listed in the *Key Terms*. Keep a glossary of terms in your notebook. Your professor may indicate through the use of course objectives which terms are important for you to know. The *Summary* will give you a quick review of the chapter's main ideas and the equations that represent those ideas mathematically. Everything else can be derived from these general equations. If your professor assigns *Problems* at the end of the chapter, make sure you work them carefully with other students. If solutions are available, do not look at the answer until you have struggled with the problem and compared your answer with someone else's. If the two of you agree on the answer, then look at the solution. If you have made a mistake, go back and rework the problem. Do not simply read the problem. You will note that the *Exercise*s are keyed to specific sections of the chapter and are easier. Work on these before you attempt the *Problems* or *Challenge Problems* which typically use multiple concepts.

Well, there you have it. We hope these suggestions will benefit your study of physics. Strive for understanding and excellence, and be persistent in your learning.

CONTENTS

WAVES/ACOUSTICS

ELECTROMAGNETISM

MODERN PHYSICS

ABOUT THE AUTHORS

Hugh D. Young is Professor of Physics at Carnegie-Mellon University in Pittsburgh, PA. He attended Carnegie-Mellon for both undergraduate and graduate study and earned his Ph.D. in fundamental particle theory under the direction of the late Richard Cutkosky. He joined the faculty of Carnegie-Mellon in 1956, and has also spent two years as a Visiting Professor at the University of California at Berkeley.

Prof. Young's career has centered entirely around undergraduate education. He has written several undergraduate-level textbooks, and in 1973 he became a co-author with Francis Sears and Mark Zemansky for their well-known introductory texts. With their deaths, he assumed full responsibility for new editions of these books, most recently the eighth edition of *University Physics.*

Prof. Young is an enthusiastic skier, climber, and hiker. He also served for several years as Associate Organist at St. Paul's Cathedral in Pittsburgh, and has played numerous organ recitals in the Pittsburgh area. Prof. Young and his wife Alice usually travel extensively in the summer, especially in Europe and in the desert canyon country of southern Utah.

Roger Freedman is a Lecturer in Physics at the University of California, Santa Barbara. He was an undergraduate at the University of California campuses in San Diego and Los Angeles, and did his doctoral research in nuclear theory at Stanford University under the direction of Professor J. Dirk Walecka. Dr. Freedman came to UCSB in 1981 after three years teaching and doing research at the University of Washington.

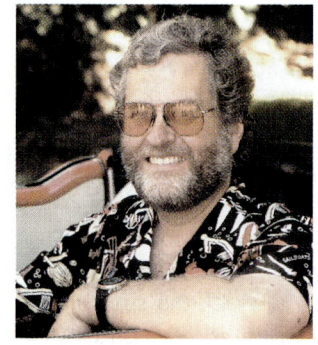

At UCSB, Dr. Freedman teaches in both the Department of Physics and the College of Creative Studies, a branch of the university intended for highly gifted and motivated undergraduates. He has published research in nuclear physics, elementary particle physics, and laser physics. In recent years he has helped to develop computer-based tools for learning introductory physics and astronomy.

Dr. Freedman holds a commercial pilot's license, and when not teaching or writing he can frequently be found flying with his wife Caroline.

T. R. Sandin is Professor of Physics at North Carolina A&T State University. He received his B.S. in physics from Santa Clara University and his M.S. and Ph.D. from Purdue University. He has received awards for excellence in teaching from both Purdue University and NC A&T. He has published research articles in low-temperature solid state physics, the Mössbauer effect, ferromagnetic anisotropy, and physics teaching, and is also the author of *Essentials of Modern Physics* (Addison-Wesley).

A. Lewis Ford is Professor of Physics at Texas A&M University. He received a B.A. from Rice University in 1968 and a Ph.D. in chemical physics from the University of Texas at Austin in 1972. After a one-year postdoc at Harvard University, he joined the Texas A&M physics faculty in 1973 and has been there ever since. Professor Ford's research area is theoretical atomic physics, with a specialization in atomic collisions. At Texas A&M he has taught a variety of undergraduate and graduate courses, but primarily introductory physics.

Units, Physical Quantities, and Vectors

The *Cassini* spacecraft, launched in October 1997, will go into orbit around the planet Saturn in 2004. Over the next four years it will closely observe Saturn, its rings, and its moons. For this ambitious mission to succeed, NASA scientists must know the spacecraft's position and motion in three dimensions at all times.

1–1 INTRODUCTION

Why study physics? For two reasons. First, physics is one of the most fundamental of the sciences. Scientists of all disciplines make use of the ideas of physics, from chemists who study the structure of molecules to paleontologists who try to reconstruct how dinosaurs walked. Physics is also the foundation of all engineering and technology. No engineer could design any kind of practical device without first understanding the basic principles involved. To design a spacecraft or a better mousetrap, you have to understand the basic laws of physics.

But there's another reason. The study of physics is an adventure. You will find it challenging, sometimes frustrating, occasionally painful, and often richly rewarding and satisfying. It will appeal to your sense of beauty as well as to your rational intelligence. Our present understanding of the physical world has been built on the foundations laid by scientific giants such as Galileo, Newton, Maxwell, and Einstein, and their influence has extended far beyond science to affect profoundly the ways in which we live and think. You can share some of the excitement of their discoveries when you learn to use physics to solve practical problems and to gain insight into everyday phenomena. If you've ever wondered why the sky is blue, how radio waves can travel through empty space, or how a satellite stays in orbit, you can find the answers by using fundamental physics. Above all, you will come to see physics as a tow-

ering achievement of the human intellect in its quest to understand our world and ourselves.

In this opening chapter, we'll go over some important preliminaries that we'll need throughout our study. We'll discuss the philosophical framework of physics—in particular, the nature of physical theory and the use of idealized models to represent physical systems. We'll introduce the systems of units used to describe physical quantities and discuss ways to describe the accuracy of a number. We'll look at examples of problems for which we can't (or don't want to) find a precise answer, but for which rough estimates can be useful and interesting. Finally, we'll study several aspects of vectors and vector algebra. Vectors will be needed throughout our study of physics to describe and analyze physical quantities, such as velocity and force, that have direction as well as magnitude.

1–2 THE NATURE OF PHYSICS

Physics is an *experimental* science. Physicists observe the phenomena of nature and try to find patterns and principles that relate these phenomena. These patterns are called physical theories or, when they are very well established and of broad use, physical laws or principles. The development of physical theory requires creativity at every stage. The physicist has to learn to ask appropriate questions, design experiments to try to answer the questions, and draw appropriate conclusions from the results. Figure 1–1 shows two famous experimental facilities.

According to legend, Galileo Galilei (1564–1642) dropped light and heavy objects from the top of the Leaning Tower of Pisa (Fig. 1–1a) to find out whether their rates of fall were the same or different. Galileo recognized that only experimental investigation could answer this question. From examining the results of his experiments (which were actually much more sophisticated than in the legend), he made the inductive leap to the principle, or theory, that the acceleration of a falling body is independent of its weight.

The development of physical theories such as Galileo's is always a two-way process that starts and ends with observations or experiments. This development often takes an indirect path, with blind alleys, wrong guesses, and the discarding of unsuccessful theories in favor of more promising ones. Physics is not simply a collection of facts and principles; it is also the *process* by which we arrive at general principles that describe how the physical universe behaves.

No theory is ever regarded as the final or ultimate truth. The possibility always exists that new observations will require that a theory be revised or discarded. It is in the nature of physical theory that we can disprove a theory by finding behavior that is inconsistent with it, but we can never prove that a theory is always correct.

Getting back to Galileo, suppose we drop a feather and a cannonball. They certainly do *not* fall at the same rate. This does not mean that Galileo was wrong; it means that his theory was incomplete. If we drop the feather and the cannonball *in a vacuum* to eliminate the effects of the air, then they do fall at the same rate. Galileo's theory has a **range of validity:** it applies only to objects for which the force exerted by the air (due to air resistance and buoyancy) is much less than the weight. Objects like feathers or parachutes are clearly outside this range.

Every physical theory has a range of validity outside of which is not applicable. Often a new development in physics extends a principle's range of validity. Galileo's analysis of falling bodies was greatly extended half a century later by Newton's laws of motion and law of gravitation.

An essential part of the interplay of theory and experiment is learning how to apply physical principles to a variety of practical problems. At various points in our study we

1–1 Two research laboratories. (a) The Leaning Tower in Pisa, Italy. According to legend, Galileo studied the motion of freely falling bodies by dropping them from the tower. He is also said to have studied pendulum motion by observing the swinging of the chandelier in the cathedral behind the tower. (b) The Hubble Space Telescope is the first major telescope to operate outside the earth's atmosphere. Its sensitive instruments measure visible, ultraviolet, and near-infrared light from astronomical objects. The Space Telescope has been used to study phenomena ranging from eruptions on the moons of Jupiter to the cores of distant galaxies. It is shown here in December 1993 after being repaired in orbit by the crew of the Space Shuttle *Endeavour.*

will discuss systematic problem-solving strategies that will help you set up and solve problems efficiently and accurately. Learning to solve problems is absolutely essential; you don't *know* physics unless you can *do* physics. This means not only learning the general principles, but also learning how to apply them in specific situations.

1–3 IDEALIZED MODELS

In everyday conversation we often use the word "model" to mean either a small-scale replica, such as a model railroad, or a person who displays articles of clothing (or the absence thereof). In physics a **model** is a simplified version of a physical system that would be too complicated to analyze in full detail.

Here's an example. Suppose we want to analyze the motion of a baseball thrown through the air. How complicated is this problem? The ball is neither perfectly spherical nor perfectly rigid; it has raised seams, and it spins as it moves through the air. Wind and air resistance influence its motion, the earth rotates beneath it, the ball's weight varies a little as its distance from the center of the earth changes, and so on. If we try to include all these things, the analysis gets hopelessly complicated. Instead, we invent a simplified version of the problem. We neglect the size and shape of the ball by representing it as a point object, or **particle.** We neglect air resistance by making the ball move in a vacuum, we forget about the earth's rotation, and we make the weight constant. Now we have a problem that is simple enough to deal with. We will analyze this model in detail in Chapter 3.

To make an idealized model of the system, we have to overlook quite a few minor effects to concentrate on the most important features of the system. Of course, we have to be careful not to neglect too much. If we ignore the effects of gravity completely, then our model predicts that when we throw the ball up, it goes in a straight line and disappears into space. We need to use some judgment and creativity to construct a model that simplifies a problem enough to make it manageable, yet keeps its essential features.

When we use a model to predict how a system will behave, the validity of our predictions is limited by the validity of the model. Going back to Galileo once more, we see that his prediction about falling bodies corresponds to an idealized model that does not include the effects of air resistance. This model works fairly well for a dropped cannonball, but not so well for a feather.

The concept of idealized models is extremely important in all physical science and technology. When we apply physical principles to complex systems, we always use idealized models, and we have to be aware of the assumptions we are making. In fact, the principles of physics themselves are stated in terms of idealized models; we speak about point masses, rigid bodies, ideal insulators, and so on. Idealized models play a crucial role throughout this book. Watch for them in discussions of physical theories and their applications to specific problems.

1–4 STANDARDS AND UNITS

Physics is an experimental science. Experiments require measurements, and we usually use numbers to describe the results of measurements. Any number that is used to describe a physical phenomenon quantitatively is called a **physical quantity.** For example, two physical quantities that describe you are your weight and your height. Some physical quantities are so fundamental that we can define them only by describing how to measure them. Such a definition is called an **operational definition.** Some examples are measuring a distance by using a ruler, and measuring a time interval by using a stopwatch. In other cases we define a physical quantity by describing how to calculate it from other quantities that we *can* measure. Thus we might define the average speed of a moving object as the distance traveled (measured with a ruler) divided by the time of travel (measured with a stopwatch).

When we measure a quantity, we always compare it with some reference standard. When we say that a Porsche 944 is 4.29 meters long, we mean that it is 4.29 times as long as a meter stick, which we define to be 1 meter long. Such a standard defines a **unit** of the quantity. The meter is a unit of distance, and the second is a unit of time. When we use a number to describe a physical quantity, we must always specify the unit that we are using; to describe a distance simply as "4.29" wouldn't mean anything.

To make accurate, reliable measurements, we need units of measurement that do not change and that can be duplicated by observers in various locations. The system of units used by scientists and engineers around the world is commonly called "the metric system," but since 1960 it has been known officially as the **International System,** or **SI** (the abbreviation for its French name, *Système International*). A list of all SI units is given in Appendix A, as are definitions of the most fundamental units.

The definitions of the basic units of the metric system have evolved over the years. When the metric system was established in 1791 by the French Academy of Sciences, the meter was defined as one ten-millionth of the distance from the North Pole to the equator (Fig. 1–2). The second was defined as the time required for a pendulum one meter long to swing from one side to the other. These definitions were cumbersome and hard to duplicate precisely, and by international agreement they have been replaced with more refined definitions.

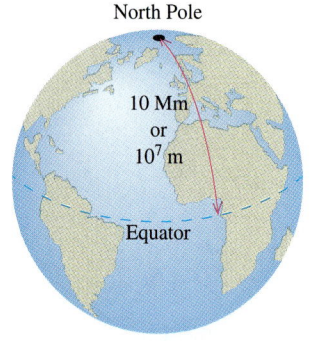

1–2 In 1791 the distance from the North Pole to the equator was defined to be exactly 10^7 m. With the modern definition of the meter this distance is about 0.02% more than 10^7 m.

TIME

From 1889 until 1967, the unit of time was defined as a certain fraction of the mean solar day, the average time between successive arrivals of the sun at its highest point in the sky. The present standard, adopted in 1967, is much more precise. It is based on an atomic clock, which uses the energy difference between the two lowest energy states of the cesium atom. When bombarded by microwaves of precisely the proper frequency, cesium atoms undergo a transition from one of these states to the other. One **second** is defined as the time required for 9,192,631,770 cycles of this radiation.

LENGTH

In 1960 an atomic standard for the meter was also established, using the wavelength of the orange-red light emitted by atoms of krypton (^{86}Kr) in a glow discharge tube. In November 1983 the length standard was changed again, in a more radical way. The speed of light in a vacuum was *defined* to be precisely 299,792,458 m/s. The meter is defined to be consistent with this number and with the above definition of the second. Hence the new definition of the **meter** is the distance that light travels in a vacuum in 1/299,792,458 second. This provides a much more precise standard of length than the one based on a wavelength of light.

MASS

The standard of mass, the **kilogram,** is defined to be the mass of a particular cylinder of platinum-iridium alloy. That cylinder is kept at the International Bureau of Weights and Measures at Sèvres, near Paris. An atomic standard of mass would be more fundamental, but at present we cannot measure masses on an atomic scale with as much accuracy as on a macroscopic scale. The *gram* (which is not a fundamental unit) is 0.001 kilogram.

UNIT PREFIXES

Once we have defined the fundamental units, it is easy to introduce larger and smaller units for the same physical quantities. In the metric system these other units are related to the fundamental units (or, in the case of mass, to the gram) by multiples of 10 or $\frac{1}{10}$. Thus one kilometer (1 km) is 1000 meters, and one centimeter (1 cm) is $\frac{1}{100}$ meter. We usually express multiples of 10 or $\frac{1}{10}$ in exponential notation: $1000 = 10^3$, $\frac{1}{1000} = 10^{-3}$, and so on. With this notation, $1 \text{ km} = 10^3 \text{ m}$ and $1 \text{ cm} = 10^{-2} \text{ m}$.

The names of the additional units are derived by adding a **prefix** to the name of the fundamental unit. For example, the prefix "kilo-," abbreviated k, always means a unit larger by a factor of 1000; thus

$$1 \text{ kilometer} = 1 \text{ km} = 10^3 \text{ meters} = 10^3 \text{ m},$$
$$1 \text{ kilogram} = 1 \text{ kg} = 10^3 \text{ grams} = 10^3 \text{ g},$$
$$1 \text{ kilowatt} = 1 \text{ kW} = 10^3 \text{ watts} = 10^3 \text{ W}.$$

A table on the inside back cover of this book lists the standard SI prefixes, with their meanings and abbreviations.

Here are several examples of the use of multiples of 10 and their prefixes with the units of length, mass, and time.

LENGTH

$$1 \text{ nanometer} = 1 \text{ nm} = 10^{-9} \text{ m (a few times the size of the largest atom)}$$
$$1 \text{ micrometer} = 1 \text{ } \mu\text{m} = 10^{-6} \text{ m (size of some bacteria and living cells)}$$
$$1 \text{ millimeter} = 1 \text{ mm} = 10^{-3} \text{ m (diameter of the point of a ballpoint pen)}$$
$$1 \text{ centimeter} = 1 \text{ cm} = 10^{-2} \text{ m (diameter of your little finger)}$$
$$1 \text{ kilometer} = 1 \text{ km} = 10^{3} \text{ m (a 10-minute walk)}$$

MASS

$$1 \text{ microgram} = 1 \text{ } \mu\text{g} = 10^{-6} \text{ g} = 10^{-9} \text{ kg (mass of a very small dust particle)}$$
$$1 \text{ milligram} = 1 \text{ mg} = 10^{-3} \text{ g} = 10^{-6} \text{ kg (mass of a grain of salt)}$$
$$1 \text{ gram} = 1 \text{ g} = 10^{-3} \text{ kg (mass of a paper clip)}$$

TIME

$$1 \text{ nanosecond} = 1 \text{ ns} = 10^{-9} \text{ s (time for light to travel 0.3 m)}$$
$$1 \text{ microsecond} = 1 \text{ } \mu\text{s} = 10^{-6} \text{ s (time for an orbiting space shuttle to travel 8 mm)}$$
$$1 \text{ millisecond} = 1 \text{ ms} = 10^{-3} \text{ s (time for sound to travel 0.35 m)}$$

THE BRITISH SYSTEM

Finally, we mention the British system of units. These units are used only in the United States and a few other countries, and in most of these they are being replaced by SI units. British units are now officially defined in terms of SI units, as follows:

Length: 1 inch = 2.54 cm (exactly)

Force: 1 pound = 4.448221615260 newtons (exactly)

The newton, abbreviated N, is the SI unit of force. The British unit of time is the second, defined the same way as in SI. In physics, British units are used only in mechanics and thermodynamics; there is no British system of electrical units.

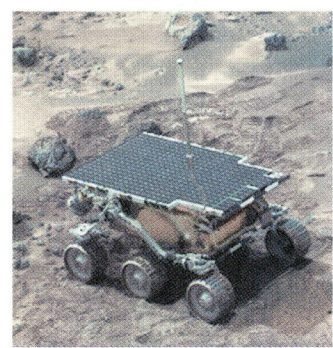

A mission to Mars involves various measurement scales. We measure the distance from the earth to Mars in hundreds of millions of kilometers, the landing spot is targeted to within tens of kilometers, and a mobile rover (like the *Sojourner* rover from the *Mars Pathfinder* mission, shown here) can move tens of meters in an hour. For a link to NASA's exploration of Mars, visit the *University Physics* web site.

Many everyday items make use of both SI and British units. An example is this speedometer from a U.S.-built automobile, which shows the speed in both kilometers per hour and miles per hour.

In this book we use SI units for all examples and problems, but we occasionally give approximate equivalents in British units. As you do problems using SI units, you may also wish to convert to the approximate British equivalents if they are more familiar to you. But you should try to think directly in SI as much as you can.

1-5 UNIT CONSISTENCY AND CONVERSIONS

We use equations to express relationships among physical quantities that are represented by algebraic symbols. Each algebraic symbol always denotes both a number and a unit. For example, d might represent a distance of 10 m, t a time of 5 s, and v a speed of 2 m/s.

An equation must always be **dimensionally consistent.** You can't add apples and automobiles; two terms may be added or equated only if they have the same units. For example, if a body moving with constant speed v travels a distance d in a time t, these quantities are related by the equation

$$d = vt. \tag{1-1}$$

If d is measured in meters, then the product vt must also be expressed in meters. Using the above numbers as an example, we may write

$$10 \text{ m} = \left(2 \, \frac{\text{m}}{\text{s}}\right)(5 \text{ s}).$$

Because the unit 1/s on the right side of the equation cancels the unit s, the product vt has units of meters, as it must. In calculations, units are treated just like algebraic symbols with respect to multiplication and division.

CAUTION ▶ When a problem requires calculations using numbers with units, *always* write the numbers with the correct units and carry the units through the calculation as in the example above. This provides a very useful check for calculations. If at some stage in a calculation you find that an equation or an expression has inconsistent units, you know you have made an error somewhere. In this book we will always carry units through all calculations, and we strongly urge you to follow this practice when you solve problems. ◀

Problem–Solving Strategy

UNIT CONVERSIONS

1. Units are multiplied and divided just like ordinary algebraic symbols. This gives us an easy way to convert a quantity from one set of units to another. The key idea is that we can express the same physical quantity in two different units and form an equality. For example, when we say that 1 min = 60 s, we don't mean that the number 1 is equal to the number 60; we mean that 1 min represents the same physical time interval as 60 s. For this reason, the ratio (1 min)/(60 s) equals 1, as does its reciprocal (60 s)/(1 min). We can multiply a quantity by either of these factors without changing that quantity's physical meaning. To find the number of seconds in 3 min, we write

$$3 \text{ min} = (3 \text{ min}) \left(\frac{60 \text{ s}}{1 \text{ min}}\right) = 180 \text{ s}.$$

2. If you do your unit conversions correctly, unwanted units will cancel, as in the example above. If instead you had multiplied 3 min by (1 min)/(60 s), your result would have been $\frac{1}{20}$ min²/s, which is a rather odd way of measuring time. To be sure you convert units properly, you must write down the units at *all* stages of the calculation.

3. As a final check, ask yourself whether your answer is reasonable. Is the result 3 min = 180 s reasonable? The answer is yes; the second is a smaller unit than the minute, so there are more seconds than minutes in the same time interval.

| EXAMPLE 1–1 |

Converting speed units The official world land speed record is 1228.0 km/h, set on October 15, 1997, by Andy Green in the jet engine car *Thrust SSC*. Express this speed in meters per second.

SOLUTION The prefix k means 10^3, so the speed 1228.0 km/h = 1228.0×10^3 m/h. We also know that there are 3600 s in 1 h. So we must combine the speed of 1228.0×10^3 m/h and a factor of 3600. But should we multiply or divide by this factor? If we treat the factor as a pure number without units, we're forced to guess how to proceed.

The correct approach is to carry the units with each factor. We then arrange the factor so that the hour unit cancels:

$$1228.0 \text{ km/h} = \left(1228.0 \times 10^3 \, \frac{\text{m}}{\text{\rlap{/}h}}\right)\left(\frac{1 \, \text{\rlap{/}h}}{3600 \text{ s}}\right) = 341.11 \text{ m/s.}$$

If you multiplied by (3600 s)/(1 h) instead of (1 h)/(3600 s), the hour unit wouldn't cancel, and you would be able to easily recognize your error. Again, the *only* way to be sure that you correctly convert units is to carry the units throughout the calculation.

| EXAMPLE 1–2 |

Converting volume units The world's largest cut diamond is the First Star of Africa (mounted in the British Royal Sceptre and kept in the Tower of London). Its volume is 1.84 cubic inches. What is its volume in cubic centimeters? In cubic meters?

SOLUTION To convert cubic inches to cubic centimeters, we multiply by $[(2.54 \text{ cm})/(1 \text{ in.})]^3$, not just (2.54 cm)/(1 in.). We find

$$1.84 \text{ in.}^3 = (1.84 \text{ in.}^3)\left(\frac{2.54 \text{ cm}}{1 \text{ in.}}\right)^3$$

$$= (1.84)(2.54)^3 \, \frac{\text{\rlap{/}in.}^3 \, \text{cm}^3}{\text{\rlap{/}in.}^3} = 30.2 \text{ cm}^3.$$

Also, 1 cm = 10^{-2} m, and

$$30.2 \text{ cm}^3 = (30.2 \text{ cm}^3)\left(\frac{10^{-2} \text{ m}}{1 \text{ cm}}\right)^3$$

$$= (30.2)(10^{-2})^3 \, \frac{\text{\rlap{/}cm}^3 \, \text{m}^3}{\text{\rlap{/}cm}^3} = 30.2 \times 10^{-6} \text{ m}^3.$$

Note: While 1 centimeter is 10^{-2} of a meter (that is, 1 cm = 10^{-2} m), a cubic centimeter (1 cm³) is *not* 10^{-2} of a cubic meter. Rather, it is the volume of a cube whose sides are 1 cm long. So 1 cm³ = $(1 \text{ cm})^3 = (10^{-2} \text{ m})^3 = (10^{-2})^3 \text{ m}^3$, or 1 cm³ = 10^{-6} m³.

1–6 UNCERTAINTY AND SIGNIFICANT FIGURES

Measurements always have uncertainties. If you measure the thickness of the cover of this book using an ordinary ruler, your measurement is reliable only to the nearest millimeter, and your result will be 3 mm. It would be *wrong* to state this result as 3.00 mm; given the limitations of the measuring device, you can't tell whether the actual thickness is 3.00 mm, 2.85 mm, or 3.11 mm. But if you use a micrometer caliper, a device that measures distances reliably to the nearest 0.01 mm, the result will be 2.91 mm. The distinction between these two measurements is in their **uncertainty.** The measurement using the micrometer caliper has a smaller uncertainty; it's a more accurate measurement. The uncertainty is also called the **error,** because it indicates the maximum difference there is likely to be between the measured value and the true value. The uncertainty or error of a measured value depends on the measurement technique used.

We often indicate the **accuracy** of a measured value—that is, how close it is likely to be to the true value—by writing the number, the symbol ±, and a second number indicating the uncertainty of the measurement. If the diameter of a steel rod is given as 56.47 ± 0.02 mm, this means that the true value is unlikely to be less than 56.45 mm or greater than 56.49 mm. In a commonly used shorthand notation, the number 1.6454(21) means 1.6454 ± 0.0021. The numbers in parentheses show the uncertainty in the final digits of the main number.

We can also express accuracy in terms of the maximum likely **fractional error** or **percent error** (also called *fractional uncertainty* and *percent uncertainty*). A resistor labeled "47 ohms ± 10%" probably has a true resistance differing from 47 ohms by no more than 10% of 47 ohms, that is, about 5 ohms. The resistance is probably between

42 and 52 ohms. For the diameter of the steel rod given above, the fractional error is (0.02 mm)/(56.47 mm), or about 0.0004; the percent error is (0.0004) (100%), or about 0.04%. Even small percent errors can sometimes be very significant (Fig. 1–3).

In many cases the uncertainty of a number is not stated explicitly. Instead, the uncertainty is indicated by the number of meaningful digits, or **significant figures,** in the measured value. We gave the thickness of the cover of this book as 2.91 mm, which has three significant figures. By this we mean that the first two digits are known to be correct, while the third digit is uncertain. The last digit is in the hundredths place, so the uncertainty is about 0.01 mm. Two values with the *same* number of significant figures may have *different* uncertainties; a distance given as 137 km also has three significant figures, but the uncertainty is about 1 km.

When you use numbers having uncertainties to compute other numbers, the computed numbers are also uncertain. It is especially important to understand this when you compare a number obtained from measurements with a value obtained from a theoretical prediction. Suppose you want to verify the value of π, the ratio of the circumference of a circle to its diameter. The true value of this ratio to ten digits is 3.141592654. To make your own calculation, you draw a large circle and measure its diameter and circumference to the nearest millimeter, obtaining the values 135 mm and 424 mm. You punch these into your calculator and obtain the quotient 3.140740741. Does this agree with the true value or not?

First, the last seven digits in this answer are meaningless; they imply a smaller uncertainty than is possible with your measurements. When numbers are multiplied or divided, the number of significant figures in the result can be no greater than in the factor with the fewest significant figures. For example, $3.1416 \times 2.34 \times 0.58 = 4.3$. Your measurements each have three significant figures, so your measured value of π, equal to (424 mm)/(135 mm), can have only three significant figures. It should be stated simply as 3.14. Within the limit of three significant figures, your value does agree with the true value.

When we add and subtract numbers, it's the location of the decimal point that matters, not the number of significant figures. For example, $123.62 + 8.9 = 132.5$. Although 123.62 has an uncertainty of about 0.01, 8.9 has an uncertainty of about 0.1. So their sum has an uncertainty of about 0.1 and should be written as 132.5, not 132.52.

In the examples and problems in this book we usually give numerical values with three significant figures, so your answers should usually have no more than three significant figures. (Many numbers that you encounter in the real world have even less accuracy. An automobile speedometer, for example, usually gives only two significant figures.) You might do the arithmetic with a calculator having a display with ten digits. But to give a ten-digit answer is not only unnecessary, it is genuinely wrong, because it misrepresents the accuracy of the results. Always round your final answer to keep only the correct number of significant figures or, in doubtful cases, one more at most. In Example 1–1 it would have been wrong to state the answer as 341.11111 m/s. Note that when you reduce such an answer to the appropriate number of significant figures, you must *round,* not *truncate.* Your calculator will tell you that the ratio of 525 m to 311 m is 1.688102894; to three significant figures, this is 1.69, not 1.68.

When we calculate with very large or very small numbers, we can show significant figures much more easily by using **scientific notation,** sometimes called **powers-of-10 notation.** The distance from the earth to the moon is about 384,000,000 m, but writing the number in this form gives no indication of the number of significant figures. Instead, we move the decimal point eight places to the left (corresponding to dividing by 10^8) and multiply by 10^8. That is,

$$384{,}000{,}000 \text{ m} = 3.84 \times 10^8 \text{ m}.$$

In this form, it is clear that we have three significant figures. The number 4.00×10^{-7} also

1–3 This spectacular mishap was the result of a very small percent error—traveling a few meters too far in a journey of hundreds of thousands of meters.

has three significant figures, even though two of them are zeros. Note that in scientific notation the usual practice is to express the quantity as a number between 1 and 10 multiplied by the appropriate power of 10.

When an integer or a fraction occurs in a general equation, we treat that number as having no uncertainty at all. For example, in the equation $v^2 = v_0^2 + 2a(x - x_0)$, which is Eq. (2–13) in Chapter 2, the coefficient 2 is *exactly* 2. We can consider this coefficient as having an infinite number of significant figures (2.000000...). The same is true of the exponent 2 in v^2 and v_0^2.

Finally, let's note that **precision** is not the same as *accuracy*. A cheap digital watch that says the time is 10:35:17 A.M. is very *precise* (the time is given to the second), but if the watch runs several minutes slow, then this value isn't very *accurate*. On the other hand, a grandfather clock might be very accurate (that is, display the correct time), but if the clock has no second hand, it isn't very precise. A high-quality measurement, like those used to define standards (Section 1–4), is both precise *and* accurate.

EXAMPLE 1–3

Significant figures in multiplication The rest energy E of an object with rest mass m is given by Einstein's equation

$$E = mc^2,$$

where c is the speed of light in a vacuum. Find E for an object with $m = 9.11 \times 10^{-31}$ kg (to three significant figures, the mass of an electron). The SI unit for E is the joule (J); $1\ \text{J} = 1\ \text{kg} \cdot \text{m}^2/\text{s}^2$.

SOLUTION From Section 1–4 the exact value of the speed of light in vacuum is $c = 299{,}792{,}458$ m/s $= 2.99792458 \times 10^8$ m/s. Substituting the values of m and c into Einstein's equation, we find

$$E = (9.11 \times 10^{-31}\ \text{kg})(2.99792458 \times 10^8\ \text{m/s})^2$$
$$= (9.11)(2.99792458)^2(10^{-31})(10^8)^2\ \text{kg} \cdot \text{m}^2/\text{s}^2$$
$$= (81.87659678)(10^{[-31+(2\times8)]})\ \text{kg} \cdot \text{m}^2/\text{s}^2$$
$$= 8.187659678 \times 10^{-14}\ \text{kg} \cdot \text{m}^2/\text{s}^2.$$

Since the value of m was given to only three significant figures, we must round this to

$$E = 8.19 \times 10^{-14}\ \text{kg} \cdot \text{m}^2/\text{s}^2 = 8.19 \times 10^{-14}\ \text{J}.$$

Most calculators use scientific notation and add exponents automatically, but you should be able to do such calculations by hand when necessary.

1–7 ESTIMATES AND ORDERS OF MAGNITUDE

We have stressed the importance of knowing the accuracy of numbers that represent physical quantities. But even a very crude estimate of a quantity often gives us useful information. Sometimes we know how to calculate a certain quantity but have to guess at the data we need for the calculation. Or the calculation might be too complicated to carry out exactly, so we make some rough approximations. In either case our result is also a guess, but such a guess can be useful even if it is uncertain by a factor of two, ten, or more. Such calculations are often called **order-of-magnitude estimates.** The great Italian-American nuclear physicist Enrico Fermi (1901–1954) called them "back-of-the-envelope calculations."

Exercises 1–16 through 1–27 at the end of this chapter are of the estimating, or "order-of-magnitude," variety. Some are silly, and most require guesswork for the needed input data. Don't try to look up a lot of data; make the best guesses you can. Even when they are off by a factor of ten, the results can be useful and interesting.

EXAMPLE 1–4

An order-of-magnitude estimate You are writing an adventure novel in which the hero escapes across the border with a billion dollars worth of gold in his suitcase. Is this possible? Would that amount of gold fit in a suitcase? Would it be too heavy to carry?

SOLUTION Gold sells for around $400 an ounce. On a particular day the price might be $200 or $600, but never mind. An ounce is about 30 grams. Actually, an ordinary (avoirdupois) ounce is 28.35 g; an ounce of gold is a troy ounce, which is 9.45% more.

Again, never mind. Ten dollars' worth of gold has a mass some- where around 1 gram, so a billion (10^9) dollars worth of gold is a hundred million (10^8) grams, or a hundred thousand (10^5) kilo- grams. This corresponds to a weight in British units of around 200,000 lb, or 100 tons. Whether the precise number is closer to 50 tons or 200 tons doesn't matter. Either way, the hero is not about to carry it across the border in a suitcase.

We can also estimate the *volume* of this gold. If its density were the same as that of water (1 g/cm^3), the volume would be 10^8 cm^3, or 100 m^3. But gold is a heavy metal; we might guess its density to be ten times that of water. Gold is actually 19.3 times as dense as water. But by guessing ten, we find a volume of 10 m^3. Visualize ten cubical stacks of gold bricks, each 1 meter on a side, and ask yourself whether they would fit in a suitcase! Clearly, your novel needs rewriting.

1–8 VECTORS AND VECTOR ADDITION

Some physical quantities, such as time, temperature, mass, density, and electric charge, can be described completely by a single number with a unit. But many other important quantities have a *direction* associated with them and cannot be described by a single number. Such quantities play an essential role in many of the central topics of physics, including motion and its causes and the phenomena of electricity and magnetism. A simple example of a quantity with direction is the motion of an airplane. To describe this motion completely, we must say not only how fast the plane is moving, but also in what direction. To fly from Chicago to New York, a plane has to head east, not south. The speed of the airplane combined with its direction of motion together constitute a quan- tity called *velocity.* Another example is *force,* which in physics means a push or pull exerted on a body. Giving a complete description of a force means describing both how hard the force pushes or pulls on the body and the direction of the push or pull.

When a physical quantity is described by a single number, we call it a **scalar quan- tity.** In contrast, a **vector quantity** has both a **magnitude** (the "how much" or "how big" part) and a direction in space. Calculations with scalar quantities use the operations of ordinary arithmetic. For example, 6 kg + 3 kg = 9 kg, or 4 × 2 s = 8 s. However, com- bining vectors requires a different set of operations.

To understand more about vectors and how they combine, we start with the simplest vector quantity, **displacement.** Displacement is simply a change in position of a point. (The point may represent a particle or a small body.) In Fig. 1–4a we represent the change of position from point P_1 to point P_2 by a line from P_1 to P_2, with an arrowhead at P_2 to represent the direction of motion. Displacement is a vector quantity because we must state not only how far the particle moves, but also in what direction. Walking 3 km north from your front door doesn't get you to the same place as walking 3 km southeast; these two displacements have the same magnitude, but different directions.

We usually represent a vector quantity such as displacement by a single letter, such as $\vec{A}$ in Fig. 1–4a. In this book we always print vector symbols in ***boldface italic type with an arrow above them.*** We do this to remind you that vector quantities have differ- ent properties from scalar quantities; the arrow is a reminder that vectors have direction. In handwriting, vector symbols are usually underlined or written with an arrow above them (Fig. 1–4). When you write a symbol for a vector, *always* write it in one of these ways. If you don't distinguish between scalar and vector quantities in your notation, you probably won't make the distinction in your thinking either, and hopeless confusion will result.

When drawing any vector, we always draw a line with an arrowhead at its tip. The length of the line shows the vector's magnitude, and the direction of the line shows the vector's direction. Displacement is always a straight-line segment, directed from the starting point to the end point, even though the actual path of the particle may be curved. In Fig. 1–4b the particle moves along the curved path shown from P_1 to P_2, but the dis- placement is still the vector $\vec{A}$. Note that displacement is not related directly to the total

Handwritten notation: $\underline{A}$ or $\vec{A}$

(a)

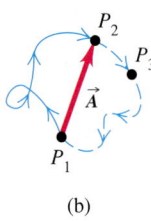

(b)

1–4 (a) Vector $\vec{A}$ is the displace- ment from point P_1 to point P_2. (b) A displacement is always a straight-line segment directed from the starting point to the end point, even if the actual path is curved. When a path ends at the same place where it started, the displacement is zero.

distance traveled. If the particle were to continue on to P_3 and then return to P_1, the displacement for the entire trip would be *zero*.

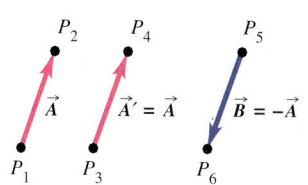

    If two vectors have the same direction, they are **parallel.** If they have the same magnitude *and* the same direction, they are *equal*, no matter where they are located in space. The vector $\vec{A}'$ from point P_3 to point P_4 in Fig. 1–5 has the same length and direction as the vector $\vec{A}$ from P_1 to P_2. These two displacements are equal, even though they start at different points. We write this as $\vec{A} = \vec{A}'$ in Fig. 1–5, using a boldface equals sign to emphasize that equality of two vector quantities is not the same relationship as equality of two scalar quantities. Two vector quantities are equal only when they have the same magnitude *and* the same direction.

    The vector $\vec{B}$ in Fig. 1–5, however, is not equal to $\vec{A}$ because its direction is *opposite* to that of $\vec{A}$. We define the **negative of a vector** as a vector having the same magnitude as the original vector but the *opposite* direction. The negative of vector quantity $\vec{A}$ is denoted as $-\vec{A}$, and we use a boldface minus sign to emphasize the vector nature of the quantities. If $\vec{A}$ is 87 m south, then $-\vec{A}$ is 87 m north. Thus the relation between $\vec{A}$ and $\vec{B}$ of Fig. 1–5 may be written as $\vec{A} = -\vec{B}$ or $\vec{B} = -\vec{A}$. When two vectors $\vec{A}$ and $\vec{B}$ have opposite directions, whether their magnitudes are the same or not, we say that they are **antiparallel.**

    We usually represent the *magnitude* of a vector quantity (its length in the case of a displacement vector) by the same letter used for the vector, but in *light italic type* with *no* arrow on top, rather than boldface italic with an arrow (which is reserved for vectors). An alternative notation is the vector symbol with vertical bars on both sides:

$$\text{(Magnitude of } \vec{A}) = A = |\vec{A}|. \tag{1–2}$$

1–5 The displacement from P_3 to P_4 is equal to that from P_1 to P_2. The displacement $\vec{B}$ from P_5 to P_6 has the same magnitude as $\vec{A}$ and $\vec{A}'$ but opposite direction; displacement $\vec{B}$ is the negative of displacement $\vec{A}$.

By definition the magnitude of a vector quantity is a scalar quantity (a number) and is *always positive*. We also note that a vector can never be equal to a scalar because they are different kinds of quantities. The expression "$\vec{A} = 6$ m" is just as wrong as "2 oranges = 3 apples" or "6 lb = 7 km"!

    When drawing diagrams with vectors, we'll usually use a scale similar to those used for maps, in which the distance on the diagram is proportional to the magnitude of the vector. For example, a displacement of 5 km might be represented in a diagram by a vector 1 cm long, since an actual-size diagram wouldn't be practical. When we work with vector quantities with units other than displacement, such as force or velocity, we *must* use a scale. In a diagram for force vectors we might use a scale in which a vector that is 1 cm long represents a force of magnitude 5 N. A 20-N force would then be represented by a vector 4 cm long, with the appropriate direction.

VECTOR ADDITION

Now suppose a particle undergoes a displacement $\vec{A}$, followed by a second displacement $\vec{B}$ (Fig. 1–6a). The final result is the same as if the particle had started at the same initial point and undergone a single displacement $\vec{C}$, as shown. We call displacement $\vec{C}$ the **vector sum**, or **resultant**, of displacements $\vec{A}$ and $\vec{B}$. We express this relationship symbolically as

$$\vec{C} = \vec{A} + \vec{B}. \tag{1–3}$$

The boldface plus sign emphasizes that adding two vector quantities requires a geometrical process and is not the same operation as adding two scalar quantities such as $2 + 3 = 5$. In vector addition we usually place the *tail* of the *second* vector at the *head*, or tip, of the *first* vector (Fig. 1–6a).

    If we make the displacements $\vec{A}$ and $\vec{B}$ in reverse order, with $\vec{B}$ first and $\vec{A}$ second, the result is the same (Fig. 1–6b). Thus

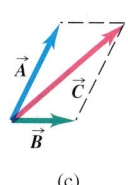

$$\vec{C} = \vec{B} + \vec{A} \quad \text{and} \quad \vec{A} + \vec{B} = \vec{B} + \vec{A}. \tag{1–4}$$

1–6 Vector $\vec{C}$ is the vector sum of vectors $\vec{A}$ and $\vec{B}$. The order in vector addition doesn't matter; vector addition is commutative.

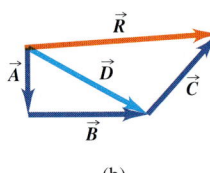

1–7 The vector sum of (a) two parallel vectors and (b) two antiparallel vectors. Note that the vectors $\vec{A}$, $\vec{B}$, and $\vec{C}$ in part (a) are not the same as the vectors $\vec{A}$, $\vec{B}$, and $\vec{C}$ in part (b).

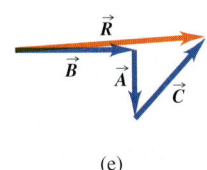

1–8 Several constructions for finding the vector sum $\vec{A} + \vec{B} + \vec{C}$.

This shows that the order of terms in a vector sum doesn't matter. In other words, vector addition obeys the commutative law.

Figure 1–6c shows an alternative representation of the vector sum. When vectors $\vec{A}$ and $\vec{B}$ are both drawn with their tails at the same point, vector $\vec{C}$ is the diagonal of a parallelogram constructed with $\vec{A}$ and $\vec{B}$ as two adjacent sides.

CAUTION ▶ A word of warning is in order about vector addition. It's a common error to conclude that if $\vec{C} = \vec{A} + \vec{B}$, the magnitude C should just equal the magnitude A plus the magnitude B. Figure 1–6 shows that in general, this conclusion is *wrong;* you can see from the figure that $C < A + B$. The magnitude of the vector sum $\vec{A} + \vec{B}$ depends on the magnitudes of $\vec{A}$ and of $\vec{B}$ *and* on the angle between $\vec{A}$ and $\vec{B}$ (see Problem 1–72). Only in the special case in which $\vec{A}$ and $\vec{B}$ are *parallel* is the magnitude of $\vec{C} = \vec{A} + \vec{B}$ equal to the sum of the magnitudes of $\vec{A}$ and $\vec{B}$ (Fig. 1–7a). By contrast, when the vectors are *antiparallel* (Fig. 1–7b) the magnitude of $\vec{C}$ equals the *difference* of the magnitudes of $\vec{A}$ and $\vec{B}$. Students who aren't careful about distinguishing between scalar and vector quantities frequently make errors about the magnitude of a vector sum. Don't let this happen to you! ◀

When we need to add more than two vectors, we may first find the vector sum of any two, add this vectorially to the third, and so on. Figure 1–8a shows three vectors $\vec{A}$, $\vec{B}$, and $\vec{C}$. In Fig. 1–8b, vectors $\vec{A}$ and $\vec{B}$ are first added, giving a vector sum $\vec{D}$; vectors $\vec{C}$ and $\vec{D}$ are then added by the same process to obtain the vector sum $\vec{R}$:

$$\vec{R} = (\vec{A} + \vec{B}) + \vec{C} = \vec{D} + \vec{C}.$$

Alternatively, we can first add $\vec{B}$ and $\vec{C}$ (Fig. 1–8c) to obtain vector $\vec{E}$, and then add $\vec{A}$ and $\vec{E}$ to obtain $\vec{R}$:

$$\vec{R} = \vec{A} + (\vec{B} + \vec{C}) = \vec{A} + \vec{E}.$$

We don't even need to draw vectors $\vec{D}$ and $\vec{E}$; all we need to do is draw the given vectors in succession, with the tail of each at the head of the one preceding it, and complete the polygon by a vector $\vec{R}$ from the tail of the first vector to the head of the last vector (Fig. 1–8d). The order makes no difference; Fig. 1–8e shows a different order, and we invite you to try others. We see that vector addition obeys the associative law.

We mentioned above that $-\vec{A}$ is a vector having the same magnitude as $\vec{A}$ but the opposite direction. This provides the basis for defining vector subtraction. We define the difference $\vec{A} - \vec{B}$ of two vectors $\vec{A}$ and $\vec{B}$ to be the vector sum of $\vec{A}$ and $-\vec{B}$:

$$\vec{A} - \vec{B} = \vec{A} + (-\vec{B}). \qquad (1\text{–}5)$$

Figure 1–9 shows an example of vector subtraction. To construct the vector difference $\vec{A} - \vec{B}$, the tail of $-\vec{B}$ is placed at the head of $\vec{A}$.

A vector quantity such as a displacement can be multiplied by a scalar quantity (an ordinary number). The displacement $2\vec{A}$ is a displacement (vector quantity) in the same direction as the vector $\vec{A}$ but twice as long; this is the same as adding $\vec{A}$ to itself. In general, when a vector $\vec{A}$ is multiplied by a scalar c, the result $c\vec{A}$ has magnitude $|c|A$ (the absolute value of c multiplied by the magnitude of the vector $\vec{A}$). If c is positive, $c\vec{A}$ is in the same direction as $\vec{A}$; if c is negative, $c\vec{A}$ is in the direction opposite to $\vec{A}$. Thus $5\vec{A}$ is parallel to $\vec{A}$, while $-5\vec{A}$ is antiparallel to $\vec{A}$.

The scalar quantity used to multiply a vector may also be a physical quantity having units. For example, you may be familiar with the relationship $\vec{F} = m\vec{a}$; the net force $\vec{F}$ (a vector quantity) that acts on a body is equal to the product of the body's mass m (a positive scalar quantity) and its acceleration $\vec{a}$ (a vector quantity). The magnitude of the net force is equal to the mass (which is positive and equals its own absolute value) multiplied by the magnitude of the acceleration. The unit of the magnitude of force is the unit of mass multiplied by the unit of the magnitude of acceleration. The direction of $\vec{F}$ is the same as that of $\vec{a}$ because m is positive.

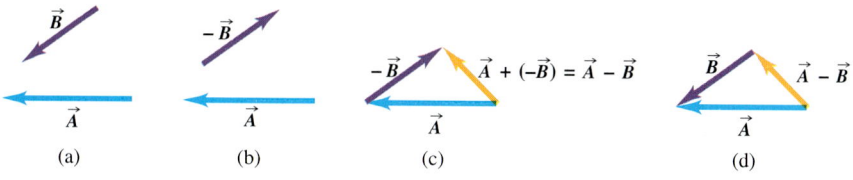

1–9 (a) Vector $\vec{A}$ and vector $\vec{B}$. (b) Vector $\vec{A}$ and vector $-\vec{B}$. (c) The vector difference $\vec{A} - \vec{B}$ is the sum of vectors $\vec{A}$ and $-\vec{B}$. The tail of $-\vec{B}$ is placed at the head of $\vec{A}$. (d) To check: $(\vec{A} - \vec{B}) + \vec{B} = \vec{A}$.

(a)　　(b)　　(c)　　(d)

EXAMPLE 1–5

Vector addition A cross-country skier skis 1.00 km north and then 2.00 km east on a horizontal snow field. a) How far and in what direction is she from the starting point? b) What are the magnitude and direction of her resultant displacement?

SOLUTION a) Figure 1–10 is a scale diagram of the skier's displacements. By careful measurement we find that the distance from the starting point is about 2.2 km and the angle ϕ (the Greek letter "phi") is about 63°. But it is much more accurate to *calculate* the result. The vectors in the diagram form a right triangle, and we can find the length of the hypotenuse by using the Pythagorean theorem:

$$\sqrt{(1.00 \text{ km})^2 + (2.00 \text{ km})^2} = 2.24 \text{ km}.$$

The angle ϕ can be found with a little simple trigonometry. If you need a review, the trigonometric functions and identities are summarized in Appendix B, along with other useful mathematical and geometrical relations. By the definition of the tangent function,

$$\tan \phi = \frac{\text{opposite side}}{\text{adjacent side}} = \frac{2.00 \text{ km}}{1.00 \text{ km}},$$

$$\phi = 63.4°.$$

b) The magnitude of the resultant displacement is just the distance that we found in part (a), 2.24 km. We can describe the direction as 63.4° east of north or 90° − 63.4° = 26.6° north of east. Take your choice!

1–10 The vector diagram, drawn to scale, for a cross-country ski trip.

1–9 COMPONENTS OF VECTORS

In Section 1–8 we added vectors by using a scale diagram and by using properties of right triangles. Measuring a diagram offers only very limited accuracy, and calculations with right triangles work only when the two vectors are perpendicular. So we need a simple but general method for adding vectors. This is called the method of *components*.

To define what we mean by the components of a vector, we begin with a rectangular (Cartesian) coordinate system of axes (Fig. 1–11). We then draw the vector we're considering with its tail at O, the origin of the coordinate system. We can represent any vector lying in the xy-plane as the sum of a vector parallel to the x-axis and a vector parallel to the y-axis. These two vectors are labeled $\vec{A}_x$ and $\vec{A}_y$ in the figure; they are called the **component vectors** of vector $\vec{A}$, and their vector sum is equal to $\vec{A}$. In symbols,

$$\vec{A} = \vec{A}_x + \vec{A}_y. \tag{1–6}$$

By definition, each component vector lies along a coordinate-axis direction. Thus we need only a single number to describe each one. When the component vector $\vec{A}_x$ points in the positive x-direction, we define the number A_x to be equal to the magnitude of $\vec{A}_x$. When the component vector $\vec{A}_x$ points in the negative x-direction, we define the number A_x to be equal to the negative of that magnitude, keeping in mind that the magnitude of a vector quantity is itself never negative. We define the number A_y in the same way. The two numbers A_x and A_y are called the **components** of $\vec{A}$.

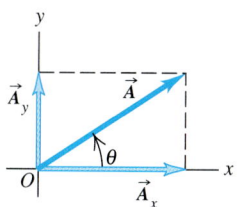

1–11 Vectors $\vec{A}_x$ and $\vec{A}_y$ are the rectangular component vectors of $\vec{A}$ in the directions of the x- and y-axes. For the vector $\vec{A}$ shown here, the components A_x and A_y are both positive.

The components A_x and A_y of a vector $\vec{A}$ are just numbers; they are *not* vectors themselves. This is why we print the symbols for components in light italic type with no arrow on top instead of the boldface italic with an arrow reserved for vectors.

We can calculate the components of the vector $\vec{A}$ if we know its magnitude A and its direction. We'll describe the direction of a vector by its angle relative to some reference direction. In Fig. 1–11 this reference direction is the positive x-axis, and the angle between vector $\vec{A}$ and the positive x-axis is θ (the Greek letter "theta"). Imagine that the vector $\vec{A}$ originally lies along the $+x$-axis and that you then rotate it to its correct direction, as indicated by the arrow in Fig. 1–11 on the angle θ. If this rotation is from the $+x$-axis toward the $+y$-axis, as shown in Fig. 1–11, then θ is *positive;* if the rotation is from the $+x$-axis toward the $-y$-axis, θ is *negative.* Thus the $+y$-axis is at an angle of $90°$, the $-x$-axis at $180°$, and the $-y$-axis at $270°$ (or $-90°$). If θ is measured in this way, then from the definition of the trigonometric functions,

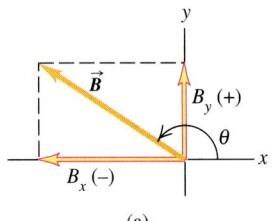

$$\frac{A_x}{A} = \cos\theta \qquad \text{and} \qquad \frac{A_y}{A} = \sin\theta;$$

$$A_x = A\cos\theta \qquad \text{and} \qquad A_y = A\sin\theta.$$

(1–7)

(θ measured from the $+x$-axis, rotating toward the $+y$-axis)

CAUTION ▶ Equations (1–7) are correct *only* when the angle θ is measured from the positive x-axis as described above. If the angle of the vector is given from a different reference direction or using a different sense of rotation, the relationships are different. Be careful! ◀

In Fig. 1–11, A_x is positive because its direction is along the positive x-axis, and A_y is positive because its direction is along the positive y-axis. This is consistent with Eqs. (1–7); θ is in the first quadrant (between $0°$ and $90°$), and both the cosine and the sine of an angle in this quadrant are positive. But in Fig. 1–12a the component B_x is negative; its direction is opposite to that of the positive x-axis. Again, this agrees with Eqs. (1–7); the cosine of an angle in the second quadrant is negative. The component B_y is positive (sin θ is positive in the second quadrant). In Fig. 1–12b, both C_x and C_y are negative (both cos θ and sin θ are negative in the third quadrant).

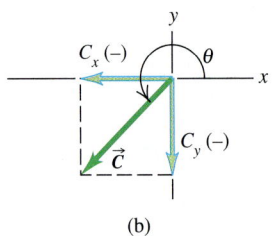

1–12 The components of a vector may be positive or negative numbers.

EXAMPLE 1-6

Finding components a) What are the x- and y-components of vector $\vec{D}$ in Fig. 1–13a? The magnitude of the vector is $D = 3.00$ m, and the angle $\alpha = 45°$. b) What are the x- and y-components of vector $\vec{E}$ in Fig. 1–13b? The magnitude of the vector is $E = 4.50$ m, and the angle $\beta = 37.0°$.

SOLUTION a) The angle between $\vec{D}$ and the positive x-axis is α (the Greek letter "alpha"), but this angle is measured toward the *negative* y-axis. So the angle we must use in Eqs. (1–7) is $\theta = -\alpha = -45°$. We find

$$D_x = D\cos\theta = (3.00 \text{ m})(\cos(-45°)) = +2.1 \text{ m},$$

$$D_y = D\sin\theta = (3.00 \text{ m})(\sin(-45°)) = -2.1 \text{ m}.$$

The vector has a positive x-component and a negative y-component, as shown in the figure. Had you been careless and substituted $+45°$ for θ in Eqs. (1–7), you would have gotten the wrong sign for D_y.

b) The x-axis isn't horizontal in Fig. 1–13b, nor is the y-axis vertical. In general, *any* orientation of the x- and y-axes is per-

missible, just so the axes are mutually perpendicular. (Later we'll use axes like these to study the motion of an object sliding on a ramp or incline; one axis will lie along the ramp, and the other will be perpendicular to the ramp.)

Here the angle β (the Greek letter "beta") is the angle between $\vec{E}$ and the positive y-axis, *not* the positive x-axis, so we

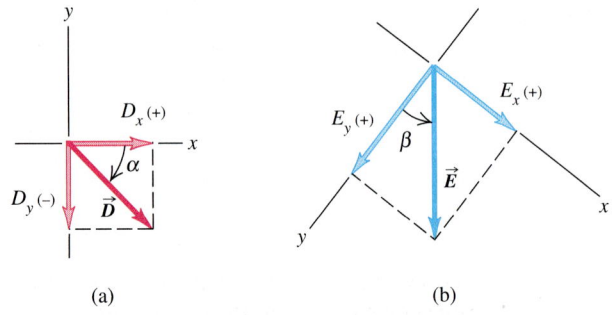

1–13 Calculating the x- and y-components of vectors.

cannot use this angle in Eqs. (1–7). Instead, note that $\vec{E}$ defines the hypotenuse of a right triangle; the other two sides of the triangle are the magnitudes of E_x and E_y, the x- and y-components of $\vec{E}$. The sine of β is the opposite side (the magnitude of E_x) divided by the hypotenuse (the magnitude E), and the cosine of β is the adjacent side (the magnitude of E_y) divided by the hypotenuse (again, the magnitude E). Both components of $\vec{E}$ are positive, so

$$E_x = E \sin \beta = (4.50 \text{ m}) (\sin 37.0°) = +2.71 \text{ m},$$
$$E_y = E \cos \beta = (4.50 \text{ m}) (\cos 37.0°) = +3.59 \text{ m}.$$

Had you used Eqs. (1–7) directly and written $E_x = E \cos 37.0°$ and $E_y = E \sin 37.0°$, your answers for E_x and E_y would have been reversed!

If you insist on using Eqs. (1–7), you must first find the angle between $\vec{E}$ and the positive x-axis, measured toward the positive y-axis; this is $\theta = 90.0° - \beta = 90.0° - 37.0° = 53.0°$. Then $E_x = E \cos \theta$ and $E_y = E \sin \theta$. You can substitute the values of E and θ into Eqs. (1–7) to show that the results for E_x and E_y are the same as given above.

Notice that the answers to part (b) have three significant figures, but the answers to part (a) have only two. Can you see why?

USING COMPONENTS

We can describe a vector completely by giving either its magnitude and direction or its x- and y-components. Equations (1–7) show how to find the components if we know the magnitude and direction. We can also reverse the process: we can find the magnitude and direction if we know the components. By applying the Pythagorean theorem to Fig. 1–11, we find that the magnitude of a vector $\vec{A}$ is

$$A = \sqrt{A_x{}^2 + A_y{}^2}, \tag{1–8}$$

where we always take the positive root. Equation (1–8) is valid for any choice of x-axis and y-axis, as long as they are mutually perpendicular. The expression for the vector direction comes from the definition of the tangent of an angle. If θ is measured from the positive x-axis, and a positive angle is measured toward the positive y-axis (as in Fig. 1–11), then

$$\tan \theta = \frac{A_y}{A_x} \quad \text{and} \quad \theta = \arctan \frac{A_y}{A_x}. \tag{1–9}$$

We will always use the notation arctan for the inverse tangent function. The notation $\tan^{-1}$ is also commonly used, and your calculator may have an INV button to be used with the TAN button. Some computer languages use atan (FORTRAN) or atn (BASIC). Pascal uses ArcTan.

CAUTION ▶ There is one slight complication in using Eqs. (1–9) to find θ. Suppose $A_x = 2$ m and $A_y = -2$ m; then $\tan \theta = -1$. But there are two angles having tangents of –1, namely, 135° and 315° (or –45°). In general, any two angles that differ by 180° have the same tangent. To decide which is correct, we have to look at the individual components. Because A_x is positive and A_y is negative, the angle must be in the fourth quadrant; thus $\theta = 315°$ (or –45°) is the correct value. Most pocket calculators give arctan (–1) = –45°. In this case that is correct; but if instead we have $A_x = -2$ m and $A_y = 2$ m, then the correct angle is 135°. Similarly, when A_x and A_y are both negative, the tangent is positive, but the angle is in the third quadrant. You should *always* draw a sketch to check which of the two possibilities is the correct one. ◀

Here's how we use components to calculate the vector sum (resultant) of two or more vectors. Figure 1–14 shows two vectors $\vec{A}$ and $\vec{B}$ and their vector sum $\vec{R}$, along with the x- and y-components of all three vectors. You can see from the diagram that the x-component R_x of the vector sum is simply the sum $(A_x + B_x)$ of the x-components of the vectors being added. The same is true for the y-components. In symbols,

$$R_x = A_x + B_x, \quad R_y = A_y + B_y \quad \text{(components of } \vec{R} = \vec{A} + \vec{B}\text{).} \tag{1–10}$$

Figure 1–14 shows this result for the case in which the components A_x, A_y, B_x, and B_y are all positive. You should draw additional diagrams to verify for yourself that Eqs. (1–10) are valid for *any* signs of the components of $\vec{A}$ and $\vec{B}$.

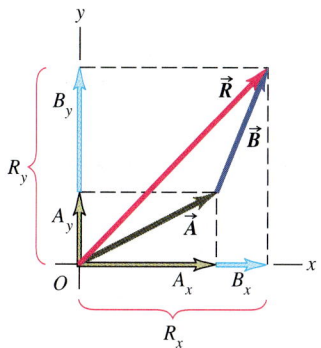

1–14 Vector $\vec{R}$ is the vector sum (resultant) of $\vec{A}$ and $\vec{B}$. Its x-component, R_x, equals the sum of the x-components of $\vec{A}$ and $\vec{B}$. The y-components are similarly related.

If we know the components of any two vectors $\vec{A}$ and $\vec{B}$, perhaps by using Eqs. (1–7), we can compute the components of the vector sum $\vec{R}$. Then if the magnitude and direction of $\vec{R}$ are needed, we can obtain them from Eqs. (1–8) and (1–9), with the A's replaced by R's.

This procedure for finding the sum of two vectors can easily be extended to any number of vectors. Let $\vec{R}$ be the vector sum of $\vec{A}, \vec{B}, \vec{C}, \vec{D}, \vec{E}, \dots$. Then the components of $\vec{R}$ are

$$R_x = A_x + B_x + C_x + D_x + E_x + \cdots,$$
$$R_y = A_y + B_y + C_y + D_y + E_y + \cdots. \tag{1–11}$$

We have talked only about vectors that lie in the xy-plane, but the component method works just as well for vectors having any direction in space. We introduce a z-axis perpendicular to the xy-plane; then in general a vector $\vec{A}$ has components A_x, A_y, and A_z in the three coordinate directions. The magnitude A is given by

$$A = \sqrt{A_x{}^2 + A_y{}^2 + A_z{}^2}. \tag{1–12}$$

Again, we always take the positive root. Also, Eqs. (1–11) for the components of the vector sum $\vec{R}$ have an additional member:

$$R_z = A_z + B_z + C_z + D_z + E_z + \cdots.$$

Finally, while our discussion of vector addition has centered on combining *displacement* vectors, the method is applicable to all other vector quantities as well. When we study the concept of force in Chapter 4, we'll find that forces are vectors that obey the same rules of vector addition that we've used with displacement. Other vector quantities will make their appearance in later chapters.

Problem–Solving Strategy

VECTOR ADDITION

1. First draw the individual vectors being summed and the coordinate axes being used. In your drawing, place the tail of the first vector at the origin of coordinates, place the tail of the second vector at the head of the first vector, and so on. Draw the vector sum $\vec{R}$ from the tail of the first vector to the head of the last vector.

2. Find the x- and y-components of each individual vector and record your results in a table. If a vector is described by its magnitude A and its angle θ, measured from the $+x$-axis toward the $+y$-axis, then the components are given by

$$A_x = A \cos \theta, \qquad A_y = A \sin \theta.$$

Some components may be positive and some may be negative, depending on how the vector is oriented (that is, what quadrant θ lies in). You can use this sign table as a check:

Quadrant	I	II	III	IV
A_x	+	–	–	+
A_y	+	+	–	–

If the angles of the vectors are given in some other way, perhaps using a different reference direction, convert them to angles measured from the $+x$-axis as described above. Be particularly careful with signs.

3. Add the individual x-components algebraically, including signs, to find R_x, the x-component of the vector sum. Do the same for the y-components to find R_y.

4. Then the magnitude R and direction θ of the vector sum are given by

$$R = \sqrt{R_x{}^2 + R_y{}^2},$$
$$\theta = \arctan \frac{R_y}{R_x}.$$

Remember that the magnitude R is *always* positive and that θ is measured from the positive x-axis. The value of θ that you find with a calculator may be the correct one, or it may be off by 180°. You can decide by examining your drawing.

EXAMPLE 1-7

Adding vectors with components The three finalists in a contest are brought to the center of a large, flat field. Each is given a meter stick, a compass, a calculator, a shovel, and (in a different order for each contestant) the following three displacements:

> 72.4 m, 32.0° east of north;
> 57.3 m, 36.0° south of west;
> 17.8 m straight south.

The three displacements lead to the point where the keys to a new Porsche are buried. Two contestants start measuring immediately, but the winner first *calculates* where to go. What does she calculate?

SOLUTION The situation is shown in Fig. 1–15. We have chosen the +x-axis as east and the +y-axis as north, the usual choice for

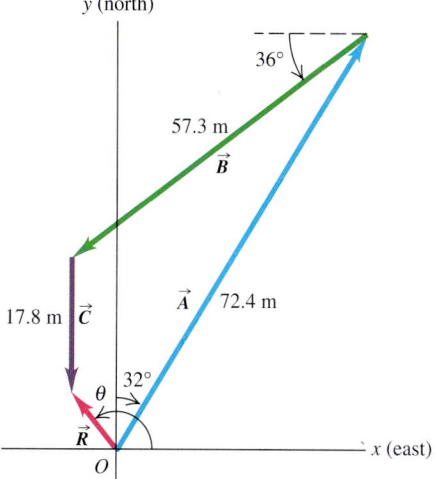

1–15 Three successive displacements $\vec{A}$, $\vec{B}$, and $\vec{C}$ and the resultant (vector sum) displacement $\vec{R} = \vec{A} + \vec{B} + \vec{C}$.

maps. Let $\vec{A}$ be the first displacement, $\vec{B}$ the second, and $\vec{C}$ the third. We can estimate from the diagram that the vector sum $\vec{R}$ is about 10 m, 40° west of north. The angles of the vectors, measured from the +x-axis toward the +y-axis, are $(90.0° - 32.0°) = 58.0°$, $(180.0° + 36.0°) = 216.0°$, and 270.0°. We have to find the components of each. Because of our choice of axes, we may use Eqs. (1–7), and so the components of $\vec{A}$ are

$$A_x = A \cos \theta_A = (72.4 \text{ m})(\cos 58.0°) = 38.37 \text{ m},$$
$$A_y = A \sin \theta_A = (72.4 \text{ m})(\sin 58.0°) = 61.40 \text{ m}.$$

Note that we have kept one too many significant figures in the components; we will wait until the end to round to the correct number of significant figures. The table below shows the components of all the displacements, the addition of components, and the other calculations. Always arrange your component calculations systematically like this.

Distance	Angle	x-component	y-component
A = 72.4 m	58.0°	38.37 m	61.40 m
B = 57.3 m	216.0°	−46.36 m	−33.68 m
C = 17.8 m	270.0°	0.00 m	−17.80 m
		$R_x = -7.99$ m	$R_y = 9.92$ m

$$R = \sqrt{(-7.99 \text{ m})^2 + (9.92 \text{ m})^2} = 12.7 \text{ m},$$

$$\theta = \arctan \frac{9.92 \text{ m}}{-7.99 \text{ m}} = 129° = 39° \text{ west of north.}$$

The losers try to measure three angles and three distances totaling 147.5 m, one meter at a time. The winner measured only one angle and one much shorter distance.

Notice that $\theta = -51°$, or 51° south of east, also satisfies the equation for θ. But since the winner has made a drawing of the displacement vectors (Fig. 1–15), she knows that $\theta = 129°$ is the only correct solution for the angle.

EXAMPLE 1-8

A vector in three dimensions After an airplane takes off, it travels 10.4 km west, 8.7 km north, and 2.1 km up. How far is it from the takeoff point?

SOLUTION Let the +x-axis be east, the +y-axis north, and the +z-axis up. Then $A_x = -10.4$ km, $A_y = 8.7$ km, and $A_z = 2.1$ km; Eq. (1–12) gives

$$A = \sqrt{(-10.4 \text{ km})^2 + (8.7 \text{ km})^2 + (2.1 \text{ km})^2} = 13.7 \text{ km.}$$

1–10 UNIT VECTORS

A **unit vector** is a vector that has a magnitude of 1, with no units. Its only purpose is to *point,* that is, to describe a direction in space. Unit vectors provide a convenient notation for many expressions involving components of vectors. We will always include a caret or "hat" (^) in the symbol for a unit vector to distinguish it from ordinary vectors whose magnitude may or may not be equal to 1.

In an *x-y* coordinate system we can define a unit vector $\hat{\imath}$ that points in the direction of the positive *x*-axis and a unit vector $\hat{\jmath}$ that points in the direction of the positive *y*-axis. Then we can express the relationship between component vectors and components,

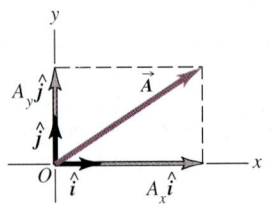

1–16 Using unit vectors, we can express a vector $\vec{A}$ in terms of its components A_x and A_y as $\vec{A} = A_x\hat{\imath} + A_y\hat{\jmath}$.

described at the beginning of Section 1–9, as follows:

$$\vec{A}_x = A_x\hat{\imath},$$
$$\vec{A}_y = A_y\hat{\jmath}. \tag{1–13}$$

Similarly, we can write a vector $\vec{A}$ in terms of its components as

$$\vec{A} = A_x\hat{\imath} + A_y\hat{\jmath}. \tag{1–14}$$

Equations (1–13) and (1–14) are vector equations; each term, such as $A_x\hat{\imath}$, is a vector quantity (Fig. 1–16). The boldface equals and plus signs denote vector equality and addition.

When two vectors $\vec{A}$ and $\vec{B}$ are represented in terms of their components, we can express the vector sum $\vec{R}$ using unit vectors as follows:

$$\vec{A} = A_x\hat{\imath} + A_y\hat{\jmath},$$
$$\vec{B} = B_x\hat{\imath} + B_y\hat{\jmath},$$
$$\vec{R} = \vec{A} + \vec{B}$$
$$= (A_x\hat{\imath} + A_y\hat{\jmath}) + (B_x\hat{\imath} + B_y\hat{\jmath})$$
$$= (A_x + B_x)\hat{\imath} + (A_y + B_y)\hat{\jmath}$$
$$= R_x\hat{\imath} + R_y\hat{\jmath}. \tag{1–15}$$

Equation (1–15) restates the content of Eqs. (1–10) in the form of a single vector equation rather than two component equations.

If the vectors do not all lie in the xy-plane, then we need a third component. We introduce a third unit vector $\hat{k}$ that points in the direction of the positive z-axis. The generalized forms of Eqs. (1–14) and (1–15) are

$$\vec{A} = A_x\hat{\imath} + A_y\hat{\jmath} + A_z\hat{k},$$
$$\vec{B} = B_x\hat{\imath} + B_y\hat{\jmath} + B_z\hat{k}, \tag{1–16}$$
$$\vec{R} = (A_x + B_x)\hat{\imath} + (A_y + B_y)\hat{\jmath} + (A_z + B_z)\hat{k}$$
$$= R_x\hat{\imath} + R_y\hat{\jmath} + R_z\hat{k}. \tag{1–17}$$

EXAMPLE 1–9

Using unit vectors   Given the two displacements

$$\vec{D} = (6\hat{\imath} + 3\hat{\jmath} - \hat{k})\text{ m} \quad \text{and} \quad \vec{E} = (4\hat{\imath} - 5\hat{\jmath} + 8\hat{k})\text{ m},$$

find the magnitude of the displacement $2\vec{D} - \vec{E}$.

SOLUTION   Letting $F = 2\vec{D} - \vec{E}$, we have

$$\vec{F} = 2(6\hat{\imath} + 3\hat{\jmath} - \hat{k})\text{ m} - (4\hat{\imath} - 5\hat{\jmath} + 8\hat{k})\text{ m}$$
$$= [(12 - 4)\hat{\imath} + (6 + 5)\hat{\jmath} + (-2 - 8)\hat{k}]\text{ m}$$
$$= (8\hat{\imath} + 11\hat{\jmath} - 10\hat{k})\text{ m}.$$

The units of the vectors $\vec{D}$, $\vec{E}$, and $\vec{F}$ are meters, so the components of these vectors are also in meters. From Eq. (1–12),

$$F = \sqrt{F_x^2 + F_y^2 + F_z^2}$$
$$= \sqrt{(8\text{ m})^2 + (11\text{ m})^2 + (-10\text{ m})^2} = 17\text{ m}.$$

1–11 PRODUCTS OF VECTORS

We have seen how addition of vectors develops naturally from the problem of combining displacements, and we will use vector addition for many other vector quantities later.

We can also express many physical relationships concisely by using *products* of vectors. Vectors are not ordinary numbers, so ordinary multiplication is not directly applicable to vectors. We will define two different kinds of products of vectors. The first, called the scalar product, yields a result that is a scalar quantity. The second, the vector product, yields another vector.

SCALAR PRODUCT

The **scalar product** of two vectors $\vec{A}$ and $\vec{B}$ is denoted by $\vec{A} \cdot \vec{B}$. Because of this notation, the scalar product is also called the **dot product.**

To define the scalar product $\vec{A} \cdot \vec{B}$ of two vectors $\vec{A}$ and $\vec{B}$, we draw the two vectors with their tails at the same point (Fig. 1–17a). The angle between their directions is ϕ as shown; the angle ϕ always lies between $0°$ and $180°$. (As usual, we use Greek letters for angles.) Figure 1–17b shows the projection of the vector $\vec{B}$ onto the direction of $\vec{A}$; this projection is the component of $\vec{B}$ parallel to $\vec{A}$ and is equal to $B \cos \phi$. (We can take components along any direction that's convenient, not just the x- and y-axes.) We define $\vec{A} \cdot \vec{B}$ to be the magnitude of $\vec{A}$ multiplied by the component of $\vec{B}$ parallel to $\vec{A}$. Expressed as an equation,

$$\vec{A} \cdot \vec{B} = AB \cos \phi = \left|\vec{A}\right|\left|\vec{B}\right| \cos \phi \quad \text{(definition of the scalar (dot) product),} \quad (1\text{–}18)$$

where ϕ ranges from $0°$ to $180°$.

Alternatively, we can define $\vec{A} \cdot \vec{B}$ to be the magnitude of $\vec{B}$ multiplied by the component of $\vec{A}$ parallel to $\vec{B}$, as in Fig. 1–17c. Hence $\vec{A} \cdot \vec{B} = B(A \cos \phi) = AB \cos \phi$, which is the same as Eq. (1–18).

The scalar product is a scalar quantity, not a vector, and it may be positive, negative, or zero. When ϕ is between $0°$ and $90°$, the scalar product is positive. When ϕ is between $90°$ and $180°$, it is negative. You should draw a diagram like Fig. 1–17, but with ϕ between $90°$ and $180°$, to confirm for yourself that the component of $\vec{B}$ parallel to $\vec{A}$ is negative in this case, as is the component of $\vec{A}$ parallel to $\vec{B}$. Finally, when $\phi = 90°$, $\vec{A} \cdot \vec{B} = 0$. *The scalar product of two perpendicular vectors is always zero.*

For any two vectors $\vec{A}$ and $\vec{B}$, $AB \cos \phi = BA \cos \phi$. This means that $\vec{A} \cdot \vec{B} = \vec{B} \cdot \vec{A}$. The scalar product obeys the commutative law of multiplication; the order of the two vectors does not matter.

We will use the scalar product in Chapter 6 to describe work done by a force. When a constant force $\vec{F}$ is applied to a body that undergoes a displacement $\vec{s}$, the work W (a scalar quantity) done by the force is given by

$$W = \vec{F} \cdot \vec{s}.$$

The work done by the force is positive if the angle between $\vec{F}$ and $\vec{s}$ is between $0°$ and $90°$, negative if this angle is between $90°$ and $180°$, and zero if $\vec{F}$ and $\vec{s}$ are perpendicular. (This is another example of a term that has a special meaning in physics; in everyday language, "work" isn't something that can be positive or negative.) In later chapters we'll use the scalar product for a variety of purposes, from calculating electric potential to determining the effects that varying magnetic fields have on electric circuits.

We can calculate the scalar product $\vec{A} \cdot \vec{B}$ directly if we know the x-, y-, and z-components of $\vec{A}$ and $\vec{B}$. To see how this is done, let's first work out the scalar products of the unit vectors. This is easy, since $\hat{\imath}, \hat{\jmath}$, and $\hat{k}$ are all perpendicular to each other. Using Eq. (1–18), we find

$$\hat{\imath} \cdot \hat{\imath} = \hat{\jmath} \cdot \hat{\jmath} = \hat{k} \cdot \hat{k} = (1)(1) \cos 0 \quad = 1,$$
$$\hat{\imath} \cdot \hat{\jmath} = \hat{\imath} \cdot \hat{k} = \hat{\jmath} \cdot \hat{k} = (1)(1) \cos 90° = 0. \quad (1\text{–}19)$$

Now we express $\vec{A}$ and $\vec{B}$ in terms of their components, expand the product, and use

(a)

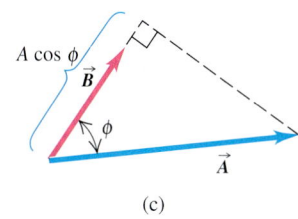

(b)

(c)

1–17 (a) Two vectors drawn from a common starting point to define their scalar product $\vec{A} \cdot \vec{B} = AB \cos \phi$. (b) $B \cos \phi$ is the component of $\vec{B}$ in the direction of $\vec{A}$, and $\vec{A} \cdot \vec{B}$ is the product of the magnitude of $\vec{A}$ and this component. (c) $\vec{A} \cdot \vec{B}$ is also the product of the magnitude of $\vec{B}$ and the component of $\vec{A}$ in the direction of $\vec{B}$.

these products of unit vectors:

$$\vec{A} \cdot \vec{B} = (A_x\hat{\imath} + A_y\hat{\jmath} + A_z\hat{k}) \cdot (B_x\hat{\imath} + B_y\hat{\jmath} + B_z\hat{k})$$

$$= A_x\hat{\imath} \cdot B_x\hat{\imath} + A_x\hat{\imath} \cdot B_y\hat{\jmath} + A_x\hat{\imath} \cdot B_z\hat{k}$$

$$+ A_y\hat{\jmath} \cdot B_x\hat{\imath} + A_y\hat{\jmath} \cdot B_y\hat{\jmath} + A_y\hat{\jmath} \cdot B_z\hat{k}$$

$$+ A_z\hat{k} \cdot B_x\hat{\imath} + A_z\hat{k} \cdot B_y\hat{\jmath} + A_z\hat{k} \cdot B_z\hat{k}$$

$$= A_xB_x\hat{\imath} \cdot \hat{\imath} + A_xB_y\hat{\imath} \cdot \hat{\jmath} + A_xB_z\hat{\imath} \cdot \hat{k}$$

$$+ A_yB_x\hat{\jmath} \cdot \hat{\imath} + A_yB_y\hat{\jmath} \cdot \hat{\jmath} + A_yB_z\hat{\jmath} \cdot \hat{k}$$

$$+ A_zB_x\hat{k} \cdot \hat{\imath} + A_zB_y\hat{k} \cdot \hat{\jmath} + A_zB_z\hat{k} \cdot \hat{k}. \tag{1-20}$$

From Eqs. (1–19) we see that six of these nine terms are zero, and the three that survive give simply

$$\vec{A} \cdot \vec{B} = A_xB_x + A_yB_y + A_zB_z \qquad \text{(scalar (dot) product in terms of components).}$$

$$\tag{1-21}$$

Thus the scalar product of two vectors is the sum of the products of their respective components.

The scalar product gives a straightforward way to find the angle ϕ between any two vectors $\vec{A}$ and $\vec{B}$ whose components are known. In this case, Eq. (1–21) can be used to find the scalar product of $\vec{A}$ and $\vec{B}$. From Eq. (1–18) the scalar product is also equal to $AB \cos \phi$. The vector magnitudes A and B can be found from the vector components with Eq. (1–12), so $\cos \phi$ and hence the angle ϕ can be determined (see Example 1–11).

EXAMPLE 1-10

Calculating a scalar product Find the scalar product $\vec{A} \cdot \vec{B}$ of the two vectors in Fig. 1–18. The magnitudes of the vectors are $A = 4.00$ and $B = 5.00$.

SOLUTION There are two ways to calculate the scalar product. The first way uses the magnitudes of the vectors and the angle between them (Eq. 1–18), and the second uses the components of the two vectors (Eq. 1–21).

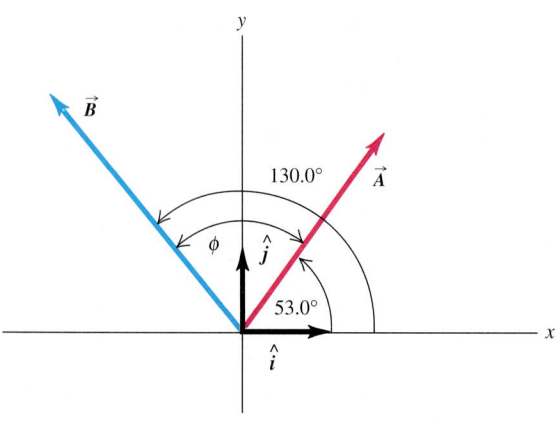

1–18 Two vectors in two dimensions.

Using the first approach, the angle between the two vectors is $\phi = 130.0° - 53.0° = 77.0°$, so

$$\vec{A} \cdot \vec{B} = AB \cos \phi = (4.00)(5.00) \cos 77.0° = 4.50.$$

This is positive because the angle between $\vec{A}$ and $\vec{B}$ is between 0° and 90°.

To use the second approach, we first need to find the components of the two vectors. Since the angles of $\vec{A}$ and $\vec{B}$ are given with respect to the +x-axis, and these angles are measured in the sense from the +x-axis to the +y-axis, we can use Eqs. (1–7):

$$A_x = (4.00) \cos 53.0° = 2.407,$$

$$A_y = (4.00) \sin 53.0° = 3.195,$$

$$A_z = 0,$$

$$B_x = (5.00) \cos 130.0° = -3.214,$$

$$B_y = (5.00) \sin 130.0° = 3.830,$$

$$B_z = 0.$$

The z-components are zero because both vectors lie in the xy-plane. As in Example 1–7, we are keeping one too many significant figures in the components; we'll round to the correct number at the end. From Eq. (1–21) the scalar product is

$$\vec{A} \cdot \vec{B} = A_xB_x + A_yB_y + A_zB_z$$

$$= (2.407)(-3.214) + (3.195)(3.830) + (0)(0) = 4.50.$$

We get the same result for the scalar product with both methods, as we should.

EXAMPLE 1–11

Finding angles with the scalar product Find the angle between the two vectors

$$\vec{A} = 2\hat{\imath} + 3\hat{\jmath} + \hat{k} \quad \text{and} \quad \vec{B} = -4\hat{\imath} + 2\hat{\jmath} - \hat{k}.$$

SOLUTION The vectors are shown in Fig. 1–19. The scalar product of two vectors is given by either Eq. (1–18) or Eq. (1–21). Equating these two and rearranging, we obtain

$$\cos\phi = \frac{A_x B_x + A_y B_y + A_z B_z}{AB}.$$

This formula can be used to find the angle between any two vectors $\vec{A}$ and $\vec{B}$. For our example the components of $\vec{A}$ are $A_x = 2$, $A_y = 3$, and $A_z = 1$, and the components of $\vec{B}$ are $B_x = -4$, $B_y = 2$, and $B_z = -1$. Thus

$$\vec{A} \cdot \vec{B} = A_x B_x + A_y B_y + A_z B_z = (2)(-4) + (3)(2) + (1)(-1) = -3,$$

$$A = \sqrt{A_x^2 + A_y^2 + A_z^2} = \sqrt{2^2 + 3^2 + 1^2} = \sqrt{14},$$

$$B = \sqrt{B_x^2 + B_y^2 + B_z^2} = \sqrt{(-4)^2 + 2^2 + (-1)^2} = \sqrt{21},$$

$$\cos\phi = \frac{A_x B_x + A_y B_y + A_z B_z}{AB} = \frac{-3}{\sqrt{14}\sqrt{21}} = -0.175,$$

$$\phi = 100°.$$

As a check on this result, note that the scalar product $\vec{A} \cdot \vec{B}$ is negative. This means that ϕ is between 90° and 180°, in agreement with our answer.

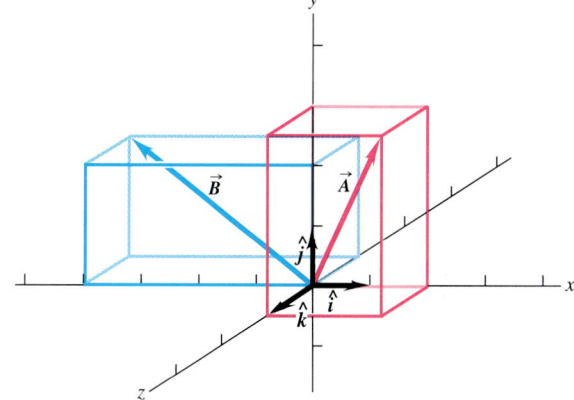

1–19 Two vectors in three dimensions.

VECTOR PRODUCT

The **vector product** of two vectors $\vec{A}$ and $\vec{B}$, also called the **cross product,** is denoted by $\vec{A} \times \vec{B}$. We will use this product in Chapter 10 to describe torque and angular momentum. Later we will also use it extensively for magnetic fields, where it will help us describe the relationships among the directions of several vector quantities.

To define the vector product $\vec{A} \times \vec{B}$ of two vectors $\vec{A}$ and $\vec{B}$, we again draw the two vectors with their tails at the same point (Fig. 1–20a). The two vectors then lie in a plane. We define the vector product to be a vector quantity with a direction perpendicular to this plane (that is, perpendicular to both $\vec{A}$ and $\vec{B}$) and a magnitude equal to $AB \sin\phi$. That is, if $\vec{C} = \vec{A} \times \vec{B}$, then

$$C = AB \sin\phi \quad \text{(magnitude of the vector (cross) product of } \vec{A} \text{ and } \vec{B}\text{).} \quad (1\text{–}22)$$

We measure the angle ϕ from $\vec{A}$ toward $\vec{B}$ and take it to be the smaller of the two possible angles, so ϕ ranges from 0° to 180°. Thus C in Eq. (1–22) is always positive, as a vector magnitude should be. Note also that when $\vec{A}$ and $\vec{B}$ are parallel or antiparallel, $\phi = 0$ or 180° and $C = 0$. That is, *the vector product of two parallel or antiparallel vectors is always zero.* In particular, *the vector product of any vector with itself is zero.* To see the contrast between the scalar product and the magnitude of the vector product, imagine that we vary the angle between $\vec{A}$ and $\vec{B}$ while keeping their magnitudes constant. When $\vec{A}$ and $\vec{B}$ are parallel, the scalar product will be maximum and the magnitude of the vector product will be zero. When $\vec{A}$ and $\vec{B}$ are perpendicular, the scalar product will be zero and the magnitude of the vector product will be maximum.

There are always *two* directions perpendicular to a given plane, one on each side of the plane. We choose which of these is the direction of $\vec{A} \times \vec{B}$ as follows. Imagine rotating vector $\vec{A}$ about the perpendicular line until it is aligned with $\vec{B}$, choosing the smaller of the two possible angles between $\vec{A}$ and $\vec{B}$. Curl the fingers of your right hand around the perpendicular line so that the fingertips point in the direction of rotation; your thumb will then point in the direction of $\vec{A} \times \vec{B}$. This **right-hand rule** is shown in Fig. 1–20a. The direction of the vector product is also the direction in which a right-hand screw advances if turned from $\vec{A}$ toward $\vec{B}$, as shown.

(a)

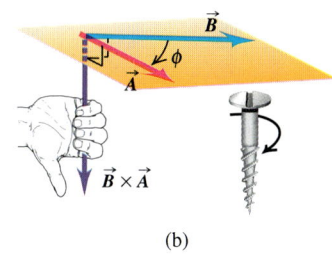

(b)

1–20 (a) Vectors $\vec{A}$ and $\vec{B}$ lie in a plane; the vector product $\vec{A} \times \vec{B}$ is perpendicular to this plane in a direction determined by the right-hand rule. (b) $\vec{B} \times \vec{A} = -\vec{A} \times \vec{B}$; the vector product is anticommutative.

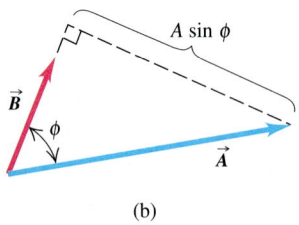

1–21 (a) $B \sin \phi$ is the component of $\vec{B}$ perpendicular to the direction of $\vec{A}$, and the magnitude of $\vec{A} \times \vec{B}$ is the product of the magnitude of $\vec{A}$ and this component. (b) The magnitude of $\vec{A} \times \vec{B}$ is also the product of the magnitude of $\vec{B}$ and the component of $\vec{A}$ perpendicular to $\vec{B}$.

Similarly, we determine the direction of $\vec{B} \times \vec{A}$ by rotating $\vec{B}$ into $\vec{A}$ in Fig. 1–20b. The result is a vector that is *opposite* to the vector $\vec{A} \times \vec{B}$. The vector product is not commutative! In fact, for any two vectors $\vec{A}$ and $\vec{B}$,

$$\vec{A} \times \vec{B} = -\vec{B} \times \vec{A}. \tag{1–23}$$

Just as we did for the scalar product, we can give a geometrical interpretation of the magnitude of the vector product. In Fig. 1–21a, $B \sin \phi$ is the component of vector $\vec{B}$ that is *perpendicular* to the direction of vector $\vec{A}$. From Eq. (1–22) the magnitude of $\vec{A} \times \vec{B}$ equals the magnitude of $\vec{A}$ multiplied by the component of $\vec{B}$ perpendicular to $\vec{A}$. Figure 1–21b shows that the magnitude of $\vec{A} \times \vec{B}$ also equals the magnitude of $\vec{B}$ multiplied by the component of $\vec{A}$ perpendicular to $\vec{B}$. Note that Fig. 1–21 shows the case in which ϕ is between $0°$ and $90°$; you should draw a similar diagram for ϕ between $90°$ and $180°$ to show that the same geometrical interpretation of the magnitude of $\vec{A} \times \vec{B}$ still applies.

If we know the components of $\vec{A}$ and $\vec{B}$, we can calculate the components of the vector product, using a procedure similar to that for the scalar product. First we work out the multiplication table for the unit vectors $\hat{\imath}, \hat{\jmath}$, and $\hat{k}$. The vector product of any vector with itself is zero, so

$$\hat{\imath} \times \hat{\imath} = \hat{\jmath} \times \hat{\jmath} = \hat{k} \times \hat{k} = \mathbf{0}.$$

The boldface zero is a reminder that each product is a zero *vector*, that is, one with all components equal to zero and an undefined direction. Using Eqs. (1–22) and (1–23) and the right-hand rule, we find

$$\hat{\imath} \times \hat{\jmath} = -\hat{\jmath} \times \hat{\imath} = \hat{k},$$

$$\hat{\jmath} \times \hat{k} = -\hat{k} \times \hat{\jmath} = \hat{\imath}, \tag{1–24}$$

$$\hat{k} \times \hat{\imath} = -\hat{\imath} \times \hat{k} = \hat{\jmath}.$$

Next we express $\vec{A}$ and $\vec{B}$ in terms of their components and the corresponding unit vectors, and we expand the expression for the vector product:

$$\begin{aligned}\vec{A} \times \vec{B} &= (A_x \hat{\imath} + A_y \hat{\jmath} + A_z \hat{k}) \times (B_x \hat{\imath} + B_y \hat{\jmath} + B_z \hat{k}) \\ &= A_x \hat{\imath} \times B_x \hat{\imath} + A_x \hat{\imath} \times B_y \hat{\jmath} + A_x \hat{\imath} \times B_z \hat{k} \\ &\quad + A_y \hat{\jmath} \times B_x \hat{\imath} + A_y \hat{\jmath} \times B_y \hat{\jmath} + A_y \hat{\jmath} \times B_z \hat{k} \\ &\quad + A_z \hat{k} \times B_x \hat{\imath} + A_z \hat{k} \times B_y \hat{\jmath} + A_z \hat{k} \times B_z \hat{k}. \end{aligned} \tag{1–25}$$

We can also rewrite the individual terms as $A_x \hat{\imath} \times B_y \hat{\jmath} = (A_x B_y) \hat{\imath} \times \hat{\jmath}$, and so on. Evaluating these by using the multiplication table for the unit vectors and then grouping the terms, we find

$$\vec{A} \times \vec{B} = (A_y B_z - A_z B_y)\hat{\imath} + (A_z B_x - A_x B_z)\hat{\jmath} + (A_x B_y - A_y B_x)\hat{k}. \tag{1–26}$$

Thus the components of $\vec{C} = \vec{A} \times \vec{B}$ are given by

$$C_x = A_y B_z - A_z B_y, \quad C_y = A_z B_x - A_x B_z, \quad C_z = A_x B_y - A_y B_x$$

$$\text{(components of } \vec{C} = \vec{A} \times \vec{B}). \tag{1–27}$$

The vector product can also be expressed in determinant form as

$$\vec{A} \times \vec{B} = \begin{vmatrix} \hat{\imath} & \hat{\jmath} & \hat{k} \\ A_x & A_y & A_z \\ B_x & B_y & B_z \end{vmatrix}.$$

If you aren't familiar with determinants, don't worry about this form.

With the axis system of Fig. 1–22a, if we reverse the direction of the z-axis, we get the system shown in Fig. 1–22b. Then, as you may verify, the definition of the vector

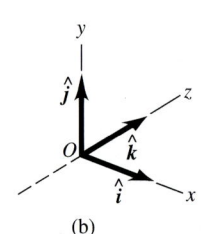

1–22 (a) A right-handed coordinate system, in which $\hat{\imath} \times \hat{\jmath} = \hat{k}$, $\hat{\jmath} \times \hat{k} = \hat{\imath}$, and $\hat{k} \times \hat{\imath} = \hat{\jmath}$. (b) A left-handed coordinate system, in which $\hat{\imath} \times \hat{\jmath} = -\hat{k}$, and so on. We'll use only right-handed systems.

product gives $\hat{\imath} \times \hat{\jmath} = -\hat{k}$ instead of $\hat{\imath} \times \hat{\jmath} = \hat{k}$. In fact, all vector products of the unit vectors $\hat{\imath}, \hat{\jmath}$, and $\hat{k}$ would have signs opposite to those in Eqs. (1–24). We see that there are two kinds of coordinate systems, differing in the signs of the vector products of unit vectors. An axis system in which $\hat{\imath} \times \hat{\jmath} = \hat{k}$, as in Fig. 1–22a, is called a **right-handed system.** The usual practice is to use *only* right-handed systems, and we will follow that practice throughout this book.

EXAMPLE 1-12

Calculating a vector product Vector $\vec{A}$ has magnitude 6 units and is in the direction of the +x-axis. Vector $\vec{B}$ has magnitude 4 units and lies in the xy-plane, making an angle of 30° with the +x-axis (Fig. 1–23). Find the vector product $\vec{A} \times \vec{B}$.

SOLUTION We can find the vector product in one of two ways. The first way is to use Eq. (1–22) to determine the magnitude of $\vec{A} \times \vec{B}$ and then use the right-hand rule to find the direction of the vector product. The second way is to use the components of $\vec{A}$ and $\vec{B}$ to find the components of the vector product $\vec{C} = \vec{A} \times \vec{B}$ using Eqs. (1–27).

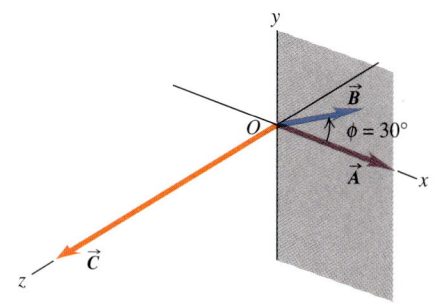

1–23 Vectors $\vec{A}$ and $\vec{B}$ and their vector product $\vec{C} = \vec{A} \times \vec{B}$. The vector $\vec{B}$ lies in the xy-plane.

With the first approach, from Eq. (1–22) the magnitude of the vector product is

$$AB \sin \phi = (6)(4)(\sin 30°) = 12.$$

From the right-hand rule, the direction of $\vec{A} \times \vec{B}$ is along the +z-axis, so we have $\vec{A} \times \vec{B} = 12\,\hat{k}$.

To use the second approach, we first write the components of $\vec{A}$ and $\vec{B}$:

$$A_x = 6, \qquad A_y = 0, \qquad A_z = 0,$$
$$B_x = 4\cos 30° = 2\sqrt{3}, \quad B_y = 4\sin 30° = 2, \quad B_z = 0.$$

Defining $\vec{C} = \vec{A} \times \vec{B}$, we have from Eqs. (1–27) that

$$C_x = (0)(0) - (0)(2) \quad = 0,$$
$$C_y = (0)(2\sqrt{3}) - (6)(0) = 0,$$
$$C_z = (6)(2) - (0)(2\sqrt{3}) = 12.$$

The vector product $\vec{C}$ has only a z-component, and it lies along the +z-axis. The magnitude agrees with the result we obtained with the first approach, as it should.

For this example the first approach was more direct because we knew the magnitudes of each vector and the angle between them, and furthermore, both vectors lay in one of the planes of the coordinate system. But often you will need to find the vector product of two vectors that are not so conveniently oriented or for which only the components are given. In such a case the second approach, using components, is the more direct one.

SUMMARY

- The fundamental physical quantities of mechanics are mass, length, and time. The corresponding basic SI units are the kilogram, the meter, and the second. Other units for these quantities, related by powers of 10, are identified by adding prefixes to the basic unit. Derived units for other physical quantities are products or quotients of the basic units. Equations must be dimensionally consistent; two terms can be added only when they have the same units.

- The accuracy of a measurement can be indicated by the number of significant figures, or by a stated uncertainty. The result of a multiplication or division usually has no more significant figures than the input data. When only crude estimates are available for input data, we can often make useful order-of-magnitude estimates.

- Scalar quantities are numbers and are combined with the usual rules of arithmetic. Vector quantities have direction as well as magnitude and are combined according to the rules of vector addition. Graphically, two vectors $\vec{A}$ and $\vec{B}$ are added by

KEY TERMS

range of validity, 2
model, 3
particle, 3
physical quantity, 3
operational definition, 3
unit, 4
International System (SI), 4
second, 4
meter, 4
kilogram, 4
prefix, 5
dimensionally consistent, 6

placing the tail of $\vec{B}$ at the head, or tip, of $\vec{A}$. The vector sum $\vec{A} + \vec{B}$ then extends from the tail of $\vec{A}$ to the head of $\vec{B}$.

■ Vector addition can be carried out by using components of vectors. If A_x and A_y are the components of vector $\vec{A}$ and B_x and B_y are the components of vector $\vec{B}$, the components of the vector sum $\vec{R} = \vec{A} + \vec{B}$ are given by

$$R_x = A_x + B_x, \qquad R_y = A_y + B_y. \tag{1-10}$$

■ Unit vectors describe directions in space. A unit vector has a magnitude of 1, with no units. We always write unit vectors with a caret or "hat" (^). The unit vectors $\hat{\imath}, \hat{\jmath},$ and $\hat{k},$ aligned with the x-, y-, and z-axes of a rectangular coordinate system, are especially useful.

■ The scalar product $C = \vec{A} \cdot \vec{B}$ of two vectors $\vec{A}$ and $\vec{B}$ is a scalar quantity, defined as

$$\vec{A} \cdot \vec{B} = AB \cos \phi = \left|\vec{A}\right|\left|\vec{B}\right| \cos \phi. \tag{1-18}$$

The scalar product can also be expressed in terms of components:

$$\vec{A} \cdot \vec{B} = A_x B_x + A_y B_y + A_z B_z. \tag{1-21}$$

The scalar product is commutative; for any two vectors $\vec{A}$ and $\vec{B}, \vec{A} \cdot \vec{B} = \vec{B} \cdot \vec{A}$. The scalar product of two perpendicular vectors is zero.

■ The vector product $\vec{C} = \vec{A} \times \vec{B}$ of two vectors $\vec{A}$ and $\vec{B}$ is another vector $\vec{C},$ with magnitude given by

$$C = AB \sin \phi. \tag{1-22}$$

The direction of the vector product is perpendicular to the plane of the two vectors being multiplied, as given by the right-hand rule. In terms of the components of the two vectors being multiplied, the components of the vector product are

$$C_x = A_y B_z - A_z B_y, \qquad C_y = A_z B_x - A_x B_z, \qquad C_z = A_x B_y - A_y B_x. \tag{1-27}$$

The vector product is not commutative; for any two vectors $\vec{A}$ and $\vec{B}, \vec{A} \times \vec{B} = -\vec{B} \times \vec{A}$. The vector product of two parallel or antiparallel vectors is zero.

DISCUSSION QUESTIONS

Q1–1 How many correct experiments do we need to disprove a theory? How many to prove a theory? Explain.

Q1–2 A guidebook describes the rate of climb of a mountain trail as 120 meters per kilometer. How can this be expressed as a number with no units?

Q1–3 Suppose you are asked to compute the tangent of 5.00 meters. Is this possible? Why or why not?

Q1–4 A highway contractor stated that in building a bridge deck he poured 250 yards of concrete. What do you think he meant?

Q1–5 What is your height in centimeters? What is your weight in newtons?

Q1–6 The U.S. National Institute of Science and Technology (NIST) maintains several accurate copies of the international standard kilogram. Even after careful cleaning, these national standard kilograms are gaining mass at an average rate of about 1 μg/y (1 y = 1 year) when compared every ten years or so to the standard international kilogram. Does this apparent change have any importance? Explain.

Q1–7 What physical phenomena (other than a pendulum or cesium clock) could be used to define a time standard?

Q1–8 Describe how you could measure the thickness of a sheet of paper with an ordinary ruler.

Q1–9 The quantity $\pi = 3.14159...$ is a number with no dimensions, since it is a ratio of two lengths. Describe two or three other geometrical or physical quantities that are dimensionless.

Q1–10 What are the units of volume? Suppose another student tells you that a cylinder of radius r and height h has volume given by $\pi r^3 h$. Explain why this cannot be right.

Q1–11 Three archers each fire four arrows at a target. Joe's four arrows hit at points 10 cm above, 10 cm below, 10 cm to the left, and 10 cm to the right of the center of the target. All four of Moe's arrows hit within 1 cm of a point 20 cm from the center, and Flo's four arrows all hit within 1 cm of the center. The contest judge says that one of the archers is precise but not accurate, another archer is accurate but not precise, and the third archer is both accurate and precise. Which description goes with which archer? Explain your reasoning.

Q1–12 A circular racetrack has radius 500 m. What is the displacement of a bicyclist when she travels around the track from the north side to the south side? When she makes one complete circle around the track? Explain your reasoning.

Q1–13 Can you find two vectors with different lengths that have a vector sum of zero? What length restrictions are required for three vectors to have a vector sum of zero? Explain your reasoning.

Q1–14 One sometimes speaks of the "direction of time," evolving from past to future. Does this mean that time is a vector quantity? Explain your reasoning.

Q1–15 Air traffic controllers give instructions to airline pilots telling them in what direction they are to fly. These instructions are called "vectors." If these are the only instructions given, is the name "vector" being used correctly? Why or why not?

Q1–16 Can you find a vector quantity that has a magnitude of zero, but components that are different from zero? Explain.

Q1–17 Let $\vec{A}$ represent any nonzero vector. Why is $\vec{A}/A$ a unit vector and what is its direction? If θ is the angle that $\vec{A}$ makes with the +x-axis, explain why $(\vec{A}/A) \cdot \hat{\imath}$ is called the *direction cosine* for that axis.

Q1–18 Which of the following are legitimate mathematical operations: a) $\vec{A} \cdot (\vec{B} - \vec{C})$; b) $(\vec{A} - \vec{B}) \times \vec{C}$; c) $\vec{A} \cdot (\vec{B} \times \vec{C})$; d) $\vec{A} \times (\vec{B} \times \vec{C})$; e) $\vec{A} \times (\vec{B} \cdot \vec{C})$? In each case, give the reason for your answer.

Q1–19 Consider the two repeated vector products $\vec{A} \times (\vec{B} \times \vec{C})$ and $(\vec{A} \times \vec{B}) \times \vec{C}$. Give an example that illustrates the general rule that these two vector products do not have the same magnitude or direction. Can you choose the vectors $\vec{A}$, $\vec{B}$, and $\vec{C}$ such that these two vector products *are* equal to each other? If so, give an example.

EXERCISES

SECTION 1–4 STANDARDS AND UNITS

SECTION 1–5 UNIT CONSISTENCY AND CONVERSIONS

1–1 Starting with the definition 1 in. = 2.54 cm, find the number of kilometers in 1.00 mile.

1–2 According to the label on a bottle of salad dressing, the volume of the contents is 0.473 liter (L). Using only the conversions 1 L = 1000 cm^3 and 1 in. = 2.54 cm, express this volume in cubic inches.

1–3 How many nanoseconds does it take light to travel 1.00 km in vacuum?

1–4 The density of lead is 11.3 g/cm^3. What is this value in kilograms per cubic meter?

1–5 The most powerful engine available for the classic 1963 Chevrolet Corvette Sting Ray developed 360 horsepower and had a displacement of 327 cubic inches. Express this displacement in liters (L) by using only the conversions 1 L = 1000 cm^3 and 1 in. = 2.54 cm.

1–6 Having been told he needs to set goals for himself, Billy Joe Bob decides to drink one cubic meter of his favorite beverage during the coming year. How many 16-fluid-ounce bottles should he drink each day? (Use Appendix E. A fluid ounce is a volume unit; 128 fluid ounces equals one gallon.)

1–7 The Concorde is the fastest airliner used for commercial service. It cruises at 1450 mi/h (about two times the speed of sound, or Mach 2). a) What is the cruise speed of the Concorde in km/h? b) What is it in m/s?

1–8 While driving in an exotic foreign land you see a speed limit sign on a highway that reads 180,000 furlongs per fortnight. How many miles per hour is this? (One furlong is $\frac{1}{8}$ mile, and a fortnight is 14 days. A furlong originally referred to the length of a plowed furrow.)

1–9 The gasoline consumption of a small car is advertised as 15.0 km/L (1 L = 1 liter). How many miles per gallon is this? Use the conversion factors in Appendix E.

SECTION 1–6 UNCERTAINTY AND SIGNIFICANT FIGURES

1–10 A useful and easy-to-remember approximate value for the number of seconds in a year is $\pi \times 10^7$. Determine the percent error in this approximate value. (There are 365.24 days in one year.)

1–11 Figure 1–3 shows the result of unacceptable error in the stopping position of a train. a) If a train travels 890 km from Berlin to Paris and then overshoots the end of the track by 10 m, what is the percentage error in the total distance covered? b) Would it be correct to write the total distance covered by the train as 890,010 m? Explain.

1–12 With a wooden ruler you measure the length of a rectangular piece of sheet metal to be 12 mm. You use a micrometer calipers to measure the width of the rectangle and obtain the value 5.98 mm. Give your answers to the following questions to the correct number of significant figures. a) What is the area of the rectangle? b) What is the ratio of the rectangle's width to its length? c) What is the perimeter of the rectangle? d) What is the difference between the length and width? e) What is the ratio of the length to the width?

1–13 Estimate the percent error in measuring a) a distance of about 75 cm with a meter stick; b) a mass of about 12 g with a chemical balance; c) a time interval of about 6 min with a stopwatch.

1–14 A rectangular piece of aluminum is 5.60 ± 0.01 cm long and 1.90 ± 0.01 cm wide. a) Find the area of the rectangle and the uncertainty in the area. b) Verify that the fractional uncertainty in the area is equal to the sum of the fractional uncertainties in the length and in the width. (This is a general result; see Challenge Problem 1–76.)

1–15 As you eat your way through a bag of chocolate chip cookies, you observe that each cookie is a circular disk of diameter 8.50 ± 0.02 cm and thickness 0.050 ± 0.005 cm. a) Find the average volume of a cookie and the uncertainty in the volume. b) Find the ratio of the diameter to the thickness and the uncertainty in this ratio.

SECTION 1–7 ESTIMATES AND ORDERS OF MAGNITUDE

1–16 How many gallons of gasoline are used in the United States in one day?

1–17 A box of typewriter paper has dimensions 11 in. × 17 in. × 9 in.; it is marked "10 M." Does that mean it contains 10,000 sheets or 10 million?

1–18 How many kernels of corn does it take to fill a 2-L soft drink bottle?

1–19 How many words are there in this book?

1–20 What total volume of air does a person breathe in a lifetime? How does that compare with the volume of the Houston Astrodome? (Estimate that a person breathes about 500 cm^3 of air with each breath.)

1–21 How many hairs do you have on your head?

1–22 How many times does a human heart beat during a lifetime? How many gallons of blood does it pump? (Estimate that the heart pumps 50 cm^3 of blood with each beat.)

1–23 In Wagner's opera *Ring of the Nibelung,* the goddess Freia is ransomed by a pile of gold just tall enough and wide enough to hide her from sight. Estimate the monetary value of this pile. (Refer to Example 1–4 for information on the price per ounce and density of gold.)

1–24 How many drops of water are in all the oceans on earth?

1–25 How many pizzas are consumed each academic year by students at your school?

1–26 How many dollar bills would have to be stacked to reach the moon? Would that be cheaper than building and launching a spacecraft?

1–27 How much would it cost to paper the entire United States (including Alaska and Hawaii) with dollar bills? What would be the cost to each person in the United States?

SECTION 1–8 VECTORS AND VECTOR ADDITION

1–28 Hearing rattles from a snake, you make two rapid displacements of magnitude 1.8 m and 2.4 m. In sketches (roughly to scale), show how your two displacements might add to give a resultant of magnitude a) 4.2 m; b) 0.6 m; c) 3.0 m.

1–29 A postal employee drives a delivery truck along the route shown in Fig. 1–24. Determine the magnitude and direction of

FIGURE 1–24 Exercises 1–29 and 1–34.

the resultant displacement by drawing a scale diagram. (See also Exercise 1–34 for a different approach to this same problem.)

1–30 For the vectors $\vec{A}$ and $\vec{B}$ in Fig. 1–25, use a scale drawing to find the magnitude and direction of a) the vector sum $\vec{A} + \vec{B}$; b) the vector difference $\vec{A} - \vec{B}$. From your answers to parts (a) and (b), find the magnitude and direction of c) $-\vec{A} - \vec{B}$; d) $\vec{B} - \vec{A}$. (See also Exercise 1–35 for a different approach to this problem.)

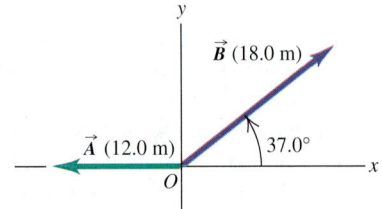

FIGURE 1–25 Exercises 1–30, 1–35, 1–40, and 1–48.

1–31 A spelunker is surveying a cave. She follows a passage 180 m straight west, then 210 m in a direction 45° east of south, then 90 m at 30° west of north. After a fourth unmeasured displacement she finds herself back where she started. Use a scale drawing to determine the magnitude and direction of the fourth displacement. (See also Problem 1–59 for a different approach to a similar problem.)

SECTION 1–9 COMPONENTS OF VECTORS

1–32 Use a scale drawing to find the *x*- and *y*-components of the following vectors. For each vector the numbers given are i) the magnitude of the vector and ii) the angle, measured in the sense from the +*x*-axis toward the +*y*-axis, that it makes with the +*x*-axis. Find a) magnitude 9.30 m, angle 60.0°; b) magnitude 22.0 km, angle 135°; c) magnitude 6.35 cm, angle 307°.

1–33 Compute the *x*- and *y*-components of the vectors $\vec{A}, \vec{B},$ and $\vec{C}$ in Fig. 1–26.

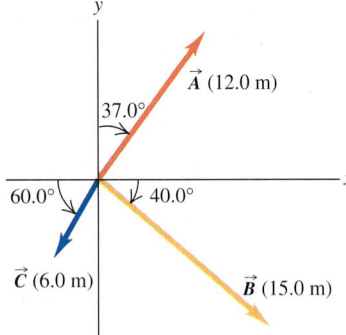

FIGURE 1–26 Exercises 1–33, 1–41, and 1–44, and Problem 1–58.

1–34 A postal employee drives a delivery truck over the route shown in Fig. 1–24. Use the method of components to determine the magnitude and direction of her resultant displacement. In a vector addition diagram (roughly to scale), show that the resultant displacement found from your diagram is in qualitative agreement with the result you obtained using the method of components.

1–35 For the vectors $\vec{A}$ and $\vec{B}$ in Fig. 1–25, use the method of components to find the magnitude and direction of a) the vector sum $\vec{A} + \vec{B}$; b) the vector sum $\vec{B} + \vec{A}$; c) the vector difference $\vec{A} - \vec{B}$; d) the vector difference $\vec{B} - \vec{A}$.

1–36 Find the magnitude and direction of the vector represented by the following pairs of components: a) $A_x = -8.60$ cm, $A_y = 5.20$ cm; b) $A_x = -9.70$ m, $A_y = -2.45$ m; c) $A_x = 7.75$ km, $A_y = -2.70$ km.

1–37 A disoriented physics professor drives 3.25 km north, then 4.75 km west, then 1.50 km south. Find the magnitude and direction of the resultant displacement, using the method of components. In a vector addition diagram (roughly to scale), show that the resultant displacement found from your diagram is in qualitative agreement with the result you obtained using the method of components.

1–38 Vector $\vec{A}$ has components $A_x = 1.30$ cm, $A_y = 2.25$ cm; vector $\vec{B}$ has components $B_x = 4.10$ cm, $B_y = -3.75$ cm. Find a) the components of the vector sum $\vec{A} + \vec{B}$; b) the magnitude and direction of $\vec{A} + \vec{B}$; c) the components of the vector difference $\vec{B} - \vec{A}$; d) the magnitude and direction of $\vec{B} - \vec{A}$.

1–39 Vector $\vec{A}$ is 2.80 cm long and is 60.0° above the x-axis in the first quadrant. Vector $\vec{B}$ is 1.90 cm long and is 60.0° below the x-axis in the fourth quadrant (Fig. 1–27). Find the magnitude and direction of a) $\vec{A} + \vec{B}$; b) $\vec{A} - \vec{B}$; c) $\vec{B} - \vec{A}$. In each case, sketch the vector addition or subtraction and show that your numerical answers are in qualitative agreement with your sketch.

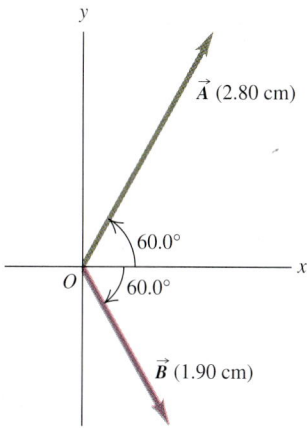

FIGURE 1–27 Exercises 1–39 and 1–50.

SECTION 1–10 UNIT VECTORS

1–40 Write each vector in Fig. 1–25 in terms of the unit vectors $\hat{\imath}$ and $\hat{\jmath}$.

1–41 Write each vector in Fig. 1–26 in terms of the unit vectors $\hat{\imath}$ and $\hat{\jmath}$.

1–42 a) Write each vector in Fig. 1–28 in terms of the unit vectors $\hat{\imath}$ and $\hat{\jmath}$. b) Use unit vectors to express the vector $\vec{C}$, where $\vec{C} = 3.00\vec{A} - 4.00\vec{B}$. c) Find the magnitude and direction of $\vec{C}$.

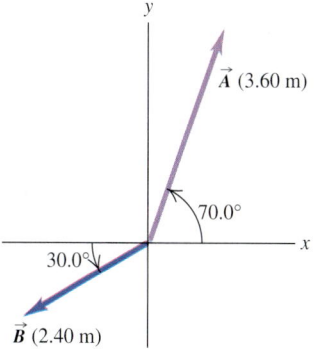

FIGURE 1–28 Exercise 1–42 and Problem 1–66.

1–43 Given two vectors $\vec{A} = 4.00\,\hat{\imath} + 3.00\,\hat{\jmath}$ and $\vec{B} = 5.00\,\hat{\imath} - 2.00\,\hat{\jmath}$, a) find the magnitude of each vector; b) write an expression for the vector difference $\vec{A} - \vec{B}$ using unit vectors; c) find the magnitude and direction of the vector difference $\vec{A} - \vec{B}$. d) In a vector diagram show $\vec{A}$, $\vec{B}$, and $\vec{A} - \vec{B}$, and also show that your diagram agrees qualitatively with your answer in part (c).

SECTION 1–11 PRODUCTS OF VECTORS

1–44 For the vectors $\vec{A}$, $\vec{B}$, and $\vec{C}$ in Fig. 1–26, find the scalar products a) $\vec{A} \cdot \vec{B}$; b) $\vec{B} \cdot \vec{C}$; c) $\vec{A} \cdot \vec{C}$.

1–45 a) Find the scalar product of the two vectors $\vec{A}$ and $\vec{B}$ given in Exercise 1–43. b) Find the angle between these two vectors.

1–46 Find the angle between each of the following pairs of vectors:

a) $\vec{A} = -2.00\hat{\imath} + 6.00\hat{\jmath}$ and $\vec{B} = 2.00\hat{\imath} - 3.00\hat{\jmath}$;
b) $\vec{A} = 3.00\hat{\imath} + 5.00\hat{\jmath}$ and $\vec{B} = 10.0\hat{\imath} + 6.00\hat{\jmath}$;
c) $\vec{A} = -4.00\hat{\imath} + 2.00\hat{\jmath}$ and $\vec{B} = 7.00\hat{\imath} + 14.0\hat{\jmath}$.

1–47 Assuming a right-handed coordinate system, find the direction of the +z-axis in a) Fig. 1–13a; b) Fig. 1–13b.

1–48 For the two vectors in Fig. 1–25, a) find the magnitude and direction of the vector product $\vec{A} \times \vec{B}$; b) find the magnitude and direction of $\vec{B} \times \vec{A}$.

1–49 Find the vector product $\vec{A} \times \vec{B}$ (expressed in unit vectors) of the two vectors given in Exercise 1–43. What is the magnitude of the vector product?

1–50 For the two vectors in Fig. 1–27, a) find the magnitude and direction of the vector product $\vec{A} \times \vec{B}$; b) find the magnitude and direction of $\vec{B} \times \vec{A}$.

PROBLEMS

1–51 An acre, a unit of land measurement still in wide use, has a length of one furlong ($\frac{1}{8}$ mi) and a width one-tenth of its length. a) How many acres are in a square mile? b) How many square feet are in an acre? See Appendix E. c) An acre-foot is the volume of water that would cover one acre of flat land to a depth of one foot. How many gallons are in an acre-foot?

1–52 An estate on the California coast was offered for sale for $4,950,000. The total area of the estate was 102 acres (see

Problem 1–51). a) Considering the price of the estate to be pro-portional to its area, what was the cost of one square meter of the estate? b) What would be the price of a portion of the estate the size of a postage stamp ($\frac{7}{8}$ in. by 1.0 in.)?

1–53 The Hydrogen Maser. The radio waves generated by a hydrogen maser can be used as a standard of frequency. The fre-quency of these waves is 1,420,405,751.786 hertz. (A hertz is another name for one cycle per second.) A clock controlled by a hydrogen maser is off by only 1 s in 100,000 years. For the fol-lowing questions, use only three significant figures. (The large number of significant figures given for the frequency simply illustrates the remarkable accuracy to which it has been mea-sured.) a) What is the time for one cycle of the radio wave? b) How many cycles occur in 1 h? c) How many cycles would have occurred during the age of the earth, which is estimated to be 4.6×10^9 years? d) By how many seconds would a hydrogen maser clock be off after a time interval equal to the age of the earth?

1–54 Estimate the number of atoms in your body. (*Hint:* Based on what you know about biology and chemistry, what are the most common types of atom in your body? What is the mass of each type of atom? Appendix D gives the atomic masses for dif-ferent elements, measured in atomic mass units; you can find the value of an atomic mass unit, or 1 u, in Appendix F.)

1–55 a) Estimate the number of dentists in your city. You will need to consider the number of people in your city, how often they need to go to the dentist, how often they actually go, how many hours a typical dental procedure (filling, root canal, and so on) takes, and how many hours a dentist works in a week. b) Using your local telephone directory, check to see whether your estimate was roughly correct.

1–56 Physicists, mathematicians, and others often deal with large numbers. The number 10^{100} has been given the whimsical name *googol* by mathematicians. Let us compare some large numbers in physics with the googol. (*Note:* This problem requires numerical values that can be found in the appendices of the book, with which you should become familiar.) a) Approximately how many atoms make up our planet? For simplicity, take the average atomic mass of the atoms to be 14 g/mol. Avogadro's number gives the number of atoms in a mole. b) Approximately how many neutrons are in a neutron star? Neutron stars are composed almost entirely of neutrons and have approximately twice the mass of the sun. c) In the leading theory of the origin of the universe, the entire universe that we can now observe occupied at a very early time a sphere whose radius was approximately equal to the present distance of the earth to the sun. At that time the universe had a density (mass divided by volume) of 10^{15} g/cm^3. Assuming that $\frac{1}{3}$ of the particles were protons, $\frac{1}{3}$ of the particles were neutrons, and the remaining $\frac{1}{3}$ were electrons, how many particles then made up the universe?

1–57 You are to program a robotic arm on an assembly line to move in the *xy*-plane. Its first displacement is $\vec{A}$; its second dis-placement is $\vec{B}$, of magnitude 6.40 cm and direction 63.0° measured in the sense from the $+x$-axis toward the $-y$-axis. The resultant $\vec{C} = \vec{A} + \vec{B}$ of the two displacements should also have a magnitude of 6.40 cm, but a direction 22.0° measured in the

sense from the $+x$-axis toward the $+y$-axis. a) Draw the vector addition diagram for these vectors, roughly to scale. b) Find the components of $\vec{A}$. c) Find the magnitude and direction of $\vec{A}$.

1–58 a) Find the magnitude and direction of the vector $\vec{R}$ that is the sum of the three vectors $\vec{A}$, $\vec{B}$, and $\vec{C}$ in Fig. 1–26. In a diagram, show how $\vec{R}$ is formed from the three vectors in Fig. 1–26. b) Find the magnitude and direction of the vector $\vec{S} = \vec{C} - \vec{A} - \vec{B}$. In a diagram, show how $\vec{S}$ is formed from the three vectors in Fig. 1–26.

1–59 As noted in Exercise 1–31, a spelunker is surveying a cave. She follows a passage 180 m straight west, then 210 m in a direction 45° east of south, then 280 m at 30° east of north. After a fourth unmeasured displacement she finds herself back where she started. Use the method of components to determine the magnitude and direction of the fourth displacement. Draw the vector addition diagram and show that it is in qualitative agree-ment with your numerical solution.

1–60 A sailor in a small sailboat encounters shifting winds. She sails 2.00 km east, then 3.50 km southeast, then an additional distance in an unknown direction. Her final position is 5.80 km directly east of the starting point (Fig. 1–29). Find the magnitude and direction of the third leg of the journey. Draw the vector addition diagram and show that it is in qualitative agreement with your numerical solution.

FIGURE 1–29 Problem 1–60.

1–61 A cross-country skier skis 2.80 km in the direction 45.0° west of south, then 7.40 km in the direction 30.0° north of east, and finally 3.30 km in the direction 22.0° south of west. a) Show these displacements on a diagram. b) How far is the skier from the starting point?

1–62 On a training flight, a student pilot flies from Lincoln, Nebraska, to Clarinda, Iowa; then to St. Joseph, Missouri; then to Manhattan, Kansas (Fig. 1–30). The directions are shown rel-ative to north: 0° is north, 90° is east, 180° is south, and 270° is west. Use the method of components to find a) the distance she has to fly from Manhattan to get back to Lincoln; b) the direc-tion (relative to north) she must fly to get there. Illustrate your solution with a vector diagram.

1–63 A graphic artist is creating a new logo for her company's web site. In the graphics program she is using, each pixel in an image file has coordinates (*x*, *y*), where the origin (0, 0) is at the upper left corner of the image, the $+x$-axis points to the right, and the $+y$-axis points down. Distances are measured in pixels.

FIGURE 1–30 Problem 1–62.

a) The artist draws a line from the pixel location (10, 20) to the location (210, 200). She wishes to draw a second line that starts at (10, 20), is 250 pixels long, and is at angle of 30° measured clockwise from the first line. At which pixel location should this second line end? Give your answer to the nearest pixel. b) The artist now draws an arrow that connects the lower right end of the first line to the lower right end of the second line. Find the length and direction of this arrow. Draw a diagram showing all three lines.

1–64 An explorer in the dense jungles of equatorial Africa leaves his hut. He takes 40 steps northeast, then 80 steps 60° north of west, then 50 steps due south. Assume his steps all have equal length. a) Sketch, roughly to scale, the three vectors and their resultant. b) Save him from becoming hopelessly lost in the jungle by giving him the displacement, calculated using the method of components, that will return him to his hut.

1–65 Vectors $\vec{A}$ and $\vec{B}$ are drawn from a common point. Vector $\vec{A}$ has magnitude A and angle θ_A measured in the sense from the $+x$-axis to the $+y$-axis. The corresponding quantities for vector $\vec{B}$ are B and θ_B. Then $\vec{A} = A \cos \theta_A \hat{\imath} + A \sin \theta_A \hat{\jmath}$, $\vec{B} = B \cos \theta_B \hat{\imath} + B \sin \theta_B \hat{\jmath}$, and $\phi = |\theta_B - \theta_A|$ is the angle between $\vec{A}$ and $\vec{B}$. a) Derive Eq. (1–18) from Eq. (1–21). b) Derive Eq. (1–22) from Eq. (1–27).

1–66 For the two vectors $\vec{A}$ and $\vec{B}$ in Fig. 1–28, a) find the scalar product $\vec{A} \cdot \vec{B}$; b) find the magnitude and direction of the vector product $\vec{A} \times \vec{B}$.

1–67 Figure 1–6c shows a parallelogram based on the two vectors $\vec{A}$ and $\vec{B}$. a) Show that the magnitude of the cross product of these two vectors is equal to the area of the parallelogram. (*Hint:* area = base × height.) b) What is the angle between the cross product and the plane of the parallelogram?

1–68 The vector $\vec{A}$ is 3.50 cm long and is directed into this page. Vector $\vec{B}$ points from the lower right corner of this page to the upper left corner of this page. Define an appropriate right-handed coordinate system and find the three components of the vector product $\vec{A} \times \vec{B}$, measured in cm². In a diagram show your coordinate system and the vectors $\vec{A}$, $\vec{B}$, and $\vec{A} \times \vec{B}$.

1–69 Given two vectors $\vec{A} = -2.00\hat{\imath} + 3.00\hat{\jmath} + 4.00\hat{k}$ and $\vec{B} = 3.00\hat{\imath} + 1.00\hat{\jmath} - 3.00\hat{k}$, do the following. a) Find the magnitude of each vector. b) Write an expression for the vector difference $\vec{A} - \vec{B}$, using unit vectors. c) Find the magnitude of

the vector difference $\vec{A} - \vec{B}$. Is this the same as the magnitude of $\vec{B} - \vec{A}$? Explain.

1–70 Bond Angle in Methane. In the methane molecule, CH_4, each hydrogen atom is at a corner of a regular tetrahedron with the carbon atom at the center. In coordinates where one of the C—H bonds is in the direction of $\hat{\imath} + \hat{\jmath} + \hat{k}$, an adjacent C—H bond is in the $\hat{\imath} - \hat{\jmath} - \hat{k}$ direction. Calculate the angle between these two bonds.

1–71 The two vectors $\vec{A}$ and $\vec{B}$ are drawn from a common point, and $\vec{C} = \vec{A} + \vec{B}$. a) Show that if $C^2 = A^2 + B^2$, the angle between the vectors $\vec{A}$ and $\vec{B}$ is 90°. b) Show that if $C^2 < A^2 + B^2$, the angle between the vectors $\vec{A}$ and $\vec{B}$ is greater than 90°. c) Show that if $C^2 > A^2 + B^2$, the angle between the vectors $\vec{A}$ and $\vec{B}$ is between 0° and 90°.

1–72 When two vectors $\vec{A}$ and $\vec{B}$ are drawn from a common point, the angle between them is ϕ. a) Using vector techniques, show that the magnitude of their vector sum is given by

$$\sqrt{A^2 + B^2 + 2AB \cos \phi}.$$

b) If $\vec{A}$ and $\vec{B}$ have the same magnitude, for what value of ϕ will their vector sum have the same magnitude as $\vec{A}$ or $\vec{B}$? c) Derive a result analogous to that in part (a) for the magnitude of the vector difference $\vec{A} - \vec{B}$. d) If $\vec{A}$ and $\vec{B}$ have the same magnitude, for what value of ϕ will $\vec{A} - \vec{B}$ have this same magnitude?

1–73 A cube is placed so that one corner is at the origin and three edges are along the x-, y-, and z-axes of a coordinate system (Fig. 1–31). Use vectors to compute a) the angle between the edge along the z-axis (line ab) and the diagonal from the origin to the opposite corner (line ad); b) the angle between line ac (the diagonal of a face) and line ad.

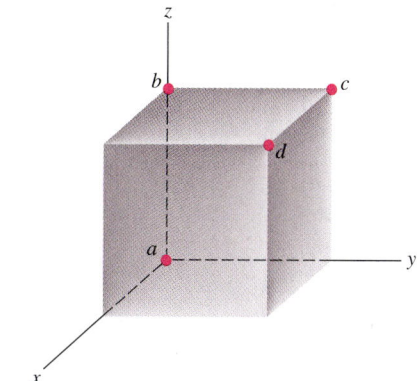

FIGURE 1–31 Problem 1–73.

1–74 Obtain a *unit vector* perpendicular to the two vectors given in Problem 1–69.

1–75 Later in our study of physics we will encounter quantities represented by $(\vec{A} \times \vec{B}) \cdot \vec{C}$. a) Prove that for any three vectors $\vec{A}$, $\vec{B}$, and $\vec{C}$, $\vec{A} \cdot (\vec{B} \times \vec{C}) = (\vec{A} \times \vec{B}) \cdot \vec{C}$. b) Calculate $(\vec{A} \times \vec{B}) \cdot \vec{C}$ for the three vectors $\vec{A}$ with magnitude $A = 5.00$ and angle $\theta_A = 26.0°$ measured in the sense from the $+x$-axis toward the $+y$-axis, $\vec{B}$ with $B = 4.00$ and $\theta_B = 63.0°$, and $\vec{C}$ with magnitude 6.00 and in the $+z$-direction. Vectors $\vec{A}$ and $\vec{B}$ are in the xy-plane.

CHALLENGE PROBLEMS

1–76 The length of a rectangle is given as $L \pm l$, and its width as $W \pm w$. a) Show that the uncertainty in its area A is $a = Lw + lW$. Assume that the uncertainties l and w are small, so that the product lw is very small and can be neglected. b) Show that the fractional uncertainty in the area is equal to the sum of the fractional uncertainty in length and the fractional uncertainty in width. c) A rectangular solid has dimensions $L \pm l$, $W \pm w$, and $H \pm h$. Find the fractional uncertainty in the volume, and show that it equals the sum of the fractional uncertainties in the length, width, and height.

1–77 At Enormous State University (ESU), the football team records its plays using vector displacements, with the origin taken to be the position of the ball before the play starts. In a certain pass play, the receiver starts at $+1.0\hat{\imath} - 5.0\hat{\jmath}$, where the units are yards, $\hat{\imath}$ is to the right, and $\hat{\jmath}$ is downfield. Subsequent displacements of the receiver are $+9.0\hat{\imath}$ (in motion before the snap), $+11.0\hat{\jmath}$ (breaks downfield), $-6.0\hat{\imath} + 4.0\hat{\jmath}$ (zigs), and $+12.0\hat{\imath} + 18.0\hat{\jmath}$ (zags). Meanwhile, the quarterback has dropped straight back to a position $-7.0\hat{\jmath}$. How far and in what direction must the quarterback throw the ball? (Like the coach, you will be well advised to diagram the situation before solving it numerically.)

1–78 Navigating in the Solar System. The *Mars Polar Lander* spacecraft was launched on January 3, 1999. On December 3, 1999, the day that *Mars Polar Lander* touched down on the Martian surface, the positions of the earth and Mars were given by these coordinates:

	x	y	z
Earth	0.3182 AU	0.9329 AU	0.0000 AU
Mars	1.3087 AU	−0.4423 AU	−0.0414 AU

In these coordinates, the sun is at the origin and the plane of the earth's orbit is the xy-plane. The earth passes through the $+x$-axis once a year on the autumnal equinox, the first day of autumn in the northern hemisphere (on or around September 22). One AU, or *astronomical unit,* is equal to 1.496×10^8 km, the average distance from the earth to the sun. a) In a diagram, show the positions of the sun, the earth, and Mars on December 3, 1999. b) Find the following distances in AU on December 3,

1999: i) from the sun to the earth; ii) from the sun to Mars; iii) from the earth to Mars. c) As seen from the earth, what was the angle between the direction to the sun and the direction to Mars on December 3, 1999? d) Explain whether Mars was visible from your location at midnight on December 3, 1999. (When it is midnight at your location, the sun is on the opposite side of the earth from you.)

1–79 Navigating in the Big Dipper. The stars of the Big Dipper (part of the constellation Ursa Major) may appear to all be at the same distance from the earth, but in fact they are very far from each other. Figure 1–32 shows the distances from the earth to each of these stars. The distances are given in light years (ly), the distance that light travels in one year. One light year equals 9.461×10^{15} m. a) Alkaid and Merak are $25.6°$ apart in the earth's sky. In a diagram, show the relative positions of Alkaid, Merak, and our sun. Find the distance in light years from Alkaid to Merak. b) To an inhabitant of a planet orbiting Merak, how many degrees apart in the sky would Alkaid and our sun be?

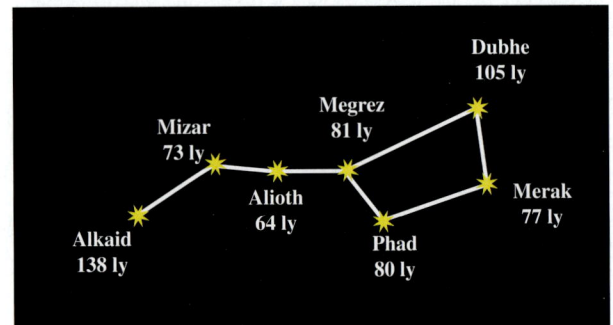

FIGURE 1–32 Challenge Problem 1–79.

1–80 The vector $\vec{r} = x\hat{\imath} + y\hat{\jmath} + z\hat{k}$, called the *position vector,* points from the origin (0, 0, 0) to an arbitrary point in space with coordinates (x, y, z). Use what you know about vectors to prove the following: All points (x, y, z) that satisfy the equation $Ax + By + Cz = 0$, where A, B, and C are constants, lie in a plane that passes through the origin and that is perpendicular to the vector $A\hat{\imath} + B\hat{\jmath} + C\hat{k}$. Sketch this vector and the plane.

Motion Along a Straight Line

Every successful space shuttle mission ends with a brief period of straight-line motion as it comes to a stop on the runway. No larger than a typical airliner, the space shuttle orbiter lands at more than 350 km/h (220 mi/h). Even with a drag parachute for braking, the fast-moving orbiter requires up to 3 km (about 2 mi) to stop.

2-1 INTRODUCTION

How do you describe the motion of a jet fighter being catapulted down the deck of an aircraft carrier? When you throw a baseball straight up in the air, how high does it go? When a glass slips from your hand, how much time do you have to catch it before it hits the floor? These are the kinds of questions you will learn to answer in this chapter. We are beginning our study of physics with *mechanics,* the study of the relationships among force, matter, and motion. The goal of this chapter and the next is to develop general methods for describing motion. The part of mechanics that deals with the description of motion is called *kinematics.* Later we will study *dynamics,* the relation of motion to its causes.

In this chapter we will study the simplest kind of motion: a single particle moving along a straight line. We will often use a particle as a model for a moving body when effects such as rotation or change of shape are not important. To describe the motion of a particle, we will introduce the physical quantities *velocity* and *acceleration.* These quantities have simple definitions in physics; however, those definitions are more precise and slightly different than the ones used in everyday language. Paying careful attention to how velocity and acceleration are defined will help you work with these and other important physical quantities.

An important part of how a physicist defines velocity and acceleration is that these quantities are *vectors.* As you learned in Chapter 1, this means that they have both magnitude and direction. Our concern in this chapter is with motion along a straight line only, so we won't need the full mathematics of vectors just yet. But in Chapter 3 we'll extend our discussion to include motion in two or three dimensions, and using vectors will be essential.

An important special case of straight-line motion is when the acceleration is constant, a situation that we will encounter many times in our study of physics. An example is the motion of a freely falling body. We'll develop simple equations to describe motion with constant acceleration. We'll also consider situations in which the acceleration varies during the motion; in this case, it's necessary to use integration to describe the motion. (If you haven't studied integration yet, this section is optional.)

2-2 DISPLACEMENT, TIME, AND AVERAGE VELOCITY

Suppose a drag racer drives her AA-fuel dragster along a straight track. To study this motion, we need a coordinate system to describe the dragster's position. We choose the *x*-axis of our coordinate system to lie along the dragster's straight-line path, with the origin O at the starting line (Fig. 2–1). We will describe the dragster's position in terms of the position of a representative point, such as its front end. By doing this, we represent the entire dragster by that point and so treat the dragster as a **particle.**

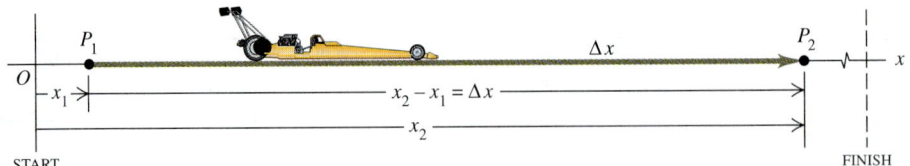

START   FINISH

2–1 Positions of an AA-fuel dragster at two times during its run.

The position of the front of the dragster—that is, the position of the particle—is given by the coordinate x, which varies with time as the dragster moves. A useful way to describe the dragster's motion is in terms of the change in x over a time interval. Suppose that 1.0 s after the start the front of the dragster is at point P_1, 19 m from the origin, and 4.0 s after the start it is at point P_2, 277 m from the origin. Then it has traveled a distance of (277 m − 19 m) = 258 m in a time interval of (4.0 s − 1.0 s) = 3.0 s. We define the dragster's **average velocity** during this time interval as a *vector* quantity whose x-component is the change in x divided by the time interval: (258 m)/(3.0 s) = 86 m/s. In general, the average velocity depends on the particular time interval chosen. For a 3.0-s time interval *before* the start of the race, the average velocity would be zero, because the dragster would be at rest at the starting line and would have zero displacement.

Let's generalize the concept of average velocity. At time t_1 the dragster is at point P_1, with coordinate x_1, and at time t_2 it is at point P_2, with coordinate x_2. The displacement of the dragster during the time interval from t_1 to t_2 is the vector from P_1 to P_2, with x-component $(x_2 - x_1)$ and with y- and z-components equal to zero. The x-component of the dragster's displacement is just the change in the coordinate x, which we write in a more compact way as

$$\Delta x = x_2 - x_1. \qquad (2\text{–}1)$$

The Greek capital letter Δ *("delta") represents a* change *in a quantity, calculated by subtracting the initial value from the final value.* Be sure you understand that Δx is *not* the product of Δ and x: it is a single symbol that means "the change in the quantity x." We likewise write the time interval from t_1 to t_2 as $\Delta t = t_2 - t_1$. Note that Δx or Δt always means the final value minus the initial value, never the reverse.

We can now define the x-component of average velocity more precisely: it is the x-component of displacement, Δx, divided by the time interval Δt during which the displacement occurs. We represent this quantity by the letter v with a subscript "av" to signify average value:

$$v_{av} = \frac{x_2 - x_1}{t_2 - t_1} = \frac{\Delta x}{\Delta t} \qquad \text{(average velocity, straight-line motion).} \qquad (2\text{–}2)$$

For the above example we had $x_1 = 19$ m, $x_2 = 277$ m, $t_1 = 1.0$ s, and $t_2 = 4.0$ s, so Eq. (2–2) gives

$$v_{av} = \frac{277\ \text{m} - 19\ \text{m}}{4.0\ \text{s} - 1.0\ \text{s}} = \frac{258\ \text{m}}{3.0\ \text{s}} = 86\ \text{m/s}.$$

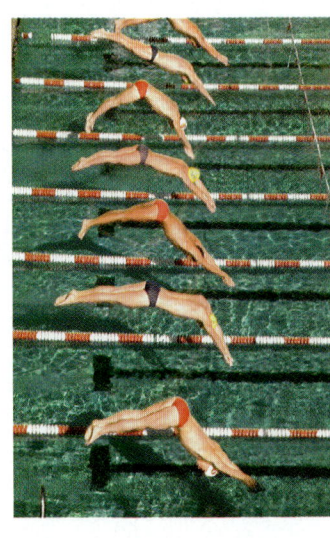

The winner of a 50-m swimming race is the swimmer whose average velocity has the greatest magnitude — that is, the swimmer who traverses a displacement Δx of 50 m in the shortest elapsed time Δt.

The average velocity of the dragster is positive. This means that during the time interval, the coordinate x increased and the dragster moved in the positive x-direction (to the right in Fig. 2–1). If a particle moves in the *negative* x-direction during a time interval, its average velocity for that time interval is negative. For example, suppose an official's truck moves to the left along the track (Fig. 2–2). The truck is at $x_1 = 277$ m at $t_1 = 16.0$ s and is at $x_2 = 19$ m at $t_2 = 25.0$ s. Then $\Delta x = (19\ \text{m} - 277\ \text{m}) = -258$ m and $\Delta t = (25.0\ \text{s} - 16.0\ \text{s}) = 9.0$ s, and the x-component of average velocity is $v_{av} = \Delta x/\Delta t = (-258\ \text{m})/(9.0\ \text{s}) = -29$ m/s. Whenever x is positive and increasing or is negative and becoming less negative, the particle is moving in the $+x$-direction and v_{av} is positive (Fig. 2–1). Whenever x is positive and decreasing or is negative and becoming more negative, the particle is moving in the $-x$-direction and v_{av} is negative (Fig. 2–2).

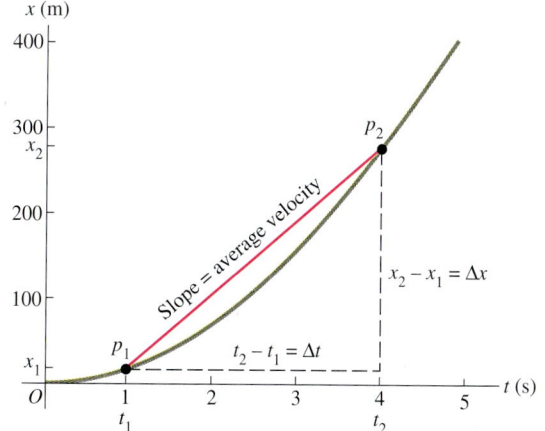

2–2 Positions of an official's truck at two times during its motion. The points P_1 and P_2 now refer to the motion of the truck, and so are different than in Fig. 2–1. The x-component of the truck's displacement is negative, so v_{av} is negative.

CAUTION ▶ You might be tempted to conclude that positive average velocity necessarily implies motion to the right, as in Fig. 2–1, and that negative average velocity necessarily implies motion to the left, as in Fig. 2–2. But these conclusions are correct *only* if the positive x-direction is to the right, as we have chosen it to be in Figs. 2–1 and 2–2. We could just as well have chosen the positive x-direction to be to the left, with the origin at the finish line. Then the dragster would have negative average velocity, and the official's truck would have positive average velocity. In most problems the direction of the coordinate axis will be yours to choose. Once you've made your choice, you *must* take it into account when interpreting the signs of v_{av} and other quantities that describe motion! ◀

With straight-line motion we will usually call Δx simply the displacement and v_{av} simply the average velocity. But be sure to remember that these are really the x-components of vector quantities that, in this special case, have *only* x-components. In Chapter 3, displacement, velocity, and acceleration vectors will have two or three nonzero components.

Figure 2–3 is a graph of the dragster's position as a function of time, that is, an **x-t graph.** The curve in the figure *does not* represent the dragster's path in space; as Fig. 2–1 shows, the path is a straight line. Rather, the graph is a pictorial way to represent how the dragster's position changes with time. The points on the graph labeled p_1 and p_2 correspond to the points P_1 and P_2 along the dragster's path. Line p_1p_2 is the hypotenuse of a right triangle with vertical side $\Delta x = x_2 - x_1$ and horizontal side $\Delta t = t_2 - t_1$. The average velocity $v_{av} = \Delta x / \Delta t$ of the dragster is then equal to the *slope* of the line p_1p_2, that is, the ratio of the vertical side Δx of the triangle to the horizontal side Δt.

Average velocity depends only on the total displacement $\Delta x = x_2 - x_1$ that occurs during the time interval $\Delta t = t_2 - t_1$, not on the details of what happens during the time interval. A second dragster might have passed through point P_1 in Fig. 2–1 at the same time t_1 as the first dragster, raced past the first dragster, then blown its engine and slowed down to pass through point P_2 at the same time t_2 as the first dragster. Both dragsters have the same displacement during the same time interval and so have the same average velocity.

2–3 The position of an AA-fuel dragster as a function of time. The average velocity v_{av} between points P_1 and P_2 in Fig. 2–1 is the slope of the line p_1p_2. This line slopes upward to the right, so the slope is positive and v_{av} is positive.

TABLE 2–1
TYPICAL VELOCITY MAGNITUDES

A snail's pace	10^{-3} m/s	Random motion of air molecules	500 m/s
A brisk walk	2 m/s	Fastest airplane	1000 m/s
Fastest human	11 m/s	Orbiting communications satellite	3000 m/s
Running cheetah	35 m/s	Electron orbiting in a hydrogen atom	2×10^6 m/s
Fastest car	341 m/s	Light traveling in a vacuum	3×10^8 m/s

If we express distance in meters and time in seconds, average velocity is measured in meters per second (m/s). Other common units of velocity are kilometers per hour (km/h), feet per second (ft/s), miles per hour (mi/h), and knots (1 knot = 1 nautical mile/h = 6080 ft/h). Table 2–1 shows a few typical velocity magnitudes.

2–3 INSTANTANEOUS VELOCITY

The average velocity of a particle during a time interval cannot tell us how fast, or in what direction, the particle was moving at any given time during the interval. To describe the motion in greater detail, we need to define the velocity at any specific instant of time or specific point along the path. Such a velocity is called **instantaneous velocity,** and it needs to be defined carefully.

Note that the word *instant* has a somewhat different definition in physics than in everyday language. You might use the phrase "It lasted just an instant" to refer to something that lasted for a very short time interval. But in physics an instant has *no* duration at all; it refers to a single value of time.

To find the instantaneous velocity of the dragster in Fig. 2–1 at the point P_1, we imagine moving the second point P_2 closer and closer to the first point P_1. We compute the average velocity $v_{av} = \Delta x / \Delta t$ over these shorter and shorter displacements and time intervals. Both Δx and Δt become very small, but their ratio does not necessarily become small. In the language of calculus the limit of $\Delta x / \Delta t$ as Δt approaches zero is called the **derivative** of x with respect to t and is written dx/dt. *The instantaneous velocity is the limit of the average velocity as the time interval approaches zero; it equals the instantaneous rate of change of position with time.* We use the symbol v, with no subscript, for instantaneous velocity:

$$v = \lim_{\Delta t \to 0} \frac{\Delta x}{\Delta t} = \frac{dx}{dt} \qquad \text{(instantaneous velocity, straight-line motion).} \qquad (2\text{–}3)$$

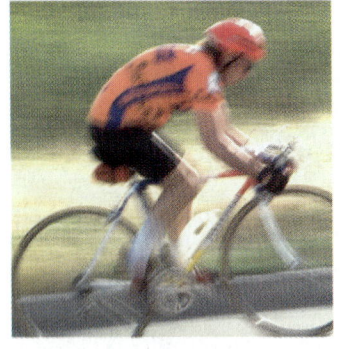

Even when moving forward, this cyclist's instantaneous velocity can be negative — if he's traveling in the negative *x*-direction. In any problem, the choice of which direction is positive and which is negative is entirely up to you.

We always assume that the time interval Δt is positive so that v has the same algebraic sign as Δx. If the positive *x*-axis points to the right, as in Fig. 2–1, a positive value of v means that x is increasing and the motion is toward the right; a negative value of v means that x is decreasing and the motion is toward the left. A body can have positive x and negative v, or the reverse; x tells us where the body is, while v tells us how it's moving.

Instantaneous velocity, like average velocity, is a vector quantity. Equation (2–3) defines its *x*-component, which can be positive or negative. In straight-line motion, all other components of instantaneous velocity are zero, and in this case we will often call v simply the instantaneous velocity. When we use the term "velocity," we will always mean instantaneous rather than average velocity, unless we state otherwise.

The terms "velocity" and "speed" are used interchangeably in everyday language, but they have distinct definitions in physics. We use the term **speed** to denote distance traveled divided by time, on either an average or an instantaneous basis. Instantaneous *speed* measures how fast a particle is moving; instantaneous *velocity* measures how fast *and* in what direction it's moving. For example, a particle with instantaneous velocity $v = 25$ m/s and a second particle with $v = -25$ m/s are moving in opposite directions at the same instantaneous speed of 25 m/s. Instantaneous speed is the magnitude of instantaneous velocity, and so instantaneous speed can never be negative.

Average speed, however, is *not* the magnitude of average velocity. When Alexander Popov set a world record in 1994 by swimming 100.0 m in 48.21 s, his average speed was (100.0 m)/(48.21 s) = 2.074 m/s. But because he swam two lengths in a 50-m pool, he started and ended at the same point, giving him zero total displacement and zero average *velocity* for his effort! Both average speed and instantaneous speed are scalars, not vectors, because these quantities contain no information about direction.

EXAMPLE 2–1

Average and instantaneous velocities A cheetah is crouched in ambush 20 m to the east of an observer's blind (Fig. 2–4). At time $t = 0$ the cheetah charges an antelope in a clearing 50 m east of the observer. The cheetah runs along a straight line. Later analysis of a videotape shows that during the first 2.0 s of the attack, the cheetah's coordinate x varies with time according to the equation $x = 20$ m $+ (5.0$ m/s$^2)t^2$. (Note that the units for the numbers 20 and 5.0 *must* be as shown to make the expression dimensionally consistent.) a) Find the displacement of the cheetah during the interval between $t_1 = 1.0$ s and $t_2 = 2.0$ s. b) Find the average velocity during the same time interval. c) Find the instantaneous velocity at time $t_1 = 1.0$ s by taking $\Delta t = 0.1$ s, then $\Delta t = 0.01$ s, then $\Delta t = 0.001$ s. d) Derive a general expression for the instantaneous velocity as a function of time, and from it find v at $t = 1.0$ s and $t = 2.0$ s.

SOLUTION a) At time $t_1 = 1.0$ s the cheetah's position x_1 is

$$x_1 = 20 \text{ m} + (5.0 \text{ m/s}^2)(1.0 \text{ s})^2 = 25 \text{ m}.$$

At time $t_2 = 2.0$ s its position x_2 is

$$x_2 = 20 \text{ m} + (5.0 \text{ m/s}^2)(2.0 \text{ s})^2 = 40 \text{ m}.$$

The displacement during this interval is

$$\Delta x = x_2 - x_1 = 40 \text{ m} - 25 \text{ m} = 15 \text{ m}.$$

b) The average velocity during this time interval is

$$v_{av} = \frac{x_2 - x_1}{t_2 - t_1} = \frac{40 \text{ m} - 25 \text{ m}}{2.0 \text{ s} - 1.0 \text{ s}} = \frac{15 \text{ m}}{1.0 \text{ s}} = 15 \text{ m/s}.$$

c) With $\Delta t = 0.1$ s, the time interval is from $t_1 = 1.0$ s to $t_2 = 1.1$ s. At time t_2, the position is

$$x_2 = 20 \text{ m} + (5.0 \text{ m/s}^2)(1.1 \text{ s})^2 = 26.05 \text{ m}.$$

The average velocity during this interval is

$$v_{av} = \frac{26.05 \text{ m} - 25 \text{ m}}{1.1 \text{ s} - 1.0 \text{ s}} = 10.5 \text{ m/s}.$$

We invite you to follow this same pattern to work out the average velocities for the 0.01-s and 0.001-s intervals. The results are 10.05 m/s and 10.005 m/s. As Δt gets smaller, the average velocity gets closer to 10.0 m/s, so we conclude that the instantaneous velocity at time $t = 1.0$ s is 10.0 m/s.

d) We find the instantaneous velocity as a function of time by taking the derivative of the expression for x with respect to t. For any n the derivative of t^n is nt^{n-1}, so the derivative of t^2 is $2t$. Therefore

$$v = \frac{dx}{dt} = (5.0 \text{ m/s}^2)(2t) = (10 \text{ m/s}^2)t.$$

At time $t = 1.0$ s, $v = 10$ m/s as we found in part (c). At time $t = 2.0$ s, $v = 20$ m/s.

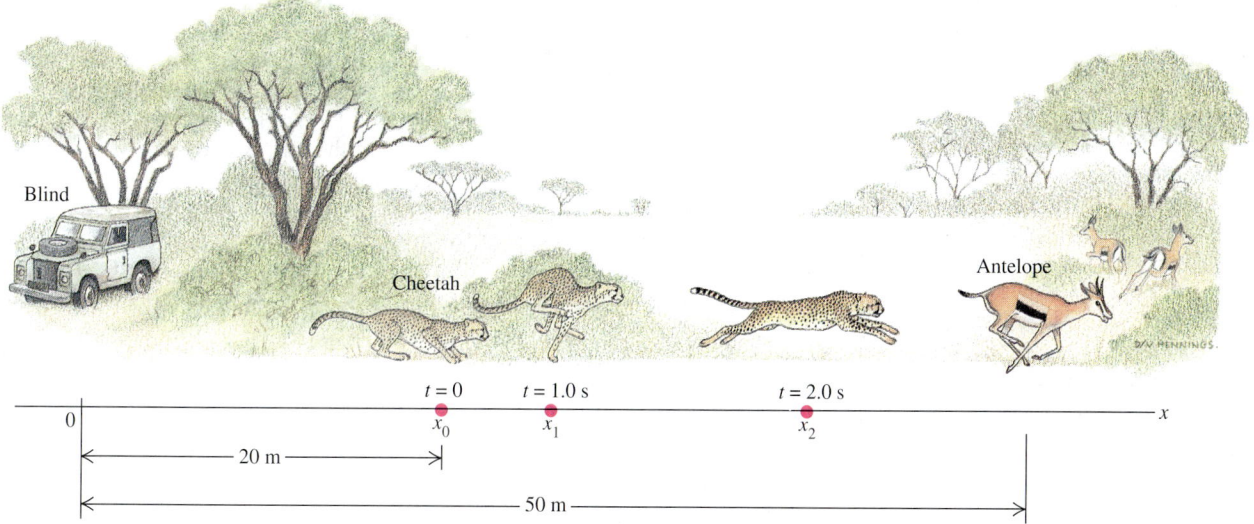

2–4 A cheetah attacking an antelope from ambush. The animals are not drawn to the same scale as the axis.

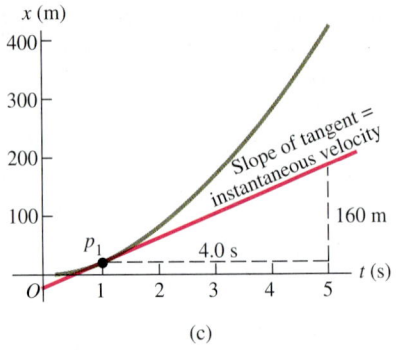

(a)   (b)   (c)

2–5 (a) and (b) As the average velocity is calculated over shorter and shorter time intervals, its value approaches the instantaneous velocity. (c) In the limit $\Delta t \to 0$ the slope of the line $p_1 p_2$ approaches the slope of the tangent to the x-t curve at point p_1. The value of this slope equals the instantaneous velocity v at point P_1 in Fig. 2–1. We find the slope of the tangent by dividing any vertical interval (with distance units) along the tangent line by the corresponding horizontal interval (with time units). Here, v at P_1 equals $(160 \text{ m})/(4.0 \text{ s}) = 40 \text{ m/s}$.

1.1

Analyzing Motion Using Diagrams

FINDING VELOCITY ON AN *x-t* GRAPH

The velocity of a particle can also be found from the graph of the particle's position as a function of time. Suppose we want to find the velocity of the dragster in Fig. 2–1 at point P_1. As point P_2 in Fig. 2–1 approaches point P_1, point p_2 in the x-t graph of Fig. 2–3 approaches point p_1. This is shown in Figs. 2–5a and 2–5b, in which the average velocity is calculated over shorter time intervals Δt. In the limit that $\Delta t \to 0$, shown in Fig. 2–5c, the slope of the line $p_1 p_2$ equals the slope of the line tangent to the curve at point p_1. *On a graph of position as a function of time for straight-line motion, the instantaneous velocity at any point is equal to the slope of the tangent to the curve at that point.*

If the tangent to the x-t curve slopes upward to the right, as in Fig. 2–5c, then its slope is positive, the velocity is positive, and the motion is in the positive x-direction. If the tangent slopes downward to the right, the slope and velocity are negative and the motion is in the negative x-direction. When the tangent is horizontal, the slope is zero and the velocity is zero. Figure 2–6 illustrates these three possibilities.

Note that Fig. 2–6 depicts the motion of a particle in two ways. Figure 2–6a is an x-t graph, and Fig. 2–6b is an example of a **motion diagram.** A motion diagram shows the position of the particle at various times during the motion, like frames from a movie or video of the particle's motion, as well as arrows to represent the particle's velocity at each instant. Both x-t graphs and motion diagrams are helpful aids to understanding motion, and we will use both frequently in this chapter. You will find it worth your while to draw *both* an x-t graph and a motion diagram as part of solving any problem involving motion.

2–6 (a) The x-t graph of the motion of a particular particle. The slope of the tangent at any point equals the velocity at that point. Between points A and B, x is negative but increasing. The slope of the graph is becoming more positive, so the particle is moving with increasing velocity in the positive x-direction. At B the slope and velocity are greatest. Between points B and C the particle is still moving in the positive x-direction, but the slope and velocity are decreasing. At point C the slope and velocity are zero. From C to E, x is positive but decreasing, so the velocity is negative; the most negative value of v occurs at D, where the slope is most negative. At point E the slope and velocity are approaching zero as x approaches a constant value, that is, the particle is coming to a stop. (b) A motion diagram showing the position and velocity of the particle at each of the five times labeled on the x-t graph.

(a)

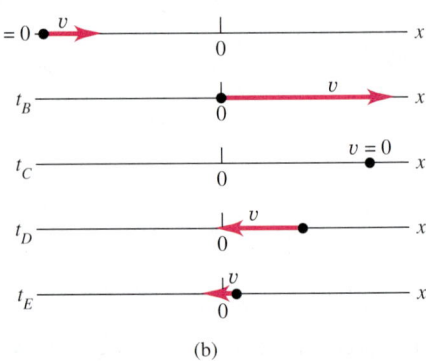

(b)

2-4 AVERAGE AND INSTANTANEOUS ACCELERATION

When the velocity of a moving body changes with time, we say that the body has an *acceleration*. Just as velocity describes the rate of change of position with time, acceleration describes the rate of change of velocity with time. Like velocity, acceleration is a vector quantity. In straight-line motion its only nonzero component is along the axis along which the motion takes place.

AVERAGE ACCELERATION

Let's consider again the motion of a particle along the *x*-axis. Suppose that at time t_1 the particle is at point P_1 and has *x*-component of (instantaneous) velocity v_1, and at a later time t_2 it is at point P_2 and has *x*-component of velocity v_2. So the *x*-component of velocity changes by an amount $\Delta v = v_2 - v_1$ during the time interval $\Delta t = t_2 - t_1$.

We define the **average acceleration** a_{av} of the particle as it moves from P_1 to P_2 to be a vector quantity whose *x*-component is Δv, the change in the *x*-component of velocity, divided by the time interval Δt:

$$a_{av} = \frac{v_2 - v_1}{t_2 - t_1} = \frac{\Delta v}{\Delta t} \quad \text{(average acceleration, straight-line motion).} \quad (2\text{-}4)$$

For straight-line motion we will usually call a_{av} simply the average acceleration, remembering that in fact it is the *x*-component of the average acceleration vector.

If we express velocity in meters per second and time in seconds, then average acceleration is in meters per second per second, or (m/s)/s. This is usually written as m/s^2 and is read "meters per second squared."

CAUTION▶ Be very careful not to confuse acceleration with velocity! Velocity describes a body's speed and direction of motion at any time; acceleration describes how the speed and direction of motion *change* over time. ◀

EXAMPLE 2-2

Average acceleration An astronaut has left an orbiting space shuttle to test a new personal maneuvering unit. As she moves along a straight line, her partner on board the shuttle measures her velocity every 2.0 s, starting at time $t = 1.0$ s, as follows:

t	v	t	v
1.0 s	0.8 m/s	9.0 s	−0.4 m/s
3.0 s	1.2 m/s	11.0 s	−1.0 m/s
5.0 s	1.6 m/s	13.0 s	−1.6 m/s
7.0 s	1.2 m/s	15.0 s	−0.8 m/s

Find the average acceleration, and describe whether the speed of the astronaut increases or decreases, for each of the following time intervals: a) $t_1 = 1.0$ s to $t_2 = 3.0$ s; b) $t_1 = 5.0$ s to $t_2 = 7.0$ s; c) $t_1 = 9.0$ s to $t_2 = 11.0$ s; d) $t_1 = 13.0$ s to $t_2 = 15.0$ s.

SOLUTION The upper part of Fig. 2–7 graphs the velocity as a function of time. On this graph, the slope of the line connecting the points at the beginning and end of each interval equals the average acceleration $a_{av} = \Delta v/\Delta t$ for that interval. The values of a_{av} are graphed in the lower part of Fig. 2–7. For each time

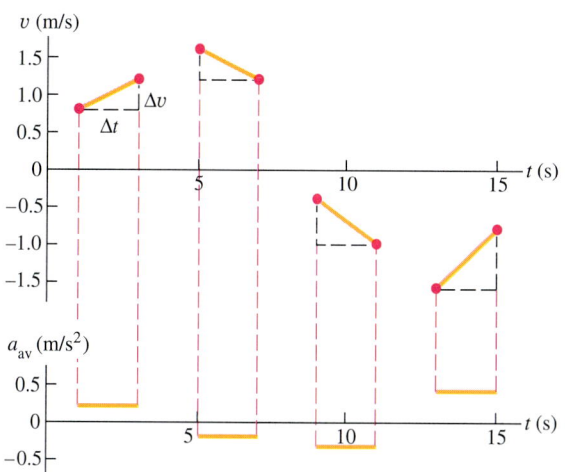

2–7 The slope of the line connecting two points on a graph of velocity versus time (upper graph) equals the average acceleration between those two points (lower graph).

interval, we have

a) $a_{av} = (1.2 \text{ m/s} - 0.8 \text{ m/s})/(7.0 \text{ s} - 5.0 \text{ s}) = 0.2 \text{ m/s}^2$. The speed (magnitude of instantaneous velocity) increases from 0.8 m/s to 1.2 m/s.

b) $a_{av} = (1.2 \text{ m/s} - 1.6 \text{ m/s})/(7.0 \text{ s} - 5.0 \text{ s}) = -0.2 \text{ m/s}^2$. The speed decreases from 1.6 m/s to 1.2 m/s.

c) $a_{av} = [-1.0 \text{ m/s} - (-0.4 \text{ m/s})]/[11.0 \text{ s} - 9.0 \text{ s})] = -0.3 \text{ m/s}^2$. The speed increases from 0.4 m/s to 1.0 m/s.

d) $a_{av} = [-0.8 \text{ m/s} - (-1.6 \text{ m/s})]/[15.0 \text{ s} - 13.0 \text{ s}] = 0.4 \text{ m/s}^2$. The speed decreases from 1.6 m/s to 0.8 m/s.

When the acceleration has the *same* direction (same algebraic sign) as the initial velocity, as in intervals a and c, the astronaut goes faster; when it has the *opposite* direction (opposite algebraic sign), as in intervals b and d, she slows down. When she moves in the negative direction with increasing speed (interval c), her velocity is algebraically decreasing (becoming more negative), and her acceleration is negative. But when she moves in the negative direction with decreasing speed (interval d), her velocity is algebraically increasing (becoming less negative) and her acceleration is positive.

INSTANTANEOUS ACCELERATION

We can now define **instantaneous acceleration,** following the same procedure that we used to define instantaneous velocity. Consider this situation: A race car driver has just entered the final straightaway at the Grand Prix. He reaches point P_1 at time t_1, moving with velocity v_1. He passes point P_2, closer to the finish line, at time t_2 with velocity v_2 (Fig. 2–8).

To define the instantaneous acceleration at point P_1, we take the second point P_2 in Fig. 2–8 to be closer and closer to the first point P_1 so that the average acceleration is computed over shorter and shorter time intervals. *The instantaneous acceleration is the limit of the average acceleration as the time interval approaches zero.* In the language of calculus, *instantaneous acceleration equals the instantaneous rate of change of velocity with time.* Thus

$$a = \lim_{\Delta t \to 0} \frac{\Delta v}{\Delta t} = \frac{dv}{dt} \quad \text{(instantaneous acceleration, straight-line motion).} \quad (2\text{–}5)$$

Note that Eq. (2–5) is really the definition of the x-component of the acceleration vector; in straight-line motion, all other components of this vector are zero. Instantaneous acceleration plays an essential role in the laws of mechanics. From now on, when we use the term "acceleration," we will always mean instantaneous acceleration, not average acceleration.

2–8 A Grand Prix car at two points on the straightaway.

EXAMPLE 2–3

Average and instantaneous accelerations Suppose the velocity v of the car in Fig. 2–8 at any time t is given by the equation

$$v = 60 \text{ m/s} + (0.50 \text{ m/s}^3)t^2.$$

(Note that the units for the numbers 60 and 0.50 *must* be as shown to make the expression dimensionally consistent.) a) Find the change in velocity of the car in the time interval between $t_1 = 1.0$ s and $t_2 = 3.0$ s. b) Find the average acceleration in this time interval. c) Find the instantaneous acceleration at time $t_1 = 1.0$ s by taking Δt to be first 0.1 s, then 0.01 s, then 0.001 s. d) Derive an expression for the instantaneous acceleration at any time, and use it to find the acceleration at $t = 1.0$ s and $t = 3.0$ s.

SOLUTION a) We first find the velocity at each time by substituting each value of t into the equation. At time $t_1 = 1.0$ s,

$$v_1 = 60 \text{ m/s} + (0.50 \text{ m/s}^3)(1.0 \text{ s})^2 = 60.5 \text{ m/s}.$$

At time $t_2 = 3.0$ s,

$$v_2 = 60 \text{ m/s} + (0.50 \text{ m/s}^3)(3.0 \text{ s})^2 = 64.5 \text{ m/s}.$$

The change in velocity Δv is

$$\Delta v = v_2 - v_1 = 64.5 \text{ m/s} - 60.5 \text{ m/s} = 4.0 \text{ m/s}.$$

The time interval is $\Delta t = 3.0 \text{ s} - 1.0 \text{ s} = 2.0 \text{ s}.$

b) The average acceleration during this time interval is

$$a_{av} = \frac{v_2 - v_1}{t_2 - t_1} = \frac{4.0 \text{ m/s}}{2.0 \text{ s}} = 2.0 \text{ m/s}^2.$$

During the time interval from $t_1 = 1.0$ s to $t_2 = 3.0$ s, the velocity and average acceleration have the same algebraic sign (in this case, positive), and the car speeds up.

c) When $\Delta t = 0.1$ s, $t_2 = 1.1$ s and

$$v_2 = 60 \text{ m/s} + (0.50 \text{ m/s}^3)(1.1 \text{ s})^2 = 60.605 \text{ m/s},$$

$$\Delta v = 0.105 \text{ m/s},$$

$$a_{av} = \frac{\Delta v}{\Delta t} = \frac{0.105 \text{ m/s}}{0.1 \text{ s}} = 1.05 \text{ m/s}^2.$$

We invite you to repeat this pattern for $\Delta t = 0.01$ s and $\Delta t = 0.001$ s; the results are $a_{av} = 1.005$ m/s^2 and $a_{av} = 1.0005$ m/s^2 respectively. As Δt gets smaller, the average acceleration gets closer to 1.0 m/s^2. We conclude that the instantaneous accelera-

tion at $t = 1.0$ s is 1.0 m/s^2.

d) The instantaneous acceleration is $a = dv/dt$, the derivative of a constant is zero, and the derivative of t^2 is $2t$. Using these, we obtain

$$a = \frac{dv}{dt} = \frac{d}{dt}[60 \text{ m/s} + (0.50 \text{ m/s}^3)t^2]$$

$$= (0.50 \text{ m/s}^3)(2t) = (1.0 \text{ m/s}^3)t.$$

When $t = 1.0$ s,

$$a = (1.0 \text{ m/s}^3)(1.0 \text{ s}) = 1.0 \text{ m/s}^2.$$

When $t = 3.0$ s,

$$a = (1.0 \text{ m/s}^3)(3.0 \text{ s}) = 3.0 \text{ m/s}^2.$$

Note that neither of these values is equal to the average acceleration found in part (b). The instantaneous acceleration of this car varies with time. Automotive engineers sometimes call the rate of change of acceleration with time the "jerk."

FINDING ACCELERATION ON A v-t GRAPH OR AN x-t GRAPH

We interpreted average and instantaneous velocity in terms of the slope of a graph of position versus time. In the same way, we can get additional insight into the concepts of average and instantaneous acceleration from a graph with instantaneous velocity v on the vertical axis and time t on the horizontal axis, that is, a **v-t graph** (Fig. 2–9). The points on the graph labeled p_1 and p_2 correspond to points P_1 and P_2 in Fig. 2–8. The average acceleration $a_{av} = \Delta v/\Delta t$ during this interval is the slope of the line p_1p_2. As point P_2 in Fig. 2–8 approaches point P_1, point p_2 in the v-t graph of Fig. 2–9 approaches point p_1, and the slope of the line $p_1 p_2$ approaches the slope of the line tangent to the curve at point p_1. Thus *on a graph of velocity as a function of time, the instantaneous acceleration at any point is equal to the slope of the tangent to the curve at that point.* In Fig. 2–9 the instantaneous acceleration varies with time.

Note that by itself, the algebraic sign of the acceleration doesn't tell you whether a body is speeding up or slowing down. You must compare the signs of the velocity and the acceleration. When v and a have the *same* sign, the body is speeding up. If both are positive, the body is moving in the positive direction with increasing speed. If both are negative, the body is moving in the negative direction with a velocity that is becoming more and more negative, and again the speed is increasing. When v and a have *opposite* signs, the body is slowing down. If v is positive and a is negative, the body is moving in the positive direction with decreasing speed; if v is negative and a is positive, the body is moving in the negative direction with a velocity that is becoming less negative, and again the body is slowing down. Figure 2–10 illustrates these various possibilities.

2–9 A v-t graph of the motion in Fig. 2–8. The average acceleration between t_1 and t_2 equals the slope of the line p_1p_2. The instantaneous acceleration at P_1 equals the slope of the tangent at p_1.

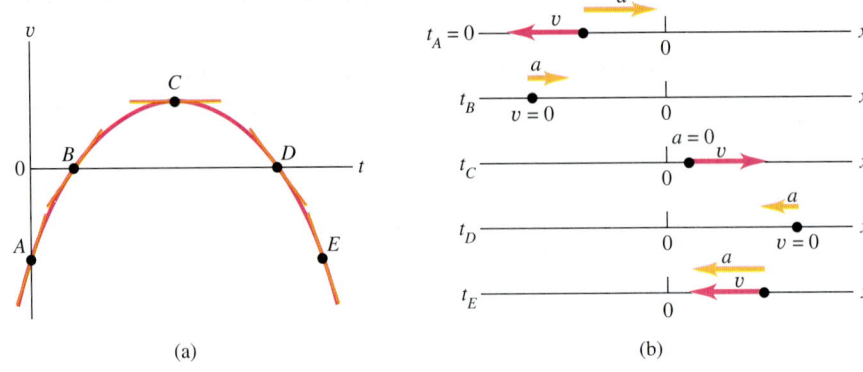

(a)   (b)

2–10 (a) A v-t graph of the motion of a particle. The slope of the tangent at any point equals the acceleration at that point. From A to B, v is negative but increasing, and the slope and acceleration are positive. The particle slows down until time B, when it is instantaneously at rest ($v = 0$); it is still accelerating, because the slope is not zero. From B to C, v is positive and increasing, and the acceleration is positive; the particle is speeding up. At point C the velocity has its maximum value, and the acceleration is instantaneously zero. From C to D the velocity is positive but decreasing, and the acceleration is negative. The body slows down again until $v = 0$ at D; the acceleration is negative at this point. From D to E, v is negative and decreasing, and the acceleration is negative; once again the particle is speeding up. (b) A motion diagram showing the position, velocity, and acceleration of the particle at each of the times labeled on the v-t graph. The positions are consistent with the v-t graph; for instance, from t_A to t_B the velocity is negative, so at t_B the particle is at a more negative value of x than at t_A.

The term *deceleration* is sometimes used for a decrease in speed. Because this may mean positive or negative a, depending on the sign of v, we avoid this term.

We can also learn about the acceleration of a body from a graph of its *position* versus time. Because $a = dv/dt$ and $v = dx/dt$, we can write

$$a = \frac{dv}{dt} = \frac{d}{dt}\left(\frac{dx}{dt}\right) = \frac{d^2x}{dt^2}. \tag{2–6}$$

2–11 (a) The same x-t graph as shown in Fig. 2–6a. The velocity is equal to the *slope* of the graph, and the acceleration is given by the *concavity* or *curvature* of the graph. Between points A and B the graph is curved upward (concave up). The velocity is increasing, and the acceleration is positive. At B the graph is not curved either upward or downward, and the acceleration is zero. At this point the velocity is maximum. From B to D the graph is curved downward (concave down), and the acceleration is negative; the velocity is decreasing from a positive value at B to zero at C and to a negative value at D. From D to E the graph is once again curved upward (concave up). The velocity is negative and approaching zero, so the acceleration is positive. (b) A motion diagram showing the position, velocity, and acceleration of the particle at each of the times labeled on the x-t graph.

That is, a is the second derivative of x with respect to t. The second derivative of any function is directly related to the *concavity* or *curvature* of the graph of that function. At a point where the x-t graph is concave up (curved upward), the acceleration is positive and v is increasing; at a point where the x-t graph is concave down (curved downward), the acceleration is negative and v is decreasing. At a point where the x-t graph has no curvature, such as an inflection point, the acceleration is zero and the velocity is constant. All three of these possibilities are shown in Fig. 2–11. Examining the curvature of an x-t graph is an easy way to decide what the *sign* of acceleration is. This technique is less helpful for determining numerical values of acceleration, because the curvature of a graph is hard to measure accurately.

(a)

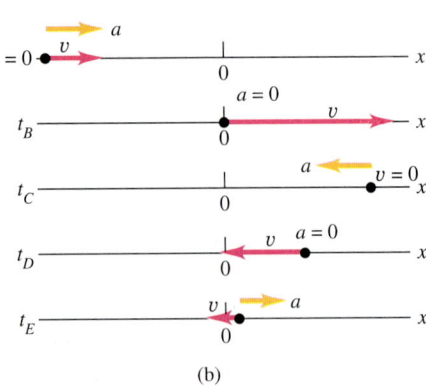

(b)

2–5 MOTION WITH CONSTANT ACCELERATION

The simplest accelerated motion is straight-line motion with *constant* acceleration. In this case the velocity changes at the same rate throughout the motion. This is a very special situation, yet one that occurs often in nature. As we will discuss in the next section, a falling body has a constant acceleration if the effects of the air are not important. The same is true for a body sliding on an incline or along a rough horizontal surface. Straight-line motion with nearly constant acceleration also occurs in technology, such as a jet fighter being catapulted from the deck of an aircraft carrier.

In this section we'll derive key equations for straight-line motion with constant acceleration. These equations will enable us to solve a wide range of problems.

Figure 2–12 is a motion diagram showing the position, velocity, and acceleration at five different times for a particle moving with constant acceleration. Figures 2–13 and 2–14 depict this same motion in the form of graphs. Since the acceleration a is constant, the *a-t* **graph** (graph of acceleration versus time) in Fig. 2–13 is a horizontal line. The graph of velocity versus time has a constant *slope* because the acceleration is constant, and so the *v-t* graph is a straight line (Fig. 2–14).

When the acceleration is constant, it's easy to derive equations for position x and velocity v as functions of time. Let's start with velocity. In Eq. (2–4) we can replace the average acceleration a_{av} by the constant (instantaneous) acceleration a. We then have

$$a = \frac{v_2 - v_1}{t_2 - t_1}. \qquad (2\text{–}7)$$

Now we let $t_1 = 0$ and let t_2 be any arbitrary later time t. We use the symbol v_0 for the velocity at the initial time $t = 0$; the velocity at the later time t is v. Then Eq. (2–7) becomes

$$a = \frac{v - v_0}{t - 0}, \qquad \text{or}$$

$$v = v_0 + at \qquad \text{(constant acceleration only)}. \qquad (2\text{–}8)$$

We can interpret this equation as follows. The acceleration a is the constant rate of change of velocity, that is, the change in velocity per unit time. The term at is the product of the change in velocity per unit time, a, and the time interval t. Therefore it equals the *total* change in velocity from the initial time $t = 0$ to the later time t. The velocity v at any time t then equals the initial velocity v_0 (at $t = 0$) plus the change in velocity at. Graphically, we can consider the height v of the graph in Fig. 2–14 at any time t to be the sum of two segments: one with length v_0 equal to the initial velocity, the other with length at equal to the change in velocity during time t. The graph of velocity as a function of time is a straight line with slope a that intercepts the vertical axis (the v-axis) at v_0.

Another interpretation of Eq. (2–8) is that the change in velocity $v - v_0$ of the particle between $t = 0$ and any later time t equals the *area* under the a-t graph between those two times. In Fig. 2–13, the area under the graph of acceleration versus time is shown as a red rectangle of vertical side a and horizontal side t. The area of this rectangle is equal to at, which from Eq. (2–8) is just equal to the change in velocity $v - v_0$. In Section 2–7 we'll show that even if the acceleration is not constant, the change in velocity during a time interval is still equal to the area under the a-t curve, although in that case Eq. (2–8) does not apply.

Next we want to derive an equation for the position x of a particle moving with constant acceleration. To do this, we make use of two different expressions for the average velocity v_{av} during the interval from $t = 0$ to any later time t. The first expression comes

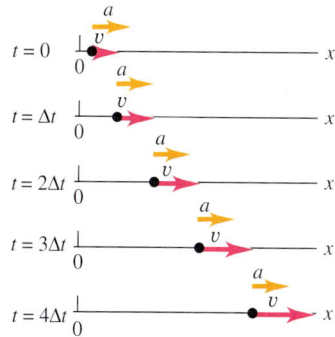

2–12 A motion diagram for a particle moving in a straight line with constant positive acceleration a. The position, velocity, and acceleration are shown at five equally spaced times. The velocity changes by equal amounts in equal time intervals because the acceleration is constant. The position changes by *different* amounts in equal time intervals because the velocity is changing.

2–13 An acceleration-time (a-t) graph for straight-line motion with constant positive acceleration a.

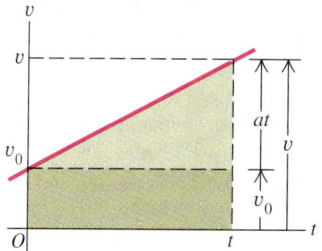

2–14 A velocity-time (v-t) graph for straight-line motion with constant positive acceleration a. The initial velocity v_0 is also positive.

from the definition of v_{av}, Eq. (2–2), which holds true whether or not the acceleration is constant. We call the position at time $t = 0$ the *initial position,* denoted by x_0. The position at the later time t is simply x. Thus for the time interval $\Delta t = t - 0$ and the corresponding displacement $\Delta x = x - x_0$, Eq. (2–2) gives

$$v_{av} = \frac{x - x_0}{t}. \tag{2–9}$$

We can also get a second expression for v_{av} that is valid only when the acceleration is constant, so that the v-t graph is a straight line (as in Fig. 2–14) and the velocity changes at a constant rate. In this case the average velocity during any time interval is simply the arithmetic average of the velocities at the beginning and end of the interval. For the time interval 0 to t,

$$v_{av} = \frac{v_0 + v}{2} \quad \text{(constant acceleration only).} \tag{2–10}$$

(This equation is *not* true if the acceleration varies and the v-t graph is a curve, as in Fig. 2–10.) We also know that with constant acceleration the velocity v at any time t is given by Eq. (2–8). Substituting that expression for v into Eq. (2–10), we find

$$v_{av} = \frac{1}{2}(v_0 + v_0 + at)$$

$$= v_0 + \frac{1}{2}at \quad \text{(constant acceleration only).} \tag{2–11}$$

Finally, we equate Eqs. (2–9) and (2–11) and simplify the result:

$$v_0 + \frac{1}{2}at = \frac{x - x_0}{t}, \quad \text{or}$$

$$x = x_0 + v_0 t + \frac{1}{2}at^2 \quad \text{(constant acceleration only).} \tag{2–12}$$

This equation states that if at the initial time $t = 0$, a particle is at position x_0 and has velocity v_0, its new position x at any later time t is the sum of three terms: its initial position x_0, plus the distance $v_0 t$ that it would move if its velocity were constant, plus an additional distance $\frac{1}{2}at^2$ caused by the change in velocity.

Just as the change in velocity of the particle equals the area under the a-t graph, the displacement—that is, the change in position—equals the area under the v-t graph. To be specific, the displacement $x - x_0$ of the particle between $t = 0$ and any later time t equals the area under the v-t graph between those two times. In Fig. 2–14 the area under the graph is divided into a dark-colored rectangle of vertical side v_0 and horizontal side t and a light-colored right triangle of vertical side at and horizontal side t. The area of the rectangle is $v_0 t$ and the area of the triangle is $\frac{1}{2}(at)(t) = \frac{1}{2}at^2$, so the total area under the curve is

$$x - x_0 = v_0 t + \frac{1}{2}at^2,$$

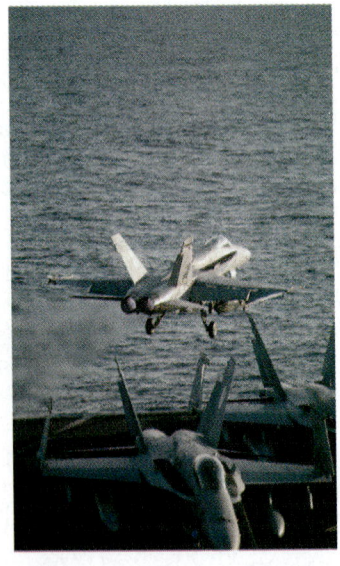

The motion of an airplane being catapulted from the deck of an aircraft carrier is an example of nearly constant acceleration. Thanks to the catapult, these airplanes attain takeoff speed in a distance of less than 100 m (see Exercise 2–18).

in agreement with Eq. (2–12).

The displacement during a time interval can always be found from the area under the v-t curve. This is true even if the acceleration is *not* constant, although in that case Eq. (2–12) does not apply. (We'll show this in Section 2–7.)

We can check whether Eqs. (2–8) and (2–12) are consistent with the assumption of constant acceleration by taking the derivative of Eq. (2–12). We find

$$v = \frac{dx}{dt} = v_0 + at,$$

which is Eq. (2–8). Differentiating again, we find simply

$$\frac{dv}{dt} = a,$$

as we should expect.

In many problems, it's useful to have a relationship between position, velocity, and acceleration that does not involve the time. To obtain this, we first solve Eq. (2–8) for t, then substitute the resulting expression into Eq. (2–12) and simplify:

$$t = \frac{v - v_0}{a},$$

$$x = x_0 + v_0\left(\frac{v - v_0}{a}\right) + \frac{1}{2}a\left(\frac{v - v_0}{a}\right)^2.$$

We transfer the term x_0 to the left side and multiply through by $2a$:

$$2a(x - x_0) = 2v_0 v - 2v_0^2 + v^2 - 2v_0 v + v_0^2.$$

Finally, simplifying gives us

$$v^2 = v_0^2 + 2a(x - x_0) \qquad \text{(constant acceleration only)}. \qquad (2\text{–}13)$$

We can get one more useful relationship by equating the two expressions for v_{av}, Eqs. (2–9) and (2–10), and multiplying through by t. Doing this, we obtain

$$x - x_0 = \left(\frac{v_0 + v}{2}\right)t \qquad \text{(constant acceleration only)}. \qquad (2\text{–}14)$$

Note that Eq. (2–14) does not contain the acceleration a. This is sometimes a useful equation when a is constant but its value is unknown.

Equations (2–8), (2–12), (2–13), and (2–14) are the *equations of motion with constant acceleration.* By using these equations, we can solve *any* kinematics problem involving straight-line motion of a particle with constant acceleration.

Figure 2–15 shows a graph of the coordinate x as a function of time for motion with constant acceleration. That is, it is a graph of Eq. (2–12); the x-t graph for constant acceleration is always a *parabola.* The curve intercepts the vertical axis (x-axis) at x_0, the position at $t = 0$. The slope of the tangent at $t = 0$ equals v_0, the initial velocity, and the slope of the tangent at any time t equals the velocity v at that time. The graph in Fig. 2–15 is concave up (it curves upward). The slope and velocity are continuously increasing, so the acceleration a is positive. If a is negative, the x-t graph is a parabola that is concave down (has a downward curvature).

For the particular case of motion with constant acceleration depicted in Fig. 2–12 and graphed in Figs. 2–13, 2–14, and 2–15, the values of x_0, v_0, and a are all positive. We invite you to redraw these figures for cases in which one, two, or all three of these quantities are negative.

A special case of motion with constant acceleration occurs when the acceleration is *zero.* The velocity is then constant, and the equations of motion become simply

$$v = v_0 = \text{constant},$$

$$x = x_0 + vt.$$

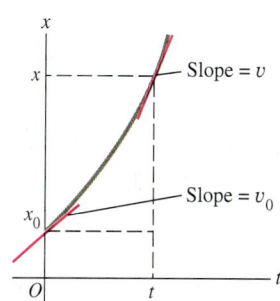

2–15 A position-time (x-t) graph for straight-line motion with constant acceleration. This graph refers to the same motion as is shown in Figs. 2–12, 2–13, and 2–14. For this motion the initial position x_0, the initial velocity v_0, and the acceleration a are all positive.

Problem–Solving Strategy

MOTION WITH CONSTANT ACCELERATION

1. You *must* decide at the beginning of a problem where the origin of coordinates is and which axis direction is positive. The choices are usually a matter of convenience. It is often easiest to place the particle at the origin at time $t = 0$; then $x_0 = 0$. It is always helpful to make a motion diagram showing these choices and some later positions of the particle.

2. Remember that your choice of the positive axis direction automatically determines the positive directions for velocity and acceleration. If x is positive to the right of the origin, then v and a are also positive toward the right.

3. Restate the problem in words first, and then translate this description into symbols and equations. *When* does the particle arrive at a certain point (that is, at what value of t)? *Where* is the particle when its velocity has a specified value (that is, what is the value of x when v has the specified value)? The following example asks "Where is the motorcyclist when his velocity is 25 m/s?" Translated

into symbols, this becomes "What is the value of x when $v = 25$ m/s?"

4. Make a list of quantities such as x, x_0, v, v_0, a, and t. In general, some of these will be known and some will be unknown. Write down the values of those that are known. Be on the lookout for implicit information. For example, "A car sits at a stoplight" usually means $v_0 = 0$.

5. Once you have identified the unknowns, try to choose an equation from Eqs. (2–8), (2–12), (2–13), and (2–14) that contains only one of the unknowns. Solve this equation for the unknown, using symbols only. Then substitute the known values and compute the value of the unknown. Sometimes you will have to solve two simultaneous equations for two unknowns.

6. Take a hard look at your results to see whether they make sense. Are they within the general range of values you expected?

EXAMPLE 2–4

Constant-acceleration calculations A motorcyclist heading east through a small Iowa city accelerates after he passes the signpost marking the city limits (Fig. 2–16). His acceleration is a constant 4.0 m/s². At time $t = 0$ he is 5.0 m east of the signpost, moving east at 15 m/s. a) Find his position and velocity at time $t = 2.0$ s. b) Where is the motorcyclist when his velocity is 25 m/s?

SOLUTION We take the signpost as the origin of coordinates ($x = 0$), and choose the positive x-axis to point east (Fig. 2–16). At the initial time $t = 0$ the initial position is $x_0 = 5.0$ m, and the initial velocity is $v_0 = 15$ m/s. The constant acceleration is $a = 4.0$ m/s².

a) The unknowns are the values of the position x and the velocity v at the later time $t = 2.0$ s. We determine the first of these using Eq. (2–12), which gives position x as a function of time t:

$$x = x_0 + v_0 t + \frac{1}{2} a t^2$$

$$= 5.0 \text{ m} + (15 \text{ m/s})(2.0 \text{ s}) + \frac{1}{2} (4.0 \text{ m/s}^2)(2.0 \text{ s})^2$$

$$= 43 \text{ m}.$$

We also use Eq. (2–8), which gives velocity v as a function of time t:

$$v = v_0 + at$$

$$= 15 \text{ m/s} + (4.0 \text{ m/s}^2)(2.0 \text{ s}) = 23 \text{ m/s}.$$

Do these results make sense? The motorcyclist accelerates

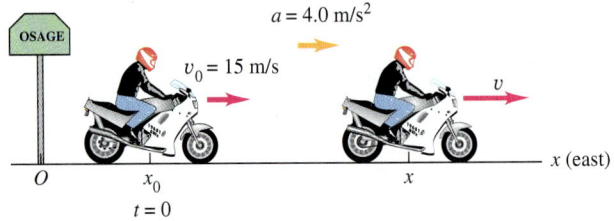

2–16 A motorcyclist traveling with constant acceleration.

from 15 m/s (about 34 mi/h, or 54 km/h) to 23 m/s (about 51 mi/h, or 83 km/h) in 2.0 s while traveling a distance of 38 m (about 125 ft). This is pretty brisk acceleration, but well within the capabilities of a high-performance bike.

b) We want to know the value of x when $v = 25$ m/s. From our solution to part (a), we can see that this occurs at a time later than 2.0 s and at a point farther than 43 m from the signpost. From Eq. (2–13) we have

$$v^2 = v_0^2 + 2a(x - x_0).$$

Solving for x and substituting in the known values, we find

$$x = x_0 + \frac{v^2 - v_0^2}{2a}$$

$$= 5.0 \text{ m} + \frac{(25 \text{ m/s})^2 - (15 \text{ m/s})^2}{2(4.0 \text{ m/s}^2)}$$

$$= 55 \text{ m}.$$

Alternatively, we may use Eq. (2–8) to first find the time when $v = 25$ m/s:

$$v = v_0 + at, \qquad \text{so}$$

$$t = \frac{v - v_0}{a} = \frac{25 \text{ m/s} - 15 \text{ m/s}}{4.0 \text{ m/s}^2}$$

$$= 2.5 \text{ s}.$$

Then from Eq. (2–12) we have

$$x = x_0 + v_0 t + \frac{1}{2} at^2$$

$$= 5.0 \text{ m} + (15 \text{ m/s})(2.5 \text{ s}) + \frac{1}{2}(4.0 \text{ m/s}^2)(2.5 \text{ s})^2$$

$$= 55 \text{ m}.$$

Again, this seems to be a plausible result.

EXAMPLE 2–5

Two bodies with different accelerations A motorist traveling with constant velocity of 15 m/s (about 34 mi/h) passes a school-crossing corner, where the speed limit is 10 m/s (about 22 mi/h). Just as the motorist passes, a police officer on a motorcycle stopped at the corner starts off in pursuit with constant acceleration of 3.0 m/s² (Fig. 2–17a). a) How much time elapses before the officer catches up with the motorist? b) What is the officer's speed at that point? c) What is the total distance each vehicle has traveled at that point?

SOLUTION The police officer and the motorist both move with constant acceleration (equal to zero for the motorist), so we can use the formulas we have developed. We take the origin at the corner, so $x_0 = 0$ for both, and take the positive direction to the right. Let x_P (for police) be the officer's position and x_M (for motorist) the motorist's position at any time. The initial velocities are $v_{P_0} = 0$ for the officer and $v_{M_0} = 15$ m/s for the motorist; the constant accelerations are $a_P = 3.0$ m/s² for the officer and $a_M = 0$ for the motorist. Then, applying Eq. (2–12) to each, we have

$$x = x_0 + v_0 t + \frac{1}{2} at^2,$$

$$x_M = v_{M_0} t,$$

$$x_P = \frac{1}{2} a_P t^2.$$

a) We want to find the time t when the officer catches the motorist. This is when the two vehicles are at the same position, so at this time, $x_M = x_P$. Equating the two expressions above, we have

$$v_{M_0} t = \frac{1}{2} a_P t^2,$$

$$t = 0 \quad \text{or} \quad t = \frac{2v_{M_0}}{a_P} = \frac{2(15 \text{ m/s})}{3.0 \text{ m/s}^2} = 10 \text{ s}.$$

There are *two* times when both the vehicles have the same x-coordinate. The first, $t = 0$, is the time when the motorist passes the parked motorcycle at the corner. The second, $t = 10$ s, is the time when the officer catches up with the motorist. b) We want the magnitude of the officer's velocity v_P at the time t found in part (a). Her velocity at any time is given by Eq. (2–8):

$$v_P = v_{P_0} + a_P t = 0 + (3.0 \text{ m/s}^2)t,$$

so when $t = 10$ s, we find $v_P = 30$ m/s. When the officer overtakes the motorist, she is traveling twice as fast as the motorist is. c) In 10 s the distance the motorist travels is

$$x_M = (15 \text{ m/s})(10 \text{ s}) = 150 \text{ m},$$

and the distance the officer travels is

$$x_P = \frac{1}{2}(3.0 \text{ m/s}^2)(10 \text{ s})^2 = 150 \text{ m}.$$

This verifies that at the time the officer catches the motorist, they have gone equal distances.

Figure 2–17b shows graphs of x versus t for each vehicle. We see again that there are two times when the two positions are the same (where the two graphs cross). At neither of these times do the two vehicles have the same velocity (i.e., where the two graphs cross, their slopes are different). At $t = 0$, the officer is at rest; at $t = 10$ s, the officer has twice the speed of the motorist.

In a real pursuit the officer would accelerate to a speed faster than that of the motorist, then slow down in order to have the same velocity as the motorist when she catches him. We haven't treated this case here because it involves a changing acceleration (but see Problem 2–56).

2–17 (a) Motion with constant acceleration overtaking motion with constant velocity. (b) Graph of x versus t for each vehicle.

(a)

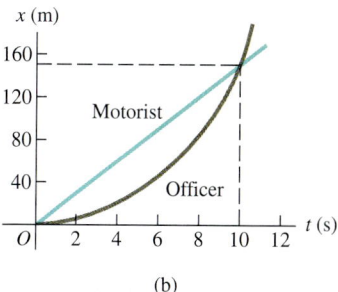

(b)

2-6 FREELY FALLING BODIES

The most familiar example of motion with (nearly) constant acceleration is that of a body falling under the influence of the earth's gravitational attraction. Such motion has held the attention of philosophers and scientists since ancient times. In the 4th century B.C., Aristotle thought (erroneously) that heavy objects fall faster than light objects, in proportion to their weight. Nineteen centuries later, Galileo argued that a body should fall with a downward acceleration that is constant and independent of its weight. We mentioned in Section 1–2 that, according to legend, Galileo experimented by dropping bullets and cannonballs from the Leaning Tower of Pisa.

The motion of falling bodies has since been studied with great precision. When the effects of the air can be neglected, Galileo is right; all bodies at a particular location fall with the same downward acceleration, regardless of their size or weight. If the distance of the fall is small compared to the radius of the earth, the acceleration is constant. In the following discussion we use an idealized model in which we neglect the effects of the air, the earth's rotation, and the decrease of acceleration with increasing altitude. We call this idealized motion **free fall,** although it includes rising as well as falling motion. (In Chapter 3 we will extend the discussion of free fall to include the motion of projectiles, which move both vertically and horizontally.)

Figure 2–18 is a photograph of a falling ball taken with a stroboscopic light source that produces a series of intense flashes at equal time intervals. Each flash is so short (a few millionths of a second) that there is little blur in the image of even a rapidly moving body. As each flash occurs, the position of the ball at that instant is recorded on the film. Because of the equal time intervals between flashes, the average velocity of the ball between any two flashes is proportional to the distance between corresponding images in the photograph. The increasing distances between images show that the velocity is continuously changing; the ball is accelerating downward. Careful measurement shows that the velocity change is the same in each time interval, so the acceleration of the freely falling ball is constant.

The constant acceleration of a freely falling body is called the **acceleration due to gravity,** and we denote its magnitude with the letter g. At or near the earth's surface the value of g is approximately 9.8 m/s^2, 980 cm/s^2, or 32 ft/s^2. The exact value varies with location, so we will often give the value of g at the earth's surface to only two significant figures. Because g is the magnitude of a vector quantity, it is always a *positive* number. On the surface of the moon the acceleration due to gravity is caused by the attractive force of the moon rather than the earth, and $g = 1.6$ m/s^2. Near the surface of the sun, $g = 270$ m/s^2.

In the following examples we use the constant-acceleration equations developed in Section 2–5. We suggest that you review the problem-solving strategies discussed in that section before you study the examples below.

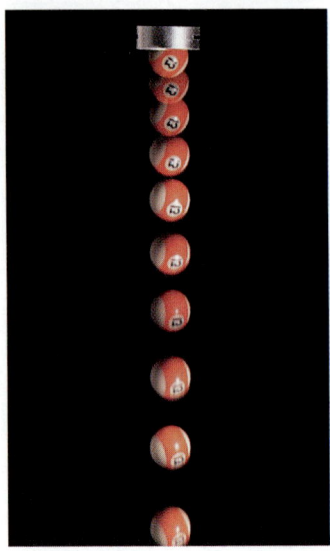

2–18 Multiflash photo of a freely falling ball.

1.7
Balloonist Drops Lemonade

1.10
Pole-Vaulter Lands

A 500-lira coin is dropped from the Leaning Tower of Pisa. It starts from rest and falls freely. Compute its position and velocity after 1.0, 2.0, and 3.0 s.

SOLUTION We'll take the origin O at the starting point, the coordinate axis as vertical, and the upward direction as positive (Fig. 2–19). Because the coordinate axis is vertical, let's call the coordinate y instead of x. We replace all the x's in the constant-acceleration equations by y's. The initial coordinate y_0

and the initial velocity v_0 are both zero. The acceleration is downward (in the negative y-direction), so $a = -g = -9.8$ m/s^2. (Remember that, by definition, g itself is *always* positive.)

The unknowns are the values of y and v at the three specified times. From Eqs. (2–12) and (2–8), with x replaced by y, we get

$$y = v_0 t + \frac{1}{2} a t^2 = 0 + \frac{1}{2}(-g)t^2 = (-4.9 \text{ m/s}^2)t^2,$$

$$v = v_0 + at = 0 + (-g)t = (-9.8 \text{ m/s}^2)t.$$

When $t = 1.0$ s, $y = (-4.9 \text{ m/s}^2)(1.0 \text{ s})^2 = -4.9$ m and $v = (-9.8 \text{ m/s}^2)(1.0 \text{ s}) = -9.8$ m/s; after one second, the coin is 4.9 m below the origin (y is negative) and has a downward velocity (v is negative) with magnitude 9.8 m/s.

The position and velocity at 2.0 and 3.0 s are found in the same way. The results are shown in Fig. 2–19; check the numerical values for yourself.

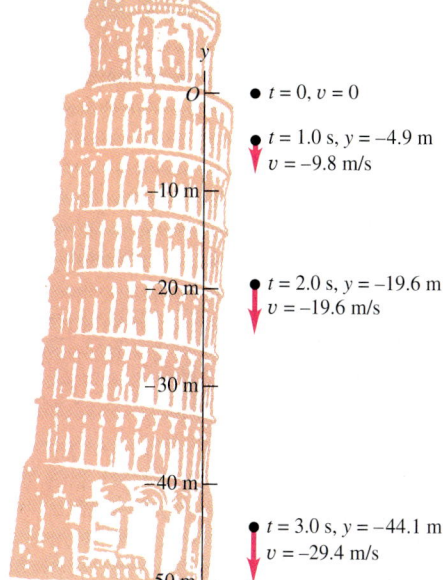

2–19 Position and velocity of a coin freely falling from rest.

EXAMPLE 2-7

You throw a ball vertically upward from the roof of a tall building. The ball leaves your hand at a point even with the roof railing with an upward speed of 15.0 m/s; the ball is then in free fall. On its way back down, it just misses the railing. At the location of the building, $g = 9.80 \text{ m/s}^2$. Find a) the position and velocity of the ball 1.00 s and 4.00 s after leaving your hand; b) the velocity when the ball is 5.00 m above the railing; c) the maximum height reached and the time at which it is reached; d) the acceleration of the ball when it is at its maximum height.

SOLUTION In Fig. 2–20 the downward path is displaced a little to the right of its actual position for clarity. Take the origin at the roof railing, at the point where the ball leaves your hand, and take the positive direction to be upward. First, let's collect our data. The initial position y_0 is zero. The initial velocity v_0 is +15.0 m/s, and the acceleration is $a = -g = -9.80 \text{ m/s}^2$.

a) The position y and velocity v at any time t after the ball leaves your hand are given by Eqs. (2–12) and (2–8) with x's replaced by y's, so

$$y = y_0 + v_0 t + \frac{1}{2} at^2 = y_0 + v_0 t + \frac{1}{2}(-g)t^2$$
$$= (0) + (15.0 \text{ m/s})t + \frac{1}{2}(-9.80 \text{ m/s}^2)t^2,$$

$$v = v_0 + at = v_0 + (-g)t$$
$$= 15.0 \text{ m/s} + (-9.80 \text{ m/s}^2)t.$$

When $t = 1.00$ s, these equations give

$$y = +10.1 \text{ m}, \qquad v = +5.2 \text{ m/s}.$$

The ball is 10.1 m above the origin (y is positive), and it is moving upward (v is positive) with a speed of 5.2 m/s. This is less than the initial speed of 15.0 m/s, as expected.

When $t = 4.00$ s, the equations for y and v as functions of time t give

$$y = -18.4 \text{ m}, \qquad v = -24.2 \text{ m/s}.$$

The ball has passed its highest point and is 18.4 m *below* the origin (y is negative). It has a *downward* velocity (v is negative) with magnitude 24.2 m/s. This speed is greater than the initial speed, as we should expect for points below the ball's launching point. Note that to get these results, we don't need to find the highest point reached or the time at which it was reached. The equations of motion give the position and velocity at any time, whether the ball is on the way up or on the way down.

b) The velocity v at any position y is given by Eq. (2–13) with x's replaced by y's:

$$v^2 = v_0{}^2 + 2a(y - y_0) = v_0{}^2 + 2(-g)(y - 0)$$
$$= (15.0 \text{ m/s})^2 + 2(-9.80 \text{ m/s}^2)y.$$

2–20 Position and velocity of a ball thrown vertically upward.

When the ball is 5.00 m above the origin, $y = +5.00$ m, so

$$v^2 = (15.0 \text{ m/s})^2 + 2(-9.80 \text{ m/s}^2)(5.00 \text{ m}) = 127 \text{ m}^2/\text{s}^2,$$

$$v = \pm 11.3 \text{ m/s}.$$

We get *two* values of v, one positive and one negative. As shown in Fig. 2–20, the ball passes this point twice, once on the way up and again on the way down. The velocity on the way up is +11.3 m/s, and on the way down it is −11.3 m/s.

c) At the highest point, the ball stops going up (positive v) and starts going down (negative v). Just at the instant when it reaches the highest point, $v = 0$. The maximum height y_1 can then be found in two ways. The first way is to use Eq. (2–13) and substitute $v = 0$, $y_0 = 0$, and $a = -g$:

$$0 = v_0^2 + 2(-g)(y_1 - 0),$$

$$y_1 = \frac{v_0^2}{2g} = \frac{(15.0 \text{ m/s})^2}{2(9.80 \text{ m/s}^2)} = +11.5 \text{ m}.$$

The second way is find the time at which $v = 0$ using Eq. (2–8), $v = v_0 + at$, and then substitute this value of t into Eq. (2–12) to find the position at this time. From Eq. (2–8), the time t_1 when the ball reaches the highest point is given by

$$v = 0 = v_0 + (-g)t_1,$$

$$t_1 = \frac{v_0}{g} = \frac{15.0 \text{ m/s}}{9.80 \text{ m/s}^2} = 1.53 \text{ s}.$$

Substituting this value of t into Eq. (2–12), we find

$$y = y_0 + v_0 t + \frac{1}{2} at^2 = (0) + (15 \text{ m/s})(1.53 \text{ s})$$

$$+ \frac{1}{2}(-9.8 \text{ m/s}^2)(1.53 \text{ s})^2 = +11.5 \text{ m}.$$

Notice that with the first way of finding the maximum height, it's not necessary to find the time first.

d) **CAUTION** ▶ It's a common misconception that at the highest point of the motion the velocity is zero *and* the acceleration is zero. If this were so, once the ball reached the highest point it would hang there in midair forever! To see why, remember that acceleration is the rate of change of velocity. If the acceleration were zero at the highest point, the ball's velocity would no longer change, and once the ball was instantaneously at rest it would remain at rest forever. ◀

The truth of the matter is that at the highest point, the acceleration is still $a = -g = -9.80 \text{ m/s}^2$, the same value as when the ball is moving up and when it's moving down. The ball stops for an instant at the highest point, but its velocity is continuously changing, from positive values through zero to negative values.

Figure 2–21 shows graphs of position and velocity as functions of time for this problem. Note that the v-t graph has a constant negative slope. The acceleration is negative (downward) on the way up, at the highest point, and on the way down.

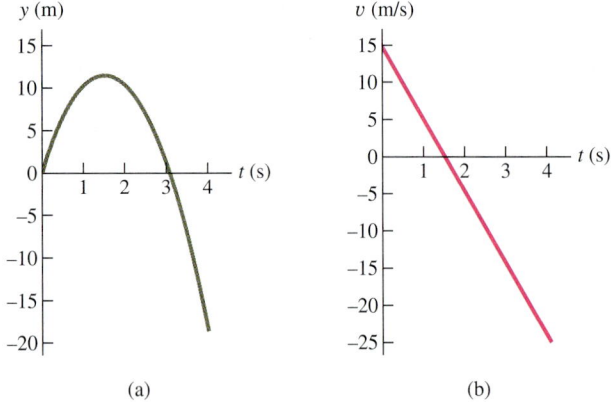

(a)   (b)

2–21 (a) Position and (b) velocity as functions of time for a ball thrown upward with an initial velocity of 15 m/s.

EXAMPLE 2–8

Find the time when the ball in Example 2–7 is 5.00 m below the roof railing.

SOLUTION We again choose the y-axis as in Fig. 2–20, so y_0, v_0, and $a = -g$ have the same values as in Example 2–7. The position y as a function of time t is again given by Eq. (2–12):

$$y = y_0 + v_0 t + \frac{1}{2} at^2 = y_0 + v_0 t + \frac{1}{2}(-g)t^2.$$

We want to solve this for the value of t when $y = -5.00$ m. Since this equation involves t^2, it is a *quadratic* equation for t. We first rearrange our equation into the same form as the standard quadratic equation for an unknown x, $Ax^2 + Bx + C = 0$:

$$\left(\frac{1}{2} g\right)t^2 + (-v_0)t + (y - y_0) = At^2 + Bt + C = 0,$$

so $A = g/2$, $B = -v_0$, and $C = y - y_0$. Using the quadratic formula (Appendix B), we find that this equation has *two* solutions:

$$t = \frac{-B \pm \sqrt{B^2 - 4AC}}{2A}$$

$$= \frac{-(-v_0) \pm \sqrt{(-v_0)^2 - 4(g/2)(y - y_0)}}{2(g/2)},$$

$$t = \frac{v_0 \pm \sqrt{v_0^2 - 2g(y - y_0)}}{g}.$$

Substituting the values $y_0 = 0$, $v_0 = +15.0$ m/s, $g = 9.80 \text{ m/s}^2$, and $y = -5.00$ m, we find

$$t = \frac{(15.0 \text{ m/s}) \pm \sqrt{(15.0 \text{ m/s})^2 - 2(9.80 \text{ m/s}^2)(-5.00 \text{ m} - 0)}}{9.80 \text{ m/s}^2},$$

$$t = +3.36 \text{ s} \qquad \text{or} \qquad t = -0.30 \text{ s}.$$

To decide which of these is the right answer, the key question to ask is "Are these answers reasonable?" The second answer, $t = -0.30$ s, is simply not reasonable; it refers to a time 0.30 s *before* the ball left your hand! The correct answer is $t = +3.36$ s. The ball is 5.00 m below the railing 3.36 s *after* it leaves your hand.

Where did the erroneous "solution" $t = -0.30$ s come from? Remember that we began with Eq. (2–12) with $a = -g$, that is, $y = y_0 + v_0 t + \frac{1}{2}(-g)t^2$. This has built into it the assumption that the acceleration is constant for *all* values of *t*, whether positive, negative, or zero. Taken at face value, this equation would tell us that the ball had been moving upward in free fall ever since the dawn of time; it eventually passes your hand at $y = 0$ at the special instant we chose to call $t = 0$, then continues in free fall. But anything that this equation describes happening before $t = 0$ is pure fiction, since the ball went into free fall only after leaving your hand at $t = 0$; the "solution" $t = -0.30$ s is part of this fiction.

We invite you to repeat these calculations to find the times at which the ball is 5.00 m *above* the origin ($y = +5.00$ m). The two answers are $t = +0.38$ s and $t = +2.68$ s; these are both positive values of *t*, and both refer to the real motion of the ball after leaving your hand. The earlier time is when the ball passes through $y = +5.00$ m moving upward, and the later time is when the ball passes through this point moving downward. (Compare this with part (b) of Example 2–7.) You should also solve for the times at which $y = +15.0$ m. In this case, both solutions involve the square root of a negative number, so there are *no* real solutions. This makes sense; we found in part (c) of Example 2–7 that the ball's maximum height is only $y = +11.5$ m, so it never reaches $y = +15.0$ m. While a quadratic equation such as Eq. (2–12) always has two solutions, in some situations one or both of the solutions will not be physically reasonable.

*2–7 VELOCITY AND POSITION BY INTEGRATION

This optional section is intended for students who have already learned a little integral calculus. In Section 2–5 we analyzed the special case of straight-line motion with constant acceleration. When *a* is not constant, as is frequently the case, the equations that we derived in that section are no longer valid. But even when *a* varies with time, we can still use the relation $v = dx/dt$ to find the velocity *v* as a function of time if the position *x* is a known function of time. And we can still use $a = dv/dt$ to find the acceleration *a* as a function of time if the velocity *v* is a known function of time.

In many physical situations, however, position and velocity are not known as functions of time, while the acceleration is. How can we find the position and velocity from the acceleration function $a(t)$? This problem arises in navigating an airliner between North America and Europe (Fig. 2–22). The crew of the airliner must know their position precisely at all times, because the airspace over the North Atlantic is very congested. But over the ocean an airliner is usually out of range of both radio navigation beacons on land and air traffic controllers' radar. To determine their position, airliners carry a device called an inertial navigation system (INS), which measures the acceleration of the moving airliner. This is done in much the same way that you can sense changes in the velocity of a car in which you're riding, even when your eyes are closed. (In Chapter 4 we'll discuss how your body detects acceleration.) Given this information, along with the airliner's initial position (say, a particular gate at Kennedy International Airport) and its initial velocity (zero when parked at the gate), the INS calculates and displays for the crew the current velocity and position of the airliner at all times during the flight. Our goal in the remainder of this section is to see how these calculations are done for the simpler case of motion in a straight line with time-varying acceleration.

We first consider a graphical approach. Figure 2–23 is a graph of acceleration versus time for a body whose acceleration is not constant but increases with time. We can divide the time interval between times t_1 and t_2 into many smaller intervals, calling a typical one Δt. Let the average acceleration during Δt be a_{av}. From Eq. (2–4) the change in velocity Δv during Δt is

$$\Delta v = a_{av}\Delta t.$$

Graphically, Δv equals the area of the shaded strip with height a_{av} and width Δt, that is, the area under the curve between the left and right sides of Δt. The total velocity change during any interval (say, t_1 to t_2) is the sum of the velocity changes Δv in the small subintervals. So the total velocity change is represented graphically by the *total* area under the *a-t* curve between the vertical lines t_1 and t_2. (In Section 2–5 we showed this to be true in the special case in which the acceleration was constant.)

In the limit that all the Δt's become very small and their number very large, the value of a_{av} for the interval from any time *t* to $t + \Delta t$ approaches the instantaneous acceleration

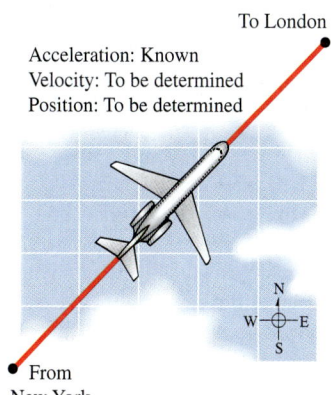

To London

Acceleration: Known
Velocity: To be determined
Position: To be determined

From New York

2–22 The position and velocity of an airliner crossing the Atlantic are found by integrating its acceleration with respect to time.

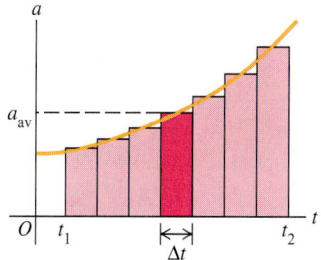

2–23 The area under an *a-t* graph between times t_1 and t_2 equals the change in velocity, $v_2 - v_1$, that occurs between those times.

a at time t. In this limit, the area under the a-t curve is the *integral* of a (which is in general a function of t) from t_1 to t_2. If v_1 is the velocity of the body at time t_1 and v_2 is the velocity at time t_2, then

$$v_2 - v_1 = \int_{v_1}^{v_2} dv = \int_{t_1}^{t_2} a\ dt. \qquad (2\text{–}15)$$

The change in velocity v is the integral of acceleration a with respect to time.

We can carry out exactly the same procedure with the curve of velocity versus time, where v is in general a function of t. If x_1 is a body's position at time t_1 and x_2 is its position at time t_2, from Eq. (2–2) the displacement Δx during a small time interval Δt is equal to $v_{av}\ \Delta t$, where v_{av} is the average velocity during Δt. The total displacement $x_2 - x_1$ during the interval $t_2 - t_1$ is given by

$$x_2 - x_1 = \int_{x_1}^{x_2} dx = \int_{t_1}^{t_2} v\ dt. \qquad (2\text{–}16)$$

The change in position x—that is, the displacement—is the time integral of velocity v. Graphically, the displacement between times t_1 and t_2 is the area under the v-t curve between those two times. (This is the same result that we obtained in Section 2–5 for the special case in which v is given by Eq. (2–8).)

If $t_1 = 0$ and t_2 is any later time t, and if x_0 and v_0 are the position and velocity, respectively, at time $t = 0$, then we can rewrite Eqs. (2–15) and (2–16) as follows:

$$v = v_0 + \int_0^t a\ dt, \qquad (2\text{–}17)$$

$$x = x_0 + \int_0^t v\ dt. \qquad (2\text{–}18)$$

Here x and v are the position and velocity at time t. If we know the acceleration a as a function of time and we know the initial velocity v_0, we can use Eq. (2–17) to find the velocity v at any time; in other words, we can find v as a function of time. Once we know this function, and given the initial position x_0, we can use Eq. (2–18) to find the position x at any time.

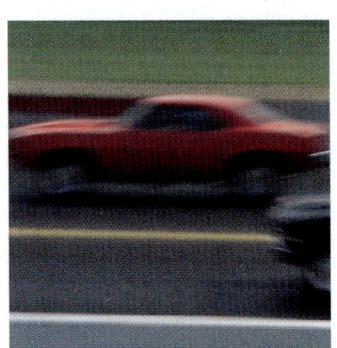

When you push your car's accelerator pedal to the floorboard, the resulting acceleration is *not* constant: the greater the car's speed, the more slowly it gains additional speed. A typical car takes twice as long to accelerate from 50 km/h to 100 km/h as it does to accelerate from 0 to 50 km/h.

EXAMPLE 2-9

Motion with changing acceleration Sally is driving along a straight highway in her classic 1965 Mustang. At time $t = 0$, when Sally is moving at 10 m/s in the positive x-direction, she passes a signpost at $x = 50$ m. Her acceleration is a function of time:

$$a = 2.0\ \text{m/s}^2 - (0.10\ \text{m/s}^3)t.$$

a) Derive expressions for her velocity and position as functions of time. b) At what time is her velocity greatest? c) What is the maximum velocity? d) Where is the car when it reaches maximum velocity?

SOLUTION a) At time $t = 0$, Sally's position is $x_0 = 50$ m, and her velocity is $v_0 = 10$ m/s. Since we are given the acceleration a as a function of time, we first use Eq. (2–17) to find the velocity v as a function of time t. The integral of t^n is $\int t^n dt = \frac{1}{n+1} t^{n+1}$ for $n \neq -1$, so

$$v = 10\ \text{m/s} + \int_0^t [2.0\ \text{m/s}^2 - (0.10\ \text{m/s}^3)t]\ dt$$

$$= 10\ \text{m/s} + (2.0\ \text{m/s}^2)t - \frac{1}{2}(0.10\ \text{m/s}^3)t^2.$$

Then we use Eq. (2–18) to find x as a function of t:

$$x = 50\ \text{m} + \int_0^t [10\ \text{m/s} + (2.0\ \text{m/s}^2)t - \frac{1}{2}(0.10\ \text{m/s}^3)t^2]\ dt$$

$$= 50\ \text{m} + (10\ \text{m/s})t + \frac{1}{2}(2.0\ \text{m/s}^2)t^2 - \frac{1}{6}(0.10\ \text{m/s}^3)t^3.$$

Figure 2–24 shows graphs of x, v, and a as functions of time. Note that for any time t the slope of the x-t graph equals the value of v, and the slope of the v-t graph equals the value of a. b) The maximum value of v occurs when v stops increasing and begins to decrease. At this instant, $dv/dt = a = 0$. Setting the expression for acceleration equal to zero, we obtain

$$0 = 2.0 \text{ m/s}^2 - (0.10 \text{ m/s}^3)t,$$

$$t = \frac{2.0 \text{ m/s}^2}{0.10 \text{ m/s}^3} = 20 \text{ s}.$$

As Fig. 2–24 shows, a is positive between $t = 0$ and $t = 20$ s and negative after that. It is zero at $t = 20$ s, the time at which v is maximum. The car speeds up until $t = 20$ s (because v and a have the same sign) and slows down after $t = 20$ s (because v and a have opposite signs).

c) We find the maximum velocity by substituting $t = 20$ s (when velocity is maximum) into the general velocity equation:

$$v_{max} = 10 \text{ m/s} + (2.0 \text{ m/s}^2)(20 \text{ s}) - \frac{1}{2}(0.10 \text{ m/s}^3)(20 \text{ s})^2$$

$$= 30 \text{ m/s}.$$

d) The maximum value of v occurs at time $t = 20$ s. We obtain the position of the car (that is, the value of x) at that time by substituting $t = 20$ s into the general expression for x:

$$x = 50 \text{ m} + (10 \text{ m/s})(20 \text{ s}) + \frac{1}{2}(2.0 \text{ m/s}^2)(20 \text{ s})^2$$

$$- \frac{1}{6}(0.10 \text{ m/s}^3)(20 \text{ s})^3$$

$$= 517 \text{ m}.$$

Since v is maximum at $t = 20$ s, the x-t graph has its maximum positive slope at this time. Note that the x-t graph is concave up (curved upward) from $t = 0$ to $t = 20$ s when a is positive and is concave down (curved downward) after $t = 20$ s when a is negative.

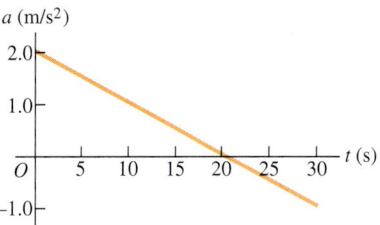

2–24 The position, velocity, and acceleration of the car in Example 2–9 as functions of time. Can you show that if this motion continues, the car will stop at $t = 44.5$ s?

EXAMPLE 2-10

Constant-acceleration formulas via integration Use Eqs. (2–17) and (2–18) to find v and x as functions of time in the case in which the acceleration is constant. Compare the results to the constant-acceleration formulas $v = v_0 + at$ (Eq. 2–8) and $x = x_0 + v_0 t + \frac{1}{2}at^2$ (Eq. 2–12), which were derived in Section 2–5 without using integration.

SOLUTION From Eq. (2–17) the velocity is given by

$$v = v_0 + \int_0^t a \ dt = v_0 + a\int_0^t dt = v_0 + at.$$

We were able to take a outside the integral because it is constant. Our result is identical to Eq. (2–8), as it should be. Substituting this expression for v into Eq. (2–18), we get

$$x = x_0 + \int_0^t v \ dt = x_0 + \int_0^t (v_0 + at) \ dt.$$

Since v_0 and a are constants, they may be taken outside the integral:

$$x = x_0 + v_0 \int_0^t dt + a\int_0^t t \ dt = x_0 + v_0 t + \frac{1}{2}at^2.$$

This result is the same as Eq. (2–12). Our expressions for v and x, Eqs. (2–17) and (2–18), which we developed to deal with cases in which acceleration depends on time, can be used just as well when the acceleration is constant.

SUMMARY

■ When a particle moves along a straight line, we describe its position with respect to an origin O by means of a coordinate such as x.

- The particle's average velocity during a time interval $\Delta t = t_2 - t_1$ is defined as

$$v_{av} = \frac{x_2 - x_1}{t_2 - t_1} = \frac{\Delta x}{\Delta t}. \tag{2-2}$$

The instantaneous velocity at any time t is defined as

$$v = \lim_{\Delta t \to 0} \frac{\Delta x}{\Delta t} = \frac{dx}{dt}. \tag{2-3}$$

- The average acceleration during a time interval $\Delta t = t_2 - t_1$ is defined as

$$a_{av} = \frac{v_2 - v_1}{t_2 - t_1} = \frac{\Delta v}{\Delta t}. \tag{2-4}$$

The instantaneous acceleration at any time t is defined as

$$a = \lim_{\Delta t \to 0} \frac{\Delta v}{\Delta t} = \frac{dv}{dt}. \tag{2-5}$$

- When the acceleration is constant, the position x and velocity v at any time t are related to the acceleration a, the initial position x_0, and the initial velocity v_0 (both at time $t = 0$) by the following equations:

$$v = v_0 + at \qquad \text{(constant acceleration only)}, \tag{2-8}$$

$$x = x_0 + v_0 t + \frac{1}{2} at^2 \qquad \text{(constant acceleration only)}, \tag{2-12}$$

$$v^2 = v_0^2 + 2a(x - x_0) \qquad \text{(constant acceleration only)}, \tag{2-13}$$

$$x - x_0 = \left(\frac{v_0 + v}{2} \right) t \qquad \text{(constant acceleration only)}. \tag{2-14}$$

- Free fall is a case of motion with constant acceleration. The magnitude of the acceleration due to gravity is a positive quantity g. The acceleration of a body in free fall is always downward.

- When the acceleration is not constant but is a known function of time, we can find the velocity and position as functions of time by integrating the acceleration function.

DISCUSSION QUESTIONS

Q2-1 Does the speedometer of a car measure speed or velocity? Explain.

Q2-2 Janet claims that a speed of 60 mi/h is the same as 88 ft/s and 27 m/s. Which relation is exact and which approximate?

Q2-3 Some European countries have highway speed limits of 110 km/h. What is the equivalent number of miles per hour?

Q2-4 Under what conditions is average velocity equal to instantaneous velocity?

Q2-5 In a given time interval, is the total displacement of a particle equal to the product of the average velocity and the time interval, even when the velocity is not constant? Explain.

Q2-6 Under what conditions does the magnitude of the average velocity equal the average speed?

Q2-7 When a Dodge Viper is at Elwood's Car Wash, a BMW Z3 is at Elm and Main. Later, when the Dodge reaches Elm and Main, the BMW reaches Elwood's Car Wash. How are the cars' average velocities between these two times related?

Q2-8 A driver in Massachusetts was sent to traffic court for speeding. The evidence against the driver was that a policewoman observed the driver's car to be alongside a second car at a certain moment, and that the policewoman had already clocked the second car as going faster than the speed limit. The driver argued that "the second car was passing me. I was not speeding." The judge ruled against the driver because, in the judge's words, "if two cars were side by side, you were both speeding." If you were a lawyer representing the accused driver, how would you argue this case?

Q2-9 Can you have a zero displacement and a nonzero average velocity? A nonzero velocity? Illustrate your answers on a x-t graph.

Q2-10 Can you have zero acceleration and nonzero velocity? Explain using a v-t graph.

Q2–11 Can you have zero velocity and nonzero average acceleration? Zero velocity and nonzero acceleration? Explain using a *v-t* graph.

Q2–12 An automobile is traveling west. Can it have a velocity toward the west and at the same time have an acceleration toward the east? Under what circumstances?

Q2–13 The official's truck in Fig. 2–2 is at $x_1 = 277$ m at $t_1 = 16.0$ s, and is at $x_2 = 19$ m at $t_2 = 25.0$ s. a) Sketch *two* different possible *x-t* graphs for the motion of the truck. b) Does the average velocity v_{av} during the time interval from t_1 to t_2 have the same value for both of your graphs? Why or why not?

Q2–14 Under constant acceleration the average velocity of a particle is half the sum of its initial and final velocities. Is this still true if the acceleration is *not* constant? Explain.

Q2–15 You throw a baseball straight up in the air so that it rises to a maximum height much greater than your height. Is the magnitude of the acceleration greater while it is being thrown or after it leaves your hand? Explain.

Q2–16 Prove these statements: i) As long as you can neglect the effects of the air, if you throw anything vertically upward, it will have the same speed when it returns to the release point as when it was released. ii) The time of flight will be twice the time it takes to get to its highest point.

Q2–17 In Example 2–7, substituting $y = -18.4$ m into Eq. (2–13) gives $v = \pm24.2$ m/s. The negative root is the velocity at $t = 4.00$ s. Explain the meaning of the positive root.

Q2–18 If the initial position and initial velocity of a vehicle are known and a record is kept of the acceleration at each instant, can the vehicle's position after a certain time be computed from these data? If so, explain how this might be done.

EXERCISES

SECTION 2–2 DISPLACEMENT, TIME, AND AVERAGE VELOCITY

2–1 A rocket carrying a satellite is accelerating straight up from the earth's surface. At 1.15 s after liftoff, the rocket clears the top of its launch platform, 63 m above the ground. After an additional 4.75 s it is 1.00 km above the ground. Calculate the magnitude of the average velocity of the rocket for a) the 4.75-s part of its flight; b) the first 5.90 s of its flight.

2–2 In an experiment, a shearwater (a seabird) was taken from its nest, flown 5150 km away, and released. It found its way back to its nest 13.5 days after release. If we place the origin in the nest and extend the +x-axis to the release point, what was the bird's average velocity in m/s a) for the return flight? b) for the whole episode, from leaving the nest to returning?

2–3 You normally drive on the freeway between San Diego and Los Angeles at an average speed of 105 km/h (65 mi/h), and the trip takes 2 h and 20 min. On a Friday afternoon, however, heavy traffic slows you down and you drive the same distance at an average speed of only 70 km/h (50 mi/h). How much longer does the trip take?

2–4 From Pillar to Post. Starting from a pillar, you run 200 m east (the +x-direction) at an average speed of 5.0 m/s, then run 280 m west at an average speed of 4.0 m/s to a post. Calculate a) your average speed from pillar to post; b) your average velocity from pillar to post.

2–5 a) Your old VW van sputters along at an average speed of 8.0 m/s for 60 s, then catches hold and zips along at an average speed of 20.0 m/s for another 60 s. Calculate your average speed for the entire 120 s. b) Suppose the speed of 8.0 m/s was maintained while you traveled 240 m, followed by the average speed of 20.0 m/s for another 240 m. Calculate the van's average speed for the entire distance. c) In which case is the average speed for the entire motion the numerical average of the two speeds?

2–6 A Honda Civic travels in a straight line along a road. Its distance x from a stop sign is given as a function of time t by the equation $x(t) = \alpha t^2 - \beta t^3$, where $\alpha = 1.50$ m/s^2 and $\beta = 0.0500$ m/s^3. Calculate the average velocity of the car for the following time intervals: a) $t = 0$ to $t = 2.00$ s; b) $t = 0$ to $t = 4.00$ s; c) $t = 2.00$ s to $t = 4.00$ s.

SECTION 2–3 INSTANTANEOUS VELOCITY

2–7 A car is stopped at a traffic light. It then travels along a straight road so that its distance from the light is given by $x(t) = bt^2 - ct^3$, where $b = 2.40$ m/s^2 and $c = 0.120$ m/s^3. a) Calculate the average velocity of the car for the time interval $t = 0$ to $t = 10.0$ s. b) Calculate the instantaneous velocity of the car at i) $t = 0$; ii) $t = 5.0$ s; iii) $t = 10.0$ s. c) How long after starting from rest is the car again at rest?

2–8 A physics professor leaves her house and walks along the sidewalk toward campus. After 5 min it starts to rain and she returns home. Her distance from her house as a function of time is shown in Fig. 2–25. At which of the labeled points is her velocity a) zero? b) constant and positive? c) constant and negative? d) increasing in magnitude? e) decreasing in magnitude?

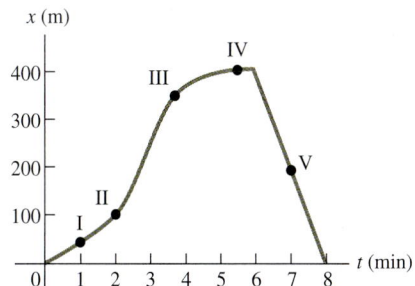

FIGURE 2–25 Exercise 2–8.

SECTION 2–4 AVERAGE AND INSTANTANEOUS ACCELERATION

2–9 A test driver at Incredible Motors, Inc., is testing a new model car having a speedometer calibrated to read m/s rather than mi/h. The following series of speedometer readings was obtained during a test run along a long, straight road:

Time (s)	0	2	4	6	8	10	12	14	16
Speed (m/s)	0	0	2	6	10	16	19	22	22

a) Compute the average acceleration during each 2-s interval. Is the acceleration constant? Is it constant during any part of the test run? b) Make a v-t graph of the data above, using scales of 1 cm = 1 s horizontally and 1 cm = 2 m/s vertically. Draw a smooth curve through the plotted points. By measuring the slope of your curve, find the instantaneous acceleration at $t = 9$ s, 13 s, and 15 s.

2–10 Figure 2–26 shows the velocity of a solar-powered car as a function of time. The driver accelerates from a stop sign, cruises for 20 s at a constant speed of 60 km/h, and then brakes to come to a stop 40 s after leaving the stop sign. Compute the average acceleration for the following time intervals: a) $t = 0$ to $t = 10$ s; b) $t = 30$ s to $t = 40$ s; c) $t = 10$ s to $t = 30$ s; d) $t = 0$ to $t = 40$ s.

FIGURE 2–26 Exercises 2–10 and 2–11.

2–11 Refer to Exercise 2–10 and Fig. 2–26. a) During what time interval does the instantaneous acceleration a have its most positive value? b) During what time interval does the instantaneous acceleration a have its most negative value? c) What is the instantaneous acceleration at $t = 20$ s? d) What is the instantaneous acceleration at $t = 35$ s? e) In a motion diagram (like Fig. 2–10b or Fig. 2–11b), show the position, velocity, and acceleration of the car at the four times $t = 5$ s, $t = 15$ s, $t = 25$ s, and $t = 35$ s.

2–12 An astronaut has left the International Space Station to test a new space scooter. Her partner measures the following velocity changes, each taking place in a 10-s interval. What are the magnitude, the algebraic sign, and the direction of the average acceleration in each interval? Assume that the positive direction is to the right. a) At the beginning of the interval the astronaut is moving toward the right along the x-axis at 15.0 m/s, and at the end of the interval she is moving toward the right at 5.0 m/s. b) At the beginning she is moving toward the left at 5.0 m/s, and at the end she is moving toward the left at 15.0 m/s. c) At the beginning she is moving toward the right at 15.0 m/s, and at the end she is moving toward the left at 15.0 m/s.

2–13 a) Based on your experiences of riding in automobiles, estimate the magnitude of a car's average acceleration when it brakes from highway speeds to a sudden stop. b) Explain why this average acceleration could be regarded as either positive or negative.

2–14 A car's velocity as a function of time is given by $v(t) = \alpha + \beta t^2$, where $\alpha = 3.00$ m/s and $\beta = 0.100$ m/s³. a) Calculate the average acceleration for the time interval $t = 0$ to

$t = 5.00$ s. b) Calculate the instantaneous acceleration for i) $t = 0$; ii) $t = 5.00$ s. c) Draw accurate v-t and a-t graphs for the car's motion between $t = 0$ and $t = 5.00$ s.

2–15 Figure 2–27 is a graph of the coordinate of a spider crawling along the x-axis. a) Graph its velocity and acceleration as functions of time. b) In a motion diagram (like Fig. 2–10b or Fig. 2–11b), show the position, velocity, and acceleration of the spider at the five times $t = 2.5$ s, $t = 10$ s, $t = 20$ s, $t = 30$ s, and $t = 37.5$ s.

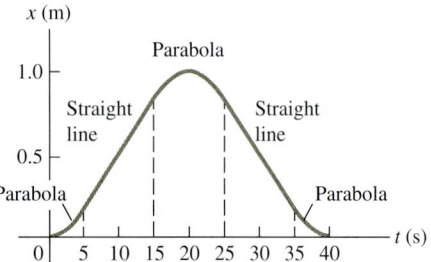

FIGURE 2–27 Exercise 2–15.

2–16 The position of the front bumper of a test car under microprocessor control is given by $x(t) = 2.17$ m + (4.80 m/s²)t^2 − (0.100 m/s⁶)t^6. Find its a) position and acceleration at the instants when the car has zero velocity. b) Draw x-t, v-t, and a-t graphs for the motion of the bumper between $t = 0$ and $t = 2.00$ s.

SECTION 2–5 MOTION WITH CONSTANT ACCELERATION

2–17 An antelope moving with constant acceleration covers the distance between two points 70.0 m apart in 7.00 s. Its speed as it passes the second point is 15.0 m/s. a) What is its speed at the first point? b) What is the acceleration?

2–18 The catapult of the aircraft carrier *USS Abraham Lincoln* accelerates an F/A-18 Hornet jet fighter from rest to a takeoff speed of 173 mi/h in a distance of 307 ft. Assume constant acceleration. a) Calculate the acceleration of the fighter in m/s². b) Calculate the time required for the fighter to be accelerated to takeoff speed.

2–19 **Automobile Airbags.** The human body can survive a negative acceleration trauma incident (sudden stop) if the magnitude of the acceleration is less than 250 m/s² (approximately 25g). If you are in an automobile accident with an initial speed of 105 km/h (65 mi/h) and are stopped by an airbag that inflates from the dashboard, over what distance must the airbag stop you for you to survive the crash?

2–20 An airplane travels 280 m down the runway before taking off. If it starts from rest, moves with constant acceleration, and becomes airborne in 8.00 s, what is its speed, in m/s, when it takes off?

2–21 A car sits in an entrance ramp to a freeway, waiting for a break in the traffic. The driver sees a small gap between a van and an 18-wheel truck and accelerates with constant acceleration along the ramp and onto the freeway. The car starts from rest, moves in a straight line, and has speed 20 m/s (45 mi/h) when it reaches the end of the 120-m-long ramp. a) What is the acceleration of the car? b) How much time does it take the car to travel the length of the ramp? c) The traffic on the freeway is moving at a constant speed of 20 m/s. What distance does the traffic travel while the car is moving the length of the ramp?

2–22 Figures 2–12, 2–13, 2–14, and 2–15 were drawn for motion with constant acceleration with positive values of x_0, v_0, and a. Redraw these four figures for the following cases: a) $x_0 < 0$, $v_0 < 0$, $a < 0$; b) $x_0 > 0$, $v_0 < 0$, $a > 0$; c) $x_0 > 0$, $v_0 > 0$, $a < 0$.

2–23 At time $t = 0$ a Corvette is traveling on a long, straight stretch of highway in Arizona at a constant speed of 30 m/s. This motion continues for 20 s. Then the driver, worried that she is behind schedule, accelerates at a constant rate for 5 s to bring the car's speed to 40 m/s. The car travels at this new speed for 10 s. But then the driver spots a police motorcycle parked behind a large cactus up ahead and slows with a constant acceleration of magnitude 4.0 m/s² until the car's speed is reduced once again to the legal limit of 30 m/s. She then maintains this speed, and waves to the officer as she passes him 5 s later. a) For the motion of the car from $t = 0$ until when the car passes the parked motorcycle, draw accurate *a-t*, *v-t*, and *x-t* graphs. b) In a motion diagram (like Fig. 2–10b or 2–11b), show the position, velocity, and acceleration of the car.

2–24 At $t = 0$ a car is stopped at a traffic light. When the light turns green, the car starts to speed up, and gains speed at a constant rate until it reaches a speed of 20 m/s 8 seconds after the light turns green. The car continues at a constant speed for 60 m. Then the driver sees a red light up ahead at the next intersection, and starts slowing down at a constant rate. The car stops at the red light, 180 m from where it was at $t = 0$. a) Draw accurate *x-t*, *v-t*, and *a-t* graphs for the motion of the car. b) In a motion diagram (like Fig. 2–10b or 2–11b), show the position, velocity, and acceleration of the car.

2–25 The graph in Fig. 2–28 shows the velocity of a motorcycle police officer plotted as a function of time. a) Find the instantaneous acceleration at $t = 3$ s, at $t = 7$ s, and at $t = 11$ s. b) How far does the officer go in the first 5 s? The first 9 s? The first 13 s?

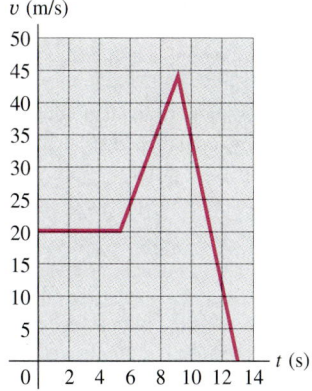

FIGURE 2–28 Exercise 2–25.

2–26 Figure 2–29 is a graph of the acceleration of a model railroad locomotive moving on the *x*-axis. Graph its velocity and coordinate as functions of time if $x = 0$ and $v = 0$ at $t = 0$.

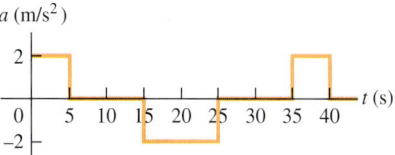

FIGURE 2–29 Exercise 2–26.

2–27 A spaceship ferrying workers to Moon Base I takes a straight-line path from the earth to the moon, a distance of 384,000 km. Suppose it accelerates at 20.0 m/s² for the first 15.0 min of the trip, then travels at constant speed until the last 15.0 min, when it accelerates at –20.0 m/s², just coming to rest as it reaches the moon. a) What is the maximum speed attained? b) What fraction of the total distance is traveled at constant speed? c) What total time is required for the trip?

2–28 A subway train starts from rest at a station and accelerates at a rate of 1.60 m/s² for 14.0 s. It runs at constant speed for 70.0 s and slows down at a rate of 3.50 m/s² until it stops at the next station. Find the *total* distance covered.

2–29 Two cars, A and B, move along the *x*-axis. Figure 2–30 is a graph of the positions of A and B versus time. a) In motion diagrams (like Figure 2–10b or Figure 2–11b), show the position, velocity, and acceleration of each of the two cars at $t = 0$, $t = 1$ s, and $t = 3$ s. b) At what time(s), if any, do A and B have the same position? c) Graph velocity versus time for both A and B. d) At what time(s), if any, do A and B have the same velocity? e) At what time(s), if any, does car A pass car B? f) At what time(s), if any, does car B pass car A?

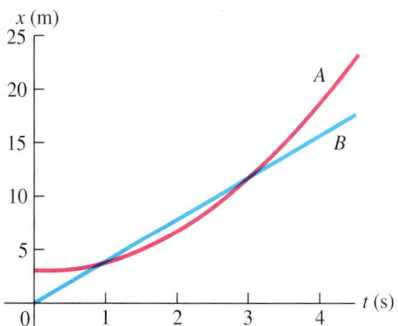

FIGURE 2–30 Exercise 2–29.

2–30 At the instant the traffic light turns green, a car that has been waiting at an intersection starts ahead with a constant acceleration of 3.20 m/s². At the same instant a truck, traveling with a constant speed of 20.0 m/s, overtakes and passes the car. a) How far beyond its starting point does the car overtake the truck? b) How fast is the car traveling when it overtakes the truck? c) Sketch an *x-t* graph of the motion of both vehicles. Take $x = 0$ at the intersection. d) Sketch a *v-t* graph of the motion of both vehicles.

2–31 As in Example 2–5, a car is traveling at a constant velocity with magnitude v_C. At the instant that the car passes a motorcycle officer, the motorcycle accelerates from rest with acceleration a_M. a) Sketch an *x-t* graph of the motion of both objects. Show that when the motorcycle overtakes the car the motorcycle has a speed twice that of the car, no matter what the value of a_M. b) Let d be the distance the motorcycle travels before catching up with the car. In terms of d, how far has the motorcycle traveled when its velocity equals the velocity of the car?

SECTION 2–6 FREELY FALLING BODIES

2–32 If the effects of the air acting on falling raindrops can be neglected, then we can treat raindrops as freely falling objects. a) Rain clouds are typically a few hundred meters above the ground. Estimate the speed with which raindrops would strike the ground if they were freely falling objects. Give your estimate in m/s, km/h, and mi/h. b) Estimate (from your own personal observations of rain) the speed with which raindrops actually strike the ground. c) Based on your answers to parts (a) and (b), is it a good approximation to neglect the effects of the air on falling raindrops? Explain.

2–33 a) If a flea can jump straight up to a height of 0.440 m, what is its initial speed as it leaves the ground? b) For how much time is it in the air?

2–34 Touchdown on the Moon. A lunar lander is making its descent to Moon Base I (Fig. 2–31). The lander descends slowly under the retro-thrust of its descent engine. The engine is cut off when the lander is 5.0 m above the surface and has a downward speed of 0.8 m/s. With the engine off, the lander is in free fall. What is the speed of the lander just before it touches the surface? The acceleration due to gravity on the moon is 1.6 m/s².

FIGURE 2–31 Exercise 2–34.

2–35 A Simple Reaction-Time Test. A meter stick is held vertically above your hand, with the lower end between your thumb and first finger. On seeing the meter stick released, you grab it with these two fingers. Your reaction time can be calculated from the distance the meter stick falls, read directly from the point where your fingers grabbed it. a) Derive a relationship for your reaction time in terms of this measured distance, d. b) If the measured distance is 17.6 cm, what is the reaction time?

2–36 A brick is dropped (zero initial speed) from the roof of a building. The brick strikes the ground in 2.50 s. Air resistance may be ignored, so the brick is in free fall. a) How tall, in

meters, is the building? b) What is the magnitude of the brick's velocity just before it reaches the ground? c) Sketch a-t, v-t, and y-t graphs for the motion of the brick.

2–37 Jillian angrily throws her engagement ring straight up from the roof of a building, 12.0 m above the ground, with an initial speed of 5.00 m/s. Air resistance may be ignored. For the motion from her hand to the ground, what are the magnitude and direction of a) the average velocity of the ring? b) the average acceleration of the ring? c) In how many seconds after being thrown does the ring strike the ground? d) What is the speed of the ring just before it strikes the ground? e) Sketch a-t, v-t, and y-t graphs for the motion.

2–38 A hot-air balloonist, rising vertically with a constant velocity of magnitude 5.00 m/s, releases a sandbag at an instant when the balloon is 40.0 m above the ground (Fig. 2–32). After it is released, the sandbag is in free fall. a) Compute the position and velocity of the sandbag at 0.250 s and 1.00 s after its release. b) How many seconds after its release will the bag strike the ground? c) With what magnitude of velocity does it strike? d) What is the greatest height above the ground that the sandbag reaches? e) Sketch a-t, v-t, and y-t graphs for the motion.

$v = 5.00$ m/s

40.0 m to ground

FIGURE 2–32 Exercise 2–38.

2–39 A student throws a water balloon vertically downward from the top of a building. The balloon leaves the thrower's hand with a speed of 6.00 m/s. Air resistance may be ignored, so the water balloon is in free fall after it leaves the thrower's hand. a) What is its speed after falling for 2.00 s? b) How far does it fall in 2.00 s? c) What is the magnitude of its velocity after falling 10.0 m? d) Sketch a-t, v-t, and y-t graphs for the motion.

2–40 An egg is thrown nearly vertically upward from a point near the cornice of a tall building. It just misses the cornice on the way down and passes a point 50.0 m below its starting point 5.00 s after it leaves the thrower's hand. Air resistance may be ignored. a) What is the initial speed of the egg? b) How high does it rise above its starting point? c) What is the magnitude of its velocity at the highest point? d) What are the magnitude and direction of its acceleration at the highest point? e) Sketch a-t, v-t, and y-t graphs for the motion of the egg.

2–41 The rocket-driven sled *Sonic Wind No. 2,* used for investigating the physiological effects of large accelerations, runs on a straight, level track 1070 m (3500 ft) long. Starting from rest, it can reach a speed of 224 m/s (500 mi/h) in 0.900 s. a) Compute the acceleration in m/s^2, assuming that it is constant. b) What is the ratio of this acceleration to that of a freely falling body (g)? c) What is the distance covered in 0.900 s? d) A magazine article states that at the end of a certain run, the speed of the sled decreased from 283 m/s (632 mi/h) to zero in 1.40 s and that during this time the magnitude of the acceleration was more than 40g. Are these figures consistent?

2–42 A large boulder is ejected vertically upward from a volcano with an initial speed of 40.0 m/s. Air resistance may be ignored. a) At what time after being ejected is the boulder moving at 20.0 m/s upward? b) At what time is it moving at 20.0 m/s downward? c) When is the displacement of the boulder from its initial position zero? d) When is the velocity of the boulder zero? e) What are the magnitude and direction of the accelera-

tion while the boulder is i) moving upward? ii) moving downward? iii) at the highest point? f) Sketch a-t, v-t, and y-t graphs for the motion.

2–43 Suppose the acceleration of gravity were only 0.98 m/s^2 instead of 9.8 m/s^2, but the initial velocities with which you can jump upward or throw a ball are unchanged. a) Calculate the height to which you could jump vertically from a standing start if you can jump to 0.75 m when g = 9.8 m/s^2. b) How high could you throw a baseball if you can throw it 18 m up when g = 9.8 m/s^2? c) Calculate the maximum height of a window from which you would care to jump to a concrete sidewalk below if for g = 9.8 m/s^2 you are willing to jump from a height of 2.0 m, which is the typical height of a first-story window.

*SECTION **2–7 VELOCITY AND POSITION BY INTEGRATION**

***2–44** The acceleration of a bus is given by $a(t) = \alpha t$, where α = 1.2 m/s^3. a) If the bus's velocity at time t = 1.0 s is 5.0 m/s, what is its velocity at time t = 2.0 s? b) If the bus's position at time t = 1.0 s is 6.0 m, what is its position at time t = 2.0 s? c) Sketch a-t, v-t, and x-t graphs for the motion.

***2–45** The acceleration of a motorcycle is given by $a(t) = At - Bt^2$, where A = 1.50 m/s^3 and B = 0.120 m/s^4. The motorcycle is at rest at the origin at time t = 0. a) Find its position and velocity as functions of time. b) Calculate the maximum velocity it attains.

PROBLEMS

2–46 On a 20-mile bike ride, you ride the first 10 miles at an average speed of 8 mi/h. What must your average speed over the next 10 miles be to have your average speed for the total 20 miles be a) 4 mi/h? b) 12 mi/h? c) Given this average speed for the first 10 miles, can you possibly attain an average speed of 16 mi/h for the total 20-mile ride? Explain.

2–47 The position of a particle between t = 0 and t = 2.00 s is given by $x(t) = (3.00 \text{ m/s}^3)t^3 - (10.0 \text{ m/s}^2)t^2 + (9.00 \text{ m/s})t$. a) Draw the x-t, v-t, and a-t graphs of this particle. b) At what time(s) between t = 0 and t = 2.00 s is the particle instantaneously at rest? Does your numerical result agree with the v-t graph in part (a)? c) At each time calculated in part (b) is the acceleration of the particle positive or negative? Show that in each case the same answer is deduced from $a(t)$ and from the v-t graph. d) At what time(s) between t = 0 and t = 2.00 s is the velocity of the particle instantaneously not changing? Locate this point on the v-t and a-t graphs of part (a). e) What is the particle's greatest distance from the origin (x = 0) between t = 0 and t = 2.00 s? f) At what time(s) between t = 0 and t = 2.00 s is the particle *speeding up* at the greatest rate? At what time(s) between t = 0 and t = 2.00 s is the particle *slowing down* at the greatest rate? Locate these points on the v-t and a-t graphs of part (a).

2–48 In a relay race, each contestant runs 25.0 m while carrying an egg balanced on a spoon, turns around, and comes back to the starting point. Edith runs the first 25.0 m in 20.0 s. On the return trip she is more confident and takes only 15.0 s. What is the magnitude of her average velocity for a) the first 25.0 m?

b) the return trip? c) What is her average velocity for the entire round trip? d) What is her average speed for the round trip?

2–49 Dan gets on Interstate Highway I–80 at Seward, Nebraska, and drives due west in a straight line and at an average velocity of magnitude 88 km/h. After traveling 76 km, he reaches the Aurora exit (Fig. 2–33). Realizing he has gone too far, he turns around and drives due east 34 km back to the York exit at an average velocity of magnitude 72 km/h. For his whole trip from Seward to the York exit, what is a) his average speed? b) the magnitude of his average velocity?

FIGURE 2–33 Problem 2–49.

2–50 Freeway Traffic. According to a *Scientific American* article (May, 1990), current freeways can sustain about 2400 vehicles per lane per hour in smooth traffic flow at 96 km/h (60 mi/h). Above that figure the traffic flow becomes "turbulent" (stop-and-go). a) If a vehicle is 4.6 m (15 ft) long on the average, what is the average spacing between vehicles at the above traffic density? b) Collision-avoidance automated control systems, which operate by bouncing radar or sonar signals off surrounding vehicles and then accelerate or brake the car when necessary, could greatly reduce the required spacing between vehicles. If the average spacing is 9.2 m (two car lengths), how many vehicles per hour can a lane of traffic carry at 96 km/h?

2–51 A world-class sprinter accelerates to his maximum speed in 4.0 s. He then maintains this speed for the remainder of a 100-m race, finishing with a total time of 9.1 s. a) What is the runner's average acceleration during the first 4.0 s? b) What is his average acceleration during the last 5.1 s? c) What is his average acceleration for the entire race? d) Explain why your answer to part (c) is not the average of the answers to parts (a) and (b).

2–52 A sled starts from rest at the top of a hill and slides down with a constant acceleration. At some later time it is 14.4 m from the top; 2.00 s after that it is 25.6 m from the top, 2.00 s later 40.0 m from the top, and 2.00 s later it is 57.6 m from the top. a) What is the magnitude of the average velocity of the sled during each of the 2.00-s intervals after passing the 14.4-m point? b) What is the acceleration of the sled? c) What is the speed of the sled when it passes the 14.4-m point? d) How much time did it take to go from the top to the 14.4-m point? e) How far did the sled go during the first second after passing the 14.4-m point?

2–53 A car 3.5 m in length and traveling at a constant speed of 20 m/s is approaching an intersection (Fig. 2–34). The width of the intersection is 20 m. The light turns yellow when the front of the car is 50 m from the beginning of the intersection. If the driver steps on the brake, the car will slow at -3.8 m/s^2. If the driver instead steps on the gas pedal, the car will accelerate at 2.3 m/s^2. The light will be yellow for 3.0 s. Ignore the reaction time of the driver. To avoid being in the intersection while the light is red, should the driver hit the brake pedal or the gas pedal?

$v_0 = 20$ m/s

50 m   20 m

3.5 m

FIGURE 2–34 Problem 2–53.

2–54 The engineer of a passenger train traveling at 25.0 m/s sights a freight train whose caboose is 200 m ahead on the same track (Fig. 2–35). The freight train is traveling at 15.0 m/s in the same direction as the passenger train. The engineer of the passenger train immediately applies the brakes, causing a constant acceleration of -0.100 m/s^2, while the freight train continues with constant speed. Take $x = 0$ at the location of the front of the passenger train when the brakes are applied. a) Will the cows

$v_{PT} = 25.0$ m/s

$a = -0.100$ m/s^2

$v_{FT} = 15.0$ m/s

200 m

FIGURE 2–35 Problem 2–54.

nearby witness a collision? b) If so, where will it take place? c) On a single graph, sketch the positions of the front of the passenger train and the back of the freight train.

2–55 Large cockroaches can run as fast as 1.50 m/s in short bursts. Suppose you turn on the light in a cheap motel and see one scurrying directly away from you at a constant 1.50 m/s as you move toward it at 0.80 m/s. If you start 0.90 m behind it, what minimum constant acceleration would you need to catch up with it when it has traveled 1.20 m, just short of safety under a counter?

2–56 Consider the situation described in Example 2–5. The example is somewhat unrealistic, because the constantly accelerating officer continues past the car. In a real pursuit the officer would accelerate to a speed faster than that of the car, then slow down in order to have the same velocity as the car when she catches up with it. Suppose that the officer in Example 2–5 accelerates from rest at $a = 2.5$ m/s^2 until her speed is 20 m/s. She then slows down at a constant rate until she comes alongside the car at $x = 360$ m, traveling with the same speed of 15 m/s as the car. a) At what time does the officer catch up with the car? b) At what time does the officer change from speeding up to slowing down? At this time, how far is the officer from the stop sign? How far behind the car is she? c) Find the acceleration of the officer while she is slowing down. d) Draw an x-t graph for the two vehicles. e) Draw a v-t graph for the two vehicles.

2–57 An automobile and a truck start from rest at the same instant, with the automobile initially at some distance behind the truck. The truck has a constant acceleration of 2.10 m/s^2, and the automobile an acceleration of 3.40 m/s^2. The automobile overtakes the truck after the truck has moved 40.0 m. a) How much time does it take the automobile to overtake the truck? b) How far was the automobile behind the truck initially? c) What is the speed of each when they are abreast? d) On a single graph, sketch the position of each vehicle as a function of time. Take $x = 0$ at the initial location of the truck.

2–58 Two stunt drivers drive directly toward each other. At time $t = 0$ the two cars are a distance D apart, car 1 is at rest, and car 2 is moving to the left with speed v_0. Car 1 begins to move at $t = 0$, speeding up with a constant acceleration a. Car 2 continues to move with a constant velocity. a) At what time do the two cars collide? b) Find the speed of car 1 just before it collides with car 2. c) Sketch x-t and v-t graphs for car 1 and car 2. Draw the curves for each car using the same axes.

2–59 In his Mustang, Juan rounds a curve onto a straight section of country road while traveling at 20 m/s (45 mi/h), only to find a fully loaded manure spreader completely blocking the road 37 m ahead of him. Startled, he travels at a constant velocity for 0.80 s of reaction time, then hits the brakes and slows at constant acceleration so that he manages to stop just short of the spreader. With the same reaction time and acceleration, if he had roared around the corner at 25 m/s (56 mi/h) instead of 20 m/s, a) how fast would he have been going when he hit the spreader? b) how much time would he have had for his life to flash before his eyes from the time he first sees the spreader until he hits it?

2–60 A police car is traveling in a straight line with constant speed v_p. A truck traveling in the same direction with speed $\frac{3}{2}v_p$ passes the police car. The truck driver realizes that she is speeding, and immediately begins to slow down at a constant rate until she comes to a stop. This is her lucky day, however, and the police car (still moving with the same constant speed) passes the truck driver without giving her a ticket. a) Show that the truck's speed at the instant that the police car passes the truck does *not* depend on the magnitude of the truck's acceleration as it slows down, and find the value of that speed. b) Sketch the *x-t* graph for the two vehicles.

2–61 The driver of a car wishes to pass a truck that is traveling at a constant speed of 20.0 m/s (about 45 mi/h). Initially, the car is also traveling at 20.0 m/s and its front bumper is 24.0 m behind the truck's rear bumper. The car accelerates at a constant 0.600 m/s², then pulls back into the truck's lane when the rear of the car is 26.0 m ahead of the front of the truck. The car is 4.5 m long and the truck is 21.0 m long. a) How much time is required for the car to pass the truck? b) What distance does the car travel during this time? c) What is the final speed of the car?

*2–62 An object's velocity is measured to be $v(t) = \alpha - \beta t^2$, where $\alpha = 4.00$ m/s and $\beta = 2.00$ m/s³. At $t = 0$ the object is at $x = 0$. a) Calculate the object's position and acceleration as functions of time. b) What is the object's maximum *positive* displacement from the origin?

*2–63 The acceleration of a particle is given by $a(t) = -2.00$ m/s² + (3.00 m/s³)t. a) Find the initial velocity v_0 such that the particle will have the same *x*-coordinate at $t = 4.00$ s as it had at $t = 0$. b) What will be the velocity at $t = 4.00$ s?

2–64 You are on the roof of the physics building, 46.0 m above the ground (Fig. 2–36). Your physics professor, who is 1.80 m tall, is walking alongside the building at a constant speed of 1.20 m/s. If you wish to drop an egg on your professor's head, where should the professor be when you release the egg? Assume that the egg is in free fall.

2–65 A physics student with too much free time drops a watermelon from the roof of a building. He hears the sound of the watermelon going "splat" 2.50 s later. How high is the building? The speed of sound is 340 m/s. Ignore air resistance.

2–66 Estimate the maximum speed and the magnitude of acceleration of an elevator. You will need to use your observations of roughly how long it takes the elevator to go from one floor to another, the approximate vertical distance from one floor to the next, and the distance over which an elevator accelerates to its maximum speed or slows to a stop.

46.0 m

$v = 1.20$ m/s

1.80 m

FIGURE 2–36 Problem 2–64.

2–67 Visitors at an amusement park watch divers step off a platform 21.3 m (70 ft) above a pool of water. According to the announcer, the divers enter the water at a speed of 56 mi/h (25 m/s). Air resistance may be ignored. a) Is the announcer correct in this claim? b) Is it possible for a diver to leap directly upward off the board so that, missing the board on the way down, she enters the water at 25.0 m/s? If so, what initial upward speed is required? Is the required initial speed physically attainable?

2–68 A flowerpot falls off a windowsill and falls past the window below. Air resistance may be ignored. It takes 0.420 s to pass this window, which is 1.90 m high. How far is the top of the window below the windowsill from which the flowerpot fell?

2–69 A football is kicked vertically upward from the ground and a student gazing out of the window sees it moving upward past her at 5.00 m/s. The window is 12.0 m above the ground. Air resistance may be ignored. a) How high does the football go above ground? b) How much time does it take to go from the ground to its highest point?

2–70 A model rocket has a constant upward acceleration of 40.0 m/s² while its engine is running. The rocket is fired vertically, and the engine runs for 2.50 s before the fuel is used up. After the engine stops, the rocket is in free fall. The motion of the rocket is purely up and down. a) Sketch the *a-t*, *v-t*, and *y-t* graphs for the rocket. b) What is the maximum height that the rocket reaches? c) What will be the speed of the rocket just before it hits the ground? d) Is the total time of flight equal to twice the time the rocket takes to reach its highest point? Why or why not? (See Question Q2–16.)

2–71 Sam heaves a 16-lb shot straight upward, giving it a constant upward acceleration from rest of 45.0 m/s² for 64.0 cm. He releases it 2.20 m above the ground. Air resistance may be ignored. a) What is the speed of the shot when he releases it? b) How high above the ground does it go? c) How much time does he have to get out of its way before it returns to the height of the top of his head, 1.83 m above the ground?

2–72 Determined to test the law of gravity for himself, a student walks off a skyscraper 180 m high, stopwatch in hand, and starts his free fall (zero initial velocity). Five seconds later, Superman arrives at the scene and dives off the roof to save the student. a) Superman leaves the roof with an initial speed v_0 that he produces by pushing himself downward from the edge of the roof with his legs of steel. He then falls with the same acceleration as any freely falling body. What must be the value of v_0 so that Superman catches the student just before the ground is reached? b) On the same graph, sketch the positions of the student and of Superman as functions of time. Take Superman's initial speed to have the value calculated in part (a). c) If the height of the skyscraper is less than some minimum value, even Superman can't reach the student before he hits the ground. What is this minimum height?

2–73 Another determined student (see Problem 2–72) walks off the top of the CN Tower in Toronto, which is 553 m high, and falls freely. His initial velocity is zero. The Rocketeer arrives at the scene five seconds later and dives off the top of the tower to save the student. The Rocketeer leaves the roof with an initial downward speed v_0. In order both to catch the student and to prevent injury to him, the Rocketeer should catch the student at a sufficiently great height above ground so that the Rocketeer and the student slow down and arrive at the ground with zero velocity. The upward acceleration that accomplishes this is provided by the Rocketeer's jet pack, which he turns on just as he catches the student; before then the Rocketeer is in free fall. To prevent discomfort to the student, the magnitude of the acceleration of the Rocketeer and the student as they move downward together should be no more than five times g. a) What is the minimum height above the ground at which the Rocketeer should catch the student? b) What must be the Rocketeer's initial downward speed so that he catches the student at the minimum height found in part (a)? c) Sketch the y-t, v-t, and a-t graphs for the student and for the Rocketeer. On each graph, draw the curves for the Rocketeer and for the student using the same axes.

2–74 A ball is thrown straight up from the ground with speed v_0. At the same instant, a second ball is dropped from rest from a height H, directly above the point where the first ball was thrown upward. There is no air resistance. a) Find the time at which the two balls collide. b) Find the value of H in terms of v_0 and g so that at the instant when the balls collide, the first ball is at the highest point of its motion.

2–75 Two cars, A and B, travel in a straight line. The distance of A from the starting point is given as a function of time by $x_A(t) = \alpha t + \beta t^2$, with $\alpha = 2.60$ m/s and $\beta = 1.20$ m/s^2. The distance of B from the starting point is $x_B(t) = \gamma t^2 - \delta t^3$, with $\gamma = 2.80$ m/s^2 and $\delta = 0.20$ m/s^3. a) Which car is ahead just after they leave the starting point? b) At what time(s) are the cars at the same point? c) At what time(s) is the distance from A to B neither increasing nor decreasing? d) At what time(s) do A and B have the same acceleration?

2–76 An apple drops from an apple tree and falls freely. The apple is originally at rest a height H above the top of the grass of a thick lawn, which is made of blades of grass of height h. When the apple enters the grass, it slows down at a constant rate so that its speed is zero when it reaches ground level. a) Find the speed of the apple just before it enters the grass. b) Find the acceleration of the apple while it is in the grass. c) Sketch the y-t, v-t, and a-t graphs for the apple's motion.

CHALLENGE PROBLEMS

2–77 A student is running at her top speed of 5.0 m/s to catch a bus, which is stopped at the bus stop. When the student is still 40.0 m from the bus, it starts to pull away, moving with a constant acceleration of 0.170 m/s^2. a) For how much time and what distance does the student have to run at 5.0 m/s before she overtakes the bus? b) When she reaches the bus, how fast is the bus traveling? c) Sketch an x-t graph for both the student and the bus. Take $x = 0$ at the initial position of the student. d) The equations you used in part (a) to find the time have a second solution, corresponding to a later time for which the student and bus are again at the same place if they continue their specified motions. Explain the significance of this second solution. How fast is the bus traveling at this point? e) If the student's top speed is 3.5 m/s, will she catch the bus? f) What is the *minimum* speed the student must have to just catch up with the bus? For what time and what distance does she have to run in that case?

2–78 In the vertical jump, an athlete starts from a crouch and jumps upward to reach as high as possible. Even the best athletes spend little more than 1.00 s in the air (their "hang time"). Treat the athlete as a particle and let y_{max} be his maximum height above the floor. To explain why he seems to hang in the air, calculate the ratio of the time he is above $y_{max}/2$ to the time it takes him to go from the floor to that height. Air resistance may be ignored.

2–79 A ball is thrown straight up from the edge of the roof of a building. A second ball is dropped from the roof 1.00 s later. Air resistance may be ignored. a) If the height of the building is 20.0 m, what must be the initial speed of the first ball if both are to hit the ground at the same time? On the same graph, sketch the position of each ball as a function of time, measured from when the first ball is thrown. Consider the same situation, but now let the initial speed v_0 of the first ball be given and treat the height h of the building as an unknown. b) What must the height of the building be for both balls to reach the ground at the same time for each of the following values of v_0: i) 6.0 m/s; ii) 9.5 m/s? c) If v_0 is greater than some value v_{max}, a value of h does not exist that allows both balls to hit the ground at the same time. Solve for v_{max}. The value v_{max} has a simple physical interpretation. What is it? d) If v_0 is less than some value v_{min}, a value of h does not exist that allows both balls to hit the ground at the same time. Solve for v_{min}. The value v_{min} also has a simple physical interpretation. What is it?

2–80 An alert hiker sees a boulder fall from the top of a distant cliff and notes that it takes 1.30 s for the boulder to fall the last third of the way to the ground. Air resistance may be ignored. a) What is the height of the cliff in meters? b) If in part (a) you get two solutions of a quadratic equation and use one for your answer, what does the other solution represent?

Motion in Two or Three Dimensions

Navigation—the problem of determining where you are, what direction you should travel to reach your destination, and how soon you will get there—has been revolutionized by the Global Positioning System (GPS). A handheld GPS unit like this one can determine its position to within a few meters by analyzing signals from several orbiting GPS satellites.

3-1 INTRODUCTION

When a baseball is hit by a bat, what determines where the baseball lands? How do you describe the motion of a roller coaster car along a curved track or the flight of a hawk circling over an open field? If you throw a water balloon horizontally from your window, will it take the same amount of time to hit the ground as a balloon that you simply drop?

We can't answer these kinds of questions using the techniques of Chapter 2, in which particles moved only along a straight line. Instead, we have to confront the reality that our world is three-dimensional. To understand the curved flight of a baseball, the orbital motion of a satellite, or the trajectory of a thrown projectile, we need to extend our descriptions of motion to two- and three-dimensional situations. We'll still use the vector quantities displacement, velocity, and acceleration, but now these quantities will have two or three components and will no longer lie along a single line. We'll find that several important and interesting kinds of motion take place in two dimensions only, that is, in a *plane*. These motions can be described with two coordinates and two components of velocity and acceleration.

We also need to consider how the motion of a particle is described by different observers who are moving relative to each other. The concept of *relative velocity* contains the seeds of the special theory of relativity, although we won't get to that until later in the book.

This chapter merges the vector language we learned in Chapter 1 with the kinematic language of Chapter 2. As before, we are concerned with describing motion, not with analyzing its causes. But the language you learn here will be an essential tool in later chapters when you use Newton's laws of motion to study the relation between force and motion.

3-2 POSITION AND VELOCITY VECTORS

To describe the *motion* of a particle in space, we first need to be able to describe the *position* of the particle. Consider a particle that is at a point P at a certain instant. The **position vector** $\vec{r}$ of the particle at this instant is a vector that goes from the origin of the coordinate system to the point P (Fig. 3–1). The figure also shows that the Cartesian coordinates x, y, and z of point P are the x-, y-, and z-components of vector $\vec{r}$. Using the unit vectors introduced in Section 1–10, we can write

$$\vec{r} = x\hat{\imath} + y\hat{\jmath} + z\hat{k}. \qquad (3-1)$$

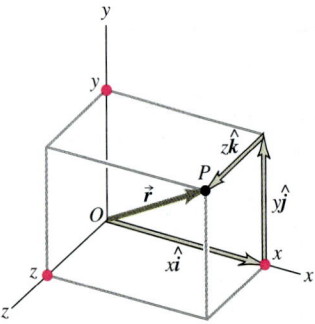

3–1 The position vector $\vec{r}$ from the origin to point P has components x, y, and z.

As the particle moves through space, the path that it follows is in general a curve (Fig. 3–2). During a time interval Δt the particle moves from P_1, where its position vector is $\vec{r}_1$, to P_2, where its position vector is $\vec{r}_2$. The change in position (the displacement) during this interval is $\Delta\vec{r} = \vec{r}_2 - \vec{r}_1$. We define the **average velocity** $\vec{v}_{av}$ during this interval in the same way we did in Chapter 2 for straight-line motion, as the displacement divided by the time interval:

$$\vec{v}_{av} = \frac{\vec{r}_2 - \vec{r}_1}{t_2 - t_1} = \frac{\Delta\vec{r}}{\Delta t} \quad \text{(average velocity vector).} \tag{3–2}$$

Note that *dividing* a vector by a scalar is really a special case of *multiplying* a vector by a scalar, described in Section 1–8; the average velocity $\vec{v}_{av}$ is equal to the displacement vector $\Delta\vec{r}$ multiplied by $1/\Delta t$, the reciprocal of the time interval.

We now define **instantaneous velocity** just as we did in Chapter 2; it is the limit of the average velocity as the time interval approaches zero, and it equals the instantaneous rate of change of position with time. The key difference is that position $\vec{r}$ and instantaneous velocity $\vec{v}$ are now both vectors:

$$\vec{v} = \lim_{\Delta t \to 0} \frac{\Delta\vec{r}}{\Delta t} = \frac{d\vec{r}}{dt} \quad \text{(instantaneous velocity vector).} \tag{3–3}$$

The *magnitude* of the vector $\vec{v}$ at any instant is the *speed* v of the particle at that instant. The *direction* of $\vec{v}$ at any instant is the same as the direction in which the particle is moving at that instant.

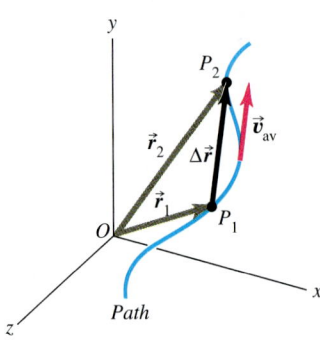

3–2 The average velocity $\vec{v}_{av}$ between points P_1 and P_2 has the same direction as the displacement $\Delta\vec{r}$.

Note that as $\Delta t \to 0$, points P_1 and P_2 in Fig. 3–2 move closer and closer together. In this limit, the vector $\Delta\vec{r}$ becomes tangent to the curve. The direction of $\Delta\vec{r}$ in the limit is also the direction of the instantaneous velocity $\vec{v}$. This leads to an important conclusion: *at every point along the path, the instantaneous velocity vector is tangent to the path at that point* (Fig. 3–3).

It's often easiest to calculate the instantaneous velocity vector using components. During any displacement $\Delta\vec{r}$, the changes Δx, Δy, and Δz in the three coordinates of the particle are the *components* of $\Delta\vec{r}$. It follows that the components v_x, v_y, and v_z of the instantaneous velocity $\vec{v}$ are simply the time derivatives of the coordinates x, y, and z. That is,

$$v_x = \frac{dx}{dt}, \qquad v_y = \frac{dy}{dt}, \qquad v_z = \frac{dz}{dt} \quad \text{(components of instantaneous velocity).} \tag{3–4}$$

We can also get this result by taking the derivative of Eq. (3–1). The unit vectors $\hat{\imath}$, $\hat{\jmath}$, and $\hat{k}$ are constant in magnitude and direction, so their derivatives are zero, and we find

$$\vec{v} = \frac{d\vec{r}}{dt} = \frac{dx}{dt}\hat{\imath} + \frac{dy}{dt}\hat{\jmath} + \frac{dz}{dt}\hat{k}. \tag{3–5}$$

This shows again that the components of $\vec{v}$ are dx/dt, dy/dt, and dz/dt.

The magnitude of the instantaneous velocity vector $\vec{v}$—that is, the speed—is given in terms of the components v_x, v_y, and v_z by the Pythagorean relation

$$|\vec{v}| = v = \sqrt{v_x{}^2 + v_y{}^2 + v_z{}^2}. \tag{3–6}$$

Figure 3–4 shows the situation when the particle moves in the xy-plane. In this case, z and v_z are zero. Then the speed (the magnitude of $\vec{v}$) is

$$v = \sqrt{v_x{}^2 + v_y{}^2},$$

and the direction of the instantaneous velocity $\vec{v}$ is given by the angle α in the figure. We see that

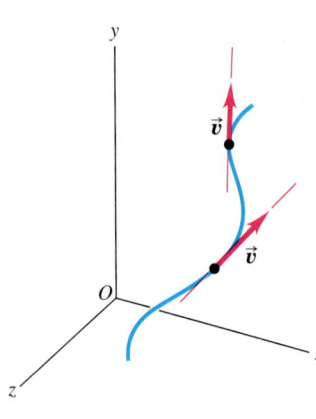

3–3 The instantaneous velocity $\vec{v}$ at any point is tangent to the path at that point.

$$\tan \alpha = \frac{v_y}{v_x}. \qquad (3-7)$$

(We always use Greek letters for angles. We use α for the direction of the instantaneous velocity vector to avoid confusion with the direction θ of the *position* vector of the particle.)

The instantaneous velocity vector is usually more interesting and useful than the average velocity vector. From now on, when we use the word "velocity," we will always mean the instantaneous velocity vector $\vec{v}$ (rather than the average velocity vector). Usually, we won't even bother to call $\vec{v}$ a vector; it's up to you to remember that velocity is a vector quantity with both magnitude and direction.

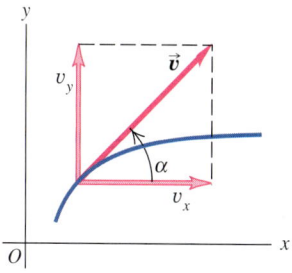

3-4 The two velocity components for motion in the xy-plane.

EXAMPLE 3-1

Calculating average and instantaneous velocity You are operating a radio-controlled model car on a vacant tennis court. Your position is the origin of coordinates, and the surface of the court lies in the xy-plane. The car, which we represent as a point, has x- and y-coordinates that vary with time according to

$$x = 2.0 \text{ m} - (0.25 \text{ m/s}^2)t^2,$$

$$y = (1.0 \text{ m/s})t + (0.025 \text{ m/s}^3)t^3.$$

a) Find the car's coordinates and its distance from you at time $t = 2.0$ s. b) Find the car's displacement and average velocity vectors during the interval from $t = 0.0$ s to $t = 2.0$ s. c) Derive a general expression for the car's instantaneous velocity vector, and find the instantaneous velocity at $t = 2.0$ s. Express the instantaneous velocity in component form and also in terms of magnitude and direction.

SOLUTION a) The car's path is shown in Fig. 3-5. At time $t = 2.0$ s the car's coordinates are

$$x = 2.0 \text{ m} - (0.25 \text{ m/s}^2)(2.0 \text{ s})^2 = 1.0 \text{ m},$$

$$y = (1.0 \text{ m/s})(2.0 \text{ s}) + (0.025 \text{ m/s}^3)(2.0 \text{ s})^3 = 2.2 \text{ m}.$$

The car's distance from the origin at this time is

$$r = \sqrt{x^2 + y^2} = \sqrt{(1.0 \text{ m})^2 + (2.2 \text{ m})^2} = 2.4 \text{ m}.$$

b) To find the displacement and average velocity, we express the position vector $\vec{r}$ as a function of time t. From Eq. (3-1), this is

$$\vec{r} = x\hat{\imath} + y\hat{\jmath}$$

$$= [2.0 \text{ m} - (0.25 \text{ m/s}^2)t^2]\hat{\imath}$$

$$+ [(1.0 \text{ m/s})t + (0.025 \text{ m/s}^3)t^3]\hat{\jmath}.$$

At time $t = 0.0$ s the position vector is

$$\vec{r}_0 = (2.0 \text{ m})\hat{\imath} + (0.0 \text{ m})\hat{\jmath}.$$

From part (a) we find that at time $t = 2.0$ s the position vector $\vec{r}_2$ is

$$\vec{r}_2 = (1.0 \text{ m})\hat{\imath} + (2.2 \text{ m})\hat{\jmath}.$$

Therefore the displacement from $t = 0.0$ s to $t = 2.0$ s is

$$\Delta\vec{r} = \vec{r}_2 - \vec{r}_0 = (1.0 \text{ m})\hat{\imath} + (2.2 \text{ m})\hat{\jmath} - (2.0 \text{ m})\hat{\imath}$$

$$= (-1.0 \text{ m})\hat{\imath} + (2.2 \text{ m})\hat{\jmath}.$$

During the time interval the car has moved 1.0 m in the negative

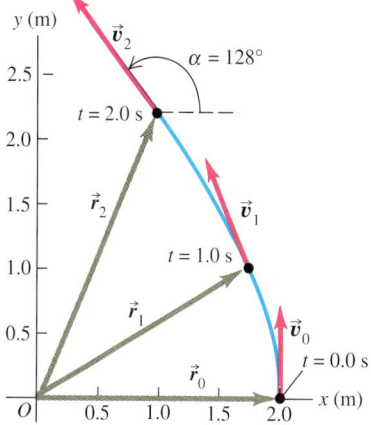

3-5 Path of a radio-controlled model car. At $t = 0$ the car has position vector $\vec{r}_0$ and instantaneous velocity vector $\vec{v}_0$. Likewise, $\vec{r}_1$ and $\vec{v}_1$ are the vectors at $t = 1.0$ s; $\vec{r}_2$ and $\vec{v}_2$ are the vectors at $t = 2.0$ s.

x-direction and 2.2 m in the positive y-direction. The average velocity during the interval from $t = 0.0$ s to $t = 2.0$ s is this displacement divided by the elapsed time (Eq. (3-2)):

$$\vec{v}_{av} = \frac{\Delta\vec{r}}{\Delta t} = \frac{(-1.0 \text{ m})\hat{\imath} + (2.2 \text{ m})\hat{\jmath}}{2.0 \text{ s} - 0.0 \text{ s}}$$

$$= (-0.50 \text{ m/s})\hat{\imath} + (1.1 \text{ m/s})\hat{\jmath}.$$

The components of this average velocity are

$$(v_{av})_x = -0.50 \text{ m/s}, \qquad (v_{av})_y = 1.1 \text{ m/s}.$$

c) From Eq. (3-4), the components of instantaneous velocity are the time derivatives of the coordinates:

$$v_x = \frac{dx}{dt} = (-0.25 \text{ m/s}^2)(2t),$$

$$v_y = \frac{dy}{dt} = 1.0 \text{ m/s} + (0.025 \text{ m/s}^3)(3t^2).$$

Then we can write the instantaneous velocity vector $\vec{v}$ as

$$\vec{v} = v_x\hat{\imath} + v_y\hat{\jmath} = (-0.50 \text{ m/s}^2)t\hat{\imath}$$

$$+ [1.0 \text{ m/s} + (0.075 \text{ m/s}^3)t^2]\hat{\jmath}.$$

At time $t = 2.0$ s, the components of instantaneous velocity are

$$v_x = (-0.50 \text{ m/s}^2)(2.0 \text{ s}) = -1.0 \text{ m/s},$$

$$v_y = 1.0 \text{ m/s} + (0.075 \text{ m/s}^3)(2.0 \text{ s})^2 = 1.3 \text{ m/s}.$$

Take a moment to compare these components of $\vec{v}$ with the components of $\vec{v}_{av}$ that we found in part (b). Just as in one dimension, the average velocity vector $\vec{v}_{av}$ over an interval is in general *not* equal to the instantaneous velocity vector $\vec{v}$ at the beginning or end of the interval (see Example 2–1).

The magnitude of the instantaneous velocity (that is, the speed) at $t = 2.0$ s is

$$v = \sqrt{v_x{}^2 + v_y{}^2} = \sqrt{(-1.0 \text{ m/s})^2 + (1.3 \text{ m/s})^2}$$

$$= 1.6 \text{ m/s}.$$

Its direction with respect to the positive x-axis is given by the angle α, where, from Eq. (3–7),

$$\tan \alpha = \frac{v_y}{v_x} = \frac{1.3 \text{ m/s}}{-1.0 \text{ m/s}} = -1.3, \quad \text{so} \quad \alpha = 128°.$$

Your calculator will tell you that the inverse tangent of -1.3 is $-52°$. But as we learned in Section 1–9, you have to examine a sketch of a vector (Fig. 3–5) to decide on its direction. The correct answer for α is $-52° + 180° = 128°$.

We invite you to calculate the position, instantaneous velocity vector, speed, and direction of motion at $t = 0.0$ s and $t = 1.0$ s. The position vectors and instantaneous velocity vectors at $t = 0.0$ s, 1.0 s, and 2.0 s are shown in Fig. 3–5. Notice that at every point, the instantaneous velocity vector $\vec{v}$ is tangent to the path. The magnitude of $\vec{v}$ increases as the car moves, which shows that the car's speed is increasing.

3-3 THE ACCELERATION VECTOR

Now let's consider the *acceleration* of a particle moving in space. Just as for motion in a straight line, acceleration describes how the velocity of the particle changes. But now we'll generalize acceleration to describe changes in the magnitude of the velocity (that is, the speed) *and* changes in the direction of the velocity (that is, the direction in which the particle is moving in space).

In Fig. 3–6a, a particle is moving along a curved path. The vectors $\vec{v}_1$ and $\vec{v}_2$ represent the particle's instantaneous velocities at time t_1, when the particle is at point P_1, and at time t_2, when the particle is at point P_2. The two velocities may differ in both magnitude and direction. We define the **average acceleration** $\vec{a}_{av}$ of the particle as it moves from P_1 to P_2 as the *vector change in velocity,* $\vec{v}_2 - \vec{v}_1 = \Delta\vec{v}$, divided by the time interval $t_2 - t_1 = \Delta t$:

$$\vec{a}_{av} = \frac{\vec{v}_2 - \vec{v}_1}{t_2 - t_1} = \frac{\Delta\vec{v}}{\Delta t} \quad \text{(average acceleration vector).} \quad (3\text{–}8)$$

Average acceleration is a *vector* quantity in the same direction as the vector $\Delta\vec{v}$ (Fig. 3–6a). Note that $\vec{v}_2$ is the vector sum of the original velocity $\vec{v}_1$ and the change $\Delta\vec{v}$ (Fig. 3–6b).

As in Chapter 2, we define the **instantaneous acceleration** $\vec{a}$ at point P_1 as the limit approached by the average acceleration when point P_2 approaches point P_1 and $\Delta\vec{v}$ and Δt both approach zero; the instantaneous acceleration is also equal to the instantaneous rate of change of velocity with time. Because we are not restricted to straight-line motion, instantaneous acceleration is now a vector:

$$\vec{a} = \lim_{\Delta t \to 0} \frac{\Delta\vec{v}}{\Delta t} = \frac{d\vec{v}}{dt} \quad \text{(instantaneous acceleration vector).} \quad (3\text{–}9)$$

The velocity vector $\vec{v}$, as we have seen, is tangent to the path of the particle. But the construction in Fig. 3–6c shows that the instantaneous acceleration vector $\vec{a}$ of a moving particle always points toward the concave side of a curved path—that is, toward the inside of any turn that the particle is making. We can also see that when a particle is moving in a curved path, it *always* has nonzero acceleration, even when it moves with constant speed. This conclusion may seem contrary to your intuition, but it's really just contrary to the everyday use of the word "acceleration" to mean that speed is increasing. The more precise definition given in Eq. (3–9) shows that there is a nonzero acceleration whenever the velocity vector changes in *any* way, whether there is a change of speed, direction, or both.

(a)

(b)

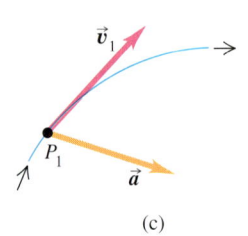

(c)

3–6 (a) The vector $\vec{a}_{av} = \Delta\vec{v}/\Delta t$ represents the average acceleration between P_1 and P_2. (b) Construction for obtaining $\Delta\vec{v} = \vec{v}_2 - \vec{v}_1$. (c) Instantaneous acceleration $\vec{a}$ at point P_1. Vector $\vec{v}$ is tangent to the path; vector $\vec{a}$ points toward the concave side of the path.

To convince yourself that a particle has a nonzero acceleration when moving on a curved path with constant speed, think of your sensations when you ride in a car. When the car accelerates, you tend to move within the car in a direction *opposite* to the car's acceleration. (We'll discover the reason for this behavior in Chapter 4.) Thus you tend to slide toward the back of the car when it accelerates forward (speeds up) and toward the front of the car when it accelerates backward (slows down). If the car makes a turn on a level road, you tend to slide toward the outside of the turn; hence the car has an acceleration toward the inside of the turn.

We will usually be interested in the instantaneous acceleration, not the average acceleration. From now on, we will use the term "acceleration" to mean the instantaneous acceleration vector $\vec{a}$.

Each component of the acceleration vector is the derivative of the corresponding component of velocity:

$$a_x = \frac{dv_x}{dt}, \qquad a_y = \frac{dv_y}{dt}, \qquad a_z = \frac{dv_z}{dt} \qquad \text{(components of instantaneous acceleration).} \tag{3–10}$$

In terms of unit vectors,

$$\vec{a} = \frac{dv_x}{dt}\hat{\imath} + \frac{dv_y}{dt}\hat{\jmath} + \frac{dv_z}{dt}\hat{k}. \tag{3–11}$$

Also, because each component of velocity is the derivative of the corresponding coordinate, we can express the components a_x, a_y, and a_z of the acceleration vector $\vec{a}$ as

$$a_x = \frac{d^2x}{dt^2}, \qquad a_y = \frac{d^2y}{dt^2}, \qquad a_z = \frac{d^2z}{dt^2}, \tag{3–12}$$

and the acceleration vector $\vec{a}$ as

$$\vec{a} = \frac{d^2x}{dt^2}\hat{\imath} + \frac{d^2y}{dt^2}\hat{\jmath} + \frac{d^2z}{dt^2}\hat{k}. \tag{3–13}$$

4.1
Determining Acceleration
Direction Graphically

4.2
Acceleration Direction of a Car

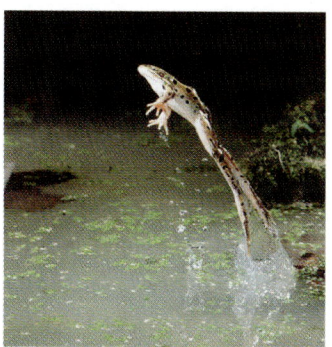

When a frog leaps, it accelerates both forward and upward. Thus its acceleration vector has both a horizontal component (a_x) and a vertical component (a_y).

EXAMPLE 3–2

Calculating average and instantaneous acceleration Let's look again at the radio-controlled model car in Example 3–1. We found that the components of instantaneous velocity at any time t are

$$v_x = \frac{dx}{dt} = (-0.25 \text{ m/s}^2)(2t),$$

$$v_y = \frac{dy}{dt} = 1.0 \text{ m/s} + (0.025 \text{ m/s}^3)(3t^2),$$

and that the velocity vector is

$$\vec{v} = v_x\hat{\imath} + v_y\hat{\jmath} = (-0.50 \text{ m/s}^2)t\hat{\imath}$$
$$+ [1.0 \text{ m/s} + (0.075 \text{ m/s}^3)t^2]\hat{\jmath}.$$

a) Find the components of the average acceleration in the interval from $t = 0.0$ s to $t = 2.0$ s. b) Find the instantaneous acceleration at $t = 2.0$ s.

SOLUTION a) From Eq. (3–8), in order to calculate the components of the average acceleration, we need the instantaneous velocity at the beginning and the end of the time interval. We find the components of instantaneous velocity at time $t = 0.0$ s

by substituting this value into the above expressions for v_x and v_y. We find that at time $t = 0.0$ s,

$$v_x = 0.0 \text{ m/s}, \qquad v_y = 1.0 \text{ m/s}.$$

We found in Example 3–1 that at $t = 2.0$ s the values of these components are

$$v_x = -1.0 \text{ m/s}, \qquad v_y = 1.3 \text{ m/s}.$$

Thus the components of average acceleration in this interval are

$$(a_{av})_x = \frac{\Delta v_x}{\Delta t} = \frac{-1.0 \text{ m/s} - 0.0 \text{ m/s}}{2.0 \text{ s} - 0.0 \text{ s}} = -0.5 \text{ m/s}^2,$$

$$(a_{av})_y = \frac{\Delta v_y}{\Delta t} = \frac{1.3 \text{ m/s} - 1.0 \text{ m/s}}{2.0 \text{ s} - 0.0 \text{ s}} = 0.15 \text{ m/s}^2.$$

b) From Eq. (3–10) the components of instantaneous acceleration are the time derivatives of the components of instantaneous velocity. We find

$$a_x = \frac{dv_x}{dt} = -0.50 \text{ m/s}^2, \qquad a_y = \frac{dv_y}{dt} = (0.075 \text{ m/s}^3)(2t).$$

We can write the instantaneous acceleration vector $\vec{a}$ as

$$\vec{a} = a_x\hat{\imath} + a_y\hat{\jmath} = (-0.50 \text{ m/s}^2)\hat{\imath} + (0.15 \text{ m/s}^3)t\hat{\jmath}.$$

At time $t = 2.0$ s, the components of instantaneous acceleration are

$$a_x = -0.50 \text{ m/s}^2, \qquad a_y = (0.15 \text{ m/s}^3)(2.0 \text{ s}) = 0.30 \text{ m/s}^2.$$

The acceleration vector at this time is

$$\vec{a} = (-0.50 \text{ m/s}^2)\hat{\imath} + (0.30 \text{ m/s}^2)\hat{\jmath}.$$

The magnitude of acceleration at this time is

$$a = \sqrt{a_x{}^2 + a_y{}^2}$$
$$= \sqrt{(-0.50 \text{ m/s}^2)^2 + (0.30 \text{ m/s}^2)^2} = 0.58 \text{ m/s}^2.$$

The direction of $\vec{a}$ with respect to the positive x-axis is given by the angle β, where

$$\tan \beta = \frac{a_y}{a_x} = \frac{0.30 \text{ m/s}^2}{-0.50 \text{ m/s}^2} = -0.60,$$

$$\beta = 180° - 31° = 149°.$$

We invite you to calculate the instantaneous acceleration at $t = 0.0$ s and $t = 1.0$ s. Figure 3–7 shows the path of the car and the velocity and acceleration vectors at $t = 0.0$ s, 1.0 s, and 2.0 s. Note that the two vectors $\vec{v}$ and $\vec{a}$ are *not* in the same direction at any of these times. The velocity vector is tangent to the path at

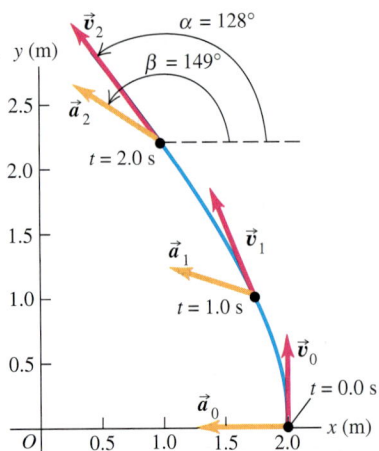

3–7 The path of the radio-controlled car, showing the velocity and acceleration at $t = 0.0$ s ($\vec{v}_0$ and $\vec{a}_0$), $t = 1.0$ s ($\vec{v}_1$ and $\vec{a}_1$), and $t = 2.0$ s ($\vec{v}_2$ and $\vec{a}_2$).

each point, and the acceleration vector points toward the concave side of the path.

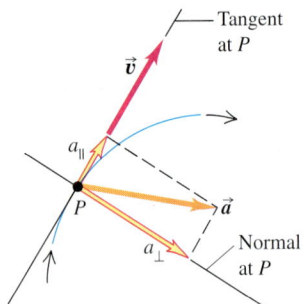

3–8 The acceleration can be resolved into a component $a_\parallel$ parallel to the path (and to the velocity) and a component $a_\perp$ perpendicular to the path (that is, along the normal to the path).

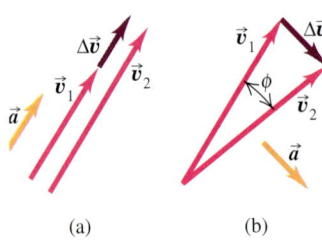

3–9 (a) When $\vec{a}$ is parallel to $\vec{v}$, the magnitude of $\vec{v}$ increases, but its direction does not change. The particle moves in a straight line with changing speed. (b) When $\vec{a}$ is perpendicular to $\vec{v}$, the direction of $\vec{v}$ changes, but its magnitude is constant. The particle moves along a curve with constant speed.

PARALLEL AND PERPENDICULAR COMPONENTS OF ACCELERATION

We can also represent the acceleration of a particle moving in a curved path in terms of components parallel and perpendicular to the velocity at each point (Fig. 3–8). In the figure these components are labeled $a_\parallel$ and $a_\perp$. To see why these components are useful, we'll consider two special cases. In Fig. 3–9a the acceleration vector is *parallel* to the velocity $\vec{v}_1$. The change in $\vec{v}$ during a small time interval Δt is a vector $\Delta\vec{v}$ having the same direction as $\vec{a}$ and hence the same direction as $\vec{v}_1$. The velocity $\vec{v}_2$ at the end of Δt, given by $\vec{v}_2 = \vec{v}_1 + \Delta\vec{v}$, is a vector having the same direction as $\vec{v}_1$ but greater magnitude. In other words, during the time interval Δt the particle in Fig. 3–9a moved in a straight line with increasing speed.

In Fig. 3–9b the acceleration $\vec{a}$ is *perpendicular* to the velocity $\vec{v}$. In a small time interval Δt, the change $\Delta\vec{v}$ is a vector very nearly perpendicular to $\vec{v}_1$, as shown. Again $\vec{v}_2 = \vec{v}_1 + \Delta\vec{v}$, but in this case $\vec{v}_1$ and $\vec{v}_2$ have different directions. As the time interval Δt approaches zero, the angle ϕ in the figure also approaches zero, $\Delta\vec{v}$ becomes perpendicular to *both* $\vec{v}_1$ and $\vec{v}_2$, and $\vec{v}_1$ and $\vec{v}_2$ have the same magnitude. In other words, the speed of the particle stays the same, but the path of the particle curves.

Thus when $\vec{a}$ is *parallel* (or antiparallel) to $\vec{v}$, its effect is to change the magnitude of $\vec{v}$ but not its direction; when $\vec{a}$ is *perpendicular* to $\vec{v}$, its effect is to change the direction of $\vec{v}$ but not its magnitude. In general, $\vec{a}$ may have components *both* parallel and perpendicular to $\vec{v}$, but the above statements are still valid for the individual components. In particular, when a particle travels along a curved path with constant speed, its acceleration is always perpendicular to $\vec{v}$ at each point.

Figure 3–10 shows a particle moving along a curved path for three different situations: constant speed, increasing speed, and decreasing speed. If the speed is constant, $\vec{a}$ is perpendicular, or *normal*, to the path and to $\vec{v}$ and points toward the concave side of the path (Fig. 3–10a). If the speed is increasing, there is still a perpendicular component of $\vec{a}$, but there is also a parallel component having the same direction as $\vec{v}$ (Fig. 3–10b). Then $\vec{a}$ points ahead of the normal to the path. (This was the case in Example 3–2.) If the speed is decreasing, the parallel component has the direction opposite to $\vec{v}$, and $\vec{a}$ points behind the normal to the path (Fig. 3–10c). We will use these ideas in Section 3–5 when we study the special case of motion in a circle.

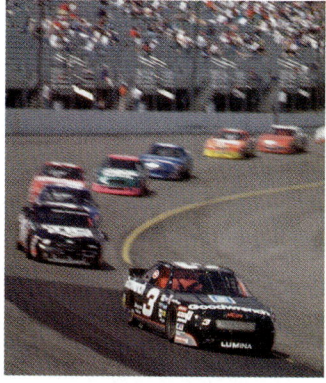

(a)   (b)   (c)

3-10 Velocity and acceleration vectors for a particle moving through a point P on a curved path with (a) constant speed, (b) increasing speed, and (c) decreasing speed.

CAUTION ▶ Note that the two quantities

$$\frac{d|\vec{v}|}{dt} \quad \text{and} \quad \left|\frac{d\vec{v}}{dt}\right|$$

are not the same. The first is the rate of change of speed; it is zero whenever a particle moves with constant speed, even when its direction of motion changes. The second is the magnitude of the vector acceleration; it is zero only when the particle's acceleration is zero, that is, when the particle moves in a straight line with constant speed. ◀

A race car typically slows down when it enters a turn and speeds up as it exits the turn. Thus the car has a component of acceleration parallel or antiparallel to the direction of motion (which describes the change in speed) and a component perpendicular to the direction of motion (which describes the change in direction).

EXAMPLE 3-3

Calculating parallel and perpendicular components of acceleration For the radio-controlled car of Examples 3-1 and 3-2, find the parallel and perpendicular components of the acceleration at $t = 2.0$ s.

SOLUTION To find the parallel and perpendicular components, we need to know the angle between the acceleration vector $\vec{a}$ and the velocity vector $\vec{v}$. In Example 3-2 we found that at $t = 2.0$ s the particle has an acceleration of magnitude 0.58 m/s² at an angle of 149° with respect to the positive x-axis. From Example 3-1, at this same time the velocity vector is at an angle of 128° with respect to the positive x-axis. So Fig. 3-7 shows that the angle between $\vec{a}$ and $\vec{v}$ is 149° − 128° = 21° (Fig. 3-11). The parallel and perpendicular components of acceleration are then

$$a_{\parallel} = a\cos 21° = (0.58 \text{ m/s}^2)\cos 21° = 0.54 \text{ m/s}^2,$$
$$a_{\perp} = a\sin 21° = (0.58 \text{ m/s}^2)\sin 21° = 0.21 \text{ m/s}^2.$$

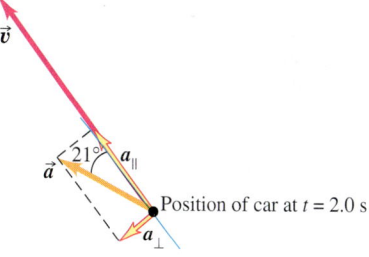

3-11 The parallel and perpendicular components of the acceleration of the radio-controlled car at $t = 2.0$ s.

The parallel component $a_{\parallel}$ is in the same direction as $\vec{v}$, which means that the speed is increasing at this instant. Because the perpendicular component $a_{\perp}$ is not zero, it follows that the car's path is curved at this instant; in other words, the car is turning.

EXAMPLE 3-4

Acceleration of a skier A skier moves along a ski-jump ramp as shown in Fig. 3-12. The ramp is straight from point A to point C and curved from point C onwards. The skier picks up speed as she moves downhill from point A to point E, where her speed is maximum. She slows down after passing point E. Draw the direction of the acceleration vector at points B, D, E, and F.

SOLUTION At point B the skier is moving in a straight line with increasing speed, so her acceleration points downhill, in the same direction as her velocity.

At point D the skier is moving along a curved path, so her acceleration has a component perpendicular to the path. There is also a component in the direction of her motion because she is

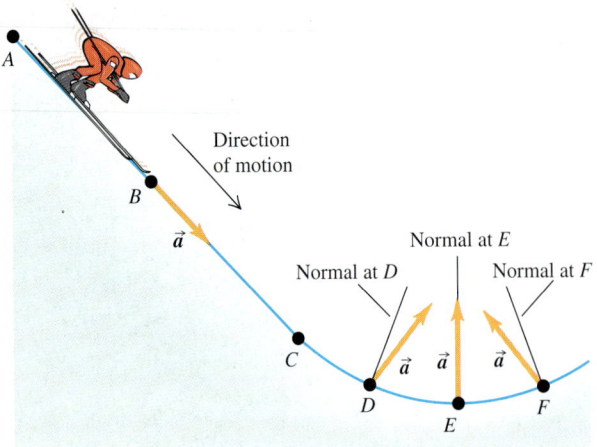

3-12 The direction of a skier's acceleration is different at different points along the path.

still speeding up at this point. So the acceleration vector points *ahead* of the normal to her path at point D.

The skier's speed is instantaneously not changing at point E; the speed is maximum at this point, so its derivative is zero. So there is no parallel component of $\vec{a}$, and the acceleration is perpendicular to her motion. The acceleration is vertical because at point E the skier's path is horizontal.

Finally, at point F the acceleration has a perpendicular component (because her path is curved at this point) and a parallel component *opposite* to the direction of her motion (because she's slowing down). So at this point, the acceleration vector points *behind* the normal to her path.

In the next section we'll examine the skier's acceleration after she flies off the ramp.

3-4 PROJECTILE MOTION

A **projectile** is any body that is given an initial velocity and then follows a path determined entirely by the effects of gravitational acceleration and air resistance. A batted baseball, a thrown football, a package dropped from an airplane, and a bullet shot from a rifle are all projectiles. The path followed by a projectile is called its **trajectory.**

To analyze this common type of motion, we'll start with an idealized model, representing the projectile as a single particle with an acceleration (due to gravity) that is constant in both magnitude and direction. We'll neglect the effects of air resistance and the curvature and rotation of the earth. Like all models, this one has limitations. Curvature of the earth has to be considered in the flight of long-range missiles, and air resistance is of crucial importance to a sky diver. Nevertheless, we can learn a lot from analysis of this simple model. For the remainder of this chapter the phrase "projectile motion" will imply that we're ignoring air resistance. In Chapter 5 we will see what happens when air resistance cannot be ignored.

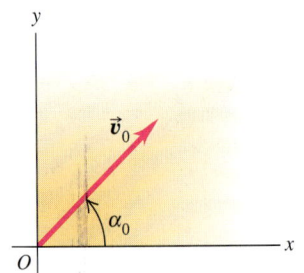

3-13 All projectile motion occurs in the vertical plane containing the initial velocity vector $\vec{v}_0$.

We first notice that projectile motion is always confined to a vertical plane determined by the direction of the initial velocity (Fig. 3–13). This is because the acceleration due to gravity is purely vertical; gravity can't move the projectile sideways. Thus projectile motion is *two-dimensional*. We will call the plane of motion the *xy*-coordinate plane, with the *x*-axis horizontal and the *y*-axis vertically upward.

The key to analysis of projectile motion is that we can treat the *x*- and *y*-coordinates separately. The *x*-component of acceleration is zero, and the *y*-component is constant and equal to $-g$. (Remember that by definition, g is always positive, while with our choice of coordinate directions, a_y is negative.) So *we can analyze projectile motion as a combination of horizontal motion with constant velocity and vertical motion with constant acceleration.* Figure 3–14 shows two projectiles with different *x*-motion but identical *y*-motion; one is dropped from rest and the other is projected horizontally, but both projectiles fall the same distance in the same time.

We can then express all the vector relationships for position, velocity, and acceleration by separate equations for the horizontal and vertical components. The actual motion of the projectile is the superposition of these separate motions. The components of $\vec{a}$ are

$$a_x = 0, \qquad a_y = -g \qquad \text{(projectile motion, no air resistance).} \qquad (3\text{–}14)$$

We will usually use $g = 9.8 \text{ m/s}^2$. Since the *x*-acceleration and *y*-acceleration are both constant, we can use Eqs. (2–8), (2–12), (2–13), and (2–14) directly. For example, suppose that at time $t = 0$ our particle is at the point (x_0, y_0) and that at this time its velocity components have the initial values v_{0x} and v_{0y}. The components of acceleration are $a_x = 0$, $a_y = -g$. Considering the x-motion first, we substitute v_x for v, v_{0x} for v_0, and 0 for a in Eqs. (2–8) and (2–12). We find

$$v_x = v_{0x}, \qquad (3\text{–}15)$$

$$x = x_0 + v_{0x}t. \qquad (3\text{–}16)$$

For the y-motion we substitute y for x, v_y for v, v_{0y} for v_0, and $-g$ for a:

$$v_y = v_{0y} - gt, \qquad (3\text{–}17)$$

$$y = y_0 + v_{0y}t - \frac{1}{2}gt^2. \qquad (3\text{–}18)$$

It's usually simplest to take the initial position (at time $t = 0$) as the origin; in this case, $x_0 = y_0 = 0$. This point might be, for example, the position of a ball at the instant it leaves the thrower's hand or the position of a bullet at the instant it leaves the gun barrel.

Figure 3–15 shows the path of a projectile that starts at (or passes through) the origin at time $t = 0$. The position, velocity, velocity components, and acceleration are shown at a series of times separated by equal intervals. The x-component of acceleration is zero, so v_x is constant. The y-component of acceleration is constant and not zero, so v_y changes by equal amounts in equal times. At the highest point in the trajectory, $v_y = 0$.

We can also represent the initial velocity $\vec{v}_0$ by its magnitude v_0 (the initial speed) and its angle α_0 with the positive x-axis. In terms of these quantities, the components v_{0x} and v_{0y} of the initial velocity are

$$v_{0x} = v_0 \cos \alpha_0, \qquad v_{0y} = v_0 \sin \alpha_0. \qquad (3\text{–}19)$$

Using these relations in Eqs. (3–15) through (3–18) and setting $x_0 = y_0 = 0$, we find

$$x = (v_0 \cos \alpha_0)t \qquad \text{(projectile motion)}, \qquad (3\text{–}20)$$

$$y = (v_0 \sin \alpha_0)t - \frac{1}{2}gt^2 \qquad \text{(projectile motion)}, \qquad (3\text{–}21)$$

$$v_x = v_0 \cos \alpha_0 \qquad \text{(projectile motion)}, \qquad (3\text{–}22)$$

$$v_y = v_0 \sin \alpha_0 - gt \qquad \text{(projectile motion)}. \qquad (3\text{–}23)$$

3–14 Independence of horizontal and vertical motion. The red ball is dropped from rest, and the yellow ball is simultaneously projected horizontally; successive images in this stroboscopic photograph are separated by equal time intervals. At any given time, both balls have the same y-position, y-velocity, and y-acceleration, despite having different x-positions and x-velocities.

3–15 The trajectory of a body projected with an initial velocity $\vec{v}_0$ at an angle α_0 above the horizontal with negligible air resistance. The distance R is the horizontal range, and h is the maximum height.

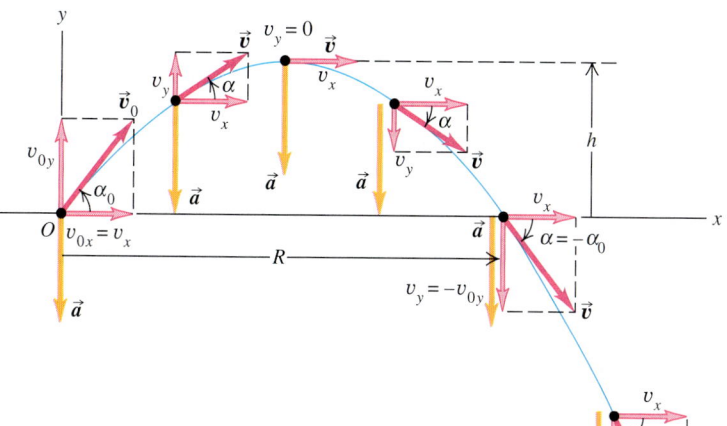

3.7
Target Practice II

3.8
Gravitational Constant on Planet Zeus

3.9
Motorcycle Daredevil

3–16 Stroboscopic photograph of a bouncing ball, showing parabolic trajectories after each bounce. Successive images are separated by equal time intervals, as in Fig. 3–14. Each peak in the trajectory is lower than the preceding one because of energy loss during the "bounce," or collision with the horizontal surface.

Rock fragments ejected from this volcano follow nearly parabolic trajectories. This *pyroclastic* eruption is caused by the sudden release of high-pressure gases from within the erupting lava. A web site at the U.S. Geological Survey describes the physics of volcanoes (for a link, visit the *University Physics* web site)

3–17 Air resistance has a large cumulative effect on the motion of a baseball.

These equations describe the position and velocity of the projectile in Fig. 3–15 at any time t.

We can get a lot of information from these equations. For example, at any time the distance r of the projectile from the origin (the magnitude of the position vector $\vec{r}$) is given by

$$r = \sqrt{x^2 + y^2}. \qquad (3\text{–}24)$$

The projectile's speed (the magnitude of its velocity) at any time is

$$v = \sqrt{v_x{}^2 + v_y{}^2}. \qquad (3\text{–}25)$$

The *direction* of the velocity, in terms of the angle α it makes with the positive x-axis, is given by

$$\tan \alpha = \frac{v_y}{v_x}. \qquad (3\text{–}26)$$

The velocity vector $\vec{v}$ is tangent to the trajectory at each point.

We can derive an equation for the trajectory's shape in terms of x and y by eliminating t. From Eqs. (3–20) and (3–21), which assume $x_0 = y_0 = 0$, we find $t = x/(v_0 \cos \alpha_0)$ and

$$y = (\tan \alpha_0)x - \frac{g}{2v_0{}^2 \cos^2 \alpha_0} x^2. \qquad (3\text{–}27)$$

Don't worry about the details of this equation; the important point is its general form. The quantities v_0, $\tan \alpha_0$, $\cos \alpha_0$, and g are constants, so the equation has the form

$$y = bx - cx^2,$$

where b and c are constants. This is the equation of a *parabola*. In projectile motion, with our simple model, the trajectory is always a parabola (Fig. 3–16).

We mentioned at the beginning of this section that air resistance isn't always negligible. When it has to be included, the calculations become a lot more complicated; the effects of air resistance depend on velocity, so the acceleration is no longer constant. Figure 3–17 shows a computer simulation of the trajectory of a baseball with $v_0 = 50$ m/s and $\alpha_0 = 53.1°$, both without air resistance and with air resistance proportional to the square of the baseball's speed. We see that air resistance has a very large effect; the maximum height and range both decrease, and the trajectory is no longer a parabola. Projectile motion with air resistance is discussed in greater detail in Section 5–7.

Problem–Solving Strategy

PROJECTILE MOTION

The strategies we used in Sections 2–5 and 2–6 for straight-line, constant-acceleration problems are also useful here.

1. Define your coordinate system and make a sketch showing your axes. Usually, it is easiest to place the origin at the initial ($t = 0$) position of the projectile, with the x-axis horizontal and the y-axis upward. Then $x_0 = 0$, $y_0 = 0$, $a_x = 0$, and $a_y = -g$.

2. List the known and unknown quantities. In some problems the components (or magnitude and direction) of initial velocity will be given, and you can use Eqs. (3–20) through (3–23) to find the coordinates and velocity components at some later time. In other problems you might be given two points on the trajectory and be asked to find the initial velocity. Be sure you know which quantities are given and which are to be found.

3. It often helps to state the problem in words and then translate into symbols. For example, *when* does the particle arrive at a certain point? (That is, at what value of t?) *Where* is the particle when its velocity has a certain value? (That is, what are the values of x and y when v_x or v_y has the specified value?)

4. At the highest point in a trajectory, $v_y = 0$. So the question "When does the projectile reach its highest point?" translates into "What is the value of t when $v_y = 0$?" Similarly, "When does the projectile return to its initial elevation?" translates into "What is the value of t when $y = y_0$?"

5. Resist the temptation to break the trajectory into segments and analyze each one separately. You don't have to start all over, with a new axis system and a new time scale, when the projectile reaches its highest point. It is usually easier to set up Eqs. (3–20) through (3–23) at the start and use the same axes and time scale throughout the problem.

EXAMPLE 3–5

Acceleration of a skier, continued Let's consider again the skier in Example 3–4. What is her acceleration at points G, H, and I in Fig. 3–18 *after* she flies off the ramp? Neglect air resistance.

SOLUTION The skier's acceleration changed from point to point while she was on the ramp. But as soon as she leaves the ramp, she becomes a projectile. So at points G, H, and I, and indeed at *all* points after she leaves the ramp, the skier's acceleration points vertically downward and has magnitude g. No matter how complicated the acceleration of a particle before it becomes a projectile, its acceleration as a projectile is given by $a_x = 0$, $a_y = -g$.

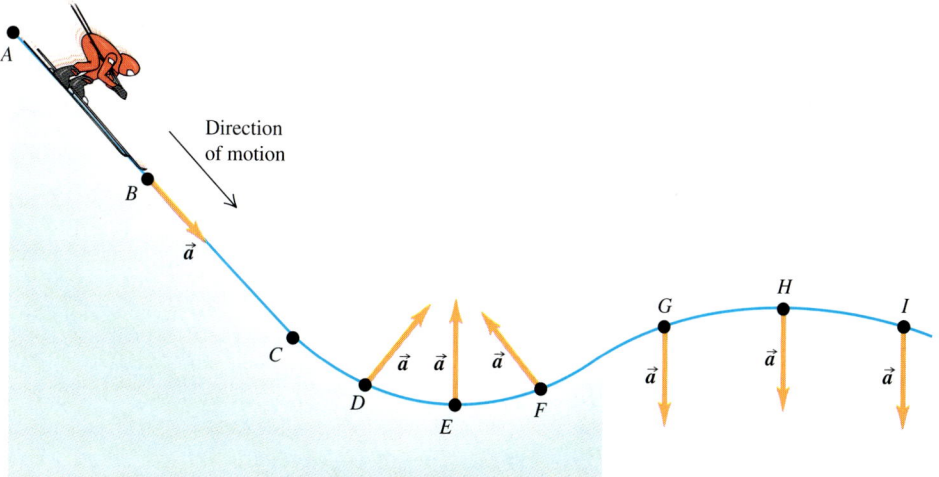

3–18 Once the skier is in projectile motion, her acceleration is constant and downward.

EXAMPLE 3–6

A body projected horizontally A motorcycle stunt rider rides off the edge of a cliff. Just at the edge his velocity is horizontal, with magnitude 9.0 m/s. Find the motorcycle's position, distance from the edge of the cliff, and velocity after 0.50 s.

SOLUTION The coordinate system is shown in Fig. 3–19. We choose the origin at the edge of the cliff, where the motorcycle first becomes a projectile, so $x_0 = 0$ and $y_0 = 0$. The initial velocity is purely horizontal (that is, $\alpha_0 = 0$), so the initial velocity components are $v_{0x} = v_0 \cos \alpha_0 = 9.0$ m/s and $v_{0y} = v_0 \sin \alpha_0 = 0$.

Where is the motorcycle at $t = 0.50$ s? The answer is contained in Eqs. (3–20) and (3–21), which give x and y as functions of time. When $t = 0.50$ s, the x- and y-coordinates are

$$x = v_{0x}t = (9.0 \text{ m/s})(0.50 \text{ s}) = 4.5 \text{ m},$$

$$y = -\frac{1}{2} gt^2 = -\frac{1}{2}(9.8 \text{ m/s}^2)(0.50 \text{ s})^2 = -1.2 \text{ m}.$$

The negative value of y shows that at this time the motorcycle is below its starting point.

What is the motorcycle's distance from the origin at this time? From Eq. (3–24),

$$r = \sqrt{x^2 + y^2} = \sqrt{(4.5 \text{ m})^2 + (-1.2 \text{ m})^2} = 4.7 \text{ m}.$$

What is the velocity at time $t = 0.50$ s? From Eqs. (3–22) and (3–23), which give v_x and v_y as functions of time, the components of velocity at time $t = 0.50$ s are

$$v_x = v_{0x} = 9.0 \text{ m/s},$$

$$v_y = -gt = (-9.8 \text{ m/s}^2)(0.50 \text{ s}) = -4.9 \text{ m/s}.$$

The motorcycle has the same horizontal velocity v_x as when it left the cliff at $t = 0$, but in addition there is a vertical velocity v_y

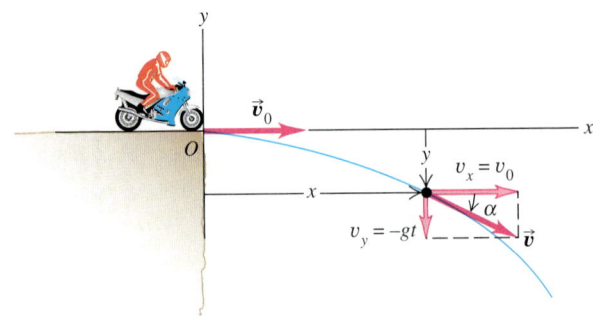

3–19 Trajectory of a body projected horizontally.

directed downward (in the negative y-direction). If we use unit vectors, the velocity at $t = 0.50$ s is

$$\vec{v} = v_x \hat{\imath} + v_y \hat{\jmath} = (9.0 \text{ m/s})\hat{\imath} + (-4.9 \text{ m/s})\hat{\jmath}.$$

We can also express the velocity in terms of magnitude and direction. From Eq. (3–25), the speed (magnitude of the velocity) at this time is

$$v = \sqrt{v_x^2 + v_y^2}$$
$$= \sqrt{(9.0 \text{ m/s})^2 + (-4.9 \text{ m/s})^2} = 10.2 \text{ m/s}.$$

From Eq. (3–26), the angle α of the velocity vector is

$$\alpha = \arctan \frac{v_y}{v_x}$$
$$= \arctan \left(\frac{-4.9 \text{ m/s}}{9.0 \text{ m/s}} \right) = -29°.$$

At this time the velocity is 29° below the horizontal.

EXAMPLE 3–7

Height and range of a baseball A batter hits a baseball so that it leaves the bat with an initial speed $v_0 = 37.0$ m/s at an initial angle $\alpha_0 = 53.1°$, at a location where $g = 9.80$ m/s^2 (Fig. 3–20). Let's see how we can predict where the ball is and how it's moving at a certain time, how we can find the ball's maximum height, and how we can find the distance from home plate to where the ball comes down. (As shown in Fig. 3–17, the effects of air resistance on the motion of a baseball aren't really negligible. We'll nonetheless ignore air resistance for this example.)

The ball is probably struck a meter or so above ground level, but we neglect this distance and assume that it starts at ground level ($y_0 = 0$). The initial velocity of the ball has components

$$v_{0x} = v_0 \cos \alpha_0 = (37.0 \text{ m/s}) \cos 53.1° = 22.2 \text{ m/s},$$

$$v_{0y} = v_0 \sin \alpha_0 = (37.0 \text{ m/s}) \sin 53.1° = 29.6 \text{ m/s}.$$

a) Find the position of the ball, and the magnitude and direction of its velocity, when $t = 2.00$ s.

Using the coordinate system shown in Fig. 3–15, we want to find x, y, v_x, and v_y at time $t = 2.00$ s. From Eqs. (3–20) through (3–23),

$$x = v_{0x}t = (22.2 \text{ m/s})(2.00 \text{ s}) = 44.4 \text{ m},$$

$$y = v_{0y}t - \frac{1}{2} gt^2$$

$$= (29.6 \text{ m/s})(2.00 \text{ s}) - \frac{1}{2}(9.80 \text{ m/s}^2)(2.00 \text{ s})^2$$

$$= 39.6 \text{ m},$$

$$v_x = v_{0x} = 22.2 \text{ m/s},$$

$$v_y = v_{0y} - gt = 29.6 \text{ m/s} - (9.80 \text{ m/s}^2)(2.00 \text{ s})$$

$$= 10.0 \text{ m/s}.$$

The y-component of velocity is positive, which means that the ball is still moving upward at this time (Fig. 3–20). The magnitude and direction of the velocity are found from Eqs. (3–25) and (3–26):

$$v = \sqrt{v_x^2 + v_y^2} = \sqrt{(22.2 \text{ m/s})^2 + (10.0 \text{ m/s})^2}$$

$$= 24.3 \text{ m/s},$$

$$\alpha = \arctan \left(\frac{10.0 \text{ m/s}}{22.2 \text{ m/s}} \right) = \arctan 0.450 = 24.2°.$$

3–20 Trajectory of a batted baseball, treated as a projectile.

The velocity is 24.2° above the horizontal.

b) Find the time when the ball reaches the highest point of its flight, and find its height h at this point.

At the highest point, the vertical velocity v_y is zero. When does this happen? Call the time t_1; then

$$v_y = 0 = v_{0y} - gt_1,$$

$$t_1 = \frac{v_{0y}}{g} = \frac{29.6 \text{ m/s}}{9.80 \text{ m/s}^2} = 3.02 \text{ s}.$$

The height h at this time is the value of y when $t = t_1 = 3.02$ s:

$$h = v_{0y}t_1 - \frac{1}{2}gt_1^2$$

$$= (29.6 \text{ m/s})(3.02 \text{ s}) - \frac{1}{2}(9.80 \text{ m/s}^2)(3.02 \text{ s})^2$$

$$= 44.7 \text{ m}.$$

Alternatively, we can apply the constant-acceleration formula Eq. (2–13) to the y-motion:

$$v_y^2 = v_{0y}^2 + 2a_y(y - y_0) = v_{0y}^2 - 2g(y - y_0).$$

At the highest point, $v_y = 0$ and $y = h$. Substituting these, along with $y_0 = 0$, we find

$$0 = v_{0y}^2 - 2gh,$$

$$h = \frac{v_{0y}^2}{2g} = \frac{(29.6 \text{ m/s})^2}{2(9.80 \text{ m/s}^2)} = 44.7 \text{ m}.$$

This is roughly half the height above the playing field of the roof of Skydome in Toronto.

c) Find the *horizontal range R*, that is, the horizontal distance from the starting point to the point at which the ball hits the ground.

First, *when* does the ball hit the ground? This occurs when $y = 0$. Call this time t_2; then

$$y = 0 = v_{0y}t_2 - \frac{1}{2}gt_2^2 = t_2\left(v_{0y} - \frac{1}{2}gt_2\right).$$

This is a quadratic equation for t_2. It has two roots,

$$t_2 = 0 \quad \text{and} \quad t_2 = \frac{2v_{0y}}{g} = \frac{2(29.6 \text{ m/s})}{9.80 \text{ m/s}^2} = 6.04 \text{ s}.$$

There are two times at which $y = 0$; $t_2 = 0$ is the time the ball *leaves* the ground, and $t_2 = 6.04$ s is the time of its return. This is exactly twice the time to reach the highest point, so the time of descent equals the time of ascent. (This is *always* true if the starting and end points are at the same elevation and air resistance can be neglected. We will prove this in Example 3–9.)

The horizontal range R is the value of x when the ball returns to the ground, that is, at $t = 6.04$ s:

$$R = v_{0x}t_2 = (22.2 \text{ m/s})(6.04 \text{ s}) = 134 \text{ m}.$$

For comparison the distance from home plate to the center field fence at Pittsburgh's Three Rivers Stadium is 125 m. Can you show that when the ball reaches this fence, it is 11.4 m above the ground? (This is more than enough to clear the fence, so this ball is a home run.)

The vertical component of velocity when the ball hits the ground is

$$v_y = v_{0y} - gt_2 = 29.6 \text{ m/s} - (9.80 \text{ m/s}^2)(6.04 \text{ s})$$

$$= -29.6 \text{ m/s}^2.$$

That is, v_y has the same magnitude as the initial vertical velocity v_{0y} but the opposite direction (down). Since v_x is constant, the angle $\alpha = -53.1°$ (below the horizontal) at this point is the negative of the initial angle $\alpha_0 = 53.1°$.

d) If the ball could continue to travel on *below* its original level (through an appropriately shaped hole in the ground), then negative values of y, corresponding to times greater than 6.04 s, would be possible. Can you compute the position and velocity at a time 7.55 s after the start, corresponding to the last position shown in Fig. 3–15? The results are

$$x = 168 \text{ m}, \qquad y = -55.8 \text{ m},$$

$$v_x = 22.2 \text{ m/s}, \qquad v_y = -44.4 \text{ m/s}.$$

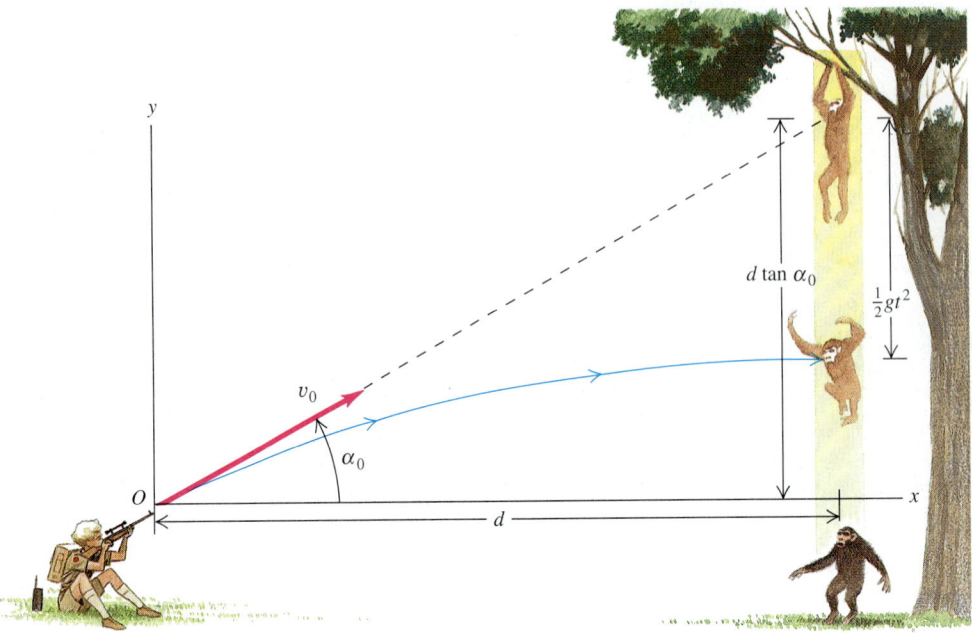

3–21 The tranquilizer dart hits the falling monkey.

EXAMPLE 3–8

The zoo keeper and the monkey A clever monkey escapes from the zoo. The zoo keeper finds him in a tree. After failing to entice the monkey down, the zoo keeper points her tranquilizer gun directly at the monkey and shoots (Fig. 3–21). The clever monkey lets go at the same instant the dart leaves the gun barrel, intending to land on the ground and escape. Show that the dart *always* hits the monkey, regardless of the dart's muzzle velocity (provided that it gets to the monkey before he hits the ground).

SOLUTION We place the origin of coordinates at the end of the barrel of the tranquilizer gun. To show that the dart hits the monkey, we have to prove that there is some time when the monkey and the dart have the same x-coordinate and the same y-coordinate. First let's ask when the x-coordinates x_{monkey} and x_{dart} are the same. The monkey drops straight down, so $x_{\text{monkey}} = d$ at *all* times. For the dart we use Eq. (3–20): $x_{\text{dart}} = (v_0 \cos \alpha_0)t$. When these x-coordinates are equal, $d = (v_0 \cos \alpha_0)t$, or

$$t = \frac{d}{v_0 \cos \alpha_0}.$$

Now we ask whether y_{monkey} and y_{dart} are also equal at this time; if they are, we have a hit. The monkey is in one-dimensional free fall; his position at any time is given by Eq. (2–12), with appropriate symbol changes. Figure 3–21 shows that the monkey's

initial height is $d \tan \alpha_0$ (the opposite side of a right triangle with angle α_0 and adjacent side d), and we find

$$y_{\text{monkey}} = d \tan \alpha_0 - \frac{1}{2} gt^2.$$

For the dart we use Eq. (3–21):

$$y_{\text{dart}} = (v_0 \sin \alpha_0)t - \frac{1}{2} gt^2.$$

So we see that if $d \tan \alpha_0 = (v_0 \sin \alpha_0)t$ at the time when the two x-coordinates are equal, then $y_{\text{monkey}} = y_{\text{dart}}$, and we have a hit. To prove that this happens, we replace t with $d/(v_0 \cos \alpha_0)$, the time when $x_{\text{monkey}} = x_{\text{dart}}$; then

$$(v_0 \sin \alpha_0)t = (v_0 \sin \alpha_0) \frac{d}{v_0 \cos \alpha_0} = d \tan \alpha_0.$$

We have proved that at the time the x-coordinates are equal, the y-coordinates are also equal; a dart aimed at the initial position of the monkey always hits it, no matter what v_0 is. This result is also independent of the value of g, the acceleration due to gravity. With no gravity ($g = 0$), the monkey would remain motionless, and the dart would travel in a straight line to hit him. With gravity, both "fall" the same distance ($\frac{1}{2}gt^2$) below their $g = 0$ positions, and the dart still hits the monkey.

EXAMPLE 3–9

Height and range of a projectile For a projectile launched with speed v_0 at initial angle α_0 (between 0 and 90°), derive general expressions for the maximum height h and horizontal range R (Fig. 3–15). For a given v_0, what value of α_0 gives maximum height? What value gives maximum horizontal range?

SOLUTION We follow the same pattern as in part (b) of Example 3–7. First, for a given α_0, when does the projectile reach its highest point? At this point, $v_y = 0$, so the time t_1 at the highest point ($y = h$) is given by Eq. (3–23):

$$v_y = v_0 \sin \alpha_0 - gt_1 = 0, \qquad t_1 = \frac{v_0 \sin \alpha_0}{g}.$$

Next, in terms of v_0 and α_0, what is the value of y at this time? From Eq. (3–21),

$$h = v_0 \sin \alpha_0 \left(\frac{v_0 \sin \alpha_0}{g} \right) - \frac{1}{2} g \left(\frac{v_0 \sin \alpha_0}{g} \right)^2$$

$$= \frac{v_0^2 \sin^2 \alpha_0}{2g}.$$

If we vary α_0, the maximum value of h occurs when $\sin \alpha_0 = 1$ and $\alpha_0 = 90°$; that is, when the projectile is launched straight up. That's what we should expect. If it is launched horizontally, as in Example 3–6, $\alpha_0 = 0$ and the maximum height is zero!

To derive a general expression for the horizontal range R, we first find an expression for the time t_2 when the projectile returns to its initial elevation. At that time, $y = 0$, and from Eq. (3–21),

$$(v_0 \sin \alpha_0) t_2 - \frac{1}{2} g t_2^2 = t_2 (v_0 \sin \alpha_0 - \frac{1}{2} g t_2) = 0.$$

The two roots of this quadratic equation for t_2 are $t_2 = 0$ (the time of launch) and $t_2 = (2v_0 \sin \alpha_0)/g$. The horizontal range R is the value of x at the second time. From Eq. (3–20),

$$R = (v_0 \cos \alpha_0) \frac{2v_0 \sin \alpha_0}{g}.$$

We can now use the trigonometric identity $2 \sin \alpha_0 \cos \alpha_0 = \sin 2\alpha_0$ to rewrite this as

$$R = \frac{v_0^2 \sin 2\alpha_0}{g}.$$

The maximum value of $\sin 2\alpha_0$ is 1; this occurs when $2\alpha_0 = 90°$, or $\alpha_0 = 45°$. This angle gives the maximum range for a given initial speed.

Figure 3–22 is based on a composite photograph of three trajectories of a ball projected from a spring gun at angles of 30°, 45°, and 60°. The initial speed v_0 is approximately the same in all three cases. The horizontal ranges are nearly the same for the 30° and 60° angles, and the range for 45° is greater than either. Can you prove that for a given value of v_0 the range is the same for an initial angle α_0 as for an initial angle $90° - \alpha_0$?

Comparing the expressions for t_1 and t_2, we see that $t_2 = 2t_1$; that is, the total flight time is twice the time up to the highest point. It follows that the time to reach the highest point equals the time to fall from there back to the initial elevation, as we asserted in Example 3–7.

CAUTION ▶ We don't recommend memorizing the above expressions for h and R. They are applicable only in the special circumstances we have described. In particular, the expression for the range R can be used *only* when launch and landing heights are equal. There are many end-of-chapter problems to which these equations are *not* applicable. ◀

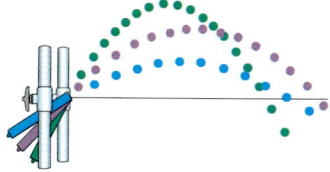

3–22 A firing angle of 45° (shown in purple) gives the maximum horizontal range. The range is less with firing angles of 30° (blue) or 60° (green).

EXAMPLE 3–10

Different initial and final heights You toss a water balloon from your window 8.0 m above the ground. When the water balloon leaves your hand, it is moving at 10.0 m/s at an angle of 20° below the horizontal. How far horizontally from your window will the balloon hit the ground? Ignore air resistance.

SOLUTION We have $v_0 = 10.0$ m/s and $\alpha_0 = -20°$; the angle is negative because the initial velocity is below the horizontal. We place the origin of coordinates at the point where the balloon leaves your hand, so that $x_0 = y_0 = 0$ and ground level is at $y = -8.0$ m. We want to find the value of x at the point where the balloon reaches the ground. The initial and final heights of the balloon are different, so we can't use the expression for the horizontal range found in Example 3–9. Instead, we first find the time t when the balloon reaches $y = -8.0$ m. We rewrite Eq. (3–21) in the standard form for a quadratic equation for t:

$$\frac{1}{2} gt^2 - (v_0 \sin \alpha_0) t + y = 0.$$

The roots of this equation are

$$t = \frac{v_0 \sin \alpha_0 \pm \sqrt{(-v_0 \sin \alpha_0)^2 - 4 \left(\frac{1}{2} g \right) y}}{2 \left(\frac{1}{2} g \right)}$$

$$= \frac{v_0 \sin \alpha_0 \pm \sqrt{v_0^2 \sin^2 \alpha_0 - 2gy}}{g}$$

$$= \left[\begin{array}{c} (10.0 \text{ m/s}) \sin (-20°) \\ \pm \sqrt{(10.0 \text{ m/s})^2 \sin^2 (-20°) - 2(9.8 \text{ m/s}^2)(-8.0 \text{ m})} \end{array} \right] \Big/ 9.8 \text{ m/s}^2$$

$$= -1.7 \text{ s} \quad \text{or} \quad 0.98 \text{ s}.$$

We can discard the negative root, since it refers to a time before the balloon left your hand (see Example 2–8 in Section 2–6 for a discussion of such "fictional" solutions to quadratic equations). The positive root tells us that the balloon takes 0.98 s to reach the ground. From Eq. (3–20), the balloon's x-coordinate at that time is

$$x = (v_0 \cos \alpha_0) t = (10.0 \text{ m/s})(\cos (-20°))(0.98 \text{ s})$$

$$= 9.2 \text{ m}.$$

The balloon hits the ground a horizontal distance of 9.2 m from your window.

(a)

(b)

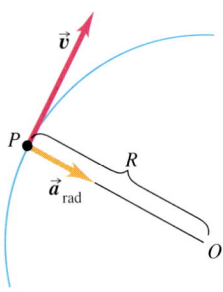

3–23 Finding the change in velocity, $\Delta\vec{v}$, of a particle moving in a circle with constant speed.

4.3

Magnitude of Centripetal Acceleration

3–5 MOTION IN A CIRCLE

When a particle moves along a curved path, the direction of its velocity changes. As we saw in Section 3–3, this means that the particle *must* have a component of acceleration perpendicular to the path, even if its speed is constant. In this section we'll calculate the acceleration for the important special case of motion in a circle.

UNIFORM CIRCULAR MOTION

When a particle moves in a circle with *constant speed,* the motion is called **uniform circular motion.** A car rounding a curve with constant radius at constant speed, a satellite moving in a circular orbit, and an ice skater skating in a circle with constant speed are all examples of uniform circular motion. There is no component of acceleration parallel (tangent) to the path; otherwise, the speed would change. The component of acceleration perpendicular (normal) to the path, which causes the direction of the velocity to change, is related in a simple way to the speed of the particle and the radius of the circle. Our next project is to derive this relation.

First we note that this is a different problem from the projectile motion considered in Section 3–4, in which the acceleration was constant in both its magnitude *(g)* and its direction (straight down). In uniform circular motion the acceleration is perpendicular to the velocity at each instant; as the direction of the velocity changes, the direction of the acceleration also changes. As we will see, the acceleration vector at each point in the circular path is directed toward the *center* of the circle.

Figure 3–23a shows a particle moving with constant speed in a circular path of radius R with center at O. The particle moves from P_1 to P_2 in a time Δt. The vector change in velocity $\Delta\vec{v}$ during this time is shown in Fig. 3–23b.

The angles labeled $\Delta\phi$ in Figs. 3–23a and 3–23b are the same because $\vec{v}_1$ is perpendicular to the line OP_1 and $\vec{v}_2$ is perpendicular to the line OP_2. Hence the triangles OP_1P_2 (Fig. 3–23a) and Op_1p_2 (Fig. 3–23b) are *similar.* Ratios of corresponding sides are equal, so

$$\frac{|\Delta\vec{v}|}{v_1} = \frac{\Delta s}{R}, \qquad \text{or} \qquad |\Delta\vec{v}| = \frac{v_1}{R}\,\Delta s.$$

The magnitude a_{av} of the average acceleration during Δt is therefore

$$a_{av} = \frac{|\Delta\vec{v}|}{\Delta t} = \frac{v_1}{R}\frac{\Delta s}{\Delta t}.$$

The magnitude a of the *instantaneous* acceleration $\vec{a}$ at point P_1 is the limit of this expression as we take point P_2 closer and closer to point P_1:

$$a = \lim_{\Delta t \to 0} \frac{v_1}{R}\frac{\Delta s}{\Delta t} = \frac{v_1}{R}\lim_{\Delta t \to 0}\frac{\Delta s}{\Delta t}.$$

But the limit of $\Delta s/\Delta t$ is the speed v_1 at point P_1. Also, P_1 can be any point on the path, so we can drop the subscript and let v represent the speed at any point. Then

$$a_{rad} = \frac{v^2}{R} \qquad \text{(uniform circular motion).} \qquad (3\text{–}28)$$

We have added the subscript "rad" as a reminder that the direction of the instantaneous acceleration at each point is always along a radius of the circle, toward its center. This is consistent with our discussion in Section 3–3: the acceleration vector points toward the *concave* side of the particle's circular path, that is, toward the *inside* of the circle (never the outside). Because the speed is constant, the acceleration is always perpendicular to the instantaneous velocity. This is shown in Fig. 3–23c; compare with Fig. 3–10a.

We conclude: *In uniform circular motion, the magnitude a of the instantaneous acceleration is equal to the square of the speed v divided by the radius R of the circle. Its direction is perpendicular to $\vec{v}$ and inward along the radius.* Because the acceleration is always directed toward the center of the circle, it is sometimes called **centripetal acceleration.** The word *centripetal* is derived from two Greek words meaning "seeking the center." Figure 3–24 shows the directions of the velocity and acceleration vectors at several points for a particle moving with uniform circular motion. Compare this with the projectile motion shown in Fig. 3–15, in which the acceleration is always directed straight down and is *not* perpendicular to the path, except at one point.

We can also express the magnitude of the acceleration in uniform circular motion in terms of the **period** T of the motion, the time for one revolution (one complete trip around the circle). In a time T the particle travels a distance equal to the circumference $2\pi R$ of the circle, so its speed is

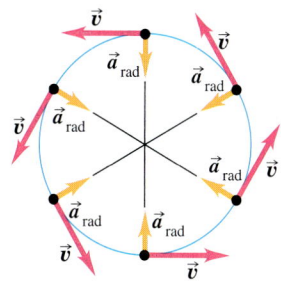

3–24 For a particle in uniform circular motion, the velocity at each point is tangent to the circle and the acceleration is directed toward the center.

$$v = \frac{2\pi R}{T}. \qquad (3\text{–}29)$$

When we substitute this into Eq. (3–28), we obtain the alternative expression

$$a_{rad} = \frac{4\pi^2 R}{T^2} \qquad \text{(uniform circular motion)}. \qquad (3\text{–}30)$$

EXAMPLE 3-11

A BMW Z3 roadster has a "lateral acceleration" of $0.87g$, which is $(0.87)(9.8\ \text{m/s}^2) = 8.5\ \text{m/s}^2$. This represents the maximum centripetal acceleration that can be attained without skidding out of the circular path. If the car is traveling at a constant 40 m/s (about 89 mi/h, or 144 km/h), what is the minimum radius of curve it can negotiate? (Assume that the curve is unbanked.)

SOLUTION We are given a_{rad} and v, so we first solve Eq. (3–28) for R:

$$R = \frac{v^2}{a_{rad}} = \frac{(40\ \text{m/s})^2}{8.5\ \text{m/s}^2} = 190\ \text{m} \qquad \text{(about 620 ft)}.$$

If the curve is banked, the radius can be smaller, as we will see in Chapter 5.

EXAMPLE 3-12

In a carnival ride, the passengers travel at constant speed in a circle of radius 5.0 m. They make one complete circle in 4.0 s. What is their acceleration?

SOLUTION The speed is constant, so this is uniform circular motion. We can use Eq. (3–30) to calculate the acceleration, since $R = 5.0$ m and the period $T = 4.0$ s (the time for one revolution) are both given:

$$a_{rad} = \frac{4\pi^2 (5.0\ \text{m})}{(4.0\ \text{s})^2} = 12\ \text{m/s}^2.$$

We could instead calculate a_{rad} using Eq. (3–28), but this approach requires an extra step to first calculate the speed v. From Eq. (3–29), the speed is the circumference of the circle divided by the period T:

$$v = \frac{2\pi R}{T} = \frac{2\pi (5.0\ \text{m})}{4.0\ \text{s}} = 7.9\ \text{m/s}.$$

The centripetal acceleration is then

$$a_{rad} = \frac{v^2}{R} = \frac{(7.9\ \text{m/s})^2}{5.0\ \text{m}} = 12\ \text{m/s}^2.$$

As in the preceding example, the direction of $\vec{a}$ is always toward the center of the circle. The magnitude of $\vec{a}$ is greater than g, the acceleration due to gravity, so this is not a ride for the faint-hearted. (Some roller coasters subject their passengers to accelerations as great as $4g$.)

NON-UNIFORM CIRCULAR MOTION

We have assumed throughout this section that the particle's speed is constant. If the speed varies, we call the motion **non-uniform circular motion.** An example is a roller coaster car that slows down and speeds up as it moves around a vertical loop. In

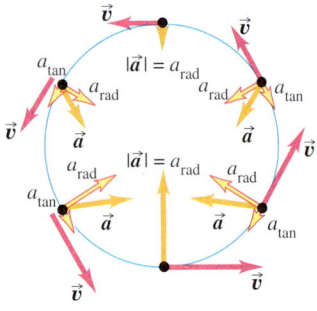

3–25 A particle moving in a vertical loop, like a roller coaster car, with a varying speed. The radial component of acceleration a_{rad} is largest where the speed is greatest (at the bottom of the loop) and smallest where the speed is least (at the top). The tangential component of acceleration a_{tan} is in the same direction as $\vec{v}$ when the particle is speeding up (going downhill) and opposite $\vec{v}$ when it is slowing down (going uphill).

non-uniform circular motion, Eq. (3–28) still gives the *radial* component of acceleration $a_{rad} = v^2/R$, which is always *perpendicular* to the instantaneous velocity and directed toward the center of the circle. But since the speed v has different values at different points in the motion, the value of a_{rad} is not constant. The radial (centripetal) acceleration is greatest at the point in the circle where the speed is greatest.

In non-uniform circular motion there is also a component of acceleration that is *parallel* to the instantaneous velocity. This is the component $a_{\parallel}$ that we discussed in Section 3–3; here we call this component a_{tan} to emphasize that it is *tangent* to the circle. From the discussion at the end of Section 3–3 we see that the tangential component of acceleration a_{tan} is equal to the rate of change of *speed*. Thus

$$a_{rad} = \frac{v^2}{R} \quad \text{and} \quad a_{tan} = \frac{d|\vec{v}|}{dt} \quad \text{(non-uniform circular motion).} \quad (3\text{–}31)$$

The vector acceleration of a particle moving in a circle with varying speed is the vector sum of the radial and tangential components of accelerations. The tangential component is in the same direction as the velocity if the particle is speeding up, and is in the opposite direction if the particle is slowing down (Fig. 3–25).

In *uniform* circular motion there is no tangential component of acceleration, but the radial component is the magnitude of $d\vec{v}/dt$. We have mentioned before that the two quantities $|d\vec{v}/dt|$ and $d|\vec{v}|/dt$ are in general *not* equal. In uniform circular motion the first is constant and equal to v^2/R; the second is zero.

3-6 RELATIVE VELOCITY

You've no doubt observed how a car that is moving slowly forward appears to be moving backward when you pass it. In general, when two observers measure the velocity of a moving body, they get different results if one observer is moving relative to the other. The velocity seen by a particular observer is called the velocity *relative* to that observer, or simply **relative velocity.** We'll first consider relative velocity along a straight line, then generalize to relative velocity in a plane. Recall that for straight-line (one-dimensional) motion we use the term *velocity* to mean the component of the velocity vector along the line of motion; this can be positive, negative, or zero.

Airshow pilots face a complicated problem involving relative velocities. They must keep track of their velocity relative to the air (to maintain enough airflow over the wings to sustain lift), their velocity relative to each other (to keep a tight formation without colliding), and their velocity relative to their audience (to remain in sight of the spectators).

RELATIVE VELOCITY IN ONE DIMENSION

A woman walks with a velocity of 1.0 m/s along the aisle of a train that is moving with a velocity of 3.0 m/s (Fig. 3–26a). What is the woman's velocity? It's a simple enough question, but it has no single answer. As seen by a passenger sitting in the train, she is moving at 1.0 m/s. A person on a bicycle standing beside the train sees the woman moving at 1.0 m/s + 3.0 m/s = 4.0 m/s. An observer in another train going in the opposite direction would give still another answer. We have to specify which observer we mean, and we speak of the velocity *relative* to a particular observer. The woman's velocity relative to the train is 1.0 m/s, her velocity relative to the cyclist is 4.0 m/s, and so on. Each observer, equipped in principle with a meter stick and a stopwatch, forms what we call a **frame of reference.** Thus a frame of reference is a coordinate system plus a time scale.

Let's call the cyclist's frame of reference (at rest with respect to the ground) A and the frame of reference of the moving train B (Fig. 3–26b). In straight-line motion the position of a point P relative to frame of reference A is given by the distance $x_{P/A}$ (the position of P with respect to A), and the position relative to frame B is given by $x_{P/B}$. The distance from the origin of A to the origin of B (position of B with respect to A) is $x_{B/A}$. We can see from the figure that

$$x_{P/A} = x_{P/B} + x_{B/A}. \quad (3\text{–}32)$$

(a)

(b)

3–26 (a) A woman walking in a train. (b) At the instant shown, the position of the woman (particle P) relative to frame of reference A is different from her position relative to frame of reference B.

This says that the total distance from the origin of A to point P is the distance from the origin of B to point P plus the distance from the origin of A to the origin of B.

The velocity of P relative to frame A, denoted by $v_{P/A}$, is the derivative of $x_{P/A}$ with respect to time. The other velocities are similarly obtained. So the time derivative of Eq. (3–32) gives us a relationship among the various velocities:

$$\frac{dx_{P/A}}{dt} = \frac{dx_{P/B}}{dt} + \frac{dx_{B/A}}{dt},$$

or

$$v_{P/A} = v_{P/B} + v_{B/A} \qquad \text{(relative velocity along a line).} \qquad (3\text{–}33)$$

Getting back to the woman on the train, A is the cyclist's frame of reference, B is the frame of reference of the train, and point P represents the woman. Using the above notation, we have

$$v_{P/B} = 1.0 \text{ m/s}, \qquad v_{B/A} = 3.0 \text{ m/s}.$$

From Eq. (3–33) the woman's velocity $v_{P/A}$ relative to the cyclist is

$$v_{P/A} = 1.0 \text{ m/s} + 3.0 \text{ m/s} = 4.0 \text{ m/s},$$

as we already knew.

In this example, both velocities are toward the right, and we have implicitly taken this as the positive direction. If the woman walks toward the *left* relative to the train, then $v_{P/B} = -1.0$ m/s, and her velocity relative to the cyclist is 2.0 m/s. The sum in Eq. (3–33) is always an algebraic sum, and any or all of the velocities may be negative.

When the woman looks out the window, the stationary cyclist on the ground appears to her to be moving backward; we can call the cyclist's velocity relative to her $v_{A/P}$. Clearly, this is just the negative of $v_{P/A}$. In general, if A and B are any two points or frames of reference,

$$v_{A/B} = -v_{B/A}. \qquad (3\text{–}34)$$

Problem-Solving Strategy

RELATIVE VELOCITY

Note the order of the double subscripts on the velocities above; $v_{A/B}$ always means "velocity of A relative to B." These subscripts obey an interesting kind of algebra, as Eq. (3–33) shows. If we regard each one as a fraction, then the fraction on the left side is the *product* of the fractions on the right sides: $P/A = (P/B)(B/A)$. This is a handy rule you can use

when applying Eq. (3–33) to any number of frames of reference. For example, if there are three different frames of reference A, B, and C, we can write immediately

$$v_{P/A} = v_{P/C} + v_{C/B} + v_{B/A}.$$

EXAMPLE 3–13

You are driving north on a straight two-lane road at a constant 88 km/h. A truck traveling at a constant 104 km/h approaches you (in the other lane, fortunately). a) What is the truck's velocity relative to you? b) What is your velocity with respect to the truck? c) How do the relative velocities change after you and the truck have passed each other?

SOLUTION Let you be Y, the truck be T, and the earth be E, and let the positive direction be north (Fig. 3–27). Then $v_{Y/E} = +88$ km/h.
a) The truck is approaching you, so it must be moving south, giving $v_{T/E} = -104$ km/h. We want to find $v_{T/Y}$. Transcribing Eq. (3–33), we have

$$v_{T/E} = v_{T/Y} + v_{Y/E},$$
$$v_{T/Y} = v_{T/E} - v_{Y/E}$$
$$= -104 \text{ km/h} - 88 \text{ km/h} = -192 \text{ km/h}.$$

The truck is moving 192 km/h south relative to you.
b) From Eq. (3–34),

$$v_{Y/T} = -v_{T/Y} = -(-192 \text{ km/h}) = +192 \text{ km/h}.$$

You are moving 192 km/h north relative to the truck.
c) The relative velocities don't change at all after you and the truck have passed each other. The relative positions of the bodies

3–27 Reference frames for you and the truck.

don't matter. The velocity of the truck relative to you is still −192 km/h, but it is now moving away from you instead of toward you.

RELATIVE VELOCITY IN TWO OR THREE DIMENSIONS

We can extend the concept of relative velocity to include motion in a plane or in space by using vector addition to combine velocities. Suppose that the woman in Fig. 3–26a is walking not down the aisle of the railroad car but from one side of the car to the other, with a speed of 1.0 m/s (Fig. 3–28a). We can again describe the woman's position P in two different frames of reference, A for the stationary ground observer and B for the moving train. But instead of coordinates x, we use position vectors $\vec{r}$ because the problem is now two-dimensional. Then, as Fig. 3–28b shows,

$$\vec{r}_{P/A} = \vec{r}_{P/B} + \vec{r}_{B/A}. \tag{3–35}$$

Just as we did before, we take the time derivative of this equation to get a relation among the various velocities; the velocity of P relative to A is $\vec{v}_{P/A} = d\vec{r}_{P/A}/dt$, and so on for the other velocities. We get

$$\vec{v}_{P/A} = \vec{v}_{P/B} + \vec{v}_{B/A} \qquad \text{(relative velocity in space).} \tag{3–36}$$

If all three of these velocities lie along the same line, then Eq. (3–36) reduces to Eq. (3–33) for the components of the velocities along that line.

(b)

(c)

If the train's velocity relative to the ground has magnitude $v_{B/A} = 3.0$ m/s and the woman's velocity relative to the railroad car has magnitude $v_{P/B} = 1.0$ m/s, then her velocity vector $\vec{v}_{P/A}$ relative to the ground is as shown in the vector diagram of Fig. 3–28c. The Pythagorean theorem then gives us

$$v_{P/A} = \sqrt{(3.0 \text{ m/s})^2 + (1.0 \text{ m/s})^2} = \sqrt{10 \text{ m}^2/\text{s}^2} = 3.2 \text{ m/s}.$$

We can also see from the diagram that the *direction* of her velocity vector relative to the ground makes an angle α with the train's velocity vector $\vec{v}_{B/A}$, where

$$\tan \alpha = \frac{v_{P/B}}{v_{B/A}} = \frac{1.0 \text{ m/s}}{3.0 \text{ m/s}}, \qquad \alpha = 18°.$$

As in the case of motion along a straight line, we have the general rule that if A and B are *any* two points or frames of reference,

$$\vec{v}_{A/B} = -\vec{v}_{B/A}. \tag{3–37}$$

The velocity of the woman relative to the train is the negative of the velocity of the train relative to the woman, and so on.

3–28 (a) A woman walking across a railroad car, seen from above. (b) The position vector depends on the frame of reference. (c) Vector diagram for the velocity of the woman relative to the ground.

EXAMPLE 3-14

Flying in a crosswind The compass of an airplane indicates that it is headed due north, and its airspeed indicator shows that it is moving through the air at 240 km/h. If there is a wind of 100 km/h from west to east, what is the velocity of the airplane relative to the earth?

SOLUTION Let subscript P refer to the plane and subscript A to the moving air (which now plays the role of the railroad car in Fig. 3–28). Subscript E refers to the earth. The information given is

$$\vec{v}_{P/A} = 240 \text{ km/h} \quad \text{due north,}$$

$$\vec{v}_{A/E} = 100 \text{ km/h} \quad \text{due east.}$$

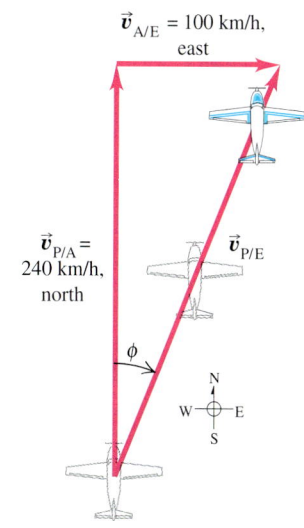

3–29 The plane is pointed north, but the wind blows east, giving the resultant velocity $\vec{v}_{P/E}$ relative to the earth.

We want to find the magnitude and direction of $\vec{v}_{P/E}$. Using Eq. (3–36), we have

$$\vec{v}_{P/E} = \vec{v}_{P/A} + \vec{v}_{A/E}.$$

The three relative velocities and their relationship are shown in Fig. 3–29; the unknowns are the speed $v_{P/E}$ and the angle ϕ. From this diagram we find

$$v_{P/E} = \sqrt{(240 \text{ km/h})^2 + (100 \text{ km/h})^2} = 260 \text{ km/h},$$

$$\phi = \arctan\left(\frac{100 \text{ km/h}}{240 \text{ km/h}}\right) = 23° \text{ E of N}.$$

The crosswind increases the speed of the airplane relative to the earth, but at the price of pushing the airplane off course.

EXAMPLE 3-15

In Example 3–14, in what direction should the pilot head to travel due north? What will then be her velocity relative to the earth? (Assume that her airspeed and the velocity of the wind are the same as in Example 3–14.)

SOLUTION Now the information given is

$$\vec{v}_{P/A} = 240 \text{ km/h} \quad \text{direction unknown,}$$

$$\vec{v}_{A/E} = 100 \text{ km/h} \quad \text{due east.}$$

Note that both this and the preceding example require us to determine two unknown quantities. In Example 3–14 these were the direction and magnitude of $\vec{v}_{P/E}$; in this example the unknowns are the direction of $\vec{v}_{P/A}$ (the direction the pilot must point the airplane) and the magnitude of $\vec{v}_{P/E}$ (her speed relative to the earth).

The three relative velocities must still satisfy the vector equation

$$\vec{v}_{P/E} = \vec{v}_{P/A} + \vec{v}_{A/E}.$$

The appropriate vector diagram is shown in Fig. 3–30. The speed $v_{P/E}$ and the angle ϕ are given by

$$v_{P/E} = \sqrt{(240 \text{ km/h})^2 - (100 \text{ km/h})^2} = 218 \text{ km/h},$$

$$\phi = \arcsin\left(\frac{100 \text{ km/h}}{240 \text{ km/h}}\right) = 25°.$$

The pilot should head 25° west of north, and her ground speed is then 218 km/h.

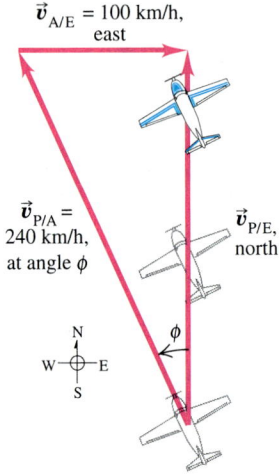

3–30 The pilot must point the plane in the direction of the vector $\vec{v}_{P/A}$ to travel due north relative to the earth.

When we derived the relative-velocity relations, we assumed that all the observers use the same time scale. This is precisely the point at which Einstein's special theory of relativity departs from the physics of Galileo and Newton. When speeds approach the speed of light, denoted by c, the velocity-addition equation has to be modified. It turns out that if the woman in Fig. 3–26a could walk down the aisle at $0.30c$ and the train could move at $0.90c$, then her speed relative to the ground would be not $1.20c$ but $0.94c$; nothing can travel faster than light! We'll return to the special theory of relativity in Chapter 39.

SUMMARY

KEY TERMS

position vector, 61
average velocity, 62
instantaneous velocity, 62
average acceleration, 64
instantaneous acceleration, 64
projectile, 68

- The position vector $\vec{r}$ of a point P in space is the vector from the origin to P. Its components are the coordinates x, y, and z.

- The average velocity vector $\vec{v}_{av}$ during the time interval Δt is the displacement $\Delta \vec{r}$ (the change in the position vector $\vec{r}$) divided by Δt:

$$\vec{v}_{av} = \frac{\vec{r}_2 - \vec{r}_1}{t_2 - t_1} = \frac{\Delta \vec{r}}{\Delta t}. \tag{3–2}$$

The instantaneous velocity vector is $\vec{v} = d\vec{r}/dt$. Its components are

$$v_x = \frac{dx}{dt}, \qquad v_y = \frac{dy}{dt}, \qquad v_z = \frac{dz}{dt}. \qquad (3\text{--}4)$$

The instantaneous speed v is the magnitude of $\vec{v}$.

■ The average acceleration vector $\vec{a}_{av}$ during the time interval Δt is the velocity change $\Delta \vec{v}$ divided by Δt:

$$\vec{a}_{av} = \frac{\vec{v}_2 - \vec{v}_1}{t_2 - t_1} = \frac{\Delta \vec{v}}{\Delta t}. \qquad (3\text{--}8)$$

The instantaneous acceleration vector is $\vec{a} = d\vec{v}/dt$. Its components are

$$a_x = \frac{dv_x}{dt}, \qquad a_y = \frac{dv_y}{dt}, \qquad a_z = \frac{dv_z}{dt}. \qquad (3\text{--}10)$$

■ Acceleration can also be represented in terms of its components parallel and perpendicular to the direction of the instantaneous velocity.

■ In projectile motion with no air resistance, $a_x = 0$ and $a_y = -g$. The coordinates and velocity components, as functions of time, are

$$x = (v_0 \cos \alpha_0)t, \qquad (3\text{--}20)$$

$$y = (v_0 \sin \alpha_0)t - \frac{1}{2} gt^2, \qquad (3\text{--}21)$$

$$v_x = v_0 \cos \alpha_0, \qquad (3\text{--}22)$$

$$v_y = v_0 \sin \alpha_0 - gt. \qquad (3\text{--}23)$$

The shape of the path in projectile motion with no air resistance is always a parabola.

■ When a particle moves in a circular path of radius R with constant speed v, it has an acceleration with magnitude

$$a_{rad} = \frac{v^2}{R}. \qquad (3\text{--}28)$$

The acceleration is always directed toward the center of the circle and perpendicular to $\vec{v}$. The period T of a circular motion is the time for one revolution. If the speed is constant, then $v = 2\pi R/T$ and

$$a_{rad} = \frac{4\pi^2 R}{T^2}. \qquad (3\text{--}30)$$

When the speed is not constant, there is still a radial component of $\vec{a}$ given by Eq. (3–28) or (3–30), but there is also a component of $\vec{a}$ parallel to the path; this component is equal to the rate of change of speed, dv/dt.

■ When a body P moves relative to a body (or reference frame) B, and B moves relative to A, we denote the velocity of P relative to B by $\vec{v}_{P/B}$, the velocity of P relative to A by $\vec{v}_{P/A}$, and the velocity of B relative to A by $\vec{v}_{B/A}$. If these velocities are all along the same line, their components along that line are related by

$$v_{P/A} = v_{P/B} + v_{B/A} \qquad \text{(relative velocity along a line).} \qquad (3\text{--}33)$$

More generally, these velocities are related by

$$\vec{v}_{P/A} = \vec{v}_{P/B} + \vec{v}_{B/A} \qquad \text{(relative velocity in space).} \qquad (3\text{--}36)$$

trajectory, 68
uniform circular motion, 76
centripetal acceleration, 77
period, 77
non-uniform circular motion, 77
relative velocity, 78
frame of reference, 78

DISCUSSION QUESTIONS

Q3–1 A simple pendulum (a mass swinging at the end of a string) swings back and forth in a circular arc. What is the direction of its acceleration at the ends of the swing? At the midpoint? In each case, explain how you obtain your answer.

Q3–2 Redraw Fig. 3–9a if $\vec{a}$ is antiparallel to $\vec{v}_1$. Does the particle move in a straight line? What happens to the speed?

Q3–3 A projectile moves in a parabolic path without air resistance. Is there any point at which $\vec{a}$ is parallel to $\vec{v}$? Perpendicular to $\vec{v}$? Explain.

Q3–4 When a rifle is fired at a distant target, the barrel is not lined up exactly on the target. Why not? Does the angle of correction depend on the distance of the target?

Q3–5 At the same instant that a bullet is fired horizontally from a gun, you drop a bullet from the height of the barrel. If there is no air resistance, which bullet hits the ground first? Explain.

Q3–6 A package falls out of an airplane that is flying in a straight line at a constant altitude and speed. If air resistance could be neglected, what would be the path of the package as observed by the pilot? As observed by a person on the ground?

Q3–7 Sketch the six graphs of the x- and y-components of position, velocity, and acceleration versus time for projectile motion with $x_0 = y_0 = 0$ and $0 < \alpha_0 < 90°$.

Q3–8 If $y_0 = 0$ and α_0 is negative, y can never be positive for a projectile. However, the expression for h in Example 3–9 seems to give a positive maximum height for a negative α_0. Explain this apparent contradiction.

Q3–9 If a jumping frog can give itself the same initial speed regardless of the direction in which it jumps (forward or straight up), how is the maximum vertical height to which it can jump related to its maximum horizontal range $R_{max} = v_0^2/g$?

Q3–10 Suppose the tranquilizer dart in Fig. 3–21 is fired at a relatively low initial speed v_0, so that the dart has passed the high point of its trajectory and is descending when it hits the monkey (the monkey is still in midair when hit). At the instant that the dart was at the high point of its trajectory, was the monkey's height above the ground the same, lower, or higher than that of the dart? Explain your answer with a drawing.

Q3–11 In uniform circular motion, what are the *average* velocity and *average* acceleration for one revolution? Explain.

Q3–12 In uniform circular motion, how does the acceleration change when the speed is increased by a factor of 3? When the radius is decreased by a factor of 2?

Q3–13 In uniform circular motion, the acceleration is perpendicular to the velocity at every instant, even though both change continuously in direction. Is there any other motion having this property, or is uniform circular motion unique?

Q3–14 Raindrops hitting the side windows of a car in motion often leave diagonal streaks. Why? Is the explanation the same or different for diagonal streaks on the windshield?

Q3–15 In a rainstorm with a strong wind, what determines the best position in which to hold an umbrella?

Q3–16 You are on the west bank of a river that is flowing north with a speed of 1.2 m/s. Your swimming speed relative to the water is 1.5 m/s and the river is 60 m wide. What is your path relative to earth that allows you to cross the river in the shortest time? Explain your reasoning

EXERCISES

SECTION 3–2 POSITION AND VELOCITY VECTORS

3–1 A squirrel has x- and y-coordinates (1.1 m, 3.4 m) at time $t_1 = 0$ and coordinates (5.3 m, –0.5 m) at time $t_2 = 3.0$ s. For this time interval, find a) the components of the average velocity; b) the magnitude and direction of the average velocity.

3–2 A rhinoceros is at the origin of coordinates at time $t_1 = 0$. For the time interval from $t_1 = 0$ to $t_2 = 12.0$ s, the rhino's average velocity has x-component –3.8 m/s and y-component 4.9 m/s. At time $t_2 = 12.0$ s, a) what are the x- and y-coordinates of the rhino? b) how far is the rhino from the origin?

3–3 A web page designer creates an animation in which a dot on a computer screen has a position of $\vec{r} = [4.0 \text{ cm} + (2.5 \text{ cm/s}^2)t^2]\hat{i} + (5.0 \text{ cm/s})t\hat{j}$. a) Find the magnitude and direction of the dot's average velocity between $t = 0$ and $t = 2.0$ s. b) Find the magnitude and direction of the instantaneous velocity at $t = 0$, $t = 1.0$ s, and $t = 2.0$ s. c) Sketch the dot's trajectory from $t = 0$ to $t = 2.0$ s, and show the velocities calculated in part (b).

3–4 If $\vec{r} = bt^2\hat{i} + ct^3\hat{j}$, where b and c are positive constants, when does the velocity vector make an angle of 45.0° with the x- and y-axes?

SECTION 3–3 THE ACCELERATION VECTOR

3–5 A jet plane is flying at a constant altitude. At time $t_1 = 0$ it has components of velocity $v_x = 90$ m/s, $v_y = 110$ m/s. At time $t_2 = 30.0$ s the components are $v_x = -170$ m/s, $v_y = 40$ m/s. a) Sketch the velocity vectors at t_1 and t_2. How do these two vectors differ? For this time interval calculate b) the components of the average acceleration; c) the magnitude and direction of the average acceleration.

3–6 A dog running in an open field has components of velocity $v_x = 2.6$ m/s and $v_y = -1.8$ m/s at $t_1 = 10.0$ s. For the time interval from $t_1 = 10.0$ s to $t_2 = 20.0$ s, the average acceleration of the dog has magnitude 0.45 m/s^2 and direction 31.0° measured from the $+x$-axis toward the $+y$-axis. At $t_2 = 20.0$ s, a) what are the x- and y-components of the dog's velocity? b) what are the magnitude and direction of the dog's velocity? c) Sketch the velocity vectors at t_1 and t_2. How do these two vectors differ?

3–7 The coordinates of a bird flying in the xy-plane are given by $x(t) = \alpha t$ and $y(t) = 3.0 \text{ m} - \beta t^2$, where $\alpha = 2.4$ m/s and $\beta = 1.2$ m/s^2. a) Sketch the path of the bird between $t = 0$ and $t = 2.0$ s. b) Calculate the velocity and acceleration vectors of the bird as functions of time. c) Calculate the magnitude and direction of the bird's velocity and acceleration at $t = 2.0$ s. d) Sketch the velocity and acceleration vectors at $t = 2.0$ s. At this instant, is the bird speeding up, is it slowing down, or is its speed instantaneously not changing? Is the bird turning? If so, in what direction?

3–8 A particle moves along a path as shown in Fig. 3–31. Between points B and D, the path is a straight line. Sketch the acceleration vectors at A, C, and E in the cases in which a) the particle moves with a constant speed; b) the particle moves with a steadily increasing speed; c) the particle moves with a steadily decreasing speed.

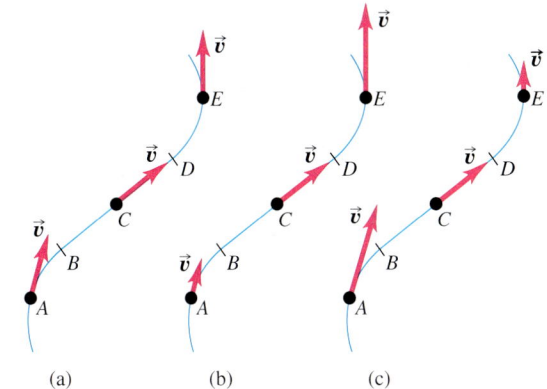

FIGURE 3–31 Exercise 3–8.

SECTION 3–4 PROJECTILE MOTION

3–9 A physics book slides off a horizontal table top with a speed of 1.10 m/s. It strikes the floor in 0.350 s. Ignore air resistance. Find a) the height of the table top above the floor; b) the horizontal distance from the edge of the table to the point where the book strikes the floor; c) the horizontal and vertical components of the book's velocity and the magnitude and direction of its velocity just before the book reaches the floor. d) Draw x-t, y-t, v_x-t, and v_y-t graphs for the motion.

3–10 A military helicopter on a training mission is flying horizontally at a speed of 60.0 m/s and accidentally drops a bomb (fortunately not armed) at an elevation of 300 m. Ignore air resistance. a) How much time is required for the bomb to reach the earth? b) How far does it travel horizontally while falling? c) Find the horizontal and vertical components of its velocity just before it strikes the earth. d) Draw x-t, y-t, v_x-t, and v_y-t graphs for the bomb's motion. e) If the velocity of the helicopter remains constant, where is the helicopter when the bomb hits the ground?

3–11 A football is thrown with an initial speed $v_0 = 15.0$ m/s at an initial angle $\alpha_0 = 45.0°$. a) Find the time T when the football is at its maximum height. b) At the three times $t_1 = T - 0.50$ s, $t_2 = T$, and $t_3 = T + 0.50$ s, find the x- and y-components of the position vector. c) At the three times t_1, t_2, and t_3, find the magnitude and direction of the velocity vector. d) At the three times t_1, t_2, and t_3, find the component of the acceleration vector that is parallel (or antiparallel) to the velocity, and find the component of the acceleration vector that is perpendicular to the velocity. e) Sketch the trajectory of the football. On your sketch, label the position of the football at the three times t_1, t_2, and t_3. At each of these positions, draw the velocity vector and the parallel and perpendicular components of the acceleration vector. f) Discuss how the speed and the direction of motion of the football are changing at the three times t_1, t_2, and t_3, and explain how the vectors in your sketch describe these changes.

3–12 A tennis ball rolls off the edge of a table top 0.750 m above the floor and strikes the floor at a point 1.40 m horizontally from the edge of the table. Ignore air resistance. a) Find the time of flight. b) Find the magnitude of the initial velocity. c) Find the magnitude and direction of the velocity of the ball just before it strikes the floor. d) Draw x-t, y-t, v_x-t, and v_y-t graphs for the motion.

3–13 A pistol that fires a signal flare gives the flare an initial speed (muzzle speed) of 120 m/s. a) If the flare is fired at an angle of 55° above the horizontal on the level salt flats of Utah, what is its horizontal range? Ignore air resistance. b) If the flare is fired at the same angle over the flat Sea of Tranquility on the moon, where $g = 1.6$ m/s^2, what is its horizontal range?

3–14 John Elway throws a football with an initial upward velocity component of 16.0 m/s and a horizontal velocity component of 20.0 m/s. Ignore air resistance. a) How much time is required for the football to reach the highest point of the trajectory? b) How high is this point? c) How much time (after being thrown) is required for the football to return to its original level? How does this compare with the time calculated in part (a)? d) How far has it traveled horizontally during this time? e) Draw x-t, y-t, v_x-t, and v_y-t graphs for the motion.

3–15 Mark McGwire hits a baseball so that it leaves the bat at a speed of 30.0 m/s and at an angle of 36.9° above the horizontal. Ignore air resistance. a) At what *two* times is the baseball at a height of 10.0 m above the point at which it left the bat? b) Calculate the horizontal and vertical components of the baseball's velocity at each of the two times calculated in part (a). c) What is the magnitude and direction of the baseball's velocity when it returns to the level at which it left the bat?

3–16 A shot putter releases the shot some distance above the level ground with a velocity of 12.0 m/s, 51.0° above the horizontal. The shot hits the ground 2.08 s later. Ignore air resistance. a) What are the components of the shot's acceleration while in flight? b) What are the components of the shot's velocity at the beginning and at the end of its trajectory? c) How far did she throw the shot horizontally? d) Why does the expression for R in Example 3–9 *not* give the correct answer for part (c)? e) How high was the shot above the ground when she released it? f) Draw x-t, y-t, v_x-t, and v_y-t graphs for the motion.

3–17 In a carnival booth you win a stuffed giraffe if you toss a quarter into a small dish. The dish is on a shelf above the point where the quarter leaves your hand and is a horizontal distance of 2.1 m from this point (Fig. 3–32; see next page). If you toss the coin with a velocity of 6.4 m/s at an angle of 60° above the horizontal, the coin lands in the dish. Ignore air resistance. a) What is the height of the shelf above the point where the quarter leaves your hand? b) What is the vertical component of the velocity of the quarter just before it lands in the dish?

FIGURE 3–32 Exercise 3–17.

3–18 Suppose the departure angle α_0 in Fig. 3–21 is 42.0° and the distance d is 3.00 m. Where will the dart and monkey meet if the initial speed of the dart is a) 12.0 m/s? b) 8.0 m/s? c) What will happen if the initial speed of the dart is 4.0 m/s? Sketch the trajectory in each case.

3–19 A man stands on the roof of a 15.0-m-tall building and throws a rock with a velocity of magnitude 30.0 m/s at an angle of 33.0° above the horizontal. Ignore air resistance. Calculate a) the maximum height above the roof reached by the rock; b) the magnitude of the velocity of the rock just before it strikes the ground; c) the horizontal distance from the base of the building to the point where the rock strikes the ground. d) Draw x-t, y-t, v_x-t, and v_y-t graphs for the motion.

SECTION 3–5 MOTION IN A CIRCLE

3–20 On your first day at work for an appliance manufacturer, you are told to figure out what to do to the period of rotation during a washer spin cycle to triple the centripetal acceleration. You impress your boss by answering immediately. What do you tell her?

3–21 The earth has a radius of 6380 km and turns around once on its axis in 24 h. a) What is the radial acceleration of an object at the earth's equator? Give your answer in m/s² and as a fraction of g. b) If a_{rad} at the equator is greater than g, objects would fly off the earth's surface and into space. (We will see the reason for this in Chapter 5.) What would the period of the earth's rotation have to be for this to occur?

3–22 A model of a helicopter rotor has four blades, each 3.40 m in length from the central shaft to the blade tip. The model is rotated in a wind tunnel at 550 rev/min. a) What is the linear speed of the blade tip, in m/s? b) What is the radial acceleration of the blade tip expressed as a multiple of the acceleration of gravity, g?

3–23 In a test of a "g-suit," a volunteer is rotated in a horizontal circle of radius 7.0 m. What is the period of rotation at which the centripetal acceleration has a magnitude of a) $3.0g$? b) $10g$?

3–24 The radius of the earth's orbit around the sun (assumed to be circular) is 1.50×10^8 km, and the earth travels around this orbit in 365 days. a) What is the magnitude of the orbital velocity of the earth in m/s? b) What is the radial acceleration of the earth toward the sun in m/s²? c) Repeat parts (a) and (b) for the motion of the planet Mercury (orbit radius = 5.79×10^7 km, orbital period = 88.0 days).

3–25 A Ferris wheel with radius 14.0 m is turning about a horizontal axis through its center (Fig. 3–33). The linear speed of a passenger on the rim is constant and equal to 7.00 m/s. What are the magnitude and direction of the passenger's acceleration as she passes through a) the lowest point in her circular motion? b) the highest point in her circular motion? c) How much time does it take the Ferris wheel to make one revolution?

FIGURE 3–33 Exercises 3–25 and 3–26.

3–26 The Ferris wheel in Fig. 3–33, which rotates counterclockwise, is just starting up. At a given instant, a passenger on the rim of the wheel and passing through the lowest point of his circular motion is moving at 3.00 m/s and is gaining speed at a rate of 0.500 m/s². a) Find the magnitude and the direction of the passenger's acceleration at this instant. b) Sketch the Ferris wheel and passenger showing his velocity and acceleration vectors.

3–27 A level racetrack has the shape of an ellipse. (See your math textbook, a dictionary, or an encyclopedia for an illustration of an ellipse.) A car drives around this track at a constant speed. a) Draw a picture of the track showing the car's velocity vector and acceleration vector at five or more different places around the track. b) Does the car's acceleration vector always point toward the geometrical center of the ellipse? Explain. c) At what point(s) on the ellipse does the car's acceleration have the greatest magnitude? Explain.

SECTION 3–6 RELATIVE VELOCITY

3–28 A railroad flatcar is traveling to the right at a speed of 13.0 m/s relative to an observer standing on the ground. A motor scooter is being ridden on the flatcar (Fig. 3–34). What is the velocity (magnitude and direction) of the motor scooter relative to the flatcar if its velocity relative to the observer on the ground is a) 18.0 m/s to the right? b) 3.0 m/s to the left? c) zero?

3–29 A "moving sidewalk" in an airport terminal building moves at 1.0 m/s and is 35.0 m long. If a woman steps on at one end and walks at 1.5 m/s relative to the moving sidewalk, how much time does she require to reach the opposite end if she walks a) in the same direction the sidewalk is moving? b) in the opposite direction?

3–30 Two piers A and B are located on a river: B is 1500 m downstream from A (Fig. 3–35). Two friends must make round

FIGURE 3–34 Exercise 3–28.

trips from pier A to pier B and return. One rows a boat at a constant speed of 4.00 km/h relative to the water; the other walks on the shore at a constant speed of 4.00 km/h. The velocity of the river is 2.80 km/h in the direction from A to B. How much time does it take each person to make the round trip?

FIGURE 3–35 Exercise 3–30.

PROBLEMS

3–31 A canoe has a velocity of 0.40 m/s southeast relative to the earth. The canoe is on a river that is flowing 0.50 m/s east relative to the earth. Find the velocity (magnitude and direction) of the canoe relative to the river.

3–32 An airplane pilot wishes to fly due west. A wind of 80.0 km/h (about 50 mi/h) is blowing toward the south. a) If the airspeed of the plane (its speed in still air) is 320.0 km/h (about 200 mi/h), in what direction should the pilot head? b) What is the speed of the plane over the ground? Illustrate with a vector diagram.

3–33 A river flows due south with a speed of 2.0 m/s. A man steers a motorboat across the river; his velocity relative to the water is 4.2 m/s due east. The river is 800 m wide. a) What is his velocity (magnitude and direction) relative to the earth? b) How much time is required to cross the river? c) How far south of his starting point will he reach the opposite bank?

3–34 a) In what direction should the motorboat in Exercise 3–33 be headed in order to reach a point on the opposite bank directly east from the starting point? (The boat's speed relative to the water remains 4.2 m/s.) b) What is the velocity of the boat relative to the earth? c) How much time is required to cross the river?

3–35 The nose of an ultralight plane is pointed south, and its air speed indicator shows 35 m/s. The plane is in a 10 m/s wind blowing toward the southwest relative to the earth. a) In a vector addition diagram, show the relation of $\vec{v}_{P/E}$ (the velocity of the plane relative to the earth) to the two given vectors. b) Letting x be east and y be north, find the components of $\vec{v}_{P/E}$. c) Find the magnitude and direction of $\vec{v}_{P/E}$.

PROBLEMS

3–36 A faulty model rocket moves in the xy-plane (the positive y-direction is vertically upward). The rocket's acceleration has components $a_x(t) = \alpha t^2$ and $a_y(t) = \beta - \gamma t$, where $\alpha = 2.50$ m/s^4, $\beta = 9.00$ m/s^2, and $\gamma = 1.40$ m/s^3. At $t = 0$ the rocket is at the origin and has velocity $\vec{v}_0 = v_{0x}\hat{i} + v_{0y}\hat{j}$ with $v_{0x} = 1.00$ m/s and $v_{0y} = 7.00$ m/s. a) Calculate the velocity and position vectors as functions of time. b) What is the maximum height reached by the rocket? c) Sketch the path of the rocket. d) What is the horizontal displacement of the rocket when it returns to $y = 0$?

3–37 A student is moving in the xy-plane in a darkened room, trying to locate a lost $20 bill. The student's coordinates are given as functions of time by $x(t) = \alpha t$ and $y(t) = 15.0$ m $- \beta t^2$, where $\alpha = 1.20$ m/s and $\beta = 0.500$ m/s^2. Unknown to the student, the $20 bill is at the origin. a) At what times is the student's velocity perpendicular to his acceleration? b) At what times is the student's speed instantaneously not changing? c) At what times is the student's velocity perpendicular to his position vector? What are the locations of the student at these times? d) What is the minimum distance from the student to the $20 bill? At what time does this minimum occur? e) Sketch the path of the hapless student.

3–38 A bird flies in the xy-plane with a velocity vector given by $\vec{v} = (\alpha - \beta t^2)\hat{i} + \gamma t\hat{j}$, with $\alpha = 2.4$ m/s, $\beta = 1.6$ m/s^3, and $\gamma = 4.0$ m/s^2. The positive y-direction is vertically upward. At $t = 0$ the bird is at the origin. a) Calculate the position and acceleration vectors of the bird as functions of time. b) What is the bird's altitude (y-coordinate) as it flies over $x = 0$ for the first time after $t = 0$?

3–39 A Piper Warrior, a small four-seat airplane, requires 300 m of runway to take off. Its liftoff speed is 88 km/h. It then climbs with a constant speed of 88 km/h along a straight-line path, just clearing a power line 15 m high at a horizontal distance of 460 m from where the Piper began its takeoff roll. a) What was the acceleration of the Piper (assumed constant) during its takeoff roll? b) After the Piper takes off, what is the angle of its flight path above the horizontal? c) What is the rate of climb (in m/s) of the Piper? d) What is the elapsed time from the beginning of the takeoff roll to when the Piper just clears the power line?

3–40 A track and field coach (who is also a physics teacher) trains an athlete to throw a javelin so that it leaves the athlete's hand at a height h above the ground, moving at a speed of $\sqrt{25gh/8}$ and at an angle of 36.9° above the horizontal. The javelin continues in flight until it strikes the ground. The ground around the athlete is level, and air resistance is negligible. a) Graph the javelin's horizontal velocity versus time and the javelin's vertical velocity versus time. b) Find the maximum height reached by the javelin. c) Find the horizontal distance that the javelin travels from where it leaves the athlete's hand to where it strikes the ground.

3–41 A demolition crew uses dynamite to blow an old building apart. Debris from the explosion flies off in all directions, and is later found at distances up to 50 m from the explosion. Estimate the maximum speed at which debris was blown outward by the explosion. Describe any assumptions that you make.

3–42 A movie stuntwoman drops from a helicopter that is 30.0 m above the ground and moving with a constant velocity whose components are 10.0 m/s upward and 15.0 m/s horizontal and toward the south. Ignore air resistance. a) Where on the ground (relative to the position of the helicopter when she drops) should the stuntwoman have placed the foam mats that break her fall? b) Draw x-t, y-t, v_x-t, and v_y-t graphs of the motion.

3–43 In fighting forest fires, airplanes work in support of ground crews by dropping water on the fires. A pilot is practicing by dropping a canister of red dye, hoping to hit a target on the ground below. If the plane is flying in a horizontal path 90.0 m above the ground and with a speed of 64.0 m/s (143 mi/h), at what horizontal distance from the target should the pilot release the canister? Ignore air resistance.

3–44 A girl throws a water-filled balloon at an angle of 50.0° above the horizontal with a speed of 12.0 m/s. The horizontal component of the balloon's velocity is directed toward a car that is approaching the girl at a constant speed of 8.00 m/s (Fig. 3–36). If the balloon is to hit the car at the same height at which it leaves her hand, what is the maximum distance the car can be from the girl when the balloon is thrown? Ignore air resistance.

FIGURE 3–36 Problem 3–44.

3–45 The Longest Home Run. According to the *Guinness Book of World Records,* the longest home run ever measured was hit by Roy "Dizzy" Carlyle in a minor league game. The ball traveled 188 m (618 ft) before landing on the ground outside the ballpark. a) Assuming the ball's initial velocity was 45° above the horizontal and neglecting air resistance, what did the initial speed of the ball need to be to produce such a home run if the ball was hit at a point 0.9 m (3.0 ft) above ground level? Assume that the ground was perfectly flat. b) How far would the ball be above a fence 3.0 m (10 ft) in height if the fence was 116 m (380 ft) from home plate?

3–46 On the day after graduation, you decide to toss a burning match into the top of a cylindrical wastebasket (diameter D and height 2D) full of old homework papers. To make this event more sporting, the bottom of the wastebasket is at the same height as the point where the match leaves your hand, and the near side of the wastebasket is a horizontal distance 6D from the point where the match leaves your hand. You launch the match at a 45.0° angle above the horizontal. Find the *minimum* and *maxi-*

mum values of the launch speed that will make the match enter the top of the wastebasket. Ignore air resistance and give your answers in terms of g and D.

3–47 You are playing catch with a friend in the hallway of your dormitory. The distance from floor to ceiling is D, and you throw the ball with an initial speed $v_0 = \sqrt{6gD}$. What is the maximum horizontal distance (in terms of D) that the ball can travel without bouncing? (Assume that the ball is launched from the floor.)

3–48 A baseball thrown at an angle of 60.0° above the horizontal strikes a building 18.0 m away at a point 8.00 m above the point from which it is thrown. Ignore air resistance. a) Find the magnitude of the initial velocity of the baseball (the velocity with which the baseball is thrown). b) Find the magnitude and direction of the velocity of the baseball just before it strikes the building. c) Draw x-t, y-t, v_x-t, and v_y-t graphs of the motion.

3–49 A projectile is launched with speed v_0 at an angle α_0 above the horizontal. The launch point is a height h above the ground. a) Show that if air resistance is neglected, the horizontal distance that the projectile travels before striking the ground is

$$x = \frac{v_0 \cos \alpha_0}{g} \left(v_0 \sin \alpha_0 + \sqrt{v_0^2 \sin^2 \alpha_0 + 2gh} \right).$$

Verify that if the launch point is at ground level so that h = 0, this is equal to the horizontal range R found in Example 3–9. b) For the case where $v_0 = 10$ m/s and h = 5.0 m, graph x as a function of launch angle α_0 for values of α_0 from 0° to 90°. Your graph should show that x is zero if $\alpha_0 = 90°$, but x is nonzero if $\alpha_0 = 0$; explain why this is so. c) We saw in Example 3–9 that for a projectile that lands at the same height from which it is launched, the horizontal range is maximum for $\alpha_0 = 45°$. For the case graphed in part (b), is the angle for maximum horizontal distance equal to, less than, or greater than 45°? (This is a general result for the situation where a projectile is launched from a point higher than where it lands.)

3–50 A snowball rolls off a barn roof that slopes downward at an angle of 40° (Fig. 3–37). The edge of the roof is 14.0 m above the ground, and the snowball has a speed of 7.00 m/s as it rolls off the roof. Ignore air resistance. a) How far from the edge of the barn does the snowball strike the ground if it doesn't strike anything else while falling? b) Draw x-t, y-t, v_x-t, and v_y-t graphs for the motion in part (a). c) A man 1.9 m tall is standing 4.0 m from the edge of the barn. Will he be hit by the snowball?

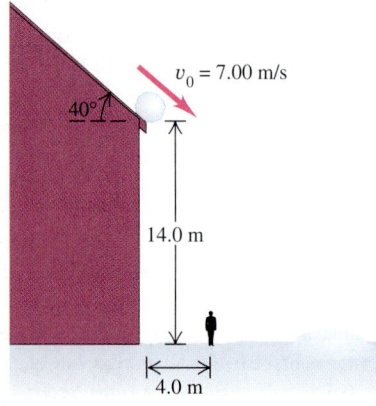

FIGURE 3–37 Problem 3–50.

3–51 a) Prove that a projectile launched at angle α_0 has the same horizontal range as one launched with the same speed at angle $(90° - \alpha_0)$. b) A frog jumps at a speed of 2.2 m/s and lands 25 cm from its starting point. At what angles above the horizontal could it have jumped?

3–52 On the Flying Trapeze. A new circus act is called the Texas Tumblers. Lovely Mary Belle swings from a trapeze, projects herself at an angle of 53°, and is supposed to be caught by Joe Bob, whose hands are 6.1 m above and 8.2 m horizontally from her launch point (Fig. 3–38). Ignore air resistance. a) What initial speed v_0 must Mary Belle have to just reach Joe Bob? b) For the initial speed calculated in part (a), what are the magnitude and direction of her velocity when Mary Belle reaches Joe Bob? c) Assuming that Mary Belle has the initial speed calculated in part (a), draw x-t, y-t, v_x-t, and v_y-t graphs showing the motion of both tumblers. Your graphs should show the motion up until the point where Mary Belle reaches Joe Bob. d) The night of their debut performance, Joe Bob misses her completely as she flies past. How far horizontally does Mary Belle travel, from her initial launch point, before landing in the safety net 8.6 m below her starting point?

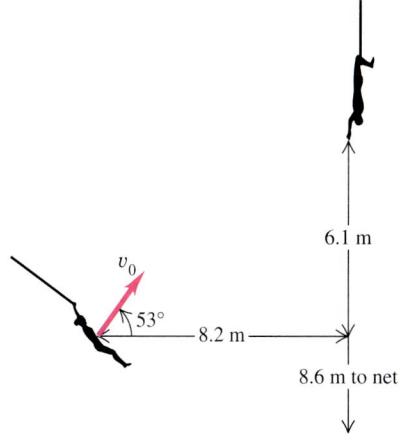

FIGURE 3–38 Problem 3–52.

3–53 A physics professor did daredevil stunts in his spare time. His last stunt was an attempt to jump across a river on a motorcycle (Fig. 3–39). The takeoff ramp was inclined at 53.0°, the river was 40.0 m wide, and the far bank was 15.0 m lower than the top of the ramp. The river itself was 100 m below the ramp. Ignore air resistance. a) What should his speed have been at the top of the ramp to have just made it to the edge of the far bank? b) If his speed was only half the value found in (a), where did he land?

3–54 A rock is thrown from the roof of a building with a velocity v_0 at an angle of α_0 from the horizontal. The building has height h. Ignore air resistance. Calculate the magnitude of the velocity of the rock just before it strikes the ground, and show that this speed is independent of α_0.

3–55 A basketball player is fouled and knocked to the floor during a layup attempt. The player is awarded two free throws. The center of the basket is a horizontal distance of 4.21 m (13.8 ft) from the foul line and is a height of 3.05 m (10.0 ft) above the floor (Fig. 3–40). On the first free-throw attempt, he shoots the ball at an angle of 35° above the horizontal and with a speed of $v_0 = 4.88$ m/s (16.0 ft/s). The ball is released 1.83 m (6.0 ft) above the floor. This

FIGURE 3–39 Problem 3–53.

shot misses badly. Ignore air resistance. a) What is the maximum height reached by the ball? b) At what distance along the floor from the free-throw line does the ball land? c) For the second throw, the ball goes through the center of the basket. For this second free throw, the player again shoots the ball at 35° above the horizontal and releases it 1.83 m above the floor. What initial speed does the player give the ball on this second attempt? d) For the second throw, what is the maximum height reached by the ball? At this point, how far horizontally is the ball from the basket?

FIGURE 3–40 Problem 3–55.

3–56 Romeo tosses a pebble at Juliet's window to wake her. Unfortunately, he throws too large a pebble too fast. Just before crashing through the glass, the pebble is moving horizontally, having traveled a horizontal distance x and a vertical distance y as a projectile. Find the magnitude and direction of the pebble's velocity as it leaves Romeo's hand.

3–57 A rocket is initially at rest on the ground. When its engines are fired, the rocket flies off in a straight line at an angle of 53.1° above the horizontal with a constant acceleration of magnitude g. The engines stop at a time T after launch, after which the rocket is in projectile motion. Ignore air resistance and assume g is independent of altitude. a) Draw the trajectory of the rocket from when its engines are first fired until the rocket hits the ground. Indicate the direction of the velocity and acceleration vectors at various points along the trajectory. b) Sketch v_x-t and v_y-t graphs for the motion of the rocket from when its engines are first fired until the rocket hits the ground. c) Find the maximum altitude reached by the rocket. Your answer will be in terms of g and T. d) Find the horizontal distance from the launch point to where the rocket hits the ground (that is, the range). Your answer will be in terms of g and T.

3–58 In an action-adventure film the hero is supposed to throw a grenade from his car, which is going 90.0 km/h, to his enemy's car, which is going 110 km/h. The enemy's car is 15.8 m in front of the hero's when he lets go of the grenade. If the hero throws the grenade so its initial velocity relative to him is at an angle of 45° above the horizontal, what should be the magnitude of the initial velocity? The cars are both traveling in the same direction on a level road. Ignore air resistance. Find the magnitude of the velocity both relative to the hero and relative to the earth.

3–59 A rock tied to a rope moves in the xy-plane. Its coordinates are given as functions of time by.

$$x(t) = R\cos\omega t, \qquad y(t) = R\sin\omega t,$$

where R and ω are constants. a) Show that the rock's distance from the origin is constant and equal to R, that is, that its path is a circle of radius R. b) Show that at every point the rock's velocity is perpendicular to its position vector. c) Show that the rock's acceleration is always opposite in direction to its position vector and has magnitude $\omega^2 R$. d) Show that the magnitude of the rock's velocity is constant and equal to ωR. e) Combine the results of parts (c) and (d) to show that the rock's acceleration has constant magnitude v^2/R.

3–60 The speed of a particle moving in a plane is equal to the magnitude of its instantaneous velocity, $v = |\vec{v}| = \sqrt{v_x^2 + v_y^2}$. a) Show that the rate of change of the speed is $dv/dt = (v_x a_x + v_y a_y)/\sqrt{v_x^2 + v_y^2}$. b) Use this expression to find the value of dv/dt at time $t = 2.0$ s for the radio-controlled car in Examples 3–1, 3–2, and 3–3. Compare your answer to the components of acceleration found in Example 3–3. Explain why your answer is *not* equal to the magnitude of the acceleration found in part (b) of Example 3–2. c) Show that the rate of change of speed can be expressed as $dv/dt = \vec{v} \cdot \vec{a}/v$, and use this result to explain why dv/dt is equal to $a_\parallel$, the component of $\vec{a}$ that is parallel to $\vec{v}$.

3–61 A particle moves in the xy-plane. Its coordinates are given as functions of time by

$$x(t) = R(\omega t - \sin\omega t), \quad y(t) = R(1 - \cos\omega t),$$

where R and ω are constants. a) Sketch the trajectory of the particle. (This is the trajectory of a point on the rim of a wheel that is rolling at a constant speed on a horizontal surface. The curve traced out by such a point as it moves through space is called a *cycloid*.) b) Determine the velocity components and the acceleration components of the particle at any time t. c) At what times is the particle momentarily at rest? What are the coordinates of the particle at these times? What are the magnitude and direction of the acceleration at these times? d) Does the magnitude of the acceleration depend on time? Compare to uniform circular motion.

3–62 You are flying in a light airplane spotting traffic for a radio station. Your flight carries you due east above a highway. Landmarks below tell you that your speed is 50 m/s relative to the ground and your air speed indicator also reads 50 m/s. However, the nose of your airplane is pointed somewhat south of east and the station's weather person tells you that a 20 m/s wind is blowing. What direction is the wind blowing?

3–63 The Carrier Pigeon Problem. Larry is driving east at 40 km/h. His twin brother Harry is driving west at 30 km/h, toward Larry in an identical car on the same straight road. When they are 42 km apart, Larry sends out a carrier pigeon, which flies at a constant speed of 50 km/h. (All speeds are relative to the earth.) The pigeon flies to Harry, gets confused and immediately returns to Larry, gets more confused and immediately flies back to Harry. This continues until the twins meet, at which time the dazed pigeon drops to the ground in exhaustion. Ignoring turnaround time, how far did the pigeon fly?

3–64 When a train's velocity is 12.0 m/s eastward, raindrops that are falling vertically with respect to the earth make traces that are inclined 30.0° to the vertical on the windows of the train. a) What is the horizontal component of a drop's velocity with respect to the earth? With respect to the train? b) What is the magnitude of the velocity of the raindrop with respect to the earth? With respect to the train?

3–65 An airplane pilot sets a compass course due west and maintains an airspeed of 220 km/h. After flying for 0.500 h, she finds herself over a town 120 km west and 20 km south of her starting point. a) Find the wind velocity (magnitude and direction). b) If the wind velocity is 40 km/h due south, in what direction should the pilot set her course to travel due west? Take the same airspeed of 220 km/h.

3–66 In an air race, an airplane flies from a point directly over Metropolis to a point directly over Gotham City, then turns around and flies back to the starting point. The airspeed (that is, speed of the airplane relative to the air) is constant throughout the flight and equal to v. Gotham City lies a distance D due east of Metropolis. a) If there is no wind, how much time is required for the round trip? How much time is required for the round trip if a steady wind of speed w is blowing b) toward the east? c) toward the south? d) If $D = 3.00 \times 10^2$ km, $v = 4.00 \times 10^2$ km/h, and $w = 1.00 \times 10^2$ km/h, calculate the round-trip flight times for parts (a), (b), and (c). Which case gives the slowest round-trip?

3–67 An elevator is moving upward at a constant speed of 2.50 m/s. A bolt in the elevator ceiling 3.00 m above the elevator floor works loose and falls. a) How long does it take for the bolt to fall to the elevator floor? What is the speed of the bolt just as it hits the elevator floor b) according to an observer in the elevator? c) according to an observer standing on one of the floor landings of the building? d) According to the observer in part (c), what distance did the bolt travel between the ceiling and floor of the elevator?

CHALLENGE PROBLEMS

3–68 A man is riding on a flatcar traveling with a constant speed of 9.10 m/s (Fig. 3–41). He wishes to throw a ball through a stationary hoop 4.90 m above the height of his hands in such a

manner that the ball will be moving horizontally as it passes through the hoop. He throws the ball with a speed of 10.8 m/s with respect to himself. a) What must be the vertical component of the initial velocity of the ball? b) How many seconds after he releases the ball will it pass through the hoop? c) At what horizontal distance in front of the hoop must he release the ball? d) When the ball leaves the man's hands, what is the direction of its velocity relative to the frame of reference of the flatcar? Relative to the frame of reference of an observer standing on the ground?

4.90 m

$v = 9.10$ m/s

FIGURE 3–41 Challenge Problem 3–68.

3–69 A shotgun fires a large number of pellets upward, with some pellets traveling very nearly vertically and others as much as $1.0°$ from the vertical. Assume that the initial speed of the pellets is uniformly 150 m/s, and ignore air resistance. a) Within what radius from the point of firing will the pellets land? b) If there are 1000 pellets, and they fall in a uniform distribution over a circle with the radius calculated in part (a), what is the probability that at least one pellet will fall on the head of the person who fires the shotgun? Assume that his head has a radius of 10 cm. c) Air resistance in fact has several effects. It slows down the rising pellets, decreases their horizontal component of velocity, and limits the speed with which they fall. Which of these effects will tend to make the radius greater than calculated in part (a), and which will tend to make it less? What do you think the overall effect of air resistance will be? (The effect of air resistance on a velocity component increases as the magnitude of the component increases.)

3–70 A projectile is thrown from a point P. It moves in such a way that its distance from P is always increasing. Find the maximum angle above the horizontal with which the projectile could have been thrown. Ignore air resistance.

3–71 A baseball is given an initial velocity with magnitude v_0 at an angle ϕ above the surface of an incline, which is in turn inclined at an angle θ above the horizontal (Fig. 3–42). a) Calculate the distance, measured along the incline, from the launch point to where the baseball strikes the incline. Your answer will be in terms of v_0, g, θ, and ϕ. b) What angle ϕ gives the maximum range, measured along the incline? (*Note:* You might be interested in the three different methods of solution presented by I. R. Lapidus in *Amer. Jour. of Phys.*, Vol. 51 (1983), pp. 806 and 847. See also H. A. Buckmaster in *Amer. Jour. of Phys.*, Vol. 53 (1985), pp. 638–641, for a thorough study of this and some similar problems.)

3–72 Refer to Challenge Problem 3–71. a) An archer on ground of constant upward slope of $30.0°$ aims at a target 60.0 m farther up the incline. The arrow in the bow and the bull's-eye at the center of the target are each 1.50 m above the ground. The initial velocity of the arrow just after it leaves the bow has magnitude 32.0 m/s. At what angle above the *horizontal* should the archer aim to hit the bull's-eye? If there are two such angles, calculate the smaller of the two. You might have to solve the equation for the angle by iteration, that is, by trial and error. How does the angle compare to that required when the ground is level, with zero slope? b) Repeat the above for ground that has a constant *downward* slope of $30.0°$.

3–73 For no apparent reason, a poodle is running at a constant speed of $v = 5.00$ m/s in a circle with radius $R = 2.50$ m. Let $\vec{v}_1$ be the velocity vector at time t_1, and let $\vec{v}_2$ be the velocity vector at time t_2. Consider $\Delta\vec{v} = \vec{v}_2 - \vec{v}_1$ and $\Delta t = t_2 - t_1$. Recall that $\vec{a}_{av} = \Delta\vec{v}/\Delta t$. For $\Delta t = 0.5$ s, 0.1 s, and 0.05 s, calculate the magnitude (to four significant figures) and direction (relative to $\vec{v}_1$) of the average acceleration $\vec{a}_{av}$. Compare your results to the general expression for the instantaneous acceleration $\vec{a}$ for uniform circular motion that is derived in the text.

3–74 A rocket designed to place small payloads into orbit is carried to an altitude of 12.0 km above sea level by a converted airliner. When the airliner is flying in a straight line at a constant speed of 850 km/h, the rocket is dropped. After the drop, the airliner maintains the same altitude and speed and continues to fly in a straight line. The rocket falls for a brief time, after which its rocket motor turns on. Once its rocket motor is on, the combined effects of thrust and gravity give the rocket a constant acceleration of magnitude $3.00g$ directed at an angle of $30.0°$ above the horizontal. For reasons of safety, the rocket should be at least 1.00 km in front of the airliner when it climbs through the airliner's altitude. Your job is to determine the minimum time that the rocket must fall before its engine starts. Ignore air resistance. Your answer should include i) a diagram showing the flight paths of both the rocket and the airliner, labeled at several points with vectors for their velocities and accelerations; ii) an x-t graph showing the motions of both the rocket and the airliner; and iii) a y-t graph showing the motions of both the rocket and the airliner. On the diagram and the graphs, indicate when the rocket is dropped, when the rocket motor turns on, and when the rocket climbs through the altitude of the airliner.

3–75 Two students are canoeing on a river. While heading upstream they accidentally drop an empty bottle overboard. They then continue paddling for 60 minutes, reaching a point 2.0 km further upstream. At this point they realize that the bottle is missing and, driven by ecological awareness, they turn around and head downstream. They catch up with and retrieve the bottle (which has been moving along with the current) 5.0 km downstream from the turn-around point. a) Assuming a constant paddling effort throughout, how fast is the river flowing? b) What would be the canoe speed in a still lake for the same paddling effort?

v_0

ϕ

θ

FIGURE 3–42 Challenge Problem 3–71.

4

Newton's Laws of Motion

To hit a home run, a baseball player requires an intuitive understanding of Newton's laws. He must swing the bat so that it hits the ball with a force of sufficient magnitude and duration and in just the right direction. If the batter does everything just right, the ball will be accelerated into an upward trajectory that sends it flying into the stands.

4-1 INTRODUCTION

How can a tugboat push a cruise ship that's much heavier than the tug? Why does it take a long distance to stop the ship once it is in motion? Why does your foot hurt more when you kick a big rock than when you kick a pebble? Why is it harder to control a car on wet ice than on dry concrete? The answers to these and similar questions take us into the subject of **dynamics,** the relationship of motion to the forces that cause it. In the two preceding chapters we studied *kinematics,* the language for *describing* motion. Now we are ready to think about what makes bodies move the way they do.

In this chapter we will use the kinematic quantities displacement, velocity, and acceleration along with two new concepts, *force* and *mass,* to analyze the principles of dynamics. These principles can be wrapped up in a neat package of three statements called **Newton's laws of motion.** The first law states that when the net force on a body is zero, its motion doesn't change. The second law relates force to acceleration when the net force is *not* zero. The third law is a relation between the forces that two interacting bodies exert on each other. These laws, based on experimental studies of moving bodies, are *fundamental* in two ways. First, they cannot be deduced or proved from other principles. Second, they make it possible to understand most familiar kinds of motion; they are the foundation of **classical mechanics** (also called **Newtonian mechanics**).

Newton's laws are not universal, however; they require modification at very high speeds (near the speed of light) and for very small sizes (such as within the atom).

The laws of motion were clearly stated for the first time by Sir Isaac Newton (1642–1727), who published them in 1687 in his *Philosophiae Naturalis Principia Mathematica* ("Mathematical Principles of Natural Philosophy"). Many other scientists before Newton contributed to the foundations of mechanics, including Copernicus, Brahe, Kepler, and especially Galileo Galilei (1564–1642), who died the same year Newton was born. Indeed, Newton himself said, "If I have been able to see a little farther than other men, it is because I have stood on the shoulders of giants." Now it's your turn to stand on the shoulders of Newton and use his laws to understand how the physical world works.

Newton's laws are very simple to state, yet many students find these laws difficult to grasp and to work with. The reason is that before studying physics, you've spent years walking, throwing balls, pushing boxes, and doing dozens of things that involve motion. Along the way, you've developed a set of "common sense" ideas about motion and its causes. But many of these "common sense" ideas, while they seem to work well enough in everyday life, don't stand up to logical analysis or comparison with experiment. A big part of the job of this chapter—and of the rest of our study of physics—is helping you to recognize when "common sense" ideas need to be replaced by other kinds of analysis.

4–2 FORCE AND INTERACTIONS

In everyday language, a **force** is a push or pull. The concept of force gives us a quantitative description of the interaction between two bodies or between a body and its environment. When you push on a car that is stuck in the snow, you exert a force on it. A locomotive exerts a force on the train it is pulling or pushing, a steel cable exerts a force on the beam it is hoisting at a construction site, and so on.

When a force involves direct contact between two bodies, we call it a **contact force.** Contact forces include the pushes or pulls you exert with your hand, the force of a rope pulling on a block to which it is tied, and the friction force that the ground exerts on a ball player sliding into home. There are also forces, called **long-range forces,** that act even when the bodies are separated by empty space. You've experienced long-range forces if you've ever played with a pair of magnets. Gravity, too, is a long-range force; the sun exerts a gravitational pull on the earth, even over a distance of 150 million kilometers, that keeps the earth in orbit. The force of gravitational attraction that the earth exerts on your body is called your **weight.**

Force is a vector quantity; you can push or pull a body in different directions. Thus to describe a force, we need to describe the *direction* in which it acts as well as its *magnitude,* the quantity that describes "how much" or "how hard" the force pushes or pulls. The SI unit of the magnitude of force is the *newton,* abbreviated N. (We'll give a precise definition of the newton in Section 4–4.) Table 4–1 lists some typical force magnitudes.

A common instrument for measuring forces is the *spring balance.* It consists of a coil spring, enclosed in a case for protection, with a pointer attached to one end. When forces are applied to the ends of the spring, it stretches; the amount of stretch depends on the force. We can make a scale for the pointer and calibrate it by using a number of identical bodies with weights of exactly 1 N each. When two, three, or more of these are suspended simultaneously from the balance, the total force stretching the spring is 2 N, 3 N, and so on, and we can label the corresponding positions of the pointer 2 N, 3 N, and so on. Then we can use this instrument to measure the magnitude of an unknown force. We can also make a similar instrument that measures pushes instead of pulls.

Suppose we slide a box along the floor, applying a force to it by pulling it with a string or pushing it with a stick (Fig. 4–1). In each case we draw a vector to represent the force applied. The labels indicate the magnitude and direction of the force, and the length of the arrow (drawn to some scale, such as 1 cm = 10 N) also shows the magnitude.

When *two* forces $\vec{F}_1$ and $\vec{F}_2$ act at the same time at a point A of a body (Fig. 4–2), experiment shows that the effect on the body's motion is the same as the effect of a single force $\vec{R}$ equal to the *vector sum* of the original forces: $\vec{R} = \vec{F}_1 + \vec{F}_2$. More generally,

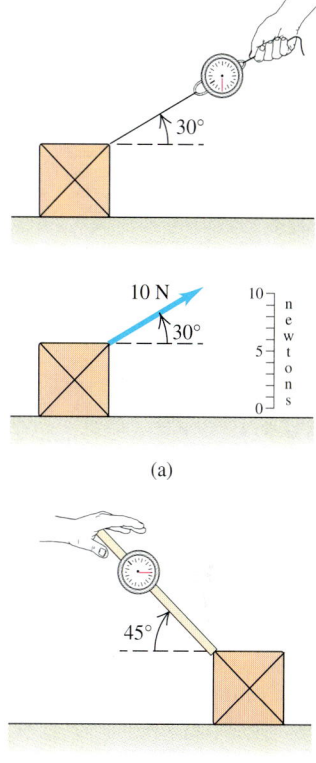

(a)

(b)

4–1 A force may be exerted on the box by either (a) pulling it or (b) pushing it. A force diagram illustrates each case.

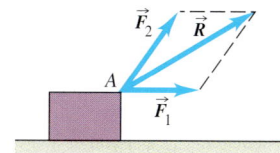

4–2 Two forces $\vec{F}_1$ and $\vec{F}_2$ acting simultaneously on a body have the same effect as a single force $\vec{R}$ equal to the vector sum (resultant) of $\vec{F}_1$ and $\vec{F}_2$.

TABLE 4–1

TYPICAL FORCE MAGNITUDES

Sun's gravitational force on the earth	3.5×10^{22} N
Thrust of the *Energia* rocket	3.9×10^{7} N
Weight of a large blue whale	1.9×10^{6} N
Maximum pulling force of a locomotive	8.9×10^{5} N
Weight of a 250-lb linebacker	1.1×10^{3} N
Weight of a medium apple	1 N
Weight of smallest insect eggs	2×10^{-6} N
Electric attraction between the proton and the electron in a hydrogen atom	8.2×10^{-8} N
Weight of a very small bacterium	1×10^{-18} N
Weight of a hydrogen atom	1.6×10^{-26} N
Weight of an electron	8.9×10^{-30} N
Gravitational attraction between the proton and the electron in a hydrogen atom	3.6×10^{-47} N

(a)

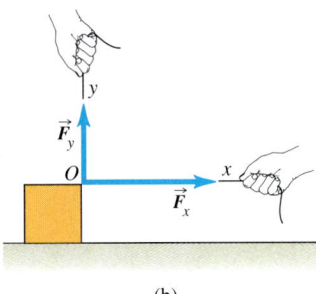

(b)

4–3 The force $\vec{F}$, which acts at an angle θ from the x-axis, may be replaced by its rectangular component vectors $\vec{F}_x$ and $\vec{F}_y$. The x- and y-components of $\vec{F}$ are $F_x = F \cos \theta$ and $F_y = F \sin \theta$.

4–4 F_x and F_y are the components of $\vec{F}$ parallel and perpendicular to the sloping surface of the inclined plane.

the effect of any number of forces applied at a point on a body is the same as the effect of a single force equal to the vector sum of the forces. This important principle goes by the name **superposition of forces.**

The experimental discovery that forces combine according to vector addition is of the utmost importance. We will use this fact many times throughout our study of physics. It allows us to replace a force by its component vectors, as we did with displacements in Section 1–9. For example, in Fig. 4–3a, force $\vec{F}$ acts on a body at point O. The component vectors of $\vec{F}$ in the directions Ox and Oy are $\vec{F}_x$ and $\vec{F}_y$. When $\vec{F}_x$ and $\vec{F}_y$ are applied simultaneously, as in Fig. 4–3b, the effect is exactly the same as the effect of the original force $\vec{F}$. *Any force can be replaced by its component vectors, acting at the same point.*

It's frequently more convenient to describe a force $\vec{F}$ in terms of its x- and y-components F_x and F_y rather than by its component vectors (recall from Section 1–9 that *component vectors* are vectors, but *components* are just numbers). For the case shown in Fig. 4–3, both F_x and F_y are positive; for other orientations of the force $\vec{F}$, either F_x or F_y can be negative or zero.

There is no law that says our coordinate axes have to be vertical and horizontal. Figure 4–4 shows a stone block being pulled up a ramp by a force $\vec{F}$, represented by its components F_x and F_y parallel and perpendicular to the sloping surface of the ramp.

CAUTION ▶ In Fig. 4–4 we draw a wiggly line through the force vector $\vec{F}$ to show that we have replaced it by its x- and y-components. Otherwise, the diagram would include the same force twice. We will draw such a wiggly line in any force diagram where a force is replaced by its components. ◀

We will often need to find the vector sum (resultant) of *all* the forces acting on a body. We will call this the **net force** acting on the body. We will use the Greek letter Σ (capital "sigma," equivalent to the Roman S) as a shorthand notation for a sum. If the forces are labeled $\vec{F}_1$, $\vec{F}_2$, $\vec{F}_3$, and so on, we abbreviate the sum as

$$\vec{R} = \vec{F}_1 + \vec{F}_2 + \vec{F}_3 + \cdots = \Sigma \vec{F}, \qquad (4\text{–}1)$$

where $\Sigma \vec{F}$ is read as "the vector sum of the forces" or "the net force." The component version of Eq. (4–1) is the pair of component equations

$$R_x = \Sigma F_x, \qquad R_y = \Sigma F_y, \qquad (4\text{–}2)$$

where ΣF_x is the sum of the x-components, and so on. Each component may be positive or negative, so be careful with signs when you evaluate the sums in Eqs. (4–2).

Once we have R_x and R_y, we can find the magnitude and direction of the net force $\vec{R} = \Sigma \vec{F}$ acting on the body. The magnitude is

$$R = \sqrt{R_x{}^2 + R_y{}^2},$$

and the angle θ between $\vec{R}$ and the +x-axis can be found from the relation $\tan\theta = R_y/R_x$. The components R_x and R_y may be positive, negative, or zero, and the angle θ may be in any of the four quadrants.

In three-dimensional problems, forces may also have z-components; then we add the equation $R_z = \Sigma F_z$ to Eqs. (4–2). The magnitude of the net force is then

$$R = \sqrt{R_x{}^2 + R_y{}^2 + R_z{}^2}.$$

EXAMPLE 4–1

Superposition of forces Three customers are fighting over the same bargain basement coat. They apply the three horizontal forces to the coat that are shown in Fig. 4–5, where the coat is located at the origin. Find the x- and y-components of the net force on the coat, and find the magnitude and direction of the net force.

SOLUTION This is just a problem in vector addition, which we attack using the component method. The angles between the three forces $\vec{F}_1$, $\vec{F}_2$, and $\vec{F}_3$ and the +x-axis are $\theta_1 = 30°$, $\theta_2 = 180° - 45° = 135°$, and $\theta_3 = 180° + 53° = 233°$. The x- and y-components of the three forces are

$$F_{1x} = (200\text{ N})\cos 30° = 173\text{ N},$$
$$F_{1y} = (200\text{ N})\sin 30° = 100\text{ N},$$
$$F_{2x} = (300\text{ N})\cos 135° = -212\text{ N},$$
$$F_{2y} = (300\text{ N})\sin 135° = 212\text{ N},$$
$$F_{3x} = (155\text{ N})\cos 233° = -93\text{ N},$$
$$F_{3y} = (155\text{ N})\sin 233° = -124\text{ N}.$$

From Eqs. (4–2) the net force $\vec{R} = \Sigma \vec{F}$ has components

$$R_x = \Sigma F_x = F_{1x} + F_{2x} + F_{3x}$$
$$= 173\text{ N} + (-212\text{ N}) + (-93\text{ N}) = -132\text{ N},$$
$$R_y = \Sigma F_y = F_{1y} + F_{2y} + F_{3y}$$
$$= 100\text{ N} + 212\text{ N} + (-124\text{ N}) = 188\text{ N}.$$

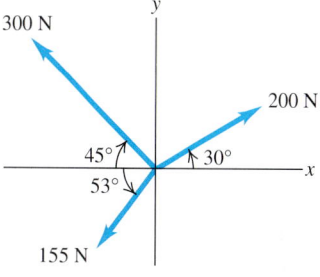

4–5 Three forces acting on a coat. What is the net force?

The net force has a negative x-component and a positive y-component, so it points to the left and toward the top of the page in Fig. 4–5 (that is, in the second quadrant).

The magnitude of the net force $\vec{R} = \Sigma \vec{F}$ is

$$R = \sqrt{R_x{}^2 + R_y{}^2} = \sqrt{(-132\text{ N})^2 + (188\text{ N})^2} = 230\text{ N}.$$

To find the angle between the net force and the +x-axis, we use the relation $\tan\theta = R_y/R_x$, or

$$\theta = \arctan\frac{R_y}{R_x} = \arctan\left(\frac{188\text{ N}}{-132\text{ N}}\right) = \arctan(-1.42).$$

The two possible solutions are $\theta = -55°$ or $\theta = -55° + 180° = 125°$. Since the net force lies in the second quadrant, as mentioned above, the correct answer is $\theta = 125°$.

4–3 NEWTON'S FIRST LAW

We have discussed some of the properties of forces, but so far have said nothing about how forces affect motion. To begin with, let's consider what happens when the net force on a body is *zero*. You would almost certainly agree that if a body is at rest, and if no net force acts on it (that is, no net push or pull), that body will remain at rest. But what if there is zero net force acting on a body in *motion?*

To see what happens in this case, suppose you slide a hockey puck along a horizontal table top, applying a horizontal force to it with your hand (Fig. 4–6a). After you stop pushing, the puck *does not* continue to move indefinitely; it slows down and stops. To keep it moving, you have to keep pushing (that is, applying a force). You might come to the "common sense" conclusions that bodies in motion naturally come to rest and that a force is required to sustain motion.

But now imagine pushing the puck across the smooth surface of a freshly waxed floor (Fig. 4–6b; see next page). After you quit pushing, the puck will slide a lot farther

(a)

4–6 A hockey puck is given an initial velocity. (a) It stops in a short distance on a table top.

(b)

(c)

4–6 (cont.) (b) A freshly waxed surface decreases the friction force, and the puck slides farther. (c) On an air-hockey table the friction force is practically zero, so the puck continues with almost constant velocity.

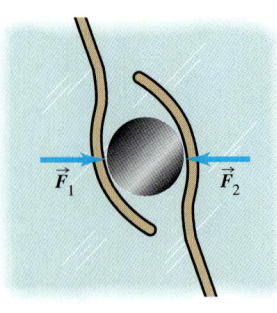

(a)

(b)

4–7 (a) A hockey puck accelerates in the direction of a net applied force $\vec{F}_1$. (b) When the net force is zero, the acceleration is zero, and the puck is in equilibrium.

before it stops. Put it on an air-hockey table, where it floats on a thin cushion of air, and it moves still farther (Fig. 4–6c). In each case, what slows the puck down is *friction,* an interaction between the lower surface of the puck and the surface on which it slides. Each surface exerts a frictional force on the puck that resists the puck's motion; the difference between the three cases is the magnitude of the frictional force. The slippery waxed floor exerts less friction than the table top, so the puck travels farther. The gas molecules of the air-hockey table exert the least friction of all. If we could eliminate friction completely, the puck would never slow down, and we would need no force at all to keep the puck moving once it had been started. Thus the "common sense" idea that a force is required to sustain motion is *incorrect.*

Experiments like the ones we've just described show that when no net force acts on a body, the body either remains at rest or moves with constant velocity in a straight line. Once a body has been set in motion, no net force is needed to keep it moving. In other words:

> **A body acted on by no net force moves with constant velocity (which may be zero) and zero acceleration.**

This is **Newton's first law of motion.**

The tendency of a body to keep moving once it is set in motion results from a property called **inertia.** You use this when you try to get ketchup out of a bottle by shaking it. First you start the bottle (and the ketchup inside) moving forward; when you jerk the bottle back, the ketchup tends to keep moving forward and, you hope, ends up on your burger. The tendency of a body at rest to remain at rest is also due to inertia. You may have seen a tablecloth yanked out from under the china without breaking anything. The force on the china isn't great enough to make it move appreciably during the short time it takes to pull the tablecloth away.

It's important to note that the *net* force is what matters in Newton's first law. For example, a physics book at rest on a horizontal table top has two forces acting on it: the downward force of the earth's gravitational attraction (a long-range force that acts even if the table top is elevated above the ground) and an upward supporting force exerted by the table top (a contact force). The upward push of the surface is just as great as the downward pull of gravity, so the *net* force acting on the book (that is, the vector sum of the two forces) is zero. In agreement with Newton's first law, if the book is at rest on the table top, it remains at rest. The same principle applies to a hockey puck sliding on a horizontal, frictionless surface: The vector sum of the upward push of the surface and the downward pull of gravity is zero. Once the puck is in motion, it continues to move with constant velocity because the *net* force acting on it is zero. (The upward supporting force of the surface is called a **normal force** because it is *normal,* or perpendicular, to the surface of contact. We'll discuss the normal force in more detail in Chapter 5.)

Here's another example. Suppose a hockey puck rests on a horizontal surface with negligible friction, such as an air-hockey table or a slab of wet ice. If the puck is initially at rest and a single horizontal force $\vec{F}_1$ acts on it (Fig. 4–7a), the puck starts to move. If the puck is in motion to begin with, the force changes its speed, its direction, or both, depending on the direction of the force. In this case the net force is equal to $\vec{F}_1$, which is *not* zero. (There are also two vertical forces: the earth's gravitational attraction and the upward contact force exerted by the surface. But as we mentioned earlier, these two forces cancel.)

Now suppose we apply a second force $\vec{F}_2$ (Fig. 4–7b), equal in magnitude to $\vec{F}_1$ but opposite in direction. The two forces are negatives of each other, $\vec{F}_2 = -\vec{F}_1$, and their vector sum is zero:

$$\Sigma \vec{F} = \vec{F}_1 + \vec{F}_2 = \vec{F}_1 + (-\vec{F}_1) = 0.$$

Again, we find that if the body is at rest at the start, it remains at rest; if it is initially moving, it continues to move in the same direction with constant speed. These results show that in Newton's first law, *zero net force is equivalent to no force at all.* This is just the principle of superposition of forces that we saw in Section 4–2.

When a body is acted on by no forces, or by several forces such that their vector sum (resultant) is zero, we say that the body is in **equilibrium.** In equilibrium, a body is either at rest or moving in a straight line with constant velocity. For a body in equilibrium, the net force is zero:

$$\Sigma \vec{F} = 0 \qquad \text{(body in equilibrium).} \tag{4–3}$$

For this to be true, each component of the net force must be zero, so

$$\Sigma F_x = 0, \qquad \Sigma F_y = 0 \qquad \text{(body in equilibrium).} \tag{4–4}$$

When Eqs. (4–4) are satisfied, the body is in equilibrium.

We are assuming that the body can be represented adequately as a point particle. When the body has finite size, we also have to consider *where* on the body the forces are applied. We will return to this point in Chapter 11.

EXAMPLE 4–2

Zero net force means constant velocity In the classic 1950 science fiction film *Rocketship X-M,* a spaceship is moving in the vacuum of outer space, far from any planet, when its engine dies. As a result, the spaceship slows down and stops. What does Newton's first law say about this event?

SOLUTION In this situation there are no forces acting on the spaceship, so according to Newton's first law, it will *not* stop. It continues to move in a straight line with constant speed. This example of science fiction contains more fiction than science.

EXAMPLE 4–3

Constant velocity means zero net force You are driving your Porsche 911 Carrera along a straight testing track at a constant speed of 150 km/h. You pass a 1971 Volkswagen Beetle doing a constant 75 km/h. For which car is the net force greater?

SOLUTION The key word in this question is "net." Both cars are in equilibrium because their velocities are both constant; therefore the *net* force on each car is *zero.*

This conclusion seems to contradict the "common sense" idea that the faster car must have a greater force pushing it. It's true that there is a forward force on both cars, and it's true that

the forward force on your Porsche is much greater than that on the Volkswagen (thanks to your Porsche's high-power engine). But there is also a *backward* force acting on each car due to friction with the road and air resistance. The only reason these cars need their engines running at all is to counteract this backward force so that the vector sum of the forward and backward forces will be zero and the car will travel with constant velocity. The backward force on your Porsche is greater because of its greater speed, which is why your engine has to be more powerful than the Volkswagen's.

INERTIAL FRAMES OF REFERENCE

In discussing relative velocity in Section 3–6, we introduced the concept of *frame of reference.* This concept is central to Newton's laws of motion. Suppose you are in an airplane that is accelerating down the runway during takeoff. If you could stand in the aisle on roller skates, you would start moving *backward* relative to the plane as the pilot opens the throttle. If instead the airplane was landing, you would start moving forward down the aisle as the airplane slowed to a stop. In either case, it looks as though Newton's first law is not obeyed; there is no net force acting on you, yet your velocity changes. What's wrong?

The point is that the airplane is accelerating with respect to the earth and is *not* a suitable frame of reference for Newton's first law. This law is valid in some frames of reference, and is not valid in others. A frame of reference in which Newton's first law *is* valid is called an **inertial frame of reference.** The earth is at least approximately an

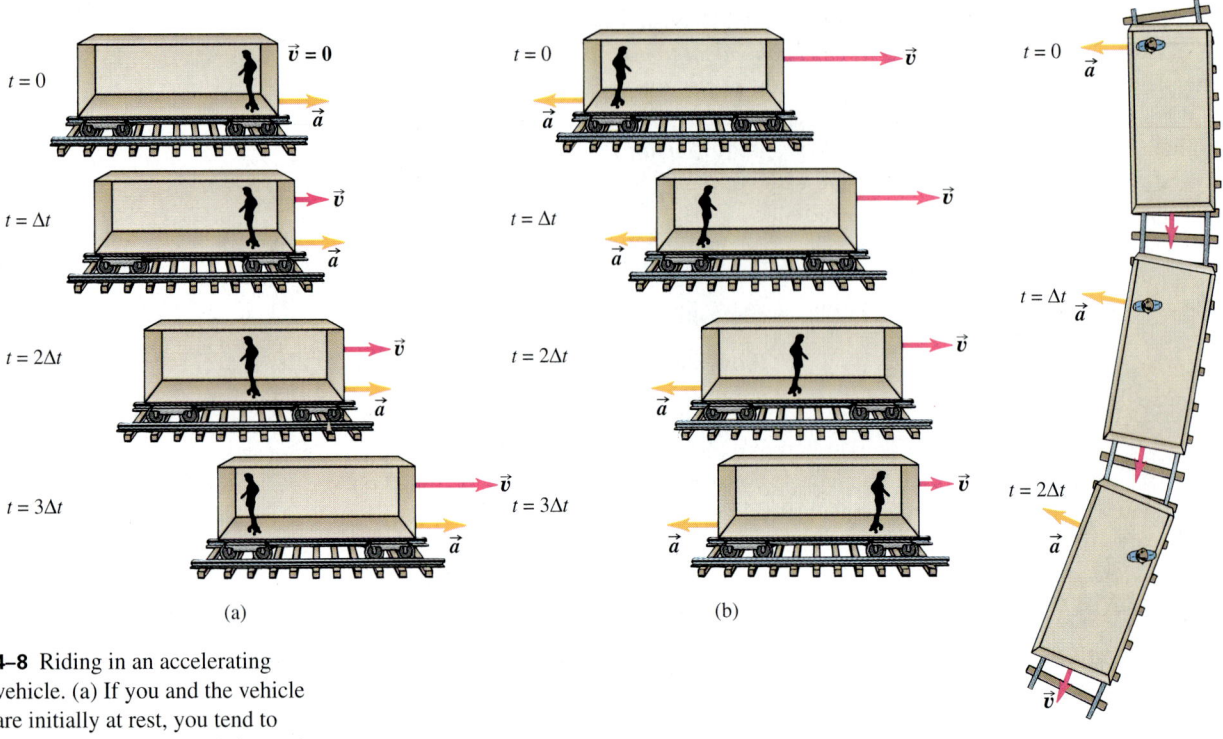

(a)

(b)

(c)

4–8 Riding in an accelerating vehicle. (a) If you and the vehicle are initially at rest, you tend to remain at rest as the vehicle accelerates around you. (b) If you and the vehicle are initially moving, you tend to continue moving with constant velocity as the vehicle slows down. (c) You tend to continue moving in a straight line as the vehicle turns.

inertial frame of reference, but the airplane is not. (The earth is not a completely inertial frame, owing to the acceleration associated with its rotation and its motion around the sun. These effects are quite small, however; see Exercises 3–21 and 3–24.) Because Newton's first law is used to define what we mean by an inertial frame of reference, it is sometimes called the *law of inertia*.

Figure 4–8 shows how we can use Newton's first law to understand what you experience when riding in a vehicle that's accelerating. In Fig. 4–8a a vehicle is initially at rest, then begins to accelerate to the right. A passenger on roller skates has virtually no net force acting on her, since the skates nearly eliminate the effects of friction; hence she tends to remain at rest relative to the inertial frame of the earth, in accordance with Newton's first law. As the vehicle accelerates around her, she moves backwards relative to the vehicle. In the same way, a passenger in a vehicle that is slowing down tends to continue moving with constant velocity relative to the earth (Fig. 4–8b). This passenger moves forward relative to the vehicle. A vehicle is also accelerating if it moves at a constant speed but is turning (Fig. 4–8c). In this case a passenger tends to continue moving relative to the earth at constant speed in a straight line; relative to the vehicle, the passenger moves to the side of the vehicle on the outside of the turn.

In each case shown in Fig. 4–8, an observer in the vehicle's frame of reference might be tempted to conclude that there *is* a net force acting on the passenger in each case, since the passenger's velocity *relative to the vehicle* changes in each case. This conclusion is simply wrong; the net force on the passenger is indeed zero. The vehicle observer's mistake is in trying to apply Newton's first law in the vehicle's frame of reference, which is *not* an inertial frame and in which Newton's first law isn't valid. In this book we will use *only* inertial frames of reference.

We've mentioned only one (approximately) inertial frame of reference: the earth's surface. But there are many inertial frames. If we have an inertial frame of reference *A*, in which Newton's first law is obeyed, then any second frame of reference *B* will also be inertial if it moves relative to A with constant velocity $\vec{v}_{B/A}$. To prove this, we use the relative velocity relation Eq. (3–36) from Section 3–6:

$$\vec{v}_{P/A} = \vec{v}_{P/B} + \vec{v}_{B/A}$$

Suppose that P is a body that moves with constant velocity $\vec{v}_{P/A}$ with respect to an inertial frame A. By Newton's first law the net force on this body is zero. The velocity of P relative to another frame B has a different value, $\vec{v}_{P/B} = \vec{v}_{P/A} - \vec{v}_{B/A}$. But if the relative velocity $\vec{v}_{B/A}$ of the two frames is constant, then $\vec{v}_{P/B}$ will be constant as well. Thus B is also an inertial frame; the velocity of P in this frame is constant, and the net force on P is zero, so Newton's first law is obeyed in B. Observers in frames A and B will disagree about the velocity of P, but they will agree that P has a constant velocity (zero acceleration) and has zero net force acting on it.

There is no single inertial frame of reference that is preferred over all others for formulating Newton's laws. If one frame is inertial, then every other frame moving relative to it with constant velocity is also inertial. Viewed in this light, the state of rest and the state of motion with constant velocity are not very different; both occur when the vector sum of forces acting on the body is zero.

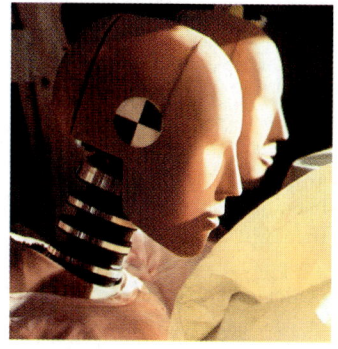

From the frame of reference of this car, it seems as though a force is pushing the crash test dummies forward as the car comes to a sudden stop. But there is really no such force: as the car stops, the dummies keep moving forward as a consequence of Newton's first law.

4–4 NEWTON'S SECOND LAW

In discussing Newton's first law, we have seen that when a body is acted on by no force or zero net force, it moves with constant velocity and zero acceleration. In Fig. 4–9a a hockey puck is sliding to the right on wet ice, so there is negligible friction. There are no horizontal forces acting on the puck; the downward force of gravity and the upward contact force exerted by the ice surface sum to zero. So the net force $\Sigma\,\vec{F}$ acting on the puck is zero, the puck has zero acceleration, and its velocity is constant.

But what happens when the net force is *not* zero? In Fig. 4–9b we apply a constant horizontal force to a sliding puck in the same direction that the puck is moving. Then $\Sigma\vec{F}$ is constant and in the same horizontal direction as $\vec{v}$. We find that during the time the force is acting, the velocity of the puck changes at a constant rate; that is, the puck moves with constant acceleration. The speed of the puck increases, so the acceleration $\vec{a}$ is in the same direction as $\vec{v}$ and as $\Sigma\,\vec{F}$.

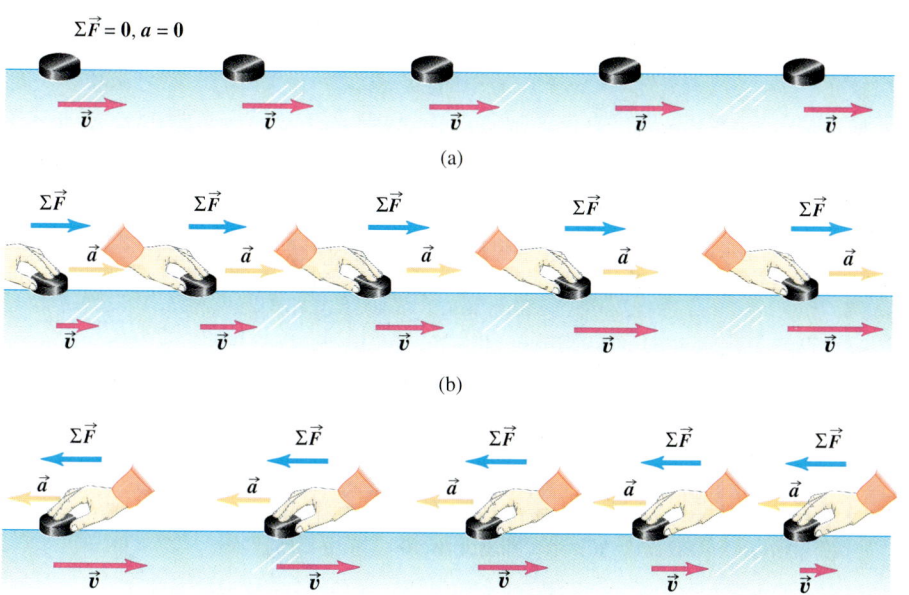

4–9 The acceleration of a body is in the same direction as the net force acting on the body (in this case, a hockey puck on a frictionless surface). (a) If $\Sigma\,\vec{F} = 0$, the puck is in equilibrium; the velocity is constant and the acceleration is zero. (b) If $\Sigma\,\vec{F}$ is to the right, the acceleration is to the right. (c) If $\Sigma\,\vec{F}$ is to the left, the acceleration is to the left.

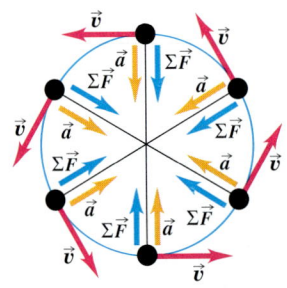

4–10 A top view of a hockey puck in uniform circular motion on a frictionless horizontal surface. At any point in the motion the acceleration $\vec{a}$ and the net force $\Sigma\vec{F}$ are in the same direction, toward the center of the circle.

(a)

(b)

(c)

4–11 (a) Acceleration $\vec{a}$ is proportional to the net force $\Sigma\vec{F}$. (b) Doubling the net force doubles the acceleration. (c) Halving the net force halves the acceleration.

Figure 4–9c shows another experiment, in which we reverse the direction of the force on the puck so that $\Sigma\vec{F}$ acts in the direction opposite to $\vec{v}$. In this case as well, the puck has an acceleration; the puck moves more and more slowly to the right. If the leftward force continues to act, the puck eventually stops and begins to move more and more rapidly to the left. The acceleration $\vec{a}$ in this experiment is to the left, in the same direction as $\Sigma\vec{F}$. As in the previous case, experiment shows that the acceleration is constant if $\Sigma\vec{F}$ is constant.

We conclude that the presence of a net force acting on a body causes the body to accelerate. The direction of the acceleration is the same as that of the net force. If the magnitude of the net force is constant, as in Figs. 4–9b and 4–9c, then so is the magnitude of the acceleration.

These conclusions about net force and acceleration also apply to a body moving along a curved path. For example, Fig. 4–10 shows a hockey puck moving in a horizontal circle on an ice surface of negligible friction. A string attaching the puck to the ice exerts a force of constant magnitude toward the center of the circle. The result is acceleration that is constant in magnitude and directed toward the center of the circle. The speed of the puck is constant, so this is uniform circular motion, as discussed in Section 3–5.

Figure 4–11a shows another experiment to explore the relationship between the acceleration of a body and the net force acting on that body. We apply a constant horizontal force to a puck on a frictionless horizontal surface, using the spring balance described in Section 4–2 with the spring stretched a constant amount. As in Figs. 4–9b and 4–9c, this horizontal force equals the net force on the puck. If we change the magnitude of the net force, the acceleration changes in the same proportion. Doubling the net force doubles the acceleration (Fig. 4–11b), halving the force halves the acceleration (Fig. 4–11c), and so on. Many such experiments show that for any given body the magnitude of the acceleration is directly proportional to the magnitude of the net force acting on the body.

For a given body the ratio of the magnitude $|\Sigma\vec{F}|$ of the net force to the magnitude $a = |\vec{a}|$ of the acceleration is constant, regardless of the magnitude of the net force. We call this ratio the inertial mass, or simply the **mass**, of the body and denote it by m. That is, $m = |\Sigma\vec{F}|/a$, or

$$|\Sigma\vec{F}| = ma. \qquad (4\text{–}5)$$

MASS AND FORCE

Mass is a quantitative measure of inertia, which we discussed in Section 4–3. The greater its mass, the more a body "resists" being accelerated. The concept is easy to relate to everyday experience. If you hit a table-tennis ball and then a basketball with the same force, the basketball has much smaller acceleration because it has much greater mass. If a force causes a large acceleration, the mass of the body is small; if the same force causes only a small acceleration, the mass of the body is large.

The SI unit of mass is the **kilogram.** We mentioned in Section 1–4 that the kilogram is officially defined to be the mass of a chunk of platinum-iridium alloy kept in a vault near Paris. We can use this standard kilogram, along with Eq. (4–5), to define the **newton:**

> **One newton is the amount of net force that gives an acceleration of one meter per second squared to a body with a mass of one kilogram.**

We can use this definition to calibrate the spring balances and other instruments used to measure forces. Because of the way we have defined the newton, it is related to the units of mass, length, and time. For Eq. (4–5) to be dimensionally consistent, it must be true that

$$1 \text{ newton} = (1 \text{ kilogram})(1 \text{ meter per second squared}),$$

or

$$1 \text{ N} = 1 \text{ kg} \cdot \text{m/s}^2.$$

We will use this relation many times in the next few chapters, so keep it in mind.

We can also use Eq. (4–5) to compare a mass with the standard mass and thus to *measure* masses. Suppose we apply a constant net force F to a body having a known mass m_1 and we find an acceleration of magnitude a_1. We then apply the same force to another body having an unknown mass m_2, and we find an acceleration of magnitude a_2. Then, according to Eq. (4–5),

$$m_1 a_1 = m_2 a_2,$$

$$\frac{m_2}{m_1} = \frac{a_1}{a_2} \qquad \text{(same net force)}. \tag{4–6}$$

The ratio of the masses is the inverse of the ratio of the accelerations. Figure 4–12 shows the inverse proportionally between mass and acceleration. In principle we could use Eq. (4–6) to measure an unknown mass m_2, but it is usually easier to determine mass indirectly by measuring the body's *weight*. We'll return to this point in Section 4–5.

When two bodies with masses m_1 and m_2 are fastened together, we find that the mass of the composite body is always $m_1 + m_2$ (Fig. 4–12c). This additive property of mass may seem obvious, but it has to be verified experimentally. Ultimately, the mass of a body is related to the number of protons, electrons, and neutrons it contains. This wouldn't be a good way to *define* mass because there is no practical way to count these particles. But the concept of mass is the most fundamental way to characterize the quantity of matter in a body.

(a)

(b)

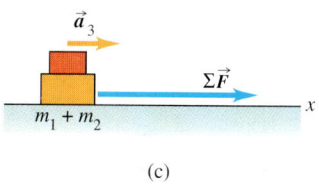

(c)

4–12 For a constant net force $\Sigma \vec{F}$, acceleration is inversely proportional to the mass of a body. Masses add like ordinary scalars. (a) $\vec{a}_1 = \Sigma \vec{F}/m_1$. (b) With a greater mass m_2, the acceleration has smaller magnitude: $\vec{a}_2 = \Sigma \vec{F}/m_2$. (c) If the two masses are combined, the acceleration is $\vec{a}_3 = \Sigma \vec{F}/(m_1 + m_2)$.

NEWTON'S SECOND LAW

We've been careful to state that the *net* force on a body is what causes that body to accelerate. Experiment shows that if a combination of forces $\vec{F}_1, \vec{F}_2, \vec{F}_3, \ldots$ is applied to a body, the body will have the same acceleration (magnitude and direction) as when only a single force is applied, if that single force is equal to the vector sum $\vec{F}_1 + \vec{F}_2 + \vec{F}_3 + \ldots$. In other words, the principle of superposition of forces also holds true when the net force is not zero and the body is accelerating.

Equation (4–5) relates the magnitude of the net force on a body to the magnitude of the acceleration that it produces. We have also seen that the direction of the net force is the same as the direction of the acceleration, whether the body's path is straight or curved. Newton wrapped up all these relationships and experimental results in a single concise statement that we now call **Newton's second law of motion**:

> **If a net external force acts on a body, the body accelerates. The direction of acceleration is the same as the direction of the net force. The net force vector is equal to the mass of the body times the acceleration of the body.**

In symbols,

$$\Sigma \vec{F} = m\vec{a} \qquad \text{(Newton's second law)}. \tag{4–7}$$

An alternative statement is that the acceleration of a body (the rate of change of its velocity) is equal to the vector sum (resultant) of all forces acting on the body, divided by its mass. The acceleration has the same direction as the net force. Newton's second law is a fundamental law of nature, the basic relation between force and motion. Most

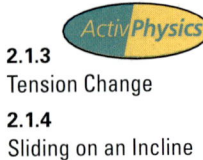

2.1.3
Tension Change
2.1.4
Sliding on an Incline

of the remainder of this chapter and all of the next are devoted to learning how to apply this principle in various situations.

Though Eq. (4–7) may be new to you, you've actually been using it all your life to measure your body's acceleration. In your inner ear, microscopic hair cells sense the magnitude and direction of the force that they must exert to cause small membranes to accelerate along with the rest of your body. By Newton's second law, the acceleration of the membranes—and hence that of your body as a whole—is proportional to this force and has the same direction. In this way, you can sense the magnitude and direction of your acceleration even with your eyes closed!

There are at least four aspects of Newton's second law that deserve special attention. First, Eq. (4–7) is a *vector* equation. Usually, we will use it in component form, with a separate equation for each component of force and the corresponding acceleration:

$$\Sigma F_x = ma_x, \qquad \Sigma F_y = ma_y, \qquad \Sigma F_z = ma_z \qquad \text{(Newton's second law).} \quad (4\text{–}8)$$

This set of component equations is equivalent to the single vector equation (4–7). Each component of total force equals the mass times the corresponding component of acceleration.

Second, the statement of Newton's second law refers to *external* forces. By this we mean forces exerted on the body by other bodies in its environment. It's impossible for a body to affect its own motion by exerting a force on itself; if it was possible, you could lift yourself to the ceiling by pulling up on your belt! That's why only external forces are included in the sum $\Sigma \vec{F}$ in Eqs. (4–7) and (4–8).

Third, Eqs. (4–7) and (4–8) are valid only when the mass m is *constant*. It's easy to think of systems whose masses change, such as a leaking tank truck, a rocket ship, or a moving railroad car being loaded with coal. But such systems are better handled by using the concept of momentum; we'll get to that in Chapter 8.

Finally, Newton's second law is valid only in inertial frames of reference, just like the first law. Thus it is not valid in the reference frame of any of the accelerating vehicles in Fig. 4–8; relative to any of these frames, the passenger accelerates even though the net force on the passenger is zero. We will usually assume that the earth is an adequate approximation to an inertial frame, although because of its rotation and orbital motion it is not precisely inertial.

CAUTION ▶ Please note that the quantity $m\vec{a}$ is *not* a force. All that Eqs. (4–7) and (4–8) say is that the vector $m\vec{a}$ is equal in magnitude and direction to the vector sum $\Sigma \vec{F}$ of all the forces acting on the body. It's incorrect to think of acceleration as a force; rather, acceleration is a result of a nonzero net force. It's "common sense" to think that there is a "force of acceleration" that pushes you back into your seat when your car accelerates forward from rest. But *there is no such force;* instead, your inertia causes you to tend to stay at rest relative to the earth, and the car accelerates around you. The "common sense" confusion arises from trying to apply Newton's second law in a frame of reference where it isn't valid, like the non-inertial reference frame of an accelerating car. We will always examine motion relative to *inertial* frames of reference only. ◀

In learning how to use Newton's second law, we will begin in this chapter with examples of straight-line motion. Then in Chapter 5 we will consider more general cases and develop more detailed problem-solving strategies for applying Newton's laws of motion.

The design of high-performance motorcycles depends fundamentally on Newton's second law. To maximize the forward acceleration, the designer makes the motorcycle as light as possible (that is, minimizes the mass) and uses the most powerful engine possible (thus maximizing the forward force).

EXAMPLE 4-4

Determining acceleration from force A worker applies a constant horizontal force with magnitude 20 N to a box with mass 40 kg resting on a level surface with negligible friction. What is the acceleration of the box?

SOLUTION The first steps in *any* problem involving forces are to choose a coordinate system and to identify all the forces acting on the body in question. We take the +x-axis in the direction of the horizontal force and the +y-axis to be upward (Fig. 4–13).

4–13 (a) A 20-N force accelerates the box to the right. (b) Vector diagram of forces acting on the box, considered as a particle. The tiles under the box are freshly waxed, so friction is negligible.

The forces acting on the box are (i) the horizontal force $\vec{F}$ exerted by the worker; (ii) the weight $\vec{w}$ of the box, that is, the downward force of the earth's gravitational attraction; and (iii) the upward supporting force $\vec{n}$ exerted by the surface. As in Section 4–3, we call the force $\vec{n}$ a *normal* force because it is normal (perpendicular) to the surface of contact. (We use a script letter n to avoid confusion with the abbreviation N for newton.) We are told that friction is negligible, so no friction force is present. The acceleration is given by Newton's second law, Eqs. (4–8). There is no vertical acceleration, so we know that the two vertical forces must sum to zero. There is only one horizontal

force, and we have

$$\Sigma F_x = ma_x,$$

$$a_x = \frac{\Sigma F_x}{m} = \frac{20 \text{ N}}{40 \text{ kg}} = \frac{20 \text{ kg} \cdot \text{m/s}^2}{40 \text{ kg}} = 0.50 \text{ m/s}^2.$$

The acceleration is in the $+x$-direction, the same direction as the net force. The net force is constant, so the acceleration is also constant. If we are given the initial position and velocity of the box, we can find the position and velocity at any later time from the equations of motion with constant acceleration we derived in Chapter 2.

EXAMPLE 4–5

Determining force from acceleration   A waitress shoves a ketchup bottle with mass 0.45 kg toward the right along a smooth, level lunch counter. As the bottle leaves her hand, it has an initial velocity of 2.8 m/s. As it slides, it slows down because of the constant horizontal friction force exerted on it by the counter top. It slides a distance of 1.0 m before coming to rest. What are the magnitude and direction of the friction force acting on it?

SOLUTION   As in the previous example, the first step is to choose a coordinate system. Suppose the bottle slides in the $+x$-direction, leaving her hand at the point $x_0 = 0$ with the given initial velocity (Fig. 4–14a). The forces acting on the bottle are shown in Fig. 4–14b. The friction force $\vec{f}$ acts to slow the bottle down, so its direction must be opposite the direction of velocity (see Fig. 4–9c).

To find the magnitude of $\vec{f}$ using Eqs. (4–8), we must first

4–14 (a) As the bottle slides to the right, the friction force slows it down. (b) Vector diagram of forces acting on the bottle, considered as a particle.

determine the acceleration of the bottle. We are given that the friction force is constant. The acceleration is then constant also, so we may use the constant-acceleration formula Eq. (2–13):

$$v^2 = v_0{}^2 + 2a(x - x_0),$$

$$a = \frac{v^2 - v_0{}^2}{2(x - x_0)} = \frac{(0 \text{ m/s})^2 - (2.8 \text{ m/s})^2}{2(1.0 \text{ m} - 0 \text{ m})} = -3.9 \text{ m/s}^2.$$

The negative sign means that the acceleration is toward the *left;* the velocity is in the opposite direction to acceleration, as it

must be, since the bottle is slowing down. The net force in the *x*-direction is the *x*-component –*f* of the friction force, so

$$\Sigma F_x = -f = ma = (0.45 \text{ kg})(-3.9 \text{ m/s}^2)$$
$$= -1.8 \text{ kg} \cdot \text{m/s}^2 = -1.8 \text{ N}.$$

Again the negative sign shows that the force on the bottle is directed toward the left. The magnitude of the friction force is $f = 1.8 \text{ N}$. Remember that magnitudes are *always* positive!

SOME NOTES ON UNITS

A few words about units are in order. In the cgs metric system (not used in this book), the unit of mass is the gram, equal to 10^{-3} kg, and the unit of distance is the centimeter, equal to 10^{-2} m. The corresponding unit of force is called the *dyne:*

$$1 \text{ dyne} = 1 \text{ g} \cdot \text{cm/s}^2.$$

One dyne is equal to 10^{-5} N. In the British system, the unit of force is the *pound* (or pound-force) and the unit of mass is the *slug*. The unit of acceleration is one foot per second squared, so

$$1 \text{ pound} = 1 \text{ slug} \cdot \text{ft/s}^2.$$

The official definition of the pound is

$$1 \text{ pound} = 4.448221615260 \text{ newtons}.$$

It is handy to remember that a pound is about 4.4 N and a newton is about 0.22 pound. Next time you want to order a "quarter-pounder," try asking for a "one-newtoner" and see what happens. Another useful fact: A body with a mass of 1 kg has a weight of about 2.2 lb at the earth's surface.

The units of force, mass, and acceleration in the three systems are summarized in Table 4–2.

Despite its name, the English unit of mass has nothing to do with the type of slug shown here. A common garden slug has a mass of about 15 grams, or about 10^{-3} slug.

TABLE 4–2

UNITS OF FORCE, MASS, AND ACCELERATION

SYSTEM OF UNITS	FORCE	MASS	ACCELERATION
SI	newton (N)	kilogram (kg)	m/s^2
cgs	dyne (dyn)	gram (g)	cm/s^2
British	pound (lb)	slug	ft/s^2

4–5 MASS AND WEIGHT

The *weight* of a body is a familiar force. It is the force of the earth's gravitational attraction for the body. We will study gravitational interactions in detail in Chapter 12, but we need some preliminary discussion now. The terms *mass* and *weight* are often misused and interchanged in everyday conversation. It is absolutely essential for you to understand clearly the distinctions between these two physical quantities.

Mass characterizes the *inertial* properties of a body. Mass is what keeps the china on the table when you yank the tablecloth out from under it. The greater the mass, the greater the force needed to cause a given acceleration; this is reflected in Newton's second law, $\Sigma \vec{F} = m\vec{a}$. Weight, on the other hand, is a *force* exerted on a body by the pull of the earth or some other large body. Everyday experience shows us that bodies having large mass also have large weight. A large stone is hard to throw because of its large *mass,* and hard to lift off the ground because of its large *weight.* On the moon the stone would be just as hard to throw horizontally, but it would be easier to lift. So what exactly *is* the relationship between mass and weight?

The answer to this question, according to legend, came to Newton as he sat under an apple tree watching the apples fall. A freely falling body has an acceleration equal to g, and because of Newton's second law, a force must act to produce this acceleration. If a 1-kg body falls with an acceleration of 9.8 m/s^2, the required force has magnitude

$$F = ma = (1 \text{ kg})(9.8 \text{ m/s}^2) = 9.8 \text{ kg} \cdot \text{m/s}^2 = 9.8 \text{ N}.$$

But the force that makes the body accelerate downward is the gravitational pull of the earth, that is, the *weight* of the body. Any body near the surface of the earth that has a mass of 1 kg *must* have a weight of 9.8 N to give it the acceleration we observe when it is in free fall. More generally, a body with mass m must have weight with magnitude w given by

$$w = mg \quad \text{(magnitude of the weight of a body of mass } m\text{).} \qquad (4\text{–}9)$$

2.11
Pole Vaulter Vaults

The weight of a body is a force, a vector quantity, and we can write Eq. (4–9) as a vector equation:

$$\vec{w} = m\vec{g}. \qquad (4\text{–}10)$$

Remember that g is the *magnitude* of $\vec{g}$, the acceleration due to gravity, so g is always a positive number, by definition. Thus w, given by Eq. (4–9), is the *magnitude* of the weight and is also always positive.

It is important to understand that the weight of a body acts on the body *all the time,* whether it is in free fall or not. When a 10-kg flowerpot hangs suspended from a chain, it is in equilibrium, and its acceleration is zero. But its weight, given by Eq. (4–10), is still pulling down on it. In this case the chain pulls up on the pot, applying an upward force. The *vector sum* of the forces is zero, and the pot is in equilibrium.

EXAMPLE 4–6

Net force and acceleration in free fall In Example 2–6 (Section 2–6) a 500-lira coin was dropped from rest from the Leaning Tower of Pisa. If the coin falls freely, so that the effects of the air are negligible, how does the net force on the coin vary as it falls?

SOLUTION In free fall, the acceleration $\vec{a}$ of the coin is constant and equal to $\vec{g}$. Hence by Newton's second law the net force $\Sigma \vec{F} = m\vec{a}$ is also constant and equal to $m\vec{g}$, which is the weight $\vec{w}$ of the coin (Fig. 4–15). The velocity of the coin changes as it falls, but the net force acting on it remains constant. If this result surprises you, it's because you still believe in the erroneous "common sense" idea that greater speed implies greater force. If so, you should reread Example 4–3.

The net force on a freely-falling coin is constant even if you initially toss it upward. The force that your hand exerts on the coin is a contact force, and it disappears the instant that the coin loses contact with your hand. From then on, the only force acting on the coin is its weight $\vec{w}$.

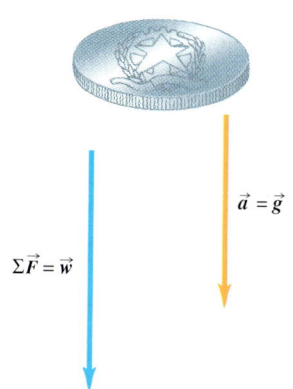

4–15 The acceleration of a freely-falling object is constant, and so is the net force acting on the object.

VARIATION OF *g* WITH LOCATION

We will use $g = 9.80$ m/s^2 for problems on the earth. In fact, the value of g varies somewhat from point to point on the earth's surface, from about 9.78 to 9.82 m/s^2, because the earth is not perfectly spherical and because of effects due to its rotation and orbital

(a)

(b)

4–16 (a) A standard kilogram weighs about 9.8 N on earth. (b) The same kilogram weighs only about 1.6 N on the moon.

4–17 An equal-arm balance determines the mass of a body by comparing its weight to a known weight.

motion. At a point where $g = 9.80 \text{ m/s}^2$, the weight of a standard kilogram is $w = 9.80$ N. At a different point, where $g = 9.78 \text{ m/s}^2$, the weight is $w = 9.78$ N but the mass is still 1 kg. The weight of a body varies from one location to another; the mass does not. If we take a standard kilogram to the surface of the moon, where the acceleration of free fall (equal to the value of g at the moon's surface) is 1.62 m/s^2, its weight is 1.62 N, but its mass is still 1 kg (Fig. 4–16). An 80.0-kg astronaut has a weight on earth of $(80.0 \text{ kg})(9.80 \text{ m/s}^2) = 784$ N, but on the moon the astronaut's weight would be only $(80.0 \text{ kg})(1.62 \text{ m/s}^2) = 130$ N.

The following excerpt from Astronaut Buzz Aldrin's book *Men from Earth,* describing his experiences during the Apollo 11 mission to the moon in 1969, offers some interesting insights into the distinction between mass and weight:

> Our portable life-support system backpacks looked simple, but they were hard to put on and tricky to operate. On Earth the portable life-support system and space suit combination weighed 190 pounds, but on the moon it was only 30. Combined with my own body weight, that brought me to a total lunar-gravity weight of around 60 pounds.
>
> One of my tests was to jog away from the lunar module to see how maneuverable an astronaut was on the surface. I remembered what Isaac Newton had taught us two centuries before: Mass and weight are not the same. I weighed only 60 pounds, but my *mass* was the same as it was on Earth. Inertia was a problem. I had to plan ahead several steps to bring myself to a stop or to turn without falling.

MEASURING MASS AND WEIGHT

In Section 4–4 we described a way to compare masses by comparing their accelerations when subjected to the same net force. Usually, however, the easiest way to measure the mass of a body is to measure its weight, often by comparing with a standard. Because of Eq. (4–9), two bodies that have the same weight at a particular location also have the same mass. We can compare weights very precisely; the familiar equal-arm balance (Fig. 4–17) can determine with great precision (up to 1 part in 10^6) when the weights of two bodies are equal and hence when their masses are equal. This method doesn't work in the apparent "zero-gravity" environment of outer space. Instead, we have to use Newton's second law directly. We apply a known force to the body, measure its acceleration, and compute the mass as the ratio of force to acceleration. This method, or a variation of it, is used to measure the masses of astronauts in orbiting space stations as well as the masses of atomic and subatomic particles.

The concept of mass plays two rather different roles in mechanics. The weight of a body (the gravitational force acting on it) is proportional to its mass; we may call the property related to gravitational interactions *gravitational mass.* On the other hand, we can call the inertial property that appears in Newton's second law the *inertial mass.* If these two quantities were different, the acceleration due to gravity might well be different for different bodies. However, extraordinarily precise experiments have established with a precision of better than one part in 10^{12} that in fact the two *are* the same.

CAUTION ▶ The SI units for mass and weight are often misused in everyday life. Incorrect expressions such as "This box weighs 6 kg" are nearly universal. What is meant is that the *mass* of the box, probably determined indirectly by *weighing,* is 6 kg. This usage is so common that there is probably no hope of straightening things out, but be sure you recognize that the term *weight* is often used when *mass* is meant. Be careful to avoid this kind of mistake in your own work! In SI units, weight (a force) is measured in newtons, while mass is measured in kilograms. ◀

EXAMPLE 4–7

A 1.96×10^4 N Lincoln Town Car traveling in the +x-direction makes a fast stop; the x-component of the net force acting on it is -1.50×10^4 N. What is its acceleration?

SOLUTION Because the newton is a unit of force, we know that 1.96×10^4 N is the weight, not the mass, of the car. Its mass m is

$$m = \frac{w}{g} = \frac{1.96 \times 10^4 \text{ N}}{9.80 \text{ m/s}^2} = \frac{1.96 \times 10^4 \text{ kg} \cdot \text{m/s}^2}{9.80 \text{ m/s}^2}$$

$$= 2000 \text{ kg}.$$

Then $\Sigma F_x = ma_x$ gives

$$a_x = \frac{\Sigma F_x}{m} = \frac{-1.50 \times 10^4 \text{ N}}{2000 \text{ kg}} = \frac{-1.50 \times 10^4 \text{ kg} \cdot \text{m/s}^2}{2000 \text{ kg}}$$

$$= -7.5 \text{ m/s}^2.$$

This acceleration can be written as $-0.77g$. Note that -0.77 is also the ratio of -1.50×10^4 N (the x-component of the net force) to 1.96×10^4 N (the weight).

4–6 NEWTON'S THIRD LAW

A force acting on a body is always the result of its interaction with another body, so forces always come in pairs. You can't pull on a doorknob without the doorknob pulling back on you. When you kick a football, the forward force that your foot exerts on the ball launches it into its trajectory, but you also feel the force the ball exerts back on your foot. If you kick a boulder, the pain you feel is due to the force that the boulder exerts on your foot.

In each of these cases the force that you exert on the other body is in the opposite direction to the force that body exerts on you. Experiments show that whenever two bodies interact, the two forces that they exert on each other are always *equal in magnitude* and *opposite in direction*. This fact is called **Newton's third law of motion.** In Fig. 4–18, $\vec{F}_{A \text{ on } B}$ is the force applied *by* body A (first subscript) *on* body B (second subscript), and $\vec{F}_{B \text{ on } A}$ is the force applied *by* body B (first subscript) *on* body A (second subscript). The mathematical statement of Newton's third law is

$$\vec{F}_{A \text{ on } B} = -\vec{F}_{B \text{ on } A}. \tag{4–11}$$

Expressed in words,

> **If body A exerts a force on body B (an "action"), then body B exerts a force on body A (a "reaction"). These two forces have the same magnitude but are opposite in direction. These two forces act on *different* bodies.**

In this statement, "action" and "reaction" are the two opposite forces; we sometimes refer to them as an **action-reaction pair.** This is not meant to imply any cause-and-effect relationship; we can consider either force as the "action" and the other as the "reaction." We often say simply that the forces are "equal and opposite," meaning that they have equal magnitudes and opposite directions.

We stress that the two forces described in Newton's third law act on *different* bodies. This is important in problems involving Newton's first or second law, which involve the forces that act *on* a body. For instance, the net force acting on the football in Fig. 4–18 is the vector sum of the weight of the football and the force $\vec{F}_{A \text{ on } B}$ exerted by the kicker. You would not include the force $\vec{F}_{B \text{ on } A}$ because this force acts on the *kicker,* not on the football.

In Fig. 4–18 the action and reaction forces are *contact* forces that are present only when the two bodies are touching. But Newton's third law also applies to *long-range* forces that do not require physical contact, such as the force of gravitational attraction. A table-tennis ball exerts an upward gravitational force on the earth that's equal in magnitude to the downward gravitational force the earth exerts on the ball. When you drop the ball, both the ball and the earth accelerate toward each other. The net force on each body has the same magnitude, but the earth's acceleration is microscopically small because its mass is so great. Nevertheless, it does move!

4–18 If body A exerts a force $\vec{F}_{A \text{ on } B}$ on body B, then body B exerts a force $\vec{F}_{B \text{ on } A}$ on body A that is equal in magnitude and opposite in direction: $\vec{F}_{A \text{ on } B} = -\vec{F}_{B \text{ on } A}$.

EXAMPLE 4-8

After your sports car breaks down, you start to push it to the nearest repair shop. While the car is starting to move, how does the force you exert on the car compare to the force the car exerts on you? How do these forces compare when you are pushing the car along at a constant speed?

SOLUTION In *both* cases, the force you exert on the car is equal in magnitude and opposite in direction to the force the car exerts on you. It's true that you have to push harder to get the car going than to keep it going. But no matter how hard you push on the car, the car pushes just as hard back on you. Newton's third law gives the same result whether the two bodies are at rest, moving with constant velocity, or accelerating.

EXAMPLE 4-9

An apple sits on a table in equilibrium. What forces act on it? What is the reaction force to each of the forces acting on the apple? What are the action-reaction pairs?

SOLUTION Figure 4–19a shows the apple on the table, and Fig. 4–19b shows the forces acting on the apple. The apple is A, the table T, and the earth E. In the diagram, $\vec{F}_{\text{E on A}}$ is the weight of the apple; that is, the downward gravitational force exerted *by* the earth E (first subscript) *on* the apple A (second subscript). Similarly, $\vec{F}_{\text{T on A}}$ is the upward force exerted *by* the table T (first subscript) *on* the apple A (second subscript).

As the earth pulls down on the apple, the apple exerts an equally strong upward pull $\vec{F}_{\text{A on E}}$ on the earth, as shown in Fig.

4–19d. $\vec{F}_{\text{A on E}}$ and $\vec{F}_{\text{E on A}}$ are an action-reaction pair, representing the mutual interaction of the apple and the earth, so

$$\vec{F}_{\text{A on E}} = -\vec{F}_{\text{E on A}}.$$

Also, as the table pushes up on the apple with force $\vec{F}_{\text{T on A}}$, the corresponding reaction is the downward force $\vec{F}_{\text{A on T}}$ exerted by the apple on the table (Fig. 4–19c). So we have

$$\vec{F}_{\text{A on T}} = -\vec{F}_{\text{T on A}}.$$

The two forces acting on the apple are $\vec{F}_{\text{T on A}}$ and $\vec{F}_{\text{E on A}}$. Are they an action-reaction pair? No, they aren't, despite being equal and opposite. They do not represent the mutual interaction of two bodies; they are two different forces acting on the *same*

(a)

(b)

(c)

(d)

(e)

4–19 The two forces in an action-reaction pair always act on different bodies.

body. *The two forces in an action-reaction pair **never** act on the same body.* Here's another way to look at it. Suppose we suddenly yank the table out from under the apple (Fig. 4–19e). The two forces $\vec{F}_{\text{A on T}}$ and $\vec{F}_{\text{T on A}}$ then become zero, but $\vec{F}_{\text{A on E}}$ and $\vec{F}_{\text{E on A}}$ are still there (the gravitational interaction is still present). Since $\vec{F}_{\text{T on A}}$ is now zero, it can't be the negative of $\vec{F}_{\text{E on A}}$, and these two forces can't be an action-reaction pair.

EXAMPLE 4–10

A stonemason drags a marble block across a floor by pulling on a rope attached to the block (Fig. 4–20a). The block may or may not be in equilibrium. How are the various forces related? What are the action-reaction pairs?

SOLUTION Figure 4–20b shows the horizontal forces acting on each body, the block (B), the rope (R), and the mason (M). We'll use subscripts on all the forces to help explain things. Vector $\vec{F}_{\text{M on R}}$ represents the force exerted *by* the mason *on* the rope. Its reaction is the equal and opposite force $\vec{F}_{\text{R on M}}$ exerted *by* the rope *on* the mason. Vector $\vec{F}_{\text{R on B}}$ represents the force exerted by the rope on the block. The reaction to it is the equal and opposite force $\vec{F}_{\text{B on R}}$ exerted by the block on the rope:

$$\vec{F}_{\text{R on M}} = -\vec{F}_{\text{M on R}} \quad \text{and} \quad \vec{F}_{\text{B on R}} = -\vec{F}_{\text{R on B}}.$$

Be sure you understand that the forces $\vec{F}_{\text{M on R}}$ and $\vec{F}_{\text{B on R}}$ are *not* an action-reaction pair. Both of these forces act on the *same* body (the rope); an action and its reaction *must* always act on *different* bodies. Furthermore, the forces $\vec{F}_{\text{M on R}}$ and $\vec{F}_{\text{B on R}}$ are not necessarily equal in magnitude. Applying Newton's second law to the rope, we get

$$\Sigma \vec{F} = \vec{F}_{\text{M on R}} + \vec{F}_{\text{B on R}} = m_{\text{rope}} \vec{a}_{\text{rope}}.$$

If the block and rope are accelerating (that is, speeding up or slowing down), the rope is not in equilibrium, and $\vec{F}_{\text{M on R}}$ must have a different magnitude than $\vec{F}_{\text{B on R}}$. By contrast, the action-reaction forces $\vec{F}_{\text{M on R}}$ and $\vec{F}_{\text{R on M}}$ are always equal in magnitude, as are $\vec{F}_{\text{R on B}}$ and $\vec{F}_{\text{B on R}}$. Newton's third law holds whether or not the bodies are accelerating.

In the special case in which the rope is in equilibrium, the forces $\vec{F}_{\text{M on R}}$ and $\vec{F}_{\text{B on R}}$ are equal in magnitude. But this is an example of Newton's *first* law, not his *third*. Another way to look at this is that in equilibrium, $\vec{a}_{\text{rope}} = \mathbf{0}$ in the above equation. Then $\vec{F}_{\text{B on R}} = -\vec{F}_{\text{M on R}}$ because of Newton's first or second law.

This is also true if the rope is accelerating but has negligibly small mass compared to the block or the mason. In this case, $m_{\text{rope}} = 0$ in the above equation, so again $\vec{F}_{\text{B on R}} = -\vec{F}_{\text{M on R}}$. Since $\vec{F}_{\text{B on R}}$ *always* equals $-\vec{F}_{\text{R on B}}$ by Newton's third law (they are an action-reaction pair), in these same special cases, $\vec{F}_{\text{R on B}}$ also equals $\vec{F}_{\text{M on R}}$. In other words, in these cases the force of the rope on the block equals the force of the mason on the rope, and we can then think of the rope as "transmitting" to the block, without change, the force the person exerts on the rope (Fig. 4–20c). This is a useful point of view, but you have to remember that it is valid *only* when the rope has negligibly small mass or is in equilibrium.

If you feel as though you're drowning in subscripts at this point, take heart. Go over this discussion again, comparing the symbols with the vector diagrams, until you're sure you see what's going on.

(a)

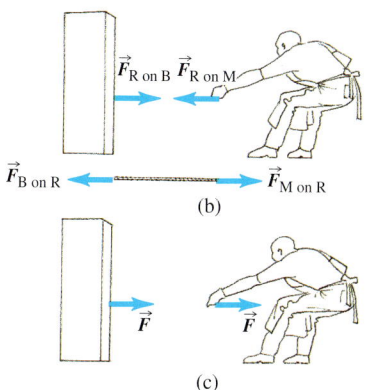

(b)

(c)

4–20 (a) A mason pulls on a rope attached to a block. (b) Separate diagrams showing the force of the rope on the block, the force of the rope on the mason, and the forces of the block and the mason on the rope. (c) If the rope is not accelerating or if its mass can be neglected, it can be considered to transmit an undiminished force from the mason to the block, and vice versa.

A body, such as the rope in Fig. 4–20, that has pulling forces applied at its ends is said to be in **tension.** The tension at any point is the magnitude of force acting at that point. In Fig. 4–20b the tension at the right-hand end of the rope is the magnitude of $\vec{F}_{\text{M on R}}$ (or of $\vec{F}_{\text{R on M}}$), and the tension at the left-hand end equals the magnitude of $\vec{F}_{\text{B on R}}$ (or of $\vec{F}_{\text{R on B}}$). If the rope is in equilibrium and if no forces act except at its ends, the tension is the same at both ends and throughout the rope. Thus if the magnitudes of $\vec{F}_{\text{B on R}}$ and $\vec{F}_{\text{M on R}}$ are 50 N each, the tension in the rope is 50 N (*not* 100 N). The *total* force vector $\vec{F}_{\text{B on R}} + \vec{F}_{\text{M on R}}$ acting on the rope in this case is zero!

Finally, we emphasize once more a fundamental truth: The two forces in an action-reaction pair *never* act on the same body. Remembering this simple fact can often help you avoid confusion about action-reaction pairs and Newton's third law.

4–7 USING NEWTON'S LAWS

Newton's three laws of motion contain all the basic principles we need to solve a wide variety of problems in mechanics. These laws are very simple in form, but the process of applying them to specific situations can pose real challenges. This section introduces some useful techniques. You'll learn others in Chapter 5, which also extends the use of Newton's laws to cover more complex situations.

When you use Newton's first law, $\Sigma \vec{F} = 0$, for an equilibrium situation, or Newton's second law, $\Sigma \vec{F} = m\vec{a}$, for a non-equilibrium situation, you must apply it to some specific body. It is absolutely essential to decide at the beginning what body you are talking about. This may sound trivial, but it isn't. Once you have chosen a body, then you have to identify all the forces acting on it. Don't get confused between the forces acting on a body and the forces exerted by that body on some other body. Only the forces acting on the body you've chosen go into $\Sigma \vec{F}$.

To help identify the relevant forces, draw a **free-body diagram.** What's that? It is a diagram showing the chosen body by itself, "free" of its surroundings, with vectors drawn to show the magnitudes and directions of all the forces applied to the body by the various other bodies that interact with it. We have already shown some free-body diagrams in Figs. 4–13, 4–14, and 4–19. Be careful to include all the forces acting *on* the body, but be equally careful *not* to include any forces that the body exerts on any other body. In particular, the two forces in an action-reaction pair must *never* appear in the same free-body diagram because they never act on the same body. Furthermore, forces that a body exerts on itself are never included, since these can't affect the body's motion.

CAUTION ▶ When you have a complete free-body diagram, you *must* be able to answer for each force the question "What other body is applying this force?" If you can't answer that question, you may be dealing with a nonexistent force. Be especially on your guard to avoid nonexistent forces such as "the force of acceleration" or "the $m\vec{a}$ force," discussed in Section 4–4. ◀

When a problem involves more than one body, you have to take the problem apart and draw a separate free-body diagram for each body. For example, Fig. 4–20b shows a separate free-body diagram for the rope in the case in which the rope is considered massless (so that no gravitational force acts on it). This figure also shows diagrams for the block and the mason, but these are *not* complete free-body diagrams because they don't show all the forces acting on each body. Which forces are missing in these diagrams?

The simple act of walking depends crucially on Newton's third law. To start moving forward, you push backward on the ground with your foot. As a reaction, the ground pushes forward on your foot (and hence on your body as a whole) with a force of the same magnitude. This *external* force provided by the ground is what accelerates your body forward.

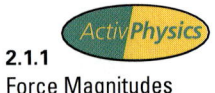

2.1.1
Force Magnitudes

Problem–Solving Strategy

NEWTON'S LAWS

1. Define your coordinate system. A diagram showing the location of the origin and the positive axis direction is always helpful. If you know the direction of the acceleration, it is often convenient to take that as your positive direction.

2. Be consistent with signs. Once you define the x-axis and its positive direction, then components in that direction of velocity, acceleration, and force are also positive.

3. In applying Newton's first or second law, always concentrate on a specific body. Draw a free-body diagram showing all the forces acting *on* this body, but do not include forces that the body exerts on any other body. The acceleration of the body is determined by the forces acting on it, not by the forces it exerts on anything else. In your free-body diagram, you can usually represent the body as a particle. Using a colored pencil or pen for the force vectors may help (Fig. 4–21). Make sure that you can answer the question "What other body is applying this force?" for each force in your diagram. To avoid confusion, the only vector quantities in your diagram should be forces; don't include other vector quantities such as velocity and acceleration.

Normal Force

Friction Force — Tension

Weight

4–21 A free-body diagram. We will use different colors to distinguish certain force vectors in free-body diagrams: dark blue for weight, purple for normal forces, and green for friction forces. All other forces will appear in bright blue. We'll use this color scheme for force vectors throughout the remainder of this book.

4. Identify the known and unknown quantities, and give each unknown quantity an algebraic symbol. If you know the direction of a force at the start, use a symbol to represent the *magnitude* of the force, always a positive quantity. Keep in mind that the *component* of this force along a particular axis direction may still be either positive or negative.

5. Write Newton's first law (for a problem with zero acceleration) or Newton's second law (for problems where there is an acceleration) in component form, using the coordinate system you defined in Step 1. Solve these equations for the unknown quantities.

6. Always check for unit consistency; when appropriate, use the conversion $1\ \text{N} = 1\ \text{kg} \cdot \text{m/s}^2$.

In Chapter 5 we will expand this strategy to deal with more complex problems, but it is important for you to use it consistently from the very start, to develop good habits in the systematic analysis of problems.

EXAMPLE 4–11

Tension in a massless chain To improve the acoustics in an auditorium, a sound reflector with a mass of 200 kg is suspended by a chain from the ceiling (Fig. 4–22a). What is the weight of the reflector? What force (magnitude and direction) does the chain exert on it? What is the tension in the chain? Assume that the mass of the chain itself is negligible.

SOLUTION The reflector is in equilibrium, so we use Newton's first law, $\Sigma \vec{F} = 0$. We draw separate free-body diagrams for the reflector (Fig. 4–22b) and the chain (Fig. 4–22c). We take the positive y-axis to be upward, as shown. Each force has only a y-component. The magnitude of the weight of the reflector is given by Eq. (4–9):

$$w = mg = (200\ \text{kg})(9.80\ \text{m/s}^2) = 1960\ \text{N}.$$

This force points in the negative y-direction, so its y-component is (-1960 N). The upward force exerted by the chain has unknown magnitude T. Because the reflector is in equilibrium, the sum of the y-components of force acting on it must be zero:

$$\Sigma F_y = T + (-1960\ \text{N}) = 0, \quad \text{so} \quad T = 1960\ \text{N}.$$

The chain pulls *up* on the reflector with a force T of magnitude 1960 N. By Newton's third law, the reflector pulls *down* on the chain with a force of magnitude 1960 N.

The chain is also in equilibrium. We have assumed that it is weightless, so an upward force of 1960 N must act on it at its

top end to make the vector sum of forces on it equal to zero. The tension in the chain is 1960 N.

Note that we have defined T to be the *magnitude* of a force, so it is always positive. But the y-component of force acting on the chain at its lower end is $-T = -1960$ N.

4–22 (a) The sound reflector and chain. (b) Free-body diagram of the reflector. (c) Free-body diagram of the chain, assuming its weight to be negligible.

EXAMPLE 4–12

Tension in a chain with mass In Example 4–11, suppose the mass of the chain is not negligible but is 10.0 kg. Find the forces at the ends of the chain.

SOLUTION The chain's weight is $w_C = m_C g =$

$(10.0\ \text{kg})(9.8\ \text{m/s}^2) = 98\ \text{N}$. Again we draw separate free-body diagrams for the reflector and the chain; each is in equilibrium. The free-body diagrams, shown in Figs. 4–23a and 4–23b, differ from those in Fig. 4–22 because the magnitudes of the forces at the two ends of the chain are no longer equal. We label them T_1

and T_2, as shown. Note that the two forces labeled T_2 form an action-reaction pair; that's how we know they have the same magnitude. The equilibrium condition $\Sigma F_y = 0$ for the reflector is

$$T_2 + (-1960\ \text{N}) = 0, \quad \text{so} \quad T_2 = 1960\ \text{N}.$$

There are now three forces acting on the chain: its weight and the forces at the two ends. The equilibrium condition $\Sigma F_y = 0$ for the chain is

$$T_1 + (-T_2) + (-w_C) = 0.$$

Note that the y-component of T_1 is positive, because it points in the $+y$-direction, but the y-components of both T_2 and w_C are negative. When we substitute the values $T_2 = 1960\ \text{N}$ and $w_C = 98\ \text{N}$ and solve for T_1, we find

$$T_1 = T_2 + w_C = 2058\ \text{N}.$$

Alternatively, we could draw a free-body diagram for the composite body consisting of the reflector and the chain together (Fig. 4–23c). The two forces on this composite object are the upward force T_1 at the top of the chain and the total weight, with magnitude $w_R + w_C = 1960\ \text{N} + 98\ \text{N} = 2058\ \text{N}$. Again we find $T_1 = 2058\ \text{N}$. Note that we cannot find T_2 by this method.

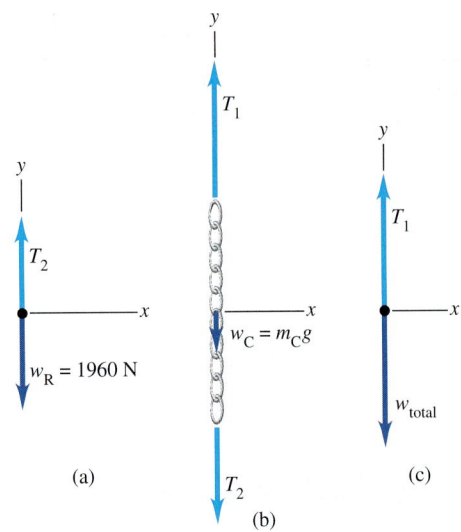

4–23 (a) Free-body diagram for the reflector (weight w_R). (b) Free-body diagram for the chain (weight w_C). (c) Free-body diagram for the reflector and chain, considered as a single particle.

EXAMPLE 4–13

The cutting-blade assembly on a radial-arm saw has a mass of 5.0 kg. It is pulled along a pair of frictionless horizontal rails aligned with the x-axis by a force F_x (Fig. 4–24a). The position of the blade assembly as a function of time is

$$x = (0.18\ \text{m/s}^2)t^2 - (0.030\ \text{m/s}^3)t^3.$$

Find the net force acting on the blade assembly as a function of time. What is the force at time $t = 3.0$ s? For what times is the force positive? Negative? Zero?

SOLUTION   Figure 4–24b shows the free-body diagram and coordinate axes. The forces are the horizontal force F_x, the weight w, and the upward force n that the rails exert to support the cutting head. There is no acceleration in the vertical direction, so the sum of the vertical components of force must be zero. The net force has only a horizontal component, equal to F_x; from Newton's second law, this is equal to ma_x. To determine F_x, we

first find the acceleration a_x by taking the second derivative of x. The second derivative of t^2 is 2, and the second derivative of t^3 is $6t$, so

$$a_x = \frac{d^2x}{dt^2} = 0.36\ \text{m/s}^2 - (0.18\ \text{m/s}^3)t.$$

Then, from Newton's second law,

$$F_x = ma_x = (5.0\ \text{kg})[0.36\ \text{m/s}^2 - (0.18\ \text{m/s}^3)t]$$

$$= 1.80\ \text{N} - (0.90\ \text{N/s})t.$$

At time $t = 3.0$ s the force is $1.80\ \text{N} - (0.90\ \text{N/s})(3.0\ \text{s}) = -0.90\ \text{N}$. The force is zero when $a_x = 0$, that is, when $0.36\ \text{m/s}^2 - (0.18\ \text{m/s}^3)t = 0$. This happens when $t = 2.0$ s. When $t < 2.0$ s, F_x is positive, and the blade assembly is being pulled to the right. When $t > 2.0$ s, F_x is negative, and the horizontal force on the assembly is to the left. Figure 4–25 shows graphs of F_x and a_x as functions of time. You shouldn't be surprised that the net force and acceleration are directly proportional; according to Newton's second law, this is *always* true.

4–24 (a) A radial-arm saw cutting blade. (b) Free-body diagram of blade assembly. Both the magnitude and direction of the net force are functions of time.

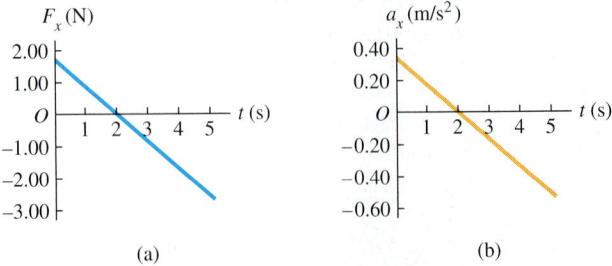

4–25 (a) The net force on the cutting-blade assembly is directly proportional to (b) its acceleration.

4–8 FREE-BODY DIAGRAMS VISUALIZED

Drawing a correct free-body diagram is the first step in analyzing almost any physics or engineering problem involving the motion of a body. In this section we present some real-life situations and the corresponding free-body diagrams. Note that while the situations depicted in the following figures occur in very different surroundings, the free-body diagrams are remarkably similar.

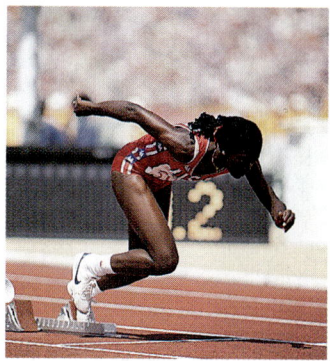

Immersed in water, a person's body experiences an upward force $\vec{B}$ due to buoyancy. This is balanced by the downward force of the diver's weight $\vec{w}$. In this situation the diver's motion depends on the force with which the water pushes on her, due either to the water currents or to reaction forces to the swimmer's arm and leg movements.

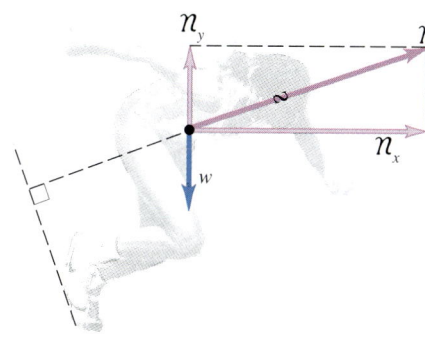

A sprinter gains a large forward acceleration at the start of a race by kicking back hard against the angled starting blocks. The blocks exert an equally large reaction force $\vec{n}$ on the sprinter. If this force has a large forward component n_x that springs her into motion, as well as a smaller upward component n_y. If this upward component has the same magnitude as the sprinter's weight w, the net vertical force acting on her is zero and she does not accelerate vertically.

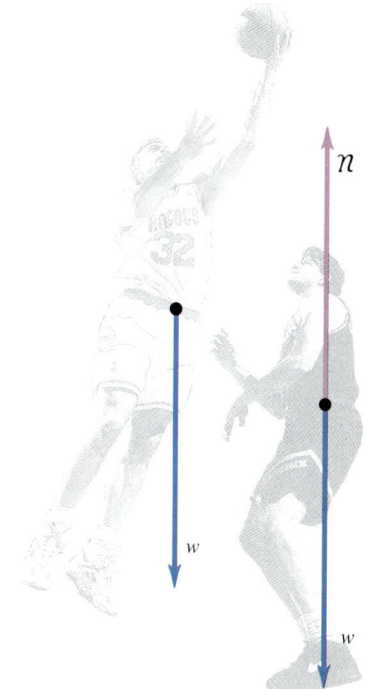

A basketball player jumps by pushing with his feet against the floor. The forces acting on him are the reaction force with which the floor pushes back up and his weight $\vec{w}$. Once the player is in the air, the only force acting on him is his weight; his acceleration is downward, even as he rises up. His stationary opponent is acted on by his own weight and by the upward normal force $\vec{n}$ exerted on him by the floor.

SUMMARY

■ Force, a vector quantity, is a quantitative measure of the interaction between two bodies. When several forces act on a body, the effect on its motion is the same as when a single force, equal to the vector sum (resultant) of the forces, acts on the body.

■ Newton's first law states that when no force acts on a body, or when the vector sum of all forces acting on it (the *net force*) is zero, the body is in equilibrium. If the body is initially at rest, it remains at rest; if it is initially in motion, it continues to move with constant velocity. This law is valid only in inertial frames of reference.

■ The inertial properties of a body are characterized by its *mass*. The acceleration of a body under the action of a given set of forces is directly proportional to the vector sum of the forces (the *net force*) and inversely proportional to the mass of the body. This relationship is Newton's second law:

$$\Sigma \vec{F} = m\vec{a}. \tag{4–7}$$

Like the first law, this is valid only in inertial frames of reference.

■ The unit of force is defined in terms of the units of mass and acceleration. In SI units the unit of force is the newton (N), equal to $1 \text{ kg} \cdot \text{m/s}^2$.

■ The weight of a body is the gravitational force exerted on it by the earth (or whatever other body exerts the gravitational force). Weight is a force and is therefore a vector quantity. The magnitude of the weight of a body at any specific location is equal to the product of its mass m and the magnitude of the acceleration due to gravity g at that location:

$$w = mg. \tag{4–9}$$

The weight of a body depends on its location, but the mass is independent of location.

■ Newton's third law states that "action equals reaction"; when two bodies interact, they exert forces on each other that are equal in magnitude and opposite in direction. Each force in an action-reaction pair acts on only one of the two bodies; the action and reaction forces never act on the same body.

DISCUSSION QUESTIONS

Q4–1 Can a body be in equilibrium when only one force acts on it? Explain.

Q4–2 A ball thrown straight up has zero velocity at its highest point. Is the ball in equilibrium at this point? Why or why not?

Q4–3 A helium balloon hovers in midair, neither ascending nor descending. Is it in equilibrium? What forces act on it?

Q4–4 When you fly in an airplane at night in smooth air, there is no sensation of motion, even though the plane may be moving at 800 km/h (500 mi/h). Why is this?

Q4–5 If the two ends of a rope in equilibrium are pulled with forces of equal magnitude and opposite direction, why is the total tension in the rope not zero?

Q4–6 You tie a brick to the end of a rope and whirl the brick around you in a horizontal circle. Describe the path of the brick after you suddenly let go of the rope.

Q4–7 When a car stops suddenly, the passengers tend to move forward relative to their seats. Why? When a car makes a sharp turn, the passengers tend to slide to one side of the car. Why?

Q4–8 Some people say that the "force of inertia" (or "force of momentum") throws the passengers forward when a car brakes sharply. What is wrong with this explanation?

Q4–9 A passenger in a moving bus with no windows notices that a ball that has been at rest in the aisle suddenly starts to move toward the rear of the bus. Think of two different possible explanations, and devise a way to decide which is correct.

Q4–10 Suppose the fundamental SI units had been chosen to be force, length, and time instead of mass, length, and time. What would be the units of mass in terms of those fundamental units?

Q4–11 Inertia is *not* a force that keeps things in place or that keeps them moving. How do we know this?

Q4–12 Why is the earth only approximately an inertial reference frame?

Q4–13 Does Newton's second law hold true for an observer in a van as it speeds up, slows down, or rounds a corner? Explain.

Q4–14 Some students refer to the quantity $m\vec{a}$ as "the force of acceleration." Is it correct to refer to this quantity as a force? If so, what exerts this force? If not, what is a better description of this quantity?

Q4–15 The acceleration of a falling body is measured in an elevator traveling upward at a constant speed of 9.8 m/s. What result is obtained?

Q4–16 You can play catch with a softball in a bus moving with constant speed on a straight road, just as though the bus were at rest. Is this still possible when the bus is making a turn at constant speed on a level road? Why or why not?

Q4–17 Is the coat of Example 4-1 in equilibrium? Explain.

Q4–18 The head of a hammer begins to come loose from its wooden handle. How should you strike the handle on a concrete sidewalk to reset the head? Why does this work?

Q4–19 Why can it hurt your foot more to kick a big rock than a small pebble? *Must* the big rock hurt more? Explain.

Q4–20 "It's not the fall that hurts you; it's the sudden stop at the bottom." Translate this saying into the language of Newton's laws of motion.

Q4–21 A person can dive into water from a height of 10 m without injury, but a person who jumps off the roof of a 10-m-tall building and lands on a concrete street is likely to be seriously injured. Why is there a difference?

Q4–22 Why are cars designed to crumple up in front and back for safety? Why not for side collisions and rollovers?

Q4–23 When a bullet is fired from a gun, what is the origin of the force that accelerates the bullet?

Q4–24 When a heavy weight is lifted by a string that is barely strong enough, the weight can be lifted by a steady pull; but if the string is jerked, it will break. Explain in terms of Newton's laws of motion.

Q4–25 A large crate is suspended from the end of a vertical rope. Is the tension in the rope larger when the crate is at rest or when it is moving upward at constant speed? If the crate is traveling upward, is the tension in the rope greatest when the crate is speeding up or when it is slowing down? In each case explain in terms of Newton's laws of motion.

Q4–26 Automotive engineers, in discussing the motion of an automobile, call the rate of change of the acceleration the "jerk." Why is this a useful quantity in characterizing the riding qualities of an automobile?

Q4–27 Why can't we correctly say that 1.0 kg *equals* 2.2 lb?

Q4–28 A horse is hitched to a wagon. Since the wagon pulls back on the horse just as hard as the horse pulls on the wagon, why doesn't the wagon remain in equilibrium, no matter how hard the horse pulls?

Q4–29 A 450-N girl from the south slaps a 800-N boy from the north. Her fingers exert a force of 30 N to the west on his cheek. There may be other reactions, but by Newton's third law, what is the reaction *force* to the slap?

Q4–30 A large truck and a small compact car have a head-on collision. During the collision the truck exerts a force $\vec{F}_{\text{T on C}}$ on the car, and the car exerts a force $\vec{F}_{\text{C on T}}$ on the truck. Which force has the larger magnitude, or are they the same? Does your answer depend on how fast each vehicle was moving before the collision? Why or why not?

Q4–31 If you ask most people what force makes a car accelerate forward, they will say "the force of the engine." But what force is *directly* responsible for making the car accelerate?

Q4–32 A small compact car is giving a push to a large van that has broken down, and they travel along the road with equal velocities and accelerations. While the car is speeding up, is the force it exerts on the van larger than, smaller than, or the same magnitude as the force the van exerts on it? Which object, the car or the van, has the largest net force on it, or are the net forces same? Explain.

Q4–33 Consider a tug-of-war between two people who pull in opposite directions on the ends of a rope. By Newton's third law, the force that *A* exerts on *B* is just as great as the force that *B* exerts on *A*. So what determines who wins? (*Hint:* Draw a free-body diagram showing all the forces that act on each person.)

Q4–34 On the moon, $g = 1.62$ m/s^2. If a 2-kg brick drops onto your foot from a height of 2 m, will this hurt more, or less, or the same if it happens on the moon instead of on the earth? Explain. If a 2-kg brick is thrown and hits you when it is moving horizontally at 6 m/s, will this hurt more, less, or the same if it happens on the moon instead of on the earth? Explain. (On the moon, assume that you are inside a pressurized structure, so you are not wearing a spacesuit.)

Q4–35 A manual for student pilots contains the following passage: "When an airplane flies at a steady altitude, neither climbing nor descending, the upward lift force from the wings equals the airplane's weight. When the airplane is climbing at a steady rate, the upward lift is greater than the weight; when the airplane is descending at a steady rate, the upward lift is less than the weight." Are these statements correct? Explain.

EXERCISES

SECTION 4–2 **FORCE AND INTERACTIONS**

4–1 Two forces have the same magnitude F. What is the angle between the two vectors if their sum has a magnitude of a) $2F$? b) $\sqrt{2}F$? c) zero? Sketch the three vectors in each case.

4–2 Instead of using the x- and y-axes of Fig. 4–5 to analyze the situation of Example 4–1, use axes rotated 30.0° counter-clockwise, so the x-axis is parallel to the 200-N force. a) For these axes find the x- and y-components of the net force on the coat. b) From the components computed in part (a) find the magnitude and direction of the net force. Compare your results to Example 4–1.

4–3 A warehouse worker pushes a crate along the floor as in Fig. 4–1b by a force of 10 N that points downward at an angle of

45° below the horizontal. Find the horizontal and vertical components of the force.

4–4 A man is dragging a trunk up the loading ramp of a mover's truck. The ramp has a slope angle of 20.0°, and the man pulls upward with a force $\vec{F}$ whose direction makes an angle of 30.0° with the ramp (Fig. 4–26). a) How large a force $\vec{F}$ is necessary for the component F_x parallel to the ramp to be 60.0 N? b) How large will the component F_y then be?

FIGURE 4–26 Exercise 4–4.

4–5 Two dogs pull horizontally on ropes attached to a post; the angle between the ropes is 60.0°. If dog A exerts a force of 270 N and dog B exerts a force of 300 N, find the magnitude of the resultant force and the angle it makes with dog A's rope.

4–6 Two forces, $\vec{F}_1$ and $\vec{F}_2$, act at a point. The magnitude of $\vec{F}_1$ is 9.00 N, and its direction is 60.0° above the x-axis in the second quadrant. The magnitude of $\vec{F}_2$ is 6.00 N, and its direction is 53.1° below the x-axis in the third quadrant. a) What are the x- and y-components of the resultant force? b) What is the magnitude of the resultant force?

SECTION 4–4 NEWTON'S SECOND LAW

4–7 If a net horizontal force of 132 N is applied to a person with mass 60 kg who is resting on the edge of a swimming pool, what horizontal acceleration is produced?

4–8 What magnitude of net force is required to give a 135-kg refrigerator an acceleration of magnitude 1.40 m/s²?

4–9 A box rests on a frozen pond, which serves as a frictionless horizontal surface. If a fisherman applies a horizontal force with magnitude 48.0 N to the box and produces an acceleration of magnitude 3.00 m/s², what is the mass of the box?

4–10 A dockworker applies a constant horizontal force of 80.0 N to a block of ice on a smooth horizontal floor. The frictional force is negligible. The block starts from rest and moves 11.0 m in 5.00 s. a) What is the mass of the block of ice? b) If the worker stops pushing at the end of 5.00 s, how far does the block move in the next 5.00 s?

4–11 A hockey puck with mass 0.160 kg is at rest at the origin ($x = 0$) on the horizontal frictionless surface of the rink. At time $t = 0$ a player applies a force of 0.250 N to the puck, parallel to

the x-axis; he continues to apply this force until $t = 2.00$ s. a) What are the position and speed of the puck at $t = 2.00$ s? b) If the same force is again applied at $t = 5.00$ s, what are the position and speed of the puck at $t = 7.00$ s?

4–12 A crate with mass 32.5 kg initially at rest on a warehouse floor is acted on by a net horizontal force of 140 N. a) What acceleration is produced? b) How far does the crate travel in 10.0 s? c) What is its speed at the end of 10.0 s?

4–13 A hockey puck moves from point A to point B at constant velocity while under the influence of several forces. a) What can you say about the forces? b) Graph the puck's path from A to B. c) On the graph, continue the path to point C if a new constant force is added to the puck at point B, where the new force is perpendicular to the puck's velocity at point B. d) Continue the path to point D if at point C the constant force added at B is replaced by one of constant magnitude but a direction always perpendicular to the path of the puck.

4–14 An electron (mass = 9.11×10^{-31} kg) leaves one end of a TV picture tube with zero initial speed and travels in a straight line to the accelerating grid, which is 1.80 cm away. It reaches the grid with a speed of 3.00×10^6 m/s. If the accelerating force is constant, compute a) the acceleration; b) the time to reach the grid; c) the net force, in newtons. (The gravitational force on the electron may be neglected.)

SECTION 4–5 MASS AND WEIGHT

4–15 Superman throws a 2400-N boulder at an adversary. What horizontal force must Superman apply to the boulder to give it a horizontal acceleration of 12.0 m/s²?

4–16 A bowling ball weighs 71.2 N (16.0 lb). The bowler applies a horizontal force of 160 N (36.0 lb) to the ball. What is the magnitude of the horizontal acceleration of the ball?

4–17 At the surface of Jupiter's moon Io the acceleration due to gravity is $g = 1.81$ m/s². A watermelon weighs 44.0 N at the surface of the earth. a) What is its mass on the earth's surface? b) What are its mass and weight on the surface of Io?

4–18 a) What is the mass of a book that weighs 3.20 N at a point where $g = 9.80$ m/s²? b) At the same location, what is the weight of a dog whose mass is 14.0 kg?

SECTION 4–6 NEWTON'S THIRD LAW

4–19 World-class sprinters can accelerate out of the starting blocks with an acceleration that is nearly horizontal and has magnitude 15 m/s². How much horizontal force must a 55-kg sprinter exert on the starting blocks during a start to produce this acceleration? Which body exerts the force that propels the sprinter: the blocks or the sprinter herself?

4–20 Imagine that you are holding a book weighing 4 N at rest on the palm of your hand. Complete the following sentences: a) A downward force of magnitude 4 N is exerted on the book by _____. b) An upward force of magnitude _____ is exerted on _____ by the hand. c) Is the upward force in part (b) the reaction to the downward force in part (a)? d) The reaction to the force in part (a) is a force of magnitude _____, exerted on _____ by _____. Its direction is _____. e) The reaction to the force in part (b) is a force of magnitude _____, exerted on _____ by _____. Its direction is _____. f) The forces in parts (a) and (b) are equal and opposite because of Newton's _____ law. g) The forces in parts (b) and (e) are equal and opposite because of Newton's _____ law. Now suppose that you exert an upward force of magnitude 5 N on the book. h) Does the book remain in equilibrium? i) Is the force exerted on the book by your hand equal and opposite to the force exerted on the book by the earth? j) Is the force exerted on the book by the earth equal and opposite to the force exerted on the earth by the book? k) Is the force exerted on the book by your hand equal and opposite to the force exerted on your hand by the book? Finally, suppose that you snatch your hand away while the book is moving upward. l) How many forces then act on the book? m) Is the book in equilibrium?

4–21 A bottle is given a push along a table top and slides off the edge of the table. Do *not* neglect air resistance. a) What forces are exerted on the bottle while it is falling from the table to the floor? b) What is the reaction to each force; that is, on what body and by what body is the reaction exerted?

4–22 The upward normal force exerted by the floor is 620 N on an elevator passenger who weighs 650 N. What are the reaction forces to these two forces? Is the passenger accelerating? If so, what are the magnitude and direction of the acceleration?

4–23 A student with mass 45 kg jumps off a high diving board. Using 6.0×10^{24} kg for the mass of the earth, what is the acceleration of the earth toward her as she accelerates toward the earth with an acceleration of 9.8 m/s²? Assume that the net force on the earth is the force of gravity she exerts on it.

SECTION **4–7** USING NEWTON'S LAWS

4–24 An astronaut is tethered by a strong cable to a spacecraft. The astronaut and her spacesuit have a total mass of 105 kg, while the mass of the cable is negligible. The mass of the spacecraft is 9.05×10^4 kg. The spacecraft is far from any large astronomical bodies, so we can neglect the gravitational forces on it and the astronaut. We also assume that both the spacecraft and the astronaut are initially at rest in an inertial reference frame. The astronaut then pulls on the cable with a force of 80.0 N. a) What force does the cable exert on the astronaut? b) Since $\Sigma \vec{F} = m\vec{a}$, how can a "massless" ($m = 0$) cable exert a force? c) What is the astronaut's acceleration? d) What force does the cable exert on the spacecraft? e) What is the acceleration of the spacecraft?

4–25 A 4.80-kg bucket of water is accelerated upward by a cord of negligible mass whose breaking strength is 75.0 N. Find the maximum upward acceleration that can be given to the bucket without breaking the cord.

4–26 An elevator with mass m is moving upward and has an acceleration of magnitude $|\vec{a}|$. The mass of the supporting cable can be neglected. What is the tension in the supporting cable a) if the elevator is speeding up as it rises? b) if the elevator is slowing down as it rises?

4–27 Two crates, one with mass 4.00 kg and the other with mass 6.00 kg, sit on the frictionless surface of a frozen pond, connected by a light rope (Fig. 4–27). A woman wearing golf shoes (so she can get traction on the ice) pulls horizontally on the 6.00-kg crate with a force F that gives the crate an acceleration of 2.50 m/s². a) What is the magnitude of the force F? b) What is the tension T in the rope connecting the two crates?

4–28 Refer to Fig. 4–27. The crates are on a horizontal frictionless surface. The woman (still wearing golf shoes for traction) applies a horizontal force $F = 50.0$ N to the 6.00-kg crate. The ropes are of negligible mass. a) Draw a free-body diagram for the 4.00-kg crate, a free-body diagram for the 6.00-kg crate, and a free-body diagram for the woman. For each force, indicate what body exerts that force. b) What is the magnitude of the acceleration of the 6.00-kg crate? c) What is the tension T in the rope connecting the two crates?

FIGURE 4–27 Exercises 4–27 and 4–28.

4–29 A parachutist relies on air resistance (mainly on her parachute) to decrease her downward velocity. If she and her parachute have a mass of 55.0 kg and air resistance exerts a total upward force of 620 N on her and her parachute, what is her acceleration?

4–30 The position of a 2.75×10^5 N training helicopter under test is given by $\vec{r} = (0.020 \text{ m/s}^3)t^3\,\hat{i} + (2.2 \text{ m/s})t\,\hat{j} - (0.060 \text{ m/s}^2)t^2\hat{k}$. Find the net force on the helicopter at $t = 5.0$ s.

4–31 An object with mass m moves along the x-axis. Its position as a function of time is given by $x(t) = At - Bt^3$, where A and B are constants. Calculate the net force on the object as a function of time.

PROBLEMS

4–32 A .22 rifle bullet, traveling at 350 m/s, strikes a block of soft wood, which it penetrates to a depth of 0.130 m. The mass of the bullet is 1.80 g. Assume a constant retarding force. a) How much time is required for the bullet to stop? b) What force, in newtons, does the wood exert on the bullet?

4–33 Two horses pull horizontally on ropes attached to a stump. The two forces $\vec{F}_1$ and $\vec{F}_2$ that they apply to the stump are such that the net (resultant) force $\vec{R}$ has a magnitude equal to that of $\vec{F}_1$ and makes an angle of 90° with $\vec{F}_1$ (Fig. 4–28; see next page). Let $F_1 = 1300$ N and $R = 1300$ N also. Find the magnitude of $\vec{F}_2$ and its direction (relative to $\vec{F}_1$).

4–34 A proud angler hangs her catch from a spring balance, which is supported from the roof of an elevator. a) If the elevator has an upward acceleration of 2.45 m/s² and the balance reads

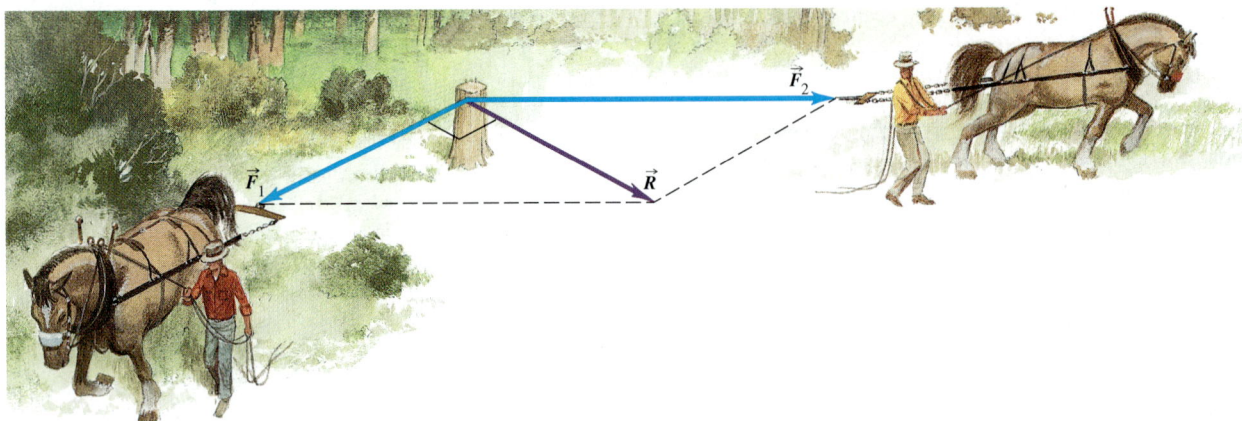

FIGURE 4–28 Problem 4–33.

50.0 N, what is the true weight of the fish? b) Under what circumstances will the balance read 30.0 N? c) What will the balance read if the elevator cable breaks?

4–35 Two adults and a child want to push a wheeled cart in the direction marked x in Fig. 4–29. The two adults push with horizontal forces $\vec{F}_1$ and $\vec{F}_2$ as shown in the figure. a) Find the magnitude and direction of the *smallest* force that the child should exert. Ignore the effects of friction. b) If the child exerts the minimum force found in part (a), the cart accelerates at 2.0 m/s^2 in the $+x$-direction. What is the weight of the cart?

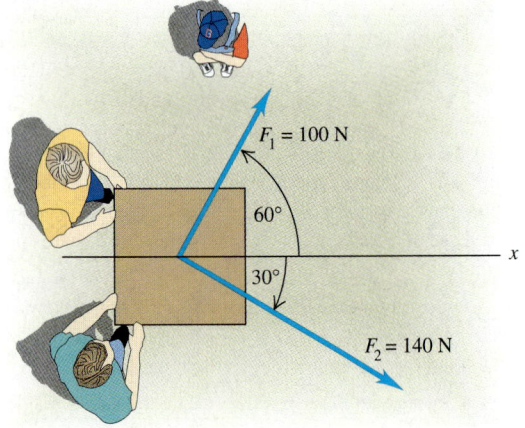

FIGURE 4–29 Problem 4–35.

4–36 An oil tanker's engines have broken down, and the wind is blowing the tanker straight toward a reef at a constant speed of 1.5 m/s (Fig. 4–30). When the tanker is 500 m from the reef, the wind dies down just as the engineer gets the engines going again. The rudder is stuck, so the only choice is to try to

FIGURE 4–30 Problem 4–36.

accelerate straight backward away from the reef. The mass of the tanker and cargo is 3.6×10^7 kg, and the engines produce a net horizontal force of 8.0×10^4 N on the tanker. Will the ship hit the reef? If it does, will the oil be safe? The hull can withstand an impact at a speed of 0.2 m/s or less. Neglect the retarding force of the water on the tanker's hull.

4–37 A Standing Vertical Jump. Basketball player Darrell Griffith is on record as attaining a standing vertical jump of 1.2 m (4 ft). (This means that he moved upward by 1.2 m after his feet left the floor.) If Griffith weighed 890 N (200 lb) and the time of the jump before his feet left the ground was 0.300 s, what was the average force he applied to the ground?

4–38 An advertisement claims that a particular automobile can "stop on a dime." What net force would actually be necessary to stop a 850-kg automobile traveling initially at 45.0 km/h in a distance equal to the diameter of a dime, which is 1.8 cm?

4–39 To study damage to aircraft that collide with large birds, you design a test gun that will accelerate chicken-sized objects so that their displacement along the gun barrel is given by $x = (9.0 \times 10^3 \text{ m/s}^2)t^2 - (8.0 \times 10^4 \text{ m/s}^3)t^3$. The object leaves the end of the barrel at $t = 0.025$ s. a) How long must the gun barrel be? b) What will be the speed of the objects as they leave the end of the barrel? c) What net force must be exerted on a 1.50-kg object at i) $t = 0$? ii) $t = 0.025$ s?

4–40 A spacecraft descends vertically near the surface of Planet X. An upward thrust of 25.0 kN from its engines slows it down at a rate of 1.20 m/s^2, but it speeds up at a rate of 0.80 m/s^2 with an upward thrust of 10.0 kN. What is the spacecraft's weight near the surface of Planet X?

4–41 A train (an engine plus four cars) running on level tracks is speeding up with an acceleration of magnitude $|\vec{a}|$. If each car has a mass m and if each car has negligible frictional forces acting on it, what is a) the force of the engine on the first car? b) the force of the first car on the second car? c) the force of the second car on the third car? d) the force of the fourth car on the third car? e) What would these same four forces be if the train were slowing down with an acceleration of magnitude $|\vec{a}|$? Your answer to each question must include a clearly labeled free-body diagram.

4–42 A gymnast of mass m climbs a vertical rope attached to the ceiling. The weight of the rope can be neglected. Calculate the tension in the rope if the gymnast a) climbs at a constant rate; b) hangs motionless on the rope; c) accelerates up the rope with an acceleration of magnitude $|\vec{a}|$; d) slides down the rope with a downward acceleration of magnitude $|\vec{a}|$.

4–43 A loaded elevator with very worn cables has a total mass of 2200 kg, and the cables can withstand a maximum tension of 28,000 N. a) What is the maximum upward acceleration for the elevator if the cables are not to break? b) What would be the answer to part (a) if the elevator were on the moon, where $g = 1.62$ m/s²?

4–44 Jumping to the Ground. A 75.0-kg man steps off a platform 3.10 m above the ground. He keeps his legs straight as he falls, but at the moment his feet touch the ground his knees begin to bend, and, treated as a particle, he moves an additional 0.60 m before coming to rest. a) What is his speed at the instant his feet touch the ground? b) Treating him as a particle, what is his acceleration as he slows down, if the acceleration is assumed to be constant? c) What force do his feet exert on the ground while he slows down? Express this force in newtons and also as a multiple of his weight.

4–45 A 4.9-N hammer head is stopped from an initial downward velocity of 3.2 m/s in a distance of 0.45 cm by a nail in a pine board. In addition to its weight, there is a 15-N downward force on the hammer head applied by the person using the hammer. Assume that the acceleration of the hammer head is constant while it is in contact with the nail and moving downward. a) Draw a free-body diagram for the hammer head. Identify the reaction force to each force in the diagram. b) Calculate the downward force $\vec{F}$ exerted by the hammer head on the nail while the hammer head is in contact with the nail and moving downward. c) Suppose the nail is in hardwood and the distance the hammer head travels in coming to rest is only 0.12 cm. The downward forces on the hammer head are the same as on part (b). What then is the force $\vec{F}$ exerted by the hammer head on the nail while the hammer head is in contact with the nail and moving downward?

4–46 A uniform cable of weight w hangs vertically downward, supported by an upward force of magnitude w at its top end. What is the tension in the cable a) at its top end? b) at its bottom end? c) at its middle? Your answer to each part must include a free-body diagram. (*Hint:* For each question choose the body to be analyzed to be a section of the cable or a point along the cable.) d) Graph the tension in the rope versus the distance from its top end.

4–47 The two blocks in Fig. 4–31 are connected by a heavy uniform rope with a mass of 4.00 kg. An upward force of 200 N is applied as shown. a) Draw three free-body diagrams, one for the 6.00-kg block, one for the 4.00-kg rope, and another one for the 5.00-kg block. For each force, indicate what body exerts that force. b) What is the acceleration of the system? c) What is the tension at the top of the heavy rope? d) What is the tension at the midpoint of the rope?

FIGURE 4–31 Problem 4–47.

4–48 If a beach ball with mass 0.0900 kg is thrown vertically upward in a vacuum, so there is no drag force on it, it reaches a height of 5.0 m. If the ball is thrown upward with the same initial velocity but in air instead of vacuum, its maximum height is 3.8 m. What is the average force exerted on the ball by the air during its upward motion?

*4–49 An object with mass m initially at rest is acted on by a force $\vec{F} = k_1\hat{\imath} + k_2 t^3\hat{\jmath}$, where k_1 and k_2 are constants. Calculate the velocity $\vec{v}(t)$ of the object as a function of time.

CHALLENGE PROBLEMS

*4–50 If we know $F(t)$, the force as a function of time, for straight-line motion, Newton's second law gives us $a(t)$, the acceleration as a function of time. We can then integrate $a(t)$ to find $v(t)$ and $x(t)$. However, suppose we know $F(v)$ instead. a) The net force on a body moving along the x-axis equals $-Cv^2$. Use Newton's second law written as $\Sigma F = m\, dv/dt$ and two integrations to show that $x - x_0 = (m/C) \ln (v_0/v)$. b) Show that Newton's second law can be written as $\Sigma F = mv\, dv/dx$. Derive

the same expression as in part (a) using this form of the second law and one integration.

*4–51 An object of mass m is at rest in equilibrium at the origin. At $t = 0$ a new force $\vec{F}(t)$ is applied that has components

$$F_x(t) = k_1 + k_2 y, \qquad F_y(t) = k_3 t,$$

where k_1, k_2, and k_3 are constants. Calculate the position $\vec{r}(t)$ and velocity $\vec{v}(t)$ vectors as functions of time.

5

Applications of Newton's Laws

A bird glides by making use of Newton's first and third laws. The wings push downward on the air as it flows past the bird, and the air reacts by pushing upward on the wings with a force—called *lift*—of equal magnitude but opposite direction. In a steady glide the aerodynamic forces on the bird exactly balance the downward force of gravity, so that the net external force on the bird is zero.

5–1 INTRODUCTION

Newton's three laws of motion, the foundation of classical mechanics, can be stated very simply, as we have seen. But *applying* these laws to situations such as an iceboat skating across a frozen lake, a toboggan sliding down a hill, a jet fighter making a steep turn, or even a weight hanging from the ceiling requires analytical skills and problem-solving technique. In this chapter we'll help you to extend the problem-solving skills you began to develop in the previous chapter.

We begin with equilibrium problems, in which a body is at rest or moving with constant velocity. Then we'll generalize our problem-solving techniques to include bodies that are not in equilibrium, for which we need to deal precisely with the relationships between forces and motion. We'll learn how to describe and analyze the contact force acting on a body when it rests or slides on a surface. Next we'll study the important case of uniform circular motion, in which a body moves in a circle with constant speed. We'll use the concept of force throughout our study of physics, and we close the chapter with a brief look at the fundamental nature of force and the classes of forces found in nature.

5–2 USING NEWTON'S FIRST LAW: PARTICLES IN EQUILIBRIUM

We learned in Chapter 4 that a body is in *equilibrium* when it is at rest or moving with constant velocity in an inertial frame of reference. A hanging lamp, a rope and pulley setup for hoisting heavy loads, a suspension bridge—all are examples of equilibrium situations. In this section we consider only equilibrium of a body that can be modeled as a particle. (In Chapter 11 we'll see what to do when the body can't be represented adequately as a particle.) The essential physical principle is Newton's first law: When a particle is at rest or is moving with constant velocity in an inertial frame of reference, the net force acting on it—that is, the vector sum of all the forces acting on it—must be zero:

$$\Sigma \vec{F} = 0 \quad \text{(particle in equilibrium).} \tag{5–1}$$

We will usually use this in component form:

$$\Sigma F_x = 0, \quad \Sigma F_y = 0 \quad \text{(particle in equilibrium).} \tag{5–2}$$

This section is about using Newton's first law to solve problems dealing with bodies in equilibrium. The important thing to remember is that *all* such problems are done in the same way. The following strategy details the steps you need to follow for any and all such problems. Study this strategy carefully, look at how it's applied in the worked-out examples, and try to apply it yourself when you solve problems.

Problem–Solving Strategy

EQUILIBRIUM OF A PARTICLE

1. Draw a simple sketch of the physical situation, showing dimensions and angles.

2. Choose some body that is in equilibrium, and draw a free-body diagram of this body. For the present we will consider it as a particle, so a large dot will do to represent the body. In your free-body diagram, *do not* include the other bodies that interact with it, such as a surface it may be resting on, or a rope pulling on it.

3. Now ask yourself what is interacting with the body by touching it or in any other way. On your free-body diagram, draw a force vector for each of the interactions. If you know the angle at which a force is directed, draw the angle accurately and label it. A surface in contact with the body exerts a normal force perpendicular to the surface and possibly a friction force parallel to the surface. Remember that a rope or chain can't push on a body, but can only pull in a direction along its length. Be sure to include the body's weight, except in the case in which the body has negligible mass (and hence negligible weight). If the mass is given, use $w = mg$ to find the weight. Label each force with a symbol representing the *magnitude* of the force and the numerical value if it's given. Make sure you can answer the question "What other body causes that force?" for each force. If you can't answer that question, you may be imagining a force that isn't there.

4. *Do not* show in the free-body diagram any forces exerted *by* the body on any other body. The sums in Eqs. (5–1) and (5–2) include only forces that act *on* the body.

5. Choose a set of coordinate axes and represent each force acting on the body by its components along these axes. Draw a wiggly line through each force vector that you replace by its components so that you don't count it twice. Often you can simplify the problem by your choice of coordinate axes. For example, when a body rests or slides on a plane surface, it's usually simplest to take the axes in the directions parallel and perpendicular to this surface, even when the plane is tilted.

6. Set the algebraic sum of all x-components of force equal to zero. In a separate equation, set the algebraic sum of all y-components equal to zero. (*Never* add x- and y-components in a single equation.) You can then solve these equations for up to two unknown quantities, which may be force magnitudes, components, or angles.

7. If there are two or more bodies, repeat Steps 2 through 6 for each body. If the bodies interact with each other, use Newton's third law to relate the forces they exert on each other. You need to find as many independent equations as the number of unknown quantities. Then solve these equations to obtain the unknowns. This part is algebra, not physics, but it's an essential step.

8. Look at your results and ask whether they make sense. When the result is a symbolic expression or formula, try to think of special cases (particular values or extreme cases for the various quantities) for which you can guess what the results ought to be. Check to see that your formula works in these particular cases.

EXAMPLE 5–1

One-dimensional equilibrium A gymnast has just begun climbing up a rope hanging from a gymnasium ceiling (Fig. 5–1a). She stops, suspended from the lower end of the rope by her hands. Her weight is 500 N, and the weight of the rope is 100 N. Analyze the forces on the gymnast and on the rope. (This example is similar to Example 4–12 in Section 4–7; our purpose is to emphasize the steps in the problem-solving strategy presented above.)

SOLUTION Figure 5–1b is a free-body diagram for the gymnast. The forces acting on her are her weight (magnitude $w_G = 500$ N) and the upward tension force exerted on her by the rope, of magnitude $T_{R \text{ on } G}$ (force exerted *by* the rope *on* the gymnast). We don't include the downward force she exerts on the rope because it isn't a force that acts *on* her. We take the y-axis vertically upward and the x-axis horizontal; there are no x-components of force. The rope pulls upward (in the positive y-direction), and the corresponding y-component of force is just the magnitude $T_{R \text{ on } G}$, a positive (scalar) quantity. But the weight acts in the negative y-direction, and its y-component is the *negative* of the magnitude, that is, $-w_G = -500$ N. The algebraic sum of

y-components is $T_{R \text{ on } G} + (-w_G)$, and from the equilibrium condition, Eq. (5–2), we have

$$\Sigma F_y = T_{R \text{ on } G} + (-w_G) = 0,$$

$$T_{R \text{ on } G} = w_G = 500 \text{ N}.$$

The tension at the bottom of the rope equals the gymnast's weight, as you probably expected.

The two forces acting on the gymnast have equal magnitude 500 N and are opposite in direction, but they are *not* an action-reaction pair. The reason is that the weight and the tension force in Fig. 5–1b both act on the gymnast, whereas action and reaction forces always act on different bodies. These two forces are equal and opposite because of Newton's first law, $\Sigma F_y = 0$, not Newton's third law. The gymnast's weight is the attractive (downward) force exerted on the gymnast by the earth. Its reaction force is the equal and opposite (upward) attractive force exerted on the *earth* by the *gymnast*. This force acts on the earth, not on the gymnast, so it does not appear in the free-body diagram for the gymnast. Compare the discussion of the apple in Example 4–9 (Section 4–6).

Figure 5–1c shows a free-body diagram for the rope. The

reaction to the upward force of magnitude 500 N exerted by the rope on the gymnast is a downward force exerted by the gymnast on the rope, as shown. According to Newton's third law, the magnitude $T_{G \, on \, R}$ of this downward force is also 500 N. The other forces on the rope are its own weight (magnitude 100 N) and the upward force (magnitude $T_{C \, on \, R}$) exerted by the ceiling on the upper end of the rope.

The y-component of force acting on the top end of the rope is $+T_{C \, on \, R}$, the y-component acting on the bottom end is $-T_{G \, on \, R}$ = −500 N, and the y-component of the weight is $-w_R$ = −100 N.

The equilibrium condition $\Sigma F_y = 0$ for the rope gives

$$\Sigma F_y = T_{C \, on \, R} + (-T_{G \, on \, R}) + (-w_R) = 0,$$

$$T_{C \, on \, R} = T_{G \, on \, R} + w_R = 500 \text{ N} + 100 \text{ N} = 600 \text{ N}.$$

This result makes sense; the tension is 100 N greater at the top of the rope (where it must support the combined weight of both the rope and the gymnast) than at the bottom (where it supports only the gymnast). Figure 5–1d is a partial free-body diagram for the ceiling, showing that the rope exerts a downward force $F_{R \, on \, C}$ on the ceiling. From Newton's third law the magnitude of this force is 600 N, the same as $T_{C \, on \, R}$.

(a)   (b)   (c)   (d)

5–1 (a) A gymnast hanging at rest from the end of a vertical rope. (b) Free-body diagram for the gymnast. (c) Free-body diagram for the rope. (d) The force exerted on the ceiling by the rope. (This is not a complete free-body diagram because it does not show the other forces acting on the ceiling.)

EXAMPLE 5-2

Two-dimensional equilibrium In Fig. 5–2a a car engine with weight w hangs from a chain that is linked at point O to two other chains, one fastened to the ceiling and the other to the wall. Find the tensions in these three chains, assuming that w is given and the weights of the chains themselves are negligible.

SOLUTION It may seem strange that we neglect the weight of the chains in this example, while in Example 5–1 we did *not* neglect the weight of a mere rope. The reason is that the weight of the chains is very small compared to the weight of the massive engine. By contrast, in Example 5–1 the weight of the rope was a reasonable fraction of the gymnast's weight (100 N compared to 500 N).

Figure 5–2b is a free-body diagram for the engine. The two forces acting on the engine are its weight and the upward force exerted by the vertical chain. Without further ceremony we can conclude that $T_1 = w$. The horizontal and slanted chains do not exert forces on the engine itself, because they are not attached to it, but they do exert forces on the ring where three chains join. So let's consider the *ring* as a particle in equilibrium; the weight of the ring itself is negligible.

In the free-body diagram for the ring (Fig. 5–2c), remember that T_1, T_2, and T_3 are the *magnitudes* of the forces; their directions are shown by the vectors on the diagram. An x-y coordinate axis system is also shown, and the force with magnitude T_3 has been resolved into its x- and y-components. Note that the vertical chain exerts forces of the same magnitude T_1 at both of its ends, upward on the engine in Fig. 5–2b and downward on the ring in

Fig. 5–2c. This is because the weight of the chain is negligible (see Example 4–11 in Section 4–7). If the weight were not negligible, these two forces would have different magnitudes, as was the case for the rope in Example 5–1.

We now apply the equilibrium conditions for the ring, writing separate equations for the x- and y-components. (Note that x- and y-components are *never* added together in a single equation.) We find

$$\Sigma F_x = 0: \quad T_3 \cos 60° + (-T_2) = 0;$$
$$\Sigma F_y = 0: \quad T_3 \sin 60° + (-T_1) = 0.$$

Because $T_1 = w$, we can rewrite the second equation as

$$T_3 = \frac{T_1}{\sin 60°} = \frac{w}{\sin 60°} = 1.155w.$$

We can now use this result in the first equation:

$$T_2 = T_3 \cos 60° = w \frac{\cos 60°}{\sin 60°} = 0.577w.$$

So we can express all three tensions as multiples of the weight w of the engine, which we assume is known. To summarize,

$$T_1 = w,$$
$$T_2 = 0.577w,$$
$$T_3 = 1.155w.$$

If the engine's weight is $w = 2200$ N (about 500 lb), then

$$T_1 = 2200 \text{ N},$$

$$T_2 = (0.577)(2200 \text{ N}) = 1270 \text{ N},$$

$$T_3 = (1.155)(2200 \text{ N}) = 2540 \text{ N}.$$

The chain attached to the ceiling exerts a force on the ring of magnitude T_3, which is greater than the weight of the engine. If this seems strange, note that the *vertical* component of this force is equal to T_1, which in turn is equal to w. But since this force also has a horizontal component, its magnitude T_3 must be somewhat *larger* than w. Hence the chain attached to the ceiling is under the greatest tension and is the one most susceptible to breaking.

(a)

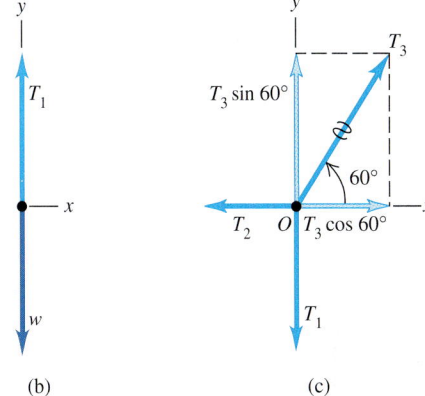

(b) (c)

5–2 (a) A car engine with weight w is suspended from a chain linked at O to two other chains. The chains and ring are considered to be massless. (b) Free-body diagram for the engine. (c) Free-body diagram for the ring with $\vec{T}_3$ replaced by its components.

EXAMPLE 5-3

An inclined plane A car rests on the slanted tracks of a ramp leading to a car-transporter trailer (Fig. 5–3a). The car's brakes and transmission lock are released; only a cable attached to the car and to the frame of the trailer prevents the car from rolling down the ramp. If the car's weight is w, find the tension in the cable and the force with which the tracks push on the car tires.

SOLUTION Figure 5–3b shows a free-body diagram for the car. The three forces acting on the car are its weight (magnitude w), the tension in the cable (magnitude T), and the forces exerted on the wheels by the tracks; we have lumped the forces on the wheels together as a single force with magnitude n. This force must be normal (perpendicular) to the tracks, and hence we call it a *normal force* (we first used this term in Section 4–3). We use a special script letter n to avoid confusion with the abbreviation N for newton. (If this force had a component *along* the tracks, it would tend to prevent the car from moving along the tracks. We have assumed that there is no such effect.)

We take our coordinate axes parallel and perpendicular to the tracks, as shown. To find the components of the weight, we first note that the angle α between the plane and the horizontal is equal to the angle α between the weight vector $\vec{w}$ and the normal

(a)

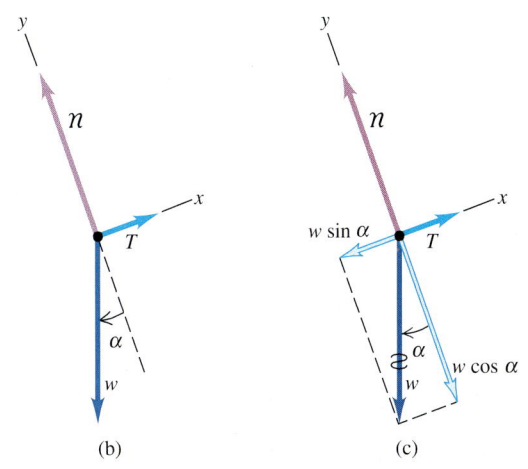

(b) (c)

5–3 (a) A cable holds the car on the ramp. (b) Free-body diagram for the car. (c) The weight component $w \sin \alpha$ acts down the plane, and $w \cos \alpha$ is perpendicular to the plane.

to the plane of the ramp. The angle α is *not* measured from the $+x$-axis toward the $+y$-axis, so we *cannot* use Eqs. (1–7) directly to find the components. (You may want to review Section 1–9 to make sure that you understand this important point.) Instead, consider the right triangles in Fig. 5–3c. The sine of α is the magnitude of the x-component of $\vec{w}$ (that is, the side of the triangle opposite the angle α) divided by the magnitude w (the hypotenuse of the triangle). Similarly, the cosine of α is the magnitude of the y-component (the side of the triangle adjacent to α) divided by w. Both components are negative, so $w_x = -w \sin \alpha$ and $w_y = -w \cos \alpha$.

We draw a wiggly line through the original vector representing the weight to remind us not to count it twice. The equilibrium conditions then give us

$$\Sigma F_x = 0: \qquad T + (-w \sin \alpha) = 0;$$
$$\Sigma F_y = 0: \qquad n + (-w \cos \alpha) = 0.$$

Be sure you understand how these signs are related to our choice of coordinates. Remember that by definition, T, w, and n are all *magnitudes* of vectors and are therefore all positive.

Solving these equations for T and n, we find

$$T = w \sin \alpha,$$
$$n = w \cos \alpha.$$

In general, the magnitude n of the normal force is *not* equal to the weight w.

To check some special cases, first note that if the angle α is zero, then $\sin \alpha = 0$ and $\cos \alpha = 1$. In this case the tracks are horizontal; no cable tension T is needed to hold the car, and the total normal force n is equal in magnitude to the weight. If the angle is 90°, then $\sin \alpha = 1$ and $\cos \alpha = 0$. Then the cable tension T equals the weight w, and the normal force n is zero. Are these the results you would expect for these particular cases?

As a final note, ask yourself how the answers for T and n would be affected if the car were not stationary but were being pulled up the ramp at a constant speed. You should immediately realize that this, too, is an equilibrium situation, since the car's velocity is constant. So the calculation is exactly the same, and T and n have the same values as when the car is at rest. (It's true that T must be greater than $w \sin \alpha$ to *start* the car moving up the ramp, but that's not what we asked.)

EXAMPLE 5–4

Tension over a frictionless pulley   Blocks of granite are being hauled up a 15° slope out of a quarry. For environmental reasons, dirt is also being dumped into the quarry to fill up old holes. You have been asked to find a way to use this dirt to move the granite out more easily. You design a system in which a granite block on a cart with steel wheels (weight w_1, including the cart) is pulled uphill on steel rails by a bucket of dirt (weight w_2, including the bucket) dropping vertically into the quarry (Fig. 5–4a). Ignoring friction in the pulley and wheels and the weight of the cable, determine how the weights w_1 and w_2 must be related in order for the system to move with constant speed.

SOLUTION   Our idealized model for the system is shown in Fig. 5–4b. Figure 5–4c is the free-body diagram for the dirt and bucket, and Fig. 5–4d is the free-body diagram for the granite block and cart. We have drawn an axis system for each body. Note that we are at liberty to orient the axes differently for each body; the choices shown are the most convenient ones. We have represented the weight of the granite block in terms of its components in the chosen axis system. These components are found in the same way as in Example 5–3. We assume the weight of the cable to be negligible so that the tension T is the same throughout the cable.

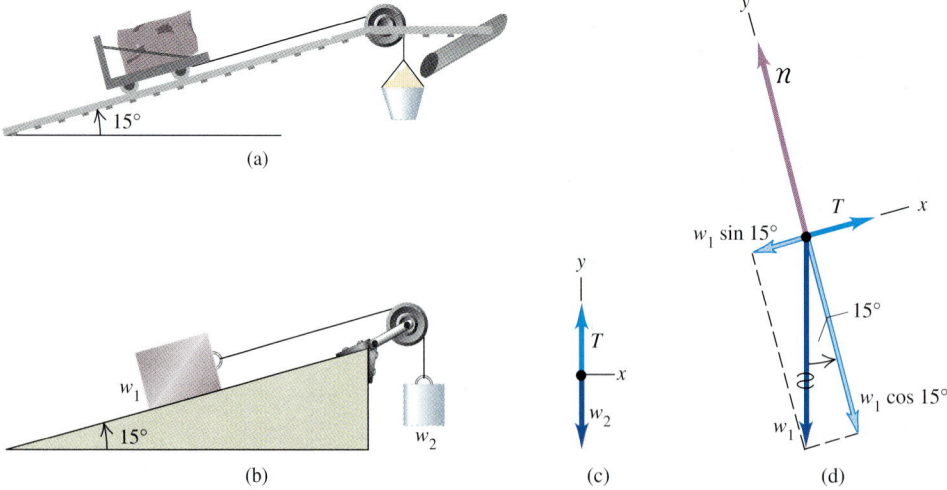

5–4 (a) A bucket of dirt pulls up a cart carrying a granite block. (b) Idealized model of the system. (c) Free-body diagram for the dirt and bucket. (d) Free-body diagram for the block and cart.

Applying $\Sigma F_y = 0$ to the dirt and bucket in Fig. 5–4c, we find

$$T + (-w_2) = 0, \qquad T = w_2.$$

Applying $\Sigma F_x = 0$ to the block and cart in Fig. 5–4d, we get

$$T + (-w_1 \sin 15°) = 0, \qquad T = w_1 \sin 15°.$$

Equating the two expressions for T, we find

$$w_2 = w_1 \sin 15° = 0.26 w_1.$$

If the weight of dirt and bucket totals 26% of the weight of the granite block and cart, the system can move with constant speed in either direction (our analysis doesn't depend on the direction of motion, only on the velocity being constant). Notice that we didn't need to apply the equation $\Sigma F_y = 0$ to the cart and block; this would be useful only if we wanted to find the value of n. Can you show that $n = w_1 \cos 15°$?

5–3 USING NEWTON'S SECOND LAW: DYNAMICS OF PARTICLES

2.1.5
Car Race

2.2
Lifting a Crate

2.3
Lowering a Crate

We are now ready to discuss *dynamics* problems, in which we apply Newton's second law to bodies that are accelerating and hence are *not* in equilibrium. In this case the net force on the body is not zero, but is equal to the mass of the body times its acceleration:

$$\Sigma \vec{F} = m\vec{a} \qquad \text{(Newton's second law)}. \tag{5–3}$$

We will usually use this relation in component form:

$$\Sigma F_x = ma_x, \qquad \Sigma F_y = ma_y \qquad \text{(Newton's second law)}. \tag{5–4}$$

The following problem-solving strategy is very similar to our strategy for equilibrium problems in Section 5–2. We urge you to study it carefully, watch how it's applied in our examples, and use it when you tackle the end-of-chapter problems. Remember that *all* dynamics problems can be solved by using this strategy.

CAUTION ▶ We emphasize again that the quantity $m\vec{a}$ is *not* a force; it's not a push or a pull exerted by anything in the body's environment. All that Eqs. (5–3) and (5–4) say is that the acceleration $\vec{a}$ is proportional to the net force $\Sigma \vec{F}$. When you draw the free-body diagram for a body not in equilibrium, make sure you *never* include "the $m\vec{a}$ force," because *there is no such force* (Fig. 5–5). Sometimes we will draw the acceleration vector $\vec{a}$ *alongside* a free-body diagram; the acceleration will *never* be drawn with its tail touching the body (a position reserved for the forces acting on the body). ◀

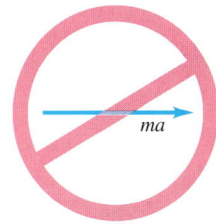

5–5 The quantity $m\vec{a}$ is *not* a force. It must never be included in free-body diagrams.

Problem–Solving Strategy

NEWTON'S SECOND LAW

1. Sketch the physical situation. Identify one or more moving bodies to which you will apply Newton's second law.

2. Draw a free-body diagram for each chosen body. Be sure to include all the forces acting *on* the body, but be equally careful *not* to include any force exerted *by* the body on some other body. Never include the quantity $m\vec{a}$ in your free-body diagram; it's not a force! Label the magnitude of each force with an algebraic symbol and the numerical value if it's given. Usually, one of the forces will be the body's weight; it is usually best to label this as $w = mg$. If a numerical value of mass is given, you can compute the corresponding weight.

3. Show your coordinate axes explicitly in the free-body diagram, and then determine components of forces with reference to these axes. If you know the direction of the acceleration, it is usually best to take that direction as one of the axes. When you represent a force in terms of its components, draw a wiggly line through the original

force vector to remind you not to include it twice. When there are two or more bodies, you can use a separate axis system for each body; you don't have to use the same axis system for all the bodies. But in the equations for each body, the signs of the components *must* be consistent with the axes you have chosen for that body.

4. Write the equations for Newton's second law, Eqs. (5–4), using a separate equation for each component.

5. If more than one body is involved, repeat Steps 2 through 4 for each body. There may be relationships among the motions of the bodies; for example, they may be connected by a rope. Express any such relationships in algebraic form as relations between the accelerations of the various bodies. Then solve the equations to find the required unknowns.

6. Check particular values or extreme cases of quantities, when possible, and compare the results with your intuitive expectations. Ask, "Does this result make sense?"

EXAMPLE 5-5

Acceleration in one dimension An iceboat is at rest on a frictionless horizontal surface (Fig. 5–6a). What constant horizontal force F do we need to apply (along the direction of the runners) to give it a velocity of 6.0 m/s (about 22 km/h, or 13 mi/h) at the end of 4.0 s? The mass of iceboat and rider is 200 kg.

SOLUTION The forces acting on the iceboat and rider are the weight, the normal force exerted by the surface, and the horizontal force F. Figure 5–6b shows a free-body diagram and a coordinate system. We can find the unknown force by using Eqs. (5–4) if we can find the acceleration, so let's start there. The y-component of acceleration is zero, and we can get the x-component from the velocity data. The forces are all constant, so a_x is also constant, and we can use one of the constant-acceleration formulas, Eq. (2–8). Since the iceboat starts from rest, we find

$$a_x = \frac{v - v_0}{t} = \frac{6.0 \text{ m/s} - 0 \text{ m/s}}{4.0 \text{ s}} = 1.5 \text{ m/s}^2.$$

The sum of the x-components of force is simply

$$\Sigma F_x = F,$$

and Newton's second law gives

$$\Sigma F_x = F = ma_x,$$
$$F = (200 \text{ kg})(1.5 \text{ m/s}^2) = 300 \text{ kg} \cdot \text{m/s}^2$$
$$= 300 \text{ N} \quad \text{(about 75 lb)}.$$

Note that we did not need the y-components at all to find F. We can use these to find the normal force n:

$$a_y = 0,$$
$$\Sigma F_y = n + (-mg) = ma_y = 0,$$
$$n = mg = (200 \text{ kg})(9.8 \text{ m/s}^2) = 1960 \text{ N} \quad \text{(about 440 lb)}.$$

In this situation the magnitude n of the normal force is equal to the weight of the iceboat and rider because the vertical acceleration is zero, the surface is horizontal, and these are the only vertical forces acting. As we will see in Examples 5–8 and 5–9, there are many common situations in which the normal force can be quite different from the weight.

(a)   (b)

5–6 (a) An iceboat starting from rest. (b) Free-body diagram for the iceboat and rider with no friction.

EXAMPLE 5-6

In the situation of Example 5–5, suppose the motion of the iceboat is opposed by a constant horizontal friction force with magnitude 100 N. Now what force F must we apply to give the iceboat a velocity of 6.0 m/s at the end of 4.0 s?

SOLUTION The acceleration is the same as before, $a_x = 1.5 \text{ m/s}^2$. The new free-body diagram is shown in Fig. 5–7; the difference between it and Fig. 5–6b is the addition of the friction force $\vec{f}$. (Note that its *magnitude*, $f = 100$ N, is a positive quantity but that its *component* in the x-direction is negative, equal to $-f$ or -100 N.) Now Newton's second law gives

$$\Sigma F_x = F + (-f) = ma_x,$$
$$F = ma_x + f = (200 \text{ kg})(1.5 \text{ m/s}^2) + (100 \text{ N}) = 400 \text{ N}.$$

We need 100 N to overcome friction and 300 N more to give the iceboat the necessary acceleration.

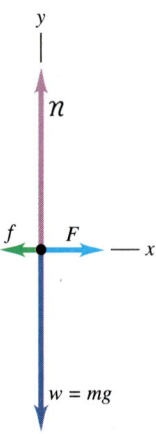

5–7 Free-body diagram for the iceboat and rider with a frictional force $\vec{f}$ opposing the motion.

EXAMPLE 5–7

Tension in an elevator cable An elevator and its load have a total mass of 800 kg (Fig. 5–8a). The elevator is originally moving downward at 10.0 m/s; it is brought to rest with constant

5–8 (a) A loaded elevator moving downward being brought to rest. (b) Free-body diagram for the elevator. Note that we draw the acceleration vector (in yellow) to one side of the free-body diagram; unlike tension and weight, acceleration is *not* a force.

acceleration in a distance of 25.0 m. Find the tension T in the supporting cable while the elevator is being brought to rest.

SOLUTION The only forces acting on the elevator are its weight and the tension force of the cable. The acceleration vector is also shown in Fig. 5–8b; we draw it off to one side because it's not a force. We can use Eqs. (5–4) to find T if we can first find the acceleration. The simplest way to do this is to use the constant-acceleration formula $v^2 = v_0^2 + 2a_y(y - y_0)$. Taking the positive y-axis upward, we have $v_0 = -10.0$ m/s, $v = 0$, and $y - y_0 = -25.0$ m. Then

$$a_y = \frac{v^2 - v_0^2}{2(y - y_0)} = \frac{(0)^2 - (-10.0 \text{ m/s})^2}{2(-25.0 \text{ m})} = +2.00 \text{ m/s}^2.$$

Note that the velocity is downward and the acceleration is upward, corresponding to downward motion with decreasing speed.

Now we are ready to use Newton's second law:

$$\Sigma F_y = T + (-w) = ma_y,$$
$$T = w + ma_y = mg + ma_y = m(g + a_y)$$
$$= (800 \text{ kg})(9.80 \text{ m/s}^2 + 2.00 \text{ m/s}^2) = 9440 \text{ N}.$$

The tension must be 1600 N *greater* than the weight ($w = mg = 7840$ N) to stop the elevator in the required distance.

Can you see that we would get the same answer for T if the elevator were moving upward and gaining speed at a rate of 2.00 m/s²?

EXAMPLE 5–8

Apparent weight in an accelerating elevator A 50.0-kg woman stands on a bathroom scale while riding in the elevator in Example 5–7 (Fig. 5–9a). What is the reading on the scale?

SOLUTION The scale reads the magnitude of the downward force exerted *by* the passenger *on* the scale; by Newton's third law, this equals the magnitude of the upward normal force exerted *by* the scale *on* the passenger. Hence we can solve the problem by finding the magnitude n of the normal force.

Figure 5–9b shows a free-body diagram for the passenger. The forces acting on her are her weight $w = mg = (50.0 \text{ kg})(9.80 \text{ m/s}^2) = 490$ N and the normal force n exerted by the scale. The passenger's acceleration is the same as the elevator's, and Newton's second law gives

$$\Sigma F_y = n + (-mg) = ma_y,$$
$$n = m(g + a_y) = (50.0 \text{ kg})(9.80 \text{ m/s}^2 + 2.00 \text{ m/s}^2)$$
$$= 590 \text{ N}.$$

While the elevator is stopping, the scale pushes up on the passenger with a force of 590 N. By Newton's third law, she pushes down on the scale with the same force; so the scale reads 590 N, which is 100 N more than her actual weight. The scale reading is called the passenger's **apparent weight.** The passen-

ger *feels* a greater strain in her legs and feet than when the elevator is stationary or moving with constant velocity because the floor is pushing up harder on her feet. You can feel this yourself; try taking a few steps in an elevator that is coming to a stop after descending.

5–9 (a) A passenger in the elevator as it slows down. (b) Free-body diagram for the passenger.

2.4
Rocket Blasts Off

2.13
Bricklayer Lowers Bricks

2.14
Modified Atwood Machine

Astronauts in orbit feel as though they are weightless because they have the same acceleration as their spacecraft—*not* because they are "outside the pull of the earth's gravity." (If they were, the astronauts and their spacecraft could not stay in orbit, but would fly off into deep space.)

Let's generalize the result of Example 5–8. A passenger with mass m rides in an elevator with y-acceleration a_y. We saw above that a scale shows the passenger's apparent weight to be

$$n = m(g + a_y).$$

When a_y is positive, the elevator is accelerating upward (moving upward with increasing speed, or moving downward with decreasing speed) and n is greater than the passenger's weight $w = mg$. When the elevator is accelerating downward (moving upward with decreasing speed, or moving downward with increasing speed), a_y is negative and n is less than the weight. If the passenger doesn't know the elevator is accelerating, she may feel as though her weight is changing; indeed, this is just what the scale shows. With upward acceleration the apparent weight is greater than the true weight, and the passenger feels heavier; with downward acceleration she feels lighter.

The extreme case occurs when the elevator has a downward acceleration $a_y = -g$, that is, when it is in free fall. In that case, $n = 0$ and the passenger *seems* to be weightless. Similarly, an astronaut orbiting the earth in a spacecraft experiences *apparent* weightlessness. In each case the person is not really weightless because there is still a gravitational force acting. But the effect of this free-fall condition is just the same as though the body were in outer space with no gravitational force at all. In both cases the person and their vehicle (elevator or spacecraft) are falling together with the same acceleration g, so nothing pushes the person against the floor or walls of the vehicle.

The physiological effect of prolonged apparent weightlessness is an interesting medical problem that is being actively explored. Gravity plays an important role in blood distribution in the body; one reaction to apparent weightlessness is a decrease in blood volume through increased excretion of water. In some cases, astronauts returning to earth have experienced temporary impairment of their sense of balance and a greater tendency toward motion sickness.

EXAMPLE 5–9

Acceleration down a hill A toboggan loaded with vacationing students (total weight w) slides down a long, snow-covered slope (Fig. 5–10a). The hill slopes at a constant angle α, and the toboggan is so well waxed that there is virtually no friction. What is the toboggan's acceleration?

SOLUTION The only forces acting on the toboggan are its weight w and the normal force n exerted by the hill (Fig. 5–10b). We take axes parallel and perpendicular to the surface of the hill and resolve the weight into x- and y-components: $w_x = w \sin \alpha$ and $w_y = -w \cos \alpha$. (Compare to Example 5–3, in which the

(a)   (b)

5–10 (a) A loaded toboggan slides down a frictionless hill. (b) The free-body diagram shows that the x-component of the weight, $w \sin \alpha$, accelerates the toboggan down the hill.

x-component of weight was $-w \sin \alpha$. The difference is that the positive *x*-axis is uphill in Example 5–3, while in Fig. 5–10b it is downhill.) Newton's second law in the *x*-direction then gives

$$\Sigma F_x = w \sin \alpha = ma_x,$$

and since $w = mg$,

$$a_x = g \sin \alpha.$$

The mass does not appear in the final result. This means that *any* toboggan, regardless of its mass or number of passengers, slides down a frictionless hill with an acceleration of $g \sin \alpha$. If the plane is horizontal, $\alpha = 0$ and $a_x = 0$ (the toboggan does not accelerate); if the plane is vertical, $\alpha = 90°$ and $a_x = g$ (the toboggan is in free fall).

Once again, we did not need the *y*-components to find the acceleration; let's use them to find the normal force n exerted on the toboggan by the surface of the hill. We know that $a_y = 0$, so $\Sigma F_y = 0$ and

$$n = mg \cos \alpha.$$

This force is *not* equal to the toboggan's weight (compare Example 5–3 in Section 5–2). The moral is that you should *never* automatically assume that the normal force is equal to the weight. Very often, as in this example, their values are different.

CAUTION ▶ Figure 5–11 shows a common *incorrect* way to draw the free-body diagram for the toboggan. This diagram is wrong for two reasons: the normal force should be drawn perpendicular to the surface, and it's completely bogus to include an "$m\vec{a}$ force." If you remember that "normal" means "perpendicular" and that the mass times the acceleration is *not* itself a force, you won't make these mistakes. ◀

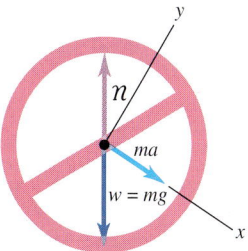

5–11 The *wrong* way to draw the free-body diagram for the toboggan. The normal force points in the wrong direction, and $m\vec{a}$ shouldn't be included because it isn't a force.

EXAMPLE 5–10

Two bodies with the same acceleration A robot arm pulls a 4.0-kg cart along a horizontal frictionless track with a 0.50-kg rope, applying a horizontal force with magnitude $F = 9.0$ N to the rope (Fig. 5–12a; see page 130). Find the acceleration of the system and the tension at the point where the rope is fastened to the cart. (On earth the rope would sag a little; to avoid this complication, suppose the robot arm is operating in a zero-gravity space station.)

SOLUTION We choose the *x*-axis in the direction of acceleration, so only *x*-components are relevant. The cart and the rope move together in the same direction and so have the same acceleration a in the *x*-direction. We can proceed in either of two ways.

Method 1: We write Newton's second law for the cart and a separate equation for Newton's second law for the rope, using a and T for the unknown acceleration and tension (Fig. 5–12b). For the cart,

$$\Sigma F_x = T = m_{cart}a,$$

and for the rope,

$$\Sigma F_x = F + (-T) = m_{rope}a.$$

Note the role of Newton's third law in equating the magnitudes of the two forces labeled T; one is the force on the cart exerted by the rope, the other is the force on the rope exerted by the cart. We now have two simultaneous equations for T and a. An easy way to solve them is to add the two equations; this eliminates T, giving

$$F = m_{cart}a + m_{rope}a = (m_{cart} + m_{rope})a,$$

and

$$a = \frac{F}{m_{cart} + m_{rope}} = \frac{9.0 \text{ N}}{4.0 \text{ kg} + 0.50 \text{ kg}} = 2.0 \text{ m/s}^2.$$

Substituting this back into the first equation,

$$T = m_{cart}a = (4.0 \text{ kg})(2.0 \text{ m/s}^2) = 8.0 \text{ N}.$$

The tension is different at the two ends of the rope; $T = 8.0$ N on the right, $F = 9.0$ N on the left. This is because the rope has mass, and so there must be a net force acting on the rope in order to accelerate it. This net force is equal to $F - T = 9.0$ N -8.0 N $= 1.0$ N, exactly enough to accelerate a 0.50-kg rope at 2.0 m/s².

Method 2: Take the cart and rope as a composite body with a total mass $m = m_{cart} + m_{rope} = 4.5$ kg (Fig. 5–12c). The only force acting on this composite body in the *x*-direction is F, so

$$a = \frac{F}{m} = \frac{9.0 \text{ N}}{4.5 \text{ kg}} = 2.0 \text{ m/s}^2.$$

Then, looking at the 4.0-kg cart by itself, we see that to give it an acceleration of 2.0 m/s² requires a force

$$T = m_{cart}a = (4.0 \text{ kg})(2.0 \text{ m/s}^2) = 8.0 \text{ N}.$$

CAUTION ▶ The method of treating the two bodies as a single composite body works here because the two bodies have the same magnitude *and* direction of acceleration. If the accelerations are different, you should treat the two bodies separately, as in the following example. ◀

(a)

(b)

(c)

5–12 (a) A robot arm pulling a cart along a frictionless track in a zero-gravity space station. (b) Individual free-body diagrams for the rope and cart. (c) Free-body diagram for the rope and cart as a composite body.

EXAMPLE 5-11

Two bodies with the same magnitude of acceleration In Fig. 5–13a an air-track glider with mass m_1 moves on a level, frictionless air track in the physics lab. It is connected to a lab weight with mass m_2 by a light, flexible, nonstretching string that passes over a small frictionless pulley. Find the acceleration of each body and the tension in the string.

SOLUTION The two bodies have different motions—one horizontal, one vertical—so we won't consider them together as we did the bodies in Example 5–10. We use a separate free-body diagram and $\Sigma \vec{F} = m\vec{a}$ equations for each. Figures 5–13b and 5–13c show a free-body diagram and a coordinate system for each body. There is no friction in the pulley, and we consider the string to be massless, so the tension T in the string is the same throughout; it applies a force of magnitude T to each body. The weights are $m_1 g$ and $m_2 g$. For the glider on the track, Newton's second law gives

$$\Sigma F_x = T = m_1 a_{1x},$$
$$\Sigma F_y = n + (-m_1 g) = m_1 a_{1y} = 0.$$

For the lab weight it's convenient to take the +y-direction as downward so that both bodies accelerate in positive axis directions. Then for the lab weight,

$$\Sigma F_y = m_2 g + (-T) = m_2 a_{2y}.$$

We note again that it is perfectly all right to use different coordinate axes for the two bodies.

If the string doesn't stretch, the two bodies must move equal distances in equal times, and so their speeds at any instant must be equal. When the speeds change, they change by equal amounts in a given time, so the accelerations of the two bodies must have the same magnitude a. We can express this relation as

$$a_{1x} = a_{2y} = a.$$

The *directions* of the two accelerations are different, but their *magnitudes* are the same. The two equations for Newton's second law are then

$$T = m_1 a \quad \text{(glider)},$$
$$m_2 g + (-T) = m_2 a \quad \text{(lab weight)}.$$

These are two simultaneous equations for the unknowns T and a. We add the two equations to eliminate T, getting

$$m_2 g = m_1 a + m_2 a = (m_1 + m_2)a$$

and

$$a = \frac{m_2}{m_1 + m_2} g.$$

The acceleration is less than g, as you might expect. Substituting this back into the first equation, we get

$$T = \frac{m_1 m_2}{m_1 + m_2} g.$$

We see that the tension T is *not* equal to the weight $m_2 g$ of mass m_2, but is *less* by a factor of $m_1/(m_1 + m_2)$. If T *were* equal to $m_2 g$, then m_2 would be in equilibrium, and it isn't.

CAUTION▶ It's a common mistake to assume that if an object is attached to a vertical string, the string tension must be equal to the object's weight. That would certainly be wrong in

this example! The only safe approach is to treat the tension as a variable, as we have done here. ◀

Now let's check some special cases. If $m_1 = 0$, then m_2 would fall freely and there would be no tension in the string. The equations do give $T = 0$ and $a = g$ when $m_1 = 0$. Also, if $m_2 = 0$, we expect no tension and no acceleration; for this case the equations give $T = 0$ and $a = 0$. Thus in these two special cases the results agree with our intuitive expectations.

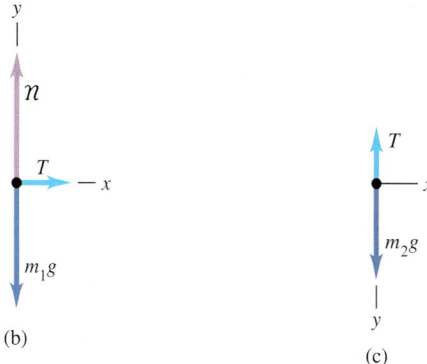

5–13 (a) A lab weight accelerates a glider along an air track. (b) Free-body diagram for the glider (mass m_1). (c) Free-body diagram for the lab weight (mass m_2).

EXAMPLE 5-12

A simple accelerometer Figure 5–14a shows a lead fish-line sinker hanging from a string attached to point P on the ceiling of a car. When the system has an acceleration a toward the right, the string makes an angle β with the vertical. In a practical instrument, some form of damping would be needed to keep the string from swinging when the acceleration changes. For example, the system might hang in a tank of oil. (You can make a primitive version of this device by taping a thread to the ceiling light in a car, tying a key or nearly any other small object to the other end, and using a protractor. Let someone else drive the car while you are doing the experiment.) The problem is this: Given m and β, what is the acceleration a?

SOLUTION As shown in the free-body diagram, Fig. 5–14b, two forces act on the body: its weight $w = mg$ and the tension T in the string. In finding the components of the tension, note that the angle β is not measured from the $+x$-axis. Referring to the right triangles in Fig. 5–14b and recalling the definitions of the sine and cosine functions, we find the components shown (compare to Example 1–6 in Section 1–9). The sum of the horizontal components of force is

$$\Sigma F_x = T \sin \beta,$$

and the sum of the vertical components is

$$\Sigma F_y = T \cos \beta + (-mg).$$

The x-acceleration is the acceleration a of the system, and the y-acceleration is zero, so

$$T \sin \beta = ma, \qquad T \cos \beta = mg.$$

When we divide the first equation by the second and use $(\sin \beta)/(\cos \beta) = \tan \beta$, we get

$$a = g \tan \beta.$$

The acceleration a is proportional to the tangent of the angle β.

When $\beta = 0$, the sinker hangs vertically and the acceleration is zero; when $\beta = 45°$, $a = g$, and so on. We note that β can never be 90° because that would require an infinite acceleration. Also note that this result doesn't depend on the mass of the sinker.

5–14 (a) A simple accelerometer. (b) Free-body diagram for the lead sinker.

5–4 FRICTIONAL FORCES

We have seen several problems where a body rests or slides on a surface that exerts forces on the body, and we have used the terms *normal force* and *friction force* to describe these forces. Whenever two bodies interact by direct contact (touching) of their surfaces, we call the interaction forces *contact forces.* Normal and friction forces are both contact forces.

Our concern in this section is with friction, an important force in many aspects of everyday life. The oil in a car engine minimizes friction between moving parts, but without friction between the tires and the road we couldn't drive or turn the car. Air drag—the frictional force exerted by the air on a body moving through it—decreases automotive fuel economy but makes parachutes work. Without friction, nails would pull out, light bulbs would unscrew effortlessly, and riding a bicycle would be hopeless.

KINETIC AND STATIC FRICTION

Let's consider a body sliding across a surface. When you try to slide a heavy box of books across the floor, the box doesn't move at all unless you push with a certain minimum force. Then the box starts moving, and you can usually keep it moving with less force than you needed to get it started. If you take some of the books out, you need less force than before to get it started or keep it moving. What general statements can we make about this behavior?

First, when a body rests or slides on a surface, we can always represent the contact force exerted by the surface on the body in terms of components of force perpendicular and parallel to the surface. We call the perpendicular component vector the *normal force,* denoted by $\vec{n}$. (Recall that *normal* is a synonym for *perpendicular.*) The component vector parallel to the surface is the **friction force,** denoted by $\vec{f}$. By definition, $\vec{n}$ and $\vec{f}$ are always perpendicular to each other. We use script symbols for these quantities to emphasize their special role in representing the contact force. If the surface is frictionless, then the contact force has *only* a normal component, and $\vec{f}$ is zero. (Frictionless surfaces are an unattainable idealization, but we can approximate a surface as frictionless if the effects of friction are negligibly small.) The direction of the friction force is always such as to oppose relative motion of the two surfaces.

The kind of friction that acts when a body slides over a surface is called a **kinetic friction force** $\vec{f}_k$. The adjective "kinetic" and the subscript "k" remind us that the two surfaces are moving relative to each other. The *magnitude* of the kinetic friction force usually increases when the normal force increases. It takes more force to slide a box full of books across the floor than to slide the same box when it is empty. This principle is also used in automotive braking systems; the harder the brake pads are squeezed against the rotating brake disks, the greater the braking effect. In many cases the magnitude of the kinetic friction force f_k is found experimentally to be approximately *proportional* to the magnitude n of the normal force. In such cases we can write

$$f_k = \mu_k n \qquad \text{(magnitude of kinetic friction force)}, \qquad (5\text{–}5)$$

where μ_k (pronounced "mu-sub-k") is a constant called the **coefficient of kinetic friction.** The more slippery the surface, the smaller the coefficient of friction. Because it is a quotient of two force magnitudes, μ_k is a pure number without units.

CAUTION ▶ Remember, the friction force and the normal force are always perpendicular. Equation (5–5) is *not* a vector equation, but a *scalar* relation between the *magnitudes* of the two perpendicular forces. ◀

Equation (5–5) is only an approximate representation of a complex phenomenon. On a microscopic level, friction and normal forces result from the intermolecular forces (fundamentally electrical in nature) between two rough surfaces at points where they come into contact (Fig. 5–15). The actual area of contact is usually much smaller than

The sport of ice hockey depends crucially on having just the right amount of friction between a player's skates and the ice. If there were too much friction, the players would move much more slowly; if there were too little friction, they could hardly keep from falling over.

5–15 The normal and friction forces arise from interactions between molecules at high points on the surfaces of the block and the floor.

the total surface area. As a box slides over the floor, bonds between the two surfaces form and break, and the total number of such bonds varies; hence the kinetic friction force is not perfectly constant. Smoothing the surfaces can actually increase friction, since more molecules are able to interact and bond; bringing two smooth surfaces of the same metal together can cause a "cold weld." Lubricating oils work because an oil film between two surfaces (such as the pistons and cylinder walls in a car engine) prevents them from coming into actual contact.

Table 5–1 shows a few representative values of μ_k. Although these values are given with two significant figures, they are only approximate, since friction forces can also depend on the *speed* of the body relative to the surface. We'll ignore this effect and assume that μ_k and f_k are independent of speed so that we can concentrate on the simplest cases. Table 5–1 also lists coefficients of *static* friction; we'll define these shortly.

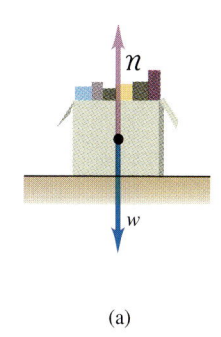

(a)

TABLE 5–1

APPROXIMATE COEFFICIENTS OF FRICTION

MATERIALS	STATIC, μ_s	KINETIC, μ_k
Steel on steel	0.74	0.57
Aluminum on steel	0.61	0.47
Copper on steel	0.53	0.36
Brass on steel	0.51	0.44
Zinc on cast iron	0.85	0.21
Copper on cast iron	1.05	0.29
Glass on glass	0.94	0.40
Copper on glass	0.68	0.53
Teflon on Teflon	0.04	0.04
Teflon on steel	0.04	0.04
Rubber on concrete (dry)	1.0	0.8
Rubber on concrete (wet)	0.30	0.25

(No sliding)

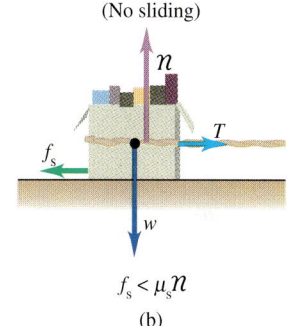

$f_s < \mu_s n$

(b)

(Just about to slide)

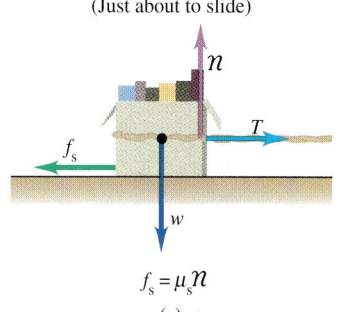

$f_s = \mu_s n$

(c)

Friction forces may also act when there is *no* relative motion. If you try to slide that box of books across the floor, the box may not move at all because the floor exerts an equal and opposite friction force on the box. This is called a **static friction force** $\vec{f}_s$. In Fig. 5–16a the box is at rest in equilibrium under the action of its weight $\vec{w}$ and the upward normal force $\vec{n}$, which is equal in magnitude to the weight and exerted on the box by the floor. Now we tie a rope to the box (Fig. 5–16b) and gradually increase the tension T in the rope. At first the box remains at rest because, as T increases, the force of static friction f_s also increases (staying equal in magnitude to T).

At some point, T becomes greater than the maximum static friction force f_s the surface can exert. Then the box "breaks loose" (the tension T is able to break the bonds between molecules in the surfaces of the box and floor) and starts to slide. Figure 5–16c is the force diagram when T is at this critical value. If T exceeds this value, the box is no longer in equilibrium. For a given pair of surfaces the maximum value of f_s depends on the normal force. Experiment shows that in many cases this maximum value, called $(f_s)_{max}$, is approximately *proportional* to n; we call the proportionality factor μ_s (pronounced "mu-sub-s") the **coefficient of static friction**. Some representative values of μ_s are shown in Table 5–1. In a particular situation, the actual force of static friction can have any magnitude between zero (when there is no other force parallel to the surface) and a maximum value given by $\mu_s n$. In symbols,

$$f_s \le \mu_s n \quad \text{(magnitude of static friction force)}. \tag{5-6}$$

Like Eq. (5–5), this is a relation between magnitudes, *not* a vector relation. The

(Now sliding)

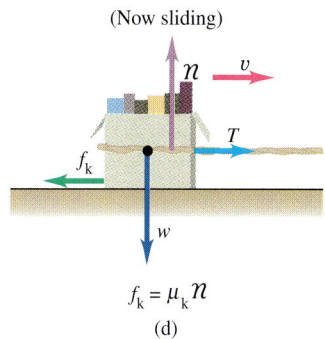

$f_k = \mu_k n$

(d)

5–16 (a), (b), (c) When there is no relative motion of the surfaces, the magnitude of the static friction force f_s is less than or equal to $\mu_s n$. (d) When there is relative motion, the magnitude of the kinetic friction force f_k equals $\mu_k n$.

equality sign holds only when the applied force T, parallel to the surface, has reached the critical value at which motion is about to start (Fig. 5–16c). When T is less than this value (Fig. 5–16b), the inequality sign holds. In that case we have to use the equilibrium conditions ($\Sigma \vec{F} = \mathbf{0}$) to find f_s. If there is no applied force ($T = 0$) as in Fig. 5–16a, then there is no static friction force either ($f_s = 0$).

As soon as sliding starts (Fig. 5–16d), the friction force usually *decreases;* it's easier to keep the box moving than to start it moving. Hence the coefficient of kinetic friction is usually *less* than the coefficient of static friction for any given pair of surfaces, as shown in Table 5–1. If we start with no applied force ($T = 0$) at time $t = 0$ and gradually increase the force, the friction force varies somewhat, as shown in Fig. 5–17.

In some situations the surfaces will alternately stick (static friction) and slip (kinetic friction). This is what causes the horrible squeak made by chalk held at the wrong angle while writing on the blackboard. Another stick-slip phenomenon is the squeaky noise your windshield-wiper blades make when the glass is nearly dry; still another is the outraged shriek of tires sliding on asphalt pavement. A more positive example is the motion of a violin bow against the string.

When a body slides on a layer of gas, friction can be made very small. In the linear air track used in physics laboratories, the gliders are supported on a layer of air. The frictional force is velocity-dependent, but at typical speeds the effective coefficient of friction is of the order of 0.001. A similar device is the air table, where the pucks are supported by an array of small air jets about 2 cm apart.

5–17 In response to an externally applied force, the friction force increases to $(f_s)_{max}$. Then the surfaces begin to slide across one another, and the frictional force drops back to a nearly constant value f_k. The kinetic friction force varies somewhat as intermolecular bonds form and break.

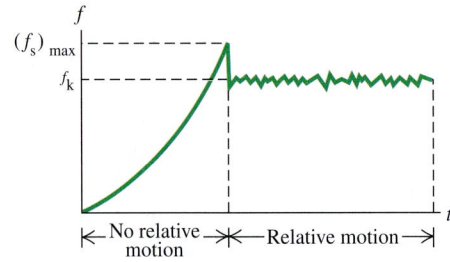

EXAMPLE 5-13

Friction in horizontal motion   A delivery company has just unloaded a 500-N crate full of home exercise equipment on the sidewalk in front of your house (Fig. 5–18a). You find that to get it started moving toward your front door, you have to pull with a horizontal force of magnitude 230 N. Once it "breaks loose" and starts to move, you can keep it moving at constant velocity with only 200 N. What are the coefficients of static and kinetic friction?

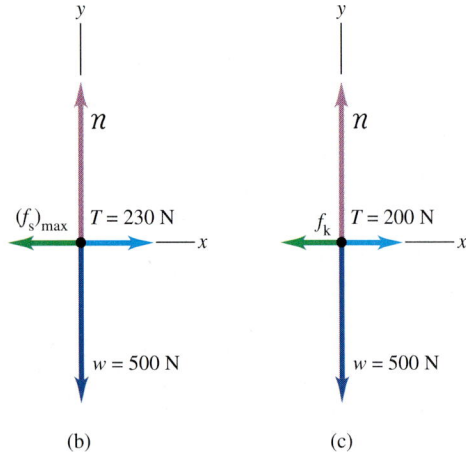

(a)   (b)   (c)

5–18 (a) Pulling a crate with a horizontal force. (b) Free-body diagram for the crate as it starts to move. (c) Free-body diagram for the crate moving at constant velocity.

SOLUTION The state of rest and the state of motion with constant velocity arc both equilibrium conditions, so we use Eqs. (5–2). An instant before the crate starts to move, the static friction force has its maximum possible value, $(f_s)_{max} = \mu_s n$. The appropriate force diagram is Fig. 5–18b. We find

$$\Sigma F_x = T + (-(f_s)_{max}) = 230\ N - (f_s)_{max} = 0, \qquad (f_s)_{max} = 230\ N,$$
$$\Sigma F_y = n + (-w) = n - 500\ N = 0, \qquad n = 500\ N,$$
$$(f_s)_{max} = \mu_s n \quad \text{(motion about to start)},$$
$$\mu_s = \frac{(f_s)_{max}}{n} = \frac{230\ N}{500\ N} = 0.46.$$

After the crate starts to move, the forces are as shown in Fig. 5–18c, and we have

$$\Sigma F_x = T + (-f_k) = 200\ N - f_k = 0, \qquad f_k = 200\ N,$$
$$\Sigma F_y = n + (-w) = n - 500\ N = 0, \qquad n = 500\ N,$$
$$f_k = \mu_k n \quad \text{(motion occurs)},$$
$$\mu_k = \frac{f_k}{n} = \frac{200\ N}{500\ N} = 0.40.$$

It's easier to keep the crate moving than to start it moving, and so the coefficient of kinetic friction is less than the coefficient of static friction.

EXAMPLE 5-14

In Example 5–13, what is the friction force if the crate is at rest on the surface and a horizontal force of 50 N is applied to it?

SOLUTION From the equilibrium conditions we have

$$\Sigma F_x = T + (-f_s) = 50\ N - f_s = 0,$$
$$f_s = 50\ N.$$

In this case, f_s is less than the maximum value $(f_s)_{max} = \mu_s n$. The frictional force can prevent motion for any horizontal applied force up to 230 N.

EXAMPLE 5-15

In Example 5–13, suppose you try to move your crate full of exercise equipment by tying a rope around it and pulling upward on the rope at an angle of 30° above the horizontal (Fig. 5–19a). How hard do you have to pull to keep the crate moving with constant velocity? Is this easier or harder than pulling horizontally? Assume that $w = 500\ N$ and $\mu_k = 0.40$.

SOLUTION Figure 5–19b is a free-body diagram showing the forces on the crate. The kinetic friction force f_k is still equal to $\mu_k n$, but now the normal force n is *not* equal in magnitude to the weight of the crate. The force exerted by the rope has an additional vertical component that tends to lift the crate off the floor.

The crate is in equilibrium, since its velocity is constant, so

$$\Sigma F_x = T \cos 30° + (-f_k) = T \cos 30° - 0.40n = 0,$$
$$\Sigma F_y = T \sin 30° + n + (-500\ N) = 0.$$

These are two simultaneous equations for the two unknown quantities T and n. To solve them, we can eliminate one unknown and solve for the other. There are many ways to do this; here is one way. Rearrange the second equation to the form

$$n = 500\ N - T \sin 30°.$$

Substitute this expression for n back into the first equation:

$$T \cos 30° - 0.40(500\ N - T \sin 30°) = 0.$$

5–19 (a) Pulling a crate with a force applied at an upward angle. (b) Free-body diagram for the crate moving at constant velocity.

Finally, solve this equation for T, then substitute the result back into either of the original equations to obtain n. The results are

$$T = 188 \text{ N}, \qquad n = 406 \text{ N}.$$

Note that the normal force is *less* than the weight of the box

($w = 500$ N) because the vertical component of tension pulls upward on the crate. Despite this, the tension required is a little less than the 200-N force needed when you pulled horizontally in Example 5–13. Try pulling at 22°; you'll find that you need even less force (see Challenge Problem 5–111).

EXAMPLE 5–16

Toboggan ride with friction I Let's go back to the toboggan that we studied in Example 5–9 (Section 5–3). The wax has worn off, and there is now a coefficient of kinetic friction μ_k. The slope has just the right angle to make the toboggan slide with constant speed. Derive an expression for the slope angle in terms of w and μ_k.

SOLUTION Figure 5–20 shows the free-body diagram. The slope angle is α. The forces on the toboggan are its weight w and the normal and frictional components of the contact force exerted on it by the sloping surface. We take axes perpendicular and parallel to the surface and represent the weight in terms of its components in these two directions, as shown. The toboggan is in equilibrium because its velocity is constant, and the equilibrium conditions are

$$\Sigma F_x = w \sin \alpha + (-f_k) = w \sin \alpha - \mu_k n = 0,$$
$$\Sigma F_y = n + (-w \cos \alpha) = 0.$$

Rearranging, we get

$$\mu_k n = w \sin \alpha, \qquad n = w \cos \alpha.$$

Just as in Example 5–9, the normal force n is *not* equal to the weight w. When we divide the first of these equations by the second, we find

$$\mu_k = \frac{\sin \alpha}{\cos \alpha} = \tan \alpha.$$

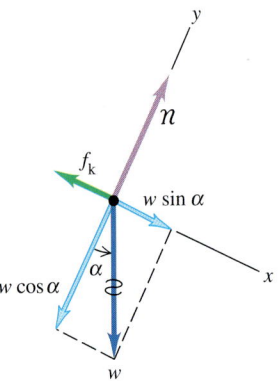

5–20 Free-body diagram for the toboggan with friction.

The weight w doesn't appear in this expression. *Any* toboggan, regardless of its weight, slides down an incline with constant speed if the coefficient of kinetic friction equals the tangent of the slope angle of the incline. The steeper the slope, the greater the coefficient of friction has to be for the toboggan to slide with constant velocity. This is just what we should expect.

EXAMPLE 5–17

Toboggan ride with friction II What if we have the same toboggan and coefficient of friction as in Example 5–16, but a steeper hill? This time the toboggan accelerates, although not as much as in Example 5–9, when there was no friction. Derive an expression for the acceleration in terms of g, α, μ_k, and w.

SOLUTION The free-body diagram (Fig. 5–21) is almost the same as for Example 5–16, but the body is no longer in equilibrium; a_y is still zero, but a_x is not. Using $w = mg$, Newton's second law gives us the two equations

$$\Sigma F_x = mg \sin \alpha + (-f_k) = ma_x,$$
$$\Sigma F_y = n + (-mg \cos \alpha) = 0.$$

From the second equation and Eq. (5–5), $f_k = \mu_k n$, we get an expression for f_k:

$$n = mg \cos \alpha,$$
$$f_k = \mu_k n = \mu_k mg \cos \alpha.$$

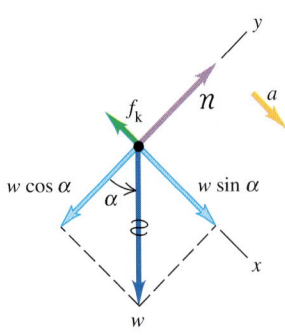

5–21 Free-body diagram for the toboggan with friction, going down a steeper hill.

We substitute this back into the x-component equation. The result is

$$mg \sin \alpha + (-\mu_k mg \cos \alpha) = ma_x,$$

$$a_x = g(\sin \alpha - \mu_k \cos \alpha).$$

Does this result make sense? Here are some special cases we can check. First, if the hill is *vertical,* $\alpha = 90°$; then $\sin \alpha = 1$, $\cos \alpha = 0$, and $a_x = g$. This is free fall, just what we would expect. Second, on a hill at angle α with *no* friction, $\mu_k = 0$. Then $a_x = g \sin \alpha$. The situation is the same as in Example 5–9, and we get the same result; that's encouraging! Next, suppose that there is just enough friction to make the toboggan move with constant velocity. In that case, $a_x = 0$, and our result gives

$$\sin \alpha = \mu_k \cos \alpha \qquad \text{and} \qquad \mu_k = \tan \alpha.$$

This agrees with our result from Example 5–16; good! Finally, note that there may be so much friction that $\mu_k \cos \alpha$ is actually greater than $\sin \alpha$. In that case, a_x is negative; if we give the

toboggan an initial downhill push to start it moving, it will slow down and eventually stop.

We have pretty much beaten the toboggan problem to death, but there is an important lesson to be learned. We started out with a simple problem and then extended it to more and more general situations. Our most general result, found in this example, includes *all* the previous ones as special cases, and that's a nice, neat package! Don't memorize this package; it is useful only for this one set of problems. But do try to understand how we obtained it and what it means.

One final variation that you may want to try out is the case in which we give the toboggan an initial push *up* the hill. The direction of the friction force is now reversed, so the acceleration is different from the downhill value. It turns out that the expression for a_x is the same as for downhill motion except that the minus sign becomes plus. Can you prove this?

ROLLING FRICTION

It's a lot easier to move a loaded filing cabinet across a horizontal floor by using a cart with wheels than to slide it. How much easier? We can define a **coefficient of rolling friction** μ_r, which is the horizontal force needed for constant speed on a flat surface divided by the upward normal force exerted by the surface. Transportation engineers call μ_r the *tractive resistance.* Typical values of μ_r are 0.002 to 0.003 for steel wheels on steel rails and 0.01 to 0.02 for rubber tires on concrete. These values show one reason why railroad trains are in general much more fuel-efficient than highway trucks.

EXAMPLE 5–18

Motion with rolling friction A typical car weighs about 12,000 N (about 2700 lb). If the coefficient of rolling friction is $\mu_r = 0.010$, what horizontal force must you apply to push the car at constant speed on a level road? Neglect air resistance.

SOLUTION The normal force n is equal to the weight w, because the road surface is horizontal and there are no other vertical forces. From the definition of μ_r, the rolling friction force f_r is

$$f_r = \mu_r n = (0.010)(12{,}000 \text{ N}) = 120 \text{ N} \qquad \text{(about 27 lb)}.$$

From Newton's first law, a forward force with this magnitude is needed to keep the car moving with constant speed.

We invite you to apply this analysis to your crate of exercise equipment (Example 5–13). If the delivery company brings it on a rubber-wheeled dolly with $\mu_r = 0.02$, only a 10-N force is needed to keep it moving at constant velocity. Can you verify this?

FLUID RESISTANCE AND TERMINAL SPEED

Sticking your hand out the window of a fast-moving car will convince you of the existence of **fluid resistance,** the force that a fluid (a gas or liquid) exerts on a body moving through it. The moving body exerts a force on the fluid to push it out of the way. By Newton's third law, the fluid pushes back on the body with an equal and opposite force.

The *direction* of the fluid resistance force acting on a body is always opposite the direction of the body's velocity relative to the fluid. The *magnitude* of the fluid resistance force usually increases with the speed of the body through the fluid. Contrast this behavior with that of the kinetic friction force between two surfaces in contact, which we can usually regard as independent of speed. For low speeds, the magnitude f of the resisting force of the fluid is approximately proportional to the body's speed v:

$$f = kv \qquad \text{(fluid resistance at low speed)}, \qquad (5\text{–}7)$$

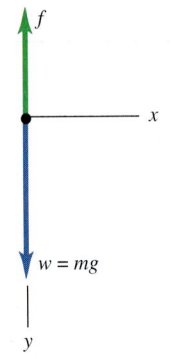

5–22 Free-body diagram for a body falling through a fluid.

where k is a proportionality constant that depends on the shape and size of the body and the properties of the fluid. In motion through air at the speed of a tossed tennis ball or faster, the resisting force is approximately proportional to v^2 rather than to v. It is then called **air drag** or simply *drag*. Airplanes, falling raindrops, and cars moving at high speed all experience air drag. In this case we replace Eq. (5–7) by

$$f = Dv^2 \qquad \text{(fluid resistance at high speed).} \qquad (5\text{–}8)$$

Because of the v^2 dependence, air drag increases rapidly with increasing speed. The air drag on a typical car is negligible at low speeds but comparable to or greater than the force of rolling resistance at highway speeds. The value of D depends on the shape and size of the body and on the density of the air.

We invite you to show that the units of the constant k in Eq. (5–7) are N · s/m or kg/s and that the units of the constant D in Eq. (5–8) are N · s²/m² or kg/m.

Because of the effects of fluid resistance, an object falling in a fluid will *not* have a constant acceleration. To describe its motion, we can't use the constant-acceleration relationships from Chapter 2; instead, we have to start over, using Newton's second law. Let's consider the following situation. You release a rock at the surface of a deep pond, and it falls to the bottom. The fluid resistance force in this situation is given by Eq. (5–7). What are the acceleration, velocity, and position of the rock as functions of time?

The free-body diagram is shown in Fig. 5–22. We take the positive direction to be downward and neglect any force associated with buoyancy in the water. There are no x-components, and Newton's second law gives

$$\Sigma F_y = mg + (-kv) = ma.$$

When the rock first starts to move, $v = 0$, the resisting force is zero, and the initial acceleration is $a = g$. As its speed increases, the resisting force also increases until finally it is equal in magnitude to the weight. At this time, $mg - kv = 0$, the acceleration becomes zero, and there is no further increase in speed. The final speed v_t, called the **terminal speed**, is given by $mg - kv_t = 0$, or

$$v_t = \frac{mg}{k} \qquad \text{(terminal speed, fluid resistance } f = kv). \qquad (5\text{–}9)$$

Figure 5–23 shows how the acceleration, velocity, and position vary with time. As time goes by, the acceleration approaches zero, and the velocity approaches v_t (remember that we chose the positive y-direction to be down). The slope of the graph of y versus t becomes constant as the velocity becomes constant.

To see how the graphs in Fig. 5–23 are derived, we must find the relation between speed and time during the interval before the terminal speed is reached. We go back to Newton's second law, which we rewrite as

$$m\frac{dv}{dt} = mg - kv.$$

After rearranging terms and replacing mg/k by v_t, we integrate both sides, noting that $v = 0$ when $t = 0$:

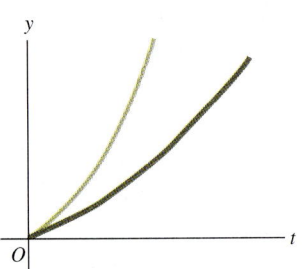

5–23 Graphs of acceleration, velocity, and position versus time for a body falling with fluid resistance proportional to v, shown as dark color curves. The light color curves show the corresponding relations if there is *no* fluid resistance.

$$\int_0^v \frac{dv}{v - v_t} = -\frac{k}{m} \int_0^t dt,$$

which integrates to

$$\ln \frac{v_t - v}{v_t} = -\frac{k}{m} t, \qquad \text{or} \qquad 1 - \frac{v}{v_t} = e^{-(k/m)t},$$

and finally

$$v = v_t\left[1 - e^{-(k/m)t}\right]. \qquad (5\text{–}10)$$

2.1.2
Skydiver

Note that v becomes equal to the terminal speed v_t only in the limit that $t \to \infty$; the rock cannot attain terminal speed in any finite length of time.

The derivative of v gives a as a function of time, and the integral of v gives y as a function of time. We leave the derivations for you to complete (see Exercise 5–40); the results are

$$a = ge^{-(k/m)t}, \qquad (5\text{–}11)$$

$$y = v_t\left[t - \frac{m}{k}(1 - e^{-(k/m)t})\right]. \qquad (5\text{–}12)$$

Now look again at Fig. 5–23, which shows graphs of these three relations.

In deriving the terminal speed in Eq. (5–9), we assumed that the fluid resistance force was proportional to the speed. For an object falling through the air at high speeds, so that the fluid resistance is proportional to v^2 as in Eq. (5–8), we invite you to show that the terminal speed v_t is given by

$$v_t = \sqrt{\frac{mg}{D}} \qquad \text{(terminal speed, fluid resistance } f = Dv^2). \qquad (5\text{–}13)$$

This expression for terminal speed explains the observation that heavy objects in air tend to fall faster than light objects. Two objects with the same physical size but different mass (say, a table-tennis ball and a lead ball with the same radius) have the same value of D but different values of m. The more massive object has a larger terminal speed and falls faster. The same idea explains why a sheet of paper falls faster if you first crumple it into a ball; the mass m is the same, but the smaller size makes D smaller (less air drag for a given speed) and v_t larger.

By changing the positions of their arms and legs while falling, sky-divers can change the value of the constant D in Eq. (5–8) and hence adjust the terminal speed of their fall (Eq. (5–13)).

EXAMPLE 5–19

Terminal speed of a sky diver For a human body falling through air in a spread-eagle position, the numerical value of the constant D in Eq. (5–8) is about 0.25 kg/m. For an 80-kg sky diver the terminal velocity is, from Eq. (5–13),

$$v_t = \sqrt{\frac{mg}{D}} = \sqrt{\frac{(80 \text{ kg})(9.8 \text{ m/s}^2)}{0.25 \text{ kg/m}}}$$

$$= 56 \text{ m/s} \qquad \text{(about 200 km/h, or 125 mi/h).}$$

When the sky diver deploys the parachute, the value of D increases greatly, and the terminal speed of the sky diver and parachute is (thankfully) much less than 56 m/s.

5–5 DYNAMICS OF CIRCULAR MOTION

We talked about uniform circular motion in Section 3–5. We showed that when a particle moves in a circular path with constant speed, the particle's acceleration is always directed toward the center of the circle (perpendicular to the instantaneous velocity). The magnitude a_{rad} of the acceleration is constant and is given in terms of the speed v and the radius R of the circle by

$$a_{rad} = \frac{v^2}{R}. \qquad (5\text{–}14)$$

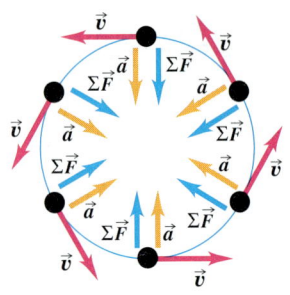

5–24 In uniform circular motion, both the acceleration and the net force are directed toward the center of the circle.

The subscript "rad" is a reminder that at each point the acceleration is radially inward toward the center of the circle, perpendicular to the instantaneous velocity. We explained in Section 3–5 why this acceleration is sometimes called *centripetal acceleration*.

We can also express the centripetal acceleration a_{rad} in terms of the *period T*, the time for one revolution:

$$T = \frac{2\pi R}{v}. \tag{5–15}$$

In terms of the period, a_{rad} is

$$a_{rad} = \frac{4\pi^2 R}{T^2}. \tag{5–16}$$

Uniform circular motion, like all other motion of a particle, is governed by Newton's second law. The particle's acceleration toward the center of the circle must be caused by a force, or several forces, such that the vector sum $\Sigma \vec{F}$ is a vector directed always toward the center (Fig. 5–24). The magnitude of the acceleration is constant, so the magnitude F_{net} of the net inward radial force must also be constant. If an ice skater whirls his partner around in a circle on the ice, he must constantly pull on her inward toward the center of the circle. If he lets go, then the inward force no longer acts, and his partner flies off in a straight line tangent to the circle (Fig. 5–25).

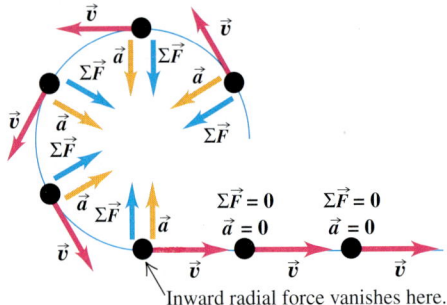

Inward radial force vanishes here.

5–25 If the inward radial force suddenly ceases to act on a body in circular motion, the body flies off in a straight line with constant velocity (as it must, since the net force acting is zero).

4.4
Circular Motion Problem Solving

4.5
Cart Goes Over Circular Path

4.6
Ball Swings on a String

4.9
Bag Hits Bystander

The magnitude of the radial acceleration is given by $a_{rad} = v^2/R$, so the magnitude of the net inward radial force F_{net} on a particle with mass m must be

$$F_{net} = ma_{rad} = m\frac{v^2}{R}. \tag{5–17}$$

Uniform circular motion can result from *any* combination of forces, just so the net force $\Sigma \vec{F}$ is always directed toward the same point at the center of the circle and has a constant magnitude.

CAUTION ▶ In doing problems involving uniform circular motion, you may be tempted to include an extra outward force of magnitude $m(v^2/R)$ to "keep the body out there" or to "keep it in equilibrium"; this outward force is usually called "*centrifugal force*" ("centrifugal" means "fleeing from the center"). Resist this temptation, because this approach is simply *wrong!* First, the body does *not* "stay out there"; it is in constant motion around its circular path. Its velocity is constantly changing in direction, so it accelerates and is *not* in equilibrium. Second, if there *were* an additional outward ("centrifugal") force to balance the inward force, there would then be *no* net inward force to cause the circular motion, and the body would move in a straight line, not a circle (Fig. 5–25). Third, the quantity $m(v^2/R)$ is *not* a force; it corresponds to the $m\vec{a}$ side of $\Sigma \vec{F} = m\vec{a}$ and *does not* appear in $\Sigma \vec{F}$. It's certainly true that a passenger riding in a car going around a circular path on a level road tends to slide to the outside of the turn, as though responding to a "centrifugal force." But such a passenger is in an accelerating, non-inertial frame of reference in which Newton's first and second laws don't apply. As we discussed in Section 4–3, what really happens is that the passenger tends to keep moving in a straight line, and the outer side of the car "runs into" the passenger as the car turns (Fig. 4–8c). *In an inertial frame of reference there is no such thing as a "centrifugal force" acting on a body.* We promise not to mention this term again, and we strongly advise you to avoid using it as well. ◄

EXAMPLE 5–20

Force in uniform circular motion A small plastic box with a mass of 0.300 kg revolves uniformly in a circle on a horizontal frictionless surface such as an air-hockey table (Fig. 5–26a). The box is attached by a cord 0.140 m long to a pin set in the surface. If the box makes two complete revolutions per second, find the force F exerted on it by the cord.

SOLUTION To calculate the force F using Newton's second law, we first need to find the magnitude of the centripetal acceleration. The period is $T = (1.00 \text{ s})/(2 \text{ rev}) = 0.500 \text{ s}$, so from Eq. (5–16) the acceleration is

$$a_{\text{rad}} = \frac{4\pi^2 R}{T^2} = \frac{4\pi^2(0.140 \text{ m})}{(0.500 \text{ s})^2} = 22.1 \text{ m/s}^2.$$

Alternatively, we can first use Eq. (5–15) to find the speed v:

$$v = \frac{2\pi R}{T} = \frac{2\pi(0.140 \text{ m})}{0.500 \text{ s}} = 1.76 \text{ m/s}.$$

Then, using Eq. (5–14),

$$a_{\text{rad}} = \frac{v^2}{R} = \frac{(1.76 \text{ m/s})^2}{0.140 \text{ m}} = 22.1 \text{ m/s}^2.$$

The body has no vertical acceleration, so the vertical forces $\vec{n}$ and $\vec{w}$ in Fig. 5–26b add to zero. The only force toward the center of the circle is the tension $\vec{F}$ in the cord, and its magnitude is

$$F = ma_{\text{rad}} = (0.300 \text{ kg})(22.1 \text{ m/s}^2)$$
$$= 6.63 \text{ kg} \cdot \text{m/s}^2 = 6.63 \text{ N}.$$

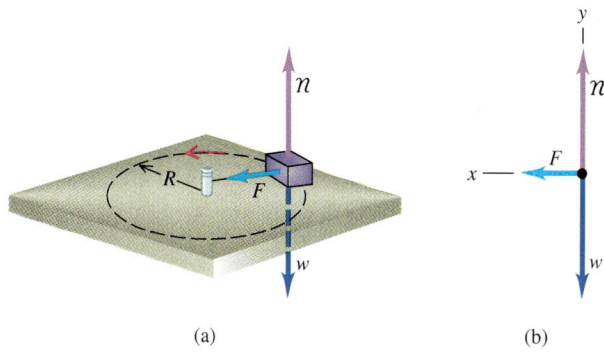

(a)   (b)

5–26 (a) A box moving in uniform circular motion on a horizontal frictionless surface. (b) Free-body diagram for the box. The $+x$-direction is toward the center of the circle.

EXAMPLE 5–21

The conical pendulum An inventor who dares to be different proposes to make a pendulum clock using a pendulum bob with mass m at the end of a thin wire of length L. Instead of swinging back and forth, the bob moves in a horizontal circle with constant speed v, with the wire making a constant angle β with the vertical direction (Fig. 5–27a). Assuming that the time T for one revolution (that is, the period) is known, find the tension F in the wire and the angle β.

SOLUTION A free-body diagram for the bob and a coordinate system are shown in Fig. 5–27b. The center of the circular path is at the center of the shaded area, *not* at the top end of the wire. The forces on the bob in the position shown are the weight mg and the tension F in the wire. The tension has a horizontal component $F \sin \beta$ and a vertical component $F \cos \beta$. (This free-body diagram is exactly like the one in Fig. 5–14b, but in this case the acceleration a is the *radial* acceleration, v^2/R.)

The system has no vertical acceleration, so the sum of the vertical components of force is zero. The horizontal force is directed toward the center of the circle; it must equal the mass m times the radial acceleration $a_{\text{rad}} = v^2/R$. Thus the $\Sigma \vec{F} = m\vec{a}$ equations are

$$\Sigma F_x = F \sin \beta = m\frac{v^2}{R}, \qquad \Sigma F_y = F \cos \beta + (-mg) = 0.$$

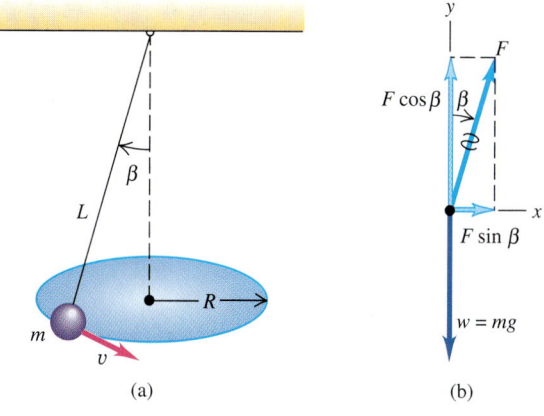

5–27 (a) The bob on the end of the wire moves in uniform circular motion. (b) Free-body diagram for the bob. The $+x$-direction is toward the center of the circle.

(a)   (b)

These are two simultaneous equations for the unknowns F and β. The equation for ΣF_y gives $F = mg/\cos\beta$; substituting this result into the equation for ΣF_x, we find

$$\tan\beta = \frac{v^2}{gR}.$$

We can rewrite this expression to relate β to the period T. The radius of the circle is $R = L\sin\beta$, and the speed is the circumference $2\pi L\sin\beta$ divided by the period T:

$$v = \frac{2\pi L\sin\beta}{T}.$$

Substituting this into $\tan\beta = v^2/gR$, we obtain

$$\cos\beta = \frac{gT^2}{4\pi^2 L},$$

or

$$T = 2\pi\sqrt{\frac{L\cos\beta}{g}}.$$

Once we know β, we can find the tension F from $F = mg/\cos\beta$. For a given length L, as the period T becomes smaller, $\cos\beta$ decreases, the angle β increases, and the tension $F = mg/\cos\beta$ increases. The angle can never be 90°, however; this would require that $T = 0$, $F = \infty$, and $v = \infty$. This system is called a *conical pendulum* because the suspending wire traces out a cone. It would not make a very good clock because the period depends on β in such a direct way.

EXAMPLE 5-22

Rounding a flat curve The BMW Z3 roadster in Example 3–11 (Section 3–5) is rounding a flat, unbanked curve with radius R (Fig. 5–28a). If the coefficient of friction between tires and road is μ_s, what is the maximum speed v_{max} at which the driver can take the curve without sliding?

SOLUTION Figure 5–28b shows a free-body diagram for the car. The acceleration v^2/R toward the center of the curve must be caused by the friction force f, and there is no vertical acceleration. Thus we have

$$\Sigma F_x = f = m\frac{v^2}{R}, \qquad \Sigma F_y = n - mg = 0.$$

The first equation shows that the friction force *needed* to keep the car moving in its circular path increases with the car's speed. But the maximum friction force *available* is $f_{max} = \mu_s n = \mu_s mg$, which is constant, and this determines the car's maximum speed. Substituting f_{max} for f and v_{max} for v in the ΣF_x equation, we find

$$\mu_s mg = m\frac{v_{max}^2}{R},$$

so the maximum speed is

$$v_{max} = \sqrt{\mu_s gR}.$$

If $\mu_s = 0.87$ and $R = 230$ m, then

$$v_{max} = \sqrt{(0.87)(9.8 \text{ m/s}^2)(230 \text{ m})} = 44 \text{ m/s},$$

or about 160 km/h (100 mi/h). This is the maximum speed for this radius. Note that the maximum centripetal acceleration (called the "lateral acceleration" in Example 3–11) is equal to $\mu_s g$. Why do we use the coefficient of *static* friction for a moving car?

If we take the curve at a slower speed than $\sqrt{\mu_s gR}$, the friction force is *less* than the maximum possible value $f_{max} = \mu_s mg$. If we try to take the curve going *faster* than the maximum speed, the car can still go in a circle without skidding, but the radius will be larger and the car will run off the road.

(a)

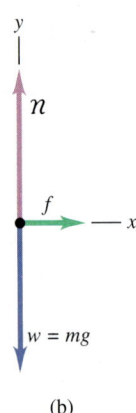

(b)

5-28 (a) A sports car rounding a curve on a level road. (b) Free-body diagram for the car. The +x-direction is toward the center of the circle.

EXAMPLE 5-23

Rounding a banked curve For a car traveling at a certain speed, it is possible to bank a curve at just the right angle so that *no* friction at all is needed. Then a car can round the curve even on wet ice with Teflon tires. Bobsled racing depends on this same idea. An engineer proposes to rebuild the curve in Example 5–22 so that a car moving at speed v can safely make the turn even if there is no friction (Fig. 5–29a). At what angle β should the curve be banked?

SOLUTION The free-body diagram is shown in Fig. 5–29b. The normal force $\vec{n}$ is no longer vertical but is perpendicular to the roadway at an angle β with the vertical. Thus it has a vertical component $n \cos \beta$ and a horizontal component $n \sin \beta$, as shown. We want to get around the curve without relying on friction, so no friction force is included. The horizontal component of $\vec{n}$ must now cause the acceleration v^2/R; there is no vertical acceleration. Thus

$$\Sigma F_x = n \sin \beta = \frac{mv^2}{R}, \qquad \Sigma F_y = n \cos \beta + (-mg) = 0.$$

From the ΣF_y equation, $n = mg/\cos \beta$. Substituting this into the ΣF_x equation gives an expression for the banking angle:

$$\tan \beta = \frac{v^2}{gR}.$$

The banking angle depends on the speed and the radius. For a given radius, no one angle is correct for all speeds. In the design of highways and railroads, curves are often banked for the *average* speed of the traffic over them. If $R = 230$ m and $v = 25$ m/s (a highway speed of 88 km/h, or 55 mi/h), then

$$\beta = \arctan \frac{(25 \text{ m/s})^2}{(9.8 \text{ m/s}^2)(230 \text{ m})} = 15°.$$

This is within the range of banking angles actually used in highways. With the same radius and $v = 44$ m/s, as in Example 5–22, $\beta = 41°$; such steeply banked curves are found at automobile raceways.

Note that the expression for the banking angle β is the same as that found for the angle of a conical pendulum in Example 5–21. In fact, the free-body diagrams of Figs. 5–27b and 5–29b are very similar; the normal force acting on the car plays the role of the tension acting on the pendulum bob.

The same considerations apply to the correct banking angle of an airplane when it makes a turn in level flight. The free-body diagram for the pilot of the airplane is exactly as shown in Fig. 5–29b; the normal force $n = mg/\cos \beta$ is exerted on the pilot by the seat. As in Example 5–8, n is equal to the *apparent weight* of the pilot, which is greater than his true weight mg. For a jet fighter making a tight turn (small R) at high speed (large v), $\tan \beta = v^2/gR$ will be large, and the required angle β will approach 90°. The pilot's apparent weight will then be tremendous, $5.8mg$ at $\beta = 80°$ and $9.6mg$ at $\beta = 84°$. Pilots black out in such tight turns because the apparent weight of their blood increases by the same factor, and the human heart isn't strong enough to pump such apparently "heavy" blood to the brain.

(a)

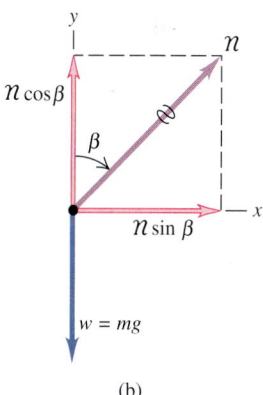

(b)

5–29 (a) A sports car rounding a banked curve. If the curve is banked at the correct angle β, no friction is needed to round the curve at speed v. (b) Free-body diagram for the car. The $+x$-direction is toward the center of the circle.

MOTION IN A VERTICAL CIRCLE

In all the preceding examples the body moved in a horizontal circle. Motion in a vertical circle is no different in principle, but the weight of the body has to be treated carefully. The following example shows what we mean.

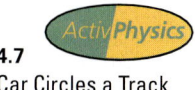

4.7
Car Circles a Track

EXAMPLE 5-24

Uniform circular motion in a vertical circle A passenger on a carnival Ferris wheel moves in a vertical circle of radius R with constant speed v. Assuming that the seat remains upright during the motion, derive expressions for the force the seat exerts on the passenger at the top of the circle and at the bottom.

SOLUTION Figure 5–30 is a diagram of the situation, with free-body diagrams for the two positions. We take the positive y-direction as upward in both cases. Let n_T be the upward normal force the seat applies to the passenger at the top of the circle, and let n_B be the normal force at the bottom. At the top the acceleration has magnitude v^2/R, but its vertical component is negative because its direction is downward, toward the center of the circle. So the $\Sigma F_y = ma_y$ equation at the top is

$$\Sigma F_y = n_T + (-mg) = m\left(-\frac{v^2}{R}\right), \qquad \text{or}$$

$$n_T = m\left(g - \frac{v^2}{R}\right).$$

This means that at the top of the Ferris wheel, the upward force the seat applies to the passenger is *smaller* in magnitude than the passenger's weight. If the ride goes fast enough that $g - v^2/R$ becomes zero, the seat applies *no* force, and the passenger is about to become airborne. If v becomes still larger, n_T becomes negative; this means that a *downward* force (such as from a seat belt) is needed to keep the passenger in the seat.

At the bottom the acceleration is upward, and Newton's second law is

$$\Sigma F_y = n_B + (-mg) = m\left(+\frac{v^2}{R}\right), \qquad \text{or}$$

$$n_B = m\left(g + \frac{v^2}{R}\right).$$

At this point the upward force provided by the seat is always *greater than* the passenger's weight. You feel the seat pushing up on you more firmly than when you are at rest.

We recognize that n_T and n_B are the values of the passenger's *apparent weight* at the top and bottom of the circle, respectively. As a specific example, suppose the passenger's mass is 60.0 kg, the radius of the circle is $R = 8.00$ m, and the wheel makes one revolution in 10.0 s. From Eq. (5–16),

$$a_{\text{rad}} = \frac{4\pi^2 R}{T^2} = \frac{4\pi^2 (8.00 \text{ m})}{(10.0 \text{ s})^2} = 3.16 \text{ m/s}^2.$$

The two values of the apparent weight are

$$n_T = (60.0 \text{ kg})(9.80 \text{ m/s}^2 - 3.16 \text{ m/s}^2) = 398 \text{ N},$$

$$n_B = (60.0 \text{ kg})(9.80 \text{ m/s}^2 + 3.16 \text{ m/s}^2) = 778 \text{ N}.$$

The passenger's weight is 588 N, and n_T and n_B are about 30% less than and greater than the weight, respectively.

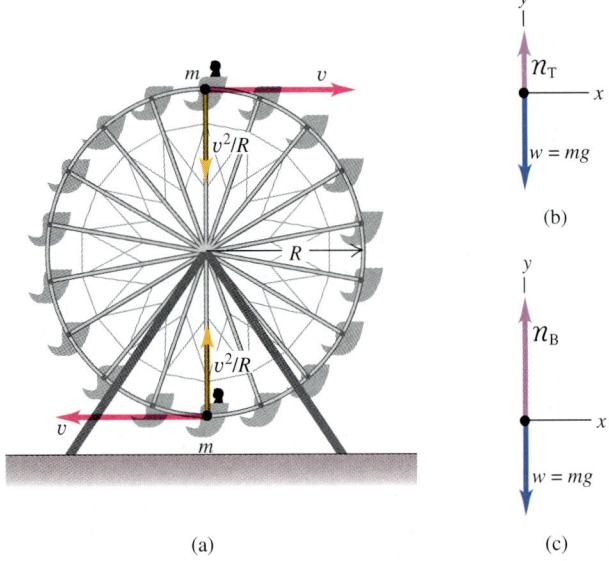

5–30 (a) The passenger's acceleration vector has the same magnitude at the top and bottom of the Ferris wheel and in both cases points toward the center of the circle. (b) Free-body diagram for a passenger at the top of the circle. (c) Free-body diagram for a passenger at the bottom of the circle.

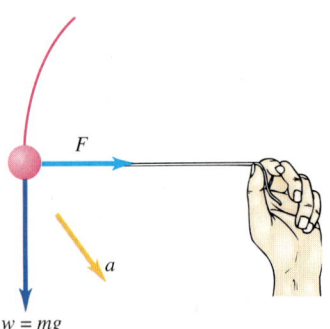

5–31 A ball moving in a vertical circle. At a typical point in the circle, there is a component of the net force (and hence a component of acceleration) tangent to the circle. The ball is in *non-uniform* circular motion.

When we tie a string to an object and whirl it in a vertical circle, the preceding analysis isn't directly applicable. The reason is that v is *not* constant in this case; at every point in the circle except the top and bottom, the net force (and hence the acceleration) does *not* point toward the center of the circle (Fig. 5–31). So both $\Sigma \vec{F}$ and $\vec{a}$ have a component tangent to the circle, and this is a case of *non-uniform* circular motion (see Section 3–5). Even worse, we can't use the constant-acceleration formulas to relate the speeds at various points because *neither* the magnitude nor the direction of the acceleration is constant. The speed relations that we need are best obtained by using the concept of energy. We'll consider such problems in Chapter 7.

*5–6 THE FUNDAMENTAL FORCES OF NATURE

We have discussed several different kinds of forces—including weight, tension, friction, fluid resistance, and the normal force—and we will encounter others as we continue our study of physics. But just how many different kinds of forces are there? Our current understanding is that all forces are expressions of just four distinct classes of *fundamental* forces, or interactions between particles (Fig. 5–32). Two are familiar in everyday experience. The other two involve interactions between subatomic particles that we cannot observe with the unaided senses.

Of the two familiar classes, **gravitational interactions** were the first to be studied in detail. The *weight* of a body results from the earth's gravitational attraction acting on it. The sun's gravitational attraction for the earth keeps the earth in its nearly circular orbit around the sun. Newton recognized that the motions of the planets around the sun and the free fall of objects on earth are both the result of gravitational forces. In Chapter 12 we will study gravitational interactions in greater detail, and we will analyze their vital role in the motions of planets and satellites.

The second familiar class of forces, **electromagnetic interactions,** includes electric and magnetic forces. If you run a comb through your hair, you can then use the comb to pick up bits of paper or fluff; this interaction is the result of electric charge on the comb. All atoms contain positive and negative electric charge, so atoms and molecules can exert electric forces on each other. Contact forces, including the normal force, friction, and fluid resistance, are the combination of all such forces exerted on the atoms of a body by atoms in its surroundings. *Magnetic* forces occur in interactions between magnets or between a magnet and a piece of iron. These may seem to form a different category, but magnetic interactions are actually the result of electric charges in motion. In an electromagnet an electric current in a coil of wire causes magnetic interactions. We will study electric and magnetic interactions in detail in the second half of this book.

These two interactions differ enormously in their strength. The electrical repulsion between two protons at a given distance is stronger than their gravitational attraction by a factor of the order of 10^{35}. Gravitational forces play no significant role in atomic or molecular structure. But in bodies of astronomical size, positive charge and negative charge are usually present in nearly equal amounts, and the resulting electrical interactions nearly cancel each other out. Gravitational interactions are thus the dominant influence in the motion of planets and in the internal structure of stars.

The other two classes of interactions are less familiar. One, the **strong interaction,** is responsible for holding the nucleus of an atom together. Nuclei contain electrically

(a)

(b)

5–32 Examples of the fundamental interactions in nature.
(a) Gravitational forces cause the orbital motions of the earth around the sun and of the moon around the earth. (b) Electromagnetic forces act between atoms to form molecules, as in this atomic force micrograph of bacterial plasmid DNA. The smallest details seen here are 3 to 4 nm in size.
(c) Strong forces between nuclear particles are responsible for thermonuclear reactions in the center of the sun; the released energy reaches us as sunlight. (d) Weak forces, characteristic of interactions involving subatomic particles called neutrinos, play a crucial role when a star (shown on the left by an arrow) explodes and becomes a supernova (shown on the right).

(c)

(d)

neutral neutrons and positively charged protons. The charged protons repel each other, and a nucleus could not be stable if it were not for the presence of an attractive force of a different kind that counteracts the repulsive electrical interactions. In this context the strong interaction is also called the *nuclear force.* It has much shorter range than electrical interactions, but within its range it is much stronger. The strong interaction is also responsible for the creation of unstable particles in high-energy particle collisions.

Finally, there is the **weak interaction.** It plays no direct role in the behavior of ordinary matter, but it is of vital importance in interactions among fundamental particles. The weak interaction is responsible for a common form of radioactivity called beta decay, in which a neutron in a radioactive nucleus is transformed into a proton while ejecting an electron and an essentially massless particle called an antineutrino. The weak interaction between the antineutrino and ordinary matter is so feeble that an antineutrino could easily penetrate a wall of lead a million kilometers thick!

During the past several decades a unified theory of the electromagnetic and weak interactions has been developed. We now speak of the *electroweak* interaction, and in a sense this reduces the number of classes of interactions from four to three. Similar attempts have been made to understand strong, electromagnetic, and weak interactions on the basis of a single unified theory called a *grand unified theory* (GUT), and the first tentative steps have been taken toward a possible unification of all interactions into a *theory of everything* (TOE). Such theories are still speculative, and there are many unanswered questions in this very active field of current research.

5-7 PROJECTILE MOTION WITH AIR RESISTANCE

A Case Study in Computer Analysis

In our study of projectile motion in Chapter 3, we assumed that air-resistance effects are negligibly small. But in fact air resistance (often called *air drag,* or simply *drag*) has a major effect on the motion of many objects, including tennis balls, bicycle riders, and airplanes. In Section 5-4 we considered how a fluid resistance force affected a body falling straight down. We'd now like to extend this analysis to a projectile moving in a plane.

It's not difficult to include the force of air resistance in the equations for a projectile, but solving them for the position and velocity as functions of time, or the shape of the path, can get quite complex. Fortunately, it is fairly easy to make quite precise numerical approximations to these solutions, using a computer. That's what this section is about.

When we omitted air drag, the only force acting on a projectile with mass m was its weight $\vec{w} = m\vec{g}$. The components of the projectile's acceleration were simply

$$a_x = 0, \qquad a_y = -g.$$

The $+x$-axis is horizontal, and the $+y$-axis is vertically upward. We must now include the air drag force acting on the projectile. At the speed of a tossed tennis ball or faster, the magnitude f of the air drag force is approximately proportional to the square of the projectile's speed relative to the air, as discussed in Section 5-4:

$$f = Dv^2, \tag{5-8}$$

where $v^2 = v_x^2 + v_y^2$. We'll assume that the air is still, so $\vec{v}$ is the velocity of the projectile relative to the ground as well as to the air (Fig. 5-33a). The direction of $\vec{f}$ is opposite the direction of $\vec{v}$, so we can write $\vec{f} = -Dv\vec{v}$ and the components of $\vec{f}$ are

$$f_x = -Dvv_x, \qquad f_y = -Dvv_y.$$

(a)

5-33 (a) A projectile in flight with velocity $\vec{v}$ relative to the ground.

Note that each component is opposite in direction to the corresponding component of velocity and that $f = \sqrt{f_x^2 + f_y^2}$. The free-body diagram is shown in Fig. 5–33b. Newton's second law gives

$$\Sigma F_x = -Dvv_x = ma_x, \qquad \Sigma F_y = -mg - Dvv_y = ma_y,$$

and the components of acceleration including the effects of both gravity and air drag are

$$a_x = -(D/m)vv_x, \qquad a_y = -g - (D/m)vv_y. \qquad (5\text{–}18)$$

The constant D depends on the density ρ of air, the silhouette area A of the body (its area as seen from the front), and a dimensionless constant C called the **drag coefficient** that depends on the shape of the body. Typical values of C for baseballs, tennis balls, and the like are in the range from 0.2 to 1.0. In terms of these quantities, D is given by

$$D = \frac{\rho CA}{2}. \qquad (5\text{–}19)$$

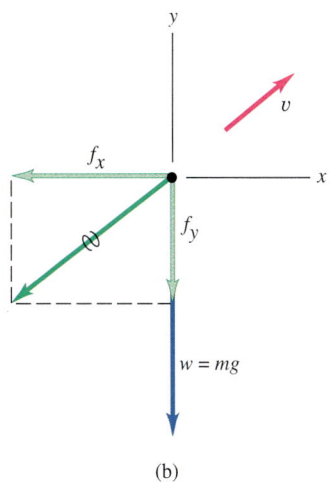

5–33 (b) Free-body diagram for the projectile. In still air, the air drag force $\vec{f}$ is always directed opposite $\vec{v}$.

Now comes the basic idea of our numerical calculation. The acceleration components a_x and a_y are constantly changing as the velocity components change. But over a sufficiently short time interval Δt, we can regard the acceleration as essentially constant. If we know the coordinates and velocity components at some time t, we can find these quantities at a slightly later time $t + \Delta t$ using the formulas for constant acceleration. Here's how we do it. During a time interval Δt, the average x-component of acceleration is $a_x = \Delta v_x / \Delta t$ and the x-velocity v_x changes by an amount $\Delta v_x = a_x \Delta t$. Similarly, v_y changes by an amount $\Delta v_y = a_y \Delta t$. So the values of the x-velocity and y-velocity at the end of the interval are

$$v_x + \Delta v_x = v_x + a_x \Delta t, \qquad v_y + \Delta v_y = v_y + a_y \Delta t. \qquad (5\text{–}20)$$

While this is happening, the projectile is moving, so the coordinates are also changing. The average x-velocity during the time interval Δt is the average of the value v_x (at the beginning of the interval) and $v_x + \Delta v_x$ (at the end of the interval), or $v_x + \Delta v_x / 2$. During Δt the coordinate x changes by an amount

$$\Delta x = (v_x + \Delta v_x / 2)\Delta t = v_x \Delta t + \frac{1}{2} a_x (\Delta t)^2,$$

and similarly for y. (Compare this to Eq. (2–12) for motion with constant acceleration; it's exactly the same expression.) So the coordinates of the projectile at the end of the interval are

$$x + \Delta x = x + v_x \Delta t + \frac{1}{2} a_x (\Delta t)^2, \qquad y + \Delta y = y + v_y \Delta t + \frac{1}{2} a_y (\Delta t)^2. \qquad (5\text{–}21)$$

We have to specify the starting conditions, that is, the initial values of x, y, v_x, and v_y. Then we can step through the calculation to find the position and velocity at the end of each interval in terms of their values at the beginning, and thus to find the values at the end of any number of intervals. It would be a lot of work to do all the calculations for 100 or more steps with a hand calculator, but the computer does it for us quickly and accurately.

Exactly how you implement this plan depends on whether you use a computer language such as C++, BASIC, FORTRAN, or Pascal or use a spreadsheet or a numerical-analysis software package such as MathCad. Here's a sketch of a general algorithm using a programming language.

Step 1: Identify the parameters of the problem, m, A, C, and ρ, and evaluate D.
Step 2: Choose the time interval Δt and the initial values of x, y, v_x, v_y and t. You may want to express the initial velocity components in terms of the magnitude v_0 and direction α_0 of the initial velocity.

Step 3: Choose the maximum number of intervals N or the maximum time $t_{max} = N\Delta t$ for which you want to get the numerical solution.

Step 4: Loop (or iterate) Steps 5 through 9 while $n < N$ or $t < t_{max}$:

    Step 5: Calculate the acceleration components.

$$a_x = -(D/m)vv_x, \qquad a_y = -g - (D/m)vv_y.$$

Step 6: Print or plot x, y, v_x, v_y, a_x, and a_y.

Step 7: Calculate the new velocity components from Eqs. (5–20):

$$v_x + \Delta v_x = v_x + a_x\Delta t, \qquad v_y + \Delta v_y = v_y + a_y\Delta t.$$

Step 8: Calculate the new coordinates from Eqs. (5–21):

$$x + \Delta x = x + v_x\Delta t + \frac{1}{2}a_x(\Delta t)^2, \qquad y + \Delta y = y + v_y\Delta t + \frac{1}{2}a_y(\Delta t)^2.$$

Step 9: Increment the time by Δt:

$$t = t + \Delta t.$$

Step 10: Stop.

If you are using a spreadsheet, the following notation may be useful. Let $x(n)$, $y(n)$, $v_x(n)$, and $v_y(n)$ be the values at the end of the nth interval. These values appear across the nth line of the spreadsheet. The initial values are $x(1)$, $y(1)$, $v_x(1)$, and $v_y(1)$. Then Eqs. (5–20) and (5–21) become

$$v_x(n + 1) = v_x(n) + a_x(n)\Delta t, \qquad x(n + 1) = x(n) + v_x(n)\Delta t + \frac{1}{2}a_x(n)(\Delta t)^2 \quad (5\text{–}22)$$

$$v_y(n + 1) = v_y(n) + a_y(n)\Delta t, \qquad y(n + 1) = y(n) + v_y(n)\Delta t + \frac{1}{2}a_y(n)(\Delta t)^2. \quad (5\text{–}23)$$

In Eqs. (5–22) and (5–23) we compute $a_x(n)$ and $a_y(n)$ by substituting the values $v_x(n)$ and $v_y(n)$ into Eqs. (5–18).

Figure 5–34 shows the trajectory of a baseball with and without air drag. The radius of the baseball is $r = 0.0366$ m, and $A = \pi r^2$. The mass is $m = 0.145$ kg, and we have chosen the drag coefficient to be $C = 0.5$ (appropriate for a batted ball or pitched fastball) and the density of air to be $\rho = 1.2$ kg/m^3 (appropriate for a ballpark at sea level). In this example the baseball was given an initial velocity of 50 m/s at an angle of 35° above the $+x$-axis. You can see that both the range of the baseball and the maximum height reached are substantially less than the zero-drag calculation would lead you to believe. Hence the baseball trajectory that we calculated in Example 3–7 (Section 3–4) is quite unrealistic, since there we ignored air drag completely. Air drag is an important part of the game of baseball!

5–34 Computer-generated trajectories of a baseball with and without drag. Note that different scales are used on the horizontal and vertical axes.

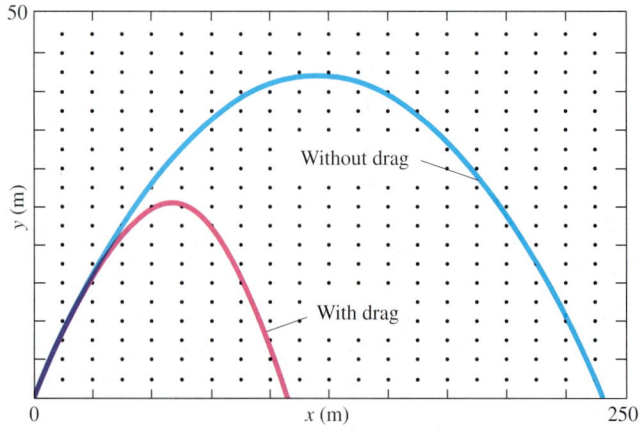

SUMMARY

- When a body is in equilibrium in an inertial frame of reference, the vector sum of forces acting on it must be zero. In component form,

$$\Sigma F_x = 0, \qquad \Sigma F_y = 0 \qquad \text{(particle in equilibrium)}. \qquad (5\text{–}2)$$

Free-body diagrams are useful in identifying the forces acting on the body being considered. Newton's third law is also frequently needed in equilibrium problems. The two forces in an action-reaction pair *never* act on the same body.

- When the vector sum of forces on a body is not zero, the body has an acceleration determined by Newton's second law:

$$\Sigma \vec{F} = m\vec{a}. \qquad (5\text{–}3)$$

In component form,

$$\Sigma F_x = ma_x, \qquad \Sigma F_y = ma_y. \qquad (5\text{–}4)$$

- The contact force between two bodies can always be represented in terms of a normal force $\vec{n}$ perpendicular to the interaction surface and a frictional force $\vec{f}$ parallel to the surface. When sliding occurs, f_k is approximately proportional to n, and the proportionality constant is μ_k, the coefficient of kinetic friction:

$$f_k = \mu_k n \qquad \text{(magnitude of kinetic friction force)}. \qquad (5\text{–}5)$$

When there is no relative motion, the maximum possible friction force is approximately proportional to the normal force. The proportionality constant is μ_s, the coefficient of static friction:

$$f_s \leq \mu_s n \qquad \text{(magnitude of static friction force)}. \qquad (5\text{–}6)$$

The actual static friction force may be anything from zero to this maximum value, depending on the situation. Usually, μ_k is less than μ_s for a given pair of surfaces.

- In uniform circular motion, the acceleration vector is directed toward the center of the circle and has magnitude v^2/R. The motion is governed by $\Sigma \vec{F} = m\vec{a}$, just as for any other dynamics problem.

- The fundamental forces of nature are the gravitational, electromagnetic, strong, and weak interactions. The electromagnetic and weak interactions have been unified into a single interaction, the electroweak interaction.

DISCUSSION QUESTIONS

Q5–1 A man sits in a seat that is suspended from a rope. The rope passes over a pulley suspended from the ceiling, and the man holds the other end of the rope in his hands. What is the tension in the rope, and what force does the seat exert on the man? Draw a free-body force diagram for the man.

Q5–2 "In general, the normal force is not equal to the weight." Give an example where the two forces are equal in magnitude, and at least two examples where they are not.

Q5–3 A clothesline is hung between two poles. No matter how tightly the line is stretched, it always sags a little at the center. Explain why.

Q5–4 A car is being driven up a steep hill at constant speed. Discuss all the forces acting on the car. What pushes it up the hill?

Q5–5 When you tighten a nut on a bolt, how are you increasing the frictional force? How does a lock washer work?

Q5–6 In pushing a box up a ramp, is the force required smaller if you push horizontally or if you push parallel to the ramp? Why?

Q5–7 A block rests on an inclined plane with enough friction to prevent its sliding down. To start the block moving, is it easier to push it up the plane, down the plane, or sideways? Why?

Q5–8 A crate of books rests on a level floor. To move it along the floor at a constant velocity, why do you exert a smaller force if you pull it at an angle θ above the horizontal than if you push it at the same angle below the horizontal?

Q5–9 In stopping a car on an icy road, is it better to push the brake pedal hard enough to "lock" the wheels and make them slide or to push gently so that the wheels continue to roll? Why?

Q5–10 Can the force of kinetic friction acting on an object ever cause that object to *speed up?* If it cannot, explain why not. If it can, give at least one example. Repeat for the force of *static* friction.

Q5–11 When you stand with bare feet in a wet bathtub, the grip feels fairly secure, and yet a catastrophic slip is quite possible. Explain this in terms of the two coefficients of friction.

Q5–12 For medical reasons it is important for astronauts in outer space to determine their body mass at regular intervals. Devise a scheme for measuring body mass in an apparently weightless environment.

Q5–13 A woman in an elevator lets go of her briefcase but it does not fall to the floor. How is the elevator moving?

Q5–14 Scales for weighing objects can be classified as those that use springs and those that use standard masses to balance unknown masses. Which group would be the more accurate when used in an accelerating elevator? When used on the moon? Does it matter whether you are trying to determine mass or weight?

Q5–15 Fast pitchers can make a baseball move faster than its 43 m/s (96 mi/h) terminal speed. How is this possible?

Q5–16 Because of air resistance, two objects of unequal mass do *not* fall at precisely the same rate. If two bodies of identical shape but unequal mass are dropped simultaneously from the same height, which one reaches the ground first? Explain.

Q5–17 A tennis ball is dropped from rest at the top of a tall glass cylinder, first with the air pumped out of the cylinder so there is no air resistance and then a second time after the air has been readmitted to the cylinder. You examine multiflash photographs (like Fig. 2–18) of the two drops. From these photos how can you tell which one is which, or can you?

Q5–18 Roller Coaster Road near Tucson, Arizona, was so named because it resembled an amusement park ride. If you drove on it at constant speed, where was the normal force the greatest and the least? Why would it have been especially poor road design to put an unbanked right or left curve at the top of one of the hills?

Q5–19 An automotive magazine calls decreasing-radius curves "the bane of the Sunday driver." Explain.

Q5–20 If you hang fuzzy dice from your rear-view mirror and drive through a banked curve, how can you tell whether you are traveling less than, equal to, or greater than the speed used to calculate the banking angle?

Q5–21 If there is a net force on a particle in uniform circular motion, why doesn't the particle's speed change?

Q5–22 A curve in a road has the banking angle calculated for 80 km/h. However, the road is covered with ice, and you plan to creep around the highest lane at 20 km/h. What may happen to your car? Why?

Q5–23 If you swing a ball on the end of a lightweight string in a horizontal circle at constant speed, the string at any instant does not lie exactly along the radius vector from the center of the path to the ball's location. Does the string lie above or below the horizontal? In relation to the direction the ball is moving, is the string in front of or behind the radius vector? Draw a free-body force diagram for the ball and use it to explain your answers. (Note that air resistance may be a factor.)

Q5–24 The centrifugal force is not included in the free-body diagrams of Figs. 5–28b and 5–29b. Explain why not.

Q5–25 A professor swings a rubber stopper in a horizontal circle on the end of a string in front of his class. He tells Caroline in the first row that he is going to let the string go when the stopper is directly in front of her face. Should Caroline worry?

Q5–26 To keep the forces on the riders within allowable limits, loop-the-loop roller coaster rides are often designed so the loop, rather than being a perfect circle, has a larger radius of curvature at the bottom than at the top. Explain.

Q5–27 You throw a baseball straight upward. If air resistance is *not* neglected, how does the time required for the ball to go from the height at which it was thrown up to its maximum height compare to the time required for it to fall from its maximum height back down to the height from which it was thrown? Explain your answer.

Q5–28 Will the frictional force always be opposite a baseball's velocity even if the wind is blowing? Explain.

Q5–29 When can a baseball in flight have an acceleration with a positive upward component? Explain in terms of the forces on the ball and also in terms of the velocity components compared to the terminal speed. Do *not* neglect air resistance.

Q5–30 When a batted baseball moves with air drag, does it spend more time climbing to its maximum height, or descending from its maximum height back to the ground? Or is the time the same for both? Explain in terms of the forces acting on the ball.

Q5–31 When a batted baseball moves with air drag, does it travel a greater horizontal distance while climbing to its maximum height, or while descending from its maximum height back to the ground? Or is the horizontal distance traveled the same for both? Explain in terms of the forces acting on the ball.

Q5–32 "A ball is thrown from the edge of a high cliff. No matter what the angle at which it is thrown, due to air resistance the ball will eventually end up moving vertically downward." Justify this statement.

EXERCISES

SECTION 5–2 USING NEWTON'S FIRST LAW: PARTICLES IN EQUILIBRIUM

5–1 Two 25.0-N weights are suspended at opposite ends of a rope that passes over a light, frictionless pulley. The pulley is attached to a chain that goes to the ceiling. a) What is the tension in the rope? b) What is the tension in the chain?

5–2 In Fig. 5–35 each of the suspended blocks has weight w. The pulleys are frictionless and the ropes have negligible weight. Calculate in each case the tension T in the rope in terms of the weight w. In each case, include the free-body diagram or diagrams you used to determine the answer.

(a)   (b)   (c)

FIGURE 5–35 Exercise 5–2.

5–3 An adventurous archaeologist crosses between two rock cliffs by slowly going hand-over-hand along a rope stretched between the cliffs. He stops to rest at the middle of the rope (Fig. 5–36). The rope will break if the tension in it exceeds 2.50×10^4 N, and our hero's mass is 90.0 kg. a) If the angle θ is 10.0°, find the tension in the rope. b) What is the smallest value the angle θ can have if the rope is not to break?

FIGURE 5–36 Exercise 5–3.

5–4 A picture frame hung against a wall is suspended by two wires attached to its upper corners. If the two wires make the same angle with the vertical, what must this angle be if the tension in each wire is equal to 0.75 of the weight of the frame? (Neglect any friction between the wall and the picture frame.)

5–5 Solve the problem in Example 5–3 using coordinate axes where the y-axis is vertical and the x-axis is horizontal. Do you get the same answers as obtained in the example using this different set of axes?

5–6 Certain streets in San Francisco make an angle of 17.5° with the horizontal. What force parallel to the street surface is required to keep a loaded 1967 Corvette of mass 1390 kg from rolling down such a street?

5–7 A large wrecking ball is held in place by two light steel cables (Fig. 5–37). If the mass m of the wrecking ball is 4090 kg, what is a) the tension T_B in the cable that makes an angle of 40° with the vertical? b) the tension T_A in the horizontal cable?

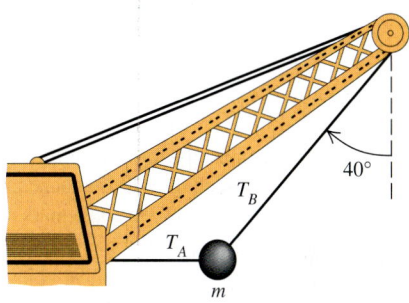

FIGURE 5–37 Exercise 5–7.

5–8 Find the tension in each cord in Fig. 5–38 if the weight of the suspended object is w.

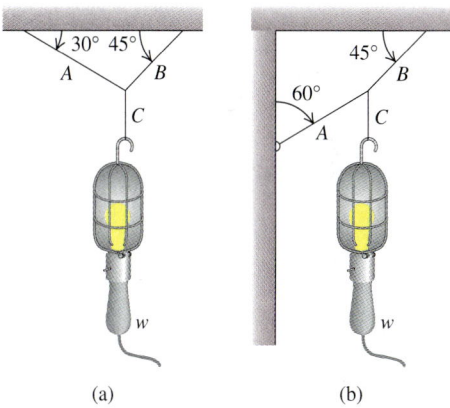

(a)   (b)

FIGURE 5–38 Exercise 5–8.

5–9 At a certain part of your drive from home to school, your car (mass 1600 kg) will coast in neutral at a constant speed of 72 km/h if there is no wind. Examination of a topographical map shows that for this stretch of straight highway the elevation decreases 200 m for each 6000 m of road. What is the total resistive force (friction plus air resistance) that acts on your car when it is traveling at 72 km/h?

5–10 A man pushes on a piano with mass 180 kg so that it slides at constant velocity down a ramp that is inclined at 11.0° above the horizontal. Neglect friction acting on the piano. If the force applied by the man is parallel to the incline, calculate the magnitude of this force.

5–11 In Fig. 5–39 the weight w is 60.0 N. a) What is the tension in the diagonal string? a) Find the magnitudes of the horizontal forces $\vec{F}_1$ and $\vec{F}_2$ that must be applied to hold the system in the position shown.

FIGURE 5–39 Exercise 5–11.

5–12 A tether ball leans against the post to which it is attached (Fig. 5–40). If the string to which the ball is attached is 1.40 m long and the ball has a radius of 0.110 m and a mass of 0.270 kg, what are the tension in the rope and the force the pole exerts on the ball? Neglect any friction between the ball and the pole. (The string is attached to the ball such that a line along the string passes through the center of the ball.)

FIGURE 5–40 Exercise 5–12.

5–13 Two blocks, each with weight w, are held in place on a frictionless incline (Fig. 5–41). In terms of w and the angle α of the incline, calculate the tension in a) the rope connecting the blocks; b) the rope that connects block A to the wall. c) Calculate the magnitude of the force that the incline exerts on each block. d) Interpret your answers for the cases $\alpha = 0$ and $\alpha = 90°$.

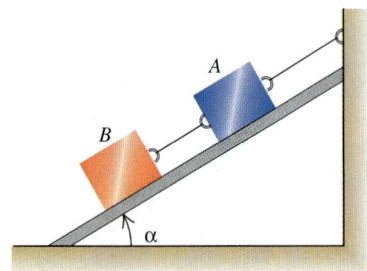

FIGURE 5–41 Exercise 5–13.

5–14 An airplane is in level flight at a constant speed. There are four forces acting on it: its weight $w = mg$, a forward force F provided by the engine (the *thrust*), the air resistance or *drag*

force f which acts opposite to the direction of flight, and a lift force L provided by the wings, which acts perpendicular to the direction of flight. The drag force f is proportional to the square of the speed. a) Show that $F = f$ and $L = w$. b) Suppose the pilot pushes the throttle forward, doubling the value of F, while maintaining a constant altitude. The airplane will eventually attain a new, faster, constant speed. At this faster speed, how does the new value of f compare to the previous value? c) How much faster than the original speed is the airplane going now?

SECTION 5–3 USING NEWTON'S SECOND LAW: DYNAMICS OF PARTICLES

5–15 Atwood's Machine. A 15.0-kg load of bricks hangs from one end of a rope that passes over a small, frictionless pulley. A 28.0-kg counterweight is suspended from the other end of the rope, as shown in Fig. 5–42. The system is released from rest. a) Draw two free-body diagrams, one for the load of bricks and one for the counterweight. b) What is the magnitude of the upward acceleration of the load of bricks? c) What is the tension in the rope while the load is moving? How does the tension compare to the weight of the load of bricks? To the weight of the counterweight?

FIGURE 5–42 Exercise 5–15.

5–16 A 8.00-kg block of ice released from rest at the top of a 1.50-m-long frictionless ramp slides downhill, reaching a speed of 2.50 m/s at the bottom. What is the angle between the ramp and the horizontal?

5–17 A light rope is attached to a block with a mass of 4.00 kg that rests on a horizontal, frictionless surface. The horizontal rope passes over a frictionless, massless pulley and a block of mass m is suspended from the other end. When the blocks are released, the tension in the rope is 10.0 N. a) Draw two free-body diagrams, one for the 4.00-kg block and one for the block of mass m. What is b) the acceleration of either block? c) the mass m of the hanging block? d) How does the tension compare to the weight of the hanging block?

5–18 A transport plane takes off from a level landing field with two gliders in tow, one behind the other. The mass of each glider is 700 kg, and the total resistance (air drag plus friction with the runway) on each may be assumed constant and equal to 2500 N. The tension in the towrope between the transport plane and the first glider is not to exceed 12,000 N. a) If a speed of 40 m/s is required for takeoff, what minimum length of runway is needed? b) What is the tension in the towrope between the two gliders while they are accelerating for the takeoff?

5–19 A 550-N physics student stands on a bathroom scale in an elevator. As the elevator starts moving, the scale reads 450 N. a) Find the acceleration of the elevator (magnitude and direction). b) What is the acceleration if the scale reads 670 N? c) If the scale reads zero, should the student worry? Explain.

5–20 A physics student playing with an air hockey table (a frictionless surface) finds that if she gives the puck a velocity of 3.80 m/s along the length (1.75 m) of the table at one end, by the time it has reached the other end the puck has drifted 2.50 cm to the right but still has a velocity component along the length of 3.80 m/s. She concludes correctly that the table is not level and correctly calculates its inclination from the above information. What is the angle of inclination?

5–21 a) Which way will the accelerometer in Fig. 5–14 deflect if the car is at rest on a sloping surface? b) Suppose the car is given a push up an icy (and hence frictionless) hill. After being pushed, the car moves uphill, slows to a halt, then moves back downhill. What is the direction of the accelerometer deflection in each stage of the motion? Explain your answers.

5–22 Which way will the accelerometer in Fig. 5–14 deflect under the following conditions? a) The car is moving toward the left and speeding up. b) The car is moving toward the left and slowing down. c) The car is moving toward the right and speeding up. Explain your answers.

SECTION 5–4 FRICTIONAL FORCES

5–23 Free-Body Diagrams. The first two steps in the solution of Newton's second law problems are to select an object for analysis and then to draw free-body diagrams for that object. Draw free-body diagrams for the following situations: a) a mass M sliding down a frictionless inclined plane of angle α; b) a mass M sliding up a frictionless inclined plane of angle α; c) a mass M sliding up an inclined plane of angle α with kinetic friction present; d) masses M and m sliding down an inclined plane of angle α with friction present, as shown in Fig. 5–43a. Here draw free-body diagrams for both m and M. Identify the forces that are action-reaction pairs. e) Draw free-body diagrams for masses m and M shown in Fig. 5–43b. Identify all action-reaction pairs. There is a frictional force between all surfaces in contact. The pulley is frictionless and massless. In all cases be sure you have the correct direction of the forces and are completely clear on what object is causing each force in your free-body diagram.

5–24 a) A large rock rests upon a rough horizontal surface. A bulldozer pushes on the rock with a horizontal force T that is slowly increased, starting from zero. In a graph with T along the x-axis and the friction force f along the y-axis, start at $T = 0$ and show the region of no motion, the point at which the rock is ready to move, and the region where the rock is moving. b) A block with weight w rests on a rough horizontal plank. The slope

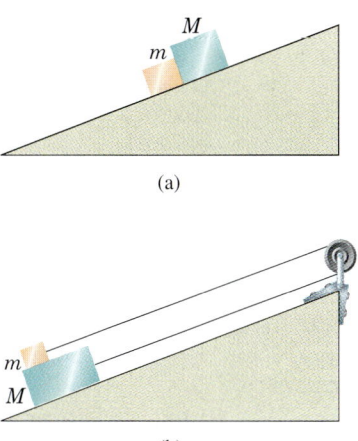

FIGURE 5–43 Exercise 5–23.

angle of the plank θ is gradually increased until the block starts to slip. Draw two graphs, both with θ along the x-axis. In one graph, show the ratio of the normal force to the weight, n/w, as a function of θ. In the second graph, show the ratio of the friction force to the weight, f/w. Indicate the region of no motion, the point at which the block is ready to move, and the region where the block is moving.

5–25 A stockroom worker pushes a box with mass 11.2 kg on a horizontal surface with a constant speed of 3.50 m/s. The coefficient of kinetic friction between the box and the surface is 0.20. a) What horizontal force must be applied by the worker to maintain the motion? b) If the force calculated in part (a) is removed, how far does the box slide before coming to rest?

5–26 A box of bananas weighing 40.0 N rests on a horizontal surface. The coefficient of static friction between the box and the surface is 0.40, and the coefficient of kinetic friction is 0.20. a) If no horizontal force is applied to the box and the box is at rest, how large is the friction force exerted on the box? b) What is the magnitude of the friction force if a monkey applies a horizontal force of 6.0 N to the box and the box is initially at rest? c) What minimum horizontal force must the monkey apply to start the box in motion? d) What minimum horizontal force must the monkey apply to keep the box moving at constant velocity once it has been started? e) If the monkey applies a horizontal force of 18.0 N, what is the magnitude of the friction force?

5–27 In a physics lab experiment, a 6.00-kg box is pushed across a flat table by a horizontal force $\vec{F}$. a) If the box is moving at a constant speed of 0.350 m/s and the coefficient of kinetic friction is 0.12, what is the magnitude of $\vec{F}$? b) What is the magnitude of $\vec{F}$ if the box is speeding up with a constant acceleration of 0.180 m/s²? c) How would your answers to parts (a) and (b) change if the experiments were performed on the moon, where $g = 1.62$ m/s²?

5–28 An 85-N box of oranges is being pushed across a horizontal floor. As it moves, it is slowing at a constant rate of 0.90 m/s each second. The push force has a horizontal component of 20 N and a vertical component of 25 N downward. Calculate the coefficient of kinetic friction between the box and floor.

5–29 A safe with mass 260 kg is to be lowered at constant speed down skids 20.0 m long, from a truck 2.00 m high. a) If the coefficient of kinetic friction between safe and skids is 0.25, does the safe need to be pulled down or held back? b) How great a force parallel to the skids is needed?

5–30 a) If the coefficient of kinetic friction between tires and dry pavement is 0.80, what is the shortest distance in which an automobile can be stopped by locking the brakes when traveling at 28.7 m/s (about 65 mi/h)? b) On wet pavement the coefficient of kinetic friction may be only 0.25. How fast should you drive on wet pavement in order to be able to stop in the same distance as in part (a)? (*Note:* Locking the brakes is *not* the safest way to stop.)

5–31 A clean brass washer slides along a horizontal clean steel surface until it stops. Using the values from Table 5–1, how many times farther would it have slid with the same initial speed if the washer had been Teflon-coated?

5–32 Consider the system shown in Fig. 5–44. Block A has weight w_A and block B has weight w_B. Once block B is set into downward motion, it descends at a constant speed. a) Calculate the coefficient of kinetic friction between block A and the table top. b) A cat, also of weight w_A, falls asleep on top of block A. If block B is now set into downward motion, what is its acceleration (magnitude and direction)?

FIGURE 5–44 Exercise 5–32, Exercise 5-35, and Problem 5-68.

5–33 Two crates connected by a rope lie on a horizontal surface (Fig. 5–45). Crate A has mass m_A, and crate B has mass m_B. The coefficient of kinetic friction between each crate and the surface is μ_k. The crates are pulled to the right at constant velocity by a horizontal force $\vec{F}$. In terms of m_A, m_B, and μ_k, calculate a) the magnitude of the force $\vec{F}$; b) the tension in the rope connecting the blocks. Include the free-body diagram or diagrams you used to determine each answer.

FIGURE 5–45 Exercise 5–33.

5–34 Two bicycle tires are set rolling with the same initial speed of 3.50 m/s along a long, straight road, and the distance each travels before its speed is reduced by half is measured. One tire is inflated to a pressure of 40 psi and goes 18.1 m; the other is at 105 psi and goes 92.9 m. What is the coefficient of rolling friction μ_r for each? Assume that the net horizontal force is due to rolling friction only.

5–35 As shown in Fig. 5–44, block A (mass 2.25 kg) rests on a table top. It is connected by a horizontal cord passing over a light, frictionless pulley to a hanging block B (mass 1.30 kg). The coefficient of kinetic friction between block A and the table top is 0.450. After the blocks are released from rest, find a) the speed of each block after moving 3.00 cm; b) the tension in the cord. Include the free-body diagram or diagrams you used to determine the answers.

5–36 A 25.0-kg box of textbooks rests on a loading ramp that makes an angle α with the horizontal. The coefficient of kinetic friction is 0.25, and the coefficient of static friction is 0.35. a) As the angle α is increased, find the minimum angle at which the box starts to slip. b) At this angle, find the acceleration once the box has begun to move. c) At this angle, how fast will the box be moving after it has slid 5.0 m along the inclined plane?

5–37 A large crate with mass m rests on a horizontal floor. The coefficients of friction between the crate and the floor are μ_s and μ_k. A woman pushes downward at an angle θ below the horizontal on the crate with a force $\vec{F}$. a) What magnitude of force $\vec{F}$ is required to keep the crate moving at constant velocity? b) If μ_s is larger than some critical value, the woman cannot start the crate moving no matter how hard she pushes. Calculate this critical value of μ_s.

5–38 A box with mass m is dragged across a level floor having a coefficient of kinetic friction μ_k by a rope that is pulled upward at an angle θ above the horizontal with a force of magnitude F. a) In terms of m, μ_k, θ, and g, obtain an expression for the magnitude of force required to move the box with constant speed. b) Knowing that you are studying physics, a CPR instructor asks you how much force it would take to slide a 90-kg patient across a floor at constant speed by pulling on him at an angle of 25° above the horizontal. By dragging some weights wrapped in an old pair of pants down the hall with a spring balance, you find that $\mu_k = 0.35$. Use the result of part (a) to answer the instructor's question.

5–39 Blocks A, B, and C are placed as in Fig. 5–46 and connected by ropes of negligible mass. Both A and B weigh 25.0 N each, and the coefficient of kinetic friction between each block and the surface is 0.35. Block C descends with constant velocity. a) Draw two separate free-body diagrams showing the forces acting on A and on B. b) Find the tension in the rope connecting blocks A and B. c) What is the weight of block C? d) If the rope connecting A and B were cut, what would be the acceleration of C?

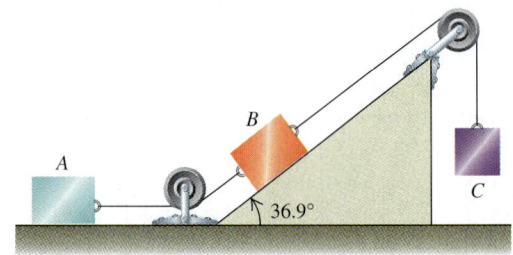

FIGURE 5–46 Exercise 5–39.

5–40 Starting from Eq. (5–10), derive Eqs. (5–11) and (5–12).

5–41 a) In Example 5–19, what value of D is required to make $v_t = 42$ m/s for this sky diver? b) If the sky diver's daughter, whose mass is 45 kg, is falling through the air and has the same D (0.25 kg/m) as her father, what is the daughter's terminal speed?

5–42 A baseball is thrown straight up. The drag force is proportional to v^2. In terms of g, what is the y-component of the ball's acceleration when its speed is half its terminal speed and a) it is moving up? b) it is moving back down?

SECTION 5–5 DYNAMICS OF CIRCULAR MOTION

5–43 A stone with a mass of 0.80 kg is attached to one end of a string 0.90 m long. The string will break if its tension exceeds 600 N. (This is called the *breaking strength* of the string.) The stone is whirled in a horizontal circle on a frictionless table top; the other end of the string is kept fixed. Find the maximum speed the stone can attain without breaking the string.

5–44 A flat (unbanked) curve on a highway has a radius of 220 m. A car rounds the curve at a speed of 25.0 m/s. What is the minimum coefficient of friction that will prevent sliding?

5–45 Aircraft experience a lift force (due to the air) that is perpendicular to the plane of the wings. Light airplanes are designed so that their wings can safely provide a lift force of 3.8 times the weight of the airplane. A greater force can damage the structure of the wing. (Aerobatic airplanes and jet fighters are designed with much higher limits.) a) What is the maximum banking angle that a pilot can safely maintain in level flight without threatening the safety of the airplane (and her own safety)? b) Does your answer to part (a) depend on the speed of the airplane? Explain why or why not.

5–46 The "Giant Swing" at a county fair consists of a vertical central shaft with a number of horizontal arms attached at its upper end (Fig. 5–47). Each arm supports a seat suspended from a cable 5.00 m long, the upper end of the cable being fastened to the arm at a point 3.00 m from the central shaft. a) Find the time of one revolution of the swing if the cable supporting a seat makes an angle of 30.0° with the vertical. b) Does the angle depend on the weight of the passenger for a given rate of revolution?

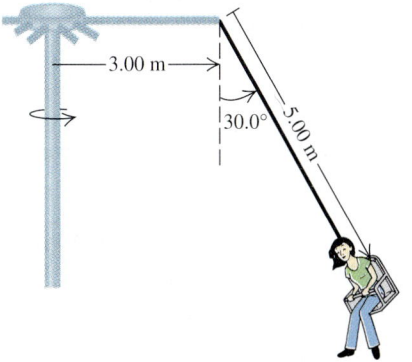

FIGURE 5–47 Exercise 5–46.

5–47 Aircraft experience a lift force (due to the air) that is perpendicular to the plane of the wings. A small airplane is flying at a constant speed of 240 km/h. At what angle from the horizontal must the wings of the airplane be tilted for the plane to execute a horizontal turn from east to north with a turning radius of 1200 m?

5–48 A small button placed on a horizontal rotating platform with diameter 0.320 m will revolve with the platform when it is brought up to a speed of 40.0 rev/min, provided the button is no more than 0.150 m from the axis. a) What is the coefficient of static friction between the button and the platform? b) How far from the axis can the button be placed, without slipping, if the platform rotates at 60.0 rev/min?

5–49 Rotating Space Stations. One problem for humans living in outer space is that they are apparently weightless. One way around this problem is to design a space station that spins about its center at a constant rate. This creates "artificial gravity" at the outside rim of the station. a) If the diameter of the space station is 800 m, how many revolutions per minute are needed in order for the "artificial gravity" acceleration to be 9.8 m/s^2? b) If the space station is a waiting area for travelers going to Mars, it might be desirable to simulate the acceleration due to gravity on the Martian surface (3.70 m/s^2). How many revolutions per minute are needed in this case?

5–50 The Cosmoclock 21 Ferris wheel in Yokohama City, Japan, has a diameter of 100 m. Its name comes from its 60 arms, each of which can function as a second hand (so that it makes one revolution every 60.0 s). a) Find the speed of the passengers when the Ferris wheel is rotating at this rate. b) A passenger weighs 882 N at the weight-guessing booth on the ground. What is his apparent weight at the highest and at the lowest point on the Ferris wheel? c) What would be the time for one revolution if the passenger's apparent weight at the highest point were zero? d) What then would be the passenger's apparent weight at the lowest point?

5–51 An airplane flies in a loop (a circular path in a vertical plane) of radius 150 m. The pilot's head always points toward the center of the loop. The speed of the airplane is not constant; the airplane goes slowest at the top of the loop and fastest at the bottom. a) At the top of the loop, the pilot feels weightless. What is the speed of the airplane at this point? b) At the bottom of the loop, the speed of the airplane is 280 km/h. What is the apparent weight of the pilot at this point? His true weight is 700 N.

5–52 A 50.0-kg stunt pilot who has been diving her airplane vertically pulls out of the dive by changing her course to a circle in a vertical plane. a) If the plane's speed at the lowest point of the circle is 95.0 m/s, what is the minimum radius of the circle for the acceleration at this point not to exceed $4.00g$? b) What is the apparent weight of the pilot at the lowest point of the pullout?

5–53 A cord is tied to a pail of water, and the pail is swung in a vertical circle of radius 0.600 m. What must be the minimum speed of the pail at the highest point of the circle if no water is to spill from it?

5–54 A bowling ball weighing 71.2 N (16.0 lb) is attached to the ceiling by a 3.80-m rope. The ball is pulled to one side and released; it then swings back and forth as a pendulum. As the rope swings through the vertical, the speed of the bowling ball is 4.20 m/s. a) What is the acceleration of the bowling ball, in magnitude and direction, at this instant? b) What is the tension in the rope at this instant?

SECTION 5–7 PROJECTILE MOTION WITH AIR RESISTANCE: A CASE STUDY IN COMPUTER ANALYSIS

5–55 a) Implement the algorithm in Section 5–7 on a computer, and use it to reproduce the graph labeled "with drag" in Fig. 5–34. b) The home field for the Colorado Rockies is in Denver (altitude 1 mile), where the air density ρ is only 1.0 kg/m^3. For $m = 0.145$ kg, $r = 0.0366$ m, and $C = 0.5$, how much farther does a ball go than at sea level (where $\rho = 1.2$ kg/m^3) if $v_0 = 50$ m/s and $\alpha_0 = 35°$?

5–56 A baseball has $m = 0.145$ kg, $r = 0.0366$ m, and $C = 0.5$. At what angle should the ball be hit or thrown to maximize the range? Assume $v_0 = 50$ m/s and $\rho = 1.2$ kg/m^3.

5–57 Bo Jackson made a flat-footed 300-ft throw (no bounces) from left field to home plate to put out a tenth-inning tying run. Assuming that $\alpha_0 = 40°$, with what speed did he throw the ball? Use $m = 0.145$ kg, $r = 0.0366$ m, $C = 0.5$, and $\rho = 1.2$ kg/m^3. How does this compare to the 100-mi/h throwing speed of a good fastball pitcher?

5–58 In tennis, 100 mi/h is a very good serving speed. How fast is the ball moving when it crosses the baseline in the opposite court (24 m distant)? Use $m = 0.055$ kg, $r = 0.031$ m, $C = 0.75$, and $\rho = 1.2$ kg/m^3. The ball leaves the racquet moving horizontally and does not hit the ground until after it crosses the opposite court baseline.

5–59 Estimate the maximum distance a human being can throw a ping-pong ball. A ping-pong ball has $r = 0.019$ m and $m = 0.0024$ kg; use $C = 0.5$ and $\rho = 1.2$ kg/m^3. The fastest pitchers can throw at about 100 mi/h. Why can a baseball be thrown much farther than a ping-pong ball?

PROBLEMS

5–60 In Fig. 5–48 a worker lifts a weight w by pulling down on a rope with a force $\vec{F}$. The upper pulley is attached to the ceiling by a chain, and the lower pulley is attached to the weight by another chain. In terms of w, find the tension in each chain and the magnitude of the force $\vec{F}$ if the weight is lifted at constant speed. Include the free-body diagram or diagrams you used to determine your answers. Assume that the rope, pulleys, and chains all have negligible weights.

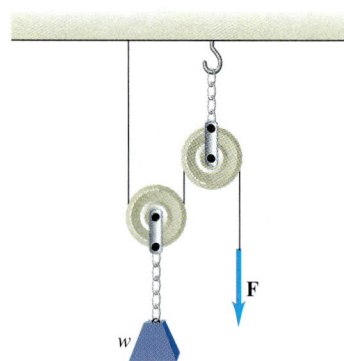

FIGURE 5–48 Problem 5–60.

5–61 A man is pushing a refrigerator up a ramp at constant speed. The ramp is inclined at an angle α above the horizontal, but the man applies a *horizontal* force $\vec{F}$. Calculate the magnitude of $\vec{F}$ in terms of α and the refrigerator mass m. Neglect friction on the refrigerator.

5–62 A Rope with Mass. In most problems in this book, the ropes, cords, or cables have so little mass compared to other objects in the problem that their mass can safely be neglected. But if the rope is the *only* object in the problem, then clearly its mass cannot be neglected. For example, suppose we have a clothesline attached to two poles (Fig. 5–49). The clothesline has a mass M, and each end makes an angle θ with the horizontal.

What is a) the tension at the ends of the clothesline? b) the tension at the lowest point? c) Why can't we have $\theta = 0$? (See Discussion Question Q5–3.) The curve of the clothesline, or of any flexible cable hanging under its own weight, is called a *catenary*. For a more advanced treatment of this curve, see K. R. Symon, *Mechanics*, 3rd ed. (Addison–Wesley, Reading, MA, 1971), pp. 237–241.

FIGURE 5–49 Problem 5–62.

5–63 A block of mass M is attached to the lower end of a vertical, uniform rope of mass m and length L. A constant upward force $\vec{F}$ is applied to the top of the rope, causing the rope and block to accelerate upward. Find the tension in the rope at a distance x from the top end of the rope, where x can have any value from 0 to L.

5–64 A block with mass m_1 is placed on an inclined plane with slope angle α and is connected to a second hanging block that has mass m_2 by a cord passing over a small, frictionless pulley (Fig. 5–50). The coefficient of static friction is μ_s, and the coefficient of kinetic friction is μ_k. a) Find the mass m_2 for which block m_1 moves up the plane at constant speed once it has been set in motion. b) Find the mass m_2 for which block m_1 moves down the plane at constant speed once it has been set in motion. c) For what range of values of m_2 will the blocks remain at rest if they are released from rest?

5–65 a) Block A in Fig. 5–51 weighs 60.0 N. The coefficient of static friction between the block and the surface on which it rests is 0.25. The weight w is 12.0 N, and the system is in equilibrium. Find the friction force exerted on block A. b) Find the maximum weight w for which the system will remain in equilibrium.

FIGURE 5–50 Problem 5–64.

FIGURE 5–51 Problem 5–65.

5–66 Block A in Fig. 5–52 weighs 1.20 N, and block B weighs 3.60 N. The coefficient of kinetic friction between all surfaces is 0.300. Find the magnitude of the horizontal force $\vec{F}$ necessary to drag block B to the left at constant speed a) if A rests on B and moves with it (Fig. 5–52a); b) if A is held at rest (Fig. 5–52b).

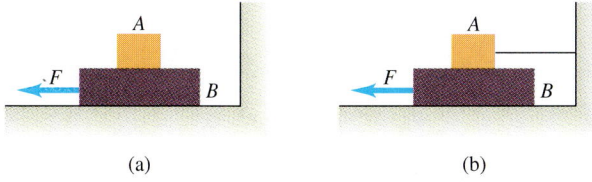

FIGURE 5–52 Problem 5–66.

5–67 A window washer pushes his scrub brush up a vertical window at constant speed by applying a force $\vec{F}$ as shown in Fig. 5–53. The brush weighs 12.0 N, and the coefficient of kinetic friction is $\mu_k = 0.150$. Calculate a) the magnitude of the force $\vec{F}$; b) the normal force exerted by the window on the brush.

FIGURE 5–53 Problem 5–67.

5–68 In the system shown in Fig. 5–44, block A has mass m_A, block B has mass m_B, and the rope connecting them has a *nonzero* mass m_{rope}. The rope has a total length L and the pulley has a very small radius. Ignore any sag in the horizontal part of the string. a) If there is no friction between block A and the table top, find the acceleration of the blocks at an instant when a length d of rope hangs vertically between the pulley and block B. As block B falls, will the magnitude of the acceleration of the system increase, decrease, or remain constant? Explain. b) Let $m_A = 2.00$ kg, $m_B = 0.400$ kg, $m_{rope} = 0.160$ kg, and $L = 1.00$ m. If there is friction between block A and the table top, with $\mu_k = 0.200$ and $\mu_s = 0.250$, find the minimum value of the distance d such that the blocks will start to move if they are initially at rest. c) Repeat part (b) for the case $m_{rope} = 0.040$ kg. Will the blocks move in this case?

5–69 If the coefficient of static friction between a table and a uniform massive rope is μ_s, what fraction of the rope can hang over the edge of the table without the rope sliding?

5–70 A woman attempts to push a box of books that has mass m up a ramp inclined at an angle α above the horizontal. The coefficients of friction between the ramp and the box are μ_s and μ_k. The force $\vec{F}$ applied by the woman is *horizontal*. a) If μ_s is greater than some critical value, the woman cannot start the box moving up the ramp no matter how hard she pushes. Calculate this critical value of μ_s. b) Assume that μ_s is less than this critical value. What magnitude of force must the woman apply to keep the box moving up the ramp at constant speed?

5–71 A 30.0-kg packing case is initially at rest on the floor of a 1500-kg pickup truck. The coefficient of static friction between the case and the truck floor is 0.30, and the coefficient of kinetic friction is 0.20. Before each acceleration given below, the truck is traveling due north at constant speed. Find the magnitude and direction of the friction force acting on the case a) when the truck accelerates at 2.20 m/s² northward; b) when it accelerates at 3.40 m/s² southward.

5–72 The truck in Problem 5–71 is traveling at a constant speed on a country road where the speed limit is 72 km/h (45 mi/h). Spotting a stop sign ahead, the driver applies the brakes and comes to a halt in a distance of 47.0 m. Without warning a policeman emerges from behind a bush and hands the driver a speeding ticket. When the driver protests that he was going less than the speed limit, the policeman replies that "I saw that packing case slide from the back to the front of your truck's cargo area. To make it slide like that, you must have had to brake very hard, which means you had to have been speeding." Will the policeman's argument hold up in traffic court? (Assume that the judge, like you, knows physics.)

5–73 Troy uses a frayed rope to pull a crate across a level floor. The maximum tension in the rope without breaking is T_{max} and the coefficient of kinetic friction is μ_k. a) Show that the maximum weight that he can pull at constant speed is $T_{max}/(\sin \theta)$, where $\theta = \arctan \mu_k$ is the angle of the rope above the horizontal. b) The answer to part (a) suggests that Troy can pull a weight that approaches infinity with a bit of spiderweb as the coefficient of kinetic friction approaches zero. Explain.

5–74 The airplane in Exercise 5–14 has an engine failure (so $F = 0$), and the airplane glides at constant speed toward a safe landing. The direction of flight is a constant angle α (called the *glide angle*) below the horizontal (Fig. 5–54). a) Find the magnitude of the lift force L (which acts perpendicular to the direction of flight) and of the drag force f in terms of w and α. b) Show that $\alpha = \arctan (f/L)$. c) A fully loaded Cessna 182 (a single-engine airplane) weighs 12,900 N and has 1300 N of drag at a speed of 130 km/h. If this airplane has an engine failure at an altitude of 2500 m, what maximum distance over the ground can it glide while searching for a safe landing spot? d) Justify the statement that "it's drag, not gravity, that makes the airplane go down."

FIGURE 5–54 Problem 5–74.

5–75 The pilot of the Cessna 182 airplane in part (c) of Problem 5–74 manages to get the engine restarted. He applies full throttle, and the airplane climbs along a straight line angled above the horizontal. The airplane is flying at a constant speed of 130 km/h, weighs 12,900 N, and has 1300 N of drag. The rate-of-climb indicator on the instrument panel shows that the airplane is gaining altitude at a steady rate of 5.00 m/s (300 m/min, or about 980 feet/min). Determine the magnitude of the thrust (the forward force produced by the engine). (*Hint:* The thrust acts in the direction that the airplane is moving.)

5–76 A 12.0-kg box rests on the flat floor of a truck. The coefficients of friction between the box and floor are $\mu_s = 0.19$ and $\mu_k = 0.15$. The truck stops at a stop sign and then starts to move with an acceleration of 2.20 m/s^2. If the box is 1.80 m from the rear of the truck when the truck starts, how much time elapses before the box falls off the truck? How far does the truck travel in this time?

5–77 Block A in Fig. 5–55 weighs 1.40 N, and block B weighs 4.20 N. The coefficient of kinetic friction between all surfaces is 0.30. Find the magnitude of the horizontal force $\vec{F}$ necessary to drag block B to the left at constant speed if A and B are connected by a light, flexible cord passing around a fixed, frictionless pulley.

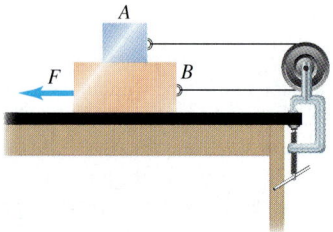

FIGURE 5–55 Problem 5–77.

5–78 A 30.0-kg crate is released from an airplane traveling due east at an altitude of 1200 m with a ground speed of 70.0 m/s. The wind applies a constant force on the crate of 180 N directed horizontally in the opposite direction to the plane's flight path. Where and when (with respect to the release location and time) does the crate hit the ground?

5–79 Block A in Fig. 5–56 has a mass of 4.00 kg, and block B has a mass of 12.0 kg. The coefficient of kinetic friction between block B and the horizontal surface is 0.25. a) What is the mass of block C if block B is moving to the right and speeding up with an acceleration 2.00 m/s^2? b) What is the tension in each cord when block B has this acceleration?

FIGURE 5–56 Problem 5–79.

5–80 Two blocks connected by a cord passing over a small, frictionless pulley rest on frictionless planes (Fig. 5–57). a) Which way will the system move when the blocks are released from rest? b) What is the acceleration of the blocks? c) What is the tension in the cord?

FIGURE 5–57 Problem 5–80.

5–81 In terms of m_1, m_2, and g, find the accelerations of each block in Fig. 5–58. There is no friction anywhere in the system.

FIGURE 5–58 Problem 5–81.

5–82 Block B, of mass m_B, rests on block A, of mass m_A, which in turn is on a horizontal table top (Fig. 5–59). The coefficient of kinetic friction between block A and the table top is μ_k and the coefficient of static friction between block A and block B is μ_s. A light string attached to block A passes over a frictionless, massless pulley and block C is suspended from the other end of the string. What is the largest mass m_C that block C can have so that blocks A and B still slide together when the system is released from rest?

FIGURE 5–59 Problem 5–82.

5–83 Two objects with masses of 5.00 kg and 2.00 kg hang 0.600 m above the floor from the ends of a cord 6.00 m long passing over a frictionless pulley. Both objects start from rest. Find the maximum height reached by the 2.00-kg object.

5–84 Friction in an Elevator. You are riding in an elevator on the way to the eighteenth floor of your dormitory. The elevator is accelerating upward with $a = 1.90$ m/s². Beside you is the box containing your new computer; box and contents have a total mass of 28.0 kg. While the elevator is accelerating upward, you push horizontally on the box to slide it at constant speed toward the elevator door. If the coefficient of kinetic friction between the box and elevator floor is $\mu_k = 0.32$, what magnitude of force must you apply?

5–85 What acceleration must the cart in Fig. 5–60 have in order that block A does not fall? The coefficient of static friction between the block and the cart is μ_s. How would the behavior of the block be described by an observer on the cart?

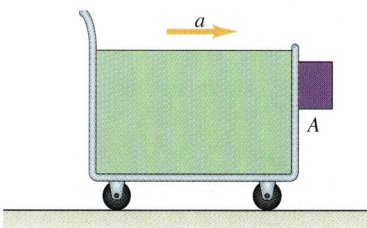

FIGURE 5–60 Problem 5–85.

5–86 Two blocks with masses of 4.00 kg and 8.00 kg are connected by a string and slide down a 30.0° inclined plane (Fig. 5–61). The coefficient of kinetic friction between the 4.00-kg block and the plane is 0.25; that between the 8.00-kg block and the plane is 0.35. a) Calculate the acceleration of each block. b) Calculate the tension in the string. c) What happens if the positions of the blocks are reversed, so the 4.00-kg block is above the 8.00-kg block?

5–87 Block A, with weight $3w$, slides down an inclined plane S of slope angle 36.9° at a constant speed while plank B, with weight w, rests on top of A. The plank is attached by a cord to the

FIGURE 5–61 Problem 5–86.

top of the plane (Fig. 5–62). a) Draw a diagram of all the forces acting on block A. b) If the coefficient of kinetic friction is the same between A and B and between S and A, determine its value.

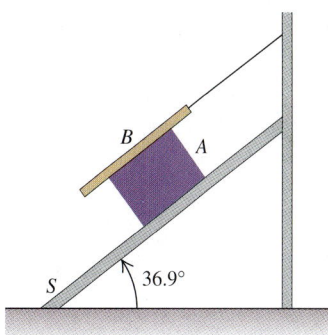

FIGURE 5–62 Problem 5–87.

5–88 A man with mass 70.0 kg stands on a platform with mass 25.0 kg. He pulls on the free end of a rope that runs over a pulley on the ceiling and has its other end fastened to the platform. The mass of the rope and the mass of the pulley can be neglected, and the pulley is frictionless. The rope is vertical on either side of the pulley. a) With what force does he have to pull so that he and the platform have an upward acceleration of 1.80 m/s²? b) What is the acceleration of the rope relative to him?

5–89 Two blocks, of mass m_1 and m_2, are stacked as shown in Fig. 5–63 and placed on a frictionless horizontal surface. There is friction between the two blocks. An external force of magnitude F is applied to the top block at an angle α below the horizontal. a) If the two blocks move together, find their acceleration. b) Show that the two blocks will move together only if

$$F \le \frac{\mu_s m_1 (m_1 + m_2)g}{m_2 \cos \alpha - \mu_s (m_1 + m_2) \sin \alpha},$$

where μ_s is the coefficient of static friction between the two blocks.

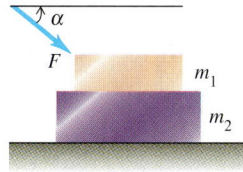

FIGURE 5–63 Problem 5–89.

5–90 A curve with a 120-m radius on a level road is banked at the correct angle for a speed of 20 m/s. If an automobile rounds this curve at 30 m/s, what is the minimum coefficient of static friction between tires and road needed to prevent skidding?

5–91 Consider a wet roadway banked as in Example 5–23 (Section 5–5), where there is a coefficient of static friction of 0.30 and a coefficient of kinetic friction of 0.25 between the tires and the roadway. The radius of the curve is $R = 50$ m. a) If the banking angle is $\beta = 25°$, what is the *maximum* speed the automobile can have before sliding *up* the banking? b) What is the *minimum* speed the automobile can have before sliding *down* the banking?

5–92 You are riding in a school bus. As the bus rounds a flat curve at constant speed, a lunch box with a mass of 0.500 kg suspended from the ceiling of the bus by a string 1.80 m long is found to hang at rest relative to the bus when the string makes an angle of 30.0° with the vertical. In this position the lunch box is 50.0 m from the center of curvature of the curve. What is the speed v of the bus?

5–93 The Monkey and Bananas Problem. A 20-kg monkey has a firm hold on a light rope that passes over a frictionless pulley and is attached to a 20-kg bunch of bananas (Fig. 5–64). The monkey looks upward, sees the bananas, and starts to climb the rope to get them. a) As the monkey climbs, do the bananas move up, down, or remain at rest? b) As the monkey climbs, does the distance between the monkey and the bananas decrease, increase, or remain constant? c) The monkey releases her hold on the rope. What happens to the distance between the monkey and the bananas while she is falling? d) Before reaching the ground, the monkey grabs the rope to stop her fall. What do the bananas do?

FIGURE 5–64 Problem 5–93.

5–94 A rock is thrown downward into water with a speed of $3mg/k$, where k is the coefficient in Eq. (5–7). Assume that the rela-

tion between fluid resistance and speed is as given in Eq. (5–7), and calculate the speed of the rock as a function of time.

5–95 A rock with mass $m = 3.00$ kg falls from rest in a viscous medium. The rock is acted on by a net constant downward force of 18.0 N (a combination of gravity and the buoyant force exerted by the medium) and by a fluid resistance force $f = kv$, where v is the speed in m/s and $k = 2.20$ N · s/m. (See Section 5–4.) a) Find the initial acceleration, a_0. b) Find the acceleration when the speed is 3.00 m/s. c) Find the speed when the acceleration equals $0.1a_0$. d) Find the terminal speed, v_t. e) Find the coordinate, speed, and acceleration 2.00 s after the start of the motion. f) Find the time required to reach a speed $0.9v_t$.

5–96 The 4.00-kg block in Fig. 5–65 is attached to a vertical rod by means of two strings. When the system rotates about the axis of the rod, the strings are extended as shown in the diagram and the tension in the upper string is 80.0 N. a) What is the tension in the lower cord? b) How many revolutions per minute does the system make? c) Find the number of revolutions per minute at which the lower cord just goes slack. d) Explain what happens if the number of revolutions per minute is less than in part (c).

FIGURE 5–65 Problem 5–96.

5–97 Two identical twins, Jena and Jackie, are playing one December on a large merry-go-round (a disk mounted parallel to the ground on a vertical axle through its center) in their school playground in northern Minnesota. Each twin has a mass of 30.0 kg. The icy coating on the merry-go-round surface makes it frictionless. The merry-go-round revolves at a constant rate as the twins ride on it. Jena, sitting 1.80 m from the center of the merry-go-round, must hold onto one of the metal posts attached to the merry-go-round with a horizontal force of 60.0 N to keep from sliding off. Jackie is sitting at the edge, 3.60 m from the center. a) With what horizontal force must Jackie hold on to keep from falling off? b) If Jackie falls off, what will be her horizontal velocity when she becomes airborne?

5–98 Consider a passenger on a Ferris wheel like that of Example 5–24. a) What is the passenger's apparent weight when the passenger's velocity is vertical, in the $\pm y$-direction? b) Where in the rotation are the passenger's apparent weight and true weight equal in magnitude?

5–99 In the ride "Spindletop" at the amusement park Six Flags Over Texas, people stand against the inner wall of a hollow vertical cylinder with radius 2.5 m. The cylinder starts to rotate, and when it reaches a constant rotation rate of 0.60 rev/s, the floor on which people are standing drops about 0.5 m. The people remain pinned against the wall. a) Draw a force diagram for a person in this ride, after the floor has dropped. b) What minimum coefficient of static friction is required if the person in the ride is not to slide downward to the new position of the floor? c) Does your answer in part (b) depend on the mass of the passenger? (*Note:* When the ride is over, the cylinder is slowly brought to rest. As it slows down, people slide down the walls to the floor.)

5–100 A physics major is working to pay his college tuition by performing in a traveling carnival. He rides a motorcycle inside a hollow transparent plastic sphere. After gaining sufficient speed, he travels in a vertical circle with a radius of 13.0 m. The physics major has a mass of 70.0 kg, and his motorcycle has a mass of 40.0 kg. a) What minimum speed must he have at the top of the circle if the tires of the motorcycle are not to lose contact with the sphere? b) At the bottom of the circle his speed is twice the value calculated in part (a). What is the magnitude of the normal force exerted on the motorcycle by the sphere at this point?

5–101 You are driving a classic 1954 Nash Ambassador with a friend who is sitting to your right on the passenger side of the front seat. The Ambassador has flat bench seats. You would like to be closer to your friend and decide to use physics to achieve your romantic goal by making a quick turn. a) Which way (to the left or to the right) should you turn the car to get your friend to slide closer to you? b) If the coefficient of static friction between your friend and the car seat is 0.35 and you keep driving at a constant speed of 20 m/s, what is the maximum radius you could make your turn and still have your friend slide your way?

5–102 A small block with mass m rests on a frictionless horizontal tabletop a distance r from a hole in the center of the table (Fig. 5–66). A string tied to the small block passes down through the hole, and a larger block with mass M is suspended from the free end of the string. The small block is set into uniform circular motion with radius r and speed v. What must v be if the large block is to remain motionless when released?

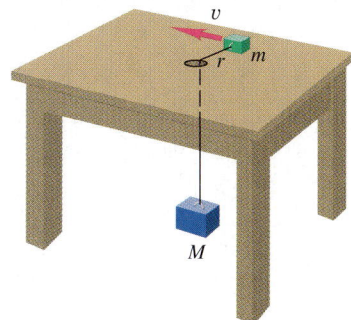

FIGURE 5–66 Problem 5–102.

5–103 A small bead can slide without friction on a circular hoop that is in a vertical plane and has a radius of 0.100 m. The hoop rotates at a constant rate of 4.00 rev/s about a vertical diameter (Fig. 5–67). a) Find the angle β at which the bead is in vertical equilibrium. (Of course, it has a radial acceleration toward the axis.) b) Is it possible for the bead to "ride" at the

same elevation as the center of the hoop? c) What will happen if the hoop rotates at 1.00 rev/s?

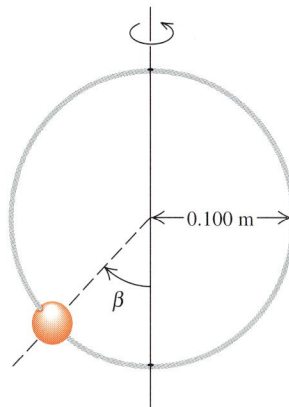

FIGURE 5–67 Problem 5–103.

5–104 A model airplane with mass 2.20 kg moves in the xy-plane such that its x- and y-coordinates vary in time according to $x(t) = \alpha - \beta t^3$ and $y(t) = \gamma t - \delta t^2$, where $\alpha = 1.50$ m, $\beta = 0.120$ m/s^3, $\gamma = 3.00$ m/s, and $\delta = 1.00$ m/s^2. a) Calculate the x- and y-components of the net force on the plane as functions of time. b) Sketch the trajectory of the airplane between $t = 0$ and $t = 3.00$ s, and draw on your sketch vectors showing the net force on the airplane at $t = 0$, $t = 1.00$ s, $t = 2.00$ s, and $t = 3.00$ s. For each of these times, relate the direction of the net force to the direction that the airplane is turning, and to whether the airplane is speeding up or slowing down (or neither). c) What are the magnitude and direction of the net force at $t = 3.00$ s?

5–105 A particle moves on a frictionless surface along a path as shown in Fig. 5–68. (The figure gives a view looking down on the surface.) The particle is initially at rest at point A, then begins to move toward B as it gains speed at a constant rate. From B to C the particle moves along a circular path at a constant speed. The speed remains constant along the straight-line path from C to D. From D to E the particle moves along a circular path, but now its speed is decreasing at a constant rate. The speed continues to decrease at a constant rate as the particle moves from E to F; the particle comes to a halt at F. (The time intervals between the marked points are not equal.) At each point marked with a dot, draw arrows to represent the velocity, the acceleration, and the net force acting on the particle. Draw a longer or shorter arrow to represent vectors of larger or smaller magnitude.

FIGURE 5–68 Problem 5–105.

5–106 A small remote-control car with a mass of 1.60 kg moves at a constant speed of $v = 12.0$ m/s in a vertical circle inside a hollow metal cylinder that has a radius of 5.00 m (Fig. 5–69). What is the magnitude of the normal force exerted on the car by the walls of the cylinder at a) point A (at the bottom of the vertical circle)? b) point B (at the top of the vertical circle)?

B

$v = 12.0$ m/s

5.00 m

$v = 12.0$ m/s

A

FIGURE 5–69 Problem 5–106.

5–107 A small block with mass m is placed inside an inverted cone that is rotating about a vertical axis such that the time for one revolution of the cone is T (Fig. 5–70). The walls of the cone make an angle β with the vertical. The coefficient of static friction between the block and the cone is μ_s. If the block is to remain at a constant height h above the apex of the cone, what are the maximum and minimum values of T?

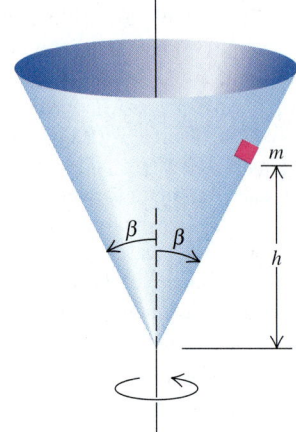

m

β β

h

FIGURE 5–70 Problem 5–107.

CHALLENGE PROBLEMS

5–108 A wedge with mass M rests on a frictionless horizontal table top. A block with mass m is placed on the wedge (Fig. 5–71a). There is no friction between the block and the wedge. The system is released from rest. a) Calculate the acceleration of the wedge and the horizontal and vertical components of the acceleration of the block. b) Do your answers to part (a) reduce to the correct results when M is very large? c) As seen by a stationary observer, what is the shape of the trajectory of the block?

5–109 A wedge with mass M rests on a frictionless horizontal table top. A block with mass m is placed on the wedge, and a horizontal force $\vec{F}$ is applied to the wedge (Fig. 5–71b). What must be the magnitude of $\vec{F}$ if the block is to remain at a constant height above the table top?

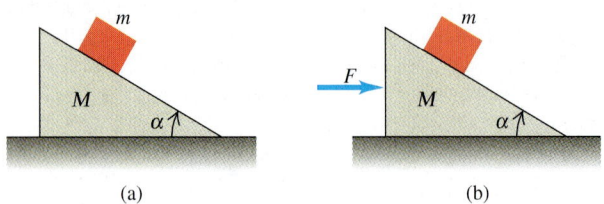

m m

M F M

α α

(a)   (b)

FIGURE 5–71 Challenge Problems 5–108 and 5–109.

5–110 A box of weight w is accelerated up a ramp by a rope that exerts a tension T. The ramp makes an angle α with the horizontal and the rope makes an angle θ above the ramp. The coefficient of kinetic friction between the box and the ramp is μ_k. Show that no matter what the value of α, the acceleration is maximum if $\theta = \arctan \mu_k$ (as long as the box remains in contact with the ramp).

5–111 A box with weight w is pulled at constant speed along a level floor by a force $\vec{F}$ that is at an angle θ above the horizontal. The coefficient of kinetic friction between the floor and box is μ_k. a) In terms of θ, μ_k, and w, calculate F. b) For $w = 400$ N and $\mu_k = 0.25$, calculate F for θ ranging from 0° to 90° in increments of 10°. Graph F versus θ. c) From the general expression in part (a), calculate the value of θ for which the value of F required to maintain constant speed is a minimum. (*Hint:* At a point where a function is minimum, what are the first and second derivatives of the function? Here F is a function of θ.) For the special case of $w = 400$ N and $\mu_k = 0.25$, evaluate this optimal θ and compare your result to the graph you constructed in part (b).

5–112 A baseball is dropped from the roof of a tall building. As the ball falls, the air exerts a drag force proportional to the square of the ball's speed ($f = Dv^2$). a) In a diagram, show the direction of motion and indicate with the aid of vectors all of the forces acting on the ball. b) Apply Newton's second law and infer from the resulting equation the general properties of the motion. c) Show that the ball acquires a terminal speed that is as given in Eq. (5–13). d) Derive the equation for the speed at any time. (*Note:*

$$\int \frac{dx}{a^2 - x^2} = \frac{1}{a} \operatorname{arctanh} \left(\frac{x}{a} \right),$$

where

$$\tanh (x) = \frac{e^x - e^{-x}}{e^x + e^{-x}} = \frac{e^{2x} - 1}{e^{2x} + 1}$$

defines the hyperbolic tangent.)

5–113 Double Atwood's Machine. In Fig. 5–72 masses m_1 and m_2 are connected by a light string A over a light frictionless pulley B. The axle of pulley B is connected by a second light string C over a second light frictionless pulley D to a mass m_3. Pulley D is suspended from the ceiling by an attachment to its axle. The system is released from rest. In terms of m_1, m_2, m_3, and g, what is a) the acceleration of block m_3? b) the acceleration of pulley B? c) the acceleration of block m_1? d) the acceleration of block m_2? e) the tension in string A? f) the tension in string C? g) What do your expressions give for the special case of $m_1 = m_2$ and $m_3 = m_1 + m_2$? Is this sensible?

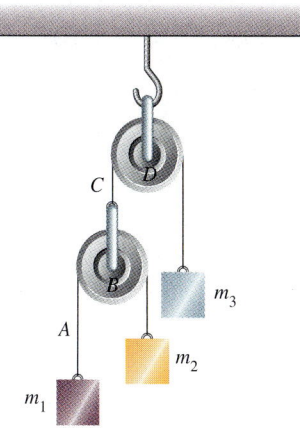

FIGURE 5–72 Challenge Problem 5–113.

5–114 The masses of blocks A and B in Fig. 5–73 are 20.0 kg and 10.0 kg, respectively. The blocks are initially at rest on the floor and are connected by a massless string passing over a massless and frictionless pulley. An upward force $\vec{F}$ is applied to the pulley. Find the accelerations $\vec{a}_1$ of block A and $\vec{a}_2$ of block B when F is a) 124 N; b) 294 N; c) 424 N.

FIGURE 5–73 Challenge Problem 5–114.

5–115 A ball is held at rest at position A in Fig. 5–74 by two light strings. The horizontal string is cut, and the ball starts swinging as a pendulum. Point B is the farthest to the right the ball goes as it swings back and forth. What is the ratio of the tension in the supporting string in position B to its value at A before the horizontal string was cut?

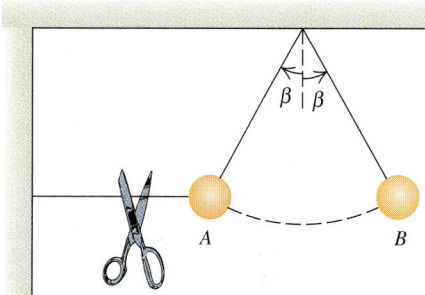

FIGURE 5–74 Challenge Problem 5–115.

6

Work and Kinetic Energy

To prepare an arrow for flight, an archer does work—that is, exerts a force over a distance—to draw back the arrow and stretch the bow and bowstring. The bow in turn does work on the arrow when it is released, giving the arrow kinetic energy, or energy of motion. This kinetic energy is used to do work on the target as the arrow penetrates into the bull's-eye.

6–1 INTRODUCTION

Some problems are harder than they look. Suppose you try to find the speed of an arrow that has been shot from a bow. You apply Newton's laws and all the problem-solving techniques that we've learned, but you run across a major stumbling block: After the archer releases the arrow, the bow string exerts a *varying* force that depends on the arrow's position. As a result, the simple methods that we've learned aren't enough to calculate the speed. Never fear; we aren't by any means finished with mechanics, and there are other methods for dealing with such problems.

The new method that we're about to introduce uses the ideas of *work* and *energy*. The applications of these ideas go far beyond mechanics, however. The importance of the energy idea stems from the *principle of conservation of energy:* Energy is a quantity that can be converted from one form to another but cannot be created or destroyed. In an automobile engine, chemical energy stored in the fuel is converted partially to the energy of the automobile's motion and partially to thermal energy. In a microwave oven, electromagnetic energy obtained from your power company is converted to thermal energy of the food being cooked. In these and all other processes, the *total* energy the sum of all energy present in all different forms— remains the same. No exception has ever been found.

We'll use the energy idea throughout the rest of this book to study a tremendous range of physical phenomena.

This idea will help you understand why a sweater keeps you warm, how a camera's flash unit can produce a short burst of light, and the meaning of Einstein's famous equation $E = mc^2$.

In this chapter, though, our concentration will be on mechanics. We'll learn about one important form of energy called *kinetic energy,* or energy of motion, and how it relates to the concept of *work*. We'll also consider *power,* which is the time rate of doing work. In Chapter 7 we'll expand the ideas of work and kinetic energy into an understanding of the general concept of energy, and we'll see how the conservation of energy arises.

6–2 WORK

You'd probably agree that it's hard work to pull a heavy sofa across the room, to lift a stack of encyclopedias from the floor to a high shelf, or to push a stalled car off the road. Indeed, all of these examples agree with the everyday meaning of *work*—any activity that requires muscular or mental effort.

In physics, work has a much more precise definition. By making use of this definition we'll find that in any motion, no matter how complicated, the total work done on a particle by all forces that act on it equals the change in its *kinetic energy*—a quantity that's related to the particle's speed. This relationship holds even when the forces acting on the particle aren't constant, a situation that can be difficult or impossible to handle with the techniques you learned in Chapters 4 and 5. So the ideas of work and kinetic energy will enable us to solve problems in mechanics that we could not have attempted before.

We'll develop the relationship between work and kinetic energy in Section 6–3, and see what to do with non-constant forces in Section 6–4. Meanwhile, let's see how work is defined and how to calculate work in a variety of situations involving *constant* forces. Even though we already know how to solve problems in which the forces are constant, the idea of work is still a useful one in such problems.

The three examples of work described above—pulling a sofa, lifting encyclopedias, and pushing a car—have something in common. In each case you do work by exerting a *force* on a body while that body *moves* from one place to another, that is, undergoes a *displacement*. You do more work if the force is greater (you pull harder on the sofa) or if the displacement is greater (you pull the sofa farther across the room).

The physicist's definition of work is based on these observations. Consider a body that undergoes a displacement of magnitude s along a straight line. (For now, we'll assume that any body we discuss can be treated as a particle so that we can ignore any rotation or changes in shape of the body.) While the body moves, a constant force with magnitude F acts on it in the same direction as the displacement (Fig. 6–1). We define the **work** W done by a constant force F acting on the body under these conditions as

$$W = Fs \quad \text{(constant force in direction of straight-line displacement).} \quad (6\text{--}1)$$

The work done on the body is greater if either the force F or the displacement s is greater, in agreement with our observations above.

6–1 When a constant force $\vec{F}$ acts in the same direction as the displacement $\vec{s}$, the work done by the force is $W = Fs$.

CAUTION ▶ Don't confuse W (work) with w (weight). Though the symbols may be almost the same, work and weight are different quantities. ◀

The SI unit of work is the **joule** (abbreviated J, pronounced "jewel," and named in honor of the nineteenth century English physicist James Prescott Joule). From Eq. (6–1) we see that in any system of units, the unit of work is the unit of force multiplied by the unit of distance. In SI units the unit of force is the newton and the unit of distance is the meter, so one joule is equivalent to one *newton-meter* (N · m):

$$1 \text{ joule} = (1 \text{ newton})(1 \text{ meter}) \quad \text{or} \quad 1 \text{ J} = 1 \text{ N} \cdot \text{m}.$$

In the British system the unit of force is the pound (lb), the unit of distance is the foot, and the unit of work is the *foot-pound* (ft · lb). The following conversions are useful:

$$1 \text{ J} = 0.7376 \text{ ft} \cdot \text{lb}, \quad 1 \text{ ft} \cdot \text{lb} = 1.356 \text{ J}.$$

Work done by a force in the direction of motion Steve is trying to impress Elayne with his new car, but the engine dies in the middle of an intersection. While Elayne steers, Steve pushes the car 19 m to clear the intersection. If he pushes in the direction of motion with a constant force of 210 N (about 47 lb), how much work does he do on the car?

SOLUTION From Eq. (6–1),

$$W = Fs = (210 \text{ N})(19 \text{ m}) = 4.0 \times 10^3 \text{ J}.$$

In Example 6–1, Steve pushed the car in the direction he wanted it to go. What if he had pushed at an angle ϕ with the car's displacement (Fig. 6–2)? Only the component of force in the direction of the car's motion, $(210 \text{ N}) \cos \phi$, would be effective in moving the car. (Other forces must act on the car so that it moves along $\vec{s}$, not in the direction of $\vec{F}$. But we're interested only in the work done by Steve, so we'll consider only the force

6–2 When a constant force $\vec{F}$ acts at an angle ϕ to the displacement $\vec{s}$, the work done by the force is $(F \cos \phi)\, s = Fs \cos \phi$.

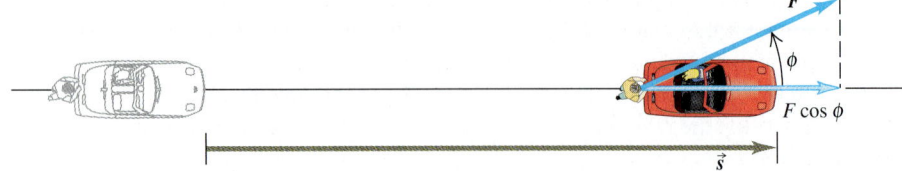

he exerts.) When the force $\vec{F}$ and the displacement $\vec{s}$ have different directions, we take the component of $\vec{F}$ in the direction of the displacement $\vec{s}$, and we define the work as the product of this component and the magnitude of the displacement. The component of $\vec{F}$ in the direction of $\vec{s}$ is $F \cos \phi$, so

$$W = Fs \cos \phi \qquad \text{(constant force, straight-line displacement).} \qquad (6\text{–}2)$$

We are assuming that F and ϕ are constant during the displacement. If $\phi = 0$, so that $\vec{F}$ and $\vec{s}$ are in the same direction, then $\cos \phi = 1$ and we are back to Eq. (6–1).

Equation (6–2) has the form of the *scalar product* of two vectors, introduced in Section 1–11: $\vec{A} \cdot \vec{B} = AB \cos \phi$. You may want to review that definition. Using this, we can write Eq. (6–2) more compactly as

$$W = \vec{F} \cdot \vec{s} \qquad \text{(constant force, straight-line displacement).} \qquad (6\text{–}3)$$

It's important to understand that work is a *scalar* quantity, even though it's calculated by using two vector quantities (force and displacement). A 5-N force toward the east acting on a body that moves 6 m to the east does exactly the same work as a 5-N force toward the north acting on a body that moves 6 m to the north.

It's also important to realize that work can be positive, negative, or zero. This is the essential way in which work as defined in physics differs from the "everyday" definition of work. When the force has a component in the *same direction* as the displacement (ϕ between zero and 90°), $\cos \phi$ in Eq. (6–2) is positive and the work W is *positive* (Fig. 6–3a). When the force has a component *opposite* to the displacement (ϕ between 90° and 180°), $\cos \phi$ is negative and the work is *negative* (Fig. 6–3b). When the force is *perpendicular* to the displacement, $\phi = 90°$ and the work done by the force is *zero* (Fig. 6–3c). The cases of zero work and negative work bear closer examination, so let's look at some examples.

There are many situations in which forces act but do zero work. You might think it's "hard work" to hold this book motionless at arm's length for five minutes, but you actually aren't doing any work at all on the book because there is no displacement. You get tired because the muscle fibers in your arm do work as they continually contract and relax. But this is work done by one part of your arm exerting a force on another part, *not* on the book. (We'll say more in Section 6–3 about work done by one part of a body on another part.) Even when you walk with constant velocity on a level floor while carrying the book, you still do no work on it. The book now has a displacement, but the

A weight lifter exerts an upward force as he raises the barbell upward, and so does *positive* work on it. But when the weight lifter lowers the barbell to the ground, he does *negative* work because the upward force he exerts on the barbell is opposite to its displacement.

6–3 (a) W is positive because $\vec{F}$ has a component in the direction of $\vec{s}$. (b) W is negative because $\vec{F}$ has a component opposite the direction of $\vec{s}$. (c) W is zero because $\vec{F}$ has no component in the direction of $\vec{s}$.

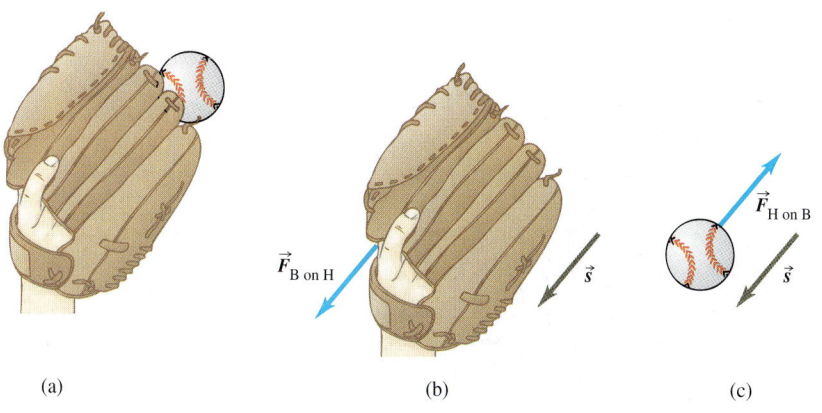

(a)   (b)   (c)

6–4 (a) When an outfielder catches a baseball, his hand and the ball have the same displacement $\vec{s}$. (b) The work done on his hand by the baseball is positive because the force $\vec{F}_{\text{B on H}}$ is in the same direction as the displacement $\vec{s}$. (c) The outfielder's hand exerts an equal and opposite force $\vec{F}_{\text{H on B}} = -\vec{F}_{\text{B on H}}$ on the baseball. The work done on the baseball by the outfielder's hand is negative because $\vec{F}_{\text{H on B}}$ and $\vec{s}$ are in opposite directions.

(vertical) supporting force that you exert on the book has no component in the direction of the (horizontal) motion. Then $\phi = 90°$ in Eq. (6–2), and $\cos\phi = 0$. When a body slides along a surface, the work done on the body by the normal force is zero; and when a ball on a string moves in uniform circular motion, the work done on the ball by the tension in the string is also zero. In both cases the work is zero because the force has no component in the direction of motion.

What does it really mean to do *negative* work? The answer comes from Newton's third law of motion. When a baseball outfielder catches a fly ball as in Fig. 6–4a, the outfielder's hand and the ball move together with the same displacement $\vec{s}$ (Fig. 6–4b). The ball exerts a force $\vec{F}_{\text{B on H}}$ on the outfielder's hand in the same direction as the hand's displacement, so the work done on the outfielder's *hand* by the *ball* is *positive*. But by Newton's third law the outfielder's hand exerts an equal and opposite force $\vec{F}_{\text{H on B}} = -\vec{F}_{\text{B on H}}$ on the ball (Fig. 6–4c). This force, which slows the ball to a stop, acts opposite to the displacement of the ball. Thus the work done on the *ball* by the outfielder's *hand* is *negative*. Since the outfielder's hand and the ball have the same displacement, the work done on the ball by the hand is just the negative of the work done on the hand by the ball. In general, when one body does negative work on a second body, the second body does an equal amount of *positive* work on the first body.

CAUTION ▶ We always speak of work done *on* a particular body *by* a specific force. Always be sure to specify exactly what force is doing the work you are talking about. When you lift a book, you exert an upward force on the book and the book's displacement is upward, so the work done by the lifting force on the book is positive. But the work done by the *gravitational* force (weight) on a book being lifted is *negative* because the downward gravitational force is opposite to the upward displacement. ◀

How do we calculate work when several forces act on a body? One way is to use Eq. (6–2) or (6–3) to compute the work done by each separate force. Then, because work is a scalar quantity, the *total* work W_{tot} done on the body by all the forces is the algebraic sum of the quantities of work done by the individual forces. An alternative way to find the total work W_{tot} is to compute the vector sum of the forces (that is, the net force) and then use this vector sum as $\vec{F}$ in Eq. (6–2) or (6–3).

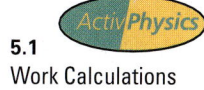

5.1
Work Calculations

EXAMPLE 6–2

Work done by several forces Farmer Johnson hitches his tractor to a sled loaded with firewood and pulls it a distance of 20 m along level frozen ground (Fig. 6–5). The total weight of sled and load is 14,700 N. The tractor exerts a constant 5000-N force at an angle of 36.9° above the horizontal, as shown. There is a 3500-N friction force opposing the motion. Find the work done

by each force acting on the sled and the total work done by all the forces.

SOLUTION Let's do the easy parts first. The work W_w done by the weight is zero because its direction is perpendicular to the displacement. (The angle between the force of gravity and the

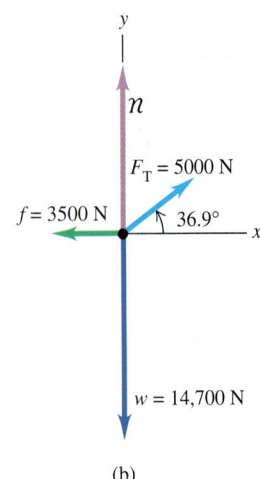

6–5 (a) A tractor pulling a sled of firewood. (b) Free-body diagram for the sled and its load, treated as a particle.

displacement is 90°, and the cosine of the angle is zero.) For the same reason the work W_n done by the normal force is also zero. So $W_w = W_n = 0$. (Incidentally, the normal force is *not* equal in magnitude to the weight; see Example 5–15 of Section 5–4, in which the free-body diagram is very similar to Fig. 6–5b.)

That leaves the force F_T exerted by the tractor and the friction force f. From Eq. (6–2) the work W_T done by the tractor is

$$W_T = F_T s \cos \phi = (5000 \text{ N})(20 \text{ m})(0.800) = 80,000 \text{ N} \cdot \text{m}$$
$$= 80 \text{ kJ}.$$

The friction force $\vec{f}$ is opposite to the displacement, so for this force $\phi = 180°$ and $\cos \phi = -1$. The work W_f done by the friction force is

$$W_f = fs \cos 180° = (3500 \text{ N})(20 \text{ m})(-1) = -70,000 \text{ N} \cdot \text{m}$$
$$= -70 \text{ kJ}.$$

The total work W_{tot} done on the sled by all forces is the *algebraic* sum of the work done by the individual forces:

$$W_{tot} = W_w + W_n + W_T + W_f = 0 + 0 + 80 \text{ kJ} + (-70 \text{ kJ})$$
$$= 10 \text{ kJ}.$$

In the alternative approach, we first find the *vector* sum of all the forces (the net force) and then use it to compute the total work. The vector sum is best found by using components. From Fig. 6–4b,

$$\Sigma F_x = F_T \cos \phi + (-f) = (5000 \text{ N}) \cos 36.9° - 3500 \text{ N}$$
$$= 500 \text{ N},$$

$$\Sigma F_y = F_T \sin \phi + n + (-w)$$
$$= (5000 \text{ N}) \sin 36.9° + n - 14,700 \text{ N}.$$

We don't really need the second equation; we know that the y-component of force is perpendicular to the displacement, so it does no work. Besides, there is no y-component of acceleration, so ΣF_y has to be zero anyway. The total work is therefore the work done by the total x-component:

$$W_{tot} = \left(\Sigma \vec{F} \right) \cdot \vec{s} = (\Sigma F_x)s = (500 \text{ N})(20 \text{ m}) = 10,000 \text{ J}$$
$$= 10 \text{ kJ}.$$

This is the same result that we found by computing the work done by each force separately.

EXAMPLE 6–3

Total work when velocity is constant An electron moves in a straight line toward the east with a constant speed of 8×10^7 m/s. It has electric, magnetic, and gravitational forces acting on it. Calculate the total work done on the electron during a 1-m displacement.

SOLUTION The electron moves with constant velocity, so its acceleration is zero, and by Newton's second law the vector sum of forces is also zero. Therefore the total work done by all the forces (equal to the work done by the vector sum of all the forces) must be *zero*. Individual forces may do nonzero work, but that's not what the problem asks for.

6–3 WORK AND KINETIC ENERGY

The total work done on a body by external forces is related to the body's displacement, that is, to changes in its position. But the total work is also related to changes in the *speed*

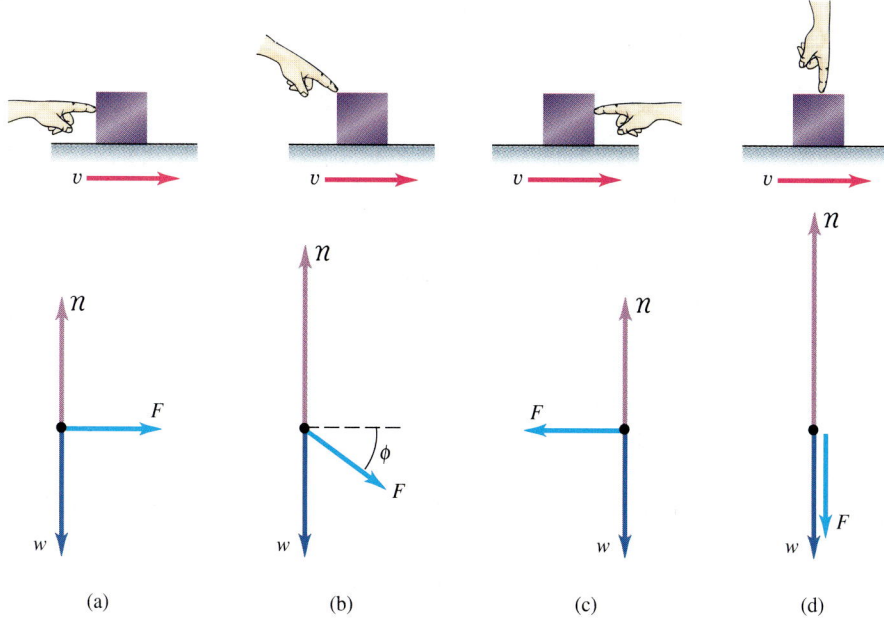

6–6 A block sliding on a friction-less table. (a) The net force causes the speed to increase and does positive work. (b) Again the net force causes the speed to increase and does positive work. (c) The net force opposes the displacement, causes the speed to decrease, and does negative work. (d) The net force is zero and does no work, and the speed is constant.

of the body. To see this, consider Fig. 6–6, which shows several examples of a block sliding on a frictionless table. The forces acting on the block are its weight $\vec{w}$, the normal force $\vec{n}$, and the force $\vec{F}$ exerted on it by the hand.

In Fig. 6–6a the net force on the block is in the direction of its motion. From Newton's second law, this means that the block speeds up; from Eq. (6–1), this also means that the total work W_{tot} done on the block is positive. The total work is also positive in Fig. 6–6b, but only the component $F \cos \phi$ contributes to W_{tot}. The block again speeds up, and this same component $F \cos \phi$ is what causes the acceleration. The total work is *negative* in Fig. 6–6c because the net force opposes the displacement; in this case the block slows down. The net force is zero in Fig. 6–6d, so the speed of the block stays the same and the total work done on the block is zero. We can conclude that when a particle undergoes a displacement, it speeds up if $W_{tot} > 0$, slows down if $W_{tot} < 0$, and maintains the same speed if $W_{tot} = 0$.

Let's make these observations more quantitative. Consider a particle with mass m moving along the x-axis under the action of a constant net force with magnitude F directed along the positive x-axis (Fig. 6–1). The particle's acceleration is constant and given by Newton's second law, $F = ma$. Suppose the speed changes from v_1 to v_2 while the particle undergoes a displacement $s = x_2 - x_1$ from point x_1 to x_2. Using a constant-acceleration equation, Eq. (2–13), and replacing v_0 by v_1, v by v_2, and $(x - x_0)$ by s, we have

$$v_2{}^2 = v_1{}^2 + 2as,$$

$$a = \frac{v_2{}^2 - v_1{}^2}{2s}.$$

When we multiply this equation by m and equate ma to the net force F, we find

$$F = ma = m\frac{v_2{}^2 - v_1{}^2}{2s}$$

and

$$Fs = \frac{1}{2}mv_2{}^2 - \frac{1}{2}mv_1{}^2. \tag{6–4}$$

The product Fs is the work done by the net force F and thus is equal to the total work W_{tot} done by all the forces acting on the particle. The quantity $\frac{1}{2}mv^2$ is called the **kinetic energy** K of the particle:

$$K = \frac{1}{2}mv^2 \qquad \text{(definition of kinetic energy)}. \qquad (6-5)$$

Like work, the kinetic energy of a particle is a scalar quantity; it depends only on the particle's mass and speed, not its direction of motion. A car (viewed as a particle) has the same kinetic energy when going north at 10 m/s as when going east at 10 m/s. Kinetic energy can never be negative, and it is zero only when the particle is at rest.

We can now interpret Eq. (6–4) in terms of work and kinetic energy. The first term on the right side of Eq. (6–4) is $K_2 = \frac{1}{2}mv_2^2$, the final kinetic energy of the particle (that is, after the displacement). The second term is the initial kinetic energy, $K_1 = \frac{1}{2}mv_1^2$, and the difference between these terms is the *change* in kinetic energy. So Eq. (6–4) says that **the work done by the net force on a particle equals the change in the particle's kinetic energy:**

$$W_{\text{tot}} = K_2 - K_1 = \Delta K \qquad \text{(work-energy theorem)}. \qquad (6-6)$$

This result is the **work-energy theorem.**

The work-energy theorem agrees with our observations about the block in Fig. 6–6. When W_{tot} is *positive*, K_2 is greater than K_1, the kinetic energy *increases,* and the particle is going faster at the end of the displacement than at the beginning. When W_{tot} is *negative,* the kinetic energy *decreases* and the speed is less after the displacement. When $W_{\text{tot}} = 0$, the initial and final kinetic energies K_1 and K_2 are the same and the speed is unchanged. We stress that the work-energy theorem by itself tells us only about changes in *speed,* not velocity, since the kinetic energy carries no information about the direction of motion.

From Eq. (6–4) or (6–6), kinetic energy and work must have the same units. Hence the joule is the SI unit of both work and kinetic energy (and, as we will see later, of all kinds of energy). To verify this, note that in SI units the quantity $K = \frac{1}{2}mv^2$ has units $\text{kg} \cdot (\text{m/s})^2$ or $\text{kg} \cdot \text{m}^2/\text{s}^2$; we recall that $1\,\text{N} = 1\,\text{kg} \cdot \text{m/s}^2$, so

$$1\,\text{J} = 1\,\text{N} \cdot \text{m} = 1\,(\text{kg} \cdot \text{m/s}^2) \cdot \text{m} = 1\,\text{kg} \cdot \text{m}^2/\text{s}^2.$$

In the British system the unit of kinetic energy and of work is

$$1\,\text{ft} \cdot \text{lb} = 1\,\text{ft} \cdot \text{slug} \cdot \text{ft/s}^2 = 1\,\text{slug} \cdot \text{ft}^2/\text{s}^2.$$

When a billiards player hits a cue ball at rest, the ball's kinetic energy after being hit is equal to the work that was done on it by the cue. The greater the force exerted by the cue and the greater the distance the ball moves while in contact with it, the greater the ball's kinetic energy.

Because we used Newton's laws in deriving the work-energy theorem, we can use it only in an inertial frame of reference. The speeds that we use to compute the kinetic energies and the distances that we use to compute work *must* be measured in an inertial frame. Note also that the work-energy theorem is valid in *any* inertial frame, but the values of W_{tot} and $K_2 - K_1$ may differ from one inertial frame to another (because the displacement and speed of a body may be different in different frames).

We have derived the work-energy theorem for the special case of straight-line motion with constant forces, and in the following examples we'll apply it to this special case only. We'll find in the next section that the theorem is valid in general, even when the forces are not constant and the particle's trajectory is curved.

Problem–Solving Strategy

CALCULATIONS WITH WORK AND KINETIC ENERGY

1. Choose the initial and final positions of the body, and draw a free-body diagram showing all the forces that act on the body. List the forces, and calculate the work done by each force. In some cases, one or more forces may be unknown; represent the unknowns by algebraic symbols. Be sure to check signs. When a force has a component in the same direction as the displacement, its work is positive; when it has a component opposite to the displacement, the work is negative. When force and displacement are perpendicular, the work is zero.

2. Add the amounts of work done by the separate forces to find the total work. Again be careful with signs. Sometimes it may be easier to calculate the vector sum of the

forces (the net force) first and then find the work done by the net force.

3. Write expressions for the initial and final kinetic energies K_1 and K_2. If a quantity such as v_1 or v_2 is unknown, express it in terms of the corresponding algebraic symbol. When you calculate kinetic energies, make sure you use the *mass* of the body, not its *weight*.

4. Use the relationship $W_{tot} = K_2 - K_1 = \Delta K$; insert the results from the above steps, and solve for whatever unknown is required. Remember that kinetic energy can never be negative. If you come up with a negative K, you've made a mistake. Maybe you interchanged subscripts 1 and 2 or made a sign error in one of the work calculations.

EXAMPLE 6–4

Using work and energy to calculate speed Let's look again at the sled in Fig. 6–5 and the numbers at the end of Example 6–2. The free-body diagram is shown again in Fig. 6–7. We found that the total work done by all the forces is 10,000 J = 10 kJ, so the kinetic energy of the sled must increase by 10 kJ. The mass of the sled and its load is $m = w/g$, equal to $(14{,}700 \text{ N})/(9.8 \text{ m/s}^2) = 1500$ kg. Suppose the initial speed v_1 is 2.0 m/s. What is the final speed?

SOLUTION Steps 1 and 2 of the problem-solving strategy were done in Example 6–2, where we found $W_{tot} = 10$ kJ. The initial kinetic energy K_1 is

$$K_1 = \frac{1}{2} m v_1^2 = \frac{1}{2}(1500 \text{ kg})(2.0 \text{ m/s})^2 = 3000 \text{ kg} \cdot \text{m}^2/\text{s}^2$$

$$= 3000 \text{ J}.$$

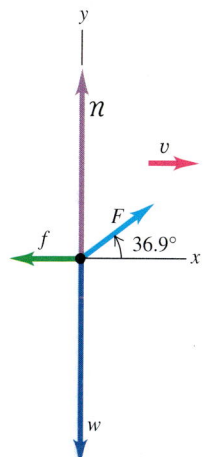

6–7 Free-body diagram for the sled and its load in Example 6–2.

The final kinetic energy K_2 is

$$K_2 = \frac{1}{2} m v_2^2 = \frac{1}{2}(1500 \text{ kg})v_2^2,$$

where v_2 is the unknown speed we want to find. Equation (6–6) gives

$$K_2 = K_1 + W_{tot} = 3000 \text{ J} + 10{,}000 \text{ J} = 13{,}000 \text{ J}.$$

Setting these two expressions for K_2 equal, substituting 1 J = 1 kg $\cdot$ m^2/s^2, and solving for v_2, we find

$$v_2 = 4.2 \text{ m/s}.$$

The total work is positive, so the kinetic energy increases ($K_2 > K_1$) and the speed increases ($v_2 > v_1$).

This problem can also be done without the work-energy theorem. We can find the acceleration from $\Sigma \vec{F} = m\vec{a}$ and then use the equations of motion for constant acceleration to find v_2. Since the acceleration is along the x-axis,

$$a = a_x = \frac{\Sigma F_x}{m} = \frac{(5000 \text{ N})\cos 36.9° - 3500 \text{ N}}{1500 \text{ kg}}$$

$$= 0.333 \text{ m/s}^2;$$

then

$$v_2^2 = v_1^2 + 2as = (2.0 \text{ m/s})^2 + 2(0.333 \text{ m/s}^2)(20 \text{ m})$$

$$= 17.3 \text{ m}^2/\text{s}^2,$$

$$v_2 = 4.2 \text{ m/s}.$$

This is the same result that we obtained with the work-energy approach, but there we avoided the intermediate step of finding the acceleration. You will find several other examples in this chapter and the next that *can* be done without using energy considerations but that are easier when energy methods are used. When a problem can be done by two different methods, doing it by both methods (as we did in this example) is a very good way to check your work.

EXAMPLE 6-5

Forces on a hammerhead In a pile driver, a steel hammerhead with mass 200 kg is lifted 3.00 m above the top of a vertical I-beam being driven into the ground (Fig. 6–8a). The hammer is then dropped, driving the I-beam 7.4 cm farther into the ground. The vertical rails that guide the hammerhead exert a constant 60-N friction force on the hammerhead. Use the work-energy theorem to find a) the speed of the hammerhead just as it hits the I-beam and b) the average force the hammerhead exerts on the I-beam. Ignore the effects of the air.

SOLUTION Figure 6–8b is a free-body diagram showing the vertical forces on the falling hammerhead. Because the displacement is vertical, any horizontal forces that may be present do no work.

a) Let point 1 be the initial position of the hammerhead, and let point 2 be where it just hits the I-beam. The vertical forces are the downward weight $w = mg = (200 \text{ kg})(9.8 \text{ m/s}^2) = 1960$ N and the upward friction force $f = 60$ N. Thus the net downward force is $w - f = 1900$ N. The displacement of the hammerhead from point 1 to point 2 is downward and equal to $s_{12} = 3.00$ m. The total work done on the hammerhead as it moves from point 1 to point 2 is then

$$W_{\text{tot}} = (w - f)s_{12} = (1900 \text{ N})(3.00 \text{ m}) = 5700 \text{ J}.$$

At point 1 the hammerhead is at rest, so its initial kinetic energy K_1 is zero. Equation (6–6) gives

$$W_{\text{tot}} = K_2 - K_1 = \frac{1}{2}mv_2^2 - 0,$$

$$v_2 = \sqrt{\frac{2W_{\text{tot}}}{m}} = \sqrt{\frac{2(5700 \text{ J})}{200 \text{ kg}}} = 7.55 \text{ m/s}.$$

This is the hammerhead's speed at point 2, just as it hits the beam.

b) Let point 3 be where the hammerhead finally comes to rest. Then $K_3 = 0$. As Fig. 6–8c shows, there is now an additional force, the upward normal force n that the beam exerts on the hammerhead during the additional downward displacement $s_{23} = 7.4$ cm. This force actually varies as the hammerhead comes to a halt, but for simplicity we'll treat n as a constant; our result for n will then be an *average* value of the upward force during the motion. The total work done on the hammerhead during the 7.4-cm displacement is

$$W_{\text{tot}} = (w - f - n)s_{23}.$$

This is equal to the change in kinetic energy $K_3 - K_2$, which is negative because the kinetic energy of the hammerhead decreases. So we have

$$(w - f - n)s_{23} = K_3 - K_2,$$

$$n = w - f - \frac{(K_3 - K_2)}{s_{23}}$$

$$= 1960 \text{ N} - 60 \text{ N} - \frac{(0 \text{ J} - 5700 \text{ J})}{0.074 \text{ m}}$$

$$= 79{,}000 \text{ N}.$$

The force that the *hammerhead* exerts on the *I-beam* during this part of the motion is the equal and opposite reaction force of 79,000 N (about 9 tons) downward—more than 40 times the weight of the hammerhead.

The total change in the hammerhead's kinetic energy during the whole process is zero; a relatively small net force does

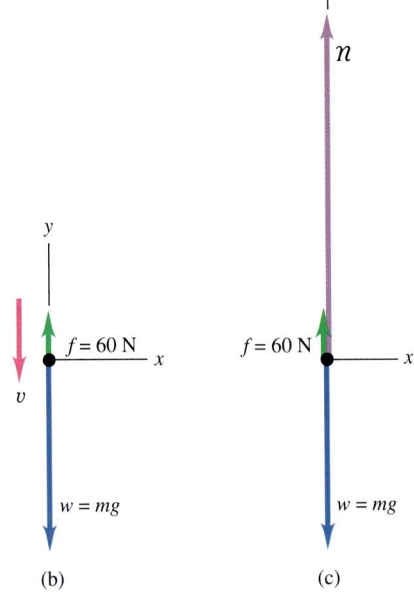

6–8 (a) A pile driver pounds an I-beam into the ground. (b) Free-body diagram for the hammerhead while falling. (c) Free-body diagram for the hammerhead while driving the I-beam. Vector lengths are not to scale.

positive work over a large distance, and then a much larger net force does negative work over a much smaller distance. The same thing happens if you speed your car up gradually and then drive it into a brick wall. The very large force needed to reduce the kinetic energy to zero over a short distance is what does the damage to your car—and possibly to you.

THE MEANING OF KINETIC ENERGY

Example 6–5 gives insight into the physical meaning of kinetic energy. The hammerhead is dropped from rest, and its kinetic energy when it hits the I-beam equals the total work done on it up to that point by the net force. This result is true in general: to accelerate a particle of mass m from rest (zero kinetic energy) up to a speed v, the total work done on it must equal the change in kinetic energy from zero to $K = \frac{1}{2}mv^2$;

$$W_{\text{tot}} = K - 0 = K.$$

So *the kinetic energy of a particle is equal to the total work that was done to accelerate it from rest to its present speed.* The definition $K = \frac{1}{2}mv^2$, Eq. (6–5), wasn't chosen at random; it's the *only* definition that agrees with this interpretation of kinetic energy.

In the second part of Example 6–5 the kinetic energy of the hammerhead was used to do work on the I-beam and drive it into the ground. This gives us another interpretation of kinetic energy: *the kinetic energy of a particle is equal to the total work that particle can do in the process of being brought to rest.* This is why a baseball outfielder pulls his hand and arm back when catching a fly ball (Fig. 6–4). As the ball comes to rest, it does an amount of work (force times distance) on the outfielder's hand equal to the ball's initial kinetic energy. By pulling his hand back, the outfielder maximizes the distance over which the force acts and so minimizes the force exerted on his hand.

EXAMPLE 6-6

Comparing kinetic energies Two iceboats like the one in Example 5–5 (Section 5–3) hold a race on a frictionless horizontal lake (Fig. 6–9). The two iceboats have masses m and $2m$, respectively. Each iceboat has an identical sail, so the wind exerts the same constant force $\vec{F}$ on each iceboat. The two iceboats start from rest and cross the finish line a distance s away. Which iceboat crosses the finish line with greater kinetic energy?

SOLUTION If you simply use the mathematical definition of kinetic energy, $K = \frac{1}{2}mv^2$ (Eq. (6–5)), the answer to this problem isn't immediately obvious. The iceboat of mass $2m$ has greater mass, so you might guess that the larger iceboat attains a greater

kinetic energy at the finish line. But the smaller iceboat, of mass m, will cross the finish line with a greater speed, and you might guess that *this* iceboat will have the greater kinetic energy. How can we decide?

The correct way to approach this problem is to remember that *the kinetic energy of a particle is just equal to the total work done to accelerate it from rest.* Both iceboats travel the same distance s, and only the horizontal force F in the direction of motion does work on either iceboat. Hence the total work done between the starting line and the finish line is the *same* for each iceboat, $W_{\text{tot}} = Fs$. At the finish line, each iceboat has a kinetic energy equal to the work W_{tot} done on it, because each iceboat started from rest. So both iceboats have the *same* kinetic energy at the finish line!

6–9 A race between iceboats.

You might think this is a "trick" question, but it isn't. If you really understand the physical meanings of quantities such as kinetic energy, you can solve problems more easily and with better insight into the physics.

Notice that we didn't need to say anything about how much time each iceboat took to reach the finish line. This is because the work-energy theorem makes no direct reference to time, only to displacement. In fact the iceboat of mass m takes less time to reach the finish line than does the larger iceboat of mass $2m$; we leave the calculation to you (Exercise 6–12).

6–10 The external forces acting on a skater pushing off a wall. The work done by these forces is zero, but his kinetic energy changes nonetheless.

WORK AND KINETIC ENERGY IN COMPOSITE SYSTEMS

You may have noticed that in this section we've been careful to apply the work-energy theorem only to bodies that we can represent as *particles,* that is, as moving point masses. The reason is that new subtleties appear for more complex systems that have to be represented in terms of many particles with different motions. We can't go into these subtleties in detail in this chapter, but here's an example.

Consider a man standing on frictionless roller skates on a level surface, facing a rigid wall (Fig. 6–10). He pushes against the wall, setting himself in motion to the right. The forces acting on him are his weight $\vec{w}$, the upward normal forces $\vec{n}_1$ and $\vec{n}_2$ exerted by the ground on his skates, and the horizontal force $\vec{F}$ exerted on him by the wall. There is no vertical displacement, so $\vec{w}$, $\vec{n}_1$, and $\vec{n}_2$ do no work. The force $\vec{F}$ is the horizontal force that accelerates him to the right, but the parts of his body where that force is applied (the man's hands) do not move. Thus the force $\vec{F}$ also does no work. So where does the man's kinetic energy come from?

The difficulty is that it's simply not adequate to represent the man as a single point mass. For the motion to occur as we've described it, different parts of the man's body must have different motions; his hands are stationary against the wall while his torso is moving away from the wall. The various parts of his body interact with each other, and one part can exert forces and do work on another part. Therefore the *total* kinetic energy of this composite system of body parts can change, even though no work is done by forces applied by bodies (such as the wall) that are outside the system. This would not be possible with a system that can be represented as a single point particle. In Chapter 8 we'll consider further the motion of a collection of particles that interact with each other. We'll discover that just as for the man in this example, the total kinetic energy of such a system can change even when no work is done on any part of the system by anything outside it.

6–4 WORK AND ENERGY WITH VARYING FORCES

So far in this chapter we've considered work done by *constant forces* only. But what happens when you stretch a spring? The more you stretch it, the harder you have to pull, so the force you exert is *not* constant as the spring is stretched. We've also restricted our discussion to *straight-line* motion. You can think of many situations in which a force that varies in magnitude, direction, or both acts on a body moving along a curved path. We need to be able to compute the work done by the force in these more general cases. Fortunately, we'll find that the work-energy theorem holds true even when varying forces are considered and when the body's path is not straight.

To add only one complication at a time, let's consider straight-line motion with a force that is directed along the line but with an x-component F that may change as the body moves. For example, imagine a train moving on a straight track with the engineer constantly changing the locomotive's throttle setting or applying the brakes. Suppose a particle moves along the x-axis from point x_1 to x_2. Figure 6–11a is a graph of the

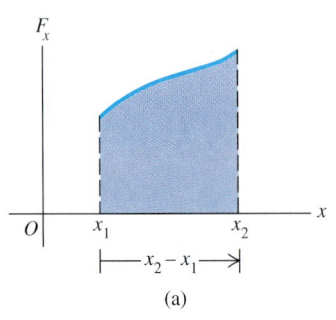

6–11 (a) Curve showing how a certain force F varies with x.

x-component of force as a function of the particle's coordinate x. To find the work done by this force, we divide the total displacement into small segments Δx_a, Δx_b, and so on (Fig. 6–11b). We approximate the work done by the force during segment Δx_a as the average force F_a in that segment multiplied by the displacement Δx_a. We do this for each segment and then add the results for all the segments. The work done by the force in the total displacement from x_1 to x_2 is approximately

$$W = F_a \Delta x_a + F_b \Delta x_b + \cdots .$$

As the number of segments becomes very large and the width of each becomes very small, this sum becomes (in the limit) the *integral* of F from x_1 to x_2:

$$W = \int_{x_1}^{x_2} F \, dx \qquad \text{(varying x-component of force, straight-line displacement)}.$$

(6–7)

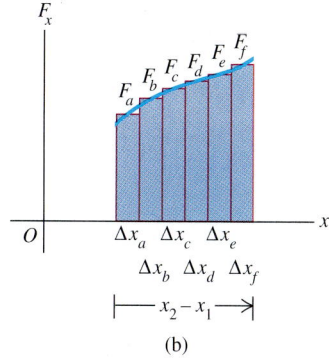

6-11 (b) If the area is partitioned into small rectangles, the sum of their areas approximates the total work done during the displacement; the greater the number of rectangles used, the closer the approximation. The total area under the curve equals the work done by the force as the particle moves from x_1 to x_2.

Note that $F_a \Delta x_a$ represents the *area* of the first vertical strip in Fig. 6–11b and that the integral in Eq. (6–7) represents the area under the curve of Fig. 6–11a between x_1 and x_2. *On a graph of force as a function of position, the total work done by the force is represented by the area under the curve between the initial and final positions.* An alternative interpretation of Eq. (6–7) is that the work W equals the average force that acts over the entire displacement, multiplied by the displacement.

Equation (6–7) also applies if F, the x-component of the force, is constant. In that case, F may be taken outside the integral:

$$W = \int_{x_1}^{x_2} F \, dx = F \int_{x_1}^{x_2} dx = F(x_2 - x_1).$$

But $x_2 - x_1 = s$, the total displacement of the particle. So in the case of a constant force F, Eq. (6–7) says that $W = Fs$, in agreement with Eq. (6–1). The interpretation of work as the area under the curve of F as a function of x also holds for a constant force; $W = Fs$ is the area of a rectangle of height F and width s (Fig. 6–12).

Now let's apply what we've learned to the stretched spring. To keep a spring stretched beyond its unstretched length by an amount x, we have to apply a force with magnitude F at each end (Fig. 6–13). If the elongation x is not too great, we find that F is directly proportional to x:

$$F = kx \qquad \text{(force required to stretch a spring)},$$

(6–8)

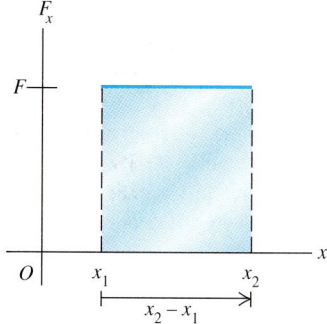

6-12 The work done by a constant force F in the x-direction as a particle moves from x_1 to x_2 is equal to the rectangular area under the graph of force versus displacement.

where k is a constant called the **force constant** (or spring constant) of the spring. Equation (6–8) shows that the units of k are force divided by distance, N/m in SI units and lb/ft in British units. A floppy toy spring such as a Slinky™ has a force constant of about 1 N/m; for the much stiffer springs in an automobile's suspension, k is about 10^5 N/m. The observation that elongation is directly proportional to force for elongations that are not too great was made by Robert Hooke in 1678 and is known as **Hooke's law.** It really shouldn't be called a "law," since it's a statement about a specific device and not a fundamental law of nature. Real springs don't always obey Eq. (6–8) precisely, but it's still a useful idealized model. We'll discuss Hooke's law more fully in Chapter 11.

To stretch a spring, we must do work. We apply equal and opposite forces to the ends of the spring and gradually increase the forces. We hold the left end stationary, so the force we apply at this end does no work. The force at the moving end *does* do work. Figure 6–14 is a graph of F as a function of x, the elongation of the spring. The work done by F when the elongation goes from zero to a maximum value X is

$$W = \int_{0}^{X} F \, dx = \int_{0}^{X} kx \, dx = \frac{1}{2} k X^2 .$$

(6–9)

6-13 The force needed to stretch an ideal spring is proportional to its elongation: $F = kx$.

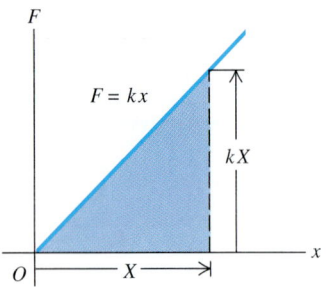

6–14 The work done to stretch a spring by a length X is equal to the triangular area under the graph of force versus displacement.

We can also obtain this result graphically. The area of the shaded triangle in Fig. 6–14, representing the total work done by the force, is equal to half the product of the base and altitude, or

$$W = \frac{1}{2}(X)(kX) = \frac{1}{2}kX^2.$$

This equation also says that the work is the *average* force $kX/2$ multiplied by the total displacement X. We see that the total work is proportional to the *square* of the final elongation X. To stretch an ideal spring by 2 cm, you must do four times as much work as is needed to stretch it by 1 cm.

Equation (6–9) assumes that the spring was originally unstretched. If initially the spring is already stretched a distance x_1, the work we must do to stretch it to a greater elongation x_2 is

$$W = \int_{x_1}^{x_2} F\, dx = \int_{x_1}^{x_2} kx\, dx = \frac{1}{2}kx_2{}^2 - \frac{1}{2}kx_1{}^2. \tag{6–10}$$

If the spring has spaces between the coils when it is unstretched, then it can also be compressed, and Hooke's law holds for compression as well as stretching. In this case the force F and displacement x are in the opposite directions from those shown in Fig. 6–13, and so F and x in Eq. (6–8) will both be negative. Since both F and x are reversed, the force again is in the same direction as the displacement, and the work done by F is again positive. So the total work is still given by Eq. (6–9) or (6–10), even when X is negative or either or both of x_1 and x_2 are negative.

EXAMPLE 6–7

Work done on a spring scale A woman weighing 600 N steps on a bathroom scale containing a stiff spring (Fig. 6–15). In equilibrium the spring is compressed 1.0 cm under her weight. Find the force constant of the spring and the total work done on it during the compression.

SOLUTION In equilibrium the net force on the woman is zero, so the woman's weight and the spring force acting on her have the same 600-N magnitude but opposite directions. We take positive values of x to correspond to elongation, so $x = -1.0$ cm $= -0.010$ m, and the force that the woman applies to the spring is $F = -600$ N. From Eq. (6–8) the force constant k is

$$k = \frac{F}{x} = \frac{-600\text{ N}}{-0.010\text{ m}} = 60{,}000\text{ N/m}.$$

Then, from Eq. (6–9),

$$W = \frac{1}{2}kX^2 = \frac{1}{2}(60{,}000\text{ N/m})(-0.010\text{ m})^2 = 3.0\text{ J}.$$

6–15 Compressing a spring in a bathroom scale. A negative force causes a negative displacement but does a positive amount of work.

WORK-ENERGY THEOREM FOR STRAIGHT-LINE MOTION, VARYING FORCES

In Section 6–3 we derived the work-energy theorem, $W_{\text{tot}} = K_2 - K_1$, for the special case of straight-line motion with a constant net force. We can now prove that this theorem is true even when the force varies with position. As in Section 6–3, let's consider a particle that undergoes a displacement x while being acted on by a net force with x-component F, which we now allow to vary. Just as in Fig. 6–11, we divide the total displacement x into a large number of small segments Δx. We can apply the work-energy theorem, Eq. (6–6), to each segment because the value of F in each small segment is approximately constant. The change in kinetic energy in segment Δx_a is equal to the work $F_a \Delta x_a$, and so on. The total change of kinetic energy is the sum of the changes in the individual segments, and thus is equal to the total work done on the particle during the entire displacement. So $W_{tot} = \Delta K$ holds for varying forces as well as for constant ones.

Here's an alternative derivation of the work-energy theorem for a force that may vary with position. It involves making a change of variable from x to v in the work integral. As a preliminary, we note that the acceleration a of the particle can be expressed in various ways, using $a = dv/dt$, $v = dx/dt$, and the chain rule for derivatives:

$$a = \frac{dv}{dt} = \frac{dv}{dx}\frac{dx}{dt} = v\frac{dv}{dx}.$$
(6–11)

Using this result, Eq. (6–7) tells us that the total work done by the *net* force F is

$$W_{tot} = \int_{x_1}^{x_2} F\,dx = \int_{x_1}^{x_2} ma\,dx = \int_{x_1}^{x_2} mv\,\frac{dv}{dx}\,dx.$$
(6–12)

Now $(dv/dx)\,dx$ is the change in velocity dv during the displacement dx, so in Eq. (6–12) we can substitute dv for $(dv/dx)\,dx$. This changes the integration variable from x to v, so we change the limits from x_1 and x_2 to the corresponding velocities v_1 and v_2 at these points. This gives us

$$W_{tot} = \int_{v_1}^{v_2} mv\,dv.$$

The integral of $v\,dv$ is just $v^2/2$. Substituting the upper and lower limits, we finally find

$$W_{tot} = \frac{1}{2}mv_2{}^2 - \frac{1}{2}mv_1{}^2.$$
(6–13)

This is the same result as Eq. (6–6) but without the assumption that the net force F is constant. So the work-energy theorem is valid even when F varies during the displacement.

EXAMPLE 6–8

Motion with a varying force An air-track glider of mass 0.100 kg is attached to the end of a horizontal air track by a spring with force constant 20.0 N/m (Fig. 6–16a). Initially the spring is unstretched and the glider is moving at 1.50 m/s to the right. Find the maximum distance d that the glider moves to the right a) if the air track is turned on so that there is no friction, and b) if the air is turned off so that there is kinetic friction with coefficient $\mu_k = 0.47$.

SOLUTION Figures 6–16b and 6–16c show the free-body diagrams for the glider without and with friction, respectively. The force exerted by the spring is not constant, so we *cannot* use the constant-acceleration formulas of Chapter 2 to solve this

(a)

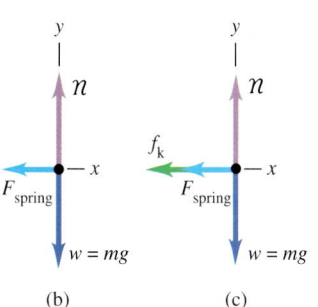

(b)   (c)

6–16 (a) A glider attached to an air track by a spring. (b) Free-body diagram for the glider without friction. (c) Free-body diagram for the glider with kinetic friction.

problem. Instead, we'll use the work-energy theorem.

a) The glider moves only horizontally, so only the horizontal spring force does work. As it moves a distance d to the right, the glider stretches the spring by d and does an amount of work on the spring equal to $\frac{1}{2}kd^2$. The *spring* does an amount of work on the *glider* equal to the negative of this, or $-\frac{1}{2}kd^2$. The spring stretches until the glider comes instantaneously to rest, so the glider's final kinetic energy is zero. The initial kinetic energy of the glider is $\frac{1}{2}mv_1^2$. Using the work-energy theorem, we find

$$-\frac{1}{2}kd^2 = 0 - \frac{1}{2}mv_1^2,$$

so the distance the glider moves is

$$d = v_1\sqrt{\frac{m}{k}} = (1.50 \text{ m/s})\sqrt{\frac{0.100 \text{ kg}}{20.0 \text{ N/m}}}$$

$$= 0.106 \text{ m} = 10.6 \text{ cm}.$$

The stretched spring pulls the glider back to the left, so the glider is at rest only instantaneously.

b) If the air is turned off, we must also include the work done by the constant force of kinetic friction. The normal force n is equal in magnitude to the weight of the glider, since the track is horizontal and there are no other vertical forces. The magnitude of the kinetic friction force is then $f_k = \mu_k n = \mu_k mg$. The friction force is directed opposite to the displacement, so the work done by friction is

$$W_{\text{fric}} = f_k d \cos 180° = -f_k d = -\mu_k mgd.$$

The total work is the sum of W_{fric} and the work done by the spring, $-\frac{1}{2}kd^2$. Hence

$$-\mu_k mgd - \frac{1}{2}kd^2 = 0 - \frac{1}{2}mv_1^2,$$

$$-(0.47)(0.100 \text{ kg})(9.8 \text{ m/s}^2)d - \frac{1}{2}(20.0 \text{ N/m})d^2$$

$$= -\frac{1}{2}(0.100 \text{ kg})(1.50 \text{ m/s})^2,$$

$$(10.0 \text{ N/m})d^2 + (0.461 \text{ N})d - (0.113 \text{ N} \cdot \text{m}) = 0.$$

This is a quadratic equation for d. The solutions are

$$d = \frac{-(0.461 \text{ N}) \pm \sqrt{(0.461 \text{ N})^2 - 4(10.0 \text{ N/m})(-0.113 \text{ N} \cdot \text{m})}}{2(10.0 \text{ N/m})}$$

$$= 0.086 \text{ m} \quad \text{or} \quad -0.132 \text{ m}.$$

We have used d as the symbol for a positive displacement, so only the positive value of d makes sense. Thus with friction the glider moves a distance

$$d = 0.086 \text{ m} = 8.6 \text{ cm}.$$

With friction present, the glider goes a shorter distance and the spring stretches less, as you might expect. Again the glider stops instantaneously, and again the spring force pulls the glider to the left; whether it moves or not depends on how great the *static* friction force is. How large would the coefficient of static friction μ_s have to be to keep the glider from springing back to the left?

WORK-ENERGY THEOREM FOR MOTION ALONG A CURVE

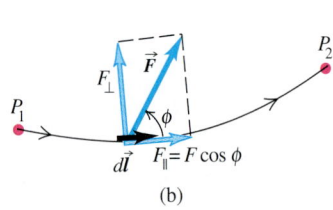

6–17 (a) A particle moves along a curved path from point P_1 to P_2, acted on by a force $\vec{F}$ that varies in magnitude and direction. During an infinitesimal displacement $d\vec{l}$ (a small segment of the path), the work dW done by the force is given by $dW = \vec{F} \cdot d\vec{l} = F \cos \phi \, dl$. (b) The force contributing to the work is the component of force parallel to the displacement, $F_\parallel = F \cos \phi$.

We can generalize our definition of work further to include a force that varies in direction as well as magnitude, and a displacement that lies along a curved path. Suppose a particle moves from point P_1 to P_2 along a curve, as shown in Fig. 6–17a. We divide the portion of the curve between these points into many infinitesimal vector displacements, and we call a typical one of these $d\vec{l}$. Each $d\vec{l}$ is tangent to the path at its position. Let $\vec{F}$ be the force at a typical point along the path, and let ϕ be the angle between $\vec{F}$ and $d\vec{l}$ at this point. Then the small element of work dW done on the particle during the displacement $d\vec{l}$ may be written as

$$dW = F \cos \phi \, dl = F_\parallel \, dl = \vec{F} \cdot d\vec{l},$$

where $F_\parallel = F \cos \phi$ is the component of $\vec{F}$ in the direction parallel to $d\vec{l}$ (Fig. 6–17b). The total work done by $\vec{F}$ on the particle as it moves from P_1 to P_2 is then

$$W = \int_{P_1}^{P_2} F \cos \phi \, dl = \int_{P_1}^{P_2} F_\parallel \, dl = \int_{P_1}^{P_2} \vec{F} \cdot d\vec{l} \quad \text{(work done on a curved path).}$$

$$(6-14)$$

We can now show that the work-energy theorem, Eq. (6–6), holds true even with varying forces and a displacement along a curved path. The force $\vec{F}$ is essentially constant over any given infinitesimal segment $d\vec{l}$ of the path, so we can apply the work-energy theorem for straight-line motion to that segment. Thus the change in the particle's kinetic energy K over that segment equals the work $dW = F_\parallel \, dl = \vec{F} \cdot d\vec{l}$ done on the particle. Adding up these infinitesimal quantities of work from all the segments along the whole path gives the total work done, Eq. (6–14), and this equals the total change in kinetic energy over the whole path. So $W_{\text{tot}} = \Delta K = K_2 - K_1$ is true *in general*, no matter what the path and no matter what the character of the forces. This can be proved more rigorously by using steps like those in Eqs. (6–11) through (6–13) (see Challenge Problem 6–90).

Note that only the component of the net force parallel to the path, $F_\parallel$, does work on the particle, so only this component can change the speed and kinetic energy of the particle. The component perpendicular to the path, $F_\perp = F \sin \phi$, has no effect on the particle's speed; it only acts to change the particle's direction.

The integral in Eq. (6–14) is called a *line integral*. To evaluate this integral in a specific problem, we need some sort of detailed description of the path and of how $\vec{F}$ varies along the path. We usually express the line integral in terms of some scalar variable, as in the following example.

EXAMPLE 6–9

Motion on a curved path At a family picnic you are appointed to push your obnoxious cousin Throckmorton in a swing (Fig. 6–18a). His weight is w, the length of the chains is R, and you push Throcky until the chains make an angle θ_0 with the vertical. To do this, you exert a varying horizontal force $\vec{F}$ that starts at zero and gradually increases just enough so that Throcky and the swing move very slowly and remain very nearly in equilibrium. What is the total work done on Throcky by all forces? What is the work done by the tension T in the chains? What is the work you do by exerting the force $\vec{F}$? (Neglect the weight of chains and seat.)

SOLUTION The free-body diagram is shown in Fig. 6–18b. We have replaced the tensions in the two chains with a single tension T. Because Throcky is in equilibrium at every point, the net force on him is zero, and the total work done on him by all

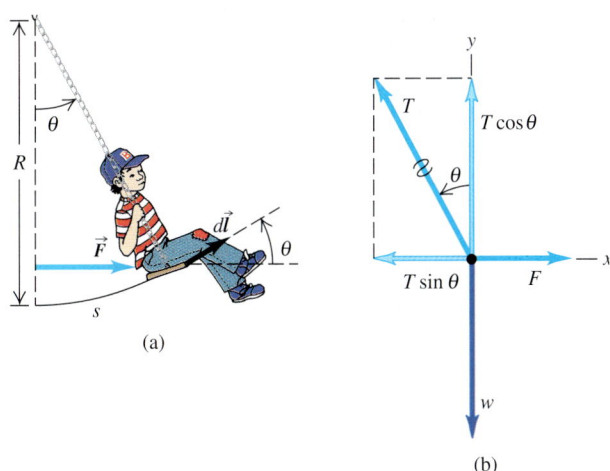

6–18 (a) Pushing cousin Throckmorton in a swing. (b) Free-body diagram for Throcky, considered as a particle and neglecting the small weight of chains and seat.

forces is zero. At any point during the motion the chain force on Throcky is perpendicular to each $d\vec{l}$, so the angle between the chain force and the displacement is always 90°. Therefore the work done by the chain tension is zero.

To compute the work done by $\vec{F}$, we have to find out how it varies with the angle θ. Throcky is in equilibrium at every point, so from $\Sigma F_x = 0$ we get

$$F + (-T \sin \theta) = 0,$$

and from $\Sigma F_y = 0$ we find

$$T \cos \theta + (-w) = 0.$$

By eliminating T from these two equations, we obtain

$$F = w \tan \theta.$$

The point where $\vec{F}$ is applied swings through the arc s. The arc length s equals the radius R of the circular path multiplied by the length θ (in radians), so $s = R\theta$. Therefore the displacement $d\vec{l}$ corresponding to a small change of angle $d\theta$ has a magnitude $dl = ds = R \, d\theta$. The work done by $\vec{F}$ is

$$W = \int \vec{F} \cdot d\vec{l} = \int F \cos \theta \; ds.$$

Now we express everything in terms of the varying angle θ:

$$W = \int_0^{\theta_0} (w \tan \theta) \cos \theta \, (R \, d\theta) = wR \int_0^{\theta_0} \sin \theta \, d\theta$$

$$= wR (1 - \cos \theta_0).$$

If $\theta_0 = 0$, there is no displacement; in that case, $\cos \theta_0 = 1$ and $W = 0$, as we should expect. If $\theta_0 = 90°$, then $\cos \theta_0 = 0$ and $W = wR$. In that case the work you do is the same as if you had lifted Throcky straight up a distance R with a force equal to his weight w. In fact, the quantity $R (1 - \cos \theta_0)$ is the increase in his height above the ground during the displacement, so for any value of θ_0 the work done by force $\vec{F}$ is the change in height multiplied by the weight. This is an example of a more general result that we'll prove in Section 7–2.

6–5 POWER

The definition of work makes no reference to the passage of time. If you lift a barbell weighing 400 N through a vertical distance of 0.5 m at constant velocity, you do

Propeller-driven airliners of the 1950s had maximum speeds of about 600 km/h (370 mi/h). By contrast, modern jet airliners are much larger and fly much faster— about 900 km/h (560 mi/h). This is due in large part to a nearly 20-fold increase in engine power, from 3400 hp (2.5×10^6 W) for a large 1950's piston engine to over 66,000 hp (4.9×10^7 W) for a modern jet engine (see Example 6–10).

(400 N)(0.5 m) = 200 J of work whether it takes you 1 second, 1 hour, or 1 year to do it. But often we need to know how quickly work is done. We describe this in terms of *power.* In ordinary conversation the word "power" is often synonymous with "energy" or "force." In physics we use a much more precise definition: **power** is the time *rate* at which work is done. Like work and energy, power is a scalar quantity.

When a quantity of work ΔW is done during a time interval Δt, the average work done per unit time or **average power** P_{av} is defined to be

$$P_{av} = \frac{\Delta W}{\Delta t} \qquad \text{(average power)}. \tag{6–15}$$

The rate at which work is done might not be constant. Even when it varies, we can define **instantaneous power** P as the limit of the quotient in Eq. (6–15) as Δt approaches zero:

$$P = \lim_{\Delta t \to 0} \frac{\Delta W}{\Delta t} = \frac{dW}{dt} \qquad \text{(instantaneous power)}. \tag{6–16}$$

The SI unit of power is the **watt** (W), named for the English inventor James Watt. One watt equals one joule per second (1 W = 1 J/s). The kilowatt (1 kW = 10^3 W) and the megawatt (1 MW = 10^6 W) are also commonly used. In the British system, work is expressed in foot-pounds, and the unit of power is the foot-pound per second. A larger unit called the *horsepower* (hp) is also used:

$$1 \text{ hp} = 550 \text{ ft} \cdot \text{lb/s} = 33{,}000 \text{ ft} \cdot \text{lb/min}.$$

That is, a 1-hp motor running at full load does 33,000 ft · lb of work every minute. A useful conversion factor is

$$1 \text{ hp} = 746 \text{ W} = 0.746 \text{ kW},$$

or 1 horsepower equals about $\frac{3}{4}$ of a kilowatt.

The watt is a familiar unit of *electrical* power; a 100-W light bulb converts 100 J of electrical energy into light and heat each second. But there's nothing inherently electrical about a watt or a kilowatt. A light bulb could be rated in horsepower, and some automobile manufacturers rate their engines in kilowatts rather than horsepower.

The units of power can be used to define new units of work or energy. The *kilowatt-hour* (kWh) is the usual commercial unit of electrical energy. One kilowatt-hour is the total work done in 1 hour (3600 s) when the power is 1 kilowatt (10^3 J/s), so

$$1 \text{ kWh} = (10^3 \text{ J/s})(3600 \text{ s}) = 3.6 \times 10^6 \text{ J} = 3.6 \text{ MJ}.$$

The kilowatt-hour is a unit of *work* or *energy,* not power.

In mechanics we can also express power in terms of force and velocity. Suppose that a force $\vec{F}$ acts on a body while it undergoes a vector displacement $\Delta \vec{s}$. If $F_{\parallel}$ is the component of $\vec{F}$ tangent to the path (parallel to $\Delta \vec{s}$), then the work done by the force is $\Delta W = F_{\parallel} \Delta s$, and the average power is

$$P_{av} = \frac{F_{\parallel} \Delta s}{\Delta t} = F_{\parallel} \frac{\Delta s}{\Delta t} = F_{\parallel} v_{av}. \tag{6–17}$$

Instantaneous power P is the limit of this as $\Delta t \to 0$:

$$P = F_{\parallel} v, \tag{6–18}$$

where v is the magnitude of the instantaneous velocity. We can also express Eq. (6–18) in terms of the scalar product:

$$P = \vec{F} \cdot \vec{v} \qquad \text{(instantaneous rate at which force } \vec{F} \text{ does work on a particle)}. \tag{6–19}$$

EXAMPLE 6-10

Each of the two jet engines on a Boeing 767 airliner develops a thrust (a forward force on the airplane) of 197,000 N (44,300 lb). When the airplane is flying at 250 m/s (900 km/h, or roughly 560 mi/h), what horsepower does each engine develop?

SOLUTION The force is in the same direction as the velocity, so $F = F_\parallel$. From Eq. (6–18),

$$P = Fv = (1.97 \times 10^5 \text{ N})(250 \text{ m/s}) = 4.93 \times 10^7 \text{ W}$$

$$= (4.93 \times 10^7 \text{ W}) \frac{1 \text{ hp}}{746 \text{ W}} = 66,000 \text{ hp}.$$

EXAMPLE 6-11

As part of a charity fund-raising drive, a Chicago marathon runner with mass 50.0 kg runs up the stairs to the top of the 443-m-tall Sears Tower, the tallest building in the United States (Fig. 6–19). In order to lift herself to the top in 15.0 minutes, what must be her average power output in watts? In kilowatts? In horsepower?

SOLUTION We'll treat the runner as a particle of mass m. As in Example 6–9, lifting a mass m against gravity requires an amount of work equal to the weight mg multiplied by the height h it is lifted. Hence the work she must do is

$$W = mgh = (50.0 \text{ kg})(9.80 \text{ m/s}^2)(443 \text{ m})$$

$$= 2.17 \times 10^5 \text{ J}.$$

The time is 15.0 min = 900 s, so from Eq. (6–15) the average power is

$$P_{av} = \frac{2.17 \times 10^5 \text{ J}}{900 \text{ s}} = 241 \text{ W} = 0.241 \text{ kW} = 0.323 \text{ hp}.$$

An alternative approach is to use Eq. (6–17). The force exerted is vertical, and the average vertical component of velocity is (443 m)/(900 s) = 0.492 m/s, so the average power is

$$P_{av} = F_\parallel v_{av} = (mg)v_{av}$$

$$= (50.0 \text{ kg})(9.80 \text{ m/s}^2)(0.492 \text{ m/s}) = 241 \text{ W}.$$

In fact the runner's *total* power output will be several times greater than we have calculated. The reason is that the runner isn't really a particle but a collection of parts that exert forces on each other and do work, such as the work done to inhale and exhale and to make her arms and legs swing. What we've calculated is only the part of her power output that goes into lifting her to the top of the building.

6–19 How much power is required to run up the stairs of Chicago's Sears Tower in 15 minutes?

6–6 **AUTOMOTIVE POWER**

A Case Study in Energy Relations

The power requirements of a gasoline-powered automobile are an important and practical example of the concepts in this chapter. If roads were flat and frictionless and air resistance didn't exist, there would be no need for an automobile to have an engine. All you'd need to go for a drive would be a few strong friends to give you a push to get started and a few other friends at your destination to stop you. (Steering on frictionless roads would be a problem, though.) In the real world, however, a moving car without an engine slows down because of forces that resist its motion. The engine's function is to continuously provide power to overcome this resistance. So to understand how much power is required from a car's engine, we must analyze the forces that act on the car.

Two forces oppose the motion of an automobile: rolling friction and air resistance. We described rolling friction in Section 5–4 in terms of a coefficient of rolling friction μ_r. A typical value of μ_r for properly inflated tires on hard pavement is 0.015. A Porsche 911 Carrera has a mass of 1251 kg and a weight of (1251 kg)(9.80 m/s²) = 12,260 N, and so the resisting force of rolling friction on a level road (where the normal force $n = mg$) is

$$F_{roll} = \mu_r n = (0.015)(12{,}260 \text{ N}) = 180 \text{ N}.$$

This force is nearly independent of car speed.

The air resistance force F_{air} is approximately proportional to the square of the speed and can be expressed by the equation

$$F_{air} = \frac{1}{2} CA\rho v^2, \tag{6–20}$$

where A is the silhouette area of the car (seen from the front), ρ is the density of air (about 1.2 kg/m^3 at sea level at ordinary temperatures), v is the car's speed, and C is a dimensionless constant called the *drag coefficient* that depends on the shape of the moving body. Typical values of C for cars range from 0.35 to 0.50; for the 911 Carrera, $C = 0.38$ and $A = 1.77$ m^2. For this car the air-resistance force is

$$F_{air} = \frac{1}{2}(0.38)(1.77 \text{ m}^2)(1.2 \text{ kg/m}^3)v^2 = (0.40 \text{ N} \cdot \text{s}^2/\text{m}^2)v^2.$$

In a residential speed zone, where $v = 10$ m/s (36 km/h, or about 22 mi/h), the air-resistance force is about

$$F_{air} = (0.40 \text{ N} \cdot \text{s}^2/\text{m}^2)(10 \text{ m/s})^2 = 40 \text{ N}.$$

At a moderate speed of 15 m/s (54 km/h or 34 mi/h), F_{air} is 90 N, and at a highway speed of 30 m/s (110 km/h or 67 mi/h) it is 360 N. Thus at slow speeds, air resistance is less important than rolling friction. At moderate speeds they are comparable, and at highway speeds air resistance dominates.

What does this mean in terms of the *power* needed from the engine? In constant-speed driving on a level road, the sum of F_{roll} and F_{air} must be just balanced by the forward force $F_{forward}$ supplied by the drive wheels. (The drive wheels push backward on the pavement, and the pavement pushes forward on the drive wheels.) The power is just this forward force multiplied by the speed v. For the 911 Carrera the power needed for constant speed v is

$$P = F_{forward}v = (F_{roll} + F_{air})v = [180 \text{ N} + (0.40 \text{ N} \cdot \text{s}^2/\text{m}^2)v^2]v.$$

For the three speeds mentioned above, you can do the arithmetic yourself to find the following results:

v (m/s)	F_{roll} (N)	F_{air} (N)	$F_{forward}$ (N)	P (kW)	P (hp)
10	180	40	220	2.2	2.9
15	180	90	270	4.1	5.5
30	180	360	540	16	22

How much fuel must be consumed in the engine to provide this power? Burning one liter (1 L) of gasoline releases about 3.5×10^7 J of energy. But not all of this is converted into useful work. The laws of thermodynamics, which we'll encounter in Chapters 17 and 18, impose fundamental limits on the efficiency of converting heat to work. In a typical car engine, about 65% of the heat released from gasoline combustion is wasted in the cooling system and the exhaust. Another 20% or so is converted to work that does nothing to propel the car; this includes work done to oppose friction in the drive train and to run accessories such as the air conditioner and power steering. This leaves only

External antenna · Driveshaft · Stand-out mirrors

Non-aerodynamic steel body

Iron V-8 engine

Flush-mounted windshield with built-in antenna

Aerodynamic composite body

Faired rear-view mirrors

Differential & rear axle

Chrome grille & bumper

Belt-driven cooling fan

Steel box frame

Transverse 4-cylinder fuel-injected computer-controlled aluminum front wheel drive engine

Air dam

Trailing arm suspension

Thermostatically controlled electric cooling fan

(a)   (b)

6–20 (a) Cars designed in the early 1970s did not feature aerodynamic styling and used heavy materials such as iron for engines and steel for body panels. Rear-wheel drive required a heavy drive train as well. (b) By the 1990s, economics forced new ideas in car designs. Body shapes had lower drag coefficients, engines were built of aluminum, and body panels were often made of plastic. These changes doubled the typical fuel efficiency of cars.

about 15% of the energy to do work against the rolling friction and air resistance we described above. The available energy per liter is then

$$(0.15)(3.5 \times 10^7 \text{ J/L}) = 5.3 \times 10^6 \text{ J/L}. \qquad (6\text{–}21)$$

Let's look at the fuel consumption in the 15-m/s case. The power required is 4.1 kW = 4100 J/s. In 1 hour (3600 s) the total energy required is

$$(4100 \text{ J/s})(3600 \text{ s}) = 1.5 \times 10^7 \text{ J},$$

and during that hour the car travels a distance of

$$(15 \text{ m/s})(3600 \text{ s}) = 5.4 \times 10^4 \text{ m} = 54 \text{ km}.$$

From Eq. (6–21) the amount of fuel consumed in 1 hour, traveling 54 km at 15 m/s, is

$$\frac{1.5 \times 10^7 \text{ J}}{5.3 \times 10^6 \text{ J/L}} = 2.8 \text{ L}.$$

That amount of gasoline moves the car 54 km, so the distance traveled per liter of fuel is (54 km)/(2.8 L) = 19 km/L, or about 45 miles per gallon. (Figure 6–20 shows some of the features of contemporary car design that have improved fuel efficiency.)

The power required for a steady 15 m/s on level ground is 4.1 kW, but the power required for acceleration and hill climbing may be much greater. The 911 Carrera is advertised as going from zero to 60 mi/h (27 m/s) in 6.1 s. The final kinetic energy is then

$$K = \frac{1}{2}mv^2 = \frac{1}{2}(1251 \text{ kg})(27 \text{ m/s})^2 = 4.6 \times 10^5 \text{ J}.$$

The average additional power required for the acceleration is

$$P_{\text{av}} = \frac{4.6 \times 10^5 \text{ J}}{6.1 \text{ s}} = 7.5 \times 10^4 \text{ W} = 75 \text{ kW} = 100 \text{ hp}.$$

This rapid acceleration requires about 18 times as much power as cruising at a steady 15 m/s (not including the power to overcome friction in the drive train). For the record, the 911 Carrera advertises a maximum horsepower of 214 hp at an engine speed of 5900 rpm.

What about hill climbing? A 5% grade, about the maximum found on most interstate highways, rises 5 meters for every 100 meters of horizontal distance. A car moving at

30 m/s up a 5% grade is gaining elevation at the rate of $(0.05)(30 \text{ m/s}) = 1.5$ m/s. The 911 Carrera weighs 12,260 N, so lifting it at this rate requires a power of

$$P = Fv = (12{,}260 \text{ N})(1.5 \text{ m/s}) = 1.8 \times 10^4 \text{ J/s} = 18 \text{ kW} = 24 \text{ hp}.$$

The *total* power required is this amount plus the 16 kW needed to maintain 30 m/s on a level road, that is,

$$P_{\text{tot}} = 18 \text{ kW} + 16 \text{ kW} = 34 \text{ kW} = 46 \text{ hp}.$$

Finally, let's compare the energy requirements of an automobile to your energy requirements while walking. A 70-kg man needs about 2.0×10^5 J of energy (released from food) to walk 1 km at 5 km/h = 1.4 m/s. If the 3.5×10^7 J obtained from 1 L of gasoline could somehow be made available to the man's body, he could travel a distance of $(3.5 \times 10^7 \text{ J})/(2.0 \times 10^5 \text{ J/km}) = 170$ km. Yet this same amount of energy can propel our car only 19 km. Of course, the car travels at over ten times the speed of a walking man (15 m/s versus 1.4 m/s). But this example shows that the speed and convenience of traveling by automobile come only at the cost of greatly increased energy consumption.

SUMMARY

KEY TERMS

work, 165

joule, 165

kinetic energy, 170

work-energy theorem, 170

force constant, 175

Hooke's law, 175

power, 180

average power, 180

instantaneous power, 180

watt, 180

■ When a constant force $\vec{F}$ acts on a particle that undergoes a straight-line displacement $\vec{s}$, the work W done by the force is defined as

$$W = Fs \cos \phi = \vec{F} \cdot \vec{s}, \qquad (6\text{–}2), (6\text{–}3)$$

where ϕ is the angle between the directions of $\vec{F}$ and $\vec{s}$. The unit of work in SI units is 1 joule = 1 newton · meter (1 J = 1 N · m). Work is a scalar quantity; it has an algebraic sign (positive or negative) but no direction in space.

■ The kinetic energy K of a particle equals the amount of work required to accelerate the particle from rest to speed v; it is also equal to the amount of work the particle can do in the process of being brought to rest. The kinetic energy of a particle with mass m and speed v is

$$K = \frac{1}{2} mv^2. \qquad (6\text{–}5)$$

Kinetic energy is a scalar quantity that has no direction in space; it is always positive or zero. Its units are the same as the units of work: 1 J = 1 N · m = 1 kg · m²/s².

■ When forces act on a particle while it undergoes a displacement, the particle's kinetic energy changes by an amount equal to the total work W_{tot} done on the particle by all the forces:

$$W_{\text{tot}} = K_2 - K_1 = \Delta K. \qquad (6\text{–}6)$$

This relation is called the work-energy theorem. It is valid whether the forces are constant or varying and whether the path followed by the particle is straight or curved. It is applicable only to bodies that can be treated as a particle.

■ When the force varies during a straight-line displacement, and the force is in the same direction as the displacement, the work done by the force is given by

$$W = \int_{x_1}^{x_2} F \, dx. \qquad (6\text{–}7)$$

If the force makes an angle ϕ with the displacement, the work done by the force is

$$W = \int_{P_1}^{P_2} F \cos \phi \, dl = \int_{P_1}^{P_2} F_\parallel \, dl = \int_{P_1}^{P_2} \vec{F} \cdot d\vec{l}. \qquad (6\text{--}14)$$

This expression is valid even if the path is curved and the angle ϕ varies during the displacement.

■ Power is the time rate of doing work. If an amount of work ΔW is done in a time Δt, the average power P_{av} is

$$P_{av} = \frac{\Delta W}{\Delta t}. \qquad (6\text{--}15)$$

The instantaneous power is defined as

$$P = \lim_{\Delta t \to 0} \frac{\Delta W}{\Delta t} = \frac{dW}{dt}. \qquad (6\text{--}16)$$

When a force $\vec{F}$ acts on a particle moving with velocity $\vec{v}$, the instantaneous power or rate at which the force does work is

$$P = \vec{F} \cdot \vec{v}. \qquad (6\text{--}19)$$

Like work and kinetic energy, power is a scalar quantity. The unit of power in SI units is 1 watt = 1 joule/second (1 W = 1 J/s).

DISCUSSION QUESTIONS

Q6–1 For a constant force in the direction of the displacement, how can twice the work be done by a force of half the magnitude?

Q6–2 An elevator is hoisted by its cables at constant speed. Is the total work done on the elevator positive, negative, or zero? Explain.

Q6–3 A rope tied to a body is pulled, causing the body to accelerate. But according to Newton's third law, the body pulls back on the rope with an equal and opposite force. Is the total work done then zero? If so, how can the body's kinetic energy change? Explain.

Q6–4 When you use an automobile jack, you lift a car by exerting a force much smaller in magnitude than the weight of the car. Does this mean that you do less work on the car than if you had lifted the car directly? Explain.

Q6–5 If there is a net force on a moving object that is nonzero and constant in magnitude, is it possible for the total work done on the object to be zero? Explain, with an example that illustrates your answer.

Q6–6 In Example 5–4 (Section 5–2), how does the work done on the bucket by the tension in the cable compare to the work done on the cart by the tension in the cable?

Q6–7 In Example 5–20 (Section 5–5), does the force $\vec{F}$ do work on the box? Do any of the forces on the box do work on it? Is the speed of the box constant? Is the velocity of the box constant? Explain your answers.

Q6–8 In Example 5–9 (Section 5–3), gravity does work on the toboggan as it travels a distance d down the hill, where d is measured parallel to the slope. The magnitude of the gravity force depends on the mass of the toboggan and its contents but is independent of the hill's slope angle α. For a constant distance d, does the *work* done by gravity depend on α? Explain.

Q6–9 A force $\vec{F}$ is in the x-direction and has a magnitude that depends on x. Sketch a possible graph of F versus x such that the force does zero work on an object that moves from x_1 to x_2, even though the force magnitude is not zero at all x in this range.

Q6–10 Does the kinetic energy of a car change more when it speeds up from 10 to 15 m/s or from 15 to 20 m/s? Explain.

Q6–11 A falling brick has a mass of 1.5 kg and is moving straight downward with a speed of 5.0 m/s. A 1.5-kg physics book is sliding across the floor with a speed of 5.0 m/s. A 1.5-kg melon is traveling with a horizontal velocity component 3.0 m/s to the right and a vertical component 4.0 m/s upward. Do these objects all have the same velocity? Do these objects all have the same kinetic energy? For each question, give the reasoning behind your answer.

Q6–12 Can the *total* work done on an object during a displacement be negative? Explain. If the total work is negative, can its magnitude be larger than the initial kinetic energy of the object? Explain.

Q6–13 A net force acts on an object and accelerates it from rest to a speed v_1. In doing so the force does an amount of work W_1. By what factor must the work done on the object be increased to produce three times the final speed, with the object again starting from rest?

Q6–14 A truck speeding down the highway has a lot of kinetic energy relative to a stopped state trooper, but no kinetic energy relative to the truck driver. In these two frames of reference, is the same amount of work required to stop the truck? Explain.

Q6–15 A vertical spring has one end attached to the floor. A force $\vec{F}$ is applied to the other end of the spring, slowly stretching it. Is the net work done on the spring equal to the change in its kinetic energy? Explain.

Q6–16 When a book slides along a table top, the force of friction does negative work on it. Can friction ever do *positive* work? Explain. (*Hint:* Think of a box in the back of an accelerating truck, with friction between the box and the truck floor.)

Q6–17 Time yourself while running up a flight of steps and compute the average rate at which you do work against the force of gravity. Express your answer in watts and in horsepower.

Q6–18 When a constant force is applied to a body moving with constant acceleration, is the power of the force constant? If not, how would the force have to vary with speed for the power to be constant?

Q6–19 An advertisement for a portable electrical generating unit claims that the unit's diesel engine produces 28,000 hp to drive an electrical generator that produces 30 MW of electrical power. Is this possible? Explain.

Q6–20 A car speeds up while the engine delivers constant power. Is the acceleration greater at the beginning of this process or at the end? Explain.

Q6–21 Consider a graph of instantaneous power versus time, with the vertical P axis starting at $P = 0$. What is the physical significance of the area under the P versus t curve between vertical lines at t_1 and t_2? How could you find the average power from the graph? Draw a P versus t curve that consists of two straight-line sections and for which the peak power is equal to twice the average power.

Q6–22 A traffic engineer claims that timing the signal lights so that motorists can travel long distances at constant speeds is one way to improve the air quality of a city. Explain the physics that supports this statement.

EXERCISES

SECTION 6–2 **WORK**

6–1 You push your physics book 1.50 m along a horizontal tabletop with a horizontal force of 2.40 N. The opposing force of friction is 0.600 N. a) How much work does your 2.40-N force do on the book? b) What is the work done on the book by the friction force? c) What is the total work done on the book?

6–2 An old oaken bucket of mass 6.75 kg hangs in a well at the end of a rope. The rope passes over a frictionless pulley at the top of the well, and you pull horizontally on the end of the rope to raise the bucket slowly a distance of 4.00 m. a) How much work do you do on the bucket in pulling it up? b) How much work is done by the gravitational force acting on the bucket? c) What is the total work done on the bucket?

6–3 A fisherman reels in 12.0 m of line while pulling in a fish that exerts a constant resisting force of 25.0 N. If the fish is pulled in at constant velocity, how much work is done on it by the tension in the line?

6–4 A factory worker pushes a 30.0-kg crate a distance of 4.5 m along a level floor at constant velocity by pushing horizontally on it. The coefficient of kinetic friction between the crate and floor is 0.25. a) What magnitude of force must the worker apply? b) How much work is done on the crate by this force? c) How much work is done on the crate by friction? d) How much work is done by the normal force? By gravity? e) What is the total work done on the crate?

6–5 Suppose the worker in Exercise 6–4 pushes downward at an angle of 30° below the horizontal. a) What magnitude of force must the worker apply to move the crate at constant velocity? b) How much work is done on the crate by this force when the crate is pushed a distance of 4.5 m? c) How much work is done on the crate by friction during this displacement? d) How much work is done by the normal force? By gravity? e) What is the total work done on the crate?

6–6 A water skier is pulled behind a boat by a horizontal tow rope. She skis off to the side, so the rope makes an angle of 15.0° with her direction of motion, and then continues in a straight line. The tension in the rope is 180 N. How much work is done on the skier by the rope during a displacement of 300 m?

6–7 Two tugboats pull a disabled supertanker. Each tug exerts a constant force of 1.80×10^6 N, one 14° west of north and the other 14° east of north, as they pull the tanker 0.75 km toward the north. What is the total work they do on the supertanker?

SECTION 6–3 **WORK AND KINETIC ENERGY**

6–8 a) Compute the kinetic energy, in joules, of a 1600-kg automobile traveling at 50.0 km/h. b) By what factor does the kinetic energy change if the speed is doubled?

6–9 The dinosaur *Tyrannosaurus rex* is thought to have had a mass of about 7000 kg. a) Treat the dinosaur as a particle and estimate its kinetic energy at a walking speed of 4.0 km/h. b) With what speed would a 70-kg person have to move to have the same kinetic energy as a walking *T. rex*?

6–10 A car is stopped by a constant friction force that is independent of the car's speed. By what factor is the stopping distance changed if the car's initial speed is doubled? (Solve using work-energy methods.)

6–11 A baseball leaves a pitcher's hand at a speed of 32.0 m/s. The mass of the baseball is 0.145 kg. Ignore air resistance. How much work has the pitcher done on the ball in throwing it?

6–12 In Example 6–6 (Section 6–3), call the iceboat with mass m boat A and the other iceboat, of mass $2m$, boat B. a) At the finish line what is the ratio v_A/v_B of the speeds of the two iceboats? b) Let t_A be the elapsed time it takes boat A to reach the finish line and t_B the elapsed time for boat B. What is the ratio t_B/t_A of these two times?

6–13 A moving electron has kinetic energy K_1. After a net amount of work W has been done on it, the electron is moving one-quarter as fast in the opposite direction. a) Find W in terms of K_1. b) Does your answer depend on the final direction of the electron's motion?

6–14 A sled with a mass of 8.00 kg moves in a straight line on a frictionless horizontal surface. At one point in its path, its speed is 4.00 m/s; after it has traveled 2.50 m beyond this point, its speed is 6.00 m/s. Use the work-energy theorem to find the force acting on the sled, assuming that this force is constant and that it acts in the direction of the sled's motion.

6–15 A soccer ball of mass 0.420 kg is initially moving with speed 2.00 m/s. A soccer player kicks the ball, exerting a constant force of magnitude 40.0 N in the same direction as the ball's motion. Over what distance must her foot be in contact with the ball to increase the ball's speed to 6.00 m/s?

6–16 A 12–pack of Omni-Cola (mass 4.30 kg) is initially at rest on a horizontal floor. It is then pushed in a straight line for 1.20 m by a trained dog who exerts a horizontal force with magnitude 36.0 N. Use the work-energy theorem to find the final speed of the 12-pack if a) there is no friction between the 12-pack and the floor; b) the coefficient of kinetic friction between the 12-pack and the floor is 0.30.

6–17 A baseball with mass 0.145 kg is thrown straight upward with an initial speed of 25.0 m/s. a) How much work has gravity done on the baseball when it reaches a height of 20.0 m above the pitcher's hand? b) Use the work-energy theorem to calculate the speed of the baseball at a height of 20.0 m above the pitcher's hand. Ignore air resistance. c) Does the answer to part (b) depend on whether the baseball is moving upward or downward at a height of 20.0 m? Explain.

6–18 A 4.80-kg watermelon is dropped (zero initial speed) from the roof of a 25.0-m-tall building. a) Calculate the work done by gravity on the watermelon during its displacement from the roof to the ground. b) What is the kinetic energy of the watermelon just before it strikes the ground? Ignore air resistance.

6–19 A little red wagon with a mass of 7.00 kg moves in a straight line on a frictionless horizontal surface. It has an initial speed of 4.00 m/s and then is pushed 3.0 m in the direction of the initial velocity by a force with a magnitude of 10.0 N. a) Use the work-energy theorem to calculate the wagon's final speed. b) Calculate the acceleration produced by the force. Use this acceleration in the kinematic relations of Chapter 2 to calculate the wagon's final speed. Compare this result to that calculated in part (a).

6–20 A block of ice with mass 2.00 kg slides 0.750 m down an inclined plane that slopes downward at an angle of 36.9° below the horizontal. If the block of ice starts from rest, what is its final speed? Friction can be neglected.

6–21 A car is traveling on a level road with speed v_0 at the instant when the brakes lock, so that the tires slide rather than roll. a) Use the work-energy theorem to calculate the minimum stopping distance of the car in terms of v_0, g, and the coefficient of kinetic friction μ_k between the tires and the road. b) The car stops in a distance of 91.2 m if $v_0 = 80.0$ km/h. What is the stopping distance if $v_0 = 60.0$ km/h? Assume that the value of μ_k remains the same.

SECTION 6–4 WORK AND ENERGY WITH VARYING FORCES

6–22 To stretch a spring 3.00 cm from its unstretched length, 12.0 J of work must be done. How much work must be done to compress this spring 4.00 cm from its unstretched length?

6–23 A force of 160 N stretches a spring 0.050 m beyond its unstretched length. a) What magnitude of force is required to stretch the spring 0.015 m beyond its unstretched length? To compress the spring 0.020 m? b) How much work must be done to stretch the spring 0.015 m beyond its unstretched length? How much work to compress the spring 0.020 m from its unstretched length?

6–24 A child applies a force $\vec{F}$ parallel to the x-axis to a 10.0-kg sled moving on the frozen surface of a small pond. As the child controls the speed of the sled, the x-component of the force she applies varies with the x-coordinate of the sled as shown in Fig. 6–21. Calculate the work done by the force $\vec{F}$ when the sled moves a) from $x = 0$ to $x = 8.0$ m; b) from $x = 8.0$ m to $x = 12.0$ m; c) from $x = 0$ to 12.0 m.

6–25 Suppose the sled in Exercise 6–24 is initially at rest at $x = 0$. Use the work-energy theorem to find the speed of the sled at a) $x = 8.0$ m; b) $x = 12.0$ m. Neglect friction between the sled and the surface of the pond.

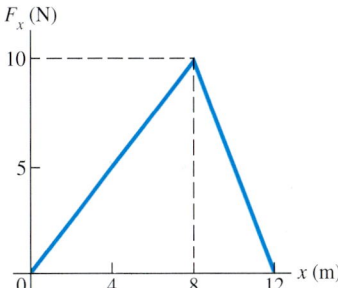

FIGURE 6–21 Exercises 6–24 and 6–25.

6–26 Leg Presses. As part of your daily workout, you lie on your back and push with your feet against a platform attached to two stiff springs arranged side by side so that they are parallel to each other. When you push the platform, you compress the springs. You do 80.0 J of work when you compress the springs 0.200 m from their uncompressed length. a) What magnitude of force must you apply to hold the platform in this position? b) How much *additional* work must you do to move the platform 0.200 m *farther,* and what maximum force must you apply?

6–27 a) In Example 6–8 (Section 6–4) it is calculated that with the air track turned off the glider travels 8.6 cm before it stops instantaneously. How large would the coefficient of static friction μ_s have to be to keep the glider from springing back to the left? b) If the coefficient of static friction between the glider and the track is $\mu_s = 0.60$, what is the maximum initial speed v_1 that the glider can be given and remain at rest after it stops instantaneously? With the air track turned off, the coefficient of kinetic friction is $\mu_k = 0.47$.

6–28 A 4.00-kg block of ice is placed against a horizontal spring that has force constant $k = 200$ N/m and is compressed 0.025 m. The spring is released and accelerates the block along a horizontal surface. Friction and the mass of the spring can be neglected. a) Calculate the work done on the block by the spring during the motion of the block from its initial position to where the spring has returned to its uncompressed length. b) What is the speed of the block after it leaves the spring?

6–29 A force $\vec{F}$ is applied to a 2.0-kg radio-controlled model car parallel to the x-axis as it moves along a straight track. The x-component of the force varies with the x-coordinate of the car as shown in Fig. 6–22. Calculate the work done by the force $\vec{F}$ when the car moves from a) $x = 0$ to $x = 3.0$ m; b) $x = 3.0$ m to $x = 4.0$ m; c) $x = 4.0$ m to $x = 7.0$ m; d) $x = 0$ to $x = 7.0$ m; e) $x = 7.0$ m to $x = 2.0$ m.

6–30 Suppose the 2.0-kg model car in Exercise 6–29 is initially at rest at $x = 0$ and $\vec{F}$ is the net force acting on the car. Use the work-energy theorem to find the speed of the car at a) $x = 3.0$ m; b) $x = 4.0$ m; c) $x = 7.0$ m.

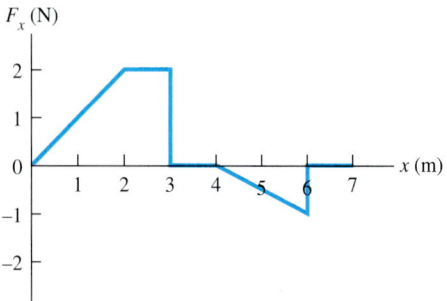

FIGURE 6–22 Exercises 6–29 and 6–30.

6–31 At a waterpark, sleds with riders are sent along a slippery horizontal surface by the release of a large compressed spring. The spring with a force constant $k = 4000$ N/m and negligible mass rests on the frictionless horizontal surface. One end is in contact with a stationary wall. A sled and rider with total mass 70.0 kg are pushed against the other end, compressing the spring 0.375 m. The sled is then released with zero initial velocity. What is the sled's speed when the spring a) returns to its uncompressed length? b) is still compressed 0.200 m?

6–32 In Example 6–9 (Section 6–4), instead of applying a varying horizontal force $\vec{F}$ that maintains Throcky very nearly in equilibrium, you apply a constant horizontal force with magnitude $F = 2w$, where w is Throcky's weight. As in Example 6–9, consider Throcky to be a particle and neglect the small weight of the chains and seat. You push Throcky until the chains make an angle θ_0 with the vertical. a) Use Eq. (6–14) to calculate the work done on Throcky by the force $\vec{F}$ that you apply. b) At the angle θ_0, how does the magnitude of $\vec{F}$ in this exercise compare to that in Example 6–9? c) How does the work done by $\vec{F}$ in this exercise compare to that in Example 6–9?

6–33 A small glider is placed against a compressed spring at the bottom of an air track that slopes upward at an angle of 40.0° above the horizontal. The glider has mass 0.0900 kg. The spring has $k = 640$ N/m and negligible mass. When the spring is

released, the glider travels a maximum distance of 1.80 m along the air track before sliding back down. Before reaching this maximum distance, the glider loses contact with the spring. a) What distance was the spring originally compressed? b) When the glider has traveled along the air track 0.80 m from its initial position against the compressed spring, is it still in contact with the spring? What is the kinetic energy of the glider at this point?

6–34 An ingenious bricklayer builds a device for shooting bricks up to the top of the wall where he is working. He places a brick on a vertical compressed spring with force constant $k = 450$ N/m and negligible mass. When the spring is released, the brick is propelled upward. If the brick has mass 1.80 kg and is to reach a maximum height of 3.6 m above its initial position on the compressed spring, what distance must the spring be compressed initially? (The brick loses contact with the spring when the spring has returned to its uncompressed length. Why?)

SECTION 6–5 POWER

6–35 A tandem (two-person) bicycle team must overcome a force of 165 N to maintain a speed of 9.00 m/s. Find the power required per rider, assuming that each contributes equally. Express your answer in watts and in horsepower.

6–36 The total consumption of electrical energy in the United States is about 1.0×10^{19} joules per year. a) What is the average rate of electrical energy consumption in watts? b) If the population of the United States is 260 million, what is the average rate of electrical energy consumption per person? c) The sun transfers energy to the earth by radiation at a rate of approximately 1.0 kW per square meter of surface. If this energy could be collected and converted to electrical energy with 40% efficiency, how great an area (in square kilometers) would be required to collect the electrical energy used by the United States?

6–37 When its 75-kW (100-hp) engine is generating full power, a small single-engine airplane with a mass of 700 kg gains altitude at a rate of 2.5 m/s (150 m/min, or 500 ft/min). What fraction of the engine power is being used to make the airplane climb? (The remainder is used to overcome the effects of air resistance and of inefficiencies in the propeller and engine.)

6–38 Your job is to lift 30-kg crates a vertical distance of 0.90 m from the ground onto the bed of a truck. a) How many crates would you have to load onto the truck in one minute for your average power output that goes into lifting the crates to be 0.50 hp? b) How many crates for an average power output of 100 W?

6–39 An elevator has a mass of 600 kg, not including passengers. The elevator is designed to ascend at constant speed a vertical distance of 20.0 m (five floors) in 16.0 s, and is driven by a motor that can provide up to 40 hp to the elevator. What is the maximum number of passengers that can ride in the elevator? Assume that an average passenger has a mass of 65.0 kg.

6–40 The hammer of a pile driver weighs 3800 N and must be lifted 2.80 m vertically at constant speed in 4.00 s. What horsepower must the engine provide to the hammer?

6–41 The aircraft carrier *John F. Kennedy* has a mass of 7.4×10^7 kg. When its engines are developing their full power of 280,000 hp, *John F. Kennedy* travels at its top speed of 35 knots (65 km/h). If 70% of the power output of the engines is applied to pushing the ship through the water, what is the magnitude of the force of water resistance that opposes the carrier's motion at this speed?

6–42 A ski tow is operated on a 15.0° slope of length 300 m. The rope moves at 12.0 km/h and power is provided for 50 riders at one time, with an average mass per rider of 70.0 kg. Estimate the power required to operate the tow.

6–43 A particle is accelerated from rest by a constant net force. a) Show that the instantaneous power supplied by the net force is ma^2t. b) To triple the acceleration at any given time, by what factor must the power be increased? c) At $t = 5.0$ s the power supplied by the net force is 36 W. What must the power be at $t = 15.0$ s to maintain constant acceleration?

6–44 Show that the instantaneous power P supplied by the net force acting on a particle is related to the particle's kinetic energy K by $P = dK/dt$.

SECTION 6–6 **AUTOMOTIVE POWER: A CASE STUDY IN ENERGY RELATIONS**

6–45 Consider the Porsche 911 Carrera of Section 6–6. a) Verify that the power needed for a constant speed of 30 m/s on a level road is 16 kW. b) If 15% of the 3.5×10^7 J of energy obtained by

burning each liter of gasoline is available to propel the car, what volume of gasoline is consumed in 1.0 h at this speed? c) Calculate the fuel consumption per unit distance in L/km and in gal/mi.

6–46 A truck engine transmits 28.0 kW (37.5 hp) to the driving wheels when the truck is traveling at a constant velocity of magnitude 60.0 km/h (37.3 mi/h) on a level road. a) What is the resisting force acting on the truck? b) Assume that 65% of the resisting force is due to rolling friction, and that the remainder is due to air resistance. If the force of rolling friction is independent of speed and the force of air resistance is proportional to the square of the speed, what power will drive the truck at 30.0 km/h? At 120.0 km/h? Give your answers in kilowatts and in horsepower.

6–47 a) If 8.00 hp are required to drive a 1800-kg automobile at 60.0 km/h on a level road, what is the total retarding force due to friction, air resistance, and so on? b) What power is necessary to drive the car at 60.0 km/h up a 10.0% grade (a hill rising 10.0 m vertically in 100 m horizontally)? c) What power is necessary to drive the car at 60.0 km/h *down* a 1.00% grade? d) Down what percent grade would the car coast at 60.0 km/h?

6–48 Adding a 75-kg passenger to the Porsche of Section 6–6 increases the mass by 6%. By what percent does this increase the required power at a) 10 m/s? b) 30 m/s?

PROBLEMS

6–49 A luggage handler pulls a 20.0-kg suitcase up a ramp inclined at 25.0° above the horizontal by a force $\vec{F}$ of magnitude 140 N that acts parallel to the ramp. The coefficient of kinetic friction between the ramp and the incline is $\mu_k = 0.300$. If the suitcase travels 3.80 m along the ramp, calculate a) the work done on the suitcase by the force $\vec{F}$; b) the work done on the suitcase by the gravitational force; c) the work done on the suitcase by the normal force; d) the work done on the suitcase by the friction force; e) the total work done on the suitcase. f) If the speed of the suitcase is zero at the bottom of the ramp, what is its speed after it has traveled 3.80 m along the ramp?

6–50 While doing a chin-up, a man's body is lifted 0.40 m. a) How much work must the man do per kilogram of body mass? b) The muscles involved in doing a chin-up can generate about 70 J of work per kilogram of muscle mass. If the man can just barely do a 0.40-m chin-up, what percentage of his body's mass do these muscles constitute? (For comparison, the *total* percentage of muscle in a typical 70-kg man with 14% body fat is about 43%.) c) Repeat part (b) for the man's young son, who has arms half as long as his father's but whose muscles can also generate 70 J of work per kilogram of muscle mass. d) Adults and children have about the same percentage of muscle in their bodies. Explain why children can commonly do chin-ups more easily than their fathers.

6–51 Ramps for the disabled are used because a large weight w can be raised by relatively small force equal to $w \sin \alpha$ plus the small friction force. Such inclined planes are an example of a class of devices called *simple machines*. An input force F_{in} is applied to the system and results in an output force F_{out} applied to the object being moved. For a simple machine the ratio of these forces, F_{out}/F_{in}, is called the actual mechanical advantage (AMA).

The inverse ratio of the distances that the points of application of these forces move through during the motion of the object, s_{in}/s_{out}, is called the ideal mechanical advantage (IMA). a) Find the IMA for an inclined plane. b) What can we say about the relation between the work supplied to the machine, W_{in}, and the work output of the machine, W_{out}, if AMA = IMA? c) Sketch a single pulley arranged to give IMA = 2. d) We define the efficiency e of a simple machine to equal the ratio of the output work to the input work, $e = W_{out}/W_{in}$. Show that $e = $ AMA/IMA.

6–52 A woman stands in an elevator that has a constant upward acceleration while the elevator travels upward a distance of 18.0 m. During the 18.0-m displacement, the normal force exerted by the elevator floor does 8.25 kJ of work on her and gravity does −7.35 kJ of work on her. a) What is the mass of the woman? b) What is the normal force that the elevator floor exerts on her? c) What is the acceleration of the elevator?

6–53 The space shuttle *Endeavour*, with mass 86,400 kg, is in a circular orbit of radius 6.66×10^6 m around the earth. It takes 90.1 min for the shuttle to complete each orbit. On a repair mission, it is cautiously moving 1.00 m closer to a disabled satellite every 3.00 s. Calculate the shuttle's kinetic energy a) relative to the earth; b) relative to the satellite.

6–54 A 5.00-kg package slides 1.50 m down a long ramp that is inclined at 12.0° below the horizontal. The coefficient of kinetic friction between the package and the ramp is $\mu_k = 0.310$. Calculate a) the work done on the package by friction; b) the work done on the package by gravity; c) the work done on the package by the normal force; d) the total work done on the package. e) If the package has a speed of 2.20 m/s at the top of the ramp, what is its speed after sliding 1.50 m down the ramp?

6–55 The package in Problem 6–54 has a speed of 2.20 m/s at the top of the ramp. Use the work-energy theorem to find how far down the ramp the package slides before it stops.

6–56 An object is attracted toward the origin with a force given by $F_x = -k/x^2$. (Gravitational and electrical forces have this distance dependence.) a) Calculate the work done by the force F_x when the object moves in the x-direction from x_1 to x_2. If $x_2 > x_1$, is the work done by F_x positive or negative? b) The only other force acting on the object is a force that you exert with your hand to move the object slowly from x_1 to x_2. How much work do you do? If $x_2 > x_1$, is the work you do positive or negative? c) Explain the similarities and differences between your answers to parts (a) and (b).

6–57 An object that can move along the x-axis is attracted toward the origin with a force of magnitude $F = \alpha x^3$, where $\alpha = 4.00$ N/m^3. What is the force F when the object is a) at $x = 1.00$ m? b) at $x = 2.00$ m? c) How much work is done by the force F when the object moves from $x = 1.00$ m to $x = 2.00$ m? Is this work positive or negative?

6–58 Consider a spring that does not obey Hooke's law very faithfully. One end of the spring is held fixed. To keep the spring stretched or compressed an amount x, a force along the x-axis with x-component $F_x = kx - bx^2 + cx^3$ must be applied to the free end. Here $k = 100$ N/m, $b = 700$ N/m^2, and $c = 12{,}000$ N/m^3. Note that $x > 0$ when the spring is stretched and $x < 0$ when it is compressed. a) How much work must be done to stretch this spring by 0.050 m from its unstretched length? b) How much work must be done to *compress* this spring by 0.050 m from its unstretched length? c) Is it easier to stretch or compress this spring? Explain why in terms of the dependence of F_x on x. (Many real springs behave qualitatively in the same way.)

6–59 A small block with a mass of 0.120 kg is attached to a cord passing through a hole in a horizontal frictionless surface (Fig. 6–23). The block is originally revolving at a distance of 0.40 m from the hole with a speed of 0.70 m/s. The cord is then pulled from below, shortening the radius of the circle in which the block revolves to 0.10 m. At this new distance the speed of the block is observed to be 2.80 m/s. a) What is the tension in the cord in the original situation when the block has speed $v = 0.70$ m/s? b) What is the tension in the cord in the final situation when the block has speed $v = 2.80$ m/s? c) How much work was done by the person who pulled on the cord?

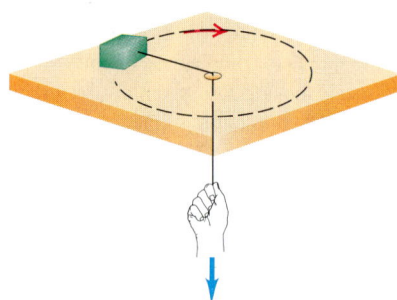

FIGURE 6–23 Problem 6–59.

6–60 Proton Bombardment. A proton with a mass of 1.67×10^{-27} kg is propelled at an initial speed of 3.00×10^5 m/s directly toward a uranium nucleus 5.00 m away. The proton is repelled by the uranium nucleus with a force of magnitude $F = \alpha/x^2$, where x is the separation between the two objects and $\alpha = 2.12 \times 10^{-26}$ N · m^2. Assume that the uranium nucleus remains at rest. a) What is the speed of the proton when it is 8.00×10^{-10} m from the uranium nucleus? b) As the proton approaches the uranium nucleus, the repulsive force slows down the proton until it comes momentarily to rest, after which the proton moves away from the uranium nucleus. How close to the uranium nucleus does the proton get? c) What is the speed of the proton when it is again 5.00 m away from the uranium nucleus?

6–61 A block of ice with mass 6.00 kg is initially at rest on a frictionless horizontal surface. A worker then applies a horizontal force $\vec{F}$ to it. As a result, the block moves along the x-axis such that its position as a function of time is given by $x(t) = \alpha t^2 + \beta t^3$, where $\alpha = 0.200$ m/s^2 and $\beta = 0.0200$ m/s^3. a) Calculate the velocity of the object when $t = 4.00$ s. b) Calculate the magnitude of $\vec{F}$ when $t = 4.00$ s. c) Calculate the work done by the force $\vec{F}$ during the first 4.00 s of the motion.

6–62 A net force with magnitude $(5.00$ N/m$^2)x^2$ and directed at a constant angle of 31.0° with the $+x$-axis acts on an object of mass 0.250 kg as the object moves parallel to the x-axis. How fast is the object moving at $x = 1.50$ m if it has a speed of 4.00 m/s at $x = 1.00$ m?

6–63 You and your bicycle have a combined mass of 80.0 kg. When you reach the base of an bridge, you are traveling along the road at 5.00 m/s (Fig. 6–24). At the top of the bridge you have climbed a vertical distance of 5.20 m and have slowed to 1.50 m/s. Ignore work done by friction and any inefficiency in the bike or your legs. a) What is the total work done on you and your bicycle when you go from the base to the top of the bridge? b) How much work have you done with the force you apply to the pedals?

FIGURE 6–24 Problem 6–63.

6–64 A force in the $+x$-direction has magnitude $F = b/x^n$, where b and n are constants. a) For $n > 1$, calculate the work done on a particle by this force when the particle moves along the x-axis from $x = x_0$ to infinity. b) Show that for $0 < n < 1$, even though F becomes zero as x becomes very large, an infinite amount of work is done by F when the particle moves from $x = x_0$ to infinity.

6–65 You are asked to design spring bumpers for the walls of a parking garage. A freely rolling 1200-kg car moving at 0.65 m/s is to compress the spring no more than 0.070 m before stopping. What should be the force constant of the spring? Assume that the spring has negligible mass.

6–66 The spring of a spring gun has force constant $k = 400$ N/m and negligible mass. The spring is compressed 6.00 cm and a ball with mass 0.0300 kg is placed in the horizontal barrel against the compressed spring. The spring is then released, and the ball is propelled out the barrel of the gun. The barrel is 6.00 cm long, so the ball leaves the barrel at the same point that it loses contact with the spring. The gun is held so the barrel is horizontal. a) Calculate the speed with which the ball leaves the barrel if friction can be neglected. b) Calculate the speed of the ball as it leaves the barrel if a constant resisting force of 6.00 N acts on the ball as it moves along the barrel. c) For the situation in part (b), at what position along the barrel does the ball have the greatest speed, and what is that speed? (In this case the maximum speed does not occur at the end of the barrel.)

6–67 A 2.50-kg textbook is forced against a horizontal spring of negligible mass and force constant 250 N/m, compressing the spring a distance of 0.250 m. When released, the textbook slides on a horizontal table top with coefficient of kinetic friction $\mu_k = 0.30$. Use the work-energy theorem to find how far the textbook moves from its initial position before coming to rest.

6–68 Your cat "Ms." (mass 7.00 kg) is trying to make it to the top of a frictionless ramp 2.00 m long and inclined upward at 30.0° above the horizontal. Since the poor cat can't get any traction on the ramp, you push her up the entire length of the ramp by exerting a constant 100-N force parallel to the ramp. If Ms. takes a running start so that she is moving at 2.40 m/s at the bottom of the ramp, what is her speed when she reaches the top of the incline? Use the work-energy theorem.

6–69 A student proposes a design for an automobile crash barrier in which a 1700-kg sport utility vehicle moving at 20.0 m/s crashes into a spring of negligible mass that slows it to a stop. To avoid injuring the passengers, the acceleration of the vehicle as it slows can be no more than $5.00g$. a) Find the required spring constant k, and find the distance the spring will compress in slowing the vehicle to a stop. In your calculation disregard any deformation or crumpling of the vehicle and the friction between the vehicle and the ground. b) What disadvantages are there to this design?

6–70 A physics professor is pushed up a ramp inclined upward at 30.0° above the horizontal as he sits in his desk chair that slides on frictionless rollers. The combined mass of the professor and chair is 85.0 kg. He is pushed 2.50 m along the incline by a group of students who together exert a constant horizontal force of 600 N. The professor's speed at the bottom of the ramp is 2.00 m/s. Use the work-energy theorem to find his speed at the top of the ramp.

6–71 A 5.00-kg block is moving at $v_0 = 6.00$ m/s along a frictionless horizontal surface toward a spring with force constant $k = 500$ N/m that is attached to a wall (Fig. 6–25). The spring has negligible mass. a) Find the maximum distance the spring

will be compressed. b) If the spring is to compress by no more than 0.150 m, what should be the maximum value of v_0?

FIGURE 6–25 Problem 6–71.

6–72 Consider the system shown in Fig. 6–26. The rope and pulley have negligible mass, and the pulley is frictionless. The coefficient of kinetic friction between the 8.00-kg block and the table top is $\mu_k = 0.250$. The blocks are released from rest. Use energy methods to calculate the speed of the 6.00-kg block after it has descended 1.50 m.

6–73 Consider the system shown in Fig. 6–26. The rope and pulley have negligible mass, and the pulley is frictionless. Initially the 6.00-kg block is moving downward and the 8.00-kg block is moving to the right, both with a speed of 0.900 m/s. The blocks come to rest after moving 2.00 m. Use the work-energy theorem to calculate the coefficient of kinetic friction between the 8.00-kg block and the table top.

FIGURE 6–26 Problems 6–72 and 6–73.

6–74 Figure 6–27 shows how the force exerted by the string of a compound bow on an arrow varies as a function of how far back the arrow is pulled (the *draw length*). Assume that the same force is exerted on the arrow as it moves forward after being released. Full draw for this bow is at a draw length of 75.0 cm. If the bow shoots a 0.0250-kg arrow from full draw, what is the speed of the arrow as it leaves the bow?

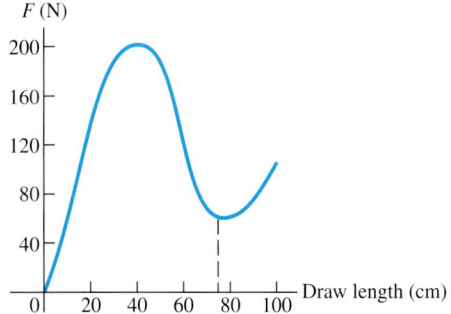

FIGURE 6–27 Problem 6–74.

6–75 A pump is required to lift 800 kg of water (about 210 gallons) per minute from a well 14.0 m deep and eject it with a speed of 18.0 m/s. a) How much work is done per minute in lifting the water? b) How much in giving the water the kinetic energy it has when ejected? c) What must be the power output of the pump?

6–76 Find the power output of the worker in Problem 6–61 as a function of time. What is the numerical value of the power (in watts) at $t = 4.00$ s?

6–77 A physics student spends part of her day walking between classes or for recreation, during which time she expends energy at an average rate of 280 W. The remainder of the day she is sitting in class, studying, or resting; during these activities, she expends energy at an average rate of 100 W. If she expends a total of 1.1×10^7 J of energy in a 24-hour day, how much of the day did she spend walking?

6–78 All birds, independent of their size, must maintain a power output of 10 to 25 watts per kilogram of body mass in order to fly by flapping their wings. (a) The Andean giant hummingbird (*Patagona gigas*) has a mass of 70 g and flaps its wings ten times per second while hovering. Estimate the amount of work done by such a hummingbird in each wingbeat. (b) A 70-kg athlete can maintain a power output of 1.4 kW for no more than a few seconds; the *steady* power output of a typical athlete is only 500 W or so. Is it possible for a human-powered aircraft to fly for extended periods by flapping its wings? Explain.

6–79 The Grand Coulee Dam is 1270 m long and 170 m high. The electrical power output from generators at its base is approximately 2000 MW. How many cubic meters of water must flow from the top of the dam per second to produce this amount of power if 92% of the work done on the water by gravity is converted to electrical energy? (Each cubic meter of water has a mass of 1000 kg.)

*6–80 The engine of a car of mass m supplies a constant power P to the wheels to accelerate the car. Rolling friction and air resistance can be neglected. The car is initially at rest. a) Show that the speed of the car is given as a function of time by $v = (2Pt/m)^{1/2}$. b) Show that the acceleration of the car is not constant, but is given as a function of time by $a = (P/2mt)^{1/2}$. c) Show that the displacement as a function of time is given by $x - x_0 = (8P/9m)^{1/2} t^{3/2}$.

6–81 Power of the Human Heart. The human heart is a powerful and extremely reliable pump. Each day it takes in and discharges about 7500 L of blood. Assume that the work done by the heart is equal to the work required to lift this amount of blood a height equal to that of the average American female (1.63 m). The density (mass per unit volume) of blood is 1.05×10^3 kg/m^3. a) How much work does the heart do in a day? b) What is its power output in watts?

6–82 Six diesel units in series can provide 13.4 MW of power to the lead car of a freight train. The diesel units have a total mass of 1.10×10^6 kg. The average car in the train has a mass of 8.2×10^4 kg and requires a horizontal pull of 2.8 kN to move at a constant 27 m/s on level tracks. a) How many cars can be in the train under these conditions? b) However, this would leave no power for accelerating or climbing hills. Show that the extra force needed to accelerate the train is about the same for a 0.10 m/s^2 acceleration or a 1.0% slope (slope angle $\alpha = $ arctan 0.010). c) With the 1.0% slope, show that an extra 2.9 MW of power is needed to maintain the 27 m/s speed of the diesel units. d) With 2.9 MW less power available, how many cars can the six diesel units pull up a 1.0% slope at a constant 27 m/s?

6–83 It takes a force of 53 kN on the lead car of a 16-car passenger train with a mass of 9.1×10^5 kg to pull it at a constant 45 m/s (101 mi/h) on level tracks. a) What power must the locomotive provide to the lead car? b) How much more power to the lead car than calculated in part (a) would be needed to give the train an acceleration of 1.5 m/s^2, at the instant that the train has a speed of 45 m/s on level tracks? c) How much more power to the lead car than that calculated in part (a) would be needed to move the train up a 1.5% grade (slope angle $\alpha = $ arctan 0.015) at a constant 45 m/s?

6–84 An object has several forces acting on it. One of these forces is $\vec{F} = \alpha x y \hat{\imath}$, a force in the x-direction whose magnitude depends on the position of the object, with $\alpha = 2.50$ N/m^2. Calculate the work done on the object by this force for the following displacements of the object: a) The object starts at the point $x = 0$, $y = 3.00$ m and moves parallel to the x-axis to the point $x = 2.00$ m, $y = 3.00$ m. b) The object starts at the point $x = 2.00$ m, $y = 0$ and moves in the y-direction to the point $x = 2.00$ m, $y = 3.00$ m. c) The object starts at the origin and moves on the line $y = 1.5x$ to the point $x = 2.00$ m, $y = 3.00$ m.

6–85 For a touring bicyclist the drag coefficient is 1.00, the frontal area is 0.463 m^2, and the coefficient of rolling friction is 0.0045. The rider has a mass of 50.0 kg, and her bike has a mass of 12.0 kg. a) To maintain a speed of 12.0 m/s (about 27 mi/h) on a level road, what must the rider's power output to the rear wheel be? b) For racing, the same rider uses a different bike with a coefficient of rolling friction 0.0030 and a mass of 9.00 kg. She also crouches down, reducing her drag coefficient to 0.88 and reducing her frontal area to 0.366 m^2. What must her power output to the rear wheel be then to maintain a speed of 12.0 m/s? c) For the situation in part (b), what power output is required to maintain a speed of 6.0 m/s? Note the great drop in power requirement when the speed is only halved. (For more on aerodynamic speed limitations for a wide variety of human-powered vehicles see "The Aerodynamics of Human-Powered Land Vehicles," *Scientific American,* December, 1983.)

CHALLENGE PROBLEMS

6–86 On a winter's day in Maine, a warehouse worker is shoving boxes up a rough plank inclined at an angle α above the horizontal. The plank is partially covered with ice, with more ice near the bottom of the plank than near the top, so that the coeffi-

cient of friction increases with the distance x along the plank: $\mu = Ax$, where A is a positive constant and the bottom of the plank is at $x = 0$. (For this plank the coefficients of kinetic and static friction are equal: $\mu_k = \mu_s = \mu$.) A box is shoved up the plank so that it leaves the bottom of the plank moving at speed v_0. Show that when the box first comes to rest, it will remain at rest if

$$v_0^2 \geq \frac{3g \sin^2 \alpha}{A \cos \alpha}$$

6–87 A Spring with Mass. We usually ignore the kinetic energy of the moving coils of a spring, but let's try to get a reasonable approximation to this. Consider a spring of mass M, equilibrium length L_0, and spring constant k. The work done to stretch or compress the spring by a distance L is $\frac{1}{2}kX^2$, where $X = L - L_0$. a) Consider a spring as described above that has one end fixed and the other end moving with speed v. Assume that the speed of points along the length of the spring varies linearly with distance l from the fixed end. Assume also that the mass M of the spring is distributed uniformly along the length of the spring. Calculate the kinetic energy of the spring in terms of M and v. (*Hint:* Divide the spring into pieces of length dl; find the speed of each piece in terms of l, v, and L; find the mass of each piece in terms of dl, M, and L; and integrate from 0 to L. The result is *not* $\frac{1}{2}Mv^2$, since not all of the spring moves with the same speed.) In a spring gun, a spring of mass 0.243 kg and force constant 3200 N/m is compressed 2.50 cm from its unstretched length. When the trigger is pulled, the spring pushes horizontally on a 0.053-kg ball. The work done by friction is negligible. Calculate the ball's speed when the spring reaches its uncompressed length b) neglecting the mass of the spring; c) including, using the results of part (a), the mass of the spring. d) In part (c), what is the final kinetic energy of the ball and of the spring?

6–88 An airplane in flight is subject to an air resistance force proportional to the square of its speed v, just as in Eq. (6–20). But there is an additional resistive force because the airplane has wings. Air flowing over the wings is pushed down and slightly forward, so from Newton's third law the air exerts a force on the wings and airplane that is up and slightly backward (Fig. 6–28). The upward force is the lift force that keeps the airplane aloft, and the backward force is called *induced drag*. At flying speeds induced drag is inversely proportional to v^2, so that the total air

FIGURE 6–28 Challenge Problem 6–88.

resistance force can be expressed by $F_{air} = \alpha v^2 + \beta/v^2$, where α and β are positive constants that depend on the shape and size of the airplane and the density of the air. For a Cessna 150, a small single-engine airplane, $\alpha = 0.30 \text{ N} \cdot \text{s}^2/\text{m}^2$ and $\beta = 3.5 \times 10^5 \text{ N} \cdot \text{m}^2/\text{s}^2$. In steady flight, the engine must provide a forward force that exactly balances the air resistance force. a) Calculate the speed (in km/h) at which this airplane will have the maximum *range* (that is, travel the greatest distance) for a given quantity of fuel. b) Calculate the speed (in km/h) for which the airplane will have the maximum *endurance* (that is, will remain in the air the longest time).

6–89 Figure 6–29 shows the oxygen consumption rate of men walking and running at different speeds. The vertical axis shows the volume of oxygen (in cm^3) that a man consumes per kilogram of body mass per minute. Note the transition from walking to running that occurs naturally at about 9 km/h. The metabolism of 1 cm^3 of oxygen releases about 20 J of energy. Using the data in the graph, calculate the energy required for a 70-kg man to travel 1 km on foot at each of the following speeds a) 5 km/h (walking); b) 10 km/h (running); c) 15 km/h (running). d) Which speed is the most efficient, that is, requires the least energy to travel 1 km?

FIGURE 6–29 Challenge Problem 6–89.

6–90 General Proof of the Work-Energy Theorem. Consider a particle that moves along a curved path in space from (x_1, y_1, z_1) to (x_2, y_2, z_2). At the initial point the particle has velocity $\vec{v} = v_{1x}\hat{i} + v_{1y}\hat{j} + v_{1z}\hat{k}$. The path that the particle follows may be divided into infinitesimal segments $d\vec{l} = dx\hat{i} + dy\hat{j} + dz\hat{k}$. As the particle moves, it is acted on by a net force $\vec{F} = F_x\hat{i} + F_y\hat{j} + F_z\hat{k}$. The force components F_x, F_y, and F_z are in general functions of position. By the same sequence of steps used in Eqs. (6–11) through (6–13), prove the work-energy theorem for this general case. That is, prove that

$$W_{tot} = K_2 - K_1,$$

where

$$W_{tot} = \int_{(x_1, y_1, z_1)}^{(x_2, y_2, z_2)} \vec{F} \cdot d\vec{l} = \int_{(x_1, y_1, z_1)}^{(x_2, y_2, z_2)} (F_x \, dx + F_y \, dy + F_z \, dz).$$

As water cascades down a waterfall, gravitational potential energy is converted into kinetic energy (so that the water gains speed) and thermal energy (manifested as an increase in the water's temperature). Careful measurements show that no energy is lost in this process—instead, the energy merely changes form.

7

Potential Energy and Energy Conservation

7–1 INTRODUCTION

When a diver jumps off a high board into a swimming pool, she hits the water moving pretty fast, with a lot of kinetic energy. Where does that energy come from? The answer we learned in Chapter 6 was that the gravitational force (her weight) does work on the diver as she falls. The diver's kinetic energy—energy associated with her *motion*—increases by an amount equal to the work done.

But there is a very useful alternative way to think about work and kinetic energy. This new approach is based on the concept of *potential energy,* which is energy associated with the *position* of a system rather than its motion. In this approach, there is *gravitational potential energy* even while the diver is standing on the high board. Energy is not added to the earth-diver system as the diver falls, but rather a storehouse of energy is *transformed* from one form (potential energy) to another (kinetic energy) as she falls. In this chapter we'll see how this transformation can be understood from the work-energy theorem.

If the diver bounces on the end of the board before she jumps, the bent board stores a second kind of potential energy called *elastic potential energy.* We'll discuss elastic potential energy of simple systems such as a stretched or compressed spring. (An important third kind of potential energy is associated with the positions of electrically charged particles relative to each other. We'll encounter this in Chapter 24.)

We will prove that in some cases the sum of a system's kinetic and potential energy, called the *total mechanical energy* of the system, is constant during the motion of the system. This will lead us to the general statement of the *law of conservation of energy,* one of the most fundamental and far-reaching principles in all of science.

7–2 GRAVITATIONAL POTENTIAL ENERGY

A particle gains or loses kinetic energy because it interacts with other objects that exert forces on it. We learned in Chapter 6 that during any interaction the change in a particle's kinetic energy is equal to the total work done on the particle by the forces that act on it.

In many situations it seems as though energy has been stored in a system, to be recovered later. It's like a savings account in a bank; you deposit money, and then you can draw it out later. For example, you must do work on the hammerhead of a pile driver to lift it. It seems reasonable that in hoisting the hammerhead into the air you are storing energy in the system, energy that is later converted into kinetic energy as the hammerhead falls. Or consider your cousin Throckmorton on a swing. Suppose you give him a shove, which gives him some kinetic energy, and then you let him swing freely back and forth. He stops momentarily when he reaches the front and back end points of his arc, so at these points he has no kinetic energy. But he regains his kinetic energy as he passes through the low point in the arc. It seems as though at the high points the energy is stored in some other form, related to his height above the ground, and is converted back to kinetic energy as he swings toward the low point.

Both these examples point to the idea of an energy associated with the *position* of bodies in a system. This kind of energy is a measure of the *potential* or *possibility* for work to be done; when a hammerhead is raised into the air, there is a potential for work to be done on it by the gravitational force, but only if the hammerhead is allowed to fall to the ground. For this reason, energy associated with position is called **potential energy.** Our discussion suggests that there is potential energy associated with a body's weight and its height above the ground. We call this *gravitational potential energy.*

When a body falls without air resistance, gravitational potential energy decreases and the body's kinetic energy increases. But in Chapter 6 we used the work-energy theorem to conclude that the kinetic energy of a falling body increases because the force of the earth's gravity (the body's weight) does work on the body. Let's use the work-energy theorem to show that these two descriptions of a falling body are equivalent and to derive the expression for gravitational potential energy.

Let's consider a body with mass m that moves along the (vertical) y-axis, as in Fig. 7–1. The forces acting on it are its weight, with magnitude $w = mg$, and possibly some other forces; we call the vector sum (resultant) of all the other forces $\vec{F}_{\text{other}}$. We'll assume that the body stays close enough to the earth's surface that the weight is constant. (We'll find in Chapter 12 that weight decreases with altitude.) We want to find the work done by the weight when the body drops from a height y_1 above the origin to a lower height y_2 (Fig. 7–1a). The weight and displacement are in the same direction, so the work W_{grav} done on the body by its weight is positive;

$$W_{\text{grav}} = Fs = w(y_1 - y_2) = mgy_1 - mgy_2. \qquad (7\text{–}1)$$

This expression also gives the correct work when the body moves upward and y_2 is greater than y_1 (Fig. 7–1b). In that case the quantity $(y_1 - y_2)$ is negative, and W_{grav} is negative because the weight and displacement are opposite in direction.

Equation (7–1) shows that we can express W_{grav} in terms of the values of the quantity mgy at the beginning and end of the displacement. This quantity, the product of the weight mg and the height y above the origin of coordinates, is called the **gravitational potential energy,** U:

$$U = mgy \qquad \text{(gravitational potential energy).} \qquad (7\text{–}2)$$

Its initial value is $U_1 = mgy_1$, and its final value is $U_2 = mgy_2$. The change in U is the final value minus the initial value, or $\Delta U = U_2 - U_1$. We can express the work W_{grav} done by the gravitational force during the displacement from y_1 to y_2 as

$$W_{\text{grav}} = U_1 - U_2 = -(U_2 - U_1) = -\Delta U. \qquad (7\text{–}3)$$

The negative sign in front of ΔU is *essential.* When the body moves up, y increases, the work done by the gravitational force is negative, and the gravitational potential energy increases ($\Delta U > 0$). When the body moves down, y decreases, the gravitational force does positive work, and the gravitational potential energy decreases ($\Delta U < 0$). It's like drawing money out of the bank (decreasing U) and spending it (doing positive work). As Eq. (7–3) shows, the unit of potential energy is the joule (J), the same unit as is used for work.

CAUTION ▶ Although you might be tempted to do so, it is *not* correct to call $U = mgy$ the "gravitational potential energy of the body." The reason is that gravitational potential energy is a *shared* property of the body and the earth. Gravitational potential energy

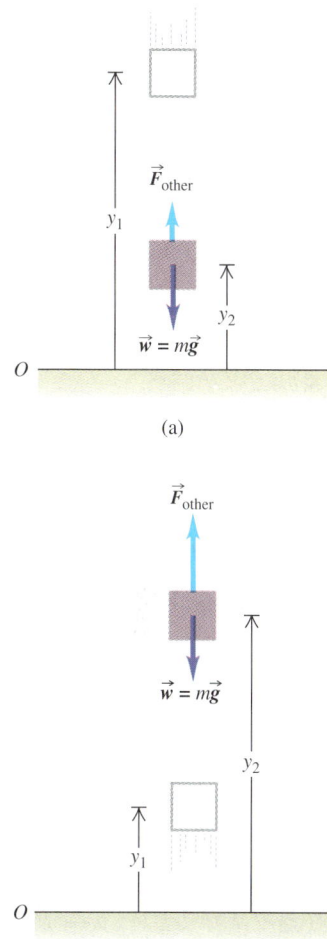

7–1 Work is done by the gravitational force $\vec{w}$ during the vertical motion of a body from an initial height y_1 to a final height y_2. The vertical displacement can be either (a) downward or (b) upward.

increases if the earth stays fixed and the height of the body increases; it also increases if the body is fixed in space and the earth is moved away from the body. Notice that the formula $U = mgy$ involves characteristics of both the body (its mass m) and the earth (the value of g). ◄

CONSERVATION OF MECHANICAL ENERGY (GRAVITATIONAL FORCES ONLY)

To see what gravitational potential energy is good for, suppose the body's weight is the *only* force acting on it, so $\vec{F}_{\text{other}} = 0$. The body is then falling freely with no air resistance, and can be moving either up or down. Let its speed at point y_1 be v_1, and let its speed at y_2 be v_2. The work-energy theorem, Eq. (6–6), says that the total work done on the body equals the change in the body's kinetic energy; $W_{\text{tot}} = \Delta K = K_2 - K_1$. If gravity is the only force that acts, then from Eq. (7–3), $W_{\text{tot}} = W_{\text{grav}} = -\Delta U = U_1 - U_2$. Putting these together, we get

$$\Delta K = -\Delta U \qquad \text{or} \qquad K_2 - K_1 = U_1 - U_2,$$

which we can rewrite as

$$K_1 + U_1 = K_2 + U_2 \qquad \text{(if only gravity does work)}, \qquad (7\text{–}4)$$

or

$$\frac{1}{2}mv_1{}^2 + mgy_1 = \frac{1}{2}mv_2{}^2 + mgy_2 \qquad \text{(if only gravity does work)}. \qquad (7\text{–}5)$$

We now define the sum $K + U$ of kinetic and potential energy to be E, the **total mechanical energy of the system.** By "system" we mean the body of mass m *and* the earth considered together, because gravitational potential energy U is a shared property of both bodies. Then $E_1 = K_1 + U_1$ is the total mechanical energy at y_1, and $E_2 = K_2 + U_2$ is the total mechanical energy at y_2. Equation (7–4) says that when the body's weight is the only force doing work on it, $E_1 = E_2$. That is, E is constant; it has the same value at y_1 and y_2. But since the positions y_1 and y_2 are arbitrary points in the motion of the body, the total mechanical energy E has the same value at *all* points during the motion;

$$E = K + U = \text{constant} \qquad \text{(if only gravity does work)}.$$

A quantity that always has the same value is called a *conserved* quantity. *When only the force of gravity does work, the total mechanical energy is constant, that is, is conserved.* This is our first example of the **conservation of mechanical energy.**

When we throw a ball into the air, its speed decreases on the way up as kinetic energy is converted to potential energy; $\Delta K < 0$ and $\Delta U > 0$. On the way back down, potential energy is converted back to kinetic energy and the ball's speed increases; $\Delta K > 0$ and $\Delta U < 0$. But the *total* mechanical energy (kinetic plus potential) is the same at every point in the motion, provided that no force other than gravity does work on the ball (that is, air resistance must be negligible). It's still true that the gravitational force does work on the body as it moves up or down, but we no longer have to calculate work directly; keeping track of changes in the value of U takes care of this completely.

An important point about gravitational potential energy is that it doesn't matter what height we choose to be $y = 0$, the origin of coordinates. If we shift the origin for y, the values of y_1 and y_2 both change, but the *difference* $y_2 - y_1$ stays the same. It follows that although U_1 and U_2 depend on where we place the origin, the difference $U_2 - U_1 = mg(y_2 - y_1)$ does not. The physically significant quantity is not the value of U at a particular point, but only the *difference* in U between two points. So we can define U to be zero at whatever point we choose without affecting the physics of the situation. This is shown in the following example.

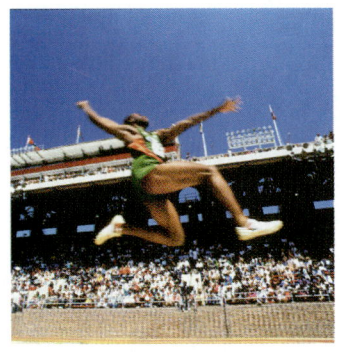

During the time this athlete is in midair, only gravity does work on him (if we neglect the minor effects of air resistance). Mechanical energy—the sum of kinetic and gravitational potential energy—is conserved.

EXAMPLE 7–1

Height of a baseball from energy conservation You throw a 0.145-kg baseball straight up in the air, giving it an initial upward velocity of magnitude 20.0 m/s. Use conservation of energy to find how high it goes, ignoring air resistance.

SOLUTION After the ball leaves your hand, the only force doing work on the ball is its weight, so we can use Eq. (7–4). Let's take the origin at the point where the ball leaves your hand (point 1). Then $y_1 = 0$ (Fig. 7–2), and the potential energy at point 1 is $U_1 = mgy_1 = 0$. At this point, $v_1 = 20.0$ m/s. We want to find the height y_2 at point 2, where it stops and begins to fall back to earth. At this point, $v_2 = 0$, and the kinetic energy is $K_2 = \frac{1}{2}mv_2{}^2 = 0$. Equation (7–4) says that $K_1 + U_1 = K_2 + U_2$; since $U_1 = 0$ and $K_2 = 0$,

$$K_1 = U_2.$$

As the energy bar graphs in Fig. 7–2 show, the kinetic energy of the ball at point 1 is completely converted into gravitational potential energy at point 2. At point 1 the kinetic energy is

$$K_1 = \frac{1}{2}mv_1{}^2 = \frac{1}{2}(0.145 \text{ kg})(20.0 \text{ m/s})^2 = 29.0 \text{ J}.$$

This equals the gravitational potential energy $U_2 = mgy_2$ at point 2, so

$$y_2 = \frac{U_2}{mg} = \frac{29.0 \text{ J}}{(0.145 \text{ kg})(9.80 \text{ m/s}^2)} = 20.4 \text{ m}.$$

We can also solve the equation $K_1 = U_2$ algebraically for y_2:

$$\frac{1}{2}mv_1{}^2 = mgy_2,$$

$$y_2 = \frac{v_1{}^2}{2g} = \frac{(20.0 \text{ m/s})^2}{2(9.80 \text{ m/s}^2)} = 20.4 \text{ m}.$$

The mass divides out, as we should expect; we learned in Chapter 2 that the motion of a body in free fall doesn't depend

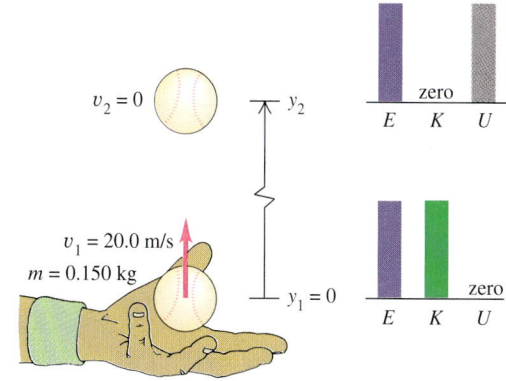

7–2 After a baseball leaves your hand, the only force acting on it is its weight (neglecting air resistance), so mechanical energy $E = K + U$ is conserved. The energy bar graphs show the values of E, K, and U at $y_1 = 0$ and y_2.

on its mass. Indeed, we could have derived the result $y_2 = v_1{}^2/2g$ using Eq. (2–13).

In carrying out the calculation above, we chose the origin to be at point 1, so $y_1 = 0$ and $U_1 = 0$. What happens if we make a different choice? As an example, suppose we choose the origin to be 5.0 m below point 1, so $y_1 = 5.0$ m. With this choice the total mechanical energy at point 1 is part kinetic and part potential, while at point 2 it's purely potential energy. If you work through the calculation again with this choice of origin, you'll find $y_2 = 25.4$ m; this is 20.4 m above point 1, just as with the first choice of origin. In every problem, it's up to you to choose the height at which $U = 0$; don't agonize over the choice, however, because the physics of the answer doesn't depend on your choice.

EFFECT OF OTHER FORCES

When other forces act on the body in addition to its weight, then $\vec{F}_{\text{other}}$ in Fig. 7–1 is *not* zero. For the pile driver described in Example 6–5 (Section 6–3) the force applied by the hoisting cable and the friction with the vertical guide rails are examples of forces that might be included in $\vec{F}_{\text{other}}$. The gravitational work W_{grav} is still given by Eq. (7–3), but the total work W_{tot} is then the sum of W_{grav} and the work done by $\vec{F}_{\text{other}}$. We will call this additional work W_{other}, so the total work done by all forces is $W_{\text{tot}} = W_{\text{grav}} + W_{\text{other}}$. Equating this to the change in kinetic energy, we have

$$W_{\text{other}} + W_{\text{grav}} = K_2 - K_1. \tag{7–6}$$

Also, from Eq. (7–3), $W_{\text{grav}} = U_1 - U_2$, so

$$W_{\text{other}} + U_1 - U_2 = K_2 - K_1,$$

which we can rearrange in the form

$$K_1 + U_1 + W_{\text{other}} = K_2 + U_2 \quad \text{(if forces other than gravity do work).} \tag{7–7}$$

5.3
Sliding on Snow and Grass

5.5
Upward-Moving Elevator Stops

5.6
Stopping a Downward-Moving Elevator

5.9
Skier Speed

5.10
Skier Rope Tow

5.13
Bricklayer Perils

Finally, using the appropriate expressions for the various energy terms, we obtain

$$\frac{1}{2}mv_1{}^2 + mgy_1 + W_{other} = \frac{1}{2}mv_2{}^2 + mgy_2 \qquad \text{(if forces other than gravity do work).}$$

(7–8)

The meaning of Eqs. (7–7) and (7–8) is this: *The work done by all forces* other than the gravitational force *equals the change in the total mechanical energy $E = K + U$ of the system, where U is the gravitational potential energy.* When W_{other} is positive, E increases, and $K_2 + U_2$ is greater than $K_1 + U_1$. When W_{other} is negative, E decreases. In the special case in which no forces other than the body's weight do work, $W_{other} = 0$. The total mechanical energy is then constant, and we are back to Eq. (7–4) or (7–5).

Problem–Solving Strategy

PROBLEMS USING MECHANICAL ENERGY

1. First decide whether the problem should be solved by energy methods, by using $\Sigma \vec{F} = m\vec{a}$ directly, or by a combination. The energy approach is particularly useful when the problem involves motion with varying forces, motion along a curved path (discussed later in this section), or both. But if the problem involves elapsed time, the energy approach is usually *not* the best choice because this approach doesn't involve time directly.

2. When using the energy approach, first decide what the initial and final states (the positions and velocities) of the system are. Use the subscript 1 for the initial state and the subscript 2 for the final state. It helps to draw sketches showing the initial and final states.

3. Define your coordinate system, particularly the level at which $y = 0$. You will use this to compute gravitational potential energies. Equation (7–2) assumes that the positive direction for y is upward; we suggest that you use this choice consistently.

4. List the initial and final kinetic and potential energies, that is, K_1, K_2, U_1, and U_2. In general, some of these will be known and some will be unknown. Use algebraic symbols for any unknown coordinates or velocities.

5. Identify all nongravitational forces that do work. A free-body diagram is always helpful. Calculate the work W_{other} done by all these forces. If some of the quantities you need are unknown, represent them by algebraic symbols.

6. Relate the kinetic and potential energies and the nongravitational work W_{other} using Eq. (7–7). If there is no nongravitational work, this becomes Eq. (7–4). It's helpful to draw bar graphs showing the initial and final values of K, U, and $E = K + U$. Then solve to find whatever unknown quantity is required.

7. Keep in mind, here and in later sections, that the work done by each force must be represented either in $U_1 - U_2 = -\Delta U$ or as W_{other}, but *never* in both places. If the gravitational work is included in ΔU, do not include it again in W_{other}.

EXAMPLE 7–2

Work and energy throwing a baseball In Example 7–1, suppose your hand moves up 0.50 m while you are throwing the ball, which leaves your hand with an upward velocity of 20.0 m/s. Again ignore air resistance. a) Assuming that your hand exerts a constant upward force on the ball, find the magnitude of that force. b) Find the speed of the ball at a point 15.0 m above the point where it leaves your hand.

SOLUTION Figure 7–3 shows a diagram of the situation, including a free-body diagram for the ball while it is being thrown. a) Let's use Eq. (7–8) to solve for W_{other}, the work done by the upward force $\vec{F}$ that your hand applies to the ball as you throw it. We'll be able to determine the magnitude F from this. Let point 1 be the point where your hand first starts to move, and let point 2 be the point where the ball leaves your hand. With the same

coordinate system as in Example 7–1, we have $y_1 = -0.50$ m and $y_2 = 0$. Then

$$K_1 = 0,$$

$$U_1 = mgy_1 = (0.145 \text{ kg})(9.80 \text{ m/s}^2)(-0.50 \text{ m}) = -0.71 \text{ J},$$

$$K_2 = \frac{1}{2}mv_2{}^2 = \frac{1}{2}(0.145 \text{ kg})(20.0 \text{ m/s})^2 = 29.0 \text{ J},$$

$$U_2 = mgy_2 = (0.145 \text{ kg})(9.80 \text{ m/s}^2)(0) = 0.$$

The initial potential energy U_1 is negative because the ball was initially below the origin. According to Eq. (7–7), $K_1 + U_1 + W_{other} = K_2 + U_2$, so

$$W_{other} = (K_2 - K_1) + (U_2 - U_1)$$
$$= (29.0 \text{ J} - 0) + (0 - (-0.71 \text{ J})) = 29.7 \text{ J}.$$

(a)

(b)

7–3 (a) Throwing a ball vertically upward. (b) The free-body diagram for the ball as the force $\vec{F}$ applied by the hand does the work W_{other} on the ball. The force $\vec{F}$ and gravity both act between y_1 and y_2. From y_2 to y_3, only gravity acts on the ball.

The kinetic energy of the ball increases by $K_2 - K_1 = 29.0$ J, and the potential energy increases by $U_2 - U_1 = 0.71$ J; the sum is $E_2 - E_1$, the change in total mechanical energy, which is equal to W_{other}.

Assuming that the upward force $\vec{F}$ that your hand applies is constant, the work W_{other} done by this force is equal to the magnitude F of the force multiplied by the upward displacement $y_2 - y_1$ over which it acts:

$$W_{other} = F(y_2 - y_1),$$

$$F = \frac{W_{other}}{y_2 - y_1} = \frac{29.7 \text{ J}}{0.50 \text{ m}} = 59 \text{ N}.$$

This is about 40 times greater than the weight of the ball.
b) Let point 3 be at the 15-m height. Then $y_3 = 15$ m, and we want to find the speed v_3 at this point. Between points 2 and 3, total mechanical energy is conserved; the force of your hand no longer acts, so $W_{other} = 0$. We can then find the kinetic energy at point 3 using Eq. (7–4):

$$K_2 + U_2 = K_3 + U_3,$$

$$U_3 = mgy_3 = (0.145 \text{ kg})(9.80 \text{ m/s}^2)(15.0 \text{ m}) = 21.3 \text{ J},$$

$$K_3 = (K_2 + U_2) - U_3 = (29.0 \text{ J} + 0 \text{ J}) - 21.3 \text{ J} = 7.7 \text{ J}.$$

Since $K_3 = \frac{1}{2}mv_3^2$, where v_3 is the y-component of the ball's velocity at point 3, we have

$$v_3 = \pm\sqrt{\frac{2K_3}{m}} = \pm\sqrt{\frac{2(7.7 \text{ J})}{0.145 \text{ kg}}} = \pm 10 \text{ m/s}.$$

The significance of the plus-or-minus sign is that the ball passes point 3 *twice*, once on the way up and again on the way down. The total mechanical energy E is constant and equal to 29.0 J while the ball is in free fall, and the potential energy at point 3 is $U_3 = 21.3$ J whether the ball is moving up or down. So at point 3, the ball's kinetic energy K_3 and *speed* don't depend on the direction the ball is moving. The velocity is positive (+10 m/s) when the ball is moving up and negative (−10 m/s) when it is moving down; the speed is 10 m/s in either case.

GRAVITATIONAL POTENTIAL ENERGY FOR MOTION ALONG A CURVED PATH

In our first two examples the body moved along a straight vertical line. What happens when the path is slanted or curved (Fig. 7–4a)? The body is acted on by the gravitational force $\vec{w} = m\vec{g}$ and possibly by other forces whose resultant we call $\vec{F}_{other}$. To find the work

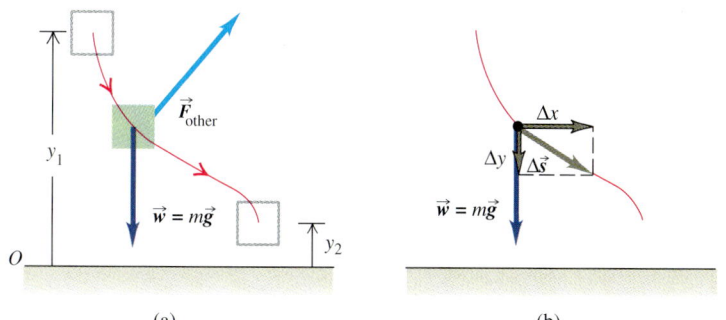

(a)   (b)

7–4 (a) A displacement along a curved path. (b) The work done by the gravitational force $\vec{w} = m\vec{g}$ depends only on the vertical component of displacement Δy (in this figure, Δy is negative).

5.14
Cart Leaves Circular Hill

done by the gravitational force during this displacement, we divide the path up into small segments $\Delta \vec{s}$; a typical segment is shown in Fig. 7–4b. The work done by the gravitational force over this segment is the scalar product of the force and the displacement. In terms of unit vectors, the force is $\vec{w} = m\vec{g} = -mg\hat{\jmath}$ and the displacement is $\Delta \vec{s} = \Delta x \hat{\imath} + \Delta y \hat{\jmath}$, so the work done by the gravitational force is

$$\vec{w} \cdot \Delta \vec{s} = -mg\hat{\jmath} \cdot (\Delta x \hat{\imath} + \Delta y \hat{\jmath}) = -mg\Delta y.$$

The work done by gravity is the same as though the body had been displaced vertically a distance Δy, with no horizontal displacement. This is true for every segment, so the *total* work done by the gravitational force is $-mg$ multiplied by the *total* vertical displacement $(y_2 - y_1)$:

$$W_{\text{grav}} = -mg(y_2 - y_1) = mgy_1 - mgy_2 = U_1 - U_2.$$

This is the same as Eq. (7–1) or (7–3), in which we assumed a purely vertical path. So even if the path a body follows between two points is curved, the total work done by the gravitational force depends only on the difference in height between the two points of the path. This work is unaffected by any horizontal motion that may occur. So *we can use the same expression for gravitational potential energy whether the body's path is curved or straight.*

EXAMPLE 7–3

Energy in projectile motion A batter hits two identical fly balls with the same initial speed but different initial angles. Prove that at a given height h, both balls have the same speed if air resistance can be neglected.

SOLUTION If there is no air resistance, the only force acting on each ball after it is hit is its weight. Hence the total mechanical energy for each ball is constant. Figure 7–5 shows the trajectories of two balls batted at the same height with the same initial speed, and thus the same total mechanical energy, but with different initial angles. At all points at the same height the potential energy is the same. Thus the kinetic energy at this height must be the same for both balls, and the speeds are the same.

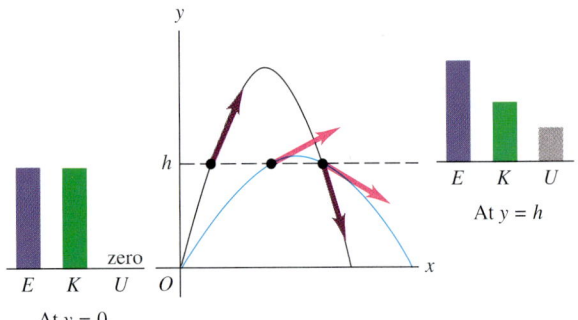

7–5 For the same initial speed and initial height the speed of a projectile at a given elevation h is always the same, neglecting air resistance.

EXAMPLE 7–4

Maximum height of a projectile, using energy methods In Example 3–9 (Section 3–4) we derived an expression for the maximum height h of a projectile launched with initial speed v_0 at initial angle α_0:

$$h = \frac{v_0^2 \sin^2 \alpha_0}{2g}.$$

Derive this expression using energy considerations.

SOLUTION We neglect air resistance, so just as in Example 7–3, total mechanical energy is conserved. We take point 1 to be the

launch point and point 2 to be the highest point on the trajectory (Fig. 7–6). We choose $y = 0$ at launch level; then $U_1 = 0$ and $U_2 = mgh$. We can express the kinetic energy at each point in terms of the components of velocity, using $v^2 = v_x^2 + v_y^2$:

$$K_1 = \frac{1}{2} m(v_{1x}^2 + v_{1y}^2),$$

$$K_2 = \frac{1}{2} m(v_{2x}^2 + v_{2y}^2).$$

Conservation of energy then gives $K_1 + U_1 = K_2 + U_2$, so

$$\frac{1}{2} m(v_{1x}^2 + v_{1y}^2) + 0 = \frac{1}{2} m(v_{2x}^2 + v_{2y}^2) + mgh.$$

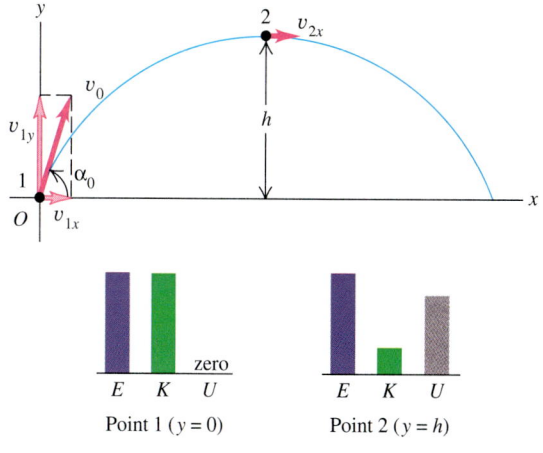

7–6 Trajectory of a projectile.

To simplify this, we multiply through by $2/m$ to obtain

$$v_{1x}^2 + v_{1y}^2 = v_{2x}^2 + v_{2y}^2 + 2gh.$$

Now for the *coup de grace:* We recall that in projectile motion the x-component of acceleration is zero, so the x-component of velocity is constant and $v_{1x} = v_{2x}$. Also, because point 2 is the highest point, the vertical component of velocity is zero at that point: $v_{2y} = 0$. Subtracting the v_x^2 terms from both sides, we get

$$v_{1y}^2 = 2gh.$$

But v_{1y} is just the y-component of initial velocity, which is equal to $v_0 \sin \alpha_0$. Making this substitution and solving for h, we find

$$h = \frac{v_0^2 \sin^2 \alpha_0}{2g},$$

which agrees with the result of Example 3–9.

EXAMPLE 7–5

Calculating speed along a vertical circle Your cousin Throckmorton skateboards down a curved playground ramp. Treating Throcky and his skateboard as a particle, his center moves through a quarter-circle with radius R (Fig. 7–7). The total mass of Throcky and his skateboard is 25.0 kg. He starts from rest, and there is no friction. (a) Find his speed at the bottom of the ramp. (b) Find the normal force that acts on him at the bottom of the curve.

SOLUTION a) We can't use the equations of motion with constant acceleration; the acceleration isn't constant because the slope decreases as Throcky descends. Instead, we'll use the energy approach. Since there is no friction, the only force other than Throcky's weight is the normal force $\vec{n}$ exerted by the ramp (Fig. 7–7b). Although this force acts all along the path, it does *zero* work because $\vec{n}$ is perpendicular to Throcky's velocity at every point. Thus $W_{\text{other}} = 0$, and total mechanical energy is conserved.

(a)

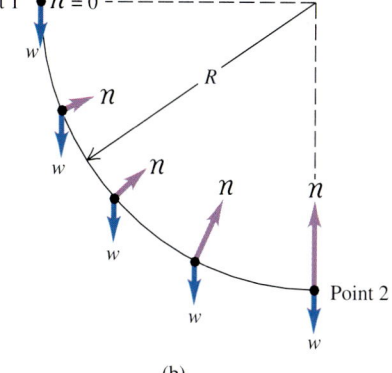

(b)

7–7 (a) Throcky skateboarding down a frictionless circular ramp. The total mechanical energy is constant. (b) Free-body diagrams for Throcky and his skateboard at various points on the ramp.

Take point 1 at the starting point and point 2 at the bottom of the curved ramp, and let $y = 0$ be at the bottom of the ramp (Fig. 7–7a). Then $y_1 = R$ and $y_2 = 0$. Throcky starts from rest at the top, so $v_1 = 0$. The various energy quantities are

$$K_1 = 0, \qquad U_1 = mgR,$$

$$K_2 = \frac{1}{2} m{v_2}^2, \qquad U_2 = 0.$$

From conservation of energy,

$$K_1 + U_1 = K_2 + U_2,$$

$$0 + mgR = \frac{1}{2} m{v_2}^2 + 0,$$

$$v_2 = \sqrt{2gR}.$$

The speed is the same as if Throcky had fallen vertically through a height R, and it is independent of his mass.

As a numerical example, let $R = 3.00$ m. Then

$$v_2 = \sqrt{2(9.80 \text{ m/s}^2)(3.00 \text{ m})} = 7.67 \text{ m/s}.$$

This example shows a general rule about the role of forces in problems in which we use energy techniques: What matters is not simply whether a force acts, but whether that force *does* work. If the force does no work, like the normal force $\vec{n}$ in this example, then it does not appear at all in Eq. (7–7), $K_1 + U_1 + W_{\text{other}} = K_2 + U_2$.

b) We want the magnitude n of the normal force at point 2. Since n doesn't appear at all in the energy equation, we'll use Newton's second law. The free-body diagram at point 2 is shown in Fig. 7–7b. At point 2, Throcky is moving at speed $v_2 = \sqrt{2gR}$ in a circle of radius R; his acceleration is radial and has magnitude

$$a_{\text{rad}} = \frac{{v_2}^2}{R} = \frac{2gR}{R} = 2g.$$

If we take the positive y-direction upward, the y-component of Newton's second law is

$$\Sigma F_y = n + (-w) = ma_{\text{rad}} = 2mg,$$

$$n = w + 2mg = 3mg.$$

At point 2 the normal force is three times Throcky's weight. This result is independent of the radius of the circular ramp. We learned in Example 5–8 (Section 5–3) and Example 5–24 (Section 5–5) that the magnitude of n is the *apparent weight,* so Throcky will feel as though he weighs three times his true weight mg. But as soon as he reaches the horizontal part of the ramp to the right of point 2, the normal force will decrease to $w = mg$, and Throcky will feel normal again. Can you see why?

Notice we had to use *both* the energy approach and Newton's second law to solve this problem; energy conservation gave us the speed, and $\Sigma \vec{F} = m\vec{a}$ gave us the normal force. For each part of the problem we used the technique that most easily led us to the answer.

EXAMPLE 7–6

A vertical circle with friction In Example 7–5, suppose that the ramp is not frictionless and that Throcky's speed at the bottom is only 6.00 m/s. What work was done by the friction force acting on him? Use $R = 3.00$ m.

SOLUTION We choose the same initial and final points and reference level as in Example 7–5 (Fig. 7–8). Again the normal force does no work, but now there is a friction force $\vec{f}$ that *does* do work. In this case, $W_{\text{other}} = W_f$. The energy quantities are

$$K_1 = 0,$$

$$U_1 = mgR = (25.0 \text{ kg})(9.80 \text{ m/s}^2)(3.00 \text{ m}) = 735 \text{ J},$$

$$K_2 = \frac{1}{2} m{v_2}^2 = \frac{1}{2}(25.0 \text{ kg})(6.00 \text{ m/s})^2 = 450 \text{ J},$$

$$U_2 = 0.$$

From Eq. (7–7),

$$W_f = K_2 + U_2 - K_1 - U_1$$

$$= 450 \text{ J} + 0 - 0 - 735 \text{ J} = -285 \text{ J}.$$

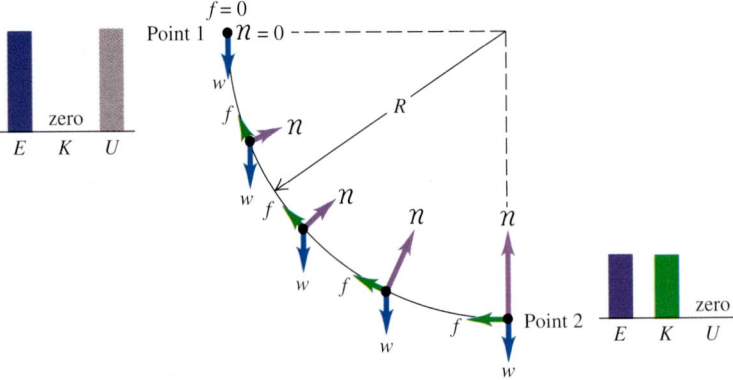

7–8 Free-body diagram for Throcky skateboarding down a ramp with friction. The total mechanical energy decreases as Throcky moves down the ramp.

The work done by the friction force is −285 J, and the total mechanical energy *decreases* by 285 J. Do you see why W_f has to be negative?

Throcky's motion is determined by Newton's second law, $\Sigma \vec{F} = m\vec{a}$. But it would be very difficult to apply the second law directly to this problem because the normal and friction forces and the acceleration are continuously changing in both magnitude and direction as Throcky skates down. The energy approach, by contrast, relates the motions at the top and bottom of the ramp without involving the details of what happens in between. Many problems are easy if energy considerations are used but very complex if we try to use Newton's laws directly.

EXAMPLE 7–7

An inclined plane with friction A 12-kg crate sits on the floor. We want to load it into a truck by sliding it up a ramp 2.5 m long, inclined at 30°. A worker, giving no thought to friction, calculates that he can get the crate up the ramp by giving it an initial speed of 5.0 m/s at the bottom and letting it go. But friction is *not* negligible; the crate slides 1.6 m up the ramp, stops, and slides back down (Fig. 7–9). a) Assuming that the friction force acting on the crate is constant, find its magnitude. b) How fast is the crate moving when it reaches the bottom of the ramp?

SOLUTION a) As in Example 7–2, we'll find the magnitude of the nongravitational force that does work (in this case, friction) by using the energy approach. Let point 1 be the bottom of the ramp, and let point 2 be where the crate stops (Fig. 7–9a). If we take $U = 0$ at floor level, we have $y_1 = 0$, $y_2 = (1.6 \text{ m}) \sin 30°$ = 0.80 m. The energy quantities are

$$K_1 = \frac{1}{2}(12 \text{ kg})(5.0 \text{ m/s})^2 = 150 \text{ J}, \qquad U_1 = 0,$$

$$K_2 = 0, \qquad U_2 = (12 \text{ kg})(9.8 \text{ m/s}^2)(0.80 \text{ m}) = 94 \text{ J},$$

$$W_{other} = -fs,$$

where f is the unknown magnitude of the friction force and $s = 1.6$ m. Using Eq. (7–7), we find

$$K_1 + U_1 + W_{other} = K_2 + U_2,$$

$$W_{other} = -fs = (K_2 + U_2) - (K_1 + U_1),$$

$$f = -\frac{(K_2 + U_2) - (K_1 + U_1)}{s}$$

$$= -\frac{[(0 + 94 \text{ J}) - (150 \text{ J} + 0)]}{1.6 \text{ m}} = 35 \text{ N}.$$

The friction force of 35 N, acting over 1.6 m, causes the mechanical energy of the crate to decrease from 150 J to 94 J. b) The crate returns to point 3 at the bottom of the ramp; $y_3 = 0$ and $U_3 = 0$ (Fig. 7–9b). On the way down, the friction force and the displacement both reverse direction but have the same magnitudes, so the frictional work has the same negative value for each half of the trip. The total work done by friction between points 1 and 3 is

$$W_{other} = W_{fric} = -2fs = -2(35 \text{ N})(1.6 \text{ m}) = -112 \text{ J}.$$

From part (a), $K_1 = 150$ J and $U_1 = 0$. Eq. (7–7) then gives

$$K_1 + U_1 + W_{other} = K_3 + U_3,$$

$$K_3 = K_1 + U_1 - U_3 + W_{other}$$

$$= 150 \text{ J} + 0 - 0 + (-112 \text{ J}) = 38 \text{ J}.$$

(a)

(b)

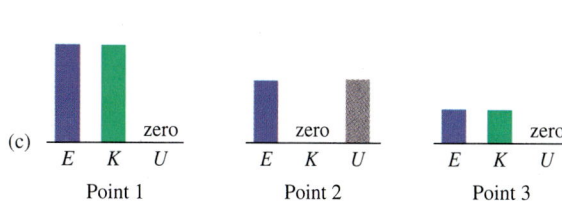

(c)

7–9 (a) A crate slides part way up the ramp, stops, and (b) slides back down. (c) Energy bar graphs for points 1, 2, and 3.

The crate returns to the bottom of the ramp with only 38 J of the original 150 J of mechanical energy. Using $K_3 = \frac{1}{2}mv_3^2$, we get

$$v_3 = \sqrt{\frac{2K_3}{m}} = \sqrt{\frac{2(38 \text{ J})}{12 \text{ kg}}} = 2.5 \text{ m/s}.$$

Alternatively, we could have applied Eq. (7–7) to points 2 and 3, considering the second half of the trip by itself. Try it and see whether you get the same result.

7–3 ELASTIC POTENTIAL ENERGY

When a railroad car runs into a spring bumper at the end of the track, the spring is compressed as the car is brought to a stop. If there is no friction, the bumper springs back and the car moves away with its original speed in the opposite direction. During the interaction with the spring, the car's kinetic energy has been "stored" in the elastic deformation of the spring. Something very similar happens in a rubber-band slingshot. Work is done on the rubber band by the force that stretches it, and that work is stored in the rubber band until you let it go. Then the rubber band gives kinetic energy to the projectile.

This is the same pattern we saw with the pile-driver in Section 7–2: do work on the system to store energy, which can later be converted to kinetic energy. We'll describe the process of storing energy in a deformable body such as a spring or rubber band in terms of *elastic potential energy*. A body is called *elastic* if it returns to its original shape and size after being deformed. To be specific, we'll consider storing energy in an ideal spring like the ones we discussed in Section 6–4. To keep such an ideal spring stretched by a distance x, we must exert a force $F = kx$, where k is the force constant of the spring. The ideal spring is a useful idealization because many other elastic bodies show this same direct proportionality between force F and displacement x, provided that x is sufficiently small.

We proceed just as we did for gravitational potential energy. We begin with the work done by the elastic (spring) force and then combine this with the work-energy theorem. The difference is that gravitational potential energy is a shared property of a body and the earth, but elastic potential energy is stored just in the spring (or other deformable body).

Figure 7–10 shows the ideal spring from Fig. 6–13, with its left end held stationary and its right end attached to a block with mass m that can move along the x-axis. In Fig. 7–10a the body is at $x = 0$ when the spring is neither stretched nor compressed. We move the block to one side, thereby stretching or compressing the spring, and then let it go. As the block moves from one position x_1 to another position x_2, how much work does the elastic (spring) force do on the block?

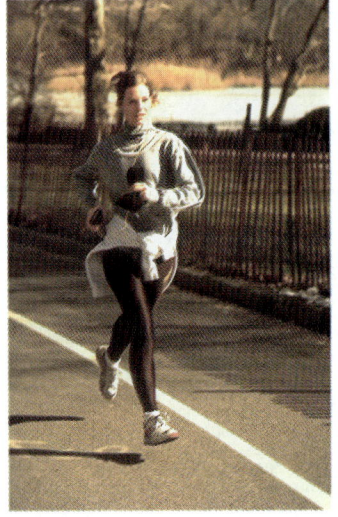

The Achilles tendon, which runs along the back of the ankle to the heel bone, acts like a natural spring. When it stretches and then relaxes, this tendon stores and then releases elastic potential energy. This spring action reduces the amount of work your leg muscles must do as you run.

7–10 (a) A block attached to a spring in equilibrium ($x = 0$) on a horizontal surface. (b) When the block moves from positive x_1 to positive $x_2 > x_1$, the spring does negative work as it stretches. (c) When the block moves from positive x_1 to positive $x_2 < x_1$, the spring relaxes and does positive work. (d) When the block moves from negative x_1 to a less negative x_2, the compressed spring relaxes and again does positive work.

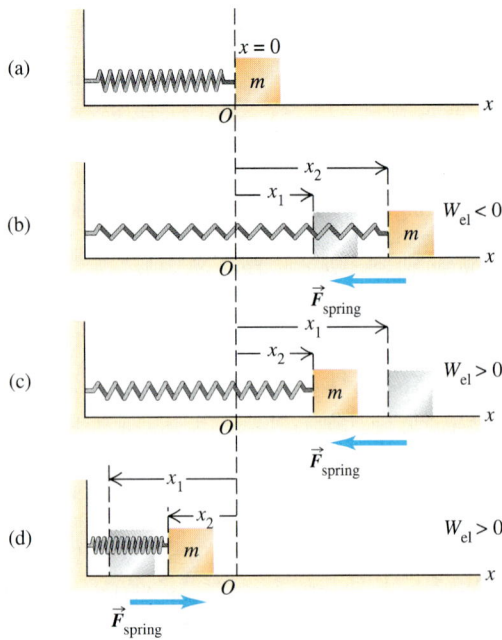

We found in Section 6–4 that the work we must do *on* the spring to move one end from an elongation x_1 to a different elongation x_2 is

$$W = \frac{1}{2}kx_2{}^2 - \frac{1}{2}kx_1{}^2 \qquad \text{(work done \emph{on} a spring),}$$

where k is the force constant of the spring. If we stretch the spring further, we do positive work on the spring; if we let the spring relax while holding one end, we do negative work on it. We also saw that this expression for work is still correct if the spring is compressed, not stretched, so that x_1 or x_2 or both are negative. Now we need to find the work done *by* the spring. From Newton's third law the two quantities of work are just negatives of each other. Changing the signs in this equation, we find that in a displacement from x_1 to x_2 the spring does an amount of work W_{el} given by

$$W_{\text{el}} = \frac{1}{2}kx_1{}^2 - \frac{1}{2}kx_2{}^2 \qquad \text{(work done \emph{by} a spring).}$$

The subscript "el" stands for *elastic*. When x_1 and x_2 are both positive and $x_2 > x_1$ (Fig. 7–10b), the spring does negative work on the block, which moves in the $+x$-direction while the spring pulls on it in the $-x$-direction. The spring stretches further, and the block slows down. When x_1 and x_2 are both positive and $x_2 < x_1$ (Fig. 7–10c), the spring does positive work as it relaxes and the block speeds up. If the spring can be compressed as well as stretched, x_1 or x_2 or both may be negative, but the expression for W_{el} is still valid. In Fig. 7–10d, both x_1 and x_2 are negative, but x_2 is less negative than x_1; the compressed spring does positive work as it relaxes, speeding the block up.

Just as for gravitational work, we can express the work done by the spring in terms of a given quantity at the beginning and end of the displacement. This quantity is $\frac{1}{2}kx^2$, and we define it to be the **elastic potential energy:**

$$U = \frac{1}{2}kx^2 \qquad \text{(elastic potential energy).} \qquad (7\text{–}9)$$

Figure 7–11 is a graph of Eq. (7–9). The unit of U is the joule (J), the unit used for *all* energy and work quantities; to see this from Eq. (7–9), recall that the units of k are N/m and that $1 \text{ N} \cdot \text{m} = 1 \text{ J}$.

We can use Eq. (7–9) to express the work W_{el} done on the block by the elastic force in terms of the change in potential energy:

$$W_{\text{el}} = \frac{1}{2}kx_1{}^2 - \frac{1}{2}kx_2{}^2 = U_1 - U_2 = -\Delta U. \qquad (7\text{–}10)$$

When a stretched spring is stretched further, as in Fig. 7–10b, W_{el} is negative and U *increases;* a greater amount of elastic potential energy is stored in the spring. When a stretched spring relaxes, as in Fig. 7–10c, x decreases, W_{el} is positive, and U *decreases;* the spring loses elastic potential energy. Negative values of x refer to a compressed spring. But, as Fig. 7–11 shows, U is positive for both positive and negative x, and Eqs. (7–9) and (7–10) are valid for both cases. Thus when a compressed spring is compressed further, $W_{\text{el}} < 0$ and U increases; when a compressed spring relaxes (Fig. 7–10d), $W_{\text{el}} > 0$ and U decreases. The more a spring is compressed *or* stretched, the greater its elastic potential energy.

CAUTION ▶ An important difference between gravitational potential energy $U = mgy$ and elastic potential energy $U = \frac{1}{2}kx^2$ is that we do *not* have the freedom to choose $x = 0$ to be wherever we wish. To be consistent with Eq. (7–9), $x = 0$ *must* be the position at which the spring is neither stretched nor compressed. At that position, its elastic potential energy and the force that it exerts are both zero. ◀

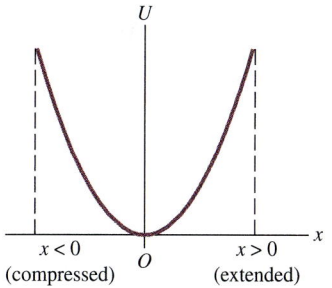

7–11 The graph of elastic potential energy for an ideal spring is a parabola: $U = \frac{1}{2}kx^2$, where x is the extension or compression of the spring. For extension (stretching), x is positive. For compression (when that is possible), x is negative. Elastic potential energy U is never negative.

The work-energy theorem says that $W_{tot} = K_2 - K_1$, no matter what kind of forces are acting on a body. If the elastic force is the *only* force that does work on the body, then

$$W_{tot} = W_{el} = U_1 - U_2.$$

The work-energy theorem $W_{tot} = K_2 - K_1$ then gives us

$$K_1 + U_1 = K_2 + U_2 \qquad \text{(if only the elastic force does work).} \qquad (7\text{--}11)$$

Here U is given by Eq. (7–9), so

$$\tfrac{1}{2}mv_1{}^2 + \tfrac{1}{2}kx_1{}^2 = \tfrac{1}{2}mv_2{}^2 + \tfrac{1}{2}kx_2{}^2 \qquad \text{(if only the elastic force does work).}$$

$$(7\text{--}12)$$

In this case the total mechanical energy $E = K + U$ (the sum of kinetic and elastic potential energy) is *conserved*. An example of this is the motion of the block in Fig. 7–10, provided that the horizontal surface is frictionless so that no force does work other than that exerted by the spring.

For Eq. (7–12) to be strictly correct, the ideal spring that we've been discussing must also be *massless*. If the spring has a mass, it will also have kinetic energy as the coils of the spring move back and forth. We can neglect the kinetic energy of the spring if its mass is much less than the mass m of the body attached to the spring. For instance, a typical automobile has a mass of 1200 kg or more. The springs in its suspension have masses of only a few kilograms, so their mass can be neglected if we want to study how a car bounces on its suspension.

If forces other than the elastic force also do work on the body, we call their work W_{other}, as before. Then the total work is $W_{tot} = W_{el} + W_{other}$, and the work-energy theorem gives

$$W_{el} + W_{other} = K_2 - K_1.$$

The work done by the spring is still $W_{el} = U_1 - U_2$, so again

$$K_1 + U_1 + W_{other} = K_2 + U_2 \qquad \text{(if forces other than the elastic force do work)}$$

$$(7\text{--}13)$$

and

$$\tfrac{1}{2}mv_1{}^2 + \tfrac{1}{2}kx_1{}^2 + W_{other} = \tfrac{1}{2}mv_2{}^2 + \tfrac{1}{2}kx_2{}^2$$

$$(7\text{--}14)$$

(if forces other than the elastic force do work).

This equation shows that *the work done by all forces* other than the elastic force *equals the change in the total mechanical energy $E = K + U$ of the system, where U is the elastic potential energy.* The "system" is made up of the body of mass m and the spring of force constant k. When W_{other} is positive, E increases; when W_{other} is negative, E decreases. You should compare Eq. (7–14) to Eq. (7–8), which describes situations in which there is gravitational potential energy but no elastic potential energy.

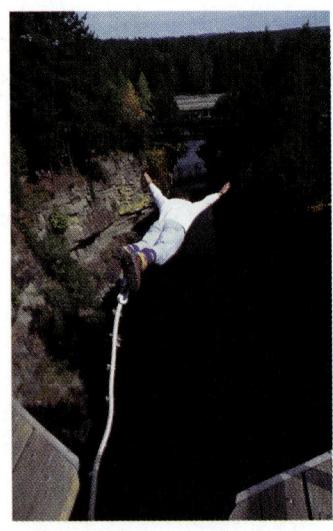

Work is done on this bungee jumper by both gravity and the elastic force of the bungee cord. Mechanical energy is not conserved, however, because work is also done by frictional forces within the bungee cord and by air resistance. This is a good thing—if mechanical energy were conserved, the bungee jumper would keep bouncing up and down forever!

SITUATIONS WITH BOTH GRAVITATIONAL AND ELASTIC POTENTIAL ENERGY

Equations (7–11), (7–12), (7–13), and (7–14) are valid when the only potential energy in the system is elastic potential energy. What happens when we have *both* gravitational and elastic forces, such as a block attached to the lower end of a vertically hanging spring? We can still use Eq. (7–13), but now U_1 and U_2 are the initial and final values of the *total* potential energy, including both gravitational and elastic potential energies. That is, $U = U_{grav} + U_{el}$. Thus *the most general statement* of the relationship between kinetic energy, potential energy, and work done by other forces is

$$K_1 + U_{grav,\,1} + U_{el,\,1} + W_{other} = K_2 + U_{grav,\,2} + U_{el,\,2} \qquad \text{(valid in general).} \quad (7\text{--}15)$$

That is, **the work done by all forces other than the gravitational force or elastic force equals the change in the total mechanical energy $E = K + U$ of the system, where U is the sum of the gravitational potential energy and elastic potential energy.** If the gravitational and elastic forces are the *only* forces that do work on the body, then $W_{other} = 0$ and the total mechanical energy (including both gravitational and elastic potential energy) is conserved. Pole vaulting is a familiar example of transformations between kinetic energy, elastic potential energy, and gravitational potential energy. The athlete's initial kinetic energy is partly stored as elastic potential energy in the flexed pole. Most of this elastic potential energy is then used to help give the increase in gravitational potential energy needed to clear the bar.

The strategy outlined in Section 7–2 is equally useful in solving problems that involve elastic forces as well as gravitational forces. The only new wrinkle is that the potential energy U now includes the elastic potential energy $U_{el} = \frac{1}{2}kx^2$, where x is the displacement of the spring *from its unstretched length.* The work done by the gravitational and elastic forces is accounted for by their potential energies; the work of the other forces, W_{other}, has to be included separately.

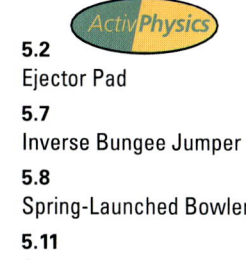

5.2
Ejector Pad

5.7
Inverse Bungee Jumper

5.8
Spring-Launched Bowler

5.11
Spring-Launched Ski Lift

EXAMPLE 7–8

Motion with elastic potential energy In Fig. 7–12a a glider with mass $m = 0.200$ kg sits on a frictionless horizontal air track, connected to a spring with force constant $k = 5.00$ N/m. You pull on the glider, stretching the spring 0.100 m, and then release it with no initial velocity (Fig. 7–12b). The glider begins to move back toward its equilibrium position ($x = 0$). What is its x-velocity when $x = 0.080$ m?

SOLUTION Because the spring force varies with position, this problem *cannot* be done by using the equations for motion with

constant acceleration. The energy method, by contrast, offers a simple and elegant solution.

As the glider starts to move, elastic potential energy is converted into kinetic energy. (The glider remains at the same height throughout the motion, so *gravitational* potential energy is not a factor.) The spring force is the only force doing work on the glider, so $W_{other} = 0$, and we may use Eq. (7–11). Let point 1 be where the glider is released (Fig. 7–12b), and let point 2 be at $x = 0.080$ m (Fig. 7–12c). The energy quantities are

7–12 (a) An air-track glider attached to a spring. (b) Elastic potential energy is added to the system by stretching the spring. (c) Elastic potential energy is transformed to kinetic energy as the glider moves back toward its equilibrium position.

$$K_1 = \frac{1}{2}(0.200 \text{ kg})(0)^2 = 0,$$

$$U_1 = \frac{1}{2}(5.00 \text{ N/m})(0.100 \text{ m})^2 = 0.0250 \text{ J},$$

$$K_2 = \frac{1}{2}mv_2{}^2,$$

$$U_2 = \frac{1}{2}(5.00 \text{ N/m})(0.080 \text{ m})^2 = 0.0160 \text{ J},$$

Then from Eq. (7–11),

$$K_2 = K_1 + U_1 - U_2 = 0 + 0.0250 \text{ J} - 0.0160 \text{ J} = 0.0090 \text{ J},$$

$$v_2 = \pm\sqrt{\frac{2K_2}{m}} = \pm\sqrt{\frac{2(0.0090 \text{ J})}{0.200 \text{ kg}}} = \pm 0.30 \text{ m/s}.$$

We choose the negative root because the glider is moving in the $-x$-direction; the answer we want is $v_2 = -0.30$ m/s.

EXAMPLE 7–9

Motion with elastic potential energy and work done by other forces For the system of Example 7–8, suppose the glider is initially at rest at $x = 0$, with the spring unstretched. You then apply a constant force $\vec{F}$ in the $+x$-direction with magnitude 0.610 N to the glider. What is the glider's speed when it has moved to $x = 0.100$ m?

SOLUTION Again, the acceleration of the glider isn't constant. (Do you see why not?) Total mechanical energy is not conserved because of the work done by the force $\vec{F}$, but we can still use the energy relation of Eq. (7–13). Now let point 1 be at $x = 0$ and point 2 at $x = 0.100$ m (these are different from the points labeled in Fig. 7–12). The energy quantities are

$$K_1 = 0, \qquad U_1 = 0,$$

$$K_2 = \frac{1}{2}mv_2{}^2, \qquad U_2 = \frac{1}{2}(5.00 \text{ N/m})(0.100 \text{ m})^2$$

$$= 0.0250 \text{ J},$$

$$W_{\text{other}} = (0.610 \text{ N})(0.100 \text{ m}) = 0.0610 \text{ J}.$$

(To calculate W_{other} we multiplied the magnitude of the force by the displacement, since both are in the $+x$-direction.) Initially, the total mechanical energy is zero; the work done by the force $\vec{F}$ increases the total mechanical energy to 0.0610 J, of which 0.0250 J is elastic potential energy. The remainder is kinetic energy. From Eq. (7–13),

$$K_1 + U_1 + W_{\text{other}} = K_2 + U_2,$$

$$K_2 = K_1 + U_1 + W_{\text{other}} - U_2$$

$$= 0 + 0 + 0.0610 \text{ J} - 0.0250 \text{ J} = 0.0360 \text{ J},$$

$$v_2 = \sqrt{\frac{2K_2}{m}} = \sqrt{\frac{2(0.0360 \text{ J})}{0.200 \text{ kg}}} = 0.60 \text{ m/s}.$$

EXAMPLE 7–10

Motion with elastic potential energy after other forces have ceased In Example 7–9, suppose the force $\vec{F}$ is removed when the glider reaches the point $x = 0.100$ m. How much farther does the glider move before coming to rest?

SOLUTION After $\vec{F}$ is removed, the spring force is the only force doing work. Hence for this part of the motion, total mechanical energy $E = K + U$ is conserved. We saw in Example 7–9 that the kinetic energy at the 0.100-m point (point 2) is $K_2 = 0.0360$ J and the potential energy at this point is $U_2 = 0.0250$ J. The total mechanical energy at and beyond this point is therefore 0.0610 J (equal to the work done by the force $\vec{F}$ between points 1 and 2).

When the body comes to rest at x_{max} (point 3), its kinetic energy K_3 is zero, and its potential energy U_3 equal to the total mechanical energy 0.0610 J. We can also see this from $K_2 + U_2 = K_3 + U_3$:

$$U_3 = K_2 + U_2 - K_3 = 0.0360 \text{ J} + 0.0250 \text{ J} - 0 = 0.0610 \text{ J}.$$

But $U_3 = \frac{1}{2}kx_{\text{max}}{}^2$, so

$$x_{\text{max}} = \sqrt{\frac{2U_3}{k}} = \sqrt{\frac{2(0.0610 \text{ J})}{5.00 \text{ N/m}}} = 0.156 \text{ m}.$$

So the body moves an additional 0.056 m after the force is removed at $x = 0.100$ m.

EXAMPLE 7–11

Motion with gravitational, elastic, and friction forces In a "worst-case" design scenario, a 2000-kg elevator with broken cables is falling at 25 m/s when it first contacts a cushioning spring at the bottom of the shaft. The spring is supposed to stop the elevator, compressing 3.00 m as it does so (Fig. 7–13). During the motion a safety clamp applies a constant 17,000-N frictional force to the elevator. As a design consultant, you are asked to determine what the force constant of the spring should be.

SOLUTION In this problem there is both gravitational and elastic potential energy, so we must use Eq. (7–15). Total mechanical energy is not conserved because the friction force does negative work W_{other} on the elevator.

Take point 1 as the elevator's position when it initially contacts the spring, and take point 2 as its position when it is at rest. We choose the origin to be at point 1, so $y_1 = 0$ and $y_2 = -3.00$ m. With this choice the coordinate of the upper end of the spring is

7-13 The fall of an elevator is stopped by a spring and by a constant friction force.

the same as the coordinate of the elevator, so the elastic potential energy at any point between point 1 and point 2 is $U_{el} = \frac{1}{2}ky^2$. The elevator's initial speed is $v_1 = 25$ m/s, so the initial kinetic energy is

$$K_1 = \frac{1}{2}mv_1^2 = \frac{1}{2}(2000 \text{ kg})(25 \text{ m/s})^2 = 625{,}000 \text{ J}.$$

The elevator stops at point 2, so $K_2 = 0$. The potential energy at point 1, U_1, is zero; U_{grav} is zero because $y_1 = 0$, and $U_{el} = 0$ because the spring is not yet compressed. At point 2 there is both gravitational and elastic potential energy, so

$$U_2 = mgy_2 + \frac{1}{2}ky_2^2;$$

the gravitational potential energy at point 2 is

$$mgy_2 = (2000 \text{ kg})(9.80 \text{ m/s}^2)(-3.00 \text{ m}) = -58{,}800 \text{ J}.$$

The other force is the 17,000-N friction force, acting opposite to the 3.00-m displacement, so

$$W_{other} = -(17{,}000 \text{ N})(3.00 \text{ m}) = -51{,}000 \text{ J}.$$

Putting these into $K_1 + U_1 + W_{other} = K_2 + U_2$, we have

$$K_1 + 0 + W_{other} = 0 + (mgy_2 + \frac{1}{2}ky_2^2),$$

so the force constant of the spring is

$$\begin{aligned} k &= \frac{2(K_1 + W_{other} - mgy_2)}{y_2^2} \\ &= \frac{2(625{,}000 \text{ J} + (-51{,}000 \text{ J}) - (-58{,}800 \text{ J}))}{(-3.00 \text{ m})^2} \\ &= 1.41 \times 10^5 \text{ N/m}. \end{aligned}$$

This is comparable to the springs in an automobile suspension.

Let's notice what might seem to be a paradox in this problem. The elastic potential energy in the spring at point 2 is

$$\frac{1}{2}ky_2^2 = \frac{1}{2}(1.41 \times 10^5 \text{ N/m})(-3.00 \text{ m})^2 = 632{,}800 \text{ J}.$$

This is *more* than the total mechanical energy at point 1,

$$E_1 = K_1 + U_1 = 625{,}000 \text{ J} + 0 = 625{,}000 \text{ J}.$$

But the friction force caused the mechanical energy of the system to *decrease* by 51,000 J between point 1 and point 2. Does this mean that energy appeared from nowhere? Don't panic; there is no paradox. At point 2 there is also *negative* gravitational potential energy $mgy_2 = -58{,}800$ J because point 2 is below the origin. The total mechanical energy at point 2 is

$$\begin{aligned} E_2 &= K_2 + U_2 = 0 + \frac{1}{2}ky_2^2 + mgy_2 \\ &= 632{,}800 \text{ J} + (-58{,}800 \text{ J}) = 574{,}000 \text{ J}. \end{aligned}$$

This is just the initial mechanical energy of 625,000 J, minus 51,000 J lost to friction.

Your next job as design consultant would be to tell your client that the elevator won't stay at the bottom of the shaft. Instead, it will bounce back up. The reason is that at point 2 the compressed spring exerts an upward force of magnitude $F_{spring} = (1.41 \times 10^5 \text{ N/m})(3.00 \text{ m}) = 422{,}000$ N. The weight of the elevator is only $w = mg = (2000 \text{ kg})(9.80 \text{ m/s}^2) = 19{,}600$ N, so the net force will be upward. The elevator bounces back upward even if the safety clamp now exerts a *downward* friction force of magnitude $f = 17{,}000$ N; the spring force is greater than the sum of f and mg. The elevator will bounce again and again until enough mechanical energy has been removed by friction for it to stop.

Can you also show that the acceleration of the falling elevator when it hits the spring is unacceptably high?

7-4 CONSERVATIVE AND NONCONSERVATIVE FORCES

In our discussions of potential energy we have talked about "storing" kinetic energy by converting it to potential energy. We always have in mind that later we may retrieve it again as kinetic energy. When you throw a ball up in the air, it slows down as kinetic energy is converted into potential energy. But on the way down, the conversion is reversed, and the ball speeds up as potential energy is converted back to kinetic energy. If there is no air resistance, the ball is moving just as fast when you catch it as when you threw it.

If a glider moving on a frictionless horizontal air track runs into a spring bumper at the end of the track, the spring compresses and the glider stops. But then it bounces back, and if there is no friction, the glider has the same speed and kinetic energy it had before the collision. Again, there is a two-way conversion from kinetic to potential energy and back. In both cases we find that we can define a potential-energy function so that the total mechanical energy, kinetic plus potential, is constant or *conserved* during the motion.

A force that offers this opportunity of two-way conversion between kinetic and potential energies is called a **conservative force.** We have seen two examples of conservative forces: the gravitational force and the spring force. An essential feature of conservative forces is that their work is always *reversible.* Anything that we deposit in the energy "bank" can later be withdrawn without loss. Another important aspect of conservative forces is that a body may move from point 1 to point 2 by various paths, but the work done by a conservative force is the same for all of these paths (Fig. 7–14). Thus if a body stays close to the surface of the earth, the gravitational force $m\vec{g}$ is independent of height, and the work done by this force depends only on the change in height. If the body moves around a closed path, ending at the same point where it started, the *total* work done by the gravitational force is always zero.

The work done by a conservative force *always* has these properties:

1. It can always be expressed as the difference between the initial and final values of a *potential energy* function.
2. It is reversible.
3. It is independent of the path of the body and depends only on the starting and ending points.
4. When the starting and ending points are the same, the total work is zero.

When the only forces that do work are conservative forces, the total mechanical energy $E = K + U$ is constant.

Not all forces are conservative. Consider the friction force acting on the crate sliding on a ramp in Example 7–7 (Section 7–2). When the body slides up and then back down to the starting point, the total work done on it by the friction force is *not* zero. When the direction of motion reverses, so does the friction force, and friction does *negative* work in *both* directions. When a car with its brakes locked skids across the pavement with decreasing speed (and decreasing kinetic energy), the lost kinetic energy cannot be recovered by reversing the motion or in any other way, and mechanical energy is *not* conserved. There is no potential-energy function for the friction force.

In the same way, the force of fluid resistance (Section 5–4) is not conservative. If you throw a ball up in the air, air resistance does negative work on the ball while it's rising *and* while it's descending. The ball returns to your hand with less speed and less

7–14 For any conservative force, the work done by that force depends only on the end points, not on the path taken. Thus the gravitational force, which is conservative, does the same work on the hummingbird no matter what path it flies from point 1 to point 2.

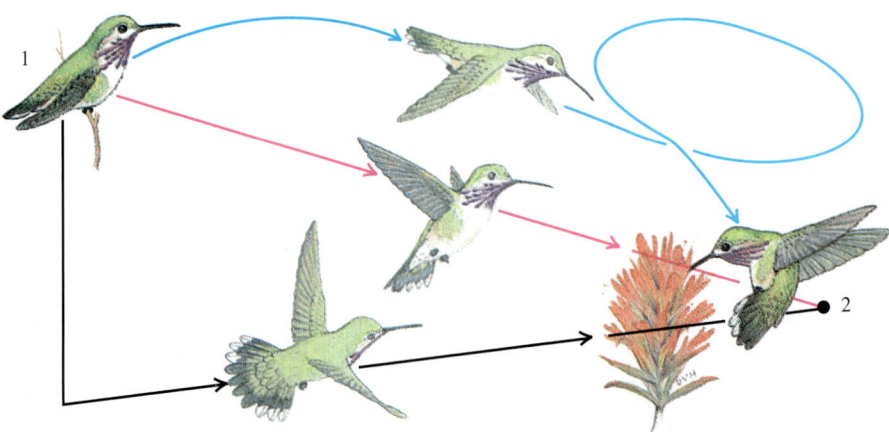

kinetic energy than when it left, and there is no way to get back the lost mechanical energy.

A force that is not conservative is called a **nonconservative force.** The work done by a nonconservative force *cannot* be represented by a potential-energy function. Some nonconservative forces, like kinetic friction or fluid resistance, cause mechanical energy to be lost or dissipated; a force of this kind is called a **dissipative force.** There are also nonconservative forces that *increase* mechanical energy. The fragments of an exploding firecracker fly off with very large kinetic energy, thanks to a chemical reaction of gunpowder with oxygen. The forces unleashed by this reaction are nonconservative because the process is not reversible. Imagine the fragments spontaneously reassembling themselves into a complete firecracker!

EXAMPLE 7–12

Frictional work depends on the path You are rearranging your furniture and wish to move a 40.0-kg futon 2.50 m across the room (Fig. 7–15). However, the straight-line path is blocked by a heavy coffee table that you don't want to move. Instead, you slide the futon in a dogleg path over the floor; the doglegs are 2.00 m and 1.50 m long. Compared to the straight-line path, how much more work must you do to push the futon in the dogleg path? The coefficient of kinetic friction is 0.200.

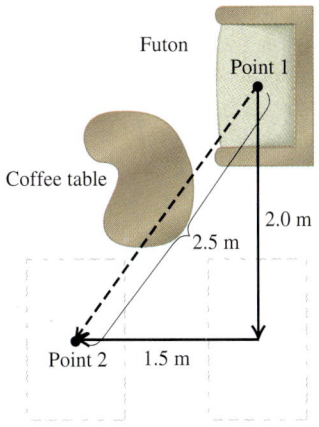

Futon
Point 1
Coffee table
2.0 m
2.5 m
Point 2   1.5 m

7–15 A view from above of furniture being moved. How much more work must you do to move the futon on a dogleg path?

SOLUTION The initial and final points are shown in Fig. 7–15. The futon is at rest at both point 1 and point 2, so $K_1 = K_2 = 0$. The gravitational potential energy does not change because the futon moves only horizontally; to be specific, we'll say $U_1 = U_2 = 0$. From Eq. (7–7) it follows that $W_{other} = 0$. The other work done on the futon is the sum of the positive work you do, W_{you}, and the negative work W_{fric} done by the kinetic friction force. Since the sum of these is zero, we have

$$W_{you} = -W_{fric}.$$

Because the floor is horizontal, the normal force on the futon equals its weight mg, and the magnitude of the friction force is $f_k = \mu_k n = \mu_k mg$. The work you must do over each path is then

$$W_{you} = -W_{fric} = -(-f_k s) = +\mu_k mgs$$
$$= (0.200)(40.0 \text{ kg})(9.80 \text{ m/s}^2)(2.50 \text{ m})$$
$$= 196 \text{ J} \quad \text{(straight-line path)},$$

$$W_{you} = -W_{fric} = (0.200)(40.0 \text{ kg})(9.80 \text{ m/s}^2)(2.00 \text{ m} + 1.50 \text{ m})$$
$$= 274 \text{ J} \quad \text{(dogleg path)}.$$

The extra work you must do is 274 J – 196 J = 78 J.

The work done by friction is $W_{fric} = -W_{you} = -196$ J on the straight-line path, and –274 J on the dogleg. Friction is a nonconservative force, so the work done by friction depends on the path taken.

EXAMPLE 7–13

Conservative or nonconservative? In a certain region of space the force on an electron is $\vec{F} = Cx\hat{j}$, where C is a positive constant. The electron moves in a counterclockwise direction around a square loop in the xy-plane (Fig. 7–16). The corners of the square are at $(x, y) = (0, 0)$, $(L, 0)$, (L, L) and $(0, L)$. Calculate the work done on the electron by the force $\vec{F}$ during one complete trip around the square. Is this force conservative or nonconservative?

SOLUTION The force $\vec{F}$ is not constant and in general is not in the same direction as the displacement. So we'll use the more general expression for work, Eq. (6–14):

$$W = \int_{P_1}^{P_2} \vec{F} \cdot d\vec{l},$$

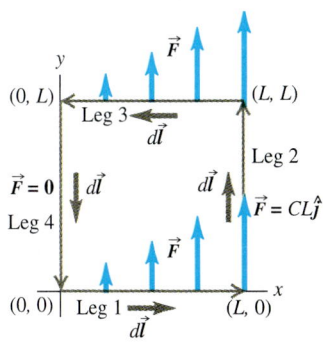

7–16 An electron moving around a square loop while being acted on by the force $\vec{F} = Cx\hat{j}$.

where $d\vec{l}$ is an infinitesimal displacement. Let's calculate the work done by the force $\vec{F}$ on each leg of the square and then add the results to find the work done on the round trip.

On the first leg, from $(0, 0)$ to $(L, 0)$, the force varies but is everywhere perpendicular to the displacement. So $\vec{F} \cdot d\vec{l} = 0$, and the work done on the first leg is $W_1 = 0$. The force has the same value $\vec{F} = CL\hat{\jmath}$ everywhere on the second leg from $(L, 0)$ to (L, L). The displacement on this leg is in the $+y$-direction, so $d\vec{l} = dy\hat{\jmath}$ and

$$\vec{F} \cdot d\vec{l} = CL\hat{\jmath} \cdot dy\hat{\jmath} = CL\,dy.$$

The work done on the second leg is then

$$W_2 = \int_{(L,0)}^{(L,L)} \vec{F} \cdot d\vec{l} = \int_{y=0}^{y=L} CL\,dy = CL \int_0^L dy = CL^2.$$

On the third leg, from (L, L) to $(0, L)$, $\vec{F}$ is again perpendicular to the displacement so $W_3 = 0$. The force is zero on the final leg, from $(0, L)$ to $(0, 0)$, so no work is done and $W_4 = 0$. The work done by the force $\vec{F}$ on the round trip is

$$W = W_1 + W_2 + W_3 + W_4 = 0 + CL^2 + 0 + 0 = CL^2.$$

The starting and ending points are the same, but the total work done by $\vec{F}$ is not zero. This is a nonconservative force; it cannot be represented by a potential energy function.

Because W is positive, the mechanical energy of the electron *increases* as it goes around the loop. This is not a mathematical curiosity; it's a description of what happens in an electrical generating plant. A loop of wire is moved through a magnetic field, which gives rise to a nonconservative force similar to the one in this example. Electrons in the wire gain energy as they move around the loop, and this energy is carried via transmission lines to the consumer. (We'll discuss how this works in detail in Chapter 30.) All the electrical energy used in the home and in industry comes from work done by nonconservative forces!

How would the value of W change if the electron went around the loop clockwise instead of counterclockwise? The force $\vec{F}$ would be unaffected, but the direction of each infinitesimal displacement $d\vec{l}$ would reverse. Thus the sign of work would also reverse, and the work for a clockwise round trip would be $W = -CL^2$. This is a different behavior than the nonconservative friction force. When a body slides over a stationary surface with friction, the work done by friction is always negative, no matter what the direction of motion (see Example 7–7).

THE LAW OF CONSERVATION OF ENERGY

Nonconservative forces cannot be represented in terms of potential energy. But we can describe the effects of these forces in terms of kinds of energy other than kinetic and potential energy. When a car with locked brakes skids to a stop, the tires and the road surface both become hotter. The energy associated with this change in the state of the materials is called **internal energy.** Raising the temperature of a body increases its internal energy; lowering the body's temperature decreases its internal energy.

To see the significance of internal energy, let's consider a block sliding on a rough surface. Friction does *negative* work on the block as it slides, and the change in internal energy of the block and surface (both of which get hotter) is *positive.* Careful experiments show that the increase in the internal energy is *exactly* equal to the absolute value of the work done by friction. In other words,

$$\Delta U_{int} = -W_{other},$$

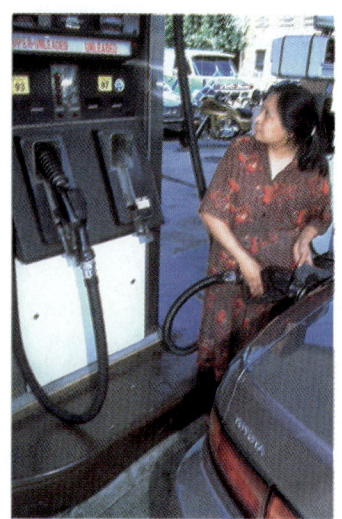

When a liter of gasoline is burned in an automotive engine, it releases 3.3×10^7 J of internal energy. Hence $\Delta U_{int} = -3.3 \times 10^7$ J, where the minus sign means that the amount of energy stored in the gasoline has decreased. This energy can be converted into kinetic energy (making the car go faster) or into potential energy (enabling the car to climb uphill).

where ΔU_{int} is the change in internal energy. If we substitute this into Eq. (7–7), (7–13), or (7–15), we find

$$K_1 + U_1 - \Delta U_{int} = K_2 + U_2.$$

Writing $\Delta K = K_2 - K_1$ and $\Delta U = U_2 - U_1$, we can finally express this as

$$\Delta K + \Delta U + \Delta U_{int} = 0 \qquad \text{(law of conservation energy).} \qquad (7\text{–}16)$$

This remarkable statement is the general form of the **law of conservation of energy.** In a given process, the kinetic energy, potential energy, and internal energy of a system may all change. But the *sum* of those changes is always zero. If there is a decrease in one form of energy, it is made up for by an increase in the other forms. When we expand our definition of energy to include internal energy, Eq. (7–16) says that *energy is never created or destroyed; it only changes form.* No exception to this rule has ever been found.

Notice that the concept of work has been banished from Eq. (7–16). Instead, this equation invites us to think purely in terms of the conversion of energy from one form to another. When you throw a baseball straight up, you convert a portion of the internal energy of your molecules into kinetic energy of the baseball. This is converted into grav-

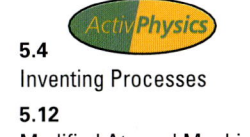

5.4
Inventing Processes
5.12
Modified Atwood Machine

itational potential energy as the ball climbs and back to kinetic energy as the ball falls. If there is air resistance, part of the energy is used to heat up the air and the ball and increase their internal energy. Energy is converted back into the kinetic form as the ball falls. If you catch the ball in your hand, whatever energy was not lost to the air once again becomes internal energy; the ball and your hand are now warmer than they were at the beginning.

In a hydroelectric generating station, falling water is used to drive turbines ("water wheels"), which in turn run electric generators. Fundamentally, gravitational potential energy is released as the water falls, and the generating station converts this into electrical energy. Even if we know nothing about the details of how this is done, we can still use the law of conservation of energy, Eq. (7–16), to draw an important conclusion: the amount of electrical energy produced can be no greater than the amount of gravitational potential energy lost. (Because of friction, some of the potential energy goes into raising the temperatures of the water and the mechanism.)

In later chapters we will study the relation of internal energy to temperature changes, heat, and work. This is the heart of the area of physics called *thermodynamics.*

EXAMPLE 7–14

Work done by friction Let's look again at Example 7–6 in Section 7–2, in which your cousin Throcky skateboards down a curved ramp. He starts with zero kinetic energy and 735 J of potential energy, and at the bottom he has 450 J of kinetic energy and zero potential energy. So $\Delta K = +450$ J, and $\Delta U = -735$ J. The work $W_{other} = W_{fric}$ done by the nonconservative friction forces is -285 J, so the change in internal energy is

$\Delta U_{int} = -W_{other} = +285$ J. The wheels, the bearings, and the ramp all get a little warmer as Throcky rolls down. In accordance with Eq. (7–16) the sum of the energy changes equals zero:

$$\Delta K + \Delta U + \Delta U_{int} = +450 \text{ J} + (-735 \text{ J}) + 285 \text{ J} = 0.$$

The total energy of the system (including nonmechanical forms of energy) is conserved.

7–5 FORCE AND POTENTIAL ENERGY

For the two kinds of conservative forces (gravitational and elastic) we have studied, we started with a description of the behavior of the force and derived from that an expression for the potential energy. For a body with mass m in a uniform gravitational field, the gravitational force is $F_y = -mg$. We found that the corresponding potential energy is $U(y) = mgy$ (Fig. 7–17a). To stretch an ideal spring by a distance x, we exert a force equal to $+kx$. By Newton's third law the force that an ideal spring exerts on a body is opposite this, or $F_x = -kx$. The corresponding potential energy function is $U(x) = \frac{1}{2}kx^2$ (Fig. 7–17b).

We can reverse this procedure: If we are given a potential-energy expression, we can find the corresponding force. Here's how we do it. First let's consider motion along a straight line, with coordinate x. We denote the x-component of force, a function of x, by $F_x(x)$, and the potential energy as $U(x)$. This notation reminds us that both F_x and U are *functions* of x. Now we recall that in any displacement the work W done by a conservative force equals the negative of the change ΔU in potential energy:

$$W = -\Delta U.$$

Let's apply this to a small displacement Δx. The work done by the force $F_x(x)$ during this displacement is approximately equal to $F_x(x)\,\Delta x$. We have to say "approximately" because $F_x(x)$ may vary a little over the interval Δx. But it is at least approximately true that

$$F_x(x)\Delta x = -\Delta U \qquad \text{and} \qquad F_x(x) = -\frac{\Delta U}{\Delta x}.$$

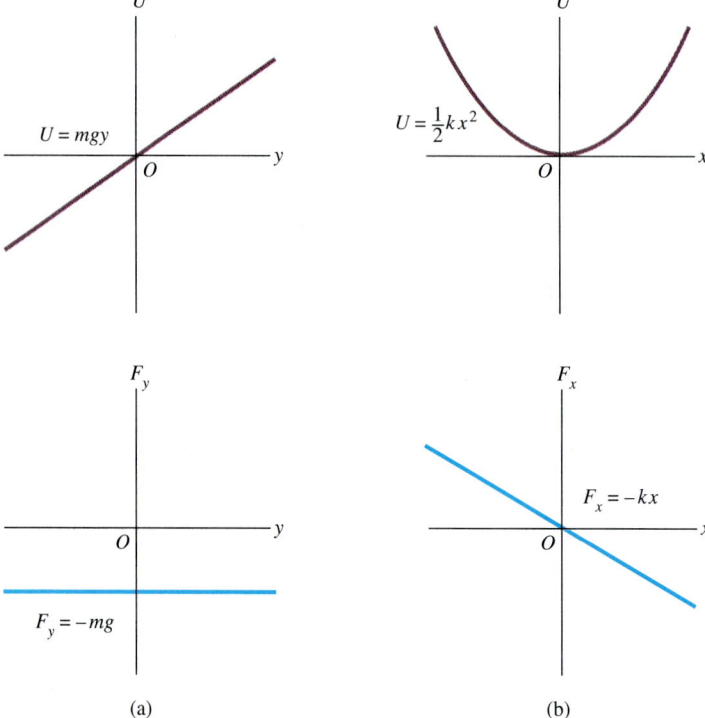

7–17 Graphs of potential energy and force versus position for (a) the gravitational force and (b) the spring force. In each case the force is the negative derivative of the potential energy.

You can probably see what's coming. We take the limit as $\Delta x \to 0$; in this limit, the variation of F_x becomes negligible, and we have the exact relation

$$F_x(x) = -\frac{dU(x)}{dx} \quad \text{(force from potential energy, one dimension).} \quad (7–17)$$

This result makes sense; in regions where $U(x)$ changes most rapidly with x (that is, where $dU(x)/dx$ is large), the greatest amount of work is done during a given displacement, and this corresponds to a large force magnitude. Also, when $F_x(x)$ is in the positive x-direction, $U(x)$ *decreases* with increasing x. So $F_x(x)$ and $dU(x)/dx$ should indeed have opposite signs. The physical meaning of Eq. (7–17) is that *a conservative force always acts to push the system toward lower potential energy.*

As a check, let's consider the function for elastic potential energy, $U(x) = \frac{1}{2}kx^2$. Substituting this into Eq. (7–17) yields

$$F_x(x) = -\frac{d}{dx}\left(\frac{1}{2}kx^2\right) = -kx.$$

which is the correct expression for the force exerted by an ideal spring.

EXAMPLE 7–15

An electrical force and its potential energy An electrically charged particle is held at rest at the point $x = 0$, while a second particle with equal charge is free to move along the positive x-axis. The potential energy of the system is

$$U(x) = \frac{C}{x},$$

where C is a positive constant that depends on the magnitude of the charges. Derive an expression for the x-component of force acting on the movable charge, as a function of its position.

SOLUTION We are given the potential-energy function $U(x)$, so we can use Eq. (7–17). The derivative with respect to x of the function $1/x$ is $-1/x^2$. So the force on the movable charge for $x > 0$ is

$$F_x(x) = -\frac{dU(x)}{dx} = -C\left(-\frac{1}{x^2}\right) = \frac{C}{x^2}.$$

The x-component of force is positive, corresponding to a repulsive interaction between like electric charges. The potential energy is very large for small x and approaches zero as x

becomes large; the force pushes the movable charge toward large positive values of x, for which the potential energy is less. The force varies as $1/x^2$; it is small when the particles are far apart (large x) but becomes large when the particles are close together (small x). This is an example of Coulomb's law for electrical interactions, which we will study in greater detail in Chapter 22.

FORCE AND POTENTIAL ENERGY IN THREE DIMENSIONS

We can extend this analysis to three dimensions, where the particle may move in the x-, y-, or z-direction, or all at once, under the action of a conservative force that has components F_x, F_y, and F_z. Each component of force may be a function of the coordinates x, y, and z. The potential-energy function U is also a function of all three space coordinates. We can now use Eq. (7–17) to find each component of force. The potential-energy change ΔU when the particle moves a small distance Δx in the x-direction is again given by $-F_x \Delta x$; it doesn't depend on F_y and F_z, which represent force components that are perpendicular to the displacement and do no work. So we again have the approximate relation

$$F_x = -\frac{\Delta U}{\Delta x}.$$

The y- and z-components of force are determined in exactly the same way:

$$F_y = -\frac{\Delta U}{\Delta y}, \qquad F_z = -\frac{\Delta U}{\Delta z}.$$

To make these relations exact, we need to take the limits $\Delta x \to 0$, $\Delta y \to 0$, and $\Delta z \to 0$ so that these ratios become derivatives. Because U may be a function of all three coordinates, we need to remember that when we calculate each of these derivatives, only one coordinate changes at a time. We compute the derivative of U with respect to x by assuming that y and z are constant and only x varies, and so on. Such a derivative is called a *partial derivative*. The usual notation is $\partial U / \partial x$ and so on; the symbol ∂ is a modified d to remind us of the nature of this operation. So we write

$$F_x = -\frac{\partial U}{\partial x}, \qquad F_y = -\frac{\partial U}{\partial y}, \qquad F_z = -\frac{\partial U}{\partial z} \qquad \text{(force from potential energy)}.$$

$$(7-18)$$

We can use unit vectors to write a single compact vector expression for the force $\vec{F}$:

$$\vec{F} = -\left(\frac{\partial U}{\partial x} \hat{\imath} + \frac{\partial U}{\partial y} \hat{\jmath} + \frac{\partial U}{\partial z} \hat{k} \right) \qquad \text{(force from potential energy).} \qquad (7-19)$$

The expression inside the parentheses represents a particular operation on the function U, in which we take the partial derivative of U with respect to each coordinate, multiply by the corresponding unit vector, and take the vector sum. This operation is called the **gradient** of U and is often abbreviated as $\vec{\nabla} U$. Thus the force is the negative of the gradient of the potential-energy function:

$$\vec{F} = -\vec{\nabla} U. \qquad (7-20)$$

As a check, let's substitute into Eq. (7–20) the function $U = mgy$ for gravitational potential energy:

$$\vec{F} = -\vec{\nabla}(mgy) = -\left(\frac{\partial (mgy)}{\partial x} \hat{\imath} + \frac{\partial (mgy)}{\partial y} \hat{\jmath} + \frac{\partial (mgy)}{\partial z} \hat{k} \right) = (-mg)\hat{\jmath}.$$

This is just the familiar expression for the gravitational force.

EXAMPLE 7-16

Force and potential energy in two dimensions A puck slides on a level, frictionless air-hockey table. The coordinates of the puck are x and y. It is acted on by a conservative force described by the potential-energy function

$$U(x, y) = \frac{1}{2}k(x^2 + y^2).$$

Derive an expression for the force acting on the puck.

SOLUTION We can use Eqs. (7–18) to find the components of force:

$$F_x = -\frac{\partial U}{\partial x} = -kx, \qquad F_y = -\frac{\partial U}{\partial y} = -ky.$$

From Eq. (7–19) this corresponds to the vector expression

$$\vec{F} = -k(x\hat{\imath} + y\hat{\jmath}).$$

Now $x\hat{\imath} + y\hat{\jmath}$ is just the position vector $\vec{r}$ of the particle, so we can rewrite this as $\vec{F} = -k\vec{r}$. This represents a force that at each point is opposite in direction to the position vector of the point, that is, a force that at each point is directed toward the origin.

The potential energy is minimum at the origin, so again the force pushes in the direction of decreasing potential energy.

Furthermore, the *magnitude* of the force at any point is

$$F = k\sqrt{x^2 + y^2} = kr,$$

where r is the particle's distance from the origin. This is the force exerted by a spring that obeys Hooke's law and has a negligibly small length (compared to the other distances in the problem) when it is not stretched. So the motion of the puck is the same as if it were attached to one end of an ideal spring of negligible unstretched length; the other end is attached to the air-hockey table at the origin.

The potential-energy function can also be expressed as $U = \frac{1}{2}kr^2$, so the radial component of force F_r can be expressed as

$$F_r = -\frac{\partial U}{\partial r} = -\frac{\partial\left(\frac{1}{2}kr^2\right)}{\partial r} = -kr.$$

The minus sign indicates that the force is radially inward (toward the origin).

7-6 ENERGY DIAGRAMS

When a particle moves along a straight line under the action of a conservative force, we can get a lot of insight into the possible motions by looking at the graph of the potential-energy function $U(x)$. Figure 7–18a shows a glider with mass m that moves along the x-axis on an air track. The spring exerts on the glider a force with x-component $F_x = -kx$. Figure 7–18b is a graph of the corresponding potential-energy function $U(x) = \frac{1}{2}kx^2$. If the elastic force of the spring is the *only* horizontal force acting on the glider, the total mechanical energy $E = K + U$ is constant, independent of x. A graph of E as a function of x is thus a straight horizontal line.

The vertical distance between the U and E graphs at each point represents the difference $E - U$, equal to the kinetic energy K at that point. We see that K is greatest at $x = 0$. It is zero at the values of x where the two graphs cross, labeled A and $-A$ in the diagram. Thus the speed v is greatest at $x = 0$, and it is zero at $x = \pm A$, the points of *maximum* possible displacement from $x = 0$ for a given value of the total energy E. The potential energy U can never be greater than the total energy E; if it were, K would be negative, and that's impossible. The motion is a back-and-forth oscillation between the points $x = A$ and $x = -A$.

At each point, the force F_x on the glider is equal to the negative of the slope of the $U(x)$ curve: $F_x = -dU/dx$. When the particle is at $x = 0$, the slope and the force are zero, so this is an *equilibrium* position. When x is positive, the slope of the $U(x)$ curve is positive and the force F_x is negative, directed toward the origin. When x is negative, the slope is negative and F_x is positive, again toward the origin. Such a force is sometimes called a *restoring force;* when the glider is displaced to either side of $x = 0$, the resulting force tends to "restore" it back to $x = 0$. An analogous situation is a marble rolling around in a round-bottomed salad bowl. We say that $x = 0$ is a point of **stable equilibrium.** More generally, *any minimum in a potential-energy curve is a stable equilibrium position.*

Figure 7–19a shows a hypothetical but more general potential-energy function $U(x)$. Figure 7–19b shows the corresponding force $F_x = -dU/dx$. Points x_1 and x_3 are stable

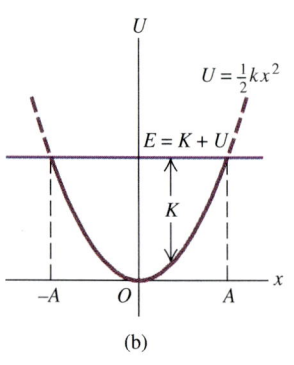

7–18 (a) A glider on an air track. The spring exerts a force $F_x = -kx$. (b) The potential-energy function. The limits of the motion are the points where the U curve intersects the horizontal line representing the total mechanical energy E.

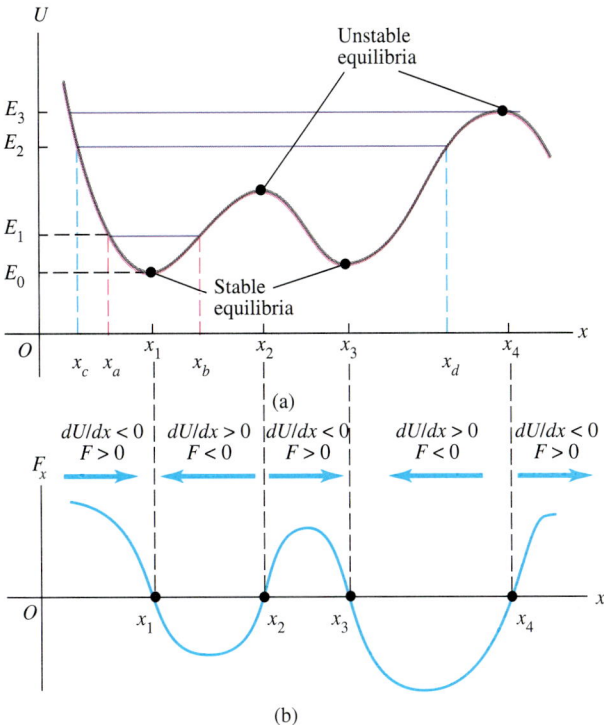

(a)

(b)

7–19 (a) A hypothetical potential-energy function $U(x)$. (b) The corresponding force $F_x = -dU/dx$. Maxima and minima of $U(x)$ correspond to points where $F_x = 0$.

equilibrium points. At each, F_x is zero because the slope of the $U(x)$ curve is zero. When the particle is displaced to either side, the force pushes back toward the equilibrium point. The slope of the $U(x)$ curve is also zero at points x_2 and x_4, and these are also equilibrium points. But when the particle is displaced a little to the right of either point, the slope of the $U(x)$ curve becomes negative, corresponding to a positive F_x that tends to push the particle still farther from the point. When the particle is displaced a little to the left, F_x is negative, again pushing away from equilibrium. This is analogous to a marble rolling on the top of a bowling ball. Points x_2 and x_4 are called **unstable equilibrium** points; *any maximum in a potential-energy curve is an unstable equilibrium position.*

CAUTION ▶ The direction of the force on a body is *not* determined by the sign of the potential energy U. Rather, it's the sign of $F_x = -dU/dx$ that matters. As we discussed in Section 7–2, the physically significant quantity is the *difference* in the value of U between two points, which is just what the derivative $F_x = -dU/dx$ measures. This means that you can always add a constant to the potential-energy function without changing the physics of the situation. ◀

If the total energy is E_1 and the particle is initially near x_1, it can move only in the region between x_a and x_b determined by the intersection of the E_1 and U graphs (Fig. 7–19a). Again, U cannot be greater than E_1 because K can't be negative. We speak of the particle as moving in a *potential well*, and x_a and x_b are the *turning points* of the particle's motion (since at these points, the particle stops and reverses direction). If we increase the total energy to the level E_2, the particle can move over a wider range, from x_c to x_d. If the total energy is greater than E_3, the particle can "escape" and move to indefinitely large values of x. At the other extreme, E_0 represents the least possible total energy the system can have.

Summary

■ The work done on a particle by a constant gravitational force can be represented in terms of a potential energy $U = mgy$:

$$W_{\text{grav}} = mgy_1 - mgy_2 = U_1 - U_2 = -\Delta U. \qquad (7\text{–}1), (7\text{–}3)$$

■ The work done by a stretched or compressed spring that exerts a force $F_x = -kx$ on a particle, where x is the amount of stretch or compression, can be represented in terms of a potential-energy function $U = \frac{1}{2}kx^2$:

$$W_{\text{el}} = \frac{1}{2}kx_1{}^2 - \frac{1}{2}kx_2{}^2 = U_1 - U_2 = -\Delta U. \qquad (7\text{–}10)$$

■ The total potential energy U is the sum of the gravitational and elastic potential energy. If no forces other than the gravitational and elastic forces do work on a particle, the sum of kinetic and potential energy is conserved:

$$K_1 + U_1 = K_2 + U_2. \qquad (7\text{–}11)$$

The sum $K + U = E$ is called the total mechanical energy.

■ When forces other than the gravitational and elastic forces do work on a particle, the work W_{other} done by these other forces equals the change in total mechanical energy (kinetic energy plus total potential energy):

$$K_1 + U_1 + W_{\text{other}} = K_2 + U_2. \qquad (7\text{–}13)$$

■ All forces are either conservative or nonconservative. A conservative force is one for which the work-kinetic energy relation is completely reversible. The work of a conservative force can always be represented by a potential-energy function, but the work of a nonconservative force cannot.

■ The work done by nonconservative forces manifests itself as changes in the internal energy of bodies. The sum of kinetic, potential, and internal energy is always conserved:

$$\Delta K + \Delta U + \Delta U_{\text{int}} = 0. \qquad (7\text{–}16)$$

■ For motion along a straight line, a conservative force $F_x(x)$ and its associated potential energy $U(x)$ are related by

$$F_x(x) = -\frac{dU}{dx}. \qquad (7\text{–}17)$$

In three dimensions, where U is a function of x, y, and z, the components of force are

$$F_x = -\frac{\partial U}{\partial x}, \qquad F_y = -\frac{\partial U}{\partial y}, \qquad F_z = -\frac{\partial U}{\partial z}. \qquad (7\text{–}18)$$

or, in vector form,

$$\vec{F} = -\left(\frac{\partial U}{\partial x}\hat{\imath} + \frac{\partial U}{\partial y}\hat{\jmath} + \frac{\partial U}{\partial z}\hat{k}\right). \qquad (7\text{–}19)$$

DISCUSSION QUESTIONS

Q7–1 A baseball is thrown straight up with initial speed v_1, as in Example 7–1 (Section 7–2). If air resistance is *not* neglected, when the ball returns to its initial height its speed is less than v_1. Explain, using energy concepts.

Q7–2 In Fig. 7–5 the projectile has the same initial kinetic energy in each case. Why does it not rise to the same maximum height in each case?

Q7–3 In Example 7–5 (Section 7–2), does Throckmorton's speed at the bottom depend on the shape of the ramp or just the difference in height of points 1 and 2? Explain. Answer this same question if the ramp is not frictionless, as in Example 7–6 (Section 7–2).

Q7–4 An egg is released from rest at the roof of a building and falls to the ground. Its fall is observed by a student on the roof of the building who uses coordinates with origin at the roof and by a student on the ground who uses coordinates with origin at the ground. Do the two students assign the same or different values to the initial gravitational potential energy, the final gravitational potential energy, the change in gravitational potential energy, and the kinetic energy of the egg just before it strikes the ground? Explain.

Q7–5 A physics teacher had a bowling ball suspended from a very long rope attached to the high ceiling of a large lecture hall. To illustrate his faith in conservation of energy, he would back up to one side of the stage, pull the ball far to one side until the taut rope brought it just to the end of his nose, and then release it. The massive ball would swing in a mighty arc across the stage and then return to stop momentarily just in front of the nose of the stationary, unflinching teacher. However, one day after the demonstration he looked up just in time to see a student at the other side of the stage *push* the ball away from his nose as he tried to duplicate the demonstration. Tell the rest of the story and explain the reason for the potentially tragic outcome.

Q7–6 When a returning space shuttle touches down on the runway, it has lost almost all the kinetic energy it had in orbit. The gravitational potential energy has also decreased considerably. Where did all that energy go?

Q7–7 Is it possible for a frictional force to *increase* the mechanical energy of a system? If so, give examples.

Q7–8 A woman bounces on a trampoline, going a little higher with each bounce. Explain how she increases the total mechanical energy.

Q7–9 In a siphon, water is lifted above its original level during its flow from one container to another. Where does it get the needed potential energy?

Q7–10 A compressed spring is clamped in its compressed position and then is dissolved in acid. What becomes of its potential energy?

Q7–11 When an object moves away from the earth the gravitational potential energy increases; when it moves closer the potential energy decreases. But the potential energy for a spring increases both when the spring is stretched and when it is compressed. Explain the reason for the difference in behavior of these two potential energies.

Q7–12 Since only changes in potential energy are important in any problem, a student decides to let the elastic potential energy of a spring be zero when the spring is stretched a distance x_1. The student decides therefore to let $U = \frac{1}{2}k(x - x_1)^2$. Is this correct? Explain.

Q7–13 Figure 7–17b shows the potential energy function for the force $F_x = -kx$. Sketch the potential energy function for the force $F_x = +kx$. For this force, is $x = 0$ a point of equilibrium? Is this equilibrium stable or unstable? Explain.

Q7–14 Figure 7–17a shows the potential energy function associated with the gravitational force between an object and the earth. Use this graph to explain why objects always fall toward the earth when they are released.

Q7–15 For a system of two particles we often let the potential energy for the force between the particles approach zero as the separation of the particles approaches infinity. If this choice is made, explain why the potential energy at noninfinite separation is positive if the particles repel one another and negative if they attract.

Q7–16 Explain why the points $x = A$ and $x = -A$ in Fig. 7–18b are called *turning points*. How are the values of E and U related at a turning point?

Q7–17 A particle is in *neutral equilibrium* if the net force on it is zero and remains zero if the particle is displaced slightly in any direction. Sketch the potential energy function near a point of neutral equilibrium, for the case of one-dimensional motion. Give an example of an object in neutral equilibrium.

Q7–18 The net force on a particle of mass m has the potential-energy function graphed in Figure 7–19a. If the total energy is E_1, graph the speed v of the particle versus its position x. At what value of x is the speed greatest? Sketch v versus x if the total energy is E_2.

EXERCISES

SECTION 7–2 GRAVITATIONAL POTENTIAL ENERGY

7–1 What is the potential energy for an 800-kg elevator at the top of Chicago's Sears Tower, 440 m above street level? Let the potential energy be zero at street level.

7–2 A 5.00-kg sack of flour is lifted vertically at a constant speed of 3.50 m/s through a height of 15.0 m. a) How great a force is required? b) How much work is done on the sack by the lifting force? What becomes of this work?

7–3 Repeat part (a) of Example 6–5 (Section 6–3) using Eq. (7–7).

7–4 A mail bag with a mass of 120 kg is suspended by a vertical rope 6.0 m long. a) What horizontal force is necessary to hold the bag in a position displaced sideways 3.0 m from its initial position? b) How much work is done by the worker in moving the bag to this position?

7–5 A baseball is thrown from the roof of a 22.0-m-tall building with an initial velocity of magnitude 12.0 m/s and directed at an angle of 53.1° above the horizontal. a) What is the speed of the ball just before it strikes the ground? Use energy methods and ignore air resistance. b) What is the answer for part (a) if the initial velocity is at an angle of 53.1° *below* the horizontal? c) If the effects of air resistance are included, will part (a) or (b) give the higher speed?

7–6 a) In Example 7–7 (Section 7–2), calculate the minimum initial speed required for the crate to reach the top of the ramp. b) If the initial speed of the crate in Example 7–7 is 11.0 m/s, what is its speed at the top of the ramp?

7–7 Answer part (b) of Example 7–7 (Section 7–2) by applying Eq. (7–7) to points 2 and 3, rather than to points 1 and 3 as was done in the example.

7–8 An empty crate slides down a ramp, starting with a speed v_0 and reaching the bottom with a speed v and kinetic energy K. Some books are placed in the box, so that the total mass is quadrupled. The coefficient of kinetic friction is constant and air resistance is negligible. Again starting with v_0 at the top of the ramp, what are the speed and kinetic energy at the bottom? Explain the reasoning behind your answers.

7–9 A small rock with a mass of 0.20 kg is released from rest at point *A,* which is at the top edge of a large hemispherical bowl with radius $R = 0.50$ m (Fig. 7–20). Assume that the size of the rock is small compared to R, so that the rock can be treated as a particle, and assume that the rock slides rather than rolling. The work done by friction on the rock when it moves from point *A* to point *B* at the bottom of the bowl is −0.22 J. What is the speed of the rock when it reaches point *B*?

FIGURE 7–20 Exercise 7–9.

7–10 Tarzan, in one tree, sights Jane in another tree. He grabs the end of a vine of length 20 m that makes an angle of 45° with the vertical, steps off his tree limb, and swings down and then up to Jane's open arms. When he arrives, his vine makes an angle of 30° with the vertical. Determine whether he gives her a tender embrace or knocks her off her limb by calculating Tarzan's speed just before he reaches Jane. Ignore air resistance and the mass of the vine.

7–11 A 10.0-kg microwave oven is pushed 8.00 m up the sloping surface of a loading ramp inclined at an angle of 36.9° above the horizontal, by a constant force $\vec{F}$ with a magnitude of 110 N and acting parallel to the ramp. The coefficient of kinetic friction between the oven and the ramp is 0.250. a) What is the work done on the oven by the force $\vec{F}$? b) What is the work done on the oven by the friction force? c) Compute the increase in potential energy for the oven. d) Use your answers to parts (a), (b), and (c) to calculate the increase in the oven's kinetic energy. e) Use $\Sigma\,\vec{F} = m\vec{a}$ to calculate the acceleration of the oven. Assuming that the oven is initially at rest, use the acceleration to calculate the oven's speed after traveling 8.00 m. Compute from this the increase in the oven's kinetic energy, and compare to the answer you got in part (d).

7–12 A small rock with mass 0.12 kg is fastened to a massless string with length 0.80 m to form a pendulum. The pendulum is swinging so as to make a maximum angle of 45° with the vertical. Air resistance is negligible. a) What is the speed of the rock when the string passes through the vertical position? b) What is the tension in the string when it makes an angle of 45° with the vertical? c) What is the tension in the string as it passes through the vertical?

SECTION 7–3 ELASTIC POTENTIAL ENERGY

7–13 A force of 800 N stretches a certain spring a distance of 0.200 m. a) What is the potential energy of the spring when it is stretched 0.200 m? b) What is its potential energy when it is compressed 5.00 cm?

7–14 A force of 720 N stretches a certain spring a distance of 0.150 m. What is the potential energy of the spring when a 60.0-kg mass hangs vertically from it?

7–15 A spring of negligible mass has force constant $k = 1600$ N/m. a) How far must the spring be compressed for 3.20 J of potential energy to be stored in it? b) You place the spring vertically with one end on the floor. You then drop a 1.20-kg book onto it from a height of 0.80 m above the top of the spring. Find the maximum distance the spring will be compressed.

7–16 A slingshot will shoot a 10-g pebble 22.0 m straight up. a) How much potential energy is stored in the slingshot's rubber band? b) With the same potential energy stored in the rubber band, how high can the slingshot shoot a 25-g pebble? c) What physical effects did you ignore in solving this problem?

7–17 A 1.20-kg piece of cheese is placed on a vertical spring of negligible mass and force constant $k = 1800$ N/m that is compressed 15.0 cm. When the spring is released, how high does the cheese rise from this initial position? (The cheese and the spring are *not* attached.)

7–18 Consider the glider of Example 7–8 (Section 7–3) and Fig. 7–12. As in the example, the glider is released from rest with the spring stretched 0.100 m. What is the displacement x of the glider from its equilibrium position when its speed is 0.20 m/s? (You should get more than one answer. Explain why.)

7–19 Consider the glider of Example 7–8 (Section 7–3) and Fig. 7–12. a) As in the example, the glider is released from rest with the spring stretched 0.100 m. What is the speed of the glider when it returns to $x = 0$? b) What must be the initial displacement of the glider if its maximum speed in the subsequent motion is to be 2.50 m/s?

7–20 Consider the glider of Example 7–8 (Section 7–3) and Fig. 7–12. As in the example, the glider is released from rest with the spring stretched 0.100 m. But now the air track is turned off so that there is a kinetic friction force acting on the glider. a) If the coefficient of kinetic friction between the air track and the glider is $\mu_k = 0.050$, what is the speed of the glider when it has traveled 0.020 m, so that $x = 0.080$ m? b) If $\mu_k = 0.050$, what is the speed of the glider when it has traveled 0.100 m, so that $x = 0$? c) What must be the value of μ_k so that the glider reaches the $x = 0$ position with zero speed?

7–21 a) For the elevator of Example 7–11 (Section 7–3), what is the speed of the elevator after it has moved downward 1.00 m from point 1 in Fig. 7–13? b) When the elevator is 1.00 m below point 1 in Fig. 7–13, what is its acceleration?

7–22 You are asked to design a spring that will give a 1160-kg satellite a speed of 2.50 m/s relative to an orbiting space shuttle. Your spring is to give the satellite a maximum acceleration of 5.00g. The spring's mass, the recoil kinetic energy of the shuttle, and changes in gravitational potential energy will all be negligible. a) What must be the force constant of the spring? b) What distance must the spring be compressed?

SECTION 7–4 CONSERVATIVE AND NONCONSERVATIVE FORCES

7–23 A 0.75-kg book moves vertically upward a distance of 16 m, and then vertically downward 16 m, returning to its initial position. a) How much work is done by gravity during the upward motion of the book? b) How much work is done by gravity during the downward motion of the book? c) What is the total work done on the book by gravity during the complete round trip? d) On the basis of your answer to part (c), would you say that the gravitational force is conservative or nonconservative? Explain.

7–24 A 0.050-kg rock moves from the origin to the point (3.0 m, 5.0 m) in a coordinate system in which the positive y-direction is upward. a) The rock first moves horizontally from the origin to the point (3.0 m, 0). Then it moves vertically from (3.0 m, 0) to (3.0 m, 5.0 m). Sketch the path of the rock in the xy-plane. How much work is done on the rock by gravity during the displacement? b) Instead of the path in part (a), let the rock first move vertically from the origin to (0, 5.0 m) and then horizontally from (0, 5.0 m) to (3.0 m, 5.0 m). Sketch the path of the rock in the xy-plane. How much work is done on the rock by gravity during the displacement? c) Compare the answers to parts (a) and (b). From your results would you say that the gravitational force is conservative or nonconservative? Explain.

7–25 In an experiment, one of the forces exerted on a proton is $\vec{F} = -\alpha x^2 \hat{\imath}$, where $\alpha = 12$ N/m². a) How much work does $\vec{F}$ do when the proton moves along the straight-line path from the point (0.10 m, 0) to the point (0.10 m, 0.40 m)? b) Along the straight-line path from the point (0.10 m, 0) to the point (0.30 m, 0)? c) Along the straight-line path from the point (0.30 m, 0) to the point (0.10 m, 0)? d) Is the force $\vec{F}$ conservative? Explain. If $\vec{F}$ is conservative, what is the potential energy function for it? Let $U = 0$ when $x = 0$.

7–26 Consider the electron and force $\vec{F}$ of Example 7–13 (Section 7–4). a) The electron travels from the point (0, 0) to the point (L, L) along the straight line from (0, 0) to (0, L) and then along the straight line from (0, L) to (L, L). What is the work done by the force $\vec{F}$ during this displacement? b) The electron travels from the point (0, 0) to the point (L, L) along the straight line from (0, 0) to (L, 0) and then along the straight line from (L, 0) to (L, L). What is the work done by the force $\vec{F}$ during this displacement? c) The electron travels from the point (0, 0) to the point (L, L) along the straight line from (0, 0) to (L, L). What is the work done by the force $\vec{F}$ during this displacement? d) Compare the answers in parts (a), (b), and (c) and explain the results of this comparison.

7–27 A 0.60-kg book slides on a horizontal table. The kinetic friction force on the book has magnitude 1.2 N. a) How much work is done on the book by friction during a displacement of 3.0 m to the left? b) The book now slides 3.0 m to the right, returning to its starting point. During this second 3.0-m displacement, how much work is done on the book by friction? c) What is the total work done on the book by friction during the complete round trip? d) On the basis of your answer to part (c), would you say that the friction force is conservative or nonconservative? Explain.

7–28 A 30.0-kg packing crate in a warehouse is pushed to the loading dock by a worker who applies a horizontal force. The coefficient of kinetic friction between the crate and the floor is 0.20. The loading dock is 15.0 m southwest of the initial position of the crate. a) If the crate is pushed 10.6 m south and then 10.6 m west, what is the total work done on the crate by friction? b) If the crate is pushed along a straight-line path to the dock, so that it travels 15.0 m southwest, what is the work done on the crate by friction? c) Sketch the paths along which the crate moves in parts (a) and (b). Based on your answers to parts (a) and (b), would you say that the friction force is conservative or nonconservative? Explain.

7–29 You and three friends stand at the corners of a square whose sides are 8.0 m long in the middle of the gym floor, as shown in Fig. 7–21. You take your physics book and push it from one person to the other. The book has a mass of 1.5 kg, and the coefficient of kinetic friction between the book and the floor is $\mu_k = 0.25$. a) The book slides from you to Beth and then from Beth to Carlos, along the lines connecting these people. What is the work done by friction during this displacement? b) You slide the book from you to Carlos along the diagonal of the square. What is the work done by friction during this displacement? c) You slide the book to Kim who then slides it back to you. What is the total work done by friction during this motion of the book? d) Is the friction force on the book conservative or nonconservative? Explain.

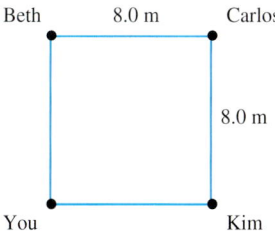

FIGURE 7–21 Exercise 7–29.

7–30 A block of mass m is attached to an ideal spring that has force constant k. a) The block moves from x_1 to x_2, where $x_2 > x_1$. How much work does the spring force do during this displacement? b) The block moves from x_1 to x_2 and then from x_2 to x_1. How much work does the spring force do during the displacement from x_2 to x_1? What is the total work done by the spring during the entire $x_1 \rightarrow x_2 \rightarrow x_1$ displacement? Explain why you obtained the answer you did. c) The block moves from x_1 to x_3, where $x_3 > x_2$. How much work does the spring force do during this displacement? The block then moves from x_3 to x_2. How much work does the spring force do during this displacement?

What is the total work done by the spring force during the $x_1 \rightarrow$ $x_3 \rightarrow x_2$ displacement? Compare your answer to the answer to part (a), where the starting and ending points are the same but the path is different.

SECTION 7–5 FORCE AND POTENTIAL ENERGY

7–31 The potential energy of a pair of hydrogen atoms separated by a large distance x is given by $U(x) = -C_6/x^6$, where C_6 is a positive constant. What is the force that one atom exerts on the other? Is this force attractive or repulsive?

7–32 A force parallel to the x-axis acts on a particle moving along the x-axis. This force produces a potential energy $U(x)$ given by $U(x) = \alpha x^4$, where $\alpha = 1.20 \text{ J/m}^4$. What is the force (magnitude and direction) when the particle is at $x = -0.800$ m?

7–33 An experimental device moving in the xy-plane is acted on by a conservative force described by the potential-energy function $U(x, y) = k(x^2 + y^2) + k'xy$, where k and k' are positive constants. Derive an expression for the force expressed in terms of the unit vectors $\hat{\imath}$ and $\hat{\jmath}$.

7–34 An object moving in the xy-plane is acted on by a conservative force described by the potential-energy function $U(x, y) = \alpha(1/x^2 + 1/y^2)$, where α is a positive constant. Derive an expression for the force expressed in terms of the unit vectors $\hat{\imath}$ and $\hat{\jmath}$.

SECTION 7–6 ENERGY DIAGRAMS

7–35 The potential energy of two atoms in a diatomic molecule is approximated by $U(r) = a/r^{12} - b/r^6$, where r is the spacing between atoms and a and b are positive constants. a) Find the

PROBLEMS

7–37 A man with mass 70.0 kg sits on a platform suspended from a movable pulley as shown in Fig. 7–23 and raises himself at constant speed by a rope passing over a fixed pulley. The platform and the pulleys have negligible mass. Assume that there are no friction losses. a) Find the force he must exert. b) Find the

FIGURE 7–23 Problem 7–37.

force $F(r)$ on one atom as a function of r. Make two graphs, one of $U(r)$ versus r and one of $F(r)$ versus r. b) Find the equilibrium distance between the two atoms. Is this equilibrium stable? c) Suppose the distance between the two atoms is equal to the equilibrium distance found in part (b). What minimum energy must be added to the molecule to *dissociate* it, that is, to separate the two atoms to an infinite distance apart? This is called the *dissociation energy* of the molecule. d) For the molecule CO, the equilibrium distance between the carbon and oxygen atoms is 1.13×10^{-10} m and the dissociation energy is 1.54×10^{-18} J per molecule. Find the values of the constants a and b.

7–36 A marble moves along the x-axis. The potential-energy function is shown in Fig. 7–22. a) At which of the labeled x-coordinates is the force on the marble zero? b) Which of the labeled x-coordinates is a position of stable equilibrium? c) Which of the labeled x-coordinates is a position of unstable equilibrium?

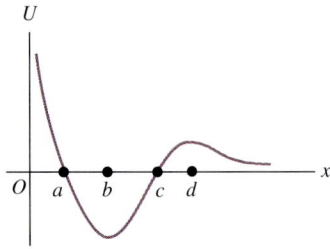

FIGURE 7–22 Exercise 7–36.

increase in the energy of the system when he raises himself 1.20 m. (Answer by calculating the increase in potential energy and also by computing the product of the force on the rope and the length of the rope passing through his hands.)

7–38 A 2.00-kg block is pushed against a spring with negligible mass and force constant $k = 400$ N/m, compressing it 0.220 m. When the block is released, it moves along a frictionless, horizontal surface and then up a frictionless incline with slope $37.0°$ (Fig. 7–24). a) What is the speed of the block as it slides along the horizontal surface after having left the spring? b) How far does the block travel up the incline before starting to slide back down?

![Figure 7-24 diagram showing spring k = 400 N/m, block m = 2.00 kg, compression 0.220 m, and incline at 37.0°]

FIGURE 7–24 Problem 7–38.

7–39 A block with mass 0.50 kg is forced against a horizontal spring of negligible mass, compressing the spring a distance of 0.20 m (Fig. 7–25). When released, the block moves on a horizontal table top for 1.00 m before coming to rest. The spring constant k is 100 N/m. What is the coefficient of kinetic friction, μ_k, between the block and the table?

FIGURE 7–25 Problem 7–39.

7–40 Riding a Loop-the-loop. A car in an amusement park ride rolls without friction around the track shown in Fig. 7–26. It starts from rest at point A at a height h above the bottom of the loop. Treat the car as a particle. a) What is the minimum value of h (in terms of R) such that the car moves around the loop without falling off at the top (point B)? b) If $h = 3.50R$ and $R = 20.0$ m, compute the speed, radial acceleration, and tangential acceleration of the passengers when the car is at point C, which is at the end of a horizontal diameter. Show these acceleration components in a diagram, approximately to scale.

FIGURE 7–26 Problem 7–40.

7–41 The Great Sandini is a 60-kg circus performer who is shot from a cannon (actually a spring gun). You don't find many men of his caliber, so you help him design a new gun. This new gun has a very large spring with a very small mass and a force constant of 1100 N/m that he will compress with a force of 4400 N. The inside of the gun barrel is coated with Teflon, so the average friction force will be only 40 N during the 4.0 m he moves in the barrel. At what speed will he emerge from the end of the barrel, 2.5 m above his initial rest position?

7–42 You are designing a delivery ramp for crates containing exercise equipment. The 1470-N crates will move at 1.8 m/s at the top of a ramp that slopes downward at 22.0°. The ramp exerts a 550-N kinetic friction force on each crate, and the maximum static friction force also has this value. Each crate will compress a spring at the bottom of the ramp and will come to rest after traveling a total distance of 8.0 m along the ramp. Once stopped, a crate must not rebound back up the ramp. Calculate the force constant of the spring that will need to be compressed the least in order to meet the design criteria.

7–43 A system of two paint buckets connected by a lightweight rope is released from rest with the 12.0-kg bucket 2.00 m above the floor (Fig. 7–27). Use the principle of conservation of energy to find the speed with which this bucket strikes the floor. Neglect friction and the inertia of the pulley.

FIGURE 7–27 Problem 7–43.

7–44 a) For the elevator of Example 7–11 (Section 7–3), how much potential energy is stored in the spring when the elevator is at point 2 in Fig. 7–13? b) What maximum distance does the elevator travel upward from point 2 before it starts to slide down the shaft again? c) As the elevator slides back down, what is its speed just before it reaches the spring? d) When the elevator compresses the spring the second time, what is the maximum energy that is stored in the spring and what is the force that the spring exerts on the elevator?

7–45 Redesign the elevator safety system of Example 7–11 (Section 7–3) so that the elevator does not bounce but stays at rest the first time its speed becomes zero. The mass of the elevator is 2000 kg and its speed when it first touches the spring is 25 m/s. There is a kinetic friction force of 17,000 N, and the maximum static friction force on the elevator is also 17,000 N. The mass of the spring can be neglected. a) What spring constant is required, and what distance is the spring compressed when the elevator is stopped? Do you think this design is practical? Explain. b) What is the maximum magnitude of the acceleration of the elevator?

7–46 A wooden rod of negligible mass and length 80.0 cm is pivoted about a horizontal axis through its center. A white rat of mass 0.500 kg clings to one end of the stick, and a mouse of mass 0.200 kg clings to the other end. The system is released from rest with the rod horizontal. If the animals can manage to hold on, what are their speeds as the rod swings through a vertical position?

7–47 A 0.100-kg potato is tied to a string with length 2.50 m, and the other end of the string is tied to a rigid support. The potato is held straight out horizontally from the point of support, with the string pulled taut, and is then released. a) What is the speed of the potato at the lowest point of its motion? b) What is the tension in the string at this point?

7–48 The following is data from a computer simulation for a batted baseball of mass 0.145 kg, including air resistance:

t	x	y	v_x	v_y
0	0	0	30.0 m/s	40.0 m/s
3.05 s	70.2 m	53.6 m	18.6 m/s	0
6.59 s	124.4 m	0	11.9 m/s	−28.7 m/s

a) How much work was done by the air on the baseball as it moved from its initial position to its maximum height? b) How much work was done by the air on the baseball as it moved from its maximum height back to the starting elevation? c) Explain why the magnitude of the answer to part (b) is smaller than the magnitude of the answer in part (a).

7–49 A fireman of mass m slides a distance d down a pole. He starts from rest. He moves as fast at the bottom as if he had stepped off a platform a distance $h \leq d$ above the ground and descended with negligible air resistance. a) What average friction force did the fireman exert on the pole? Does your answer make sense in the special cases of $h = d$ and $h = 0$? b) Find a numerical value for the average friction force a 75-kg fireman exerts, for $d = 2.5$ m and $h = 1.0$ m. c) In terms of g, h, and d, what is the speed of the fireman when he is a distance y above the bottom of the pole?

7–50 A 60.0-kg skier starts from rest at the top of a ski slope 65.0 m high. a) If frictional forces do -10.5 kJ of work on her as she descends, how fast is she going at the bottom of the slope? b) Now moving horizontally, the skier crosses a patch of soft snow, where $\mu_k = 0.20$. If the patch is 82.0 m wide and the average force of air resistance on the skier is 160 N, how fast is she going after crossing the patch? c) The skier hits a snowdrift and penetrates 2.5 m into it before coming to a stop. What is the average force exerted on her by the snowdrift as it stops her?

7–51 A skier starts at the top of a very large frictionless snowball, with a very small initial speed, and skis straight down the side (Fig. 7–28). At what point does she lose contact with the snowball and fly off at a tangent? That is, at the instant she loses contact with the snowball, what angle α does a radial line from the center of the snowball to the skier make with the vertical?

FIGURE 7–28 Problem 7–51.

7–52 A rock is tied to a cord, and the other end of the cord is held fixed. The rock is given an initial tangential velocity that causes it to rotate in a vertical circle. Prove that the tension in the cord at the lowest point exceeds that at the highest point by six times the weight of the rock.

7–53 In a truck-loading station at a post office a small 0.200-kg package is released from rest at point A on a track that is one-quarter of a circle with radius 1.60 m (Fig. 7–29). The size of the package is much less than 1.60 m, so the package can be treated as a particle. It slides down the track and reaches point B with a speed of 4.80 m/s. From point B it slides on a level surface a distance of 3.00 m to point C, where it comes to rest. a) What is the coefficient of kinetic friction on the horizontal surface? b) How much work is done on the package by friction as it slides down the circular arc from A to B?

FIGURE 7–29 Problem 7–53.

7–54 A truck of mass m has a brake failure while going down an icy mountain road of constant downward slope angle α (Fig. 7–30). Initially the truck is moving downhill at speed v_0. After careening downhill a distance L with negligible friction, the truck driver steers the runaway vehicle onto a runaway truck ramp of constant upward slope angle β. The truck ramp has a soft sand surface for which the coefficient of rolling friction is μ_r. What is the distance that the truck moves up the ramp before coming to a halt? Use energy methods.

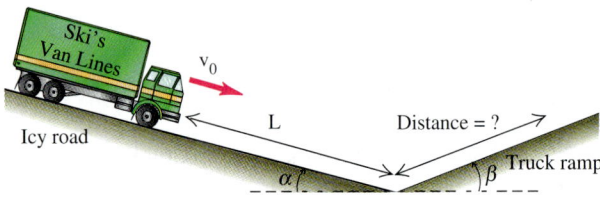

FIGURE 7–30 Problem 7–54.

7–55 A certain spring is found *not* to obey Hooke's law; it exerts a restoring force $F_x(x) = -\alpha x - \beta x^2$ if it is stretched or compressed, where $\alpha = 60.0$ N/m and $\beta = 18.0$ N/m^2. The mass of the spring is negligible. a) Calculate the potential energy function $U(x)$ for this spring. Let $U = 0$ when $x = 0$. b) An object with a mass of 0.900 kg on a frictionless horizontal surface is attached to this spring, pulled a distance 1.00 m to the right (the $+x$-direction) to stretch the spring, and released. What is the speed of the object when it is 0.50 m to the right of the $x = 0$ equilibrium position?

7–56 A variable force $\vec{F}$ is maintained tangent to a frictionless semicircular surface (Fig. 7–31). By slowly varying the force, a block with weight w is moved, and the spring to which it is attached is stretched from position 1 to position 2. The spring has negligible mass and force constant k. The end of the spring moves in an arc of radius a. Calculate the work done by the force $\vec{F}$.

FIGURE 7–31 Problem 7–56.

7–57 A 0.150-kg block of ice is placed against a horizontal compressed spring mounted on a horizontal tabletop that is 1.20 m above the floor. The spring has force constant 1900 N/m and is initially compressed 0.045 m. The mass of the spring is negligible. The spring is released, and the block slides along the table, goes off the edge, and travels to the floor. If there is negligible friction between the ice and the table, what is the speed of the block of ice when it reaches the floor?

7–58 An 80.0-kg man jumps from a height of 2.50 m onto a platform mounted on springs. As the springs compress, the platform is pushed down a maximum distance of 0.240 m below its initial position, and then it rebounds. The platform and springs have negligible mass. a) What is the man's speed at the instant the platform is depressed 0.120 m? b) If the man had just stepped gently onto the platform, what maximum distance would it have been pushed down?

7–59 An experimental apparatus of mass m is placed on a vertical spring of negligible mass and pushed down until the spring has been compressed a distance x. The apparatus is then released and reaches its maximum height at a distance h above the point where it was released. The apparatus is not attached to the spring, and at its maximum height it is no longer in contact with the spring. The maximum magnitude of acceleration the apparatus can have without being damaged is a, where $a > g$. a) What should be the force constant of the spring? b) What distance x must the spring be compressed initially?

7–60 If a fish is attached to a vertical spring and slowly lowered to its equilibrium position, it is found to stretch the spring by an amount d. If the same fish is attached to the end of the unstretched spring and then allowed to fall from rest, through what maximum distance does it stretch the spring? (*Hint:* Calculate the force constant of the spring in terms of the distance d and the mass m of the fish.)

7–61 A wooden block with mass 1.50 kg is placed against a compressed spring at the bottom of an incline of slope 30.0° (point A). When the spring is released, it projects the block up the incline. At point B, a distance of 6.00 m up the incline from A, the block is moving up the incline at 7.00 m/s and is no longer in contact with the spring. The coefficient of kinetic friction between the block and incline is $\mu_k = 0.50$. The mass of the spring is negligible. Calculate the amount of potential energy that was initially stored in the spring.

7–62 A 2.00-kg package is released on a 53.1° incline, 4.00 m from a long spring with force constant 120 N/m that is attached at the bottom of the incline (Fig. 7–32). The coefficients of friction between the package and the incline are $\mu_s = 0.40$ and $\mu_k = 0.20$. The mass of the spring is negligible. a) What is the speed of the package just before it reaches the spring? b) What is the maximum compression of the spring? c) The package rebounds back up the incline. How close does it get to its initial position?

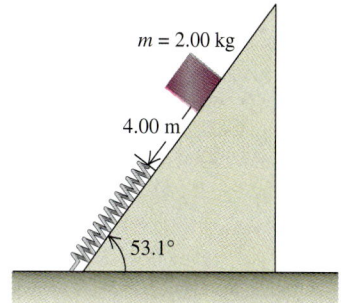

FIGURE 7–32 Problem 7–62.

7–63 A 0.500-kg block attached to a spring with length 0.60 m and force constant 40.0 N/m is at rest with the back of the block at point A on a frictionless, horizontal air table (Fig. 7–33). The mass of the spring is negligible. You move the block to the right along the surface by pulling with a constant 20.0-N horizontal force. a) What is the block's speed when the back of the block reaches point B, which is 0.25 m to the right of point A? b) When the back of the block reaches point B, you let go of the block. In the subsequent motion, how close does the block get to the wall where the left end of the spring is attached?

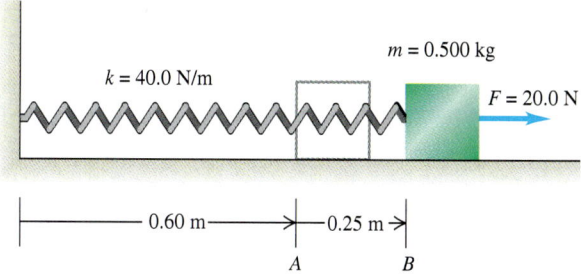

FIGURE 7–33 Problem 7–63.

7–64 A box of mass m is pushed against a spring of negligible mass and force constant k, compressing it a distance x. The box is then released and travels up a ramp that is at an angle α above the horizontal. The coefficient of kinetic friction between the box and the ramp is μ_k, where $\mu_k < 1$. The box is still moving up the ramp after traveling a distance $s > |x|$ along the ramp. Calculate the angle α for which the speed of the box after traveling distance s is a minimum. Explain why the minimum speed doesn't occur when $\alpha = 90°$, even though for this α there is the maximum increase in gravitational potential energy.

7–65 The brothers of Iota Eta Pi fraternity build a platform supported at all four corners by vertical springs in the basement of their frat house. A brave fraternity brother wearing a football helmet stands in the middle of the platform; his weight compresses the springs by 0.18 m. Then four of his fraternity brothers, pushing down at the corners of the platform, compress the springs another 0.53 m until the top of the brave brother's helmet is 0.90 m below the basement ceiling. They then simultaneously release the platform. Neglect the masses of the springs and platform. a) When the dust clears, the fraternity asks you to calculate their fraternity brother's speed just before his helmet hit the flimsy ceiling. b) Without the ceiling, how high would he have gone? c) In discussing their probation, the dean of students suggests that the next time they try this, they do it outdoors on another planet. Would the answer to part (b) be the same if this stunt were performed on a planet with a different value of g? Assume that the fraternity brothers push the platform down 0.53 m as before. Explain your reasoning.

7–66 a) Is the force $\vec{F} = Cy^2\hat{j}$, where C is a negative constant with units of N/m², conservative or nonconservative? Justify your answer. b) Is the force $\vec{F} = Cy^2\hat{i}$, where C is a negative constant with units of N/m², conservative or nonconservative? Justify your answer.

7–67 A cutting tool under microprocessor control has several forces acting on it. One force is $\vec{F} = -\alpha xy^2\hat{j}$, a force in the negative y-direction whose magnitude depends on the position of the tool. The constant is $\alpha = 2.50$ N/m³. Consider the displacement of the tool from the origin to the point $x = 3.00$ m, $y = 3.00$ m. a) Calculate the work done on the tool by $\vec{F}$ if this displacement is along the straight line $y = x$ that connects these two points. b) Calculate the work done on the tool by $\vec{F}$ if the tool is first moved out along the x-axis to the point $x = 3.00$ m, $y = 0$ and then moved parallel to the y-axis to $x = 3.00$ m, $y = 3.00$ m. c) Compare the work done by $\vec{F}$ along these two paths. Is $\vec{F}$ conservative or nonconservative? Explain.

7–68 An object has several forces acting on it. One force is $\vec{F} = \alpha xy\hat{i}$, a force in the x-direction whose magnitude depends on the position of the object. (See Problem 6–84.) The constant is $\alpha = 2.00$ N/m². The object moves along the following path: (1) it starts at the origin and moves along the y-axis to the point $x = 0$, $y = 1.50$ m; (2) it moves parallel to the x-axis to the point $x = 1.50$ m, $y = 1.50$ m; (3) it moves parallel to the y-axis to the point $x = 1.50$ m, $y = 0$; (4) it moves parallel to the x-axis back to the origin. a) Sketch this path in the xy-plane. b) Calculate the work done on the object by $\vec{F}$ for each leg of the path and for the complete round trip. c) Is $\vec{F}$ conservative or nonconservative? Explain.

7–69 A Hooke's law force $-kx$ and a constant conservative force F in the $+x$-direction act on an atomic ion. a) Show that a possible potential-energy function for this combination of forces is $U(x) = \frac{1}{2}kx^2 - Fx - F^2/2k$. Is this the *only* possible function? Explain. b) Find the stable equilibrium position. c) Graph $U(x)$ (in units of F^2/k) versus x (in units of F/k) for values of x between $-5F/k$ to $5F/k$. d) Are there any unstable equilibrium positions? e) If the total energy is $E = F^2/k$, what are the maximum and minimum values of x that the ion reaches in its motion? f) If the ion has mass m, find its maximum speed if the total energy is $E = F^2/k$. For what value of x is the speed maximum?

7–70 A particle moves along the x-axis while acted on by a single conservative force parallel to the x-axis. The force corresponds to the potential-energy function graphed in Fig. 7–34. The particle is released from rest at point A. a) What is the direction of the force on the particle when it is at point A? b) At point B? c) At what value of x is the kinetic energy of the particle a maximum? d) What is the force on the particle when it is at point C? e) What is the largest value of x reached by the particle during its motion? f) What value or values of x correspond to points of stable equilibrium? g) Of unstable equilibrium?

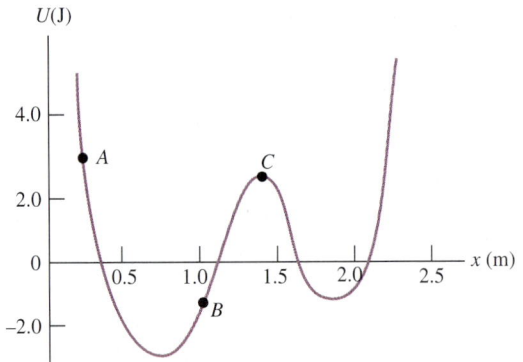

FIGURE 7–34 Problem 7–70.

CHALLENGE PROBLEM

7–71 A proton with mass m moves in one dimension. The potential-energy function is $U(x) = \alpha/x^2 - \beta/x$, where α and β are positive constants. The proton is released from rest at $x_0 = \alpha/\beta$. a) Show that $U(x)$ can be written as

$$U(x) = \frac{\alpha}{x_0{}^2}\left[\left(\frac{x_0}{x}\right)^2 - \frac{x_0}{x}\right].$$

Graph $U(x)$. Calculate $U(x_0)$ and thereby locate the point x_0 on the graph. b) Calculate $v(x)$, the speed of the proton as a func-

tion of position. Graph $v(x)$, and give a qualitative description of the motion. c) For what value of x is the speed of the proton a maximum? What is the value of that maximum speed? d) What is the force on the proton at the point in part (c)? e) Let the proton be released instead at $x_1 = 3\alpha/\beta$. Locate the point x_1 on the graph of $U(x)$. Calculate $v(x)$, and give a qualitative description of the motion. f) For each release point ($x = x_0$ and $x = x_1$), what are the maximum and minimum values of x reached during the motion?

Momentum, Impulse, and Collisions

Between July 16 and July 22, 1994, 23 kilometer-sized fragments of a comet collided with the planet Jupiter. This artist's conception shows the chain of fragments streaming toward the giant planet, as viewed from Jupiter's moon Io. Though the energy released in each impact was equivalent to more than 10^{14} tons of TNT, the fragments had so little momentum compared to Jupiter that there was no noticeable effect on the planet's orbit.

8-1 INTRODUCTION

When an eighteen-wheeler collides head-on with a compact car, what determines which way the wreckage goes after the collision? Why are the occupants of the car much more likely to be injured than those of the truck? How do you decide how to aim the cue ball in pool to knock the eight ball into the pocket?

A common theme of all these questions is that they can't be answered by directly applying Newton's second law, $\Sigma \vec{F} = m\vec{a}$, because there are forces acting about which we know very little: the forces acting between the car and the eighteen-wheeler, or between two pool balls. Remarkably, we will find in this chapter that we don't have to know *anything* about these forces in order to answer questions of this kind!

Our approach will use two new concepts, *momentum* and *impulse,* and a new conservation law, *conservation of momentum.* This conservation law is every bit as important as that of conservation of energy. The law of conservation of momentum is valid even in situations in which Newton's laws are inadequate, such as bodies moving at very high speeds (near the speed of light) or objects on a very small scale (such as the constituents of atoms). Within the domain of Newtonian mechanics, conservation of momentum enables us to analyze many situations that would be very difficult if we tried to use Newton's laws directly. Among these are *collision* problems, in which

two bodies collide and can exert very large forces on each other for a short time.

8-2 MOMENTUM AND IMPULSE

In Chapter 6 we re-expressed Newton's second law for a particle, $\Sigma \vec{F} = m\vec{a}$, in terms of the work-energy theorem. This theorem helped us to tackle a great number of physics problems and led us to the law of conservation of energy. Let's now return to $\Sigma \vec{F} = m\vec{a}$ and see yet another useful way in which this fundamental law can be restated.

Let's consider a particle of constant mass m. (Later in this chapter we'll see how to deal with situations in which the mass of a body changes.) Because $\vec{a} = d\vec{v}/dt$, we can write Newton's second law for this particle as

$$\Sigma \vec{F} = m\frac{d\vec{v}}{dt} = \frac{d}{dt}(m\vec{v}). \qquad (8-1)$$

We can take the mass m inside the derivative because it is constant. Thus Newton's second law says that the net force $\Sigma \vec{F}$ acting on a particle equals the time rate of change of the combination $m\vec{v}$, the product of the particle's mass and velocity. We'll call this combination the **momentum,** or **linear momentum,** of the particle. Using the symbol $\vec{p}$ for momentum, we have

$$\vec{p} = m\vec{v} \qquad \text{(definition of momentum)}. \qquad (8-2)$$

Momentum is a vector quantity that has a magnitude (mv) and a direction (the same as the velocity vector $\vec{v}$). The momentum of a car driving north at 20 m/s is different from the momentum of the same car driving east at the same speed. A fast ball thrown by a major-league pitcher has greater magnitude of momentum than the same ball thrown by a child because the speed is greater. An eighteen-wheeler going 65 mi/h has greater magnitude of momentum than a Saturn automobile with the same speed because the truck's mass is greater. The units of the magnitude of momentum are units of mass times speed; the SI units of momentum are kg · m/s. The plural of momentum is "momenta."

Substituting Eq. (8–2) into Eq. (8–1), we have

$$\Sigma \vec{F} = \frac{d\vec{p}}{dt} \qquad \text{(Newton's second law in terms of momentum).} \qquad (8\text{–}3)$$

The net force (vector sum of all forces) acting on a particle equals the time rate of change of momentum of the particle. This, not $\Sigma \vec{F} = m\vec{a}$, is the form in which Newton originally stated his second law (though he called momentum the "quantity of motion"). It is valid only in inertial frames of reference.

According to Eq. (8–3), a rapid change in momentum requires a large net force, while a gradual change in momentum requires less net force. This principle is used in the design of automobile safety devices such as air bags. The driver of a fast-moving automobile has a large momentum (the product of the driver's mass and velocity). If the car stops suddenly in a collision, the driver's momentum becomes zero. An air bag causes the driver to lose momentum more gradually than would an abrupt collision with the steering wheel, reducing the force exerted on the driver (and the possibility for injury). The same principle applies to the padding used to package fragile objects for shipping.

We will often express the momentum of a particle in terms of its components. If the particle has velocity components v_x, v_y, and v_z, its momentum components p_x, p_y, and p_z are given by

$$p_x = mv_x, \qquad p_y = mv_y, \qquad p_z = mv_z. \qquad (8\text{–}4)$$

These three component equations are equivalent to Eq. (8–2).

A particle's momentum $\vec{p} = m\vec{v}$ and its kinetic energy $K = \frac{1}{2}mv^2$ both depend on the mass and velocity of the particle. What is the fundamental difference between these two quantities? A purely mathematical answer is that momentum is a vector whose magnitude is proportional to speed, while kinetic energy is a scalar proportional to the speed squared. But to see the *physical* difference between momentum and kinetic energy, we must first define a quantity closely related to momentum called *impulse*.

Let's first consider a particle acted on by a *constant* net force $\Sigma \vec{F}$ during a time interval Δt from t_1 to t_2. (We'll look at the case of varying forces shortly.) The **impulse** of the net force, denoted by $\vec{J}$, is defined to be the product of the net force and the time interval:

$$\vec{J} = \Sigma \vec{F}(t_2 - t_1) = \Sigma \vec{F} \Delta t \qquad \text{(assuming constant net force).} \qquad (8\text{–}5)$$

Impulse is a vector quantity; its direction is the same as the net force $\Sigma \vec{F}$. Its magnitude is the product of the magnitude of the net force and the length of time that the net force acts. The SI unit of impulse is the newton-second (N · s). Because 1 N = 1 kg · m/s^2, an alternative set of units for impulse is kg · m/s, the same as the units of momentum.

To see what impulse is good for, let's go back to Newton's second law as restated in terms of momentum, Eq. (8–3). If the net force $\Sigma \vec{F}$ is constant, then $d\vec{p}/dt$ is also constant. In that case, $d\vec{p}/dt$ is equal to the *total* change in momentum $\vec{p}_2 - \vec{p}_1$ during the time interval $t_2 - t_1$, divided by the interval:

$$\Sigma \vec{F} = \frac{\vec{p}_2 - \vec{p}_1}{t_2 - t_1}.$$

Multiplying this equation by $(t_2 - t_1)$, we have

$$\Sigma \vec{F}(t_2 - t_1) = \vec{p}_2 - \vec{p}_1.$$

Comparing to Eq. (8–5), we end up with a result called the **impulse-momentum theorem**:

$$\vec{J} = \vec{p}_2 - \vec{p}_1 \qquad \text{(impulse-momentum theorem).} \qquad (8\text{–}6)$$

The change in momentum of a particle during a time interval equals the impulse of the net force that acts on the particle during that interval.

The impulse-momentum theorem also holds when forces are not constant. To see this, we integrate both sides of Newton's second law $\Sigma \vec{F} = d\vec{p}/dt$ over time between the limits t_1 and t_2:

$$\int_{t_1}^{t_2} \Sigma \vec{F} \, dt = \int_{t_1}^{t_2} \frac{d\vec{p}}{dt} \, dt = \int_{\vec{p}_1}^{\vec{p}_2} d\vec{p} = \vec{p}_2 - \vec{p}_1.$$

The integral on the left is defined to be the impulse $\vec{J}$ of the net force $\Sigma \vec{F}$ during this interval:

$$\vec{J} = \int_{t_1}^{t_2} \Sigma \vec{F} \, dt \qquad \text{(general definition of impulse).} \qquad (8\text{–}7)$$

With this definition the impulse-momentum theorem $\vec{J} = \vec{p}_2 - \vec{p}_1$, Eq. (8–6), is valid even when the net force $\Sigma \vec{F}$ varies with time.

We can define an *average* net force $\vec{F}_{av}$ such that even when $\Sigma \vec{F}$ is not constant, the impulse $\vec{J}$ is given by

$$\vec{J} = \vec{F}_{av}(t_2 - t_1). \qquad (8\text{–}8)$$

When $\Sigma \vec{F}$ is constant, $\Sigma \vec{F} = \vec{F}_{av}$ and Eq. (8–8) reduces to Eq. (8–5).

Figure 8–1 shows a graph of the x-component of net force ΣF_x as a function of time during a collision. This might represent the force on a soccer ball that is in contact with a player's foot from time t_1 to t_2. The x-component of impulse during this interval is represented by the blue area under the curve between t_1 and t_2. This area is equal to the rectangular area bounded by t_1, t_2, and $(F_{av})_x$, so $(F_{av})_x(t_2 - t_1)$ is equal to the impulse of the actual time-varying force during the same interval.

Impulse and momentum are both vector quantities, and Eqs. (8–5) through (8–8) are all vector equations. In specific problems it is often easiest to use them in component form:

$$J_x = \int_{t_1}^{t_2} \Sigma F_x \, dt = (F_{av})_x(t_2 - t_1) = p_{2x} - p_{1x} = mv_{2x} - mv_{1x},$$

$$(8\text{–}9)$$

$$J_y = \int_{t_1}^{t_2} \Sigma F_y \, dt = (F_{av})_y(t_2 - t_1) = p_{2y} - p_{1y} = mv_{2y} - mv_{1y},$$

and similarly for the z-component.

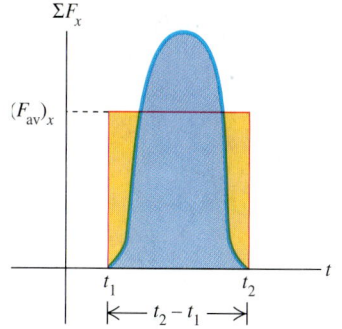

8–1 The x-component of the impulse of ΣF_x between t_1 and t_2 equals the blue area under the ΣF_x–t curve, which also equals the area under the rectangle with height $(F_{av})_x$.

MOMENTUM AND KINETIC ENERGY COMPARED

We can now see the fundamental difference between momentum and kinetic energy. The impulse-momentum theorem $\vec{J} = \vec{p}_2 - \vec{p}_1$ says that changes in a particle's momentum are due to impulse, which depends on the *time* over which the net force acts. By contrast, the work-energy theorem $W_{tot} = K_2 - K_1$ tells us that kinetic energy changes when work is done on a particle; the total work depends on the *distance* over which the net force

6.1

Momentum and Energy Change

acts. Consider a particle that starts from rest at t_1 so that $\vec{v}_1 = \mathbf{0}$. Its initial momentum is $\vec{p}_1 = m\vec{v}_1 = \mathbf{0}$, and its initial kinetic energy is $K_1 = \frac{1}{2}mv_1^2 = 0$. Now let a constant net force equal to $\vec{F}$ act on that particle from time t_1 until time t_2. During this interval the particle moves a distance s in the direction of the force. From Eq. (8–6), the particle's momentum at time t_2 is

$$\vec{p}_2 = \vec{p}_1 + \vec{J} = \vec{J},$$

where $\vec{J} = \vec{F}(t_2 - t_1)$ is the impulse that acts on the particle. So *the momentum of a particle equals the impulse that accelerated it from rest to its present speed;* impulse is the product of the net force that accelerated the particle and the *time* required for the acceleration. By comparison the kinetic energy of the particle at t_2 is $K_2 = W_{tot} = Fs$, the total *work* done on the particle to accelerate it from rest. The total work is the product of the net force and the *distance* required to accelerate the particle.

Here's an application of the distinction between momentum and kinetic energy. Suppose you have a choice between catching a 0.50-kg ball moving at 4.0 m/s or a 0.10-kg ball moving at 20 m/s. Which will be easier to catch? Both have the same magnitude of momentum, $p = mv = (0.50\text{ kg})(4.0\text{ m/s}) = (0.10\text{ kg})(20\text{ m/s}) = 2.0\text{ kg} \cdot \text{m/s}$, but the two balls have very different values of kinetic energy $K = \frac{1}{2}mv^2$; the large, slow-moving ball has $K = 4.0$ J, while the small, fast-moving ball has $K = 20$ J. Since the momentum is the same for both balls, both require the same *impulse* in order to be brought to rest. But stopping the 0.10-kg ball with your hand requires five times more *work* than stopping the 0.50-kg ball, because the smaller ball has five times as much kinetic energy. Hence for a given force that you exert with your hand, it takes the same amount of time (the duration of the catch) to stop either ball, but your hand and arm will be pushed back five times farther if you choose to catch the small, fast-moving ball. To minimize arm strain, you should choose to catch the 0.50-kg ball with its lower kinetic energy. (Kinetic energy may not be the only factor to consider, however. A pitched baseball and a .22 bullet fired from a rifle have roughly the same kinetic energy, and the bullet has less momentum than does the baseball. Nonetheless, you'd probably prefer to catch the baseball rather than the bullet. Why?)

Both the impulse-momentum and work-energy theorems are relationships between force and motion, and both rest on the foundation of Newton's laws. They are *integral* principles, relating the motion at two different times separated by a finite interval. By contrast, Newton's second law itself (in either of the forms $\Sigma\vec{F} = m\vec{a}$ or $\Sigma\vec{F} = d\vec{p}/dt$) is a *differential* principle, relating the forces to the rate of change of velocity or momentum at each instant.

The momentum of a pitched base-ball is equal to the impulse imparted to it by the pitcher, while the kinetic energy of the ball is equal to the work the pitcher does on it. Hence the momentum is related to the *time* it took to bring the ball up to speed, while the kinetic energy depends on the *distance* the ball moved while being accelerated.

EXAMPLE 8–1

Momentum vs. kinetic energy Consider again the race described in Example 6–6 (Section 6–3) between two iceboats on a frictionless frozen lake. The iceboats have masses m and $2m$ respectively, and the wind exerts the same constant horizontal force $\vec{F}$ on each iceboat (Fig. 6–9). The two iceboats start from rest and cross the finish line a distance s away. Which iceboat crosses the finish line with greater momentum?

SOLUTION In Example 6–6 we asked how the kinetic energies of the iceboats compare when they cross the finish line. The way to determine this was not by using the formula $K = \frac{1}{2}mv^2$, but rather by remembering that a body's kinetic energy is equal to the total work done to accelerate it from rest. Both iceboats started from rest, and the total work done between the starting and finish lines was the same for both iceboats (because the net

force and displacement were the same for both). Hence both iceboats cross the finish line with the same kinetic energy.

Similarly, the best way to compare the momenta of the iceboats is *not* to use the formula for momentum, $\vec{p} = m\vec{v}$. By itself this formula isn't enough to decide which iceboat has greater momentum at the finish line. The iceboat of mass $2m$ has greater mass, which suggests greater momentum; but this iceboat crosses the finish line going slower than the other one, which suggests less momentum.

Instead, we use the idea that the momentum of each iceboat equals the impulse that accelerated it from rest. For each iceboat the downward force of gravity and the upward normal force add to zero, so the net force equals the constant horizontal wind force $\vec{F}$. Let Δt be the time an iceboat takes to reach the finish line, so that the impulse on the iceboat during that time is

$\vec{J} = \vec{F}\,\Delta t$. Since the iceboat starts from rest, this just equals the iceboat's momentum $\vec{p}$ at the finish line:

$$\vec{p} = \vec{F}\,\Delta t.$$

Both iceboats are subjected to the same force $\vec{F}$, but they do *not* take the same amount of time Δt to reach the finish line. The more massive iceboat of mass $2m$ accelerates more slowly and takes a longer time to travel the distance s; thus there is a greater impulse on this iceboat between the starting and finish lines. So the iceboat of mass $2m$ crosses the finish line with a greater magnitude of momentum than the iceboat of mass m (but with the same kinetic energy). Can you show that the iceboat of mass $2m$ has $\sqrt{2}$ times as much momentum at the finish line as the iceboat of mass m?

EXAMPLE 8-2

A ball hits a wall Suppose you throw a ball with a mass of 0.40 kg against a brick wall. It hits the wall moving horizontally to the left at 30 m/s and rebounds horizontally to the right at 20 m/s. a) Find the impulse of the net force on the ball during its collision with the wall. b) If the ball is in contact with the wall for 0.010 s, find the average horizontal force that the wall exerts on the ball during the impact.

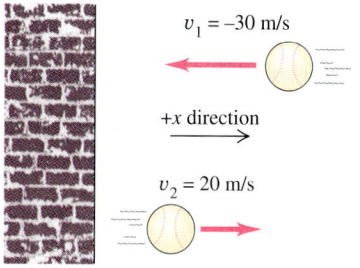

$v_1 = -30$ m/s

$+x$ direction

$v_2 = 20$ m/s

8-2 A ball moving to the left, hitting a wall, and rebounding to the right.

SOLUTION a) Take the x-axis as horizontal and the positive direction to the right (Fig. 8-2). Then the initial x-component of momentum of the ball is

$$p_1 = mv_1 = (0.40 \text{ kg})(-30 \text{ m/s}) = -12 \text{ kg} \cdot \text{m/s}.$$

The final x-component of momentum is

$$p_2 = mv_2 = (0.40 \text{ kg})(+20 \text{ m/s}) = +8.0 \text{ kg} \cdot \text{m/s}.$$

The *change* in the x-component of momentum is

$$p_2 - p_1 = mv_2 - mv_1 = 8.0 \text{ kg} \cdot \text{m/s} - (-12 \text{ kg} \cdot \text{m/s})$$
$$= 20 \text{ kg} \cdot \text{m/s}.$$

According to Eqs. (8-9), this equals the x-component of impulse of the net force on the ball, so $J_x = 20 \text{ kg} \cdot \text{m/s} = 20 \text{ N} \cdot \text{s}$.

The time variation of the net horizontal force may be similar to one of the curves in Fig. 8-3. The net horizontal force is zero before impact, rises to a maximum, and then decreases to zero when the ball loses contact with the wall. If the ball is relatively rigid, like a baseball or a golf ball, the collision lasts a short time and the maximum force is large, as in curve (a). If the ball is softer, like a tennis ball, the collision time is longer and the maximum force is less, as in curve (b). In any case the *area* under the curve represents the impulse.

b) If the collision time is $\Delta t = 0.010$ s, then from Eq. (8-9),

$$J_x = (F_{av})_x \Delta t, \qquad (F_{av})_x = \frac{J_x}{\Delta t} = \frac{20 \text{ N} \cdot \text{s}}{0.010 \text{ s}} = 2000 \text{ N}.$$

ΣF_x (a)

(b)

t

8-3 (a) A "hard" collision, such as that of a golf ball. (b) A "soft" collision, such as that of a tennis ball. The impulse is the same for both collisions, but the maximum force is less for the longer-lasting collision.

This average force is represented by the horizontal line $(F_{av})_x$ in Fig. 8-1. The horizontal force is exerted on the ball by the wall itself. The magnitude of this force has to be very large to cause the ball to change momentum in such a short time. Other forces that act on the ball during the collision are very weak by comparison; for instance, the gravitational force is only 3.9 N. Thus during the brief time that the collision lasts, we can ignore all other forces on the ball to a very good approximation. Figure 8-4 is a photograph showing the impact of a tennis ball and racket.

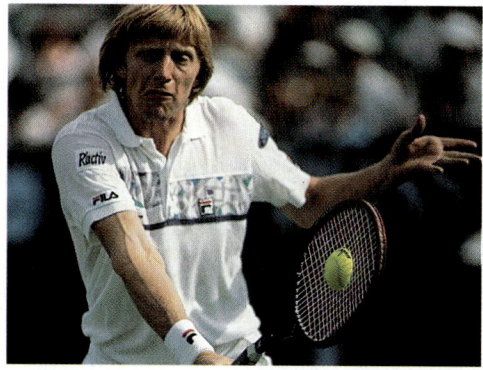

8-4 Typically, a tennis ball is in contact with the racket for approximately 0.01 s. The ball flattens noticeably due to the tremendous force exerted by the racket.

EXAMPLE 8-3

Kicking a soccer ball A soccer ball has a mass of 0.40 kg. Initially, it is moving to the left at 20 m/s, but then it is kicked and given a velocity at 45° upward and to the right, with a magnitude of 30 m/s (Fig. 8–5a). Find the impulse of the net force and the average net force, assuming a collision time $\Delta t = 0.010$ s.

SOLUTION The initial and final velocities are not along the same line, and we have to be careful to treat momentum and impulse as vector quantities, using their x- and y-components. Taking the x-axis horizontally to the right and the y-axis vertically upward, we find the following velocity components:

$$v_{1x} = -20 \text{ m/s}, \qquad v_{1y} = 0,$$
$$v_{2x} = v_{2y} = (0.707)(30 \text{ m/s}) = 21.2 \text{ m/s}$$
$$\text{(since } \cos 45° = \sin 45° = 0.707\text{)}.$$

The x-component of impulse is equal to the x-component of momentum change, and the same is true for the y-components:

$$J_x = p_{2x} - p_{1x} = m(v_{2x} - v_{1x})$$
$$= (0.40 \text{ kg})[21.2 \text{ m/s} - (-20 \text{ m/s})] = 16.5 \text{ kg} \cdot \text{m/s},$$
$$J_y = p_{2y} - p_{1y} = m(v_{2y} - v_{1y})$$
$$= (0.40 \text{ kg})(21.2 \text{ m/s} - 0) = 8.5 \text{ kg} \cdot \text{m/s}.$$

The components of the average net force on the ball are

$$(F_{av})_x = \frac{J_x}{\Delta t} = 1650 \text{ N}, \qquad (F_{av})_y = \frac{J_y}{\Delta t} = 850 \text{ N}.$$

The magnitude and direction of the average force are

$$F_{av} = \sqrt{(1650 \text{ N})^2 + (850 \text{ N})^2} = 1.9 \times 10^3 \text{ N},$$
$$\theta = \arctan \frac{850 \text{ N}}{1650 \text{ N}} = 27°,$$

where θ is measured upward from the +x-axis (Fig. 8–5b). Note that because the ball was not initially at rest, the ball's final velocity does *not* have the same direction as the average force acting on it.

The average net force $\vec{F}_{av}$ includes the effects of the force of gravity, though these are small; the weight of the ball is only 3.9 N. As in Example 8–2, the average force acting during the collision is exerted almost entirely by the object that the ball hit (in this case the soccer player's foot).

8–5 (a) A soccer ball (1) before and (2) after being kicked. (b) Finding the average force from its components.

(a)

(b)

8-3 CONSERVATION OF MOMENTUM

The concept of momentum is particularly important in situations in which we have two or more interacting bodies. Let's consider first an idealized system consisting of two bodies that interact with each other but not with anything else—for example, two astronauts who touch each other as they float freely in the zero-gravity environment of outer space (Fig. 8–6a). Think of the astronauts as particles. Each particle exerts a force on the other; according to Newton's third law, the two forces are always equal in magnitude and opposite in direction. Hence the *impulses* that act on the two particles will be equal and opposite, and the changes in momentum of the two particles will be equal and opposite.

Let's go over that again with some new terminology. For any system, the forces that the particles of the system exert on each other are called **internal forces.** Forces exerted

(a)

$\vec{F}_{B \text{ on } A}$

(b)

$\vec{F}_{A \text{ on } B}$

(c)

8–6 (a) Two astronauts push each other as they float freely in the zero-gravity environment of space. (b) Free-body diagram for astronaut A. (c) Free-body diagram for astronaut B. The only forces that act are internal forces; they don't change the total momentum of the two astronauts.

on any part of the system by some object outside it are called **external forces.** For the system we have described, the internal forces are $\vec{F}_{B \text{ on } A}$, exerted by particle B on particle A, and $\vec{F}_{A \text{ on } B}$, exerted by particle A on particle B (Figs. 8–6b, c). There are *no* external forces; when this is the case, we have an **isolated system.**

The net force on particle A is $\vec{F}_{B \text{ on } A}$, and the net force on particle B is $\vec{F}_{A \text{ on } B}$, so from Eq. (8–3) the rates of change of the momenta of the two particles are

$$\vec{F}_{B \text{ on } A} = \frac{d\vec{p}_A}{dt}, \qquad \vec{F}_{A \text{ on } B} = \frac{d\vec{p}_B}{dt}. \qquad (8-10)$$

The momentum of each particle changes, but these changes are not independent; according to Newton's third law, the two forces $\vec{F}_{B \text{ on } A}$ and $\vec{F}_{A \text{ on } B}$ are always equal in magnitude and opposite in direction. That is, $\vec{F}_{B \text{ on } A} = -\vec{F}_{A \text{ on } B}$, so $\vec{F}_{B \text{ on } A} + \vec{F}_{A \text{ on } B} = 0$. Adding together the two equations in Eq. (8–10), we have

$$\vec{F}_{B \text{ on } A} + \vec{F}_{A \text{ on } B} = \frac{d\vec{p}_A}{dt} + \frac{d\vec{p}_B}{dt} = \frac{d(\vec{p}_A + \vec{p}_B)}{dt} = 0. \qquad (8-11)$$

The rates of change of the two momenta are equal and opposite, so that the rate of change of the vector sum $\vec{p}_A + \vec{p}_B$ is zero. We now define the **total momentum $\vec{P}$** of the system of two particles as the vector sum of the momenta of the individual particles. That is,

$$\vec{P} = \vec{p}_A + \vec{p}_B. \qquad (8-12)$$

Then Eq. (8–11) becomes, finally,

$$\vec{F}_{B \text{ on } A} + \vec{F}_{A \text{ on } B} = \frac{d\vec{P}}{dt} = 0. \qquad (8-13)$$

(a)

8–7 (a) Two ice skaters push each other as they skate on a frictionless, horizontal surface. (b) Free-body diagram for skater A. (c) Free-body diagram for skater B. The normal forces and gravitational forces are external forces, but the vector sum of the external forces is zero and total momentum is conserved. (Compare to Fig. 8–6.)

(b)

(c)

6.3
Momentum Conservation and Collisions

6.7
Explosion Problems

6.8
Block Sliding Down an Incline

6.11
Pendulum Person-Projectile Bowling

The time rate of change of the *total* momentum $\vec{P}$ is zero. Hence the total momentum of the system is constant, even though the individual momenta of the particles that make up the system can change.

If external forces are also present, they must be included on the left side of Eq. (8–13) along with the internal forces. Then the total momentum is, in general, not constant. But if the vector sum of the external forces is zero, as in Fig. 8–7, these forces don't contribute to the sum, and $d\vec{P}/dt$ is again zero. Thus we have the following general result:

> **If the vector sum of the external forces on a system is zero, the total momentum of the system is constant.**

This is the simplest form of the **principle of conservation of momentum.** This principle is a direct consequence of Newton's third law. What makes this principle useful is that it doesn't depend on the detailed nature of the internal forces that act between members of the system. This means that we can apply conservation of momentum even if (as is often the case) we know very little about the internal forces. We have used Newton's second law to derive this principle, so we have to be careful to use it only in inertial frames of reference.

We can generalize this principle for a system containing any number of particles A, B, C, … interacting only with each other. The total momentum of such a system is

$$\vec{P} = \vec{p}_A + \vec{p}_B + \cdots = m_A \vec{v}_A + m_B \vec{v}_B + \cdots \qquad \text{(total momentum of a system of particles).} \qquad (8\text{–}14)$$

We make the same argument as before; the total rate of change of momentum of the system due to each action-reaction pair of internal forces is zero. Thus the total rate of change of momentum of the entire system is zero whenever the vector sum of the external forces acting on it is zero. The internal forces can change the momenta of individual particles in the system but not the *total* momentum of the system.

CAUTION ▶ When you apply the conservation of momentum to a system, it's absolutely essential to remember that momentum is a *vector* quantity. Hence you *must* use vector addition to compute the total momentum of a system. Using components is usually the simplest method. If p_{Ax}, p_{Ay}, and p_{Az} are the components of momentum of particle A, and similarly for the other particles, then Eq. (8–14) is equivalent to the component equations

$$P_x = p_{Ax} + p_{Bx} + \cdots,$$

$$P_y = p_{Ay} + p_{By} + \cdots, \qquad (8\text{–}15)$$

$$P_z = p_{Az} + p_{Bz} + \cdots.$$

If the vector sum of the external forces on the system is zero, then P_x, P_y, and P_z are all constant. ◀

In some ways the principle of conservation of momentum is more general than the principle of conservation of mechanical energy. For example, mechanical energy is conserved only when the internal forces are *conservative*—that is, when the forces allow two-way conversion between kinetic and potential energy—but conservation of momentum is valid even when the internal forces are *not* conservative. In this chapter we will analyze situations in which both momentum and mechanical energy are conserved, and others in which only momentum is conserved. These two principles play a fundamental role in all areas of physics, and we will encounter them throughout our study of physics.

Problem–Solving Strategy

CONSERVATION OF MOMENTUM

1. Before applying conservation of momentum to a problem, you must first decide whether or not momentum *is* conserved! This will be true *only* if the vector sum of the external forces acting on the system of particles is zero. If this is not the case, you can't use conservation of momentum.

2. Define a coordinate system. Make a sketch showing the coordinate axes, including the positive direction for each. Often it is easiest to choose the *x*-axis to have the direction of one of the initial velocities. Make sure you are using an inertial frame of reference. Most of the problems in this chapter deal with two-dimensional situations, in which the vectors have only *x*- and *y*-components; all the following statements can be generalized to include *z*-components when necessary.

3. Treat each body as a particle. Draw "before" and "after" sketches, and include vectors on each to represent all known velocities. Label the vectors with magnitudes, angles, components, or whatever information is given, and give each unknown magnitude, angle, or component an algebraic symbol. You may find it helpful to use the subscripts 1 and 2 for velocities before and after the inter-

action, respectively; if you use these subscripts, use letters (not numbers) to label each particle.

4. Write an equation in terms of symbols equating the total *initial x*-component of momentum (that is, before the interaction) to the total *final x*-component of momentum (that is, after the interaction), using $p_x = mv_x$ for each particle. Write another equation for the *y*-components, using $p_y = mv_y$ for each particle. Remember that the *x*- and *y*-components of velocity or momentum are *never* added together in the same equation! Even when all the velocities lie along a line (such as the *x*-axis), the components of velocity along this line can be positive or negative; be careful with signs!

5. Solve these equations to determine whatever results are required. In some problems you will have to convert from the *x*- and *y*-components of a velocity to its magnitude and direction, or the reverse.

6. In some problems, energy considerations give additional relationships among the various velocities, as we will see later in this chapter.

EXAMPLE 8-4

Recoil of a rifle A marksman holds a rifle of mass $m_R = 3.00$ kg loosely in his hands, so as to let it recoil freely when fired. He fires a bullet of mass $m_B = 5.00$ g horizontally with a velocity relative to the ground of $v_B = 300$ m/s (Fig. 8–8). What is the recoil velocity v_R of the rifle? What are the final momentum and kinetic energy of the bullet? Of the rifle?

SOLUTION We consider an idealized model in which the horizontal forces the marksman exerts on the rifle are negligible. Then there is no net horizontal force on the system (the bullet and rifle) during the firing of the rifle, and so the total horizontal momentum of the system is the same before and after the rifle is fired (i.e., is conserved).

Take the positive x-axis to be the direction the rifle is aimed. Initially, both the rifle and bullet are at rest, so the initial x-component of total momentum is zero. After the bullet is fired, its x-component of momentum is $m_B v_B$, and that of the rifle is $m_R v_R$. Conservation of the x-component of total momentum gives

$$P_x = 0 = m_B v_B + m_R v_R,$$

$$v_R = -\frac{m_B}{m_R} v_B = -\left(\frac{0.00500 \text{ kg}}{3.00 \text{ kg}} \right)(300 \text{ m/s}) = -0.500 \text{ m/s}.$$

The negative sign means that the recoil is in the direction oppo-

v_R $v_B = 300$ m/s $\longrightarrow$ $+x$

$m_B = 5.00$ g

$m_R = 3.00$ kg

8–8 The ratio of bullet speed to recoil speed is the inverse of the ratio of bullet mass to rifle mass.

site to that of the bullet. If the butt of a rifle were to hit your shoulder traveling at this speed, you'd feel it. That's what the "kick" of a rifle is all about. It's more comfortable to hold the rifle tightly against your shoulder when you fire it; then m_R is replaced by the sum of your mass and the rifle's mass, and the recoil speed is much less.

The momentum and kinetic energy of the bullet are

$$p_B = m_B v_B = (0.00500 \text{ kg})(300 \text{ m/s}) = 1.50 \text{ kg} \cdot \text{m/s},$$

$$K_B = \frac{1}{2} m_B v_B{}^2 = \frac{1}{2} (0.00500 \text{ kg})(300 \text{ m/s})^2 = 225 \text{ J};$$

for the rifle, the momentum and the kinetic energy are

$$p_R = m_R v_R = (3.00 \text{ kg})(-0.500 \text{ m/s}) = -1.50 \text{ kg} \cdot \text{m/s},$$

$$K_R = \frac{1}{2} m_R v_R{}^2 = \frac{1}{2} (3.00 \text{ kg})(-0.500 \text{ m/s})^2 = 0.375 \text{ J}.$$

The bullet and the rifle have equal and opposite momenta after the interaction because they were subjected to equal and opposite interaction forces for the same amount of *time* (that is, equal and opposite impulses). But the bullet acquires much greater kinetic energy than the rifle because the bullet travels a much greater *distance* than the rifle during the interaction. Thus the force on the bullet does more work than the force on the rifle. The ratio of the two kinetic energies, 600:1, is equal to the inverse ratio of the masses; in fact, it can be shown that this always happens in recoil situations. We leave the proof as a problem (see Exercise 8–21).

Notice that our calculation doesn't depend at all on the details of how the rifle works. In a real rifle, the bullet is propelled forward by an explosive charge; if instead a very stiff spring were used to give the same bullet velocity, the answers would have been exactly the same.

EXAMPLE 8-5

Collision along a straight line Two gliders move toward each other on a frictionless linear air track (Fig. 8–9a). After they collide, glider B moves away with a final velocity of $+2.0$ m/s (Fig. 8–9b). What is the final velocity of glider A? How do the changes in momentum and in velocity compare for the two gliders?

SOLUTION The total vertical force on each glider is zero; the net force on each glider is the horizontal force exerted on it by the other glider. The net *external* force on the two gliders together is zero, so the total momentum is constant. Take the x-axis to lie along the air track, with the positive direction to the right. The masses of the gliders and their initial x-components of velocity are shown in the figure. The x-component of total momentum before the collision is

$$P_x = m_A v_{A1} + m_B v_{B1}$$
$$= (0.50 \text{ kg})(2.0 \text{ m/s}) + (0.30 \text{ kg})(-2.0 \text{ m/s}) = 0.40 \text{ kg} \cdot \text{m/s}.$$

8–9 Two gliders (a) before and (b) after their collision.

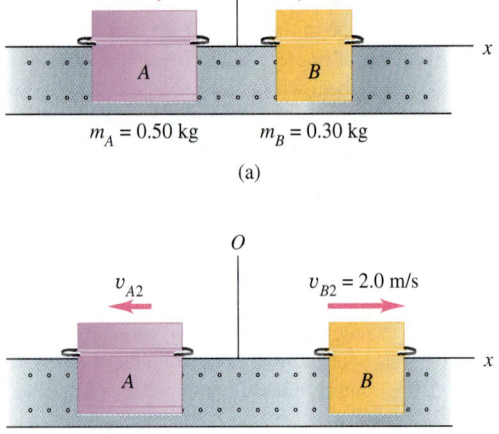

$v_{A1} = 2.0$ m/s $v_{B1} = -2.0$ m/s

A B

$m_A = 0.50$ kg $m_B = 0.30$ kg

(a)

v_{A2} $v_{B2} = 2.0$ m/s

A B

(b)

This is positive (to the right in Fig. 8–9) because glider A has a greater magnitude of momentum before the collision than does glider B. The x-component of total momentum has the same value after the collision, so

$$P_x = m_A v_{A2} + m_B v_{B2}.$$

Solving this equation for v_{A2}, the final velocity of A, we find

$$v_{A2} = \frac{P_x - m_B v_{B2}}{m_A} = \frac{0.40 \text{ kg} \cdot \text{m/s} - (0.30 \text{ kg})(2.0 \text{ m/s})}{0.50 \text{ kg}}$$

$$= -0.40 \text{ m/s}.$$

The change in momentum of glider A is

$$m_A v_{A2} - m_A v_{A1} = (0.50 \text{ kg})(-0.40 \text{ m/s}) - (0.50 \text{ kg})(2.0 \text{ m/s})$$

$$= -1.2 \text{ kg} \cdot \text{m/s},$$

and the change in momentum of glider B is

$$m_B v_{B2} - m_B v_{B1} = (0.30 \text{ kg})(2.0 \text{ m/s}) - (0.30 \text{ kg})(-2.0 \text{ m/s})$$

$$= +1.2 \text{ kg} \cdot \text{m/s}.$$

The two interacting gliders undergo changes in momentum that are equal in magnitude and opposite in direction. The same is not true of their changes in *velocity*, however. For A, $v_{A2} - v_{A1}$ = $(-0.40 \text{ m/s}) - 2.0 \text{ m/s} = -2.4 \text{ m/s}$; for B, $v_{B2} - v_{B1} = 2.0 \text{ m/s} - (-2.0 \text{ m/s}) = +4.0 \text{ m/s}$. The magnitude of the interaction force is the same for both bodies, but from Newton's second law the magnitude of acceleration (and hence of velocity change) is greater for the less massive body (glider B). In a collision between an eighteen-wheeler truck and a compact car, the car's occupants are subjected to greater acceleration (and greater chance of injury) than the truck driver is. If the truck collides with an insect, the truck driver won't notice the resulting acceleration at all, but the insect surely will!

EXAMPLE 8-6

Collision in a horizontal plane Figure 8–10a shows two chunks of ice sliding on a frictionless frozen pond. Chunk A, with mass $m_A = 5.0 \text{ kg}$, moves with initial velocity $v_{A1} = 2.0 \text{ m/s}$ parallel to the x-axis. It collides with chunk B, which has mass $m_B = 3.0 \text{ kg}$ and is initially at rest. After the collision, the velocity of chunk A is found to be $v_{A2} = 1.0 \text{ m/s}$ in a direction making an angle $\alpha = 30°$ with the initial direction. What is the final velocity of chunk B?

SOLUTION There are no horizontal (x or y) external forces, so the total horizontal momentum of the system is the same before and after the collision. The velocities are not all along a single

line, so we have to treat momentum as a *vector* quantity, using components of each momentum in the x- and y-directions. Then momentum conservation requires that the sum of the x-components *before* the collision (subscript 1) must equal the sum *after* the collision (subscript 2), and similarly for the y-components. Just as with force equilibrium problems, we write a separate equation for each component. For the x-components (before and after) we have

$$m_A v_{A1x} + m_B v_{B1x} = m_A v_{A2x} + m_B v_{B2x},$$

$$v_{B2x} = \frac{m_A v_{A1x} + m_B v_{B1x} - m_A v_{A2x}}{m_B}$$

$$= \frac{\left[\begin{array}{c}(5.0 \text{ kg})(2.0 \text{ m/s}) + (3.0 \text{ kg})(0) \\ -(5.0 \text{ kg})(1.0 \text{ m/s})(\cos 30°)\end{array}\right]}{3.0 \text{ kg}}$$

$$= 1.89 \text{ m/s}.$$

Conservation of the y-component of total momentum gives

$$m_A v_{A1y} + m_B v_{B1y} = m_A v_{A2y} + m_B v_{B2y},$$

$$v_{B2y} = \frac{m_A v_{A1y} + m_B v_{B1y} - m_A v_{A2y}}{m_B}$$

$$= \frac{\left[\begin{array}{c}(5.0 \text{ kg})(0) + (3.0 \text{ kg})(0) \\ -(5.0 \text{ kg})(1.0 \text{ m/s})(\sin 30°)\end{array}\right]}{3.0 \text{ kg}}$$

$$= -0.83 \text{ m/s}.$$

After the collision, chunk B moves in the positive x-direction and the negative y-direction (Fig. 8–10b). The magnitude of v_{B2} is

$$v_{B2} = \sqrt{(1.89 \text{ m/s})^2 + (-0.83 \text{ m/s})^2} = 2.1 \text{ m/s},$$

and the angle of its direction from the positive x-axis is

$$\beta = \arctan \frac{-0.83 \text{ m/s}}{1.89 \text{ m/s}} = -24°.$$

(a)

(b)

8–10 Views from above of the velocities (a) before and (b) after the collision.

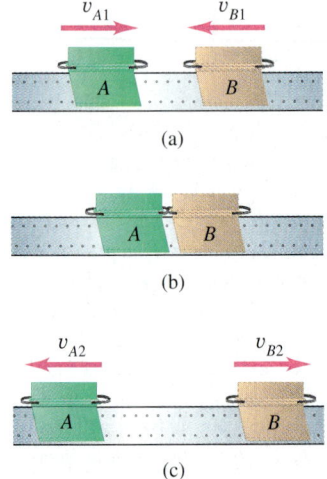

8–11 Model of an elastic collision: Glider A and glider B approach each other on a frictionless surface. Each glider has a steel spring bumper on the end to ensure an elastic collision.

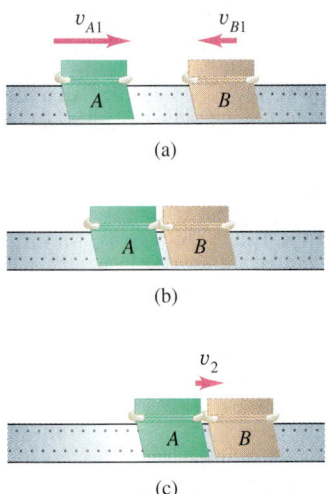

8–12 Model of a completely inelastic collision. The spring bumpers on the gliders are replaced by putty, so the gliders stick together after collision.

6.4.1
Collision Problems

6.9
Skier and Cart

8–4 INELASTIC COLLISIONS

To the average person on the street the term *collision* is likely to mean some sort of automotive disaster. We'll use it in that sense, but we'll also broaden the meaning to include any strong interaction between bodies that lasts a relatively short time. So we include not only car accidents but also balls colliding on a billiard table, neutrons hitting atomic nuclei in a nuclear reactor, a bowling ball striking the pins, the impact of a meteor on the Arizona desert, and a close encounter of a spacecraft with the planet Saturn.

If the forces between the bodies are much larger than any external forces, as is the case in most collisions, we can neglect the external forces entirely and treat the bodies as an *isolated* system. Then momentum is conserved in the collision, and the total momentum of the system has the same value before and after the collision. Two cars colliding at an icy intersection provide a good example. Even two cars colliding on dry pavement can be treated as an isolated system during the collision if the forces between the cars are much larger than the friction forces of pavement against tires.

If the forces between the bodies are also *conservative,* so that no mechanical energy is lost or gained in the collision, the total *kinetic energy* of the system is the same after the collision as before. Such a collision is called an **elastic collision.** A collision between two marbles or two billiard balls is almost completely elastic. Figure 8–11 shows a model for an elastic collision. When the bodies collide, the springs are momentarily compressed and some of the original kinetic energy is momentarily converted to elastic potential energy. Then the bodies bounce apart, the springs expand, and this potential energy is converted back to kinetic energy.

A collision in which the total kinetic energy after the collision is *less* than that before the collision is called an **inelastic collision.** A meatball landing on a plate of spaghetti and a bullet embedding itself in a block of wood are examples of inelastic collisions. An inelastic collision in which the colliding bodies stick together and move as one body after the collision is often called a **completely inelastic collision.** An example is shown in Fig. 8–12; we have replaced the spring bumper in Fig. 8–11 with a ball of putty that squashes and sticks the two bodies together.

CAUTION ▶ It's a common misconception that the *only* inelastic collisions are those in which the colliding bodies stick together. In fact, inelastic collisions include many situations in which the bodies do *not* stick. If two cars bounce off each other in a "fender-bender," the work done to deform the fenders cannot be recovered as kinetic energy of the cars, and so the collision is inelastic. ◀

Remember the following rule: **In any collision in which external forces can be neglected, momentum is conserved and the total momentum before equals the total momentum after; in elastic collisions *only*, the total kinetic energy before equals the total kinetic energy after.**

COMPLETELY INELASTIC COLLISIONS

Let's look at what happens to momentum and kinetic energy in a *completely* inelastic collision of two bodies (A and B), as in Fig. 8–12. Because the two bodies stick together after the collision, their final velocities must be equal:

$$\vec{v}_{A2} = \vec{v}_{B2} = \vec{v}_2.$$

Conservation of momentum gives the relation

$$m_A \vec{v}_{A1} + m_B \vec{v}_{B1} = (m_A + m_B)\vec{v}_2 \qquad \text{(completely inelastic collision).} \quad (8\text{–}16)$$

If we know the masses and initial velocities, we can compute the common final velocity $\vec{v}_2$.

Suppose, for example, that a body with mass m_A and initial x-component of velocity v_1 collides inelastically with a body with mass m_B that is initially at rest ($v_{B1} = 0$). From Eq. (8–16) the common x-component of velocity v_2 of both bodies after the collision is

$$v_2 = \frac{m_A}{m_A + m_B} v_1 \qquad \text{(completely inelastic collision, } B \text{ initially at rest).} \quad (8\text{–}17)$$

Let's verify that the total kinetic energy after this completely inelastic collision is less than before the collision. The kinetic energies K_1 and K_2 before and after the collision, respectively, are

$$K_1 = \frac{1}{2} m_A v_1{}^2,$$

$$K_2 = \frac{1}{2}(m_A + m_B)v_2{}^2 = \frac{1}{2}(m_A + m_B)\left(\frac{m_A}{m_A + m_B}\right)^2 v_1{}^2.$$

The ratio of final to initial kinetic energy is

$$\frac{K_2}{K_1} = \frac{m_A}{m_A + m_B} \qquad \text{(completely inelastic collision, } B \text{ initially at rest).} \quad (8\text{–}18)$$

The right side is always less than unity because the denominator is always greater than the numerator. Even when the initial velocity of m_B is not zero, it is not hard to verify that the kinetic energy after a completely inelastic collision is always less than before.

Please note: We don't recommend memorizing Eqs. (8–17) or (8–18). We derived them only to prove that kinetic energy is always lost in a completely inelastic collision.

Automobile collisions are intended to be inelastic, so that the structure of the car absorbs as much of the energy of the collision as possible. This absorbed energy cannot be recovered, since it goes into a permanent deformation of the car.

EXAMPLE 8–7

A completely inelastic collision Suppose that in the collision described in Example 8–5 (Section 8–3), the two gliders do not bounce but stick together after the collision. The masses and initial velocities are as shown in Fig. 8–9a. Find the common final velocity v_2, and compare the initial and final kinetic energies.

SOLUTION Take the x-axis to lie along the direction of motion, as before. From conservation of the x-component of momentum,

$$m_A v_{A1} + m_B v_{B1} = (m_A + m_B)v_2,$$

$$v_2 = \frac{m_A v_{A1} + m_B v_{B1}}{m_A + m_B}$$

$$= \frac{(0.50\ \text{kg})(2.0\ \text{m/s}) + (0.30\ \text{kg})(-2.0\ \text{m/s})}{(0.50\ \text{kg} + 0.30\ \text{kg})}$$

$$= 0.50\ \text{m/s}.$$

Because v_2 is positive, the gliders move together to the right (the $+x$-direction) after the collision.

Before the collision, the kinetic energies of gliders A and B are

$$K_A = \frac{1}{2} m_A v_{A1}{}^2 = \frac{1}{2}(0.50\ \text{kg})(2.0\ \text{m/s})^2 = 1.0\ \text{J},$$

$$K_B = \frac{1}{2} m_B v_{B1}{}^2 = \frac{1}{2}(0.30\ \text{kg})(-2.0\ \text{m/s})^2 = 0.60\ \text{J}.$$

(Note that the kinetic energy of glider B is positive, even though the x-components of its velocity v_{B1} and momentum mv_{B1} are both negative.) The *total* kinetic energy before the collision is 1.6 J. The kinetic energy after the collision is

$$\frac{1}{2}(m_A + m_B)v_2{}^2 = \frac{1}{2}(0.50\ \text{kg} + 0.30\ \text{kg})(0.50\ \text{m/s})^2 = 0.10\ \text{J}.$$

The final kinetic energy is only $\frac{1}{16}$ of the original; $\frac{15}{16}$ is converted from mechanical energy to various other forms. If there is a ball of chewing gum between the gliders, it squashes and becomes warmer. If the gliders couple together like two freight cars, the energy goes into elastic vibrations that eventually dissipate. If there is a spring between the gliders that is compressed as they lock together, then the energy is stored as potential energy of the spring. In all of these cases the *total* energy of the system is conserved, although *kinetic* energy is not. However, in an isolated system, momentum is *always* conserved, whether the collision is elastic or not.

EXAMPLE 8–8

The ballistic pendulum Figure 8–13 shows a ballistic pendulum, a system for measuring the speed of a bullet. The bullet, with mass m, is fired into a block of wood with mass M, suspended like a pendulum, and makes a completely inelastic collision with it. After the impact of the bullet, the block swings up to a maximum height y. Given the values of y, m, and M, what is the initial speed v of the bullet?

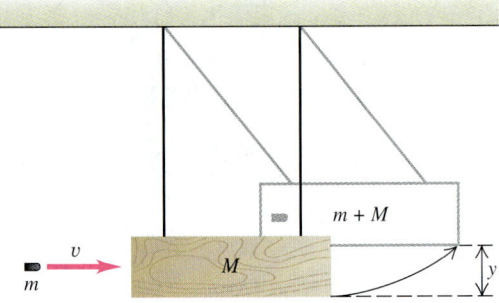

8–13 A ballistic pendulum.

SOLUTION We analyze this event in two stages. The first is the embedding of the bullet in the block, and the second is the subsequent swinging of the block on its strings. As we will see, momentum is conserved in the first stage, mechanical energy in the second.

During the first stage, the bullet embeds itself in the block so quickly that the block has no time to swing appreciably away from its straight-down position. So during the impact the supporting strings remain very nearly vertical, there is negligible external horizontal force acting on the system, and the horizontal component of momentum is conserved. Let V be the x-component of velocity of bullet and block just after the collision. Momentum conservation gives

$$mv = (m + M)V, \qquad v = \frac{m + M}{m} V.$$

The kinetic energy of the system just after the collision is $K = \frac{1}{2}(m + M)V^2$. As in Eq. (8–18), this is less than the kinetic energy before the collision; the collision is inelastic. Momentum is conserved during the first stage, but mechanical energy is not.

In the second stage, after the collision, the block and bullet move as a unit. The only forces are the weight (a conservative force) and the string tensions (which do no work). So as the pendulum swings upward and to the right, mechanical energy is conserved. Momentum is *not* conserved during this stage because there is a net external force. The pendulum comes to rest for an instant at a height y, where its kinetic energy $\frac{1}{2}(m + M)V^2$ has all become potential energy $(m + M)gy$; then it swings back down. Energy conservation gives

$$\frac{1}{2}(m + M)V^2 = (m + M)gy,$$

$$V = \sqrt{2gy}.$$

Now we substitute this into the momentum equation to find v:

$$v = \frac{m + M}{m}\sqrt{2gy}.$$

By measuring m, M, and y, we can compute the original speed v of the bullet. For example, if $m = 5.00$ g $= 0.00500$ kg, $M = 2.00$ kg, and $y = 3.00$ cm $= 0.0300$ m,

$$v = \frac{2.00 \text{ kg} + 0.00500 \text{ kg}}{0.00500 \text{ kg}}\sqrt{2(9.80 \text{ m/s}^2)(0.0300 \text{ m})}$$

$$= 307 \text{ m/s}.$$

The velocity V of the block just after impact is

$$V = \sqrt{2gy} = \sqrt{2(9.80 \text{ m/s}^2)(0.0300 \text{ m})}$$

$$= 0.767 \text{ m/s}.$$

The kinetic energy of the bullet just before impact is $\frac{1}{2}(0.00500 \text{ kg})(307 \text{ m/s})^2 = 236$ J. Just after impact the kinetic energy of the bullet and block is $\frac{1}{2}(2.005 \text{ kg})(0.767 \text{ m/s})^2 = 0.589$ J. Nearly all the kinetic energy disappears as the wood splinters and the bullet and block become hotter.

EXAMPLE 8–9

Analysis of an automobile collision A small compact car with a mass of 1000 kg is traveling north on Morewood Avenue at 15 m/s. At the intersection of Morewood and Fifth Avenue it collides with an enormous luxury car with a mass of 2000 kg, traveling east on Fifth Avenue at 10 m/s (Fig. 8–14a). Fortunately, all occupants are wearing seat belts and there are no injuries, but the two cars are thoroughly tangled and move away from the impact point as one mass. a) Treating each car as a particle, find the total momentum just before the collision. b) The insurance adjustor has asked you to find the velocity of the wreckage just after impact. What do you tell her?

SOLUTION a) In the figure, we have drawn coordinate axes and have labeled the small car A and the large car B. From Eqs.

(8–15) the components of the total momentum $\vec{P}$ are

$$P_x = p_{Ax} + p_{Bx} = m_A v_{Ax} + m_B v_{Bx}$$

$$= (1000 \text{ kg})(0) + (2000 \text{ kg})(10 \text{ m/s})$$

$$= 2.0 \times 10^4 \text{ kg} \cdot \text{m/s},$$

$$P_y = p_{Ay} + p_{By} = m_A v_{Ay} + m_B v_{By}$$

$$= (1000 \text{ kg})(15 \text{ m/s}) + (2000 \text{ kg})(0)$$

$$= 1.5 \times 10^4 \text{ kg} \cdot \text{m/s}.$$

The magnitude of $\vec{P}$ is

$$P = \sqrt{(2.0 \times 10^4 \text{ kg} \cdot \text{m/s})^2 + (1.5 \times 10^4 \text{ kg} \cdot \text{m/s})^2}$$

$$= 2.5 \times 10^4 \text{ kg} \cdot \text{m/s},$$

and its direction is given by the angle θ shown in Fig. 8–14b,

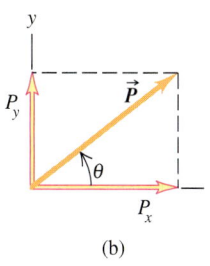

(a)

(b)

8–14 (a) View from a traffic helicopter overhead of the start of the collision at Morewood and Fifth. (b) Finding the total momentum just before the collision.

where

$$\tan \theta = \frac{P_y}{P_x} = \frac{1.5 \times 10^4 \ \text{kg} \cdot \text{m/s}}{2.0 \times 10^4 \ \text{kg} \cdot \text{m/s}} = 0.75, \qquad \theta = 37°.$$

b) To find out what happens *after* the collision, we'll assume that we can treat the cars as an isolated system during the collision. This may seem implausible, so let's look at some numbers. The mass of the large car is 2000 kg, its weight is about 20,000 N, and if the coefficient of friction is 0.5, the friction force when it slides across the pavement is about 10,000 N. Its kinetic energy just before the impact is $\frac{1}{2}(2000 \ \text{kg})(10 \ \text{m/s})^2 = 1.0 \times 10^5$ J. The car may crumple 0.2 m or so; to do -1.0×10^5 J of work on it to stop the car in 0.2 m would require a force of 5.0×10^5 N, which is 50 times greater than the friction force.

Compared to the large forces that act during such a collision, it's not unreasonable to neglect the friction forces on both cars and assume that the cars form an isolated system during the collision. In that case the total momentum just after the collision is

the same as just before, which we found to be 2.5×10^4 kg · m/s in a direction 37° north of due east. Assuming that no parts fall off, the total mass of wreckage is $M = 3000$ kg. Calling the final velocity $\vec{V}$, we have $\vec{P} = M\vec{V}$. The direction of the velocity is the same as that of the momentum, and its magnitude is

$$V = \frac{P}{M} = \frac{2.5 \times 10^4 \ \text{kg} \cdot \text{m/s}}{3000 \ \text{kg}} = 8.3 \ \text{m/s}.$$

This is an inelastic collision, so we expect the total kinetic energy to be less after the collision than before. We invite you to carry out the calculations; you will find that the initial kinetic energy is 2.1×10^5 J and the final value is 1.0×10^5 J. Over half of the initial kinetic energy is converted to other forms.

CAUTION ▶ You may have been tempted to try to find the final velocity by taking the vector sum of the initial velocities. But there is *no* "law of conservation of velocities." The conserved quantity is the total momentum of the system, and the above analysis is the only correct way to do the problem. ◀

8–5 ELASTIC COLLISIONS

6.2
Collisions and Elasticity

6.4.2 and 6.4.3
Collision Problems

6.5
Car Collision: Two Dimensions

6.10
Pendulum Bashes Box

As discussed in Section 8–4, an *elastic collision* in an isolated system is a collision in which kinetic energy (as well as momentum) is conserved. Elastic collisions occur when the forces between the colliding bodies are *conservative*. When two steel balls collide, they squash a little near the surface of contact, but then they spring back. Some of the kinetic energy is stored temporarily as elastic potential energy, but at the end it is reconverted to kinetic energy.

Let's look at an elastic collision between two bodies A and B. We start with a one-dimensional collision, in which all the velocities lie along the same line; we choose this line to be the x-axis. Each momentum and velocity then has only an x-component. We call the x-components of velocity before the collision v_{A1} and v_{B1}, and those after the collision v_{A2} and v_{B2}. From conservation of kinetic energy we have

$$\frac{1}{2} m_A v_{A1}{}^2 + \frac{1}{2} m_B v_{B1}{}^2 = \frac{1}{2} m_A v_{A2}{}^2 + \frac{1}{2} m_B v_{B2}{}^2,$$

and conservation of momentum gives

$$m_A v_{A1} + m_B v_{B1} = m_A v_{A2} + m_B v_{B2}.$$

If the masses m_A and m_B and the initial velocities v_{A1} and v_{B1} are known, these two equations can be solved simultaneously to find the two final velocities v_{A2} and v_{B2}.

The general solution is a little complicated, so we will concentrate on the particular case in which body B is at rest before the collision. Think of it as a target for body A to hit. We can then simplify the velocity notation; we let v be the initial x-component of velocity of A, and v_A and v_B be the final x-components for A and B. Then the kinetic energy and momentum conservation equations are, respectively.

$$\frac{1}{2} m_A v^2 = \frac{1}{2} m_A v_A{}^2 + \frac{1}{2} m_B v_B{}^2, \tag{8–19}$$

$$m_A v = m_A v_A + m_B v_B. \tag{8–20}$$

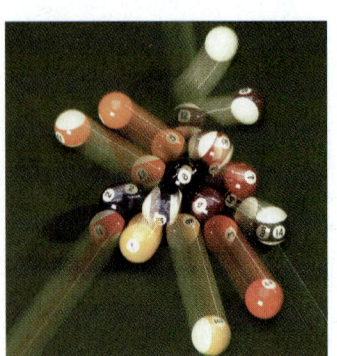

Billiard balls deform very little when they collide, and quickly spring back from any deformation they do undergo. Hence the force of interaction between the balls is almost perfectly conservative, and the collision is elastic.

We may solve for v_A and v_B in terms of the masses and the initial velocity v. This involves some fairly strenuous algebra, but it's worth it. No pain, no gain! The simplest approach is somewhat indirect, but along the way it uncovers an additional interesting feature of elastic collisions.

First rearrange Eqs. (8–19) and (8–20) as follows:

$$m_B v_B{}^2 = m_A (v^2 - v_A{}^2) = m_A (v - v_A)(v + v_A), \tag{8–21}$$

$$m_B v_B = m_A (v - v_A). \tag{8–22}$$

Now divide Eq. (8–21) by Eq. (8–22) to obtain

$$v_B = v + v_A. \tag{8–23}$$

Substitute this back into Eq. (8–22) to eliminate v_B, and then solve for v_A:

$$m_B (v + v_A) = m_A (v - v_A),$$

$$v_A = \frac{m_A - m_B}{m_A + m_B} v. \tag{8–24}$$

Finally, substitute this result back into Eq. (8–23) to obtain

$$v_B = \frac{2 m_A}{m_A + m_B} v. \tag{8–25}$$

Now we can interpret the results. Suppose body A is a ping-pong ball and body B is a bowling ball. Then we expect A to bounce off after the collision with a velocity nearly equal to its original value but in the opposite direction (Fig. 8–15a), and we expect B's velocity to be much smaller. That's just what the equations predict. When m_A is much smaller than m_B, the fraction in Eq. (8–24) is approximately equal to (-1), so v_A is approximately equal to $-v$. The fraction in Eq. (8–25) is much smaller than unity, so v_B is much less than v. Figure 8–15b shows the opposite case, in which A is the bowling ball and B the ping-pong ball and m_A is much larger than m_B. What do you expect to happen then? Check your predictions against Eqs. (8–24) and (8–25).

8–15 Collisions between a ping-pong ball and a bowling ball.

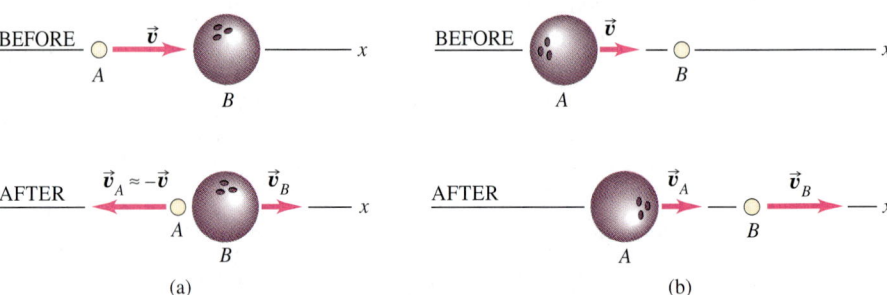

Another interesting case occurs when the masses are equal (Fig. 8–16). If $m_A = m_B$, then Eqs. (8–24) and (8–25) give $v_A = 0$ and $v_B = v$. That is, the body that was moving stops dead; it gives all its momentum and kinetic energy to the body that was at rest. This behavior is familiar to all pool players and marbles shooters.

Now comes the surprise bonus. Equation (8–23) can be rewritten as

$$v = v_B - v_A. \tag{8–26}$$

Here $v_B - v_A$ is just the velocity of B relative to A *after* the collision; according to Eq. (8–26), this equals v, which is the *negative* of the velocity of B relative to A *before* the collision. (You may want to review the discussion of relative velocity in Section 3–6.) The relative velocity has the same magnitude, but opposite sign, before and after the collision. The sign changes because the two bodies are approaching before the collision but receding from each other after the collision. If we view this collision from a second coordinate system moving with constant velocity relative to the first, the velocities of the bodies are different but the *relative* velocities are the same. Hence our statement about relative velocities holds in general for *any* straight-line elastic collision, even when neither body is at rest initially. *In a straight-line elastic collision of two bodies, the relative velocities before and after the collision have the same magnitude but opposite sign.* This can be expressed using our original notation as

$$v_{B2} - v_{A2} = -(v_{B1} - v_{A1}). \tag{8–27}$$

It turns out that a *vector* relationship similar to Eq. (8–27) is a general property of *all* elastic collisions, even when both bodies are moving initially and the velocities do not all lie along the same line. This result provides an alternative and equivalent definition of an elastic collision: *In an elastic collision the relative velocity of the two bodies has the same magnitude before and after the collision.* Whenever this condition is satisfied, the total kinetic energy is also conserved.

When an elastic two-body collision isn't head-on, the velocities don't all lie along a single line. If they all lie in a plane, each final velocity has two unknown components, and there are four unknowns in all. Conservation of energy and conservation of the x- and y-components of momentum give only three equations. To determine the final velocities uniquely, we need additional information, such as the direction or magnitude of one of the final velocities.

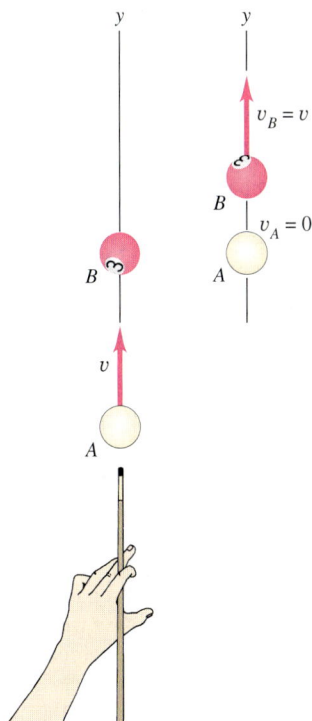

8–16 In a one-dimensional elastic collision between bodies of equal mass, if one body is at rest, it receives all the momentum and kinetic energy of the moving body.

EXAMPLE 8–10

An elastic straight-line collision The situation in Fig. 8–17 is the same as for Example 8–5 in Section 8–3, but now we've added ideal spring bumpers to the gliders so that the collision is elastic. What are the velocities of A and B after the collision?

SOLUTION From conservation of momentum,

$$m_A v_{A1} + m_B v_{B1} = m_A v_{A2} + m_B v_{B2},$$

$$(0.50 \text{ kg})(2.0 \text{ m/s}) + (0.30 \text{ kg})(-2.0 \text{ m/s})$$

$$= (0.50 \text{ kg})v_{A2} + (0.30 \text{ kg})v_{B2},$$

$$0.50v_{A2} + 0.30v_{B2} = 0.40 \text{ m/s}.$$

(In the last equation we have divided through by the unit "kg.") From Eq. (8–27), the relative velocity relation for an elastic collision,

$$v_{B2} - v_{A2} = -(v_{B1} - v_{A1})$$

$$= -(-2.0 \text{ m/s} - 2.0 \text{ m/s}) = 4.0 \text{ m/s}.$$

$v_{A1} = 2.0$ m/s $v_{B1} = -2.0$ m/s

$m_A = 0.50$ kg $m_B = 0.30$ kg

(a)

v_{A2} v_{B2}

(b)

8–17 Air-track gliders (a) before and (b) after an elastic collision.

Before the collision, the velocity of B relative to A is to the left at 4.0 m/s; after the collision, the velocity of B relative to A is to the right at 4.0 m/s. Solving these equations simultaneously, we find

$$v_{A2} = -1.0 \text{ m/s}, \qquad v_{B2} = 3.0 \text{ m/s}.$$

Both bodies reverse their directions of motion; A moves to the left at 1.0 m/s, and B moves to the right at 3.0 m/s. This is differ-

ent from the result of Example 8–5 because that collision was *not* elastic. The total kinetic energy after the elastic collision is

$$\frac{1}{2}(0.50 \text{ kg})(-1.0 \text{ m/s})^2 + \frac{1}{2}(0.30 \text{ kg})(3.0 \text{ m/s})^2 = 1.6 \text{ J}.$$

As expected, this equals the total kinetic energy before the collision (which we calculated in Example 8–7, Section 8–4).

EXAMPLE 8–11

Moderator in a nuclear reactor High-speed neutrons are produced in a nuclear reactor during nuclear fission processes. Before a neutron can trigger additional fissions, it has to be slowed down by collisions with nuclei in the *moderator* of the reactor. The first nuclear reactor (built in 1942 at the University of Chicago) and the reactor involved in the 1986 Chernobyl accident both used carbon (graphite) as the moderator material. A neutron (mass 1.0 u) traveling at 2.6×10^7 m/s undergoes a head-on elastic collision with a carbon nucleus (mass 12 u) initially at rest. The external forces during the collision are negligible. What are the velocities after the collision? (1 u is the *atomic mass unit,* equal to 1.66×10^{-27} kg.)

SOLUTION Because one body is initially at rest, we can use Eqs. (8–24) and (8–25) with

$$m_A = 1.0 \text{ u}, \qquad m_B = 12 \text{ u}, \qquad \text{and} \qquad v = 2.6 \times 10^7 \text{ m/s}.$$

We'll let you do the arithmetic; the results are

$$v_A = -2.2 \times 10^7 \text{ m/s}, \qquad v_B = 0.4 \times 10^7 \text{ m/s}.$$

The neutron ends up with $\frac{11}{13}$ of its initial speed, and the speed of the recoiling carbon nucleus is $\frac{2}{13}$ of the neutron's initial speed. Kinetic energy is proportional to speed squared, so the neutron's final kinetic energy is $(\frac{11}{13})^2$, or about 0.72 of its original value. If it makes a second such collision, its kinetic energy is $(0.72)^2$, or about half its original value, and so on. After several collisions the neutron will be moving quite slowly.

EXAMPLE 8–12

The gravitational slingshot effect Figure 8–18 shows the planet Saturn moving in the negative x-direction at its orbital speed (with respect to the sun) of 9.6 km/s. The mass of Saturn is 5.69×10^{26} kg. A spacecraft with mass 825 kg approaches Saturn, moving initially in the $+x$-direction at 10.4 km/s. The gravitational attraction of Saturn (a conservative force) causes the spacecraft to swing around it and head off in the opposite direction. Find the final speed of the spacecraft after it is far enough away to be nearly free of Saturn's gravitational pull.

SOLUTION Here the "collision" is not an impact but a gravitational interaction. Let the spacecraft be body A, and let Saturn be body B. We can assume that Saturn's speed is essentially constant during the interaction because its mass is so much greater than that of the spacecraft. So we have a one-dimensional elastic collision in which $v_{B1} = v_{B2} = -9.6$ km/s and $v_{A1} = 10.4$ km/s. The relative velocity of the two bodies has the same magnitude before and after the interaction, as shown by Eq. (8–27). Being very careful with signs, we find

$$v_{A2} = v_{B2} + v_{B1} - v_{A1}$$

$$= (-9.6 \text{ km/s}) + (-9.6 \text{ km/s}) - 10.4 \text{ km/s} = -29.6 \text{ km/s}.$$

The spacecraft's final speed (relative to the sun) is nearly three times as great as before the encounter with Saturn. Not

only does the spacecraft get a close look at Saturn, but its kinetic energy increases by a factor of $(29.6/10.4)^2 = 8.1$ in the process. The reason for this seemingly mysterious increase is that Saturn is *moving* in its orbit around the sun. If Saturn were initially at rest, the spacecraft would reverse direction but would undergo hardly any change in speed; a baseball behaves similarly when it bounces off a bat that's held stationary (a "bunt"). But when a baseball collides with a bat that's swinging toward it, the baseball leaves the bat with a much greater speed, perhaps enough to make a home run. In our example, Saturn takes the place of the swinging bat, and the spacecraft gains speed much like a batted baseball.

This is a simplified version of the *gravitational slingshot effect,* used to give an extra boost to spacecraft. (The actual motion of a spacecraft is never purely one-dimensional as we have assumed here. We'll discuss the trajectories of spacecraft and planets in detail in Chapter 12.) The *Voyager 2* spacecraft, launched in 1977, used the slingshot effect in flying past Jupiter, Saturn, and Uranus. Thanks to the kinetic energy gained in this way, *Voyager 2* reached the planet Neptune in 1989; without the slingshot effect, *Voyager 2* would not have arrived at Neptune until 2008.

In this bit of celestial gymnastics, does the spacecraft get something for nothing? Where does its extra energy come from?

8–18 The gravitational slingshot effect: a simplified model of a spacecraft making a non-impact "collision" with the planet Saturn.

EXAMPLE 8-13

A two-dimensional elastic collision Figure 8–19 shows an elastic collision of two pucks on a frictionless air table. Puck A has mass $m_A = 0.500$ kg, and puck B has mass $m_B = 0.300$ kg. Puck A has an initial velocity of 4.00 m/s in the positive x-direction and a final velocity of 2.00 m/s in an unknown direction. Puck B is initially at rest. Find the final speed v_{B2} of puck B and the angles α and β in the figure.

SOLUTION Because the collision is elastic, the initial and final kinetic energies are equal:

$$\frac{1}{2} m_A v_{A1}{}^2 = \frac{1}{2} m_A v_{A2}{}^2 + \frac{1}{2} m_B v_{B2}{}^2,$$

$$v_{B2}{}^2 = \frac{m_A v_{A1}{}^2 - m_A v_{A2}{}^2}{m_B}$$

$$= \frac{(0.500 \text{ kg})(4.00 \text{ m/s})^2 - (0.500 \text{ kg})(2.00 \text{ m/s})^2}{0.300 \text{ kg}},$$

$$v_{B2} = 4.47 \text{ m/s}.$$

Conservation of the x-component of total momentum gives

$$m_A v_{A1x} = m_A v_{A2x} + m_B v_{B2x},$$

$$(0.500 \text{ kg})(4.00 \text{ m/s}) = (0.500 \text{ kg})(2.00 \text{ m/s})(\cos \alpha)$$
$$+ (0.300 \text{ kg})(4.47 \text{ m/s})(\cos \beta),$$

and conservation of the y-component gives

$$0 = m_A v_{A2y} + m_B v_{B2y},$$

$$0 = (0.500 \text{ kg})(2.00 \text{ m/s})(\sin \alpha) - (0.300 \text{ kg})(4.47 \text{ m/s})(\sin \beta).$$

These are two simultaneous equations for α and β. The simplest solution is to eliminate β as follows: We solve the first equation for $\cos \beta$ and the second for $\sin \beta$; we then square each equation and add. Since $\sin^2 \beta + \cos^2 \beta = 1$, this eliminates β and leaves an equation that we can solve for $\cos \alpha$ and hence for α. We can then substitute this value back into either of the two equations and solve the result for β. We leave the details for you to work out as a problem; the results are

$$\alpha = 36.9°, \qquad \beta = 26.6°.$$

(a)

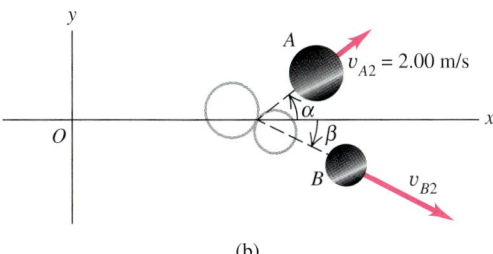

(b)

8–19 (a) Before and (b) after an elastic collision that isn't head-on.

The examples in this and the preceding sections show that we can classify collisions according to energy considerations. A collision in which kinetic energy is conserved is called *elastic*. A collision in which the total kinetic energy decreases is called *inelastic*. When the two bodies have a common final velocity, we say that the collision is *completely inelastic*. There are also cases in which the final kinetic energy is *greater* than the initial value. Rifle recoil, discussed in Example 8–4 (Section 8–3), is an example.

Finally, we emphasize again that we can sometimes use momentum conservation even when there are external forces acting on the system, if the net external force acting on the colliding bodies is small in comparison to the internal forces during the collision.

8-6 CENTER OF MASS

We can restate the principle of conservation of momentum in a useful way by using the concept of **center of mass.** Suppose we have several particles, with masses m_1, m_2, and so on. Let the coordinates of m_1 be (x_1, y_1), those of m_2 be (x_2, y_2), and so on. We define the center of mass of the system as the point having coordinates (x_{cm}, y_{cm}) given by

$$x_{cm} = \frac{m_1x_1 + m_2x_2 + m_3x_3 + \cdots}{m_1 + m_2 + m_3 + \cdots} = \frac{\sum_i m_i x_i}{\sum_i m_i},$$

$$\text{(center of mass)} \qquad (8\text{--}28)$$

$$y_{cm} = \frac{m_1y_1 + m_2y_2 + m_3y_3 + \cdots}{m_1 + m_2 + m_3 + \cdots} = \frac{\sum_i m_i y_i}{\sum_i m_i}.$$

The position vector $\vec{r}_{cm}$ of the center of mass can be expressed in terms of the position vectors $\vec{r}_1, \vec{r}_2, \ldots$ of the particles as

$$\vec{r}_{cm} = \frac{m_1\vec{r}_1 + m_2\vec{r}_2 + m_3\vec{r}_3 + \cdots}{m_1 + m_2 + m_3 + \cdots} = \frac{\sum_i m_i \vec{r}_i}{\sum_i m_i} \qquad \text{(center of mass).} \qquad (8\text{--}29)$$

In statistical language the center of mass is a *mass-weighted average* position of the particles.

EXAMPLE 8–14

Center of mass of the water molecule Figure 8–20 shows a simple model of the structure of a water molecule. The separation between atoms is $d = 9.57 \times 10^{-11}$ m. Each hydrogen atom has mass 1.0 u, and the oxygen atom has mass 16.0 u. We represent each atom as a point because nearly all the mass of each atom is concentrated in its nucleus, which is only about 10^{-5} times the overall radius of the atom. Using the coordinate system shown, find the position of the center of mass.

SOLUTION The x-coordinate of each hydrogen atom is $d \cos(105°/2)$; the y-coordinates of the upper and lower hydrogen atoms are $+d \sin(105°/2)$ and $-d \sin(105°/2)$, respectively. The coordinates of the oxygen atom are $x = 0$, $y = 0$. From Eqs. (8–28) the x-coordinate of the center of mass is

$$x_{cm} = \frac{(1.0\ \text{u})(d\cos 52.5°) + (1.0\ \text{u})(d\cos 52.5°) + (16.0\ \text{u})(0)}{1.0\ \text{u} + 1.0\ \text{u} + 16.0\ \text{u}}$$

$$= 0.068d,$$

and the y-coordinate is

$$y_{cm} = \frac{(1.0\ \text{u})(d\sin 52.5°) + (1.0\ \text{u})(-d\sin 52.5°) + (16.0\ \text{u})(0)}{1.0\ \text{u} + 1.0\ \text{u} + 16.0\ \text{u}}$$

$$= 0.$$

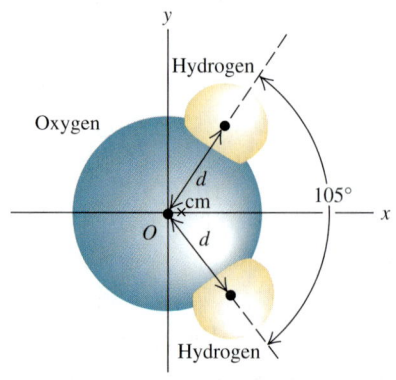

8–20 Where is the center of mass of a water molecule?

The actual value of d for the water molecule is 9.57×10^{-11} m, so we find

$$x_{cm} = (0.068)(9.57 \times 10^{-11} \text{ m}) = 6.5 \times 10^{-12} \text{ m}.$$

The center of mass is much closer to the oxygen atom than to either hydrogen atom because the oxygen atom is much more massive. Notice that the center of mass lies along the x-axis, which is the *axis of symmetry* of this molecule. If the molecule is rotated by 180° around this axis, it looks exactly the same as before. The position of the center of mass can't be affected by this rotation, so it must lie on the axis of symmetry.

For solid bodies, in which we have (at least on a macroscopic level) a continuous distribution of matter, the sums in Eqs. (8–28) have to be replaced by integrals. The calculations can get quite involved, but we can say some general things about such problems. First, whenever a homogeneous body has a geometric center, such as a billiard ball, a sugar cube, or a can of frozen orange juice, the center of mass is at the geometric center. Second, whenever a body has an axis of symmetry, such as a wheel or a pulley, the center of mass always lies on that axis. Third, there is no law that says the center of mass has to be within the body. For example, the center of mass of a doughnut is right in the middle of the hole. You can probably think of other examples.

We'll talk a little more about locating the center of mass in Chapter 11 in connection with the related concept *center of gravity.*

MOTION OF THE CENTER OF MASS

To see the significance of the center of mass of a collection of particles, we must ask what happens to the center of mass when the particles move. The x- and y-components of velocity of the center of mass, $v_{cm\text{-}x}$ and $v_{cm\text{-}y}$, are the time derivatives of x_{cm} and y_{cm}. Also, dx_1/dt is the x-component of velocity of particle 1, and so on, so $dx_1/dt = v_{1x}$, and so on. Taking time derivatives of Eqs. (8–28), we get

$$v_{cm\text{-}x} = \frac{m_1 v_{1x} + m_2 v_{2x} + m_3 v_{3x} + \cdots}{m_1 + m_2 + m_3 + \cdots},$$

$$v_{cm\text{-}y} = \frac{m_1 v_{1y} + m_2 v_{2y} + m_3 v_{3y} + \cdots}{m_1 + m_2 + m_3 + \cdots}. \tag{8–30}$$

These equations are equivalent to the single vector equation obtained by taking the time derivative of Eq. (8–29):

$$\vec{v}_{cm} = \frac{m_1 \vec{v}_1 + m_2 \vec{v}_2 + m_3 \vec{v}_3 + \cdots}{m_1 + m_2 + m_3 + \cdots} \tag{8–31}$$

We denote the *total* mass $m_1 + m_2 + \ldots$ by M. We can then rewrite Eq. (8–31) as

$$M\vec{v}_{cm} = m_1 \vec{v}_1 + m_2 \vec{v}_2 + m_3 \vec{v}_3 + \cdots = \vec{P}. \tag{8–32}$$

The right side is simply the total momentum $\vec{P}$ of the system. Thus we have proved that the *total momentum is equal to the total mass times the velocity of the center of mass.* When you catch a baseball, you are really catching a collection of a very large number of molecules of masses $m_1, m_2, m_3, \ldots$. The impulse you feel is due to the total momentum of this entire collection. But this impulse is the same as if you were catching a single particle of mass $M = m_1 + m_2 + m_3 + \cdots$ moving with velocity $\vec{v}_{cm}$, the velocity of the collection's center of mass. So Eq. (8–32) helps to justify representing an extended body as a particle.

For a system of particles on which the net external force is zero, so that the total momentum $\vec{P}$ is constant, the velocity of the center of mass $\vec{v}_{cm} = \vec{P}/M$ is also constant. Suppose we mark the center of mass of an adjustable wrench, which lies at a point partway down the handle. We then slide the wrench with a spinning motion across a smooth, horizontal table top (Fig. 8–21). The overall motion appears complicated, but the center of mass follows a straight line, as though all the mass were concentrated at that point.

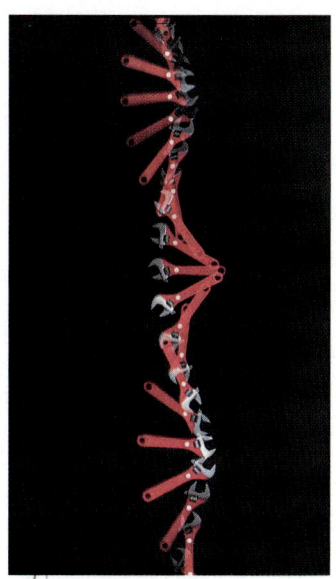

8–21 The center of mass of this wrench is marked with a white dot. The total external force acting on the wrench is almost zero. As it spins on a smooth horizontal surface, the center of mass moves in a straight line with nearly constant velocity.

EXAMPLE 8-15

A tug-of-war on the ice James and Ramon are standing 20.0 m apart on the slippery surface of a frozen pond. Ramon has mass 60.0 kg, and James has mass 90.0 kg. Midway between the two men a mug of their favorite beverage sits on the ice (Fig. 8–22). They pull on the ends of a light rope that is stretched between them. When James has moved 6.0 m toward the mug, how far and in what direction has Ramon moved?

SOLUTION The frozen surface is horizontal and essentially frictionless, so the net external force on the system of James, Ramon, and the rope is zero. Hence their total momentum is conserved. Initially, there is no motion, so the total momentum is zero; thus the velocity of the center of mass is zero, and the center of mass remains at rest. Let's take the origin at the position of the mug, and let the +x-axis extend from the mug toward Ramon. Since the rope is light, we can neglect its mass in calculating the position of the center of mass. The initial x-coordinates of James and Ramon are −10.0 m and +10.0 m, respectively, so the x-coordinate of the center of mass is

$$x_{cm} = \frac{(90.0 \text{ kg})(-10.0 \text{ m}) + (60.0 \text{ kg})(10.0 \text{ m})}{90.0 \text{ kg} + 60.0 \text{ kg}} = -2.0 \text{ m}.$$

When James moves 6.0 m toward the mug, his new x-coordinate is −4.0 m; we'll call Ramon's new x-coordinate x_2. The center of mass doesn't move, so

$$x_{cm} = \frac{(90.0 \text{ kg})(-4.0 \text{ m}) + (60.0 \text{ kg})x_2}{90.0 \text{ kg} + 60.0 \text{ kg}} = -2.0 \text{ m},$$

$$x_2 = 1.0 \text{ m}.$$

Ramon 60.0 kg

10.0 m

James
90.0 kg

10.0 m

8–22 No matter how hard James or Ramon pulls, their center of mass will not move.

James has moved 6.0 m and is still 4.0 m from the mug, but Ramon has moved 9.0 m and is only 1.0 m from it. The ratio of how far each man moved, (6.0 m)/(9.0 m) = $\frac{2}{3}$, equals the inverse ratio of their masses. Can you see why? If the two men keep moving (and if the surface is frictionless, they will!), Ramon will reach the mug first. This result is completely independent of how hard either person pulls; pulling harder just helps Ramon quench his thirst sooner.

EXTERNAL FORCES AND CENTER-OF-MASS MOTION

If the net external force on a system of particles is not zero, then total momentum is not conserved and the velocity of the center of mass will change. Let's look at the relation between the motion of the center of mass and the forces acting on the system.

Equations (8–31) and (8–32) give the *velocity* of the center of mass in terms of the velocities of the individual particles. Proceeding one additional step, we take the time derivatives of these equations to show that the accelerations are related in the same way. Let $\vec{a}_{cm} = d\vec{v}_{cm}/dt$ be the acceleration of the center of mass; then we find

$$M\vec{a}_{cm} = m_1\vec{a}_1 + m_2\vec{a}_2 + m_3\vec{a}_3 + \cdots \tag{8–33}$$

Now $m_1\vec{a}_1$ is equal to the vector sum of forces on the first particle, and so on, so the right side of Eq. (8–33) is equal to the vector sum $\Sigma \vec{F}$ of *all* the forces on *all* the particles. Just as we did in Section 8–3, we may classify each force as *internal* or *external*. The sum of all forces on all the particles is then

$$\Sigma \vec{F} = \Sigma \vec{F}_{ext} + \Sigma \vec{F}_{int} = M\vec{a}_{cm}.$$

Because of Newton's third law, the internal forces all cancel in pairs, and $\Sigma \vec{F}_{int} = 0$. What survives on the left side is only the sum of the *external* forces, and we have

$$\Sigma \vec{F}_{ext} = M\vec{a}_{cm} \qquad \text{(body or collection of particles).} \tag{8–34}$$

When a body or a collection of particles is acted on by external forces, the center of

(a)

(b)

8–23 (a) A shell explodes into two parts in flight. If air resistance is ignored, the fragments follow individual parabolic paths, but the center of mass continues on the same trajectory as the shell's path before exploding—until a fragment hits the ground. (b) The same effect occurs with exploding fireworks.

mass moves just as though all the mass were concentrated at that point and it were acted on by a net force equal to the sum of the external forces on the system.

This result may not sound very impressive, but in fact it is central to the whole subject of mechanics. In fact, we've been using this result all along; without it, we would not be able to represent an extended body as a point particle when we apply Newton's laws. It explains why only *external* forces can affect the motion of an extended body. If you pull upward on your belt, your belt exerts an equal downward force on your hands; these are *internal* forces that cancel and have no effect on the overall motion of your body.

Suppose a cannon shell traveling in a parabolic trajectory (neglecting air resistance) explodes in flight, splitting into two fragments with equal mass (Fig. 8–23a). The fragments follow new parabolic paths, but the center of mass continues on the original parabolic trajectory, just as though all the mass were still concentrated at that point. A skyrocket exploding in air (Fig. 8–23b) is a spectacular example of this effect.

This property of the center of mass is important when we analyze the motion of rigid bodies. We describe the motion of an extended body as a combination of translational motion of the center of mass and rotational motion about an axis through the center of mass. We will return to this topic in Chapter 10. This property also plays an important role in the motion of astronomical objects. It's not correct to say that the moon orbits the earth; rather, the earth and moon both move in orbits around their center of mass.

There's one more useful way to describe the motion of a system of particles. Using $\vec{a}_{cm} = d\vec{v}_{cm}/dt$, we can rewrite Eq. (8–33) as

$$M\vec{a}_{cm} = M\frac{d\vec{v}_{cm}}{dt} = \frac{d(M\vec{v}_{cm})}{dt} = \frac{d\vec{P}}{dt}. \tag{8–35}$$

The total system mass M is constant, so we're allowed to take it inside the derivative. Substituting Eq. (8–35) into Eq. (8–34), we find

$$\Sigma\vec{F}_{ext} = \frac{d\vec{P}}{dt} \qquad \text{(extended body or system of particles).} \tag{8–36}$$

This equation looks like Eq. (8–3). The difference is that Eq. (8–36) describes a *system* of particles, such as an extended body, while Eq. (8–3) describes a single particle. The interactions between the particles that make up the system can change the individual momenta of the particles, but the *total* momentum $\vec{P}$ of the system can be changed only by external forces acting from outside the system.

Finally, we note that if the net external force is zero, Eq. (8–34) shows that the acceleration $\vec{a}_{cm}$ of the center of mass is zero. So the center-of-mass velocity $\vec{v}_{cm}$ is constant, as for the wrench in Fig. 8–21. From Eq. (8–36) the total momentum $\vec{P}$ is also constant. This reaffirms our statement in Section 8–3 of the principle of conservation of momentum.

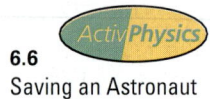

6.6
Saving an Astronaut

*8-7 ROCKET PROPULSION

Momentum considerations are particularly useful for analyzing a system in which the masses of parts of the system change with time. In such cases we can't use Newton's second law $\Sigma \vec{F} = m\vec{a}$ directly because m changes. Rocket propulsion offers a typical and interesting example of this kind of analysis. A rocket is propelled forward by rearward ejection of burned fuel that initially was in the rocket. The forward force on the rocket is the reaction to the backward force on the ejected material. The total mass of the system is constant, but the mass of the rocket itself decreases as material is ejected.

As a simple example, we consider a rocket fired in outer space, where there is no gravitational force and no air resistance. We choose our x-axis to be along the rocket's direction of motion. Figure 8–24a shows the rocket at a time t, when its mass is m and the x-component of its velocity relative to our coordinate system is v. The x-component of total momentum at this instant is $P_1 = mv$. In a short time interval dt the mass of the rocket changes by an amount dm. This is an inherently negative quantity because the rocket's mass m *decreases* with time. During dt a *positive* mass $-dm$ of burned fuel is ejected from the rocket. Let v_{ex} be the exhaust *speed* of this material *relative to the rocket;* the burned fuel is ejected opposite the direction of motion, so its x-component of *velocity* relative to the rocket is $-v_{ex}$. The x-component of velocity v_{fuel} of the burned fuel relative to our coordinate system is then

$$v_{fuel} = v + (-v_{ex}) = v - v_{ex},$$

and the x-component of momentum of the ejected mass $(-dm)$ is

$$(-dm)v_{fuel} = (-dm)(v - v_{ex}).$$

As shown in Fig. 8–24b, at the end of the time interval dt, the x-component of velocity of the rocket and unburned fuel has increased to $v + dv$, and its mass has decreased to $m + dm$ (remember that dm is negative). The rocket's momentum at this time is

$$(m + dm)(v + dv).$$

8–24 (a) A rocket moving in gravity-free outer space with mass m and x-component of velocity v at time t. (b) At time $t + dt$ the rocket's mass has decreased, so dm is negative. The mass of the rocket (including the unburned fuel) is $m + dm$, and its velocity component is $v + dv$. The mass of burned fuel in the exhaust is $-dm$, and its velocity relative to the coordinate system shown is $v_{fuel} = v - v_{ex}$.

Thus the *total* x-component of momentum P_2 of rocket plus ejected fuel at time $t + dt$ is

$$P_2 = (m + dm)(v + dv) + (-dm)(v - v_{ex}).$$

According to our initial assumption, the rocket and fuel are an isolated system. Thus momentum is conserved, and the total x-component of momentum of the system must be the same at time t and at time $t + dt$: $P_1 = P_2$. Hence

$$mv = (m + dm)(v + dv) + (-dm)(v - v_{ex}).$$

(a)   (b)

This can be simplified to

$$m\,dv = -dm\,v_{ex} - dm\,dv.$$

We can neglect the term $(-dm\,dv)$ because it is a product of two small quantities and thus is much smaller than the other terms. Dropping this term, dividing by dt, and rearranging, we find

$$m\frac{dv}{dt} = -v_{ex}\frac{dm}{dt}. \qquad (8\text{–}37)$$

Now dv/dt is the acceleration of the rocket, so the left side of this equation (mass times acceleration) equals the net force F, or *thrust*, on the rocket,

$$F = -v_{ex}\frac{dm}{dt}. \qquad (8\text{–}38)$$

The thrust is proportional both to the relative speed v_{ex} of the ejected fuel and to the mass of fuel ejected per unit time, $-dm/dt$. (Remember that dm/dt is negative because it is the rate of change of the rocket's mass.)

The x-component of acceleration of the rocket is

$$a = \frac{dv}{dt} = -\frac{v_{ex}}{m}\frac{dm}{dt}. \qquad (8\text{–}39)$$

This is positive because v_{ex} is positive (remember, it's the exhaust *speed*) and dm/dt is negative. The rocket's mass m decreases continuously while the fuel is being consumed. If v_{ex} and dm/dt are constant, the acceleration increases until all the fuel is gone.

An effective rocket burns fuel at a rapid rate (large $-dm/dt$) and ejects the burned fuel at a high relative speed (large v_{ex}), as in Fig. 8–25. In the early days of rocket propulsion, people who didn't understand conservation of momentum thought that a rocket couldn't function in outer space because "it doesn't have anything to push against." On the contrary, rockets work *best* in outer space, where there is no air resistance! The rocket in Fig. 8–25 is *not* "pushing against the ground" to get into the air.

8–25 Launch of a space shuttle, a dramatic example of rocket propulsion. In order to provide enough upward thrust to overcome gravity, space launch vehicles can consume over 10,000 kg/s of fuel and eject the burned fuel at speeds of over 4000 m/s.

If the exhaust speed v_{ex} is constant, we can integrate Eq. (8–39) to find a relation between the velocity v at any time and the remaining mass m. At time $t = 0$, let the mass be m_0 and the velocity v_0. Then we rewrite Eq. (8–39) as

$$dv = -v_{ex} \frac{dm}{m}.$$

We change the integration variables to v' and m', so we can use v and m as the upper limits (the final speed and mass). Then we integrate both sides, using limits v_0 to v and m_0 to m, and take the constant v_{ex} outside the integral:

$$\int_{v_0}^{v} dv' = -\int_{m_0}^{m} v_{ex} \frac{dm'}{m'} = -v_{ex} \int_{m_0}^{m} \frac{dm'}{m'},$$

$$v - v_0 = -v_{ex} \ln \frac{m}{m_0} = v_{ex} \ln \frac{m_0}{m}. \tag{8–40}$$

The ratio m_0/m is the original mass divided by the mass after the fuel has been exhausted. In practical spacecraft this ratio is made as large as possible to maximize the speed gain, which means that the initial mass of the rocket is almost all fuel. The final velocity of the rocket will be greater in magnitude (and is often *much* greater) than the relative speed v_{ex} if $\ln (m_0/m) > 1$, that is, if $m_0/m > e = 2.71828 \ldots$.

We've assumed throughout this analysis that the rocket is in gravity-free outer space. However, gravity must be taken into account when a rocket is launched from the surface of a planet, as in Fig. 8–25 (see Problem 8–90).

EXAMPLE 8-16

Acceleration of a rocket A rocket is in outer space, far from any planet, when the rocket engine is turned on. In the first second of firing, the rocket ejects $\frac{1}{120}$ of its mass with a relative speed of 2400 m/s. What is the rocket's initial acceleration?

SOLUTION We are given that the initial rate of change of mass is

$$\frac{dm}{dt} = -\frac{m_0/120}{1 \text{ s}} = -\frac{m_0}{120 \text{ s}},$$

where m_0 is the initial ($t = 0$) mass of the rocket. From Eq. (8–39) the initial acceleration is

$$a = -\frac{v_{ex}}{m_0} \frac{dm}{dt} = -\frac{2400 \text{ m/s}}{m_0} \left(-\frac{m_0}{120 \text{ s}}\right) = 20 \text{ m/s}^2.$$

Note that the answer didn't depend on the value of m_0. If v_{ex} is the same, the initial acceleration is the same for a 120,000-kg spacecraft that ejects 1000 kg/s as for a 60-kg astronaut equipped with a small rocket that ejects 0.5 kg/s.

EXAMPLE 8-17

Speed of a rocket Suppose that $\frac{3}{4}$ of the initial mass m_0 of the rocket in Example 8–16 is fuel, so the final mass is $m = m_0/4$, and that the fuel is completely consumed at a constant rate in a total time $t = 90$ s. If the rocket starts from rest in our coordinate system, find its speed at the end of this time.

SOLUTION We have $m_0/m = 4$, so from Eq. (8–40),

$$v = v_0 + v_{ex} \ln \frac{m_0}{m} = 0 + (2400 \text{ m/s})(\ln 4) = 3327 \text{ m/s}.$$

At the start of the flight, when the velocity of the rocket is zero, the ejected fuel is moving to the left, relative to our coordi-

nate system, at 2400 m/s. At the end of the first second ($t = 1$ s) the rocket is moving at 20 m/s, and the fuel's speed relative to our system is 2380 m/s. During the next second the acceleration, given by Eq. (8–39), is a little greater. At $t = 2$ s the rocket is moving at a little over 40 m/s, and the fuel's speed is a little less than 2360 m/s. Detailed calculation shows that at about $t = 75.6$ s, the rocket's velocity v in our coordinate system equals 2400 m/s. The burned fuel ejected after this time moves *forward*, not backward, in our system. Since the final velocity of the rocket is 3327 m/s and the relative velocity is 2400 m/s, the last portion of the ejected fuel has a forward velocity (relative to our frame) of $(3327 - 2400)$ m/s = 927 m/s. (We are using more figures than are significant to illustrate our point.)

8–8 The Neutrino

A Case Study in Modern Physics

The laws of conservation of energy and momentum, vitally important in all areas of physical science, rest on a very solid foundation of experimental evidence. Nevertheless, the decay of radioactive nuclei by a process called *beta decay* led the great physicist Niels Bohr to suggest seriously in 1930 that these laws might not be obeyed in nuclear processes. Most physicists disagreed; they believed so strongly in these conservation laws that they accepted instead an alternative hypothesis proposed by Wolfgang Pauli that an unseen "ghost particle" is emitted in beta decay. Not until 25 years later was this particle, the *neutrino,* actually observed experimentally.

To see how this challenge to the conservation laws arose, let's consider an isolated system consisting of two particles, with masses m_1 and m_2, initially at rest. The system might be the bullet and rifle of Example 8–4 (Section 8–3), or two balls tied together with a compressed spring between them, or an unstable nucleus that comes apart into two fragments. When the two bodies fly apart, with speeds v_1 and v_2, some total quantity of energy Q is divided between them:

$$Q = K_1 + K_2 = \frac{1}{2} m_1 v_1^2 + \frac{1}{2} m_2 v_2^2. \qquad (8\text{–}41)$$

In Example 8–4, the value of Q was 225 J. We noted in that example that the ratio of the kinetic energies of the two particles equals the inverse ratio of their masses. With a 600:1 mass ratio the recoiling rifle and the bullet get kinetic energies of $\frac{1}{601} Q$ and $\frac{600}{601} Q$, respectively. The energy always divides this way, with the lighter particle getting more kinetic energy than the heavier one.

To prove this, we note that the system is isolated, so the total momentum is zero both before and after it flies apart. Thus the final momenta of the two particles must be equal and opposite:

$$m_1 v_1 = -m_2 v_2.$$

When we square this equation and divide by 2, we get

$$\frac{1}{2} m_1^2 v_1^2 = \frac{1}{2} m_2^2 v_2^2,$$

$$m_1 K_1 = m_2 K_2, \qquad \text{or} \qquad \frac{K_2}{K_1} = \frac{m_1}{m_2}.$$

When we combine this with Eq. (8–41) to get expressions for K_1 and K_2, we find

$$K_1 = \frac{m_2}{m_1 + m_2} Q \qquad \text{and} \qquad K_2 = \frac{m_1}{m_1 + m_2} Q. \qquad (8\text{–}42)$$

The essential point is that when there are only two final particles, the total available energy Q *must* divide this way every time. If there are three or more final particles, the energy can be divided up in many different ways, depending on the directions in which the particles move.

EXAMPLE 8–18

Energy in radioactive decay The radioactive isotope ^{222}Rn, the inert gas radon, is naturally present in the air. It undergoes a process called *alpha decay,* the end products of which are a ^{218}Po nucleus (mass $= 3.62 \times 10^{-25}$ kg) and an alpha particle (mass $= 6.65 \times 10^{-27}$ kg). (This radiation, along with that pro-

duced by further decay events, is responsible for the health hazards of radon.) The quantity of energy released in the decay of ^{222}Rn is $Q = 9.0 \times 10^{-13}$ J. What are the kinetic energies of the emitted alpha particle and the recoiling ^{218}Po nucleus?

SOLUTION Let m_1 be the alpha particle, and let m_2 be the ^{218}Po nucleus. We can use Eqs. (8–42) with $m_1 = 6.65 \times 10^{-27}$ kg and $m_2 = 3.62 \times 10^{-25}$ kg. The energy of the alpha particle is

$$K_1 = \frac{m_2}{m_1 + m_2} Q = \frac{3.62 \times 10^{-25} \text{ kg}}{3.69 \times 10^{-25} \text{ kg}} (9.0 \times 10^{-13} \text{ J})$$

$$= 8.8 \times 10^{-13} \text{ J}.$$

The amount left for the ^{218}Po nucleus is

$$K_2 = Q - K_1 = 2 \times 10^{-14} \text{ J}.$$

About 98% of the energy released is given to the alpha particle, leaving only 2% for kinetic energy of the ^{218}Po nucleus.

Experimental measurements of ^{222}Rn decay give data on the number of alpha particles emitted with various kinetic energies (Fig. 8–26). The graph shows that *every* alpha particle emitted from *every* ^{222}Rn nucleus at rest has the same kinetic energy, within experimental error. (The alpha particles are said to be *monoenergetic*.) If there are only two final particles, then this is to be expected. If all ^{222}Rn nuclei are identical, then Q is the same for all, and K_1 must be the same. Hence these measurements confirm the two-particle picture of the decay process.

Another type of radioactive decay process, called beta decay, involves an electron or its antiparticle, the positron. Electrons are sometimes called beta-minus (β^-) particles, and positrons are called beta-plus (β^+) particles. An example of β^- decay occurs in the unstable nucleus ^{210}Bi, which decays into an electron and a ^{210}Po nucleus. Suppose this is a two-particle decay, like alpha emission. The rest mass of the ^{210}Po nucleus is 383,000 times the rest mass of the electron; a calculation similar to the one in Example 8–18 predicts that virtually all the released energy $Q = 1.86 \times 10^{-13}$ J should go to the electron and that the electrons from all decays should be monoenergetic. (The calculation requires relativistic expressions for kinetic energy and momentum, developed in Chapter 39, because the electrons emitted in ^{210}Bi decay can move as fast as 95% of the speed of light.)

Henri Becquerel, who discovered radioactivity in 1896, found indications by 1900 that the electrons emitted in β^- decay were *not* monoenergetic. Measurements made in 1914 by James Chadwick, the discoverer of the neutron, suggested this even more strongly. In the case of ^{210}Bi decay, later experimenters found a whole range of electron energies, with a maximum value of 1.86×10^{-13} J (Fig. 8–27). Since conservation of energy and momentum predicted that all the electrons should have the *same* energy, these results cast doubt on the validity of these conservation laws.

Looking for a way to save the conservation laws, Wolfgang Pauli suggested in 1931 that an unseen, electrically neutral third particle might be emitted during β^- decay. If so, it could carry off some of the energy and momentum. With three particles the energy and momentum can be shared in many different ways that are consistent with the conservation laws. With two particles (Fig. 8–28a) the momenta must be equal and opposite, and this requirement determines the energy of each particle uniquely. But with three particles, zero total momentum is represented by a triangle of momentum vectors (Figs. 8–28b, c, d) that allows various divisions of the momentum and energy, provided only that the kinetic energies add up to Q. So if there is a third particle, the electron can have a whole range of kinetic energies from zero to almost Q, as is observed.

Since Pauli's hypothetical particle was neutral, the Italian physicist Enrico Fermi first wanted to call it a neutron, but Chadwick had already claimed that name for a massive neutral particle. So Fermi christened the particle *neutrino* ("little neutral one"). (In present-day nomenclature the particle emitted in β^- decay is called an *antineutrino*, the antiparticle of the neutrino.) Subsequent measurements of electron energies in β^- decay were consistent with the existence of this unseen neutral particle. Because it interacts with other particles only extremely weakly, it is very difficult to detect. Not until 1956 were

8–26 A graph of the numbers of alpha particles (N) emitted with various energies K in the decay of ^{222}Rn to ^{218}Po. The graph consists of a single sharp peak at $K = 8.8 \times 10^{-13}$ J.

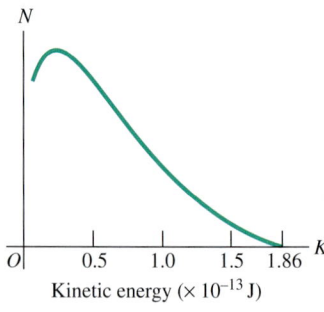

8–27 A graph of the number N of electrons emitted with various energies K in the β^- decay of ^{210}Bi to ^{210}Po. The graph shows a continuous distribution of electron energies, with a maximum value of 1.86×10^{-13} J.

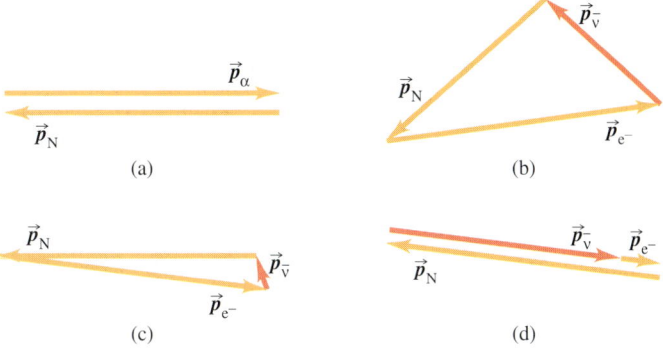

(a)

(b)

(c)

(d)

8–28 (a) Two-particle decay. The momenta are equal in magnitude and opposite in direction. (b) In the case of three particles the electron's kinetic energy will be somewhere between zero and Q. (c) When the momentum and energy of the antineutrino are small, the electron's kinetic energy is nearly Q. (d) When the antineutrino's energy is nearly Q, the electron's kinetic energy is nearly zero.

neutrinos observed directly. But in the intervening years, physicists had such confidence in the conservation laws that hardly anyone doubted that the neutrino would be found.

The neutrino's importance in modern physics goes far beyond helping to confirm conservation laws. Neutrinos have helped us understand what happens in a *supernova,* one of the most dramatic phenomena in nature, in which a star many times more massive than the sun explodes cataclysmically. The most recent nearby supernova, SN 1987A, was observed on February 24, 1987 (Fig. 8–29). At the same time, two laboratories on earth (one in Japan, the other under Lake Erie) detected a total of 19 neutrinos coming from the direction of SN 1987A. Because of the extremely weak interactions of neutrinos, even detecting this many neutrinos required enormous detectors made up of millions of kilograms of purified water. (Collisions of neutrinos with atomic nuclei in the water produce fast-moving secondary particles, which in turn produce brief flashes of light. It is these light flashes that are detected.) Analysis of the data showed that during the first second of the supernova explosion, approximately 10^{58} neutrinos

8–29 Before (left) and after (right) photographs of Supernova 1987A. Less than 0.1% of the energy released in the supernova appeared as visible light; most of the remainder was in the form of neutrinos.

were emitted from the supernova, with a total energy of 10^{46} J—over 100 times as much energy as the sun has emitted in the last five billion years! (About 5×10^{14} supernova neutrinos passed through your body on that Monday in 1987; the chance that even one of them interacted with your body is about 1 in 5,000, and even then there would have been no biological effect on you.) These neutrinos were produced by nuclear reactions deep within the core of the exploding supernova; detecting even 19 supernova neutrinos gives astrophysicists direct information about these reactions.

Spurred on by the observation of SN 1987A, a new generation of "neutrino telescopes" is coming into operation. The largest is DUMAND (Deep Underwater Muon and Neutrino Detector), which will look for light flashes generated by neutrinos passing through the clear water 4800 m below the surface of the Pacific off the island of Hawaii. By scanning 10^8 tons of seawater, DUMAND will look for ultra-high-energy neutrinos that are predicted to be emitted by material falling into supermassive black holes. Neutrino physics continues to be a vital and exciting field of research.

SUMMARY

KEY TERMS

momentum (linear momentum), 227

impulse, 228

impulse-momentum theorem, 229

internal force, 232

external force, 233

isolated system, 233

total momentum, 233

principle of conservation of momentum, 234

elastic collision, 238

inelastic collision, 238

completely inelastic collision, 238

center of mass, 246

- The momentum $\vec{p}$ of a particle with mass m moving with velocity $\vec{v}$ is defined as the vector quantity

$$\vec{p} = m\vec{v}. \tag{8-2}$$

In terms of momentum, Newton's second law for a particle may be expressed as

$$\Sigma \vec{F} = \frac{d\vec{p}}{dt}. \tag{8-3}$$

- The impulse $\vec{J}$ of a constant net force $\Sigma \vec{F}$ acting for a time interval from t_1 to t_2 is the vector quantity

$$\vec{J} = \Sigma \vec{F}(t_2 - t_1) = \Sigma \vec{F} \Delta t. \tag{8-5}$$

If the net force varies with time, the impulse is

$$\vec{J} = \int_{t_1}^{t_2} \Sigma \vec{F} \, dt. \tag{8-7}$$

The change in momentum of a particle in any time interval equals the impulse of the net force that acts on the particle during that interval:

$$\vec{J} = \vec{p}_2 - \vec{p}_1. \tag{8-6}$$

The momentum of a particle equals the impulse that accelerated it from rest to its present speed.

- An internal force is a force exerted by one part of a system on another. An external force is a force exerted on a part of a system by something outside the system. An isolated system is one with no external forces.

- The total momentum of a system of particles A, B, C, ... is the vector sum of the momenta of the individual particles:

$$\vec{P} = \vec{p}_A + \vec{p}_B + \cdots = m_A \vec{v}_A + m_B \vec{v}_B + \cdots. \tag{8-14}$$

- If the net external force on a system is zero, the total momentum of the system is constant (conserved); each component of total momentum is separately conserved.

- Collisions can be classified according to energy relations and final velocities. In an elastic collision between two bodies, the initial and final total kinetic energies are equal and the initial and final relative velocities have the same magnitude. In an inelastic two-body collision the final total kinetic energy is less than the initial total kinetic energy. If the two bodies have the same final velocity, the collision is completely inelastic.

■ The coordinates x_{cm} and y_{cm} of the center of mass of a system of particles are defined as

$$x_{cm} = \frac{m_1 x_1 + m_2 x_2 + m_3 x_3 + \cdots}{m_1 + m_2 + m_3 + \cdots} = \frac{\sum\limits_i m_i x_i}{\sum\limits_i m_i},$$

$$y_{cm} = \frac{m_1 y_1 + m_2 y_2 + m_3 y_3 + \cdots}{m_1 + m_2 + m_3 + \cdots} = \frac{\sum\limits_i m_i y_i}{\sum\limits_i m_i}. \qquad (8\text{--}28)$$

In terms of the position vectors $\vec{r}_1$, $\vec{r}_2$, ... of the particles, the position vector $\vec{r}_{cm}$ of the center of mass is

$$\vec{r}_{cm} = \frac{m_1 \vec{r}_1 + m_2 \vec{r}_2 + m_3 \vec{r}_3 + \cdots}{m_1 + m_2 + m_3 + \cdots} = \frac{\sum\limits_i m_i \vec{r}_i}{\sum\limits_i m_i}. \qquad (8\text{--}29)$$

■ The total momentum $\vec{P}$ of a system equals its total mass M multiplied by the velocity $\vec{v}_{cm}$ of its center of mass:

$$M\vec{v}_{cm} = m_1 \vec{v}_1 + m_2 \vec{v}_2 + m_3 \vec{v}_3 + \cdots = \vec{P}. \qquad (8\text{--}32)$$

■ The center of mass of a system moves as though all the mass M was concentrated at the center of mass. If the net external force on the system is zero, the center of mass velocity $\vec{v}_{cm}$ is constant. If the net external force is not zero, the center of mass accelerates:

$$\sum \vec{F}_{ext} = M\vec{a}_{cm}. \qquad (8\text{--}34)$$

■ In rocket propulsion the mass of a rocket changes as the fuel is burned and exhausted. Analysis of the motion of the rocket must include the momentum carried away by the fuel as well as the momentum of the rocket itself.

DISCUSSION QUESTIONS

Q8–1 In splitting logs with a hammer and wedge, is a heavy hammer more effective than a lighter hammer? Why?

Q8–2 Suppose you catch a baseball and then someone invites you to catch a bowling ball with either the same momentum or with the same kinetic energy as the baseball. Which would you choose? Explain.

Q8–3 When rain falls from the sky, what becomes of its momentum as it hits the ground? Is your answer also valid for Newton's famous apple?

Q8–4 A car has the same kinetic energy when it is traveling south at 30 m/s as when it is traveling northwest at 30 m/s. Is the momentum of the car the same in both cases? Explain.

Q8–5 A truck is accelerating as it speeds down the highway. One inertial frame of reference is attached to the ground with its origin at a fence post. A second frame of reference is attached to a police car that is traveling down the highway at constant velocity. Is the momentum of the truck the same in these two

reference frames? Explain. Is the rate of change of the momentum of the truck the same in these two frames? Explain.

Q8–6 When a large, heavy truck collides with a passenger car, the occupants of the car are more likely to be hurt than the truck driver. Why?

Q8–7 A woman holding a large rock stands on a frictionless horizontal sheet of ice. She throws the rock with speed v_0 at an angle α above the horizontal. Consider the system consisting of the woman plus the rock. Is the momentum of the system conserved? Why or why not? Is any component of the momentum of the system conserved? Again, why or why not?

Q8–8 In Example 8–7 (Section 8–4), where the two gliders in Fig. 8–9a stick together after the collision, the collision is inelastic since $K_2 < K_1$. In Example 8–5 (Section 8–3), is the collision inelastic? Explain.

Q8–9 In a completely inelastic collision between two objects, where the objects stick together after the collision, is it possible for the final kinetic energy of the system to be zero? If so, give an example in which this would occur. If the final kinetic energy is zero, what must be the initial momentum of the system? Is the initial kinetic energy of the system zero? Explain.

Q8–10 Since for a particle the kinetic energy is given by $K = \frac{1}{2}mv^2$ and the momentum by $\vec{p} = m\vec{v}$, it is easy to show that $K = p^2/2m$. How, then, is it possible to have an event during which the total momentum of the system is constant but the total kinetic energy changes?

Q8–11 In each of Examples 8–10, 8–11, 8–12, and 8–13 (Section 8–5), verify that the relative velocity vector of the two bodies has the same magnitude before and after the collision. In each case what happens to the *direction* of the relative velocity vector?

Q8–12 A glass dropped on the floor is more likely to break if the floor is concrete than if it is wood. Why? (Refer to Fig. 8–3.)

Q8–13 In Fig. 8–18 the kinetic energy of the spacecraft is larger after its interaction with Saturn than before. Where does the extra energy come from? Describe the event in terms of conservation of energy.

Q8–14 A machine gun is fired at a steel plate. Is the average force on the plate from the bullet impact greater if the bullets bounce off or if they are squashed and stick to the plate? Explain.

Q8–15 A net force of 4 N acts on an object initially at rest for 0.25 s and gives it a final speed of 5 m/s. How could a net force of 2 N produce the same final speed?

Q8–16 A net force with x-component ΣF_x acts on an object from time t_1 to time t_2. The x-component of the momentum of the object

is the same at t_1 as it is at t_2, but ΣF_x is not zero at all times between t_1 and t_2. What can you say about the graph of ΣF_x versus t?

Q8–17 A tennis player hits a tennis ball with a racket. Consider the system made up of the ball and the racket. Is the total momentum of the system the same just before and just after the hit? Is the total momentum just after the hit the same as two seconds later, when the ball is in midair at the high point of its trajectory? Explain any differences between the two cases.

Q8–18 In Example 8–4 (Section 8–3) consider the system consisting of the rifle plus the bullet. What is the speed of the center of mass of the system after the rifle is fired? Explain.

Q8–19 An egg is released from rest at the roof of a building and falls to the ground. As the egg falls, what happens to the momentum of the system of the egg plus the earth?

Q8–20 A woman stands in the middle of a perfectly smooth, frictionless frozen lake. She can set herself in motion by throwing things, but suppose she has nothing to throw. Can she propel herself to shore *without* throwing anything?

Q8–21 In a zero-gravity environment, can a rocket-propelled spaceship ever attain a speed greater than the relative speed with which the burnt fuel is exhausted?

Q8–22 It is estimated that Supernova 1987A, a distance of 170,000 light years from the earth, emitted 10^{58} neutrinos. But two huge detectors on earth detected only 19 of them. Give at least two reasons why the number detected was so much smaller than the number emitted.

EXERCISES

SECTION **8–2 MOMENTUM AND IMPULSE**

8–1 a) What is the magnitude of the momentum of a 10,000-kg truck whose speed is 12.0 m/s? b) What speed would a 2,000-kg sport utility vehicle have to attain in order to have i) the same momentum? ii) the same kinetic energy?

8–2 In Example 8–1 (Section 8–2), show that the iceboat of mass $2m$ has $\sqrt{2}$ times as much momentum at the finish line as does the iceboat of mass m.

8–3 a) Show that the kinetic energy K and the momentum magnitude p of a particle of mass m are related by $K = p^2/2m$. b) A 0.040-kg cardinal (*Richmondena cardinalis*) and a 0.145-kg baseball have the same kinetic energy. Which has the greater magnitude of momentum? What is the ratio of the cardinal's magnitude of momentum to the baseball's? c) A 700-N man and a 450-N woman have the same momentum. Which has the greater kinetic energy? What is the ratio of the man's kinetic energy to that of the woman?

8–4 A soccer ball of mass 0.420 kg is traveling with a speed of 4.50 m/s at an angle of 20.0° counterclockwise from the $+x$-axis (Fig. 8–30). What are the ball's x- and y-components of momentum?

8–5 A baseball of mass 0.145 kg is moving in the $+y$-direction with a speed of 1.30 m/s, and a tennis ball of mass 0.0570 kg is moving in the $-y$-direction with a speed of 7.80 m/s. What are

FIGURE **8–30** Exercise 8–4.

the magnitude and direction of the total momentum of the system consisting of the two balls?

8–6 A golf ball of mass 0.045 kg is moving in the $+x$-direction with a speed of 9.00 m/s, and a baseball of mass 0.145 kg is moving in the $-y$-direction with a speed of 7.00 m/s. What are the magnitude and direction of the total momentum of the system consisting of the two balls?

8–7 Force of a Golf Swing. A 0.0450-kg golf ball initially at rest is given a speed of 25.0 m/s when it is struck by a club. If the club and ball are in contact for 2.00 ms, what average force acts on the ball? Is the effect of the ball's weight during the time of contact significant? Why or why not?

8–8 Force of a Baseball Swing. A baseball has a mass of 0.145 kg. a) If the velocity of a pitched ball has a magnitude of 45.0 m/s and the batted ball's velocity is 55.0 m/s in the opposite direction, find the magnitude of the change in momentum of the ball and of the impulse applied to it by the bat. b) If the ball remains in contact with the bat for 2.00 ms, find the magnitude of the average force applied by the bat.

8–9 A 0.160-kg hockey puck is moving on an icy, frictionless, horizontal surface. At $t = 0$ the puck is moving to the right at 3.00 m/s. a) Calculate the velocity of the puck (magnitude and direction) after a force of 25.0 N directed to the right has been applied for 0.050 s. b) If instead a force of 12.0 N directed to the left is applied from $t = 0$ to $t = 0.050$ s, what is the final velocity of the puck?

8–10 An engine of the orbital maneuvering system (OMS) on a space shuttle exerts a force of $(26{,}700$ N$)\hat{j}$ for 3.90 s, exhausting a negligible mass of fuel relative to the 95,000-kg mass of the shuttle. a) What is the impulse of the force for this 3.90 s? b) What is the shuttle's change in momentum from this impulse? c) What is the shuttle's change in velocity from this impulse? d) Why can't we find the resulting change in kinetic energy of the shuttle?

8–11 Coach Johnson's bat exerts a horizontal force on a 0.145-kg baseball of $\vec{F} = [(1.60 \times 10^7$ N/s$)t - (6.00 \times 10^9$ N/s$^2)t^2]\hat{i}$ between $t = 0$ and $t = 2.50$ ms. At $t = 0$ the baseball's velocity is $-(40.0\hat{i} + 5.0\hat{j})$ m/s. a) Calculate the impulse exerted by the bat on the ball during the 2.50 ms that they are in contact. b) Calculate the impulse exerted by gravity on the ball during this time interval. c) Calculate the average force exerted by the bat on the ball during this time interval. d) Calculate the momentum and the velocity of the baseball at $t = 2.50$ ms.

8–12 A 0.145-kg baseball is struck by a bat. Just before impact, the ball is traveling horizontally to the right at 50.0 m/s, and it leaves the bat traveling to the left at an angle of 30° above horizontal with a speed of 65.0 m/s. If the ball and bat are in contact for 1.75 ms, find the horizontal and vertical components of the average force on the ball.

8–13 A net force $\Sigma F_x(t) = A + Bt^2$ in the +x-direction is applied to a girl on roller skates. The girl has mass m. The force starts at $t_1 = 0$ and continues until $t = t_2$. a) What is the impulse J_x of the force? b) If the girl is initially at rest, what is her speed at time t_2?

SECTION 8–3 CONSERVATION OF MOMENTUM

8–14 Frustrated by having the goalie block his shots, a 75.0-kg hockey player standing on ice throws a 0.160-kg puck horizontally at the net with a speed of 20.0 m/s. With what speed and in what direction will the hockey player begin to move if there is no friction between his feet and the ice?

8–15 You are standing on a sheet of ice that covers the football stadium parking lot in Buffalo; there is negligible friction between your feet and the ice. A friend throws you a 0.400-kg ball that is traveling horizontally at 10.0 m/s. Your mass is 70.0 kg. a) If you catch the ball, with what speed do you and the ball move afterwards? b) If the ball hits you and bounces off your chest, so afterwards it is moving horizontally at

8.0 m/s in the opposite direction, what is your speed after the collision?

8–16 On a frictionless horizontal air table, puck A (with mass 0.250 kg) is moving toward puck B (with mass 0.350 kg), which is initially at rest. After the collision, puck A has a velocity of 0.120 m/s to the left, and puck B has velocity 0.650 m/s to the right. a) What was the speed of puck A before the collision? b) Calculate the change in the total kinetic energy of the system that occurs during the collision.

8–17 Energy Change During a Hip Check. Ice hockey star Wayne Gretzky is skating at 13.0 m/s toward a defender, who in turn is skating at 5.00 m/s toward Gretzky (Fig. 8–31). Gretzky's weight is 756 N; that of the defender is 900 N. Immediately after the collision Gretzky is moving at 1.50 m/s in his original direction. Neglect external horizontal forces applied by the ice to the skaters during the collision. a) What is the velocity of the defender immediately after the collision? b) Calculate the change in the combined kinetic energy of the two players.

FIGURE 8–31 Exercise 8–17.

8–18 The expanding gases that leave the muzzle of a rifle also contribute to the recoil. A .30 caliber bullet has a mass of 0.00720 kg and a speed of 601 m/s relative to the muzzle when fired from a rifle that has a mass of 2.80 kg. The loosely held rifle recoils at a speed of 1.85 m/s relative to the earth. Find the momentum of the propellant gases in a coordinate system attached to the earth as they leave the muzzle of the rifle.

8–19 Block A in Fig. 8–32 (see next page) has a mass of 1.00 kg, and block B has a mass of 3.00 kg. The blocks are forced together, compressing a spring S between them; then the system is released from rest on a level, frictionless surface. The spring, which has negligible mass, is not fastened to either block and drops to the surface after it has expanded. Block B acquires a speed of 1.20 m/s. a) What is the final speed of block A? b) How much potential energy was stored in the compressed spring?

8–20 One of James Bond's adversaries is standing on a frozen lake; there is no friction between his feet and the ice. He throws his steel-lined hat with a velocity of 22.0 m/s at 36.9° above the horizontal, hoping to hit James. If his mass is 120 kg and that of his hat is 4.50 kg, what is the magnitude of his horizontal recoil velocity?

FIGURE 8–32 Exercise 8–19.

8–21 A ceramic penguin on top of your television set suddenly bursts apart into two pieces. One piece, mass m_A, travels off to the left with speed v_A. The other piece, mass m_B, travels off to the right with speed v_B. a) Use conservation of momentum to solve for v_B in terms of m_A, m_B, and v_A. b) Use the results of part (a) to show that $K_A/K_B = m_B/m_A$, where K_A and K_B are the kinetic energies of the two pieces.

8–22 Two ice skaters, Daniel (mass 65.0 kg) and Rebecca (mass 45.0 kg), are practicing. Daniel stops to tie his shoelace and, while at rest, is struck by Rebecca, who is moving at 13.0 m/s before she collides with him. After the collision Rebecca has a velocity of magnitude 8.00 m/s at an angle of 53.1° from her initial direction. Both skaters move on the frictionless horizontal surface of the rink. a) What are the magnitude and direction of Daniel's velocity after the collision? b) What is the change in total kinetic energy of the two skaters as a result of the collision?

8–23 Ken and Kim are skating together on a rink at 3.00 m/s. Ken keeps asking Kim how much she weighs. Annoyed, Kim pushes away from Ken so that she speeds up to 4.00 m/s and he slows down to 2.25 m/s in the same direction. Friction, in the physics sense, is negligible in this drama. If Ken weighs 700 N, what does Kim weigh?

8–24 An open-topped freight car of mass 24,000 kg is coasting without friction along a level track. It is raining very hard, and the rain is falling vertically downward. The car is originally empty and moving with a speed of 4.00 m/s. What is the speed of the car after it has collected 3000 kg of rainwater?

8–25 A hockey puck B rests on a smooth ice surface and is struck by a second puck A, which was originally traveling at 40.0 m/s and which is deflected 30.0° from its original direction (Fig. 8–33). Puck B acquires a velocity at a 45.0° angle to the original direction of A. The pucks have the same mass. a) Compute the speed of each puck after the collision. b) What fraction of the original kinetic energy of puck A is dissipated during the collision?

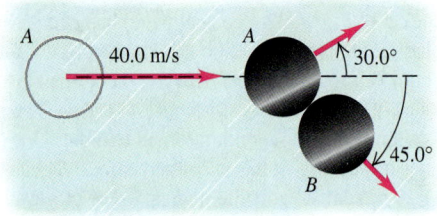

FIGURE 8–33 Exercise 8–25.

SECTION 8–4 INELASTIC COLLISIONS

8–26 On a greasy, essentially frictionless lunch counter, a 0.500-kg submarine sandwich moving 3.00 m/s to the left col-

lides with an 0.250-kg grilled cheese sandwich moving 1.20 m/s to the right. a) If the two sandwiches stick together, what is the final velocity? b) How much mechanical energy is dissipated in the collision?

8–27 Your 1050-kg sports car, parked on a hill without the parking brake being set, has rolled to the bottom of the hill and is moving westbound on a level road at a speed of 15.0 m/s. A truck driver driving east on the same road sees the sports car coming, and decides to stop the car by running his 6320-kg truck into it. The two vehicles remain locked together after the collision. a) If the truck is moving at 10.0 m/s when it hits your sports car head-on, what is the velocity (magnitude and direction) of the two vehicles just after the collision? b) At what speed should the truck be moving so that the truck and car are both stopped in the collision? c) Find the change in kinetic energy of the system of two vehicles for the situation of part (a) and the situation of part (b). For which situation is the change in kinetic energy greater in magnitude?

8–28 On a very muddy football field, a 110-kg linebacker tackles an 85-kg halfback. Immediately before the collision, the linebacker is slipping with a velocity of 8.8 m/s north and the halfback is sliding with a velocity of 7.2 m/s east. What is the velocity (magnitude and direction) at which the two players move together immediately after the collision?

8–29 In Dallas the morning after a winter ice storm, a 1400-kg automobile going westward on Chestnut Street at 35.0 km/h collides with a 2800-kg truck going southward *across* Chestnut Street at 50.0 km/h. If they become coupled on collision, what are the magnitude and direction of their velocity after colliding? Friction forces between the vehicles and the icy road can be neglected.

8–30 At the intersection of Texas Avenue and University Drive, a blue subcompact car of mass 950 kg traveling east on University collides with a maroon pickup truck of mass 1900 kg that is traveling north on Texas and ran a red light (Fig. 8–34). The two vehicles stick together as a result of the collision, and after the collision the wreckage is sliding at 16.0 m/s in the direction 24.0° east of north. Calculate the speed of each vehicle before the collision. The collision occurs during a heavy rainstorm; friction forces between the vehicles and the wet road can be neglected.

FIGURE 8–34 Exercise 8–30.

8-31 A 5.00-g bullet is fired horizontally into a 1.20-kg wooden block resting on a horizontal surface. The coefficient of kinetic friction between block and surface is 0.20. The bullet remains embedded in the block, which is observed to slide 0.230 m along the surface before stopping. What was the initial speed of the bullet?

8-32 A Ballistic Pendulum. A 12.0-g rifle bullet is fired with a speed of 380 m/s into a ballistic pendulum with a mass of 6.00 kg, suspended from a cord 70.0 cm long. (See Example 8-8 in Section 8-4.) Compute a) the vertical height through which the pendulum rises; b) the initial kinetic energy of the bullet; c) the kinetic energy of the bullet and pendulum immediately after the bullet becomes embedded in the pendulum.

SECTION 8-5 **ELASTIC COLLISIONS**

8-33 A 0.150-kg glider is moving to the right on a horizontal, frictionless air track with a speed of 0.80 m/s. It makes a head-on collision with a 0.300-kg glider that is moving to the left with a speed of 2.20 m/s. Find the final velocity (magnitude and direction) of each glider if the collision is elastic.

8-34 A 10.0-g marble slides to the left with a velocity of magnitude 0.400 m/s on the frictionless horizontal surface of an icy New York sidewalk and makes a head-on elastic collision with a larger 30.0-g marble sliding to the right with a velocity of magnitude 0.200 m/s (Fig. 8-35). a) Find the velocity of each marble (magnitude and direction) after the collision. (Since the collision is head-on, all the motion is along a line.) b) Calculate the *change in momentum* (that is, the momentum after the collision minus the momentum before the collision) for each marble. Compare the values you get for each marble. c) Calculate the *change in kinetic energy* (that is, the kinetic energy after the collision minus the kinetic energy before the collision) for each marble. Compare the values you get for each marble.

FIGURE 8-35 Exercise 8-34.

8-35 Supply the details of the calculation of α and β in Example 8-13 (Section 8-5).

8-36 Canadian nuclear reactors use *heavy water* moderators in which elastic collisions occur between the neutrons and deuterons of mass 2.0 u. (See Example 8-11 in Section 8-5.) a) What is the speed of a neutron, expressed as a fraction of its original speed, after a head-on elastic collision with a deuteron which is initially at rest? b) What is its kinetic energy, expressed as a fraction of its original kinetic energy? c) How many such successive collisions will reduce the speed of a neutron to 1/59,000 of its original value?

8-37 You are at the controls of a particle accelerator, sending a beam of 1.50×10^7 m/s protons (mass m) at a gas target of an unknown element. Your detector tells you that some protons bounce straight back after a collision with one of the nuclei of the unknown element. All such protons rebound with a speed of 1.20×10^7 m/s. Assume that the initial speed of the target nucleus is negligible and that the collision is elastic. a) Find the mass of one nucleus of the unknown element. Express your answer in terms of the proton mass m. b) What is the speed of the unknown nucleus immediately after such a collision?

SECTION 8-6 **CENTER OF MASS**

8-38 Three odd-shaped blocks of chocolate have the following masses and center-of-mass coordinates: (1) 0.300 kg, (0.200 m, 0.300 m); (2) 0.400 kg, (0.100 m, −0.400 m); (3) 0.200 kg, (−0.300 m, 0.600 m). Find the coordinates of the center of mass of the system of three chocolate blocks.

8-39 Find the position of the center of mass of the system of the sun and Jupiter. (Since Jupiter is more massive than the rest of the planets put together, this is essentially the position of the center of mass of the solar system.) Does the center of mass lie inside or outside the sun? Use the data in Appendix F.

8-40 A 1200-kg station wagon is moving along a straight highway at 12.0 m/s. Another car, with mass 1800 kg and speed 20.0 m/s, has its center of mass 40.0 m ahead of the center of mass of the station wagon (Fig. 8-36). a) Find the position of the center of mass of the system consisting of the two automobiles. b) Find the magnitude of the total momentum of the system from the above data. c) Find the speed of the center of mass of the system. d) Find the total momentum of the system, using the speed of the center of mass. Compare your result with that of part (b).

FIGURE 8-36 Exercise 8-40.

8-41 At one instant, the center of mass of a system of two particles is located on the x-axis at $x = 2.0$ m and has a velocity of $(5.0 \text{ m/s})\hat{\imath}$. One of the particles is at the origin. The other particle has a mass of 0.10 kg and is at rest on the x-axis at $x = 8.0$ m. a) What is the mass of the particle at the origin? b) Calculate the total momentum of this system. c) What is the velocity of the particle at the origin?

8-42 In Example 8-15 (Section 8-6) Ramon pulls on the rope to give himself a speed of 0.70 m/s. What is the speed of James?

8-43 A system consists of two particles. At $t = 0$ one particle is at the origin; the other, which has a mass of 0.50 kg, is on the y-axis at $y = 6.0$ m. At $t = 0$ the center of mass of the system is on the y-axis at $y = 2.4$ m. The velocity of the center of mass is given by $(0.75 \text{ m/s}^3)t^2\hat{\imath}$. a) Find the total mass of the system. b) Find the acceleration of the center of mass at any time t. c) Find the net external force acting on the system at $t = 3.0$ s.

8-44 A radio-controlled model airplane has a momentum given by $[(-0.75 \text{ kg} \cdot \text{m/s}^3)t^2 + (3.0 \text{ kg} \cdot \text{m/s})]\hat{\imath} + (0.25 \text{ kg} \cdot \text{m/s}^2)t\hat{\jmath}$. a) What are the x-, y-, and z-components of the net force on the airplane? b) At what time t is the x-component of the net force on the airplane equal to zero?

*SECTION 8–7 ROCKET PROPULSION

*8–45 A small rocket burns 0.0500 kg of fuel per second, ejecting it as a gas with a velocity relative to the rocket of magnitude 1600 m/s. a) What is the thrust of the rocket? b) Would the rocket operate in outer space where there is no atmosphere? If so, how would you steer it? Could you brake it?

*8–46 A 70-kg astronaut floating in space in a 110-kg MMU (manned maneuvering unit) experiences an acceleration of 0.029 m/s^2 when he fires one of the MMU's thrusters. a) If the speed of the escaping N_2 gas relative to the astronaut is 490 m/s, how much gas is used by the thruster in 5.0 s? b) What is the thrust of the thruster?

*8–47 A rocket is fired in deep space, where gravity is negligible. If the rocket has an initial mass of 6000 kg and ejects gas at a relative velocity of magnitude 2000 m/s, how much gas must it eject in the first second to have an initial acceleration of 25.0 m/s^2?

*8–48 A rocket is fired in deep space, where gravity is negligible. In the first second it ejects $\frac{1}{160}$ of its mass as exhaust gas and has an acceleration of 15.0 m/s^2. What is the speed of the exhaust gas relative to the rocket?

*8–49 A C 6-5 model rocket engine has an impulse of 10.0 N · s for 1.70 s, while burning 0.0125 kg of propellant. It has a maximum thrust of 13.3 N. The initial mass of the engine plus propellant is 0.0258 kg. a) What fraction of the maximum thrust is the average thrust? b) Calculate the relative speed of the exhaust gases, assuming it is constant. c) Assuming that the relative speed of the exhaust gases is constant, find the final speed of the engine if it was attached to a very light frame and fired from rest in gravity-free outer space.

*8–50 A single-stage rocket is fired from rest from a deep-space platform, where gravity is negligible. If the rocket burns its fuel in a time of 50.0 s and the relative speed of the exhaust gas is $v_{ex} =$ 2100 m/s, what must the mass ratio m_0/m for a final speed v of 8.00 km/s (about equal to the orbital speed of an earth satellite)?

*8–51 Obviously rockets can be made to go very fast, but what is a reasonable top speed? Assume that a rocket is fired from rest at a space station in deep space, where gravity is negligible. a) If the rocket ejects gas at a relative speed of 2000 m/s and you want the rocket's speed eventually to be $1.00 \times 10^{-3}\ c$, where c is the speed of light, what fraction of the initial mass of the rocket and fuel is *not* fuel? b) What is this fraction if the final speed is to be 3000 m/s?

SECTION 8–8 THE NEUTRINO: A CASE STUDY IN MODERN PHYSICS

8–52 A ^{232}Th (thorium) nucleus at rest decays to a ^{228}Ra (radon) nucleus with the emission of an alpha particle. The total kinetic energy of the decay fragments is 6.54×10^{-13} J. An alpha particle has 1.76% of the mass of a ^{228}Ra nucleus. Calculate the kinetic energy of a) the recoiling ^{228}Ra nucleus; b) the alpha particle.

8–53 In a certain alpha decay the kinetic energy of the alpha particle is 9.650×10^{-13} J and the Q value for the decay is 9.850×10^{-13} J. What is the mass of the recoiling nucleus?

8–54 A ^{210}Bi (bismuth) nucleus at rest undergoes beta decay to ^{210}Po (polonium). Suppose that the emitted electron moves to the right with a momentum (calculated using the theory of relativity) of 5.60×10^{-22} kg · m/s. The ^{210}Po nucleus, with a mass of 3.50×10^{-25} kg, recoils to the left at a speed of 1.14×10^3 m/s. Calculate the magnitude and direction of the momentum of the antineutrino that is emitted in the decay. (The ^{210}Po nucleus moves at much less than the speed of light, so the theory of relativity is not needed to calculate its momentum. Instead, the nonrelativistic expression in Eq. (8–2) can be used.)

8–55 A ^{210}Bi (bismuth) nucleus at rest undergoes β^- decay to ^{210}Po (polonium). In a particular decay event the electron and antineutrino are emitted at right angles to one another. The magnitudes of their momenta are 3.60×10^{-22} kg · m/s for the electron and 5.20×10^{-22} kg · m/s for the antineutrino. The ^{210}Po nucleus has a mass of 3.50×10^{-25} kg. Calculate a) the magnitude of the momentum of the recoiling ^{210}Po nucleus; b) the kinetic energy of the recoiling ^{210}Po nucleus.

PROBLEMS

8–56 A steel ball of mass 40.0 g is dropped from a height of 2.00 m onto a horizontal steel slab. The ball rebounds to a height of 1.60 m. a) Calculate the impulse delivered to the ball during impact. b) If the ball is in contact with the slab for 2.00 ms, find the average force on the ball during impact.

8–57 The net force acting on a 2.00-kg discus while it is being thrown is $(\alpha t^2)\hat{\imath} + (\beta + \gamma t)\hat{\jmath}$, where $\alpha = 25.0$ N/s^2, $\beta = 30.0$ N, and $\gamma = 5.0$ N/s. If the discus was originally at rest, what is its velocity after the net force has acted for 0.500 s? Express your answer in terms of the unit vectors $\hat{\imath}$ and $\hat{\jmath}$.

8–58 Just before being struck by a racket, a tennis ball weighing 0.560 N has a velocity of $(20.0$ m/s$)\hat{\imath} - (4.0$ m/s$)\hat{\jmath}$. During the 3.00 ms that the racket and ball are in contact, the net force on the ball is constant and equal to $-(380$ N$)\hat{\imath} + (110$ N$)\hat{\jmath}$. a) What are the x- and y-components of the impulse of the net force applied to the ball? b) What are the x- and y-components of the final velocity of the ball?

8–59 Three coupled railroad cars roll along and couple with a fourth car, which is initially at rest. These four roll along and couple to a fifth car initially at rest. This process continues until the speed of the final collection of railroad cars is one-fifth the speed of the initial three railroad cars. All the cars are identical. Ignoring friction, how many cars are in the final collection?

8–60 A 1500-kg blue convertible is traveling south, and a 2000-kg red sports utility vehicle is traveling west. If the total momentum of the system consisting of the two cars is 8000 kg · m/s directed at 60.0° west of south, what is the speed of each car?

8–61 Three identical pucks on a horizontal air table have repelling magnets. They are held together, then released simultaneously. Each has the same speed at any instant. One puck moves due west. What is the direction of the velocity of each of the other two pucks?

8–62 Spheres *A* (mass 0.020 kg), *B* (mass 0.030 kg), and *C* (mass 0.050 kg), are each approaching the origin as they slide on a frictionless air table (Fig. 8–37). The initial velocities of *A* and *B* are given in the figure. All three spheres arrive at the origin at the same time and stick together. a) What must be the *x*- and *y*-components of the initial velocity of *C* if all three objects are to end up moving at 0.50 m/s in the +*x*-direction after the collision? b) If *C* has the velocity found in part (a), what is the change in the kinetic energy of the system of three spheres as a result of the collision?

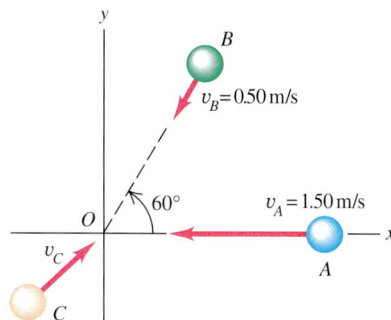

FIGURE 8–37 Problem 8–62.

8–63 A railroad handcar is moving along straight frictionless tracks with negligible air resistance. In the following cases, the car initially has a total mass (car and contents) of 200 kg and is traveling east with a velocity of magnitude 5.00 m/s. Find the *final velocity* of the car in each case, assuming that the handcar does not leave the tracks. a) A 25.0-kg mass is thrown sideways out of the car with a velocity of magnitude 2.00 m/s relative to the car's initial velocity. b) A 25.0-kg mass is thrown backward out of the car with a velocity of 5.00 m/s relative to the initial motion of the car. c) A 25.0-kg mass is thrown into the car with a velocity of 6.00 m/s relative to the ground and opposite in direction to the initial velocity of the car.

8–64 A railroad hopper car filled with sand is rolling with an initial speed of 15.0 m/s on straight horizontal tracks. Ignore frictional forces on the railroad car. The total mass of the car plus sand is 85,000 kg. The hopper door is not fully closed so sand leaks out the bottom. After 20 minutes, 13,000 kg of sand has leaked out. What then is the speed of the railroad car? (Compare your analysis to that used to solve Exercise 8–24.)

8–65 At a classic auto show, a 840-kg 1955 Nash Metropolitan motors by at 9.0 m/s, followed by a 1620-kg 1957 Packard Clipper purring past at 5.0 m/s. a) Which car has the greater kinetic energy? What is the ratio of the kinetic energy of the Nash to that of the Packard? b) Which car has the greater magnitude of momentum? What is the ratio of the magnitude of momentum of the Nash to that of the Packard? c) Let F_N be the net force required to stop the Nash in time *t*, and let F_P be the net force required to stop the Packard in the same time. Which is larger, F_N or F_P? What is the ratio F_N/F_P of these two forces? d) Now let F_N be the net force required to stop the Nash in a distance *d*, and let F_P be the net force required to stop the Packard in the same distance. Which is larger, F_N or F_P? What is the ratio F_N/F_P?

8–66 A soldier on a firing range fires an 8-shot burst from an assault weapon at a full automatic rate of 1000 rounds per minute. Each bullet has a mass of 7.45 g and a speed of 293 m/s relative to the ground as it leaves the barrel of the weapon. Calculate the average recoil force exerted on the weapon during that burst.

8–67 A 0.150-kg frame, when suspended from a coil spring, stretches the spring 0.050 m. A 0.200-kg lump of putty is dropped from rest onto the frame from a height of 30.0 cm (Fig. 8–38). Find the maximum distance the frame moves downward from its initial position.

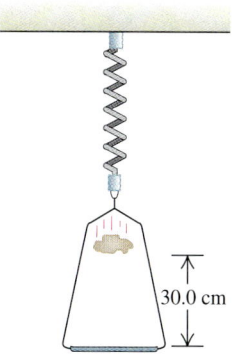

FIGURE 8–38 Problem 8–67.

8–68 A rifle bullet of mass 8.00 g strikes and embeds itself in a block of mass 0.992 kg that rests on a frictionless horizontal surface and is attached to a coil spring (Fig. 8–39). The impact compresses the spring 15.0 cm. Calibration of the spring shows that a force of 0.750 N is required to compress the spring 0.250 cm. a) Find the magnitude of the velocity of the block just after impact. b) What was the initial speed of the bullet?

FIGURE 8–39 Problem 8–68.

8–69 A Ricocheting Bullet. A 0.100-kg stone rests on a frictionless horizontal surface. A bullet of mass 6.00 g, traveling horizontally at 350 m/s, strikes the stone and rebounds horizontally at right angles to its original direction with a speed of 250 m/s. a) Compute the magnitude and direction of the velocity of the stone after it is struck. b) Is the collision perfectly elastic?

8–70 A movie stuntman (mass 80.0 kg) stands on a window ledge 5.0 m above the floor (Fig. 8–40). Grabbing a rope attached to a chandelier, he swings down to grapple with the movie's villain (mass 70.0 kg), who is standing directly under the chandelier. (Assume that the stuntman's center of mass moves downward 5.0 m. He releases the rope just as he reaches the villain.) a) With what speed do the entwined foes start to slide across the floor? b) If the coefficient of kinetic friction of their bodies with the floor is $\mu_k = 0.250$, how far do they slide?

5.0 m

$m = 80.0$ kg

$m = 70.0$ kg

FIGURE 8–40 Problem 8–70.

8–71 A 4.00-g bullet traveling horizontally with a velocity of magnitude 400 m/s is fired into a wooden block with a mass of 0.800 kg, initially at rest on a level surface. The bullet passes through the block and emerges with its speed reduced to 120 m/s. The block slides a distance of 45.0 cm along the surface from its initial position. a) What is the coefficient of kinetic friction between block and surface? b) What is the decrease in kinetic energy of the bullet? c) What is the kinetic energy of the block at the instant after the bullet passes through it?

8–72 A 5.00-g bullet is shot *through* a 1.00-kg wood block suspended on a string 2.000 m long. The center of mass of the block rises a distance of 0.45 cm. Find the speed of the bullet as it emerges from the block if its initial speed is 450 m/s.

8–73 A neutron of mass m makes a head-on elastic collision with a nucleus of mass M, which is initially at rest. a) Show that if the neutron's initial kinetic energy is K_0, the kinetic energy that it loses during the collision is $4mMK_0/(M + m)^2$. b) For what value of M does the incident neutron lose the most energy? c) When M has the value calculated in part (b), what is the speed of the neutron after the collision?

8–74 A blue puck of mass 0.0400 kg, sliding with a velocity of magnitude 0.200 m/s on a frictionless, horizontal air table, makes a perfectly elastic head-on collision with a red puck of mass m, initially at rest. After the collision the velocity of the blue puck is 0.050 m/s in the same direction as its initial velocity. Find a) the velocity (magnitude and direction) of the red puck after the collision; b) the mass m of the red puck.

8–75 Two asteroids with masses m_A and m_B are moving with velocities $\vec{v}_A$ and $\vec{v}_B$ with respect to an astronomer in a space vehicle. a) Show that the total kinetic energy as measured by the astronomer is

$$K = \frac{1}{2} M v_{cm}^2 + \frac{1}{2}(m_A v'^2_A + m_B v'^2_B),$$

with $\vec{v}_{cm}$ and M defined as in Section 8–6, $\vec{v}'_A = \vec{v}_A - \vec{v}_{cm}$ and $\vec{v}'_B = \vec{v}_B - \vec{v}_{cm}$. In this expression the total kinetic energy of the two

asteroids is the energy associated with their center of mass plus that associated with the internal motion relative to the center of mass. b) If the asteroids collide with each other, what is the *minimum* possible kinetic energy they can have after the collision, as measured by the astronomer? Explain.

8–76 Suppose you hold a small ball in contact with and directly over the center of a large ball. If you then drop the small ball a short time after dropping the large ball, the small ball rebounds with surprising speed. To show the extreme case, neglect air resistance and suppose the large ball makes an elastic collision with the floor, then rebounds to make an elastic collision with the still-descending small ball. Just before the collision between the two balls, the large ball is moving upward with velocity $\vec{v}$, and the small ball has velocity $-\vec{v}$. (Do you see why?) Assume the large ball has a much greater mass than the small ball. a) What is the velocity of the small ball immediately after its collision with the large ball? b) From the answer to part (a), what is the ratio of the small ball's rebound distance to the distance it fell before the collision?

8–77 Jack and Jill are standing on a crate at rest on the frictionless horizontal surface of a frozen pond. Jack has mass 75.0 kg, Jill has mass 45.0 kg, and the crate has mass 15.0 kg. They remember that they must fetch a pail of water, so each jumps horizontally from the top of the crate. Just after each jumps, that person is moving away from the crate with a speed of 4.00 m/s relative to the crate. a) What is the final speed of the crate if both Jack and Jill jump simultaneously and in the same direction? (*Hint:* Use an inertial coordinate system attached to the ground.) b) What is the final speed of the crate if Jack jumps first and then a few seconds later Jill jumps in the same direction? c) What is the final speed of the crate if Jill jumps first and then, again in the same direction?

8–78 A proton moving with speed v_{A1} in the +x-direction makes an elastic, off-center collision with an identical proton originally at rest. After impact the first proton moves with speed v_{A2} in the first quadrant at an angle α with the x-axis, and the second moves with speed v_{B2} in the fourth quadrant at an angle β with the x-axis (Fig. 8–10). a) Write the equations expressing conservation of linear momentum in the x- and y-directions. b) Square the equations from part (a) and add them. c) At this point, introduce the fact that the collision is elastic. d) Prove that $\alpha + \beta = \pi/2$. (You have shown that this equation is obeyed in any elastic off-center collision between objects of equal mass when one object is initially at rest.)

8–79 Hockey puck B rests on a smooth ice surface and is struck by a second puck A, which has the same mass. Puck A is initially traveling at 15.0 m/s and is deflected 25.0° from its initial direction. Assume that the collision is perfectly elastic. Find the final speed of each puck and the direction of B's velocity after the collision. (*Hint:* Use the relation derived in part (d) of Problem 8–78.)

8–80 Jonathan and Jane are sitting in a sleigh that is at rest on frictionless ice. Jonathan's weight is 800 N, Jane's weight is 600 N, and that of the sleigh is 1000 N. They see a poisonous spider on the floor of the sleigh and immediately jump off. Jonathan jumps to the left with a velocity (relative to the ice) of 5.00 m/s at 30.0° above the horizontal, and Jane jumps to the right at 7.00 m/s at 36.9° above the horizontal (relative to the ice). Calculate the sleigh's horizontal velocity (magnitude and direction) after they jump out.

8–81 The objects in Fig. 8–41 are constructed of uniform wire bent into the shapes shown. Find the position of the center of mass of each.

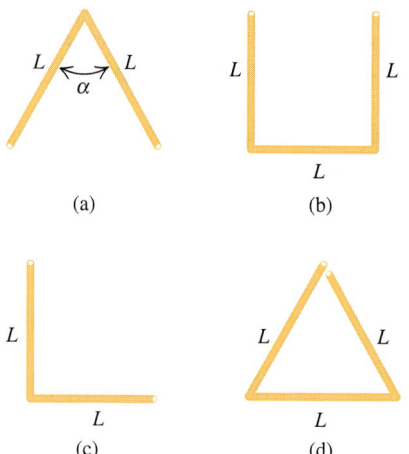

(a)   (b)

(c)   (d)

FIGURE 8–41 Problem 8–81.

8–82 A 45.0-kg woman stands up in a 60.0-kg canoe of length 5.00 m. She walks from a point 1.00 m from one end to a point 1.00 m from the other end (Fig. 8–42). If resistance to motion of the canoe in the water can be neglected, how far does the canoe move during this process?

Start   Finish

1.00 m   3.00 m   1.00 m

FIGURE 8–42 Problem 8–82.

8–83 You are standing on a concrete slab, which in turn is resting on a frozen lake. Assume that there is no friction between the slab and the ice. The slab has a weight five times your weight. If you begin walking forward at 2.00 m/s, relative to the ice, with what speed relative to the ice does the slab move?

8–84 A 20.0-kg projectile is fired at an angle of 60.0° above the horizontal and with a speed of 80.0 m/s. At the highest point of its trajectory the projectile explodes into two fragments with equal mass, one of which falls vertically with zero initial speed. Ignore air resistance. a) How far from the point of firing does the other fragment strike if the terrain is level? b) How much energy is released during the explosion?

8–85 A Nuclear Reaction. Fission, the process that supplies energy in nuclear power plants, occurs when a heavy nucleus is

split into two medium-sized nuclei. One such reaction occurs when a neutron colliding with a ^{235}U (uranium) nucleus splits that nucleus into a ^{141}Ba (barium) nucleus and a ^{92}Kr (krypton) nucleus. In this reaction, two neutrons also are split off from the original ^{235}U. Before the collision we have the arrangement in Fig. 8–43a. After the collision we have the ^{141}Ba nucleus moving in the +z-direction and the ^{92}Kr nucleus in the −z-direction. The three neutrons are moving in the xy-plane as shown in Fig. 8–43b. If the incoming neutron has an initial velocity of magnitude 3.0×10^3 m/s and a final velocity of magnitude 2.0×10^3 m/s in the directions shown, what are the speeds of the other two neutrons, and what can you say about the speeds of the ^{141}Ba and ^{92}Kr nuclei? (The mass of the ^{141}Ba nucleus is approximately 2.3×10^{-25} kg, and that of ^{92}Kr is about 1.5×10^{-25} kg.)

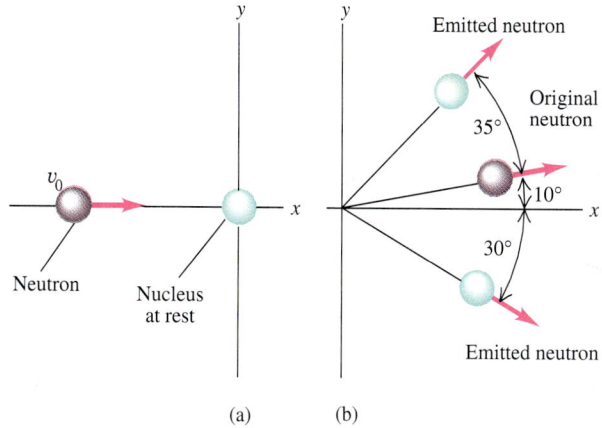

(a)   (b)

FIGURE 8–43 Problem 8–85.

8–86 Center-of-Mass Coordinate System. Puck A (mass m_A) is moving on a frictionless horizontal air table in the +x-direction with velocity $\vec{v}_{A1}$ and makes an elastic head-on collision with puck B (mass m_B) that is initially at rest. After the collision both pucks are moving along the x-axis. a) Calculate the velocity of the center of mass of the two-puck system before the collision. b) Consider a coordinate system whose origin is at the center of mass and moves with it. Is this an inertial reference frame? c) What are the initial velocities $\vec{u}_{A1}$ and $\vec{u}_{B1}$ of the two pucks in this center-of-mass reference frame? What is the total momentum in this frame? d) Use conservation of momentum and energy, applied in the center-of-mass reference frame, to relate the final momentum of each puck to its initial momentum and thus the final velocity of each puck to its initial velocity. Your results should show that a one-dimensional elastic collision has a very simple description in center-of-mass coordinates. e) Let $m_A = 0.400$ kg, $m_B = 0.200$ kg, and $v_{A1} = 6.00$ m/s. Find the center-of-mass velocities $\vec{u}_{A1}$ and $\vec{u}_{B1}$, apply the simple result found in part (d), and transform back to velocities in a stationary frame to find the final velocities of the pucks. Does your result agree with Eqs. (8–24) and (8–25)?

8–87 The *coefficient of restitution* ε for a head-on collision is defined as the ratio of the relative speed after the collision to the relative speed before. a) What is ε for a completely inelastic collision? b) What is ε for an elastic collision? c) A ball is dropped from a height h onto a stationary surface and rebounds back to a height H_1. Show that $\epsilon = \sqrt{H_1/h}$. d) A properly inflated basketball should have a coefficient of restitution 0.85. When dropped from a height of 1.2 m above a solid wood floor, to what height should a properly inflated basketball bounce? e) The height of the first bounce is H_1. If ϵ is constant, show that the height of the nth bounce is $H_n = \epsilon^{2n}h$. f) If ϵ is constant, what is the height of the eighth bounce of a properly inflated basketball dropped from 1.2 m?

8–88 Binding Energy of the Hydrogen Molecule. When two hydrogen atoms of mass m combine to form a diatomic hydrogen molecule (H_2), the potential energy of the system after they combine is $-\Delta$, where Δ is a positive quantity called the *binding energy* of the molecule. a) Show that in a collision that only involves two hydrogen atoms, it is *impossible* to form an H_2 molecule because momentum and energy cannot simultaneously be conserved. (*Hint:* If you can show this to be true in one frame of reference, then it is true in all frames of reference. Can you see why?) b) An H_2 molecule can be formed in a collision that involves *three* hydrogen atoms. Suppose that before such a collision, each of the three atoms has speed 1.00×10^3 m/s, and they are approaching at 120° angles so that at any instant, the atoms lie at the corners of an equilateral triangle. Find the speeds of the H_2 molecule and of the single hydrogen atom that remains after the collision. The binding energy of H_2 is $\Delta = 7.23 \times 10^{-19}$ J, and the mass of the hydrogen atom is 1.67×10^{-27} kg.

8–89 A wagon with two boxes of gold and having total mass 300 kg is cut loose from the horses by an outlaw when the wagon is at rest 50 m up a 6.0° slope (Fig. 8–44). The outlaw plans to have the wagon roll down the slope and across the level ground, and then fall into a canyon where his confederates wait. But in a tree 40 m from the canyon edge wait the Lone Ranger (mass 75.0 kg) and Tonto (mass 60.0 kg). They drop vertically into the wagon as it passes beneath them. a) If they require 5.0 s to grab the gold and jump out, will they make it before the wagon goes over the edge? The wagon rolls with negligible friction. b) When the two heroes drop into the wagon, is the kinetic energy of the system of the heroes plus the wagon conserved? If not, does it increase or decrease, and by how much?

FIGURE 8–44 Problem 8–89.

*8–90 In Section 8–7 we considered a rocket fired in outer space where there is no air resistance and where gravity is negligible. Suppose instead that the rocket is accelerating vertically upward from rest on the earth's surface. Continue to neglect air resistance and consider only that part of the motion

where the altitude of the rocket is small so that g may be assumed to be constant. a) How is Eq. (8–37) modified by the presence of the gravity force? b) Derive an expression for the acceleration a of the rocket, analogous to Eq. (8–39). c) What is the acceleration of the rocket in Example 8–16 (Section 8–7) if it is near the earth's surface rather than in outer space? Neglect air resistance. d) Find the speed of the rocket in Example 8–17 (Section 8–7) after 90 s if the rocket is fired from the earth's surface rather than in outer space. Neglect air resistance. How does your answer compare to the rocket speed calculated in Example 8–17?

*8–91 A Multistage Rocket. Suppose the first stage of a two-stage rocket has a total mass of 12,000 kg, of which 9000 kg is fuel. The total mass of the second stage is 1000 kg, of which 700 kg is fuel. Assume that the relative speed v_{ex} of ejected material is constant, and neglect any effect of gravity. (The latter effect is small during the firing period if the rate of fuel consumption is large.) a) Suppose the entire fuel supply carried by the two-stage rocket is utilized in a single-stage rocket of the same total mass of 13,000 kg. In terms of v_{ex}, what is the speed of the rocket, starting from rest, when its fuel is exhausted? b) For the two-stage rocket, what is the speed when the fuel of the first stage is exhausted if the first stage carries the second stage with it to this point? This speed then becomes the initial speed of the second stage. At this point the second stage separates from the first stage. c) What is the final speed of the second stage? d) What value of v_{ex} is required to give the second stage of the above rocket a speed of 7.00 km/s?

*8–92 The equation $F = -v_{ex}(dm/dt)$ for the thrust on a rocket, Eq. (8–38), can also be applied to an airplane propeller. In fact, there are *two* contributions to the thrust: one positive and one negative. The positive contribution comes from air pushed backward, away from the propeller (so $dm/dt < 0$), at a speed v_{ex} relative to the propeller. The negative contribution comes from this same quantity of air flowing into the front of the propeller (so $dm/dt > 0$) at speed v, equal to the speed of the airplane through the air. a) Write an equation for the *net* thrust developed by an aircraft propeller in terms of v, v_{ex}, and the absolute value $|dm/dt|$. b) For a Cessna 182 (a single-engine airplane) flying at 130 km/h, 150 kg of air flows through the propeller each second and the propeller develops a net thrust of 1300 N. Determine the speed increase (in km/h) that the propeller imparts to the air.

*8–93 For the rocket described in Examples 8–16 and 8–17 (Section 8–7), the mass of the rocket as a function of time is

$$m(t) = \begin{cases} m_0, & \text{for } t < 0 \\ m_0\left(1 - \dfrac{t}{120 \text{ s}}\right), & \text{for } 0 \le t \le 90 \text{ s} \\ m_0/4, & \text{for } t \ge 90 \text{ s}. \end{cases}$$

a) Calculate and graph the velocity of the rocket as a function of time from $t = 0$ s to $t = 100$ s. b) Calculate and graph the acceleration of the rocket as a function of time from $t = 0$ s to $t = 100$ s. c) A 75-kg astronaut lies on a reclined chair during the firing of the rocket. What is the maximum net force on the astronaut exerted by the chair during the firing? How does your answer compare to her weight on earth?

CHALLENGE PROBLEMS

8–94 You are going to entertain your Aunt Emma by pulling a tablecloth out from under the dishes at her birthday party. The birthday cake is resting on a tablecloth at the center of a round table of radius $r = 0.90$ m. The tablecloth is the same size as the table top. You grab the edge of the tablecloth and pull sharply. The tablecloth and cake are in contact for time t after you start pulling. Then the sliding cake is stopped (you hope) by the friction between the cake and table top. The coefficient of kinetic friction between the cake and tablecloth is $\mu_{k1} = 0.30$, and that between the cake and table top is $\mu_{k2} = 0.40$. Apply the impulse-momentum relation (Eq. 8–9) and the work-energy theorem (Eq. 6–6) to calculate the maximum value of t if the cake is not to end up on the floor. (*Hint:* Assume that the cake moves a distance d while still on the tablecloth and therefore a distance $r - d$ while sliding on the table top. Assume that the friction forces are independent of the relative speed of the sliding surfaces. You can easily try this trick yourself by pulling a sheet of paper out from under a glass of water, but have a mop handy just in case!)

8–95 In Section 8–6 we calculated the center of mass by considering objects composed of a *finite* number of point masses or objects that by symmetry could be represented by a finite number of point masses. For a solid object whose mass distribution does not allow for a simple determination of the center of mass by symmetry, the sums of Eqs. (8–28) must be generalized to integrals:

$$x_{cm} = \frac{1}{M} \int x \, dm, \qquad y_{cm} = \frac{1}{M} \int y \, dm,$$

where x and y are the coordinates of the small piece of the object that has mass dm. The integration is over the whole of the object. Consider a thin rod of length L, mass M, and cross-sectional area A. Let the origin of coordinates be at the left end of the rod and the positive x-axis lie along the rod. a) If the density $\rho = M/V$ of the object is uniform, perform the integration described above to show that the x-coordinate of the center of mass of the rod is at its geometrical center. b) If the density of the object varies linearly with x—that is, $\rho = \alpha x$, where α is a positive constant—calculate the x-coordinate of the rod's center of mass.

8–96 Use the methods of Challenge Problem 8–95 to calculate the x- and y-coordinates of the center of mass of a semicircular metal plate with uniform density ρ and thickness t. Let the radius of the plate be a. The mass of the plate is thus $M = \frac{1}{2} \rho \pi a^2 t$. Use the coordinate system indicated in Fig. 8–45.

8–97 One-fourth of a rope of length l is hanging down over the edge of a frictionless table. The rope has a uniform linear density (mass per unit length) λ ("lambda"), and the end already on the table is held by a person. How much work does the person do when she pulls on the rope to slowly raise the rest of the rope onto the table? Do the problem in two ways as follows. a) Find the force that the person must exert to raise the rope and from this the work done. Note that this is a variable force because at different times different amounts of rope are hanging over the edge. b) Suppose the segment of the rope initially hanging over the edge of the table has all of its mass concentrated at its center of mass. Find the work necessary to raise this to table height. You will probably find this approach simpler than that of part (a). How do the answers compare, and why is this so?

*8–98 A Variable-Mass Raindrop. In a rocket-propulsion problem the mass is variable. Another such problem is a raindrop falling through a cloud of small water droplets. Some of these small droplets adhere to the raindrop, thereby *increasing* its mass as it falls. The force on the raindrop is

$$F_{ext} = \frac{dp}{dt} = m\frac{dv}{dt} + v\frac{dm}{dt}.$$

Suppose the mass of the raindrop depends on the distance x that it has fallen. Then $m = kx$, where k is a constant, and $dm/dt = kv$. This gives, since $F_{ext} = mg$,

$$mg = m\frac{dv}{dt} + v(kv).$$

Or, dividing by k,

$$xg = x\frac{dv}{dt} + v^2.$$

This is a differential equation that has a solution of the form $v = at$, where a is the acceleration and is constant. Take the initial velocity of the raindrop to be zero. a) Using the proposed solution for v, find the acceleration a. b) Find the distance the raindrop has fallen in $t = 3.00$ s. c) Given that $k = 2.00$ g/m, find the mass of the raindrop at $t = 3.00$ s. For many more intriguing aspects of this problem see K. S. Krane, *Amer. Jour. Phys.*, Vol. 49 (1981), pp. 113–117.

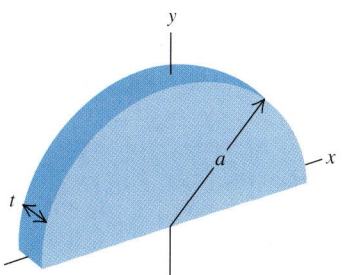

FIGURE 8–45 Challenge Problem 8–96.

9

Rotation of Rigid Bodies

No matter where you sit on a rotating amusement park ride, you have the same angular velocity and angular acceleration. But your *linear* velocity and *linear* acceleration depend on how far you sit from the axis of rotation. In general, the farther you are from the axis, the greater the linear velocity and acceleration and the wilder the ride.

9–1 INTRODUCTION

What do the motions of a compact disc, a Ferris wheel, a circular saw blade, and a ceiling fan have in common? None of these can be represented adequately as a moving *point;* each involves a body that *rotates* about an axis that is stationary in some inertial frame of reference. Rotation occurs at all scales, from the motion of electrons in atoms to the motions of entire galaxies. We need to develop some general methods for analyzing the motion of a rotating body. In this chapter and the next we consider bodies that have definite size and definite shape, and that in general can have rotational as well as translational motion.

Real-world bodies can be even more complicated; the forces that act on them can deform them—stretching, twisting, and squeezing them. We'll neglect these deformations for now and assume that the body has a perfectly definite and unchanging shape and size. We call this idealized model a **rigid body.** This chapter and the next are mostly about rotational motion of a rigid body. We begin with kinematic language for *describing* rotational motion. Next we look at the kinetic energy of rotation, the key to using energy methods for rotational motion. Then in Chapter 10 we'll develop dynamic principles that relate the forces on a body to its rotational motion.

9–2 ANGULAR VELOCITY AND ACCELERATION

In analyzing rotational motion, let's think first about a rigid body that rotates about a fixed axis. By *fixed axis* we mean an axis that is at rest in some inertial frame of reference and does not change direction relative to that frame. The rotating rigid body might be a motor shaft, a chunk of roast beef on a barbecue skewer, or a merry-go-round.

Figure 9–1 shows a rigid body rotating about a fixed axis that passes through point O and is perpendicular to the plane of the diagram, which we choose to call the xy-plane. One way to describe the rotation of this body would be to choose a particular point P on the body and to keep track of the x- and y-coordinates of this point. This isn't a terribly convenient method, since it takes two numbers (the two coordinates x and y) to specify the rotational position of the body. Instead, we notice that the line OP is fixed in the body and rotates with it. The angle θ that this line makes with the $+x$-axis describes the rotational position of the body; we will use this single quantity θ as a *coordinate* for rotation.

The angular coordinate θ of a rigid body rotating around a fixed axis can be positive or negative. If we choose positive angles to be measured counterclockwise from the positive x-axis, then the angle θ in Fig. 9–1 is positive. If we instead choose the positive rotation direction to be clockwise, then θ in Fig. 9–1 is negative. When we considered the motion of a particle along a straight line, it was essential to specify the direction of positive displacement along that line; in discussing rotation around a fixed axis, it's just as essential to specify the direction of positive rotation.

In describing rotational motion, the most natural way to measure the angle θ is not in degrees, but in **radians.** As shown in Fig. 9–2a, one radian (1 rad) is the angle subtended at the center of a circle by an arc with a length equal to the radius of the circle. In Fig. 9–2b an angle θ is subtended by an arc of length s on a circle of radius r. The value of θ (in radians) is equal to s divided by r:

$$\theta = \frac{s}{r}, \quad \text{or} \quad s = r\theta. \tag{9–1}$$

An angle in radians is the ratio of two lengths, so it is a pure number, without dimensions. If $s = 3.0$ m and $r = 2.0$ m, then $\theta = 1.5$, but we will often write this as 1.5 rad to distinguish it from an angle measured in degrees or revolutions.

The circumference of a circle (that is, the arc length all the way around the circle) is 2π times the radius, so there are 2π (about 6.283) radians in one complete revolution (360°). Therefore

$$1 \text{ rad} = \frac{360°}{2\pi} = 57.3°.$$

Similarly, $180° = \pi$ rad, $90° = \pi/2$ rad, and so on. If we had insisted on measuring the angle θ in degrees, we would have needed to include an extra factor of $(2\pi/360)$ on the right-hand side of $s = r\theta$ in Eq. (9–1). By measuring angles in radians, we keep the relationship between angle and distance along an arc as simple as possible.

ANGULAR VELOCITY

The coordinate θ shown in Fig. 9–1 specifies the rotational position of a rigid body at a given instant. We can describe the rotational *motion* of such a rigid body in terms of the rate of change of θ. We'll do this in an analogous way to our description of straight-line motion in Chapter 2. In Fig. 9–3a a reference line OP in a rotating body makes an angle θ_1 with the +x-axis at time t_1. At a later time t_2 the angle has changed to θ_2. We define the **average angular velocity** ω_{av} (the Greek letter "omega") of the body in the time interval $\Delta t = t_2 - t_1$ as the ratio of the *angular displacement* $\Delta\theta = \theta_2 - \theta_1$ to Δt:

$$\omega_{av} = \frac{\theta_2 - \theta_1}{t_2 - t_1} = \frac{\Delta\theta}{\Delta t}. \tag{9–2}$$

The **instantaneous angular velocity** ω is the limit of ω_{av} as Δt approaches zero, that is, the derivative of θ with respect to t:

$$\omega = \lim_{\Delta t \to 0} \frac{\Delta\theta}{\Delta t} = \frac{d\theta}{dt} \quad \text{(definition of angular velocity).} \tag{9–3}$$

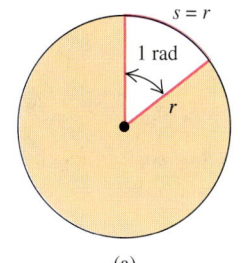

9–1 A rigid body rotating counterclockwise about a fixed axis that passes through O and is perpendicular to the page.

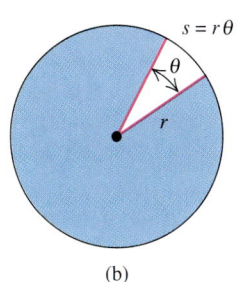

(a)

(b)

9–2 An angle θ in radians is defined as the ratio of the arc length s to the radius r.

(a)

(b)

9–3 (a) Angular displacement $\Delta\theta$ of a rotating body. (b) Every part of a rigid body has the same angular velocity.

When we refer simply to "angular velocity," we mean the instantaneous angular velocity, not the average angular velocity.

Different points on a rotating rigid body move different distances in a given time interval, depending on how far the point lies from the rotation axis. But because the body is rigid, *all* points rotate through the same angle in the same time (Fig. 9–3b). Hence *at any instant, every part of a rotating rigid body has the same angular velocity.* The angular velocity is positive if the body is rotating in the direction of increasing θ and negative if rotating in the direction of decreasing θ.

If the angle θ is in radians, the unit of angular velocity is the radian per second (rad/s). Other units, such as the revolution per minute (rev/min or rpm), are often used. Since 1 rev = 2π rad, two useful conversions are

$$1 \text{ rev/s} = 2\pi \text{ rad/s} \quad \text{and} \quad 1 \text{ rev/min} = 1 \text{ rpm} = \frac{2\pi}{60} \text{ rad/s}.$$

That is, 1 rad/s is about 10 rpm.

EXAMPLE 9–1

Calculating angular velocity Figure 9–4 shows the flywheel in a car engine under test. The angular position θ of the flywheel is given by

$$\theta = (2.0 \text{ rad/s}^3)t^3.$$

The diameter of the flywheel is 0.36 m. a) Find the angle θ, in radians and in degrees, at times $t_1 = 2.0$ s and $t_2 = 5.0$ s. b) Find the distance that a particle on the rim moves during that time interval. c) Find the average angular velocity, in rad/s and in rev/min (rpm), between $t_1 = 2.0$ s and $t_2 = 5.0$ s. d) Find the instantaneous angular velocity at time $t = 3.0$ s.

SOLUTION (a) We substitute the values of t into the given equation:

$$\theta_1 = (2.0 \text{ rad/s}^3)(2.0 \text{ s})^3 = 16 \text{ rad}$$

$$= (16 \text{ rad})\frac{360°}{2\pi \text{ rad}} = 920°,$$

$$\theta_2 = (2.0 \text{ rad/s}^3)(5.0 \text{ s})^3 = 250 \text{ rad}$$

$$= (250 \text{ rad})\frac{360°}{2\pi \text{ rad}} = 14,000°.$$

b) The flywheel turns through an angular displacement of $\theta_2 - \theta_1 = 250 \text{ rad} - 16 \text{ rad} = 234 \text{ rad}$. The radius r is half the diameter, or 0.18 m. Equation (9–1) gives

$$s = r\theta = (0.18 \text{ m})(234 \text{ rad}) = 42 \text{ m}.$$

Notice that in order to use Eq. (9–1), the angle *must* be expressed in radians. We drop "radians" from the unit for s because θ is really a dimensionless pure number; s is a distance and is measured in meters, the same unit as r.

Flywheel
$d = 0.36$ m

9–4 Rotating flywheel in a car engine.

c) In Eq. (9–2) we have

$$\omega_{av} = \frac{\theta_2 - \theta_1}{t_2 - t_1} = \frac{250 \text{ rad} - 16 \text{ rad}}{5.0 \text{ s} - 2.0 \text{ s}} = 78 \text{ rad/s}$$

$$= \left(78 \frac{\text{rad}}{\text{s}}\right)\left(\frac{1 \text{ rev}}{2\pi \text{ rad}}\right)\left(\frac{60 \text{ s}}{1 \text{ min}}\right) = 740 \text{ rev/min}.$$

d) We use Eq. (9–3):

$$\omega = \frac{d\theta}{dt} = \frac{d}{dt}[(2.0 \text{ rad/s}^3)t^3] = (2.0 \text{ rad/s}^3)(3t^2)$$

$$= (6.0 \text{ rad/s}^3)t^2.$$

At time $t = 3.0$ s,

$$\omega = (6.0 \text{ rad/s}^3)(3.0 \text{ s})^2 = 54 \text{ rad/s}.$$

ANGULAR ACCELERATION

When the angular velocity of a rigid body changes, it has an *angular acceleration.* If ω_1 and ω_2 are the instantaneous angular velocities at times t_1 and t_2, we define the **average angular acceleration** α_{av} over the interval $\Delta t = t_2 - t_1$ as the change in angular velocity divided by Δt:

$$\alpha_{av} = \frac{\omega_2 - \omega_1}{t_2 - t_1} = \frac{\Delta\omega}{\Delta t}. \tag{9–4}$$

The **instantaneous angular acceleration** α is the limit of α_{av} as $\Delta t \to 0$:

$$\alpha = \lim_{\Delta t \to 0} \frac{\Delta\omega}{\Delta t} = \frac{d\omega}{dt} \quad \text{(definition of angular acceleration)}. \tag{9–5}$$

The usual unit of angular acceleration is the radian per second per second, or rad/s^2. Henceforth we will use the term "angular acceleration" to mean the instantaneous angular acceleration rather than the average angular acceleration.

Because $\omega = d\theta/dt$, we can also express angular acceleration as the second derivative of the angular coordinate:

$$\alpha = \frac{d}{dt}\frac{d\theta}{dt} = \frac{d^2\theta}{dt^2}. \tag{9–6}$$

You have probably noticed that we are using Greek letters for angular kinematic quantities: θ for angular position, ω for angular velocity, and α for angular acceleration. These are analogous to x for position, v for velocity, and a for acceleration, respectively, in straight-line motion. In each case, velocity is the rate of change of position with respect to time and acceleration is the rate of change of velocity with respect to time. We will sometimes use the terms *linear* velocity and *linear* acceleration for the familiar quantities we defined in Chapters 2 and 3 to make sure we distinguish clearly between these and the *angular* quantities introduced in this chapter.

In rotational motion, if the angular acceleration α is positive, the angular velocity ω is increasing; if α is negative, ω is decreasing. The rotation is speeding up if α and ω have the same sign and slowing down if α and ω have opposite signs. (These relationships are exactly the same as those between *linear* acceleration a and *linear* velocity v for straight-line motion; see Section 2–4.)

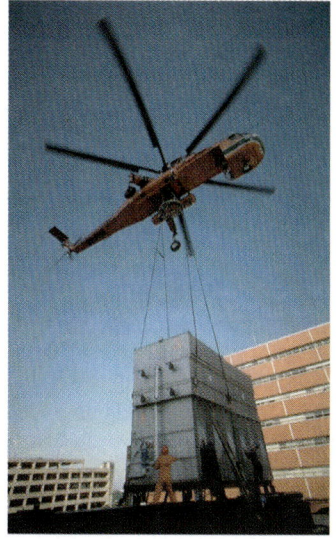

The pilot of a helicopter can change the angular velocity of the rotor—that is, give it an angular acceleration—by adjusting a control in the cockpit.

EXAMPLE 9–2

Calculating angular acceleration In Example 9–1 we found that the instantaneous angular velocity ω of the flywheel at any time t is given by

$$\omega = (6.0 \text{ rad/s}^3)t^2.$$

a) Find the average angular acceleration between $t_1 = 2.0$ s and $t_2 = 5.0$ s. b) Find the instantaneous angular acceleration at time $t = 3.0$ s.

SOLUTION (a) The values of ω at the two times are

$$\omega_1 = (6.0 \text{ rad/s}^3)(2.0 \text{ s})^2 = 24 \text{ rad/s},$$
$$\omega_2 = (6.0 \text{ rad/s}^3)(5.0 \text{ s})^2 = 150 \text{ rad/s}.$$

From Eq. (9–4) the average angular acceleration is

$$\alpha_{av} = \frac{150 \text{ rad/s} - 24 \text{ rad/s}}{5.0 \text{ s} - 2.0 \text{ s}} = 42 \text{ rad/s}^2.$$

b) From Eq. (9–5) the instantaneous angular acceleration at any time t is

$$\alpha = \frac{d\omega}{dt} = \frac{d}{dt}[(6.0 \text{ rad/s}^3)(t^2)] = (6.0 \text{ rad/s}^3)(2t)$$
$$= (12 \text{ rad/s}^3)t.$$

At time $t = 3.0$ s,

$$\alpha = (12 \text{ rad/s}^3)(3.0 \text{ s}) = 36 \text{ rad/s}^2.$$

Note that the angular acceleration is *not* constant in this situation. The angular velocity ω is always increasing because α is always positive. Furthermore, the rate at which the angular velocity increases is itself increasing, since α increases with time.

In straight-line motion, velocity v and acceleration a are the *components* of the vectors $\vec{v}$ and $\vec{a}$, respectively. In fixed-axis rotation problems it is likewise useful to think of ω and α as components of vector quantities. The angular velocity *vector* $\vec{\omega}$ is directed

(a)

(b)

Speeding up   Slowing down

(c)

9–5 (a) When the fingers of the right hand curl in the direction of rotation, the thumb position is the direction of the angular velocity vector. The direction is also the direction of advance of an ordinary screw with a right-hand thread. (b) Reversing the direction of rotation makes the angular velocity vector point in the opposite direction. (c) When the rotation axis is fixed, the angular acceleration and angular velocity vectors both lie along that axis.

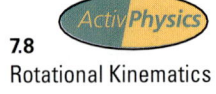

7.8
Rotational Kinematics

along the axis of rotation. As Fig. 9–5 shows, the direction is given by the right-hand rule we used to define the vector product in Section 1–11. We define the angular acceleration vector $\vec{\alpha}$ as the time derivative of the angular velocity vector. If the direction of the axis is fixed, then $\vec{\alpha}$ is also along the axis; it points in the same direction as $\vec{\omega}$ if the rotation is speeding up and in the opposite direction if the rotation is slowing down (Fig. 9–5c). The vector formulation is especially useful in situations in which the direction of the axis *changes;* we'll examine such situations briefly at the end of Chapter 10.

9–3 ROTATION WITH CONSTANT ANGULAR ACCELERATION

In Chapter 2 we found that straight-line motion is particularly simple when the acceleration is constant. This is also true of rotational motion about a fixed axis. When the angular acceleration is constant, we can derive equations for angular velocity and angular position using exactly the same procedure that we used for straight-line motion in Section 2–5. In fact, the equations we are about to derive are identical to Eqs. (2–8), (2–12), (2–13), and (2–14) if we replace x with θ, v with ω, and a with α. We suggest that you review Section 2–5 before continuing.

Let ω_0 be the angular velocity of a rigid body at time $t = 0$, and let ω be its angular velocity at any later time t. The angular acceleration α is constant and equal to the average value for any interval. Using Eq. (9–4) with the interval from 0 to t, we find

$$\alpha = \frac{\omega - \omega_0}{t - 0}, \qquad \text{or}$$

$$\omega = \omega_0 + \alpha t \qquad \text{(constant angular acceleration only).} \qquad (9\text{--}7)$$

The product αt is the total change in ω between $t = 0$ and the later time t; the angular velocity ω at time t is the sum of the initial value ω_0 and this total change.

With constant angular acceleration the angular velocity changes at a uniform rate, so its average value between 0 and t is the average of the initial and final values:

$$\omega_{av} = \frac{\omega_0 + \omega}{2}. \qquad (9\text{--}8)$$

We also know that ω_{av} is the total angular displacement $(\theta - \theta_0)$ divided by the time interval $(t - 0)$;

$$\omega_{av} = \frac{\theta - \theta_0}{t - 0}. \qquad (9\text{--}9)$$

When we equate Eqs. (9–8) and (9–9) and multiply the result by t, we get

$$\theta - \theta_0 = \frac{1}{2}(\omega_0 + \omega)t \qquad \text{(constant angular acceleration only).} \qquad (9\text{--}10)$$

To obtain a relation between θ and t that doesn't contain ω, we substitute Eq. (9–7) into Eq. (9–10):

$$\theta - \theta_0 = \frac{1}{2}[\omega_0 + (\omega_0 + \alpha t)]t,$$

$$\theta = \theta_0 + \omega_0 t + \frac{1}{2}\alpha t^2 \qquad \text{(constant angular acceleration only).} \qquad (9\text{--}11)$$

That is, if at the initial time $t = 0$ the body is at angular position θ_0 and has angular velocity ω_0, then its angular position θ at any later time t is the sum of three terms: its initial

TABLE 9–1

COMPARISON OF LINEAR AND ANGULAR MOTION WITH CONSTANT ACCELERATION

STRAIGHT-LINE MOTION WITH CONSTANT LINEAR ACCELERATION	FIXED-AXIS ROTATION WITH CONSTANT ANGULAR ACCELERATION
$a = $ constant	$\alpha = $ constant
$v = v_0 + at$	$\omega = \omega_0 + \alpha t$
$x = x_0 + v_0 t + \frac{1}{2}at^2$	$\theta = \theta_0 + \omega_0 t + \frac{1}{2}\alpha t^2$
$v^2 = v_0^2 + 2a(x - x_0)$	$\omega^2 = \omega_0^2 + 2\alpha(\theta - \theta_0)$
$x - x_0 = \frac{1}{2}(v + v_0)t$	$\theta - \theta_0 = \frac{1}{2}(\omega + \omega_0)t$

angular position θ_0, plus the rotation $\omega_0 t$ it would have if the angular velocity were constant, plus an additional rotation $\frac{1}{2}\alpha t^2$ caused by the changing angular velocity.

Following the same procedure as for straight-line motion in Section 2–5, we can combine Eqs. (9–7) and (9–11) to obtain a relation between θ and ω that does not contain t. We invite you to work out the details, following the same procedure we used to get Eq. (2–13). (See Exercise 9–10.) In fact, because of the perfect analogy between straight-line and rotational quantities, we can simply take Eq. (2–13) and replace each straight-line quantity by its rotational analog. We get

$$\omega^2 = \omega_0^2 + 2\alpha(\theta - \theta_0) \qquad \text{(constant angular acceleration only).} \qquad (9-12)$$

CAUTION▶ Keep in mind that all of these results are valid *only* when the angular acceleration α is *constant;* be careful not to try to apply them to problems in which α is *not* constant. Table 9–1 shows the analogy between Eqs. (9–7), (9–10), (9–11), and (9–12) for fixed-axis rotation with constant angular acceleration and the corresponding equations for straight-line motion with constant linear acceleration. ◀

EXAMPLE 9–3

Rotation with constant angular acceleration A bicycle wheel is being tested at a repair shop. The angular velocity of the wheel is 4.00 rad/s at time $t = 0$, and its angular acceleration is constant and equal to -1.20 rad/s^2. A spoke OP on the wheel coincides with the $+x$-axis at time $t = 0$ (Fig. 9–6). a) What is the wheel's angular velocity at $t = 3.00$ s? b) What angle does the spoke OP make with the $+x$-axis at this time?

SOLUTION We can use Eqs. (9–7) and (9–11) to find ω and θ at any time in terms of the given initial conditions.
a) From Eq. (9–7), $\omega = \omega_0 + \alpha t$. At time $t = 3.00$ s,

$$\omega = 4.00 \text{ rad/s} + (-1.20 \text{ rad/s}^2)(3.00 \text{ s}) = 0.40 \text{ rad/s}.$$

The angular velocity has decreased because α is negative.
b) The angle θ is given as a function of time by Eq. (9–11):

$$\theta = \theta_0 + \omega_0 t + \frac{1}{2}\alpha t^2$$

$$= 0 + (4.00 \text{ rad/s})(3.00 \text{ s}) + \frac{1}{2}(-1.20 \text{ rad/s}^2)(3.00 \text{ s})^2$$

$$= 6.60 \text{ rad} = 6.60 \text{ rad}\left(\frac{1 \text{ rev}}{2\pi \text{ rad}}\right) = 1.05 \text{ rev.}$$

The wheel has turned through one complete revolution plus an additional 0.05 rev, that is, through an additional angle of $(0.05 \text{ rev})(2\pi \text{ rad/rev}) = 0.32 \text{ rad} = 18°$. The spoke OP is at an angle of 18° with the $+x$-axis.

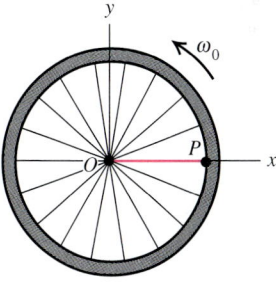

9–6 The spoke OP of a rotating wheel at $t = 0$. What angle does this spoke make with the $+x$-axis at $t = 3.00$ s?

We can also determine θ by rearranging Eq. (9–12), $\omega^2 = \omega_0^2 + 2\alpha(\theta - \theta_0)$, so that it reads

$$\theta = \theta_0 + \frac{\omega^2 - \omega_0^2}{2\alpha}$$

$$= 0 + \frac{(0.40 \text{ rad/s})^2 - (4.00 \text{ rad/s})^2}{2(-1.20 \text{ rad/s}^2)} = 6.60 \text{ rad.}$$

Can you show that the wheel will stop instantaneously at $t = 3.33$ s after having rotated through 6.67 rad? If the angular acceleration continues to remain constant after this time, the wheel will reverse direction and start rotating clockwise.

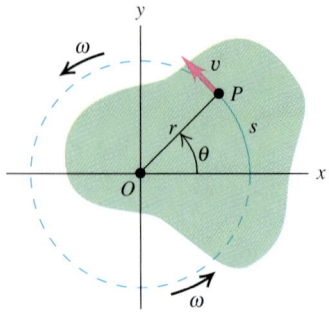

9–7 A rigid body rotating about a fixed axis through point O. The distance s that point P on the body moves through equals $r\theta$ if θ is measured in radians. The linear speed v of point P equals $r\omega$ if the angular speed ω is measured in rad/s.

9–4 RELATING LINEAR AND ANGULAR KINEMATICS

How do we find the velocity and acceleration of a particular point in a rotating rigid body? We need to answer this question in order to proceed with our study of rotation. For example, to find the kinetic energy of a rotating body, we have to start from $K = \frac{1}{2}mv^2$ for a particle, and this requires knowing v for each particle in the body. So it's worthwhile to develop general relations between the *angular* velocity and acceleration of a rigid body rotating about a fixed axis and the *linear* velocity and acceleration of a specific point or particle in the body.

When a rigid body rotates about a fixed axis, every particle in the body moves in a circular path. The circle lies in a plane perpendicular to the axis and is centered on the axis. The speed of a particle is directly proportional to the body's angular velocity; the faster the body rotates, the greater the speed of each particle. In Fig. 9–7, point P is a constant distance r from the axis of rotation, so it moves in a circle of radius r. At any time the angle θ (in radians) and the arc length s are related by

$$s = r\theta.$$

We take the time derivative of this, noting that r is constant for any specific particle, and take the absolute value of both sides:

$$\left| \frac{ds}{dt} \right| = r \left| \frac{d\theta}{dt} \right|.$$

Now $|ds/dt|$ is the absolute value of the rate of change of arc length, which is equal to the instantaneous *linear* speed v of the particle. Analogously, $|d\theta/dt|$, the absolute value of the rate of change of the angle, is the instantaneous **angular speed** ω—that is, the magnitude of the instantaneous angular velocity in rad/s. Thus

$$v = r\omega \qquad \text{(relation between linear and angular speed).} \qquad (9\text{–}13)$$

The farther a point is from the axis, the greater its linear speed. The *direction* of the linear velocity *vector* is tangent to its circular path at each point (Fig. 9–7).

We can represent the acceleration of a particle moving in a circle in terms of its centripetal and tangential components, a_{rad} and a_{tan} (Fig. 9–8), as we did in Section 3–5. It would be a good idea to review that section now. We found that the **tangential component of acceleration** a_{tan}, the component parallel to the instantaneous velocity, acts to change the *magnitude* of the particle's velocity (i.e., the speed) and is equal to the rate of change of speed. Taking the derivative of Eq. (9–13), we find

$$a_{\text{tan}} = \frac{dv}{dt} = r\frac{d\omega}{dt} = r\alpha \qquad \begin{array}{l}\text{(tangential acceleration of a} \\ \text{point on a rotating body).}\end{array} \qquad (9\text{–}14)$$

This component of a particle's acceleration is always tangent to the circular path of the particle.

The component of the particle's acceleration directed toward the rotation axis, the **centripetal component of acceleration** a_{rad}, is associated with the change of *direction* of the particle's velocity. In Section 3–5 we worked out the relation $a_{\text{rad}} = v^2/r$. We can express this in terms of ω by using Eq. (9–13):

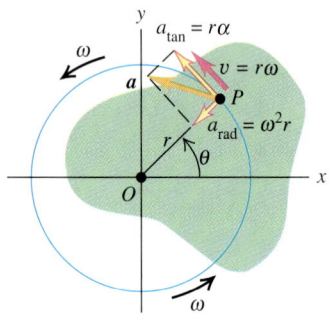

9–8 The component of acceleration of point P parallel to $\vec{v}$ is $a_{\text{tan}} = r\alpha$; the component perpendicular to $\vec{v}$, radially inward toward the rotation axis, is $a_{\text{rad}} = \omega^2 r$. The rigid body shown here is speeding up.

$$a_{\text{rad}} = \frac{v^2}{r} = \omega^2 r \qquad \begin{array}{l}\text{(centripetal acceleration of a point} \\ \text{on a rotating body).}\end{array} \qquad (9\text{–}15)$$

This is true at each instant *even when ω and v are not constant.* The centripetal component always points toward the axis of rotation.

The vector sum of the centripetal and tangential components of acceleration of a particle in a rotating body is the linear acceleration $\vec{a}$ (Fig. 9–8).

CAUTION ▶ It's important to remember that Eq. (9–1), $s = r\theta$, is valid *only* when θ is measured in radians. The same is true of any equation derived from this, including

Eqs. (9–13), (9–14), and (9–15). When you use these equations, you *must* express the angular quantities in radians, not revolutions or degrees. ◄

Equations (9–1), (9–13), and (9–14) also apply to any particle that has the same tangential velocity as a point in a rotating rigid body. For example, when a rope wound around a circular cylinder unwraps without stretching or slipping, its speed and acceleration at any instant are equal to the speed and tangential acceleration of the point at which it is tangent to the cylinder. The same principle holds for situations such as bicycle chains and sprockets, belts and pulleys that turn without slipping, and so on. We will have several opportunities to use these relations later in this chapter and in Chapter 10. Note that Eq. (9–15) for the centripetal component a_{rad} is applicable to the rope or chain *only* at points that are in contact with the cylinder or sprocket. Other points do not have the same acceleration toward the center of the circle that points on the cylinder or sprocket have.

EXAMPLE 9-4

Throwing a discus A discus thrower turns with angular acceleration $\alpha = 50$ rad/s^2, moving the discus in a circle of radius 0.80 m. We model the thrower's arm as a rigid body, so r is constant. Find the tangential and centripetal components of acceleration of the discus and the magnitude of its acceleration at the instant when the angular velocity is 10 rad/s.

SOLUTION We model the discus as a particle moving in a circular path (Fig. 9–9). The acceleration components are given by Eqs. (9–14) and (9–15):

$$a_{tan} = r\alpha = (0.80 \text{ m})(50 \text{ rad/s}^2) = 40 \text{ m/s}^2,$$

$$a_{rad} = \omega^2 r = (10 \text{ rad/s})^2(0.80 \text{ m}) = 80 \text{ m/s}^2.$$

The magnitude of the acceleration vector is

$$a = \sqrt{a_{rad}^2 + a_{tan}^2} = 89 \text{ m/s}^2,$$

or about nine times the acceleration due to gravity. Note that we have dropped the unit "radian" from our results for a_{tan}, a_{rad},

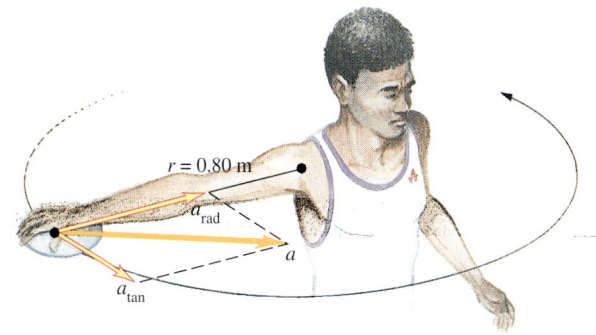

9–9 Treating a discus thrower's arm as a rigid body.

and a; we can do this because "radian" is a dimensionless quantity.

EXAMPLE 9-5

Designing a propeller You are asked to design an airplane propeller to turn at 2400 rpm. The forward air speed of the plane is to be 75.0 m/s (270 km/h, or about 168 mi/h), and the speed of the tips of the propeller blades through the air must not exceed 270 m/s. (This is about 0.80 times the speed of sound in air. If the propeller tips were to move at the speed of sound, they would produce a tremendous amount of noise. Keeping the speed comfortably below this keeps the noise level acceptably low.) a) What is the maximum radius the propeller can have? b) With this radius, what is the acceleration of the propeller tip?

SOLUTION First we convert ω to rad/s:

$$\omega = 2400 \text{ rpm} = \left(2400 \frac{\text{rev}}{\text{min}}\right)\left(\frac{2\pi \text{ rad}}{1 \text{ rev}}\right)\left(\frac{1 \text{ min}}{60 \text{ s}}\right)$$

$$= 251 \text{ rad/s}.$$

a) The tangential velocity $v_P = r\omega$ of the propeller tip is perpendicular to the forward velocity v_A through the air (Fig. 9–10; see page 276). The magnitude of the vector sum of these velocities

is v_{total}, where

$$v_{total}^2 = v_A^2 + v_P^2 = v_A^2 + r^2\omega^2,$$

so

$$r^2 = \frac{v_{total}^2 - v_A^2}{\omega^2} \quad \text{and} \quad r = \frac{\sqrt{v_{total}^2 - v_A^2}}{\omega}.$$

If $v_{total} = 270$ m/s, the propeller radius is

$$r = \frac{\sqrt{(270 \text{ m/s})^2 - (75.0 \text{ m/s})^2}}{251 \text{ rad/s}} = 1.03 \text{ m}.$$

b) The angular velocity of the propeller is constant, so the acceleration of the propeller tip is purely centripetal:

$$a_{rad} = \omega^2 r$$

$$= (251 \text{ rad/s})^2(1.03 \text{ m}) = 6.5 \times 10^4 \text{ m/s}^2.$$

From $\Sigma \vec{F} = m\vec{a}$, the propeller must exert a force of 6.5×10^4 N on each kilogram of material at its tip! This is why propellers are made out of tough material (usually aluminum alloy).

9–10 Designing an airplane propeller.

EXAMPLE 9-6

Bicycle gears How are the angular speeds of the two bicycle sprockets in Fig. 9–11 related to the number of teeth on each sprocket?

SOLUTION The chain does not slip or stretch, so it moves at the same tangential speed on both sprockets. Letting v and ω denote the linear and angular *speeds,* we have

$$v = r_1 \omega_1 = r_2 \omega_2, \qquad \text{so} \qquad \frac{\omega_2}{\omega_1} = \frac{r_1}{r_2}.$$

The angular speed is inversely proportional to the radius. This relationship also holds for pulleys connected by a belt, provided that the belt doesn't slip. For chain sprockets the teeth must be equally spaced on the circumferences of both sprockets in order for the chain to mesh properly with both. Let N_1 and N_2 be the numbers of teeth; the condition that the tooth spacing is the same on both sprockets is

$$\frac{2\pi r_1}{N_1} = \frac{2\pi r_2}{N_2}, \qquad \text{or} \qquad \frac{r_1}{r_2} = \frac{N_1}{N_2}.$$

Combining this with the other equation, we get

$$\frac{\omega_2}{\omega_1} = \frac{N_1}{N_2}.$$

The angular speed of each sprocket is inversely proportional to the number of teeth. On a multi-speed bike, you get the highest

9–11 How are the angular velocities of the two sprockets related to their number of teeth?

angular speed ω_2 of the rear wheel for a given pedaling rate ω_1 when the ratio N_1/N_2 is maximum; this means using the largest-radius front sprocket (largest N_1) and the smallest-radius rear sprocket (smallest N_2).

9-5 ENERGY IN ROTATIONAL MOTION

A rotating rigid body consists of mass in motion, so it has kinetic energy. We can express this kinetic energy in terms of the body's angular velocity and a new quantity called *moment of inertia* that we will define below. To develop this relationship, we think of the body as being made up of a large number of particles, with masses m_1, m_2, ... , at distances r_1, r_2, ... from the axis of rotation. We label the particles with the index i: The mass of the ith particle is m_i, and its distance from the axis of rotation is r_i. The particles don't necessarily all lie in the same plane, so we specify that r_i is the *perpendicular* distance from the axis to the ith particle.

When a rigid body rotates about a fixed axis, the speed v_i of the ith particle is given by Eq. (9–13), $v_i = r_i \omega$, where ω is the body's angular speed. Different particles have different values of r, but ω is the same for all (otherwise, the body wouldn't be rigid).

The kinetic energy of the ith particle can be expressed as

$$\frac{1}{2} m_i v_i^{\,2} = \frac{1}{2} m_i r_i^{\,2} \omega^2.$$

The *total* kinetic energy of the body is the sum of the kinetic energies of all its particles:

$$K = \frac{1}{2} m_1 r_1^{\,2} \omega^2 + \frac{1}{2} m_2 r_2^{\,2} \omega^2 + \cdots = \sum_i \frac{1}{2} m_i r_i^{\,2} \omega^2.$$

Taking the common factor $\omega^2/2$ out of this expression, we get

$$K = \frac{1}{2}(m_1 r_1^{\,2} + m_2 r_2^{\,2} + \cdots)\omega^2 = \frac{1}{2}\left(\sum_i m_i r_i^{\,2}\right)\omega^2.$$

The quantity in parentheses, obtained by multiplying the mass of each particle by the square of its distance from the axis of rotation and adding these products, is denoted by I and is called the **moment of inertia** of the body for this rotation axis:

$$I = m_1 r_1^{\,2} + m_2 r_2^{\,2} + \cdots = \sum_i m_i r_i^{\,2} \qquad \text{(definition of moment of inertia).} \quad (9\text{--}16)$$

(a)

The word "moment" means that I depends on how the body's mass is distributed in space; it has nothing to do with a "moment" of time. For a body with a given rotation axis and a given total mass, the greater the distance from the axis to the particles that make up the body, the greater the moment of inertia. In a rigid body, the distances r_i are all constant and I is independent of how the body is rotating around the given axis. The SI unit of moment of inertia is the kilogram-meter^2 (kg $\cdot$ m^2).

In terms of moment of inertia I, the **rotational kinetic energy** K of a rigid body is

$$K = \frac{1}{2} I \omega^2 \qquad \text{(rotational kinetic energy of a rigid body).} \quad (9\text{--}17)$$

(b)

The kinetic energy given by Eq. (9–17) is not a new form of energy; it's the sum of the kinetic energies of the individual particles that make up the rigid body, written in a compact and convenient form in terms of the moment of inertia. When using Eq. (9–17), ω *must* be measured in radians per second, not revolutions or degrees per second, to give K in joules; this is because we used $v_i = r_i \omega$ in our derivation.

Equation (9–17) gives a simple physical interpretation of moment of inertia: the greater the moment of inertia, the greater the kinetic energy of a rigid body rotating with a given angular speed ω. We learned in Chapter 6 that the kinetic energy of a body equals the amount of work done to accelerate that body from rest. So the greater a body's moment of inertia, the harder it is to start the body rotating if it's at rest and the harder it is to stop its rotation if it's already rotating (Fig. 9–12). For this reason, I is also called the *rotational inertia*.

Example 9–7 shows how *changing* the rotation axis can affect the value of I.

9-12 An apparatus free to rotate around a vertical axis. The two cylinders of mass m can be locked into any position on the horizontal shaft. (a) If the two cylinders are located close to the rotation axis, the moment of inertia is small and it's easy to start the apparatus rotating. (b) If the cylinders are further from the rotation axis, the moment of inertia is greater and it's more difficult to start or stop the rotation.

EXAMPLE 9-7

Moments of inertia for different rotation axes An engineer is designing a one-piece machine part consisting of three heavy connectors linked by light molded struts (Fig. 9–13). The connectors can be considered as massive particles connected by massless rods. a) What is the moment of inertia of this body about an axis through point A, perpendicular to the plane of the diagram? b) What is the moment of inertia about an axis coinciding with rod BC? c) If the body rotates about an axis through A perpendicular to the plane of the diagram, with angular speed $\omega = 4.0$ rad/s, what is its kinetic energy?

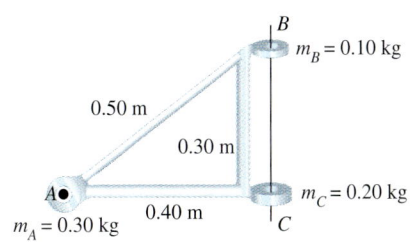

9-13 A strangely shaped machine part.

SOLUTION a) The particle at point A lies *on* the axis. Its distance r from the axis is zero, so it contributes nothing to the moment of inertia. Equation (9–16) gives

$$I = \Sigma m_i r_i^2 = (0.10 \text{ kg})(0.50 \text{ m})^2 + (0.20 \text{ kg})(0.40 \text{ m})^2$$

$$= 0.057 \text{ kg} \cdot \text{m}^2.$$

b) The particles at B and C both lie *on* the axis, so for them $r = 0$, and neither contributes to the moment of inertia. Only A contributes, and we have

$$I = \Sigma m_i r_i^2 = (0.30 \text{ kg})(0.40 \text{ m})^2 = 0.048 \text{ kg} \cdot \text{m}^2.$$

Since this moment of inertia is less than in part (a), it's easier to make the body rotate about this axis than about the axis through point A.

c) From Eq. (9–17),

$$K = \frac{1}{2} I\omega^2 = \frac{1}{2} (0.057 \text{ kg} \cdot \text{m}^2)(4.0 \text{ rad/s})^2 = 0.46 \text{ J}.$$

CAUTION ▶ The results of parts (a) and (b) of Example 9–7 show that the moment of inertia of a body depends on the location and orientation of the axis. It's not enough to just say, "The moment of inertia of this body is $0.048 \text{ kg} \cdot \text{m}^2$." We have to be specific and say, "The moment of inertia of this body *about axis BC* is $0.048 \text{ kg} \cdot \text{m}^2$." ◀

In Example 9–7 we represented the body as several point masses, and we evaluated the sum in Eq. (9–16) directly. When the body is a *continuous* distribution of matter, such as a solid cylinder or plate, the sum becomes an integral, and we need to use calculus to calculate the moment of inertia. We will give several examples of such calculations in Section 9–7; meanwhile, Table 9–2 gives moments of inertia for several familiar shapes in terms of the masses and dimensions. Each body shown in Table 9–2 is *uniform;* that is, the density has the same value at all points within the solid parts of the body.

7.7
Rotational Inertia

TABLE 9–2
MOMENTS OF INERTIA OF VARIOUS BODIES

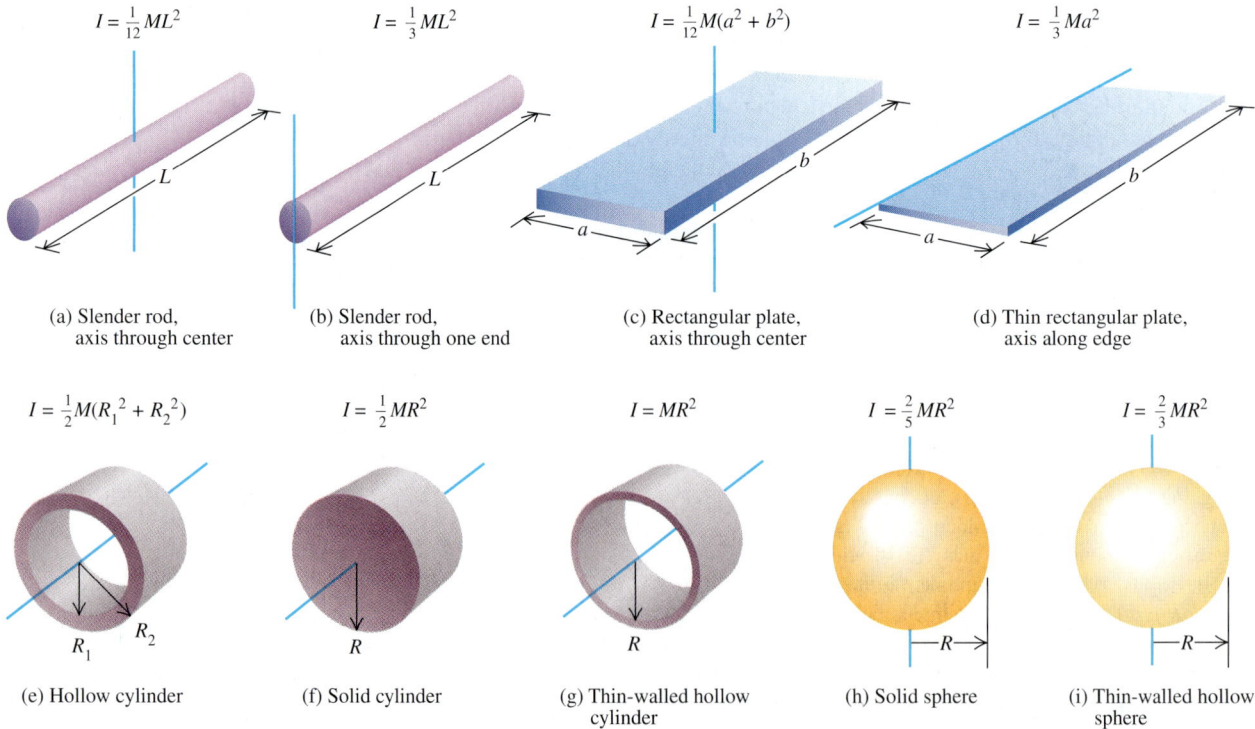

$I = \frac{1}{12} ML^2$

(a) Slender rod, axis through center

$I = \frac{1}{3} ML^2$

(b) Slender rod, axis through one end

$I = \frac{1}{12} M(a^2 + b^2)$

(c) Rectangular plate, axis through center

$I = \frac{1}{3} Ma^2$

(d) Thin rectangular plate, axis along edge

$I = \frac{1}{2} M(R_1^2 + R_2^2)$

(e) Hollow cylinder

$I = \frac{1}{2} MR^2$

(f) Solid cylinder

$I = MR^2$

(g) Thin-walled hollow cylinder

$I = \frac{2}{5} MR^2$

(h) Solid sphere

$I = \frac{2}{3} MR^2$

(i) Thin-walled hollow sphere

You may be tempted to try to compute the moment of inertia of a body by assuming that all the mass is concentrated at the center of mass and then multiplying the total mass by the square of the distance from the center of mass to the axis. Resist that temptation; it doesn't work! For example, when a uniform thin rod of length L and mass M is pivoted about an axis through one end, perpendicular to the rod, the moment of inertia is $I = ML^2/3$ (case (b) in Table 9–2). If we took the mass as concentrated at the center, a distance $L/2$ from the axis, we would obtain the *incorrect* result $I = M(L/2)^2 = ML^2/4$.

Now that we know how to calculate the kinetic energy of a rotating rigid body, we can apply the energy principles of Chapter 7 to rotational motion. Here are some points of strategy and some examples.

Problem–Solving Strategy

ROTATIONAL ENERGY

1. You can use work-energy relations and conservation of energy to find relations involving position and motion of a rotating body. The problem-solving strategy outlined for energy problems in Section 7–2 is equally useful for rotating rigid bodies; the only difference is that rotational kinetic energy is expressed in terms of the moment of inertia I and angular speed ω of the body ($K = \frac{1}{2}I\omega^2$) instead of its mass m and speed v.

2. Many problems involve a rope or cable wrapped around a rotating rigid body that functions as a pulley. In these situations, remember that the point on the pulley that contacts the rope has the same linear speed as the rope, provided that the rope doesn't slip on the pulley. You can then take advantage of Eqs. (9–13) and (9–14), which relate the linear speed and tangential acceleration of a point on a rigid body to the angular speed and angular acceleration of the body. Examples 9–8 and 9–9 illustrate this point.

EXAMPLE 9–8

An unwinding cable A light, flexible, nonstretching cable is wrapped several times around a winch drum, a solid cylinder of mass 50 kg and diameter 0.120 m, which rotates about a stationary horizontal axis held by frictionless bearings (Fig. 9–14). The free end of the cable is pulled with a constant force of magnitude 9.0 N for a distance of 2.0 m. It unwinds without slipping, turning the cylinder as it does so. If the cylinder is initially at rest, find its final angular speed and the final speed of the cable.

SOLUTION There is friction between the cable and the cylinder; this is what makes the cylinder rotate when the cable is pulled. But because the cable doesn't slip, there is no sliding of the cable relative to the cylinder, and no mechanical energy is lost due to friction. The change in kinetic energy of the cylinder is

equal to the work $W = Fs$ done by the force $F = 9.0$ N acting over a distance $s = 2.0$ m; hence $W = (9.0 \text{ N})(2.0 \text{ m}) = 18$ J. From Table 9–2 the moment of inertia is

$$I = \frac{1}{2} MR^2 = \frac{1}{2} (50 \text{ kg})(0.060 \text{ m})^2 = 0.090 \text{ kg} \cdot \text{m}^2.$$

(The radius R is just half the diameter of the cylinder.) Since the cylinder is initially at rest ($\omega_0 = 0$), the work-energy theorem gives

$$W = K_2 - K_1$$
$$= \frac{1}{2} I\omega^2 - \frac{1}{2} I\omega_0^2 = \frac{1}{2} I\omega^2,$$
$$\omega = \sqrt{\frac{2W}{I}} = \sqrt{\frac{2(18 \text{ J})}{0.090 \text{ kg} \cdot \text{m}^2}}$$
$$= 20 \text{ rad/s}.$$

The final speed of the cable is equal to the final tangential speed v of the cylinder, which is given by Eq. (9–13):

$$v = r\omega = (0.060 \text{ m})(20 \text{ rad/s}) = 1.2 \text{ m/s}.$$

M

$F = 9.0$ N

$d = 0.120$ m

9–14 A cable unwinds from a cylinder. The force F does work and increases the kinetic energy of the cylinder.

EXAMPLE 9-9

An unwinding cable II In a lab experiment to test conservation of energy in rotational motion, we wrap a light, flexible cable around a solid cylinder with mass M and radius R. The cylinder rotates with negligible friction about a stationary horizontal axis (Fig. 9–15). We tie the free end of the cable to an object of mass m and release the object with no initial velocity at a distance h above the floor. As the object falls, the cable unwinds without stretching or slipping, turning the cylinder. Find the speed of the falling object and the angular speed of the cylinder just as the object strikes the floor.

SOLUTION Initially, the system has no kinetic energy ($K_1 = 0$). We take the potential energy to be zero when the object is at

9–15 As the cylinder rotates counterclockwise, the cable unwinds, the hanging object falls, and gravitational potential energy is transformed into kinetic energy.

floor level; then $U_1 = mgh$ and $U_2 = 0$. (We can ignore the gravitational potential energy for the rotating cylinder, since its height doesn't change.) As in the previous example, no work is done by friction, so $W_{other} = 0$ and $K_1 + U_1 = K_2 + U_2$. (The cable does no net work; at one end the force and displacement are in the same direction, and at the other end they are in opposite directions. So the total work done by both ends of the cable is zero.) Just before the falling object hits the floor, both this object and the cylinder have kinetic energy. The total kinetic energy K_2 at that instant is

$$K_2 = \frac{1}{2} mv^2 + \frac{1}{2} I\omega^2.$$

From Table 9–2 the moment of inertia of the cylinder is $I = \frac{1}{2}MR^2$. Also, v and ω are related by $v = R\omega$, since the speed of the falling mass must be equal to the tangential speed at the outer surface of the cylinder. Using these relations and equating the initial and final total energies, we find

$$0 + mgh = \frac{1}{2} mv^2 + \frac{1}{2} \left(\frac{1}{2} MR^2 \right) \left(\frac{v}{R} \right)^2 = \frac{1}{2} \left(m + \frac{1}{2} M \right) v^2,$$

$$v = \sqrt{\frac{2gh}{1 + M/2m}}.$$

The final angular speed ω is obtained from $\omega = v/R$.

Let's check some particular cases. When M is much larger than m, v is very small, as we would expect. When M is much smaller than m, v is nearly equal to $\sqrt{2gh}$, which is the speed of a body that falls freely from an initial height h. Does it surprise you that v doesn't depend on the radius of the cylinder?

7.14
Woman and Flywheel Elevator—
Energy Approach

7.15
Rotoride—Energy Approach

GRAVITATIONAL POTENTIAL ENERGY FOR AN EXTENDED BODY

In Example 9–9 the cable was of negligible mass, so we could ignore its kinetic energy as well as the gravitational potential energy associated with it. If the mass is *not* negligible, we need to know how to calculate the *gravitational potential* energy associated with such an extended body. If the acceleration of gravity g is the same at all points on the body, the gravitational potential energy is the same as though all the mass were concentrated at the center of mass of the body. Suppose we take the y-axis vertically upward. Then for a body with total mass M, the gravitational potential energy U is simply

$$U = Mgy_{cm} \qquad \text{(gravitational potential energy for an extended body),} \quad (9\text{--}18)$$

where y_{cm} is the y-coordinate of the center of mass. This expression applies to any extended body, whether it is rigid or not.

To prove Eq. (9–18), we again represent the body as a collection of mass elements m_i. The potential energy for element m_i is $m_i gy_i$, so the total potential energy is

$$U = m_1 gy_1 + m_2 gy_2 + \cdots = (m_1 y_1 + m_2 y_2 + \cdots)g.$$

But from Eq. (8–28), which defines the coordinates of the center of mass,

$$m_1 y_1 + m_2 y_2 + \cdots = (m_1 + m_2 + \cdots)y_{cm} = My_{cm},$$

where $M = m_1 + m_2 + \cdots$ is the total mass. Combining this with the above expression for U, we find $U = Mgy_{cm}$, in agreement with Eq. (9–18).

We leave the application of Eq. (9–18) to the problems. We'll make use of this relationship in Chapter 10 in the analysis of rigid-body problems where the axis of rotation moves.

9–6 PARALLEL-AXIS THEOREM

We pointed out in Section 9–5 that a body doesn't have just one moment of inertia. In fact, it has infinitely many, because there are infinitely many axes about which it might rotate. But there is a simple relationship between the moment of inertia I_{cm} of a body of mass M about an axis through its center of mass and the moment of inertia I_P about any other axis parallel to the original one but displaced from it by a distance d. This relationship, called the **parallel-axis theorem,** states that

$$I_P = I_{cm} + Md^2 \qquad \text{(parallel-axis theorem).} \qquad (9\text{--}19)$$

To prove this theorem, we consider two axes, both parallel to the z-axis, one through the center of mass and the other through a point P (Fig. 9–16). First we take a very thin slice of the body, parallel to the xy-plane and perpendicular to the z-axis. We take the origin of our coordinate system to be at the center of mass of the body; the coordinates of the center of mass are then $x_{cm} = y_{cm} = z_{cm} = 0$. The axis through the center of mass passes through this thin slice at point O, and the parallel axis passes through point P, whose x- and y-coordinates are (a, b). The distance of this axis from the axis through the center of mass is d, where $d^2 = a^2 + b^2$.

We can write an expression for the moment of inertia I_P about the axis through point P. Let m_i be a mass element in our slice, with coordinates (x_i, y_i, z_i). Then the moment of inertia I_{cm} of the slice about the axis through the center of mass (at O) is

$$I_{cm} = \sum_i m_i (x_i{}^2 + y_i{}^2).$$

The moment of inertia of the slice about the axis through P is

$$I_P = \sum_i m_i [(x_i - a)^2 + (y_i - b)^2].$$

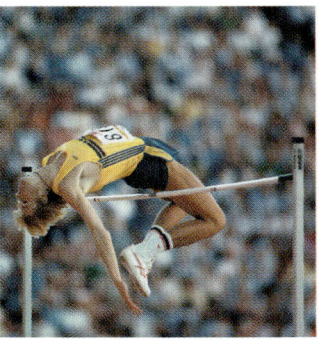

In a technique called the "Fosbury flop" after its innovator, this athlete arches his body as he passes over the bar in the high jump. As a result, his center of mass actually passes *under* the bar. This technique requires a smaller increase in gravitational potential energy (Eq. (9–18)) than the older method of straddling the bar.

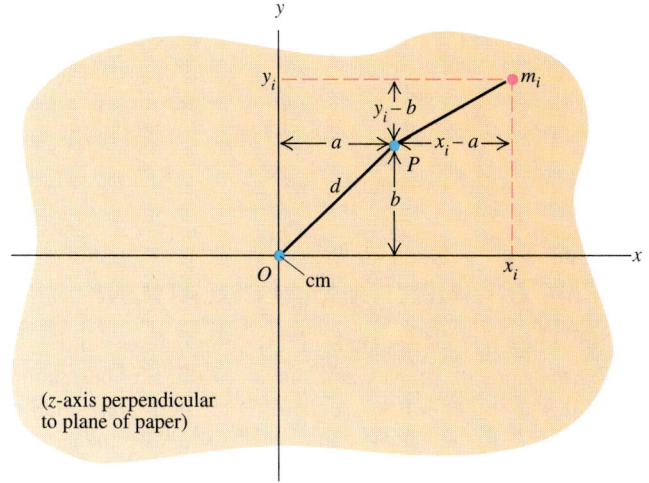

(z-axis perpendicular to plane of paper)

9–16 The mass element m_i has coordinates (x_i, y_i) with respect to an axis of rotation through the center of mass and perpendicular to the plane of the figure. The mass element has coordinates $(x_i - a, y_i - b)$ with respect to the parallel axis through point P.

These expressions don't involve the coordinates z_i measured perpendicular to the slices, so we can extend the sums to include *all* particles in *all* slices. Then I_P becomes the moment of inertia of the *entire* body for an axis through P. We then expand the squared terms and regroup, and obtain

$$I_P = \sum_i m_i (x_i^2 + y_i^2) - 2a \sum_i m_i x_i - 2b \sum_i m_i y_i + (a^2 + b^2) \sum_i m_i.$$

The first sum is I_{cm}. From Eq. (8–28), the definition of the center of mass, the second and third sums are proportional to x_{cm} and y_{cm}; these are zero because we have taken our origin to be the center of mass. The final term is d^2 multiplied by the total mass, or Md^2. This completes our proof that $I_P = I_{cm} + Md^2$.

As Eq. (9–19) shows, a rigid body has a lower moment of inertia about an axis through its center of mass than about any other parallel axis. Thus it's easier to start a body rotating if the rotation axis passes through the center of mass. This suggests that it's somehow most natural for a rotating body to rotate about an axis through its center of mass; we'll make this idea more quantitative in Chapter 10.

EXAMPLE 9-10

Using the parallel-axis theorem I A part of a mechanical linkage (Fig. 9–17) has a mass of 3.6 kg. We measure its moment of inertia about an axis 0.15 m from its center of mass to be $I_P = 0.132$ kg · m². What is the moment of inertia I_{cm} about a parallel axis through the center of mass?

SOLUTION We rearrange Eq. (9–19) and substitute in the values:

$$I_{cm} = I_P - Md^2 = 0.132 \text{ kg} \cdot \text{m}^2 - (3.6 \text{ kg})(0.15 \text{ m})^2$$
$$= 0.051 \text{ kg} \cdot \text{m}^2.$$

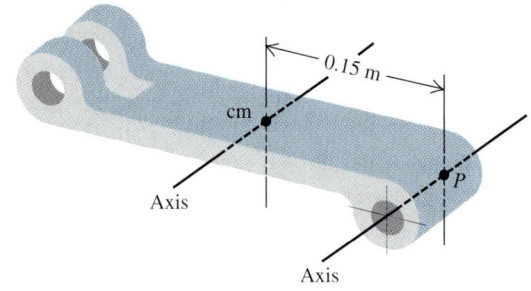

9–17 Calculating I_{cm} from a measurement of I_P.

EXAMPLE 9-11

Using the parallel-axis theorem II Find the moment of inertia of a thin uniform disk of mass M and radius R about an axis perpendicular to its plane at the edge.

SOLUTION From Table 9–2, $I_{cm} = MR^2/2$, and in this case $d = R$. The parallel-axis theorem gives

$$I_P = I_{cm} + Md^2 = \frac{MR^2}{2} + MR^2 = \frac{3MR^2}{2}.$$

This is a very simple solution of a problem that would be fairly complicated if we tried to solve it using the integration techniques to be discussed in the next section.

---

*9-7 MOMENT OF INERTIA CALCULATIONS

When a rigid body cannot be represented by a few point masses but is a continuous distribution of mass, the sum of masses and distances that defines moment of inertia (Eq. (9–16)) becomes an integral. Imagine dividing the body into small mass elements dm so that all points in a particular element are at essentially the same perpendicular distance from the axis of rotation. We call this distance r, as before. Then the moment

of inertia is

$$I = \int r^2 \, dm. \tag{9–20}$$

To evaluate the integral, we have to represent r and dm in terms of the same integration variable. When we have an effectively one-dimensional object, such as the slender rods (a) and (b) in Table 9–2, we can use a coordinate x along the length and relate dm to an increment dx. For a three-dimensional object it is usually easiest to express dm in terms of an element of volume dV and the *density* ρ of the body. Density is mass per unit volume, $\rho = dm/dV$, so we may also write Eq. (9–20) as

$$I = \int r^2 \, \rho \, dV.$$

If the body is uniform in density, then we may take ρ outside the integral:

$$I = \rho \int r^2 \, dV. \tag{9–21}$$

To use this equation, we have to express the volume element dV in terms of the differentials of the integration variables, such as $dV = dx \, dy \, dz$. The element dV must always be chosen so that all points within it are at very nearly the same distance from the axis of rotation. The limits on the integral are determined by the shape and dimensions of the body. For regularly shaped bodies this integration can often be carried out quite easily.

By measuring small variations in the orbits of satellites, geophysicists can measure the earth's moment of inertia. This tells us how our planet's mass is distributed within its interior. The same technique has been used with interplanetary spacecraft to explore the internal structure of other worlds.

EXAMPLE 9-12

Uniform thin rod, axis perpendicular to length Figure 9–18 shows a slender uniform rod with mass M and length L. It might be a baton held by a twirler in a marching band (less the rubber end caps). Compute its moment of inertia about an axis through O, at an arbitrary distance h from one end.

SOLUTION We choose as an element of mass a short section of rod with length dx at a distance x from point O. The ratio of the mass dm of this element to the total mass M is equal to the ratio of its length dx to the total length L:

$$\frac{dm}{M} = \frac{dx}{L}.$$

We solve this for dm, substitute into Eq. (9–20), and add the appropriate integration limits on x to obtain

$$I = \int x^2 \, dm = \frac{M}{L} \int_{-h}^{L-h} x^2 dx$$

$$= \left[\frac{M}{L} \left(\frac{x^3}{3} \right) \right]_{-h}^{L-h} = \frac{1}{3} M(L^2 - 3Lh + 3h^2).$$

From this general expression we can find the moment of inertia about an axis through any point on the rod. For example, if the axis is at the left end, $h = 0$ and

$$I = \frac{1}{3} ML^2.$$

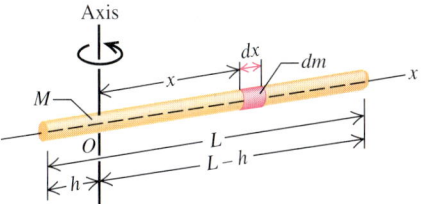

9–18 Finding the moment of inertia of a thin rod about an axis through O. The mass element is a segment with length dx.

If the axis is at the right end, we should get the same result. Putting $h = L$, we again get

$$I = \frac{1}{3} ML^2.$$

If the axis passes through the center, the usual place for a twirled baton, then $h = L/2$ and

$$I = \frac{1}{12} ML^2.$$

These results agree with the expressions in Table 9–2.

EXAMPLE 9–13

Hollow or solid cylinder, rotating about axis of symmetry Figure 9–19 shows a hollow, uniform cylinder with length L, inner radius R_1, and outer radius R_2. This might be a steel cylinder in a printing press or a sheet-steel rolling mill. Find the moment of inertia about the axis of symmetry of the cylinder.

SOLUTION We choose as a volume element a thin cylindrical shell of radius r, thickness dr, and length L. All parts of this element are at very nearly the same distance from the axis. The volume of the element is very nearly equal to that of a flat sheet with thickness dr, length L, and width $2\pi r$ (the circumference of the shell). Then

$$dm = \rho\, dV = \rho\,(2\pi rL\, dr)$$

The moment of inertia is given by

$$I = \int r^2\, dm = \int_{R_1}^{R_2} r^2\, \rho\,(2\pi rL\, dr)$$

$$= 2\pi\rho L \int_{R_1}^{R_2} r^3\, dr$$

$$= \frac{2\pi\rho L}{4}\,(R_2{}^4 - R_1{}^4)$$

$$= \frac{\pi\rho L}{2}\,(R_2{}^2 - R_1{}^2)(R_2{}^2 + R_1{}^2).$$

It is usually more convenient to express the moment of inertia in terms of the total mass M of the body, which is its density ρ multiplied by the total volume V. The volume is

$$V = \pi L(R_2{}^2 - R_1{}^2),$$

so the total mass M is

$$M = \rho V = \pi L\rho(R_2{}^2 - R_1{}^2),$$

and the moment of inertia is

$$I = \frac{1}{2} M(R_1{}^2 + R_2{}^2).$$

This is as shown in Table 9–2, case (e).

If the cylinder is solid, such as a lawn roller, $R_1 = 0$. Calling the outer radius R_2 simply R, we find that the moment of inertia

9–19 Finding the moment of inertia of a hollow cylinder about its symmetry axis. The mass element is a cylindrical shell with radius r and thickness dr.

of a solid cylinder of radius R is

$$I = \frac{1}{2} MR^2.$$

If the cylinder has a very thin wall (like a stovepipe), R_1 and R_2 are very nearly equal; if R represents this common radius,

$$I = MR^2.$$

We could have predicted this last result; in a thin-walled cylinder, all the mass is the same distance $r = R$ from the axis, so $I = \int r^2\, dm = R^2 \int dm = MR^2$.

Note that the moment of inertia of a cylinder about an axis coinciding with its axis of symmetry depends on its mass and radii but not on its length L. Two hollow cylinders with the same inner and outer radii, one of wood and one of brass, but having the same mass M have equal moments of inertia even though the length of the wood cylinder is much greater. The moment of inertia depends only on the *radial* distribution of mass, not on its distribution along the axis. Thus the above results hold also for a thin disk, such as a washer or a compact disc (CD).

EXAMPLE 9–14

Uniform sphere with radius R, axis through center The object in Fig. 9–20 might be a billiard ball, a steel ball bearing, or the world's largest ball of string. Find the moment of inertia about an axis through the center of the sphere.

SOLUTION We begin by dividing the sphere into thin disks of thickness dx, whose moment of inertia we know from Example 9–13. The radius r of the disk shown is

$$r = \sqrt{R^2 - x^2}.$$

Its volume is

$$dV = \pi r^2\, dx = \pi(R^2 - x^2)\, dx,$$

and its mass is

$$dm = \rho\, dV = \pi\rho(R^2 - x^2)\, dx.$$

From Example 9–13, the moment of inertia of a disk of radius r and mass dm is

$$dI = \frac{1}{2} r^2\, dm = \frac{1}{2}\left(\sqrt{R^2 - x^2}\right)^2 [\pi\rho(R^2 - x^2)\, dx]$$

$$= \frac{\pi\rho}{2}\,(R^2 - x^2)^2\, dx.$$

Integrating this expression from $x = 0$ to $x = R$ gives the moment of inertia of the right hemisphere. From symmetry, the total I for the entire sphere is just twice this:

$$I = (2)\frac{\pi\rho}{2}\int_0^R (R^2 - x^2)^2 dx.$$

Carrying out the integration, we obtain

$$I = \frac{8\pi\rho}{15}R^5.$$

The mass M of the sphere of volume $V = 4\pi R^3/3$ is

$$M = \rho V = \frac{4\pi\rho R^3}{3}.$$

By comparing the expressions for I and M, we find

$$I = \frac{2}{5}MR^2.$$

This agrees with the expression in Table 9–2, case (h). Note that the moment of inertia of a solid sphere of mass M and R is less than the moment of inertia of a solid *cylinder* of the same mass and radius, $I = \frac{1}{2}MR^2$. The reason is that more of the sphere's mass is located close to the axis.

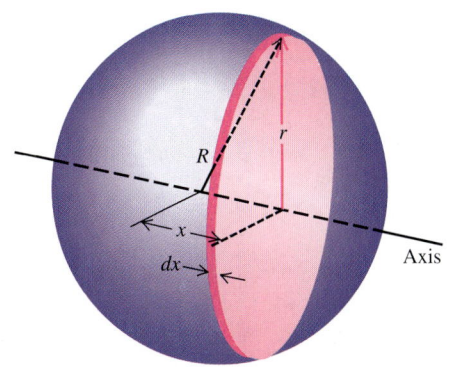

9–20 Finding the moment of inertia of a sphere about an axis through its center. The mass element is a disk with thickness dx.

SUMMARY

- When a rigid body rotates about a fixed axis, its position is described by an angular coordinate θ. The angular velocity ω is defined as the derivative of the angular coordinate θ:

$$\omega = \lim_{\Delta t \to 0}\frac{\Delta\theta}{\Delta t} = \frac{d\theta}{dt}. \tag{9–3}$$

The angular acceleration α is defined as the derivative of the angular velocity ω, or the second derivative of the angular coordinate θ:

$$\alpha = \lim_{\Delta t \to 0}\frac{\Delta\omega}{\Delta t} = \frac{d\omega}{dt} = \frac{d^2\theta}{dt^2}. \tag{9–6}$$

- When a rigid body rotates about a fixed axis with constant angular acceleration, the angular position, velocity, and acceleration are related by

$$\theta = \theta_0 + \omega_0 t + \frac{1}{2}\alpha t^2, \tag{9–11}$$

$$\theta - \theta_0 = \frac{1}{2}(\omega_0 + \omega)t, \tag{9–10}$$

$$\omega = \omega_0 + \alpha t, \tag{9–7}$$

$$\omega^2 = \omega_0^2 + 2\alpha(\theta - \theta_0), \tag{9–12}$$

where θ_0 and ω_0 are the initial values of angular position and angular velocity, respectively.

- A particle in a rigid body rotating with angular speed ω has a speed v given by

$$v = r\omega, \tag{9–13}$$

and an acceleration $\vec{a}$ with tangential component

$$a_{\text{tan}} = \frac{dv}{dt} = r\frac{d\omega}{dt} = r\alpha \tag{9–14}$$

KEY TERMS

rigid body, 268
radian, 269
average angular velocity, 269
instantaneous angular velocity, 269
average angular acceleration, 270
instantaneous angular acceleration, 271
angular speed, 274
tangential component of acceleration, 274
centripetal component of acceleration, 274
moment of inertia, 277
rotational kinetic energy, 277
parallel-axis theorem, 281

and radial (centripetal) component

$$a_{rad} = \frac{v^2}{r} = \omega^2 r. \qquad (9\text{--}15)$$

- The moment of inertia I of a body about a given axis is defined as

$$I = m_1 r_1^2 + m_2 r_2^2 + \cdots = \sum_i m_i r_i^2, \qquad (9\text{--}16)$$

  where r_i is the perpendicular distance of mass m_i from the axis of rotation. The greater the moment of inertia, the more difficult it is to change the state of the body's rotation.

- The kinetic energy of a rigid body rotating about a fixed axis (rotational kinetic energy) is given by

$$K = \frac{1}{2} I \omega^2, \qquad (9\text{--}17)$$

  where I is the moment of inertia for that rotation axis.

- The moment of inertia I_{cm} of a body of mass M about an axis through the center of mass is related to the moment of inertia I_P about a parallel axis at a distance d from the first axis by

$$I_P = I_{cm} + Md^2. \qquad (9\text{--}19)$$

DISCUSSION QUESTIONS

Q9–1 When rewinding an audio or video tape, why does the tape wind up fastest at the end?

Q9–2 Does a body rotating about a fixed axis have to be *perfectly* rigid for all points on the body to have the same angular velocity and the same angular acceleration? Explain.

Q9–3 What is the difference between tangential and radial acceleration for a point on a rotating body?

Q9–4 In Fig. 9–11, all points on the chain have the same linear speed v. Is the magnitude a of the linear acceleration also the same for all points on the chain? How are the angular accelerations of the two sprockets related? Explain.

Q9–5 In Fig. 9–11, how are the radial accelerations of points at the teeth of the two sprockets related? Explain the reasoning behind your answer.

Q9–6 A flywheel rotates with constant angular velocity. Does a point on its rim have a tangential acceleration? A radial acceleration? Are these accelerations constant in magnitude? In direction? In each case give the reasoning behind your answer.

Q9–7 What is the purpose of the spin cycle of a washing machine? Explain in terms of acceleration components.

Q9–8 Although angular velocity and angular acceleration can be treated as vectors, the angular displacement θ, despite having a magnitude and a direction, cannot. This is because θ does not follow the commutative law of vector addition (Eq. 1–4). Prove this to yourself in the following way. Lay your physics textbook flat on the desk in front of you with the cover side up and where you can read the writing on it. Rotate the farthest edge toward you by 90° about a horizontal axis. Call this angular displacement θ_1. Then rotate the left edge toward you by 90° about a vertical axis. Call this angular displacement θ_2. The spine of the

book should now face you, with the writing on it oriented so that it can be read. Now start over again but carry out the two rotations in the reverse order. Do you get a different result? That is, is $\theta_1 + \theta_2$ equal to $\theta_2 + \theta_1$? Now repeat this experiment but this time with an angle of 1° rather than 90°. Do you think that the infinitesimal displacement $d\vec{\theta}$ obeys the commutative law of addition and hence qualifies as a vector? If so, how is the direction of $d\vec{\theta}$ related to the direction of $\vec{\omega}$?

Q9–9 Can you think of a body that has the same moment of inertia for all possible axes? If so, give an example, and if not, explain why this is not possible. Can you think of a body that has the same moment of inertia for all axes passing through a certain point? If so, give an example and indicate where the point is located.

Q9–10 To maximize the moment of inertia of a flywheel while minimizing its weight, what shape and distribution of mass should it have? Explain.

Q9–11 How might you determine experimentally the moment of inertia of an irregularly shaped body about a given axis?

Q9–12 A cylindrical body has mass M and radius R. Can the mass be distributed within the body in such a way that its moment of inertia about its axis of symmetry is greater than MR^2? Explain.

Q9–13 Describe how part (b) of Table 9–2 could be used to derive the result in part (d).

Q9–14 The moment of inertia I of a rigid body for an axis through its center of mass is I_{cm}. Is there any other axis parallel to this axis for which I is less than I_{cm}? Explain.

Q9–15 For the equations for I given in parts (a) and (b) of Table 9–2 to be valid, must the rod have a circular cross section?

Is there any restriction on the size of the cross section in order for these equations to apply? Explain.

Q9–16 In part (d) of Table 9–2, the thickness of the plate must be much less than a for the expression given for I to apply. But in part (c), the expression given for I applies no matter how thick the plate is. Explain.

Q9–17 In Fig. 5–26a, use the expressions $K = \frac{1}{2}I\omega^2$ and $K = \frac{1}{2}mv^2$ to calculate the kinetic energy of the box (treated as a single particle). How do these two results compare? Explain.

Q9–18 Equation (9–18) says to use y_{cm} to calculate U for an extended rigid body. But in Example 9–9 (Section 9–5), y is measured to the bottom of the hanging mass rather than to the center of mass. Is this a mistake? Explain.

Q9–19 Any angular measure—radians, degrees, or revolutions—can be used in some of the equations in Chapter 9, but only radian measure can be used in others. Identify those for which using radians is necessary and those for which it is not, and in each case give the reasoning behind your answer.

EXERCISES

SECTION 9–2 ANGULAR VELOCITY AND ACCELERATION

9–1 a) What angle in radians is subtended by an arc 1.50 m in length on the circumference of a circle of radius 2.50 m? What is this angle in degrees? b) An arc 14.0 cm in length on the circumference of a circle subtends an angle of 128°. What is the radius of the circle? c) The angle between two radii of a circle with radius 1.50 m is 0.700 rad. What length of arc is intercepted on the circumference of the circle by the two radii?

9–2 An airplane propeller is rotating at 1900 rev/min. a) Compute the propeller's angular velocity in rad/s. b) How long in seconds does it take for the propeller to turn through 35°?

9–3 Consider the flywheel of Examples 9–1 and 9–2 (Section 9–2). a) Calculate the instantaneous angular acceleration at $t = 3.5$ s. Explain why your result is equal to the average angular acceleration for the time interval from 2.0 s to 5.0 s. b) Calculate the instantaneous angular velocity at $t = 3.5$ s. Explain why your result is *not* equal to the average angular velocity for the 2.0 s to 5.0 s time interval, even though 3.5 s is at the middle of this time interval.

9–4 A fan blade rotates with angular velocity given by $\omega(t) = \gamma - \beta t^2$, where $\gamma = 5.00$ rad/s and $\beta = 0.800$ rad/s^3. (a) Calculate the angular acceleration as a function of time. (b) Calculate the instantaneous angular acceleration α at $t = 3.00$ s and the average angular acceleration α_{av} for the time interval $t = 0$ to $t = 3.00$ s. How do these two quantities compare? If they are different, why are they different?

9–5 A child is pushing a merry-go-round. The angle through which the merry-go-round has turned varies with time according to $\theta(t) = \gamma t + \beta t^3$, where $\gamma = 0.400$ rad/s and $\beta = 0.0120$ rad/s^3. a) Calculate the angular velocity of the merry-go-round as a function of time. b) What is the initial value of the angular velocity? c) Calculate the instantaneous value of the angular velocity ω at $t = 5.00$ s and the average angular velocity ω_{av} for the time interval $t = 0$ to $t = 5.00$ s. Show that ω_{av} is *not* equal to the average of the instantaneous angular velocities at $t = 0$ and $t = 5.00$ s, and explain why it is not.

9–6 At $t = 0$ the current to a dc electric motor is reversed, resulting in an angular displacement of the motor shaft given by $\theta(t) = (250 \text{ rad/s})t - (20.0 \text{ rad/s}^2)t^2 - (1.50 \text{ rad/s}^3)t^3$. a) At what time is the angular velocity of the motor shaft zero? b) Calculate the angular acceleration at the instant that the motor shaft has zero angular velocity. c) How many revolutions does the motor shaft turn through between the time when the current is reversed and the instant when the angular velocity is zero? d) How fast was the motor shaft rotating at $t = 0$, when the current was reversed? e) Calculate the average angular velocity for the time period from $t = 0$ to the time calculated in part (a).

9–7 The angle θ through which a bicycle wheel turns is given by $\theta(t) = a + bt^2 - ct^3$, where a, b, and c are positive constants such that for t in seconds, θ will be in radians. a) Calculate the angular acceleration of the wheel as a function of time. b) At what time is the angular velocity of the wheel instantaneously not changing?

SECTION 9–3 ROTATION WITH CONSTANT ANGULAR ACCELERATION

9–8 A bicycle wheel has an initial angular velocity of 1.50 rad/s. a) If its angular acceleration is constant and equal to 0.300 rad/s^2, what is its angular velocity at $t = 2.50$ s? b) Through what angle has the wheel turned between $t = 0$ and $t = 2.50$ s?

9–9 An electric fan is turned off, and its angular velocity decreases uniformly from 500 rev/min to 200 rev/min in 4.00 s. a) Find the angular acceleration in rev/s^2 and the number of revolutions made by the motor in the 4.00-s interval. b) How many more seconds are required for the fan to come to rest if the angular acceleration remains constant at the value calculated in part (a)?

9–10 a) Derive Eq. (9–12) by combining Eqs. (9–7) and (9–11) to eliminate t. b) The angular velocity of an airplane propeller increases from 12.0 rad/s to 16.0 rad/s while turning through 7.00 rad. What is the angular acceleration in rad/s^2?

9–11 The rotating blade of a blender turns with constant angular acceleration 1.50 rad/s^2. a) How much time does it take to reach an angular velocity of 36.0 rad/s, starting from rest? b) Through how many revolutions does the blade turn in this time interval?

9–12 A flywheel requires 4.00 s to rotate through 162 rad. Its angular velocity at the end of this time is 108 rad/s. Find a) the angular velocity at the beginning of the 4.00-s interval; b) the constant angular acceleration.

9–13 Emilie's potter's wheel rotates with a constant 2.25-rad/s^2 angular acceleration. After 4.00 s the wheel has rotated through an angle of 60.0 rad. What was the angular velocity of the wheel at the beginning of the 4.00-s interval?

9–14 A circular saw blade 0.200 m in diameter starts from rest. In 6.00 s it accelerates with constant angular acceleration to an angular velocity of 140 rad/s. Find the angular acceleration and the angle through which the blade has turned.

9–15 A safety device brings the blade of a power mower from an initial angular velocity of ω_i to rest in 1.00 revolution. At the same constant acceleration, how many revolutions would it take the blade to come to rest from an initial angular velocity ω_3 that was three times as great, $\omega_3 = 3\omega_1$?

9–16 A straight piece of reflecting tape extends from the center of a wheel to its rim. You darken the room and use a camera and strobe unit that flashes once every 0.050 s to take pictures of the wheel as it rotates counterclockwise. You trigger the strobe so that the first flash ($t = 0$) occurs when the tape is horizontal to the right at an angular displacement of zero. For the following situations draw a sketch of the photo you will get for the time exposure over five flashes (at $t = 0$, 0.050 s, 0.100 s, 0.150 s, and 0.200 s) and graph θ versus t and ω versus t for $t = 0$ to $t = 0.200$ s. a) The angular velocity is constant at 10.0 rev/s. b) The wheel starts from rest with a constant angular acceleration of 25.0 rev/s^2. c) The wheel is rotating at 10.0 rev/s at $t = 0$ and changes angular velocity at a constant rate of -50.0 rev/s^2.

9–17 At $t = 0$ a grinding wheel has an angular velocity of 24.0 rad/s. It has a constant angular acceleration of 30.0 rad/s^2 until a circuit breaker trips at $t = 2.00$ s. From then on, it turns through 432 rad as it coasts to a stop at constant angular acceleration. a) Through what total angle did the wheel turn between $t = 0$ and the time it stopped? b) At what time did it stop? c) What was its acceleration as it slowed down?

9–18 a) Derive a constant-angular-acceleration equation that gives $\theta - \theta_0$ in terms of ω, α, and t (no ω_0 in the equation). b) At $t = 8.0$ s a gear is rotating about a fixed axis at 4.50 rad/s. During the preceding 8.0 s, it has turned through 40.0 rad. Use the result of part (a) to calculate the gear's constant angular acceleration. c) What was the angular velocity of the gear at $t = 0$?

SECTION 9–4 RELATING LINEAR AND ANGULAR KINEMATICS

9–19 The main rotor of a helicopter is turning in a horizontal plane at 90.0 rev/min. The distance from the center of the rotor shaft to each blade tip is 5.00 m. Calculate the speed through the air of the blade tip if a) the helicopter is sitting on the ground; b) the helicopter is rising vertically at 4.00 m/s.

9–20 A compact disc (CD) stores music in a coded pattern of tiny pits 10^{-7} m deep. The pits are arranged in a track that spirals outward toward the rim of the disc; the inner and outer radii of this spiral are 25.0 mm and 58.0 mm, respectively. As the disc spins inside a CD player, the track is scanned at a constant *linear* speed of 1.25 m/s. a) What is the angular speed of the CD when scanning the innermost part of the track? The outermost part of the track? b) The maximum playing time of a CD is 74.0 min. What would be the length of the track on such a maximum-duration CD if it was stretched out in a straight line? c) What is the average angular acceleration of a maximum-duration CD during its 74.0-min playing time? Take the direction of rotation of the disc to be positive.

9–21 A wheel rotates with a constant angular velocity of 6.00 rad/s. a) Compute the radial acceleration of a point 0.500 m from the axis, using the relation $a_{rad} = \omega^2 r$. b) Find the tangential speed of the point and compute its radial acceleration from the relation $a_{rad} = v^2/r$.

9–22 Find the required angular velocity (in rev/min) of an ultracentrifuge for the radial acceleration of a point 2.50 cm from the axis to equal 400,000g (that is, 400,000 times the acceleration due to gravity).

9–23 A flywheel with a radius of 0.300 m starts from rest and accelerates with a constant angular acceleration of 0.600 rad/s^2. Compute the magnitude of the tangential acceleration, the radial acceleration and the resultant acceleration of a point on its rim a) at the start; b) after it has turned through 60.0°; c) after it has turned through 120.0°.

9–24 An electric ceiling fan with blades 0.750 m in diameter is rotating about a fixed axis with an initial angular velocity of 0.250 rev/s. The angular acceleration is 0.900 rev/s^2. a) Compute the angular velocity after 0.200 s. b) Through how many revolutions has the blade turned in this time interval? c) What is the tangential speed of a point on the tip of the blade at $t = 0.200$ s? d) What is the magnitude of the resultant acceleration of a point on the tip of the blade at $t = 0.200$ s?

9–25 An ad claims that a centrifuge takes up only 0.127 m of bench space but can produce a radial acceleration of 3000g at 5000 rev/min. Calculate the required radius of the centrifuge. Is the claim a realistic one?

9–26 a) Derive an equation for the radial acceleration that includes v and ω but not r. b) You are designing a merry-go-round for which a point of the rim will have a radial acceleration of 0.500 m/s^2 when the tangential velocity of that point has magnitude 2.00 m/s. What angular velocity is required to achieve these values?

9–27 A Drill Problem. When drilling a 12.7-mm-diameter hole in wood, plastic, or aluminum, a shop manual recommends a drill speed of 1250 rev/min. For a 12.7-mm-diameter drill bit turning at a constant 1250 rev/min, find a) the maximum linear speed of any part of the bit; b) the maximum radial acceleration of any part of the bit.

9–28 At $t = 3.00$ s a point on the rim of a 0.200-m-radius wheel has a tangential speed of 50.0 m/s as the wheel slows down with a tangential acceleration of constant magnitude 10.0 m/s^2. a) Calculate the wheel's constant angular acceleration. b) Calculate the angular velocities at $t = 3.00$ s and $t = 0$. c) Through what angle did the wheel turn between $t = 0$ and $t = 3.00$ s? d) At what time will the radial acceleration equal g?

9–29 The spin cycles of a washing machine have two angular speeds, 423 rev/min and 640 rev/min. The internal diameter of the drum is 0.470 m. a) What is the ratio of the maximum radial force on the laundry for the higher angular speed to that for the lower speed? b) What is the ratio of the maximum tangential speed of the laundry for the higher angular speed to that for the lower speed? c) Find the laundry's maximum tangential speed and the maximum radial acceleration, in terms of g.

SECTION 9–5 ENERGY IN ROTATIONAL MOTION

9–30 Small blocks, each with mass m, are clamped at the ends and at the center of a light rod of length L. Compute the moment of inertia of the system about an axis perpendicular to the rod and passing through a point $\frac{1}{4}$ of the length from one end. Neglect the moment of inertia of the light rod.

9–31 A twirler's baton is made of a slender metal cylinder of mass M and length L. Each end has a rubber cap of mass m and each cap can be accurately treated as a particle in this problem. Find the total moment of inertia of the baton about the usual twirling axis (perpendicular to the baton through its center).

9–32 Find the moment of inertia about each of the following axes for a rod that is 0.300 cm in diameter and 1.50 m long and has a mass of 0.0420 kg. Use the formulas of Table 9–2. a) About an axis perpendicular to the rod and passing through its center. b) About an axis perpendicular to the rod and passing through one end. c) About a longitudinal axis passing through the center of the rod.

9–33 Four small spheres, each of which can be regarded as a point of mass 0.200 kg, are arranged in a square 0.400 m on a side and connected by light rods (Fig. 9–21). Find the moment of inertia of the system about an axis a) through the center of the square, perpendicular to its plane (an axis through point O in the figure); b) bisecting two opposite sides of the square (an axis along the line AB in the figure); c) that passes through the centers of the upper left and lower right spheres and through point O.

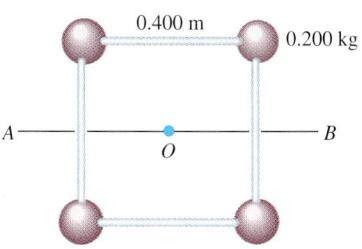

FIGURE **9–21** Exercise 9–33.

9–34 How *I* Scales. If we multiply all design dimensions of an object by a scaling factor f, its volume and mass will be multiplied by f^3. a) By what factor will its moment of inertia be multiplied? b) If a $\frac{1}{48}$-scale model has a rotational kinetic energy of 2.5 J, what will be the kinetic energy for the full-scale object of the same material rotating at the same angular velocity?

9–35 A wagon wheel is constructed as in Fig. 9–22. The radius of the wheel is 0.300 m, and the rim has a mass of 1.40 kg. Each of the eight spokes, which lie along a diameter and are 0.300 m long, has a mass of 0.280 kg. What is the moment of inertia of the wheel about an axis through its center and perpendicular to the plane of the wheel? (Use the formulas given in Table 9–2.)

FIGURE **9–22** Exercise 9–35.

9–36 An airplane propeller is 2.08 m in length (from tip to tip) and has a mass of 117 kg. The propeller is rotating at 2400 rev/min about an axis through its center. a) What is its rotational kinetic energy? Treat the propeller as a slender rod. b) If it were not rotating, how far would it have to drop in free fall to acquire the same kinetic energy?

9–37 a) Show that $\frac{1}{2}I\omega^2$ has units equivalent to joules. Explain why "rads" need not be included in the units. b) Often ω is expressed in rev/min rather than rad/s. Write an expression for rotational kinetic energy such that if I is in kg $\cdot$ m^2 and ω is in rev/min, the energy will be in joules.

9–38 An antique phonograph turntable has a kinetic energy of 0.0250 J when turning at 45.0 rev/min. What is the moment of inertia of the turntable about the rotation axis?

9–39 The flywheel of a gasoline engine is required to give up 500 J of kinetic energy while its angular velocity decreases from 650 rev/min to 520 rev/min. What moment of inertia is required?

9–40 A light flexible rope is wrapped several times around a *hollow* cylinder with a weight of 40.0 N and a radius of 0.25 m, which rotates without friction about a fixed horizontal axis. The cylinder is attached to the axle by spokes of negligible moment of inertia. The cylinder is initially at rest. The free end of the rope is pulled with a constant force P for a distance of 5.00 m, at which point the end of the rope is moving at 6.00 m/s. If the rope does not slip on the cylinder, what is the value of P?

9–41 Energy is to be stored in a 70.0-kg flywheel in the shape of a uniform solid disk with radius $R = 1.20$ m. To prevent structural failure of the flywheel, the maximum allowed radial acceleration of a point on its rim is 3500 m/s^2. What is the maximum kinetic energy that can be stored in the flywheel?

9–42 Suppose the solid cylinder in the apparatus described in Example 9–9 (Section 9–5) is replaced by a thin-walled hollow cylinder of the same mass M and radius R. The cylinder is attached to the axle by spokes of negligible moment of inertia. a) Find the speed of the hanging mass m just as it strikes the floor. b) Is the answer to part (a) greater than, equal to, or less than the speed found in Example 9–9? Explain why, using energy concepts.

9–43 Rate of Kinetic Energy Loss. A rigid body with moment of inertia I spins once every T seconds. The rotation is slowing down, so that $dT/dt > 0$. a) Express the kinetic energy of rotation of the body in terms of I and T. b) Express the rate of change of the kinetic energy of rotation in terms of I, T, and dT/dt. c) A large flywheel has $I = 8.0$ kg $\cdot$ m^2. What is the kinetic energy of the flywheel when the period of rotation is 1.5 s? d) What is the rate of change of the kinetic energy of the flywheel in part (c) at the instant when the period is 1.5 s and is changing at the rate of $dT/dt = 0.0060$?

9–44 A uniform rope 10.0 m long and with mass 3.00 kg is hanging with one end attached to a gymnasium ceiling and the other end just touching the floor. The upper end of the rope is released, and the rope falls to the floor. What is the change in the gravitational potential energy if the rope ends up flat on the floor (not coiled up)?

9–45 Center of Mass of an Extended Object. How much work must a wrestler do to raise the center of mass of his 120-kg opponent a vertical distance of 0.700 m?

SECTION **9–6 PARALLEL-AXIS THEOREM**

9–46 Find the moment of inertia of a hoop (a thin-walled hollow ring) of mass M and radius R about an axis perpendicular to the hoop's plane at an edge.

9–47 About what axis will a uniform balsa-wood sphere have the same moment of inertia as for a thin-walled hollow lead sphere of the same mass and radius with the axis along a diameter?

9–48 Use the parallel-axis theorem to show that the moments of inertia given in parts (a) and (b) of Table 9–2 are consistent.

9–49 A thin, rectangular sheet of metal has mass M and sides of length a and b. Use the parallel-axis theorem to calculate the moment of inertia of the sheet for an axis that is perpendicular to the plane of the sheet and that passes through one corner of the sheet.

9·50 a) For the thin rectangular plate shown in part (d) of Table 9–2, find the moment of inertia about an axis that lies in the plane of the plate, passes through the center of the plate, and is parallel to the axis shown in the figure. b) Find the moment of inertia of the plate for an axis that lies in the plane of the plate, passes through the center of the plate, and is perpendicular to the axis in part (a).

SECTION 9–7 MOMENT OF INERTIA CALCULATIONS

*9–51 Using the information in Table 9–2 and the parallel-axis theorem, find the moment of inertia of the slender rod with mass

M and length L shown in Fig. 9–18 about an axis through O, at an arbitrary distance h from one end. Compare your result to that found by integration in Example 9–12 (Section 9–7).

*9–52 Use Eq. (9–20) to calculate the moment of inertia of a uniform solid disk with mass M and radius R for an axis perpendicular to the plane of the disk and passing through its center.

*9–53 Use Eq. (9–20) to calculate the moment of inertia of a slender, uniform rod with mass M and length L about an axis at one end, perpendicular to the rod.

*9–54 A slender rod with length L has a mass per unit length that varies with distance from the left-hand end, where $x = 0$, according to $dm/dx = \gamma x$, where γ has units of kg/m^2. a) Calculate the total mass of the rod in terms of γ and L. b) Use Eq. (9–20) to calculate the moment of inertia of the rod for an axis at the left-hand end, perpendicular to the rod. Use the expression you derived in part (a) to express I in terms of M and L. How does your result compare to that for a uniform rod? Explain this comparison. c) Repeat part (b) for an axis at the right-hand end of the rod. How do the results for parts (b) and (c) compare? Explain this result.

PROBLEMS

9–55 Sketch a wheel lying in the plane of your paper and rotating counterclockwise. Choose a point on the rim and draw a vector $\vec{r}$ from the center of the wheel to that point. a) What is the direction of $\vec{\omega}$? b) Show that the velocity of the point is $\vec{v} = \vec{\omega} \times \vec{r}$. c) Show that the radial acceleration of the point is $\vec{a}_{rad} = \vec{\omega} \times \vec{v} = \vec{\omega} \times (\vec{\omega} \times \vec{r})$. (See Exercise 9–26.)

9–56 a) Prove that when an object starts from rest and rotates about a fixed axis with constant angular acceleration, the radial acceleration of a point in the object is directly proportional to its angular displacement. b) Through what angle has the object turned at the instant when the resultant acceleration of a point makes an angle of 36.9° with the radial direction?

9–57 A roller in a printing press turns through an angle $\theta(t)$ given by $\theta(t) = \gamma t^2 - \beta t^3$, where $\gamma = 3.20$ rad/s^2 and $\beta = 0.500$ rad/s^3. a) Calculate the angular velocity of the roller as a function of time. b) Calculate the angular acceleration of the roller as a function of time. c) What is the maximum positive angular velocity, and at what value of t does it occur?

*9–58 A bicycle wheel with a radius of 0.33 m turns with angular acceleration $\alpha(t) = \gamma - \beta t$, where $\gamma = 1.80$ rad/s^2 and $\beta = 0.25$ rad/s^3. It is at rest at $t = 0$. a) Calculate the angular velocity and angular displacement as functions of time. b) Calculate the maximum positive angular velocity and maximum positive angular displacement of the wheel. (*Hint:* See Section 2–7.)

9–59 When a toy car is rapidly scooted across the floor, it stores energy in a flywheel. The car has mass 0.180 kg, and its flywheel has a moment of inertia of 4.00×10^{-5} kg · m^2. The car is 15.0 cm long. An advertisement claims that the car can be made to travel at a scale speed of up to 700 km/h (440 mi/h). The scale speed is the speed of the toy car multiplied by the ratio of the length of an actual car to the length of the toy. Assume a

length of 3.0 m for a real car. a) For a scale speed of 700 km/h, what is the actual translational speed of the car? b) If all the kinetic energy that is initially in the flywheel is converted to the translational kinetic energy of the toy, how much energy is originally stored in the flywheel? c) What initial angular velocity of the flywheel was needed to store the amount of energy calculated in part (b)?

9–60 A classic 1957 Chevrolet Corvette of mass 1240 kg starts from rest and speeds up with a constant tangential acceleration of 3.00 m/s^2 on a circular test track of radius 60.0 m. Treat the car as a particle. a) What is its angular acceleration? b) What is its angular velocity 6.00 s after it starts? c) What is its radial acceleration at this time? d) Sketch a view from above showing the circular track, the car, the velocity vector, and the acceleration component vectors 6.00 s after the car starts. e) What are the magnitudes of the total acceleration and net force for the car at this time? f) What angle do the total acceleration and net force make with the car's velocity at this time?

9–61 The flywheel of a punch press has a moment of inertia of 16.0 kg · m^2 and runs at 300 rev/min. The flywheel supplies all the energy needed in a quick punching operation. a) Find the speed in rev/min to which the flywheel will be reduced by a sudden punching operation requiring 4000 J of work. b) What must be the constant power supply to the flywheel (in watts) to bring it back to its initial speed in a time of 5.00 s?

9–62 A totally inedible cafeteria meatball of mass 40.0 g is attached to the free end of a 2.50-m piece of string that is attached to the ceiling. The meatball is pulled to the left so that the string makes a 36.9° angle with the vertical and is then released. a) What is the magnitude and direction of the angular velocity of the meatball the first time the angular acceleration is zero? b) What is it the second time that $\alpha = 0$? c) At the times described in parts (a) and (b), what is the magnitude and direction of the meatball's radial acceleration? d) Show that the answer to part (c) is independent of the length of the string.

9–63 A vacuum cleaner belt is looped over a shaft of radius 0.45 cm and a wheel of radius 2.00 cm. The arrangement of the belt, shaft, and wheel is similar to that of the chain and sprockets in Fig. 9–11. The motor turns the shaft at 60.0 rev/s and the moving belt turns the wheel, which in turn is connected by another shaft to the roller that beats the dirt out of the rug being vacuumed. Assume that the belt doesn't slip on either the shaft or the wheel. a) What is the speed of a point on the belt? b) What is the angular velocity of the wheel in rad/s?

9–64 The motor of a table saw is rotating at 3450 rev/min. A pulley attached to the motor shaft drives a second pulley of half the diameter by means of a V-belt. A circular saw blade of diameter 0.208 m is mounted on the same rotating shaft as the second pulley. a) The operator is careless and the blade catches and throws back a small piece of wood. This piece of wood moves with linear speed equal to the tangential speed of the rim of the blade. What is this speed? b) Calculate the radial acceleration of points on the outer edge of the blade to see why sawdust doesn't stick to its teeth.

9–65 A wheel changes its angular velocity with a constant angular acceleration while rotating about a fixed axis through its center. a) Show that the change in the magnitude of the radial acceleration during any time interval of a point on the wheel is twice the product of the angular acceleration, the angular displacement, and the perpendicular distance of the point from the axis. b) The radial acceleration of a point on the wheel that is 0.250 m from the axis changes from 25.0 m/s^2 to 85.0 m/s^2 as the wheel rotates through 15.0 rad. Calculate the tangential acceleration of this point. c) Show that the change in the wheel's kinetic energy during any time interval is the product of the moment of inertia about the axis, the angular acceleration, and the angular displacement. d) During the 15.0-rad angular displacement of part (b), the kinetic energy of the wheel increases from 20.0 J to 45.0 J. What is the moment of inertia of the wheel about the rotation axis?

9–66 The three uniform objects shown in Fig. 9–23 have the same mass m. Object A is a solid cylinder with radius R. Object B is a hollow, thin cylinder with radius R. Object C is a solid cube whose sides are of length $2R$. The axis of rotation of each object is perpendicular to the page and through the object's center of mass. a) Which object has the smallest moment of inertia? Explain. b) Which object has the largest moment of inertia? Explain. c) Where would the moment of inertia of a uniform solid sphere rank if its radius is R, its mass is m, and the axis of rotation is along a diameter of the sphere? Explain.

9–67 The earth, which is not a uniform sphere, has moment of inertia of $0.3308MR^2$ about an axis through its north and south poles. It takes the earth 86,164 s to spin once about this axis. Use Appendix F to calculate a) the earth's kinetic energy due to its rotation about this axis and b) the earth's kinetic energy due to its orbital motion around the sun. c) Explain how the value of the earth's moment of inertia tells us that the mass of the earth is concentrated toward the planet's center.

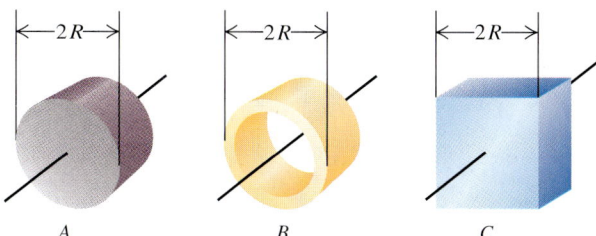

FIGURE **9–23** Problem 9–66.

9–68 A uniform solid disk with mass m and radius R is pivoted about a horizontal axis through its center. A small object of the same mass m is glued to the rim of the disk. If the disk is released from rest with the small object at the end of a horizontal radius, find the angular speed when the small object is directly below the axis.

9–69 A meter stick with a mass of 0.160 kg is pivoted about one end so it can rotate without friction about a horizontal axis. The meter stick is held in a horizontal position and released. As it swings through the vertical, calculate a) the change in gravitational potential energy that has occurred; b) the angular speed of the stick; c) the linear speed of the end of the stick opposite the axis. d) Compare the answer in part (c) to the speed of a particle that has fallen 1.00 m, starting from rest.

9–70 Exactly one turn of a flexible rope of mass m is wrapped around a uniform cylinder of mass M and radius R. The cylinder rotates without friction about a horizontal axle along the cylinder axis. One end of the rope is attached to the cylinder. The cylinder starts with an angular speed ω_0. After one revolution of the cylinder the rope has unwrapped and at this instant hangs vertically down, tangent to the cylinder. Find the angular speed of the cylinder and the linear speed of the lower end of the rope at this time. Neglect the thickness of the rope. (*Hint:* Use Eq. (9–18).)

9–71 The pulley in Fig. 9–24 has radius R and moment of inertia I. The rope does not slip over the pulley, and the pulley spins on a frictionless axle. The coefficient of kinetic friction between block A and the tabletop is μ_k. The system is released from rest, and block B descends. Block A has mass m_A and block B has mass m_B. Use energy methods to calculate the speed of block B as a function of the distance d that it has descended.

FIGURE **9–24** Problem 9–71.

9–72 The pulley in Fig. 9–25 has radius 0.160 m and moment of inertia 0.480 kg · m^2. The rope does not slip on the pulley rim. Use energy methods to calculate the speed of the 4.00-kg block just before it strikes the floor.

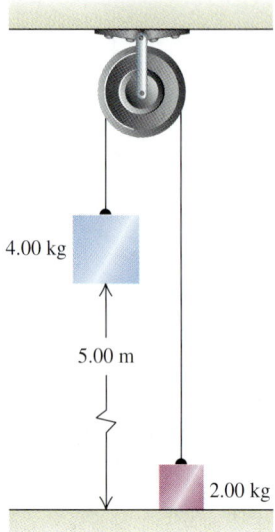

4.00 kg

5.00 m

2.00 kg

FIGURE 9–25 Problem 9–72.

9–73 You hang a thin hoop of radius R over a nail at the rim of the hoop. You displace it to the side through an angle β from its equilibrium position and let it go. What is its angular speed when it returns to its equilibrium position? (*Hint:* Use Eq. (9–18).)

9–74 A passenger bus in Zurich, Switzerland, derived its motive power from the energy stored in a large flywheel. The wheel was brought up to speed periodically, when the bus stopped at a station, by an electric motor, which could then be attached to the electric power lines. The flywheel was a solid cylinder of mass 1000 kg and diameter 1.80 m; its top angular speed was 3000 rev/min. a) At this angular speed, what is the kinetic energy of the flywheel? b) If the average power required to operate the bus is 1.86×10^4 W, how long could it operate between stops?

9–75 Two metal disks, one with radius $R_1 = 2.50$ cm and mass $M_1 = 0.80$ kg and the other with radius $R_2 = 5.00$ cm and mass $M_2 = 1.60$ kg, are welded together and mounted on a frictionless axis through their common center (Fig. 9–26). a) What is the total moment of inertia of the two disks? b) A light string is wrapped around the edge of the smaller disk, and a 1.50-kg block suspended from the free end of the string. If the block is released from rest a distance of 2.00 m above the floor, what is its speed just before it strikes the floor? c) Repeat the calculation of part (b), this time with the string wrapped around the edge of the larger disk. In which case is the final speed of the block the greatest? Explain why this should be so.

9–76 In the cylinder and mass combination described in Example 9–9 (Section 9–5), suppose the falling mass m is made of ideal rubber, so that no mechanical energy is lost when the mass hits the ground. a) If the cylinder is originally not rotating and the mass m is released from rest at a height h above the ground, to what height will this mass rebound if it bounces straight back up from the floor? b) Explain, in terms of energy, why the answer to (a) is *less* than h.

9–77 A thin flat uniform disk has mass M and radius R. A circular hole of radius $R/4$, centered at a point $R/2$ from the disk's center, is then punched in the disk. a) Find the moment

R_1 R_2

1.50 kg

FIGURE 9–26 Problem 9–75.

of inertia of the disk with the hole about an axis through the original center of the disk, perpendicular to the plane of the disk. (*Hint:* Find the moment of inertia of the piece punched from the disk.) b) Find the moment of inertia of the disk with the hole about an axis through the center of the hole, perpendicular to the plane of the disk.

9–78 A pendulum is made of a uniform solid sphere of mass M and radius R suspended from the end of a light rod. The distance from the pivot at the upper end of the rod to the center of the sphere is L. The pendulum's moment of inertia I_P for rotation about the pivot is usually approximated as ML^2. a) Use the parallel-axis theorem to show that if R is 5% of L and the mass of the rod is neglected, I_P is only 0.1% larger than ML^2. b) If the mass of the rod is 1% of M and R is 5% of L, what is the ratio of I_{rod} for an axis at the pivot to ML^2?

9–79 Perpendicular-Axis Theorem. Consider a rigid body that is a thin plane sheet of arbitrary shape. Take the body to lie in the xy-plane and let the origin O of coordinates be located at any point within or outside the body. Let I_x and I_y be the moments of inertia about the x- and y-axes and let I_O be the moment of inertia about an axis through O perpendicular to the plane. a) By considering mass elements m_i with coordinates (x_i, y_i), show that $I_x + I_y = I_O$. This is called the perpendicular-axis theorem. Note that point O does not have to be the center of mass. b) For a thin washer of mass M and with inner and outer radii R_1 and R_2, use the perpendicular-axis theorem to find the moment of inertia about an axis that is in the plane of the washer and that passes through its center. You may use the information in Table 9–2. c) Use the perpendicular-axis theorem to show that for a thin, square sheet of mass M and side L, the moment of inertia about *any* axis in the plane of the sheet that passes through the center of the sheet is $\frac{1}{12}ML^2$. You may use the information in Table 9–2.

9–80 A thin uniform rod is bent into a square of side length a. If the total mass is M, find the moment of inertia about an axis through the center and perpendicular to the plane of the square. (*Hint:* Use the parallel-axis theorem.)

*9–81 A cylinder of radius R and mass M has density that increases linearly with distance r from the cylinder axis, $\rho = \alpha r$, where α is a positive constant. a) Calculate the moment of inertia of the cylinder about a longitudinal axis through its center in terms of M and R. b) Is your answer greater or smaller than the moment of inertia of a cylinder of the same mass and radius but of uniform density? Explain why this result makes qualitative sense.

9–82 Neutron Stars and Supernova Remnants. The Crab Nebula is a cloud of glowing gas about 10 light years across, located about 6500 light years from the earth (Fig. 9–27). It is the remnant of a star that underwent a *supernova explosion*, seen on earth in 1054 A.D. Energy is released by the Crab Nebula at a rate of about 5×10^{31} W, about 10^5 times the rate at which the sun radiates energy. The Crab Nebula obtains its energy from the rotational kinetic energy of a rapidly spinning *neutron star* at its center. This object rotates once every 0.0331 s, and this period is increasing by 4.22×10^{-13} s for each second of time that elapses. a) If the rate at which energy is lost by the neutron star is equal to the rate at which energy is released by the nebula, find the moment of inertia of the neutron star. (Use the expression derived in Exercise 9–43.) b) Theories of supernovae predict that the neutron star in the Crab Nebula has a mass of about 1.4 times that of the sun. Modeling the neutron star as a solid uniform sphere, calculate its radius in kilometers. c) What is the linear speed of a point on the equator of the neutron star? Compare to the speed of light. d) Assume that the neutron star is

uniform and calculate its density. Compare to the density of ordinary rock (3000 kg/m^3) and to the density of an atomic nucleus (about 10^{17} kg/m^3). Justify the statement that a neutron star is essentially a large atomic nucleus.

FIGURE 9–27 Problem 9–82.

CHALLENGE PROBLEMS

9–83 The moment of inertia of a sphere with uniform density about an axis through its center is $\frac{2}{5}MR^2 = 0.400MR^2$. Satellite observations show that the earth's moment of inertia is $0.3308MR^2$. Geophysical data suggests the earth consists of five main regions: the inner core ($r = 0$ to $r = 1220$ km) of average density 12,900 kg/m^3, the outer core ($r = 1220$ km to $r = 3480$ km) of average density 10,900 kg/m^3, the lower mantle ($r = 3480$ km to $r = 5700$ km) of average density 4900 kg/m^3, the upper mantle ($r = 5700$ km to $r = 6350$ km) of average density 3600 kg/m^3, and the outer crust and oceans ($r = 6350$ km to $r = 6370$ km) of average density 2400 kg/m^3. a) Show that the moment of inertia about a diameter of a uniform spherical shell of inner radius R_1, outer radius R_2, and density ρ is $I = \rho(8\pi/15)(R_2^5 - R_1^5)$. (*Hint:* Form the shell by superposition of a sphere of density ρ and a smaller sphere of density $-\rho$.) b) Check the given data by using them to calculate the mass of the earth. c) Use the given data to calculate the earth's moment of inertia in terms of MR^2.

*9–84 Calculate the moment of inertia of a uniform solid cone about an axis through its center (Fig. 9–28). The cone has mass M and altitude h. The radius of its circular base is R.

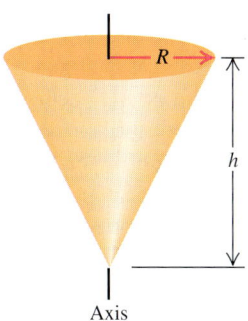

FIGURE 9–28 Challenge Problem 9–84.

9–85 On a compact disc (CD), music is coded in a pattern of tiny pits arranged in a track that spirals outward toward the rim of the disc. As the disc spins inside a CD player, the track is scanned at a constant *linear* speed of $v = 1.25$ m/s. Because the radius of the track varies as it spirals outward, the *angular* speed of the disc must change as the CD is played. (See Exercise 9–20.) Let's see what angular acceleration is required to keep v constant. The equation of a spiral is $r(\theta) = r_0 + \beta\theta$, where r_0 is the radius of the spiral at $\theta = 0$ and β is a constant. On a CD, r_0 is the inner radius of the spiral track. If we take the rotation direction of the CD to be positive, β must be positive so that r increases as the disc turns and θ increases. a) When the disc rotates through a small angle $d\theta$, the distance scanned along the track is $ds = r\, d\theta$. Using the above expression for $r(\theta)$, integrate ds to find the total distance s scanned along the track as a function of the total angle θ through which the disc has rotated. b) Since the track is scanned at a constant linear speed v, the distance s found in part (a) is equal to vt. Use this to find θ as a function of time. There will be two solutions for θ; choose the positive one, and explain why this is the solution to choose. c) Use your expression for $\theta(t)$ to find the angular velocity ω and angular acceleration α as functions of time. Is α constant? d) On a CD, the inner radius of the track is 25.0 mm, the track radius increases by 1.55 μm per revolution, and the playing time is 74.0 min. Find the values of r_0 and β, and find the total number of revolutions made during the playing time. e) Using your results from parts (c) and (d), make graphs of ω (in rad/s) versus t and α (in rad/s^2) versus t between $t = 0$ and $t = 74.0$ min.

10

Dynamics of Rotational Motion

The swirling air in a hurricane both circles around and moves inward toward the low-pressure region at the center. As the air moves inward, it gains speed in accordance with the law of conservation of angular momentum. As a result, the fastest and most destructive winds are found just outside the "eye," a region of calm air at the very center of the hurricane.

10–1 INTRODUCTION

When you use a wrench to loosen the lug nuts while changing a flat tire, the whole wheel starts to turn unless you figure out a way to hold it. What is it about the force you apply to the wrench handle that makes the wheel turn? More generally, what is it that gives a rotating body an *angular* acceleration? A force is a push or a pull, but to produce rotational motion, we need a twisting or turning action.

In this chapter we will define a new physical quantity, *torque,* that describes the twisting or turning effort of a force. We'll find that the net torque acting on a rigid body determines its angular acceleration, in the same way that the net force on a body determines its linear acceleration. We'll also look at work and power in rotational motion in order to understand such problems as how energy is transmitted by the rotating drive shaft in a car. Finally, we will develop a new conservation principle, *conservation of angular momentum,* that is tremendously useful for understanding the rotational motion of both rigid and nonrigid bodies. We'll finish this chapter by studying *gyroscopes,* rotating devices that seemingly defy common sense and don't fall over when you might think they should—but that actually behave in perfect accordance with the dynamics of rotational motion.

10–2 TORQUE

What is it about a force that determines how effective it is in causing or changing rotational motion? The magnitude and direction of the force are important, but so is the position of the point where the force is applied. When you try to swing a heavy door open, it's a lot more effective to push farther away from the axis of rotation (near the knob) than close to the axis (near the hinges). In Fig. 10–1 a wrench is being used to loosen a tight bolt. Force $\vec{F}_b$, applied near the end of the handle, is more effective than an equal force $\vec{F}_a$ applied near the bolt. Force $\vec{F}_c$ doesn't do any good at all; it's applied at the same point and has the same magnitude as $\vec{F}_b$, but it's directed along the length of the handle.

The quantitative measure of the tendency of a force to cause or change the rotational motion of a body is called *torque.* Figure 10–2 shows a body that can rotate about an axis that passes through point O and is perpendicular to the plane of the figure. The body is acted on by three forces, $\vec{F}_1$, $\vec{F}_2$, and $\vec{F}_3$, in the plane of the figure. The tendency of $\vec{F}_1$ to cause a rotation about point O depends on its magnitude F_1. It also depends on the *perpendicular* distance l_1 between the **line of action** of the force (that is, the line along which the force vector lies) and point O. We call the distance l_1 the **lever arm** (or **moment arm**) of force $\vec{F}_1$ about O. The twisting effort is directly proportional to both F_1 and l_1. We define the **torque** (or *moment*) of the force $\vec{F}_1$ with respect to point O as the product $F_1 l_1$. We will use the Greek letter τ ("tau") for torque. For a force of magnitude F whose line of action is a perpendicular distance l from

the point O, the torque is

$$\tau = Fl. \qquad (10\text{--}1)$$

Physicists usually use the term "torque," while engineers usually use "moment" (unless they are talking about a rotating shaft). Both groups use the term "lever arm" or "moment arm" for the distance l.

The lever arm of $\vec{F}_1$ in Fig. 10–2 is the perpendicular distance OA or l_1, and the lever arm of $\vec{F}_2$ is the perpendicular distance OB or l_2. The line of action of $\vec{F}_3$ passes through the reference point O, so the lever arm for $\vec{F}_3$ is zero and its torque with respect to point O is zero. In the same way, force $\vec{F}_c$ in Fig. 10–1 has zero torque with respect to point O. On the other hand, $\vec{F}_b$ has a greater torque than $\vec{F}_a$ because its lever arm is greater.

CAUTION ▶ Note that torque is always defined with reference to a specific point, often (but not always) the origin of a coordinate system. If we shift the position of this point, the torque of each force may also change. For example, the torque of force $\vec{F}_3$ in Fig. 10–2 is zero with respect to point O, but is *not* zero about point A or B. When describing the torque of a certain force, it's not enough to call it "the torque of $\vec{F}$"; you must say "the torque of $\vec{F}$ with respect to point X" or "the torque of $\vec{F}$ about point X." ◀

10–1 Which of these three forces is most likely to loosen the tight bolt?

Force $\vec{F}_1$ in Fig. 10–2 tends to cause *counterclockwise* rotation about O, while $\vec{F}_2$ tends to cause *clockwise* rotation. To distinguish between these two possibilities, we will choose a positive sense of rotation. With the choice that *counterclockwise torques are positive and clockwise torques are negative,* the torques of $\vec{F}_1$ and $\vec{F}_2$ about O are

$$\tau_1 = +F_1 l_1, \qquad \tau_2 = -F_2 l_2.$$

We will often use the symbol

to indicate our choice of the positive sense of rotation.

The SI unit of torque is the newton-meter. In our discussion of work and energy we called this combination the joule. But torque is *not* work or energy, and torque should be expressed in newton-meters, *not* joules.

Figure 10–3 shows a force $\vec{F}$ applied at a point P described by a position vector $\vec{r}$ with respect to the chosen point O. There are several ways to calculate the torque of this force. One is to find the lever arm l and use $\tau = Fl$. Or we can determine the angle ϕ between the vectors $\vec{r}$ and $\vec{F}$; the lever arm is $r \sin \phi$, so $\tau = rF \sin \phi$. A third method is to represent $\vec{F}$ in terms of a radial component F_{rad} along the direction of $\vec{r}$ and a

ActivPhysics

7.1
Calculating Torques

10–2 The torque or moment of a force about a point is the product of the force magnitude and the lever arm.

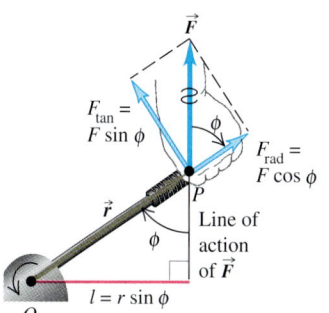

10–3 The torque of the force $\vec{F}$ about the point O is defined by $\vec{\tau} = \vec{r} \times \vec{F}$. The magnitude of $\vec{\tau}$ is $rF \sin \phi$. In this figure, $\vec{r}$ and $\vec{F}$ are in the plane of the paper; by the right-hand rule for the vector product, $\vec{\tau}$ points out of the page toward you.

10–4 The torque, $\vec{\tau} = \vec{r} \times \vec{F}$, is directed along the axis of the bolt. When the fingers of the right hand curl from the direction of $\vec{r}$ into the direction of $\vec{F}$—that is, in the direction of rotation that the torque tends to cause—the outstretched right thumb points in the direction of $\vec{\tau}$ (perpendicular to both $\vec{r}$ and $\vec{F}$).

tangential component F_{tan} at right angles, perpendicular to $\vec{r}$. (We call this a tangential component because if the body rotates, the point where the force acts moves in a circle, and this component is tangent to that circle.) Then $F_{tan} = F \sin \phi$, and $\tau = r(F \sin \phi) = F_{tan} r$. The component F_{rad} has no torque with respect to O because its lever arm with respect to that point is zero (compare to forces $\vec{F}_c$ in Fig. 10–1 and $\vec{F}_3$ in Fig. 10–2). Summarizing these three expressions for torque, we have

$$\tau = Fl = rF \sin \phi = F_{tan} r \quad \text{(magnitude of torque)}. \quad (10\text{–}2)$$

We saw in Section 9–2 that angular velocity and angular acceleration can be represented as vectors; the same is true for torque. To see how to do this, note that the quantity $rF \sin \phi$ in Eq. (10–2) is the magnitude of the *vector product* $\vec{r} \times \vec{F}$ that we defined in Section 1–11. You should review that definition. We now generalize the definition of torque as follows: When a force $\vec{F}$ acts at a point having a position vector $\vec{r}$ with respect to an origin O, as in Fig. 10–3, the torque $\vec{\tau}$ of the force with respect to O is the *vector quantity*

$$\vec{\tau} = \vec{r} \times \vec{F} \quad \text{(definition of torque vector)}. \quad (10\text{–}3)$$

The torque as defined in Eq. (10–2) is just the magnitude of the torque vector $\vec{r} \times \vec{F}$. The direction of $\vec{\tau}$ is perpendicular to both $\vec{r}$ and $\vec{F}$. In particular, if both $\vec{r}$ and $\vec{F}$ lie in a plane perpendicular to the axis of rotation, as in Fig. 10–3, then the torque vector $\vec{\tau} = \vec{r} \times \vec{F}$ is directed along the axis of rotation, with a sense given by the right-hand rule (Fig. 1–20). The direction relationships are shown in Fig. 10–4.

In the following sections we will usually be concerned with rotation of a body about an axis oriented in a specified constant direction. In that case, only the component of torque along that axis is of interest, and we often call that component the torque with respect to the specified axis.

EXAMPLE 10–1

A weekend plumber, unable to loosen a pipe fitting, slips a piece of scrap pipe (a "cheater") over his wrench handle. He then applies his full weight of 900 N to the end of the cheater by standing on it. The distance from the center of the fitting to the point where the weight acts is 0.80 m, and the wrench handle and cheater make an angle of 19° with the horizontal (Fig. 10–5a). Find the magnitude and direction of the torque he applies about the center of the pipe fitting.

SOLUTION From Fig. 10–5b, the angle ϕ between $\vec{r}$ and $\vec{F}$ is 109°, and the lever arm l is

$$l = (0.80 \text{ m})(\sin 109°) = (0.80 \text{ m})(\sin 71°) = 0.76 \text{ m}.$$

10–5 (a) A weekend plumber tries to loosen a pipe fitting by standing on a "cheater." (b) Vector diagram to find the torque about O.

(a)

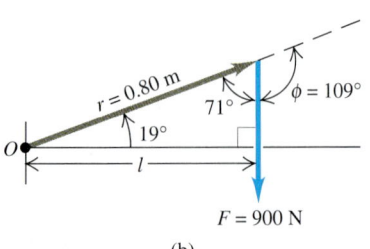

(b)

We can find the magnitude of the torque from Eq. (10–1):

$$\tau = Fl = (900 \text{ N})(0.76 \text{ m}) = 680 \text{ N} \cdot \text{m}.$$

Or, from Eq. (10–2),

$$\tau = rF \sin \phi = (0.80 \text{ m})(900 \text{ N})(\sin 109°) = 680 \text{ N} \cdot \text{m}.$$

Alternatively, we can find F_{tan}, the tangential component of $\vec{F}$. This is the component that acts perpendicular to $\vec{r}$ (that is, per-pendicular to the "cheater"). The vector $\vec{r}$ is oriented 19° from the horizontal, so the perpendicular to $\vec{r}$ is oriented 19° from the vertical. Since $\vec{F}$ is vertical, this means $F_{\text{tan}} = F(\cos 19°)$ $= (900 \text{ N})(\cos 19°) = 851 \text{ N}$. Then the torque is

$$\tau = F_{\text{tan}} r = (851 \text{ N})(0.80 \text{ m}) = 680 \text{ N} \cdot \text{m}.$$

The force tends to produce a clockwise rotation about O, and the direction of the torque vector $\vec{\tau}$ is *into* the plane of the figure.

10–3 TORQUE AND ANGULAR ACCELERATION FOR A RIGID BODY

We are now ready to develop the fundamental relation for the rotational dynamics of a rigid body. We will show that the angular acceleration of a rotating rigid body is directly proportional to the sum of the torque components along the axis of rotation. The pro-portionality factor is the moment of inertia.

To develop this relation, we again imagine the body as being made up of a large num-ber of particles. We choose the axis of rotation to be the y-axis; the first particle has mass m_1 and distance r_1 from this axis (Fig. 10–6). The *net force* acting on this particle has a component $F_{1,\text{rad}}$ along the radial direction, a component $F_{1,\text{tan}}$ that is tangent to the cir-cle of radius r_1 in which the particle moves as the body rotates, and a component F_{1y} along the axis of rotation. Newton's second law for the tangential component is

$$F_{1,\text{tan}} = m_1 a_{1,\text{tan}}. \tag{10–4}$$

We can express the tangential acceleration of the first particle in terms of the angular acceleration α, using Eq. (9–14): $a_{1,\text{tan}} = r_1 \alpha$. Using this relation and multiplying both sides of Eq. (10–4) by r_1, we obtain

$$F_{1,\text{tan}} r_1 = m_1 r_1^2 \alpha. \tag{10–5}$$

From Eq. (10–2), $F_{1,\text{tan}} r_1$ is just the magnitude of the *torque* τ_1 of the net force with respect to the rotation axis (equal to the component of the torque vector along the rota-tion axis). Neither of the components $F_{1,\text{rad}}$ or F_{1y} contributes to the torque about the y-axis, since neither tends to change the particle's rotation about that axis. So $\tau_1 = F_{1,\text{tan}} r_1$ is the total torque acting on the particle with respect to the rotation axis. Also, $m_1 r_1^2$ is I_1, the moment of inertia of the particle about the rotation axis. With this in mind, we rewrite Eq. (10–5) as

$$\tau_1 = I_1 \alpha = m_1 r_1^2 \alpha.$$

We write an equation like this for every particle in the body and then add all these equations:

$$\tau_1 + \tau_2 + \cdots = I_1 \alpha + I_2 \alpha + \cdots = m_1 r_1^2 \alpha + m_2 r_2^2 \alpha + \cdots,$$

or

$$\Sigma \tau_i = (\Sigma m_i r_i^2) \alpha.$$

The left side of this equation is the sum of all the torques about the rotation axis that act on all the particles. The right side is $I = \Sigma m_i r_i^2$, the total moment of inertia about the rota-tion axis, multiplied by the angular acceleration α. Note that α is the same for every particle because this is a *rigid* body. Thus for the entire rigid body we have the *rotational analog of Newton's second law:*

$$\Sigma \tau = I\alpha \qquad \text{(rotational analog of Newton's second law} \atop \text{for a rigid body).} \tag{10–6}$$

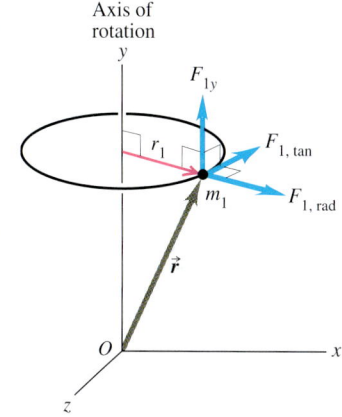

10–6 Three components of the net force acting on one of the particles of a rigid body. Only $F_{1,\text{tan}}$ has a y-component of torque about O.

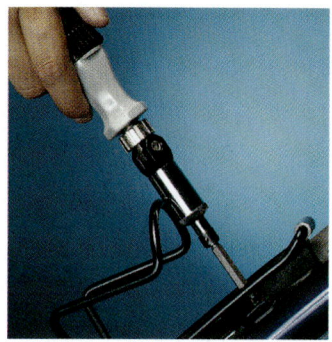

Loosening or tightening a screw requires giving it an angular accel-eration and hence applying a torque. This is made easier by using a screwdriver with a large-radius handle, which provides a large lever arm for the force you apply with your hand.

10–7 Particle 1 and particle 2 in a rigid body exert equal and opposite forces on each other. If these forces act along the line from one particle to another, the lever arms are the same, and the torques due to the two forces are equal and opposite. Only external torques affect the rotation of a rigid body.

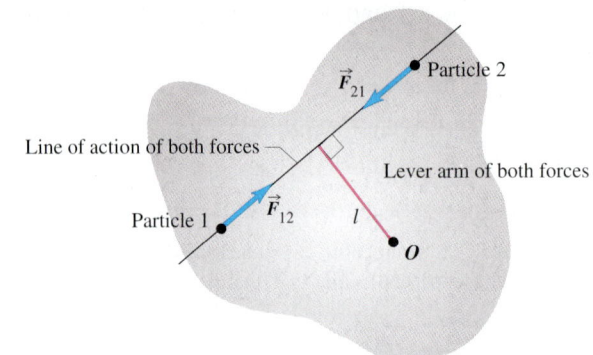

Line of action of both forces

Lever arm of both forces

Particle 1 $\vec{F}_{12}$ l Particle 2 $\vec{F}_{21}$

O

7.10
Rotoride—Dynamics Approach

7.11
Falling Ladder

7.12
Woman and Flywheel Elevator—
Dynamics Approach

Just as Newton's second law says that the net force on a particle equals the particle's mass times its acceleration, Eq. (10–6) says that the net torque on a rigid body equals the body's moment of inertia about the rotation axis times its angular acceleration.

We emphasize that Eq. (10–6) is valid *only* for *rigid* bodies. If the body is not rigid, like a rotating tank of water or a swirling tornado of air, the angular acceleration α is different for different particles in the body, and the derivation of Eq. (10–6) isn't valid. Also note that since our derivation used Eq. (9–14), $a_{tan} = r\alpha$, α must be measured in rad/s^2.

The torque on each particle is due to the net force on that particle, which is the vector sum of external and internal forces (defined in Section 8–3). According to Newton's third law, the *internal* forces that any pair of particles in the rigid body exert on each other are equal and opposite (Fig. 10–7). If these forces act along the line joining the two particles, their lever arms with respect to any axis are also equal. So the torques for each such pair are equal and opposite, and add to zero. Indeed, *all* the internal torques add to zero, so the sum $\Sigma\tau$ in Eq. (10–6) includes only the torques of the *external* forces.

Often, one of the important external forces acting on a body is its *weight*. This force is not concentrated at a single point; it acts on every particle in the entire body. Nevertheless, it turns out that if the value of $\vec{g}$ is the same at all points, we always get the correct torque (about any specified axis) if we assume that all the weight is concentrated at the *center of mass* of the body. We will prove this statement in Chapter 11, but meanwhile we will use it for some of the problems in this chapter.

Problem–Solving Strategy

ROTATIONAL DYNAMICS

Our strategy for solving problems in rotational dynamics is very similar to that used in Section 5–3 for applications of Newton's laws:

1. Draw a sketch of the situation, and select the body or bodies to be analyzed.

2. For each body, draw a free-body diagram isolating the body and including all the forces (and *only* those) that act on the body, including its weight. Label unknown quantities with algebraic symbols. A new consideration is that you must show the *shape* of the body accurately, including all dimensions and angles you will need for torque calculations.

3. Choose coordinate axes for each body, and indicate a positive sense of rotation for each rotating body. If there is a linear acceleration, it's usually simplest to pick a positive axis in its direction. If you know the sense of α in advance, picking that as the positive sense of rotation simplifies the calculations. When you represent a force in terms of its components, cross out the original force so that you don't include it twice.

4. Some problems will include bodies that have translational motion, others that have rotational motion, and some that have both. Depending on the behavior of the body in question, apply $\Sigma \vec{F} = m\vec{a}$ or $\Sigma\tau = I\alpha$ or both.

5. If more than one body is involved, carry out Steps 2 through 4 for each body. Write a separate equation of motion for each body. There may also be *geometrical relations* between the motions of two or more bodies, as with a string that unwinds from a pulley while turning it or a wheel that rolls without slipping (to be discussed in Section 10–4). Express these in algebraic form, usually as relations between two linear accelerations or a linear acceleration and an angular acceleration. Then solve the equations to find the unknown quantities. Often this involves solving a set of simultaneous equations.

6. Check the results for special cases or extreme values of quantities when possible, and compare with your intuitive expectations. Ask yourself: "Does this result make sense?"

EXAMPLE 10–2

An unwinding cable Figure 10–8a shows the same situation that we analyzed in Example 9–8 (Section 9–5) using energy methods. A cable is wrapped several times around a uniform solid cylinder that can rotate about its axis. The cylinder has diameter 0.120 m and mass 50 kg. The cable is pulled with a force of 9.0 N. Assuming that the cable unwinds without stretching or slipping, what is its acceleration?

SOLUTION Figure 10–8b shows the free-body diagram. We take the positive sense of rotation for the cylinder to be clockwise. The net force on the cylinder must be zero because its center of mass remains at rest. The weight (magnitude Mg) and the normal force (magnitude n) exerted by the bearing act along lines through the rotation axis and thus have no torque with respect to that axis. The only torque about the rotation axis is due to the force F exerted by the cable on the cylinder: $\tau = Fl$ = (9.0 N)(0.060 m) = 0.54 N · m. From Example 9–8, the moment of inertia is $I = \frac{1}{2}MR^2 = \frac{1}{2}(50\text{ kg})(0.060\text{ m})^2 = 0.090\text{ kg} \cdot \text{m}^2$. The angular acceleration α is given by Eq. (10–6):

$$\alpha = \frac{\tau}{I} = \frac{0.54\text{ N} \cdot \text{m}}{0.090\text{ kg} \cdot \text{m}^2} = 6.0\text{ rad/s}^2.$$

We invite you to check the units in this equation and make sure they come out right.

Now we need a kinematic relation. We remarked in Section 9–4 that the acceleration of a cable unwinding from a cylinder is

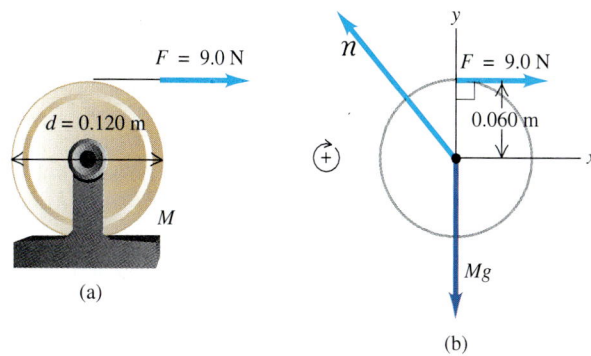

(a)

(b)

10–8 (a) Cylinder and cable. (b) The free-body diagram for the cylinder.

the same as the tangential component of acceleration of a point on the surface of the cylinder where the cable is tangent to it. This tangential acceleration is given by Eq. (9–14):

$$a = r\alpha = (0.060\text{ m})(6.0\text{ rad/s}^2) = 0.36\text{ m/s}^2.$$

Can you use this result, together with an equation from Chapter 2, to determine the speed of the cable after it has been pulled 2.0 m? Try it, and compare your result with Example 9–8, in which we found this speed using work and energy considerations.

EXAMPLE 10–3

An unwinding cable II Figure 10–9a shows the same situation that we analyzed in Example 9–9 (Section 9–5) using energy methods. Find the acceleration of the object of mass m and the angular acceleration of the cylinder.

SOLUTION We have to treat the two bodies separately. Figure 10–9b shows a free-body diagram for each body. We take the positive sense of rotation for the cylinder to be counterclockwise and the positive direction of the y-coordinate for the object to be downward. For the object, Newton's second law gives

$$\Sigma F_y = mg + (-T) = ma.$$

The weight Mg and the normal force n (exerted by the bearing) have no torques with respect to the rotation axis because they act along lines through that axis, just as in Example 10–2. The only torque is that due to the cable tension T. Applying Eq. (10–6) to the cylinder gives

$$\Sigma \tau = RT = I\alpha = \frac{1}{2}MR^2\alpha.$$

As in Example 10–2, the acceleration of the cable is the same as the tangential acceleration of a point on the cylinder rim. According to Eq. (9–14), this acceleration is given by $a = a_{\text{tan}} = R\alpha$. We use this to replace ($R\alpha$) with a in the cylinder equation above, and then divide by R; the result is

$$T = \frac{1}{2}Ma.$$

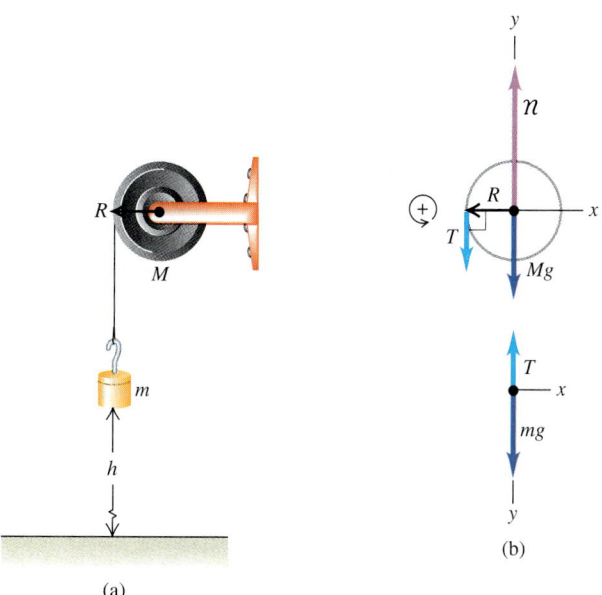

(a)

(b)

10–9 (a) Cylinder, object, and cable. (b) Free-body diagrams for the cylinder and for the object hanging from the cable. The cable is assumed to have negligible mass.

Now we substitute this expression for T into Newton's second law for the object and solve for the acceleration a:

$$mg - \frac{1}{2} Ma = ma,$$

$$a = \frac{g}{1 + M/2m}.$$

Finally, we can substitute this back into Newton's second law for the object to find T:

$$T = mg - ma = mg - m\left(\frac{g}{1 + M/2m}\right) = \frac{mg}{1 + 2m/M}.$$

We note that the tension in the cable is *not* equal to the weight

mg of the object; if it were, the object could not accelerate.

Let's check some particular cases. When M is much larger than m, the tension is nearly equal to mg, and the acceleration is correspondingly much less than g. When M is zero, $T = 0$ and $a = g$; the object then falls freely. If the object starts from a height h above the floor with an initial speed v_0, its speed v when it strikes the ground is given by $v^2 = v_0^2 + 2ah$. If it starts from rest, $v_0 = 0$ and

$$v = \sqrt{2ah} = \sqrt{\frac{2gh}{1 + M/2m}}.$$

This is the same result we obtained from energy considerations in Example 9–9.

EXAMPLE 10-4

In Fig. 10–10a a glider of mass m_1 slides without friction on a horizontal air track. It is attached to an object of mass m_2 by a massless string. The pulley is a thin cylindrical shell (with massless spokes) with mass M and radius R, and the string turns the pulley without slipping or stretching. Find the acceleration of each body, the angular acceleration of the pulley, and the tension in each part of the string.

SOLUTION Figure 10–10b shows the free-body diagrams and coordinate systems for the three bodies.

CAUTION ▶ We considered a similar situation in Example 5–11 (Section 5–3). In that example the string slid without friction over a fixed pulley, and the tension was the same throughout the massless string. But with a rotating pulley, and with friction between the pulley and string to prevent slipping, the two tensions T_1 and T_2 *cannot* be equal. If they were, the pulley could not have an angular acceleration. Labeling the tension in both parts of the string as simply T would be a serious error. Be on your guard against this error in any problem that involves a rotating pulley. ◀

The equations of motion for the glider and the object are

$$\Sigma F_x = T_1 = m_1 a_1, \tag{10–7}$$

$$\Sigma F_y = m_2 g + (-T_2) = m_2 a_2. \tag{10–8}$$

The unknown normal force n_2 acts on a line through the pulley's axis of rotation, so it has no lever arm or torque with respect to

that axis. From Table 9–2 the moment of inertia of the pulley about this axis is $I = MR^2$. We take the positive sense of rotation as clockwise (the sense of the pulley's actual angular acceleration). Then the pulley's equation of motion is

$$\Sigma\tau = T_2 R + (-T_1 R) = I\alpha = (MR^2)\alpha. \tag{10–9}$$

Because the string does not stretch or slip, we have the additional *kinematic* relations

$$a_1 = a_2 = R\alpha. \tag{10–10}$$

(The accelerations of the glider and the object have different directions but the same magnitude.)

Equations (10–7) through (10–10) are five simultaneous equations for the five unknowns a_1, a_2, α, T_1, and T_2. (Equation (10–10) is actually two equations.) This may sound like an awesome problem, but we have as many equations as unknowns, and solving them is a lot easier than you might imagine. First we use Eqs. (10–10) to eliminate a_2 and α from Eqs. (10–7) through (10–9). We then have three equations for the three unknowns T_1, T_2, and a_1:

$$T_1 = m_1 a_1,$$

$$m_2 g - T_2 = m_2 a_1,$$

$$T_2 - T_1 = Ma_1.$$

The easiest way to solve these is simply to *add* the three equations, which eliminates T_1 and T_2, and then to solve for a_1:

$$a_1 = \frac{m_2 g}{m_1 + m_2 + M}.$$

10–10 (a) An air-track glider pulled by an object hanging over a pulley. (b) Free-body diagrams for all three bodies.

(a)

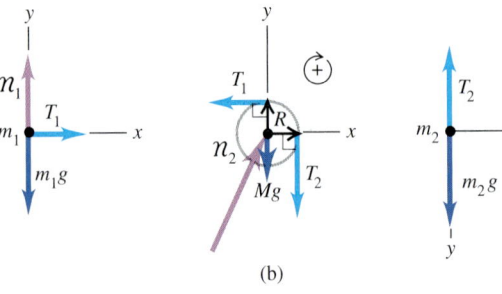

(b)

We can then substitute this back into Eqs. (10–7) and (10–8) to find the tensions. The results are

$$T_1 = \frac{m_1 m_2 g}{m_1 + m_2 + M}, \qquad T_2 = \frac{(m_1 + M)m_2 g}{m_1 + m_2 + M}.$$

Let's check some special cases to see whether these results make sense. First, if either m_1 or M is much larger than m_2, the accelerations are very small, and T_2 is approximately $m_2 g$. But if m_2 is much *larger* than either m_1 or M, the acceleration is approximately g. Both of these results are what we should expect. When $M = 0$, do we get the same result as Example 5–11 (Section 5–3)? Why or why not? Can you think of other special cases to check?

10–4 RIGID-BODY ROTATION ABOUT A MOVING AXIS

We can extend our analysis of the dynamics of rotational motion to some cases in which the axis of rotation moves. When that happens, the motion of the body is **combined translation and rotation.** The key to understanding such situations is this: every possible motion of a rigid body can be represented as a combination of *translational motion of the center of mass* and *rotation about an axis through the center of mass.* This is true even when the center of mass accelerates, so that it is not at rest in any inertial frame. A graphic example is the motion of a hammer thrown upward (Fig. 10–11). The center of mass of the hammer follows a parabolic curve, as though the hammer were a particle located at the center of mass. At the same time, the hammer rotates with constant angular velocity about the center of mass (compare the motion of the wrench in Fig. 8–21). The translation of the center of mass and the rotation about the center of mass can be treated as separate but related problems. Other examples of such motion are a ball rolling down a hill and a yo-yo unwinding at the end of a string.

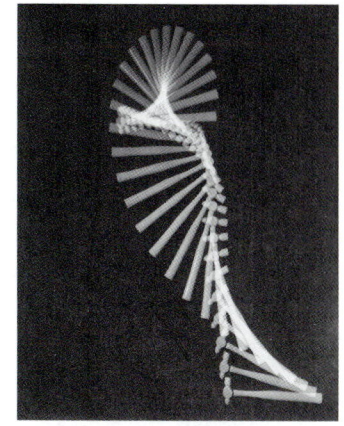

10–11 The motion of a rigid body like this thrown hammer is a combination of translational motion of the center of mass and rotation around the center of mass.

COMBINED TRANSLATION AND ROTATION: ENERGY RELATIONS

It's beyond the scope of this book to prove that the motion of a rigid body can always be divided into translation of the center of mass and rotation about the center of mass. But we can show that this is true for the *kinetic energy* of a rigid body that has both translational and rotational motion. In this case, the body's kinetic energy is the sum of a part $\frac{1}{2}Mv_{cm}^2$ associated with motion of the center of mass and a part $\frac{1}{2}I_{cm}\omega^2$ associated with rotation about an axis through the center of mass:

$$K = \frac{1}{2}Mv_{cm}^2 + \frac{1}{2}I_{cm}\omega^2 \qquad \text{(rigid body with both translation and rotation).} \qquad (10\text{–}11)$$

To prove this relation, we again imagine the rigid body to be made up of particles. Consider a typical particle with mass m_i as shown in Fig. 10–12. The velocity $\vec{v}_i$ of this particle relative to an inertial frame is the vector sum of the velocity $\vec{v}_{cm}$ of the center of mass and the velocity $\vec{v}_i'$ of the particle *relative to* the center of mass:

$$\vec{v}_i = \vec{v}_{cm} + \vec{v}_i'. \qquad (10\text{–}12)$$

The kinetic energy K_i of this particle in the inertial frame is $\frac{1}{2}m_i v_i^2$, which we can also express as $\frac{1}{2}m_i(\vec{v}_i \cdot \vec{v}_i)$. Substituting Eq. (10–12) into this, we get

$$\begin{aligned} K_i &= \frac{1}{2}m_i(\vec{v}_{cm} + \vec{v}_i') \cdot (\vec{v}_{cm} + \vec{v}_i') \\ &= \frac{1}{2}m_i(\vec{v}_{cm} \cdot \vec{v}_{cm} + 2\vec{v}_{cm} \cdot \vec{v}_i' + \vec{v}_i' \cdot \vec{v}_i') \\ &= \frac{1}{2}m_i(v_{cm}^2 + 2\vec{v}_{cm} \cdot \vec{v}_i' + v_i'^2). \end{aligned}$$

The total kinetic energy is the sum ΣK_i for all the particles making up the body.

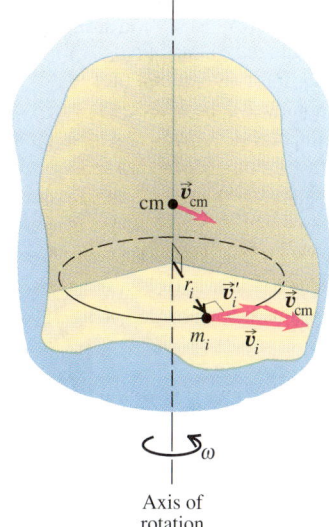

10–12 A rigid body with both translational and rotational motion. The velocity $\vec{v}_i$ of a typical particle in the rigid body is the sum of the center-of-mass velocity $\vec{v}_{cm}$ and the particle's velocity $\vec{v}_i'$ relative to the center of mass. The angular speed of the rigid body about the center of mass is ω, so $v_i' = r_i\omega$.

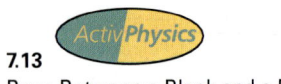

7.13
Race Between a Block and a Disk

Expressing the three terms in this equation as separate sums, we get

$$K = \Sigma K_i = \Sigma\left(\frac{1}{2}m_i v_{cm}{}^2\right) + \Sigma(m_i \vec{\boldsymbol{v}}_{cm} \cdot \vec{\boldsymbol{v}}_i{}') + \Sigma\left(\frac{1}{2}m_i v_i{}'^2\right).$$

The first and second terms have common factors that can be taken outside the sum:

$$K = \frac{1}{2}(\Sigma m_i)v_{cm}{}^2 + \vec{\boldsymbol{v}}_{cm}\cdot(\Sigma m_i\vec{\boldsymbol{v}}_i{}') + \Sigma\left(\frac{1}{2}m_i v_i{}'^2\right). \qquad (10\text{–}13)$$

Now comes the reward for our effort. In the first term, Σm_i is the total mass M. The second term is zero because $\Sigma m_i\vec{\boldsymbol{v}}_i{}'$ is M times the velocity of the center of mass *relative to the center of mass,* and this is zero by definition. The last term is the sum of kinetic energies of the particles computed by using their speeds with respect to the center of mass; this is just the kinetic energy of rotation around the center of mass. Using the same steps that led to Eq. (9–17) for the rotational kinetic energy of a rigid body, we can write this last term as $\frac{1}{2}I_{cm}\omega^2$, where I_{cm} is the moment of inertia with respect to the axis through the center of mass and ω is the angular speed. So Eq. (10–13) becomes Eq. (10–11):

$$K = \frac{1}{2}Mv_{cm}{}^2 + \frac{1}{2}I_{cm}\omega^2.$$

Rolling without slipping is the norm for automobile tires, but is not always the case. If the car slides on ice, the tires will spin too fast for the car's speed and $R\omega$ will be greater than v_{cm}. If the driver applies the brakes too heavily so that the car skids, the tires will spin too slowly (if at all) and $R\omega$ will be less than v_{cm}.

An important case of combined translation and rotation is **rolling without slipping,** such as the motion of the wheel shown in Fig. 10–13. The wheel is symmetrical, so its center of mass is at its geometric center. We view the motion in an inertial frame of reference in which the surface on which the wheel rolls is at rest. In this frame, the point on the wheel that contacts the surface must be instantaneously *at rest* so that it does not slip. Hence the velocity $\vec{\boldsymbol{v}}_1{}'$ of the point of contact relative to the center of mass must have the same magnitude but opposite direction as the center-of-mass velocity $\vec{\boldsymbol{v}}_{cm}$. If the radius of the wheel is R and its angular speed about the center of mass is ω, then the magnitude of $\vec{\boldsymbol{v}}_1{}'$ is $R\omega$; hence we must have

$$v_{cm} = R\omega \qquad \text{(condition for rolling without slipping).} \qquad (10\text{–}14)$$

As Fig. 10–13 shows, the velocity of a point on the wheel is the vector sum of the velocity of the center of mass and the velocity of the point relative to the center of mass. Thus while point 1, the point of contact, is instantaneously at rest, point 3 at the top of the wheel is moving forward *twice as fast* as the center of mass, and points 2 and 4 at the sides have velocities at 45° to the horizontal.

At any instant we can think of the wheel as rotating about an "instantaneous axis" of rotation that passes through the point of contact with the ground. The angular velocity ω is the same for this axis as for an axis through the center of mass; an observer at the center of mass sees the rim make the same number of revolutions per second as does an observer at the rim watching the center of mass spin around him. If we think of the

10–13 As seen from the inertial frame of the surface, the velocity $\vec{\boldsymbol{v}}$ of a point on the rim of a rolling wheel is the vector sum of the velocity $\vec{\boldsymbol{v}}_{cm}$ of the center of mass and the velocity $\vec{\boldsymbol{v}}'$ of the point relative to the center of mass due to the rotation of the wheel. The magnitude of $\vec{\boldsymbol{v}}'$ is equal to $R\omega$. When the wheel rolls without slipping, $v' = v_{cm}$. The rolling wheel is instantaneously at rest at the point where it contacts the ground. The instantaneous velocity is greatest at the top of the wheel.

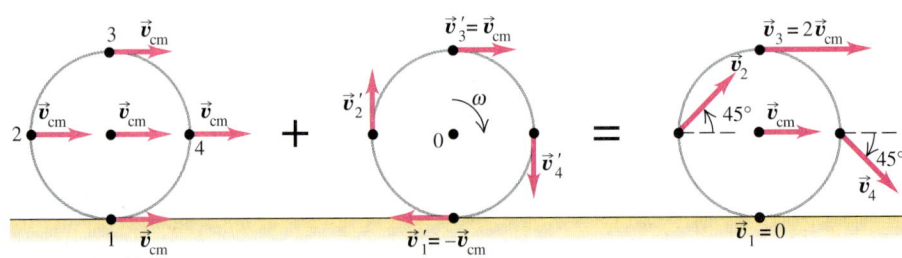

motion of the rolling wheel in Fig. 10–13 in this way, the kinetic energy of the wheel is $K = \frac{1}{2}I_1\omega^2$, where I_1 is the moment of inertia of the wheel about an axis through point 1. But by the parallel-axis theorem, Eq. (9–19), $I_1 = I_{cm} + MR^2$, where M is the total mass of the wheel and I_{cm} is the moment of inertia with respect to an axis through the center of mass. Using Eq. (10–14), the kinetic energy of the wheel is

$$K = \frac{1}{2}I_1\omega^2 = \frac{1}{2}I_{cm}\omega^2 + \frac{1}{2}MR^2\omega^2 = \frac{1}{2}I_{cm}\omega^2 + \frac{1}{2}Mv_{cm}^2,$$

which is the same as Eq. (10–11).

If a rigid body changes height as it moves, we must also consider gravitational potential energy. As we discussed in Section 9–5, the gravitational potential energy associated with any extended body of mass M, rigid or not, is the same as if we replace the body by a particle of mass M located at the body's center of mass. That is,

$$U = Mgy_{cm}.$$

EXAMPLE 10–5

A rolling cylindrical shell A hollow cylindrical shell with mass M and radius R rolls without slipping with speed v_{cm} on a flat surface. What is its kinetic energy?

SOLUTION The moment of inertia is $I_{cm} = MR^2$, and from Eq. (10–14) the angular speed ω is equal to v_{cm}/R. Substituting these expressions into Eq. (10–11), we get

$$K = \frac{1}{2}Mv_{cm}^2 + \frac{1}{2}(MR^2)\left(\frac{v_{cm}}{R}\right)^2$$

$$= Mv_{cm}^2.$$

The kinetic energy is twice as great as it would be if the cylindrical shell were sliding at speed v_{cm} without rolling. Half of the total kinetic energy is translational, and half is rotational.

EXAMPLE 10–6

Speed of a primitive yo-yo A primitive yo-yo is made by wrapping a string several times around a solid cylinder with mass M and radius R (Fig. 10–14). You hold the end of the string stationary while releasing the cylinder with no initial motion. The string unwinds but does not slip or stretch as the cylinder drops and rotates. Use energy considerations to find the speed v_{cm} of the center of mass of the solid cylinder after it has dropped a distance h.

SOLUTION The upper end of the string is held fixed, not pulled upward, so the hand in Fig. 10–14 does no work on the system of string and cylinder. As in Example 9–8 (Section 9–5), there is friction between the string and the cylinder, but because the string never slips on the surface of the cylinder, no mechanical energy is lost. Thus we can use conservation of mechanical energy. The potential energies are $U_1 = Mgh$ and $U_2 = 0$. The string has no kinetic energy because it's massless. The initial kinetic energy of the cylinder is $K_1 = 0$, and its final kinetic energy K_2 is given by Eq. (10–11):

$$K_2 = \frac{1}{2}Mv_{cm}^2 + \frac{1}{2}I_{cm}\omega^2.$$

But $\omega = v_{cm}/R$ from Eq. (10–14), and $I_{cm} = \frac{1}{2}MR^2$, so

$$K_2 = \frac{1}{2}Mv_{cm}^2 + \frac{1}{2}\left(\frac{1}{2}MR^2\right)\left(\frac{v_{cm}}{R}\right)^2$$

$$= \frac{3}{4}Mv_{cm}^2.$$

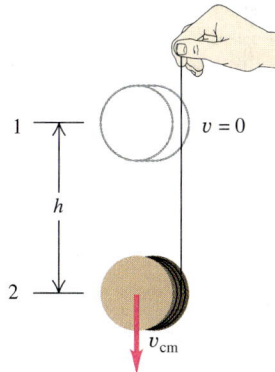

10–14 Calculating the speed of a primitive yo-yo.

Finally, conservation of energy gives

$$K_1 + U_1 = K_2 + U_2,$$

$$0 + Mgh = \frac{3}{4}Mv_{cm}^2 + 0,$$

and

$$v_{cm} = \sqrt{\frac{4}{3}gh}.$$

This is less than the speed $\sqrt{2gh}$ that a dropped object would have, because one third of the potential energy released as the cylinder falls appears as rotational kinetic energy.

EXAMPLE 10-7

Race of the rolling bodies In a physics lecture demonstration, an instructor "races" various round rigid bodies by releasing them from rest at the top of an inclined plane (Fig. 10–15). What shape should a body have to reach the bottom of the incline first?

SOLUTION We can again use conservation of energy because there is no sliding of the rigid bodies over the inclined plane. No work is done by kinetic friction if the bodies roll without slipping. We can also ignore the effects of *rolling friction,* introduced in Section 5–4, provided that the bodies and the surface on which they roll are perfectly rigid. (Later in this section we'll explain why this is so.) Each body starts from rest at the top of an incline with height h, so $K_1 = 0$, $U_1 = Mgh$, and $U_2 = 0$. From Eq. (10–11),

$$K_2 = \frac{1}{2} M v_{cm}^2 + \frac{1}{2} I_{cm} \omega^2.$$

If the bodies roll without slipping, $\omega = v_{cm}/R$. The moments of inertia of all the round bodies in Table 9–2 (about axes through their centers of mass) can be expressed as $I_{cm} = cMR^2$, where c is a pure number less than or equal to 1 that depends on the shape of the body. Then from conservation of energy,

$$K_1 + U_1 = K_2 + U_2,$$

$$0 + Mgh = \frac{1}{2} M v_{cm}^2 + \frac{1}{2} cMR^2 \left(\frac{v_{cm}}{R} \right)^2$$

$$= \frac{1}{2} (1 + c) M v_{cm}^2,$$

10–15 Which body rolls down the plane fastest, and why?

so the speed at the bottom of the incline is

$$v_{cm} = \sqrt{\frac{2gh}{1 + c}}.$$

This is a fairly amazing result; the speed doesn't depend on either the mass M of the body or its radius R. All uniform solid cylinders have the same speed at the bottom, even if their masses and radii are different, because they have the same c. All solid spheres have the same speed, and so on. The smaller the value of c, the faster the body is moving at the bottom (and at any point on the way down). Small-c bodies always beat large-c bodies, because they have less of their kinetic energy tied up in rotation and have more available for translation. Reading the values of c from Table 9–2, we see that the order of finish is as follows: any solid sphere, any solid cylinder, any thin-walled hollow sphere, and any thin-walled hollow cylinder.

COMBINED TRANSLATION AND ROTATION: DYNAMICS

We can also analyze the combined translational and rotational motion of a rigid body from the standpoint of dynamics. We showed in Section 8–6 that for a body with total mass M, the acceleration $\vec{a}_{cm}$ of the center of mass is the same as that of a point mass M acted on by all the external forces on the actual body:

$$\Sigma \vec{F}_{ext} = M\vec{a}_{cm}. \tag{10–15}$$

The rotational motion about the center of mass is described by the rotational analog of Newton's second law, Eq. (10–6):

$$\Sigma \tau = I_{cm} \alpha, \tag{10–16}$$

where I_{cm} is the moment of inertia with respect to an axis through the center of mass and the sum $\Sigma \tau$ includes all external torques with respect to this axis. It's not immediately obvious that Eq. (10–16) should apply to the motion of a translating rigid body; after all, our derivation of $\Sigma \tau = I\alpha$ in Section 10–3 assumed that the axis of rotation was stationary. But in fact, Eq. (10–16) is valid *even when the axis of rotation moves,* provided the following two conditions are met:

1. The axis through the center of mass must be an axis of symmetry.
2. The axis must not change direction.

Note that in general this moving axis of rotation is *not* at rest in an inertial frame of reference.

We can now solve dynamics problems involving a rigid body that undergoes translational and rotational motion at the same time, provided that the rotation axis satisfies

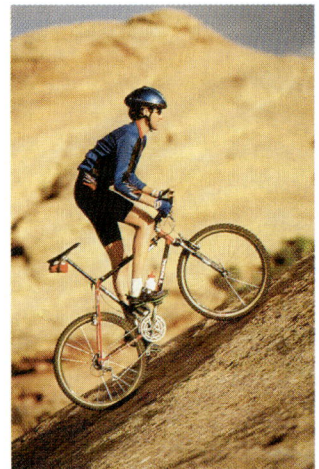

The axle of a bicycle wheel passes through the wheel's center of mass and is an axis of symmetry. Hence the rotation of the wheel is described by Eq. (10–16), provided the bicycle doesn't turn or tilt to one side (which would change the orientation of the axle).

the two conditions just mentioned. The problem-solving strategy outlined in Section 10–3 is equally useful here, and you should review it now. Keep in mind that when a body undergoes translational and rotational motion at the same time, we need two separate equations of motion *for the same body*. One of these, Eq. (10–15), describes the translational motion of the center of mass. The other equation of motion, Eq. (10–16), describes the rotational motion about the axis through the center of mass.

EXAMPLE 10–8

Acceleration of a primitive yo-yo For the primitive yo-yo in Example 10–6, find the downward acceleration of the cylinder and the tension in the string.

SOLUTION Figure 10–16 shows a free-body diagram for the yo-yo, including the choice of positive coordinate directions. The equation for the translational motion of the center of mass is

$$\Sigma F_y = Mg + (-T) = Ma_{cm}. \qquad (10\text{–}17)$$

The moment of inertia for an axis through the center of mass is $I_{cm} = \frac{1}{2}MR^2$. Only the tension force has a torque with respect to the axis through the center of mass, so the equation for rotational motion about this axis is

$$\Sigma \tau = TR = I_{cm}\alpha = \frac{1}{2}MR^2\alpha. \qquad (10\text{–}18)$$

The string unwinds without slipping, so $v_{cm} = R\omega$ from Eq. (10–14); the derivative of this relation with respect to time is

$$a_{cm} = R\alpha. \qquad (10\text{–}19)$$

We now use Eq. (10–19) to eliminate α from Eq. (10–18) and then solve Eqs. (10–17) and (10–18) simultaneously for T and a_{cm}. The results are amazingly simple:

$$a_{cm} = \frac{2}{3}g, \qquad T = \frac{1}{3}Mg.$$

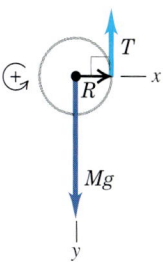

10–16 Free-body diagram for a primitive yo-yo.

Using the constant-acceleration formula $v_{cm}^2 = v_0^2 + 2a_{cm}h$, you can show that the speed of the yo-yo after it has fallen a distance h is $v_{cm} = \sqrt{\frac{4}{3}gh}$, just as we found in Example 10–6.

From the standpoint of dynamics, the tension force is essential; it causes the yo-yo's acceleration to be less than g, and its torque is what causes the yo-yo to turn. Yet when we analyzed this situation using energy methods in Example 10–6, we didn't have to consider the tension force at all! Because no mechanical energy was lost or gained, from the energy standpoint the string is significant only as a way of helping to convert some of the gravitational potential energy into rotational kinetic energy.

EXAMPLE 10–9

Acceleration of a rolling sphere A solid bowling ball rolls without slipping down the return ramp at the side of the alley (Fig. 10–17a). The ramp is inclined at an angle β to the horizontal. What is the ball's acceleration? Treat the ball as a uniform solid sphere, ignoring the finger holes.

SOLUTION Figure 10–17b shows a free-body diagram, with positive coordinate directions indicated. From Table 9–2 the moment of inertia of a solid sphere is $I_{cm} = \frac{2}{5}MR^2$. The equations of motion for translation and for rotation about the axis through

(a)   (b)

10–17 (a) A bowling ball rolling down a ramp. (b) The free-body diagram for the bowling ball.

the center of mass, respectively, are

$$\Sigma F_x = Mg \sin \beta + (-f) = Ma_{cm}, \qquad (10\text{--}20)$$

$$\Sigma \tau = fR = I_{cm}\alpha = \left(\frac{2}{5}MR^2\right)\alpha. \qquad (10\text{--}21)$$

Note that only the friction force f has a torque with respect to the axis through the center of mass. If the ball rolls without slipping, we have the same kinematic relation $a_{cm} = R\alpha$ as in Example 10–8. We use this to eliminate α from Eq. (10–21):

$$fR = \frac{2}{5}MRa_{cm}.$$

This equation and Eq. (10–20) are two equations for two unknowns, a_{cm} and f. We solve Eq. (10–20) for f, substitute the expression into the above equation to eliminate f, and then solve for a_{cm} to obtain

$$a_{cm} = \frac{5}{7} g \sin \beta.$$

The acceleration is just $\frac{5}{7}$ as large as it would be if the ball could *slide* without friction down the slope, like the toboggan in Example 5–9 (Section 5–3). Finally, we substitute this back into Eq. (10–20) and solve for f:

$$f = \frac{2}{7} Mg \sin \beta.$$

Because the ball does not slip at the instantaneous point of contact with the ramp, the friction force f is a *static* friction force; it prevents slipping and gives the ball its angular acceleration. We can derive an expression for the minimum coefficient of static friction μ_s needed to prevent slipping. The normal force is $n = Mg \cos \beta$. The maximum force of static friction equals $\mu_s n$, so the coefficient of friction must be at least as great as

$$\mu_s = \frac{f}{n} = \frac{\frac{2}{7} Mg \sin \beta}{Mg \cos \beta} = \frac{2}{7} \tan \beta.$$

If the plane is tilted only a little, β is small, and only a small value of μ_s is needed to prevent slipping. But as the angle increases, the required value of μ_s increases, as we might expect intuitively. If the ball begins to slip, Eqs. (10–20) and (10–21) are both still valid, but it's no longer true that $v_{cm} = R\omega$ or $a_{cm} = R\alpha$; we have only two equations for three unknowns (a_{cm}, α, and f). To solve the problem of rolling *with* slipping requires taking *kinetic* friction into account (see Challenge Problem 10–85).

If the bowling ball descends a vertical distance h as it moves down the ramp, the displacement along the ramp is $h/\sin \beta$. You should be able to show that the speed of the ball at the bottom of the ramp would be $v_{cm} = \sqrt{\frac{10}{7}gh}$, which is just the result you found in Example 10–7 with $c = \frac{2}{5}$.

If the ball were rolling *uphill*, the force of friction would still be directed uphill as in Fig. 10–17b. Can you see why?

7.9
Car "Peels Out"

9.14
The Sweet Spot

ROLLING FRICTION

In Example 10–7 we said that we can ignore rolling friction if both the rolling body and the surface over which it rolls are perfectly rigid. In Fig. 10–18a a perfectly rigid sphere is rolling down a perfectly rigid incline. The line of action of the normal force passes through the center of the sphere, so its torque is zero; there is no sliding at the point of contact, so the friction force does no work. Figure 10–18b shows a more realistic situation, in which the surface "piles up" in front of the sphere and the sphere rides in a shallow trench. Because of these deformations, the contact forces on the sphere no longer act along a single point, but over an area; the forces are concentrated on the front

10–18 (a) Forces on a perfectly rigid sphere rolling down a perfectly rigid incline. (b) If the sphere or incline is deformable, the contact forces act at different positions. The normal force produces a counterclockwise torque about the rotation axis, opposing the clockwise rotation. The deformation is shown greatly exaggerated.

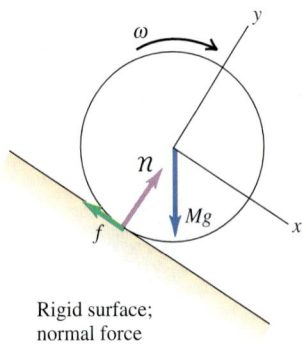

Rigid surface;
normal force
produces no torque

(a)

Deformable surface;
normal force produces
torque opposing rotation

(b)

of the sphere as shown. As a result, the normal force now exerts a torque that opposes the rotation. In addition, there is some sliding of the sphere over the surface due to the deformation, causing mechanical energy to be lost. The combination of these two effects is the phenomenon of *rolling friction*. Rolling friction also occurs if the rolling body is deformable, such as an automobile tire. Often the rolling body and surface are rigid enough that rolling friction can be ignored, as we have assumed in all the examples in this section.

10–5 WORK AND POWER IN ROTATIONAL MOTION

When you pedal a bicycle, you apply forces to a rotating body and do work on it. Similar things happen in many other real-life situations, such as a rotating motor shaft driving a power tool or a car engine propelling the vehicle. We can express this work in terms of torque and angular displacement.

Suppose a tangential force $\vec{F}_{tan}$ acts at the rim of a pivoted disk—for example, a child running while pushing on a playground merry-go-round (Fig. 10–19a). The disk rotates through an infinitesimal angle $d\theta$ about a fixed axis during an infinitesimal time interval dt (Fig. 10–19b). The work dW done by the force $\vec{F}_{tan}$ while a point on the rim moves a distance ds is $dW = F_{tan}\, ds$. If $d\theta$ is measured in radians, then $ds = R\, d\theta$ and

$$dW = F_{tan} R\, d\theta.$$

Now $F_{tan}R$ is the *torque* τ due to the force $\vec{F}_{tan}$, so

$$dW = \tau\, d\theta. \tag{10–22}$$

The total work W done by the torque during an angular displacement from θ_1 to θ_2 is

$$W = \int_{\theta_1}^{\theta_2} \tau\, d\theta \qquad \text{(work done by a torque)}. \tag{10–23}$$

If the torque is *constant* while the angle changes by a finite amount $\Delta\theta = \theta_2 - \theta_1$, then

$$W = \tau(\theta_2 - \theta_1) = \tau\,\Delta\theta \qquad \text{(work done by a constant torque).} \tag{10–24}$$

The work done by a *constant* torque is the product of torque and the angular displacement. If torque is expressed in newton-meters (N · m) and angular displacement in radians, the work is in joules. Equation (10–24) is the rotational analog of Eq. (6–1), $W = Fs$, and Eq. (10–23) is the analog of Eq. (6–7), $W = \int F\, dx$, for the work done by a force in a straight-line displacement.

If the force in Fig. 10–19 had an axial or radial component, that component would do no work because the displacement of the point of application has only a tangential component. An axial or radial component of force would also make no contribution to the torque about the axis of rotation, so Eqs. (10–23) and (10–24) are correct for any force, no matter what its components.

When a torque does work on a rotating rigid body, the kinetic energy changes by an amount equal to the work done. We can prove this by using exactly the same procedure that we used in Eqs. (6–11) through (6–13) for a particle. We first let τ represent the *net* torque on the body so that from Eq. (10–6), $\tau = I\alpha$. By using this equation, we are assuming that the body is rigid so that the moment of inertia I is constant. We then transform the integrand in Eq. (10–23) into an integral on ω as follows:

$$\tau\, d\theta = (I\alpha)\, d\theta = I\frac{d\omega}{dt}\, d\theta = I\frac{d\theta}{dt}\, d\omega = I\omega\, d\omega.$$

Since τ is the net torque, the integral in Eq. (10–23) is the *total* work done on the

(a)

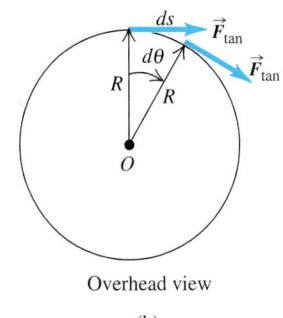

Overhead view

(b)

10–19 A tangential force acting on a rotating body does work.

rotating rigid body. This equation then becomes

$$W_{tot} = \int_{\omega_1}^{\omega_2} I\omega \, d\omega = \frac{1}{2} I\omega_2^2 - \frac{1}{2} I\omega_1^2. \tag{10-25}$$

The change in the rotational kinetic energy of a *rigid* body equals the work done by forces exerted from outside the body. This equation is analogous to Eq. (6–13), the work-energy theorem for a particle.

What about the *power* associated with work done by a torque acting on a rotating body? When we divide both sides of Eq. (10–22) by the time interval dt during which the angular displacement occurs, we find

$$\frac{dW}{dt} = \tau \frac{d\theta}{dt}.$$

But dW/dt is the rate of doing work, or *power P*, and $d\theta/dt$ is angular velocity ω, so

$$P = \tau\omega. \tag{10-26}$$

When a torque τ (with respect to the axis of rotation) acts on a body that rotates with angular velocity ω, its power (rate of doing work) is the product of τ and ω. This is the analog of the relation $P = \vec{F} \cdot \vec{v}$ that we developed in Section 6–5 for particle motion.

EXAMPLE 10–10

The power output of an automobile engine is advertised to be 200 hp at 6000 rpm. What is the corresponding torque?

SOLUTION First we have to convert the power to watts and the angular velocity to rad/s:

$$P = 200 \text{ hp} = 200 \text{ hp} \left(\frac{746 \text{ W}}{1 \text{ hp}} \right) = 1.49 \times 10^5 \text{ W},$$

$$\omega = 6000 \text{ rev/min} = \left(\frac{6000 \text{ rev}}{1 \text{ min}} \right)\left(\frac{2\pi \text{ rad}}{1 \text{ rev}} \right)\left(\frac{1 \text{ min}}{60 \text{ s}} \right)$$

$$= 628 \text{ rad/s}.$$

From Eq. (10–26),

$$\tau = \frac{P}{\omega} = \frac{1.49 \times 10^5 \text{ N} \cdot \text{m/s}}{628 \text{ rad/s}} = 237 \text{ N} \cdot \text{m}.$$

You could apply this much torque by using a wrench 0.25 m long and applying a force of 948 N (213 lb) to the end of its handle. Could you do it?

EXAMPLE 10–11

An electric motor exerts a constant torque of $\tau = 10 \text{ N} \cdot \text{m}$ on a grindstone mounted on its shaft. The moment of inertia of the grindstone is $I = 2.0 \text{ kg} \cdot \text{m}^2$. If the system starts from rest, find the work done by the motor in 8.0 seconds and the kinetic energy at the end of this time. What was the average power delivered by the motor?

SOLUTION From $\tau = I\alpha$, the angular acceleration is 5.0 rad/s^2. The angular velocity and kinetic energy after 8.0 s are

$$\omega = \alpha t = (5.0 \text{ rad/s}^2)(8.0 \text{ s}) = 40 \text{ rad/s},$$

$$K = \frac{1}{2} I\omega^2 = \frac{1}{2}(2.0 \text{ kg} \cdot \text{m}^2)(40 \text{ rad/s})^2 = 1600 \text{ J}.$$

The total angle through which the system turns in 8.0 s is

$$\Delta\theta = \frac{1}{2}\alpha t^2 = \frac{1}{2}(5.0 \text{ rad/s}^2)(8.0 \text{ s})^2 = 160 \text{ rad},$$

and the total work done by the torque is

$$W = \tau\Delta\theta = (10 \text{ N} \cdot \text{m})(160 \text{ rad}) = 1600 \text{ J}$$

This equals the total kinetic energy, as it must.

The average power is the total work divided by the time interval:

$$P_{av} = \frac{1600 \text{ J}}{8.0 \text{ s}} = 200 \text{ J/s} = 200 \text{ W}.$$

The instantaneous power, given by $P = \tau\omega$, is not constant because ω increases continuously. But we can compute the total work by taking the time integral of P, as follows:

$$W = \int_{t_1}^{t_2} P \, dt = \int_{t_1}^{t_2} \tau\omega \, dt = \int_{t_1}^{t_2} \tau(\alpha t) dt$$

$$= \int_0^{8.0 \text{ s}} (10 \text{ N} \cdot \text{m})(5 \text{ rad/s}^2)t \, dt = \left[(50 \text{ J/s}^2)\frac{t^2}{2} \right]_0^{8.0 \text{ s}}$$

$$= 1600 \text{ J},$$

as before. The instantaneous power P increases from zero at $t = 0$ to $(10 \text{ N} \cdot \text{m})(40 \text{ rad/s}) = 400 \text{ W}$ at $t = 8.0 \text{ s}$. The angular velocity and the power increase uniformly with time, so the *average* power is just half this maximum value, or 200 W.

10–6 ANGULAR MOMENTUM

Every rotational quantity that we have encountered in Chapters 9 and 10 is the analog of some quantity in the translational motion of a particle. The analog of *momentum* of a particle is **angular momentum,** a vector quantity denoted as $\vec{L}$. Its relation to momentum $\vec{p}$ (which we will sometimes call *linear momentum* for clarity) is exactly the same as the relation of torque to force, $\vec{\tau} = \vec{r} \times \vec{F}$. For a particle with constant mass m, velocity $\vec{v}$, momentum $\vec{p} = m\vec{v}$, and position vector $\vec{r}$ relative to the origin O of an inertial frame, we define angular momentum $\vec{L}$ as

$$\vec{L} = \vec{r} \times \vec{p} = \vec{r} \times m\vec{v} \quad \text{(angular momentum of a particle).} \quad (10\text{–}27)$$

The value of $\vec{L}$ depends on the choice of origin O, since it involves the position vector of the particle relative to the origin. The units of angular momentum are kg $\cdot$ m²/s.

In Fig. 10–20 a particle moves in the xy-plane; its position vector $\vec{r}$ and momentum $\vec{p} = m\vec{v}$ are shown. The angular momentum vector $\vec{L}$ is perpendicular to the xy-plane. The right-hand rule for vector products shows that its direction is along the $+z$-axis, and its magnitude is

$$L = mvr \sin \phi = mvl, \quad (10\text{–}28)$$

where l is the perpendicular distance from the line of $\vec{v}$ to O. This distance plays the role of "lever arm" for the momentum vector.

When a net force $\vec{F}$ acts on a particle, its velocity and momentum change, so its angular momentum may also change. We can show that the *rate of change* of angular momentum is equal to the torque of the net force. We take the time derivative of Eq. (10–27), using the rule for the derivative of a product:

$$\frac{d\vec{L}}{dt} = \left(\frac{d\vec{r}}{dt} \times m\vec{v} \right) + \left(\vec{r} \times m \frac{d\vec{v}}{dt} \right) = (\vec{v} \times m\vec{v}) + (\vec{r} \times m\vec{a}).$$

The first term is zero because it contains the vector product of the vector $\vec{v} = d\vec{r}/dt$ with itself. In the second term we replace $m\vec{a}$ with the net force $\vec{F}$, obtaining

$$\frac{d\vec{L}}{dt} = \vec{r} \times \vec{F} = \vec{\tau} \quad \text{(for a particle acted on by net force } \vec{F}). \quad (10\text{–}29)$$

The rate of change of angular momentum of a particle equals the torque of the net force acting on it. Compare this result to Eq. (8–3), which states that the rate of change $d\vec{p}/dt$ of the *linear* momentum of a particle equals the net force that acts on it.

We can use Eq. (10–28) to find the total angular momentum of a rigid body rotating about the z-axis with angular speed ω. First consider a thin slice of the body lying in the xy-plane (Fig. 10–21). Each particle in the slice moves in a circle centered at the origin, and at each instant its velocity $\vec{v}_i$ is perpendicular to its position vector $\vec{r}_i$, as shown. Hence in Eq. (10–28), $\phi = 90°$ for every particle. A particle with mass m_i at a distance r_i from O has a speed v_i equal to $r_i\omega$. From Eq. (10–28) the magnitude L_i of its angular momentum is

$$L_i = m_i(r_i\omega)r_i = m_i r_i{}^2 \omega. \quad (10\text{–}30)$$

The direction of each particle's angular momentum, as given by the right-hand rule for the vector product, is along the $+z$-axis.

The *total* angular momentum of the slice of the body lying in the xy-plane is the sum ΣL_i of the angular momenta L_i of the particles. Summing Eq. (10–30), we have

$$L = \Sigma L_i = (\Sigma m_i r_i{}^2)\omega = I\omega,$$

where I is the moment of inertia of the slice about the z-axis.

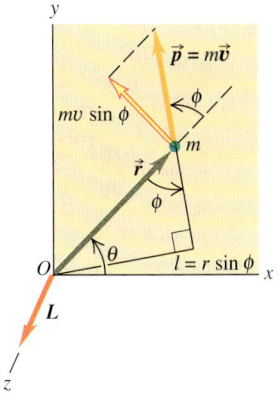

10–20 A particle with mass m moving in the xy-plane. The angular momentum of the particle is $\vec{L} = \vec{r} \times m\vec{v} = \vec{r} \times \vec{p}$, and its magnitude is $L = mvl$.

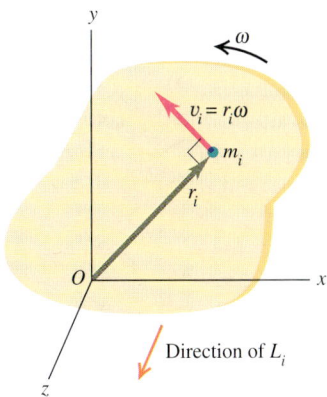

10–21 Calculating the angular momentum of a rotating rigid body. Every particle in the body moves in a circle about the rotation axis (in this case, the z-axis) with the same angular speed.

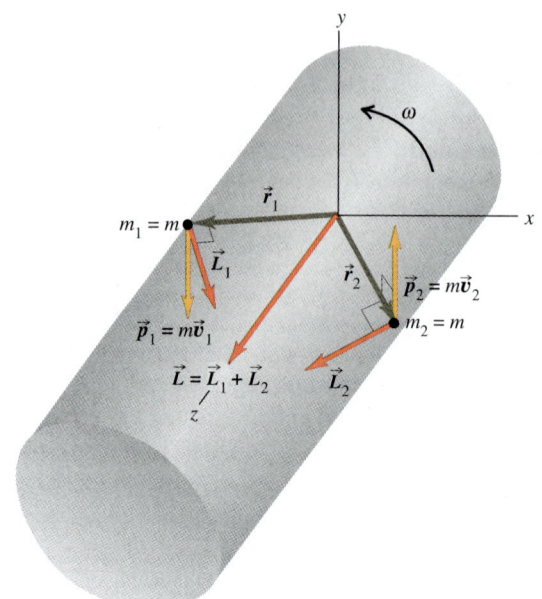

10–22 Two particles of the same mass ($m_1 = m_2 = m$) located symmetrically on either side of the axis of rotation (the z-axis). The angular momentum vectors $\vec{L}_1$ and $\vec{L}_2$ of the individual particles do not lie along the rotation axis, but their vector sum $\vec{L} = \vec{L}_1 + \vec{L}_2$ does.

We can do this same calculation for the other slices of the body, all parallel to the xy-plane. For points that do not lie in the xy-plane, a complication arises because the $\vec{r}$ vectors have components in the z-direction as well as the x- and y-directions; this gives the angular momentum of each particle a component perpendicular to the z-axis. But *if the z-axis is an axis of symmetry,* the perpendicular components for particles on opposite sides of this axis add up to zero (Fig. 10–22). So when a body rotates about an axis of symmetry, its angular momentum vector $\vec{L}$ lies along the symmetry axis, and its magnitude is $L = I\omega$.

The angular velocity vector $\vec{\omega}$ also lies along the rotation axis, as we discussed at the end of Section 9–2. Hence for a rigid body rotating around an axis of symmetry, $\vec{L}$ and $\vec{\omega}$ are in the same direction (Fig. 10–23). So we have the *vector* relationship

$$\vec{L} = I\vec{\omega} \qquad \text{(for a rigid body rotating around a symmetry axis).} \qquad (10\text{--}31)$$

From Eq. (10–29) the rate of change of angular momentum of a particle equals the torque of the net force acting on the particle. For any system of particles (including both rigid and nonrigid bodies) the rate of change of the *total* angular momentum equals the sum of the torques of all forces acting on all the particles. The torques of the *internal* forces add to zero if these forces act along the line from one particle to another, as in Fig. 10–7, and so the sum of torques includes only the torques of the *external* forces. (A similar cancellation occurred in our discussion of center-of-mass motion in Section 8–6.) If the total angular momentum of the system of particles is $\vec{L}$ and the sum of external torques is $\Sigma\vec{\tau}$, then

$$\Sigma\vec{\tau} = \frac{d\vec{L}}{dt} \qquad \text{(for any system of particles).} \qquad (10\text{--}32)$$

10–23 For rotation about an axis of symmetry, $\vec{\omega}$ and $\vec{L}$ are parallel and along the axis. The directions are given by the same right-hand rule as in Fig. 10–4.

Finally, if the system of particles is a rigid body rotating about a symmetry axis, then $L = I\omega$ and I is constant. If this axis has a fixed direction in space, then the vectors $\vec{L}$ and $\vec{\omega}$ change only in magnitude, not in direction. In that case, $dL/dt = I\, d\omega/dt = I\alpha$, or

$$\Sigma\tau = I\alpha,$$

which is again our basic relation for the dynamics of rigid-body rotation. If the body is not rigid, I may change, and in that case, L changes even when ω is constant. For a non-rigid body, Eq. (10–32) is still valid, even though Eq. (10–6) is not.

When the axis of rotation is *not* a symmetry axis, the angular momentum is in general *not* parallel to the axis (Fig. 10–24). As the body turns, the angular momentum vector $\vec{L}$ traces out a cone around the rotation axis. Because $\vec{L}$ changes, there must be a net external torque acting on the body even though the angular velocity magnitude ω may be constant. If the body is an unbalanced wheel on a car, this torque is provided by friction in the bearings, which causes the bearings to wear out. "Balancing" a wheel means distributing the mass so that the rotation axis is an axis of symmetry; then $\vec{L}$ points along the rotation axis, and no net torque is required to keep the wheel turning.

In fixed-axis rotation we often use the term "angular momentum of the body" to refer just to the *component* of $\vec{L}$ along the rotation axis of the body (the z-axis in Fig. 10–24), with a sign to indicate the sense of rotation just as with angular velocity.

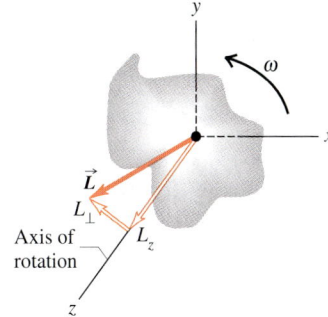

10–24 If the rotation axis of a rigid body is not a symmetry axis, the angular momentum vector $\vec{L}$ does not in general lie along the rotation axis. Even if ω is constant, the direction of $\vec{L}$ changes, and a net torque is required to maintain rotation.

EXAMPLE 10–12

A turbine fan in a jet engine (Fig. 10–25) has a moment of inertia of 2.5 kg · m^2 about its axis of rotation. As the turbine is starting up, its angular velocity as a function of time is

$$\omega = (400 \text{ rad/s}^3)t^2.$$

a) Find the fan's angular momentum as a function of time, and find its value at time $t = 3.0$ s. b) Find the net torque acting on the fan as a function of time, and find the torque at time $t = 3.0$ s.

SOLUTION a) Just like an electric fan, the turbine fan rotates around an axis of symmetry, so we may use Eq. (10–31) to relate the magnitudes L and ω:

$$L = I\omega = (2.5 \text{ kg} \cdot \text{m}^2)(400 \text{ rad/s}^3)t^2 = (1000 \text{ kg} \cdot \text{m}^2/\text{s}^3)t^2.$$

At time $t = 3.0$ s,

$$L = (1000 \text{ kg} \cdot \text{m}^2/\text{s}^3)(3.0 \text{ s})^2 = 9000 \text{ kg} \cdot \text{m}^2/\text{s}.$$

b) From Eq. (10–32),

$$\tau = \frac{dL}{dt} = (1000 \text{ kg} \cdot \text{m}^2/\text{s}^2)(2t) = (2000 \text{ kg} \cdot \text{m}^2/\text{s}^2)t.$$

10–25 A turbine fan is used in turbojet engines to force air into the engine at high speeds.

At time $t = 3.0$ s,

$$\tau = (2000 \text{ kg} \cdot \text{m}^2/\text{s}^3)(3.0 \text{ s}) = 6000 \text{ kg} \cdot \text{m}^2/\text{s}^2 = 6000 \text{ N} \cdot \text{m}.$$

10–7 CONSERVATION OF ANGULAR MOMENTUM

We have just seen that angular momentum can be used for an alternative statement of the basic dynamic principle for rotational motion. It also forms the basis for the **principle of conservation of angular momentum.** Like conservation of energy and of linear momentum, this principle is a universal conservation law, valid at all scales from atomic and nuclear systems to the motions of galaxies. This principle follows directly from Eq. (10–32): $\Sigma\vec{\tau} = d\vec{L}/dt$. If $\Sigma\vec{\tau} = 0$, then $d\vec{L}/dt = 0$, and $\vec{L}$ is constant. **When the net external torque acting on a system is zero, the total angular momentum of the system is constant (conserved).**

A circus acrobat, a diver, and an ice skater pirouetting on the toe of one skate all take advantage of this principle. Suppose an acrobat has just left a swing with arms and legs extended and rotating counterclockwise about her center of mass. When she pulls her arms and legs in, her moment of inertia I_{cm} with respect to her center of mass changes from a large value I_1 to a much smaller value I_2. The only external force acting on her is her weight, which has no torque with respect to an axis through her center of mass. So her angular momentum $L = I_{cm}\omega$ remains constant, and her angular velocity ω increases

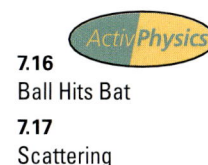

ActivPhysics

7.16
Ball Hits Bat

7.17
Scattering

as I_{cm} decreases. That is,

$$I_1\omega_1 = I_2\omega_2 \qquad \text{(zero net external torque).} \qquad (10-33)$$

When a skater or ballerina spins with arms outstretched and then pulls her arms in, her angular velocity increases as her moment of inertia decreases. In each case there is conservation of angular momentum in a system in which the net external torque is zero.

When a system has several parts, the internal forces that the parts exert on each other cause changes in the angular momenta of the parts, but the *total* angular momentum doesn't change. Here's an example. Consider two bodies A and B that interact with each other but not with anything else, such as the astronauts we discussed in Section 8–3 (Fig. 8–6). Suppose body A exerts a force $\vec{F}_{A \text{ on } B}$ on body B; the corresponding torque (with respect to whatever point we choose) is $\vec{\tau}_{A \text{ on } B}$. According to Eq. (10–32), this torque is equal to the rate of change of angular momentum of B:

$$\vec{\tau}_{A \text{ on } B} = \frac{d\vec{L}_B}{dt}.$$

At the same time, body B exerts a force $\vec{F}_{B \text{ on } A}$ on body A, with a corresponding torque $\vec{\tau}_{B \text{ on } A}$, and

$$\vec{\tau}_{B \text{ on } A} = \frac{d\vec{L}_A}{dt}.$$

From Newton's third law, $\vec{F}_{B \text{ on } A} = -\vec{F}_{A \text{ on } B}$. Furthermore, if the forces act along the same line, as in Fig. 10–7, their lever arms with respect to the chosen axis are equal. Thus the *torques* of these two forces are equal and opposite, and $\vec{\tau}_{B \text{ on } A} = -\vec{\tau}_{A \text{ on } B}$. So if we add the two previous equations, we find

$$\frac{d\vec{L}_A}{dt} + \frac{d\vec{L}_B}{dt} = 0,$$

or, because $\vec{L}_A + \vec{L}_B$ is the *total* angular momentum $\vec{L}$ of the system,

$$\frac{d\vec{L}}{dt} = 0 \qquad \text{(zero net external torque).} \qquad (10-34)$$

That is, the total angular momentum of the system is constant. The torques of the internal forces can transfer angular momentum from one body to the other, but they can't change the *total* angular momentum of the system.

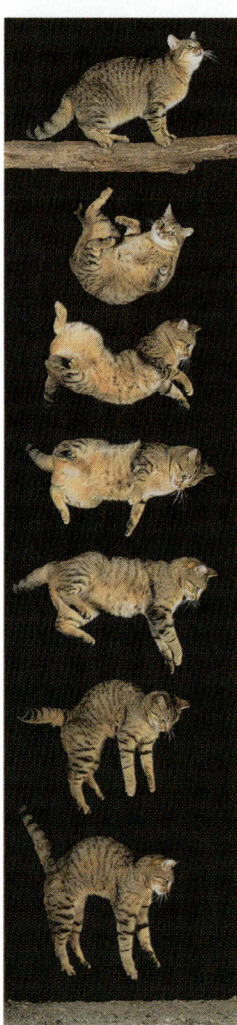

A falling cat twists different parts of its body in different directions so that it lands feet first. At all times during this process the angular momentum of the cat as a whole remains zero.

Anyone can be a ballerina An acrobatic physics professor stands at the center of a turntable, holding his arms extended horizontally with a 5.0-kg dumbbell in each hand (Fig. 10–26). He is set rotating about a vertical axis, making one revolution in 2.0 s. Find the prof's new angular velocity if he pulls the dumbbells in to his stomach, and discuss how this affects the kinetic energy. His moment of inertia (without the dumbbells) is 3.0 kg · m^2 when his arms are outstretched, dropping to 2.2 kg · m^2 when his hands are at his stomach. The dumbbells are 1.0 m from the axis initially and 0.20 m from it at the end. Treat the dumbbells as particles.

SOLUTION If we neglect friction in the turntable, no external torques act about the vertical axis, so the angular momentum

about this axis is constant, and we can use Eq. (10–33). The moment of inertia of the system is $I = I_{prof} + I_{dumbbells}$, so

$$I_1 = 3.0 \text{ kg} \cdot \text{m}^2 + 2(5.0 \text{ kg})(1.0 \text{ m})^2 = 13 \text{ kg} \cdot \text{m}^2,$$

$$I_2 = 2.2 \text{ kg} \cdot \text{m}^2 + 2(5.0 \text{ kg})(0.20 \text{ m})^2 = 2.6 \text{ kg} \cdot \text{m}^2,$$

$$\omega_1 = \frac{1 \text{ rev}}{2.0 \text{ s}} = 0.5 \text{ rev/s}.$$

From conservation of angular momentum the new angular velocity is

$$\omega_2 = \frac{I_1}{I_2}\omega_1 = \frac{13 \text{ kg} \cdot \text{m}^2}{2.6 \text{ kg} \cdot \text{m}^2}(0.5 \text{ rev/s}) = 2.5 \text{ rev/s}.$$

That is, the angular velocity increases by a factor of five while the angular momentum remains constant. Note that we didn't have to change "revolutions" to "radians" in this calculation. Why not?

To calculate the kinetic energy, we must express ω_1 and ω_2 in rad/s. (Why?) We have $\omega_1 = (0.5 \text{ rev/s})(2\pi \text{ rad/rev}) = 3.14 \text{ rad/s}$ and $\omega_2 = (2.5 \text{ rev/s})(2\pi \text{ rad/rev}) = 15.7 \text{ rad/s}$. The initial kinetic energy is

$$K_1 = \frac{1}{2} I_1 \omega_1^2 = \frac{1}{2}(13 \text{ kg} \cdot \text{m}^2)(3.14 \text{ rad/s})^2 = 64 \text{ J},$$

and the final kinetic energy is

$$K_2 = \frac{1}{2} I_2 \omega_2^2 = \frac{1}{2}(2.6 \text{ kg} \cdot \text{m}^2)(15.7 \text{ rad/s})^2 = 320 \text{ J}.$$

Where did the extra energy come from?

10–26 Fun with conservation of angular momentum—if you don't get dizzy.

EXAMPLE 10–14

Figure 10–27 shows two disks, one an engine flywheel, the other a clutch plate attached to a transmission shaft. Their moments of inertia are I_A and I_B; initially, they are rotating with constant angular velocities ω_A and ω_B, respectively. We then push the disks together with forces acting along the axis, so as not to apply any torque on either disk. The disks rub against each other and eventually reach a common final angular speed ω. Derive an expression for ω.

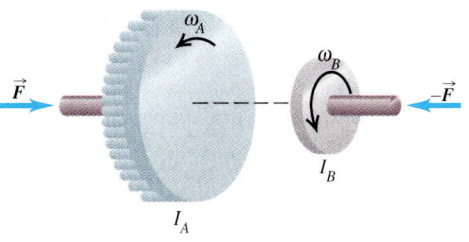

SOLUTION The only torque acting on either disk is the torque applied by the other disk; there are no external torques. Thus the total angular momentum of the system of two disks is the same before and after they are pushed together. At the end they rotate together as one body with total moment of inertia $I = I_A + I_B$ and angular speed ω. Conservation of angular momentum gives

$$I_A \omega_A + I_B \omega_B = (I_A + I_B)\omega,$$

$$\omega = \frac{I_A \omega_A + I_B \omega_B}{I_A + I_B}.$$

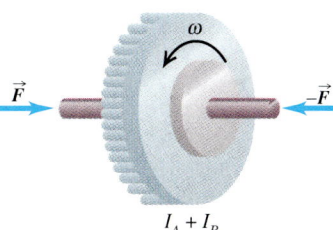

10–27 When the net external torque is zero, angular momentum is conserved. The forces shown are along the axis of rotation and thus exert no torque about any point on it.

EXAMPLE 10–15

In Example 10–14, suppose flywheel A has a mass of 2.0 kg, a radius of 0.20 m, and an initial angular speed of 50 rad/s (about 500 rpm) and that clutch plate B has a mass of 4.0 kg, a radius of 0.10 m, and an initial angular speed of 200 rad/s. Find the common final angular speed ω after the disks are pushed into contact. Is kinetic energy conserved during this process?

SOLUTION The moments of inertia of the two disks are

$$I_A = \frac{1}{2} m_A r_A^2 = \frac{1}{2}(2.0 \text{ kg})(0.20 \text{ m})^2 = 0.040 \text{ kg} \cdot \text{m}^2,$$

$$I_B = \frac{1}{2} m_B r_B^2 = \frac{1}{2}(4.0 \text{ kg})(0.10 \text{ m})^2 = 0.020 \text{ kg} \cdot \text{m}^2.$$

From Example 10–14 we have

$$\omega = \frac{I_A \omega_A + I_B \omega_B}{I_A + I_B}$$

$$= \frac{(0.040 \text{ kg} \cdot \text{m}^2)(50 \text{ rad/s}) + (0.020 \text{ kg} \cdot \text{m}^2)(200 \text{ rad/s})}{0.040 \text{ kg} \cdot \text{m}^2 + 0.020 \text{ kg} \cdot \text{m}^2}$$

$$= 100 \text{ rad/s}.$$

The initial kinetic energy is

$$K_1 = \frac{1}{2} I_A \omega_A^2 + \frac{1}{2} I_B \omega_B^2$$

$$= \frac{1}{2}(0.040 \text{ kg} \cdot \text{m}^2)(50 \text{ rad/s})^2 + \frac{1}{2}(0.020 \text{ kg} \cdot \text{m}^2)(200 \text{ rad/s})^2$$

$$= 450 \text{ J}.$$

The final kinetic energy is

$$K_2 = \frac{1}{2}(I_A + I_B)\omega^2$$

$$= \frac{1}{2}(0.040 \text{ kg} \cdot \text{m}^2 + 0.020 \text{ kg} \cdot \text{m}^2)(100 \text{ rad/s})^2 = 300 \text{ J}.$$

One third of the initial kinetic energy was lost during this "angular collision," the rotational analog of a completely inelastic collision. We shouldn't expect kinetic energy to be conserved, even though the resultant external force and torque are zero, because nonconservative (frictional) internal forces act while the two disks rub together and gradually approach a common angular velocity.

EXAMPLE 10-16

Angular momentum in a crime bust A door 1.0 m wide, of mass 15 kg, is hinged at one side so that it can rotate without friction about a vertical axis. It is unlatched. A police officer fires a bullet with a mass of 10 g and a speed of 400 m/s into the exact center of the door, in a direction perpendicular to the plane of the door (Fig. 10–28). Find the angular speed of the door just after the bullet imbeds itself in the door. Is kinetic energy conserved?

SOLUTION We consider the door and bullet together as a system. There is no external torque about the axis defined by the hinges, so angular momentum about this axis is conserved. The initial angular momentum of the bullet is given by Eq. (10–28):

$$L = mvl = (0.010 \text{ kg})(400 \text{ m/s})(0.50 \text{ m}) = 2.0 \text{ kg} \cdot \text{m}^2/\text{s}.$$

The final angular momentum is $I\omega$, where $I = I_{\text{door}} + I_{\text{bullet}}$. From Table 9–2, for a door of width d,

$$I_{\text{door}} = \frac{Md^2}{3} = \frac{(15 \text{ kg})(1.0 \text{ m})^2}{3} = 5.0 \text{ kg} \cdot \text{m}^2.$$

The moment of inertia of the bullet (with respect to the axis along the hinges) is

$$I_{\text{bullet}} = ml^2 = (0.010 \text{ kg})(0.50 \text{ m})^2 = 0.0025 \text{ kg} \cdot \text{m}^2.$$

Conservation of angular momentum requires that $mvl = I\omega$, or

$$\omega = \frac{mvl}{I} = \frac{2.0 \text{ kg} \cdot \text{m}^2/\text{s}}{5.0 \text{ kg} \cdot \text{m}^2 + 0.0025 \text{ kg} \cdot \text{m}^2} = 0.40 \text{ rad/s}.$$

10-28 Shooting open a door (top view). The bullet imbeds itself in the center of the door.

The collision of bullet and door is inelastic because nonconservative forces act during the impact. Thus we do not expect kinetic energy to be conserved. To check, we calculate initial and final kinetic energies:

$$K_1 = \frac{1}{2}mv^2 = \frac{1}{2}(0.010 \text{ kg})(400 \text{ m/s})^2 = 800 \text{ J},$$

$$K_2 = \frac{1}{2}I\omega^2 = \frac{1}{2}(5.0025 \text{ kg} \cdot \text{m}^2)(0.40 \text{ rad/s})^2$$

$$= 0.40 \text{ J}.$$

The final kinetic energy is only 1/2000 of the initial value!

10-8 GYROSCOPES AND PRECESSION

In all of the situations we've looked at so far in this chapter, the axis of rotation either has stayed fixed or has moved and kept the same direction (such as rolling without slipping). But a variety of new physical phenomena, some quite unexpected, can occur when the axis of rotation can change direction. For example, consider a toy gyroscope that's supported at one end (Fig. 10–29). If we hold it with the flywheel axis horizontal and let go, the free end of the axis simply drops owing to gravity—*if* the flywheel isn't spinning. But if the flywheel *is* spinning, what happens is quite different. One possible motion is a steady circular motion of the axis in a horizontal plane, combined with the spin motion of the flywheel about the axis. This surprising, non-intuitive motion of the axis is called **precession.** Precession is found in nature, as well as in rotating machines such as gyroscopes. As you read these words, the earth itself is precessing; its spin axis (through the north and south poles) slowly changes direction, going through a complete cycle of precession every 26,000 years.

To study this strange phenomenon of precession, we must remember that angular velocity, angular momentum, and torque are all *vector* quantities. In particular, we need the general relation between the net torque $\Sigma \vec{\tau}$ that acts on a body and the rate of change

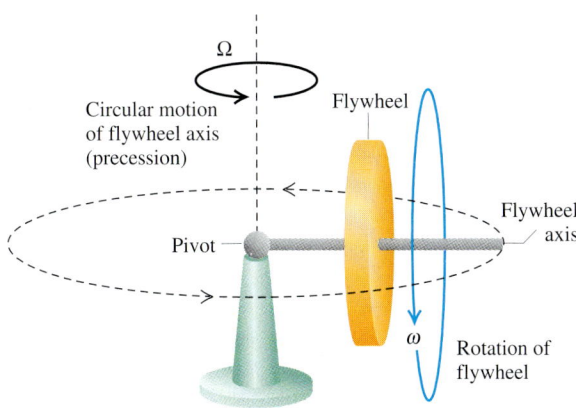

10–29 A gyroscope spinning with angular speed ω and supported at one end. The flywheel and axis do not drop, but move in a horizontal circular motion called precession. The angular speed of precession is Ω.

of the body's angular momentum $\vec{L}$, given by Eq. (10–32), $\Sigma\vec{\tau} = d\vec{L}/dt$. Let's first apply this equation to the case in which the flywheel is *not* spinning (Fig. 10–30a). We take the origin O at the pivot and assume that the flywheel is symmetrical, with mass M and moment of inertia I about the flywheel axis. The flywheel axis is initially along the x-axis. The only external forces on the gyroscope are the normal force $\vec{n}$ acting at the pivot (assumed to be frictionless) and the weight $\vec{w}$ of the flywheel that acts at its center of mass, a distance r from the pivot. The normal force has zero torque with respect to the pivot, and the weight has a torque $\vec{\tau}$ in the y-direction as shown in Fig. 10–30a. Initially, there is no rotation, and the initial angular momentum $\vec{L}_i$ is zero. From Eq. (10–32) the *change $d\vec{L}$* in angular momentum in a short time interval dt following this is

$$d\vec{L} = \vec{\tau}\,dt. \qquad (10\text{–}35)$$

This is in the y-direction because $\vec{\tau}$ is. As each additional time interval dt elapses, the angular momentum changes by additional increments $d\vec{L}$ in the y-direction because the direction of the torque is constant (Fig. 10–30b). The steadily increasing horizontal angular momentum means that the gyroscope rotates downward faster and faster around the y-axis until it hits either the stand or the table on which it sits.

Now let's see what happens if the flywheel *is* spinning initially, so the initial angular momentum $\vec{L}_i$ is not zero (Fig. 10–31a). Since the flywheel rotates around its symmetry axis, $\vec{L}_i$ lies along the axis. But each change in angular momentum $d\vec{L}$ is perpendicular to the axis, because the torque $\vec{\tau} = \vec{r} \times \vec{w}$ is perpendicular to the axis (Fig. 10–31b). This causes the *direction* of $\vec{L}$ to change, but not its magnitude. The changes $d\vec{L}$ are always in the horizontal xy-plane, so the angular momentum vector and the flywheel axis with which it moves are always horizontal. In other words, the axis doesn't fall—it just precesses.

If this still seems mystifying to you, think about a ball attached to a string. If the ball is initially at rest and you pull the string toward you, the ball will move toward you also.

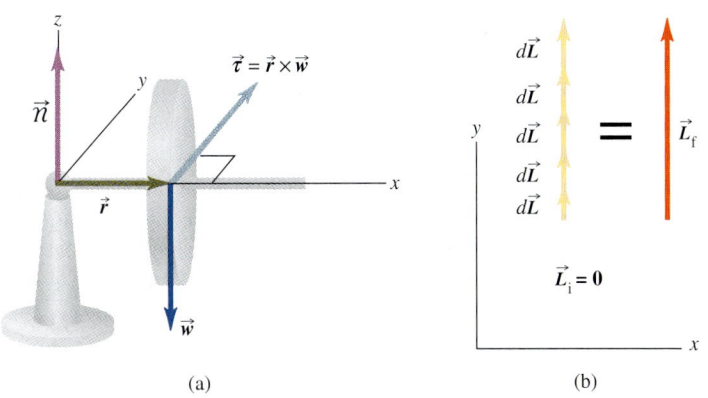

(a)   (b)

10–30 (a) The flywheel is initially not spinning. The torque $\vec{\tau}$ is due to the weight $\vec{w}$. (b) Looking down from directly over the gyroscope. In each successive time interval dt the torque produces a change $d\vec{L}$ in the angular momentum; this change is in the same direction as $\vec{\tau}$. The flywheel has zero initial angular momentum, so all the $d\vec{L}$'s are parallel, and the final angular momentum $\vec{L}_f$ also has the same direction as $\vec{\tau}$.

10–31 (a) The flywheel is spinning initially with angular momentum $\vec{L}_i$. The free-body diagram is the same as in Fig. 10–30, so the forces are not shown.
(b) With an initial angular momentum, each change $d\vec{L}$ is perpendicular to $\vec{L}$; this continuously changes the direction of $\vec{L}$, causing precession. The magnitude of the angular momentum is constant: $|\vec{L}_i| = |\vec{L}_f|$.

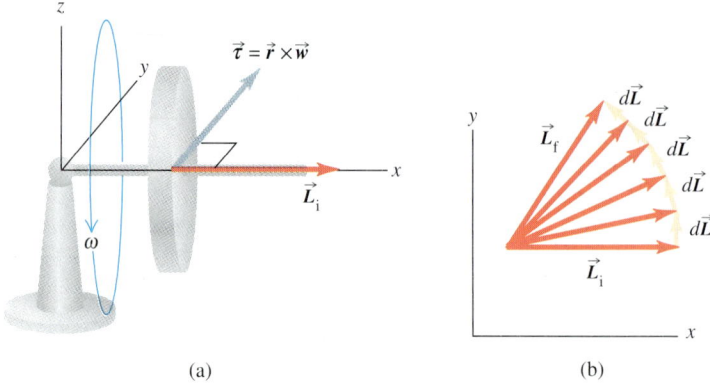

(a)   (b)

But if the ball is initially moving and you continuously pull the string in a direction perpendicular to the ball's motion, the ball moves in a circle around your hand; it does not approach your hand at all. In the first case the ball has zero linear momentum $\vec{p}$ to start with; when you apply a force $\vec{F}$ toward you for a time dt, the ball acquires a momentum $d\vec{p} = \vec{F}\,dt$, which is also toward you. But if the ball already has linear momentum $\vec{p}$, a change in momentum $d\vec{p}$ that's perpendicular to $\vec{p}$ will change the direction of motion, not the speed. Replace $\vec{p}$ with $\vec{L}$ and $\vec{F}$ with $\vec{\tau}$ in this argument, and you'll see that precession is simply the rotational analog of uniform circular motion.

At the instant shown in Fig. 10–31a, the gyroscope has angular momentum $\vec{L}$. A short time interval dt later, the angular momentum is $\vec{L} + d\vec{L}$; the infinitesimal change in angular momentum is $d\vec{L} = \vec{\tau}\,dt$, which is perpendicular to $\vec{L}$. As the vector diagram in Fig. 10–32 shows, this means that the flywheel axis of the gyroscope has turned through a small angle $d\phi$ given by $d\phi = |d\vec{L}|/|\vec{L}|$. The rate at which the axis moves, $d\phi/dt$, is called the **precession angular speed;** denoting this quantity by Ω, we find

$$\Omega = \frac{d\phi}{dt} = \frac{|d\vec{L}|/|\vec{L}|}{dt} = \frac{\tau}{L} = \frac{wr}{I\omega}. \qquad (10\text{–}36)$$

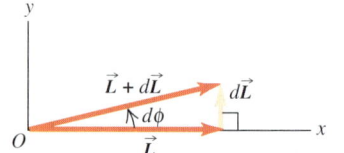

10–32 In a time dt the angular momentum vector and the flywheel axis precess together through an angle $d\phi$.

Thus the precession angular speed is *inversely* proportional to the angular speed of spin about the axis. A rapidly spinning gyroscope precesses slowly; if friction in its bearings causes the flywheel to slow down, the precession angular speed *increases!* The precession angular speed of the earth is very slow (1 rev/26,000 yr) because its spin angular momentum L is large and the torque τ, due to the gravitational influences of the moon and sun, is relatively small.

As a gyroscope precesses, its center of mass moves in a circle with radius r in a horizontal plane. Its vertical component of acceleration is zero, so the upward normal force $\vec{n}$ exerted by the pivot must be just equal in magnitude to the weight. The circular motion of the center of mass with angular speed Ω requires a force $\vec{F}$ directed toward the center of the circle, with magnitude $F = M\Omega^2 r$. This force must also be supplied by the pivot.

One key assumption that we made in our analysis of the gyroscope was that the angular momentum vector $\vec{L}$ is associated only with the spin of the flywheel and is purely horizontal. But there will also be a vertical component of angular momentum associated with the precessional motion of the gyroscope. By ignoring this, we've tacitly assumed that the precession is *slow,* that is, that the precession angular speed Ω is very much less than the spin angular speed ω. As Eq. (10–36) shows, a large value of ω automatically gives a small value of Ω, so this approximation is reasonable. When the precession is not slow, additional effects show up, including an up-and-down wobble or *nutation* of the flywheel axis that's superimposed on the precessional motion. You can see nutation occurring in a gyroscope as its spin slows down, so that Ω increases and the vertical component of $\vec{L}$ can no longer be ignored.

EXAMPLE 10-17

Figure 10–33a shows a top view of a cylindrical gyroscope wheel that has been set spinning by an electric motor. The pivot is at O, and the mass of the axle is negligible. a) As seen from above, is the precession clockwise or counterclockwise? b) If the gyro takes 4.0 s for one revolution of precession, at what angular speed does the wheel spin?

SOLUTION a) The right-hand rule shows that $\vec{\omega}$ and $\vec{L}$ are to the left (Fig. 10–33b). The weight $\vec{w}$ points into the page in this top view and acts at the center of mass (denoted by an ×); the torque $\vec{\tau} = \vec{r} \times \vec{w}$ is toward the top of the page; and $d\vec{L}/dt$ is also toward the top of the page. Adding a small $d\vec{L}$ to the $\vec{L}$ that we have initially changes the direction of $\vec{L}$ as shown, so the precession is clockwise as seen from above.

b) Be careful not to confuse ω and Ω! We are given $\Omega = $ (1 rev)/(4.0 s) = $(2\pi$ rad)/(4.0 s) = 1.57 rad/s. The weight is equal to mg, and the moment of inertia about its symmetry axis of a solid cylinder with radius R is $I = \frac{1}{2}mR^2$. Solving Eq. (10–36) for ω, we find

$$\omega = \frac{wr}{I\Omega} = \frac{mgr}{(mR^2/2)\Omega} = \frac{2gr}{R^2\Omega}$$

$$= \frac{2(9.8 \text{ m/s}^2)(2.0 \times 10^{-2} \text{ m})}{(3.0 \times 10^{-2} \text{ m})^2(1.57 \text{ rad/s})} = 280 \text{ rad/s} = 2600 \text{ rev/min}.$$

The precession angular speed Ω is very much less than the spin angular speed ω, so this is an example of slow precession.

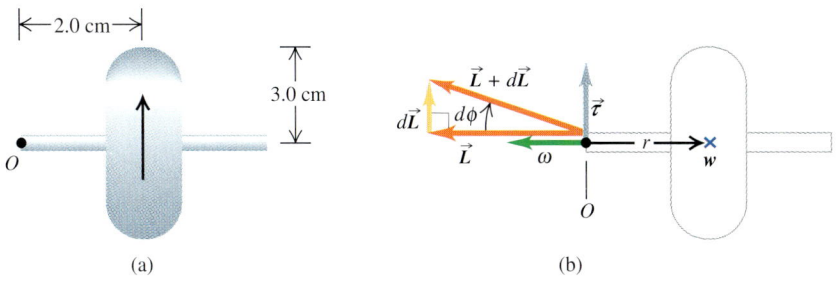

(a)   (b)

10–33 In which direction does this gyroscope precess?

SUMMARY

■ When a force $\vec{F}$ acts on a body, the torque τ of that force with respect to a point O is given by

$$\tau = Fl, \qquad (10-2)$$

where l is the lever arm. A common sign convention is "counterclockwise positive, clockwise negative." A generalized definition of torque as a vector quantity $\vec{\tau}$ is

$$\vec{\tau} = \vec{r} \times \vec{F}, \qquad (10-3)$$

where $\vec{r}$ is the position vector of the point at which the force acts.

■ The angular acceleration α of a rigid body rotating about a stationary axis is related to the net torque $\Sigma\tau$ and the moment of inertia I of the body by

$$\Sigma\tau = I\alpha. \qquad (10-6)$$

■ If a rigid body has both translational and rotational motion, its kinetic energy may be expressed as the sum of translational kinetic energy due to motion of the center of mass and rotational kinetic energy about an axis through the center of mass:

$$K = \frac{1}{2}Mv_{cm}^2 + \frac{1}{2}I_{cm}\omega^2. \qquad (10-11)$$

In the special case of a rigid body with radius R that rolls without slipping, $v_{cm} = R\omega$. Combined translation and rotation may also be described by $\Sigma\vec{F}_{ext} = M\vec{a}_{cm}$ (for translation of the center of mass) and $\Sigma\tau = I_{cm}\alpha$ (for rotation about the center of mass), provided that the rotation axis is a symmetry axis that does not change direction as the body moves.

KEY TERMS

line of action, 294
lever arm (moment arm), 294
torque, 294
combined translation and rotation, 301
rolling without slipping, 302
angular momentum, 309
principle of conservation of angular momentum, 311
precession, 314
precession angular speed, 316

■ When a torque τ acts on a rigid body that undergoes an angular displacement from θ_1 to θ_2, the work W done by the torque is

$$W = \int_{\theta_1}^{\theta_2} \tau \, d\theta. \tag{10–23}$$

If the torque is *constant*, then

$$W = \tau(\theta_2 - \theta_1) = \tau \, \Delta\theta. \tag{10–24}$$

The work-energy theorem for rotational motion of a rigid body is

$$W_{\text{tot}} = \frac{1}{2} I\omega_2{}^2 - \frac{1}{2} I\omega_1{}^2. \tag{10–25}$$

When the body rotates with angular velocity ω, the power P (rate at which the torque does work) is

$$P = \tau\omega. \tag{10–26}$$

■ The angular momentum $\vec{L}$, with respect to point O, of a particle with mass m, velocity $\vec{v}$, and linear momentum $\vec{p} = m\vec{v}$, is

$$\vec{L} = \vec{r} \times \vec{p} = \vec{r} \times m\vec{v}, \tag{10–27}$$

where $\vec{r}$ is the particle's position vector with respect to point O. When a symmetric rigid body with moment of inertia I rotates with angular velocity $\vec{\omega}$ about a stationary axis of symmetry, its angular momentum is given by

$$\vec{L} = I\vec{\omega}. \tag{10–31}$$

If the body is not symmetric or the rotation axis is not an axis of symmetry, the component of angular momentum along the axis of rotation is equal to $I\omega$.

■ In terms of vector torque $\vec{\tau}$ and angular momentum $\vec{L}$, the basic dynamic relation for the rotational motion of any system can be restated as

$$\Sigma\vec{\tau} = \frac{d\vec{L}}{dt}. \tag{10–32}$$

■ If the net torque associated with the external forces acting on a system is zero, the total angular momentum of the system is constant (conserved).

DISCUSSION QUESTIONS

Q10–1 In tightening cylinder-head bolts in an automobile engine, the critical quantity is the *torque* applied to the bolts. Why is this more important than the actual *force* applied to the wrench handle?

Q10–2 Can a single force applied to a body change both its translational and rotational motion? Explain.

Q10–3 Suppose you could use wheels of any type in the design of a soapbox-derby racer (an unpowered four-wheel vehicle that coasts from rest down a hill). Within the rules on the total weight of the vehicle and rider, should you design with large massive wheels or small light wheels? Should you use solid wheels or wheels with most of the mass at the rim? Explain.

Q10–4 Unless $\vec{r}$ and $\vec{F}$ are perpendicular, there are always two angles between their directions that give the same torque for given magnitudes of $\vec{r}$ and $\vec{F}$. Explain why. Illustrate your answer with a sketch.

Q10–5 The crankshaft of an automobile engine has a flywheel to increase the moment of inertia about the rotation axis. Why is this desirable?

Q10–6 The harder you hit the brakes while driving forward, the more the front end of your car will move down (and the rear end move up). Why? What happens when accelerating? Why don't drag racers use front-wheel drive only?

Q10–7 When an acrobat walks on a tightrope, she extends her arms straight out from her sides. She does this to make it easier for her to catch herself if she should tip to one side or the other. Explain how this works. (*Hint:* Think about Eq. (10–6).)

Q10–8 When an electric motor is turned on, it takes longer to come up to final speed if there is a grinding wheel attached to the shaft. Why?

Q10–9 Experienced cooks can tell whether an egg is raw or hard-boiled by rolling it down a slope (taking care to catch it at the bottom). How is this possible? What are they looking for?

Q10–10 An electric grinding wheel coasts for a minute or more after the power is turned off, but an electric drill coasts for only a few seconds. Why is there a difference?

Q10–11 The hammer in Fig. 10–11 is acted on by the force of gravity, and forces produce torques that cause a body's angular velocity to change. Why, then, is the angular velocity of the hammer in the figure constant?

Q10–12 In Example 10–6 (Section 10–4) suppose you pulled your hand and the end of the string upward. Would mechanical energy be conserved in this situation? Why or why not?

Q10–13 A wheel is rolling without slipping on a horizontal surface. In an inertial frame of reference in which the surface is at rest, is there any point on the wheel that has a velocity that is purely vertical? Is there any point that has a horizontal velocity component opposite to the velocity of the center of mass? Explain. Do your answers change if the wheel is slipping as it rolls? Why or why not?

Q10–14 Part of the kinetic energy of a moving automobile is in the rotational motion of its wheels. When the brakes are applied hard on an icy street, the wheels "lock," and the car starts to slide. What becomes of the rotational kinetic energy?

Q10–15 You are standing at the center of a large horizontal turntable in a carnival funhouse. The turntable is set rotating on frictionless bearings, and rotates freely (that is, there is no motor driving the turntable). As you walk toward the edge of the turntable, what happens to the combined angular momentum of you and the turntable? What happens to the rotation speed of the turntable? Explain your answer.

Q10–16 A point particle travels in a circle at constant speed. With respect to the origin at the center of the circle, is there a net torque acting on the particle? A net force? What if the speed of the particle is changing? In each case, explain your answers.

Q10–17 A point particle travels in a straight line at constant speed and the closest distance it comes to the origin of coordinates is a distance l. With respect to this origin, does the particle have nonzero angular momentum? As the particle moves along its straight-line path, does its angular momentum with respect to origin change?

Q10–18 In Example 10–13 (Section 10–7) the angular speed ω changes and this must mean that there is nonzero angular acceler-

ation. But there is no torque about the rotation axis if the forces the professor applies to the weights are directly radially inward. Then by Eq. (10–6) α must be zero. Explain what is wrong with this reasoning that leads to this apparent contradiction.

Q10–19 In Example 10–13 (Section 10–7) the rotational kinetic energy of the professor and dumbbells increases. But since there are no external torques, no work is being done to change the rotational kinetic energy. Then, by Eq. (10–25) the kinetic energy must remain the same! Explain what is wrong with this reasoning that leads to this apparent contradiction. Where *does* the extra kinetic energy come from?

Q10–20 As discussed in Section 10–7, the angular momentum of a circus acrobat is conserved as she tumbles through the air. Is her *linear* momentum conserved? Why or why not?

Q10–21 If you stop a spinning raw egg for the shortest instant you can and then release it, the egg will start spinning again. If you do the same to a hard-boiled egg, it will remain stopped. Try it. Explain it.

Q10–22 A helicopter has a large main rotor that rotates in a horizontal plane and provides lift. There is also a small rotor on the tail that rotates in a vertical plane. What is the purpose of the tail rotor? (*Hint:* if there were no tail rotor, what would happen when the pilot changed the angular speed of the main rotor?) Some helicopters have no tail rotor, but instead have two large main rotors that rotate in a horizontal plane. Why is it important that the two main rotors rotate in opposite directions?

Q10–23 In a common design for a gyroscope, the flywheel and flywheel axis are enclosed in a light, spherical frame with the flywheel at the center of the frame. The gyroscope is then balanced on top of a pivot so that the flywheel is directly above the pivot. Does the gyroscope precess if it is released while the flywheel is spinning? Explain.

Q10–24 A gyroscope takes 3.8 s to precess 1.0 revolution about a vertical axis. Two minutes later it takes only 1.9 s to precess 1.0 revolution. No one has touched the gyroscope. Explain.

Q10–25 A gyroscope is precessing as in Fig. 10–29. What happens if you gently add some weight to the end of the flywheel axis farthest from the pivot?

EXERCISES

SECTION 10–2 TORQUE

10–1 Calculate the torque (magnitude and direction) about point O due to the force $\vec{F}$ in each of the situations sketched in Fig. 10–34. In each case, the force $\vec{F}$ and the yellow rod both lie in the plane of the page, the rod has length 4.00 m, and the force has magnitude $F = 10.0$ N.

10–2 Calculate the net torque about point O for the two forces applied as in Fig. 10–35. The rod and both forces are in the plane of the page.

10–3 A square metal plate 0.180 m on each side is pivoted about an axis through point O at its center and perpendicular to the plate (Fig. 10–36). Calculate the net torque about this axis due to the three forces shown in the figure if the magnitudes of

the forces are $F_1 = 18.0$ N, $F_2 = 26.0$ N, and $F_3 = 14.0$ N. The plate and all forces are in the plane of the page.

10–4 Forces $F_1 = 7.50$ N and $F_2 = 5.30$ N are applied tangentially to a wheel with radius 0.330 m, as shown in Fig. 10–37. What is the net torque on the wheel due these two forces for an axis perpendicular to the wheel and passing through its center?

10–5 One force acting on a machine part is $\vec{F} = (-5.00 \text{ N})\hat{\imath} + (4.00 \text{ N})\hat{\jmath}$. The vector from the origin to the point where the force is applied is $\vec{r} = (-0.450 \text{ m})\hat{\imath} + (0.150 \text{ m})\hat{\jmath}$. a) In a sketch show $\vec{r}, \vec{F},$ and the origin. b) Use the right-hand rule to determine the direction of the torque. c) Calculate the vector torque produced by this force. Verify that the direction of the torque is the same as you obtained in (b).

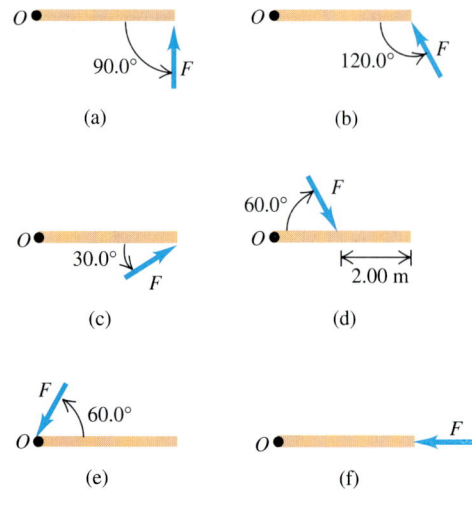

(a)   (b)

(c)   (d)

(e)   (f)

FIGURE 10–34 Exercise 10–1.

FIGURE 10–35 Exercise 10–2.

FIGURE 10–36 Exercise 10–3.

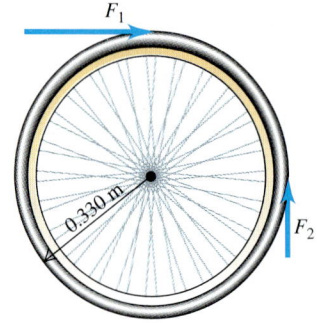

FIGURE 10–37 Exercise 10–4.

SECTION 10–3 TORQUE AND ANGULAR ACCELERATION FOR A RIGID BODY

10–6 The flywheel of an engine has moment of inertia $2.50 \text{ kg} \cdot \text{m}^2$ about its rotation axis. a) What constant torque

is required to bring it up to an angular speed of 400 rev/min in 8.00 s, starting from rest? b) What is its final kinetic energy?

10–7 Using the value of a calculated in Example 10–2 (Section 10–3), what is the speed of the cable after it has been pulled 2.0 m? Compare your result with Example 9–8 (Section 9–5).

10–8 A cord is wrapped around the rim of a wheel 0.250 m in radius, and a steady pull of 40.0 N is exerted on the cord. The wheel is mounted on frictionless bearings on a horizontal shaft through its center. The moment of inertia of the wheel about this shaft is $5.00 \text{ kg} \cdot \text{m}^2$. Compute the angular acceleration of the wheel.

10–9 a) Find the magnitude n of the normal force for the situation described in Example 10–3 (Section 10–3). b) Is your answer to part (a) less than, equal to, or greater than the total weight $(M + m)g$ of the cylinder and mass? Explain why this should be so. c) Suppose the cylinder is initially rotating clockwise so that the hanging mass m is initially moving upward at speed v_0 (the cable remains taut). What effect does this have on the tension T and normal force n? Explain.

10–10 a) In the situation described in Example 10–2 of Section 10–3 (Fig. 10–8), the normal force $\vec{n}$ exerted on the cylinder by the bearing is directed up and to the left. Why must the normal force be in this direction? b) Calculate the magnitude and direction of $\vec{n}$.

10–11 A grindstone in the shape of a solid disk with a diameter of 0.520 m and a mass of 50.0 kg is rotating at 850 rev/min. You press an ax against the rim with a normal force of 160 N (Fig. 10–38), and the grindstone comes to rest in 7.50 s. Find the coefficient of friction between the ax and the grindstone. Neglect friction in the bearings.

FIGURE 10–38 Exercise 10–11 and Problem 10–44.

10–12 A 15.0-kg bucket of water is suspended by a rope wrapped around a windlass, which is a solid cylinder 0.300 m in diameter with mass 12.0 kg. The cylinder is pivoted on a frictionless axle through its center. The bucket is released from rest at the top of a well and falls 10.0 m to the water. Neglect the weight of the rope. a) What is the tension in the rope while the bucket is falling? b) With what speed does the bucket strike the water? c) What is the time of fall? d) While the bucket is falling, what is the force exerted on the cylinder by the axle?

10–13 A 2.00-kg textbook rests on a frictionless horizontal surface. A cord attached to the book passes over a pulley whose diameter is 0.150 m to a hanging book of mass 3.00 kg. The system is released from rest, and the books are observed to move 1.20 m in 0.800 s. a) What is the tension in each part of the cord? b) What is the moment of inertia of the pulley about its rotation axis?

10–14 A thin, horizontal rod with length l and mass M is pivoted about a vertical axis at one end. A force with constant magnitude F is applied to the other end, causing the rod to rotate in a horizontal plane. The force is maintained perpendicular to the rod and to the axis of rotation. Calculate the magnitude of the angular acceleration of the rod.

SECTION 10–4 RIGID-BODY ROTATION ABOUT A MOVING AXIS

10–15 A string is wrapped several times around the rim of a small hoop of radius 0.0800 m and mass 0.180 kg. If the free end of the string is held in place and the hoop is released from rest (Fig. 10–39), calculate a) the tension in the string while the hoop descends as the string unwinds; b) the time it takes the hoop to descend 0.750 m; c) the angular speed of the rotating hoop after it has descended 0.750 m.

0.0800 m

FIGURE 10–39 Exercise 10–15 and Problem 10–63.

10–16 Repeat part (c) of Exercise 10–15, this time using energy considerations.

10–17 In Example 10–5 (Section 10–4) we found that for a hollow cylindrical shell rolling without slipping on a horizontal surface, half of the total kinetic energy is translational and half is rotational. What fraction of the total kinetic energy is rotational for the following objects rolling without slipping on a horizontal surface? a) A uniform solid cylinder. b) A uniform sphere. c) A thin-walled hollow sphere. d) A hollow cylinder of outer radius R and inner radius $R/2$.

10–18 A hollow spherical shell of mass 2.00 kg rolls without slipping down a 38.0° slope. a) Find the acceleration, the friction force, and the minimum coefficient of friction needed to prevent slipping. b) How would your answers to part (a) change if the mass were doubled to 4.00 kg?

10–19 A 392-N wheel comes off a moving truck and rolls without slipping along a highway. At the bottom of a hill it is rotating at 25.0 rad/s. The radius of the wheel is 0.600 m and its moment of inertia about its rotation axis is $0.800MR^2$. Friction does work on the wheel as it rolls up the hill to a stop, a height h above the bottom of the hill; this work has absolute value 3500 J. Calculate h.

10–20 A Ball Rolling Uphill. A bowling ball rolls without slipping up a ramp that slopes upward at an angle β to the horizontal. (See Example 10–9 in Section 10–4.) Treat the ball as a uniform solid sphere, ignoring the finger holes. a) Draw the free-body diagram for the ball. Explain why the friction force must be directed *uphill*. b) What is the acceleration of the center of mass of the ball? c) What minimum coefficient of static friction is needed to prevent slipping?

SECTION 10–5 WORK AND POWER IN ROTATIONAL MOTION

10–21 A playground merry-go-round has radius 2.40 m and moment of inertia 2100 kg · m² about a vertical axle through its center, and turns with negligible friction. a) A child applies an 18.0-N force tangentially to the edge of the merry-go-round for 15.0 s. If the merry-go-round is initially at rest, what is its angular speed after this 15.0-s interval? b) How much work did the child do on the merry-go-round? c) What is the average power supplied by the child?

10–22 Example 9–5 (Section 9–4) described the design of an aircraft propeller. The engine delivers 175 hp to the propeller at 2400 rev/min. a) How much torque does the aircraft engine provide? b) How much work does the engine do in one revolution of the propeller?

10–23 A 1.50-kg grinding wheel is in the form of a solid cylinder of radius 0.100 m. a) What constant torque will bring it from rest to an angular speed of 1200 rev/min in 2.5 s? b) Through what angle has it turned during that time? c) Use Eq. (10–24) to calculate the work done by the torque. d) What is the grinding wheel's kinetic energy when it is rotating at 1200 rev/min? Compare your answer to the result in part (c).

10–24 What is the power output in horsepower of an electric motor turning at 4800 rev/min and developing a torque of 4.30 N · m?

10–25 The carbide tips of the cutting teeth of a circular saw are 8.6 cm from the axis of rotation. a) The no-load speed of the saw, when it is not cutting anything, is 4800 rev/min. Why is its no-load power output negligible? b) While cutting lumber, the angular speed of the saw slows to 2400 rev/min and the power output is 1.9 hp. What is the tangential force that the wood exerts on the carbide tips?

10–26 An airplane propeller is 2.08 m in length (from tip to tip) and has a mass of 117 kg. When the airplane's engine is first started, it applies a constant torque of 1950 N · m to the propeller, which starts from rest. a) What is the angular acceleration of the propeller? Treat the propeller as a slender rod. (*Hint:* See Table 9–2.) b) What is the propeller's angular speed after making 5.00 rev? c) How much work is done by the engine during the first 5.00 rev? d) What is the average power output of the engine during the first 5.00 rev? e) What is the instantaneous power output of the motor at the instant that the propeller has turned through 5.00 rev?

10–27 a) Compute the torque developed by an industrial motor whose output is 150 kW at an angular speed of 4000 rev/min. b) A drum of negligible mass, 0.400 m in diameter, is attached to the motor shaft and the power output of the motor is used to raise a weight hanging from a rope wrapped around the drum. How heavy a weight can be lifted at constant speed? c) At what constant speed will the weight rise?

SECTION 10-6 ANGULAR MOMENTUM

10-28 A woman of mass 50 kg is standing on the rim of a large disk that is rotating at 0.50 rev/s about an axis through its center. The disk has a mass of 110 kg and a radius of 4.0 m. Calculate the magnitude of the total angular momentum of the woman-plus-disk system. (Assume that the woman can be treated as a point.)

10-29 A 2.00-kg rock has a horizontal velocity of magnitude 12.0 m/s when it is at point P in Fig. 10-40. a) At this instant, what is the magnitude and direction of its angular momentum relative to point O? b) If the only force acting on the rock is its weight, what is the rate of change (magnitude and direction) of its angular momentum at this instant?

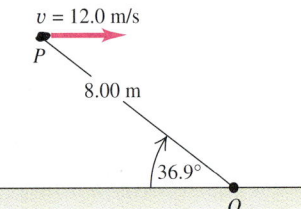

FIGURE 10-40 Exercise 10-29.

10-30 a) Calculate the magnitude of the angular momentum of the earth considered as a particle orbiting the sun. The mass of the earth is 5.97×10^{24} kg. Treat it as moving in a circular orbit of radius 1.50×10^{11} m at a speed of 2.98×10^4 m/s. b) Calculate the magnitude of the angular momentum of the earth due its rotation around an axis through the north and south poles. Treat the earth as a uniform sphere of radius 6.38×10^6 m that makes one revolution in 24.0 hours.

10-31 Find the magnitude of the angular momentum of the second hand on a clock about an axis through the center of the clock face. The clock hand has a length of 15.0 cm and a mass of 6.00 g. Take the second hand to be a slender rod rotating with constant angular velocity about one end.

SECTION 10-7 CONSERVATION OF ANGULAR MOMENTUM

10-32 Under some circumstances a star can collapse into an extremely dense object made mostly of neutrons and called a *neutron star*. The density of a neutron star is roughly 10^{14} times as great as that of ordinary solid matter. Suppose we represent the star as a uniform solid rigid sphere, both before and after the collapse. The star's initial radius was 7.0×10^5 km (comparable to our sun); its final radius is 16 km. If the original star rotated once in 30 days, find the angular speed of the neutron star.

10-33 A small block on a frictionless horizontal surface has a mass of 0.0250 kg. It is attached to a massless cord passing through a hole in the surface (Fig. 10-41). The block is originally revolving at a distance of 0.300 m from the hole with an angular speed of 1.75 rad/s. The cord is then pulled from below, shortening the radius of the circle in which the block revolves to 0.150 m. The block may be treated as a particle. a) Is angular momentum conserved? Why or why not? b) What is the new angular speed? c) Find the change in kinetic energy of the block. d) How much work was done in pulling the cord?

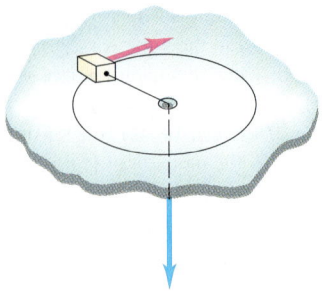

FIGURE 10-41 Exercise 10-33, Problem 10-76, and Challenge Problem 10-87.

10-34 The Spinning Figure Skater. The outstretched hands and arms of a figure skater preparing for a spin can be considered a slender rod pivoting about an axis through its center (Fig. 10-42). When his hands and arms are brought in and wrapped around his body to execute the spin, the hands and arms can be considered a thin-walled hollow cylinder. His hands and arms have a combined mass of 8.0 kg. When outstretched, they span 1.8 m; when wrapped, they form a cylinder of radius 25 cm. The moment of inertia about the rotation axis of the remainder of his body is constant and equal to 0.40 kg · m². If his original angular speed is 0.40 rev/s, what is his final angular speed?

FIGURE 10-42 Exercise 10-34.

10-35 A diver comes off a board with arms straight up and legs straight down, giving her a moment of inertia about her rotation axis of 18 kg · m². She then tucks into a small ball, decreasing this moment of inertia to 3.6 kg · m². While tucked, she makes two complete revolutions in 1.0 s. If she hadn't tucked at all, how many revolutions would she have made in the 1.5 s from board to water?

10-36 A large turntable rotates about a fixed vertical axis, making one revolution in 6.00 s. The moment of inertia of the turntable about this axis is 1200 kg · m². A child of mass 40.0 kg, initially standing at the center of the turntable, runs out along a radius. What is the angular speed of the turntable when the child is 2.00 m from the center? (Assume that the child can be treated as a particle.)

10-37 A large wooden turntable in the shape of a flat disk has a radius of 2.00 m and a total mass of 120 kg. The turntable is initially rotating at 3.00 rad/s about a vertical axis through its center. Suddenly, a 70-kg parachutist makes a soft landing on the turntable at a point near the outer edge. a) Find the angular speed of the turntable after the parachutist lands. (Assume that the parachutist can be treated as a particle.) b) Compute the kinetic energy of the system before and after the parachutist lands. Why are these kinetic energies not equal?

10–38 A solid wood door 1.00 m wide and 2.00 m high is hinged along one side and has a total mass of 40.0 kg. Initially open and at rest, the door is struck at its center by a handful of sticky mud of mass 0.500 kg, traveling perpendicular to the door at 12.0 m/s just before impact. Find the final angular speed of the door. Does the mud make a significant contribution to the moment of inertia?

SECTION 10–8 GYROSCOPES AND PRECESSION

10–39 Draw a top view of the gyroscope in Fig. 10–29. a) Draw labeled arrows on your sketch for $\vec{\omega}$, $\vec{L}$, and $\vec{\tau}$. Draw $d\vec{L}$ produced by $\vec{\tau}$. Draw $\vec{L} + d\vec{L}$. Determine the sense of precession by examining the directions of $\vec{L}$ and $\vec{L} + d\vec{L}$. b) Reverse the direction of the spin angular velocity of the rotor and repeat all steps in part (a). c) Move the pivot to the other end of the shaft, with the same direction of spin angular velocity as in part (b), and repeat all the steps. d) Keeping the pivot as in part (c), reverse the spin angular velocity of the rotor and repeat all steps.

10–40 The rotor (flywheel) of a toy gyroscope has a mass of 0.140 kg. Its moment of inertia about its axis is 1.20×10^{-4} kg $\cdot$ m^2. The mass of the frame is 0.0250 kg. The gyroscope is supported on a single pivot (Fig. 10–43) with its center of mass a horizontal distance of 4.00 cm from the pivot. The gyroscope is precessing in a horizontal plane at the rate of one revolution in 2.20 s. a) Find the upward force exerted by the pivot. b) Find the angular speed with which the rotor is spinning about its axis, expressed in rev/min. c) Copy the diagram, and draw vectors to show the angular momentum of the rotor and the torque acting on it.

FIGURE 10–43 Exercise 10–40.

10–41 A Gyro Stabilizer. The stabilizing gyroscope of a ship is a solid disk of mass 60,000 kg; its radius is 2.00 m, and it rotates about a vertical axis with an angular speed of 500 rev/min. a) How much time is required to bring it up to speed, starting from rest, with a constant power input of 7.46×10^4 W? b) Find the torque needed to cause the axis to precess in a vertical fore-and-aft plane at the rate of 1.00°/s.

10–42 A gyroscope is precessing about a vertical axis. Describe what happens to the precession angular speed if the following changes in the variables are made, with all other variables remaining the same: a) the angular speed of the spinning flywheel is doubled; b) the total weight is doubled; c) the moment of inertia about the axis of the spinning flywheel is doubled; d) the distance from the pivot to the center of gravity is doubled. e) What happens if all four of the variables in parts (a) through (d) are doubled?

10–43 The earth precesses once every 26,000 years, and spins on its axis once a day. Estimate the magnitude of the torque that causes the precession of the earth. You may need some data from Appendix F. Make the estimate by assuming i) the earth is a uniform sphere and ii) the precession of the earth is like that of the gyroscope shown in Fig. 10–30. In this model, the precession axis and rotation axis are perpendicular. Actually, the angle between these two axes for the earth is only $23\frac{1}{2}°$; this affects the calculated torque by about a factor of 2.

PROBLEMS

10–44 A 55.0-kg grindstone is a solid disk 0.520 m in diameter. You press an ax down on the rim with a normal force of 160 N (Fig. 10–38). The coefficient of kinetic friction between the blade and the stone is 0.60, and there is a constant friction torque of 6.50 N $\cdot$ m between the axle of the stone and its bearings. a) How much force must be applied tangentially at the end of a crank handle 0.500 m long to bring the stone from rest to 120 rev/min in 9.00 s? b) After the grindstone attains an angular speed of 120 rev/min, what tangential force at the end of the handle is needed to maintain a constant angular speed of 120 rev/min? c) How much time does it take the grindstone to come from 120 rev/min to rest if it is acted on by the axle friction alone?

10–45 An experimental bicycle wheel is placed on a test stand so that it is free to turn on its axle. If a constant net torque of 5.00 N $\cdot$ m is applied to the tire for 2.00 s, the angular speed of the tire increases from zero to 100 rev/min. The external torque is then removed, and the wheel is brought to rest by friction in its bearings in 125 s. Compute a) the moment of inertia of the wheel

about the rotation axis; b) the friction torque; c) the total number of revolutions made by the wheel in the 125-s time interval.

10–46 A flywheel 0.600 m in diameter is pivoted on a horizontal axis. A rope is wrapped around the outside of the flywheel, and a steady pull of 40.0 N is exerted on the rope. The flywheel starts from rest, and 5.00 m of rope are unwound in 2.00 s. a) What is the angular acceleration of the flywheel? b) What is its final angular speed? c) What is its final kinetic energy? d) What is its moment of inertia about the rotation axis?

10–47 A wheel starts from rest and rotates with constant angular acceleration about a fixed axis. a) Prove that the power at any given time is proportional to the square of the net torque about the axis. b) If the power at $t = 3.00$ s is 500 W with a constant net torque of 20.0 N $\cdot$ m, what would the power have been at $t = 3.00$ s with a constant net torque of 60.0 N $\cdot$ m? c) Prove that the power at any given angular displacement is proportional to the $\frac{3}{2}$ power of the total torque about the axis at that angular displacement. d) If the power after having rotated 37.5 rad with a torque of 20.0 N $\cdot$ m is 500 W, what would the power have been after having rotated

37.5 rad with a torque of 60.0 N • m? e) Do the answers to parts (a) and (b) contradict the answers to parts (c) and (d)? Why or why not?

10–48 A beam of length l lies on the +x-axis with its left end at the origin. A cable pulls on the beam in the +y-direction with a force $\vec{F}$ with a magnitude that depends on where along the beam it is exerted, $F = F_0(1 - x/l)$, where F_0 is a constant equal to the force magnitude when it is applied at the left end of the beam. a) What is the direction of the torque due to $\vec{F}$? b) Graph F versus x from $x = 0$ to $x = l$. Express F in terms of F_0 and x in terms of l. c) Graph the torque versus x from $x = 0$ to $x = l$. Express the torque in terms of $F_0 l$, and express x in terms of l. d) Where along the beam should the force be applied to produce maximum torque and what is that maximum torque?

10–49 While exploring a castle, Exena the Exterminator is spotted by a dragon who chases her down a hallway. Exena runs into a room and attempts to swing the heavy door shut before the dragon gets her. The door is initially perpendicular to the wall, so it must be turned through 90° to close. The door is 3.00 m tall and 1.25 m wide, and it weighs 750 N. The friction at the hinges can be neglected. If Exena applies a force of 220 N at the edge of the door and perpendicular to it, how much time does it take her to close the door?

10–50 A thin rod of length l lies on the +x-axis with its left end at the origin. A string pulls on the rod with a force $\vec{F}$ directed toward a point P a distance h above the rod. Where along the rod should you attach the string to get the greatest torque about the origin if point P is a) above the right end of the rod? b) above the left end of the rod? c) above the center of the rod?

10–51 Balancing Act. Attached to one end of a long, thin, uniform rod of length L and mass M is a small blob of clay of the same mass M. a) Locate the position of the center of mass of the system of rod and clay. Note this position on a drawing of the rod. b) You carefully balance the rod on a frictionless table top so that it is standing vertically, with the end without the clay touching the table. If the rod is now tipped so that it is a small angle θ away from the vertical, determine its angular acceleration at this instant. Assume that the end without the clay remains in contact with the table top. (*Hint:* See Table 9–2.) c) You again balance the rod on the frictionless table top so that it is standing vertically, but now the end of the rod *with* the clay is touching the table. If the rod is again tipped so that it is a small angle θ away from the vertical, determine its angular acceleration at this instant. Assume that the end with the clay remains in contact with the table top. How does this compare to the angular acceleration in part (b)? d) A pool cue is a tapered wooden rod that is thick at one end and thin at the other. You can easily balance a pool cue vertically on one finger if the thin end is in contact with your finger; this is quite a bit harder to do if the thick end is in contact with your finger. Explain why there is a difference.

10–52 You connect a light string to a point on the edge of a uniform vertical disk of radius R and mass M. The disk is free to rotate without friction about a stationary horizontal axis through its center. Initially, the disk is at rest with the string connection at the highest point on the disk. You pull the string with a constant horizontal force $\vec{F}$ until the wheel has made exactly one-quarter revolution about a horizontal axis through its center, then let go. a) Use Eq. (10–23) to find the work done by the

string. b) Use Eq. (6–14) to find the work done by the string. Do you obtain the same result as in part (a)? c) Find the final angular speed of the disk. d) Find the maximum tangential acceleration of a point on the disk. e) Find the maximum radial (centripetal) acceleration of a point on the disk.

10–53 The mechanism shown in Fig. 10–44 is used to raise a crate of supplies from a ship's hold. The crate has a total mass of 50 kg. A rope is wrapped around a wooden cylinder that turns on a metal axle. The cylinder has radius 0.25 m and moment of inertia $I = 2.9 \text{ kg} \cdot \text{m}^2$ about the axle. The crate is suspended from the free end of the rope. One end of the axle is pivoted on frictionless bearings; a crank handle is attached to the other end. When the crank is turned, the end of the handle rotates about the axle in a vertical circle of radius 0.12 m, the cylinder turns, and the crate is raised. What magnitude of the force $\vec{F}$ applied tangentially to the rotating crank is required to raise the crate with an acceleration of 0.80 m/s²? (The mass of the rope can be neglected, as can the moments of inertia of the axle and the crank.)

FIGURE 10–44 Problem 10–53.

10–54 A large 16.0-kg roll of paper with radius $R = 18.0$ cm rests against the wall and is held in place by a bracket attached to a rod through the center of the roll (Fig. 10–45). The rod turns without friction in the bracket, and the moment of inertia of the paper and rod about the axis is $0.260 \text{ kg} \cdot \text{m}^2$. The other end of the bracket is attached by a frictionless hinge to the wall such that the bracket makes an angle of 30.0° with the wall. The weight of the bracket is negligible. The coefficient of kinetic friction between the paper and the wall is $\mu_k = 0.25$. A constant vertical force $F = 40.0$ N is applied to the paper, and the paper unrolls. a) What is the magnitude of the force that the rod exerts on the paper as it unrolls? b) What is the angular acceleration of the roll?

10–55 A block with mass $m = 5.00$ kg slides down a surface inclined 36.9° to the horizontal (Fig. 10–46). The coefficient of kinetic friction is 0.25. A string attached to the block is wrapped around a flywheel on a fixed axis at O. The flywheel has mass 25.0 kg and moment of inertia $0.500 \text{ kg} \cdot \text{m}^2$ with respect to the axis of rotation. The string pulls without slipping at a perpendicular distance of 0.200 m from that axis. a) What is the acceleration of the block down the plane? b) What is the tension in the string?

10–56 Two metal disks, one with radius $R_1 = 2.50$ cm and mass $M_1 = 0.80$ kg and the other with radius $R_2 = 5.00$ cm and mass

FIGURE 10–45 Problem 10–54.

FIGURE 10–46 Problem 10–55.

FIGURE 10–47 Problem 10–58.

$M_2 = 1.60$ kg, are welded together and mounted on a frictionless axis through their common center, as in Problem 9–75. a) A light string is wrapped around the edge of the smaller disk and a 1.50-kg block is suspended from the free end of the string. What is the magnitude of the downward acceleration of the block after it is released? b) Repeat the calculation of part (a), this time with the string wrapped around the edge of the larger disk. In which case is the acceleration of the block the greatest? Does your answer make sense?

10–57 A lawn roller in the form of a thin-walled hollow cylinder of mass M is pulled horizontally with a constant horizontal force F applied by a handle attached to the axle. If it rolls without slipping, find the acceleration and the friction force.

10–58 Atwood's Machine. Figure 10–47 shows an Atwood's machine. Find the linear accelerations of blocks A and B, the angular acceleration of the wheel C, and the tension in each side of the cord if there is no slipping between the cord and the surface of the wheel. The masses of the blocks A and B are m_A and m_B, respectively, the moment of inertia of the wheel about its axis is I, and the radius of the semicircle in which the string moves is R.

10–59 A solid disk is rolling without slipping on a level surface at a constant speed of 2.50 m/s. a) If the disk rolls up a 30.0° ramp, how far along the ramp will it move before it stops? b) Explain why your answer to (a) does not depend on either the mass or the radius of the disk.

10–60 The Yo-yo. A yo-yo is made from two uniform disks, each with mass m and radius R, connected by a light axle of radius b. A light, thin string is wound several times around the axle and

then held stationary while the yo-yo is released from rest, dropping as the string unwinds. Find the linear acceleration and angular acceleration of the yo-yo and the tension in the string.

10–61 A uniform marble of radius r starts from rest with its center of mass a height h above the lowest point of a loop-the-loop track of radius R, a track shaped like the one in Fig. 7–26. It rolls without slipping. Rolling friction and air resistance are negligible. a) What is the minimum value of h such that the marble will not leave the track at the top of the loop? (*Note:* The radius r is *not* negligible compared to the radius R.) b) What is the answer to part (a) if the track is well lubricated, making friction negligible?

10–62 Figure 10–48 shows three identical yo-yos initially at rest on a horizontal surface. For each yo-yo the string is pulled in the direction shown. In each case there is sufficient friction for the yo-yo to roll without slipping. Draw the free-body diagram for each yo-yo. In what direction will each yo-yo rotate? (Try it!) Explain your answers.

FIGURE 10–48 Problem 10–62.

10–63 As shown in Fig. 10–39, a string is wrapped several times around the rim of a small hoop of radius 0.0800 m and mass 0.180 kg. The free end of the string is pulled upward in just the right way so that the hoop does not move vertically as the string unwinds. a) Find the tension in the string as the string unwinds. b) Find the angular acceleration of the hoop as the string unwinds. c) Find the upward acceleration of the hand that pulls on the free end of the string. d) How would your answers be different if the hoop were replaced by a solid disk of the same mass and radius?

10–64 In a lab experiment you let a uniform ball roll down a curved track. The ball starts from rest and rolls without slipping. While on the track, the ball descends a vertical distance h. The lower end of the track is horizontal and extends over the edge of the lab table; the ball leaves the track traveling horizontally. While free falling after leaving the track, the ball moves a horizontal distance x and a vertical distance y. a) Calculate x in terms of h and y, ignoring the work done by friction. b) Would the answer to part (a) be any different on the moon? c) In doing the experiment very carefully, your measured value of x is consistently a bit smaller than that calculated in part (a). Why? d) What would x be for the same h and y as in part (a) if you let a silver dollar roll down the track? Neglect the work done by friction.

10–65 In a spring gun, a spring of force constant 400 N/m is compressed 0.15 m. When fired, 80.0% of the elastic potential energy stored in the spring is eventually converted into kinetic energy of a 0.0590-kg uniform ball that is rolling without slipping at the base of a ramp. The ball continues to roll without slipping up the ramp with 90.0% of the kinetic energy at the bottom converted into an increase in gravitational potential energy at the instant it stops. a) What is the speed of the ball's center of mass at the base of the ramp? b) At this position, what is the speed of a point at the top of the ball? c) At this position, what is the speed of a point at the bottom of the ball? d) What maximum vertical height up the ramp does the ball move?

10–66 If a wheel rolls along a horizontal surface at constant speed, the coordinates of a certain point on the rim of the wheel are $x(t) = R[(2\pi t/T) - \sin(2\pi t/T)]$ and $y(t) = R[1 - \cos(2\pi t/T)]$, where R and T are constants. a) Sketch the trajectory of the point from $t = 0$ to $t = 2T$. A curve with this shape is called a *cycloid*. b) What are the meanings of the constants R and T? c) Find the x- and y-components of the velocity and of the acceleration of the point at any time t. d) Find the times at which the point is instantaneously at rest. What are the x- and y-components of the acceleration at these times? e) Find the magnitude of the acceleration of the point. Does it depend on time? Compare to the magnitude of the acceleration of a particle in uniform circular motion, $a_{\text{rad}} = 4\pi^2 R/T^2$. Explain your result for the magnitude of the acceleration of the point on the rolling wheel using the idea that rolling is a combination of rotational and translational motion.

10–67 A child rolls a 0.600-kg basketball up a long ramp. The basketball can be considered a thin-walled hollow sphere. When the child releases the basketball at the bottom of the ramp, it has a speed of 8.0 m/s. When the ball returns to her after rolling up the ramp and then rolling back down, it has a speed of 4.0 m/s. Assume the work done by friction on the basketball is the same when the ball moves up or down the ramp and that the basketball rolls without slipping. Find the maximum vertical height increase of the ball as it rolls up the ramp.

10–68 A wheel rotates from rest about a fixed axis through its center of mass such that $\theta = bt^3$, where b is a positive constant with units rad/s^3. a) Use Eq. (10–23) to show that the work done by the net torque on the wheel when it has turned through an angle θ is $\frac{9}{2}I_{\text{cm}}b^{2/3}\theta^{4/3}$. b) Use Eq. (9–3) to calculate the angular speed of the wheel when it has turned through an angle θ. c) Use the result of part (b) to calculate the kinetic energy of the wheel when it has turned through an angle θ. Is the work-energy theorem, Eq. (10–25), obeyed? Explain.

10–69 A uniform solid cylinder of mass M and radius $2R$ rests on a horizontal table top. A string is attached by a yoke to a frictionless axle through the center of the cylinder so that the cylinder can rotate about the axle. The string runs over a disk-shaped pulley of mass M and radius R that is mounted on a

frictionless axle through its center. A block of mass M is suspended from the free end of the string (Fig. 10–49). The string doesn't slip over the pulley surface, and the cylinder rolls without slipping on the table top. Find the magnitude of the acceleration of the block after the system is released from rest.

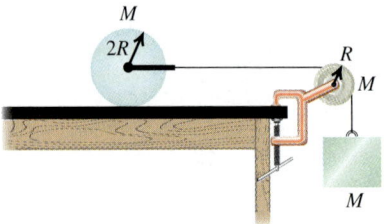

FIGURE 10–49 Problem 10–69.

10–70 A uniform 0.0300-kg rod of length 0.400 m rotates in a horizontal plane about a fixed axis through its center and perpendicular to the rod. Two small rings, each of mass 0.0200 kg, are mounted so that they can slide along the rod. They are initially held by catches at positions 0.0500 m on each side of the center of the rod, and the system is rotating at 30.0 rev/min. Without otherwise changing the system, the catches are released, and the rings slide outward along the rod and fly off at the ends. a) What is the angular speed of the system at the instant when the rings reach the ends of the rod? b) What is the angular speed of the rod after the rings leave it?

10–71 A uniform rod of length L rests on a frictionless horizontal surface. The rod is pivoted about a fixed frictionless axis at one end. The rod is initially at rest. A bullet traveling parallel to the horizontal surface and perpendicular to the rod with speed v strikes the rod at its center and becomes embedded in it. The mass of the bullet is one-fourth the mass of the rod. a) What is the final angular speed of the rod? b) What is the ratio of the kinetic energy of the system after the collision to the kinetic energy of the bullet before the collision?

10–72 The solid wood door of a gymnasium is 1.00 m wide and 2.00 m high, has a total mass of 35.0 kg, and is hinged along one side. The door is open and at rest when a stray basketball hits the center of the door head-on, applying an average force of 1500 N to the door for 8.00 ms. Find the angular speed of the door after the impact. (*Hint:* Integrating Eq. (10–32) yields $\Delta L = \int_{t_1}^{t_2} (\Sigma \tau)\, dt = (\Sigma \tau)_{\text{av}}\Delta t$. The quantity $\int_{t_1}^{t_2} (\Sigma \tau)\, dt$ is called the *angular impulse.*)

10–73 A target in a shooting gallery consists of a vertical square wooden board, 0.250 m on a side and of mass 0.750 kg, that is pivoted on a horizontal axis along its top edge. The board is struck face-on at its center by a bullet of mass 1.90 g that is traveling at 360 m/s and that remains embedded in the board. a) What is the angular speed of the board just after the bullet's impact? b) What maximum height above the equilibrium position does the center of the board reach before starting to swing down again? c) What minimum bullet speed would be required for the board to swing all the way over after impact?

10–74 Neutron Star Glitches. Occasionally a rotating neutron star (Exercise 10–32) undergoes a sudden and unexpected speedup called a *glitch*. One explanation is that a glitch occurs when the crust of the neutron star settles slightly, decreasing the moment of inertia about the rotation axis. A neutron star with angular speed $\omega_0 = 70.4$ rad/s underwent such a glitch in October 1975 that increased its angular speed to $\omega = \omega_0 + \Delta\omega$, where $\Delta\omega/\omega_0 = 2.01 \times 10^{-6}$. If the radius of the neutron star before the glitch was 11 km, by how much did its radius decrease in the starquake? Assume that the neutron star is a uniform sphere.

10–75 Disks A and B are mounted on a shaft SS and may be connected or disconnected by a clutch C (Fig. 10–50). Disk A is made of a lighter material than disk B so that the moment of inertia of disk A about the shaft is one-third that of disk B. The moments of inertia of the shaft and clutch are negligible. With the clutch disconnected, A is brought up to an angular speed ω_0. The accelerating torque is then removed from A, and A is coupled to disk B by the clutch. (Bearing friction may be neglected.) It is found that 2400 J of thermal energy is developed in the clutch when the connection is made. What was the original kinetic energy of disk A?

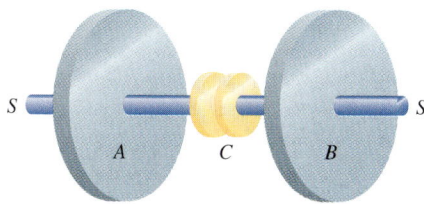

FIGURE 10–50 Problem 10–75.

10–76 A small block of mass 0.250 kg is attached to a string passing through a hole in a frictionless horizontal surface (see Fig. 10–41). The block is originally revolving in a circle with a radius of 0.800 m about the hole with a tangential speed of 4.00 m/s. The string is then pulled slowly from below, shortening the radius of the circle in which the block revolves. The breaking strength of the string is 30.0 N. What is the radius of the circle when the string breaks?

10–77 A horizontal plywood disk of mass 7.00 kg and diameter 1.00 m is pivoted on frictionless bearings about a vertical axis through its center. You attach a circular model railroad track of negligible mass and average diameter 0.95 m to the disk. A 1.20-kg, battery-driven model train rests on the tracks. To demonstrate conservation of angular momentum, you switch on the train's engine. The train moves counterclockwise, soon attaining a constant speed of 0.600 m/s relative to the tracks. Find the magnitude and direction of the angular velocity of the disk relative to the earth.

10–78 One particle of mass m moves at a constant speed v in a circle of radius R that is a distance R above the xz-plane. Another particle of mass m moves the same way at the same speed in another circle of radius R that is a distance R below the xz-plane. The two particles are one-half a revolution apart, so that when one is at (x, R, z), the other is at $(-x, -R, -z)$. Thus their center of mass is at the origin but their rotation axis (the y-axis) is not an axis of symmetry. a) Sketch these two particles at the instant they are at $(R, R, 0)$ and $(-R, -R, 0)$, and show their position, velocity, and angular momentum vectors with respect to the origin. b) Show that at any time the two particles have equal angular momentum. c) What angle do $\vec{\omega}$ (the angular velocity vector of the system of two particles) and the total angular momentum of the system make

with each other? d) Show that the y-component of the system's total angular momentum is constant and equal to $L_y = 2mvR$. e) What is the y-component of the total torque acting on the system? f) Find the magnitude of the net force acting on each particle and the magnitude of the total torque acting on the system. g) Show, using your sketch from part (a), the direction of the net torque on the system and that this torque rotates parallel to the xz-plane.

10–79 In a physics laboratory you do the following ballistic pendulum experiment. You shoot a ball of mass m horizontally from a spring gun with a speed v. The ball is immediately caught a distance r below a frictionless pivot by a pivoted catcher assembly of mass M. The moment of inertia of this assembly about its rotation axis through the pivot is I. The distance r is much greater than the radius of the ball. a) Use conservation of angular momentum to show that the angular speed of the ball and catcher just after the ball is caught is $\omega = mvr/(mr^2 + I)$. b) After the ball is caught, the center of mass of the ball-catcher assembly system swings up with a maximum height increase h. Use conservation of energy to show that $\omega = \sqrt{2(M + m)gh/(mr^2 + I)}$. c) Your lab partner says that linear momentum is conserved in the collision and derives the expression $mv = (m + M)V$, where V is the speed of the ball immediately after the collision. She then uses conservation of energy to derive that $V = \sqrt{2gh}$, so that $mv = (m + M)\sqrt{2gh}$. Use the results of parts (a) and (b) to show that this equation is satisfied only for the special case when r is given by $I = Mr^2$.

10–80 A 55-kg runner runs around the edge of a horizontal turntable mounted on a vertical frictionless axis through its center. The runner's velocity relative to the earth has magnitude 2.8 m/s. The turntable is rotating in the opposite direction with an angular velocity of magnitude 0.20 rad/s relative to the earth. The radius of the turntable is 3.0 m, and its moment of inertia about the axis of rotation is 80 kg · m². Find the final angular velocity of the system if the runner comes to rest relative to the turntable. (The runner may be treated as a particle.)

10–81 A Falling Bicycle. The moment of inertia of the front wheel of a bicycle about its axle is 0.085 kg · m², its radius is 0.33 m, and the forward speed of the bicycle is 6.00 m/s. With what angular velocity must the front wheel be turned about a vertical axis to counteract the capsizing torque due to a mass of 50.0 kg located 0.040 m horizontally to the right or left of the line of contact of wheels and ground? (Bicycle riders: Compare your own experience and decide whether your answer seems reasonable.)

10–82 Center of Percussion. A baseball bat rests on a horizontal frictionless surface. The bat has a length of 0.900 m, a mass of 0.800 kg, and its center of mass is 0.600 m from the handle end of the bat (Fig. 10–51). The moment of inertia of the bat about its center of mass is 0.0530 kg · m². The bat is struck by a baseball traveling perpendicular to the bat. The impact applies an impulse $J = \int_{t_1}^{t_2} F \, dt$ at a point a distance x from the handle end of the bat. What must x be so that the handle end of the bat remains at rest as the bat begins to move? (*Hint:* Consider the motion of the center of mass and the rotation about the center of mass. Find x so that these two motions combine to give $v = 0$ for the end of the bat just after the collision. Also, note that integration of Eq. (10–32) gives $\Delta L = \int_{t_1}^{t_2} (\Sigma \, \tau) \, dt$ (Problem 10–72).) The point on the bat you have located is called the *center of percussion*. Hitting a pitched ball at the center of percussion of the bat minimizes the "sting" the batter experiences on the hands.

FIGURE 10-51 Problem 10–82.

10–83 Consider a gyroscope with an axis that is not horizontal, but is inclined from the horizontal by an angle β. Show that the precession angular frequency does not depend on the value of β, but is given by Eq. (10–36).

CHALLENGE PROBLEMS

10–84 A uniform ball of radius R rolls without slipping between two rails such that the horizontal distance is d between the two contact points of the rails to the ball. a) In a sketch, show that at any instant $v_{cm} = \omega\sqrt{R^2 - d^2/4}$. Discuss this expression in the limits $d = 0$ and $d = 2R$. b) For a uniform ball starting from rest and descending a vertical distance h while rolling without slipping down a ramp, $v_{cm} = \sqrt{10gh/7}$. Replacing the ramp with the two rails, show that

$$v_{cm} = \sqrt{\frac{10gh}{5 + 2/(1 - d^2/4R^2)}}.$$

In each case, the work done by friction has been neglected. c) Which speed in part (b) is smaller? Why? Answer in terms of how the loss of potential energy is shared between the gain in translational and rotational kinetic energies. d) For what value of the ratio d/R do the two expressions for the speed in part (b) differ by 5.0%? By 0.50%?

10–85 When an object is rolling without slipping, the rolling friction force is much less than the friction force when the object is sliding; a silver dollar will roll on its edge much farther than it will slide on its flat side. (See Section 5–4.) When an object is rolling without slipping on a horizontal surface, we can approximate the friction force to be zero, so that a and α are approximately zero and v and ω are approximately constant. Rolling without slipping means $v = r\omega$ and $a = r\alpha$. If an object is set into motion on a surface *without* these equalities, sliding (kinetic) friction will act on the object as it slips until rolling without slipping is established. A solid cylinder with mass M and radius R, rotating with angular velocity ω_0 about an axis through its center, is set on a horizontal surface for which the kinetic friction coefficient is μ_k. a) Draw a free-body diagram for the cylinder on the surface. Think carefully about the direction of the kinetic friction force on the cylinder. Calculate the accelerations a of the center of mass and α of rotation about the center of mass. b) The cylinder is initially slipping completely, as initially $\omega = \omega_0$ but $v = 0$. Rolling without slipping sets in when $v = R\omega$. Calculate the *distance* the cylinder rolls before slipping stops. c) Calculate the work done by the friction force on the cylinder as it moves from where it was set down to where it begins to roll without slipping.

10–86 A demonstration gyroscope wheel is constructed by removing the tire from a bicycle wheel 0.650 m in diameter, wrapping lead wire around the rim, and taping it in place. The shaft projects 0.200 m at each side of the wheel, and a woman holds the ends of the shaft in her hands. The mass of the system is 8.00 kg; its entire mass may be assumed to be located at its rim. The shaft is horizontal and the wheel is spinning about the shaft at 5.00 rev/s. Find the magnitude and direction of the force each hand exerts on the shaft a) when the shaft is at rest; b) when the shaft is rotating in a horizontal plane about its center at 0.050 rev/s; c) when the shaft is rotating in a horizontal plane about its center at 0.300 rev/s. d) At what rate must the shaft rotate in order that it may be supported at one end only?

10–87 A block with mass m is revolving with linear speed v_1 in a circle of radius r_1 on a frictionless horizontal surface (see Fig. 10–41). The string is slowly pulled from below until the radius of the circle in which the block is revolving is reduced to r_2. a) Calculate the tension T in the string as a function of r, the distance of the block from the hole. Your answer will be in terms of the initial velocity v_1 and the radius r_1. b) Use $W = \int_{r_1}^{r_2} \vec{T}(r) \cdot d\vec{r}$ to calculate the work done by $\vec{T}$ when r changes from r_1 to r_2. c) Compare the results of part (b) to the change in the kinetic energy of the block.

11

Equilibrium and Elasticity

To help submarines maintain structural integrity while diving deep in the ocean, their hulls are built with a cylindrical cross section. This shape stabilizes the hull against compressive stresses due to the inward water pressure. The rounded shape also minimizes the possibility of stress fractures in the hull, which would tend to form at sharp corners.

11–1 INTRODUCTION

We've devoted a good deal of effort to understanding why and how bodies accelerate in response to the forces that act on them. But very often we're interested in making sure that bodies *don't* accelerate. Any building, from the World Trade Center to the humblest shed, must be designed so that it won't topple over. Similar concerns occur with a suspension bridge, a ladder leaning against a wall, or a crane hoisting a bucket full of concrete.

A body that can be modeled as a *particle* is in equilibrium whenever the vector sum of the forces acting on it is zero. But for the situations we've just described, that condition isn't enough. If forces act at different points on an extended body, an additional requirement must be satisfied to ensure that the body has no tendency to *rotate*: the sum of the *torques* about any point must be zero. This requirement is based on the principles of rotational dynamics developed in Chapter 10. We can compute the torque due to the weight of a body using the concept of center of mass from Section 8–6 and the related concept of center of gravity, which we introduce in this chapter.

Rigid bodies don't bend, stretch, or squash when forces act on them. But the rigid body is an idealization; all real materials are *elastic* and do deform to some extent. Elastic properties of materials are tremendously important. You want the wings of an airplane to be able to bend a little, but you'd rather not have them break off. The steel frame of an earthquake-resistant building has to be able to flex, but not

too much. Many of the necessities of everyday life, from rubber bands to suspension bridges, depend on the elastic properties of materials. In this chapter we'll introduce the concepts of *stress, strain,* and *elastic modulus* and a simple principle called *Hooke's law* that helps us predict what deformations will occur when forces are applied to a real (not perfectly rigid) body.

11–2 CONDITIONS FOR EQUILIBRIUM

We learned in Sections 4–3 and 5–2 that a particle is in equilibrium—that is, the particle does not accelerate—in an inertial frame of reference if the vector sum of all the forces acting on the particle is zero, $\Sigma \vec{F} = 0$. The equivalent statement for an *extended* body is that the center of mass of the body has zero acceleration if the vector sum of all external forces acting on the body is zero, as discussed in Section 8–6. This is often called the **first condition for equilibrium.** In terms of components,

$$\Sigma F_x = 0, \quad \Sigma F_y = 0, \quad \Sigma F_z = 0 \quad \begin{array}{l}\text{(first condition} \\ \text{for equilibrium),}\end{array} \quad (11\text{–}1)$$

where the sum includes *external* forces only.

A second condition for an extended body to be in equilibrium is that the body must have no tendency to *rotate.* This condition is based on the dynamics of rotational motion in exactly the same way that the first condition is based on Newton's first law. A rigid body that, in an inertial frame, is not rotating about a certain point has zero angular momentum about that point, $\vec{L} = \mathbf{0}$. If it is not to start rotating about that point, the rate of change of angular momentum $d\vec{L}/dt$ must *also* be zero. From the discussion in Section 10–6, particularly Eq. (10–32), this means that the sum of torques $\Sigma \vec{\tau}$ due to all the external forces acting on the body must be zero. A rigid body in equilibrium can't have any tendency to start rotating about *any* point, so the sum of external torques must be zero about any point. This is the **second condition for equilibrium:**

$$\Sigma \vec{\tau} = \mathbf{0} \quad \text{about any point} \qquad \text{(second condition for equilibrium).} \qquad (11-2)$$

The sum of the torques due to all external forces acting on the body, with respect to any specified point, must be zero.

CAUTION ▶ Although the choice of reference point is arbitrary, once you choose a point you must use the *same* point to calculate *all* of the torques on a body. An important element of problem-solving strategy is to pick the point so as to simplify the calculations as much as possible. ◀

In this chapter we will apply the first and second conditions for equilibrium to situations in which a rigid body is at rest (no translation or rotation). Such a body is said to be in **static equilibrium.** But the same conditions apply to a rigid body in uniform *translational* motion (without rotation), such as an airplane in flight with constant speed, direction, and altitude. Such a body is in equilibrium but is not static.

11–3 CENTER OF GRAVITY

In most equilibrium problems, one of the forces acting on the body is its weight. We need to be able to calculate the *torque* of this force. The weight doesn't act at a single point; it is distributed over the entire body. But we can always calculate the torque due to the body's weight by assuming that the entire force of gravity (weight) is concentrated at a point called the **center of gravity** (abbreviated "cg"). The acceleration due to gravity $\vec{g}$ decreases with altitude; but if we can ignore this variation over the vertical dimension of the body, then the body's center of gravity is identical with its *center of mass,* which we defined in Section 8–6. We stated this result without proof in Section 10–3, and now we'll prove it.

First let's review the definition of the center of mass. For a collection of particles with masses $m_1, m_2, \ldots$ and coordinates $(x_1, y_1, z_1), (x_2, y_2, z_2), \ldots$, the coordinates x_{cm}, y_{cm}, and z_{cm} of the center of mass are given by

$$x_{cm} = \frac{m_1 x_1 + m_2 x_2 + m_3 x_3 + \cdots}{m_1 + m_2 + m_3 + \ldots} = \frac{\sum\limits_i m_i x_i}{\sum\limits_i m_i},$$

$$y_{cm} = \frac{m_1 y_1 + m_2 y_2 + m_3 y_3 + \cdots}{m_1 + m_2 + m_3 + \cdots} = \frac{\sum\limits_i m_i y_i}{\sum\limits_i m_i}, \qquad \text{(center of mass)} \qquad (11-3)$$

$$z_{cm} = \frac{m_1 z_1 + m_2 z_2 + m_3 z_3 + \cdots}{m_1 + m_2 + m_3 + \cdots} = \frac{\sum\limits_i m_i z_i}{\sum\limits_i m_i}.$$

Also, x_{cm}, y_{cm}, and z_{cm} are the components of the position vector $\vec{r}_{cm}$ of the center of mass, so Eqs. (11–3) are equivalent to the vector equation

$$\vec{r}_{cm} = \frac{m_1 \vec{r}_1 + m_2 \vec{r}_2 + \cdots}{m_1 + m_2 + \cdots} = \frac{\sum\limits_i m_i \vec{r}_i}{\sum\limits_i m_i}. \qquad (11-4)$$

Now let's consider the gravitational torque on a body of arbitrary shape (Fig. 11–1). We assume that the acceleration due to gravity $\vec{g}$ has the same magnitude and direction

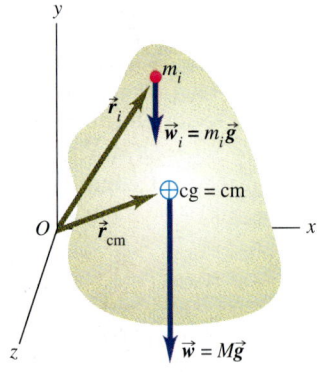

11–1 The gravitational torque about any point can be found by assuming that all the weight of the body acts as its center of gravity (cg), which is identical to its center of mass (cm) if $\vec{g}$ is the same at all points on the body.

at every point in the body. Every particle in the body experiences a gravitational force, and the total weight of the body is the vector sum of a large number of parallel forces. A typical particle has mass m_i and weight $\vec{w}_i = m_i\vec{g}$. If $\vec{r}_i$ is the position vector of this particle with respect to an arbitrary origin O, the torque vector $\vec{\tau}_i$ of the weight $\vec{w}_i$ with respect to O is, from Eq. (10–3),

$$\vec{\tau}_i = \vec{r}_i \times \vec{w}_i = \vec{r}_i \times m_i\vec{g}.$$

The *total* torque due to the gravitational forces on all the particles is

$$\vec{\tau} = \Sigma\vec{\tau}_i = \vec{r}_1 \times m_1\vec{g} + \vec{r}_2 \times m_2\vec{g} + \cdots$$
$$= (m_1\vec{r}_1 + m_2\vec{r}_2 + \cdots) \times \vec{g}$$
$$= \left(\sum_i m_i\vec{r}_i\right) \times \vec{g}.$$

When we multiply and divide this by the total mass of the body,

$$M = m_1 + m_2 + \cdots = \sum_i m_i,$$

we get

$$\vec{\tau} = \frac{m_1\vec{r}_1 + m_2\vec{r}_2 + \cdots}{m_1 + m_2 + \cdots} \times M\vec{g} = \frac{\sum\limits_i m_i\vec{r}_i}{\sum\limits_i m_i} \times M\vec{g}.$$

The fraction in this equation is just the position vector $\vec{r}_{cm}$ of the center of mass, with components x_{cm}, y_{cm}, and z_{cm}, as given by Eq. (11–4), and $M\vec{g}$ is equal to the total weight $\vec{w}$ of the body. Thus

$$\vec{\tau} = \vec{r}_{cm} \times M\vec{g} = \vec{r}_{cm} \times \vec{w}. \qquad (11\text{–}5)$$

The total gravitational torque, given by Eq. (11–5), is the same as though the total weight $\vec{w}$ were acting on the position $\vec{r}_{cm}$ of the center of mass, which we also call the *center of gravity*. **If $\vec{g}$ has the same value at all points on a body, its center of gravity is identical to its center of mass.** Note, however, that the center of mass is defined independently of any gravitational effect. In this chapter we'll assume that the center of mass and the center of gravity are identical unless explicitly stated otherwise.

FINDING AND USING THE CENTER OF GRAVITY

We can often use symmetry considerations to locate the center of gravity of a body, just as we did for center of mass. The center of gravity of a homogeneous sphere, cube, circular sheet, or rectangular plate is at its geometric center. The center of gravity of a right circular cylinder or cone is on its axis of symmetry.

    For a body of more complex shape, we can sometimes locate the center of gravity by thinking of the body as being made of symmetrical pieces. For example, we could approximate the human body as a collection of solid cylinders, with a sphere for the head. Then we can compute the coordinates of the center of gravity of the combination from Eqs. (11–3), letting m_1, m_2, ... be the masses of the individual pieces and (x_1, y_1, z_1), (x_2, y_2, z_2), ... be the coordinates of their centers of gravity.

    When a body acted on by gravity is supported or suspended at a single point, the center of gravity is always at or directly above or below the point of suspension. If it were anywhere else, the weight would have a torque with respect to the point of suspension, and the body could not be in rotational equilibrium. This fact can be used to determine experimentally the location of the center of gravity of an irregular body. To see how this works, hold one corner of a piece of paper gently between your thumb and forefinger so that the paper hangs below your hand. Draw a line on the paper that extends vertically downward from the point of suspension; the center of gravity must lie somewhere along this line. Then hold the paper by a different corner and draw a line vertically downward

The center of gravity and center of mass of an orbiting space station are not at exactly the same place. This is because $\vec{g}$, the acceleration due to gravity, can vary measurably over the dimensions of the station. On the earth's surface this variation is so small that the center of mass and center of gravity do coincide for all but the largest structures.

11–2 In (a) the center of gravity is within the area bounded by the supports, and the car is in equilibrium. The car in (b) and the truck in (c) will tip over because their centers of gravity lie outside the area of support.

Area of support (a)   Area of support (b)   Area of support (c)

7.2
A Tilted Beam: Torques and Equilibrium

7.3
Arm Levers

from the new point of suspension. The point where the lines intersect is the paper's center of gravity. (Try this with a coffee cup suspended first from the upper part of its handle, then from the lower part of its handle.)

Using the same reasoning, we can see that a body supported at several points must have its center of gravity somewhere within the area bounded by the supports. This explains why a car can drive on a straight but slanted road if the slant angle is relatively small (Fig. 11–2a) but will tip over if the angle is too great (Fig. 11–2b). The truck in Fig. 11–2c has a higher center of gravity than the car and will tip over on a shallower incline; when a truck overturns on a highway and blocks traffic for hours, it's the high center of gravity that's to blame.

The lower the center of gravity and the larger the area of support, the more difficult it is to overturn a body. Four-legged animals such as deer and horses have a large area of support bounded by their legs; hence they are naturally stable and need only small feet or hooves. Animals that walk erect on two legs, such as humans and birds, need relatively large feet to give them a reasonable area of support. If a two-legged animal holds its body approximately horizontal, like a chicken or the dinosaur *Tyrannosaurus rex,* it must perform a delicate balancing act as it walks to keep its center of gravity over the foot that is on the ground. A chicken does this by moving its head; *T. rex* probably did it by moving its massive tail.

EXAMPLE 11–1

Walking the plank A massive uniform plank of length $L = 6.0$ m and mass $M = 90$ kg rests on top of two sawhorses separated by $D = 1.5$ m, located equal distances from the center of the plank (Fig. 11–3). Your cousin Throckmorton tries to stand on the right-hand end of the plank. If the plank is to remain at rest, how massive can Throckmorton be?

SOLUTION If the system of plank and Throckmorton is just in balance, the center of gravity of this system will be directly over the right-hand sawhorse. We take the origin at C, the geometrical center and center of gravity of the uniform plank. Then the x-coordinates of the centers of gravity of the plank and Throckmorton are $x_p = 0$ and $x_T = L/2 = 3.0$ m, respectively. If we let m be Throcky's mass, then from Eqs. (11–3) the center of gravity of the system of plank and Throcky is at

$$x_{cg} = \frac{M(0) + m(L/2)}{M + m} = \frac{m}{M + m}\frac{L}{2}.$$

$L = 6.0$ m

$D = 1.5$ m

$\frac{L}{2} - \frac{D}{2}$

Throckmorton, mass m

Plank, mass $M = 90$ kg

$\oplus$ cg

C S

11–3 Throckmorton and the plank, just barely in equilibrium.

Setting this equal to $D/2$, the x-coordinate of the right-hand sawhorse, we have

$$\frac{m}{M+m}\frac{L}{2} = \frac{D}{2},$$

$$mL = (M+m)D,$$

$$m = M\frac{D}{L-D} = (90\text{ kg})\frac{1.5\text{ m}}{6.0\text{ m} - 1.5\text{ m}}$$

$$= 30\text{ kg}.$$

We can get the same result even more easily if we take the origin at S, the position of the right-hand sawhorse, so that $x_{cg} = 0$. The centers of gravity of the plank and Throcky are now at $x_P = -D/2$ and $x_T = (L/2) - (D/2)$ respectively, so

$$x_{cg} = \frac{M(-D/2) + m((L/2) - (D/2))}{M+m} = 0,$$

$$m = \frac{MD/2}{(L/2) - (D/2)} = M\frac{D}{L-D} = 30\text{ kg}.$$

A 60-kg child could only stand halfway between the right-hand sawhorse and the end of the plank. Can you see why?

11–4 SOLVING RIGID-BODY EQUILIBRIUM PROBLEMS

There are just two key principles of rigid-body equilibrium: the vector sum of the forces on the body must be zero, and the sum of torques about any point must be zero. To keep things simple, we'll restrict our attention to situations in which we can treat all forces as acting in a single plane, which we'll call the xy-plane. Then we can ignore the condition $\Sigma F_z = 0$ in Eq. (11–1), and in Eq. (11–2) we need only consider the z-components of torque (perpendicular to the plane). The first and second conditions for equilibrium are then

$$\Sigma F_x = 0 \quad \text{and} \quad \Sigma F_y = 0 \qquad \text{(first condition for equilibrium, forces in xy-plane),}$$

$$\Sigma \tau_z = 0 \qquad\qquad\qquad \text{(second condition for equilibrium, forces in xy-plane).}$$

(11–6)

The challenge is to apply these simple principles to specific problems. The following strategy is very similar to the suggestions in Section 5–3 for equilibrium of a particle. You should compare it with the strategy for rotational dynamics problems given in Section 10–3.

Problem–Solving Strategy

EQUILIBRIUM OF A RIGID BODY

1. Draw a sketch of the physical situation, including dimensions, and select the body in equilibrium to be analyzed.

2. Draw a free-body diagram showing the forces acting *on* the selected body and no others. *Do not* include forces exerted *by* this body on other bodies. Be careful to show correctly the point at which each force acts; this is crucial for correct torque calculations. You can't represent a rigid body as a point.

3. Choose coordinate axes and specify a positive sense of rotation for torques. Represent forces in terms of their components with respect to the axes you have chosen; when you do this, cross out the original force so that you don't include it twice.

4. In choosing a point to compute torques, note that if a force has a line of action that goes *through* a particular point, the torque of the force with respect to that point is zero. You can often eliminate unknown forces or components from the torque equation by a clever choice of point for your calculation. The body doesn't actually have to be pivoted about an axis through the chosen point.

5. Write equations expressing the equilibrium conditions. Remember that $\Sigma F_x = 0$, $\Sigma F_y = 0$, and $\Sigma \tau_z = 0$ are always separate equations; *never* add x- and y-components in a single equation. Also remember that when a force is represented in terms of its components, you can compute the torque of that force by finding the torque of each component separately, each with its appropriate lever arm and sign, and adding the results. This is often easier than determining the lever arm of the original force.

6. You always need as many equations as you have unknowns. Depending on the number of unknowns, you may need to compute torques with respect to two or more axes to obtain enough equations. Often, there are several equally good sets of force and torque equations for a particular problem; there is usually no single "right" combination of equations. When you have as many independent equations as unknowns, you can solve the equations simultaneously. We will illustrate this point in some of the following examples by using various sets of equations.

EXAMPLE 11-2

Weight distribution for a car An auto magazine reports, "The Nissan 240SX (Fig. 11–4a) has 53% of its weight on its front wheels and 47% on its rear wheels, with a 2.46-m wheelbase." This means that the total normal force on the front wheels is $0.53w$ and that on the rear wheels is $0.47w$, where w is the total weight. The wheelbase is the distance between front and rear axles. How far in front of the rear axle is the 240SX's center of gravity?

SOLUTION Figure 11–4b shows a free-body diagram for the car, along with the x- and y-axes and our convention that counterclockwise torques are positive. We can see that the first condition for equilibrium in Eqs. (11–6) is satisfied; $\Sigma F_x = 0$ because there aren't any x-components of force, and $\Sigma F_y = 0$ because $0.47w + 0.53w + (-w) = 0$. So far, so good. We have drawn the weight w as acting at the center of gravity, and the distance we want is L_{cg}. This is the lever arm of the weight with respect to the rear axle R, so it is reasonable to take torques with respect to R. Note that the torque due to the weight is negative because it tends to cause a clockwise rotation about R. The torque due to the upward normal force at the front wheels is positive because it tends to cause a counterclockwise rotation about R. The torque equation is

$$\Sigma \tau_R = 0.47w(0) - wL_{cg} + 0.53w(2.46 \text{ m}) = 0,$$

$$L_{cg} = 1.30 \text{ m}.$$

Can you show that if f is the fraction of the weight on the front

(a)

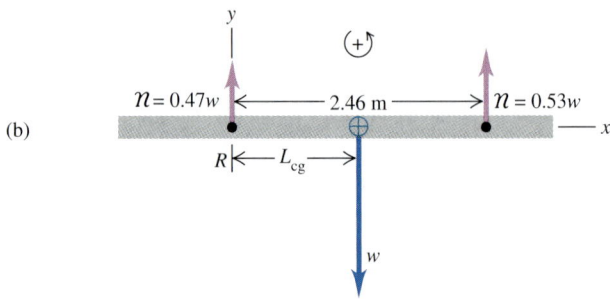
(b)

11–4 (a) The Nissan 240SX at rest on a horizontal surface. (b) Free-body diagram for the car.

wheels and d is the wheelbase, then the center of gravity is a distance fd in front of the rear wheels? In terms of the center of gravity, why do owners of rear-wheel-drive vehicles put bags of sand in their trunks to improve traction on snow and ice? Would this help with a front-wheel-drive car? With a four-wheel-drive car?

EXAMPLE 11-3

A heroic rescue Sir Lancelot is trying to rescue the Lady Elayne from Castle Von Doom by climbing a uniform ladder that is 5.0 m long and weighs 180 N. Lancelot, who weighs 800 N, stops a third of the way up the ladder (Fig. 11–5a). The bottom of the ladder rests on a horizontal stone ledge and leans across the moat in equilibrium against a vertical wall that is frictionless

because of a thick layer of moss. The ladder makes an angle of 53.1° with the horizontal, conveniently forming a 3–4–5 right triangle. a) Find the normal and friction forces on the ladder at its base. b) Find the minimum coefficient of static friction needed to prevent slipping at the base. c) Find the magnitude and direction of the contact force on the ladder at the base.

(a)

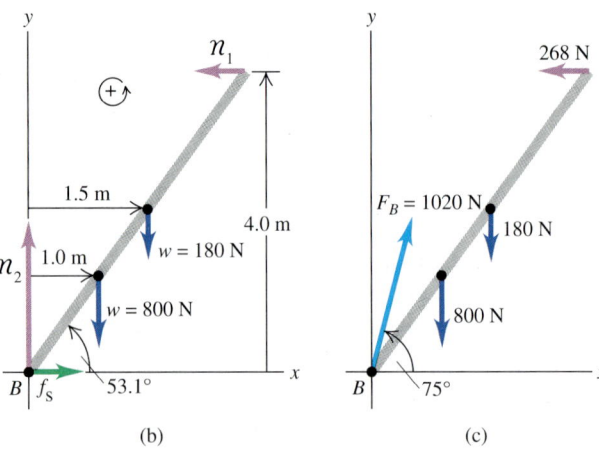
(b)   (c)

11–5 (a) Sir Lancelot pauses a third of the way up the ladder, fearing it will slip. (b) Free-body diagram for the ladder.

SOLUTION a) Figure 11–5b shows the free-body diagram for the ladder. The ladder is described as "uniform," so its center of gravity is at its geometrical center, halfway between the base and the wall. Lancelot's weight pushes down on the ladder at a point one-third of the way from the base toward the wall. The forces at the base are the upward normal force n_2 and the static friction force f_s, which must point to the right to prevent slipping. The frictionless wall exerts only a normal force n_1 at the top of the ladder.

From Eqs. (11–6), the first condition for equilibrium gives

$$\Sigma F_x = f_s + (-n_1) = 0,$$
$$\Sigma F_y = n_2 + (-800 \text{ N}) + (-180 \text{ N}) = 0.$$

These are two equations for the three unknowns n_1, n_2, and f_s. The first equation tells us that the two horizontal forces must be equal and opposite, and the second equation gives

$$n_2 = 980 \text{ N}.$$

The ground pushes up with a force of 980 N to balance the total (downward) weight (800 N + 180 N).

We don't yet have enough equations, but now we can use the second condition for equilibrium. We can take torques about any point we choose. The smart choice is the point that will give us the fewest terms and fewest unknowns in the torque equation. With this in mind, we choose point B, at the base of the ladder. The two forces n_2 and f_s have no torque about that point. From Fig. 11–5b we see that the lever arm for the ladder's weight is 1.5 m, the lever arm for Lancelot's weight is 1.0 m, and the lever arm for n_1 is 4.0 m. The torque equation for point B is

$$\Sigma \tau_B = n_1(4.0 \text{ m}) - (180 \text{ N})(1.5 \text{ m}) - (800 \text{ N})(1.0 \text{ m})$$
$$+ n_2(0) + f_s(0) = 0.$$

Solving for n_1, we get $n_1 = 268$ N. We now substitute this back into the $\Sigma F_x = 0$ equation to get

$$f_s = 268 \text{ N}.$$

b) The static friction force f_s cannot exceed $\mu_s n_2$, so the *minimum* coefficient of static friction to prevent slipping is

$$(\mu_s)_{\min} = \frac{f_s}{n_2} = \frac{268 \text{ N}}{980 \text{ N}} = 0.27.$$

c) The components of the contact force $\vec{F}_B$ at the base are the static friction force f_s and the normal force n_2, so

$$\vec{F}_B = f_s \hat{i} + n_2 \hat{j} = (268 \text{ N})\hat{i} + (980 \text{ N})\hat{j}.$$

The magnitude and direction of $\vec{F}_B$ (Fig. 11–5c) are then

$$F_B = \sqrt{(268 \text{ N})^2 + (980 \text{ N})^2} = 1020 \text{ N},$$
$$\theta = \arctan \frac{980 \text{ N}}{268 \text{ N}} = 75°.$$

As Fig. 11–5c shows, the contact force $\vec{F}_B$ is *not* directed along the length of the ladder. You may be surprised by this, but there's really no good reason why the two directions should be the same. Can you show that if $\vec{F}_B$ were directed along the ladder, there would be a net counterclockwise torque with respect to the top of the ladder, and equilibrium would be impossible?

Here are a few final comments. First, as Lancelot climbs higher on the ladder, the lever arm and torque of his weight about B increase; this increases the values of n_1, f_s, and $(\mu_s)_{\min}$. At the top, his lever arm would be nearly 3 m, giving a minimum coefficient of static friction of nearly 0.7. The value of μ_s would not be this large for a present-day aluminum ladder on a wood floor, which is why such ladders are usually equipped with nonslip rubber pads. Lancelot's medieval ladder is unlikely to have this feature, and the ladder is likely to slip as he climbs.

Second, a larger ladder angle would decrease the lever arms with respect to B of the weights of the ladder and Lancelot and increase the lever arm of n_1, all of which would decrease the required friction force. The R. D. Werner Ladder Co. recommends that its ladders be used at an angle of 75°. (Why not 90°?)

Finally, if we had assumed friction on the wall as well as on the floor, the problem would be impossible to solve by using the equilibrium conditions alone. (Try it!) Such a problem is said to be *statically indeterminate*. The difficulty is that it's no longer adequate to treat the body as being perfectly rigid. Another simple example of such a problem is a four-legged table; there is no way to use the equilibrium conditions alone to find the force on each separate leg. In a sense, four legs are one too many; three, properly located, are sufficient for stability.

EXAMPLE 11–4

Another rescue attempt After Lancelot falls into the moat, Sir Gawain tosses a grappling hook through an open upstairs window. In Fig. 11–6a (page 336) he is resting partway up the rope. Gawain weighs 700 N, his body makes an angle of 60° with the wall, and his center of gravity is 0.85 m from his feet. The rope force acts 1.30 m from his feet, and the rope makes an angle of 20° with the vertical. Find the tension in the rope and the forces exerted on his feet by the moss-free region of the wall.

SOLUTION Figure 11–6b shows the free-body diagram for Gawain. The forces on his feet include a (horizontal) normal component n and a (vertical) frictional component f. The rope tension is T. The angles needed to find the torques are shown.

Because two unknown forces (n and f) act at point B at Gawain's feet, we take torques about B, obtaining a torque equa-

tion in which T is the only unknown. The easiest way to find the torque of each force is to use $\tau = rF \sin \phi$, where ϕ is the angle between the position vector $\vec{r}$ (from B to the point of application of the force) and the force vector $\vec{F}$. Taking counterclockwise torques as positive, we obtain the torque equation

$$\Sigma \tau_B = (1.30 \text{ m})T (\sin 140°)$$
$$- (0.85 \text{ m})(700 \text{ N}) (\sin 60°) + (0)n + (0)f = 0.$$

We solve this equation for T; the result is

$$T = 617 \text{ N}.$$

We note that the rope tension is *less than* Gawain's weight.

To find the components of force n and f at Gawain's feet, we use the first equilibrium condition. From Fig. 11–6b, T_x is negative and T_y is positive, so $T_x = -T \sin 20°$ and $T_y = T \cos 20°$.

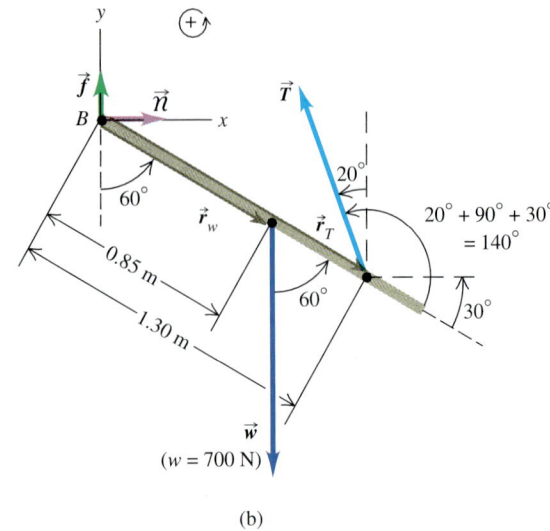

11–6 (a) Gawain pausing to catch his breath. (b) Free-body diagram for Gawain, considered as a rod.

Then

$$\Sigma F_x = n + (-T \sin 20°) = 0,$$

$$\Sigma F_y = f + T \cos 20° + (-700 \text{ N}) = 0.$$

When we substitute the value of T into these equations, we get

$$n = 211 \text{ N} \quad \text{and} \quad f = 120 \text{ N}.$$

The sum of the vertical friction force ($f = 120$ N) and the vertical component of the tension ($T \cos 20° = 580$ N) equals Gawain's 700-N weight.

EXAMPLE 11–5

Equilibrium and pumping iron Figure 11–7a shows a human arm lifting a dumbbell, and Fig. 11–7b is a free-body diagram for the forearm. The forearm is in equilibrium under the action of the weight w of the dumbbell, the tension T in the tendon connected to the biceps muscle, and the force E exerted on the forearm by the upper arm at the elbow joint. For clarity the point A where the tendon is attached is drawn farther away from the elbow than its actual position. The weight w and the angle θ are given; we want to find the tendon tension and the two components of force at the elbow (three unknown scalar quantities in all). We neglect the weight of the forearm itself.

SOLUTION First we represent the tendon force in terms of its components T_x and T_y, using the given angle θ and the unknown magnitude T:

$$T_x = T \cos \theta, \qquad T_y = T \sin \theta.$$

We also represent the force at the elbow in terms of its components E_x and E_y. We'll guess that the directions of these components are as shown in Fig. 11–7b; there's no need to agonize over this guess, since the results for E_x and E_y will tell us the actual directions. Next we note that if we take torques about the elbow joint, the resulting torque equation does not contain

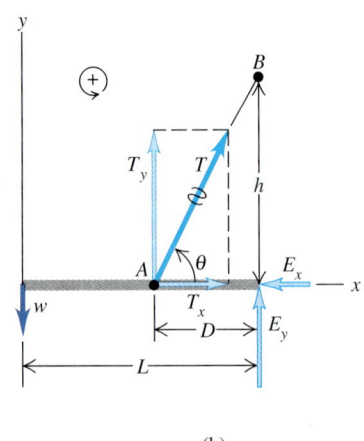

11–7 (a) Building up the biceps. (b) Free-body diagram for the forearm. The weight of the forearm is neglected, and the distance D is greatly exaggerated for clarity.

E_x, E_y, or T_x because the lines of action of all these forces pass through this point. The torque equation is then simply

$$\Sigma \tau_E = Lw - DT_y = 0.$$

From this we find

$$T_y = \frac{Lw}{D} \quad \text{and} \quad T = \frac{Lw}{D \sin \theta}.$$

To find E_x and E_y, we use the first conditions for equilibrium, $\Sigma F_x = 0$ and $\Sigma F_y = 0$:

$$\Sigma F_x = T_x + (-E_x) = 0,$$

$$E_x = T_x = T \cos \theta = \frac{Lw}{D \sin \theta} \cos \theta = \frac{Lw}{D} \cot \theta$$

$$= \frac{Lw}{D} \frac{D}{h} = \frac{Lw}{h};$$

$$\Sigma F_y = T_y + E_y + (-w) = 0,$$

$$E_y = w - \frac{Lw}{D} = -\frac{(L-D)w}{D}.$$

The negative sign shows that our guess for the direction of E_y, shown in Fig. 11–7b, was wrong; it is actually vertically *downward*.

An alternative way to find E_x and E_y is to use two more torque equations. We take torques about the tendon attach point, A:

$$\Sigma \tau_A = (L-D)w + DE_y = 0 \quad \text{and} \quad E_y = -\frac{(L-D)w}{D}.$$

Finally, we take torques about point B in the figure:

$$\Sigma \tau_B = Lw - hE_x = 0 \quad \text{and} \quad E_x = \frac{Lw}{h}.$$

Notice how much we have simplified these calculations by choosing the point for calculating torques so as to eliminate one or more of the unknown quantities. In the last step, the force T has no torque about point B; thus when the torques of T_x and T_y are computed separately, they must add to zero. We invite you to check out this statement.

As a specific example, suppose $w = 200$ N, $D = 0.050$ m, $L = 0.30$ m, and $\theta = 80°$. Then from $\tan \theta = h/D$, we find

$$h = D \tan \theta = (0.050 \text{ m})(5.67) = 0.28 \text{ m}.$$

From the previous general results we find

$$T = \frac{Lw}{D \sin \theta} = \frac{(0.30 \text{ m})(200 \text{ N})}{(0.050 \text{ m})(0.98)} = 1220 \text{ N},$$

$$E_y = -\frac{(L-D)w}{D} = -\frac{(0.30 \text{ m} - 0.050 \text{ m})(200 \text{ N})}{0.050 \text{ m}}$$

$$= -1000 \text{ N},$$

$$E_x = \frac{Lw}{h} = \frac{(0.30 \text{ m})(200 \text{ N})}{0.28 \text{ m}} = 210 \text{ N}.$$

The magnitude of the force at the elbow is

$$E = \sqrt{E_x{}^2 + E_y{}^2} = 1020 \text{ N}.$$

In our alternative determination of E_x and E_y, we didn't explicitly use the first condition for equilibrium (that the vector sum of the forces is zero). As a check, compute ΣF_x and ΣF_y to verify that they really *are* zero. Consistency checks are always a good idea!

In view of the magnitudes of our results, neglecting the weight of the forearm itself, which may be 20 N or so, will cause only relatively small errors in our results.

11–5 STRESS, STRAIN, AND ELASTIC MODULI

The rigid body is a useful idealized model, but the stretching, squeezing, and twisting of real bodies when forces are applied are often too important to ignore. Figure 11–8 shows three examples. We want to study the relation between the forces and deformations for each case.

11–8 Three types of stress. (a) A cable under *tension*, being stretched by forces acting at its ends. (b) A submersible under *compression*, being squeezed from all sides by the force of the water pressure. (c) A drive shaft under *shear*, being twisted by forces at its ends that cause torques about its axis.

(a)   (b)   (c)

For each kind of deformation we will introduce a quantity called **stress** that characterizes the strength of the forces causing the stretch, squeeze, or twist, usually on a "force per unit area" basis. Another quantity, **strain,** describes the resulting deformation. When the stress and strain are small enough, we often find that the two are directly proportional, and we call the proportionality constant an **elastic modulus.** The harder you pull on something, the more it stretches; and the more you squeeze it, the more it compresses. The general pattern that emerges can be formulated as

$$\frac{\text{Stress}}{\text{Strain}} = \text{Elastic modulus} \qquad \text{(Hooke's law)}. \tag{11-7}$$

The proportionality of stress and strain (under certain conditions) is called **Hooke's law,** after Robert Hooke (1635–1703), a contemporary of Newton. We used one form of Hooke's law in Sections 6–4 and 7–3: the elongation of an ideal spring is proportional to the stretching force. Remember that Hooke's law is not really a general law but an experimental finding that is valid only over a limited range. The last section of this chapter discusses what this limited range is.

TENSILE AND COMPRESSIVE STRESS AND STRAIN

The simplest elastic behavior to understand is the stretching of a bar, rod, or wire when its ends are pulled. Figure 11–9a shows a bar with uniform cross-section area A, with equal and opposite forces F pulling at its ends. We say that the bar is in **tension.** We've already talked a lot about tension in ropes and strings; it's the same concept here. Figure 11–9b shows a cross section through the bar. The part of the bar to the right of the section pulls on the part to the left with a force $F_\perp$, and vice versa. We use the notation $F_\perp$ as a reminder that the force acts in a direction perpendicular to the cross section. We will assume that the forces at every cross section are distributed uniformly over the section, as shown by the short arrows in Fig. 11–9b. (This is the case if the bar is uniform and the forces at the *ends* are uniformly distributed.)

We define the **tensile stress** at the cross section as the ratio of the force $F_\perp$ to the cross-section area A:

$$\text{Tensile stress} = \frac{F_\perp}{A}. \tag{11-8}$$

This is a *scalar* quantity because $F_\perp$ is the *magnitude* of the force. The SI unit of stress is the **pascal** (abbreviated Pa, and named for the 17th-century French scientist and philosopher Blaise Pascal). Equation (11–8) shows that one pascal equals one newton per square meter (N/m^2):

$$1 \text{ pascal} = 1 \text{ Pa} = 1 \text{ N/m}^2.$$

In the British system the logical unit of stress would be the pound per square foot, but the pound per square inch (lb/in.^{2} or psi) is more commonly used. The conversion factors are

$$1 \text{ psi} = 6895 \text{ Pa} \qquad \text{and} \qquad 1 \text{ Pa} = 1.450 \times 10^{-4} \text{ psi}.$$

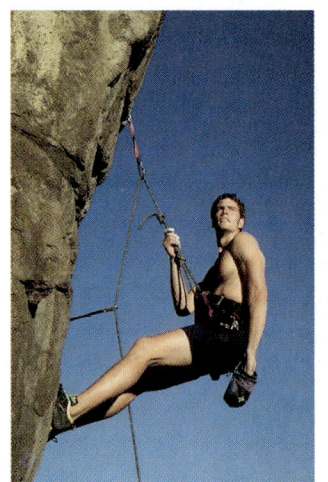

This rock climber exerts a stress on the climbing rope due to his weight. The strain of the rope is the fractional amount that it stretches as a result of this stress.

11–9 (a) A bar in tension. (b) The stress at a perpendicular section equals $F_\perp/A$, the ratio of the force to the cross-section area.

(a)

(b)

The units of stress are the same as those of *pressure*, which we will encounter often in later chapters. Air pressure in automobile tires is typically around 3×10^5 Pa = 300 kPa, and steel cables are commonly required to withstand tensile stresses of the order of 10^8 Pa.

The fractional change of length (stretch) of a body under a tensile stress is called the **tensile strain.** Figure 11–10 shows a bar with unstretched length l_0 that stretches to a length $l = l_0 + \Delta l$ when equal and opposite forces F are applied to its ends. The elongation Δl does not occur only at the ends; every part of the bar stretches in the same proportion. The tensile strain is defined as the ratio of the elongation Δl to the original length l_0:

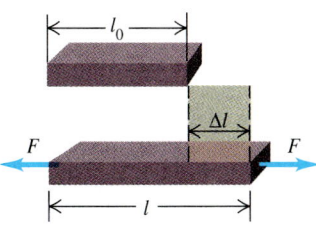

11–10 The tensile strain is defined as $\Delta l / l_0$. The elongation Δl is exaggerated for clarity.

$$\text{Tensile strain} = \frac{l - l_0}{l_0} = \frac{\Delta l}{l_0}. \qquad (11–9)$$

Tensile strain is stretch per unit length. It is a ratio of two lengths, always measured in the same units, and so is a pure (dimensionless) number with no units.

Experiment shows that for a sufficiently small tensile stress, stress and strain are proportional, as in Eq. (11–7). The corresponding elastic modulus is called **Young's modulus,** denoted by Y:

$$Y = \frac{\text{Tensile stress}}{\text{Tensile strain}},$$

$$Y = \frac{F_\perp / A}{\Delta l / l_0} = \frac{F_\perp}{A} \frac{l_0}{\Delta l} \qquad \text{(Young's modulus)}. \qquad (11–10)$$

Since strain is a pure number, the units of Young's modulus are the same as those of stress: force per unit area. Some typical values are listed in Table 11–1. (This table also gives values of two other elastic moduli that we will discuss later in this chapter.) A material with a large value of Y is relatively unstretchable; a large stress is required for a given strain. For example, the value of Y for cast steel (2×10^{11} Pa) is much larger than that for rubber (5×10^8 Pa).

TABLE 11–1

APPROXIMATE ELASTIC MODULI

MATERIAL	YOUNG'S MODULUS, Y (Pa)	BULK MODULUS, B (Pa)	SHEAR MODULUS, S (Pa)
Aluminum	7.0×10^{10}	7.5×10^{10}	2.5×10^{10}
Brass	9.0×10^{10}	6.0×10^{10}	3.5×10^{10}
Copper	11×10^{10}	14×10^{10}	4.4×10^{10}
Crown glass	6.0×10^{10}	5.0×10^{10}	2.5×10^{10}
Iron	21×10^{10}	16×10^{10}	7.7×10^{10}
Lead	1.6×10^{10}	4.1×10^{10}	0.6×10^{10}
Nickel	21×10^{10}	17×10^{10}	7.8×10^{10}
Steel	20×10^{10}	16×10^{10}	7.5×10^{10}

When the forces on the ends of a bar are pushes rather than pulls (Fig. 11–11), the bar is in **compression,** and the stress is a **compressive stress.** At the cross section shown,

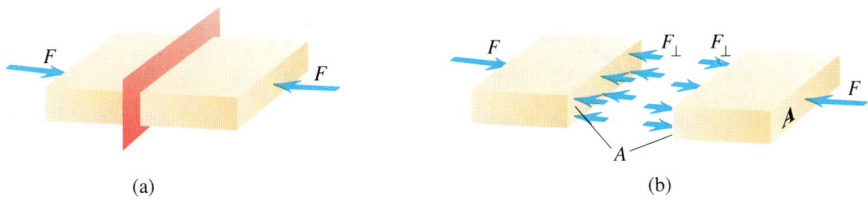

(a)   (b)

11–11 (a) A bar in compression. (b) The stress at a perpendicular section also equals $F_\perp/A$.

(d)

11–12 (a) A beam supported at both ends is under both compression and tension. (b) The cross-section shape of an I-beam minimizes both stress and weight. (c) A hollow pole or tube is resistant to bending in all directions. (d) At a height of 553 m, the CN Tower in Toronto is the world's tallest self-supporting structure. In cross section the tower forms a Y, which gives the 1.3×10^8-kg structure tremendous rigidity against twisting or bending. In a 200-km/h wind, the Skypod (the seven-story building two-thirds of the way up the tower) would sway back and forth only 25 cm.

each side pushes rather than pulls on the other. The **compressive strain** of a bar in compression is defined in the same way as tensile strain, but Δl has the opposite direction. Hooke's law and Eq. (11–10) are valid for compression as well as tension if the compressive stress is not too great. For many materials, Young's modulus has the same value for both tensile and compressive stresses; composite materials such as concrete are an exception.

In many situations, bodies can experience both tensile and compressive stresses at the same time. As an example, a horizontal beam supported at each end sags under its own weight. As a result, the top of the beam is under compression, while the bottom of the beam is under tension (Fig. 11–12a). To minimize the stress and hence the bending strain, the top and bottom of the beam are given a large cross-section area. There is neither compression nor tension along the centerline of the beam, so this part can have a small cross section; this helps to keep the weight of the bar to a minimum and further helps to reduce the stress. The result is an I-beam of the familiar shape used in building construction (Fig. 11–12b).

The vertical pole that supports a traffic light or a highway sign has a circular cross section because it must be resistant to bending in all directions due to wind or earthquakes. A hollow circular pole is more resistant to bending than a solid circular pole of the same mass but smaller radius, since it has a greater effective cross-section area (Fig. 11–12c). The CN Tower in Toronto (Fig. 11–12d) is basically a hollow circular pole with the sides pinched in to give some of the natural stability of a tripod. Sections of the tower closer to the ground must support a larger fraction of the tower's weight, so the cross-section area is greater near the ground to keep the stress relatively constant.

Bridges are under tremendous stress due to the weight they must support. The design of bridges is an exercise in imparting this stress to the supporting ground. A suspension bridge supports its load primarily through tension in the cables and compression in the

11–13 (a) In a suspension bridge, the cables are under tension and the towers are under compression. Tension structures are most efficient for supporting light or moderate loads over long distances; the world's longest suspension bridge, the Humber Bridge in England, carries cars and trucks (but no trains) across a center span of 1410 m. (b) An arched bridge is under compression. Compression structures are very efficient for supporting large loads over short distances. An example is the Sydney Harbour Bridge in Australia, which has a 500-m span and was built to carry four railway lines and a 17-m-wide road.

(a)   (b)

towers (Fig. 11–13a). The downward force on the towers due to tension is balanced by the upward force provided by the ground under the towers. The ground must be strong enough to bear this stress without fracturing. An arched bridge supports its load primarily through compression (Fig. 11–13b). The compressive stress is borne by the ground at the ends of the arch.

EXAMPLE 11–6

A steel rod 2.0 m long has a cross-section area of 0.30 cm^2. A 550-kg milling machine is now hung from the rod. Determine the stress, the strain, and the elongation of the rod.

SOLUTION We use the definitions of stress, strain, and Young's modulus given by Eqs. (11–8), (11–9), and (11–10), respectively, and the value of Y for steel from Table 11–1:

$$\text{Stress} = \frac{F_\perp}{A} = \frac{(550 \text{ kg})(9.8 \text{ m/s}^2)}{3.0 \times 10^{-5} \text{ m}^2} = 1.8 \times 10^8 \text{ Pa};$$

$$\text{Strain} = \frac{\Delta l}{l_0} = \frac{\text{Stress}}{Y} = \frac{1.8 \times 10^8 \text{ Pa}}{20 \times 10^{10} \text{ Pa}} = 9.0 \times 10^{-4};$$

$$\text{Elongation} = \Delta l = (\text{Strain}) \times l_0 = (9.0 \times 10^{-4})(2.0 \text{ m})$$
$$= 0.0018 \text{ m} = 1.8 \text{ mm}.$$

The small size of this elongation, which results from a load of over half a ton, is a testament to the stiffness of steel.

11–6 BULK STRESS AND STRAIN

When a submersible plunges deep into the ocean, the water exerts nearly uniform pressure everywhere on its surface and squeezes the submersible to a slightly smaller volume (Fig. 11–8b). This is a different situation from the tensile and compressive stresses and strains we have discussed. The stress is now a uniform pressure on all sides, and the resulting deformation is a volume change. We use the terms **bulk stress** (or **volume stress**) and **bulk strain** (or **volume strain**) to describe these quantities. Another familiar example is the compression of a gas under pressure, such as the air in a car's tire.

If we choose an arbitrary cross section within a fluid (liquid or gas) at rest, the force acting on each side of the section is always *perpendicular* to it. If we tried to exert a force parallel to a section, the fluid would slip sideways to counteract the effort. When a solid is immersed in a fluid and both are at rest, the forces that the fluid exerts on the surface of the solid are always perpendicular to the surface at each point. The force $F_\perp$ per unit area A on such a surface is called the **pressure** p in the fluid:

$$p = \frac{F_\perp}{A} \qquad \text{(pressure in a fluid).} \qquad (11\text{–}11)$$

When we apply pressure to the surface of a fluid in a container, such as the cylinder and piston shown in Fig. 11–14, the pressure is transmitted through the fluid and also acts on the surface of any body immersed in the fluid. This principle is called *Pascal's law*. If pressure differences due to differences in depth within the fluid can be neglected, the pressure is the same at every point in the fluid and at every point on the surface of any submerged body.

Pressure has the same units as stress; commonly used units include 1 Pa (= 1 N/m^2) and 1 lb/in.^{2} (1 psi). Also in common use is the **atmosphere,** abbreviated atm. One atmosphere is the approximate average pressure of the earth's atmosphere at sea level:

$$1 \text{ atmosphere} = 1 \text{ atm} = 1.013 \times 10^5 \text{ Pa} = 14.7 \text{ lb/in.}^2.$$

Pressure is a scalar quantity, not a vector quantity; it has no direction.

11–14 Every element dA of the surface of a submerged body experiences an inwardly normal force $dF_\perp = p \, dA$ from the surrounding fluid.

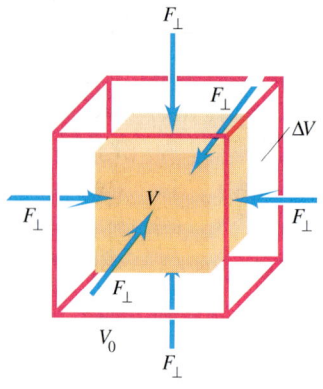

11–15 Under bulk stress, the cube occupies the shaded volume. Without the stress, the cube would occupy the outlined volume. The bulk strain is the fractional change in volume, $\Delta V/V_0$. The volume change ΔV is exaggerated for clarity.

TABLE 11–2

COMPRESSIBILITIES OF LIQUIDS

LIQUID	COMPRESSIBILITY, k	
	Pa^{-1}	atm^{-1}
Carbon disulfide	93×10^{-11}	94×10^{-6}
Ethyl alcohol	110×10^{-11}	111×10^{-6}
Glycerine	21×10^{-11}	21×10^{-6}
Mercury	3.7×10^{-11}	3.8×10^{-6}
Water	45.8×10^{-11}	46.4×10^{-6}

Pressure plays the role of stress in a volume deformation. The corresponding strain is the fractional change in volume (Fig. 11–15), that is, the ratio of the volume change ΔV to the original volume V_0:

$$\text{Bulk (volume) strain} = \frac{\Delta V}{V_0}. \qquad (11\text{–}12)$$

Volume strain is the change in volume per unit volume. Like tensile or compressive strain, it is a pure number, without units.

When Hooke's law is obeyed, an increase in pressure (bulk stress) produces a *proportional* bulk strain (fractional change in volume). The corresponding elastic modulus (ratio of stress to strain) is called the **bulk modulus,** denoted by B. When the pressure on a body changes by a small amount Δp, from p_0 to $p_0 + \Delta p$, and the resulting bulk strain is $\Delta V/V_0$, Hooke's law takes the form

$$B = -\frac{\Delta p}{\Delta V/V_0} \qquad \text{(bulk modulus).} \qquad (11\text{–}13)$$

We include a minus sign in this equation because an *increase* of pressure always causes a *decrease* in volume. In other words, if Δp is positive, ΔV is negative. The bulk modulus B itself is a positive quantity.

For small pressure changes in a solid or a liquid, we consider B to be constant. The bulk modulus of a *gas,* however, depends on the initial pressure p_0. Table 11–1 includes values of the bulk modulus for several solid materials. Its units, force per unit area, are the same as those of pressure (and of tensile or compressive stress).

The reciprocal of the bulk modulus is called the **compressibility** and is denoted by k. From Eq. (11–13),

$$k = \frac{1}{B} = -\frac{\Delta V/V_0}{\Delta p} = -\frac{1}{V_0}\frac{\Delta V}{\Delta p} \qquad \text{(compressibility).} \qquad (11\text{–}14)$$

Compressibility is the fractional decrease in volume, $-\Delta V/V_0$, per unit increase Δp in pressure. The units of compressibility are those of *reciprocal pressure,* Pa^{-1} or atm^{-1}.

Values of compressibility k for several liquids are listed in Table 11–2. From this table the compressibility of water is 46.4×10^{-6} atm^{-1}. This means that for each atmosphere increase in pressure, the volume of water decreases by 46.4 parts per million. Materials with small bulk modulus and large compressibility are easy to compress; those with larger bulk modulus and smaller compressibility compress less with the same pressure increase.

EXAMPLE 11–7

A hydraulic press contains 0.25 m³ (250 L) of oil. Find the decrease in volume of the oil when it is subjected to a pressure increase $\Delta p = 1.6 \times 10^7$ Pa (about 160 atm or 2300 psi). The bulk modulus of the oil is $B = 5.0 \times 10^9$ Pa (about 5.0×10^4 atm), and its compressibility is $k = 1/B = 20 \times 10^{-6}$ atm^{-1}.

SOLUTION To find the volume change ΔV, we solve Eq. (11–13) for ΔV and then substitute in the given values:

$$\Delta V = -\frac{V_0 \Delta p}{B} = -\frac{(0.25 \text{ m}^3)(1.6 \times 10^7 \text{ Pa})}{5.0 \times 10^9 \text{ Pa}}$$

$$= -8.0 \times 10^{-4} \text{ m}^3 = -0.80 \text{ L.}$$

Alternatively, we can use Eq. (11–14). Solving for ΔV and using the approximate unit conversions given above, we get

$$\Delta V = -kV_0 \, \Delta p = -(20 \times 10^{-6} \text{ atm}^{-1})(0.25 \text{ m}^3)(160 \text{ atm})$$

$$= -8.0 \times 10^{-4} \text{ m}^3.$$

Even though the pressure increase is very large, the fractional change in volume is very small:

$$\Delta V/V_0 = (-8.0 \times 10^{-4} \text{ m}^3)/(0.25 \text{ m}^3) = -0.0032, \text{ or } -0.32\%.$$

11-7 SHEAR STRESS AND STRAIN

The third kind of stress-strain situation is called *shear*. The drive shaft in Fig. 11–8c and the body in Fig. 11–16 are under **shear stress;** the result will be a twisting or deformation of the body. We define the shear stress as the force $F_\parallel$ *tangent* to a material surface, divided by the area A on which it acts:

$$\text{Shear stress} = \frac{F_\parallel}{A}. \qquad (11\text{–}15)$$

This is the sort of stress provided by a pair of scissors or shears. Shear stress, like the other two types of stress, is a force per unit area.

A shear deformation is shown in Fig. 11–17. The black outline *abcd* represents an unstressed block of material, such as a book, a wall, or a chunk of the earth's crust. The area *a′ b′ c′ d′* shaded in yellow shows the same block under shear stress. In Fig. 11–17a the centers of the stressed and unstressed block coincide. The deformation in Fig. 11–17b is the same as that in Fig. 11–17a, but the figure has been shifted and rotated to make the edges *ad* and *a′ d′* coincide. In shear stress the lengths of the faces remain nearly constant; all dimensions parallel to the diagonal *ac* increase in length, and those parallel to the diagonal *bd* decrease in length. We define **shear strain** as the ratio of the displacement *x* of corner *b* to the transverse dimension *h*:

$$\text{Shear strain} = \frac{x}{h} = \tan\phi. \qquad (11\text{–}16)$$

In real-life situations, x is nearly always much smaller than h, $\tan\phi$ is very nearly equal to ϕ (in radians), and the strain is simply the angle ϕ measured in radians. Like all strains, shear strain is a dimensionless number; it is a ratio of two lengths.

If the forces are small enough that Hooke's law is obeyed, the shear strain is *proportional* to the shear stress. The corresponding elastic modulus (ratio of shear stress to shear strain) is called the **shear modulus,** denoted by S:

$$S = \frac{\text{Shear stress}}{\text{Shear strain}} = \frac{F_\parallel/A}{x/h} = \frac{F_\parallel}{A}\frac{h}{x} = \frac{F_\parallel/A}{\phi} \quad \text{(shear modulus)}, \qquad (11\text{–}17)$$

with x and h defined as in Fig. 11–17.

Several values of shear modulus are given in Table 11–1. For a given material, S is usually $\frac{1}{3}$ to $\frac{1}{2}$ as large as Young's modulus Y for tensile stress. Keep in mind that the concepts of shear stress, shear strain, and shear modulus apply to *solid* materials only. Here's why: the shear forces in Fig. 11–17a are required to deform the solid block, and the block tends to return to its original shape if the shear forces are removed. By contrast, gases and liquids do not have a definite shape.

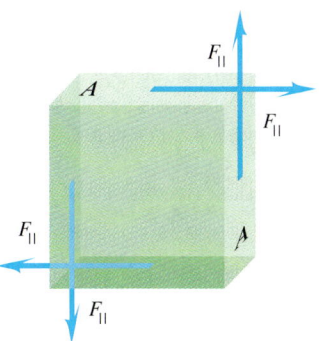

11–16 A body under shear stress. The area A is the edge area on which each force $F_\parallel$ acts.

Scissors and knives work by applying shear stress on the object to be cut. Cutting takes place when the strain caused by the stress is so large that chemical bonds between molecules in the object break apart.

(a)

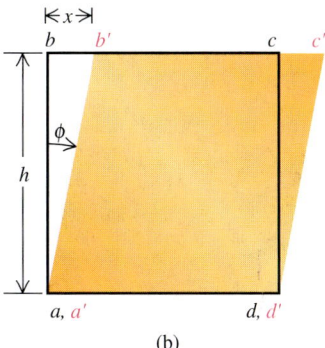

(b)

11–17 (a) A block changes shape under shear stress. (b) The deformed block has been shifted and rotated so that its bottom edge coincides with that of the undeformed block to show the angle ϕ clearly. The shear strain, defined as x/h, is very nearly equal to ϕ in radians.

EXAMPLE 11-8

Suppose the body in Fig. 11–17 is the brass base plate of an out-door sculpture; it experiences shear forces as a result of an earthquake. The frame is 0.80 m square and 0.50 cm thick. How large a force $F_\parallel$ must be exerted on each of its edges if the displacement x (see Fig. 11–17b) is 0.16 mm?

SOLUTION From Table 11–1 the shear modulus of brass is 3.5×10^{10} Pa. The shear strain is

$$\text{Shear strain} = \frac{x}{h} = \frac{1.6 \times 10^{-4}\,\text{m}}{0.80\,\text{m}} = 2.0 \times 10^{-4}.$$

From Eq. (11–17) the shear stress equals the shear strain multiplied by the shear modulus S,

$$\text{Stress} = (\text{Shear strain}) \times S$$
$$= (2.0 \times 10^{-4})(3.5 \times 10^{10}\,\text{Pa}) = 7.0 \times 10^{6}\,\text{Pa},$$

and the force at each edge is the shear stress multiplied by the area of the edge:

$$F_\parallel = (\text{Shear stress}) \times A$$
$$= (7.0 \times 10^{6}\,\text{Pa})(0.80\,\text{m})(0.0050\,\text{m}) = 2.8 \times 10^{4}\,\text{N}.$$

This is a force of over three tons.

11-8 ELASTICITY AND PLASTICITY

Hooke's law, the proportionality of stress and strain in elastic deformations, has a limited range of validity. In the preceding three sections we used phrases such as "provided that the forces are small enough that Hooke's law is obeyed." Just what *are* the limitations of Hooke's law? We know that if you pull, squeeze, or twist *anything* hard enough, it will bend or break. Can we be more precise than that?

Let's look at tensile stress and strain again. Suppose we plot a graph of stress as a function of strain. If Hooke's law is obeyed, the graph is a straight line with a slope equal to Young's modulus. Figure 11–18 shows a typical stress-strain graph for a metal such as copper or soft iron. The strain is shown as the *percent* elongation; the horizontal scale is not uniform beyond the first portion of the curve, up to a strain of less than 1%. The first portion is a straight line, indicating Hooke's law behavior with stress directly proportional to strain. This straight-line portion ends at point *a;* the stress at this point is called the *proportional limit.*

From *a* to *b*, stress and strain are no longer proportional, and Hooke's law is *not* obeyed. If the load is gradually removed, starting at any point between *O* and *b*, the curve is retraced until the material returns to its original length. The deformation is *reversible,* and the forces are conservative; the energy put into the material to cause the deformation is recovered when the stress is removed. In region *Ob* we say that the material shows *elastic behavior.* Point *b,* the end of this region, is called the *yield point;* the stress at the yield point is called the *elastic limit.*

11–18 Typical stress-strain diagram for a ductile metal under tension.

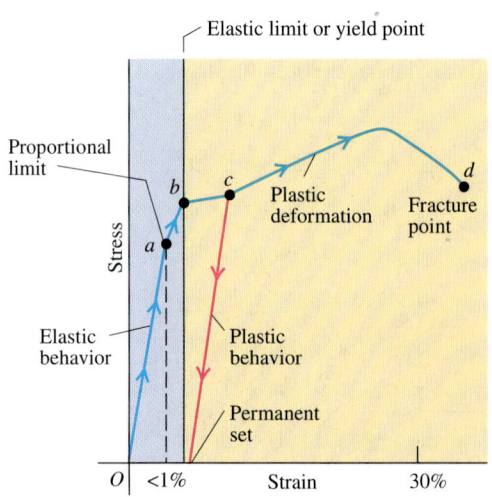

When we increase the stress beyond point b, the strain continues to increase. But now when we remove the load at some point beyond b, say c, the material does not come back to its original length. Instead, it follows the red line in Fig. 11–18. The length at zero stress is now greater than the original length; the material has undergone an irreversible deformation and has acquired what we call a *permanent set*. Further increase of load beyond c produces a large increase in strain for a relatively small increase in stress, until a point d is reached at which *fracture* takes place. The behavior of the material from b to d is called *plastic flow* or *plastic deformation*. A plastic deformation is irreversible; when the stress is removed, the material does not return to its original state.

For some materials a large amount of plastic deformation takes place between the elastic limit and the fracture point. Such a material is said to be *ductile*. But if fracture occurs soon after the elastic limit is passed, the material is said to be *brittle*. A soft iron wire that can have considerable permanent stretch without breaking is ductile, while a steel piano string that breaks soon after its elastic limit is reached is brittle.

Figure 11–19 shows a stress-strain curve for vulcanized rubber that has been stretched by over seven times its original length. The stress is not proportional to the strain, but the behavior is elastic because when the load is removed, the material returns to its original length. However, the material follows different curves for increasing and decreasing stress. This is called *elastic hysteresis*. The work done by the material when it returns to its original shape is less than the work required to deform it; there are nonconservative forces associated with internal friction. Rubber with large elastic hysteresis is very useful for absorbing vibrations, such as in engine mounts and shock-absorber bushings for cars.

The stress required to cause actual fracture of a material is called the *breaking stress,* the *ultimate strength,* or (for tensile stress) the *tensile strength.* Two materials, such as two types of steel, may have very similar elastic constants but vastly different breaking stresses. Table 11–3 gives a few typical values of breaking stress for several materials in tension.

The conversion factor 6.9×10^8 Pa = 100,000 psi will help put these numbers in perspective. For example, if the breaking stress of a particular steel is 6.9×10^8 Pa, then a bar with 1-in.^{2} cross section has a breaking strength of 100,000 lb.

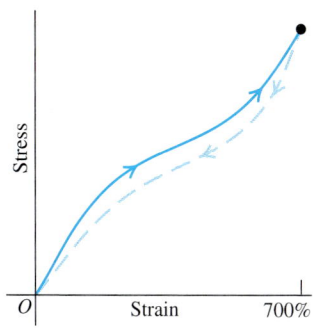

11–19 Typical stress-strain diagram for vulcanized rubber, showing elastic hysteresis.

TABLE 11–3

APPROXIMATE BREAKING STRESSES OF MATERIALS

MATERIAL	BREAKING STRESS (Pa or N/m^2)
Aluminum	2.2×10^8
Brass	4.7×10^8
Glass	10×10^8
Iron	3.0×10^8
Phosphor bronze	5.6×10^8
Steel	$5–20 \times 10^8$

Summary

- For a rigid body to be in equilibrium, two conditions must be satisfied. First, the vector sum of forces must be zero. Second, the sum of torques about any point must be zero. The first condition, in terms of components, is

$$\Sigma F_x = 0, \qquad \Sigma F_y = 0, \qquad \Sigma F_z = 0. \tag{11–1}$$

The second condition is

$$\Sigma \vec{\tau} = 0 \qquad \text{about } any \text{ point.} \tag{11–2}$$

- The torque due to the weight of a body can be obtained by assuming the entire weight to be concentrated at the center of gravity. If $\vec{g}$ has the same value at all points, the coordinates of the center of gravity are the same as those of the center of mass:

$$x_{cm} = \frac{m_1 x_1 + m_2 x_2 + \cdots}{m_1 + m_2 + \cdots} = \frac{\sum_i m_i x_i}{\sum_i m_i}, \qquad y_{cm} = \frac{m_1 y_1 + m_2 y_2 + \cdots}{m_1 + m_2 + \cdots} = \frac{\sum_i m_i y_i}{\sum_i m_i},$$

$$z_{cm} = \frac{m_1 z_1 + m_2 z_2 + \cdots}{m_1 + m_2 + \cdots} = \frac{\sum_i m_i z_i}{\sum_i m_i}. \tag{11–3}$$

KEY TERMS

first condition for equilibrium, 329

second condition for equilibrium, 330

static equilibrium, 330

center of gravity, 330

stress, 338

strain, 338

elastic modulus, 338

Hooke's law, 338

tension, 338

tensile stress, 338

pascal, 338

tensile strain, 339

Young's modulus, 339

compression, 339

compressive stress, 339

■ The torque due to a force can be computed by finding the torque of each force component, using its appropriate lever arm and sign, and adding these.

■ Hooke's law states that in elastic deformations, stress is proportional to strain:

$$\frac{\text{Stress}}{\text{Strain}} = \text{Elastic modulus.} \qquad (11-7)$$

■ Tensile stress is tensile force per unit area, $F_\perp/A$. Tensile strain is fractional change in length, $\Delta l/l_0$. Young's modulus Y is the ratio of tensile stress to tensile strain:

$$Y = \frac{F_\perp/A}{\Delta l/l_0} = \frac{F_\perp}{A}\frac{l_0}{\Delta l}. \qquad (11-10)$$

Compressive stress and strain are defined the same way as tensile stress and strain. For many materials, Young's modulus has the same value for both tension and compression.

■ Pressure in a fluid is force per unit area:

$$p = \frac{F_\perp}{A}. \qquad (11-11)$$

■ The bulk modulus B is the negative of the ratio of pressure change Δp (bulk stress) to fractional volume change $\Delta V/V_0$:

$$B = -\frac{\Delta p}{\Delta V/V_0}. \qquad (11-13)$$

Compressibility k is the reciprocal of bulk modulus: $k = 1/B$.

■ Shear stress is force per unit area $F_\parallel/A$ for a force applied parallel to a surface. Shear strain is the angle ϕ shown in Fig. 11–17. The shear modulus S is the ratio of shear stress to shear strain:

$$S = \frac{\text{Shear stress}}{\text{Shear strain}} = \frac{F_\parallel/A}{x/h} = \frac{F_\parallel}{A}\frac{h}{x} = \frac{F_\parallel/A}{\phi}. \qquad (11-17)$$

■ The proportional limit is the maximum stress for which stress and strain are proportional. Beyond the proportional limit, Hooke's law is not valid. The elastic limit is the stress beyond which irreversible deformation occurs. The breaking stress, or ultimate strength, is the stress at which the material breaks.

DISCUSSION QUESTIONS

Q11–1 Does a rigid body in uniform rotation about a fixed axis satisfy the first and second conditions for equilibrium? Why or why not? An alternative definition of equilibrium states: "For a rigid body to be in equilibrium, every part of the body must be in equilibrium." On the basis of this definition, is a rigid body in uniform rotation about a fixed axis in equilibrium? Explain.

Q11–2 Car tires are sometimes "balanced" on a machine that pivots the tire and wheel about the center. Weights are placed around the wheel rim until it does not tip from the horizontal plane. Discuss this procedure in terms of the center of gravity.

Q11–3 Does the center of gravity of a solid body always lie within the material of the body? If not, give a counterexample.

Q11–4 In Section 11–3 we always assumed that the value of g was the same at all points on the body. This is *not* a good approximation if the dimensions of the body are great enough, because the value of g decreases with altitude. If this is taken into account, will the center of gravity of a long vertical rod be above, below, or at its center of mass? Explain how this can be used to keep the long axis of an orbiting spacecraft pointed toward the earth. (This would be useful for a weather satellite that must always keep its camera lens trained on the earth.) The moon is not exactly spherical, but is somewhat elongated. Explain why this same effect is responsible for keeping the same face of the moon pointed toward the earth at all times.

Q11–5 You are balancing a wrench by suspending it at a single point. Is the equilibrium stable, unstable, or neutral if the point is above, at, or below the wrench's center of gravity? In each case give the reasoning behind your answer. (For rotation a rigid body is in *stable* equilibrium if a small rotation of the body produces a torque that tends to return the body to equilibrium, it is in *unstable* equilibrium if a small rotation produces a torque that tends to take the body farther from equilibrium, and it is in *neutral* equilibrium if a small rotation produces no torque.)

Q11–6 You can probably stand flatfooted on the floor, then rise up and balance on your tiptoes. Why can't you do it if your toes are touching the wall of your room? (Try it!)

Q11–7 You freely pivot a horseshoe from a horizontal nail through one of its nail holes. You then hang a long string with a weight at its bottom from the same nail, so that the string hangs vertically in front of the horseshoe without touching it. How do you know that the horseshoe's center of gravity is along the line behind the string? How can you locate the center of gravity by repeating the process at another nail hole? Will the center of gravity be within the solid material of the horseshoe?

Q11–8 A stunt driver drives a car in a straight line at constant speed while balanced on the two right-hand wheels. Where is the car's center of gravity?

Q11–9 When four-legged animals walk, they always have three of their legs on the ground at any instant. Explain why, using the concept of center of gravity.

Q11–10 Why must a water skier moving with constant velocity lean backwards? What determines how far back she must lean? Draw a free-body diagram for the water skier to justify your answers.

Q11–11 In pioneer days, when a Conestoga wagon was stuck in the mud, people would grasp the wheel spokes and try to turn the wheels, rather than simply pushing the wagon. Why?

Q11–12 A physical therapist says a physicist-patient has weak stomach muscles because he can't do a partial sit-up without his feet being held down. The physicist, who has short slender legs and a long heavy upper torso, correctly disagrees. Explain, using the concept of center of gravity.

Q11–13 Why is it easier to hold a 10-kg dumbbell in your hand at your side than to hold it with your arm extended horizontally?

Q11–14 Figure 11–2 shows that an object at rest on an incline will tip over if its center of gravity is not within the area of support. Explain why this result does *not* depend on the value of the coefficient of static friction between the object and the surface of the incline. Draw a free-body diagram as part of your explanation.

Q11–15 In Example 11–3 (Section 11–4), explain why n_1 is negative in the $\Sigma F_x = 0$ equation but gives a positive torque in the $\Sigma \tau_B = 0$ equation.

Q11–16 Why is a tapered water glass with a narrow base easier to tip over than one with straight sides? Does it matter whether the glass is full or empty?

Q11–17 When a tall, heavy refrigerator is pushed across a rough floor, what determines whether it slides or tips?

Q11–18 Climbing ropes used by mountaineers are usually made of nylon. Would a steel cable of equal strength be just as good? What advantages and disadvantages would it have in comparison to nylon?

Q11–19 Why is concrete with steel reinforcing rods embedded in it stronger than plain concrete?

Q11–20 Electric power lines are sometimes made by using wires with steel core and copper jacket or with strands of copper and steel twisted together. Why?

Q11–21 Compare the mechanical properties of a steel cable, made by twisting many thin wires together, with those of a solid steel rod of the same diameter. What advantages does each have?

Q11–22 Two vases have flat bottoms, the same weight, and centers of gravity at the same distance above their bottoms. The one that exerts the greater pressure on the table is likely to be the one that tips over easier. Why?

Q11–23 Is the bulk modulus of steel greater or less than that of air? By roughly what factor?

Q11–24 Is the shear modulus of steel greater or less than that of jelly? By roughly what factor?

Q11–25 Why does Fig. 11–16 show two pairs of antiparallel forces, all of the same magnitude? Wouldn't just one pair result in shear strain and equilibrium?

Q11–26 The material in human bones and elephant bones is essentially the same, but an elephant has much thicker legs. Explain why, in terms of breaking stress.

Q11–27 There is a small but appreciable amount of elastic hysteresis in the large tendon at the back of a horse's leg. Explain how this can cause damage to the tendon if a horse runs too hard for too long a time.

Q11–28 When rubber mounting blocks are used to absorb machine vibrations through elastic hysteresis, as mentioned in Section 11–8, what becomes of the energy associated with the vibrations?

EXERCISES

SECTION **11–3 CENTER OF GRAVITY**

11–1 A ball with radius $r_1 = 0.080$ m and mass 1.00 kg is attached by a light rod 0.400 m in length to a second ball with radius $r_2 = 0.100$ m and mass 2.00 kg (Fig. 11–20). Where is the center of gravity of this system?

11–2 Suppose the rod in Exercise 11–1 is uniform and has a mass of 1.50 kg. Where is the system's center of gravity?

11–3 For the plank of Example 11–1 (Section 11–3), show

FIGURE 11–20 Exercise 11–1.

that a 60-kg child could stand no farther to the right than halfway between the right-hand sawhorse and the end of the plank without the plank tipping over.

SECTION 11–4 SOLVING RIGID-BODY EQUILIBRIUM PROBLEMS

11–4 A uniform 300-N trapdoor in a ceiling is hinged at one side. Find the net upward force needed to begin to open it and the total force exerted on the door by the hinges, a) if the upward force is applied at the center; b) if the upward force is applied at the center of the edge opposite the hinges.

11–5 **Raising a Ladder.** A ladder carried by a fire truck is 20.0 m long. The ladder weighs 2800 N and its center of gravity is at its center. The ladder is pivoted at one end (*A*) about a pin (Fig. 11–21); the friction torque at the pin can be neglected. The ladder is raised into position by a force applied by a hydraulic piston at *C*. Point *C* is 8.0 m from *A*, and the force $\vec{F}$ exerted by the piston makes an angle of 40° with the ladder. What magnitude must $\vec{F}$ have to just lift the ladder off the support bracket at *B*?

FIGURE 11–21 Exercise 11–5.

11–6 Two people are carrying a uniform wooden board that is 3.00 m long and weighs 160 N. If one person applies an upward force equal to 60 N at one end, at what point does the other person lift?

11–7 Two people carry a heavy electric motor by placing it on a light board 2.00 m long. One person lifts at one end with a force of 400 N, and the other lifts the opposite end with a force of 600 N. What is the weight of the motor, and where along the board is its center of gravity located?

11–8 Suppose the board in Exercise 11–7 is not light but weighs 200 N, with its center of gravity at its center. The two people each exert the same forces as before. What is the weight of the motor in this case, and where is its center of gravity located?

11–9 Show that the torques about point *B* due to T_x and T_y in Example 11–5 (Section 11–4) sum to zero. Do this for the general case, rather than for the specific numerical values given at the end of the example.

11–10 A uniform ladder 5.0 m long rests against a frictionless vertical wall with its lower end 3.0 m from the wall. The ladder weighs 160 N. The coefficient of static friction between the foot of the ladder and the ground is 0.40. A man weighing 740 N climbs slowly up the ladder. a) What is the maximum frictional force that the ground can exert on the ladder at its lower end? b) What is the actual frictional force when the man has climbed 1.0 m along the ladder? c) How far along the ladder can the man climb before the ladder starts to slip?

11–11 A diving board 3.00 m long is supported at a point 1.00 m from the end, and a diver weighing 500 N stands at the free end (Fig. 11–22). The diving board is of uniform cross section and weighs 280 N. Find a) the force at the support point; b) the force at the end that is held down.

FIGURE 11–22 Exercise 11–11.

11–12 A uniform aluminum beam 9.00 m long, weighing 300 N, rests symmetrically on two supports 5.00 m apart (Fig. 11–23). A boy weighing 600 N starts at point *A* and walks toward the right. a) In the same diagram construct two graphs showing the upward forces F_A and F_B exerted on the beam at points *A* and *B*, as functions of the coordinate *x* of the boy. Let 1 cm = 100 N vertically, and 1 cm = 1.00 m horizontally. b) From your diagram, how far beyond point *B* can the boy walk before the beam tips? c) How far from the right end of the beam should support *B* be placed so that the boy can walk just to the end of the beam without causing it to tip?

FIGURE 11–23 Exercise 11–12.

11–13 Find the tension *T* in each cable and the magnitude and direction of the force exerted on the strut by the pivot in each of the arrangements in Fig. 11–24. In each case let *w* be the weight of the suspended crate full of priceless art objects. The strut is uniform and also has weight *w*.

FIGURE 11–24 Exercise 11–13.

11–14 The horizontal beam in Fig. 11–25 weighs 150 N, and its center of gravity is at its center. Find a) the tension in the cable; b) the horizontal and vertical components of the force exerted on the beam at the wall.

FIGURE 11–25 Exercise 11–14.

11–15 A door 1.00 m wide and 2.00 m high weighs 280 N and is supported by two hinges, one 0.50 m from the top and the other 0.50 m from the bottom. Each hinge supports half the total weight of the door. Assuming that the door's center of gravity is at its center, find the horizontal components of force exerted on the door by each hinge.

11–16 The boom in Fig. 11–26 weighs 2600 N. The boom is attached with a frictionless pivot at its lower end. It is not uniform; the distance of its center of gravity from the pivot is 35% of its length. a) Find the tension in the guy wire and the horizontal and vertical components of the force exerted on the boom at its lower end. b) Does the line of action of the force exerted on the boom at its lower end lie along the boom?

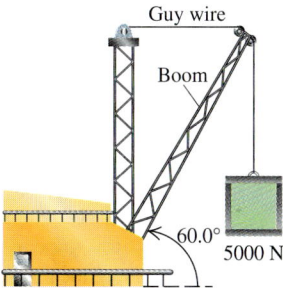

FIGURE 11–26 Exercise 11–16.

11–17 A 3.00-m-long, 240-N uniform rod at the zoo is held in a horizontal position by two ropes at its ends (Fig. 11–27). The left rope makes an angle of 150° with the rod and the right rope

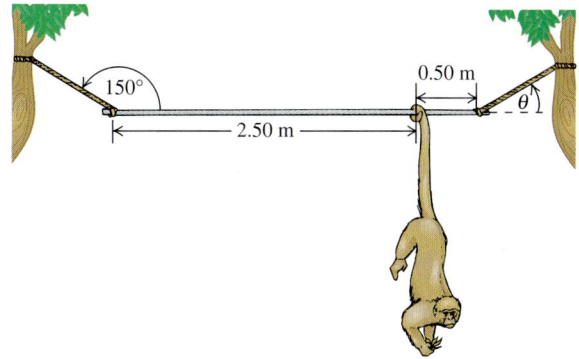

FIGURE 11–27 Exercise 11–17.

makes an angle θ with the horizontal. A 90-N howler monkey (*Alouatta seniculus*) hangs motionless 0.50 m from the right end of the rod as he carefully studies you. Calculate the tensions in the two ropes and the angle θ.

11–18 A non-uniform beam 4.50 m long and weighing 1.00 kN makes an angle of 25.0° with the horizontal. It is held in position by a frictionless pivot at its upper right end and by a cable 3.00 m farther down the beam and perpendicular to it. The center of gravity of the beam is 2.00 m down the beam from the pivot. Lighting equipment exerts a 5.00-kN downward force on the lower left end of the beam. Find the tension T in the cable and the horizontal and vertical components of the force exerted on the beam by the pivot.

11–19 A Couple. Two forces equal in magnitude and opposite in direction acting on an object at two different points form what is called a *couple*. Two antiparallel forces with equal magnitudes $F_1 = F_2 = 8.00$ N are applied to a rod as shown in Fig. 11–28. a) What should be the distance l between the forces if they are to provide a net torque of 6.40 N · m about the left-hand end of the rod? b) Is the sense of this torque clockwise or counterclockwise? c) Repeat parts (a) and (b) for a pivot at the point on the rod where $\vec{F}_2$ is applied.

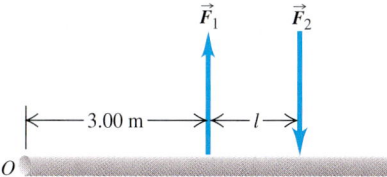

FIGURE 11–28 Exercise 11–19.

SECTION **11–5 STRESS, STRAIN, AND ELASTIC MODULI**

11–20 A relaxed biceps muscle requires a force of 25.0 N for an elongation of 3.0 cm; the same muscle under maximum tension requires a force of 500 N for the same elongation. Find Young's modulus for the muscle tissue under each of these conditions if the muscle is assumed to be a uniform cylinder with length 0.200 m and cross-section area 50.0 cm^2.

11–21 A circular steel wire 2.00 m long is to stretch no more than 0.25 cm when a tensile force of 400 N is applied to each end of the wire. What minimum diameter is required for the wire?

11–22 Two round rods, one steel and the other copper, are joined end to end. Each rod is 0.750 m long and 1.50 cm in diameter. The combination is subjected to a tensile force with magnitude 4000 N. For each rod, what is a) the strain? b) the elongation?

11–23 A metal rod that is 4.00 m long and 0.50 cm^2 in cross-section area is found to stretch 0.20 cm under a tension of 5000 N. What is Young's modulus for this metal?

11–24 Stress on a Mountaineer's Rope. A nylon rope used by mountaineers elongates 1.10 m under the weight of an 65.0-kg climber. If the rope is 45.0 m in length and 7.0 mm in diameter, what is Young's modulus for this material?

11–25 In constructing a large mobile, an artist hangs an aluminum sphere of mass 6.0 kg from a vertical steel wire 0.50 m long and 2.5×10^{-3} cm^2 in cross-section area. On the bottom of the sphere he attaches a similar steel wire, from which he hangs a brass cube of mass 10.0 kg. For each wire, compute a) the tensile strain; b) the elongation.

11–26 A vertical solid steel post 25 cm in diameter and 2.50 m long is required to support a load of 8000 kg. The weight of the post can be neglected. What is a) the stress in the post? b) the strain in the post? c) the change in post's length when the load is applied?

SECTION **11–6** BULK STRESS AND STRAIN

11–27 Outside a house 1.0 km from ground zero of a 100-kiloton nuclear bomb explosion, the pressure will rapidly rise to as high as 2.8 atm while the pressure inside the house remains 1.0 atm. If the area of the front of the house is 50 m^2, what is the resulting net force exerted by the air on the front of the house?

11–28 A solid gold bar is pulled up from the hold of the sunken RMS *Titanic*. a) What happens to its volume as it goes from the pressure at the ship to the lower pressure at the ocean's surface? b) The pressure difference is proportional to the depth. How many times greater would the volume change have been had the ship been twice as deep? c) The bulk modulus of lead is one-fourth that of gold. Find the ratio of the volume change of a solid lead bar to that of an equal volume gold bar for the same pressure change.

11–29 A petite young woman distributes her 500 N weight equally over the heels of her high-heel shoes. Each heel has an area of 0.750 cm^2. a) What is the pressure exerted on the floor by each heel? b) With the same pressure, how much weight could be supported by two flat-bottomed sandals, each of area 200 cm^2?

11–30 In the Challenger Deep of the Marianas Trench the depth of seawater is 10.9 km and the pressure is 1.16×10^8 Pa (about 1.15×10^3 atm). a) If a cubic meter of water is taken from the surface to this depth, what is the change in its volume? (Normal atmospheric pressure is about 1.0×10^5 Pa. Assume that k for seawater is the same as the freshwater value given in Table 11–2.) b) What is the density of seawater at this depth? (At the surface, seawater has a density of 1.03×10^3 kg/m^3.)

11–31 A specimen of oil having an initial volume of 600 cm^3 is subjected to a pressure increase of 3.6×10^6 Pa, and the

volume is found to decrease by 0.45 cm^3. What is the bulk modulus of the material? The compressibility?

SECTION **11–7** SHEAR STRESS AND STRAIN

11–32 Suppose the object in Fig. 11–17 is a square steel plate, 10.0 cm on a side and 0.500 cm thick. a) Find the shear strain that results if a force of magnitude 9.0×10^5 N is applied to each of the four sides as shown in Fig. 11–17a. b) Find the deformation angle ϕ in degrees and the displacement x in centimeters (see Fig. 11–17b).

11–33 Shear forces are applied to a rectangular solid. The same forces are applied to another rectangular solid of the same material, but with three times each edge length. In each case the forces are small enough that Hooke's law is obeyed. What is the ratio of the shear strain for the larger object to that of the smaller object?

11–34 Two strips of brass, each 3.00 cm wide and 45.0 cm long, are placed with their ends overlapping by 1.00 cm. The overlapping ends are then riveted together by four rivets, each with diameter 0.250 cm. Tests show that when a tension of at least 1.20×10^4 N is applied to the ends of the riveted strip, the rivets fail and shear apart. What is the shearing stress on each rivet when it fails? Assume that each rivet carries one quarter of the load.

SECTION **11–8** ELASTICITY AND PLASTICITY

11–35 In a materials testing laboratory, a metal wire made from a new alloy is found to break when a tensile force of 90.8 N is applied perpendicular to each end. If the diameter of the wire is 1.84 mm, what is the breaking stress of the alloy?

11–36 A 4.0-m-long steel wire has a cross-section area of 0.050 cm^2. Its proportional limit has a value of 0.0016 times its Young's modulus (Table 11–1). Its breaking stress has a value of 0.0065 times its Young's modulus. The wire is fastened at its upper end and hangs vertically. a) How great a weight can be hung from the wire without exceeding the proportional limit? b) How much will the wire stretch under this load? c) What is the maximum weight that can be supported?

11–37 A steel cable with cross-section area 3.00 cm^2 has an elastic limit of 2.40×10^8 Pa. Find the maximum upward acceleration that can be given a 1200-kg elevator supported by the cable if the stress is not to exceed one-third of the elastic limit.

11–38 A brass wire is to withstand a tensile force of 350 N without breaking. What minimum diameter must the wire have?

PROBLEMS

11–39 **Center of Gravity of a Car.** For an automobile with weight w the front wheels support a fraction f of the weight, so the normal force exerted at the front wheels is fw, and at the rear wheels the normal force is $(1 - f)w$. The distance between the front and real axles is d. a) Show that the center of gravity of the car is a distance fd in front of the rear wheels. b) Show that the general result derived in part (a) reproduces the numerical answer of Example 11–2 (Section 11–4).

11–40 Sir Lancelot rides slowly out of the castle at Camelot and onto the 12.0-m-long drawbridge that passes over the moat (Fig. 11–29). Unbeknownst to him, his enemies have partially severed the vertical cable holding up the front end of the bridge so that it will break under a tension of 5.80×10^3 N. The bridge has mass 200 kg and its center of gravity is at its center. Lancelot, his lance, his armor, and his horse together have a combined mass of 600 kg. Will the cable break before

Lancelot reaches the end of the drawbridge? If so, how far from the castle end of the bridge will the center of gravity of the horse plus rider be when the cable breaks?

FIGURE 11–29 Problem 11–40.

11–41 Three vertical forces act on an airplane when it is flying at a constant altitude and with a constant velocity. These are the weight of the airplane, an aerodynamic force on the wing of the airplane, and an aerodynamic force on the airplane's horizontal tail. (The aerodynamic forces are exerted by the surrounding air, and are reactions to the forces that the wing and tail exert on the air as the airplane flies through it.) For a particular light airplane with a weight of 6700 N, the center of gravity is 0.30 m in front of the point where the wing's vertical aerodynamic force acts and 3.66 m in front of the point where the tail's vertical aerodynamic force acts. Determine the magnitude and direction (upward or downward) of each of the two vertical aerodynamic forces.

11–42 A pickup truck has a wheelbase of 3.00 m. Ordinarily, 10,780 N rests on the front wheels and 8820 N on the rear wheels when the truck is parked on a level road. a) A box weighing 3600 N is now placed on the tailgate, 1.00 m behind the rear axle. How much total weight now rests on the front wheels? On the rear wheels? b) How much weight would need to be placed on the tailgate to make the front wheels come off the ground?

11–43 A uniform seesaw is 4.00 m long and weighs 240 N. The seesaw is pivoted about a point on its bottom surface and the location of this pivot along the length of the seesaw can be adjusted as far as 0.20 m from the center of the seesaw. Little Karen, who weighs 420 N, has her center of gravity over the right end, with the seesaw set so she exerts the most torque about the pivot. The seesaw balances horizontally when Elwood has his center of gravity 0.15 m from the left end. Both have their feet off the ground. a) What does Elwood weigh? b) Is this equilibrium stable, unstable, or neutral? (*Hint:* See Question Q11–5.)

11–44 A piece of steel will not balance with the tip of one end just barely resting on the edge of a table (point *O* in Fig. 11–30). But suppose we fasten on two other pieces as shown in Fig. 11–30. All pieces have the same square cross section and density. Neglect the weight of the fasteners. The system can now balance if we find the right length *l* of the third piece. a) What is the length *l* for which the system will balance with the first and third pieces horizontal, as shown in the figure?

b) How would your answer change if the pieces were made of copper rather than steel? Explain.

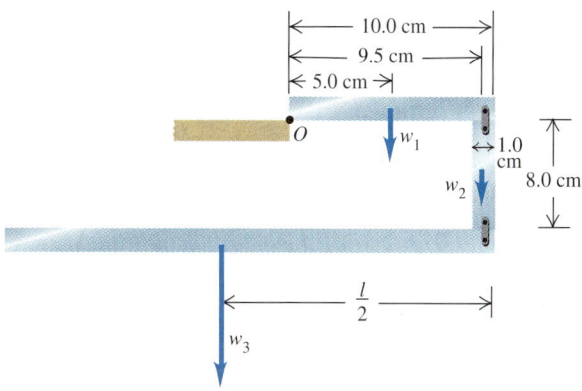

FIGURE 11–30 Problem 11–44.

11–45 Several external forces $\vec{F}_i$ are applied to a rigid body. With respect to an origin *O*, force $\vec{F}_1$ is applied at the point located at $\vec{r}_1$, force $\vec{F}_2$ at $\vec{r}_2$, and so on. Point *P* is located at $\vec{r}_P$. If the body is in translational equilibrium, so that $\Sigma\,\vec{F}_i = \mathbf{0}$, prove that the sum of the torques about point *P* equals the sum of torques about point *O*. (A special case of this result is when $\Sigma\,\vec{\tau}_O = \mathbf{0}$; for a body in translational equilibrium, if the resultant torque about one point is zero, the resultant torque about any point is zero.)

11–46 A claw hammer is used to pull a nail out of a board (Fig. 11–31). The nail is at an angle of 60° to the board, and a force $\vec{F}_1$ of magnitude 500 N applied to the nail is required to pull it from the board. The hammer head contacts the board at point *A*, which is 0.080 m from where the nail enters the board. A horizontal force $\vec{F}_2$ is applied to the hammer handle at a distance of 0.300 m above the board. What magnitude of force $\vec{F}_2$ is required to apply the required 500-N force (F_1) to the nail? (Neglect the weight of the hammer.)

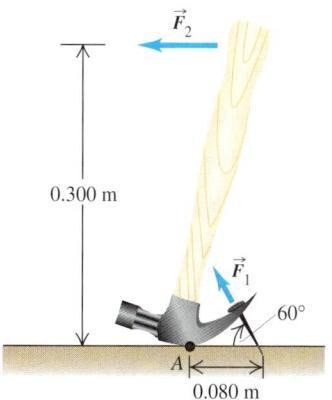

FIGURE 11–31 Problem 11–46.

11–47 End *A* of the bar *AB* in Fig. 11–32 rests on a frictionless horizontal surface, and end *B* is hinged. A horizontal force $\vec{F}$ of magnitude 120 N is exerted on end *A*. Neglect the weight of the

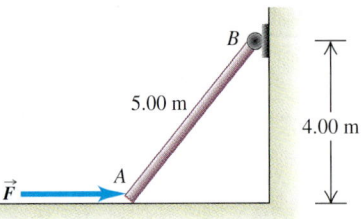

FIGURE 11–32 Problem 11–47.

bar. What are the horizontal and vertical components of the force exerted by the bar on the hinge at B?

11–48 A single additional force is to be applied to the bar in Fig. 11–33 to maintain it in equilibrium in the position shown. The weight of the bar can be neglected. a) What are the horizontal and vertical components of the required force? b) What is the angle the force must make with the bar? c) What is the magnitude of the required force? d) Where should the force be applied?

FIGURE 11–33 Problem 11–48.

11–49 Your dog Nikita is 0.90 m long (nose to hindlegs). Her forelegs are 0.18 m horizontally from her nose, her center of gravity is 0.28 m horizontally from her hindlegs, and she weighs 190 N. a) How much force does a level floor exert on each of Nikita's forefeet, and on each of her hindfeet? b) If Nikita picks up a 25-N bone and holds it in her mouth (directly under her nose), what is the force exerted by the floor on each of her forefeet and each of her hindfeet?

11–50 A Truck on a Drawbridge. A loaded cement mixer drives onto an old drawbridge, where it stalls with its center of gravity three-quarters of the way across the span. The truck driver radios for help, sets the hand-brake, and waits. Meanwhile, a boat approaches, so the drawbridge is raised by means of a cable attached to its end opposite the hinge (Fig. 11–34). The drawbridge span is 40.0 m long and has a mass of 12,000 kg; its center of gravity is at its midpoint. The cement mixer, with driver, has a mass of 30,000 kg. When the drawbridge has been raised to an angle of 30° above the horizontal, the cable makes an angle of 70° with the surface of the bridge. a) What is the tension T in the cable when the drawbridge is held in this position? b) What are the horizontal and vertical components of the force the hinge exerts on the span?

FIGURE 11–34 Problem 11–50.

11–51 Couples. a) A force $\vec{F}_1$ is applied to a rod at a distance x from some point P on the rod. A second force $\vec{F}_2$ is applied at a distance $x + l$ from point P. The forces are equal in magnitude and opposite in direction and hence form a couple (Exercise 11–19). The angle between the rod and the direction of each force is ϕ. Derive an expression for the net torque about P produced by these two forces and show that it is independent of x. This shows that a couple produces the same torque about any pivot point. b) Two forces with magnitude $F_1 = F_2 = 14.0$ N are applied to a rod as shown in Fig. 11–35, with $\phi = 37°$. Calculate the net torque about point O due to these two forces by calculating the torque due to each separate force. Calculate the net torque about point P due to these two forces by calculating the torque due to each separate force. Compare your results to the general result derived in part (a).

FIGURE 11–35 Problem 11–51.

11–52 A non-uniform fire escape ladder is 6.0 m long when extended to the icy alley below. It is held at the top by a frictionless pivot, and there is negligible frictional force from the icy surface at the bottom. The ladder weighs 250 N, and its center of gravity is 2.0 m along the ladder from its bottom. A mother and child of total weight 750 N are on the ladder 1.5 m from the pivot. The ladder makes an angle θ with the horizontal. Find the magnitude and direction of a) the force exerted by the icy alley on the ladder and b) the force exerted by the ladder on the pivot.

11–53 A uniform strut of mass m makes an angle θ with the horizontal. It is supported by a frictionless pivot located at one-third its length from its lower left end and a horizontal rope at its upper right end. A cable and package of total weight w hang from its upper right end. a) Find the vertical and horizontal components V and H of the pivot's force on the strut as well as the tension T in the rope. b) If the maximum safe tension in the rope is 700 N and the mass of the strut is 20.0 kg, find the maximum safe weight of the cable and package when the strut makes an angle of 55.0° with the horizontal. c) For what angle θ can no weight be safely suspended from the right end of the strut?

11–54 a) In Fig. 11–36 a 6.00-m-long uniform beam is hanging from a point 1.00 m to the right of its center. The beam weighs 140 N and makes an angle of 30.0° with the vertical. At the right-hand end of the beam a 100-N weight is hung; an unknown weight w hangs at the other end. If the system is in equilibrium, what is w? Ignore the thickness of the beam. b) If the beam makes instead an angle of 45.0° with the vertical, what is w?

FIGURE 11–36 Problem 11–54.

11–55 A uniform horizontal flagpole 5.00 m in length and with a weight of 200 N is hinged to a vertical wall at one end. A 600-N stuntwoman hangs from its other end. The flagpole is supported by a guy wire running from its outer end to a point on the wall directly above the pole. a) If the tension in this wire is not to exceed 1000 N, what is the minimum height above the pole at which it may be fastened to the wall? b) If the flagpole remains horizontal, by how many newtons would the tension be increased if the wire were fastened 0.50 m below this point?

11–56 A holiday decoration consists of two shiny glass spheres with masses 0.0240 kg and 0.0360 kg suspended as shown in Fig. 11–37 from a uniform rod with mass 0.120 kg and length 1.00 m. The rod is suspended from the ceiling by a vertical cord at each end, so that it is horizontal. Calculate the tension in each of the cords A through F.

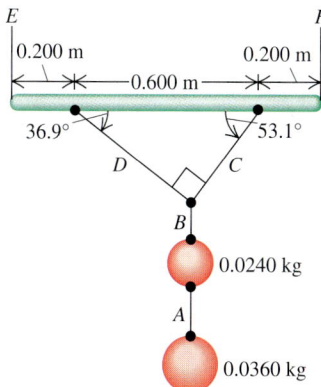

FIGURE 11–37 Problem 11–56.

11–57 The sign for Lew's Bait Shop and Sushi Restaurant is a uniform rectangle, 1.50 m wide and 0.800 m tall, weighing 300 N. It is supported by a wire at its upper-right corner and a pivot at its lower-left corner. a) Late one night Lew's competitor from across the street cuts the wire. The sign stays in place because the pivot has rusted tight. After the wire has been cut, what are the magnitude and direction of the force and of the torque exerted by the pivot on the sign? b) When the sign was brand new, the pivot was frictionless and the wire was at the angle that minimized the tension in it. What was the tension in the wire and the magnitude and direction of the force exerted on the sign by the pivot?

11–58 When you stretch a wire, rope, or rubber band, it gets thinner as well as longer. When Hooke's law holds, the fractional decrease in width is proportional to the tensile strain. If w_0 is the original width and Δw is the change in width, then $\Delta w/w_0 = -\sigma \Delta l/l_0$, where the minus sign reminds us that width decreases when length increases. The dimensionless constant, σ, different for different materials, is called *Poisson's ratio*. a) If the steel cable of Example 11–6 (Section 11–5) has a circular cross section and a Poisson's ratio of 0.23, what is its change in diameter when the milling machine is hung from it? b) A cylinder made of nickel (Poisson's ratio = 0.42) has radius 2.0 cm. What tensile force $F_\perp$ must be applied perpendicular to each end of the cylinder to cause its radius to decrease by 0.10 mm? Assume that the breaking stress and proportional limit for the metal are extremely large and are not exceeded.

11–59 A circular disk 0.500 m in diameter, pivoted about a horizontal axis through its center, has a cord wrapped around its rim. The cord passes over a frictionless pulley P and is attached to an object that weighs 240 N. A uniform rod 2.00 m long is fastened to the disk, with one end at the center of the disk. The apparatus is in equilibrium, with the rod horizontal (Fig. 11–38). a) What is the weight of the rod? b) What is the new equilibrium direction of the rod when a second object weighing 20.0 N is suspended from the other end of the rod, as shown by the broken line? That is, what angle does the rod then make with the horizontal?

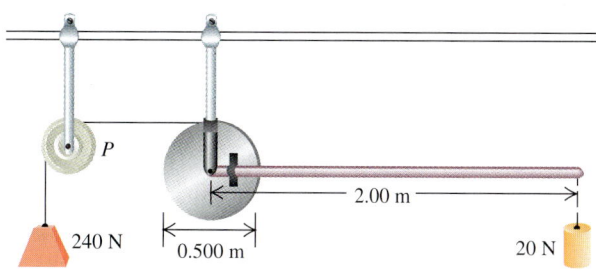

FIGURE 11–38 Problem 11–59.

11–60 One end of a uniform meter stick is placed against a vertical wall (Fig. 11–39). The other end is held by a lightweight cord making an angle θ with the stick. The coefficient of static friction between the end of the meter stick and the wall is 0.40. a) What is the maximum value the angle θ can have if the stick is to remain in equilibrium? b) Let the angle θ be 15°. A block of the same weight as the meter stick is suspended from the stick as

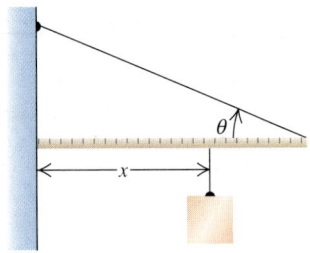

FIGURE 11–39 Problem 11–60.

shown, at a distance x from the wall. What is the minimum value of x for which the stick will remain in equilibrium? c) When $\theta = 15°$, how large must the coefficient of static friction be so that the block can be attached 10 cm from the left end of the stick without causing it to slip?

11–61 Two friends are carrying a 200-kg crate up a flight of stairs. The crate is 1.25 m long and 0.50 m high, and its center of gravity is at its center. The stairs make a 45.0° angle with respect to the floor. The crate also is carried at a 45.0° angle, so that its bottom side is parallel to the slope of the stairs (Fig. 11–40). If the force each person applies is vertical, what is the magnitude of each of these forces? Is it best to be the person above or below on the stairs?

FIGURE 11–40 Problem 11–61.

11–62 You are trying to raise a bicycle wheel of mass m and radius R up over a curb of height h. To do this, you apply a horizontal force $\vec{F}$. What is the least magnitude of the force $\vec{F}$ that

FIGURE 11–41 Problem 11–62.

will succeed in raising the wheel onto the curb when the force is applied a) at the center of the wheel (Fig. 11–41)? b) at the top of the wheel? c) In which case is less force required?

11–63 The Farmyard Gate. A gate 4.00 m wide and 2.00 m high weighs 500 N. Its center of gravity is at its center, and it is hinged at A and B. To relieve the strain on the top hinge, a wire CD is connected as shown in Fig. 11–42. The tension in CD is increased until the horizontal force at hinge A is zero. a) What is the tension in the wire CD? b) What is the magnitude of the horizontal component of the force at hinge B? c) What is the combined vertical force exerted by hinges A and B?

FIGURE 11–42 Problem 11–63.

11–64 If you put a uniform block at the edge of a table, the center of the block must be over the table for the block not to fall off. a) If you stack two identical blocks at the table edge, the center of the top block must be over the bottom block, and the center of gravity of the two blocks together must be over the table. In terms of the length L of each block, what is the maximum overhang possible (Fig. 11–43)? b) Repeat part (a) for three identical blocks and for four identical blocks. c) Is it possible to make a stack of blocks such that the uppermost block is not directly over the table at all? How many blocks would it take to do this? (Try this with your friends using copies of this book.)

FIGURE 11–43 Problem 11–64.

11–65 An engineer is designing a conveyor system for loading hay bales into a wagon (Fig. 11–44). Each bale is 0.25 m wide, 0.50 m high, and 0.80 m long (the dimension perpendicular to the plane of the figure), and has a mass of 30.0 kg. The center

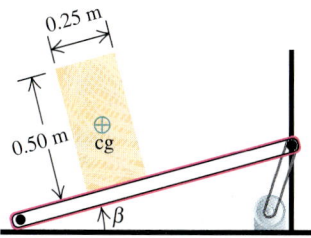

FIGURE 11–44 Problem 11–65.

of gravity of each bale is at its geometrical center. The coefficient of static friction between a bale and the conveyor belt is 0.60, and the belt moves with constant speed. a) The angle β of the conveyor is slowly increased. At some critical angle a bale will tip (if it doesn't slip first), and at some different critical angle it will slip (if it doesn't tip first). Find the two critical angles, and determine which happens at the smaller angle. b) Would the outcome of part (a) be different if the coefficient of friction were 0.40?

11–66 The hay bale of Problem 11–65 is dragged along a horizontal surface with constant speed by a force $\vec{F}$ (Fig. 11–45). The coefficient of kinetic friction is 0.35. a) Find the magnitude of the force $\vec{F}$. b) Find the value of h for which the bale just begins to tip.

FIGURE 11–45 Problem 11–66.

11–67 A garage door is mounted on an overhead rail (Fig. 11–46). The wheels at A and B have rusted so that they do not roll, but rather slide along the track. The coefficient of kinetic friction is 0.52. The distance between the wheels is 2.00 m, and each is 0.50 m from the vertical sides of the door. The door is uniform and weighs 950 N. It is pushed to the left at constant speed by a horizontal force $\vec{F}$. a) If the distance h is 1.60 m, what is the vertical component of the force exerted on each wheel by the track? b) Find the maximum value h can have without causing one wheel to leave the track.

FIGURE 11–46 Problem 11–67.

11–68 A horizontal boom is supported at its left end by a frictionless pivot. It is held in place by a cable attached to the right-hand end of the boom. A chain and crate of total weight w hang from somewhere along the boom. The boom's weight w_b cannot be neglected and the boom may or may not be uniform. a) Show that the tension in the cable is the same whether the cable makes an angle θ or an angle $180° - \theta$ with the horizontal, and that the horizontal force component exerted on the boom by the pivot has equal magnitude but opposite direction for the two angles. b) Show that the cable cannot be horizontal. c) Show that the tension in the cable is a minimum when the cable is vertical, pulling upward on the right end of the boom. d) Show that when the cable is vertical, the force exerted by the pivot on the boom is vertical.

11–69 Prior to being placed in its hole, a 5700-N, 9.0-m-long uniform utility pole makes some nonzero angle with the vertical. A vertical cable attached 2.0 m below its upper end holds it in place while its lower end rests on the ground. a) Find the tension in the cable and the magnitude and direction of the force exerted by the ground on the pole. b) Why don't we need the angle the pole makes with the vertical, as long as it is not zero?

11–70 Hooke's law for a tensile stress can be written as $F = kx$, where x is the object's change in length and k is the force constant. a) What is the force constant of a rod of length l_0, cross-section area A, and Young's modulus Y? b) In terms of l_0, A, and Y, how much work is required to stretch the object a distance x?

11–71 A 12.0-kg mass, fastened to the end of an aluminum wire with an unstretched length of 0.50 m, is whirled in a vertical circle with a constant angular speed of 120 rev/min. The cross-section area of the wire is 0.014 cm^2. Calculate the elongation of the wire when the mass is a) at the lowest point of the path; b) at the highest point of its path.

11–72 A metal wire 3.50 m long and 0.70 mm in diameter was given the following test. A load weighing 20 N was originally hung from the wire to keep it taut. The position of the lower end of the wire was read on a scale as load was added.

Added load (N)	Scale reading (cm)
0	3.02
10	3.07
20	3.12
30	3.17
40	3.22
50	3.27
60	3.32
70	4.27

a) Graph these values, plotting the increase in length horizontally and the added load vertically. b) Calculate the value of Young's modulus. c) The proportional limit occurred at a scale reading of 3.34 cm. What was the stress at this point?

11–73 A 1.05-m-long rod of negligible weight is supported at its ends by wires A and B of equal length (Fig. 11–47). The cross-section area of A is 2.00 mm^2 and that of B is 4.00 mm^2. Young's modulus for wire A is 1.80×10^{11} Pa; that for B is 1.20×10^{11} Pa. At what point along the rod should a weight w

be suspended to produce a) equal stresses in A and B? b) equal strains in A and B?

FIGURE 11–47 Problem 11–73.

11–74 An amusement park ride consists of airplane-shaped cars attached to steel rods (Fig. 11–48). Each rod has a length of 15.0 m and a cross-section area of 8.00 cm^2. a) How much is the rod stretched when the ride is at rest? (Assume that each car plus two people seated in it has a total weight of 1900 N.) b) When operating, the ride has a maximum angular speed of 8.0 rev/min. How much is the rod stretched then?

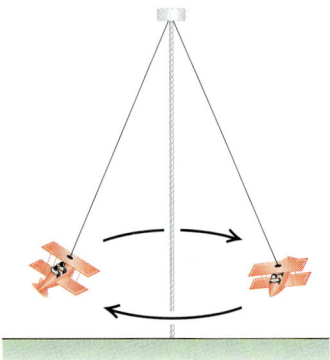

FIGURE 11–48 Problem 11–74.

11–75 A brass rod with a length of 1.40 m and a cross-section area of 2.00 cm^2 is fastened end to end to a nickel rod with length L and cross-section area 1.00 cm^2. The compound rod is subjected to equal and opposite pulls of magnitude 4.00×10^4 N at its ends. a) Find the length L of the nickel rod if the elongations of the two rods are equal. b) What is the stress in each rod? c) What is the strain in each rod?

11–76 Stress on the Shin Bone. Compressive strength of our bones is important in everyday life. Young's modulus for bone is about 1.4×10^{10} Pa. Bone can take only about a 1.0% change in its length before fracturing. a) What is the maximum force that can be applied to a bone whose minimum cross-section area is 3.0 cm^2? (This is approximately the cross-section area of a tibia, or shin bone, at its narrowest point.) b) Estimate the maximum

height from which a 70-kg man could jump and not fracture the tibia. Take the time between when he first touches the floor and when he has stopped to be 0.030 s, and assume that the stress is distributed equally between his two legs.

11–77 You hang a floodlamp from the end of a vertical steel wire. The floodlamp stretches the wire 0.18 mm and the stress is proportional to the strain. How much would it have stretched a) if the wire were twice as long? b) if the wire had the same length but twice the diameter? c) for a copper wire of the original length and diameter?

11–78 A moonshiner produces pure ethanol (ethyl alcohol) late at night and stores it in a stainless steel tank in the form of a cylinder 0.300 m in diameter with a tight-fitting piston at the top. The total volume of the tank is 250 L (0.250 m^3). In an attempt to squeeze a little more into the tank, the moonshiner piles 1420 kg of lead bricks on top of the piston. What additional volume of ethanol can the moonshiner squeeze into the tank? (Assume that the wall of the tank is perfectly rigid.)

11–79 A bar with cross-section area A is subjected to equal and opposite tensile forces $\vec{F}$ at its ends. Consider a plane through the bar making an angle θ with a plane at right angles to the bar (Fig. 11–49). a) What is the tensile (normal) stress at this plane in terms of F, A, and θ? b) What is the shear (tangential) stress at the plane in terms of F, A, and θ? c) For what value of θ is the tensile stress a maximum? d) For what value of θ is the shear stress a maximum?

FIGURE 11–49 Problem 11–79.

11–80 A horizontal uniform copper rod has an original length l_0, a cross-section area A, a Young's modulus Y, and a mass m. It is supported by a frictionless pivot at its right end and by a cable at its left end. Both pivot and cable are attached so that they exert their forces uniformly over the rod's cross section. The cable makes an angle θ with the rod and compresses it. a) Find the stress exerted by the cable and pivot on the rod. b) Find the change in length of the rod due to this stress. c) The mass of the rod equals $\rho A l_0$, where ρ is the density. Show that the answers to parts (a) and (b) are independent of the cross-section area of the rod. d) The density of copper is 8900 kg/m^3. Take Y for compression as given for copper in Table 11–1. Find the stress and change in length for an original length of 1.8 m and an angle of 30°. e) By how much would you multiply the answers of part (d) if the rod were twice as long?

CHALLENGE PROBLEMS

11–81 A bookcase weighing 1500 N rests on a horizontal surface for which the coefficient of static friction is $\mu_s = 0.40$. The bookcase is 1.80 m tall and 2.00 m wide; its center of gravity is

at its geometrical center. The bookcase rests on four short legs that are each 0.10 m from the edge of the bookcase. A person pulls on a rope attached to an upper corner of the bookcase with

FIGURE 11–50 Challenge Problem 11–81.

a force $\vec{F}$ that makes an angle θ with the bookcase (Fig. 11–50). a) If $\theta = 90°$, so $\vec{F}$ is horizontal, show that as F is increased from zero, the bookcase will start to slide before it tips, and calculate the magnitude of $\vec{F}$ that will start the bookcase sliding. b) If $\theta = 0°$, so $\vec{F}$ is vertical, show that the bookcase will tip over rather than slide, and calculate the magnitude of $\vec{F}$ that will cause the bookcase to start to tip. c) Calculate as a function of θ the magnitude of $\vec{F}$ that will cause the bookcase to start to slide and the magnitude that will cause it to start to tip. What is the smallest value that θ can have so that the bookcase will still start to slide before it starts to tip?

11–82 Knocking Over a Post. One end of a post weighing 400 N and with height h rests on a rough horizontal surface with $\mu_s = 0.30$. The upper end is held by a rope fastened to the surface and making an angle of 36.9° with the post (Fig. 11–51). A horizontal force $\vec{F}$ is exerted on the post as shown. a) If the force $\vec{F}$ is applied at the midpoint of the post, what is the largest value it can have without causing the post to slip? b) How large can the force be without causing the post to slip if its point of application is six-tenths of the way from the ground to the top of the post? c) Show that if the point of application of the force is too high, the post cannot be made to slip, no matter how great the force. Find the critical height for the point of application.

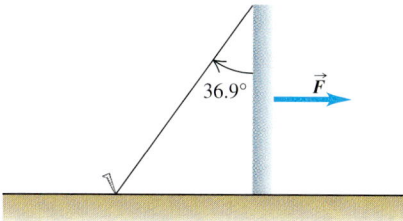

FIGURE 11–51 Challenge Problem 11–82.

11–83 Minimizing the Tension. A heavy horizontal girder of length L has several objects suspended from it. It is supported by a frictionless pivot at its left end and a cable of negligible weight that is attached to an I-beam at a point a distance h directly above the girder's center. Where should the other end of the cable be attached to the girder so that the cable's tension is a minimum? (*Hint:* In evaluating and presenting your answer, don't forget that the maximum distance of the point of attachment from the pivot is the length L of the beam.)

11–84 Two ladders, 4.00 m and 3.00 m long, are hinged at point A and tied together by a horizontal rope 0.90 m above the floor (Fig. 11–52). The ladders weigh 480 N and 360 N, respectively, and the center of gravity of each is at its center. Assume that the floor is freshly waxed and frictionless. a) Find the upward force at the bottom of each ladder. b) Find the tension in the rope. c) Find the magnitude of the force one ladder exerts on the other at point A. d) If an 800-N painter stands at point A, find the tension in the horizontal rope.

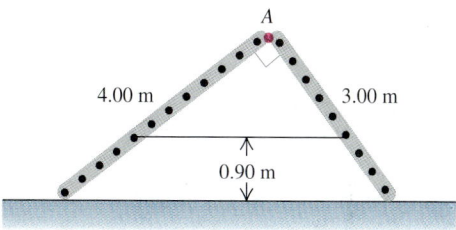

FIGURE 11–52 Challenge Problem 11–84.

11–85 The compressibility of sodium is to be measured by observing the displacement of the piston in Fig. 11–14 when a force is applied. The sodium is immersed in an oil that fills the cylinder below the piston. Assume that the piston and walls of the cylinder are perfectly rigid and that there is no friction and no oil leak. Compute the compressibility of the sodium in terms of the applied force F, the piston displacement x, the piston area A, the initial volume of the oil V_O, the initial volume of the sodium V_S, and the compressibility of the oil k_O.

11–86 Bulk Modulus of an Ideal Gas. The equation of state (the equation relating pressure, volume, and temperature) for an ideal gas is $pV = nRT$, where n and R are constants. a) Show that if the gas is compressed while the temperature T is held constant, the bulk modulus is equal to the pressure. b) When an ideal gas is compressed without the transfer of any heat into or out of it, the pressure and volume are related by $pV^\gamma = $ constant, where γ is a constant having different values for different gases. Show that in this case the bulk modulus is given by $B = \gamma p$.

11–87 An angler hangs a 4.50-kg fish from a vertical steel wire 1.50 m long and 5.00×10^{-3} cm^2 in cross-section area. The upper end of the wire is securely fastened to a support. a) Calculate the amount the wire is stretched by the hanging fish. The angler now applies a force $\vec{F}$ to the fish, pulling it very slowly downward by 0.500 mm from its equilibrium position. For this downward motion, calculate b) the work done by gravity; c) the work done by the force $\vec{F}$; d) the work done by the force the wire exerts on the fish; and e) the change in the elastic potential energy (the potential energy associated with the tensile stress in the wire). Compare the answers in parts (d) and (e). (*Hint:* See Problem 11–70.)

12

Gravitation

The great galaxy M31 in the constellation Andromeda is more than 10^{18} km (100,000 light years) across. It lies some 2.9×10^6 light years from earth. All the matter in this galaxy—stars, glowing gas, interstellar dust, and other material not visible in this image—is kept in orbit around the galaxy's center by its mutual gravitational attraction.

12-1 INTRODUCTION

Why are planets, moons, and the sun all nearly spherical? Why do some earth satellites circle the earth in 90 minutes, while the moon takes 27 days for the trip? And why don't satellites fall back to earth? The study of gravitation provides the answers for these and many related questions.

As we remarked in Chapter 5, gravitation is one of the four classes of interactions found in nature, and it was the earliest of the four to be studied extensively. Newton discovered in the seventeenth century that the same interaction that makes an apple fall out of a tree also keeps the planets in their orbits around the sun. This was the beginning of *celestial mechanics,* the study of the dynamics of objects in space. Today, our knowledge of celestial mechanics allows us to determine how to put a satellite into any desired orbit around the earth or to choose just the right trajectory to send a spacecraft to another planet.

In this chapter you will learn the basic law that governs gravitational interactions. This law is *universal:* gravity acts in the same fundamental way between the earth and your body, between the sun and a planet, and between a planet and one of its moons. We'll apply the law of gravitation to phenomena such as the variation of weight with altitude, the orbits of satellites around the earth, and the orbits of planets around the sun.

12-2 NEWTON'S LAW OF GRAVITATION

The example of gravitational attraction that's probably most familiar to you is your *weight,* the force that attracts you toward the earth. During his study of the motions of the planets and of the moon, Newton discovered the fundamental character of the gravitational attraction between *any* two bodies. Along with his three laws of motion, Newton published the **law of gravitation** in 1687. It may be stated as follows:

> **Every particle of matter in the universe attracts every other particle with a force that is directly proportional to the product of the masses of the particles and inversely proportional to the square of the distance between them.**

Translating this into an equation, we have

$$F_g = \frac{Gm_1m_2}{r^2} \qquad \text{(law of gravitation)}, \qquad (12-1)$$

where F_g is the magnitude of the gravitational force on either particle, m_1 and m_2 are their masses, r is the distance between them (Fig. 12–1), and G is a fundamental physical constant called the **gravitational constant.** The numerical value of G depends on the system of units used.

CAUTION ▶ Because the symbols g and G are almost the same, it's common to confuse the two gravitational quantities that these symbols represent. Lowercase g is the acceleration due to gravity, which relates the weight w of a body to its mass m: $w = mg$. The value of g is different at different locations on the earth's surface and on the surfaces of different planets. By contrast, capital G relates the gravitational force between any two bodies to their masses and the distance between them. We call G a *universal* constant because it has the same value for any two bodies, no matter where in space they are located. In the next section we'll see how the values of g and G are related, but remember that they're two very different quantities! ◄

Gravitational forces always act along the line joining the two particles, and they form an action-reaction pair. Even when the masses of the particles are different, the two interaction forces have equal magnitude. The attractive force that your body exerts on the earth has the same magnitude as the force that it exerts on you. When you fall from a diving board into a swimming pool, the entire earth rises up to meet you! (Why don't you notice this? The earth's mass is greater than yours by a factor of about 10^{23}, so its acceleration is only 10^{-23} as great as yours.)

We have stated the law of gravitation in terms of the interaction between two *particles*. It turns out that the gravitational interaction of any two bodies having *spherically symmetric* mass distributions (such as solid spheres or spherical shells) is the same as though we concentrated all the mass of each at its center, as in Fig. 12–2. Thus if we model the earth as a spherically symmetric body with mass m_E, the force it exerts on a particle or a spherically symmetric body with mass m, at a distance r between centers, is

$$F_g = \frac{Gm_E m}{r^2}, \qquad (12-2)$$

provided that the body lies outside the earth. A force of the same magnitude is exerted *on* the earth by the body. (We will prove these statements in Section 12–7.)

At points *inside* the earth the situation is different. If we could drill a hole to the center of the earth and measure the gravitational force on a body at various depths, we would find that toward the center of the earth the force *decreases*, rather than increasing as $1/r^2$. As the body enters the interior of the earth (or other spherical body), some of the earth's mass is on the side of the body opposite from the center and pulls in the opposite direction. Exactly at the center, the earth's gravitational force on the body is zero.

Spherically symmetric bodies are an important case because moons, planets, and stars all tend to be spherical (Fig. 12–3a). Since all particles in a body gravitationally attract each other, the particles tend to move to minimize the distance between them. As a result, the body naturally tends to assume a spherical shape, just as a lump of clay forms into a sphere if you squeeze it with equal forces on all sides. This effect is greatly reduced in celestial bodies of low mass since the gravitational attraction is less, and these bodies tend *not* to be spherical (Fig. 12–3b).

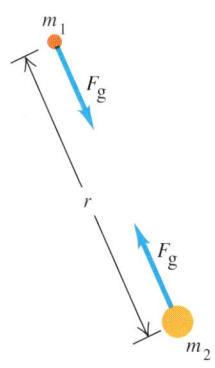

12–1 Two particles, separated by a distance r, exert attractive gravitational forces on each other. The forces on each particle are of equal magnitude even if their masses are quite different.

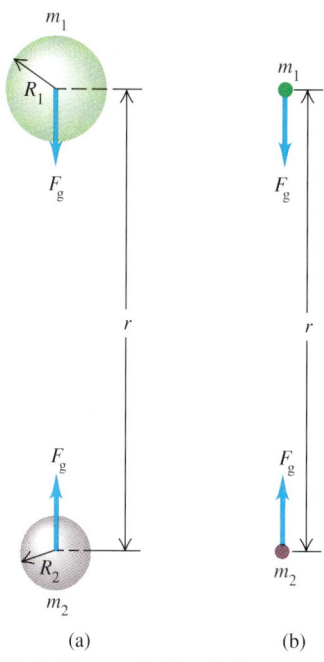

(a)   (b)

12–2 The gravitational effect *outside* any spherically symmetric mass distribution is the same as though all the mass of the sphere were concentrated at its center.

12–3 (left) The sun's mass of 1.99×10^{30} kg is great enough that gravitation pulls it into a nearly spherical shape of radius 696,000 km. (right) The asteroid Gaspra is an irregularly shaped body about 20 km long. The mass of Gaspra is only about 10^{16} kg, and the gravitational attraction between its parts isn't great enough to overcome the interatomic forces that maintain Gaspra's solid shape.

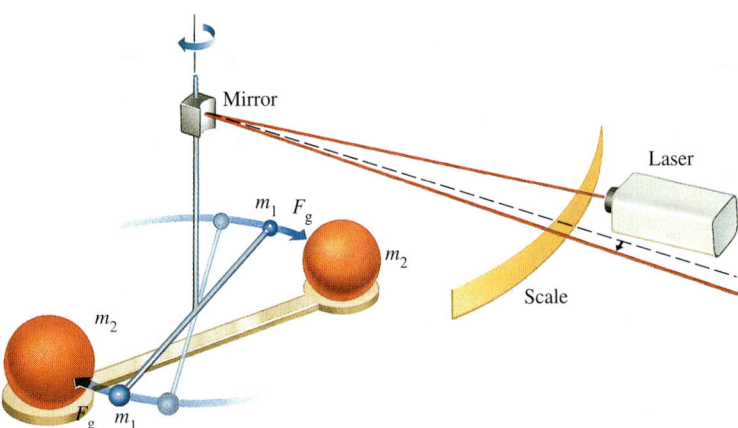

12–4 The principle of the Cavendish balance, used for determining the value of G. The angle of deflection has been exaggerated for clarity.

DETERMINING THE VALUE OF G

To determine the value of the gravitational constant G, we have to *measure* the gravitational force between two bodies of known masses m_1 and m_2 at a known distance r. The force is extrememly small for bodies that are small enough to be brought into the laboratory, but it can be measured with an instrument called a *torsion balance*, which Sir Henry Cavendish used in 1798 to determine G.

A modern version of the Cavendish torsion balance is shown in Fig. 12–4. A light, rigid rod shaped like an inverted T is supported by a very thin vertical quartz fiber. Two small spheres, each of mass m_1, are mounted at the ends of the horizontal arms of the T. When we bring two large spheres, each of mass m_2, to the positions shown, the attractive gravitational forces twist the T through a small angle. To measure this angle, we shine a beam of light on a mirror fastened to the T. The reflected beam strikes a scale, and as the T twists, the reflected beam moves along the scale.

After calibrating the Cavendish balance, we can measure gravitational forces and thus determine G. The presently accepted value (in SI units) is

$$G = 6.67259(85) \times 10^{-11} \text{ N} \cdot \text{m}^2/\text{kg}^2.$$

To three significant figures, $G = 6.67 \times 10^{-11}$ N $\cdot$ m^2/kg^2. Because 1 N = 1 kg $\cdot$ m/s^2, the units of G can also be expressed (in fundamental SI units) as m^3/(kg $\cdot$ s^2).

Gravitational forces combine vectorially. If each of two masses exerts a force on a third, the *total* force on the third mass is the vector sum of the individual forces of the first two. Example 12–3 makes use of this property, which is often called *superposition of forces.*

EXAMPLE 12-1

Calculating gravitational force The mass m_1 of one of the small spheres of a Cavendish balance is 0.0100 kg, the mass m_2 of one of the large spheres is 0.500 kg, and the center-to-center distance between each large sphere and the nearer small one is 0.0500 m. Find the gravitational force F_g on each sphere due to the nearest other sphere.

SOLUTION The magnitude of each force is

$$F_g = \frac{(6.67 \times 10^{-11} \text{ N} \cdot \text{m}^2/\text{kg}^2)(0.0100 \text{ kg})(0.500 \text{ kg})}{(0.0500 \text{ m})^2}$$

$$= 1.33 \times 10^{-10} \text{ N}.$$

This is a very small force. *Reminder:* The two bodies experience the *same* magnitude of force, even though their masses are very different.

EXAMPLE 12-2

Acceleration due to gravitational attraction Suppose one large sphere and one small sphere are detached from the apparatus in Example 12–1 and placed 0.0500 m (between centers) from each other at a point in space far removed from all other bodies. What is the magnitude of the acceleration of each, relative to an inertial system?

SOLUTION The force on each sphere has the same magnitude that we found in Example 12–1. The acceleration a_1 of the smaller sphere has magnitude

$$a_1 = \frac{F_g}{m_1} = \frac{1.33 \times 10^{-10}\ \text{N}}{0.0100\ \text{kg}} = 1.33 \times 10^{-8}\ \text{m/s}^2.$$

The acceleration a_2 of the larger sphere has magnitude

$$a_2 = \frac{F_g}{m_2} = \frac{1.33 \times 10^{-10}\ \text{N}}{0.500\ \text{kg}} = 2.66 \times 10^{-10}\ \text{m/s}^2.$$

Although the forces on the bodies are equal in magnitude, the magnitudes of the two accelerations are *not* equal. Also, the accelerations are not constant; the gravitational forces increase as the spheres start to move toward each other.

EXAMPLE 12-3

Superposition of gravitational forces Three spheres are arranged at the vertices of a 45° right triangle (Fig. 12–5). Find the magnitude and direction of the total gravitational force exerted on the small sphere by both large ones.

SOLUTION We use the principle of superposition: the total force on the small sphere is the vector sum of the two forces due to each large sphere. We can compute the vector sum by using components, but first we find the magnitudes of the forces. The magnitude F_1 of the force on the small mass due to the upper large one is

$$F_1 = \frac{(6.67 \times 10^{-11}\ \text{N} \cdot \text{m}^2/\text{kg}^2)(0.500\ \text{kg})(0.0100\ \text{kg})}{(0.200\ \text{m})^2 + (0.200\ \text{m})^2}$$

$$= 4.17 \times 10^{-12}\ \text{N}.$$

The magnitude F_2 of the force due to the lower large mass is

$$F_2 = \frac{(6.67 \times 10^{-11}\ \text{N} \cdot \text{m}^2/\text{kg}^2)(0.500\ \text{kg})(0.0100\ \text{kg})}{(0.200\ \text{m})^2}$$

$$= 8.34 \times 10^{-12}\ \text{N}.$$

The x- and y-components of these forces are

$$F_{1x} = (4.17 \times 10^{-12}\ \text{N})(\cos 45°) = 2.95 \times 10^{-12}\ \text{N},$$
$$F_{1y} = (4.17 \times 10^{-12}\ \text{N})(\sin 45°) = 2.95 \times 10^{-12}\ \text{N},$$
$$F_{2x} = 8.34 \times 10^{-12}\ \text{N},$$
$$F_{2y} = 0.$$

The components of the total force on the small mass are

$$F_x = F_{1x} + F_{2x} = 11.3 \times 10^{-12}\ \text{N},$$
$$F_y = F_{1y} + F_{2y} = 2.95 \times 10^{-12}\ \text{N}.$$

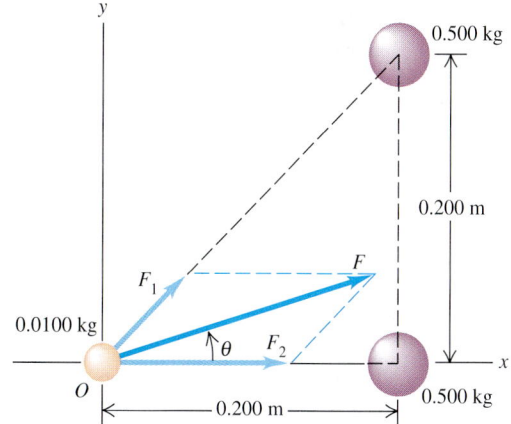

12–5 The total gravitational force on the 0.0100-kg sphere is the vector sum of the forces exerted on it by the two 0.500-kg spheres.

The magnitude of this force is

$$F = \sqrt{F_x^2 + F_y^2} = \sqrt{(11.3 \times 10^{-12}\ \text{N})^2 + (2.95 \times 10^{-12}\ \text{N})^2}$$
$$= 1.17 \times 10^{-11}\ \text{N},$$

and its direction relative to the x-axis is

$$\theta = \arctan \frac{F_y}{F_x} = \arctan \frac{2.95 \times 10^{-12}\ \text{N}}{11.3 \times 10^{-12}\ \text{N}} = 14.6°.$$

Can you show that this force is *not* directed toward the center of mass of the two large masses? (See Problem 12–42.)

Perhaps the most remarkable aspect of the gravitational force of one body on another is that it acts *at a distance,* without direct contact between the two bodies. Electric and magnetic forces have this same remarkable property, as do the strong and weak interactions that we discussed in Section 5–6. A useful way to describe forces that act at a distance is in terms of a *field*. One body sets up a disturbance or field at all points in space, and the force that acts on a second body at a particular point is its response to the first body's field at that point. There is a field associated with each force that acts at a distance, and so we refer to gravitational fields, electric fields, magnetic fields, and so on. We won't need the field concept for our study of gravitation in this chapter, so we won't discuss it further here. But in later chapters we'll find that the field description is an extraordinarily powerful tool for describing electric and magnetic interactions.

12–3 WEIGHT

We defined the *weight* of a body in Section 4–5 as the attractive gravitational force exerted on it by the earth. We can now broaden our definition. **The weight of a body is the total gravitational force exerted on the body by all other bodies in the universe.** When the body is near the surface of the earth, we can neglect all other gravitational forces and consider the weight as just the earth's gravitational attraction. At the surface of the *moon* we consider a body's weight to be the gravitational attraction of the moon, and so on.

If we again model the earth as a spherically symmetric body with radius R_E and mass m_E, the weight w of a small body of mass m at the earth's surface (distance R_E from its center) is

$$w = F_g = \frac{Gm_E m}{R_E^2} \qquad \text{(weight of a body of mass } m \text{ at the earth's surface).} \qquad (12\text{–}3)$$

But we also know from Section 4–5 that the weight w of a body is the force that causes the acceleration g of free fall, so by Newton's second law, $w = mg$. Equating this with Eq. (12–3) and dividing by m, we find

$$g = \frac{Gm_E}{R_E^2} \qquad \text{(acceleration due to gravity at the earth's surface).} \qquad (12\text{–}4)$$

The acceleration due to gravity g is independent of the mass m of the body because m doesn't appear in this equation. We already knew that, but we can now see how it follows from the law of gravitation.

We can *measure* all the quantities in Eq. (12–4) except for m_E, so this relation allows us to compute the mass of the earth. Solving Eq. (12–4) for m_E and using $R = 6380$ km $= 6.38 \times 10^6$ m and $g = 9.80$ m/s^2, we find

$$m_E = \frac{gR_E^2}{G} = 5.98 \times 10^{24} \text{ kg},$$

In an airliner at high altitude, you are farther from the center of the earth than when on the ground and you actually weigh less. The effect is quite small, though measurable. Can you show that at an altitude of 10 km above the surface, you weigh just 0.3% less than you do on the ground?

very close to the currently accepted value of 5.974×10^{24} kg. Once Cavendish had measured G, he computed the mass of the earth in just this way.

At a point above the earth's surface a distance r from the center of the earth (a distance $r - R_E$ above the surface), the weight of a body is given by Eq. (12–3) with R_E replaced by r:

$$w = F_g = \frac{Gm_E m}{r^2}. \qquad (12\text{–}5)$$

The weight of a body decreases inversely with the square of its distance from the earth's center. Figure 12–6 shows how the weight varies with height above the earth for an astronaut who weighs 700 N at the earth's surface.

The *apparent* weight of a body on earth differs slightly from the earth's gravitational force because the earth rotates and is therefore not precisely an inertial frame of refer-

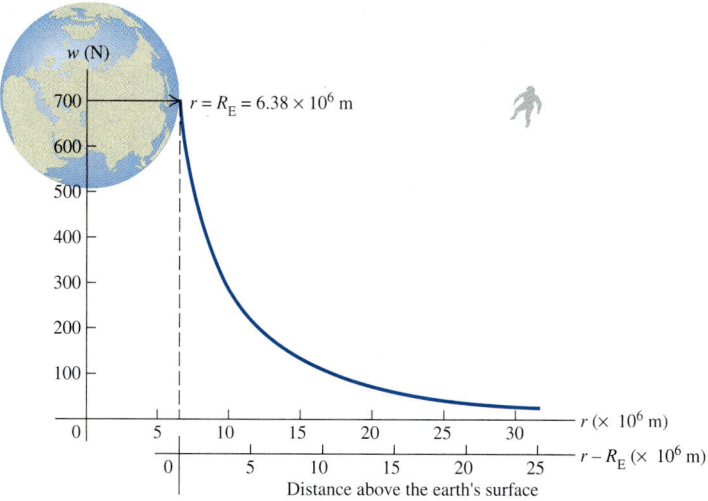

12-6 An astronaut weighing 700 N at the earth's surface experiences less gravitational attraction when above that surface (10^6 m = 1000 km = 620 mi). The astronaut's distance from the *center* of the earth is r; her distance from the *surface* of the earth is $r - R_E = r - 6.38 \times 10^6$ m.

ence. We have ignored this effect in the above discussion and have assumed that the earth *is* an inertial system. We will return to the effect of the earth's rotation in Section 12–8.

In our discussion of weight, we've used the fact that the earth is an approximately spherically symmetric distribution of mass. But this does *not* mean that the earth is uniform. To demonstrate that it cannot be uniform, let's first calculate the average *density*, or mass per unit volume, of the earth. If we assume a spherical earth, the volume is

$$V_E = \frac{4}{3}\pi R_E{}^3 = \frac{4}{3}\pi (6.38 \times 10^6 \text{ m})^3 = 1.09 \times 10^{21} \text{ m}^3.$$

The average density ρ (the Greek letter "rho") of the earth is the total mass divided by the total volume:

$$\rho = \frac{m_E}{V_E} = \frac{5.97 \times 10^{24} \text{ kg}}{1.09 \times 10^{21} \text{ m}^3}$$

$$= 5500 \text{ kg/m}^3 = 5.5 \text{ g/cm}^3.$$

(For comparison, the density of water is $1000 \text{ kg/m}^3 = 1.00 \text{ g/cm}^3$.) If the earth were uniform, we would expect the density of individual rocks near the earth's surface to have this same value. In fact, the density of igneous surface rocks such as granite or gneiss is about $3000 \text{ kg/m}^3 = 3 \text{ g/cm}^3$, although some basaltic rock has a density of about $5000 \text{ kg/m}^3 = 5 \text{ g/cm}^3$. So the earth *cannot* be uniform, and the interior of the earth must be much more dense than the surface in order that the *average* density be $5500 \text{ kg/m}^3 = 5.5 \text{ g/cm}^3$. According to geophysical models of the earth's interior, the maximum density at the center is about $13{,}000 \text{ kg/m}^3 = 13 \text{ g/cm}^3$. Figure 12–7 is a graph of density as a function of distance from the center.

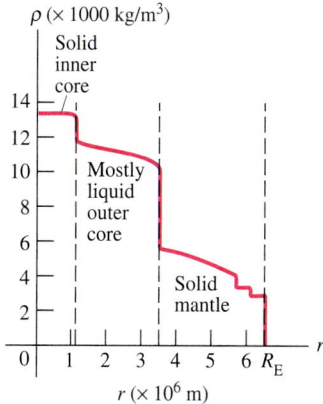

12-7 The density of the earth decreases with increasing distance from its center.

EXAMPLE 12-4

Gravity on Mars You're involved in the design of a mission carrying humans to the surface of the planet Mars, which has a radius $r_M = 3.40 \times 10^6$ m and a mass $m_M = 6.42 \times 10^{23}$ kg. The earth weight of the Mars lander is 39,200 N. Calculate its weight F_g and the acceleration g_M due to the gravity of Mars: a) 6.0×10^6 m above the surface of Mars (the distance at which the moon Phobos orbits Mars); b) at the surface of Mars.

Neglect the gravitational effects of the (very small) moons of Mars.

SOLUTION a) In Eq. (12–5) we replace m_E with m_M. The value of G is the same everywhere in the universe; it is a fundamental physical constant. The distance r from the *center* of Mars is

$$r = (6.0 \times 10^6 \text{ m}) + (3.40 \times 10^6 \text{ m}) = 9.4 \times 10^6 \text{ m}.$$

The mass m of the lander is its earth weight w divided by the acceleration of gravity g on earth:

$$m = \frac{w}{g} = \frac{39,200 \text{ N}}{9.8 \text{ m/s}^2} = 4000 \text{ kg}.$$

The mass is the same whether the lander is on the earth, on Mars, or in between. From Eq. (12–5),

$$F_g = \frac{Gm_Mm}{r^2}$$
$$= \frac{(6.67 \times 10^{-11} \text{ N} \cdot \text{m}^2/\text{kg}^2)(6.42 \times 10^{23} \text{ kg})(4000 \text{ kg})}{(9.4 \times 10^6 \text{ m})^2}$$
$$= 1940 \text{ N}.$$

The acceleration due to the gravity of Mars at this point is

$$g_M = \frac{F_g}{m} = \frac{1940 \text{ N}}{4000 \text{ kg}} = 0.48 \text{ m/s}^2.$$

This is also the acceleration experienced by Phobos in its orbit, 6.0×10^6 m above the surface of Mars.

b) To find F_g and g_M at the surface, we repeat the above calculations, replacing $r = 9.4 \times 10^6$ m with $R_M = 3.40 \times 10^6$ m. Alternatively, because F_g and g_M are inversely proportional to $1/r^2$ (at any point outside the planet), we can multiply the results of (a) by the factor

$$\left(\frac{9.4 \times 10^6 \text{ m}}{3.40 \times 10^6 \text{ m}} \right)^2.$$

We invite you to complete the calculation by both methods and show that at the surface $F_g = 15,000$ N and $g_M = 3.7$ m/s². That is, on the surface of Mars, F_g and g are roughly 40% as large as at the surface of the earth.

12–4 GRAVITATIONAL POTENTIAL ENERGY

When we first developed the concept of gravitational potential energy in Section 7–2, we assumed that the gravitational force on a body is constant in magnitude and direction. This led to the expression $U = mgy$. But we now know that the earth's gravitational force on a body of mass m at any point outside the earth is given more generally by Eq. (12–2), $F_g = Gm_Em/r^2$, where m_E is the mass of the earth and r is the distance of the body from the earth's center. For problems in which r changes enough that the gravitational force can't be considered constant, we need a more general expression for gravitational potential energy.

To find this expression, we follow the same basic sequence of steps as in Section 7–2. We consider a body of mass m outside the earth, and first compute the work W_{grav} done by the gravitational force when the body moves directly away from or toward the center of the earth from $r = r_1$ to $r = r_2$ as in Fig. 12–8. This is given by

$$W_{grav} = \int_{r_1}^{r_2} F_r \, dr, \tag{12–6}$$

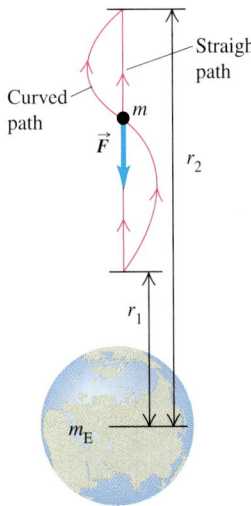

12–8 Work done by the gravitational force $\vec{F}$ as a body moves from radial coordinate r_1 to r_2. The work done by $\vec{F}$ is the same whether the body moves on a straight path or a curved one.

where F_r is the radial component of the gravitational force $\vec{F}$, that is, the component in the direction *outward* from the center of the earth. Because $\vec{F}$ points directly *inward* toward the center of the earth, F_r is negative. It differs from Eq. (12–2), the magnitude of the gravitational force, by a minus sign:

$$F_r = -\frac{Gm_Em}{r^2}. \tag{12–7}$$

Substituting Eq. (12–7) into Eq. (12–6), we see that W_{grav} is given by

$$W_{grav} = -Gm_Em \int_{r_1}^{r_2} \frac{dr}{r^2} = \frac{Gm_Em}{r_2} - \frac{Gm_Em}{r_1}. \tag{12–8}$$

The path doesn't have to be a straight line; it could also be a curve like the one in Fig. 12–8. By an argument similar to that in Section 7–2, this work depends only on the initial and final values of r, not on the path taken. This also proves that the gravitational force is always *conservative*.

We now define the corresponding potential energy U so that $W_{grav} = U_1 - U_2$, as in Eq. (7–3). Comparing this with Eq. (12–8), we see that the appropriate definition for

gravitational potential energy is

$$U = -\frac{Gm_{\mathrm{E}}m}{r} \qquad \text{(gravitational potential energy).} \qquad (12\text{–}9)$$

Figure 12–9 shows how the gravitational potential energy depends on the distance r between the body of mass m and the center of the earth. When the body moves away from the earth, r increases, the gravitational force does negative work, and U increases (i.e., becomes less negative). When the body "falls" toward earth, r decreases, the gravitational work is positive, and the potential energy decreases (i.e., becomes more negative).

You may be troubled by Eq. (12–9) because it states that gravitational potential energy is always negative. But in fact you've seen negative values of U before. In using the formula $U = mgy$ in Section 7–2, we found that U was negative whenever the body of mass m was at a value of y below the arbitrary height we chose to be $y = 0$—that is, whenever the body and the earth were closer together than some certain arbitrary distance. (See, for instance, Example 7–2 in Section 7–2.) In defining U by Eq. (12–9), we have chosen U to be zero when the body of mass m is infinitely far away from the earth ($r = \infty$). As the body moves toward the earth, gravitational potential energy decreases and so becomes negative. If we wanted, we could make $U = 0$ at the surface of the earth, where $r = R_{\mathrm{E}}$, by simply adding the quantity $Gm_{\mathrm{E}}m/R_{\mathrm{E}}$ to Eq. (12–9). This would make U positive when $r > R_{\mathrm{E}}$, but at the price of making the expression for U more complicated. This added term would not affect the *difference* in U between any two points, which is the only physically significant quantity; that is why we omit this term and use Eq. (12–9) for potential energy.

CAUTION ▶ Be careful not to confuse the expressions for gravitational force, Eq. (12–7), and gravitational potential energy, Eq. (12–9). The force F_r is proportional to $1/r^2$, while potential energy U is proportional to $1/r$. ◀

Armed with Eq. (12–9), we can now use general energy relations for problems in which the $1/r^2$ behavior of the earth's gravitational force has to be included. If the gravitational force on the body is the only force that does work, the total mechanical energy of the system is constant, or *conserved*. In the following example we'll use this principle to calculate **escape speed,** the speed required for a body to escape completely from a planet.

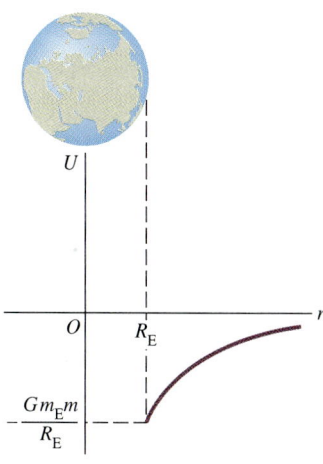

12–9 Graph of gravitational potential energy U versus distance from the center of the earth r. Note that U is always negative, but U becomes less negative with increasing r.

EXAMPLE 12–5

"From the earth to the moon" In Jules Verne's 1865 story with this title, three men were shot to the moon in a shell fired from a giant cannon sunk in the earth in Florida. a) Find the muzzle speed needed to shoot the shell straight up to a height above the earth equal to the earth's radius. (b) Find the *escape speed*—that is, the muzzle speed that would allow the shell to escape from the earth completely. Neglect air resistance, the earth's rotation, and the gravitational pull of the moon. The earth's radius is $R_{\mathrm{E}} = 6380$ km $= 6.38 \times 10^6$ m, and its mass is $m_{\mathrm{E}} = 5.97 \times 10^{24}$ kg (see Appendix F).

SOLUTION a) The only force doing work is the conservative gravitational force, so mechanical energy is conserved: $K_1 + U_1 = K_2 + U_2$. Let point 1 be the starting point, and let point 2 be the point of maximum height, so the speed at point 2 is $v_2 = 0$. If the radius of the earth is R_{E}, then $r_1 = R_{\mathrm{E}}$ and $r_2 = 2R_{\mathrm{E}}$ (Fig. 12–10). The mass of the shell (with passengers) is m. We can determine v_1 from the energy-conservation equation

$K_1 + U_1 = K_2 + U_2$:

$$\frac{1}{2}mv_1^2 + \left(-\frac{Gm_{\mathrm{E}}m}{R_{\mathrm{E}}}\right) = 0 + \left(-\frac{Gm_{\mathrm{E}}m}{2R_{\mathrm{E}}}\right).$$

Rearranging this, we find that

$$v_1 = \sqrt{\frac{Gm_{\mathrm{E}}}{R_{\mathrm{E}}}}$$

$$= \sqrt{\frac{(6.67 \times 10^{-11}\ \mathrm{N \cdot m^2/kg^2})(5.97 \times 10^{24}\ \mathrm{kg})}{6.38 \times 10^6\ \mathrm{m}}}$$

$$= 7900\ \mathrm{m/s}\ (= 28,400\ \mathrm{km/h} = 17,700\ \mathrm{mi/h}).$$

b) Again mechanical energy is conserved, so $K_1 + U_1 = K_2 + U_2$. We want the shell barely to be able to "reach" $r_2 = \infty$, with no kinetic energy left over, so $K_2 = 0$. When the shell is infinitely far from the earth, the potential energy $U_2 = 0$ as well, so the total mechanical energy $K_2 + U_2$ must be zero if the shell is just barely to escape to infinity. When the shell is fired, its positive kinetic energy K_1 and negative potential energy U_1 must also add to zero:

$$\frac{1}{2}mv_1^2 + \left(-\frac{Gm_{\mathrm{E}}m}{R_{\mathrm{E}}}\right) = 0,$$

$$v_1 = \sqrt{\frac{2Gm_{\mathrm{E}}}{R_{\mathrm{E}}}}$$

$$= \sqrt{\frac{2(6.67 \times 10^{-11}\ \mathrm{N \cdot m^2/kg^2})(5.97 \times 10^{24}\ \mathrm{kg})}{6.38 \times 10^6\ \mathrm{m}}}$$

$$= 1.12 \times 10^4\ \mathrm{m/s} \quad (= 40{,}200\ \mathrm{km/h} = 25{,}000\ \mathrm{mi/h}).$$

This result does not depend on the mass of the shell, nor does it depend on the direction in which the shell is launched. Modern spacecraft launched from Florida must attain essentially this same speed to escape the earth. A spacecraft on the ground at Cape Canaveral is already moving at 410 m/s to the east because of the earth's rotation; by launching to the east, the spacecraft takes advantage of this "free" contribution toward escape speed.

Generalizing our result, the initial speed v_1 needed for a body to escape from the surface of a spherical mass M with radius R (ignoring air resistance) is

$$v_1 = \sqrt{\frac{2GM}{R}} \quad \text{(escape speed)}.$$

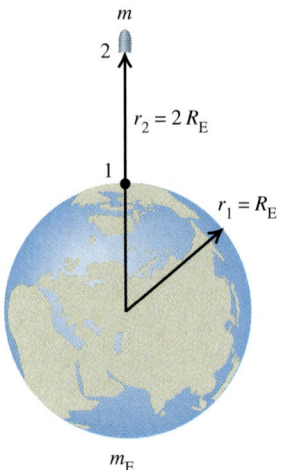

12–10 A projectile fired from the surface of the earth to a height equal to the earth's radius.

You can use this result to compute the escape speed for other bodies. You will find 5.02×10^3 m/s for Mars, 5.95×10^4 m/s for Jupiter, and 6.18×10^5 m/s for the sun.

MORE ON GRAVITATIONAL FORCE AND POTENTIAL ENERGY

The relationship between the gravitational force on a body, given by Eq. (12–7), and its gravitational potential energy, given by Eq. (12–9), can be expressed in a different way by using the methods of Section 7–5. Equations (7–17) and (7–18) show that the component of force in a given direction equals the negative of the derivative of U with respect to the corresponding coordinate. For motion along the x-axis,

$$F_x = -\frac{dU}{dx}. \tag{12–10}$$

The gravitational force has a component only in the radial direction, so we replace x by r in Eq. (12–10). We find

$$F_r = -\frac{dU}{dr} = -\frac{d}{dr}\left(-\frac{Gm_{\mathrm{E}}m}{r}\right) = -\frac{Gm_{\mathrm{E}}m}{r^2}. \tag{12–11}$$

This agrees with the F_r we started with in Eq. (12–7). As we remarked before, F_r is negative, showing that the force points in the opposite direction from the direction of increasing r.

As a final note, let's show that when we are close to the earth's surface, Eq. (12–9) reduces to the familiar $U = mgy$ from Chapter 7. We first rewrite Eq. (12–8) as

$$W_{\mathrm{grav}} = Gm_{\mathrm{E}}m\,\frac{r_1 - r_2}{r_1 r_2}.$$

If the body stays close to the earth, then in the denominator we may replace r_1 and r_2 by R_{E}, the earth's radius, so

$$W_{\mathrm{grav}} = Gm_{\mathrm{E}}m\,\frac{r_1 - r_2}{R_{\mathrm{E}}^2}.$$

According to Eq. (12–4), $g = Gm_E/R_E^2$, so

$$W_{grav} = mg(r_1 - r_2).$$

If we replace the r's by y's, this is just Eq. (7–1) for the work done by a constant gravitational force. In Section 7–2 we used this equation to derive Eq. (7–2), $U = mgy$, so we may consider this expression for gravitational potential energy to be a special case of the more general Eq. (12–9).

12–5 THE MOTION OF SATELLITES

Artificial satellites orbiting the earth are a familiar fact of contemporary life. But how do they stay in orbit, and what determines the properties of their orbits? We can use Newton's laws and the law of gravitation to provide the answers. We'll see in the next section that the motion of planets can be analyzed in the same way.

To begin, think back to the discussion of projectile motion in Section 3–4. In Example 3–6 a motorcycle rider rides horizontally off the edge of a cliff, launching himself into a parabolic path that ends on the flat ground at the base of the cliff. If he survives and repeats the experiment with increased launch speed, he will land farther from the starting point. We can imagine him launching himself with great enough speed that the earth's curvature becomes significant. As he falls, the earth curves away beneath him. If he is going fast enough, and if his launch point is high enough that he clears the mountain tops, he may be able to go right on around the earth without ever landing.

Figure 12–11 shows a variation on this theme. We launch a projectile from point A in the direction AB, tangent to the earth's surface. Trajectories (1) to (7) show the effect of increasing the initial speed. In trajectories (3) through (5) the projectile misses the earth and becomes an earth satellite. If there is no retarding force, the projectile's speed when it returns to point A is the same as its initial speed and it repeats its motion indefinitely.

Trajectories (1) through (5) close on themselves and are called **closed orbits.** All closed orbits are ellipses or segments of ellipses; trajectory (4) is a circle, a special case

Many satellites, including this gamma-ray telescope, have been placed in orbit by NASA's fleet of space shuttles.

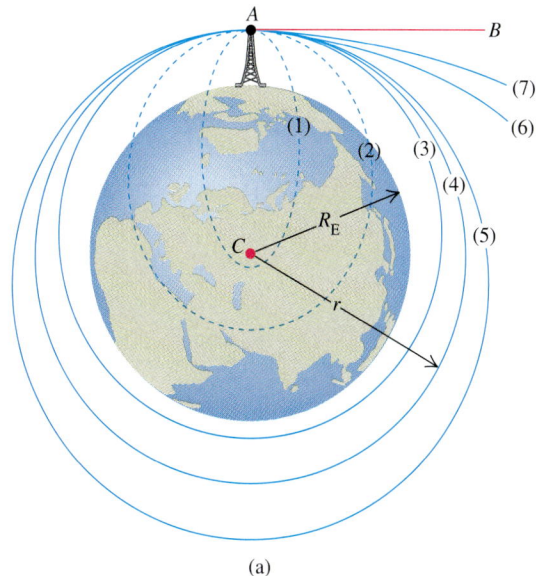

(a)

12–11 Trajectories of a body projected from point A in the direction AB with different initial velocities. Orbits (1) and (2) would be completed as shown if the earth were a point mass at C.

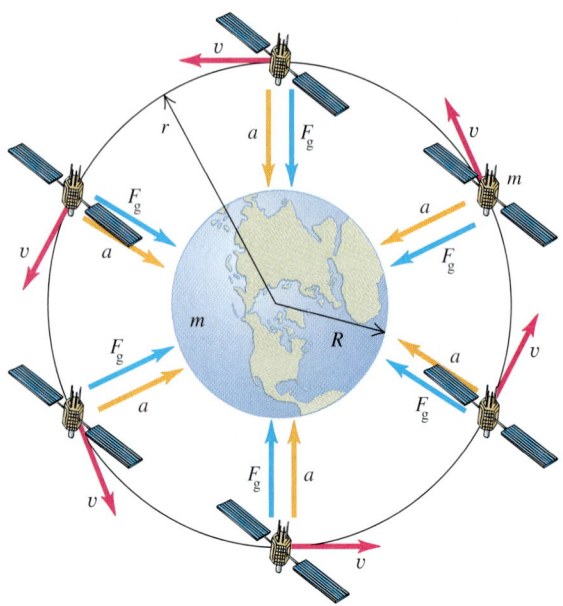

12–12 The force F_g due to the earth's gravitational attraction provides the centripetal acceleration that keeps a satellite in orbit. Compare this figure to Fig. 5–24.

4.8

Satellites Orbit

of an ellipse. Trajectories (6) and (7) are **open orbits.** For these the projectile never returns to its starting point but travels ever farther away from the earth.

A circular orbit like trajectory (4) in Fig. 12–11 is the simplest case, so we'll analyze it in detail. It is also an important case, since many artificial satellites have nearly circular orbits and the orbits of the planets around the sun are also fairly circular. The only force acting on a satellite in circular orbit around the earth is the earth's gravitational attraction, which is directed toward the center of the earth and hence toward the center of the orbit (Fig. 12–12). As we discussed in Section 5–5, this means that the satellite is in *uniform* circular motion and its speed is constant. The satellite isn't falling *toward* the earth; rather, it's constantly falling *around* the earth, and in a circular orbit the speed is just right to keep the distance from the satellite to the center of the earth constant.

How can we find the constant speed v of a satellite in a circular orbit? The radius of the orbit is r, measured from the *center* of the earth; the acceleration of the satellite has magnitude $a_{rad} = v^2/r$ and is always directed toward the center of the circle. By the law of gravitation, the net force (gravitational force) on the satellite of mass m has magnitude $F_g = Gm_E m/r^2$ and is in the same direction as the acceleration. Newton's second law ($\Sigma \vec{F} = m\vec{a}$) then tells us that

$$\frac{Gm_E m}{r^2} = \frac{mv^2}{r}.$$

Solving this for v, we find

$$v = \sqrt{\frac{Gm_E}{r}} \qquad \text{(circular orbit).} \qquad (12\text{–}12)$$

This relation shows that we can't choose the orbit radius r and the speed v independently; for a given radius r, the speed v for a circular orbit is determined.

Equation (12–12) also shows that the satellite's motion does not depend on its mass m because m doesn't appear in the equation. If we could cut a satellite in half without changing its speed, each half would continue on with the original motion. An astronaut on board a space shuttle is herself a satellite of the earth, held by the earth's gravitational attraction in the same orbit as the shuttle. The astronaut has the same velocity and acceleration as the shuttle, so nothing is pushing her against the floor or walls of the shuttle. She is in a state of *apparent weightlessness,* as in a freely falling elevator; see the discussion following Example 5–8 in Section 5–3. (*True* weightlessness would occur only if the astronaut were infinitely far from any other masses, so that the gravitational force on her would be zero.) Indeed, every part of her body is apparently weightless; she feels nothing pushing her stomach against her intestines or her head against her shoulders.

Apparent weightlessness is not just a feature of circular orbits; it occurs whenever gravity is the only force acting on a spacecraft. Hence it occurs for orbits of any shape, including open orbits such as (6) and (7) in Fig. 12–11.

We can derive a relation between the radius r of a circular orbit and the period T, the time for one revolution. The speed v is the distance $2\pi r$ traveled in one revolution, divided by the period:

$$v = \frac{2\pi r}{T}. \qquad (12\text{–}13)$$

To get an expression for T, we solve Eq. (12–13) for T and substitute v from Eq. (12–12):

$$T = \frac{2\pi r}{v} = 2\pi r \sqrt{\frac{r}{Gm_E}} = \frac{2\pi r^{3/2}}{\sqrt{Gm_E}} \qquad \text{(circular orbit).} \qquad (12\text{–}14)$$

Equations (12–12) and (12–14) show that larger orbits correspond to slower speeds and longer periods.

It's interesting to compare Eq. (12–12) to the calculation of escape speed in Example 12–5. We see that the escape speed from a spherical body with radius R is just $\sqrt{2}$ times as large as the speed of a satellite in a circular orbit at that radius. If our spacecraft is in circular orbit around *any* planet, we have to multiply our speed by a factor of $\sqrt{2}$ to escape to infinity, regardless of the planet's mass.

Since the speed v in a circular orbit is determined by Eq. (12–12) for a given orbit radius r, the total mechanical energy $E = K + U$ is determined as well. Using Eqs. (12–9) and (12–12), we have

$$E = K + U = \frac{1}{2} mv^2 + \left(-\frac{Gm_E m}{r}\right) = \frac{1}{2} m \left(\frac{Gm_E}{r}\right) - \frac{Gm_E m}{r},$$

$$(12\text{–}15)$$

$$E = -\frac{Gm_E m}{2r} \qquad \text{(circular orbit).}$$

The total mechanical energy in a circular orbit is negative and equal to one-half the potential energy. Increasing the orbit radius r means increasing the mechanical energy (that is, making E less negative). If the satellite is in a relatively low orbit that encounters the outer fringes of earth's atmosphere, mechanical energy will decrease due to negative work done by the force of air resistance; as a result, the orbit radius will decrease until the satellite hits the ground or burns up in the atmosphere.

We have talked mostly about earth satellites, but we can apply the same analysis to the circular motion of *any* body under its gravitational attraction to a stationary body. Other examples include the earth's moon, the moons of other planets, and the rings of Saturn. If Saturn's rings were a rotating rigid body, all parts of the rings would have the same orbital period T, and by Eq. (12–13) the outer regions of the rings (large r) would have a greater speed v than the inner regions (small r). But since the rings are actually composed of many individual orbiting particles, Eqs. (12–12) and (12–14) tell us that particles in the outer regions of the rings move at a slower speed and have a longer period than do particles in the inner regions.

The rings of Saturn are made up of countless numbers of icy particles, each of which is an independent satellite of Saturn.

EXAMPLE 12-6

Suppose you want to place a 1000-kg weather satellite into a circular orbit 300 km above the earth's surface. a) What speed, period, and radial acceleration must it have? b) How much work had to be done to place this satellite in orbit? c) How much additional work would have to be done to make this satellite escape the earth? The earth's radius is $R_E = 6380$ km, and its mass is $m_E = 5.97 \times 10^{24}$ kg.

SOLUTION a) The radius of the satellite's orbit is

$$r = 6380 \text{ km} + 300 \text{ km} = 6680 \text{ km} = 6.68 \times 10^6 \text{ m}.$$

From Eq. (12–12),

$$v = \sqrt{\frac{Gm_E}{r}} = \sqrt{\frac{(6.67 \times 10^{-11} \text{ N} \cdot \text{m}^2/\text{kg}^2)(5.97 \times 10^{24} \text{ kg})}{6.68 \times 10^6 \text{ m}}}$$

$$= 7720 \text{ m/s}.$$

From Eq. (12–14),

$$T = \frac{2\pi r}{v} = \frac{2\pi(6.68 \times 10^6 \text{ m})}{7720 \text{ m/s}}$$

$$= 5440 \text{ s} = 90.6 \text{ min}.$$

The radial acceleration is

$$a_{rad} = \frac{v^2}{r} = \frac{(7720 \text{ m/s})^2}{6.68 \times 10^6 \text{ m}}$$

$$= 8.92 \text{ m/s}^2.$$

This is the value of g at a height of 300 km above the earth's surface; it is somewhat less than the value of g at the surface. b) The work required is the difference between E_2, the total mechanical energy when the satellite is in orbit, and E_1, the orig-

inal mechanical energy when the satellite was at rest on the launch pad back on earth. In orbit, if we use Eq. (12–15), the energy is

$$E_2 = -\frac{Gm_E m}{2r} = -\frac{(6.67 \times 10^{-11} \text{ N} \cdot \text{m}^2/\text{kg}^2)(5.97 \times 10^{24} \text{ kg})(1000 \text{ kg})}{2(6.68 \times 10^6 \text{ m})}$$

$$= -2.99 \times 10^{10} \text{ J}.$$

At rest on the earth's surface the energy is purely potential:

$$E_1 = K_1 + U_1 = 0 + \left(-\frac{Gm_E m}{R_E}\right)$$

$$= -\frac{(6.67 \times 10^{-11} \text{ N} \cdot \text{m}^2/\text{kg}^2)(5.97 \times 10^{24} \text{ kg})(1000 \text{ kg})}{6.38 \times 10^6 \text{ m}}$$

$$= -6.25 \times 10^{10} \text{ J},$$

and so

$$W_{required} = E_2 - E_1 = -2.99 \times 10^{10} \text{ J} - (-6.25 \times 10^{10} \text{ J})$$

$$= 3.26 \times 10^{10} \text{ J}.$$

Notice that we ignored the initial kinetic energy due to the rotation of the earth. How much difference does this make? (See Example 12–5.)

c) We saw in part (b) of Example 12–5 that for a satellite to escape to infinity, the total mechanical energy must be zero. The total mechanical energy in the circular orbit is $E_2 = -2.99 \times 10^{10}$ J; to increase this to zero, an amount of work equal to 2.99×10^{10} J would have to be done. This extra energy could be supplied by rocket engines attached to the satellite.

12-6 THE MOTION OF PLANETS

The name *planet* comes from a Greek word meaning "wanderer," and indeed the planets continually change their positions in the sky relative to the background of stars. One of the great intellectual accomplishments of the sixteenth and seventeenth centuries was the threefold realization that the earth is also a planet, that all planets orbit the sun, and that the apparent motions of the planets as seen from the earth can be used to precisely determine the orbits of the planets.

The first and second of these ideas were published by Nicolaus Copernicus in Poland in 1543. The determination of planetary orbits was carried out between 1601 and 1619 by the German astronomer and mathematician Johannes Kepler, using a voluminous set of precise data on apparent planetary motions compiled by his mentor, the Danish astronomer Tycho Brahe. By trial and error, Kepler discovered three empirical laws that accurately described the motions of the planets:

1. **Each planet moves in an elliptical orbit, with the sun at one focus of the ellipse.**
2. **A line from the sun to a given planet sweeps out equal areas in equal times.**
3. **The periods of the planets are proportional to the $\frac{3}{2}$ powers of the major axis lengths of their orbits.**

Kepler did not know *why* the planets moved in this way. Three generations later, when Newton turned his attention to the motion of the planets, he discovered that each of Kepler's laws can be *derived;* they are consequences of Newton's laws of motion and the law of gravitation. Let's examine each of Kepler's laws in turn and see how each law arises.

Let's first consider the elliptical orbits described in Kepler's first law. Figure 12–13 shows the geometry of the ellipse. The longest dimension is the *major axis,* with half-length a; this half-length is called the **semi-major axis.** The sum of the distances from S to P and from S' to P is the same for all points on the curve. S and S' are the *foci* (plural of focus). The sun is at S, and the planet is at P; we think of them both as points because the size of each is very small in comparison to the distance between them. There is nothing at the other focus S'.

The distance of each focus from the center of the ellipse is ea, where e is a dimensionless number between 0 and 1 called the **eccentricity.** If $e = 0$, the ellipse is a circle. The actual orbits of the planets are fairly circular; their eccentricities range from 0.007 for Venus to 0.248 for Pluto (the earth's orbit has $e = 0.017$). The point in the planet's orbit closest to the sun is the *perihelion,* and the point most distant from the sun is the *aphelion.*

Newton was able to show that for a body acted on by an attractive force proportional to $1/r^2$, the only possible closed orbits are a circle or an ellipse; he also showed that open orbits (trajectories (6) and (7) in Fig. 12–11) must be parabolas or hyperbolas. These results can be derived by a straightforward application of Newton's laws and the law of gravitation, together with a lot more differential equations than we're ready for.

Kepler's second law is shown in Fig. 12–14. In a small time interval dt, the line from the sun S to the planet P turns through an angle $d\theta$. The area swept out is the colored triangle with height r, base length $r\,d\theta$, and area $dA = \frac{1}{2}r^2\,d\theta$. The rate at which area is swept out, dA/dt, is called the *sector velocity:*

$$\frac{dA}{dt} = \frac{1}{2}r^2\frac{d\theta}{dt}. \tag{12–16}$$

The essence of Kepler's second law is that the sector velocity has the same value at all points in the orbit. When the planet is close to the sun, r is small and $d\theta/dt$ is large; when the planet is far from the sun, r is large and $d\theta/dt$ is small.

To see how Kepler's second law follows from Newton's laws, we express dA/dt in terms of the velocity vector $\vec{v}$ of the planet P. The component of $\vec{v}$ perpendicular to the radial line is $v_\perp = v \sin\phi$. From Fig. 12–14b the displacement along the direction of $v_\perp$ during time dt is $r\,d\theta$, so we also have $v_\perp = r\,d\theta/dt$. Using this relation in Eq. (12–16), we find

$$\frac{dA}{dt} = \frac{1}{2}rv\sin\phi \qquad \text{(sector velocity).} \tag{12–17}$$

Now $rv\sin\phi$ is the magnitude of the vector product $\vec{r}\times\vec{v}$, which in turn is $1/m$ times the angular momentum $\vec{L} = \vec{r}\times m\vec{v}$ of the planet with respect to the sun. So we have

$$\frac{dA}{dt} = \frac{1}{2m}\left|\vec{r}\times m\vec{v}\right| = \frac{L}{2m}. \tag{12–18}$$

Thus Kepler's second law, that sector velocity is constant, means that angular momentum is constant!

It is easy to see why the angular momentum of the planet *must* be constant. According to Eq. (10–29), the rate of change of $\vec{L}$ equals the torque of the gravitational force $\vec{F}$ acting on the planet:

$$\frac{d\vec{L}}{dt} = \vec{\tau} = \vec{r}\times\vec{F}.$$

In our situation, $\vec{r}$ is the vector from the sun to the planet, and the force $\vec{F}$ is directed from the planet to the sun. So these vectors always lie along the same line, and their vector

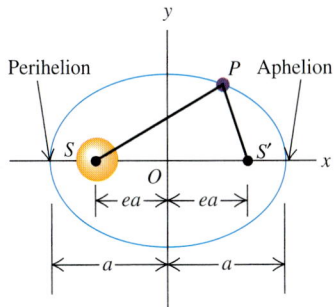

12–13 Geometry of an ellipse. The sum of the distances SP and $S'P$ is the same for every point on the curve. The sizes of the sun and planet are exaggerated for clarity.

(a)

(b)

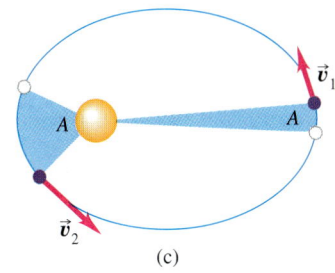

(c)

12–14 (a) The planet P moves about the sun (at S) in an elliptical orbit. (b) In a time dt the line from the sun to the planet sweeps out a triangular area $dA = \frac{1}{2}(r\,d\theta)\,r = \frac{1}{2}r^2\,d\theta$. (c) The planet's velocity varies so that the ratio of area dA to time interval dt is constant, regardless of the planet's position in its orbit.

product $\vec{r} \times \vec{F}$ is zero. Hence $d\vec{L}/dt = 0$. This conclusion does not depend on the $1/r^2$ behavior of the force; angular momentum is conserved for *any* force that acts always along the line joining the particle to a fixed point. Such a force is called a *central force*. (Kepler's first and third laws are valid *only* for a $1/r^2$ force.)

Conservation of angular momentum also explains why the orbit lies in a plane. The vector $\vec{L} = \vec{r} \times m\vec{v}$ is always perpendicular to the plane of the vectors $\vec{r}$ and $\vec{v}$; since $\vec{L}$ is constant in magnitude *and* direction, $\vec{r}$ and $\vec{v}$ always lie in the same plane, which is just the plane of the planet's orbit.

We have already derived Kepler's third law for the particular case of circular orbits. Equation (12–14) shows that the period of a satellite or planet in a circular orbit is proportional to the $\frac{3}{2}$ power of the orbit radius. Newton was able to show that this same relationship holds for an elliptical orbit, with the orbit radius r replaced by the semimajor axis a:

$$T = \frac{2\pi a^{3/2}}{\sqrt{Gm_S}} \qquad \text{(elliptical orbit around the sun).} \qquad (12\text{–}19)$$

Since the planet orbits the sun, not the earth, we have replaced the earth's mass m_E in Eq. (12–14) by the sun's mass m_S. Note that the period does not depend on the eccentricity e. An asteroid in an elongated elliptical orbit with semi-major axis a will have the same orbital period as a planet in a circular orbit of radius a. The key difference is that the asteroid moves at different speeds at different points in its elliptical orbit (Fig. 12–14c), while the planet's speed is constant around its circular orbit.

EXAMPLE 12-7

Kepler's third law Using the data in Appendix F, check whether the orbital semi-major axes and periods of Uranus and Saturn obey Kepler's third law.

SOLUTION The ratio of the orbital semi-major axis of Uranus to that of Saturn is

$$\frac{a_U}{a_S} = \frac{2.88 \times 10^{12} \text{ m}}{1.43 \times 10^{12} \text{ m}} = 2.01.$$

According to Kepler's third law, the ratio of the periods should be the $\frac{3}{2}$ power of this, or 2.85. The actual ratio is

$$\frac{T_U}{T_S} = \frac{83.75 \text{ y}}{29.42 \text{ y}} = 2.85.$$

As a historical note, Uranus wasn't discovered until 1781, a century and a half after Kepler's publication of his third law. While Kepler deduced his three laws from the motions of the five planets (other than the earth) known in his time, these laws have proven to apply equally well to all of the planets, asteroids, and comets subsequently discovered to be orbiting the sun.

EXAMPLE 12-8

Halley's comet Halley's comet moves in an elongated elliptical orbit around the sun (Fig. 12–15). At perihelion, Halley's comet is 8.75×10^7 km from the sun; at aphelion it is 5.26×10^9 km from the sun. a) At what point in the orbit does the comet have the greatest speed? b) Find the semi-major axis, eccentricity, and period of the orbit.

SOLUTION a) Mechanical energy is conserved as the comet moves around its orbit. The comet's kinetic energy $K = \frac{1}{2}mv^2$ will be maximum when the potential energy $U = -Gm_S m/r$ is minimum (that is, the most negative), which occurs when r is a minimum. Hence the speed v is maximum at perihelion. Your intuition about falling bodies is helpful here; as the comet falls inward toward the sun, it picks up speed, and its speed is maximum when closest to the sun. By the same reasoning, the comet

slows down as it moves away from the sun, and its speed is minimum at aphelion.

b) From Fig. 12–13 the length of the major axis equals the sum of the comet-sun distance at perihelion plus the comet-sun distance at aphelion. The length of the major axis is $2a$, so

$$a = \frac{8.75 \times 10^7 \text{ km} + 5.26 \times 10^9 \text{ km}}{2} = 2.67 \times 10^9 \text{ km}.$$

Further inspection of Fig. 12–13 shows that the comet-sun distance at perihelion is given by

$$a - ea = a(1 - e).$$

Since we are given that this distance is 8.75×10^7 km, the eccentricity is

$$e = 1 - \frac{8.75 \times 10^7 \text{ km}}{a} = 1 - \frac{8.75 \times 10^7 \text{ km}}{2.67 \times 10^9 \text{ km}} = 0.967.$$

12–15 (a) The orbit of Halley's comet. (b) Halley's comet as it appeared in 1986. At the heart of the comet is an icy body, called the nucleus, that is about 10 km across. When the comet's orbit carries it close to the sun, the heat of sunlight causes the nucleus to partially evaporate. The evaporated material forms the tail, which can be tens of millions of kilometers long.

This is very close to 1, so this is a very elongated orbit.

The period is given by Eq. (12–19):

$$T = \frac{2\pi a^{3/2}}{\sqrt{Gm_S}} = \frac{2\pi(2.67 \times 10^9 \text{ km})^{3/2}}{\sqrt{(6.67 \times 10^{-11} \text{ N} \cdot \text{m}^2/\text{kg}^2)(1.99 \times 10^{30} \text{ kg})}}$$

$$= 2.38 \times 10^9 \text{ s} = 75.5 \text{ years.}$$

Halley's comet was at perihelion in early 1986; it will next reach perihelion one period later, in 2061.

We have assumed that as a planet or comet orbits the sun, the sun remains absolutely stationary. Of course, this can't be correct; because the sun exerts a gravitational force on the planet, the planet exerts a gravitational force on the sun of the same magnitude but opposite direction. In fact, *both* the sun and the planet orbit around their common center of mass. We've made only a small transgression by ignoring this effect, however; the sun's mass is about 750 times the total mass of all the planets combined, so the center of mass of the solar system is not far from the center of the sun. But in binary star systems, in which two stars of comparable mass orbit each other, these effects are not small, and we must account for the motions of both stars about the center of mass.

Newton's analysis of planetary motions was a complete success, and his results are used on a daily basis by modern-day astronomers. But the most remarkable result of Newton's work is that the motions of bodies in the heavens obey the *same* laws of motion as do bodies on the earth. This *Newtonian synthesis,* as it has come to be called, is one of the great unifying principles of science. It has had profound effects on the way that humanity looks at the universe—not as a realm of impenetrable mystery, but as a direct extension of our everyday world, subject to scientific study and calculation. The Newtonian synthesis was an astonishing leap in understanding, made by a giant intellect.

*12–7 SPHERICAL MASS DISTRIBUTIONS

We have used without proof the statement that the gravitational interaction between two spherically symmetric mass distributions is the same as though all the mass of each were concentrated at its center. Now we're ready to prove this statement. Newton searched for

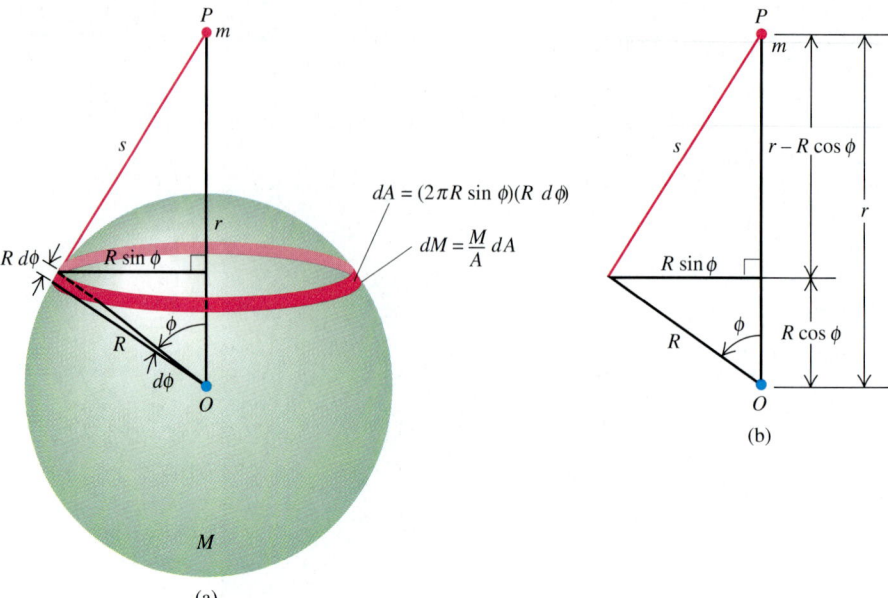

12–16 (a) To calculate the gravitational effects *outside* a spherical shell, all the mass M can be considered concentrated at the center. (b) The distance s is the hypotenuse of a right triangle with sides $(r - R \cos \phi)$ and $R \sin \phi$.

a proof for several years, and he delayed publication of the law of gravitation until he found one.

Here's our program. Rather than starting with two spherically symmetric masses, we'll tackle the simpler problem of a point mass m interacting with a thin spherical shell with total mass M. We will show that when m is outside the sphere, the *potential energy* associated with this gravitational interaction is the same as though M were all concentrated at the center of the sphere. According to Eq. (12–10), the force is the negative derivative of the potential energy, so the *force* on m is also the same as for a point mass M. Any spherically symmetric mass distribution can be thought of as being made up of many concentric spherical shells, so our result will also hold for *any* spherically symmetric M.

We start by considering a ring on the surface of the shell (Fig. 12–16a), centered on the line from the center of the shell to m. The reason we do this is that all of the particles that make up the ring are the same distance s from the point mass m. From Eq. (12–9) the potential energy of interaction between the earth (mass m_E) and a point mass m, separated by a distance r, is $U = -Gm_E m/r$. By changing notation in this expression, we see that in the situation shown in Fig. 12–16a, the potential energy of interaction between the point mass m and a particle of mass m_i within the ring is given by

$$U_i = -\frac{Gmm_i}{s}.$$

To find the potential energy of interaction between m and the entire ring of mass $dM = \sum_i m_i$, we sum this expression for U_i over all particles in the ring. Calling this potential energy dU, we find

$$dU = \sum_i U_i = \sum_i \left(-\frac{Gmm_i}{s} \right) = -\frac{Gm}{s} \sum_i m_i = -\frac{Gm \, dM}{s}. \qquad (12\text{–}20)$$

To proceed, we need to know the mass dM of the ring. We can find this with the aid of a little geometry. The radius of the shell is R, so in terms of the angle ϕ shown in the figure, the radius of the ring is $R \sin \phi$, and its circumference is $2\pi R \sin \phi$. The width of the ring is $R \, d\phi$, and its area dA is approximately equal to its width times its circumference:

$$dA = 2\pi R^2 \sin \phi \, d\phi.$$

The ratio of the ring mass dM to the total mass M of the shell is equal to the ratio of the area dA of the ring to the total area $A = 4\pi R^2$ of the shell:

$$\frac{dM}{M} = \frac{2\pi R^2 \sin\phi \, d\phi}{4\pi R^2} = \frac{1}{2} \sin\phi \, d\phi. \qquad (12\text{-}21)$$

Now we solve Eq. (12–21) for dM and substitute the result into Eq. (12–20) to find the potential energy of interaction between the point mass m and the ring:

$$dU = -\frac{GMm \sin\phi \, d\phi}{2s}. \qquad (12\text{-}22)$$

The total potential energy of interaction between the point mass and the *shell* is the integral of Eq. (12–22) over the whole sphere as ϕ varies from 0 to π (*not* 2π!) and s varies from $r - R$ to $r + R$. To carry out the integration, we have to express the integrand in terms of a single variable; we choose s. To express ϕ and $d\phi$ in terms of s, we have to do a little more geometry. Figure 12–16b shows that s is the hypotenuse of a right triangle with sides $(r - R\cos\phi)$ and $R\sin\phi$, so the Pythagorean theorem gives

$$s^2 = (r - R\cos\phi)^2 + (R\sin\phi)^2$$
$$= r^2 - 2rR\cos\phi + R^2. \qquad (12\text{-}23)$$

We take differentials of both sides:

$$2s \, ds = 2rR \sin\phi \, d\phi.$$

Next we divide this by $2rR$ and substitute the result into Eq. (12–22):

$$dU = -\frac{GMm}{2s} \frac{s \, ds}{rR} = -\frac{GMm}{2rR} \, ds. \qquad (12\text{-}24)$$

We can now integrate Eq. (12–24), recalling that s varies from $r - R$ to $r + R$:

$$U = -\frac{GMm}{2rR} \int_{r-R}^{r+R} ds = -\frac{GMm}{2rR} [(r + R) - (r - R)]. \qquad (12\text{-}25)$$

Finally, we have

$$U = -\frac{GmM}{r} \qquad \text{(point mass } m \text{ outside spherical shell } M\text{).} \qquad (12\text{-}26)$$

This is equal to the potential energy of two point masses m and M at a distance r. So we have proved that the gravitational potential energy of the spherical shell M and the point mass m at any distance r is the same as though they were point masses. Because the force is given by $F_r = -dU/dr$, the force is also the same.

Any spherically symmetric mass distribution can be thought of as a combination of concentric spherical shells. Because of the principle of superposition of forces, what is true of one shell is also true of the combination. So we have proved half of what we set out to prove, that the gravitational interaction between any spherically symmetric mass distribution and a point mass is the same as though all the mass of the spherically symmetric distribution were concentrated at its center.

The other half is to prove that *two* spherically symmetric mass distributions interact as though they were both points. That's easier. In Fig. 12–16 the forces the two bodies exert on each other are an action-reaction pair, and they obey Newton's third law. So we have also proved that the force that m exerts *on* the sphere M is the same as though M were a point. But now if we replace m with a spherically symmetric mass distribution

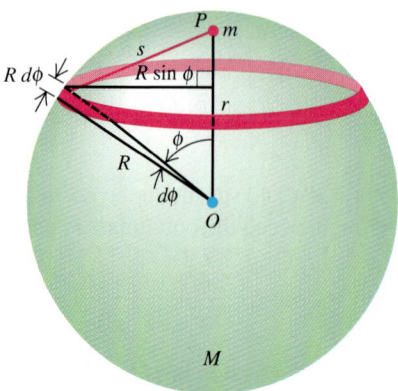

12–17 When a point mass m is *inside* a uniform spherical shell of mass M, the potential energy is the same no matter where inside the shell the point mass is located. The force from the masses' mutual gravitational interaction is zero.

centered at m's location, the resulting gravitational force on any part of M is the same as before, and so is the total force. This completes our proof.

A POINT MASS INSIDE A SPHERICAL SHELL

We assumed at the beginning that the point mass m was outside the spherical shell, so our proof is valid only when m is outside a spherically symmetric mass distribution. When m is *inside* a spherical shell, the geometry is as shown in Fig. 12–17. The entire analysis goes just as before; Eqs. (12–20) through (12–24) are still valid. But when we get to Eq. (12–25), the limits of integration have to be changed to $R - r$ and $R + r$. We then have

$$U = -\frac{GMm}{2rR} \int_{R-r}^{R+r} ds = -\frac{GMm}{2rR}[(R + r) - (R - r)], \qquad (12\text{–}27)$$

and the final result is

$$U = -\frac{GmM}{R} \qquad \text{(point mass } m \text{ inside spherical shell } M\text{).} \qquad (12\text{–}28)$$

Compare this result to Eq. (12–26): instead of having r, the distance between m and the center of M, in the denominator, we have R, the radius of the shell. This means that U in Eq. (12–28) doesn't depend on r and thus has the same value everywhere inside the shell. When m moves around inside the shell, no work is done on it, so the force on m at any point inside the shell must be zero.

More generally, at any point in the interior of any spherically symmetric mass distribution (not necessarily a shell), at a distance r from its center, the gravitational force on a point mass m is the same as though we removed all the mass at points farther than r from the center and concentrated all the remaining mass at the center.

EXAMPLE 12–9

"Journey to the center of the earth" Suppose we drill a hole through the earth (radius R_E, mass m_E) along a diameter and drop a mail pouch (mass m) down the hole. Derive an expression for the gravitational force on the pouch as a function of its distance r from the center. Assume that the density of the earth is uniform (not a very realistic model; see Fig. 12–7).

SOLUTION According to the statements above, the gravitational force at distance r from the center is determined only by the mass M within a sphere of radius r (Fig. 12–18). With uniform

density the mass is proportional to the volume of the sphere, which is $\frac{4}{3}\pi r^3$ for this sphere and $\frac{4}{3}\pi R_E{}^3$ for the entire earth. So we have

$$\frac{M}{m_E} = \frac{\frac{4}{3}\pi r^3}{\frac{4}{3}\pi R_E{}^3} = \frac{r^3}{R_E{}^3}.$$

The magnitude of the gravitational force on m is given by

$$F_g = \frac{GMm}{r^2} = \frac{Gm_Em}{r^2}\frac{r^3}{R_E{}^3} = \frac{Gm_Em}{R_E{}^3}r.$$

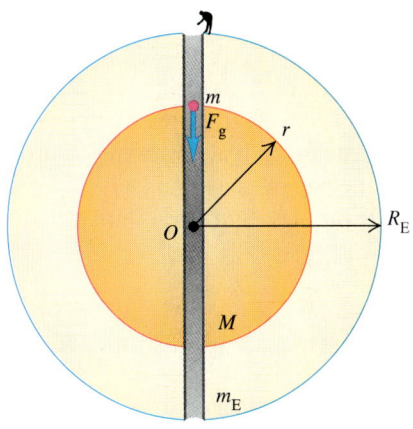

At points inside the earth, F_g is *directly proportional* to the distance r from the center, rather than proportional to $1/r^2$ as it is outside the sphere. Right at the surface, where $r = R_E$, the above expression gives $F_g = Gm_E m/R_E^2$, as we should expect. In the next chapter we'll learn how to compute the time it would take for the mail pouch to emerge on the other side of the earth. Meanwhile, keep in mind that the assumption of uniform density of the earth is, like the hole, pure fantasy. We discussed the variation of density in the earth in Section 12–3.

12–18 A hole through the center of the earth (assumed to be uniform). When an object is at a distance r from the center, only the mass inside a sphere of radius r exerts a net gravitational force on it.

*12–8 APPARENT WEIGHT AND THE EARTH'S ROTATION

Because the earth rotates on its axis, it is not precisely an inertial frame of reference. For this reason the apparent weight of a body on earth is not precisely equal to the earth's gravitational attraction, which we will call the **true weight** $\vec{w}_0$ of the body. Figure 12–19 is a cutaway view of the earth, showing three observers. Each one holds a spring balance with a body of mass m hanging from it. Each balance applies a tension force $\vec{F}$ to the body hanging from it, and the reading on each balance is the magnitude F of this force. If the observers are unaware of the earth's rotation, each one *thinks* that the scale reading equals the weight of the body because he thinks the body on his spring balance is in equilibrium. So each observer thinks that the tension $\vec{F}$ must be opposed by an equal and opposite force $\vec{w}$, which we call the **apparent weight.** But if the bodies are rotating with the earth, they are *not* precisely in equilibrium. Our problem is to find the relation between the apparent weight $\vec{w}$ and the true weight $\vec{w}_0$.

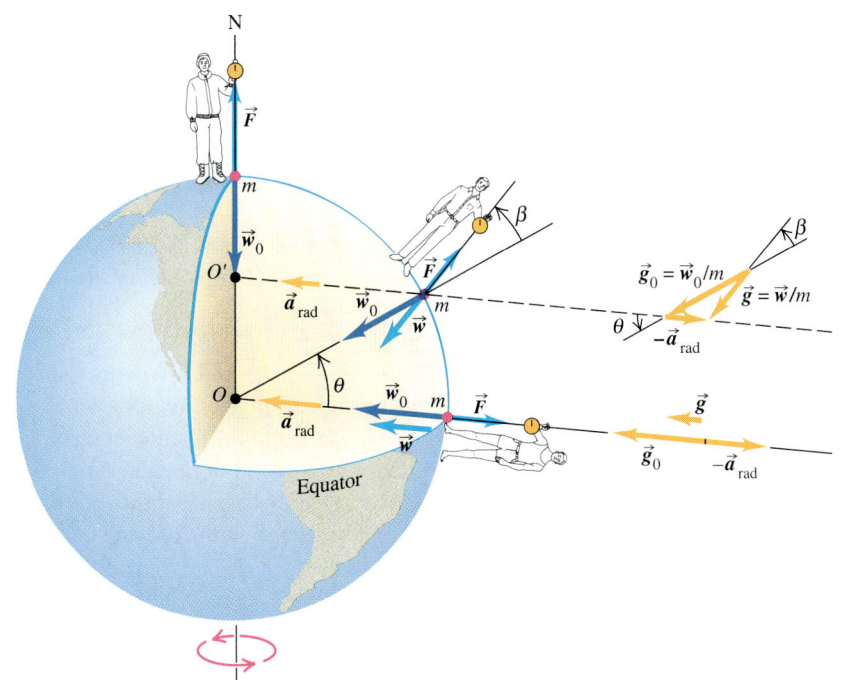

12–19 Except at the poles, the scale reads less than the gravitational force of attraction because a net force is needed to provide the centripetal acceleration. The sketch exaggerates the difference for clarity. On the earth the maximum angle β between the true and apparent weight vectors is 0.1°, and the magnitude of the apparent weight is smaller than that of the true weight by less than 0.35%.

If we assume that the earth is spherically symmetric, then the true weight $\vec{w}_0$ has magnitude $Gm_E m/R_E^2$, where m_E and R_E are the mass and radius of the earth. This value is the same for all points on the earth's surface. If the center of the earth can be taken as the origin of an inertial coordinate system, then the body at the north pole really *is* in equilibrium in an inertial system, and the reading on that observer's spring balance is equal to w_0. But the body at the equator is moving in a circle of radius R_E with speed v, and there must be a net inward force equal to the mass times the centripetal acceleration:

$$w_0 - F = \frac{mv^2}{R_E}.$$

So the magnitude of the apparent weight (equal to the magnitude of F) is

$$w = w_0 - \frac{mv^2}{R_E} \qquad \text{(at the equator)}. \tag{12–29}$$

If the earth were not rotating, the body when released would have a free-fall acceleration $g_0 = w_0/m$. Since the earth *is* rotating, the falling body's actual acceleration relative to the observer at the equator is $g = w/m$. Dividing Eq. (12–29) by m and using these relations, we find

$$g = g_0 - \frac{v^2}{R_E} \qquad \text{(at the equator)}.$$

To evaluate v^2/R_E, we note that in 86,164 s a point on the equator moves a distance equal to the earth's circumference, $2\pi R_E = 2\pi(6.38 \times 10^6 \text{ m})$. (The solar day, 86,400 s, is $\frac{1}{365}$ longer than this because in one day the earth also completes $\frac{1}{365}$ of its orbit around the sun.) Thus we find

$$v = \frac{2\pi(6.38 \times 10^6 \text{ m})}{86,164 \text{ s}} = 465 \text{ m/s},$$

$$\frac{v^2}{R_E} = \frac{(465 \text{ m/s})^2}{6.38 \times 10^6 \text{ m}} = 0.0339 \text{ m/s}^2.$$

So for a spherically symmetric earth the acceleration due to gravity should be about 0.03 m/s^2 less at the equator than at the poles.

At locations intermediate between the equator and the poles, the true weight $\vec{w}_0$ and the centripetal acceleration are not along the same line, and we need to write a vector equation corresponding to Eq. (12–29). From Fig. 12–19 we see that the appropriate equation is

$$\vec{w} = \vec{w}_0 - m\vec{a}_{\text{rad}} = m\vec{g}_0 - m\vec{a}_{\text{rad}}. \tag{12–30}$$

The difference in the magnitudes of g and g_0 lies between zero and 0.0339 m/s^2. As shown in Fig. 12–19, the *direction* of the apparent weight differs from the direction toward the center of the earth by a small angle β, which is $0.1°$ or less.

Table 12–1 gives the values of g at several locations, showing variations with latitude. There are also small additional variations due to the lack of perfect spherical symmetry of the earth, local variations in density, and differences in elevation.

Our discussion of apparent weight can also be applied to the phenomenon of apparent weightlessness in orbiting spacecraft, which we described in Section 12–5. Bodies in an orbiting spacecraft are *not* weightless; the earth's gravitational attraction continues to act on them just as though they were at rest relative to the earth. The apparent weight of a body in a spacecraft is again given by Eq. (12–30):

$$\vec{w} = \vec{w}_0 - m\vec{a}_{\text{rad}} = m\vec{g}_0 - m\vec{a}_{\text{rad}}.$$

But for a spacecraft in orbit, as well as any body inside the spacecraft, the acceleration

TABLE 12–1
VARIATIONS OF g WITH LATITUDE AND ELEVATION

STATION	NORTH LATITUDE	ELEVATION (m)	$g(\text{m/s}^2)$	$g(\text{ft/s}^2)$
Canal Zone	9°	0	9.78243	32.0944
Jamaica	18°	0	9.78591	32.1059
Bermuda	32°	0	9.79806	32.1548
Denver	40°	1638	9.79609	32.1393
Pittsburgh, Pa.	40.5°	235	9.80118	32.1561
Cambridge, Mass.	42°	0	9.80398	32.1652
Greenland	70°	0	9.82534	32.2353

$\vec{a}_{\text{rad}}$ toward the earth's center is equal to the value of the acceleration of gravity $\vec{g}_0$ at the position of the spacecraft. Hence

$$\vec{g}_0 = \vec{a}_{\text{rad}},$$

and the apparent weight is

$$\vec{w} = 0.$$

This is what we mean when we say that an astronaut or other body in the spacecraft is apparently weightless. Note that we didn't make any assumptions about the shape of the orbit; as we mentioned in Section 12–5, an astronaut will be apparently weightless no matter what the orbit.

This orbiting astronaut is acted on by the earth's gravity, but *feels* weightless because his acceleration is equal to $\vec{g}$.

12–9 BLACK HOLES

A Case Study in Modern Physics

The concept of a black hole is one of the most interesting and startling products of modern gravitational theory, yet the basic idea can be understood on the basis of Newtonian principles. Think first about the properties of our own sun. Its mass $M = 1.99 \times 10^{30}$ kg and radius $R = 6.96 \times 10^8$ m are much larger than those of any planet, but compared to other stars, our sun is not exceptionally massive.

What's the average *density* ρ of the sun? You can find it the same way we found the average density of the earth in Section 12–3:

$$\rho = \frac{M}{V} = \frac{M}{\frac{4}{3}\pi R^3} = \frac{1.99 \times 10^{30} \text{ kg}}{\frac{4}{3}\pi(6.96 \times 10^8 \text{ m})^3}$$

$$= 1410 \text{ kg/m}^3.$$

The sun's temperatures range from 5800 K (about 5500°C, or 10,000°F) at the surface up to 1.5×10^7 K (about 2.7×10^7 °F) in the interior, so it surely contains no solids or liquids. Yet gravitational attraction pulls the sun's gas atoms together until the sun is, on average, 41% denser than water and about 1200 times as dense as the air we breathe.

Now think about the escape speed for a body at the surface of the sun. In Example 12–5 (Section 12–4) we found that the escape speed from the surface of a spherical mass M with radius R is $v = \sqrt{2GM/R}$. We can relate this to the average density. Substituting $M = \rho V = \rho(\frac{4}{3}\pi R^3)$ into the expression for escape speed gives

$$v = \sqrt{\frac{2GM}{R}} = \sqrt{\frac{8\pi G\rho}{3}} R. \qquad (12\text{–}31)$$

Using either form of this equation, you can show that the escape speed for a body at the surface of our sun is $v = 6.18 \times 10^5$ m/s (about 2.2 million km/h, or 1.4 million mi/h).

This value, roughly 1/500 the speed of light, is independent of the mass of the escaping body; it depends only on the mass and radius (or average density and radius) of the sun.

Now consider various stars with the same average density ρ and different radii R. Equation (12–31) shows that for a given value of density ρ, the escape speed v is directly proportional to R. In 1783 the Rev. John Mitchell, an amateur astronomer, noted that if a body with the same average density as the sun had about 500 times the radius of the sun, the magnitude of its escape velocity would be greater than the speed of light c. With his statement that "all light emitted from such a body would be made to return toward it," Mitchell became the first person to suggest the existence of what we now call a **black hole.**

The first expression for escape speed in Eq. (12–31) also suggests that a body of mass M will act as a black hole if its radius R is less than or equal to a certain critical radius. How can we determine this critical radius? You might think that you can find the answer by simply setting $v = c$ in Eq. (12–31). As a matter of fact, this does give the correct result, but only because of two compensating errors. The kinetic energy of light is *not* $mc^2/2$, and the gravitational potential energy near a black hole is *not* given by Eq. (12–9). In 1916, Karl Schwarzschild used Einstein's general theory of relativity (in part a generalization and extension of Newtonian gravitation theory) to derive an expression for the critical radius R_S, now called the **Schwarzschild radius.** The result turns out to be the same as though we had set $v = c$ in Eq. (12–31), so

$$c = \sqrt{\frac{2GM}{R_S}}.$$

Solving for the Schwarzschild radius R_S, we find

$$R_S = \frac{2GM}{c^2}. \tag{12–32}$$

If a spherical, nonrotating body with mass M has a radius less than R_S, then *nothing* (not even light) can escape from the surface of the body, and the body functions as a black hole (Fig. 12–20). In this case, any other body within a distance R_S of the center of the black hole is trapped by the gravitational attraction of the black hole and cannot escape from it.

The surface of the sphere with radius R_S surrounding a black hole is called the **event horizon** because, since light can't escape from within that sphere, we can't see events occurring inside. All that an observer outside the event horizon can know about a black hole is its mass (from its gravitational effects on other bodies), its electric charge (from

12–20 (a) When the radius R of a body is greater than the Schwarzschild radius R_S, light can escape from the surface of the body. As it travels away, it is "redshifted" to longer wavelengths. (b) When the body is within the event horizon (radius R_S), it is a black hole with an escape speed greater than the speed of light. In that case we can obtain very little information about it.

(a)

(b)

the electric forces it exerts on other charged bodies), and its angular momentum (because a rotating black hole tends to drag space—and everything in that space—around with it). All other information about the body is irretrievably lost when it collapses inside its event horizon.

EXAMPLE 12–10

Current astrophysical theory suggests that a burned-out star can collapse under its own gravity to form a black hole when its mass is as small as two solar masses. If it does, what is the radius of its event horizon?

SOLUTION The radius is R_S, and "two solar masses" means $M = 2(1.99 \times 10^{30} \text{ kg}) = 4.0 \times 10^{30} \text{ kg}$. From Eq. (12–32),

$$R_S = \frac{2GM}{c^2} = \frac{2(6.67 \times 10^{-11} \text{ N} \cdot \text{m}^2/\text{kg}^2)(4.0 \times 10^{30} \text{ kg})}{(3.00 \times 10^8 \text{ m/s})^2}$$

$$= 5.9 \times 10^3 \text{ m} = 5.9 \text{ km},$$

or less than 4 miles.

If the radius of such an object is just equal to the Schwarzschild radius, the average density has the incredibly large value

$$\rho = \frac{M}{\frac{4}{3}\pi R^3} = \frac{4.0 \times 10^{30} \text{ kg}}{\frac{4}{3}\pi(5.9 \times 10^3 \text{ m})^3}$$

$$= 4.6 \times 10^{18} \text{ kg/m}^3.$$

This is of the order of 10^{15} times as great as the density of familiar matter on earth and is comparable to the densities of atomic nuclei. In fact, once the body collapses to a radius of R_S, nothing can prevent it from collapsing further. All of the mass ends up being crushed down to a single point called a *singularity* at the center of the event horizon. This point has zero volume and so has *infinite* density.

At points far from a black hole, its gravitational effects are the same as those of any normal body with the same mass. If the sun collapsed to form a black hole, the orbits of the planets would be unaffected. But things get dramatically different close to the black hole. If you decided to become a martyr for science and jump into a black hole, those you left behind would notice several odd effects as you moved toward the event horizon, most of them associated with effects of general relativity. If you carried a radio transmitter to send back your comments on what was happening, they would have to retune their receiver continuously to lower and lower frequencies, an effect called the *gravitational red shift*. Consistent with this shift, they would observe that your clocks (electronic or biological) would appear to run more and more slowly, an effect called *time dilation*. In fact, during their lifetimes they would never see you make it to the event horizon. In your frame of reference, you would make it to the event horizon in a rather short time but in a rather disquieting way. As you fell feet first into the black hole, the gravitational pull on your feet would be greater than that on your head, which would be slightly farther away from the black hole. The *differences* in gravitational force on different parts of your body would be great enough to stretch you along the direction toward the black hole and compress you perpendicular to it. These effects (called *tidal forces*) would rip you to atoms, and then rip your atoms apart, before you reached the event horizon.

If light cannot escape from a black hole and if black holes are as small as Example 12–10 suggests, how can we know that such things exist in space? Here's the idea. Any gas or dust near the black hole tends to be pulled into an *accretion disk* that swirls around and into the black hole, rather like a whirlpool (Fig. 12–21). Friction within the accretion disk's material causes it to lose mechanical energy and spiral into the black hole; as it moves inward, it is compressed together. This causes heating of the material, just as air compressed in a bicycle pump gets hotter. Temperatures in excess of 10^6 K can occur in the accretion disk, so hot that the disk emits not just visible light (as do bodies that are "red-hot" or "white-hot") but x rays. Astronomers look for these x rays (emitted *before* the material crosses the event horizon) to signal the presence of a black hole. Several promising candidates have been found, and astronomers now express considerable confidence in the existence of black holes.

Black holes in binary star systems like the one depicted in Fig. 12–21 have masses a few times greater than the sun's mass. There is also mounting evidence for the

12–21 A binary star system in which an ordinary star and a black hole orbit each other. Matter is pulled from the ordinary star to form an accretion disk around the black hole. The gas in the accretion disk is compressed and heated to such high temperatures that it becomes an intense source of x rays.

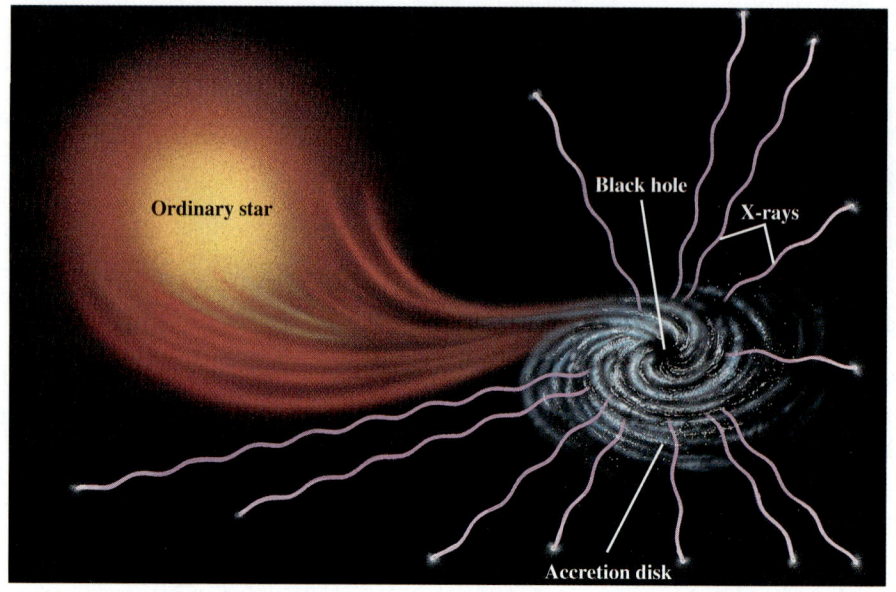

existence of much larger *supermassive black holes*. One example is believed to occur in the galaxy M87, a collection of about 10^{12} stars located 50 million light years away. Figure 12–22 shows a 6000-light-year-long jet of material being ejected from the center of M87 at speeds of about one tenth the speed of light. At the galaxy's center itself, shown in the inset, a spiral-shaped disk of hot gas 300 light years across is swirling about a massive unseen object at the center of the disk. The most plausible, albeit bizarre, explanation for both of these phenomena is that there is a black hole at the center of the disk, with about 3×10^9 times the mass of the sun. In this model, material in the disk spirals in to form a hot accretion disk within about 10^{11} km of the black hole. But the Schwarzschild radius of the black hole is only about 10^{10} km (roughly the same size as the orbit of the planet Pluto around the sun), and not all of the infalling material hits this relatively small target. The material that misses is swirled outward at high speed

12–22 Two Hubble Space Telescope images of the galaxy M87. At the center of M87, shown in the inset, gas clouds orbit at speeds in excess of 550 km/s around an invisible object that is presumed to be a supermassive black hole.

Gas Disk in Nucleus of
Active Galaxy M87

Hubble Space Telescope
Wide Field Planetary Camera 2

along the rotation axis of the accretion disk, and magnetic forces cause this material to form jets like the one shown in Fig. 12–22.

Astronomical observations like those for M87 indicate that there are supermassive black holes at the center of a number of galaxies. In fact, there is now compelling evidence that a black hole with a mass more than 2×10^6 times that of the sun lies at the center of our own Milky Way galaxy. Observational and theoretical studies of such black holes continue to be an exciting area of research in contemporary physics and astronomy.

SUMMARY

- Newton's law of gravitation states that two particles with masses m_1 and m_2, a distance r apart, attract each other with forces of magnitude

$$F_g = G \frac{m_1 m_2}{r^2}.$$ (12–1)

These forces form an action-reaction pair and obey Newton's third law. When two or more bodies exert gravitational forces on a particular body, the total gravitational force on that particular body is the vector sum of the forces exerted by the other bodies individually.

- The weight w of a body is the total gravitational force exerted on it by all other bodies in the universe. Near the surface of the earth (mass m_E and radius R_E), this is essentially equal to the gravitational force of the earth alone. The weight w of a body with mass m is then

$$w = F_g = \frac{G m_E m}{R_E^2},$$ (12–3)

and the acceleration due to gravity g is

$$g = \frac{G m_E}{R_E^2}.$$ (12–4)

- The gravitational potential energy U of two spherically symmetric bodies of masses m and m_E with their centers separated by a distance r is

$$U = -\frac{G m_E m}{r}.$$ (12–9)

This potential energy is never positive; it is zero only when the two bodies are infinitely far apart.

- When a satellite moves in a circular orbit, the centripetal acceleration is provided by the gravitational attraction of the earth. The speed v and period T of a satellite in a circular orbit with radius r are

$$v = \sqrt{\frac{G m_E}{r}},$$ (12–12)

$$T = \frac{2\pi r}{v} = 2\pi r \sqrt{\frac{r}{G m_E}} = \frac{2\pi r^{3/2}}{\sqrt{G m_E}}.$$ (12–14)

- Kepler's three laws describe characteristics of the elliptical orbits of planets around the sun or satellites around a planet.

- The gravitational interaction of any spherically symmetric mass distribution, at points outside the distribution, is the same as though all the mass were concentrated at the center.

- Because of the earth's rotation, a body's apparent weight at the equator is about 0.3% less than its true weight, and the free-fall acceleration g is about the same percentage less than the value it would have without rotation.
- If a nonrotating spherical mass distribution with total mass M has a radius less than $R_S = 2\,GM/c^2$, called the Schwarzschild radius, then the gravitational interaction prevents anything, including light, from escaping from within a sphere with radius R_S. Such a body is called a black hole.

DISCUSSION QUESTIONS

Q12–1 A student wrote: "The only reason an apple falls downward to meet the earth instead of the earth rising upward to meet the apple is that the earth is much more massive and so exerts a much greater pull." Please comment.

Q12–2 To use the Cavendish balance (Fig. 12–4) to determine G, you must first know the masses m_1 and m_2. How could you determine these masses *without* weighing them?

Q12–3 If all planets had the same average density, how would the acceleration due to gravity at the surface of a planet depend on its radius?

Q12–4 Is a pound of butter on earth the same amount as a pound of butter on Mars? What about a kilogram of butter? Explain.

Q12–5 Example 12–2 (Section 12–2) shows that the acceleration of each sphere caused by the gravitational force is inversely proportional to the mass of that sphere. So why does the force of gravity give all masses the same acceleration when dropped near the surface of the earth?

Q12–6 When will you attract the sun more: today at noon, or tonight at midnight? Explain.

Q12–7 Since the moon is constantly attracted toward the earth by the gravitational interaction, why doesn't it crash into the earth?

Q12–8 Imagine that the sun somehow became twice as massive as it is now. What effect would this have on your weight on earth (as measured by standing on a scale)? Explain your answer.

Q12–9 The sun pulls on the moon with a force that is more than twice the magnitude of the force with which the earth attracts the moon. Why, then, doesn't the sun take the moon away from the earth?

Q12–10 As defined in Chapter 7, gravitational potential energy is $U = mgy$ and is positive for a body of mass m above the earth's surface (which is at $y = 0$). But in this chapter, gravitational potential energy is $U = -Gm_Em/r^2$, which is *negative* for a body of mass m above the earth's surface (which is at $r = R_E$). How can you reconcile these seemingly incompatible descriptions of gravitational potential energy?

Q12–11 A planet is moving at constant speed in a circular orbit around a star. In one complete orbit, what is the net amount of work done on the planet by the star's gravitational force: positive, negative, or zero? What if the planet's orbit is an ellipse, so that the speed is not constant? Explain your answers.

Q12–12 Does the escape speed for an object at the earth's surface depend on the direction in which it is launched? Explain.

Does your answer depend on whether or not you include the effects of air resistance?

Q12–13 If a projectile is fired straight up from the earth's surface, what would happen if the total mechanical energy (kinetic plus potential) is a) less than zero? c) greater than zero? In each case, ignore air resistance and the gravitational effects of the sun, the moon and the other planets.

Q12–14 Discuss whether the following statement is correct: "In the absence of air resistance, the trajectory of a projectile thrown near the earth's surface is an *ellipse*, not a parabola."

Q12–15 The earth is closer to the sun in November than in May. In which month does it move faster in its orbit? Explain.

Q12–16 A communications firm wants to place a satellite in orbit so that it is always directly above the earth's 45th parallel (latitude 45° north). This means that the plane of the orbit will not pass through the center of the earth. Is such an orbit possible? Why or why not?

Q12–17 At what point in an elliptical orbit is the acceleration maximum? At what point is it minimum? Justify your answers.

Q12–18 Which takes more fuel, a voyage from the earth to the moon or from the moon to the earth? Explain.

Q12–19 What would Kepler's third law be for circular orbits if an amendment to Newton's law of gravitation made the gravitational force inversely proportional to r^3? Would this change affect Kepler's other two laws? Explain.

Q12–20 In the elliptical orbit of Halley's comet, shown in Fig. 12–15a, the sun's gravity is responsible for making the comet fall inward from aphelion to perihelion. But what is responsible for making the comet move from perihelion back outward to aphelion?

Q12–21 Many people believe that orbiting astronauts feel weightless because they are "beyond the pull of the earth's gravity." How far from the earth would a spacecraft have to travel to be truly beyond the earth's gravitational influence? If a spacecraft were really unaffected by the earth's gravity, would it remain in orbit? Explain. What is the real reason why astronauts in orbit feel weightless?

Q12–22 As part of their training before going into orbit, astronauts ride in an airliner that is flown along the same parabolic trajectory as a freely falling projectile. Explain why this gives the same experience of apparent weightlessness as being in orbit.

EXERCISES

SECTION 12–2 NEWTON'S LAW OF GRAVITATION

12–1 What is the ratio of the gravitational pull of the sun on the moon to that of the earth on the moon? Use the data in Appendix F. Is it more accurate to say that the moon orbits the earth, or that the moon orbits the sun?

12–2 A 2150-kg satellite used in a cellular telephone network is in a circular orbit at a height of 780 km above the surface of the earth. What is the gravitational force on the satellite? What fraction is this of its weight at the surface of the earth?

12–3 A set of two uniform spheres of masses m_1 and m_2 have their centers separated by a distance r_{12}. They exert a gravitational force of magnitude F_{12} on each other. What magnitude of gravitational force does a second set of uniform spheres exert on one another if they have masses nm_1 and nm_2 and a center-to-center separation of nr_{12}, where n is any positive number?

12–4 Two uniform spheres, each of mass M and radius R, touch one another. What is the magnitude of their gravitational force of attraction?

12–5 An interplanetary spaceship passes through the point in space where the gravitational forces from the sun and the earth on the ship exactly cancel. a) How far from the center of the earth is it? Use the data in Appendix F. b) What, if anything, happens to the spaceship when it passes through the point described in part (a)? Explain.

12–6 a) In Fig. 12–23 what are the magnitude and direction of the net gravitational force exerted on the 0.100-kg uniform sphere by the other two uniform spheres? The centers of all three spheres are on the same line. b) According to Newton's third law, does the 0.100-kg sphere exert forces of the same magnitude as your answer to part (a), but opposite direction, on *each* of the other two spheres?

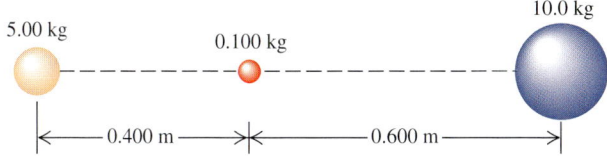

FIGURE 12–23 Exercise 12–6.

12–7 The typical adult human has a mass of about 70 kg. What force does a full moon exert on such a human when it is directly above with its center 378,000 km away?

12–8 The sun has a mass 333,000 times that of the earth. For a person on earth, the average distance to the center of the sun is 23,500 times the distance to the center of the earth. In magnitude, what is the ratio of the sun's gravitational force on you to the earth's gravitational force on you?

12–9 Calculate the magnitude and direction of the net gravitational force on the moon due to the earth and the sun when the moon is in each of the positions shown in Fig. 12–24. (Note that the figure is *not* drawn to scale. Assume that the sun is in the

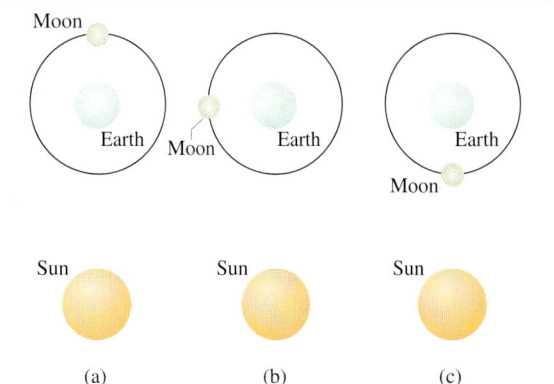

FIGURE 12–24 Exercise 12–9

plane of the earth-moon orbit, even though this is usually not the case.) Use the data in Appendix F.

12–10 Determine the magnitude and direction of the gravitational force on a particle of mass m that is midway between two spherically symmetric objects, one of mass m_1 and the other of mass m_2, where $m_2 > m_1$. The centers of the two spherically symmetric objects are a distance d apart.

12–11 Two uniform spheres, each of mass 0.260 kg, are fixed at points A and B (Fig. 12–25). Find the magnitude and direction of the initial acceleration of a uniform sphere with mass 0.010 kg if released from rest at point P and acted on only by forces of gravitational attraction of the spheres at A and B.

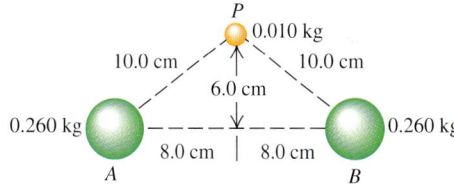

FIGURE 12–25 Exercise 12–11.

SECTION 12–3 WEIGHT

12–12 Use the mass and radius of the planet Pluto given in Appendix F to calculate the acceleration due to gravity at the surface of Pluto.

12–13 At what distance above the surface of the earth is the acceleration due to the earth's gravity 0.980 m/s² if the acceleration due to gravity at the surface has magnitude 9.80 m/s²?

12–14 The mass of Venus is 81.5% that of the earth, and its radius is 94.9% that of the earth. a) Compute the acceleration due to gravity on the surface of Venus from these data. b) What is the weight of a 5.00-kg rock on the surface of Venus?

12–15 Titania, the largest moon of the planet Uranus, has $\frac{1}{8}$ the radius of the earth and $\frac{1}{1700}$ the mass of the earth. a) What is the acceleration due to gravity at the surface of Titania? b) What is the average density of Titania? (This is less than the density of rock, which is one piece of evidence that Titania is made primarily of ice.)

12–16 Rhea, one of Saturn's moons, has a radius of 765 km and an acceleration due to gravity of 0.278 m/s^2 at its surface. Calculate its mass and average density.

12–17 An experiment using the Cavendish balance to measure the gravitational constant G found that a uniform 0.400-kg sphere attracts another 0.00300-kg sphere with a force of 8.00×10^{-10} N when the distance between the centers of the spheres is 0.0100 m. The acceleration due to gravity at the earth's surface is 9.80 m/s^2, and the radius of the earth is 6380 km. Compute the mass of the earth from these data.

12–18 In Example 12–4 (Section 12–3) we neglected the gravitational effects of the moons of Mars. The larger of the two moons, Phobos, has a radius of approximately 12 km and an average density of 2000 kg/m^3. a) Calculate the gravitational force of Phobos on the Mars lander of Example 12–4 if the Mars lander were on the surface of Phobos. b) Discuss whether the approximation made in Example 12–4 was appropriate.

SECTION 12–4 GRAVITATIONAL POTENTIAL ENERGY

12–19 The asteroid Dactyl, discovered in 1993, has a radius of only about 700 m and a mass of about 3.6×10^{12} kg. Use the results of Example 12–5 (Section 12–4) to calculate the escape speed for an object at the surface of Dactyl. Could a person reach this speed just by walking?

12–20 When in orbit, a communication satellite attracts the earth with a force of 19.0 kN and the earth-satellite gravitational potential energy (relative to zero at infinite separation) is -1.39×10^{11} J. a) Find the satellite's altitude above the earth's surface. b) Find the mass of the satellite.

12–21 Use the results of Example 12–5 (Section 12–4) to calculate the escape speed for a spacecraft a) from the surface of Mars; b) from the surface of Jupiter. Use the data in Appendix F. c) Why is the escape speed for a spacecraft independent of the spacecraft's mass?

12–22 Ten days after it was launched toward Mars in December 1998, the *Mars Climate Orbiter* spacecraft (mass 629 kg) was 2.87×10^6 km from the earth and traveling at 1.20×10^4 km/h relative to the earth. At this time, what were a) the spacecraft's kinetic energy relative to the earth and b) the potential energy of the earth-spacecraft system?

SECTION 12–5 THE MOTION OF SATELLITES

12–23 An earth satellite moves in a circular orbit with an orbital speed of 6200 m/s. a) Find the time of one revolution. b) Find the radial acceleration of the satellite in its orbit.

12–24 What is the period of revolution of a satellite of mass m that orbits the earth in a circular path of radius 7880 km (about 1500 km above the surface of the earth)?

12–25 For a satellite to be in a circular orbit 780 km above the surface of the earth, what must be its orbital speed?

SECTION 12–6 THE MOTION OF PLANETS

12–26 For a time in the nineteenth century it was thought (incorrectly) that there was a planet called Vulcan orbiting the

sun in a circular orbit with a period of 19.7 days. If Vulcan had turned out to be real, what would have been the radius of its orbit? (The name "Vulcan" may have been the inspiration for the planet of the same name in *Star Trek*.)

12–27 The star Rho^1 Cancri is 57 light years from the earth and has a mass 0.85 times that of our sun. A planet has been detected in a circular orbit around Rho^1 Cancri with an orbital radius equal to 0.11 times the radius of the earth's orbit around the sun. What is a) the orbital speed and b) the orbital period of the planet of Rho^1 Cancri?

12–28 Venus orbits the sun in a nearly circular orbit. Use the data in Appendix F for the orbital radius and period of Venus to calculate the mass of the sun.

12–29 a) Use Fig. 12–13 to show that the sun-planet distance at perihelion is $(1 - e)a$, the sun-planet distance at aphelion is $(1 + e)a$, and therefore that the sum of these two distances is $2a$. b) Pluto is usually called the outermost planet, but at perihelion in 1989 it was almost 100 million km closer to the sun than Neptune. The semi-major axes of the orbits of Pluto and Neptune are 5.92×10^{12} m and 4.50×10^{12} m, respectively, and the eccentricities are 0.248 and 0.010. Find Pluto's closest distance and Neptune's farthest distance from the sun. c) How many years after being at perihelion in 1989 will Pluto again be at perihelion?

12–30 The star 70 Virginis is 59 light years from earth and has a mass of 1.9×10^{30} kg. a) A large planet of mass 1.3×10^{28} kg is known to orbit this star. The planet is attracted to the star by a force of 3.3×10^{26} N when their centers are separated by a distance equal to the semi-major axis of the planet's orbit. Calculate this distance (in km). b) What is the period (in days) of this planet's orbit? c) Other as yet undiscovered planets may orbit 70 Virginis. If a second planet is in a circular orbit around 70 Virginis with an orbital period equal to eight times the value found in part (b), what is the radius (in km) of the second planet's orbit?

12–31 The *Helios B* spacecraft had a speed of 71 km/s when it was 4.3×10^7 km from the sun. a) Prove that it was not in a circular orbit about the sun. b) Prove that its orbit about the sun was closed and therefore elliptical.

*SECTION 12–7 SPHERICAL MASS DISTRIBUTIONS

***12–32** A thin, uniform rod has length L and mass M. A small uniform sphere of mass m is placed a distance x from one end of the rod, along the axis of the rod (Fig. 12–26). a) Calculate the gravitational potential energy of the rod-sphere system. Take the potential energy to be zero when the rod and sphere are infinitely far apart. Show that your answer reduces to the expected result when x is much larger than L. (*Hint:* Use the power series expansion for $\ln(1 + x)$ given in Appendix B.) b) Use $F_x = -dU/dx$ to find the magnitude and direction of the gravitational force exerted on the sphere by the rod. Show that your answer reduces to the expected result when x is much larger than L.

FIGURE 12–26 Exercise 12–32, Problem 12–71.

*12–33 Consider the ring-shaped body of Fig. 12–27. A particle of mass m is placed a distance x from the center of the ring, along the line through the center of the ring and perpendicular to its plane. a) Calculate the gravitational potential energy U of this system. Take the potential energy to be zero when the two objects are far apart. b) Show that your answer to part (a) reduces to the expected result when x is much larger than the radius a of the ring. c) Use $F_x = -dU/dx$ to find the magnitude and direction of the force on the particle. d) Show that your answer to part (c) reduces to the expected result when x is much larger than a. e) What are the values of U and F_x when $x = 0$? Explain why these results make sense.

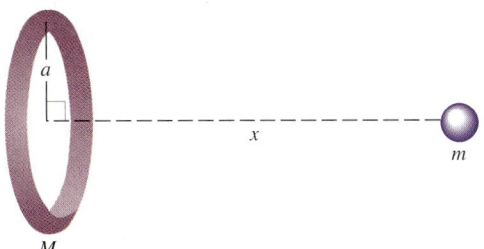

FIGURE 12–27 Exercise 12–33, Problem 12–70.

*SECTION 12–8 APPARENT WEIGHT AND THE EARTH'S ROTATION

*12–34 The weight of a person at the equator, as determined by a spring balance, is 690 N. By how much does this differ from the true force of gravitational attraction at the same point? Assume that the earth is spherically symmetric.

*12–35 The acceleration due to gravity at the north pole of Neptune is approximately 10.7 m/s^2. Neptune has mass 1.0×10^{26} kg and radius 2.5×10^4 km and rotates once around its axis in about 16 h. a) What is the gravitational force on a 5.0-kg object at the north pole of Neptune? b) What is the apparent weight of this same object at Neptune's equator? (Note that Neptune's "surface" is gaseous, not solid, so it is impossible to stand on it.)

SECTION 12–9 BLACK HOLES: A CASE STUDY IN MODERN PHYSICS

12–36 a) Show that a black hole attracts an object of mass m with a force of $mc^2 R_S/(2r^2)$, where r is the distance between the object and the center of the black hole. b) Calculate the magnitude of the gravitational force exerted by a black hole of Schwarzschild radius 14.0 mm on a 5.00-kg mass 3000 km from it. c) What is the mass of this black hole?

12–37 At the Galaxy's Core. Astronomers have observed a small, massive object at the center of our Milky Way galaxy. A ring of material orbits this massive object; the ring has a diameter of about 15 light years and an orbital speed of about 200 km/s. a) Determine the mass of the massive object at the center of the Milky Way galaxy. Give your answer both in kilograms and in solar masses (one solar mass is the mass of the sun). b) Observations of stars, as well as theories of the structure of stars, suggest that it is impossible for a single star to have a mass of more than about 50 solar masses. Can this massive object be a single, ordinary star? c) Many astronomers believe that the massive object at the center of the Milky Way galaxy is a black hole. If so, what must the Schwarzschild radius of this black hole be? Would a black hole of this size fit inside the earth's orbit around the sun?

12–38 Derive the relation $R_S = (3.0 \text{ km/solar mass})M$, where M is in solar masses. (One solar mass is the mass of the sun.) How accurate is this relation?

12–39 To what fraction of its current radius would the earth have to be compressed to become a black hole?

PROBLEMS

12–40 Three uniform spheres are fixed at positions shown in Fig. 12–28. a) What is the magnitude and direction of the force on a 0.0150-kg particle placed at P? b) If the spheres are in deep outer space and a 0.0150-kg particle is released from rest 300 m from the origin along a line 45° below the $-x$-axis, what will the particle's speed be when it reaches the origin?

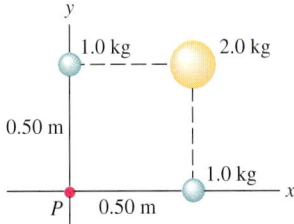

FIGURE 12–28 Problem 12–40.

12–41 A uniform sphere with mass 60.0 kg is held with its center at the origin and a second uniform sphere with mass 80.0 kg is held with its center at the point $x = 0$, $y = 3.00$ m. a) What are the magnitude and direction of the net gravitational force due to these objects on a third uniform sphere with mass 0.500 kg placed at the point $x = 4.00$ m, $y = 0$? b) Where, other than infinitely far away, could the third sphere be placed such that the net gravitational force acting on it from the other two spheres is equal to zero?

12–42 a) Show that the gravitational force on the small mass due to the two large masses in Example 12–3 (Section 12–2) is *not* directed toward the point midway between the two large masses. b) Consider the two large masses as making up a single rigid body. (That is, the two masses are joined by a rod of negligible mass). Calculate the torque exerted by the small mass on the rigid body for a pivot at its center of mass. c) Explain how the result in part (b) shows that the center of mass does not coincide with the center of gravity. Why is this the case in this situation?

12–43 At a certain instant the earth, the moon, and a stationary 1250-kg spacecraft lie at the vertices of an equilateral triangle whose sides are 3.84×10^5 km in length. a) Find the magnitude and direction of the net gravitational force exerted on the spacecraft by the earth and moon. State the direction as an angle measured from a line connecting the earth and the spacecraft. In a sketch, show the earth, the moon, the spacecraft, and the force vector. b) What is the minimum amount of work that you would have to do to move the spacecraft to a point far from the earth and moon? Ignore any gravitational effects due to the other planets or the sun.

12–44 An experiment is performed in deep space with two uniform spheres of equal radii, one with mass 25.0 kg and the other with mass 100.0 kg. The spheres are released from rest with their centers 40.0 m apart. They accelerate toward each other because of their mutual gravitational attraction. Neglect all gravitational forces other than that between the two spheres. a) Explain why linear momentum is conserved. b) When their centers are 20.0 m apart, find i) the speed of each sphere and ii) the magnitude of the relative velocity with which one sphere is approaching the other. c) How far from the initial position of the center of the 25.0-kg sphere do the surfaces of the two spheres collide?

12–45 Assume that the moon orbits the earth in a circular orbit. From the observed orbital period of 27.3 d, calculate the distance of the moon from the center of the earth. Assume that the motion of the moon is determined solely by the gravitational force exerted on it by the earth, and use the mass of the earth given in Appendix F.

12–46 Gravitational Field. The *gravitational field* $\vec{g}$ produced by an object is defined as the gravitational force $\vec{F}_g$ that the object exerts on a small test particle of mass m divided by m. That is, $\vec{g} = \vec{F}_g/m$. What are the magnitude and direction of the gravitational field produced by a 20.0-kg spherically symmetric object at a point 1.50 m directly above the center of the object?

12–47 Geosynchronous Satellites. Many satellites are moving in a circle in the earth's equatorial plane. They are at such a height above the earth's surface that they remain always above the same point. a) Find the altitude of these satellites above the earth's surface. (Such an orbit is said to be *geosynchronous*.) b) Explain, with a sketch, why the radio signals from these satellites cannot directly reach receivers on earth that are north of 81.3° N latitude.

12–48 Suppose an object is to be placed in a circular orbit around the asteroid Dactyl (see Exercise 12–19), with the radius of its orbit just slightly larger than the asteroid's radius. What is the speed of the object? Could you launch yourself into orbit around Dactyl just by walking?

12–49 What is the escape speed from a 300-km-diameter asteroid with a density of 2500 kg/m³?

12–50 a) Asteroids have average densities of about 2500 kg/m³ and radii from 470 km down to less than a kilometer. Assuming that the asteroid has a spherically symmetric mass distribution, estimate the radius of the largest asteroid from which you could escape simply by jumping off. (*Hint:* You can estimate your jump speed by relating it to the maximum height that you can jump on earth.) b) Europa, one of Jupiter's four large moons, has a radius of 1570 km. The acceleration due to gravity at its surface is 1.33 m/s². Calculate its average density.

12–51 There are two equations from which a change in the gravitational potential energy U of the system of a mass m and the earth can be calculated. One is $U = mgy$ (Eq. 7–2). The other

is $U = -Gm_E m/r$ (Eq. 12–9). As shown in Section 12–4, the first equation is correct only if the gravitational force is a constant over the change in height Δy. The second is always correct. Actually, the gravitational force is never exactly constant over any change in height, but if the variation is small we can ignore it. Consider the difference in U between a mass at the earth's surface and a distance h above it using both equations, and find the value of h for which Eq. (7–2) is in error by 1%. Express this value of h as a fraction of the earth's radius, and also obtain a numerical value for it.

12–52 In Example 12–5 (Section 12–4) we ignored the gravitational effects of the moon on a spacecraft en route from the earth to the moon. In fact, we must include the gravitational potential energy due to the moon as well. For this problem, ignore the motion of the earth and moon. a) If the moon has radius R_M and the distance between the centers of the earth and the moon is R_{EM}, find the total gravitational potential energy of the particle-earth and particle-moon systems when a particle of mass m is between the earth and the moon and a distance r from the center of the earth. Take the gravitational potential energy to be zero when the objects are far from each other. b) There is a point along a line between the earth and the moon where the net gravitational force is zero. Use the expression derived in part (a) and numerical values from Appendix F to find the distance of this point from the center of the earth. With what speed must a spacecraft be launched from the surface of the earth to just barely reach this point? c) If a spacecraft were launched from the earth's surface toward the moon with an initial speed of 11.2 km/s, with what speed would it impact the moon?

12–53 An unmanned spacecraft is in a circular orbit around the moon, observing the lunar surface from an altitude of 50.0 km. (See Appendix F.) To the dismay of scientists on earth, an electrical fault causes an on-board thruster to fire, decreasing the speed of the spacecraft by 20.0 m/s. If nothing is done to correct its orbit, with what speed (in km/h) will the spacecraft crash into the lunar surface?

*12–54 What would be the length of a day (that is, the time required for one rotation of the earth on its axis) if the rate of rotation of the earth were such that $g = 0$ at the equator?

12–55 A hammer with mass m is dropped from rest from a height h above the earth's surface. This height is not necessarily small compared with the radius R_E of the earth. If air resistance is neglected, derive an expression for the speed v of the hammer when it reaches the surface of the earth. Your expression should involve h, R_E, and m_E, the mass of the earth.

12–56 a) Calculate how much work is required to launch a spacecraft of mass m from the surface of the earth (mass m_E, radius R_E) and place it in a circular *low earth orbit*, that is, an orbit whose altitude above the earth's surface is much less than R_E. (As an example, the International Space Station is in low earth orbit at an altitude of about 400 km, much less than $R_E = 6380$ km.) Ignore the kinetic energy that the spacecraft has on the ground due to the earth's rotation. b) Calculate the minimum amount of additional work required to move the spacecraft from low earth orbit to a very great distance from the earth. Ignore the gravitational effects of the sun, the moon, and the other planets. c) Justify the statement that "in terms of energy, low earth orbit is halfway to the edge of the universe."

12–57 A spacecraft is to be launched from the surface of the earth so that it will escape from the solar system altogether. a) Find the speed relative to the center of the earth with which it must be launched. Take into consideration the gravitational effects of both the earth and the sun and include the effects of the earth's orbital speed, but ignore air resistance. b) The rotation of the earth can help this spacecraft achieve escape speed. Find the speed that the spacecraft must have relative to the earth's *surface* if the spacecraft is launched from Florida at the point shown in Fig. 12–29. The rotation and orbital motions of the earth are in the same direction. The launch facilities at Cape Canaveral are 28.5° north of the equator. c) The European Space Agency (ESA) uses launch facilities in French Guiana (immediately north of Brazil), 5.15° north of the equator. What speed relative to the earth's surface would a spacecraft need to escape the solar system if launched from French Guiana?

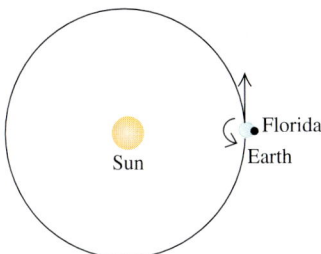

FIGURE 12–29 Problem 12–57.

12–58 We can define a useful quantity called the *gravitational potential* of the earth (denoted by ϕ) by requiring that the product of its value at any point and the mass of a particle placed at that point gives us the gravitational potential energy of that earth-particle system. (We will use an analogous quantity, the electric potential V, in the discussion of electrical interactions in Chapter 24.) a) Find the SI units for the gravitational potential and show they are equivalent to m^2/s^2. b) What is the equation for the gravitational potential at any point outside the earth, if it is taken to be zero infinitely far from the earth? c) What is the increase in gravitational potential between the earth's surface and the International Space Station's altitude of 400 km? d) Use the result of part (c) to calculate the amount of work that must be done against the earth's gravitational force to move a 15,000-kg payload from Cape Canaveral to an altitude of 400 km.

12–59 Kirkwood Gaps. Hundreds of thousands of asteroids orbit the sun within the *asteroid belt*, which extends from about 3×10^8 km to about 5×10^8 km from the sun. a) Find the orbital period (in years) of i) an asteroid at the inside of the belt and ii) an asteroid at the outside of the belt. Assume circular orbits. b) In 1867 the American astronomer Daniel Kirkwood pointed out that several gaps exist in the asteroid belt where relatively few asteroids are found. It is now understood that these *Kirkwood gaps* are caused by the gravitational attraction of Jupiter, largest of the planets, which orbits the sun once every

11.86 years. As an example, if an asteroid has an orbital period half that of Jupiter, or 5.93 years, on every other orbit this asteroid would be at its closest to Jupiter and feel a strong attraction toward the planet. This attraction, acting over and over on successive orbits, could sweep asteroids out of the Kirkwood gap. Use this hypothesis to determine the orbital radius for this Kirkwood gap. c) One of several other Kirkwood gaps appears at a distance from the sun where the orbital period is 0.400 that of Jupiter. Explain why this happens, and find the orbital radius for this Kirkwood gap.

12–60 If a satellite is in a sufficiently low orbit, it will encounter air drag from the earth's atmosphere. Since air drag does negative work (the force of air drag is directed opposite the motion), the mechanical energy will decrease. According to Eq. (12–15), if E decreases (becomes more negative), the radius r of the orbit will decrease. If air drag is relatively small, the satellite can be considered to be in a circular orbit of continually decreasing radius. a) According to Eq. (12–12), if the radius of a satellite's circular orbit decreases, the satellite's orbital speed v *increases*. How can you reconcile this with the statement that the mechanical energy *decreases*? (*Hint:* Is air drag the only force that does work on the satellite as the orbital radius decreases?) b) Due to air drag, the radius of a satellite's circular orbit decreases from r to $r - \Delta r$, where the positive quantity Δr is much less than r. The mass of the satellite is m. Show that the increase in orbital speed is $\Delta v = +(\Delta r/2)\sqrt{Gm_E/r^3}$; that the change in kinetic energy is $\Delta K = -W = +(Gm_E m/2r^2)\Delta r$; that the change in gravitational potential energy is $\Delta U = 2W = -2\Delta K = -(Gm_E m/r^2)\Delta r$; and that the amount of work done by the force of air drag is $W = -(Gm_E m/2r^2)\Delta r$. Interpret these results in light of your comments in part (a). c) A satellite of mass 3000 kg is initially in a circular orbit 300 km above the earth's surface. Due to air drag, the satellite's altitude decreases to 250 km. Calculate the initial orbital speed; the increase in orbital speed; the initial mechanical energy; the change in kinetic energy; the change in gravitational potential energy; the change in mechanical energy; and the work done by the force of air drag. d) Eventually a satellite will descend to a low enough altitude in the atmosphere that the satellite burns up and the debris falls to earth. What becomes of the initial mechanical energy?

12–61 Binary Star—Equal Masses. Two identical stars of mass M orbit around their center of mass. Each orbit is circular and has radius R, so that the two stars are always on opposite sides of the circle. a) Find the gravitational force of one star on the other. b) Find the orbital speed of each star and the period of the orbit. c) How much energy would be required to separate the two stars to infinity?

12–62 Binary Star—Different Masses. Two stars, of mass M_1 and M_2, are in circular orbits around their center of mass. The star of mass M_1 has an orbit of radius R_1; the star of mass M_2 has an orbit of radius R_2. a) Show that the ratio of the orbital radii of the two stars equals the reciprocal of the ratio of their masses, that is, $R_1/R_2 = M_2/M_1$. b) Explain why the two stars have the same orbital period, and show that the period T is given by $T = 2\pi(R_1 + R_2)^{3/2}/\sqrt{G(M_1 + M_2)}$. c) The two stars in a certain binary star system move in circular orbits. The first star, Alpha, has an orbital speed of 36.0 km/s. The other star, Beta, has an orbital speed of 12.0 km/s. The orbital period is 137 d. What are the masses of each of the two stars? d) One of the best candidates for a black hole is found in the binary system called A0620-0090. The two objects in the binary system are an orange star, V616 Monocerotis, and a compact object believed to be a black hole (Fig. 12–21). The orbital period of A0620-0090 is 7.75 hours, the mass of V616 Monocerotis is estimated to be 0.67 times the mass of the sun, and the mass of the black hole is estimated to be 3.8 times the mass of the sun. Assuming that the orbits are circular, find the radius of each object's orbit and the orbital speed of each object. Compare these answers to the orbital radius and orbital speed of the earth in its orbit around the sun.

12–63 Comets travel around the sun in elliptical orbits with large eccentricities. If a comet has speed 2.0×10^4 m/s when at a distance of 2.5×10^{11} m from the center of the sun, what is its speed when at a distance of 5.0×10^{10} m?

12–64 As Mars orbits the sun in its elliptical orbit, its distance of closest approach to the center of the sun (at perihelion) is 2.067×10^{11} m, and its maximum distance from the center of the sun (at aphelion) is 2.492×10^{11} m. If the orbital speed of Mars at aphelion is 2.198×10^4 m/s, what is its orbital speed at perihelion? (Neglect the influence of the other planets.)

12–65 Consider a spacecraft in an elliptical orbit around the earth. At the low point, or perigee of its orbit, it is 400 km above the earth's surface; at the high point, or apogee, it is 4000 km above the earth's surface. a) What is the period of the spacecraft's orbit? b) Using conservation of angular momentum, find the ratio of the spacecraft's speed at perigee to its speed at apogee. c) Using conservation of energy, find the speed at perigee and the speed at apogee. d) It is desired to have the spacecraft escape from the earth completely. If the spacecraft's rockets are fired at perigee, by how much would the speed have to be increased to achieve this? What if the rockets were fired at apogee? Which point in the orbit is the more efficient to use?

12–66 The planet Uranus has a radius of 25,560 km and a surface acceleration due to gravity of 11.1 m/s^2 at its poles. Its moon Miranda (discovered by Kuiper in 1948) is in a circular orbit about Uranus at an altitude of 104,000 km above the planet's surface. Miranda has a mass of 6.6×10^{19} kg and a radius of 235 km. a) Calculate the mass of Uranus from the above data. b) Calculate the magnitude of Miranda's acceleration due to its orbital motion about Uranus. c) Calculate the acceleration due to Miranda's gravity at the surface of Miranda. d) Do the answers to parts (b) and (c) mean that an object released 1 m above Miranda's surface on the side toward Uranus will fall *up* relative to Miranda? Explain.

12–67 A 3000-kg spacecraft is in a circular orbit 2000 km above the surface of Mars. How much work must the spacecraft engines perform to move the spacecraft to a circular orbit with radius 4000 km?

12–68 In Example 12–8 (Section 12–6) the orbital period of Halley's comet was calculated to be 75.5 years. Many other comets have orbits that extend much farther from the sun, and hence have much longer periods. a) At aphelion, a typical long-period comet is about 8×10^{12} km from the sun; at perihelion, it passes inside the earth's orbit. Estimate the orbital period of such a comet. Give your answer in years. b) Estimate the speed of a typical long-period comet at perihelion. (*Hint:* At aphelion, the comet is moving very slowly.) c) The massive nucleus of a comet contains about 10^{15} kg of material. If the earth were to be struck by a long-period comet, estimate the kinetic energy the comet would have just before impact. (*Hint:* The motion of the comet is due primarily to the gravitational attraction of the sun, not the earth. Why?) Compare this to the energy released in a very large volcanic eruption, about 6×10^{18} J, and to the energy that would be released by burning all the fossil fuel on earth, about 2×10^{23} J. An impact of this kind apparently occurred 65 million years ago in the Yucatan, and is implicated in the demise of the dinosaurs as well as many other ancient species.

12–69 One of the brightest comets of the twentieth century was Comet Hyakutake, which passed close to the sun in early 1996. The orbital period of this comet is estimated to be about 30,000 years. Find the semi-major axis of this comet's orbit. Compare to the average sun-Pluto distance and to the distance to Alpha Centauri, the nearest star to the sun, which is 4.3 light years distant.

*12–70 An object in the shape of a thin ring has radius a and mass M. A uniform sphere with a mass m and radius R is placed with its center at a distance x to the right of the center of the ring along a line through the center of the ring and perpendicular to its plane (Fig. 12–27). What is the gravitational force that the sphere exerts on the ring-shaped object? Show that your result reduces to the expected result when x is much larger than a.

*12–71 A thin uniform rod has length L and mass M. Calculate the magnitude of the gravitational force the rod exerts on a particle of mass m that is at a point along the axis of the rod a distance x from one end (Fig. 12–26). Show that your result reduces to the expected result when x is much larger than L.

*12–72 A shaft is drilled from the surface to the center of the earth (Fig. 12–18). As in Example 12–9 (Section 12–7), make the rather unrealistic assumption that the density of the earth is uniform. With this approximation, the gravitational force on an object of mass m that is inside the earth at a distance r from the center has magnitude $F_g = Gm_E mr/R_E^3$ (as shown in Example 12–9) and points toward the center of the earth. a) Derive an expression for the gravitational potential energy $U(r)$ of the object-and-earth system as a function of the object's distance from the center of the earth. Take the potential energy to be zero when the object is at the center of the earth. b) If an object is released in the shaft at the surface of the earth, what speed will it have when it reaches the center of the earth?

CHALLENGE PROBLEMS

12–73 a) When an object is in a circular orbit of radius r around the earth (mass m_E), the period of the orbit is T (given by Eq. (12–14)) and the orbital speed is v (given by Eq. (12–12)). Show that when the object is moved into a circular orbit of slightly larger radius $r + \Delta r$, where $\Delta r \ll r$, its new period is $T + \Delta T$ and its new orbital speed is $v - \Delta v$, where Δr, ΔT, and Δv are all positive quantities and

$$\Delta T = \frac{3\pi \Delta r}{v} \quad \text{and} \quad \Delta v = \frac{\pi \Delta r}{T}$$

(*Hint:* Use the expression $(1 + x)^n \approx 1 + nx$, valid for $|x| \ll 1$.) b) The International Space Station (ISS) is in a nearly circular orbit at an altitude of 398.00 km above the surface of the earth. A maintenance crew is about to arrive on a space shuttle that is also in a circular orbit in the same orbital plane as the ISS, but with an altitude of 398.10 km. The crew has come to remove a faulty 125-m electrical cable, one end of which is attached to the ISS and the other end of which is floating free in space. The plan is for the shuttle to snag the free end just at the moment that the shuttle, the ISS, and the center of the earth all lie along the same line. The cable will then break free from the ISS when it becomes taut. How long after the free end is caught by the space shuttle will it detach from the ISS? Give your answer in minutes. c) If the shuttle misses catching the cable, show that the crew must wait a time $t \approx T^2/\Delta T$ before they have a second chance. Find the numerical value of t and explain whether it would be worth the wait.

12–74 Interplanetary Navigation. The most efficient way to send a spacecraft from the earth to another planet is by using a *Hohmann transfer orbit* (Fig. 12–30). If the orbits of the departure and destination planets are circular, the Hohmann transfer orbit is an elliptical orbit whose perihelion and aphelion are tangent to the orbits of the two planets. The rockets are fired briefly at the departure planet to put the spacecraft into the transfer orbit; the spacecraft then coasts until it reaches the destination planet. The rockets are then fired again to put the spacecraft into the same orbit about the sun as the destination planet. a) For a flight from earth to Mars, in what direction must the rockets be fired at the earth and at Mars: in the direction of motion, or opposite the direction of motion? What about from a flight from Mars to the earth? b) How long does a one-way trip from the

earth to Mars take, between the firings of the rockets? c) To reach Mars from earth, the launch must be timed so that Mars will be at the right spot when the spacecraft reaches Mars's orbit around the sun. At launch, what must be the angle between a sun-Mars line and a sun-earth line? Use data from Appendix F.

12–75 Tidal Forces Near a Black Hole. An astronaut inside a spacecraft, which protects her from harmful radiation, is orbiting a black hole at a distance of 120 km from its center. The black hole is 5.00 times the mass of the sun and has a Schwarzschild radius of 15.0 km. The astronaut is positioned inside the spaceship such that one of her 0.030-kg ears is 6.0 cm farther from the black hole than the center of mass of the spacecraft and the other ear is 6.0 cm closer. a) What is the tension between her ears? Would the astronaut find it difficult to keep from being torn apart by the gravitational forces? (Since her whole body orbits with the same angular velocity, one ear is moving too slowly for the radius of its orbit and the other is moving too fast. Hence her head must exert forces on her ears to keep them in their orbits.) b) Is the center of gravity of her head at the same point as the center of mass? Explain.

*12–76 Mass M is distributed uniformly over a disk of radius a. Find the gravitational force (magnitude and direction) between this disk-shaped mass and a particle of mass m located a distance x above the center of the disk (Fig. 12–31). Does your result reduce to the correct expression as x becomes very large? (*Hint:* Divide the disk into infinitesimally thin concentric rings, use the expression derived in Problem 12–70 for the gravitational force due to each ring, and integrate to find the total force.)

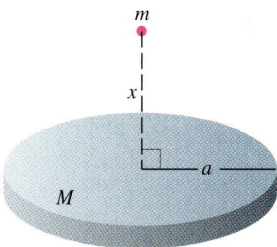

FIGURE 12–31 Challenge Problem 12–76.

*12–77 Mass M is distributed uniformly along a line of length $2L$. A particle of mass m is at a point that is a distance a above the center of the line on its perpendicular bisector (point P in Fig. 12–32). For the gravitational force that the line exerts on the particle, calculate the components perpendicular and parallel to the line. Does your result reduce to the correct expression as a becomes very large?

FIGURE 12–32 Challenge Problem 12–77.

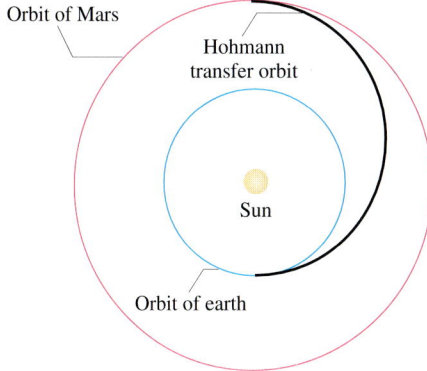

FIGURE 12–30 Challenge Problem 12–74.

13

Periodic Motion

An electric toothbrush is an everyday example of periodic motion, or motion that repeats itself. The brush oscillates through 31,000 up-and-down motions per second, dislodging harmful plaque that an ordinary manual toothbrush might not remove. Using such a toothbrush can make it less likely that your teeth will have to encounter another rapidly moving object—a dentist's drill.

13–1 INTRODUCTION

The vibration of a quartz crystal in a watch, the swinging pendulum of a grandfather clock, the sound vibrations produced by a clarinet or an organ pipe, the back-and-forth motion of the pistons in a car engine—all these are examples of motion that repeats itself over and over. We call this **periodic motion** or **oscillation,** and it is the subject of this chapter. Understanding periodic motion will be essential for our later study of waves, sound, alternating electric currents, and light.

A body that undergoes periodic motion always has a stable equilibrium position. When it is moved away from this position and released, a force or torque comes into play to pull it back toward equilibrium. But by the time it gets there, it has picked up some kinetic energy, so it overshoots, stopping somewhere on the other side, and is again pulled back toward equilibrium. Picture a ball rolling back and forth in a round bowl or a pendulum that swings back and forth past its straight-down position.

In this chapter we will concentrate on two simple examples of systems that can undergo periodic motions: spring-mass systems and pendulums. We will also study why oscillations often tend to die out with time and why some oscillations can build up to greater and greater displacements from equilibrium when periodically varying forces act.

13–2 THE CAUSES OF OSCILLATION

One of the simplest systems that can have periodic motion is shown in Fig. 13–1a. A body with mass m moves on a frictionless horizontal guide system, such as a linear air track, so it moves only along the x-axis. The body is attached to a spring of negligible mass that can be either stretched or compressed. The left end of the spring is held fixed, and the right end is attached to the body. The spring force is the only horizontal force acting on the body; the vertical normal and gravitational forces always add to zero. The quantities x, v, a, and F refer to the x-components of the position, velocity, acceleration, and force vectors, respectively. Like all components, they can be positive, negative, or zero.

It's simplest to define our coordinate system so that the origin O is at the equilibrium position, where the spring is neither stretched nor compressed. Then x is the x-component of the **displacement** of the body from equilibrium and is also the change in length of the spring. The x-component of acceleration a is given by $a = F/m$.

Figure 13–1b shows free-body diagrams for three different positions of the body. Whenever the body is displaced from its equilibrium position, the spring force tends to restore it to the equilibrium position. We call a force with this character a **restoring force.** Oscillation can occur only when there is a restoring force tending to return the system to equilibrium.

13-1 Model for periodic motion. (a) When the body is to the left of the equilibrium position, the compressed spring exerts a force to the right. When the body is at the equilibrium position, the spring is neither compressed nor stretched and exerts no horizontal force. When the body is to the right of the equilibrium position, the stretched spring exerts a force to the left. (b) Free-body diagrams for the three positions. If the spring force F is proportional to the displacement from equilibrium, the motion is simple harmonic.

Let's analyze how oscillation occurs in this system. If we displace the body to the right to $x = A$ and then let go, the net force and the acceleration are to the left. The speed increases as the body approaches the equilibrium position O. When the body is at O, the net force acting on it is zero, but because of its motion it *overshoots* the equilibrium position. On the other side of the equilibrium position the velocity is to the left but the acceleration is to the right; the speed decreases until the body comes to a stop. We will show later that with an ideal spring, the stopping point is at $x = -A$. The body then accelerates to the right, overshoots equilibrium again, and stops at the starting point $x = A$, ready to repeat the whole process. The body is oscillating! If there is no friction or other force to remove mechanical energy from the system, this motion repeats itself forever; the restoring force perpetually draws the body back toward the equilibrium position, only to have the body overshoot time after time.

In different situations the force may depend on the displacement x from equilibrium in different ways. But oscillation will *always* occur if the force is a *restoring* force that tends to return the system to equilibrium.

Here are some terms that we'll use in discussing periodic motions of all kinds:

The **amplitude** of the motion, denoted by A, is the maximum magnitude of displacement from equilibrium; that is, the maximum value of $|x|$. It is always positive. If the spring in Fig. 13-1 is an ideal one, the total overall range of the motion is $2A$. The SI unit of A is the meter. A complete vibration, or **cycle,** is one complete round trip, say from A to $-A$ and back to A, or from O to A, back through O to $-A$, and back to O. Note that motion from one side to the other (say, $-A$ to A) is a half-cycle, not a whole cycle.

The **period,** T, is the time for one cycle. It is always positive. The SI unit is the second, but it is sometimes expressed as "seconds per cycle."

The **frequency,** f, is the number of cycles in a unit of time. It is always positive. The SI unit of frequency is the hertz:

$$1 \text{ hertz} = 1 \text{ Hz} = 1 \text{ cycle/s} = 1 \text{ s}^{-1}.$$

This unit is named in honor of the German physicist Heinrich Hertz (1857–1894), a pioneer in investigating electromagnetic waves.

The **angular frequency**, ω, is 2π times the frequency:

$$\omega = 2\pi f.$$

We'll learn shortly why ω is a useful quantity. It represents the rate of change of an angular quantity (not necessarily related to a rotational motion) that is always measured in radians, so its units are rad/s. Since f is in cycle/s, we may regard the number 2π as having units rad/cycle.

From the definitions of period T and frequency f we see that each is the reciprocal of the other:

$$f = \frac{1}{T}, \qquad T = \frac{1}{f} \qquad \text{(relationships between frequency and period).} \quad (13\text{--}1)$$

Also, from the definition of ω,

$$\omega = 2\pi f = \frac{2\pi}{T} \qquad \text{(angular frequency).} \qquad (13\text{--}2)$$

EXAMPLE 13-1

An ultrasonic transducer (a kind of loudspeaker) used for medical diagnosis oscillates at a frequency of 6.7 MHz = 6.7×10^6 Hz. How much time does each oscillation take, and what is the angular frequency?

SOLUTION The period T is given by Eq. (13–1):

$$T = \frac{1}{f} = \frac{1}{6.7 \times 10^6 \text{ Hz}} = 1.5 \times 10^{-7} \text{ s} = 0.15 \text{ } \mu s.$$

We get ω from Eq. (13–2):

$$\omega = 2\pi f = 2\pi (6.7 \times 10^6 \text{ Hz})$$
$$= (2\pi \text{ rad/cycle})(6.7 \times 10^6 \text{ cycle/s})$$
$$= 4.2 \times 10^7 \text{ rad/s.}$$

A very rapid vibration corresponds to large f and ω and small T; a slow vibration corresponds to small f and ω and large T.

13-3 SIMPLE HARMONIC MOTION

The very simplest kind of oscillation occurs when the restoring force F is *directly proportional* to the displacement from equilibrium x. This happens if the spring in Fig. 13–1 is an ideal one that obeys Hooke's law. The constant of proportionality between F and x is the force constant k. (You may want to review Hooke's law and the definition of the force constant in Section 6–4.) On either side of the equilibrium position, F and x always have opposite signs. In Section 6–4 we represented the force acting *on* a stretched ideal spring as $F = kx$. The x-component of force the spring exerts *on the body* is the negative of this, so the x-component of force F on the body is

$$F = -kx \qquad \text{(restoring force exerted by an ideal spring).} \qquad (13\text{--}3)$$

This equation gives the correct magnitude and sign of the force, whether x is positive, negative, or zero. The force constant k is always positive and has units of N/m (a useful alternative set of units is kg/s^2). We are assuming that there is no friction, so Eq. (13–3) gives the *net* force on the body.

When the restoring force is directly proportional to the displacement from equilibrium, as given by Eq. (13–3), the oscillation is called **simple harmonic motion,** abbreviated **SHM.** The acceleration $a = d^2x/dt^2 = F/m$ of a body in SHM is given by

$$a = \frac{d^2x}{dt^2} = -\frac{k}{m}x \qquad \text{(simple harmonic motion).} \qquad (13\text{--}4)$$

The minus sign means the acceleration and displacement always have opposite signs. This acceleration is *not* constant, so don't even think of using the constant-acceleration equations from Chapter 2. We'll see shortly how to solve this equation to find the displacement x as a function of time. A body that undergoes simple harmonic motion is called a **harmonic oscillator.**

Why is simple harmonic motion important? Keep in mind that not all periodic motions are simple harmonic; in periodic motion in general, the restoring force depends on displacement in a more complicated way than in Eq. (13–3). But in many systems the restoring force is *approximately* proportional to displacement if the displacement is sufficiently small (Fig. 13–2). That is, if the amplitude is small enough, the oscillations of such systems will be approximately simple harmonic and therefore approximately described by Eq. (13–4). Thus we can use SHM as an approximate model for many different periodic motions, such as the vibration of the quartz crystal in a watch, the motion of a tuning fork, the electric current in an alternating-current circuit, and the vibrations of atoms in molecules and solids.

EQUATIONS OF SIMPLE HARMONIC MOTION

To explore the properties of simple harmonic motion, we must express the displacement x of the oscillating body as a function of time, $x(t)$. The second derivative of this function, d^2x/dt^2, must be equal to $(-k/m)$ times the function itself, as required by Eq. (13–4). As we mentioned above, the formulas for constant acceleration from Section 2–5 are no help because the acceleration changes constantly as the displacement x changes. Instead, we'll find $x(t)$ by noticing a striking similarity between SHM and another form of motion that we've already studied in detail.

Figure 13–3 shows a top view of a horizontal disk of radius A with a ball attached to its rim at point Q. The disk rotates with constant angular speed ω (measured in rad/s), so the ball moves in uniform circular motion. A horizontal light beam shines on the rotating disk and casts a shadow of the ball on a screen. The shadow at point P oscillates back and forth as the ball moves in a circle. We then arrange a body attached to an ideal spring, like the combination shown in Fig. 13–1, so that the body oscillates parallel to the shadow. We will prove that the motion of the body and the motion of the ball's shadow are *identical* if the amplitude of the body's oscillation is equal to the disk radius A, and if the angular frequency $2\pi f$ of the oscillating body is equal to the angular speed ω of the rotating disk. That is, *simple harmonic motion is the projection of uniform circular motion onto a diameter.*

We can verify this remarkable statement by finding the acceleration of the shadow at P and comparing it to the acceleration of a body undergoing SHM, given by Eq. (13–4). The circle in which the ball moves so that its projection matches the motion of the oscillating body is called the **circle of reference;** we will call the point Q the *reference point.* We take the circle of reference to lie in the xy-plane, with the origin O at the center of the circle (Fig. 13–4a). At time t the vector OQ from the origin to the reference point Q makes an angle θ with the positive x-axis. As the point Q moves around the

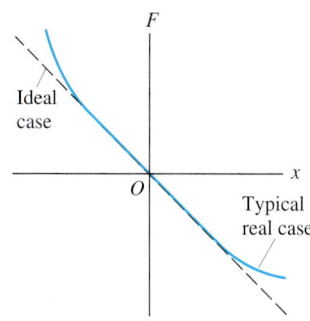

13–2 In most real oscillations the restoring force is not directly proportional to the displacement, and is different in magnitude for displacements on either side of the equilibrium position. But $F = -kx$ is often a good approximation to the force if the displacement x is sufficiently small.

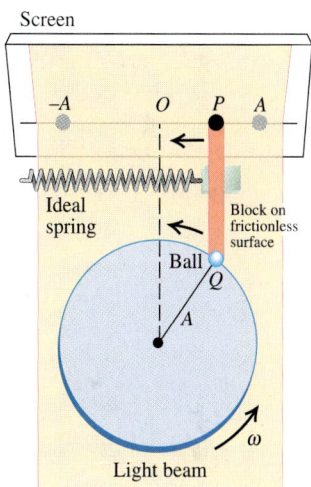

13–3 The ball at point Q rotates counterclockwise in uniform circular motion. Its shadow at point P, projected onto the screen by a beam of light, moves in exactly the same way as a body oscillating on an ideal spring, that is, in simple harmonic motion.

13–4 (a) The coordinate of the ball's shadow P along the x-axis changes with time as the ball Q rotates counterclockwise in uniform circular motion. (b) and (c) The velocity and acceleration of point P are the x-components of the velocity and acceleration vectors, respectively, of point Q.

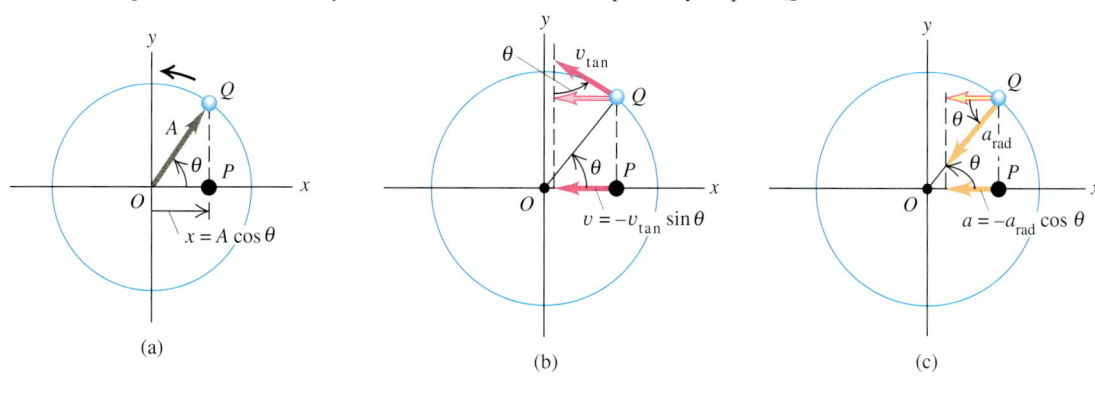

(a) (b) (c)

reference circle with constant angular speed ω, the vector OQ rotates with the same angular speed. Such a rotating vector is called a **phasor.** (This term was in use long before the invention of the similarly named Star Trek stun gun. The phasor method for analyzing oscillations is useful in many areas of physics. We'll use phasors when we study alternating-current circuits in Chapter 32 and the interference of light in Chapters 37 and 38.)

The x-component of the phasor at time t is just the x-coordinate of the point Q:

$$x = A \cos \theta. \qquad (13\text{--}5)$$

This is also the x-coordinate of the shadow P, which is the *projection* of Q onto the x-axis. Hence the acceleration of the shadow P along the x-axis is equal to the x-component of the acceleration vector of the reference point Q (Fig. 13–4c). Since point Q is in uniform circular motion, its acceleration vector $\vec{a}_Q$ is always directed toward O. Furthermore, the magnitude of $\vec{a}_Q$ is constant and given by the angular velocity squared times the radius of the circle (see Section 3–5):

$$a_Q = \omega^2 A. \qquad (13\text{--}6)$$

Figure 13–4c show that the x-component of $\vec{a}_Q$ is $a = -a_Q \cos \theta$. Combining this with Eqs. (13–5) and (13–6), we get that the acceleration of point P is

$$a = -a_Q \cos \theta = -\omega^2 A \cos \theta, \qquad (13\text{--}7)$$

or

$$a = -\omega^2 x. \qquad (13\text{--}8)$$

The acceleration of the point P is directly proportional to the displacement x and always has the opposite sign. These are precisely the hallmarks of simple harmonic motion.

Equation (13–8) will be *exactly* the same as Eq. (13–4) for the acceleration of a harmonic oscillator, provided that the angular speed ω of the reference point Q is related to the force constant k and mass m of the oscillating body by

$$\omega^2 = \frac{k}{m}, \qquad \text{or} \qquad \omega = \sqrt{\frac{k}{m}}. \qquad (13\text{--}9)$$

We have been using the same symbol ω for the angular *speed* of the reference point Q and the angular *frequency* of the oscillating point P. The reason is that these quantities are equal! If point Q makes one complete revolution in time T, then point P goes through one complete cycle of oscillation in the same time; hence T is the period of the oscillation. During time T the point Q moves through 2π radians, so its angular speed is $\omega = 2\pi/T$. But this is just the same as Eq. (13–2) for the angular frequency of the point P, which verifies our statement about the two interpretations of ω. This is why we introduced angular frequency in Section 13–2; it's this quantity that makes the connection between oscillation and circular motion. So we re-interpret Eq. (13–9) as an expression for the angular frequency of simple harmonic motion for a body of mass m, acted on by a restoring force with force constant k:

$$\omega = \sqrt{\frac{k}{m}} \qquad \text{(simple harmonic motion).} \qquad (13\text{--}10)$$

When you start a body oscillating in SHM, the value of ω is not yours to choose; it is predetermined by the values of k and m. The units of k are N/m or kg/s^2, so k/m is in (kg/s^2)/kg $=$ s^{-2}. When we take the square root in Eq. (13–10), we get s^{-1}, or more properly rad/s because this is an *angular* frequency (recall that a radian is not a true unit).

According to Eqs. (13–1) and (13–2), the frequency f and period T are

$$f = \frac{\omega}{2\pi} = \frac{1}{2\pi} \sqrt{\frac{k}{m}} \qquad \text{(simple harmonic motion),} \qquad (13\text{--}11)$$

$$T = \frac{1}{f} = \frac{2\pi}{\omega} = 2\pi\sqrt{\frac{m}{k}} \qquad \text{(simple harmonic motion).} \qquad (13-12)$$

We see from Eq. (13–12) that a larger mass m, with its greater inertia, will have less acceleration, move more slowly, and take a longer time for a complete cycle. In contrast, a stiffer spring (one with a larger force constant k) exerts a greater force at a given deformation x, causing greater acceleration, higher speeds, and a shorter time T per cycle.

CAUTION▶ You can run into trouble if you don't make the distinction between frequency f and angular frequency $\omega = 2\pi f$. Frequency tells you how many cycles of oscillation occur per second, while angular frequency tells you how many radians per second this corresponds to on the reference circle. In solving problems, pay careful attention to whether the goal is to find f or ω. ◀

Equations (13–11) and (13–12) show that the period and frequency of simple harmonic motion are completely determined by the mass m and the force constant k. In simple harmonic motion the period and frequency do not depend on the amplitude A. For given values of m and k the time of one complete oscillation is the same whether the amplitude is large or small. Equation (13–3) shows why we should expect this. Larger A means that the body reaches larger values of $|x|$ and is subjected to larger restoring forces. This increases the average speed of the body over a complete cycle; this exactly compensates for having to travel a larger distance, so the same total time is involved.

The oscillations of a tuning fork are essentially simple harmonic motion, which means that it always vibrates with the same frequency, independent of amplitude. This is why a tuning fork can be used as a standard for musical pitch. If it were not for this characteristic of simple harmonic motion, it would be impossible to make familiar types of mechanical and electronic clocks run accurately or to play most musical instruments in tune. If you encounter an oscillating body with a period that *does* depend on the amplitude, the oscillation is *not* simple harmonic motion.

Equation (13–11) explains why tuning forks of different sizes produce sounds of different pitch. The smaller the mass m in the tuning fork's tines, the higher the frequency of oscillation and the higher the pitch of the sound the tuning fork produces.

EXAMPLE 13–2

Angular frequency, frequency, and period in SHM A spring is mounted horizontally, with its left end held stationary. By attaching a spring balance to the free end and pulling toward the right (Fig. 13–5), we determine that the stretching force is proportional to the displacement and that a force of 6.0 N causes a displacement of 0.030 m. We remove the spring balance and attach a 0.50-kg body to the end, pull it a distance of 0.020 m, release it, and watch it oscillate in SHM. a) Find the force constant of the spring. b) Find the angular frequency, frequency, and period of the oscillation.

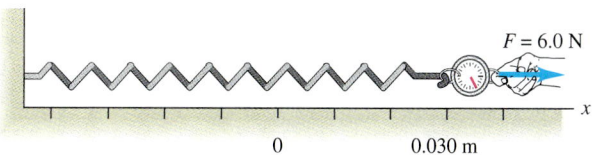

13–5 The force exerted *on* the spring is 6.0 N. The force exerted *by* the spring is –6.0 N.

SOLUTION a) When $x = 0.030$ m, the force the spring exerts on the spring balance is $F = -6.0$ N. From Eq. (13–3),

$$k = -\frac{F}{x} = -\frac{-6.0 \text{ N}}{0.030 \text{ m}} = 200 \text{ N/m}.$$

b) We are given $m = 0.50$ kg and $k = 200$ N/m $= 200$ kg/s². Using Eq. (13–10), we find

$$\omega = \sqrt{\frac{k}{m}} = \sqrt{\frac{200 \text{ kg/s}^2}{0.50 \text{ kg}}} = 20 \text{ rad/s}.$$

The frequency f is

$$f = \frac{\omega}{2\pi} = \frac{20 \text{ rad/s}}{2\pi \text{ rad/cycle}} = 3.2 \text{ cycle/s} = 3.2 \text{ Hz}.$$

The period T is the reciprocal of the frequency f:

$$T = \frac{1}{f} = \frac{1}{3.2 \text{ cycle/s}} = 0.31 \text{ s}.$$

A period is usually stated in "seconds" rather than "seconds per cycle."

The amplitude of the oscillation is 0.020 m, the distance to the right that we pulled the body attached to the spring before releasing it. We didn't need to use this information to find the angular frequency, frequency, or period because in SHM, none of these quantities depend on the amplitude.

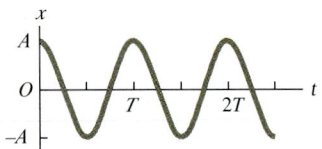

13–6 Graph of x versus t [Eq. (13–13)] for simple harmonic motion. The case shown has $\phi = 0$.

(a)

(b)

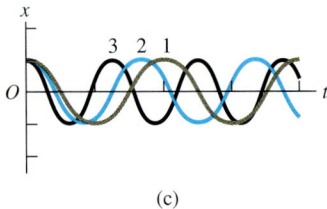

(c)

13–7 Variations of simple harmonic motion. All cases shown have $\phi = 0$. (a) Amplitude A increases from curve 1 to 2 to 3, while mass m and force constant k are unchanged. Changing the amplitude has no effect on the period. (b) Mass m increases from 1 to 2 to 3, while A and k are unchanged. (c) Force constant k increases from 1 to 2 to 3, while A and m are unchanged.

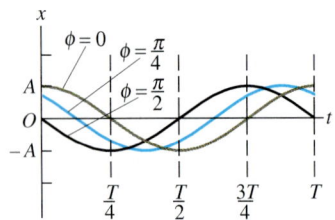

13–8 Position as a function of time for simple harmonic motion with the same frequency and amplitude but with three different phase angles: $\phi = 0$ (green), $\phi = \pi/4$ (blue), and $\phi = \pi/2$ (black).

DISPLACEMENT, VELOCITY, AND ACCELERATION IN SHM

We still need to find the displacement x as a function of time for a harmonic oscillator. Equation (13–4) for a body in simple harmonic motion along the x-axis is identical to Eq. (13–8) for the x-coordinate of the reference point in uniform circular motion with constant angular speed $\omega = \sqrt{k/m}$. It follows that Eq. (13–5), $x = A \cos \theta$, describes the coordinate x for both of these situations. If at $t = 0$ the phasor OQ makes an angle ϕ ("phi") with the positive x-axis, then at any later time t this angle is $\theta = \omega t + \phi$. We substitute this into Eq. (13–5) to obtain

$$x = A \cos(\omega t + \phi) \quad \text{(displacement in SHM),} \quad (13\text{–}13)$$

where $\omega = \sqrt{k/m}$. Figure 13–6 shows a graph of Eq. (13–13) for the particular case $\phi = 0$. The displacement x is a periodic function of time, as expected for SHM. We could also have written Eq. (13–13) in terms of a sine function rather than a cosine by using the identity $\cos \alpha = \sin(\alpha + \pi/2)$. *In simple harmonic motion the position is a periodic, sinusoidal function of time.* There are many other periodic functions, but none so smooth and simple as a sine or cosine function.

The value of the cosine function is always between -1 and 1, so in Eq. (13–13), x is always between $-A$ and A. This confirms that A is the amplitude of the motion. Figure 13–7a shows the graph of x versus t for various values of A.

The period T is the time for one complete cycle of oscillation. The cosine function repeats itself whenever the quantity in parentheses in Eq. (13–13) increases by 2π radians. Thus if we start at time $t = 0$, the time T to complete one cycle is given by

$$\omega T = \sqrt{\frac{k}{m}}\, T = 2\pi, \quad \text{or} \quad T = 2\pi\sqrt{\frac{m}{k}},$$

which is just Eq. (13–12). Changing either m or k changes the period of oscillation, as shown in Figs. 13–7b and 13–7c.

The constant ϕ in Eq. (13–13) is called the **phase angle.** It tells us at what point in the cycle the motion was at $t = 0$ (equivalent to where around the circle the point Q was at $t = 0$). We denote the position at $t = 0$ by x_0. Putting $t = 0$ and $x = x_0$ in Eq. (13–13), we get

$$x_0 = A \cos \phi. \quad (13\text{–}14)$$

If $\phi = 0$, then $x_0 = A \cos 0 = A$, and the body starts at its maximum positive displacement. If $\phi = \pi$, then $x_0 = A \cos \pi = -A$, and the particle starts at its maximum *negative* displacement. If $\phi = \pi/2$, then $x_0 = A \cos(\pi/2) = 0$, and the particle is initially at the origin. Figure 13–8 shows the displacement x versus time for different phase angles.

We find the velocity v and acceleration a as functions of time for a harmonic oscillator by taking derivatives of Eq. (13–13) with respect to time:

$$v = \frac{dx}{dt} = -\omega A \sin(\omega t + \phi) \quad \text{(velocity in SHM),} \quad (13\text{–}15)$$

$$a = \frac{dv}{dt} = \frac{d^2x}{dt^2} = -\omega^2 A \cos(\omega t + \phi) \quad \text{(acceleration in SHM).} \quad (13\text{–}16)$$

The velocity v oscillates between $v_{max} = +\omega A$ and $-v_{max} = -\omega A$, and the acceleration a oscillates between $a_{max} = +\omega^2 A$ and $-a_{max} = -\omega^2 A$ (Fig. 13–9). Comparing Eq. (13–16) with Eq. (13–13) and recalling that $\omega^2 = k/m$ from Eq. (13–9), we see that

$$a = -\omega^2 x = -\frac{k}{m}\, x,$$

which is just Eq. (13–4) for simple harmonic motion. This confirms that Eq. (13–13) for x as a function of time is correct.

We actually derived Eq. (13–16) earlier in a geometrical way by taking the x-component of the acceleration vector of the reference point Q. This was done in Fig. 13–4c and Eq. (13–7) (recall that $\theta = \omega t + \phi$). In the same way, we could have derived Eq. (13–15) by taking the x-component of the velocity vector of Q, as shown in Fig. 13–4b. We'll leave the details for you to work out (see Problem 13–67).

Note that the sinusoidal graph of displacement versus time (Fig. 13–9a) is shifted by one-quarter period from the graph of velocity versus time (Fig. 13–9b) and by one-half period from the graph of acceleration versus time (Fig. 13–9c). When the body is passing through the equilibrium position so that the displacement is zero, the velocity equals either v_{max} or $-v_{max}$ (depending on which way the body is moving) and the acceleration is zero. When the body is at either its maximum positive displacement, $x = +A$, or its maximum negative displacement, $x = -A$, the velocity is zero and the body is instantaneously at rest. At these points, the restoring force $F = -kx$ and the acceleration of the body have their maximum magnitudes. At $x = +A$ the acceleration is negative and equal to $-a_{max}$. At $x = -A$ the acceleration is positive: $a = +a_{max}$.

If we are given the initial position x_0 and initial velocity v_0 for the oscillating body, we can determine the amplitude A and the phase angle ϕ. Here's how to do it. The initial velocity v_0 is the velocity at time $t = 0$; putting $v = v_0$ and $t = 0$ in Eq. (13–15), we find

$$v_0 = -\omega A \sin \phi. \qquad (13\text{–}17)$$

To find ϕ, divide Eq. (13–17) by Eq. (13–14). This eliminates A and gives an equation that we can solve for ϕ:

$$\frac{v_0}{x_0} = \frac{-\omega A \sin \phi}{A \cos \phi} = -\omega \tan \phi,$$

$$\phi = \arctan\left(-\frac{v_0}{\omega x_0}\right) \qquad \text{(phase angle in SHM).} \qquad (13\text{–}18)$$

It is also easy to find the amplitude A if we are given x_0 and v_0. We'll sketch the derivation, and you can fill in the details. Square Eq. (13–14); divide Eq. (13–17) by ω, square it, and add to the square of Eq. (13–14). The right side will be $A^2(\sin^2 \phi + \cos^2 \phi)$, which is equal to A^2. The final result is

$$A = \sqrt{x_0^2 + \frac{v_0^2}{\omega^2}} \qquad \text{(amplitude in SHM).} \qquad (13\text{–}19)$$

Note that when the body has both an initial displacement x_0 and a nonzero initial velocity v_0, the amplitude A is *not* equal to the initial displacement. That's reasonable; if you start the body at a positive x_0 but give it a positive velocity v_0, it will go *farther* than x_0 before it turns and comes back.

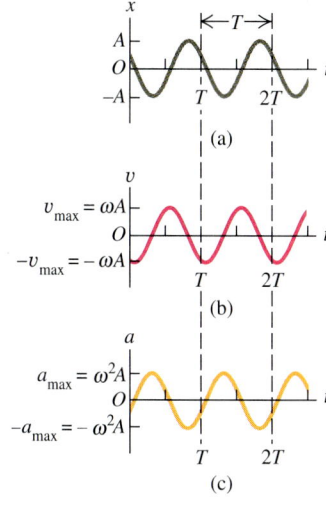

13–9 (a) Graph of x versus t for simple harmonic motion. In this graph, $\phi = \pi/3$. (b) Graph of v versus t for the same motion. Note that this graph is shifted by $\frac{1}{4}$ cycle from the graph of x versus t. (c) Graph of a versus t for the same motion. This graph is shifted by $\frac{1}{4}$ cycle from the graph of v versus t and by $\frac{1}{2}$ cycle from the graph of x versus t.

9.1
Position Graphs and Equations
9.2
Describing Vibrational Motion
9.5
Ape Drops Tarzan

Problem-Solving Strategy

SIMPLE HARMONIC MOTION I

1. Be careful to distinguish between two different kinds of quantities: those that represent basic physical properties of the system and those that describe a particular motion that occurs when the system is set in motion in a specific way. The physical properties include the mass m, the force constant k, and quantities derived from these, including the period T, the frequency f, and the angular frequency $\omega = 2\pi f$. In some problems, m or k or both can be determined from other information given about the

system. Quantities that describe a particular motion include the amplitude A, the maximum velocity v_{max}, the phase angle ϕ, and the displacement, velocity, or acceleration at a particular time.

2. When detailed information is required about displacements, velocities, and accelerations at various times, you will need to use Eqs. (13–13), (13–15), and (13–16). If the initial position x_0 and initial velocity v_0 are both given,

you can determine the phase angle and amplitude from Eqs. (13–18) and (13–19). If the body is given an initial positive displacement x_0 but no initial velocity ($v_0 = 0$), then the amplitude is $A = x_0$ and the phase angle

is $\phi = 0$. If it has an initial positive velocity v_0 but no initial displacement ($x_0 = 0$), the amplitude is $A = v_0/\omega$ and the phase angle is $\phi = -\pi/2$.

EXAMPLE 13–3

Let's return to the horizontal spring we considered in Example 13–2. The force constant is $k = 200$ N/m, and the spring is attached to a body with mass $m = 0.50$ kg. This time we give the body an initial displacement of $+0.015$ m and an initial velocity of $+0.40$ m/s. a) Find the period, amplitude, and phase angle of the motion. b) Write equations for the displacement, velocity, and acceleration as functions of time.

SOLUTION a) The period is exactly the same as in Example 13–2, $T = 0.31$ s. In simple harmonic motion the period does not depend on the amplitude, only on the values of k and m.

In Example 13–2 we found that $\omega = 20$ rad/s. So from Eq. (13–19),

$$A = \sqrt{x_0{}^2 + \frac{v_0{}^2}{\omega^2}}$$

$$= \sqrt{(0.015 \text{ m})^2 + \frac{(0.40 \text{ m/s})^2}{(20 \text{ rad/s})^2}}$$

$$= 0.025 \text{ m}.$$

To find the phase angle ϕ, we use Eq. (13–18):

$$\phi = \arctan\left(\frac{-v_0}{\omega x_0}\right)$$

$$= \arctan\left(\frac{-0.40 \text{ m/s}}{(20 \text{ rad/s})(0.015 \text{ m})}\right) = -53° = -0.93 \text{ rad.}$$

b) The displacement at any time is given by Eq. (13–13); substituting the appropriate values, we find

$$x = (0.025 \text{ m}) \cos [(20 \text{ rad/s})t - 0.93 \text{ rad}].$$

The velocity at any time is given by Eq. (13–15); substituting the values, we find

$$v = -(0.50 \text{ m/s}) \sin [(20 \text{ rad/s})t - 0.93 \text{ rad}].$$

The velocity varies sinusoidally between -0.50 m/s and $+0.50$ m/s. The acceleration is given by Eq. (13–16); substituting the values, we get

$$a = -(10 \text{ m/s}^2) \cos [(20 \text{ rad/s})t - 0.93 \text{ rad}].$$

The acceleration varies between -10 m/s^2 and $+10$ m/s^2.

13–4 ENERGY IN SIMPLE HARMONIC MOTION

9.3
Vibrational Energy

9.4
Two Ways to Measure Young
Tarzan's Mass

9.6
Releasing a Vibrating Skier I

9.7
Releasing a Vibrating Skier II

We can learn even more about simple harmonic motion by using energy considerations. Take another look at the body oscillating on the end of a spring in Fig. 13–1. We've already noted that the spring force is the only horizontal force on the body. The force exerted by an ideal spring is a conservative force, and the vertical forces do no work, so the total mechanical energy of the system is *conserved*. We will also assume that the mass of the spring itself is negligible.

The kinetic energy of the body is $K = \frac{1}{2}mv^2$, and the potential energy of the spring is $U = \frac{1}{2}kx^2$, just as in Section 7–3. (It would be helpful to review that section.) There are no nonconservative forces that do work, so the total mechanical energy $E = K + U$ is conserved:

$$E = \frac{1}{2}mv^2 + \frac{1}{2}kx^2 = \text{constant.} \tag{13–20}$$

The total mechanical energy E is also directly related to the amplitude A of the motion. When the body reaches the point $x = A$, its maximum displacement from equilibrium, it momentarily stops as it turns back toward the equilibrium position. That is, when $x = A$ (or $-A$), $v = 0$. At this point the energy is entirely potential, and $E = \frac{1}{2}kA^2$. Because E is constant, this quantity equals E at any other point. Combining this expression with Eq. (13–20), we get

$$E = \frac{1}{2}mv^2 + \frac{1}{2}kx^2 = \frac{1}{2}kA^2 = \text{constant} \qquad \text{(total mechanical energy in SHM).} \tag{13–21}$$

13-10 In simple harmonic motion the total mechanical energy E is constant, continuously transforming from potential energy U to kinetic energy K and back again as the body oscillates.

9.8
One- and Two-Spring Vibrating Systems

9.9
Vibro-Ride

We can verify this equation by substituting x and v from Eqs. (13–13) and (13–15) and using $\omega^2 = k/m$ from Eq. (13–9):

$$E = \frac{1}{2} mv^2 + \frac{1}{2} kx^2 = \frac{1}{2} m(-\omega A \sin (\omega t + \phi))^2 + \frac{1}{2} k(A \cos (\omega t + \phi))^2$$

$$= \frac{1}{2} kA^2 \sin^2 (\omega t + \phi) + \frac{1}{2} kA^2 \cos^2 (\omega t + \phi)$$

$$= \frac{1}{2} kA^2.$$

(Recall that $\sin^2 \alpha + \cos^2 \alpha = 1$.) Hence our expressions for displacement and velocity in SHM are consistent with energy conservation, as they must be.

We can use Eq. (13–21) to solve for the velocity v of the body at a given displacement x:

$$v = \pm \sqrt{\frac{k}{m}} \sqrt{A^2 - x^2}. \qquad (13\text{–}22)$$

The $\pm$ sign means that at a given value of x the body can be moving in either direction. For example, when $x = \pm A/2$,

$$v = \pm \sqrt{\frac{k}{m}} \sqrt{A^2 - \left(\pm \frac{A}{2}\right)^2} = \pm \sqrt{\frac{3}{4}} \sqrt{\frac{k}{m}} A.$$

Equation (13–22) also shows that the *maximum* speed v_{max} occurs at $x = 0$. Using Eq. (13–10), $\omega = \sqrt{k/m}$, we find that

$$v_{max} = \sqrt{\frac{k}{m}} A = \omega A. \qquad (13\text{–}23)$$

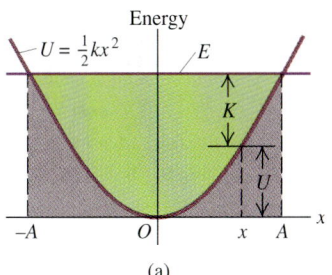

(a)

This agrees with Eq. (13–15), which showed that v oscillates between $-\omega A$ and $+\omega A$.

Figure 13–10 shows the energy quantities E, K, and U at $x = 0$, $x = \pm A/2$, and $x = \pm A$. Figure 13–11 is a graphical display of Eq. (13–21); energy (kinetic, potential, and total) is plotted vertically and the coordinate x is plotted horizontally. The parabolic curve in Fig. 13–11a represents the potential energy $U = \frac{1}{2} kx^2$. The horizontal line represents the total mechanical energy E, which is constant and does not vary with x. This line intersects the potential-energy curve at $x = -A$ and $x = A$, where the energy is entirely potential. At any value of x between $-A$ and A, the vertical distance from the x-axis to the parabola is U; since $E = K + U$, the remaining vertical distance up to the horizontal line is K. Figure 13–11b shows both K and U as functions of x. As the body oscillates between $-A$ and A, the energy is continuously transformed from potential to kinetic and back again.

Figure 13–11a shows the connection between the amplitude A and the corresponding total mechanical energy $E = \frac{1}{2} kA^2$. If we tried to make x greater than A (or less than $-A$), U would be greater than E, and K would have to be negative. But K can never be negative, so x can't be greater than A or less than $-A$.

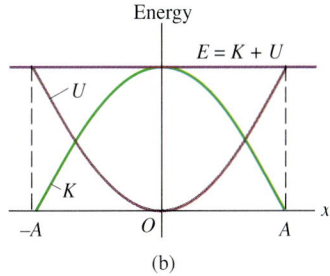

(b)

13-11 Kinetic energy K, potential energy U, and total mechanical energy E as functions of position for SHM. At each value of x the sum of the values of K and U equals the constant value of E.

Problem-Solving Strategy

SIMPLE HARMONIC MOTION II

The energy equation, Eq. (13–21), is a useful alternative relation between velocity and position, especially when energy quantities are also required. If the problem involves a relation among position, velocity, and acceleration without reference to time, it is usually easier to use Eq. (13–4) (from Newton's second law) or (13–21) (from energy conservation) than to use the general expression for x, v, and a as functions of time

[Eqs. (13–13), (13–15), and (13–16), respectively]. Because the energy equation involves x^2 and v^2, it cannot tell you the sign of x or of v; you have to infer the sign from the situation. For instance, if the body is moving from the equilibrium position toward the point of greatest positive displacement, then x is positive and v is positive.

EXAMPLE 13–4

In the oscillation described in Example 13–2, $k = 200$ N/m, $m = 0.50$ kg, and the oscillating mass is released from rest at $x = 0.020$ m. a) Find the maximum and minimum velocities attained by the oscillating body. b) Compute the maximum acceleration. c) Determine the velocity and acceleration when the body has moved halfway to the center from its original position. d) Find the total energy, potential energy, and kinetic energy at this position.

SOLUTION a) The velocity v at any displacement x is given by Eq. (13–22):

$$v = \pm \sqrt{\frac{k}{m}} \sqrt{A^2 - x^2}.$$

The maximum velocity occurs when the body is moving to the right through the equilibrium position, where $x = 0$:

$$v = v_{max} = \sqrt{\frac{k}{m}} A = \sqrt{\frac{200 \text{ N/m}}{0.50 \text{ kg}}} (0.020 \text{ m}) = 0.40 \text{ m/s}.$$

The minimum (i.e., most negative) velocity occurs when the body is moving to the left through $x = 0$; its value is $-v_{max} = -0.40$ m/s.
b) From Eq. (13–4),

$$a = -\frac{k}{m} x.$$

The maximum (most positive) acceleration occurs at the most negative value of x, that is, $x = -A$; therefore

$$a_{max} = -\frac{k}{m}(-A) = -\frac{200 \text{ N/m}}{0.50 \text{ kg}}(-0.020 \text{ m}) = 8.0 \text{ m/s}^2.$$

The minimum (most negative) acceleration is -8.0 m/s², occurring at $x = +A = +0.020$ m.
c) At a point halfway to the center from the initial position, $x = A/2 = 0.010$ m. From Eq. (13–22),

$$v = -\sqrt{\frac{200 \text{ N/m}}{0.50 \text{ kg}}} \sqrt{(0.020 \text{ m})^2 - (0.010 \text{ m})^2} = -0.35 \text{ m/s}.$$

We choose the negative square root because the body is moving from $x = A$ toward $x = 0$. From Eq. (13–4),

$$a = -\frac{200 \text{ N/m}}{0.50 \text{ kg}}(0.010 \text{ m}) = -4.0 \text{ m/s}^2.$$

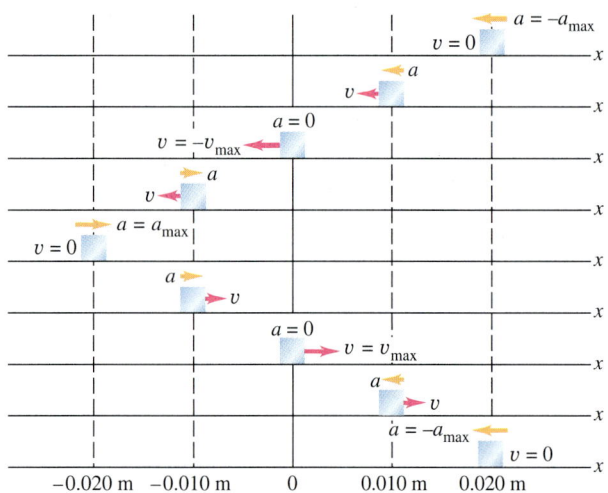

13–12 A body is attached to a spring, pulled 0.020 m from the equilibrium position, and released. The velocity and acceleration are shown at $x = 0$, $x = \pm A/2$, and $x = \pm A$. The velocity of the body is not constant, so these images of the body at equally spaced points in its motion are *not* equally spaced in time.

At this point the velocity and the acceleration have the same sign, so the speed is increasing. The conditions at $x = 0$, $\pm A/2$, and $\pm A$ are shown in Fig. 13–12.
d) The total energy has the same value at all points during the motion:

$$E = \frac{1}{2} kA^2 = \frac{1}{2}(200 \text{ N/m})(0.020 \text{ m})^2 = 0.040 \text{ J}.$$

The potential energy is

$$U = \frac{1}{2} kx^2 = \frac{1}{2}(200 \text{ N/m})(0.010 \text{ m})^2 = 0.010 \text{ J},$$

and the kinetic energy is

$$K = \frac{1}{2} mv^2 = \frac{1}{2}(0.50 \text{ kg})(-0.35 \text{ m/s})^2 = 0.030 \text{ J}.$$

At this point, E is one fourth potential energy and three fourths kinetic energy.

EXAMPLE 13-5

Energy and momentum in SHM A block with mass M attached to a horizontal spring with force constant k is moving with simple harmonic motion having amplitude A_1. At the instant when the block passes through its equilibrium position, a lump of putty with mass m is dropped vertically onto the block from a very small height and sticks to it (Fig. 13–13a). a) Find the new amplitude and period. b) Repeat part (a) for the case in which the putty is dropped on the block when it is at one end of its path (Fig. 13–13b).

SOLUTION a) The problem involves the motion at a given position, not a given time, so let's use energy methods. Before the putty lands on the block, the mechanical energy of the oscillating block and spring is constant. When the putty lands on the block, it's a completely inelastic collision (Section 8–4); the x-component of momentum is conserved, but mechanical energy decreases. Once the collision is over, the mechanical energy remains constant at its new, lower value. Let's look at these three stages in turn—before, during, and after the collision.

Before the collision the total mechanical energy of the block and spring is $E_1 = \frac{1}{2}kA_1{}^2$. Since the block is at the equilibrium position, $U = 0$, and the energy is purely kinetic. If we let v_1 be the speed of the block at the equilibrium position, we have

$$E_1 = \frac{1}{2}Mv_1{}^2 = \frac{1}{2}kA_1{}^2, \qquad \text{so} \qquad v_1 = \sqrt{\frac{k}{M}}A_1.$$

During the collision the x-component of momentum of the system of block and putty is conserved. (Why?) Just before the collision this component is the sum of Mv_1 (for the block) and zero (for the putty). Just after the collision the block and putty move together with speed v_2, and their combined x-component of momentum is $(M + m)v_2$. From conservation of momentum,

$$Mv_1 + 0 = (M + m)v_2, \qquad \text{so} \qquad v_2 = \frac{M}{M + m}v_1.$$

The collision lasts a very short time, so just after the collision the block and putty are still at the equilibrium position. The energy is still purely kinetic, but is *less* than before the collision:

$$E_2 = \frac{1}{2}(M + m)v_2{}^2 = \frac{1}{2}\frac{M^2}{M + m}v_1{}^2 = \frac{M}{M + m}\left(\frac{1}{2}Mv_1{}^2\right)$$

$$= \left(\frac{M}{M + m}\right)E_1.$$

(The lost mechanical energy goes into raising the temperature of the block and the putty.) Since E_2 equals $\frac{1}{2}kA_2{}^2$, where A_2 is the amplitude after the collision, we have

$$\frac{1}{2}kA_2{}^2 = \left(\frac{M}{M + m}\right)\frac{1}{2}kA_1{}^2,$$

$$A_2 = A_1\sqrt{\frac{M}{M + m}}.$$

The larger the putty mass m, the smaller the final amplitude.

Finding the period of oscillation after the collision is the easy

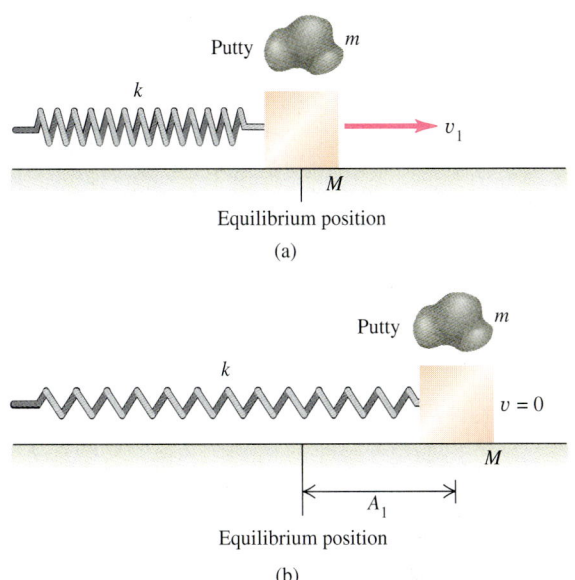

13–13 (a) Dropping a lump of putty on an oscillating block as it passes through equilibrium. (b) Dropping a lump of putty on the block at $x = A_1$.

part. Using Eq. (13–12), we have

$$T_2 = 2\pi\sqrt{\frac{M + m}{k}}.$$

Dropping the putty on the block at the equilibrium position makes each cycle take a longer time and cover a smaller total distance.

b) When the putty drops on the block, the block is instantaneously at rest; all the mechanical energy is stored in the spring as potential energy. Again the x-component of momentum is conserved during the collision, but now this component is zero both before and after the collision. The block had zero kinetic energy just before the collision, and the block and putty have zero kinetic energy just after the collision. So in this case, adding the extra mass of the putty has *no effect* on the mechanical energy. That is,

$$E_2 = E_1 = \frac{1}{2}kA_1{}^2,$$

and the amplitude is still A_1. The period still changes when the putty is added, though; its value doesn't depend on how the mass is added, only on what the total mass is. So T_2 is the same as we found in part (a), $T_2 = 2\pi\sqrt{(M + m)/k}$. Adding the mass of the putty in this way slows down the oscillation but doesn't change the amplitude.

13–5 APPLICATIONS OF SIMPLE HARMONIC MOTION

So far, we've looked at a grand total of *one* situation in which simple harmonic motion (SHM) occurs: a body attached to an ideal horizontal spring. But SHM can occur in any system in which there is a restoring force that is directly proportional to the displacement from equilibrium, as given by Eq. (13–3), $F = -kx$. The restoring force will originate in different ways in different situations, so the force constant k has to be found for each case by examining the net force on the system. Once this is done, it's straightforward to find the angular frequency ω, frequency f, and period T; we just substitute the value of k into Eqs. (13–10), (13–11), and (13–12). Let's use these ideas to examine several examples of simple harmonic motion.

VERTICAL SHM

Suppose we hang a spring with force constant k (Fig. 13–14a) and suspend from it a body with mass m. Oscillations will now be vertical; will they still be SHM? In Fig.

13–14 (a) A hanging spring. (b) A body suspended from the spring is in equilibrium when the upward spring force has the same magnitude as the body's weight. (c) If the body is displaced from equilibrium, the restoring force is proportional to the coordinate measured from the equilibrium position. Hence the oscillations of the body are simple harmonic.

13–14b the body hangs at rest, in equilibrium. In this position the spring is stretched an amount Δl just great enough that the spring's upward vertical force $k\Delta l$ on the body balances its weight mg:

$$k\Delta l = mg.$$

Take $x = 0$ to be this equilibrium position, and take the positive x-direction to be upward. When the body is a distance x *above* its equilibrium position (Fig. 13–14c), the extension of the spring is $\Delta l - x$. The upward force it exerts on the body is then $k(\Delta l - x)$, and the net x-component of force on the body is

$$F_{\text{net}} = k(\Delta l - x) + (-mg) = -kx,$$

that is, a net downward force of magnitude kx. Similarly, when the body is *below* the equilibrium position, there is a net upward force with magnitude kx. In either case there is a restoring force with magnitude kx. If the body is set in vertical motion, it oscillates in SHM with the same angular frequency as though it were horizontal, $\omega = \sqrt{k/m}$. So vertical SHM doesn't differ in any essential way from horizontal SHM. The only real change is that the equilibrium position $x = 0$ no longer corresponds to the point at which the spring is unstretched. The same ideas hold if a body with weight mg is placed atop a compressible spring (Fig. 13–15) and compresses it a distance Δl.

13–15 If the weight mg compresses the spring a distance Δl, the force constant is $k = mg/\Delta l$ and the angular frequency for vertical SHM is $\omega = \sqrt{k/m}$.

EXAMPLE 13–6

Vertical SHM in an old car The shock absorbers in an old car with mass 1000 kg are completely worn out. When a 980-N person climbs slowly into the car to its center of gravity, the car sinks 2.8 cm. When the car, with the person aboard, hits a bump, the car starts oscillating up and down in SHM. Model the car and person as a single body on a single spring, and find the period and frequency of the oscillation.

SOLUTION When the force increases by 980 N, the spring compresses an additional 0.028 m, and the coordinate x of the car changes by -0.028 m. Hence the effective force constant (including the effect of the entire suspension) is

$$k = -\frac{F}{x} = -\frac{980 \text{ N}}{-0.028 \text{ m}} = 3.5 \times 10^4 \text{ kg/s}^2.$$

The person's mass is $w/g = (980 \text{ N})/(9.8 \text{ m/s}^2) = 100$ kg. The *total* oscillating mass is $m = 1000$ kg $+ 100$ kg $= 1100$ kg. The period T is

$$T = 2\pi\sqrt{\frac{m}{k}} = 2\pi\sqrt{\frac{1100 \text{ kg}}{3.5 \times 10^4 \text{ kg/s}^2}} = 1.11 \text{ s},$$

and the frequency is

$$f = \frac{1}{T} = \frac{1}{1.11 \text{ s}} = 0.90 \text{ Hz}.$$

If you have ridden in a car suffering from this affliction, the persistent oscillation may have left your stomach in an altered state.

ANGULAR SHM

Figure 13–16 shows the balance wheel of a mechanical watch. The wheel has a moment of inertia I about its axis. A coil spring (called the *hairspring*) exerts a restoring torque τ that is proportional to the angular displacement θ from the equilibrium position. We write $\tau = -\kappa\theta$, where κ (the Greek letter "kappa") is a constant called the *torsion constant*. Using the rotational analog of Newton's second law for a rigid body, $\Sigma\tau = I\alpha = I\, d^2\theta/dt^2$, the equation of motion is

$$-\kappa\theta = I\alpha, \qquad \text{or} \qquad \frac{d^2\theta}{dt^2} = -\frac{\kappa}{I}\theta.$$

The form of this equation is exactly the same as Eq. (13–4) for the acceleration in simple harmonic motion, with x replaced by θ and k/m replaced by κ/I. So we are dealing with a form of *angular* simple harmonic motion. The angular frequency ω and frequency f are given by Eqs. (13–10) and (13–11), respectively, with the same replacement:

$$\omega = \sqrt{\frac{\kappa}{I}} \qquad \text{and} \qquad f = \frac{1}{2\pi}\sqrt{\frac{\kappa}{I}} \qquad \text{(angular SHM).} \qquad (13\text{–}24)$$

The motion is described by the function

$$\theta = \Theta \cos(\omega t + \phi),$$

where Θ (capital "theta") plays the role of an angular amplitude.

It's a good thing that the motion of a balance wheel *is* simple harmonic. If it weren't, the frequency might depend on the amplitude, and the watch would run too fast or too slow as the spring ran down.

13–16 The balance wheel of a mechanical watch. The spring exerts a restoring torque that is proportional to the angular displacement from the equilibrium position. Therefore the motion is angular SHM.

*VIBRATIONS OF MOLECULES

The following discussion of the vibrations of molecules uses the binomial theorem. If you aren't familiar with this theorem, you should read about it in the appropriate section of your math textbook.

When two atoms are separated from each other by a few atomic diameters, they can exert attractive forces on each other. But if the atoms are so close to each other that their electron shells overlap, the forces between the atoms are repulsive. Between these limits, there can be an equilibrium separation distance at which two atoms form a *molecule*. If these atoms are displaced slightly from equilibrium, they will oscillate. Let's see whether these oscillations can be simple harmonic.

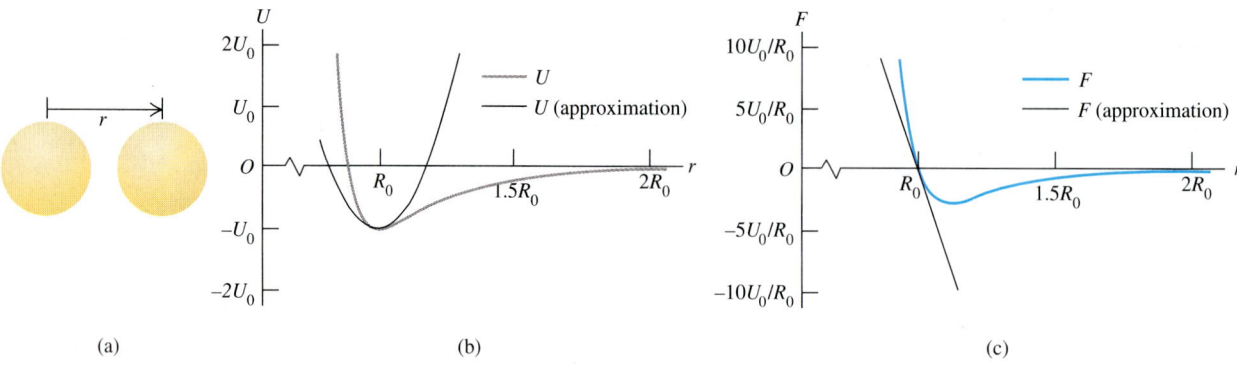

(a)   (b)   (c)

13–17 (a) Two atoms with centers separated by r. (b) Potential energy U in the van der Waals interaction as a function of r. At the equilibrium separation $r = R_0$, U is minimum. Near $r = R_0$, U can be approximated by a parabola. (c) Force F on the right-hand atom as function of r. At the equilibrium separation $r = R_0$, F is zero. Near $r = R_0$, F can be approximated by a straight line.

As an example, we'll consider one type of interaction between atoms called the *van der Waals interaction*. Our immediate task here is specifically to study oscillations, so we won't go into the details of how this interaction arises. Let the center of one atom be at the origin, and let the center of the other atom be a distance r away (Fig. 13–17a); the equilibrium distance between centers is $r = R_0$. Experiment shows that the van der Waals interaction can be described by the potential energy function

$$U = U_0\left[\left(\frac{R_0}{r}\right)^{12} - 2\left(\frac{R_0}{r}\right)^{6}\right], \qquad (13\text{–}25)$$

where U_0 is a positive constant with units of joules. When the two atoms are very far apart, $U = 0$; when they are separated by the equilibrium distance $r = R_0$, $U = -U_0$. The force on the second atom is the negative derivative of Eq. (13–25),

$$F = -\frac{dU}{dr} = U_0\left[\frac{12R_0{}^{12}}{r^{13}} - 2\frac{6R_0{}^{6}}{r^{7}}\right] = 12\frac{U_0}{R_0}\left[\left(\frac{R_0}{r}\right)^{13} - \left(\frac{R_0}{r}\right)^{7}\right]. \quad (13\text{–}26)$$

The potential energy and force are plotted in Figs. 13–17b and 13–17c, respectively. The force is positive for $r < R_0$ and negative for $r > R_0$, so this is a *restoring* force.

To study small-amplitude oscillations around the equilibrium separation $r = R_0$, we introduce the quantity x to represent the displacement from equilibrium:

$$x = r - R_0, \qquad \text{so} \qquad r = R_0 + x.$$

In terms of x, the force F in Eq. (13–26) becomes

$$F = 12\frac{U_0}{R_0}\left[\left(\frac{R_0}{R_0 + x}\right)^{13} - \left(\frac{R_0}{R_0 + x}\right)^{7}\right] = 12\frac{U_0}{R_0}\left[\frac{1}{(1 + x/R_0)^{13}} - \frac{1}{(1 + x/R_0)^{7}}\right].$$

$$(13\text{–}27)$$

This looks nothing like Hooke's law, $F = -kx$, so we might be tempted to conclude that molecular oscillations cannot be SHM. But let us restrict ourselves to *small-amplitude* oscillations so that the absolute value of the displacement x will be small in comparison to R_0 and the absolute value of the ratio x/R_0 will be much less than 1. We can then simplify Eq. (13–27) by using the *binomial theorem*:

$$(1 + u)^{n} = 1 + nu + \frac{n(n-1)}{2!}u^{2} + \frac{n(n-1)(n-2)}{3!}u^{3} + \cdots. \quad (13\text{–}28)$$

If $|u|$ is much less than 1, each successive term in Eq. (13–28) is much smaller than the one it follows, and we can safely approximate $(1 + u)^{n}$ by just the first two terms. In Eq.

(13-27), u is replaced by x/R_0 and n equals -13 or -7, so

$$\frac{1}{(1 + x/R_0)^{13}} = (1 + x/R_0)^{-13} \approx 1 + (-13)\frac{x}{R_0},$$

$$\frac{1}{(1 + x/R_0)^{7}} = (1 + x/R_0)^{-7} \approx 1 + (-7)\frac{x}{R_0},$$

$$F \approx 12\frac{U_0}{R_0}\left[\left(1 + (-13)\frac{x}{R_0}\right) - \left(1 + (-7)\frac{x}{R_0}\right)\right] = -\left(\frac{72U_0}{R_0^2}\right)x. \qquad (13-29)$$

This is just Hooke's law, with force constant $k = 72U_0/R_0^2$. (Note that k has the correct units, J/m^2 or N/m.) So oscillations of molecules bound by the van der Waals interaction can be simple harmonic motion, provided that the amplitude is small in comparison to R_0 so that the approximation $|x/R_0| \ll 1$ used in the derivation of Eq. (13-29) is valid.

You can also show that the potential energy U, Eq. (13-25), can be written as $U \approx \frac{1}{2}kx^2 + C$, where $C = -U_0$ and k is again equal to $72U_0/R_0^2$. Adding a constant to the potential energy has no effect on the physics, so the system of two atoms is fundamentally no different from a mass attached to a horizontal spring for which $U = \frac{1}{2}kx^2$. The proof is left as an exercise.

EXAMPLE 13-7

Two argon atoms can form a weakly bound molecule, Ar$_2$, held together by a van der Waals interaction with $U_0 = 1.68 \times 10^{-21}$ J and $R_0 = 3.82 \times 10^{-10}$ m. Find the frequency for small oscillations of one of the atoms about its equilibrium position.

SOLUTION Since the oscillations are small, we can use Eq. (13-11) for the frequency of simple harmonic motion. From Eq. (13-29) the force constant is

$$k = \frac{72U_0}{R_0^2} = \frac{72(1.68 \times 10^{-21}\ \text{J})}{(3.82 \times 10^{-10}\ \text{m})^2} = 0.829\ \text{J/m}^2 = 0.829\ \text{N/m}.$$

This is comparable to the force constant of a loose, floppy toy spring like a Slinky™.

From the periodic table of the elements (Appendix D), the average atomic mass of argon is $(39.948\ \text{u})(1.66 \times 10^{-27}\ \text{kg/1 u}) = 6.63 \times 10^{-26}$ kg. If one of the argon atoms is fixed and the other atom oscillates, the frequency of oscillation is

$$f = \frac{1}{2\pi}\sqrt{\frac{k}{m}} = \frac{1}{2\pi}\sqrt{\frac{0.829\ \text{N/m}}{6.63 \times 10^{-26}\ \text{kg}}} = 5.63 \times 10^{11}\ \text{Hz}.$$

The oscillating mass is very small, so even a loose spring causes very rapid oscillations.

Our answer for f isn't quite right, however. If there is no net external force acting on the molecule, the center of mass of the molecule (located halfway between the two atoms) doesn't accelerate. To ensure this, *both* atoms must oscillate with the same amplitude in opposite directions. It turns out that we can account for this by replacing m with $m/2$ in the expression for f. (See Problem 13-69.) This makes f larger by a factor of $\sqrt{2}$, so $f = \sqrt{2}(5.63 \times 10^{11}\ \text{Hz}) = 7.96 \times 10^{11}$ Hz. An additional complication is that on the atomic scale we must use *quantum mechanics*, not Newtonian mechanics, to describe oscillation and other motions; happily, the frequency has the same value in quantum mechanics.

13-6 THE SIMPLE PENDULUM

A **simple pendulum** is an idealized model consisting of a point mass suspended by a massless, unstretchable string. When the point mass is pulled to one side of its straight-down equilibrium position and released, it oscillates about the equilibrium position. Familiar situations such as a wrecking ball on a crane's cable, the plumb bob on a surveyor's transit, and a child on a swing can be modeled as simple pendulums.

The path of the point mass (sometimes called a pendulum bob) is not a straight line but the arc of a circle with radius L equal to the length of the string (Fig. 13-18). We use as our coordinate the distance x measured along the arc. If the motion is simple harmonic, the restoring force must be directly proportional to x or (because $x = L\theta$) to θ. Is it?

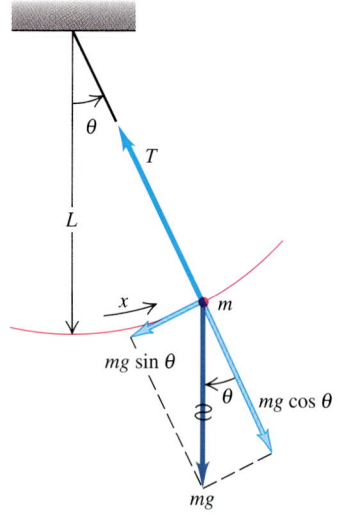

13–18 The forces on the bob of a simple pendulum.

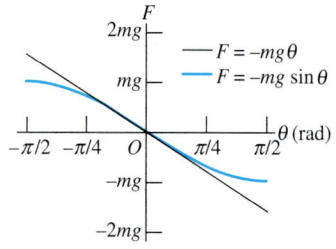

13–19 The restoring force for a simple pendulum, $F = -mg \sin \theta$ (shown in blue), can be approximated by $F = -mg\theta$ (shown in black) for small values of θ. Hence for small angles the restoring force is approximately proportional to θ, and the oscillations are simple harmonic.

9.10
Pendulum Frequency

9.11
Pendulum Timer

9.12
Risky Pendulum Walk

In Fig. 13–18 we represent the forces on the mass in terms of tangential and radial components. The restoring force F is the tangential component of the net force:

$$F = -mg \sin \theta. \qquad (13\text{–}30)$$

The restoring force is provided by gravity; the tension T merely acts to make the point mass move in an arc. The restoring force is proportional *not* to θ but to sin θ, so the motion is *not* simple harmonic. However, *if the angle θ is small,* sin θ is very nearly equal to θ in radians (Fig. 13–19). For example, when $\theta = 0.1$ rad (about 6°), sin $\theta = 0.0998$, a difference of only 0.2%. With this approximation, Eq. (13–30) becomes

$$F = -mg\theta = -mg\frac{x}{L}, \qquad \text{or}$$
$$\qquad\qquad\qquad\qquad\qquad\qquad (13\text{–}31)$$
$$F = -\frac{mg}{L}x.$$

The restoring force is then proportional to the coordinate *for small displacements,* and the force constant is $k = mg/L$. From Eq. (13–10) the angular frequency ω of a simple pendulum with small amplitude is

$$\omega = \sqrt{\frac{k}{m}} = \sqrt{\frac{mg/L}{m}} = \sqrt{\frac{g}{L}} \qquad \text{(simple pendulum, small amplitude)}. \quad (13\text{–}32)$$

The corresponding frequency and period relations are

$$f = \frac{\omega}{2\pi} = \frac{1}{2\pi}\sqrt{\frac{g}{L}} \qquad \text{(simple pendulum, small amplitude)}, \qquad (13\text{–}33)$$

$$T = \frac{2\pi}{\omega} = \frac{1}{f} = 2\pi\sqrt{\frac{L}{g}} \qquad \text{(simple pendulum, small amplitude)}. \quad (13\text{–}34)$$

Note that these expressions do not involve the *mass* of the particle. This is because the restoring force, a component of the particle's weight, is proportional to m. Thus the mass appears on *both* sides of $\Sigma \vec{F} = m\vec{a}$ and cancels out. (This is the same physics that explains why bodies of different masses fall with the same acceleration in a vacuum.) For small oscillations the period of a pendulum for a given value of g is determined entirely by its length.

The dependence on L and g in Eqs. (13–32) through (13–34) is just what we should expect. A long pendulum has a longer period than a shorter one. Increasing g increases the restoring force, causing the frequency to increase and the period to decrease.

We emphasize again that the motion of a pendulum is only *approximately* simple harmonic. When the amplitude is not small, the departures from simple harmonic motion can be substantial. But how small is "small"? The period can be expressed by an infinite series; when the maximum angular displacement is Θ, the period T is given by

$$T = 2\pi\sqrt{\frac{L}{g}}\left(1 + \frac{1^2}{2^2}\sin^2\frac{\Theta}{2} + \frac{1^2 \cdot 3^2}{2^2 \cdot 4^2}\sin^4\frac{\Theta}{2} + \cdots\right). \qquad (13\text{–}35)$$

We can compute the period to any desired degree of precision by taking enough terms in the series. We invite you to check that when $\Theta = 15°$ (on either side of the central position), the true period is longer than that given by the approximate Eq. (13–34) by less than 0.5%.

The usefulness of the pendulum as a timekeeper depends on the period being *very nearly* independent of amplitude, provided that the amplitude is small. Thus as a pendulum clock runs down and the amplitude of the swings decreases a little, the clock still keeps very nearly correct time.

A simple pendulum, or a variation thereof, is also a precise and convenient method for measuring the acceleration of gravity g, since L and T can be measured easily and precisely. Such measurements are often used in geophysics. Local deposits of ore or oil affect the local value of g because their density differs from that of their surroundings. Precise measurements of this quantity over an area being surveyed often furnish valuable information about the nature of underlying deposits.

EXAMPLE 13-8

Find the period and frequency of a simple pendulum 1.000 m long at a location where $g = 9.800$ m/s^2.

Then

$$f = \frac{1}{T} = 0.4983 \text{ Hz}.$$

SOLUTION From Eq. (13–34),

$$T = 2\pi \sqrt{\frac{L}{g}} = 2\pi \sqrt{\frac{1.000 \text{ m}}{9.800 \text{ m/s}^2}} = 2.007 \text{ s}.$$

The period is almost exactly 2 s. In fact, when the metric system was first established, the second was defined as half the period of a one-meter pendulum. This wasn't a very good standard for time, however, because the value of g varies from place to place. We discussed more modern time standards in Section 1–4.

13-7 THE PHYSICAL PENDULUM

A **physical pendulum** is any *real* pendulum, using a body of finite size, as contrasted to the idealized model of the *simple* pendulum with all the mass concentrated at a single point. For small oscillations, analyzing the motion of a real, physical pendulum is almost as easy as for a simple pendulum. Figure 13–20 shows a body of irregular shape pivoted so that it can turn without friction about an axis through point O. In the equilibrium position the center of gravity is directly below the pivot; in the position shown in the figure, the body is displaced from equilibrium by an angle θ, which we use as a coordinate for the system. The distance from O to the center of gravity is d, the moment of inertia of the body about the axis of rotation through O is I, and the total mass is m. When the body is displaced as shown, the weight mg causes a restoring torque

$$\tau = -(mg)(d \sin \theta). \tag{13–36}$$

The negative sign shows that the restoring torque is clockwise when the displacement is counterclockwise, and vice versa.

When the body is released, it oscillates about its equilibrium position. The motion is not simple harmonic because the torque τ is proportional to $\sin \theta$ rather than to θ itself. However, if θ is small, we can again approximate $\sin \theta$ by θ in radians, and the motion is *approximately* simple harmonic. With this approximation,

$$\tau = -(mgd)\theta.$$

The equation of motion is $\Sigma \tau = I\alpha$, so

$$-(mgd)\theta = I\alpha = I\frac{d^2\theta}{dt^2},$$

$$\frac{d^2\theta}{dt^2} = -\frac{mgd}{I}\theta. \tag{13–37}$$

Comparing this with Eq. (13–4), we see that the role of (k/m) for the mass-spring system is played here by the quantity (mgd/I). Thus the angular frequency is given by

$$\omega = \sqrt{\frac{mgd}{I}} \quad \text{(physical pendulum, small amplitude)}. \tag{13–38}$$

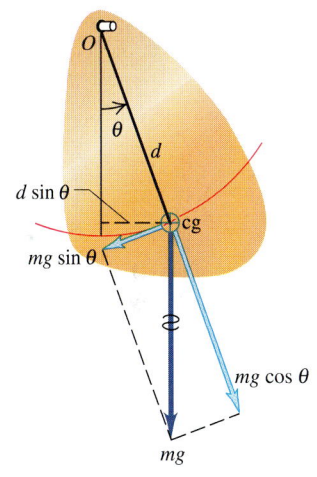

13-20 Dynamics of a physical pendulum.

9.13

Physical Pendulum

The frequency f is $1/2\pi$ times this, and the period T is

$$T = 2\pi \sqrt{\frac{I}{mgd}} \qquad \text{(physical pendulum, small amplitude).} \qquad (13\text{-}39)$$

Equation (13-39) is the basis of a common method for experimentally determining the moment of inertia of a body with a complicated shape. The center of gravity of the body is first located by balancing. The body is then suspended so that it is free to oscillate about an axis, and the period T of small-amplitude oscillations is measured. By using Eq. (13-39), the moment of inertia I of the body about this axis can then be calculated from T, the body's mass m, and the distance d from the axis to the center of gravity (see Exercise 13-42). Biomechanics researchers use this method to find the moments of inertia of an animal's limbs. This information is important for analyzing how an animal walks, as we'll see in the second of the two following examples.

EXAMPLE 13-9

Physical pendulum versus simple pendulum Suppose the body in Fig. 13-20 is a uniform rod with length L, pivoted at one end. Find the period of its motion.

SOLUTION From Chapter 9 the moment of inertia of a uniform rod about an axis through one end is $I = \frac{1}{3}ML^2$. The distance from the pivot to the center of gravity is $d = L/2$. From Eq. (13-39),

$$T = 2\pi \sqrt{\frac{\frac{1}{3}ML^2}{MgL/2}} = 2\pi \sqrt{\frac{2L}{3g}}.$$

If the rod is a meter stick ($L = 1.00$ m) and $g = 9.80$ m/s^2, then

$$T = 2\pi \sqrt{\left(\frac{2}{3}\right)(1.00 \text{ m})/(9.80 \text{ m/s}^2)} = 1.64 \text{ s}.$$

The period is smaller by a factor of $\sqrt{2/3} = 0.816$ than the period of a simple pendulum with the same length, calculated in Example 13-8 (Section 13-6).

EXAMPLE 13-10

Tyrannosaurus rex and the physical pendulum All walking animals, including humans, have a natural walking pace, a number of steps per minute that is more comfortable than a faster or slower pace. Suppose this natural pace is equal to the period of the leg, viewed as a uniform rod pivoted at the hip joint. a) How does the natural walking pace depend on the length L of the leg, measured from hip to foot? b) Fossil evidence shows that *Tyrannosaurus rex*, a two-legged dinosaur that lived about 65 million years ago at the end of the Cretaceous period, had a leg length $L = 3.1$ m and a stride length (the distance from one footprint to the next print of the same foot) $S = 4.0$ m (Fig. 13-21). Estimate the walking speed of *Tyrannosaurus rex*.

SOLUTION a) The period of oscillation of the leg is given by the expression found in Example 13-9, $T = 2\pi\sqrt{2L/3g}$, which is proportional to $\sqrt{L}$. Each period (a complete back-and-forth swing of the leg) corresponds to *two* steps, so the walking pace in steps per unit time is just twice the oscillation frequency $f = 1/T$. Hence the walking pace is proportional to $1/\sqrt{L}$. Animals with short legs (small values of L), such as mice or Chihuahuas, have rapid walking paces; humans, giraffes, and other animals with long legs (large values of L) walk at slower paces.

b) According to our model for the natural walking pace, the elapsed time for one stride of a walking *Tyrannosaurus rex* is

$$T = 2\pi \sqrt{\frac{2L}{3g}} = 2\pi \sqrt{\frac{2(3.1 \text{ m})}{3(9.8 \text{ m/s}^2)}} = 2.9 \text{ s}.$$

The distance moved during this time is the stride length S, so the walking speed is

$$v = \frac{S}{T} = \frac{4.0 \text{ m}}{2.9 \text{ s}} = 1.4 \text{ m/s} = 5.0 \text{ km/h} = 3.1 \text{ mi/h}.$$

This is about the same as a typical human walking speed! Our estimate must be somewhat in error, however, because a uniform rod isn't a very good model for a leg. The legs of many animals, including *T. rex* as well as humans, are tapered; there is a lot more mass between the knee and the hip than between the knee and the foot. Thus the center of mass is less than $L/2$ from the hip; a reasonable guess would be about $L/4$. The moment of inertia is *considerably* less than $ML^2/3$, probably somewhere around $ML^2/15$. Try these numbers out with the analysis of Example 13-9; you'll get a shorter oscillation period and an even faster walking speed for *T. rex*.

13-21 The walking speed of *Tyrannosaurus rex* can be estimated from its leg length L and its stride length S.

13-8 DAMPED OSCILLATIONS

The idealized oscillating systems we have discussed thus far are frictionless. There are no nonconservative forces, the total mechanical energy is constant, and a system set into motion continues oscillating forever with no decrease in amplitude.

Real-world systems always have some dissipative forces, however, and oscillations do die out with time unless we provide some means for replacing the dissipated mechanical energy. A mechanical pendulum clock continues to run because potential energy stored in the spring or a hanging weight system replaces the mechanical energy lost due to friction in the pivot and the gears. But eventually the spring runs down or the weights reach the bottom of their travel. Then no more energy is available, and the pendulum swings decrease in amplitude and stop.

The decrease in amplitude caused by dissipative forces is called **damping,** and the corresponding motion is called **damped oscillation.** The simplest case to analyze in detail is a simple harmonic oscillator with a frictional damping force that is directly proportional to the *velocity* of the oscillating body. This behavior occurs in friction involving viscous fluid flow, such as in shock absorbers or sliding between oil-lubricated surfaces. We then have an additional force on the body due to friction, $F = -bv$, where $v = dx/dt$ is the velocity and b is a constant that describes the strength of the damping force. The negative sign shows that the force is always opposite in direction to the velocity. The *net* force on the body is then

$$\Sigma F = -kx - bv, \tag{13-40}$$

and Newton's second law for the system is

$$-kx - bv = ma, \quad \text{or} \quad -kx - b\frac{dx}{dt} = m\frac{d^2x}{dt^2}. \tag{13-41}$$

A disturbance such as an earthquake would make this chandelier oscillate back and forth. The oscillations would be damped, however, by air resistance and friction at the point of suspension, and would eventually die out.

13–22 Graph of Eq. (13–42), showing damped harmonic motion, in the case in which the phase angle $\phi = 0$. The period when there is no damping ($b = 0$) is T_0. The blue curve shows the motion when $b = 0.1\sqrt{km}$, and the red curve shows the motion when $b = 0.4\sqrt{km}$. The amplitude decreases more rapidly for the larger value of b. Close inspection of the points where the curves cross the t-axis also reveals that the period increases slightly with increasing b. The critical-damping condition is $b = 2\sqrt{km}$.

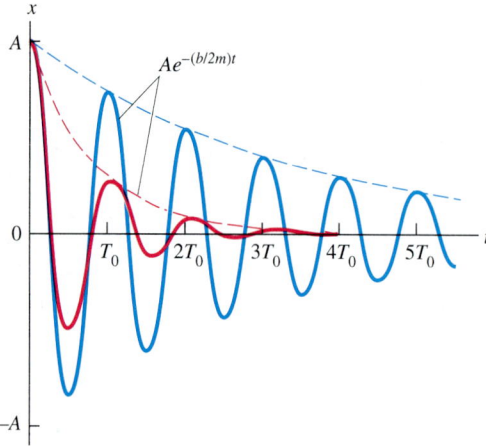

Equation (13–41) is a differential equation for x; it would be the same as Eq. (13–4), the equation for the acceleration in SHM, except for the added term $-b\,dx/dt$. Solving this equation is a straightforward problem in differential equations, but we won't go into the details here. If the damping force is relatively small, the motion is described by

$$x = Ae^{-(b/2m)t}\cos(\omega't + \phi) \qquad \text{(oscillator with little damping).} \qquad (13\text{–}42)$$

The angular frequency of oscillation ω' is given by

$$\omega' = \sqrt{\frac{k}{m} - \frac{b^2}{4m^2}} \qquad \text{(oscillator with little damping).} \qquad (13\text{–}43)$$

You can verify that Eq. (13–42) is a solution of Eq. (13–41) by calculating the first and second derivatives of x, substituting them into Eq. (13–41), and checking whether the left and right sides are equal. This is a straightforward but slightly tedious procedure.

The motion described by Eq. (13–42) differs from the undamped case in two ways. First, the amplitude $Ae^{-(b/2m)t}$ is not constant but decreases with time because of the decreasing exponential factor $e^{-(b/2m)t}$. Figure 13–22 is a graph of Eq. (13–42) for the case $\phi = 0$; it shows that the larger the value of b, the more quickly the amplitude decreases.

Second, the angular frequency ω', given by Eq. (13–43), is no longer equal to $\omega = \sqrt{k/m}$ but is somewhat smaller. It becomes zero when b becomes so large that

$$\frac{k}{m} - \frac{b^2}{4m^2} = 0, \qquad \text{or} \qquad b = 2\sqrt{km}. \qquad (13\text{–}44)$$

When Eq. (13–44) is satisfied, the condition is called **critical damping.** The system no longer oscillates but returns to its equilibrium position without oscillation when it is displaced and released.

If b is greater than $2\sqrt{km}$, the condition is called **overdamping.** Again there is no oscillation, but the system returns to equilibrium more slowly than with critical damping. For the overdamped case the solutions of Eq. (13–41) have the form

$$x = C_1e^{-a_1t} + C_2e^{-a_2t},$$

where C_1 and C_2 are constants that depend on the initial conditions and a_1 and a_2 are constants determined by m, k, and b.

When b is less than the critical value, as in Eq. (13–42), the condition is called **underdamping.** The system oscillates with steadily decreasing amplitude.

In a vibrating tuning fork or guitar string, it is usually desirable to have as little damping as possible. By contrast, damping plays a beneficial role in the oscillations of an automobile's suspension system. The shock absorbers provide a velocity-dependent damping force so that when the car goes over a bump, it doesn't continue bouncing forever (Fig. 13–23). For optimal passenger comfort, the system should be critically damped or slightly underdamped. As the shocks get old and worn, the value of b decreases and the bouncing persists longer. Not only is this nauseating, it is bad for steering because the front wheels have less positive contact with the ground. Thus damping is an advantage in this system. Too much damping would be counterproductive, however; if b is excessively large, the system is overdamped and the suspension returns to equilibrium more slowly. If the suspension is overdamped and the car hits a second bump just after the first one, the springs in the suspension will still be compressed somewhat from the first bump and will be unable to fully absorb the impact.

In damped oscillations the damping force is nonconservative; the mechanical energy of the system is not constant but decreases continuously, approaching zero after a long time. To derive an expression for the rate of change of energy, we first write an expression for the total mechanical energy E at any instant:

$$E = \frac{1}{2} mv^2 + \frac{1}{2} kx^2.$$

To find the rate of change of this quantity, we take its time derivative:

$$\frac{dE}{dt} = mv \frac{dv}{dt} + kx \frac{dx}{dt}.$$

But $dv/dt = a$ and $dx/dt = v$, so

$$\frac{dE}{dt} = v(ma + kx).$$

From Eq. (13–41), $ma + kx = -b \, dx/dt = -bv$, so

$$\frac{dE}{dt} = v(-bv) = -bv^2 \qquad \text{(damped oscillations).} \qquad (13\text{–}45)$$

The right side of Eq. (13–45) is always negative, whether v is positive or negative. This shows that E continuously decreases, though not at a uniform rate. The term $-bv^2 = (-bv)v$ (force times velocity) is the rate at which the damping force does (negative) work on the system (that is, the damping *power*). This equals the rate of change of the total mechanical energy of the system.

Similar behavior occurs in electric circuits containing inductance, capacitance, and resistance. There is a natural frequency of oscillation, and the resistance plays the role of the damping constant b. In such circuits it is usually desirable to minimize damping, but damping can never be prevented completely. We will study these circuits in detail in Chapters 31 and 32.

Upper cylinder remains relatively stationary

Piston

Viscous fluid

Lower cylinder moves with wheel and axle

13–23 A car shock absorber. The top part, connected to the piston, is attached to the car's frame; the bottom part, connected to the lower cylinder, is attached to the axle. The viscous fluid causes a damping force that depends on the relative velocity of the two ends of the unit. This helps to control wheel bounce and jounce.

13–9 FORCED OSCILLATIONS, RESONANCE, AND CHAOS

A damped oscillator left to itself will eventually stop moving altogether. But we can maintain a constant-amplitude oscillation by applying a force that varies with time in a periodic or cyclic way, with a definite period and frequency. As an example, consider your cousin Throckmorton on a playground swing. You can keep him swinging with constant amplitude by giving him a little push once each cycle. We call this additional force a **driving force.**

If we apply a periodically varying driving force with angular frequency ω_d to a damped harmonic oscillator, the motion that results is called a **forced oscillation** or a *driven oscillation.* It is different from the motion that occurs when the system is simply displaced from equilibrium and then left alone, in which case the system oscillates with a **natural angular frequency** ω' determined by m, k, and b, as in Eq. (13–43). In a forced oscillation, however, the angular frequency with which the mass oscillates will be equal to the driving angular frequency ω_d. This does *not* have to be equal to the angular frequency ω' with which the system would oscillate without a driving force. If you grab the ropes of Throckmorton's swing, you can force the swing to oscillate with any frequency you like.

Suppose we force the oscillator to vibrate with an angular frequency ω_d that is nearly *equal* to the angular frequency ω' it would have with no driving force. What happens? The oscillator is naturally disposed to oscillate at $\omega = \omega'$, so we expect the amplitude of the resulting oscillation to be larger than when the two frequencies are very different. Detailed analysis and experiment shows that this is just what happens. The easiest case to analyze is a *sinusoidally* varying force, say, $F(t) = F_{max} \cos \omega_d t$. If we vary the frequency ω_d of the driving force, the amplitude of the resulting forced oscillation varies in an interesting way (Fig. 13–24). When there is very little damping (small b), the amplitude goes through a sharp peak as the driving angular frequency ω_d nears the natural oscillation angular frequency ω'. When the damping is increased (larger b), the peak becomes broader and smaller in height and shifts toward lower frequencies.

We could work out an expression that shows how the amplitude A of the forced oscillation depends on the frequency of a sinusoidal driving force, with maximum value F_{max}. That would involve more differential equations than we're ready for, but here is the result:

$$A = \frac{F_{max}}{\sqrt{(k - m\omega_d^2)^2 + b^2 \omega_d^2}} \qquad \text{(amplitude of a driven oscillator).} \quad (13\text{–}46)$$

When $k - m\omega_d^2 = 0$, the first term under the radical is zero, so A has a maximum near $\omega_d = \sqrt{k/m}$. The height of the curve at this point is proportional to $1/b$; the less damping, the higher the peak. At the low-frequency extreme, when $\omega_d = 0$, we get $A = F_{max}/k$. This corresponds to a *constant* force F_{max} and a constant displacement $A = F_{max}/k$ from equilibrium, as we might expect.

13–24 Graph of the amplitude A of forced oscillation of a damped harmonic oscillator as a function of the frequency ω_d of the driving force. The horizontal axis shows the ratio of ω_d to the angular frequency $\omega = \sqrt{k/m}$ of an undamped oscillator. Each curve is labeled with the value of the dimensionless quantity $b/\sqrt{km}$, which characterizes the amount of damping. The highest curve has $b = 0.2\sqrt{km}$, the next has $b = 0.4\sqrt{km}$, and so on. As b increases, the peak becomes broader and less sharp and shifts toward lower frequencies. When b is larger than $\sqrt{2km}$, the peak disappears completely.

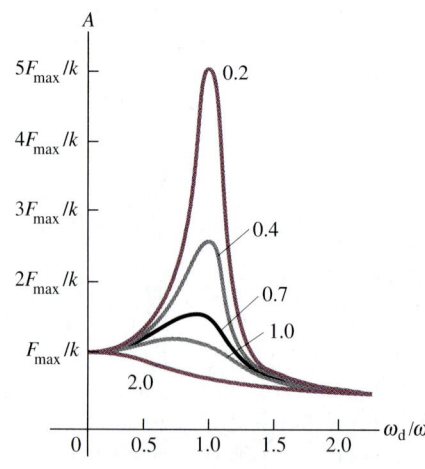

The fact that there is an amplitude peak at driving frequencies close to the natural frequency of the system is called **resonance.** Physics is full of examples of resonance; building up the oscillations of a child on a swing by pushing with a frequency equal to the swing's natural frequency is one. A vibrating rattle in a car that occurs only at a certain engine speed or wheel-rotation speed is an all too familiar example. Inexpensive loudspeakers often have an annoying boom or buzz when a musical note happens to coincide with the resonant frequency of the speaker cone or the speaker housing. A tuned circuit in a radio or television receiver responds strongly to waves having frequencies near its resonant frequency, and this fact is used to select a particular station and reject the others. We will study resonance in electric circuits in detail in Chapter 32.

Resonance in mechanical systems can be destructive. A company of soldiers once destroyed a bridge by marching across it in step; the frequency of their steps was close to a natural vibration frequency of the bridge, and the resulting oscillation had large enough amplitude to tear the bridge apart. Ever since, marching soldiers have been ordered to break step before crossing a bridge. Some years ago, vibrations of the engines of a particular airplane had just the right frequency to resonate with the natural frequencies of its wings. Large oscillations built up, and occasionally the wings fell off.

Nearly everyone has seen the film of the collapse of the Tacoma Narrows suspension bridge in 1940 (Fig. 13–25). This is usually cited as an example of resonance driven by the wind, but there's some doubt as to whether it should be called that. The wind didn't have to vary *periodically* with a frequency close to a natural frequency of the bridge. The air flow past the bridge was turbulent, and vortices were formed in the air with a regular frequency that depended on the flow speed. It is conceivable that this frequency may have coincided with a natural frequency of the bridge. But the cause may well have been something more subtle called a *self-excited oscillation*, in which the aerodynamic forces caused by a *steady* wind blowing on the bridge tended to displace it farther from equilibrium at times when it was already moving away from equilibrium. It is as though we had a damping force such as the $-bv$ term in Eq. (13–40) but with the sign reversed. Instead of draining mechanical energy away from the system, this anti-damping force pumps energy into the system, building up the oscillations to destructive amplitudes. The approximate differential equation is Eq. (13–41) with the sign of the b term reversed, and the oscillating solution is Eq. (13–42) with a *positive* sign in the exponent. You can see that we're headed for trouble. Engineers have learned how to stabilize suspension bridges, both structurally and aerodynamically, to prevent such disasters.

13–25 The Tacoma Narrows Bridge collapsed four months and six days after it was opened for traffic. The main span was 2800 ft long and 39 ft wide, with 8-ft-high steel stiffening girders on both sides. The maximum amplitude of the torsional vibrations was 35°; the frequency was about 0.2 Hz.

CHAOTIC OSCILLATIONS

If a sinusoidally varying driving force $F(t) = F_{max} \cos \omega_d t$ is applied to a harmonic oscillator, the resulting forced oscillation has the same angular frequency ω_d as the driving force. But what happens if the oscillator is *not* a harmonic one?

Here's an example. Suppose you have a pendulum that is made up of a mass m attached to a lightweight but rigid rod of length L and is free to rotate around a pivot at the top of the rod, as in Fig. 13–18. We saw in Section 13–6 that large-amplitude oscillations of a simple pendulum are *not* simple harmonic motion, because the period depends on the amplitude. To apply a sinusoidally varying driving force to the pendulum, you hold the pivot in your hand and oscillate it up and down with angular frequency ω_d. To be specific, let's suppose that ω_d is twice the angular frequency $\omega = \sqrt{g/L}$ at which the pendulum would oscillate by itself and that there is friction in the pivot.

As you increase the amplitude F_{max} of the driving force that you apply to the pivot, the amplitude of oscillation of the pendulum increases as well. If F_{max} isn't too large, the pendulum oscillates at the same angular frequency as the driving force, just like a driven harmonic oscillator. But if you increase F_{max} to a value close to the weight mg of the pendulum, after the transient behavior has died down, you will notice something odd: the motion of the pendulum during one up-and-down cycle of the pivot is slightly different from the oscillation during the next up-and-down cycle. The pendulum's motion repeats itself only after *two* up-and-down cycles. So if the period of the driving force is $T_d = 2\pi/\omega_d$, the period of the actual motion of the pendulum is $T = 2T_d$, a phenomenon called *period doubling*. If you increase the value of F_{max} to about $1.5mg$, the pendulum will make complete circles around the pivot, moving somewhat differently during one circle than during the following circle, so again $T = 2T_d$. Further increases in F_{max} will cause the period of the pendulum to double again, to $T = 4T_d$; the pendulum's motion repeats on only every fourth cycle of the pivot's motion. As you continue to increase F_{max}, T becomes a larger and larger multiple of T_d. At a finite value of F_{max} (around $2mg$ for the pendulum we've described) the period of the motion becomes infinite. This means that the motion *never* repeats!

Although the motion of the pendulum is completely determined by its initial position, its initial velocity, and the forces that act on it (through Newton's second law, $\Sigma \vec{F} = m\vec{a}$), the motion appears to be completely random and unpredictable. Such *seemingly* random motion is called **chaotic motion,** and its behavior is called **chaos.** Many physical systems in nature and technology can display chaotic behavior. Examples include water dripping from a faucet, a mass bouncing on top of a vertically oscillating piston, and turbulent fluid flow. Chaos also occurs in such situations outside of physics such as the growth and decay of animal populations and fluctuations in stock prices. The study of chaotic motion in mechanical systems can give tremendous insight into chaotic behavior in biology, economics, and other fields. Chaos is an active field of research in many fields of natural and social science.

SUMMARY

KEY TERMS

periodic motion, 392
oscillation, 392
displacement, 392
restoring force, 392
amplitude, 393
cycle, 393

■ Periodic motion is motion that repeats itself in a definite cycle. It occurs whenever a body has a stable equilibrium position and a restoring force or torque that acts when it is displaced from equilibrium. Period T is the time for one cycle. Frequency f is the number of cycles per unit time:

$$f = \frac{1}{T}, \qquad T = \frac{1}{f}. \tag{13–1}$$

Angular frequency ω is

$$\omega = 2\pi f = \frac{2\pi}{T}. \tag{13-2}$$

- If the net force is a restoring force F that is directly proportional to the displacement x, the motion is called simple harmonic motion (SHM). In that case,

$$F = -kx, \tag{13-3}$$

$$a = \frac{F}{m} = -\frac{k}{m}x. \tag{13-4}$$

In many cases the restoring force F is given approximately by Eq. (13–3) if the displacement from equilibrium is small.

- The circle of reference construction uses a rotating vector called a phasor, having a length equal to the amplitude of the motion. Its projection on the horizontal axis represents the actual motion of a body in simple harmonic motion.

- The angular frequency, frequency, and period in SHM do not depend on the amplitude, but depend only on the mass m and force constant k:

$$\omega = \sqrt{\frac{k}{m}}, \tag{13-10}$$

$$f = \frac{\omega}{2\pi} = \frac{1}{2\pi}\sqrt{\frac{k}{m}}, \tag{13-11}$$

$$T = \frac{1}{f} = 2\pi\sqrt{\frac{m}{k}}. \tag{13-12}$$

- In SHM the displacement, velocity, and acceleration are sinusoidal functions of time. The displacement x is given as a function of time t by

$$x = A\cos(\omega t + \phi), \tag{13-13}$$

where $\omega = \sqrt{k/m}$. The amplitude A and phase angle ϕ are determined by the initial position and velocity of the body.

- Conservation of energy in SHM leads to the relation

$$E = \frac{1}{2}mv^2 + \frac{1}{2}kx^2 = \frac{1}{2}kA^2 = \text{constant}. \tag{13-21}$$

- In angular simple harmonic motion the frequency and angular frequency are related to the moment of inertia I and the torsion constant κ by

$$\omega = \sqrt{\frac{\kappa}{I}} \quad \text{and} \quad f = \frac{1}{2\pi}\sqrt{\frac{\kappa}{I}}. \tag{13-24}$$

- A simple pendulum consists of a point mass m at the end of a massless string of length L. Its motion is approximately simple harmonic for sufficiently small amplitude; the angular frequency, frequency, and period are then given by

$$\omega = \sqrt{\frac{g}{L}}, \tag{13-32}$$

$$f = \frac{\omega}{2\pi} = \frac{1}{2\pi}\sqrt{\frac{g}{L}}, \tag{13-33}$$

$$T = \frac{2\pi}{\omega} = \frac{1}{f} = 2\pi\sqrt{\frac{L}{g}}. \tag{13-34}$$

- A physical pendulum is a body suspended from an axis of rotation a distance d from its center of gravity. If the moment of inertia about the axis of rotation is I, the angular frequency and period for small-amplitude oscillations are

$$\omega = \sqrt{\frac{mgd}{I}}, \qquad (13\text{–}38)$$

$$T = 2\pi\sqrt{\frac{I}{mgd}}. \qquad (13\text{–}39)$$

- When a damping force $F = -bv$ proportional to velocity is added to a simple harmonic oscillator, the motion is called a damped oscillation. For a relatively small damping force,

$$x = Ae^{-(b/2m)t}\cos\omega' t. \qquad (13\text{–}42)$$

The angular frequency of oscillation ω' is given by

$$\omega' = \sqrt{\frac{k}{m} - \frac{b^2}{4m^2}}. \qquad (13\text{–}43)$$

This motion occurs when $b < 2\sqrt{km}$, a condition called underdamping. When $b = 2\sqrt{km}$, the system is critically damped and no longer oscillates. When $b > 2\sqrt{km}$, the system is overdamped.

- When a sinusoidally varying driving force is added to a damped harmonic oscillator, the resulting motion is called a forced oscillation. The amplitude is given as a function of driving angular frequency ω_d by

$$A = \frac{F_{max}}{\sqrt{(k - m\omega_d^2)^2 + b^2\omega_d^2}}. \qquad (13\text{–}46)$$

The amplitude reaches a peak at a driving frequency close to the natural oscillation frequency of the system. This behavior is called resonance.

DISCUSSION QUESTIONS

Q13–1 Your schedule of classes, study time, meals, and recreation is essentially the same from one week to the next. Is this a form of periodic motion? Of simple harmonic motion? Explain.

Q13–2 Think of several examples in everyday life of motion that is at least approximately simple harmonic. In what respects does each differ from SHM?

Q13-3 Does a tuning fork or similar tuning instrument undergo simple harmonic motion? Why is this a crucial question for musicians?

Q13–4 A highly elastic "superball" bouncing on a hard floor has a motion that is approximately periodic. In what ways is the motion similar to SHM? In what ways is it different?

Q13–5 If a uniform spring is cut in half, what is the force constant of each half? Justify your answer. How would the frequency of SHM using a half-spring differ from that using the same mass and the entire spring?

Q13–6 The analysis of SHM in this chapter neglected the mass of the spring. How does the spring mass change the characteristics of the motion?

Q13–7 Suppose the stationary end of the spring in Fig. 13–1 is connected instead to another identical body that is free to slide along the same line. Could such a system undergo SHM? Explain. How would the period compare with that of the original system? Explain.

Q13–8 For the system shown in Fig. 13–1, is there any point during the motion at which the body is in equilibrium? Explain.

Q13–9 The system shown in Fig. 13–14 is mounted in an elevator, which accelerates upward with constant acceleration. Does the period increase, decrease, or remain the same? Justify your answer.

Q13–10 Why is the "springiness" of a diving board adjusted for different dives and different weights of divers? How is the adjustment made?

Q13–11 A pendulum is mounted in an elevator that accelerates upward with constant acceleration. Does the period increase, decrease, or remain the same? Explain your answer.

Q13–12 What should you do to the length of the string of a simple pendulum to a) double its frequency? b) double its period? c) double its angular frequency?

Q13–13 If a pendulum clock is taken to a mountain top, does it gain or lose time, assuming it is correct at a lower elevation? Explain your answer.

Q13–14 When the amplitude of a simple pendulum increases, should its period increase or decrease? Give a qualitative argument; do not rely on Eq. (13–35). Is your argument also valid for a physical pendulum?

Q13–15 Why do short dogs (like Chihuahuas) walk with quicker strides than do tall dogs (like Great Danes)?

Q13–16 At what point in the motion of a simple pendulum is the string tension greatest? Least? In each case give the reasoning behind your answer.

Q13–17 Could a standard of time be based on the period of a certain standard pendulum? What advantages and disadvantages would such a standard have compared to the actual present-day standard discussed in Section 1–4?

Q13–18 The frequency with which dogs pant is the natural frequency of their respiratory system. Why do they choose this frequency?

Q13–19 In designing structures in an earthquake-prone region, how should the natural frequencies of oscillation of a structure relate to typical earthquake frequencies? Why? Should the structure have a large or small amount of damping?

EXERCISES

SECTION 13–2 THE CAUSES OF OSCILLATION

13–1 A piano string sounds a middle A by vibrating primarily at 220 Hz. a) Calculate its period and angular frequency. b) Calculate the period and angular frequency for a soprano singing a "high A," two octaves up, which is four times the frequency of the piano string.

13–2 The body in Fig. 13–1 is displaced 0.120 m from its equilibrium position and released with zero initial speed. After 0.800 s its displacement is found to be 0.120 m on the opposite side, and it has passed the equilibrium position once during this interval. Find a) the amplitude; b) the period; c) the frequency.

13–3 The tip of a tuning fork goes through 440 complete vibrations in 0.500 s. Find the angular frequency and the period of the motion.

SECTION 13–3 SIMPLE HARMONIC MOTION

13–4 In physics lab, you attach a 0.200-kg air-track glider to the end of an ideal spring of negligible mass as in Fig. 13–1 and start it oscillating. The elapsed time from when the glider first moves through the equilibrium point to the second time it moves through that point is 2.60 s. Find the spring's force constant.

13–5 A body of unknown mass is attached to an ideal spring with force constant 120 N/m in the arrangement shown in Fig. 13–1. It is found to vibrate with a frequency of 6.00 Hz. Find a) the period; b) the angular frequency; c) the mass of the body.

13–6 A harmonic oscillator is made by using a 0.600-kg frictionless block and an ideal spring of unknown force constant. The oscillator is found to have a period of 0.150 s. Find the force constant of the spring.

13–7 A harmonic oscillator has a mass of 0.500 kg and an ideal spring with force constant 140 N/m. Find a) the period; b) the frequency; c) the angular frequency.

13–8 Substitute the following equations, in which A, ω, and β are constant, into Eq. (13–4) to see if they describe SHM. If so, what must ω equal? a) $x = A \sin(\omega t + \beta)$. b) $x = A\omega t^2 + \beta$. c) $x = Ae^{i(\omega t + \beta)}$, where $i = \sqrt{-1}$.

13–9 A guitar string vibrates at a frequency of 440 Hz. A point at its center moves in SHM with an amplitude of 3.0 mm and a phase angle of zero. a) Write an equation for the position of the center of the string as a function of time. b) What are the maximum values of the magnitudes of the velocity and acceleration of the center of the string? c) The derivative of the acceleration with respect to time is a quantity called the *jerk*. Write an equation for the jerk of the center of the string as a function of time, and find the maximum value of the magnitude of the jerk.

13–10 A 2.00-kg frictionless block is attached to an ideal spring with force constant 300 N/m. At $t = 0$ the spring is neither stretched nor compressed and the block is moving in the negative direction at 12.0 m/s. Find a) the amplitude and b) the phase angle. c) Write an equation for the position as a function of time.

13–11 Repeat Exercise 13–10, but assume that at $t = 0$ the block has velocity –4.00 m/s and displacement +0.200 m.

13–12 The point of the needle of a sewing machine moves in SHM along the x-axis with a frequency of 2.5 Hz. At $t = 0$ its position and velocity components are +1.1 cm and –15 cm/s. a) Find the acceleration component of the needle at $t = 0$. b) Write equations giving the position, velocity, and acceleration components of the point as a function of time.

13–13 An object is undergoing simple harmonic motion with period 1.200 s and amplitude 0.600 m. At $t = 0$ the object is at $x = 0$. How far is the object from the equilibrium position when $t = 0.480$ s?

13–14 An object is undergoing simple harmonic motion with period 0.300 s and amplitude 6.00 cm. At $t = 0$ the object is instantaneously at rest at $x = 6.00$ cm. Calculate the time it takes the object to go from $x = 6.00$ cm to $x = -1.50$ cm.

SECTION 13–4 ENERGY IN SIMPLE HARMONIC MOTION

13–15 Derive Eq. (13–19) a) from Eqs. (13–14) and (13–17); b) from conservation of energy, Eq. (13–21).

13–16 A tuning fork labeled 392 Hz has the tip of each of its two prongs vibrating with an amplitude of 0.600 mm. a) What is the maximum speed of the tip of a prong? b) A housefly (*Musca domestica*) of mass 0.0270 g is holding on to the tip of one of the prongs. As the prong vibrates, what is the fly's maximum kinetic energy? Assume that the fly's mass has a negligible effect on the frequency of oscillation.

13–17 A harmonic oscillator has angular frequency ω and amplitude A. a) What are the magnitudes of the displacement and velocity when the elastic potential energy is equal to the kinetic energy? (Assume that $U = 0$ at equilibrium.) b) How often does this occur each cycle? What is the time between occurrences? c) At an instant when the displacement is equal to $A/2$, what fraction of the total energy of the system is kinetic and what fraction is potential?

13–18 An ideal spring with force constant $k = 450$ N/m is mounted as in Fig. 13–1. A 0.500-kg glider attached to the end undergoes simple harmonic motion with an amplitude 0.040 m. Compute a) the maximum speed of the glider; b) the speed of the glider when it is at $x = -0.015$ m; c) the magnitude of the maximum acceleration of the glider; d) the acceleration of the glider at $x = -0.015$ m; e) the total mechanical energy of the glider at any point in its motion.

13–19 A cheerleader waves her pom-pom in SHM with an amplitude of 18.0 cm and a frequency of 0.850 Hz. Find a) the maximum magnitude of the acceleration and of the velocity; b) the acceleration and speed when the pom-pom's coordinate is $x = +9.0$ cm; c) the time required to move from the equilibrium position directly to a point 12.0 cm away. d) Which of the quantities asked for in parts (a), (b), and (c) can be found using the energy approach used in Section 13–4, and which cannot? Explain.

13–20 For the situation described in part (a) of Example 13–5, what should be the value of the putty mass m so that the amplitude after the collision is one-half the original amplitude? For this value of m, what fraction of the original mechanical energy is converted into heat?

13–21 A 0.150-kg toy is undergoing simple harmonic motion on the end of a horizontal spring with force constant $k = 300$ N/m. When the object is 0.012 m from its equilibrium position, it is observed to have a speed of 0.300 m/s. What is a) the total energy of the object at any point of its motion? b) the amplitude of the motion? c) the maximum speed attained by the object during its motion?

SECTION 13–5 APPLICATIONS OF SIMPLE HARMONIC MOTION

13–22 The scale of a spring balance reading from zero to 200 N is 12.5 cm long. A fish hanging from the bottom of the spring oscillates vertically at 2.60 Hz. What is the mass of the fish? Neglect the mass of the spring.

13–23 A proud deep-sea fisherman hangs a 65.0-kg fish from an ideal spring having negligible mass. The fish stretches the spring 0.120 m. a) What is the force constant of the spring? b) What is the period of oscillation of the fish if it is pulled down and released?

13–24 A thrill-seeking cat with a mass of 4.00 kg is attached by a harness to an ideal spring of negligible mass and oscillates vertically in simple harmonic motion. The amplitude is 0.050 m, and at the highest point of the motion the spring has its natural unstretched length. Calculate the elastic potential energy of the spring (take it to be zero for the unstretched spring), the kinetic energy of the cat, the gravitational potential energy of the system relative to the lowest point of the motion, and the sum of these three energies when the cat is a) at its highest point; b) at its lowest point; c) at its equilibrium position.

13–25 A 2.00-kg block of Cheddar cheese hangs from an ideal spring of negligible mass. When displaced from equilibrium and released, the cheese oscillates with a period of 0.400 s. How much is the spring stretched when the cheese hangs in equilibrium (at rest)?

13–26 Your boss at the Cut-Rate Cuckoo Clock Company asks you what would happen to the frequency of the angular SHM of the balance wheel if it had the same density and the same coil spring (thus the same torsion constant), but all the balance wheel dimensions were made one-third as great to save material. a) What do you tell her? b) By what factor would the torsion constant need to be changed to make the smaller balance wheel oscillate at the original frequency?

13–27 A certain alarm clock ticks four times each second, each tick representing half a period. The balance wheel consists of a thin rim with radius 0.55 cm, connected to the balance staff by thin spokes of negligible mass. The total mass of the balance wheel is 0.90 g. a) What is the moment of inertia of the balance wheel about its shaft? b) What is the torsion constant of the hairspring?

13–28 A thin metal disk of mass 2.00×10^{-3} kg and radius 2.20 cm is attached at its center to a long fiber (Fig. 13–26). The disk, when twisted and released, oscillates with a period of 1.00 s. Find the torsion constant of the fiber.

FIGURE 13–26 Exercise 13–28.

13–29 You want to find the moment of inertia of a complicated machine part about an axis through its center of mass. You suspend it from a wire along this axis. The wire has a torsion constant of 0.450 N · m/rad. You twist the part a small amount about this axis and let it go, timing 125 oscillations in 265 s. What is the desired moment of inertia?

13–30 The balance wheel of a watch vibrates with an angular amplitude Θ, angular frequency ω, and phase angle $\phi = 0$. a) Find expressions for the angular velocity $d\theta/dt$ and angular acceleration $d^2\theta/dt^2$ as functions of time. b) Find its angular velocity and angular acceleration when its angular displacement is Θ; when its angular displacement is $\Theta/2$ and θ is decreasing. (*Hint:* Sketch a graph of θ versus t.)

*13–31 For the van der Waals interaction with potential energy function given by Eq. (13–25), show that when the magnitude of the displacement x from equilibrium ($r = R_0$) is small the potential energy can be written approximately as $U \approx \frac{1}{2}kx^2 - U_0$. (*Hint*: In Eq. (13–25), let $r = R_0 + x$ and $u = x/R_0$. Then approximate $(1 + u)^n$ by the first *three* terms in Eq. (13–28).) How does k in this equation compare with the force constant in Eq. (13–29) for the force?

*13–32 When displaced from equilibrium, the two hydrogen atoms in an H_2 molecule are acted on by a restoring force $F = -kx$ with $k = 580$ N/m. Calculate the oscillation frequency of the H_2 molecule. (*Hint:* The mass of a hydrogen atom is 1.008 atomic mass units, or 1 u; see Appendix E. As in Example 13–7 of Section 13–5, use $m/2$ instead of m in the expression for f.)

SECTION 13–6 THE SIMPLE PENDULUM

13–33 A Pendulum on Mars. A certain simple pendulum has a period on earth of 1.60 s. What is its period on the surface of Mars, where $g = 3.71$ m/s²?

13–34 You pull a simple pendulum of length 0.240 m to the side through an angle of 3.50° and release it. a) How much time does it take the pendulum bob to reach its highest speed? b) How much time does it take if the pendulum is relased at an angle of 1.75° instead of 3.50°?

13–35 An apple weighs 1.00 N. When you hang it from the end of a long spring of force constant 1.50 N/m and negligible mass, it bounces up and down in SHM. If you stop the bouncing and let the apple swing from side to side through a small angle, the frequency of this simple pendulum is half the bounce frequency. (Because the angle is small, the back and forth swings do not cause any appreciable change in the length of the spring.) What is the unstretched length of the spring?

13–36 A small sphere of mass m is attached to a massless rod of length L that is pivoted at the top, forming a simple pendulum. The pendulum is pulled to one side so that the rod is at an angle Θ from the vertical, and released from rest. a) In a diagram, show the pendulum just after it is released. On the diagram, draw vectors representing the *forces* acting on the small sphere and the *acceleration* of the sphere. Accuracy counts! At this point, what is the linear acceleration of the sphere? b) Repeat part (a) for the instant when the pendulum rod is at an angle $\Theta/2$ from the vertical. c) Repeat part (a) for the instant when the pendulum rod is vertical. At this point, what is the linear speed of the sphere?

13–37 After landing on an unfamiliar planet, a space explorer constructs a simple pendulum of length 50.0 cm. She finds that the pendulum makes 100 complete swings in 136 s. What is the value of g on this planet?

SECTION 13–7 THE PHYSICAL PENDULUM

13–38 We want to support a thin hoop by a horizontal nail and have the hoop make one complete small-angle oscillation each 2.0 s. What must the hoop's radius be?

13–39 Show that the expression for the period of a physical pendulum reduces to that of a simple pendulum if the physical pendulum consists of a particle of mass m on the end of a massless string of length L.

13–40 A 1.80-kg monkey wrench is pivoted 0.250 m from its center of mass and allowed to swing as a physical pendulum. The period for small-angle oscillations is 0.940 s. a) What is the moment of inertia of the wrench about an axis through the pivot? b) If the wrench is initially displaced 0.400 rad from its equilibrium position, what is the angular speed of the wrench as it passes through the equilibrium position?

13–41 A holiday ornament in the shape of a hollow sphere with mass $M = 0.015$ kg and radius $R = 0.050$ m is hung from a tree limb by a small loop of wire attached to the surface of the sphere. If the ornament is displaced a small distance and released, it swings back and forth as a physical pendulum. Calculate its period. (Neglect friction at the pivot. The moment of inertia of the sphere about the pivot at the tree limb is $5MR^2/3$.)

13–42 A 1.80-kg connecting rod from a car engine is pivoted about a horizontal knife edge as shown in Fig. 13–27. The center of gravity of the rod was located by balancing and is 0.200 m from the pivot. When it is set into small amplitude oscillation, the rod makes 100 complete swings in 120 s. Calculate the moment of inertia of the rod about the rotation axis through the pivot.

FIGURE 13–27 Exercise 13–42.

SECTION 13–8 DAMPED OSCILLATIONS

13–43 An unhappy 0.300-kg mouse moving on the end of a spring with force constant $k = 2.50$ N/m is acted on by a damping force $F_x = -bv$. a) If the constant b has the value 0.900 kg/s, what is the frequency of oscillation of the mouse? b) For what value of the constant b will the motion be critically damped?

13–44 A 50.0-g hard-boiled egg moves on the end of a spring with force constant $k = 25.0$ N/m. Its initial displacement is 0.300 m. A damping force $F_x = -bv$ acts on the egg, and the amplitude of the motion decreases to 0.100 m in 5.00 s. Calculate the magnitude of the damping constant b.

13–45 The motion of an underdamped oscillator is described by Eq. (13–42). Let the phase angle ϕ be zero. a) According to this equation, what is the value of x at $t = 0$? b) What are the magnitude and direction of the velocity at $t = 0$? What does the result tell you about the slope of the graph of x versus t near $t = 0$? c) Obtain an expression for the acceleration a at $t = 0$. For what value or range of values of the damping constant b (in terms of k and m) is the acceleration at $t = 0$ negative, zero, and positive? Discuss each case in terms of the shape of the graph of x versus t near $t = 0$.

SECTION 13–9 FORCED OSCILLATIONS, RESONANCE, AND CHAOS

13–46 A sinusoidally varying driving force is applied to a damped harmonic oscillator of force constant k and mass m. If the damping constant has a value b_1, the amplitude is A_1 when the driving angular frequency equals $\sqrt{k/m}$. In terms of A_1, what is the amplitude for the same driving frequency and the same driving force amplitude F_{max} if the damping constant is a) $3b_1$? b) $b_1/2$?

13-47 A sinusoidally varying driving force is applied to a damped harmonic oscillator. a) What are the units of the damping constant b? b) Show that the quantity $\sqrt{km}$ has the same units as b. c) In terms of F_{max} and k, what is the amplitude for $\omega_d = \sqrt{k/m}$ when i) $b = 0.2\sqrt{km}$? ii) $b = 0.4\sqrt{km}$? Compare your results to Fig. 13-24.

13-48 An experimental package and its support structure to be placed on board the International Space Station act as an underdamped spring-mass system with a force constant of 2.1×10^6 N/m and a mass of 108 kg. A NASA requirement is that resonance for forced oscillations not occur for any frequency below 35 Hz. Does this package meet the requirement?

PROBLEMS

13-49 SHM in a Car Engine. The motion of the piston of an automobile engine (Fig. 13-28) is approximately simple harmonic. a) If the stroke of an engine (twice the amplitude) is 0.100m and the engine runs at 3500 rev/min, compute the acceleration of the piston at the endpoint of its stroke. b) If the piston has a mass of 0.450 kg, what net force must be exerted on it at this point? c) What are the speed and kinetic energy of the piston at the midpoint of its stroke? d) What average power is required to accelerate the piston from rest to the speed found in part (c)? e) If the engine runs at 7000 rev/min, what are the answers to parts (b), (c), and (d)?

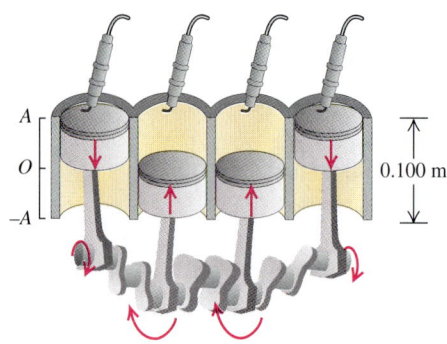

FIGURE 13-28 Problem 13-49.

13-50 Four passengers whose combined mass is 250 kg compress the springs of a car with worn-out shock absorbers by 4.00 cm when they enter it. Model the car and passengers as a single body on a single ideal spring. If the loaded car has a period of vibration of 1.08 s, what is the period of vibration of the empty car?

13-51 A glider is oscillating in SHM on an air track with an amplitude A_1. You slow it so that its amplitude is halved. What happens to its a) period, frequency, and angular frequency? b) total mechanical energy? c) maximum speed? d) speed at $x = \pm A_1/4$? e) potential and kinetic energies at $x = \pm A_1/4$?

13-52 You hang an unknown weight from the end of a spring, moving your hand down slowly so that when the weight comes to equilibrium it has stretched the spring a distance L. If the spring has negligible mass, prove that the weight can oscillate in SHM with the same period as a simple pendulum of length L.

13-53 A child with poor table manners is sliding his 250-g dinner plate back and forth in SHM with an amplitude of 0.100 m on a horizontal surface. At a point 0.060 m away from equilibrium, the speed of the plate is 0.300 m/s. a) What is the period?

b) What is the displacement when the speed is 0.160 m/s? c) In the center of the dinner plate is a 10.0-g carrot slice. If the carrot slice is just on the verge of slipping at the end point of the path, what is the coefficient of static friction between the carrot slice and the plate?

13-54 A mechanical bucking horse moves vertically in SHM with an amplitude of 0.250 m and a frequency of 1.50 Hz, that are the same whether there is a rider or not. A drugstore cowboy on the horse decides the macho way to ride is to not hold on in any way. a) He leaves the saddle when it is moving upward. What is the magnitude of the downward acceleration of the saddle when he loses contact with it? b) How high is the top surface of the saddle above its equilibrium position when he first gets airborne? c) How fast is he moving upward when he leaves the saddle? d) He is in free fall until he returns to the saddle. Show that this happens 0.538 s later. e) What is his speed relative to the saddle at the instant he returns?

13-55 A block of mass M rests on a frictionless surface and is connected to a horizontal spring of force constant k. The other end of the spring is attached to a wall (Fig. 13-29). A second block of mass m rests on top of the first block. The coefficient of static friction between the blocks is μ_s. Find the *maximum* amplitude of oscillation such that the top block will not slip on the bottom block.

FIGURE 13-29 Problem 13-55.

13-56 An object with a mass of 0.200 kg is acted on by an elastic restoring force with force constant 10.0 N/m. a) Graph elastic potential energy U as a function of displacement x over a range of x from -0.300 m to $+0.300$ m. On your graph, let 1 cm = 0.05 J vertically and 1 cm = 0.05 m horizontally. The object is set into oscillation with an initial potential energy of 0.140 J and an initial kinetic energy of 0.060 J. Answer the following questions by referring to the graph. b) What is the amplitude of oscillation? c) What is the potential energy when the displacement is one-half the amplitude? d) At what displacement are the kinetic and potential energies equal? e) What is the value of the phase angle ϕ if the initial velocity is positive and the initial displacement is negative?

13-57 Figure 13-14 shows a body of mass m hanging from a spring of force constant k. The positive x-axis is upward and $x = 0$

is the equilibrium position of the body. a) Show that when the body is at coordinate x, the elastic potential energy in the spring is $U_{el} = \frac{1}{2}k(\Delta l - x)^2$. b) Let $x = x_0$ be the coordinate at which the gravitational potential energy is zero. Show that the total potential energy is $U = \frac{1}{2}kx^2 + \frac{1}{2}k(\Delta l)^2 - mgx_0$. c) The expression for the total potential energy in part (b) is of the form $U = \frac{1}{2}kx^2 + C$, where the constant C equals $\frac{1}{2}k(\Delta l)^2 - mgx_0$. Explain why the behavior of the system does not depend on the value of this constant, so that vertical SHM is fundamentally no different than horizontal SHM, for which $U = \frac{1}{2}kx^2$.

13-58 A hanging wire is 1.80 m in length. When a 60.0-kg steel ball is suspended from the wire, the wire stretches by 2.00 mm. If the ball is pulled down a small additional distance and released, at what frequency will it vibrate? Assume that the stress on the wire is less than the proportional limit (see Section 11-8).

13-59 A 5.00-kg partridge is suspended from a pear tree by an ideal spring of negligible mass. When the partridge is pulled down 0.100 m below its equilibrium position and released, it vibrates with a period of 4.20 s. a) What is its speed as it passes through the equilibrium position? b) What is its acceleration when it is 0.050 m above the equilibrium position? c) When it is moving upward, how much time is required for it to move from a point 0.050 m below its equilibrium position to a point 0.050 m above it? d) The motion of the partridge is stopped, and then it is removed from the spring. How much does the spring shorten?

13-60 A 0.0200-kg bolt moves with simple harmonic motion that has an amplitude of 0.240 m and a period of 1.500 s. The displacement of the bolt is +0.240 m when $t = 0$. Compute a) the displacement of the bolt when $t = 0.500$ s; b) the magnitude and direction of the force acting on the bolt when $t = 0.500$ s; c) the minimum time required for the bolt to move from its initial position to the point where $x = -0.180$ m; d) the speed of the bolt when $x = -0.180$ m.

13-61 SHM of a Butcher's Scale. A spring of negligible mass and force constant $k = 400$ N/m is hung vertically, and a 0.200-kg pan is suspended from its lower end. A butcher drops a 2.2-kg steak onto the pan from a height of 0.40 m. The steak makes a totally inelastic collision with the pan and sets the system into vertical simple harmonic motion. What is a) the speed of the pan and steak immediately after the collision? b) the amplitude of the subsequent motion? c) the period of that motion?

13-62 A 40.0-N force stretches a vertical spring 0.250 m. a) What mass must be suspended from the spring so that the system will oscillate with a period of 1.00 s? b) If the amplitude of the motion is 0.050 m and the period is that specified in part (a), where is the object and in what direction is it moving 0.35 s after it has passed the equilibrium position, moving downward? c) What force (magnitude and direction) does the spring exert on the object when it is 0.030 m below the equilibrium position, moving upward?

13-63 A small excursion boat with a flat deck bobs up and down executing simple harmonic motion due to the waves on a lake. The amplitude of the motion is 0.200 m and the period is 2.80 s. A stable dock is next to the boat at a level equal to the highest level of the deck. People wish to step off the boat onto the dock but can do so comfortably only if the level of the deck is within 0.100 m of the dock level. How much time do the people have to get off comfortably during each period of the simple harmonic motion?

13-64 An interesting though highly impractical example of oscillation is the motion of an object dropped down a hole that extends from one side of the earth, through its center, to the other side. With the assumption (not very realistic) that the earth is a sphere of uniform density, prove that the motion is simple harmonic and find the period. (*Note:* The gravitational force on the object as a function of the object's distance r from the center of the earth was derived in Example 12-9 (Section 12-7). The motion is simple harmonic if the acceleration a and the displacement from equilibrium x are related by Eq. (13-8), and the period is then $T = 2\pi/\omega$.)

13-65 Let t_1 be the time it takes a body undergoing SHM to move from $x = 0$ (when $t = 0$) to $x = A$. Obtain an equation for t_1 in the following way. In Eq. (13-22) replace v by dx/dt. Separate variables by writing all factors containing x on one side and all those containing t on the other side so that each side can be integrated. Integrate the equation with limits on t of 0 to t_1 and limits on x of 0 to A and thereby obtain an equation for t_1. How does t_1 compare to the period T?

13-66 For a certain oscillator the net force on the body of mass m is given by $F = -cx^3$. a) What is the potential energy function for this oscillator, if we take $U = 0$ at $x = 0$? b) One-quarter of a period is the time for the body to move from $x = 0$ to $x = A$. Calculate this time and hence the period. (*Hint:* Begin with Eq. (13-20), modified to include the potential energy function you found in part (a), and solve for the velocity v as a function of x. Then replace v by dx/dt and follow the steps outlined in Problem 13-65. In the x-integral make the change of variable $u = x/A$. The resulting integral can be evaluated by numerical methods on a computer and has the value $\int_0^1 du/\sqrt{1 - u^4} = 1.31$.) c) According to the result you obtained in part (b), does the period depend on the amplitude A of the motion? Are the oscillations simple harmonic?

13-67 Consider the circle of reference shown in Fig. 13-4. The x-component of the velocity of Q is the velocity of P. Compute this component and show that the velocity of P is as given by Eq. (13-15).

13-68 To measure g in an unorthodox manner, a student places a ball bearing on the concave side of a lens (Fig. 13-30). She attaches the lens to a simple harmonic oscillator (actually a small stereo speaker) whose amplitude is A and whose frequency f can be varied. She can measure both A and f with a strobe light. a) If the ball bearing has mass m, find the normal force exerted by the lens on the ball bearing as a function of time. Your result should be in terms of A, f, m, g, and a phase angle ϕ. b) The frequency is slowly increased. When it reaches a value f_b, the ball is heard to bounce. What is g in terms of A and f_b?

*13-69 Two identical atoms in a diatomic molecule vibrate as harmonic oscillators. However, their center of mass, midway between them, remains at rest. a) Show that at any instant the momenta of the atoms relative to the center of mass are $\vec{p}$ and $-\vec{p}$. b) Show that the total kinetic energy K of the two atoms

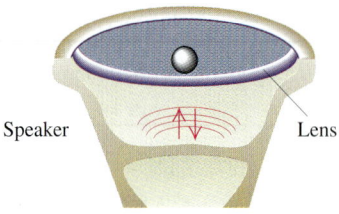

FIGURE 13–30 Problem 13–68.

at any instant is the same as that of a single object of mass $m/2$ with a momentum of magnitude p. (Use $K = p^2/2m$.) This result shows why $m/2$ should be used in the expression for f in Example 13–7 (Section 13–5). c) If the atoms are not identical, but have masses m_1 and m_2, show that the result of part (a) still holds and the single object's mass in part (b) is $m_1m_2/(m_1 + m_2)$. The quantity $m_1m_2/(m_1 + m_2)$ is called the *reduced mass* of the system.

*13–70 An approximation for the potential energy of a KCl molecule is $U = A[(R_0^7/8r^8) - 1/r]$, where $R_0 = 2.67 \times 10^{-10}$ m and $A = 2.31 \times 10^{-28}$ J · m. Using this approximation: a) Show that the radial component of the force on each atom is $F = A[(R_0^7/r^9) - 1/r^2]$. b) Show that R_0 is the equilibrium separation. c) Find the minimum potential energy. d) Use $r = R_0 + x$ and the first two terms of the binomial theorem (Eq. 13–28) to show that $F \approx -(7A/R_0^3)x$ so that the molecule's force constant is $k = 7A/R_0^3$. e) With both the K and Cl atoms vibrating in opposite directions on opposite sides of the molecule's center of mass, $m_1m_2/(m_1 + m_2) = 3.06 \times 10^{-26}$ kg is the mass to use in calculating the frequency. (See Problem 13–69.) Calculate the frequency of small-amplitude vibrations.

*13–71 Three positively charged particles are maintained in a straight line. The end particles are identical and are held fixed at a distance $2R_0$ apart. The potential energy for the force acting on the center particle can be written as $U = A[1/r - 1/(r - 2R_0)]$, where A is a constant and r is the distance from the left-hand particle to the center particle. a) Show that the force on the center particle is $F = A[1/r^2 - 1/(r - 2R_0)^2]$. b) Show that $r = R_0$ at equilibrium. c) Use $r = R_0 + x$ and the first two terms of the binomial theorem (Eq. 13–28) to show that $F \approx -(4A/R_0^3)x$ so that the force constant is $k = 4A/R_0^3$ d) What is the frequency of small-amplitude oscillation of the center particle if it has mass m?

13–72 Many real springs are more easily stretched than compressed. We can represent this by using different spring constants for $x > 0$ and for $x < 0$. As an example, consider a spring that exerts the following restoring force:

$$F = \begin{cases} -kx & \text{for } x > 0, \\ -2kx & \text{for } x < 0. \end{cases}$$

A mass m on a frictionless horizontal surface is attached to this spring, displaced to $x = A$ by stretching the spring, and released. a) Find the period of the motion. Does the period depend on A? Are the oscillations simple harmonic? b) What is the most negative value of x that the mass m reaches? Is the oscillation symmetric about the point $x = 0$?

13–73 Two solid cylinders connected along their common axis by a short, light rod have a radius R and a total mass M and rest

on a horizontal table top. A spring with force constant k has one end attached to a clamp and the other end attached to a frictionless ring at the center of mass of the cylinders (Fig. 13–31). The cylinders are pulled to the left a distance x, which stretches the spring, and released. There is sufficient friction between the table top and the cylinders for the cylinders to roll without slipping as they move back and forth on the end of the spring. Show that the motion of the center of mass of the cylinders is simple harmonic, and calculate its period in terms of M and k. (*Hint:* The motion is simple harmonic if a and x are related by Eq. (13–8), and the period then is $T = 2\pi/\omega$. Apply $\Sigma \tau = I_{cm}\alpha$ and $\Sigma F = Ma_{cm}$ to the cylinders in order to relate a_{cm} and the displacement x of the cylinders from their equilibrium position.)

FIGURE 13–31 Problem 13–73.

13–74 A slender, uniform metal rod with mass M is pivoted without friction about an axis through its midpoint and perpendicular to the rod. A horizontal spring with force constant k is attached to the lower end of the rod, with the other end of the spring attached to a rigid support. If the rod is displaced by a small angle Θ from the vertical (Fig. 13–32) and released, show that it moves in angular simple harmonic motion and calculate the period. (*Hint:* Assume that the angle Θ is small enough for the approximations $\sin \Theta \approx \Theta$ and $\cos \Theta \approx 1$ to be valid. The motion is simple harmonic if $d^2\theta/dt^2 = -\omega^2\theta$, and the period is then $T = 2\pi/\omega$.)

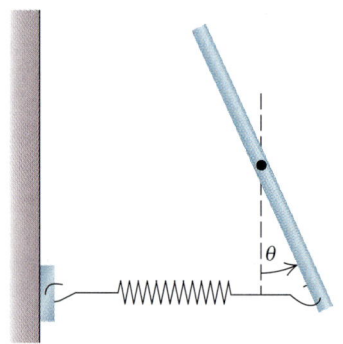

FIGURE 13–32 Problem 13–74.

13–75 The Silently Ringing Bell Problem. A large bell is hung from a wooden beam so it can swing back and forth with negligible friction. The center of mass of the bell is 0.60 m below the pivot, the bell's mass is 34.0 kg, and the moment of inertia of the bell about an axis at the pivot is 18.0 kg · m². The clapper is a small 1.8-kg mass attached to one end of a slender rod that has length L and negligible mass. The other end of the rod is attached to the inside of the bell so it can swing freely about the same axis as the bell. What should be the length L of the clapper rod for the bell to ring silently, that is, for the period of oscillation for the bell to equal that for the clapper?

13–76 Two identical thin rods, each of mass m and length L, are joined at right angles to form an L-shaped object. This object is balanced on top of a sharp edge (Fig. 13–33). If the L-shaped object is deflected slightly, it oscillates. Find the frequency of oscillation.

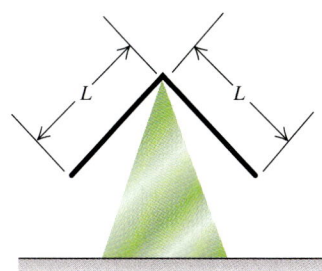

FIGURE 13–33 Problem 13–76.

13–77 You want to construct a pendulum with a period of 4.00 s at a location where $g = 9.80$ m/s^2. a) What is the length of a *simple* pendulum having this period? b) Suppose the pendulum must be mounted in a case that is not over 0.50 m high. Can you devise a pendulum having a period of 4.00 s that will satisfy this requirement?

13–78 Show that $x(t)$ as given in Eq. (13–42) is a solution of Newton's second law for damped oscillations, Eq. (13–41), if ω' is as defined in Eq. (13–43).

13–79 A uniform rod of length L oscillates through small angles about a point a distance x from its center. a) Prove that its angular frequency is $\sqrt{gx/[(L^2/12) + x^2]}$. b) Show that its maximum angular frequency occurs when $x = L/\sqrt{12}$. c) What is the length of the rod if the maximum angular frequency is 2π rad/s?

13–80 When Is It Simple? To make a pendulum, you hang a uniform solid sphere of radius R on the end of a light string. The distance from the pivot to the center of the sphere is L. a) Prove that the period of your pendulum is $T = T_{sp} \sqrt{1 + 2R^2/5L^2}$, where T_{sp} is the period of a simple pendulum of length L. b) At what value of L is T only 0.10% larger than T_{sp}? c) For a sphere of diameter 2.540 cm, what is the value of L in part (b)?

CHALLENGE PROBLEMS

13–81 Two springs, each with unstretched length 0.200 m but with different force constants k_1 and k_2, are attached to opposite ends of a block with mass m on a level, frictionless surface. The outer ends of the springs are now attached to two pins P_1 and P_2, 0.100 m from the original positions of the ends of the springs (Fig. 13–34). Let $k_1 = 2.00$ N/m, $k_2 = 6.00$ N/m, and $m = 0.100$ kg. a) Find the length of each spring when the block is in its new equilibrium position after the springs have been attached to the pins. b) Find the period of vibration of the block if it is slightly displaced from its new equilibrium position and released.

FIGURE 13–34 Challenge Problem 13–81.

13–82 The Effective Force Constant of Two Springs. Two springs with the same unstretched length but different force constants k_1 and k_2 are attached to a block of mass m on a level, frictionless surface. Calculate the effective force constant k_{eff} in each of the three cases (a), (b), and (c) depicted in Fig. 13–35. (The effective force constant is defined by $\Sigma F = -k_{eff}x$.) d) An object with mass m, suspended from a uniform spring with a force constant k, vibrates with a frequency f_1. When the spring is cut in half and the same object is suspended from one of the halves, the frequency is f_2. What is the ratio f_2/f_1?

FIGURE 13–35 Challenge Problem 13–82.

13–83 a) What is the change ΔT in the period of a simple pendulum when the acceleration of gravity g changes by Δg? (*Hint:* The new period $T + \Delta T$ is obtained by substituting $g + \Delta g$ for g:

$$T + \Delta T = 2\pi \sqrt{\frac{L}{g + \Delta g}}.$$

To obtain an approximate expression, expand the factor $(g + \Delta g)^{-1/2}$ using the binomial theorem (Appendix B) and keeping only the first two terms:

$$(g + \Delta g)^{-1/2} = g^{-1/2} - \frac{1}{2} g^{-3/2} \Delta g + \cdots.$$

The other terms contain higher powers of Δg and are very small if Δg is small.) Express your result as the *fractional* change in period $\Delta T/T$ in terms of the fractional change $\Delta g/g$. b) A pendulum clock keeps correct time at a point where $g = 9.8000$ m/s^2, but is found to lose 4.0 s each day at a higher elevation. Use the result of part (a) to find the approximate value of g at this new location.

13–84 A Spring with Mass. The preceding problems in this chapter have assumed that the springs had negligible mass. But of course no spring is completely massless. To find the effect of the spring's mass, consider a spring of mass M, equilibrium length L_0, and spring constant k. When stretched or compressed to a length L, the potential energy is $\frac{1}{2}kx^2$, where $x = L - L_0$. a) Consider a spring as described above that has one end fixed and the other end moving with speed v. Assume that the speed of points along the length of the spring varies linearly with distance l from the fixed end. Assume also that the mass M of the spring is distributed uniformly along the length of the spring. Calculate the kinetic energy of the spring in terms of M and v. (*Hint:* Divide the spring into pieces of length dl; find the speed of each piece in terms of l, v, and L; find the mass of each piece in terms of dl, M, and L; and integrate from 0 to L. The result is *not* $\frac{1}{2}Mv^2$, since not all of the spring moves with the same speed.) b) Take the time derivative of the conservation of energy equation, Eq. (13–21), for a mass m moving on the end of a *massless* spring. By comparing your results to Eq. (13–8), which defines ω, show that the angular frequency of oscillation is $\omega = \sqrt{k/m}$. c) Apply the procedure of part (b) to obtain the angular frequency of oscillation ω of the spring considered in part (a). If the *effective mass* M' of the spring is defined by $\omega = \sqrt{k/M'}$, what is M' in terms of M?

13–85 A uniform 1.00-m stick hangs from a horizontal axis at one end and oscillates as a physical pendulum. An object of small dimensions and of mass equal to that of the meter stick can be clamped to the stick at a distance y below the axis. Let T represent the period of the system with the body attached and T_0 the period of the meter stick alone. a) Find the ratio T/T_0. Evaluate your expression for y ranging from 0 to 1.0 m in steps

of 0.1 m, and graph T/T_0 versus y. b) Is there any value of y, other than $y = 0$, for which $T = T_0$? If so, find it and explain why the period is unchanged when y has this value.

13–86 You measure the period of a physical pendulum about one pivot point to be T. Then you find another pivot point on the opposite side of the center of mass that gives the same period. The two points are separated by a distance L. Use the parallel-axis theorem to show that $g = L(2\pi/T)^2$. (This result shows a way that you can measure g without knowing the mass or any moments of inertia of the physical pendulum.)

13–87 Resonance in a Mechanical System. A mass m is attached to one end of a massless spring with a force constant k and an unstretched length l_0. The other end of the spring is free to turn about a nail driven into a frictionless horizontal surface (Fig. 13–36). The mass is made to revolve in a circle with an angular frequency of revolution ω'. a) Calculate the length l of the spring as a function of ω'. b) What happens to the result in part (a) when ω' approaches the natural frequency $\omega = \sqrt{k/m}$ of the mass-spring system? (If your result bothers you, remember that massless springs and frictionless surfaces don't exist as such, but are only approximate descriptions of real springs and surfaces. Also, Hooke's law is itself only an approximation to the way real springs behave; the greater the elongation of the spring, the greater the deviation from Hooke's law.)

FIGURE 13–36 Challenge Problem 13–87.

*13–88 Vibration of a Covalently Bonded Molecule. Many diatomic (two-atom) molecules are bound together by *covalent bonds* that are much stronger than the van der Waals interaction. Examples include H_2, O_2, and N_2. Experiment shows that for many such molecules, the interaction can be described by a force of the form

$$F = A\left[e^{-2b(r - R_0)} - e^{-b(r - R_0)}\right],$$

where A and b are positive constants, r is the center-to-center separation of the atoms, and R_0 is the equilibrium separation. For the hydrogen molecule (H_2), $A = 2.97 \times 10^{-8}$ N, $b = 1.95 \times 10^{10}$ m^{-1}, and $R_0 = 7.4 \times 10^{-11}$ m. Find the force constant for small oscillations around equilibrium. (*Hint:* Use the expansion for e^x given in Appendix B.) Compare your result to the value given in Exercise 13–32.

Fluid Mechanics

The beach is a wonderful place to watch fluid mechanics in action. Water moves in response to gravity and the pressure of the surrounding fluid; the flow of water changes from laminar to turbulent as the waves break; and surface tension makes the water in a breaking wave form into drops.

14-1 INTRODUCTION

Fluids play a vital role in many aspects of everyday life. We drink them, breathe them, swim in them. They circulate through our bodies and control our weather. Airplanes fly through them; ships float in them. A fluid is any substance that can flow; we use the term for both liquids and gases. We usually think of a gas as easily compressed and a liquid as nearly incompressible, although there are exceptional cases.

We begin our study with **fluid statics,** the study of fluids at rest in equilibrium situations. Like other equilibrium situations, it is based on Newton's first and third laws. We will explore the key concepts of density, pressure, buoyancy, and surface tension. **Fluid dynamics,** the study of fluids in motion, is much more complex; indeed, it is one of the most complex branches of mechanics. Fortunately, we can analyze many important situations using simple idealized models and familiar principles such as Newton's laws and conservation of energy. Even so, we will barely scratch the surface of this broad and interesting topic.

14-2 DENSITY

An important property of any material is its **density,** defined as its mass per unit volume. A homogeneous mate-

rial such as ice or iron has the same density throughout. We use the Greek letter ρ ("rho") for density. If a mass m of homogeneous material has volume V, the density ρ is

$$\rho = \frac{m}{V} \quad \text{(definition of density)}. \quad (14\text{--}1)$$

The density of some materials varies from point to point within the material; some examples are the earth's atmosphere (which is less dense at high altitude) and oceans (which are denser at greater depths). For these materials, Eq. (14–1) describes the *average* density. In general, the density of a material depends on environmental factors such as temperature and pressure.

The SI unit of density is the kilogram per cubic meter (1 kg/m^3). The cgs unit, the gram per cubic centimeter (1 g/cm^3), is also widely used. The conversion factor

$$1 \text{ g/cm}^3 = 1000 \text{ kg/m}^3$$

is useful. Densities of several common substances at ordinary temperatures are given in Table 14–1. Note the wide range of magnitudes. The densest material found on earth is the metal osmium ($\rho = 22{,}500$ kg/m^3), but this pales by comparison to the densities of exotic astronomical objects such as white dwarf stars and neutron stars.

The **specific gravity** of a material is the ratio of its density to the density of water at 4.0°C, 1000 kg/m^3; it is a pure number without units. For example, the specific

428   CHAPTER 14   FLUID MECHANICS

TABLE 14–1
DENSITIES OF SOME COMMON SUBSTANCES

MATERIAL	DENSITY (kg/m³)*	MATERIAL	DENSITY (kg/m³)*
Air (1 atm, 20° C)	1.20	Iron, steel	7.8×10^3
Ethanol	0.81×10^3	Brass	8.6×10^3
Benzene	0.90×10^3	Copper	8.9×10^3
Ice	0.92×10^3	Silver	10.5×10^3
Water	1.00×10^3	Lead	11.3×10^3
Seawater	1.03×10^3	Mercury	13.6×10^3
Blood	1.06×10^3	Gold	19.3×10^3
Glycerin	1.26×10^3	Platinum	21.4×10^3
Concrete	2×10^3	White dwarf star	10^{10}
Aluminum	2.7×10^3	Neutron star	10^{18}

*To obtain the densities in grams per cubic centimeter, simply divide by 10^3.

The price of gold is quoted by weight (say, in dollars per ounce). Because gold is one of the densest of the metals, a small fortune in gold can be stored in a small volume.

gravity of aluminum is 2.7. "Specific gravity" is a poor term, since it has nothing to do with gravity; "relative density" would have been better.

Measuring density is an important analytical technique. For example, we can determine the charge condition of a storage battery by measuring the density of its electrolyte, a sulfuric acid solution. As the battery discharges, the sulfuric acid (H_2SO_4) combines with lead in the battery plates to form insoluble lead sulfate ($PbSO_4$), decreasing the concentration of the solution. The density decreases from about 1.30×10^3 kg/m³ for a fully charged battery to 1.15×10^3 kg/m³ for a discharged battery.

Another automotive example is permanent-type antifreeze, which is usually a solution of ethylene glycol ($\rho = 1.12 \times 10^3$ kg/m³) and water. The freezing point of the solution depends on the glycol concentration, which can be determined by measuring the density. Such measurements of fluid density are performed routinely in service stations with a device called a hydrometer, which we'll discuss in Section 14–4.

EXAMPLE 14–1

The weight of a roomful of air Find the mass and weight of the air in a living room with a 4.0 m × 5.0 m floor and a ceiling 3.0 m high. What is the mass and weight of an equal volume of water?

SOLUTION The densities of air and water are given in Table 14–1. The volume of the room is $V = (3.0 \text{ m})(4.0 \text{ m})(5.0 \text{ m}) = 60$ m³. The mass m_{air} of air is given by Eq. (14–1):

$$m_{air} = \rho_{air}V = (1.2 \text{ kg/m}^3)(60 \text{ m}^3) = 72 \text{ kg}.$$

The weight of the air is

$$w_{air} = m_{air}g = (72 \text{ kg})(9.8 \text{ m/s}^2) = 700 \text{ N} = 160 \text{ lb}.$$

Does it surprise you that a roomful of air weighs this much? The mass of an equal volume of water is

$$m_{water} = \rho_{water}V = (1000 \text{ kg/m}^3)(60 \text{ m}^3) = 6.0 \times 10^4 \text{ kg}.$$

The weight is

$$w_{water} = m_{water}g = (6.0 \times 10^4 \text{ kg})(9.8 \text{ m/s}^2) = 5.9 \times 10^5 \text{ N}$$
$$= 1.3 \times 10^5 \text{ lb} = 66 \text{ tons}.$$

This much weight would certainly collapse the floor of an ordinary house.

14–3 PRESSURE IN A FLUID

When a fluid (either liquid or gas) is at rest, it exerts a force perpendicular to any surface in contact with it, such as a container wall or a body immersed in the fluid. This is the force that you feel pressing on your legs when you dangle them in a swimming pool. While the fluid as a whole is at rest, the molecules that make up the fluid are in motion; the force exerted by the fluid is due to molecules colliding with their surroundings.

If we think of an imaginary surface *within* the fluid, the fluid on the two sides of the surface exerts equal and opposite forces on the surface. (Otherwise, the surface would accelerate and the fluid would not remain at rest.) Consider a small surface of area dA centered on a point in the fluid; the normal force exerted by the fluid on each side is $dF_\perp$ (Fig. 14–1). We define the **pressure** p at that point as the normal force per unit area, that is, the ratio of $dF_\perp$ to dA:

$$p = \frac{dF_\perp}{dA} \quad \text{(definition of pressure).} \quad (14\text{–}2)$$

If the pressure is the same at all points of a finite plane surface with area A, then

$$p = \frac{F_\perp}{A}, \quad (14\text{–}3)$$

where $F_\perp$ is the net normal force on one side of the surface. The SI unit of pressure is the **pascal,** where

$$1 \text{ pascal} = 1 \text{ Pa} = 1 \text{ N/m}^2.$$

We introduced the pascal in Chapter 11. Two related units, used principally in meteorology, are the *bar,* equal to 10^5 Pa, and the *millibar,* equal to 100 Pa.

Atmospheric pressure p_a is the pressure of the earth's atmosphere, the pressure at the bottom of this sea of air in which we live. This pressure varies with weather changes and with elevation. Normal atmospheric pressure at sea level (an average value) is 1 *atmosphere* (atm), defined to be exactly 101,325 Pa. To four significant figures,

$$(p_a)_{av} = 1 \text{ atm} = 1.013 \times 10^5 \text{ Pa}$$

$$= 1.013 \text{ bar} = 1013 \text{ millibar} = 14.70 \text{ lb/in.}^2$$

CAUTION ▶ In everyday language the words "pressure" and "force" mean pretty much the same thing. In fluid mechanics, however, these words describe distinct quantities with different characteristics. Fluid pressure acts perpendicular to any surface in the fluid, no matter how that surface is oriented. Hence pressure has no intrinsic direction of its own; it's a scalar. By contrast, force is a vector with a definite direction. Remember, too, that pressure is force per unit area. ◀

14–1 An imaginary small area dA within a fluid has equal normal forces $dF_\perp$ exerted on each side. The pressure acting over the area is $p = dF_\perp/dA$.

EXAMPLE 14–2

In the room described in Example 14–1, what is the total downward force on the floor surface due to air pressure of 1.00 atm?

SOLUTION The floor area is $A = (4.0 \text{ m})(5.0 \text{ m}) = 20 \text{ m}^2$. The pressure is uniform, so from Eq. (14–3) the total downward force is

$$F = pA = (1.013 \times 10^5 \text{ N/m}^2)(20 \text{ m}^2)$$

$$= 2.0 \times 10^6 \text{ N} = 4.5 \times 10^5 \text{ lb} = 225 \text{ tons.}$$

As in Example 14–1, this is more than enough force to collapse the floor. So why doesn't it collapse? Because there is an upward force on the underside of the floor. If the house has a basement, this upward force is exerted by the air underneath the floor. In this case, if we neglect the thickness of the floor, the *net* force due to air pressure is zero.

PRESSURE, DEPTH, AND PASCAL'S LAW

If the weight of the fluid can be neglected, the pressure in a fluid is the same throughout its volume. We used that approximation in our discussion of bulk stress and strain in Section 11–6. But often the fluid's weight is *not* negligible. Atmospheric pressure is less at high altitude than at sea level, which is why an airplane cabin has to be pressurized when flying at 35,000 feet. When you dive into deep water, your ears tell you that the pressure increases rapidly with increasing depth below the surface.

We can derive a general relation between the pressure p at any point in a fluid at rest and the elevation y of the point. We'll assume that the density ρ and the acceleration due to gravity g are the same throughout the fluid. If the fluid is in equilibrium, every

(a)

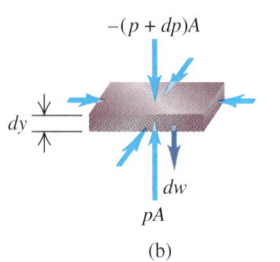

(b)

14-2 The forces on an element of fluid in equilibrium.

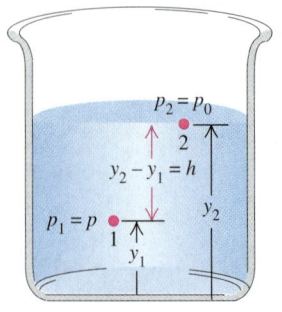

14-3 The pressure at a depth h in a fluid is greater than the surface pressure p_0 by $\rho g h$.

14-4 The pressure in the fluid is the same at all points having the same elevation. The shape of the container does not affect the pressure.

volume element is in equilibrium. Consider a thin element of fluid, with height dy (Fig. 14–2). The bottom and top surfaces each have area A, and they are at elevations y and $y + dy$ above some reference level where $y = 0$. The volume of the fluid element is $dV = A \, dy$, its mass is $dm = \rho \, dV = \rho A \, dy$, and its weight is $dw = dm \, g = \rho g A \, dy$.

What are the other forces on this fluid element? Call the pressure at the bottom surface p; the total y-component of upward force on this surface is pA. The pressure at the top surface is $p + dp$, and the total y-component of (downward) force on the top surface is $-(p + dp)A$. The fluid element is in equilibrium, so the total y-component of force, including the weight and the forces at the bottom and top surfaces, must be zero:

$$\Sigma F_y = 0, \quad \text{so} \quad pA - (p + dp)A - \rho g A \, dy = 0.$$

When we divide out the area A and rearrange, we get

$$\frac{dp}{dy} = -\rho g. \tag{14-4}$$

This equation shows that when y increases, p decreases; that is, as we move upward in the fluid, pressure decreases, as we expect. If p_1 and p_2 are the pressures at elevations y_1 and y_2, respectively, and if ρ and g are constant, then

$$p_2 - p_1 = -\rho g(y_2 - y_1) \quad \text{(pressure in a fluid of uniform density).} \tag{14-5}$$

It's often convenient to express Eq. (14–5) in terms of the *depth* below the surface of a fluid (Fig. 14–3). Take point 1 at any level in the fluid and let p represent the pressure at this point. Take point 2 at the *surface* of the fluid, where the pressure is p_0 (subscript zero for zero depth). The depth of point 1 below the surface is $h = y_2 - y_1$, and Eq. (14–5) becomes

$$p_0 - p = -\rho g(y_2 - y_1) = -\rho g h,$$
$$p = p_0 + \rho g h \quad \text{(pressure in a fluid of uniform density).} \tag{14-6}$$

The pressure p at a depth h is greater than the pressure p_0 at the surface by an amount $\rho g h$. Note that the pressure is the same at any two points at the same level in the fluid. The *shape* of the container does not matter (Fig. 14–4).

Equation (14–6) shows that if we increase the pressure p_0 at the top surface, possibly by using a piston that fits tightly inside the container to push down on the fluid surface, the pressure p at any depth increases by exactly the same amount. This fact was recognized in 1653 by the French scientist Blaise Pascal (1623–1662) and is called **Pascal's law: Pressure applied to an enclosed fluid is transmitted undiminished to every portion of the fluid and the walls of the containing vessel.**

The hydraulic lift shown schematically in Fig. 14–5 illustrates Pascal's law. A piston with small cross-section area A_1 exerts a force F_1 on the surface of a liquid such as oil. The applied pressure $p = F_1/A_1$ is transmitted through the connecting pipe to a larger piston of area A_2. The applied pressure is the same in both cylinders, so

$$p = \frac{F_1}{A_1} = \frac{F_2}{A_2} \quad \text{and} \quad F_2 = \frac{A_2}{A_1} F_1. \tag{14-7}$$

The hydraulic lift is a force-multiplying device with a multiplication factor equal to the ratio of the areas of the two pistons. Dentist's chairs, car lifts and jacks, many elevators, and hydraulic brakes all use this principle.

For gases the assumption that the density ρ is uniform is realistic only over short vertical distances. In a room with a ceiling height of 3.0 m filled with air of uniform density 1.2 kg/m^3, the difference in pressure between floor and ceiling, given by Eq. (14–6), is

$$\rho g h = (1.2 \text{ kg/m}^3)(9.8 \text{ m/s}^2)(3.0 \text{ m}) = 35 \text{ Pa},$$

or about 0.00035 atm, a very small difference. But between sea level and the summit of Mount Everest (8882 m) the density of air changes by nearly a factor of three, and in this case we cannot use Eq. (14–6). Liquids, by contrast, are nearly incompressible, and it is usually a very good approximation to regard their density as independent of pressure. A pressure of several hundred atmospheres will cause only a few percent increase in the density of most liquids.

14–5 Principle of the hydraulic lift, an application of Pascal's law. The size of the fluid-filled container is exaggerated for clarity.

ABSOLUTE PRESSURE, GAUGE PRESSURE, AND PRESSURE GAUGES

If the pressure inside a car tire is equal to atmospheric pressure, the tire is flat. The pressure has to be *greater* than atmospheric to support the car, so the significant quantity is the *difference* between the inside and outside pressures. When we say that the pressure in a car tire is "32 pounds" (actually 32 lb/in.^{2}, equal to 220 kPa or 2.2×10^5 Pa), we mean that it is *greater* than atmospheric pressure (14.7 lb/in.^{2} or 1.01×10^5 Pa) by this amount. The *total* pressure in the tire is then 47 lb/in.^{2} or 320 kPa. The excess pressure above atmospheric pressure is usually called **gauge pressure,** and the total pressure is called **absolute pressure.** Engineers use the abbreviations psig and psia for "pounds per square inch gauge" and "pounds per square inch absolute," respectively. If the pressure is *less* than atmospheric, as in a partial vacuum, the gauge pressure is negative.

EXAMPLE 14-3

Finding absolute and gauge pressures A solar water-heating system uses solar panels on the roof, 12.0 m above the storage tank. The water pressure at the level of the panels is one atmosphere. What is the absolute pressure in the tank? The gauge pressure?

SOLUTION From Eq. (14–6), the absolute pressure is

$$p = p_0 + \rho g h$$

$$= (1.01 \times 10^5 \text{ Pa}) + (1000 \text{ kg/m}^3)(9.80 \text{ m/s}^2)(12.0 \text{ m})$$

$$= 2.19 \times 10^5 \text{ Pa} = 2.16 \text{ atm} = 31.8 \text{ lb/in.}^2.$$

The gauge pressure is

$$p - p_0 = (2.19 - 1.01) \times 10^5 \text{ Pa} = 1.18 \times 10^5 \text{ Pa}$$

$$= 1.16 \text{ atm} = 17.1 \text{ lb/in.}^2.$$

If such a tank has a pressure gauge, it is usually calibrated to read gauge pressure rather than absolute pressure. As we have mentioned, the variation in *atmospheric* pressure over this height is negligibly small.

The simplest pressure gauge is the open-tube *manometer* (Fig. 14–6a). The U-shaped tube contains a liquid of density ρ, often mercury or water. One end of the tube

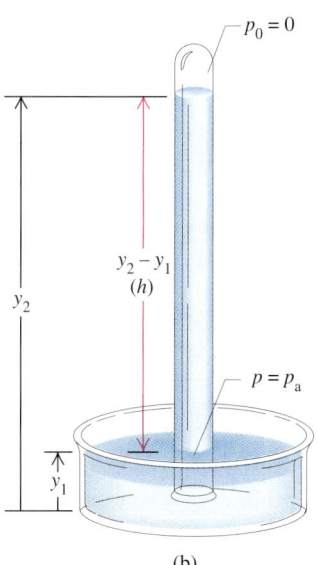

14–6 Pressure gauges. (a) The open-tube manometer. (b) The mercury barometer.

(a)

(b)

is connected to the container where the pressure p is to be measured, and the other end is open to the atmosphere at pressure $p_0 = p_a$. The pressure at the bottom of the tube due to the fluid in the left column is $p + \rho g y_1$, and the pressure at the bottom due to the fluid in the right column is $p_a + \rho g y_2$. These pressures are measured at the same point, so they must be equal:

$$p + \rho g y_1 = p_a + \rho g y_2,$$
$$p - p_a = \rho g(y_2 - y_1) = \rho g h. \tag{14-8}$$

In Eq. (14-8), p is the *absolute pressure,* and the difference $p - p_a$ between absolute and atmospheric pressure is the gauge pressure. Thus the gauge pressure is proportional to the difference in height $(y_2 - y_1)$ of the liquid columns.

Another common pressure gauge is the **mercury barometer.** It consists of a long glass tube, closed at one end, that has been filled with mercury and then inverted in a dish of mercury (Fig. 14-6b). The space above the mercury column contains only mercury vapor; its pressure is negligibly small, so the pressure p_0 at the top of the mercury column is practically zero. From Eq. (14-6),

$$p_a = p = 0 + \rho g(y_2 - y_1) = \rho g h. \tag{14-9}$$

Thus the mercury barometer reads the atmospheric pressure p_a directly from the height of the mercury column.

In many applications, pressures are still commonly described in terms of the height of the corresponding mercury column, as so many "inches of mercury" or "millimeters of mercury" (abbreviated mm Hg). A pressure of 1 mm Hg is called *one torr,* after Evangelista Torricelli, inventor of the mercury barometer. But these units depend on the density of mercury, which varies with temperature, and on the value of g, which varies with location, so the pascal is the preferred unit of pressure.

EXAMPLE 14-4

Compute the atmospheric pressure p_a on a day when the height of mercury in a barometer is 760 mm.

SOLUTION We use Eq. (14-9). Assuming that $g = 9.80$ m/s^2 and

$\rho = 13.6 \times 10^3$ kg/m^3 for mercury, we find

$$p_a = \rho g h = (13.6 \times 10^3 \text{ kg/m}^3)(9.80 \text{ m/s}^2)(0.760 \text{ m})$$
$$= 1.01 \times 10^5 \text{ N/m}^2 = 1.01 \times 10^5 \text{ Pa} = 1.01 \text{ bar.}$$

Pressure p being measured

14-7 A Bourdon pressure gauge. The spiral metal tube is attached to a pointer. When the pressure inside the tube increases, the tube straightens out a little, deflecting the pointer on the scale.

One common type of blood-pressure gauge, called a *sphygmomanometer*, uses a mercury-filled manometer. Blood-pressure readings, such as 130/80, refer to the maximum and minimum gauge pressures in the arteries, measured in mm Hg or torrs. Blood pressure varies with height; the standard reference point is the upper arm, level with the heart.

Many types of pressure gauges use a flexible sealed vessel (Fig. 14-7). A change in the pressure either inside or outside the vessel causes a change in its dimensions. This change is detected optically, electrically, or mechanically.

14-4 BUOYANCY

Buoyancy is a familiar phenomenon: a body immersed in water seems to weigh less than when it is in air. When the body is less dense than the fluid, it floats. The human body usually floats in water, and a helium-filled balloon floats in air.

Archimedes' principle states: **When a body is completely or partially immersed in a fluid, the fluid exerts an upward force on the body equal to the weight of the fluid displaced by the body.** To prove this principle, we consider an arbitrary portion of

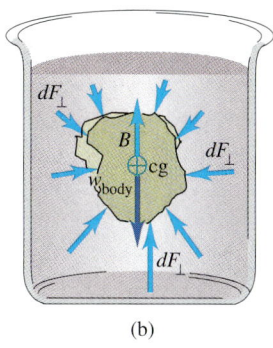

(a)   (b)

14–8 (a) An element of fluid in equilibrium under its weight and the buoyant force of the surrounding fluid. (b) The element is replaced by a body with identical shape. The buoyant force is the same and is equal to the weight of the fluid that the body replaced, no matter what the weight of the body that replaces the fluid.

fluid at rest. In Fig. 14–8a the irregular outline is the surface bounding this portion of fluid. The arrows represent the forces exerted on the boundary surface by the surrounding fluid.

The entire fluid is in equilibrium, so the sum of all the y-components of force on this portion of fluid is zero. Hence the sum of the y-components of the *surface* forces must be an upward force equal in magnitude to the weight mg of the fluid inside the surface. Also, the sum of the torques on the portion of fluid must be zero, so the line of action of the resultant y-component of surface force must pass through the center of gravity of this portion of fluid.

Now we remove the fluid inside the surface and replace it with a solid body having exactly the same shape (Fig. 14–8b). The pressure at every point is exactly the same as before. So the total upward force exerted on the body by the fluid is also the same, again equal in magnitude to the weight mg of the fluid displaced to make way for the body. We call this upward force the **buoyant force** on the solid body. The line of action of the buoyant force again passes through the center of gravity of the displaced fluid (which doesn't necessarily coincide with the center of gravity of the body).

When a balloon floats in equilibrium in air, its weight (including the gas inside it) must be the same as the weight of the air displaced by the balloon. When a submerged submarine is in equilibrium, its weight must equal the weight of the water it displaces. A body whose average density is less than that of a liquid can float partially submerged at the free upper surface of the liquid. The greater the density of the liquid, the less of the body is submerged. When you swim in seawater (density 1030 kg/m³), your body floats higher than in fresh water (1000 kg/m³). Unlikely as it may seem, lead floats in mercury. Very flat-surfaced "float glass" is made by floating molten glass on molten tin and then letting it cool.

Another familiar example is the hydrometer, used to measure the density of liquids (Fig. 14–9a). The calibrated float sinks into the fluid until the weight of the fluid it displaces is exactly equal to its own weight. The hydrometer floats *higher* in denser liquids than in less dense liquids. It is weighted at its bottom end so that the upright position is stable, and a scale in the top stem permits direct density readings. Figure 14–9b shows a type of hydrometer that is commonly used to measure the density of battery acid or antifreeze. The bottom of the large tube is immersed in the liquid; the bulb is squeezed to expel air and is then released, like a giant medicine dropper. The liquid rises into the outer tube, and the hydrometer floats in this sample of the liquid.

(a)   (b)

14–9 (a) A simple hydrometer. (b) Use of a hydrometer as a tester for battery acid or antifreeze.

EXAMPLE 14–5

A 15.0-kg solid gold statue is being raised from a sunken ship. What is the tension in the hoisting cable when the statue is at rest and a) completely immersed; b) out of the water?

SOLUTION a) When the statue is immersed, it experiences an upward buoyant force equal in magnitude to the weight of the water displaced (Fig. 14–10a). To find this force, we first find

(a)   (b)

14–10 (a) The completely immersed statue at rest. (b) Free-body diagram for the immersed statue.

the volume of the statue, using the density of gold from Table 14–1:

$$V = \frac{m}{\rho_{gold}} = \frac{15.0 \text{ kg}}{19.3 \times 10^3 \text{ kg/m}^3} = 7.77 \times 10^{-4} \text{ m}^3.$$

EXAMPLE 14–6

You place a container of seawater on a scale and note the reading on the scale. You now suspend the statue of Example 14–5 in the water (Fig. 14–11). How does the scale reading change?

SOLUTION Consider the water, the statue, and the container together as a system; the total weight of the system does not depend on whether or not the statue is immersed. The total supporting force, including the tension T and the upward force F of the scale on the container (equal to the scale reading), is the same in both cases. But we saw in Example 14–5 that T decreases by 7.84 N when the statue is immersed, so the scale reading F must *increase* by 7.84 N. An alternative viewpoint is that the water exerts an upward buoyant force of 7.84 N on the statue, so the statue must exert an equal downward force on the water, making the scale reading 7.84 N greater than the weight of water and container.

14–11 How does the scale reading change when the statue is immersed in the water?

Using Table 14–1 again, we find the weight of this volume of seawater:

$$w_{sw} = m_{sw}g = \rho_{sw}Vg$$
$$= (1.03 \times 10^3 \text{ kg/m}^3)(7.77 \times 10^{-4} \text{ m}^3)(9.80 \text{ m/s}^2)$$
$$= 7.84 \text{ N}.$$

This equals the buoyant force B.

To find the cable tension T, we draw a free-body diagram for the statue (Figure 14–10b). Since the statue is at rest, the net external force acting on it is zero:

$$\Sigma F_y = B + T + (-mg) = 0,$$
$$T = mg - B = (15.0 \text{ kg})(9.80 \text{ m/s}^2) - 7.84 \text{ N}$$
$$= 147 \text{ N} - 7.84 \text{ N} = 139 \text{ N}.$$

The submerged statue seems to weigh 139 N, about 5% less than its actual weight of 147 N.

b) The density of air is about 1.2 kg/m³, so the buoyant force of air on the statue is

$$B = \rho_{air}Vg = (1.2 \text{ kg/m}^3)(7.77 \times 10^{-4} \text{ m}^3)(9.80 \text{ m/s}^2)$$
$$= 9.1 \times 10^{-3} \text{ N}.$$

This is only 62 parts per million of the statue's actual weight. This effect is not within the precision of our data, and we ignore it. Thus the tension in the cable with the statue in air is equal to the statue's weight, 147 N.

14–5 SURFACE TENSION

A paper clip can rest atop a water surface even though its density is several times that of water. Some insects can walk on the surface of water; their feet make indentations in the surface but do not penetrate it. Such phenomena are examples of *surface tension:* the surface of the liquid behaves like a membrane under tension. The molecules of the liquid exert attractive forces on each other; there is zero net force on a molecule inside the volume of the liquid, but a surface molecule is drawn into the volume (Fig. 14–12). Thus the liquid tends to minimize its surface area, just as does a stretched membrane. Freely falling raindrops are spherical (*not* teardrop-shaped) because a sphere has a smaller surface area for a given volume than any other shape. Figure 14–13 is a beautiful example of the formation of a spherical droplet.

Figure 14–14 shows how we can make quantitative measurements of surface tension. A piece of wire is bent into a U shape, and a second piece of wire slides on the arms of the U. When the apparatus is dipped into a soap solution and removed, creating a liquid film, the film exerts a surface-tension force on the slider that quickly pulls it up toward the top of the inverted U (if the slider's weight w is not too great). When we pull the slider down, increasing the area of the film, molecules move from the interior of the liquid (which is many molecules thick, even in a thin film) into the surface layers. The surface layers do not simply stretch like a rubber sheet. Instead, more surface is created by molecules moving from the bulk liquid.

To hold the slider in equilibrium, we need a total downward force $F = w + T$. In equilibrium, F is also equal to the surface-tension force exerted by the soap film on the slider. Let l be the length of the wire slider. The film has both front and back surfaces, so the force F acts along a total length $2l$. The **surface tension** γ (Greek "gamma") in the film is defined as the ratio of the surface-tension force F to the length d along which the force acts:

$$\gamma = \frac{F}{d} \qquad \text{(definition of surface tension).} \qquad (14\text{–}10)$$

In this case, $d = 2l$ and

$$\gamma = \frac{F}{2l}.$$

Surface tension is a *force per unit length*. The SI unit is the newton per meter (N/m), but the cgs unit, the dyne per centimeter (dyn/cm), is more commonly used:

$$1 \text{ dyn/cm} = 10^{-3} \text{ N/m} = 1 \text{ mN/m}.$$

Table 14–2 shows some typical values of surface tension. The lowest values of γ occur in the liquified noble gases neon and helium, in which the attraction between atoms is very weak. (For the same reason, these elements do not form compounds.)

14–12 A molecule of a liquid is attracted by its fellow molecules. A molecule on the surface is attracted into the bulk liquid, which tends to reduce the surface area of the liquid.

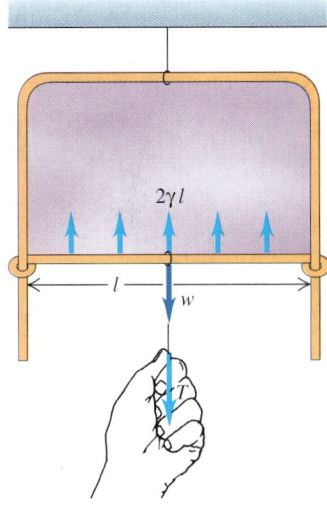

14–14 Measuring the surface tension of a soap film (shown in purple). The horizontal slide wire is in equilibrium under the action of the upward surface force $2\gamma l$ and downward pull $w + T$.

14–13 The splash from a drop of water falling into a liquid produces a column called a Rayleigh jet. Surface tension pulls the top of the jet into a spherical droplet.

TABLE 14–2

EXPERIMENTAL VALUES OF SURFACE TENSION

LIQUID IN CONTACT WITH AIR	TEMPERATURE (°C)	SURFACE TENSION (mN/m, or dyn/cm)
Benzene	20	28.9
Carbon tetrachloride	20	26.8
Ethanol	20	22.3
Glycerin	20	63.1
Mercury	20	465.0
Olive oil	20	32.0
Soap solution	20	25.0
Water	0	75.6
Water	20	72.8
Water	60	66.2
Water	100	58.9
Oxygen	−193	15.7
Neon	−247	5.15
Helium	−269	0.12

Water pressure p

R Fibers

Air pressure p_a

14–15 Surface tension makes it difficult to force water through small crevices. The required water pressure can be reduced by heating the water and adding soap, both of which lower the surface tension γ.

Liquid between surfaces

R

O

Air fills interior

(a)

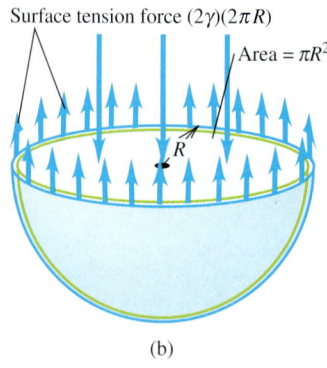

Surface tension force $(2\gamma)(2\pi R)$

Area $= \pi R^2$

R

(b)

14–16 (a) Cross section of a soap bubble showing the two surfaces, the thin layer of liquid between them, and the air inside the bubble. (b) Equilibrium of half of a soap bubble. The surface-tension force exerted by the other half is $(2\gamma)(2\pi R)$, and the force exerted by the air inside the bubble is the pressure p times the area πR^2.

The surface tension of a particular liquid usually decreases as temperature increases; the table shows this behavior for water. In Chapter 16 we will learn that temperature is related to the energy of molecular motion in a material. As temperature increases and the molecules of a liquid move faster, the interactions between molecules have less effect on their motion and the surface tension decreases.

To wash clothing thoroughly, water must be forced through the tiny spaces between the fibers (Fig. 14–15). This requires increasing the surface area of the water, which is difficult to do because of surface tension. The job is made easier by lowering the value of γ. Hence very hot water ($\gamma = 58.9$ mN/m at 100°C) is better for washing than tepid water ($\gamma = 72.8$ mN/m at 20°C), and soapy water ($\gamma = 25.0$ mN/m at 20°C) is better still.

PRESSURE INSIDE A BUBBLE

Surface tension causes a pressure difference between the inside and outside of a soap bubble or a liquid drop. A soap bubble consists of two spherical surface films with a thin layer of liquid between (Fig. 14–16a). Because of surface tension, the films tend to contract in an attempt to minimize their surface area. But as the bubble contracts, it compresses the inside air, eventually increasing the interior pressure to a level that prevents further contraction.

We can derive an expression for the excess pressure inside a bubble in terms of its radius R and the surface tension γ of the liquid. Assume first that there is no external pressure. Each half of the soap bubble is in equilibrium; the lower half is shown in Fig. 14–16b. The forces at the flat circular surface where this half joins the upper half are the upward force of surface tension and the downward force due to the pressure of air in the upper half. The circumference of the circle along which the surface tension acts is $2\pi R$. (We neglect the small difference between inner and outer radii.) The total surface-tension force for each surface (inner and outer) is $\gamma(2\pi R)$, for a total of $(2\gamma)(2\pi R)$. Air pressure pushes both downward and outward on the lower half of the bubble, but the *resultant* force due to air pressure is downward only; its magnitude is the pressure p times πR^2, the area of the circle where the two half-bubbles meet. For the sum of these forces to be zero, we must have

$$(2\gamma)(2\pi R) = p(\pi R^2),$$

$$p = \frac{4\gamma}{R}.$$

(14–11)

In general, the pressure outside the bubble is *not* zero. But Eq. (14–11) still gives the *difference* between inside and outside pressure. If the outside pressure is atmospheric pressure p_a, then

$$p - p_a = \frac{4\gamma}{R} \qquad \text{(soap bubble).} \qquad (14\text{–}12)$$

A liquid drop has only *one* surface film. Hence the surface tension force is $\gamma(2\pi R)$, half that for a soap bubble. In equilibrium the difference between the pressure in the liquid and that in the outside air is also half that for a soap bubble:

$$p - p_a = \frac{2\gamma}{R} \qquad \text{(liquid drop).} \qquad (14\text{–}13)$$

The smaller the radius of the bubble or drop, the greater the pressure difference. A large pressure is required to force water through small crevices, since the water must form droplets of small radius R (Fig. 14–15).

EXAMPLE 14–7

Calculate the excess pressure inside a drop of water at 20°C if the diameter is 2.00 mm, 20.0 μm, or 0.200 μm.

SOLUTION From Table 14–2, $\gamma = 72.8$ mN/m $= 72.8 \times 10^{-3}$ N/m. From Eq. (14–13), with a diameter $d = 2.00$ mm,

$$p - p_a = \frac{2\gamma}{R} = \frac{2(72.8 \times 10^{-3}\ \text{N/m})}{1.00 \times 10^{-3}\ \text{m}}$$

$$= 146\ \text{N/m}^2 = 146\ \text{Pa} = 0.00144\ \text{atm.}$$

For $d = 20.0\ \mu$m $= 20.0 \times 10^{-6}$ m, $p - p_a = 14{,}600$ Pa $= 0.144$ atm. If $d = 0.200\ \mu$m, $p - p_a = 1.46 \times 10^6$ Pa, or 14.4 atmospheres! The pressure inside such a minuscule droplet can be gigantic.

CAPILLARITY

When a gas-liquid interface meets a solid surface, such as the wall of a container (Fig. 14–17), the gas-liquid interface usually curves up or down near the solid surface. The angle θ at which it meets the surface is called the **contact angle.** When the molecules of the liquid are attracted to each other less strongly than to the solid, as with water and glass, the liquid "wets" or adheres to the solid surface. The gas-liquid interface curves up, and θ is less than 90°. The liquid is nonwetting when the attraction between liquid molecules is the stronger force, as with mercury and glass; the gas-liquid interface curves down, and θ is greater than 90°.

Surface tension causes elevation or depression of the liquid in a narrow tube (Fig. 14–18). This effect is called **capillarity.** When the contact angle is less than 90° (Fig. 14–18a), the total surface tension force along the line of contact with the tube wall acts upward; the liquid *rises* until it reaches an equilibrium height y at which the total surface tension force just balances the extra weight of the liquid in the tube. The curved liquid surface is called a *meniscus.* For a nonwetting liquid such as mercury (Figure 14–18b) the contact angle is greater than 90°. The meniscus curves down, and the surface is depressed, pulled *down* by the surface-tension forces.

Capillarity is responsible for the absorption of water by paper towels, the rise of melted wax in a candle wick, and many other everyday phenomena. Blood is pumped through the arteries and veins of your body, but capillarity is important in causing flow through the smallest blood vessels, which indeed are called capillaries.

Related to surface tension is the phenomenon of *negative pressure*. The stress in a liquid is ordinarily *compressive,* but in some circumstances, liquids can sustain *tensile* stresses. Consider a cylindrical tube, closed at one end, with a tight-fitting piston at the other. We fill the tube completely with a liquid that wets both the inner surface of the

Water: $\theta < 90°$ Mercury: $\theta > 90°$

14–17 When a gas-liquid interface meets a solid surface, the interface usually curves up or down near the solid surface.

14–18 Surface-tension forces on a liquid in a capillary tube. The liquid (a) rises if $\theta < 90°$ or (b) is depressed if $\theta > 90°$. The diameter of the tube is greatly exaggerated for clarity.

(a) (b)

tube and the piston face; the molecules of liquid adhere to all the surfaces. If the surfaces are very clean and the liquid very pure, then when we pull the piston face, we observe a *tensile* stress and a slight *increase* in volume; we are *stretching* the liquid. Adhesive forces prevent it from pulling away from the walls of the container.

With water, tensile stresses as large as 300 atm have been observed in the laboratory. This situation is highly unstable; a liquid under tension tends to break up into many small droplets. In tall trees, however, negative pressures are a regular occurrence. Negative pressure is believed to be an important mechanism for transport of water and nutrients from the roots to the leaves in the small xylem tubes (diameter of the order of 0.1 mm) in the growing layers of the tree.

14–6 FLUID FLOW

We are now ready to consider *motion* of a fluid. Fluid flow can be extremely complex, as shown by the currents in river rapids or the swirling flames of a campfire. But some situations can be represented by relatively simple idealized models. An **ideal fluid** is a fluid that is *incompressible* (that is, its density cannot change) and has no internal friction (called *viscosity*). Liquids are approximately incompressible in most situations, and we may also treat a gas as incompressible if the pressure differences from one region to another are not too great. Internal friction in a fluid causes shear stresses when two adjacent layers of fluid move relative to each other, as when fluid flows inside a tube or around an obstacle. In some cases we can neglect these shear forces in comparison with forces arising from gravitation and pressure differences.

The path of an individual particle in a moving fluid is called a **flow line.** If the overall flow pattern does not change with time, the flow is called **steady flow.** In steady flow, every element passing through a given point follows the same flow line. In this case the "map" of the fluid velocities at various points in space remains constant, although the velocity of a particular particle may change in both magnitude and direction during its motion. A **streamline** is a curve whose tangent at any point is in the direction of the fluid velocity at that point. When the flow pattern changes with time, the streamlines do not coincide with the flow lines. We will consider only steady-flow situations, for which flow lines and streamlines are identical.

The flow lines passing through the edge of an imaginary element of area, such as the area A in Fig. 14–19, form a tube called a **flow tube.** From the definition of a flow line,

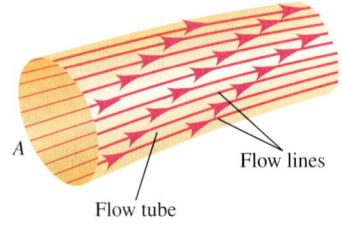

14–19 A flow tube bounded by flow lines. In steady flow, fluid cannot cross the walls of a flow tube.

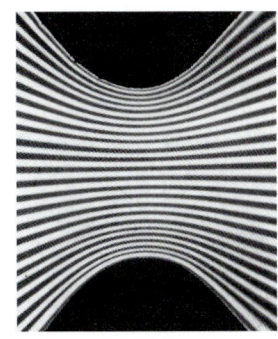

in steady flow no fluid can cross the side walls of a flow tube; the fluids in different flow tubes cannot mix.

Figure 14–20 shows patterns of fluid flow from left to right around a number of obstacles and in a channel of varying cross section. The photographs were made by injecting dye into water flowing between two closely spaced glass plates. These patterns are typical of **laminar flow,** in which adjacent layers of fluid slide smoothly past each other and the flow is steady. (A *lamina* is a thin sheet.) At sufficiently high flow rates, or when boundary surfaces cause abrupt changes in velocity, the flow can become irregular and chaotic. This is called **turbulent flow.** In turbulent flow there is no steady-state pattern; the flow pattern continuously changes.

14–20 (a) (b) (c) Laminar flow around obstacles of different shapes. (d) Flow through a channel of varying cross-section area.

THE CONTINUITY EQUATION

The mass of a moving fluid doesn't change as it flows. This leads to an important quantitative relationship called the **continuity equation.** Consider a portion of a flow tube between two stationary cross sections with areas A_1 and A_2 (Fig. 14–21). The fluid speeds at these sections are v_1 and v_2, respectively. No fluid flows in or out across the sides of the tube because the fluid velocity is tangent to the wall at every point on the wall. During a small time interval dt the fluid at A_1 moves a distance $v_1\,dt$, so a cylinder of fluid with height $v_1\,dt$ and volume $dV_1 = A_1 v_1\,dt$ flows into the tube across A_1. During this same interval, a cylinder of volume $dV_2 = A_2 v_2\,dt$ flows out of the tube across A_2.

Let's first consider the case of an incompressible fluid so that the density ρ has the same value at all points. The mass dm_1 flowing into the tube across A_1 in time dt is $dm_1 = \rho A_1 v_1\,dt$. Similarly, the mass dm_2 that flows out across A_2 in the same time is $dm_2 = \rho A_2 v_2\,dt$. In steady flow the total mass in the tube is constant, so $dm_1 = dm_2$ and

$$\rho A_1 v_1 dt = \rho A_2 v_2 dt, \quad \text{or}$$

$$A_1 v_1 = A_2 v_2 \quad \text{(continuity equation, incompressible fluid).} \quad (14\text{–}14)$$

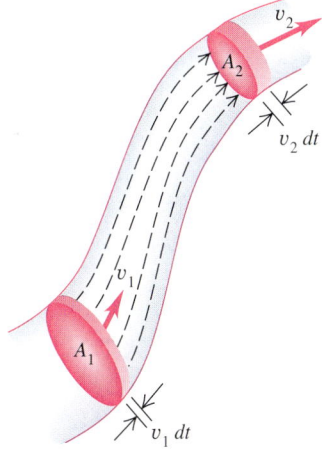

14–21 A flow tube with changing cross-section area. If the fluid is incompressible, the product Av has the same value at all points along the tube.

The product Av is the *volume flow rate* dV/dt, the rate at which volume crosses a section of the tube:

$$\frac{dV}{dt} = Av \quad \text{(volume flow rate).} \quad (14\text{–}15)$$

The *mass* flow rate is the mass flow per unit time through a cross section. This is equal to the density ρ times the volume flow rate dV/dt.

Equation (14–14) shows that the volume flow rate has the same value at all points along any flow tube. When the cross section of a flow tube decreases (Fig. 14–20d), the speed increases, and vice versa. The deep part of a river has larger cross section and slower current than the shallow part, but the volume flow rates are the same in both. This is the essence of the familiar maxim "Still waters run deep." The stream of water from a faucet narrows as it gains speed during its fall, but dV/dt is the same everywhere along the stream. If a water pipe with 2-cm diameter is connected to a pipe with 1-cm diameter, the flow speed is four times as great in the 1-cm part as in the 2-cm part.

We can generalize Eq. (14–14) for the case in which the fluid is *not* incompressible. If ρ_1 and ρ_2 are the densities at sections 1 and 2, then

$$\rho_1 A_1 v_1 = \rho_2 A_2 v_2 \qquad \text{(continuity equation, compressible fluid).} \qquad (14\text{–}16)$$

We leave the details for an exercise. If the fluid is incompressible so that ρ_1 and ρ_2 are always equal, Eq. (14–16) reduces to Eq. (14–14).

14-7 BERNOULLI'S EQUATION

According to the continuity equation, the speed of fluid flow can vary along the paths of the fluid. The pressure can also vary; it depends on height as in the static situation (Section 14–3), and it also depends on the speed of flow. We can derive an important relationship called *Bernoulli's equation* that relates the pressure, flow speed, and height for flow of an ideal, incompressible fluid. Bernoulli's equation is an essential tool in analyzing plumbing systems, hydroelectric generating stations, and the flight of airplanes.

The dependence of pressure on speed follows from the continuity equation, Eq. (14–14). When an incompressible fluid flows along a flow tube with varying cross section, its speed *must* change, and so an element of fluid must have an acceleration. If the tube is horizontal, the force that causes this acceleration has to be applied by the surrounding fluid. This means that the pressure *must* be different in regions of different cross section; if it were the same everywhere, the net force on every fluid element would be zero. When a horizontal flow tube narrows and a fluid element speeds up, it must be moving toward a region of lower pressure in order to have a net forward force to accelerate it. If the elevation also changes, this causes an additional pressure difference.

To derive Bernoulli's equation, we apply the work-energy theorem to the fluid in a section of a flow tube. In Fig. 14–22 we consider the element of fluid that at some initial time lies between the two cross sections a and c. The speeds at the lower and upper ends are v_1 and v_2. In a small time interval dt the fluid that is initially at a moves to b, a distance $ds_1 = v_1 dt$, and the fluid that is initially at c moves to d, a distance $ds_2 = v_2 dt$. The cross-section areas at the two ends are A_1 and A_2, as shown. The fluid is incompressible; hence by the continuity equation, Eq. (14–14), the volume of fluid dV passing *any* cross section during time dt is the same. That is, $dV = A_1 ds_1 = A_2 ds_2$.

Let's compute the *work* done on this fluid element during dt. The pressures at the two ends are p_1 and p_2; the force on the cross section at a is $p_1 A_1$, and the force at c is $p_2 A_2$. The net work dW done on the element by the surrounding fluid during this displacement is therefore

$$dW = p_1 A_1 ds_1 - p_2 A_2 ds_2 = (p_1 - p_2)\,dV. \qquad (14\text{–}17)$$

The second term has a negative sign because the force at c opposes the displacement of the fluid.

The work dW is due to forces other than the conservative force of gravity, so it equals the change in the total mechanical energy (kinetic energy plus gravitational potential energy) associated with the fluid element. The mechanical energy for the fluid between sections b and c does not change. At the beginning of dt the fluid between a and b has volume $A_1 ds_1$, mass $\rho A_1 ds_1$, and kinetic energy $\frac{1}{2}\rho(A_1 ds_1)v_1^2$. At the end of dt the fluid between c and d has kinetic energy $\frac{1}{2}\rho(A_2 ds_2)v_2^2$. The net change in kinetic energy dK during time dt is

$$dK = \frac{1}{2}\rho\,dV(v_2^2 - v_1^2). \qquad (14\text{–}18)$$

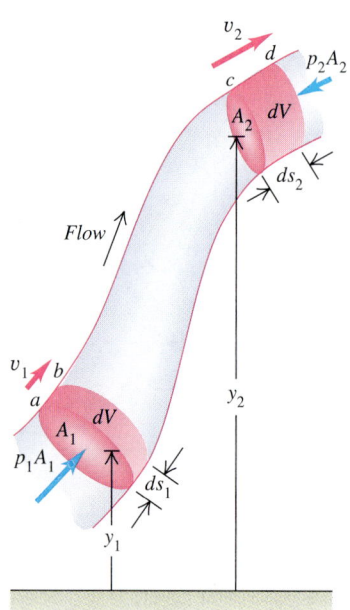

14–22 The net work done on a fluid element by the pressure of the surrounding fluid equals the change in the kinetic energy plus the change in the gravitational potential energy.

What about the change in gravitational potential energy? At the beginning of dt the potential energy for the mass between a and b is $dm\, gy_1 = \rho\, dV\, gy_1$. At the end of dt the potential energy for the mass between c and d is $dm\, gy_2 = \rho\, dV\, gy_2$. The net change in potential energy dU during dt is

$$dU = \rho\, dV\, g(y_2 - y_1). \tag{14-19}$$

Combining Eqs. (14–17), (14–18), and (14–19) in the energy equation $dW = dK + dU$, we obtain

$$(p_1 - p_2)\, dV = \frac{1}{2}\rho\, dV(v_2{}^2 - v_1{}^2) + \rho\, dVg(y_2 - y_1),$$

$$p_1 - p_2 = \frac{1}{2}\rho\,(v_2{}^2 - v_1{}^2) + \rho g(y_2 - y_1). \tag{14-20}$$

This is **Bernoulli's equation.** It states that the work done on a unit volume of fluid by the surrounding fluid is equal to the sum of the changes in kinetic and potential energies per unit volume that occur during the flow. We may also interpret Eq. (14–20) in terms of pressures. The first term on the right is the pressure difference associated with the change of speed of the fluid. The second term on the right is the additional pressure difference caused by the weight of the fluid and the difference in elevation of the two ends.

We can also express Eq. (14–20) in a more convenient form as

$$p_1 + \rho gy_1 + \frac{1}{2}\rho v_1{}^2 = p_2 + \rho gy_2 + \frac{1}{2}\rho v_2{}^2 \qquad \text{(Bernoulli's equation)}.$$
$$\tag{14-21}$$

The subscripts 1 and 2 refer to *any* two points along the flow tube, so we can also write

$$p + \rho gy + \frac{1}{2}\rho v^2 = \text{constant}. \tag{14-22}$$

Note that when the fluid is *not* moving (so $v_1 = v_2 = 0$), Eq. (14–21) reduces to the pressure relation we derived for a fluid at rest, Eq. (14–5).

CAUTION▶ We stress again that Bernoulli's equation is valid *only* for incompressible, steady flow of a fluid with no viscosity. It's a simple equation that's easy to use; don't let this tempt you into using it in situations in which it doesn't apply! ◀

Problem-Solving Strategy

BERNOULLI'S EQUATION

Bernoulli's equation is derived from the work-energy relationship, so it isn't surprising that much of the problem-solving strategy suggested in Sections 7–2 and 7–3 is equally applicable here. In particular:

1. Always begin by identifying clearly the points 1 and 2 referred to in Bernoulli's equation.

2. Make lists of the known and unknown quantities in Eq. (14–21). The variables are p_1, p_2, v_1, v_2, y_1, and y_2, and the constants are ρ and g. What is given? What do you need to determine?

3. In some problems you will need to use the continuity equation, Eq. (14–14), to get a relation between the two speeds in terms of cross-section areas of pipes or containers. Or perhaps you will know both speeds and will need to determine one of the areas. You may also need to use Eq. (14–15) to find the volume flow rate.

4. As always, consistent units are essential. In SI units, pressure is in pascals, density in kilograms per cubic meter, and speed in meters per second. Also note that the pressures must be either all absolute pressures or all gauge pressures.

EXAMPLE 14–8

Water pressure in the home Water enters a house through a pipe with an inside diameter of 2.0 cm at an absolute pressure of 4.0×10^5 Pa (about 4 atm). A 1.0-cm-diameter pipe leads to the second-floor bathroom 5.0 m above (Fig. 14–23). When the flow speed at the inlet pipe is 1.5 m/s, find the flow speed, pressure, and volume flow rate in the bathroom.

SOLUTION Let points 1 and 2 be at the inlet pipe and at the bathroom, respectively. The speed v_2 at the bathroom is obtained from the continuity equation, Eq. (14–14):

$$v_2 = \frac{A_1}{A_2} v_1 = \frac{\pi (1.0 \text{ cm})^2}{\pi (0.50 \text{ cm})^2} (1.5 \text{ m/s}) = 6.0 \text{ m/s}.$$

We take $y_1 = 0$ (at the inlet) and $y_2 = 5.0$ m (at the bathroom). We are given p_1 and v_1, and can find p_2 from Bernoulli's equation:

$$p_2 = p_1 - \frac{1}{2} \rho (v_2{}^2 - v_1{}^2) - \rho g (y_2 - y_1)$$

$$= 4.0 \times 10^5 \text{ Pa} - \frac{1}{2} (1.0 \times 10^3 \text{ kg/m}^3)(36 \text{ m}^2/\text{s}^2 - 2.25 \text{ m}^2/\text{s}^2)$$

$$- (1.0 \times 10^3 \text{ kg/m}^3)(9.8 \text{ m/s}^2)(5.0 \text{ m})$$

$$= 4.0 \times 10^5 \text{ Pa} - 0.17 \times 10^5 \text{ Pa} - 0.49 \times 10^5 \text{ Pa}$$

$$= 3.3 \times 10^5 \text{ Pa} = 3.3 \text{ atm} = 48 \text{ psia}.$$

The volume flow rate is

$$\frac{dV}{dt} = A_2 v_2 = \pi (0.50 \times 10^{-2} \text{ m})^2 (6.0 \text{ m/s})$$

$$= 4.7 \times 10^{-4} \text{ m}^3/\text{s} = 0.47 \text{ L/s}.$$

Note that after the water is turned off, the term $\frac{1}{2} \rho (v_2{}^2 - v_1{}^2)$ in the equation for pressure vanishes, and the pressure rises to 3.5×10^5 Pa.

14–23 What is the water pressure in the second-story bathroom of this house?

EXAMPLE 14–9

Speed of efflux Figure 14–24 shows a gasoline storage tank with cross-section area A_1, filled to a depth h. The space above the gasoline contains air at pressure p_0, and the gasoline flows out through a short pipe with area A_2. Derive expressions for the flow speed in the pipe and the volume flow rate.

SOLUTION We can consider the entire volume of moving liquid as a single flow tube; v_1 and v_2 are the speeds at points 1 and 2, respectively, in Fig. 14–24. The pressure at point 2 is atmospheric pressure, p_a. Applying Bernoulli's equation to points 1 and 2 and taking $y = 0$ at the bottom of the tank, we find

$$p_0 + \frac{1}{2} \rho v_1{}^2 + \rho g h = p_a + \frac{1}{2} \rho v_2{}^2,$$

$$v_2{}^2 = v_1{}^2 + 2 \frac{p_0 - p_a}{\rho} + 2gh.$$

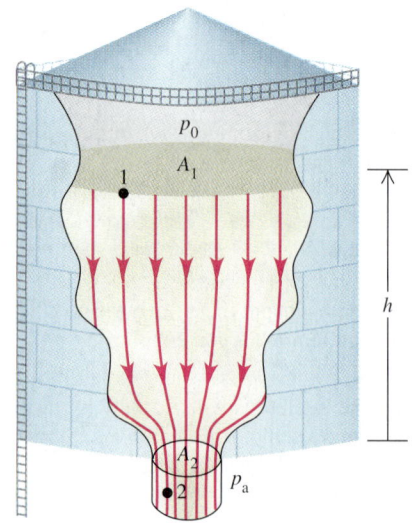

14–24 Calculating the speed of efflux for gasoline flowing out the bottom of a storage tank.

Because A_2 is much smaller than A_1, $v_1{}^2$ is very much smaller than $v_2{}^2$ and can be neglected. We then find

$$v_2{}^2 = 2\,\frac{p_0 - p_a}{\rho} + 2gh.$$

The speed v_2, sometimes called the *speed of efflux*, depends on both the pressure difference ($p_0 - p_a$) and the height h of the liquid level in the tank. If the top of the tank is vented to the atmosphere, there is no excess pressure: $p_0 = p_a$ and $p_0 - p_a = 0$. In that case,

$$v_2 = \sqrt{2gh}.$$

That is, the speed of efflux from an opening at a distance h below the top surface of the liquid is the *same* as the speed a body would acquire in falling freely through a height h. This result is called *Torricelli's theorem*. It is valid not only for an opening in the bottom of a container, but also for a hole in a side wall at a depth h below the surface. We find the volume flow rate from Eq. (14–15):

$$\frac{dV}{dt} = A_2\sqrt{2gh}.$$

EXAMPLE 14–10

The Venturi meter Figure 14–25 shows a *Venturi meter*, used to measure flow speed in a pipe. The narrow part of the pipe is called the *throat*. Derive an expression for the flow speed v_1 in terms of the cross-section areas A_1 and A_2 and the difference in height h of the liquid levels in the two vertical tubes.

SOLUTION We apply Bernoulli's equation to the wide (point 1) and narrow (point 2) parts of the pipe, with $y_1 = y_2$:

$$p_1 + \frac{1}{2}\rho v_1{}^2 = p_2 + \frac{1}{2}\rho v_2{}^2.$$

From the continuity equation, $v_2 = (A_1/A_2)v_1$. Substituting this and rearranging, we get

$$p_1 - p_2 = \frac{1}{2}\rho v_1{}^2\left(\frac{A_1{}^2}{A_2{}^2} - 1\right).$$

Because A_1 is greater than A_2, v_2 is greater than v_1 and the pressure p_2 in the throat is *less* than p_1. A net force to the right accelerates the fluid as it enters the throat, and a net force to the

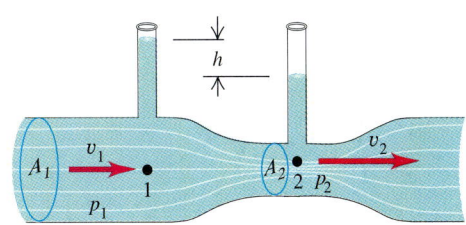

14–25 The Venturi meter.

left slows it as it leaves. The pressure difference $p_1 - p_2$ is also equal to ρgh, where h is the difference in liquid level in the two tubes. Combining this with the above result and solving for v_1, we get

$$v_1 = \sqrt{\frac{2gh}{(A_1/A_2)^2 - 1}}.$$

EXAMPLE 14–11

Lift on an airplane wing Figure 14–26 shows flow lines around a cross section of an airplane wing. The flow lines crowd together above the wing, corresponding to increased flow speed and reduced pressure in this region, just as in the Venturi throat. The upward force on the underside of the wing is greater than the downward force on the top side; there is a net upward force, or *lift*. Lift is not simply due to the impulse of air striking the underside of the wing; in fact, it turns out that the reduced pressure on the upper wing surface makes the greatest contribution to the lift. (This highly simplified discussion ignores the formation of vortices; a more complete discussion would take these into account.)

We can also understand the lift force on the basis of momentum changes. Figure 14–26 shows that there is a net *downward* change in the vertical component of momentum of the air flowing past the wing, corresponding to the downward force the wing exerts on the air. The reaction force *on* the wing is *upward*, as we concluded above.

14–26 Flow lines around an airplane wing. The momentum of a parcel of air (relative to the wing) is $\vec{p}_i$ before encountering the wing and $\vec{p}_f$ afterwards.

14–8 TURBULENCE

When the speed of a flowing fluid exceeds a certain critical value, the flow is no longer laminar. The flow pattern becomes extremely irregular and complex, and it changes continuously with time; there is no steady-state pattern. This irregular, chaotic flow is called **turbulence.** The contrast between laminar and turbulent flow is shown in Fig. 14–27 for two familiar systems: a stream of water and rising smoke in air. Bernoulli's equation is *not* applicable to regions where there is turbulence, because the flow is not steady.

The transition from laminar to turbulent flow is often very sudden. A flow pattern that is stable at low speeds suddenly becomes unstable when a critical speed is reached. Irregularities in the flow pattern can be caused by roughness in the pipe wall, variations in density of the fluid, and many other factors. At small flow speeds these disturbances damp out; the flow pattern is *stable* and tends to maintain its laminar nature. But when the critical speed is reached, the flow pattern becomes unstable. The disturbances no longer damp out, but grow until they destroy the entire laminar-flow pattern. Normal blood flow in the human aorta is laminar, but a small disturbance such as a heart pathology can cause the flow to become turbulent. Turbulence makes noise, which is why listening to blood flow with a stethoscope is a useful diagnostic tool.

Turbulence poses some profound questions for theoretical physics. The motion of any particular particle in a fluid is presumably determined by Newton's laws. If we know the motion of the fluid at some initial time, shouldn't we be able to predict the motion at any later time? After all, when we launch a projectile, we can compute the entire trajectory if we know the initial position and velocity. How is it that a system that obeys well-defined physical laws can have unpredictable, chaotic behavior? There is as yet no simple answer to these questions. Indeed, the study of *chaos*—seemingly unpredictable behavior in systems that should be entirely predictable—is a very active field of research. A particularly significant result of chaos research has been the discovery of similar behavior in a broad range of systems, including the evolution of biological populations, the growth of crystals, the shapes of coastlines, and many others.

14–27 (a) Laminar flow of water. (b) Turbulent flow of water. (c) First laminar, then turbulent flow of smoke.

(a)

(b)

(c)

EXAMPLE 14–12

The curve ball Does a curve ball *really* curve? Yes, it certainly does, and the reason is turbulence. Figure 14–28a shows a ball moving through the air from left to right. To an observer moving with the center of the ball, the air stream appears to move from right to left, as shown by the flow lines in the figure. Because of the large speeds that are ordinarily involved (near 160 km/h, or 100 mi/h), there is a region of *turbulent* flow behind the ball.

Figure 14–28b shows a *spinning* ball with "top spin." Layers of air near the ball's surface are pulled around in the direction of spin by friction between the ball and air and by the air's internal friction (viscosity). The speed of air relative to the ball's surface becomes smaller at the top of the ball than at the bottom, and turbulence occurs farther forward on the top side than on the bottom. This asymmetry causes a pressure difference; the average pressure at the top of the ball is now greater than that at the bottom. The net force deflects the ball downward, as shown in Fig. 14–28c. This is why "top spin" is used in tennis to keep a

very fast serve in the court (Fig. 14–28d). In a baseball curve pitch, the ball spins about a nearly vertical axis, and the actual deflection is sideways. In that case, Fig. 14–28c is a *top* view of the situation. A curve ball thrown by a left-handed pitcher curves *toward* a right-handed batter, making it harder to hit (Fig. 14–28e).

A similar effect occurs with golf balls, which always have "back spin" from impact with the slanted face of the golf club. The resulting pressure difference between the top and bottom of the ball causes a lift force that keeps the ball in the air considerably longer than would be possible without spin. A well-hit drive appears from the tee to "float" or even curve *upward* during the initial portion of its flight. This is a real effect, not an illusion. The dimples on the ball play an essential role; the viscosity of air gives an undimpled ball a much shorter trajectory than a dimpled one with the same initial velocity and spin. Figure 14–29 shows the backspin of a golf ball just after it is struck by a club.

14–28 Motion of the air from right to left relative to the ball corresponds to motion of a ball through still air from left to right. (a) A nonspinning ball has a symmetric region of turbulence behind it. (b) A spinning ball drags along layers of air near its surface. (c) The resulting asymmetric region of turbulence and the deflection of the air stream by the spinning ball. The force shown is exerted on the ball by the air stream; it pushes the ball in the direction of the tangential velocity of the front of the ball. The force can (d) push a tennis ball downward or (e) make a baseball curve.

14–29 Stroboscopic photograph of a golf ball being struck by a club. The picture was taken at 1000 flashes per second. The ball rotates about once in eight pictures, corresponding to an angular speed of 125 rev/s, or 7500 rpm.

*14–9 VISCOSITY

Viscosity is internal friction in a fluid. Viscous forces oppose the motion of one portion of a fluid relative to another. Viscosity is the reason it takes effort to paddle a canoe through calm water, but it is also the reason the paddle works. Viscous effects are important in the flow of fluids in pipes, the flow of blood, the lubrication of engine parts, and many other situations.

A viscous fluid tends to cling to a solid surface in contact with it. There is a thin *boundary layer* of fluid near the surface, in which the fluid is nearly at rest with respect to the surface. That's why dust particles can cling to a fan blade even when it is rotating rapidly and why you can't get all the dirt off your car by just squirting a hose at it.

The simplest example of viscous flow is motion of a fluid between two parallel plates (Fig. 14–30). The bottom plate is stationary, and the top plate moves with constant velocity $\vec{v}$. The fluid in contact with each surface has the same velocity as that surface. The flow speeds of intermediate layers of fluid increase uniformly from one surface to the other, as shown by the arrows, so the fluid layers slide smoothly over one another; the flow is laminar.

A portion of the fluid that has the shape *abcd* at some instant has the shape *abc′d′* a moment later and becomes more and more distorted as the motion continues. That is, the fluid is in a state of continuously increasing *shear strain*. To maintain this motion, we have to apply a constant force F to the right on the upper plate to keep it moving and a force of equal magnitude toward the left on the lower plate to hold it stationary. If A is the surface area of each plate, the ratio F/A is the *shear stress* exerted on the fluid.

In Section 11–7 we defined shear strain as the ratio of the displacement *dd′* to the length l. In a solid, shear strain is proportional to shear stress. In a fluid the shear strain *increases continuously and without limit* as long as the stress is applied. The stress depends not on the shear strain, but on its *rate of change*. The rate of change of strain, also called the *strain rate,* equals the rate of change of *dd′* (the speed v of the moving surface) divided by l. That is,

$$\text{Rate of change of shear strain} = \text{strain rate} = \frac{v}{l}.$$

We define the **viscosity** of the fluid, denoted by η ("eta"), as the ratio of the shear stress, F/A, to the strain rate:

$$\eta = \frac{\text{Shear stress}}{\text{Strain rate}} = \frac{F/A}{v/l} \qquad \text{(definition of viscosity).} \tag{14–23}$$

Rearranging Eq. (14–23), we see that the force required for the motion in Fig. 14–30 is directly proportional to the speed:

$$F = \eta A \frac{v}{l}. \tag{14–24}$$

Fluids that flow readily, such as water or gasoline, have smaller viscosities than "thick" liquids such as honey or motor oil. Viscosities of all fluids are strongly temper-

Lava is an example of a viscous fluid. The viscosity decreases with increasing temperature: the hotter the lava, the more easily it can flow.

14–30 Laminar flow of a viscous fluid. The plates above and below the fluid each have area A. Viscosity is the ratio of shear stress F/A to strain rate v/l.

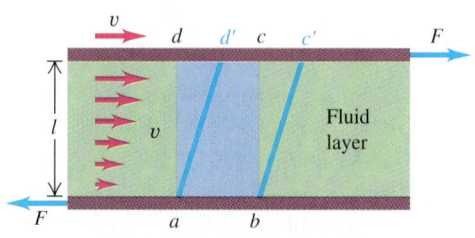

ature dependent, increasing for gases and decreasing for liquids as the temperature increases. An important goal in the design of oils for engine lubrication is to *reduce* the temperature variation of viscosity as much as possible.

From Eq. (14–23) the unit of viscosity is that of force times distance, divided by area times speed. The SI unit is

$$1 \text{ N} \cdot \text{m}/[\text{m}^2 \cdot (\text{m/s})] = 1 \text{ N} \cdot \text{s/m}^2 = 1 \text{ Pa} \cdot \text{s}.$$

The corresponding cgs unit, $1 \text{ dyn} \cdot \text{s/cm}^2$, is the only viscosity unit in common use; it is called a **poise,** in honor of the French scientist Jean Louis Marie Poiseuille (pronounced "pwa-*zoo*-yuh"):

$$1 \text{ poise} = 1 \text{ dyn} \cdot \text{s/cm}^2 = 10^{-1} \text{ N} \cdot \text{s/m}^2.$$

The centipoise and the micropoise are also used. The viscosity of water is 1.79 centipoise at 0°C and 0.28 centipoise at 100°C. Viscosities of lubricating oils are typically 1 to 10 poise, and the viscosity of air at 20°C is 181 micropoise.

For a *Newtonian fluid* the viscosity η is independent of the speed v, and from Eq. (14–24) the force F is directly proportional to the speed. Fluids that are suspensions or dispersions are often *non*-Newtonian in their viscous behavior. One example is blood, which is a suspension of corpuscles in a liquid. As the strain rate increases, the corpuscles deform and become preferentially oriented to facilitate flow, causing η to decrease. The fluids that lubricate human joints show similar behavior.

Figure 14–31 shows the flow speed profile for laminar flow of a viscous fluid in a long cylindrical pipe. The speed is greatest along the axis and zero at the pipe walls. The motion is like a lot of concentric tubes sliding relative to one another, with the central tube moving fastest and the outermost tube at rest. By applying Eq. (14–23) to a cylindrical fluid element, we could derive an equation describing the speed profile. We'll omit the details; the flow speed v at a distance r from the axis of a pipe with radius R is

$$v = \frac{p_1 - p_2}{4\eta L} (R^2 - r^2), \tag{14–25}$$

14-31 Velocity profile for a viscous fluid in a cylindrical pipe.

where p_1 and p_2 are the pressures at the two ends of a pipe with length L. The speed at any point is proportional to the change of pressure per unit length, $(p_2 - p_1)/L$ or dp/dx, called the *pressure gradient*. (The flow is always in the direction of decreasing pressure.) To find the total volume flow rate through the pipe, we consider a ring with inner radius r, outer radius $r + dr$, and cross-section area $dA = 2\pi r \, dr$. The volume flow rate through this element is $v dA$; the total volume flow rate is found by integrating from $r = 0$ to $r = R$. The result is

$$\frac{dV}{dt} = \frac{\pi}{8}\left(\frac{R^4}{\eta}\right)\left(\frac{p_1 - p_2}{L}\right) \quad \text{(Poiseuille's equation).} \tag{14–26}$$

This relation was first derived by Poiseuille and is called **Poiseuille's equation.** The volume flow rate is inversely proportional to viscosity, as we would expect. It is also proportional to the pressure gradient $(p_2 - p_1)/L$, and it varies as the *fourth power* of the radius R. If we double R, the flow rate increases by a factor of 16. This relation is important for the design of plumbing systems and hypodermic needles. Needle size is much more important than thumb pressure in determining the flow rate from the needle; doubling the needle diameter has the same effect as increasing the thumb force sixteenfold. Similarly, blood flow in arteries and veins can be controlled over a wide range by relatively small changes in diameter, an important temperature-control mechanism in warm-blooded animals. Relatively slight narrowing of arteries due to arteriosclerosis can result in elevated blood pressure and added strain on the heart muscle.

One more useful relation in viscous fluid flow is the expression for the force F exerted on a sphere of radius r moving with speed v through a fluid with viscosity η.

When the flow is laminar, the relationship is simple:

$$F = 6\pi\eta r v \quad \text{(Stokes's law).} \tag{14–27}$$

We encountered this kind of velocity-proportional force in Section 5–4. Equation (14–27) is called **Stokes's law.**

EXAMPLE 14-13

Terminal speed in a viscous fluid Derive an expression for the terminal speed v_t of a sphere falling in a viscous fluid in terms of the sphere's radius r and density ρ and the fluid viscosity η, assuming that the flow is laminar so that Stokes's law is valid.

SOLUTION As in Example 5–19, the terminal speed is reached when the total force is zero, including the weight of the sphere, the viscous retarding force, and the buoyant force (Fig. 14–32). Let ρ and ρ' be the densities of the sphere and of the fluid, respectively. The weight of the sphere is then $\frac{4}{3}\pi r^3\rho g$, and the buoyant force is $\frac{4}{3}\pi r^3\rho' g$. At terminal speed,

$$\Sigma F_y = F_{\text{buoyancy}} + F_{\text{visc}} + (-mg)$$

$$= \frac{4}{3}\pi r^3\rho' g + 6\pi\eta r v_t - \frac{4}{3}\pi r^3\rho g = 0,$$

$$v_t = \frac{2}{9}\frac{r^2 g}{\eta}(\rho - \rho').$$

We can use this formula to *measure* the viscosity of a fluid. Or, if we know the viscosity, we can determine the radius of the sphere (or the approximate sizes of other small particles) by measuring the terminal speed. Robert Millikan used this method to determine the radii of very small electrically charged oil drops by observing their terminal speeds in air. He used these drops to measure the charge of an individual electron.

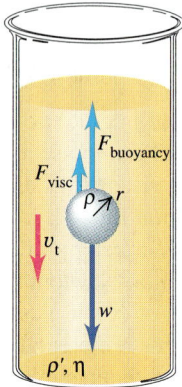

14–32 A sphere falling in a viscous fluid reaches a terminal speed when the sum of the forces acting on it is zero.

The terminal speed for a water droplet of radius 10^{-5} m in air is of the order of 1 cm/s; the Stokes's-law force is what keeps clouds in the air. At higher speeds the flow often becomes turbulent, and Stokes's law is no longer valid. In air the drag force at highway speeds is approximately proportional to v^2. We discussed the applications of this relation to skydiving in Section 5–4 and to air drag of a moving automobile in Section 6–6.

SUMMARY

KEY TERMS

■ Density is mass per unit volume. If a mass m of homogeneous material has volume V, its density ρ is

$$\rho = \frac{m}{V}. \tag{14–1}$$

Specific gravity is the ratio of the density of a material to the density of water.

■ Pressure is normal force per unit area. Pascal's law states that pressure applied to the surface of an enclosed fluid is transmitted undiminished to every portion of the fluid. Absolute pressure is the total pressure in a fluid; gauge pressure is the difference between absolute pressure and atmospheric pressure. The SI unit of pressure is the pascal (Pa): 1 Pa = 1 N/m^2.

■ If a fluid of uniform density ρ (an incompressible fluid) is at rest, the pressure difference between points 1 and 2 at elevations y_1 and y_2 is

$$p_2 - p_1 = -\rho g(y_2 - y_1). \tag{14–5}$$

If the pressure at the surface of an incompressible fluid at rest is p_0, the pressure at a depth h is

$$p = p_0 + \rho gh. \tag{14–6}$$

- Archimedes' principle states that when a body is immersed in a fluid, the fluid exerts an upward buoyant force on the body equal to the weight of the fluid the body displaces.

- The surface of a liquid behaves like a membrane under tension; the force per unit length across a line on the surface is called the surface tension, denoted by γ.

- An ideal fluid is incompressible and has no viscosity. A flow line is the path of a fluid particle; a streamline is a curve tangent at each point to the velocity vector at that point. A flow tube is a tube bounded at its sides by flow lines. In laminar flow, layers of fluid slide smoothly past each other. In turbulent flow there is great disorder and a constantly changing flow pattern.

- Conservation of mass in an incompressible fluid is expressed by the equation of continuity: for two cross sections A_1 and A_2 in a flow tube, the flow speeds v_1 and v_2 are related by

$$A_1 v_1 = A_2 v_2. \tag{14-14}$$

The product Av is the *volume flow rate*, dV/dt, the rate at which volume crosses a section of the tube:

$$\frac{dV}{dt} = Av. \tag{14-15}$$

- Bernoulli's equation relates the pressure p, flow speed v, and elevation y for steady flow in an ideal fluid. For any two points, denoted by subscripts 1 and 2,

$$p_1 + \rho g y_1 + \frac{1}{2}\rho v_1{}^2 = p_2 + \rho g y_2 + \frac{1}{2}\rho v_2{}^2. \tag{14-21}$$

- The viscosity of a fluid characterizes its resistance to shear strain. In a Newtonian fluid the viscous force is proportional to strain rate. When such a fluid flows in a cylindrical pipe of inner radius R, and length L is the length of pipe, the total volume rate is given by Poiseuille's equation:

$$\frac{dV}{dt} = \frac{\pi}{8}\left(\frac{R^4}{\eta}\right)\left(\frac{p_1 - p_2}{L}\right). \tag{14-26}$$

where p_1 and p_2 are the pressures at the two ends and η is the viscosity.

- A sphere of radius r moving with speed v through a fluid having viscosity η experiences a viscous resisting force F given by Stokes's law:

$$F = 6\pi\eta r v. \tag{14-27}$$

DISCUSSION QUESTIONS

Q14–1 If the weight of a roomful of water is so great (see Example 14–1 in Section 14–2), why don't the first floors of houses built over basements or crawl spaces all collapse when flooded to the ceiling?

Q14–2 A rubber hose is attached to a funnel, and the free end is bent around to point upward. When water is poured into the funnel, it rises in the hose to the same level as in the funnel, even though the funnel has a lot more water in it than the hose. Why?

Q14–3 Comparing Examples 14–1 and 14–2 in Sections 14–2 and 14–3, it seems that 700 N of air is exerting a downward force of 2.0×10^6 N on the floor. How is this possible?

Q14–4 Equation (14–7) shows that an area ratio of 100 to 1 can give 100 times more output force than input force. Doesn't this violate conservation of energy? Explain.

Q14–5 You have probably noticed that the lower the tire pressure, the greater the contact area between the tire and the pavement. Why?

Q14–6 In hot-air ballooning a large balloon is filled with air heated by a gas burner at the bottom. Why must the air be heated? How does the balloonist control ascent and descent?

Q14–7 In describing the size of a large ship, one uses such expressions as "it displaces 20,000 tons." What does this mean? Can the weight of the ship be obtained from this information?

Q14–8 You drop a solid sphere of aluminum in a bucket of water that sits on the ground. The buoyant force equals the weight of water displaced; this is less than the weight of the sphere, so the sphere sinks to the bottom. If you take the bucket with you on an elevator that accelerates upward, the apparent weight of the water increases and the buoyant force on the sphere increases. Could the acceleration of the elevator be great enough to make the sphere pop up out of the water? Explain.

Q14–9 A rigid lighter-than-air dirigible filled with helium cannot continue to rise indefinitely. Why not? What determines the maximum height it can attain?

Q14–10 Air pressure decreases with increasing altitude. So why isn't air near the surface continuously drawn upward toward the lower-pressure regions above?

Q14–11 The purity of gold can be tested by weighing it in air and in water. How? Do you think you could get away with making a fake gold brick by gold-plating some cheaper material?

Q14–12 During the Great Mississippi Flood of 1993, the levees in St. Louis tended to rupture first at the bottom. Why?

Q14–13 A cargo ship travels from the Atlantic Ocean (salt water) to Lake Ontario (fresh water) via the St. Lawrence River. The ship rides several centimeters lower in the water in Lake Ontario than it did in the ocean. Explain why.

Q14–14 A submarine is more compressible than water. Why, then, can a submarine completely surrounded by water be only in unstable equilibrium?

Q14–15 An old question is "Which weighs more, a pound of feathers or a pound of lead?" If the weight in pounds is the gravitational force, will a pound of feathers balance a pound of lead on opposite pans of an equal-arm balance? Explain, taking into account buoyant forces.

Q14–16 Suppose the door of a room makes an airtight but frictionless fit in its frame. Do you think you could open the door if the air pressure on one side were standard atmospheric pressure and that on the other side differed from standard by 1%? Explain.

Q14–17 A balloon is less compressible than air. Why, then, is there a height at which a helium balloon is in stable equilibrium?

Q14–18 A piece of iron is glued to the top of a block of wood. When the block is placed in a bucket of water with the iron on top, the block floats. The block is now turned over so that the iron is submerged beneath the wood. Does the block float or sink? Does the water level in the bucket rise, drop, or stay the same? Explain your answers.

Q14–19 You take an empty glass jar and push it into a tank of water with the open mouth of the jar downward, so that the air inside the jar is trapped and cannot get out. If you push the jar deeper into the water, does the buoyant force on the jar stay the same? If not, does it increase or decrease? Explain your answer.

Q14–20 An ice cube floats in a glass of water. The ice cube contains many air bubbles. When the ice melts, will the water level in the glass rise, fall, or remain unchanged? Explain.

Q14–21 A small drop on the surface of a freshly waxed car is nearly spherical, but a larger drop is more flattened. Why?

Q14–22 Water drops easily roll off a duck's back. What can you conclude about the contact angle between water and the duck?

Q14–23 Can mercury in a clean glass tube with a clean glass piston exhibit the tensile stress of negative pressure? Explain.

Q14–24 You are told, "Bernoulli's equation tells us where there is higher fluid speed, there is lower fluid pressure, and vice versa." Is this statement always true, even for an idealized fluid? Explain.

Q14–25 As you walk past a cup of coffee sitting on a desk, the coffee has a velocity relative to you. Would you describe the coffee with fluid statics or fluid dynamics? Why?

Q14–26 If the velocity at each point in space in steady-state fluid flow is constant, how can a fluid particle accelerate?

Q14–27 Why do jet airplanes usually fly at altitudes above 30,000 ft, though it takes a lot of fuel to climb that high?

Q14–28 In a store-window vacuum cleaner display, a table-tennis ball is suspended in midair in a jet of air blown from the outlet hose of a tank-type vacuum cleaner. The ball bounces around a little but always moves back toward the center of the jet, even if the jet is tilted from the vertical. How does this behavior illustrate Bernoulli's equation?

Q14–29 How does a baseball pitcher give the ball the spin that makes it curve? Can he make it curve in either direction? Does it matter whether he is right-handed or left-handed?

Q14–30 What causes the sharp hammering sound that is sometimes heard from a water pipe when an open faucet is suddenly turned off?

Q14–31 A tornado consists of a rapidly whirling air vortex. Why is the pressure always much lower in the center than at the outside? How does this condition account for the destructive power of a tornado?

Q14–32 Airports at high elevations have longer runways for takeoffs and landings than do airports at sea level. One reason is that aircraft engines develop less power in the thin air well above sea level. What is another reason?

Q14–33 Why would a fire department add a viscosity-decreasing substance to its water for fighting fires in tall buildings?

Q14–34 When a smooth-flowing stream of water comes out of a faucet, it narrows as it falls; if it falls far enough it eventually breaks up into drops. Why does it narrow? Why does it break up?

Q14–35 Spreading margarine with a knife is an example of viscous flow. Is margarine a Newtonian fluid, or is it non-Newtonian? Explain.

EXERCISES

SECTION 14–2 DENSITY

14–1 On a part-time job, you are asked to bring a cylindrical iron rod of length 85.8 cm and diameter 2.85 cm from a storage room to a machinist. Will you need a cart? (To answer, calculate the weight of the rod.)

14–2 The moon has a radius of 1740 km and a mass of 7.35×10^{22} kg. What is its average density?

14–3 You purchase a rectangular piece of metal that has dimensions $5.0 \times 15.0 \times 30.0$ mm and a mass of 0.0158 kg. The seller tells you that the metal is gold. To check this, you compute the average density of the piece. What value do you get? Were you cheated?

14–4 A kidnapper demands a 40.0-kg cube of platinum as a ransom. What is the length of a side?

SECTION 14–3 PRESSURE IN A FLUID

14–5 A barrel contains a 0.120-m layer of oil floating on water that is 0.250 m deep. The density of the oil is 600 kg/m^3 a) What is the gauge pressure at the oil-water interface? b) What is the gauge pressure at the bottom of the barrel?

14–6 An empty sport utility vehicle weighs 16.5 kN. Each of its tires has a gauge pressure of 205 kPa (29.7 lb/in.^{2}). a) What is the total contact area of the four tires with the pavement? (Assume that the tire walls are flexible so that the pressure exerted by the tire on the pavement equals the air pressure inside the tire.) b) For the same tire pressure, what is that area when the vehicle is loaded with 9.1 kN of passengers and cargo?

14–7 You are designing a diving bell to withstand the pressure of seawater at a depth of 250 m. a) What is the gauge pressure at this depth? (Neglect changes in the density of the water with depth.) b) At this depth, what is the net force due to the water outside and the air inside the bell on a circular glass window 30.0 cm in diameter if the pressure inside the diving bell equals the pressure at the surface of the water? (Neglect the small variation of pressure over the surface of the window.)

14–8 What gauge pressure must a pump produce to pump water from the bottom of the Grand Canyon (elevation 730 m) to Indian Gardens (1370 m)? Express your results in pascals and in atmospheres.

14–9 The liquid in the open-tube manometer in Fig. 14–6a is mercury, $y_1 = 3.00$ cm, and $y_2 = 7.00$ cm. Atmospheric pressure is 980 millibars. a) What is the absolute pressure at the bottom of the U-shaped tube? b) What is the absolute pressure in the open tube at a depth of 4.00 cm below the free surface? c) What is the absolute pressure of the gas in the tank? d) What is the gauge pressure of the gas in pascals?

14–10 There is a maximum depth at which a diver can breathe through a snorkel tube (Fig. 14–33) because as the depth increases, so does the pressure difference, tending to collapse the diver's lungs. Since the snorkel connects the air in the lungs to the atmosphere at the surface, the pressure inside the lungs is atmospheric pressure. What is the external-internal pressure difference when the diver's lungs are at a depth of 6.1 m (about 20 ft)? Assume that the diver is in fresh water. (A scuba diver breathing from compressed air tanks can operate at greater depths than can a snorkeler, since the pressure of the air inside the scuba diver's lungs increases to match the external pressure of the water.)

14–11 An electrical short cuts off all power to a submersible when it is 30 m below the surface of the ocean. The crew must push out a hatch of area 0.75 m^2 and weight 300 N on the bottom to escape. If the pressure inside is 1.0 atm, what downward force must they exert on the hatch to open it?

14–12 You are assigned the design of a cylindrical pressurized water tank for a future colony on Mars, where the acceleration due to gravity is 3.71 m/s^2. The pressure at the surface of the water will be 130 kPa and the depth of the water will be 14.2 m.

FIGURE 14–33 Exercise 14–10.

The pressure of the air in the building outside the tank will be 93 kPa. Find the net downward force on the tank's flat bottom, of area 2.00 m^2, exerted by the water and air inside the tank and the air outside the tank.

14–13 A tapered, pressurized tank for a rocket contains 0.250 m^3 of kerosene, which has a mass of 205 kg. The pressure at the top of the kerosene is 2.01×10^5 Pa. The kerosene exerts a force of 16.4 kN on the tank's bottom, which has an area of 0.0700 m^2. Find the depth of the kerosene.

14–14 The piston of a hydraulic automobile lift is 0.30 m in diameter. What gauge pressure, in pascals, is required to lift a car with a mass of 1200 kg? Also express this pressure in atmospheres.

SECTION 14–4 BUOYANCY

14–15 A slab of ice floats on a freshwater lake. What minimum volume must the slab have for a 45.0-kg woman to be able to stand on it without getting her feet wet?

14–16 An ore sample weighs 17.50 N in air. When the sample is suspended by a light cord and totally immersed in water, the tension in the cord is 11.20 N. Find the total volume and density of the sample.

14–17 An object of average density ρ floats at the surface of a fluid of density ρ_{fluid}. a) How must the two densities be related? b) In view of the answer to part (a), how can steel ships float in water? c) In terms of ρ and ρ_{fluid}, what fraction of the object is submerged and what fraction is above the fluid? Check that your answers give the correct limiting behavior as $\rho \to \rho_{fluid}$ and as $\rho \to 0$. d) While on board your yacht, your cousin Throckmorton cuts a rectangular piece (dimensions $5.0 \times 4.0 \times 3.0$ cm) out of a life preserver and throws it into the ocean. The piece has a mass of 42 g. As it floats in the ocean, what percentage of its volume is above the surface?

14–18 A hollow plastic sphere is held below the surface of a freshwater lake by a cord anchored to the bottom of the lake.

The sphere has a volume of 0.650 m^3 and the tension in the cord is 900 N. a) Calculate the buoyant force exerted by the water on the sphere. b) What is the mass of the sphere? c) The cord breaks and the sphere rises to the surface. When the sphere comes to rest, what fraction of its volume will be submerged?

14–19 A cubical block of wood 10.0 cm on a side floats at the interface between oil and water with its lower surface 1.50 cm below the interface (Fig. 14–34). The density of the oil is 790 kg/m^3. a) What is the gauge pressure at the upper face of the block? b) What is the gauge pressure at the lower face of the block? c) What is the mass and density of the block?

FIGURE 14–34 Exercise 14–19.

14–20 A solid aluminum ingot weighs 89 N in air. a) What is its volume? b) The ingot is suspended from a rope and totally immersed in water. What is the tension in the rope (the *apparent* weight of the ingot in water)?

SECTION 14–5 SURFACE TENSION

14–21 Find the gauge pressure in pascals in a soap bubble 3.00 cm in diameter. The surface tension is 25.0×10^{-3} N/m.

14–22 Find the excess pressure at 20°C a) inside a large raindrop with a radius of 1.00 mm; b) inside a water drop with a radius of 0.0100 mm (typical water droplets in fog).

14–23 How to Stand on Water. Estimate the upward surface-tension force that would be exerted on your feet if you attempted to stand on water. (You will need to measure your feet.) What maximum body mass could the water support this way?

14–24 Why Trees Don't Suck Air. The negative pressures that occur in a column of sap in a tall tree have been measured to be as great as −20 atm. These columns of sap are open at the top to the outside air, and water evaporates from the leaves. But if the pressure is negative, why doesn't air get sucked into the leaves? To answer this question, estimate the pressure difference required to force air into the interstices of the cell walls within the leaves (about 10^{-8} m across) and explain why outside air cannot in fact get into the tree. (Take the surface tension of the sap to be that of water at 20°C. The situation is the reverse of that in Fig. 14–15, with air displacing the sap in the interstices.)

14–25 The soap film in Fig. 14–14 is from the soap solution of Table 14–2. It is 22.0 cm wide and is at 20°C. The slide wire has a mass of 0.700 g. What is the magnitude of the downward pull T required to keep the slide wire in equilibrium?

SECTION 14–6 FLUID FLOW

14–26 Water is flowing in a pipe with varying cross-section area, and at all points the water completely fills the pipe. At point 1 the cross-section area of the pipe is 0.070 m^2, and the magnitude of the fluid velocity is 3.50 m/s. a) What is the fluid speed at points in the pipe where the cross-section area is i) 0.105 m^2? ii) 0.047 m^2? b) Calculate the volume of water discharged from the open end of the pipe in 1.00 hour.

14–27 Water is flowing in a circular pipe of varying cross-section area, and at all points the water completely fills the pipe. a) At one point in the pipe the radius is 0.150 m. What is the speed of the water at this point if the volume flow rate in the pipe is 1.20 m^3/s? b) At a second point in the pipe the water speed is 3.80 m/s. What is the radius of the pipe at this point?

14–28 a) Derive Eq. (14–16). b) If the density increases by 1.50% from point 1 to point 2, what happens to the volume flow rate?

SECTION 14–7 BERNOULLI'S EQUATION

14–29 A sealed tank containing seawater to a height of 11.0 m also contains air above the water at a gauge pressure of 3.00 atm. Water flows out from the bottom through a small hole. Calculate the efflux speed of the water.

14–30 A small circular hole 6.00 mm in diameter is cut in the side of a large water tank, 14.0 m below the water level in the tank. The top of the tank is open to the air. Find a) the speed of efflux; b) the volume discharged per unit time.

14–31 What gauge pressure is required in the city mains for a stream from a fire hose connected to the mains to reach a vertical height of 15.0 m? (Assume that the mains have a much larger diameter than the fire hose.)

14–32 At one point in a pipeline the water's speed is 3.00 m/s and the gauge pressure is 5.00×10^4 Pa. Find the gauge pressure at a second point in the line, 11.0 m lower than the first, if the pipe diameter at the second point is twice that at the first.

14–33 Lift on an Airplane. Air streams horizontally past a small airplane's wings such that the speed is 70.0 m/s over the top surface and 60.0 m/s past the bottom surface. If the plane has a mass of 1340 kg and a wing area of 16.2 m^2, what is the net vertical force (including the effects of gravity) on the airplane? The density of the air is 1.20 kg/m^3.

14–34 A soft drink (mostly water) flows in a pipe at a beverage plant with a mass flow rate that would fill 220 0.355-L cans per minute. At point 2 in the pipe, the gauge pressure is 152 kPa and the cross-section area is 8.00 cm^2. At point 1, 1.35 m above point 2, the cross-section area is 2.00 cm^2. Find the a) mass flow rate; b) volume flow rate; c) flow speeds at points 1 and 2; d) gauge pressure at point 1.

14–35 Water discharges from a horizontal cylindrical pipe at the rate of 465 cm^3/s. At a point in the pipe where the radius is 2.05 cm the absolute pressure is 1.60×10^5 Pa. What is the pipe radius at a constriction if the pressure there is reduced to 1.20×10^5 Pa?

14–36 At a certain point in a horizontal pipeline the water's speed is 2.50 m/s, and the gauge pressure is 1.80×10^4 Pa. Find the gauge pressure at a second point in the line if the cross-section area at the second point is twice that at the first.

*SECTION 14–9 VISCOSITY

*14–37 Water at 20°C is flowing in a pipe of radius 10.0 cm. The viscosity of water at 20°C is 1.005 centipoise. If the water's speed in the center of the pipe is 2.50 m/s, what is the water's speed a) 5.0 cm from the center of the pipe (halfway between the center and the walls)? b) at the walls of the pipe?

*14–38 Water at 20°C flows through a pipe of radius 8.50 mm. The viscosity of water at 20°C is 1.005 centipoise. If the flow speed at the center is 0.200 m/s and the flow is laminar, find the pressure drop due to viscosity along a 3.00-m section of pipe.

*14–39 Water at 20°C is flowing in a horizontal pipe that is 15.0 m long; the flow is laminar, and the water completely fills the pipe. A pump maintains a gauge pressure of 1200 Pa at a large tank at one end of the pipe. The other end of the pipe is open to the air. The viscosity of water at 20°C is 1.005 centipoise. a) If the pipe has diameter 9.00 cm, what is the volume flow rate? b) What gauge pressure must the pump provide to achieve the same volume flow rate for a pipe with a diameter of 3.00 cm? c) For the pipe in part (a) and the same gauge pressure maintained by the pump, what does the volume flow rate become if the water is at a temperature of 60°C? (The viscosity of water at 60°C is 0.469 centipoise.)

*14–40 The South American insect *Rhodinus prolixus* sucks the blood of mammals. Its mouthparts are like a very fine hypodermic needle (which allows it to draw blood from its victims painlessly, and so without being noticed). The narrowest part of the "needle" is 0.20 mm long and is only 10 μm in diameter. a) What must be the gauge pressure in the insect's mouth cavity if it is to take a large meal of 0.25 cm^3 of blood in 15 minutes? Express your answer in Pa and in atm. (The viscosity of human blood in such a narrow tube is 1.0 centipoise. To get an approximate answer, apply Poiseuille's equation to the blood even though it is a non-Newtonian fluid.) b) Why is it a good approximation to ignore the dimensions of the rest of the insect's mouthparts?

*14–41 What speed must an aluminum sphere with radius 2.00 mm have in castor oil at 20°C for the viscous drag force to be one-fourth the weight of the sphere? (The viscosity of castor oil at this temperature is 9.86 poise.)

*14–42 **Measuring Viscosity.** A brass sphere with a mass of 0.35 g falls with a terminal speed of 5.0 cm/s in an unknown liquid. If the density of the liquid is 2900 kg/m^3, what is the viscosity of the liquid?

*14–43 Keeping all other variables constant, what happens to the volume flow rate in laminar flow if you double a) the pipe diameter? b) the viscosity? c) the pressure difference? d) the pressure gradient? e) the pipe length?

*14–44 For normal shots in basketball (not the desperation half-court-line ones) the force of air resistance is negligible. To show this, consider the ratio of the Stokes's law force to the weight of a 0.600-kg basketball. The basketball has radius 0.124 m and is moving through air of density 1.20 kg/m^3 at 5.00 m/s.

*14–45 A very narrow, high-intensity laser beam drills a cylindrical hole through the hull of a Federation spaceship; the hole is 0.180 m long and only 50.0 μm in radius. Air at 20°C with a viscosity of 181 micropoise begins to rush out in laminar flow from the 1.00-atm interior to the vacuum outside. a) What is the air speed along the axis of the cylinder, at the edge, and halfway between? b) How many days will it take for a cubic meter of air to flow through the hole? (Assume that the interior pressure remains 1.00 atm.) c) By what factor are the answers to parts (a) and (b) changed if the radius of the hole is twice as much and the flow remains laminar?

PROBLEMS

14–46 In a lecture demonstration, a professor pulls apart two hemispherical steel shells (diameter D) with ease using their attached handles. She then places them together, pumps out the air to an absolute pressure of p, and hands them to a bodybuilder in the back row to pull apart. a) If atmospheric pressure is p_0, how much force must the bodybuilder exert on each shell? b) Evaluate your answer for the case $p = 0.025$ atm, $D = 10.0$ cm.

14–47 The deepest point known in any of the earth's oceans is in the Marianas Trench, 10.92 km deep. a) Assuming water to be incompressible, what is the pressure at this depth? Use the density of seawater. b) The actual pressure is 1.16×10^8 Pa; your calculated value will be less because the density actually varies with depth. Using the compressibility of water and the actual pressure, find the density of the water at the bottom of the Marianas Trench. What is the percent change in the density of the water?

14–48 A swimming pool measures 5.0 m long × 4.0 m wide × 3.0 m deep. Compute the force exerted by the water against a) the bottom; b) either end. (*Hint:* Calculate the force on a thin horizontal strip at a depth h, and integrate this over the end of the pool.) Do not include the force due to air pressure.

14–49 The upper edge of a gate in a dam runs along the water surface. The gate is 2.00 m high and 4.00 m wide and is hinged along a horizontal line through its center (Fig. 14–35). Calculate the torque about the hinge arising from the force due to the water. (*Hint:* Use a procedure similar to that used in Problem 14–48; calculate the torque of a thin horizontal strip at a depth h and integrate this over the gate.)

FIGURE 14–35 Problem 14–49.

14–50 Force and Torque on a Dam. A dam has the shape of a rectangular solid. The side facing the lake has area A and height H. The surface of the freshwater lake behind the dam is at the top of the dam. a) Show that the net horizontal force exerted by the water on the dam equals $\frac{1}{2}\rho g H A$, that is, the average gauge pressure across the face of the dam times the area. (See Problem 14–48.) b) Show that the torque exerted by the water about an axis along the bottom of the dam is $\rho g H^2 A/6$. c) How do the force and torque depend on the size of the lake?

14–51 An astronaut is standing at the north pole of a newly discovered spherically symmetric planet of radius R. In his hands he holds a container full of a liquid of mass m and volume V. At the surface of the liquid, the pressure is p_0; at a depth d below the surface, the pressure has a greater value p. From this information, determine the mass of the planet.

14–52 To find the density at a given point within a material, consider a small volume dV centered on that point. If the mass within that volume is dm, the density at the point is $\rho = dm/dV$. Consider a cylindrical rod of mass M, radius R, and length L with a density that is proportional to the square of the distance from one end, $\rho = Cx^2$. a) Show that $C = 3M/\pi R^2 L^3$. b) Show that the average density, given by Eq. (14–1), is one-third of the density at the $x = L$ end.

14–53 The earth does not have a uniform density; it is most dense at its center and least dense at its surface. An approximation to its density is $\rho(r) = A - Br$, where $A = 12,700$ kg/m^3 and $B = 1.50 \times 10^{-3}$ kg/m^4. Use $R = 6.37 \times 10^6$ m for the radius of the earth approximated as a sphere. a) Geological evidence indicates that the densities are 13,100 kg/m^3 and 2,400 kg/m^3 at the earth's center and surface. What values does the linear approximation model give for the densities at these two locations? b) Imagine dividing the earth into concentric spherical shells. Each shell has radius r, thickness dr, volume $dV = 4\pi r^2 dr$, and mass $dm = \rho(r)\,dV$. By integrating from $r = 0$ to $r = R$, show that the mass of the earth in this model is $M = \frac{4}{3}\pi R^3(A - \frac{3}{4}BR)$. c) Show that the given values of A and B give the correct mass of the earth to within 0.4%. d) We saw in Section 12–7 that a uniform spherical shell gives no contribution to g inside it. Show that $g(r) = \frac{4}{3}\pi Gr(A - \frac{3}{4}Br)$ inside the earth in this model. e) Verify that the expression of part (d) gives $g = 0$ at the center of the earth and $g = 9.85$ m/s^2 at the surface. f) Show that in this model g does *not* decrease uniformly with depth but rather has a maximum of $4\pi GA^2/9B = 10.01$ m/s^2 at $r = 2A/3B = 5640$ km.

14–54 In Example 12–9 (Section 12–7) we saw that inside a planet of uniform density (not a very realistic assumption for the earth) the acceleration due to gravity increases uniformly with distance from the center of the planet. That is, $g(r) = g_s r/R$, where g_s is the acceleration due to gravity at the surface, r is the distance from the center of the planet, and R is the radius of the planet. The interior of the planet can be treated approximately as an incompressible fluid of density ρ. a) Replace the height y in Eq. (14–4) with the radial coordinate r and integrate to find the pressure inside a uniform planet as a function of r. Let the pressure at the surface be zero. (This means ignoring the pressure of the planet's atmosphere.) b) Using this model, calculate the pressure at the center of the earth. (Use a value of ρ equal to the average density of the earth, calculated from the mass and radius given in Appendix F.) c) Geologists estimate the pressure at the center of the earth to be approximately 4×10^{11} Pa. Does this agree with your calculation for the pressure at $r = 0$? What might account for any differences?

14–55 A U-shaped tube open to the air at both ends contains some mercury. A quantity of water is carefully poured into the left arm of the U-shaped tube until the vertical height of the water column is 15.0 cm (Fig. 14–36). a) What is the gauge pressure at the water-mercury interface? b) Calculate the vertical distance h from the top of the mercury in the right-hand arm of the tube to the top of the water in the left-hand arm.

FIGURE 14–36 Problem 14–55.

14–56 The Great Molasses Flood. On the afternoon of January 15, 1919, an unusually warm day in Boston, a 27.4-m-high (90 ft), 27.4-m-diameter cylindrical metal tank used for storing molasses ruptured. Molasses flooded into the streets in a 9-m-deep stream, killing pedestrians and horses and knocking down buildings. The molasses had a density of 1600 kg/m^3. If the tank was full before the accident, what was the total outward force the molasses exerted on its sides? (*Hint:* Consider the outward force on a circular ring of the tank wall of width dy and at a depth y below the surface. Integrate to find the total outward force. Assume that before the tank ruptured, the pressure at the surface of the molasses was equal to the air pressure outside the tank.)

14–57 An open barge has the dimensions shown in Fig. 14–37. If the barge is made out of 4.0-cm-thick steel plate on each of its four sides and its bottom, what mass of coal can the barge carry in fresh water without sinking? Is there enough room in the barge to hold this amount of coal? (The density of coal is about 1500 kg/m^3.)

FIGURE 14–37 Problem 14–57.

14–58 A hot-air balloon has a volume of 2200 m^3. The balloon fabric (the envelope) weighs 900 N. The basket with gear and full propane tanks weighs 1700 N. If the balloon can barely lift an additional 3200 N of passengers, breakfast, and champagne when the outside air density is 1.23 kg/m^3, what is the average density of the heated gases in the envelope?

14–59 Advertisements for a certain small car claim it floats in water. a) If the car's mass is 900 kg and its interior volume 3.0 m^3, what fraction of the car is immersed when it floats? The volume of steel and other materials may be neglected. b) Water gradually leaks in and displaces the air in the car. What fraction of the interior volume is filled with water when the car sinks?

14–60 A single ice cube of mass 9.70 g floats in a glass completely full of 420 cm^3 of water. Ignore the water's surface tension and its variation in density with temperature (as long as it remains a liquid). a) What volume of water does the ice cube displace? b) When the ice cube has completely melted, has any water overflowed? If so, how much? If not, explain why this is so. c) Suppose the water in the glass had been very salty water of density 1050 kg/m^3. What volume of salt water would the 9.70-g ice cube displace? d) Redo part (b) for the freshwater ice cube in the salty water.

14–61 A piece of wood is 0.600 m long, 0.250 m wide, and 0.080 m thick. Its density is 600 kg/m^3. What volume of lead must be fastened underneath to sink the wood in calm water so that its top is just even with the water level? What is the mass of this volume of lead?

14–62 A hydrometer consists of a spherical bulb and a cylindrical stem with a cross-section area of 0.400 cm^2 (Fig. 14–9a). The total volume of bulb and stem is 13.2 cm^3. When immersed in water, the hydrometer floats with 8.00 cm of the stem above the water surface. When immersed in an organic fluid, 3.20 cm of the stem is above the surface. Find the density of the organic fluid. (*Note:* This illustrates the precision of such a hydrometer. Relatively small density differences give rise to relatively large differences in hydrometer readings.)

14–63 The densities of air, helium, and hydrogen (at $p = 1.0$ atm and $T = 20°C$) are 1.20 kg/m^3, 0.166 kg/m^3, and 0.0899 kg/m^3, respectively. a) What is the volume in cubic meters displaced by a hydrogen-filled airship that has a total "lift" of 120 kN? (The "lift" is the amount by which the buoyant force exceeds the weight of the gas that fills the airship.) b) What would be the "lift" if helium were used instead of hydrogen? In view of your answer, why is helium used in modern airships like advertising blimps?

14–64 SHM of a Floating Object. An object with height h, mass M, and uniform cross-section area A floats upright in a liquid with density ρ. a) Calculate the vertical distance from the surface of the liquid to the bottom of the floating object at equilibrium. b) A downward force with magnitude F is applied to the top of the object. At the new equilibrium position, how much farther below the surface of the liquid is the bottom of the object than it was in part (a)? (Assume that some of the object remains above the surface of the liquid.) c) Your result in part (b) shows that if the force is suddenly removed, the object will oscillate up and down in simple harmonic motion. Calculate the period of this motion in terms of the density ρ of the liquid and the mass

M and cross-section area A of the object. Neglect the damping due to fluid friction (Section 13–8).

14–65 A 950-kg cylindrical can buoy floats vertically in salt water. The diameter of the buoy is 0.900 m. a) Calculate the additional distance the buoy will sink when a 70.0-kg man stands on top. (Use the expression derived in part (b) of Problem 14–64.) b) Calculate the period of the resulting vertical simple harmonic motion when the man dives off. (Use the expression derived in part (c) of Problem 14–64, and as in that problem neglect the damping due to fluid friction.)

14–66 In seawater a life preserver with a volume of 0.0400 m^3 will support a 75.0-kg person (average density 980 kg/m^3) with 20% of the person's volume above water when the life preserver is fully submerged. What is the density of the material composing the life preserver?

14–67 A block of balsa wood placed in one scale pan of an equal-arm balance is exactly balanced by a 0.0950-kg brass mass in the other scale pan. Find the true mass of the balsa wood if its density is 150 kg/m^3. Explain why it is accurate to neglect the buoyancy in air of the brass but *not* the buoyancy in air of the balsa wood.

14–68 Block A in Fig. 14–38 hangs by a cord from spring balance D and is submerged in a liquid C contained in beaker B. The mass of the beaker is 1.00 kg; the mass of the liquid is 1.80 kg. Balance D reads 3.50 kg and balance E reads 7.50 kg. The volume of block A is 3.80×10^{-3} m^3. a) What is the density of the liquid? b) What will each balance read if block A is pulled up out of the liquid?

FIGURE 14–38 Problem 14–68.

14–69 A hunk of aluminum is completely covered with a gold shell to form an ingot of weight 45.0 N. When you suspend the ingot from a spring balance and submerge the ingot in water, the balance reads 39.0 N. What is the weight of the gold in the shell?

14–70 Untethered helium balloons floating in a car that has all the windows rolled up and outside air vents closed move in the direction of the car's acceleration, but loose balloons filled with air move opposite. To show why, consider only the horizontal forces acting on the balloons. Let a be the magnitude of the car's forward acceleration. Consider a horizontal tube of cross-section area A that extends from the windshield, where $x = 0$ and $p = p_0$, back along the x-axis. Now consider a volume element of thickness dx in this tube. The pressure on its front surface is p and the pressure on its rear surface is $p + dp$. Assume the air has a constant density ρ. a) Apply Newton's second law to the volume element to show $dp = \rho a\, dx$. b) Integrate the result of part (a) to find the pressure at the front surface in terms of a and x. c) To show that considering ρ constant is reasonable, calculate the pressure difference in atm for a distance as long as 2.5 m and a large acceleration of 5.0 m/s^2. d) Show that the net horizontal force on a balloon of volume V is ρVa. e) For negligible friction forces, show that the acceleration of the balloon (average density ρ_{bal}) is $(\rho/\rho_{bal})a$, so that the acceleration relative to the car is $a_{rel} = [(\rho/\rho_{bal}) - 1]a$. f) Use the expression for a_{rel} in part (e) to explain the movement of the balloons.

14–71 The weight of a king's solid crown is w. When suspended by a light rope and completely immersed in water, the tension in the rope (the crown's apparent weight) is fw. a) Prove that the crown's relative density (specific gravity) is $1/(1 - f)$. Discuss the meaning of the limits as f approaches 0 and 1. b) If the crown is solid gold and weighs 12.9 N in air, what is its apparent weight when completely immersed in water? c) Repeat part (b) if the crown is solid lead with a very thin gold plating, but still has a weight in air of 12.9 N.

14–72 A piece of steel has a weight w, an apparent weight (see Problem 14–71) w_{water} when completely immersed in water, and an apparent weight w_{fluid} when completely immersed in an unknown fluid. a) Prove that the fluid's density relative to water (specific gravity) is $(w - w_{fluid})/(w - w_{water})$. b) Is this result reasonable for the three cases of w_{fluid} greater than, equal to, or less than w_{water}? c) The apparent weight of the piece of steel in water of density 1000 kg/m^3 is 87.2% of its weight. What percentage of its weight will its apparent weight be in formic acid (density 1220 kg/m^3)?

14–73 You cast some metal of density ρ_m in a mold, but you are worried that there might be cavities within the casting. You measure the weight of the casting to be w, and the buoyant force when it is completely surrounded by water to be B. a) Show that $V_0 = B/(\rho_{water}g) - w/(\rho_m g)$ is the total volume of any enclosed cavities. b) If your metal is copper, the casting's weight is 156 N, and the buoyant force is 20 N, what is the total volume of any enclosed cavities in your casting? What fraction is this of the total volume of the casting?

14–74 A cubical block of wood 0.100 m on a side and with a density of 550 kg/m^3 floats in a jar of water. Oil with a density of 750 kg/m^3 is poured on the water until the top of the oil layer is 0.035 m below the top of the block. a) How deep is the oil layer? b) What is the gauge pressure at the block's lower face?

14–75 Dropping Anchor. An iron anchor with mass 35.0 kg and density 7860 kg/m^3 lies on the deck of a small barge that has vertical sides and that is floating in a freshwater river. The area of the bottom of the barge is 8.00 m^2. The anchor is thrown overboard but is suspended above the bottom of the river by a rope; the mass and volume of the rope are small enough to ignore. After the anchor is overboard and the barge has finally stopped bobbing up and down, has the barge risen or sunk down in the water? By what vertical distance?

14–76 Assume that crude oil from a supertanker has density 750 kg/m^3. The tanker runs aground on a sandbar. To refloat the tanker, its oil cargo is pumped out into steel barrels, each of

which has a mass of 15.0 kg when empty and holds 0.120 m^3 of oil. Neglect the volume occupied by the steel from which the barrel is made. a) If a salvage worker accidentally drops a filled, sealed barrel overboard, will it float or sink in the seawater? b) If the barrel floats, what fraction of its volume will be above the water surface? If it sinks, what minimum tension would have to be exerted by a rope to haul the barrel up from the ocean floor? c) Repeat parts (a) and (b) if the density of the oil is 910 kg/m^3 and the mass of each empty barrel is 32.0 kg.

14–77 A cubical block of density ρ_B and with sides of length L floats in a liquid of greater density ρ_L. a) What fraction of the block's volume is above the surface of the liquid? b) The liquid is denser than water (density ρ_W) and does not mix with it. If water is poured on the surface of the liquid, how deep must the water layer be so that the water surface just rises to the top of the block? Express your answer in terms of L, ρ_B, ρ_L, and ρ_W. c) Find the depth of the water layer in part (b) if the liquid is mercury, the block is made of iron, and the side length is 10.0 cm.

14–78 A barge is in a rectangular lock on a freshwater river. The lock is 60.0 m long and 20.0 m wide, and the steel doors on each end are closed. With the barge floating in the lock, a 2.50×10^6 N load of scrap metal is put onto the barge. The metal has density 9000 kg/m^3. a) When the load of scrap metal, initially on the bank, is placed onto the barge, what vertical distance does the water in the lock rise? b) The scrap metal is now pushed overboard into the water. Does the water level in the lock rise, fall, or remain the same? If it rises or falls, by what vertical distance does it change?

14–79 A U-shaped tube with a horizontal portion of length l (Fig. 14–39) contains a liquid. What is the difference in height between the liquid columns in the vertical arms a) if the tube has an acceleration a toward the right? b) if the tube is mounted on a horizontal turntable rotating with an angular speed ω with one of the vertical arms on the axis of rotation? c) Explain why the difference in height does not depend on the density of the liquid or on the cross-section area of the tube. Would it be the same if the vertical tubes did not have equal cross-section areas? Would it be the same if the horizontal portion were tapered from one end to the other? Explain.

FIGURE 14–39 Problem 14–79.

14–80 A cylindrical container of an incompressible liquid with density ρ rotates with constant angular speed ω about its axis of symmetry, which we take to be the y-axis (Fig. 14–40). a) Show that the pressure at a given height within the fluid increases in the radial direction (outward from the axis of rotation) according to $\partial p/\partial r = \rho \omega^2 r$. b) Integrate this partial differential equation to find the pressure as a function of distance from the axis of rotation along a horizontal line at $y = 0$. c) Combine the result of part (b) with Eq. (14–5) to show that the surface of the rotating liquid has a *parabolic* shape, that is, the height of the liquid is given by $h(r) = \omega^2 r^2/2g$. (This technique is used for making parabolic telescope mirrors; liquid glass is rotated and allowed to solidify while rotating.)

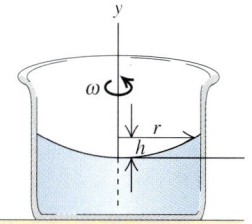

FIGURE 14–40 Problem 14–80.

14–81 An incompressible fluid with density ρ is in a horizontal test tube of inner cross-section area A. The test tube spins in a horizontal circle in an ultracentrifuge at an angular speed ω. Gravitational forces are negligible. Consider a volume element of the fluid of area A and thickness dr' a distance r' from the rotation axis. The pressure on its inner surface is p and on its outer surface is $p + dp$. a) Apply Newton's second law to the volume element to show $dp = \rho\omega^2 r' \, dr'$. b) If the surface of the fluid is at a radius r_0 where the pressure is p_0, show that the pressure p at a distance $r \geq r_0$ is $p = p_0 + \rho\omega^2(r^2 - r_0^2)/2$. c) An object of volume V and density ρ_{ob} has its center of mass at a distance R_{cmob} from the axis. Show that the net horizontal force on the object is $\rho V\omega^2 R_{cm}$, where R_{cm} is the distance from the axis to the center of mass of the displaced fluid. d) Explain why the object will move inward if $\rho R_{cm} > \rho_{ob}R_{cmob}$ and outward if $\rho R_{cm} < \rho_{ob}R_{cmob}$. e) For small objects of uniform density, $R_{cm} = R_{cmob}$. What happens to a mixture of small objects of this kind with different densities in an ultracentrifuge?

14–82 What radius must a water drop have for the difference between inside and outside pressures to be 0.0250 atm? Assume $T = 20°C$.

14–83 A cubical block of wood 0.30 m on a side is weighted so that its center of gravity is at the point shown in Fig. 14–41a and it floats in water with one-half its volume submerged. If the block is "heeled" at an angle of 45.0° as in Fig. 14–41b, compute the net torque about a horizontal axis perpendicular to the block and passing through its geometrical center.

(a)   (b)

FIGURE 14–41 Problem 14–83.

14–84 Water stands at a depth H in a large open tank whose side walls are vertical (Fig. 14–42). A hole is made in one of the walls at a depth h below the water surface. a) At what distance R from the foot of the wall does the emerging stream strike the floor? b) How far above the bottom of the tank could a second hole be cut so that the stream emerging from it could have the same range as for the first hole?

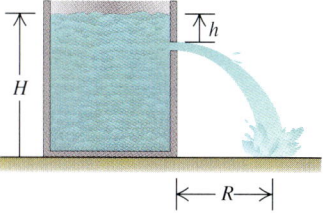

FIGURE 14–42 Problem 14–84.

14–85 A cylindrical bucket, open at the top, is 25.0 cm high and 10.0 cm in diameter. A circular hole with cross-section area 1.50 cm² is cut in the center of the bottom of the bucket. Water flows into the bucket from a tube above it at the rate of 2.40×10^{-4} m³/s. How high will the water in the bucket rise?

14–86 Water flows steadily from an open tank as in Fig. 14–43. The elevation of point 1 is 10.0 m, and the elevation of points 2 and 3 is 2.00 m. The cross-section area at point 2 is 0.0480 m²; at point 3 it is 0.0160 m². The area of the tank is very large compared with the cross-section area of the pipe. Assuming that Bernoulli's equation applies, compute a) the discharge rate in cubic meters per second; b) the gauge pressure at point 2.

FIGURE 14–43 Problem 14–86.

14–87 Modern airplane design calls for a lift due to the net force of the moving air on the wing of about 2000 N per square meter of wing area. Assume that air flows past the wing of an aircraft with streamline flow. If the speed of flow past the lower wing surface is 120 m/s, what is the required speed over the upper surface to give a lift of 2000 N/m²? The density of the air is 1.20 kg/m³.

14–88 The radius of 1993's Hurricane Emily was about 350 km. The wind speed near the center ("eye") of the hurricane, whose radius was about 30 km, reached about 200 km/h. As air swirls in from the rim of the hurricane toward the eye, its angular momentum remains roughly constant. a) Estimate the wind speed at the rim of the hurricane. b) Estimate the pressure difference at the earth's surface between the eye and the rim. (*Hint:* See Table 14–1.). Where is the pressure greater? c) If the kinetic energy of the swirling air in the eye could be converted completely to gravitational potential energy, how high would the air go? d) In fact the air in the eye is lifted to heights of several kilometers. How can you reconcile this with your answer to part (c)?

14–89 Two very large open tanks A and F (Fig. 14–44) contain the same liquid. A horizontal pipe BCD, having a constriction at C and open to the air at D, leads out of the bottom of tank A, and a vertical pipe E opens into the constriction at C and dips into the liquid in tank F. Assume streamline flow and no viscosity. If the cross-section area at C is one half that at D and if D is a distance h_1 below the level of the liquid in A, to what height h_2 will liquid rise in pipe E? Express your answer in terms of h_1.

FIGURE 14–44 Problem 14–89.

14–90 The horizontal pipe shown in Fig. 14–45 has a cross-section area of 40.0 cm^2 at the wider portions and 10.0 cm^2 at the constriction. Water is flowing in the pipe, and the discharge from the pipe is 6.00×10^{-3} m^3/s (6.00 L/s). Find a) the flow speeds at the wide and the narrow portions; b) the pressure difference between these portions; c) the difference in height between the mercury columns in the U-shaped tube.

FIGURE 14–45 Problem 14–90.

14–91 Figure 14–27a shows a liquid flowing from a vertical pipe. Notice that the vertical stream of the liquid has a very definite shape as it flows from the pipe. To get the equation for this shape, assume that the liquid is in free fall once it leaves the pipe. Just as it leaves the pipe, the liquid has speed v_0 and the radius of the stream of liquid is r_0. a) Find an equation for the speed of the liquid as a function of the distance y it has fallen. Combining this with the equation of continuity, find an expression for the radius of the stream as a function of y. b) If water flows out of a vertical pipe at a speed of 1.20 m/s, how far below the outlet will the radius be one-half the original radius of the stream?

*14–92 a) With what speed is a solid brass sphere 2.50 mm in radius falling in a tank of glycerin at an instant when its acceleration is one half that of a freely falling body? The viscosity of the glycerin is 8.30 poise. b) What is the terminal speed of the sphere?

*14–93 Speed of a Bubble in Liquid. a) With what terminal speed does an air bubble 2.00 mm in diameter rise in a liquid with viscosity 1.50 poise and density 900 kg/m^3? (Assume that the density of the air is 1.20 kg/m^3 and that the bubble diameter remains constant.) b) What is the terminal speed of the same bubble in water at 20°C, which has viscosity 1.005 centipoise?

*14–94 Oil of viscosity 3.00 poise and density 860 kg/m^3 is to be pumped from one large open tank to another through a smooth, horizontal steel pipe 1.50 km in length and 0.110 m in diameter. The pipe discharges into the air. a) What gauge pressure, in pascals and in atmospheres, must the pump exert in order to maintain a flow of 0.0600 m^3/s? b) Explain why the power consumed by the pump is equal to the volume flow rate multiplied by the gauge pressure that the pump maintains. What is the numerical value of the power?

*14–95 The tank at the left in Fig. 14–46a has a very large cross-section area and is open to the atmosphere. The depth $y = 0.600$ m. The cross-section areas of the horizontal tubes leading out of the tank are 1.00 cm^2, 0.40 cm^2, and 0.20 cm^2, respectively. The liquid is ideal, having zero viscosity. a) What is the volume flow rate out of the tank? b) What is the speed in each portion of the horizontal tube? c) What is the height of the liquid in each of the five vertical side tubes? d) Suppose that the liquid in Fig. 14–46b has a viscosity of 0.0600 poise and a density of 800 kg/m^3 and that the depth of liquid in the large tank is such that the volume rate of flow is the same as that in part (a). The distance between the side tubes at c and d, and between those at e and f, is 0.200 m. The cross-section areas of the horizontal tubes are the same in both diagrams. What is the difference in level between the tops of the liquid columns in tubes c and d? e) In tubes e and f? f) What is the flow speed on the axis of each part of the horizontal tube?

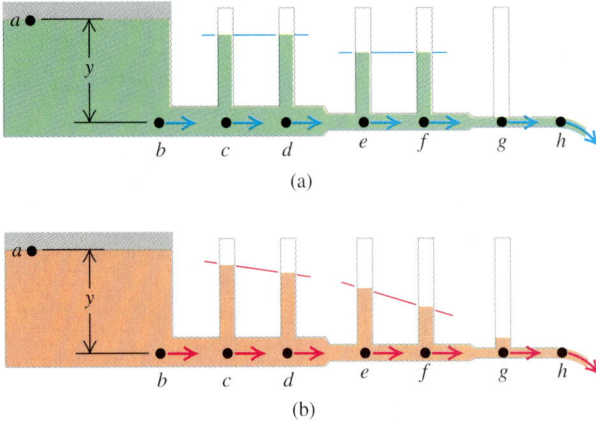

FIGURE 14–46 Problem 14–95.

CHALLENGE PROBLEMS

14–96 A rock with mass $m = 3.00$ kg is suspended from the roof of an elevator by a light cord. The rock is totally immersed in a bucket of water that sits on the floor of the elevator, but the rock doesn't touch the bottom or sides of the bucket. a) When the elevator is at rest, the tension in the cord is 21.0 N. Calculate the volume of the rock. b) Derive an expression for the tension in the cord when the elevator is accelerating *upward* with an acceleration of magnitude a. Calculate the tension when $a = 2.50$ m/s^2 upward. c) Derive an expression for the tension in the cord when the elevator is accelerating *downward* with an acceleration of magnitude a. Calculate the tension when $a = 2.50$ m/s^2 downward. d) What is the tension when the elevator is in free fall with a downward acceleration equal to g?

14–97 Suppose a piece of styrofoam, $\rho = 180$ kg/m^3, is held completely submerged in water (Fig. 14–47). a) What is the tension in the cord? Find this using Archimedes's principle. b) Use $p = p_0 + \rho gh$ to directly calculate the force exerted by the water on the two sloped sides and the bottom of the styrofoam; then show the vector sum of these forces is the buoyant force.

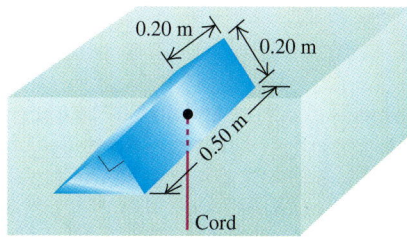

FIGURE 14–47 Challenge Problem 14–97.

14–98 A large tank with diameter D, open to the air, contains water to a height H. A small hole with diameter d ($d \ll D$) is made at the base of the tank. Neglecting any effects of viscosity, calculate the time it takes for the tank to drain completely.

14–99 A *siphon*, as depicted in Fig. 14–48, is a convenient device for removing liquids from containers. To establish the flow, the tube must be initially filled with fluid. Let the fluid have density ρ and let the atmospheric pressure be p_a. Assume that the cross-section area of the tube is the same at all points along it. a) If the lower end of the siphon is at a distance h below the surface of the liquid in the container, what is the speed of the fluid as it flows out the lower end of the siphon? (Assume that the container has a very large diameter, and neglect any effects of viscosity.) b) A curious feature of a siphon is that the fluid initially flows "uphill." What is the greatest height H that the high point of the tube can have if flow is still to occur?

FIGURE 14–48 Challenge Problem 14–99.

14–100 The following is quoted from a letter. *It is the practice of carpenters hereabouts, when laying out and leveling up the foundations of relatively long buildings, to use a garden hose filled with water, into the ends of the hose being thrust glass tubes 10 to 12 inches long. The theory is that water, seeking a common level, will be of the same height in both the tubes and thus effect a level. Now the question rises as to what happens if a bubble of air is left in the hose. Our greybeards contend the air will not affect the reading from one end to the other. Others say that it will cause important inaccuracies.* Can you give a relatively simple answer to this question, together with an explanation? Fig. 14–49 gives a rough sketch of the situation that caused the dispute.

FIGURE 14–49 Challenge Problem 14–100.

15

Temperature and Heat

Perhaps the highest-temperature material you will ever see is the sun's outer atmosphere, or corona. At a temperature of about 2,000,000°C (3,600,000°F), the corona glows with a light that is literally unearthly. But because the corona is also very thin, its light is rather faint. You can only see the corona during a total solar eclipse when the sun's disk is covered by the moon, as shown here.

15-1 INTRODUCTION

Whether it's a sweltering summer day or a frozen midwinter night, your body needs to be kept at a nearly constant temperature. It has effective temperature-control mechanisms, but sometimes it needs help. On a hot day you wear less clothing to improve heat transfer from your body to the air and for better cooling by evaporation of perspiration. You probably drink cold beverages, possibly with ice in them, and sit near a fan or in an air-conditioned room. On a cold day you wear more clothes or stay indoors where it's warm. When you're outside, you keep active and drink hot liquids to stay warm. The concepts in this chapter will help you understand the basic physics of keeping warm or cool.

First, we need to define *temperature,* including temperature scales and ways to measure temperature. Next we discuss how the dimensions and volume of a body are affected by temperature changes. We'll encounter the concept of *heat,* which describes energy transfer caused by temperature differences, and we'll learn how to calculate the *rates* of such energy transfers.

Our emphasis in this chapter is on the concepts of temperature and heat as they relate to *macroscopic* objects such as cylinders of gas, ice cubes, and the human body. In Chapter 16 we'll look at these same concepts from a *microscopic* viewpoint in terms of the behavior of individual atoms and molecules. These two chapters lay the groundwork for the subject of **thermodynamics,** the study of energy transformations involving heat, mechanical work,

and other aspects of energy and how these transformations relate to the properties of matter. Thermodynamics forms an indispensable part of the foundation of physics, chemistry, and the life sciences, and its applications turn up in such places as car engines, refrigerators, biochemical processes, and the structure of stars. We'll explore the key ideas of thermodynamics in Chapters 17 and 18.

15-2 TEMPERATURE AND THERMAL EQUILIBRIUM

The concept of **temperature** is rooted in qualitative ideas of "hot" and "cold" based on our sense of touch. A body that feels hot usually has a higher temperature than a similar body that feels cold. That's pretty vague, and the senses can be deceived. But many properties of matter that we can *measure* depend on temperature. The length of a metal rod, steam pressure in a boiler, the ability of a wire to conduct an electric current, and the color of a very hot glowing object—all these depend on temperature.

Temperature is also related to the kinetic energies of the molecules of a material. In general this relationship is fairly complex, so it's not a good place to start in *defining* temperature. In Chapter 16 we will look at the relationship between temperature and the energy of molecular motion for an ideal gas. It is important to understand, however, that temperature and heat are inherently *macroscopic* concepts. They can and must be defined independently of any detailed molecular picture. In this section we'll develop a macroscopic definition of temperature.

To use temperature as a measure of hotness or coldness, we need to construct a temperature scale. To do this, we can use any measurable property of a system that varies with its "hotness" or "coldness." Figure 15–1a shows a familiar system that is used to measure temperature. When the system becomes hotter, the colored liquid (usually mercury or ethanol) expands and rises in the tube, and the value of L increases. Another simple system is a quantity of gas in a constant-volume container (Fig. 15–1b). The pressure p, measured by the gauge, increases or decreases as the gas becomes hotter or colder. A third example is the electrical resistance R of a conducting wire, which also varies when the wire becomes hotter or colder. Each of these properties gives us a number (L, p, or R) that varies with hotness and coldness, so each property can be used to make a **thermometer.**

To measure the temperature of a body, you place the thermometer in contact with the body. If you want to know the temperature of a cup of hot coffee, you stick the thermometer in the coffee; as the two interact, the thermometer becomes hotter and the coffee cools off a little. After the thermometer settles down to a steady value, you read the temperature. The system has reached an *equilibrium* condition, in which the interaction between the thermometer and the coffee causes no further change in the system. We call this a state of **thermal equilibrium.**

If two systems are separated by an insulating material or **insulator** such as wood, plastic foam, or fiberglass, they influence each other more slowly. Camping coolers are made with insulating materials to delay the ice and cold food inside from warming up and attaining thermal equilibrium with the hot summer air outside. An *ideal insulator* is a material that permits no interaction at all between the two systems. It prevents the systems from attaining thermal equilibrium if they aren't in thermal equilibrium at the start. An ideal insulator is just that, an idealization; real insulators, like those in camping coolers, aren't ideal, so the contents of the cooler will warm up eventually.

We can discover an important property of thermal equilibrium by considering three systems, A, B, and C, that initially are not in thermal equilibrium (Fig. 15–2). We surround them with an ideal insulating box so that they cannot interact with anything except each other. We separate systems A and B with an ideal insulating wall (the blue slab in Fig. 15–2a), but we let system C interact with both systems A and B. This interaction is shown in the figure by an orange slab representing a thermal **conductor,** a material that *permits* thermal interactions through it. We wait until thermal equilibrium is attained; then A and B are each in thermal equilibrium with C. But are they in thermal equilibrium *with each other?*

To find out, we separate system C from systems A and B with an ideal insulating wall (Fig. 15–2b), and then we replace the insulating wall between A and B with a *conducting* wall that lets A and B interact. What happens? Experiment shows that *nothing*

(a)

(b)

15–1 (a) A system whose temperature is specified by the value of length L. (b) A system whose temperature is given by the value of the pressure p.

(a)

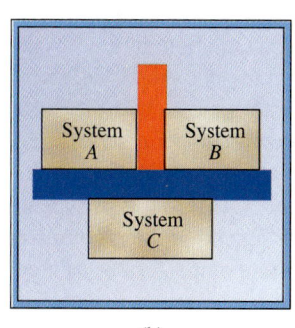

(b)

15–2 The zeroth law of thermodynamics. (a) If systems A and B are each in thermal equilibrium with system C, then (b) A and B are in thermal equilibrium with each other. Blue slabs represent insulating walls; orange slabs represent conducting walls.

happens; there are no additional changes to *A* or *B*. We conclude that **if *C* is initially in thermal equilibrium with both *A* and *B*, then *A* and *B* are also in thermal equilibrium with each other.** This result is called the **zeroth law of thermodynamics.** (The importance of this law was recognized only after the first, second, and third laws of thermodynamics had been named. Since it is fundamental to all of them, the name "zeroth" seemed appropriate.) It may seem trivial and obvious, but even so it needs to be verified by experiment.

Now suppose system *C* is a thermometer, such as the tube-and-liquid system of Fig. 15–1a. In Fig. 15–2a the thermometer *C* is in contact with both *A* and *B*. In thermal equilibrium, when the thermometer reading reaches a stable value, the thermometer measures the temperature of both *A* and *B*; hence *A* and *B* both have the *same* temperature. Experiment shows that thermal equilibrium isn't affected by adding or removing insulators, so the reading of thermometer *C* wouldn't change if it were in contact only with *A* or only with *B*. We conclude that **two systems are in thermal equilibrium if and only if they have the same temperature.** This is what makes a thermometer useful; a thermometer actually measures *its own* temperature, but when a thermometer is in thermal equilibrium with another body, the temperatures must be equal. When the temperatures of two systems are different, they *cannot* be in thermal equilibrium.

15–3 THERMOMETERS AND TEMPERATURE SCALES

To make the liquid-in-tube device shown in Fig. 15–1a into a useful thermometer, we need to mark a scale on the tube wall with numbers on it. These numbers are arbitrary, and historically many different schemes have been used. Suppose we label the thermometer's liquid level at the freezing temperature of pure water "zero" and the level at the boiling temperature "100," and divide the distance between these two points into 100 equal intervals called *degrees*. The result is the **Celsius temperature scale** (formerly called the *centigrade* scale in English-speaking countries). The Celsius temperature for a state colder than freezing water is a negative number. The Celsius scale is used, both in everyday life and in science and industry, almost everywhere in the world.

Another common type of thermometer uses a *bimetallic strip,* made by bonding strips of two different metals together (Fig. 15–3a). When the system gets hotter, one metal expands more than the other, so the composite strip bends when the temperature changes. This strip is usually formed into a spiral, with the outer end anchored to the thermometer case and the inner end attached to a pointer (Fig. 15–3b). The pointer rotates in response to temperature changes.

In a *resistance thermometer* the changing electrical resistance of a coil of fine wire, a carbon cylinder, or a germanium crystal is measured. Because resistance can be measured very precisely, resistance thermometers are usually more precise than most other types.

To measure very high temperatures, an *optical pyrometer* can be used. It measures the intensity of radiation emitted by a red-hot or white-hot substance. The instrument does not touch the hot substance, so the optical pyrometer can be used at temperatures that would destroy most other thermometers.

In the **Fahrenheit temperature scale,** still used in everyday life in the United States, the freezing temperature of water is 32°F (thirty-two degrees Fahrenheit) and the boiling temperature 212°F, both at standard atmospheric pressure. There are 180 degrees between freezing and boiling, compared to 100 on the Celsius scale, so one Fahrenheit degree represents only $\frac{100}{180}$, or $\frac{5}{9}$, as great a temperature change as one Celsius degree.

To convert temperatures from Celsius to Fahrenheit, note that a Celsius temperature T_C is the number of Celsius degrees above freezing; the number of Fahrenheit degrees above freezing is $\frac{9}{5}$ of this. But freezing on the Fahrenheit scale is at 32°F, so to obtain

(a)

(b)

15–3 (a) A bimetallic strip bends when heated because one metal (shown in red) expands more than the other. (b) A bimetallic strip, usually in the form of a spiral, may be used as a thermometer.

the actual Fahrenheit temperature T_F, multiply the Celsius value by $\frac{9}{5}$ and then add $32°$. Symbolically,

$$T_F = \frac{9}{5} T_C + 32° . \qquad (15-1)$$

To convert Fahrenheit to Celsius, solve this equation for T_C:

$$T_C = \frac{5}{9} (T_F - 32°) . \qquad (15-2)$$

In words, subtract $32°$ to get the number of Fahrenheit degrees above freezing, and then multiply by $\frac{5}{9}$ to obtain the number of Celsius degrees above freezing, that is, the Celsius temperature.

We don't recommend memorizing Eqs. (15–1) and (15–2). Instead, try to understand the reasoning that led to them so that you can derive them on the spot when you need them, checking your reasoning with the relation $100°C = 212°F$.

It is useful to distinguish between an actual temperature and a temperature *interval* (a difference or change in temperature). An actual temperature of $20°$ is stated as $20°C$ (twenty degrees Celsius), and a temperature *interval* of $10°$ is $10 \, C°$ (ten Celsius degrees). A beaker of water heated from $20°C$ to $30°C$ has a temperature change of $10 \, C°$.

15–4 GAS THERMOMETERS AND THE KELVIN SCALE

When we calibrate two thermometers, such as a liquid-in-tube system and a resistance thermometer, so that they agree at $0°C$ and $100°C$, they may not agree exactly at intermediate temperatures. Any temperature scale defined in this way always depends somewhat on the specific properties of the material used. Ideally, we would like to define a temperature scale that *doesn't* depend on the properties of a particular material. To establish a truly material-independent scale, we first need to develop some principles of thermodynamics. We'll return to this fundamental problem in Chapter 18. Here we'll discuss a thermometer that comes close to the ideal, the *gas thermometer*.

The principle of a gas thermometer is that the pressure of a gas at constant volume increases with temperature. A quantity of gas is placed in a constant-volume container (Fig. 15–4a), and its pressure is measured by one of the devices described in Section

(a)

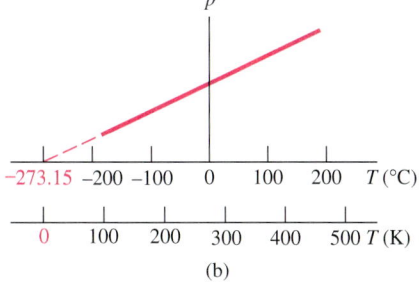

(b)

15–4 (a) A constant-volume gas thermometer. (b) Graph of absolute pressure versus temperature for a constant-volume low-density gas thermometer. A straight-line extrapolation predicts that the pressure would become zero at $-273.15°C$ if that temperature could be reached and the proportionality of pressure to temperature held exactly.

14–3. To calibrate a constant-volume gas thermometer, we measure the pressure at two temperatures, say 0°C and 100°C, plot these points on a graph, and draw a straight line between them (Fig. 15–4b). Then we can read from the graph the temperature corresponding to any other pressure.

By extrapolating this graph, we see that there is a hypothetical temperature, −273.15°C, at which the absolute pressure of the gas would become zero. We might expect that this temperature would be different for different gases, but it turns out to be the same for many different gases (at least in the limit of very low gas density). We can't actually observe this zero-pressure condition. Gases liquefy and solidify at very low temperatures, and the proportionality of pressure to temperature no longer holds.

We use this extrapolated zero-pressure temperature as the basis for a temperature scale with its zero at this temperature. This is the **Kelvin temperature scale,** named for the British physicist Lord Kelvin (1824–1907). The units are the same size as those on the Celsius scale, but the zero is shifted so that 0 K = −273.15°C and 273.15 K = 0°C; that is,

$$T_K = T_C + 273.15. \qquad (15\text{–}3)$$

This scale is shown in Fig. 15–4b. A common room temperature, 20°C (= 68°F), is 20 + 273.15, or about 293 K.

In SI nomenclature, "degree" is not used with the Kelvin scale; the temperature mentioned above is read "293 kelvins," not "degrees Kelvin." We capitalize Kelvin when it refers to the temperature scale; however, the *unit* of temperature is the *kelvin,* which is not capitalized (but is nonetheless abbreviated as a capital K).

Body temperature You place a small piece of melting ice in your mouth. Eventually, the water all converts from ice at $T_1 = 32.00°$F to body temperature, $T_2 = 98.60°$F. Express these temperatures as °C and K, and find $\Delta T = T_2 - T_1$ in both cases.

SOLUTION First we find the Celsius temperatures. We know that $T_1 = 32.00°$F = 0.00°C, and 98.60°F is 98.60 − 32.00 = 66.60 F° above freezing; multiply this by (5 C°/9 F°) to find 37.00 C° above freezing, or $T_2 = 37.00°$C.

To get the Kelvin temperatures, we just add 273.15 to each Celsius temperature: $T_1 = 273.15$ K and $T_2 = 310.15$ K. "Normal" body temperature is 37.0°C, but if your doctor says that your temperature is 310 K, don't be alarmed.

The temperature *difference* $\Delta T = T_2 - T_1$ is 37.00 C° = 37.00 K. The Celsius and Kelvin scales have different zero points but the same size degrees. Therefore any temperature difference is the *same* on the Celsius and Kelvin scales but not the same on the Fahrenheit scale.

The Celsius scale has two fixed points, the normal freezing and boiling temperatures of water. But we can define the Kelvin scale using a gas thermometer with only a single reference temperature. We define the ratio of any two temperatures T_1 and T_2 on the Kelvin scale as the ratio of the corresponding gas-thermometer pressures p_1 and p_2:

$$\frac{T_2}{T_1} = \frac{p_2}{p_1} \qquad \text{(constant-volume gas thermometer, } T \text{ in kelvins).} \qquad (15\text{–}4)$$

The pressure p is directly proportional to the Kelvin temperature, as shown in Fig. 15–4b. To complete the definition of T, we need only specify the Kelvin temperature of a single specific state. For reasons of precision and reproducibility the state chosen is the *triple point* of water. This is the unique combination of temperature and pressure at which solid water (ice), liquid water, and water vapor can all coexist. This occurs at a temperature of 0.01°C and a water-vapor pressure of 610 Pa (about 0.006 atm). (This is the pressure of the *water;* it has nothing to do directly with the gas pressure in the *thermometer.*) The triple-point temperature T_{triple} of water is *defined* to have the value $T_{\text{triple}} = 273.16$ K, corresponding to 0.01°C. From Eq. (15–4), if p_{triple} is the pressure in a gas thermometer at temperature T_{triple} and p is the pressure at some other temperature T,

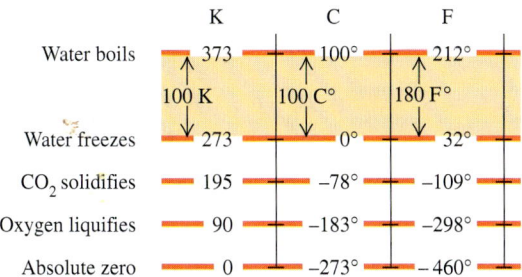

	K	C	F
Water boils	373	100°	212°
	100 K	100 C°	180 F°
Water freezes	273	0°	32°
CO_2 solidifies	195	−78°	−109°
Oxygen liquifies	90	−183°	−298°
Absolute zero	0	−273°	−460°

15–5 Relations among Kelvin, Celsius, and Fahrenheit temperature scales. Temperatures have been rounded off to the nearest degree.

then T is given on the Kelvin scale by

$$T = T_{triple} \frac{p}{p_{triple}} = (273.16 \text{ K}) \frac{p}{p_{triple}}. \qquad (15\text{–}5)$$

Low-pressure gas thermometers using various gases are found to agree very closely, but they are large, bulky, and very slow to come to thermal equilibrium. They are used principally to establish high-precision standards and to calibrate other thermometers.

EXAMPLE 15-2

Suppose a constant-volume gas thermometer has a pressure of 1.50×10^4 Pa at temperature T_{triple} and a pressure of 1.95×10^4 Pa at some unknown temperature T. What is T?

SOLUTION From Eq. (15–5) the Kelvin temperature is

$$T = (273.16 \text{ K}) \frac{1.95 \times 10^4 \text{ Pa}}{1.50 \times 10^4 \text{ Pa}} = 355 \text{ K } (= 82°\text{C}).$$

The relationships among the three temperature scales we have discussed are shown graphically in Fig. 15–5. The Kelvin scale is called an **absolute temperature scale,** and its zero point ($T = 0$ K $= -273.15°$C, the temperature at which $p = 0$ in Eq. (15–5)) is called **absolute zero.** At absolute zero a system of molecules (such as a quantity of a gas, a liquid, or a solid) has its *minimum* possible total energy (kinetic plus potential); because of quantum effects, however, it is *not* correct to say that all molecular motion ceases at absolute zero. To define more completely what we mean by absolute zero, we need to use the thermodynamic principles developed in the next several chapters. We will return to this concept in Chapter 18.

15–5 THERMAL EXPANSION

Most materials expand when their temperatures increase. The decks of bridges need special joints and supports to allow for expansion. A completely filled and tightly capped bottle of water cracks when it is heated, but you can loosen a metal jar lid by running hot water over it. These are all examples of *thermal expansion.*

LINEAR EXPANSION

Suppose a rod of material has a length L_0 at some initial temperature T_0. When the temperature changes by ΔT, the length changes by ΔL. Experiment shows that if ΔT is not too large (say, less than 100 C° or so), ΔL is *directly proportional* to ΔT. If two rods made of the same material have the same temperature change, but one is twice as long as the other, then the *change* in its length is also twice as great. Therefore ΔL must also be proportional to L_0. Introducing a proportionality constant α (which is different for different materials), we may express these relations in an equation:

$$\Delta L = \alpha L_0 \, \Delta T \qquad \text{(linear thermal expansion).} \qquad (15\text{–}6)$$

(a)

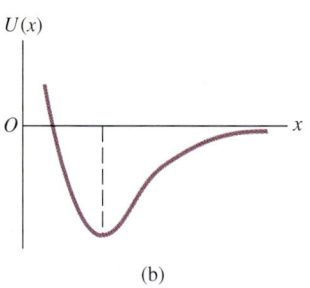

(b)

15–6 (a) We can visualize the forces between neighboring atoms in a solid by imagining them to be connected by springs that are easier to stretch than to compress. (b) A graph of potential energy versus distance between neighboring atoms shows that the forces are not symmetric. Compare this figure to Fig. 13–17b.

If a body has length L_0 at temperature T_0, then its length L at a temperature $T = T_0 + \Delta T$ is

$$L = L_0 + \Delta L = L_0 + \alpha L_0 \Delta T = L_0(1 + \alpha \Delta T). \qquad (15\text{–}7)$$

The constant α, which describes the thermal expansion properties of a particular material, is called the **coefficient of linear expansion.** The units of α are K^{-1} or $(C°)^{-1}$. (Remember that a temperature *interval* is the same in the Kelvin and Celsius scales.) For many materials, every linear dimension changes according to Eq. (15–6) or (15–7). Thus L could be the thickness of a rod, the side length of a square sheet, or the diameter of a hole. Some materials, such as wood or single crystals, expand differently in different directions. We won't consider this complication.

We can understand thermal expansion qualitatively on a molecular basis. Picture the interatomic forces in a solid as springs, as in Fig. 15–6 (we explored the analogy between spring forces and interatomic forces in Section 13–5). Each atom vibrates about its equilibrium position. When the temperature increases, the energy and amplitude of the vibration also increase. The interatomic spring forces are not symmetric about the equilibrium position; they usually behave like a spring that is easier to stretch than to compress. As a result, when the amplitude of vibration increases, the *average* distance between molecules also increases. As the atoms get farther apart, every dimension increases.

CAUTION ▶ If a solid object has a hole in it, what happens to the size of the hole when the temperature of the object increases? A common misconception is that if the object expands, the hole will shrink because material expands into the hole. But the truth of the matter is that if the object expands, the hole will expand too; as we stated above, *every* linear dimension of an object changes in the same way when the temperature changes. If you're not convinced, think of the atoms in Fig. 15–6a as outlining a cubical hole. When the object expands, the atoms move apart and the hole increases in size. ◀

The direct proportionality expressed by Eq. (15–6) is not exact; it is *approximately* correct only for sufficiently small temperature changes. For a given material, α varies somewhat with the initial temperature T_0 and the size of the temperature interval. We'll ignore this complication here, however. Average values of α for several materials are listed in Table 15–1. Within the precision of these values we don't need to worry whether T_0 is 0°C or 20°C or some other temperature. Note that typical values of α are very small; even for a temperature change of 100 C°, the fractional length change $\Delta L/L_0$ is only of the order of 1/1000 for the metals in the table.

VOLUME EXPANSION

Increasing temperature usually causes increases in *volume* for both solid and liquid materials. Just as with linear expansion, experiments show that if the temperature change ΔT is not too great (less than 100 C° or so), the increase in volume ΔV is approximately

TABLE 15–1

COEFFICIENTS OF LINEAR EXPANSION

MATERIAL	$\alpha\,[K^{-1}\ \text{or}\ (C°)^{-1}]$
Aluminum	2.4×10^{-5}
Brass	2.0×10^{-5}
Copper	1.7×10^{-5}
Glass	$0.4\text{–}0.9 \times 10^{-5}$
Invar (nickel-iron alloy)	0.09×10^{-5}
Quartz (fused)	0.04×10^{-5}
Steel	1.2×10^{-5}

TABLE 15–2

COEFFICIENTS OF VOLUME EXPANSION

SOLIDS	$\beta[\text{K}^{-1} \text{ or } (\text{C}°)^{-1}]$	LIQUIDS	$\beta[\text{K}^{-1} \text{ or } (\text{C}°)^{-1}]$
Aluminum	7.2×10^{-5}	Ethanol	75×10^{-5}
Brass	6.0×10^{-5}	Carbon disulfide	115×10^{-5}
Copper	5.1×10^{-5}	Glycerin	49×10^{-5}
Glass	$1.2–2.7 \times 10^{-5}$	Mercury	18×10^{-5}
Invar	0.27×10^{-5}		
Quartz (fused)	0.12×10^{-5}		
Steel	3.6×10^{-5}		

proportional to both the temperature change ΔT and the initial volume V_0:

$$\Delta V = \beta V_0 \Delta T \qquad \text{(volume thermal expansion)}. \qquad (15\text{–}8)$$

The constant β characterizes the volume expansion properties of a particular material; it is called the **coefficient of volume expansion.** The units of β are K^{-1} or $(\text{C}°)^{-1}$. As with linear expansion, β varies somewhat with temperature, and Eq. (15–8) is an approximate relation that is valid only for small temperature changes. For many substances, β decreases at low temperatures. Several values of β in the neighborhood of room temperature are listed in Table 15–2. Note that the values for liquids are generally much larger than those for solids.

For solid materials there is a simple relation between the volume expansion coefficient β and the linear expansion coefficient α. To derive this relation, we consider a cube of material with side length L and volume $V = L^3$. At the initial temperature the values are L_0 and V_0. When the temperature increases by dT, the side length increases by dL and the volume increases by an amount dV given by

$$dV = \frac{dV}{dL} dL = 3L^2 \, dL.$$

Now we replace L and V by the initial values L_0 and V_0. From Eq. (15–6), dL is

$$dL = \alpha L_0 \, dT;$$

since $V_0 = L_0^3$, this means that dV can also be expressed as

$$dV = 3L_0^2 \alpha L_0 \, dT = 3\alpha V_0 \, dT.$$

This is consistent with the infinitesimal form of Eq. (15–8), $dV = \beta V_0 \, dT$, only if

$$\beta = 3\alpha. \qquad (15\text{–}9)$$

You should check this relation for some of the materials listed in Tables 15–1 and 15–2.

When this SR-71 aircraft is sitting on the ground, its wing panels fit together so loosely that fuel leaks out of the wings onto the ground. But once it is in flight at over three times the speed of sound, air friction heats the panels so much that they expand to make a perfect fit.

Problem–Solving Strategy

THERMAL EXPANSION

1. Identify which quantities in Eq. (15–6) or (15–8) are known and which are unknown. Often you will be given two temperatures and will have to compute ΔT. Or you may be given an initial temperature T_0 and have to find a final temperature corresponding to a given length or volume change. In this case, find ΔT first; then the final temperature is $T_0 + \Delta T$.

2. Unit consistency is crucial, as always. L_0 and ΔL (or V_0 and ΔV) must have the same units, and if you use a value of α or β in K^{-1} or $(\text{C}°)^{-1}$, then ΔT must be in kelvins or Celsius degrees (C°). But you can use K and C° interchangeably.

3. Remember that the sizes of holes in a material expand with temperature just like any other linear dimension, and the volume of a hole (such as the volume of a container) expands in the same way as the corresponding solid shape.

EXAMPLE 15-3

Length change due to temperature change A surveyor uses a steel measuring tape that is exactly 50.000 m long at a temperature of 20°C. What is its length on a hot summer day when the temperature is 35°C?

SOLUTION We have $L_0 = 50.000$ m, $T_0 = 20°C$, $T = 35°C$, and $\Delta T = T - T_0 = 15$ C° = 15 K. From Eq. (15–6),

$$\Delta L = \alpha L_0 \, \Delta T = (1.2 \times 10^{-5} \text{ K}^{-1})(50 \text{ m})(15 \text{ K})$$
$$= 9.0 \times 10^{-3} \text{ m} = 9.0 \text{ mm,}$$
$$L = L_0 + \Delta L = 50.000 \text{ m} + 0.009 \text{ m} = 50.009 \text{ m.}$$

Thus the length at 35°C is 50.009 m. Note that L_0 is given to five significant figures but that we need only two of them to compute ΔL.

EXAMPLE 15-4

In Example 15–3 the surveyor uses the measuring tape to measure a distance when the temperature is 35°C; the value that she reads off the tape is 35.794 m. What is the actual distance?

SOLUTION As we saw in Example 15–3, at 35°C the tape has expanded slightly; the distance between two successive meter marks on the tape is a little more than a meter, in the ratio (50.009 m)/(50.000 m). Hence the true distance is greater than the value read off the tape by this same factor:

$$\frac{50.009 \text{ m}}{50.000 \text{ m}} (35.794 \text{ m}) = 35.800 \text{ m.}$$

EXAMPLE 15-5

Volume change due to temperature change A glass flask with volume 200 cm³ is filled to the brim with mercury at 20°C. How much mercury overflows when the temperature of the system is raised to 100°C? The coefficient of linear expansion of the glass is 0.40×10^{-5} K⁻¹.

SOLUTION Mercury overflows because β is much larger for mercury than for glass. From Table 15–2 the coefficient of volume expansion for mercury is $\beta_{\text{mercury}} = 18 \times 10^{-5}$ K⁻¹; from Eq. (15–9) the coefficient of volume expansion for the glass is

$$\beta_{\text{glass}} = 3\alpha_{\text{glass}} = 3(0.40 \times 10^{-5} \text{ K}^{-1}) = 1.2 \times 10^{-5} \text{ K}^{-1}.$$

The increase in volume of the glass flask is

$$\Delta V_{\text{glass}} = \beta_{\text{glass}} V_0 \Delta T$$
$$= (1.2 \times 10^{-5} \text{ K}^{-1})(200 \text{ cm}^3)(100°C - 20°C) = 0.19 \text{ cm}^3.$$

The increase in volume of the mercury is

$$\Delta V_{\text{mercury}} = \beta_{\text{mercury}} V_0 \Delta T$$
$$= (18 \times 10^{-5} \text{ K}^{-1})(200 \text{ cm}^3)(100°C - 20°C) = 2.9 \text{ cm}^3.$$

The volume of mercury that overflows is

$$\Delta V_{\text{mercury}} - \Delta V_{\text{glass}} = 2.9 \text{ cm}^3 - 0.19 \text{ cm}^3 = 2.7 \text{ cm}^3.$$

This is basically how a mercury-in-glass thermometer works, except that instead of letting the mercury overflow and run all over the place, the thermometer has it rise inside a sealed tube as T increases.

As Tables 15–1 and 15–2 show, glass has smaller coefficients of expansion α and β than do most metals. This is why you can use hot water to loosen a metal lid on a glass jar; the metal expands more than the glass does.

THERMAL EXPANSION OF WATER

Water, in the temperature range from 0°C to 4°C, *decreases* in volume with increasing temperature. In this range its coefficient of volume expansion is *negative*. Above 4°C, water expands when heated (Fig. 15–7). Hence water has its greatest density at 4°C. Water also expands when it freezes, which is why ice humps up in the middle of the compartments in an ice cube tray. By contrast, most materials contract when they freeze.

This anomalous behavior of water has an important effect on plant and animal life in lakes. A lake cools from the surface down; above 4°C, the cooled water at the surface flows to the bottom because of its greater density. But when the surface temperature drops below 4°C, the water near the surface is less dense than the warmer water below. Hence the downward flow ceases, and the water near the surface remains colder than that at the bottom. As the surface freezes, the ice floats because it is less dense than water.

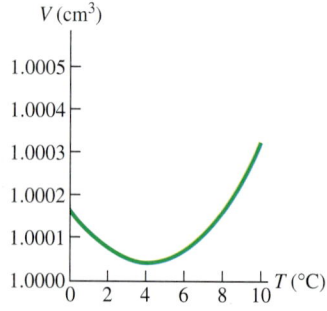

15-7 The volume of one gram of water in the temperature range from 0°C to 10°C. By 100°C the volume has increased to 1.034 cm³. If the coefficient of volume expansion were constant, the curve would be a straight line.

The water at the bottom remains at 4°C until nearly the entire lake is frozen. If water behaved like most substances, contracting continuously on cooling and freezing, lakes would freeze from the bottom up. Circulation due to density differences would continuously carry warmer water to the surface for efficient cooling, and lakes would freeze solid much more easily. This would destroy all plant and animal life that cannot withstand freezing. If water did not have this special property, the evolution of life would have taken a very different course.

THERMAL STRESS

If we clamp the ends of a rod rigidly to prevent expansion or contraction and then change the temperature, tensile or compressive stresses called **thermal stresses** develop. The rod would like to expand or contract, but the clamps won't let it. The resulting stresses may become large enough to strain the rod irreversibly or even break it. Concrete highways and bridge decks usually have gaps between sections, filled with a flexible material or bridged by interlocking teeth (Fig. 15–8), to permit expansion and contraction of the concrete. Long steam pipes have expansion joints or U-shaped sections to prevent buckling or stretching with temperature changes. If one end of a steel bridge is rigidly fastened to its abutment, the other end usually rests on rollers.

15–8 The interlocking teeth of an expansion joint on a bridge. These joints are needed to accommodate changes in length that result from thermal expansion.

To calculate the thermal stress in a clamped rod, we compute the amount the rod *would* expand (or contract) if not held and then find the stress needed to compress (or stretch) it back to its original length. Suppose that a rod with length L_0 and cross-section area A is held at constant length while the temperature is reduced (negative ΔT), causing a tensile stress. The fractional change in length if the rod were free to contract would be

$$\left(\frac{\Delta L}{L_0}\right)_{\text{thermal}} = \alpha \, \Delta T. \qquad (15\text{--}10)$$

Both ΔL and ΔT are negative. The tension must increase by an amount F that is just enough to produce an equal and opposite fractional change in length $(\Delta L/L_0)_{\text{tension}}$. From the definition of Young's modulus, Eq. (11–10),

$$Y = \frac{F/A}{\Delta L/L_0}, \qquad \text{so} \qquad \left(\frac{\Delta L}{L_0}\right)_{\text{tension}} = \frac{F}{AY}. \qquad (15\text{--}11)$$

If the length is to be constant, the *total* fractional change in length must be zero. From Eqs. (15–10) and (15–11), this means that

$$\left(\frac{\Delta L}{L_0}\right)_{\text{thermal}} + \left(\frac{\Delta L}{L_0}\right)_{\text{tension}} = \alpha \, \Delta T + \frac{F}{AY} = 0.$$

Solving for the tensile stress F/A required to keep the rod's length constant, we find

$$\frac{F}{A} = -Y\alpha \, \Delta T \qquad \text{(thermal stress)}. \qquad (15\text{--}12)$$

For a decrease in temperature, ΔT is negative, so F and F/A are positive; this means that a *tensile* force and stress are needed to maintain the length. If ΔT is positive, F and F/A are negative, and the required force and stress are *compressive*.

If there are temperature differences within a body, non-uniform expansion or contraction will result and thermal stresses can be induced. You can break a glass bowl by pouring very hot water into it; the thermal stress between the hot and cold parts of the bowl exceeds the breaking stress of the glass, causing cracks. The same phenomenon makes ice cubes crack when dropped into warm water. Heat-resistant glasses such as Pyrex™ have exceptionally low expansion coefficients and high strength.

EXAMPLE 15-6

An aluminum cylinder 10 cm long, with a cross-section area of 20 cm², is to be used as a spacer between two steel walls. At 17.2°C it just slips in between the walls. When it warms to 22.3°C, calculate the stress in the cylinder and the total force it exerts on each wall, assuming that the walls are perfectly rigid and a constant distance apart.

SOLUTION Equation (15–12) relates the stress to the temperature change. From Table 11–1 we find $Y = 7.0 \times 10^{10}$ Pa, and from Table 15–1, $\alpha = 2.4 \times 10^{-5}$ K⁻¹. The temperature change is $\Delta T = 22.3°C - 17.2°C = 5.1$ C° $= 5.1$ K. The stress is F/A; from Eq. (15–12),

$$\frac{F}{A} = -Y\alpha \, \Delta T = -(0.70 \times 10^{11} \text{ Pa})(2.4 \times 10^{-5} \text{ K}^{-1})(5.1 \text{ K})$$

$$= -8.6 \times 10^{6} \text{ Pa (or } -1200 \text{ lb/in.}^{2}).$$

The negative sign indicates that compressive rather than tensile stress is needed to keep the cylinder's length constant. This stress is independent of the length and cross-section area of the cylinder. The total force F is the cross-section area times the stress:

$$F = A\left(\frac{F}{A}\right) = (20 \times 10^{-4} \text{ m}^2)(-8.6 \times 10^6 \text{ Pa})$$

$$= -1.7 \times 10^4 \text{ N},$$

or nearly two tons. The negative sign indicates compression.

15–6 QUANTITY OF HEAT

When you put a cold spoon into a cup of hot coffee, the spoon warms up and the coffee cools down as they approach thermal equilibrium. The interaction that causes these temperature changes is fundamentally a transfer of *energy* from one substance to another. Energy transfer that takes place solely because of a temperature difference is called *heat flow* or *heat transfer,* and energy transferred in this way is called **heat.**

An understanding of the relation between heat and other forms of energy emerged gradually during the eighteenth and nineteenth centuries. Sir James Joule (1818–1889) studied how water can be warmed by vigorous stirring with a paddle wheel (Fig. 15–9a). The paddle wheel adds energy to the water by doing *work* on it, and Joule found that *the temperature rise is directly proportional to the amount of work done.* The same temperature change can also be caused by putting the water in contact with some hotter body (Fig. 15–9b); hence this interaction must also involve an energy exchange. We will explore the relation between heat and mechanical energy in greater detail in Chapters 17 and 18.

(a)

(b)

15–9 The same temperature change of the same system may be accomplished by (a) doing work on it or (b) adding heat to it.

CAUTION ▶ It is absolutely essential for you to keep clearly in mind the distinction between *temperature* and *heat.* Temperature depends on the physical state of a material and is a quantitative description of its hotness or coldness. In physics the term "heat" always refers to energy in transit from one body or system to another because of a temperature difference, never to the amount of energy contained within a particular system. We can change the temperature of a body by adding heat to it or taking heat away, or by adding or subtracting energy in other ways such as mechanical work (Fig. 15–9a). If we cut a body in half, each half has the same temperature as the whole; but to raise the temperature of each half by a given interval, we add *half* as much heat as for the whole. ◀

We can define a *unit* of quantity of heat based on temperature changes of some specific material. The **calorie** (abbreviated cal) is defined as *the amount of heat required to raise the temperature of one gram of water from* 14.5°C *to* 15.5°C. The kilocalorie (kcal), equal to 1000 cal, is also used; a food-value calorie is actually a kilocalorie. A corresponding unit of heat using Fahrenheit degrees and British units is the **British thermal unit,** or Btu. One Btu is the quantity of heat required to raise the temperature of one pound (weight) of water 1 F° from 63°F to 64°F.

Because heat is energy in transit, there must be a definite relation between these units and the familiar mechanical energy units such as the joule. Experiments similar in concept to Joule's have shown that

$$1 \text{ cal} = 4.186 \text{ J},$$

$$1 \text{ kcal} = 1000 \text{ cal} = 4186 \text{ J},$$

$$1 \text{ Btu} = 778 \text{ ft} \cdot \text{lb} = 252 \text{ cal} = 1055 \text{ J}.$$

The calorie is not a fundamental SI unit. The International Committee on Weights and Measures recommends using the joule as the basic unit of energy in all forms, including heat. We will follow that recommendation in this book.

SPECIFIC HEAT CAPACITY

We use the symbol Q for quantity of heat. When it is associated with an infinitesimal temperature change dT, we call it dQ. The quantity of heat Q required to increase the temperature of a mass m of a certain material from T_1 to T_2 is found to be approximately proportional to the temperature change $\Delta T = T_2 - T_1$. It is also proportional to the mass m of material. When you're heating water to make tea, you need twice as much heat for two cups as for one if the temperature interval is the same. The quantity of heat needed also depends on the nature of the material; raising the temperature of one kilogram of water by 1 C° requires 4190 J of heat, but only 910 J are needed to raise the temperature of one kilogram of aluminum by 1 C°.

Putting all these relationships together, we have

$$Q = mc\,\Delta T \qquad \text{(heat required for temperature change of mass } m\text{)}, \qquad (15\text{–}13)$$

where c is a quantity, different for different materials, called the **specific heat capacity** (or sometimes *specific heat*) of the material. For an infinitesimal temperature change dT and corresponding quantity of heat dQ,

$$dQ = mc\,dT, \qquad (15\text{–}14)$$

$$c = \frac{1}{m}\frac{dQ}{dT} \qquad \text{(specific heat capacity)}. \qquad (15\text{–}15)$$

In Eqs. (15–13), (15–14), and (15–15), Q (or dQ) and ΔT (or dT) can be either positive or negative. When they are positive, heat enters the body and its temperature increases; when they are negative, heat leaves the body and its temperature decreases.

CAUTION▶ Remember that dQ does not represent a change in the amount of heat *contained* in a body; this is a meaningless concept. Heat is always energy *in transit* as a result of a temperature difference. There is no such thing as "the amount of heat in a body." The term *specific heat capacity* is unfortunate because it tends to suggest the erroneous idea that a body *contains* a certain amount of heat. ◀

The specific heat capacity of water is approximately

$$4190 \text{ J/kg} \cdot \text{K}, \qquad 1 \text{ cal/g} \cdot \text{C}°, \qquad \text{or } 1 \text{ Btu/lb} \cdot \text{F}°.$$

The specific heat capacity of a material always depends somewhat on the initial temperature and the temperature interval. Figure 15–10 shows this variation for water. In the problems and examples in this chapter we will usually ignore this small variation.

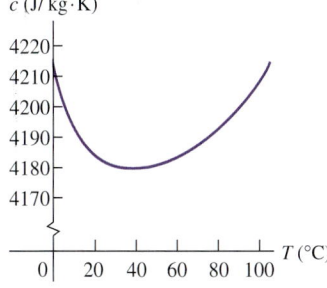

15–10 Specific heat capacity of water as a function of temperature. The value of c varies by less than 1% between 0°C and 100°C.

EXAMPLE 15–7

Feed a cold, starve a fever During a bout with the flu an 80-kg man ran a fever of 2.0 C° above normal, that is, a body temperature of 39.0°C (102.2°F) instead of the normal 37.0°C (98.6°F). Assuming that the human body is mostly water, how much heat is required to raise his temperature by that amount?

SOLUTION The temperature change is $\Delta T = 39.0°\text{C} - 37.0°\text{C}$ $= 2.0 \text{ C}° = 2.0 \text{ K}$. From Eq. (15–13),

$$Q = mc\,\Delta T = (80 \text{ kg})(4190 \text{ J/kg} \cdot \text{K})(2.0 \text{ K}) = 6.7 \times 10^5 \text{ J}.$$

This corresponds to 160 kcal, or 160 food-value calories. In fact the specific heat capacity of the human body is more nearly equal to 3480 J/kg · K, about 83% that of water. The difference is due to the presence of protein, fat, and minerals, which have lower specific heat capacities. With this value of c, the required heat is 5.6×10^5 J = 133 kcal.

EXAMPLE 15–8

Overheating electronics You are designing an electronic circuit element made of 23 mg of silicon. The electric current through it adds energy at the rate of 7.4 mW = 7.4×10^{-3} J/s. If your design doesn't allow any heat transfer out of the element, at what rate does its temperature increase? The specific heat capacity of silicon is 705 J/kg · K.

SOLUTION In one second, $Q = (7.4 \times 10^{-3} \text{ J/s})(1 \text{ s}) = 7.4 \times 10^{-3}$ J. From Eq. (15–13), $Q = mc \, \Delta T$, the temperature change in one second is

$$\Delta T = \frac{Q}{mc} = \frac{7.4 \times 10^{-3} \text{ J}}{(23 \times 10^{-6} \text{ kg})(705 \text{ J/kg} \cdot \text{K})} = 0.46 \text{ K}.$$

Alternatively, we can divide both sides of Eq. (15–14) by dt and rearrange:

$$\frac{dT}{dt} = \frac{dQ/dt}{mc}$$

$$= \frac{7.4 \times 10^{-3} \text{ J/s}}{(23 \times 10^{-6} \text{ kg})(705 \text{ J/kg} \cdot \text{K})} = 0.46 \text{ K/s}.$$

At this rate of temperature rise (27 K every minute) the circuit element would soon self-destruct. Heat transfer is an important design consideration in electronic circuit elements (see Section 15–9).

MOLAR HEAT CAPACITY

Sometimes it's more convenient to describe a quantity of substance in terms of the number of *moles n* rather than the *mass m* of material. Recall from your study of chemistry that a mole of any pure substance always contains the same number of molecules. (We will discuss this point in more detail in Chapter 16.) The *molar mass* of any substance, denoted by M, is the mass per mole. (The quantity M is sometimes called *molecular weight,* but *molar mass* is preferable; the quantity depends on the mass of a molecule, not its weight.) For example, the molar mass of water is 18.0 g/mol = 18.0×10^{-3} kg/mol; one mole of water has a mass of 18.0 g = 0.0180 kg. The total mass m of material is equal to the mass per mole M times the number of moles n:

$$m = nM. \tag{15–16}$$

Replacing the mass m in Eq. (15–13) by the product nM, we find

$$Q = nMc \, \Delta T. \tag{15–17}$$

The product Mc is called the **molar heat capacity** (or *molar specific heat*) and is denoted by C (capitalized). With this notation we rewrite Eq. (15–17) as

$$Q = nC \, \Delta T \qquad \text{(heat required for temperature change of n moles).} \tag{15–18}$$

Comparing to Eq. (15–15), we can express the molar heat capacity C (heat per mole per temperature change) in terms of the specific heat capacity c (heat per mass per temperature change) and the molar mass M (mass per mole):

$$C = \frac{1}{n} \frac{dQ}{dT} = Mc \qquad \text{(molar heat capacity).} \tag{15–19}$$

For example, the molar heat capacity of water is

$$C = Mc = (0.0180 \text{ kg/mol})(4190 \text{ J/kg} \cdot \text{K}) = 75.4 \text{ J/mol} \cdot \text{K}.$$

Values of specific and molar heat capacities for several substances are given in Table 15–3.

Precise measurements of heat capacities require great experimental skill. Usually, a measured quantity of energy is supplied by an electric current in a heater wire wound around the specimen. The temperature change ΔT is measured with a resistance thermometer or thermocouple embedded in the specimen. This sounds simple, but great care is needed to avoid or compensate for unwanted heat transfer between the sample and its surroundings. Measurements for solid materials are usually made at constant atmo-

TABLE 15–3
APPROXIMATE SPECIFIC AND MOLAR HEAT CAPACITIES (CONSTANT PRESSURE)

SUBSTANCE	SPECIFIC HEAT CAPACITY, c $(J/kg \cdot K)$	M (kg/mol)	MOLAR HEAT CAPACITY, C $(J/mol \cdot K)$
Aluminum	910	0.0270	24.6
Beryllium	1970	0.00901	17.7
Copper	390	0.0635	24.8
Ethanol	2428	0.0461	111.9
Ethylene glycol	2386	0.0620	148.0
Ice (near 0°C)	2100	0.0180	37.8
Iron	470	0.0559	26.3
Lead	130	0.207	26.9
Marble ($CaCO_3$)	879	0.100	87.9
Mercury	138	0.201	27.7
Salt (NaCl)	879	0.0585	51.4
Silver	234	0.108	25.3
Water (liquid)	4190	0.0180	75.4

Water has a much higher specific heat capacity than the glass or metals used to make cookware. This helps explain why it takes several minutes to boil water on a stove, even though the pot or kettle reaches a high temperature very quickly.

spheric pressure; the corresponding values are called *heat capacities at constant pressure,* denoted by c_p and C_p. For a gas it is usually easier to keep the substance in a container with constant *volume;* the corresponding values are called *heat capacities at constant volume,* denoted by c_V and C_V. For a given substance, C_V and C_p are different. If the system can expand while heat is added, there is additional energy exchange through the performance of *work* by the system on its surroundings. If the volume is constant, the system does no work. For gases the difference between C_p and C_V is substantial. We will study heat capacities of gases in detail in Section 17–8.

The last column of Table 15–3 shows something interesting. The molar heat capacities for most elemental solids are about the same, about 25 J/mol · K. This correlation, named the *rule of Dulong and Petit* (for its discoverers), forms the basis for a very important idea. The number of atoms in one mole is the same for all elemental substances. This means that on a *per atom* basis, about the same amount of heat is required to raise the temperature of each of these elements by a given amount, even though the *masses* of the atoms are very different. The heat required for a given temperature increase depends only on *how many* atoms the sample contains, not on the mass of an individual atom. We will see the reason why the rule of Dulong and Petit works so well when we study the molecular basis of heat capacities in greater detail in Chapter 16.

15–7 CALORIMETRY AND PHASE CHANGES

Calorimetry means "measuring heat." We have discussed energy transfer (heat) involved in temperature changes. Heat is also involved in *phase changes,* such as the melting of ice or boiling of water. Once we understand these additional heat relationships, we can analyze a variety of problems involving quantity of heat.

PHASE CHANGES

We use the term **phase** to describe a specific state of matter, such as a solid, liquid, or gas. The compound H_2O exists in the *solid phase* as ice, in the *liquid phase* as water, and in the *gaseous phase* as steam. A transition from one phase to another is called a **phase change** or *phase transition.* For any given pressure a phase change takes place at a definite temperature, usually accompanied by absorption or emission of heat and a change of volume and density.

A familiar example of a phase change is the melting of ice. When we add heat to ice at 0°C and normal atmospheric pressure, the temperature of the ice *does not* increase. Instead, some of it melts to form liquid water. If we add the heat slowly, to maintain the system very close to thermal equilibrium, the temperature remains at 0°C until all the ice is melted. The effect of adding heat to this system is not to raise its temperature but to change its *phase* from solid to liquid.

To change 1 kg of ice at 0°C to 1 kg of liquid water at 0°C and normal atmospheric pressure requires 3.34×10^5 J of heat. The heat required per unit mass is called the **heat of fusion** (or sometimes *latent heat of fusion*), denoted by L_f. For water at normal atmospheric pressure the heat of fusion is

$$L_f = 3.34 \times 10^5 \text{ J/kg} = 79.6 \text{ cal/g} = 143 \text{ Btu/lb}.$$

More generally, to melt a mass m of material that has a heat of fusion L_f requires a quantity of heat Q given by

$$Q = mL_f.$$

This process is *reversible*. To freeze liquid water to ice at 0°C, we have to *remove* heat; the magnitude is the same, but in this case, Q is negative because heat is removed rather than added. To cover both possibilities and to include other kinds of phase changes, we write

$$Q = \pm mL \qquad \text{(heat transfer in a phase change)}. \tag{15–20}$$

The plus sign (heat entering) is used when the material melts; the minus sign (heat leaving) is used when it freezes. The heat of fusion is different for different materials, and it also varies somewhat with pressure.

For any given material at any given pressure the freezing temperature is the same as the melting temperature. At this unique temperature the liquid and solid phases (liquid water and ice, for example) can coexist in a condition called **phase equilibrium.**

We can go through this whole story again for *boiling* or *evaporation,* a phase transition between liquid and gaseous phases. The corresponding heat (per unit mass) is called the **heat of vaporization** L_v. At normal atmospheric pressure the heat of vaporization L_v for water is

$$L_v = 2.256 \times 10^6 \text{ J/kg} = 539 \text{ cal/g} = 970 \text{ Btu/lb}.$$

That is, it takes 2.256×10^6 J to change 1 kg of liquid water at 100°C to 1 kg of water vapor at 100°C. By comparison, to raise the temperature of a kilogram of water from 0°C to 100°C requires $Q = mc\Delta T = (1.00 \text{ kg})(4190 \text{ J/kg} \cdot \text{C}°)(100 \text{ C}°) = 4.19 \times 10^5$ J, less than one-fifth as much heat as is required for vaporization at 100°C. This agrees with everyday kitchen experience; a pot of water may reach boiling temperature in a few minutes, but it takes a much longer time to completely evaporate all the water away.

Like melting, boiling is a reversible transition. When heat is removed from a gas at the boiling temperature, the gas returns to the liquid phase, or *condenses,* giving up to its surroundings the same quantity of heat (heat of vaporization) that was needed to vaporize it. At a given pressure the boiling and condensation temperatures are always the same; at this temperature the liquid and gaseous phases can coexist in phase equilibrium.

Both L_v and the boiling temperature of a material depend on pressure. Water boils at a lower temperature (about 95°C) in Denver than in Pittsburgh because Denver is at higher elevation and the average atmospheric pressure is less. The heat of vaporization is somewhat greater at this lower pressure, about 2.27×10^6 J/kg.

Table 15–4 lists heats of fusion and vaporization for several materials and their melting and boiling temperatures at normal atmospheric pressure. Very few *elements* have melting temperatures in the vicinity of ordinary room temperatures; one of the few is the metal gallium, shown in Fig. 15–11.

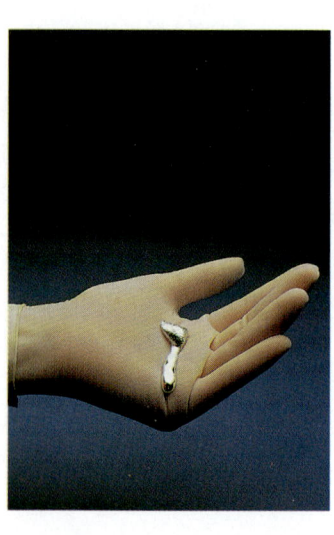

15–11 The metal gallium, shown here melting in a person's hand, is one of the few elements that melt in the vicinity of room temperature. Its melting temperature is 29.8°C, and its heat of fusion is 8.04×10^4 J/kg.

TABLE 15–4
HEATS OF FUSION AND VAPORIZATION

SUBSTANCE	NORMAL MELTING POINT K	NORMAL MELTING POINT °C	HEAT OF FUSION, L_f (J/kg)	NORMAL BOILING POINT K	NORMAL BOILING POINT °C	HEAT OF VAPORIZATION, L_v (J/kg)
Helium	*	*	*	4.216	−268.93	20.9×10^3
Hydrogen	13.84	−259.31	58.6×10^3	20.26	−252.89	452×10^3
Nitrogen	63.18	−209.97	25.5×10^3	77.34	−195.8	201×10^3
Oxygen	54.36	−218.79	13.8×10^3	90.18	−183.0	213×10^3
Ethanol	159	−114	104.2×10^3	351	78	854×10^3
Mercury	234	−39	11.8×10^3	630	357	272×10^3
Water	273.15	0.00	334×10^3	373.15	100.00	2256×10^3
Sulfur	392	119	38.1×10^3	717.75	444.60	326×10^3
Lead	600.5	327.3	24.5×10^3	2023	1750	871×10^3
Antimony	903.65	630.50	165×10^3	1713	1440	561×10^3
Silver	1233.95	960.80	88.3×10^3	2466	2193	2336×10^3
Gold	1336.15	1063.00	64.5×10^3	2933	2660	1578×10^3
Copper	1356	1083	134×10^3	1460	1187	5069×10^3

*A pressure in excess of 25 atmospheres is required to make helium solidify. At 1 atmosphere pressure, helium remains a liquid down to absolute zero.

Figure 15–12 shows how the temperature varies when we add heat continuously to a specimen of ice with an initial temperature below 0°C (point *a*). The temperature rises until we reach the melting point (point *b*). Then as more heat is added, the temperature remains constant until all the ice has melted (point *c*). Then the temperature starts to rise again until the boiling temperature is reached (point *d*). At that point the temperature again is constant until all the water is transformed into the vapor phase (point *e*). If the rate of heat input is constant, the slope of the line for the solid phase (ice) has a steeper slope than does the line for the liquid phase (water). Do you see why? (See Table 15–3.)

A substance can sometimes change directly from the solid to the gaseous phase. This process is called *sublimation,* and the solid is said to *sublime*. The corresponding heat is called the *heat of sublimation, L_s*. Liquid carbon dioxide cannot exist at a pressure lower than about 5×10^5 Pa (about 5 atm), and "dry ice" (solid carbon dioxide) sublimes at atmospheric pressure. Sublimation of water from frozen food causes freezer burn. The reverse process, a phase change from gas to solid, occurs when frost forms on cold bodies such as refrigerator cooling coils.

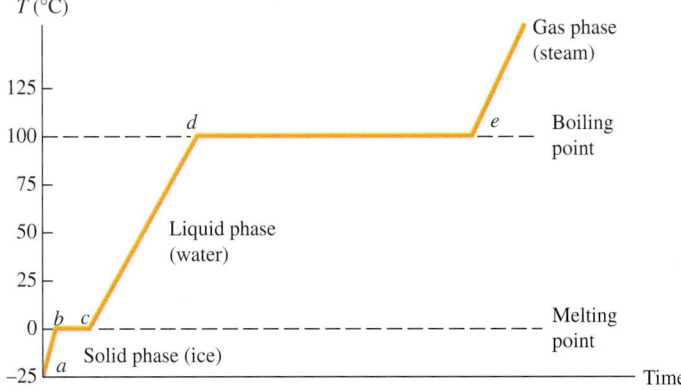

15–12 Graph of temperature versus time for a specimen of water initially in the solid phase (ice). Heat is added to the specimen at a constant rate. The temperature remains constant during each change of phase, provided that the pressure remains constant.

Very pure water can be cooled several degrees below the freezing temperature without freezing; the resulting unstable state is described as *supercooled*. When a small ice crystal is dropped in or the water is agitated, it crystallizes within a second or less. Supercooled water *vapor* condenses quickly into fog droplets when a disturbance, such as dust particles or ionizing radiation, is introduced. This principle is used in "seeding" clouds, which often contain supercooled water vapor, to cause condensation and rain.

A liquid can sometimes be *superheated* above its normal boiling temperature. Again, any small disturbance such as agitation or the passage of a charged particle through the material causes local boiling with bubble formation.

Steam heating systems for buildings use a boiling-condensing process to transfer heat from the furnace to the radiators. Each kilogram of water that is turned to steam in the boiler absorbs over 2×10^6 J (the heat of vaporization L_v of water) from the boiler and gives it up when it condenses in the radiators. Boiling-condensing processes are also used in refrigerators, air conditioners, and heat pumps. We will discuss these systems in Chapter 18.

The temperature-control mechanisms of many warm-blooded animals make use of heat of vaporization, removing heat from the body by using it to evaporate water from the tongue (panting) or from the skin (sweating). Evaporative cooling enables humans to maintain normal body temperature in hot, dry desert climates where the air temperature may reach 55°C (about 130°F). The skin temperature may be as much as 30°C cooler than the surrounding air. Under these conditions a normal person may perspire several liters per day. Unless this lost water is replaced, dehydration, heat stroke, and death result. Old-time desert rats (such as one of the authors) state that in the desert, any canteen that holds less than a gallon should be viewed as a toy!

Evaporative cooling is also used to cool buildings in hot, dry climates and to condense and recirculate "used" steam in coal-fired or nuclear-powered electric-generating-plants. That's what is going on in the large, tapered concrete towers that you see at such plants.

Chemical reactions such as combustion are analogous to phase changes in that they involve definite quantities of heat. Complete combustion of one gram of gasoline produces about 46,000 J or about 11,000 cal, so the **heat of combustion** L_c of gasoline is

$$L_c = 46,000 \text{ J/g} = 4.6 \times 10^7 \text{ J/kg}.$$

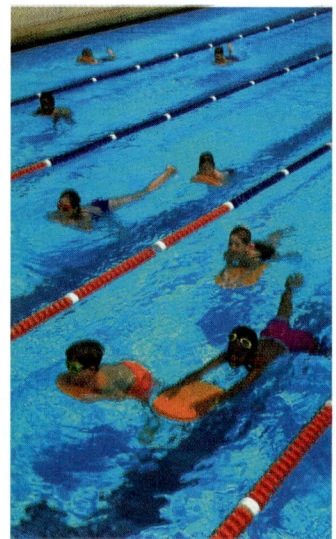

The water may be warm and it may be a hot day, but these children will feel cold when they first step out of the swimming pool. That's because as water evaporates from their skin, it removes the heat of vaporization from their bodies. To stay warm, they will need to dry off immediately.

Energy values of foods are defined similarly; the unit of food energy, although called a calorie, is a *kilocalorie,* equal to 1000 cal or 4186 J. When we say that a gram of peanut butter "contains 6 calories," we mean that 6 kcal of heat (6,000 cal or 25,000 J) is released when the carbon and hydrogen atoms in the peanut butter react with oxygen (with the help of enzymes) and are completely converted to CO_2 and H_2O. Not all of this energy is directly useful for mechanical work. We will study the *efficiency* of energy utilization in Chapter 18.

HEAT CALCULATIONS

Let's look at some examples of calorimetry calculations (calculations with heat). The basic principle is very simple: when heat flow occurs between two bodies that are isolated from their surroundings, the amount of heat lost by one body must equal the amount gained by the other. Heat is energy in transit, so this principle is really just conservation of energy. We take each quantity of heat *added to* a body as *positive* and each quantity *leaving* a body as *negative*. When several bodies interact, the *algebraic sum* of the quantities of heat transferred to all the bodies must be zero. Calorimetry, dealing entirely with one conserved quantity, is in many ways the simplest of all physical theories!

Problem-Solving Strategy

CALORIMETRY PROBLEMS

1. To avoid confusion about algebraic signs when calculating quantities of heat, use Eqs. (15–13) and (15–20) consistently for each body, noting that each Q is positive when heat enters a body and negative when it leaves. The algebraic sum of all the Q's must be zero.

2. Often you will need to find an unknown temperature. Represent it by an algebraic symbol such as T. Then if a body has an initial temperature of 20°C and an unknown final temperature T, the temperature change for the body is $\Delta T = T_{final} - T_{initial} = T - 20°C$ (not $20°C - T$).

3. In problems in which a phase change takes place, as when ice melts, you may not know in advance whether *all* the material undergoes a phase change or only part of it. You can always assume one or the other, and if the resulting calculation gives an absurd result (such as a final temperature that is higher or lower than *all* of the initial temperatures), you know that the initial assumption was wrong. Back up and try again!

EXAMPLE 15-9

A temperature change with no phase change A geologist working in the field drinks her morning coffee out of an aluminum cup. The cup has a mass of 0.120 kg and is initially at 20.0°C when she pours in 0.300 kg of coffee initially at 70.0°C. What is the final temperature after the coffee and the cup attain thermal equilibrium? (Assume that coffee has the same specific heat capacity as water and that there is no heat exchange with the surroundings.)

SOLUTION No phase changes occur in this situation, so we need use only Eq. (15–13). Call the final temperature T. By using Table 15–3, the (negative) heat gained by the coffee is

$$Q_{coffee} = m_{coffee} c_{water}\, \Delta T_{coffee}$$
$$= (0.300\ \text{kg})(4190\ \text{J/kg} \cdot \text{K})(T - 70.0°C).$$

The (positive) heat gained by the aluminum cup is

$$Q_{aluminum} = m_{aluminum} c_{aluminum}\, \Delta T_{aluminum}$$
$$= (0.120\ \text{kg})(910\ \text{J/kg} \cdot \text{K})(T - 20.0°C).$$

We equate the sum of these two quantities of heat to zero, obtaining an algebraic equation for T:

$$Q_{coffee} + Q_{aluminum} = 0, \quad \text{or}$$
$$(0.300\ \text{kg})(4190\ \text{J/kg} \cdot \text{K})(T - 70.0°C)$$
$$+ (0.120\ \text{kg})(910\ \text{J/kg} \cdot \text{K})(T - 20.0°C) = 0.$$

Solution of this equation gives $T = 66.0°C$. The final temperature is much closer to the initial temperature of the coffee than to that of the cup; water has a much larger specific heat capacity than aluminum, and we have over twice as much mass of water. We can also find the quantities of heat by substituting this value for T back into the original equations. We find that $Q_{coffee} = -5.0 \times 10^3$ J and $Q_{aluminum} = +5.0 \times 10^3$ J; Q_{coffee} is negative, which means that the coffee loses heat.

EXAMPLE 15-10

Changes in both temperature and phase A physics student wants to cool 0.25 kg of Diet Omni-Cola (mostly water), initially at 25°C, by adding ice initially at --20°C. How much ice should she add so that the final temperature will be 0°C with all the ice melted if the heat capacity of the container may be neglected?

SOLUTION The Omni-Cola loses heat, so the heat added to it is negative:

$$Q_{Omni} = m_{Omni} c_{water}\, \Delta T_{Omni}$$
$$= (0.25\ \text{kg})(4190\ \text{J/kg} \cdot \text{K})(0°C - 25°C)$$
$$= -26{,}000\ \text{J}.$$

From Table 15–3, the specific heat capacity of ice (not the same as for liquid water) is 2.1×10^3 J/kg · K. Let the mass of ice be m_{ice}; then the heat Q_1 needed to warm it from −20°C to 0°C is

$$Q_1 = m_{ice} c_{ice}\, \Delta T_{ice}$$
$$= m_{ice}(2.1 \times 10^3\ \text{J/kg} \cdot \text{K})(0°C - (-20°C))$$
$$= m_{ice}(4.2 \times 10^4\ \text{J/kg}).$$

From Eq. (15–20) the additional heat Q_2 needed to melt this mass of ice is the mass times the heat of fusion. Using Table 15–4, we find

$$Q_2 = m_{ice} L_f$$
$$= m_{ice}(3.34 \times 10^5\ \text{J/kg}).$$

The sum of these three quantities must equal zero:

$$Q_{Omni} + Q_1 + Q_2 = -26{,}000\ \text{J} + m_{ice}(42{,}000\ \text{J/kg})$$
$$+ m_{ice}(334{,}000\ \text{J/kg}) = 0.$$

Solving this for m_{ice}, we get $m_{ice} = 0.069$ kg = 69 g (three or four medium-size ice cubes).

EXAMPLE 15–11

What's cooking? A heavy copper pot of mass 2.0 kg (including the copper lid) is at a temperature of 150°C. You pour 0.10 kg of water at 25°C into the pot, then quickly close the lid of the pot so that no steam can escape. Find the final temperature of the pot and its contents, and determine the phase (liquid or gas) of the water. Assume that no heat is lost to the surroundings.

SOLUTION There are three conceivable outcomes in this situation. One, none of the water boils, and the final temperature is less than 100°C; two, a portion of the water boils, giving a mixture of water and steam at 100°C; or three, all the water boils, giving 0.10 kg of steam at a temperature of 100°C or greater.

The simplest case to calculate is the first possibility, so let's try that first. Let the common final temperature of the liquid water and the copper pot be T. Since we are assuming that no phase changes take place, the sum of the quantities of heat added to the two materials is

$$Q_{water} + Q_{copper} = m_{water}c_{water}(T - 25°C)$$
$$+ m_{copper}c_{copper}(T - 150°C)$$
$$= (0.10 \text{ kg})(4190 \text{ J/kg} \cdot \text{K})(T - 25°C)$$
$$+ (2.0 \text{ kg})(390 \text{ J/kg} \cdot \text{K})(T - 150°C)$$
$$= 0.$$

Solving this for T, we find $T = 106°C$. But this is above the boiling point of water, which contradicts our assumption that none of the water boils! So this assumption can't be correct; at least some of the water undergoes a phase change.

If we try the second possibility, in which the final temperature is 100°C, we have to find the fraction of water that changes to the gaseous phase. Let this fraction be x; the (positive)

amount of heat needed to vaporize this water is $(xm_{water}) L_v$. Setting the final temperature T equal to 100°C, we have

$$Q_{water} = m_{water}c_{water}(100°C - 25°C) + xm_{water}L_v$$
$$= (0.10 \text{ kg})(4190 \text{ J/kg} \cdot \text{K})(75 \text{ K})$$
$$+ x(0.10 \text{ kg})(2.256 \times 10^6 \text{ J/kg})$$
$$= 3.14 \times 10^4 \text{ J} + x(2.256 \times 10^5 \text{ J}),$$
$$Q_{copper} = m_{copper}c_{copper}(100°C - 150°C)$$
$$= (2.0 \text{ kg})(390 \text{ J/kg} \cdot \text{K})(-50 \text{ K}) = -3.90 \times 10^4 \text{ J}.$$

The requirement that the sum of all the quantities of heat be zero then gives

$$Q_{water} + Q_{copper} = 3.14 \times 10^4 \text{ J} + x(2.256 \times 10^5 \text{ J})$$
$$- 3.90 \times 10^4 \text{ J} = 0,$$
$$x = \frac{3.90 \times 10^4 \text{ J} - 3.14 \times 10^4 \text{ J}}{2.256 \times 10^5 \text{ J}} = 0.034.$$

This makes sense, and we conclude that the final temperature of the water and copper is 100°C. Of the original 0.10 kg of water, 0.034(0.10 kg) = 0.0034 kg = 3.4 g has been converted to steam at 100°C.

Had x turned out to be greater than 1, we would have again had a contradiction (the fraction of water that vaporized can't be greater than 1). In this case the third possibility would have been the correct description, all the water would have vaporized, and the final temperature would have been greater than 100°C. Can you show that this would have been the case if we had originally poured less than 15 g of 25°C water into the pot?

EXAMPLE 15–12

Combustion, temperature change, and phase change In a particular gasoline camp stove, 30% of the energy released in burning the fuel actually goes to heating the water in the pot on the stove. If we heat 1.00 L (1.00 kg) of water from 20°C to 100°C and boil 0.25 kg of it away, how much gasoline do we burn in the process?

SOLUTION The heat required to raise the temperature of the water from 20°C to 100°C is

$$Q_1 = mc \, \Delta T = (1.00 \text{ kg})(4190 \text{ J/kg} \cdot \text{K})(80 \text{ K})$$
$$= 3.35 \times 10^5 \text{ J}.$$

To boil 0.25 kg of water at 100°C requires

$$Q_2 = mL_v = (0.25 \text{ kg})(2.256 \times 10^6 \text{ J/kg}) = 5.64 \times 10^5 \text{ J}.$$

The total energy needed is the sum of these, or 8.99×10^5 J. This is only 0.30 of the total heat of combustion, so that energy is $(8.99 \times 10^5 \text{ J})/0.30 = 3.00 \times 10^6$ J. As we mentioned above, each gram of gasoline releases 46,000 J, so the mass of gasoline required is

$$\frac{3.00 \times 10^6 \text{ J}}{46,000 \text{ J/g}} = 65 \text{ g},$$

or a volume of about 0.09 L of gasoline.

15–8 MECHANISMS OF HEAT TRANSFER

We have talked about *conductors* and *insulators,* materials that permit or prevent heat transfer between bodies. Now let's look in more detail at *rates* of energy transfer. In the kitchen you use an aluminum pot for good heat transfer from the stove to whatever you're cooking, but your refrigerator is insulated with a material that *prevents* heat from flowing into the food inside the refrigerator. How do we describe the difference between these two materials?

The three mechanisms of heat transfer are conduction, convection, and radiation. *Conduction* occurs within a body or between two bodies in contact. *Convection* depends on motion of mass from one region of space to another. *Radiation* is heat transfer by electromagnetic radiation, such as sunshine, with no need for matter to be present in the space between bodies.

CONDUCTION

If you hold one end of a copper rod and place the other end in a flame, the end you are holding gets hotter and hotter, even though it is not in direct contact with the flame. Heat reaches the cooler end by **conduction** through the material. On the atomic level, the atoms in the hotter regions have more kinetic energy, on the average, than their cooler neighbors. They jostle their neighbors, giving them some of their energy. The neighbors jostle *their* neighbors, and so on through the material. The atoms themselves do not move from one region of material to another, but their energy does.

Most metals also use another, more effective mechanism to conduct heat. Within the metal, some electrons can leave their parent atoms and wander through the crystal lattice. These "free" electrons can rapidly carry energy from the hotter to the cooler regions of the metal, so metals are generally good conductors of heat. A metal rod at 20°C feels colder than a piece of wood at 20°C because heat can flow more easily from your hand into the metal. The presence of "free" electrons also causes most metals to be good electrical conductors.

Heat transfer occurs only between regions that are at different temperatures, and the direction of heat flow is always from higher to lower temperature. Figure 15–13 shows a rod of conducting material with cross-section area A and length L. The left end of the rod is kept at a temperature T_H and the right end at a lower temperature T_C, and heat flows from left to right. The sides of the rod are covered by an ideal insulator, so no heat transfer occurs at the sides.

When a quantity of heat dQ is transferred through the rod in a time dt, the rate of heat flow is dQ/dt. We call this rate the **heat current,** denoted by H. That is, $H = dQ/dt$. Experiments show that the heat current is proportional to the cross-section area A of the rod and to the temperature difference $(T_H - T_C)$ and is inversely proportional to the rod length L. Introducing a proportionality constant k called the **thermal conductivity** of the material, we have

$$H = \frac{dQ}{dt} = kA\frac{T_H - T_C}{L} \quad \text{(heat current in conduction).} \quad (15\text{–}21)$$

The quantity $(T_H - T_C)/L$ is the temperature difference *per unit length;* it is called the magnitude of the **temperature gradient.** The numerical value of k depends on the material of the rod. Materials with large k are good conductors of heat; materials with small k are poor conductors or insulators. Equation (15–21) also gives the heat current through a slab or through *any* homogeneous body with uniform cross section A perpendicular to the direction of flow; L is the length of the heat-flow path.

The units of heat current H are units of energy per time, or power; the SI unit of heat current is the watt (1 W = 1 J/s). We can find the units of k by solving Eq. (15–21) for k. We invite you to verify that the SI units of k are W/m · K. Some numerical values of k are given in Table 15–5.

The thermal conductivity of "dead" (that is, nonmoving) air is very small. A wool sweater keeps you warm because it traps air between the fibers. In fact, many insulating

15–13 Steady-state heat flow due to conduction in a uniform rod.

TABLE 15–5
THERMAL CONDUCTIVITIES

SUBSTANCE	k (W/m · K)
Metals	
Aluminum	205.0
Brass	109.0
Copper	385.0
Lead	34.7
Mercury	8.3
Silver	406.0
Steel	50.2
Various solids (representative values)	
Brick, insulating	0.15
Brick, red	0.6
Concrete	0.8
Cork	0.04
Felt	0.04
Fiberglass	0.04
Glass	0.8
Ice	1.6
Rock wool	0.04
Styrofoam	0.01
Wood	0.12–0.04
Gases	
Air	0.024
Argon	0.016
Helium	0.14
Hydrogen	0.14
Oxygen	0.023

materials such as styrofoam and fiberglass are mostly dead air. Figure 15–14 shows a ceramic material with very unusual thermal properties, including very small conductivity.

If the temperature varies in a non-uniform way along the length of the conducting rod, we introduce a coordinate x along the length and generalize the temperature gradient to be dT/dx. The corresponding generalization of Eq. (15–21) is

$$H = \frac{dQ}{dt} = -kA\frac{dT}{dx}. \qquad (15\text{–}22)$$

The negative sign shows that heat always flows in the direction of *decreasing* temperature.

For thermal insulation in buildings, engineers use the concept of **thermal resistance,** denoted by R. The thermal resistance R of a slab of material with area A is defined so that the heat current H through the slab is

$$H = \frac{A(T_H - T_C)}{R}, \qquad (15\text{–}23)$$

where T_H and T_C are the temperatures on the two sides of the slab. Comparing this with Eq. (15–21), we see that R is given by

$$R = \frac{L}{k}, \qquad (15\text{–}24)$$

15–14 This protective tile, developed for use in the space shuttle, has extraordinary thermal properties. The extremely small thermal conductivity and small heat capacity of the material make it possible to hold the tile by its edges, even though its temperature is high enough to emit the light for this photograph.

where L is the thickness of the slab. The SI unit of R is $1\ \text{m}^2 \cdot \text{K/W}$. In the units used for commercial insulating materials in the United States, H is expressed in Btu/h, A is in ft^2, and $T_H - T_C$ in F°. (1 Btu/h = 0.293 W.) The units of R are then $\text{ft}^2 \cdot \text{F}° \cdot \text{h/Btu}$, though values of R are usually quoted without units; a 6-inch-thick layer of fiberglass has an R value of 19 (that is, $R = 19\ \text{ft}^2 \cdot \text{F}° \cdot \text{h/Btu}$), a two-inch-thick slab of polyurethane foam has a value of 12, and so on. Doubling the thickness doubles the R-value. Common practice in new construction in severe northern climates is to specify R values of around 30 for exterior walls and ceilings. When the insulating material is in layers, such as a plastered wall, fiberglass insulation, and wood exterior siding, the R values are additive. Do you see why? (See Problem 15–100.)

Electronics engineers concerned with cooling microprocessor chips use a different definition of thermal resistance. This subject is discussed in Section 15–9.

Problem–Solving Strategy

HEAT CONDUCTION

1. Identify the direction of heat flow in the problem (from hot to cold). In Eq. (15–21), L is always measured along this direction, and A is always an area perpendicular to this direction. Often when a box or other container has an irregular shape but uniform wall thickness, you can approximate it as a flat slab with the same thickness and total wall area.

2. In some problems the heat flows through two different materials in succession. The temperature at the interface between the two materials is then intermediate between T_H and T_C; represent it by a symbol such as T. The temperature differences for the two materials are then $(T_H - T)$ and $(T - T_C)$. In steady-state heat flow, the same heat has to pass through both materials in succession, so

the heat current H must be *the same* in both materials. It's like a series electric circuit (to be discussed in Chapter 25).

3. If there are two *parallel* heat flow paths, so that some heat flows through each, then the total H is the sum of the quantities H_1 and H_2 for the separate paths. An example is heat flow from inside to outside a house, both through the glass in a window and through the surrounding frame. In this case the temperature difference is the same for the two paths, but L, A, and k may be different for the two paths. This is like a parallel electric circuit.

4. As always, it is essential to use a consistent set of units. If you use a value of k expressed in W/m · K, don't use distances in cm, heat in calories, or T in °F!

EXAMPLE 15–13

Conduction through a picnic cooler A Styrofoam box used to keep drinks cold at a picnic has total wall area (including the lid) of 0.80 m² and wall thickness 2.0 cm. It is filled with ice, water, and cans of Omni-Cola at 0°C. What is the rate of heat flow into the box if the temperature of the outside wall is 30°C? How much ice melts in one day?

SOLUTION We assume that the total heat flow is approximately the same as it would be through a flat slab of area 0.80 m² and thickness 2.0 cm = 0.020 m (Fig. 15–15). We find k from Table 15–5. From Eq. (15–21) the heat current (rate of heat flow) is

$$H = kA \frac{T_H - T_C}{L} = (0.010 \text{ W/m} \cdot \text{K})(0.80 \text{ m}^2) \frac{30°C - 0°C}{0.020 \text{ m}}$$

$$= 12 \text{ W} = 12 \text{ J/s.}$$

The total heat flow Q in one day (86,400 s) is

$$Q = Ht = (12 \text{ J/s})(86,400 \text{ s}) = 1.04 \times 10^6 \text{ J.}$$

The heat of fusion of ice is 3.34×10^5 J/kg, so the quantity of ice melted by this quantity of heat is

$$m = \frac{Q}{L_f}$$

$$= \frac{1.04 \times 10^6 \text{ J}}{3.34 \times 10^5 \text{ J/kg}} = 3.1 \text{ kg.}$$

15–15 Conduction of heat. We can approximate heat flow through the walls of a picnic cooler by heat flow through a single flat slab of styrofoam.

EXAMPLE 15–14

A steel bar 10.0 cm long is welded end-to-end to a copper bar 20.0 cm long (Fig. 15–16). Both bars are insulated perfectly on their sides. Each bar has a square cross section, 2.00 cm on a side. The free end of the steel bar is maintained at 100°C by placing it in contact with steam, and the free end of the copper bar is maintained at 0°C by placing it in contact with ice. Find the temperature at the junction of the two bars and the total rate of heat flow.

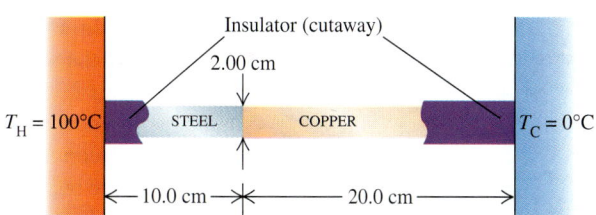

15–16 Heat flow along two metal bars, one of steel and one of copper, connected end-to-end.

Rearranging and solving for T, we find

$$T = 20.7°C.$$

Even though the steel bar is shorter, the temperature drop across it is much greater than across the copper bar because steel is a much poorer conductor.

We can find the total heat current by substituting this value for T back into either of the above expressions:

$$H_{steel} = \frac{(50.2 \text{ W/m} \cdot \text{K})(0.0200 \text{ m})^2(100°C - 20.7°C)}{0.100 \text{ m}} = 15.9 \text{ W,}$$

or

$$H_{copper} = \frac{(385 \text{ W/m} \cdot \text{K})(0.0200 \text{ m})^2(20.7°C)}{0.200 \text{ m}} = 15.9 \text{ W.}$$

SOLUTION As we discussed in the Problem-Solving Strategy, the heat currents in the two bars must be equal; this is the key to the solution. Let T be the unknown junction temperature; we use Eq. (15–21) for each bar and equate the two expressions:

$$H_{steel} = \frac{k_{steel}A(100°C - T)}{L_{steel}} = H_{copper} = \frac{k_{copper}A(T - 0°C)}{L_{copper}}.$$

The areas A are equal and may be divided out. Substituting $L_{steel} = 0.100$ m, $L_{copper} = 0.200$ m, and numerical values of k from Table 15–5, we find

$$\frac{(50.2 \text{ W/m} \cdot \text{K})(100°C - T)}{0.100 \text{ m}} = \frac{(385 \text{ W/m} \cdot \text{K})(T - 0°C)}{0.200 \text{ m}}.$$

EXAMPLE 15–15

In Example 15–14, suppose the two bars are separated. One end of each bar is maintained at 100°C, and the other end of each bar is maintained at 0°C (Fig. 15–17). What is the *total* rate of heat flow in the two bars?

SOLUTION In this case the bars are in parallel rather than in series. The total heat current is now the *sum* of the currents in the two bars, and for each bar, $T_H - T_C = 100°C - 0°C = 100$ K:

$$H = H_{\text{steel}} + H_{\text{copper}} = \frac{k_{\text{steel}} A (T_H - T_C)}{L_{\text{steel}}} + \frac{k_{\text{copper}} A (T_H - T_C)}{L_{\text{copper}}}$$

$$= \frac{(50.2 \text{ W/m} \cdot \text{K})(0.0200 \text{ m})^2 (100 \text{ K})}{0.100 \text{ m}}$$

$$+ \frac{(385 \text{ W/m} \cdot \text{K})(0.0200 \text{ m})^2 (100 \text{ K})}{0.200 \text{ m}}$$

$$= 20.1 \text{ W} + 77.0 \text{ W} = 97.1 \text{ W}.$$

The heat flow in the copper bar is much greater than that in the steel bar, even though it is longer, because the thermal conductivity of copper is much larger. The total heat flow is much greater than in Example 15–14, partly because the total cross

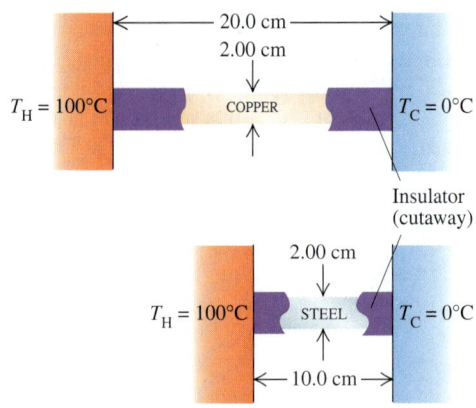

15–17 Heat flow along two metal bars, one of steel and one of copper, parallel to and separated from each other.

section for heat flow is greater and partly because the full 100-K temperature difference appears across each bar.

CONVECTION

Convection is transfer of heat by mass motion of a fluid from one region of space to another. Familiar examples include hot-air and hot-water home heating systems, the cooling system of an automobile engine, and the flow of blood in the body. If the fluid is circulated by a blower or pump, the process is called *forced convection;* if the flow is caused by differences in density due to thermal expansion, such as hot air rising, the process is called *natural convection* or *free convection.*

Free convection in the atmosphere plays a dominant role in determining the daily weather (Fig. 15–18), and convection in the oceans is an important global heat-transfer mechanism. On a smaller scale, soaring hawks and glider pilots make use of thermal updrafts from the warm earth. Sometimes the updrafts are strong enough to form a thunderstorm (Fig. 15–19). The most important mechanism for heat transfer within the human body (needed to maintain nearly constant temperature in various environments) is *forced* convection of blood, with the heart serving as the pump.

Convective heat transfer is a very complex process, and there is no simple equation to describe it. Here are a few experimental facts:

1. The heat current due to convection is directly proportional to the surface area. This is the reason for the large surface areas of radiators and cooling fins.
2. The viscosity of fluids slows natural convection near a stationary surface, giving a surface film that on a vertical surface typically has about the same insulating value as 1.3 cm of plywood (*R* value 0.7). Forced convection decreases the thickness of this film, increasing the rate of heat transfer. This is the reason for the "wind-chill factor"; you get cold faster in a cold wind than in still air with the same temperature.
3. The heat current due to convection is found to be approximately proportional to the $\frac{5}{4}$ power of the temperature difference between the surface and the main body of fluid.

A heating element in the tip of this submerged tube warms the surrounding water, producing a complex pattern of free convection.

RADIATION

Radiation is the transfer of heat by electromagnetic waves such as visible light, infrared, and ultraviolet radiation. Everyone has felt the warmth of the sun's radiation

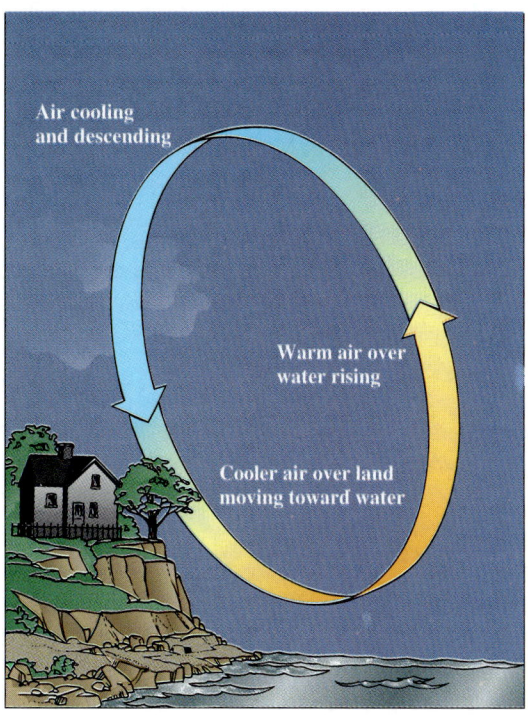

(a)　　　　　　　　　　　　　　　　　　(b)

and the intense heat from a charcoal grill or the glowing coals in a fireplace. Most of the heat from these very hot bodies reaches you not by conduction or convection in the intervening air but by *radiation*. This heat transfer would occur even if there were nothing but vacuum between you and the source of heat.

Every body, even at ordinary temperatures, emits energy in the form of electromagnetic radiation. At ordinary temperatures, say 20°C, nearly all the energy is carried by infrared waves with wavelengths much longer than those of visible light. As the temperature rises, the wavelengths shift to shorter values. At 800°C a body emits enough visible radiation to be self-luminous and appears "red-hot," although even at this temperature most of the energy is carried by infrared waves. At 3000°C, the temperature of an incandescent lamp filament, the radiation contains enough visible light so the body appears "white-hot."

The rate of energy radiation from a surface is proportional to the surface area A. The rate increases very rapidly with temperature, depending on the fourth power of the absolute (Kelvin) temperature. The rate also depends on the nature of the surface; this

15–18 Water has a higher specific heat capacity than does land. The heat of the sun has comparatively little effect on the temperature of the sea; by contrast, the land heats up during the day and cools off rapidly at night. Near the shore the temperature difference between land and sea gives rise to daytime sea breezes and nighttime land breezes.

15–19 Convection in a thunderstorm. Hot, moist air is less dense than cool, dry air and hence is lifted by buoyancy. The lifting can continue to altitudes of 12,000 to 18,000 m (40,000 to 60,000 ft). A typical thunderstorm 5 km in diameter may contain 5×10^8 kg of water; when this lifted moisture condenses to form raindrops or hailstones, about 10^{15} J of heat is released into the upper atmosphere (equal to the electrical energy used by a city of 100,000 people in a month). Air is pulled along with the falling rain and hail, creating strong downdrafts and violent surface winds. The downdrafts weaken as the energy supplied by the lifted moisture is used up.

dependence is described by a quantity e called the **emissivity.** This is a dimensionless number between 0 and 1, representing the ratio of the rate of radiation from a particular surface to the rate of radiation from an equal area of an ideal radiating surface at the same temperature. Emissivity also depends somewhat on temperature. Thus the heat current $H = dQ/dt$ due to radiation from a surface area A with emissivity e at absolute temperature T can be expressed as

$$H = Ae\sigma T^4 \qquad \text{(heat current in radiation)}, \qquad (15\text{--}25)$$

where σ is a fundamental physical constant called the **Stefan-Boltzmann constant.** This relation is called the **Stefan-Boltzmann law** in honor of its late nineteenth century discoverers. The current best numerical value of σ is

$$\sigma = 5.67051(19) \times 10^{-8} \text{ W/m}^2 \cdot \text{K}^4.$$

We invite you to check unit consistency in Eq. (15–25). Emissivity (e) is often larger for dark surfaces than for light ones. The emissivity of a smooth copper surface is about 0.3, but e for a dull black surface can be close to unity.

EXAMPLE 15–16

A thin square steel plate, 10 cm on a side, is heated in a black-smith's forge to a temperature of 800°C. If the emissivity is 0.60, what is the total rate of radiation of energy?

SOLUTION The total surface area, including both sides, is $2(0.10 \text{ m})^2 = 0.020 \text{ m}^2$. We must convert the temperature to the

Kelvin scale; 800°C = 1073 K. Then Eq. (15–25) gives

$$H = Ae\sigma T^4$$
$$= (0.020 \text{ m}^2)(0.60)(5.67 \times 10^{-8} \text{ W/m}^2 \cdot \text{K}^4)(1073 \text{ K})^4$$
$$= 900 \text{ W}.$$

While a body at absolute temperature T is radiating, its surroundings at temperature T_s are also radiating, and the body *absorbs* some of this radiation. If it is in thermal equilibrium with its surroundings, $T = T_s$ and the rates of radiation and absorption must be equal. For this to be true, the rate of absorption must be given in general by $H = Ae\sigma T_s^4$. Then the *net* rate of radiation from a body at temperature T with surroundings at temperature T_s is

$$H_{net} = Ae\sigma T^4 - Ae\sigma T_s^4 = Ae\sigma(T^4 - T_s^4). \qquad (15\text{--}26)$$

In this equation a positive value of H means a net heat flow *out of* the body. Equation (15–26) shows that for radiation, as for conduction and convection, the heat current depends on the temperature *difference* between two bodies.

EXAMPLE 15–17

Radiation from the human body If the total surface area of the human body is 1.20 m^2 and the surface temperature is 30°C = 303 K, find the total rate of radiation of energy from the body. If the surroundings are at a temperature of 20°C, what is the *net* rate of heat loss from the body by radiation? The emissivity of the body is very close to unity, irrespective of skin pigmentation.

SOLUTION The rate of radiation of energy per unit area is given by Eq. (15–25). Taking $e = 1$, we find

$$H = Ae\sigma T^4$$
$$= (1.20 \text{ m}^2)(1)(5.67 \times 10^{-8} \text{ W/m}^2 \cdot \text{K}^4)(303 \text{ K})^4$$
$$= 574 \text{ W}.$$

This loss is partly offset by *absorption* of radiation, which depends on the temperature of the surroundings. The *net* rate of radiative energy transfer is given by Eq. (15–26):

$$H = Ae\sigma(T^4 - T_s^4)$$
$$= (1.20 \text{ m}^2)(1)(5.67 \times 10^{-8} \text{ W/m}^2 \cdot \text{K}^4)[(303 \text{ K})^4 - (293 \text{ K})^4]$$
$$= 72 \text{ W}.$$

This is positive because the body is losing heat to its colder surroundings.

Heat transfer by radiation is important in some surprising places. A premature baby in an incubator can be cooled dangerously by radiation if the walls of the incubator happen to be cold, even when the *air* in the incubator is warm. Some incubators regulate the air temperature by measuring the baby's skin temperature.

A body that is a good absorber must also be a good emitter. An ideal radiator, with an emissivity of unity, is also an ideal absorber, absorbing *all* of the radiation that strikes it. Such an ideal surface is called an ideal black body or simply a **blackbody.** Conversely, an ideal *reflector,* which absorbs *no* radiation at all, is also a very ineffective radiator.

This is the reason for the silver coatings on vacuum ("Thermos") bottles, invented by Sir James Dewar (1842–1923). A vacuum bottle has double glass walls. The air is pumped out of the spaces between the walls; this eliminates nearly all heat transfer by conduction and convection. The silver coating on the walls reflects most of the radiation from the contents back into the container, and the wall itself is a very poor emitter. Thus a vacuum bottle can keep coffee or soup hot for several hours. The Dewar flask, used to store very cold liquefied gases, is exactly the same in principle.

15–9 INTEGRATED CIRCUITS

A Case Study in Heat Transfer

Integrated circuits (ic's) and VLSI (very large scale integration) chips are at the heart of most modern electronic devices, including computers, stereo systems, and fuel-injection systems in car engines (Fig. 15–20). The widespread use of these devices has posed new and interesting problems in heat transfer. Some of the electrical energy associated with the electric current in a chip is dissipated in the chip as heat; if a chip gets too hot as a result, the circuits become unreliable or are permanently destroyed. Keeping chips cool during operation is vitally important.

To introduce the problem, let's think about what happens when you turn on a light bulb. Electrical energy is dissipated as heat in the filament; the light bulb warms up until it reaches a temperature at which the rate of transfer of energy from the light bulb to its surroundings (by radiation as well as by conduction and convection to the surrounding air) just balances the rate of electrical energy input. A standard 60-W light bulb has a power input of 60 W and a surface area of about 120 cm^2. In equilibrium its power loss to its surroundings per unit surface area, or *power density,* is (60 W)/(120 cm^2) = 0.5 W/cm^2.

15–20 (a) Modern ic chips are usually placed in plastic or ceramic packages for protection. The packages contain conducting paths that lead from the chip to outside pins, which connect the circuit to the overall circuit board in the final product. (b) This ceramic package was specifically designed to ensure r_{th} of less than 1 K/W.

(a)

(b)

The heating element on an electric range has a greater power loss relative to its surface area, so it has a larger power density of 2–3 W/cm^2.

Now consider the *much larger* power dissipation values that occur in microelectronics. Many thousands of electronic elements are packed tightly together on a silicon chip only a few millimeters on a side. Power densities for present-day VLSI chips range up to 40 W/cm^2. For comparison, the ceramic tiles that protect the space shuttle when it reenters the earth's atmosphere typically have to dissipate 100 W/cm^2 (see Fig. 15–14).

For an ic chip in a plastic package, the maximum safe temperature is about 100°C. A chip in a ceramic package can operate reliably up to about 120°C. To see whether the temperature of an operating chip is within limits, we use the same principle as for a light bulb: power input equals power output. The power output H (the rate of heat transfer out of the chip) is approximately proportional to the difference $T_{ic} - T_{amb}$ between the temperature T_{ic} of the chip and the ambient temperature T_{amb} (temperature of the surroundings). Using a proportionality constant r_{th} that depends on the shape and size of the ic, we express the rate of heat loss H as

$$H = \frac{T_{ic} - T_{amb}}{r_{th}}. \tag{15–27}$$

(Note that r_{th} is the same as R/A in Eq. (15–23).) When the ic reaches its final operating temperature, the rate of heat loss must equal the electrical power P dissipated in the device. Equating H and P and solving for T_{ic}, we find

$$T_{ic} = T_{amb} + r_{th}P. \tag{15–28}$$

Values of r_{th} for common ic packages in still air vary from 30 to 70 K/W (30 to 70 C°/W). For example, one watt of electrical power ($P = 1$ W) into an ic will raise the temperature of a 40-pin plastic package to about 62 C° = 62 K above the surrounding air temperature. For this unit, $r_{th} = 62$ K/W.

EXAMPLE 15-18

A 40-pin ceramic package has $r_{th} = 40$ K/W. If the maximum temperature the circuit can safely reach is 120°C, what is the highest power level at which the circuit can safely operate in an ambient temperature of 75°C?

This power level is adequate in many typical ic applications, but chips used in high-speed computing applications often require considerably higher power levels.

SOLUTION We use Eq. (15–27), replacing H by P:

$$P = \frac{T_{ic} - T_{amb}}{r_{th}}$$

$$= \frac{120°C - 75°C}{40 K/W} = 1.1 \text{ W}.$$

To remove heat from the chip more efficiently, we can force air past the circuit, improving the heat transfer by convection. Blowing air through the system at a rate of about 20 m^3/min can reduce the effective value of r_{th} for the chip and package by about 10 to 15 K/W. This is an improvement, but in many cases it is still not enough cooling for a high-performance ic.

One method of cooling that is currently being investigated is direct immersion of the ic package in a fluorocarbon fluid. These fluids are electrical insulators and are chemically inert, making them compatible with operation of electronic components. Their thermal transport properties are not very favorable because of small values of thermal conductivity and heat of vaporization. To improve heat transfer in these fluids, design engineers have tried forced convection, structural modification of packages, and boiling.

In forced convection, microscopic channels are cut into the back of the silicon base of the ic chip. Typical channel dimensions are 50 μm wide and 300 μm deep. Experiments with water (not suitable for cooling an actual ic) have achieved heat transfers as large as 790 W per square centimeter of surface area.

Adding fins to the basic cylindrical pin form of an ic package can increase the heated surface area of the package by a factor of 8 to 12. The larger surface area increases heat transfer by convection, conduction, and radiation (see Eqs. (15–23) and (15–25)), reducing the effective value of r_{th} by as much as a factor of 20.

Heat flow values of 45 W/cm^2 have been obtained by allowing the cooling liquid to boil so that heat transfer by convection involves both liquid and vapor at the same time. However, the sudden drop in temperature at the surface of the ic package when boiling begins results in thermal stress on the package. Further work in package design is underway to try to minimize this problem.

Chip packaging has become both an art and a science. Careful layout of electronic components on the circuit board can minimize the overall thermal resistance of the chip, allowing the heat generated to dissipate more readily. Much recent attention has been devoted to research and development of new packages, so there will be rapid changes over the next few years in the heat limitations of ic technology.

SUMMARY

■ A thermometer measures temperature. Two bodies in thermal equilibrium must have the same temperature. A conducting material between two bodies permits interaction leading to thermal equilibrium; an insulating material prevents or impedes this interaction.

■ The Celsius and Fahrenheit temperature scales are based on the freezing temperature (0°C = 32°F) and boiling temperature (100°C = 212°F) of water. They are related by

$$T_F = \frac{9}{5} T_C + 32°, \tag{15–1}$$

$$T_C = \frac{5}{9}(T_F - 32°), \tag{15–2}$$

and 1 C° = $\frac{9}{5}$F°.

■ The Kelvin scale has its zero at the extrapolated zero-pressure temperature for a constant-volume gas thermometer, which is −273.15°C. Thus 0 K = −273.15°C, and

$$T_K = T_C + 273.15. \tag{15–3}$$

In the gas-thermometer scale, the ratio of two temperatures is defined to be equal to the ratio of the two corresponding gas-thermometer pressures:

$$\frac{T_2}{T_1} = \frac{p_2}{p_1}. \tag{15–4}$$

The triple-point temperature of water (0.01°C) is defined to be 273.16 K.

■ Under a temperature change ΔT, any linear dimension L_0 of a solid body changes by an amount ΔL given approximately by

$$\Delta L = \alpha L_0 \Delta T, \tag{15–6}$$

where α is the coefficient of linear expansion. Under a temperature change ΔT, the change ΔV in the volume V_0 of any solid or liquid material is given

KEY TERMS

thermodynamics, 460

temperature, 460

thermometer, 461

thermal equilibrium, 461

insulator, 461

conductor, 461

zeroth law of thermodynamics, 462

Celsius temperature scale, 462

Fahrenheit temperature scale, 462

Kelvin temperature scale, 464

absolute temperature scale, 465

absolute zero, 465

coefficient of linear expansion, 466

coefficient of volume expansion, 467

thermal stress, 469

heat, 470

calorie, 470

British thermal unit, 470

specific heat capacity, 471

molar heat capacity, 472

phase, 473

phase change, 473

heat of fusion, 474

phase equilibrium, 474

heat of vaporization, 474

approximately by

$$\Delta V = \beta V_0 \, \Delta T, \tag{15-8}$$

where β is the coefficient of volume expansion. For solids, $\beta = 3\alpha$.

■ When a material is cooled or heated and held so that it cannot contract or expand, the tensile stress F/A is given by

$$\frac{F}{A} = -Y\alpha \, \Delta T. \tag{15-12}$$

■ Heat is energy in transit from one place to another as a result of a temperature difference. The quantity of heat Q required to raise the temperature of a mass m of material by a small amount ΔT is

$$Q = mc \, \Delta T, \tag{15-13}$$

where c is the specific heat capacity of the material. When the quantity of material is given by the number of moles n, the corresponding relation is

$$Q = nC \, \Delta T, \tag{15-18}$$

where $C = Mc$ is the molar heat capacity (M is the molar mass). The number of moles n and the mass m of material are related by $m = nM$.

■ The molar heat capacities of many solid elements are approximately 25 J/mol · K; this is the rule of Dulong and Petit.

■ To change a mass m of a material to a different phase at the same temperature (such as liquid to vapor or liquid to solid) requires the addition or subtraction of a quantity of heat Q given by

$$Q = \pm mL, \tag{15-20}$$

where L is the heat of fusion, vaporization, or sublimation.

■ When heat is added to a body, the corresponding Q is positive; when it is removed, Q is negative. The basic principle of calorimetry comes from conservation of energy. In an isolated system whose parts interact by heat exchange, the algebraic sum of the Q's for all parts of the system must be zero.

■ The three mechanisms of heat transfer are conduction, convection, and radiation. Conduction is transfer of energy of molecular motion within materials without bulk motion of the materials. Convection involves mass motion from one region to another, and radiation is energy transfer through electromagnetic radiation.

■ The heat current H for conduction depends on the area A through which the heat flows, the length L of the heat path, the temperature difference $(T_H - T_C)$, and the thermal conductivity k of the material:

$$H = \frac{dQ}{dt} = kA \frac{T_H - T_C}{L}. \tag{15-21}$$

■ Heat transfer by convection is a complex process, depending on surface area, orientation, and the temperature difference between a body and its surroundings.

■ The heat current H due to radiation is given by

$$H = Ae\sigma T^4, \tag{15-25}$$

where A is the surface area, e the emissivity of the surface (a pure number between 0 and 1), T the absolute temperature, and σ a fundamental constant called the Stefan-Boltzmann constant. When a body at temperature T is surrounded by material at temperature T_s, the net heat current from the body to its surroundings is

$$H_{net} = Ae\sigma T^4 - Ae\sigma T_s^{\,4} = Ae\sigma(T^4 - T_s^{\,4}). \tag{15-26}$$

DISCUSSION QUESTIONS

Q15–1 A thermometer is laid out in direct sunlight. Does it measure the temperature of the air, or of the sun, or of something else? Explain.

Q15–2 What is the temperature of vacuum? Explain.

Q15–3 Does it make sense to say that one body is twice as hot as another? Why or why not?

Q15–4 Many automobile engines have cast-iron cylinders and aluminum pistons. What kinds of problems could occur if the engine gets too hot? (The coefficient of volume expansion of cast iron is approximately the same as that of steel.)

Q15–5 A piece of copper with a hole through it is placed in a bell jar and the air is pumped out. A student tells you that if the piece of copper warms up, the hole cannot expand because there are no longer any atoms in the hole to move apart. How would you explain the student's error to him?

Q15–6 Telescope mirrors must hold their curved shape precisely so that the images they produce will be in focus. Explain why this is a reason for not making the mirrors of large telescopes from ordinary glass, but from glasslike materials that have very low coefficients of thermal expansion.

Q15–7 Why do frozen water pipes burst? Would a mercury thermometer break if the temperature went below the freezing temperature of mercury? Why or why not?

Q15–8 Two bodies made of the same material have the same external dimensions and appearance, but one is solid and the other is hollow. When their temperature is increased, is the overall volume expansion the same or different? Why?

Q15–9 The inside of an oven is at a temperature of 200°C (392°F). You can put your hand in the oven without injury as long as you don't touch anything. But since the air inside the oven is also at 200°C, why isn't your hand burned just the same?

Q15–10 A newspaper article about the weather states that "the temperature of a body measures how much heat the body contains." Is this description correct? Why or why not?

Q15–11 To raise the temperature of an object, must you add heat to it? If you add heat to an object, must you raise its temperature? Explain.

Q15–12 A student asserted that a suitable unit for specific heat capacity is $1 \ \mathrm{m^2/s^2 \cdot C°}$. Is she correct? Why or why not?

Q15–13 Why do you think that the heat of vaporization for water is so much larger than the heat of fusion?

Q15–14 In some household air conditioners used in dry climates, air is cooled by blowing it through a water-soaked filter, evaporating some of the water. How does this cool the air? Would such a system work well in a high-humidity climate? Why or why not?

Q15–15 The units of specific heat capacity c are $\mathrm{J/kg \cdot K}$, but the units of heat of fusion L_f or heat of vaporization L_v are simply J/kg. Why don't the units of L_f and L_v include a factor of $(\mathrm{K})^{-1}$ to account for a temperature change?

Q15–16 If you set an ice cube on your kitchen table, it will stay frozen longer if it is wrapped in a wet paper towel. Why?

Q15–17 Why does food cook faster in a pressure cooker than in an open pot of boiling water?

Q15–18 Desert travelers sometimes keep water in a canvas bag. Some water seeps through the bag and evaporates. How does this cool the water inside?

Q15–19 When you first step out of the shower, you feel cold. But as soon as you are dry you feel warmer, even though the room temperature does not change. Why?

Q15–20 The climate of regions adjacent to large bodies of water (like the Pacific or Atlantic coasts) is usually more moderate than that of regions far from large bodies of water (like the prairies). Why?

Q15–21 When water is placed in ice-cube trays in a freezer, why does the water not freeze all at once when the temperature has reached 0°C? In fact the water freezes first in a layer adjacent to the sides of the tray. Why?

Q15–22 Before giving you an injection, a physician swabs your arm with isopropyl alcohol at room temperature. Why does this make your arm feel cold? (*Hint:* The reason is *not* the fear of the injection! The boiling point of isopropyl alcohol is 82.4°C.)

Q15–23 A cold block of metal feels colder than a block of wood at the same temperature. Why? A *hot* block of metal feels hotter than a block of wood at the same temperature. Again, why? Is there any temperature at which the two blocks feel equally hot or cold? What temperature is this?

Q15–24 A person pours a cup of hot coffee, intending to drink it five minutes later. To keep the coffee as hot as possible, should she put cream in it now, or wait until just before she drinks it? Explain.

Q15–25 When a freshly baked apple pie has just been removed from the oven, the crust and filling are both at the same temperature. Yet if you sample the pie, the filling will burn your tongue but the crust will not. Why is there a difference? (*Hint:* The filling is moist while the crust is dry.)

Q15–26 Old-time kitchen lore suggests that things cook better (evenly and without burning) in heavy cast-iron pots. What desirable characteristics do such pots have?

Q15–27 Figure 15–18 shows that the land has a higher temperature than the ocean during the day, but a lower temperature at night. Explain why. (*Hint:* The specific heat capacity of soil is only 0.2–0.8 times as great as that of water.)

Q15–28 Why does moisture condense on the insides of windows on cold days? If it is cold enough outside, frost will form, but only when the outside temperature is far below freezing. Why is this so? Why are the bottoms of windows frostier than the tops? How do storm windows change the situation?

Q15–29 Why do glasses containing cold liquids sweat in warm weather? Glasses containing alcoholic drinks sometimes form frost on the outside. How can this happen?

Q15–30 It is well known that a potato bakes faster if a large nail is stuck through it. Why? Is an aluminum nail better than a steel one? Why or why not? (*Note:* Don't try this in a microwave oven!) There is also a gadget on the market to hasten roasting of meat, consisting of a hollow metal tube containing a wick and some water; this is claimed to be much better than a solid metal rod. How does this work?

Q15–31 Glider pilots in the Midwest know that thermal updrafts are likely to occur in the vicinity of freshly plowed fields. Why?

Q15–32 Some folks claim that ice cubes freeze faster if the trays are filled with hot water, because hot water cools off faster than cold water. What do you think?

Q15–33 We're lucky that the earth isn't in thermal equilibrium with the sun (which has a surface temperature of 5800 K). But why isn't it?

Q15–34 A microwave oven works by producing electromagnetic radiation that is absorbed by water molecules, raising the temperature of the water. This radiation can easily penetrate nonmetallic objects. Explain why a microwave oven can cook massive objects (such as potatoes and beef roasts) much faster than can a conventional oven. What disadvantages are associated with the reasons for this greater speed?

EXERCISES

SECTION 15–3 THERMOMETERS AND TEMPERATURE SCALES

15–1 Convert the following Celsius temperatures to Fahrenheit: a) −62.8°C, the lowest temperature ever recorded in North America (February 3, 1947, Snag, Yukon); b) 56.7°C, the highest temperature ever recorded in the United States (July 10, 1913, Death Valley, California); c) 31.1°C, the world's highest average annual temperature (Lugh Ferrandi, Somalia).

15–2 Find the Celsius temperatures corresponding to a) a winter night in Seattle (41.0°F); b) a hot summer day in Palm Springs (107.0°F); c) a cold winter day in northern Manitoba (−18.0°F).

15–3 One day in Saskatoon, Saskatchewan, the Fahrenheit and Celsius temperatures are the same. What are the temperatures and the season?

15–4 a) On January 22, 1943, the temperature in Spearfish, South Dakota, rose from −4.0°F to 45.0°F in just two minutes. What was the temperature change in Celsius degrees? b) The temperature in Browning, Montana, was 44.0°F on January 23, 1916. The next day the temperature plummeted to −56.0°F. What was the temperature change in Celsius degrees?

15–5 a) You feel sick and are told that you have a temperature of 40.2°C. What is your temperature in °F? Should you be concerned? b) The morning weather report in Sydney gives a current temperature of 12°C. What is this temperature in °F?

15–6 A "blue norther" passes through Lubbock, Texas, one September afternoon and the temperature drops 11.8 C° in one hour. What is this temperature change in F°?

SECTION 15–4 GAS THERMOMETERS AND THE KELVIN SCALE

15–7 Convert the following record-setting temperatures to the Kelvin scale: a) the lowest temperature recorded in the 48 contiguous states (−70.0°F at Rogers Pass, Montana, on January 20, 1954); b) Australia's highest temperature (127.0°F at Cloncurry, Queensland, on January 16, 1889); c) the lowest temperature recorded in the northern hemisphere (−90.0°F at Verkhoyansk, Siberia, in 1892).

15–8 Convert the following Kelvin temperatures to the Celsius and Fahrenheit scales: a) the midday temperature at the surface of the moon (400 K); b) the temperature at the tops of the clouds in the atmosphere of Saturn (95 K); c) the temperature at the center of the sun (1.55×10^7 K).

15–9 The normal boiling point of liquid neon is −245.92°C. What is this temperature on the Kelvin scale?

15–10 The ratio of the pressures of a gas at the melting point of platinum and its pressure at the triple point of water, when the gas is kept at constant volume, is found to be 7.476. What is the Celsius temperature of the melting point of platinum?

15–11 A gas thermometer registers an absolute pressure corresponding to 325 mm of mercury when in contact with water at the triple point. What pressure does it read when in contact with water at the normal boiling point?

15–12 Like the Kelvin scale, the *Rankine scale* is an absolute temperature scale: absolute zero is zero degrees Rankine (0°R). However, the units of this scale are the same size as those on the Fahrenheit scale rather than the Celsius scale. What is the numerical value of the triple-point temperature of water on the Rankine scale?

15–13 A Constant-Volume Gas Thermometer. An experimenter using a gas thermometer found the pressure at the triple point of water (0.01°C) to be 4.80×10^4 Pa and the pressure at the normal boiling point (100°C) to be 6.50×10^4 Pa. a) Assuming that the pressure varies linearly with temperature, use these two data points to find the Celsius temperature at which the gas pressure would be zero (that is, find the Celsius temperature of absolute zero). b) Does the gas in this thermometer obey Eq. (15–4) precisely? If that equation were precisely obeyed, and the pressure at 100°C were 6.50×10^4 Pa, what pressure would the experimenter have measured at 0.01°C? (As we will learn in Section 16–2, Eq. (15–4) is accurate only for gases at very low density.)

SECTION 15–5 THERMAL EXPANSION

15–14 Air Friction and Thermal Expansion. The supersonic airliner Concorde is 62.1 m long when sitting on the ground on a typical (15°C) day. It is primarily made of aluminum. In flight at twice the speed of sound, friction with the air warms Concorde's skin and causes the aircraft to lengthen by 25 cm. (The passenger cabin is on rollers, and the airplane expands around the passengers.) What is the temperature of Concorde's skin in flight?

15–15 The Humber Bridge in England has the world's longest single span, 1410 m in length. Calculate the change in length of the steel deck of the span when the temperature increases from −5.0°C to 18.0°C.

15–16 Ensuring a Tight Fit. Aluminum rivets used in airplane construction are made slightly larger than the rivet holes and cooled by "dry ice" (solid CO_2) before being driven. If the diameter of a hole is 4.500 mm, what should be the diameter of a rivet at 23.0°C if its diameter is to equal that of the hole when the rivet is cooled to −78.0°C, the temperature of dry ice? Assume that the expansion coefficient remains constant at the value given in Table 15–1.

15–17 A U.S. one-cent coin has a diameter of 1.9000 cm at 20.0°C. The coin is made of a metal alloy (mostly zinc) for which the coefficient of linear expansion is 2.6×10^{-5} K^{-1}. What would its diameter be a) on a hot day in Death Valley (48.0°C); b) on a cold night in the mountains of Greenland (−53.0°C)?

15–18 The pendulum shaft of a clock is made of brass. What is the fractional change in length of the shaft when it is cooled from 19.50°C to 5.00°C?

15–19 A metal rod is 40.125 cm long at 20.0°C and 40.148 cm long at 45.0°C. Calculate the average coefficient of linear expansion of the rod for this temperature range.

15–20 A copper cylinder is initially at 20.0°C. At what temperature will its volume be 0.150% larger than it is at 20.0°C?

15–21 An underground tank with a capacity of 1700 L (1.70 m^3) is filled with ethanol that has an initial temperature of 19.0°C. After the ethanol has cooled off to the temperature of the tank and ground, which is 10.0°C, how much air space will there be above the ethanol in the tank? (Assume that the volume of the tank doesn't change.)

15–22 Determine the coefficient of volume expansion for water at a temperature of 9°C. (*Hint:* You will need to make measurements on Fig. 15–7.)

15–23 A glass flask whose volume is 1000.00 cm^3 at 0.0°C is completely filled with mercury at this temperature. When flask and mercury are warmed to 55.0°C, 8.95 cm^3 of mercury overflow. If the coefficient of volume expansion of mercury is 18.0×10^{-5} K^{-1}, compute the coefficient of volume expansion of the glass.

15–24 a) If an area measured on the surface of a solid body is A_0 at some initial temperature and then changes by ΔA when the temperature changes by ΔT, show that

$$\Delta A = (2\alpha)A_0\Delta T.$$

where α is the coefficient of linear expansion. b) A circular sheet of aluminum is 55.0 cm in diameter at 15.0°C. By how much does the area of one side of the sheet change when the temperature increases to 27.5°C?

15–25 A machinist bores a hole of diameter 1.350 cm in a steel plate at a temperature of 25°C. What is the cross-section area of the hole a) at 25°C; b) when the temperature of the plate is increased to 175°C? Assume that the coefficient of linear expansion remains constant over this temperature range. (*Hint:* See Exercise 15–24.)

15–26 A brass rod is 185 cm long and 1.60 cm in diameter. What force must be applied to each end of the rod to prevent it from contracting when it is cooled from 120°C to 10°C?

15–27 a) A wire that is 1.50 m long at 20°C is found to increase in length by 1.9 cm when warmed to 420°C. Compute its average coefficient of linear expansion for this temperature range. b) The wire is stretched just taut (zero tension) at 420°C.

Find the stress in the wire if it is cooled to 20°C without being allowed to contract. Young's modulus for the wire is 2.0×10^{11} Pa.

15–28 Steel train rails are laid in 12.0-m-long segments placed end-to-end. The rails are laid on a winter day when their temperature is −2.0°C. a) How much space must be left between adjacent rails if they are to just touch on a summer day when their temperature is 33.0°C? b) If the rails are originally laid in contact, what is the stress in them on a summer day when their temperature is 33.0°C?

SECTION 15–6 QUANTITY OF HEAT

15–29 Heat Loss During Breathing. In very cold weather a significant mechanism for heat loss by the human body is energy expended in warming the air taken into the lungs with each breath. a) On a cold winter day when the temperature is −20°C, what is the amount of heat needed to warm to body temperature (37°C) the 0.50 L of air exchanged with each breath? Assume that the specific heat capacity of air is 1020 J/kg · K and that 1.0 L of air has mass 1.3×10^{-3} kg. b) How much heat is lost per hour if the respiration rate is 20 breaths per minute?

15–30 While running, a 70-kg student generates thermal energy at a rate of 1200 W. To maintain a constant body temperature of 37°C, this energy must be removed by perspiration or other mechanisms. If these mechanisms failed and the heat could not flow out of the student's body, for what amount of time could a student run before irreversible body damage occurs? (Protein structures in the body are irreversibly damaged if body temperature rises to 44°C or above. The specific heat capacity of a typical human body is 3480 J/kg · K, slightly less than that of water. The difference is due to the presence of protein, fat, and minerals, which have lower specific heat capacities).

15–31 While painting the top of an antenna 225 m in height, a worker accidentally lets a 1.00-L water bottle fall from his lunchbox. The bottle lands in some bushes at ground level and does not break. If a quantity of heat equal to the magnitude of the change in mechanical energy of the water goes into the water, what is its increase in temperature?

15–32 A crate of fruit of mass 35.0 kg and specific heat capacity 3650 J/kg · K slides down a ramp inclined at 36.9° below the horizontal. The ramp is 8.00 m long. a) If the crate was at rest at the top of the incline and has a speed of 2.50 m/s at the bottom, how much work was done on the crate by friction? b) If an amount of heat equal to the magnitude of the work done by friction goes into the crate of fruit and the fruit reaches a uniform final temperature, what is its temperature change?

15–33 An engineer is working on a new engine design. One of the moving parts contains 1.60 kg of aluminum and 0.30 kg of iron and is designed to operate at 210°C. How much heat is required to raise its temperature from 20° to 210°C?

15–34 A nail being driven into a board increases in temperature. If we assume that 60% of the kinetic energy delivered by a 1.80-kg hammer with a speed of 7.80 m/s is transformed into heat that flows into the nail and does not flow out, what is the temperature increase of an 8.00-g aluminum nail after it is struck ten times?

15–35 An aluminum tea kettle with a mass of 1.50 kg and containing 1.80 kg of water is placed on a stove. If no heat is lost to the surroundings, how much heat must be added to raise the temperature from 20.0°C to 85.0°C?

15–36 In an effort to stay awake for an all-night study session, a student makes a cup of coffee by first placing a 200-W electric immersion heater in 0.320 kg of water. a) How much heat must be added to the water to raise its temperature from 20.0°C to 80.0°C? b) How much time is required? Assume that all of the heater power goes into heating the water.

15–37 A technician measures the specific heat capacity of an unidentified liquid by immersing an electrical resistor in it. Electrical energy is converted to heat transferred to the liquid for 120 s at a constant rate of 65.0 W. The mass of the liquid is 0.780 kg, and its temperature increases from 18.55°C to 22.54°C. a) Find the average specific heat capacity of the liquid in this temperature range. Assume that negligible heat is transferred to the container that holds the liquid and that no heat is lost to the surroundings. b) Suppose that in this experiment heat transfer from the liquid to the container or surroundings cannot be neglected. Is the result calculated in part (a) an *overestimate* or an *underestimate* of the average specific heat capacity? Explain.

SECTION **15–7 CALORIMETRY AND PHASE CHANGES**

15–38 Before going in for his annual physical, a 70.0-kg man whose body temperature is 37.0°C consumes an entire 0.355-L can of a soft drink (mostly water) at 12.0°C. a) What will be his body temperature after equilibrium is attained? Ignore any heating by the man's metabolism. The specific heat capacity of the man's body is 3480 J/kg · K. b) Is the change in his body temperature great enough to be measured by a medical thermometer?

15–39 In the situation described in Exercise 15–38, the man's metabolism will eventually return the temperature of his body (and of the soft drink that he consumed) to 37.0°C. If his body releases energy at a rate of 7.00×10^3 kJ/day (the *basal metabolic rate,* or BMR), how long does this take? Assume that all of the released energy goes into raising the temperature.

15–40 An ice-cube tray of negligible mass contains 0.350 kg of water at 18.0°C. How much heat must be removed to cool the water to 0.0°C and freeze it? Express your answer in joules, calories, and Btu.

15–41 How much heat is required to convert 12.0 g of ice at −10.0°C to steam at 100.0°C? Express your answer in joules, calories, and Btu.

15–42 An open container has in it 0.550 kg of ice at −15.0°C. The mass of the container can be neglected. Heat is supplied to the container at the constant rate of 800 J/min for 500 min. a) After how many minutes does the ice *start* to melt? b) After how many minutes, from the time when the heating is first started, does the temperature start to rise above 0°C? c) Plot a curve showing the temperature as a function of the elapsed time.

15–43 The capacity of commercial air conditioners is sometimes expressed in "tons," the number of tons of ice (1 ton = 2000 lb) that can be frozen from water at 0°C in 24 h by the unit. Express the capacity of a 2-ton air conditioner in Btu/h and in watts.

15–44 Steam Burns vs. Water Burns. What is the amount of heat input to your skin when it receives the heat released a) by 25.0 g of steam initially at 100.0°C when it is cooled to skin temperature (34.0°C)? b) by 25.0 g of water initially at 100.0°C when it is cooled to 34.0°C? c) What does this tell you about the relative severity of steam and hot water burns?

15–45 What must be the initial speed of a lead bullet at a temperature of 25°C so that the heat developed when it is brought to rest will be just sufficient to melt it? Assume that all the initial mechanical energy of the bullet is converted to heat and that no heat flows from the bullet to its surroundings. (Typical rifles have muzzle speeds that exceed the speed of sound in air, which is 347 m/s at 25°C.)

15–46 Evaporation of sweat is an important mechanism for temperature control in some warm-blooded animals. a) What mass of water must evaporate from the skin of a 70.0-kg man to cool his body 1.00 C°? The heat of vaporization of water at body temperature (37°C) is 2.42×10^6 J/kg · K. The specific heat capacity of a typical human body is 3480 J/kg · K. (See Exercise 15–30.) b) What volume of water must the man drink to replenish the evaporated water? Compare to the volume of a soft-drink can (355 cm³).

15–47 "The Ship of the Desert." Camels require very little water because they are able to tolerate relatively large changes in their body temperature. While humans keep their body temperatures constant to within one or two Celsius degrees, a dehydrated camel permits its body temperature to drop to 34.0°C overnight and rise to 40.0°C during the day. To see how effective this mechanism is for saving water, calculate how many liters of water a 400-kg camel would have to drink if it instead attempted to keep its body temperature at a constant 34.0°C by evaporation of sweat during the day (12 hours) instead of rising to 40.0°C. (The specific heat capacity of a camel or other mammal is about the same as that of a typical human, 3480 J/kg · K. The heat of vaporization of water at 34°C is 2.42×10^6 J/kg.)

15–48 In a physics lab experiment a student immersed 200 one-cent coins (each having a mass of 3.00 g) in boiling water. After they reached thermal equilibrium, she fished them out and dropped them into 0.240 kg of water at 20.0°C in an insulated container of negligible mass. What was the final temperature of the coins? (One-cent coins are made of a metal alloy—mostly zinc—with a specific heat capacity of 390 J/kg · K.)

15–49 A copper pot with a mass of 0.500 kg contains 0.170 kg of water at a temperature of 20.0°C. A 0.250-kg block of iron at 85.0°C is dropped into the pot. Find the final temperature, assuming no heat loss to the surroundings.

15–50 A laboratory technician drops a 0.0850-kg sample of unknown material, at a temperature of 100.0°C, into a calorimeter. The calorimeter can, initially at 19.0°C, is made of 0.150 kg of copper and contains 0.200 kg of water. The final temperature of the calorimeter can is 26.1°C. Compute the specific heat capacity of the sample.

15–51 An insulated beaker with negligible mass contains 0.250 kg of water at a temperature of 75.0°C. How many kilograms of ice at a temperature of −20.0°C must be dropped in the water so that the final temperature of the system will be 30.0°C?

15–52 A glass vial containing a 16.0-g sample of an enzyme is cooled in an ice bath. The bath contains water and 0.120 kg of ice. The sample has specific heat capacity 2250 J/kg · K; the glass vial has mass 6.0 g and specific heat capacity 2800 J/kg · K. How much ice melts in cooling the enzyme sample from room temperature (19.5°C) to the temperature of the ice bath?

15–53 A 4.00-kg silver ingot is taken from a furnace where its temperature is 750°C and placed on a large block of ice at 0°C. Assuming that all the heat given up by the silver is used to melt the ice, how much ice is melted?

15–54 A copper calorimeter can with a mass of 0.100 kg contains 0.160 kg of water and 0.018 kg of ice in thermal equilibrium at atmospheric pressure. If 0.750 kg of lead at a temperature of 255°C is dropped into the calorimeter can, what is the final temperature? Assume that no heat is lost to the surroundings.

15–55 A vessel whose walls are thermally insulated contains 2.40 kg of water and 0.450 kg of ice, all at a temperature of 0.0°C. The outlet of a tube leading from a boiler in which water is boiling at atmospheric pressure is inserted into the water. How many grams of steam must condense inside the vessel (also at atmospheric pressure) to raise the temperature of the system to 28.0°C? Neglect the heat transferred to the container.

SECTION 15–8 MECHANISMS OF HEAT TRANSFER

15–56 Use Eq. (15–21) to show that the SI units of thermal conductivity are W/m · K.

15–57 Suppose that the rod in Fig. 15–13 is made of copper, is 45.0 cm long, and has a cross-section area of 1.25 cm². Let $T_H = 100.0$°C and $T_C = 0.0$°C. a) What is the final steady-state temperature gradient along the rod? b) What is the heat current in the rod in the final steady state? c) What is the final steady-state temperature at a point in the rod 12.0 cm from its left end?

15–58 One end of an insulated metal rod is maintained at 100°C, and the other end is maintained at 0°C by an ice-water mixture. The rod is 60.0 cm long and has a cross-section area of 1.25 cm². The heat conducted by the rod melts 8.50 g of ice in 10.0 min. Find the thermal conductivity k of the metal.

15–59 A carpenter builds an outer house wall with a layer of wood 3.0 cm thick on the outside and a layer of styrofoam insulation 2.2 cm thick as the inside wall surface. The wood has $k = 0.080$ W/m · K and the styrofoam has $k = 0.010$ W/m · K. The interior surface temperature is 19.0°C, and the exterior surface temperature is −10.0°C. a) What is the temperature at the plane where the wood meets the styrofoam? b) What is the rate of heat flow per square meter through this wall?

15–60 An electric kitchen range has total wall area of 1.40 m² and is insulated with a layer of fiberglass 4.0 cm thick. The inside surface of the fiberglass has a temperature of 175°C and its outside surface is at 35°C. The fiberglass has a thermal conductivity of 0.040 W/m · K. a) What is the heat current through the insulation, assuming it may be treated as a flat slab with an area of 1.40 m²? b) What electric-power input to the heating element is required to maintain this temperature?

15–61 The ceiling of a room has an area of 125 ft². The ceiling is insulated to an R value of 30 (in units of ft² · F° · h/Btu). The surface in the room is maintained at 69°F, and the surface in the attic has a temperature of 35°F. What is the heat flow through the ceiling into the attic in 5.0 h? Express your answer in Btu and in joules.

15–62 A long rod, insulated to prevent heat loss along its sides, is in perfect thermal contact with boiling water (at atmospheric pressure) at one end with an ice-water mixture at the other (Fig. 15–21). The rod consists of 1.00-m section of copper (one end in steam) joined end-to-end to a length L_2 of steel (one end in ice). Both sections of the rod have cross-section areas of 4.00 cm². The temperature of the copper-steel junction is 65.0°C after a steady state has been set up. a) How much heat per second flows from the steam bath to the ice-water mixture? b) What is the length L_2 of the steel section?

FIGURE 15–21 Exercise 15–62.

15–63 A pot with a steel bottom 8.50 mm thick rests on a hot stove. The area of the bottom of the pot is 0.150 m². The water inside the pot is at 100.0°C, and 0.390 kg are evaporated every 3.00 min. Find the temperature of the lower surface of the pot, which is in contact with the stove.

15–64 What is the rate of energy radiation per unit area of a blackbody at a temperature of a) 273 K? b) 2730 K?

15–65 What is the net rate of heat loss by radiation in Example 15–17 (Section 15–8) if the temperature of the surroundings is 5.0°C?

15–66 The emissivity of tungsten is 0.35. A tungsten sphere with a radius of 1.50 cm is suspended within a large evacuated enclosure whose walls are at 290 K. What power input is required to maintain the sphere at a temperature of 3000 K if heat conduction along the supports is neglected?

15–67 Size of a Light-Bulb Filament. The operating temperature of a tungsten filament in an incandescent light bulb is 2450 K, and its emissivity is 0.35. Find the surface area of the filament of a 150-W bulb if all the electrical energy consumed by the bulb is radiated by the filament as electromagnetic waves. (Only a fraction of the radiation appears as visible light.)

15–68 The Sizes of Stars. The hot, glowing surfaces of stars emit energy in the form of electromagnetic radiation. It is a good approximation to assume $e = 1$ for these surfaces. Find the radii of the following stars (assumed to be spherical): a) Rigel, the bright blue star in the constellation Orion, which radiates energy at a rate of 2.7×10^{32} W and has surface temperature 11,000 K; b) Procyon B (only visible using a telescope), which radiates energy at a rate of 2.1×10^{23} W and has surface temperature 10,000 K. c) Compare your answers to the radius of the earth, the radius of the sun, and the distance between the earth and the sun. (Rigel is an example of a *supergiant* star, and Procyon B is an example of a *white dwarf* star.)

SECTION 15–9 INTEGRATED CIRCUITS: A CASE STUDY IN HEAT TRANSFER

15–69 A night light has a 12-W bulb. The glass envelope of the bulb has surface area 20 cm². For this bulb, what is the power dissipation per unit surface area (by radiation, conduction, and convection to the surrounding air)?

15–70 The light bulb in Exercise 15–69 has $r_{th} = 5.0$ K/W. What is the operating temperature of the bulb surface if the ambient temperature is 18°C?

15–71 Maximum Temperature for an Operating Chip. One method for controlling the temperature of an operating VLSI chip is to decrease the ambient temperature by air conditioning or by immersing the chip in a cold liquid. A VLSI chip has a heat power output of 28 W and a value of $r_{th} = 4.5$ K/W. If the chip is to operate at temperatures up to 120°C, what is the maximum allowable ambient temperature?

15–72 The value of r_{th} for a computer chip running at 300 MHz is measured to determine the effectiveness of forced air cooling. Without the forced air, when the electrical power dissipation (by conversion to thermal energy) in the chip is 22 W, the chip rises to 88 C° above the ambient temperature. With the forced air system operating, the chip rises 58 C° above the ambient temperature for the same 22-W power dissipation. What is the value of r_{th} for the chip both a) without the forced air cooling and b) with the forced air cooling?

15–73 A certain computer chip has an average power consumption of 14.0 W when operating at a clock speed of 300 MHz. If the ambient air temperature is 20.0°C and the maximum allowable temperature of the chip is 100.0°C, what is the minimum value of r_{th} that will ensure safe operation of the chip?

PROBLEMS

15–74 Suppose that a steel hoop could be constructed to just fit around the earth's equator at a temperature of 20.0°C. What would be the thickness of space between the hoop and the earth if the temperature of the hoop were increased by 0.50 C°?

15–75 At an absolute temperature T_0, a cube has sides of length L_0 and has density ρ_0. The cube is made of a material with coefficient of volume expansion β. a) Show that if the temperature increases to $T_0 + \Delta T$, the density of the cube becomes, approximately,

$$\rho \approx \rho_0(1 - \beta \Delta T).$$

(*Hint:* Use the expression $(1 + x)^n \approx 1 + nx$, valid for $|x| \ll 1$.) Explain why this approximate result is only valid if ΔT is much less than $1/\beta$, and explain why you would expect this to be the case in most situations. b) A copper cube has sides of length 1.25 cm at 20.0°C. Find the change in its volume and density when its temperature is increased to 70.0°C.

15–76 At a temperature of 20.0°C the volume of a certain glass flask, up to a reference mark on the long stem of the flask, is exactly 100 cm³. The flask is filled to this point with a liquid whose coefficient of volume expansion is 8.00×10^{-4} K⁻¹, with both flask and liquid at 20.0°C. The coefficient of volume expansion of the glass is 2.00×10^{-5} K⁻¹. The cross-section area of the stem is 50.0 mm² and can be considered constant. a) Explain why it is a good approximation to ignore the change in the cross-section area of the stem. b) How far will the liquid rise or fall in the stem when the temperature is raised to 50.0°C?

15–77 A metal rod that is 30.0 cm long expands by 0.0650 cm when its temperature is raised from 0°C to 100°C. A rod of a different metal and of the same length expands by 0.0350 cm for the same rise in temperature. A third rod, also 30.0 cm long, is made up of pieces of each of the above metals placed end-to-end and expands 0.0580 cm between 0°C and 100°C. Find the length of each portion of the composite bar.

15–78 On a cool (4°C) Saturday morning, a pilot fills the fuel tanks of her Pitts S-2C (a two-seat aerobatic airplane) to their full capacity of 106.0 L. Before flying on Sunday morning, when the temperature is again 4°C, she checks the fuel level and finds only 103.4 L of gasoline in the tanks. She realizes that it was hot on Saturday afternoon, and that thermal expansion of the gasoline caused the missing fuel to empty out of the tank's vent. a) What was the maximum temperature (in °C) reached by the fuel and tank on Saturday afternoon? The coefficient of volume expansion of gasoline is 9.5×10^{-4} K⁻¹, and the tank is made of aluminum. b) In order to have the maximum amount of fuel available for flight, when should the pilot have filled the fuel tanks?

15–79 a) Equation (15–12) gives the stress required to keep the length of a rod constant as its temperature changes. Show that if the length is permitted to change by an amount ΔL when its temperature changes by ΔT, the stress is equal to

$$\frac{F}{A} = Y\left(\frac{\Delta L}{L_0} - \alpha \Delta T\right),$$

where F is the tension on the rod, L_0 is the original length of the rod, A its cross-section area, α its coefficient of linear expansion, and Y its Young's modulus. b) A heavy brass bar has projections at its ends, as in Fig. 15–22. Two fine steel wires fastened between the projections are just taut (zero tension) when the whole system is at 20°C. What is the tensile stress in the steel wires when the temperature of the system is raised to 140°C? Make any simplifying assumptions you think are justified, but state what they are.

FIGURE 15–22 Problem 15–79.

15–80 A steel rod with a length of 0.350 m and an aluminum rod with a length of 0.250 m, both with the same diameter, are placed end-to-end between rigid supports with no initial stress in the rods. The temperature of the rods is now raised by 60.0 C°. What is the stress in each rod? (*Hint:* The length of the combined rod remains the same, but the lengths of the individual rods do not. See Problem 15–79.)

15–81 A steel ring with 2.5000-in. inside diameter at 20.0°C is to be warmed and slipped over a brass shaft with 2.5020-in. outside diameter at 20.0°C. a) To what temperature should the ring be warmed? b) If the ring and shaft together are cooled by some means such as liquid air, at what temperature will the ring just slip off the shaft?

15–82 Bulk Stress Due to a Temperature Increase. a) Prove that if an object under pressure is raised in temperature but not allowed to expand, the increase in pressure is

$$\Delta p = B\beta\Delta T,$$

where the bulk modulus B and the average coefficient of volume expansion β are both assumed positive and constant. b) What pressure is necessary to prevent a steel block from expanding when its temperature is increased from 20.0°C to 35.0°C?

15–83 A liquid is enclosed in a metal cylinder that is provided with a piston of the same metal. The system is originally at a pressure of 1.00 atm (1.013×10^5 Pa) and at a temperature of 30.0°C. The piston is forced down until the pressure on the liquid is increased by 50.0 atm, and then clamped in this position. Find the new temperature at which the pressure of the liquid is again 1.00 atm. Assume that the cylinder is sufficiently strong so that its volume is not altered by changes in pressure, but only by changes in temperature. Use the result derived in Problem 15–82. (*Hint:* See Section 11–6.)

Compressibility of liquid: $k = 8.50 \times 10^{-10}$ Pa^{-1}.

Coefficient of volume expansion of liquid: $\beta = 4.80 \times 10^{-4}$ K^{-1}.

Coefficient of volume expansion of metal: $\beta = 3.90 \times 10^{-5}$ K^{-1}.

15–84 A thirsty mechanic cools a 2.00-L bottle of a soft drink (mostly water) by pouring it into a large aluminum mug with a mass of 0.257 kg and adding 0.120 kg of ice initially at −15.0°C. If the soft drink and mug are initially at 20.0°C, what is the final temperature of the system, assuming no heat losses?

15–85 Spacecraft Reentry. A spacecraft made of aluminum circles the earth at a speed of 7700 m/s. a) Find the ratio of its kinetic energy to the energy required to raise its temperature from 0°C to 600°C. (The melting point of aluminum is 660°C. Assume a constant specific heat capacity of 910 J/kg · K.) b) Discuss the bearing of your answer on the problem of reentry of a manned space vehicle into the earth's atmosphere.

15–86 A capstan is a rotating drum or cylinder over which a rope or cord slides in order to provide a great amplification of the rope's tension while keeping both ends free (Fig. 15–23). Since the added tension in the rope is due to friction, the capstan generates thermal energy. a) If the difference in tension between the two ends of the rope is 520 N and the capstan has a diameter of 10.0 cm and turns once in 0.900 s, find the rate at which thermal energy is being generated. Why does the number of turns not matter? b) If the capstan is made of iron and has a mass of 6.00 kg, at what rate does its temperature rise? Assume that the temperature in the capstan is uniform and that all the thermal energy generated flows into it.

FIGURE 15–23 Problem 15–86.

15–87 Debye's T^3 Law. At very low temperatures the molar heat capacity of rock salt varies with temperature according to Debye's T^3 law:

$$C = k\frac{T^3}{\Theta^3},$$

where $k = 1940$ J/mol · K and $\Theta = 281$ K. a) How much heat is required to raise the temperature of 1.50 mol of rock salt from 10.0 K to 40.0 K? (*Hint:* Use Eq. (15–18) in the form $dQ = nCdT$ and integrate.) b) What is the average molar heat capacity in this range? c) What is the true molar heat capacity at 40.0 K?

15–88 A person of mass 70.0 kg is sitting in the bathtub. The bathtub is 190 cm by 80 cm; before the person got in, the water was 10 cm deep. The water is at a temperature of 37.0°C. Suppose that the water were spontaneously to cool down to form ice at 0.0°C, and that all the energy released was used to launch the hapless bather vertically into the air. How high would the bather go? (As you will see in Chapter 18, this event is allowed by energy conservation but is prohibited by the second law of thermodynamics.)

15–89 Hot Air in a Physics Lecture. a) A typical student listening attentively to a physics lecture has a heat output of 100 W. How much heat energy does a class of 90 physics students release into a lecture hall over the course of a 50-min lecture? b) Assume that all the heat energy in part (a) is transferred to the 3200 m^3 of air in the room. The air has specific heat capacity 1020 J/kg · K and density 1.20 kg/m^3. If none of the heat escapes and the air conditioning system is off, how much will the temperature of the air in the room rise during the 50-min lecture? c) If the class is taking an exam, the heat output per student rises to 280 W. What is the temperature rise during 50 min in this case?

15-90 The molar heat capacity of a certain substance varies with temperature according to the empirical equation

$$C = 29.5 \text{ J/mol} \cdot \text{K} + (8.20 \times 10^{-3} \text{ J/mol} \cdot \text{K}^2)T.$$

How much heat is necessary to change the temperature of 3.00 mol of this substance from 27°C to 227°C? (*Hint:* Use Eq. (15–18) in the form $dQ = nC \, dT$ and integrate.)

15-91 a) An ice cube with a mass of 0.075 kg is taken from a freezer, where the cube's temperature was −10.0°C, and dropped into a glass of water at 0.0°C. If no heat is gained or lost from outside, how much water will freeze onto the cube? b) Is it possible to add so much ice to the glass that all of the water freezes? Explain.

15-92 Hot Water Versus Steam Heating. In a household hot-water heating system, water is delivered to the radiators at 70.0°C (158.0°F) and leaves at 28.0°C (82.4°F). The system is to be replaced by a steam system in which steam at atmospheric pressure condenses in the radiators and the condensed steam leaves the radiators at 35.0°C (95.0°F). How many kilograms of steam will supply the same heat as was supplied by 1.00 kg of hot water in the first system?

15-93 A copper calorimeter can with a mass of 0.446 kg contains 0.0950 kg of ice. The system is initially at 0.0°C. a) If 0.0350 kg of steam at 100.0°C and 1.00 atm pressure is added to the can, what is the final temperature of the calorimeter can and its contents? b) At the final temperature, how many kilograms are there of ice, how many of liquid water, and how many of steam?

15-94 In a container of negligible mass, 0.140 kg of ice initially at −15.0°C is added to 0.190 kg of water that has a temperature of 35.0°C. a) If no heat is lost to the surroundings, what is the final temperature of the system? b) At the final temperature, how many kilograms are there of ice and how many of liquid water?

15-95 In a container of negligible mass, 0.0400 kg of steam at 100°C and atmospheric pressure is added to 0.200 kg of water at 50.0°C. a) If no heat is lost to the surroundings, what is the final temperature of the system? b) At the final temperature, how many kilograms are there of steam and how many of liquid water?

15-96 A tube leads from a 0.150-kg calorimeter to a flask in which water is boiling under atmospheric pressure. The calorimeter has specific heat capacity 420 J/kg · K, and it originally contains 0.340 kg of water at 15.0°C. Steam is allowed to condense in the calorimeter at atmospheric pressure until the temperature of the calorimeter and contents reaches 71.0°C, at which point the total mass of calorimeter and contents is found to be 0.525 kg. Compute the heat of vaporization of water from these data.

15-97 In a container of negligible mass, 0.150 kg of ice at 0°C and 0.0950 kg of steam at 100°C are added to 0.200 kg of water that has a temperature of 50.0°C. a) If no heat is lost to the surroundings and the pressure in the container is a constant 1.00 atm, what is the final temperature of the system? b) At the final temperature, how many kilograms are there of ice, how many of liquid water, and how many of steam? c) Repeat parts (a) and

(b) if 0.350 kg of ice at 0°C and 0.012 kg of steam at 100°C are added to 0.200 kg of water that has a temperature of 40.0°C.

15-98 One experimental method of measuring an insulating material's thermal conductivity is to construct a box of the material and measure the power input to an electric heater inside the box that maintains the interior at a measured temperature above the outside surface. Suppose that in such an apparatus a power input of 180 W is required to keep the interior surface of the box 65.0 C° (about 120 F°) above the temperature of the outer surface. The total area of the box is 2.18 m², and the wall thickness is 3.9 cm. Find the thermal conductivity of the material in SI units.

15-99 Effect of a Window in a Door. A carpenter builds a solid wood door with dimensions 2.00 m × 0.95 m × 5.0 cm. Its thermal conductivity is $k = 0.120$ W/m · K. The air film on the inner and outer surfaces of the door have the same combined thermal resistance as an additional 1.8 cm thickness of solid wood. The inside air temperature is 20.0°C and the outside air temperature is −8.0°C. a) What is the rate of heat flow through the door? b) By what factor is the heat flow increased if a window 0.50 m on a side is inserted in the door? The glass is 0.45 cm thick and the glass has a thermal conductivity of 0.80 W/m · K. The air films on the two sides of the glass have a total thermal resistance that is the same as an additional 12.0 cm of glass.

15-100 A wood ceiling with thermal resistance R_1 is covered with a layer of insulation with thermal resistance R_2. Prove that the effective thermal resistance of the combination is $R = R_1 + R_2$.

15-101 Compute the ratio of the rate of heat loss through a single-pane window with area 0.15 m² to that for a double-pane window with the same area. The glass of a single pane is 4.2 mm thick, and the air space between the two panes of the double-pane window is 7.0 mm thick. The glass has thermal conductivity 0.80 W/m · K. The air films on the room and outdoor surfaces of either window have a combined thermal resistance of 0.15 m² · K/W.

15-102 Rods of copper, brass, and steel are welded together to form a Y-shaped figure. The cross-section area of each rod is 2.00 cm². The free end of the copper rod is maintained at 100.0°C and the free ends of the brass and steel rods at 0.0°C. Assume that there is no heat loss from the surfaces of the rods. The lengths of the rods are as follows: copper, 13.0 cm; brass, 18.0 cm; steel, 24.0 cm. a) What is the temperature of the junction point? b) What is the heat current in each of the three rods?

15-103 Time Needed for a Lake to Freeze Over. a) When the air temperature is below 0°C, the water at the surface of a lake freezes to form an ice sheet. Why doesn't freezing occur throughout the entire volume of the lake? b) Show that the thickness of the ice sheet formed on the surface of a lake is proportional to the square root of the time if the heat of fusion of the water freezing on the underside of the ice sheet is conducted through the sheet. c) Assuming that the upper surface of the ice sheet is at −10°C and that the bottom surface is at 0°C, calculate the time it will take to form an ice sheet 25 cm thick. d) If the lake in part (c) is uniformly 40 m deep, how long would it take to freeze all the water in the lake? Is this likely to occur?

15–104 A rod is initially at a uniform temperature of 0°C throughout. One end is kept at 0°C, and the other is brought into contact with a steam bath at 100°C. The surface of the rod is insulated so that heat can flow only lengthwise along the rod. The cross-section area of the rod is 2.50 cm², its length is 120 cm, its thermal conductivity is 380 W/m · K, its density is 1.00×10^4 kg/m³, and its specific heat capacity is 520 J/kg · K. Consider a short cylindrical element of the rod 1.00 cm in length. a) If the temperature gradient at the cooler end of this element is 140 C°/m, how many joules of heat energy flow across this end per second? b) If the average temperature of the element is increasing at the rate of 0.250 C°/s, what is the temperature gradient at the other end of the element?

15–105 If the solar radiation energy incident per second on the frozen surface of a lake is 600 W/m² and 70% of this energy is absorbed by the ice, how much time will it take for a 2.50-cm-thick layer of ice to melt? The ice and the water beneath it are at a temperature of 0°C.

15–106 The rate at which radiant energy from the sun reaches the earth's upper atmosphere is about 1.50 kW/m². The distance from earth to sun is 1.50×10^{11} m, and the radius of the sun is 6.96×10^8 m. a) What is the rate of radiation of energy per unit area from the sun's surface? b) If the sun radiates as an ideal blackbody, what is the temperature of its surface?

15–107 A Thermos for Liquid Helium. A physicist uses a cylindrical metal can 0.250 m high and 0.090 m in diameter to store liquid helium at 4.22 K; at that temperature the heat of vaporization of helium is 2.09×10^4 J/kg. Completely surrounding the metal can are walls maintained at the temperature of liquid nitrogen, 77.3 K, with vacuum between the can and the surrounding walls. How much helium is lost per hour? The emissivity of the metal can is 0.200. The only heat transfer between the metal can and the surrounding walls is by radiation.

15–108 Thermal Expansion of an Ideal Gas. a) The pressure p, volume V, number of moles n, and Kelvin temperature T of an ideal gas are related by the equation $pV = nRT$, where R is a constant. Prove that the coefficient of volume expansion for an ideal gas is equal to the reciprocal of the Kelvin temperature if the expansion occurs at constant pressure. b) Compare the coefficients of volume expansion of copper and air at a temperature of 20°C. Assume that air may be treated as an ideal gas and that the pressure remains constant.

15–109 An engineer is developing an electric water heater to provide a continuous supply of hot water. One trial design is shown in Fig. 15–24. Water is flowing at the rate of 0.500 kg/min, the inlet thermometer registers 18.0°C, the voltmeter reads 120 V, and the ammeter reads 15.0 A (corresponding to a power input of (120 V)(15.0 A) = 1800 W). a) When a steady state is finally reached, what is the reading of the outlet thermometer? b) Why is it unnecessary to take into account the heat capacity mc of the apparatus itself?

FIGURE 15–24 Problem 15–109.

15–110 Food Intake of a Hamster. The energy output of an animal engaged in an activity is called the basal metabolic rate (BMR) and is a measure of the conversion of food energy into other forms of energy. A simple calorimeter to measure the BMR consists of an insulated box with a thermometer to measure the temperature of the air. The air has density 1.20 kg/m³ and specific heat capacity 1020 J/kg · K. A 50.0-g hamster is placed in a calorimeter that contains 0.0500 m³ of air at room temperature. a) When the hamster is running in a wheel, the temperature of the air in the calorimeter rises 1.60 C° per hour. How much heat does the running hamster generate in an hour? Assume that all this heat goes into the air in the calorimeter. Neglect the heat that goes into the walls of the box and into the thermometer, and assume that no heat is lost to the surroundings. b) Assuming that the hamster converts seed into heat with an efficiency of 10% and that hamster seed has a food energy value of 24 J/g, how many grams of seed must the hamster eat per hour to supply this energy?

CHALLENGE PROBLEMS

15–111 a) A spherical shell has inner and outer radii a and b, respectively, and the temperatures at the inner and outer surfaces are T_2 and T_1. The thermal conductivity of the material of which the shell is made is k. Derive an equation for the total heat current through the shell. b) Derive an equation for the temperature variation within the shell in part (a); that is, calculate T as a function of r, the distance from the center of the shell. c) A hollow cylinder has length L, inner radius a, and outer radius b, and the temperatures at the inner and outer surfaces are T_2 and T_1. (The cylinder could represent an insulated hot-water pipe, for example.) The thermal conductivity of the material of which the cylinder is made is k. Derive an equation for the total heat current through the walls of the cylinder. d) For the cylinder of part (c), derive an equation for the temperature variation inside the cylinder walls. e) For the spherical shell of part (a) and the hollow cylinder of part (c), show that the equation for the total heat current in each case reduces to Eq. (15–21) for linear heat flow when the shell or cylinder is very thin.

15–112 A steam pipe with a radius of 2.00 cm, carrying steam at 140°C, is surrounded by a cylindrical jacket with inner and outer radii 2.00 cm and 4.00 cm and made of a type of cork with thermal conductivity 4.00×10^{-2} W/m · K. This in turn is surrounded by a cylindrical jacket made of a brand of styrofoam with thermal conductivity 1.00×10^{-2} W/m · K and having inner and outer radii 4.00 cm and 6.00 cm (Fig. 15–25). The outer surface of the styrofoam is in contact with air at 15°C. Assume that this outer surface has a temperature of 15°C. a) What is the temperature at a radius of 4.00 cm, where the two insulating layers meet? b) What is the total rate of transfer of heat out of a 2.00-m length of pipe? (*Hint:* Use the expression derived in part (c) of Challenge Problem 15–111.)

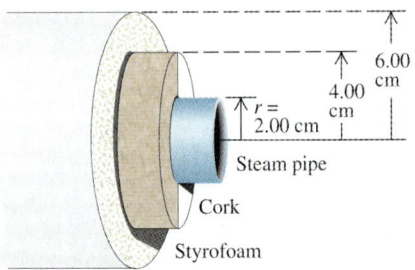

FIGURE 15–25 Challenge Problem 15–112.

15–113 Suppose that both ends of the rod in Fig. 15–13 are kept at a temperature of 0°C, and that the initial temperature distribution along the rod is given by $T = (100°C) \sin \pi x/L$, where x is measured from the left end of the rod. Let the rod be copper, with length $L = 0.100$ m and cross-section area 1.00 cm^2. a) Show the initial temperature distribution in a diagram. b) What is the final temperature distribution after a very long time has elapsed? c) Sketch curves that you think would represent the temperature distribution at intermediate times. d) What is the initial temperature gradient at the ends of the rod? e) What is the initial heat current from the ends of the rod into the bodies making contact with its ends? f) What is the initial heat current at the center of the rod? Explain. What is the heat current at this point at any later time? g) What is the value of the *thermal diffusivity* $k/\rho c$ for copper, and in what unit is it expressed? (Here k is the thermal conductivity, $\rho = 8.9 \times 10^3$ kg/m^3 is the density, and c is the specific heat capacity.) h) What is the initial time rate of change of temperature at the center of the rod? i) How much time would be required for the center of the rod to reach its final temperature if the temperature continued to decrease at this rate? (This time is called the *relaxation time* of the rod.) j) From the graphs in part (c), would you expect the magnitude of the rate of change of temperature at the midpoint to remain constant, increase, or decrease as a function of time? k) What is the initial rate of change of temperature at a point in the rod 2.5 cm from its left end?

15–114 Temperature Change in a Clock. A pendulum clock is designed to tick off one second on each side-to-side swing of the pendulum (two ticks per complete period). a) Will a pendulum clock gain time in hot weather and lose in cold, or the reverse? Explain your reasoning. b) A particular pendulum clock keeps correct time at 20.0°C. The pendulum shaft is of steel, and its mass may be neglected compared with that of the bob. What is the fractional change in length of the shaft when it is cooled to 10.0°C? c) How many seconds per day will the clock gain or lose at 10.0°C? d) How closely must the temperature be controlled if the clock is not to gain or lose more than 1.00 s a day? Does the answer depend on the period of the pendulum?

15–115 A solid cylindrical copper rod 0.200 m long has one end maintained at a temperature of 20.00 K. The other end is blackened and exposed to thermal radiation from surrounding walls at 500 K. The sides of the rod are insulated, so no energy is lost or gained except at the ends of the rod. When equilibrium is reached, what is the temperature of the blackened end? (*Hint:* Since copper is a very good conductor of heat at low temperature, with $k = 1670$ W/m · K at 20 K, the temperature of the blackened end is only slightly greater than 20.00 K.)

15–116 A Walk in the Sun. Consider a poor lost soul walking at 5 km/h on a hot day in the desert, wearing only a bathing suit. This person's skin temperature tends to rise due to four mechanisms: i) energy is generated by metabolic reactions in the body at a rate of 280 W, and almost all of this energy is converted to heat that flows to the skin; ii) heat is delivered to the skin by convection from the outside air at a rate equal to $k'A_{skin}(T_{air} - T_{skin})$, where k' is 54 J/h · C° · m^2, the exposed skin area A_{skin} is 1.5 m^2, the air temperature T_{air} is 47°C, and the skin temperature T_{skin} is 36°C; iii) the skin absorbs radiant energy from the sun at a rate of 1400 W/m^2; iv) the skin absorbs radiant energy from the environment, which has temperature 47°C. a) Calculate the net rate (in watts) at which the person's skin is heated by all four of these mechanisms. Assume that the emissivity of the skin is $e = 1$ and that the skin temperature is initially 36°C. Which mechanism is the most important? b) At what rate (in L/h) must perspiration evaporate from this person's skin to maintain a constant skin temperature? (The heat of vaporization of water at 36°C is 2.42×10^6 J/kg.) c) Suppose instead the person is protected by light-colored clothing ($e \approx 0$) so that the exposed skin area is only 0.45 m^2. What rate of perspiration is required now? Discuss the usefulness of the traditional clothing worn by desert peoples.

Hot-air balloons use the principle that a gas tends to expand when its temperature is increased. When the gas expands so much that the average density of the balloon—including its envelope, gondola, and crew—is the same as that of the surrounding air, the balloon floats. The first successful flight of a human-carrying hot-air balloon was in 1783, more than a century before the first airplane.

16

Thermal Properties of Matter

16–1 INTRODUCTION

The kitchen is a great place to learn about how the properties of matter depend on temperature. When you boil water in a tea kettle, the increase in temperature produces steam that whistles out of the spout at high pressure. If you forget to poke holes in a potato before baking it, the high-pressure steam produced inside the potato can cause it to explode messily. Water vapor in the air can condense into droplets of liquid on the sides of a glass of ice water; if the glass is just out of the freezer, frost will form on the sides as water vapor changes to a solid.

All of these examples show the interrelations between the large-scale or *macroscopic* properties of a substance such as pressure, volume, temperature, and mass of substance. But we can also describe a substance using a *microscopic* perspective. This means investigating small-scale quantities such as the masses, speeds, kinetic energies, and momenta of the individual molecules that make up a substance.

The macroscopic and microscopic descriptions are intimately related. For example, the (microscopic) collision forces that occur when air molecules strike a solid surface (such as your skin) cause (macroscopic) atmospheric pressure. Standard atmospheric pressure is 1.01×10^5 Pa; to produce this pressure, 10^{32} molecules strike your skin every day with an average speed of over 1700 km/h (1000 mi/h)!

In this chapter we'll use both macroscopic and microscopic approaches to gain an understanding of the thermal properties of matter. One of the simplest kinds of matter to understand is the *ideal gas*. For such a gas, we'll be able to relate pressure, volume, temperature, and amount of substance to one another and to the speeds and masses of individual molecules. We will explore the molecular basis of heat capacities of both gases and solids, and we will take a look at the various *phases* of matter—gas, liquid, and solid—and the conditions under which each occurs.

16–2 EQUATIONS OF STATE

The conditions in which a particular material exists are described by physical quantities such as pressure, volume, temperature, and amount of substance. For example, a tank of oxygen in a welding outfit has a pressure gauge and a label stating its volume. We could add a thermometer and place the tank on a scale to determine its mass. These variables describe the *state* of the material and are called **state variables.**

The volume V of a substance is usually determined by its pressure p, temperature T, and amount of substance, described by the mass m or number of moles n. Ordinarily, we can't change one of these variables without causing a change in another. When the tank of oxygen gets hotter, the pressure increases. If the tank gets too hot, it explodes; this happens occasionally with overheated steam boilers.

In a few cases the relationship among p, V, T, and m (or n) is simple enough that we can express it as an equation called the **equation of state.** When it's too complicated for that, we can use graphs or numerical tables. Even then, the relation among the variables still exists; we call it an equation of state even when we don't know the actual equation.

Here's a simple (though approximate) equation of state for a solid material. The temperature coefficient of volume expansion β is the fractional volume change $\Delta V/V_0$ per unit temperature change, and the compressibility k is the negative of the fractional volume change $\Delta V/V_0$ per unit pressure change. If a certain amount of material has volume V_0 when the pressure is p_0 and the temperature is T_0, the volume V at slightly differing pressure p and temperature T is approximately

$$V = V_0[1 + \beta(T - T_0) - k(p - p_0)]. \tag{16-1}$$

(There is a negative sign in front of the term $k(p - p_0)$ because an *increase* in pressure causes a *decrease* in the volume.) Equation (16–1) is called an *equation of state* for the material.

THE IDEAL GAS EQUATION

Another simple equation of state is the one for an *ideal gas*. Figure 16–1 shows an experimental setup to study the behavior of a gas. The cylinder has a movable piston to vary the volume, the temperature can be varied by heating, and we can pump any desired amount of any gas into the cylinder. We then measure the pressure, volume, temperature, and amount of gas. Note that *pressure* refers both to the force per unit area exerted by the cylinder on the gas and to the force per unit area exerted by the gas on the cylinder; by Newton's third law, these must be equal.

It is usually easiest to describe the amount of gas in terms of the number of moles n, rather than the mass. We did this when we defined molar heat capacity in Section 15–6; you may want to review that section. The **molar mass** M of a compound (sometimes called *molecular weight*) is the mass per mole, and the total mass m_{tot} of a given quantity of that compound is the number of moles n times the mass per mole M:

$$m_{tot} = nM \qquad \text{(total mass, number of moles, and molar mass).} \tag{16-2}$$

We are calling the total mass m_{tot} because later in the chapter we will use m for the mass of one molecule.

Measurements of the behavior of various gases lead to several conclusions. First, the volume V is proportional to the number of moles n. If we double the number of moles, keeping pressure and temperature constant, the volume doubles.

Second, the volume varies *inversely* with the absolute pressure p. If we double the pressure while holding the temperature T and number of moles n constant, the gas compresses to one half of its initial volume. In other words, $pV = \text{constant}$ when n and T are constant.

Third, the pressure is proportional to the *absolute* temperature. If we double the absolute temperature, keeping the volume and number of moles constant, the pressure doubles. In other words, $p = (\text{constant})T$ when n and V are constant.

These three relationships can be combined neatly into a single equation, called the **ideal-gas equation:**

$$pV = nRT \qquad \text{(ideal-gas equation),} \tag{16-3}$$

where R is a proportionality constant. An **ideal gas** is one for which Eq. (16–3) holds precisely for *all* pressures and temperatures. This is an idealized model; it works best at very low pressures and high temperatures, when the gas molecules are far apart and in rapid motion. It is reasonably good (within a few percent) at moderate pressures (such as a few atmospheres) and at temperatures well above those at which the gas liquefies.

We might expect that the constant R in the ideal-gas equation would have different values for different gases, but it turns out to have the same value for *all* gases, at least at

16–1 A hypothetical setup for studying the behavior of gases. The pressure p, volume V, temperature T, and number of moles of a gas can be controlled and measured.

8.4
State Variables and Ideal Gas Law

sufficiently high temperature and low pressure. It is called the **gas constant** (or *ideal-gas constant*). The numerical value of R depends on the units of p, V, and T. In SI units, in which the unit of p is Pa ($1 \text{ Pa} = 1 \text{ N/m}^2$) and the unit of V is m^3, the current best numerical value of R is

$$R = 8.314510(70) \text{ J/mol} \cdot \text{K}.$$

Note that the units of pressure times volume are the same as units of work or energy (for example, N/m^2 times m^3); that's why R has units of energy per mole per unit of absolute temperature. In chemical calculations, volumes are often expressed in liters (L) and pressures in atmospheres (atm). In this system, to four significant digits,

$$R = 0.08206 \, \frac{\text{L} \cdot \text{atm}}{\text{mol} \cdot \text{K}}.$$

We can express the ideal-gas equation, Eq. (16–3), in terms of the mass m_{tot} of gas, using $m_{tot} = nM$ from Eq. (16–2):

$$pV = \frac{m_{tot}}{M} RT. \tag{16–4}$$

From this we can get an expression for the density $\rho = m_{tot}/V$ of the gas:

$$\rho = \frac{pM}{RT}. \tag{16–5}$$

CAUTION ▶ When using Eq. (16–5), be certain that you distinguish between the Greek letter ρ ("rho") for density and the letter p for pressure. ◀

For a *constant mass* (or constant number of moles) of an ideal gas the product nR is constant, so the quantity pV/T is also constant. If the subscripts 1 and 2 refer to any two states of the same mass of a gas, then

$$\frac{p_1 V_1}{T_1} = \frac{p_2 V_2}{T_2} = \text{constant} \qquad \text{(ideal gas, constant mass).} \tag{16–6}$$

Notice that you don't need the value of R to use this equation.

The proportionality of pressure to absolute temperature is familiar; in fact, in Chapter 15 we *defined* a temperature scale in terms of pressure in a constant-volume gas thermometer. That may make it seem that the pressure-temperature relation in the ideal-gas equation, Eq. (16–3), is just a result of the way we define temperature. But the equation also tells us what happens when we change the volume or the amount of substance. Also, the gas-thermometer scale turns out to correspond closely to a temperature scale that we will define in Chapter 18 that doesn't depend on the properties of any particular material. For now, consider this equation as being based on this genuinely material-independent temperature scale, even though we haven't defined it yet.

Problem–Solving Strategy

IDEAL GASES

1. In some problems you will be concerned with only one state of the system; some of the quantities in Eq. (16–3) will be known, some unknown. Make a list of what you know and what you have to find. For example, $p = 1.0 \times 10^6$ Pa, $V = 4 \text{ m}^3$, $T = ?$, $n = 2$ mol, or something comparable.

2. In other problems you will compare two different states of the same amount of gas. Decide which is state 1 and

which is state 2, and make a list of the quantities for each: $p_1, p_2, V_1, V_2, T_1, T_2$. If all but one of these quantities are known, you can use Eq. (16–6). Otherwise, you have to use Eq. (16–3). For example, if only p_1, V_1, and n are given, you can't use Eq. (16–6) to find T_1.

3. As always, be sure to use a consistent set of units. If you need the value of the gas constant R, choose its units and then convert the units of the other quantities accordingly.

You may have to convert atmospheres to pascals or liters to cubic meters ($1 \text{ m}^3 = 10^3 \text{ L} = 10^6 \text{ cm}^3$). Sometimes the problem statement will make one system of units clearly more convenient than others. Decide on your system, and stick to it.

4. Don't forget that T must always be an *absolute* temperature. If you are given temperatures in °C, be sure to convert them to Kelvin temperatures by adding 273.15 (to three significant figures, 273). Likewise, p is always the absolute pressure, never the gauge pressure.

5. You may sometimes have to convert between mass m_{tot} and number of moles n. The relationship is $m_{tot} = Mn$, where M is the molar mass. Here's a tricky point: If you use Eq. (16–4), you *must* use the same mass units for m_{tot} and M. So if M is in grams per mole (the usual units for molar mass), then m_{tot} must also be in grams. If you want to use m_{tot} in kg, then you must convert M to kg/mol. For example, the molar mass of oxygen is 32 g/mol or 32×10^{-3} kg/mol. Be careful!

EXAMPLE 16–1

Volume of a gas at STP The condition called **standard temperature and pressure** (STP) for a gas is defined to be a temperature of 0°C = 273.15 K and a pressure of 1 atm = 1.013×10^5 Pa. If you want to keep a mole of an ideal gas in your room at STP, how big a container do you need?

SOLUTION From Eq. (16–3), using R in J/mol · K,

$$V = \frac{nRT}{p} = \frac{(1 \text{ mol})(8.315 \text{ J/mol} \cdot \text{K})(273.15 \text{ K})}{1.013 \times 10^5 \text{ Pa}}$$

$$= 0.0224 \text{ m}^3 = 22.4 \text{ L}.$$

This is almost exactly the volume of three basketballs. A cube 0.282 m on a side would also do the job.

EXAMPLE 16–2

Compressing gas in an automobile engine In an automobile engine, a mixture of air and gasoline is compressed in the cylinders before being ignited. A typical engine has a compression ratio of 9.00 to 1; this means that the gas in the cylinders is compressed to 1/(9.00) of its original volume (Fig. 16–2). The initial pressure is 1.00 atm, and the initial temperature is 27°C. If the pressure after compression is 21.7 atm, find the temperature of the compressed gas.

SOLUTION Let state 1 be the uncompressed gas, and let state 2 be the fully compressed gas. Then $p_1 = 1.00$ atm, $p_2 = 21.7$ atm, and $V_1 = 9.00 \, V_2$. Converting temperature to the Kelvin scale by adding 273, $T_1 = 300$ K; the final temperature T_2 is unknown. The intake and exhaust valves at the top of the cylinder in Fig. 16–2 stay closed during the compression, so the number of moles of gas n is constant. Hence we can use Eq. (16–6):

$$\frac{p_1 V_1}{T_1} = \frac{p_2 V_2}{T_2}.$$

The temperature of the compressed gas is

$$T_2 = T_1 \frac{p_2 V_2}{p_1 V_1} = (300 \text{ K}) \frac{(21.7 \text{ atm}) V_2}{(1.00 \text{ atm})(9.00 \, V_2)}$$

$$= 723 \text{ K} = 450°\text{C}.$$

16–2 Cutaway of an automobile engine. While the air-gasoline mixture is being compressed prior to ignition, the intake and exhaust valves are both in the closed (up) position.

We didn't need to know the values of V_1 or V_2, only their ratio. Note that T_2 is the temperature of the air-gasoline mixture *before* the mixture is ignited; when burning starts, the temperature becomes higher still.

EXAMPLE 16–3

Mass of air in a scuba tank A typical tank used for scuba diving has a volume of 11.0 L (about 0.4 ft³) and a gauge pressure, when full, of 2.10×10^7 Pa (about 3000 psig). The "empty" tank contains 11.0 L of air at 21°C and 1 atm (1.013×10^5 Pa). When the tank is filled with hot air from a compressor, the temperature is 42°C and the gauge pressure is 2.10×10^7 Pa. What mass of air was added? (Air is a mixture of gases, about 78% nitrogen, 21% oxygen, and 1% miscellaneous; its average molar mass is 28.8 g/mol = 28.8×10^{-3} kg/mol.)

SOLUTION Let's first find the number of moles in the tank at the beginning (state 1) and at the end (state 2). We must remember to convert the temperatures to the Kelvin scale by adding 273 and to convert the pressure to absolute by adding 1.013×10^5 Pa. From Eq. (16–3), the number of moles n_1 in the "empty" tank is

$$n_1 = \frac{p_1 V_1}{RT_1} = \frac{(1.013 \times 10^5 \text{ Pa})(11.0 \times 10^{-3} \text{ m}^3)}{(8.315 \text{ J/mol} \cdot \text{K})(294 \text{ K})} = 0.46 \text{ mol}.$$

The volume of the metal tank is hardly affected by the increased pressure, so $V_1 = V_2$. The number of moles in the full tank is

$$n_2 = \frac{p_2 V_2}{RT_2} = \frac{(21.1 \times 10^6 \text{ Pa})(11.0 \times 10^{-3} \text{ m}^3)}{(8.315 \text{ J/mol} \cdot \text{K})(315 \text{ K})} = 88.6 \text{ mol}.$$

We added $n_2 - n_1 = 88.6$ mol $- 0.46$ mol $= 88.1$ mol to the tank. The added mass is not insubstantial: $m(n_2 - n_1)$ = (88.1 mol)(28.8×10^{-3} kg/mol) = 2.54 kg.

Could this problem have been solved in the same way as Example 16–2? The volume is constant, so $p/nT = R/V$ is constant and $p_1/n_1 T_1 = p_2/n_2 T_2$; this can be solved for n_2/n_1, the ratio of the final and initial number of moles. But we need the *difference* of these two numbers, not the ratio, so this equation by itself isn't enough to solve the problem.

EXAMPLE 16–4

Variation of atmospheric pressure with elevation Find the variation of atmospheric pressure with elevation in the earth's atmosphere, assuming that the temperature is 0°C at all elevations. Ignore the variation of g with elevation.

SOLUTION We begin with the pressure relation that we derived in Section 14–3, Eq. (14–4): $dp/dy = -\rho g$. The density ρ is given by Eq. (16–5): $\rho = pM/RT$. We substitute this into Eq. (14–4), separate variables, and integrate, letting p_1 be the pressure at elevation y_1 and p_2 be the pressure at y_2:

$$\frac{dp}{dy} = -\frac{pM}{RT} g,$$

$$\int_{p_1}^{p_2} \frac{dp}{p} = -\frac{Mg}{RT} \int_{y_1}^{y_2} dy,$$

$$\ln \frac{p_2}{p_1} = -\frac{Mg}{RT}(y_2 - y_1),$$

$$\frac{p_2}{p_1} = e^{-Mg(y_2 - y_1)/RT}.$$

Now let $y_1 = 0$ be at sea level, and let the pressure at that point be $p_0 = 1.013 \times 10^5$ Pa. Then our final expression for the pressure p at any height y is

$$p = p_0 e^{-Mgy/RT}.$$

At the summit of Mount Everest, where $y = 8863$ m,

$$\frac{Mgy}{RT} = \frac{(28.8 \times 10^{-3} \text{ kg/mol})(9.80 \text{ m/s}^2)(8863 \text{ m})}{(8.315 \text{ J/mol} \cdot \text{K})(273 \text{ K})} = 1.10,$$

$$p = (1.013 \times 10^5 \text{ Pa})e^{-1.10} = 0.337 \times 10^5 \text{ Pa}$$

$$= 0.33 \text{ atm}.$$

The assumption of constant temperature isn't realistic, and g decreases a little with increasing elevation (see Challenge Problem 16–76). Even so, this example shows why mountaineers need to carry oxygen on Mount Everest.

The ability of the human body to absorb oxygen from the atmosphere depends critically on atmospheric pressure. Absorption drops sharply when the pressure is less than about 0.65×10^5 Pa, corresponding to an elevation above sea level of about 4700 m (15,000 ft). There is no permanent human habitation on earth above 6000 m (about 20,000 ft), although survival for short periods of time is possible at higher elevations. Jet airplanes, which typically fly at altitudes of 8000 to 12,000 m, *must* have pressurized cabins for passenger comfort and health.

THE VAN DER WAALS EQUATION

The ideal-gas equation, Eq. (16–3), can be obtained from a simple molecular model that ignores the volumes of the molecules themselves and the attractive forces between them. We'll examine that model in Section 16–4. Meanwhile, we mention another equation of state, the **van der Waals equation,** that makes approximate corrections for these two omissions. This equation was developed by the 19th-century Dutch physicist J. D. van der Waals; the interaction between atoms that we discussed in Section 13–5 was named the *van der Waals interaction* after him. The van der Waals equation is

$$\left(p + \frac{an^2}{V^2}\right)(V - nb) = nRT. \tag{16–7}$$

The constants a and b are empirical constants, different for different gases. Roughly speaking, b represents the volume of a mole of molecules; the total volume of the molecules is then nb, and the net volume available for the molecules to move around in is $V - nb$. The constant a depends on the attractive intermolecular forces, which reduce the pressure of the gas for given values of n, V, and T by *pulling* the molecules together as they *push* on the walls of the container. The decrease in pressure is proportional to the number of molecules per unit volume in a layer near the wall (which are exerting the pressure on the wall) and is also proportional to the number per unit volume in the next layer beyond the wall (which are doing the attracting). Hence the decrease in pressure due to intermolecular forces is proportional to n^2/V^2.

When n/V is small (that is, when the gas is *dilute*), the average distance between molecules is large, the corrections in the van der Waals equation become insignificant, and Eq. (16–7) reduces to the ideal-gas equation. As an example, for carbon dioxide gas (CO_2) the constants in the van der Waals equation are $a = 0.364$ J $\cdot$ m^3/mol^2 and $b = 4.27 \times 10^{-5}$ m^3/mol. We found in Example 16–1 that one mole of an ideal gas at $T = 0°C = 273.15$ K and $p = 1$ atm $= 1.013 \times 10^5$ Pa occupies a volume $V = 0.0224$ m^3; according to Eq. (16–7), one mole of CO_2 occupying this volume at this temperature would be at a pressure 532 Pa less than 1 atm, a difference of only 0.5% from the ideal-gas value.

pV-DIAGRAMS

We could in principle represent the p-V-T relationship graphically as a *surface* in a three-dimensional space with coordinates p, V, and T. This representation sometimes helps in grasping the overall behavior of the substance, but ordinary two-dimensional graphs are usually more convenient. One of the most useful of these is a set of graphs of pressure as a function of volume, each for a particular constant temperature. Such a diagram is called a **pV-diagram.** Each curve, representing behavior at a specific temperature, is called an **isotherm,** or a *pV-isotherm*.

Figure 16–3 shows *pV*-isotherms for a constant amount of an ideal gas. The highest temperature is T_4; the lowest is T_1. This is a graphical representation of the ideal-gas equation of state. We can read off the volume V corresponding to any given pressure p and temperature T in the range shown.

Figure 16–4 shows a *pV*-diagram for a material that *does not* obey the ideal-gas equation. At temperatures below T_c the isotherms develop flat regions in which we can compress the material without an increase in pressure. Observation of the gas shows that it is *condensing* from the vapor (gas) to the liquid phase. The flat parts of the isotherms in the shaded area of Fig. 16–4 represent conditions of liquid-vapor *phase equilibrium*. As the volume decreases, more and more material goes from vapor to liquid, but the pressure does not change. (To keep the temperature constant during condensation, we have to remove the heat of vaporization, discussed in Section 15–7.)

When we compress such a gas at a constant temperature T_2 in Fig. 16–4, it is vapor until point a is reached. Then it begins to liquefy; as the volume decreases further, more material liquefies, and *both* the pressure and the temperature remain constant. At point b, all the material is in the liquid state. After this, any further compression results in a very rapid rise of pressure, because liquids are in general much less compressible than gases. At a lower constant temperature T_1, similar behavior occurs, but the condensation begins at lower pressure and greater volume than at the constant temperature T_2. At temperatures greater than T_c, *no* phase transition occurs as the material is compressed; at the highest temperatures, such as T_4, the curves resemble the ideal-gas curves of Fig. 16–3. We call T_c the *critical temperature* for this material. In Section 16–7 we'll discuss what happens to the phase of the gas above the critical temperature.

We will use *pV*-diagrams often in the next two chapters. We will show that the *area* under a *pV*-curve (whether or not it is an isotherm) represents the *work* done by the system during a volume change. This work, in turn, is directly related to heat transfer and changes in the *internal energy* of the system, which we'll get to in Chapter 17.

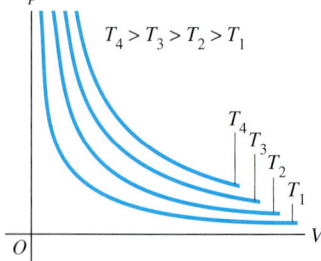

16–3 Isotherms, or constant-temperature curves, for a constant amount of an ideal gas. For each curve, the product $pV = nRT$ is constant, so p is proportional to $1/V$; the proportionality constant increases with increasing T.

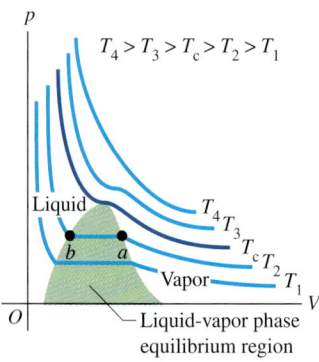

16–4 A *pV*-diagram for a non-ideal gas, showing isotherms for temperatures above and below the critical temperature T_c. The liquid-vapor equilibrium region is shown as a green shaded area. At still lower temperatures the material might undergo phase transitions from liquid to solid or from gas to solid; these are not shown in this diagram.

16-3 MOLECULAR PROPERTIES OF MATTER

We have studied several properties of matter in bulk, including elasticity, density, surface tension, heat capacities, and equations of state, with only passing references to molecular structure. Now we want to look in more detail at the relation of bulk behavior to microscopic structure. We begin with a general discussion of the molecular structure of matter. Then in the next two sections we develop the kinetic-molecular model of an ideal gas, obtaining from this molecular model the equation of state and an expression for heat capacity.

All familiar matter is made up of **molecules.** For any specific chemical compound, all the molecules are identical. The smallest molecules contain one atom each and are of the order of 10^{-10} m in size; the largest contain many atoms and are at least 10,000 times that large. In gases the molecules move nearly independently; in liquids and solids they are held together by intermolecular forces that are electrical in nature, arising from interactions of the electrically charged particles that make up the molecules. Gravitational forces between molecules are negligible in comparison with electrical forces.

The interaction of two *point* electric charges is described by a force (repulsive for like charges, attractive for unlike charges) with a magnitude proportional to $1/r^2$, where r is the distance between the points. We will study this relationship, called *Coulomb's law,* in Chapter 22. Molecules are *not* point charges but complex structures containing both positive and negative charge, and their interactions are more complex. The force between molecules in a gas varies with the distance r between molecules somewhat as shown in Fig. 16-5, where a positive F corresponds to a repulsive force and a negative F to an attractive force. When molecules are far apart, the intermolecular forces are very small and usually attractive. As a gas is compressed and its molecules are brought closer together, the attractive forces increase. The intermolecular force becomes zero at an equilibrium spacing r_0, corresponding roughly to the spacing between molecules in the liquid and solid states. In liquids and solids, relatively large pressures are needed to compress the substance appreciably. This shows that at molecular distances slightly *less* than the equilibrium spacing, the forces become *repulsive* and relatively large.

Figure 16-5 also shows the potential energy as a function of r. This function has a *minimum* at r_0, where the force is zero. The two curves are related by $F(r) = -dU/dr$, as we showed in Section 7-5. Such a potential energy function is often called a **potential well.** A molecule at rest at a distance r_0 from a second molecule would need an additional energy $|U_0|$, the "depth" of the potential well, to "escape" to an indefinitely large value of r.

Molecules are always in motion; their kinetic energies usually increase with temperature. At very low temperatures the average kinetic energy of a molecule may be much *less* than the depth of the potential well. The molecules then condense into the liquid or solid phase with average intermolecular spacings of about r_0. But at higher temperatures the average kinetic energy becomes larger than the depth $|U_0|$ of the potential well. Molecules can then escape the intermolecular force and become free to move independently, as in the gaseous phase of matter.

In *solids,* molecules vibrate about more or less fixed points. In a crystalline solid these points are arranged in a recurring *crystal lattice.* Figure 16-6 shows the cubic crystal structure of sodium chloride (ordinary salt). A scanning tunneling microscope image of individual silicon atoms on the surface of a crystal is shown in Fig. 16-7.

The vibration of molecules in a solid about their equilibrium positions may be nearly simple harmonic if the potential well is approximately parabolic in shape at distances close to r_0. (We discussed this kind of simple harmonic motion in Section 13-5.) But if the potential-energy curve rises more gradually for $r > r_0$ than for $r < r_0$, as in Fig. 16-5, the average position shifts to larger r with increasing amplitude. As we pointed out in

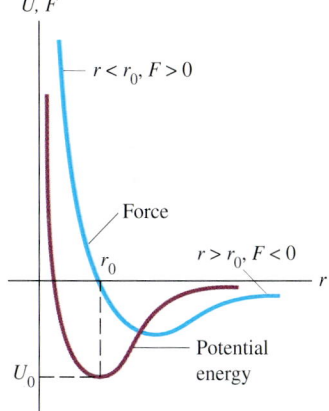

16-5 The force between two molecules (blue curve) is zero at a separation $r = r_0$, where the potential energy (dark red curve) is a minimum. The force is attractive when the separation is greater than r_0 and repulsive when the separation is less than r_0. (Compare this figure to Fig. 13-17.)

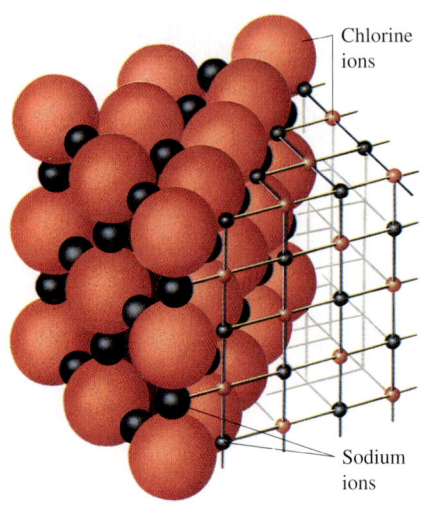

Chlorine ions

Sodium ions

16–6 Schematic representation of the cubic crystal structure of sodium chloride.

Section 15–5, this is the basis of thermal expansion.

In a *liquid,* the intermolecular distances are usually only slightly greater than in the solid phase of the same substance, but the molecules have much greater freedom of movement. Liquids show regularity of structure only in the immediate neighborhood of a few molecules. This is called *short-range order,* in contrast with the *long-range order* of a solid crystal.

The molecules of a *gas* are usually widely separated and so have only very small attractive forces. A gas molecule moves in a straight line until it collides with another molecule or with a wall of the container. In molecular terms, an *ideal gas* is a gas whose molecules exert *no* attractive forces on each other and therefore have no *potential* energy.

At low temperatures, most common substances are in the solid phase. As the temperature rises, a substance melts and then vaporizes. From a molecular point of view, these transitions are in the direction of increasing molecular kinetic energy. Thus temperature and molecular kinetic energy are closely related.

We have used the mole as a measure of quantity of substance. One **mole** of any pure chemical element or compound contains a definite number of molecules, the same number for all elements and compounds. The official SI definition is that **one mole is the amount of substance that contains as many elementary entities as there are atoms in 0.012 kilogram of carbon 12.** In our discussion, the "elementary entities" are molecules. (In a monatomic substance such as carbon or helium, each molecule is a single

16–7 A scanning tunneling microscope image of the surface of a silicon crystal. The area shown is only 9.0 nm (9.0×10^{-9} m) wide. Each blue "bead" is an individual silicon atom; you can clearly see how these atoms are arranged in a (nearly) perfect array of hexagons.

atom, but we'll still call it a molecule here.) Note that atoms of a given element may occur in any of several isotopes, which are chemically identical but have different atomic masses; "carbon 12" refers to a specific isotope of carbon.

The number of molecules in a mole is called **Avogadro's number** and denoted by N_A. The current best numerical value of N_A is

$$N_A = 6.0221367(36) \times 10^{23} \text{ molecules/mol} \qquad \text{(Avogadro's number)}.$$

The *molar mass M* of a compound is the mass of one mole. This is equal to the mass m of a single molecule multiplied by Avogadro's number.

$$M = N_A m \qquad \text{(molar mass, Avogadro's number, and mass}$$
$$\text{of a molecule)}. \qquad (16\text{–}8)$$

When the molecule consists of a single atom, the term *atomic mass* is often used instead of molar mass or molecular weight.

EXAMPLE 16–5

Find the mass of a single hydrogen atom and the mass of an oxygen molecule.

SOLUTION We use Eq. (16–8). From the periodic table of the elements (Appendix D) the mass per mole of atomic hydrogen (that is, the atomic mass) is 1.008 g/mol. Therefore the mass m_H of a single hydrogen atom is

$$m_H = \frac{1.008 \text{ g/mol}}{6.022 \times 10^{23} \text{ molecules/mol}} = 1.674 \times 10^{-24} \text{ g/molecule}.$$

From Appendix D, the atomic mass of oxygen is 16.0 g/mol, so the molar mass of oxygen, which has diatomic (two-atom)

molecules, is 32.0 g/mol. The mass of a single molecule of O_2 is

$$m_{O_2} = \frac{32.0 \text{ g/mol}}{6.022 \times 10^{23} \text{ molecules/mol}} = 53.1 \times 10^{-24} \text{ g/molecule}.$$

We note that the values in Appendix D are for the *average* atomic masses of a natural sample of each element. Such a sample may contain several different isotopes of the element, each with a different atomic mass. Natural samples of hydrogen and oxygen are almost entirely made up of just one isotope; this is not the case for all elements, however.

16–4 KINETIC-MOLECULAR MODEL OF AN IDEAL GAS

The goal of any molecular theory of matter is to understand the *macroscopic* properties of matter in terms of its atomic or molecular structure and behavior. Such theories are of tremendous practical importance; once we have this understanding, we can design materials to have specific desired properties. Such analysis has led to the development of high-strength steels, glasses with special optical properties, semiconductor materials for electronic devices, and countless other materials essential to contemporary technology.

In this section and the ones that follow, we will consider a simple molecular model of an ideal gas. This *kinetic-molecular model* represents the gas as a large number of particles bouncing around in a closed container. In this section we use the kinetic-molecular model to understand how the ideal-gas equation of state, Eq. (16–3), is related to Newton's laws. In the following section we'll use the kinetic-molecular model to predict the molar heat capacity of an ideal gas. We'll go on to elaborate the model to include "particles" that are not points but have a finite size. We will be able to see why polyatomic gases have larger molar heat capacities than monatomic gases.

The following discussion of the kinetic-molecular model has several steps, and you may need to go over it several times to grasp how it all goes together. Don't get discouraged!

Here are the assumptions of our model:

8.1
Characteristics of a Gas

1. A container with volume V contains a very large number N of identical molecules, each with mass m.
2. The molecules behave as point particles; their size is small in comparison to the average distance between particles and to the dimensions of the container.

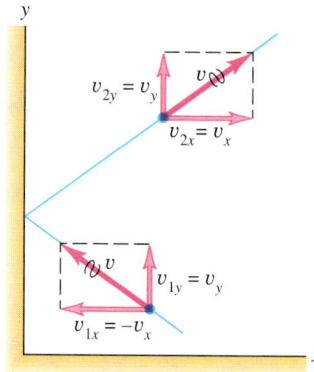

16–8 Elastic collision of a molecule with an idealized container wall. The velocity component parallel to the wall does not change; the velocity component perpendicular to the wall reverses direction. The speed v does not change.

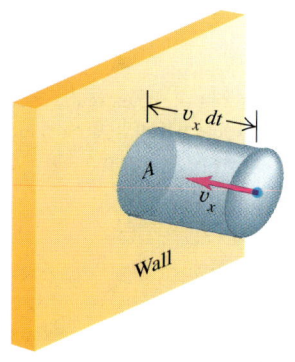

16–9 A molecule moving toward the wall with speed v_x collides with the area A during the time interval dt only if it is within a distance $v_x\,dt$ of the wall at the beginning of the interval. All such molecules are contained within a volume $Av_x\,dt$.

3. The molecules are in constant motion; they obey Newton's laws of motion. Each molecule collides occasionally with a wall of the container. These collisions are perfectly elastic.

4. The container walls are perfectly rigid and infinitely massive and do not move.

CAUTION ▶ Make sure you don't confuse N, the number of *molecules* in the gas, with n, the number of *moles*. ◀

During collisions the molecules exert *forces* on the walls of the container; this is the origin of the *pressure* that the gas exerts. In a typical collision (Fig. 16–8) the velocity component parallel to the wall is unchanged, and the component perpendicular to the wall reverses direction but does not change in magnitude.

Here is our program. First we determine the *number* of collisions per unit time for a certain wall area A. Then we find the total momentum change associated with these collisions and the force needed to cause this momentum change. Then we can determine the pressure, which is force per unit area, and compare the result to the ideal-gas equation. We'll find a direct connection between the temperature of the gas and the kinetic energy of the gas molecules.

To begin, let v_x be the *magnitude* of the x-component of velocity of a molecule of the gas. For now we assume that all molecules have the same v_x. This isn't right, but making this temporary assumption helps to clarify the basic ideas. We will show later that this assumption isn't really necessary.

As shown in Fig. 16–8, for each collision the x-component of velocity changes from $-v_x$ to $+v_x$. So the x-component of momentum changes from $-mv_x$ to $+mv_x$, and the *change* in the x-component of momentum is $mv_x - (-mv_x) = 2mv_x$.

If a molecule is going to collide with a given wall area A during a small time interval dt, then at the beginning of dt it must be within a distance $v_x\,dt$ from the wall (Fig. 16–9) and it must be headed toward the wall. So the number of molecules that collide with A during dt is equal to the number of molecules within a cylinder with base area A and length $v_x\,dt$ that have their x-velocity aimed toward the wall. The volume of such a cylinder is $A\,v_x\,dt$. Assuming that the number of molecules per unit volume (N/V) is uniform, the *number* of molecules in this cylinder is $(N/V)(Av_x\,dt)$. On the average, half of these molecules are moving toward the wall and half are moving away from it. So the number of collisions with A during dt is

$$\frac{1}{2}\left(\frac{N}{V}\right)(Av_x\,dt).$$

For the system of all molecules in the gas, the total momentum change dP_x during dt is the *number* of collisions multiplied by $2mv_x$:

$$dP_x = \frac{1}{2}\left(\frac{N}{V}\right)(Av_x\,dt)(2mv_x) = \frac{NAmv_x^2\,dt}{V}. \tag{16–9}$$

(We are using capital P for total momentum and small p for pressure. Be careful!) The *rate* of change of momentum component P_x is

$$\frac{dP_x}{dt} = \frac{NAmv_x^2}{V}. \tag{16–10}$$

According to Newton's second law, this rate of change of momentum equals the force exerted by the wall area A on the gas molecules. From Newton's *third* law this is equal and opposite to the force exerted *on* the wall *by* the molecules. Pressure p is the magnitude of the force exerted on the wall per unit area, and we obtain

$$p = \frac{F}{A} = \frac{Nmv_x^2}{V}. \tag{16–11}$$

The pressure exerted by the gas depends on the number of molecules per volume (N/V), the mass m per molecule, and the speed of the molecules.

We mentioned that v_x is really *not* the same for all the molecules. But we could have sorted the molecules into groups having the same v_x within each group, then added up the resulting contributions to the pressure. The net effect of all this is just to replace v_x^2 in Eq. (16–11) by the *average* value of v_x^2, which we denote by $(v_x^2)_{av}$. Furthermore, $(v_x^2)_{av}$ is related simply to the *speeds* of the molecules. The speed v of any molecule is related to the velocity components v_x, v_y, and v_z by

$$v^2 = v_x^2 + v_y^2 + v_z^2.$$

We can average this relation over all molecules:

$$(v^2)_{av} = (v_x^2)_{av} + (v_y^2)_{av} + (v_z^2)_{av}.$$

But there is no real difference in our model between the x-, y-, and z-directions. (Molecular speeds are very fast in a typical gas, so the effects of gravity are negligibly small.) It follows that $(v_x^2)_{av}$, $(v_y^2)_{av}$, and $(v_z^2)_{av}$ must all be *equal*. Hence $(v^2)_{av}$ is equal to $3(v_x^2)_{av}$, and

$$(v_x^2)_{av} = \frac{1}{3}(v^2)_{av},$$

so Eq. (16–11) becomes

$$pV = \frac{1}{3}Nm(v^2)_{av} = \frac{2}{3}N\left[\frac{1}{2}m(v^2)_{av}\right]. \tag{16–12}$$

We notice that $\frac{1}{2}m(v^2)_{av}$ is the average translational kinetic energy of a single molecule. The product of this and the total number of molecules N equals the total random kinetic energy K_{tr} of translational motion of all the molecules. (The notation K_{tr} reminds us that this energy is associated with *translational* motion. There may be additional energies associated with rotational and vibrational motion of molecules.) The product pV equals two thirds of the total translational kinetic energy:

$$pV = \frac{2}{3}K_{tr}. \tag{16–13}$$

Now we compare this with the ideal-gas equation,

$$pV = nRT,$$

which is based on experimental studies of gas behavior. For the two equations to agree, we must have

$$K_{tr} = \frac{3}{2}nRT \qquad \text{(average translational kinetic energy of } n \text{ moles} \atop \text{of ideal gas).} \tag{16–14}$$

This remarkably simple result shows that K_{tr} is *directly proportional* to the absolute temperature T. We will use this important result several times in the following discussion.

The average translational kinetic energy of a single molecule is the total translational kinetic energy K_{tr} of all molecules divided by the number of molecules, N:

$$\frac{K_{tr}}{N} = \frac{1}{2}m(v^2)_{av} = \frac{3nRT}{2N}.$$

Also, the total number of molecules N is the number of moles n multiplied by Avogadro's number N_A, so

$$N = nN_A, \qquad \frac{n}{N} = \frac{1}{N_A},$$

and

$$\frac{K_{tr}}{N} = \frac{1}{2} m(v^2)_{av} = \frac{3}{2}\left(\frac{R}{N_A}\right)T. \tag{16-15}$$

The ratio R/N_A occurs frequently in molecular theory. It is called the **Boltzmann constant,** k:

$$k = \frac{R}{N_A} = \frac{8.315 \text{ J/mol} \cdot \text{K}}{6.022 \times 10^{23} \text{ molecules/mol}}$$

$$= 1.381 \times 10^{-23} \text{ J/molecule} \cdot \text{K}.$$

In terms of k we can rewrite Eq. (16–15) as

$$\frac{1}{2} m(v^2)_{av} = \frac{3}{2} kT \qquad \text{(average translational kinetic energy of} \atop \text{a gas molecule).} \tag{16-16}$$

This shows that the average translational kinetic energy *per molecule* depends only on the temperature, not on the pressure, volume, or kind of molecule. We can obtain the average translational kinetic energy *per mole* by multiplying Eq. (16–16) by Avogadro's number and using the relation $M = N_A m$:

$$N_A \frac{1}{2} m(v^2)_{av} = \frac{1}{2} M(v^2)_{av} = \frac{3}{2} RT \qquad \text{(average translational kinetic} \atop \text{energy per mole of gas).} \tag{16-17}$$

The translational kinetic energy of a mole of ideal-gas molecules depends only on T.

Finally, it is sometimes convenient to rewrite the ideal-gas equation on a molecular basis. We use $N = N_A n$ and $R = N_A k$ to obtain the alternative form of the ideal-gas equation:

$$pV = NkT. \tag{16-18}$$

This shows that we can think of the Boltzmann constant k as a gas constant on a "per-molecule" basis instead of the usual "per-mole" basis for R.

MOLECULAR SPEEDS

From Eqs. (16–16) and (16–17) we can obtain expressions for the square root of $(v^2)_{av}$, called the **root-mean-square speed** v_{rms}:

$$v_{rms} = \sqrt{(v^2)_{av}} = \sqrt{\frac{3kT}{m}} = \sqrt{\frac{3RT}{M}} \qquad \text{(root-mean-square speed of a gas molecule).} \tag{16-19}$$

It might seem more natural to characterize molecular speeds by their *average* value rather than by v_{rms}, but we see that v_{rms} evolves more directly from Eqs. (16–16) and (16–17). To compute the rms speed, we square each molecular speed, add, divide by the number of molecules, and take the square root; v_{rms} is the *root* of the *mean* of the *squares*. Example 16–7 illustrates this procedure.

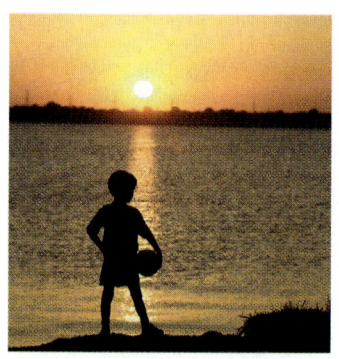

When the sun goes down and the air temperature drops, the average speed of air molecules decreases as well.

Equations (16–16) and (16–19) show that at a given temperature T, gas molecules of different mass m have the same average kinetic energy but different root-mean-square speeds. On average, the nitrogen molecules ($M = 28$ g/mol) in the air around you are moving faster than are the oxygen molecules ($M = 32$ g/mol). Hydrogen molecules ($M = 2$ g/mol) are fastest of all; this is why there is hardly any hydrogen in the earth's atmosphere, despite its being the most common element in the universe. A sizable fraction of any H_2 molecules in the atmosphere would have speeds greater than the earth's "escape speed" of 1.12×10^4 m/s (calculated in Example 12–5, Section 12–4) and would escape into space. The heavier, slower-moving gases cannot escape so easily, which is why they dominate our atmosphere.

The assumption that individual molecules undergo perfectly elastic collisions with the container wall is actually a little too simple. More detailed investigation has shown that in most cases, molecules actually adhere to the wall for a short time and then leave again with speeds that are characteristic of the temperature *of the wall*. However, the gas and the wall are ordinarily in thermal equilibrium and have the same temperature. So there is no net energy transfer between gas and wall, and this discovery does not alter the validity of our conclusions.

Problem–Solving Strategy

KINETIC-MOLECULAR THEORY

As usual, using a consistent set of units is essential. The following are several places where caution is needed.

1. The usual units for molar mass M are grams per mole; the molar mass of oxygen (O_2) is 32 g/mol, for example. These units are often omitted in tables. In equations such as Eq. (16–19), when you use SI units you *must* express M in kilograms per mole by multiplying the table value by (1 kg/10^3 g). Thus in SI units, M for oxygen molecules is 32×10^{-3} kg/mol.

2. Are you working on a "per-molecule" basis or a "per-mole" basis? Remember that m is the mass of a single molecule and M is the mass of a mole of molecules; N is the number of molecules, n is the number of moles; k is the gas constant per molecule, and R is the gas constant per mole. Although N, the number of molecules, is in one sense a dimensionless number, you can do a complete unit check if you think of N as having the unit "molecules"; then m has units "mass per molecule," and k has units "joules per molecule per kelvin."

3. Remember that T is always *absolute* (Kelvin) temperature.

EXAMPLE 16-6

Calculating molecular kinetic energy and v_{rms} a) What is the average translational kinetic energy of a molecule of an ideal gas at a temperature of 27°C? b) What is the total random translational kinetic energy of the molecules in one mole of this gas? c) What is the root-mean-square speed of oxygen molecules at this temperature?

SOLUTION a) To use Eq. (16–16), we first convert the temperature to the Kelvin scale: 27°C = 300 K. Then

$$\frac{1}{2} m(v^2)_{av} = \frac{3}{2} kT = \frac{3}{2}(1.38 \times 10^{-23} \text{ J/K})(300 \text{ K})$$

$$= 6.21 \times 10^{-21} \text{ J}.$$

This answer does not depend on the mass of the molecule.
b) The total translational kinetic energy K_{tr} is just the average translational kinetic energy of a molecule (which we just calculated) multiplied by the number of molecules, in this case Avogadro's number:

$$K_{tr} = N_A \left(\frac{1}{2} m(v^2)_{av} \right)$$

$$= (6.022 \times 10^{23} \text{ molecules})(6.21 \times 10^{-21} \text{ J/molecule})$$

$$= 3740 \text{ J}.$$

We can also get this result from Eq. (16–14):

$$K_{tr} = \frac{3}{2} nRT = \frac{3}{2}(1 \text{ mol})(8.315 \text{ J/mol} \cdot \text{K})(300 \text{ K})$$

$$= 3740 \text{ J}.$$

This is about the same kinetic energy as that of a sprinter in a 100-m dash.
c) From Example 16–5 (Section 16–3) the mass of an oxygen molecule is

$$m_{O_2} = (53.1 \times 10^{-24} \text{ g})(1 \text{ kg}/10^3 \text{ g}) = 5.31 \times 10^{-26} \text{ kg}.$$

From Eq. (16–19),

$$v_{rms} = \sqrt{\frac{3kT}{m}} = \sqrt{\frac{3(1.38 \times 10^{-23} \text{ J/K})(300 \text{ K})}{5.31 \times 10^{-26} \text{ kg}}}$$

$$= 484 \text{ m/s}.$$

This is 1740 km/h, or 1080 mi/h! Alternatively,

$$v_{rms} = \sqrt{\frac{3RT}{M}} = \sqrt{\frac{3(8.315 \text{ J/mol} \cdot \text{K})(300 \text{ K})}{32.0 \times 10^{-3} \text{ kg/mol}}}$$

$$= 484 \text{ m/s}.$$

Note again that when we use Eq. (16–19) with R in SI units, we have to express M in *kilograms* per mole, not grams per mole. In this example we use $M = 32.0 \times 10^{-3}$ kg/mol, *not* 32.0 g/mol.

EXAMPLE 16–7

Five gas molecules chosen at random are found to have speeds of 500, 600, 700, 800, and 900 m/s. Find the rms speed. Is it the same as the *average* speed?

SOLUTION The average value of v^2 for the five molecules is

$$(v^2)_{av} = \frac{\left[\begin{array}{c}(500 \text{ m/s})^2 + (600 \text{ m/s})^2 + (700 \text{ m/s})^2 \\ +(800 \text{ m/s})^2 + (900 \text{ m/s})^2\end{array}\right]}{5}$$

$$= 5.10 \times 10^5 \text{ m}^2/\text{s}^2.$$

The square root of this is v_{rms}:

$$v_{rms} = 714 \text{ m/s}.$$

The *average* speed v_{av} is given by

$$v_{av} = \frac{500 \text{ m/s} + 600 \text{ m/s} + 700 \text{ m/s} + 800 \text{ m/s} + 900 \text{ m/s}}{5}$$

$$= 700 \text{ m/s}.$$

We see that in general, v_{rms} and v_{av} are *not* the same. Roughly speaking, v_{rms} gives greater weight to the larger speeds than does v_{av}.

EXAMPLE 16–8

Volume of a gas at STP, revisited Find the number of molecules and the number of moles in one cubic meter of air at atmospheric pressure and 0°C.

SOLUTION From Eq. (16–18),

$$N = \frac{pV}{kT} = \frac{(1.013 \times 10^5 \text{ Pa})(1 \text{ m}^3)}{(1.38 \times 10^{-23} \text{ J/K})(273 \text{ K})}$$

$$= 2.69 \times 10^{25} \text{ molecules}.$$

The number of moles n is

$$n = \frac{N}{N_A} = \frac{2.69 \times 10^{25} \text{ molecules}}{6.022 \times 10^{23} \text{ molecules/mol}} = 44.7 \text{ mol}.$$

To check: The total volume is 1.00 m³, so the volume of one mole is

$$\frac{1.00 \text{ m}^3}{44.7 \text{ mol}} = 0.0224 \text{ m}^3/\text{mol} = 22.4 \text{ L/mol}.$$

This is the same result that we obtained in Example 16–1 (Section 16–2) using only macroscopic quantities.

COLLISIONS BETWEEN MOLECULES

We have ignored the possibility that two gas molecules might collide. If they are really points, they *never* collide. But consider a more realistic model in which the molecules are rigid spheres with radius r. How often do they collide with other molecules? How far do they travel, on average, between collisions? We can get approximate answers from the following rather primitive model.

Consider N spherical molecules with radius r in a volume V. Suppose only one molecule is moving. When it collides with another molecule, the distance between centers is $2r$. Suppose we draw a cylinder with radius $2r$, with its axis parallel to the velocity of the molecule (Fig. 16–10). The moving molecule collides with any other molecule whose center is inside this cylinder. In a short time dt a molecule with speed v travels a distance $v\,dt$; during this time it collides with any molecule that is in the cylindrical volume of radius $2r$ and length $v\,dt$. The volume of the cylinder is $4\pi r^2 v\,dt$. There are N/V molecules per unit volume, so the number dN with centers in this cylinder is

$$dN = 4\pi r^2 v\,dt\,N/V.$$

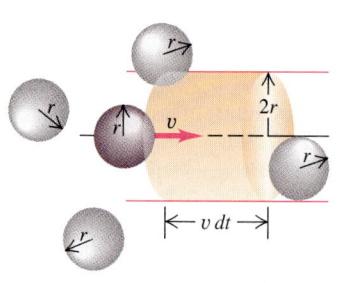

16–10 In a time dt a molecule with radius r will collide with any other molecule within a cylindrical volume of radius $2r$ and length $v\,dt$.

Thus the number of collisions *per unit time* is

$$\frac{dN}{dt} = \frac{4\pi r^2 vN}{V}.$$

This result assumes that only one molecule is moving. The analysis is quite a bit more involved when all the molecules move at once. It turns out that in this case the collisions are more frequent, and the above equation has to be multiplied by a factor of $\sqrt{2}$:

$$\frac{dN}{dt} = \frac{4\pi\sqrt{2}\,r^2 vN}{V}.$$

The average time t_{mean} between collisions, called the *mean free time,* is the reciprocal of this expression:

$$t_{mean} = \frac{V}{4\pi\sqrt{2}r^2 v N}. \qquad (16\text{-}20)$$

The average distance traveled between collisions is called the **mean free path,** denoted by λ (the Greek letter "lambda"). In our simple model, this is just the molecule's speed v multiplied by t_{mean}:

$$\lambda = v t_{mean} = \frac{V}{4\pi\sqrt{2}r^2 N} \qquad \text{(mean free path of a gas molecule).} \qquad (16\text{-}21)$$

The mean free path is inversely proportional to the number of molecules per unit volume (N/V) and inversely proportional to the cross-section area πr^2 of a molecule; the more molecules there are and the larger the size of a molecule, the shorter the mean distance between collisions. Note that the mean free path *does not* depend on the speed of the molecule.

We can express Eq. (16-21) in terms of macroscopic properties of the gas, using the ideal-gas equation in the form of Eq. (16-18), $pV = NkT$. We find

$$\lambda = \frac{kT}{4\pi\sqrt{2}r^2 p}. \qquad (16\text{-}22)$$

If the temperature is increased at constant pressure, the gas expands, the average distance between molecules increases, and λ increases. If the pressure is increased at constant temperature, the gas compresses and λ decreases.

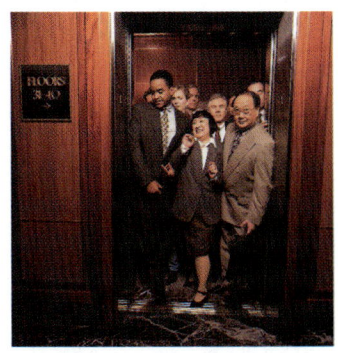

If you try to walk through a crowd, your mean free path—the distance you can travel on average without running into another person—depends on how large the people are and how closely they are spaced.

EXAMPLE 16-9

Calculating mean free path a) Estimate the mean free path of a molecule of air at 27°C and 1 atm. Model the molecules as spheres with radius $r = 2.0 \times 10^{-10}$ m. b) Estimate the mean free time of an oxygen molecule with $v = v_{rms}$.

SOLUTION a) From Eq. (16-22),

$$\lambda = \frac{kT}{4\pi\sqrt{2}r^2 p}$$

$$= \frac{(1.38 \times 10^{-23} \text{ J/K})(300 \text{ K})}{4\pi\sqrt{2}(2.0 \times 10^{-10} \text{ m})^2 (1.01 \times 10^5 \text{ Pa})}$$

$$= 5.8 \times 10^{-8} \text{ m}.$$

The molecule doesn't get very far between collisions, but the distance is still several hundred times the radius of the molecule. To get a mean free path of one meter, the pressure must be about 5.8×10^{-8} atm. Pressures this low are found 100 km or so above

the earth's surface, at the outer fringe of our atmosphere.
b) We could use Eq. (16-20) to find the mean free time t_{mean}, but it's more convenient to use the basic relationship in Eq. (16-21) between t_{mean} and the mean free path: $\lambda = v t_{mean}$, so $t_{mean} = \lambda/v$. From Example 16-6, for oxygen at 27°C the root-mean-square speed is $v_{rms} = 484$ m/s, so the mean free time for a molecule with this speed is

$$t_{mean} = \frac{\lambda}{v} = \frac{5.8 \times 10^{-8} \text{ m}}{484 \text{ m/s}} = 1.2 \times 10^{-10} \text{ s}.$$

This molecule undergoes about 10^{10} collisions per second!

Note that the mean free path calculated in part (a) doesn't depend on the molecule's speed, but the mean free time does. Slower molecules have a longer average time interval t_{mean} between collisions than do fast ones, but the average *distance* λ between collisions is the same no matter what the molecule's speed.

16-5 HEAT CAPACITIES

When we introduced the concept of heat capacity in Section 15-6, we talked about ways to *measure* the specific or molar heat capacity of a particular material. Now we'll see how these numbers can be *predicted* on theoretical grounds. That's a significant step forward.

(a)

(b)

16–11 When an amount of heat dQ is added to (a) a constant volume of monatomic ideal gas molecules, (b) the total translational kinetic energy increases by $dK_{tr} = dQ$, and the temperature increases by $dT = dQ/nC_V$.

HEAT CAPACITIES OF GASES

The basis of our analysis is that heat is *energy* in transit. When we add heat to a substance, we are increasing its molecular energy. In this discussion we will keep the volume of the gas constant so that we don't have to worry about energy transfer through mechanical work. If we were to let the gas expand, it would do work by pushing on moving walls of its container, and this additional energy transfer would have to be included in our calculations. We'll return to this more general case in Chapter 17. For now, with the volume held constant, we are concerned with C_V, the molar heat capacity *at constant volume*.

In the simple kinetic-molecular model of Section 16–4 the molecular energy consists only of the translational kinetic energy K_{tr} of the pointlike molecules. This energy is directly proportional to the absolute temperature T, as shown by Eq. (16–14), $K_{tr} = \frac{3}{2}nRT$. When the temperature changes by a small amount dT, the corresponding change in kinetic energy is

$$dK_{tr} = \frac{3}{2} nR\, dT. \tag{16–23}$$

From the definition of molar heat capacity at constant volume, C_V (Section 15–6), we also have

$$dQ = nC_V\, dT, \tag{16–24}$$

where dQ is the heat input needed for a temperature change dT. Now if K_{tr} represents the total molecular energy, as we have assumed, then dQ and dK_{tr} must be *equal* (Fig. 16–11). Equating the expressions given by Eqs. (16–23) and (16–24), we get

$$nC_V\, dT = \frac{3}{2} nR\, dT,$$

$$C_V = \frac{3}{2} R \qquad \text{(ideal gas of point particles).} \tag{16–25}$$

This surprisingly simple result says that the molar heat capacity (at constant volume) of *every* gas whose molecules can be represented as points is equal to $3R/2$.

To see whether this makes sense, let's first check the units. The gas constant *does* have units of energy per mole per kelvin, the correct units for a molar heat capacity. But more important is whether Eq. (16–25) agrees with *measured* values of molar heat capacities. In SI units, Eq. (16–25) gives

$$C_V = \frac{3}{2}(8.315\ \text{J/mol} \cdot \text{K}) = 12.47\ \text{J/mol} \cdot \text{K}.$$

For comparison, Table 16–1 gives measured values of C_V for several gases. We see that for *monatomic* gases our prediction is right on the money, but that it is way off for diatomic and polyatomic gases.

This comparison tells us that our point-molecule model is good enough for monatomic gases but that for diatomic and polyatomic molecules we need something more sophisticated. For example, we can picture a diatomic molecule as *two* point masses, like a little elastic dumbbell, with an interaction force between the atoms of the kind shown in Fig. 16–5. Such a molecule can have additional kinetic energy associated with *rotation* about axes through its center of mass. The atoms may also have a back-and-forth *vibrating* motion along the line joining them, with additional kinetic and potential energies. These possibilities are shown in Fig. 16–12.

When heat flows into a *monatomic* gas at constant volume, *all* of the added energy goes into an increase in random *translational* molecular kinetic energy. Equation

TABLE 16–1
MOLAR HEAT CAPACITIES OF GASES

TYPE OF GAS	GAS	C_V (J/mol · K)
Monatomic	He	12.47
	Ar	12.47
Diatomic	H_2	20.42
	N_2	20.76
	O_2	21.10
	CO	20.85
Polyatomic	CO_2	28.46
	SO_2	31.39
	H_2S	25.95

(a) Translational motion

(b) Rotational motion

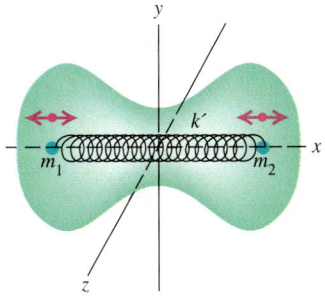

(c) Vibrational motion

16–12 A diatomic molecule. Almost all the mass of each atom is in its tiny nucleus. (a) The center of mass has three independent velocity components. (b) The molecule has two independent axes of rotation through its center of mass. (c) The atoms and "spring" have additional kinetic and potential energies of vibration.

(16–23) shows that this gives rise to an increase in temperature. But when the temperature is increased by the same amount in a *diatomic* or *polyatomic* gas, additional heat is needed to supply the increased rotational and vibrational energies. Thus polyatomic gases have *larger* molar heat capacities than monatomic gases, as Table 16–1 shows.

But how do we know how much energy is associated with each additional kind of motion of a complex molecule, compared to the translational kinetic energy? The new principle that we need is called the principle of **equipartition of energy.** It can be derived from sophisticated statistical-mechanics considerations; that derivation is beyond our scope, and we will treat the principle as an axiom.

The principle of equipartition of energy states that each velocity component (either linear or angular) has, on average, an associated kinetic energy per molecule of $\frac{1}{2}kT$, or one-half the product of the Boltzmann constant and the absolute temperature. The number of velocity components needed to describe the motion of a molecule completely is called the number of **degrees of freedom.** For a monatomic gas, there are three degrees of freedom (for the velocity components v_x, v_y, and v_z); this gives a total average kinetic energy per molecule of $3(\frac{1}{2}kT)$, consistent with Eq. (16–16).

For a *diatomic* molecule there are two possible axes of rotation, perpendicular to each other and to the molecule's axis. (We don't include rotation about the molecule's own axis because in ordinary collisions there is no way for this rotational motion to change.) If we assign five degrees of freedom to a diatomic molecule, the average total kinetic energy per molecule is $\frac{5}{2}kT$ instead of $\frac{3}{2}kT$. The total kinetic energy of n moles is $K_{tot} = nN_A(\frac{1}{2}kT) = \frac{5}{2}n(kN_A)T = \frac{5}{2}nRT$, and the molar heat capacity (at constant volume) is

$$C_V = \frac{5}{2}R \quad \text{(diatomic gas, including rotation).} \quad (16–26)$$

In SI units,

$$C_V = \frac{5}{2}(8.315 \text{ J/mol} \cdot \text{K}) = 20.79 \text{ J/mol} \cdot \text{K}.$$

This agrees within a few percent with the measured values for diatomic gases given in Table 16–1.

Vibrational motion can also contribute to the heat capacities of gases. Molecular bonds are not rigid; they can stretch and bend, and the resulting vibrations lead to additional degrees of freedom and additional energies. For most diatomic gases, however, vibrational motion does *not* contribute appreciably to heat capacity. The reason for this is a little subtle and involves some concepts of quantum mechanics. Briefly, vibrational energy can change only in finite steps. If the energy change of the first step is much larger than the energy possessed by most molecules, then nearly all the molecules remain in the minimum-energy state of motion. In that case, changing the temperature

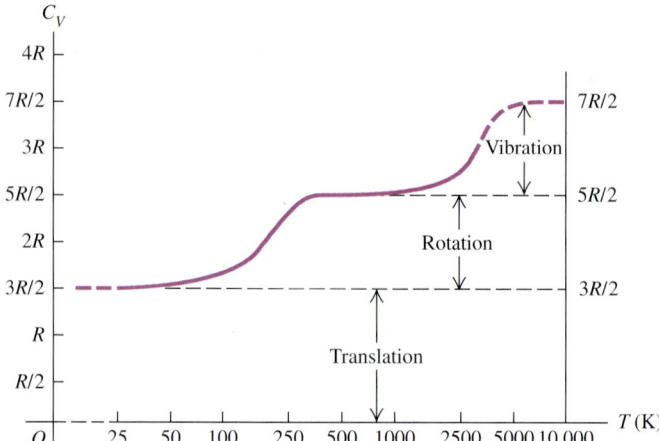

16–13 Experimental values of C_V for hydrogen gas (H_2). The temperature is plotted on a logarithmic scale. Appreciable rotational motion begins to occur above 50 K, and above 600 K the molecule begins to appreciably increase its vibrational motion.

does not change their average vibrational energy appreciably, and the vibrational degrees of freedom are said to be "frozen out." In more complex molecules the gaps between permitted energy levels are sometimes much smaller, and then vibration *does* contribute to heat capacity. The rotational energy of a molecule also changes by finite steps, but they are usually much smaller; the "freezing out" of rotational degrees of freedom occurs only in rare instances, such as for the hydrogen molecule below about 100 K.

In Table 16–1 the large values of C_V for some polyatomic molecules show the contributions of vibrational energy. In addition, a molecule with three or more atoms that are not in a straight line has three, not two, rotational degrees of freedom.

From this discussion we expect heat capacities to be temperature-dependent, generally increasing with increasing temperature. Figure 16–13 is a graph of the temperature dependence of C_V for hydrogen gas (H_2), showing the temperatures at which the rotational and vibrational energies begin to contribute to the heat capacity.

HEAT CAPACITIES OF SOLIDS

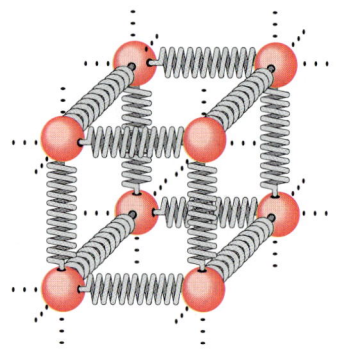

16–14 The forces between neighboring particles in a crystal may be visualized by imagining every particle as being connected to its neighbors by springs.

We can carry out a similar heat-capacity analysis for a crystalline solid. Consider a crystal consisting of N identical atoms (a *monatomic solid*). Each atom is bound to an equilibrium position by interatomic forces. The elasticity of solid materials shows us that these forces must permit stretching and bending of the bonds. We can think of a crystal as an array of atoms connected by little springs (Fig. 16–14). Each atom can *vibrate* about its equilibrium position.

Each atom has three degrees of freedom, corresponding to its three components of velocity. According to the equipartition principle, each atom has an average kinetic energy of $\frac{1}{2}kT$ for each degree of freedom. In addition, each atom has *potential* energy associated with the elastic deformation. For a simple harmonic oscillator (discussed in Chapter 13) it is not hard to show that the average kinetic energy of an atom is *equal* to its average potential energy. In our model of a crystal, each atom is essentially a three-dimensional harmonic oscillator; it can be shown that the equality of average kinetic and potential energies also holds here, provided that the "spring" forces obey Hooke's law.

Thus we expect each atom to have an average kinetic energy $\frac{3}{2}kT$ and an average potential energy $\frac{3}{2}kT$, or an average total energy $3kT$ per atom. If the crystal contains N atoms or n moles, its total energy is

$$E_{\text{tot}} = 3NkT = 3nRT. \qquad (16\text{–}27)$$

16–15 Experimental values of C_V for lead, aluminum, silicon, and diamond. At high temperatures, C_V for each solid approaches about $3R$, in agreement with the rule of Dulong and Petit. At low temperatures, C_V is much less than $3R$.

From this we conclude that the molar heat capacity of a crystal should be

$$C_V = 3R \qquad \text{(ideal monatomic solid)}. \qquad (16–28)$$

In SI units,

$$C_V = (3)(8.315 \text{ J/mol} \cdot \text{K}) = 24.9 \text{ J/mol} \cdot \text{K}.$$

This is the **rule of Dulong and Petit,** which we encountered as an *empirical* finding in Section 15–6: elemental solids all have molar heat capacities of about 25 J/mol · K. Now we have *derived* this rule from kinetic theory. The agreement is only approximate, to be sure, but considering the very simple nature of our model, it is quite significant.

At low temperatures, the heat capacities of most solids *decrease* with decreasing temperature (Fig. 16–15) for the same reason that vibrational degrees of freedom of molecules are frozen out at low temperatures. At very low temperatures the quantity kT is much *smaller* than the smallest energy step the vibrating atoms can take. Hence most of the atoms remain in their lowest energy states because the next higher energy level is out of reach. The average vibrational energy per atom is then *less* than $3kT$, and the heat capacity per molecule is *less* than $3k$. At higher temperatures when kT is *large* in comparison to the minimum energy step, the equipartition principle holds, and the total heat capacity is $3k$ per molecule or $3R$ per mole as the Dulong and Petit relation predicts. Quantitative understanding of the temperature variation of heat capacities was one of the triumphs of quantum mechanics during its initial development in the 1920s.

*16–6 MOLECULAR SPEEDS

As we mentioned in Section 16–4, the molecules in a gas don't all have the same speed. Figure 16–16 shows one experimental scheme for measuring the distribution of molecular speeds. A substance is vaporized in a hot oven; molecules of the vapor escape through an aperture in the oven wall and into a vacuum chamber. A series of slits blocks all molecules except those in a narrow beam, which is aimed at a pair of rotating disks. A molecule passing through the slit in the first disk is blocked by the second disk unless it arrives just as the slit in the second disk is lined up with the beam. The disks function as a speed selector that passes only molecules within a certain narrow speed range. This range can be varied by changing the disk rotation speed, and we can measure how many molecules lie within each of various speed ranges.

To describe the results of such measurements, we define a function $f(v)$ called a *distribution function*. If we observe a total of N molecules, the number dN having speeds in

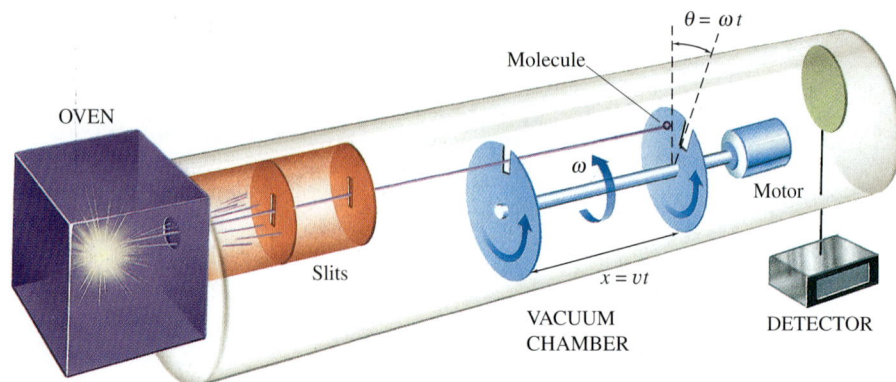

16–16 A molecule with a speed v is passing through the first slit. When it reaches the second slit, the slits have rotated through the offset angle θ. If $v = \omega x/\theta$, the molecule passes through the second slit and reaches the detector.

(a)

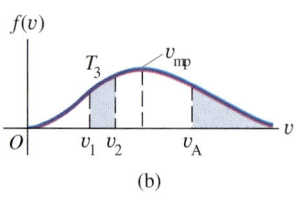

(b)

16–17 (a) Curves of the Maxwell-Boltzmann distribution function $f(v)$ for various temperatures. As the temperature increases, the curve becomes flatter, and its maximum shifts to higher speeds. (b) At temperature T_3 the fraction of molecules having speeds in the range v_1 to v_2 is shown by the shaded area under the T_3 curve. The fraction with speeds greater than v_A is shown by the area from v_A to infinity.

the range between v and $v + dv$ is given by

$$dN = Nf(v)\ dv. \tag{16–29}$$

We can also say that the *probability* that a randomly chosen molecule will have a speed in the interval v to $v + dv$ is $f(v)\ dv$. Hence $f(v)$ is the probability per unit speed *interval*; it is *not* equal to the probability that a molecule has speed exactly equal to v. Since a probability is a pure number, $f(v)$ has units of reciprocal speed (s/m).

Figure 16–17a shows distribution functions for several different temperatures. At each temperature the height of the curve for any value of v is proportional to the number of molecules with speeds near v. The peak of the curve represents the *most probable speed* v_{mp} for the corresponding temperature. As the temperature increases, the average molecular kinetic energy increases, and so the peak of $f(v)$ shifts to higher and higher speeds.

Figure 16–17b shows that the area under a curve between any two values of v represents the fraction of all the molecules having speeds in that range. Every molecule must have *some* value of v, so the integral of $f(v)$ over all v must be unity for any T.

If we know $f(v)$, we can calculate the most probable speed v_{mp}, the average speed v_{av}, and the rms speed v_{rms}. To find v_{mp}, we simply find the point where $df/dv = 0$; this gives the value of the speed where the curve has its peak. To find v_{av}, we take the number $Nf(v)\ dv$ having speeds in each interval dv, multiply each number by the corresponding speed v, add all these products (by integrating over all v from zero to infinity), and finally divide by N. That is,

$$v_{av} = \int_0^\infty vf(v)\ dv. \tag{16–30}$$

The rms speed is obtained similarly; the average of v^2 is given by

$$(v^2)_{av} = \int_0^\infty v^2 f(v)\ dv, \tag{16–31}$$

and v_{rms} is the square root of this.

The function $f(v)$ describing the actual distribution of molecular speeds is called the **Maxwell-Boltzmann distribution.** It can be derived from statistical-mechanics considerations, but that derivation is beyond our scope. Here is the result:

$$f(v) = 4\pi \left(\frac{m}{2\pi kT}\right)^{3/2} v^2 e^{-mv^2/2kT} \quad \text{(Maxwell-Boltzmann distribution).}$$

$$\tag{16–32}$$

We can also express this function in terms of the translational kinetic energy of a molecule, which we denote by ϵ. That is, $\epsilon = \frac{1}{2}mv^2$. We invite you to verify that when this is substituted into Eq. (16–32), the result is

$$f(v) = \frac{8\pi}{m}\left(\frac{m}{2\pi kT}\right)^{3/2} \epsilon e^{-\epsilon/kT}. \tag{16–33}$$

This form shows that the exponent in the Maxwell-Boltzmann distribution function is $-\epsilon/kT$ and that the shape of the curve is determined by the relative magnitude of ϵ and kT at any point. We leave it as an exercise to prove that the *peak* of each curve occurs where $\epsilon = kT$, corresponding to a most probable speed v_{mp} given by

$$v_{mp} = \sqrt{\frac{2kT}{m}}. \tag{16–34}$$

8.2
Maxwell-Boltzmann Distribution—
Conceptual Analysis

8.3
Maxwell-Boltzmann Distribution—
Quantitative Analysis

To find the average speed, we substitute Eq. (16–32) into Eq. (16–30) and carry out the integration, making a change of variable $v^2 = x$ and then integrating by parts. We leave the details as an exercise; the result is

$$v_{av} = \sqrt{\frac{8kT}{\pi m}}. \tag{16–35}$$

Finally, to find the rms speed, we substitute Eq. (16–32) into Eq. (16–31). Evaluating the resulting integral takes some mathematical acrobatics, but we can find it in a table of integrals. The result is

$$v_{rms} = \sqrt{\frac{3kT}{m}}. \tag{16–36}$$

This result agrees with Eq. (16–19); it *must* agree if the Maxwell-Boltzmann distribution is to be consistent with the equipartition theorem and our other kinetic-theory calculations.

Table 16–2 shows the fraction of all the molecules in an ideal gas that have speeds *less than* various multiples of v_{rms}. These numbers were obtained by numerical integration; they are the same for all ideal gases.

The distribution of molecular speeds in liquids is similar, although not identical, to that for gases. We can understand the vapor pressure of a liquid and the phenomenon of boiling on this basis. Suppose a molecule must have a speed at least as great as v_A in Fig. 16–17b to escape from the surface of a liquid into the adjacent vapor. The number of such molecules, represented by the area under the "tail" of each curve (to the right of

TABLE 16–2

FRACTIONS OF MOLECULES IN AN IDEAL GAS WITH SPEEDS LESS THAN VARIOUS MULTIPLES OF v/v_{rms}

v/v_{rms}	FRACTION
0.20	0.011
0.40	0.077
0.60	0.218
0.80	0.411
1.00	0.608
1.20	0.771
1.40	0.882
1.60	0.947
1.80	0.979
2.00	0.993

v_A), increases rapidly with temperature. Thus the rate at which molecules can escape is strongly temperature-dependent. This process is balanced by another one in which molecules in the vapor phase collide inelastically with the surface and are trapped back into the liquid phase. The number of molecules suffering this fate per unit time is proportional to the pressure in the vapor phase. Phase equilibrium between liquid and vapor occurs when these two competing processes proceed at exactly the same rate. So if the molecular speed distributions are known for various temperatures, we can make a theoretical prediction of vapor pressure as a function of temperature. When liquid evaporates, it's the high-speed molecules that escape from the surface. The ones that are left have less energy on average; this gives us a molecular view of evaporative cooling.

Rates of chemical reactions are often strongly temperature-dependent, and the reason is contained in the Maxwell-Boltzmann distribution. When two reacting molecules collide, the reaction can occur only when the molecules are close enough for the electric-charge distributions of their electrons to interact strongly. This requires a minimum energy, called the *activation energy,* and thus a certain minimum molecular speed. Figure 16–17a shows that the number of molecules in the high-speed tail of the curve increases rapidly with temperature. Thus we expect the rate of any reaction that depends on an activation energy to increase rapidly with temperature. Similarly, many plant growth processes have strongly temperature-dependent rates, as can be seen by the rapid and diverse growth in tropical rain forests.

16–7 PHASES OF MATTER

We've talked a lot about ideal gases in the last few sections. An ideal gas is the simplest system to analyze from a molecular viewpoint because we ignore the interactions between molecules. But those interactions are the very thing that makes matter condense into the liquid and solid phases under some conditions. So it's not surprising that theoretical analysis of liquid and solid structure and behavior is a lot more complicated than that for gases. We won't try to go far here with a microscopic picture, but we can talk in general about phases of matter, phase equilibrium, and phase transitions.

In Section 15–7 we learned that each phase is stable only in certain ranges of temperature and pressure. A transition from one phase to another ordinarily takes place under conditions of **phase equilibrium** between the two phases, and for a given pressure this occurs at only one specific temperature. We can represent these conditions on a graph with axes p and T, called a **phase diagram;** Fig. 16–18 shows an example. Each point on the diagram represents a pair of values of p and T. Only a single phase can exist at each point, except for points on the solid lines, where two phases can coexist in phase equilibrium.

16–18 A typical pT phase diagram, showing regions of temperature and pressure at which the various phases exist and where phase changes occur.

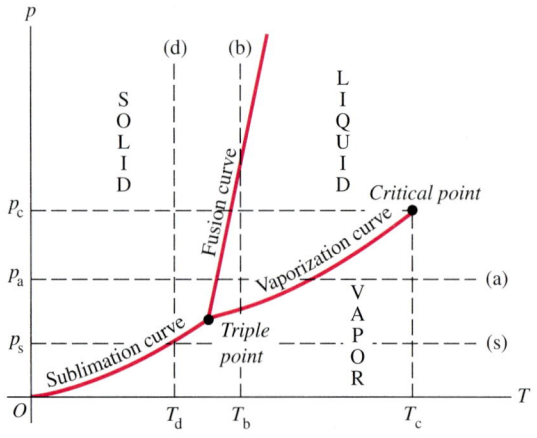

TABLE 16–3

TRIPLE-POINT DATA

SUBSTANCE	TEMPERATURE (K)	PRESSURE (Pa)
Hydrogen	13.80	0.0704×10^5
Deuterium	18.63	0.171×10^5
Neon	24.56	0.432×10^5
Nitrogen	63.18	0.125×10^5
Oxygen	54.36	0.00152×10^5
Ammonia	195.40	0.0607×10^5
Carbon dioxide	216.55	5.17×10^5
Sulfur dioxide	197.68	0.00167×10^5
Water	273.16	0.00610×10^5

These lines separate the diagram into solid, liquid, and vapor regions. For example, the fusion curve separates the solid and liquid areas and represents possible conditions of solid-liquid phase equilibrium. Similarly, the vaporization curve separates the liquid and vapor areas, and the sublimation curve separates the solid and vapor areas. The three curves meet at the **triple point,** the only condition under which all three phases can coexist. In Section 15–4 we used the triple-point temperature of water to define the Kelvin temperature scale. Triple-point data for several substances are given in Table 16–3.

If we add heat to a substance at a constant pressure p_a, it goes through a series of states represented by the horizontal line (a) in Figure 16–18. The melting and boiling temperatures at this pressure are the temperatures at which the line intersects the fusion and vaporization curves, respectively. When the pressure is p_s, constant-pressure heating transforms a substance from solid directly to vapor. This process is called *sublimation;* the intersection of line (s) with the sublimation curve gives the temperature T_s at which it occurs for a pressure p_s. At any pressure less than the triple-point pressure, no liquid phase is possible. The triple-point pressure for carbon dioxide is 5.1 atm. At normal atmospheric pressure, solid carbon dioxide ("dry ice") undergoes sublimation; there is no liquid phase at this pressure.

Line (b) in Fig. 16–18 represents compression at a constant temperature T_b. The material passes from vapor to liquid and then to solid at the points where line (b) crosses the vaporization curve and fusion curve, respectively. Line (d) shows constant-temperature compression at a lower temperature T_d; the material passes from vapor to solid at the point where line (d) crosses the sublimation curve.

We saw in the pV-diagram of Fig. 16–4 that a liquid-vapor phase transition occurs only when the temperature and pressure are less than those at the point lying at the top of the green shaded area labeled "Liquid-vapor phase equilibrium region." This point corresponds to the end point at the top of the vaporization curve in Fig. 16–18. It is called the **critical point,** and the corresponding values of p and T are called the critical pressure and temperature, p_c and T_c. A gas at a pressure *above* the critical pressure does not separate into two phases when it is cooled at constant pressure (along a horizontal line above the critical point in Fig. 16–18). Instead, its properties change gradually and continuously from those we ordinarily associate with a gas (low density, large compressibility) to those of a liquid (high density, small compressibility) *without a phase transition.*

If this stretches credibility, think about liquid-phase transitions at successively higher points on the vaporization curve. As we approach the critical point, the *differences* in physical properties (such as density, bulk modulus, and viscosity) between the liquid and vapor phases become smaller and smaller. Exactly *at* the critical point they all become zero, and at this point the distinction between liquid and vapor disappears. The

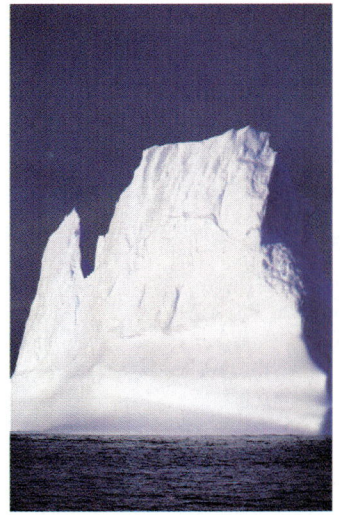

The average temperature and pressure at the earth's surface are close to the triple point of water. That's why water can be found on earth in all three phases: vapor (in the atmosphere), liquid (in the oceans), and solid (like this iceberg).

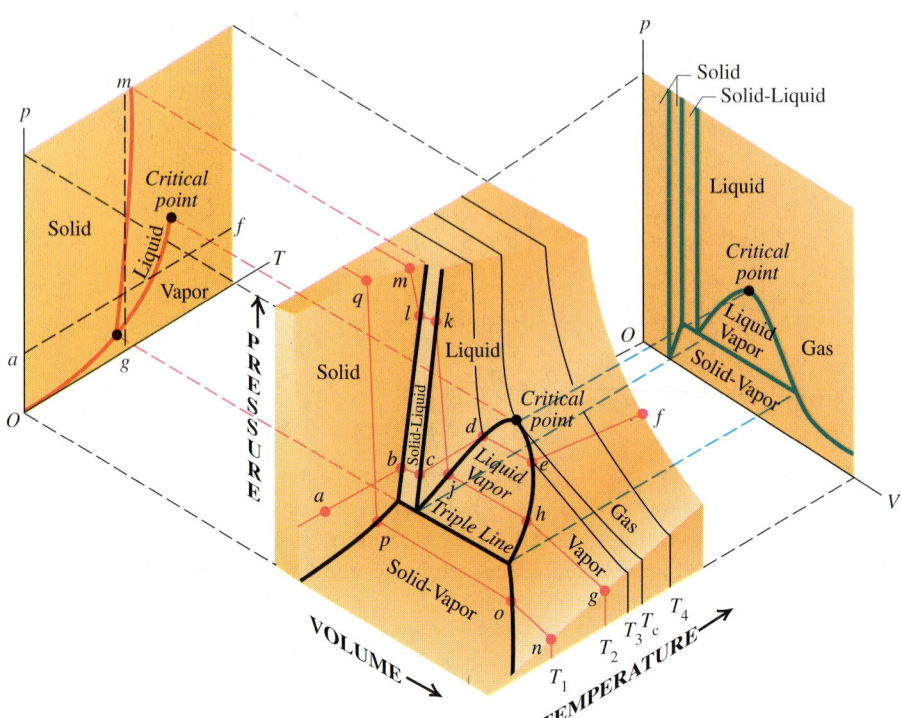

16–19 A *pVT*-surface for a substance that expands on melting. Projections of the boundaries on the surface on the *pT*- and *pV*-planes are also shown.

heat of vaporization also grows smaller and smaller as we approach the critical point, and it too becomes zero at the critical point.

For nearly all familiar materials the critical pressures are much greater than atmospheric pressure, so we don't observe this behavior in everyday life. For example, the critical point for water is at 647.4 K and 221.2×10^5 Pa (about 218 atm or 3210 psi). But high-pressure steam boilers in electric generating plants regularly run at pressures and temperatures well above the critical point.

Many substances can exist in more than one solid phase. A familiar example is carbon, which exists as noncrystalline soot and crystalline graphite and diamond. Water is another example; at least eight types of ice, differing in crystal structure and physical properties, have been observed at very high pressures.

We remarked in Section 16–2 that the equation of state of any material can be represented graphically as a surface in a three-dimensional space with coordinates *p*, *V*, and *T*. Such a surface is seldom useful in representing detailed quantitative information, but it can add to our general understanding of the behavior of materials at various temperatures and pressure. Figure 16–19 shows a typical *pVT*-surface. The light lines represent *pV*-isotherms; projecting them onto the *pV*-plane would give a diagram similar to Fig. 16–4. The *pV*-isotherms represent contour lines on the *pVT*-surface, just as contour lines on a topographic map represent the elevation (the third dimension) at each point. The projections of the edges of the surface onto the *pT*-plane gives the *pT*-phase diagram of Fig. 16–18.

Line *abcdef* represents constant-pressure heating, with melting along *bc* and vaporization along *de*. Note the volume changes that occur as *T* increases along this line. Line *ghjklm* corresponds to an isothermal (constant temperature) compression, with liquefaction along *hj* and solidification along *kl*. Between these, segments *gh* and *jk* represent isothermal compression with increase in pressure; the pressure increases are much greater in the liquid region *jk* and the solid region *lm* than in the vapor region *gh*. Finally, *nopq* represents isothermal solidification directly from the vapor phase; this is the process involved in growth of crystals directly from vapor, as in the formation of

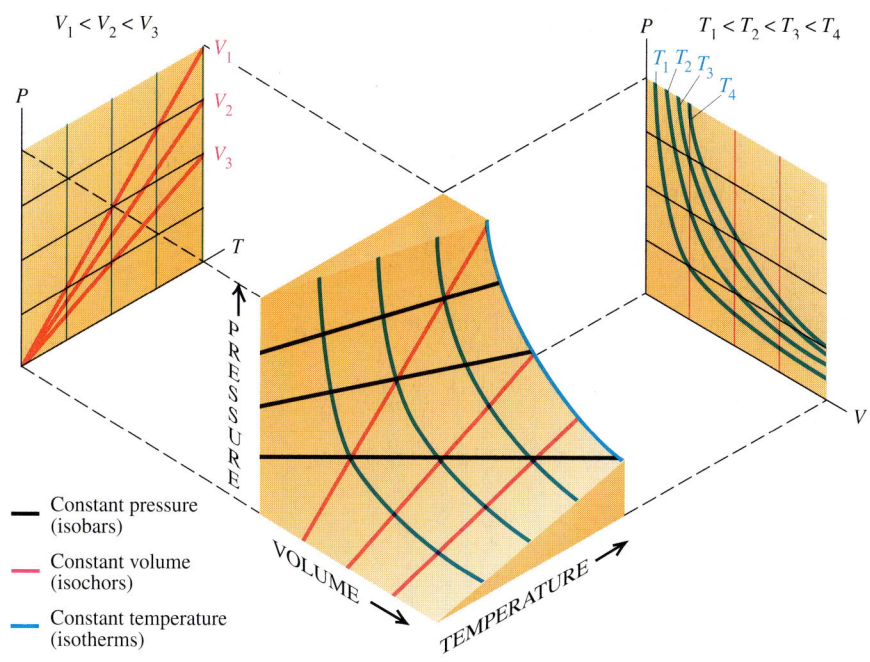

$V_1 < V_2 < V_3$

$T_1 < T_2 < T_3 < T_4$

— Constant pressure (isobars)

— Constant volume (isochors)

— Constant temperature (isotherms)

16–20 A pVT-surface for an ideal gas. At the left, each red line corresponds to a certain constant volume; at the right, each blue-green line corresponds to a certain constant temperature.

snowflakes or frost and in the fabrication of some solid-state electronic devices. These three lines on the pVT-surface are worth careful study.

For contrast, Fig. 16–20 shows the much simpler pVT-surface for a substance that obeys the ideal-gas equation of state under all conditions. The projections of the constant-temperature curves onto the pV-plane correspond to the curves of Fig. 16–3, and the projections of the constant-volume curves onto the pT-plane show the direct proportionality of pressure to absolute temperature.

SUMMARY

- The pressure p, volume V, and temperature T of a given quantity of a substance are called state variables. They are related by an equation of state. The ideal-gas equation of state is

$$pV = nRT, \qquad (16\text{–}3)$$

where n is the number of moles of gas and T is the absolute temperature. The constant R is the same for all gases for conditions in which this equation is applicable.

- A pV-diagram is a set of graphs, called isotherms, each showing pressure as a function of volume for a constant temperature.

- The molar mass M of a pure substance is the mass per mole. The total mass m_{tot} is related to the number of moles n by

$$m_{tot} = nM. \qquad (16\text{–}2)$$

Avogadro's number N_A is the number of molecules in a mole. The mass m of an individual molecule is related to M and N_A by

$$M = N_A m. \qquad (16\text{–}8)$$

- The average translational kinetic energy of the molecules of an ideal gas is directly proportional to absolute temperature:

$$K_{tr} = \frac{3}{2} nRT. \tag{16-14}$$

By using the Boltzmann constant, $k = R/N_A$, this can be expressed in terms of the average translational kinetic energy per molecule:

$$\frac{1}{2} m(v^2)_{av} = \frac{3}{2} kT. \tag{16-16}$$

The root-mean-square speed of molecules in an ideal gas is

$$v_{rms} = \sqrt{(v^2)_{av}} = \sqrt{\frac{3kT}{m}} = \sqrt{\frac{3RT}{M}}. \tag{16-19}$$

- The mean free path λ of molecules in an ideal gas is

$$\lambda = v t_{mean} = \frac{V}{4\pi\sqrt{2}r^2 N}. \tag{16-21}$$

- The molar heat capacity C_V at constant volume for an ideal monatomic gas is

$$C_V = \frac{3}{2} R. \tag{16-25}$$

For an ideal diatomic gas, including rotational kinetic energy,

$$C_V = \frac{5}{2} R. \tag{16-26}$$

For an ideal monatomic solid,

$$C_V = 3R. \tag{16-28}$$

- The speeds of molecules in an ideal gas are distributed according to the Maxwell-Boltzmann distribution:

$$f(v) = 4\pi\left(\frac{m}{2\pi kT}\right)^{3/2} v^2 e^{-mv^2/2kT}. \tag{16-32}$$

- Ordinary matter exists in the solid, liquid, and gas phases. A phase diagram shows conditions under which two phases can coexist in phase equilibrium. All three phases can coexist at the triple point. The vaporization curve ends at the critical point, above which the distinction between liquid and gas phases disappears.

DISCUSSION QUESTIONS

Q16-1 Section 16-2 states that ordinarily, pressure, volume, and temperature cannot change individually without one affecting the others. Yet when a liquid evaporates, its volume changes, even though its pressure and temperature are constant. Is this inconsistent? Why or why not?

Q16-2 In the ideal-gas equation, could an equivalent Celsius temperature be used instead of the Kelvin one if an appropriate numerical value of the constant R is used? Why or why not?

Q16-3 On a chilly morning you can "see your breath." Can you really? What are you really seeing? Does this phenomenon depend on the temperature of the air, the humidity, or both? Explain.

Q16-4 When a car is driven some distance, the air pressure in the tires increases. Why? Should you let out some air to reduce the pressure? Why or why not?

Q16-5 The coolant in an automobile radiator is kept at a pressure higher than atmospheric pressure. Why is this desirable? The radiator cap will release coolant when the gauge pressure of the coolant reaches a certain value, typically 15 lb/in.^{2} or so. Why not just seal the system completely?

Q16-6 Unwrapped food placed in a freezer experiences dehydration, known as "freezer burn." Why?

Q16-7 "Freeze-drying" food involves the same process as "freezer burn," referred to in the previous question. For freeze-drying, the food is usually frozen first, then placed in a vacuum chamber and irradiated with infrared radiation. What is the purpose of the vacuum? The radiation? What advantages might freeze-drying have in comparison to ordinary drying?

Q16–8 A group of students drove from their university (near sea level) up into the mountains for a skiing weekend. Upon arriving at the slopes, they discovered that the bags of potato chips they had brought for snacks had all burst open. What caused this to happen?

Q16–9 An old or burned-out incandescent light bulb usually has a dark gray area on part of the inside surfaces. What is this? Why is it not a uniform deposit all over the inside of the bulb?

Q16–10 Helium is a mixture of two isotopes, one having molar mass 3 g/mol and the other 4 g/mol. In a sample of helium gas which atoms move faster, on the average? Explain why.

Q16–11 Which has more atoms, a kilogram of hydrogen or a kilogram of lead? Which has more mass? Explain.

Q16–12 In the kinetic-molecular model of an ideal gas, is the pressure of the gas due to collisions of molecules with the container walls, collisions of molecules with each other, or both? How does the kinetic-molecular model deal with collisions between molecules?

Q16–13 The proportion of various gases in the earth's atmosphere changes somewhat with altitude. Would you expect the proportion of oxygen at high altitude to be greater or less than at sea level compared to the proportion of nitrogen? Why?

Q16–14 Comment on the following statement: *When two gases are mixed, if they are to be in thermal equilibrium they must have the same average molecular speed.* Is the statement correct? Why or why not?

Q16–15 The kinetic-molecular model contains a hidden assumption about the temperature of the container walls. What is this assumption? What would happen if this assumption was not valid?

Q16–16 The temperature of an ideal gas is directly proportional to the average kinetic energy of its molecules. If a container of ideal gas is moving past you at 2000 m/s, is the temperature of the gas higher than if the container was at rest? Defend your answer.

Q16–17 You are sitting at the back of a chemistry lecture hall when your instructor opens a gas valve. You can *hear* the sound of the escaping gas almost immediately, but it takes several seconds before you can *smell* the gas. What is the reason for the "smell delay"?

Q16–18 In deriving the ideal-gas equation from the kinetic-molecular model, we ignored potential energy due to the earth's gravity. Is this omission justified? Why or why not?

Q16–19 The derivation of the ideal-gas equation included the assumption that the number of molecules is very large, so that we could compute the average force due to many collisions. However, the ideal-gas equation holds accurately only at low pressures, where the molecules are few and far between. Is this inconsistent? Why or why not?

Q16–20 A gas storage tank has a small leak. The pressure in the tank drops more quickly if the gas is hydrogen or helium than if it is oxygen. Why?

Q16–21 Consider two specimens of ideal gas at the same temperature. Specimen A has the same total mass as specimen B, but the molecules in specimen A have greater molar mass than they do in specimen B. In which specimen is the total kinetic energy of the gas greater? Does your answer depend on the molecular structure of the gases? Why or why not?

Q16–22 Some elements that form solid crystals have molar heat capacities *greater* than $3R$. What could account for this?

*Q16–23 In a gas that contains N molecules, would it be accurate to say that the number of molecules with speed v is equal to $f(v)$? Or would it be accurate to say that this number is given by $N f(v)$? Explain your answers.

*Q16–24 Figure 16–16 shows a velocity selector. The two disks are a distance x apart, the slits in the two disks are at an angle θ to each other, and the disks rotate together with angular speed ω. Explain why a molecule that passes through the first slit of the selector will only be able to pass through the second slit if its speed is $v = \omega x/\theta$.

*Q16–25 Imagine a special air filter placed in a window of a house. The tiny holes in the filter only allow air molecules moving faster than a certain speed to exit the house, and only allow air molecules moving slower than that speed to enter the house from outside. What effect would this filter have on the temperature inside the house? (It turns out that the second law of thermodynamics—which we will discuss in Chapter 18—tells us that such a wonderful air filter would be quite impossible to make.)

Q16–26 A beaker of water at room temperature is placed in an enclosure, and the air pressure in the enclosure is slowly reduced. When the air pressure is reduced sufficiently, the water begins to boil. The temperature of the water does not rise when it boils; in fact, the temperature *drops* slightly. Explain these phenomena.

Q16–27 Ice is slippery to walk on, and especially slippery if you wear ice skates. What does this tell you about how the melting temperature of ice depends on pressure? Explain.

Q16–28 Hydrothermal vents are openings in the ocean floor that discharge very hot water. The water emerging from one such vent off the Oregon coast, 2400 m below the surface, has a temperature of 279°C. Despite its high temperature, the water doesn't boil. Why not?

Q16–29 The dark areas on the moon's surface are called *maria*, Latin for "seas," and were once thought to be bodies of water. In fact the maria are not "seas" at all, but plains of solidified lava. Given that there is no atmosphere on the moon, how can you explain the absence of liquid water on the moon's surface?

Q16–30 In addition to the normal cooking directions printed on the back of a box of rice, there are also "high altitude directions." The only difference is that the "high altitude directions" suggest increasing the cooking time and using a greater volume of boiling water in which to cook the rice. Why should the directions depend on the altitude in this way?

EXERCISES

SECTION 16–2 EQUATIONS OF STATE

16–1 A 20.0-L tank contains 0.225 kg of helium at 18.0°C. The molar mass of helium is 4.00 g/mol. a) How many moles of helium are in the tank? b) What is the pressure in the tank, in pascals and in atmospheres?

16–2 Helium gas with a volume of 2.60 L, under a pressure of 1.30 atm and at a temperature of 41.0°C, is warmed until both pressure and volume are doubled. a) What is the final temperature? b) How many grams of helium are there? The molar mass of helium is 4.00 g/mol.

16–3 A cylindrical tank has a tight-fitting piston that allows the volume of the tank to be changed. The tank originally contains 0.110 m^3 of air at a pressure of 3.40 atm. The piston is slowly pulled out until the volume of the gas is increased to 0.390 m^3. If the temperature remains constant, what is the final value of the pressure?

16–4 A 3.00-L tank contains air at 3.00 atm and 20.0°C. The tank is sealed and cooled until the pressure is 1.00 atm. a) What is the temperature then in degrees Celsius? Assume that the volume of the tank is constant. b) If the temperature is kept at the value found in part (a) and the gas is compressed, what is the volume when the pressure again becomes 3.00 atm?

16–5 A Jaguar XK8 convertible has an eight-cylinder engine. At the beginning of its compression stroke, one of the cylinders contains 499 cm^3 of air at atmospheric pressure (1.01×10^5 Pa) and a temperature of 27.0°C. At the end of the stroke the air has been compressed to a volume of 46.2 cm^3 and the gauge pressure has increased to 2.72×10^6 Pa. Compute the final temperature.

16–6 A welder using a tank of volume 0.0750 m^3 fills it with oxygen (molar mass = 32.0 g/mol) at a gauge pressure of 3.00×10^5 Pa and temperature of 37.0°C. The tank has a small leak, and in time some of the oxygen leaks out. On a day when the temperature is 22.0°C, the gauge pressure of the oxygen in the tank is 1.80×10^5 Pa. Find a) the initial mass of oxygen; b) the mass of oxygen that has leaked out.

16–7 A large cylindrical tank contains 0.750 m^3 of nitrogen gas at 27°C and 1.50×10^5 Pa (absolute pressure). The tank has a tight-fitting piston that allows the volume to be changed. What will be the pressure if the volume is decreased to 0.480 m^3 and the temperature is increased to 157°C?

16–8 A room with dimensions 7.00 m × 8.00 m × 2.50 m is filled with pure oxygen at 22.0°C and 1.00 atm. The molar mass of oxygen is 32.0 g/mol. a) How many moles of oxygen are required? b) What is the mass of this oxygen, in kilograms?

16–9 The gas inside a balloon will always have a pressure nearly equal to atmospheric pressure, since that is the pressure applied to the outside of the balloon. You fill a balloon with helium (a nearly ideal gas) to a volume of 0.600 L at a temperature of 19.0°C. What is the volume of the balloon if you cool it to the boiling point of liquid nitrogen (77.3 K)?

16–10 Deviations from the Ideal-Gas Equation. For carbon dioxide gas (CO_2), the constants in the van der Waals equation are $a = 0.364$ J · m^3/mol^2 and $b = 4.27 \times 10^{-5}$ m^3/mol. a) If 1.00 mol of CO_2 gas at 350 K is confined to a volume of 400 cm^3, find the pressure of the gas using the ideal-gas equation and the van der Waals equation. b) Which equation gives a lower pressure? Why? What is the percentage difference of the van der Waals result from the ideal-gas equation result? c) The gas is kept at the same temperature as it expands to a volume of 4000 cm^3. Repeat the calculations of parts (a) and (b). d) Explain how your calculations show that the van der Waals equation is equivalent to the ideal-gas equation if n/V is small.

16–11 The total lung volume for a typical physics student is 6.00 L. A physics student fills her lungs with air at an absolute pressure of 1.00 atm. Then, holding her breath, she compresses her chest cavity, decreasing her lung volume to 5.70 L. What is then the pressure of the air in her lungs? Assume that the temperature of the air remains constant.

16–12 A diver observes a bubble of air rising from the bottom of a lake (where the absolute pressure is 3.50 atm) to the surface (where the pressure is 1.00 atm). The temperature at the bottom is 4.0°C, and the temperature at the surface is 23.0°C. a) What is the ratio of the volume of the bubble as it reaches the surface to its volume at the bottom? b) Would it be safe for the diver to hold his breath while ascending from the bottom of the lake to the surface? Why or why not?

16–13 A metal tank with volume 3.10 L will burst if the absolute pressure of the gas it contains exceeds 100 atm. a) If 11.0 mol of an ideal gas is put into the tank at a temperature of 23.0°C, to what temperature can the gas be warmed before the tank ruptures? Neglect the thermal expansion of the tank. b) Based on your answer to part (a), is it reasonable to neglect the thermal expansion of the tank? Explain.

16–14 With the assumption (not very realistic) that the air temperature is a uniform 0°C, what is the atmospheric pressure at an altitude of 5000 m? This is roughly the maximum altitude usually attained by aircraft without pressurized cabins.

16–15 At an altitude of 11,000 m (a typical cruising altitude for a jet airliner), the air temperature is −56.5°C and the air density is 0.364 kg/m^3. What is the pressure of the atmosphere at that altitude? (Note that the temperature at this altitude is not the same as at the surface of the earth, so that the calculation of Example 16–4 (Section 16–2) doesn't apply.)

SECTION 16–3 MOLECULAR PROPERTIES OF MATTER

16–16 Modern vacuum pumps make it easy to attain pressures of the order of 10^{-13} atm in the laboratory. At a pressure of 9.00×10^{-14} atm and an ordinary temperature (say $T = 300$ K), how many molecules are present in a volume of 1.00 cm^3?

16–17 The Lagoon Nebula (Fig. 16–21) is a cloud of hydrogen gas located 3900 light years from earth. The cloud is about 45 light years in diameter, and glows because of its high temperature of 7500 K. (The gas is raised to this temperature by the stars that lie within the nebula). The cloud is also very thin; there are only 80 molecules per cubic centimeter. a) Find the gas pressure (in atmospheres) in the Lagoon Nebula. Compare to the laboratory pressure referred to in Exercise 16–16. b) Science fiction films sometimes show starships being buffeted by turbulence as they fly through gas clouds such as the Lagoon Nebula. Does this seem realistic? Why or why not?

FIGURE 16–21 Exercise 16–17.

16–18 In a gas at standard conditions, what is the length of the side of a cube that contains a number of molecules equal to the population of the earth (about 6×10^9)?

16–19 How many moles are there in a 1.00-kg bottle of water? How many molecules? The molar mass of water is 18.0 g/mol.

16–20 Consider an ideal gas at 27°C and 1.00 atm pressure. Imagine the molecules to be, on average, uniformly spaced, with each molecule at the center of a small cube. a) What is the length of an edge of each small cube if adjacent cubes touch but don't overlap? b) How does this distance compare with the diameter of a molecule?

16–21 Consider 5.00 mol of liquid water. a) What volume is occupied by this amount of water? The molar mass of water is 18.0 g/mol. b) Imagine the molecules to be, on average, uniformly spaced, with each molecule at the center of a small cube. What is the length of an edge of each small cube if adjacent cubes touch but don't overlap? c) How does this distance compare with the diameter of a molecule?

SECTION 16–4 KINETIC-MOLECULAR MODEL OF AN IDEAL GAS

16–22 A flask contains a mixture of neon (Ne), krypton (Kr), and radon (Rn) gases. Compare a) the average kinetic energies of the three types of atoms; b) the root-mean-square speeds. (*Hint:* The periodic table in Appendix D shows the molar mass (in g/mol) of each element under the chemical symbol for that element.)

16–23 **Gaseous Diffusion of Uranium.** a) A process called *gaseous diffusion* is often used to separate isotopes of uranium, that is, atoms of the elements that have different masses, such as ^{235}U and ^{238}U. The only gaseous compound of uranium at ordinary temperatures is uranium hexafluoride, UF_6. Speculate on how $^{235}UF_6$ and $^{238}UF_6$ molecules might be separated by diffusion. b) The molar masses for $^{235}UF_6$ and $^{238}UF_6$ molecules are

0.349 kg/mol and 0.352 kg/mol, respectively. If uranium hexafluoride acts as an ideal gas, what is the ratio of the root-mean-square speed of $^{235}UF_6$ molecules to that of $^{238}UF_6$ molecules if the temperature is uniform?

16–24 The ideas of average and root-mean-square value can be applied to any distribution. A class of 150 students had the following scores on a 100-point quiz:

Score	Number of Students
10	11
20	12
30	24
40	15
50	19
60	10
70	12
80	20
90	17
100	10

a) Find the average score for the class. b) Find the root-mean-square score for the class.

16–25 a) Oxygen (O_2) has a molar mass of 32.0 g/mol. What is the average translational kinetic energy of an oxygen molecule at a temperature of 300 K? b) What is the average value of the square of its speed? c) What is the root-mean-square speed? d) What is the momentum of an oxygen molecule traveling at this speed? e) Suppose an oxygen molecule traveling at this speed bounces back and forth between opposite sides of a cubical vessel 0.10 m on a side. What is the average force it exerts on one of the walls of the container? (Assume that the molecule's velocity is perpendicular to the two sides that it strikes.) f) What is the average force per unit area? g) How many oxygen molecules traveling at this speed are necessary to produce an average pressure of 1 atm? h) Compute the number of oxygen molecules that are actually contained in a vessel of this size at 300 K and atmospheric pressure. i) Your answer for part (h) should be three times as large as the answer for part (g). Where does this discrepancy arise?

16–26 Calculate the mean free path of air molecules at a pressure of 3.50×10^{-13} atm and a temperature of 300 K. (This pressure is readily attainable in the laboratory; see Exercise 16–16.) As in Example 16–9, model the air molecules as spheres of radius 2.0×10^{-10} m.

16–27 At what temperature is the root-mean-square speed of nitrogen molecules equal to the root-mean-square speed of hydrogen molecules at 20.0°C? (*Hint:* The periodic table in Appendix D shows the molar mass (in g/mol) of each element under the chemical symbol for that element. The molar mass of H_2 is twice the molar mass of hydrogen atoms, and similarly for N_2.)

16–28 Smoke particles in the air typically have masses of the order of 10^{-16} kg. The Brownian motion (rapid, irregular movement) of these particles, resulting from collisions with air molecules, can be observed with a microscope. a) Find the root-mean-square speed of Brownian motion for a particle with a mass of 3.00×10^{-16} kg in air at 300 K. b) Would the root-mean-square speed be different if the particle were in hydrogen gas at the same temperature? Explain.

SECTION 16–5 HEAT CAPACITIES

16–29 a) Calculate the specific heat capacity at constant volume of water vapor, assuming the nonlinear triatomic molecule has three translational and three rotational degrees of freedom and that vibrational motion does not contribute. The molar mass of water is 18.0 g/mol. b) The actual specific heat capacity of water vapor at low pressures is about 2000 J/kg · K. Compare this with your calculation and comment on the actual role of vibrational motion.

16–30 a) The specific heat of ice at constant volume is 833 J/kg · K at $-180°C$, 1640 J/kg · K at $-60°C$, and 2060 J/kg · K at $-5.0°C$. Calculate C_V, the molar heat capacity at constant volume for ice, at each of these temperatures. The molar mass of H_2O is 18.0 g/mol. b) Why do you suppose the value of C_V increases with increasing temperature? c) Do the values that you calculate approach the value of $3R$ (given in the rule of Dulong and Petit) as the temperature increases? Speculate about why this should be so.

16–31 a) How much heat does it take to increase the temperature of 2.50 mol of a diatomic ideal gas by 30.0 K near room temperature if the gas is held at constant volume? b) What is the answer to the question in part (a) if the gas is monatomic rather than diatomic?

16–32 a) Compute the specific heat capacity at constant volume of nitrogen (N_2) gas, and compare with the specific heat capacity of liquid water. The molar mass of N_2 is 28.0 g/mol. b) You warm 1.00 kg of water at a constant volume of 1.00 L from $20.0°C$ to $30.0°C$ in a kettle. For the same amount of heat, how many kilograms of $20.0°C$ air would you be able to warm to $30.0°C$? What volume (in liters) would this air occupy at $20.0°C$ and a pressure of 1.00 atm? Make the simplifying assumption that air is 100% N_2.

SECTION 16–6 MOLECULAR SPEEDS

*16–33 For a gas of nitrogen molecules (N_2), what must be the temperature if 94.7% of all the molecules have speeds less than

a) 1500 m/s; b) 1000 m/s; c) 500 m/s? Use Table 16–2. The molar mass of N_2 is 28.0 g/mol.

*16–34 Derive Eq. (16–33) from Eq. (16–32).

*16–35 Prove that $f(v)$ as given by Eq. (16–33) is maximum for $\epsilon = kT$. Use this result to obtain Eq. (16–34).

*16–36 For diatomic carbon dioxide gas (CO_2, molar mass = 44.0 g/mol) at $T = 300$ K, calculate a) the most probable speed v_{mp}; b) the average speed v_{av}; c) the root-mean-square speed v_{rms}.

SECTION 16–7 PHASES OF MATTER

16–37 Puffy cumulus clouds, which are made of water droplets, occur at lower altitudes in the atmosphere. Wispy cirrus clouds, which are made of ice crystals, occur only at higher altitudes. Find the altitude y (measured from sea level) above which only cirrus clouds can occur. On a typical day and at altitudes less than 11 km, the temperature at an altitude y is given by $T = T_0 - \alpha y$, where $T_0 = 15.0°C$ and $a = 6.0$ C°/1000 m.

16–38 Solid water (ice) is slowly warmed from a very low temperature. a) What minimum external pressure p_1 must be applied to the solid if a melting phase transition is to be observed? Describe the sequence of phase transitions that occur if the applied pressure p is such that $p < p_1$. b) Above a certain maximum pressure p_2, no boiling transition is observed. What is this pressure? Describe the sequence of phase transitions that occur if $p_1 < p < p_2$.

16–39 A physicist places a piece of ice at $0.00°C$ and a beaker of water at $0.00°C$ inside a glass box and closes the lid of the box. All the air is then removed from the box. If the ice, water, and beaker are all maintained at a temperature of $0.00°C$, describe the final equilibrium state inside the box.

16–40 The atmosphere of the planet Mars is 95.3% carbon dioxide (CO_2) and about 0.03% water vapor. The atmospheric pressure is only 600 Pa, and the surface temperature varies from $-30°C$ to $-100°C$. The polar ice caps contain both CO_2 ice and water ice. Could there be *liquid* CO_2 on the surface of Mars? Could there be liquid water? Why or why not?

PROBLEMS

16–41 a) Use Eq. (16–1) to estimate the change in the volume of a solid steel sphere of volume 11 L when the temperature and pressure increase from $21°C$ and 1.013×10^5 Pa to $42°C$ and 2.10×10^7 Pa. (*Hint:* Consult Chapters 11 and 15 to determine the values of β and k.) b) In Example 16–3 the change in volume of an 11-L steel scuba tank was neglected. Was this a good approximation? Explain.

16–42 At an absolute pressure of 2.00×10^{-13} atm, a partial vacuum easily obtained in laboratories (see Exercise 16–16), calculate the mass of nitrogen present in a volume of 3000 cm^3 if the temperature of the gas is $22.0°C$. The molar mass of nitrogen (N_2) is 28.0 g/mol.

16–43 A cylinder 1.00 m tall with inside diameter 0.120 m is used to hold propane gas (molar mass 44.1 g/mol) for use in a barbecue. It is initially filled with gas until the gauge pressure is 1.30×10^6 Pa and the temperature is $22.0°C$. The temperature of the gas remains constant as it is partially emptied out of the tank,

until the gauge pressure is 2.50×10^5 Pa. Calculate the mass of propane that has been used.

16–44 During a test dive in 1939 prior to being accepted by the U.S. Navy, the submarine *Squalus* sank at a point where the depth of water was 73.0 m. The temperature at the surface was $27.0°C$ and at the bottom it was $7.0°C$. The density of seawater is 1030 kg/m^3. a) A diving bell was used to rescue 33 trapped crewmen from the *Squalus*. The diving bell was in the form of a circular cylinder 2.30 m high, open at the bottom and closed at the top. When the diving bell was lowered to the bottom of the sea, to what height did water rise within the diving bell? (*Hint:* You may ignore the relatively small variation in water pressure between the bottom of the bell and the surface of the water within the bell.) b) At what gauge pressure must compressed air have been supplied to the bell while on the bottom to expel all the water from it?

16–45 A glassblower makes a barometer using a tube 0.900 m long and with a cross-section area of 0.620 cm^2. Mercury stands in this tube to a height of 0.750 m. The room temperature is 20.0°C. A small amount of air is accidentally introduced into the evacuated space above the mercury, and the column drops to a height of 0.690 m. How many grams of air were introduced? The molar mass of air is 28.8 g/mol.

16–46 A hot-air balloon stays aloft because hot air at atmospheric pressure is less dense than cooler air at the same pressure; the calculation of the buoyant force is discussed in Chapter 14. If the volume of the balloon is 500 m^3 and the surrounding air is at 15.0°C, what must be the temperature of the air in the balloon for it to lift a total load of 290 kg (in addition to the mass of the hot air)? The density of air at 15.0°C and atmospheric pressure is 1.23 kg/m^3.

16–47 An automobile tire has a volume of 0.0150 m^3 on a cold day when the temperature of the air in the tire is 5.0°C and atmospheric pressure is 1.02 atm. Under these conditions the gauge pressure is measured to be 1.70 atm (about 25 lb/in.^{2}). After the car is driven on the highway for 30 min, the temperature of the air in the tires has risen to 45.0°C and the volume to 0.0159 m^3. What then is the gauge pressure?

16–48 A flask with a volume of 1.50 L, provided with a stopcock, contains ethane gas (C$_2$H$_6$) at 300 K and atmospheric pressure (1.013 × 10^5 Pa). The molar mass of ethane is 30.1 g/mol. The system is warmed to a temperature of 380 K, with the stopcock open to the atmosphere. The stopcock is then closed, and the flask cooled to its original temperature. a) What is the final pressure of the ethane in the flask? b) How many grams of ethane remain in the flask?

16–49 A balloon whose volume is 750 m^3 is to be filled with hydrogen at atmospheric pressure (1.01 × 10^5 Pa). a) If the hydrogen is stored in cylinders with a volume of 1.90 m^3 at a gauge pressure of 1.20 × 10^6 Pa, how many cylinders are required? Assume that the temperature of the hydrogen remains constant. b) What is the total weight (in addition to the weight of the gas) that can be supported by the balloon if the gas in the balloon and the surrounding air are both at 15.0°C? The molar mass of hydrogen (H$_2$) is 2.02 g/mol. The density of air at 15.0°C and atmospheric pressure is 1.23 kg/m^3. See Chapter 14 for a discussion of buoyancy. c) What weight could be supported if the balloon were filled with helium (with a molar mass of 4.00 g/mol) instead of hydrogen, again at 15.0°C?

16–50 A vertical cylindrical tank 0.900 m high has its top end closed by a tightly fitting frictionless piston of negligible weight. The air inside the cylinder is at an absolute pressure of 1.00 atm. The piston is depressed by pouring mercury on it slowly (Fig. 16–22). How far will the piston descend before mercury spills over the top of the cylinder? The temperature of the air is kept constant.

16–51 A large tank of water has a hose connected to it, as shown in Fig. 16–23. The tank is sealed at the top and has compressed air between the water surface and the top. When the water height h has the value 3.50 m, the absolute pressure p of the compressed air is 4.20 × 10^5 Pa. Assume that the air above the water expands at constant temperature, and take the atmospheric pressure to be 1.00 × 10^5 Pa. a) What is the speed with which water flows out of the hose when h = 3.50 m? b) As water flows out of

FIGURE 16–22 Problem 16–50.

FIGURE 16–23 Problem 16–51.

the tank, h decreases. Calculate the speed of flow for h = 3.00 m and for h = 2.00 m. c) At what value of h does the flow stop?

16–52 In one hour, an average person at rest consumes 14.5 L of oxygen at a pressure of 1.00 atm and a temperature of 20.0°C. a) Express this rate of oxygen consumption in terms of number of molecules per second. b) A person at rest inhales and exhales 0.5 L of air with each breath. The inhaled air is 21.0% oxygen, and the exhaled air is 16.3% oxygen. How many breaths per minute will satisfy this person's oxygen requirements? c) Repeat part (b) for a resting person at an elevation of 3000 m, where the pressure is 0.72 atm and the temperature is 0.0°C. Assume that the oxygen percentages and volume per inhalation are the same as stated in part (b). (To maintain its functions, the body still requires the same number of oxygen molecules per second as at sea level.) Explain why some people report "shortness of breath" at such elevations.

16–53 How Many Atoms Are You? Estimate the number of atoms in the body of a 50-kg physics student. Note that the human body is mostly water, which has molar mass 18.0 g/mol, and that each water molecule contains three atoms.

16–54 The size of an oxygen molecule is about 2.0 × 10^{-10} m. Make a rough estimate of the pressure at which the finite volume of the molecules should cause noticeable deviations from ideal-gas behavior at ordinary temperatures (T = 300 K).

16–55 Successive Approximations and the van der Waals Equation. In the ideal-gas equation, the number of moles per volume n/V is simply equal to p/RT. In the van der Waals equation, solving for n/V in terms of the pressure p and temperature

T is somewhat more involved. a) Show that the van der Waals equation can be written as

$$\frac{n}{V} = \left(\frac{p + an^2/V^2}{RT} \right)\left(1 - \frac{bn}{V}\right).$$

b) The van der Waals parameters for hydrogen sulfide gas (H_2S) are $a = 0.448$ J $\cdot$ m^3/mol^2 and $b = 4.29 \times 10^{-5}$ m^3/mol. Determine the number of moles per volume of H_2S gas at $127°C$ and an absolute pressure of 9.80×10^5 Pa as follows: i) Calculate a first approximation using the ideal-gas equation, $n/V = p/RT$. ii) Substitute this approximation for n/V into the right-hand side of the equation in part (a). The result is a new, improved approximation for n/V. iii) Substitute the new approximation for n/V into the right-hand side of the equation in (a). The result is a further-improved approximation for n/V. iv) Repeat step (iii) until successive approximations agree to the desired level of accuracy (in this case, to three significant figures). c) Compare your final result in part (b) to the result p/RT obtained using the ideal-gas equation. Which result gives a larger value of n/V? Why?

16–56 a) Compute the increase in gravitational potential energy for a nitrogen molecule (molar mass 28.0 g/mol) for an increase in elevation of 400 m near the earth's surface. b) At what temperature is this equal to the average kinetic energy of a nitrogen molecule? c) Is it possible that a nitrogen molecule near sea level where $T = 15.0°C$ could rise to an altitude of 400 m? Is it likely that it could do so without hitting any other molecules along the way? Explain.

16–57 **The Lennard-Jones Potential.** A commonly used potential-energy function for the interaction of two molecules (Fig. 16–5) is the Lennard-Jones 6-12 potential

$$U(r) = U_0\left[\left(\frac{R_0}{r}\right)^{12} - 2\left(\frac{R_0}{r}\right)^6\right],$$

where r is the distance between the centers of the molecules and U_0 and R_0 are positive constants. The corresponding force $F(r)$ is given in Eq. (13–26). a) Graph $U(r)$ and $F(r)$ versus r. b) Let r_1 be the value of r at which $U(r) = 0$, and let r_2 be the value of r at which $F(r) = 0$. Show the locations of r_1 and r_2 on your graphs of $U(r)$ and $F(r)$. Which of these values represents the equilibrium separation between the molecules? c) Find the values of r_1 and r_2 in terms of R_0, and find the ratio r_1/r_2. d) If the molecules are located a distance r_2 apart (as calculated in part (c)), how much work must be done to pull them apart so that $r \to \infty$?

16–58 a) What is the total random translational kinetic energy of 5.00 L of hydrogen gas, with pressure 1.01×10^5 Pa and temperature 300 K? b) If the tank containing the gas is moved with a speed of 30.0 m/s, by what percentage is the total kinetic energy of the gas increased? The molar mass of hydrogen (H_2) is 2.016 g/mol.

16–59 The speed of propagation of a sound wave in air at $27°C$ is about 350 m/s. Calculate, for comparison, a) v_{rms} for nitrogen molecules; b) the rms value of v_x at this temperature. The molar mass of nitrogen (N_2) is 28.0 g/mol.

16–60 The surface of the sun has a temperature of about 5800 K and consists largely of hydrogen atoms. a) Find the rms speed of

a hydrogen atom at this temperature. (The mass of a single hydrogen atom is 1.67×10^{-27} kg.) b) The escape speed for a particle to leave the gravitational influence of the sun is given by $(2GM/R)^{1/2}$, where M is the sun's mass, R its radius, and G the gravitational constant (Example 12–5 of Section 12–4). Use the data in Appendix F to calculate this escape speed. c) Can appreciable quantities of hydrogen escape from the sun? Can *any* hydrogen escape? Explain.

16–61 a) Show that a projectile with mass m can "escape" from the surface of a planet if it is launched vertically upward with a kinetic energy greater than mgR_p, where g is the acceleration due to gravity at the planet's surface and R_p is the planet's radius. Ignore air resistance. (See Problem 16–60.) b) If the planet in question is the earth, at what temperature does the average translational kinetic energy of a nitrogen molecule (molar mass = 28.0 g/mol) equal that required to escape? What about a hydrogen molecule (molar mass = 2.02 g/mol)? c) Repeat part (b) for the moon, for which $g = 1.63$ m/s^2 and $R_p = 1740$ km. d) While the earth and the moon have similar average surface temperatures, the moon has essentially no atmosphere. Use your results from parts (b) and (c) to explain why.

16–62 **Planetary Atmospheres.** a) The temperature near the top of Jupiter's multicolored cloud layer is about 140 K. The temperature at the top of the earth's troposphere, at an altitude of about 20 km, is about 220 K. Calculate the rms speed of hydrogen molecules in both these environments. Give your answers in m/s and as a fraction of the escape speed from the respective planet. (See Problem 16–60.) b) Hydrogen gas (H_2) is a rare element in the earth's atmosphere. In the atmosphere of Jupiter, by contrast, 89% of all molecules are H_2. Explain why, using your results from part (a). c) Suppose an astronomer claims to have discovered an oxygen (O_2) atmosphere on the asteroid Ceres. How likely is this? Ceres has a mass equal to 0.014 times the mass of the moon, a density of 2400 kg/m^3, and a surface temperature of about 200 K.

16–63 a) For what mass of molecule or particle is v_{rms} equal to 1.00 mm/s at 300 K? b) If the particle is an ice crystal, how many molecules does it contain? The molar mass of water is 18.0 g/mol. c) Calculate the diameter of the particle if it is a spherical piece of ice. Would it be visible to the naked eye?

16–64 In describing the heat capacities of solids in Section 16–5, we stated that the potential energy $U = \frac{1}{2}kx^2$ of a harmonic oscillator averaged over one period of the motion is equal to the kinetic energy $K = \frac{1}{2}mv^2$ averaged over one period. Prove this result using Eqs. (13–13) and (13–15) for the position and velocity of a simple harmonic oscillator. For simplicity, assume that the initial position and velocity make the phase angle ϕ equal to zero. (*Hint:* Use the trigonometric identities $\cos^2(\theta) = [1 + \cos(2\theta)]/2$ and $\sin^2(\theta) = [1 - \cos(2\theta)]/2$. What is the average value of $\cos(2\omega t)$ averaged over one period?)

16–65 It is possible to make crystalline solids that are only one layer of atoms thick. Such "two-dimensional" crystals can be created by depositing atoms on a very flat surface. a) If the atoms in such a two-dimensional crystal can only move within the plane of the crystal, what will be its molar heat capacity near room temperature? Give your answer as a multiple of R and in J/mol · K. b) At very low temperatures, will the molar heat capacity of a two-dimensional crystal be greater, less, or equal to the result you found in part (a)? Explain why.

16–66 a) Calculate the total *rotational* kinetic energy of the molecules in 1.00 mol of a diatomic gas at 300 K. b) Calculate the moment of inertia of an oxygen molecule (O_2) for rotation about either the y- or z-axis shown in Fig. 16–12. Treat the molecule as two massive points (representing the oxygen atoms) separated by a distance of 1.21×10^{-10} m. The molar mass of oxygen *atoms* is 16.0 g/mol. c) Find the rms angular velocity of rotation of an oxygen molecule about either the y- or z-axis shown in Fig. 16–12. How does your answer compare to the angular velocity of a typical piece of rapidly rotating machinery (10,000 rev/min)?

16–67 For each polyatomic gas in Table 16–1, compute the value of the molar heat capacity at constant volume, C_V, on the assumption that there is no vibrational energy. Compare with the measured values in the table, and compute the fraction of the total heat capacity that is due to vibration for each of the three gases. (*Note:* CO_2 is linear; SO_2 and H_2S are not. Recall that a linear polyatomic molecule has two rotational degrees of freedom, and a nonlinear molecule has three.)

*16–68 a) Show that $\int_0^\infty f(v)\, dv = 1$, where $f(v)$ is the Maxwell-Boltzmann distribution of Eq. (16–32). b) In terms of the physical definition of $f(v)$, explain why the integral in part (a) *must* have this value.

*16–69 Calculate the integral in Eq. (16–31), $\int_0^\infty v^2 f(v)\, dv$, and compare this result to $(v^2)_{av}$ as given by Eq. (16–16). (*Hint:* You may use the tabulated integral

$$\int_0^\infty x^{2n} e^{-\alpha x^2}\, dx = \frac{1 \cdot 3 \cdot 5 \cdots (2n-1)}{2^{n+1}\alpha^n}\sqrt{\frac{\pi}{\alpha}},$$

where n is a positive integer and α is a positive constant.)

*16–70 Calculate the integral in Eq. (16–30), $\int_0^\infty v f(v)\, dv$, and compare this result to v_{av} as given by Eq. (16–35). (*Hint:* Make the change of variable $v^2 = x$ and use the tabulated integral

$$\int_0^\infty x^n e^{-\alpha x}\, dx = \frac{n!}{\alpha^{n+1}},$$

where n is a positive integer and α is a positive constant.)

*16–71 a) Explain why in a gas of N molecules, the number of molecules having speeds in the *finite* interval v to $v + \Delta v$ is $\Delta N = N\int_v^{v+\Delta v} f(v)\, dv$. b) If Δv is small, then $f(v)$ is approximately constant over the interval and $\Delta N \approx Nf(v)\, \Delta v$. For oxygen gas ($O_2$, molar mass = 32.0 g/mol) at $T = 300$ K, use this approximation to calculate the number of molecules with speeds within $\Delta v = 20$ m/s of v_{mp}. Express your answer as a multiple of N. c) Repeat part (b) for speeds within $\Delta v = 20$ m/s of $7v_{mp}$.

CHALLENGE PROBLEMS

16–75 Dark Nebulae and the Interstellar Medium. The dark area in Fig. 16–24 that appears devoid of stars is a *dark nebula*, a cold gas cloud in interstellar space that contains enough material to block out light from the stars behind it. A typical dark

d) Repeat parts (b) and (c) for a temperature of 600 K. e) Repeat parts (b) and (c) for a temperature of 150 K. f) What do your results tell you about the shape of the distribution as a function of temperature? Do your conclusions agree with what is shown in Fig. 16–17?

16–72 The *vapor pressure* is the pressure of the vapor phase of a substance when it is in equilibrium with the solid or liquid phase of the substance. The *relative humidity* is the partial pressure of water vapor in the air divided by the vapor pressure of water at that same temperature, expressed as a percentage. The air is saturated when the humidity is 100%. a) The vapor pressure of water at 20.0°C is 2.34×10^3 Pa. If the air temperature is 20.0°C and the relative humidity is 60%, what is the partial pressure of water vapor in the atmosphere (that is, the pressure due to water vapor alone)? b) Under the conditions of part (a), what is the mass of water in 1.00 m^3 of air? (The molar mass of water is 18.0 g/mol. Assume that water vapor can be treated as an ideal gas.)

16–73 The Dew Point. The vapor pressure of water (Problem 16–72) decreases as the temperature decreases. If the amount of water vapor in the air is kept constant as the air is cooled, a temperature is reached, called the *dew point*, at which the partial pressure and vapor pressure coincide and the vapor is saturated. If the air is cooled further, vapor condenses to liquid until the partial pressure again equals the vapor pressure at that temperature. The temperature in a room is 30.0°C. A meteorologist cools a metal can by gradually adding cold water. When the can temperature reaches 16.0°C, water droplets form on its outside surface. What is the relative humidity of the 30.0°C air in the room? The table below lists the vapor pressure of water at various temperatures:

Temperature (°C)	Vapor Pressure (Pa)
10.0	1.23×10^3
12.0	1.40×10^3
14.0	1.60×10^3
16.0	1.81×10^3
18.0	2.06×10^3
20.0	2.34×10^3
22.0	2.65×10^3
24.0	2.99×10^3
26.0	3.36×10^3
28.0	3.78×10^3
30.0	4.25×10^3

16–74 Altitude at Which Clouds Form. On a spring day in the midwestern United States, the air temperature at the surface is 28.0°C. Puffy cumulus clouds form at an altitude where the air temperature equals the dew point (Problem 16–73). If the air temperature decreases with altitude at a rate of 0.6 C°/100 m, at approximately what height above the ground will clouds form if the relative humidity at the surface is a) 35% b) 80%? (*Hint:* Use the table in Problem 16–73.)

nebula is about 20 light years in diameter and contains about 50 hydrogen atoms per cubic centimeter (monatomic hydrogen, *not* H_2) at a temperature of about 20 K. (A light year is the distance light travels in vacuum in one year and is equal to

9.46×10^{15} m.) a) Estimate the mean free path for a hydrogen atom in a dark nebula. The radius of a hydrogen atom is 5.0×10^{-11} m. b) Estimate the rms speed of a hydrogen atom and the mean free time (the average time between collisions for a given atom). Based on this result, do you think that atomic collisions, such as those leading to H_2 molecule formation, are very important in determining the composition of the nebula? c) Estimate the pressure inside a dark nebula. d) Compare the rms speed of a hydrogen atom to the escape speed at the surface of the nebula (assumed spherical). If the space around the nebula were a vacuum, would such a cloud be stable or would it tend to evaporate? e) The stability of dark nebulae is explained by the presence of the *interstellar medium* (ISM), an even thinner gas that permeates space and into which the dark nebulae are embedded. Show that for dark nebulae to be in equilibrium with the ISM, the numbers of atoms per volume (N/V) and the temperatures (T) of dark nebulae and the ISM must be related by

$$\frac{(N/V)_{\text{nebula}}}{(N/V)_{\text{ISM}}} = \frac{T_{\text{ISM}}}{T_{\text{nebula}}}.$$

f) In the vicinity of the sun, the ISM contains about 1 hydrogen atom per 200 cm³. Estimate the temperature of the ISM in the vicinity of the sun. Compare to the temperature of the sun's surface, about 5800 K. Would a spacecraft coasting through interstellar space burn up? Why or why not?

FIGURE 16–24 Challenge Problem 16–75.

16–76 In the *troposphere*, the part of the atmosphere that extends from the surface to an altitude of about 11 km, the temperature is not uniform but decreases with increasing elevation. a) Show that if the temperature variation is approximated by the linear relation

$$T = T_0 - \alpha y,$$

where T_0 is the temperature at the earth's surface and T is the temperature at height y, the pressure p at height y is given by

$$\ln\left(\frac{p}{p_0}\right) = \frac{Mg}{R\alpha} \ln\left(\frac{T_0 - \alpha y}{T_0}\right),$$

where p_0 is the pressure at the earth's surface and M is the molar mass for air. The coefficient α is called the lapse rate of temperature. It varies with atmospheric conditions, but an average value

is about 0.6 C°/100 m. b) Show that the above result reduces to the result of Example 16–4 (Section 16–2) in the limit that $\alpha \to 0$. c) With $\alpha = 0.6$ C°/100 m, calculate p for $y = 8863$ m and compare your answer to the result of Example 16–4. Take $T_0 = 288$ K and $p_0 = 1.00$ atm.

16–77 Van der Waals Equation and Critical Points. a) In pV-diagrams the slope $\partial p/\partial V$ along an isotherm is never positive. Explain why. b) Regions where $\partial p/\partial V = 0$ represent equilibrium between two phases; volume can change with no change in pressure, as when water boils at atmospheric pressure. We can use this to determine the temperature, pressure, and volume per mole at the critical point using the equation of state $p = p(V, T, n)$. If $T > T_c$, then $p(V)$ has no maximum along an isotherm, but if $T < T_c$, then $p(V)$ has a maximum. Explain how this leads to the following condition for determining the critical point:

$$\frac{\partial p}{\partial V} = 0 \quad \text{and} \quad \frac{\partial^2 p}{\partial V^2} = 0 \quad \text{at the critical point.}$$

c) Solve the van der Waals equation (Eq. 16–7) for p. That is, find $p(V, T, n)$. Find $\partial p/\partial V$ and $\partial^2 p/\partial V^2$. Set these equal to zero to obtain two equations for V, T, and n. d) Simultaneous solution of the two equations obtained in part (c) gives the temperature and volume per mole at the critical point, T_c and $(V/n)_c$. Find these constants in terms of a and b. (*Hint:* Divide one equation by the other to eliminate T.) e) Substitute these values into the equation of state to find p_c, the pressure at the critical point. f) Use the results from parts (d) and (e) to find the ratio $RT_c/p_c(V/n)_c$. This should not contain either a or b, and so should have the same value for all gases. g) Compute the ratio $RT_c/p_c(V/n)_c$ for the gases H_2, N_2, and H_2O using the critical point data given below.

Gas	T_c (K)	p_c (Pa)	$(V/n)_c$ (m³/mol)
H_2	33.3	13.0×10^5	65.0×10^{-6}
N_2	126.2	33.9×10^5	90.1×10^{-6}
H_2O	647.4	221.2×10^5	56.0×10^{-6}

h) Discuss how well the results of part (g) compare to the prediction of part (f) based on the van der Waals equation. What do you conclude about the accuracy of the van der Waals equation as a description of the behavior of gases near the critical point?

*16–78 In Example 16–7 (Section 16–4) we saw that $v_{\text{rms}} > v_{\text{av}}$. It is not difficult to show that this is *always* the case. (The only exception is when the particles have the same speed, in which case $v_{\text{rms}} = v_{\text{av}}$.) a) For two particles with speeds v_1 and v_2, show that $v_{\text{rms}} \geq v_{\text{av}}$, regardless of the numerical values of v_1 and v_2. Then show that $v_{\text{rms}} > v_{\text{av}}$ if $v_1 \neq v_2$. b) Suppose that for a collection of N particles you know that $v_{\text{rms}} > v_{\text{av}}$. Another particle, with speed u, is added to the collection of particles. If the new rms and average speeds are denoted as v'_{rms} and v'_{av}, show that

$$v'_{\text{rms}} = \sqrt{\frac{N v_{\text{rms}}^2 + u^2}{N + 1}} \quad \text{and} \quad v'_{\text{av}} = \frac{N v_{\text{av}} + u}{N + 1}.$$

c) Use the expressions in part (b) to show that $v'_{\text{rms}} > v'_{\text{av}}$, regardless of the numerical value of u. d) Explain why your results for (a) and (c) together show that $v_{\text{rms}} > v_{\text{av}}$ for any collection of particles if the particles do not all have the same speed.

This steam locomotive operates using the first law of thermodynamics. When heat is generated by burning wood or coal in the locomotive's engine, part of the energy goes into increasing the temperature of the engine's water (which boils and produces steam). The remainder is used by the expanding steam to do work and propel the locomotive forward.

17–1 INTRODUCTION

Every time you drive a car, turn on an air conditioner, or use an electrical appliance, you reap the practical benefits of *thermodynamics,* the study of relationships involving heat, mechanical work, and other aspects of energy and energy transfer. If you've seen demonstrations that use liquid nitrogen, you may have wondered how gases are liquefied. One method is by first compressing the gas to very high pressure while keeping the temperature constant, then insulating it and allowing it to expand. The gas cools so much during the expansion that it liquefies. This is an example of a *thermodynamic process.*

The first law of thermodynamics, central to the understanding of such processes, is an extension of the principle of conservation of energy. It broadens this principle to include energy exchange by both heat transfer and mechanical work and introduces the concept of the *internal energy* of a system. Conservation of energy plays a vital role in every area of physical science, and the first law has extremely broad usefulness. To state energy relationships precisely, we need the concept of a *thermodynamic system,* and we discuss *heat* and *work* as two means of transferring energy into or out of such a system.

17–2 THERMODYNAMIC SYSTEMS

We have studied energy transfer through mechanical work (Chapter 6) and through heat transfer (Chapters 15 and 16).

Now we are ready to combine and generalize these principles. We will always talk about energy transfer to or from some specific *system.* The system might be a mechanical device, a biological organism, or a specified quantity of material such as the refrigerant in an air conditioner or steam expanding in a turbine. A **thermodynamic system** is a system that can interact (and exchange energy) with its surroundings, or environment, in at least two ways, one of which is heat transfer. A familiar example is a quantity of popcorn kernels in a pot with a lid. When the pot is placed on a stove, energy is added to the popcorn by conduction of heat; as the popcorn pops and expands, it does work as it exerts an upward force on the lid and moves it through a displacement (Fig. 17–1). The *state* of the popcorn changes in this process, since the volume, temperature, and pressure of the popcorn all change as it pops. A process such as this one, in which there are changes in the state of a thermodynamic system, is called a **thermodynamic process.**

In mechanics we used the concept of *system* on a regular basis in connection with free-body diagrams and conservation of energy and momentum. With *thermodynamic* systems, as with all others, it is essential to define clearly at the start exactly what is and is not included in the system. Only then can we describe unambiguously the energy transfers into and out of that system. For instance, in our popcorn example we defined the system to include the popcorn but not the pot, lid, or stove.

17–1 The popcorn in the pot is a thermodynamic system. In the thermodynamic process shown here, heat is added to the system, and the system does work on its surroundings to lift the lid of the pot.

17–2 (a) A jet engine uses the heat of combustion of its fuel to do work propelling the plane. (b) Humans and other biological organisms are more complicated systems than we can analyze fully in this book, but the same basic principles of thermodynamics apply to them.

Thermodynamics has its roots in many practical problems other than popping popcorn (Fig. 17–2). The engine in an automobile and the jet engines in an airplane use the heat of combustion of their fuel to perform mechanical work in propelling the vehicle. Muscle tissue in living organisms metabolizes chemical energy in food and performs mechanical work on the organism's surroundings. A steam engine or steam turbine uses the heat of combustion of coal or other fuel to perform mechanical work such as driving an electric generator or pulling a train.

SIGNS FOR HEAT AND WORK IN THERMODYNAMICS

We describe the energy relations in any thermodynamic process in terms of the quantity of heat Q added *to* the system and the work W done *by* the system. Both Q and W may be positive, negative, or zero (Fig. 17–3). A positive value of Q represents heat flow *into* the system, with a corresponding input of energy to it; negative Q represents heat flow *out of* the system. A positive value of W represents work done *by* the system against its surroundings, such as work done by an expanding gas, and hence corresponds to energy *leaving* the system. Negative W, such as work done during compression of a gas in which work is done *on the gas* by its surroundings, represents energy *entering* the system. We will use these conventions consistently in the examples in this chapter and the next.

 CAUTION ▶ Note that our sign rule for work is *opposite* to the one we used in mechanics, in which we always spoke of the work done by the forces acting *on* a body. In thermodynamics it is usually more convenient to call W the work done *by* the system so that when a system expands, the pressure, volume change, and work are all positive. Take care to use the sign rules for work and heat consistently! ◀

17–3 WORK DONE DURING VOLUME CHANGES

A gas in a cylinder with a movable piston is a simple example of a thermodynamic system. Internal-combustion engines, steam engines, and compressors in refrigerators and air conditioners all use some version of such a system. In the next several sections we will use the gas-in-cylinder system to explore several kinds of processes involving energy transformations.

(a)

(b)

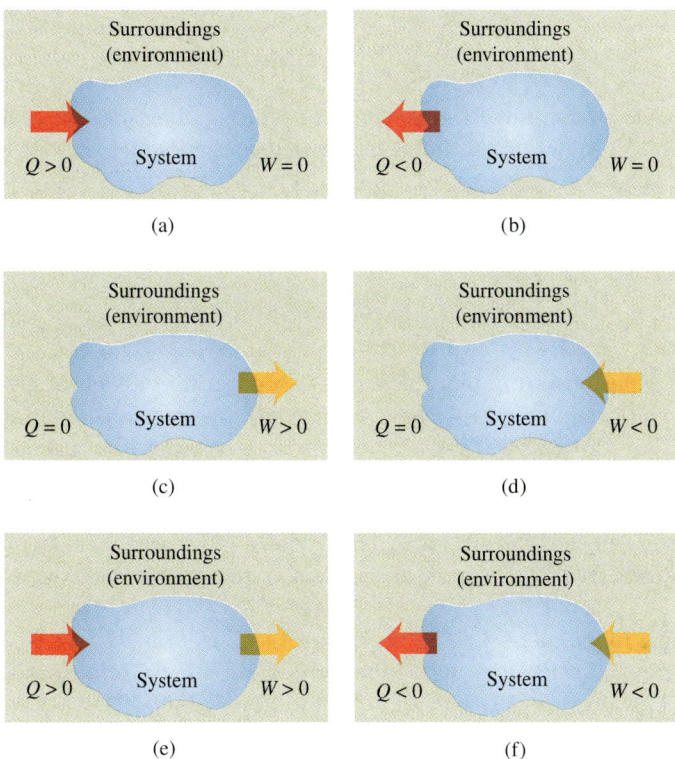

(a)   (b)

(c)   (d)

(e)   (f)

17–3 A thermodynamic system may exchange energy with its surroundings (environment) by means of heat and work. (a) When heat is added *to* the system, Q is positive. (b) When heat is transferred *out of* the system, Q is negative. (c) When work is done *by* the system, W is positive. (d) When work is done *on* the system, W is negative. Energy transfer by both heat and work can occur simultaneously; in (e), heat is added to the system and work is done by the system, and in (f), heat is transferred out of the system and work is done on the system.

(a)

(b)

17–4 (a) When a molecule strikes a wall moving away from it, the molecule does work on the wall; the molecule's speed and kinetic energy decrease. The gas does positive work on the piston. (b) When a molecule strikes a wall moving toward it, the wall does work on the molecule; the molecule's speed and kinetic energy increase. The gas does negative work on the piston.

We'll use a microscopic viewpoint, based on the kinetic and potential energies of individual molecules in a material, to develop intuition about thermodynamic quantities. But it is important to understand that the central principles of thermodynamics can be treated in a completely *macroscopic* way, without reference to microscopic models. Indeed, part of the great power and generality of thermodynamics is that it does *not* depend on details of the structure of matter.

First we consider the *work* done by the system during a volume change. When a gas expands, it pushes outward on its boundary surfaces as they move outward. Hence an expanding gas always does positive work. The same thing is true of any solid or fluid material that expands under pressure, such as the popcorn in Fig. 17–1.

We can understand the work done by a gas in a volume change by considering the molecules that make up the gas. When one such molecule collides with a stationary surface, it exerts a momentary force on the wall but does no work because the wall does not move. But if the surface is moving, such as a piston in a gasoline engine, the molecule *does* do work on the surface during the collision. If the piston in Fig. 17–4a moves to the right, so that the volume of the gas increases, the molecules that strike the piston exert a force through a distance and do *positive* work on the piston. If the piston moves toward the left as in Fig. 17–4b, so the volume of the gas decreases, then positive work is done *on* the molecule during the collision. Hence the gas molecules do *negative* work on the piston.

Figure 17–5 shows a solid or fluid in a cylinder with a movable piston. Suppose that the cylinder has a cross-section area A and that the pressure exerted by the system at the piston face is p. The total force F exerted by the system on the piston is $F = pA$. When the piston moves out an infinitesimal distance dx, the work dW done by this force is

$$dW = F \, dx = pA \, dx.$$

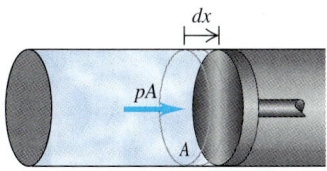

17–5 The infinitesimal work done by the system during the small expansion dx is $dW = pA \, dx$.

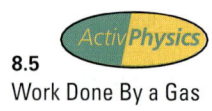

8.5
Work Done By a Gas

But

$$A \, dx = dV,$$

where dV is the infinitesimal change of volume of the system. Thus we can express the work done by the system in this infinitesimal volume change as

$$dW = p \, dV. \tag{17–1}$$

In a finite change of volume from V_1 to V_2,

$$W = \int_{V_1}^{V_2} p \, dV \qquad \text{(work done in a volume change).} \tag{17–2}$$

In general the pressure of the system may vary during the volume change. To evaluate the integral in Eq. (17–2), we have to know how the pressure varies as a function of volume. We can represent this relationship as a graph of p as a function of V (a pV-diagram, described at the end of Section 16–2). Figure 17–6a shows a simple example. In this figure, Eq. (17–2) is represented graphically as the *area* under the curve of p versus V between the limits V_1 and V_2. (In Section 6–4 we used a similar interpretation of the work done by a force F as the area under the curve of F versus x between the limits x_1 and x_2.)

According to the rule we stated in Section 17–2, work is *positive* when a system *expands*. In an expansion from state 1 to state 2 in Fig. 17–6a the area under the curve and the work are positive. A *compression* from 1 to 2 in Fig. 17–6b gives a *negative* area; when a system is compressed, its volume decreases and it does *negative* work on its surroundings (see also Fig. 17–4b).

If the pressure p remains constant while the volume changes from V_1 to V_2 (Fig. 17–6c), the work done by the system is

When fuel is burned in one of the cylinders of an automotive engine, the heated fuel-air mixture expands and does an amount of work given by Eq. (17–2).

$$W = p(V_2 - V_1) \qquad \begin{array}{l}\text{(work done in a volume change at}\\\text{constant pressure).}\end{array} \tag{17–3}$$

In any process in which the volume is *constant*, the system does no work because there is no displacement.

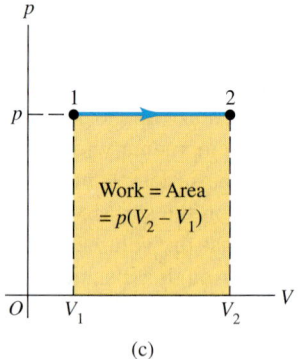

(a)   (b)   (c)

17–6 The work done equals the area under the curve on a pV-diagram. (a) In this change from state 1 to state 2 the volume increases and the work and area are positive. (b) In this change from state 1 to state 2 the volume decreases; the area is taken to be negative to agree with the sign of W. The two states are reversed from part (a). (c) The area of the rectangle gives the work done for a constant-pressure process. In this case, volume increases and $W > 0$.

EXAMPLE 17–1

Isothermal expansion of an ideal gas An ideal gas undergoes an *isothermal* (constant-temperature) *expansion* at temperature T, during which its volume changes from V_1 to V_2. How much work does the gas do?

SOLUTION From Eq. (17–2),

$$W = \int_{V_1}^{V_2} p \, dV.$$

From Eq. (16–3) the pressure p of n moles of ideal gas occupying volume V at absolute temperature T is

$$p = \frac{nRT}{V},$$

where R is the gas constant. We substitute this into the integral,

take the constants n, R, and T outside, and evaluate the integral:

$$W = nRT \int_{V_1}^{V_2} \frac{dV}{V} = nRT \ln \frac{V_2}{V_1} \qquad \begin{array}{l} \text{(ideal gas,} \\ \text{isothermal process).} \end{array}$$

In an expansion, $V_2 > V_1$ and W is positive. Also, when T is constant,

$$p_1 V_1 = p_2 V_2 \qquad \text{or} \qquad \frac{V_2}{V_1} = \frac{p_1}{p_2},$$

so the isothermal work may also be expressed as

$$W = nRT \ln \frac{p_1}{p_2} \qquad \text{(ideal gas, isothermal process).}$$

In an isothermal expansion the volume increases and the pressure drops, so $p_1 > p_2$ and again the work is positive.

17–4 PATHS BETWEEN THERMODYNAMIC STATES

We've seen that if a thermodynamic process involves a change in volume, the system undergoing the process does work (either positive or negative) on its surroundings. Heat also flows into or out of the system during the process if there is a temperature difference between the system and its surroundings. Let's now examine how the work done by and heat added to the system during a thermodynamic process depend on the details of how the process takes place.

When a thermodynamic system changes from an initial state to a final state, it passes through a series of intermediate states. We call this series of states a **path.** There are always infinitely many different possibilities for these intermediate states. When they are all equilibrium states, the path can be plotted on a pV-diagram (Fig. 17–7a). Point 1 represents an initial state with pressure p_1 and volume V_1, and point 2 represents a final state with pressure p_2 and volume V_2. To pass from state 1 to state 2, we could keep the pressure constant at p_1 while the system expands to volume V_2 (point 3 in Fig. 17–7b), then reduce the pressure to p_2 (probably by decreasing the temperature) while keeping the volume constant at V_2 (to point 2 on the diagram). The work done by the system during this process is the area under the line $1 \rightarrow 3$; no work is done during the constant-volume process $3 \rightarrow 2$. Or the system might traverse the path $1 \rightarrow 4 \rightarrow 2$ (Fig. 17–7c); in that case the work is the area under the line $4 \rightarrow 2$, since no work is done during the constant-volume process $1 \rightarrow 4$. The smooth curve from 1 to 2 is another

17–7 (a) Three different paths between state 1 and state 2. (b)–(d) The work done by the system during a transition between two states depends on the path chosen.

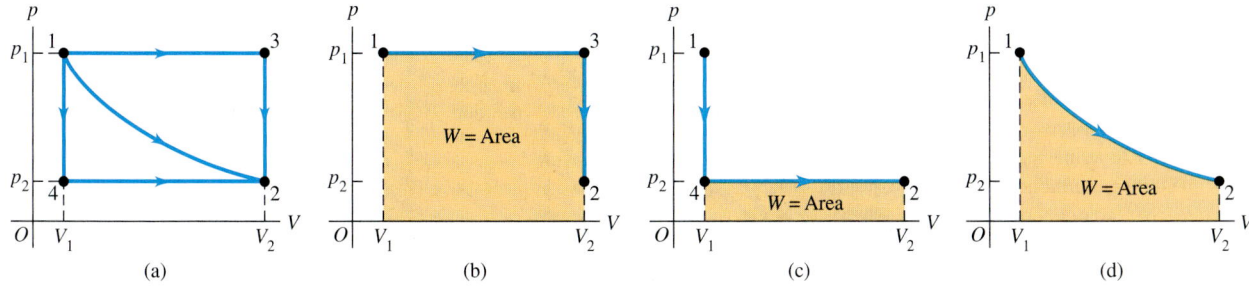

possibility (Fig. 17–7d), and the work for this path is different from that for either of the other paths.

We conclude that *the work done by the system depends not only on the initial and final states, but also on the intermediate states, that is, on the path.* Furthermore, we can take the system through a series of states forming a closed loop, such as $1 \rightarrow 3 \rightarrow 2 \rightarrow 4 \rightarrow 1$. In this case the final state is the same as the initial state, but the total work done by the system is *not* zero. (In fact, it is represented on the graph by the area enclosed by the loop; can you prove that? See Exercise 17–6.) It follows that it doesn't make sense to talk about the amount of work *contained in* a system. In a particular state, a system may have definite values of the state coordinates p, V, and T, but it wouldn't make sense to say that it has a definite value of W.

Like work, the *heat* added to a thermodynamic system when it undergoes a change of state depends on the path from the initial state to the final state. Here's an example. Suppose we want to change the volume of a certain quantity of an ideal gas from 2.0 L to 5.0 L while keeping the temperature constant at $T = 300$ K. Figure 17–8 shows two different ways in which we can do this. In Fig. 17–8a the gas is contained in a cylinder with a piston, with an initial volume of 2.0 L. We let the gas expand slowly, supplying heat from the electric heater to keep the temperature at 300 K. After expanding in this slow, controlled, isothermal manner, the gas reaches its final volume of 5.0 L; it absorbs a definite amount of heat in the process.

Figure 17–8b shows a different process leading to the same final state. The container is surrounded by insulating walls and is divided by a thin, breakable partition into two compartments. The lower part has volume 2.0 L, and the upper part has volume 3.0 L. In the lower compartment we place the same amount of the same gas as in Fig. 17–8a, again at $T = 300$ K. The initial state is the same as before. Now we break the partition; the gas undergoes a rapid, uncontrolled expansion, with no heat passing through the insulating walls. The final volume is 5.0 L, the same as in Fig. 17–8a. The gas does no work during this expansion because it doesn't push against anything that moves. This uncontrolled expansion of a gas into vacuum is called a **free expansion;** we will discuss it further in Section 17–7.

Experiments have shown that when an ideal gas undergoes a free expansion, there is no temperature change. Therefore the final state of the gas is the same as in Fig. 17–8a. The intermediate states (pressures and volumes) during the transition from state 1 to state 2 are entirely different in the two cases; Figs. 17–8a and 17–8b represent *two different paths* connecting the *same states* 1 and 2. For the path in Fig. 17–8b, *no* heat is transferred into the system, and the system does no work. Like work, *heat depends not only on the initial and final states but also on the path.*

Because of this path dependence, it would not make sense to say that a system "contains" a certain quantity of heat. To see this, suppose we assign an arbitrary value to "the

17–8 (a) Slow, controlled isothermal expansion of a gas from an initial state 1 to a final state 2 with the same temperature but lower pressure. (b) Rapid, uncontrolled expansion of the same gas starting at the same state 1 and ending at the same state 2.

heat in a body" in some reference state. Then presumably the "heat in the body" in some other state would equal the heat in the reference state plus the heat added when the body goes to the second state. But that's ambiguous, as we have just seen; the heat added depends on the *path* we take from the reference state to the second state. We are forced to conclude that there is *no* consistent way to define "heat in a body"; it is not a useful concept.

While it doesn't make sense to talk about "work in a body" or "heat in a body," it *does* make sense to speak of the amount of *internal energy* in a body. This important concept is our next topic.

17–5 INTERNAL ENERGY AND THE FIRST LAW OF THERMODYNAMICS

Internal energy is one of the most important concepts in thermodynamics. In Section 7–4, when we discussed energy changes for a body sliding with friction, we stated that warming a body increased its internal energy and that cooling the body decreased its internal energy. But what *is* internal energy? We can look at it in various ways; let's start with one based on the ideas of mechanics. Matter consists of atoms and molecules, and these are made up of particles having kinetic and potential energies. We *tentatively* define the **internal energy** of a system as the sum of the kinetic energies of all of its constituent particles, plus the sum of all the potential energies of interaction among these particles.

Note that internal energy does *not* include potential energy arising from the interaction between the system and its surroundings. If the system is a glass of water, placing it on a high shelf increases the gravitational potential energy arising from the interaction between the glass and the earth. But this has no effect on the interaction between the molecules of the water, and so the internal energy of the water does not change.

We use the symbol U for internal energy. (We used this same symbol in our study of mechanics to represent potential energy. You may have to remind yourself occasionally that U has a different meaning in thermodynamics.) During a change of state of the system the internal energy may change from an initial value U_1 to a final value U_2. We denote the change in internal energy as $\Delta U = U_2 - U_1$.

We know that heat transfer is energy transfer. When we add a quantity of heat Q to a system and it does no work during the process, the internal energy increases by an amount equal to Q; that is, $\Delta U = Q$. When a system does work W by expanding against its surroundings and no heat is added during the process, energy leaves the system and the internal energy decreases. That is, when W is positive, ΔU is negative, and conversely. So $\Delta U = -W$. When *both* heat transfer and work occur, the *total* change in internal energy is

$$U_2 - U_1 = \Delta U = Q - W \quad \text{(first law of thermodynamics).} \quad (17\text{–}4)$$

We can rearrange this to the form

$$Q = \Delta U + W. \quad (17\text{–}5)$$

The message of Eq. (17–5) is that in general when heat Q is added to a system, some of this added energy remains within the system, changing its internal energy by an amount ΔU; the remainder leaves the system again as the system does work W against its surroundings. Because W and Q may be positive, negative, or zero, ΔU can be positive, negative, or zero for different processes (Fig. 17–9).

Equation (17–4) or (17–5) is the **first law of thermodynamics.** It is a generalization of the principle of conservation of energy to include energy transfer through heat as

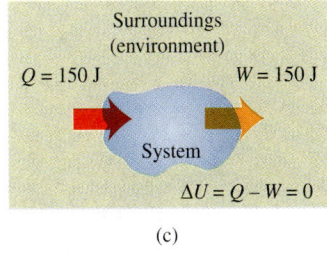

17–9 In a thermodynamic process, the internal energy of a system may increase, decrease, or stay the same. (a) If more heat is added to the system than the system does work, ΔU is positive and the internal energy increases. (b) If more heat flows out of the system than work is done on the system, ΔU is negative and the internal energy decreases. (c) If the heat added to the system equals the work done by the system, $\Delta U = 0$ and the internal energy is unchanged.

well as mechanical work. As you will see in later chapters, this principle can be extended to ever broader classes of phenomena by identifying additional forms of energy and energy transfer. In every situation in which it seems that the total energy in all known forms is not conserved, it has been possible to identify a new form of energy such that the total energy, including the new form, *is* conserved. There is energy associated with electric fields, with magnetic fields, and, according to the theory of relativity, even with mass itself.

At the beginning of this discussion we tentatively defined internal energy in terms of microscopic kinetic and potential energies. This has drawbacks, however. Actually *calculating* internal energy in this way for any real system would be hopelessly complicated. Furthermore, this definition isn't an *operational* one because it doesn't describe how to determine internal energy from physical quantities that we can measure directly.

So let's look at internal energy in another way. Starting over, we define the *change* in internal energy ΔU during any change of a system as the quantity given by Eq. (17–4), $\Delta U = Q - W$. This *is* an operational definition, because we can measure Q and W. It does not define U itself, only ΔU. This is not a shortcoming, because we can *define* the internal energy of a system to have a specified value in some reference state, and then use Eq. (17–4) to define the internal energy in any other state. This is analogous to our treatment of potential energy in Chapter 7, in which we arbitrarily defined the potential energy of a mechanical system to be zero at a certain position.

This new definition trades one difficulty for another. If we define ΔU by Eq. (17–4), then when the system goes from state 1 to state 2 by two different paths, how do we know that ΔU is the same for the two paths? We have already seen that Q and W are, in general, *not* the same for different paths. If ΔU, which equals $Q - W$, is also path-dependent, then ΔU is ambiguous. If so, the concept of internal energy of a system is subject to the same criticism as the erroneous concept of quantity of heat in a system, as we discussed at the end of Section 17–4.

The only way to answer this question is through *experiment*. For various materials we measure Q and W for various changes of state and various paths to learn whether ΔU is or is not path-dependent. The results of many such investigations are clear and unambiguous: While Q and W depend on the path, $\Delta U = Q - W$ *is independent of path. The change in internal energy of a system during any thermodynamic process depends only on the initial and final states, not on the path leading from one to the other.*

Experiment, then, is the ultimate justification for believing that a thermodynamic system in a specific state has a unique internal energy that depends only on that state. An equivalent statement is that the internal energy U of a system is a function of the state coordinates p, V, and T (actually, any two of these, since the three variables are related by the equation of state).

To say that the first law of thermodynamics, given by Eq. (17–4) or (17–5), represents conservation of energy for thermodynamic processes is correct, as far as it goes. But an important *additional* aspect of the first law is the fact that internal energy depends only on the state of a system. In changes of state, the change in internal energy is path-independent.

All this may seem a little abstract if you are satisfied to think of internal energy as microscopic mechanical energy. There's nothing wrong with that view, and we will make use of it at various times during our discussion. But in the interest of precise *operational* definitions, internal energy, like heat, can and must be defined in a way that is independent of the detailed microscopic structure of the material.

Two special cases of the first law of thermodynamics are worth mentioning. A process that eventually returns a system to its initial state is called a *cyclic* process. For such a process, the final state is the same as the initial state, and so the *total* internal

8.6
Heat, Internal Energy, and First Law of Thermodynamics

The internal energy of a cup of coffee just depends on its thermodynamic state—how much water and ground coffee it contains, and what its temperature is. It does not depend on the history of how the coffee was prepared—that is, the thermodynamic path that led to its current state.

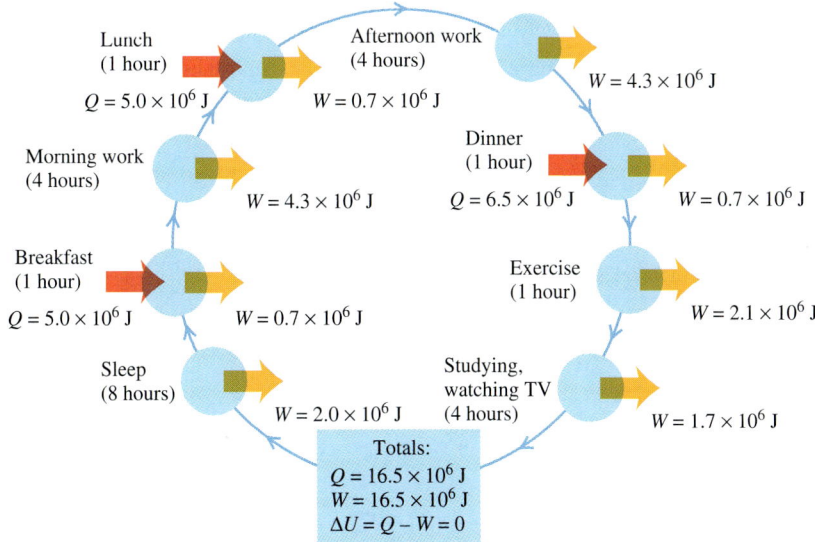

17–10 Every day, your body (a thermodynamic system) goes through a cyclic thermodynamic process like this one. Heat Q is added by metabolizing food, and your body does work W in breathing, walking, and other activities. If you return to the same state at the end of the day, $Q = W$ and the net change in your internal energy is zero.

energy change must be zero. Then

$$U_2 = U_1 \quad \text{and} \quad Q = W.$$

If a net quantity of work W is done by the system during this process, an equal amount of energy must have flowed into the system as heat Q. But there is no reason why either Q or W individually has to be zero (Fig. 17–10).

Another special case occurs in an *isolated* system, one that does no work on its surroundings and has no heat flow to or from its surroundings. For any process taking place in an isolated system,

$$W = Q = 0,$$

and therefore

$$U_2 - U_1 = \Delta U = 0.$$

In other words, *the internal energy of an isolated system is constant.*

Problem–Solving Strategy

THE FIRST LAW OF THERMODYNAMICS

1. The internal energy change ΔU in any thermodynamic process or series of processes is independent of the path, whether the substance is an ideal gas or not. This is of the utmost importance in the problems in this chapter and the next. Sometimes you will be given enough information about one path between given initial and final states to calculate ΔU for that path. Since ΔU is the same for every other path between the same two states, you can then relate the various energy quantities for other paths. For a cyclic process, $\Delta U = 0$.

2. As usual, consistent units are essential. If p is in Pa and V in m^3, then W is in joules. Otherwise, you may want to convert the pressure and volume units into Pa and m^3. If a

heat capacity is given in terms of calories, the simplest procedure is usually to convert it to joules. Be especially careful with moles. When you use $n = m_{tot}/M$ to convert between total mass and number of moles, remember that if m_{tot} is in kilograms, M must be in *kilograms* per mole. The usual units for M are *grams* per mole; be careful!

3. When a process consists of several distinct steps, it often helps to make a chart showing Q, W, and ΔU for each step. Put these quantities for each step on a different line, and arrange them so that the Q's, W's, and ΔU's form columns. Then you can apply the first law to each line; in addition, you can add each column and apply the first law to the sums. Do you see why?

EXAMPLE 17-2

Working off your dessert You propose to eat a 900-calorie hot fudge sundae (with whipped cream) and then run up several flights of stairs to work off the energy you have taken in. How high do you have to climb? Assume that your mass is 60 kg.

SOLUTION The system consists of you and the earth. Remember that one food-value calorie is 1 kcal = 1000 cal = 4190 J. The energy intake is

$$Q = 900 \text{ kcal } (4190 \text{ J/1 kcal}) = 3.77 \times 10^6 \text{ J}.$$

The energy output required to climb a height h is

$$W = mgh = (60 \text{ kg})(9.8 \text{ m/s}^2)h = (588 \text{ N})h.$$

If the final state of the system is the same as the initial state (that is, no fatter, no leaner), these two energy quantities must be equal: $Q = W$. Then

$$h = \frac{Q}{mg} = \frac{3.77 \times 10^6 \text{ J}}{588 \text{ N}} = 6410 \text{ m} \quad \text{(about 21,000 ft).}$$

Good luck! We have assumed 100% efficiency in the conversion of food energy into mechanical work; this isn't very realistic. We'll talk more about efficiency later.

EXAMPLE 17-3

Figure 17–11 shows a *pV*-diagram for a *cyclic* process, one in which the initial and final states are the same. It starts at point *a* and proceeds counterclockwise in the *pV*-diagram to point *b*, then back to *a*, and the total work is $W = -500$ J. a) Why is the work negative? b) Find the change in internal energy and the heat added during this process.

SOLUTION a) The work done equals the area under the curve, with the area taken as positive for increasing volume and negative for decreasing volume. The area under the lower curve from *a* to *b* is positive, but it is smaller than the absolute value of the negative area under the upper curve from *b* back to *a*. Therefore the net area (the area enclosed by the path, shown with red stripes) and the work are negative. In other words, 500 more joules of work are done *on* the system than *by* the system. b) For this and any other cyclic process (in which the beginning and end points are the same), $\Delta U = 0$, so $Q = W = -500$ J. That is, 500 joules of heat must come *out of* the system.

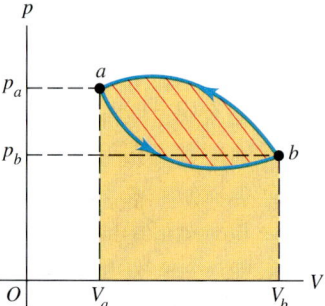

17-11 The net work done by the system in the process *aba* is –500 J. What would it have been if the process had proceeded clockwise in this *pV*-diagram?

EXAMPLE 17-4

A series of thermodynamic processes is shown in the *pV*-diagram of Fig. 17–12. In process *ab*, 150 J of heat are added to the system, and in process *bd*, 600 J of heat are added. Find a) the internal energy change in process *ab*; b) the internal energy change in process *abd* (shown in light blue); and c) the total heat added in process *acd* (shown in dark blue).

SOLUTION a) No volume change occurs during process *ab*, so $W_{ab} = 0$ and $\Delta U_{ab} = Q_{ab} = 150$ J.
b) Process *bd* occurs at constant pressure, so the work done by the system during this expansion is

$$W_{bd} = p(V_2 - V_1)$$
$$= (8.0 \times 10^4 \text{ Pa})(5.0 \times 10^{-3} \text{ m}^3 - 2.0 \times 10^{-3} \text{ m}^3)$$
$$= 240 \text{ J}.$$

The total work for process *abd* is

$$W_{abd} = W_{ab} + W_{bd} = 0 + 240 \text{ J} = 240 \text{ J},$$

17-12 A *pV*-diagram showing the various thermodynamic processes.

and the total heat is

$$Q_{abd} = Q_{ab} + Q_{bd} = 150 \text{ J} + 600 \text{ J} = 750 \text{ J}.$$

Applying Eq. (17–4) to *abd*, we find

$$\Delta U_{abd} = Q_{abd} - W_{abd} = 750 \text{ J} - 240 \text{ J} = 510 \text{ J}.$$

c) Because ΔU is independent of path, the internal energy change is the same for path *acd* as for path *abd*; that is,

$$\Delta U_{acd} = \Delta U_{abd} = 510 \text{ J}.$$

The total work for the path *acd* is

$$W_{acd} = W_{ac} + W_{cd} = p(V_2 - V_1) + 0$$
$$= (3.0 \times 10^4 \text{ Pa})(5.0 \times 10^{-3} \text{ m}^3 - 2.0 \times 10^{-3} \text{ m}^3)$$
$$= 90 \text{ J}.$$

Now we apply Eq. (17–5) to process *acd*:

$$Q_{acd} = \Delta U_{acd} + W_{acd} = 510 \text{ J} + 90 \text{ J} = 600 \text{ J}.$$

We see that although ΔU is the same (510 J) for *abd* and *acd*, W (240 J versus 90 J) and Q (750 J versus 600 J) have different values for the two processes.

Here is a tabulation of the various quantities:

STEP	Q	W	$\Delta U = Q - W$	STEP	Q	W	$\Delta U = Q - W$
ab	150 J	0 J	150 J	*ac*	?	90 J	?
bd	600 J	240 J	360 J	*cd*	?	0 J	?
abd	750 J	240 J	510 J	*acd*	600 J	90 J	510 J

EXAMPLE 17–5

Thermodynamics of boiling water One gram of water (1 cm^3) becomes 1671 cm^3 of steam when boiled at a constant pressure of 1 atm (1.013×10^5 Pa). The heat of vaporization at this pressure is $L_v = 2.256 \times 10^6$ J/kg. Compute a) the work done by the water when it vaporizes; b) its increase in internal energy.

SOLUTION a) For a constant-pressure process we may use Eq. (17–3) to compute the work done by the vaporizing water:

$$W = p(V_2 - V_1)$$
$$= (1.013 \times 10^5 \text{ Pa})(1671 \times 10^{-6} \text{ m}^3 - 1 \times 10^{-6} \text{ m}^3)$$
$$= 169 \text{ J}.$$

b) The heat added to the water is the heat of vaporization:

$$Q = mL_v = (10^{-3} \text{ kg})(2.256 \times 10^6 \text{ J/kg}) = 2256 \text{ J}.$$

From the first law of thermodynamics, Eq. (17–4), the change in internal energy is

$$\Delta U = Q - W = 2256 \text{ J} - 169 \text{ J} = 2087 \text{ J}.$$

To vaporize one gram of water, we have to add 2256 J of heat. Most (2087 J) of this added energy remains in the system as an increase in internal energy. The remaining 169 J leaves the system again as it does work against the surroundings while expanding from liquid to vapor. The increase in internal energy is associated mostly with the intermolecular forces that hold the molecules together in the liquid state. These forces are attractive, so the associated potential energies are greater after work has been done to pull the molecules apart, forming the vapor state. It's like increasing gravitational potential energy by pulling an elevator farther from the center of the earth.

INFINITESIMAL CHANGES OF STATE

In the preceding examples the initial and final states differ by a finite amount. Later we will consider *infinitesimal* changes of state in which a small amount of heat dQ is added to the system, the system does a small amount of work dW, and its internal energy changes by an amount dU. For such a process we state the first law in differential form as

$$dU = dQ - dW \qquad \text{(first law of thermodynamics, infinitesimal process).} \qquad (17\text{–}6)$$

For the systems we will discuss, the work dW is given by $dW = p\, dV$, so we can also state the first law as

$$dU = dQ - p\, dV. \qquad (17\text{–}7)$$

17–6 KINDS OF THERMODYNAMIC PROCESSES

In this section we describe four specific kinds of thermodynamic processes that occur often in practical situations. These can be summarized briefly as "no heat transfer" or *adiabatic*, "constant volume" or *isochoric*, "constant pressure" or *isobaric*, and "constant temperature" or *isothermal*. For some of these we can use a simplified form of the first law of thermodynamics.

ADIABATIC PROCESS

An **adiabatic process** (pronounced "ay-dee-ah-*bat*-ic") is defined as one with no heat transfer into or out of a system; $Q = 0$. We can prevent heat flow either by surrounding the system with thermally insulating material or by carrying out the process so quickly that there is not enough time for appreciable heat flow. From the first law we find that for every adiabatic process,

$$U_2 - U_1 = \Delta U = -W \qquad \text{(adiabatic process)}. \qquad (17\text{–}8)$$

8.4
State Variables and Ideal Gas Law

When a system expands adiabatically, W is positive (the system does work on its surroundings), so ΔU is negative and the internal energy decreases. When a system is *compressed* adiabatically, W is negative (work is done on the system by its surroundings) and U increases. In many (but not all) systems an increase of internal energy is accompanied by a rise in temperature.

The compression stroke in an internal-combustion engine is an approximately adiabatic process. The temperature rises as the air-fuel mixture in the cylinder is compressed. The expansion of the burned fuel during the power stroke is also an approximately adiabatic expansion with a drop in temperature.

ISOCHORIC PROCESS

An **isochoric process** (pronounced "eye-so-*kor*-ic") is a *constant-volume* process. When the volume of a thermodynamic system is constant, it does no work on its surroundings. Then $W = 0$, and

$$U_2 - U_1 = \Delta U = Q \qquad \text{(isochoric process)}. \qquad (17\text{–}9)$$

In an isochoric process, all the energy added as heat remains in the system as an increase in internal energy. Heating a gas in a closed constant-volume container is an example of an isochoric process. (Note that there are types of work that do not involve a volume change. For example, we can do work on a fluid by stirring it. In some literature, "isochoric" is used to mean that no work of any kind is done.)

ISOBARIC PROCESS

An **isobaric process** (pronounced "eye-so-*bear*-ic") is a *constant-pressure* process. In general, none of the three quantities ΔU, Q, and W is zero in an isobaric process, but calculating W is easy nonetheless. From Eq. (17–3),

$$W = p(V_2 - V_1) \qquad \text{(isobaric process)}. \qquad (17\text{–}10)$$

Example 17–5 concerns an isobaric process, boiling water at constant pressure.

ISOTHERMAL PROCESS

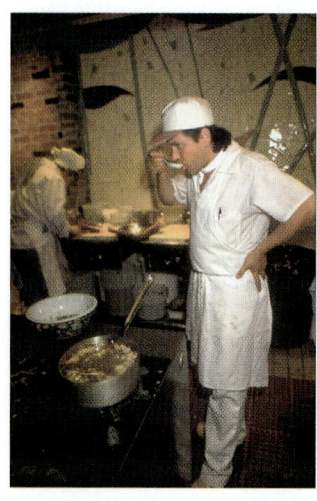

Most cooking involves isobaric processes. That's because the air pressure above a saucepan or frying pan, or inside a microwave oven, remains essentially constant while the food is being heated. (But see page 550 for an example of an *adiabatic* process in the kitchen.)

An **isothermal process** is a *constant-temperature* process. For a process to be isothermal, any heat flow into or out of the system must occur slowly enough that thermal equilibrium is maintained. In general, none of the quantities ΔU, Q, or W is zero in an isothermal process.

In some special cases the internal energy of a system depends *only* on its temperature, not on its pressure or volume. The most familiar system having this special property is an ideal gas, as we'll discuss in the next section. For such systems, if the temperature is constant, the internal energy is also constant; $\Delta U = 0$ and $Q = W$. That is, any energy entering the system as heat Q must leave it again as work W done by the system. Example 17–1, involving an ideal gas, is an example of an isothermal process in which U is also constant. For most systems other than ideal gases the internal energy depends on pressure as well as temperature, so U may vary even when T is constant.

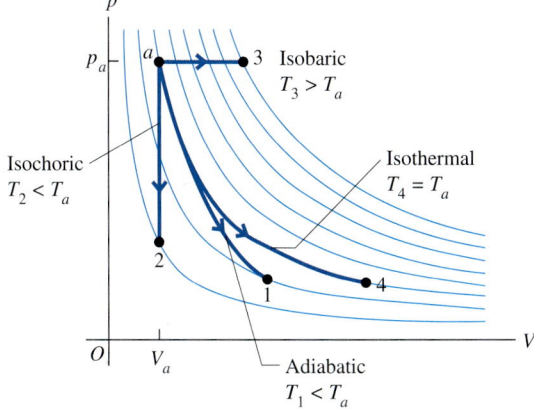

Figure 17–13 shows a *pV*-diagram for each of these four processes for a constant amount of an ideal gas. The path followed in an adiabatic process (*a* to 1) is called an **adiabat.** A vertical line (constant volume) is an **isochor,** a horizontal line (constant pressure) is an **isobar,** and a curve of constant temperature (shown as light blue lines in Fig. 17–13) is an **isotherm.**

17–7 INTERNAL ENERGY OF AN IDEAL GAS

We will now show that for an ideal gas, the internal energy U depends only on temperature, not on pressure or volume. Let's think again about the free-expansion experiment described in Section 17–4. A thermally insulated container with rigid walls is divided into two compartments by a partition (Fig. 17–14). One compartment has a quantity of an ideal gas, and the other is evacuated.

When the partition is removed or broken, the gas expands to fill both parts of the container. The gas does no work on its surroundings because the walls of the container don't move, and there is no heat flow through the insulation. So both Q and W are zero, and the internal energy U is constant. This is true of any substance, whether it is an ideal gas or not.

Does the *temperature* change during a free expansion? Suppose it *does* change, while the internal energy stays the same. In that case we have to conclude that the internal energy depends on both the temperature and the volume or on both the temperature and the pressure, but certainly not on the temperature alone. But if T is constant during a free expansion, for which we know that U is constant even though both p and V change, then we have to conclude that U depends only on T, not on p or V.

Many experiments have shown that when a low-density gas undergoes a free expansion, its temperature *does not* change. Such a gas is essentially an ideal gas. The conclusion is: **The internal energy of an ideal gas depends only on its temperature, not on its pressure or volume.** This property, in addition to the ideal-gas equation of state, is part of the ideal-gas model. Make sure that you understand that U depends only on T for an ideal gas, for we will make frequent use of this fact.

For non-ideal gases, some temperature change occurs during free expansions, even though the internal energy is constant. This shows that the internal energy cannot depend *only* on temperature; it must depend on pressure as well. From the microscopic viewpoint, in which internal energy U is the sum of the kinetic and potential energies for all the particles that make up the system, this is not surprising. Non-ideal gases usually have attractive intermolecular forces, and when molecules move farther apart, the associated potential energies increase. If the total internal energy is constant, the kinetic energies

17–14 The partition is broken (or removed) to start the free expansion of gas into the vacuum region.

must decrease. Temperature is directly related to molecular *kinetic* energy, and for such a gas a free expansion is usually accompanied by a *drop* in temperature.

17–8 HEAT CAPACITIES OF AN IDEAL GAS

We defined specific heat capacity and molar heat capacity in Section 15–6. We also remarked at the end of that section that the specific or molar heat capacity of a substance depends on the conditions under which the heat is added. It is usually easiest to measure the heat capacity of a gas in a closed container under constant-volume conditions. The corresponding heat capacity is the **molar heat capacity at constant volume,** denoted by C_V. Heat capacity measurements for solids and liquids are usually carried out in the atmosphere under constant atmospheric pressure, and we call the corresponding heat capacity the **molar heat capacity at constant pressure,** C_p. If neither p nor V is constant, we have an infinite number of possible heat capacities.

Let's consider C_V and C_p for an ideal gas. To measure C_V, we raise the temperature of an ideal gas in a rigid container with constant volume (neglecting its thermal expansion). To measure C_p, we let the gas expand just enough to keep the pressure constant as the temperature rises.

Why should these two molar heat capacities be different? The answer lies in the first law of thermodynamics. In a constant-volume temperature increase, the system does no work, and the change in internal energy ΔU equals the heat added Q. In a constant-pressure temperature increase, on the other hand, the volume *must* increase; otherwise, the pressure (given by the ideal-gas equation of state $p = nRT/V$) could not remain constant. As the material expands, it does an amount of work W. According to the first law,

$$Q = \Delta U + W. \tag{17–11}$$

For a given temperature increase, the internal energy change ΔU of an ideal gas has the same value no matter what the process (remember that the internal energy of an ideal gas depends only on temperature, not pressure or volume). Equation (17–11) then shows that the heat input for a constant-pressure process must be *greater* than that for a constant-volume process because additional energy must be supplied to account for the work W done during the expansion. So C_p is greater than C_V for an ideal gas. The pV-diagram in Fig. 17–15 shows this relationship. For air, C_p is 40% greater than C_V.

For a very few substances (one of which is water between 0°C and 4°C) the volume *decreases* during heating. In this case, W is negative, the heat input is *less* than in the constant-volume case, and C_p is *less* than C_V.

We can derive a simple relation between C_p and C_V for an ideal gas. First consider the constant-*volume* process. We place n moles of an ideal gas at temperature T in a constant-volume container. We place it in thermal contact with a hotter body; an infinitesimal quantity of heat dQ flows into the gas, and its temperature increases by an infinitesimal amount dT. By the definition of C_V, the molar heat capacity at constant volume,

$$dQ = nC_V \, dT. \tag{17–12}$$

The pressure increases during this process, but the gas does no work ($dW = 0$) because the volume is constant. The first law in differential form, Eq. (17–6), is $dQ = dU + dW$. Since $dW = 0$, $dQ = dU$ and Eq. (17–12) can also be written as

$$dU = nC_V \, dT. \tag{17–13}$$

Now consider a constant-*pressure* process with the same temperature change dT. We place the same gas in a cylinder with a piston that we can allow to move just enough to

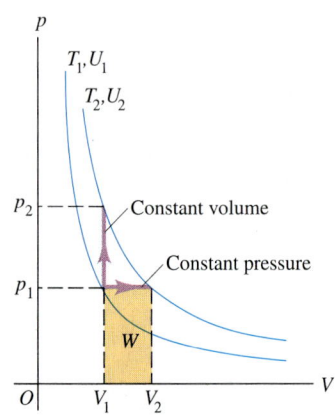

17–15 Raising the temperature of an ideal gas from T_1 to T_2 by a constant-volume or a constant-pressure process. For an ideal gas, U depends only on T, so ΔU is the same for both processes. In the constant-volume process, no work is done, so $Q = \Delta U$. But for the constant-pressure process, Q is greater, since it must include both ΔU and $W = p_1(V_2 - V_1)$. Thus $C_p > C_V$.

maintain constant pressure. Again we bring the system into contact with a hotter body. As heat flows into the gas, it expands at constant pressure and does work. By the definition of C_p, the molar heat capacity at constant pressure, the amount of heat dQ entering the gas is

$$dQ = nC_p \, dT. \tag{17-14}$$

The work dW done by the gas in this constant-pressure process is

$$dW = p \, dV.$$

8.7
Heat Capacity

8.8
Isochoric Process

8.9
Isobaric Process

8.10
Isothermal Process

We can also express dW in terms of the temperature change dT by using the ideal-gas equation of state, $pV = nRT$. Because p is constant, the change in V is proportional to the change in T:

$$dW = p \, dV = nR \, dT. \tag{17-15}$$

Now we substitute Eqs. (17–14) and (17–15) into the first law, $dQ = dU + dW$. We obtain

$$nC_p \, dT = dU + nR \, dT. \tag{17-16}$$

Now here comes the crux of the calculation. The internal energy change dU for the constant-pressure process is again given by Eq. (17–13), $dU = nC_V \, dT$, *even though now the volume is not constant.* Why is this so? Recall the discussion of Section 17–7; one of the special properties of an ideal gas is that its internal energy depends *only* on temperature. Thus the *change* in internal energy during any process must be determined only by the temperature change. If Eq. (17–13) is valid for an ideal gas for one particular kind of process, it must be valid for an ideal gas for *every* kind of process with the same dT. So we may replace dU in Eq. (17–16) by $nC_V \, dT$:

$$nC_p \, dT = nC_V \, dT + nR \, dT.$$

When we divide each term by the common factor $n \, dT$, we get

$$C_p = C_V + R \qquad \text{(molar heat capacities of an ideal gas).} \tag{17-17}$$

As we predicted, the molar heat capacity of an ideal gas at constant pressure is *greater* than the molar heat capacity at constant volume; the difference is the gas constant R. (Of course, R must be expressed in the same units as C_p and C_V, such as J/mol · K.)

We have used the ideal-gas model to derive Eq. (17–17), but it turns out to be obeyed to within a few percent by many real gases at moderate pressures. Measured values of C_p and C_V are given in Table 17–1 for several real gases at low pressures; the difference in most cases is approximately $R = 8.315$ J/mol · K.

TABLE 17–1

MOLAR HEAT CAPACITIES OF GASES AT LOW PRESSURE

TYPE OF GAS	GAS	C_V (J/mol · K)	C_p (J/mol · K)	$C_p - C_V$ (J/mol · K)	$\gamma = C_p/C_V$
Monatomic	He	12.47	20.78	8.31	1.67
	Ar	12.47	20.78	8.31	1.67
Diatomic	H_2	20.42	28.74	8.32	1.41
	N_2	20.76	29.07	8.31	1.40
	O_2	20.85	29.17	8.31	1.40
	CO	20.85	29.16	8.31	1.40
Polyatomic	CO_2	28.46	36.94	8.48	1.30
	SO_2	31.39	40.37	8.98	1.29
	H_2S	25.95	34.60	8.65	1.33

The table also shows that the molar heat capacity of a gas is related to its molecular structure, as we discussed in Section 16–5. In fact, the first two columns of Table 17–1 are the same as Table 16–1.

The last column of Table 17–1 lists the values of the dimensionless **ratio of heat capacities,** C_p/C_V, denoted by the Greek letter γ ("gamma"):

$$\gamma = \frac{C_p}{C_V} \qquad \text{(ratio of heat capacities).} \qquad (17\text{–}18)$$

(This is sometimes called the "ratio of specific heats.") For gases, C_p is always greater than C_V and γ is always greater than unity. This quantity plays an important role in *adiabatic* processes for an ideal gas, which we will study in the next section.

We can use our kinetic-theory discussion of the molar heat capacity of an ideal gas (Section 16–5) to predict values of γ. As an example, an ideal monatomic gas has $C_V = \frac{3}{2}R$. From Eq. (17–17),

$$C_p = C_V + R = \frac{3}{2}R + R = \frac{5}{2}R,$$

so

$$\gamma = \frac{C_p}{C_V} = \frac{\frac{5}{2}R}{\frac{3}{2}R} = \frac{5}{3} = 1.67.$$

As Table 17–1 shows, this agrees well with values of γ computed from measured heat capacities. For most diatomic gases near room temperature, $C_V = \frac{5}{2}R$, $C_p = C_V + R = \frac{7}{2}R$, and

$$\gamma = \frac{C_p}{C_V} = \frac{\frac{7}{2}R}{\frac{5}{2}R} = \frac{7}{5} = 1.40,$$

also in good agreement with measured values.

Here's a final reminder: For an ideal gas the internal energy change in *any* process is given by $\Delta U = nC_V \Delta T$, *whether the volume is constant or not.* This relation, which comes in handy in the following example, holds for other substances *only* when the volume is constant.

EXAMPLE 17–6

Cooling your room A typical dorm room or bedroom contains about 2500 moles of air. Find the change in the internal energy of this much air when it is cooled from 23.9°C to 11.6°C at a constant pressure of 1.00 atm. Treat the air as an ideal gas with $\gamma = 1.400$.

SOLUTION This is a constant-pressure process. Your first impulse may be to find C_p and then calculate Q from $Q = nC_p \Delta T$; find the volume change and find the work done by the gas from $W = p\,\Delta V$; then finally use the first law to find ΔU. This would be perfectly correct, but there's a much easier way. For an ideal gas the internal-energy change is $\Delta U = nC_V \Delta T$ for *every* process, *whether the volume is constant or not.* So all we have to do is find C_V and use this expression for ΔU. From Eqs.

(17–17) and (17–18),

$$\gamma = \frac{C_p}{C_V} = \frac{C_V + R}{C_V} = 1 + \frac{R}{C_V},$$

$$C_V = \frac{R}{\gamma - 1} = \frac{8.315 \text{ J/mol} \cdot \text{K}}{1.400 - 1} = 20.79 \text{ J/mol} \cdot \text{K}.$$

Then

$$\Delta U = nC_V \Delta T = (2500 \text{ mol})(20.79 \text{ J/mol} \cdot \text{K})(11.6 \text{ K} - 23.9 \text{ K})$$

$$= -6.39 \times 10^5 \text{ J}.$$

A room air conditioner must extract this much internal energy from the air in your room and transfer it to the air outside. We'll discuss how this is done in Chapter 18.

17–9 ADIABATIC PROCESSES FOR AN IDEAL GAS

An adiabatic process, defined in Section 17–6, is a process in which no heat transfer takes place between a system and its surroundings. Zero heat transfer is an idealization, but a process is approximately adiabatic if the system is well insulated or if the process takes place so quickly that there is not enough time for appreciable heat flow to occur.

In an adiabatic process, $Q = 0$, so from the first law, $\Delta U = -W$. An adiabatic process for an ideal gas is shown in the pV-diagram of Fig. 17–16. As the gas expands from volume V_a to V_b, it does positive work, so its internal energy decreases and its temperature drops. If point a, representing the initial state, lies on an isotherm at temperature $T + dT$, then point b for the final state is on a different isotherm at a lower temperature T. For an ideal gas an adiabatic curve (adiabat) at any point is always *steeper* than the isotherm passing through the same point. For an adiabatic *compression* from V_b to V_a the situation is reversed and the temperature rises.

The air in the output pipes of air compressors used in gasoline stations and in paint-spraying equipment and to fill scuba tanks is always warmer than the air entering the compressor; this is because the compression is rapid and hence approximately adiabatic. Adiabatic *cooling* occurs when you open a bottle of your favorite carbonated beverage. The gas just above the beverage surface expands rapidly in a nearly adiabatic process; the temperature of the gas drops so much that water vapor in the gas condenses, forming a miniature cloud.

CAUTION ▶ Keep in mind that when we talk about "adiabatic heating" and "adiabatic cooling," we really mean "raising the temperature" and "lowering the temperature," respectively. In an adiabatic process, the temperature change is due to work done by or on the system; there is *no* heat flow at all. ◀

We can derive a relation between volume and temperature changes for an infinitesimal adiabatic process in an ideal gas. Equation (17–13) gives the internal energy change dU for *any* process for an ideal gas, adiabatic or not, so we have $dU = nC_V \, dT$. Also, the work done by the gas during the process is given by $dW = p \, dV$. Then, since $dU = -dW$ for an adiabatic process, we have

$$nC_V \, dT = -p \, dV. \tag{17–19}$$

To obtain a relation containing only the volume V and temperature T, we eliminate p using the ideal-gas equation in the form $p = nRT/V$. Substituting this into Eq. (17–19) and rearranging, we get

$$nC_V \, dT = -\frac{nRT}{V} \, dV,$$

$$\frac{dT}{T} + \frac{R}{C_V}\frac{dV}{V} = 0.$$

The coefficient R/C_V can be expressed in terms of $\gamma = C_p/C_V$. We have

$$\frac{R}{C_V} = \frac{C_p - C_V}{C_V} = \frac{C_p}{C_V} - 1 = \gamma - 1,$$

$$\frac{dT}{T} + (\gamma - 1)\frac{dV}{V} = 0. \tag{17–20}$$

Because γ is always greater than unity for a gas, $(\gamma - 1)$ is always positive. This means that in Eq. (17–20), dV and dT always have opposite signs. An adiabatic *expansion* of an ideal gas ($dV > 0$) always occurs with a *drop* in temperature ($dT < 0$), and an adiabatic *compression* ($dV < 0$) always occurs with a *rise* in temperature ($dT > 0$); this confirms our earlier prediction.

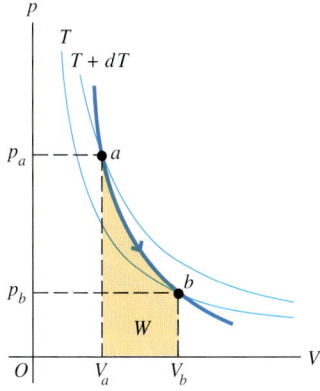

17–16 A pV-diagram of an adiabatic process for an ideal gas. As the gas expands from V_a to V_b, its temperature drops from $T + dT$ to T, corresponding to the decrease in internal energy due to the work W done by the gas (indicated by shaded area). For an ideal gas, when an isotherm and an adiabat pass through the same point on a pV-diagram, the adiabat is always steeper. (An adiabatic process is also shown in Fig. 17–13.)

8.11
Adiabatic Process

For finite changes in temperature and volume we integrate Eq. (17–20), obtaining

$$\ln T + (\gamma - 1)\ln V = \text{constant},$$

$$\ln T + \ln V^{\gamma-1} = \text{constant},$$

$$\ln\left(TV^{\gamma-1}\right) = \text{constant},$$

and finally,

$$TV^{\gamma-1} = \text{constant}. \tag{17–21}$$

Thus for an initial state (T_1, V_1) and a final state (T_2, V_2),

$$T_1V_1^{\gamma-1} = T_2V_2^{\gamma-1} \qquad \text{(adiabatic process, ideal gas)}. \tag{17–22}$$

Because we have used the ideal-gas equation in our derivation of Eqs. (17–21) and (17–22), the T's must always be *absolute* (Kelvin) temperatures.

We can also convert Eq. (17–21) into a relation between pressure and volume by eliminating T, using the ideal-gas equation in the form $T = pV/nR$. Substituting this into Eq. (17–21), we find

$$\frac{pV}{nR} V^{\gamma-1} = \text{constant},$$

or, because n and R are constant,

$$pV^{\gamma} = \text{constant}. \tag{17–23}$$

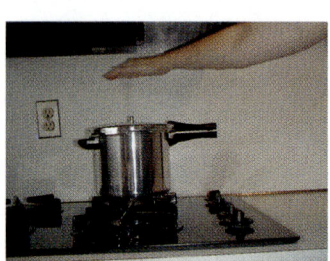

Hot steam escapes from the top of this pressure cooker at high speed. Hence it has no time to exchange heat with its surroundings, and its expansion is nearly adiabatic. As the steam's volume increases, its temperature drops so much (see Eq. (17–22)) that it feels cool on this chef's hand.

For an initial state (p_1, V_1) and a final state (p_2, V_2), Eq. (17–23) becomes

$$p_1V_1^{\gamma} = p_2V_2^{\gamma} \qquad \text{(adiabatic process, ideal gas)}. \tag{17–24}$$

We can also calculate the *work* done by an ideal gas during an adiabatic process. We know that $Q = 0$ and $W = -\Delta U$ for *any* adiabatic process. For an ideal gas, $\Delta U = nC_V(T_2 - T_1)$. If the number of moles n and the initial and final temperatures T_1 and T_2 are known, we have simply

$$W = nC_V(T_1 - T_2) \qquad \text{(adiabatic process, ideal gas)}. \tag{17–25}$$

We may also use $pV = nRT$ in this equation to obtain

$$W = \frac{C_V}{R}(p_1V_1 - p_2V_2) = \frac{1}{\gamma - 1}(p_1V_1 - p_2V_2) \qquad \substack{\text{(adiabatic process,} \\ \text{ideal gas).}} \tag{17–26}$$

(We used the result $C_V = R/(\gamma - 1)$ from Example 17–6.) If the process is an expansion, the temperature drops, T_1 is greater than T_2, p_1V_1 is greater than p_2V_2, and the work is *positive,* as we should expect. If the process is a compression, the work is negative.

Throughout this analysis of adiabatic processes we have used the ideal-gas equation of state, which is valid only for *equilibrium* states. Strictly speaking, our results are valid only for a process that is fast enough to prevent appreciable heat exchange with the surroundings (so that $Q = 0$ and the process is adiabatic), yet slow enough that the system does not depart very much from thermal and mechanical equilibrium. Even when these conditions are not strictly satisfied, though, Eqs. (17–22), (17–24), and (17–26) give useful approximate results.

EXAMPLE 17–7

Adiabatic compression in a diesel engine The compression ratio of a diesel engine is 15 to 1; this means that air in the cylinders is compressed to $\frac{1}{15}$ of its initial volume (Fig. 17–17). If the initial pressure is 1.01×10^5 Pa and the initial temperature is 27°C (300 K), find the final pressure and the temperature after compression. Air is mostly a mixture of diatomic oxygen and nitrogen; treat it as an ideal gas with $\gamma = 1.40$.

SOLUTION We have $p_1 = 1.01 \times 10^5$ Pa, $T_1 = 300$ K, and $V_1/V_2 = 15$. From Eq. (17–22),

$$T_2 = T_1\left(\frac{V_1}{V_2}\right)^{\gamma-1} = (300 \text{ K})(15)^{0.40} = 886 \text{ K} = 613°\text{C}.$$

From Eq. (17–24),

$$p_2 = p_1\left(\frac{V_1}{V_2}\right)^{\gamma} = (1.01 \times 10^5 \text{ Pa})(15)^{1.40}$$

$$= 44.8 \times 10^5 \text{ Pa} = 44 \text{ atm}.$$

If the compression had been isothermal, the final pressure would have been 15 atm, but because the temperature also increases during an adiabatic compression, the final pressure is much greater. When fuel is injected into the cylinders near the end of the compression stroke, the high temperature of the air attained during compression causes the fuel to ignite spontaneously without the need for spark plugs.

17–17 Adiabatic compression of air in a cylinder of a diesel engine.

EXAMPLE 17-8

Work done in an adiabatic process In Example 17–7, how much work does the gas do during the compression if the initial volume of the cylinder is $1.00 \text{ L} = 1.00 \times 10^{-3} \text{ m}^3$? Assume that C_V for air is 20.8 J/mol · K and $\gamma = 1.40$.

SOLUTION We may determine the number of moles using the ideal-gas equation $pV = nRT$ and then use Eq. (17–25), or we may use Eq. (17–26). With the first method we have

$$n = \frac{p_1 V_1}{RT_1} = \frac{(1.01 \times 10^5 \text{ Pa})(1.00 \times 10^{-3} \text{ m}^3)}{(8.315 \text{ J/mol} \cdot \text{K})(300 \text{ K})}$$

$$= 0.0405 \text{ mol},$$

and Eq. (17–25) gives

$$W = nC_V(T_1 - T_2)$$

$$= (0.0405 \text{ mol})(20.8 \text{ J/mol} \cdot \text{K})(300 \text{ K} - 886 \text{ K})$$

$$= -494 \text{ J}.$$

With the second method,

$$W = \frac{1}{\gamma - 1}(p_1 V_1 - p_2 V_2)$$

$$= \frac{1}{1.40 - 1}\left[\begin{array}{l}(1.01 \times 10^5 \text{ Pa})(1.00 \times 10^{-3} \text{ m}^3)\\-(44.8 \times 10^5 \text{ Pa})\left(\dfrac{1.00 \times 10^{-3} \text{ m}^3}{15}\right)\end{array}\right]$$

$$= -494 \text{ J}.$$

The work is negative because the gas is compressed.

SUMMARY

- A thermodynamic system can exchange energy with its surroundings by heat transfer or by mechanical work and in some cases by other mechanisms. When a system at pressure p expands from volume V_1 to V_2, it does an amount of work W given by

$$W = \int_{V_1}^{V_2} p \, dV. \tag{17–2}$$

If the pressure p is constant during the expansion,

$$W = p(V_2 - V_1) \quad \text{(constant pressure only)}. \tag{17–3}$$

- In any thermodynamic process, the heat added to the system and the work done by the system depend not only on the initial and final states but also on the path (the series of intermediate states through which the system passes).

KEY TERMS

thermodynamic system, 533
thermodynamic process, 533
path, 537
free expansion, 538
internal energy, 539
first law of thermodynamics, 539
adiabatic process, 544
isochoric process, 544
isobaric process, 544
isothermal process, 544
adiabat, 545
isochor, 545

■ The first law of thermodynamics states that when heat Q is added to a system while it does work W, the internal energy U changes by an amount

$$U_2 - U_1 = \Delta U = Q - W. \tag{17-4}$$

In an infinitesimal process,

$$dU = dQ - dW. \tag{17-6}$$

The internal energy of any thermodynamic system depends only on its state. The change in internal energy in any process depends only on the initial and final states, not on the path. The internal energy of an isolated system is constant.

■ Adiabatic process: No heat transfer into or out of a system; $Q = 0$.

■ Isochoric process: Constant volume; $W = 0$.

■ Isobaric process: Constant pressure; $W = p(V_2 - V_1)$.

■ Isothermal process: Constant temperature.

■ The internal energy of an ideal gas depends only on its temperature, not on its pressure or volume. For other substances the internal energy generally depends on both pressure and temperature.

■ The molar heat capacities C_V and C_p of an ideal gas are related by

$$C_p = C_V + R. \tag{17-17}$$

The ratio of heat capacities, C_p/C_V, is denoted by γ:

$$\gamma = \frac{C_p}{C_V}. \tag{17-18}$$

■ For an adiabatic process for an ideal gas the quantities $TV^{\gamma-1}$ and pV^γ are constant. For an initial state (p_1, V_1, T_1) and a final state (p_2, V_2, T_2),

$$T_1 V_1^{\gamma-1} = T_2 V_2^{\gamma-1}, \tag{17-22}$$

$$p_1 V_1^\gamma = p_2 V_2^\gamma. \tag{17-24}$$

The work done by an ideal gas during an adiabatic expansion is

$$W = nC_V(T_1 - T_2) = \frac{C_V}{R}(p_1 V_1 - p_2 V_2)$$

$$= \frac{1}{\gamma - 1}(p_1 V_1 - p_2 V_2). \tag{17-25}, (17-26)$$

DISCUSSION QUESTIONS

Q17-1 In Chapter 6 the relationship between work and changes in kinetic energy was derived from Newton's laws of motion. Would it be possible to derive the first law of thermodynamics, which relates work, heat, and changes in internal energy, from Newton's laws of motion? Why or why not?

Q17-2 It is not correct to say that a body contains a certain amount of heat, yet a body can transfer heat to another body. How can a body give away something it does not have in the first place?

Q17-3 If you are told the initial and final states of a system and the associated change in internal energy, can you determine whether the internal energy change was due to work or to heat transfer?

Q17-4 Discuss the application of the first law of thermodynamics to a mountaineer who eats food, gets warm and perspires a lot during a climb, and does a lot of mechanical work in raising herself to the summit. What about the descent? The mountaineer also gets warm during the descent. Is the source of this energy the same as during the ascent?

Q17-5 When ice melts at 0°C, its volume decreases. Is the internal energy change greater than, less than, or equal to the heat added? How can you tell?

Q17-6 Household refrigerators always have tubing on the outside, usually at the back or bottom. When the refrigerator is running, the tubing becomes quite hot. Where does the heat come from?

Q17-7 Example 17-5 (Section 17-5) describes a process that occurs at a constant temperature of 100°C. Why, then, can't we use in this example the expressions for W derived for an isothermal process in Example 17-1 (Section 17-3)?

Q17–8 Translated literally from the Greek, "isothermal" means "same heat." Give as many reasons as you can why such a literal translation is misleading.

Q17–9 Imagine a gas made up entirely of negatively charged electrons. Like charges repel, so the electrons exert repulsive forces on each other. Would you expect that the temperature of such a gas would rise, fall, or stay the same in a free expansion? Why?

Q17–10 There are a few materials that contract when their temperature is increased, such as water between 0°C and 4°C. Would you expect C_p for such materials to be greater or less than C_V? Why or why not?

Q17–11 In a constant-volume process, $dU = nC_V \, dT$. But in a constant-pressure process, it is *not* true that $dU = nC_p \, dT$. Why not?

Q17–12 When you blow on the back of your hand with your mouth wide open, your breath feels warm. But if you partially close your mouth to form an "o" and then blow on your hand, your breath feels cool. Why?

Q17–13 On a warm summer day a large cylinder of compressed gas (propane or butane) was used to supply several large gas burners at a cookout. After a while, frost formed on the outside of the tank. Why?

Q17–14 In the carburetor of an aircraft or automobile engine, air flows through a relatively small aperture, then expands. In cool, foggy weather, ice sometimes forms in this aperture even though the outside air temperature is above freezing. Why?

Q17–15 Air escaping from an air hose at a gas station always feels cold. Why?

Q17–16 When you use a hand pump to inflate the tires of your bicycle, the pump gets warm after a while. Why? What happens to the temperature of the air in the pump as you compress it? Why does this happen? When you raise the pump handle to draw outside air into the pump, what happens to the temperature of the air taken in? Again, why does this happen?

Q17–17 The prevailing winds on the Hawaiian island of Kauai blow from the northeast. The winds cool as they go up the slope of Mt. Waialeale (elevation 1523 m), causing water vapor to condense and rain to fall. There is much more precipitation at the summit than at the base of the mountain. In fact, Mt. Waialeale is the rainiest spot on earth, averaging 11.7 m of rainfall a year. But what makes the winds cool?

Q17–18 Applying the same considerations as in Question 17–17, explain why the island of Niihau, a few kilometers to the southwest of Kauai, is almost a desert, and farms there need to be irrigated.

Q17–19 When a gas is compressed adiabatically into the surrounding air, its temperature rises even though there is no heat input to the gas. Where does the energy come from to raise the temperature?

Q17–20 When a gas expands adiabatically, it does work on its surroundings. But if there is no heat input to the gas, where does the energy come from to do the work?

Q17–21 As described in Section 17–9, when you open a bottle of a carbonated beverage, a miniature cloud can form just above the surface of the beverage. This is more likely to happen if the bottle and its contents are cold before you open the bottle. Why? This effect does not occur with bottles of non-carbonated beverages such as fruit juice. Why not?

EXERCISES

SECTION **17–3** **WORK DONE DURING VOLUME CHANGES**
SECTION **17–4** **PATHS BETWEEN THERMODYNAMIC STATES**

17–1 Two moles of an ideal gas are heated at constant pressure from $T = 27°C$ to $107°C$. a) Draw a pV-diagram for this process. b) Calculate the work done by the gas.

17–2 Three moles of an ideal gas have an initial temperature of 127.0°C. While the temperature is kept constant, the volume is increased until the pressure drops to 40.0% of its original value. a) Draw a pV-diagram for this process. b) Calculate the work done by the gas.

17–3 A metal cylinder with rigid walls contains 2.50 mol of oxygen gas. The gas is cooled until the pressure increases to 30.0% of its original value. Neglect the thermal contraction of the cylinder. a) Draw a pV-diagram for this process. b) Calculate the work done by the gas.

17–4 A gas under a constant pressure of 1.50×10^5 Pa and with an initial volume of 0.0900 m^3 is cooled until its volume becomes 0.0600 m^3. a) Draw a pV-diagram for this process. b) Calculate the work done by the gas.

17–5 A gas undergoes two processes. In the first, the volume remains constant at 0.200 m^3 and the pressure increases from 2.00×10^5 Pa to 5.00×10^5 Pa. The second process is a compression to a volume of 0.120 m^3 at a constant pressure of 5.00×10^5 Pa. a) In a pV-diagram, show both processes. b) Find the total work done by the gas during both processes.

17–6 Work Done in a Cyclic Process. a) In Fig. 17–7a, consider the closed loop $1 \rightarrow 3 \rightarrow 2 \rightarrow 4 \rightarrow 1$. This is a *cyclic* process in which the initial and final states are the same. Find the total work done by the system in this cyclic process, and show that it is equal to the area enclosed by the loop. b) How is the work done for the process in part (a) related to the work done if the loop is traversed in the opposite direction, $1 \rightarrow 4 \rightarrow 2 \rightarrow 3 \rightarrow 1$? Explain.

SECTION **17–5** **INTERNAL ENERGY AND THE FIRST LAW OF THERMODYNAMICS**

17–7 In a certain chemical process, a lab technician supplies 254 J of heat to a system. At the same time, 73 J of work are done on the system by its surroundings. What is the increase in the internal energy of the system?

17–8 A gas in a cylinder expands from a volume of 0.110 m^3 to 0.320 m^3. Heat flows into the gas just rapidly enough to keep the pressure constant at 1.80×10^5 Pa during the expansion. The total heat added is 1.15×10^5 J. a) Find the work done by the gas. b) Find the change in internal energy of the gas. c) Does it matter whether or not the gas is ideal? Why or why not?

17–9 A gas in a cylinder is held at a constant pressure of 2.30×10^5 Pa and is cooled and compressed from 1.70 m³ to 1.20 m³. The internal energy of the gas decreases by 1.40×10^5 J. a) Find the work done by the gas. b) Find the absolute value $|Q|$ of the heat flow into or out of the gas, and state the direction of heat flow. c) Does it matter whether or not the gas is ideal? Why or why not?

17–10 A system is taken from state a to state b along the three paths shown in Fig. 17–18. a) Along which path is the work done by the system the greatest? The least? b) If $U_b > U_a$, along which path is the absolute value $|Q|$ of the heat transfer the greatest? For this path, is heat absorbed or liberated by the system?

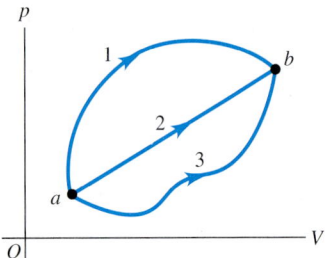

FIGURE 17–18 Exercise 17–10.

17–11 Doughnuts: Breakfast of Champions! A typical doughnut contains 2.0 g of protein, 17.0 g of carbohydrates, and 7.0 g of fat. The average food-energy values of these substances are 4.0 kcal/g for protein and carbohydrates and 9.0 kcal/g for fat. a) During heavy exercise, an average person uses energy at a rate of 510 kcal/h. How long would you have to exercise to "work off" one doughnut? b) If the energy in the doughnut could somehow be converted into kinetic energy of your body as a whole, how fast could you move after eating the doughnut? Take your mass to be 60 kg, and express your answer in m/s and km/h.

17–12 A liquid is irregularly stirred in a well-insulated container and thereby undergoes a rise in temperature. Regard the liquid as the system. a) Has heat been transferred? How can you tell? b) Has work been done? How can you tell? Why is it important that the stirring is irregular? c) What is the sign of ΔU? How can you tell?

17–13 A system is taken around the cycle shown in Fig. 17–19, from state a to state b and then back to state a. The absolute value of the heat transfer during one cycle is 7200 J. a) Does the system absorb or liberate heat when it goes around the cycle in the direction shown in the figure? How can you tell? b) What is the work W done by the system in one cycle? c) If the system

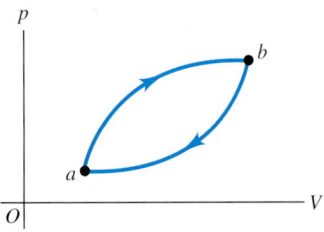

FIGURE 17–19 Exercise 17–13.

goes around the cycle in a counterclockwise direction, does it absorb or liberate heat in one cycle? What is the magnitude of the heat absorbed or liberated in one counterclockwise cycle?

17–14 A thermodynamic system undergoes a cyclic process as shown in Fig. 17–20. The cycle consists of two closed loops, loop I and loop II. a) Over one complete cycle, does the system do positive or negative work? b) In each of loops I and II, is the net work done by the system positive or negative? c) Over one complete cycle, does heat flow into or out of the system? d) In each of loops I and II, does heat flow into or out of the system?

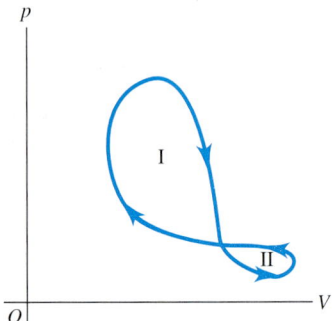

FIGURE 17–20 Exercise 17–14.

17–15 A student performs a combustion experiment by burning a mixture of fuel and oxygen in a constant-volume metal can surrounded by a water bath. During the experiment the temperature of the water is observed to rise. Regard the mixture of fuel and oxygen as the system. a) Has heat been transferred? How can you tell? b) Has work been done? How can you tell? c) What is the sign of ΔU? How can you tell?

17–16 Boiling Water at High Pressure. When water is boiled under a pressure of 2.00 atm, the heat of vaporization is 2.20×10^6 J/kg and the boiling point is 120°C. At this pressure, 1.00 kg of water has a volume of 1.00×10^{-3} m³, and 1.00 kg of steam a volume of 0.824 m³. a) Compute the work done when 1.00 kg of steam is formed at this temperature. b) Compute the increase in internal energy of the water.

SECTION **17–6 KINDS OF THERMODYNAMIC PROCESSES**
SECTION **17–7 INTERNAL ENERGY OF AN IDEAL GAS**
SECTION **17–8 HEAT CAPACITIES OF AN IDEAL GAS**

17–17 In an experiment to simulate conditions within an automobile engine, 645 J of heat is transferred to 0.185 mol of air contained within a cylinder of volume 40.0 cm³. Initially the nitrogen is at a pressure of 3.00×10^6 Pa and a temperature of 780 K. a) If the volume of the cylinder is held fixed, what is the final temperature of the air? Assume that the air is essentially all nitrogen gas, and use the data in Table 17–1. Draw a pV-diagram for this process. b) Find the final temperature of the air if the volume of the cylinder is allowed to increase while the pressure remains constant. Draw a pV-diagram for this process.

17–18 A cylinder contains 0.0100 mol of helium at $T = 27.0$°C. a) How much heat is needed to raise the temperature to 67.0°C while keeping the volume constant? Draw a pV-diagram for this process. b) If instead the pressure of the helium is kept constant, how much heat is needed to raise the temperature from 27.0°C to 67.0°C? Draw a pV-diagram for this process. c) What accounts for the difference between your answers to parts (a) and (b)? In which case is more heat required? What becomes of the additional heat? d) If the gas is ideal, what is the change in its internal energy in part (a)? In part (b)? How do the two answers compare? Why?

17–19 The temperature of 0.150 mol of an ideal gas is held constant at 77.0°C while its volume is reduced to 25.0% of its initial volume. The initial pressure of the gas is 1.25 atm. a) Determine the work done by the gas. b) What is the change in its internal energy? c) Does the gas exchange heat with its surroundings? If so, how much? Does the gas absorb or liberate heat?

17–20 During an isothermal compression of an ideal gas, 335 J of heat must be removed from the gas to maintain constant temperature. How much work is done by the gas during the process?

17–21 Propane gas (C_3H_8) behaves like an ideal gas with $\gamma = 1.127$. Determine the molar heat capacity at constant volume and the molar heat capacity at constant pressure.

17–22 A cylinder contains 0.250 mol of carbon dioxide (CO_2) gas at a temperature of 27.0°C. The cylinder is provided with a frictionless piston, which maintains a constant pressure of 1.00 atm on the gas. The gas is heated until its temperature increases to 127.0°C. Assume that the CO_2 can be treated as an ideal gas. a) Draw a pV-diagram for this process. b) How much work is done by the gas in this process? c) On what is this work done? d) What is the change in internal energy of the gas? e) How much heat was supplied to the gas? f) How much work would have been done if the pressure had been 0.50 atm?

17–23 Gaseous ethane (C_2H_6) has $\gamma = 1.220$ and may be treated as an ideal gas. a) If 2.40 mol of ethane is to be heated from 20.0°C to 25.0°C at a constant pressure of 1.00 atm, how much heat is required? b) What will be the change in internal energy of the ethane?

SECTION **17–9** ADIABATIC PROCESSES FOR AN IDEAL GAS

17–24 A monatomic ideal gas that is initially at a pressure of 1.50×10^5 Pa and with a volume of 0.0800 m³ is compressed adiabatically to a volume of 0.0400 m³. a) What is the final pressure? b) How much work is done by the gas? c) What is the ratio of the final temperature of the gas to its initial temperature? Is the gas heated or cooled by this compression?

17–25 The engine of a Ferrari F355 F1 sports car takes in air at 20.0°C and 1.00 atm and compresses it adiabatically to 0.0900 times the original volume. The air may be treated as an ideal gas with $\gamma = 1.40$. a) Draw a pV-diagram for this process. b) Find the final temperature and pressure.

17–26 An ideal gas initially at 4.00 atm and 350 K is permitted to expand adiabatically to 1.50 times its initial volume. Find the final pressure and temperature if the gas is a) monatomic; b) diatomic with $C_V = \frac{5}{2}R$.

17–27 During an adiabatic expansion the temperature of 0.450 mol of argon (Ar) drops from 50.0°C to 10.0°C. The argon may be treated as an ideal gas. a) Draw a pV-diagram for this process. b) How much work does the gas do? c) Does heat flow into or out of the gas? If so, what is the direction and absolute value of this heat flow? d) What is the change in internal energy of the gas?

17–28 A cylinder contains 0.100 mol of an ideal monatomic gas. Initially the gas is at a pressure of 1.00×10^5 Pa and occupies a volume of 2.50×10^{-3} m³. a) Find the initial temperature of the gas in kelvins. b) If the gas is allowed to expand to twice the initial volume, find the final temperature (in kelvins) and pressure of the gas if the expansion is i) isothermal; ii) isobaric; iii) adiabatic.

17–29 A quantity of sulfur dioxide gas (SO_2) occupies a volume of 5.00×10^{-3} m³ at a pressure of 1.10×10^5 Pa. The gas expands adiabatically to a volume of 1.00×10^{-2} m³, doing 285 J of work on its surroundings. Assume that the gas may be treated as ideal. a) Find the final pressure of the gas. (*Hint*: See Table 17–1.) b) How much work does the gas do on its surroundings? c) What is the ratio of the final temperature of the gas to its initial temperature?

PROBLEMS

17–30 A quantity of air is taken from state a to state b along a path that is a straight line in the pV-diagram (Fig. 17–21). a) In this process, does the temperature of the gas increase, decrease, or stay the same? Explain. b) If $V_a = 0.0700$ m³, $V_b = 0.1100$ m³, $p_a = 1.00 \times 10^5$ Pa, and $p_b = 1.40 \times 10^5$ Pa, what is the work W

done by the gas in this process? Assume that the gas may be treated as ideal.

17–31 When a system is taken from state a to state b in Fig. 17–22 along the path acb, 90.0 J of heat flows into the system and 60.0 J of work is done by the system. a) How much heat

FIGURE 17–21 Problem 17–30.

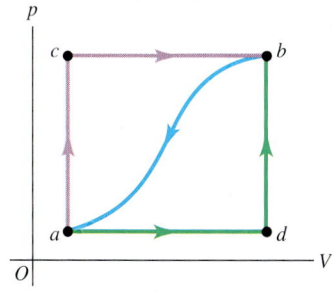

FIGURE 17–22 Problem 17–31.

flows into the system along path *adb* if the work done by the system is 15.0 J? b) When the system is returned from *b* to *a* along the curved path, the absolute value of the work done by the system is 35.0 J. Does the system absorb or liberate heat? How much heat? c) If $U_a = 0$ and $U_d = 8.0$ J, find the heat absorbed in the processes *ad* and *db*.

17–32 A thermodynamic system is taken from state *a* to state *c* in Fig. 17–23 along either path *abc* or path *adc*. Along path *abc* the work *W* done by the system is 450 J. Along path *adc*, *W* is 120 J. The internal energies of each of the four states shown in the figure are $U_a = 150$ J, $U_b = 240$ J, $U_c = 680$ J, and $U_d = 330$ J. Calculate the heat flow *Q* for each of the four processes *ab*, *bc*, *ad*, and *dc*. In each process, does the system absorb or liberate heat?

17–33 Fig. 17–23 shows four states of a thermodynamic system, *a*, *b*, *c*, and *d*. The volume of the system is V_a for both state *a* and state *b*, and is V_c for both state *c* and state *d*. The pressure of the system is p_a for both state *a* and state *d*, and is p_c for both state *b* and state *c*. The internal energies of the four states are U_a, U_b, U_c, and U_d. For each of the processes *ab*, *bc*, *ad*, and *dc*, find a) the work done by the system during the process and b) the heat flow into the system during the process. c) The system can be taken from state *a* to state *c* along either path *abc* or path *adc*. Find the net heat flow into the system and the net work done by the system for each path. For which path is the net heat flow greater? For which path is the net work greater? d) A friend tells you that the amounts of heat flow should be the same for both path *abc* and path *adc*, since the initial state (*a*) and final state (*c*) of the system are the same for both paths. How would you respond?

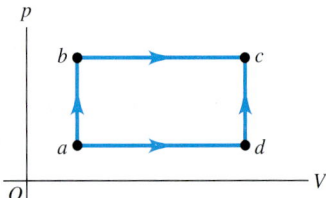

FIGURE 17–23 Problems 17–32 and 17–33.

17–34 Nitrogen gas in an expandable container is cooled from 50.0°C to 10.0°C, with the pressure held constant at 3.00×10^5 Pa. The total heat liberated by the gas is 2.50×10^4 J. Assume that the gas may be treated as ideal. a) Find the number of moles of gas. b) Find the change in internal energy of the gas. c) Find the work done by the gas. d) How much heat would be liberated by the gas for the same temperature change if the volume were constant?

17–35 In a certain process, 2.15×10^5 J of heat is liberated by a system, and at the same time the system contracts under a constant external pressure of 9.50×10^5 Pa. The internal energy of the system is the same at the beginning and end of the process. Find the change in volume of the system. (The system is *not* an ideal gas.)

17–36 A cylinder with a frictionless movable piston, like that shown in Fig. 17–5, contains a quantity of helium gas. Initially the gas is at a pressure of 1.00×10^5 Pa, has a temperature of

300 K, and occupies a volume of 1.50 L. The gas then undergoes two processes. In the first, the gas is heated and the piston is allowed to move to keep the temperature equal to 300 K. This continues until the pressure reaches 2.50×10^4 Pa. In the second process, the gas is compressed at constant pressure until it returns to its original volume of 1.50 L. Assume that the gas may be treated as ideal. a) In a *pV*-diagram, show both processes. b) Find the volume of the gas at the end of the first process, and find the pressure and temperature at the end of the second process. c) Find the total work done by the gas during both processes. d) What would you have to do to the gas to return it to its original pressure and temperature?

17–37 A Thermodynamic Process in a Liquid. A chemical engineer is studying the properties of liquid methanol (CH_3OH). She uses a steel cylinder with a cross-section area of 0.0200 m^2 and containing 1.20×10^{-2} m^3 of methanol. The cylinder is equipped with a tightly fitting piston that supports a load of 3.00×10^4 N. The temperature of the system is increased from 20.0°C to 50.0°C. For methanol, the coefficient of volume expansion is 1.20×10^{-3} K^{-1}, the density is 791 kg/m^3, and the specific heat capacity at constant pressure is $c_p = 2.51 \times 10^3$ J/kg $\cdot$ K. Neglect the expansion of the steel cylinder. Find a) the increase in volume of the methanol; b) the mechanical work done by the methanol against the 3.00×10^4 N force; c) the amount of heat added to the methanol; d) the change in internal energy of the methanol. e) Based on your results, explain whether there is any substantial difference between the specific heat capacities c_p (at constant pressure) and c_V (at constant volume) for methanol under these conditions.

17–38 A Thermodynamic Process in a Solid. A cube of copper 2.00 cm on a side is suspended by a string. (The physical properties of copper are given in Tables 14–1, 15–2, and 15–3.) The cube is heated with a burner from 20.0°C to 90.0°C. The air surrounding the cube is at atmospheric pressure (1.01×10^5 Pa). Find a) the increase in volume of the cube; b) the mechanical work done by the cube to expand against the pressure of the surrounding air; c) the amount of heat added to the cube; d) the change in internal energy of the cube. e) Based on your results, explain whether there is any substantial difference between the specific heat capacities c_p (at constant pressure) and c_V (at constant volume) for copper under these conditions.

17–39 A Thermodynamic Process in an Insect. The African bombardier beetle *Stenaptinus insignis* can emit a jet of defensive spray from the moveable tip of its abdomen (Fig. 17–24). The beetle's body has reservoirs of two different chemicals; when the beetle is disturbed, these chemicals are combined in a reaction chamber, producing a compound that is warmed from 20°C to 100°C by the heat of reaction. The high pressure produced allows the compound to be sprayed out at speeds up to 19 m/s (68 km/h), scaring away predators of all kinds. (The beetle shown in the figure is tethered to a wire fastened to its back by wax. It is responding to having its left foreleg pinched by forceps. The beetle is 2 cm long.) Calculate the heat of reaction of the two chemicals (in J/kg). Assume that the specific heat capacity of the two chemicals and the spray is the same as that of water, 4.19×10^3 J/kg $\cdot$ K, and that the initial temperature of the chemicals is 20°C.

FIGURE 17–24 Problem 17–39.

17–40 A Compressed-Air Engine. You are designing an engine that runs on compressed air. Air enters the engine at a pressure of 1.60×10^6 Pa and leaves at a pressure of 2.80×10^5 Pa. What must be the temperature of the compressed air for there to be no possibility of frost forming in the exhaust ports of the engine? Assume that the expansion is adiabatic. (*Note:* Frost will form if the moist air is cooled below 0°C in the expansion.)

17–41 During certain seasons strong winds called "chinooks" blow from the west across the eastern slopes of the Rockies and downhill into Denver and adjoining areas. Although the mountains are cool, the wind in Denver is very hot; within a few minutes after the chinook wind arrives, the temperature can climb 20 C° ("chinook" is a Native American word meaning "snow eater"). Similar winds occur in the Alps (called "foehns") and in southern California (called "Santa Anas"). a) Explain why the temperature of the chinook wind rises as it descends the slopes. Why is it important that the wind be fast-moving? b) Suppose a strong wind is blowing toward Denver (elevation 1630 m) from Grays Peak (80 km west of Denver, at an elevation of 4350 m), where the air pressure is 5.60×10^4 Pa and the air temperature is −15.0°C. The temperature and pressure in Denver before the wind arrives are 2.0°C and 8.12×10^4 Pa. By how many Celsius degrees will the temperature in Denver rise when the chinook arrives?

17–42 A certain ideal gas has molar heat capacity at constant volume C_V. A sample of this gas initially occupies a volume V_0 at pressure p_0 and absolute temperature T_0. The gas expands isobarically to a volume $2V_0$, then expands further adiabatically to a final volume of $4V_0$. a) Draw a pV-diagram for this sequence of processes. b) Compute the total work done by the gas for this sequence of processes. c) Find the final temperature of the gas. d) Find the absolute value $|Q|$ of the total heat flow into or out of the gas for this sequence of processes, and state the direction of heat flow.

17–43 An air pump has a cylinder 0.250 m long with a movable piston. The pump is used to compress air from the atmosphere (at absolute pressure 1.01×10^5 Pa) into a very large tank at 4.20×10^5 Pa gauge pressure. (For air, $C_V = 20.8$ J/mol · K.) a) The piston begins the compression stroke at the open end of the cylinder. How far down the length of the cylinder has the piston moved when air first begins to flow from the cylinder into the tank? Assume that the compression is adiabatic. b) If the air is taken into the pump at 27.0°C, what is the temperature of the compressed air? c) How much work does the pump do in putting 20.0 mol of air into the tank?

17–44 Engine Turbochargers and Intercoolers. The power output of an automobile engine is directly proportional to the mass of air that can be forced into the volume of the engine's cylinders to react chemically with gasoline. Many cars have a *turbocharger* which compresses the air before it enters the engine, giving a greater mass of air per volume. This rapid, essentially adiabatic compression also heats the air. To compress it further, the air then passes through an *intercooler* in which the air exchanges heat with its surroundings at essentially constant pressure. The air is then drawn into the cylinders. In a typical installation, air is taken into the turbocharger at atmospheric pressure (1.01×10^5 Pa), density $\rho = 1.23$ kg/m^3, and temperature 15.0°C. It is compressed adiabatically to 1.45×10^5 Pa. In the intercooler, the air is cooled to the original temperature of 15.0°C at a constant pressure of 1.45×10^5 Pa. a) Draw a pV-diagram for this sequence of processes. b) If the volume of one of the engine's cylinders is 575 cm^3, what mass of air exiting from the intercooler will fill the cylinder at 1.45×10^5 Pa? Compared to the power output of an engine that takes in air at 1.01×10^5 Pa at 15.0°C, what percentage increase in power is obtained by using the turbocharger and intercooler? c) If the intercooler is not used, what mass of air exiting from the turbocharger will fill the cylinder at 1.45×10^5 Pa? Compared to the power output of an engine that takes in air at 1.01×10^5 Pa at 15.0°C, what percentage increase in power is obtained by using the turbocharger alone?

17–45 A monatomic ideal gas expands slowly to twice its original volume, doing 300 J of work in the process. Find the heat added to the gas and the change in internal energy of the gas if the process is a) isothermal; b) adiabatic; c) isobaric.

17–46 A cylinder with a piston contains 0.250 mol of oxygen at 2.40×10^5 Pa and 355 K. The oxygen may be treated as an ideal gas. The gas first expands isobarically to twice its original volume. It is then compressed isothermally back to its original volume, and finally it is cooled isochorically to its original pressure. a) Show the series of processes on a pV-diagram. b) Compute the temperature during the isothermal compression. c) Compute the maximum pressure. d) Compute the total work done by the piston on the gas during the series of processes.

17–47 Use the conditions and processes of Problem 17–46 to compute a) the work done by the gas, the heat added to it, and its internal-energy change during the initial expansion; b) the work done, the heat added, and the internal-energy change during the final cooling; c) the internal-energy change during the isothermal compression.

17–48 A cylinder with a piston contains 0.150 mol of nitrogen at 1.80×10^5 Pa and 300 K. The nitrogen may be treated as an ideal gas. The gas is first compressed isobarically to one-half its original volume. It then expands adiabatically back to its original volume, and finally it is heated isochorically to its original pressure. a) Show the series of processes on a pV-diagram. b) Compute the temperatures at the beginning and end of the adiabatic expansion. c) Compute the minimum pressure.

17–49 Use the conditions and processes of Problem 17–48 to compute a) the work done by the gas, the heat added to it, and its internal-energy change during the initial compression; b) the work done by the gas, the heat added to it, and its internal-energy change during the adiabatic expansion; c) the work done, the heat added, and the internal-energy change during the final heating.

17–50 Comparing Thermodynamic Processes. In a cylinder, 1.20 mol of an ideal monatomic gas initially at 3.60×10^5 Pa and 300 K expands until its volume triples. Compute the work done by the gas if the expansion is a) isothermal; b) adiabatic; c) iso-

baric. d) Show each process on a pV-diagram. In which case is the absolute value of the work done by the gas greatest? Least? e) In which case is the absolute value of the heat transfer greatest? Least? f) In which case is the absolute value of the change in internal energy of the gas greatest? Least?

17–51 A flexible balloon contains 0.350 mol of hydrogen sulfide gas (H_2S). Initially the H_2S has a volume of 7.00×10^3 cm^3 and a temperature of 27.0°C. The H_2S first expands isobarically until the volume doubles. Then it expands adiabatically until the temperature returns to its initial value. Assume that the H_2S can be treated as an ideal gas. a) Draw a diagram of the process in the pV-plane. b) What is the total heat supplied to the H_2S in the process? c) What is the total change in the internal energy of the H_2S? d) What is the total work done by the H_2S? e) What is the final volume?

CHALLENGE PROBLEMS

17–52 Oscillations of a Piston. A vertical cylinder of radius r contains a quantity of ideal gas, and is fitted with a piston of mass m that is free to move (Fig. 17–25). The piston and the walls of the cylinder are frictionless and are made of a perfect thermal insulator. The outside air pressure is p_0. In equilibrium, the piston sits a height h above the bottom of the cylinder. a) Find the absolute pressure of the gas trapped below the piston when in equilibrium. b) The piston is pulled up by a small distance and released. Find the net force acting on the piston when its base is a distance $h + y$ above the bottom of the cylinder, where y is much less than h. c) After the piston is displaced from equilibrium and released, it oscillates up and down. Find the frequency of these small oscillations. If the displacement is not small, are the oscillations simple harmonic? How can you tell?

17–53 The van der Waals equation of state, an approximate representation of the behavior of gases at high pressure, is given by Eq. (16–7):

$$\left(p + \frac{an^2}{V^2} \right)(V - nb) = nRT,$$

where a and b are constants having different values for different gases. (In the special case of $a = b = 0$, this is the ideal-gas equation.) a) Calculate the work done by a gas with this equation of state in an isothermal expansion from V_1 to V_2. Show that your answer agrees with the ideal-gas result found in Example 17–1 (Section 17–3) when you set $a = b = 0$. b) For ethane gas (C_2H_6), $a = 0.554$ J · m^3/mol^2 and $b = 6.38 \times 10^{-5}$ m^3/mol. Calculate the work W done by 1.80 mol of ethane when it expands from 2.00×10^{-3} m^3 to 4.00×10^{-3} m^3 at a constant temperature of 300 K. Do the calculation using i) the van der Waals equation of state and ii) the ideal-gas equation of state. c) How large is the difference between the two results for W in part (b)? For which equation of state is W larger? Use the interpretation of the terms a and b given in Section 16–2 to explain why this should be so. Are the differences between the two equations of state important in this case?

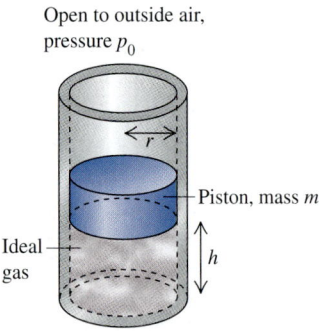

Open to outside air, pressure p_0

Piston, mass m

Ideal gas

h

FIGURE 17–25 Challenge Problem 17–52.

Could heat spontaneously flow out of these fried eggs, thus "unfrying" them and making them cooler, and flow back into the hot frying pan? Such a process would conserve energy and is allowed by the first law of thermodynamics. But such a process is in fact impossible because it would violate the *second* law of thermodynamics: heat cannot by itself flow from one body into a hotter body.

18–1 INTRODUCTION

Many thermodynamic processes proceed naturally in one direction but not the opposite. For example, heat always flows from a hot body to a cooler body, never the reverse. Heat flow from a cool body to a hot body would not violate the first law of thermodynamics; energy would be conserved. But it doesn't happen in nature. Why not? It is easy to convert mechanical energy completely into heat; this happens every time we use a car's brakes to stop it. Going in the reverse direction, there are plenty of devices that convert heat *partially* into mechanical energy. (An automobile engine is an example.) But even the cleverest would-be inventors have never succeeded in building a machine that converts heat *completely* into mechanical energy. Again, why not?

The answer to both of these questions has to do with the *directions* of thermodynamic processes and is called the *second law of thermodynamics.* This law places fundamental limitations on the efficiency of an engine or a power plant. It also places limitations on the minimum energy input needed to operate a refrigerator. So the second law is directly relevant for many important practical problems.

We can also state the second law in terms of the concept of *entropy,* a quantitative measure of the degree of disorder or randomness of a system. The idea of entropy helps explain why ink mixed with water never spontaneously unmixes and why a host of other seemingly possible processes are never observed to occur.

18–2 DIRECTIONS OF THERMODYNAMIC PROCESSES

Thermodynamic processes that occur in nature are all **irreversible processes.** These are processes that proceed spontaneously in one direction but not the other. The flow of heat from a hot body to a cooler body is irreversible, as is the free expansion of a gas discussed in Sections 17–4 and 17–7. Sliding a book across a table converts mechanical energy into heat by friction; this process is irreversible, for no one has ever observed the reverse process (in which a book initially at rest on the table would spontaneously start moving and the table and book would cool down). Our main topic for this chapter is the *second law of thermodynamics,* which determines the preferred direction for such processes.

Despite this preferred direction for every natural process, we can think of a class of idealized processes that *would* be reversible. A system that undergoes such an idealized **reversible process** is always very close to being in thermodynamic equilibrium within itself and with its surroundings. Any change of state that takes place can then be reversed (made to go the other way) by making only an infinitesimal change in the conditions of the system. For example, heat flow between two bodies whose temperatures differ only infinitesimally can be reversed by making only a very small change in one temperature or the other.

Reversible processes are thus **equilibrium processes,** with the system always in thermodynamic equilibrium. Of course, if a system were *truly* in thermodynamic equilibrium, no change of state would take place. Heat would not flow into or out of a system with truly uniform temperature throughout, and a system that is truly in mechanical equilibrium would not expand and do work against its surroundings. A reversible process is an idealization that can never be precisely attained in the real world. But by making the temperature gradients and the pressure differences in the substance very small, we can keep the system very close to equilibrium states and make the process nearly reversible. That's why we call a reversible process a *quasi-equilibrium process.*

By contrast, heat flow with finite temperature difference, free expansion of a gas, and conversion of work to heat by friction are all *irreversible* processes; no small change in conditions could make any of them go the other way. They are also all *nonequilibrium* processes, in that the system is not in thermodynamic equilibrium at any point until the end of the process.

There is a relationship between the direction of a process and the *disorder* or *randomness* of the resulting state. For example, imagine a tedious sorting job, such as alphabetizing a thousand book titles written on file cards. Throw the alphabetized stack of cards into the air. Do they come down in alphabetical order? No, their tendency is to come down in a random or disordered state. In the free expansion of a gas discussed in Sections 17–4 and 17–7, the air is more disordered after it has expanded into the entire box than when it was confined in one side, just as your clothes are more disordered when scattered all over your floor than when confined to your closet.

Similarly, macroscopic kinetic energy is energy associated with organized, coordinated motions of many molecules, but heat transfer involves changes in energy of random, disordered molecular motion. Therefore conversion of mechanical energy into heat involves an increase of randomness or disorder.

In the following sections we will introduce the second law of thermodynamics by considering two broad classes of devices: *heat engines,* which are partly successful in converting heat into work, and *refrigerators,* which are partly successful in transporting heat from cooler to hotter bodies.

18–3 HEAT ENGINES

The essence of our technological society is the ability to use sources of energy other than muscle power. Sometimes, mechanical energy is directly available; water power is an example. But most of our energy comes from the burning of fossil fuels (coal, oil, and gas) and from nuclear reactions. These supply energy that is transferred as *heat.* This is directly useful for heating buildings, for cooking, and for chemical processing, but to operate a machine or propel a vehicle, we need *mechanical* energy.

Thus it's important to know how to take heat from a source and convert as much of it as possible into mechanical energy or work. This is what happens in gasoline engines in automobiles, jet engines in airplanes, steam turbines in electric power plants, and many other systems. Closely related processes occur in the animal kingdom; food energy is "burned" (that is, carbohydrates combine with oxygen to yield water, carbon dioxide, and energy) and partly converted to mechanical energy as an animal's muscles do work on its surroundings.

Any device that transforms heat partly into work or mechanical energy is called a **heat engine.** Usually, a quantity of matter inside the engine undergoes inflow and outflow of heat, expansion and compression, and sometimes change of phase. We call this matter the **working substance** of the engine. In internal-combustion engines the working substance is a mixture of air and fuel; in a steam turbine it is water.

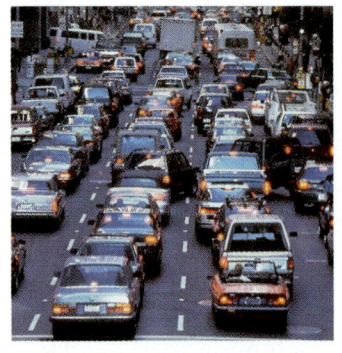

All motorized vehicles (except electric-powered vehicles) use heat engines for propulsion.

The simplest kind of engine to analyze is one in which the working substance undergoes a **cyclic process,** a sequence of processes that eventually leaves the substance in the same state in which it started. In a steam turbine the water is recycled and used over and over. Internal-combustion engines do not use the same air over and over, but we can still analyze them in terms of cyclic processes that approximate their actual operation.

8.12
Cyclic Process—Strategies

8.13
Cyclic Process—Problems

All heat engines *absorb* heat from a source at a relatively high temperature, perform some mechanical work, and *discard* or *reject* some heat at a lower temperature. As far as the engine is concerned, the discarded heat is wasted. In internal-combustion engines the waste heat is that discarded in the hot exhaust gases and the cooling system; in a steam turbine it is the heat that must flow out of the used steam to condense and recycle the water.

When a system is carried through a cyclic process, its initial and final internal energies are equal. For any cyclic process, the first law of thermodynamics requires that

$$U_2 - U_1 = 0 = Q - W, \quad \text{so} \quad Q = W.$$

That is, the net heat flowing into the engine in a cyclic process equals the net work done by the engine.

When we analyze heat engines, it helps to think of two bodies with which the working substance of the engine can interact. One of these, called the *hot reservoir,* represents the heat source; it can give the working substance large amounts of heat at a constant temperature T_H without appreciably changing its own temperature. The other body, called the *cold reservoir,* can absorb large amounts of discarded heat from the engine at a constant lower temperature T_C. In a steam-turbine system the flames and hot gases in the boiler are the hot reservoir, and the cold water and air used to condense and cool the used steam are the cold reservoir.

We denote the quantities of heat transferred from the hot and cold reservoirs as Q_H and Q_C, respectively. A quantity of heat Q is positive when heat is transferred *into* the working substance and is negative when heat leaves the working substance. Thus in a heat engine, Q_H is positive but Q_C is negative, representing heat *leaving* the working substance. This sign convention is consistent with the rules we stated in Section 17–2; we will continue to use those rules here. Frequently, it clarifies the relationships to state them in terms of the absolute values of the Q's and W's because absolute values are always positive. When we do this, our notation will show it explicitly.

We can represent the energy transformations in a heat engine by the *energy-flow diagram* of Fig. 18–1. The engine itself is represented by the circle. The amount of heat Q_H supplied to the engine by the hot reservoir is proportional to the width of the incoming "pipeline" at the top of the diagram. The width of the outgoing pipeline at the bottom is proportional to the magnitude $|Q_C|$ of the heat rejected in the exhaust. The branch line to the right represents the portion of the heat supplied that the engine converts to mechanical work, W.

When an engine repeats the same cycle over and over, Q_H and Q_C represent the quantities of heat absorbed and rejected by the engine *during one cycle;* Q_H is positive, and Q_C is negative. The *net* heat Q absorbed per cycle is

$$Q = Q_H + Q_C = |Q_H| - |Q_C|. \tag{18–1}$$

The useful output of the engine is the net work W done by the working substance. From the first law,

$$W = Q = Q_H + Q_C = |Q_H| - |Q_C|. \tag{18–2}$$

Ideally, we would like to convert *all* the heat Q_H into work; in that case we would have $Q_H = W$ and $Q_C = 0$. Experience shows that this is impossible; there is always some heat wasted, and Q_C *is never zero.* We define the **thermal efficiency** of an engine, denoted by

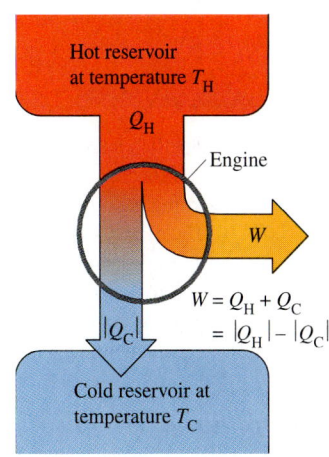

Hot reservoir at temperature T_H

Q_H

Engine

W

$W = Q_H + Q_C$
$= |Q_H| - |Q_C|$

$|Q_C|$

Cold reservoir at temperature T_C

18–1 Schematic energy-flow diagram for a heat engine.

e, as the quotient

$$e = \frac{W}{Q_H}.$$ (18–3)

The thermal efficiency e represents the fraction of Q_H that *is* converted to work. To put it another way, e is what you get divided by what you pay for. This is always less than unity, an all-too-familiar experience! In terms of the flow diagram of Fig. 18–1, the most efficient engine is the one for which the branch pipeline representing the work output is as wide as possible and the exhaust pipeline representing the heat thrown away is as narrow as possible.

When we substitute the two expressions for W given by Eq. (18–2) into Eq. (18–3), we get the following equivalent expressions for e:

$$e = \frac{W}{Q_H} = 1 + \frac{Q_C}{Q_H} = 1 - \left|\frac{Q_C}{Q_H}\right| \qquad \text{(thermal efficiency of an engine).}$$ (18–4)

Note that e is a quotient of two energy quantities and thus is a pure number, without units. Of course, we must always express W, Q_H, and Q_C in the same units.

Problem–Solving Strategy

HEAT ENGINES

1. The suggestions in the Problem-Solving Strategy in Section 17–5 are equally useful throughout the present chapter, and we suggest that you re-read them.

2. Be very careful with the sign conventions for W and the various Q's. W is positive when the system expands and does work and is negative when the system is compressed. Each Q is positive if it represents heat entering the system and is negative if it represents heat leaving the system. When in doubt, use the first law when possible,

to check consistency. When you know that a quantity is negative, such as Q_C in the above discussion, it sometimes helps to write it as $Q_C = -|Q_C|$.

3. Some problems deal with power rather than energy quantities. Power is work per unit time ($P = W/t$), and rate of heat transfer (heat current) H is heat transfer per unit time ($H = Q/t$). Sometimes it helps to ask, "What is W or Q in one second (or one hour)?"

EXAMPLE 18-1

A gasoline engine in a large truck takes in 10,000 J of heat and delivers 2000 J of mechanical work per cycle. The heat is obtained by burning gasoline with heat of combustion $L_c = 5.0 \times 10^4$ J/g. a) What is the thermal efficiency of this engine? b) How much heat is discarded in each cycle? c) How much gasoline is burned in each cycle? d) If the engine goes through 25 cycles per second, what is its power output in watts? In horsepower? e) How much gasoline is burned per second? Per hour?

SOLUTION We have $Q_H = 10,000$ J and $W = 2000$ J.
a) From Eq. (18–3) the thermal efficiency is

$$e = \frac{W}{Q_H} = \frac{2000 \text{ J}}{10,000 \text{ J}} = 0.20 = 20\%.$$

This is a fairly typical figure for cars and trucks if W includes only the work actually delivered to the wheels.

b) From Eq. (18–2), $W = Q_H + Q_C$ so

$$Q_C = W - Q_H = 2000 \text{ J} - 10,000 \text{ J}$$
$$= -8000 \text{ J}.$$

That is, 8000 J of heat leaves the engine during each cycle.
c) Let m be the mass of gasoline burned during each cycle. Then

$$Q_H = mL_c,$$

$$m = \frac{Q_H}{L_c} = \frac{10,000 \text{ J}}{5.0 \times 10^4 \text{ J/g}} = 0.20 \text{ g}.$$

d) The power P (rate of doing work) is the work per cycle multiplied by the number of cycles per second:

$$P = (2000 \text{ J/cycle})(25 \text{ cycles/s}) = 50,000 \text{ W} = 50 \text{ kW}$$

$$= (50,000 \text{ W})\frac{1 \text{ hp}}{746 \text{ W}} = 67 \text{ hp}.$$

e) The mass of gasoline burned per second is the mass per cycle multiplied by the number of cycles per second:

$$(0.20 \text{ g/cycle})(25 \text{ cycles/s}) = 5.0 \text{ g/s}.$$

The mass burned per hour is

$$(5.0 \text{ g/s}) \frac{3600 \text{ s}}{1 \text{ h}} = 18,000 \text{ g/h} = 18 \text{ kg/h}.$$

The density of gasoline is about 0.70 g/cm^3, so this is about $25,700 \text{ cm}^3$, 25.7 L, or 6.8 gallons of gasoline per hour. If the truck is traveling at 55 mi/h (88 km/h), this represents fuel consumption of 8.1 miles/gallon (3.4 km/L).

18-4 INTERNAL-COMBUSTION ENGINES

The gasoline engine, used in automobiles and many other types of machinery, is a familiar example of a heat engine. Let's look at its thermal efficiency. Figure 18–2 shows the operation of one type of gasoline engine. First a mixture of air and gasoline vapor flows into a cylinder through an open intake valve while the piston descends, increasing the volume of the cylinder from a minimum of V (when the piston is all the way up) to a maximum of rV (when it is all the way down). The quantity r is called the **compression ratio;** for present-day automobile engines its value is typically 8–10. At the end of this *intake stroke,* the intake valve closes and the mixture is compressed, approximately adiabatically, to volume V during the *compression stroke.* The mixture is then ignited by the spark plug, and the heated gas expands, approximately adiabatically, back to volume rV, pushing on the piston and doing work; this is the *power stroke.* Finally, the exhaust valve opens, and the combustion products are pushed out (during the *exhaust stroke*), leaving the cylinder ready for the next intake stroke.

THE OTTO CYCLE

Figure 18–3 (page 564) is a pV-diagram for an idealized model of the thermodynamic processes in a gasoline engine. This model is called the **Otto cycle.** At point a the gasoline-air mixture has entered the cylinder. The mixture is compressed adiabatically to point b and is then ignited. Heat Q_{H} is added to the system by the burning gasoline along

18-2 Cycle of a four-stroke internal-combustion engine. (a) *Intake stroke:* Piston moves down, causing a partial vacuum in cylinder; gasoline-air mixture flows through open intake valve into cylinder. (b) *Compression stroke:* Intake valve closes, and mixture is compressed as piston moves up. (c) *Ignition:* Spark plug ignites mixture. (d) *Power stroke:* Hot burned mixture pushes piston down, doing work. (e) *Exhaust stroke:* Exhaust valve opens and piston moves up, pushing burned mixture out of cylinder. Engine is now ready for next intake stroke, and the cycle repeats.

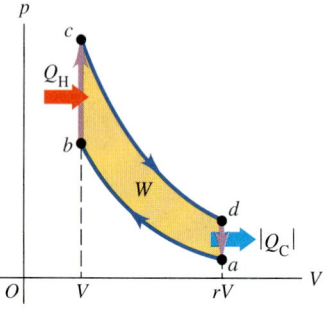

18-3 The pV-diagram for the Otto cycle, an idealized model of the thermodynamic processes in a gasoline engine.

line bc, and the power stroke is the adiabatic expansion to d. The gas is cooled to the temperature of the outside air along line da; during this process, heat $|Q_C|$ is rejected. In practice, this gas leaves the engine as exhaust and does not enter the engine again. But since an equivalent amount of gasoline and air enters, we may consider the process to be cyclic.

We can calculate the efficiency of this idealized cycle. Processes bc and da are constant-volume, so the heats Q_H and Q_C are related simply to the temperatures:

$$Q_H = nC_V(T_c - T_b) > 0,$$

$$Q_C = nC_V(T_a - T_d) < 0.$$

The thermal efficiency is given by Eq. (18–4). Inserting the above expressions and canceling out the common factor nC_V, we find

$$e = \frac{Q_H + Q_C}{Q_H} = \frac{T_c - T_b + T_a - T_d}{T_c - T_b}. \qquad (18–5)$$

To simplify this further, we use the temperature-volume relation for adiabatic processes for an ideal gas, Eq. (17–22). For the two adiabatic processes ab and cd,

$$T_a(rV)^{\gamma-1} = T_b V^{\gamma-1} \qquad \text{and} \qquad T_d(rV)^{\gamma-1} = T_c V^{\gamma-1}.$$

We divide each of these equations by the common factor $V^{\gamma-1}$ and substitute the resulting expressions for T_b and T_c back into Eq. (18–5). The result is

$$e = \frac{T_d r^{\gamma-1} - T_a r^{\gamma-1} + T_a - T_d}{T_d r^{\gamma-1} - T_a r^{\gamma-1}} = \frac{(T_d - T_a)(r^{\gamma-1} - 1)}{(T_d - T_a)r^{\gamma-1}}.$$

Dividing out the common factor $(T_d - T_a)$, we get

$$e = 1 - \frac{1}{r^{\gamma-1}} \qquad \text{(thermal efficiency in Otto cycle)}. \qquad (18–6)$$

The thermal efficiency given by Eq. (18–6) is always less than unity, even for this idealized model. With $r = 8$ and $\gamma = 1.4$ (the value for air) the theoretical efficiency is $e = 0.56$, or 56%. The efficiency can be increased by increasing r. However, this also increases the temperature at the end of the adiabatic compression of the air-fuel mixture. If the temperature is too high, the mixture explodes spontaneously during compression instead of burning evenly after the spark plug ignites it. This is called *pre-ignition* or *detonation;* it causes a knocking sound and can damage the engine. The octane rating of a gasoline is a measure of its antiknock qualities. The maximum practical compression ratio for high-octane, or "premium," gasoline is about 10. Higher ratios can be used with more exotic fuels.

The Otto cycle, which we have just described, is a highly idealized model. It assumes that the mixture behaves as an ideal gas; it neglects friction, turbulence, loss of heat to cylinder walls, and many other effects that combine to reduce the efficiency of a real engine. Another source of inefficiency is incomplete combustion. A mixture of gasoline vapor with just enough air for complete combustion of the hydrocarbons to H_2O and CO_2 does not ignite readily. Reliable ignition requires a mixture that is "richer" in gasoline. The resulting incomplete combustion leads to CO and unburned hydrocarbons in the exhaust. The heat obtained from the gasoline is then less than the total heat of combustion; the difference is wasted, and the exhaust products contribute to air pollution. Efficiencies of real gasoline engines are typically around 35%.

THE DIESEL CYCLE

The Diesel engine is similar in operation to the gasoline engine. The most important difference is that there is no fuel in the cylinder at the beginning of the compression stroke.

A little before the beginning of the power stroke, the injectors start to inject fuel directly into the cylinder, just fast enough to keep the pressure approximately constant during the first part of the power stroke. Because of the high temperature developed during the adiabatic compression, the fuel ignites spontaneously as it is injected; no spark plugs are needed.

The idealized **Diesel cycle** is shown in Fig. 18–4. Starting at point a, air is compressed adiabatically to point b, heated at constant pressure to point c, expanded adiabatically to point d, and cooled at constant volume to point a. Because there is no fuel in the cylinder during most of the compression stroke, pre-ignition cannot occur, and the compression ratio r can be much higher than for a gasoline engine. This improves efficiency and ensures reliable ignition when the fuel is injected (because of the high temperature reached during the adiabatic compression). Values of r of 15–20 are typical; with these values and $\gamma = 1.4$, the theoretical efficiency of the idealized Diesel cycle is about 0.65 to 0.70. As with the Otto cycle, the efficiency of any actual engine is substantially less than this. Diesel engines are usually more efficient than gasoline engines. They are also heavier (per unit power output) and often harder to start. They need no carburetor or ignition system, but the fuel-injection system requires expensive high-precision machining.

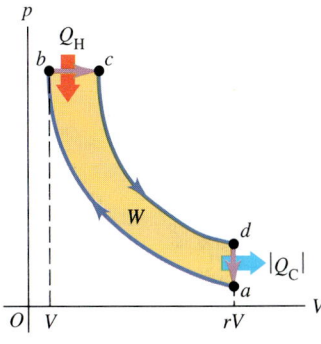

18–4 The pV-diagram for the idealized Diesel cycle.

18–5 REFRIGERATORS

We can think of a **refrigerator** as a heat engine operating in reverse. A heat engine takes heat from a hot place and gives off heat to a colder place. A refrigerator does the opposite; it takes heat from a cold place (the inside of the refrigerator) and gives it off to a warmer place (usually the air in the room where the refrigerator is located). A heat engine has a net *output* of mechanical work; the refrigerator requires a net *input* of mechanical work. Using the sign conventions from Section 18–3, for a refrigerator Q_C is positive but both W and Q_H are negative; hence $|W| = -W$ and $|Q_H| = -Q_H$.

A flow diagram for a refrigerator is shown in Fig. 18–5. From the first law for a cyclic process,

$$Q_H + Q_C - W = 0, \quad \text{or} \quad -Q_H = Q_C - W,$$

or, because both Q_H and W are negative,

$$|Q_H| = Q_C + |W|. \tag{18–7}$$

Thus, as the diagram shows, the heat $|Q_H|$ leaving the working substance and given to the hot reservoir is always *greater* than the heat Q_C taken from the cold reservoir. Note that the absolute-value relation

$$|Q_H| = |Q_C| + |W| \tag{18–8}$$

is valid for both heat engines and refrigerators.

From an economic point of view, the best refrigeration cycle is one that removes the greatest amount of heat $|Q_C|$ from the inside of the refrigerator for the least expenditure of mechanical work, $|W|$. The relevant ratio is therefore $|Q_C|/|W|$; the larger this ratio, the better the refrigerator. We call this ratio the **coefficient of performance,** denoted by K. From Eq. (18–8), $|W| = |Q_H| - |Q_C|$, so

$$K = \frac{|Q_C|}{|W|} = \frac{|Q_C|}{|Q_H| - |Q_C|} \quad \begin{array}{l}\text{(coefficient of performance of} \\ \text{a refrigerator).}\end{array} \tag{18–9}$$

As always, we measure Q_H, Q_C, and W all in the same energy units; K is then a dimensionless number.

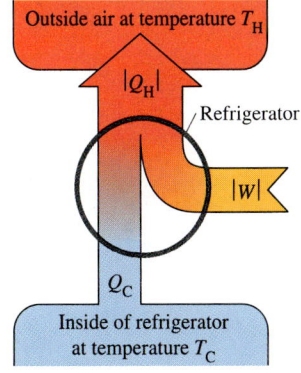

18–5 Schematic energy-flow diagram of a refrigerator.

18–6 Principle of the mechanical refrigeration cycle.

The principles of the common refrigeration cycle are shown schematically in Fig. 18–6. The fluid "circuit" contains a refrigerant fluid (the working substance). In the past this has usually been CCl_2F_2 or another member of the Freon family; because the release of such substances into the atmosphere depletes the ozone layer, alternative refrigerants are now in use. The left side of the circuit (including the cooling coils inside the refrigerator) is at low temperature and low pressure; the right side (including the condenser coils outside the refrigerator) is at high temperature and high pressure. Ordinarily, both sides contain liquid and vapor in phase equilibrium.

The compressor takes in fluid, compresses it adiabatically, and delivers it to the condenser coil at high pressure. The fluid temperature is then higher than that of the air surrounding the condenser, so the refrigerant gives off heat $|Q_H|$ and partially condenses to liquid. The fluid then expands adiabatically into the evaporator at a rate controlled by the expansion valve. As the fluid expands, it cools considerably, enough that the fluid in the evaporator coil is colder than its surroundings. It absorbs heat $|Q_C|$ from its surroundings, cooling them and partially vaporizing. The fluid then enters the compressor to begin another cycle. The compressor, usually driven by an electric motor, requires energy input and does work $|W|$ *on* the working substance during each cycle.

An air conditioner operates on exactly the same principle. In this case the refrigerator box becomes a room or an entire building. The evaporator coils are inside, the condenser is outside, and fans circulate air through these (Fig. 18–7). In large installations the condenser coils are often cooled by water. For air conditioners the quantities of greatest practical importance are the *rate* of heat removal (the heat current H from the region being cooled) and the *power* input $P = W/t$ to the compressor. If heat $|Q_C|$ is removed in time t, then $H = |Q_C|/t$. Then we can express the coefficient of performance as

$$K = \frac{|Q_C|}{|W|} = \frac{Ht}{Pt} = \frac{H}{P}.$$

Typical room air conditioners have heat removal rates H of 5000 to 10,000 Btu/h, or about 1500–3000 W, and require electric power input of about 600 to 1200 W. Typical coefficients of performance are about 2.5, with somewhat larger values for larger-capacity units. The actual values of K in operation depend on the inside and outside temperatures.

Unfortunately, K is often expressed commercially in mixed units, with H in Btu per hour and P in watts. In these units, H/P is called the **energy efficiency rating** (EER); the

Hot air

Fan

Blower

Cool air

Condenser

Expansion
valve

*Warm,
outside
air*

Hot, humid air

Compressor

Evaporator

18–7 An air conditioner works on
the same principle as a refrigerator.

units, customarily omitted, are (Btu/h)/W. Because 1 W = 3.413 Btu/h, the EER is
numerically 3.413 times as large as the dimensionless K. Room air conditioners typically have an EER of 7 to 10.

A variation on this theme is the **heat pump,** used to heat buildings by cooling the
outside air. It functions like a refrigerator turned inside out. The evaporator coils are outside, where they take heat from cold air, and the condenser coils are inside, where they
give off heat to the warmer air. With proper design, the heat $|Q_H|$ delivered to the inside
per cycle can be considerably greater than the work $|W|$ required to get it there.

Work is *always* needed to transfer heat from a colder to a hotter body. Heat flows
spontaneously from hotter to colder, and to reverse this flow requires the addition of
work from the outside. Experience shows that it is impossible to make a refrigerator that
transports heat from a colder body to a hotter body without the addition of work. If no
work were needed, the coefficient of performance would be infinite. We call such a
device a *workless refrigerator;* it is a mythical beast, like the unicorn and the free lunch.

18–6 THE SECOND LAW OF THERMODYNAMICS

Experimental evidence suggests strongly that it is *impossible* to build a heat engine that
converts heat completely to work, that is, an engine with 100% thermal efficiency. This
impossibility is the basis of one statement of the **second law of thermodynamics,** as
follows:

> **It is impossible for any system to undergo a process in which it absorbs heat
> from a reservoir at a single temperature and converts the heat completely into
> mechanical work, with the system ending in the same state in which it began.**

We will call this the "engine" statement of the second law.

The basis of the second law of thermodynamics lies in the difference between the
nature of internal energy and that of macroscopic mechanical energy. In a moving body
the molecules have random motion, but superimposed on this is a coordinated motion of
every molecule in the direction of the body's velocity. The kinetic energy associated with
this *coordinated* macroscopic motion is what we call the kinetic energy of the moving
body. The kinetic and potential energies associated with the *random* motion constitute
the internal energy.

When a body sliding on a surface comes to rest as a result of friction, the organized motion of the body is converted to random motion of molecules in the body and in the surface. Since we cannot control the motions of individual molecules, we cannot convert this random motion completely back to organized motion. We can convert *part* of it, and this is what a heat engine does.

If the second law were *not* true, we could power an automobile or run a power plant by cooling the surrounding air. Neither of these impossibilities violates the *first* law of thermodynamics. The second law, therefore, is not a deduction from the first but stands by itself as a separate law of nature. The first law denies the possibility of creating or destroying energy; the second law limits the *availability* of energy and the ways in which it can be used and converted.

Our analysis of refrigerators in Section 18–5 forms the basis for an alternative statement of the second law of thermodynamics. Heat flows spontaneously from hotter to colder bodies, never the reverse. A refrigerator does take heat from a colder to a hotter body, but its operation requires an input of mechanical energy or work. Generalizing this observation, we state:

It is impossible for any process to have as its sole result the transfer of heat from a cooler to a hotter body.

We'll call this the "refrigerator" statement of the second law. It may not seem to be very closely related to the "engine" statement. In fact, though, the two statements are completely equivalent. For example, if we could build a workless refrigerator, violating the second or "refrigerator" statement of the second law, we could use it in conjunction with a heat engine, pumping the heat rejected by the engine back to the hot reservoir to be reused. This composite machine (Fig. 18–8a) would violate the "engine" statement of the second law because its net effect would be to take a net quantity of heat $Q_H - |Q_C|$ from the hot reservoir and convert it completely to work W.

Alternatively, if we could make an engine with 100% thermal efficiency, in violation of the first statement, we could run it using heat from the hot reservoir and use the work output to drive a refrigerator that pumps heat from the cold reservoir to the hot (Fig. 18–8b). This composite device would violate the "refrigerator" statement because its net effect would be to take heat Q_C from the cold reservoir and deliver it to the hot reservoir without requiring any input of work. Thus any device that violates one form of the second law can be used to make a device that violates the other form. If violations of the first form are impossible, so are violations of the second!

The conversion of work to heat, as in friction or viscous fluid flow, and heat flow from hot to cold across a finite temperature gradient, are *irreversible* processes. The "engine" and "refrigerator" statements of the second law state that these processes can be only partially reversed. We could cite other examples. Gases always seep sponta-

18–8 Energy-flow diagrams for equivalent forms of the second law. (a) A workless refrigerator (left), if it existed, could be used in combination with an ordinary heat engine (right) to form a composite device that functions as an engine with 100% efficiency, converting heat $Q_H - |Q_C|$ completely to work. (b) An engine with 100% efficiency (left), if it existed, would be used in combination with an ordinary refrigerator (right) to form a workless refrigerator, transferring heat Q_C from the cold reservoir to the hot with no net input of work. Since either of these is impossible, the other must be also.

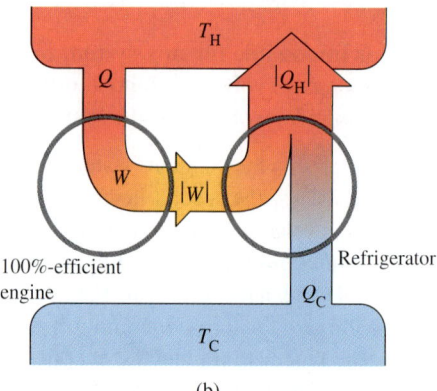

(a)   (b)

neously through an opening from a region of high pressure to a region of low pressure; gases and miscible liquids left by themselves always tend to mix, not to unmix. The second law of thermodynamics is an expression of the inherent one-way aspect of these and many other irreversible processes. Energy conversion is an essential aspect of all plant and animal life and of human technology, so the second law of thermodynamics is of the utmost fundamental importance in the world we live in.

18–7 THE CARNOT CYCLE

According to the second law, no heat engine can have 100% efficiency. How great an efficiency *can* an engine have, given two heat reservoirs at temperatures T_H and T_C? This question was answered in 1824 by the French engineer Sadi Carnot (1796–1832), who developed a hypothetical, idealized heat engine that has the maximum possible efficiency consistent with the second law. The cycle of this engine is called the **Carnot cycle.**

8.14
Carnot Cycle

To understand the rationale of the Carnot cycle, we return to a recurrent theme in this chapter: *reversibility* and its relation to directions of thermodynamic processes. Conversion of work to heat is an irreversible process; the purpose of a heat engine is a *partial* reversal of this process, the conversion of heat to work with as great an efficiency as possible. For maximum heat-engine efficiency, therefore, *we must avoid all irreversible processes.* This requirement turns out to be enough to determine the basic sequence of steps in the Carnot cycle, as we will show next.

Heat flow through a finite temperature drop is an irreversible process. Therefore, during heat transfer in the Carnot cycle there must be *no* finite temperature difference. When the engine takes heat from the hot reservoir at temperature T_H, the working substance of the engine must also be at T_H; otherwise, irreversible heat flow would occur. Similarly, when the engine discards heat to the cold reservoir at T_C, the engine itself must be at T_C. That is, every process that involves heat transfer must be *isothermal* at either T_H or T_C.

Conversely, in any process in which the temperature of the working substance of the engine is intermediate between T_H and T_C, there must be *no* heat transfer between the engine and either reservoir because such heat transfer could not be reversible. Therefore any process in which the temperature T of the working substance changes must be *adiabatic.* The bottom line is that every process in our idealized cycle must be either isothermal or adiabatic. In addition, thermal and mechanical equilibrium must be maintained at all times so that each process is completely reversible.

THE CARNOT CYCLE

The Carnot cycle consists of two reversible isothermal and two reversible adiabatic processes. Figure 18–9 (page 570) shows a Carnot cycle using as its working substance an ideal gas in a cylinder with a piston. It consists of the following steps:

1. The gas expands isothermally at temperature T_H, absorbing heat Q_H (*ab*).
2. It expands adiabatically until its temperature drops to T_C (*bc*).
3. It is compressed isothermally at T_C, rejecting heat $|Q_C|$ (*cd*).
4. It is compressed adiabatically back to its initial state at temperature T_H (*da*).

We can calculate the thermal efficiency e of a Carnot engine in the special case in which the working substance is an *ideal gas.* To carry out this calculation, we will first find the ratio Q_C/Q_H of the quantities of heat transferred in the two isothermal processes and then use Eq. (18–4) to find e.

For an ideal gas the internal energy U depends only on temperature and is thus constant in any isothermal process. For the isothermal expansion *ab*, $\Delta U_{ab} = 0$, and Q_H is

18–9 The Carnot cycle for an ideal gas. The light-blue lines in the pV-diagram are isotherms (curves of constant temperature); the dark-blue lines are adiabats (curves of zero heat flow).

equal to the work W_{ab} done by the gas during its isothermal expansion at temperature T_H. We calculated this work in Example 17–1 (Section 17–3); using that result, we have

$$Q_H = W_{ab} = nRT_H \ln \frac{V_b}{V_a}. \qquad (18\text{–}10)$$

Similarly,

$$Q_C = W_{cd} = nRT_C \ln \frac{V_d}{V_c} = -nRT_C \ln \frac{V_c}{V_d}. \qquad (18\text{–}11)$$

Because V_d is less than V_c, Q_C is negative ($Q_C = -|Q_C|$); heat flows out of the gas during the isothermal compression at temperature T_C.

The ratio of the two quantities of heat is thus

$$\frac{Q_C}{Q_H} = -\left(\frac{T_C}{T_H}\right)\frac{\ln (V_c/V_d)}{\ln (V_b/V_a)}. \qquad (18\text{–}12)$$

This can be simplified further by use of the temperature-volume relation for an adiabatic process, Eq. (17–22). We find for the two adiabatic processes:

$$T_H V_b{}^{\gamma-1} = T_C V_c{}^{\gamma-1} \qquad \text{and} \qquad T_H V_a{}^{\gamma-1} = T_C V_d{}^{\gamma-1}.$$

Dividing the first of these by the second, we find

$$\frac{V_b{}^{\gamma-1}}{V_a{}^{\gamma-1}} = \frac{V_c{}^{\gamma-1}}{V_d{}^{\gamma-1}} \qquad \text{and} \qquad \frac{V_b}{V_a} = \frac{V_c}{V_d}.$$

Thus the two logarithms in Eq. (18–12) are equal, and that equation reduces to

$$\frac{Q_C}{Q_H} = -\frac{T_C}{T_H}, \qquad \text{or} \qquad \frac{|Q_C|}{|Q_H|} = \frac{T_C}{T_H} \qquad \begin{array}{l}\text{(heat transfer in a} \\ \text{Carnot engine).}\end{array} \qquad (18\text{–}13)$$

The ratio of the heat rejected at T_C to the heat absorbed at T_H is just equal to the ratio T_C/T_H. Then from Eq. (18–4) the efficiency of the Carnot engine is

$$e_{\text{Carnot}} = 1 - \frac{T_C}{T_H} = \frac{T_H - T_C}{T_H} \qquad \text{(efficiency of a Carnot engine).} \qquad (18\text{–}14)$$

This simple result says that the efficiency of a Carnot engine depends only on the temperatures of the two heat reservoirs. The efficiency is large when the temperature *difference* is large, and it is very small when the temperatures are nearly equal. The efficiency can never be exactly unity unless $T_C = 0$; we'll see later that this, too, is impossible.

CAUTION ▶ In all calculations involving the Carnot cycle, you must make sure that you use *absolute* (Kelvin) temperatures only. That's because Eqs. (18–10) through (18–14) come from the ideal-gas equation $pV = nRT$, in which T is absolute temperature. ◀

EXAMPLE 18-2

A Carnot engine takes 2000 J of heat from a reservoir at 500 K, does some work, and discards some heat to a reservoir at 350 K. How much work does it do, how much heat is discarded, and what is the efficiency?

SOLUTION From Eq. (18–13) the heat Q_C discarded by the engine is

$$Q_C = -Q_H \frac{T_C}{T_H} = -(2000 \text{ J}) \frac{350 \text{ K}}{500 \text{ K}}$$

$$= -1400 \text{ J}.$$

Then from the first law, the work W done by the engine is

$$W = Q_H + Q_C = 2000 \text{ J} + (-1400 \text{ J})$$

$$= 600 \text{ J}.$$

From Eq. (18–14) the thermal efficiency is

$$e = 1 - \frac{T_C}{T_H} = 1 - \frac{350 \text{ K}}{500 \text{ K}} = 0.30 = 30\%.$$

Alternatively, from the basic definition of thermal efficiency,

$$e = \frac{W}{Q_H} = \frac{600 \text{ J}}{2000 \text{ J}} = 0.30 = 30\%.$$

EXAMPLE 18-3

Suppose 0.200 mol of an ideal diatomic gas ($\gamma = 1.40$) undergoes a Carnot cycle with temperatures 227°C and 27°C. The initial pressure is $p_a = 10.0 \times 10^5$ Pa, and during the isothermal expansion at the higher temperature the volume doubles. a) Find the pressure and volume at each of points a, b, c, and d in Fig. 18–9. b) Find Q, W, and ΔU for each step and for the entire cycle. c) Determine the efficiency directly from the results of part (b), and compare it with the result from Eq. (18–14).

SOLUTION a) We first remember to convert the Celsius temperatures to absolute temperatures by adding 273.15. The higher temperature is $T_H = (227 + 273.15)$ K = 500 K, and the lower temperature is $T_C = (27 + 273.15)$ K = 300 K. We then use the ideal-gas equation to find V_a:

$$V_a = \frac{nRT_H}{p_a} = \frac{(0.200 \text{ mol})(8.315 \text{ J/mol} \cdot \text{K})(500 \text{ K})}{10.0 \times 10^5 \text{ Pa}}$$

$$= 8.31 \times 10^{-4} \text{ m}^3.$$

The volume doubles during the isothermal expansion $a \rightarrow b$, so

$$V_b = 2V_a = 2(8.31 \times 10^{-4} \text{ m}^3)$$

$$= 16.6 \times 10^{-4} \text{ m}^3.$$

Also, during the isothermal expansion $a \rightarrow b$, $p_a V_a = p_b V_b$, so

$$p_b = \frac{p_a V_a}{V_b} = 5.00 \times 10^5 \text{ Pa}.$$

For the adiabatic expansion $b \rightarrow c$, $T_H V_b{}^{\gamma-1} = T_C V_c{}^{\gamma-1}$, so

$$V_c = V_b \left(\frac{T_H}{T_C} \right)^{1/(\gamma-1)} = (16.6 \times 10^{-4} \text{ m}^3) \left(\frac{500 \text{ K}}{300 \text{ K}} \right)^{2.5}$$

$$= 59.6 \times 10^{-4} \text{ m}^3.$$

Using the ideal-gas equation again for point c, we find

$$p_c = \frac{nRT_C}{V_c} = \frac{(0.200 \text{ mol})(8.315 \text{ J/mol} \cdot \text{K})(300 \text{ K})}{59.6 \times 10^{-4} \text{ m}^3}$$

$$= 0.837 \times 10^5 \text{ Pa}.$$

For the adiabatic compression $d \rightarrow a$, $T_C V_d{}^{\gamma-1} = T_H V_a{}^{\gamma-1}$, and

$$V_d = V_a \left(\frac{T_H}{T_C} \right)^{1/(\gamma-1)} = (8.31 \times 10^{-4} \text{ m}^3) \left(\frac{500 \text{ K}}{300 \text{ K}} \right)^{2.5}$$

$$= 29.8 \times 10^{-4} \text{ m}^3,$$

$$p_d = \frac{nRT_C}{V_d} = \frac{(0.200 \text{ mol})(8.315 \text{ J/mol} \cdot \text{K})(300 \text{ K})}{29.8 \times 10^{-4} \text{ m}^3}$$

$$= 1.67 \times 10^5 \text{ Pa}.$$

b) For the isothermal expansion $a \to b$, $\Delta U_{ab} = 0$. To find W_{ab} ($= Q_H$), we use Eq. (18–10):

$$W_{ab} = Q_H = nRT_H \ln \frac{V_b}{V_a}$$

$$= (0.200 \text{ mol})(8.315 \text{ J/mol} \cdot \text{K})(500 \text{ K})(\ln 2) = 576 \text{ J}.$$

For the adiabatic expansion $b \to c$, $Q_{bc} = 0$. From the first law of thermodynamics, $\Delta U_{bc} = Q_{bc} - W_{bc} = -W_{bc}$; hence the work W_{bc} done by the gas in this process equals the negative of the change in internal energy of the gas. From Eq. (17–13) we have $\Delta U = nC_V \Delta T$, where $\Delta T = T_C - T_H$ (final temperature minus initial temperature). Using $C_V = 20.8 \text{ J/mol} \cdot \text{K}$ for an ideal diatomic gas, we find

$$W_{bc} = -\Delta U_{bc} = -nC_V(T_C - T_H) = nC_V(T_H - T_C)$$

$$= (0.200 \text{ mol})(20.8 \text{ J/mol} \cdot \text{K})(500 \text{ K} - 300 \text{ K}) = 832 \text{ J}.$$

For the isothermal compression $c \to d$, $\Delta U_{cd} = 0$; Eq. (18–11) gives

$$W_{cd} = Q_C = nRT_C \ln \frac{V_d}{V_c}$$

$$= (0.200 \text{ mol})(8.315 \text{ J/mol} \cdot \text{K})(300 \text{ K}) \left(\ln \frac{29.8 \times 10^{-4} \text{ m}^3}{59.6 \times 10^{-4} \text{ m}^3} \right)$$

$$= -346 \text{ J}.$$

For the adiabatic compression $d \to a$, $Q_{da} = 0$, and

$$W_{da} = -\Delta U_{da} = -nC_V(T_H - T_C) = nC_V(T_C - T_H)$$

$$= (0.200 \text{ mol})(20.8 \text{ J/mol} \cdot \text{K})(300 \text{ K} - 500 \text{ K})$$

$$= -832 \text{ J}.$$

We can tabulate the results as follows:

PROCESS	Q	W	ΔU
$a \to b$	576 J	576 J	0
$b \to c$	0	832 J	−832 J
$c \to d$	−346 J	−346 J	0
$d \to a$	0	−832 J	832 J
Total	230 J	230 J	0

Note that for the entire cycle, $Q = W$ and $\Delta U = 0$. Also note that the quantities of work in the two adiabatic processes are negatives of each other; it is easy to prove from the analysis leading to Eq. (18–13) that this must *always* be the case.

c) From the table, $Q_H = 576 \text{ J}$ and the total work is 230 J. Thus

$$e = \frac{W}{Q_H} = \frac{230 \text{ J}}{576 \text{ J}} = 0.40 = 40\%.$$

We can compare this with the result from Eq. (18–14):

$$e = \frac{T_H - T_C}{T_H} = \frac{500 \text{ K} - 300 \text{ K}}{500 \text{ K}} = 0.40 = 40\%.$$

THE CARNOT REFRIGERATOR

Because each step in the Carnot cycle is reversible, the *entire cycle* may be reversed, converting the engine into a refrigerator. The coefficient of performance of the Carnot refrigerator is obtained by combining the general definition of K, Eq. (18–9), with Eq. (18–13) for the Carnot cycle. We first rewrite Eq. (18–9) as

$$K = \frac{|Q_C|}{|Q_H| - |Q_C|} = \frac{|Q_C|/|Q_H|}{1 - |Q_C|/|Q_H|}.$$

Then we substitute Eq. (18–13), $|Q_C|/|Q_H| = T_C/T_H$, into this. The result is

$$K_{\text{Carnot}} = \frac{T_C}{T_H - T_C} \qquad \text{(coefficient of performance of a Carnot refrigerator).} \qquad (18\text{–}15)$$

When the temperature difference $T_H - T_C$ is small, K is much larger than unity; in this case a lot of heat can be "pumped" from the lower to the higher temperature with only a little expenditure of work. But the greater the temperature difference, the smaller the value of K and the more work is required to transfer a given quantity of heat.

EXAMPLE 18–4

If the cycle described in Example 18–3 is run backward as a refrigerator, what is its coefficient of performance?

SOLUTION From Eq. (18–9),

$$K = \frac{|Q_C|}{|W|} = \frac{346 \text{ J}}{230 \text{ J}} = 1.50.$$

Because the cycle is a Carnot cycle, we may also use

Eq. (18–15):

$$K = \frac{T_C}{T_H - T_C} = \frac{300 \text{ K}}{500 \text{ K} - 300 \text{ K}} = 1.50.$$

For a Carnot cycle, e and K depend only on the temperatures, as shown by Eqs. (18–14) and (18–15), and we don't need to calculate Q and W. For cycles containing irreversible processes, however, these two equations are not valid, and more detailed calculations are necessary.

THE CARNOT CYCLE AND THE SECOND LAW

We can prove that **no engine can be more efficient than a Carnot engine operating between the same two temperatures.** The key to the proof is the above observation that since each step in the Carnot cycle is reversible, the *entire cycle* may be reversed. Run backward, the engine becomes a refrigerator. Suppose we have an engine that is more efficient than a Carnot engine (Fig. 18–10). Let the Carnot engine, run backward as a refrigerator by negative work $-|W|$, take in heat Q_C from the cold reservoir and expel heat $|Q_H|$ to the hot reservoir. The superefficient engine expels heat $|Q_C|$, but to do this, it takes in a greater amount of heat $Q_H + \Delta$. Its work output is then $W + \Delta$, and the net effect of the two machines together is to take a quantity of heat Δ and convert it completely into work. This violates the engine statement of the second law. We could construct a similar argument that a superefficient engine could be used to violate the refrigerator statement of the second law. Note that we don't have to assume that the superefficient engine is reversible. In a similar way we can show that *no refrigerator can have a greater coefficient of performance than a Carnot refrigerator operating between the same two temperatures.*

Thus the statement that no engine can be more efficient than a Carnot engine is yet another equivalent statement of the second law of thermodynamics. It also follows directly that **all Carnot engines operating between the same two temperatures have the same efficiency, irrespective of the nature of the working substance.** Although we derived Eq. (18–14) for a Carnot engine using an ideal gas as its working substance, it is in fact valid for *any* Carnot engine, no matter what its working substance.

Equation (18–14), the expression for the efficiency of a Carnot engine, sets an upper limit to the efficiency of a real engine such as a steam turbine. To maximize this upper limit and the actual efficiency of the real engine, the designer must make the intake temperature T_H as high as possible and the exhaust temperature T_C as low as possible.

The exhaust temperature cannot be lower than the lowest temperature available for cooling the exhaust. For a steam turbine at an electric power plant, T_C may be the temperature of river or lake water; then we want the boiler temperature T_H to be as high as possible. The vapor pressures of all liquids increase rapidly with temperature, so we are limited by the mechanical strength of the boiler. At 500°C the vapor pressure of water is about 240×10^5 Pa (235 atm); this is about the maximum practical pressure in large present-day steam boilers.

The unavoidable exhaust heat loss in electric power plants creates a serious environmental problem. When a lake or river is used for cooling, the temperature of the body of water may be raised several degrees. This has a severely disruptive effect on the overall ecological balance, since relatively small temperature changes can have significant effects on metabolic rates in plants and animals. **Thermal pollution,** as this effect is called, is an inevitable consequence of the second law of thermodynamics, so careful planning is essential to minimize the ecological impact of new power plants.

To maximize efficiency, the temperatures inside a jet engine are made as high as possible. Exotic ceramic materials are used that can withstand temperatures in excess of 1000°C without melting or becoming soft.

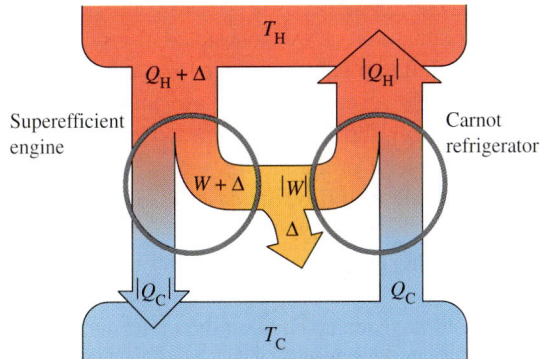

18–10 If there were a more efficient engine than a Carnot engine, it could be used in conjunction with a Carnot refrigerator to convert the heat Δ completely into work with no net heat transfer to the cold reservoir. By the second law of thermodynamics this cannot happen.

*18–8 THE KELVIN TEMPERATURE SCALE

In Chapter 15 we expressed the need for a temperature scale that doesn't depend on the properties of any particular material. We can now use the Carnot cycle to define such a scale. The thermal efficiency of a Carnot engine operating between two heat reservoirs at temperatures T_H and T_C is independent of the nature of the working substance and depends only on the temperatures. From Eq. (18–4), this thermal efficiency is

$$e = \frac{Q_H + Q_C}{Q_H} = 1 + \frac{Q_C}{Q_H}.$$

Therefore the ratio Q_C/Q_H is the same for *all* Carnot engines operating between two given temperatures T_H and T_C.

Kelvin proposed that we *define* the ratio of the temperatures, T_C/T_H, to be equal to the magnitude of the ratio Q_C/Q_H of the quantities of heat absorbed and rejected:

$$\frac{T_C}{T_H} = \frac{|Q_C|}{|Q_H|} = -\frac{Q_C}{Q_H} \qquad \text{(definition of Kelvin temperature).} \qquad (18\text{–}16)$$

Equation (18–16) looks identical to Eq. (18–13), but there is a subtle and crucial difference. The temperatures in Eq. (18–13) are based on an ideal-gas thermometer, as defined in Section 15–4, while Eq. (18–16) *defines* a temperature scale based on the Carnot cycle and the second law of thermodynamics and independent of the behavior of any particular substance. Thus the **Kelvin temperature scale** is truly *absolute*. To complete the definition of the Kelvin scale, we assign, as in Section 15–4, the arbitrary value of 273.16 K to the temperature of the triple point of water. When a substance is taken around a Carnot cycle, the ratio of the heats absorbed and rejected, $|Q_H|/|Q_C|$, is equal to the ratio of the temperatures of the reservoirs *as expressed on the gas-thermometer scale* defined in Section 15–4. Since the triple point of water is chosen to be 273.16 K in both scales, it follows that *the Kelvin and ideal-gas scales are identical.*

The zero point on the Kelvin scale is called **absolute zero.** Absolute zero can be interpreted on a molecular level; at absolute zero the system has its *minimum* possible total internal energy (kinetic plus potential). Because of quantum effects, however, it is *not* true that at $T = 0$, all molecular motion ceases. There are theoretical reasons for believing that absolute zero cannot be attained experimentally, although temperatures below 10^{-7} K have been achieved. The more closely we approach absolute zero, the more difficult it is to get closer. One statement of the *third law of thermodynamics* is that it is impossible to reach absolute zero in a finite number of thermodynamic steps.

*18–9 ENTROPY

The second law of thermodynamics, as we have stated it, is rather different in form from many familiar physical laws. It is not an equation or a quantitative relationship but rather a statement of *impossibility.* However, the second law *can* be stated as a quantitative relation with the concept of *entropy,* the subject of this section.

We have talked about several processes that proceed naturally in the direction of increasing disorder. Irreversible heat flow increases disorder because the molecules are initially sorted into hotter and cooler regions; this sorting is lost when the system comes to thermal equilibrium. Adding heat to a body increases its disorder because it increases average molecular speeds and therefore the randomness of molecular motion. Free expansion of a gas increases its disorder because the molecules have greater randomness of position after the expansion than before.

ENTROPY AND DISORDER

Entropy provides a *quantitative* measure of disorder. To introduce this concept, let's consider an infinitesimal isothermal expansion of an ideal gas. We add heat dQ and let the gas expand just enough to keep the temperature constant. Because the internal energy of an ideal gas depends only on its temperature, the internal energy is also constant; thus from the first law, the work dW done by the gas is equal to the heat dQ added. That is,

$$dQ = dW = p\,dV = \frac{nRT}{V}\,dV, \qquad \text{so} \qquad \frac{dV}{V} = \frac{dQ}{nRT}.$$

The gas is in a more disordered state after the expansion than before because the molecules are moving in a larger volume and have more randomness of position. Thus the fractional volume change dV/V is a measure of the increase in disorder, and the above equation shows that it is proportional to the quantity dQ/T. We introduce the symbol S for the entropy of the system, and we define the infinitesimal entropy change dS during an infinitesimal reversible process at absolute temperature T as

$$dS = \frac{dQ}{T} \qquad \text{(infinitesimal reversible process).} \qquad (18\text{--}17)$$

If a total amount of heat Q is added during a reversible isothermal process at absolute temperature T, the total entropy change $\Delta S = S_2 - S_1$ is given by

$$\Delta S = S_2 - S_1 = \frac{Q}{T} \qquad \text{(reversible isothermal process).} \qquad (18\text{--}18)$$

Entropy has units of energy divided by temperature; the SI unit of entropy is 1 J/K.

We can see how the quotient Q/T is related to the increase in disorder. Higher temperature means greater randomness of motion. If the substance is initially cold, with little molecular motion, adding heat Q causes a substantial fractional increase in molecular motion and randomness. But if the substance is already hot, the same quantity of heat adds relatively little to the greater molecular motion already present. So the quotient Q/T is an appropriate characterization of the increase in randomness or disorder when heat flows into a system.

EXAMPLE 18–5

One kilogram of ice at 0°C is melted and converted to water at 0°C. Compute its change in entropy, assuming that the melting is done reversibly. The heat of fusion of water is $L_f = 3.34 \times 10^5$ J/kg.

SOLUTION The temperature T is constant at 273 K. The heat needed to melt the ice is $Q = mL_f = 3.34 \times 10^5$ J. From Eq. (18–18) the increase in entropy of the system is

$$\Delta S = S_2 - S_1 = \frac{Q}{T} = \frac{3.34 \times 10^5 \text{ J}}{273 \text{ K}} = 1.22 \times 10^3 \text{ J/K}.$$

This increase corresponds to the increase in disorder when the water molecules go from the highly ordered state of a crystalline solid to the much more disordered state of a liquid (Fig. 18–11).

In any *isothermal* reversible process, the entropy change equals the heat transferred divided by the absolute temperature. When we refreeze the water, Q has the opposite sign, and the entropy change of the water is $\Delta S = -1.22 \times 10^3$ J/K. The water molecules rearrange themselves into a crystal to form ice, so disorder and entropy both decrease.

18–11 Water molecules are arranged in a regular, ordered way in an ice crystal. When the ice melts, the hydrogen bonds between molecules are broken, increasing the water's disorder and its entropy.

We can generalize the definition of entropy change to include *any* reversible process leading from one state to another, whether it is isothermal or not. We represent the process as a series of infinitesimal reversible steps. During a typical step, an infinitesimal quantity of heat dQ is added to the system at absolute temperature T. Then we sum (integrate) the quotients dQ/T for the entire process; that is,

$$\Delta S = \int_1^2 \frac{dQ}{T} \qquad \text{(entropy change in a reversible process).} \qquad (18\text{--}19)$$

The limits 1 and 2 refer to the initial and final states.

Because entropy is a measure of the disorder of a system in any specific state, it must depend only on the current state of the system, not on its past history. We will show later that this is indeed the case. When a system proceeds from an initial state with entropy S_1 to a final state with entropy S_2, the change in entropy $\Delta S = S_2 - S_1$ defined by Eq. (18–19) does not depend on the path leading from the initial to the final state but is the same for *all possible* processes leading from state 1 to state 2. Thus the entropy of a system must also have a definite value for any given state of the system. We recall that *internal energy,* introduced in Chapter 17, also has this property, although entropy and internal energy are very different quantities.

Since entropy is a function only of the state of a system, we can also compute entropy changes in *irreversible* (non-equilibrium) processes for which Eqs. (18–17) and (18–19) are not applicable. We simply invent a path connecting the given initial and final states that *does* consist entirely of reversible, equilibrium processes and compute the total entropy change for that path. It is not the actual path, but the entropy change must be the same as for the actual path.

As with internal energy, the above discussion does not tell us how to calculate entropy itself, but only the change in entropy in any given process. Just as with internal energy, we may arbitrarily assign a value to the entropy of a system in a specified reference state and then calculate the entropy of any other state with reference to this.

EXAMPLE 18-6

One kilogram of water at 0°C is heated to 100°C. Compute its change in entropy.

SOLUTION To use Eq. (18–19), we imagine that the temperature of the water is increased reversibly in a series of infinitesimal steps, in each of which the temperature is raised by an infinitesimal amount dT. From Eq. (15–14) the heat required to carry out each such step is $dQ = mc\,dT$. Substituting this into Eq. (18–19) and integrating, we find

$$\Delta S = S_2 - S_1 = \int_1^2 \frac{dQ}{T} = \int_{T_1}^{T_2} mc\,\frac{dT}{T} = mc\,\ln\frac{T_2}{T_1}$$

$$= (1.00\text{ kg})(4190\text{ J/kg}\cdot\text{K})\left(\ln\frac{373\text{ K}}{273\text{ K}}\right)$$

$$= 1.31 \times 10^3\text{ J/K}.$$

In practice, this heating would be done irreversibly, perhaps by setting a pan of water on a stove with a finite temperature difference between the burner and pan. But the entropy change of the water is the same whether the process is reversible or irreversible.

In this calculation we assumed that the specific heat capacity c doesn't depend on temperature. That's a pretty good approximation, since c for water increases by only 1% between 0°C and 100°C.

EXAMPLE 18-7

A gas expands adiabatically and reversibly. What is its change in entropy?

SOLUTION In an adiabatic process, no heat enters or leaves the system. Hence $dQ = 0$, and there is *no* change in entropy in this reversible process: $\Delta S = 0$. Every *reversible* adiabatic process is

a constant-entropy process. (For this reason, reversible adiabatic processes are also called *isentropic* processes.) The increase in disorder resulting from the gas occupying a greater volume is exactly balanced by the decrease in disorder associated with the lowered temperature and reduced molecular speeds.

EXAMPLE 18-8

A thermally insulated box is divided by a partition into two compartments, each having volume V (Fig. 18–12). Initially, one compartment contains n moles of an ideal gas at temperature T, and the other compartment is evacuated. We then break the partition, and the gas expands to fill both compartments. What is the entropy change in this free-expansion process?

SOLUTION For this process, $Q = 0$, $W = 0$, $\Delta U = 0$, and therefore (because the system is an ideal gas) $\Delta T = 0$. We might think that the entropy change is zero because there is no heat exchange. But Eq. (18–19) can be used to calculate entropy changes for *reversible* processes only; this free expansion is *not* reversible, and there *is* an entropy change. The process is adiabatic, but it is not isentropic.

To calculate ΔS, we recall that the entropy change depends only on the initial and final states. We can devise a *reversible* process having the same endpoints, use Eq. (18–19) to calculate its entropy change, and thus determine the entropy change in the original process. An appropriate reversible process in this case is an isothermal expansion from V to $2V$ at temperature T. The gas does work W during this substitute expansion, so an equal amount of heat Q must be supplied to keep the internal energy constant. We found in Example 17–1 (Section 17–3) that the work done by n moles of ideal gas in an isothermal expansion from V_1 to V_2 is $W = nRT \ln (V_2/V_1)$. Using $V_1 = V$ and $V_2 = 2V$, we have

$$Q = W = nRT \ln \frac{2V}{V} = nRT \ln 2.$$

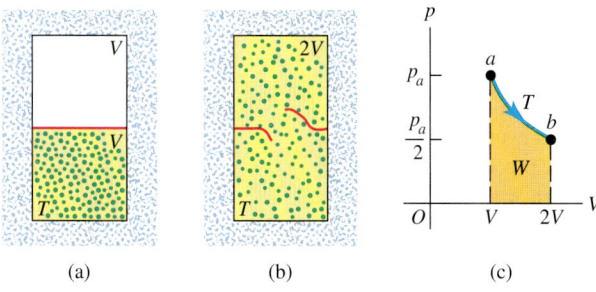

18–12 (a, b) Free expansion of an insulated ideal gas. (c) The free-expansion process doesn't pass through equilibrium states from a to b. However, the entropy change $S_b - S_a$ can be calculated by using the isothermal path shown or *any* reversible path from a to b.

Thus the entropy change is

$$\Delta S = \frac{Q}{T} = nR \ln 2,$$

and this is also the entropy change for the free expansion with the same initial and final states. For one mole,

$$\Delta S = (1 \text{ mol})(8.315 \text{ J/mol} \cdot \text{K})(\ln 2) = 5.76 \text{ J/K}.$$

EXAMPLE 18-9

Entropy and the Carnot cycle For the Carnot engine in Example 18–2 (Section 18–7), find the total entropy change in the engine during one cycle.

SOLUTION There is no entropy change during the adiabatic expansion or adiabatic compression. During the isothermal expansion at $T_H = 500$ K the engine takes in 2000 J of heat, and its entropy change is

$$\Delta S_H = \frac{Q_H}{T_H} = \frac{2000 \text{ J}}{500 \text{ K}} = 4.0 \text{ J/K}.$$

During the isothermal compression at $T_C = 350$ K the engine gives off 1400 J of heat, and its entropy change is

$$\Delta S_C = \frac{Q_C}{T_C} = \frac{-1400 \text{ J}}{350 \text{ K}} = -4.0 \text{ J/K}.$$

The total entropy change in the engine during one cycle is $\Delta S_H + \Delta S_C = 4.0 \text{ J/K} + (-4.0 \text{ J/K}) = 0$. The total entropy change of the two heat reservoirs is also zero, although each individual reservoir has an entropy change. This cycle contains no irreversible processes, and the total entropy change of the system and its surroundings is zero.

ENTROPY IN CYCLIC PROCESSES

Example 18–9 showed that the total entropy change for a cycle of a particular Carnot engine, which uses an ideal gas as its working substance, is zero. This result follows directly from Eq. (18–13), which we can rewrite as

$$\frac{Q_H}{T_H} + \frac{Q_C}{T_C} = 0. \tag{18–20}$$

The quotient Q_H/T_H equals ΔS_H, the entropy change of the engine that occurs at $T = T_H$. Likewise, Q_C/T_C equals ΔS_C, the (negative) entropy change of the engine that occurs at $T = T_C$. Hence Eq. (18–20) says that $\Delta S_H + \Delta S_C = 0$, that is, there is zero net entropy change in one cycle.

18–13 (a) A reversible cyclic process for an ideal gas, shown as a red closed path on a pV-diagram. Several ideal-gas isotherms are shown in blue. (b) The path can be approximated by a series of long, thin Carnot cycles; one cycle is highlighted in gold. The total entropy change is zero for each Carnot cycle and for the actual cyclic process. (c) The entropy change $S_b - S_a$ is the same for all paths, such as paths 1 and 2, connecting points a and b.

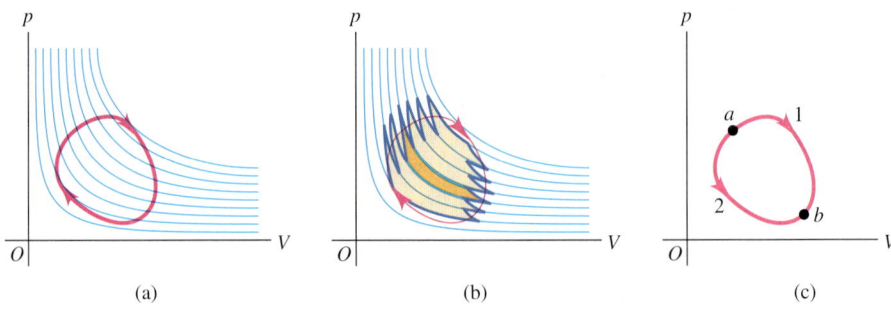
(a)   (b)   (c)

What about Carnot engines that use a different working substance? According to the second law, *any* Carnot engine operating between given temperatures T_H and T_C has the same efficiency $e = 1 - T_C/T_H$ (Eq. (18–14)). Combining this expression for e with Eq. (18–4), $e = 1 + Q_C/Q_H$, just reproduces Eq. (18–20). So Eq. (18–20) is valid for any Carnot engine working between these temperatures, whether its working substance is an ideal gas or not. We conclude that *the total entropy change in one cycle of any Carnot engine is zero.*

This result can be generalized to show that the total entropy change during *any* reversible cyclic process is zero. A reversible cyclic process appears on a pV-diagram as a closed path (Fig. 18–13a). We can approximate such as path as closely as we like by a sequence of isothermal and adiabatic processes forming parts of many long, thin Carnot cycles (Fig. 18–13b). The total entropy change for the full cycle is the sum of the entropy changes for each small Carnot cycle, each of which is zero. So **the total entropy change during *any* reversible cycle is zero:**

$$\int \frac{dQ}{T} = 0 \qquad \text{(reversible cyclic process)}. \qquad (18–21)$$

It follows that when a system undergoes a reversible process leading from any state a to any other state b, *the entropy change is independent of the path* (Fig. 18–13c). If the entropy change for path 1 were different from the change for path 2, the system could be taken along path 1 and then backward along path 2 to the starting point, with a nonzero net change in entropy. This would violate the conclusion that the total entropy change in such a cyclic process must be zero. Because the entropy change in such processes is independent of path, we conclude that in any given state, the system has a definite value of entropy that depends only on the state, not on the processes that led to that state.

ENTROPY IN IRREVERSIBLE PROCESSES

In an idealized, reversible process involving only equilibrium states, the total entropy change of the system and its surroundings is zero. But all *irreversible* processes involve an increase in entropy. Unlike energy, *entropy is not a conserved quantity.* The entropy of an isolated system *can* change, but as we shall see, it can never decrease. The free expansion of a gas, described in Example 18–8, is an irreversible process in an isolated system in which there is an entropy increase.

<hr>

EXAMPLE 18–10

Suppose 1.00 kg of water at 100°C is placed in thermal contact with 1.00 kg of water at 0°C. What is the total change in entropy? Assume that the specific heat capacity of water is constant at 4190 J/kg · K over this temperature range.

SOLUTION This process involves irreversible heat flow because of the temperature differences. The final temperature is 50°C = 323 K. The first 4190 J of heat transferred cools the hot water to 99°C and warms the cold water from 0°C to 1°C. The

net change of entropy is approximately

$$\Delta S = \frac{-4190 \text{ J}}{373 \text{ K}} + \frac{4190 \text{ J}}{273 \text{ K}} = 4.1 \text{ J/K}.$$

Further increases in entropy occur as the system approaches thermal equilibrium at 50°C (323 K). The *total* increase in entropy can be calculated in the same way as in Example 18–6 by assuming that the process occurs reversibly. The entropy change of the hot water is

$$\Delta S_{\text{hot}} = mc \int_{T_1}^{T_2} \frac{dT}{T} = (1.00 \text{ kg})(4190 \text{ J/kg} \cdot \text{K}) \int_{373 \text{ K}}^{323 \text{ K}} \frac{dT}{T}$$

$$= (4190 \text{ J/K}) \left(\ln \frac{323 \text{ K}}{373 \text{ K}} \right) = -603 \text{ J/K}.$$

The entropy change of the cold water is

$$\Delta S_{\text{cold}} = (4190 \text{ J/K}) \left(\ln \frac{323 \text{ K}}{273 \text{ K}} \right) = +705 \text{ J/K}.$$

The *total* entropy change of the system is

$$\Delta S_{\text{total}} = \Delta S_{\text{hot}} + \Delta S_{\text{cold}} = (-603 \text{ J/K}) + 705 \text{ J/K} = +102 \text{ J/K}.$$

An irreversible heat flow in an isolated system is accompanied by an increase in entropy. We could have reached the same end state by simply mixing the two quantities of water. This, too, is an irreversible process; because the entropy depends only on the state of the system, the total entropy change would be the same, 102 J/K.

ENTROPY AND THE SECOND LAW

The results of Example 18–10 about the flow of heat from a higher to a lower temperature, or the mixing of substances at different temperatures, are characteristic of *all* natural (that is, irreversible) processes. When we include the entropy changes of all the systems taking part in the process, the increases in entropy are always greater than the decreases. In the special case of a *reversible* process, the increases and decreases are equal. Hence we can state the general principle: **When all systems taking part in a process are included, the entropy either remains constant or increases.** In other words, **no process is possible in which the total entropy decreases, when all systems taking part in the process are included.** This is an alternative statement of the second law of thermodynamics in terms of entropy. Thus it is equivalent to the "engine" and "refrigerator" statements discussed earlier. Figure 18–14 shows a specific example of this general principle.

The increase of entropy in every natural, irreversible process measures the increase of disorder or randomness in the universe associated with that process. Consider again the example of mixing hot and cold water. We *might* have used the hot and cold water as the high- and low-temperature reservoirs of a heat engine. While removing heat from the hot water and giving heat to the cold water, we could have obtained some mechanical work. But once the hot and cold water have been mixed and have come to a uniform temperature, this opportunity to convert heat to mechanical work is lost irretrievably. The lukewarm water will never *unmix* itself and separate into hotter and colder portions. No decrease in *energy* occurs when the hot and cold water are mixed. What has been lost is not *energy,* but *opportunity,* the opportunity to convert part of the heat from the hot water into mechanical work. Hence when entropy increases, energy becomes less *available,* and the universe becomes more random or "run down."

18–14 The mixing of colored ink and water starts from a state of relative order (low entropy) in which each fluid is separate and distinct from the other. The final state after mixing is more disordered (has greater entropy) because particles of each fluid are randomly surrounded by particles of both fluids. Spontaneous unmixing of the ink and water, a process in which there would be a net decrease in entropy, is never observed.

*18–10 MICROSCOPIC INTERPRETATION OF ENTROPY

We described in Section 17–5 how the internal energy of a system could be calculated, at least in principle, by adding up all the kinetic energies of its constituent particles and all the potential energies of interaction among the particles. This is called a *microscopic* calculation of the internal energy. We can also make a microscopic calculation of the entropy S of a system. Unlike energy, however, entropy is not something that belongs to each individual particle or pair of particles in the system. Rather, entropy is a measure of the disorder of the system as a whole. To see how to calculate entropy microscopically, we first have to introduce the idea of *macroscopic* and *microscopic states.*

Suppose you toss N identical coins on the floor, and half of them show heads and half show tails. This is a description of the large-scale or **macroscopic state** of the

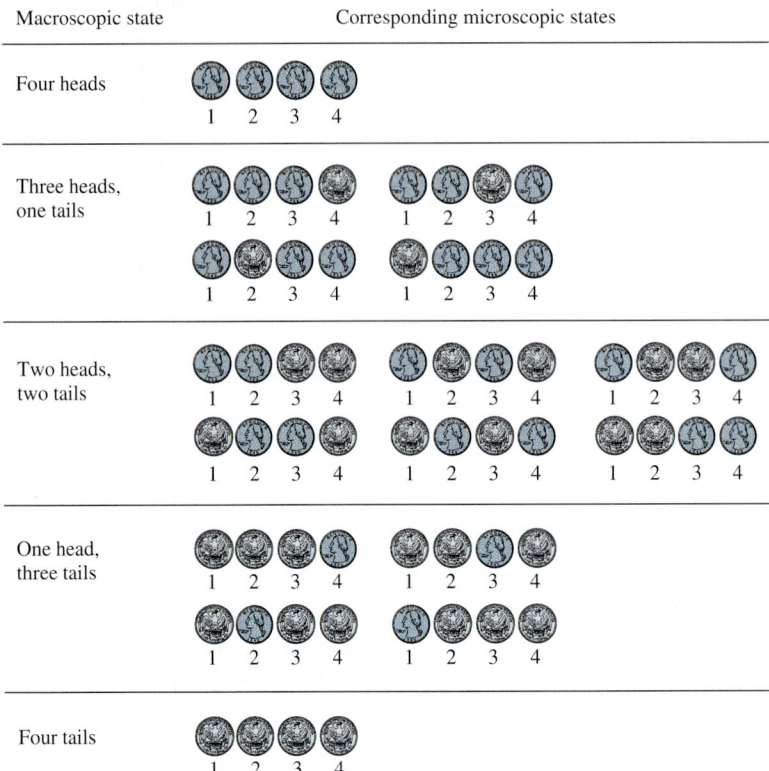

| Macroscopic state | Corresponding microscopic states |

18–15 All possible microscopic states of four coins. There can be several possible microscopic states for each macroscopic state.

system of N coins. A description of the **microscopic state** of the system includes information about each individual coin: Coin 1 was heads, coin 2 was tails, coin 3 was tails, and so on. There can be many microscopic states that correspond to the same macroscopic description. For instance, with $N = 4$ coins there are six possible states in which half are heads and half are tails (Fig. 18–15). The number of microscopic states grows rapidly with increasing N; for $N = 100$ there are $2^{100} = 1.27 \times 10^{30}$ microscopic states, of which 1.01×10^{29} are half heads and half tails.

The least probable outcomes of the coin toss are the states that are either all heads or all tails. It is certainly possible that you could throw 100 heads in a row, but don't bet on it; the probability of doing this is only 1 in 1.27×10^{30}. The most probable outcome of tossing N coins is that half are heads and half are tails. The reason is that this *macroscopic* state has the greatest number of corresponding *microscopic* states, as shown in Fig. 18–15.

To make the connection to the concept of entropy, note that N coins that are all heads constitutes a completely ordered macroscopic state; the description "all heads" completely specifies the state of each one of the N coins. The same is true if the coins are all tails. But the macroscopic description "half heads, half tails" by itself tells you very little about the state (heads or tails) of each individual coin. We say that the system is *disordered* because we know so little about its microscopic state. Compared to the state "all heads" or "all tails," the state "half heads, half tails" has a much greater number of possible microscopic states, much greater disorder, and hence much greater entropy (which is a quantitative measure of disorder).

Now instead of N coins, consider a mole of an ideal gas containing Avogadro's number of molecules. The macroscopic state of this gas is given by its pressure p, volume V, and temperature T; a description of the microscopic state involves stating the position and velocity for each molecule in the gas. At a given pressure, volume, and temperature the gas may be in any one of an astronomically large number of microscopic states, depending on the positions and velocities of its 6.02×10^{23} molecules. If the gas undergoes a free expansion into a greater volume, the range of possible positions increases, as does the number of possible microscopic states. The system becomes more disordered, and the entropy increases as calculated in Example 18–8 (Section 18–9).

We can draw the following general conclusion: *For any system the most probable macroscopic state is the one with the greatest number of corresponding microscopic states, which is also the macroscopic state with the greatest disorder and the greatest entropy.*

Let w represent the number of possible microscopic states for a given macroscopic state. (For the four coins shown in Fig. 18–15 the state of four heads has $w = 1$, the state of three heads and one tails has $w = 4$, and so on.) Then the entropy S of a macroscopic state can be shown to be given by

$$S = k \ln w \quad \text{(microscopic expression for entropy),} \quad (18\text{--}22)$$

where $k = R/N_A$ is the Boltzmann constant (gas constant per molecule) introduced in Section 16–4. As Eq. (18–22) shows, increasing the number of possible microscopic states w increases the entropy S.

What matters in a thermodynamic process is not the absolute entropy S but the *difference* in entropy between the initial and final states. Hence an equally valid and useful definition would be $S = k \ln w + C$, where C is a constant, since C cancels in any calculation of an entropy difference between two states. But it's convenient to set this constant equal to zero and use Eq. (18–22). With this choice, since the smallest possible value of w is unity, the smallest possible value of S for any system is $k \ln 1 = 0$. Entropy can *never* be negative.

In practice, calculating w is a difficult task, so Eq. (18–22) is typically used only to calculate the absolute entropy S of certain special systems. But we can use this relation to calculate *differences* in entropy between one state and another. Consider a system that undergoes a thermodynamic process that takes it from macroscopic state 1, for which there are w_1 possible microscopic states, to macroscopic state 2, with w_2 associated microscopic states. The change in entropy in this process is

$$\Delta S = S_2 - S_1 = k \ln w_2 - k \ln w_1 = k \ln \frac{w_2}{w_1}. \quad (18\text{--}23)$$

The *difference* in entropy between the two macroscopic states depends on the *ratio* of the numbers of possible microscopic states.

As the following example shows, using Eq. (18–23) to calculate a change in entropy from one macroscopic state to another gives the same results as considering a reversible process connecting those two states and using Eq. (18–19).

EXAMPLE 18–11

Use Eq. (18–23) to calculate the entropy change in the free expansion of n moles of gas at temperature T described in Example 18–8.

SOLUTION The situation is shown in Fig. 18–16. Let w_1 be the number of microscopic states when the gas occupies volume V. When the partition is broken, the velocities of the molecules are unaffected, since no work is done. But each molecule now has twice as much volume in which it can move and hence has twice the number of possible positions. The number of molecules is $N = nN_A$, so the number of possible microscopic states of the system as a whole increases by 2^N. Hence the number of microscopic states when the gas occupies volume $2V$ is $w_2 = 2^N w_1$.

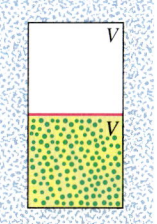
Gas occupies volume V
number of microstates w_1

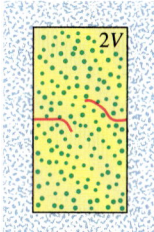
Gas occupies volume $2V$
number of microstates $w_2 = 2^N w_1$

18–16 In a free expansion of N molecules in which the volume doubles, the number of possible microscopic states increases by 2^N.

The change in entropy in this process is

$$\Delta S = k \ln \frac{w_2}{w_1} = k \ln \frac{2^N w_1}{w_1} = k \ln 2^N$$

$$= Nk \ln 2.$$

Since $N = nN_A$ and $k = R/N_A$, this becomes

$$\Delta S = (nN_A)(R/N_A) \ln 2 = nR \ln 2.$$

Thus we have found the same result as in Example 18–8, but without any reference to the thermodynamic path taken.

The relationship between entropy and the number of microscopic states gives us new insight into the entropy statement of the second law of thermodynamics, that the entropy of a closed system can never decrease. From Eq. (18–22) this means that a closed system can never spontaneously undergo a process that decreases the number of possible microscopic states.

An example of such a forbidden process would be if all of the air in your room spontaneously moved to one half of the room, leaving a vacuum in the other half. Such a "free compression" would be the reverse of the free expansion of Examples 18–8 and 18–11. This would decrease the number of possible microscopic states by a factor of 2^N. Strictly speaking, this process is not impossible! The probability of finding a given molecule in one half of the room is 1/2, so the probability of finding all of the molecules in one half of the room at once is $(1/2)^N$. (This is exactly the same as the probability of having a tossed coin come up heads N times in a row.) This probability is *not* zero. But lest you worry about suddenly finding yourself gasping for breath in the evacuated half of your room, consider that a typical room might hold 1000 moles of air, and so $N = 1000N_A = 6.02 \times 10^{26}$ molecules. The probability of all the molecules being in the same half of the room is therefore $(1/2)^{6.02 \times 10^{26}}$. Expressed as a decimal, this number has more than 10^{26} zeros to the right of the decimal point!

Because the probability of such a "free compression" taking place is so vanishingly small, it has almost certainly never occurred anywhere in the universe since the beginning of time. We conclude that for all practical purposes the second law of thermodynamics is never violated.

18–11 ENERGY RESOURCES

A Case Study in Thermodynamics

The laws of thermodynamics place very general limitations on conversion of energy from one form to another. In this time of increasing energy demand and diminishing resources, these matters are of the utmost practical importance. We conclude this chapter with a brief discussion of a few energy-conversion systems, present and proposed.

Over half of the electric power generated in the United States is obtained from coal-fired steam-turbine generating plants. Modern boilers can transfer about 80 to 90 percent of the heat of combustion of coal into steam. The maximum Carnot efficiency of a steam turbine is given by Eq. (18–14); T_H is the temperature of the boiler, and T_C is the temperature of the condenser. The highest practical value of T_H is about 800 K, and the condenser temperature is determined by the temperature of its surroundings, about 300 K. Hence $e_{Carnot} = 1 - (300 \text{ K})/(800 \text{ K}) = 0.63$, or 63%. In practice, the turbine efficiency is less than this ideal maximum, about 50%. The efficiency of large electrical generators in converting mechanical power to electrical is very high, typically 99%. Since efficiency is the ratio of the useful energy output to the energy input, the overall thermal efficiency of such a plant is the product of the individual efficiencies; this is roughly (0.85)(0.50)(0.99), or about 42%.

EXAMPLE 18–12

The Tennessee Valley Authority's Paradise power plant has a generating capacity of about 1.3 gigawatt (1.3×10^9 W) of electric power. The plant burns 10,500 tons of coal per day; the heat of combustion of coal is about 6.2×10^6 J/kg. What is the overall thermal efficiency of the plant?

SOLUTION To calculate the thermal efficiency e, we need to know the heat input and energy output per time interval. One ton is 2000 lb, and one pound has a mass of 2.2 kg, so (10,500 tons)(2000 lb/ton)(2.2 kg/lb) = 4.62×10^7 kg of coal is used daily. Hence the heat supplied to the plant each day by burning coal is $(6.2 \times 10^6$ J/kg)$(4.62 \times 10^7$ kg) = 2.86×10^{14} J. The energy output in 1 d = 86,400 s is $(1.3 \times 10^9$ J/s)(86,400 s) = 1.12×10^{14} J. Hence the overall efficiency is

$$e = \frac{W}{Q_H} = \frac{1.12 \times 10^{14} \text{ J}}{2.86 \times 10^{14} \text{ J}} = 0.39 = 39\%.$$

The efficiency of the steam turbine itself is the main limiting factor on the overall efficiency.

In a nuclear power plant, the heat to generate steam is supplied by a nuclear reaction, the fission of uranium, rather than the chemical reaction of burning coal. The steam turbines in nuclear power plants have the same theoretical efficiency limit as in coal-fired plants. At present it is not practical to run nuclear reactors at temperatures and pressures as high as those in coal boilers, so the net thermal efficiency of a nuclear plant is somewhat lower, typically 30%.

In 1990, 18% of the world's electric power supply came from nuclear power plants (20% in the United States). High construction costs, questions of public safety, and the problem of disposal of radioactive waste have slowed the development of additional nuclear power facilities. However, there is also increasing concern about the environmental impact of conventional coal-burning power plants. Coal smoke is responsible for serious and well-documented environmental hazards such as acid rain, and it contributes to global warming through the greenhouse effect. It is important to manage our development of energy resources so as to minimize the risks to human life and health, and nuclear power continues to receive attention as a future energy resource. We noted at the end of Section 18–7 that thermal pollution from both coal-fired and nuclear plants poses serious environmental problems.

Solar energy has promise for a variety of purposes. The power per unit surface area in the sun's radiation (before it passes through the earth's atmosphere) is about 1.4 kW/m^2. A maximum of about 1.0 kW/m^2 reaches the surface of the earth on a clear day, and the time average, over a 24-hour period, is about 0.2 kW/m^2. This radiation can be collected and used to heat water (Fig. 18–17a). A different solar-energy scheme is to use large banks of photovoltaic cells (Fig. 18–17b) for direct conversion of solar energy to electricity. A photovoltaic cell is not a heat engine in the usual sense and is not limited by the Carnot efficiency. There are other fundamental limitations on photocell efficiency, however; 25% is currently attainable, and 50% may be practicable with multilayer semiconductor photocells. Because of high capital costs, the price of photovoltaic-produced energy is about 25 cents per kilowatt-hour (compared with about 5 cents for conventional coal-fired steam-turbine plants). But such a system offers many

Absorber plate covered by glass

Tubes carry heated fluid

Insulation

(a)

(b)

18–17 (a) A passive solar collecting panel for home water heating. (b) An array of photovoltaic cells for direct conversion of sun power to electric power.

potential advantages, such as lack of noise, lack of moving parts, minimal maintenance requirements, and reduced pollution. Capital costs are likely to drop with improved technology, and photovoltaic systems continue to be studied and developed.

EXAMPLE 18-13

A solar water heater system is to be installed in a single-family dwelling to provide 300 L of 55°C hot water per day (normal use for a family of four). Estimate the required surface area of the collecting panels, assuming that water is supplied to the house at 15°C, that the average power per unit area from the sun is 130 W/m^2 (typical of much of the central United States), and that the collecting panels are 60% efficient.

SOLUTION The energy that must be supplied to heat 300 L of water (mass 300 kg) from 15°C to 55°C is

$$Q = mc\,\Delta T = (300\text{ kg})(4190\text{ J/kg}\cdot\text{C}°)(55°\text{C} - 15°\text{C})$$

$$= 5.03 \times 10^7\text{ J}.$$

The solar energy received by the collecting panels in one day is the average power P per unit area, multiplied by the time t (one day, or 86,400 s) and multiplied by the (unknown) panel area A. The heat Q delivered to the water is a fraction $e = 0.60$ of this, so

$$Q = ePtA,$$

$$A = \frac{Q}{ePt} = \frac{5.03 \times 10^7\text{ J}}{(0.60)(130\text{ W/m}^2)(86,400\text{ s})} = 7.5\text{ m}^2.$$

This area will fit handily onto the roof of the house.

EXAMPLE 18-14

The top of a small car (total area 3.0 m^2) is covered with photovoltaic cells that supply energy to an electric motor. It is a cloudless day, so the solar power received per unit area is 1.0 kW/m^2. If the efficiency of the photovoltaic cells is 25%, what is the power delivered to the motor?

SOLUTION The delivered power is the solar power per unit area,

multiplied by the collecting area and by the efficiency:

$$P = (1000\text{ W/m}^2)(3.0\text{ m}^2)(0.25)$$

$$= 750\text{ W} = 1.0\text{ hp}.$$

This is a very small amount of power for a car, so a solar-powered vehicle must be lightweight and have very low air and rolling resistance. The power is even less on an overcast day.

About 1% of the solar energy reaching the earth is transformed into wind energy. This can be gathered and converted by "forests" of wind turbines (Fig. 18–18). A typical turbine may have a blade diameter of 40 m; in an 8-m/s wind, such a turbine can extract about 400 kW from the air, of which about one-fourth can be converted to electric power. This is enough to provide the electric power needs of about 30 homes. About 7500 wind turbines (85% of the U.S. total) are located in California, where the state government encourages their use. As with photovoltaic systems, the capital costs of such systems are higher than for coal-fired generating plants of equal capacity, although some installations are producing energy at a cost of about 7 cents per kilowatt-hour.

Biomass is yet another mechanism for using solar energy. Plants absorb solar energy and store it; the energy can be utilized by burning or by fermentation and distillation to make alcohols. Ethanol made from fermented corn is used in the United States as a gasoline additive, although its cost is not quite competitive at present crude-oil price levels. An important feature of biomass materials is that they take as much carbon dioxide out of the atmosphere when they grow as is released when they burn; thus they don't contribute to the greenhouse effect.

An indirect scheme for collection and conversion of solar energy would use the temperature gradient in the ocean. In the Gulf of Mexico, for example, the water temperature near the surface is about 25°C, while at a depth of a few hundred meters it is about 5°C. The second law of thermodynamics forbids cooling the ocean and converting the extracted heat completely into work, but there is nothing to forbid running a heat engine between these two temperatures. The thermodynamic efficiency would be very low (less than 7%), but since two-thirds of the earth's surface is ocean, a vast supply of energy would be available.

18–18 An array of wind turbines to collect and convert wind and energy. The propellerlike blades turn electric generators, converting the kinetic energy of moving air into electrical energy.

The present level of activity in energy-conversion research is rather low in comparison to the 1970s, when the industrialized countries experienced a serious energy shortage. As we use up our available fossil-fuel resources, we will certainly have to become more active in developing alternative energy sources. The principles of thermodynamics will play a central role in this development.

Summary

KEY TERMS

- A reversible process is one whose direction can be reversed by an infinitesimal change in the conditions of the process. A reversible process is also an equilibrium process.

- A heat engine takes heat Q_H from a source, converts part of it to work W, and discards the remainder $|Q_C|$ at a lower temperature. The thermal efficiency e of a heat engine is

$$e = \frac{W}{Q_H} = 1 + \frac{Q_C}{Q_H} = 1 - \left| \frac{Q_C}{Q_H} \right|. \qquad (18\text{--}4)$$

- A gasoline engine operating on the Otto cycle with compression ratio r has a theoretical maximum thermal efficiency e given by

$$e = 1 - \frac{1}{r^{\gamma-1}}. \qquad (18\text{--}6)$$

- A refrigerator takes heat Q_C from a colder place, has a work input $|W|$, and discards heat $|Q_H|$ at a warmer place. The coefficient of performance K is defined as

$$K = \frac{|Q_C|}{|W|} = \frac{|Q_C|}{|Q_H| - |Q_C|}. \qquad (18\text{--}9)$$

- The second law of thermodynamics describes the directionality of natural thermodynamic processes. It can be stated in several equivalent forms. The *engine* statement is that no cyclic process can convert heat completely into work. The *refrigerator* statement is that no cyclic process can transfer heat from a colder place to a hotter place with no input of mechanical work.

- The Carnot cycle operates between two heat reservoirs at temperatures T_H and T_C and uses only reversible processes. Its thermal efficiency is

$$e_{\text{Carnot}} = 1 - \frac{T_C}{T_H} = \frac{T_H - T_C}{T_H}. \qquad (18\text{--}14)$$

Two additional equivalent statements of the second law are that no engine operating between the same two temperatures can be more efficient than a Carnot engine and that all Carnot engines operating between the same two temperatures have the same efficiency.

- A Carnot engine run backward is a Carnot refrigerator. Its coefficient of performance is

$$K_{\text{Carnot}} = \frac{T_C}{T_H - T_C}. \qquad (18\text{--}15)$$

Another form of the second law is that no refrigerator operating between the same two temperatures can have a larger coefficient of performance than a Carnot refrigerator. All Carnot refrigerators operating between the same two temperatures have the same coefficient of performance.

KEY TERMS

irreversible process, 559
reversible process, 559
equilibrium process, 560
heat engine, 560
working substance, 560
cyclic process, 561
thermal efficiency, 561
compression ratio, 563
Otto cycle, 563
Diesel cycle, 565
refrigerator, 565
coefficient of performance, 565
energy efficiency rating, 566
heat pump, 567
second law of thermodynamics, 567
Carnot cycle, 569
thermal pollution, 573
Kelvin temperature scale, 574
absolute zero, 574
entropy, 575
macroscopic state, 579
microscopic state, 580

- The Kelvin temperature scale is based on the efficiency of the Carnot cycle and is independent of the properties of any specific material. The zero point on the Kelvin scale is called absolute zero.

- Entropy is a quantitative measure of the disorder of a system. The entropy change in any reversible process is

$$\Delta S = \int_1^2 \frac{dQ}{T} \qquad \text{(reversible process)}, \qquad (18\text{--}19)$$

where T is absolute temperature. Entropy depends only on the state of the system, and the change in entropy between given initial and final states is the same for all processes leading from one to the other. This fact can be used to find the entropy change in an irreversible process.

- The second law can be stated in terms of entropy S: The entropy of an isolated system may increase but can never decrease. When a system interacts with its surroundings, the total entropy change of system and surroundings can never decrease. When the interaction involves only reversible processes, the total entropy is constant and $\Delta S = 0$; when there is any irreversible process, the total entropy increases and $\Delta S > 0$.

- When a system is in a particular macroscopic state, the particles that make up the system may be in any of w possible microscopic states. The greater the number w, the greater the entropy:

$$S = k \ln w. \qquad (18\text{--}22)$$

DISCUSSION QUESTIONS

Q18-1 A pot is half-filled with water, and a lid is placed on the pot, forming a tight seal so that no water vapor can escape. The pot is heated on a stove, forming water vapor inside the pot. The heat is then turned off and the water vapor condenses back to liquid. Is this cycle reversible or irreversible? Why?

Q18-2 Give two examples of reversible processes and two examples of irreversible processes in purely mechanical systems, such as blocks sliding on planes, springs, pulleys, and strings. Explain what makes each process reversible or irreversible.

Q18-3 What irreversible processes occur in a gasoline engine? Why are they irreversible?

Q18-4 Suppose you try to cool the kitchen of your house by leaving the refrigerator door open. What happens? Why? Would the result be the same if you left open a picnic cooler full of ice? Explain the reason for any differences.

Q18-5 A member of the U.S. Congress proposed a scheme to produce energy as follows. Water molecules (H_2O) are to be broken apart to produce hydrogen and oxygen. The hydrogen is then burned (that is, combined with oxygen), releasing energy in the process. The only product of this combustion is water, so there is no pollution. In light of the second law of thermodynamics, what do you think of this energy-producing scheme?

Q18-6 Some critics of biological evolution claim that it violates the second law of thermodynamics, since evolution involves simple life forms developing into more complex, more highly ordered organisms. Explain why this is not a valid argument against evolution.

Q18-7 Why must a room air conditioner be placed in a window? Why can't it just be set on the floor and plugged in?

Q18-8 Is it a violation of the second law of thermodynamics to convert mechanical energy completely into heat? To convert heat completely into work? Explain your answers.

Q18-9 A growing plant creates a highly complex and organized structure out of simple materials such as air, water, and trace minerals. Does this violate the second law of thermodynamics? Why or why not? What is the plant's ultimate source of energy? Explain your reasoning.

Q18-10 Imagine a special air filter placed in a window of a house. The tiny holes in the filter only allow air molecules moving faster than a certain speed to exit the house, and only allows air molecules moving slower than that speed to enter the house from outside. Explain why such an air filter would cool the house, and why the second law of thermodynamics makes building such a filter an impossible task.

Q18-11 An electric motor has its shaft coupled to that of an electric generator. The motor drives the generator, and some current from the generator is used to run the motor. The excess current is used to light a home. What is wrong with this scheme?

Q18-12 When the sun shines on a glass-roofed greenhouse, the temperature becomes higher inside than outside. Does this phenomenon violate the second law of thermodynamics? Why or why not?

Q18-13 When a wet cloth is hung up in a hot wind in the desert, it is cooled by evaporation to a temperature that may be

20 C° or so below that of the air. Discuss this process in light of the second law of thermodynamics.

Q18–14 If no real engine can be as efficient as a Carnot engine operating between the same two temperatures, what is the point of developing and using Eq. (18–14)?

Q18–15 Suppose you want to increase the efficiency of a heat engine. Would it be better to increase T_H or to decrease T_C by an equal amount? Why?

Q18–16 What would be the efficiency of a Carnot engine operating with $T_H = T_C$? What would be the efficiency if $T_C = 0$ K and T_H were any temperature above 0 K? Interpret your answers.

Q18–17 Real heat engines, like the gasoline engine in a car, always have some friction between their moving parts, although lubricants keep the friction to a minimum. Would a heat engine with completely frictionless parts be 100% efficient? Why or why not? Does the answer depend on whether or not the engine runs on the Carnot cycle? Again, why or why not?

Q18–18 Does a refrigerator full of food consume more power if the room temperature is 20°C than if it is 15°C?

Or is the power consumption the same? Explain your reasoning.

Q18–19 Materials scientists are developing ceramics that are as strong as metals but can withstand much higher temperatures before melting. Explain the usefulness of such ceramics in engine design.

*Q18–20 Explain why each of the following processes are examples of increasing disorder or randomness: mixing hot and cold water; free expansion of a gas; irreversible heat flow; and developing heat by mechanical friction. Are entropy increases involved in all of these? Why or why not?

*Q18–21 Are the earth and sun in thermal equilibrium? Are there entropy changes associated with the transmission of energy from the sun to the earth? Does radiation differ from other modes of heat transfer with respect to entropy changes? Explain your reasoning.

*Q18–22 Discuss the entropy changes involved in the preparation and consumption of a hot-fudge sundae.

EXERCISES

SECTION 18–3 HEAT ENGINES

18–1 A diesel engine performs 2200 J of mechanical work and discards 4300 J of heat each cycle. a) How much heat must be supplied to the engine in each cycle? b) What is the thermal efficiency of the engine?

18–2 An aircraft engine takes in 9000 J of heat and discards 6400 J each cycle. a) What is the mechanical work output of the engine during one cycle? b) What is the thermal efficiency of the engine?

18–3 A Gasoline Engine. A gasoline engine takes in 16,100 J of heat and delivers 3700 J of work per cycle. The heat is obtained by burning gasoline with a heat of combustion of 4.60×10^4 J/g. a) What is the thermal efficiency? b) How much heat is discarded in each cycle? c) What mass of fuel is burned in each cycle? d) If the engine goes through 60.0 cycles per second, what is its power output in kilowatts? In horsepower?

18–4 A gasoline engine has a power output of 180 kW (about 241 hp). Its thermal efficiency is 28.0%. a) How much heat must be supplied to the engine per second? b) How much heat is discarded by the engine per second?

18–5 A certain nuclear-power plant has a mechanical-power output (used to drive an electric generator) of 330 MW. Its rate of heat input from the nuclear reactor is 1300 MW. a) What is the thermal efficiency of the system? b) At what rate is heat discarded by the system?

SECTION 18–4 INTERNAL-COMBUSTION ENGINES

18–6 What compression ratio r must an Otto cycle have to achieve an ideal efficiency of 65.0% if $\gamma = 1.40$?

18–7 For an Otto cycle with $\gamma = 1.40$ and $r = 9.50$, the temperature of the gasoline-air mixture when it enters the cylinder is 22.0°C (point a of Fig. 18–3). a) What is the temperature at the end of the compression stroke (point b)? b) The initial pressure of the gasoline-air mixture (point a) is 8.50×10^4 Pa, slightly

below atmospheric pressure. What is the pressure at the end of the compression stroke?

18–8 The Otto-cycle engine in a Mercedes-Benz SLK230 has a compression ratio of 8.8. a) What is the ideal efficiency of the engine? Use $\gamma = 1.40$. b) The engine in a Dodge Viper GT2 has a slightly higher compression ratio of 9.6. How much increase in the ideal efficiency results from this increase in the compression ratio?

SECTION 18–5 REFRIGERATORS

18–9 A refrigerator has a coefficient of performance of 2.10. Each cycle it absorbs 3.40×10^4 J of heat from the cold reservoir. a) How much mechanical energy is required each cycle to operate the refrigerator? b) During each cycle, how much heat is discarded to the high-temperature reservoir?

18–10 Liquid refrigerant at a pressure of 1.34×10^5 Pa leaves the expansion valve of a refrigerator at −23.0°C. It then flows through the evaporation coils inside the refrigerator and leaves as vapor at the same pressure and at −20.5°C, the same temperature as the inside of the refrigerator. The boiling point of the refrigerant at this pressure is −23.0°C, the heat of vaporization is 1.60×10^5 J/kg, and the specific heat capacity of the vapor at constant pressure is 485 J/kg · K. The coefficient of performance of the refrigerator is $K = 2.8$. If 8.00 kg of refrigerant flows through the refrigerator each hour, find the electric power that must be supplied to the refrigerator.

18–11 A window air conditioner unit absorbs 9.80×10^4 J of heat per minute from the room being cooled and in the same time period deposits 1.44×10^5 J of heat into the outside air. a) What is the power consumption of the unit in watts? b) What is the energy efficiency rating of the unit?

18–12 A freezer has a coefficient of performance of 2.40. The freezer is to convert 1.80 kg of water at 25.0°C to 1.80 kg of ice at −5.0°C in one hour. a) What amount of heat must be removed from the water at 25.0°C to convert it to ice at –5.0°C? b) How much electrical energy is consumed by the freezer during this hour? c) How much wasted heat is rejected to the room in which the freezer sits?

SECTION 18–7 THE CARNOT CYCLE

18–13 A Carnot engine whose high-temperature reservoir is at 620 K takes in 550 J of heat at this temperature in each cycle and gives up 335 J to the low-temperature reservoir. a) How much mechanical work does the engine perform during each cycle? b) What is the temperature of the low-temperature reservoir? c) What is the thermal efficiency of the cycle?

18–14 A Carnot engine is operated between two heat reservoirs at temperatures of 520 K and 300 K. a) If the engine receives 6.45 kJ of heat energy from the reservoir at 520 K in each cycle, how many joules per cycle does it reject to the reservoir at 300 K? b) How much mechanical work is performed by the engine during each cycle? c) What is the thermal efficiency of the engine?

18–15 An ice-making machine operates in a Carnot cycle. It takes heat from water at 0.0°C and rejects heat to a room at 24.0°C. Suppose that 85.0 kg of water at 0.0°C are converted to ice at 0.0°C. a) How much heat is rejected to the room? b) How much energy must be supplied to the device?

18–16 A Carnot refrigerator is operated between two heat reservoirs at temperatures of 320 K and 270 K. a) If in each cycle the refrigerator receives 415 J of heat energy from the reservoir at 270 K, how many joules of heat energy does it deliver to the reservoir at 320 K? b) If the refrigerator goes through 165 cycles each minute, what power input is required to operate the refrigerator? c) What is the coefficient of performance of the refrigerator?

18–17 A Carnot device extracts 5.00 kJ of heat from a body at –10.0°C. How much work is done if the device exhausts heat into the environment at a) 25.0°C; b) 0.0°C; c) –25.0°C? In each case, is the device acting as an engine or as a refrigerator?

18–18 An inventor claims to have developed an engine that in each cycle takes in 2.60×10^8 J of heat at a temperature of 400 K, does 42.0 kWh of mechanical work, and rejects heat at a temperature of 250 K. Would you advise investing money to put this engine on the market? Why or why not?

18–19 a) Show that the efficiency e of a Carnot engine and the coefficient of performance K of a Carnot refrigerator are related by $K = (1 - e)/e$. The engine and refrigerator operate between the same hot and cold reservoirs. b) What is K for the limiting values $e \to 1$ and $e \to 0$? Explain.

*SECTION 18–9 ENTROPY

*18–20 A sophomore with nothing better to do adds heat to 0.350 kg of ice at 0.0°C until it is all melted. a) What is the change in entropy of the water? b) The source of heat is a very massive body at a temperature of 25.0°C. What is the change in entropy of this body? c) What is the total change in entropy of the water and the heat source?

*18–21 Calculate the entropy change that occurs when 1.00 kg of water at 20.0°C is mixed with 2.00 kg of water at 80.0°C.

*18–22 Three moles of an ideal gas undergo a reversible isothermal compression at 20.0°C. During this compression, 1850 J of work is done on the gas. What is the change of entropy of the gas?

*18–23 What is the entropy change of 0.130 kg of helium gas at the normal boiling point of helium when it all condenses isothermally to 1.00 L of liquid helium? (*Hint:* See Table 15–4 in Section 15–7.)

*18–24 a) Calculate the change in entropy when 1.00 kg of water at 100°C is vaporized and converted to steam at 100°C. (See Table 15–4.) b) Compare your answer to the change in entropy when 1.00 kg of ice is melted at 0°C, calculated in Example 18–5 (Section 18–9). Is the entropy change greater for melting or for vaporization? Interpret your answer, using the idea that entropy is a measure of the randomness of a system.

*18–25 a) Calculate the change in entropy when 1.00 mol of water (molecular mass 18.0 g/mol) at 100°C evaporates to form water vapor at 100°C. b) Repeat the calculation of part (a) for 1.00 mol of liquid nitrogen, 1.00 mol of silver, and 1.00 mol of mercury when each is vaporized at its normal boiling point. (See Table 15–4 for the heats of vaporization, and Appendix D for the molar masses. Note that the nitrogen molecule is N_2.) c) Your results in parts (a) and (b) should be in relatively close agreement. (This is called the *rule of Drepez and Trouton.*) Explain why this should be so, using the idea that entropy is a measure of the randomness of a system.

*18–26 A block of copper with a mass of 3.50 kg, initially at 100.0°C, is dropped into 0.800 kg of water initially at 0.0°C. a) What is the final temperature of the system? b) What is the total change in entropy of the system?

*18–27 Two moles of an ideal gas undergo a reversible isothermal expansion from 0.0280 m³ to 0.0420 m³ at a temperature of 25.0°C. What is the change in entropy of the gas?

*SECTION 18–10 MICROSCOPIC INTERPRETATION OF ENTROPY

*18–28 A box is separated by a partition into two parts of equal volume. The left side of the box contains 500 molecules of nitrogen gas; the right side contains 100 molecules of oxygen gas. The two gases are at the same temperature. The partition is punctured, and equilibrium is eventually attained. Assume that the volume of the box is large enough for each gas to undergo a free expansion and not change temperature. a) On average, how many molecules of each type will there be in either half of the box? b) What is the change in entropy of the system when the partition is punctured? c) What is the probability that the molecules will be found in the same distribution as before the partition was punctured, that is, 500 nitrogen molecules in the left half and 100 oxygen molecules in the right half?

*18–29 Two moles of an ideal gas occupies a volume V. The gas expands isothermally and reversibly to a volume $3V$. a) Is the velocity distribution changed by the isothermal expansion? Explain. b) Use Eq. (18–23) to calculate the change in entropy of the gas. c) Use Eq. (18–18) to calculate the change in entropy of the gas. Compare this result to that obtained in part (b).

*18–30 You toss four identical coins on the floor. Each coin has an equal probability of showing heads or tails. a) What is the probability of all four coins showing heads? Of all four showing tails? b) What is the probability of three coins showing heads and one tails? What is the probability of three coins showing tails and one heads? c) What is the probability of two coins showing heads and two showing tails? d) What is the sum of the five probabilities calculated in parts (a) through (c)? Explain.

SECTION 18–11 ENERGY RESOURCES: A CASE STUDY IN THERMODYNAMICS

18–31 Solar Heating in Winter. A well-insulated house in Columbus, Ohio, has 150 m^2 of floor space and requires 1.50×10^{10} J of heat for the month of January. This heat is to be supplied by a solar collector with a collection efficiency of 60.0%. The average (night and day) solar energy input in Columbus in January is 65.7 W/m^2. What area of solar collector is required? Would the collector fit on the roof of the house?

18–32 a) A home owner in a cold climate has a coal-burning furnace that burns 4500 kg (about 5 tons) of coal during a winter. The coal used has a heat of combustion of 2.70×10^7 J/kg. If stack losses (the amount of heat energy lost up the chimney) are 20%, how many joules were actually used to heat the house? b) The home owner proposes to install a solar heating system, heating large tanks of water by solar radiation during the summer and using the stored energy for heating during the winter. Find the required dimensions of the storage tank to store a quantity of energy equal to that computed in part (a). Assume that the tank is a cube and that the water is raised to 49.0°C (120°F) in the summer and cooled to 27.0°C (81°F) in the winter.

18–33 The roof of a suburban house is equipped with 8.0 m^2 of passive solar collecting panels with collection efficiency 60%, used to heat water from 15.0°C to 55.0°C for household use. a) If the average solar-energy input is 150 W/m^2, what volume of water can be heated per hour? b) During an average day, a typical household uses about 75 L of 55.0°C hot water for each person. How many inhabitants can this water heating system accommodate?

18–34 A solar-power plant is to be built with a power output capacity of 850 MW. What land area must the solar-energy collectors occupy if they are a) photocells with 60% efficiency? b) mirrors that generate steam for a turbine-generator unit with overall efficiency of 30%? Take the average power in the sun's radiation to be 200 W/m^2 at the earth's surface. Express your answers in square kilometers and square miles.

18–35 A "solar house" has storage facilities for 4.00×10^9 J (about 4 million Btu). Compare the space requirements (in m^3) for this storage on the assumption a) that the energy is stored in water heated from a minimum temperature of 21.0°C (70°F) to a maximum temperature of 49.0°C (120°F); b) that the energy is stored in Glauber salt heated in the same temperature range.

PROPERTIES OF GLAUBER SALT ($Na_2SO_4 \cdot 10H_2O$)

Specific heat capacity	
Solid	1930 J/kg · K
Liquid	2850 J/kg · K
Density	1600 kg/m^3
Melting point	32.0°C
Heat of fusion	2.42×10^5 J/kg

c) What is the advantage of using Glauber salt?

18–36 A coal-fired steam-turbine power plant has a mechanical-power output of 1100 MW and a thermal efficiency of 35.0%. a) At what rate must heat be supplied by burning coal? b) If West Virginia coal is used, which has a heat of combustion of 3.00×10^4 J/g, what mass of coal is burned per second? Per day? c) At what rate is heat discarded by the system? d) If the discarded heat is given to water in a river, and the water temperature is to rise by no more than 4.0 C°, what volume of water is needed per second? e) In part (d), if the river is of rectangular cross section with depth 5.0 m and width 100 m, what must be the speed of flow of the water?

18–37 Electric vs. Conventional Automobiles. a) Consider an energy-conversion process that involves two steps, such as using heat to boil water and using the steam to turn an electric generator. Each step has its own efficiency. Is the efficiency of the overall process equal to the sum of the two efficiencies, or the difference, or the product, or what? Explain your reasoning. b) Conventional automobiles are about 15% efficient; that is, only about 15% of the energy released by burning fuel is eventually converted to kinetic energy of the automobile. In an electric automobile with a battery-powered motor, the energy comes from a power plant that supplies the electricity to charge the batteries. Calculate the efficiency of an electric automobile using the following data: i) A typical electric power plant has an efficiency of 40%; ii) 10% of the energy supplied by the power plant is lost in transmission to the city; iii) 10% of the energy received by the city is lost in distribution within the city; iv) the efficiency of the battery-charging process is 80%; v) the efficiency of the electric motor is 90%. c) How does the overall efficiency compare to that of a conventional automobile?

PROBLEMS

18–38 A Carnot engine whose low-temperature reservoir is at −90.0°C has an efficiency of 40.0%. An engineer is assigned the problem of increasing this to 45.0%. a) By how many Celsius degrees must the temperature of the high-temperature reservoir be increased if the temperature of the low-temperature reservoir remains constant? b) By how many Celsius degrees must the temperature of the low-temperature reservoir be decreased if that of the high-temperature reservoir remains constant?

18–39 A heat engine takes 0.350 mol of a diatomic ideal gas around the cycle shown in the pV-diagram of Fig. 18–19. Process $1 \rightarrow 2$ is at constant volume, process $2 \rightarrow 3$ is adiabatic, and process $3 \rightarrow 1$ is at a constant pressure of 1.00 atm. The value of γ for this gas is 1.40. a) Find the pressure and volume at points 1, 2, and 3. b) Calculate Q, W, and ΔU for each of the three processes. c) Find the net work done by the gas in the cycle. d) Find the net heat flow into the engine in one cycle. e) What is the thermal efficiency of the engine? How does this compare to the efficiency of a Carnot-cycle engine operating between the same minimum and maximum temperatures T_1 and T_2?

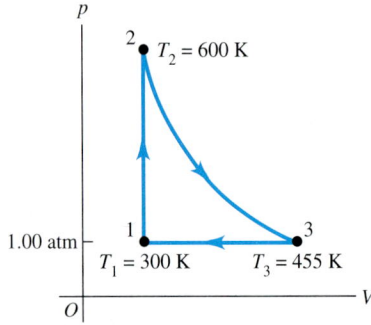

FIGURE **18–19** Problem 18–39.

18–40 An experimental power plant at the Natural Energy Laboratory of Hawaii generates electricity from the temperature gradient of the ocean. The surface and deep-water temperatures are 27°C and 6°C, respectively. a) What is the maximum theoretical efficiency of this power plant? b) If the power plant is to produce 210 kW of power, at what rate must heat be extracted from the warm water? At what rate must heat be absorbed by the cold water? Assume the maximum theoretical efficiency. c) The cold water that enters the plant leaves at a temperature of 10°C. What must be the flow rate of cold water through the system? Give your answer in kg/h and L/h.

18–41 What is the thermal efficiency of an engine that operates by taking n moles of diatomic ideal gas through the cycle $1 \rightarrow 2 \rightarrow 3 \rightarrow 4 \rightarrow 1$ shown in Fig. 18–20?

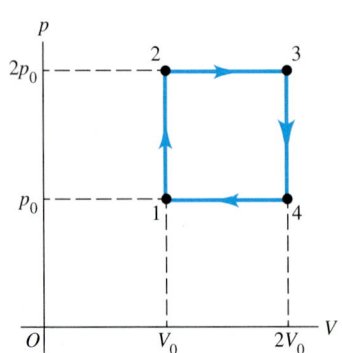

FIGURE **18–20** Problem 18–41.

18–42 A cylinder contains oxygen at a pressure of 2.00 atm. The volume is 4.00 L, and the temperature is 300 K. Assume that the oxygen may be treated as an ideal gas. The oxygen is carried through the following processes:

(i) Heated at constant pressure from the initial state (state 1) to state 2, which has $T = 450$ K.
(ii) Cooled at constant volume to 250 K (state 3).
(iii) Compressed at constant temperature to a volume of 4.00 L (state 4).
(iv) Heated at constant volume to 300 K, which takes the system back to state 1.

a) Show these four processes in a pV-diagram, giving the numerical values of p and V in each of the four states. b) Calculate Q and W for each of the four processes. c) Calculate the net work done by the oxygen. d) What is the efficiency of this device as a heat engine? How does this compare to the efficiency of a Carnot-cycle engine operating between the same minimum and maximum temperatures of 250 K and 450 K?

18–43 Thermodynamic Processes for a Refrigerator. A refrigerator operates on the cycle shown in Fig. 18–21. The compression ($d \rightarrow a$) and expansion ($b \rightarrow c$) steps are adiabatic. The temperature, pressure, and volume of the coolant in each of the four states a, b, c, and d are given in the table below.

State	T (°C)	P (kPa)	V (m³)	U (kJ)	Percentage That Is Liquid
a	80	2305	0.0682	1969	0
b	80	2305	0.00946	1171	100
c	5	363	0.2202	1005	54
d	5	363	0.4513	1657	5

a) In each cycle, how much heat is taken from inside the refrigerator into the coolant while the coolant is in the evaporator? b) In each cycle, how much heat is exhausted from the coolant into the air outside the refrigerator while the coolant is in the condenser? c) In each cycle, how much work is done by the motor that operates the compressor? d) Calculate the coefficient of performance of the refrigerator.

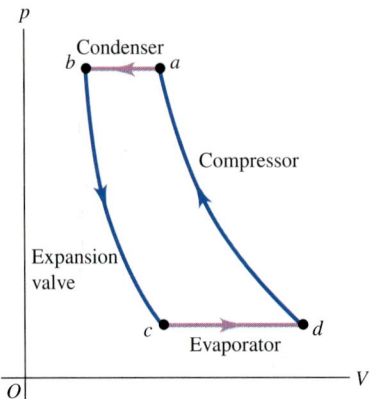

FIGURE **18–21** Problem 18–43.

18–44 A monatomic ideal gas is taken around the cycle shown in Fig. 18–22 in the direction shown in the figure. The path for process $c \to a$ is a straight line in the pV-diagram. a) Calculate Q, W, and ΔU for each process $a \to b$, $b \to c$, and $c \to a$. b) What are Q, W, and ΔU for one complete cycle? c) What is the efficiency of the cycle?

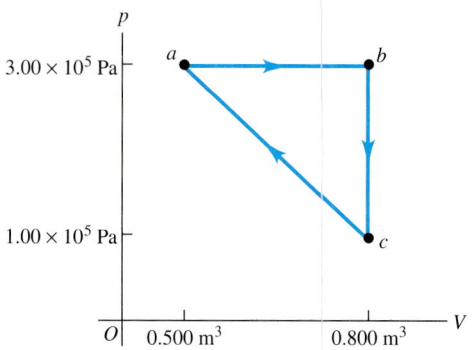

FIGURE 18–22 Problem 18–44.

18–45 A Stirling-cycle Engine. The *Stirling cycle* is similar to the Otto cycle, except that the compression and expansion of the gas is done at constant temperature, not adiabatically as in the Otto cycle. The Stirling cycle is used in *external* combustion engines, which means that the gas inside the cylinder is not used in the combustion process. Heat is supplied by burning fuel steadily outside the cylinder, instead of explosively inside the cylinder as in the Otto cycle. For this reason Stirling-cycle engines are quieter than Otto-cycle engines, since there are no intake and exhaust valves (a major source of engine noise). While small Stirling engines are used for a variety of purposes, Stirling engines for automobiles have not been successful because they are larger, heavier, and more expensive than conventional automobile engines. In the cycle, the working fluid goes through the following sequence of steps (Fig. 18–23):

(i) Compressed isothermally at temperature T_1 from the initial state a to state b, with a compression ratio r.
(ii) Heated at constant volume to state c at temperature T_2.
(iii) Expanded isothermally at T_2 to state d.
(iv) Cooled at constant volume back to the initial state a.

Assume that the working fluid is n moles of an ideal gas (for which C_V is independent of temperature). a) Calculate Q, W, and ΔU for each of the processes $a \to b$, $b \to c$, $c \to d$, and $d \to a$. b) In the Stirling cycle, the heat transfers in the processes $b \to c$ and $d \to a$ do not involve external heat sources, but rather use *regeneration*: the same substance that transfers heat to the gas inside the cylinder in the process $b \to c$ also absorbs heat back from the gas in the process $d \to a$. Hence the heat transfers $Q_{b \to c}$ and $Q_{d \to a}$ do not play a role in determining the efficiency of the engine. Explain this last statement by comparing the expressions for $Q_{b \to c}$ and $Q_{d \to a}$ calculated in part (a). c) Calculate the efficiency of a Stirling-cycle engine in terms of the temperatures T_1 and T_2. How does this compare to the efficiency of a Carnot-cycle engine operating between these same two temperatures? (Historically, the Stirling cycle was devised before the

Carnot cycle.) Does this result violate the second law of thermodynamics? Explain. Unfortunately, actual Stirling-cycle engines cannot achieve this efficiency, due to problems with the heat-transfer processes and pressure losses in the engine.

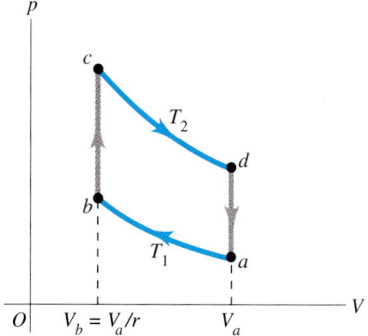

FIGURE 18–23 Problem 18–45.

18–46 A Carnot engine operates between two heat reservoirs at temperatures T_H and T_C. An inventor proposes to increase the efficiency by running one engine between T_H and an intermediate temperature T' and a second engine between T' and T_C using as input the heat expelled by the first engine. Compute the efficiency of this composite system and compare it to that of the original engine.

18–47 The maximum power that can be extracted by a wind turbine from an air stream is approximately

$$P = kd^2v^3,$$

where d is the blade diameter, v is the wind speed, and the constant $k = 0.5 \ W \cdot s^3/m^5$. a) Explain the dependence of P on d and on v by considering a cylinder of air that passes over the turbine blades in time t (Fig. 18–24). This cylinder has diameter d, length $L = vt$, and density ρ. b) The Mod–5B wind turbine at Kahaku on the Hawaiian island of Oahu has a blade diameter of 97 m (slightly longer than a football field) and sits atop a 58-m tower. It can produce 3.2 MW of electric power. Assuming 25% efficiency, what wind speed is required to produce this amount of power? Give your answer in m/s and in km/h. c) Commercial wind turbines are commonly located in or downwind of mountain passes. Why?

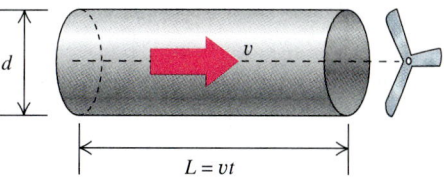

FIGURE 18–24 Problem 18–47.

18–48 Fuel Economy and Automobile Performance. The Otto-cycle engine of a Volvo V70 has a compression ratio $r = 8.5$. The U.S. Environmental Protection Agency (EPA) fuel economy rating of this car is 25 miles per gallon on the highway (105 km/h, or 65 mi/h). Gasoline has a heat of combustion of

4.60×10^7 J/kg, and its density is 740 kg/m³. a) At 105 km/h, what is the rate of gasoline consumption in L/h? b) What is the theoretical efficiency of the engine? Use $\gamma = 1.40$. c) How much power is the engine producing at 105 km/h? Assume that the engine is operating at its theoretical efficiency, and give your answer in watts and in horsepower. By comparison, the engine of the Volvo V70 has a maximum power of 236 hp. d) Due to heat losses and friction in the drive train, the actual efficiency is approximately 15%. Repeat part (c) using this information. What fraction of the maximum possible power is used for highway driving?

18–49 Automotive Thermodynamics. A Volkswagen Passat has a six-cylinder Otto-cycle engine with compression ratio $r = 10.6$. The diameter of each cylinder, called the *bore* of the engine, is 82.5 mm. The distance that the piston moves during the compression in Fig. 18–2, called the *stroke* of the engine, is 86.4 mm. The initial pressure of the air-fuel mixture (at point a in Fig. 18–3) is 8.50×10^4 Pa, and the initial temperature is 300 K (the same as the outside air). Assume that 200 J of heat is added to each cylinder in each cycle by the burning gasoline, and that the gas has $C_V = 20.5$ J/mol · K and $\gamma = 1.40$. a) Calculate the total work done in one cycle in each cylinder of the engine, and the heat released when the gas is cooled to the temperature of the outside air. b) Calculate the volume of the air-fuel mixture at point a in the cycle. c) Calculate the pressure, volume, and temperature of the gas at points b, c, and d in the cycle. In a pV-diagram, show the numerical values of p, V, and T for each of the four states. d) Compare the efficiency of this engine with the efficiency of a Carnot-cycle engine operating between the same maximum and minimum temperatures.

*18–50 Recalculate the entropy change between points a and b of Fig. 18–12c if the reversible path is a) an isobaric expansion to $2V$ followed by an isochoric process; b) an isochoric cooling to $p_a/2$ followed by an isobaric expansion.

*18–51 A 0.0500-kg cube of ice at an initial temperature of $-15.0°C$ is placed in 0.600 kg of water at $T = 45.0°C$ in an insulated container of negligible mass. Calculate the entropy change of the system.

*18–52 a) For the Otto cycle shown in Fig. 18–3, calculate the entropy changes of the gas in each of the constant–volume processes $b \rightarrow c$ and $d \rightarrow a$ in terms of the temperatures T_a, T_b, T_c, and T_d and the number of moles n and the heat capacity C_V of the gas. b) What is the total entropy change in the engine during one cycle? (*Hint:* Use the relation between T_a and T_b and between T_d and T_c.) c) The processes $b \rightarrow c$ and $d \rightarrow a$ occur

irreversibly in a real Otto engine. Explain how can this be reconciled with your result in part (b).

*18–53 A *TS*-Diagram. a) Graph a Carnot cycle, plotting Kelvin temperature vertically and entropy horizontally. This is called a temperature-entropy diagram, or *TS*-diagram. b) Show that the area under any curve representing a reversible path in a temperature-entropy diagram represents the heat absorbed by the system. c) Derive from your diagram the expression for the thermal efficiency of a Carnot cycle. d) Draw a temperature-entropy diagram for the Stirling cycle, described in Problem 18–45. Use this diagram to relate the efficiency of the Carnot and Stirling cycles.

*18–54 A physics student immerses one end of a copper rod in boiling water at 100°C and the other end in an ice-water mixture at 0°C. The sides of the rod are insulated. After steady-state conditions have been achieved in the rod, 0.160 kg of ice melts in a certain time interval. For this time interval find a) the entropy change of the boiling water; b) the entropy change of the ice-water mixture; c) the entropy change of the copper rod; d) the total entropy change of the entire system.

*18–55 To heat one cup of water (250 cm³) to make coffee, you place an electric heating element in the cup. As the water temperature increases from 20°C to 65°C, the temperature of the heating element remains at a constant 120°C. Calculate the entropy change of a) the water; b) the heating element; c) the system of water and heating element. (Make the same assumption about the specific heat of water as in Example 18–10 (Section 18–9), and neglect the heat that flows into the ceramic coffee cup itself.) d) Is this process reversible or irreversible? Explain.

*18–56 An object of mass m_1, specific heat capacity c_1, and temperature T_1 is placed in contact with a second object of mass m_2, specific heat capacity c_2, and temperature $T_2 > T_1$. As a result, the temperature of the first object increases to T and the temperature of the second object decreases to T'. a) Show that the entropy increase of the system is

$$\Delta S = m_1 c_1 \ln \frac{T}{T_1} + m_2 c_2 \ln \frac{T'}{T_2},$$

and show that energy conservation requires that

$$m_1 c_1 (T - T_1) = m_2 c_2 (T_2 - T').$$

b) Show that the entropy change ΔS, considered as a function of T, is a *maximum* if $T = T'$, which is just the condition of thermodynamic equilibrium. c) Discuss the result of part (b) in terms of the idea of entropy as a measure of disorder.

CHALLENGE PROBLEM

18–57 Consider a diesel cycle that starts (at point a in Fig. 18–4) with air at temperature T_a. The air may be treated as an ideal gas. a) If the temperature at point c is T_c, derive an expres-

sion for the efficiency of the cycle in terms of the compression ratio r. b) What is the efficiency if $T_a = 300$ K, $T_c = 950$ K, $\gamma = 1.40$, and $r = 21.0$?

When an earthquake strikes, the news of the event travels through the body of the earth in the form of seismic waves. By studying such waves, geophysicists learn about the internal structure of the earth and where future earthquake activity is likely to occur.

19

Mechanical
Waves

19–1 INTRODUCTION

When you're at the beach, enjoying the ocean surf, you're experiencing a wave motion. Ripples on a pond, musical sounds we can hear, other sounds we *can't* hear, the wiggles of a Slinky™ stretched out on the floor—all these are *wave* phenomena. Waves can occur whenever a system is disturbed from its equilibrium position and when the disturbance can travel or *propagate* from one region of the system to another. Sound, light, ocean waves, radio and television transmission, and earthquakes are all wave phenomena. Waves are important in all branches of physical and biological science; indeed, the wave concept is one of the most important unifying threads running through the entire fabric of the natural sciences.

This chapter and the next two are about mechanical waves, waves that travel within some material called a *medium.* We'll begin by deriving the basic equations for describing waves, including the important special case of *periodic* waves in which the pattern of the wave repeats itself as the wave propagates. Chapter 20 deals with what happens when two or more waves occupy the same space, giving rise to *interference,* and Chapter 21 is concerned with a particularly important type of mechanical wave called *sound.*

Not all waves are mechanical in nature. Another broad class is *electromagnetic* waves, including light, radio waves, infrared and ultraviolet radiation, x-rays, and gamma rays. *No* medium is needed for electromagnetic waves; they can travel through empty space. Yet another class of wave phenomena is the wavelike behavior of atomic and subatomic particles. This behavior forms part of the foundation of quantum mechanics, the basic theory that is used for the analysis of atomic and molecular structure. We will return to electromagnetic waves in later chapters. Meanwhile, we can learn the essential language of waves in the context of mechanical waves.

19–2 TYPES OF MECHANICAL WAVES

A **mechanical wave** is a disturbance that travels through some material or substance called the **medium** for the wave. As the wave travels through the medium, the particles that make up the medium undergo displacements of various kinds, depending on the nature of the wave.

Figure 19–1 shows three varieties of mechanical waves. In Fig. 19–1a the medium is a string or rope under tension. If we give the left end a small upward shake or wiggle, the wiggle travels along the length of the string. Successive sections of string go through the same motion that we gave to the end, but at successively later times. Because the displacements of the medium are perpendicular or *transverse* to the direction of travel of the wave along the medium, this is called a **transverse wave.**

In Fig. 19–1b the medium is a liquid or gas in a tube with a rigid wall at the right end and a movable piston at the left end. If we give the piston a single back-and-forth

19–1 (a) The hand moves the string up and then returns, producing a transverse wave. (b) The piston compresses the liquid or gas to the right, then returns, producing a longitudinal wave. (c) The board pushes to the right, then returns, producing a sum of transverse and longitudinal waves. In all three cases the solitary wave propagates to the right.

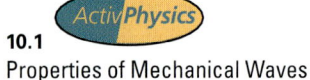

10.1

Properties of Mechanical Waves

motion, displacement and pressure fluctuations travel down the length of the medium. This time the motions of the particles of the medium are back and forth along the *same* direction that the wave travels, and we call this a **longitudinal wave.**

In Fig. 19–1c the medium is water in a channel, such as an irrigation ditch or canal. When we move the flat board at the left end forward and back once, a wave disturbance travels down the length of the channel. In this case the displacements of the water have *both* longitudinal and transverse components.

Each of these systems has an equilibrium state. For the stretched string it is the state in which the system is at rest, stretched out along a straight line. For the fluid in a tube it is a state in which the fluid is at rest with uniform pressure, and for the water in a trough it is a smooth, level water surface. In each case the wave motion is a disturbance from the equilibrium state that travels from one region of the medium to another. In each

19–2 The block of mass m is attached to a spring and undergoes simple harmonic motion, producing a sinusoidal wave that travels to the right on the string. In a real-life system a driving force would have to be applied to the mass m to replace the energy carried away by the wave.

case there are forces that tend to restore the system to its equilibrium position when it is displaced, just as the force of gravity tends to pull a pendulum toward its straight-down equilibrium position when it is displaced.

These examples have three things in common. First, in each case the disturbance travels or *propagates* with a definite speed through the medium. This speed is called the speed of propagation, or simply the **wave speed.** It is determined in each case by the mechanical properties of the medium. We will use the symbol v for wave speed. (The wave speed is *not* the same as the speed with which particles move when they are disturbed by the wave. We'll return to this point in Section 19–3.) Second, the medium itself does not travel through space; its individual particles undergo back-and-forth or up-and-down motions around their equilibrium positions. The overall pattern of the wave disturbance is what travels. Third, to set any of these systems into motion, we have to put in energy by doing mechanical work on the system. The wave motion transports this energy from one region of the medium to another. *Waves transport energy, but not matter, from one region to another.*

19-3 PERIODIC WAVES

The transverse wave on a stretched string in Fig. 19–1a is an example of a *wave pulse.* The hand shakes the string up and down just once, exerting a transverse force on it as it does so. The result is a single "wiggle," or pulse, that travels along the length of the string. The tension in the string restores its straight-line shape once the pulse has passed.

A more interesting situation develops when we give the free end of the string a repetitive, or *periodic,* motion. (You may want to review the discussion of periodic motion in Chapter 13 before going ahead.) Then each particle in the string will also undergo periodic motion as the wave propagates, and we have a **periodic wave.** In particular, suppose we move the string up and down in *simple harmonic motion* (SHM) with amplitude A, frequency f, angular frequency $\omega = 2\pi f$, and period $T = 1/f = 2\pi/\omega$. A possible experimental setup is shown in Fig. 19–2. As we will see, periodic waves with simple harmonic motion are particularly easy to analyze; we call them **sinusoidal waves.** It also turns out that *any* periodic wave can be represented as a combination of sinusoidal waves. So this particular kind of wave motion is worth special attention.

In Fig. 19–2 the wave that advances along the string is a *continuous succession* of transverse sinusoidal disturbances. Figure 19–3 shows the shape of a part of the string near the left end at time intervals of $\frac{1}{8}$ of a period, for a total time of one period. The wave shape advances steadily toward the right, as indicated by the short red arrow pointing to a particular wave crest. As the wave moves, any point on the string (the blue dot, for example) oscillates up and down about its equilibrium position with simple harmonic motion. *When a sinusoidal wave passes through a medium, every particle in the medium undergoes simple harmonic motion with the same frequency.*

CAUTION ▶ Be very careful to distinguish between the motion of the *transverse wave* along the string and the motion of a *particle* of the string. The wave moves with constant speed v *along* the length of the string, while the motion of the particle is simple harmonic and *transverse* (perpendicular) to the length of the string. ◀

For a periodic wave, the shape of the string at any instant is a repeating pattern. The length of one complete wave pattern is the distance from one crest to the next, or from one trough to the next, or from any point to the corresponding point on the next repetition of the wave shape. We call this distance the **wavelength** of the wave, denoted by λ (the Greek letter "lambda"). The wave pattern travels with constant speed v and advances a distance of one wavelength λ in a time interval of one period T. So the wave speed v is given by $v = \lambda/T$, or, because $f = 1/T$,

$$v = \lambda f \quad \text{(periodic wave)}. \tag{19-1}$$

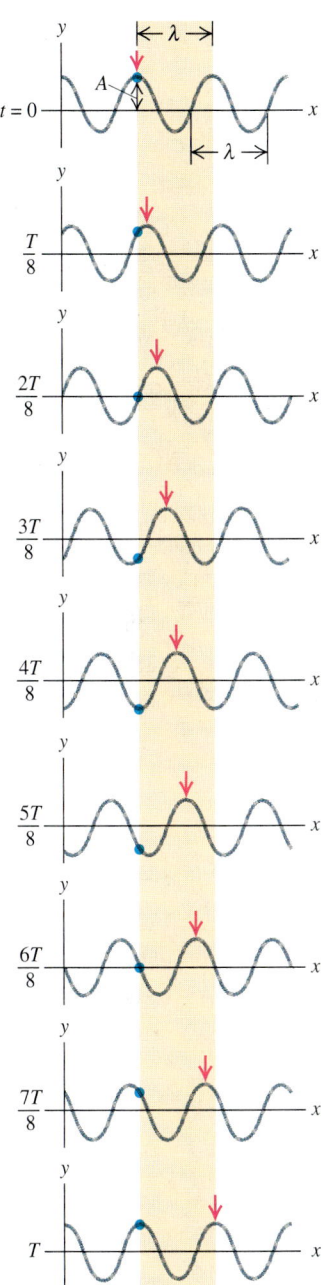

19-3 A sinusoidal transverse wave traveling to the right along a string. The shape of the string is shown at intervals of $\frac{1}{8}$ of a period; the vertical scale is exaggerated. All points on the string move up and down in SHM with amplitude A. The wave travels a distance of one wavelength λ in one period T. Any two particles in the string whose x-coordinates differ by λ, such as the blue dot at $x = \lambda$ and a particle at $x = 0$, oscillate in phase with each other.

The speed of propagation equals the product of wavelength and frequency. The frequency is a property of the *entire* periodic wave because all points on the string oscillate with the same frequency f.

In many important situations the wave speed v is determined entirely by the mechanical properties of the medium. In this case, increasing f causes λ to decrease so that the product $v = \lambda f$ remains the same, and waves of *all* frequencies propagate with the same wave speed. In this chapter we will consider *only* waves of this kind. (In later chapters we will study the propagation of light waves in matter for which the wave speed depends on frequency; this turns out to be the reason why prisms break white light into a spectrum and why raindrops create a rainbow.)

To understand the mechanics of a periodic *longitudinal* wave, we consider a long tube filled with a fluid, with a piston at the left end as in Fig. 19–1b. If we push the piston in, we compress the fluid near the piston, increasing the pressure in this region. This region then pushes against the neighboring region of fluid, and so on, and a wave pulse moves along the tube.

Now suppose we move the piston back and forth with simple harmonic motion along a line parallel to the axis of the tube (Fig. 19–4). This motion forms regions in the fluid where the pressure and density are greater or less than the equilibrium values. We call a region of increased pressure a *compression.* In the figure, compressions are represented by darkly shaded areas. A region of reduced pressure is an *expansion;* in the figure, expansions are represented by lightly shaded areas. The short red arrow shows the position of one particular compression; the compressions and expansions move to the right with constant speed v.

The motion of a single particle of the medium, such as the one shown by a blue dot in Fig. 19–4, is simple harmonic motion parallel to the direction of wave propagation. The wavelength is the distance from one compression to the next or from one expansion to the next. The fundamental equation $v = \lambda f$ holds for longitudinal waves as well as for transverse waves and indeed for *all* types of periodic waves. As for transverse waves, in this chapter we will consider only situations in which the speed of longitudinal waves does not depend on the frequency.

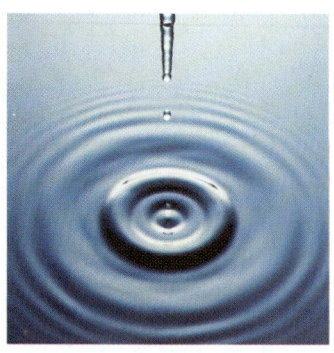

A series of drops falling vertically into water produces a periodic wave that spreads radially outward from the source. The wave crests and troughs are concentric circles. The distance along a radial line from one crest to the next, or from one trough to the next, is the wavelength.

EXAMPLE 19–1

Sound waves are longitudinal waves in air. The speed of sound depends on temperature; at 20°C it is 344 m/s (1130 ft/s). What is the wavelength of a sound wave in air at 20°C if the frequency is $f = 262$ Hz (the approximate frequency of middle C on a piano)?

SOLUTION From Eq. (19–1),

$$\lambda = \frac{v}{f} = \frac{344 \text{ m/s}}{262 \text{ s}^{-1}} = 1.31 \text{ m}.$$

The "high C" sung by coloratura sopranos (also designated C_6) is two octaves above middle C (designated C_4). Each octave corresponds to a factor of two in frequency, so the frequency of high C is four times that of middle C, $f = 4(262 \text{ Hz}) = 1048$ Hz. The speed of sound waves is unaffected by changes in frequency, so the wavelength is one-fourth as large, $\lambda = (1.31 \text{ m})/4 = 0.328$ m.

19–4 MATHEMATICAL DESCRIPTION OF A WAVE

Many characteristics of periodic waves can be described by using the concepts of wave speed, period, frequency, and wavelength. Often, though, we need a more detailed description of the positions and motions of individual particles of the medium at particular times during wave propagation. For this description we need the concept of a *wave function,* a function that describes the position of any particle in the medium at any time. We will concentrate on *sinusoidal* waves, in which each particle undergoes simple harmonic motion about its equilibrium position.

As a specific example, let's look at waves on a stretched string. If we ignore the sag of the string due to gravity, the equilibrium position of the string is along a straight line. We take this to be the x-axis of a coordinate system. Waves on a string are *transverse;* during wave motion a particle with equilibrium position x is displaced some distance y in the direction perpendicular to the x-axis. The value of y depends on which particle we are talking about (that is, y depends on x) and also on the time t when we look at it. Thus y is a *function* of x and t; $y = y(x, t)$. We call $y(x, t)$ the **wave function** that describes the wave. If we know this function for a particular wave motion, we can use it to find the displacement (from equilibrium) of any particle at any time. From this we can find the velocity and acceleration of any particle, the shape of the string, and anything else we want to know about the behavior of the string at any time.

WAVE FUNCTION FOR A SINUSOIDAL WAVE

Let's see how to determine the form of the wave function for a sinusoidal wave. Suppose a sinusoidal wave travels from left to right (the direction of increasing x) along the string, as in Fig. 19–3. Every particle of the string oscillates with simple harmonic motion with the same amplitude and frequency. But the oscillations of particles at different points on the string are *not* all in step with each other. The particle marked by the blue dot in Fig. 19–3 is at its maximum positive value of y at $t = 0$ and returns to $y = 0$ at $t = 2T/8$; these same events occur for a particle at the *center* of the colored band at $t = 4T/8$ and $t = 6T/8$, exactly one half-period later. For any two particles of the string, the motion of the particle on the right (in terms of the wave, the "downstream" particle) lags behind the motion of the particle on the left by an amount proportional to the distance between the particles.

Hence the cyclic motions of various points on the string are out of step with each other by various fractions of a cycle. We call these differences *phase differences,* and we say that the *phase* of the motion is different for different points. For example, if one point has its maximum positive displacement at the same time that another has its maximum negative displacement, the two are a half-cycle out of phase. (This is the case for the blue dot in Fig. 19–3 and a point at the center of the colored band.)

Suppose that the displacement of a particle at the left end of the string ($x = 0$), where the wave originates, is given by

$$y(x = 0, t) = A \sin \omega t = A \sin 2\pi f t. \qquad (19-2)$$

That is, the particle oscillates in simple harmonic motion with amplitude A, frequency f, and angular frequency $\omega = 2\pi f$. The notation $y(x = 0, t)$ reminds us that the motion of this particle is a special case of the wave function $y(x, t)$ that describes the entire wave. At $t = 0$ the particle at $x = 0$ has zero displacement ($y = 0$) and is moving in the $+y$-direction (the value of y is increasing with time).

The wave disturbance travels from $x = 0$ to some point x to the right of the origin in an amount of time given by x/v, where v is the wave speed. So the motion of point x at time t is the same as the motion of point $x = 0$ at the earlier time $t - x/v$. Hence we can find the displacement of point x at time t by simply replacing t in Eq. (19–2) by $(t - x/v)$. When we do that, we find the following expression for the wave function:

$$y(x, t) = A \sin \omega \left(t - \frac{x}{v} \right) = A \sin 2\pi f \left(t - \frac{x}{v} \right) \qquad (19-3)$$

(sinusoidal wave moving in $+x$-direction).

The displacement $y(x, t)$ is a function of both the location x of the point and the time t. We could make Eq. (19–3) more general by allowing for different values of the phase angle, as we did for simple harmonic motion in Section 13–3, but for now we omit this.

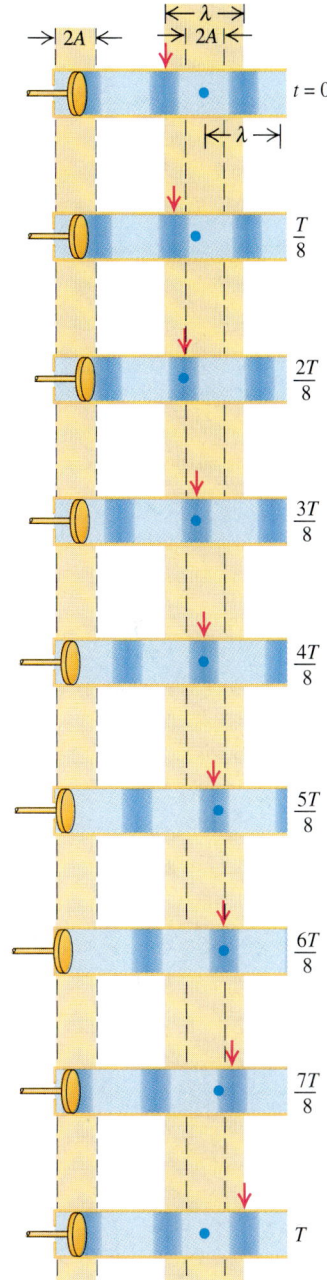

19–4 A sinusoidal longitudinal wave traveling to the right in a fluid. The disturbance of the fluid is shown at intervals of $\frac{1}{8}$ of a period. The piston and all points in the fluid move back and forth in SHM with amplitude A. The wave travels a distance of one wavelength λ in one period T. In equilibrium, the particle shown by the blue dot is $\frac{3}{2}$ wavelengths from the piston; its motion is one half-cycle out of phase with that of the piston.

We can rewrite the wave function given by Eq. (19–3) in several different but useful forms. We can express it in terms of the period $T = 1/f$ and the wavelength $\lambda = v/f$:

$$y(x, t) = A \sin 2\pi \left(\frac{t}{T} - \frac{x}{\lambda} \right)$$

(sinusoidal wave moving in +x-direction). (19–4)

We get another convenient form of the wave function if we define a quantity k, called the **wave number:**

$$k = \frac{2\pi}{\lambda} \qquad \text{(wave number).}$$

(19–5)

Substituting $\lambda = 2\pi/k$ and $f = \omega/2\pi$ into the wavelength-frequency relation $v = \lambda f$ gives

$$\omega = vk \qquad \text{(periodic wave).}$$

(19–6)

We can then rewrite Eq. (19–4) as

$$y(x, t) = A \sin (\omega t - kx) \qquad \text{(sinusoidal wave moving in +x-direction).}$$

(19–7)

Which of these various forms for the wave function $y(x, t)$ we use in any specific problem is a matter of convenience. Note that ω has units rad/s, so for unit consistency in Eqs. (19–6) and (19–7) the wave number k must have the units rad/m. (Some physicists define the wave number as $1/\lambda$ rather than $2\pi/\lambda$. When reading other texts, be sure to determine how this term is defined.)

The wave function $y(x, t)$ is graphed as a function of x for a specific time t in Fig. 19–5a. This graph gives the displacement y of a particle from its equilibrium position as a function of the coordinate x of the particle. If the wave is a transverse wave on a string, the graph in Fig. 19–5a represents the shape of the string at that instant, like a flash photograph of the string. In particular, at time $t = 0$,

$$y(x, t = 0) = A \sin (-kx) = -A \sin kx = -A \sin 2\pi \frac{x}{\lambda}.$$

A graph of the wave function versus time t for a specific coordinate x is shown in Fig. 19–5b. This graph gives the displacement y of the particle at that coordinate as a function of time. That is, it describes the motion of that particle. In particular, at the position $x = 0$,

$$y(x = 0, t) = A \sin \omega t = A \sin 2\pi \frac{t}{T}.$$

This is consistent with our original statement about the motion at $x = 0$, Eq. (19–2).

CAUTION ▶ Make sure that you understand the difference between Figs. 19–5a and 19–5b. In particular, note that Fig. 19–5b is *not* a picture of the shape of the string; it is a graph of the position y of a particle at $x = 0$ as a function of time. ◀

We can modify Eqs. (19–3) through (19–7) to represent a wave traveling in the *negative* x-direction. In this case the displacement of point x at time t is the same as the

19–5 (a) The graph of $y(x, t)$ versus coordinate x for a specific time, in this case $t = 0$, describes the shape of the wave at that time. (b) The graph of $y(x, t)$ versus time t for a specific coordinate, in this case $x = 0$, describes the motion of a particle at that coordinate as a function of time. The vertical scale is exaggerated in both (a) and (b).

(a)

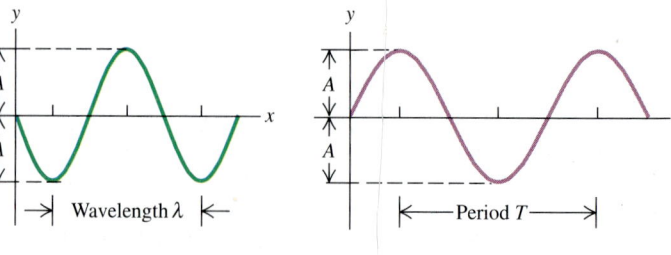

(b)

motion of point $x = 0$ at the *later* time $(t + x/v)$, so in Eq. (19–2) we replace t by $(t + x/v)$. For a wave traveling in the negative x-direction,

$$y(x, t) = A \sin 2\pi f \left(t + \frac{x}{v} \right) = A \sin 2\pi \left(\frac{t}{T} + \frac{x}{\lambda} \right) \tag{19-8}$$

$$= A \sin (\omega t + kx) \quad \text{(sinusoidal wave moving in } -x\text{-direction)}.$$

In the expression $y(x, t) = A \sin (\omega t \pm kx)$ for a wave traveling in the $-x$- or $+x$-direction, the quantity $(\omega t \pm kx)$ is called the **phase**. It plays the role of an angular quantity (always measured in radians) in Eq. (19–7) or (19–8), and its value for any values of x and t determines what part of the sinusoidal cycle is occurring at a particular point and time. For a positive crest (where $y = A$ and the sine function has the value 1), the phase could be $\pi/2$, $5\pi/2$, and so on; for a point of zero displacement it could be 0, π, 2π, and so on. The wave speed is the speed with which we have to move along with the wave to keep alongside a point of a given phase, such as a particular crest of a wave on a string. For a wave traveling in the $+x$-direction, that means $\omega t - kx = \text{constant}$. Taking the derivative with respect to t, we find $\omega = k \, dx/dt$, or

$$\frac{dx}{dt} = \frac{\omega}{k}.$$

Comparing this with Eq. (19–6), we see that dx/dt is equal to the speed v of the wave. Because of this relationship, v is sometimes called the *phase velocity* of the wave. (*Phase speed* would be a better term.)

Problem–Solving Strategy

MECHANICAL WAVES

1. It helps to make a distinction between *kinematics* problems and *dynamics* problems. In kinematics problems we are concerned only with describing motion; the relevant quantities are wave speed, wavelength (or wave number), frequency (or angular frequency), amplitude, and the position, velocity, and acceleration of individual particles. In dynamics problems, concepts such as force and mass enter; the relation of wave speed to the mechanical properties of a system is an example. We'll get into these relations in the next few sections.

2. If f is given, you can find $T = 1/f$, and vice versa. If λ is given, you can find $k = 2\pi/\lambda$, and vice versa. If any two of the quantities v, λ, and f (or v, k, and ω) are known, you can find the third, using $v = \lambda f$ or $\omega = vk$. In some problems that's all you need. To determine the wave function completely, you need to know A and any two of v, λ, and f (or v, k, and ω). Once you have this information, you can use it in Eq. (19–3), (19–4), or (19–7) to get the specific wave function for the problem at hand. Once you have that, you can find the value of y at any point (value of x) and at any time by substituting into the wave function.

3. If the wave speed v is not given, you may be able to find it in one of two ways: Use the frequency-wavelength relation $v = \lambda f$, or use relations between v and the mechanical properties of the system, such as tension and mass per unit length for a string. We will develop these relations in the next three sections. Which method you use to find v will depend on what information you are given.

EXAMPLE 19-2

Wave on a clothesline Your cousin Throckmorton is playing with the clothesline. He unties one end, holds it taut, and wiggles the end up and down sinusoidally with frequency 2.00 Hz and amplitude 0.075 m. The wave speed is $v = 12.0$ m/s. At time $t = 0$ the end has zero displacement and is moving in the $+y$-direction. Assume that no wave bounces back from the far end to muddle up the pattern. a) Find the amplitude, angular fre-

quency, period, wavelength, and wave number of the wave. b) Write a wave function describing the wave. c) Write equations for the displacement as a function of time of Throckmorton's end of the string and of a point 3.00 m from his end.

SOLUTION a) The amplitude A of the wave is just the amplitude of the motion of the end of the clothesline, $A = 0.075$ m. The angular frequency is

$$\omega = 2\pi f = (2\pi\,\text{rad/cycle})(2.00\,\text{cycles/s}) = 4.00\pi\,\text{rad/s}$$

$$= 12.6\,\text{rad/s}.$$

The period is $T = 1/f = 0.500$ s. We get the wavelength from Eq. (19–1):

$$\lambda = \frac{v}{f} = \frac{12.0\,\text{m/s}}{2.00\,\text{s}^{-1}} = 6.00\,\text{m}.$$

We find the wave number from Eq. (19–5) or (19–6):

$$k = \frac{2\pi}{\lambda} = \frac{2\pi\,\text{rad}}{6.00\,\text{m}} = 1.05\,\text{rad/m}, \qquad \text{or}$$

$$k = \frac{\omega}{v} = \frac{4.00\pi\,\text{rad/s}}{12.0\,\text{m/s}} = 1.05\,\text{rad/m}.$$

b) We take the coordinate of Throckmorton's end of the string to be $x = 0$, and choose the direction in which the wave propagates along the string to be the positive x-direction. Then the wave function is given by Eq. (19–4):

$$y(x, t) = A \sin 2\pi\!\left(\frac{t}{T} - \frac{x}{\lambda}\right)$$

$$= (0.075\,\text{m}) \sin 2\pi\!\left(\frac{t}{0.500\,\text{s}} - \frac{x}{6.00\,\text{m}}\right)$$

$$= (0.075\,\text{m}) \sin [(12.6\,\text{rad/s})t - (1.05\,\text{rad/m})x].$$

We can also get this same equation from Eq. (19–7) by using the values of ω and k obtained above. The quantity $(12.6\,\text{rad/s})t - (1.05\,\text{rad/m})x$ is the phase of a point x on the string at time t.

c) With our choice of the positive x-direction the two points in question are at $x = 0$ and $x = +3.00$ m. For each point, we can obtain an expression for the displacement as a function of time by substituting these values of x into the wave function found in part (b):

$$y(x = 0, t) = (0.075\,\text{m}) \sin 2\pi\!\left(\frac{t}{0.500\,\text{s}} - \frac{0}{6.00\,\text{m}}\right)$$

$$= (0.075\,\text{m}) \sin (12.6\,\text{rad/s})t,$$

$$y(x = +3.00\,\text{m}, t) = (0.075\,\text{m}) \sin 2\pi\!\left(\frac{t}{0.500\,\text{s}} - \frac{3.00\,\text{m}}{6.00\,\text{m}}\right)$$

$$= (0.075\,\text{m}) \sin [(12.6\,\text{rad/s})t - \pi\,\text{rad}].$$

The phases of these two points separated by one half-wavelength ($\lambda/2 = (6.00\,\text{m})/2 = 3.00$ m) differ by π radians. Both points oscillate in SHM with the same frequency and amplitude, but their oscillations are one half-cycle out of phase.

Using the above expression for $y(x = 0, t)$, can you show that the end of the string at $x = 0$ is moving in the positive direction at $t = 0$ as stated at the beginning of this example?

PARTICLE VELOCITY AND ACCELERATION IN A SINUSOIDAL WAVE

From the wave function we can get an expression for the transverse velocity of any *particle* in a transverse wave. We call this v_y to distinguish it from the wave propagation speed v. To find the transverse velocity v_y at a particular point x, we take the derivative of the wave function $y(x, t)$ with respect to t, keeping x constant. If the wave function is

$$y(x, t) = A \sin (\omega t - kx),$$

then

$$v_y(x, t) = \frac{\partial y(x, t)}{\partial t} = \omega A \cos (\omega t - kx). \tag{19–9}$$

The ∂ in this expression is a modified d, used to remind us that $y(x, t)$ is a function of *two* variables and that we are allowing only one (t) to vary. The other (x) is constant because we are looking at a particular point on the string. This derivative is called a *partial derivative*. If you haven't reached this point yet in your study of calculus, don't fret; it's a simple idea.

Equation (19–9) shows that the transverse velocity of a particle varies with time, as we expect for simple harmonic motion. The maximum particle speed is ωA; this can be greater than, less than, or equal to the wave speed v, depending on the amplitude and frequency of the wave.

The *acceleration* of any particle is the *second* partial derivative of $y(x, t)$ with respect to t:

$$a_y(x, t) = \frac{\partial^2 y(x, t)}{\partial t^2} = -\omega^2 A \sin (\omega t - kx) = -\omega^2 y(x, t). \tag{19–10}$$

The acceleration of a particle equals $-\omega^2$ times its displacement, which is the result we obtained in Section 13–3 for simple harmonic motion.

We can also compute partial derivatives of $y(x, t)$ with respect to x, holding t constant. This corresponds to studying the shape of the string at one instant of time, like a flash photo. The first derivative $\partial y(x, t)/\partial x$ is the *slope* of the string at any point. The second partial derivative with respect to x is the *curvature* of the string:

$$\frac{\partial^2 y(x, t)}{\partial x^2} = -k^2 A \sin(\omega t - kx) = -k^2 y(x, t). \qquad (19\text{--}11)$$

From Eqs. (19–10) and (19–11) and the relation $\omega = vk$ we see that

$$\frac{\partial^2 y(x, t)/\partial t^2}{\partial^2 y(x, t)/\partial x^2} = \frac{\omega^2}{k^2} = v^2,$$

$$\frac{\partial^2 y(x, t)}{\partial x^2} = \frac{1}{v^2}\frac{\partial^2 y(x, t)}{\partial t^2} \qquad \text{(wave equation).} \qquad (19\text{--}12)$$

The wave function $y = A \sin(\omega t + kx)$ also satisfies this relationship; we invite you to verify this.

Equation (19–12), called the **wave equation,** is one of the most important equations in all of physics. Whenever it occurs, we know that a disturbance can propagate as a wave along the x-axis with wave speed v. The disturbance need not be a sinusoidal wave; we'll see in the next section that *any* wave on a string obeys Eq. (19–12), whether the wave is periodic or not (see also Problems 19–36 and 19–39). In Chapter 33 we will find that electric and magnetic fields satisfy the wave equation; the wave speed will turn out to be the speed of light, which will lead us to the conclusion that light is an electromagnetic wave.

Figure 19–6a shows the velocity v_y and acceleration a_y, given by Eqs. (19–9) and (19–10), for several points on a string as a sinusoidal wave passes along it. Note that at points where the string has an upward curvature ($\partial^2 y/\partial x^2 > 0$), the acceleration of that point is positive ($a_y = \partial^2 y/\partial t^2 > 0$); this follows from the wave equation, Eq. (19–12). For the same reason the acceleration is negative ($a_y = \partial^2 y/\partial t^2 < 0$) at points where the string has a downward curvature ($\partial^2 y/\partial x^2 < 0$), and the acceleration is zero ($a_y = \partial^2 y/\partial t^2 = 0$) at points of inflection where the curvature is zero ($\partial^2 y/\partial x^2 = 0$). We

(a)

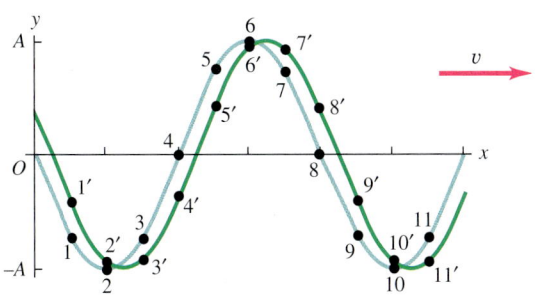

(b)

19–6 (a) Another view of the wave at $t = 0$ in Fig. 19–5a, showing the transverse velocity v_y and transverse acceleration a_y of several points on the string. The acceleration of each point is proportional to the string displacement at that point. (b) Light curve: wave at $t = 0$. Dark curve: the same wave at $t = 0.05T$. For each point, you can relate the magnitude and direction of the displacement during this time interval to v_y and a_y shown in (a).

emphasize again that v_y and a_y are the *transverse* velocity and acceleration of points on the string; these points move along the y-direction, not along the propagation direction of the wave. The transverse motions of several points on the string can be seen in Fig. 19–6b.

The concept of wave function is equally useful with *longitudinal* waves, and everything we have said about wave functions can be adapted to this case. The quantity y still measures the displacement of a particle of the medium from its equilibrium position; the difference is that for a longitudinal wave, this displacement is *parallel* to the x-axis instead of perpendicular to it. We'll discuss longitudinal waves in detail in Section 19–6.

19–5 SPEED OF A TRANSVERSE WAVE

One of the key properties of any wave is the wave speed. Light waves in air have a much greater speed of propagation than do sound waves in air (3.00×10^8 m/s versus 344 m/s); that's why you see the flash from a bolt of lightning before you hear the clap of thunder. In this section we'll see what determines the speed of propagation of one particular kind of wave: transverse waves on a string. The speed of these waves is important to understand in its own right because it is an essential part of analyzing stringed musical instruments, as we'll discuss in Chapter 20. Furthermore, the speeds of many kinds of mechanical waves turn out to have the same basic mathematical expression, as does the speed of waves on a string.

The physical quantities that determine the speed of transverse waves on a string are the *tension* in the string and its *mass per unit length* (also called *linear mass density*). We might guess that increasing the tension should increase the restoring forces that tend to straighten the string when it is disturbed, thus increasing the wave speed. We might also guess that increasing the mass should make the motion more sluggish and decrease the speed. Both these guesses turn out to be right. We'll develop the exact relationship between wave speed, tension, and mass per unit length by two different methods. The first is simple in concept and considers a specific wave shape; the second is more general but also more formal. Choose whichever you like better.

WAVE SPEED ON A STRING: FIRST METHOD

We consider a perfectly flexible string (Fig. 19–7). In the equilibrium position the tension is F, and the linear mass density (mass per unit length) is μ. (When portions of the string are displaced from equilibrium, the mass per unit length decreases a little, and the

19–7 Propagation of a transverse wave on a string. (a) String in equilibrium; (b) part of the string in motion.

tension increases a little.) We ignore the weight of the string so that when the string is at rest in the equilibrium position, the string forms a perfectly straight line as in Fig. 19–7a.

Starting at time $t = 0$, we apply a constant transverse force F_y at the left end of the string. We might expect that the end would move with constant acceleration; that would happen if the force were applied to a *point* mass. But here the effect of the force F_y is to set successively more and more mass in motion. As shown in Fig. 19–7b, the wave travels with constant speed v, so the division point P between moving and nonmoving portions moves with the same constant speed v.

Figure 19–7b shows that all particles in the moving portion of the string move upward with constant *velocity* v_y, not constant acceleration. To see why this is so, we note that the *impulse* of the force F_y up to time t is $F_y t$. According to the impulse-momentum theorem (Section 8–2), the impulse is equal to the change in the total transverse component of momentum ($mv_y - 0$) of the moving part of the string. Because the system started with *no* transverse momentum, this is equal to the total momentum at time t:

$$F_y t = mv_y.$$

The total momentum thus must increase proportionately with time. But since the division point P moves with constant speed, the length of string that is in motion and hence the total mass m in motion are also proportional to the time t that the force has been acting. So the *change* of momentum must be associated entirely with the increasing amount of mass in motion, not with an increasing velocity of an individual mass element. That is, mv_y changes because m, not v_y, changes.

At time t, the left end of the string has moved up a distance $v_y t$, and the boundary point P has advanced a distance vt. The total force at the left end of the string has components F and F_y. Why F? There is no motion in the direction along the length of the string, so there is no unbalanced horizontal force. Therefore F, the magnitude of the horizontal component, does not change when the string is displaced. In the displaced position the tension is $(F^2 + F_y^2)^{1/2}$ (greater than F), and the string stretches somewhat.

To derive an expression for the wave speed v, we again apply the impulse-momentum theorem to the portion of the string in motion at time t, that is, the portion to the left of P in Fig. 19–7b. The transverse *impulse* (transverse force times time) is equal to the change of transverse *momentum* of the moving portion (mass times transverse component of velocity). The impulse of the transverse force F_y in time t is $F_y t$. In Fig. 19–7b the right triangle whose vertex is at P, with sides $v_y t$ and vt, is similar to the right triangle whose vertex is at the position of the hand, with sides F_y and F. Hence

$$\frac{F_y}{F} = \frac{v_y t}{vt}, \qquad F_y = F\frac{v_y}{v},$$

and

$$\text{Transverse impulse} = F_y t = F\frac{v_y}{v}t.$$

The mass of the moving portion of the string is the product of the mass per unit length μ and the length vt, or μvt. The transverse momentum is the product of this mass and the transverse velocity v_y:

$$\text{Transverse momentum} = (\mu vt)v_y.$$

We note again that the momentum increases with time *not* because mass is moving faster, as was usually the case in Chapter 8, but because *more mass* is brought into motion. But the impulse of the force F_y is still equal to the total change in momentum of the system. Applying this relation, we obtain

$$F\frac{v_y}{v}t = \mu vt v_y.$$

Solving this for v, we find

$$v = \sqrt{\frac{F}{\mu}} \qquad \text{(speed of a transverse wave on a string).} \qquad (19\text{--}13)$$

Equation (19–13) confirms our prediction that the wave speed v should increase when the tension F increases but decrease when the mass per unit length μ increases.

Note that v_y does not appear in Eq. (19–13); thus the wave speed doesn't depend on v_y. Our calculation considered only a very special kind of pulse, but we can consider *any* shape of wave disturbance as a series of pulses with different values of v_y. So even though we derived Eq. (19–13) for a special case, it is valid for *any* transverse wave motion on a string, including the sinusoidal and other periodic waves we discussed in Section 19–3. Note that the wave speed doesn't depend on the amplitude or frequency of the wave, in accordance with our assumptions in Section 19–3.

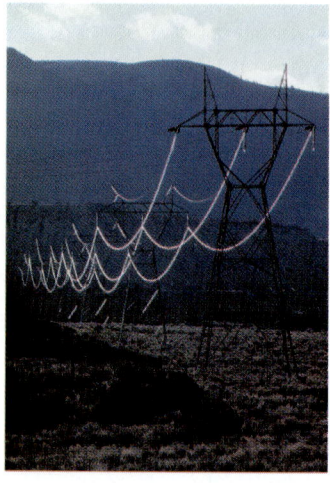

These cables have a relatively large amount of mass per unit length (μ) and a relatively low tension (F). Hence, if the cables are disturbed—say, by a bird landing on them—transverse waves will travel along them quite slowly.

WAVE SPEED ON A STRING: SECOND METHOD

Here is an alternative derivation of Eq. (19–13). If you aren't comfortable with partial derivatives, it can be omitted. We apply Newton's second law, $\Sigma \vec{F} = m\vec{a}$, to a small segment of string whose length in the equilibrium position is Δx (Fig. 19–8). The mass of the segment is $m = \mu\,\Delta x$; the forces at the ends are represented in terms of their x- and y-components. The x-components have equal magnitude F and add to zero because the motion is transverse and there is no component of acceleration in the x-direction. To obtain F_{1y} and F_{2y}, we note that the ratio F_{1y}/F is equal in magnitude to the *slope* of the string at point x and that F_{2y}/F is equal to the slope at point $x + \Delta x$. Taking proper account of signs, we find

$$\frac{F_{1y}}{F} = -\left(\frac{\partial y}{\partial x}\right)_x, \qquad \frac{F_{2y}}{F} = \left(\frac{\partial y}{\partial x}\right)_{x+\Delta x}. \qquad (19\text{--}14)$$

The notation reminds us that the derivatives are evaluated at points x and $x + \Delta x$, respectively. From Eq. (19–14) we find that the net y-component of force is

$$F_y = F_{1y} + F_{2y} = F\left[\left(\frac{\partial y}{\partial x}\right)_{x+\Delta x} - \left(\frac{\partial y}{\partial x}\right)_x\right]. \qquad (19\text{--}15)$$

We now equate F_y from Eq. (19–15) to the mass $\mu\,\Delta x$ times the y-component of acceleration $\partial^2 y/\partial t^2$. We obtain

$$F\left[\left(\frac{\partial y}{\partial x}\right)_{x+\Delta x} - \left(\frac{\partial y}{\partial x}\right)_x\right] = \mu\Delta x\,\frac{\partial^2 y}{\partial t^2}, \qquad (19\text{--}16)$$

19–8 Free-body diagram for a segment of string whose length in its equilibrium position is Δx. The force at each end of the string is tangent to the string at the point of application; each force is represented in terms of its x- and y-components.

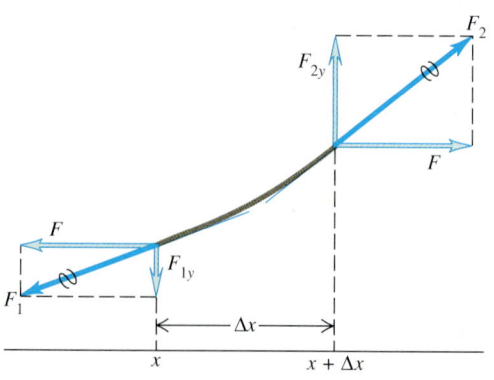

or, dividing by $F \Delta x$,

$$\frac{\left(\dfrac{\partial y}{\partial x}\right)_{x+\Delta x} - \left(\dfrac{\partial y}{\partial x}\right)_{x}}{\Delta x} = \frac{\mu}{F}\frac{\partial^2 y}{\partial t^2}. \tag{19-17}$$

We now take the limit as $\Delta x \to 0$. In this limit, the left side of Eq. (19–17) becomes the derivative of $\partial y/\partial x$ with respect to x (at constant t), that is, the *second* (partial) derivative of y with respect to x:

$$\frac{\partial^2 y}{\partial x^2} = \frac{\mu}{F}\frac{\partial^2 y}{\partial t^2}. \tag{19-18}$$

Now, finally, comes the punch line of our story. Equation (19–18) has exactly the same form as the *wave equation,* Eq. (19–12), that we derived at the end of Section 19–4. That equation and Eq. (19–18) describe the very same wave motion, so they must be identical. Comparing the two equations, we see that for this to be so, we must have

$$v = \sqrt{\frac{F}{\mu}}, \tag{19-19}$$

which is the same expression as Eq. (19–13).

In going through this derivation, we didn't make any special assumptions about the shape of the wave. Since our derivation led us to rediscover Eq. (19–12), the wave equation, we conclude that the wave equation is valid for waves on a string that have *any* shape.

EXAMPLE 19-3

In Example 19–2 the linear mass density of the clothesline is 0.250 kg/m. How much tension does Throcky have to apply to produce the observed wave speed of 12.0 m/s?

SOLUTION We use Eq. (19–13); solving for F, we find

$$F = \mu v^2 = (0.250 \text{ kg/m})(12.0 \text{ m/s})^2 = 36.0 \text{ kg} \cdot \text{m/s}^2$$
$$= 36.0 \text{ N} = 8.09 \text{ lb}.$$

This much force is easily within Throcky's capability.

EXAMPLE 19-4

One end of a nylon rope is tied to a stationary support at the top of a vertical mine shaft 80.0 m deep (Fig. 19–9). The rope is stretched taut by a box of mineral samples with mass 20.0 kg attached at the lower end. The mass of the rope is 2.00 kg. The geologist at the bottom of the mine signals to his colleague at the top by jerking the rope sideways. a) What is the speed of a transverse wave on the rope? b) If a point on the rope is given a transverse simple harmonic motion with a frequency of 2.00 Hz, what is the wavelength of the wave?

SOLUTION a) Let's neglect the variation in tension between the bottom and top of the rope due to the rope's own weight. The tension F at the bottom of the rope equals the weight of the 20.0-kg load:

$$F = (20.0 \text{ kg})(9.80 \text{ m/s}^2) = 196 \text{ N}.$$

The mass per unit length is

$$\mu = \frac{m}{L} = \frac{2.00 \text{ kg}}{80.0 \text{ m}} = 0.0250 \text{ kg/m}.$$

$m_{\text{rope}} = 2.00 \text{ kg}$

80.0m

$m_{\text{samples}} = 20.0 \text{ kg}$

19-9 Sending signals by way of transverse waves on a vertical rope.

The wave speed is given by Eq. (19–13):

$$v = \sqrt{\frac{F}{\mu}} = \sqrt{\frac{196 \text{ N}}{0.0250 \text{ kg/m}}} = 88.5 \text{ m/s}.$$

b) From Eq. (19–1),

$$\lambda = \frac{v}{f} = \frac{88.5 \text{ m/s}}{2.00 \text{ s}^{-1}} = 44.3 \text{ m}.$$

If we account for the weight of the rope, the wave speed will increase and the wavelength will decrease as the wave travels up the rope because the tension increases. Can you verify that the wave speed at the top is 92.9 m/s?

19–6 SPEED OF A LONGITUDINAL WAVE

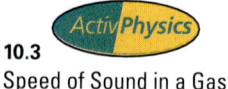

10.3

Speed of Sound in a Gas

Propagation speeds of longitudinal as well as transverse waves depend on the mechanical properties of the medium. We can derive relations for longitudinal waves that are analogous to Eq. (19–13) for transverse waves on a string. As in the wave-function discussion in Section 19–4, x is still the coordinate measured along the length of the wave medium. But for a longitudinal wave the displacement y is along that same direction rather than perpendicular to it as in a transverse wave.

Here is a derivation for the speed of longitudinal waves in a fluid in a pipe. This is a subject of some importance, since when the frequency of a longitudinal wave is within the range of human hearing, we call it **sound.** All musical wind instruments are fundamentally pipes in which a longitudinal wave (sound) propagates in a fluid (air). Human speech works on the same principle; sound waves propagate in your vocal tract, which is basically an air-filled pipe connected to the lungs at one end (your larynx) and to the outside air at the other end (your mouth). The steps in our derivation are completely parallel to the derivation of Eq. (19–13), and we invite you to compare the two developments.

Figure 19–10 shows a fluid (either liquid or gas) with density ρ in a pipe with cross-section area A. In the equilibrium state, the fluid is under a uniform pressure p. In Fig. 19–10a the fluid is at rest. At time $t = 0$ we start the piston at the left end moving toward the right with constant speed v_y. This initiates a wave motion that travels to the right along the length of the pipe, in which successive sections of fluid begin to move and become compressed at successively later times.

Figure 19–10b shows the fluid at time t. All portions of fluid to the left of point P are moving to the right with speed v_y, and all portions to the right of P are still at rest. The boundary between the moving and stationary portions travels to the right with a speed equal to the speed of propagation or wave speed v. At time t the piston has moved a distance $v_y t$, and the boundary has advanced a distance vt. As with a transverse disturbance in a string, we can compute the speed of propagation from the impulse-momentum theorem.

The quantity of fluid set in motion in time t is the amount that originally occupied a section of the cylinder with length vt, cross-section area A, and volume vtA. The mass of this fluid is ρvtA, and its longitudinal momentum (that is, momentum along the length of the pipe) is

$$\text{Longitudinal momentum} = (\rho vtA)v_y.$$

Next we compute the increase of pressure, Δp, in the moving fluid. The original volume of the moving fluid, Avt, has decreased by an amount $Av_y t$. From the definition of the bulk modulus B, Eq. (11–13) in Section 11–6,

$$B = \frac{-\text{Pressure change}}{\text{Fractional volume change}} = \frac{-\Delta p}{-Av_y t/Avt},$$

$$\Delta p = B \frac{v_y}{v}.$$

19–10 Propagation of a longitudinal wave in a fluid confined in a tube. (a) Fluid in equilibrium. (b) Part of the fluid in motion. The net force on the moving fluid is to the left and is equal to $(p + \Delta p)A - pA = \Delta pA$.

The pressure in the moving fluid is $p + \Delta p$, and the force exerted on it by the piston is $(p + \Delta p)A$. The net force on the moving fluid (see Fig. 19–10b) is ΔpA, and the longitudinal impulse is

$$\text{Longitudinal impulse} = \Delta pAt = B\frac{v_y}{v}At.$$

Because the fluid was at rest at time $t = 0$, the change in momentum up to time t is equal to the momentum at that time. Applying the impulse-momentum theorem, we find

$$B\frac{v_y}{v}At = \rho v t A v_y. \qquad (19\text{--}20)$$

When we solve this for v, we get

$$v = \sqrt{\frac{B}{\rho}} \qquad \text{(speed of a longitudinal wave in a fluid).} \qquad (19\text{--}21)$$

Thus the speed of propagation of a longitudinal pulse in a fluid depends only on the bulk modulus B and the density ρ of the medium.

While we derived Eq. (19–21) for waves in a pipe, it also applies to longitudinal waves in a bulk fluid. Thus the speed of sound waves traveling in air or water is determined by this equation. We'll work out the details in the next section.

When a longitudinal wave propagates in a *solid* rod or bar, the situation is somewhat different. The rod expands sideways slightly when it is compressed longitudinally, while a fluid in a pipe with constant cross section cannot move sideways. Using the same kind of reasoning that led us to Eq. (19–21), we can show that the speed of a longitudinal pulse in the rod is given by

$$v = \sqrt{\frac{Y}{\rho}} \qquad \text{(speed of a longitudinal wave in a solid rod),} \qquad (19\text{--}22)$$

where Y is Young's modulus, defined in Section 11–5.

CAUTION ▶ Equation (19–22) applies only to a rod or bar whose sides are free to bulge and shrink a little as the wave travels. It does not apply to longitudinal waves in a *bulk* solid, since in these materials, sideways motion in any element of material is prevented by the surrounding material. The speed of longitudinal waves in a bulk solid depends on the density, the bulk modulus, and the *shear* modulus; a full discussion is beyond the scope of this book. ◀

Note the similarity of form of Eqs. (19–13), (19–19), (19–21), and (19–22). In all these equations for wave speed, whether for transverse or longitudinal waves, the numerator is an elastic property describing the restoring force and the denominator is an inertial property of the medium. As with the derivation for a transverse wave on a string, Eqs. (19–21) and (19–22) are valid for sinusoidal and other periodic waves, not just for the special case discussed here.

Visualizing the relation between particle motion and wave motion is not as easy for longitudinal waves as for transverse waves on a string. Figure 19–11 will help you understand these motions. To use this figure, tape two index cards edge-to-edge with a gap of 1 mm or so between them, forming a thin slit. Place the cards over the figure with the slit horizontal at the top of the diagram, and move them downward with constant speed. The portions of the sine curves that are visible through the slit correspond to a row of particles in a medium in which a longitudinal sinusoidal wave is traveling. Each particle undergoes SHM about its equilibrium position, with delays or phase shifts that increase continuously along the slit. The regions of maximum compression and expansion move from left to right with constant speed. Moving the card upward simulates a wave traveling from right to left.

19–11 Diagram for illustrating longitudinal traveling waves.

Table 19–1 lists the speed of sound in several bulk materials. Sound waves travel slower in lead than in aluminum or steel because lead has a lower bulk modulus and shear modulus and a higher density.

TABLE 19–1
SPEED OF SOUND IN VARIOUS BULK MATERIALS

MATERIAL	SPEED OF SOUND (m/s)
Gases	
Air (20°C)	344
Helium (20°C)	999
Hydrogen (20°C)	1330
Liquids	
Liquid helium (4 K)	211
Mercury (20°C)	1451
Water (0°C)	1402
Water (20°C)	1482
Water (100°C)	1543
Solids	
Aluminum	6420
Lead	1960
Steel	5941

EXAMPLE 19–5

Wavelength of sonar waves A ship uses a sonar system to detect underwater objects (Fig. 19–12). The system emits underwater sound waves and measures the time interval for the reflected wave (echo) to return to the detector. Determine the speed of sound waves in water, and find the wavelength of a wave having a frequency of 262 Hz.

SOLUTION We use Eq. (19–21) to find the wave speed. From Table 11–2 we find that the compressibility of water, which is the reciprocal of the bulk modulus, is $k = 45.8 \times 10^{-11}$ Pa^{-1}. Thus $B = (1/45.8) \times 10^{11}$ Pa. The density of water is $\rho = 1.00 \times 10^3$ kg/m^3. We obtain

$$v = \sqrt{\frac{B}{\rho}} = \sqrt{\frac{(1/45.8) \times 10^{11} \text{ Pa}}{1.00 \times 10^3 \text{ kg/m}^3}}$$

$$= 1480 \text{ m/s}.$$

This value agrees well with the experimental value in Table 19–1; it is over four times the speed of sound in air at ordinary temperatures. The wavelength is given by

$$\lambda = \frac{v}{f} = \frac{1480 \text{ m/s}}{262 \text{ s}^{-1}} = 5.65 \text{ m}.$$

A wave of this frequency in air has a wavelength of 1.31 m, as we found in Example 19–1 (Section 19–3).

Dolphins emit high-frequency sound waves (typically 100,000 Hz) and use the echoes for guidance and for hunting. The corresponding wavelength in water is 1.48 cm. With this high-frequency "sonar" system they can sense objects that are roughly as small as the wavelength (but not much smaller). *Ultrasonic imaging* is a medical technique that uses exactly the same physical principle; sound waves of very high frequency and very short wavelength, called *ultrasound,* are scanned over the human body, and the "echoes" from interior organs are used to create an image. With ultrasound of frequency 5 MHz = 5×10^6 Hz, the wavelength in water (the primary constituent of the body) is 0.3 mm, and features as small as this can be discerned in the image. Ultrasound is used for the study of heart-valve action, detection of tumors, and prenatal examinations. Ultrasound is more sensitive than x rays in distinguishing various kinds of tissues and does not have the radiation hazards associated with x rays.

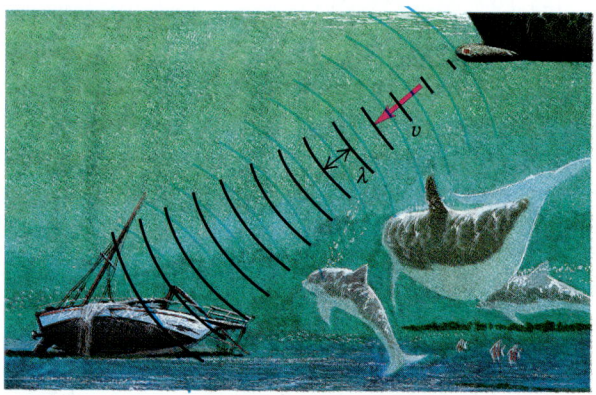

19–12 A sonar system uses underwater sound waves to detect and locate underwater objects.

EXAMPLE 19–6

What is the speed of longitudinal sound waves in a lead rod?

SOLUTION This is the situation for which Eq. (19–22) is applicable. From Table 11–1, $Y = 1.6 \times 10^{10}$ Pa, and from Table 14–1, $\rho = 11.3 \times 10^3$ kg/m^3. We find

$$v = \sqrt{\frac{Y}{\rho}} = \sqrt{\frac{1.6 \times 10^{10} \text{ Pa}}{11.3 \times 10^3 \text{ kg/m}^3}} = 1.2 \times 10^3 \text{ m/s}.$$

This is more than three times the speed of sound in air. Note that our result is the speed with which a sound wave travels along a lead *rod*. You can see from Table 19–1 that sound waves in *bulk* lead travel faster still; the reason is that for lead the bulk modulus is greater than Young's modulus.

19–7 SOUND WAVES IN GASES

In the preceding section we derived Eq. (19–21), $v = \sqrt{B/\rho}$, for the speed of longitudinal waves in a fluid of bulk modulus B and density ρ. We can use this to find the speed of sound in an ideal gas.

The bulk modulus is defined in general as in Eq. (11–13); for infinitesimal pressure and volume changes, $B = -V\, dp/dV$. So we need to know how p varies with V for an ideal gas. If the temperature is constant, then according to the ideal gas law the product pV is constant, and we can use this to evaluate dp/dV. But when a gas is compressed *adiabatically* so that there is no heat flow, its temperature rises, and when it expands adiabatically, its temperature drops. In an adiabatic process for an ideal gas, Eq. (17–24) states that pV^γ is constant (recall that $\gamma = C_p/C_V$ is the dimensionless ratio of heat capacities), and we get a different result for B. When a wave travels through a gas, are the compressions and expansions adiabatic, or is there enough heat conduction between adjacent layers of gas to maintain a nearly constant temperature throughout?

Because thermal conductivities of gases are very small, it turns out that for ordinary sound frequencies, say 20 to 20,000 Hz, propagation of sound is very nearly adiabatic. Thus in Eq. (19–21) we use the **adiabatic bulk modulus** B_{ad}, derived from the assumption

$$pV^\gamma = \text{constant.} \tag{19–23}$$

We take the derivative of Eq. (19–23) with respect to V:

$$\frac{dp}{dV} V^\gamma + \gamma p V^{\gamma-1} = 0.$$

Dividing by $V^{\gamma-1}$ and rearranging, we find

$$B_{ad} = -V \frac{dp}{dV} = \gamma p. \tag{19–24}$$

For an *isothermal* process, $pV = $ constant; we invite you to prove that the *isothermal* bulk modulus is

$$B_{iso} = p. \tag{19–25}$$

The adiabatic modulus is *larger* than the isothermal modulus by a factor γ.

Combining Eqs. (19–21) and (19–24), we find

$$v = \sqrt{\frac{\gamma p}{\rho}} \qquad \text{(speed of sound in an ideal gas).} \tag{19–26}$$

We can get a useful alternative form by using Eq. (16–5) for the density ρ of an ideal gas:

$$\rho = \frac{pM}{RT},$$

where R is the gas constant, M is the molar mass, and T is the absolute temperature. Combining this with Eq. (19–26), we find

$$v = \sqrt{\frac{\gamma RT}{M}} \qquad \text{(speed of sound in an ideal gas).} \qquad (19\text{–}27)$$

For any particular gas, γ, R, and M are constants, and the wave speed is proportional to the square root of the absolute temperature. Except for the numerical factor of 3 in one and γ in the other, this expression is identical to Eq. (16–19), which gives the root-mean-square speed of molecules in an ideal gas. This shows that sound speeds and molecular speeds are closely related; exploring that relationship in detail would be beyond our scope.

EXAMPLE 19–7

Compute the speed of sound waves in air at room temperature ($T = 20°C$).

SOLUTION From Example 16–3 (Section 16–2) the mean molar mass of air is 28.8×10^{-3} kg/mol. Also, $\gamma = 1.40$ for air, and $R = 8.315$ J/mol · K. At $T = 20°C = 293$ K we find

$$v = \sqrt{\frac{\gamma RT}{M}}$$

$$= \sqrt{\frac{(1.40)(8.315 \text{ J/mol} \cdot \text{K})(293 \text{ K})}{28.8 \times 10^{-3} \text{ kg/mol}}} = 344 \text{ m/s.}$$

This agrees with the measured speed of sound at this temperature to within 0.3 percent.

The ear is sensitive to a range of sound frequencies from about 20 Hz to about 20,000 Hz. From the relation $v = \lambda f$ the corresponding wavelength range at 20°C is from about 17 m, corresponding to a 20-Hz note, to about 1.7 cm, corresponding to 20,000 Hz.

Bats can hear much higher frequencies. Like dolphins, bats use high-frequency sound waves for navigation. A typical frequency is 100 kHz; the corresponding wavelength in air at 20°C is about 3.4 mm, small enough for them to detect the flying insects they eat.

In this discussion we have ignored the *molecular* nature of a gas and have treated it as a continuous medium. We know that a gas is actually composed of molecules in random motion, separated by distances that are large in comparison with their diameters. The vibrations that constitute a wave in a gas are superposed on the random thermal motion. At atmospheric pressure, a molecule travels an average distance (the mean free path) of about 10^{-7} m between collisions, while the displacement amplitude of a faint sound may be only 10^{-9} m. We can think of a gas with a sound wave passing through as being comparable to a swarm of bees; the swarm as a whole oscillates slightly while individual insects move about through the swarm, apparently at random.

19–8 ENERGY IN WAVE MOTION

Every wave motion has *energy* associated with it. The energy we receive from sunlight and the destructive effects of ocean surf and earthquakes bear this out. To produce any of the wave motions we have discussed in this chapter, we have to apply a force to a portion of the wave medium; the point where the force is applied moves, so we do *work* on the system. As the wave propagates, each portion of the medium exerts a force and does work on the adjoining portion. In this way a wave can transport energy from one region of space to another.

As an example of energy considerations in wave motion, let's look again at transverse waves on a string. How is energy transferred from one portion of string to another? Picture a wave traveling from left to right (the positive x-direction) on the string, and consider a particular point a on the string (Fig. 19–13a). The string to the left of a exerts a force on the string to the right of it, and vice versa. In Fig. 19–13b the string to the left of a has been removed, and the force it exerts at a is represented by the components F

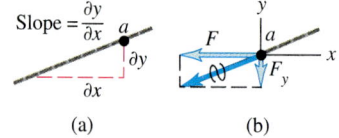

19–13 (a) Point a on a string carrying a wave from left to right. (b) The components of the force exerted on the right part of the string by the left part at point a.

and F_y, as we did in Figs. 19–7 and 19–8. We note again that F_y/F is equal to the negative of the *slope* of the string at a, which is also given by $\partial y/\partial x$. Putting these together, we have

$$F_y(x,t) = -F \frac{\partial y(x,t)}{\partial x}. \qquad (19\text{–}28)$$

We need the negative sign because F_y is negative when the slope is positive. We write the vertical force as $F_y(x, t)$ as a reminder that its value may be different at different points along the string and at different times.

When point a moves in the y-direction, the force F_y does *work* on this point and therefore transfers energy into the part of the string to the right of a. The corresponding power P (rate of doing work) at the point a is the transverse force $F_y(x, t)$ at a times the transverse velocity $v_y(x, t) = \partial y(x, t)/\partial t$ of that point:

$$P(x,t) = F_y(x,t)v_y(x,t) = -F \frac{\partial y(x,t)}{\partial x} \frac{\partial y(x,t)}{\partial t}. \qquad (19\text{–}29)$$

This power is the *instantaneous* rate at which energy is transferred along the string. Its value depends on the position x on the string and on the time t. Note that energy is being transferred only at points where the string has a nonzero slope ($\partial y/\partial x$ is nonzero), so that there is a transverse component of the tension force, and where the string has a nonzero transverse velocity ($\partial y/\partial t$ is nonzero) so that the transverse force can do work.

Equation (19–29) is valid for *any* wave on a string, sinusoidal or not. For a sinusoidal wave with wave function given by Eq. (19–7), we have

$$y(x,t) = A \sin(\omega t - kx),$$

$$\frac{\partial y(x,t)}{\partial x} = -kA \cos(\omega t - kx),$$

$$\frac{\partial y(x,t)}{\partial t} = \omega A \cos(\omega t - kx),$$

$$P(x,t) = Fk\omega A^2 \cos^2(\omega t - kx). \qquad (19\text{–}30)$$

By using the relations $\omega = vk$ and $v^2 = F/\mu$, we can also express Eq. (19–30) in the alternative form

$$P(x,t) = \sqrt{\mu F}\,\omega^2 A^2 \cos^2(\omega t - kx). \qquad (19\text{–}31)$$

The $\cos^2$ function is never negative, so the instantaneous power in a sinusoidal wave is either positive (so that energy flows in the positive x-direction) or zero (at points where there is no energy transfer). Energy is never transferred in the direction opposite to the direction of wave propagation.

The maximum value of the instantaneous power $P(x, t)$ occurs when the $\cos^2$ function has the value unity:

$$P_{\text{max}} = \sqrt{\mu F}\,\omega^2 A^2. \qquad (19\text{–}32)$$

To obtain the *average* power from Eq. (19–31), we note that the *average* value of the $\cos^2$ function, averaged over any whole number of cycles, is $\frac{1}{2}$. Hence the average power is

$$P_{\text{av}} = \frac{1}{2}\sqrt{\mu F}\,\omega^2 A^2 \qquad \text{(average power, sinusoidal wave on a string).} \qquad (19\text{–}33)$$

The rate of energy transfer is proportional to the square of the amplitude and to the square of the frequency.

Power relationships analogous to Eq. (19–33) can be worked out for longitudinal waves. We won't go into detail; the results are most easily stated in terms of the average power *per unit cross-section area* in the wave motion. The average power per unit area is called the **intensity,** denoted by *I* For fluids in a pipe it is given by

$$I = \frac{1}{2}\sqrt{\rho B}\omega^2 A^2, \qquad (19\text{--}34)$$

and for a solid rod it is given by

$$I = \frac{1}{2}\sqrt{\rho Y}\omega^2 A^2. \qquad (19\text{--}35)$$

Again the power is proportional to A^2 and to ω^2.

The proportionality of the rate of energy transfer to the square of the amplitude is a general result for waves of all types; the proportionality constant is different for different types of waves. However, the energy transfer rate is proportional to ω^2 only for mechanical waves. For electromagnetic waves, the intensity is independent of the value of ω.

EXAMPLE 19–8

a) In Examples 19–2 and 19–3, at what maximum rate does Throcky put energy into the rope? That is, what is his maximum instantaneous power? b) What is his average power? c) As Throcky tires, the amplitude decreases. What is the average power when the amplitude has dropped to 7.50 mm?

SOLUTION a) From Eq. (19–32),

$$P_{max} = \sqrt{\mu F}\omega^2 A^2$$
$$= \sqrt{(0.250 \text{ kg/m})(36.0 \text{ N})}(4.00\pi \text{ rad/s})^2(0.075 \text{ m})^2$$
$$= 2.66 \text{ W}.$$

b) From Eqs. (19–32) and (19–33), $P_{av} = \frac{1}{2}P_{max}$, so

$$P_{av} = \frac{1}{2}(2.66 \text{ W}) = 1.33 \text{ W}.$$

c) This amplitude is $\frac{1}{10}$ of the value we used in parts (a) and (b). The average power is proportional to the *square* of the amplitude, so now the average power is

$$P_{av} = \frac{1}{10^2}(1.33 \text{ W}) = 0.0133 \text{ W} = 13.3 \text{ mW}.$$

EXAMPLE 19–9

Sound waves in the outer ear Sound waves entering the human ear first pass through the auditory canal before reaching the eardrum (Fig. 19–14). The auditory canal of a typical adult is 2.5 cm long and 7.0 mm in diameter. When you listen to ordinary conversation, the intensity of sound waves in the auditory canal is about 3.2×10^{-6} W/m^2. a) What is the average power delivered to the eardrum? b) Find the amplitude of these sound waves. Assume a frequency $f = 100$ Hz (typical of many of the sounds used in human speech).

SOLUTION a) The intensity I of the wave equals the average power divided by the cross-section area of the wave, so average power equals intensity *times* this area. The wave moves over the entire auditory canal of radius $r = (7.0 \text{ mm})/2 = 3.5$ mm, so the cross-section area is $\pi r^2 = \pi(3.5 \times 10^{-3} \text{ m})^2 = 3.8 \times 10^{-5}$ m^2. The average power is therefore

$$P_{av} = I\pi r^2 = (3.2 \times 10^{-6} \text{ W/m}^2)(3.8 \times 10^{-5} \text{ m}^2)$$
$$= 1.2 \times 10^{-10} \text{ W}.$$

This minuscule amount of power is absorbed by the eardrum, whose oscillations are transformed in the middle and inner ears into an electrical signal that is sent to the brain. A healthy ear can detect sound intensities as low as 10^{-12} W/m^2, in which case the average power delivered is only 3.8×10^{-17} W. Human hearing is sensitive indeed!

b) For waves propagating in a fluid in a pipe, the relationship between intensity I and amplitude A is given by Eq. (19–34), $I = \frac{1}{2}\sqrt{\rho B}\omega^2 A^2$. Solving this equation for A, we find

$$A = \frac{1}{\omega}\frac{\sqrt{2I}}{(\rho B)^{1/4}}.$$

The fluid is air, so we can use the results of Section 19–7 for sound waves in gases; the bulk modulus to be used is $B_{ad} = \gamma p$. For air, $\gamma = 1.40$, and normal atmospheric pressure is $p = 1.013 \times 10^5$ Pa, so $B_{ad} = (1.40)(1.013 \times 10^5 \text{ Pa}) = 1.42 \times 10^5$ Pa. We can find the density of the air using the ideal-gas law in the form of Eq. (16–5), $\rho = pM/RT$. Using room temperature (20°C,

19–14 The anatomy of the human ear.

or 293 K), we have

$$\rho = \frac{(1.013 \times 10^5 \text{ Pa})(28.8 \times 10^{-3} \text{ kg/mol})}{(8.315 \text{ J/mol} \cdot \text{K})(293 \text{ K})}$$

$$= 1.20 \text{ kg/m}^3.$$

Substituting into the above expression for the amplitude A, we get

$$A = \frac{1}{2\pi(100 \text{ s}^{-1})} \frac{\sqrt{2(3.2 \times 10^{-6} \text{ W/m}^2)}}{((1.20 \text{ kg/m}^3)(1.42 \times 10^5 \text{ Pa}))^{1/4}}$$

$$= 2.0 \times 10^{-7} \text{ m}.$$

This is a very tiny distance, yet this much displacement of air molecules is perfectly detectable by a person with normal hearing. In fact, a sound of this intensity would still be audible if we increased the frequency by a factor of 100 to $f = 10,000$ Hz. Our calculations show that the amplitude A is inversely proportional to the frequency, so the amplitude would be 0.01 as great, or 2.0×10^{-9} m. Your ear can detect the motion of air molecules over a distance that is only a few times larger than the size of the molecules themselves! We will explore the sensitivity of the ear further in Chapter 21.

SUMMARY

- A wave is any disturbance from an equilibrium condition that propagates from one region to another. A mechanical wave always travels within some material called the medium. In a periodic wave, the motion of each point of the medium is periodic. When the motion is sinusoidal, the wave is called a sinusoidal wave. The frequency f of a periodic wave is the number of repetitions per unit time, and the period T is the time for one cycle. The wavelength λ is the distance over which the wave pattern repeats. The speed of propagation v is the speed with which the wave disturbance travels. For any periodic wave, these quantities are related by

$$v = \lambda f. \tag{19–1}$$

- A wave function $y(x, t)$ describes the displacements of individual particles in the medium. For a sinusoidal wave moving in the $+x$-direction,

$$y(x,t) = A \sin \omega \left(t - \frac{x}{v} \right) = A \sin 2\pi f \left(t - \frac{x}{v} \right) \tag{19–3}$$

$$= A \sin 2\pi \left(\frac{t}{T} - \frac{x}{\lambda} \right), \tag{19–4}$$

or

$$y(x,t) = A \sin (\omega t - kx), \tag{19–7}$$

where the wave number k is defined as $k = 2\pi/\lambda$ and $\omega = vk$. In all three forms, A is

KEY TERMS

mechanical wave, 593
medium, 593
transverse wave, 593
longitudinal wave, 594
wave speed, 595
periodic wave, 595
sinusoidal wave, 595
wavelength, 595
wave function, 597
wave number, 598
phase, 599
wave equation, 601
sound, 606
adiabatic bulk modulus, 609
intensity, 612

the amplitude, the maximum displacement of a particle from its equilibrium position. If the wave is moving in the $-x$-direction, the minus sign in Eqs. (19–3), (19–4), and (19–7) is replaced by a plus sign.

- The wave function must obey a partial differential equation called the wave equation,

$$\frac{\partial^2 y(x,t)}{\partial x^2} = \frac{1}{v^2} \frac{\partial^2 y(x,t)}{\partial t^2}. \tag{19–12}$$

- The speed of a transverse wave on a string with tension F and mass per unit length μ is

$$v = \sqrt{\frac{F}{\mu}} \qquad \text{(transverse wave on a string).} \tag{19–13}$$

- The speed of a longitudinal wave in a fluid with bulk modulus B and density ρ is

$$v = \sqrt{\frac{B}{\rho}} \qquad \text{(longitudinal wave in a fluid).} \tag{19–21}$$

- The speed of a longitudinal wave in a solid rod with Young's modulus Y and density ρ is

$$v = \sqrt{\frac{Y}{\rho}} \qquad \text{(longitudinal wave in a solid rod).} \tag{19–22}$$

- Sound propagation in gases is ordinarily an adiabatic process. The speed of sound in an ideal gas is

$$v = \sqrt{\frac{\gamma p}{\rho}} = \sqrt{\frac{\gamma R T}{M}} \qquad \text{(sound wave in an ideal gas).} \quad \text{(19–26), (19–27)}$$

- Wave motion conveys energy from one region to another. For a sinusoidal wave on a stretched string, the average power is

$$P_{\text{av}} = \frac{1}{2}\sqrt{\mu F}\,\omega^2 A^2. \tag{19–33}$$

For a longitudinal wave, the average power per unit cross-section area in the wave motion is called the intensity, I.

DISCUSSION QUESTIONS

Q19–1 A common pastime at sporting events is "doing the wave," in which the spectators in one section raise and lower their arms, signaling the spectators in the next section to do the same, and so on. In the language of this chapter, is this truly a wave? Why or why not?

Q19–2 Explain why it is necessary to include the factor 2π in Equation (19–4).

Q19–3 What kinds of energy are associated with waves on a stretched string? How could such energy be detected experimentally?

Q19–4 The amplitude of a wave decreases gradually as the wave travels down a long stretched string. What happens to the energy of the wave when this happens?

Q19–5 For the wave motions discussed in this chapter, does the speed of propagation depend on the amplitude? How can you tell?

Q19–6 The speed of ocean waves depends on the depth of the water; the deeper the water, the faster the wave travels. Use this to explain why ocean waves crest and "break" as the wave nears the shore.

Q19–7 Is it possible to have a longitudinal wave on a stretched string? Why or why not? Is it possible to have a transverse wave on a steel rod? Again, why or why not? If your answer is yes in either case, explain how you would create such a wave.

Q19–8 An echo is sound reflected from a distant object, such as a wall or a cliff. Explain how you can determine how far away the object is by timing the echo.

Q19–9 Why do you see lightning before you hear the thunder? A familiar rule of thumb is to start counting slowly, once per second, when you see the lightning; when you hear the thunder, divide the number you have reached by 3 to obtain your distance from the lightning in kilometers (or by 5 to obtain your distance in miles). Why does this work? Or does it?

Q19–10 For transverse waves on a string, is the wave speed the same as the maximum speed of any part of the string? Or is there a different relation between these two speeds? What about for longitudinal waves on a rod? Explain your reasoning.

Q19–11 Children make toy telephones by sticking each end of a long string through a hole in the bottom of a paper cup and knotting it so it will not pull out. When the spring is pulled taut, sound can be transmitted from one cup to the other. How does this work? Why is the transmitted sound louder than the sound traveling through air for the same distance?

Q19–12 The four strings on a violin have different thicknesses, but are all under approximately the same tension. Do waves travel faster on the thick strings or the thin strings? Why?

Q19–13 When a rock is thrown into a pond and the resulting ripples spread in ever-widening circles, the amplitude decreases with increasing distance from the center. Why?

Q19–14 When sound travels from air into water, does the frequency of the wave change? The speed? The wavelength? Explain your reasoning.

Q19–15 The hero of a western movie listens for an oncoming train by putting his ear to the track. Why does this method give early warning of the approach of a train?

Q19–16 In a transverse wave on a string, the motion of the string is perpendicular to the length of the string. How, then, is it possible for energy to move along the length of the string?

Q19–17 Justify the following statement: "A wave transfers energy fastest where the particles of the medium are moving fastest." In particular, explain how can this can be true for a transverse wave, in which the particles of the medium move in a direction perpendicular to that of wave propagation.

Q19–18 In speaker systems designed for high-fidelity music reproduction, the "tweeters" that reproduce the high frequencies are always much smaller than the "woofers" used for low frequencies. Explain why, using Eq. (19–34).

Q19–19 At very high frequencies, above 5×10^8 Hz or so, the propagation of sound in air is isothermal, not adiabatic. Explain why this should be so. (*Hint:* Very high frequencies mean very short wavelengths.)

EXERCISES

SECTION 19–3 PERIODIC WAVES

19–1 A fisherman notices that his boat is moving up and down periodically owing to waves on the surface of the water. It takes 2.5 s for the boat to travel from its highest point to its lowest, a total distance of 0.62 m. The fisherman sees that the wave crests are spaced 6.0 m apart. a) How fast are the waves traveling? b) What is the amplitude of each wave? c) If the total vertical distance traveled by the boat were 0.30 m, but the other data remained the same, how would the answers to parts (a) and (b) be affected? d) Would you expect that the motion of the boat would be purely vertical? Why or why not?

19–2 The sound waves from a loudspeaker spread out nearly uniformly in all directions when their wavelengths are large in comparison with the diameter of the speaker. When the wavelength is small in comparison with the diameter of the speaker, much of the sound energy is concentrated forward. For a speaker with a diameter of 12.0 cm, compute the frequency for which the wavelength of the sound waves in air ($v = 344$ m/s) is a) ten times the diameter of the speaker; b) equal to the diameter of the speaker; c) one-tenth the diameter of the speaker. (For a second factor in the optimal size of the speaker, see Question Q19–18.)

19–3 The speed of sound in air at 20°C is 344 m/s. a) What is the wavelength of a sound wave with a frequency of 784 Hz, corresponding to the note G_5 on a piano? b) What is the frequency of a sound wave with a wavelength of 0.0655 mm? (This frequency is too high for you to hear.)

19–4 The speed of radio waves in vacuum (equal to the speed of light) is 3.00×10^8 m/s. Find the wavelength for a) an AM radio station with frequency 540 kHz; b) an FM radio station with frequency 104.5 MHz.

19–5 Provided that the amplitude is sufficiently great, the human ear can respond to longitudinal waves over a range of frequencies from about 20.0 Hz to about 20,000 Hz. Compute the wavelengths corresponding to these frequencies a) for waves in air ($v = 344$ m/s); b) for waves in water ($v = 1480$ m/s).

SECTION 19–4 MATHEMATICAL DESCRIPTION OF A WAVE

19–6 A certain transverse wave is described by

$$y(x, t) = (6.50 \text{ mm}) \sin 2\pi \left(\frac{t}{0.0360 \text{ s}} - \frac{x}{28.0 \text{ cm}} \right)$$

Determine the wave's a) amplitude; b) wavelength; c) frequency; d) speed of propagation; e) direction of propagation.

19–7 Transverse waves on a string have wave speed 8.00 m/s, amplitude 0.0700 m, and wavelength 0.320 m. The waves travel in the $-x$-direction, and at $t = 0$ the $x = 0$ end of the string has zero displacement and is moving in the $+y$-direction. a) Find the frequency, period, and wave number of these waves. b) Write a wave function describing the wave. c) Find the transverse displacement of a particle at $x = 0.360$ m at time $t = 0.150$ s. d) How much time must elapse from the instant in part (c) until the particle at $x = 0.360$ m next has zero displacement? e) How much additional time must elapse from the instant in part (d) until the next time the particle has zero displacement?

19–8 Redraw Figs. 19–6a and 19–6b for the case in which the wave is propagating to the *left*.

19–9 Show that the following functions satisfy the wave equation, Eq. (19–12): a) $y(x, t) = A \sin (\omega t + kx)$; b) $y(x, t) = A \cos (\omega t + kx)$. c) In what directions are these waves traveling? How can you tell? d) For the wave of part (b), write the equations for the transverse velocity and transverse acceleration of a particle at point x.

19–10 a) For a wave described by $y(x, t) = A \sin(\omega t - kx)$, graph y, v_y, and a_y as functions of x for time $t = 0$. b) Consider the following points on the string: i) $x = 0$; ii) $x = \pi/4k$; iii) $x = \pi/2k$; iv) $x = 3\pi/4k$; v) $x = \pi/k$; vi) $x = 5\pi/4k$; vii) $x = 3\pi/2k$; and viii) $x = 7\pi/4k$. For a particle at each of these points at $t = 0$, describe in words whether the particle is moving and in what direction, and whether the particle is speeding up, slowing down, or instantaneously not accelerating.

19–11 A sinusoidal wave is propagating along a stretched string that lies along the x-axis. The displacement of the string as a function of time is graphed in Fig. 19–15 for particles at $x = 0$ and at $x = 0.0900$ m. a) What is the amplitude of the wave? b) What is the period of the wave? c) You are told that the two points $x = 0$ and $x = 0.0900$ m are within one wavelength of each other. If the wave is moving in the $+x$-direction, determine the wavelength and the wave speed. d) If instead the wave is moving in the $-x$-direction, determine the wavelength and the wave speed. e) Would it be possible to unambiguously determine the wavelength in parts (c) and (d) if you were not told that the two points were within one wavelength of each other? Why or why not?

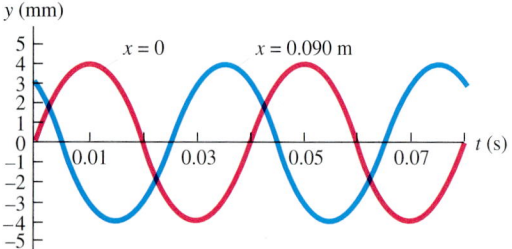

FIGURE 19–15 Exercise 19–11.

19–12 Speed of Propagation vs. Particle Speed. a) Show that Eq. (19–3) may be written as

$$y(x, t) = -A \sin\left[\frac{2\pi}{\lambda}(x - vt)\right].$$

b) Use $y(x, t)$ to find an expression for the transverse velocity v_y of a particle in the string on which the wave travels. c) Find the maximum speed of a particle of the string. Under what circumstances is this equal to the propagation speed v? Less than v? Greater than v?

19–13 A transverse wave on a string has amplitude 0.300 cm, wavelength 12.0 cm, and speed 6.00 cm/s. It is represented by $y(x, t)$ as given in Exercise 19–12. a) At time $t = 0$, compute y at 1.5-cm intervals of x (that is, at $x = 0$, $x = 1.5$ cm, $x = 3.0$ cm, and so on) from $x = 0$ to $x = 12.0$ cm. Graph the results. This is the shape of the string at time $t = 0$. b) Repeat the calculations for the same values of x at times $t = 0.400$ s and $t = 0.800$ s. Graph the shape of the string at these instants. In what direction is the wave traveling?

SECTION 19–5 SPEED OF A TRANSVERSE WAVE

19–14 With what tension must a rope of length 2.50 m and mass 0.120 kg be stretched for transverse waves of frequency 40.0 Hz to have a wavelength of 0.750 m?

19–15 One end of a horizontal rope is attached to a prong of an electrically driven tuning fork that vibrates at 120 Hz. The other end passes over a pulley and supports a 1.50-kg mass. The linear mass density of the rope is 0.0550 kg/m. a) What is the speed of a transverse wave on the rope? b) What is the wavelength? c) How would your answers to parts (a) and (b) be changed if the mass were increased to 3.00 kg?

19–16 A cowgirl ties one end of a 10.0-m-long rope to a fence post and pulls on the other end so the rope is stretched horizontally with a tension of 140 N. The mass of the rope is 0.800 kg. a) What is the speed of transverse waves on the rope? b) If the cowgirl moves the free end up and down with a frequency of 1.20 Hz, what is the wavelength of the transverse waves on the rope ? c) The cowgirl pulls harder on the rope so that the tension is doubled, to 280 N. With what frequency must she move the free end of the rope up and down to produce transverse waves of the same wavelength as in part (a)?

19–17 One end of a 14.0-m-long rubber tube, with a total mass of 0.800 kg, is fastened to a fixed support. A cord attached to the other end passes over a pulley and supports an object with a mass of 7.50 kg. The tube is struck a transverse blow at one end. Find the time required for the pulse to reach the other end.

19–18 If in Example 19–4 (Section 19–5) we do *not* neglect the weight of the rope, what is the wave speed a) at the bottom of the rope? b) at the middle of the rope? c) at the top of the rope?

SECTION 19–6 SPEED OF A LONGITUDINAL WAVE
SECTION 19–7 SOUND WAVES IN GASES

19–19 a) In a liquid with density 1300 kg/m^3, longitudinal waves with frequency 400 Hz are found to have wavelength 8.00 m. Calculate the bulk modulus of the liquid. b) A metal bar with a length of 1.50 m has density 6400 kg/m^3. Longitudinal sound waves take 3.90×10^{-4} s to travel from one end of the bar to the other. What is Young's modulus for this metal?

19–20 A major earthquake centered on Loma Prieta, California, near San Francisco, occurred at 5:04 P.M. local time on October 17, 1989 (in UTC, Coordinated Universal Time, 0h 4m 15s on October 18, 1989). The primary seismic waves (P waves) from such an earthquake are longitudinal waves that travel through the earth's crust. P waves were detected at Caracas, Venezuela, at 0h 13m 54s UTC; at Kevo, Finland, at 0h 15m 35s UTC; and at Vienna, Austria, at 0h 17m 02s UTC. The distances the P waves traveled from Loma Prieta were 6280 km to Caracas, 8690 km to Kevo, and 9650 km to Vienna. a) Use the arrival times to calculate the average speed of the P waves that traveled to these three cities. How can you account for any differences between the average speeds? b) The average density of the earth's crust is about 3.3 g/cm^3. Use this value to calculate the bulk modulus of the earth's crust along the path traveled by the P waves to each of the three cities. How do your answers compare to the bulk moduli in Table 11–1?

19–21 A submerged scuba diver hears the sound of a boat horn directly above her on the surface of the lake. At the same time, a friend on dry land 22.0 m from the boat also hears the horn (Fig. 19–16). The horn is 1.2 m above the surface of the water. What is the distance (labeled by "?" in Fig. 19–16) from the horn to the diver? Both air and water are at 20°C.

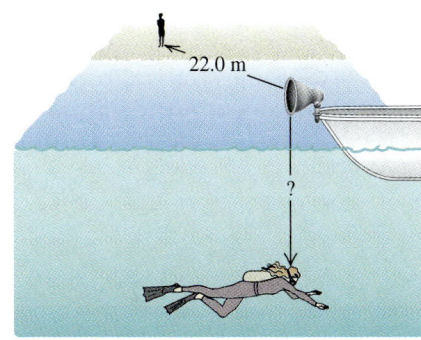

FIGURE 19–16 Exercise 19–21.

19–22 At a temperature of 27.0°C, what is the speed of longitudinal waves in a) hydrogen (molar mass 2.02 g/mol)? b) helium (molar mass 4.00 g/mol)? c) argon (molar mass 39.9 g/mol)? See Table 17–1 for values of γ. d) Compare your answers for parts (a), (b) and (c) with the speed in air at the same temperature.

19–23 A jet airliner is cruising at high altitude at a speed of 850 km/h (about 530 mi/h). This is equal to 0.85 times the speed of sound at that altitude (also called "Mach 0.85"). a) What is the air temperature at this altitude? b) With this data, can you determine the air pressure at this altitude? Explain.

19–24 Use the definition $B = -V\, dp/dV$ and the relation between p and V for an isothermal process to derive Eq. (19–25).

19–25 An 80.0-m-long brass rod is struck at one end. A person at the other end hears two sounds as a result of two longitudinal waves, one traveling in the metal rod and the other traveling in the air. What is the time interval between the two sounds? The speed of sound in air is 344 m/s; relevant information about brass can be found in Table 11–1 and Table 14–1.

19–26 What is the difference between the speed of longitudinal waves in air at 27.0°C and their speed at −13.0°C?

19–27 What must be the stress (F/A) in a stretched wire of a material whose Young's modulus is Y for the speed of longitudinal waves to equal 30 times the speed of transverse waves?

SECTION 19–8 ENERGY IN WAVE MOTION

19–28 A piano wire with mass 3.00 g and length 80.0 cm is stretched with a tension of 25.0 N. A wave with frequency 120.0 Hz and amplitude 1.6 mm travels along the wire. a) Calculate the average power carried by the wave. b) What happens to the average power if the wave amplitude is halved?

19–29 A longitudinal wave of frequency 220 Hz travels down a copper rod of radius 8.00 mm. The average power in the wave is 6.50 μW. (See Tables 11–1 and 14–1 for the necessary data about copper.) a) Find the wavelength of the wave. b) Find the amplitude A of the wave. c) Find the maximum longitudinal velocity of a particle in the rod.

19–30 Longitudinal Waves in Different Fluids. a) A longitudinal wave propagating in a water-filled pipe has intensity 3.00×10^{-6} W/m^2 and frequency 3400 Hz. Find the amplitude A and wavelength λ of the wave. Water has density 1000 kg/m^3 and bulk modulus 2.18×10^9 Pa. b) If the pipe is filled with air at pressure 1.00×10^5 Pa and density 1.20 kg/m^3, what will be the amplitude A and wavelength λ of a longitudinal wave with the same intensity and frequency as in part (a)? c) In which fluid is the amplitude larger, water or air? What is the ratio of the two amplitudes? Why is this ratio so different from one?

PROBLEMS

19–31 A transverse sine wave with an amplitude of 2.50 mm and a wavelength of 1.80 m travels from left to right along a long horizontal stretched string with a speed of 36.0 m/s. Take the origin at the left end of the undisturbed string. At time $t = 0$ the left end of the string is at the origin and is moving upward. a) What are the frequency, angular frequency, and wave number of the wave? b) What is the function $y(x, t)$ that describes the wave? c) What is $y(t)$ for a particle at the left end of the string? d) What is $y(t)$ for a particle 1.35 m to the right of the origin? e) What is the maximum magnitude of transverse velocity of any particle of the string? f) Find the transverse displacement and the transverse velocity of a particle 1.35 m to the right of the origin at time $t = 0.0625$ s.

19–32 A transverse wave on a rope is given by

$$y(x, t) = (0.750 \text{ cm}) \sin \pi[(250 \text{ s}^{-1})t + (0.400 \text{ cm}^{-1})x].$$

a) Find the amplitude, period, frequency, wavelength, and speed of propagation. b) Sketch the shape of the rope at the following values of t: 0, 0.0005 s, and 0.0010 s. c) Is the wave traveling in the $+x$- or $-x$-direction? d) The mass per unit length of the rope is 0.500 kg/m. Find the tension. e) Find the average power of this wave.

19–33 Three pieces of string, each of length L, are joined together end to end to make a combined string of length $3L$.

The first piece of string has mass per unit length μ_1, the second piece has mass per unit length $\mu_2 = 4\mu_1$, and the third piece has mass per unit length $\mu_3 = \mu_1/4$. a) If the combined string is under tension F, how much time does it take a transverse wave to travel the entire length $3L$? Give your answer in terms of L, F, and μ_1. b) Does your answer to part (a) depend on the order in which the three pieces are joined together? Explain.

19–34 An ant of mass m is standing peacefully on top of a horizontal stretched rope. The rope has mass per unit length μ and is under tension F. Without warning, Throckmorton starts a sinusoidal transverse wave of wavelength λ propagating along the rope. The motion of the rope is in a vertical plane. What minimum wave amplitude will make the ant become momentarily weightless? Assume that m is so small that the presence of the ant has no effect on the propagation of the wave.

19–35 When a transverse sinusoidal wave is present on a string, the particles of the string undergo simple harmonic motion. This is the same motion as that of a mass m attached to an ideal spring of force constant k', for which the angular frequency of oscillation was found in Chapter 13 to be $\omega = \sqrt{k'/m}$. Consider a string with tension F and mass per unit length μ along which is propagating a sinusoidal wave with amplitude A and wavelength λ. a) Find the "force constant" k' of the restoring force that acts on a short segment of the string of length Δx (where $\Delta x \ll \lambda$). b) How does the "force constant" calculated in part (b) depend on F, μ, A, and λ? Explain the physical reasons why this should be so.

19–36 Motion of a Wave Pulse. The sinusoidal waves described in Section 19–4 can be produced on a string by continually oscillating one end of the string. If instead the end of the string is given a single shake, a *wave pulse* propagates down the string. A particular wave pulse is described by the function

$$y(x, t) = \frac{A^3}{A^2 + (x - vt)^2},$$

where $A = 1.00$ cm and $v = 20.0$ m/s. a) Sketch the pulse as a function of x at $t = 0$. How far along the string does the pulse extend? b) Sketch the pulse as a function of x at $t = 0.001$ s. c) At the point $x = 4.50$ cm, at what time t is the displacement maximum? At what two times is the displacement at $x = 4.50$ cm equal to half its maximum value? d) Show that $y(x, t)$ satisfies the wave equation, Eq. (19–12).

19–37 A Non-Sinusoidal Wave. The shape of a wave on a string at a specific instant is shown in Fig. 19–17. The wave is propagating to the right, in the $+x$-direction. a) Determine the direction of the transverse *velocity* of each of the six labeled points on the string. If the velocity is zero, state so. Explain your reasoning. b) Determine the direction of the transverse *acceleration* of each of the six labeled points on the string. Explain your reasoning. c) How would your answers be affected if the wave was propagating to the left, in the $-x$-direction?

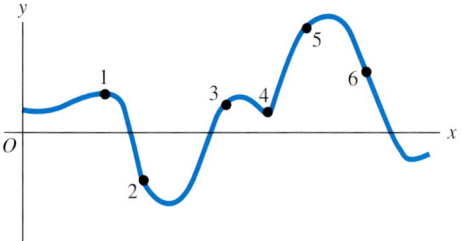

FIGURE 19–17 Problem 19–37.

19–38 Locating Lightning by Radio. While doing her physics homework, a student in the Boston area is listening in her dorm room to the radio broadcast of a Red Sox–Yankees baseball game at Boston's Fenway Park. In the bottom of the fourth inning a thunderstorm moving from west to east makes its presence known in three ways: (1) The student sees a lightning flash (and hears the electromagnetic pulse on her radio receiver); (2) 3.00 s later, she hears the thunder over the radio. (3) 4.43 s after the lightning flash, the thunder rattles her window. By a previous careful measurement she knows that she is 1.12 km due north of the broadcast booth at the ballpark. The speed of sound is 344 m/s. Where did the lightning flash occur in relation to the ballpark?

19–39 Waves of Arbitrary Shape. a) Explain why *any* wave described by a function of the form $y(x, t) = f(t - x/v)$ moves in the $+x$-direction with speed v. b) Show that $y(x, t) = f(x - vt)$ satisfies the wave equation, no matter what the functional form of f. To do this, write $y(x, t) = f(u)$, where $u = t - x/v$. Then, to take partial derivatives of $y(x, t)$, use the chain rule:

$$\frac{\partial y(x, t)}{\partial t} = \frac{df(u)}{du}\frac{\partial u}{\partial t} = \frac{df(u)}{du},$$

$$\frac{\partial y(x, t)}{\partial x} = \frac{df(u)}{du}\frac{\partial u}{\partial x} = \frac{df(u)}{du}\left(-\frac{1}{v}\right).$$

c) A wave pulse is described by the function $y(x, t) = De^{-(Bx - Ct)^2}$, where B, C, and D are all positive constants. What is the speed of this wave?

19–40 Equation (19–7) for a sinusoidal wave can be made more general by including a phase angle ϕ, where $0 \le \phi \le 2\pi$ (in radians) Then the wave function $y(x, t)$ becomes

$$y(x, t) = A \sin (\omega t - kx + \phi).$$

a) Sketch the wave as a function of x at $t = 0$ for $\phi = 0$, $\phi = \pi/4$, $\phi = \pi/2$, $\phi = 3\pi/4$, and $\phi = 3\pi/2$. b) Calculate the transverse velocity $v_y = \partial y/\partial t$. c) At $t = 0$, a particle on the string at $x = 0$ has displacement $y = A/\sqrt{2}$. Is this enough information to determine the value of ϕ? If you are told in addition that a particle at $x = 0$ is moving toward $y = 0$ at $t = 0$, what is the value of ϕ? d) Explain in general what you must know about the wave's behavior at a given instant to determine the value of ϕ.

19–41 a) Show that Eq. (19–33) can also be written as $P_{av} = \frac{1}{2} Fk\omega A^2$, where k is the wave number of the wave. b) If the tension F in the string is quadrupled while the amplitude A is kept the same, how must k and ω each change to keep the average power constant? (*Hint:* Recall Eq. (19–6).)

19–42 Instantaneous Power in a Wave. a) Graph $y(x, t)$ as given by Eq. (19–7) as a function of x for a given time t (say, $t = 0$). On the same axes, make a graph of the instantaneous power $P(x, t)$ as given by Eq. (19–31). b) Explain the connection between the slope of the graph of $y(x, t)$ versus x and the value of $P(x, t)$. In particular, explain what is happening at points where $P = 0$, where there is no instantaneous energy transfer. c) The quantity $P(x, t)$ always has the same sign. What does this imply about the direction of energy flow? d) Consider a wave moving in the $-x$-direction, for which $y(x, t) = A \sin (\omega t + kx)$. Calculate $P(x, t)$ for this wave, and make a graph of $y(x, t)$ and $P(x, t)$ as functions of x for a given time t (say, $t = 0$). What differences arise from reversing the direction of the wave?

19–43 A metal wire, with a density ρ and a Young's modulus Y, is stretched between rigid supports. At temperature T the speed of a transverse wave is found to be v_1. When the temperature is increased to $T + \Delta T$, the speed decreases to $v_2 < v_1$. Determine the coefficient of linear expansion of the wire.

CHALLENGE PROBLEMS

19–44 Longitudinal Waves on a Spring. A long spring such as a Slinky™ is often used to demonstrate longitudinal waves.

a) Show that if a spring that obeys Hooke's law has mass m, length L, and force constant k', the speed of longitudinal waves on the spring is $v = L\sqrt{k'/m}$. b) Evaluate v for a spring with $m = 0.250$ kg, $L = 2.00$ m, and $k' = 1.50$ N/m.

19–45 a) Show that for a wave on a string, the kinetic energy *per unit length of string* is

$$u_k(x, t) = \frac{1}{2}\mu v_y^2(x, t) = \frac{1}{2}\mu\left(\frac{\partial y(x, t)}{\partial t}\right)^2,$$

where μ is the mass per unit length. b) Calculate $u_k(x, t)$ for a sinusoidal wave given by Eq. (19–7). c) There is also elastic potential energy in the string, associated with the work required to deform and stretch the string. Consider a short segment of string at position x that has unstretched length Δx, as in Fig. 19–8. Neglecting the (small) curvature of the segment, its slope is $\partial y(x, t)/\partial x$. Assume that the displacement of the string from equilibrium is small, so that $\partial y/\partial x$ has a magnitude much less than unity. Show that the stretched length of the segment is approximately

$$\Delta x\left[1 + \frac{1}{2}\left(\frac{\partial y(x, t)}{\partial x}\right)^2\right].$$

(*Hint:* Use the relationship $\sqrt{1 + u} \approx 1 + \frac{1}{2}u$, valid for $|u| \ll 1$.) d) The potential energy stored in the segment equals the work done by the string tension F (which acts along the string) to stretch the segment from its unstretched length Δx to the length calculated in part (c). Calculate this work, and show that the potential energy *per unit length of string* is

$$u_p(x, t) = \frac{1}{2}F\left(\frac{\partial y(x, t)}{\partial x}\right)^2.$$

e) Calculate $u_p(x, t)$ for a sinusoidal wave given by Eq. (19–7). f) Show that $u_k(x, t) = u_p(x, t)$ for all x and t. g) Show $y(x, t)$, $u_k(x, t)$, and $u_p(x, t)$ as functions of x for $t = 0$ in one graph with all three functions on the same axes. Explain why u_k and u_p are maximum where y is zero, and vice versa. h) Show that the instantaneous power in the wave, given by Eq. (19–30), is equal to the total energy per unit length multiplied by the wave speed v. Explain why this result is reasonable.

19–46 A deep-sea diver is suspended beneath the surface of Loch Ness by a 100-m long cable that is attached to a boat on the surface (Fig. 19–18). The diver and his suit have a total mass of 120 kg and a volume of 0.0800 m³. The cable has a diameter of 2.00 cm and a linear mass density of $\mu = 1.10$ kg/m. The diver thinks he sees something moving in the murky depths and jerks the end of the cable back and forth to send transverse waves up the cable as a signal to his companions in the boat. a) What is

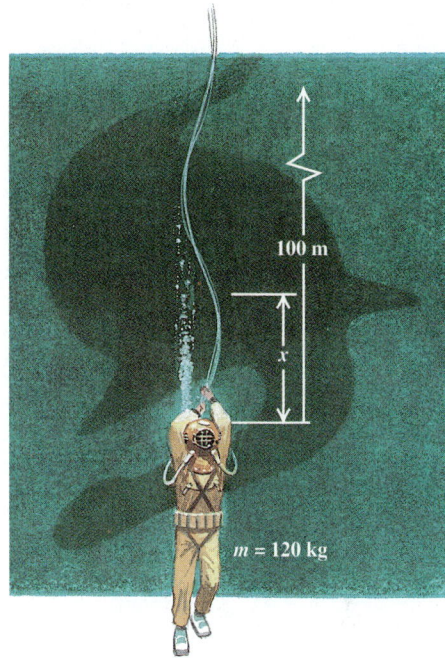

FIGURE 19–18 Challenge Problem 19–46.

the tension in the cable at its lower end, where it is attached to the diver? Do not forget to include the buoyant force that the water (density 1000 kg/m³) exerts on him. b) Calculate the tension in the cable a distance x above the diver. The buoyant force on the cable must be included in your calculation. c) The speed of transverse waves on the cable is given by $v = \sqrt{F/\mu}$ (Eq. 19–13) . The speed therefore varies along the cable, since the tension is not constant. (This expression neglects the damping force that the water exerts on the moving cable.) Integrate to find the time required for the first signal to reach the surface.

19–47 A uniform rope of length L and mass m is held at one end and whirled in a horizontal circle with angular velocity ω. Ignore the force of gravity on the rope. Find the time required for a transverse wave to travel from one end of the rope to the other.

19–48 For longitudinal waves in a solid rod, prove that the power per unit cross-section area in the wave motion (the *intensity*) is given by Eq. (19–35).

20

Wave Interference and Normal Modes

As these circular waves on the surface of a pond pass through each other, they combine in the simplest possible way: how much the water is disturbed at any point is simply the sum of the disturbances from each individual wave. This same principle of superposition applies to waves of many different kinds, including waves on a string and sound waves in a pipe.

20-1 INTRODUCTION

When a wave strikes the boundaries of its medium, all or part of the wave is *reflected*. When you yell at a building wall or a cliff face some distance away, the sound wave is reflected from the rigid surface, and an *echo* comes back. When you flip the end of a rope whose far end is tied to a rigid support, a pulse travels the length of the rope and is reflected back to you. In both cases the initial and reflected waves overlap in the same region of the medium. This overlapping of waves is called *interference*.

When there are *two* boundary points or surfaces, such as a guitar string that's tied down at both ends, we get repeated reflections. In such situations we find that sinusoidal waves can occur only for certain special frequencies, which are determined by the properties and dimensions of the medium. These special frequencies and their associated wave patterns are called *normal modes*. The pitches of most musical instruments are determined by the normal-mode frequencies of the vibrating part of the instrument. Normal-mode motion also explains why your singing voice sounds better in the shower and why a trained singer's amplified voice can break a crystal goblet if the correct note is sung.

In this chapter our focus will be on the interference of mechanical waves. But interference is also important in nonmechanical waves. It explains the colors seen in soap bubbles and is the basic principle behind holograms and the use of x rays to explore crystal structure. Interference and normal modes also appear in determining the energy levels of atoms. We'll explore these aspects of interference in later chapters.

20-2 BOUNDARY CONDITIONS FOR A STRING AND THE PRINCIPLE OF SUPERPOSITION

As a simple example of wave reflections and the role of the boundary of a wave medium, let's look again at transverse waves on a stretched string. What happens when a wave pulse or a sinusoidal wave arrives at the *end* of the string?

If the end is fastened to a rigid support, it is a *fixed* end that cannot move. The arriving wave exerts a force on the support; the reaction to this force, exerted *by* the support *on* the string, "kicks back" on the string and sets up a *reflected* pulse or wave traveling in the reverse direction. Figure 20–1 is a series of photographs showing the reflection of a pulse at the fixed end of a long coiled spring. The reflected pulse moves in the opposite direction from the initial, or *incident*, pulse, and its displacement is also opposite. This situation is illustrated for a wave pulse on a string in Fig. 20–2a.

The opposite situation from an end that is held stationary is a *free* end, one that is perfectly free to move in the direction perpendicular to the length of the string. For example, the string might be tied to a light ring that slides on a frictionless rod perpendicular to the rope, as in

20-1 A series of images of a wave pulse, equally spaced in time from top to bottom. The pulse starts at the right in the top image, travels to the left, and is reflected from the fixed end at the left.

(a)   (b)

20-2 Reflection of a wave pulse (a) at a fixed end of a string and (b) at a free end. Time increases from top to bottom in each figure.

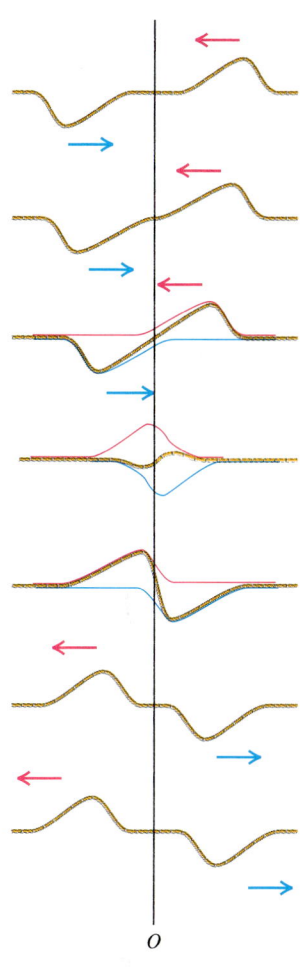

O

20-3 Overlap of two wave pulses traveling in opposite directions with one pulse inverted with respect to the other. Time increases from top to bottom.

Fig. 20–2b. The ring and rod maintain the tension but exert no transverse force. When a wave arrives at this free end, the ring slides along the rod. The ring reaches a maximum displacement, and both it and the string come momentarily to rest, as in the fourth drawing from the top in Fig. 20–2b. But the string is now stretched, giving increased tension, so the free end of the string is pulled back down, and again a reflected pulse is produced. As for a fixed end, the reflected pulse moves in the opposite direction from the initial pulse, but now the direction of the displacement is the same as for the initial pulse. The conditions at the end of the string, such as a rigid support or the complete absence of transverse force, are called **boundary conditions.**

The formation of the reflected pulse is similar to the overlap of two pulses traveling in opposite directions. Figure 20–3 shows two pulses with the same shape, one inverted with respect to the other, traveling in opposite directions. As the pulses overlap and pass each other, the total displacement of the string is the *algebraic sum* of the displacements at that point in the individual pulses. Because these two pulses have the same shape, the total displacement at point O in the middle of the figure is zero at all times. Thus the motion of the right half of the string would be the same if we cut the string at point O,

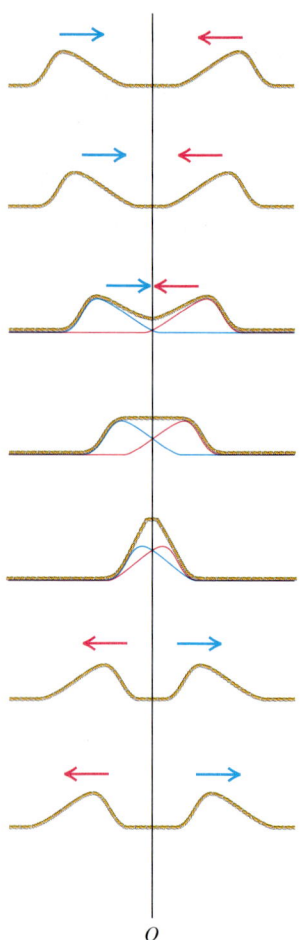

O

20–4 Overlap of two wave pulses traveling in opposite directions with no inversion of one pulse. Time increases from top to bottom.

threw away the left side, and held the end at O fixed. The two pulses on the right side then correspond to the incident and reflected pulses, combining so that the total displacement at O is *always* zero. For this to occur, the reflected pulse must be inverted relative to the incident pulse.

Figure 20–4 shows two pulses with the same shape, traveling in opposite directions but *not* inverted relative to each other. The displacement at point O in the middle of the figure is not zero, but the slope of the string at this point is always zero. According to Eq. (19–28), this corresponds to the absence of any transverse force at this point. In this case the motion of the right half of the string would be the same as if we cut the string at point O and anchored the end with a frictionless sliding ring (Fig. 20–2b) that maintains tension without exerting any transverse force. In other words, this situation corresponds to reflection of a pulse at a free end of a string at point O. In this case the reflected pulse is *not* inverted.

THE PRINCIPLE OF SUPERPOSITION

Combining the displacements of the separate pulses at each point to obtain the actual displacement is an example of the **principle of superposition:** when two waves overlap, the actual displacement of any point on the string at any time is obtained by adding the displacement the point would have if only the first wave were present and the displacement it would have if only the second wave were present. In other words, the wave function $y(x, t)$ that describes the resulting motion in this situation is obtained by *adding* the two wave functions for the two separate waves.

Mathematically, this additive property of wave functions follows from the form of the wave equation, Eq. (19–12) or (19–18), which every physically possible wave function must satisfy. Specifically, the wave equation is *linear;* that is, it contains the function $y(x, t)$ only to the first power (there are no terms involving $y(x, t)^2$, $y(x, t)^{1/2}$, etc.). As a result, if any two functions $y_1(x, t)$ and $y_2(x, t)$ satisfy the wave equation separately, their sum $y_1(x, t) + y_2(x, t)$ also satisfies it and is therefore a physically possible motion. Because this principle depends on the linearity of the wave equation and the corresponding linear-combination property of its solutions, it is also called the *principle of linear superposition.* For some physical systems, such as a medium that does not obey Hooke's law, the wave equation is *not* linear; this principle does not hold for such systems.

The principle of superposition is of central importance in all types of waves. When a friend talks to you while you are listening to music, you can distinguish the sound of speech and the sound of music from each other. This is precisely because the total sound wave reaching your ears is the algebraic sum of the wave produced by your friend's voice and the wave produced by the speakers of your stereo. If two sound waves did *not* combine in this simple linear way, the sound you would hear in this situation would be a hopeless jumble. Superposition also applies to electromagnetic waves (such as light) and many other types of waves.

20–3 STANDING WAVES ON A STRING

We have talked about the reflection of a wave *pulse* on a string when it arrives at a boundary point (either a fixed end or a free end). Now let's look at what happens when a *sinusoidal* wave is reflected by a fixed end of a string. We will again approach the problem by considering the superposition of two waves propagating through the string, one representing the original or incident wave and the other representing the wave reflected at the fixed end. The general term **interference** is used to describe the result of two or more waves passing through the same region at the same time.

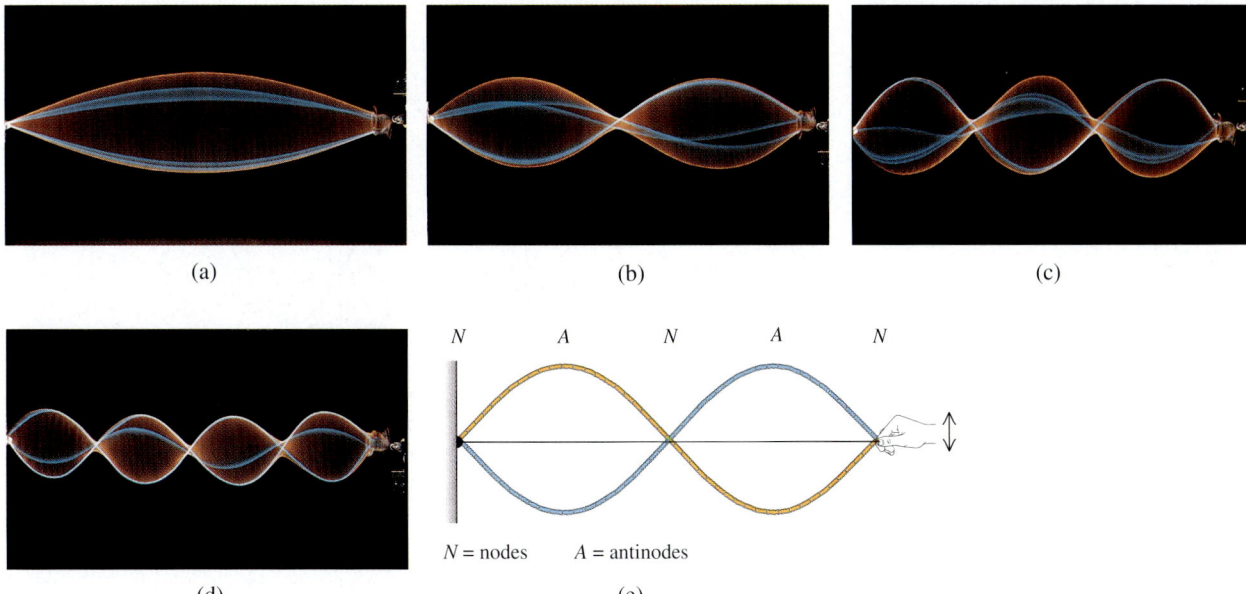

(a)   (b)   (c)

(d)   (e)

N = nodes A = antinodes

20–5 (a)–(d) Time exposures of standing waves in a stretched string. From (a) to (d) the frequency of oscillation of the right-hand end increases, and the wavelength of the standing wave
decreases. (e) The extremes of the motion of the standing wave in (b), with nodes at the center
and at the ends. The right-hand end of the string moves very little in comparison to the antinodes
and so is essentially a node.

Figure 20–5 shows a string that is fixed at its left-hand end. Its right-hand end is
moved up and down in simple harmonic motion to produce a wave that travels to the left;
the wave reflected from the fixed end travels to the right. The resulting motion when the
two waves combine no longer looks like two waves traveling in opposite directions. The
string appears to be subdivided into a number of segments, as in the time-exposure
photographs of Figs. 20–5a, 20–5b, 20–5c, and 20–5d. Figure 20–5e shows two instantaneous shapes of the string in Fig. 20–5b. Let's compare this behavior with the waves
we studied in Chapter 19. In a wave that travels along the string, the amplitude is constant, and the wave pattern moves with a speed equal to the wave speed. Here, instead,
the wave pattern remains in the same position along the string, and its amplitude fluctuates. There are particular points called **nodes** (labeled N in Fig. 20–5e) that never move
at all. Midway between the nodes are points called **antinodes** (labeled A in Fig. 20–5e)
where the amplitude of motion is greatest. Because the wave pattern doesn't appear to
be moving in either direction along the string, it is called a **standing wave.** (To emphasize the difference, a wave that *does* move along the string is called a **traveling wave.**)

The principle of superposition explains how the incident and reflected wave combine
to form a standing wave. In Fig. 20–6 the red curves show a wave traveling to the left.
The blue curves show a wave traveling to the right with the same propagation speed,
wavelength, and amplitude. The waves are shown at nine instants, one sixteenth of a
period apart. At each point along the string, we add the displacements (the values of y)
for the two separate waves; the result is the total wave on the string, shown in black.

At certain instants, such as $t = 8T/16$, the two wave patterns are exactly in phase with
each other, and the shape of the string is a sine curve with twice the amplitude of either
individual wave. At other instants, such as $t = 4T/16$, the two waves are exactly out of
phase with each other, and the total wave at that instant is zero. The resultant displacement is *always* zero at those places marked N at the bottom of Fig. 20–6. These are the
nodes. At a node the displacements of the two waves in red and blue are always equal

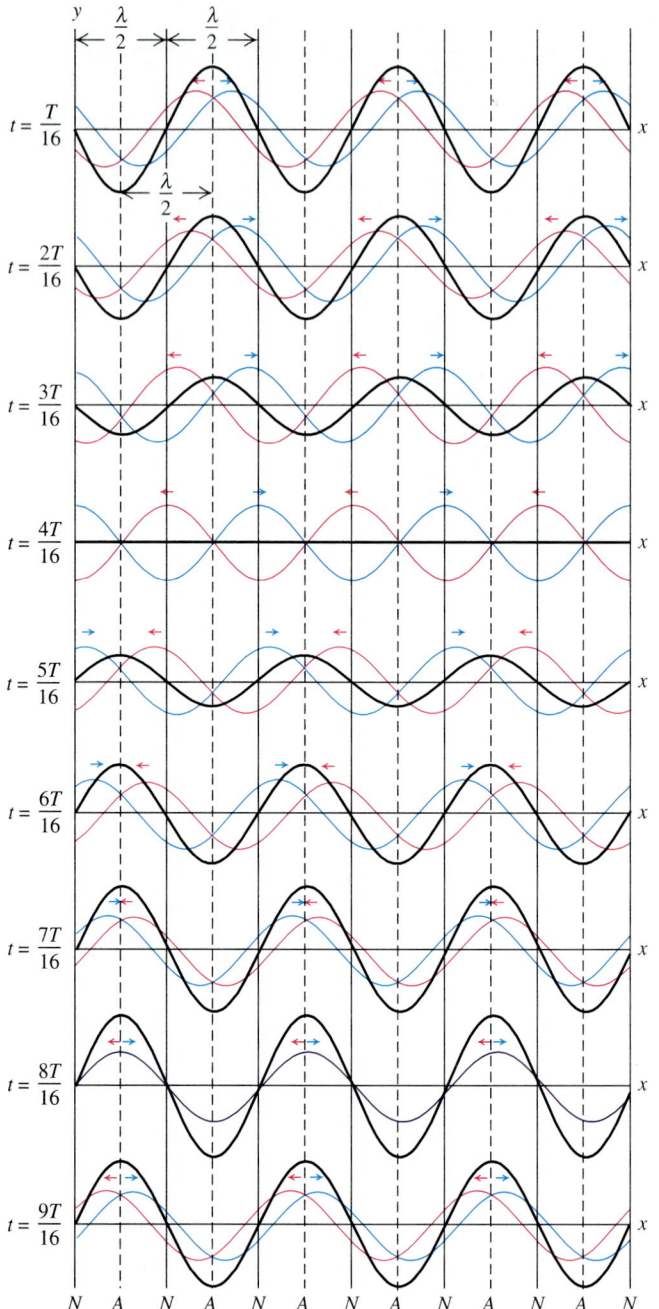

20–6 Formation of a standing wave. A wave traveling to the left (red curves) combines with a wave traveling to the right (blue curves) to form a standing wave (black curves). The horizontal x-axis in each part shows the equilibrium position of the string.

and opposite and cancel each other out. This cancellation is called **destructive interference.** Midway between the nodes are the points of *greatest* amplitude or *antinodes,* marked A. At the antinodes the displacements of the two waves in red and blue are always identical, giving a large resultant displacement; this phenomenon is called **constructive interference.** We can see from the figure that the distance between successive nodes or between successive antinodes is one half-wavelength, or $\lambda/2$.

We can derive a wave function for the standing wave of Fig. 20–6 by adding the wave functions $y_1(x, t)$ and $y_2(x, t)$ for two waves with equal amplitude, period, and wavelength traveling in opposite directions. Here $y_1(x, t)$ (the red curves in Fig. 20–6) represents an *incident* wave traveling to the left along the +*x*-axis, arriving at the point *x* = 0 and being reflected; $y_2(x, t)$ (the blue curves in Fig. 20–6) represents the *reflected* wave traveling to the right from *x* = 0. We noted in Section 20–2 that the wave reflected from a fixed end of a string is inverted, so we give a negative sign to one of the waves:

$$y_1(x, t) = A \sin(\omega t + kx) \qquad \text{(traveling to the left),}$$

$$y_2(x, t) = -A \sin(\omega t - kx) \qquad \text{(traveling to the right).}$$

Note also that the change of sign corresponds to a shift in *phase* of 180° or π radians. At *x* = 0 the motion from the incident wave is $A \sin \omega t$, and the motion from the reflected wave is $-A \sin \omega t$, which we can also write as $A \sin(\omega t + \pi)$. The wave function for the standing wave is the sum of the individual wave functions:

$$y(x, t) = y_1(x, t) + y_2(x, t) = A[\sin(\omega t + kx) - \sin(\omega t - kx)].$$

We can rearrange this by using the identities for the sine of the sum and difference of two angles: $\sin(a \pm b) = \sin a \cos b \pm \cos a \sin b$. Using these and combining terms, we obtain the wave function for the standing wave:

$$y(x, t) = y_1(x, t) + y_2(x, t) = (2A \sin kx) \cos \omega t,$$

or

$$y(x, t) = (A_{\text{SW}} \sin kx) \cos \omega t \qquad \begin{array}{l} \text{(standing wave on a string,} \\ \text{fixed end at } x = 0). \end{array} \qquad (20\text{–}1)$$

The standing wave amplitude A_{SW} is twice the amplitude A of either of the original traveling waves:

$$A_{\text{SW}} = 2A.$$

Equation (20–1) has two factors: a function of *x* and a function of *t*. The factor $2A \sin kx$ shows that at each instant the shape of the string is a sine curve. But unlike a wave traveling along a string, the wave shape stays in the same position, oscillating up and down as described by the $\cos \omega t$ factor. This behavior is shown graphically by the black curves in Fig. 20–6. Each point in the string still undergoes simple harmonic motion, but all the points between any successive pair of nodes oscillate *in phase*. This is in contrast to the phase differences between oscillations of adjacent points that we see with a wave traveling in one direction.

We can use Eq. (20–1) to find the positions of the nodes; these are the points for which $\sin kx = 0$, so the displacement is *always* zero. This occurs when $kx = 0$, π, 2π, 3π, ..., or, using $k = 2\pi/\lambda$,

$$x = 0, \frac{\pi}{k}, \frac{2\pi}{k}, \frac{3\pi}{k}, \ldots$$

$$= 0, \frac{\lambda}{2}, \frac{2\lambda}{2}, \frac{3\lambda}{2}, \ldots \qquad \begin{array}{l} \text{(nodes of a standing wave on a} \\ \text{string, fixed end at } x = 0). \end{array} \qquad (20\text{–}2)$$

In particular, there is a node at *x* = 0, as there should be, since this point is a fixed end of the string.

A standing wave, unlike a traveling wave, *does not* transfer energy from one end to the other. The two waves that form it would individually carry equal amounts of power in opposite directions. There is a local flow of energy from each node to the adjacent antinodes and back, but the *average* rate of energy transfer is zero at every point. We invite you to evaluate the wave power given by Eq. (19–29) using the wave function of Eq. (20–1) and show that the average power is zero (see Challenge Problem 20–44).

EXAMPLE 20-1

A long string lies along the x-axis when in equilibrium. The end of the string at $x = 0$ is tied down. An incident sinusoidal wave (corresponding to the red curves in Fig. 20–6) travels along the string in the $-x$-direction at 84.0 m/s with an amplitude of 1.50 mm and a frequency of 120 Hz. This wave is reflected from the fixed end at $x = 0$, and the superposition of the incident traveling wave and the reflected traveling wave forms a standing wave. a) Find the equation giving the displacement of a point on the string as a function of position and time. b) Locate the points on the string that don't move at all. c) Find the amplitude, maximum transverse velocity, and maximum transverse acceleration at points of maximum oscillation.

SOLUTION a) Since there is a fixed end at $x = 0$, we may use Eq. (20–1) to describe this standing wave. The amplitude of the incident wave is $A = 1.50$ mm $= 1.50 \times 10^{-3}$ m; the reflected wave has the same amplitude, and the standing wave amplitude is $A_{sw} = 2A = 3.00 \times 10^{-3}$ m. The angular frequency ω and wave number k are

$$\omega = 2\pi f = (2\pi \text{ rad})(120 \text{ s}^{-1}) = 754 \text{ rad/s},$$

$$k = \frac{\omega}{v} = \frac{754 \text{ rad/s}}{84.0 \text{ m/s}} = 8.98 \text{ rad/m}.$$

Then Eq. (20–1) gives

$$y(x, t) = (A_{sw} \sin kx) \cos \omega t$$

$$= [(3.00 \times 10^{-3} \text{ m}) \sin (8.98 \text{ rad/m})x] \cos (754 \text{ rad/s})t.$$

b) The *nodes* are the points on the string that don't move. If we use our expression for $y(x, t)$ from part (a), these are the points at which $\sin (8.98 \text{ rad/m})x = 0$, so $(8.98 \text{ rad/m})x = 0, \pi, 2\pi, 3\pi, \ldots$, or

$$x = 0, \frac{\pi}{8.98 \text{ rad/m}}, \frac{2\pi}{8.98 \text{ rad/m}}, \frac{3\pi}{8.98 \text{ rad/m}}, \ldots$$

$$= 0, \ 0.350 \text{ m}, \ 0.700 \text{ m}, \ 1.050 \text{ m}, \ldots.$$

We could also find the positions of the nodes using Eq. (20–2). The wavelength is $\lambda = v/f = (84.0 \text{ m/s})/(120 \text{ s}^{-1}) = 0.700$ m, and Eq. (20–2) again gives $x = 0, 0.350$ m, 0.700 m, 1.050 m, ...

c) From the expression in part (a) for $y(x, t)$ we see that the maximum displacement from equilibrium is 3.00×10^{-3} m $= 3.00$ mm, which is just twice the amplitude of the incident wave. This maximum occurs at the antinodes, where $\sin (8.98 \text{ rad/m})x = \pm 1$, $(8.98 \text{ rad/m})x = \pi/2, 3\pi/2, 5\pi/2, \ldots$, and $x = 0.175$ m, 0.525 m, 0.875 m, ... Each antinode is midway between two nodes.

To find the transverse velocity (that is, velocity in the y-direction) of a particle on the string at any point x, we take the first partial derivative of $y(x, t)$ with respect to time:

$$v_y(x, t) = \frac{\partial y(x, t)}{\partial t} = [(3.00 \times 10^{-3} \text{ m}) \sin (8.98 \text{ rad/m})x]$$

$$\times [(-754 \text{ rad/s}) \sin (754 \text{ rad/s})t]$$

$$= [(-2.26 \text{ m/s}) \sin (8.98 \text{ rad/m})x] \sin (754 \text{ rad/s})t.$$

This is largest at an antinode, where $\sin (8.98 \text{ rad/m})x = \pm 1$ and the transverse velocity varies in value between 2.26 m/s and -2.26 m/s. As is always the case in simple harmonic motion, the maximum velocity occurs when the particle is passing through the equilibrium position ($y = 0$).

The transverse acceleration $a_y(x, t)$ is the first partial derivative of $v_y(x, t)$ with respect to time (that is, the *second* partial derivative of $y(x, t)$ with respect to time). We leave the calculation to you; the result is

$$a_y(x, t) = \frac{\partial v_y(x, t)}{\partial t} = \frac{\partial^2 y(x, t)}{\partial t^2}$$

$$= [(-1.71 \times 10^3 \text{ m/s}^2) \sin (8.98 \text{ rad/m})x] \cos (754 \text{ rad/s})t.$$

This is also largest at the antinodes; at these points the transverse acceleration varies in value between $+1.71 \times 10^3$ m/s^2 and -1.71×10^3 m/s^2. This maximum value is tremendous, 174 times the acceleration due to gravity!

EXAMPLE 20-2

Cousin Throckmorton is once again playing with the clothesline in Example 19–2 (Section 19–4). One end of the clothesline is attached to a vertical post; Throckmorton holds the other end loosely in his hand so that the speed of waves on the clothesline is a relatively slow 0.720 m/s. By experiment, Throckmorton finds several frequencies at which he can oscillate his end of the clothesline so that a light clothespin 45.0 cm from the post doesn't move. What are these frequencies?

SOLUTION There will be a standing wave on the clothesline formed by the superposition of the wave that Throcky produces and the wave reflected from the post. For the clothespin not to move, it must be at a node of the standing wave. From Eq. (20–2) the nodes are located at distances $\lambda/2, 2\lambda/2$,

$3\lambda/2, \ldots$ from the post; one of these must equal the clothespin distance $d = 45.0$ cm $= 0.450$ m. So $d = n\lambda/2$, where n could be any positive integer 1, 2, 3, ..., and so the wavelength is $\lambda = 2d/n$. Solving for the frequency, we get

$$f = \frac{v}{\lambda} = \frac{nv}{2d}$$

$$= n\frac{0.720 \text{ m/s}}{2(0.450 \text{ m})} = n(0.800 \text{ Hz})$$

$$= 0.800 \text{ Hz}, \ 1.60 \text{ Hz}, \ 2.40 \text{ Hz}, \ldots.$$

With $f = 0.800$ Hz the clothespin is at the first node beyond the post; with $f = 1.60$ Hz it is at the second node; with $f = 2.40$ Hz it is at the third node; and so on.

20–4 NORMAL MODES OF A STRING

We have described standing waves on a string rigidly held at one end, as in Fig. 20–5. We made no assumptions about the length of the string or about what was happening at the other end. Let's now consider a string of a definite length L, rigidly held at *both* ends. Such strings are found in many musical instruments, including pianos, violins, and guitars. When a guitar string is plucked, a wave is produced in the string; this wave is reflected and re-reflected from the ends of the string, making a standing wave. This standing wave on the string in turn produces a sound wave in the air, with a frequency determined by the properties of the string. This is what makes stringed instruments so useful in making music.

To understand these properties of standing waves on a string fixed at both ends, let's first examine what happens when we set up a sinusoidal wave on such a string. The standing wave that results must have a node at *both* ends of the string. We saw in the preceding section that adjacent nodes are one half-wavelength ($\lambda/2$) apart, so the length of the string must be $\lambda/2$, or $2(\lambda/2)$, or $3(\lambda/2)$, or in general some integer number of half-wavelengths:

$$L = n\frac{\lambda}{2} \qquad (n = 1, 2, 3, \dots) \qquad \text{(string fixed at both ends).} \qquad (20\text{–}3)$$

That is, if a string with length L is fixed at both ends, a standing wave can exist only if its wavelength satisfies Eq. (20–3).

Solving this equation for λ and labeling the possible values of λ as λ_n, we find

$$\lambda_n = \frac{2L}{n} \qquad (n = 1, 2, 3, \dots) \qquad \text{(string fixed at both ends).} \qquad (20\text{–}4)$$

Waves can exist on the string if the wavelength is *not* equal to one of these values, but there cannot be a steady wave pattern with nodes and antinodes, and the total wave cannot be a standing wave. Equation (20–4) is illustrated by the standing waves shown in Figs. 20–5a, 20–5b, 20–5c, and 20–5d; these represent $n = 1, 2, 3,$ and 4, respectively.

Corresponding to the series of possible standing-wave wavelengths λ_n is a series of possible standing-wave frequencies f_n, each related to its corresponding wavelength by $f_n = v/\lambda_n$. The smallest frequency f_1 corresponds to the largest wavelength (the $n = 1$ case), $\lambda_1 = 2L$:

$$f_1 = \frac{v}{2L} \qquad \text{(string fixed at both ends).} \qquad (20\text{–}5)$$

This is called the **fundamental frequency.** The other standing-wave frequencies are $f_2 = 2v/2L$, $f_3 = 3v/2L$, and so on. These are all integer multiples of the fundamental frequency f_1, such as $2f_1$, $3f_1$, $4f_1$, and so on, and we can express *all* the frequencies as

$$f_n = n\frac{v}{2L} = nf_1 \qquad (n = 1, 2, 3, \dots) \qquad \text{(string fixed at both ends).} \quad (20\text{–}6)$$

These frequencies are called **harmonics,** and the series is called a **harmonic series.** Musicians sometimes call f_2, f_3, and so on **overtones;** f_2 is the second harmonic or the first overtone, f_3 is the third harmonic or the second overtone, and so on. The first harmonic is the same as the fundamental frequency.

For a string with fixed ends at $x = 0$ and $x = L$, the wave function $y(x, t)$ of the nth standing wave is given by Eq. (20–1) (which satisfies the condition that there is a node at $x = 0$), with $\omega = \omega_n = 2\pi f_n$ and $k = k_n = 2\pi/\lambda_n$:

$$y_n(x, t) = A_{sw} \cos \omega_n t \sin k_n x. \qquad (20\text{–}7)$$

We invite you to show that this wave function has nodes at both $x = 0$ and $x = L$, as it must.

10.4
Standing Waves on Strings

10.5
Tuning a Stringed Instrument: Standing Waves

10.6
String Mass and Standing Waves

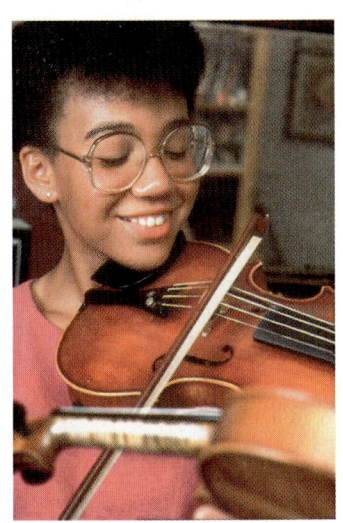

Each string of a violin naturally oscillates at one or more of its harmonic frequencies. These oscillations produce sound waves in the air with the same frequencies.

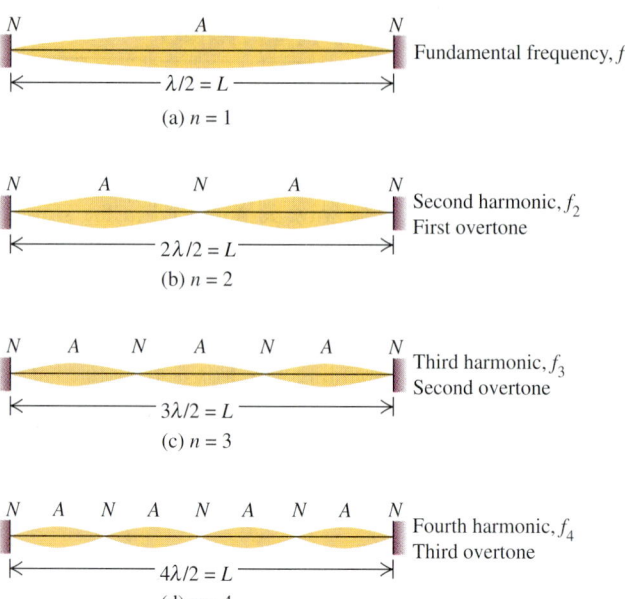

20–7 The first four normal modes of a string fixed at both ends. (Compare these to the photographs in Fig. 20–5.)

Astronomers have discovered that the sun oscillates in several different normal modes. In this theoretical model of one such mode, blue denotes places where solar material is moving outward and red denotes where it is moving inward.

A **normal mode** of an oscillating system is a motion in which all particles of the system move sinusoidally with the same frequency. For a system made up of a string of length L fixed at both ends, each of the wavelengths given by Eq. (20–4) corresponds to a possible normal mode pattern and frequency. There are infinitely many normal modes, each with its characteristic frequency and vibration pattern. Figure 20–7 shows the first four normal-mode patterns and their associated frequencies and wavelengths; these correspond to Eq. (20–7) with $n = 1$, 2, 3, and 4. By contrast, a harmonic oscillator, which has only one oscillating particle, has only one normal mode and one characteristic frequency. The string fixed at both ends has infinitely many normal modes because it is made up of a very large (effectively infinite) number of particles.

If we could displace a string so that its shape is the same as one of the normal-mode patterns and then release it, it would vibrate with the frequency of that mode. Such a vibrating string would displace the surrounding air with the same frequency, producing a traveling sinusoidal sound wave that your ears would perceive as a pure tone. But when a string is struck (as in a piano) or plucked (as is done to guitar strings), the shape of the displaced string is *not* as simple as one of the patterns in Fig. 20–7. The fundamental as well as many overtones are present in the resulting vibration. This motion is therefore a combination or *superposition* of many normal modes. Several simple-harmonic motions of different frequencies are present simultaneously, and the displacement of any point on the string is the sum (or superposition) of displacements associated with the individual modes. The sound produced by the vibrating string is likewise a superposition of traveling sinusoidal sound waves, which you perceive as a rich, complex tone with the fundamental frequency f_1. The standing wave on the string and the traveling sound wave in the air have similar **harmonic content** (the extent to which frequencies higher than the fundamental are present). The harmonic content depends on how the string is initially set into motion. If you pluck the strings of an acoustic guitar in the normal location over the sound hole, the sound that you hear has a different harmonic content than if you pluck the strings next to the fixed end on the guitar body.

It is possible to represent every possible motion of the string as some superposition of normal-mode motions. Finding this representation for a given vibration pattern is called *harmonic analysis*. The sum of sinusoidal functions that represents a complex wave is called a *Fourier series*. Figure 20–8 shows how a standing wave that is produced

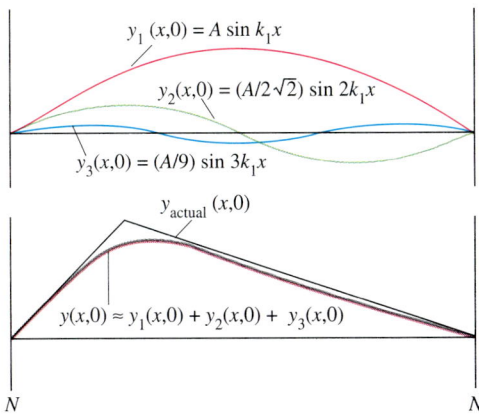

$y_1(x,0) = A \sin k_1 x$

$y_2(x,0) = (A/2\sqrt{2}) \sin 2k_1 x$

$y_3(x,0) = (A/9) \sin 3k_1 x$

$y_{\text{actual}}(x,0)$

$y(x,0) \approx y_1(x,0) + y_2(x,0) + y_3(x,0)$

N N

20-8 The standing wave that results from plucking a guitar string into a triangular shape is well represented (except at the sharp maximum point) by the sum of only three sinusoidal functions. Including additional sinusoidal functions further improves the representation.

by plucking a guitar string of length L at a point $L/4$ from one end can be represented as a combination of sinusoidal functions.

As we have seen, the fundamental frequency of a vibrating string is $f_1 = v/2L$. The speed v of waves on the string is determined by Eq. (19–13), $v = \sqrt{F/\mu}$. Combining these, we find

10.10 ActivPhysics
Complex Waves: Fourier Analysis

$$f_1 = \frac{1}{2L}\sqrt{\frac{F}{\mu}} \qquad \text{(string fixed at both ends).} \qquad (20\text{–}8)$$

This is also the fundamental frequency of the sound wave created in the surrounding air by the vibrating string. Familiar musical instruments show how f_1 depends on the properties of the string. The inverse dependence of frequency on length L is illustrated by the long strings of the bass (low-frequency) section of the piano or the bass viol compared with the shorter strings on the treble section of the piano or the violin (Fig. 20–9). The pitch of a violin or guitar is usually varied by pressing a string against the fingerboard with the fingers to change the length L of the vibrating portion of the string. Increasing the tension F increases the wave speed v and thus increases the frequency (and the pitch). All string instruments are "tuned" to the correct frequencies by varying the tension; you tighten the string to raise the pitch. Finally, increasing the mass per unit length μ decreases the wave speed and thus the frequency. The lower notes on a steel guitar are

20-9 Comparison of ranges of a piano, a violin, a viola, a cello, and a string bass. In all cases, longer strings provide bass notes and shorter strings produce treble notes.

Bass

Cello

Viola

Violin

produced by thicker strings, and one reason for winding the bass strings of a piano with wire is to obtain the desired low frequency without resorting to a string that is inconveniently long.

Problem–Solving Strategy

STANDING WAVES

1. As with the problems in Chapter 19, it is useful to distinguish between the purely kinematic quantities, such as wave speed v, wavelength λ, and frequency f, and the dynamic quantities involving the properties of the medium, including F and μ (for transverse waves on a string) and B and ρ (for longitudinal waves in a pipe, described in the next section). You can compute the wave speed if you know either λ and f (or, equivalently, $k = 2\pi/\lambda$ and $\omega = 2\pi f$) or the properties of the medium. Try to determine whether the problem is only kinematic

in nature or whether the properties of the medium are also involved.

2. In visualizing nodes and antinodes in standing waves, it is always helpful to draw diagrams. For a string you can draw the shape at one instant and label the nodes N and antinodes A. For longitudinal waves (discussed in the next section), it is not so easy to draw the shape, but you can still label the nodes and antinodes. The distance between two adjacent nodes or two adjacent antinodes is always $\lambda/2$, and the distance between a node and the adjacent antinode is always $\lambda/4$.

EXAMPLE 20-3

A giant bass viol In an effort to get your name in the *Guinness Book of World Records,* you set out to build a bass viol with strings that have a length of 5.00 m between fixed points. One string has a linear mass density of 40.0 g/m and a fundamental frequency of 20.0 Hz (the lowest frequency that the human ear can hear). a) Calculate the tension of this string. b) Calculate the frequency and wavelength on the string of the second harmonic. c) Calculate the frequency and wavelength on the string of the second overtone.

SOLUTION a) This problem involves the properties of the wave medium (the string). We are given the values of μ, L, and the fundamental frequency f_1, so we solve Eq. (20–8) for the string tension F:

$$F = 4\mu L^2 f_1^{\,2} = 4(40.0 \times 10^{-3}\ \text{kg/m})(5.00\ \text{m})^2(20.0\ \text{s}^{-1})^2$$
$$= 1600\ \text{N} = 360\ \text{lb}.$$

(In a real bass viol the string tension is typically a few hundred newtons.)

b) From Eq. (20–6) the second harmonic frequency ($n = 2$) is

$$f_2 = 2f_1 = 2(20.0\ \text{Hz}) = 40.0\ \text{Hz}.$$

From Eq. (20–4) the wavelength on the string of the second harmonic is

$$\lambda_2 = \frac{2L}{2} = 5.00\ \text{m}.$$

c) The second overtone is the "second tone over" (above) the fundamental, that is, $n = 3$. Its frequency is

$$f_3 = 3f_1 = 3(20.0\ \text{Hz}) = 60.0\ \text{Hz}.$$

The wavelength on the string is

$$\lambda_3 = \frac{2L}{3} = 3.33\ \text{m}.$$

EXAMPLE 20-4

What are the frequency and wavelength of the sound waves that are produced in the air when the string in Example 20–3 is vibrating at its fundamental frequency?

SOLUTION When the string vibrates at a particular frequency, the surrounding air is forced to vibrate at the same frequency. So the frequency of the sound wave is the same as that of the standing wave on the string, $f_1 = 20.0$ Hz. But the relationship $\lambda = v/f$ shows that the *wavelength* of the sound wave will be different from the wavelength of the standing wave on the string if the two waves have different speeds. From Example 19–7 (Section 19–7) the speed of sound in air at 20°C is $v_{\text{sound}} = 344$ m/s; for the string on the giant bass viol, for

which $\lambda_1 = 2L = 2(5.00\ \text{m}) = 10.0$ m, the wave speed is $v_{\text{string}} = \lambda_1 f_1 = (10.0\ \text{m})(20.0\ \text{s}^{-1}) = 200$ m/s. The wavelength of the sound wave is

$$\lambda_{1\ (\text{sound})} = \frac{v_{\text{sound}}}{f_1} = \frac{344\ \text{m/s}}{20.0\ \text{Hz}} = 17.2\ \text{m}.$$

The sound wave has a greater speed and longer wavelength than the standing wave on the string that produces the sound. In fact, for any normal mode on this string, the sound wave that is produced will have the same frequency as the wave on the string but will have a wavelength that is greater by a factor of $v_{\text{sound}}/v_{\text{string}} = (344\ \text{m/s})/(200\ \text{m/s}) = 1.72.$

Gas inlet tube

Vibrating diaphragm

Powder collects at displacement nodes

Speaker

20–10 Kundt's tube for determining the velocity of sound in a gas. The N's and A's are displacement nodes and antinodes. The blue shading represents the density of the gas at an instant when the gas pressure at the displacement nodes is a maximum or a minimum.

20–5 LONGITUDINAL STANDING WAVES AND NORMAL MODES

When longitudinal waves propagate in a fluid in a pipe with finite length, the waves are reflected from the ends in the same way that transverse waves on a string are reflected at its ends. The superposition of the waves traveling in opposite directions again forms a standing wave. Just as for transverse standing waves on a string, longitudinal standing waves (normal modes) in a pipe can be used to create sound waves in the surrounding air. This is the operating principle of the human voice as well as many musical instruments, including woodwinds, brasses, and pipe organs.

Transverse waves on a string, including standing waves, are usually described only in terms of the displacement of the string. But longitudinal waves in a fluid may be described either in terms of the displacement of the fluid or in terms of the pressure variation in the fluid. To avoid confusion, we'll use the terms **displacement node** and **displacement antinode** to refer to points where particles of the fluid have zero displacement and maximum displacement, respectively.

We can demonstrate longitudinal standing waves in a column of gas and measure the wave speed using an apparatus called Kundt's tube (Fig. 20–10). A horizontal glass tube a meter or so long is closed at one end and has a flexible diaphragm at the other end that can transmit vibrations. A nearby loudspeaker is driven by an audio oscillator and amplifier; this produces sound waves that force the diaphragm to vibrate sinusoidally with a frequency that we can vary. The sound waves within the tube are reflected at the other, closed end of the tube. We spread a small amount of light powder uniformly along the bottom of the tube. As we vary the frequency of the sound, we pass through frequencies at which the amplitude of the standing waves becomes large enough for the powder to be swept along the tube at those points where the gas is in motion. The powder therefore collects at the displacement nodes (where the gas is not moving). Adjacent nodes are separated by a distance equal to $\lambda/2$, and we can measure this distance. We read the frequency f from the oscillator dial, and we can then calculate the speed v of the waves from the relation $v = \lambda f$.

Figure 20–11 will help you to visualize longitudinal standing waves; it is analogous to Fig. 19–11 for longitudinal traveling waves. Again tape two index cards together edge-to-edge, with a gap of a millimeter or two forming a thin slit. Place the card over the diagram with the slit horizontal and move it vertically with constant velocity. The portions of the sine curves that appear in the slit correspond to the oscillations of the particles in a longitudinal standing wave. Each particle moves with longitudinal simple harmonic motion about its equilibrium position.

Figure 20–12 is an enlarged version of a portion of Fig. 20–11 centered on a displacement node. Note that particles on opposite sides of a displacement node vibrate in

20–11 Diagram for illustrating longitudinal standing waves. The straight vertical lines mark the locations of displacement nodes and pressure antinodes.

20–12 Particles on opposite sides of a displacement node (shown in green) vibrate in opposite phase, creating a pressure antinode. Particles on opposite sides of a displacement antinode vibrate in phase, creating a pressure node (a point of constant pressure).

opposite phase. When these particles approach each other, the gas between them is compressed and the pressure rises; when they recede from each other, there is an expansion and the pressure drops. Hence at a displacement *node* the gas undergoes the maximum amount of compression and expansion, and the variations in pressure and density above and below the average have their maximum value. By contrast, particles on opposite sides of a displacement *antinode* vibrate *in phase;* the distance between the particles is nearly constant, and there is *no* variation in pressure or density at a displacement antinode.

We use the term **pressure node** to describe a point in a longitudinal standing wave at which the pressure and density do not vary and the term **pressure antinode** to describe a point at which the variations in pressure and density are greatest. Using these terms, we can summarize our observations about longitudinal standing waves as follows: **a pressure node is always a displacement antinode, and a pressure antinode is always a displacement node.** Figure 20–10 depicts a longitudinal standing wave at an instant at which the pressure variations are greatest; the blue shading shows that the density and pressure of the gas have their maximum and minimum values at the displacement nodes (labeled *N*).

When reflection takes place at a *closed* end of a pipe (an end with a rigid barrier or plug), the displacement of the particles at this end must always be zero, analogous to a fixed end of a string. Thus a closed end of a pipe is a displacement node and a pressure antinode; the particles do not move, but the pressure variations are maximum. An *open* end of a pipe is a pressure node because it is open to the atmosphere, where the pressure is constant. Because of this, an open end is always a displacement *antinode,* in analogy to a free end of a string; the particles oscillate with maximum amplitude, but the pressure does not vary. (Strictly speaking, the pressure node actually occurs somewhat beyond an open end of a pipe. But if the diameter of the pipe is small in comparison to the wavelength, which is true for most musical instruments, this effect can safely be neglected.) Thus longitudinal waves in a column of fluid are reflected at the closed and open ends of a pipe in the same way that transverse waves in a string are reflected at fixed and free ends, respectively.

EXAMPLE 20–5

A directional loudspeaker aims a sound wave of frequency 200 Hz at a wall (Fig. 20–13). At what distances from the wall could you stand and hear no sound at all?

SOLUTION Your ear detects *pressure* variations in the air; increases or decreases in the pressure outside your eardrum

cause it to move slightly in or out, a motion that generates an electrical signal that is sent to the brain. (If you've ever had trouble getting your ears to "pop" on a drive up into the mountains or on an airline flight, you're familiar with just how sensitive your ears are to pressure changes.) Hence you will hear no sound if your ear is at a pressure node, which is a displacement

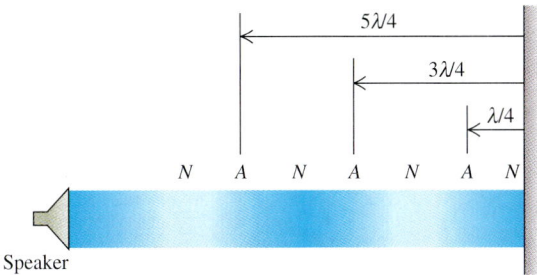

20–13 When a sound wave is directed at a wall, it interferes with the reflected wave to create a standing wave. The N's and A's are *displacement* nodes and antinodes.

antinode. The wall is a displacement node; the distance from a node to an adjacent antinode is $\lambda/4$, and the distance from one antinode to the next is $\lambda/2$. From Example 19–7 (Section 19–7),

the speed of sound waves in air at 20°C is 344 m/s; hence the wavelength of a 200-Hz sound wave in air is 1.72 m, and the distances d from the wall at which no sound will be heard are

$$d = \lambda/4 = (1.72\ \text{m})/4$$
$$= 0.43\ \text{m} \qquad \text{(first displacement antinode and pressure node)},$$

$$d = \lambda/4 + \lambda/2 = 3\lambda/4$$
$$= 1.29\ \text{m} \qquad \text{(second displacement antinode and pressure node)},$$

$$d = 3\lambda/4 + \lambda/2 = 5\lambda/4$$
$$= 2.15\ \text{m} \qquad \text{(third displacement antinode and pressure node)},$$

and so on. If the loudspeaker is not highly directional, this effect is hard to notice because of multiple reflections of sound waves from the floor, ceiling, and other points on the walls.

ORGAN PIPES AND WIND INSTRUMENTS

The most important application of longitudinal standing waves is the production of musical tones by wind instruments. Organ pipes are one of the simplest examples. Air is supplied by a blower, at a gauge pressure typically of the order of 10^3 Pa (10^{-2} atm), to the bottom end of the pipe (Fig. 20–14). A stream of air emerges from the narrow opening at the edge of the horizontal surface and is directed against the top edge of the opening, which is called the *mouth* of the pipe. The column of air in the pipe is set into vibration, and there is a series of possible normal modes, just as with the stretched string. The mouth always acts as an open end; thus it is a pressure node and a displacement antinode. The other end of the pipe (at the top in Fig. 20–14) may be either open or closed.

In Fig. 20–15, both ends of the pipe are open, so both ends are pressure nodes and displacement antinodes. An organ pipe that is open at both ends is called an *open pipe*. The fundamental frequency f_1 corresponds to a standing-wave pattern with a displacement antinode at each end and a displacement node in the middle (Fig. 20–15a). The distance between adjacent antinodes is always equal to one half-wavelength, and in this case that is equal to the length L of the pipe; $\lambda/2 = L$. The corresponding frequency, obtained from the relation $f = v/\lambda$, is

$$f_1 = \frac{v}{2L} \qquad \text{(open pipe)}. \qquad (20\text{–}9)$$

Figures 20–15b and 20–15c show the second and third harmonics (first and second overtones); their vibration patterns have two and three displacement nodes, respectively. For these, a half-wavelength is equal to $L/2$ and $L/3$, respectively, and the frequencies are twice and three times the fundamental, respectively. That is, $f_2 = 2f_1$ and $f_3 = 3f_1$. For *every* normal mode of an open pipe the length L must be an integer number of half-wavelengths, and the possible wavelengths λ_n are given by

$$L = n\frac{\lambda_n}{2} \quad \text{or} \quad \lambda_n = \frac{2L}{n} \quad (n = 1, 2, 3, \ldots) \quad \text{(open pipe)}. \quad (20\text{–}10)$$

The corresponding frequencies f_n are given by $f_n = v/\lambda_n$, so all the normal-mode frequencies for a pipe that is open at both ends are given by

$$f_n = \frac{nv}{2L} \qquad (n = 1, 2, 3, \ldots) \qquad \text{(open pipe)}. \qquad (20\text{–}11)$$

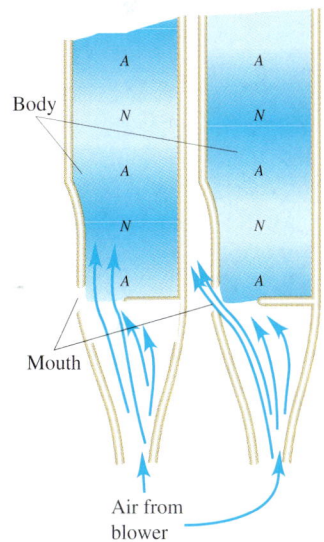

20–14 Cross sections of an organ pipe at two instants one half-period apart. Vibrations from the turbulent air flow set up standing waves in the pipe. The N's and A's are *displacement* nodes and antinodes; as the blue shading shows, these are points of maximum pressure variation and zero pressure variation, respectively.

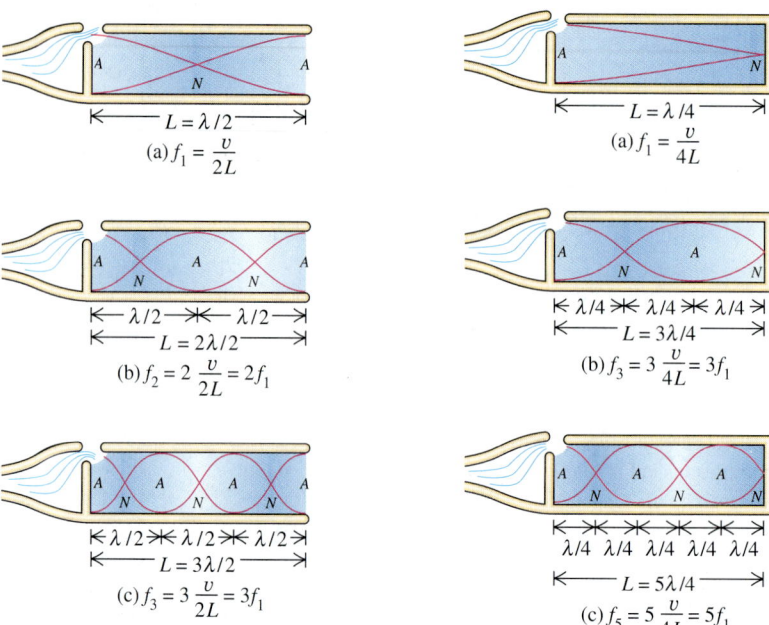

20–15 A cross section of an open pipe showing the first three normal modes. The shading indicates the pressure variations. The red curves are graphs of the displacement along the pipe axis at two instants separated in time by one half-period. The N's and A's are the *displacement* nodes and antinodes; interchange these to show the *pressure* nodes and antinodes.

20–16 A cross section of a stopped pipe showing the first three normal modes as well as the *displacement* nodes and antinodes. Only odd harmonics are possible.

The value $n = 1$ corresponds to the fundamental frequency, $n = 2$ to the second harmonic (or first overtone), and so on. Alternatively, we can say

$$f_n = nf_1 \qquad (n = 1, 2, 3, \ldots) \qquad \text{(open pipe)}, \qquad (20\text{–}12)$$

with f_1 given by Eq. (20–9).

Figure 20–16 shows a pipe that is open at the left end but closed at the right end. This is called a *stopped pipe*. The left (open) end is a displacement antinode (pressure node), but the right (closed) end is a displacement node (pressure antinode). The distance between a node and the adjacent antinode is always one quarter-wavelength. Figure 20–16a shows the lowest-frequency mode; the length of the pipe is a quarter-wavelength ($L = \lambda_1/4$). The fundamental frequency is $f_1 = v/\lambda_1$, or

$$f_1 = \frac{v}{4L} \qquad \text{(stopped pipe)}. \qquad (20\text{–}13)$$

This is one-half the fundamental frequency for an *open* pipe of the same length. In musical language, the *pitch* of a closed pipe is one octave lower (a factor of two in frequency) than that of an open pipe of the same length. Figure 20–16b shows the next mode, for which the length of the pipe is *three quarters* of a wavelength, corresponding to a frequency $3f_1$. For Fig. 20–16c, $L = 5\lambda/4$ and the frequency is $5f_1$. The possible wavelengths are given by

$$L = n\frac{\lambda_n}{4} \quad \text{or} \quad \lambda_n = \frac{4L}{n} \qquad (n = 1, 3, 5, \ldots) \qquad \text{(stopped pipe)}. \qquad (20\text{–}14)$$

The normal-mode frequencies are given by $f_n = v/\lambda_n$, or

$$f_n = \frac{nv}{4L} \qquad (n = 1, 3, 5, \ldots) \qquad \text{(stopped pipe)}, \qquad (20\text{–}15)$$

or

$$f_n = nf_1 \qquad (n = 1, 3, 5, \ldots) \qquad \text{(stopped pipe)}, \qquad (20\text{–}16)$$

with f_1 given by Eq. (20–13). We see that the second, fourth, and all *even* harmonics are missing. In a pipe that is closed at one end, the fundamental frequency is $f_1 = v/4L$, and only the odd harmonics in the series $(3f_1, 5f_1, \ldots)$ are possible.

A final possibility is a pipe that is closed at *both* ends, with displacement nodes and pressure antinodes at both ends. This wouldn't be of much use as a musical instrument because there would be no way for the vibrations to get out of the pipe.

EXAMPLE 20-6

On a day when the speed of sound is 345 m/s, the fundamental frequency of a stopped organ pipe is 220 Hz. a) How long is this stopped pipe? b) The second *overtone* of this pipe has the same wavelength as the third *harmonic* of an open pipe. How long is the open pipe?

SOLUTION a) For a stopped pipe, $f_1 = v/4L$, so the length of the stopped pipe is

$$L_{\text{stopped}} = \frac{v}{4f_1} = \frac{345 \text{ m/s}}{4(220 \text{ s}^{-1})} = 0.392 \text{ m}.$$

b) The frequency of the first overtone of a stopped pipe is $f_3 = 3f_1$, and the frequency of the second overtone is $f_5 = 5f_1$:

$$f_5 = 5f_1 = 5(220 \text{ Hz}) = 1100 \text{ Hz}.$$

If the wavelengths are the same, the frequencies are the same, so the frequency of the third harmonic of the open pipe is also 1100 Hz. The third harmonic of an open pipe is at $3f_1 = 3(v/2L)$. If this equals 1100 Hz, then

$$1100 \text{ Hz} = 3\left(\frac{345 \text{ m/s}}{2L_{\text{open}}}\right) \qquad \text{and} \qquad L_{\text{open}} = 0.470 \text{ m}.$$

The stopped pipe is 0.392 m long and has a fundamental frequency of 220 Hz; the open pipe is longer, 0.470 m, but has a higher fundamental frequency of (1100 Hz)/3 = 367 Hz. If this seems like a contradiction, you should again compare Figs. 20–15a and 20–16a.

In an organ pipe in actual use, several modes are always present at once; the motion of the air is a superposition of these modes. This situation is analogous to a string that is struck or plucked, as in Fig. 20–8. Just as for a vibrating string, a complex standing wave in the pipe produces a traveling sound wave in the surrounding air with a harmonic content similar to that of the standing wave. A very narrow pipe produces a sound wave rich in higher harmonics, which we hear as a thin and "stringy" tone; a fatter pipe produces mostly the fundamental mode, heard as a softer, more flute-like tone. The harmonic content also depends on the shape of the pipe's mouth.

We have talked about organ pipes, but this discussion is also applicable to other wind instruments. The flute and the recorder are directly analogous. The most significant difference is that those instruments have holes along the pipe. Opening and closing the holes with the fingers changes the effective length L of the air column and thus changes the pitch. Any individual organ pipe, by comparison, can play only a single note. The flute and recorder behave as *open* pipes, while the clarinet acts as a *stopped* pipe (closed at the reed end, open at the bell).

Equations (20–11) and (20–15) show that the frequencies of any wind instrument are proportional to the speed of sound v in the air column inside the instrument. As Eq. (19–27) shows, v depends on temperature; it increases when temperature increases. Thus the pitch of all wind instruments rises with increasing temperature. An organ that has some of its pipes at one temperature and others at a different temperature is bound to sound out of tune.

Each pipe in a pipe organ has a different length and diameter, and emits sound waves with a different pitch and different harmonic content.

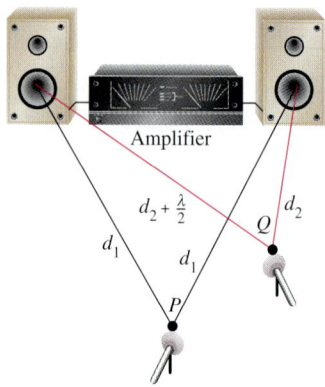

20–17 Two speakers driven by the same amplifier. The waves emitted by the speakers are in phase; they arrive at point P in phase because the two path lengths are the same. They arrive at point Q a half-cycle out of phase because the path lengths differ by $\lambda/2$.

20–6 INTERFERENCE OF WAVES

Wave phenomena that occur when two or more waves overlap in the same region of space are grouped under the heading *interference*. As we have seen, standing waves are a simple example of an interference effect: two waves traveling in opposite directions in a medium combine to produce a standing wave pattern with nodes and antinodes that do not move.

Figure 20–17 shows an example of another type of interference that involves waves that spread out in space. Two speakers, driven in phase by the same amplifier, emit identical sinusoidal sound waves with the same constant frequency. We place a microphone at point P in the figure, equidistant from the speakers. Wave crests emitted from the two speakers at the same time travel equal distances and arrive at point P at the same time; hence the waves arrive in phase, and there is constructive interference. The total wave amplitude at P is twice the amplitude from each individual wave, and we can measure this combined amplitude with the microphone.

Now let's move the microphone to point Q, where the distances from the two speakers to the microphone differ by a half-wavelength. Then the two waves arrive a half-cycle out of step, or *out of phase;* a positive crest from one speaker arrives at the same time as a negative crest from the other. Destructive interference takes place, and the amplitude measured by the microphone is much *smaller* than when only one speaker is present. If the amplitudes from the two speakers are equal, the two waves cancel each other out completely at point Q, and the total amplitude there is zero.

CAUTION ▶ Although this situation bears some resemblance to standing waves in a pipe, the total wave in Fig. 20–17 is a *traveling* wave, not a standing wave. To see why, recall that in a standing wave there is no net flow of energy in any direction. By contrast, in Fig. 20–17 there is an overall flow of energy from the speakers into the surrounding air; this is characteristic of a traveling wave. The interference between the waves from the two speakers simply causes the energy flow to be *channeled* into certain directions (for example, toward P) and away from other directions (for example, away from Q). You can see another difference between Fig. 20–17 and a standing wave by considering a point, such as Q, where destructive interference occurs. Such a point is *both* a displacement node *and* a pressure node because there is no wave at all at this point. Compare this to a standing wave, in which a pressure node is a displacement antinode and vice versa. ◀

20–18 Two speakers driven by the same amplifier, emitting waves in phase. Only the waves directed toward the microphone are shown, and they are separated for clarity. (a) Constructive interference occurs when the path difference is $0, \lambda, 2\lambda, 3\lambda, \ldots$. (b) Destructive interference occurs when the path difference is $\lambda/2, 3\lambda/2, 5\lambda/2, \ldots$; in this case little or no sound energy flows toward the microphone directly in front of the speakers. The energy is instead directed to the sides, where constructive interference occurs.

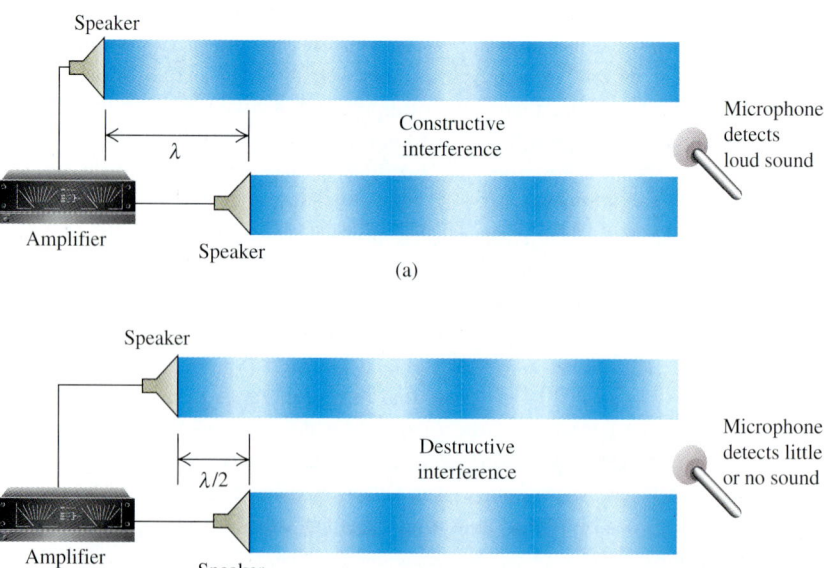

Constructive interference occurs wherever the distances traveled by the two waves differ by a whole number of wavelengths, $0, \lambda, 2\lambda, 3\lambda, \ldots$; in all these cases the waves arrive at the microphone in phase (Fig. 20–18a). If the distances from the two speakers to the microphone differ by any half-integer number of wavelengths, $\lambda/2, 3\lambda/2, 5\lambda/2, \ldots$, the waves arrive at the microphone out of phase and there will be destructive interference (Fig. 20–18b).

EXAMPLE 20-7

Two small loudspeakers, A and B (Fig. 20–19), are driven by the same amplifier and emit pure sinusoidal waves in phase. If the speed of sound is 350 m/s, a) for what frequencies does constructive interference occur at point P; b) for what frequencies does destructive interference occur at point P?

SOLUTION The nature of the interference at point P depends on the difference in path lengths from points A and B to point P; let's calculate this first. The distance from speaker A to point P is $[(2.00 \text{ m})^2 + (4.00 \text{ m})^2]^{1/2} = 4.47$ m, and the distance from speaker B to point P is $[(1.00 \text{ m})^2 + (4.00 \text{ m})^2]^{1/2} = 4.12$ m. The path difference is $d = 4.47 \text{ m} - 4.12 \text{ m} = 0.35$ m.

a) Constructive interference occurs when the path difference d is an integer number of wavelengths, $d = 0, \lambda, 2\lambda, \ldots$ or $d = 0, v/f, 2v/f, \ldots = nv/f$. So the possible frequencies are

$$f_n = \frac{nv}{d} = n\frac{350 \text{ n/s}}{0.35 \text{ m}} \qquad (n = 1, 2, 3, \ldots)$$

$$= 1000 \text{ Hz}, 2000 \text{ Hz}, 3000 \text{ Hz}, \ldots.$$

b) Destructive interference occurs when the path difference is a half-integer number of wavelengths, $d = \lambda/2, 3\lambda/2, 5\lambda/2, \ldots$ or $d = v/2f, 3v/2f, 5v/2f, \ldots$. The possible frequencies are

$$f_n = \frac{nv}{2d} = n\frac{350 \text{ m/s}}{2(0.35 \text{ m})} \qquad (n = 1, 3, 5, \ldots)$$

$$= 500 \text{ Hz}, 1500 \text{ Hz}, 2500 \text{ Hz}, \ldots.$$

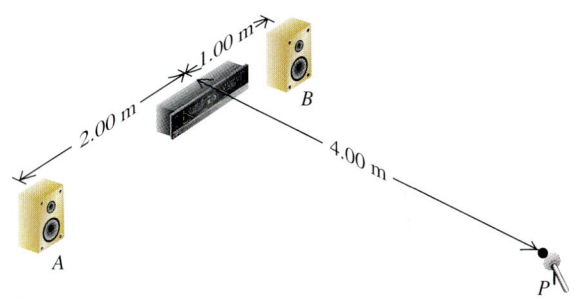

20–19 The distances from A to P and from B to P are hypotenuses of right triangles.

As we increase the frequency, the sound at point P alternates between large and small amplitudes; the maxima and minima occur at the frequencies we have found. It can be hard to notice this effect in an ordinary room because of multiple reflections from the walls, floor, and ceiling. Such an experiment is best done either outdoors or in an anechoic chamber, which has walls that absorb almost all sound and thereby eliminate reflections.

Experiments closely analogous to the one in Example 20–7, but using light, have provided both strong evidence for the wave nature of light and a means of measuring its wavelengths. We will discuss these experiments in detail in Chapter 37.

Interference effects are used to control noise from very loud sound sources such as gas-turbine power plants or jet engine test cells. The idea is to use additional sound sources that in some regions of space interfere destructively with the unwanted sound and cancel it out. Microphones in the controlled area feed signals back to the sound sources, which are continuously adjusted for optimum cancellation of noise in the controlled area.

20-7 RESONANCE

We have discussed several mechanical systems that have normal modes of oscillation. In each mode, every particle of the system oscillates with simple harmonic motion with the same frequency as the frequency of this mode. The systems we discussed in Sections 20–4 and 20–5 have an infinite series of normal modes, but the basic concept is closely

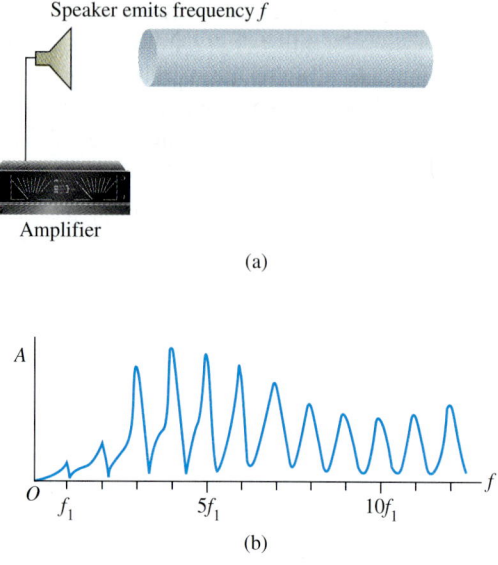

Speaker emits frequency f

Amplifier

(a)

A

O f_1 $5f_1$ $10f_1$ f

(b)

20–20 (a) The air in an open pipe is forced to oscillate at the frequency f of the sound waves coming from the loudspeaker. (b) The graph of amplitude of oscillation of the air in the open pipe as a function of the driving frequency f. This graph is called the resonance curve of the open pipe. The peaks occur at the normal-mode frequencies f_1, $f_2 = 2f_1, f_3 = 3f_1, \ldots.$

related to the simple harmonic oscillator, discussed in Chapter 13, which has only a single normal mode (that is, only one frequency at which it will oscillate after being disturbed).

Suppose we apply a periodically varying force to a system that can oscillate. The system is then forced to oscillate with a frequency equal to the frequency of the applied force (called the *driving frequency*). This motion is called a *forced oscillation*. We talked about forced oscillations of the harmonic oscillator in Section 13–9, and we suggest that you review that discussion. In particular, we described the phenomenon of mechanical **resonance.** A simple example of resonance is pushing Cousin Throckmorton on a swing. The swing is a pendulum; it has only a single normal mode, with a frequency determined by its length. If we push the swing periodically with this frequency, we can build up the amplitude of the motion. But if we push with a very different frequency, the swing hardly moves at all.

Resonance also occurs when a periodically varying force is applied to a system with many normal modes. An example is shown in Fig. 20–20a. An open organ pipe is placed next to a loudspeaker that is driven by an amplifier and emits pure sinusoidal sound waves of frequency f, which can be varied by adjusting the amplifier. The air in the pipe is forced to vibrate with the same frequency f as the *driving force* provided by the loudspeaker. In general the amplitude of this motion is relatively small, and the motion of the air inside the pipe will not be any of the normal-mode patterns shown in Fig. 20–15. But if the frequency f of the force is close to one of the normal-mode frequencies, the air in the pipe will move in the normal-mode pattern for that frequency, and the amplitude can become quite large. Figure 20–20b shows the amplitude of oscillation of the air in the pipe as a function of the driving frequency f. The shape of this graph is called the **resonance curve** of the pipe; it has peaks where f equals the normal-mode frequencies of the pipe. The detailed shape of the resonance curve depends on the geometry of the pipe.

If the frequency of the force is precisely *equal* to a normal-mode frequency, the system is in resonance, and the amplitude of the forced oscillation is maximum. If there were no friction or other energy-dissipating mechanism, a driving force at a normal-mode frequency would continue to add energy to the system, and the amplitude would increase indefinitely. In such an idealized case the peaks in the resonance curve of Fig. 20–20b would be infinitely high. But in any real system there is always some dissipation of energy, or damping, as we discussed in Section 13–9; the amplitude of oscillation in resonance may be large, but it cannot be infinite.

The "sound of the ocean" you hear when you put your ear next to a large seashell is due to resonance. The noise of the outside air moving past the seashell is a mixture of sound waves of almost all audible frequencies, which forces the air inside the seashell to oscillate. The seashell behaves like an organ pipe, with a set of normal-mode frequencies; hence the inside air oscillates most strongly at those frequencies, producing the seashell's characteristic sound. To hear a similar phenomenon, uncap a full bottle of your favorite beverage and blow across the open top. The noise is provided by your breath blowing across the top, and the "organ pipe" is the column of air inside the bottle above the surface of the liquid. If you take a drink and repeat the experiment, you will hear a lower tone because the "pipe" is longer and the normal-mode frequencies are lower.

Resonance also occurs when a stretched string is forced to oscillate. Suppose that one end of a stretched string is held fixed while the other is given a transverse sinusoidal motion with small amplitude, setting up standing waves. If the frequency of the driving mechanism is *not* equal to one of the normal-mode frequencies of the string, the amplitude at the antinodes is fairly small. However, if the frequency is equal to any one of the normal-mode frequencies, the string is in resonance, and the amplitude at the antinodes is very much larger than that at the driven end. The driven end is not precisely a node, but it lies much closer to a node than to an antinode when the string is in resonance. The photographs in Fig. 20–5 were made this way, with the left end of the string fixed and the right end oscillating vertically with small amplitude; large-amplitude standing waves resulted when the frequency of oscillation of the right end was equal to the fundamental frequency or to one of the first three overtones.

It is easy to demonstrate resonance with a piano. Push down the damper pedal (the right-hand pedal) so that the dampers are lifted and the strings are free to vibrate, and then sing a steady tone into the piano. When you stop singing, the piano seems to continue to sing the same note. The sound waves from your voice excite vibrations in the strings that have natural frequencies close to the frequencies (fundamental and harmonics) present in the note you sang.

A more spectacular example is a singer breaking a wine glass with her amplified voice. A good-quality wine glass has normal-mode frequencies that you can hear by tapping it. If the singer emits a loud note with a frequency corresponding exactly to one of these normal-mode frequencies, large-amplitude oscillations can build up and break the glass.

Resonance is a very important concept, not only in mechanical systems but in all areas of physics. In Chapter 32 we will see examples of resonance in electric circuits.

The frequency of the sound from this trumpet exactly matches one of the normal-mode frequencies of the goblet. The resonant vibrations of the goblet have such large amplitude that the goblet tears itself apart.

EXAMPLE 20–8

A stopped organ pipe is sounded near a guitar, causing one of the strings to vibrate with large amplitude. We vary the tension of the string until we find the maximum amplitude. The string is 80% as long as the stopped pipe. If both the pipe and the string vibrate at their fundamental frequency, calculate the ratio of the wave speed on the string to the speed of sound in air.

SOLUTION The large response of the string is an example of resonance; it occurs because the organ pipe and the guitar string have the same fundamental frequency. Letting the subscripts a and s stand for air and string, respectively, we have $f_{1a} = f_{1s}$.

We also know from Eq. (20–13) that $f_{1a} = v_a/4L_a$ and from Eq. (20–5) that $f_{1s} = v_s/2L_s$. Putting these together, we find

$$\frac{v_a}{4L_a} = \frac{v_s}{2L_s}.$$

Substituting $L_s = 0.80\, L_a$ and rearranging, we get

$$\frac{v_s}{v_a} = 0.40.$$

For example, if the speed of sound in air is 345 m/s, the wave speed on the string is (0.40)(345 m/s) = 138 m/s.

SUMMARY

■ A wave that reaches a boundary of the medium in which it propagates is reflected. The principle of superposition states that the total wave displacement at any point where two or more waves overlap is the sum of the displacements of the individual waves.

■ When a wave is reflected from a fixed or free end of a stretched string, the incident and reflected waves combine to form a standing wave containing nodes and antinodes. Adjacent nodes are spaced a distance $\lambda/2$ apart, as are adjacent antinodes. If the string has a fixed end at $x = 0$, the wave function of a standing wave with angular frequency $\omega = 2\pi f$, wave number $k = 2\pi/\lambda$, and amplitude A_{sw} is

$$y(x, t) = (A_{sw} \sin kx) \cos \omega t \qquad \text{(standing wave on a string, fixed end at } x = 0). \qquad (20\text{–}1)$$

■ When both ends of a string with length L are held fixed, standing waves can occur only when L is an integer multiple of $\lambda/2$. The corresponding possible frequencies are

$$f_n = n \frac{v}{2L} = nf_1$$

$$(n = 1, 2, 3, \ldots) \qquad \text{(string fixed at both ends).} \qquad (20\text{–}6)$$

Each frequency and its associated vibration pattern is called a normal mode. The lowest frequency f_1 is called the fundamental frequency. In terms of the mechanical properties F and μ of the string, the fundamental frequency is

$$f_1 = \frac{1}{2L} \sqrt{\frac{F}{\mu}} \qquad \text{(string fixed at both ends).} \qquad (20\text{–}8)$$

■ For sound waves in a pipe or tube, a closed end is a displacement node and a pressure antinode; an open end is a displacement antinode and a pressure node. For a pipe open at both ends, the normal-mode frequencies are

$$f_n = \frac{nv}{2L} \qquad (n = 1, 2, 3, \ldots) \qquad \text{(open pipe).} \qquad (20\text{–}11)$$

For a pipe open at one end and closed at the other, the normal-mode frequencies are

$$f_n = \frac{nv}{4L} \qquad (n = 1, 3, 5, \ldots) \qquad \text{(stopped pipe).} \qquad (20\text{–}15)$$

■ When two or more waves overlap in the same region of space, the resulting effects are called interference. The resulting amplitude can be either larger or smaller than the amplitude of each individual wave, depending on whether the waves are in phase or out of phase. When the waves are in phase, the result is called constructive interference; when they are one half-cycle out of phase, it is called destructive interference.

■ When a periodically varying force is applied to a system having normal modes, the system vibrates with the same frequency as that of the force; this is called a forced oscillation. If the force frequency is equal to or close to one of the normal-mode frequencies, the amplitude of the resulting forced oscillation can become very large; this phenomenon is called resonance.

DISCUSSION QUESTIONS

Q20–1 Energy can be transferred along a string by wave motion. However, in a standing wave on a string, no energy can ever be transferred past a node. Why not?

Q20–2 Can a standing wave be produced on a string by superposing two waves traveling in opposite directions with the same frequency but different amplitudes? Why or why not? Can a standing wave be produced by superposing two waves traveling in opposite directions with different frequencies but the same amplitude? Why or why not?

Q20–3 If you stretch a rubber band and pluck it, you hear a (somewhat) musical tone. How does the frequency of this tone change as you stretch the rubber band further? (Try it!) Does this agree with Eq. (20–8) for a string fixed at both ends? Explain.

Q20–4 A musical interval of an *octave* corresponds to a factor of two in frequency. By what factor must the tension in a guitar or violin string be increased to raise its pitch one octave? To raise it two octaves? Explain your reasoning.

Q20–5 By touching a string lightly at its center while bowing, a violinist can produce a note exactly one octave above the note to which the string is tuned, that is, a note with exactly twice the frequency. Why is this possible?

Q20–6 When an earthquake hits, tall buildings tend to oscillate from side to side. Is the top of the building a displacement node or antinode? What about the bottom of the building? Explain.

Q20–7 Does the pitch (or frequency) of an organ pipe increase or decrease with increasing temperature? Explain.

Q20–8 In most modern wind instruments the pitch is changed by using keys or valves to change the length of the vibrating air column. The bugle, however, has no valves or keys, yet it can play many notes. How is this possible? Are there restrictions on what notes a bugle can play?

Q20–9 Symphonic musicians always "warm up" their wind instruments by blowing them before a performance. What purpose does this serve?

Q20–10 When you inhale helium, your voice becomes high and squeaky. Why? (*Warning:* Inhaling too much helium can cause unconsciousness or death.)

Q20–11 The pitch of a kettledrum is determined by the frequencies of the normal modes of oscillation of the drumhead. How can a kettledrum be tuned?

Q20–12 As we discussed in Section 19–2, water waves are a combination of longitudinal and transverse waves. Defend the following statement: "When water waves hit a vertical wall, the wall is a node of the longitudinal displacement but an antinode of the transverse displacement."

Q20–13 A glass of water sits on a kitchen counter just above a dishwasher that is running. The surface of the water has a stationary pattern of concentric circular ripples. What causes this? Why is the pattern stationary? How does the speed of the dishwasher affect the number of concentric ripples in the water?

Q20–14 The rate at which exhaust gases flow out of a gasoline engine can be increased by using a "tuned" exhaust pipe of just the right length. (This improves the performance of the engine, since fuel can be forced into the engine at a faster rate.) How does this work? What determines the optimum length to use?

Q20–15 Some stringed instruments (such as the sitar of India) have two of the same string. When one string is plucked, the other one starts vibrating at the same frequency, even though it has not been touched. How is this possible?

Q20–16 Figure 12–13 shows the geometry of an *ellipse*. The points S and S' are called the *foci* of the ellipse (plural of *focus*); the sum of the distances SP and $S'P$ has the same value for any point P on the ellipse. If a room is built with curved walls in the shape of an ellipse, a person standing at one focus of the room can easily hear a person at the other focus, even if that person whispers and the overall dimensions of the room are quite large. Explain why, using the ideas of interference in space.

EXERCISES

SECTION 20–2 BOUNDARY CONDITIONS FOR A STRING AND THE PRINCIPLE OF SUPERPOSITION

20–1 A wave pulse on a string has the dimensions shown in Fig. 20–21 at $t = 0$. The wave speed is 5.0 m/s. a) If point O is a fixed end, draw the total wave on the string at $t = 1.0$ ms, 2.0 ms, 3.0 ms, 4.0 ms, 5.0 ms, 6.0 ms, and 7.0 ms. b) Repeat part (a) for the case in which point O is a free end.

20–2 A wave pulse on a string has the dimensions shown in Fig. 20–22 at $t = 0$. The wave speed is 40 cm/s. a) If point O is a fixed end, draw the total wave on the string at $t = 15$ ms, 20 ms, 25 ms, 30 ms, 35 ms, 40 ms, and 45 ms. b) Repeat part (a) for the case in which point O is a free end.

FIGURE 20–21 Exercise 20–1.

FIGURE 20–22 Exercise 20–2.

SECTION 20-3 **STANDING WAVES ON A STRING**

20-3 Standing waves on a wire are described by Eq. (20–1), with $A_{SW} = 2.50$ mm, $\omega = 942$ rad/s, and $k = 0.750\pi$ rad/m. The left-hand end of the wire is at $x = 0$. At what distances from the left-hand end are a) the nodes of the standing wave? b) the antinodes of the standing wave?

20-4 a) Prove by direct substitution that $y(x,\ t) = (A_{SW} \cos \omega t) \sin kx$ is a solution of the wave equation, Eq. (19–12), for $v = \omega/k$. b) Explain why the relation $v = \omega/k$ for *traveling* waves also applies to *standing* waves.

20-5 Adjacent antinodes of a standing wave on a string are 15.0 cm apart. A particle at an antinode oscillates in simple harmonic motion with amplitude 0.850 cm and period 0.0750 s. The string lies along the +x-axis and is fixed at $x = 0$. a) Find the displacement of a point on the string as a function of position and time. b) Find the speed of propagation of a transverse wave in the string. c) Find the amplitude at a point 3.0 cm to the right of an antinode.

20-6 Give the details of the derivation of Eq. (20–1) from $y_1(x,\ t) + y_2(x,\ t) = A[\sin (\omega t + kx) - \sin (\omega t - kx)]$.

20-7 Let $y_1(x,\ t) = A_1 \sin (\omega_1 t - k_1 x)$ and $y_2(x,\ t) = A_2 \sin (\omega_2 t - k_2 x)$ be two solutions to the wave equation, Eq. (19–12), for the same v. Show that $y(x,\ t) = y_1(x,\ t) + y_2(x,\ t)$ is also a solution to the wave equation.

SECTION 20-4 **NORMAL MODES OF A STRING**

20-8 A piano tuner stretches a steel piano wire with a tension of 800 N. The steel wire is 0.400 m long and has a mass of 3.00 g. a) What is the frequency of its fundamental mode of vibration? b) What is the number of the highest harmonic that could be heard by a person who is capable of hearing frequencies up to 10,000 Hz?

20-9 A wire of mass 40.0 g is stretched so that its ends are tied down at points 80.0 cm apart. The wire vibrates in its fundamental mode with frequency 60.0 Hz and with an amplitude at the antinodes of 0.300 cm. a) What is the speed of propagation of transverse waves in the wire? b) Compute the tension in the wire.

20-10 In deriving Eq. (20–2) for a string with a fixed end at $x = 0$, we found that nodes occur at positions x that satisfy $kx = n\pi$, where $n = 0,\ 1,\ 2,\ 3,\ \dots$. Apply this condition to a string with fixed ends at $x = 0$ and $x = L$ to rederive Eq. (20–4) for the possible standing-wave wavelengths.

20-11 **Playing a Cello.** The portion of a cello string between the bridge and upper end of the finger board (that part of the string that is free to vibrate) is 60.0 cm long, and this length of the string has a mass of 2.00 g. The string sounds an A_4 note (440 Hz) when played. a) Where must the cellist put a finger (what distance x from bridge) to play a D_5 note (587 Hz)? (See Fig. 20–23.) For both the A_4 and D_5 notes the string vibrates in its fundamental mode. b) Without retuning, is it possible to play a G_4 note (392 Hz) on this string? Why or why not?

20-12 a) A horizontal string tied at both ends is vibrating in its fundamental mode. The standing wave has speed v, frequency f, amplitude A, and wavelength λ. Calculate the maximum trans-

FIGURE 20-23 Exercise 20–11.

verse velocity and maximum transverse acceleration of points located at i) $x = \lambda/2$, ii) $x = \lambda/4$, and iii) $x = \lambda/8$ from the left-hand end of the string. b) At each of the points in part (a), what is the amplitude of the motion? c) At each of the points in part (a), how much time does it take the string to go from its largest upward displacement to its largest downward displacement?

20-13 A 1.50-m-long rope is stretched between two supports with a tension that makes the speed of transverse waves 48.0 m/s. What are the wavelength and frequency of a) the fundamental? b) the second overtone? c) the fourth harmonic?

20-14 One of the 63.5-cm-long strings of an ordinary guitar is tuned to produce the note B_3 (frequency 245 Hz) when vibrating in its fundamental mode. a) Find the speed of transverse waves on this string. b) If the tension in this string is increased by 1.0%, what will be the new fundamental frequency of the string? c) If the speed of sound in the surrounding air is 344 m/s, find the frequency and wavelength of the sound wave produced in the air by the vibration of the B_3 string. How do these compare to the frequency and wavelength of the standing wave on the string?

SECTION 20-5 **LONGITUDINAL STANDING WAVES AND NORMAL MODES**

20-15 Standing sound waves are produced in a pipe that is 1.20 m long. For the fundamental and first two overtones, determine the locations along the pipe (measured from the left end) of the displacement nodes and the pressure nodes if a) the pipe is open at both ends; b) the pipe is closed at the left end and open at the right end.

20-16 Find the fundamental frequency and the frequency of the first three overtones of a pipe 45.0 cm long a) if the pipe is open at both ends; b) if the pipe is closed at one end. Use $v = 344$ m/s. c) For each of the above cases, what is the number of the highest harmonic that may be heard by a person who can hear frequencies from 20 Hz to 20,000 Hz?

20-17 The human vocal tract is a pipe that extends about 17 cm from the lips to the vocal folds (also called "vocal cords") near the middle of your throat. The vocal folds behave rather like the reed of a clarinet, and the vocal tract acts as a stopped pipe. Estimate the first three standing-wave frequencies of the vocal tract. Use $v = 344$ m/s. (The answers are only an estimate, since the position of lips and tongue affect the motion of air in the vocal tract.)

20–18 A certain pipe produces a fundamental frequency of 262 Hz in air. a) If the pipe is filled with helium at the same temperature, what fundamental frequency does it produce? (The molar mass of air is 28.8 g/mol, and the molar mass of helium is 4.00 g/mol.) b) Does your answer to part (a) depend on whether the pipe is open or stopped? Why or why not?

SECTION 20–6 INTERFERENCE OF WAVES

20–19 Figure 20–24 shows two rectangular wave pulses on a stretched string traveling toward each other. Each pulse is traveling with a speed of 1.00 mm/s and has the height and width shown in the figure. If the leading edges of the pulses are 8.00 mm apart at $t = 0$, sketch the shape of the string at $t = 4.00$ s, $t = 6.00$ s, and $t = 10.0$ s.

FIGURE 20–24 Exercise 20–19.

20–20 Two triangular wave pulses are traveling toward each other on a stretched string as shown in Fig. 20–25. Each pulse is identical to the other and travels at 2.00 cm/s. The leading edges of the pulses are 1.00 cm apart at $t = 0$. Sketch the shape of the string at $t = 0.250$ s, $t = 0.500$ s, $t = 0.750$ s, $t = 1.000$ s, and $t = 1.250$ s.

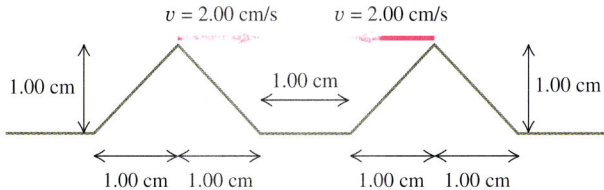

FIGURE 20–25 Exercise 20–20.

PROBLEMS

20–25 As shown in Exer   , a standing wave given by Eq. (20–1) satisfies the w: on, Eq. (19–12). a) Show that a standing wave giv   0–1) also satisfies the equation

$$\frac{\partial^2}{\quad} \qquad y(x,t)$$

Interpret this equation   iat you know about simple harmonic motion. b) D   g wave given by $y(x, t) =$

20–21 Two loudspeakers, A and B (Fig. 20–26), are driven by the same amplifier and emit sinusoidal waves in phase. Speaker B is 2.00 m to the right of speaker A. The speed of sound in the air is 344 m/s. Consider point Q along the extension of the line connecting the speakers, 1.00 m to the right of speaker B. Both speakers emit sound waves that travel directly from the speaker to point Q. a) What is the lowest frequency for which *constructive* interference occurs at point Q? b) What is the lowest frequency for which *destructive* interference occurs at point Q?

FIGURE 20–26 Exercises 20–21 and 20–22.

20–22 Two loudspeakers, A and B (Fig. 20–26), are driven by the same amplifier and emit sinusoidal waves in phase. Speaker B is 2.00 m to the right of speaker A. The frequency of the sound waves produced by the loudspeakers is 206 Hz, and their speed in the air is 344 m/s. Consider point P between the speakers and along the line connecting them, a distance x to the right of speaker A. Both speakers emit sound waves that travel directly from the speaker to point P. a) For what values of x will *destructive* interference occur at point P? b) For what values of x will *constructive* interference occur at point P? c) Interference effects like those in parts (a) and (b) are almost never a factor in listening to home stereo equipment. Why not?

SECTION 20–7 RESONANCE

20–23 You blow across the open mouth of an empty test tube and produce the fundamental standing wave of the air column inside the test tube. The speed of sound in air is 344 m/s and the test tube acts as a stopped pipe. a) If the length of the air column in the test tube is 14.0 cm, what is the frequency of this standing wave? b) What is the frequency of the fundamental standing wave in the air column if the test tube is half-filled with water?

20–24 The D note (588 Hz) from a trumpet causes a guitar string to vibrate in its second overtone with large amplitude. The vibrating portion of the guitar string has length 63.5 cm. What is the speed of transverse waves on the guitar string?

$A \sin(\omega t - kx)$ also satisfy the equation in part (a)? Interpret this result.

20–26 a) The red and blue waves in Fig. 20–3 combine so that the displacement of the string at O is always zero. To show this mathematically for a wave of arbitrary shape, consider a wave moving to the right along the string in Fig. 20–3 (shown in blue) that at time T is given by $y_1(x, T) = f(x)$, where f is some function of x. (The form of $f(x)$ determines the shape of the wave.) If the point O corresponds to $x = 0$, explain why at time T the wave moving to the left in Fig. 20–3 (shown in red) is given by the function $y_2(x, T) = -f(-x)$. b) Show that the total wave function $y(x, T) = y_1(x, T) + y_2(x, T)$ is zero at O, independent of the form of the function $f(x)$. c) The red and blue waves in Fig. 20–4 combine so that the slope of the string at O is always zero. To show this mathematically for a wave of arbitrary shape, again let the wave moving to the right in Fig. 20–4 (shown in blue) be given by $y_1(x, T) = f(x)$ at time T. Explain why the wave moving to the left (shown in red) at this same time T is given by $y_2(x, T) = f(-x)$. d) Show that the total wave function $y(x, T) = y_1(x, T) + y_2(x, T)$ has zero slope at O, independent of the form of the function $f(x)$, as long as $f(x)$ has a finite first derivative.

20–27 A string that lies along the $+x$-axis has a free end at $x = 0$. a) By using steps similar to those used to derive Eq. (20–1), show that an incident traveling wave $y_1(x, t) = A \sin(\omega t + kx)$ gives rise to a standing wave $y(x, t) = 2A \sin \omega t \cos kx$. b) Show that the standing wave has an antinode at its free end ($x = 0$). c) Find the maximum displacement, maximum speed, and maximum acceleration of the free end of the string.

20–28 A string with both ends held fixed is vibrating in its third harmonic. The waves have a speed of 192 m/s and a frequency of 240 Hz. The amplitude of the standing wave at an antinode is 0.400 cm. a) Calculate the amplitude at points on the string a distance of i) 40.0 cm; ii) 20.0 cm; and iii) 10.0 cm from the left-hand end of the string. b) At each point in part (a), how much time does it take the string to go from its largest upward displacement to its largest downward displacement? c) Calculate the maximum transverse velocity and the maximum transverse acceleration of the string at each of the points in part (a).

20–29 A wooden plank is placed over a pit which is 5.00 m wide. A physics student stands in the middle of the plank and begins to jump up and down such that she jumps upward from the plank two times each second. The plank oscillates with a large amplitude, with maximum amplitude at its center. a) What is the speed of transverse waves on the plank? b) At what rate does the student have to jump to produce large-amplitude oscillations if she is standing 1.25 m from the edge of the pit? (*Note:* The transverse standing waves of the plank have nodes at the two ends that rest on the ground on either side of the pit.)

20–30 A new musical instrument consists of a metal can with length L and diameter $L/10$. The top of the can is cut out, and a string is stretched across this open end of the can. a) The tension in the string is adjusted so that the fundamental frequency for longitudinal sound waves in the air column in the can equals the frequency of the third-harmonic standing wave on the string. What is the relationship between the speed v_t of transverse waves on the string and the speed v_a of sound waves in the air? b) What happens to the sound produced by the instrument if the tension in the string is doubled?

20–31 An organ pipe has two successive harmonics with frequencies 1372 and 1764 Hz. The speed of sound in air is 344 m/s. a) Is this an open or a stopped pipe? Explain. b) What two harmonics are these? c) What is the length of the pipe?

20–32 Longitudinal Standing Waves in a Solid. Longitudinal standing waves can be produced in a solid rod by holding it at some point between the fingers of one hand and stroking it with

the other hand. The rod oscillates with antinodes at both ends. a) Why are the ends antinodes and not nodes? b) The fundamental frequency can be obtained by stroking the rod while it is held at its center. Explain why this is the *only* place to hold the rod to obtain the fundamental. c) Calculate the fundamental frequency of a steel rod of length 1.50 m (see Table 19–1). d) What is the next possible standing-wave frequency of this rod? Where should the rod be held to excite a standing wave of this frequency?

20–33 a) Determine the first three normal-mode frequencies for a pipe of length L that is closed at *both* ends. Explain your reasoning. b) Use the results of part (a) to estimate the normal-mode frequencies of a shower stall. Use $v = 344$ m/s. Explain the connection between these frequencies and the observation that your singing voice probably sounds better in the shower, especially when you sing at certain frequencies.

20–34 One type of steel has a density of 7.8×10^3 kg/m^3 and a breaking stress of 7.0×10^8 N/m^2. A cylindrical guitar string is to be made of 4.00 g of this steel. a) What are the length and radius of the longest and thinnest string that can be placed under a tension of 900 N without breaking? b) What is the highest fundamental frequency that this string could have?

20–35 Combining Standing Waves. A guitar string of length L is plucked in such a way that the total wave produced is the sum of the fundamental and the second harmonic. That is, the standing wave is given by

$$y(x, t) = y_1(x, t) + y_2(x, t),$$

where

$$y_1(x, t) = C \cos \omega_1 t \sin k_1 x,$$
$$y_2(x, t) = C \cos \omega_2 t \sin k_2 x,$$

with $\omega_1 = vk_1$ and $\omega_2 = vk_2$. a) At what values of x are the nodes of y_1? b) At what values of x are the nodes of y_2? c) Graph the total wave at $t = 0$, $t = 1/8f_1$, $t = 1/4f_1$, $t = 3/8f_1$, and $t = 1/2f_1$. d) Does the sum of the two standing waves y_1 and y_2 produce a standing wave? Explain.

20–36 A massive aluminum sculpture is hung from a steel wire. The fundamental frequency for transverse standing waves on the wire is 200 Hz. The sculpture is then immersed in water so that one-third of its volume is submerged. a) What is the new fundamental frequency? b) Why is it a good approximation to treat the wire as being fixed at both ends?

20–37 A long tube contains air at a pressure of 1.00 atm and a temperature of 77.0°C. The tube is open at one end and closed at the other by a movable piston. A tuning fork near the open end is vibrating with a frequency of 500 Hz. Resonance is produced when the piston is at distances 18.0, 55.5, and 93.0 cm from the open end. a) From these measurements, what is the speed of sound in air at 77.0°C? b) From the result of part (a), what is the ratio γ of the molar heat capacities at constant pressure and constant volume for air at this temperature? (See Section 19–7. The molar mass of air is 28.8 g/mol.) c) These data show that a displacement antinode is slightly outside of the open end of the tube. How far outside is it?

20–38 The auditory canal of the human ear (Fig. 19–14) extends about 2.5 cm from the outside ear to the eardrum. a) Explain why the human ear is especially sensitive to sounds at frequencies around 3500 Hz. Use $v = 344$ m/s. b) Would you expect the ear to be especially sensitive to frequencies around 7000 Hz? Around 10,500 Hz? Why or why not?

20–39 A cellist tunes the C-string of her instrument to a fundamental frequency of 65.4 Hz. The vibrating portion of the string is 0.600 m long and has a mass of 14.4 g. a) With what tension must it be stretched? b) What percent increase in tension is needed to increase the frequency from 65.4 Hz to 73.4 Hz, corresponding to a rise in pitch from C to D?

20–40 A platinum wire (density 21.4 g/cm^3) is 225 μm in diameter and 0.450 m long. One end of the wire is attached to the ceiling, while a 420-g mass is attached to the other end so that the wire hangs vertically under tension. If a vibrating tuning fork of just the right frequency is held next to the wire, the wire begins to vibrate as well. a) What tuning-fork frequencies will cause this to happen? You may assume that the bottom end of the wire (to which the mass is attached) is essentially stationary, and that the tension in the wire is essentially constant along its length. b) Justify the assumptions made in part (a).

20–41 The frequency of the note F$_4$ is 349 Hz. a) If an organ pipe is open at one end and closed at the other, what length must it have for its fundamental mode to produce this note at 20.0°C? b) At what air temperature will the frequency be 370 Hz, corresponding to a rise in pitch from F to F-sharp? (See Section 19–7. Ignore the change in length of the pipe due to the temperature change.)

CHALLENGE PROBLEMS

20–44 Instantaneous Power in a Standing Wave. From Eq. (19–29), the instantaneous rate at which a wave transmits energy along a string (instantaneous power) is

$$P(x, t) = -F \frac{\partial y(x, t)}{\partial x} \frac{\partial y(x, t)}{\partial t},$$

where F is the tension. a) Evaluate $P(x, t)$ for a standing wave of the form given by Eq. (20–1). b) Show that for all values of x, the *average* power P_{av} carried by the standing wave is zero. (Equation (19–33) does not apply here. Can you see why?) c) For a standing wave given by Eq. (20–1), graph $P(x, t)$ and the displacement $y(x, t)$ as functions of x for $t = 0$, $t = \pi/4\omega$, $t = \pi/2\omega$, and $t = 3\pi/4\omega$. (Positive $P(x, t)$ means energy is flowing in the $+x$-direction; negative $P(x, t)$ means the flow is in the $-x$-direction.) d) The *kinetic* energy per unit length of string is greatest where the string has the greatest transverse speed, and the *potential* energy per unit length of string is greatest where the string has the steepest slope (because there the string is stretched the

20–42 A standing wave with a frequency of 1100 Hz in a column of methane (CH$_4$) at 20.0°C produces nodes that are 0.200 m apart. What is the ratio γ of the molar heat capacity at constant pressure to that at constant volume for methane? (See Section 19–7. The molar mass of methane is 16.0 g/mol.)

20–43 Two identical loudspeakers are located at points A and B, 2.00 m apart. The loudspeakers are driven by the same amplifier and produce sound waves with a frequency of 784 Hz. Take the speed of sound in air to be 344 m/s. A small microphone is moved out from point B along a line perpendicular to the line connecting A and B (line BC in Fig. 20–27). a) At what distances from B will there be *destructive* interference? b) At what distances from B will there be *constructive* interference? c) If the frequency is made low enough, there will be no positions along the line BC at which destructive interference occurs. How low must the frequency be for this to be the case?

FIGURE 20–27 Problem 20–43.

most). (See Challenge Problem 19–45.) Using these ideas, discuss the flow of energy along the string.

20–45 The B-string of a guitar is made of steel (density 7800 kg/m^3), is 63.5 cm long, and has diameter 0.406 mm. The fundamental frequency is $f = 247.0$ Hz. a) Find the string tension. b) If the tension F is changed by a small amount ΔF, the frequency f changes by a small amount Δf. Show that

$$\frac{\Delta f}{f} = \frac{1}{2} \frac{\Delta f}{f}.$$

c) The string is tuned as in part (a) when its temperature is 18.5°C. Strenuous playing can make the temperature of the string rise, changing its vibration frequency. Find Δf if the temperature of the string rises to 29.5°C. The steel string has a Young's modulus of 2.00×10^{11} Pa and a coefficient of linear expansion of 1.20×10^{-5} (C°)$^{-1}$. Assume that the temperature of the body of the guitar remains constant. Will the pitch rise or fall?

The human ear is an incredibly sensitive instrument, able to detect sounds so faint that they move the eardrum only a microscopic distance. But this sensitivity can be ruined by overly loud sounds. Excessive exposure to loud noises can destroy the hair cells in the inner ear, blocking the passage of signals along the auditory nerve from the ear to the brain. Hearing protection is essential for anyone working in a high-noise environment.

21–1 INTRODUCTION

Of all the mechanical waves that occur in nature, the most important in our everyday lives are longitudinal waves in a medium, usually air, called *sound* waves. The reason is that the human ear is tremendously sensitive and can detect sound waves even of very low intensity. Besides their use in spoken communication, our ears allow us to pick up a myriad of cues about our environment, from the welcome sound of a meal being prepared to the warning sound of an approaching car. The ability to hear an unseen nocturnal predator was essential to the survival of our ancestors, so it is no exaggeration to say that we humans owe our existence to our highly evolved sense of hearing.

In this chapter we will discuss several important properties of sound waves, including frequency, amplitude, and intensity. Up to this point we have described mechanical waves primarily in terms of displacement; however, a description of sound waves in terms of *pressure* fluctuations is often more appropriate, largely because the ear is primarily sensitive to changes in pressure. We'll study the relations among displacement, pressure fluctuation, and intensity and the connections between these quantities and human sound perception. We will find that interference of two sound waves differing slightly in frequency causes a phenomenon called *beats*. When a source of sound or a listener moves through the air, the listener may hear a different frequency than the one emitted by the source. This is the Doppler effect, another topic of this chapter. We'll also look briefly at shock waves and "sonic booms."

21–2 SOUND WAVES

The most general definition of **sound** is that it is a longitudinal wave in a medium. Our main concern in this chapter is with sound waves in air, but sound can travel through any gas, liquid, or solid. You may be all too familiar with the propagation of sound through a solid if your neighbor's stereo speakers are right next to your wall.

The simplest sound waves are sinusoidal waves, which have definite frequency, amplitude, and wavelength. The human ear is sensitive to waves in the frequency range from about 20 to 20,000 Hz, called the **audible range,** but we also use the term *sound* for similar waves with frequencies above (**ultrasonic**) and below (**infrasonic**) the range of human hearing.

Sound waves usually travel out in all directions from the source of sound, with an amplitude that depends on the direction and distance from the source. We'll return to this point in the next section. For now, we concentrate on the idealized case of a sound wave that propagates in the positive x-direction only. As we discussed in Section 19–4, such a wave is described by a wave function $y(x, t)$, which gives the instantaneous displacement y of a particle in the medium at position x at time t. If the wave is sinusoidal, we can express it using Eq. (19–7):

$$y(x, t) = A \sin(\omega t - kx) \quad \text{(sound wave propagating in the } +x\text{-direction).}$$

(21–1)

Remember that in a longitudinal wave the displacements are *parallel* to the direction of travel of the wave, so distances x and y are measured parallel to each other, not perpendicular as in a transverse wave. The amplitude A is the maximum displacement of a particle in the medium from its equilibrium position.

Sound waves may also be described in terms of variations of *pressure* at various points. In a sinusoidal sound wave in air, the pressure fluctuates above and below atmospheric pressure p_a in a sinusoidal variation with the same frequency as the motions of the air particles. The human ear operates by sensing such pressure variations. A sound wave entering the ear canal exerts a fluctuating pressure on one side of the eardrum; the air on the other side of the eardrum, vented to the outside by the Eustachian tube, is at atmospheric pressure. The pressure difference on the two sides of the eardrum sets it into motion. Microphones and similar devices also usually sense pressure differences, not displacements, so it is very useful to develop a relation between these two descriptions.

Let $p(x, t)$ be the instantaneous pressure fluctuation in a sound wave at any point x at time t. That is, $p(x, t)$ is the amount by which the pressure *differs* from normal atmospheric pressure p_a. Think of $p(x, t)$ as the *gauge pressure* defined in Section 14–3; it can be either positive or negative. The *absolute* pressure at a point is then $p_a + p(x, t)$.

To see the connection between the pressure fluctuation $p(x, t)$ and the displacement $y(x, t)$ in a sound wave propagating in the +x-direction, consider an imaginary cylinder of air with cross-section area S and axis along the direction of propagation (Fig. 21–1). When no sound wave is present, the cylinder has length Δx and volume $V = S \Delta x$, as shown by the shaded volume in Fig. 21–1. When a wave is present, at time t the end of the cylinder that is initially at x is displaced by $y_1 = y(x, t)$, and the end that is initially at $x + \Delta x$ is displaced by $y_2 = y(x + \Delta x, t)$; this is shown by the red lines. If $y_2 > y_1$, as shown in Fig. 21–1, the cylinder's volume has increased, which causes a decrease in pressure. If $y_2 < y_1$, the cylinder's volume has decreased and the pressure has increased. If $y_2 = y_1$, the cylinder is simply shifted to the left or right; there is no volume change and no pressure fluctuation. The pressure fluctuation depends on the *difference* between the displacement at neighboring points in the medium.

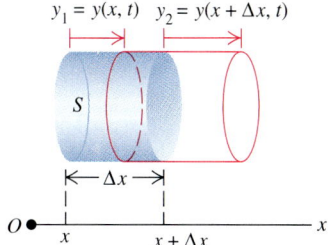

Quantitatively, the change in volume ΔV of the cylinder is

$$\Delta V = S(y_2 - y_1) = S[y(x + \Delta x, t) - y(x, t)].$$

In the limit as $\Delta x \to 0$, the fractional change in volume dV/V is

$$\frac{dV}{V} = \lim_{\Delta x \to 0} \frac{y(x + \Delta x, t) - y(x, t)}{\Delta x} = \frac{\partial y(x, t)}{\partial x}.$$

(21–2)

21–1 A cylindrical volume of gas with cross-section area S. The length in the undisplaced position is Δx. During wave propagation along the axis, the left end is displaced to the right a distance y_1 and the right end is displaced a different distance y_2. The resulting change in volume is $S(y_2 - y_1)$.

The fractional volume change is related to the pressure fluctuation by the bulk modulus B, which by definition [Eq. (11–13)] is $B = -p(x, t)/(dV/V)$. Solving for $p(x, t)$, we have

$$p(x, t) = -B \frac{\partial y(x, t)}{\partial x}.$$

(21–3)

The negative sign arises because when $\partial y(x, t)/\partial x$ is positive, the displacement is greater at $x + \Delta x$ than at x, corresponding to an increase in volume and a *decrease* in pressure.

When we evaluate $\partial y(x, t)/\partial x$ for the sinusoidal wave of Eq. (21–1), we find

$$p(x, t) = BkA \cos(\omega t - kx).$$

(21–4)

Figure 21–2 shows $y(x, t)$ and $p(x, t)$ for a sinusoidal sine wave at $t = 0$. While $y(x, t)$ and $p(x, t)$ describe the same wave, these two functions are one quarter-cycle out of phase; at a given time, the displacement is greatest where the pressure fluctuation is zero, and vice versa.

21–2 Pressure fluctuation $p(x, t)$ and displacement $y(x, t)$ in a sinusoidal traveling sound wave, shown as functions of x at a particular time t. The pressure fluctuation and displacement are a quarter-cycle out of phase; $p(x, t)$ is maximum where $y(x, t)$ is zero, and vice versa.

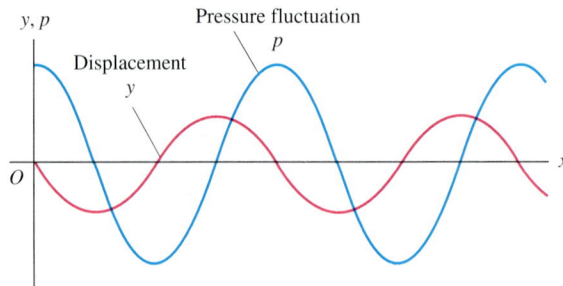

CAUTION ▶ Take care not to confuse the *traveling* sound wave described by Eqs. (21–1) and (21–2) with the *standing* sound waves in a pipe that we discussed in Section 20–5. In the traveling wave shown in Fig. 21–2, the patterns of the pressure fluctuation and the displacement move together in the $+x$-direction (to the right in Fig. 21–2); at any point x, particles oscillate in simple harmonic motion, and the pressure varies sinusoidally. By contrast, in a standing sound wave, the patterns of $p(x, t)$ and $y(x, t)$ do *not* move; there are points at which the displacement is always zero (the displacement nodes) and other points at which the pressure fluctuation is always zero (the pressure nodes). ◀

Equation (21–4) shows that the quantity BkA represents the maximum pressure fluctuation. We call this the **pressure amplitude,** denoted by p_{max}:

$$p_{max} = BkA \qquad \text{(sinusoidal sound wave).} \qquad (21–5)$$

The pressure amplitude is directly proportional to the displacement amplitude A, as we might expect, and it also depends on wavelength. Waves of shorter wavelength λ (larger wave number $k = 2\pi/\lambda$) have greater pressure variations for a given amplitude because the maxima and minima are squeezed closer together. A medium with a large value of bulk modulus B requires a relatively large pressure amplitude for a given displacement amplitude because large B means a less compressible medium, that is, greater pressure change required for a given volume change.

EXAMPLE 21–1

Amplitude of a sound wave in air In a sinusoidal sound wave of moderate loudness the maximum pressure variations are of the order of 3.0×10^{-2} Pa above and below atmospheric pressure p_a (nominally 1.013×10^5 Pa at sea level). Find the corresponding maximum displacement if the frequency is 1000 Hz and $v = 344$ m/s.

SOLUTION From Eq. (21–5) the maximum displacement is $A = p_{max}/Bk$. To determine k, we use Eq. (19–6), $\omega = vk$. We have $\omega = 2\pi f = (2\pi \text{ rad})(1000 \text{ Hz}) = 6283$ rad/s and

$$k = \frac{\omega}{v} = \frac{6283 \text{ rad/s}}{344 \text{ m/s}} = 18.3 \text{ rad/m.}$$

As described in Section 19–7, the appropriate value of B for ordinary sound waves in air is the *adiabatic* bulk modulus. From Eq. (19–24), for air at normal atmospheric pressure,

$$B = \gamma p_a = (1.40)(1.013 \times 10^5 \text{ Pa}) = 1.42 \times 10^5 \text{ Pa.}$$

Then

$$A = \frac{p_{max}}{Bk} = \frac{3.0 \times 10^{-2} \text{ Pa}}{(1.42 \times 10^5 \text{ Pa})(18.3 \text{ rad/m})}$$

$$= 1.2 \times 10^{-8} \text{ m.}$$

This displacement amplitude is only about $\frac{1}{100}$ the size of a human cell. Remember that the ear actually senses pressure fluctuations; it detects these minuscule displacements only indirectly.

EXAMPLE 21–2

Amplitude of a sound wave in the inner ear When a sound wave enters the ear, it sets the eardrum into oscillation, which in turn causes oscillation of the three tiny bones in the middle ear called the *ossicles* (Fig. 21–3). This oscillation is finally transmitted to the fluid-filled inner ear; the motion of the fluid disturbs hair cells within the inner ear, which transmit nerve impulses to the brain with the information that a sound is present. The moving part of the eardrum has an area of about 43 mm^2, and the area of

21–3 The anatomy of the human ear. The middle ear is the size of a small marble; the ossicles (hammer, anvil, and stirrup) are the smallest bones in the human body.

the stirrup (the smallest of the ossicles) where it connects to the inner ear is about 3.2 mm². For the sound in the previous example, determine a) the pressure amplitude and b) the displacement amplitude of the wave in the fluid of the inner ear.

SOLUTION a) We can safely neglect the mass of the ossicles (about 58 mg = 5.8×10^{-5} kg), so the force exerted by the ossicles on the fluid in the inner ear (mostly water) is the same as the force exerted on the eardrum and ossicles by the sound wave in air. Hence the pressure amplitude p_{max} is greater in the inner ear than in the outside air because the same force F is exerted on a smaller area S. Pressure equals force divided by area, so

$$p_{max \, (inner \, ear)} = \frac{F}{S_{inner \, ear}} = \frac{p_{max \, (air)} \, S_{eardrum}}{S_{inner \, ear}}$$

$$= \frac{(3.0 \times 10^{-2} \text{ Pa})(43 \text{ mm}^2)}{3.2 \text{ mm}^2} = 0.40 \text{ Pa}.$$

The effect of the ossicles is to increase the pressure amplitude in the inner ear by a factor of (43 mm²)/(3.2 mm²) = 13. This amplification factor helps to give the human ear its great sensitivity.

b) To find the maximum displacement, we again use the relationship $A = p_{max}/Bk$ as in Example 21–1. The values of B, k, and p_{max} are different in the inner ear than in the air, however.

The fluid in the inner ear is mostly water, which has a much greater bulk modulus than air because water is much more difficult to compress. From Table 11–2 the compressibility of water (unfortunately also called k) equals 45.8×10^{-11} Pa⁻¹, so $B_{fluid} = 1/(45.8 \times 10^{-11} \text{ Pa}^{-1}) = 2.18 \times 10^9$ Pa.

The wave in the inner ear has the same angular frequency ω as the wave in the air because the air, eardrum, ossicles, and inner ear fluid all oscillate together. But since the wave speed v is greater in water than in air, the wave number $k = \omega/v$ is less in the inner ear. From Table 19–1 the speed of sound in water is 1482 m/s at 20°C and increases somewhat with temperature; at human body temperature (37.0°C, or 98.6°F), v is about 1500 m/s. Using the value of ω from the previous example, we have

$$k_{inner \, ear} = \frac{\omega}{v_{inner \, ear}} = \frac{6283 \text{ rad/s}}{1500 \text{ m/s}} = 4.2 \text{ rad/m}.$$

When we put everything together, the maximum displacement of the fluid in the inner ear is

$$A_{inner \, ear} = \frac{p_{max \, (inner \, ear)}}{B_{fluid} \, k_{inner \, ear}} = \frac{0.40 \text{ Pa}}{(2.18 \times 10^9 \text{ Pa})(4.2 \text{ rad/m})}$$

$$= 4.4 \times 10^{-11} \text{ m}.$$

This is an even smaller displacement amplitude than in the air. What really matters in the inner ear, however, is the *pressure* amplitude, since the pressure variations within the fluid cause the forces that set the hair cells into motion.

PERCEPTION OF SOUND WAVES

The physical characteristics of a sound wave are directly related to the perception of that sound by a listener. For a given frequency, the greater the pressure amplitude of a sinusoidal sound wave, the greater the perceived **loudness.** The relationship between pressure amplitude and loudness is not a simple one, and it varies from one person to another. One important factor is that the ear is not equally sensitive to all frequencies in the audible range. A sound at one frequency may seem louder than one of equal pressure amplitude at a different frequency. At 1000 Hz the minimum pressure amplitude that can be perceived with normal hearing is about 3×10^{-5} Pa; to produce the same loudness at 200 Hz or 15,000 Hz requires about 3×10^{-4} Pa. Perceived loudness also depends on the health of the ear. A loss of sensitivity at the high-frequency end usually happens naturally with age but can be further aggravated by excessive noise levels. Studies have shown that many young rock musicians have suffered permanent ear damage and have hearing that is typical of persons 65 years of age. Portable stereo headsets used at high volume pose similar threats to hearing. Be careful!

The frequency of a sound wave is the primary factor in determining the **pitch** of a sound, the quality that lets us classify the sound as "high" or "low." The higher the frequency of a sound (within the audible range), the higher the pitch that a listener will perceive. Pressure amplitude also plays a role in determining pitch. When a listener compares two sinusoidal sound waves with the same frequency but different pressure

21–4 (a) Graph of pressure fluctuation versus time from a clarinet with fundamental frequency $f_1 = 233$ Hz. (b) Harmonic content of the sound of the clarinet. (c) Graph of pressure fluctuation versus time from an alto recorder with fundamental frequency $f_1 = 523$ Hz. (d) Harmonic content of the sound of the alto recorder.

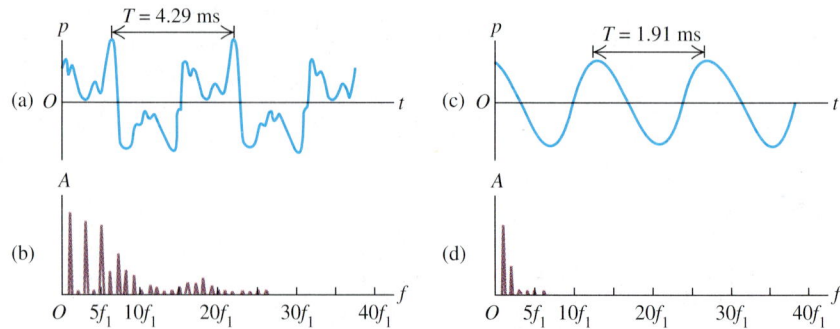

amplitudes, the one with the greater pressure amplitude is usually perceived as louder but also as slightly lower in pitch.

Musical sounds have wave functions that are more complicated than a simple sine function. The pressure fluctuation in the sound wave produced by a clarinet is shown in Fig. 21–4a. As we discussed in Section 20–5, the reason for the complexity is that the column of air in a wind instrument vibrates with the fundamental frequency and at many harmonics at the same time. The same occurs for a string that has been plucked, bowed, or struck. The sound wave produced in the surrounding air will have a similar amount of each harmonic, that is, a similar *harmonic content*. Figure 21–4b shows the harmonic content of the sound of a clarinet. Two tones produced by different instruments might have the same fundamental frequency (and thus the same pitch) but sound different because of the presence of different amounts of the various harmonics. The difference is called *tone color, quality,* or **timbre** and is often described in subjective terms such as reedy, golden, round, mellow, and tinny. A tone that is rich in harmonics, like the clarinet tone in Figs. 21–4a and 21–4b, usually sounds thin and "stringy" or "reedy," while a tone containing mostly a fundamental, like the alto recorder tone in Figs. 21–4c and 21–4d, is more mellow and flutelike. The same principle applies to the human voice, which is another example of a wind instrument; the vowels "a" and "e" sound different because of differences in harmonic content.

Another factor in determining tone quality is the behavior at the beginning (*attack*) and end (*decay*) of a tone. A piano tone begins with a thump and then dies away gradually. A harpsichord tone, in addition to having different harmonic content, begins much more quickly with a click, and the higher harmonics begin before the lower ones. When the key is released, the sound also dies away much more rapidly with a harpsichord than with a piano. Similar effects are present in other musical instruments. With wind and string instruments the player has considerable control over the attack and decay of the tone, and these characteristics help to define the unique characteristics of each instrument.

Unlike the tones made by musical instruments or the vowels in human speech, **noise** is a combination of *all* frequencies, not just frequencies that are integer multiples of a fundamental frequency. (An extreme case is "white noise," which contains equal amounts of all frequencies across the audible range.) Examples include the sound of the wind and the hissing sound you make in saying the consonant "s."

21–3 SOUND INTENSITY

Traveling sound waves, like all other traveling waves, transfer energy from one region of space to another. We define the **intensity** of a wave (denoted by I) to be *the time average rate at which energy is transported by the wave, per unit area*, across a surface perpendicular to the direction of propagation. That is, intensity I is average *power* per unit area.

In Section 6–5 we saw that power equals the product of force and velocity; see Eq. (6–18). So the power per unit area in a sound wave propagating in the x-direction equals the product of the pressure fluctuation $p(x, t)$ (force per unit area) and the *particle* velocity $v_y(x, t)$. For the sinusoidal wave described by Eq. (21–1), $p(x, t)$ is given by Eq. (21–4) and $v_y(x, t)$ is obtained by taking the time derivative of Eq. (21–1). We find

$$v_y(x, t) = \frac{\partial y(x, t)}{\partial t} = \omega A \cos(\omega t - kx),$$

$$p(x, t)v_y(x, t) = (BkA \cos(\omega t - kx))(\omega A \cos(\omega t - kx)) = B\omega k A^2 \cos^2(\omega t - kx).$$

The intensity is, by definition, the average value of $p(x, t)v_y(x, t)$. For any value of x the average value of the function $\cos^2(\omega t - kx)$ over one period $T = 2\pi/\omega$ is 1/2, so

$$I = \frac{1}{2} B\omega k A^2. \tag{21–6}$$

By using the relations $\omega = vk$ and $v^2 = B/\rho$, we can transform Eq. (21–6) into the form

$$I = \frac{1}{2} \sqrt{\rho B} \omega^2 A^2 \quad \text{(intensity of a sinusoidal sound wave)}, \tag{21–7}$$

which we cited at the end of Section 19–8. This equation shows why in a stereo system a low-frequency woofer has to vibrate with much larger amplitude than a high-frequency tweeter to produce the same sound intensity.

It is usually more useful to express I in terms of the pressure amplitude p_{max}. Using Eq. (21–5) and the relation $\omega = vk$, we find

$$I = \frac{\omega p_{max}{}^2}{2Bk} = \frac{v p_{max}{}^2}{2B}. \tag{21–8}$$

By using the wave speed relation $v^2 = B/\rho$, we can also write Eq. (21–8) in the alternative forms

$$I = \frac{p_{max}{}^2}{2\rho v} = \frac{p_{max}{}^2}{2\sqrt{\rho B}} \quad \text{(intensity of a sinusoidal sound wave)}. \tag{21–9}$$

We invite you to verify these expressions (see Exercise 21–4). Comparison of Eqs. (21–7) and (21–9) shows that sinusoidal sound waves of the same intensity but different frequency have different displacement amplitudes A but the *same* pressure amplitude p_{max}. This is another reason why it is usually more convenient to describe a sound wave in terms of pressure fluctuations, not displacement.

The *total* average power carried across a surface by a sound wave equals the product of the intensity at the surface and the surface area if the intensity over the surface is uniform. The average total sound power emitted by a person speaking in an ordinary conversational tone is about 10^{-5} W, while a loud shout corresponds to about 3×10^{-2} W. If all the residents of New York City were to talk at the same time, the total sound power would be about 100 W, equivalent to the electric power requirement of a medium-sized light bulb. On the other hand, the power required to fill a large auditorium or stadium with loud sound is considerable (see Example 21–5 below.)

Problem–Solving Strategy

SOUND INTENSITY

Quite a few quantities are involved in characterizing the amplitude and intensity of a sound wave, and it's easy to get lost in the maze of relationships. It helps to put them in categories: The amplitude is described by A or p_{max}, and the frequency f can be determined from ω, k, or λ. These quanti-ties are related through the wave speed v, which in turn is determined by the properties of the medium, B and ρ. Take a hard look at the problem at hand, identify which of these quantities are given and which you have to find; then start looking for relationships that take you where you want to go.

EXAMPLE 21–3

Intensity of a sound wave in air Find the intensity of the sound wave in Example 21–1, with $p_{max} = 3.0 \times 10^{-2}$ Pa, if the temperature is 20°C so that the density of air is $\rho = 1.20$ kg/m^3 and the speed of sound is $v = 344$ m/s.

SOLUTION From Eq. (21–9),

$$I = \frac{p_{max}^2}{2\rho v} = \frac{(3.0 \times 10^{-2} \text{ Pa})^2}{2(1.20 \text{ kg/m}^3)(344 \text{ m/s})}$$

$$= 1.1 \times 10^{-6} \text{ J/(s} \cdot \text{m}^2) = 1.1 \times 10^{-6} \text{ W/m}^2.$$

A very loud sound wave at the threshold of pain has a pressure amplitude of about 30 Pa and an intensity of about 1 W/m^2. The pressure amplitude of the faintest sound wave that can be heard is about 3×10^{-5} Pa, and the corresponding intensity is about 10^{-12} W/m^2. We invite you to verify these correspondences.

EXAMPLE 21–4

Same intensity, different frequencies A 20-Hz sound wave has the same intensity as the 1000-Hz sound wave in Examples 21–1 and 21–3. What are the amplitude A and pressure amplitude p_{max} of the 20-Hz sound wave?

SOLUTION In Eq. (21–7), ρ and B depend on the medium, not the amplitude or frequency of the wave. For I to be constant, the product ωA must be constant. That is,

$$(20 \text{ Hz})A_{20} = (1000 \text{ Hz})(1.2 \times 10^{-8} \text{ m}),$$

$$A_{20} = 6.0 \times 10^{-7} \text{ m} = 0.60 \ \mu\text{m}.$$

Do you understand why we didn't have to convert the frequencies to angular frequencies?

Since the intensity is the same for both frequencies, Eq. (21–9) shows that the *pressure* amplitude p_{max} must also be the same for both. Hence $p_{max} = 3.0 \times 10^{-2}$ Pa for $f = 20$ Hz. Note that using Eq. (21–5) and $k = \omega/v$, we get $p_{max} = BkA = (B/v)\omega A$; the bulk modulus B and wave speed v depend only on the medium, so we again conclude that the product ωA must have the same value for both frequencies.

EXAMPLE 21–5

"Play it loud!" For an outdoor concert we want the sound intensity at a distance of 20 m from the speaker array to be 1 W/m^2. Assuming that the sound waves have the same intensity in all directions, what acoustic power output is needed from the speaker array?

SOLUTION We make the assumptions that the speakers are near ground level and that the acoustic power is spread uniformly over a hemisphere 20 m in radius (that is, we assume that none

of the acoustic power is directed into the ground). The surface area of this hemisphere is equal to $(1/2)(4\pi)(20 \text{ m})^2$, or about 2500 m^2. The acoustic power needed is

$$(1 \text{ W/m}^2)(2500 \text{ m}^2) = 2500 \text{ W} = 2.5 \text{ kW}.$$

The electrical power input to the speaker would need to be considerably larger because the efficiency of such devices is not very high (typically a few percent for ordinary speakers, and up to 25% for horn-type speakers).

VARIATION OF INTENSITY WITH DISTANCE

If a source of sound can be considered as a point, the intensity at a distance r from the source is inversely proportional to r^2. This follows directly from energy conservation; if the power output of the source is P, then the average intensity I_1 through a sphere with radius r_1 and surface area $4\pi r_1^2$ is

$$I_1 = \frac{P}{4\pi r_1^2}.$$

The average intensity I_2 through a sphere with a different radius r_2 is given by a similar expression. If no energy is absorbed between the two spheres, the power P must be the same for both, and

$$4\pi r_1^2 I_1 = 4\pi r_2^2 I_2,$$

$$\frac{I_1}{I_2} = \frac{r_2^2}{r_1^2}. \tag{21–10}$$

TABLE 21–1

SOUND INTENSITY LEVELS FROM VARIOUS SOURCES (REPRESENTATIVE VALUES)

SOURCE OR DESCRIPTION OF SOUND	SOUND INTENSITY LEVEL, β (dB)	INTENSITY, I (W/m^2)
Threshold of pain	120	1
Riveter	95	3.2×10^{-3}
Elevated train	90	10^{-3}
Busy street traffic	70	10^{-5}
Ordinary conversation	65	3.2×10^{-6}
Quiet automobile	50	10^{-7}
Quiet radio in home	40	10^{-8}
Average whisper	20	10^{-10}
Rustle of leaves	10	10^{-11}
Threshold of hearing at 1000 Hz	0	10^{-12}

The intensity I at any distance r is therefore inversely proportional to r^2. This "inverse-square" relationship also holds for various other energy-flow situations with a point source, such as light emitted by a point source.

The inverse-square relationship between sound intensity and distance from the source does not apply indoors because sound energy can also reach a listener by reflection from the walls and ceiling. Indeed, part of the architect's job in designing an auditorium is to tailor these reflections so that the intensity is as nearly constant as possible over the auditorium.

THE DECIBEL SCALE

Because the ear is sensitive over such a broad range of intensities, a *logarithmic* intensity scale is usually used. The **sound intensity level** β of a sound wave is defined by the equation

$$\beta = (10 \text{ dB}) \log \frac{I}{I_0} \qquad \text{(definition of sound intensity level).} \qquad (21\text{–}11)$$

In this equation, I_0 is a reference intensity, chosen to be 10^{-12} W/m^2, approximately the threshold of human hearing at 1000 Hz. Recall that "log" means the logarithm to the base 10. Sound intensity levels are expressed in **decibels**, abbreviated dB. A decibel is $\frac{1}{10}$ of a *bel,* a unit named for Alexander Graham Bell (the inventor of the telephone). The bel is inconveniently large for most purposes, and the decibel is the usual unit of sound intensity level.

If the intensity of a sound wave equals I_0 or 10^{-12} W/m^2, its sound intensity level is 0 dB. An intensity of 1 W/m^2 corresponds to 120 dB. Table 21–1 gives the sound intensity levels in decibels of several familiar sounds. You can use Eq. (21–11) to check the value of sound intensity level β given for each intensity in the table.

Because the ear is not equally sensitive to all frequencies in the audible range, some sound-level meters weight the various frequencies unequally. One such scheme leads to the so-called dBA scale; this scale de-emphasizes the low and very high frequencies, where the ear is less sensitive than at midrange frequencies.

EXAMPLE 21–6

Temporary deafness A ten-minute exposure to 120-dB sound will typically shift your threshold of hearing at 1000 Hz from 0 dB up to 28 dB for a while. Ten years of exposure to 92-dB sound will cause a *permanent* shift up to 28 dB. What intensities correspond to 28 dB and 92 dB?

SOLUTION We rearrange Eq. (21–11) by dividing both sides by 10 dB and then using the relationship $10^{\log x} = x$:

$$I = I_0 10^{\beta/(10 \text{ dB})}.$$

When $\beta = 28$ dB,

$$I = (10^{-12} \text{ W/m}^2)10^{(28 \text{ dB}/10 \text{ dB})}$$

$$= (10^{-12} \text{ W/m}^2)10^{2.8} = 6.3 \times 10^{-10} \text{ W/m}^2.$$

Similarly, for $\beta = 92$ dB,

$$I = (10^{-12} \text{ W/m}^2)10^{(92 \text{ dB}/10 \text{ dB})} = 1.6 \times 10^{-3} \text{ W/m}^2.$$

If your answers are a factor of 10 too large, you may have entered 10×10^{-12} in your calculator instead of 1×10^{-12}. Be careful!

EXAMPLE 21-7

A bird sings in a meadow Consider an idealized model with a bird (treated as a point source) emitting constant sound power, with intensity inversely proportional to the square of the distance from the bird. By how many dB does the sound intensity level drop when you move twice as far away from the bird?

SOLUTION We label the two points 1 and 2 (Fig. 21–5), and we use Eq. (21–11) twice. The difference in sound intensity level, $\beta_2 - \beta_1$, is given by

$$\beta_2 - \beta_1 = (10 \text{ dB}) \left(\log \frac{I_2}{I_0} - \log \frac{I_1}{I_0} \right)$$

$$= (10 \text{ dB}) [(\log I_2 - \log I_0) - (\log I_1 - \log I_0)]$$

$$= (10 \text{ dB}) \log \frac{I_2}{I_1}.$$

Now we use the reciprocal of Eq. (21–10); $I_2/I_1 = r_1^2/r_2^2$, so

$$\beta_2 - \beta_1 = (10 \text{ dB}) \log \frac{r_1^2}{r_2^2} = (10 \text{ dB}) \log \frac{r_1^2}{(2r_1)^2}$$

$$= (10 \text{ dB}) \log \frac{1}{4} = -6.0 \text{ dB}.$$

A decrease in intensity of a factor of four corresponds to a 6-dB decrease in sound intensity level.

We invite you to prove that an intensity increase of a factor of two corresponds to a 3-dB increase in sound intensity level. This change is barely perceptible to the human ear. Most people usually interpret an increase of 8 to 10 dB in sound intensity level as a doubling of loudness.

21–5 When you double your distance from a point source of sound, by how much does the sound intensity level decrease?

P_2 P_1 Point source

D/V MENNINGS.

21-4 BEATS

In Section 20–6 we talked about *interference* effects that occur when two different waves with the same frequency overlap in the same region of space. Now let's look at what happens when we have two waves with equal amplitude but slightly different frequencies. This occurs, for example, when two tuning forks with slightly different frequencies are sounded together or when two organ pipes that are supposed to have exactly the same frequency are slightly "out of tune."

Consider a particular point in space where the two waves overlap. The displacements of the individual waves at this point are plotted as functions of time in Fig. 21–6a. The total length of the time axis represents one second, and the frequencies are 16 Hz (blue graph) and 18 Hz (red graph). Applying the principle of superposition, we add the two displacements at each instant of time to find the total displacement at that time. The result is the graph of Fig. 21–6b. At certain times ($t = 0.25$ s and $t = 0.75$ s in Fig. 21–6) the two waves are in phase; their maxima coincide and their amplitudes add. But

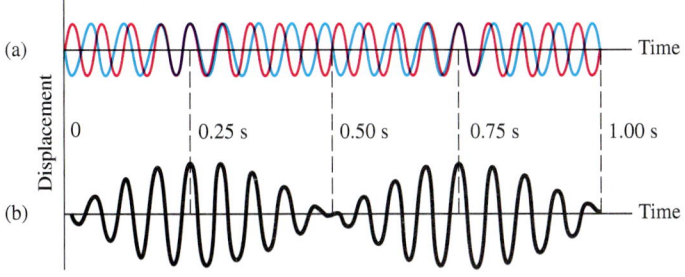

(a)

Displacement

0   0.25 s   0.50 s   0.75 s   1.00 s   Time

(b)   Time

21-6 Beats are fluctuations in amplitude produced by two sound waves of slightly different frequency, 16 Hz and 18 Hz. (a) Individual waves. (b) Resultant wave formed by superposition of the two waves. The beat frequency is 18 Hz − 16 Hz = 2 Hz.

because of their slightly different frequencies, the two waves cannot be in phase at all times. Indeed, at certain times (like $t = 0.50$ s in Fig. 21–6) the two waves are exactly *out of phase*. The two waves then cancel each other, and the total amplitude is zero.

The resultant wave in Fig. 21–6b looks like a single sinusoidal wave with a varying amplitude that goes from a maximum to zero and back. In this example the amplitude goes through two maxima and two minima in one second, so the frequency of this amplitude variation is 2 Hz. The amplitude variation causes variations of loudness called **beats,** and the frequency with which the loudness varies is called the **beat frequency.** In this example the beat frequency is the *difference* of the two frequencies. If the beat frequency is a few hertz, we hear it as a waver or pulsation in the tone.

We can prove that the beat frequency is *always* the difference of the two frequencies f_a and f_b. Suppose f_a is larger than f_b; the corresponding periods are T_a and T_b, with $T_a < T_b$. If the two waves start out in phase at time $t = 0$, they will again be in phase when the first wave has gone through exactly one more cycle than the second. This will happen at a value of t equal to T_{beat}, the *period* of the beat. Let n be the number of cycles of the first wave in time T_{beat}; then the number of cycles of the second wave in the same time is $(n - 1)$, and we have the relations

$$T_{\text{beat}} = nT_a \qquad \text{and} \qquad T_{\text{beat}} = (n - 1)T_b.$$

Eliminating n between these two equations, we find

$$T_{\text{beat}} = \frac{T_a T_b}{T_b - T_a}.$$

The reciprocal of the beat period is the beat *frequency*, $f_{\text{beat}} = 1/T_{\text{beat}}$, so

$$f_{\text{beat}} = \frac{T_b - T_a}{T_a T_b} = \frac{1}{T_a} - \frac{1}{T_b},$$

and finally

$$f_{\text{beat}} = f_a - f_b \qquad \text{(beat frequency).} \qquad (21\text{–}12)$$

As claimed, the beat frequency is the difference of the two frequencies. In using Eq. (21–12), remember that f_a is the higher frequency.

An alternative way to derive Eq. (21–12) is to write functions to describe the curves in Fig. 21–6a and then add them. Suppose that at a certain position the two waves are given by $y_a(t) = A \sin 2\pi f_a t$ and $y_b(t) = -A \sin 2\pi f_b t$. We use the trigonometric identity

$$\sin a - \sin b = 2 \sin \frac{1}{2}(a - b) \cos \frac{1}{2}(a + b).$$

We can then express the total wave $y(t) = y_a(t) + y_b(t)$ as

$$y_a(t) + y_b(t) = \left[2A \sin \frac{1}{2}(2\pi)(f_a - f_b)t\right] \cos \frac{1}{2}(2\pi)(f_a + f_b)t.$$

ActivPhysics

10.7

Beats and Beat Frequency

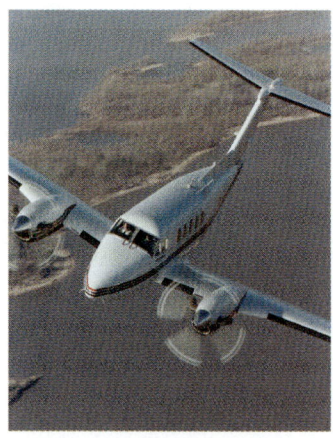

If the two propellers on this airplane are not precisely synchronized, the pilots, passengers, and listeners on the ground will hear beats.

The amplitude factor (the quantity in brackets) varies slowly with frequency $\frac{1}{2}(f_a - f_b)$. The cosine factor varies with a frequency equal to the *average* frequency $\frac{1}{2}(f_a + f_b)$. The *square* of the amplitude factor, which is proportional to the intensity that the ear hears, goes through two maxima and two minima per cycle. So the beat frequency f_{beat} that is heard is twice the quantity $\frac{1}{2}(f_a - f_b)$, or just $f_a - f_b$, in agreement with Eq. (21–12).

Beats between two tones can be heard up to a beat frequency of about 6 or 7 Hz. Two piano strings or two organ pipes differing in frequency by 2 or 3 Hz sound wavery and "out of tune," although some organ stops contain two sets of pipes deliberately tuned to beat frequencies of about 1 to 2 Hz for a gently undulating effect. Listening for beats is an important technique in tuning all musical instruments.

At frequency differences greater than about 6 or 7 Hz, we no longer hear individual beats, and the sensation merges into one of *consonance* or *dissonance,* depending on the frequency ratio of the two tones. In some cases the ear perceives a tone called a *difference tone,* with a pitch equal to the beat frequency of the two tones. For example, if you listen to a whistle that produces sounds at 1800 Hz and 1900 Hz when blown, you will hear not only these tones but also a much lower 100-Hz tone.

The engines on multiengine propeller aircraft have to be synchronized so that the sounds don't cause annoying beats, which are heard as loud throbbing sounds. On some planes this is done electronically; on others the pilot does it by ear, just like tuning a piano.

21–5 THE DOPPLER EFFECT

You've probably noticed that when a car approaches you with its horn sounding, the pitch seems to drop as the car passes. This phenomenon, first described by the 19th-century Austrian scientist Christian Doppler, is called the **Doppler effect.** When a source of sound and a listener are in motion relative to each other, the frequency of the sound heard by the listener is not the same as the source frequency. A similar effect occurs for light and radio waves; we'll return to this later in this section.

To analyze the Doppler effect for sound, we'll work out a relation between the frequency shift and the velocities of source and listener relative to the medium (usually air) through which the sound waves propagate. To keep things simple, we consider only the special case in which the velocities of both source and listener lie along the line joining them. Let v_S and v_L be the velocity components along this line for the source and the listener, respectively, relative to the medium. We choose the positive direction for both v_S and v_L to be the direction from the listener L to the source S. The speed of sound relative to the medium, v, is always considered positive.

MOVING LISTENER

Let's think first about a listener L moving with velocity v_L toward a stationary source S (Fig. 21–7). The source emits a sound wave with frequency f_S and wavelength $\lambda = v/f_S$. The figure shows several wave crests, separated by equal distances λ. The wave crests approaching the moving listener have a speed of propagation *relative to the listener* of $(v + v_L)$. So the frequency f_L with which the crests arrive at the listener's position (that is, the frequency the listener hears) is

$$f_L = \frac{v + v_L}{\lambda} = \frac{v + v_L}{v/f_S}, \qquad (21\text{–}13)$$

or

$$f_L = \left(\frac{v + v_L}{v}\right) f_S = \left(1 + \frac{v_L}{v}\right) f_S \qquad \begin{array}{l}\text{(moving listener,} \\ \text{stationary source).}\end{array} \qquad (21\text{–}14)$$

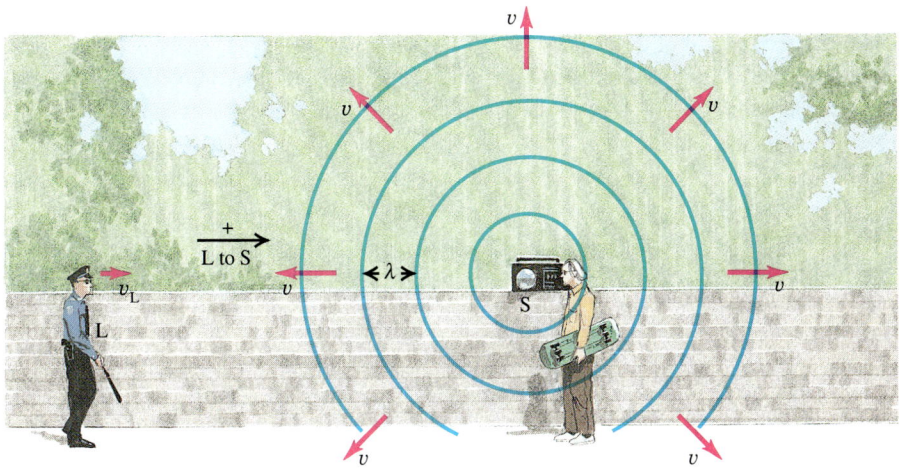

21–7 A listener moving toward a stationary source hears a frequency that is higher than the source frequency because the relative speed of listener and wave is greater than the wave speed v.

So a listener moving toward a source ($v_L > 0$), as in Fig. 21–7, hears a higher frequency (higher pitch) than does a stationary listener. A listener moving away from the source ($v_L < 0$) hears a lower frequency (lower pitch).

MOVING SOURCE AND MOVING LISTENER

Now suppose the source is also moving, with velocity v_S (Fig. 21–8). The wave speed relative to the wave medium (air) is still v; it is determined by the properties of the medium and is not changed by the motion of the source. But the wavelength is no longer equal to v/f_S. Here's why. The time for emission of one cycle of the wave is the period $T = 1/f_S$. During this time, the wave travels a distance $vT = v/f_S$ and the source moves a distance $v_S T = v_S/f_S$. The wavelength is the distance between successive wave crests, and this is determined by the *relative* displacement of source and wave. As Fig. 21–8 shows, this is different in front of and behind the source. In the region to the right of the source in Fig. 21–8 (that is, in front of the source) the wavelength is

$$\lambda = \frac{v}{f_S} - \frac{v_S}{f_S} = \frac{v - v_S}{f_S} \qquad \text{(wavelength in front of a moving source). (21–15)}$$

In the region to the left of the source (that is, behind the source) it is

$$\lambda = \frac{v + v_S}{f_S} \qquad \text{(wavelength behind a moving source).} \qquad \text{(21–16)}$$

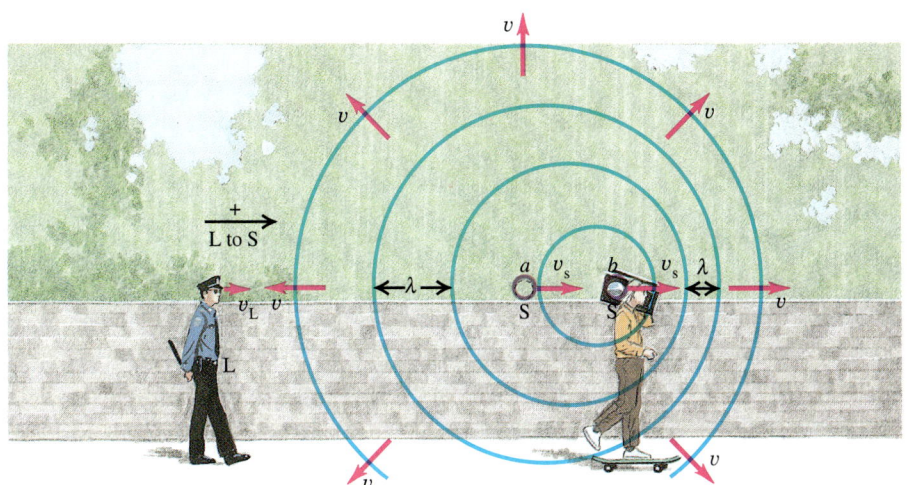

21–8 Wave crests emitted by a moving source are crowded together in front of the source (to the right of this source) and stretched out behind it (to the left of this source).

The waves in front of and behind the source are compressed and stretched out, respectively, by the motion of the source.

To find the frequency heard by the listener behind the source, we substitute Eq. (21–16) into the first form of Eq. (21–13):

$$f_L = \frac{v + v_L}{\lambda} = \frac{v + v_L}{(v + v_S)/f_S},$$

$$f_L = \frac{v + v_L}{v + v_S} f_S \qquad \text{(Doppler effect, moving source and moving listener).} \qquad (21\text{–}17)$$

This expresses the frequency f_L heard by the listener in terms of the frequency f_S of the source.

Equation (21–17) includes all possibilities for motion of source and listener (relative to the medium) along the line joining them. If the listener happens to be at rest in the medium, v_L is zero. When both source and listener are at rest or have the same velocity relative to the medium, then $v_L = v_S$ and $f_L = f_S$. Whenever the direction of the source or listener velocity is opposite to the direction from the listener toward the source (which we have defined as positive), the corresponding velocity to be used in Eq. (21–17) is negative.

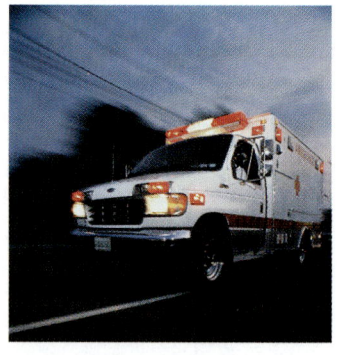

The Doppler effect explains why the siren on a fire engine or ambulance has a high pitch ($f_L > f_S$) when it is approaching you ($v_S < 0$) and a low pitch ($f_L < f_S$) when it is moving away ($v_S > 0$).

Problem–Solving Strategy

DOPPLER EFFECT

1. Draw a sketch and establish a coordinate system. Define the positive direction to be the direction from listener to source, and make sure you know the signs of all the relevant velocities. A velocity in the direction from listener toward source is positive; a velocity in the opposite direction is negative. Also, the velocities must all be measured relative to the air in which the sound is traveling.

2. Use consistent notation to identify the various quantities: subscript S for source, L for listener.

3. When a wave is reflected from a surface, either stationary or moving, the analysis can be carried out in two steps. In

the first, the surface plays the role of listener; the frequency with which the wave crests arrive at the surface is f_L. Then think of the surface as a new source, emitting waves with this same frequency f_L. Finally, determine what frequency is heard by a listener detecting this new wave.

4. Ask whether your final result makes sense. If the source and listener are moving toward each other, f_L is always greater than f_S; if they are moving apart, $f_L < f_S$. If the source and listener have no relative motion, $f_L = f_S$.

EXAMPLE 21-8

A police siren emits a sinusoidal wave with frequency $f_S = 300$ Hz. The speed of sound is 340 m/s. a) Find the wavelength of the waves if the siren is at rest in the air. b) If the siren is moving at 30 m/s (108 km/h, or 67 mi/h), find the wavelengths of the waves ahead of and behind the source.

21–9 Wavelengths ahead of and behind the police siren when the siren is moving through the air at 30 m/s.

SOLUTION a) When the source is at rest,

$$\lambda = \frac{v}{f_S} = \frac{340 \text{ m/s}}{300 \text{ Hz}} = 1.13 \text{ m}.$$

b) The situation is shown in Fig. 21–9. We use Eqs. (21–15) and (21–16). In front of the siren,

$$\lambda = \frac{v - v_S}{f_S} = \frac{340 \text{ m/s} - 30 \text{ m/s}}{300 \text{ Hz}} = 1.03 \text{ m}.$$

Behind the siren,

$$\lambda = \frac{v + v_S}{f_S} = \frac{340 \text{ m/s} + 30 \text{ m/s}}{300 \text{ Hz}} = 1.23 \text{ m}.$$

The wavelength is less in front of the siren and greater behind the siren.

EXAMPLE 21–9

If a listener L is at rest and the siren in Example 21–8 is moving away from L at 30 m/s (Fig. 21–10), what frequency does the listener hear?

SOLUTION We have $v_L = 0$ and $v_S = 30$ m/s. (The source velocity v_S is positive because the siren is moving in the same direction as the direction from listener to source.) From Eq. (21–17),

$$f_L = \frac{v}{v + v_S} f_S = \frac{340 \text{ m/s}}{340 \text{ m/s} + 30 \text{ m/s}} (300 \text{ Hz}) = 276 \text{ Hz}.$$

The source and listener are moving apart, so the frequency f_L heard by the listener is less than the frequency f_S emitted by the source.

Alternatively, we can use the result of Example 21–8 for the wavelength behind the source (which is where the listener in

21–10 The listener is at rest, and the siren moves away from the listener at 30 m/s.

Fig. 21–10 is located):

$$f_L = \frac{v}{\lambda} = \frac{340 \text{ m/s}}{1.23 \text{ m}} = 276 \text{ Hz}.$$

Even though the source is moving, the wave speed v relative to the stationary listener is unchanged.

EXAMPLE 21–10

If the siren is at rest and the listener is moving toward the left at 30 m/s (Fig. 21–11), what frequency does the listener hear?

SOLUTION Now $v_S = 0$. The positive direction (from listener to source) is still from left to right, so $v_L = -30$ m/s. Then Eq. (21–17) gives

$$f_L = \frac{v + v_L}{v} f_S = \frac{340 \text{ m/s} + (-30 \text{ m/s})}{340 \text{ m/s}} (300 \text{ Hz}) = 274 \text{ Hz}.$$

Again the frequency heard by the listener is less than the source frequency. Note that the *relative velocity* of source and listener is

21–11 The siren is at rest, and the listener moves away from it at 30 m/s.

the same as in the previous example, but the Doppler shift is different because the velocities relative to the *air* are different.

EXAMPLE 21–11

If the siren is moving away from the listener with a speed of 45 m/s relative to the air and the listener is moving toward the siren with a speed of 15 m/s relative to the air (Fig. 21–12), what frequency does the listener hear?

SOLUTION In this case, $v_L = 15$ m/s and $v_S = 45$ m/s. (Both velocities are positive because both velocity vectors point in the direction from listener to source.) From Eq. (21–17),

$$f_L = \frac{v + v_L}{v + v_S} f_S = \frac{340 \text{ m/s} + 15 \text{ m/s}}{340 \text{ m/s} + 45 \text{ m/s}} (300 \text{ Hz})$$

$$= 277 \text{ Hz}.$$

The frequency heard by the listener is again less than the source frequency, but the value is different than in the previous two examples, even though the source and listener move away from each other at 30 m/s in all three cases. The *sign* of the Doppler shift of frequency (that is, whether f_L is less than or greater than f_S) depends on how the source and listener are moving relative to each other; to determine the *value* of the Doppler shift of frequency, you must know the velocities of source and listener relative to the air.

21–12 The velocities relative to the air when the siren is moving away from the listener with a relative velocity of 30 m/s.

EXAMPLE 21-12

The police car with its 300-Hz siren is moving toward a warehouse at 30 m/s, intending to crash through the door. What frequency does the driver of the police car hear reflected from the warehouse?

SOLUTION In this situation there are *two* Doppler shifts, as shown in Fig. 21–13. In the first shift, the warehouse is the stationary "listener." The frequency of sound reaching the warehouse, which we call f_w, is greater than 300 Hz because the source is approaching. In the second shift, the warehouse acts as a source of sound with frequency f_w, and the listener is the driver of the police car; she hears a frequency greater than f_w because she is approaching the source.

To determine f_w, we use Eq. (21–17) with f_L replaced by f_w. The warehouse is at rest, so $v_L = 0$; the siren is moving in the negative direction from source to listener, so $v_S = -30$ m/s. Then

$$f_w = \frac{v}{v + v_S} f_s = \frac{340 \text{ m/s}}{340 \text{ m/s} + (-30 \text{ m/s})} (300 \text{ Hz}) = 329 \text{ Hz}.$$

To determine the frequency heard by the driver, we again use Eq. (21–17), but now with f_s replaced by f_w. The warehouse is now the source, so $v_S = 0$, and the velocity of the listener is $v_L = +30$ m/s; v_L is positive because it is in the direction from listener to source. Hence

$$f_L = \frac{v + v_L}{v} f_w = \frac{340 \text{ m/s} + 30 \text{ m/s}}{340 \text{ m/s}} (329 \text{ Hz}) = 358 \text{ Hz}.$$

21–13 (a) The police car with siren (source) moves at 30 m/s toward the stationary warehouse (listener). (b) The sound wave is reflected from the stationary warehouse (source) toward the police car (listener).

The reflected sound heard by the driver has an even higher frequency than the sound heard by a stationary listener in the warehouse.

DOPPLER EFFECT FOR ELECTROMAGNETIC WAVES

In the Doppler effect for sound, the velocities v_L and v_S are always measured relative to the *air* or whatever medium we are considering. There is also a Doppler effect for *electromagnetic* waves in empty space, such as light waves or radio waves. In this case there is no medium that we can use as a reference to measure velocities, and all that matters is the *relative* velocity of source and receiver. (By contrast, the Doppler effect for sound does not depend simply on this relative velocity, as discussed in Example 21–11.)

To derive the expression for the Doppler frequency shift for light, we have to use the special theory of relativity. We will discuss this in Chapter 39, but for now we quote the result without derivation. The wave speed is the speed of light, usually denoted by c, and it is the same for both source and receiver. In the frame of reference in which the receiver is at rest, the source is moving away from the receiver with velocity v. (If the source is *approaching* the receiver, v is negative.) The source frequency is again f_s. The frequency f_R measured by the receiver R (the frequency of arrival of the waves at the receiver) is then given by

$$f_R = \sqrt{\frac{c - v}{c + v}} f_s \qquad \text{(Doppler effect for light).} \qquad (21–18)$$

When v is positive, the source is moving directly *away* from the receiver and f_R is always less than f_s; when v is negative, the source is moving directly *toward* the receiver and f_R is *greater* than f_s. The qualitative effect is the same as for sound, but the quantitative relationship is different.

A familiar application of the Doppler effect for radio waves is the radar device mounted on the side window of a police car to check other cars' speeds. The electromagnetic wave emitted by the device is reflected from a moving car, which acts as a moving source, and the wave reflected back to the device is Doppler-shifted in frequency. The transmitted and reflected signals are combined to produce beats, and the speed can be computed from the frequency of the beats. Similar techniques ("Doppler radar") are used to measure wind velocities in the atmosphere.

The Doppler effect is also used to track satellites and other space vehicles. In Fig. 21–14 a satellite emits a radio signal with constant frequency f_S. As the satellite orbits past, it first approaches and then moves away from the receiver; the frequency f_R of the signal received on earth changes from a value greater than f_S to a value less than f_S as the satellite passes overhead.

The Doppler effect for electromagnetic waves, including visible light, is important in astronomy. Astronomers compare wavelengths of light from distant stars to those emitted by the same elements on earth. In a binary star system, in which two stars orbit about their common center of mass, the light is Doppler-shifted to higher frequencies when a star is moving toward an observer on earth and to lower frequencies when it's moving away. Measurements of the frequency shifts reveal information about the orbits and masses of the stars that comprise the binary system.

Light from most galaxies is shifted toward the longer-wavelength or red end of the visible spectrum, an effect called the *red shift*. This is often described as a Doppler shift resulting from motion of these galaxies away from us. But from the point of view of the general theory of relativity it is something much more fundamental, associated with the expansion of space itself. Distant galaxies have large red shifts because their light has been in transit for a long time and has shared in the expansion of all the space through which it moved. Extrapolating this expansion backward to over 10^{10} years ago leads to the "big bang" picture. From this point of view, the big bang was not an explosion in space but the initial rapid expansion of space itself.

21–14 Change of velocity component along the line of sight of a satellite passing a tracking station. The frequency received at the tracking station changes from high to low as the satellite passes overhead.

*21–6 SHOCK WAVES

You may have experienced "sonic booms" caused by an airplane flying over faster than the speed of sound. We can see qualitatively why this happens from Fig. 21–15. Let v_S denote the *speed* of the airplane relative to the air, so that it is always positive. The motion of the airplane through the air produces sound; if v_S is less than the speed of sound v, the waves in front of the airplane are crowded together with a wavelength given by Eq. (21–15):

$$\lambda = \frac{v - v_S}{f_S}.$$

As the speed v_S of the airplane approaches the speed of sound v, the wavelength approaches zero and the wave crests pile up on each other (Fig. 21–15a). The airplane must exert a large force to compress the air in front of it; by Newton's third law, the air exerts an equally large force back on the airplane. Hence there is a large increase in aerodynamic drag (air resistance) as the airplane approaches the speed of sound, a phenomenon known as the "sound barrier."

When v_S is greater in magnitude than v, the source of sound is **supersonic,** and Eqs. (21–15) and (21–17) for the Doppler effect no longer describe the sound wave in front of the source. Figure 21–15b shows a cross section of what happens. As the airplane moves, it displaces the surrounding air and produces sound. A series of wave crests is emitted from the nose of the airplane; each spreads out in a circle centered at the position of the airplane when it emitted the crest. After a time t the crest emitted from point S_1 has spread to a circle with radius vt, and the airplane has moved a greater distance $v_S t$ to position S_2. You can see that the circular crests interfere constructively at points along the green line that makes an angle α with the direction of the airplane velocity, leading

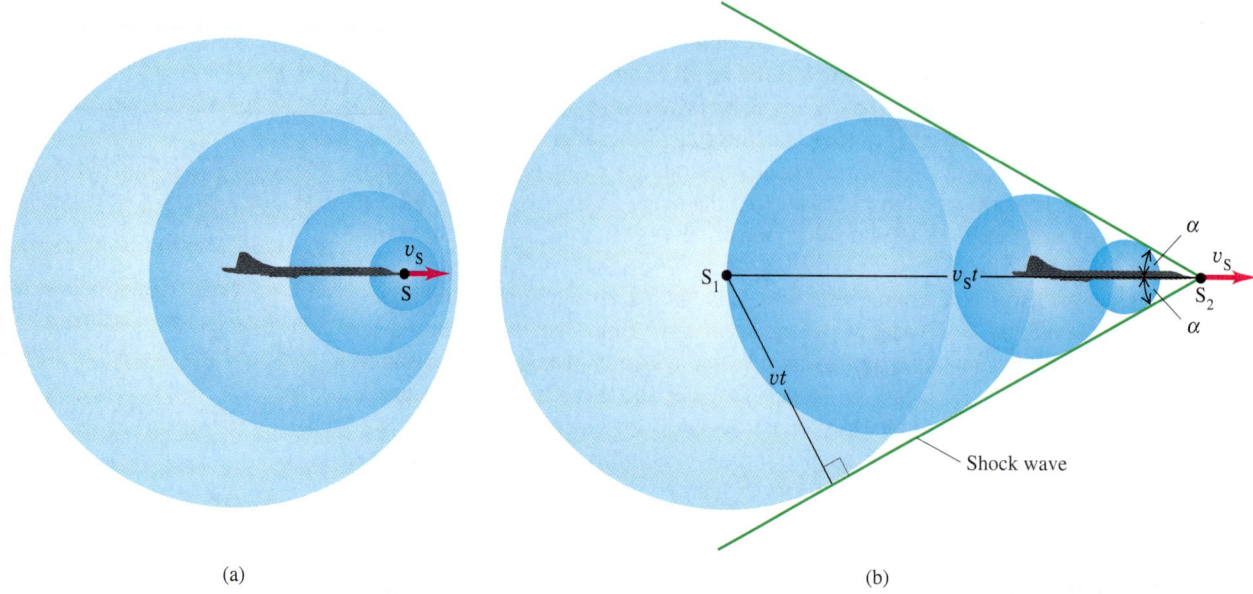

(a)   (b)

21–15 (a) As the speed of the source of sound S approaches the speed of sound, the wave crests begin to pile up in front of S. (b) A shock wave forms when the speed of the source is greater than the speed of sound. (c) Photograph of shock waves produced by a T-38 jet aircraft moving at 1.1 times the speed of sound. Separate shock waves are generated by the nose, wings, and tail. The angles of these waves vary because the air is accelerated and decelerated as it moves relative to the airplane, so the relative speed of the airplane and air is different at different points on the airplane.

to a very large-amplitude wave crest along this line. This large-amplitude crest is called a **shock wave** (Fig. 21–15c).

From the right triangle in Fig. 21–15b we can see that the angle α is given by

$$\sin \alpha = \frac{vt}{v_S t} = \frac{v}{v_S} \qquad \text{(shock wave).} \qquad (21-19)$$

In this relation, v_S is the *speed* of the source (the magnitude of its velocity) relative to the air and is always positive. The ratio v_S/v is called the **Mach number.** It is greater than unity for all supersonic speeds, and $\sin \alpha$ in Eq. (21–19) is the reciprocal of the Mach number. The first person to break the sound barrier was Capt. Chuck Yeager of the U. S. Air Force, flying the Bell X-1 at Mach 1.06 on October 14, 1947.

The actual situation is three-dimensional; the shock wave forms a *cone* around the direction of motion of the source. If the source (possibly a supersonic jet airplane or a rifle bullet) moves with constant velocity, the angle α is constant, and the shock-wave cone moves along with the source. It's the arrival of this shock wave that causes the sonic boom you hear after a supersonic airplane has passed by. The larger the airplane, the stronger the sonic boom; the shock wave produced at ground level by the Concorde supersonic airliner flying at 12,000 m (40,000 ft) causes a sudden jump in air pressure of about 20 Pa. In front of the shock-wave cone, there is no sound. Inside the cone a stationary listener hears the Doppler-shifted sound of the airplane moving away.

CAUTION ▶ We emphasize that a shock wave is produced *continuously* by any object that moves through the air at supersonic speed, not only at the instant that it "breaks the sound barrier." The sound waves that combine to form the shock wave, as in Fig. 21–15b, are created by the motion of the object itself, not by any sound source that the object may carry. The cracking noises of a bullet and of the tip of a circus whip are due to their supersonic motion. A supersonic jet airplane may have very loud engines, but these do not cause the shock wave. Indeed, a space shuttle makes a very loud sonic boom when coming in for a landing; its engines are out of fuel at this point, so it is a supersonic glider. ◀

Shock waves have applications outside of aviation. They are used to break up kidney stones and gallstones without invasive surgery, using a technique with the impressive name *extracorporeal shock-wave lithotripsy*. A shock wave produced outside the body is focused by a reflector or acoustic lens so that as much of it as possible converges on the stone. When the resulting stresses in the stone exceed its tensile strength, it breaks into small pieces and can be eliminated. This technique requires accurate determination of the location of the stone, which may be done using ultrasonic imaging techniques (see Example 19–5 in Section 19–6).

The first supersonic airplane, the Bell X-1, was shaped much like a .50-caliber bullet—which was known to be able to travel faster than sound.

EXAMPLE 21-13

Sonic boom of the Concorde The Concorde is flying at Mach 1.75 at an altitude of 8000 m, where the speed of sound is 320 m/s. How long after the plane passes directly overhead will you hear the sonic boom?

SOLUTION Figure 21–16 shows the situation just as the shock wave reaches you. From Eq. (21–19) the angle α is

$$\alpha = \arcsin \frac{1}{1.75} = 34.8°.$$

The speed of the plane is the speed of sound multiplied by the Mach number:

$$v_S = (1.75)(320 \text{ m/s}) = 560 \text{ m/s}.$$

From Fig. 21–16 we have

$$\tan \alpha = \frac{8000 \text{ m}}{v_S t},$$

$$t = \frac{8000 \text{ m}}{(560 \text{ m/s})(\tan 34.8°)} = 20.5 \text{ s}.$$

You hear the boom 20.5 s after the Concorde passes overhead, and at that time it has traveled (560 m/s)(20.5 s) = 11.5 km past the straight-overhead point.

In this calculation we assumed that the speed of sound is the same at all altitudes, so $\alpha = \arcsin v/v_S$ is a constant and the shock wave forms a perfect cone. In fact, the speed of sound decreases with increasing altitude. How would this affect the result?

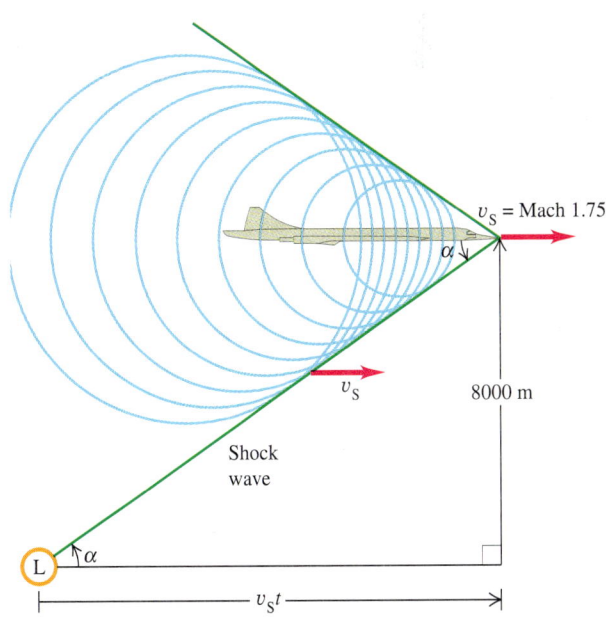

21–16 You hear a sonic boom when the shock wave reaches you at L (*not* just when the plane breaks the sound barrier). A listener to the right of L has not yet heard the sonic boom but will shortly; a listener to the left of L has already heard the sonic boom and now hears the Doppler-shifted sound of the airplane.

SUMMARY

■ Sound consists of longitudinal waves in a medium. A sinusoidal sound wave is characterized by its frequency f and wavelength λ (or angular frequency ω and wave number k) and by its amplitude A. The amplitude is also related to the pressure amplitude p_{max} by

$$p_{max} = BkA, \qquad (21\text{--}5)$$

where B is the bulk modulus of the wave medium.

■ The loudness of a sound depends on its amplitude and frequency, while the pitch depends primarily on its frequency alone. The tone quality or timbre depends on the harmonic content and the attack and decay characteristics.

■ The intensity I of a wave is the time average rate at which energy is transported by the wave, per unit area. For a sinusoidal wave of amplitude A and pressure amplitude p_{max},

$$I = \frac{1}{2}\sqrt{\rho B}\,\omega^2 A^2 = \frac{p_{max}^2}{2\rho v} = \frac{p_{max}^2}{2\sqrt{\rho B}}. \qquad (21\text{--}7),\ (21\text{--}9)$$

■ The sound intensity level β of a sound wave of intensity I is defined as

$$\beta = (10\ \mathrm{dB})\log\frac{I}{I_0}, \qquad (21\text{--}11)$$

where I_0 is a reference intensity defined to be $10^{-12}\ \mathrm{W/m^2}$. Sound intensity levels are expressed in decibels (dB).

■ Beats are heard when two tones with slightly different frequencies f_a and f_b are sounded together. With $f_a > f_b$, the beat frequency f_{beat} is

$$f_{beat} = f_a - f_b. \qquad (21\text{--}12)$$

■ The Doppler effect for sound is the frequency shift that occurs when there is motion of a source of sound or a listener, or both, relative to the medium. The source and listener frequencies f_S and f_L and the source and listener velocities v_S and v_L relative to the medium (in which the speed of sound is v) are related by

$$f_L = \frac{v + v_L}{v + v_S}\,f_S. \qquad (21\text{--}17)$$

■ A sound source moving with a speed v_S greater than the speed of sound v creates a shock wave. The wave front is a cone with angle α given by

$$\sin\alpha = \frac{v}{v_S}. \qquad (21\text{--}19)$$

DISCUSSION QUESTIONS

Q21–1 Lane dividers on highways sometimes have regularly spaced ridges or ripples. When the tires of a moving car roll along such a divider, a musical note is produced. Why? Explain how this phenomenon could be used to measure the car's speed.

Q21–2 The tone quality of an acoustic guitar is different when the strings are plucked near the bridge (the lower end of the strings) than when they are plucked near the sound hole (close to the center of the strings). Why?

Q21–3 Which has a more direct influence on the loudness of a sound wave: the *displacement* amplitude or the *pressure* amplitude? Explain your reasoning.

Q21–4 If the pressure amplitude of a sound wave is halved, by what factor does the intensity of the wave decrease? By what factor must the pressure amplitude of a sound wave be increased in order to increase the intensity by a factor of 16? Explain.

Q21–5 Defend the following statement: "*Adding* a certain number of decibels to the sound intensity level corresponds to *multiplying* the intensity by a corresponding factor."

Q21–6 A small fraction of the energy in a sound wave is absorbed by the air through which the sound passes. How does this modify the inverse-square relationship between intensity and distance from the source? Explain your reasoning.

Q21–7 According to Eq. (21–10), the intensity of sound at the back of a concert hall should be very much less than at the front of the hall, since the back row of seats is much farther from the stage. But in fact the intensity at the back is only slightly less than at the front. Why?

Q21–8 An organist in a cathedral plays a loud chord and then releases the keys. The sound persists for a few seconds and gradually dies away. Why does it persist? What happens to the sound energy when the sound dies away?

Q21–9 When you are shouting to someone a fair distance away, it is easier for her to hear you if the wind is blowing from you to her than if it is in the opposite direction. Explain the physical reason for this difference.

Q21–10 Two vibrating tuning forks have identical frequencies, but one is stationary and the other is mounted at the rim of a rotating platform. What does a listener hear? Explain.

Q21–11 A large church has part of the organ in the front of the church and part in the back. A person walking rapidly down the aisle while both segments are playing at once reports that the two segments sound out of tune. Why?

Q21–12 A sound source and a listener are both at rest on the earth, but a strong wind is blowing from the source toward the listener. Is there a Doppler effect? Why or why not?

Q21–13 Can you think of circumstances in which a Doppler effect would be observed for surface waves in water? For elastic waves propagating in a body of water deep below the surface?

If so, describe the circumstances and explain your reasoning. If not, explain why not.

*Q21–14 Does an aircraft make a sonic boom only at the instant its speed exceeds Mach 1? Explain your reasoning.

*Q21–15 If you are riding in a supersonic aircraft, what do you hear? Explain your reasoning. In particular, do you hear a continuous sonic boom? Why or why not?

*Q21–16 The speed of sound is not the same at all altitudes, but decreases with increasing altitude. How does this affect the calculation in Example 21–13 (Section 21–6)? Is the shape of the shock wave still a cone? Explain your reasoning.

*Q21–17 A jet airplane is flying at a constant altitude at a steady speed v_s greater than the speed of sound. Describe what is being heard by observers at points A, B, and C at the instant shown in Fig. 21–17, when the shock wave has just reached point B. Explain your reasoning.

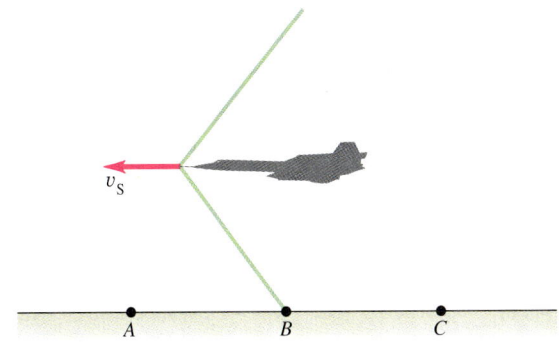

FIGURE 21–17 Question Q21–17.

EXERCISES

Unless indicated otherwise, assume the speed of sound in air to be $v = 344$ m/s.

SECTION 21–2 SOUND WAVES

21–1 Example 21–1 (Section 21–2) showed that for sound waves in air with frequency 1000 Hz, a displacement amplitude of 1.2×10^{-8} m produces a pressure amplitude of 3.0×10^{-2} Pa. a) What is the wavelength of these waves? b) For 1000-Hz waves in air, what displacement amplitude would be needed for the pressure amplitude to be at the pain threshold, which is 30 Pa? c) For what wavelength and frequency will waves with a displacement amplitude of 1.2×10^{-8} m produce a pressure amplitude of 1.5×10^{-3} Pa?

21–2 Example 21–1 (Section 21–2) showed that for sound waves in air with frequency 1000 Hz, a displacement amplitude of 1.2×10^{-8} m produces a pressure amplitude of 3.0×10^{-2} Pa. Water at 20°C has a bulk modulus of 2.2×10^9 Pa, and the speed of sound in water at this temperature is 1480 m/s. For 1000-Hz sound waves in 20°C water, what displacement amplitude is produced if the pressure amplitude is 3.0×10^{-2} Pa? Explain why your answer is much less than 1.2×10^{-8} Pa.

21–3 Consider a sound wave in air that has displacement amplitude 0.0200 mm. Calculate the pressure amplitude for frequencies of a) 150 Hz; b) 1500 Hz; c) 15,000 Hz. In each case compare the results to the pain threshold, which is 30 Pa.

SECTION 21–3 SOUND INTENSITY

21–4 Derive Eq. (21–9) from the equations that precede it.

21–5 A sound wave in air at 20°C has a frequency of 150 Hz and a displacement amplitude of 5.00×10^{-3} mm. For this sound wave calculate the a) pressure amplitude (in Pa); b) intensity (in W/m²); c) sound intensity level (in decibels).

21–6 a) What is the sound intensity level in a car when the sound intensity is 0.500 μW/m²? b) What is the sound intensity level in the air near a jackhammer when the pressure amplitude of the sound is 0.150 Pa and the temperature is 20.0°C?

21–7 For a person with normal hearing, the faintest sound that can be heard at a frequency of 400 Hz has a pressure amplitude of about 6.0×10^{-5} Pa. Calculate the corresponding intensity and sound intensity level at 20°C.

21–8 The intensity due to a number of independent sound sources is the sum of the individual intensities. a) When four quadruplets cry simultaneously, how many decibels greater is the sound intensity level than when a single one cries? b) To increase the sound intensity level again by the same number of decibels as in part (a), how many more crying babies are required?

21–9 A baby's mouth is 30 cm from her father's ear and 1.50 m from her mother's ear. What is the difference between the sound intensity levels heard by the father and by the mother?

21–10 A tornado warning siren on top of a tall pole radiates sound waves uniformly in all directions. At a distance of 15.0 m the intensity of the sound is 0.250 W/m². Neglect any effects from the reflection of the sound waves from the ground. a) At what distance from the siren is the intensity 0.010 W/m²? b) What is the total acoustical power output of the siren?

21–11 a) By what factor must the sound intensity be increased to increase the sound intensity level by 13.0 dB? b) Explain why you don't need to know the original sound intensity.

SECTION 21–4 BEATS

21–12 Two identical violin strings, when in tune and stretched with the same tension, have a fundamental frequency of 440.0 Hz. One of the strings is retuned by adjusting its tension. When this is done, 1.5 beats per second are heard when both strings are plucked simultaneously. a) What are the possible fundamental frequencies of the retuned string? b) By what fractional amount was the string tension changed if it was i) increased; ii) decreased?

21–13 Two sinusoidal sound waves with frequencies 108 Hz and 112 Hz arrive at your ear simultaneously. Each wave has an amplitude of 1.5×10^{-8} m when it reaches your ear. a) Describe in detail what you hear. b) What is the maximum amplitude of the total sound wave? What is the minimum amplitude?

SECTION 21–5 THE DOPPLER EFFECT

21–14 On the planet Arrakis a male ornithoid is flying toward his mate at 25.0 m/s while singing at a frequency of 1200 Hz. If the stationary female hears a tone of 1240 Hz, what is the speed of sound in the atmosphere of Arrakis?

21–15 In Example 21–12 (Section 21–5), suppose the police car is moving away from the warehouse at 20 m/s. What fre-

quency does the driver of the police car hear reflected from the warehouse?

21–16 Two train whistles, A and B, each have a frequency of 392 Hz. A is stationary and B is moving toward the right (away from A) at a speed of 35.0 m/s. A listener is between the two whistles and is moving toward the right with a speed of 15.0 m/s (Fig. 21–18). No wind is blowing. a) What is the frequency from A as heard by the listener? b) What is the frequency from B as heard by the listener? c) What is the beat frequency detected by the listener?

FIGURE 21–18 Exercise 21–16.

21–17 A railroad train is traveling at 25.0 m/s in still air. The frequency of the note emitted by the locomotive whistle is 400 Hz. What is the wavelength of the sound waves a) in front of the locomotive? b) behind the locomotive? What is the frequency of the sound heard by a stationary listener c) in front of the locomotive? d) behind the locomotive?

21–18 A swimming duck paddles the water with its feet once every 1.6 s, producing surface waves with this period. The duck is moving at constant speed in a pond where the speed of surface waves is 0.32 m/s, and the crests of the waves ahead of the duck are spaced 0.12 m apart. a) What is the duck's speed? b) How far apart are the crests behind the duck?

21–19 A railroad train is traveling at 30.0 m/s in still air. The frequency of the note emitted by the train whistle is 262 Hz. What frequency is heard by a passenger on a train moving in the opposite direction to the first at 18.0 m/s and a) approaching the first? b) receding from the first?

*SECTION 21–6 SHOCK WAVES

***21–20** A jet plane flies overhead at Mach 1.70 and at a constant altitude of 950 m. a) What is the angle α of the shock-wave cone? b) How much time after the plane passes directly overhead do you hear the sonic boom? Neglect the variation of the speed of sound with altitude.

PROBLEMS

21–21 a) Defend the following statement: "In a sinusoidal sound wave, the pressure variation given by Eq. (21–4) is greatest where the displacement given by Eq. (21–1) is zero." b) For a sinusoidal sound wave given by Eq. (21–1) with amplitude $A = 10.0$ μm and wavelength $\lambda = 0.250$ m, graph the displacement y and pressure fluctuation p as functions of x at time $t = 0$. Show at least two wavelengths of the wave on your graphs. c) The displacement y in a *non*sinusoidal sound wave is shown in Fig. 21–19 as a function of x for $t = 0$. Draw a graph showing the pressure fluctuation p in this wave as a function of x at $t = 0$. This sound wave has the same 10.0-μm amplitude as the wave in part (b). Does it have the same pressure amplitude? Why or why not?

d) Is the statement in part (a) necessarily true if the sound wave is *not* sinusoidal? Explain your reasoning.

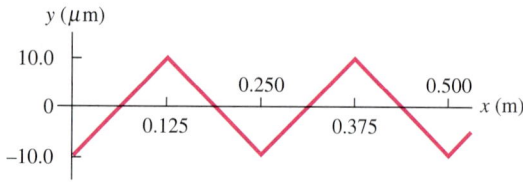

FIGURE 21–19 Problem 21–21.

21–22 Many airports have noise ordinances restricting the maximum sound intensity that an aircraft may produce when it takes off. At one California airport the maximum allowable sound intensity level is 98.5 dB as measured by a microphone at the end of the 1740-m-long runway. A certain airliner produces a sound intensity level of 100.0 dB on the ground when it flies over at an altitude of 100 m. On takeoff, this airliner rolls for 1200 m along the runway before leaving the ground, at which point it climbs at a 15° angle. Does this airliner violate the noise ordinance? Neglect any effects due to reflection of the sound waves from the ground.

21–23 The sound from a trumpet radiates uniformly in all directions in 20°C air. At a distance of 5.00 m from the trumpet the sound intensity level is 52.0 dB. The frequency is 587 Hz. a) What is the pressure amplitude at this distance? b) What is the displacement amplitude? c) At what distance is the sound intensity level 30.0 dB?

21–24 A sound wave entering a window of area A has a sound intensity level of β. a) How much acoustic power enters the window? b) Find the value of the acoustic power for the case of a 1.20-m^2 window and 55.0-dB sound.

21–25 A small sphere of radius R is arranged to pulsate so that its radius varies in simple harmonic motion between a minimum of $R - \Delta R$ and a maximum of $R + \Delta R$ with frequency f. This produces sound waves in the surrounding air of density ρ and bulk modulus B. a) Find the intensity of sound waves at the surface of the sphere. (The amplitude of oscillation of the sphere is the same as that of the air at the surface of the sphere.) b) Find the total acoustic power radiated by the sphere. c) At a distance $d \gg R$ from the center of the sphere, find the amplitude, pressure amplitude, and intensity of the sound wave.

21–26 A 2.00-MHz sound wave travels through a pregnant woman's abdomen and is reflected from the fetal heart wall of her unborn baby. The heart wall is moving toward the sound receiver as the heart beats. The reflected sound is then mixed with the transmitted sound, and 85 beats per second are detected. The speed of sound in body tissue is 1500 m/s. Calculate the speed of the fetal heart wall at the instant this measurement is made.

21–27 The sound source of a ship's sonar system operates at a frequency of 22.0 kHz. The speed of sound in water (assumed to be at a uniform 20°C) is 1482 m/s. a) What is the wavelength of the waves emitted by the source? b) What is the difference in frequency between the directly radiated waves and the waves reflected from a whale traveling directly toward the ship at 4.95 m/s? The ship is at rest in the water.

21–28 A police siren of frequency f_{siren} is attached to a vibrating platform. The platform and siren oscillate up and down in simple harmonic motion with amplitude A_P and frequency f_P. a) Find the maximum and minimum sound frequencies that you would hear at a position directly above the siren. b) At what point in the motion of the platform is the maximum frequency heard? The minimum frequency? Explain.

21–29 Horseshoe bats (genus *Rhinolophus*) emit sounds from their nostrils, then listen to the frequency of the sound reflected from their prey to determine the prey's speed. (The "horseshoe"

that gives the bat its name is a depression around the nostrils that acts like a focusing mirror, so that the bat emits sound in a narrow beam like a flashlight.) A *Rhinolophus* flying at speed v_{bat} emits sound of frequency f_{bat}; the sound it hears reflected from an insect flying toward it has a higher frequency f_{refl}. a) Show that the speed of the insect is

$$v_{\text{insect}} = v \left[\frac{f_{\text{refl}}(v - v_{\text{bat}}) - f_{\text{bat}}(v + v_{\text{bat}})}{f_{\text{refl}}(v - v_{\text{bat}}) + f_{\text{bat}}(v + v_{\text{bat}})} \right].$$

where v is the speed of sound. b) If $f_{\text{bat}} = 80.7$ kHz, $f_{\text{refl}} = 83.5$ kHz, and $v_{\text{bat}} = 3.9$ m/s, calculate the speed of the insect.

21–30 A sound wave with frequency f_0 and wavelength λ_0 travels horizontally toward the right. It strikes and is reflected from a large, rigid, vertical plane surface, perpendicular to the direction of propagation of the wave and moving toward the left with a speed v_1. a) How many positive wave crests strike the surface in a time interval t? b) At the end of this time interval, how far to the left of the surface is the wave that was reflected at the beginning of the time interval? c) What is the wavelength of the reflected waves in terms of λ_0? d) What is the frequency of the reflected waves in terms of f_0? Is your result consistent with the assertion made in point 3 of the Problem-Solving Strategy in Section 21–5? e) A listener is at rest at the left of the moving surface. How many beats per second does she detect as a result of the combined effect of the incident and reflected waves?

21–31 a) Show that Eq. (21–18) can be written as

$$f_R = f_S \left(1 - \frac{v}{c} \right)^{1/2} \left(1 + \frac{v}{c} \right)^{-1/2}.$$

b) Use the binomial theorem to show that if $v \ll c$, this is approximately equal to

$$f_R = f_S \left(1 - \frac{v}{c} \right).$$

c) A pilotless reconnaissance aircraft emits a radio signal with a frequency of 243 MHz. It is flying directly toward a test engineer on the ground. The engineer detects beats between the received signal and a local signal also of frequency 243 MHz. The beat frequency is 46.0 Hz. What is the speed of the aircraft? (Radio waves travel at the speed of light, $c = 3.00 \times 10^8$ m/s.)

21–32 The gas cloud known as the Crab Nebula (Fig. 9–27) can be seen with even a small telescope. It is the remnant of a *supernova*, a cataclysmic explosion of a star. The explosion was seen on the earth on July 4, 1054 A.D. The streamers in Fig. 9–27 glow with the characteristic red color of heated hydrogen gas. In a laboratory on the earth, heated hydrogen produces red light with frequency 4.568×10^{14} Hz; the red light received from streamers in the Crab Nebula pointed toward the earth has frequency 4.586×10^{14} Hz. a) Estimate the speed with which the outer edges of the Crab Nebula are expanding. Assume that the speed of the center of the nebula relative to the earth is negligible. (You may use the formulas derived in Problem 21–31. The speed of light is 3.00×10^8 m/s.) b) Assuming that the expansion speed has been constant since the supernova explosion, estimate the diameter of the Crab Nebula. Give your answer in meters and in light years. c) The angular diameter of the Crab Nebula as seen from earth is about 5 arc minutes (1 arc minute = $\frac{1}{60}$ of a degree). Estimate the distance (in light years) to the Crab Nebula, and estimate the year in which the supernova explosion actually took place.

21–33 A radar installation used for monitoring thunderstorms emits radio waves at a frequency of 2800 MHz. A line of thunderstorms is approaching the installation at 42.0 km/h. a) In the frame of reference of the thunderstorms, is the frequency of the radio waves greater than or less than 2800 MHz? Why? By what amount does the frequency differ from 2800 MHz? (You may use the formulas derived in Problem 21–31. Radio waves travel at the speed of light, $c = 3.00 \times 10^8$ m/s.) b) Radio waves are reflected from the water drops in the thunderstorms, and the reflected waves are detected back at the installation. As measured by the receiver at the installation, is the frequency of these reflected waves greater than or less than 2800 MHz? Why? By what amount does the frequency of the reflected waves differ from 2800 MHz?

21–34 A woman stands at rest in front of a large, smooth wall. She holds a vibrating tuning fork of frequency f_0 directly in front of her (between her and the wall). a) She now runs toward the wall with a speed v_w. She detects beats due to the interference between the sound waves reaching her directly from the fork and those reaching her after being reflected from the wall. How many beats per second will she detect? (*Note:* If the beat frequency is too large, the woman may have to use some instrumentation other than her ears to detect and count the beats.) b) If she instead runs away from the wall, holding the tuning fork at her back so it is between her and the wall, how many beats per second will she detect?

*21–35 On a clear day you see a jet plane flying overhead. From the apparent size of the plane, you determine that it is flying at a constant altitude h. You hear the sonic boom a time T after the plane passes directly overhead. Show that if the speed of sound v is the same at all altitudes, the speed of the plane is

$$v_s = \frac{hv}{\sqrt{h^2 - v^2 T^2}}$$

(*Hint:* Trigonometric identities will be useful.)

CHALLENGE PROBLEMS

21–36 Figure 21–20 shows the pressure fluctuation p of a nonsinusoidal sound wave as a function of x for $t = 0$. The wave is traveling in the $+x$-direction. a) Graph the pressure fluctuation p as a function of t for $x = 0$. Show at least two cycles of oscillation. b) Graph the displacement y in this sound wave as a function of x at $t = 0$. At $x = 0$, the displacement at $t = 0$ is zero. Show at least two wavelengths of the wave. c) Graph the displacement y as a function of t for $x = 0$. Show at least two cycles of oscillation. d) Calculate the maximum velocity and the maximum acceleration of an element of the air through which this sound wave is traveling. e) Describe how the cone of a loudspeaker must move as a function of time to produce the sound wave in this problem.

21–37 Two loudspeakers, A and B, radiate sound uniformly in all directions in air at 20°C. The acoustic power output from A is 8.00×10^{-4} W, and from B it is 6.00×10^{-5} W. Both loudspeakers are vibrating in phase at a frequency of 172 Hz. a) Determine the difference in phase of the two signals at a point C along the line joining A and B, 3.00 m from B and 4.00 m from A (Fig. 21–21). b) Determine the intensity and sound intensity level at C from speaker A if speaker B is turned off and the intensity and sound intensity level at point C from speaker B if speaker A is turned off. c) With both speakers on, what are the intensity and sound intensity level at C?

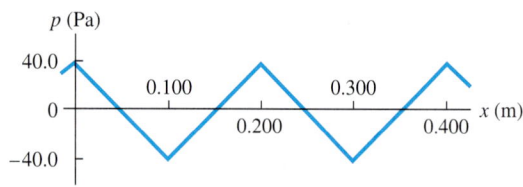

FIGURE 21–20 Challenge Problem 21–36.

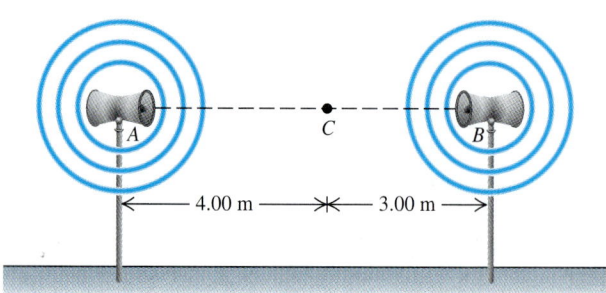

FIGURE 21–21 Challenge Problem 21–37.

22

Electric Charge and Electric Field

Most of the forces on this water skier are electric. Electric interactions between adjacent molecules give rise to the force of the water on the ski, the tension in the tow rope, and the resistance of the air on the skier's body. They also hold the skier's body together! Only one wholly nonelectric force acts on the skier: the force of gravity.

22–1 INTRODUCTION

When you scuff your shoes across a carpet and then reach for a metal doorknob, you can get zapped by an annoying spark of static electricity. Why does this happen, and why is it more likely to happen on a dry day than on a humid one? The atoms in your body hold together and don't break apart, even though the particles that make up those atoms can be moving at very high speeds. Why? What really happens in an electric circuit? How do electric motors and generators work? And what is light, anyway?

The answers to all these questions come from a fundamental branch of physics known as *electromagnetism,* the study of electric and magnetic interactions. These interactions involve particles that have a property called *electric charge,* an attribute of matter that is as fundamental as mass. Our exploration of electromagnetic phenomena will occupy our attention for most of the remainder of this book.

We begin our study of electromagnetism in this chapter by examining the nature of electric charge. We'll find that electric charge is quantized and that it obeys a conservation principle. We then turn to a discussion of the interactions of electric charges that are at rest in our frame of reference, called *electrostatic* interactions. Such interactions are exceedingly important: They hold atoms, molecules, and our bodies together and have numerous technological applications. Electrostatic interactions are governed by a simple relationship known as *Coulomb's law* and are most

conveniently described by using the concept of *electric field.* We'll explore all these concepts in this chapter and expand on them in the three chapters that follow. In later chapters we'll expand our discussion to include electric charges in motion.

While the key ideas of electromagnetism are conceptually simple, applying them to practical problems will make use of many of your mathematical skills, especially your knowledge of geometry and integral calculus. For this reason you may find this chapter and those that follow to be more mathematically demanding than previous chapters. The reward for your extra effort will be a deeper understanding of principles that are at the heart of modern physics and technology.

22–2 ELECTRIC CHARGE

We can't say what electric charge *is;* we can only describe its properties and its behavior. The ancient Greeks discovered as early as 600 B.C. that when they rubbed amber with wool, the amber could then attract other objects. Today we say that the amber has acquired a net **electric charge**, or has become *charged.* The word "electric" is derived from the Greek word *elektron,* meaning amber. When you scuff your shoes across a nylon carpet, you become electrically charged, and you can charge a comb by passing it through dry hair.

22-1 (a, b) After being rubbed with fur, two plastic rods repel each other. (c, d) After being rubbed with silk, two glass rods repel each other. (e) The charged plastic rod from (b) attracts the charged glass rod from (d). (f) The fur attracts the charged plastic rod, and the silk attracts the charged glass rod.

Plastic rods and fur (real or fake) are particularly good for demonstrating the phenomena of **electrostatics,** the interactions between electric charges that are at rest (or nearly so). Figure 22–1a shows two plastic rods and a piece of fur. After we charge each rod by rubbing it with the piece of fur, we find that the rods repel each other (Fig. 22–1b). When we rub glass rods (Fig. 22–1c) with silk, the glass rods also become charged and repel each other (Fig. 22–1d). But a charged plastic rod *attracts* a charged glass rod (Fig. 22–1e). Furthermore, the plastic rod and the fur attract each other, and the glass rod and the silk attract each other (Fig. 22–1f).

These experiments and many others like them have shown that there are exactly two kinds of electric charge: the kind on the plastic rod rubbed with fur and the kind on the glass rod rubbed with silk. Benjamin Franklin (1706–1790) suggested calling these two kinds of charge *negative* and *positive,* respectively, and these names are still used. The plastic rod and the silk have negative charge; the glass rod and the fur have positive charge. **Two positive charges or two negative charges repel each other. A positive charge and a negative charge attract each other.**

CAUTION ▶ The attraction and repulsion of two charged objects is sometimes summarized as "Like charges repel, and opposite charges attract." But keep in mind that the phrase "like charges" does *not* mean that the two charges are exactly identical, only that both charges have the same algebraic *sign* (both positive or both negative). "Opposite charges" means that both objects have an electric charge, and those charges have different signs (one positive and the other negative). ◀

One technological application of forces between charged bodies is in a photocopy machine. Positively charged regions of the machine's imaging drum attract negatively charged particles of toner. When a piece of paper is placed in contact with the drum, the toner particles stick to the paper and form an image.

22–3 ELECTRIC CHARGE AND THE STRUCTURE OF MATTER

Electric charge, like mass, is one of the fundamental attributes of the particles of which matter is made. The interactions responsible for the structure and properties of atoms and molecules are primarily *electric* interactions between electrically charged particles. The same is true for the structure and properties of ordinary matter, which is made up of atoms and molecules. The normal force exerted on you by the chair in which you're sitting, the tension force in a stretched string, and the adhesive force of glue are all fundamentally electric in nature, arising from the electric forces between charged particles in adjacent atoms.

The structure of atoms can be described in terms of three particles: the negatively charged **electron,** the positively charged **proton,** and the uncharged **neutron.** The proton and neutron are combinations of other entities called *quarks,* which have charges of $\pm\frac{1}{3}$ and $\pm\frac{2}{3}$ times the electron charge. Isolated quarks have not been observed, and there are theoretical reasons to believe that it is impossible in principle to observe a quark in isolation.

The protons and neutrons in an atom make up a small, very dense core called the **nucleus,** with dimensions of the order of 10^{-15} m. Surrounding the nucleus are the electrons, extending out to distances of the order of 10^{-10} m from the nucleus. If an atom were a few kilometers across, its nucleus would be the size of a tennis ball. The negatively charged electrons are held within the atom by the attractive electric forces exerted on them by the positively charged nucleus. (The protons and neutrons are held within the stable atomic nuclei by an attractive interaction, called the *nuclear force,* that overcomes the electric repulsion of the protons. The nuclear force has a short range, of the order of nuclear dimensions, and its effects do not extend far beyond the nucleus.)

The masses of the individual particles, to the precision that they are presently known, are

$$\text{Mass of electron} = m_{\text{e}} = 9.1093897(54) \times 10^{-31} \text{ kg},$$

$$\text{Mass of proton} = m_{\text{p}} = 1.6726231(10) \times 10^{-27} \text{ kg},$$

$$\text{Mass of neutron} = m_{\text{n}} = 1.6749286(10) \times 10^{-27} \text{ kg}.$$

The numbers in parentheses are the uncertainties in the last two digits. Note that the masses of the proton and neutron are nearly equal and are roughly 2000 times the mass of the electron. Over 99.9% of the mass of any atom is concentrated in its nucleus.

The negative charge of the electron has (within experimental error) *exactly* the same magnitude as the positive charge of the proton. In a neutral atom the number of electrons equals the number of protons in the nucleus, and the net electric charge (the algebraic sum of all the charges) is exactly zero (Fig. 22–2a; see page 672). The number of protons or electrons in a neutral atom of an element is called the **atomic number** of the element. If one or more electrons are removed, the remaining positively charged structure is called a **positive ion** (Fig. 22–2b). A **negative ion** is an atom that has *gained* one or more electrons (Fig. 22–2c). This gaining or losing of electrons is called **ionization.**

When the total number of protons in a macroscopic body equals the total number of electrons, the total charge is zero and the body as a whole is electrically neutral. To give a body an excess negative charge, we may either *add negative* charges to a neutral body or *remove positive* charges from that body. Similarly, we can create an excess positive charge by either *adding positive* charge or *removing negative* charge. In most cases, negatively charged (and highly mobile) electrons are added or removed, and a "positively charged body" is one that has lost some of its normal complement of electrons. When we speak of the charge of a body, we always mean its *net* charge. The net charge is

The electron—the first constituent of the atom to be isolated—was discovered in 1897 by the English physicist J. J. Thomson. This discovery revolutionized our understanding of the structure of matter, and led to the later discoveries of the proton and neutron. Thomson was awarded the 1906 Nobel Prize in physics and was knighted in 1908.

22–2 (a, left) The neutral lithium atom has three protons in its nucleus and three electrons surrounding the nucleus. (b, center) A positive lithium ion is made by removing an electron from a neutral atom. (c, right) A negative lithium ion is made by adding an electron to a neutral atom. (The electron "orbits" are a schematic representation of the actual electron distribution, which is a diffuse cloud many times larger than the nucleus.)

always a very small fraction (typically no more than 10^{-12}) of the total positive charge or negative charge in the body.

Implicit in the foregoing discussion are two very important principles. First is the **principle of conservation of charge: The algebraic sum of all the electric charges in any closed system is constant.** If we rub together a plastic rod and a piece of fur, both initially uncharged, the rod acquires a negative charge (since it takes electrons from the fur) and the fur acquires a positive charge of the *same* magnitude (since it has lost as many electrons as the rod has gained). Hence the total electric charge on the two bodies together does not change. In any charging process, charge is not created or destroyed; it is merely *transferred* from one body to another.

Conservation of charge is believed to be a *universal* conservation law, and there is no experimental evidence for any violation of this principle. Even in high-energy interactions in which particles are created and destroyed, such as the creation of electron-positron pairs, the total charge of any closed system is exactly constant.

The second important principle is that the magnitude of charge of the electron or proton is a natural unit of charge. Every observable amount of electric charge is always an integer multiple of this basic unit. We say that charge is *quantized*. A familiar example of quantization is money. When you pay cash for an item in a store, you have to do it in one-cent increments. Cash can't be divided into amounts smaller than one cent, and electric charge can't be divided into amounts smaller than the charge of one electron or proton. (The quark charges, $\pm\frac{1}{3}$ and $\pm\frac{2}{3}$ of the electron charge, are probably not observable as isolated charges.) Thus the charge on any macroscopic body is always either zero or an integer multiple (negative or positive) of the electron charge.

22–4 CONDUCTORS, INSULATORS, AND INDUCED CHARGES

Some materials permit electric charge to move easily from one region of the material to another, while others do not. For example, Fig. 22–3a shows a copper wire supported by a glass rod. Suppose you touch one end of the wire to a charged plastic rod and attach the other end to a metal ball that is initially uncharged; you then remove the charged rod and the wire. When you bring another charged body up close to the ball (Figs. 22–3b and 22–3c), the ball is attracted or repelled, showing that the ball has become electrically charged. Electric charge has been transferred through the copper wire between the ball and the surface of the plastic rod.

The wire is called a **conductor** of electricity. If you repeat the experiment using a rubber band or nylon thread in place of the wire, you find that *no* charge is transferred to the ball. These materials are called **insulators.** Conductors permit the easy movement of charge through them, while insulators do not. As an example, carpet fibers on a dry day are good insulators. As you walk across a carpet, the rubbing of your shoes against the fibers causes charge to build up on you, and this charge will remain on you because it can't flow through the insulating fibers. If you then touch a conducting object such as a doorknob, a rapid charge transfer takes place between your finger and the doorknob, and you feel a shock. One way to prevent this is to wind some of the carpet fibers around conducting cores so that any charge that builds up on you can be transferred harmlessly to the carpet. Another solution is to coat the carpet fibers with an antistatic layer that does not easily transfer electrons to or from your shoes; this prevents any charge from building up on you in the first place.

Most metals are good conductors, while most nonmetals are insulators. Within a solid metal such as copper, one or more outer electrons in each atom become detached and can move freely throughout the material, just as the molecules of a gas can move through the spaces between the grains in a bucket of sand. The motion of these negatively charged electrons carries charge through the metal. The other electrons remain bound to the positively charged nuclei, which themselves are bound in nearly fixed positions within the material. In an insulator there are no, or very few, free electrons, and electric charge cannot move freely through the material. Some materials called *semiconductors* are intermediate in their properties between good conductors and good insulators.

We can charge a metal ball by touching it with an electrically charged plastic rod, as in Fig. 22–3a. In this process, some of the excess electrons on the rod are transferred from it to the ball, leaving the rod with a smaller negative charge. There is a different technique in which the plastic rod can give another body a charge of *opposite* sign without losing any of its own charge. This process is called charging by **induction.**

Figure 22–4a shows an example of charging by induction. A metal sphere is supported on an insulating stand. When you bring a negatively charged rod near it, without actually touching it (Fig. 22–4b), the free electrons in the metal sphere are repelled by the excess electrons on the rod, and they shift toward the right, away from the rod. They cannot escape from the sphere because the supporting stand and the surrounding air are insulators. So we get excess negative charge at the right surface of the sphere and a deficiency of negative charge (that is, a net positive charge) at the left surface. These excess charges are called **induced charges.**

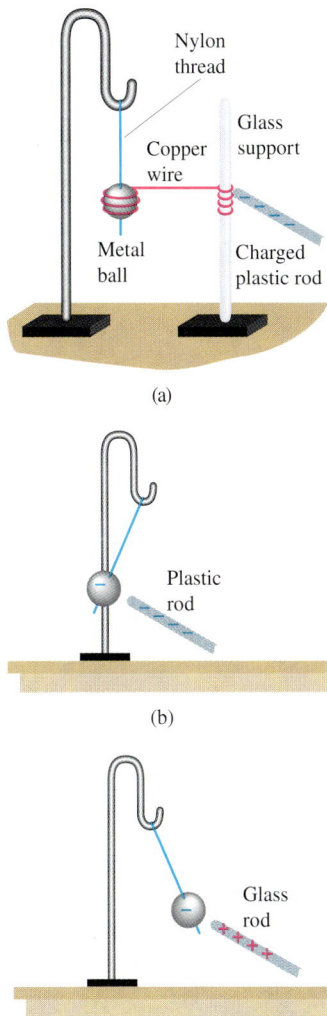

(a)

(b)

(c)

22–3 Copper is a good conductor of electricity; glass and nylon are good insulators. (a) The wire conducts charge between the metal ball and charged plastic rod to charge the ball negatively. (b) Afterward, the metal ball is repelled by a negatively charged plastic rod and (c) attracted to a positively charged glass rod.

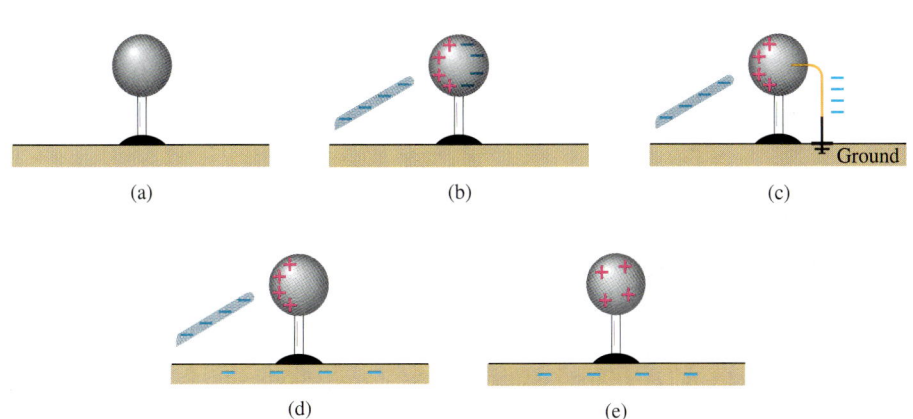

(a)   (b)   (c)

(d)   (e)

22–4 Charging a metal sphere by induction.

(a)

(b)

22–5 (a) The charges within the molecules of an insulating material can shift slightly. As a result, a charged comb attracts a neutral insulator. By Newton's third law the neutral insulator exerts an equal-magnitude attractive force on the charged comb. (b) If the sign of the charge on the comb is reversed, the charges within the insulator shift in the opposite direction, and the insulator is again attracted to the comb.

Not all of the free electrons move to the right surface of the sphere. As soon as any induced charge develops, it exerts forces toward the *left* on the other free electrons. These electrons are repelled by the negative induced charge on the right and attracted toward the positive induced charge on the left. The system reaches an equilibrium state in which the force toward the right on an electron, due to the charged rod, is just balanced by the force toward the left due to the induced charge. If we remove the charged rod, the free electrons shift back to the left, and the original neutral condition is restored.

What happens if, while the plastic rod is nearby, you touch one end of a conducting wire to the right surface of the sphere and the other end to the earth (Fig. 22–4c)? The earth is a conductor, and it is so large that it can act as a practically infinite source of extra electrons or sink of unwanted electrons. Some of the negative charge flows through the wire to the earth. Now suppose you disconnect the wire (Fig. 22–4d) and then remove the rod (Fig. 22–4e); a net positive charge is left on the sphere. The charge on the negatively charged rod has not changed during this process. The earth acquires a negative charge that is equal in magnitude to the induced positive charge remaining on the sphere.

Charging by induction would work just as well if the mobile charges in the spheres were positive charges instead of negatively charged electrons, or even if both positive and negative mobile charges were present. In a metallic conductor the mobile charges are always negative electrons, but it is often convenient to describe a process *as though* the moving charges were positive. In ionic solutions and ionized gases, both positive and negative charges are mobile.

Finally, we note that a charged body can exert forces even on objects that are *not* charged themselves. If you rub a balloon on the rug and then hold the balloon against the ceiling, it sticks, even though the ceiling has no net electric charge. After you electrify a comb by running it through your hair, you can pick up uncharged bits of paper with the comb. How is this possible?

This interaction is an induced-charge effect. In Fig. 22–4b the plastic rod exerts a net attractive force on the conducting sphere even though the total charge on the sphere is zero, because the positive charges are closer to the rod than the negative charges. Even in an insulator, electric charges can shift back and forth a little when there is charge nearby. This is shown in Fig. 22–5a; the negatively charged plastic comb causes a slight shifting of charge within the molecules of the neutral insulator, an effect called *polarization*. The positive and negative charges in the material are present in equal amounts, but the positive charges are closer to the plastic comb and so feel an attraction that is stronger than the repulsion felt by the negative charges, giving a net attractive force. (In Section 22–5 we will study how electric forces depend on distance.) Note that a neutral insulator is also attracted to a *positively* charged comb (Fig. 22–5b). Now the charges in the insulator shift in the opposite direction; the negative charges in the insulator are closer to the comb and feel an attractive force that is stronger than the repulsion felt by the positive charges in the insulator. Hence a charged object of *either* sign exerts an attractive force on an uncharged insulator. The attraction between a charged object and an uncharged one has many important practical applications, including electrostatic dust precipitators and attracting droplets of sprayed paint to a car body.

22–5 Coulomb's Law

Charles Augustin de Coulomb (1736–1806) studied the interaction forces of charged particles in detail in 1784. He used a torsion balance (Fig. 22–6a) similar to the one used 13 years later by Cavendish to study the much weaker gravitational interaction, as we discussed in Section 12–2. For **point charges,** charged bodies that are very small in comparison with the distance r between them, Coulomb found that the electric force is proportional to $1/r^2$. That is, when the distance r doubles, the force decreases to $\frac{1}{4}$ of its initial value; when the distance is halved, the force increases to four times its initial value.

The electric force between two point charges also depends on the quantity of charge on each body, which we will denote by q or Q. To explore this dependence, Coulomb divided a charge into two equal parts by placing a small charged spherical conductor into contact with an identical but uncharged sphere; by symmetry, the charge is shared equally between the two spheres. (Note the essential role of the principle of conservation of charge in this procedure.) Thus he could obtain one half, one quarter, and so on, of any initial charge. He found that the forces that two point charges q_1 and q_2 exert on each other are proportional to each charge and therefore are proportional to the *product* $q_1 q_2$ of the two charges.

Thus Coulomb established what we now call **Coulomb's law:**

> **The magnitude of the electric force between two point charges is directly proportional to the product of the charges and inversely proportional to the square of the distance between them.**

In mathematical terms, the magnitude F of the force that each of two point charges q_1 and q_2 a distance r apart exerts on the other can be expressed as

$$F = k\frac{|q_1 q_2|}{r^2}, \tag{22–1}$$

where k is a proportionality constant whose numerical value depends on the system of units used. The absolute value bars are used in Eq. (22–1) because the charges q_1 and q_2 can be either positive or negative, while the force magnitude F is always positive.

The directions of the forces the two charges exert on each other are always along the line joining them. When the charges q_1 and q_2 have the same sign, either both positive or both negative, the forces are repulsive (Fig. 22–6b); when the charges have opposite signs, the forces are attractive (Fig. 22–6c). The two forces obey Newton's third law; they are always equal in magnitude and opposite in direction, even when the charges are not equal.

The proportionality of the electric force to $1/r^2$ has been verified with great precision. There is no reason to suspect that the exponent is anything different from precisely 2. Thus the form of Eq. (22–1) is the same as that of the law of gravitation. But electric and gravitational interactions are two distinct classes of phenomena. Electric interactions depend on electric charges and can be either attractive or repulsive, while gravitational interactions depend on mass and are always attractive (because there is no such thing as negative mass).

The value of the proportionality constant k in Coulomb's law depends on the system of units used. In our study of electricity and magnetism we will use SI units exclusively. The SI electric units include most of the familiar units such as the volt, the ampere, the ohm, and the watt. (There is *no* British system of electric units.) The SI unit of electric charge is called one **coulomb** (1 C). In SI units the constant k in Eq. (22–1) is

$$k = 8.987551787 \times 10^9 \ \text{N} \cdot \text{m}^2/\text{C}^2 \cong 8.988 \times 10^9 \ \text{N} \cdot \text{m}^2/\text{C}^2.$$

The value of k is known to such a large number of significant digits because this value is closely related to the speed of light in vacuum. (We will show this in Chapter 33 when we study electromagnetic radiation.) As we discussed in Section 1–4, this speed is *defined* to be exactly $c = 2.99792458 \times 10^8$ m/s. The numerical value of k is defined in terms of c to be precisely

$$k = (10^{-7} \ \text{N} \cdot \text{s}^2/\text{C}^2)c^2.$$

You may want to check this expression to confirm that k has the right units.

(a)

(b)

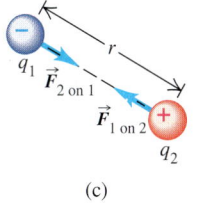

(c)

22–6 (a) A torsion balance of the type used by Coulomb to measure the electric force. (b) Electric charges of the same sign repel each other. (c) Electric charges of opposite sign attract each other. In each case the magnitude of the force on each point charge is proportional to the product of the charges and is inversely proportional to the square of the distance between them. The forces obey Newton's third law: $\vec{F}_{1 \ \text{on} \ 2} = -\vec{F}_{2 \ \text{on} \ 1}$.

11.1
Electric Force: Coulomb's Law
11.2
Electric Force: Superposition
Principle
11.3
Electric Force: Superposition
(Quantitative)

In principle we can measure the electric force F between two equal charges q at a measured distance r and use Coulomb's law to determine the charge. Thus we could regard the value of k as an operational definition of the coulomb. For reasons of experimental precision it is better to define the coulomb instead in terms of a unit of electric *current* (charge per unit time), the *ampere,* equal to one coulomb per second. We will return to this definition in Chapter 29.

In SI units we usually write the constant k in Eq. (22–1) as $1/4\pi\epsilon_0$, where ϵ_0 ("epsilon-nought" or "epsilon-zero") is another constant. This appears to complicate matters, but it actually simplifies many formulas that we will encounter in later chapters. From now on, we will usually write Coulomb's law as

$$F = \frac{1}{4\pi\epsilon_0}\frac{|q_1q_2|}{r^2}$$ (Coulomb's law: force between two   (22–2)
point charges).

The constants in Eq. (22–2) are approximately

$$\epsilon_0 = 8.854 \times 10^{-12} \text{ C}^2/\text{N} \cdot \text{m}^2 \qquad \text{and} \qquad \frac{1}{4\pi\epsilon_0} = k = 8.988 \times 10^9 \text{ N} \cdot \text{m}^2/\text{C}^2.$$

In examples and problems we will often use the approximate value

$$\frac{1}{4\pi\epsilon_0} = 9.0 \times 10^9 \text{ N} \cdot \text{m}^2/\text{C}^2,$$

which is within about 0.1% of the correct value.

As we mentioned in Section 22–3, the most fundamental unit of charge is the magnitude of the charge of an electron or a proton, which is denoted by e. The most precise value available as of the writing of this book is

$$e = 1.60217733(49) \times 10^{-19} \text{ C}.$$

One coulomb represents the negative of the total charge of about 6×10^{18} electrons. For comparison, a copper cube 1 cm on a side contains about 2.4×10^{24} electrons. About 10^{19} electrons pass through the glowing filament of a flashlight bulb every second.

In electrostatics problems (that is, problems that involve charges at rest) it's very unusual to encounter charges as large as one coulomb. Two 1-C charges separated by 1 m would exert forces on each other of magnitude 9×10^9 N (about a million tons)! The total charge of all the electrons in a copper penny is even greater, about 1.4×10^5 C, which shows that we can't disturb electric neutrality very much without using enormous forces. More typical values of charge range from about 10^{-9} to about 10^{-6} C. The microcoulomb (1 μC = 10^{-6} C) and the nanocoulomb (1 nC = 10^{-9} C) are often used as practical units of charge.

Coulomb's law as we have stated it describes only the interaction of two *point* charges. Experiment shows that when two charges exert forces simultaneously on a third charge, the total force acting on that charge is the *vector sum* of the forces that the two charges would exert individually. This important property, called the **principle of superposition of forces,** holds for any number of charges. By using this principle, we can apply Coulomb's law to *any* collection of charges. Several of the examples at the end of this section show applications of the superposition principle.

Strictly speaking, Coulomb's law as we have stated it should be used only for point charges *in vacuum.* If matter is present in the space between the charges, the net force acting on each charge is altered because charges are induced in the molecules of the intervening material. We will describe this effect later. As a practical matter, though, we can use Coulomb's law unaltered for point charges in air. At normal atmospheric pressure the presence of air changes the electric force from its vacuum value by only about one part in 2000.

The value of e was first measured during 1909–1913 by the American physicist Robert A. Millikan at the University of Chicago. (Challenge Problem 24–83 describes how his experiments worked.) Millikan won the 1923 Nobel Prize in physics for this research.

Problem-Solving Strategy

COULOMB'S LAW

1. As always, consistent units are essential. With the value of $k = 1/4\pi\epsilon_0$ given above, distances *must* be in meters, charge in coulombs, and force in newtons. If you are given distances in centimeters, inches, or furlongs, don't forget to convert! When a charge is given in microcoulombs or nanocoulombs, remember that $1 \ \mu\text{C} = 10^{-6} \ \text{C}$ and $1 \ \text{nC} = 10^{-9} \ \text{C}$.

2. Remember that the electric force, like any force, is a *vector.* When the forces acting on a charge are caused by two or more other charges, the total force on the charge is the *vector sum* of the individual forces. You may want to go back and review the vector algebra in Sections 1–8 through 1–10. It's often useful to use components in an

xy-coordinate system. Be sure to use correct vector notation; if a symbol represents a vector quantity, put an arrow over it. If you get sloppy with your notation, you will also get sloppy with your thinking.

3. Some examples and problems in this and later chapters involve a continuous distribution of charge along a line or over a surface. In these cases the vector sum described in Step 2 becomes a vector integral, usually carried out by use of components. We divide the total charge distribution into infinitesimal pieces, use Coulomb's law for each piece, and then integrate to find the vector sum. Sometimes this can be done without explicit use of integration.

EXAMPLE 22–1

Electric force vs. gravitational force An α particle ("alpha") is the nucleus of a helium atom. It has a mass $m = 6.64 \times 10^{-27}$ kg and a charge $q = +2e = 3.2 \times 10^{-19}$ C. Compare the force of the electric repulsion between two α particles with the force of gravitational attraction between them.

SOLUTION The magnitude F_e of the electric force is given by Eq. (22–2),

$$F_e = \frac{1}{4\pi\epsilon_0} \frac{q^2}{r^2},$$

and the magnitude F_g of the gravitational force is given by Eq. (12–1),

$$F_g = G\frac{m^2}{r^2}.$$

To compare these two magnitudes, we construct their ratio:

$$\frac{F_e}{F_g} = \frac{1}{4\pi\epsilon_0 G} \frac{q^2}{m^2} = \frac{9.0 \times 10^9 \ \text{N} \cdot \text{m}^2/\text{C}^2}{6.67 \times 10^{-11} \ \text{N} \cdot \text{m}^2/\text{kg}^2} \frac{(3.2 \times 10^{-19} \ \text{C})^2}{(6.64 \times 10^{-27} \ \text{kg})^2}$$

$$= 3.1 \times 10^{35}.$$

This astonishingly large number shows that the gravitational force in this situation is completely negligible in comparison to the electric force. This is always true for interactions of atomic and subatomic particles. (Notice that this result doesn't depend on the distance r between the two α particles.) But within objects the size of a person or a planet, the positive and negative charges are nearly equal in magnitude, and the net electric force is usually much *smaller* than the gravitational force.

EXAMPLE 22–2

Force between two point charges Two point charges, $q_1 = +25$ nC and $q_2 = -75$ nC, are separated by a distance of 3.0 cm (Fig. 22–7a). Find the magnitude and direction of a) the electric force that q_1 exerts on q_2; b) the electric force that q_2 exerts on q_1.

SOLUTION a) The magnitude of the force that q_1 exerts on q_2 is given by Coulomb's law, Eq. (22–2). Converting charge to coulombs and distance to meters, we have

$$F_{1 \text{ on } 2} = \frac{1}{4\pi\epsilon_0} \frac{|q_1 q_2|}{r^2}$$

$$= (9.0 \times 10^9 \ \text{N} \cdot \text{m}^2/\text{C}^2) \frac{|(+25 \times 10^{-9} \ \text{C})(-75 \times 10^{-9} \ \text{C})|}{(0.030 \ \text{m})^2}$$

$$= 0.019 \ \text{N}.$$

Since the two charges have opposite signs, the force is attractive; that is, the force that acts on q_2 is directed toward q_1 along the line joining the two charges, as shown in Fig. 22–7b.

22–7 What force does q_1 exert on q_2, and what force does q_2 exert on q_1? Gravitational forces are negligible. (a) The two charges. (b) Free-body diagram for charge q_2. (c) Free-body diagram for charge q_1.

b) Remember that Newton's third law applies to the electric force. Even though the charges have different magnitudes, the magnitude of the force that q_2 exerts on q_1 is the *same* as the magnitude of the force that q_1 exerts on q_2:

$$F_{2 \text{ on } 1} = 0.019 \text{ N.}$$

Newton's third law also states that the direction of the force that q_2 exerts on q_1 is exactly opposite the direction of the force that q_1 exerts on q_2; this is shown in Fig. 22–7c. Note that the force on q_1 is directed toward q_2, as it must be, since charges of opposite sign attract each other.

EXAMPLE 22–3

Vector addition of electric forces on a line Two point charges are located on the positive x-axis of a coordinate system (Fig. 22–8). Charge $q_1 = 1.0$ nC is 2.0 cm from the origin, and charge $q_2 = -3.0$ nC is 4.0 cm from the origin. What is the total force exerted by these two charges on a charge $q_3 = 5.0$ nC located at the origin? Gravitational forces are negligible.

SOLUTION The total force on q_3 is the vector sum of the forces due to q_1 and q_2 individually. Converting charge to coulombs and distance to meters, we use Eq. (22–2) to find the magnitude $F_{1 \text{ on } 3}$ of the force of q_1 on q_3:

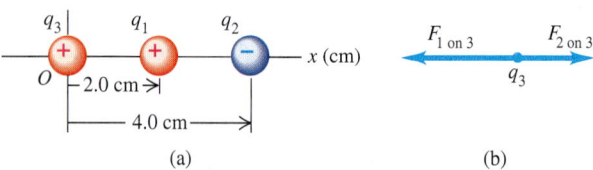

22–8 What is the total force exerted on the point charge q_3 by the other two point charges? (a) The three charges. (b) Free-body diagram for charge q_3.

$$F_{1 \text{ on } 3} = \frac{1}{4\pi\epsilon_0} \frac{|q_1 q_3|}{r^2}$$

$$= (9.0 \times 10^9 \text{ N} \cdot \text{m}^2/\text{C}^2)\frac{(1.0 \times 10^{-9} \text{ C})(5.0 \times 10^{-9} \text{ C})}{(0.020 \text{ m})^2}$$

$$= 1.12 \times 10^{-4} \text{ N} = 112 \ \mu\text{N.}$$

This force has a negative x-component because q_3 is repelled (that is, pushed in the negative x-direction) by q_1, which has the same sign.

The magnitude $F_{2 \text{ on } 3}$ of the force of q_2 on q_3 is

$$F_{2 \text{ on } 3} = \frac{1}{4\pi\epsilon_0} \frac{|q_2 q_3|}{r^2}$$

$$= (9.0 \times 10^9 \text{ N} \cdot \text{m}^2/\text{C}^2)\frac{(3.0 \times 10^{-9} \text{ C})(5.0 \times 10^{-9} \text{ C})}{(0.040 \text{ m})^2}$$

$$= 8.4 \times 10^{-5} \text{ N} = 84 \ \mu\text{N.}$$

This force has a positive x-component because q_3 is attracted (that is, pulled in the positive x-direction) by the opposite charge q_2. The sum of the x-components is

$$F_x = -112 \ \mu\text{N} + 84 \ \mu\text{N} = -28 \ \mu\text{N.}$$

There are no y- or z-components. Thus the total force on q_3 is directed to the left, with magnitude $28 \ \mu\text{N} = 2.8 \times 10^{-5}$ N.

EXAMPLE 22–4

Vector addition of electric forces in a plane In Fig. 22–9, two equal positive point charges $q_1 = q_2 = 2.0 \ \mu\text{C}$ interact with a third point charge $Q = 4.0 \ \mu\text{C}$. Find the magnitude and direction of the total (net) force on Q.

SOLUTION As in Example 22–3, we have to compute the force each charge exerts on Q and then find the vector sum of the forces. The easiest way to do this is to use components. The figure shows the force on Q due to the upper charge q_1. From Coulomb's law the magnitude F of this force is

$$F_{1 \text{ on } Q} = (9.0 \times 10^9 \text{ N} \cdot \text{m}^2/\text{C}^2)\frac{(4.0 \times 10^{-6} \text{ C})(2.0 \times 10^{-6} \text{ C})}{(0.50 \text{ m})^2}$$

$$= 0.29 \text{ N.}$$

The angle α is below the x-axis, so the components of this force

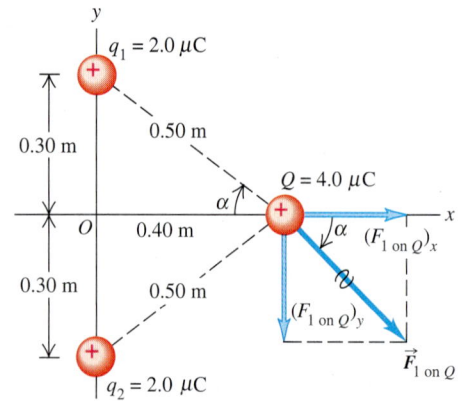

22–9 $\vec{F}_{1 \text{ on } Q}$ is the force on Q due to the upper charge q_1.

are given by

$$(F_{1\ on\ Q})_x = (F_{1\ on\ Q}) \cos \alpha = (0.29\ N)\frac{0.40\ m}{0.50\ m} = 0.23\ N,$$

$$(F_{1\ on\ Q})_y = -(F_{1\ on\ Q}) \sin \alpha = -(0.29\ N)\frac{0.30\ m}{0.50\ m} = -0.17\ N.$$

The lower charge q_2 exerts a force with the same magnitude but at an angle α *above* the x-axis. From symmetry we see that its x-component is the same as that due to the upper charge, but its

y-component has the opposite sign. So the components of the total force $\vec{F}$ on Q are

$$F_x = 0.23\ N + 0.23\ N = 0.46\ N,$$

$$F_y = -0.17\ N + 0.17\ N = 0.$$

The total force on Q is in the +x-direction, with magnitude 0.46 N. Can you show that if the lower charge were $-2.0\ \mu C$, the total force on Q would have components $F_x = 0$ and $F_y = -0.34\ N$?

22–6 ELECTRIC FIELD AND ELECTRIC FORCES

When two electrically charged particles in empty space interact, how does each one know the other is there? What goes on in the space between them to communicate the effect of each one to the other? We can begin to answer these questions, and at the same time reformulate Coulomb's law in a very useful way, by using the concept of *electric field*.

To introduce this concept, let's look at the mutual repulsion of two positively charged bodies A and B (Fig. 22–10a). Suppose B has charge q_0, and let $\vec{F}_0$ be the electric force of A on B. One way to think about this force is as an "action-at-a-distance" force, that is, as a force that acts across empty space without needing any matter (such as a push rod or a rope) to transmit it through the intervening space. (Gravity can also be thought of as an "action-at-a-distance" force.) But a more fruitful way to visualize the repulsion between A and B is as a two-stage process. We first envision that body A, as a result of the charge that it carries, somehow *modifies the properties of the space around it*. Then body B, as a result of the charge that *it* carries, senses how space has been modified at its position. The response of body B is to experience the force $\vec{F}_0$.

To elaborate how this two-stage process occurs, we first consider body A by itself: We remove body B and label its former position as point P (Fig. 22–10b). We say that the charged body A produces or causes an **electric field** at point P (and at all other points in the neighborhood). This electric field is present at P even if there is no other charge at P; it is a consequence of the charge on body A only. If a point charge q_0 is then placed at point P, it experiences the force $\vec{F}_0$. We take the point of view that this force is exerted on q_0 *by the field* at P (Fig. 22–10c). Thus the electric field is the intermediary through which A communicates its presence to q_0. Because the point charge q_0 would experience a force at *any* point in the neighborhood of A, the electric field that A produces exists at all points in the region around A.

We can likewise say that the point charge q_0 produces an electric field in the space around it and that this electric field exerts the force $-\vec{F}_0$ on body A. For each force (the force of A on q_0 and the force of q_0 on A), one charge sets up an electric field that exerts a force on the second charge. We emphasize that this is an *interaction* between *two* charged bodies. A single charge produces an electric field in the surrounding space, but this electric field cannot exert a net force on the charge that created it; this is an example of the general principle that a body cannot exert a net force on itself, as discussed in Section 4–4. (If this principle wasn't valid, you would be able to lift yourself to the ceiling by pulling up on your belt!) **The electric force on a charged body is exerted by the electric field created by *other* charged bodies.**

To find out experimentally whether there is an electric field at a particular point, we place a charged body, which we call a **test charge,** at the point (Fig. 22–10c). If the test charge experiences an electric force, then there is an electric field at that point. This field is produced by charges other than q_0.

Force is a vector quantity, so electric field is also a vector quantity. (Note the use of vector signs, as well as boldface letters and plus, minus, and equals signs in the

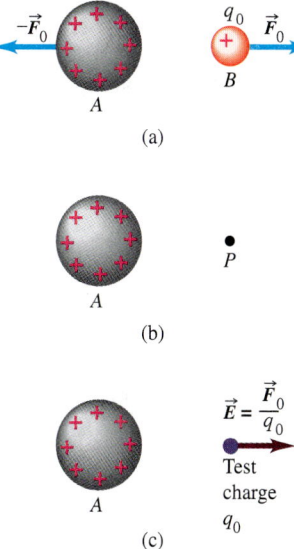

22–10 A charged body creates an electric field in the space around it.

following discussion.) We define the *electric field* $\vec{E}$ at a point as the electric force $\vec{F}_0$ experienced by a test charge q_0 at the point, divided by the charge q_0. That is, the electric field at a certain point is equal to the *electric force per unit charge* experienced by a charge at that point:

$$\vec{E} = \frac{\vec{F}_0}{q_0} \qquad \text{(definition of electric field as electric force per unit charge).} \quad (22\text{–}3)$$

In SI units, in which the unit of force is 1 N and the unit of charge is 1 C, the unit of electric field magnitude is 1 newton per coulomb (1 N/C).

If the field $\vec{E}$ at a certain point is known, rearranging Eq. (22–3) gives the force $\vec{F}_0$ experienced by a point charge q_0 placed at that point. This force is just equal to the electric field $\vec{E}$ produced at that point by charges other than q_0, multiplied by the charge q_0:

$$\vec{F}_0 = q_0\vec{E} \qquad \text{(force exerted on a point charge } q_0 \text{ by an electric field } \vec{E}). \quad (22\text{–}4)$$

The charge q_0 can be either positive or negative. If q_0 is *positive,* the force $\vec{F}_0$ experienced by the charge is the same direction as $\vec{E}$; if q_0 is *negative,* $\vec{F}_0$ and $\vec{E}$ are in opposite directions (Fig. 22–11).

While the electric field concept may be new to you, the basic idea—that one body sets up a field in the space around it, and a second body responds to that field—is one that you've actually used before. Compare Eq. (22–4) to the familiar expression for the gravitational force $\vec{F}_g$ that the earth exerts on a mass m_0:

$$\vec{F}_g = m_0\vec{g}. \quad (22\text{–}5)$$

In this expression, $\vec{g}$ is the acceleration due to gravity. If we divide both sides of Eq. (22–5) by the mass m_0, we obtain

$$\vec{g} = \frac{\vec{F}_g}{m_0}.$$

Thus $\vec{g}$ can be regarded as the gravitational force per unit mass. By analogy to Eq. (22–3), we can interpret $\vec{g}$ as the *gravitational field.* Thus we treat the gravitational interaction between the earth and the mass m_0 as a two-stage process: The earth sets up a gravitational field $\vec{g}$ in the space around it, and this gravitational field exerts a force given by Eq. (22–5) on the mass m_0 (which we can regard as a *test mass*). In this sense, you've made use of the field concept every time you've used Eq. (22–5) for the force of gravity. The gravitational field $\vec{g}$, or gravitational force per unit mass, is a useful concept because it does not depend on the mass of the body on which the gravitational force is exerted; likewise, the electric field $\vec{E}$, or electric force per unit charge, is useful because it does not depend on the charge of the body on which the electric force is exerted.

CAUTION ▶ The electric force experienced by a test charge q_0 can vary from point to point, so the electric field can also be different at different points. For this reason, Eq. (22–4) can be used only to find the electric force on a *point* charge. If a charged body is large enough in size, the electric field $\vec{E}$ may be noticeably different in magnitude and direction at different points on the body, and calculating the net electric force on the body can become rather complicated. ◀

Since the electric field $\vec{E}$ can vary from point to point, it is not a single vector quantity but an infinite set of vector quantities, one associated with each point in space. This is an example of a **vector field.** If we use a rectangular (*xyz*) coordinate system, each component of $\vec{E}$ at any point is in general a function of the coordinates (*x, y, z*) of the point. We can represent the components as $E_x(x, y, z)$, $E_y(x, y, z)$, and $E_z(x, y, z)$. In some situations the magnitude and direction of the field (and hence its vector components) are *constant* throughout a certain region; we then say that the field is *uniform* in this region. Vector fields are an important part of the language of physics, particularly in electricity and magnetism. Another example of a vector field is the velocity $\vec{v}$ of wind currents; the

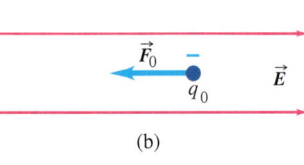

22–11 The force $\vec{F}_0$ exerted on a charge q_0 by an electric field $\vec{E}$. (a) If q_0 is positive, then $\vec{F}_0$ and $\vec{E}$ are in the same direction. (b) If q_0 is negative, $\vec{F}_0$ and $\vec{E}$ are in opposite directions.

11.4
Electric Field: Point Charge

11.9
Motion of a Charge in an Electric Field: Introduction

11.10
Motion in an Electric Field: Problems

magnitude and direction of $\vec{v}$, and hence its vector components, vary from point to point in the atmosphere.

We have so far ignored a subtle but important difficulty with our definition of electric field: In Fig. 22–10 the force exerted by the test charge q_0 on the charge distribution on body A may cause this distribution to shift around. This is especially true if body A is a conductor, on which charge is free to move. So the electric field around A when q_0 is present may not be the same as when q_0 is absent. But if q_0 is very small, the redistribution of charge on body A is also very small. So to make a completely correct definition of electric field, we take the *limit* of Eq. (22–3) as the test charge q_0 approaches zero and as the disturbing effect of q_0 on the charge distribution becomes negligible:

$$\vec{E} = \lim_{q_0 \to 0} \frac{\vec{F_0}}{q_0}.$$

In practical calculations of the electric field $\vec{E}$ produced by a charge distribution, we will consider the charge distribution to be fixed, and so we will not need this limiting process.

If the source distribution is a point charge q, it is easy to find the electric field that it produces. We call the location of the charge the **source point,** and we call the point P where we are determining the field the **field point.** It is also useful to introduce a *unit vector* $\hat{r}$ that points along the line from source point to field point (Fig. 22–12a). This unit vector is equal to the displacement vector $\vec{r}$ from the source point to the field point, divided by the distance $r = |\vec{r}|$ between these two points; that is, $\hat{r} = \vec{r}/r$. If we place a small test charge q_0 at the field point P, at a distance r from the source point, the magnitude F_0 of the force is given by Coulomb's law, Eq. (22–2):

$$F_0 = \frac{1}{4\pi\epsilon_0} \frac{|qq_0|}{r^2}.$$

From Eq. (22–3) the magnitude E of the electric field at P is

$$E = \frac{1}{4\pi\epsilon_0} \frac{|q|}{r^2} \qquad \text{(magnitude of electric field of a point charge).} \qquad (22\text{–}6)$$

Using the unit vector $\hat{r}$, we can write a *vector* equation that gives both the magnitude and direction of the electric field $\vec{E}$:

$$\vec{E} = \frac{1}{4\pi\epsilon_0} \frac{q}{r^2} \hat{r} \qquad \text{(electric field of a point charge).} \qquad (22\text{–}7)$$

By definition the electric field of a point charge always points *away from* a positive charge (that is, in the same direction as $\hat{r}$; see Fig. 22–12b) but *toward* a negative charge (that is, in the direction opposite $\hat{r}$; see Fig. 22–12c).

Another situation in which it is easy to find the electric field is inside a *conductor.* If there is an electric field within a conductor, the field exerts a force on every charge in the conductor, giving the free charges a net motion. By definition an electrostatic situation is one in which the charges have *no* net motion. We conclude that *in electrostatics the electric field at every point within the material of a conductor must be zero.* (Note that we are not saying that the field is necessarily zero in a *hole* inside a conductor.)

With the concept of electric field, our description of electric interactions has two parts. First, a given charge distribution acts as a source of electric field. Second, the electric field exerts a force on any charge that is present in the field. Our analysis often has two corresponding steps: first, calculating the field caused by a source charge distribution; second, looking at the effect of the field in terms of force and motion. The second step often involves Newton's laws as well as the principles of electric interactions. In the next section we show how to calculate fields caused by various source distributions, but first here are some examples of calculating the field due to a point charge and of finding the force on a charge due to a given field $\vec{E}$.

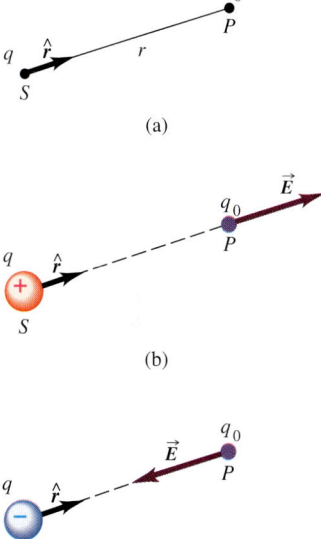

22–12 (a) The unit vector $\hat{r}$ points from the source point S to the field point P. (b) At each point, the electric field vector set up by an isolated *positive* point charge points directly away from the charge. (c) At each point, the electric field vector set up by an isolated *negative* point charge points directly toward the charge. Note that in both (b) and (c), $\vec{E}$ is *produced* by q (see Eq. (22–7)) but *acts* on q_0 (see Eq. (22–4)).

EXAMPLE 22-5

Electric-field magnitude for a point charge What is the magnitude of the electric field at a field point 2.0 m from a point charge $q = 4.0$ nC? (The point charge could represent any small charged object with this value of q, provided the dimensions of the object are much less than the distance from the object to the field point.)

SOLUTION From Eq. (22–6),

$$E = \frac{1}{4\pi\epsilon_0} \frac{|q|}{r^2} = (9.0 \times 10^9 \text{ N} \cdot \text{m}^2/\text{C}^2) \frac{4.0 \times 10^{-9} \text{C}}{(2.0 \text{ m})^2}$$

$$= 9.0 \text{ N/C}.$$

Alternatively, we can first use Coulomb's law, Eq. (22–2), to find the magnitude F_0 of the force on a test charge q_0 placed 2.0 m from q:

$$F_0 = \frac{1}{4\pi\epsilon_0} \frac{|qq_0|}{r^2} = (9.0 \times 10^9 \text{ N} \cdot \text{m}^2/\text{C}^2) \frac{(4.0 \times 10^{-9} \text{ C})|q_0|}{(2.0 \text{ m})^2}$$

$$= (9.0 \text{ N/C}) |q_0|.$$

Then, from Eq. (22–3), the magnitude of $\vec{E}$ is

$$E = \frac{F_0}{|q_0|} = 9.0 \text{ N/C}.$$

Because q is positive, the *direction* of $\vec{E}$ at this point is along the line from q toward q_0, as shown in Fig. 22–12b. However, the magnitude and direction of $\vec{E}$ do not depend on the sign of q_0. Do you see why not?

EXAMPLE 22-6

Electric-field vector for a point charge A point charge $q = -8.0$ nC is located at the origin. Find the electric-field vector at the field point $x = 1.2$ m, $y = -1.6$ m (Fig. 22–13).

SOLUTION The electric field is given in vector form by Eq. (22–7). The distance from the charge at the source point S (which in this example is at the origin O) to the field point P is

$$r = \sqrt{x^2 + y^2} = \sqrt{(1.2 \text{ m})^2 + (-1.6 \text{ m})^2} = 2.0 \text{ m}.$$

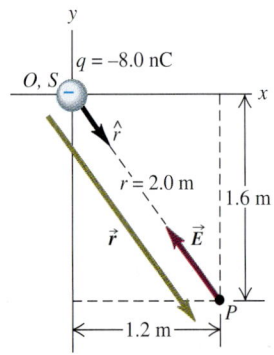

22–13 The vectors $\vec{r}$, $\hat{r}$, and $\vec{E}$ for a point charge.

The unit vector $\hat{r}$ is directed from the source point to the field point. This is equal to the displacement vector $\vec{r}$ from the source point to the field point (shown shifted to one side in Fig. 22–13 so as not to obscure the other vectors), divided by its magnitude r:

$$\hat{r} = \frac{\vec{r}}{r} = \frac{x\hat{i} + y\hat{j}}{r}$$

$$= \frac{(1.2 \text{ m})\hat{i} + (-1.6 \text{ m})\hat{j}}{2.0 \text{ m}} = 0.60\hat{i} - 0.80\hat{j}.$$

Hence the electric field vector is

$$\vec{E} = \frac{1}{4\pi\epsilon_0} \frac{q}{r^2} \hat{r}$$

$$= (9.0 \times 10^9 \text{ N} \cdot \text{m}^2/\text{C}^2) \frac{(-8.0 \times 10^{-9} \text{ C})}{(2.0 \text{ m})^2} (0.60\hat{i} - 0.80\hat{j})$$

$$= (-11 \text{ N/C})\hat{i} + (14 \text{ N/C})\hat{j}.$$

Since q is negative, $\vec{E}$ points from the field point to the charge (the source point), in the direction opposite to $\hat{r}$ (compare Fig. 22–12c). We leave the calculation of the magnitude and direction of $\vec{E}$ as an exercise.

EXAMPLE 22-7

Electron in a uniform field When the terminals of a battery are connected to two large parallel conducting plates, the resulting charges on the plates cause an electric field $\vec{E}$ in the region between the plates that is very nearly uniform. (We will see the reason for this uniformity in the next section. Charged plates of this kind are used in common electrical devices called *capacitors*, to be discussed in Chapter 25.) If the plates are horizontal and separated by 1.0 cm and the plates are connected to a 100-volt battery, the magnitude of the field is $E = 1.00 \times 10^4$ N/C. Suppose the direction of $\vec{E}$ is vertically upward, as shown by the

vectors in Fig. 22–14. a) If an electron is released from rest at the upper plate, what is its acceleration? b) What speed and kinetic energy does it acquire while traveling 1.0 cm to the lower plate? c) How much time is required for it to travel this distance? An electron has a charge $-e = -1.60 \times 10^{-19}$ C and a mass $m = 9.11 \times 10^{-31}$ kg.

SOLUTION a) A coordinate system is shown in Fig. 22–14. Note that $\vec{E}$ is upward (in the +y-direction) but $\vec{F}$ is downward because the charge of the electron is negative. Thus F_y is negative.

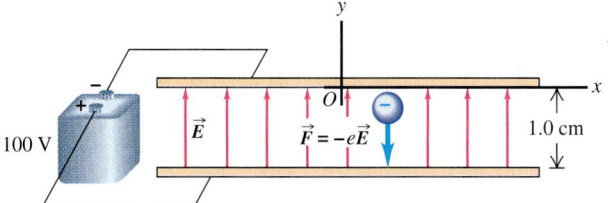

22–14 A uniform electric field between two parallel conducting plates connected to a 100-V battery. (The separation of the plates is exaggerated in this figure relative to the dimensions of the plates.)

Because F_y is constant, the electron moves with constant acceleration a_y given by

$$a_y = \frac{F_y}{m} = \frac{-eE}{m} = \frac{(-1.60 \times 10^{-19} \text{ C})(1.00 \times 10^4 \text{ N/C})}{9.11 \times 10^{-31} \text{ kg}}$$

$$= -1.76 \times 10^{15} \text{ m/s}^2.$$

This is an enormous acceleration! To give a 1000-kg car this acceleration, we would need a force of about 2×10^{18} N (about 2×10^{14} tons). The gravitational force on the electron is completely negligible compared to the electric force.

b) The electron starts from rest, so its motion is in the y-direction only (the direction of the acceleration). We can find the electron's speed at any position using the constant-acceleration formula $v_y{}^2 = v_{0y}{}^2 + 2a_y(y - y_0)$. We have $v_{0y} = 0$ and $y_0 = 0$, so the speed $|v_y|$ when $y = -1.0$ cm $= -1.0 \times 10^{-2}$ m is

$$|v_y| = \sqrt{2a_y y} = \sqrt{2(-1.76 \times 10^{15} \text{ m/s}^2)(-1.0 \times 10^{-2} \text{ m})}$$

$$= 5.9 \times 10^6 \text{ m/s}.$$

The velocity is downward, so its y-component is $v_y = -5.9 \times 10^6$ m/s. The electron's kinetic energy is

$$K = \frac{1}{2}mv^2 = \frac{1}{2}(9.11 \times 10^{-31} \text{ kg})(5.9 \times 10^6 \text{ m/s})^2$$

$$= 1.6 \times 10^{-17} \text{ J}.$$

c) From the constant-acceleration formula $v_y = v_{0y} + a_y t$, we find that the time required is very brief:

$$t = \frac{v_y - v_{0y}}{a_y} = \frac{(-5.9 \times 10^6 \text{ m/s}) - (0 \text{ m/s})}{-1.76 \times 10^{15} \text{ m/s}^2}$$

$$= 3.4 \times 10^{-9} \text{ s}.$$

(We could also have found the time by solving the equation $y = y_0 + v_{0y}t + \frac{1}{2}a_y t^2$ for t.)

EXAMPLE 22–8

An electron trajectory If we launch an electron into the electric field of Example 22–7 with an initial horizontal velocity v_0 (Fig. 22–15), what is the equation of its trajectory?

SOLUTION The force and acceleration are the same as in Example 22–7. The field is upward, but the force on the (negatively charged) electron is downward. The x-acceleration is zero; the y-acceleration is $(-e)E/m$. At $t = 0$, $x_0 = y_0 = 0$, $v_{0x} = v_0$, and $v_{0y} = 0$; hence at time t,

$$x = v_0 t \quad \text{and} \quad y = \frac{1}{2}a_y t^2 = -\frac{1}{2}\frac{eE}{m}t^2.$$

Eliminating t between these equations, we get

$$y = -\frac{1}{2}\frac{eE}{mv_0{}^2}x^2.$$

This is the equation of a parabola, just like the trajectory of a projectile launched horizontally in the earth's gravitational field (discussed in Section 3–4). For a given initial velocity of the electron, the curvature of the trajectory depends on the field

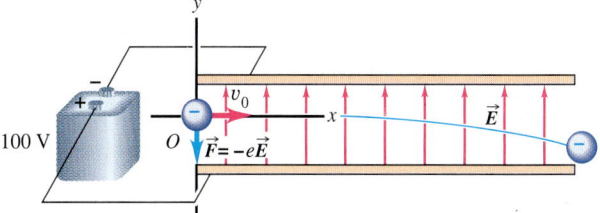

22–15 The parabolic trajectory of an electron in a uniform electric field.

magnitude E. If we reverse the signs of the charges on the two plates in Fig. 22–15, the direction of $\vec{E}$ reverses, and the electron trajectory will curve up, not down. Hence we can "steer" the electron by varying the charges on the plates. The electric field between charged conducting plates is used in this way to control the trajectory of electron beams in oscilloscopes and to control the direction of a flying stream of ink droplets in an inkjet printer.

22–7 ELECTRIC-FIELD CALCULATIONS

Equation (22–7) gives the electric field caused by a single point charge. But in most realistic situations that involve electric fields and forces, we encounter charge that is *distributed* over space. The charged plastic and glass rods in Fig. 22–1 have electric charge distributed over their surfaces, as do the drum and toner brush of an electrostatic photocopy machine. In this section we'll learn to calculate electric fields caused by various distributions of electric charge. Calculations of this kind are of tremendous importance for technological applications of electric forces. To determine the trajectories of electrons in a TV tube, of atomic nuclei in an accelerator for cancer radio-therapy, or of charged particles in a semiconductor electronic device, you have to know the detailed nature of the electric field acting on the charges.

To find the field caused by a charge distribution, we imagine the distribution to be made up of many point charges q_1, q_2, q_3, ... (This is actually quite a realistic description, since we have seen that charge is carried by electrons and protons that are so small as to be almost pointlike.) At any given point P, each point charge produces its own electric field $\vec{E}_1$, $\vec{E}_2$, $\vec{E}_3$, ..., so a test charge q_0 placed at P experiences a force $\vec{F}_1 = q_0\vec{E}_1$ from charge q_1, a force $\vec{F}_2 = q_0\vec{E}_2$ from charge q_2, and so on. From the principle of superposition of forces discussed in Section 22–5, the *total* force $\vec{F}_0$ that the charge distribution exerts on q_0 is the vector sum of these individual forces:

$$\vec{F}_0 = \vec{F}_1 + \vec{F}_2 + \vec{F}_3 + \cdots = q_0\vec{E}_1 + q_0\vec{E}_2 + q_0\vec{E}_3 + \cdots.$$

The combined effect of all the charges in the distribution is described by the *total* electric field $\vec{E}$ at point P. From the definition of electric field, Eq. (22–3), this is

$$\vec{E} = \frac{\vec{F}_0}{q_0} = \vec{E}_1 + \vec{E}_2 + \vec{E}_3 + \cdots.$$

The total electric field at P is the vector sum of the fields at P due to each point charge in the charge distribution. This is the **principle of superposition of electric fields.**

When charge is distributed along a line, over a surface, or through a volume, a few additional terms are useful. For a line charge distribution (such as a long, thin, charged plastic rod) we use λ ("lambda") to represent the **linear charge density** (charge per unit length, measured in C/m). When charge is distributed over a surface (such as the surface of the imaging drum of a photocopy machine), we use σ ("sigma") to represent the **surface charge density** (charge per unit area, measured in C/m^2). And when charge is distributed through a volume, we use ρ ("rho") to represent the **volume charge density** (charge per unit volume, C/m^3).

Some of the calculations in the following examples may look fairly intricate; in electric-field calculations a certain amount of mathematical complexity is in the nature of things. After you've worked through the examples one step at a time, the process will seem less formidable. Many of the calculational techniques in these examples will be used again in Chapter 29 to calculate the *magnetic* fields caused by charges in motion.

11.5
Electric Field Due to a Dipole

11.6
Electric Field: Problems

Problem-Solving Strategy

ELECTRIC-FIELD CALCULATIONS

1. Be sure to use a consistent set of units. Distances must be in meters and charge must be in coulombs. If you are given cm or nC, don't forget to convert.

2. When adding up the electric fields caused by different parts of the charge distribution, remember that electric field is a vector, so you *must* use vector addition. Don't simply add together the magnitudes of the individual fields; the directions are important too.

3. Usually, you will use components to compute vector sums. Use the methods you learned in Chapter 1; review them if you need to. Use proper vector notation; distinguish carefully between scalars, vectors, and components of vectors. Indicate your coordinate axes clearly on your diagram, and be certain the components are consistent with your choice of axes.

4. In finding the direction of $\vec{E}$, remember that the $\vec{E}$ vector produced by a *positive* point charge points *away* from that charge. A *negative* point charge produces an $\vec{E}$ that points *toward* that charge.

5. In some situations you will have a continuous distribution of charge along a line, over a surface, or through a volume. Then you have to define a small element of charge dQ that can be considered as a point, find its electric field $d\vec{E}$ at point P, and find a way to add the fields of all the charge elements. Usually, it is easiest to do this for each component of $\vec{E}$ separately. You will often need to evaluate one or more integrals. Make certain the limits on your integrals are correct; especially when the situation has symmetry, make sure you don't count the charge twice.

EXAMPLE 22–9

Field of an electric dipole Point charges q_1 and q_2 of +12 nC and −12nC, respectively, are placed 0.10 m apart (Fig. 22–16). This combination of two charges with equal magnitude and opposite sign is called an *electric dipole*. (Such combinations occur frequently in nature. For example, in Fig. 22–5, each molecule in the neutral insulator is an electric dipole. We'll study dipoles in more detail in Section 22–9.) Compute the electric field caused by q_1, the field caused by q_2, and the total field a) at point a; b) at point b; and c) at point c.

SOLUTION a) At point a the field $\vec{E}_1$ caused by the positive charge q_1 and the field $\vec{E}_2$ caused by the negative charge q_2 are both directed toward the right. The magnitudes of $\vec{E}_1$ and $\vec{E}_2$ are

$$E_1 = \frac{1}{4\pi\epsilon_0}\frac{|q_1|}{r^2} = (9.0\times10^9 \text{ N}\cdot\text{m}^2/\text{C}^2)\frac{12\times10^{-9}\text{ C}}{(0.060\text{ m})^2}$$

$$= 3.0\times10^4 \text{ N/C},$$

$$E_2 = \frac{1}{4\pi\epsilon_0}\frac{|q_2|}{r^2} = (9.0\times10^9 \text{ N}\cdot\text{m}^2/\text{C}^2)\frac{12\times10^{-9}\text{ C}}{(0.040\text{ m})^2}$$

$$= 6.8\times10^4 \text{ N/C}.$$

The components of $\vec{E}_1$ and $\vec{E}_2$ are

$$E_{1x} = 3.0\times10^4 \text{ N/C}, \qquad E_{1y} = 0,$$

$$E_{2x} = 6.8\times10^4 \text{ N/C}, \qquad E_{2y} = 0.$$

Hence at point a the total electric field $\vec{E}_a = \vec{E}_1 + \vec{E}_2$ has components

$$(E_a)_x = E_{1x} + E_{2x} = (3.0 + 6.8)\times10^4 \text{ N/C},$$

$$(E_a)_y = E_{1y} + E_{2y} = 0.$$

At point a the total field has magnitude $E_a = 9.8\times10^4$ N/C and is directed toward the right, so

$$\vec{E}_a = (9.8\times10^4 \text{ N/C})\hat{\imath}.$$

b) At point b the field $\vec{E}_1$ due to q_1 is directed toward the left, while the field $\vec{E}_2$ due to q_2 is directed toward the right. The magnitudes of $\vec{E}_1$ and $\vec{E}_2$ are

$$E_1 = \frac{1}{4\pi\epsilon_0}\frac{|q_1|}{r^2} = (9.0\times10^9 \text{ N}\cdot\text{m}^2/\text{C}^2)\frac{12\times10^{-9}\text{ C}}{(0.040\text{ m})^2}$$

$$= 6.8\times10^4 \text{ N/C},$$

$$E_2 = \frac{1}{4\pi\epsilon_0}\frac{|q_2|}{r^2} = (9.0\times10^9 \text{ N}\cdot\text{m}^2/\text{C}^2)\frac{12\times10^{-9}\text{ C}}{(0.140\text{ m})^2}$$

$$= 0.55\times10^4 \text{ N/C}.$$

The components of $\vec{E}_1$, $\vec{E}_2$, and the total field $\vec{E}_b$ at point b are

$$E_{1x} = -6.8\times10^4 \text{ N/C}, \qquad\qquad E_{1y} = 0,$$

$$E_{2x} = 0.55\times10^4 \text{ N/C}, \qquad\qquad E_{2y} = 0,$$

$$(E_b)_x = E_{1x} + E_{2x} = (-6.8 + 0.55)\times10^4 \text{ N/C},$$

$$(E_b)_y = E_{1y} + E_{2y} = 0.$$

That is, the electric field at b has magnitude $E_b = 6.2\times10^4$ N/C and is directed toward the left, so

$$\vec{E}_b = (-6.2\times10^4 \text{ N/C})\hat{\imath}.$$

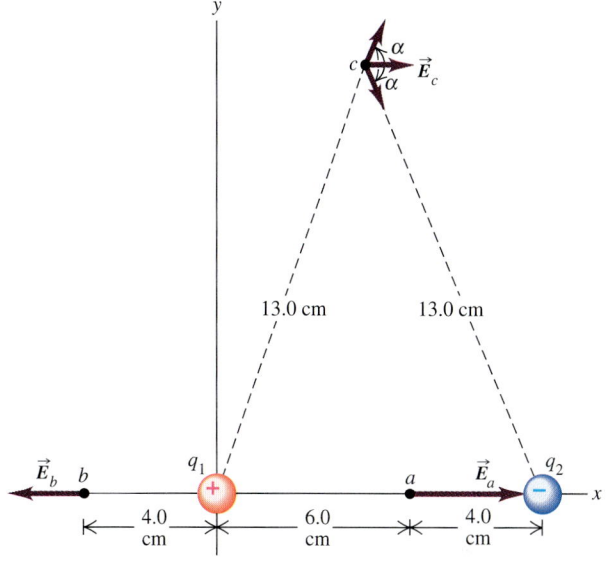

22–16 Electric field at three points, a, b, and c, set up by charges q_1 and q_2, which form an electric dipole.

c) At point c, both $\vec{E}_1$ and $\vec{E}_2$ have the same magnitude, since this point is equidistant from both charges and the charge magnitudes are the same:

$$E_1 = E_2 = \frac{1}{4\pi\epsilon_0}\frac{|q|}{r^2} = (9.0\times10^9 \text{ N}\cdot\text{m}^2/\text{C}^2)\frac{12\times10^{-9}\text{ C}}{(0.13\text{ m})^2}$$

$$= 6.39\times10^3 \text{ N/C}.$$

The directions of $\vec{E}_1$ and $\vec{E}_2$ are shown in the figure. The x-components of both vectors are the same:

$$E_{1x} = E_{2x} = E_1 \cos\alpha = (6.39\times10^3 \text{ N/C})\left(\frac{5}{13}\right)$$

$$= 2.46\times10^3 \text{ N/C}.$$

From symmetry the y-components E_{1y} and E_{2y} are equal and opposite and so add to zero. Hence the components of the total field $\vec{E}_c$ are

$$(E_c)_x = E_{1x} + E_{2x} = 2(2.46\times10^3 \text{ N/C}) = 4.9\times10^3 \text{ N/C},$$

$$(E_c)_y = E_{1y} + E_{2y} = 0.$$

So at point c the total electric field has magnitude $E_c = 4.9\times10^3$ N/C and is directed toward the right, so

$$\vec{E}_c = (4.9\times10^3 \text{ N/C})\hat{\imath}.$$

Does it surprise you that the field at point c is parallel to the line between the two charges?

An alternative way to find the electric field at c is to use the vector expression for the field of a point charge, Eq. (22–7). The displacement vector $\vec{r}_1$ from q_1 to point c, a distance $r = 13.0$ cm away, is

$$\vec{r}_1 = r\cos\alpha\,\hat{\imath} + r\sin\alpha\,\hat{\jmath}.$$

Hence the unit vector that points from q_1 to c is

$$\hat{r}_1 = \frac{\vec{r}_1}{r} = \cos\alpha\,\hat{\imath} + \sin\alpha\,\hat{\jmath},$$

and the field due to q_1 at point c is

$$\vec{E}_1 = \frac{1}{4\pi\epsilon_0}\frac{q_1}{r^2}\hat{r}_1 = \frac{1}{4\pi\epsilon_0}\frac{q_1}{r^2}(\cos\alpha\,\hat{\imath} + \sin\alpha\,\hat{\jmath}).$$

By symmetry the unit vector $\hat{r}_2$ that points from q_2 to point c has the opposite x-component but the same y-component, so the field at c due to q_2 is

$$\vec{E}_2 = \frac{1}{4\pi\epsilon_0}\frac{q_2}{r^2}\hat{r}_2 = \frac{1}{4\pi\epsilon_0}\frac{q_2}{r^2}(-\cos\alpha\,\hat{\imath} + \sin\alpha\,\hat{\jmath}).$$

Since $q_2 = -q_1$, the total field at c is

$$\vec{E}_c = \vec{E}_1 + \vec{E}_2$$

$$= \frac{1}{4\pi\epsilon_0}\frac{q_1}{r^2}(\cos\alpha\,\hat{\imath} + \sin\alpha\,\hat{\jmath}) + \frac{1}{4\pi\epsilon_0}\frac{(-q_1)}{r^2}(-\cos\alpha\,\hat{\imath} + \sin\alpha\,\hat{\jmath})$$

$$= \frac{1}{4\pi\epsilon_0}\frac{q_1}{r^2}(2\cos\alpha\,\hat{\imath})$$

$$= (9.0\times10^9\ \text{N}\cdot\text{m}^2/\text{C}^2)\frac{12\times10^{-9}\ \text{C}}{(0.13\ \text{m})^2}\left(2\left(\frac{5}{13}\right)\right)\hat{\imath}$$

$$= (4.9\times10^3\ \text{N/C})\hat{\imath},$$

as before.

EXAMPLE 22-10

Field of a ring of charge A ring-shaped conductor with radius a carries a total charge Q uniformly distributed around it (Fig. 22–17). Find the electric field at a point P that lies on the axis of the ring at a distance x from its center.

SOLUTION As shown in the figure, we imagine the ring divided into infinitesimal segments of length ds. Each segment has charge dQ and acts as a point-charge source of electric field. Let $d\vec{E}$ be the electric field from one such segment; the net electric field at P is then the sum of all contributions $d\vec{E}$ from all the segments that make up the ring. (This same technique works for any situation in which charge is distributed along a line or a curve.)

The calculation of $\vec{E}$ is greatly simplified because the field point P is on the symmetry axis of the ring. If we consider two ring segments at the top and bottom of the ring, we see that the contributions $d\vec{E}$ to the field at P from these segments have the same x-component but opposite y-components. Hence the total y-component of field due to this pair of segments is zero. When we add up the contributions from all such pairs of segments, the total field $\vec{E}$ will have only a component along the ring's symmetry axis (the x-axis), with no component perpendicular to that axis (that is, no y-component or z-component). So the field at P is described completely by its x-component E_x.

To calculate E_x, note that the square of the distance r from a ring segment to the point P is $r^2 = x^2 + a^2$. Hence the magnitude of this segment's contribution $d\vec{E}$ to the electric field at P is

$$dE = \frac{1}{4\pi\epsilon_0}\frac{dQ}{x^2 + a^2}.$$

Using $\cos\alpha = x/r = x/(x^2 + a^2)^{1/2}$, the component dE_x of this field along the x-axis is

$$dE_x = dE\cos\alpha = \frac{1}{4\pi\epsilon_0}\frac{dQ}{x^2 + a^2}\frac{x}{\sqrt{x^2 + a^2}}$$

$$= \frac{1}{4\pi\epsilon_0}\frac{x\,dQ}{(x^2 + a^2)^{3/2}}.$$

To find the *total* x-component E_x of the field at P, we integrate this expression over all segments of the ring:

$$E_x = \int\frac{1}{4\pi\epsilon_0}\frac{x\,dQ}{(x^2 + a^2)^{3/2}}.$$

Since x does not vary as we move from point to point around the

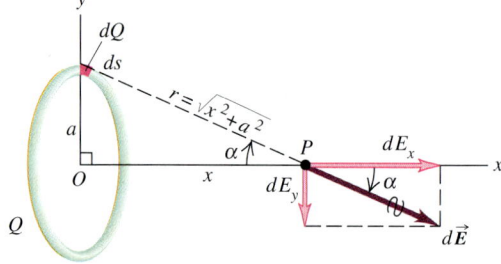

22–17 Calculating the electric field on the axis of a ring of charge. In this figure, the charge is assumed to be positive.

ring, all the factors on the right side except dQ are constant and can be taken outside the integral. The integral of dQ is just the total charge Q, and we finally get

$$\vec{E} = E_x\hat{\imath} = \frac{1}{4\pi\epsilon_0}\frac{Qx}{(x^2 + a^2)^{3/2}}\hat{\imath}. \qquad (22\text{-}8)$$

Our result for $\vec{E}$ shows that at the center of the ring ($x = 0$) the field is zero. We should expect this; charges on opposite sides of the ring would push in opposite directions on a test charge at the center, and the forces would add to zero. When the field point P is much farther from the ring than its size (that is, $x \gg a$), the denominator in Eq. (22–8) becomes approximately equal to x^3, and the expression becomes approximately

$$\vec{E} = \frac{1}{4\pi\epsilon_0}\frac{Q}{x^2}\hat{\imath}.$$

In other words, when we are so far from the ring that its size a is negligible in comparison to the distance x, its field is the same as that of a point charge. To an observer far from the ring, the ring would appear like a point, and the electric field reflects this.

In this example we used a *symmetry argument* to conclude that $\vec{E}$ had only an x-component at a point on the ring's axis of symmetry. We'll use symmetry arguments many times in this and subsequent chapters. Keep in mind, however, that such arguments can be used only in special cases. At a point in the xy-plane that is not on the x-axis in Fig. 22–17, the symmetry argument doesn't apply, and the field has in general both x- and y-components.

EXAMPLE 22–11

Field of a line of charge Positive electric charge Q is distributed uniformly along a line with length $2a$, lying along the y-axis between $y = -a$ and $y = +a$, as in Fig. 22–18. (This might represent one of the charged rods in Fig. 22–1.) Find the electric field at point P on the x-axis at a distance x from the origin.

SOLUTION We divide the line charge into infinitesimal segments, each of which acts as a point charge; let the length of a typical segment at height y be dy. If the charge is distributed uniformly, the linear charge density λ at any point on the line is equal to $Q/2a$ (the total charge divided by the total length). Hence the charge dQ in a segment of length dy is

$$dQ = \lambda \, dy = \frac{Q \, dy}{2a}.$$

The distance r from this segment to P is $(x^2 + y^2)^{1/2}$, so the magnitude of field dE at P due to this segment is

$$dE = \frac{Q}{4\pi\epsilon_0} \frac{dy}{2a(x^2 + y^2)}.$$

We represent this field in terms of its x- and y-components:

$$dE_x = dE \cos\alpha, \qquad dE_y = -dE \sin\alpha.$$

We note that $\sin\alpha = y/(x^2 + y^2)^{1/2}$ and $\cos\alpha = x/(x^2 + y^2)^{1/2}$; combining these with the expression for dE, we find

$$dE_x = \frac{Q}{4\pi\epsilon_0} \frac{x \, dy}{2a(x^2 + y^2)^{3/2}},$$

$$dE_y = -\frac{Q}{4\pi\epsilon_0} \frac{y \, dy}{2a(x^2 + y^2)^{3/2}}.$$

To find the total field components E_x and E_y, we integrate these expressions, noting that to include all of Q, we must integrate from $y = -a$ to $y = +a$. We invite you to work out the details of the integration; an integral table is helpful. The final results are

$$E_x = \frac{1}{4\pi\epsilon_0} \frac{Qx}{2a} \int_{-a}^{a} \frac{dy}{(x^2 + y^2)^{3/2}} = \frac{Q}{4\pi\epsilon_0} \frac{1}{x\sqrt{x^2 + a^2}},$$

$$E_y = -\frac{1}{4\pi\epsilon_0} \frac{Q}{2a} \int_{-a}^{a} \frac{y \, dy}{(x^2 + y^2)^{3/2}} = 0,$$

or, in vector form,

$$\vec{E} = \frac{1}{4\pi\epsilon_0} \frac{Q}{x\sqrt{x^2 + a^2}} \hat{\imath}. \qquad (22\text{–}9)$$

Using a symmetry argument as in Example 22–10, we could have guessed that E_y would be zero; if we place a positive test charge at P, the upper half of the line of charge pushes downward on it, and the lower half pushes up with equal magnitude.

To explore our result, let's first see what happens in the limit that x is much larger than a. Then we can neglect a in the denominator of Eq. (22–9), and our result becomes

$$\vec{E} = \frac{1}{4\pi\epsilon_0} \frac{Q}{x^2} \hat{\imath}.$$

This means that if point P is very far from the line charge in comparison to the length of the line, the field at P is the same as that of a point charge. We found a similar result for the charged

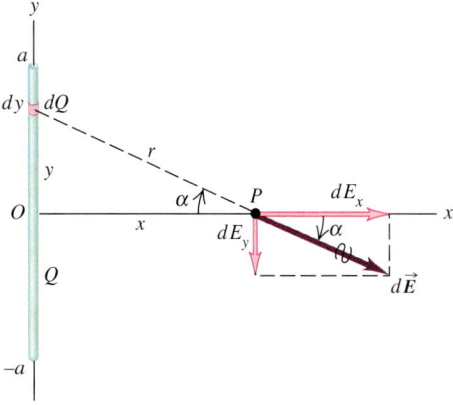

22–18 Finding the electric field at point P on the perpendicular bisector of a line of charge with length $2a$ and total charge Q. In this figure, the charge is assumed to be positive.

ring in Example 22–10.

We get an added dividend from our exact result for $\vec{E}$, Eq. (22–9), if we express it in terms of the linear charge density $\lambda = Q/2a$. Substituting $Q = 2a\lambda$ into Eq. (22–9) and simplifying, we get

$$\vec{E} = \frac{1}{2\pi\epsilon_0} \frac{\lambda}{x\sqrt{(x^2/a^2) + 1}} \hat{\imath}. \qquad (22\text{–}10)$$

Now what happens if we make the line of charge longer and longer, adding charge in proportion to the total length so that λ, the charge per unit length, remains constant? What is $\vec{E}$ at a distance x from a *very* long line of charge? To answer the question, we take the *limit* of Eq. (22–10) as a becomes very large. In this limit, the term x^2/a^2 in the denominator becomes much smaller than unity and can be thrown away. We are left with

$$\vec{E} = \frac{\lambda}{2\pi\epsilon_0 x} \hat{\imath}.$$

The field magnitude depends only on the distance of point P from the line of charge. So at any point P at a perpendicular distance r from the line in any direction, $\vec{E}$ has magnitude

$$E = \frac{\lambda}{2\pi\epsilon_0 r} \qquad \text{(infinite line of charge)}.$$

Thus the electric field due to an infinitely long line of charge is proportional to $1/r$ rather than to $1/r^2$ as for a point charge. The direction of $\vec{E}$ is radially outward from the line if λ is positive and radially inward if λ is negative.

Of course, there's really no such thing in nature as an infinite line of charge. But when the field point is close enough to the line, there's very little difference between the result for an infinite line and the real-life finite case. For example, if the distance r of the field point from the center of the line is 1% of the length of the line, the value of E differs from the infinite-length value by less than 0.02%.

EXAMPLE 22–12

Field of a uniformly charged disk Find the electric field caused by a disk of radius R with a uniform positive surface charge density (charge per unit area) σ, at a point along the axis of the disk a distance x from its center. Assume that x is positive.

SOLUTION The situation is shown in Fig. 22–19. We can represent this charge distribution as a collection of concentric rings of charge. We already know from Example 22–10 how to find the field of a single ring on its axis of symmetry, so all we have to do is add the contributions of all the rings.

As shown in the figure, a typical ring has charge dQ, inner radius r, and outer radius $r + dr$. Its area dA is approximately equal to its width dr times its circumference $2\pi r$, or $dA = 2\pi r\, dr$. The charge per unit area is $\sigma = dQ/dA$, so the charge of the ring is $dQ = \sigma\, dA = \sigma\,(2\pi r\, dr)$, or

$$dQ = 2\pi\sigma r\, dr.$$

We use this in place of Q in the expression for the field due to a ring found in Example 22–10, Eq. (22–8), and also replace the ring radius a with r. The field component dE_x at point P due to charge dQ is

$$dE_x = \frac{1}{4\pi\epsilon_0}\frac{(2\pi\sigma r\, dr)x}{(x^2 + r^2)^{3/2}}.$$

To find the total field due to all the rings, we integrate dE_x over r. To include the whole disk, we must integrate from 0 to R (*not* from $-R$ to R):

$$E_x = \int_0^R \frac{1}{4\pi\epsilon_0}\frac{(2\pi\sigma r\, dr)x}{(x^2 + r^2)^{3/2}} = \frac{\sigma x}{2\epsilon_0}\int_0^R \frac{r\, dr}{(x^2 + r^2)^{3/2}}.$$

Remember that x is a constant during the integration and that the integration variable is r. The integral can be evaluated by use of the substitution $z = x^2 + r^2$. We'll let you work out the details; the result is

$$E_x = \frac{\sigma x}{2\epsilon_0}\left[-\frac{1}{\sqrt{x^2 + R^2}} + \frac{1}{x} \right]$$

$$= \frac{\sigma}{2\epsilon_0}\left[1 - \frac{1}{\sqrt{(R^2/x^2) + 1}} \right]. \qquad (22\text{–}11)$$

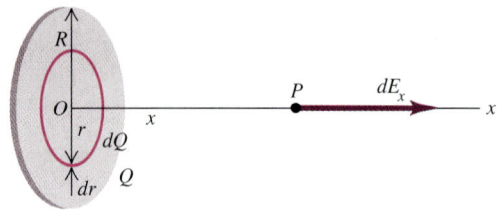

22–19 Finding the electric field on the axis of a uniformly charged disk. In this figure, the charge is assumed to be positive.

We saw in Example 22–10 that at a point on the symmetry axis of a uniformly charged ring, the electric field due to the ring has no components perpendicular to the axis. Hence at point P in Fig. 22–19, $dE_y = dE_z = 0$ for each ring, and the total field has $E_y = E_z = 0$.

Again we can ask what happens if the charge distribution gets very large. Suppose we keep increasing the radius R of the disk, simultaneously adding charge so that the surface charge density σ (charge per unit area) is constant. In the limit that R is much larger than the distance x of the field point from the disk, the term $1/\sqrt{(R^2/x^2) + 1}$ in Eq. (22–11) becomes negligibly small, and we get

$$E = \frac{\sigma}{2\epsilon_0}. \qquad (22\text{–}12)$$

Our final result does not contain the distance x from the plane. This correct but rather surprising result means that the electric field produced by an *infinite* plane sheet of charge is *independent of the distance from the sheet.* Thus the field is *uniform;* its direction is everywhere perpendicular to the sheet, away from it. Again, there is no such thing as an infinite sheet of charge, but if the dimensions of the sheet are much larger than the distance x of the field point P from the sheet, the field is very nearly the same as for an infinite sheet.

If P is to the left of the plane ($x < 0$) instead of to the right, the result is the same except that the direction of $\vec{E}$ is to the left instead of to the right. Also, if the surface charge density is negative, the directions of the fields on both sides of the plane are toward it rather than away from it.

EXAMPLE 22–13

Field of two oppositely charged infinite sheets Two infinite plane sheets are placed parallel to each other, separated by a distance d (Fig. 22–20). The lower sheet has a uniform positive surface charge density σ, and the upper sheet has a uniform negative surface charge density $-\sigma$ with the same magnitude. Find the electric field between the two sheets, above the upper sheet, and below the lower sheet.

SOLUTION The situation described in this example is an idealization of two *finite,* oppositely charged sheets, like the plates shown in Figs. 22–14 and 22–15. If the dimensions of the sheets are large in comparison to the separation d, then we can to good approximation consider the sheets to be infinite in extent.

In Example 22–12 we found the field due to a single infinite plane sheet of charge. We can then find the total field by using the principle of superposition of electric fields. Let sheet 1 be the lower sheet of positive charge, and let sheet 2 be the upper sheet of negative charge; the fields due to each sheet are $\vec{E}_1$ and $\vec{E}_2$, respectively. From Eq. (22–12) of Example 22–12, both $\vec{E}_1$ and $\vec{E}_2$ have the same magnitude at all points, no matter how far from either sheet:

$$E_1 = E_2 = \frac{\sigma}{2\epsilon_0}.$$

At all points, the direction of $\vec{E}_1$ is away from the positive charge of sheet 1, and the direction of $\vec{E}_2$ is towards the negative charge

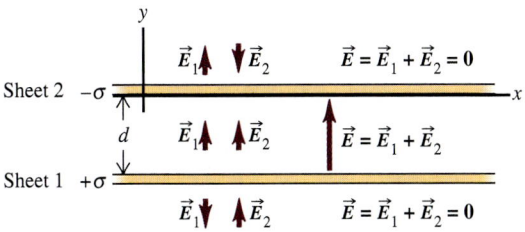

22–20 Finding the electric field due to two oppositely charged infinite sheets. The sheets are seen edge-on; only a portion of the infinite sheets can be shown!

of sheet 2. These fields, as well as the *x*- and *y*-axes, are shown in Fig. 22–20.

CAUTION ▶ You may be surprised that $\vec{E}_1$ is unaffected by the presence of sheet 2 and that $\vec{E}_2$ is unaffected by the presence of sheet 1. Indeed, you may have thought that the field of one sheet would be unable to "penetrate" the other sheet. This is a common misconception, based on the erroneous idea that the electric field represents some kind of physical substance that "flows" into or out of charges. In fact there is no such substance, and the electric fields $\vec{E}_1$ and $\vec{E}_2$ depend only on the individual charge distributions that create them. The *total* field is just the vector sum of $\vec{E}_1$ and $\vec{E}_2$. ◀

At points between the sheets, $\vec{E}_1$ and $\vec{E}_2$ reinforce each other; at points above the upper sheet or below the lower sheet, $\vec{E}_1$ and $\vec{E}_2$ cancel each other. Thus the total field is

$$\vec{E} = \vec{E}_1 + \vec{E}_2 = \begin{cases} 0 & \text{above the upper sheet,} \\ \dfrac{\sigma}{\epsilon_0}\,\hat{\jmath} & \text{between the sheets,} \\ 0 & \text{below the lower sheet.} \end{cases}$$

Because we considered the sheets to be infinite, our result does not depend on the separation *d*.

Note that the field between the oppositely charged sheets is uniform. We used this in Examples 22–7 and 22–8, in which two large parallel conducting plates were connected to the terminals of a battery. The battery causes the two plates to become oppositely charged, giving a field between the plates that is essentially uniform if the plate separation is much smaller than the dimensions of the plates. In Chapter 23 we will discuss the behavior of the field near the edges of charged sheets that are not infinite, and in Chapter 24 we will examine how a battery can produce such separation of positive and negative charge. An arrangement of two oppositely charged conducting plates is called a *capacitor;* these devices prove to be of tremendous practical utility and are the principal subject of Chapter 25.

22–8 ELECTRIC FIELD LINES

The concept of an electric field can be a little elusive because you can't see an electric field directly. Electric field *lines* can be a big help for visualizing electric fields and making them seem more real. An **electric field line** is an imaginary line or curve drawn through a region of space so that its tangent at any point is in the direction of the electric-field vector at that point. The basic idea is shown in Fig. 22–21. (We used a similar concept in our discussion of fluid flow in Section 14–6. A *streamline* is a line or curve whose tangent at any point is in the direction of the velocity of the fluid at that point. However, the similarity between electric field lines and fluid streamlines is a mathematical one only; there is nothing "flowing" in an electric field.) The English scientist Michael Faraday (1791–1867) first introduced the concept of field lines. He called them "lines of force," but the term "field lines" is preferable.

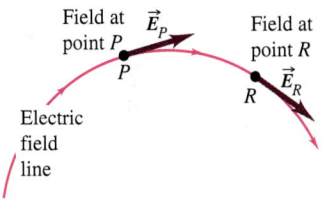

22–21 The direction of the electric field at any point is tangent to the field line through that point.

Electric field lines show the direction of $\vec{E}$ at each point, and their spacing gives a general idea of the *magnitude* of $\vec{E}$ at each point. Where $\vec{E}$ is strong, we draw lines bunched closely together; where $\vec{E}$ is weaker, they are farther apart. At any particular point, the electric field has a unique direction, so only one field line can pass through each point of the field. In other words, *field lines never intersect.*

Figure 22–22 (on page 690) shows some of the electric field lines in a plane containing (a) a single positive charge; (b) two equal-magnitude charges, one positive and one negative (a dipole); and (c) two equal positive charges. Diagrams such as this are sometimes called *field maps.* These are cross sections of the actual three-dimensional patterns. The direction of the total electric field at every point in each diagram is along the tangent to the electric field line passing through the point. Arrowheads indicate the direction of the $\vec{E}$-field vector along each field line. The actual field vectors have been drawn at several points in each pattern. Notice that in general, the magnitude of the electric field is different at different points on a given field line; a field line is *not* a curve of constant electric field magnitude! The figure shows that field lines are directed *away*

(a)

(b)

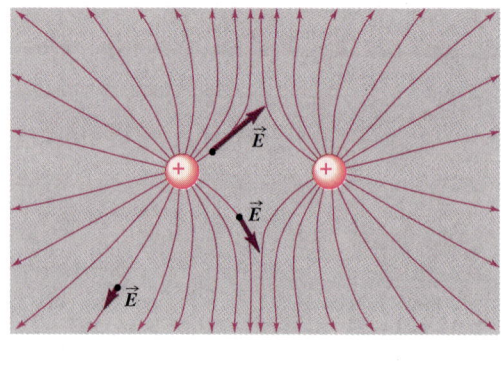

(c)

22–22 Electric field lines for several charge distributions. (a) A single positive charge; (b) two charges of equal magnitude and opposite sign (an electric dipole); (c) two equal positive charges. In general, the magnitude of $\vec{E}$ is different at different points along a given field line.

(a)

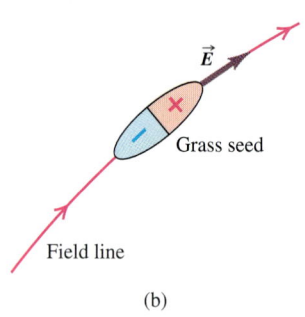

(b)

22–23 (a) Electric field lines produced by two equal point charges. The pattern is formed by grass seeds floating on a liquid above two charged wires. Compare this pattern with Fig. 22–22c. (b) The electric field causes polarization of the grass seeds, which in turn causes the seeds to align with the field.

from positive charges (since close to a positive point charge, $\vec{E}$ points away from the charge) and *toward* negative charges (since close to a negative point charge, $\vec{E}$ points toward the charge). In regions where the field magnitude is large, such as between the positive and negative charges in Fig. 22–22b, the field lines are drawn close together. In regions where the field magnitude is small, such as between the two positive charges in Fig. 22–22c, the lines are widely separated. In a *uniform* field, the field lines are straight, parallel, and uniformly spaced, as in Fig. 22–14.

Figure 22–23a is a view from above of a demonstration setup for visualizing electric field lines. In the arrangement shown here, the tips of two positively charged wires are inserted in a container of insulating liquid, and some grass seeds are floated on the liquid. The grass seeds are electrically neutral insulators, but the electric field of the two charged wires causes *polarization* of the grass seeds; there is a slight shifting of the positive and negative charges within the molecules of each seed, like that shown in Fig. 22–5. The positively charged end of each grass seed is pulled in the direction of $\vec{E}$, and the negatively charged end is pulled opposite $\vec{E}$. Hence the long axis of each grass seed tends to orient parallel to the electric field, in the direction of the field line that passes through the position of the seed (Fig. 22–23b).

CAUTION ▶ It's a common misconception that if a charged particle of charge q is in motion where there is an electric field, the particle must move along an electric field line. Because $\vec{E}$ at any point is tangent to the field line that passes through that point, it is indeed true that the *force* $\vec{F} = q\vec{E}$ on the particle, and hence the particle's acceleration, are tangent to the field line. But we learned in Chapter 3 that when a particle moves on a curved path, its acceleration *cannot* be tangent to the path. So in general, the trajectory of a charged particle is *not* the same as a field line. (The only exception is if the field lines are straight lines and the particle is released from rest, as in Fig. 22–14. Can you see why?) ◀

22–9 ELECTRIC DIPOLES

An **electric dipole** is a pair of point charges with equal magnitude and opposite sign (a positive charge q and a negative charge $-q$ separated by a distance d. We introduced electric dipoles in Example 22–9 (Section 22–7); the concept is worth exploring further because many physical systems, from molecules to TV antennas, can be described as electric dipoles. We will also use this concept extensively in our discussion of dielectrics in Chapter 25.

Figure 22–24 shows a molecule of water (H_2O), which in many ways behaves like an electric dipole. The water molecule as a whole is electrically neutral, but the chemical bonds within the molecule cause a displacement of charge; the result is a net negative charge on the oxygen end of the molecule and a net positive charge on the hydrogen end, forming an electric dipole. The effect is equivalent to shifting one electron only about 4×10^{-11} m (about the radius of a hydrogen atom), but the consequences of this shift are profound. Water is an excellent solvent for ionic substances such as table salt (sodium chloride, NaCl) precisely because the water molecule is an electric dipole. When dissolved in water, salt dissociates into a positive sodium ion (Na^+) and a negative chlorine ion (Cl^-), which tend to be attracted to the negative and positive ends, respectively, of water molecules; this holds the ions in solution. If water molecules were not electric dipoles, water would be a poor solvent, and almost all of the chemistry that occurs in aqueous solutions would be impossible. This includes all of the biochemical reactions that occur in all of the life on earth. In a very real sense, your existence as a living being depends on electric dipoles!

We will ask two questions about electric dipoles. First, what forces and torques does an electric dipole experience when placed in an external electric field (that is, a field set up by charges outside the dipole)? Second, what electric field does an electric dipole itself produce?

22–24 A water molecule is an example of an electric dipole.

FORCE AND TORQUE ON AN ELECTRIC DIPOLE

To start with the first question, let's place an electric dipole in a *uniform* external electric field $\vec{E}$, as shown in Fig. 22–25. The forces $\vec{F}_+$ and $\vec{F}_-$ on the two charges both have magnitude qE, but their directions are opposite, and they add to zero. *The net force on an electric dipole in a uniform external electric field is zero.*

However, the two forces don't act along the same line, so their *torques* don't add to zero. We calculate torques with respect to the center of the dipole. Let the angle between the electric field $\vec{E}$ and the dipole axis be ϕ; then the lever arm for both $\vec{F}_+$ and $\vec{F}_-$ is $(d/2) \sin \phi$. The torque of $\vec{F}_+$ and the torque of $\vec{F}_-$ both have the same magnitude of $(qE)(d/2) \sin \phi$, and both torques tend to rotate the dipole clockwise (that is, $\vec{\tau}$ is directed into the page in Fig. 22–25). Hence the magnitude of the net torque is just twice the magnitude of either individual torque:

$$\tau = (qE)(d \sin \phi), \qquad (22-13)$$

where $d \sin \phi$ is the perpendicular distance between the lines of action of the two forces.

The product of the charge q and the separation d is the magnitude of a quantity called the **electric dipole moment,** denoted by p:

$$p = qd \quad \text{(magnitude of electric dipole moment).} \qquad (22-14)$$

The units of p are charge times distance (C · m). For example, the magnitude of the electric dipole moment of a water molecule is $p = 6.13 \times 10^{-30}$ C · m. (Be careful not to confuse dipole moment with momentum or pressure. There aren't as many letters in the alphabet as there are physical quantities, so some letters are used several times. The context usually makes it clear what we mean, but be careful.) We further define the electric dipole moment to be a *vector* quantity $\vec{p}$ with magnitude given by Eq. (22–14) and direction along the dipole axis from the negative charge to the positive charge, as shown in Fig. 22–25.

In terms of p, Eq. (22–13) for the magnitude τ of the torque exerted by the field becomes

$$\tau = pE \sin \phi \quad \text{(magnitude of the torque on an electric dipole).} \qquad (22-15)$$

Since the angle ϕ in Fig. 22–25 is the angle between the directions of the vectors $\vec{p}$ and $\vec{E}$, this is reminiscent of the expression for the magnitude of the *vector product* discussed

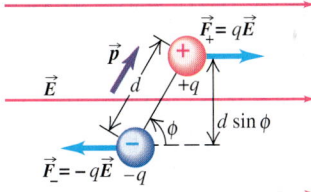

22–25 The net force on this electric dipole is zero, but there is a torque directed into the page that tends to rotate the dipole clockwise.

in Section 1–11. (You may want to review that discussion.) Hence we can write the torque on the dipole in vector form as

$$\vec{\tau} = \vec{p} \times \vec{E} \quad \text{(torque on an electric dipole, in vector form).} \quad (22\text{–}16)$$

You can use the right-hand rule for the vector product to verify that in the situation shown in Fig. 22–25, $\vec{\tau}$ is directed into the page. The torque is greatest when $\vec{p}$ and $\vec{E}$ are perpendicular and is zero when they are parallel or antiparallel. The torque always tends to turn $\vec{p}$ to line it up with $\vec{E}$. The position $\phi = 0$, with $\vec{p}$ parallel to $\vec{E}$, is a position of stable equilibrium, and the position $\phi = \pi$, with $\vec{p}$ and $\vec{E}$ antiparallel, is a position of unstable equilibrium. The polarization of a grass seed in the apparatus of Fig. 22–23 gives it an electric dipole moment; the torque exerted by $\vec{E}$ then causes the seed to align with $\vec{E}$ and hence with the field lines.

When a dipole changes direction in an electric field, the electric-field torque does *work* on it, with a corresponding change in potential energy. The work dW done by a torque τ during an infinitesimal displacement $d\phi$ is given by Eq. (10–22): $dW = \tau \, d\phi$. Because the torque is in the direction of *decreasing* ϕ, we must write the torque as $\tau = -pE \sin \phi$, and

$$dW = \tau \, d\phi = -pE \sin \phi \, d\phi.$$

In a finite displacement from ϕ_1 to ϕ_2 the total work done on the dipole is

$$W = \int_{\phi_1}^{\phi_2} (-pE \sin \phi) \, d\phi$$

$$= pE \cos \phi_2 - pE \cos \phi_1.$$

The work is the negative of the change of potential energy, just as in Chapter 7: $W = U_1 - U_2$. So we see that a suitable definition of potential energy U for this system is

$$U(\phi) = -pE \cos \phi. \quad (22\text{–}17)$$

In this expression we recognize the *scalar product* $\vec{p} \cdot \vec{E} = pE \cos \phi$, so we can also write

$$U = -\vec{p} \cdot \vec{E} \quad \text{(potential energy for a dipole in an electric field).} \quad (22\text{–}18)$$

The potential energy has its minimum value $U = -pE$ (i.e., its most negative value) at the stable equilibrium position, where $\phi = 0$ and $\vec{p}$ is parallel to $\vec{E}$. The potential energy is maximum when $\phi = \pi$ and $\vec{p}$ is antiparallel to $\vec{E}$; then $U = +pE$. At $\phi = \pi/2$, where $\vec{p}$ is perpendicular to $\vec{E}$, U is zero. We could of course define U differently so that it is zero at some other orientation of $\vec{p}$, but our definition is simplest.

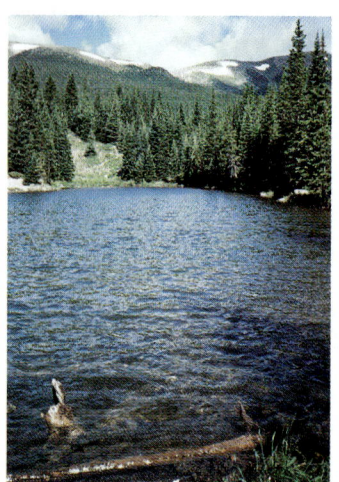

Any body of water is really a sea of electric dipoles (see Figure 22–24).

EXAMPLE 22–14

Force and torque on an electric dipole Figure 22–26a shows an electric dipole in a uniform electric field with magnitude 5.0×10^5 N/C directed parallel to the plane of the figure. The charges are $\pm 1.6 \times 10^{-19}$ C; both lie in the plane and are separated by 0.125 nm = 0.125×10^{-9} m. (Both the charge magnitude and the distance are typical of molecular quantities.) Find a) the net force exerted by the field on the dipole; b) the magnitude and direction of the electric dipole moment; c) the magnitude and direction of the torque; d) the potential energy of the system in the position shown.

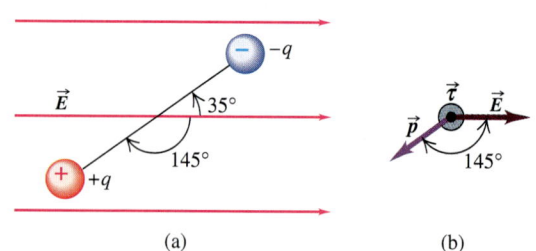

22–26 (a) An electric dipole. (b) Directions of the electric dipole moment and torque.

SOLUTION a) Since the field is uniform, the forces on the two charges are equal and opposite, and the total force is zero.

b) The magnitude p of the electric dipole moment $\vec{p}$ is

$$p = qd = (1.6 \times 10^{-19} \text{ C})(0.125 \times 10^{-9} \text{ m})$$

$$= 2.0 \times 10^{-29} \text{ C} \cdot \text{m}.$$

The direction of $\vec{p}$ is from the negative to the positive charge, 145° clockwise from the electric-field direction (Fig. 22–26b).
c) The magnitude of the torque is

$$\tau = pE \sin \phi = (2.0 \times 10^{-29} \text{ C})(5.0 \times 10^{5} \text{ N/C})(\sin 145°)$$

$$= 5.7 \times 10^{-24} \text{ N} \cdot \text{m}.$$

From the right-hand rule for vector products (Section 1–11) the direction of the torque $\vec{\tau} = \vec{p} \times \vec{E}$ is out of the page. This corresponds to a counterclockwise torque that tends to align $\vec{p}$ with $\vec{E}$.
d) The potential energy is

$$U = -pE \cos \phi$$

$$= -(2.0 \times 10^{-29} \text{ C} \cdot \text{m})(5.0 \times 10^{5} \text{ N/C})(\cos 145°)$$

$$= 8.2 \times 10^{-24} \text{ J}.$$

In this discussion we have assumed that $\vec{E}$ is uniform, so there is no net force on the dipole. If $\vec{E}$ is not uniform, the forces at the ends may not cancel completely, and the net force may not be zero. Thus a body with zero net charge but an electric dipole moment can experience a net force in a nonuniform electric field. As we mentioned in Section 22–3, an uncharged body can be polarized by an electric field, giving rise to a separation of charge and an electric dipole moment. This is how uncharged bodies can experience electrostatic forces (see Fig. 22–5).

FIELD OF AN ELECTRIC DIPOLE

Now let's think of an electric dipole as a *source* of electric field. What does the field look like? The general shape of things is shown by the field map of Fig. 22–22b. At each point in the pattern the total $\vec{E}$ field is the vector sum of the fields from the two individual charges, as in Example 22–9. We suggest you try drawing diagrams showing this vector sum for several points.

To get quantitative information about the field of an electric dipole, we have to do some calculating, as illustrated in Example 22–15 below. Notice the use of the principle of superposition of electric fields to add up the contributions to the field of the individual charges. Also notice that we need to use approximation techniques even for the relatively simple case of a field due to two charges. Field calculations often become very complicated, and computer analysis is typically used to determine the field due to an arbitrary charge distribution.

EXAMPLE 22-15

Field of an electric dipole, revisited In Fig. 22–27 (page 694) an electric dipole is centered at the origin, with $\vec{p}$ in the direction of the +y-axis. Derive an approximate expression for the electric field at a point on the y-axis for which y is much larger than d. Use the binomial expansion of $(1 + x)^{n}$, that is, $(1 + x)^{n} \cong 1 + nx + n(n - 1)x^{2}/2 + \cdots$, for the case $|x| < 1$. (This problem illustrates a useful calculational technique.)

SOLUTION The total y-component E_{y} of electric field from the two charges is

$$E_{y} = \frac{q}{4\pi\epsilon_{0}} \left[\frac{1}{(y - d/2)^{2}} - \frac{1}{(y + d/2)^{2}} \right]$$

$$= \frac{q}{4\pi\epsilon_{0}y^{2}} \left[\left(1 - \frac{d}{2y}\right)^{-2} - \left(1 + \frac{d}{2y}\right)^{-2} \right].$$

We used this exact approach in Example 22–9 (Section 22–7). Now comes the approximation. When y is much greater than d, that is, when we are far away from the dipole compared to its

size, the quantity $d/2y$ is much smaller than 1. With $n = -2$ and $d/2y$ playing the role of x in the binomial expansion, we keep only the first two terms. The terms we discard are much smaller than those we keep, and we have

$$\left(1 - \frac{d}{2y}\right)^{-2} \cong 1 + \frac{d}{y} \quad \text{and} \quad \left(1 + \frac{d}{2y}\right)^{-2} \cong 1 - \frac{d}{y}.$$

Hence E_{y} is given approximately by

$$E \cong \frac{q}{4\pi\epsilon_{0}y^{2}} \left[1 + \frac{d}{y} - \left(1 - \frac{d}{y}\right) \right]$$

$$= \frac{qd}{2\pi\epsilon_{0}y^{3}}$$

$$= \frac{p}{2\pi\epsilon_{0}y^{3}}.$$

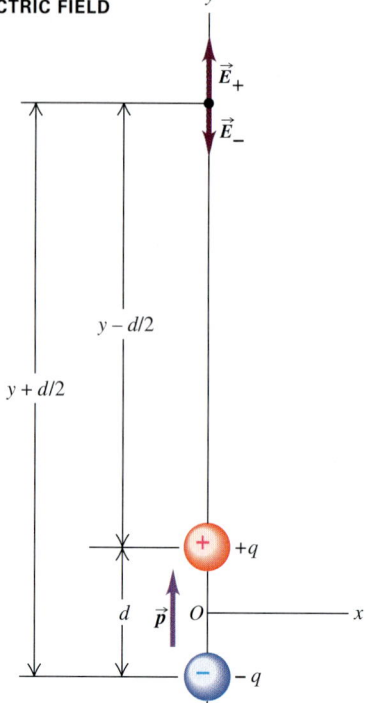

An alternative route to this expression is to put the fractions in the E_y expression over a common denominator and combine, then approximate the denominator $(y - d/2)^2(y + d/2)^2$ as y^4. We leave the details as a problem.

For points P off the coordinate axes, the expressions are more complicated, but at *all* points far away from the dipole (in any direction) the field drops off as $1/r^3$. We can compare this with the $1/r^2$ behavior of a point charge, the $1/r$ behavior of a long line charge, and the independence of r for a large sheet of charge. There are charge distributions for which the field drops off even more quickly. An *electric quadrupole* consists of two equal dipoles with opposite orientation, separated by a small distance. The field of a quadrupole at large distances drops off as $1/r^4$.

22–27 Finding the electric field of an electric dipole at a point on its axis.

SUMMARY

KEY TERMS

electric charge, 669

electrostatics, 670

electron, 671

proton, 671

neutron, 671

nucleus, 671

atomic number, 671

positive ion, 671

negative ion, 671

ionization, 671

principle of conservation of
  charge, 672

conductor, 673

insulator, 673

induction, 673

induced charge, 673

point charge, 674

Coulomb's law, 675

coulomb, 675

principle of superposition of
  forces, 676

electric field, 679

test charge, 679

vector field, 680

source point, 681

field point, 681

- The fundamental entity in electrostatics is electric charge. There are two kinds of charge: positive and negative. Charges of the same sign repel each other; charges of opposite sign attract. Charge is conserved; the total charge in an isolated system is constant.

- All ordinary matter is made of protons, neutrons, and electrons. The positive protons and electrically neutral neutrons in the nucleus of an atom are bound together by the nuclear force; the negative electrons surround it at distances much greater than the nuclear size. Electric interactions are chiefly responsible for the structure of atoms, molecules, and solids.

- Conductors are materials that permit electric charge to move easily within them. Insulators permit charge to move much less readily. Most metals are good conductors; most nonmetals are insulators.

- Coulomb's law is the basic law of interaction for point electric charges. For charges q_1 and q_2 separated by a distance r, the magnitude of the force on either charge is

$$F = \frac{1}{4\pi\epsilon_0} \frac{|q_1 q_2|}{r^2}. \qquad (22\text{--}2)$$

The force on each charge is along the line joining the two charges: repulsive if q_1 and q_2 have the same sign, attractive if they have opposite signs. The forces form an action-reaction pair and obey Newton's third law. In SI units the unit of electric charge is the coulomb, abbreviated C, and

$$\frac{1}{4\pi\epsilon_0} = 8.988 \times 10^9 \text{ N} \cdot \text{m}^2/\text{C}^2.$$

- The principle of superposition of forces states that when two or more charges each exert a force on another charge, the total force on that charge is the vector sum of the forces exerted by the individual charges.

■ Electric field, a vector quantity, is the force per unit charge exerted on a test charge at any point, provided that the test charge is small enough that it does not disturb the charges that cause the field. From Coulomb's law the electric field produced by a point charge is

$$\vec{E} = \frac{1}{4\pi\epsilon_0} \frac{q}{r^2} \hat{r}. \qquad (22\text{--}7)$$

■ The principle of superposition of electric fields states that the electric field of any combination of charges is the vector sum of the fields caused by the individual charges. To calculate the electric field caused by a continuous distribution of charge, divide the distribution into small elements, calculate the field caused by each element, and then carry out the vector sum or each component sum, usually by integrating. Charge distributions are described by linear charge density λ, surface charge density σ, and volume charge density ρ.

■ Field lines provide a graphical representation of electric fields. At any point on a field line, the tangent to the line is in the direction of $\vec{E}$ at that point. Where field lines are close together, E is larger; where field lines are far apart, E is smaller.

■ An electric dipole is a pair of electric charges of equal magnitude q but opposite sign, separated by a distance d. The electric dipole moment $\vec{p}$ is defined to have magnitude $p = qd$. The direction of $\vec{p}$ is from negative toward positive charge. An electric dipole in an electric field experiences a torque of magnitude

$$\tau = pE \sin\phi, \qquad (22\text{--}15)$$

where ϕ is the angle between the directions of $\vec{p}$ and $\vec{E}$. Vector torque $\vec{\tau}$ is

$$\vec{\tau} = \vec{p} \times \vec{E}. \qquad (22\text{--}16)$$

■ The potential energy for an electric dipole in an electric field $\vec{E}$ depends on the orientation of the dipole moment $\vec{p}$ with respect to the field:

$$U = -\vec{p} \cdot \vec{E}. \qquad (22\text{--}18)$$

principle of superposition of electric fields, 684
linear charge density, 684
surface charge density, 684
volume charge density, 684
electric field line, 689
electric dipole, 690
electric dipole moment, 691

DISCUSSION QUESTIONS

Q22–1 Two metal spheres are hanging from nylon threads. When you bring the spheres close to each other, they tend to attract. Based on this information alone, discuss all the possible ways that the spheres could be charged. Is it possible that after the spheres touch, they will cling together? Explain.

Q22–2 The electric force between two charged particles becomes weaker with increasing distance. Suppose that instead the electric force were *independent* of distance. In this case, would a charged comb still cause a neutral insulator to become polarized as in Fig. 22–5? Why or why not? Would the neutral insulator still be attracted to the comb? Again, why or why not?

Q22–3 Your clothing tends to cling together after going through the dryer. Why? Would you expect more or less clinging if all your clothing is made of the same material (say, cotton) than if you dry different kinds of clothing together? Again, why? (You may want to experiment with your next load of laundry.)

Q22–4 An uncharged metal sphere hangs from a nylon thread. When a positively charged glass rod is brought close to the metal sphere, the sphere is drawn toward the rod. But if the sphere touches the rod, it suddenly flies away from the rod. Explain why the sphere is first attracted, then repelled.

Q22–5 The free electrons in a metal are gravitationally attracted toward the earth. Why, then, don't they all settle to the bottom of the conductor, like sediment settling to the bottom of a river?

Q22–6 Some of the free electrons in a good conductor (such as a piece of copper) move at speeds of 10^6 m/s or more. Why don't these electrons fly out of the conductor completely?

Q22–7 Good electrical conductors, such as metals, are typically good conductors of heat; electrical insulators, such as wood, are typically poor conductors of heat. Explain why there should be a relationship between electrical conduction and heat conduction in these materials.

Q22–8 Defend the following statement: "If there were only one electrically charged particle in the entire universe, the concept of electric charge would be meaningless."

Q22–9 Two identical metal objects are mounted on insulating stands. Describe how you could place charges of opposite sign but exactly equal magnitude on the two objects.

Q22–10 Plastic food wrap can be used to cover a container by stretching the material across the top and pressing the overhanging material against the sides. What makes it stick? (*Hint:* The answer involves the electric force.) Does the food wrap stick to itself with equal tenacity? Why or why not? Does it work with metallic containers? Again, why or why not?

Q22–11 If you walk across a nylon rug and then touch a large metal object such as a doorknob, you may get a spark and a shock. Why does this tend to happen more on dry days than on humid days? (*Hint:* See Fig. 22–24.) Why are you less likely to get a shock if you touch a *small* metal object, such as a paper clip?

Q22–12 The electric force between an electron and a proton, between two electrons, or between two protons is much stronger than the gravitational force between any of these pairs of particles. Yet even though the sun and planets contain electrons and protons, it is the gravitational force that holds the planets in their orbits around the sun. Explain this seeming contradiction.

Q22–13 What similarities do electrical forces have to gravitational forces? What are the most significant differences?

Q22–14 Atomic nuclei are made of protons and neutrons. This shows that there must be another kind of interaction in addition to gravitational and electric forces. Explain.

Q22–15 Sufficiently strong electric fields can cause atoms to become positively ionized, that is, to lose one or more electrons. Explain how this can happen. What determines how strong the field must be to make this happen?

Q22–16 When you pull transparent plastic tape off a roll and try to position it precisely on a piece of paper, the tape often jumps over and sticks where it is not wanted. Why?

Q22–17 A particle with positive charge Q is held fixed at the origin. A second particle with positive charge q is fired at the first particle, and follows a trajectory as shown in Fig. 22–28. Is the angular momentum of the second particle constant? Why or

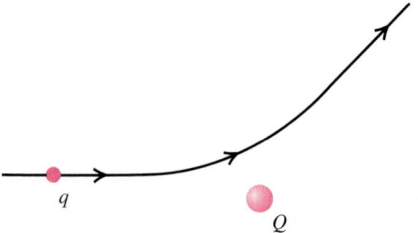

FIGURE 22–28 Question Q22–17.

why not? (*Hint:* How much torque does the first particle exert on the second particle?)

Q22–18 The air temperature and the velocity of the air have different values at different places in the earth's atmosphere. Is the air velocity a vector field? Why or why not? Is the air temperature a vector field? Again, why or why not?

Q22–19 Suppose the charge shown in Fig. 22–22a is fixed in position. A small, positively charged particle is then placed at some point in the figure and released. Will the trajectory of the particle follow an electric field line? Why or why not? Suppose instead that the particle is placed at some point in Fig. 22–22b and released (the positive and negative charges shown in the figure are fixed in position). Will its trajectory follow an electric field line? Again, why or why not? Explain any differences between your answers for the two different situations.

Q22–20 The water molecule (H_2O) has a large dipole moment, while the benzene molecule (C_6H_6) has zero dipole moment. Use these facts to explain why ordinary table salt (NaCl, or sodium chloride) dissolves very easily in water, but dissolves poorly in benzene.

EXERCISES

SECTION **22–5 COULOMB'S LAW**

22–1 Excess electrons are placed on a small lead sphere of mass 8.00 g so that its net charge is -3.20×10^{-9} C. a) Find the number of excess electrons on the sphere. b) How many excess electrons are there per lead atom? The atomic number of lead is 82, and its atomic mass is 207 g/mol.

22–2 Lightning occurs when there is a flow of electric charge (principally electrons) between the ground and a thundercloud. The maximum rate of charge flow in a lightning bolt is about 20,000 C/s; this lasts for 100 μs or less. How much charge flows between the ground and the cloud in this time? How many electrons flow during this time?

22–3 Estimate how many electrons there are in your body. Make any assumptions you feel are necessary, but clearly state what they are. (*Hint:* Most of the atoms in your body have equal numbers of electrons, protons, and neutrons.) What is the combined charge of all these electrons?

22–4 Particles in a Gold Ring. You have a pure (24-karat) gold ring with a mass of 17.7 g. Gold has an atomic mass of 197 g/mol and an atomic number of 79. a) How many protons are in the ring, and what is their total positive charge? b) If the ring carries no net charge, how many electrons are in it?

22–5 What is the total charge, in coulombs, of all the electrons in 1.80 mol of hydrogen atoms?

22–6 Two small spheres spaced 20.0 cm apart have equal charge. How many excess electrons must be present on each sphere if the magnitude of the force of repulsion between them is 4.57×10^{-21} N?

22–7 Two small plastic spheres are given positive electrical charges. When they are 15.0 cm apart, the repulsive force between them has magnitude 0.220 N. What is the charge on each sphere a) if the two charges are equal? b) if one sphere has four times the charge of the other?

22–8 Two small aluminum spheres, each having a mass of 0.0250 kg, are separated by 80.0 cm. a) How many electrons does each sphere contain? (The atomic mass of aluminum is 26.982 g/mol, and its atomic number is 13.) b) How many electrons would have to be removed from one sphere and added to the other to cause an attractive force between the spheres of magnitude 1.00×10^4 N (roughly one ton)? Assume the spheres may be treated as point charges. c) What fraction of all the electrons in each sphere does this represent?

22–9 How far does the electron of a hydrogen atom have to be removed from the nucleus for the force of attraction to equal the weight of the electron at the surface of the earth?

22–10 A negative charge $-0.550\ \mu C$ exerts an upward 0.200-N force on an unknown charge 0.300 m directly below it. a) What is the unknown charge (magnitude and sign)? b) What are the magnitude and direction of the force that the unknown charge exerts on the $-0.550\ \mu C$ charge?

22–11 A point charge of $+3.50\ \mu C$ is placed 0.800 m to the left of a second, identical point charge. What are the magnitudes and directions of the forces that each charge exerts on the other?

22–12 In Example 22–4 (Section 22–5), suppose the point charge on the y-axis at $y = -0.30$ m has negative charge $-2.0\ \mu C$, the other charges remaining the same. Find the magnitude and direction of the net force on Q. How does your answer differ from that in Example 22–4? Explain the differences.

22–13 Two point charges are located on the y-axis as follows: charge $q_1 = -1.50$ nC at $y = -0.600$ m, and charge $q_2 = +3.20$ nC at the origin ($y = 0$). What is the total force (magnitude and direction) exerted by these two charges on a third charge $q_3 = +5.00$ nC located at $y = -0.400$ m?

22–14 Two point charges are placed on the x-axis as follows: Charge $q_1 = +4.00$ nC is located at $x = 0.200$ m, and charge $q_2 = +5.00$ nC is at $x = -0.300$ m. What are the magnitude and direction of the total force exerted by these two charges on a negative point charge $q_3 = -6.00$ nC that is placed at the origin?

22–15 A positive point charge q is placed on the +y-axis at $y = a$, and a negative point charge $-q$ is placed on the $-y$-axis at $y = -a$. A negative point charge $-Q$ is located at some point on the +x-axis. a) In a free-body diagram, show the forces that act on the charge $-Q$. b) Find the x- and y-components of the net force that the two charges q and $-q$ exert on $-Q$. (Your answer should only involve k, q, Q, a, and the coordinate x of the third charge.) c) What is the net force on the charge $-Q$ when it is at the origin ($x = 0$)? d) Graph the y-component of the net force on the charge $-Q$ as a function of x for values of x between $-4a$ and $+4a$.

22–16 Two positive point charges q are placed on the y-axis at $y = a$ and $y = -a$. A negative point charge $-Q$ is located at some point on the +x-axis. a) In a free-body diagram, show the forces that act on the charge $-Q$. b) Find the x- and y-components of the net force that the two positive charges exert on $-Q$. (Your answer should only involve k, q, Q, a, and the coordinate x of the third charge.) c) What is the net force on the charge $-Q$ when it is at the origin ($x = 0$)? d) Graph the x-component of the net force on the charge $-Q$ as a function of x for values of x between $-4a$ and $+4a$.

22–17 Four identical charges q are placed at the corners of a square of side L. a) In a free-body diagram, show all of the forces that act on one of the charges. b) Find the magnitude and direction of the total force exerted on one charge by the other three charges.

SECTION 22–6 ELECTRIC FIELD AND ELECTRIC FORCES

22–18 A particle has charge -3.00 nC. a) Find the magnitude and direction of the electric field due to this particle at a point 0.250 m directly above it. b) At what distance from this particle does its electric field have a magnitude of 12.0 N/C?

22–19 a) What must be the charge (sign and magnitude) of a 1.45-g particle for it to remain stationary when placed in a

downward-directed electric field of magnitude 650 N/C? b) What is the magnitude of an electric field in which the electric force on a proton is equal in magnitude to its weight?

22–20 a) What is the electric field of an iron nucleus at a distance of 6.00×10^{-10} m from the nucleus? The atomic number of iron is 26. Assume that the nucleus may be treated as a point charge. b) What is the electric field of a proton at a distance of 5.29×10^{-11} m from the proton? (This is the radius of the electron orbit in the Bohr model for the ground state of the hydrogen atom.)

22–21 A small object carrying a charge of $-55.0\ \mu C$ experiences a downward force of 6.20×10^{-9} N when placed at a certain point in an electric field. a) What are the magnitude and direction of the electric field at this point? b) What would be the magnitude and direction of the force acting on a copper nucleus (atomic number = 29, atomic mass = 63.5 g/mol) placed at this same point in the electric field?

22–22 Electric Field of the Earth. The earth has a net electric charge that causes a field at points near its surface equal to 150 N/C and directed in toward the center of the earth. a) What magnitude and sign of charge would a 60-kg human have to acquire to overcome his or her weight by the force exerted by the earth's electric field? b) What would be the force of repulsion between two people each with the charge calculated in part (a) and separated by a distance of 100 m? Is use of the earth's electric field a feasible means of flight? Why or why not?

22–23 An electron is projected with an initial speed $v_0 = 1.60 \times 10^6$ m/s into the uniform field between the parallel plates in Fig. 22–29. Assume that the field between the plates is uniform and directed vertically downward, and that the field outside the plates is zero. The electron enters the field at a point midway between the plates. a) If the electron just misses the upper plate as it emerges from the field, find the magnitude of the electric field. b) Suppose that in Fig. 22–29 the electron is replaced by a proton with the same initial speed v_0. Would the proton hit one of the plates? If the proton does not hit one of the plates, what is the magnitude and direction of its vertical displacement as it exits the region between the plates? c) Compare the paths traveled by the electron and the proton and explain the differences.

FIGURE 22–29 Exercise 22–23.

22–24 a) Calculate the magnitude and direction (relative to the +x-axis) of the electric field in Example 22–6 (Section 22–6). b) A -2.5-nC point charge is placed at the point P in Fig. 22–13. Find the magnitude and direction of i) the force that the -8.0-nC charge at the origin exerts on this charge and ii) the force that this charge exerts on the -8.0-nC charge at the origin.

22–25 a) For the electron in Examples 22–7 and 22–8 (Section 22–6), compare the weight of the electron to the magnitude of the electric force on the electron. Is it appropriate to neglect the gravitational force on the electron in these examples? Explain. b) A particle with charge $+e$ is placed at rest between the charged plates in Fig. 22–14. What must be the mass of this object if it is to remain at rest? Give your answer in kilograms and in multiples of the electron mass. c) Does the answer to part (b) depend on where between the plates the object is placed? Why or why not?

22–26 A uniform electric field exists in the region between two oppositely charged plane parallel plates. A proton is released from rest at the surface of the positively charged plate and strikes the surface of the opposite plate, 1.60 cm distant from the first, in a time interval of 1.50×10^{-6} s. a) Find the magnitude of the electric field. b) Find the speed of the proton when it strikes the negatively charged plate.

22–27 A point charge is at the origin. With this point charge as the source point, what is the unit vector $\hat{r}$ in the direction of a) the field point at $x = 0$, $y = -1.35$ m; b) the field point at $x = 12.0$ cm, $y = 12.0$ cm; c) the field point at $x = -1.10$ m, $y = 2.60$ m? Express your results in terms of the unit vectors $\hat{i}$ and $\hat{j}$.

22–28 According to the safety standards of the Institute of Electrical and Electronics Engineers (IEEE), humans should avoid prolonged exposure to electric fields of magnitudes greater than 614 N/C. a) At a point where $E = 614$ N/C, what is the magnitude of the electric force on a single electron? b) Atomic and molecular dimensions are on the order of 10^{-10} m. What is the magnitude of the electric force on an electron that is 1.0×10^{-10} m from a proton? c) How do the answers in parts (a) and (b) compare? What do you think would happen to a person placed in an electric field that produced a force equal to that calculated in part (b)?

SECTION 22–7 ELECTRIC-FIELD CALCULATIONS

22–29 A +2.00-nC point charge is at the origin, and a second −5.00-nC point charge is on the x-axis at $x = 0.800$ m. a) Find the electric field (magnitude and direction) at each of the following points on the x-axis: i) $x = 0.200$ m; ii) $x = 1.20$ m; iii) $x = -0.200$ m. b) Find the net electric force that the two charges would exert on an electron placed at each point in part (a).

22–30 Two particles having charges $q_1 = 0.500$ nC and $q_2 = 8.00$ nC are separated by a distance of 1.20 m. At what point along the line connecting the two charges is the total electric field due to the two charges equal to zero?

22–31 Two positive point charges q are placed on the x-axis, one at $x = a$ and one at $x = -a$. a) Find the magnitude and direction of the electric field at $x = 0$. b) Derive an expression for the electric field at points on the x-axis. Use your result to graph the x-component the electric field as a function of x, for values of x between $-4a$ and $+4a$.

22–32 A positive point charge q is placed at $x = a$, and a negative point charge $-q$ is placed at $x = -a$. a) Find the magnitude and direction of the electric field at $x = 0$. b) Derive an expression for the electric field at points on the x-axis. Use your result to graph the x-component of the electric field as a function of x, for values of x between $-4a$ and $+4a$.

22–33 In a rectangular coordinate system a positive point charge $q = 6.00 \times 10^{-9}$ C is placed at the point $x = +0.150$ m, $y = 0$, and an identical point charge is placed at $x = -0.150$ m,

$y = 0$. Find the x- and y-components, the magnitude, and the direction of the electric field at the following points: a) the origin; b) $x = 0.300$ m, $y = 0$; c) $x = 0.150$ m, $y = -0.400$ m; d) $x = 0$, $y = 0.200$ m.

22–34 A point charge $q_1 = -4.00$ nC is at the point $x = 0.600$ m, $y = 0.800$ m, and a second point charge $q_2 = +6.00$ nC is at the point $x = 0.600$ m, $y = 0$. Calculate the magnitude and direction of the net electric field at the origin due to these two point charges.

22–35 Repeat Exercise 22–33 for the case where the point charge at $x = +0.150$ m, $y = 0$ is positive and the other is negative.

22–36 A very long, straight wire has charge per unit length 1.50×10^{-10} C/m. At what distance from the wire is the electric field magnitude equal to 2.50 N/C?

22–37 Positive electric charge is distributed along the y-axis with charge per unit length λ. a) Consider the case where charge is distributed only between the points $y = a$ and $y = -a$. For points on the +x-axis, graph the x-component of the electric field as a function of x for values of x between $x = a/2$ and $x = 4a$. b) Consider instead the case where charge is distributed along the entire y-axis with the same charge per unit length λ. Using the same graph as in part (a), plot the x-component of the electric field as a function of x for values of x between $x = a/2$ and $x = 4a$. Label which graph refers to which situation.

22–38 A ring-shaped conductor with radius $a = 2.50$ cm has a total positive charge $Q = +0.125$ nC uniformly distributed around it, as shown in Fig. 22–17. The center of the ring is at the origin of coordinates O. a) What is the electric field (magnitude and direction) at point P, which is on the x-axis at $x = 40.0$ cm? b) A point charge $q = -2.50$ μC is placed at the point P described in part (a). What are the magnitude and direction of the force exerted *by* the charge q *on* the ring?

22–39 A uniformly charged disk of radius R carries positive charge per unit area σ, as in Fig. 22–19. For points on the +x-axis, graph the x-component of the electric field as a function of x for values of x between $x = 0$ and $x = 4R$.

22–40 Near the earth's surface, the electric field in the open air has magnitude 150 N/C and is directed down toward the ground. If this is regarded as being due to a large sheet of charge lying on the earth's surface, calculate the charge per unit area in the sheet. What is the sign of the charge?

22–41 Each square centimeter of the surface of an infinite plane sheet of paper has 2.50×10^6 excess electrons. Find the magnitude and direction of the electric field at a point 5.00 cm from the surface of the sheet.

22–42 Two horizontal infinite plane sheets of charge are separated by a distance d. The lower sheet has negative charge with uniform surface charge density $-\sigma < 0$. The upper sheet has positive charge with uniform surface charge density $\sigma > 0$. What is the electric field (magnitude, and direction if the field is nonzero) a) above the upper sheet? b) below the lower sheet? c) between the sheets?

SECTION 22–8 ELECTRIC FIELD LINES

22–43 Two large parallel sheets are separated by a distance d. One plate has positive surface charge density $\sigma > 0$, and the other has negative surface charge density $-\sigma < 0$. Sketch the electric field lines at points near the center of the sheets and therefore well away from the edges.

22–44 Sketch the electric field lines for a disk of radius R with a positive uniform surface charge density σ. Use what you know about the electric field very close to the disk and very far from the disk to make your sketch.

22–45 a) Sketch the electric field lines for an infinite line of charge. You may find it helpful to show the field lines in a plane containing the line of charge in one sketch and the field lines in a plane perpendicular to the line of charge in a second sketch. b) Explain how your sketches show i) that the magnitude E of the electric field depends only on the distance r from the line of charge and ii) that E decreases like $1/r$.

22–46 Figure 22–30 shows some of the electric field lines due to three point charges arranged along the vertical axis. All three charges have the same magnitude. a) What are the signs of each of the three charges? Explain your reasoning. b) At what point(s) is the magnitude of the electric field the smallest? Explain your reasoning. Explain how the fields produced by each individual point charge combine to give a small net field at this point or points.

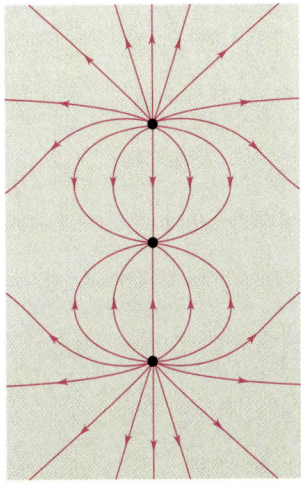

FIGURE 22–30 Exercise 22–46.

SECTION 22–9 ELECTRIC DIPOLES

22–47 Point charges $q_1 = -4.5$ nC and $q_2 = +4.5$ nC are separated by 3.1 mm, forming an electric dipole. a) Find the electric dipole moment. b) The charges are in a uniform electric field whose direction makes an angle of 36.9° with the line connecting the charges. What is the magnitude of this field if the torque exerted on the dipole has magnitude 7.2×10^{-9} N · m?

22–48 The potassium chloride molecule (KCl) has a dipole moment of 8.9×10^{-30} C · m. a) Assuming that this dipole moment arises from two charges $\pm 1.6 \times 10^{-19}$ C separated by distance d, calculate d. b) What is the maximum magnitude of the torque that a uniform electric field with magnitude 6.0×10^5 N/C can exert on a KCl molecule? Sketch the relative orientations of the electric dipole moment $\vec{p}$ and the electric field $\vec{E}$ when the torque is a maximum.

22–49 The ammonia molecule (NH_3) has a dipole moment of 5.0×10^{-30} C · m. Ammonia molecules in the gas phase are placed in a uniform electric field $\vec{E}$ with magnitude 1.6×10^6 N/C. a) What is the change in electric potential energy when the dipole moment of a molecule changes its orientation with respect to $\vec{E}$ from parallel to perpendicular? b) At what absolute temperature T is the average translational kinetic energy $\frac{3}{2}kT$ of a molecule equal to the change in potential energy calculated in part (a)? (Above this temperature, thermal agitation prevents the dipoles from aligning with the electric field.)

22–50 The dipole moment of the water molecule (H_2O) is 6.17×10^{-30} C · m. Consider a water molecule located at the origin whose dipole moment $\vec{p}$ points in the $+x$-direction. A chlorine ion (Cl^-), of charge -1.60×10^{-19} C, is located at $x = 3.00 \times 10^{-9}$ m. Find the magnitude and direction of the electric force that the water molecule exerts on the chlorine ion. Is this force attractive or repulsive? Assume that x is much larger than the separation d between the charges in the dipole, so that the approximate expression for the electric field along the dipole axis derived in Example 22–15 (Section 22–9) can be used.

22–51 In Example 22–15 (Section 22–9), the approximate result $E \cong p/2\pi\epsilon_0 y^3$ was derived for the electric field of a dipole at points on the dipole axis. a) Rederive this result by putting the fractions in the expression for E_y over a common denominator, as described in Example 22–15. b) Explain why the approximate result also gives the correct approximate expression for E_y for $y < 0$.

PROBLEMS

22–52 A charge $q_1 = +5.00$ nC is placed at the origin of an xy-coordinate system, and a charge $q_2 = -2.00$ nC is placed on the positive x-axis at $x = 4.00$ cm. a) If a third charge $q_3 = +6.00$ nC is now placed at the point $x = 4.00$ cm, $y = 3.00$ cm, find the x- and y-components of the total force exerted on this charge by the other two. b) Find the magnitude and direction of this force.

22–53 Two positive point charges Q are held fixed on the x-axis at $x = a$ and $x = -a$. A third positive point charge q, of mass m, is placed on the x-axis away from the origin at a coordinate x such that $|x| \ll a$. The charge q, which is free to move

along the x axis, is then released. a) Find the frequency of oscillation of the charge q. (*Hint:* Review the definition of simple harmonic motion in Section 13–3. Use the binomial expansion $(1 + z)^n = 1 + nz + n(n - 1)z^2/2 + \cdots$, valid for the case $|z| < 1$.) b) Suppose instead the charge q were placed on the y-axis at a coordinate y such that $|y| \ll a$, then released. If this charge is free to move anywhere in the xy-plane, what will happen to it? Explain your answer.

22–54 Two identical spheres of mass m are hung from silk threads of length L, as shown in Fig. 22–31. Each sphere has the same charge, so $q_1 = q_2 = q$. The radius of each sphere is very small compared to the distance between the spheres, so they may be treated as point charges. Show that if the angle θ is small, the equilibrium separation d between the spheres is $d = (q^2L/2\pi\epsilon_0 mg)^{1/3}$. (*Hint:* If θ is small, then $\tan\theta \cong \sin\theta$.)

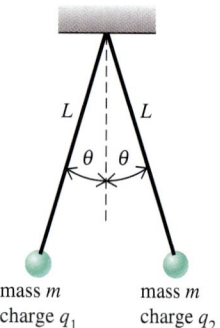

L L

θ θ

mass m mass m
charge q_1 charge q_2

FIGURE 22–31 Problems 22–54, 22–55, and 22–56.

22–55 Two small spheres of mass $m = 15.0$ g are hung by silk threads of length $L = 1.20$ m from a common point (Fig. 22–31). When the spheres are given equal quantities of negative charge, so that $q_1 = q_2 = q$, each thread hangs at $\theta = 25.0°$ from the vertical. a) Draw a diagram showing the forces on each sphere. Treat the spheres as point charges. b) Find the magnitude of q. c) Both threads are now shortened to length $L = 0.600$ m, while the charges q_1 and q_2 remain unchanged. What new angle will each thread make with the vertical? (*Hint:* This part of the problem can be solved numerically by using trial values for θ and adjusting the values of θ until a self-consistent answer is obtained.)

22–56 Two identical spheres are each attached to silk threads of length $L = 0.500$ m and hung from a common point (Fig. 22–31). Each sphere has mass $m = 8.00$ g. The radius of each sphere is very small compared to the distance between the spheres, so they may be treated as point charges. One sphere is given charge q_1, and the other a different charge q_2; this causes the spheres to separate so that when the spheres are in equilibrium, each thread makes an angle $\theta = 20.0°$ with the vertical. a) Draw a free-body diagram for each sphere when in equilibrium, and label all the forces that act on each sphere. b) Determine the magnitude of the electrostatic force that acts on each sphere, and determine the tension in each thread. c) Based on the information you have been given, what can you say about the magnitudes and the signs of q_1 and q_2? Explain your answers. d) A small wire is now connected between the spheres, allowing charge to be transferred from one sphere to the other until the two spheres have equal charges; the wire is then removed. Each thread now makes an angle of $30.0°$ with the vertical. Determine the original charges q_1 and q_2. (*Hint:* The total charge on the pair of spheres is conserved.)

22–57 Sodium chloride (NaCl, ordinary table salt) is made up of positive sodium ions (Na$^+$) and negative chlorine ions (Cl$^-$). a) If a point charge with the same charge and mass as all the Na$^+$ ions in 0.100 mol of NaCl is 2.00 cm from a point charge with the same charge and mass as all the Cl$^-$ ions, what is the magnitude of the attractive force between these two point charges? b) If the positive point charge in part (a) is held in place and the negative point charge is released from rest, what is its initial acceleration? (See Appendix D for atomic masses.) c) Does it seem reasonable that the ions in NaCl could be separated in this way? Why or why not? (In fact, when sodium chloride dissolves in water it breaks up into Na$^+$ and Cl$^-$ ions. However, in this situation there are additional electric forces exerted by the water molecules on the ions.)

22–58 Three point charges are arranged along the x-axis. Charge $q_1 = -4.50$ nC is located at $x = 0.200$ m, and charge $q_2 = +2.50$ nC is at $x = -0.300$ m. A positive point charge q_3 is located at the origin. a) What must be the value of q_3 for the net force on this point charge to have magnitude 4.00×10^{-6} N? b) What is the direction of the net force on q_3? c) Where along the x-axis can q_3 be placed and the net force on it be zero, other than the trivial answers of $x = \pm\infty$?

22–59 Three identical point charges q are placed at each of three corners of a square of side L. Find the magnitude and direction of the net force on a point charge $-3q$ placed a) at the center of the square; b) at the vacant corner of the square. In each case, draw a free-body diagram showing the forces exerted on the $-3q$ charge by each of the other three charges.

22–60 Three point charges are placed on the y-axis: a charge q at $y = a$, a charge $-2q$ at the origin, and a charge q at $y = -a$. Such an arrangement is called an *electric quadrupole*. a) Find the magnitude and direction of the electric field at points on the y-axis for $y > a$. b) Use the binomial expansion to show that very far from the quadrupole, so that $y \gg a$, the electric field is proportional to y^{-4}. Contrast this behavior to that of the electric field of a point charge and that of the electric field of a dipole.

22–61 a) For the arrangement of charges described in Problem 22–60, find the magnitude and direction of the electric field at points on the positive x-axis. b) Use the binomial expansion to find an approximate expression for the electric field valid for $x \gg a$. Contrast this behavior to that of the electric field of a point charge and that of the electric field of a dipole.

22–62 a) Suppose all the electrons in 20.0 g of carbon atoms could be located at the North Pole of the earth and all the protons at the South Pole. What would be the total force of attraction exerted on each group of charges by the other? The atomic number of carbon is 6, and the atomic mass of carbon is 12.0 g/mol. b) What would be the magnitude and direction of the force exerted by the charges in part (a) on a third charge that is equal to the charge at the South Pole, and located at a point on the surface of the earth at the equator? Draw a diagram showing the locations of the charges and the forces on the charge at the equator.

22–63 If Atoms Were Not Neutral... Because the charges on the electron and proton have the same absolute value, atoms are electrically neutral. Suppose this were not precisely true, and the absolute value of the charge of the electron was less than the charge of the proton by 0.00100%. a) Estimate what the net charge of this textbook would be under these circumstances. Make any assumptions you feel are justified, but state clearly what they are. (*Hint:* Most of the atoms in this textbook have equal numbers of electrons, protons, and neutrons.) b) What would be the magnitude of the electric force between two textbooks placed 5.0 m apart? Would this force be attractive or repulsive? Estimate what the acceleration of each book would be if they were 5.0 m apart and there are no non-electrical forces on them. c) Discuss how the fact that ordinary matter is stable shows that the absolute values of the charges on the electron and proton must be identical to a *very* high level of accuracy.

22–64 Operation of an Inkjet Printer. In an inkjet printer, letters are built up by squirting drops of ink at the paper from a rapidly moving nozzle. The pattern on the paper is controlled by an electrostatic valve that determines at each nozzle position whether ink is squirted onto the paper or not. The ink drops, 15 μm in radius, leave the nozzle and travel toward the paper at 20 m/s. The drops pass through a charging unit that gives each drop a positive charge q when the drop loses some electrons. The drops then pass between parallel deflecting plates 2.0 cm in length where there is a uniform vertical electric field with magnitude 8.0×10^4 N/C. If a drop is to be deflected 0.30 mm by the time it reaches the end of the deflection plate, what magnitude of charge must be given to the drop? (Assume that the density of the ink drop is the same as that of water, 1000 kg/m^3.)

22–65 A proton is projected into a uniform electric field that points vertically upward and has magnitude E. The initial velocity of the proton has a magnitude v_0 and is directed at an angle α below the horizontal. a) Find the maximum distance h_{max} that the proton descends vertically below its initial elevation. Ignore gravitational forces. b) After what horizontal distance d does the proton return to its original elevation? c) Sketch the trajectory of the proton. d) Find the numerical values of h_{max} and d if $E = 500$ N/C, $v_0 = 4.00 \times 10^5$ m/s, and $\alpha = 30.0°$.

22–66 A negative point charge $q_1 = -4.00$ nC is on the x-axis at $x = 0.60$ m. A second point charge q_2 is on the x-axis at $x = -1.20$ m. What must be the sign and magnitude of q_2 for the net electric field at the origin to be a) 50.0 N/C in the $+x$-direction? b) 50.0 N/C in the $-x$-direction?

22–67 A charge of 12.0 nC is at the origin; a second, unknown charge is at $x = 3.00$ m, $y = 0$; and a third charge of -16.0 nC is at $x = 5.00$ m, $y = 0$. What are the sign and magnitude of the unknown charge if the net field at $x = 8.00$ m, $y = 0$ has magnitude 12.0 N/C and is in the $+x$-direction?

22–68 Positive charge Q is distributed uniformly along the x-axis from $x = 0$ to $x = a$. A positive point charge q is located on the positive x-axis at $x = a + r$, a distance r to the right of the end of Q (Fig. 22–32). a) Calculate the x- and y-components of the electric field produced by the charge distribution Q at points on the positive x-axis where $x > a$. b) Calculate the force (magnitude and direction) that the charge distribution Q exerts on q. c) Show that if $r \gg a$, the magnitude of the force in part (b) is approximately $Qq/4\pi\epsilon_0 r^2$. Explain why this result is obtained.

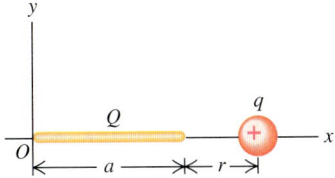

FIGURE 22–32 Problem 22–68.

22–69 Positive charge Q is distributed uniformly along the positive y-axis between $y = 0$ and $y = a$. A negative point charge $-q$ lies on the positive x-axis, a distance x from the origin (Fig. 22–33). a) Calculate the x- and y-components of the electric field produced by the charge distribution Q at points on the positive x-axis. b) Calculate the x- and y-components of the force that the charge distribution Q exerts on q. c) Show that if $x \gg a$, $F_x \cong -Qq/4\pi\epsilon_0 x^2$ and $F_y \cong +Qqa/8\pi\epsilon_0 x^3$. Explain why this result is obtained.

FIGURE 22–33 Problem 22–69.

22–70 A charged line like that shown in Fig. 22–18 extends from $y = 2.50$ cm to $y = -2.50$ cm. The total charge distributed uniformly along the line is -9.00 nC. a) Find the electric field (magnitude and direction) on the x-axis at $x = 0.25$ cm. b) Is the magnitude of the electric field you calculated in part (a) larger or smaller than the electric field 0.25 cm from an infinite line of charge that has the same charge per unit length as this finite line of charge? In terms of the approximation used to derive $E = \lambda/2\pi\epsilon_0 r$ for an infinite line from Eq. (22–9), explain why this is so. c) At what distance x does the result for the infinite line of charge differ by 1.0% from that for the finite line?

22–71 A charged line like that shown in Fig. 22–18 extends from $y = 2.50$ cm to $y = -2.50$ cm. The total charge distributed uniformly along the line is -9.00 nC. a) Find the electric field (magnitude and direction) on the x-axis at $x = 10.0$ cm. b) Is the magnitude of the electric field you calculated in part (a) larger or smaller than the electric field 10.0 cm from a point charge that has the same total charge as this finite line of charge? In terms of the approximation used to derive $E = Q/4\pi\epsilon_0 x^2$ for a point charge from Eq. (22–9), explain why this is so. c) At what distance x does the result for the finite line of charge differ by 1.0% from that for the point charge?

22–72 A uniformly charged disk like that in Fig. 22–19 has radius 2.50 cm and carries a total charge of 4.0×10^{-12} C. a) Find the electric field (magnitude and direction) on the x-axis at $x = 0.20$ cm. b) Is the magnitude of the electric field you calculated in part (a) larger or smaller than the electric field 0.20 cm from an infinite sheet of charge with the same charge per unit area as the disk? In terms of the approximation used to derive Eq. (22–12) from Eq. (22–11), explain why this is so. c) What is the percent difference between the electric field produced by the finite disk and by an infinite sheet with the same charge per unit area at i) $x = 0.20$ cm? ii) $x = 0.40$ cm?

22–73 A uniformly charged disk like that in Fig. 22–19 has radius 2.50 cm and carries a total charge of 4.0×10^{-12} C. a) Find the electric field (magnitude and direction) on the x-axis at $x = 20.0$ cm. b) Show that for $x \gg R$, Eq. (22–11) becomes $E = Q/4\pi\epsilon_0 x^2$, where Q is the total charge on the disk. c) Is the magnitude of the electric field you calculated in part (a) larger or smaller than the electric field 20.0 cm from a point charge that has the same total charge as this disk? In terms of the approximation used in part (b) to derive $E = Q/4\pi\epsilon_0 x^2$ for a point charge from Eq. (22–11), explain why this is so. d) What is the percent difference between the electric field produced by the finite disk and by a point charge with the same charge at $x = 20.0$ cm and at $x = 10.0$ cm?

22–74 a) Let $f(x)$ be an even function of x so that $f(x) = f(-x)$. Show that $\int_{-a}^{a} f(x)\,dx = 2\int_{0}^{a} f(x)\,dx$. (*Hint:* Write the integral from $-a$ to a as the sum of the integral from $-a$ to 0 and the integral from 0 to a. In the first integral, make the change of variable $x' = -x$.) b) Let $g(x)$ be an odd function of x so that $g(x) = -g(-x)$. Use the method given in the hint for part (a) to show that $\int_{-a}^{a} g(x)\,dx = 0$. c) Use the result of part (b) to show why E_y in Example 22–11 (Section 22–7) is zero.

22–75 Positive charge $+Q$ is distributed uniformly along the $+x$-axis from $x = 0$ to $x = a$. Negative charge $-Q$ is distributed uniformly along the $-x$-axis from $x = 0$ to $x = -a$. a) A positive point charge q lies on the positive y-axis, a distance y from the origin. a) Find the force (magnitude and direction) that the positive and negative charge distributions together exert on q. Show that this force is proportional to y^{-3} for $y \gg a$. b) Suppose instead that the positive point charge q lies on the positive x-axis, a distance $x > a$ from the origin. Find the force (magnitude and direction) that the charge distribution exerts on q. Show that this force is proportional to x^{-3} for $x \gg a$.

22–76 Positive charge Q is uniformly distributed around a semicircle of radius a (Fig. 22–34). Find the electric field (magnitude and direction) at the center of curvature P.

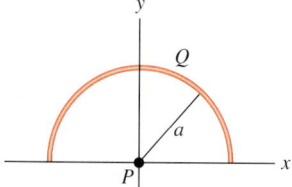

FIGURE 22–34 Problem 22–76.

22–77 Negative charge $-Q$ is distributed uniformly around a quarter-circle of radius a that lies in the first quadrant, with the center of curvature at the origin. Find the x- and y-components of the net electric field at the origin.

22–78 A small sphere of mass m carries a positive charge q and is attached to one end of a silk fiber of length L. The other end of the fiber is attached to a large vertical insulating sheet that has a positive surface charge density σ. Show that when the sphere is in equilibrium, the fiber makes an angle equal to arctan $(q\sigma/2mg\epsilon_0)$ with the vertical sheet.

22–79 The imaging drum of a photocopier is positively charged to attract negatively charged toner particles. Near the surface of the drum, its electric field has magnitude 1.40×10^5 N/C. A toner particle is to be attracted to the drum with a force that is 10 times the

weight of the particle. a) What must be the ratio of the mass of a toner particle to the magnitude of its net charge? b) If the toner particles are made of carbon (atomic number 6, atomic mass 12.0 g/mol), how many carbon atoms are there for each excess electron on a toner particle?

22–80 Electric charge is distributed uniformly along each side of a square. Two adjacent sides have positive charge with total charge $+Q$ on each. a) If the other two sides have negative charge with total charge $-Q$ on each (Fig. 22–35), what are the x- and y-components of the net electric field at the center of the square? Each side of the square has length a. b) Repeat the calculation of part (a) if all four sides have positive charge $+Q$.

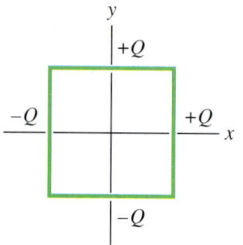

FIGURE 22–35 Problem 22–80.

22–81 Three large parallel insulating sheets have surface charge densities $+0.0200$ C/m^2, $+0.0100$ C/m^2, and -0.0200 C/m^2, respectively (Fig. 22–36). Adjacent sheets are a distance of 0.300 m from each other. Calculate the net electric field (magnitude and direction) due to all three sheets at a) point P (0.150 m to the left of sheet I); b) point R (midway between sheets I and II); c) point S (midway between sheets II and III); d) point T (0.150 m to the right of sheet III).

FIGURE 22–36 Problems 22–81 and 22–82.

22–82 For the situation described in Problem 22–81 (Fig. 22–36), find the force per unit area (magnitude and direction) exerted on each of sheets I, II, and III by the other two sheets.

22–83 An infinite sheet with positive charge per unit area σ lies in the xy-plane. A second infinite sheet with negative charge per unit area $-\sigma$ lies in the yz-plane. Find the net electric field at all points that do not lie in either of these planes. Express your answer in terms of the unit vectors $\hat{\imath}, \hat{\jmath}$, and $\hat{k}$.

22–84 A thin disk with a circular hole at its center, called an *annulus,* has inner radius R_1 and outer radius R_2 (Fig. 22–37).

The disk has a uniform positive surface charge density σ on its surface. a) Determine the total electric charge on the annulus. b) The annulus lies in the yz-plane, with its center at the origin. For an arbitrary point on the x-axis (the axis of the annulus), find the magnitude and direction of the electric field $\vec{E}$. Consider points both above and below the annulus in Fig. 22–37. c) Show that at points on the x-axis that are sufficiently close to the origin, the magnitude of the electric field is approximately proportional to the distance between the center of the annulus and the point. How close is "sufficiently close"? d) A point particle with mass m and negative charge $-q$ is free to move along the x-axis (but cannot move off the axis). The particle is originally placed at rest at $x = 0.01R_1$ and released. Find the frequency of oscillation of the particle. (*Hint:* Review Section 13–3. The annulus is held stationary.)

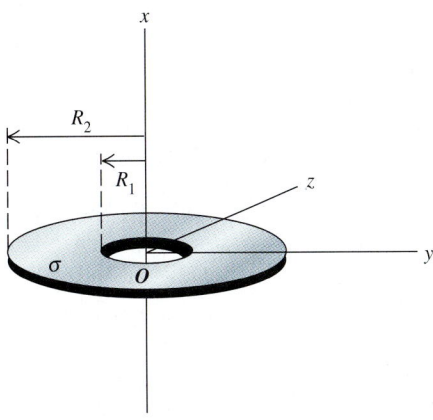

FIGURE 22–37 Problem 22–84.

CHALLENGE PROBLEMS

22–85 Three charges are placed as shown in Fig. 22–38. The magnitude of q_1 is 2.00 μC, but its sign and the value of the charge q_2 are not known. Charge q_3 is +4.00 μC, and the net force $\vec{F}$ on q_3 is entirely in the negative x-direction a) Considering the different possible signs of q_1 and q_2, there are four possible force diagrams representing the forces $\vec{F}_1$ and $\vec{F}_2$ that q_1 and q_2 exert on q_3. Sketch these four possible force configurations. b) Using the sketches from part (a) and the direction of $\vec{F}$, deduce the signs of the charges q_1 and q_2. c) Calculate the magnitude of q_2. d) Determine F, the magnitude of the net force on q_3.

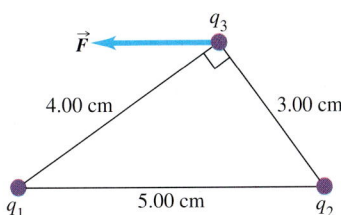

FIGURE 22–38 Challenge Problem 22–85.

22–86 Two charges are placed as shown in Fig. 22–39. The magnitude of q_1 is 3.00 μC, but its sign and the value of the charge q_2 are not known. The direction of the net electric field $\vec{E}$ at point P is entirely in the negative y-direction a) Considering the different possible signs of q_1 and q_2, there are four possible diagrams that could represent the electric fields $\vec{E}_1$ and $\vec{E}_2$ produced by q_1 and q_2.

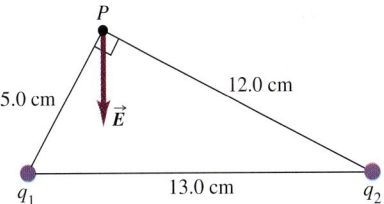

FIGURE 22–39 Challenge Problem 22–86.

Sketch the four possible electric field configurations. b) Using the sketches from part (a) and the direction of $\vec{E}$, deduce the signs of q_1 and q_2. c) Determine the magnitude of $\vec{E}$.

22–87 Two thin rods of length L lie along the x-axis, one between $x = a/2$ and $x = a/2 + L$ and the other between $x = -a/2$ and $x = -a/2 - L$. Each rod has positive charge Q distributed uniformly along its length. a) Calculate the electric field produced by the second rod at points along the positive x-axis. b) Show that the magnitude of the force that one rod exerts on the other is

$$F = \frac{Q^2}{4\pi\epsilon_0 L^2} \ln\left[\frac{(a + L)^2}{a(a + 2L)}\right].$$

c) Show that if $a \gg L$, the magnitude of this force reduces to $F = Q^2/4\pi\epsilon_0 a^2$. (*Hint:* Use the expansion $\ln (1 + z) = z - z^2/2 + z^3/3 - \cdots$, valid for $|z| \ll 1$. Carry *all* expansions to at least order L^2/a^2.) Interpret this result.

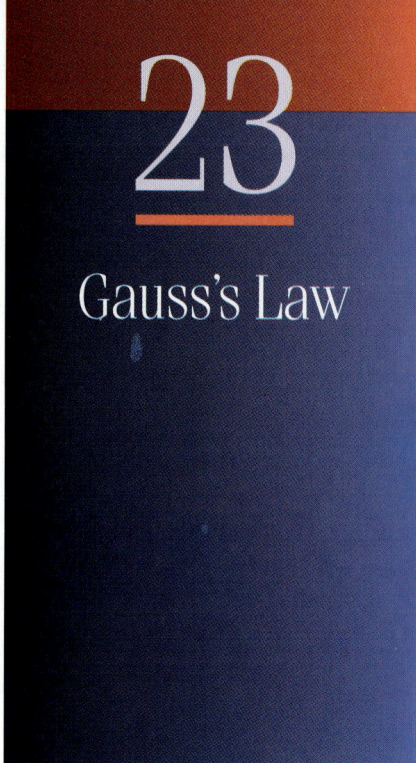

23

Gauss's Law

Gauss's law is another way to describe how electric charges and fields behave. One consequence of this law is that static charges on a conductor are found on the conductor's surface, not its interior. That's why this child acquires an electric charge by touching the charged metal sphere. The charged hairs on the child's head repel and stand out.

23-1 INTRODUCTION

Often, there's both an easy way and a hard way to do a job; the easy way may involve nothing more than using the right tools. In physics, an important tool for simplifying problems is the use of *symmetry properties* of systems. Many physical systems have symmetry; for example, a cylindrical body doesn't look any different after you've rotated it around its axis, and a charged metal sphere looks just the same after you've turned it about any axis through its center.

Gauss's law is part of the key to using symmetry considerations to simplify electric-field calculations. For example, the field of a straight-line or plane-sheet charge distribution, which we derived in Section 22–7 using some fairly strenuous integrations, can be obtained in a few lines with the help of Gauss's law. In addition to making certain calculations easier, Gauss's law will also give us insight into how electric charge distributes itself over conducting bodies.

Here's what Gauss's law is all about. Given any general distribution of charge, we surround it with an imaginary surface that encloses the charge. Then we look at the electric field at various points on this imaginary surface. Gauss's law is a relation between the field at *all* the points on the surface and the total charge enclosed within the surface. This may sound like a rather indirect way of expressing things, but it turns out to be a tremendously useful relationship. Above and beyond its use as a calculational tool, Gauss's law will help us gain deeper insights

into electric fields. We will make use of these insights repeatedly in the next several chapters as we pursue our study of electromagnetism.

23-2 CHARGE AND ELECTRIC FLUX

In Chapter 22 we asked the question "Given a charge distribution, what is the electric field produced by that distribution at a point P?" We saw that the answer could be found by representing the distribution as an assembly of point charges, each of which produces an electric field $\vec{E}$ given by Eq. (22–7). The total field at P is then the vector sum of the fields due to all the point charges.

But there is an alternative relationship between charge distributions and electric fields. To discover this relationship, let's stand the question of Chapter 22 on its head and ask, "If the electric field pattern is known in a given region, what can we determine about the charge distribution in that region?"

Here's an example. Consider the box shown in Fig. 23–1a, which may or may not contain electric charge. We'll imagine that the box is made of a material that has no effect on any electric fields; it's of the same breed as the massless rope, the frictionless incline, and the free college education. Better still, let the box represent an *imaginary* surface that may or may not enclose some charge. We'll refer to the box as a **closed surface** because it completely

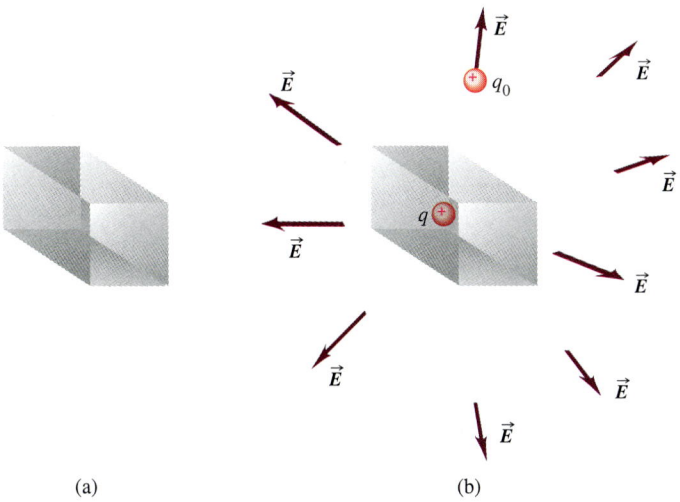

(a)   (b)

23-1 (a) A box containing an unknown amount of charge. (b) The charge inside the box can be probed by using a test charge q_0 to measure the electric field outside the box.

encloses a volume. How can you determine how much (if any) electric charge lies within the box?

Knowing that a charge distribution produces an electric field and that an electric field exerts a force on a test charge, you move a test charge q_0 around the vicinity of the box. By measuring the force $\vec{F}$ experienced by the test charge at different positions, you make a three-dimensional map of the electric field $\vec{E} = \vec{F}/q_0$ outside the box. In the case shown in Fig. 23–1b, the map turns out to be the same as that of the electric field produced by a positive point charge (Fig. 22–22a). From the details of the map, you can find the exact value of the point charge inside the box.

To determine the contents of the box, we actually only need to measure $\vec{E}$ on the *surface* of the box. In Fig. 23–2a there is a single positive point charge inside the box, and in Fig. 23–2b there are two such charges. The field patterns on the surfaces of the boxes are different in detail, but in both cases the electric field points out of the box. Figures 23–2c and 23–2d show cases with one and two negative point charges, respectively, inside the box. Again, the details of $\vec{E}$ on the surface of the box are different, but in both cases the field points into the box.

In Section 22–6 we mentioned the analogy between electric field vectors and the velocity vectors of a fluid in motion. This analogy can be helpful, even though an electric field does not actually "flow." Using this analogy, in Figs. 23–2a and 23–2b, in which the electric field vectors point out of the surface, we say that there is an outward **electric flux.** (The word *flux* comes from a Latin word meaning "flow.") In Figs. 23–2c and 23–2d the $\vec{E}$ vectors point into the surface, and the electric flux is *inward.*

23-2 The electric field on the surface of boxes containing (a) a single positive point charge, (b) two positive point charges, (c) a single negative point charge, or (d) two negative point charges.

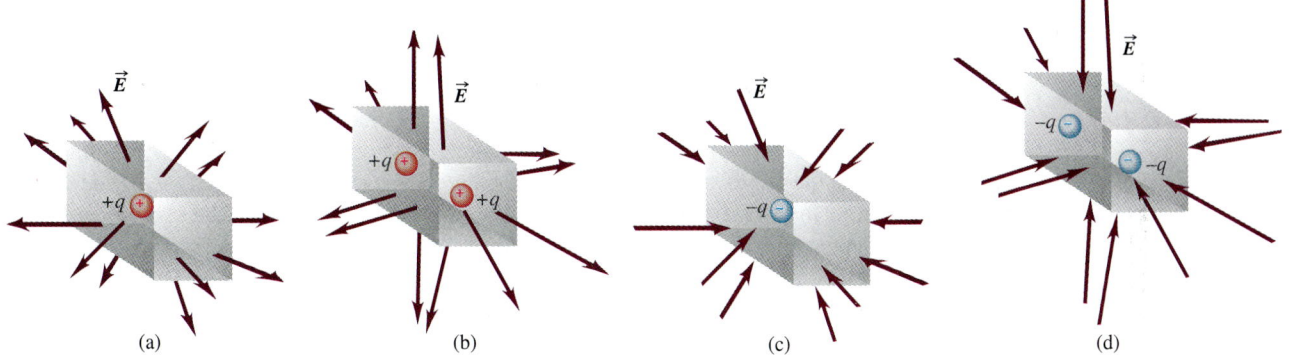

(a)   (b)   (c)   (d)

23–3 Three cases in which there is zero *net* charge inside a box and no net electric flux through the surface of the box. (a) An empty box with $\vec{E} = 0$. (b) A box containing one positive and one equal-magnitude negative point charge. (c) An empty box immersed in a uniform electric field.

(a)   (b)   (c)

11.7
Electric Flux

Figure 23–2 suggests a simple relationship: Positive charge inside the box goes with an outward electric flux through the box's surface, and negative charge inside goes with an inward electric flux. What happens if there is *zero* charge inside the box? In Fig. 23–3a the box is empty and $\vec{E} = 0$ everywhere, so there is no electric flux into or out of the box. In Fig. 23–3b, one positive and one negative point charge of equal magnitude are enclosed within the box, so the *net* charge inside the box is zero. There is an electric field, but it "flows into" the box on half of its surface and "flows out of" the box on the other half. Hence there is no *net* electric flux into or out of the box.

The box is again empty in Fig. 23–3c. However, there is charge present *outside* the box; the box has been placed with one end parallel to a uniformly charged infinite sheet, which produces a uniform electric field perpendicular to the sheet (as we learned in Example 22–12 of Section 22–7). On one end of the box, $\vec{E}$ points into the box; on the opposite end, $\vec{E}$ points out of the box; and on the sides, $\vec{E}$ is parallel to the surface and so points neither into nor out of the box. As in Fig. 23–3b, the inward electric flux on one part of the box exactly compensates for the outward electric flux on the other part. So in all of the cases shown in Fig. 23–3, there is no *net* electric flux through the surface of the box, and no *net* charge is enclosed in the box.

Figures 23–2 and 23–3 demonstrate a connection between the *sign* (positive, negative, or zero) of the *net* charge enclosed by a closed surface and the sense (outward, inward, or none) of the net electric flux through the surface. There is also a connection between the *magnitude* of the net charge inside the closed surface and the *strength* of the net "flow" of $\vec{E}$ over the surface. In both Figs. 23–4a and 23–4b there is a single point charge inside the box, but in Fig. 23–4b the magnitude of the charge is twice as great, and so $\vec{E}$ is everywhere twice as great in magnitude as in Fig. 23–4a. Keeping in mind the fluid flow analogy, this means that the net outward electric flux is also twice as great in Fig. 23–4b as in Fig. 23–4a. This suggests that the net electric flux through the surface of the box is *directly proportional* to the magnitude of the net charge enclosed by the box.

This conclusion is independent of the size of the box. In Fig. 23–4c the point charge $+q$ is enclosed by a box with twice the linear dimensions of the box in Fig. 23–4a. The magnitude of the electric field of a point charge decreases with distance according to $1/r^2$, so the average magnitude of $\vec{E}$ on each face of the large box in Fig. 23–4c is just $\frac{1}{4}$ of the average magnitude on the corresponding face in Fig. 23–4a. But each face of the large box has exactly four times the area of the corresponding face of the small box. Hence the outward electric flux will be the *same* for the two boxes if we *define* electric flux as follows: For each face of the box, take the product of the average perpendicular component of $\vec{E}$ and the area of that face; then add up the results from all faces of the box. With this definition the net electric flux due to a single point charge inside the box

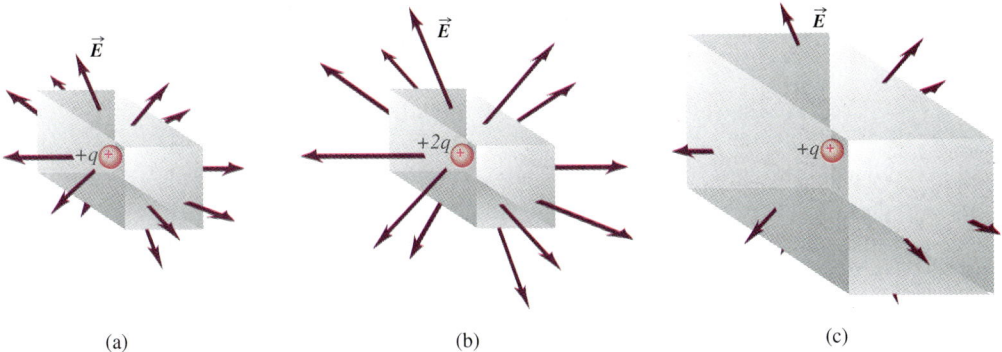

23–4 (a) A box enclosing a positive point charge +q. (b) An identical box enclosing a positive point charge +2q. Doubling the charge causes the magnitude of $\vec{E}$ to double and doubles the electric flux through the surface. (c) The point charge +q enclosed by a box of twice the dimensions of the box in (a). The magnitude of $\vec{E}$ on the surface is reduced by a factor of $\frac{1}{4}$, but the area through which $\vec{E}$ "flows" is increased by a factor of four.

is independent of the size of the box and depends only on the net charge inside the box.

We have seen that there is a relationship between the net amount of charge inside a closed surface and the electric flux through that surface. For the special cases of a closed surface in the shape of a rectangular box and charge distributions made up of point charges or infinite charged sheets, we have found the following:

1. Whether there is a net outward or inward electric flux through a closed surface depends on the sign of the enclosed charge.
2. Charges *outside* the surface do not give a net electric flux through the surface.
3. The net electric flux is directly proportional to the net amount of charge enclosed within the surface but is otherwise independent of the size of the closed surface.

These observations are a qualitative statement of *Gauss's law*.

Do these observations hold true for other kinds of charge distributions and for closed surfaces of arbitrary shape? The answer to these questions will prove to be "yes." But to explain why this is so, we need a precise mathematical statement of what we mean by electric flux. This is developed in the next section.

(a)

23–3 CALCULATING ELECTRIC FLUX

In the preceding section we introduced the concept of *electric flux*. Qualitatively, the electric flux through a surface is a description of whether the electric field $\vec{E}$ points into or out of the surface. We used this to give a rough qualitative statement of Gauss's law: The net electric flux through a closed surface is directly proportional to the net charge inside that surface. To be able to make full use of this law, we need to know how to *calculate* electric flux. To do this, let's again make use of the analogy between an electric field $\vec{E}$ and the field of velocity vectors $\vec{v}$ in a flowing fluid. (Again, keep in mind that this is only an analogy; an electric field is *not* a flow.)

Figure 23–5 shows a fluid flowing steadily from left to right. Let's examine the volume flow rate dV/dt (in, say, cubic meters per second) through the wire rectangle with area A. When the area is perpendicular to the flow velocity $\vec{v}$ (Fig. 23–5a) and the flow

(b)

23–5 The volume flow rate of fluid through the wire rectangle (a) is vA when the area of the rectangle is perpendicular to $\vec{v}$ and (b) is $vA \cos \phi$ when the rectangle is tilted at an angle ϕ.

velocity is the same at all points in the fluid, the volume flow rate dV/dt is the area A multiplied by the flow speed v:

$$\frac{dV}{dt} = vA.$$

When the rectangle is tilted at an angle ϕ (Fig. 23–5b) so that its face is not perpendicular to $\vec{v}$, the area that counts is the silhouette area that we see when we look in the direction of $\vec{v}$. This area, which is outlined in red and labeled $A_\perp$ in Fig. 23–5b, is the *projection* of the area A onto a surface perpendicular to $\vec{v}$. Two sides of the projected rectangle have the same length as the original one, but the other two are foreshortened by a factor of $\cos\phi$, so the projected area $A_\perp$ is equal to $A\cos\phi$. Then the volume flow rate through A is

$$\frac{dV}{dt} = vA\cos\phi.$$

If $\phi = 90°$, $dV/dt = 0$; the wire rectangle is edge-on to the flow, and no fluid passes through the rectangle.

Also, $v\cos\phi$ is the component of the vector $\vec{v}$ perpendicular to the plane of the area A. Calling this component $v_\perp$, we can rewrite the volume flow rate as

$$\frac{dV}{dt} = v_\perp A.$$

We can express the volume flow rate more compactly by using the concept of *vector area* $\vec{A}$, a vector quantity with magnitude A and a direction perpendicular to the plane of the area we are describing. The vector area $\vec{A}$ describes both the size of an area and its orientation in space. In terms of $\vec{A}$, we can write the volume flow rate of fluid through the rectangle in Fig. 23–5b as

$$\frac{dV}{dt} = \vec{v}\cdot\vec{A}.$$

Using the analogy between electric field and fluid flow, we now define electric flux in the same way as we have just defined the volume flow rate of a fluid; we simply replace the fluid velocity $\vec{v}$ by the electric field $\vec{E}$. The symbol that we use for electric flux is Φ_E (the capital Greek letter "phi"; the subscript E is a reminder that this is *electric* flux). Consider first a flat area A perpendicular to a uniform electric field $\vec{E}$ (Fig. 23–6a). We define the electric flux through this area to be the product of the field magnitude E and the area A:

$$\Phi_E = EA.$$

Roughly speaking, we can picture Φ_E in terms of the field lines passing through A. Increasing the area means that more lines of $\vec{E}$ pass through the area, increasing the flux; stronger field means more closely spaced lines of $\vec{E}$ and therefore more lines per unit area, so again the flux increases.

If the area A is flat but not perpendicular to the field $\vec{E}$, then fewer field lines pass through it. In this case the area that counts is the silhouette area that we see when looking in the direction of $\vec{E}$. This is the area $A_\perp$ in Fig. 23–6b and is equal to $A\cos\phi$ (compare to Fig. 23–5b). We generalize our definition of electric flux for a uniform electric field to

$$\Phi_E = EA\cos\phi \qquad \text{(electric flux for uniform } \vec{E}, \text{flat surface).} \qquad (23\text{–}1)$$

(a)

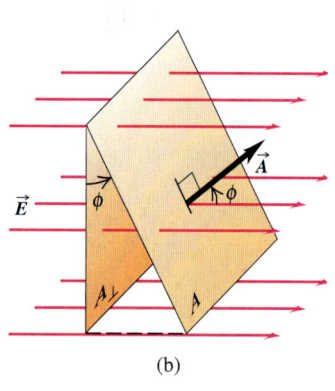

(b)

23–6 A flat surface in a uniform electric field. (a) The electric flux through the surface equals EA. (b) When the area vector makes an angle ϕ with $\vec{E}$, $A_\perp = A\cos\phi$. The flux is zero when $\phi = 90°$.

Since $E \cos \phi$ is the component of $\vec{E}$ perpendicular to the area, we can rewrite Eq. (23–1) as

$$\Phi_E = E_\perp A \qquad \text{(electric flux for uniform } \vec{E}, \text{ flat surface).} \qquad (23\text{–}2)$$

In terms of the vector area $\vec{A}$ perpendicular to the area, we can write the electric flux as the scalar product of $\vec{E}$ and $\vec{A}$:

$$\Phi_E = \vec{E} \cdot \vec{A} \qquad \text{(electric flux for uniform } \vec{E}, \text{ flat surface).} \qquad (23\text{–}3)$$

Equations (23–1), (23–2), and (23–3) express the electric flux for a *flat* surface and a *uniform* electric field in different but equivalent ways. The SI unit for electric flux is $1 \text{ N} \cdot \text{m}^2/\text{C}$.

We can represent the direction of a vector area $\vec{A}$ by using a *unit vector* $\hat{n}$ perpendicular to the area; $\hat{n}$ stands for "normal." Then

$$\vec{A} = A\hat{n}. \qquad (23\text{–}4)$$

A surface has two sides, so there are two possible directions for $\hat{n}$ and $\vec{A}$. We must always specify which direction we choose. In Section 23–2 we related the charge inside a *closed* surface to the electric flux through the surface. With a closed surface we will always choose the direction of $\hat{n}$ to be *outward,* and we will speak of the flux *out of* a closed surface. Thus what we called "outward electric flux" in Section 23–2 corresponds to a *positive* value of Φ_E, and what we called "inward electric flux" corresponds to a *negative* value of Φ_E.

What happens if the electric field $\vec{E}$ isn't uniform but varies from point to point over the area A? Or what if A is part of a curved surface? Then we divide A into many small elements dA, each of which has a unit vector $\hat{n}$ perpendicular to it and a vector area $d\vec{A} = \hat{n} \, dA$. We calculate the electric flux through each element and integrate the results to obtain the total flux:

$$\Phi_E = \int E \cos \phi \, dA = \int E_\perp dA = \int \vec{E} \cdot d\vec{A} \qquad \begin{array}{l}\text{(general definition} \\ \text{of electric flux).}\end{array} \qquad (23\text{–}5)$$

We call this integral the **surface integral** of the component $E_\perp$ over the area, or the surface integral of $\vec{E} \cdot d\vec{A}$. The various forms of the integral all express the same thing in different terms. In specific problems, one form is sometimes more convenient than another. Example 23–3 at the end of this section illustrates the use of Eq. (23–5).

In Eq. (23–5) the electric flux $\int E_\perp \, dA$ is just equal to the *average* value of the perpendicular component of the electric field, multiplied by the area of the surface. This is the same definition of electric flux that we were led to in Section 23–2, now expressed more mathematically. In the next section we will see the connection between the total electric flux through *any* closed surface, no matter what its shape, and the amount of charge enclosed within that surface.

Sunbathers at the beach orient themselves face-on toward the sun to maximize the amount of sunlight falling on their bodies. In the same way, a surface oriented face-on to an electric field (as in Fig. 23–6a) receives the maximum electric flux.

EXAMPLE 23–1

Electric flux through a disk A disk with radius 0.10 m is oriented with its normal unit vector $\hat{n}$ at an angle of 30° to a uniform electric field $\vec{E}$ with magnitude 2.0×10^3 N/C (Fig. 23–7). (Since this isn't a closed surface, it has no "inside" or "outside." That's why we have to specify the direction of $\hat{n}$ in the figure.) a) What is the electric flux through the disk? b) What is the flux through the disk if it is turned so that its normal is perpendicular to $\vec{E}$? c) What is the flux through the disk if its normal is parallel to $\vec{E}$?

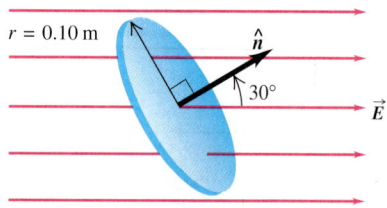

23-7 The electric flux Φ_E through a disk depends on the angle between its normal $\hat{n}$ and the electric field $\vec{E}$.

SOLUTION a) The area is $A = \pi(0.10 \text{ m})^2 = 0.0314 \text{ m}^2$. From Eq. (23–1),

$$\Phi_E = EA \cos \phi = (2.0 \times 10^3 \text{ N/C})(0.0314 \text{ m}^2)(\cos 30°)$$

$$= 54 \text{ N} \cdot \text{m}^2/\text{C}.$$

b) The normal to the disk is now perpendicular to $\vec{E}$, so $\phi = 90°$, $\cos \phi = 0$, and $\Phi_E = 0$. In this case there is no flux through the disk. c) The normal to the disk is parallel to $\vec{E}$, so $\phi = 0$, $\cos \phi = 1$, and the flux has its maximum possible value. From Eq. (23–1),

$$\Phi_E = EA \cos \phi = (2.0 \times 10^3 \text{ N/C})(0.0314 \text{ m}^2)(1)$$

$$= 63 \text{ N} \cdot \text{m}^2/\text{C}.$$

EXAMPLE 23-2

Electric flux through a cube A cube of side L is placed in a region of uniform electric field $\vec{E}$. Find the electric flux through each face of the cube and the total flux through the cube when a) it is oriented with two of its faces perpendicular to the field $\vec{E}$, as in Fig. 23–8a; b) when the cube is turned by an angle θ, as in Fig. 23–8b.

SOLUTION a) Since $\vec{E}$ is uniform, we can find the flux through each face of the cube using Eqs. (23–3) and (23–4). The unit vectors for each face ($\hat{n}_1$ through $\hat{n}_6$) are shown in the figure; the direction of each unit vector is *outward* from the closed surface of the cube. The angle between $\vec{E}$ and $\hat{n}_1$ is $180°$; the angle between $\vec{E}$ and $\hat{n}_2$ is $0°$; and the angle between $\vec{E}$ and each of the other four unit vectors is $90°$. Each face of the cube has area L^2, so the fluxes through each of the faces are

$$\Phi_{E1} = \vec{E} \cdot \hat{n}_1 A = EL^2 \cos 180° = -EL^2;$$

$$\Phi_{E2} = \vec{E} \cdot \hat{n}_2 A = EL^2 \cos 0° = +EL^2;$$

$$\Phi_{E3} = \Phi_{E4} = \Phi_{E5} = \Phi_{E6} = EL^2 \cos 90° = 0.$$

The flux is negative on face 1, where $\vec{E}$ is directed into the cube, and positive on face 2, where $\vec{E}$ is directed out of the cube. The *total* flux through the cube is the sum of the fluxes through the six faces:

$$\Phi_E = \Phi_{E1} + \Phi_{E2} + \Phi_{E3} + \Phi_{E4} + \Phi_{E5} + \Phi_{E6}$$

$$= -EL^2 + EL^2 + 0 + 0 + 0 + 0 = 0.$$

b) The fluxes through faces 1 and 3 are negative, since $\vec{E}$ is directed into those faces; the field is directed out of faces 2 and

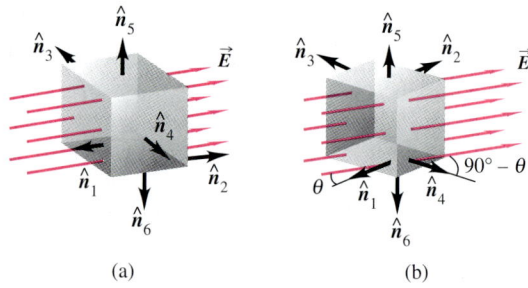

(a)   (b)

23-8 Electric flux of a uniform field $\vec{E}$ through a cubical box of side L in two orientations.

4, so the fluxes through those faces is positive. We find

$$\Phi_{E1} = \vec{E} \cdot \hat{n}_1 A = EL^2 \cos (180° - \theta) = -EL^2 \cos \theta;$$

$$\Phi_{E2} = \vec{E} \cdot \hat{n}_2 A = +EL^2 \cos \theta;$$

$$\Phi_{E3} = \vec{E} \cdot \hat{n}_3 A = EL^2 \cos (90° + \theta) = -EL^2 \sin \theta;$$

$$\Phi_{E4} = \vec{E} \cdot \hat{n}_4 A = EL^2 \cos (90° - \theta) = +EL^2 \sin \theta;$$

$$\Phi_{E5} = \Phi_{E6} = EL^2 \cos 90° = 0.$$

The total flux $\Phi_E = \Phi_{E1} + \Phi_{E2} + \Phi_{E3} + \Phi_{E4} + \Phi_{E5} + \Phi_{E6}$ through the surface of the cube is again zero. We came to this same conclusion in our discussion of Fig. 23–3c in Section 23–2; there we observed that there was zero net flux of a uniform electric field through a closed surface that contains no electric charge.

EXAMPLE 23-3

Electric flux through a sphere A positive point charge $q = 3.0 \text{ } \mu\text{C}$ is surrounded by a sphere with radius 0.20 m centered on the charge (Fig. 23–9). Find the electric flux through the sphere due to this charge.

SOLUTION At any point on the sphere the magnitude of $\vec{E}$ is

$$E = \frac{q}{4\pi\epsilon_0 r^2} = (9.0 \times 10^9 \text{ N} \cdot \text{m}^2/\text{C}^2) \frac{3.0 \times 10^{-6} \text{ C}}{(0.20 \text{ m})^2}$$

$$= 6.75 \times 10^5 \text{ N/C}.$$

From symmetry the field is perpendicular to the spherical surface at every point. The positive direction for both $\hat{n}$ and $E_\perp$ is outward, so $E_\perp = +E$ and the flux through a surface element dA is $E \, dA$. In Eq. (23–5), E is the same at every point and can be taken outside the integral; what remains is the integral $\int dA$, which is just the total area $A = 4\pi r^2$ of the spherical surface. Thus the total flux out of the sphere is

$$\Phi_E = EA = (6.75 \times 10^5 \text{ N/C})(4\pi)(0.20 \text{ m})^2$$

$$= 3.4 \times 10^5 \text{ N} \cdot \text{m}^2/\text{C}.$$

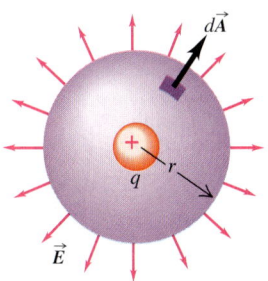

23–9 Electric flux through a sphere centered on a point charge.

Notice that we divided by $r^2 = (0.20 \text{ m})^2$ to find E, then multiplied by $r^2 = (0.20 \text{ m})^2$ to find Φ_E; hence the radius r of the sphere cancels out of the result for Φ_E. We would have obtained the same flux with a sphere of radius 2.0 m or 200 m. We came to essentially the same conclusion in our discussion of Fig. 23–4 in Section 23–2, where we considered rectangular closed surfaces of two different sizes enclosing a point charge. There we found that the flux of $\vec{E}$ was independent of the size of the surface; the same result holds true for a spherical surface. Indeed, the flux through *any* surface enclosing a single point charge is independent of the shape or size of the surface, as we'll soon see.

23–4 GAUSS'S LAW

11.8
Gauss's Law

Gauss's law is an alternative to Coulomb's law for expressing the relationship between electric charge and electric field. It was formulated by Carl Friedrich Gauss (1777–1855), one of the greatest mathematicians of all time. Many areas of mathematics bear the mark of his influence, and he made equally significant contributions to theoretical physics.

Gauss's law states that the total electric flux through any closed surface (a surface enclosing a definite volume) is proportional to the total (net) electric charge inside the surface. In Section 23–2 we observed this relationship qualitatively for certain special cases; now we'll develop it more rigorously. We'll start with the field of a single positive point charge q. The field lines radiate out equally in all directions. We place this charge at the center of an imaginary spherical surface with radius R. The magnitude E of the electric field at every point on the surface is given by

$$E = \frac{1}{4\pi\epsilon_0} \frac{q}{R^2}.$$

At each point on the surface, $\vec{E}$ is perpendicular to the surface, and its magnitude is the same at every point, just as in Example 23–3 (Section 23–3). The total electric flux is just the product of the field magnitude E and the total area $A = 4\pi R^2$ of the sphere:

$$\Phi_E = EA = \frac{1}{4\pi\epsilon_0} \frac{q}{R^2}(4\pi R^2) = \frac{q}{\epsilon_0}. \qquad (23\text{--}6)$$

The flux is independent of the radius R of the sphere. It depends only on the charge q enclosed by the sphere.

We can also interpret this result in terms of field lines. Figure 23–10 shows two spheres with radii R and $2R$, respectively, centered on the point charge q. Every field line that passes through the smaller sphere also passes through the larger sphere, so the total flux through each sphere is the same.

What is true of the entire sphere is also true of any portion of its surface. In Fig. 23–10 an area dA is outlined on a sphere of radius R and then projected onto the sphere of radius $2R$ by drawing lines from the center through points on the boundary of dA. The area projected on the larger sphere is clearly $4\,dA$. But since the electric field due to a point charge is inversely proportional to r^2, the field magnitude is $\frac{1}{4}$ as great on the sphere of radius $2R$ as on the sphere of radius R. Hence the electric flux is the same for both areas and is independent of the radius of the sphere.

This projection technique shows us how to extend this discussion to nonspherical surfaces. Instead of a second sphere, let us surround the sphere of radius R by a surface of irregular shape, as in Fig. 23–11a. Consider a small element of area dA on the irregular surface; we note that this area is *larger* than the corresponding element on a spherical surface at the same distance from q. If a normal to dA makes an angle ϕ with

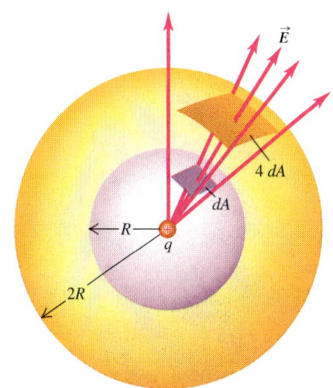

23–10 Projection of an element of area dA of a sphere of radius R onto a concentric sphere of radius $2R$. The projection multiplies each linear dimension by two, so the area element on the larger sphere is $4\,dA$. The same number of lines and the same flux pass through each area element.

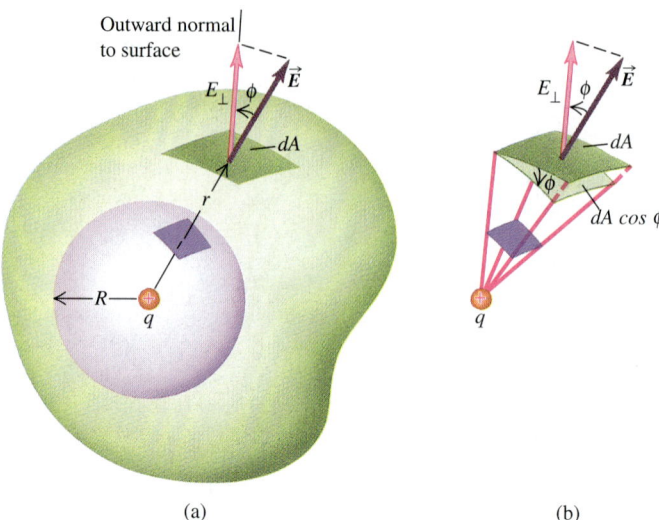

Outward normal to surface

(a)   (b)

23–11 (a) The outward normal to the surface makes an angle ϕ with the direction of $\vec{E}$. (b) The projection of the area element dA onto the spherical surface is $dA \cos \phi$.

a radial line from q, two sides of the area projected onto the spherical surface are fore-shortened by a factor $\cos \phi$ (Fig. 23–11b). The other two sides are unchanged. Thus the electric flux through the spherical surface element is equal to the flux $E\, dA \cos \phi$ through the corresponding irregular surface element.

We can divide the entire irregular surface into elements dA, compute the electric flux $E\, dA \cos \phi$ for each, and sum the results by integrating, as in Eq. (23–5). Each of the area elements projects onto a corresponding spherical surface element. Thus the *total* electric flux through the irregular surface, given by any of the forms of Eq. (23–5), must be the same as the total flux through a sphere, which Eq. (23–6) shows is equal to q/ϵ_0. Thus for the irregular surface,

$$\Phi_E = \oint \vec{E} \cdot d\vec{A} = \frac{q}{\epsilon_0}. \qquad (23\text{–}7)$$

Equation (23–7) holds for a surface of *any* shape or size, provided only that it is a *closed* surface enclosing the charge q. The circle on the integral sign reminds us that the integral is always taken over a *closed* surface.

The area elements $d\vec{A}$ and the corresponding unit vectors $\hat{n}$ always point *out of* the volume enclosed by the surface. The electric flux is then positive in areas where the electric field points out of the surface and negative where it is inward. Also, $E_\perp$ is positive at points where $\vec{E}$ points out of the surface and negative at points where $\vec{E}$ points into the surface.

If the point charge in Fig. 23–11 is negative, the $\vec{E}$ field is directed radially *inward;* the angle ϕ is then greater than $90°$, its cosine is negative, and the integral in Eq. (23–7) is negative. But since q is also negative, Eq. (23–7) still holds.

For a closed surface enclosing *no* charge,

$$\Phi_E = \oint \vec{E} \cdot d\vec{A} = 0.$$

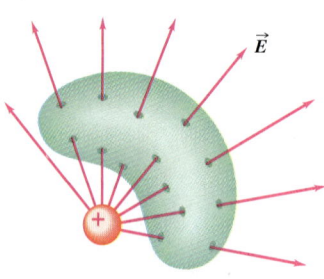

23–12 A point charge *outside* a closed surface that encloses no charge. If an electric field line from the external charge enters the surface at one point, it will leave at another.

This is the mathematical statement that when a region contains no charge, any field lines caused by charges *outside* the region that enter on one side must leave again on the other side. (In Section 23–2 we came to the same conclusion by considering the special case of a rectangular box in a uniform field.) Figure 23–12 illustrates this point. *Electric field lines can begin or end inside a region of space only when there is charge in that region.*

Now comes the final step in obtaining the general form of Gauss's law. Suppose the surface encloses not just one point charge q but several charges $q_1, q_2, q_3, \ldots$. The total (resultant) electric field $\vec{E}$ at any point is the vector sum of the $\vec{E}$ fields of the individual charges. Let Q_{encl} be the *total* charge enclosed by the surface: $Q_{encl} = q_1 + q_2 + q_3 + \cdots$. Also let $\vec{E}$ be the *total* field at the position of the surface area element $d\vec{A}$, and let $E_\perp$ be its component perpendicular to the plane of that element (that is, parallel to $d\vec{A}$). Then we can write an equation like Eq. (23–7) for each charge and its corresponding field and add the results. When we do, we obtain the general statement of Gauss's law:

$$\Phi_E = \oint \vec{E} \cdot d\vec{A} = \frac{Q_{encl}}{\epsilon_0} \qquad \text{(Gauss's law).} \qquad (23\text{–}8)$$

The total electric flux through a closed surface is equal to the total (net) electric charge inside the surface, divided by ϵ_0.

CAUTION ▶ Remember that the closed surface in Gauss's law is *imaginary;* there need not be any material object at the position of the surface. We often refer to a closed surface used in Gauss's law as a **Gaussian surface.** ◀

Using the definition of Q_{encl} and the various ways to express electric flux given in Eq. (23–5), we can express Gauss's law in the following equivalent forms:

$$\Phi_E = \oint E \cos\phi \, dA = \oint E_\perp \, dA = \oint \vec{E} \cdot d\vec{A} = \frac{Q_{encl}}{\epsilon_0} \qquad \begin{array}{l}\text{(various forms} \\ \text{of Gauss's law).}\end{array} \qquad (23\text{–}9)$$

As in Eq. (23–5), the various forms of the integral all express the same thing, the total electric flux through the Gaussian surface, in different terms. One form is sometimes more convenient than another.

In Eqs. (23–8) and (23–9), Q_{encl} is always the algebraic sum of all the positive and negative charges enclosed by the Gaussian surface, and $\vec{E}$ is the *total* field at each point on the surface. Also note that in general, this field is caused partly by charges inside the surface and partly by charges outside. But as is shown in Fig. 23–12, the outside charges do *not* contribute to the total (net) flux through the surface. So Eqs. (23–8) and (23–9) are correct even when there are charges outside the surface that contribute to the electric field at the surface. When $Q_{encl} = 0$, the total flux through the Gaussian surface must be zero, even though some areas may have positive flux and others may have negative flux.

Gauss's law is the definitive answer to the question we posed at the beginning of Section 23–2: "If the electric field pattern is known in a given region, what can we determine about the charge distribution in that region?" It provides a relationship between the electric field on a closed surface and the charge distribution within that surface. But in some cases we can use Gauss's law to answer the reverse question: "If the charge distribution is known, what can we determine about the electric field that the charge distribution produces?" Gauss's law may seem like an unappealing way to address this question, since it may look as though evaluating the integral in Eq. (23–8) is a hopeless task. Sometimes it is, but other times it is surprisingly easy. Here's an example in which *no* integration is involved at all; we'll work out several more examples in the next section.

Carl Friedrich Gauss helped develop several branches of mathematics, including differential geometry, real analysis, and number theory. The "bell curve" of statistics is one of his inventions. Gauss also made state-of-the-art investigations of the earth's magnetism and calculated the orbit of the first asteroid to be discovered.

EXAMPLE 23-4

Electric flux and enclosed charge Figure 23–13 shows the field produced by two point charges $+q$ and $-q$ of equal magnitude but opposite sign (an electric dipole). Find the electric flux through each of the closed surfaces A, B, C, and D.

SOLUTION The definition of electric flux given in Eq. (23–5) involves a surface integral, and so it might seem that integration is called for. But Gauss's law says that the total electric flux through a closed surface is equal to the total enclosed charge

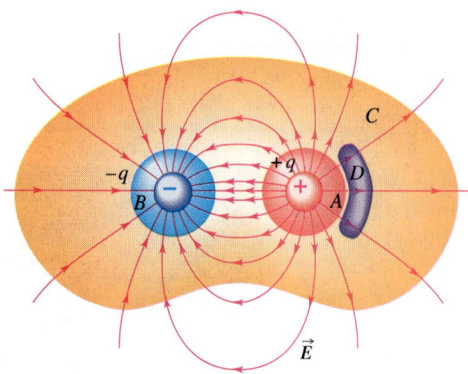

23-13 The net number of field lines leaving a closed surface is proportional to the total charge enclosed by that surface.

divided by ϵ_0. By inspection of Fig. 23-13, surface A (shown in red) encloses the positive charge, so $Q_{encl} = +q$; surface B (shown in blue) encloses the negative charge, so $Q_{encl} = -q$; surface C (shown in yellow), which encloses *both* charges, has $Q_{encl} = +q + (-q) = 0$; and surface D (shown in purple), which has no charges enclosed within it, also has $Q_{encl} = 0$. Hence without having to do any integration, we can conclude that the total fluxes for the various surfaces are $\Phi_E = +q/\epsilon_0$ for surface A,

$\Phi_E = -q/\epsilon_0$ for surface B, and $\Phi_E = 0$ for both surface C and surface D.

These results depend only on the charges enclosed within each Gaussian surface, not on the precise shapes of the surfaces. For example, compare surface C to the rectangular surface shown in Fig. 23-3b, which also encloses both charges of an electric dipole. In that case as well, we concluded that the net flux of $\vec{E}$ was zero; the inward flux on one part of the surface exactly compensates for the outward flux on the remainder of the surface.

We can draw similar conclusions by examining the electric field lines. Surface A encloses only the positive charge; in Fig. 23-13, 18 lines are depicted crossing A in an outward direction. Surface B encloses only the negative charge; it is crossed by these same 18 lines, but in an inward direction. Surface C encloses *both* charges. It is intersected by lines at 16 points; at 8 intersections the lines are outward, and at 8 they are inward. The *net* number of lines crossing in an outward direction is zero, and the net charge inside the surface is also zero. Surface D is intersected at 6 points; at 3 points the lines are outward, and at the other 3 they are inward. The net number of lines crossing in an outward direction and the total charge enclosed are both zero. There are points on the surfaces where $\vec{E}$ is not perpendicular to the surface, but this doesn't affect the counting of the field lines.

23-5 APPLICATIONS OF GAUSS'S LAW

Gauss's law is valid for *any* distribution of charges and for *any* closed surface. Gauss's law can be used in two ways. If we know the charge distribution, and if it has enough symmetry to let us evaluate the integral in Gauss's law, we can find the field. Or if we know the field, we can use Gauss's law to find the charge distribution, such as charges on conducting surfaces.

In this section we present examples of both kinds of applications. As you study them, watch for the role played by the symmetry properties of each system. We will use Gauss's law to calculate the electric fields caused by several simple charge distributions; the results are collected in a table in the chapter summary.

In practical problems we often encounter situations in which we want to know the electric field caused by a charge distribution on a conductor. These calculations are aided by the following remarkable fact: *When excess charge is placed on a solid conductor and is at rest, it resides entirely on the surface, not in the interior of the material.* (By *excess* we mean charges other than the ions and free electrons that make up the neutral conductor.) Here's the proof. We know from Section 22-6 that in an electrostatic situation (with all charges at rest) the electric field $\vec{E}$ at every point in the interior of a conducting material is zero. If $\vec{E}$ were *not* zero, the charges would move. Suppose we construct a Gaussian surface inside the conductor, such as surface A in Fig. 23-14. Because $\vec{E} = \mathbf{0}$ everywhere on this surface, Gauss's law requires that the net charge inside the surface is zero. Now imagine shrinking the surface like a collapsing balloon until it encloses a region so small that we may consider it as a point P; then the charge at that point must be zero. We can do this anywhere inside the conductor, so *there can be no excess charge at any point within a solid conductor; any excess charge must reside on the conductor's surface.* (This result is for a *solid* conductor. In the next section we'll discuss what can happen if the conductor has cavities in its interior.) We will make use of this fact frequently in the examples that follow.

23-14 Under electrostatic conditions, any excess charge resides entirely on the surface of a solid conductor.

Problem–Solving Strategy

GAUSS'S LAW

1. The first step is to select the surface that you are going to use with Gauss's law. We often call this a *Gaussian surface*. If you are trying to find the field at a particular point, then that point must lie on your Gaussian surface.

2. The Gaussian surface does not have to be a real physical surface, such as a surface of a solid body. Often, the appropriate surface is an imaginary geometric surface; it may be in empty space, embedded in a solid body, or partly both.

3. Usually, you can evaluate the integral in Gauss's law (without using a computer) only if the Gaussian surface and the charge distribution have some symmetry property. If the charge distribution has cylindrical or spherical symmetry, choose the Gaussian surface to be a coaxial cylinder or a concentric sphere, respectively.

4. Often, you can think of the closed Gaussian surface as being made up of several separate surfaces, such as the sides and ends of a cylinder. The integral $\oint E_\perp \, dA$ over the entire closed surface is always equal to the sum of the integrals over all the separate surfaces. Some of these integrals may be zero, as in points 6 and 7 below.

5. If $\vec{E}$ is *perpendicular* (normal) at every point to a surface with area A, if it points *outward* from the interior of the surface, and if it also has the same *magnitude* at every point on the surface, then $E_\perp = E =$ constant, and $\int E_\perp \, dA$ over that surface is equal to EA. If instead $\vec{E}$ is perpendicular and *inward*, $E_\perp = -E$ and $\int E_\perp \, dA = -EA$.

6. If $\vec{E}$ is *tangent* to a surface at every point, then $E_\perp = 0$, and the integral over that surface is zero.

7. If $\vec{E} = 0$ at every point on a surface, the integral is zero.

8. Finally, in the integral $\oint E_\perp \, dA$, $E_\perp$ is always the perpendicular component of the *total* electric field at each point on the closed Gaussian surface. In general, this field may be caused partly by charges within the surface and partly by charges outside. Even when there is *no* charge within the surface, the field at points on the Gaussian surface is not necessarily zero; in that case, however, the *integral* over the Gaussian surface—that is, the total electric flux through the Gaussian surface—is always zero.

EXAMPLE 23–5

Field of a charged conducting sphere We place positive charge q on a solid conducting sphere with radius R (Fig. 23–15). Find $\vec{E}$ at any point inside or outside the sphere.

SOLUTION As we discussed earlier in this section, all the charge must be on the surface of the sphere. We can use the spherical symmetry to show that the charge must be distributed *uniformly* over the surface and that the direction of the electric field at any point P outside the sphere must be along a *radial* line between the center and point P.

The role of symmetry deserves careful discussion. When we say that the system is spherically symmetric, we mean that if we rotate it through any angle about any axis through the center, the system after rotation is indistinguishable from the original unrotated system. There is nothing about the system that distinguishes one direction or orientation in space from another. The charge is free to move on the conductor, and there is nothing about the conductor that would make it tend to concentrate more in some regions than others. If it were nonuniform, then when we rotate the system, the sphere would look the same but the charge distribution would look different, and there is no property of the sphere that can make this happen. So we conclude that the surface charge distribution must be uniform.

A similar argument shows that the field must be *radial*. If we again rotate the system, the field pattern of the rotated system must be identical to that of the original system. If the field had a component at some point that was perpendicular to the radial direction, that component would have to be different after at least some rotations. Thus there can't be such a component, and

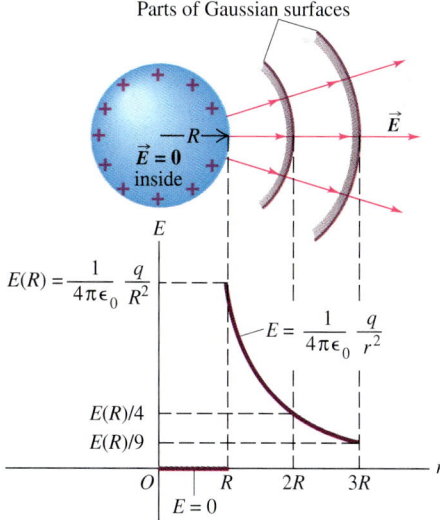

Parts of Gaussian surfaces

23–15 Under electrostatic conditions the electric field inside a solid conducting sphere is zero. Outside the sphere the electric field drops off as $1/r^2$, as though all of the excess charge on the sphere were concentrated at its center.

the field must be radial. For the same reason the magnitude E of the field can depend only on the distance r from the center and must have the same value at all points on a spherical surface concentric with the conductor.

To take advantage of these symmetry properties, we take as our Gaussian surface an imaginary sphere centered on the conductor, with radius r greater than the radius R of the conductor so that the enclosed charge is q. The area of the Gaussian surface is $4\pi r^2$; $\vec{E}$ is uniform over the surface and perpendicular to it at each point. The flux integral $\oint E_\perp \, dA$ in Gauss's law is therefore just $E(4\pi r^2)$, and Eq. (23–8) gives

$$E(4\pi r^2) = \frac{q}{\epsilon_0} \quad \text{and}$$

$$E = \frac{1}{4\pi\epsilon_0} \frac{q}{r^2} \quad \text{(outside a charged conducting sphere).}$$

This expression for the field at any point *outside* the sphere $(r > R)$ is the same as for a point charge; the field due to the charged sphere is the same as though the entire charge were concentrated at its center. Just outside the surface of the sphere, where $r = R$,

$$E = \frac{1}{4\pi\epsilon_0} \frac{q}{R^2} \quad \text{(at the surface of a charged conducting sphere).}$$

CAUTION ▶ Remember that we have chosen the charge q to be *positive*. If the charge is negative, the electric field is radially *inward* instead of radially outward, and the electric flux through

the Gaussian surface is negative. The electric field magnitudes outside and at the surface of the sphere are given by the same expressions as above, except that q denotes the *magnitude* (absolute value) of the charge. ◀

Inside the sphere $(r < R)$ the electric field is zero, as with any solid conductor when the charges are at rest. Figure 23–15 shows E as a function of the distance r from the center of the sphere. Note that in the limit as $R \to 0$, the sphere becomes a point charge; there is then only an "outside," and the field is everywhere given by $E = q/4\pi\epsilon_0 r^2$. Thus we have deduced Coulomb's law from Gauss's law. (In Section 23–4 we deduced Gauss's law from Coulomb's law, so this completes the demonstration of their logical equivalence.)

We can also use this method for a conducting spherical *shell* (a spherical conductor with a concentric spherical hole in the center) if there is no charge inside the hole. We use a spherical Gaussian surface with radius r less than the radius of the hole. If there *were* a field inside the hole, it would have to be radial and spherically symmetric as before, so $E = Q_{encl}/4\pi\epsilon_0 r^2$. But now there is no enclosed charge, so $Q_{encl} = 0$ and $E = 0$ inside the hole.

Can you use this same technique to find the electric field in the interspace between a charged sphere and a concentric hollow conducting sphere that surrounds it?

EXAMPLE 23-6

Field of a line charge Electric charge is distributed uniformly along an infinitely long, thin wire. The charge per unit length is λ (assumed positive). Find the electric field. (This is an approximate representation of the field of a uniformly charged *finite* wire, provided that the distance from the field point to the wire is much less than the length of the wire.)

SOLUTION The field must point away from the positive charges on the wire, but in what direction? Again we ask, "What is the symmetry?" We can rotate the system through any angle about its axis, and we can shift it by any amount along the axis; in each case the resulting system is indistinguishable from the original. Using the same argument as in Example 23–5, we conclude that $\vec{E}$ at each point doesn't change when either of these operations is carried out. The field can't have any component parallel to the wire; if it did, we would have to explain why the field lines that begin on the wire pointed in one direction parallel to the wire and not the other. Also, the field can't have any component tangent to a circle in a plane perpendicular to the wire with its center on the wire. If it did, we would have to explain why the component pointed in one direction around the wire rather than the other. All that's left is a component radially outward from the wire at each point. So the field lines outside a uniformly charged, infinite wire are *radial* and lie in planes perpendicular to the wire. The field *magnitude* can depend only on the radial distance from the wire.

These symmetry properties suggest that we use as a Gaussian surface a *cylinder* with arbitrary radius r and arbitrary length l, with its ends perpendicular to the wire (Fig. 23–16). We break the surface integral for the flux Φ_E into an integral over each flat end and one over the curved side walls. There is no flux through

23–16 A coaxial cylindrical Gaussian surface is used to find the electric field outside an infinitely long charged wire.

the ends because $\vec{E}$ lies in the plane of the surface and $E_\perp = 0$. To find the flux through the side walls, note that $\vec{E}$ is perpendicular to the surface at each point, so $E = E_\perp$; by symmetry, E has the same value everywhere on the walls. The area of the side walls is $2\pi rl$. (To make a paper cylinder with radius r and height l, you need a paper rectangle with width $2\pi r$, height l, and area $2\pi rl$.) Hence the total flux Φ_E through the entire cylinder is the sum of the flux through the side walls, which is $(E)(2\pi rl)$, and the zero flux through the two ends. Finally, we need the total enclosed charge, which is the charge per unit length multiplied by the length of wire inside the Gaussian surface, or $Q_{encl} = \lambda l$. From Gauss's law, Eq. (23–8),

$$\Phi_E = (E)(2\pi rl) = \frac{\lambda l}{\epsilon_0} \quad \text{and}$$

$$E = \frac{1}{2\pi\epsilon_0} \frac{\lambda}{r} \quad \text{(field of an infinite line of charge).}$$

This is the same result that we found in Example 22–11 (Section 22–7) by much more laborious means.

We have assumed that λ is *positive*. If it is *negative*, $\vec{E}$ is directed radially inward toward the line of charge, and in the above expression for the field magnitude E we must interpret λ as the *magnitude* (absolute value) of the charge per unit length.

Note that although the *entire* charge on the wire contributes to the field, only the part of the total charge that is within the Gaussian surface is considered when we apply Gauss's law. This may seem strange; it looks as though we have somehow obtained the right answer by ignoring part of the charge and that the field of a *short* wire of length l would be the same as that of a very long wire. But we *do* include the entire charge on the wire when we make use of the *symmetry* of the problem. If the wire is short, the symmetry with respect to shifts along the axis is not present, and the field is not uniform in magnitude over our Gaussian surface. Gauss's law is then no longer useful and *cannot* be used to find the field; the problem is best handled by the integration technique used in Example 22–11.

We can use a Gaussian surface like that in Fig. 23–16 to show that the field at points outside a long, uniformly charged cylinder is the same as though all the charge were concentrated on a line along its axis. We can also calculate the electric field in the space between a charged cylinder and a coaxial hollow conducting cylinder surrounding it. (This is a model of a coaxial cable, like those used to connect your TV to a cable television "feed.") We leave these calculations as problems.

EXAMPLE 23–7

Field of an infinite plane sheet of charge Find the electric field caused by a thin, flat, infinite sheet on which there is a uniform positive charge per unit area σ.

SOLUTION The field must point away from the positively charged sheet. To say more, we must again ask, "What is the symmetry?" The charge distribution doesn't change if we slide it in any direction parallel to the sheet. From this we conclude that at each point, $\vec{E}$ is perpendicular to the sheet. The symmetry also tells us that the field must have the same magnitude E at any given distance on either side of the sheet.

To take advantage of these symmetry properties, we use as our Gaussian surface a cylinder with its axis perpendicular to the sheet of charge, with ends of area A (Fig. 23–17). The charged sheet passes through the middle of the cylinder's length, so the cylinder ends are equidistant from the sheet. At each end of the cylinder, $\vec{E}$ is perpendicular to the surface, and $E_\perp$ is equal to E; hence the flux through each end is $+EA$. Because $\vec{E}$ is perpendicular to the charged sheet, it is parallel to the curved *side* walls of the cylinder, so $E_\perp$ at these walls is zero, and there is no flux through these walls. The total flux integral in Gauss's law is then $2EA$ (EA from each end and zero from the side walls). The net charge within the Gaussian surface is the charge per unit area multiplied by the sheet area enclosed by the surface, or $Q_{encl} = \sigma A$. Hence Gauss's law, Eq. (23–8), gives

$$2EA = \frac{\sigma A}{\epsilon_0} \quad \text{and}$$

$$E = \frac{\sigma}{2\epsilon_0} \quad \text{(field of an infinite sheet of charge)}.$$

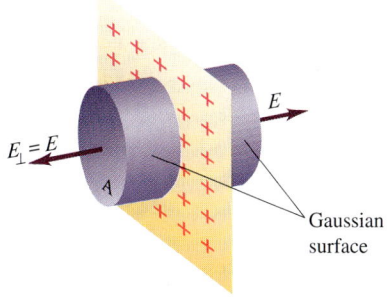

23–17 A cylindrical Gaussian surface is used to find the field of an infinite plane sheet of charge.

This is the same result that we found in Example 22–12 (Section 22–7) using a much more complex calculation. The field is uniform and directed perpendicular to the plane of the sheet. Its magnitude is *independent* of the distance from the sheet. The field lines are therefore straight, parallel to each other, and perpendicular to the sheet.

If the charge density is negative, $\vec{E}$ is directed *toward* the sheet, the flux through the Gaussian surface in Fig. 23–17 is negative, and σ in the expression $E = \sigma/2\epsilon_0$ denotes the magnitude (absolute value) of the charge density.

The assumption that the sheet is infinitely large is an idealization; nothing in nature is really infinitely large. But the result $E = \sigma/2\epsilon_0$ is a good approximation for points that are close to the sheet (compared to the sheet's dimensions) and not too near its edges. At such points, the field is very nearly uniform and perpendicular to the plane.

EXAMPLE 23–8

Field between oppositely charged parallel conducting plates
Two large plane parallel conducting plates are given charges of equal magnitude and opposite sign; the charge per unit area is $+\sigma$ for one and $-\sigma$ for the other. Find the electric field in the region between the plates.

SOLUTION The field between and around the plates is approximately as shown in Fig. 23–18a. Because opposite charges attract, most of the charge accumulates at the opposing faces of the plates. A small amount of charge resides on the *outer* surfaces of the plates, and there is some spreading or "fringing" of

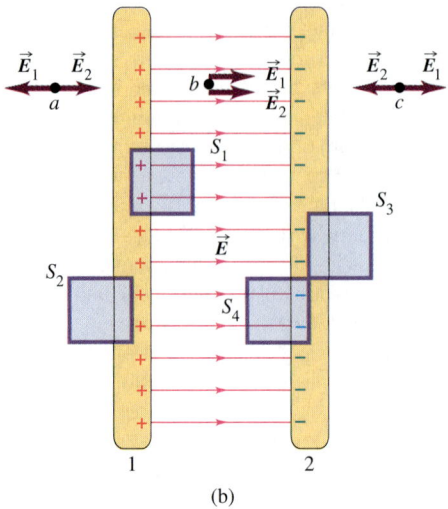

23–18 Electric field between oppositely charged parallel plates. (a) The field between the plates is quite uniform except near the edges. (b) Idealized case in which "fringing" at the edges is ignored.

the field at the edges. But if the plates are very large in comparison to the distance between them, the amount of charge on the outer surfaces is negligibly small, and the fringing can be neglected except near the edges. In this case we can assume that the field is uniform in the interior region between plates, as in Fig. 23–18b, and that the charges are distributed uniformly over the opposing surfaces.

To exploit this symmetry, we can use the shaded Gaussian surfaces S_1, S_2, S_3, and S_4. These surfaces are cylinders with ends of area A like the one shown in perspective in Fig. 23–17; they are shown in a side view in Fig. 23–18b. One end of each surface lies within one of the conducting plates. For the surface labeled S_1, the left-hand end is within plate 1 (the positive plate). Since the field is zero within the volume of any solid conductor under electrostatic conditions, there is no electric flux through this end. The electric field between the plates is perpendicular to the right-hand end, so on that end, $E_\perp$ is equal to E and the flux is EA; this is positive, since $\vec{E}$ is directed out of the Gaussian surface. There is no flux through the side walls of the cylinder, since these walls are parallel to $\vec{E}$. So the total flux integral in Gauss's law is EA. The net charge enclosed by the cylinder is

σA, so Eq. (23–8) yields

$$EA = \frac{\sigma A}{\epsilon_0} \quad \text{and}$$

$$E = \frac{\sigma}{\epsilon_0} \quad \text{(field between oppositely charged conducting plates).}$$

The field is uniform and perpendicular to the plates, and its magnitude is independent of the distance from either plate. This same result can be obtained by using the Gaussian surface S_4; furthermore, the surfaces S_2 and S_3 can be used to show that $E = 0$ to the left of plate 1 and to the right of plate 2. We leave these calculations as a problem.

We obtained the above results in Example 22–13 (Section 22–7) by using the principle of superposition of electric fields. The fields due to the two sheets of charge (one on each plate) are $\vec{E}_1$ and $\vec{E}_2$; from Example 23–7, both of these have magnitude $\sigma/2\epsilon_0$. The total (resultant) electric field at any point is the vector sum $\vec{E} = \vec{E}_1 + \vec{E}_2$. At points a and c in Fig. 23–18b, $\vec{E}_1$ and $\vec{E}_2$ have opposite directions, and their resultant is zero. This is also true at every point within the material of each plate, consistent with the requirement that with charges at rest there can be no field within a solid conductor. At any point b between the plates, $\vec{E}_1$ and $\vec{E}_2$ have the same direction; their resultant has magnitude $E = \sigma/\epsilon_0$, just as we found above using Gauss's law.

<hr>

EXAMPLE 23-9

Field of a uniformly charged sphere Positive electric charge Q is distributed uniformly *throughout the volume* of an *insulating* sphere with radius R. Find the magnitude of the electric field at a point P a distance r from the center of the sphere.

SOLUTION As in Example 23–5, the system is spherically symmetric. To make use of this symmetry, we choose as our Gaussian surface a sphere with radius r, concentric with the charge distribution. From symmetry the magnitude E of the elec-

tric field has the same value at every point on the Gaussian surface, and the direction of $\vec{E}$ is radial at every point on the surface, so $E_\perp = E$. Hence the total electric flux through the Gaussian surface is the product of E and the total area of the surface $A = 4\pi r^2$, that is, $\Phi_E = 4\pi r^2 E$.

The amount of charge enclosed within the Gaussian surface depends on the radius r. Let's first find the field magnitude *inside* the charged sphere of radius R; the magnitude E is evaluated at the radius of the Gaussian surface, so we choose $r < R$.

The volume charge density ρ is the charge Q divided by the volume of the entire charged sphere of radius R:

$$\rho = \frac{Q}{4\pi R^3/3}.$$

The volume V_{encl} enclosed by the Gaussian surface is $\frac{4}{3}\pi r^3$, so the total charge Q_{encl} enclosed by that surface is

$$Q_{encl} = \rho V_{encl} = \left(\frac{Q}{4\pi R^3/3}\right)\left(\frac{4}{3}\pi r^3\right) = Q\frac{r^3}{R^3}.$$

Then Gauss's law, Eq. (23–8), becomes

$$4\pi r^2 E = \frac{Q}{\epsilon_0}\frac{r^3}{R^3}, \qquad \text{or}$$

$$E = \frac{1}{4\pi\epsilon_0}\frac{Qr}{R^3} \qquad \text{(field inside a uniformly charged sphere).}$$

The field magnitude is proportional to the distance r of the field point from the center of the sphere. At the center ($r = 0$), $E = 0$.

To find the field magnitude *outside* the charged sphere, we use a spherical Gaussian surface of radius $r > R$. This surface encloses the entire charged sphere, so $Q_{encl} = Q$, and Gauss's law gives

$$4\pi r^2 E = \frac{Q}{\epsilon_0}, \qquad \text{or}$$

$$E = \frac{1}{4\pi\epsilon_0}\frac{Q}{r^2} \qquad \text{(field outside a uniformly charged sphere).}$$

For *any* spherically symmetric charged body the electric field outside the body is the same as though the entire charge were concentrated at the center. (We made this same observation in Example 23–5.)

Figure 23–19 shows a graph of E as a function of r for this problem. For $r < R$, E is directly proportional to r, and for $r > R$, E varies as $1/r^2$. If the charge is negative instead of positive, $\vec{E}$ is radially *inward* and Q in the expressions for E is interpreted as the magnitude (absolute value) of the charge.

Notice that if we set $r = R$ in either of the two expressions

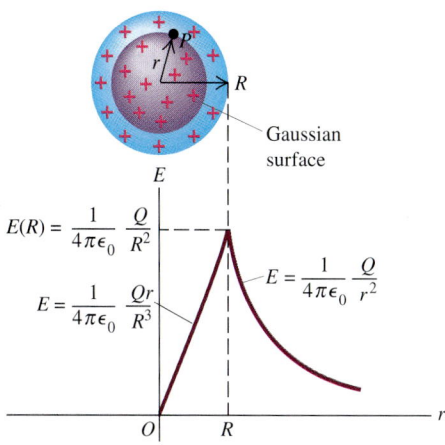

23–19 The magnitude of the electric field of a uniformly charged insulating sphere. Compare this with the field for a conducting sphere (Fig. 23–15).

for E (inside or outside the sphere), we get the same result $E = Q/4\pi\epsilon_0 R^2$ for the magnitude of the field at the surface of the sphere; this is because the magnitude E is a *continuous* function of r. By contrast, for the charged conducting sphere of Example 23–5 the electric-field magnitude is *discontinuous* at $r = R$ (it jumps from $E = 0$ just inside the sphere to $E = Q/4\pi\epsilon_0 R^2$ just outside the sphere). In general, the electric field $\vec{E}$ is discontinuous in magnitude, direction, or both wherever there is a *sheet* of charge, such as at the surface of a charged conducting sphere (Example 23–5), at the surface of an infinite charged sheet (Example 23–7), or at the surface of a charged conducting plate (Example 23–8).

The general technique used in this example can be applied to *any* spherically symmetric distribution of charge, whether it is uniform or not. Such charge distributions occur within many atoms and atomic nuclei, which is why Gauss's law is a useful tool in atomic and nuclear physics.

23–6 CHARGES ON CONDUCTORS

We have learned that in an electrostatic situation (in which there is no net motion of charge) the electric field at every point within a conductor is zero and that any excess charge on a solid conductor is located entirely on its surface (Fig. 23–20a). But what if there is a *cavity* inside the conductor (Fig. 23–20b)? If there is no charge within the cavity, we can use a Gaussian surface such as A (which lies completely within the material of the conductor) to show that the *net* charge on the *surface of the cavity* must be zero,

(a)

(b)

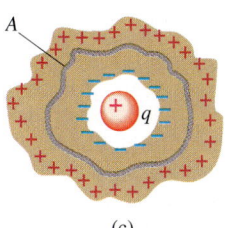

(c)

23–20 (a) Charge on a solid conductor resides entirely on its outer surface. (b) If there is no charge inside the conductor's cavity, the net charge on the surface of the cavity is zero. (c) If there is a charge q inside the cavity, the total charge on the cavity surface is $-q$.

because $\vec{E} = 0$ everywhere on the Gaussian surface. In fact, we can prove in this situation that there can't be any charge *anywhere* on the cavity surface. We will postpone detailed proof of this statement until Chapter 24.

Suppose we place a small body with a charge q inside a cavity within a conductor (Fig. 23–20c). The conductor is uncharged and is insulated from the charge q. Again $\vec{E} = 0$ everywhere on surface A, so according to Gauss's law the *total* charge inside this surface must be zero. Therefore there must be a charge $-q$ distributed on the surface of the cavity, drawn there by the charge q inside the cavity. The *total* charge on the conductor must remain zero, so a charge $+q$ must appear either on its outer surface or inside the material. But we showed in Section 23–5 that in an electrostatic situation there can't be any excess charge within the material of a conductor. So we conclude that the charge $+q$ must appear on the outer surface. By the same reasoning, if the conductor originally had a charge q_C, then the total charge on the outer surface must be $q + q_C$ after the charge q is inserted into the cavity.

EXAMPLE 23–10

The conductor shown in cross section in Fig. 23–21 carries a total charge of +3 nC. The charge within the cavity, insulated from the conductor, is –5 nC. How much charge is on each surface (inner and outer) of the conductor?

SOLUTION If the charge in the cavity is $q = -5$ nC, the charge on the inner cavity surface must be $-q = -(-5 \text{ nC}) = +5 \text{ nC}$. The conductor carries a *total* charge of +3 nC, none of which is in the interior of the material. If +5 nC is on the inner surface of the cavity, then there must be $(+3 \text{ nC}) - (+5 \text{ nC}) = -2 \text{ nC}$ on the outer surface of the conductor.

23–21 There is no excess charge in the bulk material of this conductor. Charge is located only on the inner and outer surfaces.

TESTING GAUSS'S LAW EXPERIMENTALLY

We can now consider a historic experiment, shown in Fig. 23–22. We mount a conducting container, such as a metal pail with a lid, on an insulating stand. The container is initially uncharged. Then we hang a charged metal ball from an insulating thread (Fig. 23–22a), lower it into the pail, and put the lid on (Fig. 23–22b). Charges are induced on

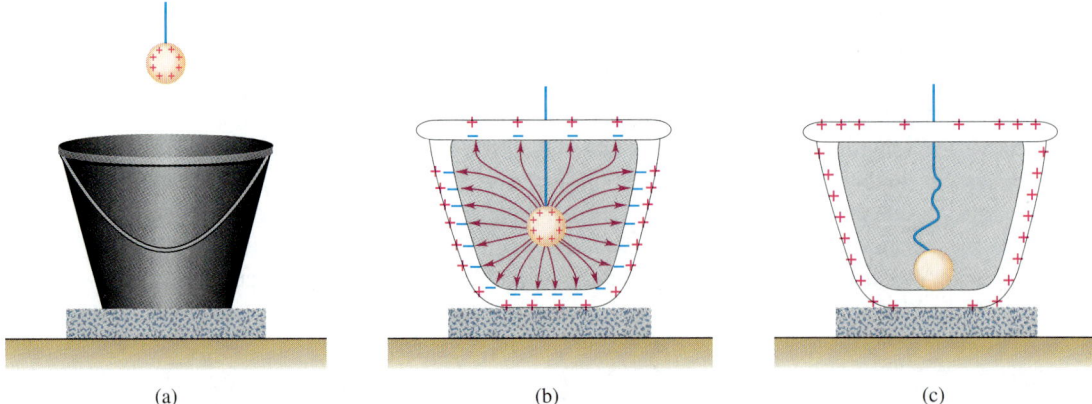

(a)   (b)   (c)

23–22 (a) A charged conducting ball suspended by an insulating thread outside a conducting container on an insulating stand. (b) The ball is lowered into the container, and the lid is put on. Charges are induced on the walls of the container. (c) When the ball is touched to the inner surface of the container, all its charge is transferred to the container and appears on the container's outer surface.

the walls of the container, as shown. But now we let the ball *touch* the inner wall (Fig. 23–22c). The surface of the ball becomes, in effect, part of the cavity surface. The situation is now the same as Fig. 23–20b; if Gauss's law is correct, the net charge on the cavity surface must be zero. Thus the ball must lose all its charge. Finally, we pull the ball out; we find that it has indeed lost all its charge.

This experiment was performed in the nineteenth century by the English scientist Michael Faraday, using a metal icepail with a lid, and it is called **Faraday's icepail experiment.** (Similar experiments were carried out in the eighteenth century by Benjamin Franklin in America and Joseph Priestley in England, although with much less precision.) The result confirms the validity of Gauss's law and therefore of Coulomb's law. Faraday's result was significant because Coulomb's experimental method, using a torsion balance and dividing of charges, was not very precise; it is very difficult to confirm the $1/r^2$ dependence of the electrostatic force with great precision by direct force measurements. By contrast, experiments like Faraday's test the validity of Gauss's law, and therefore of Coulomb's law, with much greater precision.

A modern version of Faraday's experiment is shown in Fig. 23–23. The details of the box labeled "power supply" aren't important; its job is to place charge on the outer sphere and remove it, on demand. The inner box with a dial is a sensitive *electrometer,* an instrument that can detect motion of extremely small amounts of charge between the outer and inner spheres. If Gauss's law is correct, there can never be any charge on the inner surface of the outer sphere. If so, there should be no flow of charge between spheres while the outer sphere is being charged and discharged. The fact that no flow is actually observed is a very sensitive confirmation of Gauss's law and therefore of Coulomb's law. The precision of the experiment is limited mainly by the electrometer, which can be astonishingly sensitive. Experiments have shown that the exponent 2 in the $1/r^2$ of Coulomb's law does not differ from precisely 2 by more than 10^{-16}. So there is no reason to suspect that it is anything other than exactly 2.

The same principle behind Faraday's icepail experiment is used in a *Van de Graaff electrostatic generator* (Fig. 23–24). The charged conducting ball of Fig. 23–22 is replaced by a charged belt that continuously carries charge to the inside of a conducting shell, only to have it carried away to the outside surface of the shell. As a result, the charge on the shell and the electric field around it can become very large very rapidly. The Van de Graaff generator is used as an accelerator of charged particles and for physics demonstrations.

This principle also forms the basis for *electrostatic shielding.* Suppose we have a very sensitive electronic instrument that we want to protect from stray electric fields that might cause erroneous measurements. We surround the instrument with a conducting box, or we line the walls, floor, and ceiling of the room with a conducting material such as sheet copper. The external electric field redistributes the free electrons in the conductor, leaving a net positive charge on the outer surface in some regions and a net negative charge in others (Fig. 23–25). This charge distribution causes an additional electric field such that the *total* field at every point inside the box is zero, as Gauss's law says it must be. The charge distribution on the box also alters the shapes of the field lines near the box, as the figure shows. Such a setup is often called a *Faraday cage.* The same physics tells you that one of the safest places to be in a lightning storm is inside an automobile; if the car is struck by lightning, the charge tends to remain on the metal skin of the vehicle, and little or no electric field is produced inside the passenger compartment.

FIELD AT THE SURFACE OF A CONDUCTOR

Finally, we note that there is a direct relation between the $\vec{E}$ field at a point just outside any conductor and the surface charge density σ at that point. In general, σ varies from

23–23 The outer spherical shell can be alternately charged and discharged by the power supply. If there were any flow of charge between the inner and outer shells, it would be detected by the electrometer inside the inner shell.

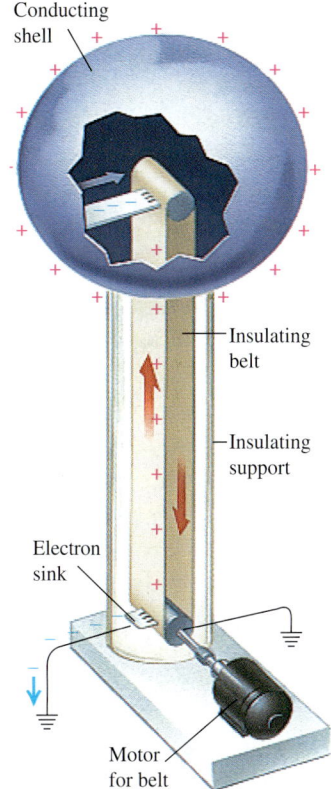

23–24 Cutaway view of the essential parts of a Van de Graaff electrostatic generator. The electron sink at the bottom draws electrons from the belt, giving it a positive charge; at the top the belt attracts electrons away from the conducting shell, giving the shell a positive charge.

23-25 (a) A conducting box (a Faraday cage) immersed in a uniform electric field. The field pushes electrons toward the left, leaving a net negative charge on the left side and a net positive charge on the right side. The result is that the net electric field at any point *inside* the box is zero; the shapes of the electric field lines near the box are somewhat distorted. (b) A Van de Graaff generator at the Boston Museum of Science produces a tremendous electrical discharge, but the person inside the conducting metal cage is shielded from its effects.

(a)   (b)

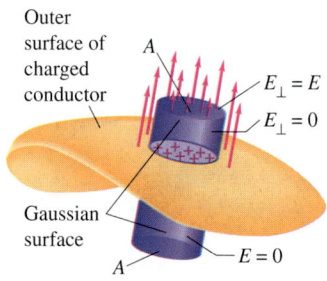

23-26 The field just outside a charged conductor is perpendicular to the surface, and its perpendicular component $E_\perp$ is equal to σ/ϵ_0.

point to point on the surface. We will show in Chapter 24 that at any such point, the direction of $\vec{E}$ is always perpendicular to the surface.

To find a relation between σ at any point on the surface and the perpendicular component of the electric field at that point, we construct a Gaussian surface in the form of a small cylinder (Fig. 23-26). One end face, with area A, lies within the conductor, and the other lies just outside. The electric field is zero at all points within the conductor. Outside the conductor the component of $\vec{E}$ perpendicular to the side walls of the cylinder is zero, and over the end face the perpendicular component is equal to $E_\perp$. (If σ is positive, the electric field will point out of the conductor and $E_\perp$ will be positive; if σ is negative, the field will point inward and $E_\perp$ will be negative.) Hence the total flux through the surface is $E_\perp A$. The charge enclosed within the Gaussian surface is σA, so from Gauss's law,

$$E_\perp A = \frac{\sigma A}{\epsilon_0}, \quad \text{and} \quad E_\perp = \frac{\sigma}{\epsilon_0} \quad \text{(field at the surface of} \quad (23-10) \quad \text{a conductor).}$$

We can check this with the results we have obtained for spherical, cylindrical, and plane surfaces.

We showed in Example 23-8 that the field magnitude between two infinite flat oppositely charged conducting plates also equals σ/ϵ_0. In this case the field magnitude is the same at *all* distances from the plates, but in all other cases it decreases with increasing distance from the surface.

EXAMPLE 23-11

Verify Eq. (23-10) for a conducting sphere with radius R and total charge q.

SOLUTION In Example 23-5 (Section 23-5) we showed that the electric field just outside the surface is

$$E = \frac{1}{4\pi\epsilon_0} \frac{q}{R^2}.$$

The surface charge density is uniform and equal to q divided by the surface area of the sphere:

$$\sigma = \frac{q}{4\pi R^2}.$$

Comparing these two expressions, we see that $E = \sigma/\epsilon_0$, as Eq. (23-10) states.

EXAMPLE 23-12

Electric field of the earth The earth has a net electric charge. The resulting electric field near the surface can be measured with sensitive electronic instruments; its average value is about

150 N/C, directed toward the center of the earth. a) What is the corresponding surface charge density? b) What is the *total* surface charge of the earth?

SOLUTION a) We know from the direction of the field that σ is negative (corresponding to $\vec{E}$ being directed *into* the surface, so $E_\perp$ is negative). From Eq. (23–10),

$$\sigma = \epsilon_0 E_\perp = (8.85 \times 10^{-12} \text{ C}^2/\text{N} \cdot \text{m}^2)(-150 \text{ N/C})$$

$$= -1.33 \times 10^{-9} \text{ C/m}^2 = -1.33 \text{ nC/m}^2.$$

b) The total charge Q is the product of the surface area $4\pi R_E^2$, where $R_E = 6.38 \times 10^6$ m is the radius of the earth, and the charge density σ:

$$Q = 4\pi(6.38 \times 10^6 \text{ m})^2(-1.33 \times 10^{-9} \text{ C/m}^2)$$

$$= -6.8 \times 10^5 \text{ C} = -680 \text{ kC}.$$

We could also use the result of Example 23–5. Solving for Q,

we find

$$Q = 4\pi\epsilon_0 R^2 E_\perp$$

$$= \frac{1}{9.0 \times 10^9 \text{ N} \cdot \text{m}^2/\text{C}^2}(6.38 \times 10^6 \text{ m})^2(-150 \text{ N/C})$$

$$= -6.8 \times 10^5 \text{ C}.$$

One electron has a charge of -1.60×10^{-19} C. Hence this much excess negative electric charge corresponds to there being $(-6.8 \times 10^5 \text{ C})/(-1.60 \times 10^{-19} \text{ C}) = 4.2 \times 10^{24}$ excess electrons on the earth, or about 7 moles of excess electrons. This is compensated by an equal *deficiency* of electrons in the earth's upper atmosphere, so the combination of the earth and its atmosphere is electrically neutral.

SUMMARY

- Electric flux is a measure of the "flow" of electric field through a surface. It is equal to the product of an area element and the perpendicular component of $\vec{E}$, integrated over a surface:

$$\Phi_E = \int E \cos\phi \, dA = \int E_\perp dA = \int \vec{E} \cdot d\vec{A}. \qquad (23\text{–}5)$$

- Gauss's law is logically equivalent to Coulomb's law, but its use greatly simplifies problems that have a high degree of symmetry. It states that the total electric flux through a closed surface, which can be written as the surface integral of the component of $\vec{E}$ normal to the surface, equals a constant times the total charge Q_{encl} enclosed by the surface:

$$\Phi_E = \oint \vec{E} \cdot d\vec{A} = \frac{Q_{\text{encl}}}{\epsilon_0}. \qquad (23\text{–}8)$$

Gauss's law can be expressed in several equivalent forms:

$$\Phi_E = \oint E \cos\phi \, dA = \oint E_\perp dA = \oint \vec{E} \cdot d\vec{A} = \frac{Q_{\text{encl}}}{\epsilon_0}. \qquad (23\text{–}9)$$

- When excess charge is placed on a conductor and is at rest, it resides entirely on the surface, and $\vec{E} = 0$ everywhere in the material of the conductor.

- The following table lists electric fields caused by several symmetric charge distributions. In the table, q, Q, λ, and σ refer to the *magnitudes* of the quantities.

KEY TERMS

closed surface, 704
electric flux, 705
surface integral, 709
Gauss's law, 711
Gaussian surface, 713
Faraday's icepail experiment, 721

CHARGE DISTRIBUTION	POINT IN ELECTRIC FIELD	ELECTRIC FIELD MAGNITUDE	CHARGE DISTRIBUTION	POINT IN ELECTRIC FIELD	ELECTRIC FIELD MAGNITUDE
Single point charge q	Distance r from q	$E = \dfrac{1}{4\pi\epsilon_0}\dfrac{q}{r^2}$	Solid insulating sphere with radius R, charge Q distributed uniformly throughout volume	Outside sphere, $r > R$	$E = \dfrac{1}{4\pi\epsilon_0}\dfrac{Q}{r^2}$
Charge q on surface of conducting sphere with radius R	Outside sphere, $r > R$	$E = \dfrac{1}{4\pi\epsilon_0}\dfrac{q}{r^2}$		Inside sphere, $r < R$	$E = \dfrac{1}{4\pi\epsilon_0}\dfrac{Qr}{R^3}$
	Inside sphere, $r < R$	$E = 0$			
Infinite wire, charge per unit length λ	Distance r from wire	$E = \dfrac{1}{2\pi\epsilon_0}\dfrac{\lambda}{r}$	Infinite sheet of charge with uniform charge per unit area σ	Any point	$E = \dfrac{\sigma}{2\epsilon_0}$
Infinite conducting cylinder with radius R, charge per unit length λ	Outside cylinder, $r > R$	$E = \dfrac{1}{2\pi\epsilon_0}\dfrac{\lambda}{r}$	Two oppositely charged conducting plates with surface charge densities $+\sigma$ and $-\sigma$	Any point between plates	$E = \dfrac{\sigma}{\epsilon_0}$
	Inside cylinder, $r < R$	$E = 0$			

DISCUSSION QUESTIONS

Q23–1 A spherical Gaussian surface encloses a point charge q. If the point charge is moved from the center of the sphere to a point away from the center, does the electric field at a point on the surface change? Does the total flux through the Gaussian surface change? Explain.

Q23–2 A certain region of space bounded by an imaginary closed surface contains no charge. Is the electric field always zero everywhere on the surface? If not, under what circumstances is it zero on the surface?

Q23–3 A rubber balloon has a single point charge in its interior. Does the electric flux through the balloon depend on whether or not it is fully inflated? Explain your reasoning.

Q23–4 Are Coulomb's law and Gauss's law *completely* equivalent? Are there any situations in electrostatics in which one is valid and the other is not? Explain your reasoning.

Q23–5 In Fig. 23–13, suppose a third point charge were placed outside the yellow Gaussian surface C. Would this affect the electric flux through any of the surfaces A, B, C, or D in the figure? Why or why not?

Q23–6 If the electric field of a point charge were proportional to $1/r^3$ instead of $1/r^2$, would Gauss's law still be valid? Explain your reasoning. (*Hint:* Consider a spherical Gaussian surface centered on a single point charge.)

Q23–7 A solid, right circular cylinder of radius R and height $h = R$ has charge distributed uniformly throughout its volume. Can Gauss's law be used to calculate the electric field at all points inside the cylinder? What about at all points outside the cylinder? Explain your reasoning.

Q23–8 It was shown in the text that the electric field inside an empty cavity in a conductor is zero. Is this statement true no matter what the shape of the cavity? Why or why not?

Q23–9 The electric field $\vec{E}$ is uniform throughout a certain region of space. A small conducting sphere that carries a net charge Q is then placed in this region. What is the electric field inside the sphere? Explain your reasoning.

Q23–10 In a conductor, one or more electrons from each atom are free to roam throughout the volume of the conductor. Does this contradict the statement that any excess charge on a solid conductor must reside on its surface? Why or why not?

Q23–11 Explain the following statement: "In a static situation, the electric field at the surface of a conductor can have no com-ponent parallel to the surface, because this would violate the condition that the charges on the surface are at rest." Would this same statement be a valid one for the electric field at the surface of an *insulator*? Explain your answer, and the reason for any differences between the cases of a conductor and an insulator.

Q23–12 The magnitude of $\vec{E}$ at the surface of an irregularly shaped solid conductor must be greatest in regions where the surface curves most sharply, such as point A in Fig. 23–27, and must be least in flat regions such as point B in Fig. 23–27. Explain why this must be so by considering how electric field lines must be arranged near a conducting surface. How does the surface charge density compare at points A and B? Explain.

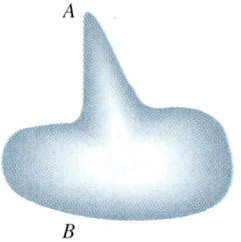

FIGURE 23–27 Question Q23–12.

Q23–13 A lightning rod is a pointed copper rod mounted on top of a building and welded to a heavy copper cable running down into the ground. Lightning rods are used to protect houses and barns from lightning; the lightning current runs through the copper rather than through the building. Why? Why should the end of the rod be pointed? (*Hint:* The answer to Q23–12 may be helpful.)

Q23–14 A solid conductor has a cavity in its interior. Would the presence of a point charge inside the cavity affect the electric field outside the conductor? Why or why not? Would the presence of a point charge outside the conductor affect the electric field inside the cavity? Again, why or why not?

Q23–15 Some modern aircraft are made primarily of composite materials that do not conduct electricity. The U.S. Federal Aviation Administration requires that such aircraft have conducting wires imbedded in their surfaces to provide protection when flying near thunderstorms. Explain the physics behind this requirement.

EXERCISES

SECTION 23–3 CALCULATING ELECTRIC FLUX

23–1 A flat sheet of paper of area 0.250 m^2 is oriented so that the normal to the sheet is at an angle of $60°$ to a uniform electric field of magnitude 14 N/C. a) Find the magnitude of the electric flux through the sheet. b) Does the answer to part (a) depend on the shape of the sheet? Why or why not? c) For what angle ϕ between the normal to the sheet and the electric field is the magnitude of the flux through the sheet i) largest? ii) smallest? Explain your answers.

23–2 The cube in Fig. 23–28 has sides of length $L = 10.0$ cm. The electric field is uniform, has a magnitude $E = 4.00 \times 10^3$ N/C, and is parallel to the xy-plane at an angle of $36.9°$ measured from the $+x$-axis toward the $+y$-axis. a) What is the electric flux through each of the six cube faces S_1, S_2, S_3, S_4, S_5, and S_6? b) What is the total electric flux through all faces of the cube?

23–3 A cube has sides of length L. It is placed with one corner at the origin as shown in Fig. 23–28. The electric field is uniform and given by $\vec{E} = -B\hat{i} + C\hat{j} - D\hat{k}$, where B, C, and D are positive constants. a) Find the electric flux through each of the six cube faces S_1, S_2, S_3, S_4, S_5, and S_6. b) Find the electric flux through the entire cube.

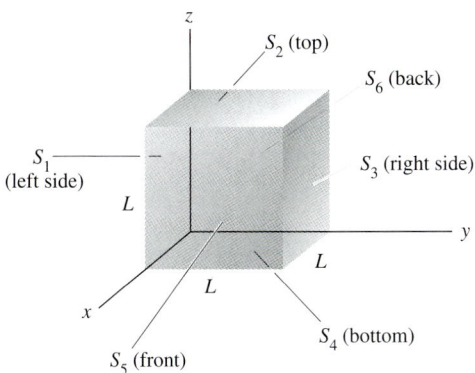

FIGURE 23–28 Exercises 23–2 and 23–3, and Problem 23–20.

23–4 A flat sheet is in the shape of a rectangle with sides of length 0.400 m and 0.600 m. The sheet is immersed in a uniform electric field of magnitude 75.0 N/C that is directed at 20° from the plane of the sheet (Fig. 23–29). Find the magnitude of the electric flux through the sheet.

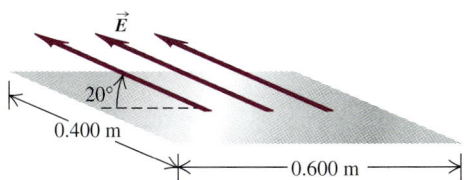

FIGURE 23–29 Exercise 23–4.

23–5 It was shown in Example 22–11 (Section 22–7) that the electric field due to an infinite line of charge is perpendicular to the line and has magnitude $E = \lambda/2\pi\epsilon_0 r$. Consider an imaginary cylinder with radius $r = 0.250$ m and length $l = 0.400$ m that has an infinite line of positive charge running along its axis. The charge per unit length on the line is $\lambda = 6.00$ μC/m. a) What is the electric flux through the cylinder due to this infinite line of charge? b) What is the flux through the cylinder if its radius is increased to $r = 0.500$ m? c) What is the flux through the cylinder if its length is increased to $l = 0.800$ m?

SECTION 23–4 GAUSS'S LAW

23–6 The three small spheres shown in Fig. 23–30 carry charges $q_1 = 4.00$ nC, $q_2 = -7.80$ nC, and $q_3 = 2.40$ nC. Find the net electric flux through each of the following closed surfaces shown in cross section in the figure: a) S_1; b) S_2; c) S_3; d) S_4; e) S_5. f) Do your answers to parts (a) through (e) depend on how the charge is distributed over each small sphere? Why or why not?

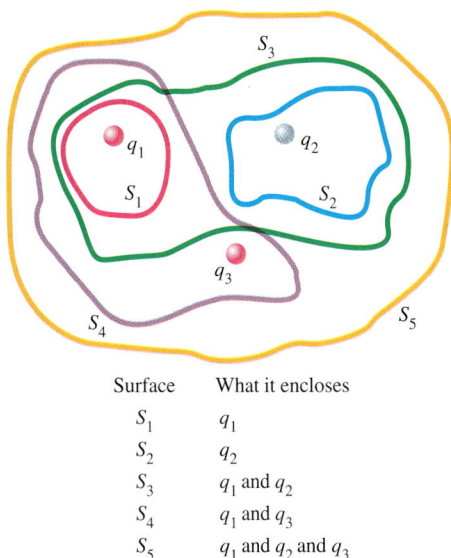

Surface	What it encloses
S_1	q_1
S_2	q_2
S_3	q_1 and q_2
S_4	q_1 and q_3
S_5	q_1 and q_2 and q_3

FIGURE 23–30 Exercise 23–6.

23–7 a) A closed surface encloses a net charge of -3.60 μC. What is the net electric flux through the surface? b) The electric flux through a closed surface is found to be 780 N · m²/C. What quantity of charge is enclosed by the surface? c) The closed surface in part (b) is a cube with sides of length 2.50 cm. From the information given in part (b), is it possible to tell where within the cube the charge is located? Explain.

23–8 A point charge $q_1 = 4.00$ nC is located on the x-axis at $x = 2.00$ m, and a second point charge $q_2 = -6.00$ nC is on the y-axis at $y = 1.00$ m. What is the total electric flux due to these two point charges through a spherical surface centered at the origin and with radius a) 0.500 m? b) 1.50 m? c) 2.50 m?

23–9 In a certain region of space, the electric field $\vec{E}$ is uniform. a) Use Gauss's law to prove that this region of space must be electrically neutral; that is, the volume charge density ρ must be zero. b) Is the converse true? That is, in a region of space where there is no charge, must $\vec{E}$ be uniform? Explain.

23–10 a) In a certain region of space, the volume charge density ρ has a uniform positive value. Can $\vec{E}$ be uniform in this region? Explain. b) Suppose that in this region of uniform positive ρ there is a "bubble" within which $\rho = 0$. Can $\vec{E}$ be uniform within this bubble? Explain.

23–11 A 9.60-μC point charge is at the center of a cube with sides of length 0.500 m. a) What is the electric flux through one of the six faces of the cube? b) How would your answer to part (a) change if the sides were length 0.250 m? Explain.

SECTION 23–5 APPLICATIONS OF GAUSS'S LAW
SECTION 23–6 CHARGES ON CONDUCTORS

23–12 A solid metal sphere of radius 0.450 m carries a net charge of 0.250 nC. Find the magnitude of the electric field a) at a point 0.100 m outside the surface of the sphere; b) at a point inside the sphere, 0.100 m below the surface.

23–13 In a physics lecture demonstration a charge of $-0.180 \mu C$ is placed on the spherical dome of a Van de Graaff generator. a) How far away from the center of the dome should you sit for the electric field at your location to be within the recommended maximum of 614 N/C (Exercise 22–28)? b) The answer to part (a) doesn't depend on the radius of the dome. Why not?

23–14 The cylindrical imaging drum of a photocopier is to have an electric field just outside its surface of 1.40×10^5 N/C. a) If the drum has a surface area of 0.0610 m^2 (the area of a $8\frac{1}{2} \times 11$ in. sheet of paper), what total quantity of charge must reside on the surface of the drum? b) If the surface area of the drum is increased to 0.122 m^2 so that larger sheets of paper can be used, what total quantity of charge is required to produce the same 1.40×10^5 N/C electric field just above the surface?

23–15 How many excess electrons must be added to an isolated spherical conductor 32.0 cm in diameter to produce an electric field of 1150 N/C just outside the surface?

23–16 A conductor with an inner cavity, like that shown in Fig. 23–20c, carries a total charge of +5.00 nC. The charge within the cavity, insulated from the conductor, is −6.00 nC. How much charge is on a) the inner surface of the conductor? b) the outer surface of the conductor?

23–17 Apply Gauss's law to the Gaussian surfaces S_2, S_3, and S_4 in Fig. 23–18b to calculate the electric field between and outside the plates.

23–18 A square insulating sheet 80.0 cm on a side is held horizontally. The sheet has 7.50 nC of charge spread uniformly over its area. a) Calculate the electric field at a point 0.100 mm above the center of the sheet. b) Estimate the electric field at a point 100 m above the center of the sheet. c) Would the answers to parts (a) and (b) be different if the sheet were made of a conducting material? Why or why not?

23–19 An infinitely long cylindrical conductor has radius R and uniform surface charge density σ. a) In terms of σ and R, what is the charge per unit length λ for the cylinder? b) In terms of σ, what is the magnitude of the electric field produced by the charged cylinder at a distance $r > R$ from its axis? c) Express the result of part (b) in terms of λ and show that the electric field outside the cylinder is the same as if all the charge were on the axis. Compare your result to the result for a line of charge in Example 23–6 (Section 23–5).

PROBLEMS

23–20 A cube has sides of length $L = 0.300$ m. It is placed with one corner at the origin as shown in Fig. 23–28. The electric field is not uniform, but is given by $\vec{E} = (-5.00 \text{ N/C} \cdot \text{m})x\hat{\imath} + (3.00 \text{ N/C} \cdot \text{m})z\hat{k}$. a) Find the electric flux through each of the six cube faces S_1, S_2, S_3, S_4, S_5, and S_6. b) Find the total electric charge inside the cube.

23–21 A flat, square surface with sides of length L is described by the equations

$$x = L, \qquad 0 \le y \le L, \qquad 0 \le z \le L.$$

a) Draw this square and show the x-, y-, and z-axes. b) Find the electric flux through the square due to a positive point charge q located at the origin ($x = 0$, $y = 0$, $z = 0$). (Hint: Think of the square as part of a cube centered on the origin.)

23–22 The electric field $\vec{E}$ in Fig. 23–31 is everywhere parallel to the x-axis, so the components E_y and E_z are zero. The x-component of the field E_x depends on x but not on y and z. At points in the yz-plane (where $x = 0$), $E_x = 125$ N/C. a) What is the electric flux through surface I in Fig. 23–31? b) What is the electric flux through surface II? c) The volume shown in the figure is a small section of a very large insulating slab 1.0 m thick. If there is a total charge −24.0 nC within the volume shown, what are the magnitude and direction of $\vec{E}$ at the face opposite surface I? d) Is the electric field produced only by charges within the slab, or is the field also due to charges outside the slab? How can you tell?

23–23 The electric field $\vec{E}_1$ at one face of a parallelepiped is uniform over the entire face and is directed out of the face. At the opposite face, the electric field $\vec{E}_2$ is also uniform over the entire face and is directed into that face (Fig. 23–32). The two faces in question are inclined at 30.0° from the horizontal, while $\vec{E}_1$ and $\vec{E}_2$ are both horizontal; $\vec{E}_1$ has a magnitude of 2.50×10^4 N/C,

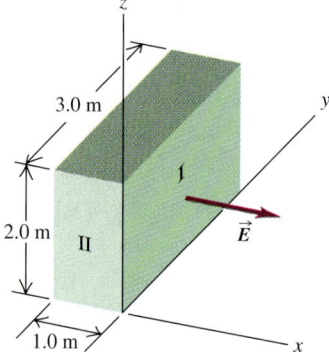

FIGURE 23–31 Problem 23–22.

and $\vec{E}_2$ a magnitude of 7.00×10^4 N/C. a) Assuming that no other electric field lines cross the surfaces of the parallelepiped, determine the net charge contained within. b) Is the electric field produced only by the charges within the parallelepiped, or is the field also due to charges outside the parallelepiped? How can you tell?

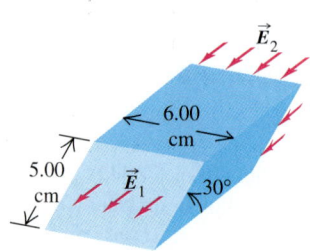

FIGURE 23–32 Problem 23–23.

23–24 A Sphere in a Sphere. A solid conducting sphere carrying charge q has radius a. It is inside a concentric hollow conducting sphere of inner radius b and outer radius c. The hollow sphere has no net charge. a) Derive expressions for the electric field magnitude in terms of the distance r from the center for the regions $r < a$, $a < r < b$, $b < r < c$, and $r > c$. b) Graph the magnitude of the electric field as a function of r from $r = 0$ to $r = 2c$. c) What is the charge on the inner surface of the hollow sphere? d) On the outer surface? e) Represent the charge of the small sphere by four plus signs. Sketch the field lines of the system within a spherical volume of radius $2c$.

23–25 A solid conducting sphere of radius R that carries positive charge Q is concentric with a very thin insulating shell of radius $2R$ that also carries charge Q. The charge Q is distributed uniformly over the insulating shell. a) Find the electric field (magnitude and direction) in each of the regions $0 < r < R$, $R < r < 2R$, and $r > 2R$. b) Graph the electric field magnitude as a function of r.

23–26 A conducting spherical shell with inner radius a and outer radius b has a positive point charge Q located at its center. The total charge on the shell is $-3Q$, and it is insulated from its surroundings (Fig. 23–33). a) Derive expressions for the electric field magnitude in terms of the distance r from the center for the regions $r < a$, $a < r < b$, and $r > b$. b) What is the surface charge density on the inner surface of the conducting shell? c) What is the surface charge density on the outer surface of the conducting shell? d) Sketch the electric field lines and the location of all charges. e) Graph the electric field magnitude as a function of r.

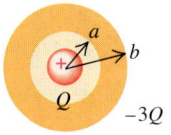

FIGURE 23–33 Problem 23–26.

23–27 Concentric Spherical Shells. A small conducting spherical shell with inner radius a and outer radius b is concentric with a larger conducting spherical shell with inner radius c and outer radius d (Fig. 23–34). The inner shell has total charge $+2q$, and the outer shell has charge $+4q$. a) Calculate the electric field (magnitude and direction) in terms of q and the distance r from the common center of the two shells for i) $r < a$; ii) $a < r < b$; iii) $b < r < c$; iv) $c < r < d$; v) $r > d$. Show your results in a graph of the radial component of $\vec{E}$ as a function of r. b) What is the total charge on the i) inner surface of the small shell; ii) outer surface of the small shell; iii) inner surface of the large shell; iv) outer surface of the large shell?

23–28 Repeat Problem 23–27, but now let the outer shell have charge $-2q$. As in Problem 23–27, the inner shell has charge $+2q$.

23–29 Repeat Problem 23–27, but now let the outer shell have charge $-4q$. As in Problem 23–27, the inner shell has charge $+2q$.

23–30 A solid conducting sphere of radius R carries a positive total charge Q. The sphere is surrounded by an insulating shell

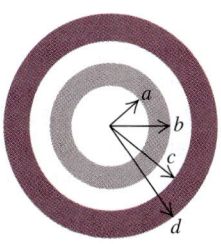

FIGURE 23–34 Problems 23–27, 23–28, 23–29, and 23–32.

with inner radius R and outer radius $2R$. The insulating shell has a uniform charge density ρ. a) Find the value of ρ so that the net charge of the entire system is zero. b) If ρ has the value found in part (a), find the electric field (magnitude and direction) in each of the regions $0 < r < R$, $R < r < 2R$, and $r > 2R$. Show your results in a graph of the radial component of $\vec{E}$ as a function of r. c) As a general rule, the electric field is only discontinuous at locations where there is a thin sheet of charge. Explain how your results in part (b) agree with this rule.

23–31 Negative charge $-Q$ is distributed uniformly over the surface of a thin spherical insulating shell with radius R. Calculate the force (magnitude and direction) that the shell exerts on a positive point charge q located a) a distance $r > R$ from the center of the shell (outside the shell); b) a distance $r < R$ from the center of the shell (inside the shell).

23–32 A small *insulating* spherical shell with inner radius a and outer radius b is concentric with a larger *insulating* spherical shell with inner radius c and outer radius d (Fig. 23–34). The inner shell has total charge $+q$ distributed uniformly over its volume, and the outer shell has charge $-q$ distributed uniformly over its volume. a) Calculate the charge densities in the inner shell and outer shell. b) Calculate the electric field (magnitude and direction) in terms of q and the distance r from the common center of the two shells for i) $r < a$; ii) $a < r < b$; iii) $b < r < c$; iv) $c < r < d$; v) $r > d$. Show your results in a graph of the radial component of $\vec{E}$ as a function of r.

23–33 The Coaxial Cable. A long coaxial cable consists of an inner cylindrical conductor with radius a and an outer coaxial cylinder with inner radius b and outer radius c. The outer cylinder is mounted on insulating supports and has no net charge. The inner cylinder has a uniform positive charge per unit length λ. Calculate the electric field a) at any point between the cylinders, a distance r from the axis; b) at any point outside the outer cylinder. c) Graph the magnitude of the electric field as a function of the distance r from the axis of the cable, from $r = 0$ to $r = 2c$. d) Find the charge per unit length on the inner surface and on the outer surface of the outer cylinder.

23–34 A very long conducting tube (hollow cylinder) has inner radius a and outer radius b. It carries charge per unit length $+\alpha$, where α is a positive constant with units of C/m. A line of charge lies along the axis of the tube. The line of charge has charge per unit length $+\alpha$. a) Calculate the electric field in terms of α and the distance r from the axis of the tube for i) $r < a$; ii) $a < r < b$; iii) $r > b$. Show your results in a graph of E as a function of r. b) What is the charge per unit length on i) the inner surface of the tube; ii) the outer surface of the tube?

23–35 Repeat Problem 23–34, but now let the conducting tube have charge per unit length $-\alpha$. As in Problem 23–34, the line of charge has charge per unit length $+\alpha$.

23–36 A very long solid cylinder of radius R has positive charge uniformly distributed throughout it, with charge per unit volume ρ. a) Derive the expression for the electric field inside the volume at a distance r from the axis of the cylinder in terms of the charge density ρ. b) What is the electric field at a point outside the volume in terms of the charge per unit length λ in the cylinder? c) Compare the answers to parts (a) and (b) for $r = R$. d) Graph the electric field magnitude as a function of r from $r = 0$ to $r = 3R$.

23–37 A single, isolated, large conducting plate (Fig. 23–35) has a charge per unit area σ on its surface. Because the plate is a conductor, the electric field at its surface is perpendicular to the surface and has magnitude $E = \sigma/\epsilon_0$. a) In Example 23–7 (Section 23–5) it was shown that the field caused by a large, uniformly charged sheet with charge per unit area σ has magnitude $E = \sigma/2\epsilon_0$, exactly *half* as much as for a charged conducting plate. Why is there a difference? b) By regarding the charge distribution on the conducting plate as being two sheets of charge (one on each surface), each with charge per unit area σ, use the result of Example 23–7 and the principle of superposition to show that $E = 0$ inside the plate and $E = \sigma/\epsilon_0$ outside the plate.

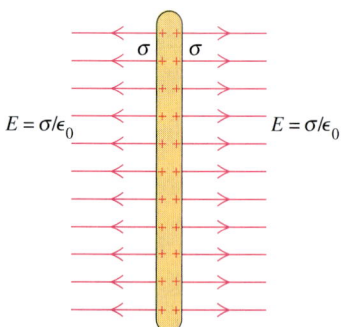

FIGURE 23–35 Problem 23–37.

23–38 Thomson's Model of the Atom. In the early years of the twentieth century, a leading model of the structure of the atom was that of the English physicist J. J. Thomson (the discoverer of the electron). In Thomson's model, an atom consisted of a sphere of positively charged material into which were imbedded negatively charged electrons, like chocolate chips in a ball of cookie dough. Consider such an atom consisting of one electron of mass m and charge $-e$, which may be regarded as a point charge, and a uniformly charged sphere of charge $+e$ and radius R. a) Explain why the equilibrium position of the electron is at the center of the nucleus. b) In Thomson's model, it was assumed that the positive material provided little or no resistance to the motion of the electron. If the electron is displaced from equilibrium by a distance less than R, show that the resulting motion of the electron will be simple harmonic, and calculate the frequency of oscillation. (*Hint:* Review the definition of simple harmonic motion in Section 13–3. If it can be shown that the net force on the electron is of this form, then it follows that the

motion is simple harmonic. Conversely, if the net force on the electron does not follow this form, the motion is not simple harmonic.) c) By Thomson's time, it was known that excited atoms emit light waves of only certain frequencies. In his model, the frequency of emitted light is the same as the oscillation frequency of the electron or electrons in the atom. What would have to be the radius of a Thomson-model atom for it to produce red light of frequency 4.57×10^{14} Hz? Compare your answer to the radii of real atoms, which are of order 10^{-10} m. (See Appendix F for data about the electron.) d) If the electron were displaced from equilibrium by a distance greater than R, would the electron oscillate? Would its motion be simple harmonic? Explain your reasoning. (*Historical note:* In 1910, the atomic nucleus was discovered, proving the Thomson model to be incorrect. An atom's positive charge is not spread over its volume as Thomson supposed, but is concentrated in the tiny nucleus of radius 10^{-14} to 10^{-15} m.)

23–39 Thomson's Model of the Atom, Continued. Using Thomson's (outdated) model of the atom described in Problem 23–38, consider an atom consisting of two electrons, each of charge $-e$, imbedded in a sphere of charge $+2e$ and radius R. In equilibrium, each electron is a distance d from the center of the atom (Fig. 23–36). Find the distance d in terms of the other properties of the atom.

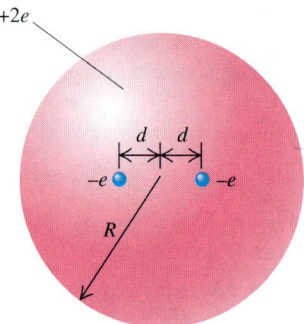

FIGURE 23–36 Problem 23–39.

23–40 Electric Field Inside a Hydrogen Atom. A hydrogen atom is made up of a proton of charge $+Q = 1.60 \times 10^{-19}$ C and an electron of charge $-Q = -1.60 \times 10^{-19}$ C. The proton may be regarded as a point charge at $r = 0$, the center of the atom. The motion of the electron causes its charge to be "smeared out" into a spherical distribution around the proton, so that the electron is equivalent to a charge per unit volume of

$$\rho(r) = -\frac{Q}{\pi a_0^3} e^{-2r/a_0},$$

where $a_0 = 5.29 \times 10^{-11}$ m is called the *Bohr radius.* a) Find the total amount of the hydrogen atom's charge that is enclosed within a sphere of radius r centered on the proton. Show that as $r \to \infty$, the enclosed charge goes to zero. Explain this result. b) Find the electric field (magnitude and direction) caused by the charge of the hydrogen atom as a function of r. c) Graph the electric field magnitude E as a function of r.

23–41 Probing Atomic Nuclei with Electron Scattering. To study the structure of the lead nucleus, electrons (charge $-e = -1.60 \times 10^{-19}$ C, mass 9.11×10^{-31} kg) are fired at a lead target. Some of the electrons actually enter the nuclei of the target, and the deflection of these electrons is measured. The deflection is caused by the charge of the nucleus, which is distributed approximately uniformly over the spherical volume of the nucleus. A lead nucleus has a charge of $+82e$ and a radius of $R = 7.1 \times 10^{-15}$ m. Find the acceleration of an electron at the following distances from the center of a calcium nucleus: a) $2R$; b) R; c) $R/2$; d) zero (at the center).

23–42 A Uniformly Charged Slab. A slab of insulating material has thickness $2d$ and is oriented so that its faces are parallel to the yz-plane and given by the planes $x = d$ and $x = -d$. The y- and z-dimensions of the slab are very large compared to d and may be treated as essentially infinite. The slab has a uniform positive charge density ρ. a) Explain why the electric field due to the slab at the center of the slab ($x = 0$) is zero. b) Using Gauss's law, find the electric field due to the slab (magnitude and direction) at all points in space.

23–43 A Non–Uniformly Charged Slab. Repeat Problem 23–42, but now let the charge density of the slab be given by $\rho(x) = \rho_0 (x/d)^2$, where ρ_0 is a positive constant.

23–44 Can Electric Forces Alone Give Stable Equilibrium? In Chapter 22, several examples were given of calculating the force exerted on a point charge by other point charges in its surroundings. a) Consider a positive point charge $+q$. Give an example of how you would place two other point charges of your choosing so that the net force on charge $+q$ will be zero. b) If the net force on charge $+q$ is zero, then that charge is in equilibrium. The equilibrium will be *stable* if, when the charge $+q$ is displaced slightly in *any* direction from its position of equilibrium, the net force on the charge pushes it back towards the equilibrium position. For this to be the case, what must be the direction of the electric field $\vec{E}$ due to the other charges at points surrounding the equilibrium position of $+q$? c) Imagine that the charge $+q$ is moved very far away, and imagine a small Gaussian surface centered on the position where $+q$ was in equilibrium. By applying Gauss's law to this surface, show that it is *impossible* to satisfy the condition for stability described in part (b). In other words, a charge $+q$ cannot be held in stable equilibrium by electrostatic forces alone. This result is known as *Earnshaw's theorem.* d) Parts (a) through (c) referred to the equilibrium of a positive point charge $+q$. Prove that Earnshaw's theorem also applies to a negative point charge $-q$.

23–45 A non-uniform but spherically symmetric distribution of charge has a charge density $\rho(r)$ given as follows:

$$\rho(r) = \rho_0(1 - r/R) \qquad \text{for} \qquad r \leq R,$$
$$\rho(r) = 0 \qquad \text{for} \qquad r \geq R,$$

where $\rho_0 = 3Q/\pi R^3$ is a positive constant. a) Show that the total charge contained in the charge distribution is Q. b) Show that the electric field in the region $r \geq R$ is identical to that produced by a point charge Q at $r = 0$. c) Obtain an expression for the electric field in the region $r \leq R$. d) Graph the electric field magnitude E as a function of r. e) Find the value of r at which the

electric field is maximum, and find the value of that maximum field.

23–46 A non-uniform but spherically symmetric distribution of charge has a charge density $\rho(r)$ given as follows:

$$\rho(r) = \rho_0(1 - 4r/3R) \qquad \text{for} \qquad r \leq R,$$
$$\rho(r) = 0 \qquad \text{for} \qquad r \geq R,$$

where ρ_0 is a positive constant. a) Find the total charge contained in the charge distribution. b) Obtain an expression for the electric field in the region $r \geq R$. c) Obtain an expression for the electric field in the region $r \leq R$. d) Graph the electric field magnitude E as a function of r. e) Find the value of r at which the electric field is maximum, and find the value of that maximum field.

23–47 Gauss's Law for Gravitation. The gravitational force between two point masses separated by a distance r is proportional to $1/r^2$, just like the electric force between two point charges. Because of this similarity between gravitational and electric interactions, there is also a Gauss's law for gravitation. a) Let $\vec{g}$ be the acceleration due to gravity caused by a point mass m at the origin, so that $\vec{g} = -(Gm/r^2)\hat{r}$. Consider a spherical Gaussian surface of radius r centered on this point mass, and show that the flux of $\vec{g}$ through this surface is given by

$$\oint \vec{g} \cdot d\vec{A} = -4\pi Gm.$$

b) By following the same logical steps used in Section 23–4 to obtain Gauss's law for the electric field, show that the flux of $\vec{g}$ through *any* closed surface is given by

$$\oint \vec{g} \cdot d\vec{A} = -4\pi GM_{encl},$$

where M_{encl} is the total mass enclosed within the closed surface.

23–48 Applying Gauss's Law for Gravitation. Using Gauss's law for gravitation (derived in part (b) of Problem 23–47), show that the following statements are true: a) For any spherically symmetric mass distribution of total mass M, the acceleration due to gravity outside the distribution is the same as though all the mass were concentrated at the center. (*Hint:* See Example 23–5 in Section 23–5.) b) At any point inside a spherically symmetric shell of mass, the acceleration due to gravity is zero. (*Hint:* See Example 23–5.) c) If we could drill a hole through a spherically symmetric planet to its center, and if the density were uniform, we would find that the magnitude of $\vec{g}$ is directly proportional to the distance r from the center. (*Hint:* See Example 23–9 in Section 23–5.) We proved these results in Section 12–7 using some fairly strenuous analysis; the proofs using Gauss's law for gravitation are *much* easier.

23–49 a) An insulating sphere of radius a has a uniform charge density ρ. The sphere is not centered at the origin, but at $\vec{r} = \vec{b}$. Show that the electric field inside the sphere is given by $\vec{E} = \rho(\vec{r} - \vec{b})/3\epsilon_0$. b) An insulating sphere of radius R has a spherical hole of radius a located within its volume and centered a distance b from the center of the sphere, where $a < b < R$ (a cross section of the sphere is shown in Fig. 23–37). The solid part of the sphere has a uniform volume charge density ρ. Find the magnitude and direction of the electric field $\vec{E}$ inside the hole, and show that $\vec{E}$ is uniform over the entire hole. (*Hint:* Use the principle of superposition and the result of part (a).)

FIGURE 23–37 Problem 23–49.

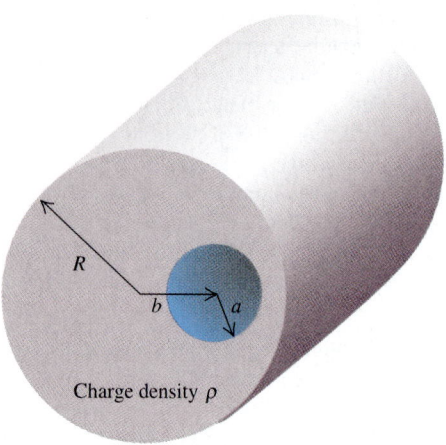

FIGURE 23–38 Problem 23–50.

23–50 A very long, solid insulating cylinder of radius R has a cylindrical hole of radius a bored along its entire length. The axis of the hole is a distance b from the axis of the cylinder, where $a < b < R$ (Fig. 23–38). The solid material of the cylinder has a uniform volume charge density ρ. Find the magnitude and direction of the electric field $\vec{E}$ inside the hole, and show that $\vec{E}$ is uniform over the entire hole. (*Hint:* See Problem 23–49.)

23–51 Positive charge Q is distributed uniformly over each of two spherical volumes of radius R. One sphere of charge is centered at the origin and the other at $x = 2R$ (Fig. 23–39). Find the magnitude and direction of the net electric field due to these two distributions of charge at the following points on the x–axis: a) $x = 0$; b) $x = R/2$; c) $x = R$; d) $x = 3R$.

23–52 Repeat Problem 23–51, but now let the left-hand sphere have positive charge Q and let the right-hand sphere have negative charge $-Q$.

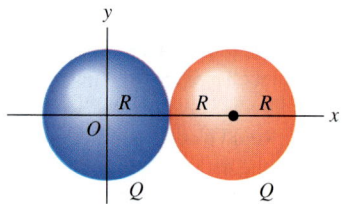

FIGURE 23–39 Problems 23–51 and 23–52.

CHALLENGE PROBLEMS

23–53 A region in space contains a total positive charge Q that is distributed spherically such that the volume charge density $\rho(r)$ is given by

$$\rho(r) = \alpha \qquad \text{for} \qquad r \le R/2,$$
$$\rho(r) = 2\alpha(1 - r/R) \qquad \text{for} \qquad R/2 \le r \le R,$$
$$\rho(r) = 0 \qquad \text{for} \qquad r \ge R.$$

Here α is a positive constant having units of C/m³. a) Determine α in terms of Q and R. b) Using Gauss's law, derive an expression for the magnitude of $\vec{E}$ as a function of r. Do this separately for all three regions. Express your answers in terms of the total charge Q. Be sure to check that your results agree on the boundaries of the regions. c) What fraction of the total charge is contained within the region $r \le R/2$? d) If an electron with charge $q' = -e$ is oscillating back and forth about $r = 0$ (the center of the distribution) with an amplitude less than $R/2$, show that the motion is simple harmonic. (*Hint:* Review the discussion of simple harmonic motion in Section 13–3. If, and only if, the net force on the electron is proportional to its displacement from equilibrium, then the motion is simple harmonic.) e) What is the period of the motion in part (d)? f) If the amplitude of the

motion described in part (e) is greater than $R/2$, is the motion still simple harmonic? Why or why not?

23–54 A region in space contains a total positive charge Q that is distributed spherically such that the volume charge density $\rho(r)$ is given by

$$\rho(r) = 3\alpha r/(2R) \qquad \text{for} \qquad r \le R/2,$$
$$\rho(r) = \alpha[1 - (r/R)^2] \qquad \text{for} \qquad R/2 \le r \le R$$
$$\rho(r) = 0 \qquad \text{for} \qquad r \ge R.$$

Here α is a positive constant having units of C/m³. a) Determine α in terms of Q and R. b) Using Gauss's law, derive an expression for the magnitude of the electric field as a function of r. Do this separately for all three regions. Express your answers in terms of the total charge Q. c) What fraction of the total charge is contained within the region $R/2 \le r \le R$? d) What is the magnitude of $\vec{E}$ at $r = R/2$? e) If an electron with charge $q' = -e$ is released from rest at any point in any of the three regions, the resulting motion will be oscillatory but not simple harmonic. Why? (See Challenge Problem 23–53.)

A lightning bolt releases a staggering amount of electric potential energy. Charge flows between a cloud and the ground at a rate that can exceed 10^5 coulombs per second. The electric potential energy per coulomb—that is, the electric potential—released in this process can be as great as 10^7 volts, or 10^7 joules per coulomb.

24-1 INTRODUCTION

This chapter is about energy associated with electrical interactions. Every time you turn on a light, a CD player, or an electric appliance, you are making use of electrical energy, an indispensable ingredient of our technological society. In Chapters 6 and 7 we introduced the concepts of *work* and *energy* in the context of mechanics; now we'll combine these concepts with what we've learned about electric charge, electric forces, and electric fields.

When a charged particle moves in an electric field, the field exerts a force that can do *work* on the particle. This work can always be expressed in terms of electric potential energy. Just as gravitational potential energy depends on the height of a mass above the earth's surface, electric potential energy depends on the position of the charged particle in the electric field. We'll describe electric potential energy using a new concept called *electric potential,* or simply *potential.* In circuits a difference in potential from one point to another is often called *voltage.* The concepts of potential and voltage are crucial to understanding how electric circuits work and have equally important applications to electron beams in TV picture tubes, high-energy particle accelerators, and many other devices.

24-2 ELECTRIC POTENTIAL ENERGY

The concepts of work, potential energy, and conservation of energy proved to be extremely useful in our study of

mechanics. In this section we'll show that these concepts are just as useful for understanding and analyzing electrical interactions.

Let's begin by reviewing several essential points from Chapters 6 and 7. First, when a force $\vec{F}$ acts on a particle that moves from point a to point b, the work $W_{a \to b}$ done by the force is given by a *line integral:*

$$W_{a \to b} = \int_a^b \vec{F} \cdot d\vec{l} = \int_a^b F \cos \phi \, dl \quad \text{(work done by a force),}$$

(24–1)

where $d\vec{l}$ is an infinitesimal displacement along the particle's path and ϕ is the angle between $\vec{F}$ and $d\vec{l}$ at each point along the path.

Second, if the force $\vec{F}$ is *conservative,* as we defined the term in Section 7–4, the work done by $\vec{F}$ can always be expressed in terms of a **potential energy** U. When the particle moves from a point where the potential energy is U_a to a point where it is U_b, the change in potential energy is $\Delta U = U_b - U_a$, and the work $W_{a \to b}$ done by the force is

$$W_{a \to b} = U_a - U_b = -(U_b - U_a) = -\Delta U \quad \text{(work done by a conservative force).}$$

(24–2)

When $W_{a \to b}$ is positive, U_a is greater than U_b, ΔU is negative, and the potential energy *decreases.* That's what happens when a baseball falls from a high point (a) to a

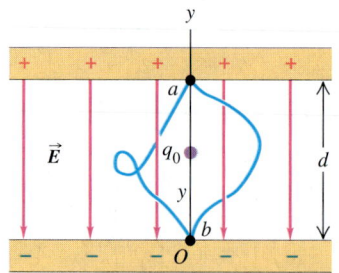

24–1 A test charge q_0 moving from a to b experiences a force of magnitude q_0E. The work done by this force is $W_{a \to b} = q_0Ed$ and is independent of the particle's path.

(a) Decreasing U

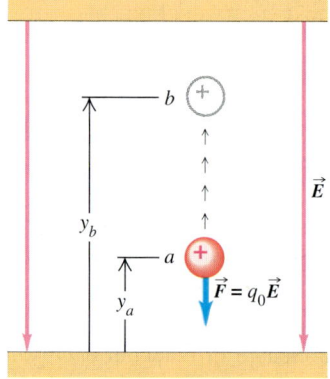

(b) Increasing U

24–2 (a) When a positive charge moves in the direction of an electric field, the field does positive work and the potential energy decreases. (b) When a positive charge moves in the direction opposite to an electric field, the field does negative work and the potential energy increases.

lower point (b) under the influence of the earth's gravity; the force of gravity does positive work, and the gravitational potential energy decreases. When a ball is thrown upward, the gravitational force does negative work during the ascent, and the potential energy increases.

Third, the work-energy theorem says that the change in kinetic energy $\Delta K = K_b - K_a$ during any displacement is equal to the *total* work done on the particle. If the only work done on the particle is done by conservative forces, then Eq. (24–2) gives the total work, and $K_b - K_a = -(U_b - U_a)$. We usually write this as

$$K_a + U_a = K_b + U_b. \tag{24–3}$$

This is, the total mechanical energy (kinetic plus potential) is *conserved* under these circumstances.

ELECTRIC POTENTIAL ENERGY IN A UNIFORM FIELD

Let's look at an electrical example of these basic concepts. In Fig. 24–1 a pair of charged parallel metal plates sets up a uniform, downward electric field with magnitude E. The field exerts a downward force with magnitude $F = q_0E$ on a positive test charge q_0. As the charge moves downward a distance d from point a to point b, the force on the test charge is constant and independent of its location. So the work done by the electric field is the product of the force magnitude and the component of displacement in the (downward) direction of the force:

$$W_{a \to b} = Fd = q_0Ed. \tag{24–4}$$

This is positive, since the force is in the same direction as the net displacement of the test charge.

The y-component of the electric force, $F_y = -q_0E$, is constant, and there is no x- or z-component. This is exactly analogous to the gravitational force on a mass m near the earth's surface; for this force, there is a constant y-component $F_y = -mg$ and the x- and z-components are zero. Because of this analogy, we can conclude that the force exerted on q_0 by the uniform electric field in Fig. 24–1 is *conservative,* just as is the gravitational force. This means that the work $W_{a \to b}$ done by the field is independent of the path the particle takes from a to b. We can represent this work with a *potential energy* function U, just as we did for gravitational potential energy in Section 7–2. The potential energy for the gravitational force $F_y = -mg$ was $U = mgy$; hence the potential energy for the electric force $F_y = -q_0E$ is

$$U = q_0Ey. \tag{24–5}$$

When the test charge moves from height y_a to height y_b, the work done on the charge by the field is given by

$$W_{a \to b} = -\Delta U = -(U_b - U_a) = -(q_0Ey_b - q_0Ey_a) = q_0E(y_a - y_b). \tag{24–6}$$

When y_a is greater than y_b (Fig. 24–2a), the positive test charge q_0 moves downward, in the same direction as $\vec{E}$; the displacement is in the same direction as the force $\vec{F} = q_0\vec{E}$, so the field does positive work and U decreases. (In particular, if $y_a - y_b = d$ as in Fig. 24–1, Eq. (24–6) gives $W_{a \to b} = q_0Ed$, in agreement with Eq. (24–4).) When y_a is less than y_b (Fig. 24–2b), the positive test charge q_0 moves upward, in the opposite direction to $\vec{E}$; the displacement is opposite the force, the field does negative work, and U increases.

If the test charge q_0 is *negative,* the potential energy increases when it moves with the field and decreases when it moves against the field (Fig. 24–3).

Whether the test charge is positive or negative, the following general rules apply: *U increases* if the test charge q_0 moves in the direction *opposite* to the electric force

(a) Increasing U

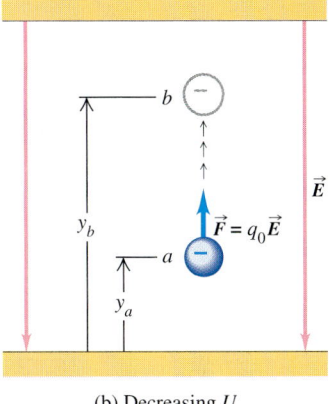

(b) Decreasing U

24-3 (a) When a negative charge moves in the direction of an electric field, the field does negative work and the potential energy increases.
(b) When a negative charge moves in the direction opposite to an electric field, the field does positive work and the potential energy decreases.

$\vec{F} = q_0\vec{E}$ (Figs. 24–2b and 24–3a); U *decreases* if q_0 moves in the *same* direction as $\vec{F} = q_0\vec{E}$ (Figs. 24–2a and 24–3b). This is the same behavior as for gravitational potential energy, which increases if a mass m moves upward (opposite the direction of the gravitational force) and decreases if m moves downward (in the same direction as the gravitational force).

ELECTRIC POTENTIAL ENERGY OF TWO POINT CHARGES

The idea of electric potential energy isn't restricted to the special case of a uniform electric field. Indeed, we can apply this concept to a point charge in *any* electric field caused by a static charge distribution. Recall from Chapter 22 that we can represent any charge distribution as a collection of point charges. Therefore it's useful to calculate the work done on a test charge q_0 moving in the electric field caused by a single, stationary point charge q. We'll consider first a displacement along the *radial* line in Fig. 24–4, from point a to point b. The force on q_0 is given by Coulomb's law, and its radial component is

$$F_r = \frac{1}{4\pi\epsilon_0}\frac{qq_0}{r^2}. \qquad (24\text{–}7)$$

If q and q_0 have the same sign (+ or −), the force is repulsive and F_r is positive; if the two charges have opposite signs, the force is attractive and F_r is negative. The force is *not* constant during the displacement, and we have to integrate to calculate the work $W_{a\to b}$ done on q_0 by this force as q_0 moves from a to b. We find

$$W_{a\to b} = \int_{r_a}^{r_b} F_r \, dr = \int_{r_a}^{r_b}\frac{1}{4\pi\epsilon_0}\frac{qq_0}{r^2}\, dr = \frac{qq_0}{4\pi\epsilon_0}\left(\frac{1}{r_a}-\frac{1}{r_b}\right). \qquad (24\text{–}8)$$

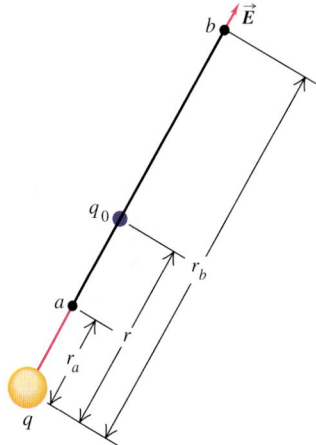

24-4 Charge q_0 moves along a straight line extending radially from charge q. As it moves from a to b, the distance varies from r_a to r_b.

The work done by the electric force for this particular path depends only on the end points.

In fact, the work is the same for *all possible* paths from a to b. To prove this, we consider a more general displacement (Fig. 24–5) in which a and b do not lie on the same radial line. From Eq. (24–1) the work done on q_0 during this displacement is given by

$$W_{a\to b} = \int_{r_a}^{r_b} F\cos\phi \, dl = \int_{r_a}^{r_b}\frac{1}{4\pi\epsilon_0}\frac{qq_0}{r^2}\cos\phi\, dl.$$

But the figure shows that $\cos\phi \, dl = dr$. That is, the work done during a small displacement $d\vec{l}$ depends only on the change dr in the distance r between the charges, which is the *radial component* of the displacement. Thus Eq. (24–8) is valid even for this more general displacement; the work done on q_0 by the electric field $\vec{E}$ produced by q depends

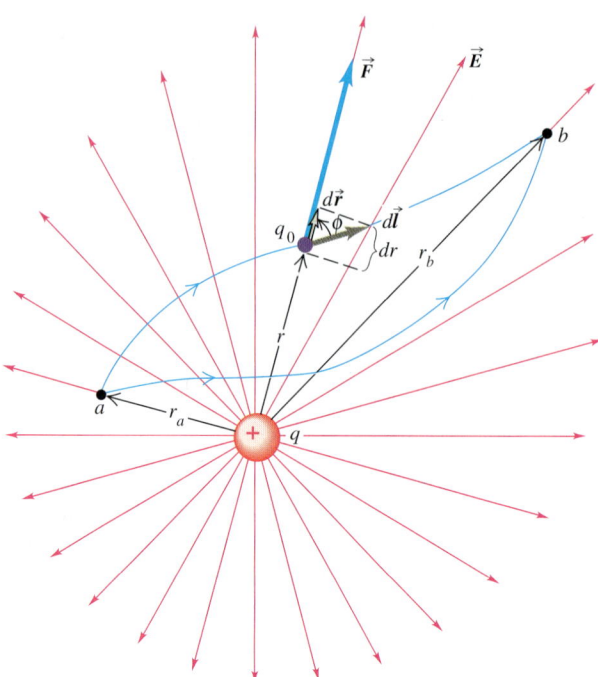

24–5 The work done on charge q_0 by the electric field produced by charge q depends only on the distances r_a and r_b.

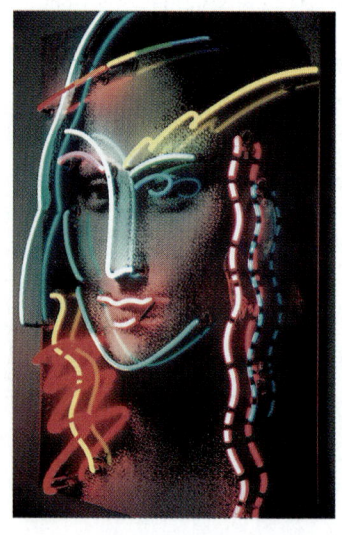

In neon lights, electric potential energy is transferred to atoms of neon gas. This added energy enables negative electrons in the atoms to move away from the positive nuclei. Within a fraction of a second the electrons fall back toward the nuclei, releasing the energy in the form of light.

only on r_a and r_b, not on the details of the path. Also, if q_0 returns to its starting point a by a different path, the total work done in the round-trip displacement is zero (the integral in Eq. (24–8) is from r_a back to r_a). These are the needed characteristics for a conservative force, as we defined it in Section 7–4. Thus the force on q_0 is a *conservative* force.

We see that Eqs. (24–2) and (24–8) are consistent if we define $qq_0/4\pi\epsilon_0 r_a$ to be the potential energy U_a when q_0 is at point a, a distance r_a from q, and we define $qq_0/4\pi\epsilon_0 r_b$ to be the potential energy U_b when q_0 is at point b, a distance r_b from q. Thus the potential energy U when the test charge q_0 is at *any* distance r from charge q is

$$U = \frac{1}{4\pi\epsilon_0}\frac{qq_0}{r} \qquad \text{(electric potential energy of two point charges } q \text{ and } q_0 \text{).} \qquad (24\text{–}9)$$

Note that we have *not* assumed anything about the signs of q and q_0; Eq. (24–9) is valid for any combination of signs. The potential energy is positive if the charges q and q_0 have the same sign and negative if they have opposite signs.

CAUTION ▶ Be careful not to confuse Eq. (24–9) for the potential energy of two point charges with the similar expression in Eq. (24–7) for the radial component of the electric force that one charge exerts on the other. The potential energy U is proportional to $1/r$, while the force component F_r is proportional to $1/r^2$. ◀

Potential energy is always defined relative to some reference point where $U = 0$. In Eq. (24–9), U is zero when q and q_0 are infinitely far apart and $r = \infty$. Therefore U represents the work that would be done on the test charge q_0 by the field of q if q_0 moved from an initial distance r to infinity. If q and q_0 have the same sign, the interaction is repulsive, this work is positive, and U is positive at any finite separation. If they have opposite sign, the interaction is attractive and U is negative.

We emphasize that the potential energy U given by Eq. (24–9) is a *shared* property of the two charges q and q_0; it is a consequence of the *interaction* between these two bodies. If the distance between the two charges is changed from r_a to r_b, the change in potential energy is the same whether q is held fixed and q_0 is moved or q_0 is held fixed and q is moved. For this reason we will never use the phrase "the electric potential energy *of* a point charge." (Likewise, if a mass m is at a height h above the earth's surface, the gravitational potential energy is a shared property of the mass m and the earth. We emphasized this in Sections 7–2 and 12–4.)

Gauss's law tells us that the electric field outside any spherically symmetric charge distribution is the same as though all the charge were concentrated at the center. Therefore Eq. (24–9) also holds if the test charge q_0 is outside any spherically symmetric charge distribution with total charge q at a distance r from the center.

EXAMPLE 24–1

Conservation of energy with electric forces The positron (the antiparticle of the electron) has a mass of 9.11×10^{-31} kg and a charge $+e = +1.60 \times 10^{-19}$ C. Suppose a positron moves in the vicinity of an alpha particle, which has a charge $+2e = 3.20 \times 10^{-19}$ C. The alpha particle is over 7000 times as massive as the positron, so we assume that it is at rest in some inertial frame of reference. When the positron is 1.00×10^{-10} m from the alpha particle, it is moving directly away from the alpha particle at a speed of 3.00×10^6 m/s. a) What is the positron's speed when the two particles are 2.00×10^{-10} m apart? b) What is the positron's speed when it is very far away from the alpha particle? c) How would the situation change if the moving particle were an electron (same mass as the positron but opposite charge)?

SOLUTION a) The electric force is conservative, so mechanical energy (kinetic plus potential) is conserved. Hence the initial and final kinetic and potential energies, K_a, K_b, U_a, and U_b, are related by Eq. (24–3), $K_a + U_a = K_b + U_b$. We want to find the final speed v_b of the positron. This appears in the expression for the final kinetic energy, $K_b = \frac{1}{2} m v_b^2$; solving the energy-conservation equation for K_b, we have

$$K_b = K_a + U_a - U_b.$$

The values of the energies on the right-hand side of this expression are

$$K_a = \frac{1}{2} m v_a^2 = \frac{1}{2} (9.11 \times 10^{-31} \text{ kg})(3.00 \times 10^6 \text{ m/s})^2$$

$$= 4.10 \times 10^{-18} \text{ J},$$

$$U_a = \frac{1}{4\pi\epsilon_0} \frac{q q_0}{r_a}$$

$$= (9.0 \times 10^9 \text{ N} \cdot \text{m}^2/\text{C}^2) \frac{(3.20 \times 10^{-19} \text{ C})(1.60 \times 10^{-19} \text{ C})}{1.00 \times 10^{-10} \text{ m}}$$

$$= 4.61 \times 10^{-18} \text{ J},$$

$$U_b = (9.0 \times 10^9 \text{ N} \cdot \text{m}^2/\text{C}^2) \frac{(3.20 \times 10^{-19} \text{ C})(1.60 \times 10^{-19} \text{ C})}{2.00 \times 10^{-10} \text{ m}}$$

$$= 2.30 \times 10^{-18} \text{ J}.$$

Hence the final kinetic energy is

$$K_b = \frac{1}{2} m v_b^2 = 4.10 \times 10^{-18} \text{ J} + 4.61 \times 10^{-18} \text{ J} - 2.30 \times 10^{-18} \text{ J}$$

$$= 6.41 \times 10^{-18} \text{ J},$$

and the final speed of the positron is

$$v_b = \sqrt{\frac{2K_b}{m}} = \sqrt{\frac{2(6.41 \times 10^{-18} \text{ J})}{9.11 \times 10^{-31} \text{ kg}}} = 3.8 \times 10^6 \text{ m/s}.$$

The force is repulsive, so the positron speeds up as it moves away from the stationary alpha particle.

b) When the final positions of the positron and alpha particle are very far apart, the separation r_b approaches infinity and the final potential energy U_b approaches zero. Then the final kinetic energy of the positron is

$$K_b = K_a + U_a - U_b = 4.10 \times 10^{-18} \text{ J} + 4.61 \times 10^{-18} \text{ J} - 0$$

$$= 8.71 \times 10^{-18} \text{ J},$$

and its final speed is

$$v_b = \sqrt{\frac{2K_b}{m}} = \sqrt{\frac{2(8.71 \times 10^{-18} \text{ J})}{9.11 \times 10^{-31} \text{ kg}}} = 4.4 \times 10^6 \text{ m/s}.$$

Comparing to part (a), we see that as the positron moves from $r = 2.00 \times 10^{-10}$ m to infinity, the additional work done on it by the electric field of the alpha particle increases the speed by only about 16%. This is because the electric force decreases rapidly with distance.

c) If the moving charge is negative, the force on it is attractive rather than repulsive, and we expect it to slow down rather than speed up. The only difference in the above calculations is that both potential-energy quantities are negative. From part (a), at a distance $r_b = 2.00 \times 10^{-10}$ m we have

$$K_b = K_a + U_a - U_b$$

$$= 4.10 \times 10^{-18} \text{ J} + (-4.61 \times 10^{-18} \text{ J}) - (-2.30 \times 10^{-18} \text{ J})$$

$$= 1.79 \times 10^{-18} \text{ J},$$

$$v_b = \sqrt{\frac{2K_b}{m}} = 2.0 \times 10^6 \text{ m/s}.$$

From part (b), at $r_b = \infty$ the kinetic energy of the electron would seem to be

$$K_b = K_a + U_a - U_b = 4.10 \times 10^{-18} \text{ J} + (-4.61 \times 10^{-18} \text{ J}) - 0$$

$$= -5.1 \times 10^{-19} \text{ J}.$$

But kinetic energies can *never* be negative! This result means that the electron can never reach $r_b = \infty$; the attractive force will bring the electron to a halt at a finite distance from the alpha particle. The electron will then begin to move back toward the alpha particle. You can solve for the distance r_b at which the electron comes momentarily to rest by setting K_b equal to zero in the equation for conservation of mechanical energy (see Exercise 24–8).

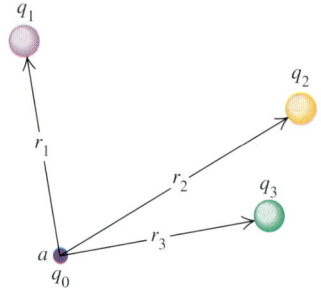

24–6 The potential energy associated with a charge q_0 at point a depends on charges q_1, q_2, and q_3 and on their distances r_1, r_2, and r_3 from point a.

ELECTRIC POTENTIAL ENERGY WITH SEVERAL POINT CHARGES

Suppose the electric field $\vec{E}$ in which charge q_0 moves is caused by *several* point charges q_1, q_2, q_3, ... at distances r_1, r_2, r_3, ... from q_0, as in Fig. 24–6. The total electric field at each point is the *vector sum* of the fields due to the individual charges, and the total work done on q_0 during any displacement is the sum of the contributions from the individual charges. From Eq. (24–9) we conclude that the potential energy associated with the test charge q_0 at point a in Fig. 24–6 is the *algebraic* sum (*not* a vector sum)

$$U = \frac{q_0}{4\pi\epsilon_0}\left(\frac{q_1}{r_1} + \frac{q_2}{r_2} + \frac{q_3}{r_3} + \cdots\right) = \frac{q_0}{4\pi\epsilon_0}\sum_i \frac{q_i}{r_i} \qquad \begin{array}{l}\text{(point charge } q_0 \\ \text{and collection} \\ \text{of charges } q_i).\end{array} \quad (24\text{–}10)$$

When q_0 is at a different point b, the potential energy is given by the same expression, but r_1, r_2, ... are the distances from q_1, q_2, ... to point b. The work done on charge q_0 when it moves from a to b along any path is equal to the difference $U_a - U_b$ between the potential energies when q_0 is at a and at b.

We can represent *any* charge distribution as a collection of point charges, so Eq. (24–10) shows that we can always find a potential-energy function for *any* static electric field. It follows that **for every electric field due to a static charge distribution the force exerted by that field is conservative.**

Equations (24–9) and (24–10) define U to be zero when all the distances r_1, r_2, ... are infinite, that is, when the test charge q_0 is very far away from all the charges that produce the field. As with any potential-energy function, the point where $U = 0$ is arbitrary; we can always add a constant to make U equal zero at any point we choose. In electrostatics problems it's usually simplest to choose this point to be at infinity. When we analyze electric circuits in Chapters 26 and 27, other choices will be more convenient.

Equation (24–10) gives the potential energy associated with the presence of the test charge q_0 in the $\vec{E}$ field produced by q_1, q_2, q_3, ... But there is also potential energy involved in assembling these charges. If we start with charges q_1, q_2, q_3, ... all separated from each other by infinite distances and then bring them together so that the distance between q_i and q_j is r_{ij}, the *total* potential energy U is the sum of the potential energies of interaction for each pair of charges. We can write this as

$$U = \frac{1}{4\pi\epsilon_0}\sum_{i<j}\frac{q_i q_j}{r_{ij}}. \qquad (24\text{–}11)$$

This sum extends over all *pairs* of charges; we don't let $i = j$ (because that would be an interaction of a charge with itself), and we include only terms with $i < j$ to make sure that we count each pair only once. Thus to account for the interaction between q_3 and q_4, we include a term with $i = 3$ and $j = 4$ but not a term with $i = 4$ and $j = 3$.

INTERPRETING ELECTRIC POTENTIAL ENERGY

As a final comment, here are two viewpoints on electric potential energy. We have defined it in terms of the work done *by the electric field* on a charged particle moving in the field, just as in Chapter 7 we defined potential energy in terms of the work done by gravity or by a spring. When a particle moves from point a to point b, the work done on it by the electric field is $W_{a\rightarrow b} = U_a - U_b$. Thus the potential-energy difference $U_a - U_b$ equals *the work that is done by the electric force when the particle moves from a to b.* When U_a is greater than U_b, the field does positive work on the particle as it "falls" from a point of higher potential energy (a) to a point of lower potential energy (b).

An alternative but equivalent viewpoint is to consider how much work we would have to do to "raise" a particle from a point b where the potential energy is U_b to a point a where it has a greater value U_a (pushing two positive charges closer together, for exam-

A crystal of table salt is made of ions of sodium (Na^+) and chlorine (Cl^-) and has a net negative potential energy. To dissolve salt in water, energy must be added to separate the ions. This energy comes from interactions between the ions and the charged poles of H_2O molecules (see Fig. 22–24).

ple). To move the particle slowly (so as not to give it any kinetic energy), we need to exert an additional external force $\vec{F}_{\text{ext}}$ that is equal and opposite to the electric-field force and does positive work. The potential-energy difference $U_a - U_b$ is then defined as *the work that must be done by an external force to move the particle slowly from b to a against the electric force*. Because $\vec{F}_{\text{ext}}$ is the negative of the electric-field force and the displacement is in the opposite direction, this definition of the potential difference $U_a - U_b$ is equivalent to that given above. This alternative viewpoint also works if U_a is less than U_b, corresponding to "lowering" the particle; an example is moving two positive charges away from each other. In this case, $U_a - U_b$ is again equal to the work done by the external force, but now this work is negative.

We will use both of these viewpoints in the next section to interpret what is meant by electric *potential,* or potential energy per unit charge.

EXAMPLE 24-2

A system of point charges Two point charges are located on the x-axis, $q_1 = -e$ at $x = 0$ and $q_2 = +e$ at $x = a$. a) Find the work that must be done by an external force to bring a third point charge $q_3 = +e$ from infinity to $x = 2a$. b) Find the total potential energy of the system of three charges.

SOLUTION a) The work that must be done on q_3 by an external force $\vec{F}_{\text{ext}}$ is equal to the difference between two quantities: the potential energy U associated with q_3 when it is at $x = 2a$ and the potential energy when it is infinitely far away. The second of these is zero, so the work that must be done is equal to U. The distances between the charges are $r_{13} = 2a$ and $r_{23} = a$, so from Eq. (24–10),

$$W = U = \frac{q_3}{4\pi\epsilon_0}\left(\frac{q_1}{r_{13}} + \frac{q_2}{r_{23}}\right) = \frac{+e}{4\pi\epsilon_0}\left(\frac{-e}{2a} + \frac{+e}{a}\right) = \frac{+e^2}{8\pi\epsilon_0 a}.$$

If q_3 is brought in from infinity along the $+x$-axis, it is attracted by q_1 but is repelled more strongly by q_2; hence positive work must be done to push q_3 to the position at $x = 2a$.

b) The total potential energy of the assemblage of three charges is given by Eq. (24–11):

$$U = \frac{1}{4\pi\epsilon_0}\sum_{i<j}\frac{q_i q_j}{r_{ij}} = \frac{1}{4\pi\epsilon_0}\left(\frac{q_1 q_2}{r_{12}} + \frac{q_1 q_3}{r_{13}} + \frac{q_2 q_3}{r_{23}}\right)$$

$$= \frac{1}{4\pi\epsilon_0}\left(\frac{(-e)(e)}{a} + \frac{(-e)(e)}{2a} + \frac{(e)(e)}{a}\right) = \frac{-e^2}{8\pi\epsilon_0 a}.$$

Since $U < 0$, the system has lower potential energy than it would if the three charges were infinitely far apart. An external force would have to do *negative* work to bring the three charges from infinity to assemble this entire arrangement and would have to do *positive* work to move the three charges back to infinity.

24-3 ELECTRIC POTENTIAL

In the preceding section we looked at the potential energy U associated with a test charge q_0 in an electric field. Now we want to describe this potential energy on a "per unit charge" basis, just as electric field describes the force per unit charge on a charged particle in the field. This leads us to the concept of *electric potential*, often called simply *potential*. This concept is very useful in calculations involving energies of charged particles. It also facilitates many electric-field calculations because electric potential is closely related to the electric field $\vec{E}$. When we need to determine an electric field, it is often easier to determine the potential first and then find the field from it.

Potential is *potential energy per unit charge*. We define the potential V at any point in an electric field as the potential energy U *per unit charge* associated with a test charge q_0 at that point:

$$V = \frac{U}{q_0}, \qquad \text{or} \qquad U = q_0 V. \tag{24–12}$$

Potential energy and charge are both scalars, so potential is a scalar quantity. From Eq. (24–12) its units are found by dividing the units of energy by those of charge. The SI unit of potential, called one **volt** (1 V) in honor of the Italian scientist and electrical experimenter Alessandro Volta (1745–1827), equals 1 joule per coulomb:

$$1 \text{ V} = 1 \text{ volt} = 1 \text{ J/C} = 1 \text{ joule/coulomb}.$$

Let's put Eq. (24–2), which equates the work done by the electric force during a displacement from a to b to the quantity $-\Delta U = -(U_b - U_a)$, on a "work per unit charge" basis. We divide this equation by q_0, obtaining

$$\frac{W_{a \to b}}{q_0} = -\frac{\Delta U}{q_0} = -\left(\frac{U_b}{q_0} - \frac{U_a}{q_0}\right) = -(V_b - V_a) = V_a - V_b, \qquad (24\text{–}13)$$

where $V_a = U_a/q_0$ is the potential energy per unit charge at point a, and similarly for V_b. We call V_a and V_b the *potential at point a* and *potential at point b*, respectively. Thus the work done per unit charge by the electric force when a charged body moves from a to b is equal to the potential at a minus the potential at b.

The difference $V_a - V_b$ is called the *potential of a with respect to b;* we sometimes abbreviate this difference as $V_{ab} = V_a - V_b$ (note the order of the subscripts). This is often called the potential difference between a and b, but that's ambiguous unless we specify which is the reference point. In electric circuits, which we will analyze in later chapters, the potential difference between two points is often called **voltage.** Equation (24–13) then states that V_{ab}, the potential of a with respect to b, equals the work done by the electric force when a UNIT charge moves from a to b.

Another way to interpret the potential difference V_{ab} in Eq. (24–13) is to use the alternative viewpoint mentioned at the end of Section 24–2. In that viewpoint, $U_a - U_b$ is the amount of work that must be done by an *external* force to move a particle of charge q_0 slowly from b to a against the electric force. The work that must be done *per unit charge* by the external force is then $(U_a - U_b)/q_0 = V_a - V_b = V_{ab}$. In other words, V_{ab}, *the potential of a with respect to b, equals the work that must be done to move a UNIT charge slowly from b to a against the electric force.*

An instrument that measures the difference of potential between two points is called a *voltmeter.* The principle of the common type of moving-coil voltmeter will be described later. There are also much more sensitive potential-measuring devices that use electronic amplification. Instruments that can measure a potential difference of $1\,\mu\text{V}$ are common, and sensitivities down to 10^{-12} V can be attained.

To find the potential V due to a single point charge q, we divide Eq. (24–9) by q_0:

$$V = \frac{U}{q_0} = \frac{1}{4\pi\epsilon_0}\frac{q}{r} \qquad \text{(potential due to a point charge),} \qquad (24\text{–}14)$$

where r is the distance from the point charge q to the point at which the potential is evaluated. If q is positive, the potential that it produces is positive at all points; if q is negative, it produces a potential that is negative everywhere. In either case, V is equal to zero at $r = \infty$, an infinite distance from the point charge. Note that potential, like electric field, is independent of the test charge q_0 that we use to define it.

Similarly, we divide Eq. (24–10) by q_0 to find the potential due to a collection of point charges:

$$V = \frac{U}{q_0} = \frac{1}{4\pi\epsilon_0}\sum_i \frac{q_i}{r_i} \qquad \begin{array}{l}\text{(potential due to a collection}\\ \text{of point charges).}\end{array} \qquad (24\text{–}15)$$

In this expression, r_i is the distance from the ith charge, q_i, to the point at which V is evaluated. Just as the electric field due to a collection of point charges is the *vector* sum of the fields produced by each charge, the electric potential due to a collection of point charges is the *scalar* sum of the potentials due to each charge. When we have a continuous distribution of charge along a line, over a surface, or through a volume, we divide the charge into elements dq, and the sum in Eq. (24–15) becomes an integral:

$$V = \frac{1}{4\pi\epsilon_0}\int \frac{dq}{r} \qquad \begin{array}{l}\text{(potential due to a continuous distribution}\\ \text{of charge),}\end{array} \qquad (24\text{–}16)$$

The potential energy for a small amount of charge at high potential may be less than that for a larger amount of charge at low potential. In the same way, the potential energy for a 1-kg melon lifted to the top of the Eiffel Tower (elevation 300 m) is less than that for a 1000-kg automobile lifted 3 m.

where r is the distance from the charge element dq to the field point where we are finding V. We'll work out several examples of such cases. The potential defined by Eqs. (24–15) and (24–16) is zero at points that are infinitely far away from *all* the charges. Later we'll encounter cases in which the charge distribution itself extends to infinity. We'll find that in such cases we cannot set $V = 0$ at infinity, and we'll need to exercise care in using and interpreting Eqs. (24–15) and (24–16).

When we are given a collection of point charges, Eq. (24–15) is usually the easiest way to calculate the potential V. But in some problems in which the electric field is known or can be found easily, it is easier to determine V from $\vec{E}$. The force $\vec{F}$ on a test charge q_0 can be written as $\vec{F} = q_0\vec{E}$, so from Eq. (24–1) the work done by the electric force as the test charge moves from a to b is given by

$$W_{a \to b} = \int_a^b \vec{F} \cdot d\vec{l} = \int_a^b q_0 \vec{E} \cdot d\vec{l}.$$

If we divide this by q_0 and compare the result with Eq. (24–13), we find

$$V_a - V_b = \int_a^b \vec{E} \cdot d\vec{l} = \int_a^b E \cos\phi \, dl \qquad \text{(potential difference as an integral of } \vec{E}\text{).} \qquad (24\text{–}17)$$

The value of $V_a - V_b$ is independent of the path taken from a to b, just as the value of $W_{a \to b}$ is independent of the path. To interpret Eq. (24–17), remember that $\vec{E}$ is the electric force per unit charge on a test charge. If the line integral $\int_a^b \vec{E} \cdot d\vec{l}$ is positive, the electric field does positive work on a positive test charge as it moves from a to b. In this case the electric potential energy decreases as the test charge moves, and so the potential energy per unit charge decreases as well; hence V_b is less than V_a, and $V_a - V_b$ is positive.

Figure 24–7a shows a positive point charge. The electric field is directed away from the charge, and $V = q/4\pi\epsilon_0 r$ is positive at any finite distance from the charge. If you move away from the charge, in the direction of $\vec{E}$, you move toward lower values of V; if you move toward the charge, in the direction opposite $\vec{E}$, you move toward greater values of V. For the negative point charge in Fig. 24–7b, $\vec{E}$ is directed toward the charge and $V = q/4\pi\epsilon_0 r$ is negative at any finite distance from the charge. In this case, if you move toward the charge, you are moving in the direction of $\vec{E}$ and in the direction of decreasing (more negative) V. Moving away from the charge, in the direction opposite $\vec{E}$, moves you toward increasing (less negative) values of V. Here's a general rule, valid for *any* electric field: moving *with* the direction of $\vec{E}$ means moving in the direction of *decreasing* V, and moving *against* the direction of $\vec{E}$ means moving in the direction of *increasing* V.

Also, a positive test charge q_0 experiences an electric force in the direction of $\vec{E}$, toward lower values of V; a negative test charge experiences a force opposite to $\vec{E}$, toward higher values of V. Thus a positive charge tends to "fall" from a high-potential region to a lower-potential one. The opposite is true for a negative charge.

Notice that Eq. (24–17) can be rewritten as follows:

$$V_a - V_b = -\int_b^a \vec{E} \cdot d\vec{l}. \qquad (24\text{–}18)$$

This has a negative sign compared to the integral in Eq. (24–17), and the limits are reversed; hence Eqs. (24–17) and (24–18) are equivalent. But Eq. (24–18) has a slightly different interpretation. To move a unit charge slowly against the electric force, we must apply an *external* force per unit charge equal to $-\vec{E}$, equal and opposite to the electric force per unit charge $\vec{E}$. Equation (24–18) says that $V_a - V_b = V_{ab}$, the potential of a with respect to b, equals the work done per unit charge by this external force to move a unit

11.13
Electrical Potential Energy
and Potential

(a)

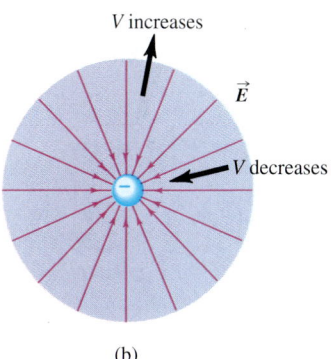

(b)

24–7 (a) A positive point charge. (b) A negative point charge. In both cases, if you move in the direction of $\vec{E}$, electric potential V decreases; if you move in the direction opposite $\vec{E}$, V increases.

charge from b to a. This is the same alternative interpretation we discussed under Eq. (24–13).

Equations (24–17) and (24–18) show that the unit of potential difference (1 V) is equal to the unit of electric field (1 N/C) multiplied by the unit of distance (1 m). Hence the unit of electric field can be expressed as 1 *volt per meter* (1 V/m), as well as 1 N/C:

$$1 \text{ V/m} = 1 \text{ volt/meter} = 1 \text{ N/C} = 1 \text{ newton/coulomb.}$$

In practice, the volt per meter is the usual unit of electric field magnitude.

The magnitude e of the electron charge can be used to define a unit of energy that is useful in many calculations with atomic and nuclear systems. When a particle with charge q moves from a point where the potential is V_b to a point where it is V_a, the change in the potential energy U of the change is

$$U_a - U_b = q(V_a - V_b) = qV_{ab}.$$

If the charge q equals the magnitude e of the electron charge, 1.602×10^{-19} C, and the potential difference is $V_{ab} = 1$ V, the change in energy is

$$U_a - U_b = (1.602 \times 10^{-19} \text{ C})(1 \text{ V}) = 1.602 \times 10^{-19} \text{ J.}$$

This quantity of energy is defined to be 1 **electron volt** (1 eV):

$$1 \text{ eV} = 1.602 \times 10^{-19} \text{ J.}$$

The multiples meV, keV, MeV, GeV, and TeV are often used.

CAUTION ▶ Remember that the electron volt is a unit of energy, *not* a unit of potential or potential difference! ◀

When a particle with charge e moves through a potential difference of one volt, the change in potential *energy* is 1 eV. If the charge is some multiple of e, say Ne, the change in potential energy in electron volts is N times the potential difference in volts. For example, when an alpha particle, which has charge $2e$, moves between two points with a potential difference of 1000 V, the change in potential energy is 2(1000 eV) = 2000 eV. To confirm this, we write

$$U_a - U_b = qV_{ab} = (2e)(1000 \text{ V}) = (2)(1.602 \times 10^{-19} \text{ C})(1000 \text{ V})$$

$$= 3.204 \times 10^{-16} \text{ J} = 2000 \text{ eV.}$$

Although we have defined the electron volt in terms of *potential* energy, we can use it for *any* form of energy, such as the kinetic energy of a moving particle. When we speak of a "one-million-electron-volt proton," we mean a proton with a kinetic energy of one million electron volts (1 MeV), equal to $(10^6)(1.602 \times 10^{-19} \text{ J}) = 1.602 \times 10^{-13}$ J.

This accelerator at the Fermi National Accelerator Laboratory in Illinois gives protons a kinetic energy of 400 MeV (4×10^8 eV). Additional acceleration stages increase their kinetic energy to 950 GeV, or 0.95 TeV (9.5×10^{11} eV).

EXAMPLE 24-3

Electric force and electric potential A proton (charge $+e$ $= 1.602 \times 10^{-19}$ C) moves in a straight line from point a to point b inside a linear accelerator, a total distance $d = 0.50$ m. The electric field is uniform along this line, with magnitude $E = 1.5 \times 10^7$ V/m $= 1.5 \times 10^7$ N/C in the direction from a to b. Determine (a) the force on the proton; (b) the work done on it by the field; (c) the potential difference $V_a - V_b$.

SOLUTION a) The force is in the same direction as the electric field, and its magnitude is

$$F = qE = (1.602 \times 10^{-19} \text{ C})(1.5 \times 10^7 \text{ N/C}) = 2.4 \times 10^{-12} \text{ N.}$$

b) The force is constant and in the same direction as the dis-

placement, so the work done is

$$W_{a \to b} = Fd = (2.4 \times 10^{-12} \text{ N})(0.50 \text{ m}) = 1.2 \times 10^{-12} \text{ J}$$

$$= (1.2 \times 10^{-12} \text{ J}) \frac{1 \text{ eV}}{1.602 \times 10^{-19} \text{ J}}$$

$$= 7.5 \times 10^6 \text{ eV} = 7.5 \text{ MeV.}$$

c) From Eq. (24–13) the potential difference is the work per unit charge, which is

$$V_a - V_b = \frac{W_{a \to b}}{q} = \frac{1.2 \times 10^{-12} \text{ J}}{1.602 \times 10^{-19} \text{ C}} = 7.5 \times 10^6 \text{ J/C}$$

$$= 7.5 \times 10^6 \text{ V} = 7.5 \text{ MV.}$$

We can get this same result even more easily by remembering that one electron volt equals one volt multiplied by the charge e. Since the work done is 7.5×10^6 eV and the charge is e, the potential difference is $(7.5 \times 10^6 \text{ eV})/e = 7.5 \times 10^6$ V.

An alternative approach is to use Eq. (24–17) or (24–18) and calculate an integral of the electric field. The angle ϕ between the constant field $\vec{E}$ and the displacement is zero, so Eq. (24–17)

becomes

$$V_a - V_b = \int_a^b E \cos \phi \, dl = \int_a^b E \, dl = E \int_a^b dl.$$

The integral of dl from a to b is just the distance d, so we again find

$$V_a - V_b = Ed = (1.5 \times 10^7 \text{ V/m})(0.50 \text{ m}) = 7.5 \times 10^6 \text{ V}.$$

EXAMPLE 24–4

Potential due to two point charges An electric dipole consists of two point charges, $q_1 = +12$ nC and $q_2 = -12$ nC, placed 10 cm apart (Fig. 24–8). Compute the potentials at points a, b, and c by adding the potentials due to either charge, as in Eq. (24–15).

SOLUTION This is the same arrangement of charges as in Example 22–9 (Section 22–7). This time we have to evaluate at each point the *algebraic* sum in Eq. (24–15),

$$V = \frac{1}{4\pi\epsilon_0} \sum_i \frac{q_i}{r_i}.$$

At point a the potential due to the positive charge q_1 is

$$\frac{1}{4\pi\epsilon_0} \frac{q_1}{r_1} = (9.0 \times 10^9 \text{ N} \cdot \text{m}^2/\text{C}^2) \frac{12 \times 10^{-9} \text{ C}}{0.060 \text{ m}}$$

$$= 1800 \text{ N} \cdot \text{m/C}$$

$$= 1800 \text{ J/C} = 1800 \text{ V},$$

and the potential due to the negative charge q_2 is

$$\frac{1}{4\pi\epsilon_0} \frac{q_2}{r_2} = (9.0 \times 10^9 \text{ N} \cdot \text{m}^2/\text{C}^2) \frac{(-12 \times 10^{-9} \text{ C})}{0.040 \text{ m}}$$

$$= -2700 \text{ N} \cdot \text{m/C}$$

$$= -2700 \text{ J/C} = -2700 \text{ V}.$$

The potential V_a at point a is the sum of these:

$$V_a = 1800 \text{ V} + (-2700 \text{ V}) = -900 \text{ V}.$$

By similar calculations you can show that at point b the potential due to the positive charge is $+2700$ V, the potential due to the

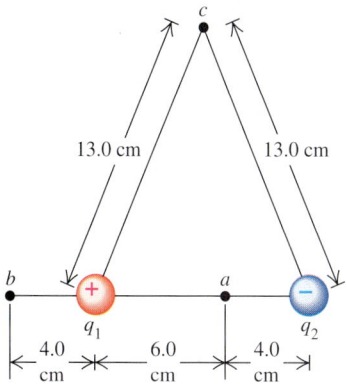

24–8 What are the potentials at points a, b, and c due to this electric dipole?

negative charge is -770 V, and

$$V_b = 2700 \text{ V} + (-770 \text{ V}) = 1930 \text{ V}.$$

At point c the potential due to the positive charge is

$$\frac{1}{4\pi\epsilon_0} \frac{q_1}{r_1} = (9.0 \times 10^9 \text{ N} \cdot \text{m}^2/\text{C}^2) \frac{12 \times 10^{-9} \text{ C}}{0.13 \text{ m}} = 830 \text{ V},$$

the potential due to the negative charge is -830 V, and the total potential is zero:

$$V_c = 830 \text{ V} + (-830 \text{ V}) = 0.$$

The potential is also equal to zero at infinity (infinitely far from both charges).

EXAMPLE 24–5

Potential and potential energy Compute the potential energy associated with a point charge of $+4.0$ nC if it is placed at points a, b, and c in Fig. 24–8.

SOLUTION For any point charge q, $U = qV$. We use the values of V from Example 24–4. At point a,

$$U_a = qV_a = (4.0 \times 10^{-9} \text{ C})(-900 \text{ J/C}) = -3.6 \times 10^{-6} \text{ J}.$$

At point b,

$$U_b = qV_b = (4.0 \times 10^{-9} \text{ C})(1930 \text{ J/C}) = 7.7 \times 10^{-6} \text{ J}.$$

At point c,

$$U_c = qV_c = 0.$$

All of these values correspond to U and V being zero at infinity. Note that *no* net work is done on the 4.0-nC charge if it moves from point c to infinity *by any path*. In particular, let the path be along the perpendicular bisector of the line joining the other two charges q_1 and q_2 in Fig. 24–8. As shown in Example 22–9 (Section 22–7), at points on the bisector the direction of $\vec{E}$ is perpendicular to the bisector. Hence the force on the 4.0-nC charge is perpendicular to the path, and no work is done in any displacement along it.

EXAMPLE 24-6

Finding potential by integration By integrating the electric field as in Eq. (24–17), find the potential at a distance r from a point charge q.

SOLUTION Let V be the potential at a distance r from the point charge. As usual, we choose the potential to be zero at an infinite distance from the charge. To find V, we let point a in Eq. (24–17) be at distance r, and let point b be at infinity. To carry out the integral, we can choose any path we like between these two points; the most convenient path is a straight radial line as shown in Fig. 24–9 so that $d\vec{l}$ is in the radial direction and has magnitude dr. If q is positive, $\vec{E}$ and $d\vec{l}$ are always parallel, so $\phi = 0$ and Eq. (24–17) becomes

$$V - 0 = \int_r^\infty E\, dr = \int_r^\infty \frac{q}{4\pi\epsilon_0 r^2}\, dr$$

$$= -\frac{q}{4\pi\epsilon_0 r}\bigg|_r^\infty = 0 - \left(-\frac{q}{4\pi\epsilon_0 r}\right),$$

$$V = \frac{q}{4\pi\epsilon_0 r}.$$

This agrees with Eq. (24–14). If q is negative, $\vec{E}$ is radially inward while $d\vec{l}$ is still radially outward, so $\phi = 180°$. Since $\cos 180° = -1$, this adds a minus sign to the above result. However, the field magnitude E is always positive, and since q is negative, we must write $E = |q|/4\pi\epsilon_0 r = -q/4\pi\epsilon_0 r$, giving another minus sign. The two minus signs cancel, and the above result for V is valid for point charges of either sign.

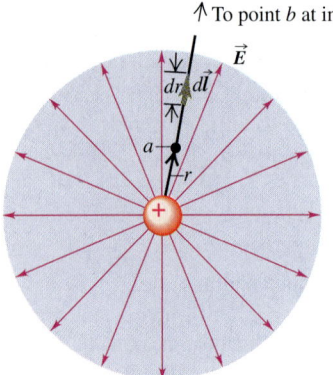

24–9 Calculating the potential by integrating $\vec{E}$ for a single point charge.

We can get the same result by using Eq. (22–7) for the electric field, which is valid for either sign of q, and writing $d\vec{l} = \hat{r}\, dr$:

$$V - 0 = V = \int_r^\infty \vec{E} \cdot d\vec{l} = \int_r^\infty \frac{1}{4\pi\epsilon_0}\frac{q}{r^2}\hat{r} \cdot \hat{r}\, dr = \int_r^\infty \frac{q}{4\pi\epsilon_0 r^2}\, dr,$$

$$V = \frac{q}{4\pi\epsilon_0 r}.$$

EXAMPLE 24-7

Moving through a potential difference In Fig. 24–10 a dust particle with mass $m = 5.0 \times 10^{-9}$ kg $= 5.0\ \mu$g and charge $q_0 = 2.0$ nC starts from rest at point a and moves in a straight line to point b. What is its speed v at point b?

SOLUTION This problem would be difficult to solve without using energy techniques, since the force that acts on the particle varies in magnitude as the particle moves from a to b. Only the conservative electric force acts on the particle, so mechanical energy is conserved:

$$K_a + U_a = K_b + U_b.$$

For this situation, $K_a = 0$ and $K_b = \frac{1}{2}mv^2$. We get the potential energies (U) from the potentials (V) using Eq. (24–12): $U_a = q_0 V_a$ and $U_b = q_0 V_b$. Substituting these into the energy-conservation equation and solving for v, we find

$$0 + q_0 V_a = \frac{1}{2}mv^2 + q_0 V_b,$$

$$v = \sqrt{\frac{2q_0(V_a - V_b)}{m}}.$$

We calculate the potentials using Eq. (24–15), just as we did in

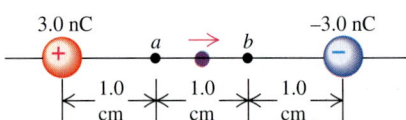

24–10 The particle moves from point a to point b; its acceleration is not constant.

Example 24–4:

$$V_a = (9.0 \times 10^9\ \text{N} \cdot \text{m}^2/\text{C}^2)\left(\frac{3.0 \times 10^{-9}\ \text{C}}{0.010\ \text{m}} + \frac{(-3.0 \times 10^{-9}\ \text{C})}{0.020\ \text{m}}\right)$$

$$= 1350\ \text{V},$$

$$V_b = (9.0 \times 10^9\ \text{N} \cdot \text{m}^2/\text{C}^2)\left(\frac{3.0 \times 10^{-9}\ \text{C}}{0.020\ \text{m}} + \frac{(-3.0 \times 10^{-9}\ \text{C})}{0.010\ \text{m}}\right)$$

$$= -1350\ \text{V},$$

$$V_a - V_b = (1350\ \text{V}) - (-1350\ \text{V}) = 2700\ \text{V}.$$

Finally,

$$v = \sqrt{\frac{2(2.0 \times 10^{-9}\ \text{C})(2700\ \text{V})}{5.0 \times 10^{-9}\ \text{kg}}} = 46\ \text{m/s}.$$

We can check unit consistency by noting that 1 V = 1 J/C, so the numerator under the radical has units of J or kg • m^2/s^2.

We can use exactly this same method to find the speed of an electron accelerated across a potential difference of 500 V in an oscilloscope tube or 20 kV in a TV picture tube. The end-of-chapter problems include several examples of such calculations.

24–4 CALCULATING ELECTRIC POTENTIAL

When calculating the potential due to a charge distribution, we usually follow one of two routes. If we know the charge distribution, we can use Eq. (24–15) or (24–16). Or if we know how the electric field depends on position, we can use Eq. (24–17), defining the potential to be zero at some convenient place. Some problems require a combination of these approaches.

Problem–Solving Strategy

CALCULATING ELECTRIC POTENTIAL

1. Remember that potential is *potential energy per unit charge.* Understanding this statement can get you a long way.

2. To find the potential due to a collection of point charges, use Eq. (24–15). If you are given a continuous charge distribution, devise a way to divide it into infinitesimal elements and then use Eq. (24–16). Carry out the integration, using appropriate limits to include the entire charge distribution. In the integral, be careful about which geometric quantities vary and which are constant.

3. If you are given the electric field, or if you can find it using any of the methods of Chapter 22 or 23, it may be easier to use Eq. (24–17) or (24–18) to calculate the potential difference between point *a* and point *b*. When appropriate, make use of your freedom to define *V* to be zero at some convenient place, and choose this place to

be point *b*. (For point charges, this will usually be at infinity, but for other distributions of charge—especially those that themselves extend to infinity—it may be convenient or necessary to define V_b to be zero at some finite distance from the charge distribution. This is just like defining *U* to be zero at ground level in gravitational problems.) Then the potential at any other point, say *a*, can be found from Eq. (24–17) or (24–18) with $V_b = 0$.

4. Remember that potential is a *scalar* quantity, not a *vector*. It doesn't have components! However, you may have to use components of the vectors $\vec{E}$ and $d\vec{l}$ when you use Eq. (24–17) or (24–18).

5. If you know the electric field, you can always get a rough check of your result for V by verifying that V decreases if you move in the direction of $\vec{E}$.

EXAMPLE 24–8

A charged conducting sphere A solid conducting sphere of radius R has a total charge q. Find the potential everywhere, both outside and inside the sphere.

SOLUTION We used Gauss's law in Example 23–7 (Section 23–5) to show that at all points *outside* the sphere the field is the same as if the sphere were removed and replaced by a point charge q. We take $V = 0$ at infinity, as we did for a point charge. Then the potential at a point outside the sphere at a distance r from its center is the same as the potential due to a point charge q at the center:

$$V = \frac{1}{4\pi\epsilon_0} \frac{q}{r}.$$

The potential at the surface of the sphere is $V_{surface} = q/4\pi\epsilon_0 R$.

Inside the sphere, $\vec{E}$ is zero everywhere; otherwise, charge would move within the sphere. Hence if a test charge moves from any point to any other point inside the sphere, no work is done on that charge. This means that the potential is the same at every point inside the sphere and is equal to its value $q/4\pi\epsilon_0 R$ at the surface.

The field and potential are shown as functions of r in Fig. 24–11 for a positive charge q. In this case the electric field points radially away from the sphere. As you move away from the sphere, in the direction of $\vec{E}$, V decreases (as it should). The electric field at the surface has magnitude $E_{surface} = |q|/4\pi\epsilon_0 R^2$.

Here is a practical application of these results. The maximum potential to which a conductor in air can be raised is limited because air molecules become *ionized,* and air becomes a conductor, at an electric field magnitude of about 3×10^6 V/m. Assume for the moment that q is positive. Comparing the above expressions for the potential $V_{surface}$ and field magnitude $E_{surface}$ at the surface of a charged conducting sphere, we note that $V_{surface} = RE_{surface}$. Thus if E_{max} represents the electric field magnitude above which air becomes conductive (known as the *dielectric strength* of air), the maximum potential V_{max} to which

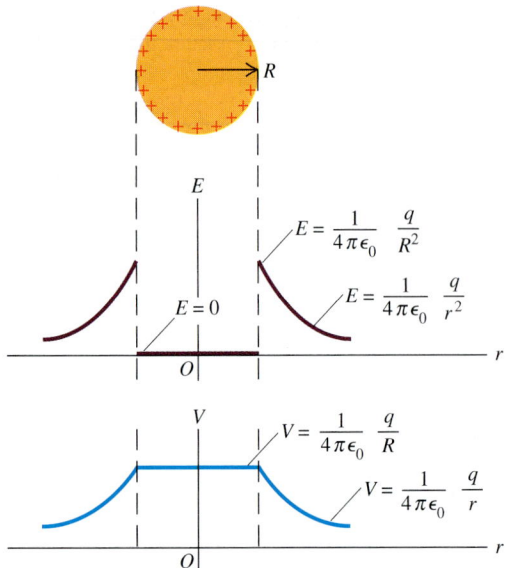

24–11 Electric field magnitude E and potential V at points inside and outside a positively charged spherical conductor.

a spherical conductor can be raised is

$$V_{max} = RE_{max}.$$

For a conducting sphere 1 cm in radius in air, $V_{max} = (10^{-2} \text{ m})(3 \times 10^6 \text{ V/m}) = 30{,}000$ V. No amount of "charging" could raise the potential of a conducting sphere of this size in air

higher than about 30,000 V; attempting to raise the potential further by adding extra charge would cause the surrounding air to become ionized and conductive, and the extra added charge would leak into the air. To attain even higher potentials, high-voltage machines such as Van de Graaff generators use spherical terminals with very large radii (see Fig. 23–24). For example, a terminal of radius $R = 2$ m has a maximum potential $V_{max} = (2 \text{ m})(3 \times 10^6 \text{ N/C}) = 6 \times 10^6 \text{ V} = 6$ MV. Such machines are sometimes placed in pressurized tanks filled with a gas such as sulfur hexafluoride (SF_6) that can withstand even larger fields without becoming conductive (that is, SF_6 has a larger value of E_{max} than does air).

At the other extreme is the effect produced by a surface of very *small* radius of curvature, such as a sharp point or the tip of a thin wire. Since the maximum potential decreases as the radius decreases, even relatively small potentials applied to sharp points in air produce sufficiently high fields just outside the point to ionize the surrounding air, making it a conductor. The resulting current and its associated glow (visible in a dark room) is called *corona*. (Along with the ionization comes ozone, which is unhealthful in large quantities.) Copying machines use corona from fine wires to charge the imaging drum, and some paper-handling machines use corona to help drain off static charges that would otherwise cause paper jams. On a larger scale, a lightning rod has a sharp end so that lightning bolts will pass through a conducting path in the air that leads to the rod and not to other nearby structures that could be damaged. Car radio antennas have a ball on the end to help *prevent* corona that would cause static.

EXAMPLE 24–9

Oppositely charged parallel plates Find the potential at any height y between the two oppositely charged parallel plates discussed in Section 24–2.

SOLUTION The potential energy U for a test charge q_0 at a point a distance y above the bottom plate (Fig. 24–12) is given by Eq. (24–5), $U = q_0 E y$. The potential V_y at this point is the potential energy per unit charge, $V_y = U/q_0$:

$$V_y = Ey.$$

We have chosen U, and therefore V_y, to be zero at point b, where $y = 0$. Even if we choose the potential to be different from zero at b, it is still true that

$$V_y - V_b = Ey.$$

The potential decreases as we move from the upper to the lower plate in the direction of $\vec{E}$. At point a, where $y = d$ and $V_y = V_a$,

$$V_a - V_b = Ed \quad \text{and} \quad E = \frac{V_a - V_b}{d} = \frac{V_{ab}}{d},$$

where V_{ab} is the potential of the positive plate with respect to the negative plate. That is, the electric field equals the potential difference between the plates divided by the distance between them. (This relation between E and V_{ab} holds *only* for the planar

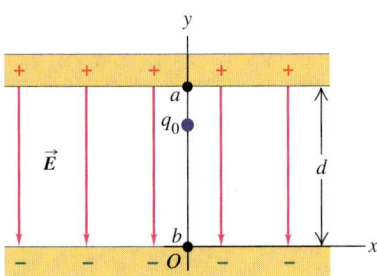

24–12 The charged parallel plates from Fig. 24–1.

geometry we have described. It does *not* work for situations such as concentric cylinders or spheres in which the electric field is not uniform.)

In Example 23–8 (Section 23–5) we derived the expression $E = \sigma/\epsilon_0$ for the electric field E between two conducting plates having surface charge densities $+\sigma$ and $-\sigma$ respectively. In dealing with conducting plates, the expression $E = V_{ab}/d$ is generally more useful; the potential difference V_{ab} can be measured easily with a voltmeter, while there are no instruments that read surface charge density directly. In fact, measuring V_{ab} allows σ to be determined; setting the two expressions for E equal to each other gives

$$\sigma = \frac{\epsilon_0 V_{ab}}{d}.$$

The surface charge density on the positive plate is directly proportional to the potential difference between the plates. On the negative plate the surface charge density is $-\sigma$.

CAUTION ▶ You might think that if a conducting body has zero potential, it must necessarily also have zero net charge. But that just isn't so! As an example, the plate at $y = 0$ in Fig. 24–12 has zero potential ($V = 0$) but has a nonzero charge per unit area $-\sigma$. Remember that there's nothing particularly special about the place where potential is zero; we can *define* this place to be wherever we want it to be. ◀

EXAMPLE 24-10

An infinite line charge or charged conducting cylinder Find the potential at a distance r from a very long line of charge with linear charge density (charge per unit length) λ.

SOLUTION We found in Example 23–6 (Section 23–5) that the electric *field* at a distance r from a long straight-line charge (Fig. 24–13a) or outside a long charged conducting cylinder (Fig. 24–13b) has only a radial component, given by

$$E_r = \frac{1}{2\pi\epsilon_0} \frac{\lambda}{r}.$$

We can find the potential by integrating $\vec{E}$ as in Eq. (24–17). Since the field has only a radial component, the scalar product $\vec{E} \cdot d\vec{l}$ is equal to $E_r \, dr$. Hence the potential of any point a with respect to any other point b, at radial distances r_a and r_b from the line of charge, is

$$V_a - V_b = \int_a^b \vec{E} \cdot d\vec{l} = \int_a^b E_r \, dr = \frac{\lambda}{2\pi\epsilon_0} \int_{r_a}^{r_b} \frac{dr}{r} = \frac{\lambda}{2\pi\epsilon_0} \ln \frac{r_b}{r_a}.$$

If we take point b at infinity and set $V_b = 0$, we find that V_a is *infinite*:

$$V_a = \frac{\lambda}{2\pi\epsilon_0} \ln \frac{\infty}{r_a} = \infty.$$

This shows that if we try to define V to be zero at infinity, then V must be infinite at *any* finite distance from the line charge. Hence this is *not* a useful way to define V for this problem! The difficulty, as we mentioned earlier, is that the charge distribution itself extends to infinity.

To get around this difficulty, remember that we can define V to be zero at any point we like. We set $V_b = 0$ at point b at an arbitrary radial distance r_0. Then the potential $V = V_a$ at point a

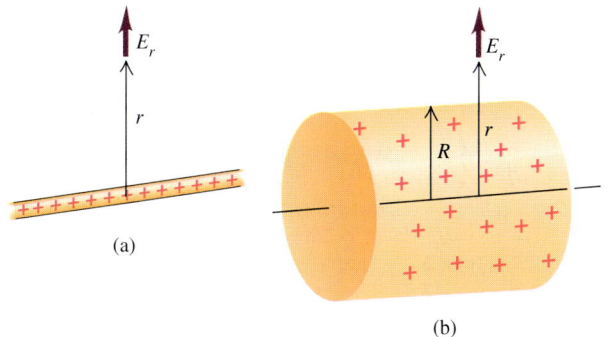

24–13 Electric field outside (a) a long positively charged wire; (b) a long positively charged cylinder.

at a radial distance r is given by $V - 0 = (\lambda/2\pi\epsilon_0) \ln (r_0/r)$, or

$$V = \frac{\lambda}{2\pi\epsilon_0} \ln \frac{r_0}{r}.$$

This equation also gives the potential in the field of a charged conducting cylinder, but only for values of r equal to or greater than the radius R of the cylinder. If we choose r_0 to be the cylinder radius R, so that $V = 0$ when $r = R$, then at any point for which $r > R$,

$$V = \frac{\lambda}{2\pi\epsilon_0} \ln \frac{R}{r},$$

where r is the distance from the axis of the cylinder. Inside the cylinder, $\vec{E} = 0$, and V has the same value (zero) as on the cylinder's surface.

EXAMPLE 24-11

A ring of charge Electric charge is distributed uniformly around a thin ring of radius a, with total charge Q (Fig. 24–14). Find the potential at a point P on the ring axis at a distance x from the center of the ring.

SOLUTION We calculated the electric field on the axis of this ring in Example 22–10 (Section 22–7). Figure 24–14 shows that the entire charge is at a distance $r = \sqrt{x^2 + a^2}$ from point P. We conclude immediately that the potential at point P is

$$V = \frac{1}{4\pi\epsilon_0} \frac{Q}{\sqrt{x^2 + a^2}}.$$

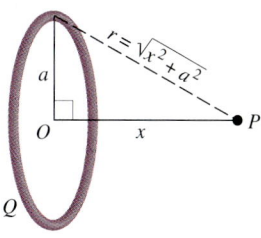

24–14 All the charge in a ring of charge Q is the same distance r from a point P on the ring axis.

Potential is a *scalar* quantity; there is no need to consider components of vectors in this calculation, as we had to do when we found the electric field at P. So the potential calculation is a lot simpler than the field calculation.

When x is much larger than a, the above expression for V becomes approximately equal to $V = Q/4\pi\epsilon_0 x$. This corresponds to the potential of a point charge Q at distance x. When we are very far away from a charged ring, it looks like a point charge. (We drew a similar conclusion about the electric field of a ring in Example 22–10.)

These results for V can also be found by integrating the expression for E_x found in Example 22–10 (see Problem 24–65).

EXAMPLE 24–12

A line of charge Electric charge Q is distributed uniformly along a line or thin rod of length 2a. Find the potential at a point P along the perpendicular bisector of the rod at a distance x from its center.

SOLUTION This is the same situation as in Example 22–11 (Section 22–7). As in that example, the element of charge dQ corresponding to an element of length dy on the rod is given by $dQ = (Q/2a)dy$ (Fig. 24–15). The distance from dQ to P is $\sqrt{x^2 + y^2}$, and the contribution dV that it makes to the potential at P is

$$dV = \frac{1}{4\pi\epsilon_0} \frac{Q}{2a} \frac{dy}{\sqrt{x^2 + y^2}}.$$

To get the potential at P due to the entire rod, we integrate dV over the length of the rod from $y = -a$ to $y = a$:

$$V = \frac{1}{4\pi\epsilon_0} \frac{Q}{2a} \int_{-a}^{a} \frac{dy}{\sqrt{x^2 + y^2}}.$$

You can look up the integral in a table. The final result is

$$V = \frac{1}{4\pi\epsilon_0} \frac{Q}{2a} \ln\left(\frac{\sqrt{a^2 + x^2} + a}{\sqrt{a^2 + x^2} - a} \right).$$

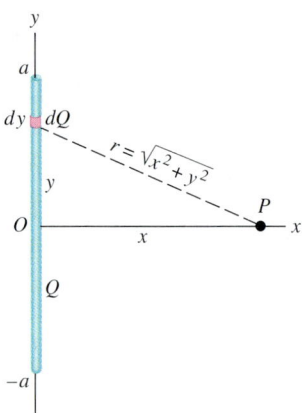

24–15 Finding the electric potential on the perpendicular bisector of a uniformly charged rod of length 2a.

When x is very large, we expect V to approach zero; we invite you to verify that it does so.

As in Example 24–11, this problem is simpler than finding $\vec{E}$ at point P because potential is a scalar quantity and no vector calculations are involved.

24–5 EQUIPOTENTIAL SURFACES

Field lines (Section 22–8) help us visualize electric fields. In a similar way the potential at various points in an electric field can be represented graphically by *equipotential surfaces*. These use the same fundamental idea as topographic maps like those used by hikers and mountain climbers (Fig. 24–16). On a topographic map, contour lines are drawn through points that are all at the same elevation. Any number of these could be drawn, but typically only a few contour lines are shown at equal spacings of elevation. If a mass m is moved over the terrain along such a contour line, the gravitational potential energy mgy does not change because the elevation y is constant. Thus contour lines on a topographic map are really curves of constant gravitational potential energy. Contour lines are close together in regions where the terrain is steep and there are large changes in elevation over a small horizontal distance; the contour lines are farther apart where the terrain is gently sloping. A ball allowed to roll downhill will experience the greatest downhill gravitational force where contour lines are closest together.

By analogy to contour lines on a topographic map, an **equipotential surface** is a three-dimensional surface on which the *electric potential V* is the same at every point. If a test charge q_0 is moved from point to point on such a surface, the *electric* potential energy $q_0 V$ remains constant. In a region where an electric field is present, we can construct an equipotential surface through any point. In diagrams we usually show only a

24–16 Contour lines on a topographic map are curves of constant elevation and hence of constant gravitational potential energy.

(a)

(b)

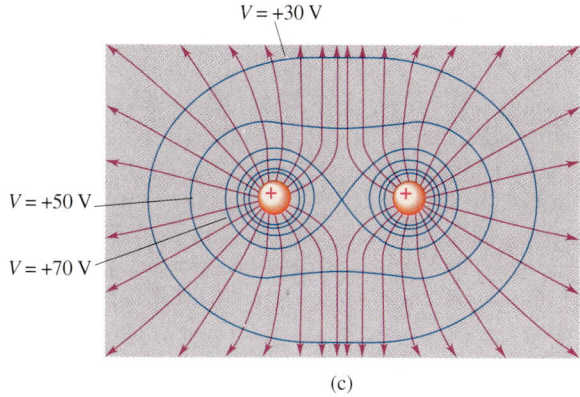

(c)

24–17 Cross sections of equipotential surfaces (blue lines) and electric field lines (red lines) for assemblies of point charges. There are equal potential differences between adjacent surfaces. (a) A single isolated positive charge. (b) An electric dipole. (c) Two equal positive charges. Compare these diagrams to those in Fig. 22–22, in which only the electric field lines were shown. How would the diagrams change if the signs of the charges were reversed?

11.11
Electric Potential: Qualitative Introduction

11.12.1 and 11.12.2
Electric Potential, Field, and Force

few representative equipotentials, often with equal potential differences between adjacent surfaces. No point can be at two different potentials, so equipotential surfaces for different potentials can never touch or intersect.

Because potential energy does not change as a test charge moves over an equipotential surface, the electric field can do no work on such a charge. It follows that $\vec{E}$ must be perpendicular to the surface at every point so that the electric force $q_0\vec{E}$ will always be perpendicular to the displacement of a charge moving on the surface. **Field lines and equipotential surfaces are always mutually perpendicular.** In general, field lines are curves, and equipotentials are curved surfaces. For the special case of a *uniform* field, in which the field lines are straight, parallel, and equally spaced, the equipotentials are parallel *planes* perpendicular to the field lines.

Figure 24–17 shows several arrangements of charges. The field lines in the plane of the charges are represented by red lines, and the intersections of the equipotential surfaces with this plane (that is, cross sections of these surfaces) are shown as blue lines. The actual equipotential surfaces are three-dimensional. At each crossing of an equipotential and a field line, the two are perpendicular.

In Fig. 24–17 we have drawn equipotentials so that there are equal potential differences between adjacent surfaces. In regions where the magnitude of $\vec{E}$ is large, the equipotential surfaces are close together because the field does a relatively large amount of work on a test charge in a relatively small displacement. This is the case near the point charge in Fig. 24–17a or between the two point charges in Fig. 24–17b; note that in these regions the field lines are also closer together. This is directly analogous to the downhill force of gravity being greatest in regions on a topographic map where contour lines are close together. Conversely, in regions where the field is weaker, the equipotential surfaces are farther apart; this happens at larger radii in Fig. 24–17a, to the left of

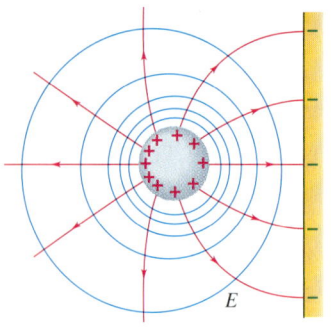

24–18 When charges are at rest, a conducting surface is always an equipotential surface. Field lines (shown in red) are perpendicular to a conducting surface.

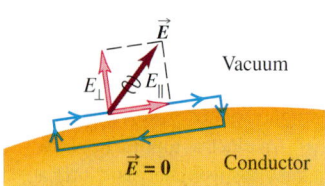

24–19 Inside the conductor, $\vec{E} = \mathbf{0}$. If $\vec{E}$ just outside the conductor had a component $E_{\parallel}$ parallel to the conductor surface, then a test charge moving around the rectangular loop and returning to the starting point would have nonzero total work done on it by the field. Because the $\vec{E}$ field is conservative, this is impossible. Therefore $E_{\parallel}$ must be zero, and $\vec{E}$ just outside the surface must be perpendicular to it.

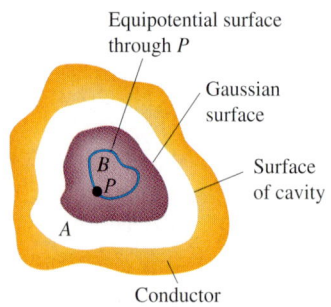

24–20 A cavity in a conductor. If the cavity contains no charge, every point in the cavity is at the same potential, the electric field is zero everywhere in the cavity, and there is no charge anywhere on the surface of the cavity.

the negative charge or the right of the positive charge in Fig. 24–17b, and at greater distances from both charges in Fig. 24–17c. (It may appear that two equipotential surfaces intersect at the center of Fig. 24–17c, in violation of the rule that this can never happen. In fact this is a single figure-8-shaped equipotential surface.)

On a given equipotential surface the potential V has the same value at every point, but in general the electric-field magnitude E does not. For instance, on the equipotential surface labeled "$V = -30$ V" in Fig. 24–17b, the magnitude E is less to the left of the negative charge than it is between the two charges. On the figure-8-shaped equipotential surface in Fig. 24–17c, $E = 0$ at the middle point halfway between the two charges; at any other point on this surface, E is nonzero.

EQUIPOTENTIALS AND CONDUCTORS

Here's an important statement about equipotential surfaces: **When all charges are at rest, the surface of a conductor is always an equipotential surface.** Since the electric field $\vec{E}$ is always perpendicular to an equipotential surface, we can prove this statement by proving that **when all charges are at rest, the electric field just outside a conductor must be perpendicular to the surface at every point** (Fig. 24–18). We know that $\vec{E} = \mathbf{0}$ everywhere inside the conductor; otherwise, charges would move. In particular, at any point just inside the surface the component of $\vec{E}$ tangent to the surface is zero. It follows that the tangential component of $\vec{E}$ is also zero just *outside* the surface. If it were not, a charge could move around a rectangular path partly inside and partly outside (Fig. 24–19) and return to its starting point with a net amount of work having been done on it. This would violate the conservative nature of electrostatic fields, so the tangential component of $\vec{E}$ just outside the surface must be zero at every point on the surface. Thus $\vec{E}$ is perpendicular to the surface at each point, proving our statement (Fig. 24–19).

Finally, we can now prove a theorem that we quoted without proof in Section 23–6. The theorem is as follows: In an electrostatic situation, if a conductor contains a cavity and if no charge is present inside the cavity, then there can be no net charge *anywhere* on the surface of the cavity. This means that if you're inside a charged conducting box, you can safely touch any point on the inside walls of the box without being shocked. To prove this theorem, we first prove that *every point in the cavity is at the same potential.* In Fig. 24–20 the conducting surface A of the cavity is an equipotential surface, as we have just proved. Suppose point P in the cavity is at a different potential; then we can construct a different equipotential surface B including point P.

Now consider a Gaussian surface, shown in Fig. 24–20, between the two equipotential surfaces. Because of the relation between $\vec{E}$ and the equipotentials, we know that the field at every point between the equipotentials is from A toward B, or else at every point it is from B toward A, depending on which equipotential surface is at higher potential. In either case the flux through this Gaussian surface is certainly not zero. But then Gauss's law says that the charge enclosed by the Gaussian surface cannot be zero. This contradicts our initial assumption that there is *no* charge in the cavity. So the potential at P *cannot* be different from that at the cavity wall.

The entire region of the cavity must therefore be at the same potential. But for this to be true, *the electric field inside the cavity must be zero everywhere.* Finally, Gauss's law shows that the electric field at any point on the surface of a conductor is proportional to the surface charge density σ at that point. We conclude that *the surface charge density on the wall of the cavity is zero at every point.* This chain of reasoning may seem tortuous, but it is worth careful study.

CAUTION ▶ Don't confuse equipotential surfaces with the Gaussian surfaces we encountered in Chapter 23. Gaussian surfaces have relevance only when we are using Gauss's law, and we can choose *any* Gaussian surface that's convenient. We are *not* free to choose the shape of equipotential surfaces; the shape is determined by the charge distribution. ◀

24–6 POTENTIAL GRADIENT

Electric field and potential are closely related. Equation (24–17), restated below, expresses one aspect of that relationship:

$$V_a - V_b = \int_a^b \vec{E} \cdot d\vec{l}.$$

If we know $\vec{E}$ at various points, we can use this equation to calculate potential differences. We should be able to turn this around; if we know the potential V at various points, we can use it to determine $\vec{E}$. Regarding V as a function of the coordinates (x, y, z) of a point in space, we will show that the components of $\vec{E}$ are directly related to the *partial derivatives* of V with respect to x, y, and z.

In Eq. (24–17), $V_a - V_b$ is the potential of a with respect to b, that is, the change of potential when a point moves from b to a. We can write this as

$$V_a - V_b = \int_b^a dV = -\int_a^b dV,$$

where dV is the infinitesimal change of potential accompanying an infinitesimal element $d\vec{l}$ of the path from b to a. Comparing to Eq. (24–17), we have

$$-\int_a^b dV = \int_a^b \vec{E} \cdot d\vec{l}.$$

These two integrals must be equal for *any* pair of limits a and b, and for this to be true the *integrands* must be equal. Thus for *any* infinitesimal displacement $d\vec{l}$,

$$-dV = \vec{E} \cdot d\vec{l}.$$

To interpret this expression, we write $\vec{E}$ and $d\vec{l}$ in terms of their components: $\vec{E} = \hat{\imath} E_x + \hat{\jmath} E_y + \hat{k} E_z$ and $d\vec{l} = \hat{\imath}\, dx + \hat{\jmath}\, dy + \hat{k}\, dz$. Then we have

$$-dV = E_x\, dx + E_y\, dy + E_z\, dz.$$

Suppose the displacement is parallel to the x-axis, so $dy = dz = 0$. Then $-dV = E_x\, dx$ or $E_x = -(dV/dx)_{y,z \text{ constant}}$, where the subscript reminds us that only x varies in the derivative; recall that V is in general a function of x, y, and z. But this is just what is meant by the partial derivative $\partial V/\partial x$. The y- and z-components of $\vec{E}$ are related to the corresponding derivatives of V in the same way, so we have

$$E_x = -\frac{\partial V}{\partial x}, \qquad E_y = -\frac{\partial V}{\partial y}, \qquad E_z = -\frac{\partial V}{\partial z} \qquad \begin{array}{l}\text{(components of } \vec{E} \ \ (24\text{–}19) \\ \text{in terms of } V\text{).}\end{array}$$

This is consistent with the units of electric field being V/m. In terms of unit vectors we can write $\vec{E}$ as

$$\vec{E} = -\left(\hat{\imath}\, \frac{\partial V}{\partial x} + \hat{\jmath}\, \frac{\partial V}{\partial y} + \hat{k}\, \frac{\partial V}{\partial z} \right) \qquad (\vec{E} \text{ in terms of } V). \qquad (24\text{–}20)$$

In vector notation the following operation is called the **gradient** of the function f:

$$\vec{\nabla} f = \left(\hat{\imath}\, \frac{\partial}{\partial x} + \hat{\jmath}\, \frac{\partial}{\partial y} + \hat{k}\, \frac{\partial}{\partial z} \right) f. \qquad (24\text{–}21)$$

The operator denoted by the symbol $\vec{\nabla}$ is called "grad" or "del." Thus in vector notation,

ActivPhysics

11.12.3
Electric Potential, Field, and Force

$$\vec{E} = -\vec{\nabla}V. \qquad (24\text{--}22)$$

This is read "$\vec{E}$ is the negative of the gradient of V" or "$\vec{E}$ equals negative grad V." The quantity $\vec{\nabla}V$ is called the *potential gradient*.

At each point, the potential gradient points in the direction in which V *increases* most rapidly with a change in position. Hence at each point the direction of $\vec{E}$ is the direction in which V *decreases* most rapidly and is always perpendicular to the equipotential surface through the point. This agrees with our observation in Section 24–3 that moving in the direction of the electric field means moving in the direction of decreasing potential.

Equation (24–22) doesn't depend on the particular choice of the zero point for V. If we were to change the zero point, the effect would be to change V at every point by the same amount; the derivatives of V would be the same.

If $\vec{E}$ is radial with respect to a point or an axis and r is the distance from the point or the axis, the relation corresponding to Eqs. (24–19) is

$$E_r = -\frac{\partial V}{\partial r} \qquad \text{(radial electric field).} \qquad (24\text{--}23)$$

Often we can compute the electric field caused by a charge distribution in either of two ways: directly, by adding the $\vec{E}$ fields of point charges, or by first calculating the potential and then taking its gradient to find the field. The second method is often easier because potential is a *scalar* quantity, requiring at worst the integration of a scalar function. Electric field is a *vector* quantity, requiring computation of components for each element of charge and a separate integration for each component. Thus, quite apart from its fundamental significance, potential offers a very useful computational technique in field calculations. Below, we present several examples in which a knowledge of V is used to find the electric field.

We stress once more that if we know $\vec{E}$ as a function of position, we can calculate V using Eq. (24–17) or (24–18), and if we know V as a function of position, we can calculate $\vec{E}$ using Eq. (24–19), (24–20), or (24–23). Deriving V from $\vec{E}$ requires integration, and deriving $\vec{E}$ from V requires differentiation.

EXAMPLE 24-13

Potential and field of a point charge From Eq. (24–14) the potential at a radial distance r from a point charge q is $V = q/4\pi\epsilon_0 r$. Find the vector electric field from this expression for V.

SOLUTION By symmetry the electric field has only a radial component, so we use Eq. (24–23):

$$E_r = -\frac{\partial V}{\partial r} = -\frac{\partial}{\partial r}\left(\frac{1}{4\pi\epsilon_0}\frac{q}{r}\right) = \frac{1}{4\pi\epsilon_0}\frac{q}{r^2},$$

so the vector electric field is

$$\vec{E} = \hat{r}\,E_r = \frac{1}{4\pi\epsilon_0}\frac{q}{r^2}\,\hat{r},$$

in agreement with Eq. (22–7).

EXAMPLE 24-14

Potential and field outside a charged conducting cylinder In Example 24–10 (Section 24–4) we found that the potential outside a charged conducting cylinder with radius R and charge per unit length λ is

$$V = \frac{\lambda}{2\pi\epsilon_0}\ln\frac{R}{r} = \frac{\lambda}{2\pi\epsilon_0}(\ln R - \ln r).$$

Find the components of the electric field outside the cylinder.

SOLUTION As in the previous example the symmetry tells us that $\vec{E}$ has only a radial component E_r. The derivative of $\ln R$ (a constant) is zero, and the derivative of $\ln r$ is $1/r$, so

$$E_r = -\frac{\partial V}{\partial r} = \frac{\lambda}{2\pi\epsilon_0}\frac{\partial(\ln r)}{\partial r} = \frac{1}{2\pi\epsilon_0}\frac{\lambda}{r}.$$

This is the same expression for E_r that we found in Example 23–6 (Section 23–5).

EXAMPLE 24-15

Potential and field of a ring of charge In Example 24–11 (Section 24–4) we found that for a ring of charge with radius a and total charge Q, the potential at a point P on the ring axis a distance x from the center is

$$V = \frac{1}{4\pi\epsilon_0} \frac{Q}{\sqrt{x^2 + a^2}}.$$

Find the electric field at P.

SOLUTION From Eq. (24–19),

$$E_x = -\frac{\partial V}{\partial x} = \frac{1}{4\pi\epsilon_0} \frac{Qx}{(x^2 + a^2)^{3/2}}.$$

This agrees with the result that we obtained in Example 22–10 (Section 22–7).

In this example, V does not appear to be a function of y or z, but it would *not* be correct to conclude that $\partial V/\partial y = \partial V/\partial z = 0$ and $E_y = E_z = 0$ everywhere. The reason is that our expression for V is valid *only for points on the x-axis,* where $y = z = 0$. If we had the complete expression for V valid at *all* points in space, then we could use it to find the components of $\vec{E}$ at any point using Eq. (24–19).

24-7 THE CATHODE-RAY TUBE

Let's look at how the concept of potential is applied to an important class of devices called **cathode-ray tubes,** or "CRTs" for short. Cathode-ray tubes are found in oscilloscopes, and similar devices are used in TV picture tubes and computer displays. The name goes back to the early 1900s. Cathode-ray tubes use an electron beam; before the basic nature of the beam was understood, it was called a cathode ray because it emanated from the cathode (negative electrode) of a vacuum tube.

Figure 24–21 is a schematic diagram of the principal elements of a cathode-ray tube. The interior of the tube is a very good vacuum, with a pressure of around 0.01 Pa (10^{-7} atm) or less. At any greater pressure, collisions of electrons with air molecules would scatter the electron beam excessively. The *cathode,* at the left end in the figure, is raised to a high temperature by the *heater,* and electrons evaporate from the surface of the cathode. The *accelerating anode,* with a small hole at its center, is maintained at a high positive potential V_1, of the order of 1 to 20 kV, relative to the cathode. This potential difference gives rise to an electric field directed from right to left in the region between the accelerating anode and the cathode. Electrons passing through the hole in the anode form a narrow beam and travel with constant horizontal velocity from the anode to the *fluorescent screen.* The area where the electrons strike the screen glows brightly.

The *control grid* regulates the number of electrons that reach the anode and hence the brightness of the spot on the screen. The *focusing anode* ensures that electrons

24–21 Basic elements of a cathode-ray tube. The width of the electron beam is exaggerated for clarity.

leaving the cathode in slightly different directions are focused down to a narrow beam and all arrive at the same spot on the screen. We won't need to worry about the control grid or focusing anode in the following analysis. The assembly of cathode, control grid, focusing anode, and accelerating electrode is called the *electron gun.*

The beam of electrons passes between two pairs of *deflecting plates.* An electric field between the first pair of plates deflects the electrons horizontally, and an electric field between the second pair deflects them vertically. If no deflecting fields are present, the electrons travel in a straight line from the hole in the accelerating anode to the center of the screen, where they produce a bright spot.

To analyze the electron motion, let's first calculate the speed v of the electrons as they leave the electron gun. We can use the same method as in Example 24–7 (Section 24–3). The initial speeds at which the electrons are emitted from the cathode are very small in comparison to their final speeds, so we assume that the initial speeds are zero. Then the speed v_x of the electrons as they leave the electron gun is given by

$$v_x = \sqrt{\frac{2eV_1}{m}}. \tag{24-24}$$

The kinetic energy of an electron leaving the anode depends only on the *potential difference* between anode and cathode, not on the details of the fields or the electron trajectories within the electron gun. As a numerical example, if $V_1 = 2000$ V,

$$v_x = \sqrt{\frac{2(1.60 \times 10^{-19} \text{ C})(2.00 \times 10^3 \text{ V})}{9.11 \times 10^{-31} \text{ kg}}} = 2.65 \times 10^7 \text{ m/s}.$$

This is about 9% of the speed of light in a vacuum.

If there is no electric field between the horizontal-deflection plates, the electrons enter the region between the vertical-deflection plates (shown in Fig. 24–22) with speed v_x. If there is a potential difference (voltage) V_2 between these plates, with the upper plate at higher potential, there is a *downward* electric field with magnitude $E = V_2/d$ between the plates. A constant upward force with magnitude eE then acts on the electrons, and their upward (y-component) acceleration is

$$a_y = \frac{eE}{m} = \frac{eV_2}{md}. \tag{24-25}$$

The *horizontal* component of velocity v_x is constant. The path of the electrons in the region between the plates is a parabolic trajectory, just as in Example 22–8 (Section 22–6) and in projectile motion (Section 3–4); in all of these cases a particle moves with constant x-velocity and constant y-acceleration. After the electrons emerge from this region, their paths again become straight lines, and they strike the screen at a point a distance y above its center. We are going to prove that this distance is *directly proportional* to the deflecting potential difference V_2.

24–22 Electrostatic deflection of an electron beam in a cathode-ray tube.

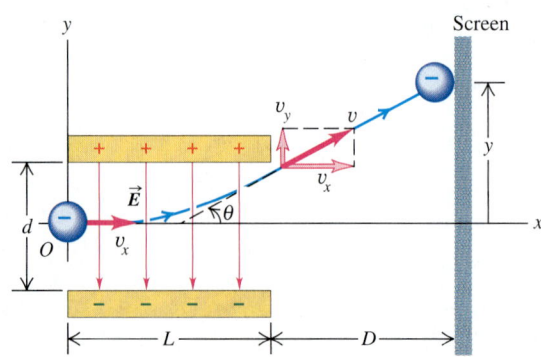

We first note that the time t required for the electrons to travel the length L of the plates is

$$t = \frac{L}{v_x}. \tag{24–26}$$

During this time, they acquire an upward velocity component v_y given by

$$v_y = a_y t. \tag{24–27}$$

Combining Eqs. (24–25), (24–26), and (24–27), we find

$$v_y = \frac{eV_2}{md}\frac{L}{v_x}. \tag{24–28}$$

When the electrons emerge from the deflecting field, their velocity $\vec{v}$ makes an angle θ with the x-axis given by

$$\tan \theta = \frac{v_y}{v_x}. \tag{24–29}$$

Ordinarily, the length L of the deflection plates is much smaller than the distance D from the plates to the screen. In this case the angle θ is also given approximately by $\tan \theta = y/D$. Combining this with Eq. (24–29), we find

$$\frac{y}{D} = \frac{v_y}{v_x}, \tag{24–30}$$

or, using Eq. (24–28) to eliminate v_y, we get

$$y = \frac{DeV_2 L}{mdv_x{}^2}. \tag{24–31}$$

Finally, we substitute the expression for v_x given by Eq. (24–24) into Eq. (24–31); the result is

$$y = \left(\frac{LD}{2d}\right)\frac{V_2}{V_1}. \tag{24–32}$$

The factor in parentheses depends only on the dimensions of the system, which are all constants. So we have proved that the deflection y is proportional to the *deflecting* voltage V_2, as claimed. It is also *inversely* proportional to the *accelerating* voltage V_1. This isn't surprising; the faster the electrons are going, the less they are deflected by the deflecting voltage.

If there is also a field between the *horizontal* deflecting plates, the beam is also deflected in the horizontal direction, perpendicular to the plane of Fig. 24–22. The coordinates of the luminous spot on the screen are then proportional to the horizontal and vertical deflecting voltages, respectively. This is the principle of the cathode-ray oscilloscope. If the horizontal deflection voltage sweeps the beam from left to right at a uniform rate, the beam traces out a graph of the vertical voltage as a function of time. Oscilloscopes are extremely useful laboratory instruments in many areas of pure and applied science.

The picture tube in a television set is similar, but the beam is deflected by *magnetic* fields (to be discussed in Chapter 28) rather than electric fields. In the system currently used in the United States and Canada, the electron beam traces out the area of the picture 30 times per second in an array of 525 horizontal lines, and the intensity of the beam is varied to make bright and dark areas on the screen. (In a color set, the screen is an array of dots of phosphors with three different colors.) The accelerating voltage in TV picture tubes (V_1 in the above discussion) is typically about 20 kV. Computer displays and monitors operate on the same principle, using a magnetically deflected electron beam to trace out images on a fluorescent screen. In this context the device is called a CRT display or a VDT (video display terminal).

24-8 CALCULATING POTENTIALS DUE TO CHARGED CONDUCTORS:

A Case Study in Computer Analysis

The techniques we have used in this chapter for calculating potentials and electric fields are useful only if the charge distribution is known. In many practical situations, however, the charge distribution is *not* known; instead, the value of the potential is known on the boundaries of a region. For instance, in an electrostatic situation the surface of a conductor is always an equipotential surface, but the distribution of charge on the surface is in general not uniform and is not readily calculated by the techniques we've encountered so far. Consider a region of space enclosed by one or more conductors maintained at fixed potentials (for example, by a battery). How can we determine the potential as a function of position in this region?

The key to solving this problem is to use the following fact about the electric potential: *In a region where there is no charge, the value of the potential at a given point is equal to the average of the potential values at surrounding points.* We'll prove this statement by using Gauss's law in conjunction with Eq. (24–19), which gives the electric field components in terms of partial derivatives of the potential.

We'll confine our discussion to situations in which the potential depends only on two coordinates, x and y. An example is the potential due to a long charged cylinder, as discussed in Example 24–10 (Section 24–4); the potential at a point depends only on the point's coordinates in a plane perpendicular to the axis of the cylinder, not on the coordinate z along the axis. For such a two-dimensional situation, consider a point P at coordinates (x, y, z), and enclose it by a Gaussian surface in the shape of a cubical box of side $2\Delta l$ centered on P (Fig. 24–23). If there is no charge in the volume enclosed by the box, the total electric flux Φ_E through the box is equal to zero. From Eq. (24–19) the z-component of the electric field $E_z = -\partial V/\partial z$ equals zero because the potential V is not a function of z. Hence there is no flux through the two faces of the Gaussian surface that are parallel to the xy-plane. Since the box is small, to a good approximation the flux through each of the other four faces of the box is equal to the product of the normal component of $\vec{E}$ at the center of each face and the area $(2\Delta l)^2$ of each face. The total flux

24–23 In a region where there is no charge, the value of the potential at a point P equals the average of the potential values at points surrounding P.

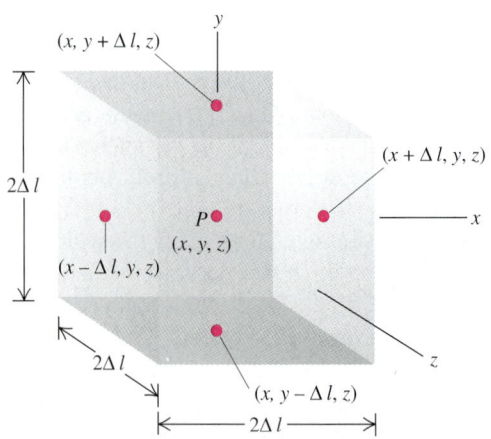

(equal to zero) through the box can then be expressed as

$$\Phi_E = E_x(x + \Delta l, \, y, \, z)(2\Delta l)^2 + (-E_x(x - \Delta l, \, y, \, z))(2\Delta l)^2$$
$$+ E_y(x, \, y + \Delta l, \, z)(2\Delta l)^2 + (-E_y(x, \, y - \Delta l, \, z))(2\Delta l)^2 = 0. \quad (24\text{–}33)$$

Using Eq. (24–19), we can write the electric-field components to the same approximation as

$$E_x(x + \Delta l, \, y, \, z) = -\frac{\partial V(x + \Delta l, \, y)}{\partial x} = -\frac{V(x + \Delta l, \, y) - V(x, \, y)}{\Delta l},$$

$$E_x(x - \Delta l, \, y, \, z) = -\frac{\partial V(x - \Delta l, \, y)}{\partial x} = -\frac{V(x, \, y) - V(x - \Delta l, \, y)}{\Delta l},$$

$$E_y(x, \, y + \Delta l, \, z) = -\frac{\partial V(x, \, y + \Delta l)}{\partial y} = -\frac{V(x, \, y + \Delta l) - V(x, \, y)}{\Delta l},$$

$$E_y(x, \, y - \Delta l, \, z) = -\frac{\partial V(x, \, y - \Delta l)}{\partial y} = -\frac{V(x, \, y) - V(x, \, y - \Delta l)}{\Delta l}. \quad (24\text{–}34)$$

Substituting Eqs. (24–34) into Eq. (24–33) and dividing through by $4 \, \Delta l$, we obtain

$$-[V(x + \Delta l, \, y) - V(x, \, y)] + [V(x, \, y) - V(x - \Delta l, \, y)]$$
$$-[V(x, \, y + \Delta l) - V(x, \, y)] + [V(x, \, y) - V(x, \, y - \Delta l)] = 0.$$

If we solve this for $V(x, y)$, the potential at point P, we find

$$V(x, \, y) = \frac{1}{4}[V(x + \Delta l, \, y) + V(x - \Delta l, \, y) + V(x, \, y + \Delta l) + V(x, \, y - \Delta l)].$$
$$(24\text{–}35)$$

In words, the value of the potential at P is the average of the potential values at the points surrounding P. This statement becomes exact in the limit that Δl becomes infinitesimally small.

To see how to use Eq. (24–35) to calculate the potential due to a set of charged conductors, let's consider a specific situation. Figure 24–24a shows a hollow conducting box with a square cross section and with a long axis parallel to the z-axis. The length of the box is very much greater than the dimension L. The top of the box, labeled a, is insulated from the other three sides, collectively labeled b; this is done by having the top be a separate piece of metal with a small gap between it and the vertical sides of the box (Fig. 24–24b). A fixed potential difference V_0 is maintained between segments a and b of the box. We choose the potentials of the lower segments to be $V_b = 0$, so the potential of the upper segment is $V_a = V_0$. As a result of the potential difference, there is a positive charge on a (the higher-potential conductor) and a negative charge on b (the lower-potential conductor).

Our goal is to find the potential V at all points in the interior volume of the box. Because the box is long, the potential inside the box is to a good approximation a function of x and y only. We imagine making a rectangular grid of points inside the box, separated by a distance Δl (Fig. 24–25). The outermost points of the grid are on the conductor surfaces themselves. Equation (24–35) then relates the potential at different grid points to each other; since the potentials of the conductors are specified, we can determine the potential at each grid point in the empty interior of the box.

Complications arise because Eq. (24–35) relates the values of the potential at four different grid points. The value of V is unknown at each grid point in the interior of the box, and each such point is surrounded by two or three other interior points at which V is also unknown. So Eq. (24–35) can't be used to solve for the values of the potential at these interior points in a single step. Instead, we have to use an *iterative* method; we will

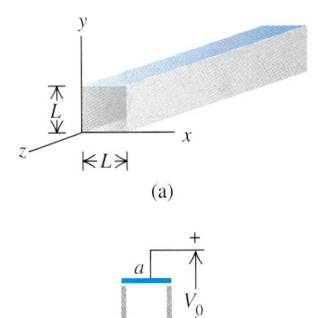

24–24 (a) A conducting box whose z-dimension is much longer than the dimension L along the x- and y-axes. (b) A view down the z-axis of the box, showing the two segments into which the box is divided and the potential difference between the segments.

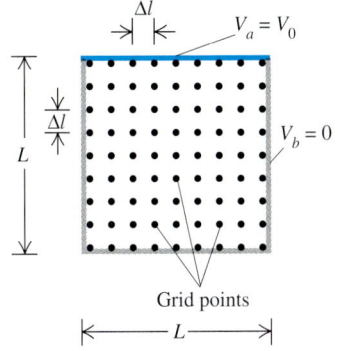

24–25 A view down the long axis of the conducting box showing a rectangular grid of points separated by Δl.

make a series of successive approximations to find a set of values of V at the interior grid points such that Eq. (24–35) is satisfied at every point. The procedure that we follow to do this is called the *relaxation method.*

Here's a skeleton of a computer program to carry out this calculation.

Step 1: Choose a *positive* value of the potential difference V_0. (The procedure described below runs into trouble if $V_0 < 0$.)

Step 2: Choose the number m of grid points across or down the region shown in Fig. 24–25. The total number of grid points is then m^2, and the total number of grid points in the interior of the box is $(m - 2)^2$. Values of m between 10 and 40 work well. Large values of m require lengthy calculation; small values of m give low resolution.

Step 3: Let (j, k) be a pair of integer indexes that identify a particular grid point and its location in the grid (the jth column and kth row in the grid); each ranges from 1 to m.

Step 4: Begin a loop on k, from $k = 1$ to $k = m$ (successive rows of grid points).

Step 5: Begin a loop on j, from $j = 1$ to $j = m$ (successive columns of grid points).

Step 6: Assign an initial value to the potential $V(j, k)$ for each grid point. If $k = 1$ (the top row, corresponding to the surface of the upper conductor a in Fig. 24–25), set $V(j, k)$ equal to V_0. If $j = 1$, $j = m$, or $k = m$, set $V(j, k)$ equal to zero; these correspond to the left, right, and bottom surfaces, respectively, of the lower conductor b in Fig. 24–25. For all other values of (j, k), corresponding to grid points in the empty interior of the box, assign an arbitrary value of $V(j, k)$. The closer this arbitrary value is to the actual value, the fewer iterations will be required to obtain a good solution. But any value greater than 0 and less than V_0 will do. A good choice for the arbitrary value might be $0.5V_0$ for all interior grid points. Do *not* use $V = 0$; this choice will cause problems in Step 13.

Step 7: End of loops over j and k.

Step 8: Specify the desired accuracy of the results, expressed as a fractional uncertainty. The program is about to begin iterating to find a solution for $V(j, k)$ at grid points inside the box, and the iteration will stop when the fractional change in the values from one iteration to the next is less than the desired accuracy. Reasonable values are from 0.01 to 0.001; the smaller the value chosen, the more iterations will be required.

Step 9: Define a quantity δ (lowercase "delta") and set it equal to zero.

Step 10: Again begin a loop on k, this time from $k = 2$ to $k = m - 1$ (that is, over interior points only).

Step 11: Begin a loop on j, this time from $j = 2$ to $j = m - 1$ (interior points only).

Step 12: Compute new values of $V_{new}(j, k)$ for each interior grid point, using Eq. (24–35). (The loops over j and k do not include the values 1 and m because the values of the potential on the conductor surfaces are fixed.) That is, let

$$V_{new}(j, k) = \frac{1}{4}[V(j + 1, k) + V(j - 1, k) + V(j, k + 1) + V(j, k - 1)].$$

Referring to Fig. 24–25, this means that the new value of the potential at (j, k) is the average of the old values of V at the four grid points immediately to the right, the left, below, and above.

Step 13: Calculate the fractional change in potential $\Delta V/V$ from the old value to the new value, and take its absolute value:

$$\frac{\Delta V}{V} = \left| \frac{V_{new}(j, k) - V(j, k)}{V(j, k)} \right|.$$

A division-by-zero error will result if $V(j, k) = 0$; this is the reason for our cautionary note at the end of Step 6. If $\Delta V/V$ is greater than δ, set δ equal to $\Delta V/V$.

Step 14: Replace the old value of V by the new value. That is, set $V(j, k)$ equal to $V_{new}(j, k)$.

Step 15: End of loops over j and k.

Step 16: If the value of δ is greater than the desired accuracy specified in Step 8, there is at least one interior grid point at which the change in V during the iteration just completed was greater than the desired accuracy. Hence one or more additional iterations are required, and the program must return to Step 10. If the value of δ is less than or equal to the desired accuracy, the current values of $V(j, k)$ are adequate, and the program may proceed to Step 17.

Step 17: Print out the potential for each grid point.

Step 18: END.

Figure 24–26 shows the result of this calculation.

The program described above is actually easier to implement with a spreadsheet than with a programming language such as BASIC or Pascal. Most popular spreadsheets permit iterative calculations and will automatically go through the steps described above. To carry out the calculation, choose a square grid of spreadsheet cells, enter the values of the conductor potentials in the cells on the periphery of the grid, and enter the formula corresponding to Eq. (24–35) into each interior cell. The results shown in Fig. 24–26 were obtained with a spreadsheet.

It's easy to modify the above program to treat other types of conducting boundaries. The values of the potential on the conducting surfaces can be chosen at will. The box can be changed from a square to a rectangle by having a different number of rows than columns. The box can also have shapes other than rectangular. Various suggestions are included in the exercises.

Once the values of the potential have been calculated for all grid points, the electric field components E_x and E_y can be calculated as well using equations similar to Eq. (24–34). At a point (x, y) in the interior of the box, we can find E_x and E_y using the approximate results

$$E_x(x, y) = -\frac{\partial V(x, y)}{\partial x} \approx -\frac{V(x + \Delta l, y) - V(x - \Delta l, y)}{2\Delta l},$$

$$E_y(x, y) = -\frac{\partial V(x, y)}{\partial y} \approx -\frac{V(x, y + \Delta l) - V(x, y - \Delta l)}{2\Delta l}. \tag{24–36}$$

1.00	1.00	1.00	1.00	1.00	1.00	1.00	1.00	1.00	1.00	1.00	1.00
0.00	0.49	0.68	0.76	0.80	0.82	0.82	0.80	0.76	0.68	0.49	0.00
0.00	0.28	0.46	0.57	0.62	0.65	0.65	0.62	0.57	0.46	0.28	0.00
0.00	0.18	0.33	0.42	0.48	0.50	0.50	0.48	0.42	0.32	0.18	0.00
0.00	0.13	0.23	0.31	0.36	0.38	0.38	0.36	0.31	0.23	0.13	0.00
0.00	0.09	0.17	0.23	0.27	0.29	0.29	0.27	0.23	0.17	0.09	0.00
0.00	0.06	0.12	0.17	0.20	0.21	0.21	0.20	0.17	0.12	0.06	0.00
0.00	0.05	0.09	0.12	0.14	0.15	0.15	0.14	0.12	0.09	0.05	0.00
0.00	0.03	0.06	0.08	0.10	0.11	0.11	0.10	0.08	0.06	0.03	0.00
0.00	0.02	0.04	0.05	0.06	0.07	0.07	0.06	0.05	0.04	0.02	0.00
0.00	0.01	0.02	0.02	0.03	0.03	0.03	0.03	0.02	0.02	0.01	0.00
0.00	0.00	0.00	0.00	0.00	0.00	0.00	0.00	0.00	0.00	0.00	0.00

24–26 Potentials after 75 iterations for the situation shown in Fig. 24–25, with $V_0 = 1.00$ V. The grid points on the conductor surfaces are shown shaded for clarity. For this grid, $m = 12$. The potential decreases as you move away from the higher-potential top surface and toward the lower-potential surfaces of the sides and bottom.

The magnitude of the field is

$$E = \sqrt{E_x^2 + E_y^2} \qquad (24\text{--}37)$$

To indicate the direction of $\vec{E}$ at each interior grid point, the program can be instructed to draw a line of length c between the two points with coordinates

$$(x,\ y) \quad \text{and} \quad \left(x + \frac{cE_x}{E},\ y + \frac{cE_y}{E}\right). \qquad (24\text{--}38)$$

You can verify that the distance between these points is c and that the direction of the line is the same as that of $\vec{E}$ at (x, y).

To draw lines indicating the direction of $\vec{E}$ at each interior grid point, insert the following additional steps between Steps 17 and 18 in the above program skeleton. (*Note:* This may be easier to implement by using BASIC or Pascal than with a spreadsheet.)

Step 17A: Choose the screen coordinates (addresses of pixels on the screen) within which the grid points and direction lines of $\vec{E}$ will be drawn, that is, values of $x_{\min}$, $y_{\min}$, $x_{\max}$, and $y_{\max}$. These values are determined by the characteristics of your computer. For example, a Macintosh with a 14-inch monitor or a PC with a VGA video adapter has $x_{\min} = 0$, $x_{\max} = 639$, $y_{\min} = 0$, and $y_{\max} = 479$. Some languages let you change the range of the screen coordinates. In BASIC, for example, the WINDOW command lets you choose any range of x and y you like; it does the conversion to screen coordinates for you.

Step 17B: Choose the distance Δl (the spacing, in pixels, between grid points on the screen) so that all the grid points will fit on the screen. For instance, if the number of grid rows and columns is $m = 20$, choose Δl to be 1/20 of the *smaller* of $(x_{\max} - x_{\min})$ and $(y_{\max} - y_{\min})$. Set the line length c equal to Δl.

Step 17C: Begin a loop on k from $k = 2$ to $k = m - 1$ (interior points only).

  Step 17D: Begin a loop on j from $j = 2$ to $j = m - 1$ (interior points only).

    Step 17E: Calculate the x- and y-coordinates of the screen position corresponding to the current grid point:

$$x = x_{\min} + j\,\Delta l \qquad \text{and} \qquad y = y_{\min} + k\,\Delta l.$$

    Step 17F: Calculate the electric-field components at the interior grid point (j, k) using Eq. (24–36):

$$E_x = -\frac{V(j+1,\ k) - V(j-1,\ k)}{2\Delta l} \qquad \text{and} \qquad E_y = -\frac{V(j,\ k+1) - V(j,\ k-1)}{2\Delta l}.$$

    Step 17G: If $E_x = 0$ and $E_y = 0$, do not draw a line for the direction of $\vec{E}$ (since the field is zero); instead, go on to the next grid point. Otherwise, find E using Eq. (24–37).

    Step 17H: Draw a line between the points given by Eq. (24–38).

Step 17I: End of loops over j and k.

Figure 24–27 shows the result of this calculation for the situation depicted in Fig. 24–25.

You may want to refine the calculation to draw at each grid point a vector with length proportional to the field magnitude E; this shows both the magnitude and the direction of the field. To do this, simply replace the parameter c in Step 17H by a multiple of E. You may have to experiment to find the appropriate scale factor for the best picture.

By a simple modification of the technique used to create the field map, we can map the equipotential lines in the xy-plane (the cross section in this plane of the equipotential *surfaces*). We use the idea that when an equipotential crosses a field line, the two are perpendicular. To map the equipotentials, we draw a line at each grid point in a direction

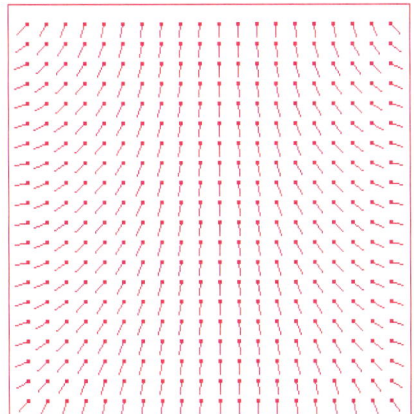

24–27 Electric-field map for the situation shown in Fig. 24–25. Each grid point is marked with a dot. For this grid, $m = 22$. The field is directed away from the positive charges on the top surface at $V = V_0$ and toward the negative charges on the sides and bottom at $V = 0$.

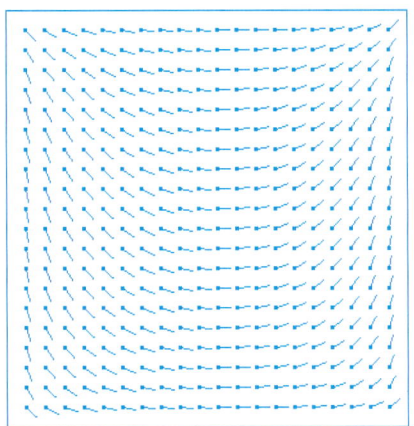

24–28 Equipotential line segments for the situation shown in Fig. 24–25. Each grid point is marked with a dot. For this grid, $m = 22$. As you approach the conducting surfaces at the top, left, bottom, and right, the equipotentials become nearly parallel to the surfaces.

perpendicular to the field line at that point. From analytic geometry, if two lines in a plane are perpendicular, their slopes are negative reciprocals of each other. The slope of the field line at each point is E_y/E_x, so the slope of the equipotential at the same point is $-E_x/E_y$. Referring to Eq. (24–38), which we used to draw lines with length c parallel to the field, we see that we need only replace E_x by E_y and E_y by $-E_x$ in this expression to get a line perpendicular to $\vec{E}$ at the point. That is, in Step 17H we use, instead of Eq. (24–38), the following:

$$(x, \; y) \quad \text{and} \quad \left(x + \frac{cE_y}{E}, \; y - \frac{cE_x}{E}\right). \tag{24–39}$$

Figure 24–28 shows the result of such a calculation. Compare this to the field map shown in Fig. 24–27. Note that this procedure doesn't plot actual equipotential curves; rather, it plots segments of various curves. But these give us a good idea of the shapes of the equipotentials.

Finally, we note that the relaxation-method approach is just another case in which it is easiest to find the potential first, then determine the electric field from the potential.

Summary

- The electric force caused by any collection of charges at rest is a conservative force. The work W done by the electric force on a charged particle moving in an electric field can be represented by a potential-energy function U:

$$W_{a \to b} = U_a - U_b. \tag{24–2}$$

The potential energy for two point charges q and q_0 separated by a distance r is

$$U = \frac{1}{4\pi\epsilon_0} \frac{qq_0}{r}, \tag{24–9}$$

and the potential energy for a charge q_0 in the electric field of a collection of charges

KEY TERMS

(electric) potential energy, 731
(electric) potential, 737
volt, 737
voltage, 738
electron volt, 740
equipotential surface, 746
gradient, 749
cathode-ray tube, 751

q_i is given by

$$U = \frac{q_0}{4\pi\epsilon_0}\left(\frac{q_1}{r_1} + \frac{q_2}{r_2} + \frac{q_3}{r_3} + \cdots\right) = \frac{q_0}{4\pi\epsilon_0}\sum_i \frac{q_i}{r_i}, \tag{24-10}$$

where r_i is the distance from q_i to q_0. If q_0 is infinitely far from all the other charges, $U = 0$.

- Potential, denoted by V, is potential energy per unit charge. The potential due to a single point charge q at a distance r from the charge is

$$V = \frac{U}{q_0} = \frac{1}{4\pi\epsilon_0}\frac{q}{r}. \tag{24-14}$$

The potential due to a collection of point charges q_i is

$$V = \frac{U}{q_0} = \frac{1}{4\pi\epsilon_0}\sum_i \frac{q_i}{r_i}, \tag{24-15}$$

and the potential due to a continuous distribution of charge is

$$\dot{V} = \frac{1}{4\pi\epsilon_0}\int \frac{dq}{r}. \tag{24-16}$$

- The potential difference between two points a and b, also called the potential of a with respect to b, is given by the line integral of $\vec{E}$:

$$V_a - V_b = \int_a^b \vec{E}\cdot d\vec{l} = \int_a^b E\cos\phi\,dl. \tag{24-17}$$

Potentials can be calculated either by summing or integrating over the charges, as in Eqs. (24–15) and (24–16), or by first finding $\vec{E}$ and then using Eq. (24–17).

- Two equivalent sets of units for electric-field magnitude are volts per meter (V/m) and newtons per coulomb (N/C). One volt is one joule per coulomb (1 V = 1 J/C).

- The electron volt, abbreviated eV, is the energy corresponding to a particle with a charge equal to that of the electron moving through a potential difference of one volt. The conversion factor is $1\text{ eV} = 1.602 \times 10^{-19}$ J.

- An equipotential surface is a surface on which the potential has the same value at every point. At a point where a field line crosses an equipotential surface, the two are perpendicular. When all charges are at rest, the surface of a conductor is always an equipotential surface, and all points in the material of a conductor are at the same potential. When a cavity within a conductor contains no charge, the entire cavity is an equipotential region, and there is no surface charge anywhere on the surface of the cavity.

- If the potential is known as a function of the coordinates x, y, and z, the components of electric field $\vec{E}$ at any point are given by

$$E_x = -\frac{\partial V}{\partial x}, \qquad E_y = -\frac{\partial V}{\partial y}, \qquad E_z = -\frac{\partial V}{\partial z}. \tag{24-19}$$

In vector form,

$$\vec{E} = -\left(\hat{\imath}\frac{\partial V}{\partial x} + \hat{\jmath}\frac{\partial V}{\partial y} + \hat{k}\frac{\partial V}{\partial z}\right). \tag{24-20}$$

- The cathode-ray tube uses an electron beam created by a set of electrodes called an electron gun. The beam is deflected by two sets of deflecting plates, then strikes a fluorescent screen and forms an image on it.

DISCUSSION QUESTIONS

Q24–1 A student asked, "Since electrical potential is always proportional to potential energy, why bother with the concept of potential at all?" How would you respond?

Q24–2 The potential (relative to a point at infinity) midway between two charges of equal magnitude and opposite sign is zero. Is it possible to bring a test charge from infinity to this midpoint in such a way that no work is done in any part of the displacement? If so, describe how it can be done. If it is not possible, explain why.

Q24–3 Is it possible to have an arrangement of two point charges separated by a finite distance such that the electric potential energy of the arrangement is the same as if the two charges were infinitely far apart? Why or why not? What if there are three charges? Explain your reasoning.

Q24–4 On a physics exam once given at a famous university, students were asked to calculate the potential energy of a particular distribution of point charges. One student did no calculations for this problem, but simply answered "The potential energy can have any value you want." Although this was not the answer that the instructor had in mind when writing the problem, the student nonetheless received full credit for her answer. Why was this answer correct? (*Note:* Don't try this on *your* exams!)

Q24–5 If $\vec{E}$ is zero everywhere along a certain path that leads from point A to point B, what is the potential difference between those two points? Does this mean that $\vec{E}$ is zero everywhere along *any* path from A to B? Explain.

Q24–6 If $\vec{E}$ is zero throughout a certain region of space, is the potential necessarily also zero in this region? Why or why not? If not, what *can* be said about the potential?

Q24–7 If you carry out the integral of the electric field $\int \vec{E} \cdot d\vec{l}$ for a *closed* path like that shown in Fig. 24–29, the integral will *always* be equal to zero, independent of the shape of the path and independent of where charges may be located relative to the path. Explain why.

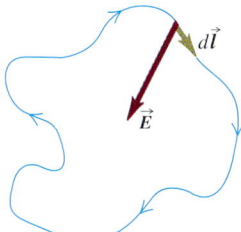

FIGURE 24–29 Question Q24–7.

Q24–8 The potential difference between the two terminals of an AA battery (used in flashlights and portable stereos) is 1.5 V. If two AA batteries are placed end-to-end with the positive terminal of one battery touching the negative terminal of the other, what is the potential difference between the terminals at the exposed ends of the combination? What if the two positive terminals are touching each other? Explain your reasoning.

Q24–9 It is easy to produce a potential difference of several thousand volts between your body and the floor by scuffing your shoes across a nylon carpet. When you touch a metal doorknob, you get a mild shock. Yet contact with a power line of comparable voltage would probably be fatal. Why is there a difference?

Q24–10 If the electric potential at a single point is known, can $\vec{E}$ at that point be determined? If so, how? If not, why not?

Q24–11 Because electric field lines and equipotential surfaces are always perpendicular, two equipotential surfaces can never cross; if they did, the direction of $\vec{E}$ would be ambiguous at the crossing points. Yet two equipotential surfaces appear to cross at the center of Fig. 24–17c. Explain why there is no ambiguity about the direction of $\vec{E}$ in this particular case.

Q24–12 As you walk through a certain region of space, the potential at your position becomes more and more positive as you move from north to south. The potential doesn't change when you move east, west, up, or down. Is there necessarily an electric field in this region? If so, what is its direction? Explain your reasoning. Suggest a way in which you could cause the potential to vary with position in the way described.

Q24–13 Is potential gradient a scalar quantity or a vector quantity? How can you tell?

Q24–14 A conducting sphere is to be charged by bringing in positive charge a little at a time until the total charge is Q. The total work required for this process is alleged to be proportional to Q^2. Is this correct? Why or why not?

Q24–15 Are there cases in electrostatics in which a conducting surface is *not* an equipotential surface? If so, give an example. If not, explain why not.

Q24–16 A conducting sphere is placed between two charged parallel plates such as those shown in Fig. 24–1. Does the electric field inside the sphere depend on precisely where between the plates the sphere is placed? What about the electric potential inside the sphere? Do the answers to these questions depend on whether or not there is a net charge on the sphere? Explain your reasoning.

Q24–17 A conductor that carries a net charge Q has a hollow, empty cavity in its interior. Does the potential vary from point to point within the material of the conductor? What about within the cavity? How does the potential inside the cavity compare to the potential within the material of the conductor?

Q24–18 A high-voltage dc power line falls on a car, so the entire metal body of the car is at a potential of 10,000 V with respect to the ground. What happens to the occupants a) when they are sitting in the car? b) when they step out of the car? Explain your reasoning.

Q24–19 When a thunderstorm is approaching, sailors at sea sometimes observe a phenomenon called "St. Elmo's fire," a bluish flickering light at the tips of masts. What causes this? Why does it occur at the tips of masts? Why is the effect most pronounced when the masts are wet? (*Hint:* Seawater is a good conductor of electricity.)

Q24–20 A positive point charge is placed near a very large conducting plane. A professor of physics asserted that the field caused by this configuration is the same as would be obtained by removing the plane and placing a negative point charge of equal magnitude in the mirror-image position behind the initial position of the plane. Is this correct? Why or why not? (*Hint:* Inspect Fig. 24–17b.)

Q24–21 In electronics it is customary to define the potential of ground (thinking of the earth as a large conductor) as zero. Is

this consistent with the fact that the earth has a net electric charge that is not zero? (Refer to Exercise 22–22.)

Q24–22 In applying the relaxation method to the box in Fig. 24–25, the values of the potential at the grid points at the *corners* of the box are irrelevant. Why?

EXERCISES

SECTION 24–2 ELECTRIC POTENTIAL ENERGY

24–1 A point charge $q_1 = +2.40\ \mu C$ is held stationary at the origin. A second point charge $q_2 = -4.30\ \mu C$ moves from the point $x = 0.150$ m, $y = 0$ to the point $x = 0.250$ m, $y = 0.250$ m. How much work is done by the electric force on q_2?

24–2 A point charge q_1 is held stationary at the origin. A second charge q_2 is placed at point a, and the electric potential energy of the pair of charges is $+5.4 \times 10^{-8}$ J. When the second charge is moved to point b, the electric force on the charge does -1.9×10^{-8} J of work. What is the electric potential energy of the pair of charges when the second charge is at point b?

24–3 A small metal sphere, carrying a net charge of $q_1 = -2.80\ \mu C$, is held in a stationary position by insulating supports. A second small metal sphere, with a net charge of $q_2 = -7.80\ \mu C$ and mass 1.50 g, is projected toward q_1. When the two spheres are 0.800 m apart, q_2 is moving toward q_1 with speed 22.0 m/s (Fig. 24–30). Assume that the two spheres can be treated as point charges. Neglect the force of gravity. a) What is the speed of q_2 when the spheres are 0.400 m apart? b) How close does q_2 get to q_1?

FIGURE 24–30 Exercise 24–3.

24–4 How far from a -7.20-μC point charge must a $+2.30$-μC point charge be placed for the electric potential energy U of the pair of charges to be -0.400 J? (Take U to be zero when the charges have infinite separation.)

24–5 A point charge $Q = +4.60\ \mu C$ is held fixed at the origin. A second point charge $q = +1.20\ \mu C$ with mass of 2.80×10^{-4} kg is placed on the x-axis, 0.250 m from the origin. a) What is the electric potential energy U of the pair of charges? (Take U to be zero when the charges have infinite separation.) b) The second point charge is released from rest. What is its speed when its distance from the origin is i) 0.500 m? ii) 5.00 m? iii) 50.0 m?

24–6 Three equal 1.20-μC point charges are placed at the corners of an equilateral triangle whose side is 0.500 m. What is the potential energy of the system? (Take as zero the potential energy of the three charges when they are infinitely far apart.)

24–7 A point charge $q_1 = 4.00$ nC is placed at the origin, and a second point charge $q_2 = -3.00$ nC is placed on the x-axis at $x = +20.0$ cm. A third point charge $q_3 = 2.00$ nC is to be placed on the x-axis between q_1 and q_2. Let the potential energy of the three charges be zero when they are infinitely far apart. a) What is the potential energy of the system of the three charges if q_3 is placed at $x = +10.0$ cm? b) Where should q_3 be placed to make the potential energy of the system equal to zero?

24–8 In part (c) of Example 24–1 (Section 24–2), calculate the distance between the positron and alpha particle when the positron comes momentarily to rest.

24–9 Three point charges, which initially are infinitely far apart, are placed at the corners of an equilateral triangle of side d. Two of the point charges are identical and have charge q. If zero net work is required to place the three charges at the corners of the triangle, what must be the value of the third charge?

SECTION 24–3 ELECTRIC POTENTIAL

24–10 A particle with a charge of $+4.20$ nC is in a uniform electric field $\vec{E}$ directed to the left. It is released from rest and moves to the left; after it has moved 6.00 cm, its kinetic energy is found to be $+1.50 \times 10^{-6}$ J. a) What work was done by the electric force? b) What is the potential of the starting point with respect to the endpoint? c) What is the magnitude of $\vec{E}$?

24–11 A charge of 28.0 nC is placed in a uniform electric field that is directed vertically upward and that has a magnitude of 4.00×10^4 V/m. What work is done by the electric force when the charge moves a) 0.450 m to the right; b) 0.670 m upward; c) 2.60 m at an angle of 45.0° downward from the horizontal?

24–12 Two stationary point charges $+3.00$ nC and $+2.00$ nC are separated by a distance of 50.0 cm. An electron is released from rest at a point midway between the two charges and moves along the line connecting the two charges. What is the speed of the electron when it is 10.0 cm from the $+3.00$-nC charge?

24–13 A point charge has a charge of 2.50×10^{-11} C. At what distance from the point charge is the electric potential a) 90.0 V? b) 30.0 V? Take the potential to be zero at an infinite distance from the charge.

24–14 The potential V at a distance of 25.0 cm from a very small charged sphere is 48.0 V, with V taken to be zero at an infinite distance from the sphere. a) If the sphere is treated as a point charge, what is its charge? b) What is the potential at a distance of 75.0 cm from the sphere?

24–15 Two point charges $q_1 = +2.40$ nC and $q_2 = -6.50$ nC are 0.100 m apart. Point A is midway between them; point B is 0.080 m from q_1 and 0.060 m from q_2 (Fig. 24–31). Take the electric potential to be zero at infinity. Find a) the potential at point A; b) the potential at point B; c) the work done by the electric field on a charge of 2.50 nC that travels from point B to point A.

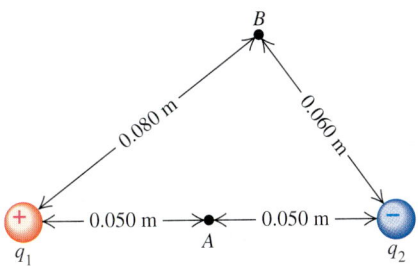

FIGURE 24–31 Exercise 24–15.

24–16 Two positive point charges, each of magnitude q, are fixed on the y-axis at the points $y = +a$ and $y = -a$. Take the potential to be zero at an infinite distance from the charges. a) Show the positions of the charges in a diagram. b) What is the potential V_0 at the origin? c) Show that the potential at any point on the x-axis is

$$V = \frac{1}{4\pi\epsilon_0} \frac{2q}{\sqrt{a^2 + x^2}}.$$

d) Graph the potential on the x-axis as a function of x over the range from $x = -4a$ to $x = +4a$. e) What is the potential when $x \gg a$? Explain why this result is obtained.

24–17 A positive charge $+q$ is located at the point $x = 0$, $y = -a$, and a negative charge $-q$ is located at the point $x = 0$, $y = +a$. a) Show the positions of the charges in a diagram. b) Derive an expression for the potential V at points on the x-axis as a function of the coordinate x. Take V to be zero at an infinite distance from the charges. c) Graph V at points on the x–axis as a function of x over the range from $x = -4a$ to $x = +4a$. d) What is the answer to part (b) if the two charges are interchanged so that $+q$ is at $y = +a$ and $-q$ is at $y = -a$?

24–18 Consider the arrangement of charges described in Exercise 24–17. a) Derive an expression for the potential V at points on the y-axis as a function of the coordinate y. Take V to be zero at an infinite distance from the charges. b) Graph V at points on the y-axis as a function of y over the range from $y = -4a$ to $y = +4a$. c) Show that for $y \gg a$, the potential at a point on the positive y-axis is given by $V = -\left(\frac{1}{4\pi\epsilon_0}\right)2qa/y^2$. d) What are the answers to parts (a) and (c) if the two charges are interchanged so that $+q$ is at $y = +a$ and $-q$ is at $y = -a$?

24–19 A positive charge q is fixed at the point $x = 0$, $y = 0$, and a negative charge $-2q$ is fixed at the point $x = a$, $y = 0$. a) Show the positions of the charges in a diagram. b) Derive an expression for the potential V at points on the x-axis as a function of the coordinate x. Take V to be zero at an infinite distance from the charges. c) At what positions on the x-axis is $V = 0$? d) Graph V at points on the x-axis as a function of x in the range from $x = -2a$ to $x = +2a$. e) What does the answer to part (b) become when $x \gg a$? Explain why this result is obtained.

24–20 Consider the arrangement of point charges described in Exercise 24–19. a) Derive an expression for the potential V at points on the y-axis as a function of the coordinate y. Take V to be zero at an infinite distance from the charges. b) At what positions on the y-axis is $V = 0$? c) Graph V at points on the y-axis as a function of y in the range from $y = -2a$ to $y = +2a$. d) What does the answer to part (a) become when $y \gg a$? Explain why this result is obtained.

24–21 A simple type of vacuum tube known as a *diode* consists essentially of two electrodes within a highly evacuated enclosure. One electrode, the *cathode,* is maintained at a high temperature and emits electrons from its surface. A potential difference of a few hundred volts is maintained between the cathode and the other electrode, known as the *anode,* with the anode at the higher potential. Suppose a diode consists of a cylindrical cathode with radius 0.062 cm, mounted coaxially within a cylindrical anode 0.558 cm in radius. The potential of the anode is 295 V higher than that of the cathode. An electron leaves the surface of the cathode with zero initial speed. Find its speed when it strikes the anode.

24–22 At a certain distance from a point charge, the potential and electric field magnitude due to that charge are 4.98 V and 12.0 V/m, respectively. (The potential is taken to be zero at infinity.) a) What is the distance to the point charge? b) What is the magnitude of the charge? c) Is the electric field directed toward or away from the point charge?

24–23 A uniform electric field has magnitude E and is directed in the negative x-direction. The potential difference between point a (at $x = 0.60$ m) and point b (at $x = 0.90$ m) is 240 V. a) Which point, a or b, is at the higher potential? b) Calculate the value of E. c) A negative point charge $q = -0.200$ μC is moved from b to a. Calculate the work done on the point charge by the electric field.

SECTION 24–4 CALCULATING ELECTRIC POTENTIAL

24–24 A total electric charge of 3.50 nC is distributed uniformly over the surface of a metal sphere with a radius of 24.0 cm. If the potential is zero at a point at infinity, find the value of the potential at the following distances from the center of the sphere: a) 48.0 cm; b) 24.0 cm; 12.0 cm.

24–25 A uniformly charged thin ring has radius 15.0 cm and has total charge +24.0 nC. An electron is placed on the ring's axis a distance 30.0 cm from the center of the ring and is constrained to stay on the axis of the ring. The electron is then released from rest. a) Describe the subsequent motion of the electron. b) Find the speed of the electron when it reaches the center of the ring.

24–26 An infinitely long line of charge has linear charge density 5.00×10^{-12} C/m. A proton (mass 1.67×10^{-27} kg, charge $+1.60 \times 10^{-19}$ C) is 18.0 cm from the line and moving directly toward the line at 1.50×10^3 m/s. How close does the proton get to the line of charge?

24–27 Two large parallel metal plates carry opposite charges of equal magnitude. They are separated by 45.0 mm, and the potential difference between them is 360 V. a) What is the magnitude of the electric field (assumed to be uniform) in the region between the plates? b) What is the magnitude of the force this field exerts on a particle with charge +2.40 nC? c) Use the results of part (b) to compute the work done by the field on the particle as it moves from the higher-potential plate to the lower. d) Compare the result of part (c) to the change of potential energy of the same charge, computed from the electric potential.

24–28 Two large parallel metal sheets carrying opposite electric charges of equal magnitude are separated by a distance of 38.0 mm. The electric field between them is uniform and has magnitude 480 N/C. a) What is the potential difference between the sheets? b) Which sheet is at higher potential, the one with positive charge or the one with negative charge? c) What is the surface charge density σ on the positive sheet?

24–29 A potential difference of 4.75 kV is established between parallel plates in air. a) If the air becomes electrically conducting when the electric field exceeds 3.00×10^6 V/m, what is the minimum separation of the plates? b) When the separation has the minimum value calculated in part (a), what is the surface charge density on each plate?

24–30 Two large parallel conducting plates carrying opposite charges of equal magnitude are separated by 2.20 cm. a) If the surface charge density for each plate has magnitude 47.0 nC/m^2, what is the magnitude of $\vec{E}$ in the region between the plates? b) What is the potential difference between the two plates? c) If the separation between the plates is doubled while the surface charge density is kept constant at the value in part (a), what happens to the magnitude of the electric field and to the potential difference?

SECTION 24–5 EQUIPOTENTIAL SURFACES
SECTION 24–6 POTENTIAL GRADIENT

24–31 In a certain region of space, the electric potential is $V(x, y, z) = Axy - Bx^2 + Cy$, where A, B, and C are positive constants. a) Calculate the x-, y-, and z-components of the electric field. b) At what points is the electric field equal to zero?

24–32 The potential due to a point charge Q at the origin may be written as

$$V = \frac{Q}{4\pi\epsilon_0 r} = \frac{Q}{4\pi\epsilon_0 \sqrt{x^2 + y^2 + z^2}}.$$

a) Calculate E_x, E_y, and E_z using Eq. (24–19). b) Show that the result of part (a) agrees with Eq. (22–7) for the electric field of a point charge.

24–33 The potential is given by $V = 0$ for $z < 0$, $V = Cz$ for $0 < z < d$, and $V = Cd$ for $z > d$, where C and d are positive constants. The potential does not depend on x or y. a) Find the electric field (magnitude and direction) at all points. b) What charge distribution could be the source of this field? Explain.

24–34 A metal sphere with radius r_a is supported on an insulating stand at the center of a hollow metal spherical shell with radius r_b. There is charge $+q$ on the inner sphere and charge $-q$ on the outer spherical shell. a) Calculate the potential $V(r)$ for i) $r < r_a$; ii) $r_a < r < r_b$; iii) $r > r_b$. (*Hint:* The net potential is the sum of the potentials due to the individual spheres.) Take V to be

zero when r is infinite. b) Show that the potential of the inner sphere with respect to the outer is

$$V_{ab} = \frac{q}{4\pi\epsilon_0}\left(\frac{1}{r_a} - \frac{1}{r_b}\right).$$

c) Use Eq. (24–23) and the result from part (a) to show that the electric field at any point between the spheres has magnitude

$$E(r) = \frac{V_{ab}}{(1/r_a - 1/r_b)}\frac{1}{r^2}.$$

d) Use Eq. (24–23) and the result from part (a) to find the electric field at a point outside the larger sphere at a distance r from the center, where $r > r_b$. e) Suppose the charge on the outer sphere is not $-q$ but a negative charge of different magnitude, say $-Q$. Show that the answers for parts (b) and (c) are the same as before but the answer for part (d) is different.

24–35 A metal sphere with radius $r_a = 1.20$ cm is supported on an insulating stand at the center of a hollow metal spherical shell with radius $r_b = 9.60$ cm. Charge $+q$ is put on the inner sphere and charge $-q$ on the outer spherical shell The magnitude of q is chosen to make the potential difference between the spheres 500 V, with the inner sphere at higher potential. a) Use the result of Exercise 24–34(b) to calculate q. b) With the help of the result of Exercise 24–34(a), sketch the equipotential surfaces that correspond to 500, 400, 300, 200, 100, and 0 V. c) On your sketch, show the electric field lines. Are the electric field lines and equipotential surfaces mutually perpendicular? Are the equipotential surfaces closer together when the magnitude of $\vec{E}$ is largest?

24–36 A charge Q is uniformly distributed along a rod with length $2a$. In Example 24–12 (Section 24–4), we derived an expression for the potential V at a point on the perpendicular bisector of the rod at a distance x from its center. a) By differentiating this expression, find the x-component E_x of the electric field at such a point. Show that the result is the same expression we found in Example 22–11 (Section 22–7) by direct integration. (*Hint:* Writing the natural logarithm of the fraction as the difference of two logarithms simplifies the calculation.) b) By symmetry, at a point on the perpendicular bisector of the rod, $E_y = 0$ and $E_z = 0$. Explain why it would *not* be correct to draw these conclusions by using Eq. (24–19).

24–37 A potential difference of 480 V is established between large parallel metal plates. Let the potential of one plate be 480 V and the other be 0 V. The plates are separated by $d = 1.70$ cm. a) Sketch the equipotential surfaces that correspond to 0, 120, 240, 360, and 480 V. b) On your sketch, show the electric field lines. Does your sketch confirm that the field lines and equipotential surfaces are mutually perpendicular?

SECTION 24–7 THE CATHODE-RAY TUBE

24–38 a) An electron is to be accelerated from 3.00×10^6 m/s to 8.00×10^6 m/s. Through what potential difference must the electron pass to accomplish this? b) Through what potential difference must the electron pass if it is to be slowed from 8.00×10^6 m/s to a halt?

24–39 Deflection in a CRT. In Fig. 24–32 an electron is projected along the axis midway between the deflection plates of a cathode–ray tube with an initial speed of 6.50×10^6 m/s. The uniform electric field between the plates has a magnitude of 1.10×10^3 V/m and is upward. a) What is the force (magnitude and direction) on the electron when it is between the plates? b) What is the acceleration of the electron (magnitude and direction) when acted on by the force in part (a)? c) How far below the axis has the electron moved when it reaches the end of the plates? d) At what angle with the axis is it moving as it leaves the plates? e) How far below the axis will it strike the fluorescent screen S?

FIGURE 24–32 Exercise 24–39.

SECTION 24–8 CALCULATING POTENTIALS DUE TO CHARGED CONDUCTORS: A CASE STUDY IN COMPUTER ANALYSIS

24–40 Repeat the derivation of Eq. (24–35), but look at the case in which V depends on all three coordinates (x, y, z). Show that in a region where there is no charge, the potential at (x, y, z) is given by

$$V(x, y, z) = \frac{1}{6}[V(x + \Delta l, y, z) + V(x - \Delta l, y, z)$$

$$+ V(x, y + \Delta l, z) + V(x, y - \Delta l, z)$$

$$+ V(x, y, z + \Delta l) + V(x, y, z - \Delta l)].$$

Use this result to show that in a region where there is no charge, there can be no points at which V has a local minimum or maximum, and hence there can be no points at which a point charge will be in static equilibrium.

24–41 Write a computer code or spreadsheet to implement the procedure outlined in Steps 1–18 in the text (not including Steps 17A through 17I). a) Run your code or spreadsheet for the situation shown in Fig. 24–24 with $V_a = V_0 = 1.00$ V and $V_b = 0$. Compare your results to Fig. 24–26. b) If using a computer code, include Steps 17A through 17I in your code and draw the field map. Compare your drawing with Fig. 24–27. c) Modify Step 17H to draw equipotential line segments, using Eq. (24–39). Compare your drawing with Fig. 24–28.

24–42 The terminals of the battery or voltage source shown in Fig. 24–24 are reconnected so that the top of the box is at $V = 0$ and the bottom and sides are at $V = V_0 = 1.00$ V. a) Modify the program or spreadsheet to find the potential inside this box. Compare your results to Fig. 24–26 and discuss any differences. b) If using a computer code, draw the field map and equipotential line segments. Compare your drawings with Figs. 24–27 and 24–28 and discuss any differences.

24–43 The sides of the box shown in Fig. 24–24 are reconnected so that the top and left sides of the box are at $V = V_0 = 1.00$ V, and the bottom and right sides are at $V = 0$. a) Modify the program or spreadsheet to find the potential inside

this box. Compare your results to Fig. 24–26 and discuss any differences. b) If using a computer code, draw the field map and equipotential line segments. Compare your drawings with Figs. 24–27 and 24–28 and discuss any differences.

24–44 The sides of the box shown in Fig. 24–24 are reconnected so that the top and bottom sides of the box are at $V = V_0 = 1.00$ V, and the left and right sides are at $V = 0$. a) Modify the program or spreadsheet to find the potential inside this box. Compare your results to Fig. 24–26 and discuss any differences. b) If using a computer code, draw the field map and equipotential line segments. Compare your drawings with Figs. 24–27 and 24–28 and discuss any differences.

24–45 The box shown in Fig. 24–24 is modified so that the potential of the top is $V = V_0$, the potential of the bottom is $V = 0$, and the potential of the sides varies uniformly from V_0 to 0. This is shown in Fig. 24–33 for the case $V_0 = 1.00$ V and $m = 11$. Modify the program or spreadsheet to find the potential inside this box. Compare your results to the result for two oppositely charged parallel plates, as discussed in Example 24–9 (Section 24–4).

FIGURE 24–33 Exercise 24–45.

24–46 The box shown in Fig. 24–24 is modified so that the upper surface has a "finger" that projects down into the interior of the box as shown in Fig. 24–34. a) Modify the program or spreadsheet to find the potential inside this box. Compare your results to Fig. 24–26 and discuss any differences. b) If using a computer code, draw the field map and equipotential line segments. Compare your drawings with Figs. 24–27 and 24–28 and discuss any differences.

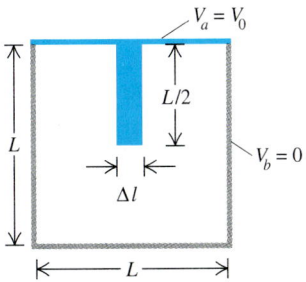

FIGURE 24–34 Exercise 24–46.

24–47 In your computer program from Exercise 24–41, make the modification described in the text to let the electric field lines be drawn with a length proportional to the field magnitude at each point. Run the program for the situation shown in Fig. 24–24 with $V_a = V_0 = 1.00$ V and $V_b = 0$.

PROBLEMS

24–49 A particle with charge +7.60 nC is in a uniform electric field directed to the left. Another force in addition to the electric force acts on the particle so that when it is released from rest it moves to the right. After it has moved 8.00 cm, the additional force has done 6.50×10^{-5} J of work and the particle has 4.35×10^{-5} J of kinetic energy. a) What work was done by the electric force? b) What is the potential of the starting point with respect to the endpoint? c) What is the magnitude of the electric field?

24–50 In the *Bohr model* of the hydrogen atom, a single electron revolves around a single proton in a circle of radius r. Assume that the proton remains at rest. a) By equating the electric force to the electron mass times its acceleration, derive an expression for the electron's speed. b) Obtain an expression for the electron's kinetic energy, and show that its magnitude is just half that of the electric potential energy. c) Obtain an expression for the total energy, and evaluate it using $r = 5.29 \times 10^{-11}$ m. Give your numerical result in joules and in electron volts.

24–51 A vacuum tube diode was described in Exercise 24–21. Because of the accumulation of charge near the cathode, the electric potential between the electrodes is not a linear function of the position, even with planar geometry, but is given by

$$V(x) = Cx^{4/3},$$

where x is the distance from the cathode (negative electrode) and C is a constant, characteristic of a particular diode and operating conditions. Assume that the distance between the cathode and anode (positive electrode) is 13.0 mm and the potential difference between electrodes is 240 V. a) Determine the value of C. b) Obtain a formula for the electric field between the electrodes as a function of x. c) Determine the force on an electron when the electron is halfway between the electrodes.

24–52 Refer to Problem 23–39. For the situation shown in Fig. 23–36, calculate the energy required to move the two electrons of charge $-e$ infinitely far from each other and infinitely far from the uniformly charged sphere. Your answer will be in terms of the distances d and R.

24–53 An Ionic Crystal. Figure 24–35 shows eight point charges arranged at the corners of a cube with sides of length d. The values of the charges are $+q$ and $-q$, as shown. This is a model of one cell of a cubic ionic crystal. In sodium chloride (NaCl), for instance, the positive ions are Na^+ and the negative ions are Cl^-. a) Calculate the potential energy U of this arrangement. (Take as zero the potential energy of the eight charges when they are infinitely far apart.) b) In part (a), you should have found that $U < 0$. Explain the relationship between this result and the observation that such ionic crystals exist in nature.

24–54 A One-Dimensional "Crystal." While real crystals are three-dimensional, much can be learned from simple one-

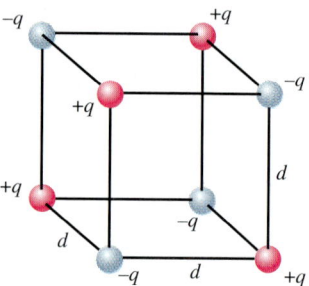

FIGURE 24–35 Problem 24–53.

dimensional models for which calculations can be carried out much more easily. As a one-dimensional model of an ionic crystal such as sodium chloride (NaCl), consider alternating positive and negative ions of charge $+e$ and $-e$, respectively, which are uniformly spaced along the x-axis with spacing d (Fig. 24–36). The charges may be considered to extend to infinity in both directions. a) Consider the potential energy of the interaction between the positive ion at $x = 0$ and all the other ions. This represents the *potential energy per ion* in this one-dimensional "crystal." Write an expression for this potential energy (your expression will be an infinite series). b) Evaluate the infinite series in part (a), using the expansion $\ln(1 + z) = z - z^2/2 + z^3/3 - z^4/4 + \cdots$, valid for the case $|z| \leq 1$. c) Does the potential energy per ion have the same value for negative ions in the "crystal" as for positive ions? Explain your reasoning. d) In the real, three-dimensional NaCl crystal, the spacing between adjacent ions is 2.82×10^{-10} m. Using this spacing for the value of d in Fig. 24–36, calculate the potential energy per ion in the one-dimensional crystal. e) For most real (three-dimensional) ionic crystal substances such as NaCl, the potential energy per ion is about -8×10^{-19} J/ion. How does this compare to your result in part (d)? How good a model is the one-dimensional "crystal" shown in Fig. 24–36?

FIGURE 24–36 Problem 24–54.

24–55 The H_2^+ Ion. The H_2^+ ion is comprised of two protons, each of charge $+e = 1.60 \times 10^{-19}$ C, and an electron of charge $-e$ and mass 9.11×10^{-31} kg. The separation between the protons is 1.07×10^{-10} m. The protons and the electron may be treated as point charges. a) Suppose the electron is located at the point midway between the two protons. What is the potential energy of the interaction between the electron and the two protons? (Do not include the potential energy due to the interaction between the two protons.) b) Suppose the electron in part (a) has a velocity of magnitude 1.50×10^6 m/s in a direction along the perpendicular bisector of the line connecting the two protons. How far from the point midway between the two protons can the electron move? Because the masses of the protons are much greater than the electron mass, the motions of the protons are very slow and can be ignored. (*Note:* A realistic description of the electron motion requires the use of quantum mechanics, not Newtonian mechanics.)

24–56 A small sphere with a mass of 1.50 g hangs by a thread between two parallel vertical plates 5.00 cm apart. The charge on the sphere is $q = 8.90 \times 10^{-6}$ C. What potential difference between the plates will cause the thread to assume an angle of $30.0°$ with the vertical (Fig. 24–37)?

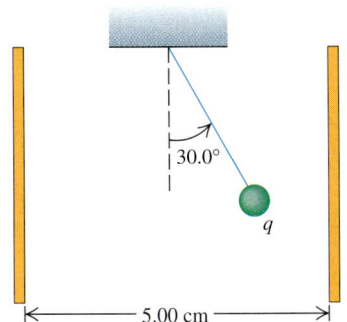

FIGURE 24–37 Problem 24–56.

24–57 Coaxial Cylinders. A long metal cylinder with radius a is supported on an insulating stand on the axis of a long hollow metal tube with radius b. The positive charge per unit length on the inner cylinder is λ, and there is an equal negative charge per unit length on the outer cylinder. a) Calculate the potential $V(r)$ for i) $r < a$; ii) $a < r < b$; iii) $r > b$. (*Hint:* The net potential is the sum of the potentials due to the individual conductors.) Take $V = 0$ at $r = b$. b) Show that the potential of the inner cylinder with respect to the outer is

$$V_{ab} = \frac{\lambda}{2\pi\epsilon_0} \ln \frac{b}{a}.$$

c) Use Eq. (24–23) and the result from part (a) to show that the electric field at any point between the cylinders has magnitude

$$E(r) = \frac{V_{ab}}{\ln(b/a)} \cdot \frac{1}{r}.$$

d) What is the potential difference between the two cylinders if the outer cylinder has no net charge?

24–58 A *Geiger counter* detects radiation such as alpha particles by using the fact that the radiation ionizes the air along its path. A thin wire lies on the axis of a hollow metal cylinder and is insulated from it (Fig. 24–38). A large potential difference is established between the wire and the outer cylinder, with the wire at higher potential; this sets up a strong electric field directed radially outward. When ionizing radiation enters the device, it ionizes a few air molecules. The free electrons produced are accelerated by the electric field toward the wire and, on the way there, ionize many more air molecules. Thus a current pulse is produced that can be detected by appropriate

electronic circuitry and converted to an audible "click." Suppose the radius of the central wire is 145 μm and the radius of the hollow cylinder is 1.80 cm. What potential difference between the wire and the cylinder produces an electric field of 2.00×10^4 V/m at a distance of 1.20 cm from the wire? (The wire and cylinder are both very long in comparison to their radii, so the results of Problem 24–57 apply.)

FIGURE 24–38 Problem 24–58.

24–59 *Electrostatic precipitators* use electric forces to remove pollutant particles from smoke, in particular in the smokestacks of coal-burning power plants. One form of precipitator consists of a vertical hollow metal cylinder with a thin wire, insulated from the cylinder, running along its axis (Fig. 24–39). A large potential difference is established between the wire and the outer cylinder, with the wire at lower potential. This sets up a strong radial electric field directed inward. The field produces a region of ionized air near the wire. Smoke enters the precipitator at the bottom, ash and dust in it pick up electrons, and the charged pollutants are accelerated toward the outer cylinder wall by the electric field. Suppose the radius of the central wire is 90.0 μm, the radius of the cylinder is 14.0 cm, and a potential difference of 50.0 kV is established between the wire and the cylinder. Also assume that the wire and cylinder are both very long in comparison to the cylinder radius, so the results of Problem 24–57 apply. a) What is the magnitude of the electric field midway between the wire and the cylinder wall? b) What magnitude of charge must a 30.0-μg ash particle have if the electric field computed in part (a) is to exert a force ten times the weight of the particle?

FIGURE 24–39 Problem 24–59.

24–60 Four line segments of charge are arranged to form a square with sides of length a. The potential is zero at infinity. Calculate the potential at the center of the square if a) two opposite sides are positively charged with charge $+Q$ each and the other two sides are negatively charged with charge $-Q$ each; b) if each side has positive charge $+Q$. (*Hint:* Use the result of Example 24–12 in Section 24–4.)

24–61 A disk of radius R has uniform surface charge density σ. a) By regarding the disk as a series of thin concentric rings, calculate the electric potential V at a point on the disk's axis a distance x from the center of the disk. Assume that the potential is zero at infinity. (*Hint:* Use the result of Example 24–11 in Section 24–4.) b) Calculate $-\partial V/\partial x$. Show that the result agrees with the expression for E_x calculated in Example 22–12 (Section 22–7).

24–62 A line of charge of length $2a$ has total charge Q. The potential at a point along the perpendicular bisector of the rod, at a distance x from its center, was calculated in Example 24–12 (Section 24–4). a) Show that if $x >> 0$, $V(x)$ approaches $Q/4\pi\epsilon_0 x$. Interpret this result. (*Hint:* For $|z| << 1$, $\ln(1 + z) \cong z$.) b) Consider points very close to the rod, such that $x << a$. Simplify the expression for $V(x)$ for this special case. Compare to the result $V = (\lambda/2\pi\epsilon_0) \ln(R/r)$ obtained in Example 24–10 (Section 24–4) for an infinitely long charged cylinder. What corresponds to R in this limit? Can you explain this?

24–63 a) From the expression for E obtained in Problem 23–36, find the expressions for the electric potential V as a function of r, both inside and outside the cylinder. Let $V = 0$ at the surface of the cylinder. In each case, express your result in terms of the charge per unit length λ of the charge distribution. b) Graph V and E as functions of r from $r = 0$ to $r = 3R$.

24–64 Refer to Problem 22–52. a) Calculate the potential at the point $x = 3.00$ cm, $y = 0$ and at the point $x = 3.00$ cm, $y = 5.00$ cm due to the first two charges. Let the potential be zero far from the charges. b) If the third charge moves from the point $x = 3.00$ cm, $y = 0$ to the point $x = 3.00$ cm, $y = 5.00$ cm, calculate the work done on it by the field of the first two charges. Comment on the *sign* of this work. Is your result reasonable?

24–65 For the ring of charge described in Example 24–11 (Section 24–4), integrate the expression for E_x found in Example 22–10 (Section 22–7) to find the potential at point P on the ring's axis. Assume that $V = 0$ at infinity. Compare your result to that obtained in Example 24–11 using Eq. (24–16).

24–66 A thin insulating rod is bent into a semicircular arc of radius a, and a total electric charge Q is distributed uniformly along the rod. Calculate the potential at the center of curvature of the arc if the potential is assumed to be zero at infinity.

24–67 For the situation described in Problem 23–20, calculate the magnitude of the potential difference between the following pairs of cube faces (Fig. 23–28): a) S_1 and S_3; b) S_2 and S_4; c) S_5 and S_6. In each case where the potential difference is nonzero, specify which face is at higher potential.

24–68 a) From the expression for $E(r)$ obtained in Example 23–9 (Section 23–5), find the expression for the electric potential $V(r)$ as a function of r both inside and outside the uniformly

charged sphere. Assume that $V = 0$ at infinity. b) Graph V and E as functions of r from $r = 0$ to $r = 3R$.

24–69 Use the electric field calculated in Problem 23–25 to calculate the potential difference between the solid conducting sphere and the thin insulating shell.

24–70 Consider a solid conducting sphere inside a hollow conducting sphere, with radii and charges specified in Problem 23–24. Take $V = 0$ as $r \to \infty$. Use the electric field calculated in Problem 23–24 to calculate the potential V at the following values of r: a) $r = c$ (at the outer surface of the hollow sphere); b) $r = b$ (at the inner surface of the hollow sphere); c) $r = a$ (at the surface of the solid sphere); d) $r = 0$ (at the center of the solid sphere).

24–71 Use the electric field calculated in Problem 23–42 to calculate the potential difference between the two faces of the uniformly charged slab.

24–72 a) If a spherical raindrop of radius 0.650 mm carries a charge of -1.20 pC uniformly distributed over its volume, what is the potential at its surface? (Take the potential to be zero at an infinite distance from the raindrop.) b) Two identical raindrops, each with radius and charge specified in part (a), collide and merge into one larger raindrop. What is the radius of this larger drop, and what is the potential at its surface if its charge is uniformly distributed over its volume?

24–73 Electric charge is distributed uniformly along a thin rod of length a, with total charge Q. Take the potential to be zero at infinity. Find the potential at the following points (see Fig. 24–40): a) point P, a distance x to the right of the rod; b) point R, a distance y above the right-hand end of the rod. c) In part (a) and part (b), what does your result reduce to as x or y becomes much larger than a?

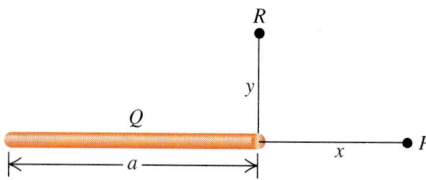

FIGURE 24–40 Problem 24–73.

24–74 An alpha particle with kinetic energy 11.0 MeV makes a head-on collision with a lead nucleus at rest. What is the distance of closest approach of the two particles? (Assume that the lead nucleus remains stationary and that it may be treated as a point charge. The atomic number of lead is 82. The alpha particle is a helium nucleus, with atomic number 2.)

24–75 Two metal spheres of different sizes are charged such that the electric potential is the same at the surface of each. Sphere A has a radius three times that of sphere B. Let Q_A and Q_B be the charges on each sphere and E_A and E_B be the electric field magnitudes at the surface of each sphere. What is a) the ratio Q_B/Q_A; b) the ratio E_B/E_A?

24–76 Use the charge distribution and electric field calculated in Problem 23–45. a) Show that for $r \geq R$ the potential is identical to that produced by a point charge Q. (Take the potential to be zero at infinity.) b) Obtain an expression for the electric potential valid in the region $r \leq R$.

24–77 A metal sphere of radius R_1 has a charge Q_1. Take the electric potential to be zero at an infinite distance from the sphere. a) What are the electric field and electric potential at the surface of the sphere? This sphere is now connected by a long, thin conducting wire to another sphere of radius R_2 that is several meters from the first sphere. Before the connection is made, this second sphere is uncharged. After electrostatic equilibrium has been reached, what is b) the total charge on each sphere? c) the electric potential at the surface of each sphere? d) the electric field at the surface of each sphere? Assume that the amount of charge on the wire is much less than the charge on each sphere.

24–78 The electric potential V in a region of space is given by

$$V(x, y, z) = A(x^2 - 3y^2 + z^2),$$

where A is a constant. a) Derive an expression for the electric field $\vec{E}$ at any point in this region. b) The work done by the field when a 1.50-μC test charge moves from the point $(x, y, z) = (0, 0, 0.250 \text{ m})$ to the origin is measured to be 6.00×10^{-5} J. Determine A. c) Determine the electric field at the point $(0, 0, 0.250 \text{ m})$. d) Show that in every plane parallel to the xz-plane the equipotential contours are circles. e) What is the radius of the equipotential contour corresponding to $V = 1280$ V and $y = 2.00$ m?

24–79 Nuclear Fusion in the Sun. The source of the sun's energy is a sequence of nuclear reactions that occur in its core. The first of these reactions involves the collision of two protons, which fuse together to form a heavier nucleus and release energy. For this process, called *nuclear fusion*, to occur, the two protons must first approach until their surfaces are essentially in contact. a) Assume both protons are moving with the same speed and that they collide head-on. If the radius of the proton is 1.2×10^{-15} m, what is the minimum speed that will allow fusion to occur? The charge distribution within a proton is spherically symmetric, so the electric field and potential outside a proton is the same as if it were a point charge. The mass of the proton is given in Appendix F. b) Another nuclear fusion reaction that occurs in the sun's core involves a collision between two helium nuclei, each of which has 2.99 times the mass of the proton, charge $+2e$, and radius 1.7×10^{-15} m. Assuming the same collision geometry as in part (a), what minimum speed is required for this fusion reaction to take place if the nuclei must approach to a center-to-center distance of about 3.5×10^{-15} m? As for the proton, the charge of the helium nucleus is uniformly distributed throughout its volume. c) In Section 16–4 it was shown that the average translational kinetic energy of a particle of mass m in a gas at absolute temperature T is $\frac{3}{2}kT$, where k is the Boltzmann constant (given in Appendix F). For two protons with kinetic energy equal to this average value to be able to undergo the process described in part (a), what absolute temperature is required? What absolute temperature is required for two average helium nuclei to be able to undergo the process described in part (b)? (At these temperatures, atoms are completely ionized, so nuclei and electrons move separately.) d) The temperature in the sun's core is about 1.5×10^7 K. How does this compare to the temperatures calculated in part (c)? How can the reactions described in parts (a) and (b) occur at all in the interior of the sun? (*Hint:* See the discussion of the distribution of molecular speeds in Section 16–6.)

24–80 Nuclear Fission. The unstable nucleus uranium-236 can be regarded as a uniformly charged sphere of charge $Q = +92e$ and radius $R = 7.4 \times 10^{-15}$ m. In nuclear fission, this can divide into two smaller nuclei, each of one-half the charge and one-half the volume of the original uranium-236 nucleus. This is one of the reactions that occurred in the nuclear weapon exploded over Hiroshima, Japan in August 1945. a) Find the radii of the two "daughter" nuclei of charge $+46e$. b) In a simple model for the fission process, immediately after the uranium-236 nucleus has undergone fission the "daughter" nuclei are at rest and just touching, as shown in Fig. 24–41. Calculate the kinetic energy that each of the "daughter" nuclei will have when they are very far apart. c) In this model the sum of the kinetic energies of the two "daughter" nuclei, calculated in part (b), is the energy released by the fission of one uranium-236 nucleus. Calculate the energy released by the fission of 10.0 kg of uranium-236. The atomic mass of uranium-236 is 236 u, where 1 u = 1 atomic mass unit = 1.66×10^{-27} kg. Express your answer both in joules and in kilotons of TNT (1 kiloton of TNT releases 4.18×10^{12} J when it explodes). d) In terms of this model, discuss why an atomic bomb could just as well be called an "electric bomb."

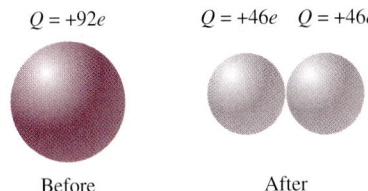

$Q = +92e$ $Q = +46e$ $Q = +46e$

Before   After

FIGURE 24–41 Problem 24–80.

CHALLENGE PROBLEMS

24–81 In experiments in which atomic nuclei collide, head-on collisions like that described in Problem 24–74 do happen, but "near misses" are more common. Suppose the alpha particle in Problem 24–74 was not "aimed" at the center of the lead nucleus, but had an initial nonzero angular momentum (with respect to the stationary lead nucleus) of magnitude $L = p_0 b$, where p_0 is the magnitude of the initial momentum of the alpha particle and $b = 1.00 \times 10^{-12}$ m. What is the distance of closest approach? Repeat for $b = 1.00 \times 10^{-13}$ m and $b = 1.00 \times 10^{-14}$ m.

24–82 In a certain region a charge distribution exists that is spherically symmetric but non-uniform. That is, the volume charge density $\rho(r)$ depends on the distance r from the center of the distribution but not on the spherical polar angles θ and ϕ. The electric potential $V(r)$ due to this charge distribution is

$$V(r) = \begin{cases} \dfrac{\rho_0 a^2}{18\epsilon_0}\left[1 - 3\left(\dfrac{r}{a}\right)^2 + 2\left(\dfrac{r}{a}\right)^3\right] & \text{for } r \leq a, \\ 0 & \text{for } r \geq a, \end{cases}$$

where ρ_0 is a constant having units of C/m^3 and a is a constant having units of meters. a) Derive expressions for $\vec{E}$ for the regions $r \le a$ and $r \ge a$. (*Hint:* Use Eq. (24–23).) Explain why $\vec{E}$ has only a radial component. b) Derive an expression for $\rho(r)$ in each of the two regions $r \le a$ and $r \ge a$. (*Hint:* Use Gauss's law for two spherical shells, one of radius r and the other of radius $r + dr$. The charge contained in the infinitesimal spherical shell of radius dr is $dq = 4\pi r^2 \rho(r)\, dr$.) c) Show that the net charge contained in the volume of a sphere of radius greater than or equal to a is zero. (*Hint:* Integrate the expressions derived in part (b) for $\rho(r)$ over a spherical volume of radius greater than or equal to a.) Is this result consistent with the electric field for $r > a$ that you calculated in part (a)?

24–83 The Millikan Oil-Drop Experiment. The charge of an electron was first measured by the American physicist Robert Millikan during 1909–1913. In his experiment, oil is sprayed in very fine drops (around 10^{-4} mm in diameter) into the space between two parallel horizontal plates separated by a distance d. A potential difference V_{AB} is maintained between the parallel plates, causing a downward electric field between them. Some of the oil drops acquire a negative charge because of frictional effects or because of ionization of the surrounding air by x rays or radioactivity. The drops are observed through a microscope. a) Show that an oil drop of radius r at rest between the plates will remain at rest if the magnitude of its charge is

$$q = \frac{4\pi}{3} \frac{\rho r^3 gd}{V_{AB}},$$

where ρ is the density of the oil. (Neglect the buoyant force of the air.) By adjusting V_{AB} to keep a given drop at rest, the charge on that drop can be determined, provided its radius is known. b) Millikan's oil drops were much too small to measure their radii directly. Instead, Millikan determined r by cutting off the electric field and measuring the *terminal speed* v_t of the drop as it fell. (We discussed the concept of terminal speed in Section 5–4.) The viscous force F on a sphere of radius r moving with speed v through a fluid with viscosity η is given by Stokes's law, Eq. (14–27): $F = 6\pi\eta r v$. When the drop is falling at v_t, the viscous force just balances the weight $w = mg$ of the drop. Show that the magnitude of the charge on the drop is

$$q = 18\pi \frac{d}{V_{AB}} \sqrt{\frac{\eta^3 v_t^3}{2\rho g}}.$$

Within the limits of their experimental error, every one of the thousands of drops that Millikan and his co-workers measured had a charge equal to some small integer multiple of a basic charge e. That is, they found drops with charges of $\pm 2e$, $\pm 5e$, and so on, but none with values such as $0.76e$ or $2.49e$. A drop with charge $-e$ has acquired one extra electron; if its charge is $-2e$, it has acquired two extra electrons, and so on. c) A charged oil drop in a Millikan oil-drop apparatus is observed to fall 1.00 mm at constant speed in 39.3 s if $V_{AB} = 0$. The same drop can be held at rest between two plates separated by 1.00 mm if $V_{AB} = 9.16$ V. How many excess electrons has the drop acquired, and what is the radius of the drop? The viscosity of air is 1.81×10^{-5} N $\cdot$ s/m^2, and the density of the oil is 824 kg/m^3.

24–84 A hollow, thin-walled insulating cylinder of radius R and length L (like the cardboard tube in a roll of toilet paper) has charge Q uniformly distributed over its surface. a) Calculate the electric potential at all points along the axis of the tube. Take the origin to be at the center of the tube, and take the potential to be zero at infinity. b) Show that if $L \ll R$, the result of part (a) reduces to the potential on the axis of a ring of charge of radius R (Example 24–11 in Section 24–4). c) Use the result of part (a) to find the electric field at all points along the axis of the tube.

24–85 Two point charges are moving to the right along the x-axis. Point charge 1 has charge $q_1 = 2.00$ μC, mass $m_1 = 6.00 \times 10^{-5}$ kg, and speed v_1. Point charge 2 is to the right of q_1 and has charge $q_2 = -5.00$ μC, mass $m_2 = 3.00 \times 10^{-5}$ kg, and speed v_2. At a particular instant the charges are separated by a distance of 9.00 mm and have speeds $v_1 = 400$ m/s and $v_2 = 1300$ m/s. The only forces on the particles are the forces they exert on each other. a) Determine the speed v_{cm} of the center of mass of the system. b) The *relative energy* E_{rel} of the system is defined as the total energy minus the kinetic energy contributed by the motion of the center of mass:

$$E_{rel} = E - \frac{1}{2}(m_1 + m_2)v_{cm}^2,$$

where $E = \frac{1}{2}m_1 v_1^2 + \frac{1}{2}m_2 v_2^2 + q_1 q_2/4\pi\epsilon_0 r$ is the total energy of the system and r is the distance between the charges. Show that $E_{rel} = \frac{1}{2}\mu v^2 + q_1 q_2/4\pi\epsilon_0 r$, where $\mu = m_1 m_2/(m_1 + m_2)$ is called the *reduced mass* of the system and $v = v_2 - v_1$ is the relative speed of the moving particles. c) For the numerical values given above, calculate the numerical value of E_{rel}. d) Based on the result of part (c), for the conditions given above, will the particles escape from one another? Explain. e) If the particles do escape, what will be their final relative speed when $r \to \infty$? If the particles do not escape, what will be their distance of maximum separation? That is, what will be the value of r when $v = 0$? f) Repeat parts (c) through (e) for $v_1 = 400$ m/s and $v_2 = 1800$ m/s when the separation is 9.00 mm.

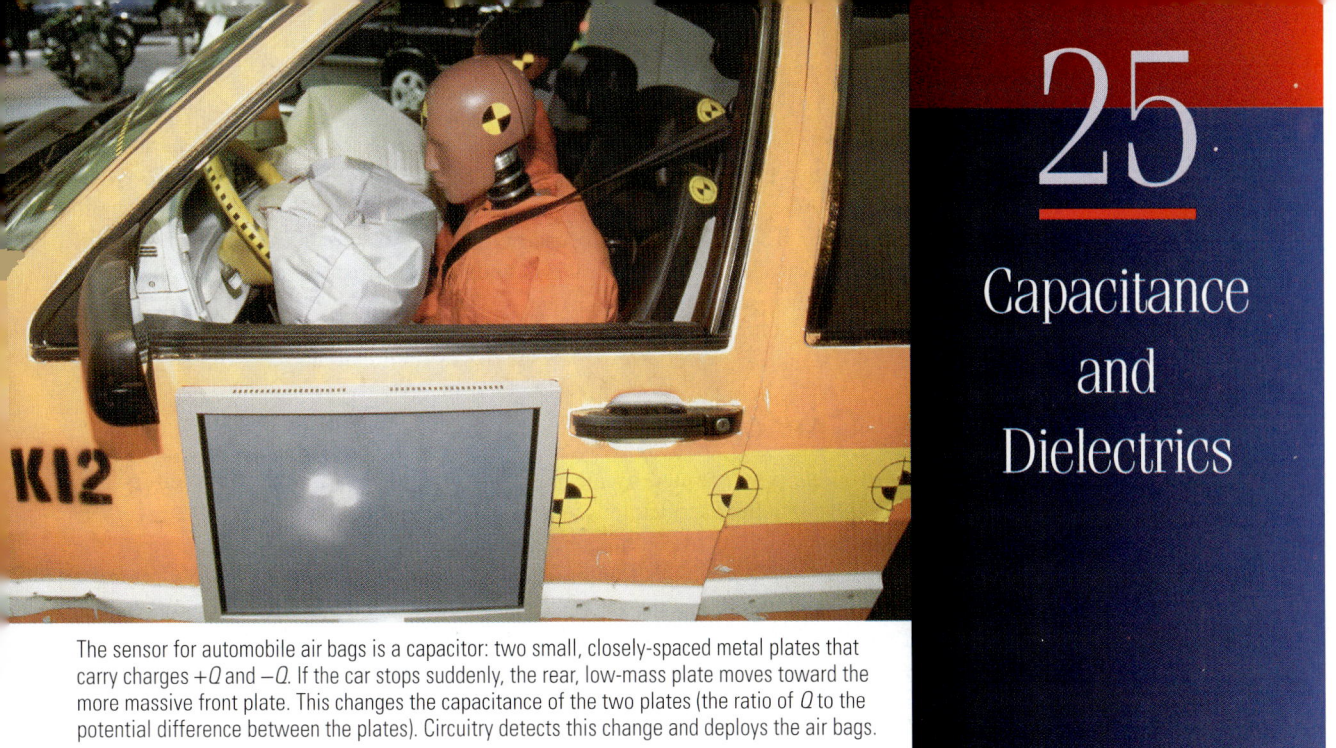

The sensor for automobile air bags is a capacitor: two small, closely-spaced metal plates that carry charges $+Q$ and $-Q$. If the car stops suddenly, the rear, low-mass plate moves toward the more massive front plate. This changes the capacitance of the two plates (the ratio of Q to the potential difference between the plates). Circuitry detects this change and deploys the air bags.

25

Capacitance and Dielectrics

25–1 INTRODUCTION

When you set an old-fashioned spring mousetrap or pull back the string of an archer's bow, you are storing mechanical energy as elastic potential energy. A capacitor is a device that stores *electric* potential energy and electric charge. To make a capacitor, just insulate two conductors from each other. To store energy in this device, transfer charge from one conductor to the other so that one has a negative charge and the other has an equal amount of positive charge. Work must be done to move the charges through the resulting potential difference between the conductors, and the work done is stored as electric potential energy.

Capacitors have a tremendous number of applications. In electronic flash units used by photographers or in energy-storage units for pulsed lasers, the energy and charge stored in a capacitor are recovered quickly. In other applications the energy is released more slowly. Springs in the suspension of an automobile help smooth out the ride by absorbing the energy from sudden jolts and releasing that energy gradually; in an analogous way, a capacitor in an electronic circuit can protect sensitive components by smoothing out variations in voltage due to power surges. And just as the presence of a spring in a mechanical system gives that system a natural frequency at which it responds most strongly to an applied periodic force, the presence of a capacitor in an electric circuit gives that circuit a natural frequency for current oscillations. This idea is used in tuned circuits such as those in radio and television receivers, which respond to broadcast signals at one particular frequency and ignore signals at other frequencies.

Capacitors also have practical effects that are not desirable. Adjacent pins on the underside of a computer chip act like a capacitor, and the property that makes capacitors useful for smoothing out voltage variations acts to retard the rate at which the potentials of the chip's pins can be changed. This limits the maximum speed at which the chip can perform computations, an unpleasant effect that becomes more important as computer chips become smaller and are pushed to operate at faster speeds.

In this chapter we'll study the fundamental properties of capacitors. We'll encounter many of their most important applications in later chapters (particularly Chapter 32, in which we'll see the crucial role played by capacitors in the *alternating-current* circuits that pervade our technological society). For a particular capacitor the ratio of the charge on each conductor to the potential difference between the conductors is a constant, called the *capacitance*. The capacitance depends on the sizes and shapes of the conductors and on the insulating material (if any) between them. Compared to the case in which there is only vacuum between the conductors, the capacitance increases when an insulating material (a *dielectric*) is present. This happens because a redistribution of charge, called *polarization*, takes place within the insulating material. Studying polarization will give us added insight into the electrical properties of matter.

Capacitors also give us a new way to think about electric potential energy. The energy stored in a charged capacitor is related to the electric field in the space between the conductors. We will see that electric potential energy can be regarded as being stored *in the field itself.* The idea that the electric field is itself a storehouse of energy is at the heart of the theory of electromagnetic waves and our modern understanding of the nature of light, to be discussed in Chapter 33.

25-2 CAPACITORS AND CAPACITANCE

Any two conductors separated by an insulator (or a vacuum) form a **capacitor** (Fig. 25–1). In most practical applications, each conductor initially has zero net charge, and electrons are transferred from one conductor to the other; this is called *charging* the capacitor. Then the two conductors have charges with equal magnitude and opposite sign, and the *net* charge on the capacitor as a whole remains zero. We will assume throughout this chapter that this is the case. When we say that a capacitor has charge Q, or that a charge Q is *stored* on the capacitor, we mean that the conductor at higher potential has charge $+Q$ and the conductor at lower potential has charge $-Q$ (assuming that Q is positive). Keep this in mind in the following discussion and examples.

In circuit diagrams a capacitor is represented by either of these symbols:

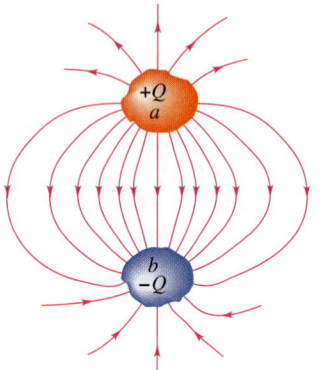

25-1 Any two conductors insulated from one another form a capacitor.

In either symbol the vertical lines (straight or curved) represent the conductors, and the horizontal lines represent wires connected to either conductor. One common way to charge a capacitor is to connect these two wires to opposite terminals of a battery. Once the charges Q and $-Q$ are established on the conductors, the battery is disconnected. This gives a fixed *potential difference* V_{ab} between the conductors (that is, the potential of the positively charged conductor a with respect to the negatively charged conductor b) that is just equal to the voltage of the battery.

The electric field at any point in the region between the conductors is proportional to the magnitude Q of charge on each conductor. It follows that the potential difference V_{ab} between the conductors is also proportional to Q. If we double the magnitude of charge on each conductor, the charge density at each point doubles, the electric field at each point doubles, and the potential difference between conductors doubles; however, the *ratio* of charge to potential difference does not change. This ratio is called the **capacitance** C of the capacitor:

$$C = \frac{Q}{V_{ab}} \qquad \text{(definition of capacitance).} \qquad (25\text{-}1)$$

The SI unit of capacitance is called one **farad** (1 F), in honor of the nineteenth century English physicist Michael Faraday. From Eq. (25–1), one farad is equal to one *coulomb per volt* (1 C/V):

$$1 \text{ F} = 1 \text{ farad} = 1 \text{ C/V} = 1 \text{ coulomb/volt.}$$

CAUTION ▶ Don't confuse the symbol C for capacitance (which is always in italics) with the abbreviation C for coulombs (which is never italicized). ◀

The greater the capacitance C of a capacitor, the greater the magnitude Q of charge on either conductor for a given potential difference V_{ab} and hence the greater the amount of stored energy. (Remember that potential is potential energy per unit charge.) Thus *capacitance is a measure of the ability of a capacitor to store energy.* We will see that the value of the capacitance depends only on the shapes and sizes of the conductors and on the nature of the insulating material between them. (The above remarks about capacitance being independent of Q and V_{ab} do not apply to certain special types of insulating materials. We won't discuss these materials in this book, however.)

CALCULATING CAPACITANCE – CAPACITORS IN VACUUM

We can calculate the capacitance C of a given capacitor by finding the potential difference V_{ab} between the conductors for a given magnitude of charge Q and then using Eq. (25–1). For now we'll consider only *capacitors in vacuum;* that is, we'll assume that the conductors that make up the capacitor are separated by empty space.

The simplest form of capacitor consists of two parallel conducting plates, each with area A, separated by a distance d that is small in comparison with their dimensions (Fig. 25–2a). When the plates are charged, the electric field is almost completely localized in the region between the plates (Fig. 25–2b). As we discussed in Example 23–8 (Section 23–5), the field between such plates is essentially *uniform,* and the charges on the plates are uniformly distributed over their opposing surfaces. We call this arrangement a **parallel-plate capacitor.**

We worked out the electric-field magnitude E for this arrangement in Example 22–13 (Section 22–7) using the principle of superposition of electric fields and again in Example 23–8 (Section 23–5) using Gauss's law. It would be a good idea to review those examples. We found that $E = \sigma/\epsilon_0$, where σ is the magnitude (absolute value) of the surface charge density on each plate. This is equal to the magnitude of the total charge Q on each plate divided by the area A of the plate, or $\sigma = Q/A$, so the field magnitude E can be expressed as

$$E = \frac{\sigma}{\epsilon_0} = \frac{Q}{\epsilon_0 A}.$$

The field is uniform, and the distance between the plates is d, so the potential difference (voltage) between the two plates is

$$V_{ab} = Ed = \frac{1}{\epsilon_0} \frac{Qd}{A}.$$

From this we see that the capacitance C of a parallel-plate capacitor in vacuum is

$$C = \frac{Q}{V_{ab}} = \epsilon_0 \frac{A}{d} \qquad \text{(capacitance of a parallel-plate capacitor in vacuum).} \qquad (25-2)$$

The capacitance depends only on the geometry of the capacitor; it is directly proportional to the area A of each plate and inversely proportional to their separation d. The quantities A and d are constants for a given capacitor, and ϵ_0 is a universal constant. Thus in a vacuum the capacitance C is a constant independent of the charge on the capacitor or the potential difference between the plates.

When matter is present between the plates, its properties affect the capacitance. We will return to this topic in Section 25–5. Meanwhile, we remark that if the space contains air at atmospheric pressure instead of vacuum, the capacitance differs from the prediction of Eq. (25–2) by less than 0.06%.

In Eq. (25–2), if A is in square meters and d in meters, C is in farads. The units of ϵ_0 are $C^2/N \cdot m^2$, so we see that

$$1 \text{ F} = 1 \text{ C}^2/\text{N} \cdot \text{m} = 1 \text{ C}^2/\text{J}.$$

Because $1 \text{ V} = 1 \text{ J/C}$ (energy per unit charge), this is consistent with our definition $1 \text{ F} = 1 \text{ C/V}$. Finally, the units of ϵ_0 can be expressed as $1 \text{ C}^2/\text{N} \cdot \text{m}^2 = 1 \text{ F/m}$, so

$$\epsilon_0 = 8.85 \times 10^{-12} \text{ F/m}.$$

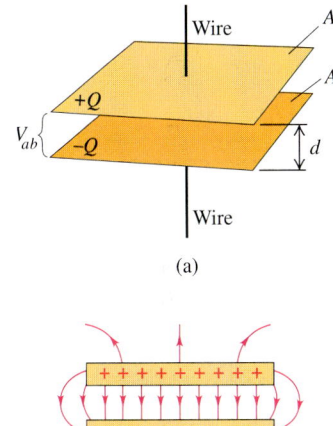

25–2 (a) A charged parallel-plate capacitor. (b) When the separation of the plates is small in comparison to their size, the fringing of the electric field at the edges is slight.

11.11.6
Electric Potential: Qualitative Introduction

11.12.1 and 11.12.3
Electric Potential, Field, and Force

This relation is useful in capacitance calculations, and it also helps us to verify that Eq. (25–2) is dimensionally consistent.

One farad is a very large capacitance, as is shown in Example 25–1. In many applications the most convenient units of capacitance are the *microfarad* ($1~\mu\text{F} = 10^{-6}~\text{F}$) and the *picofarad* ($1~\text{pF} = 10^{-12}~\text{F}$). For example, the power supply of an alternating-current AM radio contains several capacitors with capacitances of the order of 10 or more microfarads, and the capacitances in the tuning circuits are of the order of 10 to 100 picofarads.

For *any* capacitor in vacuum the capacitance C depends only on the shapes, dimensions, and separation of the conductors that make up the capacitor. If the conductor shapes are more complex than those of the parallel-plate capacitor, the expression for capacitance is more complicated than in Eq. (25–2). In the following examples we show how to calculate C for two other conductor geometries.

EXAMPLE 25–1

Size of a 1-F capacitor A parallel-plate capacitor has a capacitance of 1.0 F. If the plates are 1.0 mm apart, what is the area of the plates?

SOLUTION Solving Eq. (25–2) for the area A, we get

$$A = \frac{Cd}{\epsilon_0} = \frac{(1.0~\text{F})(1.0 \times 10^{-3}~\text{m})}{8.85 \times 10^{-12}~\text{F/m}}$$

$$= 1.1 \times 10^{8}~\text{m}^2.$$

This corresponds to a square about 10 km (about 6 miles) on a side! This area is about a third larger than Manhattan Island.

It used to be considered a good joke to send a newly graduated engineer to the stockroom for a 1-farad capacitor. That's not as funny as it used to be; recently developed technology makes it possible to make 1-F capacitors a few centimeters on a side. One type uses activated carbon granules, for which 1 gram has a surface area of about 1000 m^2.

EXAMPLE 25–2

Properties of a parallel-plate capacitor The plates of a parallel-plate capacitor in vacuum are 5.00 mm apart and 2.00 m^2 in area. A potential difference of 10,000 V (10.0 kV) is applied across the capacitor. Compute a) the capacitance; b) the charge on each plate; and c) the magnitude of the electric field in the space between them.

SOLUTION a) From Eq. (25–2),

$$C = \epsilon_0 \frac{A}{d} = \frac{(8.85 \times 10^{-12}~\text{F/m})(2.00~\text{m}^2)}{5.00 \times 10^{-3}~\text{m}}$$

$$= 3.54 \times 10^{-9}~\text{F} = 0.00354~\mu\text{F}.$$

b) The charge on the capacitor is

$$Q = CV_{ab} = (3.54 \times 10^{-9}~\text{C/V})(1.00 \times 10^{4}~\text{V})$$

$$= 3.54 \times 10^{-5}~\text{C} = 35.4~\mu\text{C}.$$

The plate at higher potential has charge $+35.4~\mu\text{C}$, and the other plate has charge $-35.4~\mu\text{C}$.

c) The electric-field magnitude is

$$E = \frac{\sigma}{\epsilon_0} = \frac{Q}{\epsilon_0 A} = \frac{3.54 \times 10^{-5}~\text{C}}{(8.85 \times 10^{-12}~\text{C}^2/\text{N} \cdot \text{m}^2)(2.00~\text{m}^2)}$$

$$= 2.00 \times 10^{6}~\text{N/C}.$$

We can get this same result by recalling that the electric field is equal in magnitude to the potential gradient (Eq. (24–22)). Since the field between the plates is uniform,

$$E = \frac{V_{ab}}{d} = \frac{1.00 \times 10^{4}~\text{V}}{5.00 \times 10^{-3}~\text{m}} = 2.00 \times 10^{6}~\text{V/m}.$$

(Remember that the newton per coulomb and the volt per meter are equivalent units.)

EXAMPLE 25–3

A spherical capacitor Two concentric spherical conducting shells are separated by vacuum. The inner shell has total charge $+Q$ and outer radius r_a, and the outer shell has charge $-Q$ and inner radius r_b (Fig. 25–3). Find the capacitance of this spherical capacitor.

SOLUTION This isn't a parallel-plate capacitor, so we can't use Eq. (25–2). Instead, we go back to the fundamental definition of

capacitance, Eq. (25–1). We'll use Gauss's law to find the electric field between the spherical conductors and from this determine the potential difference V_{ab} between the conductors.

Using the same procedure as in Example 23–5 (Section 23–5), we take as our Gaussian surface a sphere with radius r between the two spheres and concentric with them. Gauss's law, Eq. (23–8), states that the electric flux through this surface is equal to the total charge enclosed within the surface, divided

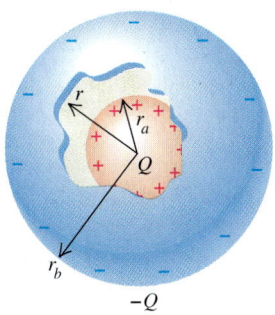

25–3 A spherical capacitor.

by ϵ_0:

$$\oint \vec{E} \cdot d\vec{A} = \frac{Q_{\text{encl}}}{\epsilon_0}.$$

By symmetry, $\vec{E}$ is constant in magnitude and parallel to $d\vec{A}$ at every point on this surface, so the integral in Gauss's law is equal to $(E)(4\pi r^2)$. The total charge enclosed is $Q_{\text{encl}} = Q$, so we have

$$(E)(4\pi r^2) = \frac{Q}{\epsilon_0},$$

$$E = \frac{Q}{4\pi\epsilon_0 r^2}.$$

The electric field between the spheres is just that due to the charge on the inner sphere; the outer sphere has no effect. We found in Example 23–5 that the charge on a conducting sphere produces zero field *inside* the sphere, which also tells us that the

outer conductor makes no contribution to the field between the conductors.

The above expression for E is the same as that for a point charge Q, so the expression for the potential can also be taken to be the same as for a point charge, $V = Q/4\pi\epsilon_0 r$. Hence the potential of the inner (positive) conductor at $r = r_a$ with respect to that of the outer (negative) conductor at $r = r_b$ is

$$V_{ab} = V_a - V_b = \frac{Q}{4\pi\epsilon_0 r_a} - \frac{Q}{4\pi\epsilon_0 r_b}$$

$$= \frac{Q}{4\pi\epsilon_0}\left(\frac{1}{r_a} - \frac{1}{r_b}\right) = \frac{Q}{4\pi\epsilon_0}\frac{r_b - r_a}{r_a r_b}.$$

Finally, the capacitance is

$$C = \frac{Q}{V_{ab}} = 4\pi\epsilon_0 \frac{r_a r_b}{r_b - r_a}.$$

As an example, if $r_a = 9.5$ cm and $r_b = 10.5$ cm,

$$C = 4\pi(8.85 \times 10^{-12} \text{ F/m}) \frac{(0.095 \text{ m})(0.105 \text{ m})}{0.010 \text{ m}}$$

$$= 1.1 \times 10^{-10} \text{ F} = 110 \text{ pF}.$$

We can relate this result to the capacitance of a parallel-plate capacitor. The quantity $4\pi r_a r_b$ is intermediate between the areas $4\pi r_a^2$ and $4\pi r_b^2$ of the two spheres; in fact, it's the *geometric mean* of these two areas, which we can denote by A_{gm}. The distance between spheres is $d = r_b - r_a$, so we can rewrite the above result as $C = \epsilon_0 A_{\text{gm}}/d$. This is exactly the same form as for parallel plates: $C = \epsilon_0 A/d$. The point is that if the distance between spheres is very small in comparison to their radii, they behave like parallel plates with the same area and spacing.

EXAMPLE 25–4

A cylindrical capacitor A long cylindrical conductor has a radius r_a and a linear charge density $+\lambda$. It is surrounded by a coaxial cylindrical conducting shell with inner radius r_b and linear charge density $-\lambda$ (Fig. 25–4). Calculate the capacitance per unit length for this capacitor, assuming that there is vacuum in the space between cylinders.

SOLUTION We again go back to the fundamental definition of capacitance, Eq. (25–1). To find the potential difference between the cylinders, we use a result that we worked out in Example 24–10 (Section 24–4). There we found that at a point outside a charged cylinder a distance r from the axis, the potential due to the cylinder is

$$V = \frac{\lambda}{2\pi\epsilon_0} \ln \frac{r_0}{r},$$

where r_0 is the (arbitrary) radius at which $V = 0$. We can use this same result for the potential *between* the cylinders in the present problem because, according to Gauss's law, the charge on the outer cylinder doesn't contribute to the field between cylinders (see Example 25–3). In our case we take the radius r_0 to be r_b, the radius of the inner surface of the outer cylinder, so that the outer conducting cylinder is at $V = 0$. Then the potential at the outer surface of the inner cylinder (where $r = r_a$) is just equal to

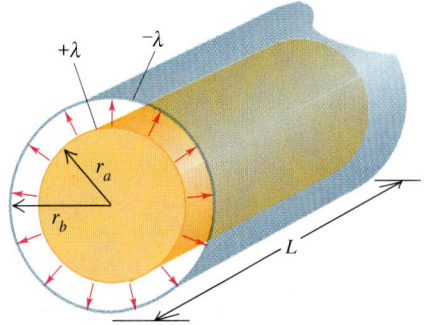

25–4 A long cylindrical capacitor. The linear charge density λ is assumed to be positive in this figure. The magnitude of charge in a length L of either cylinder is λL.

the potential V_{ab} of the inner (positive) cylinder a with respect to the outer (negative) cylinder b, or

$$V_{ab} = \frac{\lambda}{2\pi\epsilon_0} \ln \frac{r_b}{r_a}.$$

This potential difference is positive (assuming that λ is positive, as in Fig. 25–4) because the inner cylinder is at higher potential than the outer.

Capacitance is charge Q divided by potential difference V_{ab}. The total charge Q in a length L is $Q = \lambda L$, so the capacitance C of a length L is

$$C = \frac{Q}{V_{ab}} = \frac{\lambda L}{\dfrac{\lambda}{2\pi\epsilon_0}\ln\dfrac{r_b}{r_a}} = \frac{2\pi\epsilon_0 L}{\ln(r_b/r_a)}.$$

The capacitance per unit length is

$$\frac{C}{L} = \frac{2\pi\epsilon_0}{\ln(r_b/r_a)}.$$

Substituting $\epsilon_0 = 8.85 \times 10^{-12}$ F/m $= 8.85$ pF/m, we get

$$\frac{C}{L} = \frac{55.6 \text{ pF/m}}{\ln(r_b/r_a)}.$$

We see that the capacitance of the coaxial cylinders is determined entirely by the dimensions, just as for the parallel-plate case. Ordinary coaxial cables are made like this but with an insulating material instead of vacuum between the inner and outer conductors. A typical cable for TV antennas and VCR connections has a capacitance per unit length of 69 pF/m.

25–3 CAPACITORS IN SERIES AND PARALLEL

Capacitors are manufactured with certain standard capacitances and working voltages. However, these standard values may not be the ones you actually need in a particular application. You can obtain the values you need by combining capacitors; many combinations are possible, but the simplest combinations are a series connection and a parallel connection.

CAPACITORS IN SERIES

An assortment of commercially available capacitors.

Figure 25–5a is a schematic diagram of a **series connection.** Two capacitors are connected in series (one after the other) by conducting wires between points a and b. Both capacitors are initially uncharged. When a constant positive potential difference V_{ab} is applied between points a and b, the capacitors become charged; the figure shows that the charge on *all* conducting plates has the same magnitude. To see why, note first that the top plate of C_1 acquires a positive charge Q. The electric field of this positive charge pulls negative charge up to the bottom plate of C_1 until all of the field lines that begin on the top plate end on the bottom plate. This requires that the bottom plate have charge $-Q$. These negative charges had to come from the top plate of C_2, which becomes positively charged with charge $+Q$. This positive charge then pulls negative charge $-Q$ from the connection at point b onto the bottom plate of C_2. The total charge on the lower plate of C_1 and the upper plate of C_2 together must always be zero because these plates aren't connected to anything except each other. Thus *in a series connection the magnitude of charge on all plates is the same.*

Referring again to Fig. 25–5a, we can write the potential differences between points a and c, c and b, and a and b as

$$V_{ac} = V_1 = \frac{Q}{C_1}, \qquad V_{cb} = V_2 = \frac{Q}{C_2},$$

$$V_{ab} = V = V_1 + V_2 = Q\left(\frac{1}{C_1} + \frac{1}{C_2}\right),$$

25–5 (a) Two capacitors in series and (b) the equivalent capacitor.

(a)

(b)

and so

$$\frac{V}{Q} = \frac{1}{C_1} + \frac{1}{C_2}. \qquad (25\text{–}3)$$

Following a common convention, we use the symbols V_1, V_2, and V to denote the potential *differences* V_{ac} (across the first capacitor), V_{cb} (across the second capacitor), and V_{ab} (across the entire combination of capacitors), respectively.

The **equivalent capacitance** C_{eq} of the series combination is defined as the capacitance of a *single* capacitor for which the charge Q is the same as for the combination, when the potential difference V is the same. In other words, the combination can be replaced by an *equivalent capacitor* of capacitance C_{eq}. For such a capacitor, shown in Fig. 25–5b,

$$C_{eq} = \frac{Q}{V}, \qquad \text{or} \qquad \frac{1}{C_{eq}} = \frac{V}{Q}. \qquad (25\text{–}4)$$

Combining Eqs. (25–3) and (25–4), we find

$$\frac{1}{C_{eq}} = \frac{1}{C_1} + \frac{1}{C_2}.$$

We can extend this analysis to any number of capacitors in series. We find the following result for the *reciprocal* of the equivalent capacitance:

$$\frac{1}{C_{eq}} = \frac{1}{C_1} + \frac{1}{C_2} + \frac{1}{C_3} + \cdots \qquad \text{(capacitors in series)}. \qquad (25\text{–}5)$$

The reciprocal of the equivalent capacitance of a series combination equals the sum of the reciprocals of the individual capacitances. In a series connection the equivalent capacitance is always *less than* any individual capacitance.

CAUTION ▶ The magnitude of charge is the same on all plates of all the capacitors in a series combination; however, the potential differences of the individual capacitors are *not* the same unless their individual capacitances are the same. The potential differences of the individual capacitors add to give the total potential difference across the series combination. ◀

CAPACITORS IN PARALLEL

The arrangement shown in Fig. 25–6a is called a **parallel connection.** Two capacitors are connected in parallel between points a and b. In this case the upper plates of the two capacitors are connected together by conducting wires to form an equipotential surface, and the lower plates form another. Hence *in a parallel connection the potential difference for all individual capacitors is the same* and is equal to $V_{ab} = V$. The charges Q_1 and Q_2 are not necessarily equal, however, since charges can reach each capacitor independently from the source (such as a battery) of the voltage V_{ab}. The charges are

$$Q_1 = C_1 V \qquad \text{and} \qquad Q_2 = C_2 V.$$

The *total* charge Q of the combination, and thus the total charge on the equivalent

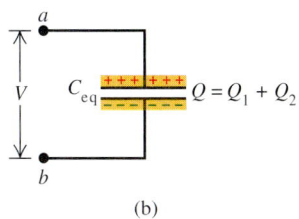

25–6 (a) Two capacitors in parallel and (b) the equivalent capacitor.

(a)   (b)

capacitor, is

$$Q = Q_1 + Q_2 = (C_1 + C_2)V,$$

so

$$\frac{Q}{V} = C_1 + C_2. \qquad (25\text{–}6)$$

The parallel combination is equivalent to a single capacitor with the same total charge $Q = Q_1 + Q_2$ and potential difference V as the combination (Fig. 25–6b). The equivalent capacitance of the combination, C_{eq}, is the same as the capacitance Q/V of this single equivalent capacitor. So from Eq. (25–6),

$$C_{eq} = C_1 + C_2.$$

In the same way we can show that for any number of capacitors in parallel,

$$C_{eq} = C_1 + C_2 + C_3 + \cdots \qquad \text{(capacitors in parallel)}. \qquad (25\text{–}7)$$

The equivalent capacitance of a parallel combination equals the *sum* of the individual capacitances. In a parallel connection the equivalent capacitance is always *greater than* any individual capacitance.

  CAUTION ▶ The potential differences are the same for all the capacitors in a parallel combination; however, the charges on individual capacitors are *not* the same unless their individual capacitances are the same. The charges on the individual capacitors add to give the total charge on the parallel combination. (Compare these statements to those in the "caution" paragraph following Eq. (25–5).) ◀

Problem–Solving Strategy

EQUIVALENT CAPACITANCE

1. Keep in mind that when we say that a capacitor has charge Q, we always mean that the plate at higher potential has charge $+Q$ and the other plate has charge $-Q$.

2. When capacitors are connected in series, as in Fig. 25–5a, they always have the same charge, assuming that they were uncharged before they were connected. The potential differences are *not* equal unless the capacitances are equal. The total potential difference across the combination is the sum of the individual potential differences.

3. When capacitors are connected in parallel, as in Fig. 25–6a, the potential difference V is always the same for all of the individual capacitors. The charges on the individual capacitors are *not* equal unless the capacitances are equal. The total charge on the combination is the sum of the individual charges.

4. To find the equivalent capacitance of more complicated combinations, you can sometimes identify parts that are simple series or parallel connections and replace these by their equivalent capacitances in a step-by-step reduction. Then, if you need to find the charge or potential difference for an individual capacitor, you may have to retrace your path to the original capacitors.

EXAMPLE 25–5

Capacitors in series and in parallel In Figs. 25–5 and 25–6, let $C_1 = 6.0\ \mu\text{F}$, $C_2 = 3.0\ \mu\text{F}$, and $V_{ab} = 18\ \text{V}$. Find the equivalent capacitance, and find the charge and potential difference for each capacitor when the two capacitors are connected a) in series; b) in parallel.

SOLUTION a) The equivalent capacitance of the series combination (Fig. 25–5a) is given by Eq. (25–5):

$$\frac{1}{C_{eq}} = \frac{1}{C_1} + \frac{1}{C_2} = \frac{1}{6.0\ \mu\text{F}} + \frac{1}{3.0\ \mu\text{F}}, \qquad C_{eq} = 2.0\ \mu\text{F}.$$

This is less than either C_1 or C_2. The charge Q on each capacitor in series is the same as the charge on the equivalent capacitor:

$$Q = C_{eq}V = (2.0\ \mu\text{F})(18\ \text{V}) = 36\ \mu\text{C}.$$

The potential difference across each capacitor is inversely proportional to its capacitance:

$$V_{ac} = V_1 = \frac{Q}{C_1} = \frac{36\ \mu\text{C}}{6.0\ \mu\text{F}} = 6.0\ \text{V},$$

$$V_{cb} = V_2 = \frac{Q}{C_2} = \frac{36\ \mu\text{C}}{3.0\ \mu\text{F}} = 12\ \text{V}.$$

For two capacitors in series the *larger* potential difference appears across the capacitor with the *smaller* capacitance. As a check, note that $V_{ac} + V_{cb} = V_{ab} = 18$ V, as it must.

b) The equivalent capacitance of the parallel combination (Fig. 25–6a) is given by Eq. (25–7):

$$C_{eq} = C_1 + C_2 = 6.0 \ \mu F + 3.0 \ \mu F = 9.0 \ \mu F.$$

This is greater than either C_1 or C_2. The potential difference across each of the two capacitors in parallel is the same as that across the equivalent capacitor, 18 V. The charges Q_1 and Q_2 are

directly proportional to the capacitances C_1 and C_2, respectively:

$$Q_1 = C_1 V = (6.0 \ \mu F)(18 \ V) = 108 \ \mu C,$$

$$Q_2 = C_2 V = (3.0 \ \mu F)(18 \ V) = 54 \ \mu C.$$

For two capacitors in parallel the *larger* charge appears on the capacitor with the *larger* capacitance. Can you show that the total charge $Q_1 + Q_2$ on the parallel combination is equal to the charge $Q = C_{eq} V$ on the equivalent capacitor?

EXAMPLE 25–6

A capacitor network Find the equivalent capacitance of the combination shown in Fig. 25–7a.

SOLUTION We first replace the 12-μF and 6-μF series combination by its equivalent capacitance; calling that C', we use Eq. (25–5):

$$\frac{1}{C'} = \frac{1}{12 \ \mu F} + \frac{1}{6 \ \mu F}, \qquad C' = 4 \ \mu F.$$

This gives us the equivalent combination shown in Fig. 25–7b. Next we find the equivalent capacitance of the three capacitors in parallel, using Eq. (25–7). Calling their equivalent capacitance C'', we have

$$C'' = 3 \ \mu F + 11 \ \mu F + 4 \ \mu F = 18 \ \mu F.$$

This gives us the simpler equivalent combination shown in Fig. 25–7c. Finally, we find the equivalent capacitance C_{eq} of these

two capacitors in series (Fig. 25–7d):

$$\frac{1}{C_{eq}} = \frac{1}{18 \ \mu F} + \frac{1}{9 \ \mu F}, \qquad C_{eq} = 6 \ \mu F.$$

The equivalent capacitance of the network is 6 μF; that is, if a potential difference V_{ab} is applied across the terminals of the network, the net charge on the network is 6 μF times V_{ab}. How is this net charge related to the charges on the individual capacitors in Fig. 25–7a? (See Exercise 25–14.)

25–7 (a) A capacitor network between points a and b. (b) The 12-μF and 6-μF capacitors in series in (a) are replaced by an equivalent 4-μF capacitor. (c) The 3-μF, 11-μF, and 4-μF capacitors in parallel in (b) are replaced by an equivalent 18-μF capacitor. (d) Finally, the 18-μF and 9-μF capacitors in series in (c) are replaced by an equivalent 6-μF capacitor.

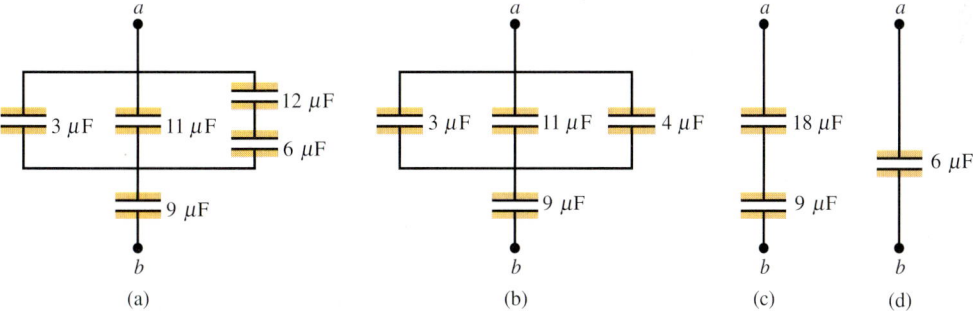

25–4 ENERGY STORAGE IN CAPACITORS AND ELECTRIC-FIELD ENERGY

Many of the most important applications of capacitors depend on their ability to store energy. The electric potential energy stored in a charged capacitor is just equal to the amount of work required to charge it, that is, to separate opposite charges and place them on different conductors. When the capacitor is discharged, this stored energy is recovered as work done by electrical forces.

We can calculate the potential energy U of a charged capacitor by calculating the work W required to charge it. Suppose that when we are done charging the capacitor, the final charge is Q and the final potential difference is V. From Eq. (25–1) these

are related by

$$V = \frac{Q}{C}.$$

Let q and v be the charge and potential difference, respectively, at an intermediate stage during the charging process; then $v = q/C$. At this stage the work dW required to transfer an additional element of charge dq is

$$dW = v\, dq = \frac{q\, dq}{C}.$$

The total work W needed to increase the capacitor charge q from zero to a final value Q is

$$W = \int_0^W dW = \frac{1}{C}\int_0^Q q\, dq = \frac{Q^2}{2C} \qquad \text{(work to charge a capacitor).} \qquad (25\text{--}8)$$

This is also equal to the total work done by the electric field on the charge when the capacitor discharges. Then q *decreases* from an initial value Q to zero as the elements of charge dq "fall" through potential differences v that vary from V down to zero.

If we define the potential energy of an *uncharged* capacitor to be zero, then W in Eq. (25–8) is equal to the potential energy U of the charged capacitor. The final stored charge is $Q = CV$, so we can express U (which is equal to W) as

$$U = \frac{Q^2}{2C} = \frac{1}{2}CV^2 = \frac{1}{2}QV \qquad \begin{array}{l}\text{(potential energy stored} \\ \text{in a capacitor).}\end{array} \qquad (25\text{--}9)$$

When Q is in coulombs, C in farads (coulombs per volt), and V in volts (joules per coulomb), U is in joules.

The last form of Eq. (25–9), $U = \frac{1}{2}QV$, shows that the total work W required to charge the capacitor is equal to the total charge Q multiplied by the *average* potential difference $\frac{1}{2}V$ during the charging process.

The expression $U = \frac{1}{2}(Q^2/C)$ in Eq. (25–9) shows that a charged capacitor is the electrical analog of a stretched spring with elastic potential energy $U = \frac{1}{2}kx^2$. The charge Q is analogous to the elongation x, and the *reciprocal* of the capacitance, $1/C$, is analogous to the force constant k. The energy supplied to a capacitor in the charging process is analogous to the work we do on a spring when we stretch it.

Equations (25–8) and (25–9) show that capacitance measures the ability of a capacitor to store both energy and charge. If a capacitor is charged by connecting it to a battery or other source that provides a fixed potential difference V, then increasing the value of C gives a greater charge $Q = CV$ and a greater amount of stored energy $U = \frac{1}{2}CV^2$. If instead the goal is to transfer a given quantity of charge Q from one conductor to another, Eq. (25–8) shows that the work W required is inversely proportional to C; the greater the capacitance, the easier it is to give a capacitor a fixed amount of charge.

An electronic flash unit stores potential energy in a capacitor. Depressing the camera's shutter button provides a conducting path from one capacitor plate to the other through the flash tube. Once this path is established and the capacitor discharges, the stored energy is converted into a brief but intense flash of light.

ELECTRIC-FIELD ENERGY

We can charge a capacitor by moving electrons directly from one plate to another. This requires doing work against the electric field between the plates. Thus we can think of the energy as being stored *in the field* in the region between the plates. To develop this relation, let's find the energy *per unit volume* in the space between the plates of a parallel-plate capacitor with plate area A and separation d. We call this the **energy density**, denoted by u. From Eq. (25–9) the total stored potential energy is $\frac{1}{2}CV^2$, and the volume between the plates is just Ad; hence the energy density is

$$u = \text{Energy density} = \frac{\frac{1}{2}CV^2}{Ad}. \qquad (25\text{-}10)$$

From Eq. (25–2) the capacitance C is given by $C = \epsilon_0 A/d$. The potential difference V is related to the electric field magnitude E by $V = Ed$. If we use these expressions in Eq. (25–10), the geometric factors A and d cancel, and we find

$$u = \frac{1}{2}\epsilon_0 E^2 \qquad \text{(electric energy density in a vacuum).} \qquad (25\text{-}11)$$

Although we have derived this relation only for a parallel-plate capacitor, it turns out to be valid for any capacitor in vacuum and indeed *for any electric field configuration in vacuum.* This result has an interesting implication. We think of vacuum as space with no matter in it, but vacuum can nevertheless have electric fields and therefore energy. Thus "empty" space need not be truly empty after all. We will use this idea and Eq. (25–11) in Chapter 33 in connection with the energy transported by electromagnetic waves.

CAUTION ▶ It's a common misconception that electric-field energy is a new kind of energy, different from the electric potential energy described before. This is *not* the case; it is simply a different way of interpreting electric potential energy. We can regard the energy of a given system of charges as being a shared property of all the charges, or we can think of the energy as being a property of the electric field that the charges create. Either interpretation leads to the same value of the potential energy. ◀

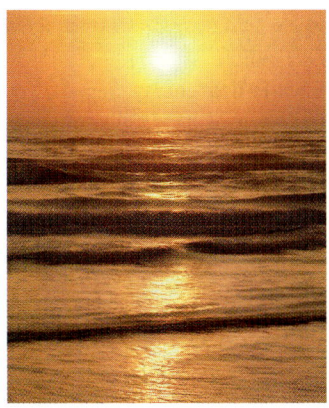

Half the energy in sunlight is in the form of electric-field energy. The other half is magnetic-field energy, which we will learn about in Chapters 31 and 33.

EXAMPLE 25-7

Transferring charge and energy between capacitors In Fig. 25–8 we charge a capacitor of capacitance $C_1 = 8.0\ \mu F$ by connecting it to a source of potential difference $V_0 = 120$ V (not shown in the figure). The switch S is initially open. Once C_1 is charged, the source of potential difference is disconnected. a) What is the charge Q_0 on C_1 if switch S is left open? b) What is the energy stored in C_1 if switch S is left open? c) The capacitor of capacitance $C_2 = 4.0\ \mu F$ is initially uncharged. After we close switch S, what is the potential difference across each capacitor, and what is the charge on each capacitor? d) What is the total energy of the system after we close switch S?

SOLUTION a) The charge Q_0 on C_1 is

$$Q_0 = C_1 V_0 = (8.0\ \mu F)(120\ V) = 960\ \mu C.$$

b) The energy initially stored in the capacitor is

$$U_\text{initial} = \frac{1}{2}Q_0 V_0 = \frac{1}{2}(960 \times 10^{-6}\ C)(120\ V) = 0.058\ J.$$

c) When the switch is closed, the positive charge Q_0 becomes distributed over the upper plates of both capacitors and the negative charge $-Q_0$ is distributed over the lower plates of both capacitors. Let Q_1 and Q_2 be the magnitudes of the final charges on the two capacitors. From conservation of charge,

$$Q_1 + Q_2 = Q_0.$$

In the final state, when the charges are no longer moving, both upper plates are at the same potential; they are connected by a conducting wire and so form a single equipotential surface. Both lower plates are also at the same potential, different from that of

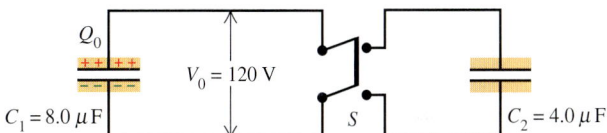

25–8 When the switch S is closed, the charged capacitor C_1 is connected to an uncharged capacitor C_2. The center part of the switch is an insulating handle; charge can flow only between the two upper terminals and between the two lower terminals.

the upper plates. The final potential difference V between the plates is therefore the same for both capacitors; in other words, this is a parallel connection. The capacitor charges are

$$Q_1 = C_1 V, \qquad Q_2 = C_2 V.$$

When we combine this with the preceding equation for conservation of charge, we find

$$V = \frac{Q_0}{C_1 + C_2} = \frac{960\ \mu C}{8.0\ \mu F + 4.0\ \mu F} = 80\ V,$$

$$Q_1 = 640\ \mu C, \qquad Q_2 = 320\ \mu C.$$

d) The final energy of the system is the sum of the energies stored in each capacitor:

$$U_\text{final} = \frac{1}{2}Q_1 V + \frac{1}{2}Q_2 V = \frac{1}{2}Q_0 V$$

$$= \frac{1}{2}(960 \times 10^{-6}\ C)(80\ V) = 0.038\ J.$$

This is less than the original energy $U_{initial} = 0.058$ J; the difference has been converted to energy of some other form. The conductors become a little warmer because of their resistance,

and some energy is radiated as electromagnetic waves. We'll study the circuit behavior of capacitors in detail in Chapters 27 and 32.

EXAMPLE 25-8

Electric-field energy Suppose you want to store 1.00 J of electric potential energy in a volume of 1.00 m^3 in vacuum. a) What is the magnitude of the required electric field? b) If the field magnitude is ten times larger, how much energy is stored per cubic meter?

SOLUTION a) The desired energy density is
$u = (1.00$ J$)/(1.00$ m$^3) = 1.00$ J/m^3. We solve Eq. (25–11) for E:

$$E = \sqrt{\frac{2u}{\epsilon_0}} = \sqrt{\frac{2(1.00 \text{ J/m}^3)}{8.85 \times 10^{-12} \text{ C}^2/\text{N} \cdot \text{m}^2}}$$
$$= 4.75 \times 10^5 \text{ N/C} = 4.75 \times 10^5 \text{ V/m}.$$

This is a sizable field, corresponding to a potential difference of nearly a half million volts over a distance of one meter.
b) Equation (25–11) shows that u is proportional to E^2. If E increases by a factor of 10, u increases by a factor of $10^2 = 100$, and the energy density is 100 J/m^3.

EXAMPLE 25-9

Two ways to calculate energy stored in a capacitor The spherical capacitor described in Example 25–3 (Section 25–2) has charges $+Q$ and $-Q$ on its inner and outer conductors. Find the electric potential energy stored in the capacitor a) by using the capacitance C found in Example 25–3; b) by integrating the electric-field energy density.

SOLUTION a) In Example 25–3 we found that the spherical capacitor has capacitance

$$C = 4\pi\epsilon_0 \frac{r_a r_b}{r_b - r_a},$$

where r_a and r_b are the radii of the inner and outer conducting spheres. From Eq. (25–9) the energy stored in this capacitor is

$$U = \frac{Q^2}{2C} = \frac{Q^2}{8\pi\epsilon_0} \frac{r_b - r_a}{r_a r_b}.$$

b) The electric field in the volume between the two conducting spheres has magnitude $E = Q/4\pi\epsilon_0 r^2$. The electric field is zero inside the inner sphere and is also zero outside the inner surface of the outer sphere, because a Gaussian surface with radius $r < r_a$ or $r > r_b$ encloses zero net charge. Hence the energy density is nonzero only in the space between the spheres ($r_a < r < r_b$). In this region,

$$u = \frac{1}{2}\epsilon_0 E^2 = \frac{1}{2}\epsilon_0 \left(\frac{Q}{4\pi\epsilon_0 r^2}\right)^2 = \frac{Q^2}{32\pi^2\epsilon_0 r^4}.$$

The energy density is *not* uniform; it decreases rapidly with increasing distance from the center of the capacitor. To find the total electric-field energy, we integrate u (the energy per unit volume) over the volume between the inner and outer conducting spheres. Dividing this volume up into spherical shells of radius r, surface area $4\pi r^2$, thickness dr, and volume $dV = 4\pi r^2 dr$, we have

$$U = \int u \, dV = \int_{r_a}^{r_b} \left(\frac{Q^2}{32\pi^2\epsilon_0 r^4}\right) 4\pi r^2 \, dr$$
$$= \frac{Q^2}{8\pi\epsilon_0} \int_{r_a}^{r_b} \frac{dr}{r^2} = \frac{Q^2}{8\pi\epsilon_0}\left(-\frac{1}{r_b} + \frac{1}{r_a}\right)$$
$$= \frac{Q^2}{8\pi\epsilon_0} \frac{r_b - r_a}{r_a r_b}.$$

This is the same result as we obtained in part (a). We emphasize that electric potential energy can be regarded as being associated with either the *charges,* as in part (a), or the *field,* as in part (b); regardless of which viewpoint you choose, the amount of stored energy is the same.

25-5 DIELECTRICS

Most capacitors have a nonconducting material, or **dielectric,** between their conducting plates. A common type of capacitor uses long strips of metal foil for the plates, separated by strips of plastic sheet such as Mylar. A sandwich of these materials is rolled up, forming a unit that can provide a capacitance of several microfarads in a compact package.

Placing a solid dielectric between the plates of a capacitor serves three functions.

First, it solves the mechanical problem of maintaining two large metal sheets at a very small separation without actual contact.

Second, using a dielectric increases the maximum possible potential difference between the capacitor plates. As we described in Example 24–8 (Section 24–4), any insulating material, when subjected to a sufficiently large electric field, experiences **dielectric breakdown,** a partial ionization that permits conduction through it. Many dielectric materials can tolerate stronger electric fields without breakdown than can air. Thus using a dielectric allows a capacitor to sustain a higher potential difference V and so store greater amounts of charge and energy.

Third, the capacitance of a capacitor of given dimensions is *greater* when there is a dielectric material between the plates than when there is a vacuum. We can demonstrate this effect with the aid of a sensitive *electrometer,* a device that measures the potential difference between two conductors without letting any appreciable charge flow from one to the other. Figure 25–9a shows an electrometer connected across a charged capacitor, with magnitude of charge Q on each plate and potential difference V_0. When we insert an uncharged sheet of dielectric, such as glass, paraffin, or polystyrene, between the plates, experiment shows that the potential difference *decreases* to a smaller value V (Fig. 25–9b). When we remove the dielectric, the potential difference returns to its original value V_0, showing that the original charges on the plates have not changed.

The original capacitance C_0 is given by $C_0 = Q/V_0$, and the capacitance C with the dielectric present is $C = Q/V$. The charge Q is the same in both cases, and V is less than V_0, so we conclude that the capacitance C with the dielectric present is *greater* than C_0. When the space between plates is completely filled by the dielectric, the ratio of C to C_0 (equal to the ratio of V_0 to V) is called the **dielectric constant** of the material, K:

$$K = \frac{C}{C_0} \qquad \text{(definition of dielectric constant).} \qquad (25\text{–}12)$$

When the charge is constant, $Q = C_0V_0 = CV$ and $C/C_0 = V_0/V$. In this case, Eq. (25–12) can be rewritten as

$$V = \frac{V_0}{K} \qquad \text{(when } Q \text{ is constant).} \qquad (25\text{–}13)$$

With the dielectric present, the potential difference for a given charge Q is *reduced* by a factor K.

The dielectric constant K is a pure number. Because C is always greater than C_0, K is always greater than unity. A few representative values of K are given in Table 25–1. For vacuum, $K = 1$ by definition. For air at ordinary temperatures and pressures, K is about 1.0006; this is so nearly equal to 1 that for most purposes an air capacitor is equivalent to one in vacuum. Note that while water has a very large value of K, it is usually not a very practical dielectric for use in capacitors. The reason is that while pure water

Vacuum

Q $-Q$

V_0

Electrometer

(a)

Dielectric

Q $-Q$

V

Electrometer

(b)

25–9 Effect of a dielectric between the plates of a parallel-plate capacitor. The electrometer measures potential difference. (a) With a given charge, the potential difference is V_0. (b) With the same charge but with a dielectric between the plates, the potential difference V is smaller than V_0.

TABLE 25–1

VALUES OF DIELECTRIC CONSTANT K AT 20°C

MATERIAL	K	MATERIAL	K
Vacuum	1	Polyvinyl chloride	3.18
Air (1 atm)	1.00059	Plexiglas	3.40
Air (100 atm)	1.0548	Glass	5–10
Teflon	2.1	Neoprene	6.70
Polyethylene	2.25	Germanium	16
Benzene	2.28	Glycerin	42.5
Mica	3–6	Water	80.4
Mylar	3.1	Strontium titanate	310

25–10 (a) Electric field lines with vacuum between the plates. (b) The induced charges on the faces of the dielectric decrease the electric field.

is a very poor conductor, it is also an excellent ionic solvent. Any ions that are dissolved in the water will cause charge to flow between the capacitor plates, so the capacitor discharges.

No real dielectric is a perfect insulator. Hence there is always some *leakage current* between the charged plates of a capacitor with a dielectric. We tacitly ignored this effect in Section 25–3 when we derived expressions for the equivalent capacitances of capacitors in series, Eq. (25–5), and in parallel, Eq. (25–7). But if a leakage current flows for a long enough time to substantially change the charges from the values we used to derive Eqs. (25–5) and (25–7), those equations may no longer be accurate.

INDUCED CHARGE AND POLARIZATION

When a dielectric material is inserted between the plates while the charge is kept constant, the potential difference between the plates decreases by a factor K. Therefore the electric field between the plates must decrease by the same factor. If E_0 is the vacuum value and E is the value with the dielectric, then

$$E = \frac{E_0}{K} \qquad \text{(when } Q \text{ is constant).} \qquad (25\text{–}14)$$

Since the electric-field magnitude is smaller when the dielectric is present, the surface charge density (which causes the field) must be smaller as well. The surface charge on the conducting plates does not change, but an *induced* charge of the opposite sign appears on each surface of the dielectric (Fig. 25–10). The dielectric was originally electrically neutral, and is still neutral; the induced surface charges arise as a result of *redistribution* of positive and negative charge within the dielectric material, a phenomenon called **polarization.** We first encountered polarization in Section 22–4, and we suggest that you reread the discussion of Fig. 22–5. We will assume that the induced surface charge is *directly proportional* to the electric-field magnitude E in the material; this is indeed the case for many common dielectrics. (This direct proportionality is analogous to Hooke's law for a spring.) In that case, K is a constant for any particular material. When the electric field is very strong or if the dielectric is made of certain crystalline materials, the relationship between induced charge and the electric field can be more complex; we won't consider such cases here.

We can derive a relation between this induced surface charge and the charge on the plates. Let's denote the magnitude of the charge per unit area induced on the surfaces of the dielectric (the induced surface charge density) by σ_i. The magnitude of the surface charge density on the capacitor plates is σ, as usual. Then the *net* surface charge on each side of the capacitor has magnitude $(\sigma - \sigma_i)$, as shown in Fig. 25–10b. As we found in Example 22–13 (Section 22–7) and in Example 23–8 (Section 23–5), the field between the plates is related to the net surface charge density by $E = \sigma_{net}/\epsilon_0$. Without and with the dielectric, respectively, we have

$$E_0 = \frac{\sigma}{\epsilon_0}, \qquad E = \frac{\sigma - \sigma_i}{\epsilon_0}.$$

Using these expressions in Eq. (25–14) and rearranging the result, we find

$$\sigma_i = \sigma\left(1 - \frac{1}{K}\right) \qquad \text{(induced surface charge density).} \qquad (25\text{–}15)$$

This equation shows that when K is very large, σ_i is nearly as large as σ. In this case, σ_i nearly cancels σ, and the field and potential difference are much smaller than their values in vacuum.

The product $K\epsilon_0$ is called the **permittivity** of the dielectric, denoted by ϵ:

$$\epsilon = K\epsilon_0 \quad \text{(definition of permittivity)}. \tag{25–16}$$

In terms of ϵ we can express the electric field within the dielectric as

$$E = \frac{\sigma}{\epsilon}. \tag{25–17}$$

The capacitance when the dielectric is present is given by

$$C = KC_0 = K\epsilon_0 \frac{A}{d} = \epsilon \frac{A}{d} \quad \begin{array}{l}\text{(parallel-plate capacitor,} \\ \text{dielectric between plates).}\end{array} \tag{25–18}$$

We can repeat the derivation of Eq. (25–11) for the energy density u in an electric field for the case in which a dielectric is present. The result is

$$u = \frac{1}{2} K\epsilon_0 E^2 = \frac{1}{2} \epsilon E^2 \quad \begin{array}{l}\text{(electric energy density} \\ \text{in a dielectric).}\end{array} \tag{25–19}$$

In empty space, where $K = 1$, $\epsilon = \epsilon_0$ and Eqs. (25–18) and (25–19) reduce to Eqs. (25–2) and (25–11), respectively, for a parallel-plate capacitor in a vacuum. For this reason, ϵ_0 is sometimes called the "permittivity of free space" or the "permittivity of vacuum." Because K is a pure number, ϵ and ϵ_0 have the same units, $C^2/N \cdot m^2$ or F/m.

Problem–Solving Strategy

DIELECTRICS

1. Always check for consistency of units; it's a bit more complex with electrical quantities than it was in mechanics. Distances must always be in meters. Remember that a microfarad is 10^{-6} farad, and so on. Don't confuse the numerical value of ϵ_0 with the value of $1/4\pi\epsilon_0$. There are two alternative sets of SI units for electric field magnitude, N/C and V/m. The units of ϵ_0 are $C^2/N \cdot m^2$ or F/m.

2. In problems such as Example 25–10 it is easy to get lost in a blizzard of formulas. Ask yourself at each step what kind of quantity each symbol represents. For example, distinguish clearly between charges and charge densities, between electric fields and electric potential differences, and between energy and energy density.

3. When you check numerical values, remember that with a dielectric present, (a) the capacitance is always greater than without a dielectric; (b) for a given amount of charge on the capacitor, the electric field and potential difference are less than without a dielectric; and (c) the induced surface charge density σ_i on the dielectric is always less in magnitude than the charge density σ on the capacitor plates.

EXAMPLE 25–10

A capacitor with and without a dielectric Suppose the parallel plates in Fig. 25–10 each have an area of 2000 cm^2 (2.00×10^{-1} m^2) and are 1.00 cm (1.00×10^{-2} m) apart. The capacitor is connected to a power supply and charged to a potential difference $V_0 = 3000$ V $= 3.00$ kV. It is then disconnected from the power supply, and a sheet of insulating plastic material is inserted between the plates, completely filling the space between them. We find that the potential difference decreases to 1000 V while the charge on each capacitor plate remains constant. Compute a) the original capacitance C_0; b) the magnitude of charge Q on each plate; c) the capacitance C after the dielectric is inserted; d) the dielectric constant K of the dielectric; e) the permittivity ϵ of the dielectric; f) the magnitude of the induced charge Q_i on each face of the dielectric; g) the original electric field E_0 between the plates; and h) the electric field E after the dielectric is inserted.

SOLUTION Most of these quantities can be obtained in several different ways. Here is a representative sample; try to think of others and compare them.

a) $C_0 = \epsilon_0 \dfrac{A}{d} = (8.85 \times 10^{-12} \text{ F/m}) \dfrac{2.00 \times 10^{-1} \text{ m}^2}{1.00 \times 10^{-2} \text{ m}}$

$\quad\quad = 1.77 \times 10^{-10}$ F $= 177$ pF.

b) $Q = C_0 V_0 = (1.77 \times 10^{-10} \text{ F})(3.00 \times 10^3 \text{ V})$

$\quad\quad = 5.31 \times 10^{-7}$ C $= 0.531$ μC.

c) $C = \dfrac{Q}{V} = \dfrac{5.31 \times 10^{-7}\ \text{C}}{1.00 \times 10^{3}\ \text{V}} = 5.31 \times 10^{-10}\ \text{F} = 531\ \text{pF}.$

d) From Eq. (25–12),

$$K = \frac{C}{C_0} = \frac{5.31 \times 10^{-10}\ \text{F}}{1.77 \times 10^{-10}\ \text{F}} = \frac{531\ \text{pF}}{177\ \text{pF}} = 3.00.$$

Alternatively, from Eq. (25–13),

$$K = \frac{V_0}{V} = \frac{3000\ \text{V}}{1000\ \text{V}} = 3.00.$$

e) $\epsilon = K\epsilon_0 = (3.00)(8.85 \times 10^{-12}\ \text{C}^2/\text{N} \cdot \text{m}^2)$

$= 2.66 \times 10^{-11}\ \text{C}^2/\text{N} \cdot \text{m}^2.$

f) Multiplying Eq. (25–15) by the area of each plate gives the induced charge $Q_i = \sigma_i A$ in terms of the charge $Q = \sigma A$ on each plate:

$$Q_i = Q\left(1 - \frac{1}{K}\right) = (5.31 \times 10^{-7}\ \text{C})\left(1 - \frac{1}{3.00}\right)$$

$= 3.54 \times 10^{-7}\ \text{C}.$

g) $E_0 = \dfrac{V_0}{d} = \dfrac{3000\ \text{V}}{1.00 \times 10^{-2}\ \text{m}} = 3.00 \times 10^{5}\ \text{V/m}.$

h) $E = \dfrac{V}{d} = \dfrac{1000\ \text{V}}{1.00 \times 10^{-2}\ \text{m}} = 1.00 \times 10^{5}\ \text{V/m},$

or

$$E = \frac{\sigma}{\epsilon} = \frac{Q}{\epsilon A} = \frac{5.31 \times 10^{-7}\ \text{C}}{(2.66 \times 10^{-11}\ \text{C}^2/\text{N} \cdot \text{m}^2)(2.00 \times 10^{-1}\ \text{m}^2)}$$

$= 1.00 \times 10^{5}\ \text{V/m},$

or

$$E = \frac{\sigma - \sigma_i}{\epsilon_0} = \frac{Q - Q_i}{\epsilon_0 A}$$

$$= \frac{(5.31 - 3.54) \times 10^{-7}\ \text{C}}{(8.85 \times 10^{-12}\ \text{C}^2/\text{N} \cdot \text{m}^2)(2.00 \times 10^{-1}\ \text{m}^2)}$$

$= 1.00 \times 10^{5}\ \text{V/m};$

or, from Eq. (25–14),

$$E = \frac{E_0}{K} = \frac{3.00 \times 10^{5}\ \text{V/m}}{3.00} = 1.00 \times 10^{5}\ \text{V/m}.$$

EXAMPLE 25-11

Energy storage with and without a dielectric Find the total energy stored in the electric field of the capacitor in Example 25–10 and the energy density, both before and after the dielectric sheet is inserted.

SOLUTION Let the original energy be U_0, and let the energy with the dielectric in place be U. From Eq. (25–9),

$$U_0 = \frac{1}{2} C_0 V_0{}^2 = \frac{1}{2}(1.77 \times 10^{-10}\ \text{F})(3000\ \text{V})^2 = 7.97 \times 10^{-4}\ \text{J},$$

$$U = \frac{1}{2} CV^2 = \frac{1}{2}(5.31 \times 10^{-10}\ \text{F})(1000\ \text{V})^2 = 2.66 \times 10^{-4}\ \text{J}.$$

The final energy is one third of the original energy.

The volume is $(0.200\ \text{m}^2)(0.0100\ \text{m}) = 0.00200\ \text{m}^3$. The original energy density (energy per unit volume) is

$$u_0 = \frac{7.97 \times 10^{-4}\ \text{J}}{0.00200\ \text{m}^3} = 0.398\ \text{J/m}^3.$$

Alternatively,

$$u_0 = \frac{1}{2}\epsilon_0 E_0{}^2 = \frac{1}{2}(8.85 \times 10^{-12}\ \text{C}^2/\text{N} \cdot \text{m}^2)(3.0 \times 10^{5}\ \text{N/C})^2$$

$= 0.398\ \text{J/m}^3.$

With the dielectric in place, from Eq. (25–19),

$$u = \frac{1}{2}\epsilon E^2 = \frac{1}{2}(2.66 \times 10^{-11}\ \text{C}^2/\text{N} \cdot \text{m}^2)(1.00 \times 10^{5}\ \text{N/C})^2$$

$= 0.133\ \text{J/m}^3.$

The energy density with the dielectric is one third of the original energy density. Do you get the same result for u if you divide U, the total energy with the dielectric, by the volume 0.00200 m³?

We can generalize the results of this example. When a dielectric is inserted into a capacitor while the charge on each plate remains the same, the permittivity ϵ increases by a factor of K

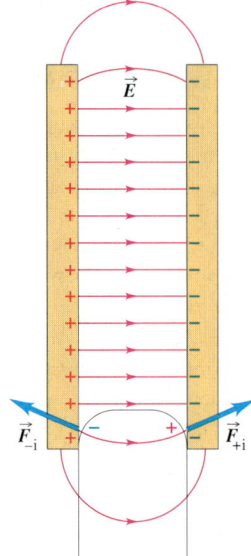

25–11 The fringing field at the edges of the capacitor exerts forces $\vec{F}_{-i}$ and $\vec{F}_{+i}$ on the negative and positive induced surface charges of a dielectric, pulling the dielectric into the capacitor.

(the dielectric constant), the electric field decreases by a factor of $1/K$, and the energy density $u = \frac{1}{2}\epsilon E^2$ decreases by a factor of $1/K$. Where did the energy go? The answer lies in the fringing field at the edges of a real parallel-plate capacitor. As Fig. 25–11 shows, that field exerts an attractive force on the dielectric, tending to pull the dielectric into the space between the plates and doing work on it as it does so. We could attach a spring to the bottom of the dielectric in Fig. 25–11 and use this force to stretch the spring. Because work is done by the field, the field energy density decreases.

TABLE 25–2
DIELECTRIC CONSTANT AND DIELECTRIC STRENGTH OF SOME INSULATING MATERIALS

MATERIAL	DIELECTRIC CONSTANT, K	DIELECTRIC STRENGTH, E_{max} (V/m)
Polycarbonate	2.8	3×10^7
Polyester	3.3	6×10^7
Polypropylene	2.2	7×10^7
Polystyrene	2.6	2×10^7
Pyrex glass	4.7	1×10^7

DIELECTRIC BREAKDOWN

We mentioned earlier that when any dielectric material is subjected to a sufficiently strong electric field, *dielectric breakdown* takes place and the dielectric becomes a conductor. This occurs when the electric field is so strong that electrons are ripped loose from their molecules and crash into other molecules, liberating even more electrons. This avalanche of moving charge, forming a spark or arc discharge, often starts quite suddenly.

Because of dielectric breakdown, capacitors always have maximum voltage ratings. When a capacitor is subjected to excessive voltage, an arc may form through a layer of dielectric, burning or melting a hole in it. This arc creates a conducting path (a short circuit) between the conductors. If a conducting path remains after the arc is extinguished, the device is rendered permanently useless as a capacitor.

The maximum electric-field magnitude that a material can withstand without the occurrence of breakdown is called its **dielectric strength.** This quantity is affected significantly by temperature, trace impurities, small irregularities in the metal electrodes, and other factors that are difficult to control. For this reason we can give only approximate figures for dielectric strengths. The dielectric strength of dry air is about 3×10^6 V/m. A few typical values of dielectric strength for common insulating materials are shown in Table 25–2. Note that the values are all substantially greater than the value for air. For example, a layer of polycarbonate 0.01 mm thick (about the smallest practical thickness) has ten times the dielectric strength of air and can withstand a maximum voltage of about $(3 \times 10^7 \text{ V/m})(1 \times 10^{-5} \text{ m}) = 300$ V.

A very strong electric field caused dielectric breakdown in a block of Plexiglas. The resulting flow of charge etched this pattern into the block.

*25–6 MOLECULAR MODEL OF INDUCED CHARGE

In Section 25–5 we discussed induced surface charges on a dielectric in an electric field. Now let's look at how these surface charges can come about. If the material were a *conductor,* the answer would be simple. Conductors contain charge that is free to move, and when an electric field is present, some of the charge redistributes itself on the surface so that there is no electric field inside the conductor. But an ideal dielectric has *no* charges that are free to move, so how can a surface charge occur?

To understand this, we have to look again at rearrangement of charge at the *molecular* level. Some molecules, such as H_2O and N_2O, have equal amounts of positive and negative charge but a lopsided distribution, with excess positive charge concentrated on one side of the molecule and negative charge on the other. As we described in Section 22–9, such an arrangement is called an *electric dipole,* and the molecule is called a *polar molecule.* When no electric field is present in a gas or liquid with polar molecules, the molecules are oriented randomly (Fig. 25–12a). When they are placed in an electric field, however, they tend to orient themselves as in Fig. 25–12b, as a result of the electric-field torques described in Section 22–9. Because of thermal agitation, the alignment of the molecules with $\vec{E}$ is not perfect.

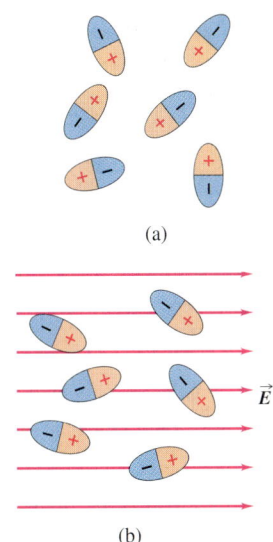

25–12 (a) Polar molecules have random orientations when there is no applied electric field. (b) The molecules tend to line up with an applied electric field $\vec{E}$.

(a)

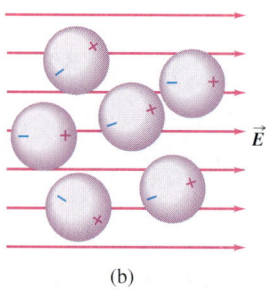

(b)

25–13 (a) Nonpolar molecules have their positive and negative charge centers at the same point. (b) These centers become separated slightly by the applied electric field $\vec{E}$.

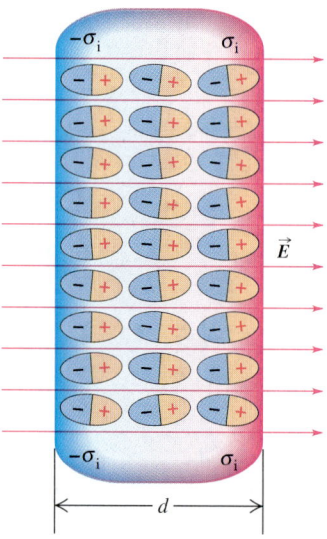

25–14 Polarization of a dielectric in an electric field gives rise to thin layers of bound charges on the surfaces, creating surface charge densities σ_i and $-\sigma_i$. The sizes of the molecules are greatly exaggerated for clarity.

Even a molecule that is *not* ordinarily polar *becomes* a dipole when it is placed in an electric field because the field pushes the positive charges in the molecules in the direction of the field and pushes the negative charges in the opposite direction. This causes a redistribution of charge within the molecule (Fig. 25–13). Such dipoles are called *induced* dipoles.

With either polar or nonpolar molecules the redistribution of charge caused by the field leads to the formation of a layer of charge on each surface of the dielectric material (Fig. 25–14). These layers are the surface charges described in Section 25–5; their surface charge density is denoted by σ_i. The charges are *not* free to move indefinitely, as they would be in a conductor, because each charge is bound to a molecule. They are in fact called **bound charges** to distinguish them from the **free charges** that are added to and removed from the conducting capacitor plates. In the interior of the material the net charge per unit volume remains zero. As we have seen, this redistribution of charge is called *polarization,* and we say that the material is *polarized.*

The four parts of Fig. 25–15 show the behavior of a slab of dielectric when it is inserted in the field between a pair of oppositely charged capacitor plates. Figure 25–15a shows the original field. Figure 25–15b is the situation after the dielectric has been inserted but before any rearrangement of charges has occurred. Figure 25–15c shows by thinner arrows the additional field set up in the dielectric by its induced surface charges. This field is *opposite* to the original field, but it is not great enough to cancel the original field completely because the charges in the dielectric are not free to move indefinitely. The resultant field in the dielectric, shown in Fig. 25–15d, is therefore decreased in magnitude. In the field-line representation, some of the field lines leaving the positive plate go through the dielectric, while others terminate on the induced charges on the faces of the dielectric.

As we discussed in Section 22–4, polarization is also the reason a charged body, such as an electrified plastic rod, can exert a force on an *uncharged* body such as a bit of paper or a pith ball. Figure 25–16 shows an uncharged dielectric sphere B in the radial field of a positively charged body A. The induced positive charges on B experience a force toward the right, while the force on the induced negative charges is toward the left. The negative charges are closer to A, and thus are in a stronger field, than are the positive charges. The force toward the left is stronger than that toward the right, and B is attracted toward A, even though its net charge is zero. The attraction occurs whether the

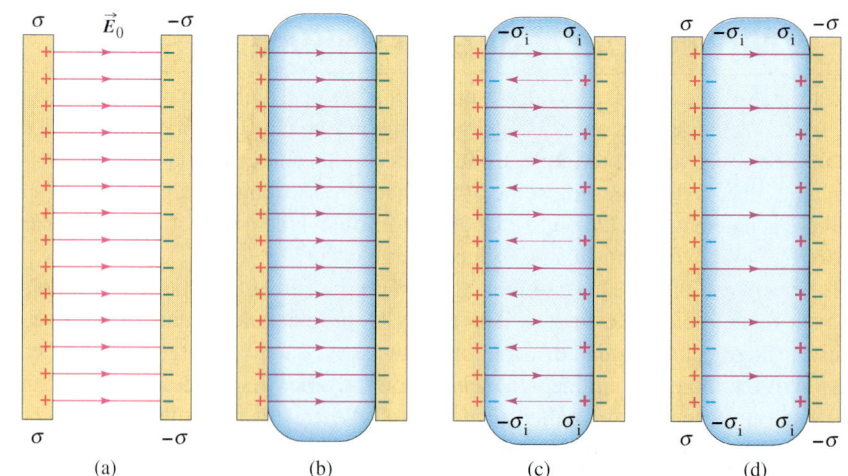

25–15 (a) Electric field of magnitude E_0 between two charged plates. (b) Introduction of a dielectric of dielectric constant K. (c) The induced surface charges and their field (thinner lines). (d) Resultant field of magnitude E_0/K when a dielectric is between charged plates.

sign of A's charge is positive or negative (see Fig. 22–5). Furthermore, the effect is not limited to dielectrics; an uncharged conducting body would be attracted in the same way.

*25-7 GAUSS'S LAW IN DIELECTRICS

We can extend the analysis of Section 25–5 to reformulate Gauss's law in a form that is particularly useful for dielectrics. Figure 25–17 is a close-up view of the left capacitor plate and left surface of the dielectric in Fig. 25–10. Let's apply Gauss's law to the rectangular box shown in cross section by the purple line; the surface area of the left and right sides is A. The left side is embedded in the conductor that forms the left capacitor plate, and so the electric field everywhere on that surface is zero. The right side is embedded in the dielectric, where the electric field has magnitude E, and $E_\perp = 0$ everywhere on the other four sides. The total charge enclosed, including both the charge on the capacitor plate and the induced charge on the dielectric surface, is $Q_{encl} = (\sigma - \sigma_i)A$, so Gauss's law gives

$$EA = \frac{(\sigma - \sigma_i)A}{\epsilon_0}. \qquad (25\text{--}20)$$

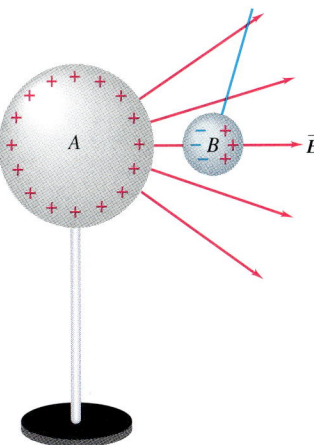

25–16 A neutral sphere B in the radial electric field of a positively charged sphere A is attracted to the charge because of polarization.

This equation is not very illuminating as it stands because it relates two unknown quantities, E inside the dielectric and the induced surface charge density σ_i. But now we can use Eq. (25–15), developed for this same situation, to simplify this equation by eliminating σ_i. Equation (25–15) is

$$\sigma_i = \sigma\left(1 - \frac{1}{K}\right), \qquad \text{or} \qquad \sigma - \sigma_i = \frac{\sigma}{K}.$$

Combining this with Eq. (25–20), we get

$$EA = \frac{\sigma A}{K\epsilon_0}, \qquad \text{or} \qquad KEA = \frac{\sigma A}{\epsilon_0}. \qquad (25\text{--}21)$$

Equation (25–21) says that the flux of $K\vec{E}$, not $\vec{E}$, through the Gaussian surface in Fig. 25–17 is equal to the enclosed *free* charge σA divided by ϵ_0. It turns out that for *any* Gaussian surface, whenever the induced charge is proportional to the electric field in the material, we can rewrite Gauss's law as

$$\oint K\vec{E} \cdot d\vec{A} = \frac{Q_{encl\text{-}free}}{\epsilon_0} \qquad \text{(Gauss's law in a dielectric)}, \qquad (25\text{--}22)$$

where $Q_{encl\text{-}free}$ is the total *free* charge (not bound charge) enclosed by the Gaussian surface. The significance of these results is that the right sides contain only the *free* charge on the conductor, not the bound (induced) charge. In fact, although we have not proved it, Eq. (25–22) remains valid even when different parts of the Gaussian surface are embedded in dielectrics having different values of K, provided that the value of K in each dielectric is independent of the electric field (usually the case for electric fields that are not too strong) and that we use the appropriate value of K for each point on the Gaussian surface.

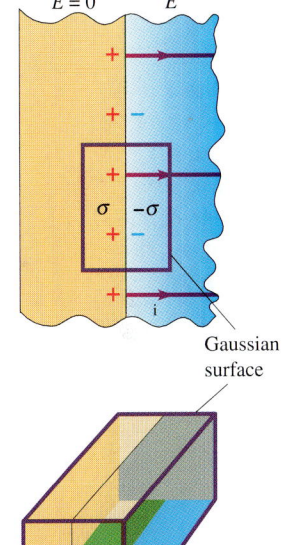

25–17 Gauss's law with a dielectric.

EXAMPLE 25-12

A spherical capacitor with dielectric In the spherical capacitor of Example 25–3 (Section 25–2), the volume between the concentric spherical conducting shells is filled with an insulating oil with dielectric constant K. Use Gauss's law to find the capacitance.

SOLUTION As we did in Example 25–3, we use a spherical Gaussian surface of radius r between the two spheres. Since a dielectric is present, we use Gauss's law in the form of Eq. (25–22). The spherical symmetry of the problem is not changed by the presence of the dielectric, so we have

$$\oint K\vec{E} \cdot d\vec{A} = \oint KE\, dA = KE \oint dA = (KE)\,(4\pi r^2) = \frac{Q}{\epsilon_0},$$

$$E = \frac{Q}{4\pi K\epsilon_0 r^2} = \frac{Q}{4\pi\epsilon r^2},$$

where $\epsilon = K\epsilon_0$ is the permittivity of the dielectric (introduced in Section 25–5). Compared to the case in which there is vacuum between the conducting shells, the electric field is reduced by a factor of $1/K$. The potential difference V_{ab} between the shells is likewise reduced by a factor of $1/K$, and so the capacitance $C = Q/V_{ab}$ is *increased* by a factor of K, just as for a parallel-plate capacitor when a dielectric is inserted. Using the result for the vacuum case in Example 25–3, we find that the capacitance with the dielectric is

$$C = \frac{4\pi K\epsilon_0 r_a r_b}{r_b - r_a} = \frac{4\pi\epsilon r_a r_b}{r_b - r_a}.$$

SUMMARY

- A capacitor is any pair of conductors separated by an insulating material. When the capacitor is charged, there are charges of equal magnitude Q and opposite sign on the two conductors, and the potential V_{ab} of the positively charged conductor with respect to the negatively charged conductor is proportional to Q. The capacitance C is defined as

$$C = \frac{Q}{V_{ab}}. \tag{25–1}$$

- A parallel-plate capacitor is made with two parallel conducting plates, each with area A, separated by a distance d. If they are separated by vacuum, the capacitance is

$$C = \frac{Q}{V_{ab}} = \epsilon_0 \frac{A}{d}. \tag{25–2}$$

The SI unit of capacitance is the farad, abbreviated F. One farad is one coulomb per volt: $1\ F = 1\ C/V$. Alternative units are

$$1\ F = 1\ C^2/N \cdot m = 1\ C^2/J.$$

The microfarad ($1\ \mu F = 10^{-6}\ F$) and the picofarad ($1\ pF = 10^{-12}\ F$) are commonly used.

- When capacitors with capacitances $C_1, C_2, C_3, \ldots$ are connected in series, the equivalent capacitance C_{eq} is

$$\frac{1}{C_{eq}} = \frac{1}{C_1} + \frac{1}{C_2} + \frac{1}{C_3} + \cdots \qquad \text{(capacitors in series).} \tag{25–5}$$

When capacitors are connected in parallel, the equivalent capacitance C_{eq} is

$$C_{eq} = C_1 + C_2 + C_3 + \cdots \qquad \text{(capacitors in parallel).} \tag{25–7}$$

- The energy U required to charge a capacitor C to a potential difference V and a charge Q is equal to the energy stored in the capacitor and is given by

$$U = \frac{Q^2}{2C} = \frac{1}{2}CV^2 = \frac{1}{2}QV. \tag{25–9}$$

This energy can be thought of as residing in the electric field between the conductors; the energy density u (energy per unit volume) is

$$u = \frac{1}{2}\epsilon_0 E^2. \tag{25–11}$$

- When the space between the conductors is filled with a dielectric material, the capacitance increases by a factor K, called the dielectric constant of the material.

For a parallel-plate capacitor with such a dielectric filling the space between its plates, the capacitance is

$$C = KC_0 = K\epsilon_0 \frac{A}{d} = \epsilon \frac{A}{d}, \qquad (25\text{–}18)$$

where $\epsilon = K\epsilon_0$ is called the permittivity of the dielectric. For a fixed amount of charge on the capacitor plates, induced charges on the surface of the dielectric decrease the the electric field and potential difference between the plates by the same factor K. The surface charge results from polarization, a microscopic rearrangement of charge in the dielectric due to the reorientation of polar molecules in an applied electric field or the creation of induced dipole moments in a nonpolar material. Under sufficiently strong fields, dielectrics become conductors; this is called dielectric breakdown. The maximum electric field magnitude that a material can withstand without breakdown is called its dielectric strength.

- The energy density u in an electric field in a dielectric is

$$u = \frac{1}{2} K\epsilon_0 E^2 = \frac{1}{2} \epsilon E^2. \qquad (25\text{–}19)$$

- Gauss's law can be reformulated for dielectrics as follows:

$$\oint K\vec{E} \cdot d\vec{A} = \frac{Q_{\text{encl-free}}}{\epsilon_0}, \qquad (25\text{–}22)$$

where $Q_{\text{encl-free}}$ includes only the free charge (not bound charge) enclosed by the Gaussian surface.

DISCUSSION QUESTIONS

Q25–1 Equation (25–2) shows that the capacitance of a parallel-plate capacitor becomes larger as the plate separation d decreases. However, there is a practical limit to how small d can be made, which places limits on how large C can be. Explain what sets the limit on d. (*Hint:* What happens to the magnitude of the electric field as $d \rightarrow 0$?)

Q25–2 A solid slab of metal is placed between the plates of a capacitor without touching either plate. Does the capacitance increase, decrease, or stay the same? Explain your reasoning.

Q25–3 Suppose the two plates of a capacitor have different areas. When the capacitor is charged by connecting it to a battery, do the charges on the two plates have equal magnitude, or may they be different? Explain your reasoning.

Q25–4 At the Fermi National Accelerator Laboratory (Fermilab) in Illinois, protons are accelerated around a ring 2 km in radius to speeds that approach that of light. The energy for this is stored in capacitors the size of a house. When these capacitors are being charged, they make a very loud creaking sound. What is the origin of this sound?

Q25–5 In the parallel-plate capacitor of Fig. 25–2, suppose the plates are pulled apart so that the separation d is much larger than the size of the plates. a) Is it still accurate to say that the electric field between the plates is uniform? Why or why not? b) In the situation shown in Fig. 25–2, the potential difference between the plates is $V_{ab} = Qd/\epsilon_0 A$. If the plates are pulled apart

as described above, is V_{ab} more or less than this formula would say? Explain your reasoning. c) With the plates pulled apart as described above, is the capacitance more than, less than, or the same as that given by Eq. (25–2)? Explain your reasoning.

Q25–6 A parallel-plate capacitor is charged by being connected to a battery and is kept connected to the battery. The separation between the plates is then doubled. How does the electric field change? The charge on the plates? The total energy? Explain your reasoning.

Q25–7 A parallel-plate capacitor is charged by being connected to a battery and is then disconnected from the battery. The separation between the plates is then doubled. How does the electric field change? The potential difference? The total energy? Explain your reasoning.

Q25–8 According to the text, we can consider the energy in a charged capacitor to be located in the field between the plates. But suppose there is vacuum between the plates; can there be energy in a vacuum? Why or why not? What form could this energy take?

Q25–9 The charged plates of a capacitor attract each other, so to pull the plates farther apart requires work by some external force. What becomes of the energy added by this work? Explain your reasoning.

Q25–10 The two plates of a capacitor are given charges $\pm Q$. The capacitor is then disconnected from the charging device so that the charges on the plates can't change, and the capacitor is immersed in a tank of oil. Does the electric field between the plates increase, decrease, or stay the same? Explain your reasoning. How may this field be measured?

Q25–11 As shown in Table 25–1, water has a very large dielectric constant $K = 80.4$. Why do you think water is not commonly used as a dielectric in capacitors?

Q25–12 Is dielectric strength the same thing as dielectric constant? Explain any differences between the two quantities. Is there a simple relationship between dielectric strength and dielectric constant? (See Table 25–2.)

Q25–13 A capacitor made of aluminum foil strips separated by Mylar film was subjected to excessive voltage, and the resulting dielectric breakdown melted holes in the Mylar. After this the capacitance was found to be about the same as before, but the breakdown voltage was much less. Why?

Q25–14 Two capacitors have equal capacitance, but one has a higher maximum voltage rating than the other. Which capacitor is likely to be bulkier? Why?

Q25–15 The freshness of fish can be measured by placing a fish between the plates of a capacitor and measuring the capacitance. How does this work? (*Hint:* As time passes, the fish dries out. See Table 25–1.)

Q25–16 *Electrolytic* capacitors use as their dielectric an extremely thin layer of nonconducting oxide between a metal plate and a conducting solution. Discuss the advantage of such a capacitor over one constructed using a solid dielectric between the metal plates.

Q25–17 A parallel-plate capacitor is connected to a power supply that maintains a fixed potential difference between the plates. If a sheet of dielectric is then slid between the plates, what happens to i) the electric field between the plates, ii) the magnitude of charge on each plate, and iii) the energy stored in the capacitor? Now suppose that before the dielectric is inserted, the charged capacitor is disconnected from the power supply. In this case, what happens to i) the electric field between the plates, ii) the magnitude of charge on each plate, and iii) the energy stored in the capacitor? Explain any differences between the two situations.

Q25–18 Liquid dielectrics having polar molecules (such as water) always have dielectric constants that decrease with increasing temperature. Why?

Q25–19 A conductor is an extreme case of a dielectric, since if an electric field is applied to a conductor, charges are free to move within the conductor to set up "induced charges." What is the dielectric constant of a perfect conductor? Is it $K = 0$? $K \to \infty$? Or something in between? Explain your reasoning.

EXERCISES

SECTION 25–2 CAPACITORS AND CAPACITANCE

25–1 A capacitor has a capacitance of 7.28 μF. What amount of charge must be placed on each of its plates to make the potential difference between its plates equal to 25.0 V?

25–2 The plates of a parallel-plate capacitor are 3.28 mm apart, and each has an area of 12.2 cm^2. Each plate carries a charge of magnitude 4.35×10^{-8} C. The plates are in vacuum. a) What is the capacitance? b) What is the potential difference between the plates? c) What is the magnitude of the electric field between the plates?

25–3 A parallel-plate air capacitor of capacitance 245 pF has a charge of magnitude 0.148 μC on each plate. The plates are 0.328 mm apart. a) What is the potential difference between the plates? b) What is the area of each plate? c) What is the electric field magnitude between the plates? d) What is the surface-charge density on each plate?

25–4 A cylindrical capacitor like that in Example 25–4 (Section 25–2) has a capacitance per length of 31.5 pF/m. a) Find the ratio of the radii of the inner and outer conductors. b) If the potential difference between the inner and outer conductors is 2.60 V, what is the magnitude of the charge per unit length on the conductors?

25–5 A cylindrical capacitor has an inner conductor of radius 1.5 mm and an outer conductor of radius 3.5 mm. The two conductors are separated by vacuum, and the entire capacitor is 2.8 m long. a) What is the capacitance per unit length? b) The potential of the inner conductor is 350 mV higher than that of the

outer conductor. Find the charge (magnitude and sign) on both conductors.

25–6 A spherical capacitor is formed from two concentric spherical conducting shells separated by vacuum. The inner sphere has radius 15.0 cm and the capacitance is 116 pF. a) What is the radius of the outer sphere? b) If the potential difference between the two spheres is 220 V, what is the magnitude of charge on each sphere?

25–7 A spherical capacitor is formed from two concentric spherical conducting shells separated by vacuum. The inner sphere has radius 12.5 cm and the outer sphere has radius 14.8 cm. A potential difference of 120 V is applied to the capacitor. a) What is the capacitance of the capacitor? b) What is the magnitude of $\vec{E}$ at $r = 12.6$ cm, just outside the inner sphere? c) What is the magnitude of $\vec{E}$ at $r = 14.7$ cm, just inside the outer sphere? d) For a parallel-plate capacitor $\vec{E}$ is uniform in the region between the plates, except near the edges of the plates. Is this also true for a spherical capacitor?

SECTION 25–3 CAPACITORS IN SERIES AND PARALLEL

25–8 In the circuit shown in Fig. 25–18, $C_1 = 3.00$ μF, $C_2 = 5.00$ μF, and $C_3 = 6.00$ μF. The applied potential is $V_{ab} = +24.0$ V. Calculate a) the charge on each capacitor; b) the potential difference across each capacitor; c) the potential difference between points a and d.

25–9 In Fig. 25–19, each capacitor has $C = 4.00$ μF and $V_{ab} = +28.0$ V. Calculate a) the charge on each capacitor; b) the potential difference across each capacitor; c) the potential difference between points a and d.

FIGURE 25-18 Exercise 25-8.

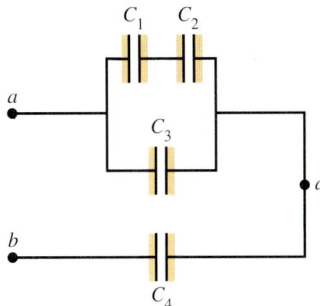

FIGURE 25-19 Exercise 25-9.

25-10 In Fig. 25-5a, let $C_1 = 3.00$ μF, $C_2 = 5.00$ μF, and $V_{ab} = +52.0$ V. Calculate a) the charge on each capacitor; b) the potential difference across each capacitor.

25-11 In Fig. 25-6a, let $C_1 = 3.00$ μF, $C_2 = 5.00$ μF, and $V_{ab} = +52.0$ V. Calculate a) the charge on each capacitor; b) the potential difference across each capacitor.

25-12 Two parallel-plate vacuum capacitors have plate spacings d_1 and d_2 and equal plate areas A. Show that when the capacitors are connected in series, the equivalent capacitance is the same as for a single capacitor with plate area A and spacing $d_1 + d_2$.

25-13 Two parallel-plate vacuum capacitors have areas A_1 and A_2 and equal plate spacings d. Show that when the capacitors are connected in parallel, the equivalent capacitance is the same as for a single capacitor with plate area $A_1 + A_2$ and spacing d.

25-14 For the situation shown in Fig. 25-7a, suppose $V_{ab} = +36$ V. Calculate a) the charge on each capacitor; b) the potential difference across each capacitor. c) The charge on the entire network is equal to $Q = C_{eq}V_{ab}$, where $C_{eq} = 6$ μF from Example 25-5 (Section 25-3). Explain why Q is equal to the charge on the 9-μF capacitor. Explain why Q is also equal to the sum of the charge on the 3-μF capacitor, the 11-μF capacitor, and $either$ the 12-μF capacitor or the 6-μF capacitor.

25-15 Suppose the 3-μF capacitor in Fig. 25-7a were removed and replaced by a different one, and that this changes the equivalent capacitance between points a and b to 8 μF. What is the capacitance of the replacement capacitor?

SECTION 25-4 ENERGY STORAGE IN CAPACITORS AND ELECTRIC-FIELD ENERGY

25-16 A parallel-plate air capacitor has a capacitance of 920 pF. The charge on each plate is 2.55 μC. a) What is the potential difference between the plates? b) If the charge is kept constant, what will be the potential difference between the plates if the separation is doubled? c) How much work is required to double the separation?

25-17 A 5.80-μF parallel-plate air capacitor has a plate separation of 5.00 mm and is charged to a potential difference of 400 V. Calculate the energy density in the region between the plates, in units of J/m^3.

25-18 An air capacitor is made from two flat parallel plates 1.50 mm apart. The magnitude of charge on each plate is 0.0180 μC when the potential difference is 200 V. a) What is the capacitance? b) What is the area of each plate? c) What maximum voltage can be applied without dielectric breakdown? (Dielectric breakdown for air occurs at an electric field strength of 3.0×10^6 V/m.) d) When the charge is 0.0180 μC, what total energy is stored?

25-19 A 450-μF capacitor is charged to 295 V. Then a wire is connected between the plates. How many joules of thermal energy are produced as the capacitor discharges if all of the energy that was stored goes into heating the wire?

25-20 A capacitor of capacitance C is charged to a potential difference V_0. The terminals of the charged capacitor are then connected to those of an uncharged capacitor of capacitance $C/2$. Compute a) the original charge of the system; b) the final potential difference across each capacitor; c) the final energy of the system; d) the decrease in energy when the capacitors are connected. e) Where did the "lost" energy go?

25-21 A parallel-plate vacuum capacitor with plate area A and separation x has charges $+Q$ and $-Q$ on its plates. The capacitor is disconnected from the source of charge, so the charge on each plate remains fixed. a) What is the total energy stored in the capacitor? b) The plates are pulled apart an additional distance dx. What is the change in the stored energy? c) If F is the force with which the plates attract each other, then the change in the stored energy must equal the work $dW = Fdx$ done in pulling the plates apart. Find an expression for F. d) Explain why F is not equal to QE, where E is the electric field between the plates.

25-22 A parallel-plate vacuum capacitor has 8.38 J of energy stored in it. The separation between the plates is 2.30 mm. If the separation is decreased to 1.15 mm, what is the energy stored if a) the capacitor is disconnected from the potential source so the charge on the plates remains constant? b) the capacitor remains connected to the potential source so the potential difference between the plates remains constant?

25-23 A cylindrical air capacitor of length 15.0 m stores 3.20×10^{-9} J of energy when the potential difference between the two conductors is 4.00 V. a) Calculate the magnitude of the charge on each conductor. b) Calculate the ratio of the radii of the inner and outer conductors.

25-24 A spherical capacitor is formed from two concentric spherical conducting shells separated by vacuum. The inner sphere has radius 10.0 cm, and the separation between the spheres is 1.50 cm. The magnitude of the charge on each sphere is 3.30 nC. a) What is the potential difference between the two spheres? b) What is the electric-field energy stored in the capacitor?

25–25 Consider the spherical capacitor of Exercise 25–7. a) What is the energy density at $r = 12.6$ cm, just outside the inner sphere? b) What is the energy density at $r = 14.7$ cm, just inside the outer sphere? c) For a parallel-plate capacitor the energy density is uniform in the region between the plates, except near the edges of the plates. Is this also true for a spherical capacitor?

25–26 a) Find the electric-field energy density at a point 12.0 cm from an isolated point charge $q = 8.00$ nC. b) If the point charge in part (a) were instead $q = -8.00$ nC, what effect would this have on the electric-field energy density? Explain your reasoning.

SECTION 25–5 DIELECTRICS

25–27 Two parallel plates have equal and opposite charges. When the space between the plates is evacuated, the electric field is $E = 3.20 \times 10^5$ V/m. When the space is filled with dielectric, the electric field is $E = 2.50 \times 10^5$ V/m. a) What is the charge density on each surface of the dielectric? b) What is the dielectric constant?

25–28 Two identical, oppositely charged conducting plates of area 2.50 cm^2 are separated by a dielectric 1.80 mm thick, with a dielectric constant of 3.60. The resultant electric field in the dielectric is 1.20×10^6 V/m. Compute a) the charge per unit area on the conducting plate; b) the charge per unit area on the surfaces of the dielectric; c) the total electric-field energy stored in the capacitor.

25–29 The dielectric to be used in a parallel-plate capacitor has a dielectric constant of 3.60 and a dielectric strength of 1.60×10^7 V/m. The capacitor is to have a capacitance of 1.25×10^{-9} F and must be able to withstand a maximum potential difference of 5500 V. What is the minimum area the plates of the capacitor may have?

25–30 Show that Eq. (25–19) holds for a parallel-plate capacitor with a dielectric material between the plates. Use a derivation analogous to that used for Eq. (25–11).

25–31 A capacitor has parallel plates of area 12 cm^2 separated by 2.0 mm. The space between the plates is filled with poly-

styrene (see Table 25–2). a) Find the permittivity of polystyrene. b) Find the maximum permissible voltage across the capacitor to avoid dielectric breakdown. c) When the voltage equals the value found in part (b), find the surface-charge density on each plate and the induced surface-charge density on the surface of the dielectric.

25–32 A constant potential difference of 12 V is maintained between the terminals of a 0.25-μF parallel-plate air capacitor. a) A sheet of Mylar is inserted between the plates of the capacitor, completely filling the space between the plates. When this is done, how much additional charge flows onto the positive plate of the capacitor? See Table 25–1. b) What is the total induced charge on either face of the Mylar sheet? c) What effect does the Mylar sheet have on the electric field between the plates? Explain how you can reconcile this with the increase in charge on the plates, which acts to *increase* the electric field.

25–33 When a 360-nF air capacitor (1 nF = 10^{-9} F) is connected to a power supply, the energy stored in the capacitor is 1.85×10^{-5} J. While the capacitor is kept connected to the power supply, a slab of dielectric is inserted that completely fills the space between the plates. This increases the stored energy by 2.32×10^{-5} J. a) What is the potential difference between the capacitor plates? b) What is the dielectric constant of the slab?

*SECTION 25–7 GAUSS'S LAW IN DIELECTRICS

*25–34 A point charge q is imbedded in a solid material of dielectric constant K. a) Use Gauss's law as stated in Eq. (25–22) to find the magnitude of the electric field due to the point charge at a distance d from the charge. b) Use your result from part (a) and Gauss's law in its original form as given in Eq. (23–8) to determine the *total* charge (free *and* bound) within a sphere of radius d centered on the point charge q. c) Find the total bound charge within the sphere described in part (b).

*25–35 A parallel-plate capacitor has the volume between its plates filled with plastic with dielectric constant K. The magnitude of the charge on each plate is Q. Each plate has area A, and the distance between the plates is d. a) Use Gauss's law as stated in Eq. (25–22) to calculate the magnitude of the electric field in the dielectric. b) Use the electric field determined in part (a) to calculate the potential difference between the two plates. c) Use the result of part (b) to determine the capacitance of the capacitor. Compare your result to Eq. (25–12).

PROBLEMS

25–36 A parallel-plate air capacitor is made by using two plates 16 cm square, spaced 4.7 mm apart. It is connected to a 12-V battery. a) What is the capacitance? b) What is the charge on each plate? c) What is the electric field between the plates? d) What is the energy stored in the capacitor? e) If the battery is disconnected and then the plates are pulled apart to a separation of 9.4 mm, what are the answers to parts (a), (b), (c), and (d)?

25–37 Suppose the battery in Problem 25–36 remains connected while the plates are pulled apart. What are the answers

then to parts (a), (b), (c), and (d) after the plates have been pulled apart?

25–38 Four identical, square conducting plates have sides of length L. All four plates lie parallel to the xy-plane, with corners at $(x, y) = (L/2, L/2), (-L/2, L/2), (L/2, -L/2)$, and $(-L/2, -L/2)$. Plate 1 is in the plane $z = 0$, plate 2 in the plane $z = d$, plate 3 in the plane $z = 2d$, and plate 4 in the plane $z = 3d$, where $d \ll L$. Plates 1 and 3 each carry a positive charge Q, and plates 2 and 4 each carry a negative charge $-Q$. There is vacuum between the plates. a) Find the total energy stored in this arrangement of plates and charges. b) Plates 2 and 3 are now interchanged without altering the charges that they carry. How much work has to be done to carry out this interchange? (*Hint:* Review the discussion of superposition in Example 22–13 (Section 22–7).)

25–39 Electronic flash units for cameras contain a capacitor for storing the energy used to produce the flash. In one such unit the flash lasts for $\frac{1}{675}$ s with an average light power output of 2.70×10^5 W. a) If the conversion of electrical energy to light is 95% efficient (the rest of the energy goes to thermal energy), how much energy must be stored in the capacitor for one flash? b) The capacitor has a potential difference between its plates of 125 V when the stored energy equals the value calculated in part (a). What is the capacitance?

25–40 In one type of computer keyboard, each key holds a small metal plate that serves as one plate of a parallel-plate air-filled capacitor. When the key is depressed, the plate separation decreases and the capacitance increases. Electronic circuitry detects the change in capacitance and thus detects that the key has been pressed. In one particular keyboard the area of each metal plate is 42.0 mm^2, and the separation between the plates is 0.700 mm before the key is depressed. If the circuitry can detect a change in capacitance of 0.250 pF, how far must the key be depressed before the circuitry detects its depression?

25–41 Consider a cylindrical capacitor like that shown in Fig. 25–4. Let $d = r_b - r_a$ be the spacing between the inner and outer conductors. a) Let the radii of the two conductors be only slightly different, so that $d \ll r_a$. Show that the result derived in Example 25–4 (Section 25–2) for the capacitance of a cylindrical capacitor then reduces to Eq. (25–2), the equation for the capacitance of a parallel-plate capacitor, with A being the surface area of each cylinder. Use the result that $\ln(1 + z) \cong z$ for $|z| \ll 1$. b) Even though the earth is essentially spherical, its surface appears flat to us because its radius is so large. Use this idea to explain why the result of part (a) makes sense from a purely geometrical standpoint.

25–42 In Fig. 25–6a, let $C_1 = 9.0\ \mu$F, $C_2 = 4.0\ \mu$F, and $V_{ab} = 28$ V. Suppose the charged capacitors are disconnected from the source and from each other, then reconnected to each other with plates of *opposite* sign together. By how much does the energy of the system decrease?

25–43 A 4.00-μF capacitor and a 6.00-μF capacitor are connected in parallel across a 660-V supply line. a) Find the charge on each capacitor and the voltage across each. b) The charged capacitors are disconnected from the line and from each other, then reconnected to each other with terminals of unlike sign together. Find the final charge on each and the voltage across each.

25–44 Several 0.25-μF capacitors are available. The voltage across each is not to exceed 600 V. You need to make a capacitor with capacitance 0.25 μF to be connected across a potential difference of 960 V. a) Show in a diagram how an equivalent capacitor having the desired properties can be obtained. b) No dielectric is a perfect insulator, which would not permit the flow of any charge through its volume. Suppose that the dielectric in one of the capacitors in your diagram is a moderately good con-

ductor. What will happen in this case when your combination of capacitors is connected across the 960-V potential difference?

25–45 In Fig. 25–20, $C_1 = C_5 = 8.4\ \mu$F and $C_2 = C_3 = C_4 = 4.2\ \mu$F. The applied potential is $V_{ab} = 220$ V. a) What is the equivalent capacitance of the network between points a and b? b) Calculate the charge on each capacitor and the potential difference across each capacitor.

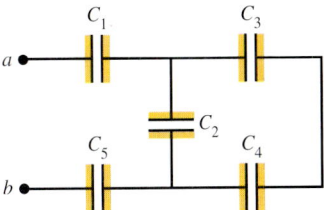

FIGURE 25–20 Problem 25–45.

25–46 The capacitors in Fig. 25–21 are initially uncharged and are connected as in the diagram with switch S open. The applied potential difference is $V_{ab} = +210$ V. a) What is the potential difference V_{cd}? b) What is the potential difference across each capacitor after switch S is closed? c) How much charge flowed through the switch when it was closed?

FIGURE 25–21 Problem 25–46.

25–47 Three capacitors having capacitances of 8.4, 8.4, and 4.2 μF are connected in series across a 36-V potential difference. a) What is the charge on the 4.2-μF capacitor? b) What is the total energy stored in all three capacitors? c) The capacitors are disconnected from the potential difference without allowing them to discharge. They are then reconnected in parallel with each other, with the positively charged plates connected together. What is the voltage across each capacitor in the parallel combination? d) What is the total energy now stored in the capacitors?

25–48 A 4.00-μF capacitor and a 6.00-μF capacitor are connected in series across a 660-V supply line. a) Find the charge on each capacitor and the voltage across each. b) The charged capacitors are disconnected from the line and from each other and then reconnected to each other, with terminals of like sign together. Find the final charge on each and the voltage across each.

25–49 In Fig. 25–22, each capacitance C_1 is 6.9 μF, and each capacitance C_2 is 4.6 μF. a) Compute the equivalent capacitance of the network between points a and b. b) Compute the charge on each of the three capacitors nearest a and b when $V_{ab} = 420$ V. c) With 420 V across a and b, compute V_{cd}.

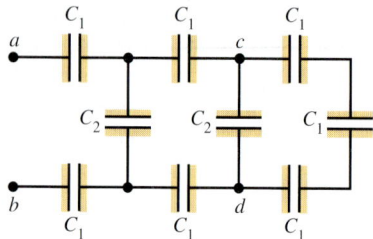

FIGURE 25–22 Problem 25–49.

25–50 An air capacitor is made by using two flat plates, each with area A, separated by a distance d. Then a metal slab having thickness a (less than d) and the same shape and size as the plates is inserted between them, parallel to the plates and not touching either plate. (See Fig. 25–23.) a) What is the capacitance of this arrangement? b) Express the capacitance as a multiple of the capacitance C_0 when the metal slab is not present. c) Discuss what happens to the capacitance in the limits $a \rightarrow 0$ and $a \rightarrow d$.

FIGURE 25–23 Problem 25–50.

25–51 a) Discuss how the concept of capacitance can also be applied to a *single* conductor. (*Hint:* In the relationship $C = Q/V_{ab}$, think of the second conductor as being located at infinity.) b) Use Eq. (25–1) to show that $C = 4\pi\epsilon_0 R$ for a solid conducting sphere of radius R. c) Use your result in part (b) to calculate the capacitance of the earth, which is a good conductor of radius 6380 km. Compare to typical capacitors used in electronic circuits, which have capacitances that range from 10 pF to 100 μF.

25–52 A solid conducting sphere of radius R carries a charge Q. Calculate the electric-field energy density at a point a distance r from the center of the sphere for a) $r < R$; b) $r > R$. c) Calculate the total electric-field energy associated with the charged sphere. (*Hint:* Consider a spherical shell of radius r and thickness dr, which has volume $dV = 4\pi r^2\, dr$, and find the energy stored in this volume. Then integrate from $r = 0$ to $r \rightarrow \infty$.) d) Explain why the result of part (c) can be interpreted as the amount of work required to assemble the charge Q on the sphere. e) By using Eq. (25–9) and the result of part (c), show that the capacitance of the sphere is as given in Problem 25–51.

25–53 A uniformly charged sphere of radius R has total charge Q, as described in Example 23–9 (Section 23–5). Calculate the electric-field energy density at a point a distance r from the center of the sphere for a) $r < R$; b) $r > R$. c) Calculate the total electric-field energy. See the hint in part (c) of Problem 25–52.

25–54 The inner cylinder of a long cylindrical capacitor has radius r_a and a linear charge density $+\lambda$. It is surrounded by a coaxial cylindrical conducting shell with inner radius r_b and lin-

ear charge density $-\lambda$ (Fig. 25–4). a) What is the energy density in the region between the conductors at a distance r from the axis? b) Integrate the energy density calculated in part (a) over the volume between the conductors in a length L of the capacitor to obtain the total electric-field energy per unit length. c) Use Eq. (25–9) and the capacitance per unit length calculated in Example 25–4 (Section 25–2) to calculate U/L. Does your result agree with that obtained in part (b)?

25–55 A parallel-plate capacitor has the space between the plates filled with two slabs of dielectric, one with constant K_1 and one with constant K_2. (See Fig. 25–24.) Each slab has thickness $d/2$, where d is the plate separation. Show that the capacitance is

$$C = \frac{2\epsilon_0 A}{d}\left(\frac{K_1 K_2}{K_1 + K_2}\right).$$

FIGURE 25–24 Problem 25–55.

25–56 A parallel-plate capacitor has the space between the plates filled with two slabs of dielectric, one with constant K_1 and one with constant K_2. (See Fig. 25–25.) The thickness of each slab is the same as the plate separation d, and each slab fills half of the volume between the plates. Show that the capacitance is

$$C = \frac{\epsilon_0 A(K_1 + K_2)}{2d}.$$

FIGURE 25–25 Problem 25–56.

25–57 Some cell walls in the human body have a layer of negative charge on the inside surface and a layer of positive charge of equal magnitude on the outside surface. Suppose that the surface charge densities are $\pm 0.50 \times 10^{-3}$ C/m^2, the cell wall is 5.0×10^{-9} m thick, and the cell-wall material has a dielectric constant of $K = 5.4$. a) Find the magnitude of $\vec{E}$ in the wall between the two charge layers. b) Find the potential difference between the inside and the outside of the cell. Which is at higher potential? c) A typical cell in the human body has volume 10^{-16} m^3. Estimate the total electric-field energy stored in the wall of a cell of this size. (*Hint:* Assume that the cell is spherical, and calculate the volume of the cell wall.)

25–58 Suppose that the parallel-plate capacitor described in Examples 25–10 and 25–11 (Section 25–5) remains connected to the 3000-V power supply while an insulating plastic sheet with $K = 2.50$ is inserted between the plates, completely filling the space between them. Compute a) the magnitude of charge Q

on each plate after the dielectric is inserted; b) the magnitude of induced charge Q_i on each face of the dielectric; c) the electric field E after the dielectric is inserted; d) the total energy stored in the electric field after the dielectric is inserted; e) the energy density after the dielectric is inserted. f) You should have found

in part (d) that the stored energy *increased* when the dielectric was inserted, whereas in Example 25–11 the stored energy *decreased* when this was done. Why is there a difference between the two cases? In the present case, where does the extra energy come from?

CHALLENGE PROBLEMS

25–59 Capacitors in networks cannot always be grouped into simple series or parallel combinations. As an example, Fig. 25–26a shows three capacitors C_x, C_y, and C_z in a *delta network*, so called because of its triangular shape. This network has *three* terminals a, b, and c and hence cannot be transformed into a single equivalent capacitor. It can be shown that as far as any effect on the external circuit is concerned, a delta network is equivalent to what is called a *Y network*. For example, the delta network of Fig. 25–26a can be replaced by the Y network of Fig. 25–26b. (The name "Y network" also refers to the shape of the network.) a) Show that the transformation equations that give C_1, C_2, and C_3 in terms of C_x, and C_y, and C_z are

$$C_1 = (C_xC_y + C_yC_z + C_zC_x)/C_x,$$
$$C_2 = (C_xC_y + C_yC_z + C_zC_x)/C_y,$$
$$C_3 = (C_xC_y + C_yC_z + C_zC_x)/C_z.$$

(*Hint:* The potential difference V_{ac} must be the same in both circuits, as V_{bc} must be. Also, the charge q_1 that flows from point a along the wire as indicated must be the same in both circuits, as must q_2. Obtain a relationship for V_{ac} as a function of q_1 and q_2 and the capacitances for each network, and obtain a separate relationship for V_{bc} as a function of the charges for each network. The coefficients of corresponding charges in corresponding equations must be the same for both networks.) b) For the network shown in Fig. 25–26c, determine the equivalent capacitance between the terminals at the left end of the network . (*Hint:* Use the delta-Y transformation derived in part (a). Use points a, b, and c to form the delta, and transform the delta into a Y. The capacitors can then be combined using the relationships for series and parallel combinations of capacitors.) c) Determine the charges of, and the potential differences across, each capacitor in Fig. 25–26c.

25–60 The parallel-plate air capacitor in Fig. 25–27 consists of two horizontal conducting plates of equal area A. The bottom plate rests on a fixed support, and the top plate is suspended by four springs with spring constant k, positioned at each of the four corners of the top plate as shown in the figure. When uncharged, the plates are separated by a distance z_0. A battery is connected to the plates and produces a potential difference V between them. This causes the plate separation to decrease to z. Neglect any fringing effects. a) Show that the electrostatic force between the charged plates has a magnitude $\epsilon_0 A V^2/2z^2$. (*Hint:* See Exercise 25–21.) b) Obtain an expression that relates the plate separation z to the potential difference V. The resulting equation will be cubic in z. c) Given the values $A = 0.300$ m^2, $z_0 = 1.20$ mm, $k = 25.0$ N/m, and $V = 120$ V, find the two values of z for which the top plate will be in equilibrium. (*Hint:* You can solve the cubic equation by plugging a trial value of z into the equation and then adjusting your guess until the equation is

satisfied to three significant figures. Locating the roots of the cubic equation graphically can help you in picking starting values of z for this trial-and-error procedure. One root of the cubic equation has a nonphysical negative value.) d) For each of the two values of z found in part (c), is the equilibrium stable or unstable? For stable equilibrium a small displacement of the object will give rise to a net force tending to return the object to the equilibrium position. For unstable equilibrium a small displacement gives rise to a net force that takes the object farther away from equilibrium.

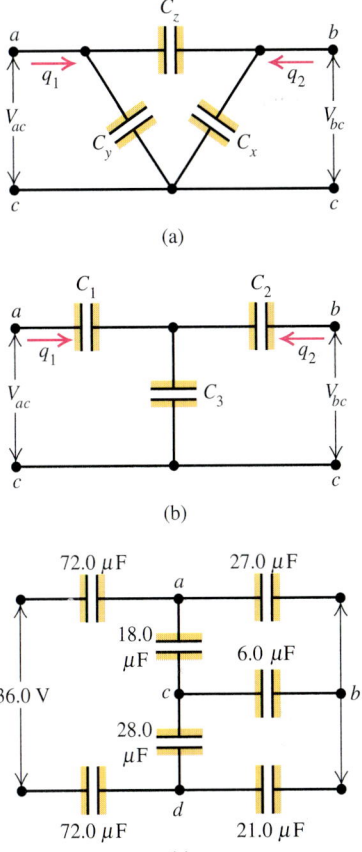

FIGURE 25–26 Challenge Problem 25–59.

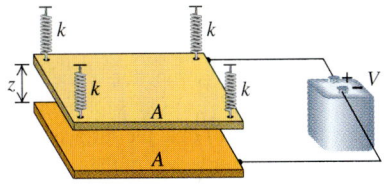

FIGURE 25–27 Challenge Problem 25–60.

25–61 Two square conducting plates with sides of length L are separated by a distance D. A dielectric slab with constant K with dimensions $L \times L \times D$ is inserted a distance x into the space between the plates, as shown in Fig. 25–28. a) Find the capacitance C of this system. (See Problem 25–56.) b) Suppose that the capacitor is connected to a battery that maintains a constant potential difference V between the plates. If the dielectric slab is inserted an additional distance dx into the space between the plates, show that the change in stored energy is

$$dU = +\frac{(K-1)\epsilon_0 V^2 L}{2D}\,dx.$$

c) Suppose that before the slab is moved by dx, the plates are disconnected from the battery, so that the charges on the plates remain constant. Determine the magnitude of the charge on each plate, and then show that when the slab is moved dx farther into the space between the plates, the stored energy changes by an amount that is the *negative* of the expression for dU given in part (b). d) If F is the force exerted on the slab by the charges on the plates, then dU should equal the work done *against* this force to move the slab a distance dx. Thus $dU = -F\,dx$. Show that applying this expression to the result of part (b) suggests that the electric force on the slab pushes it *out* of the capacitor, while the result of part (c) suggests that the force pulls the slab *into* the capacitor. e) Figure 25–11 shows that the force in fact pulls the slab into the capacitor. Explain why the result of part (b) gives an incorrect answer for the direction of this force, and calculate the magnitude of the force. (This method does not require knowledge of the nature of the fringing field.)

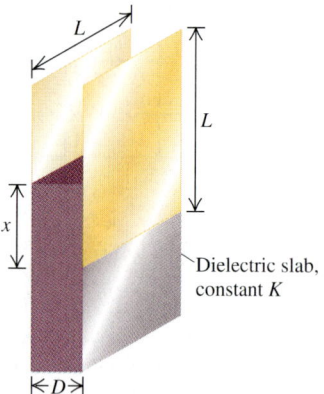

FIGURE 25–28 Challenge Problem 25–61.

25–62 An isolated spherical capacitor has charge $+Q$ on its inner conductor (radius r_a) and charge $-Q$ on its outer conductor (radius r_b). Half of the volume between the two conductors is then filled with a liquid dielectric of constant K, as shown in cross section in Fig. 25–29. a) Find the capacitance of the half-filled capacitor. b) Find the magnitude of $\vec{E}$ in the volume between the two conductors as a function of the distance r from the center of the capacitor. Give answers for both the upper and lower halves of this volume. c) Find the surface density of free charge on the upper and lower halves of the inner and outer conductors. d) Find the surface density of bound charge on the inner ($r = r_a$) and outer ($r = r_b$) surfaces of the dielectric. e) What is the

surface density of bound charge on the flat surface of the dielectric? Explain.

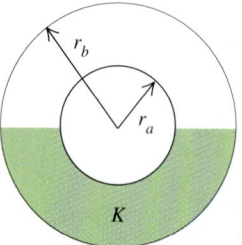

FIGURE 25–29 Challenge Problem 25–62.

25–63 Three square metal plates A, B, and C, each 12.0 cm on a side and 1.50 mm thick, are arranged as in Fig. 25–30. The plates are separated by sheets of paper 0.45 mm thick and with dielectric constant 4.2. The outer plates are connected together and connected to point b. The inner plate is connected to point a. a) Copy the diagram and show by plus and minus signs the charge distribution on the plates when point a is maintained at a positive potential relative to point b. b) What is the capacitance between points a and b?

FIGURE 25–30 Challenge Problem 25–63.

25–64 A fuel gauge uses a capacitor to determine the height of the fuel in a tank. The effective dielectric constant K_{eff} changes from a value of 1 when the tank is empty to a value of K, the dielectric constant of the fuel, when the tank is full. The appropriate electronic circuitry can determine the effective dielectric constant of the combined air and fuel between the capacitor plates. Each of the two rectangular plates has a width w and a length L (Fig. 25–31). The height of the fuel between the plates is h. Neglect any fringing effects. a) Derive an expression for K_{eff} as a function of h. b) What is the effective dielectric constant for a tank one-quarter full, one-half full, and three-quarters full if the fuel is gasoline ($K = 1.95$)? c) Repeat part (b) for methanol ($K = 33.0$). d) For which fuel is this fuel gauge more practical?

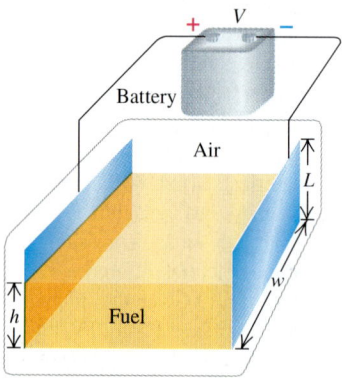

FIGURE 25–31 Challenge Problem 25–64.

26

Current, Resistance, and Electromotive Force

Pathfinder-Plus is an unmanned solar-powered aircraft. Solar cells on its 36-m-long wing provide more than 12 kilowatts of electric power to its motors, enough to lift the aircraft to 24,000 m (80,000 ft) and to maintain flight all day. Future solar aircraft with on-board energy storage will remain aloft for weeks at a time for environmental research or for relaying telecommunications.

26–1 INTRODUCTION

In the past four chapters we studied the interactions of electric charges *at rest;* now we're ready to study charges *in motion.* An *electric current* consists of charges in motion from one region to another. When this motion takes place within a conducting path that forms a closed loop, the path is called an *electric circuit.*

Fundamentally, electric circuits are a means for conveying *energy* from one place to another. As charged particles move within a circuit, electric potential energy is transferred from a source (such as a battery or generator) to a device in which that energy is either stored or converted to another form: into sound in a stereo system or into heat and light in a toaster or light bulb. From a technological standpoint, electric circuits are useful because they allow energy to be transported without any moving parts (other than the moving charged particles themselves). Electric circuits are at the heart of flashlights, CD players, computers, radio and television transmitters and receivers, and household and industrial power distribution systems. The nervous systems of animals and humans are specialized electric circuits that carry vital signals from one part of the body to another.

In this chapter we will study the basic properties of electric currents. To understand the behavior of currents in electric circuits, we'll describe the properties of conductors and how they depend on temperature. We'll learn why a short, fat, cold copper wire is a better conductor than a long, skinny, hot steel wire. We'll study the properties of batteries and how they cause current and energy transfer in a circuit. In this analysis we will use the concepts of current, potential difference (or voltage), resistance, and electromotive force. Finally, we'll look at electric current in a material from a microscopic viewpoint.

26–2 CURRENT

A **current** is any motion of charge from one region to another. In this section we'll discuss currents in conducting materials. The vast majority of technological applications of charges in motion involve currents of this kind.

In electrostatic situations (discussed in Chapters 22 through 25) the electric field is zero everywhere within the conductor, and there is *no* current. However, this does not mean that all charges within the conductor are at rest. In an ordinary metal such as copper or aluminum, some of the electrons are free to move within the conducting material. These free electrons move randomly in all directions, somewhat like the molecules of a gas but with much greater speeds, of the order of 10^6 m/s. The electrons nonetheless do not escape from the conducting material, because they are attracted to the positive ions of the material. The motion of the electrons is random, so there is no *net* flow of charge in any direction and hence no current.

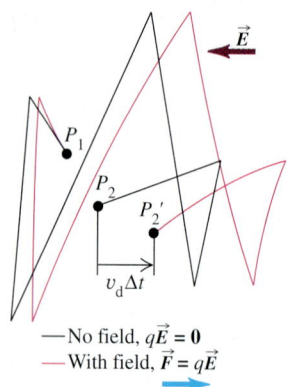

26–1 If there is no electric field inside a conducting material, charged particles move randomly within the material; the electron whose trajectory is shown in black may move from point P_1 to point P_2 after a time Δt. Because the motions of different electrons are random, there is no net flow of charge in any direction. If a field $\vec{E}$ is present, the electric force $\vec{F} = q\vec{E}$ imposes a small drift (greatly exaggerated here) on an electron's random motion. With the field the electron ends up at point P_2', a distance $v_d \Delta t$ from P_2 in the direction of the force. The electron has a negative charge q, so the force $\vec{F} = q\vec{E}$ is in the direction opposite that of the field $\vec{E}$.

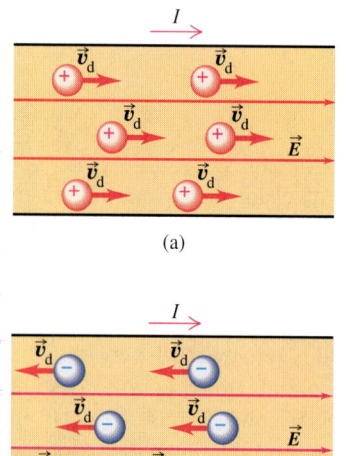

26–2 Positive charges moving in the direction of the electric field $\vec{E}$ produce the same current as the same number of negative charges of the same magnitude moving at the same speed in the direction opposite to the field.

Now consider what happens if a constant, steady electric field $\vec{E}$ is established inside a conductor. (We'll see later how this can be done.) A charged particle (such as a free electron) inside the conducting material is then subjected to a steady force $\vec{F} = q\vec{E}$. If the charged particle were moving in *vacuum,* this steady force would cause a steady acceleration in the direction of $\vec{F}$, and after a time the charged particle would be moving in that direction at high speed. But a charged particle moving in a *conductor* undergoes frequent collisions with the massive, nearly stationary ions of the material. In each such collision the particle's direction of motion undergoes a random change. The net effect of the electric field $\vec{E}$ is that in addition to the random motion of the charged particles within the conductor, there is also a very slow net motion or *drift* of the moving charged particles as a group in the direction of the electric force $\vec{F} = q\vec{E}$ (Fig. 26–1). This motion is described in terms of the **drift velocity** $\vec{v}_d$ of the particles. As a result, there is a net current in the conductor. As mentioned above, the random motion has a very large average speed; by contrast, the drift speed is very slow, often of the order of 10^{-4} m/s.

The drift of moving charges through a conductor can be interpreted in terms of work and energy. The electric field $\vec{E}$ does work on the moving charges. The resulting kinetic energy is transferred to the material of the conductor by means of collisions with the ions, which vibrate about their equilibrium positions in the crystalline structure of the conductor. This energy transfer increases the average vibrational energy of the ions and therefore the temperature of the material. Thus much of the work done by the electric field goes into heating the conductor, *not* into making the moving charges move ever faster and faster. This heating is sometimes useful, as in an electric toaster, but in many situations is simply an unavoidable by-product of current flow.

In different current-carrying materials the charges of the moving particles may be positive or negative. In metals the moving charges are always (negative) electrons, while in an ionized gas (plasma) or an ionic solution the moving charges may include both electrons and positively charged ions. In a semiconductor material such as germanium or silicon, conduction is partly by electrons and partly by motion of *vacancies,* also known as *holes;* these are sites of missing electrons and act like positive charges.

Figure 26–2 shows segments of two different current-carrying materials. In Fig. 26–2a the moving charges are positive, the electric force is in the same direction as $\vec{E}$, and the drift velocity $\vec{v}_d$ is from left to right. In Fig. 26–2b the charges are negative, the electric force is opposite to $\vec{E}$, and the drift velocity $\vec{v}_d$ is from right to left. In either case there is a net flow of positive charge from left to right, and positive charges end up to the right of negative ones. We *define* the current, denoted by I, to be in the direction in which there is a flow of *positive* charge. Thus we describe currents as though they consisted entirely of positive charge flow, even in cases in which we know that the actual current is due to electrons. Hence the current is to the right in both Fig. 26–2a and Fig. 26–2b. This choice or convention for the direction of current flow is called **conventional current.** While the direction of the conventional current is *not* necessarily the same as the direction in which charged particles are actually moving, we'll find that the sign of the moving charges is of little importance in analyzing electric circuits.

Figure 26–3 shows a segment of a conductor in which a current is flowing. We consider the moving charges to be *positive,* so they are moving in the same direction as the current. We define the current through the cross-section area A to be *the net charge flowing through the area per unit time.* Thus if a net charge dQ flows through an area in a time dt, the current I through the area is

$$I = \frac{dQ}{dt} \qquad \text{(definition of current)}. \qquad (26\text{–}1)$$

CAUTION ▶ Although we refer to the *direction* of a current, current as defined by Eq. (26–1) is *not* a vector quantity. In a current-carrying wire, the current is always

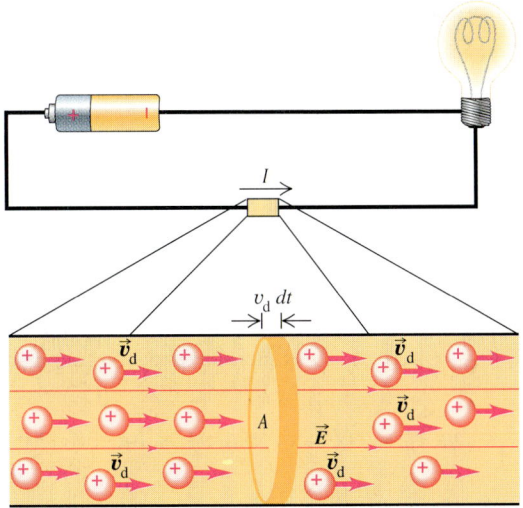

26–3 The current through the cross-section area A is the time rate of charge transfer through A. The random motion of each moving charged particle averages to zero, and the current is in the same direction as $\vec{E}$. If the moving charges are positive, as shown here, the drift velocity $\vec{v}_d$ is in the same direction as the current and $\vec{E}$; if the moving charges are negative, the drift velocity is in the opposite direction.

along the length of the wire, regardless of whether the wire is straight or curved. No single vector could describe motion along a curved path, which is why current is not a vector. We'll usually describe the direction of current either in words (as in "the current flows clockwise around the circuit") or by choosing a current to be positive if it flows in one direction along a conductor and negative if it flows in the other direction. ◄

The SI unit of current is the **ampere;** one ampere is defined to be *one coulomb per second* (1 A = 1 C/s). This unit is named in honor of the French scientist André Marie Ampère (1775–1836). When an ordinary flashlight (D-cell size) is turned on, the current in the flashlight is about 0.5 to 1 A; the current in the wires of a starter motor used to start a car engine is around 200 A. Currents in radio and television circuits are usually expressed in *milliamperes* (1 mA = 10^{-3} A) or *microamperes* (1 μA = 10^{-6} A), and currents in computer circuits are expressed in *nanoamperes* (1 nA = 10^{-9} A) or *picoamperes* (1 pA = 10^{-12} A).

CURRENT, DRIFT VELOCITY, AND CURRENT DENSITY

We can express current in terms of the drift velocity of the moving charges. Let's consider again the situation of Fig. 26–3, a conductor with cross-section area A and an electric field $\vec{E}$ directed from left to right. To begin with, we'll assume that the free charges in the conductor are positive; then the drift velocity is in the same direction as the field.

Suppose there are n charged particles per unit volume. We call n the **concentration** of particles; its SI unit is m^{-3}. Assume that all the particles move with the same drift velocity with magnitude v_d. In a time interval dt, each particle moves a distance $v_d\,dt$. The particles that flow out of the right end of the shaded cylinder with length $v_d\,dt$ during dt are the particles that were within this cylinder at the beginning of the interval dt. The volume of the cylinder is $Av_d\,dt$, and the number of particles within it is $nAv_d\,dt$. If each particle has a charge q, the charge dQ that flows out of the end of the cylinder during time dt is

$$dQ = q(nAv_d\,dt) = nqv_dA\,dt,$$

and the current is

$$I = \frac{dQ}{dt} = nqv_dA.$$

The current *per unit cross-section area* is called the **current density** J:

$$J = \frac{I}{A} = nqv_d.$$

The units of current density are amperes per square meter (A/m^2).

If the moving charges are negative rather than positive, as in Fig. 26–2b, the drift velocity is opposite to $\vec{E}$. But the *current* is still in the same direction as $\vec{E}$ at each point in the conductor. Hence the current I and current density J don't depend on the sign of the charge, and so in the above expressions for I and J we replace the charge q by its absolute value $|q|$:

$$I = \frac{dQ}{dt} = n|q|v_d A \qquad \text{(general expression for current)}, \qquad (26\text{–}2)$$

$$J = \frac{I}{A} = n|q|v_d \qquad \text{(general expression for current density)}. \qquad (26\text{–}3)$$

The current in a conductor is the product of the concentration of moving charged particles, the magnitude of charge of each such particle, the magnitude of the drift velocity, and the cross-section area of the conductor.

We can also define a *vector* current density $\vec{J}$ that includes the direction of the drift velocity:

$$\vec{J} = nq\vec{v}_d \qquad \text{(vector current density)}. \qquad (26\text{–}4)$$

There are *no* absolute value signs in Eq. (26–4). If q is positive, $\vec{v}_d$ is in the same direction as $\vec{E}$, and if q is negative, $\vec{v}_d$ is opposite to $\vec{E}$; in either case, $\vec{J}$ is in the same direction as $\vec{E}$. Equation (26–3) gives the *magnitude J* of the vector current density $\vec{J}$.

In general, a conductor may contain several different kinds of moving charged particles having charges $q_1, q_2, \ldots$, concentrations $n_1, n_2, \ldots$, and drift velocities with magnitudes $v_{d1}, v_{d2}, \ldots$. An example is current flow in an ionic solution. In a sodium chloride solution, current can be carried by both positive sodium ions and negative chlorine ions; the total current I is found by adding up the currents due to each kind of charged particle, using Eq. (26–2). Likewise, the total vector current density $\vec{J}$ is found by using Eq. (26–4) for each kind of charged particle and adding up the results.

We will see in Section 26–5 that it is possible to have a current that is *steady* (that is, one that is constant in time) only if the conducting material forms a closed loop, called a *complete circuit.* In such a steady situation, the total charge in every segment of the conductor is constant. Hence the rate of flow of charge *out* at one end of a segment at any instant equals the rate of flow of charge *in* at the other end of the segment, and *the current is the same at all cross sections of the circuit.* We'll make use of this observation when we analyze electric circuits later in this chapter.

In many simple circuits, such as flashlights or electric drills, the direction of the current is always the same; this is called *direct current.* But home appliances such as toasters, refrigerators, and televisions use *alternating current,* in which the current continuously changes direction. In this chapter we'll consider direct current only. Alternating current has many special features worthy of detailed study, which we'll examine in Chapter 32.

Part of the electric circuit that includes this light bulb passes through a beaker with a solution of sodium chloride. The current in the solution is carried by both positive charges (Na^+ ions) and negative charges (Cl^- ions).

Current density and drift velocity in a wire An 18-gauge copper wire (the size usually used for lamp cords) has a nominal diameter of 1.02 mm. This wire carries a constant current of 1.67 A to a 200-watt lamp. The density of free electrons is 8.5×10^{28} electrons per cubic meter. Find the magnitudes of a) the current density and b) the drift velocity.

SOLUTION a) The cross-section area is

$$A = \frac{\pi d^2}{4} = \frac{\pi(1.02 \times 10^{-3} \text{ m})^2}{4} = 8.17 \times 10^{-7} \text{ m}^2.$$

The magnitude of the current density is

$$J = \frac{I}{A} = \frac{1.67 \text{ A}}{8.17 \times 10^{-7} \text{ m}^2} = 2.04 \times 10^6 \text{ A/m}^2.$$

b) Solving Eq. (26–3) for the drift velocity magnitude v_d, we find

$$v_d = \frac{J}{n|q|} = \frac{2.04 \times 10^6 \text{ A/m}^2}{(8.5 \times 10^{28} \text{ m}^{-3})|-1.60 \times 10^{-19} \text{ C}|}$$

$$= 1.5 \times 10^{-4} \text{ m/s} = 0.15 \text{ mm/s}.$$

At this speed an electron would require 6700 s, or about 1 hr 50 min, to travel the length of a wire 1 m long. The speeds of random motion of the electrons are of the order of 10^6 m/s. So in this example the drift speed is around 10^{10} times slower than the speed of random motion. Picture the electrons as bouncing around frantically, with a very slow and sluggish drift!

If the electrons move this slowly, you may wonder why the light comes on right away when you turn on the switch. The reason is that the electric field is set up in the wire with a speed approaching the speed of light, and electrons start to move all along the wire at very nearly the same time. The time that it takes any individual electron to get from the switch to the light bulb isn't really relevant. A good analogy is a group of soldiers standing at attention when the sergeant orders them to start marching; the order reaches the soldiers' ears at the speed of sound, which is much faster than their marching speed, so all the soldiers start to march essentially in unison.

26-3 RESISTIVITY

The current density $\vec{J}$ in a conductor depends on the electric field $\vec{E}$ and on the properties of the material. In general, this dependence can be quite complex. But for some materials, especially metals, at a given temperature, $\vec{J}$ is nearly *directly proportional* to $\vec{E}$, and the ratio of the magnitudes E and J is constant. This relationship, called **Ohm's law,** was discovered in 1826 by the German physicist Georg Simon Ohm (1787–1854). The word "law" should actually be in quotes, since Ohm's law, like the ideal gas equation and Hooke's law, is an *idealized model* that describes the behavior of some materials quite well but is not a general description of *all* matter. In the following discussion we'll assume that Ohm's law is valid, even though there are many situations in which it is not. The situation is comparable to our representation of the behavior of the static and kinetic friction forces; we treated these friction forces as being directly proportional to the normal force, even though we knew that this was at best an approximate description.

We define the **resistivity** ρ of a material as the ratio of the magnitudes of electric field and current density:

$$\rho = \frac{E}{J} \qquad \text{(definition of resistivity).} \qquad (26–5)$$

The greater the resistivity, the greater the field needed to cause a given current density, or the smaller the current density caused by a given field. From Eq. (26–5) the units of ρ are $(\text{V/m})/(\text{A/m}^2) = \text{V} \cdot \text{m/A}$. As we will discuss in the next section, 1 V/A is called one *ohm* (1 Ω; we use the Greek letter Ω, or "omega," which is alliterative with "ohm"). So the SI units for ρ are $\Omega \cdot \text{m}$ (ohm-meters). Representative values of resistivity are given in Table 26–1. A perfect conductor would have zero resistivity, and a perfect insulator would have an infinite resistivity. Metals and alloys have the smallest resistivities and are the best conductors. The resistivities of insulators are greater than those of the metals by an enormous factor, of the order of 10^{22}.

The reciprocal of resistivity is **conductivity.** Its units are $(\Omega \cdot \text{m})^{-1}$. Good conductors of electricity have larger conductivity than insulators. Conductivity is the direct electrical analog of thermal conductivity. Comparing Table 26–1 with Table 15–5 (Thermal Conductivities), we note that good electrical conductors, such as metals, are usually also good conductors of heat. Poor electrical conductors, such as ceramic and plastic materials, are also poor thermal conductors. In a metal the free electrons that carry charge in electrical conduction also provide the principal mechanism for heat conduction, so we should expect a correlation between electrical and thermal conductivity. Because of the enormous difference between resistivity of electrical conductors and insulators, it is easy

TABLE 26–1

RESISTIVITIES AT ROOM TEMPERATURE (20°C)

	SUBSTANCE	$\rho \ (\Omega \cdot m)$	SUBSTANCE	$\rho \ (\Omega \cdot m)$
Conductors			*Semiconductors*	
Metals:	Silver	1.47×10^{-8}	Pure Carbon (graphite)	3.5×10^{-5}
	Copper	1.72×10^{-8}	Pure Germanium	0.60
	Gold	2.44×10^{-8}	Pure Silicon	2300
	Aluminum	2.75×10^{-8}	*Insulators*	
	Tungsten	5.25×10^{-8}	Amber	5×10^{14}
	Steel	20×10^{-8}	Glass	10^{10}–10^{14}
	Lead	22×10^{-8}	Lucite	$>10^{13}$
	Mercury	95×10^{-8}	Mica	10^{11}–10^{15}
Alloys:	Manganin (Cu 84%, Mn 12%, Ni 4%)	44×10^{-8}	Quartz (fused)	75×10^{16}
	Constantan (Cu 60%, Ni 40%)	49×10^{-8}	Sulfur	10^{15}
	Nichrome	100×10^{-8}	Teflon	$>10^{13}$
			Wood	10^{8}–10^{11}

to confine electric currents to well-defined paths or circuits. The variation in *thermal* conductivity is much less, only a factor of 10^3 or so, and it is usually impossible to confine heat currents to that extent.

Semiconductors have resistivities intermediate between those of metals and those of insulators. These materials are important because of the way their resistivities are affected by temperature and by small amounts of impurities.

A material that obeys Ohm's law reasonably well is called an *ohmic* conductor or a *linear* conductor. For such materials, at a given temperature, ρ is a *constant* that does not depend on the value of E. Many materials show substantial departures from Ohm's-law behavior; they are *nonohmic,* or *nonlinear.* In these materials, J depends on E in a more complicated manner.

Analogies with fluid flow can be a big help in developing intuition about electric current and circuits. For example, in making wine or maple syrup, the product is sometimes filtered to remove sediments. A pump forces the fluid through the filter under pressure; if the flow rate (analogous to J) is proportional to the pressure difference between the upstream and downstream sides (analogous to E), the behavior is analogous to Ohm's law.

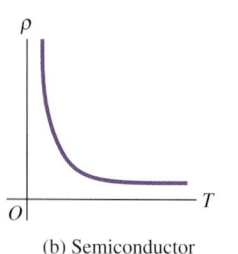

(a) Metal

RESISTIVITY AND TEMPERATURE

The resistivity of a *metallic* conductor nearly always increases with increasing temperature, as shown in Fig. 26–4a. As temperature increases, the ions of the conductor vibrate with greater amplitude, making it more likely that a moving electron will collide with an ion as in Fig. 26–1; this impedes the drift of electrons through the conductor and hence the current. Over a small temperature range (up to 100 C° or so), the resistivity of a metal can be represented approximately by the equation

$$\rho(T) = \rho_0[1 + \alpha(T - T_0)] \quad \text{(temperature dependence of resistivity),} \quad (26\text{–}6)$$

where ρ_0 is the resistivity at a reference temperature T_0 (often taken as 0°C or 20°C) and $\rho(T)$ is the resistivity at temperature T, which may be higher or lower than T_0. The factor α is called the **temperature coefficient of resistivity.** Some representative values are given in Table 26–2. The resistivity of the alloy manganin is practically independent of temperature.

The resistivity of graphite (a nonmetal) *decreases* with increasing temperature, since at higher temperatures, more electrons are "shaken loose" from the atoms and become mobile; hence the temperature coefficient of resistivity of graphite is negative. This same behavior occurs for semiconductors (Fig. 26–4b). Measuring the resistivity of a small

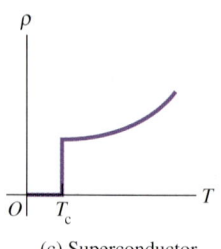

(b) Semiconductor

(c) Superconductor

26–4 Variation of resistivity with absolute temperature for (a) a normal metal; (b) a semiconductor; (c) a superconductor. In (a) the linear approximation to ρ as a function of T is shown as a green line; the approximation agrees exactly at $T = T_0$, where $\rho = \rho_0$.

TABLE 26–2

TEMPERATURE COEFFICIENTS OF RESISTIVITY (APPROXIMATE VALUES NEAR ROOM TEMPERATURE)

MATERIAL	$\alpha \ [(C°)^{-1}]$	MATERIAL	$\alpha \ [(C°)^{-1}]$
Aluminum	0.0039	Lead	0.0043
Brass	0.0020	Manganin	0.00000
Carbon (graphite)	−0.0005	Mercury	0.00088
Constantan	0.00001	Nichrome	0.0004
Copper	0.00393	Silver	0.0038
Iron	0.0050	Tungsten	0.0045

semiconductor crystal is therefore a sensitive measure of temperature; this is the principle of a type of thermometer called a *thermistor.*

Some materials, including several metallic alloys and oxides, show a phenomenon called *superconductivity.* As the temperature decreases, the resistivity at first decreases smoothly, like that of any metal. But then at a certain critical temperature T_c a phase transition occurs, and the resistivity suddenly drops to zero, as shown in Fig. 26–4c. Once a current has been established in a superconducting ring, it continues indefinitely without the presence of any driving field.

Superconductivity was discovered in 1911 by the Dutch physicist Heike Kamerlingh Onnes (1853–1926). He discovered that at very low temperatures, below 4.2 K, the resistivity of mercury suddenly dropped to zero. For the next 75 years, the highest T_c attained was about 20 K. This meant that superconductivity occurred only when the material was cooled using expensive liquid helium, with a boiling point temperature of 4.2 K, or explosive liquid hydrogen, with a boiling point of 20.3 K. But in 1986 Karl Muller and Johannes Bednorz discovered an oxide of barium, lanthanum, and copper with a T_c of nearly 40 K, and the race was on to develop "high-temperature" superconducting materials.

By 1987 a complex oxide of yttrium, copper, and barium had been found that has a value of T_c well above the 77-K boiling temperature of liquid nitrogen, a refrigerant that is both inexpensive and safe. The current (1999) record for T_c is about 160 K, and materials that are superconductors at room temperature may well become a reality. The implications of these discoveries for power-distribution systems, computer design, and transportation are enormous. Meanwhile, superconducting electromagnets cooled by liquid helium are used in particle accelerators and some experimental magnetic-levitation railroads. Superconductors have other exotic properties that require an understanding of magnetism to explore; we will discuss these further in Chapter 30.

The resistivity of wood is 10^{16} to 10^{18} times greater than that of copper. Hence, while electrons travel through a piece of copper with ease, they can hardly move at all through pieces of wood like these.

26–4 RESISTANCE

For a conductor with resistivity ρ, the current density $\vec{J}$ at a point where the electric field is $\vec{E}$ is given by Eq. (26–5), which we can write as

$$\vec{E} = \rho\vec{J}. \qquad (26\text{–}7)$$

When Ohm's law is obeyed, ρ is constant and independent of the magnitude of the electric field, so $\vec{E}$ is directly proportional to $\vec{J}$. Often, however, we are more interested in the total current in a conductor than in $\vec{J}$ and more interested in the potential difference between the ends of the conductor than in $\vec{E}$. This is so largely because current and potential difference are much easier to measure than are $\vec{J}$ and $\vec{E}$.

Suppose our conductor is a wire with uniform cross-section area A and length L, as shown in Fig. 26–5. Let V be the potential difference between the higher-potential and lower-potential ends of the conductor, so that V is positive. The *direction* of the current is always from the higher-potential end to the lower-potential end. That's because

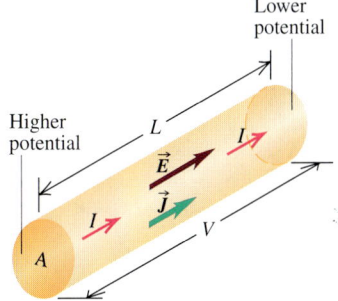

26–5 A conductor with uniform cross section. The current density is uniform over any cross section, and the electric field is constant along the length. The current flows from higher to lower electric potential.

current in a conductor flows in the direction of $\vec{E}$, no matter what the sign of the moving charges (Fig. 26–2), and because $\vec{E}$ points in the direction of *decreasing* electric potential (see Section 24–3). As the current flows through the potential difference, electric potential energy is lost; this energy is transferred to the ions of the conducting material during collisions.

We can also relate the *value* of the current I to the potential difference between the ends of the conductor. If the magnitudes of the current density $\vec{J}$ and the electric field $\vec{E}$ are uniform throughout the conductor, the total current I is given by $I = JA$, and the potential difference V between the ends is $V = EL$. When we solve these equations for J and E, respectively, and substitute the results in Eq. (26–7), we obtain

$$\frac{V}{L} = \frac{\rho I}{A} \quad \text{or} \quad V = \frac{\rho L}{A} I. \tag{26–8}$$

This shows that when ρ is constant, the total current I is proportional to the potential difference V.

The ratio of V to I for a particular conductor is called its **resistance** R:

$$R = \frac{V}{I}. \tag{26–9}$$

Comparing this definition of R to Eq. (26–8), we see that the resistance R of a particular conductor is related to the resistivity ρ of its material by

$$R = \frac{\rho L}{A} \quad \text{(relationship between resistance and resistivity).} \tag{26–10}$$

If ρ is constant, as is the case for ohmic materials, then so is R.

The equation

$$V = IR \quad \text{(relationship between voltage, current, and resistance)} \tag{26–11}$$

is often called Ohm's law, but it is important to understand that the real content of Ohm's law is the direct proportionality (for some materials) of V to I or of J to E. Equation (26–9) or (26–11) *defines* resistance R for *any* conductor, whether or not it obeys Ohm's law, but only when R is constant can we correctly call this relationship Ohm's law.

Equation (26–10) shows that the resistance of a wire or other conductor of uniform cross section is directly proportional to its length and inversely proportional to its cross-section area. It is also proportional to the resistivity of the material of which the conductor is made.

The flowing-fluid analogy is again useful. In analogy to Eq. (26–10), a thin water hose offers more resistance to flow than a fat one, and a long hose has more resistance than a short one. We can increase the resistance to flow by stuffing the hose with cotton or sand; this corresponds to increasing the resistivity. The flow rate is approximately proportional to the pressure difference between the ends. Flow rate is analogous to current, and pressure difference is analogous to potential difference ("voltage"). Let's not stretch this analogy too far, though; the water flow rate in a pipe is usually *not* proportional to its cross-section area (see Eq. (14–26) in Section 14–9).

The SI unit of resistance is the **ohm,** equal to one volt per ampere (1 $\Omega = 1$ V/A). The *kilohm* (1 k$\Omega = 10^3$ Ω) and the *megohm* (1 M$\Omega = 10^6$ Ω) are also in common use. A 100-m length of 12-gauge copper wire, the size usually used in household wiring, has a resistance at room temperature of about 0.5 Ω. A 100-W, 120-V light bulb has a resistance (at operating temperature) of 140 Ω. If the same current I flows in both the copper wire and the light bulb, the potential difference $V = IR$ is much greater across the light bulb, and much more potential energy is lost per charge in the light bulb. This lost energy is converted by the light bulb filament into light and heat. You don't want your house-

TABLE 26-3

COLOR CODES FOR RESISTORS

COLOR	VALUE AS DIGIT	VALUE AS MULTIPLIER
Black	0	1
Brown	1	10
Red	2	10^2
Orange	3	10^3
Yellow	4	10^4
Green	5	10^5
Blue	6	10^6
Violet	7	10^7
Gray	8	10^8
White	9	10^9

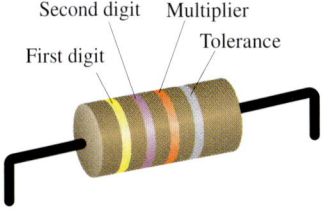

26-6 This resistor has a value of $47\ \text{k}\Omega \pm 10\%$.

hold wiring to glow white-hot, so its resistance is kept low by using wire of low resistivity and large cross-section area.

Because the resistivity of a material varies with temperature, the resistance of a specific conductor also varies with temperature. For temperature ranges that are not too great, this variation is approximately a linear relation, analogous to Eq. (26-6):

$$R(T) = R_0[1 + \alpha(T - T_0)]. \qquad (26\text{-}12)$$

In this equation, $R(T)$ is the resistance at temperature T and R_0 is the resistance at temperature T_0, often taken to be 0°C or 20°C. The temperature coefficient of *resistance* α is the same constant that appears in Eq. (26-6) if the dimensions L and A in Eq. (26-10) do not change appreciably with temperature; this is indeed the case for most conducting materials (see Problem 26-51). Within the limits of validity of Eq. (26-12), the *change* in resistance resulting from a temperature change $T - T_0$ is given by $R_0\alpha(T - T_0)$.

A circuit device made to have a specific value of resistance between its ends is called a **resistor.** Resistors in the range 0.01 to $10^7\ \Omega$ can be bought off the shelf. Individual resistors used in electronic circuitry are often cylindrical in shape, a few millimeters in diameter and length, with wires coming out of the ends. The resistance may be marked with a standard code using three or four color bands near one end (Fig. 26-6), according to the scheme shown in Table 26-3. The first two bands (starting with the band nearest an end) are digits, and the third is a power-of-ten multiplier, as shown in Fig. 26-6. For example, yellow-violet-orange means $47 \times 10^3\ \Omega$, or $47\ \text{k}\Omega$. The fourth band, if present, indicates the precision of the value; no band means ±20%, a silver band ±10%, and a gold band ±5%. Another important characteristic of a resistor is the maximum *power* it can dissipate without damage. We'll return to this point in Section 26-6.

For a resistor that obeys Ohm's law, a graph of current as a function of potential difference (voltage) is a straight line (Fig. 26-7a). The slope of the line is $1/R$. If the sign of the potential difference changes, so does the sign of the current produced; in Fig. 26-5 this corresponds to interchanging the higher- and lower-potential ends of the conductor, so the electric field, current density, and current all reverse direction. In devices that do not obey Ohm's law, the relation of voltage to current may not be a direct proportion, and it may be different for the two directions of current. Figure 26-7b shows the behavior of a vacuum *diode,* a vacuum tube used to convert high-voltage alternating current to direct current. For positive potentials of anode with respect to cathode (see Section 24-8), I is approximately proportional to $V^{3/2}$; for negative potentials the current is extremely small. The behavior of semiconductor diodes (Fig. 26-7c) is somewhat different but still strongly asymmetric. For a diode of either type, a positive potential difference V will cause a current to flow in the positive direction, but a potential difference of the other sign will cause little or no current. Hence a diode acts like a one-way

(a)

(b)

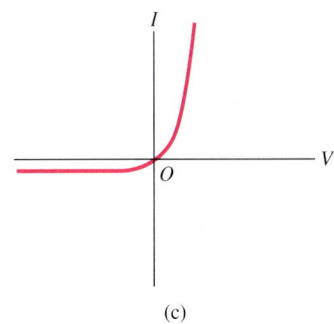

(c)

26-7 Current-voltage relations for (a) a resistor that obeys Ohm's law; (b) a vacuum diode; (c) a semiconductor diode. Only for a resistor that obeys Ohm's law is I proportional to V.

valve in a circuit. Diodes are used to perform a wide variety of logic functions in computer circuitry.

EXAMPLE 26-2

Electric field, potential difference, and resistance in a wire The 18-gauge copper wire in Example 26–1 (Section 26–2) has a diameter of 1.02 mm and a cross-section area $A = 8.2 \times 10^{-7}$ m^2. It carries a current $I = 1.67$ A. Find a) the electric-field magnitude in the wire; b) the potential difference between two points in the wire 50 m apart; c) the resistance of a 50-m length of this wire.

SOLUTION a) From Eq. (26–5) the electric-field magnitude E is equal to ρJ. The current density is $J = I/A$, and the resistivity of copper is 1.72×10^{-8} $\Omega \cdot$ m from Table 26–1. So

$$E = \rho J = \frac{\rho I}{A} = \frac{(1.72 \times 10^{-8} \ \Omega \cdot \text{m})(1.67 \ \text{A})}{8.2 \times 10^{-7} \ \text{m}^2}$$

$$= 0.035 \ \text{V/m}.$$

b) The potential difference is given by

$$V = EL = (0.035 \ \text{V/m})(50 \ \text{m}) = 1.7 \ \text{V}.$$

c) From Eq. (26–11) the resistance of a 50-m length of this wire is

$$R = \frac{V}{I} = \frac{1.7 \ \text{V}}{1.67 \ \text{A}} = 1.0 \ \Omega.$$

We can also obtain this result directly from Eq. (26–10):

$$R = \frac{\rho L}{A} = \frac{(1.72 \times 10^{-8} \ \Omega \cdot \text{m})(50 \ \text{m})}{8.2 \times 10^{-7} \ \text{m}^2} = 1.0 \ \Omega.$$

EXAMPLE 26-3

Temperature dependence of resistance Suppose the resistance of the wire in Example 26–2 is 1.05 Ω at a temperature of 20°C. Find the resistance at 0°C and at 100°C.

SOLUTION We use Eq. (26–12), with $T_0 = 20$°C and $R_0 = 1.05$ Ω. From Table 26–2 the temperature coefficient of resistivity of copper is $\alpha = 0.00393$ (C°)$^{-1}$. At $T = 0$°C,

$$R = R_0[1 + \alpha(T - T_0)]$$

$$= (1.05 \ \Omega)(1 + [0.00393 \ (\text{C}°)^{-1}][0°\text{C} - 20°\text{C}])$$

$$= 0.97 \ \Omega.$$

At $T = 100$°C,

$$R = (1.05 \ \Omega)(1 + [0.00393 \ (\text{C}°)^{-1}][100°\text{C} - 20°\text{C}])$$

$$= 1.38 \ \Omega.$$

This substantial variation in resistance of ordinary copper wire must be taken into account in designing electric circuits that are to operate over a wide range of temperatures.

EXAMPLE 26-4

The hollow cylinder shown in Fig. 26–8 has length L and inner and outer radii a and b. It is made of a material with resistivity ρ. A potential difference is set up between the inner and outer surfaces of the cylinder (each of which is an equipotential surface) so that current flows radially through the cylinder. What is the resistance to this radial current flow?

SOLUTION We can't use Eq. (26–10) directly because the cross section through which the charge travels is *not* constant; it varies from $2\pi a L$ at the inner surface to $2\pi b L$ at the outer surface. Instead, we consider a thin cylindrical shell of inner radius r and thickness dr. The area A is then $2\pi r L$, and the length of the current path through the shell is dr. The resistance dR of this shell, between inner and outer surfaces, is that of a conductor with length dr and area $2\pi r L$, that is,

$$dR = \frac{\rho \, dr}{2\pi r L}.$$

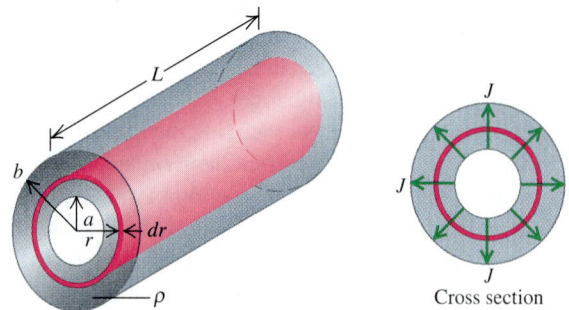

26–8 Finding the resistance for radial current flow.

The current has to pass successively through all such shells between the inner and outer radii a and b. From Eq. (26–11) the potential difference across one shell is $dV = I \, dR$, and the total potential difference between the inner and outer surfaces is the sum of the potential differences for all shells. The total current is the same through each shell, so the total resistance is the sum of the resistances of all the shells. If the area $2\pi rL$ were constant, we could just integrate dr from $r = a$ to $r = b$ to get the total length of the current path. But the area increases as the current passes through shells of greater radius, so we have to integrate the above expression for dR. The total resistance is thus given by

$$R = \int dR = \frac{\rho}{2\pi L} \int_a^b \frac{dr}{r} = \frac{\rho}{2\pi L} \ln \frac{b}{a}.$$

26–5 ELECTROMOTIVE FORCE AND CIRCUITS

For a conductor to have a steady current, it must be part of a path that forms a closed loop or **complete circuit.** Here's why. If you establish an electric field $\vec{E}_1$ inside an isolated conductor with resistivity ρ that is *not* part of a complete circuit, a current begins to flow with current density $\vec{J} = \vec{E}_1/\rho$ (Fig. 26–9a). As a result a net positive charge quickly accumulates at one end of the conductor, and a net negative charge accumulates at the other end (Fig. 26–9b). These charges themselves produce an electric field $\vec{E}_2$ in the direction opposite to $\vec{E}_1$, causing the total electric field and hence the current to decrease. Within a very small fraction of a second, enough charge builds up on the conductor ends that the total electric field $\vec{E} = \vec{E}_1 + \vec{E}_2 = 0$ inside the conductor. Then $\vec{J} = 0$ as well, and the current stops altogether. So there can be no steady motion of charge in such an *incomplete* circuit.

To see how to maintain a steady current in a *complete* circuit, we recall a basic fact about electric potential energy: If a charge q goes around a complete circuit and returns to its starting point, the potential energy must be the same at the end of the round trip as at the beginning. As described in Section 26–4, there is always a *decrease* in potential energy when charges move through an ordinary conducting material with resistance. So there must be some part of the circuit in which the potential energy *increases.*

The problem is analogous to an ornamental water fountain that recycles its water. The water pours out of openings at the top, cascades down over the terraces and spouts (moving in the direction of decreasing gravitational potential energy), and collects in a basin in the bottom. A pump then lifts it back to the top (increasing the potential energy) for another trip. Without the pump, the water would just fall to the bottom and stay there.

ELECTROMOTIVE FORCE

In an electric circuit there must be a device somewhere in the loop that acts like the water pump in a water fountain. In this device a charge travels "uphill," from lower to higher potential energy, even though the electrostatic force is trying to push it from higher to lower potential energy. The direction of current in such a device is from lower to higher potential, just the opposite of what happens in an ordinary conductor. The influence that makes current flow from lower to higher potential is called **electromotive force** (abbreviated **emf** and pronounced "ee-em-eff"). This is a poor term because emf is *not* a force but an energy-per-unit-charge quantity, like potential. The SI unit of emf is the same as for potential, the volt (1 V = 1 J/C). A typical flashlight battery has an emf of 1.5 V; this means that the battery does 1.5 J of work on every coulomb of charge that passes through it. We'll use the symbol $\mathcal{E}$ (a script capital E) for emf.

Every complete circuit with a steady current must include some device that provides emf. Such a device is called a **source of emf.** Batteries, electric generators, solar cells, thermocouples, and fuel cells are all examples of sources of emf. All such devices convert energy of some form (mechanical, chemical, thermal, and so on) into electric potential energy and transfer it into the circuit to which the device is connected. An *ideal* source of emf maintains a constant potential difference between its terminals, independent of the current through it. We define electromotive force quantitatively as the magnitude of this potential difference. As we will see, such an ideal source is a

(a)

(b)

26–9 (a) If an electric field $\vec{E}_1$ is produced inside a conductor that is *not* part of a complete circuit, a current will flow at least temporarily. (b) The current causes charge to build up at the ends of the conductor, producing an additional field $\vec{E}_2$ that opposes $\vec{E}_1$. The total field $\vec{E} = \vec{E}_1 + \vec{E}_2$ is reduced, and the current decreases. After a very short time, $\vec{E}_2$ becomes equal in magnitude to $\vec{E}_1$, so that the total field $\vec{E}$ inside the conductor equals zero; the current then stops completely.

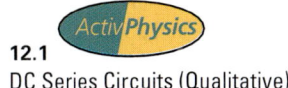

12.1
DC Series Circuits (Qualitative)

26–10 Schematic diagram of a source of emf in an "open-circuit" situation in which the source is not connected in a circuit. The electric-field force $\vec{F}_e = q\vec{E}$ and the non-electrostatic force $\vec{F}_n$ on a positive charge q are shown. The work done by $\vec{F}_n$ on a positive charge q moving from b to a is equal to $q\mathcal{E}$, where $\mathcal{E}$ is the emf. In the open-circuit situation, $\vec{F}_e$ and $\vec{F}_n$ have equal magnitude.

mythical beast, like the frictionless plane and the massless rope. We will discuss later how real-life sources of emf differ in their behavior from this idealized model.

Figure 26–10 is a schematic diagram of an ideal source of emf that maintains a potential difference between conductors a and b, called the *terminals* of the device. Terminal a, marked +, is maintained at *higher* potential than terminal b, marked –. Associated with this potential difference is an electric field $\vec{E}$ in the region around the terminals, both inside and outside the source. The electric field inside the device is directed from a to b, as shown. A charge q within the source experiences an electric force $\vec{F}_e = q\vec{E}$. But the source also provides an additional influence, which we represent as a non-electrostatic force $\vec{F}_n$. This force, operating inside the device, pushes charge from b to a in an "uphill" direction against the electric force $\vec{F}_e$. Thus $\vec{F}_n$ maintains the potential difference between the terminals. If $\vec{F}_n$ were not present, charge would flow between the terminals until the potential difference was zero. The origin of the additional influence $\vec{F}_n$ depends on the kind of source. In a generator it results from magnetic-field forces on moving charges. In a battery or fuel cell it is associated with diffusion processes and varying electrolyte concentrations resulting from chemical reactions. In an electrostatic machine such as a Van de Graaff generator (Fig. 23–24), an actual mechanical force is applied by a moving belt or wheel.

If a positive charge q is moved from b to a inside the source, the non-electrostatic force $\vec{F}_n$ does a positive amount of work $W_n = q\mathcal{E}$ on the charge. This displacement is *opposite* to the electrostatic force $\vec{F}_e$, so the potential energy associated with the charge *increases* by an amount equal to qV_{ab}, where $V_{ab} = V_a - V_b$ is the (positive) potential of point a with respect to point b. For the ideal source of emf that we've described, $\vec{F}_e$ and $\vec{F}_n$ are equal in magnitude but opposite in direction, so the total work done on the charge q is zero; there is an increase in potential energy but *no* change in the kinetic energy of the charge. It's like lifting a book from the floor to a high shelf at constant speed. The increase in potential energy is just equal to the non-electrostatic work W_n, so $q\mathcal{E} = qV_{ab}$, or

$$V_{ab} = \mathcal{E} \qquad \text{(ideal source of emf)}. \qquad (26\text{–}13)$$

Now let's make a complete circuit by connecting a wire with resistance R to the terminals of a source (Fig. 26–11). The potential difference between terminals a and b sets up an electric field within the wire; this causes current to flow around the loop from a toward b, from higher to lower potential. Notice that where the wire bends, equal amounts of positive and negative charge persist on the "inside" and "outside" of the bend. These charges exert the forces that cause the current to follow the bends in the wire.

From Eq. (26–11) the potential difference between the ends of the wire in Fig. 26–11 is given by $V_{ab} = IR$. Combining with Eq. (26–13), we have

$$\mathcal{E} = V_{ab} = IR \qquad \text{(ideal source of emf)}. \qquad (26\text{–}14)$$

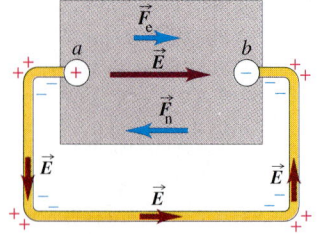

26–11 Schematic diagram of an ideal source in a complete circuit. The vectors $\vec{F}_n$ and $\vec{F}_e$ are the forces on a positive charge q inside the source. The current is in the direction from a to b in the external circuit and from b to a within the source.

That is, when a positive charge q flows around the circuit, the potential *rise* $\mathcal{E}$ as it passes through the ideal source is numerically equal to the potential *drop* $V_{ab} = IR$ as it passes through the remainder of the circuit. Once $\mathcal{E}$ and R are known, this relation determines the current in the circuit.

CAUTION ▶ It's a common misconception that in a closed circuit, current is something that squirts out of the positive terminal of a battery and is consumed or "used up" by the time it reaches the negative terminal. But in fact the current is the *same* at every point in a simple loop circuit like that in Fig. 26–11, even if the thickness of the wires is different at different points in the circuit. This happens because charge is conserved (that is, it can neither be created nor destroyed) and because charge cannot accumulate in the circuit devices we have described. If charge did accumulate, the potential differences would change with time. It's like the flow of water in an ornamental fountain;

water flows out of the top of the fountain at the same rate at which it reaches the bottom, no matter what the dimensions of the fountain. None of the water is "used up" along the way! ◄

INTERNAL RESISTANCE

Real sources in a circuit don't behave in exactly the way we have described; the potential difference across a real source in a circuit is *not* equal to the emf as in Eq. (26–14). The reason is that charge moving through the material of any real source encounters *resistance*. We call this the **internal resistance** of the source, denoted by r. If this resistance behaves according to Ohm's law, r is constant and independent of the current I. As the current moves through r, it experiences an associated drop in potential equal to Ir. Thus when a current is flowing through a source from the negative terminal b to the positive terminal a, the potential difference V_{ab} between the terminals is

$$V_{ab} = \varepsilon - Ir \quad \text{(terminal voltage, source with internal resistance).} \quad (26\text{–}15)$$

The potential V_{ab}, called the **terminal voltage,** is less than the emf ε because of the term Ir representing the potential drop across the internal resistance r. Expressed another way, the increase in potential energy qV_{ab} as a charge q moves from b to a within the source is now less than the work $q\varepsilon$ done by the non-electrostatic force $\vec{F}_n$, since some potential energy is lost in traversing the internal resistance.

A 1.5-V battery has an emf of 1.5 V, but the terminal voltage V_{ab} of the battery is equal to 1.5 V only if no current is flowing through it so that $I = 0$ in Eq. (26–15). If the battery is part of a complete circuit through which current is flowing, the terminal voltage will be less than 1.5 V. *For a real source of emf, the terminal voltage equals the emf only if no current is flowing through the source.* Thus we can describe the behavior of a source in terms of two properties: an emf ε, which supplies a constant potential difference independent of current, in series with an internal resistance r.

The current in the external circuit connected to the source terminals a and b is still determined by $V_{ab} = IR$. Combining this with Eq. (26–15), we find

$$\varepsilon - Ir = IR,$$

or

$$I = \frac{\varepsilon}{R + r} \quad \text{(current, source with internal resistance).} \quad (26\text{–}16)$$

That is, the current equals the source emf divided by the *total* circuit resistance $(R + r)$.

CAUTION ▶ You might have thought that a battery or other source of emf always produces the same current, no matter what circuit it's used in. But as Eq. (26–16) shows, the current that a source of emf produces in a given circuit depends on the resistance R of the external circuit (as well as on the internal resistance r of the source). The greater the resistance, the less current the source will produce. It's analogous to pushing an object through a thick, viscous liquid such as oil or molasses; if you exert a certain steady push (emf), you can move a small object at high speed (small R, large I) or a large object at low speed (large R, small I). ◄

SYMBOLS FOR CIRCUIT DIAGRAMS

An important part of analyzing any electric circuit is drawing a schematic *circuit diagram.* Table 26–4 shows the usual symbols used in circuit diagrams. We will use these symbols extensively in this chapter and the next. We usually assume that the wires that connect the various elements of the circuit have negligible resistance; from Eq. (26–11), $V = IR$, the potential difference between the ends of such a wire is zero. Table 26–4 includes two *meters* that are used to measure the properties of circuits. Idealized meters do not disturb the circuit in which they are connected. A **voltmeter,** introduced in

The emf of this battery—that is, the terminal voltage when it's not connected to anything—is 12 V. But because the battery has internal resistance, the terminal voltage of the battery is less than 12 V when it is supplying current to a light bulb.

TABLE 26–4

SYMBOLS FOR CIRCUIT DIAGRAMS

———	Conductor with negligible resistance
R —⌁⌁—	Resistor
+‖$\mathcal{E}$	Source of emf (Longer vertical line always represents the positive terminal, usually the terminal at higher potential)
+‖$\mathcal{E}$ —⌁⌁—r or —⌁⌁—r +‖$\mathcal{E}$	Source of emf with internal resistance r (r can be placed on either side)
—(V)—	Voltmeter (measures potential difference between its terminals)
—(A)—	Ammeter (measures current through it)

Section 24–3, measures the potential difference between its terminals; an idealized voltmeter has infinitely large resistance and measures potential difference without having any current diverted through it. An **ammeter** measures the current passing through it; an idealized ammeter has zero resistance and has no potential difference between its terminals. Because meters act as part of the circuit in which they are connected, these properties are important to remember.

EXAMPLE 26–5

A source on open circuit Figure 26–12 shows a source (a battery) with an emf $\mathcal{E}$ of 12 V and an internal resistance r of 2 Ω. (For comparison, the internal resistance of a commercial 12-V lead storage battery is only a few thousandths of an ohm.) The wires to the left of a and to the right of the ammeter A are not connected to anything. Determine the readings of the idealized voltmeter V and the idealized ammeter A.

SOLUTION There is no current because there is no complete circuit. (There is no current through our idealized voltmeter, with its infinitely large resistance.) Hence the ammeter A reads $I = 0$. Because there is no current through the battery, there is no potential difference across its internal resistance. From Eq. (26–15) with $I = 0$, the potential difference V_{ab} across the battery terminals is equal to the emf. So the voltmeter reads

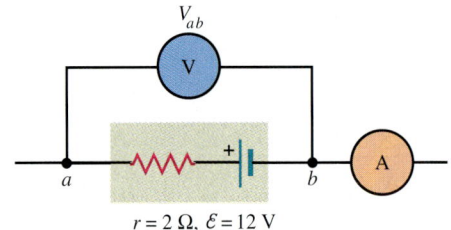

26–12 A source on open circuit.

$V_{ab} = \mathcal{E} = 12$ V. The terminal voltage of a real, non-ideal source equals the emf *only* if there is no current flowing through the source, as in this example.

EXAMPLE 26–6

A source in a complete circuit Using the battery in Example 26–5, we add a 4-Ω resistor to form the complete circuit shown in Fig. 26–13. What are the voltmeter and ammeter readings now?

SOLUTION We now have a complete circuit. The current I through the resistor R is determined by Eq. (26–16):

$$I = \frac{\mathcal{E}}{R + r} = \frac{12 \text{ V}}{4 \,\Omega + 2 \,\Omega} = 2 \text{ A}.$$

The ammeter A reads $I = 2$ A.

Our idealized conducting wires have zero resistance, and the idealized ammeter A also has zero resistance. So there is no potential difference between points a and a' or between points b

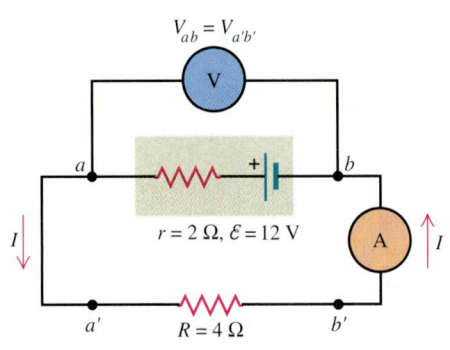

26–13 A source in a complete circuit.

and b'. That is, $V_{ab} = V_{a'b'}$. We can find V_{ab} by considering a and b either as the terminals of the resistor or as the terminals of the source. Considering them as terminals of the resistor, we use Ohm's law ($V = IR$):

$$V_{a'b'} = IR = (2 \text{ A})(4 \, \Omega) = 8 \text{ V}.$$

Considering them as the terminals of the source, we have

$$V_{ab} = \mathcal{E} - Ir = 12 \text{ V} - (2 \text{ A})(2 \, \Omega) = 8 \text{ V}.$$

Either way, we conclude that the voltmeter reads $V_{ab} = 8$ V. With a current flowing through the source, the terminal voltage V_{ab} is less than the emf. The smaller the internal resistance r, the less the difference between V_{ab} and $\mathcal{E}$.

EXAMPLE 26-7

Using voltmeters and ammeters The voltmeter and ammeter in Example 26–6 are moved to different positions in the circuit. What are the voltmeter and ammeter readings in the situations shown in a) Fig. 26–14a; b) Fig. 26–14b?

SOLUTION a) The voltmeter now measures the potential difference between points a' and b'. But as mentioned in Example 26–6, $V_{ab} = V_{a'b'}$, so the voltmeter reads the same as in Example 26–6: $V_{a'b'} = 8$ V.

CAUTION ▶ You might be tempted to conclude that the ammeter in Fig. 26–14a, which is located "upstream" of the resistor, would have a higher reading than the one located "downstream" of the resistor in Fig. 26–13. But this conclusion is based on the misconception that current is somehow "used up" as it moves through a resistor. As charges move through a resistor, there is a decrease in electric potential energy, but there is *no* change in the current. *The current in a simple loop is the same at every point.* An ammeter placed as in Fig. 26–14a reads the same as one placed as in Fig. 26–13: $I = 2$ A. ◀

b) There is no current through the voltmeter because it has infinitely large resistance. Since the voltmeter is now part of the circuit, there is no current at all in the circuit, and the ammeter reads $I = 0$.

The voltmeter measures the potential difference $V_{bb'}$ between points b and b'. Since $I = 0$, the potential difference across the resistor is $V_{a'b'} = IR = 0$, and the potential difference between the ends a and a' of the idealized ammeter is also zero. So $V_{bb'}$ is equal to V_{ab}, the terminal voltage of the source. As in Example 26–5, there is no current flowing, so the terminal voltage equals the emf, and the voltmeter reading is $V_{ab} = \mathcal{E} = 12$ V.

This example shows that ammeters and voltmeters are circuit elements, too. Moving the voltmeter from the position in Fig. 26–14a to that in Fig. 26–14b changes the current and potential

26–14 Different placements of voltmeter and ammeter in a complete circuit.

differences in the circuit, in this case rather dramatically. If you want to measure the potential difference between two points in a circuit without disturbing the circuit, use a voltmeter as in Figs. 26–13 or 26–14a, *not* as in Fig. 26–14b.

EXAMPLE 26-8

A source with a short circuit Using the same battery as in the preceding three examples, we now replace the 4-Ω resistor with a zero-resistance conductor, as shown in Fig. 26–15. What are the meter readings now?

SOLUTION Now we have a zero-resistance path between points a and b, so we must have $V_{ab} = IR = I(0) = 0$, no matter what the current. Knowing this, we can find the current I from Eq.

(26–15):

$$V_{ab} = \mathcal{E} - Ir = 0,$$

$$I = \frac{\mathcal{E}}{r} = \frac{12 \text{ V}}{2 \, \Omega} = 6 \text{ A}.$$

The ammeter reads $I = 6$ A, and the voltmeter reads $V_{ab} = 0$. This is a different value of I than in Example 26–6, even though the same battery is used. A source does *not* deliver the same current

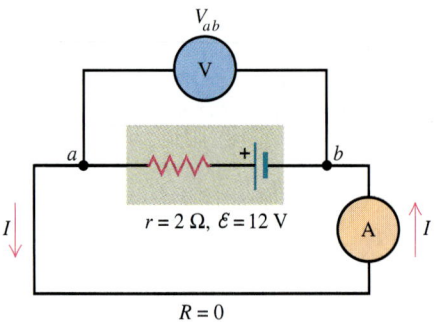

26–15 A source on short circuit. (*Warning:* See the text for the dangers of this arrangement.)

in all situations; the amount of current depends on the internal resistance r and on the resistance of the external circuit.

The situation in this example is called a *short circuit.* The terminals of the battery are connected directly to each other, with no external resistance. The short-circuit current is equal to the emf $\mathcal{E}$ divided by the internal resistance r. *Warning:* A short circuit can be an extremely dangerous situation. An automobile battery or a household power line has very small internal resistance (much less than in these examples), and the short-circuit current can be great enough to cause a small wire or a storage battery actually to explode. Don't try it!

The net change in potential energy for a charge q making a round trip around a complete circuit must be zero. Hence the net change in *potential* around the circuit must also be zero; in other words, the algebraic sum of the potential differences and emf's around the loop is zero. We can see this by rewriting Eq. (26–16) in the form

$$\mathcal{E} - Ir - IR = 0.$$

A potential gain of $\mathcal{E}$ is associated with the emf and potential drops of Ir and IR are associated with the internal resistance of the source and the external circuit, respectively. Figure 26–16 is a graph showing how the potential varies as we go around the complete circuit of Fig. 26–13. The horizontal axis doesn't necessarily represent actual distances, but rather various points in the loop. If we take the potential to be zero at the negative terminal of the battery, then we have a rise $\mathcal{E}$ and a drop Ir in the battery and an additional drop IR in the external resistor, and as we finish our trip around the loop, the potential is back where it started.

We have mentioned graphical representations of current-voltage relations for circuit elements that don't obey Ohm's law; two examples are shown in Figs. 26–7b and 26–7c. The current-voltage relation for a *source* may also be represented graphically. For a source represented by Eq. (26–15), that is, for which

$$V_{ab} = \mathcal{E} - Ir,$$

26–16 Potential rises and drops in a circuit.

the graph of current versus terminal voltage appears as in Fig. 26–17a. The intercept on the V-axis, corresponding to an open-circuit condition ($I = 0$), is at $V_{ab} = \mathcal{E}$; the intercept on the I-axis, corresponding to a short-circuit situation ($V_{ab} = 0$), is at $I = \mathcal{E}/r$.

We can use a graph of I versus V_{ab} to find the current in a circuit containing a nonlinear device, such as a diode (Fig. 26–17b). The current-voltage relation for a semiconductor diode is shown in Fig. 26–17c. Equation (26–15), the current-voltage relation for the source, is also plotted on this graph. Each of the curves represents a current-voltage relation that must be satisfied, so the *intersection* of the two curves represents the only possible values of V_{ab} and I. This amounts to a graphical solution of two simultaneous equations for V_{ab} and I, one of which is nonlinear.

Finally, we remark that Eq. (26–15) is not always an adequate representation of the behavior of a source. The emf may not be constant, and what we have described as an internal resistance may actually be a more complex voltage-current relation that doesn't obey Ohm's law. Nevertheless, the concept of internal resistance frequently provides an adequate description of batteries, generators, and other energy converters. The principal difference between a fresh flashlight battery and an old one is not in the emf, which decreases only slightly with use, but in the internal resistance, which may increase from less than an ohm when the battery is fresh to as much as $1000\ \Omega$ or more after long use. Similarly, a car battery can deliver less current to the starter motor on a cold morning than when the battery is warm, not because the emf is appreciably less but because the internal resistance is temperature-dependent, increasing with decreasing temperature. Cold-climate dwellers take a number of measures to avoid this loss, from using special battery warmers to soaking the battery in warm water on very cold mornings.

(a)

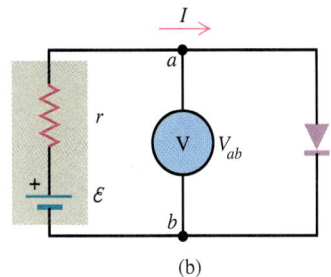

(b)

26–6 ENERGY AND POWER IN ELECTRIC CIRCUITS

Let's now look at some energy and power relations in electric circuits. The box in Fig. 26–18 represents a circuit element with potential difference $V_a - V_b = V_{ab}$ between its terminals and current I passing through it in the direction from a toward b. This element might be a resistor, a battery, or something else; the details don't matter. As charge passes through the circuit element, the electric field does work on the charge. In a source of emf, additional work is done by the force $\vec{F}_n$ that we mentioned in Section 26–5.

The total work done on a charge q passing through the circuit element is equal to the product of q and the potential difference V_{ab} (work per unit charge). When V_{ab} is positive, the electric force (and possibly a non-electric force) does a positive amount of work qV_{ab} on the charge as it "falls" from potential V_a to lower potential V_b. If the current is I, then in a time interval dt, an amount of charge $dQ = I\,dt$ passes through. The work dW done on this charge is

$$dW = V_{ab}\,dQ = V_{ab}I\,dt.$$

This work represents electrical energy transferred *into* this circuit element. The time rate of energy transfer is *power,* denoted by P. Dividing the above equation by dt, we obtain the *rate* at which the rest of the circuit delivers electrical energy to this circuit element:

$$\frac{dW}{dt} = P = V_{ab}I \quad \text{(rate of delivering electrical energy to a circuit element).} \tag{26–17}$$

It may happen that the potential at b is higher than that at a; then V_{ab} is negative, and there is a net transfer of energy *out of* the circuit element. The element is then acting as a source, delivering electrical energy into the circuit in which it is connected. This is the usual situation for a battery, which converts chemical energy into electrical energy and delivers it to the external circuit. Similarly, a generator converts mechanical energy to electrical energy and delivers it to the external circuit. Thus Eq. (26–17) can represent

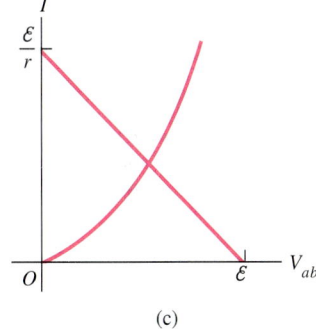

(c)

26–17 (a) Current-voltage relation for a source with emf $\mathcal{E}$ and internal resistance r. (b) A circuit containing a source and a diode (a nonlinear element). (c) Simultaneous solution of I-V_{ab} equations for this circuit.

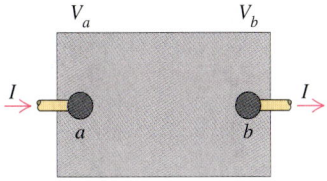

26–18 The power input P to the portion of the circuit between a and b is $P = V_{ab}I$.

either the rate at which energy is *delivered* to a circuit element or the rate at which energy is *extracted* from that element.

The unit of V_{ab} is one volt, or one joule per coulomb, and the unit of I is one ampere, or one coulomb per second. Hence the unit of $V_{ab}I$ is one watt, as it should be:

$$(1 \text{ J/C})(1 \text{ C/s}) = 1 \text{ J/s} = 1 \text{ W}.$$

Let's consider a few special cases.

PURE RESISTANCE

If the circuit element in Fig. 26–18 is a resistor, the potential difference is $V_{ab} = IR$. From Eq. (26–17) the electrical power delivered to the resistor by the circuit is

$$P = V_{ab}I = I^2R = \frac{V_{ab}^2}{R} \qquad \text{(power delivered to a resistor).} \qquad (26\text{–}18)$$

In this case the potential at a (where the current enters the resistor) is always higher than at b (where the current exits). Current enters the higher-potential terminal of the device, and Eq. (26–18) represents the rate of transfer of electric potential energy *into* the circuit element.

What becomes of this energy? The moving charges collide with atoms in the resistor and transfer some of their energy to these atoms, increasing the internal energy of the material. Either the temperature of the resistor increases or there is a flow of heat out of it, or both. In any of these cases we say that energy is *dissipated* in the resistor at a rate I^2R. Every resistor has a *power rating,* the maximum power the device can dissipate without becoming overheated and damaged. In practical applications the power rating of a resistor is often just as important a characteristic as its resistance value. Of course, some devices, such as electric heaters, are designed to get hot and transfer heat to their surroundings. But if the power rating is exceeded, even such a device may melt or even explode.

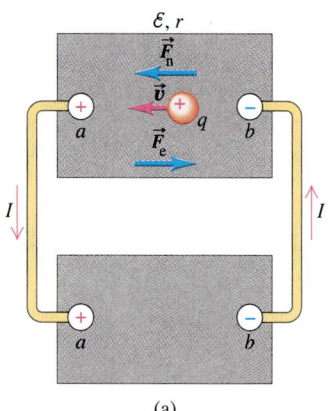

POWER OUTPUT OF A SOURCE

The upper rectangle in Fig. 26–19a represents a source with emf ε and internal resistance r, connected by ideal (resistanceless) conductors to an external circuit represented by the lower box. This could describe a car battery connected to one of the car's headlights (Fig. 26–19b). Point a is at higher potential than point b, so $V_a > V_b$ and V_{ab} is positive. Note that the current I is *leaving* the source at the higher-potential terminal (rather than entering there). Energy is being delivered to the external circuit, and the rate of its delivery to the circuit is given by Eq. (26–17):

$$P = V_{ab}I.$$

For a source that can be described by an emf ε and an internal resistance r, we may use Eq. (26–15):

$$V_{ab} = \varepsilon - Ir.$$

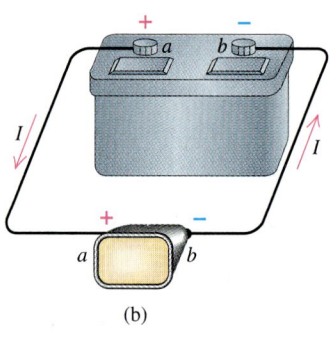

Multiplying this equation by I, we find

$$P = V_{ab}I = \varepsilon I - I^2r. \qquad (26\text{–}19)$$

26–19 (a) The rate of conversion of non-electrical to electrical energy in the source equals εI. The rate of energy dissipation in the source is I^2r. The difference $\varepsilon I - I^2r$ is the power output of the source. (b) A car battery connected to a headlight is a real-life example of the generic circuit in part (a).

What do the terms εI and I^2r mean? In Section 26–5 we defined the emf ε as the work per unit charge performed on the charges by the non-electrostatic force as the charges are pushed "uphill" from b to a in the source. In a time dt, a charge $dQ = I\,dt$ flows through the source; the work done on it by this non-electrostatic force is $\varepsilon\,dQ = \varepsilon I\,dt$. Thus εI is the *rate* at which work is done on the circulating charges by whatever agency causes the non-electrostatic force in the source. This term represents the rate of conversion of non-electrical energy to electrical energy within the source. The term I^2r is the rate at which electrical energy is *dissipated* in the internal resistance of the source. The difference $\mathcal{E}I - I^2r$ is the *net* electrical power output of the source, that is, the rate at which the source delivers electrical energy to the remainder of the circuit.

POWER INPUT TO A SOURCE

Suppose that the lower rectangle in Fig. 26–19a is itself a source, with an emf *larger* than that of the upper source and with its emf opposite to that of the upper source. Figure 26–20 shows a practical example, an automobile battery (the upper circuit element) being charged by the car's alternator (the lower element). The current I in the circuit is then *opposite* to that shown in Fig. 26–19; the lower source is pushing current backward through the upper source. Because of this reversal of current, instead of Eq. (26–15) we have for the upper source

$$V_{ab} = \mathcal{E} + Ir,$$

and instead of Eq. (26–19), we have

$$P = V_{ab}I = \mathcal{E}I + I^2r. \qquad (26\text{–}20)$$

Work is being done *on*, rather than *by*, the agent that causes the non-electrostatic force in the upper source. There is a conversion of electrical energy into non-electrical energy in the upper source at a rate $\mathcal{E}I$. The term I^2r in Eq. (26–20) is again the rate of dissipation of energy in the internal resistance of the upper source, and the sum $\mathcal{E}I + I^2r$ is the total electrical power *input* to the upper source. This is what happens when a rechargeable battery (a storage battery) is connected to a charger. The charger supplies electrical energy to the battery; part of it is converted to chemical energy, to be reconverted later, and the remainder is dissipated (wasted) in the battery's internal resistance, warming the battery and causing a heat flow out of it. If you have a power tool or laptop computer with a rechargeable battery, you may have noticed that it gets warm while it is charging.

26–20 When two sources are connected in a simple loop circuit, the source with the larger emf delivers energy to the other source.

Problem–Solving Strategy

POWER AND ENERGY IN CIRCUITS

1. A set of sign rules is useful in calculating the energy balance in a circuit. A source of emf $\mathcal{E}$ puts *positive* power $\mathcal{E}I$ into a circuit when the current I runs through the source from $-$ to $+$; the energy is converted from chemical energy in a battery, from mechanical energy in a generator, or whatever. But when current passes through a source in the direction from $+$ to $-$, the source is taking power out of the circuit (as in charging a storage battery, when electrical energy is converted back to chemical energy). In that case we count $\mathcal{E}I$ as negative. The internal resistance r of a source always removes energy from the

circuit, dissipating it at a rate I^2r, irrespective of the direction of the current. This represents a *positive* power input to the device and a *negative* power input to the circuit. Similarly, a resistor R always removes energy from the circuit at a rate given by $VI = I^2R = V^2/R$.

2. The energy balance can be expressed in one of two forms: either "net power input = net power output" or "the algebraic sum of the power inputs to the circuit is zero."

EXAMPLE 26-9

Power input and output in a complete circuit Figure 26–21 shows the same situation that we analyzed in Example 26–6. Find the rate of energy conversion (chemical to electrical) and the rate of dissipation of energy in the battery and the net power output of the battery.

SOLUTION The rate of energy conversion in the battery is

$$\mathcal{E}I = (12\text{ V})(2\text{ A}) = 24\text{ W}.$$

The rate of dissipation of energy in the battery is

$$I^2r = (2\text{ A})^2(2\ \Omega) = 8\text{ W}.$$

26–21 Power relationships in a simple circuit.

The electrical power *output* of the source is the difference between these: $\mathcal{E}I - I^2r = 16$ W. The terminal voltage is $V_{ab} = 8$ V. The power output is also given by the terminal voltage $V_{ab} = 8$ V (calculated in Example 26–6) multiplied by the current:

$$V_{ab}I = (8 \text{ V})(2 \text{ A}) = 16 \text{ W}.$$

The electrical power input to the resistor is

$$V_{a'b'}I = (8 \text{ V})(2 \text{ A}) = 16 \text{ W}.$$

This equals the rate of dissipation of electrical energy in the resistor:

$$I^2R = (2 \text{ A})^2(4 \text{ }\Omega) = 16 \text{ W}.$$

Note that our results agree with Eq. (26–19), which states that $V_{ab}I = \mathcal{E}I - I^2R$; the left side of this equation equals 16 W, and the right side equals 24 W − 8 W = 16 W. This verifies the consistency of the various power quantities.

EXAMPLE 26–10

Suppose the 4-Ω resistor in Fig. 26–21 is replaced by an 8-Ω resistor. How does this affect the electrical power dissipated in the resistor?

SOLUTION According to Eq. (26–18), the power dissipated in the resistor is given by $P = I^2R$. If you were in a hurry, you might conclude that since R now has twice the value that it had in Example 26–9, the power should also be twice as great, or $2(16 \text{ W}) = 32$ W. Or you might instead try to use the formula $P = V_{ab}^2/R$; this formula would lead you to conclude that the power should be one half as great as in the preceding example, or $(16 \text{ W})/2 = 8$ W. Which answer is correct?

In fact, *both* of these conclusions are *incorrect*. The first is incorrect because changing the resistance R also changes the current in the circuit (remember, a source of emf does *not* generate the same current in all situations). The second conclusion is also incorrect because the potential difference V_{ab} across the resistor changes when the current changes. To get the correct answer, we first use the same technique as in Example 26–6 to find the current:

$$I = \frac{\mathcal{E}}{R + r} = \frac{12 \text{ V}}{8 \text{ }\Omega + 2 \text{ }\Omega} = 1.2 \text{ A}.$$

The greater resistance causes the current to decrease. The potential difference across the resistor is

$$V_{ab} = IR = (1.2 \text{ A})(8 \text{ }\Omega) = 9.6 \text{ V},$$

which is greater than with the 4-Ω resistor. We can then find the power dissipated in the resistor in either of two ways:

$$P = I^2R = (1.2 \text{ A})^2(8 \text{ }\Omega) = 12 \text{ W},$$

or

$$P = \frac{V_{ab}^2}{R} = \frac{(9.6 \text{ V})^2}{8 \text{ }\Omega} = 12 \text{ W}.$$

Increasing the resistance R causes a *reduction* in the power input to the resistor. In the expression $P = I^2R$ the decrease in current is more important than the increase in resistance; in the expression $P = V_{ab}^2/R$ the increase in resistance is more important than the increase in V_{ab}. This same principle applies to ordinary light bulbs; a 50-W light bulb has a greater resistance than does a 100-W light bulb.

Can you show that replacing the 4-Ω resistor by an 8-Ω resistor decreases both the rate of energy conversion (chemical to electrical) in the battery and the rate of energy dissipation in the battery?

EXAMPLE 26–11

Power in a short circuit Figure 26–22 shows the same short-circuited battery that we analyzed in Example 26–8. Find the rates of energy conversion and energy dissipation in the battery and the net power output of the battery.

SOLUTION The rate of energy conversion (chemical to electrical) in the battery is

$$\mathcal{E}I = (12 \text{ V})(6 \text{ A}) = 72 \text{ W}.$$

26–22 Power relationships in a short circuit. With ideal wires and ammeter ($R = 0$), no electrical power leaves the battery.

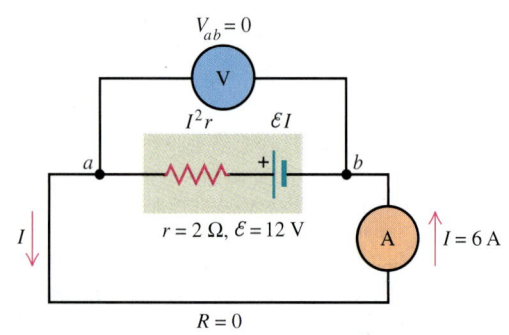

The rate of dissipation of energy in the battery is

$$I^2r = (6\ \text{A})^2(2\ \Omega) = 72\ \text{W}.$$

The net power output of the source, given by $V_{ab}I$, is zero

because the terminal voltage V_{ab} is zero. *All* of the converted energy is dissipated within the source. This is why a short-circuited battery is quickly ruined and in some cases may even explode.

*26-7 THEORY OF METALLIC CONDUCTION

We can gain additional insight into electrical conduction by looking at the microscopic origin of conductivity. We'll consider a very simple model that treats the electrons as classical particles and ignores their quantum-mechanical, wavelike behavior in solids. Using this model, we'll derive an expression for the resistivity of a metal. Even though this model is not entirely correct conceptually, it will still help you to develop an intuitive idea of the microscopic basis of conduction.

In the simplest microscopic model of conduction in a metal, each atom in the metallic crystal gives up one or more of its outer electrons. These electrons are then free to move through the crystal, colliding at intervals with the stationary positive ions. The motion of the electrons is analogous to the motion of molecules of a gas moving through a porous bed of sand, and they are often referred to as an "electron gas."

If there is no electric field, the electrons move in straight lines between collisions, the directions of their velocities are random, and on average they never get anywhere (Fig. 26–23a). But if an electric field is present, the paths curve slightly because of the acceleration caused by electric-field forces. Figure 26–23b shows a few paths of an electron in an electric field directed from right to left. As we mentioned in Section 26–2, the average speed of random motion is of the order of 10^6 m/s, while the average drift speed is *much* smaller, of the order of 10^{-4} m/s. The average time between collisions is called the **mean free time,** denoted by τ. Figure 26–24 (page 820) shows a mechanical analog of this electron motion.

We would like to derive from this model an expression for the resistivity ρ of a material, defined by Eq. (26–5):

$$\rho = \frac{E}{J}, \qquad (26\text{–}21)$$

where E and J are the magnitudes of electric field and current density. The current density $\vec{J}$ is in turn given by Eq. (26–4):

$$\vec{J} = nq\vec{v}_{\text{d}}, \qquad (26\text{–}22)$$

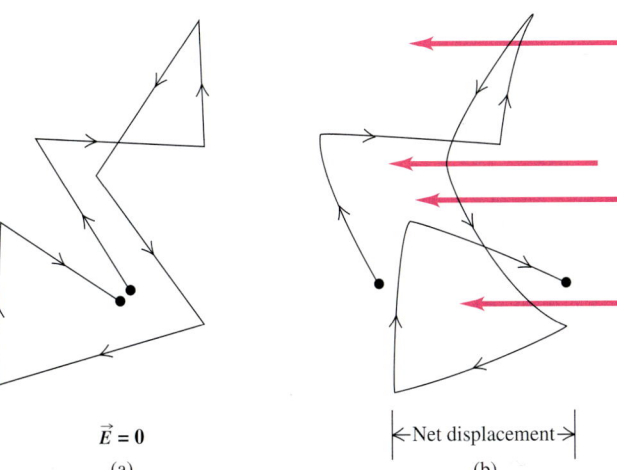

$\vec{E} = 0$

(a)

⊢Net displacement→⊣

(b)

26–23 (a) Random motion of electrons in a metallic crystal. (b) Motion with drift caused by electric-field forces. The curvatures of the paths are greatly exaggerated.

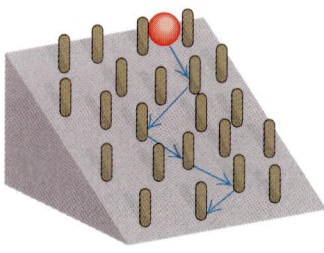

26-24 The motion of a ball rolling down an inclined plane and bouncing off pegs in its path is analogous to the motion of an electron in a metallic conductor with an electric field present.

where n is the number of free electrons per unit volume, q is the charge of each, and $\vec{v}_d$ is their average drift velocity. (We also know that $q = -e$ in an ordinary metal; we'll use that fact later.)

We need to relate the drift velocity $\vec{v}_d$ to the electric field $\vec{E}$. The value of $\vec{v}_d$ is determined by a steady-state condition in which, on average, the velocity *gains* of the charges due to the force of the $\vec{E}$ field are just balanced by the velocity *losses* due to collisions.

To clarify this process, let's imagine turning on the two effects one at a time. Suppose that before time $t = 0$ there is no field. The electron motion is then completely random. A typical electron has velocity $\vec{v}_0$ at time $t = 0$, and the value of $\vec{v}_0$ averaged over many electrons (that is, the initial velocity of an average electron) is zero: $(\vec{v}_0)_{av} = \mathbf{0}$. Then at time $t = 0$ we turn on a constant electric field $\vec{E}$. The field exerts a force $\vec{F} = q\vec{E}$ on each charge, and this causes an acceleration $\vec{a}$ in the direction of the force, given by

$$\vec{a} = \frac{\vec{F}}{m} = \frac{q\vec{E}}{m},$$

where m is the electron mass. *Every* electron has this acceleration.

We wait for a time τ, the average time between collisions, and then "turn on" the collisions. An electron that has velocity $\vec{v}_0$ at time $t = 0$ has a velocity at time $t = \tau$ equal to

$$\vec{v} = \vec{v}_0 + \vec{a}\tau.$$

The velocity $\vec{v}_{av}$ of an *average* electron at this time is the sum of the averages of the two terms on the right. As we have pointed out, the initial velocity $\vec{v}_0$ is zero for an average electron, so

$$\vec{v}_{av} = \vec{a}\tau = \frac{q\tau}{m}\vec{E}. \tag{26-23}$$

After time $t = \tau$, the tendency of the collisions to decrease the velocity of an average electron (by means of randomizing collisions) just balances the tendency of the $\vec{E}$ field to increase this velocity. Thus the velocity of an average electron, given by Eq. (26-23), is maintained over time and is equal to the drift velocity $\vec{v}_d$:

$$\vec{v}_d = \frac{q\tau}{m}\vec{E}.$$

Now we substitute this equation for the drift velocity $\vec{v}_d$ into Eq. (26-22):

$$\vec{J} = nq\vec{v}_d = \frac{nq^2\tau}{m}\vec{E}.$$

Comparing this with Eq. (26-21), which we can rewrite as $\vec{J} = \vec{E}/\rho$, and substituting $q = -e$, we see that the resistivity ρ is given by

$$\rho = \frac{m}{ne^2\tau}. \tag{26-24}$$

If n and τ are independent of $\vec{E}$, then the resistivity is independent of $\vec{E}$ and the conducting material obeys Ohm's law.

Turning the interactions on one at a time may seem artificial. But a little thought shows that the derivation would come out the same if each electron had its own clock and the $t = 0$ times were different for different electrons. If τ is the average time between collisions, then $\vec{v}_d$ is still the average electron drift velocity, even though the motions of the various electrons aren't actually correlated in the way we postulated.

What about the temperature dependence of resistivity? In a perfect crystal with no atoms out of place, a correct quantum-mechanical analysis would let the free electrons move through the crystal with no collisions at all. But the atoms vibrate about their equilibrium positions. As the temperature increases, the amplitudes of these vibrations increase, collisions become more frequent, and the mean free time τ decreases. So this theory predicts that the resistivity of a metal increases with temperature. In a superconductor, roughly speaking, there are no inelastic collisions, τ is infinite, and the resistivity ρ is zero.

In a pure semiconductor such as silicon or germanium, the number of charge carriers per unit volume, n, is not constant but increases very rapidly with increasing temperature. This increase in n far outweighs the decrease in the mean free time, and in a semiconductor the resistivity always decreases rapidly with increasing temperature. At low temperatures, n is very small, and the resistivity becomes so large that the material can be considered an insulator.

Electrons gain energy between collisions through the work done on them by the electric field. During collisions they transfer some of this energy to the atoms of the material of the conductor. This leads to an increase in the material's internal energy and temperature; that's why wires carrying current get warm. If the electric field in the material is large enough, an electron can gain enough energy between collisions to knock off electrons that are normally bound to atoms in the material. These can then knock off more electrons, and so on, possibly leading to an avalanche of current. This is the microscopic basis of dielectric breakdown in insulators.

EXAMPLE 26-12

Mean free time in copper Calculate the mean free time between collisions in copper at room temperature.

SOLUTION From Example 26–1 and Table 26–1, for copper $n = 8.5 \times 10^{28}$ m^{-3} and $\rho = 1.72 \times 10^{-8}$ $\Omega \cdot$ m. Also, $e = 1.60 \times 10^{-19}$ C and $m = 9.11 \times 10^{-31}$ kg for electrons. Rearranging Eq. (26–24), we get

$$\tau = \frac{m}{ne^2\rho}$$

$$= \frac{9.11 \times 10^{-31} \text{ kg}}{(8.5 \times 10^{28} \text{ m}^{-3})(1.60 \times 10^{-19} \text{ C})^2(1.72 \times 10^{-8} \ \Omega \cdot \text{m})}$$

$$= 2.4 \times 10^{-14} \text{ s.}$$

Taking the reciprocal of this time, we find that each electron averages about 4×10^{13} collisions every second!

*26–8 PHYSIOLOGICAL EFFECTS OF CURRENTS

Electrical potential differences and currents play a vital role in the nervous systems of animals. Conduction of nerve impulses is basically an electrical process, although the mechanism of conduction is much more complex than in simple materials such as metals. A nerve fiber, or *axon*, along which an electrical impulse can travel has a cylindrical membrane with one conducting fluid inside and another outside (Fig. 26–25a). In the resting condition the only charged bodies that can cross the membrane are positively charged potassium ions, which leak out of the axon into the surroundings. This leaves the inside of the axon with a net negative charge and a potential of about -70 mV (-0.07 V) with respect to the outer fluid.

When an electrical stimulus is applied to the axon, the membrane temporarily becomes more permeable to other ions in the fluids, leading to a local change in potential difference (Fig. 26–25b). This disturbance, called the *action potential*, propagates along the membrane as a pulse with a speed of the order of 30 m/s. The entire pulse passes a given point along the axon in a few milliseconds, after which the membrane recovers and the potential difference returns to its initial value (Fig. 26–25c).

Potassium ions (K$^+$) play an essential role in nerve conduction. A healthy diet includes foods rich in potassium, such as bananas, tomatoes, spinach, and oranges.

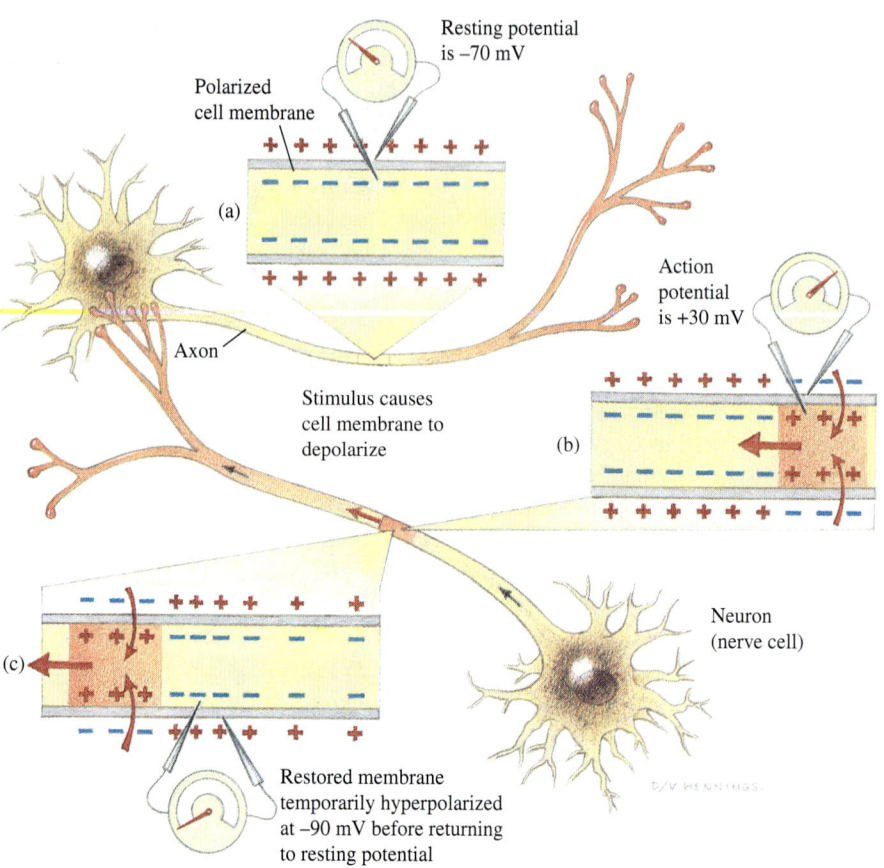

26–25 (a) The cell membrane around an axon maintains a resting potential difference of about −70 mV between inner and outer fluids. (b) An electrical stimulus depolarizes the membrane, and the potential difference suddenly rises to about +30 mV. (c) The change in potential difference, called the *action potential*, propagates along the axon. After the pulse passes, the potential difference first over-shoots to about −90 mV, then recovers its original value.

The electrical nature of nerve impulses explains why the body is sensitive to exter-nally supplied electrical currents. Currents through the body as small as 0.1 A can be fatal because they interfere with essential nerve processes such as those in the heart. Even smaller currents can also be very dangerous. A current of 0.01 A through an arm or leg causes strong, convulsive muscle action and considerable pain; with a current of 0.02 A, a person holding the conductor that is inflicting the shock is typically unable to release it. Currents of this magnitude through the chest can cause ventricular fibrillation, a disorganized twitching of heart muscles that pumps very little blood. Surprisingly, very large currents (over 0.1 A) are somewhat *less* likely to cause fatal fibrillation because the heart muscle is "clamped" in one position. The heart actually stops beating and is more likely to resume normal beating when the current is removed. The electric defibrillators used for medical emergencies apply a large current pulse to stop the heart (and the fib-rillation) to give the heart a chance to restore normal rhythm.

Body fluids are usually quite good conductors because of their substantial ion concentrations. By comparison the resistance of skin is relatively high, ranging from 500 kΩ for very dry skin to 1000 Ω or so for wet skin, depending also on the area of contact. If $R = 1000\ \Omega$, a current of 0.1 A requires a potential difference of $V = IR = (0.1\ \text{A})(1000\ \Omega) = 100\ \text{V}$. Were it not for the high resistance of skin, even an ordinary 1.5-V flashlight battery could produce a hazardous shock.

In summary, electric current poses three different kinds of hazards: Interference with the nervous system, injury caused by convulsive muscle action, and burns from I^2R heat-ing (described in Section 26–6). The moral of this rather morbid story is that under certain conditions, voltages as small as 10 V can be dangerous. *All* electrical circuits and electrical equipment should always be approached with respect and caution.

On the positive side, alternating currents with frequencies of the order of 10^6 Hz do not interfere appreciably with nerve processes and can be used for therapeutic heating for arthritic conditions, sinusitis, and other disorders. If one electrode is made very small, the resulting concentrated heating can be used for local destruction of tissue such as tumors or for cutting tissue in certain surgical procedures.

The study of particular nerve impulses is an important diagnostic tool in medicine. The most familiar examples are electrocardiography (EKG) and electroencephalography (EEG). Electrocardiograms, obtained by attaching electrodes to the chest and back and recording the regularly varying potential differences, are used to study heart function. Electrodes attached to the scalp permit study of potentials in the brain, and the resulting patterns can be helpful in diagnosing disorders such as epilepsy or brain tumors.

SUMMARY

- Current is the amount of charge flowing through a specified area, per unit time. The SI unit of current is the ampere, equal to one coulomb per second (1 A = 1 C/s). If the charge carriers in a material have concentration n, charge q, and drift velocity $\vec{v}_d$ (of magnitude v_d), the current I through an area A is

$$I = \frac{dQ}{dt} = n|q|v_d A. \qquad (26-2)$$

Current density $\vec{J}$ is current per unit cross-section area. In terms of the above quantities,

$$\vec{J} = nq\vec{v}_d. \qquad (26-4)$$

Current is conventionally described in terms of a flow of positive charge, even when the actual charge carriers are negative or of both signs.

- The resistivity ρ of a material is defined by the ratio of the magnitudes of electric field and current density:

$$\rho = \frac{E}{J}. \qquad (26-5)$$

The SI unit of resistivity is the ohm-meter (1 $\Omega \cdot$ m). Good conductors have small resistivity; good insulators have large resistivity. Ohm's law, obeyed approximately by many materials, states that ρ is a constant independent of the value of E. Resistivity usually increases with temperature; for small temperature changes, this variation can be represented approximately by

$$\rho(T) = \rho_0[1 + \alpha(T - T_0)], \qquad (26-6)$$

where α is the temperature coefficient of resistivity.

- For materials that obey Ohm's law, the potential difference V across a particular sample of material is proportional to the current I through the material:

$$V = IR, \qquad (26-11)$$

where R is the resistance of the sample. In terms of resistivity ρ, length L, and cross-section area A,

$$R = \frac{\rho L}{A}. \qquad (26-10)$$

The SI unit of resistance is 1 ohm (1 Ω = 1 V/A).

- A complete circuit is a conducting loop that provides a continuous current-carrying path. A complete circuit carrying a steady current must contain a source of electromotive force (emf) $\mathcal{E}$. The SI unit of electromotive force is 1 volt (1 V). An ideal source of emf maintains a constant potential difference, independent of current through the device, but every real source of emf has some internal resistance r. The terminal voltage V_{ab} then depends on current:

$$V_{ab} = \mathcal{E} - Ir \qquad \text{(source with internal resistance).} \qquad (26\text{–}15)$$

- Many devices do not obey Ohm's law but show a more complicated voltage-current relationship; often, this must be represented graphically.

- A circuit element with a potential difference V_{ab} and a current I puts energy into a circuit if the current direction is from lower to higher potential in the device and takes energy out of the circuit if the current is opposite. The power P (rate of energy transfer) is given by

$$P = V_{ab}I. \qquad (26\text{–}17)$$

A resistor R always takes electrical energy out of a circuit, converting it to thermal energy at a rate given by

$$P = V_{ab}I = I^2R = \frac{V_{ab}^{\,2}}{R}. \qquad (26\text{–}18)$$

- The microscopic basis of conduction in metals is the motion of electrons that move freely through the metallic crystal, bumping into ion cores in the crystal. In a crude classical model of this motion, the resistivity of the material can be related to the electron mass, charge, density, and mean free time between collisions.

DISCUSSION QUESTIONS

Q26–1 How would you expect the resistivity of a good insulator such as glass or polystyrene to vary with temperature? Explain your reasoning.

Q26–2 Two copper wires of different diameter are joined end-to-end. If a current flows in the wire combination, what happens to electrons when they move from the larger-diameter wire into the smaller-diameter wire? Does their drift speed increase, decrease, or stay the same? If the drift speed changes, what is the force that causes the change? Explain your reasoning.

Q26–3 What is the difference between an emf and a potential difference? Under what circumstances are the potential difference between the terminals of a battery and the emf of the battery equal to each other? Under what circumstances are they not equal?

Q26–4 Can the potential difference between the terminals of a battery ever be opposite in direction to the emf? If it can, give an example. If it cannot, explain why not.

Q26–5 A rule of thumb used to determine the internal resistance of a source is that it is the open-circuit voltage divided by the short-circuit current. Is this correct? Why or why not?

Q26–6 Batteries are always labeled with their emf; for instance, a AA flashlight battery is labeled "1.5 volts." Would it be appropriate to also put a label on batteries stating how much current they provide? Why or why not?

Q26–7 Is there an electric field in the space around an ordinary flashlight battery if it is not connected to anything? Explain.

Q26–8 In circuit analysis it is often assumed that a wire connecting two circuit elements has no potential difference between its ends. Yet there must be an electric field within the wire to make the charges move, and so there must be a potential difference. How do you resolve this discrepancy?

Q26–9 Electrons in an electric circuit pass through a resistor. The wire on either side of the resistor has the same diameter. a) How does the drift speed of the electrons before entering the resistor compare to the speed after leaving the resistor? Explain your reasoning. b) How does the potential energy for an electron before entering the resistor compare to the potential energy after leaving the resistor? Explain your reasoning.

Q26–10 When an electric current passes through a resistor, the current loses energy, transferring thermal energy to the resistor. Does the current lose kinetic energy, potential energy, or a combination of the two? Explain your reasoning.

Q26–11 A light bulb glows because it has resistance. The brightness of a light bulb increases with the electrical power dissipated in the bulb. a) In the circuit shown in Fig. 26–26a, the two bulbs A and B are identical. Compared to bulb A, does bulb

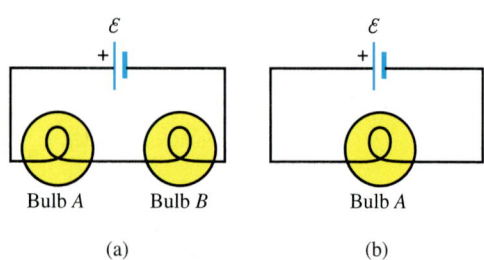

Bulb A Bulb B Bulb A

(a)   (b)

FIGURE 26–26 Question Q26–11.

B glow more brightly, just as brightly, or less brightly? Explain your reasoning. b) Bulb *B* is removed from the circuit and the circuit is completed as shown in Fig. 26–26b. Compared to the brightness of bulb *A* in Fig. 26–26a, does bulb *A* now glow more brightly, just as brightly, or less brightly? Explain your reasoning.

Q26–12 (See Question Q26–11.) An ideal ammeter A is placed in a circuit with a battery and a light bulb as shown in Fig. 26–27a, and the ammeter reading is noted. The circuit is then reconnected as in Fig. 26–27b, so that the positions of the ammeter and light bulb are reversed. a) How does the ammeter reading in the situation shown in Fig. 26–27a compare to the reading in the situation shown in Fig. 26–27b? Explain your reasoning. b) In which situation does the light bulb glow more brightly? Explain your reasoning.

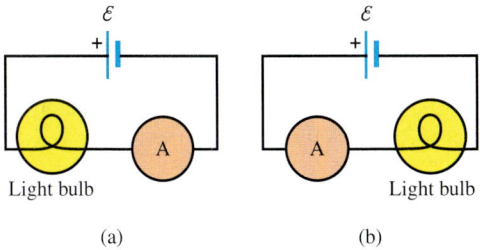

FIGURE 26–27 Question Q26–12.

Q26–13 (See Question Q26–11.) Will a light bulb glow more brightly if it is connected to a battery as shown in Fig. 26–28a, in which an ideal ammeter A is placed in the circuit, or if it is connected as shown in Fig. 26–28b, in which an ideal voltmeter V is placed in the circuit? Explain your reasoning.

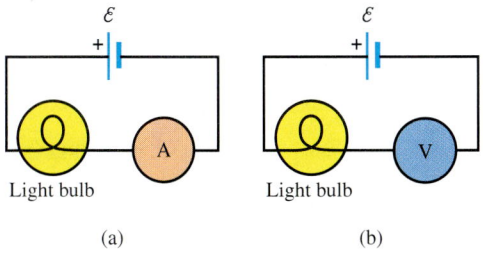

FIGURE 26–28 Question Q26–13.

Q26–14 The energy that can be extracted from a storage battery is always less than the energy that goes into it while it is being charged. Why?

Q26–15 Eight flashlight batteries in series have an emf of about 12 V, similar to that of a car battery. Could they be used to start a car with a dead battery? Why or why not?

Q26–16 A would-be inventor proposed to increase the power supplied by a battery to a light bulb by using thick wire near the battery and thinner wire near the bulb. In the thin wire the electrons from the thick wire would become more densely packed, more electrons per second would reach the bulb, and the bulb would receive more power than if a wire of constant cross section were used. What do you think of this scheme?

Q26–17 Small aircraft often have 24-V electrical systems rather than the 12-V systems in automobiles, even though the electrical power requirements are roughly the same in both applications. The explanation given by aircraft designers is that a 24-V system weighs less than a 12-V system, because thinner wires can be used. Explain why this is so.

Q26–18 Long-distance electric-power transmission lines always operate at very high voltage, sometimes as much as 750 kV. What are the advantages of such high voltages? What are the disadvantages?

Q26–19 Ordinary household electric lines in North America usually operate at 120 V. Why is this a desirable voltage, rather than a value considerably larger or smaller? On the other hand, automobiles usually have 12-V electrical systems. Why is this a desirable voltage?

Q26–20 Electric power for household and commercial use always uses *alternating current,* which reverses direction 120 times each second. A student claimed that the power conveyed by such a current would have to average out to zero, since the current is going one way half the time and the other way the rest of the time. What is your response to this claim?

Q26–21 A fuse is a device designed to break a circuit, usually by melting when the current exceeds a certain value. What characteristics should the material of the fuse have?

Q26–22 High-voltage power supplies are sometimes designed intentionally to have rather large internal resistance as a safety precaution. Why is such a power supply with a large internal resistance safer than one with the same voltage but lower internal resistance?

Q26–23 The text states that good thermal conductors are also good electrical conductors. If so, why don't the cords used to connect toasters, irons, and similar heat-producing appliances get hot by conduction of heat from the heating element?

EXERCISES

SECTION 26–2 **CURRENT**

26–1 A current of 3.6 A flows through an automobile headlight. How many coulombs of charge flow through the headlight in 3.0 h?

26–2 A silver wire 2.6 mm in diameter transfers a charge of 420 C in 80 min. Silver contains 5.8×10^{28} free electrons per cubic meter. a) What is the current in the wire? b) What is the magnitude of the drift velocity of the electrons in the wire?

26–3 Copper has 8.5×10^{28} free electrons per cubic meter. A 71.0-cm length of 12-gauge copper wire, which is 2.05 mm in diameter, carries 4.85 A of current. a) How much time does it take for an electron to travel the length of the wire? b) Repeat part (a) for 6-gauge copper wire (diameter 4.12 mm) of the same length that carries the same current. c) Generally speaking, how does changing the diameter of a wire that carries a given amount of current affect the drift velocity of the electrons in the wire?

26–4 Current passes through a solution of sodium chloride. In 1.00 s, 2.68×10^{16} Na$^+$ ions arrive at the negative electrode and 3.92×10^{16} Cl$^-$ ions arrive at the positive electrode. What is the current passing between the electrodes? b) What is the direction of the current?

26–5 The current in a wire varies with time according to the relation $I = 55$ A $- (0.65$ A/s$^2)t^2$. a) How many coulombs of charge pass a cross section of the wire in the time interval between $t = 0$ and $t = 8.0$ s? b) What constant current would transport the same charge in the same time interval?

SECTION 26–3 RESISTIVITY
SECTION 26–4 RESISTANCE

26–6 A copper wire has a square cross section 2.3 mm on a side. The wire is 4.0 m long and carries a current of 3.6 A. The density of free electrons is 8.5×10^{28}/m^3. Find the magnitudes of a) the current density in the wire; b) the electric field in the wire. c) How much time is required for an electron to travel the length of the wire?

26–7 In household wiring, copper wire 2.05 mm in diameter is often used. Find the resistance of a 24.0-m length of this wire.

26–8 What length of copper wire 0.462 mm in diameter has a resistance of 1.00 Ω?

26–9 In an experiment conducted at room temperature, a current of 0.820 A flows through a wire 3.26 mm in diameter. Find the magnitude of the electric field in the wire if the wire is made of a) tungsten; b) aluminum.

26–10 What diameter must a copper wire have if its resistance is to be the same as that of an equal length of aluminum wire with diameter 3.26 mm?

26–11 As part of a lecture demonstration a physics professor plans to hold an uninsulated current-carrying wire in her hands. For safety's sake, the potential difference between her hands is to be no more than 1.50 V. She holds her hands 1.20 m apart, with wire stretched tightly between them. The wire is to carry 6.00 A of current and is to be made of aluminum. What is the minimum wire radius that is consistent with safety?

26–12 You apply a potential difference of 4.50 V between the ends of a wire that is 2.50 m in length and 0.654 mm in radius. The resulting current through the wire is 17.6 A. What is the resistivity of the wire?

26–13 A current-carrying gold wire has diameter 0.84 mm. The electric field in the wire is 0.49 V/m. What is a) the current carried by the wire? b) the potential difference between two points in the wire 6.4 m apart? c) the resistance of a 6.4-m length of this wire?

26–14 A wire of length L and cross-section area A has resistance R. What will be the resistance of the wire if it is stretched to twice its original length? Assume that the density and resistivity of the material do not change when the wire is stretched.

26–15 The potential difference between points in a wire 75.0 cm apart is 0.938 V when the current density is 4.40×10^7 A/m^2. What is a) the magnitude of $\vec{E}$ in the wire; b) the resistivity of the material of which the wire is made?

26–16 A certain resistor has a resistance of 1.484 Ω at 20.0°C and a resistance of 1.512 Ω at 34.0°C. What is its temperature coefficient of resistivity?

26–17 a) What is the resistance of a Nichrome wire at 0.0°C if its resistance is 100.00 Ω at 11.5°C? b) What is the resistance of a carbon rod at 25.8°C if its resistance is 0.0160 Ω at 0.0°C?

26–18 A carbon resistor is to be used as a thermometer. On a winter day when the temperature is 4.0°C, the resistance of the carbon resistor is 217.3 Ω. What is the temperature on a spring day when the resistance is 215.8 Ω? (Take the reference temperature T_0 to be 4.0°C.)

26–19 A strand of wire has resistance 5.60 μΩ. Find the net resistance of 120 such strands if they are a) placed side by side to form a cable of the same length as a single strand; b) connected end to end to form a wire 120 times as long as a single strand.

SECTION 26–5 ELECTROMOTIVE FORCE AND CIRCUITS

26–20 Consider the circuit shown in Fig. 26–29. The terminal voltage of the 24.0-V battery is 21.2 V. What is a) the internal resistance r of the battery; b) the resistance R of the circuit resistor?

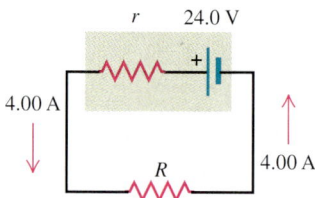

FIGURE 26–29 Exercise 26–20.

26–21 An ideal voltmeter V is connected to a 2.0-Ω resistor and a battery with emf 5.0 V and internal resistance 0.5 Ω as shown in Fig. 26–30. a) What is the current in the 2.0-Ω resistor? b) What is the terminal voltage of the battery? c) What is the reading on the voltmeter? Explain your answers.

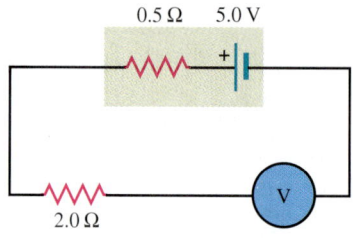

FIGURE 26–30 Exercise 26–21.

26–22 A complete circuit consists of a 24.0-V battery, a 5.60-Ω resistor, and a switch. The internal resistance of the battery is 0.28 Ω. The switch is opened. What does an ideal voltmeter read when placed a) across the terminals of the battery? b) across the resistor? c) across the switch? d) Repeat parts (a), (b), and (c) for the case when the switch is closed.

26–23 When switch S in Fig. 26–31 is open, the voltmeter V of the battery reads 3.08 V. When the switch is closed, the voltmeter reading drops to 2.97 V, and the ammeter A reads 1.65 A. Find the emf, the internal resistance of the battery, and the circuit resistance R. Assume that the two meters are ideal, so they don't affect the circuit.

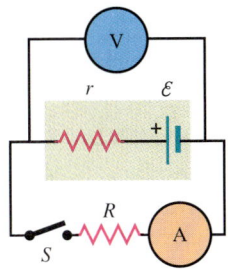

FIGURE 26–31 Exercise 26–23.

26–24 The circuit shown in Fig. 26–32 contains two batteries, each with an emf and an internal resistance, and two resistors. Find a) the current in the circuit (magnitude *and* direction); b) the terminal voltage V_{ab} of the 16.0-V battery; c) the potential difference V_{ac} of point a with respect to point c. d) Using Fig. 26–16 as a model, graph the potential rises and drops in this circuit.

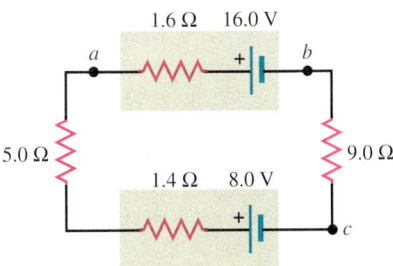

FIGURE 26–32 Exercises 26–24, 26–25, 26–26, and 26–34.

26–25 In the circuit shown in Fig. 26–32, the 16.0-V battery is removed and reinserted with the opposite polarity, so that its negative terminal is now next to point a. Find a) the current in the circuit (magnitude *and* direction); b) the terminal voltage V_{ab} of the 16.0-V battery; c) the potential difference V_{ac} of point a with respect to point c. d) Graph the potential rises and drops in this circuit (see Fig. 26–16).

26–26 In the circuit of Fig. 26–32, the 5.0-Ω resistor is removed and replaced by a resistor of unknown resistance R. When this is done, an ideal voltmeter connected across the points b and c reads 1.9 V. Find a) the current in the circuit; b) the resistance R. c) Graph the potential rises and drops in this circuit (see Fig. 26–16).

26–27 The following measurements of current and potential difference were made on a resistor constructed of Nichrome wire:

I (A)	0.50	1.00	2.00	4.00
V_{ab} (V)	1.94	3.88	7.76	15.52

a) Graph V_{ab} as a function of I. b) Does Nichrome obey Ohm's law? How can you tell? c) What is the resistance of the resistor in ohms?

26–28 The following measurements were made on a Thyrite resistor:

I (A)	0.50	1.00	2.00	4.00
V_{ab} (V)	2.55	3.11	3.77	4.58

a) Graph V_{ab} as a function of I. b) Does Thyrite obey Ohm's law? How can you tell? c) Graph the resistance $R = V_{ab}/I$ as a function of I.

26–29 The internal resistance of a flashlight battery increases gradually with age, even though the battery is not used. The emf, however, remains fairly constant at about 1.5 V. Batteries may be tested for age at the time of purchase by connecting an ammeter directly across the terminals of the battery and reading the current. The resistance of the ammeter is so small that the battery is practically short-circuited. a) The short-circuit current of a fresh flashlight battery (1.50-V emf) is 14.8 A. What is the internal resistance? b) What is the internal resistance if the short-circuit current is only 6.8 A? c) The short-circuit current of a 12.6-V car battery may be as great as 1000 A. What is its internal resistance?

SECTION 26–6 ENERGY AND POWER IN ELECTRIC CIRCUITS

26–30 A resistor with a 15.0-V potential difference across its ends develops thermal energy at a rate of 327 W. a) What is its resistance? b) What is the current in the resistor?

26–31 To stun its prey, the electric eel *Electrophorus electricus* generates 0.80-A pulses of current along its skin. This current flows across a 650-V potential difference. At what rate does *Electrophorus* deliver energy to its prey?

26–32 A battery-powered global positioning system (GPS) receiver operating on 9.0 V draws a current of 0.13 A. How much electrical energy does it consume during $1\frac{1}{2}$ h?

26–33 Consider a resistor with length L, uniform cross-section area A, and uniform resistivity ρ that is carrying a current with uniform current density J. Use Eq. (26–18) to find the electrical power dissipated per unit volume, p. Express your result in terms of a) E and J; b) J and ρ; c) E and ρ.

26–34 Consider the circuit of Fig. 26–32. a) What is the total rate at which electrical energy is being dissipated in the 5.00-Ω and 9.00-Ω resistors? b) What is the power output of the 16.0-V battery? c) At what rate is electrical energy being converted to other forms in the 8.0-V battery? d) Show that the overall rate of conversion of non-electrical (chemical) energy to electrical energy equals the overall rate of dissipation of electrical energy in the circuit.

26–35 The capacity of a storage battery, such as those used in automobile electrical systems, is rated in ampere-hours (A · h). A 50-A · h battery can supply a current of 50 A for 1.0 h, or 25 A for 2.0 h, and so on. a) What total energy can be supplied by a 12-V, 60-A · h battery if its internal resistance is negligible? b) What volume (in liters) of gasoline has a total heat of combustion equal to the energy obtained in part (a)? (See Section 15–7. The density of gasoline is 900 kg/m^3.) c) If a generator with an average electrical power output of 0.45 kW is connected to the battery, how much time will be required for it to charge the battery fully?

26–36 In the circuit analyzed in Example 26–9 (Section 26–6) the 4.0-Ω resistor is replaced by a 8.0-Ω resistor, as in Example 26–10 (Section 26–9). a) Calculate the rate of conversion of chemical energy to electrical energy in the battery. How does your answer compare to the result calculated in Example 26–9? b) Calculate the rate of electrical energy dissipation in the internal resistance of the battery. How does your answer compare to the result calculated in Example 26–9? c) Use the results of parts (a) and (b) to calculate the net power output of the battery. How does your result compare to the electrical power dissipated in the 8.0-Ω resistor as calculated for this circuit in Example 26–10?

26–37 In the circuit in Fig. 26–33, find a) the rate of conversion of internal (chemical) energy to electrical energy within the battery; b) the rate of dissipation of electrical energy in the battery; c) the rate of dissipation of electrical energy in the external resistor.

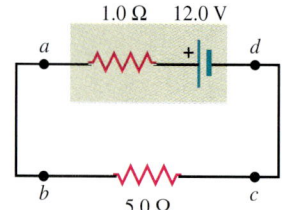

FIGURE 26–33 Exercise 26–37.

26–38 A typical small flashlight contains two batteries, each having an emf of 1.5 V, connected in series with a bulb having resistance 17 Ω. a) If the internal resistance of the batteries is negligible, what power is delivered to the bulb? b) If the batteries last for 5.0 h, what is the total energy delivered to the bulb? c) The resistance of real batteries increases as they run down. If the initial internal resistance is negligible, what is the combined internal resistance of both batteries when the power to the bulb has decreased to half its initial value? (Assume that the resistance of the bulb is constant. Actually, it will change somewhat when the current through the filament changes, because this changes the temperature of the filament and hence the resistivity of the filament wire.)

26–39 A "540-W" electric heater is designed to operate from 120-V lines. a) What is its resistance? b) What current does it draw? c) If the line voltage drops to 110 V, what power does the heater take? (Assume that the resistance is constant. Actually, it will change because of the change in temperature.) d) The heater coils are metallic so that the resistance of the heater decreases with decreasing temperature. If the change of resistance with temperature is taken into account, will the electrical power consumed by the heater be larger or smaller than what you calculated in part (c)? Explain.

*SECTION **26–7** THEORY OF METALLIC CONDUCTION

***26–40** Pure silicon contains approximately 1.0×10^{16} free electrons per cubic meter. a) At room temperature, what mean free time τ gives a value of the resistivity that agrees with the value given in Table 26–1? b) Your answer in part (a) is much greater than the mean free time for copper given in Example 26–12 (Section 26–7). Why, then, does pure silicon have such a high resistivity compared to copper?

*SECTION **26–8** PHYSIOLOGICAL EFFECTS OF CURRENTS

***26–41** A person with body resistance between his hands of 10 kΩ accidentally grasps the terminals of a 14-kV power supply. a) If the internal resistance of the power supply is 2000 Ω, what is the current through the person's body? b) What is the power dissipated in his body? c) If the power supply is to be made safe by increasing its internal resistance, what should the internal resistance be for the maximum current in the above situation to be 1.00 mA or less?

***26–42** The average bulk resistivity of the human body (apart from surface resistance of the skin) is about 5.0 $\Omega \cdot$ m. The conducting path between the hands can be represented approximately as a cylinder 1.6 m long and 0.10 m in diameter. The skin resistance can be made negligible by soaking the hands in salt water. a) What is the resistance between the hands if the skin resistance is negligible? b) What potential difference between the hands is needed for a lethal shock current of 100 mA? (Note that your result shows that small potential differences produce dangerous currents when the skin is damp.) c) With the current in part (b), what power is dissipated in the body?

PROBLEMS

26–43 An electrical conductor designed to carry large currents has a circular cross section 2.50 mm in diameter and is 14.0 m long. The resistance between its ends is 0.104 Ω. a) What is the resistivity of the material? b) If the electric field magnitude in the conductor is 1.28 V/m, what is the total current? c) If the material has 8.5×10^{28} free electrons per cubic meter, find the average drift speed under the conditions of part (b).

26–44 A 2.0-m length of wire is made by welding the end of a 120-cm-long silver wire to the end of an 80-cm-long copper wire. Each piece of wire is 0.60 mm in diameter. The wire is at room temperature, so the resistivities are as given in Table 26–1. A potential difference of 5.0 V is maintained between the ends of the 2.0-m composite wire. a) What is the current in the copper section? b) What is the current in the silver section? c) What is

the magnitude of $\vec{E}$ in the copper? d) What is the magnitude of $\vec{E}$ in the silver? e) What is the potential difference between the ends of the silver section of wire?

26–45 A 3.00-m length of copper wire at 20°C has a 1.20-m-long section with diameter 1.60 mm and a 1.80-m-long section with diameter 0.80 mm. There is a current of 2.5 mA in the 1.60-mm diameter section. a) What is the current in the 0.80-mm-diameter section? b) What is the magnitude of $\vec{E}$ in the 1.60-mm-diameter section? c) What is the magnitude of $\vec{E}$ in the 0.80-mm-diameter section? d) What is the potential difference between the ends of the 3.00-m length of wire?

26–46 a) The available kinetic energy per unit volume due to the drift velocity of the conduction electrons in a current-carrying conductor can be defined as $K/\text{volume} = n(\frac{1}{2}mv_d^2)$.

Use this expression to evaluate K/volume for the copper wire and current of Example 26–1 (Section 26–2). b) Calculate the total change in electrical potential energy for the conduction electrons in 1.0 cm^3 of copper if they fall through a potential drop of 1.0 V. How does your answer compare to the available kinetic energy in 1.0 cm^3 due to the drift velocity?

26–47 A material of resistivity ρ is formed into a solid, truncated cone of height h and radii r_1 and r_2 at either end (Fig. 26–34). a) Calculate the resistance of the cone between the two flat end faces. (*Hint:* Imagine slicing the cone into very many thin disks, and calculate the resistance of one such disk.) b) Show that your result agrees with Eq. (26–10) when $r_1 = r_2$.

FIGURE 26–34 Problem 26–47.

26–48 The region between two concentric conducting spheres with radii a and b is filled with a conducting material with resistivity ρ. a) Show that the resistance between the spheres is given by

$$R = \frac{\rho}{4\pi}\left(\frac{1}{a} - \frac{1}{b}\right).$$

b) Derive an expression for the current density as a function of radius, in terms of the potential difference V_{ab} between the spheres. c) Show that the result in part (a) reduces to Eq. (26–10) when the separation $L = b - a$ between the spheres is small.

26–49 Leakage in a Dielectric. Two parallel plates of a capacitor have equal and opposite charges Q. The dielectric has a dielectric constant K and a resistivity ρ. Show that the "leakage" current I carried by the dielectric is given by $I = Q/K\epsilon_0 \rho$.

26–50 A rectangular block of metal of resistivity ρ has dimensions $d \times 2d \times 3d$. A potential difference V is to be applied between two opposite faces of the block. a) To which two faces of the block should the potential difference be applied to give the maximum *current density*? What is this maximum current density? b) To which two faces of the block should the potential difference be applied to give the maximum *current*? What is this maximum current?

26–51 The temperature coefficient of resistance α in Eq. (26–12) equals the temperature coefficient of resistivity α in Eq. (26–6) only if the coefficient of thermal expansion is small. A cylindrical column of mercury is in a vertical glass tube. At 20°C, the length of the mercury column is 12.0 cm. The diameter of the mercury column is 1.6 mm and doesn't change with temperature because glass has a small coefficient of thermal expansion. The coefficient of volume expansion of the mercury is given in Table 15–2, its resistivity at 20°C is given in Table 26–1, and its temperature coefficient of resistivity is given in Table 26–2. a) At 20°C, what is the resistance between the ends of the mercury

column? b) The mercury column is heated to 60°C. What is the change in its resistivity? c) What is the change in its length? Explain why the coefficient of volume expansion rather than the coefficient of linear expansion determines the change in length. d) What is the change in its resistance? (*Hint:* Since the percentage changes in ρ and L are small you may find it helpful to derive from Eq. (26–10) an equation for ΔR in terms of $\Delta\rho$ and ΔL.) e) What is the temperature coefficient of resistance α for the mercury column, as defined in Eq. (26–12)? How does this value compare with the temperature coefficient of resistivity? Is the effect of the change in length important?

26–52 a) What is the potential difference V_{ad} in the circuit of Fig. 26–35? b) What is the terminal voltage of the 4.00-V battery? c) A battery with emf 10.30 V and internal resistance 0.50 Ω is inserted in the circuit at d, with its negative terminal connected to the negative terminal of the 8.00-V battery. What is now the difference of potential V_{bc} between the terminals of the 4.00-V battery?

FIGURE 26–35 Problem 26–52.

26–53 The potential difference across the terminals of a battery is 8.4 V when there is a current of 1.50 A in the battery from the negative to the positive terminal. When the current is 3.50 A in the reverse direction, the potential difference becomes 9.4 V. a) What is the internal resistance of the battery? b) What is the emf of the battery?

26–54 A semiconductor device that does not obey Ohm's law has the current-voltage relation $V = \alpha I + \beta I^2$, with $\alpha = 2.50$ Ω and $\beta = 0.360$ Ω/A. a) If the device is connected across a potential difference of 4.00 V, what is the current through the device? b) What potential difference is required to produce a current through the device twice that calculated in part (a)?

26–55 A 12.6-V car battery with negligible internal resistance is connected to a series combination of a 3.2-Ω resistor that obeys Ohm's law and a thermistor that does not obey Ohm's law, but instead has a current-voltage relation $V = \alpha I + \beta I^2$, with $\alpha = 3.8$ Ω and $\beta = 1.3$ Ω/A. What is the current through the 3.2-Ω resistor?

26–56 The open-circuit terminal voltage of a source is 7.86 V, and its short-circuit current is 9.25 A. a) What is the current when a resistor with resistance 2.4 Ω is connected to the terminals of the source? The resistor obeys Ohm's law. b) What is the current in the semiconductor device of Problem 26–54 when connected across the terminals of this source? c) What is the terminal voltage of the source at the current calculated in part (b)?

26–57 A Non-Ideal Ammeter. Unlike the idealized ammeter described in Section 26–4, any real ammeter has a nonzero resistance. a) An ammeter with resistance R_A is connected in series with a resistor R and a battery of emf $\mathcal{E}$ and internal resistance r. The current measured by the ammeter is I_A. Find the current through the circuit if the ammeter is removed so that the battery and the resistor form a complete circuit. The more "ideal" the ammeter, the smaller the difference between this current and the current I_A. b) If $R = 3.80\ \Omega$, $\mathcal{E} = 7.50$ V, and $r = 0.45\ \Omega$, find the maximum value of the ammeter resistance R_A so that I_A is within 1.0% of the current in the circuit when the ammeter is absent. c) Explain why your answer in part (b) represents a *maximum* value.

26–58 A Non-Ideal Voltmeter. Unlike the idealized voltmeter described in Section 26–4, any real voltmeter has a resistance that is not infinitely large. a) A voltmeter with resistance R_V is connected across the terminals of a battery of emf $\mathcal{E}$ and internal resistance r. Find the potential difference measured by the voltmeter. b) If $\mathcal{E} = 7.50$ V and $r = 0.45\ \Omega$, find the minimum value of the voltmeter resistance R_V so that the voltmeter reading is within 1.0% of the emf of the battery. c) Explain why your answer in part (b) represents a *minimum* value.

26–59 According to the U.S. National Electrical Code, copper wire used for interior wiring of houses, hotels, office buildings, and industrial plants is permitted to carry no more than a specified maximum amount of current. The table below shows the maximum current I_{max} for several common sizes of wire with varnished cambric insulation. The "wire gauge" is a standard method used to describe the diameter of wires. Note that the larger the diameter of the wire, the *smaller* the wire gauge.

Wire gauge	Diameter (cm)	I_{max} (A)
14	0.163	18
12	0.205	25
10	0.259	30
8	0.326	40
6	0.412	60
5	0.462	65
4	0.519	85

a) What considerations determine the maximum current-carrying capacity of household wiring? b) A total of 4200 W of power is to be supplied through the wires of a house to the household electrical appliances. If the potential difference across the group of appliances is 120 V, determine the gauge of the thinnest permissible wire that can be used. c) Suppose the wire used in this house is of the gauge found in part (b) and has total length 42.0 m. At what rate is energy dissipated in the wires? d) The house is built in a community where the consumer cost of electric energy is $0.11 per kilowatt-hour. If the house were built with wire of the next larger diameter than that found in part (b), what would be the savings in electricity costs in one year? Assume that the appliances are kept on for an average of 12 hours a day.

26–60 A toaster using a Nichrome heating element operates on 120 V. When it is switched on at 20°C, the heating element carries an initial current of 1.35 A. A few seconds later the current reaches the steady value of 1.23 A. a) What is the final

temperature of the element? The average value of the temperature coefficient of resistivity for Nichrome over the temperature range is $4.5 \times 10^{-4}\ (\text{C}°)^{-1}$. b) What is the power dissipated in the heating element i) initially; ii) when the current reaches a steady value?

26–61 In the circuit of Fig. 26–36, find a) the current through the 8.0-Ω resistor; b) the total rate of dissipation of electrical energy in the 8.0-Ω resistor and in the internal resistance of the batteries. c) In one of the batteries, chemical energy is being converted into electrical energy. In which one is this happening, and at what rate? d) In one of the batteries, electrical energy is being converted into chemical energy. In which one is this happening, and at what rate? e) Show that the overall rate of production of electrical energy equals the overall rate of consumption of electrical energy in the circuit.

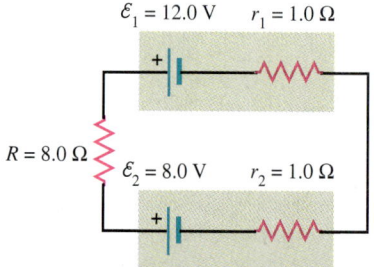

FIGURE 26–36 Problem 26–61.

26–62 A lightning bolt strikes one end of a steel lightning rod, producing a 15,000-A current burst that lasts for 65 μs. The rod is 2.0 m long and 1.8 cm in diameter and its other end is connected to the ground by 35 m of 8.0-mm-diameter copper wire. a) Find the potential difference between the top of the steel rod and the lower end of the copper wire during the current burst. b) Find the total energy deposited in the rod and wire by the current burst.

26–63 The *Tolman-Stewart experiment* (1916) demonstrated that the free charges in a metal have negative charge and provided a quantitative measurement of their charge to mass ratio, $|q|/m$. The experiment consisted of abruptly stopping a rapidly rotating spool of wire and measuring the potential difference that this produced between the ends of the wire. In a simplified model of this experiment, consider a metal rod of length L that is given a uniform acceleration $\vec{a}$ to the right. Initially the free charges in the metal lag behind the rod's motion, thus setting up an electric field $\vec{E}$ in the rod. In the steady state this field exerts a force on the free charges that makes them accelerate along with the rod. a) Apply $\Sigma \vec{F} = m\vec{a}$ to the free charges to obtain an expression for $|q|/m$ in terms of the magnitudes of the induced electric field $\vec{E}$ and the acceleration $\vec{a}$. b) If all the free charges in the metal rod have the same acceleration, the electric field $\vec{E}$ is the same at all points in the rod. Use this fact to rewrite the expression for $|q|/m$ in terms of the potential V_{bc} between the ends of the rod (Fig. 26–37). c) If the free charges have negative charge, which end of the rod, b or c, is at higher potential? d) If the rod is 0.50 m long and the free charges are electrons (charge $q = -1.60 \times 10^{-19}$ C, mass 9.11×10^{-31} kg), what magnitude of acceleration is required to produce a potential difference of 1.0 mV between the ends of the rod? e) Discuss why the actual

FIGURE 26–37 Problem 26–63.

experiment used a rotating spool of thin wire rather than a moving bar as in our simplified analysis.

26–64 A coil is to be used as an immersion heater for boiling water. The coil is to operate at 120 V and is to heat 250 cm^3 of water from 20°C to 100°C in 6.0 minutes. The specific heat capacity of water is 4190 J/kg · C°. a) What must be the resistance of the coil (assumed temperature-independent)? b) The coil is to be made of 25.0 cm^3 of Nichrome, which will be formed into a wire. What must be the total length of the wire used to make the coil, and what must be the radius of the wire? Assume that the wire has a circular cross section and that the volume of the Nichrome is unaffected by the process that forms it into a wire.

26–65 A 12.0-V battery has an internal resistance of 0.24 Ω and a capacity of 50.0 A · h. (See Exercise 26–35.) The battery is charged by passing a 10-A current through it for 5.0 h. a) What is the terminal voltage during charging? b) What total electrical energy is supplied to the battery during charging? c) What electrical energy is dissipated in the internal resistance during charging? d) The battery is now completely discharged through a resistor, again with a constant current of 10 A. What is the external circuit resistance? e) What total electrical energy is supplied to the external resistor? f) What total electrical energy is dissipated in the internal resistance? g) Why are the answers to parts (b) and (e) not equal?

26–66 Repeat Problem 26–65 with charge and discharge currents of 30 A. The charging and discharging times will now be 1.7 h rather than 5.0 h. What differences in performance do you see?

CHALLENGE PROBLEMS

26–67 The temperature coefficient of resistivity α is given by

$$\alpha = \frac{1}{\rho}\frac{d\rho}{dT},$$

where ρ is the resistivity at the temperature T. Eq. (26–6) then follows if α is assumed constant and much smaller than $(T - T_0)^{-1}$. a) If α is not constant but is given by $\alpha = -n/T$, where T is the Kelvin temperature and n is a constant, show that the resistivity is given by $\rho = a/T^n$ where a is a constant. b) From Fig. 26–7c, you can see that such a relation might be used as a rough approximation for a semiconductor. Using the values of ρ and α for carbon from Table 26–1 and Table 26–2, determine a and n. (In Table 26–1, assume that "room temperature" means 293 K.) c) Using your result from part (b), determine the resistivity of carbon at −196°C and 300°C. (Remember to express T in kelvins.)

26–68 The current-voltage relationship of a semiconductor diode is given by

$$I = I_S\left[\exp\left(\frac{eV}{kT}\right) - 1\right],$$

where I and V are the current through and the voltage across the diode, respectively, I_S is a constant characteristic of the device, e is the magnitude of the electron charge, k is the Boltzmann constant, and T is the Kelvin temperature. Such a diode is connected in series with a resistor with $R = 1.00$ Ω and a battery with $\mathcal{E} = 2.00$ V. The polarity of the battery is such that the current through the diode is in the forward direction (Fig. 26–38). The battery has negligible internal resistance. a) Obtain an equation for V. Note that you cannot solve for V algebraically. b) The value of V must be obtained by using a numerical method. One approach is to try a value of V, see how the left- and right-hand

sides of the equation compare for this V, and use this to refine your guess for V. Using $I_S = 1.50$ mA and $T = 293$ K, obtain a solution (accurate to three significant figures) for the voltage drop V across the diode and the current I through it.

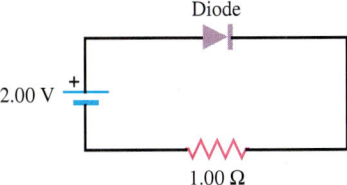

FIGURE 26–38 Challenge Problem 26–68.

26–69 The resistivity of a semiconductor can be modified by adding different amounts of impurities. A rod of semiconducting material of length L and cross-section area A lies along the x-axis between $x = 0$ and $x = L$. The material obeys Ohm's law, and its resistivity varies along the rod according to $\rho(x) = \rho_0 \exp(-x/L)$. The end of the rod at $x = 0$ is at a potential V_0 greater than the end at $x = L$. a) Find the total resistance of the rod and the current in the rod. b) Find the electric field magnitude $E(x)$ in the rod as a function of x. c) Find the electric potential $V(x)$ in the rod as a function of x. d) Graph the functions $\rho(x)$, $E(x)$, and $V(x)$ for values of x between $x = 0$ and $x = L$.

26–70 A source with emf $\mathcal{E}$ and internal resistance r is connected to an external circuit. a) Show that the power output of the source is maximum when the current in the circuit is one-half the short-circuit current of the source. b) If the external circuit consists of a resistance R, show that the power output is maximum when $R = r$ and that the maximum power is $\mathcal{E}^2/4r$.

Complex circuits are at the heart of all modern electronic devices. The conducting paths in this printed circuit are thin films (shown in blue-green) deposited on an insulating board. The operation of any such circuit, no matter how complex, can be understood using Kirchhoff's rules—a central topic of this chapter.

27 Direct-Current Circuits

27–1 INTRODUCTION

If you look inside your TV, your computer, or your stereo receiver or under the hood of a car, you will find circuits of much greater complexity than the simple circuits we studied in Chapter 26. Whether connected by wires or integrated in a semiconductor chip, these circuits often include several sources, resistors, and other circuit elements such as capacitors, transformers, and motors, interconnected in a *network*.

In this chapter we study general methods for analyzing such networks, including how to find unknown voltages, currents, and properties of circuit elements. We'll learn how to determine the equivalent resistance for several resistors connected in series or in parallel. For more general networks we need two rules called *Kirchhoff's rules*. One is based on the principle of conservation of charge applied to a junction; the other is derived from energy conservation for a charge moving around a closed loop. We'll discuss instruments for measuring various electrical quantities. We also look at a circuit containing resistance and capacitance, in which the current varies with time.

Our principal concern in this chapter is with **direct-current** (dc) circuits, in which the direction of the current does not change with time. Flashlights and automobile wiring systems are examples of direct-current circuits. Household electrical power is supplied in the form of **alternating current** (ac), in which the current oscillates back and forth. The same principles for analyzing networks apply to both kinds of circuits, and we conclude this chapter with a look at household wiring systems. We'll discuss alternating-current circuits in detail in Chapter 32.

27–2 RESISTORS IN SERIES AND PARALLEL

Resistors turn up in all kinds of circuits, ranging from hair dryers and space heaters to circuits that limit or divide current or reduce or divide a voltage. Such circuits often contain several resistors, so it's appropriate to consider *combinations* of resistors. A simple example is a string of light bulbs used for holiday decorations; each bulb acts as a resistor, and from a circuit-analysis perspective the string of bulbs is simply a combination of resistors.

Suppose we have three resistors with resistances R_1, R_2, and R_3. Figure 27–1 shows four different ways in which they might be connected between points a and b. When several circuit elements such as resistors, batteries, and motors are connected in sequence as in Fig. 27–1a, with only a single current path between the points, we say that they are connected in **series.** We studied *capacitors* in series in Section 25–3; we found that, because of conservation of charge, capacitors in series all have the same charge if they are initially uncharged. In circuits we're often more interested in the *current*, which is charge flow per unit time.

The resistors in Fig. 27–1b are said to be connected in **parallel** between points a and b. Each resistor provides an alternative path between the points. For circuit elements that are connected in parallel, the *potential difference* is the same across each element. We studied capacitors in parallel in Section 25–3.

In Fig. 27–1c, resistors R_2 and R_3 are in parallel, and this combination is in series with R_1. In Fig. 27–1d, R_2 and R_3 are in series, and this combination is in parallel with R_1.

For any combination of resistors we can always find a *single* resistor that could replace the combination and result in the same total current and potential difference. For example, a string of holiday light bulbs could be replaced by a single, appropriately chosen light bulb that would draw the same current and have the same potential difference between its terminals as the original string of bulbs. The resistance of this single resistor is called the **equivalent resistance** of the combination. If any one of the networks in Fig. 27–1 were replaced by its equivalent resistance R_{eq}, we could write

$$V_{ab} = IR_{eq}, \quad \text{or} \quad R_{eq} = \frac{V_{ab}}{I},$$

where V_{ab} is the potential difference between terminals a and b of the network and I is the current at point a or b. To compute an equivalent resistance, we assume a potential difference V_{ab} across the actual network, compute the corresponding current I, and take the ratio V_{ab}/I.

RESISTORS IN SERIES

We can derive general equations for the equivalent resistance of a series or parallel combination of resistors. If the resistors are in *series,* as in Fig. 27–1a, the current I must be the same in all of them. (As we discussed in Section 26–5, current is *not* "used up" as it passes through a circuit.) Applying $V = IR$ to each resistor, we have

$$V_{ax} = IR_1, \quad V_{xy} = IR_2, \quad V_{yb} = IR_3.$$

The potential differences across each resistor need not be the same (except for the special case in which all three resistances are equal). The potential difference V_{ab} across the entire combination is the sum of these individual potential differences:

$$V_{ab} = V_{ax} + V_{xy} + V_{yb} = I(R_1 + R_2 + R_3),$$

and so

$$\frac{V_{ab}}{I} = R_1 + R_2 + R_3.$$

The ratio V_{ab}/I is, by definition, the equivalent resistance R_{eq}. Therefore

$$R_{eq} = R_1 + R_2 + R_3.$$

It is easy to generalize this to any number of resistors:

$$R_{eq} = R_1 + R_2 + R_3 + \cdots \quad \text{(resistors in series).} \quad (27\text{–}1)$$

The equivalent resistance of *any number* of resistors in series equals the sum of their individual resistances. The equivalent resistance is *greater than* any individual resistance.

Let's compare this result with Eq. (25–5) for *capacitors* in series. Resistors in series add directly because the voltage across each is directly proportional to its resistance and to the common current. Capacitors in series add reciprocally because the voltage across each is directly proportional to the common charge but *inversely* proportional to the individual capacitance.

(a)

(b)

(c)

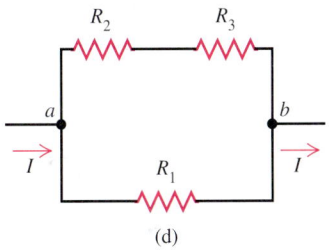

(d)

27–1 Four different ways of connecting three resistors.

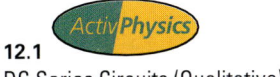

12.1
DC Series Circuits (Qualitative)

RESISTORS IN PARALLEL

If the resistors are in *parallel,* as in Fig. 27–1b, the current through each resistor need not be the same. But the potential difference between the terminals of each resistor must be the same and equal to V_{ab}. (Remember that the potential difference between any two points does not depend on the path taken between the points.) Let's call the currents in the three resistors I_1, I_2, and I_3. Then from $I = V/R$,

$$I_1 = \frac{V_{ab}}{R_1}, \qquad I_2 = \frac{V_{ab}}{R_2}, \qquad I_3 = \frac{V_{ab}}{R_3}.$$

In general, the current is different through each resistor. Because charge is not accumulating or draining out of point a, the total current I must equal the sum of the three currents in the resistors:

$$I = I_1 + I_2 + I_3 = V_{ab}\left(\frac{1}{R_1} + \frac{1}{R_2} + \frac{1}{R_3}\right),$$

or

$$\frac{I}{V_{ab}} = \frac{1}{R_1} + \frac{1}{R_2} + \frac{1}{R_3}.$$

But by definition of the equivalent resistance R_{eq}, $I/V_{ab} = 1/R_{eq}$, so

$$\frac{1}{R_{eq}} = \frac{1}{R_1} + \frac{1}{R_2} + \frac{1}{R_3}.$$

Again it is easy to generalize to *any number* of resistors in parallel:

$$\frac{1}{R_{eq}} = \frac{1}{R_1} + \frac{1}{R_2} + \frac{1}{R_3} + \cdots \qquad \text{(resistors in parallel)}. \tag{27–2}$$

For *any number* of resistors in parallel, the *reciprocal* of the equivalent resistance equals the *sum of the reciprocals* of their individual resistances. The equivalent resistance is always *less than* any individual resistance.

We can compare this result with Eq. (25–7) for *capacitors* in parallel. Resistors in parallel add reciprocally because the current in each is proportional to the common voltage across them and *inversely* proportional to the resistance of each. Capacitors in parallel add directly because the charge on each is proportional to the common voltage across them and *directly* proportional to the capacitance of each.

For the special case of *two* resistors in parallel,

$$\frac{1}{R_{eq}} = \frac{1}{R_1} + \frac{1}{R_2} = \frac{R_1 + R_2}{R_1 R_2},$$

and

$$R_{eq} = \frac{R_1 R_2}{R_1 + R_2} \qquad \text{(two resistors in parallel)}. \tag{27–3}$$

Because $V_{ab} = I_1 R_1 = I_2 R_2$, it follows that

$$\frac{I_1}{I_2} = \frac{R_2}{R_1} \qquad \text{(two resistors in parallel)}. \tag{27–4}$$

This shows that the currents carried by two resistors in parallel are *inversely proportional* to their resistances. More current goes through the path of least resistance.

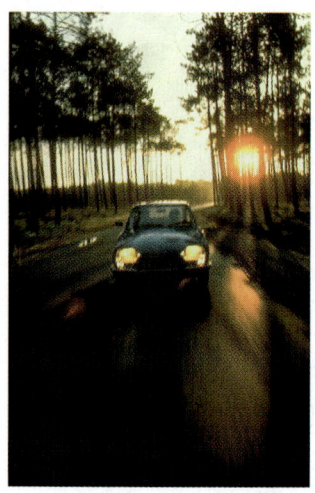

A car's headlights are connected in parallel. Hence each headlight is exposed to the full potential difference supplied by the car's electrical system, giving maximum brightness. Another advantage is that if one headlight burns out, the other one will keep shining (see Example 27–2 on page 836).

12.2
DC Parallel Circuits

Problem–Solving Strategy

RESISTORS IN SERIES AND PARALLEL

1. It helps to remember that when resistors are connected in series, the total potential difference across the combination is the sum of the individual potential differences. When they are connected in parallel, the potential difference is the same for every resistor and is equal to the potential difference across the parallel combination.

2. Also keep in mind the analogous statements for current. When resistors are connected in series, the current is the same through every resistor and is equal to the current through the series combination. When resistors are con-

nected in parallel, the total current through the combination is equal to the sum of currents through the individual resistors.

3. We can often consider networks such as those in Figs. 27–1c and 27–1d as combinations of series and parallel arrangements. In Fig. 27–1c we first replace the parallel combination of R_2 and R_3 by its equivalent resistance; this then forms a series combination with R_1. In Fig. 27–1d the combination of R_2 and R_3 in series forms a parallel combination with R_1.

EXAMPLE 27–1

Compute the equivalent resistance of the network in Fig. 27–2a, and find the current in each resistor. The source of emf has negligible internal resistance.

SOLUTION Figures 27–2b and 27–2c show successive stages in reducing the network to a single equivalent resistance. From Eq. (27–2) the 6-Ω and 3-Ω resistors in parallel in Fig. 27–2a are equivalent to the single 2-Ω resistor in Fig. 27–2b:

$$\frac{1}{R_{eq}} = \frac{1}{6\,\Omega} + \frac{1}{3\,\Omega} = \frac{1}{2\,\Omega}.$$

(You can find the same result using Eq. (27–3).) From Eq. (27–1) the series combination of this 2-Ω resistor with the 4-Ω resistor is equivalent to the single 6-Ω resistor in Fig. 27–2c.

To find the current in each resistor of the original network, we reverse the steps by which we reduced the network. In the

circuit shown in Fig. 27–2d (identical to Fig. 27–2c), the current is $I = V_{ab}/R = (18\text{ V})/(6\,\Omega) = 3$ A. So the current in the 4-Ω and 2-Ω resistors in Fig. 27–2e (identical to Fig. 27–2b) is also 3 A. The potential difference V_{cb} across the 2-Ω resistor is therefore $V_{cb} = IR = (3\text{ A})(2\,\Omega) = 6$ V. This potential difference must also be 6 V in Fig. 27–2f (identical to Fig. 27–2a). Using $I = V_{cb}/R$, the currents in the 6-Ω and 3-Ω resistors in Fig. 27–2f are (6 V)/(6 Ω) = 1 A and (6 V)/(3 Ω) = 2 A, respectively.

Note that for the two resistors in parallel between points c and b in Fig. 27–2f, there is twice as much current through the 3-Ω resistor as through the 6-Ω resistor; more current goes through the path of least resistance, in accordance with Eq. (27–4). Note also that the total current through these two resistors is 3 A, the same as it is through the 4-Ω resistor between points a and c.

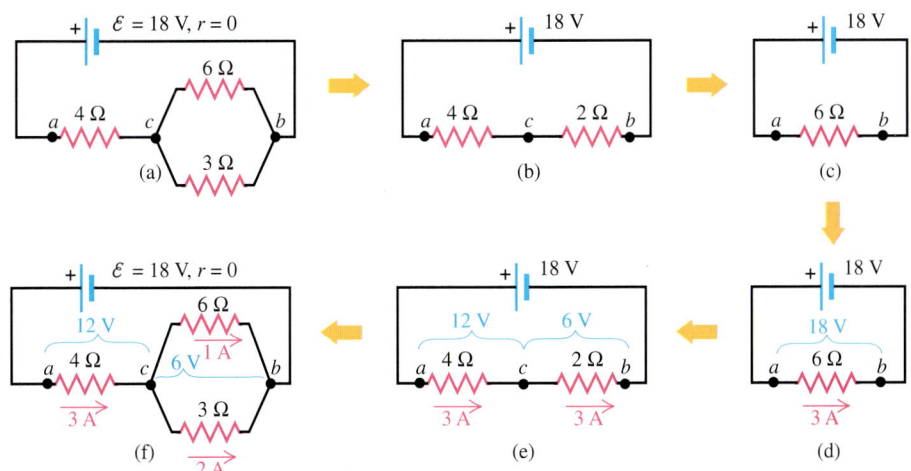

27–2 Steps in reducing a combination of resistors to a single equivalent resistor and finding the current in each resistor.

EXAMPLE 27-2

Series versus parallel combinations Two identical light bulbs are to be connected to a source with $\mathcal{E} = 8$ V and negligible internal resistance. Each light bulb has a resistance $R = 2\ \Omega$. Find the current through each bulb, the potential difference across each bulb, and the power delivered to each bulb and to the entire network if the bulbs are connected a) in series, as in Fig. 27–3a; b) in parallel, as in Fig. 27–3b. c) Suppose one of the bulbs burns out; that is, its filament breaks and current can no longer flow through it. What happens to the other bulb in the series case? In the parallel case?

SOLUTION a) From Eq. (27–1) the equivalent resistance of the two bulbs between points a and c in Fig. 27–3a is the sum of their individual resistances, or

$$R_{eq} = 2R = 2(2\ \Omega) = 4\ \Omega.$$

The current is the same through either light bulb in series:

$$I = \frac{V_{ac}}{R_{eq}} = \frac{8\ \text{V}}{4\ \Omega} = 2\ \text{A}.$$

Since the bulbs have the same resistance, the potential difference is the same across each bulb:

$$V_{ab} = V_{bc} = IR = (2\ \text{A})(2\ \Omega) = 4\ \text{V}.$$

This is one half of the 8-V terminal voltage of the source. We can find the power delivered to each light bulb using either of two formulas from Eq. (26–18):

$$P = I^2 R = (2\ \text{A})^2(2\ \Omega) = 8\ \text{W} \qquad \text{or}$$

$$P = \frac{V^2}{R} = \frac{(4\ \text{V})^2}{2\ \Omega} = 8\ \text{W},$$

where $V = 4$ V is the potential difference across a single bulb.
    The total power delivered to both bulbs is $P_{total} = 2P = 16$ W. Alternatively, we can find the total power by using the equivalent resistance $R_{eq} = 4\ \Omega$, through which the current is $I = 2$ A and across which the potential difference is $V_{ac} = 8$ V:

$$P_{total} = I^2 R_{eq} = (2\ \text{A})^2(4\ \Omega) = 16\ \text{W} \qquad \text{or}$$

$$P_{total} = \frac{V_{ac}^{\ 2}}{R_{eq}} = \frac{(8\ \text{V})^2}{4\ \Omega} = 16\ \text{W}.$$

CAUTION ▶ We emphasize that the same current I passes through both light bulbs. The bulbs are identical, so the same power $P = I^2R$ is delivered to both bulbs, and both bulbs will glow equally brightly. The current is *not* "used up" as it travels through the circuit. ◀
b) If the light bulbs are in parallel, as in Fig. 27–3b, the potential difference V_{de} across each bulb is the same and equal to 8 V, the terminal voltage of the source. Hence the current through each light bulb is

$$I = \frac{V_{de}}{R} = \frac{8\ \text{V}}{2\ \Omega} = 4\ \text{A},$$

and the power delivered to each bulb is

$$P = I^2 R = (4\ \text{A})^2(2\ \Omega) = 32\ \text{W} \qquad \text{or}$$

$$P = \frac{V^2}{R} = \frac{(8\ \text{V})^2}{2\ \Omega} = 32\ \text{W}.$$

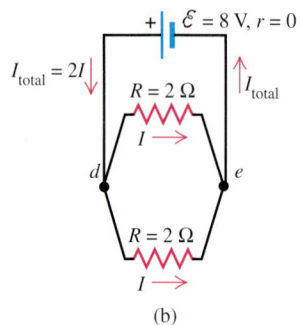

27–3 Circuit diagrams for two light bulbs (a) in series and (b) in parallel.

Both the potential difference across each bulb and the current through each bulb are twice as great as in the series case. Hence the power delivered to each bulb is *four* times greater, and each bulb glows more brightly than in the series case. If the goal is to produce the maximum amount of light from each bulb, a parallel arrangement is superior to a series arrangement.
    The total power delivered to the parallel network is $P_{total} = 2P = 64$ W, four times greater than in the series case. The increased power compared to the series case isn't obtained "for free"; energy is extracted from the source four times more rapidly in the parallel case than in the series case. If the source is a battery, it will be used up four times as fast.
    We can also find the total power by using the equivalent resistance R_{eq}, which is given by Eq. (27–2):

$$\frac{1}{R_{eq}} = 2\left(\frac{1}{2\ \Omega}\right) = 1\ \Omega^{-1}, \qquad \text{or} \qquad R_{eq} = 1\ \Omega.$$

The total current through the equivalent resistor is $I_{total} = 2I = 2(4\ \text{A}) = 8$ A, and the potential difference across the equivalent resistor is 8 V. Hence the total power is

$$P_{total} = I^2 R_{eq} = (8\ \text{A})^2(1\ \Omega) = 64\ \text{W} \qquad \text{or}$$

$$P_{total} = \frac{V^2}{R_{eq}} = \frac{(8\ \text{V})^2}{1\ \Omega} = 64\ \text{W}.$$

The potential difference across the equivalent resistance is the same for both the series and parallel cases, but for the parallel case the value of R_{eq} is less, and so $P_{total} = V^2/R_{eq}$ is greater.
    Our calculation isn't completely accurate; real light bulbs are *not* ohmic, and the bulb resistance $R = V/I$ is *not* a constant independent of the potential difference V across the bulb. (The resistance of the filament increases with increasing operating temperature and hence with increasing V.) But it is indeed true that light bulbs connected in series across a source glow less

brightly than when connected in parallel across the same source (Fig. 27–4).

c) In the series case the same current flows through both bulbs. If one of the bulbs burns out, there will be no current at all in the circuit, and neither bulb will glow.

In the parallel case the potential difference across either bulb remains equal to 8 V even if one of the bulbs burns out. Hence the current through the functional bulb remains equal to 4 A, and the power delivered to that bulb remains equal to 32 W, the same as before the other bulb burned out. This is another of the merits of a parallel arrangement of light bulbs: If one fails, the other bulbs are unaffected. This principle is used in household wiring systems, which we'll discuss in Section 27–6.

27–4 When connected to the same source, two light bulbs in series (shown on the left) draw less power and glow less brightly than when they are in parallel (shown on the right).

27–3 KIRCHHOFF'S RULES

Many practical resistor networks cannot be reduced to simple series-parallel combinations. Figure 27–5a shows a dc power supply with emf ε_1 charging a battery with a smaller emf ε_2 and feeding current to a light bulb with resistance R. Fig. 27–5b is a "bridge" circuit, used in many different types of measurement and control systems. (One important application of a "bridge" circuit is described in Problem 27–59.) We don't need any new principles to compute the currents in these networks, but there are some techniques that help us handle such problems systematically. We will describe the techniques developed by the German physicist Gustav Robert Kirchhoff (1824–1887).

First, here are two terms that we will use often. A **junction** in a circuit is a point where three or more conductors meet. Junctions are also called *nodes* or *branch points*. A **loop** is any closed conducting path. The circuit in Fig. 27–5a has two junctions, a and b. In Fig. 27–5b, points a, b, c, and d are junctions, but points e and f are not. Some possible loops in Fig. 27–5b are the closed paths *acdba*, *acdefa*, *abdefa*, and *abcdefa*.

Kirchhoff's rules consist of the following two statements:

Kirchhoff's junction rule: *The algebraic sum of the currents into any junction is zero.* That is,

$$\Sigma I = 0 \quad \text{(junction rule, valid at any junction).} \quad (27\text{–}5)$$

Kirchhoff's loop rule: *The algebraic sum of the potential differences in any loop, including those associated with emf's and those of resistive elements, must equal zero.* That is,

$$\Sigma V = 0 \quad \text{(loop rule, valid for any closed loop).} \quad (27\text{–}6)$$

The junction rule is based on *conservation of electric charge*. No charge can accumulate at a junction, so the total charge entering the junction per unit time must equal the total charge leaving per unit time. Charge per unit time is current, so if we consider the currents entering to be positive and those leaving to be negative, the algebraic sum of currents into a junction must be zero. It's like a T branch in a water pipe; if you have one liter per minute coming in one pipe, you can't have three liters per minute going out the other two pipes. We may as well confess that we used the junction rule (without saying so) in Section 27–2 in the derivation of Eq. (27–2) for resistors in parallel.

The loop rule is a statement that the electrostatic force is *conservative*. Suppose we go around a loop, measuring potential differences across successive circuit elements as

(a)

(b)

27–5 Two networks that cannot be reduced to simple series-parallel combinations of resistors.

we go. When we return to the starting point, we must find that the *algebraic sum* of these differences is zero; otherwise, we could not say that the potential at this point has a definite value.

In applying the loop rule, we need some sign conventions. The following Problem-Solving Strategy describes in detail how to use these, but here's a quick overview. We first assume a direction for the current in each branch of the circuit and mark it on a diagram of the circuit. Then, starting at any point in the circuit, we imagine traveling around a loop, adding emf's and *IR* terms as we come to them. When we travel through a source in the direction from − to +, the emf is considered to be *positive;* when we travel from + to −, the emf is considered to be *negative*. When we travel through a resistor in the *same* direction as the assumed current, the *IR* term is *negative* because the current goes in the direction of decreasing potential. When we travel through a resistor in the direction *opposite* to the assumed current, the *IR* term is *positive* because this represents a rise of potential.

Kirchhoff's two rules are all we need to solve a wide variety of network problems. Usually, some of the emf's, currents, and resistances are known, and others are unknown. We must always obtain from Kirchhoff's rules a number of independent equations equal to the number of unknowns so that we can solve the equations simultaneously. Often the hardest part of the solution is not in understanding the basic principles but in keeping track of algebraic signs!

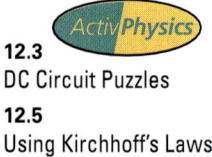

12.3
DC Circuit Puzzles

12.5
Using Kirchhoff's Laws

Problem–Solving Strategy

KIRCHHOFF'S RULES

1. Draw a *large* circuit diagram so that you have plenty of room for labels. Label all quantities, known and unknown, including an assumed direction for each unknown current and emf. Often you will not know in advance the actual direction of an unknown current or emf, but this doesn't matter. Carry out your solution, using the assumed direction. If the actual direction of a particular quantity is opposite to your assumption, the result will come out with a negative sign. If you use Kirchhoff's rules correctly, they give you the directions as well as the magnitudes of unknown currents and emf's. We will illustrate this point in the following examples.

2. When you label currents, it is usually best to use the junction rule immediately to express the currents in terms of as few quantities as possible. For example, Fig. 27–6a shows a circuit that is correctly labeled; Fig. 27–6b

shows the same circuit, relabeled by applying the junction rule to point *a* to eliminate I_3.

3. Choose any closed loop in the network, and designate a direction (clockwise or counterclockwise) to travel around the loop in applying the loop rule. The direction doesn't have to be the same as any assumed current direction.

4. Travel around the loop in the designated direction, adding potential differences as you cross them. Remember that a positive potential difference corresponds to an increase in potential, and a negative potential difference corresponds to a decrease. An emf is counted as positive when you traverse it from − to + and negative when you traverse it from + to −. An *IR* term is negative if you travel through the resistor in the same direction as the assumed current

(a)

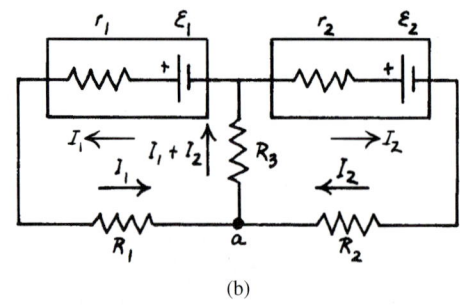

(b)

27–6 Application of the junction rule to point *a* reduces the number of unknown currents from three to two.

and positive if you travel in the opposite direction. Figure 27–7 summarizes these sign conventions. In each part of the figure, "travel" is the direction in which we imagine going around a loop in using Kirchhoff's loop law, not necessarily the direction of current.

5. Equate the sum in Step 4 to zero.

6. If necessary, choose another loop to get a different relation among the unknowns, and continue until you have as many independent equations as unknowns or until every circuit element has been included in at least one of the chosen loops.

7. Finally, solve the equations simultaneously to determine the unknowns. This step is algebra, not physics, but it can be fairly complex. Be careful with algebraic manipulations; one sign error is fatal to the entire solution. Symbolic computer programs such as Mathematica or Maple can be useful for this step.

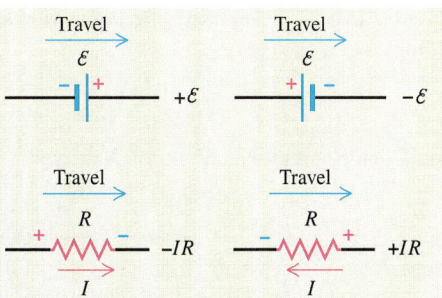

27–7 Sign conventions to use in traveling around a circuit loop when using Kirchhoff's rules.

8. You can use this same bookkeeping system to find the potential V_{ab} of any point a with respect to any other point b. Start at b and add the potential changes you encounter in going from b to a, using the same sign rules as in Step 4. The algebraic sum of these changes is $V_{ab} = V_a - V_b$.

EXAMPLE 27-3

A single-loop circuit The circuit shown in Fig. 27–8a contains two batteries, each with an emf and an internal resistance, and two resistors. Find a) the current in the circuit, b) the potential difference V_{ab}, and c) the power output of the emf of each battery.

SOLUTION a) This is a single-loop circuit with no junctions, so we don't need Kirchhoff's junction rule. To apply the loop rule to the single loop, we first assume a direction for the current; let's assume a counterclockwise direction, as shown. Then, starting at a and going counterclockwise, we add potential increases and decreases and equate the sum to zero, as in Eq. (27–6). The resulting equation is

$$-I(4\,\Omega) - 4\text{ V} - I(7\,\Omega) + 12\text{ V} - I(2\,\Omega) - I(3\,\Omega) = 0.$$

Collecting terms containing I and solving for I, we find

$$8\text{ V} = I(16\,\Omega) \quad\text{and}\quad I = 0.5\text{ A}.$$

The result for I is positive, showing that our assumed current direction is correct. For an exercise, try assuming the opposite

direction for I; you should then get $I = -0.5$ A, indicating that the actual current is opposite to this assumption.

b) To find V_{ab}, the potential at a with respect to b, we start at b and add potential changes as we go toward a. There are two possible paths from b to a; taking the lower one first, we find

$$V_{ab} = (0.5\text{ A})(7\,\Omega) + 4\text{ V} + (0.5\text{ A})(4\,\Omega) = 9.5\text{ V}.$$

Point a is at 9.5 V higher potential than b. All the terms in this sum, including the IR terms, are positive because each represents an *increase* in potential as we go from b toward a. If we use the upper path instead, the resulting equation is

$$V_{ab} = 12\text{ V} - (0.5\text{ A})(2\,\Omega) - (0.5\text{ A})(3\,\Omega) = 9.5\text{ V}.$$

Here the IR terms are negative because our path goes in the direction of the current, with potential decreases through the resistors. The result is the same as for the lower path, as it must be in order for the total potential change around the complete

27–8 (a) In this example we travel around the loop in the same direction as the assumed current, so all the IR terms are negative. The potential decreases as we travel from + to − through the bottom emf but increases as we travel from − to + through the top emf. (b) A real-life example of a circuit of this kind.

(a)

(b)

loop to be zero. In each case, potential rises are taken to be positive and drops are taken to be negative.

c) The power output of the emf of the 12-V battery is

$$P = \mathcal{E}I = (12 \text{ V})(0.5 \text{ A}) = 6 \text{ W},$$

and the power output of the emf of the 4-V battery is

$$P = \mathcal{E}I = (-4 \text{ V})(0.5 \text{ A}) = -2 \text{ W}.$$

The negative sign in $\mathcal{E}$ for the 4-V battery appears because the current actually runs from the higher-potential side of the battery to the lower-potential side. The negative value of P means that we are *storing* energy in that battery, and it is being *recharged* by the 12-V battery. The circuit shown in Fig. 27–8a is very

much like that used when a 12-V automobile battery is used to recharge a run-down battery in another automobile (Fig. 27–8b). The 3-Ω and 7-Ω resistors in Fig. 27–8a represent the resistances of the jumper cables and of the conducting path through the red automobile with the run-down battery (though the values of the resistances are different in actual automobiles and jumper cables than in this example).

By applying the expression $P = I^2R$ to each of the four resistors in Fig. 27–8a, you should be able to show that the total power dissipated in all four resistors is 4 W. Of the 6 W provided by the emf of the 12-V battery, 2 W goes into storing energy in the 4-V battery, and 4 W is dissipated in the resistances.

EXAMPLE 27–4

Charging a battery In the circuit shown in Fig. 27–9, a 12-V power supply with unknown internal resistance r is connected to a run-down rechargeable battery with unknown emf $\mathcal{E}$ and internal resistance 1 Ω and to an indicator light bulb of resistance 3 Ω carrying a current of 2 A. The current through the run-down battery is 1 A in the direction shown. Find the unknown current I, the internal resistance r, and the emf $\mathcal{E}$.

SOLUTION First we apply the junction rule, Eq. (27–5), to point a. We find

$$-I + 1 \text{ A} + 2 \text{ A} = 0, \qquad \text{so} \qquad I = 3 \text{ A}.$$

To determine r, we apply the loop rule, Eq. (27–6), to the outer loop labeled (1); we find

$$12 \text{ V} - (3 \text{ A})r - (2 \text{ A})(3 \Omega) = 0, \qquad \text{so} \qquad r = 2 \Omega.$$

The terms containing the resistances r and 3 Ω are negative because our loop traverses those elements in the same direction as the current and hence finds potential *drops*. If we had chosen to traverse loop (1) in the opposite direction, every term would have had the opposite sign, and the result for r would have been the same.

To determine $\mathcal{E}$, we apply the loop rule to loop (2):

$$-\mathcal{E} + (1 \text{ A})(1 \Omega) - (2 \text{ A})(3 \Omega) = 0, \qquad \text{so} \qquad \mathcal{E} = -5 \text{ V}.$$

The term for the 1-Ω resistor is positive because in traversing it in the direction opposite to the current we find a potential *rise*. The negative value for $\mathcal{E}$ shows that the actual polarity of this emf is opposite to the assumption made in the figure; the positive terminal of this source is really on the right side. As in Example 27–3, the battery is being recharged. Alternatively, we could use loop (3), obtaining the equation

$$12 \text{ V} - (3 \text{ A})(2 \Omega) - (1 \text{ A})(1 \Omega) + \mathcal{E} = 0,$$

from which we again find $\mathcal{E} = -5 \text{ V}$.

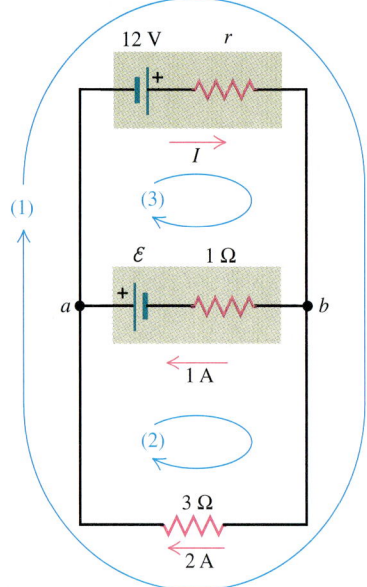

27–9 In this circuit a power supply charges a run-down battery and lights a bulb. An assumption has been made about the polarity of the emf $\mathcal{E}$ of the run-down battery; is this assumption correct?

As an additional consistency check, we note that $V_{ba} = V_b - V_a$ equals the voltage across the 3-Ω resistance, which is $(2 \text{ A})(3 \Omega) = 6 \text{ V}$. Going from a to b by the top branch, we encounter potential differences $+12 \text{ V} - (3 \text{ A})(2 \Omega) = +6 \text{ V}$, and going by the middle branch we find $-(-5 \text{ V}) + (1 \text{ A})(1 \Omega) = +6 \text{ V}$. The three ways of getting V_{ba} give the same results. Make sure that you understand all the signs in these calculations.

EXAMPLE 27–5

Power in a battery-charging circuit In the circuit of Example 27–4 (shown in Fig. 27–9), find the power delivered by the 12-V power supply and by the battery being recharged, and find the power dissipated in each resistor.

SOLUTION The power output from the emf of the power supply is

$$P_{\text{supply}} = \mathcal{E}_{\text{supply}} I_{\text{supply}} = (12 \text{ V})(3 \text{ A}) = 36 \text{ W}.$$

The power dissipated in the power supply's internal resistance r is

$$P_{r\text{-supply}} = I_{\text{supply}}^2 r_{\text{supply}} = (3\ A)^2(2\ \Omega) = 18\ W,$$

so the power supply's *net* power output is $P_{\text{net}} = 36\ W - 18\ W = 18\ W$. Alternatively, from Example 27–4 the terminal voltage of the battery is $V_{ba} = 6\ V$, so the net power output is

$$P_{\text{net}} = V_{ba} I_{\text{supply}} = (6\ V)(3\ A) = 18\ W.$$

The power output of the emf ε of the battery being charged is

$$P_{\text{emf}} = \varepsilon I_{\text{battery}} = (-5\ V)(1\ A) = -5\ W.$$

This is negative because the 1-A current runs through the battery from the higher-potential side to the lower-potential side. (As we mentioned in Example 27–4, the polarity assumed for this bat-tery in Fig. 27–9 was wrong.) We are storing energy in the battery as we charge it. Additional power is dissipated in the battery's internal resistance; this power is

$$P_{r\text{-battery}} = I_{\text{battery}}^2 r_{\text{battery}} = (1\ A)^2(1\ \Omega) = 1\ W.$$

The total power input to the battery is thus $1\ W + |-5\ W| = 6\ W$. Of this, 5 W represent useful energy stored in the battery; the remainder is wasted in its internal resistance.

The power dissipated in the light bulb is

$$P_{\text{bulb}} = I_{\text{bulb}}^2 R_{\text{bulb}} = (2\ A)^2(3\ \Omega) = 12\ W.$$

Of the 18 W of net power from the power supply, 5 W goes to recharge the battery, 1 W is dissipated in the battery's internal resistance, and 12 W is dissipated in the light bulb.

EXAMPLE 27–6

A complex network Figure 27–10 shows a "bridge" circuit of the type described at the beginning of this section (see Fig. 27–5b). Find the current in each resistor and the equivalent resistance of the network of five resistors.

SOLUTION This network cannot be represented in terms of series and parallel combinations. There are five different cur-rents to determine, but by applying the junction rule to junctions a and b, we can represent them in terms of three unknown cur-rents, as shown in the figure. The current in the battery is $I_1 + I_2$.

We apply the loop rule to the three loops shown, obtaining the following three equations:

$$13\ V - I_1(1\ \Omega) - (I_1 - I_3)(1\ \Omega) = 0; \qquad (1)$$

$$-I_2(1\ \Omega) - (I_2 + I_3)(2\ \Omega) + 13\ V = 0; \qquad (2)$$

$$-I_1(1\ \Omega) - I_3(1\ \Omega) + I_2(1\ \Omega) = 0. \qquad (3)$$

This is a set of three simultaneous equations for the three unknown currents. They may be solved by various methods; one straightforward procedure is to solve the third equation for I_2, obtaining $I_2 = I_1 + I_3$, and then substitute this expression into the first two equations to eliminate I_2. When this is done, we are left with the two equations

$$13\ V = I_1(2\ \Omega) - I_3(1\ \Omega), \qquad (1')$$

$$13\ V = I_1(3\ \Omega) + I_3(5\ \Omega). \qquad (2')$$

Now we can eliminate I_3 by multiplying Eq. $(1')$ by 5 and

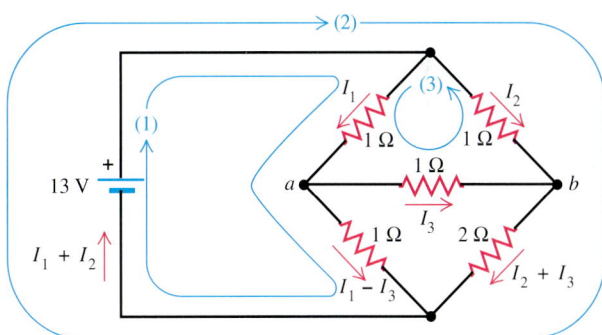

27–10 A network circuit with several resistors.

adding the two equations. We obtain

$$78\ V = I_1(13\ \Omega), \qquad I_1 = 6\ A.$$

We substitute this result back into Eq. $(1')$ to obtain $I_3 = -1\ A$, and finally, from Eq. (3) we find $I_2 = 5\ A$. The negative value of I_3 tells us that its direction is opposite to our initial assumption.

The total current through the network is $I_1 + I_2 = 11\ A$, and the potential drop across it is equal to the battery emf, namely, 13 V. The equivalent resistance of the network is

$$R_{\text{eq}} = \frac{13\ V}{11\ A} = 1.2\ \Omega.$$

EXAMPLE 27–7

A potential difference within a complex network In the circuit of Example 27–6 (Fig. 27–10), find the potential difference V_{ab}.

SOLUTION To find $V_{ab} = V_a - V_b$, we start at point b and follow a path to point a, adding potential rises and drops as we go. The simplest path is through the center 1-Ω resistor. We have found $I_3 = -1\ A$, showing that the actual current direction in this branch is from right to left. Thus as we go from b to a, there is a *drop* of potential with magnitude $IR = (1\ A)(1\ \Omega) = 1\ V$, and $V_{ab} = -1\ V$. That is, the potential at point a is 1 V less than that at point b.

Alternatively, we may go through the lower two resistors. The currents through these are

$$I_2 + I_3 = 5\ A + (-1\ A) = 4\ A \qquad \text{and}$$

$$I_1 - I_3 = 6\ A - (-1\ A) = 7\ A,$$

and so

$$V_{ab} = -(4\ A)(2\ \Omega) + (7\ A)(1\ \Omega) = -1\ V.$$

We suggest that you try some other paths from b to a to verify that they also give this result.

27–4 ELECTRICAL MEASURING INSTRUMENTS

We've been talking about potential difference, current, and resistance for two chapters, so it's about time we said something about how to *measure* these quantities. Many common devices, including car instrument panels, battery chargers, and inexpensive electrical instruments, measure potential difference (voltage), current, or resistance using a **d'Arsonval galvanometer.** In the following discussion we'll often just call it a *meter.* A pivoted coil of fine wire is placed in the magnetic field of a permanent magnet (Fig. 27–11). Attached to the coil is a spring, similar to the hairspring on the balance wheel of a watch. In the equilibrium position, with no current in the coil, the pointer is at zero. When there is a current in the coil, the magnetic field exerts a torque on the coil that is proportional to the current. (We'll discuss this magnetic interaction in detail in Chapter 28.) As the coil turns, the spring exerts a restoring torque that is proportional to the angular displacement.

Thus the angular deflection of the coil and pointer is directly proportional to the coil current, and the device can be calibrated to measure current. The maximum deflection, typically 90° or so, is called *full-scale deflection.* The essential electrical characteristics of the meter are the current I_{fs} required for full-scale deflection (typically of the order of 10 μA to 10 mA) and the resistance R_c of the coil (typically of the order of 10 to 1000 Ω).

The meter deflection is proportional to the *current* in the coil. If the coil obeys Ohm's law, the current is proportional to the *potential difference* between the terminals of the coil, and the deflection is also proportional to this potential difference. For example, consider a meter whose coil has a resistance $R_c = 20.0\ \Omega$ and that deflects full scale when the current in its coil is $I_{fs} = 1.00$ mA. The corresponding potential difference for full-scale deflection is

$$V = I_{fs}R_c = (1.00 \times 10^{-3}\ \text{A})(20.0\ \Omega) = 0.0200\ \text{V}.$$

AMMETERS

A current-measuring instrument is usually called an **ammeter** (or milliammeter, microammeter, etc., depending on the range). *An ammeter always measures the current passing through it.* An *ideal* ammeter, discussed in Section 26–5, would have *zero* resistance, so including it in a branch of a circuit would not affect the current in that branch. Real ammeters always have some finite resistance, but it is always desirable for an ammeter to have as little resistance as possible.

We can adapt any meter to measure currents that are larger than its full-scale read-

This ammeter (top) and voltmeter (bottom) are both d'Arsonval galvanometers. The difference has to do with their internal connections (see Figure 27–12 on the opposite page).

27–11 (a) A d'Arsonval galvanometer, showing a pivoted coil with attached pointer, a permanent magnet supplying a magnetic field that is uniform in magnitude, and a spring to provide restoring torque, which opposes magnetic-field torque. (b) Pivoted coil around a soft-iron core, with supports removed.

(a)   (b)

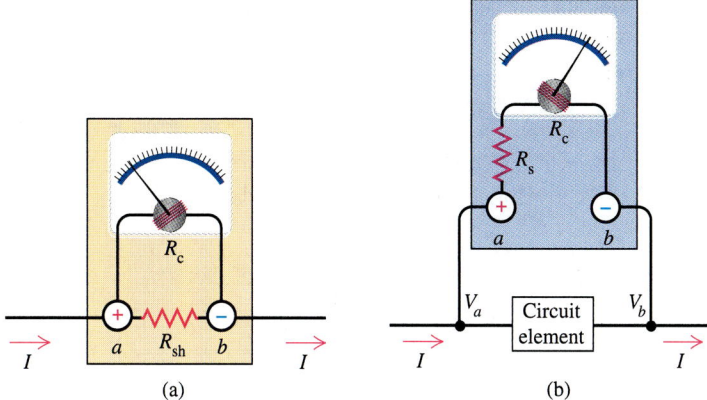

27–12 (a) Internal connections of a moving-coil ammeter. (b) Internal connections of a moving-coil voltmeter.

ing by connecting a resistor in parallel with it (Fig. 27–12a) so that some of the current bypasses the meter coil. The parallel resistor is called a **shunt resistor** or simply a *shunt,* denoted as R_{sh}.

Suppose we want to make a meter with full-scale current I_{fs} and coil resistance R_c into an ammeter with full-scale reading I_a. To determine the shunt resistance R_{sh} needed, note that at full-scale deflection the total current through the parallel combination is I_a, the current through the coil of the meter is I_{fs}, and the current through the shunt is the difference $I_a - I_{fs}$. The potential difference V_{ab} is the same for both paths, so

$$I_{fs}R_c = (I_a - I_{fs})R_{sh} \qquad \text{(for an ammeter).} \qquad (27\text{–}7)$$

EXAMPLE 27–8

Designing an ammeter What shunt resistance is required to make the 1.00-mA, 20.0-Ω meter described above into an ammeter with a range of 0 A to 50.0 mA?

SOLUTION We have $I_{fs} = 1.00$ mA $= 1.00 \times 10^{-3}$ A and $R_c = 20.0\ \Omega$, and we wish the ammeter to be able to handle a maximum current $I_a = 50.0 \times 10^{-3}$ A. Solving Eq. (27–7) for R_{sh}, we find

$$R_{sh} = \frac{I_{fs}R_c}{I_a - I_{fs}} = \frac{(1.00 \times 10^{-3}\ \text{A})(20.0\ \Omega)}{50.0 \times 10^{-3}\ \text{A} - 1.00 \times 10^{-3}\ \text{A}}$$

$$= 0.408\ \Omega.$$

The equivalent resistance R_{eq} of the instrument is given by

$$\frac{1}{R_{eq}} = \frac{1}{R_c} + \frac{1}{R_{sh}} = \frac{1}{20.0\ \Omega} + \frac{1}{0.408\ \Omega},$$

$$R_{eq} = 0.400\ \Omega.$$

The shunt resistance is so small in comparison to the meter resistance that the equivalent resistance is very nearly equal to the shunt resistance. This shunt resistance gives us a low-resistance instrument with the desired range of 0 to 50.0 mA. At full-scale deflection, $I = I_a = 50.0$ mA, the current through the galvanometer is 1.00 mA, the current through the shunt resistor is 49.0 mA, and $V_{ab} = 0.0200$ V. If the current I is *less* than 50.0 mA, the coil current and the deflection are proportionally less, but the resistance R_{eq} is still 0.400 Ω.

VOLTMETERS

This same basic meter may also be used to measure potential difference or *voltage.* A voltage-measuring device is called a **voltmeter** (or millivoltmeter, etc., depending on the range). A voltmeter always measures the potential difference between two points, and its terminals must be connected to these points. (Example 26–7 in Section 26–5 described what can happen if a voltmeter is connected incorrectly.) As we discussed in Section 26–5, an ideal voltmeter would have *infinite* resistance, so connecting it between two points in a circuit would not alter any of the currents. Real voltmeters always have finite

12.4
Using Ammeters and Voltmeters

resistance, but a voltmeter should have large enough resistance that connecting it in a circuit does not change the other currents appreciably.

For the meter described in Example 27–8 the voltage across the meter coil at full-scale deflection is only $I_{fs}R_c = (1.00 \times 10^{-3}\ \text{A})(20.0\ \Omega) = 0.0200\ \text{V}$. We can extend this range by connecting a resistor R_s in *series* with the coil (Fig. 27–12b). Then only a fraction of the total potential difference appears across the coil itself, and the remainder appears across R_s. For a voltmeter with full-scale reading V_V, we need a series resistor R_s in Fig. 27–12b such that

$$V_V = I_{fs}(R_c + R_s) \qquad \text{(for a voltmeter).} \qquad (27\text{–}8)$$

EXAMPLE 27–9

Designing a voltmeter How can we make a galvanometer with $R_c = 20.0\ \Omega$ and $I_{fs} = 0.00100\ \text{A}$ into a voltmeter with a maximum range of 10.0 V?

SOLUTION Solving Eq. (27–8) for R_s, we have

$$R_s = \frac{V_V}{I_{fs}} - R_c = \frac{10.0\ \text{V}}{0.00100\ \text{A}} - 20.0\ \Omega = 9980\ \Omega.$$

At full-scale deflection, $V_{ab} = 10.0\ \text{V}$, the voltage across the meter is 0.0200 V, the voltage across R_s is 9.98 V, and the current

through the voltmeter is 0.00100 A. In this case most of the voltage appears across the series resistor. The equivalent meter resistance is $R_{eq} = 20.0\ \Omega + 9980\ \Omega = 10,000\ \Omega$. Such a meter is described as a "1,000 ohms-per-volt meter," referring to the ratio of resistance to full-scale deflection. In normal operation the current through the circuit element being measured (I in Fig. 27–12b) is much greater than 0.00100 A, and the resistance between points a and b in the circuit is much less than 10,000 Ω. So the voltmeter draws off only a small fraction of the current and disturbs only slightly the circuit being measured.

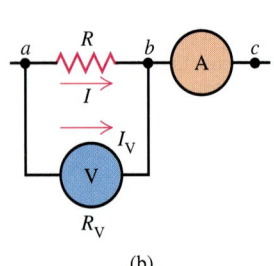

(a)

(b)

27–13 Ammeter-voltmeter method for measuring resistance.

AMMETERS AND VOLTMETERS IN COMBINATION

A voltmeter and an ammeter can be used together to measure *resistance* and *power*. The resistance R of a resistor equals the potential difference V_{ab} between its terminals divided by the current I; that is, $R = V_{ab}/I$. The power input P to any circuit element is the product of the potential difference across it and the current through it: $P = V_{ab}I$. In principle, the most straightforward way to measure R or P is to measure V_{ab} and I simultaneously.

With practical ammeters and voltmeters this isn't quite as simple as it seems. In Fig. 27–13a, ammeter A reads the current I in the resistor R. Voltmeter V, however, reads the *sum* of the potential difference V_{ab} across the resistor and the potential difference V_{bc} across the ammeter. If we transfer the voltmeter terminal from c to b, as in Fig. 27–13b, then the voltmeter reads the potential difference V_{ab} correctly, but the ammeter now reads the *sum* of the current I in the resistor and the current I_V in the voltmeter. Either way, we have to correct the reading of one instrument or the other unless the corrections are small enough to be negligible.

EXAMPLE 27–10

Measuring resistance I Suppose we want to measure an unknown resistance R using the circuit of Fig. 27–13a. The meter resistances are $R_V = 10,000\ \Omega$ (for the voltmeter) and $R_A = 2.00\ \Omega$ (for the ammeter). If the voltmeter reads 12.0 V and the ammeter reads 0.100 A, what are the resistance R and the power dissipated in the resistor?

SOLUTION If the meters were ideal (i.e., $R_V = \infty$ and $R_A = 0$), the resistance would be $R = V/I = (12.0\ \text{V})/(0.100\ \text{A}) = 120\ \Omega$. But the voltmeter reading includes the voltage V_{bc} across the

ammeter as well as the voltage V_{ab} across the resistor. We have $V_{bc} = IR_A = (0.100\ \text{A})(2.00\ \Omega) = 0.200\ \text{V}$, so the actual potential drop V_{ab} across the resistor is 12.0 V − 0.200 V = 11.8 V, and the resistance is

$$R = \frac{V_{ab}}{I} = \frac{11.8\ \text{V}}{0.100\ \text{A}} = 118\ \Omega.$$

The power dissipated in this resistor is

$$P = V_{ab}I = (11.8\ \text{V})(0.100\ \text{A}) = 1.18\ \text{W}.$$

EXAMPLE 27–11

Measuring resistance II Suppose the meters of Example 27–10 are connected to a different resistor, in the circuit shown in Fig. 27–13b, and the readings obtained on the meters are the same as in Example 27–10. What is the value of this new resistance R, and what is the power dissipated in the resistor?

SOLUTION In this case the voltmeter measures the potential across the resistor correctly; the complication is that the ammeter measures the voltmeter current I_V as well as the current I in

the resistor. We have $I_V = V/R_V = (12.0 \text{ V})/(10,000 \ \Omega) = 1.20$ mA. The actual current I in the resistor is $I = 0.100 \text{ A} - 0.0012 \text{ A} = 0.0988 \text{ A}$, and the resistance is

$$R = \frac{V_{ab}}{I} = \frac{12.0 \text{ V}}{0.0988 \text{ A}} = 121 \ \Omega.$$

The power dissipated in the resistor is

$$P = V_{ab}I = (12.0 \text{ V})(0.0988 \text{ A}) = 1.19 \text{ W}.$$

OHMMETERS

An alternative method for measuring resistance is to use a d'Arsonval meter in an arrangement called an **ohmmeter.** It consists of a meter, a resistor, and a source (often a flashlight battery) connected in series (Fig. 27–14). The resistance R to be measured is connected between terminals x and y.

The series resistance R_s is variable; it is adjusted so that when terminals x and y are short-circuited (that is, when $R = 0$), the meter deflects full-scale. When nothing is connected to terminals x and y, so that the circuit between x and y is *open* (that is, when $R \to \infty$), there is no current and no deflection. For any intermediate value of R the meter deflection depends on the value of R, and the meter scale can be calibrated to read the resistance R directly. Larger currents correspond to smaller resistances, so this scale reads backward compared to the scale showing the current.

You have probably seen multiple-range meters, or "multimeters," that use d'Arsonval galvanometers. Such a device uses a single-range moving-coil meter; various ranges are provided by switching different resistances in parallel and series with the meter coil. By using appropriate resistances, a multimeter can be used as a voltmeter or as an ammeter. Multimeters also include a battery; placing this in series with the coil makes the meter function as an ohmmeter.

In situations in which high precision is required, instruments containing d'Arsonval meters have been supplanted by electronic instruments with direct digital readouts. These are more precise, stable, and mechanically rugged than d'Arsonval meters. Digital voltmeters can be made with extremely high internal resistance, of the order of 100 MΩ.

27–14 Ohmmeter circuit. The resistor R_s has a variable resistance, as is indicated by the arrow through the resistor symbol. To use the ohmmeter, first connect x directly to y and adjust R_s until the meter reads zero. Then connect x and y across the resistor R and read the scale.

THE POTENTIOMETER

The *potentiometer* is an instrument that can be used to measure the emf of a source without drawing any current from the source; it also has a number of other useful applications. Essentially, it balances an unknown potential difference against an adjustable, measurable potential difference.

The principle of the potentiometer is shown schematically in Fig. 27–15. A resistance wire ab of total resistance R_{ab} is permanently connected to the terminals of a source of known emf $\mathcal{E}_1$. A sliding contact c is connected through the galvanometer G to a second source whose emf $\mathcal{E}_2$ is to be measured. As contact c is moved along the resistance wire, the resistance R_{cb} between points c and b varies; if the resistance wire is uniform, R_{cb} is proportional to the length of wire between c and b. To determine the value of $\mathcal{E}_2$, contact c is moved until a position is found at which the galvanometer shows no deflection; this corresponds to zero current passing through $\mathcal{E}_2$. With $I_2 = 0$, Kirchhoff's loop rule gives

$$\mathcal{E}_2 = IR_{cb}.$$

With $I_2 = 0$, the current I produced by the emf $\mathcal{E}_1$ has the same value no matter what the

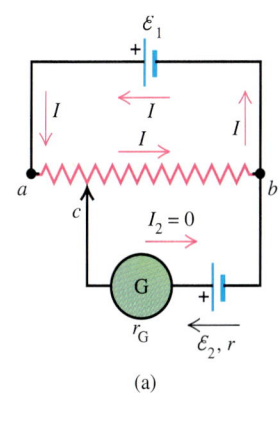

27–15 (a) Potentiometer circuit. (b) Circuit symbol for a potentiometer (variable resistor).

value of the emf $\mathcal{E}_2$. We calibrate the device by replacing $\mathcal{E}_2$ by a source of known emf; then any unknown emf $\mathcal{E}_2$ can be found by measuring the length of wire cb for which $I_2 = 0$ (see Exercise 27–25). Note that for this to work, V_{ab} must be greater than $\mathcal{E}_2$.

The term *potentiometer* is also used for any variable resistor, usually having a circular resistance element and a sliding contact controlled by a rotating shaft and knob. The circuit symbol for a potentiometer is shown in Fig. 27–15b.

27–5 RESISTANCE-CAPACITANCE CIRCUITS

In the circuits we have analyzed up to this point, we have assumed that all the emf's and resistances are *constant* (time-independent) so that all the potentials, currents, and powers are also independent of time. But in the simple act of charging or discharging a capacitor we find a situation in which the currents, voltages, and powers *do* change with time. As we described in Chapter 25, capacitors have many applications that make use of their ability to store charge and energy, so understanding what happens when they are charged or discharged is of great practical importance.

CHARGING A CAPACITOR

Figure 27–16 shows a simple circuit for charging a capacitor. A circuit such as this that has a resistor and a capacitor in series is called an **R-C circuit.** We idealize the battery (or power supply) to have a constant emf $\mathcal{E}$ and zero internal resistance ($r = 0$), and we neglect the resistance of all the connecting conductors.

We begin with the capacitor initially uncharged (Fig. 27–16a); then at some initial time $t = 0$ we close the switch, completing the circuit and permitting current around the loop to begin charging the capacitor (Fig. 27–16b). For all practical purposes, the current begins at the same instant in every conducting part of the circuit, and at each instant the current is the same in every part.

CAUTION ▶ Up this point we have been working with constant potential differences (voltages), currents, and charges and have used *capital* letters V, I, and Q, respectively, to denote these quantities. To distinguish between quantities that vary with time and those that are constant, we will use *lowercase* letters v, i, and q for time-varying voltages, currents, and charges, respectively. We suggest that you follow this same convention in your own work. ◀

Because the capacitor in Fig. 27–16 is initially uncharged, the potential difference v_{bc} across it is zero at $t = 0$. At this time, from Kirchhoff's loop law, the voltage v_{ab} across the resistor R is equal to the battery emf $\mathcal{E}$. The initial ($t = 0$) current through the resistor, which we will call I_0, is given by Ohm's law: $I_0 = v_{ab}/R = \mathcal{E}/R$.

As the capacitor charges, its voltage v_{bc} increases and the potential difference v_{ab} across the resistor decreases, corresponding to a decrease in current. The sum of these two voltages is constant and equal to $\mathcal{E}$. After a long time the capacitor becomes fully charged, the current decreases to zero, and the potential difference v_{ab} across the resistor becomes zero. Then the entire battery emf $\mathcal{E}$ appears across the capacitor, and $v_{bc} = \mathcal{E}$.

Let q represent the charge on the capacitor, and i the current in the circuit at some time t after the switch has been closed. We choose the positive direction for the current to correspond to positive charge flowing onto the left-hand capacitor plate, as in Fig. 27–16b. The instantaneous potential differences v_{ab} and v_{bc} are

$$v_{ab} = iR, \qquad v_{bc} = \frac{q}{C}.$$

Using these in Kirchhoff's loop rule, we find

$$\mathcal{E} - iR - \frac{q}{C} = 0. \tag{27–9}$$

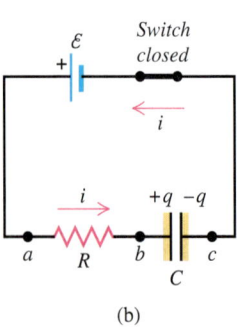

27–16 Charging a capacitor. (a) Just before the switch is closed, the charge is zero. When the switch closes (at $t = 0$), the current jumps from zero to $\mathcal{E}/R$. (b) At some later time t, as $t \rightarrow \infty$, $q \rightarrow Q_f$ and $i \rightarrow 0$.

The potential drops by an amount iR as we travel from a to b and by q/C as we travel from b to c. Solving Eq. (27–9) for i, we find

$$i = \frac{\mathcal{E}}{R} - \frac{q}{RC}. \qquad (27\text{–}10)$$

At time $t = 0$, when the switch is first closed, the capacitor is uncharged, and so $q = 0$. Substituting $q = 0$ into Eq. (27–10), we find that the *initial* current I_0 is given by $I_0 = \mathcal{E}/R$, as we have already noted. If the capacitor were not in the circuit, the last term in Eq. (27–10) would not be present; then the current would be *constant* and equal to $\mathcal{E}/R$.

As the charge q increases, the term q/RC becomes larger, and the capacitor charge approaches its final value, which we will call Q_f. The current decreases and eventually becomes zero. When $i = 0$, Eq. (27–10) gives

$$\frac{\mathcal{E}}{R} = \frac{Q_f}{RC}, \qquad Q_f = C\mathcal{E}. \qquad (27\text{–}11)$$

Note that the final charge Q_f does not depend on R.

The current and the capacitor charge are shown as functions of time in Fig. 27–17. At the instant the switch is closed ($t = 0$), the current jumps from zero to its initial value $I_0 = \mathcal{E}/R$; after that, it gradually approaches zero. The capacitor charge starts at zero and gradually approaches the final value given by Eq. (27–11), $Q_f = C\mathcal{E}$.

We can derive general expressions for the charge q and current i as functions of time. With our choice of the positive direction for current (Fig. 27–16b), i equals the rate at which positive charge arrives at the left-hand (positive) plate of the capacitor, so $i = dq/dt$. Making this substitution in Eq. (27–10), we have

$$\frac{dq}{dt} = \frac{\mathcal{E}}{R} - \frac{q}{RC} = -\frac{1}{RC}(q - C\mathcal{E}).$$

We can rearrange this to

$$\frac{dq}{q - C\mathcal{E}} = -\frac{dt}{RC}$$

and then integrate both sides. We change the integration variables to q' and t' so that we can use q and t for the upper limits. The lower limits are $q' = 0$ and $t' = 0$:

$$\int_0^q \frac{dq'}{q' - C\mathcal{E}} = -\int_0^t \frac{dt'}{RC}.$$

When we carry out the integration, we get

$$\ln\left(\frac{q - C\mathcal{E}}{-C\mathcal{E}}\right) = -\frac{t}{RC}.$$

Exponentiating both sides (that is, taking the inverse logarithm) and solving for q, we find

$$\frac{q - C\mathcal{E}}{-C\mathcal{E}} = e^{-t/RC},$$

$$q = C\mathcal{E}(1 - e^{-t/RC}) = Q_f(1 - e^{-t/RC}) \quad \substack{(R\text{-}C \text{ circuit, charging} \\ \text{capacitor}).} \qquad (27\text{–}12)$$

The instantaneous current i is just the time derivative of Eq. (27–12):

$$i = \frac{dq}{dt} = \frac{\mathcal{E}}{R} e^{-t/RC} = I_0 e^{-t/RC} \quad (R\text{-}C \text{ circuit, charging capacitor}). \qquad (27\text{–}13)$$

The charge and current are both *exponential* functions of time. Figure 27–17a is a graph of Eq. (27–13), and Fig. 27–17b is a graph of Eq. (27–12).

12.6
Capacitance

(a)

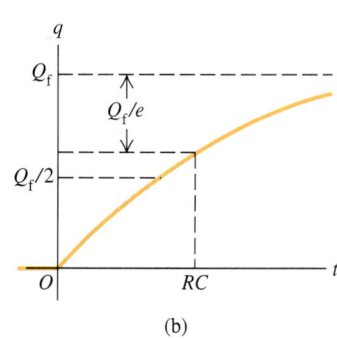

(b)

27–17 Current i and capacitor charge q as functions of time for the circuit of Fig. 27–16. The initial current is I_0, and the initial capacitor charge is zero. The current asymptotically approaches zero, and the capacitor charge asymptotically approaches a final value of Q_f.

12.7
Series and Parallel Capacitors
12.8
Circuit Time Constants

TIME CONSTANT

After a time equal to RC, the current in the R-C circuit has decreased to $1/e$ (about 0.368) of its initial value. At this time, the capacitor charge has reached $(1 - 1/e) = 0.632$ of its final value $Q_f = C\mathcal{E}$. The product RC is therefore a measure of how quickly the capacitor charges. We call RC the **time constant,** or the **relaxation time,** of the circuit, denoted by τ:

$$\tau = RC \qquad \text{(time constant for } R\text{-}C \text{ circuit).} \qquad (27\text{--}14)$$

When τ is small, the capacitor charges quickly; when it is larger, the charging takes more time. If the resistance is small, it's easier for current to flow, and the capacitor charges more quickly. If R is in ohms and C in farads, τ is in seconds.

In Fig. 27–17a the horizontal axis is an *asymptote* for the curve. Strictly speaking, i never becomes precisely zero. But the longer we wait, the closer it gets. After a time equal to $10\,RC$, the current has decreased to 0.000045 of its initial value. Similarly, the curve in Fig. 27–17b approaches the horizontal dashed line labeled Q_f as an asymptote. The charge q never attains precisely this value, but after a time equal to $10RC$ the difference between q and Q_f is only 0.000045 of Q_f. We invite you to verify that the product RC has units of time.

DISCHARGING A CAPACITOR

Now suppose that after the capacitor in Fig. 27–16b has acquired a charge Q_0, we remove the battery from our R-C circuit and connect points a and c to an open switch (Fig. 27–18a). We then close the switch and at the same instant reset our stopwatch to $t = 0$; at that time, $q = Q_0$. The capacitor then *discharges* through the resistor, and its charge eventually decreases to zero.

Again let i and q represent the time-varying current and charge at some instant after the connection is made. In Fig. 27–18b we make the same choice of the positive direction for current as in Fig. 27–16b. Then Kirchhoff's loop rule gives Eq. (27–10) but with $\mathcal{E} = 0$; that is,

$$i = \frac{dq}{dt} = -\frac{q}{RC}. \qquad (27\text{--}15)$$

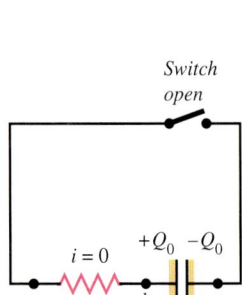

The current i is now negative; this is because positive charge q is leaving the left-hand capacitor plate in Fig. 27–18b, so the current is in the direction opposite to that shown in the figure. At time $t = 0$, when $q = Q_0$, the initial current is $I_0 = -Q_0/RC$.

To find q as a function of time, we rearrange Eq. (27–15), again change the names of the variables to q' and t', and integrate. This time the limits for q' are Q_0 to q. We get

$$\int_{Q_0}^{q} \frac{dq'}{q'} = -\frac{1}{RC} \int_{0}^{t} dt',$$

$$\ln \frac{q}{Q_0} = -\frac{t}{RC},$$

$$q = Q_0 e^{-t/RC} \qquad (R\text{-}C \text{ circuit, discharging capacitor).} \qquad (27\text{--}16)$$

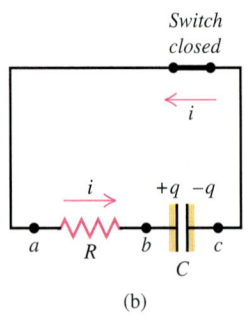

The instantaneous current i is the derivative of this with respect to time:

$$i = \frac{dq}{dt} = -\frac{Q_0}{RC} e^{-t/RC} = I_0 e^{-t/RC} \qquad (R\text{-}C \text{ circuit, discharging capacitor).}$$

$$(27\text{--}17)$$

27–18 (a) Before the switch is closed at time $t = 0$, the capacitor charge is Q_0, and the current is zero. (b) At time t after the switch is closed, the capacitor charge is q and the current is i. The actual current direction is opposite to the direction shown; i is negative. After a long time, q and i both approach zero.

The current and the charge are graphed in Fig. 27–19; both quantities approach zero exponentially with time. Comparing these results with Eqs. (27–12) and (27–13), we note that the expressions for the current are identical, apart from the sign of I_0. The capacitor charge approaches zero asymptotically in Eq. (27–16), while the *difference* between q and Q approaches zero asymptotically in Eq. (27–12).

Energy considerations give us additional insight into the behavior of an RC circuit. While the capacitor is charging, the instantaneous rate at which the battery delivers energy to the circuit is $P = \mathcal{E}i$. The instantaneous rate at which electrical energy is dissipated in the resistor is i^2R, and the rate at which energy is stored in the capacitor is $iv_{bc} = iq/C$. Multiplying Eq. (27–9) by i, we find

$$\mathcal{E}i = i^2R + iq/C. \qquad (27\text{–}18)$$

This means that of the power $\mathcal{E}i$ supplied by the battery, part (i^2R) is dissipated in the resistor and part (iq/C) is stored in the capacitor.

The *total* energy supplied by the battery during charging of the capacitor equals the battery emf $\mathcal{E}$ multiplied by the total charge Q_f, or $\mathcal{E}Q_f$. The total energy stored in the capacitor, from Eq. (25–9), is $Q_f\mathcal{E}/2$. Thus of the energy supplied by the battery, *exactly half* is stored in the capacitor, and the other half is dissipated in the resistor. It is a little surprising that this half-and-half division of energy doesn't depend on C, R, or $\mathcal{E}$. This result can also be verified in detail by taking the integral over time of each of the power quantities in Eq. (27–18). We leave this calculation for your amusement (see Problem 27–69).

(a)

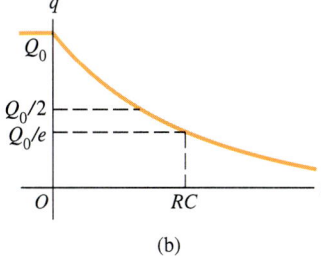

(b)

27–19 Current i and capacitor charge q as functions of time for the circuit of Fig. 27–18. The initial current is I_0, and the initial capacitor charge is Q_0; both i and q asymptotically approach zero.

EXAMPLE 27-12

Charging a capacitor A resistor with resistance 10 MΩ is connected in series with a capacitor with capacitance 1.0 μF and a battery with emf 12.0 V, as in Fig. 27–16. Before the switch is closed at time $t = 0$, the capacitor is uncharged. a) What is the time constant? b) What fraction of the final charge is on the plates at time $t = 46$ s? c) What fraction of the initial current remains at $t = 46$ s?

SOLUTION a) From Eq. (27–14) the time constant is

$$\tau = RC = (10 \times 10^6 \ \Omega)(1.0 \times 10^{-6} \ \text{F}) = 10 \ \text{s}.$$

b) The fraction of the final capacitor charge is q/Q_f. From Eq.

(27–12),

$$\frac{q}{Q_f} = 1 - e^{-t/RC} = 1 - e^{-(46 \ \text{s})/(10 \ \text{s})} = 0.99.$$

The capacitor is 99% charged after a time equal to 4.6 RC, or 4.6 time constants.
c) From Eq. (27–13),

$$\frac{i}{I_0} = e^{-4.6} = 0.010.$$

After 4.6 time constants the current has decreased to 1.0% of its initial value.

EXAMPLE 27-13

Discharging a capacitor The resistor and capacitor described in Example 27–12 are reconnected as shown in Fig. 27–18. The capacitor is originally given a charge of 5.0 μC, then discharged by closing the switch at $t = 0$. a) At what time will the charge be equal to 0.50 μC? b) What is the current at this time?

SOLUTION a) Solving Eq. (27–16) for the time t gives

$$t = -RC \ln \frac{q}{Q_0}$$

$$= -(10 \times 10^6 \ \Omega)(1.0 \times 10^{-6} \ \text{F}) \ln \frac{0.50 \ \mu\text{C}}{5.0 \ \mu\text{C}} = 23 \ \text{s}.$$

This is 2.3 times the time constant $\tau = RC = 10$ s.
b) From Eq. (27–17), with $Q_0 = 5.0 \ \mu\text{C} = 5.0 \times 10^{-6}$ C,

$$i = -\frac{Q_0}{RC} e^{-t/RC} = -\frac{5.0 \times 10^{-6} \ \text{C}}{10 \ \text{s}} e^{-2.3} = -5.0 \times 10^{-8} \ \text{A}.$$

The current has the opposite sign when the capacitor is discharging than when it is charging. We could have saved the effort required to calculate $e^{-t/RC}$ by noticing that at the time in question, $q = 0.10 \ Q_0$; from Eq. (27–16) this means $e^{-t/RC} = 0.10$.

27–6 POWER DISTRIBUTION SYSTEMS

A Case Study in Circuit Analysis

We conclude this chapter with a brief discussion of practical household and automotive electric-power distribution systems. Automobiles use direct-current (dc) systems, while nearly all household, commercial, and industrial systems use alternating current (ac) because of the ease of stepping voltage up and down with transformers. Most of the same basic wiring concepts apply to both. We'll talk about alternating-current circuits in greater detail in Chapter 32.

The various lamps, motors, and other appliances to be operated are always connected in *parallel* to the power source (the wires from the power company for houses, or from the battery and alternator for a car). If appliances were connected in series, shutting one appliance off would shut them all off (see Example 27–2 in Section 27–2). The basic idea of house wiring is shown in Fig. 27–20. One side of the "line," as the pair of conductors is called, is called the *neutral* side; it is always connected to "ground" at the entrance panel. For houses, *ground* is an actual electrode driven into the earth (which is usually a good conductor) or sometimes connected to the household water pipes. Electricians speak of the "hot" side and the "neutral" side of the line. Most modern house wiring systems have *two* hot lines with opposite polarity with respect to the neutral. We'll return to this detail later.

Household voltage is nominally 120 V in the United States and Canada, often 240 V in Europe. (For alternating current, which varies sinusoidally with time, these numbers represent the *root-mean-square* voltage, which is $1/\sqrt{2}$ times the peak voltage. We'll discuss this further in Section 32–2.) The amount of current I drawn by a given device is determined by its power input P, given by Eq. (26–17): $P = VI$. Hence $I = P/V$. For example, the current in a 100-W light bulb is

$$I = \frac{P}{V} = \frac{100 \text{ W}}{120 \text{ V}} = 0.83 \text{ A}.$$

The power input to this bulb is actually determined by its resistance R. Using Eq. (26–18), which states that $P = VI = I^2 R = V^2/R$ for a resistor, the resistance of this bulb at operating temperature is

$$R = \frac{V}{I} = \frac{120 \text{ V}}{0.83 \text{ A}} = 144 \ \Omega, \qquad \text{or} \qquad R = \frac{V^2}{P} = \frac{(120 \text{ V})^2}{100 \text{ W}} = 144 \ \Omega.$$

Similarly, a 1500-W waffle iron draws a current of $(1500 \text{ W})/(120 \text{ V}) = 12.5$ A and has a resistance, at operating temperature, of 9.6 Ω. Because of the temperature dependence of resistivity, the resistances of these devices are considerably less when they are cold. If you measure the resistance of a 100-W light bulb with an ohmmeter (whose small cur-

27–20 Schematic diagram of part of a house wiring system. Only two branch circuits are shown; an actual system might have four to thirty branch circuits. Lamps and appliances may be plugged into the outlets. The grounding wires, which normally carry no current, are not shown. (Note that actual wires have a different color-coding system.)

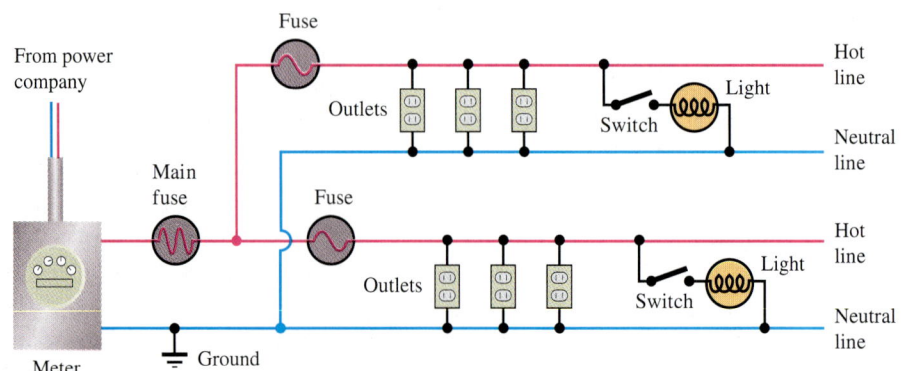

rent causes very little temperature rise), you will probably get a value of about 10 Ω. When a light bulb is turned on, this low resistance causes there to be an initial surge of current until the filament heats up. That's why a light bulb that's ready to burn out nearly always does so just when you turn it on.

The maximum current available from an individual circuit is limited by the resistance of the wires. As we discussed in Section 26–6, the I^2R power loss in the wires causes them to become hot, and in extreme cases this can cause a fire or melt the wires. Ordinary lighting and outlet wiring in houses usually uses 12-gauge wire. This has a diameter of 2.05 mm and can carry a maximum current of 20 A safely (without overheating). Larger sizes such as 8-gauge (3.26 mm) or 6-gauge (4.11 mm) are used for high-current appliances such as electric ranges and clothes dryers, and 2-gauge (6.54 mm) or larger is used for the main power lines entering a house.

Protection against overloading and overheating of circuits is provided by fuses or circuit breakers. A *fuse* contains a link of lead-tin alloy with a very low melting temperature; the link melts and breaks the circuit when its rated current is exceeded. A *circuit breaker* is an electromechanical device that performs the same function, using an electromagnet or a bimetallic strip to "trip" the breaker and interrupt the circuit when the current exceeds a specified value. Circuit breakers have the advantage that they can be reset after they are tripped, while a blown fuse must be replaced. However, fuses are somewhat more reliable in operation than circuit breakers are.

If your system has fuses and you plug too many high-current appliances into the same outlet, the fuse blows. *Do not* replace the fuse with one of larger rating; if you do, you risk overheating the wires and starting a fire. The only safe solution is to distribute the appliances among several circuits. Modern kitchens often have three or four separate 20-A circuits.

Contact between the hot and neutral sides of the line causes a *short circuit.* Such a situation, which can be caused by faulty insulation or by any of a variety of mechanical malfunctions, provides a very low-resistance current path, permitting a very large current that would quickly melt the wires and ignite their insulation if the current were not interrupted by a fuse or circuit breaker. An equally dangerous situation is a broken wire that interrupts the current path, creating an *open circuit.* This is hazardous because of the sparking that can occur at the point of intermittent contact.

In approved wiring practice, a fuse or breaker is placed *only* in the hot side of the line, never in the neutral side. Otherwise, if a short circuit should develop because of faulty insulation or other malfunction, the ground-side fuse could blow. The hot side would still be live and would pose a shock hazard if you touched the live conductor and a grounded object such as a water pipe. For similar reasons the wall switch for a light fixture is always in the hot side of the line, never the neutral side.

Further protection against shock hazard is provided by a third conductor called the *grounding wire,* included in all present-day wiring. This conductor corresponds to the long round or U-shaped prong of the three-prong connector plug on an appliance or power tool. It is connected to the neutral side of the line at the entrance panel. The grounding wire normally carries no current, but it connects the metal case or frame of the device to ground. If a conductor on the hot side of the line accidentally contacts the frame or case, the grounding conductor provides a current path, and the fuse blows. Without the ground wire, the frame could become "live," that is, at a potential 120 V above ground. Then if you touched it and a water pipe (or even a damp basement floor) at the same time, you could get a dangerous shock (Fig. 27–21). In some situations, especially outlets located outdoors or near a sink or other water pipes, a special kind of circuit breaker called a *ground-fault interrupter* (GFI or GFCI) is used. This device senses the difference in current between the hot and neutral conductors (which is normally zero) and trips when this difference exceeds some very small value, typically 5 mA.

27–21 (a) If a malfunctioning electric drill is connected to a wall socket via a two-prong plug, a person may receive a shock. (b) If the drill malfunctions when it is connected via a three-prong plug, a person touching it receives no shock, since electric charge flows through the third prong and into the ground rather than into the person's body. If the ground current is appreciable, the fuse blows.

(a)   (b)

Most modern household wiring systems actually use a slight elaboration of the system described above. The power company provides *three* conductors (Fig. 27–22). One is neutral; the other two are both at 120 V with respect to the neutral but with opposite polarity, giving a voltage between them of 240 V. The power company calls this a *three-wire line,* in contrast to the 120-V two-wire (plus ground wire) line described above. With a three-wire line, 120-V lamps and appliances can be connected between neutral and either hot conductor, and high-power devices requiring 240 V, such as electric ranges and clothes dryers, are connected between the two hot lines.

27–22 Diagram of a typical 120-240-V wiring system in a kitchen. Grounding wires are not shown. For each line, the hot side is shown in red, and the neutral line is shown in blue. As described in the text, different colors are used in actual household wiring.

To help prevent wiring errors, household wiring uses a standardized color code in which the hot side of a line has black insulation (black and red for the two sides of a 240-V line), the neutral side has white insulation, and the grounding conductor is bare or has green insulation. But in electronic devices and equipment the ground or neutral side of the line is usually black. Beware! (Our illustrations do not follow this standard code but use red for the hot line and blue for neutral.)

All of the above discussion can be applied directly to automobile wiring. The voltage is about 13 V (direct current); the power is supplied by the battery and by the alternator, which charges the battery when the engine is running. The neutral side of each circuit is connected to the body and frame of the vehicle. For this low voltage a separate grounding conductor is not required for safety. The fuse or circuit breaker arrangement is the same in principle as in household wiring. Because of the lower voltage (less energy per charge), more current (a greater number of charges per second) is required for the same power; a 100-W headlight bulb requires a current of about $(100 \text{ W})/(13 \text{ V}) = 8 \text{ A}$.

Although we spoke of *power* in the above discussion, what we buy from the power company is *energy*. Power is energy transferred per unit time, so energy is average power multiplied by time. The usual unit of energy sold by the power company is the kilowatt-hour (1 kWh):

$$1 \text{ kWh} = (10^3 \text{ W})(3600 \text{ s}) = 3.6 \times 10^6 \text{ W} \cdot \text{s} = 3.6 \times 10^6 \text{ J}.$$

One kilowatt-hour typically costs 2 to 10 cents, depending on the location and quantity of energy purchased. To operate a 1500-W (1.5-kW) waffle iron continuously for one hour requires 1.5 kWh of energy; at 10 cents per kilowatt-hour, the energy cost is 15 cents. The cost of operating any lamp or appliance for a specified time can be calculated in the same way if the power rating is known. However, many electric cooking utensils (including waffle irons) cycle on and off to maintain a constant temperature, so the average power may be less than the power rating marked on the device.

EXAMPLE 27-14

An 1800-W toaster, a 1.3-kW electric frying pan, and a 100-W lamp are plugged into the same 20-A, 120-V circuit. a) What current is drawn by each device, and what is the resistance of each device? b) Will this combination blow the fuse?

SOLUTION a) When plugged into the same circuit, the three devices are in parallel, and the voltage across each is $V = 120$ V. The power delivered to each device is $P = VI = V^2/R$, where I is the current drawn by the device and R is its resistance. Hence $I = P/V$ and $R = V^2/P$. We find

$$I_{\text{toaster}} = \frac{1800 \text{ W}}{120 \text{ V}} = 15 \text{ A}, \qquad R_{\text{toaster}} = \frac{(120 \text{ V})^2}{1800 \text{ W}} = 8 \text{ }\Omega;$$

$$I_{\text{frying pan}} = \frac{1300 \text{ W}}{120 \text{ V}} = 11 \text{ A}, \qquad R_{\text{frying pan}} = \frac{(120 \text{ V})^2}{1300 \text{ W}} = 11 \text{ }\Omega;$$

$$I_{\text{lamp}} = \frac{100 \text{ W}}{120 \text{ V}} = 0.83 \text{ A}, \qquad R_{\text{lamp}} = \frac{(120 \text{ V})^2}{100 \text{ W}} = 144 \text{ }\Omega.$$

For constant voltage the device with the *least* resistance (in this case the toaster) draws the most current and receives the most power.

b) The total current through the line is the sum of the currents drawn by the three devices:

$$I = I_{\text{toaster}} + I_{\text{frying pan}} + I_{\text{lamp}} = 15 \text{ A} + 11 \text{ A} + 0.83 \text{ A} = 27 \text{ A}.$$

This exceeds the 20-A rating of the line, and the fuse will indeed blow. We could also find the current by first finding the equivalent resistance of the three devices in parallel:

$$\frac{1}{R_{\text{eq}}} = \frac{1}{R_{\text{toaster}}} + \frac{1}{R_{\text{frying pan}}} + \frac{1}{R_{\text{lamp}}}$$

$$= \frac{1}{8 \text{ }\Omega} + \frac{1}{11 \text{ }\Omega} + \frac{1}{144 \text{ }\Omega} = 0.22 \text{ }\Omega^{-1},$$

$$R_{\text{eq}} = 4.5 \text{ }\Omega;$$

the total current is then $I = V/R_{\text{eq}} = (120 \text{ V})/(4.5 \text{ }\Omega) = 27 \text{ A}$, as before. A third way to determine I is to use $I = P/V$ and simply divide the total power delivered to all three devices by the voltage:

$$I = \frac{P_{\text{toaster}} + P_{\text{frying pan}} + P_{\text{lamp}}}{V} = \frac{1800 \text{ W} + 1300 \text{ W} + 100 \text{ W}}{120 \text{ V}}$$

$$= 27 \text{ A}.$$

Current demands like these are encountered in everyday life in kitchens, which is why modern kitchens have more than one 20-A circuit. In actual practice, the toaster and frying pan should be connected to different circuits; the current in each circuit would then be safely below the 20-A rating.

SUMMARY

- When several resistors R_1, R_2, R_3, ... , are connected in series, the equivalent resistance R_{eq} is the sum of the individual resistances:

$$R_{\text{eq}} = R_1 + R_2 + R_3 + \cdots \qquad \text{(resistors in series)}. \qquad (27\text{-}1)$$

The same *current* flows through all the resistors in a series connection. When

KEY TERMS

direct current, 832
alternating current, 832
series, 832
parallel, 832

several resistors are connected in parallel, the equivalent resistance R_{eq} is given by

$$\frac{1}{R_{eq}} = \frac{1}{R_1} + \frac{1}{R_2} + \frac{1}{R_3} + \cdots \qquad \text{(resistors in parallel).} \qquad (27\text{-}2)$$

All resistors in a parallel connection have the same *potential difference* between their terminals.

■ Kirchhoff's junction rule is based on conservation of charge. It states that the algebraic sum of the currents into any junction must be zero. Kirchhoff's loop rule is based on conservation of energy and the conservative nature of electrostatic fields. It states that the algebraic sum of potential differences around any loop must be zero. Careful use of consistent sign rules is essential in applications of Kirchhoff's rules.

■ In a d'Arsonval galvanometer the deflection is proportional to the current in the coil. For a larger current range, a shunt resistor is added, so some of the current bypasses the meter coil. Such an instrument is called an ammeter. If a resistor is added in series, the meter can also be calibrated to read potential difference or voltage. The instrument is then called a voltmeter. A good ammeter has very low resistance; a good voltmeter has very high resistance.

■ When a capacitor is charged by a battery in series with a resistor, the current and capacitor charge are not constant. The charge approaches its final value asymptotically, and the current approaches zero asymptotically. The charge and current in the circuit are

$$q = C\mathcal{E}(1 - e^{-t/RC}) = Q_f(1 - e^{-t/RC}), \qquad (27\text{-}12)$$

$$i = \frac{dq}{dt} = \frac{\mathcal{E}}{R} e^{-t/RC} = I_0 e^{-t/RC}. \qquad (27\text{-}13)$$

After a time $\tau = RC$ the charge has approached within $1/e$ of its final value. This time is called the time constant or relaxation time of the circuit. When the capacitor discharges, the charge and current are given as functions of time by

$$q = Q_0 e^{-t/RC}, \qquad (27\text{-}16)$$

$$i = \frac{dq}{dt} = -\frac{Q_0}{RC} e^{-t/RC} = I_0 e^{-t/RC}. \qquad (27\text{-}17)$$

The time constant is the same for charging and discharging.

■ In household wiring systems the various electrical devices are connected in parallel across the power line, which consists of a pair of conductors, one "hot" and the other "neutral." An additional "ground" wire is included for safety. The maximum permissible current in a circuit is determined by the size of the wires and the maximum temperature they can tolerate. Protection against excessive current and the resulting fire hazard is provided by fuses or circuit breakers.

DISCUSSION QUESTIONS

Q27–1 In which 120-V light bulb does the filament have greater resistance, a 60-W bulb or a 120-W bulb? If the two bulbs are connected to a 120-V line in series, through which bulb will there be the greater voltage drop? What if they are connected in parallel? Explain your reasoning.

Q27–2 Two 120-V light bulbs, one 25-W and one 200-W, were connected in series across a 240-V line. It seemed like a good idea at the time, but one bulb burned out almost instantaneously. Which one burned out, and why?

Q27–3 You connect a number of identical light bulbs to a flashlight battery. What happens to the brightness of each bulb as more and more bulbs are added to the circuit if you connect them in i) series; ii) parallel? Will the battery last longer if the bulbs are in series or in parallel? Explain your reasoning.

Q27–4 In the circuit shown in Fig. 27–23, three identical light bulbs are connected to a flashlight battery. How do the brightnesses of the bulbs compare? Which light bulb has the greatest current passing through it? Which light bulb has the greatest potential difference between its terminals? What happens if bulb A is unscrewed? Bulb B? Bulb C? Explain your reasoning.

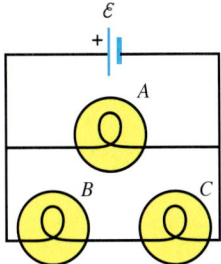

FIGURE 27–23 Question Q27–4.

Q27–5 Why do the lights on a car become dimmer when the starter is operated?

Q27–6 Lights in a house often dim momentarily when a motor, such as a washing machine or power saw, is turned on. Why does this happen?

Q27–7 Compare the formulas for resistors in series and in parallel with those for capacitors in series and in parallel. What similarities and differences do you see? Sometimes circuit analysis uses the quantity *conductance,* denoted as G and defined as the reciprocal of resistance: $G = 1/R$. Make the corresponding comparison for conductance and capacitance.

Q27–8 Is it possible to connect resistors together in a way that cannot be reduced to some combination of series and parallel combinations? If so, give examples. If not, state why not.

Q27–9 The direction of current in a battery can be reversed by connecting it to a second battery of greater emf with the positive terminals of the two batteries together. When the direction of current is reversed in a battery, does its emf also reverse? Why or why not?

Q27–10 In a two-cell flashlight, the batteries are usually connected in series. Why not connect them in parallel? What possible advantage could there be in connecting several identical batteries in parallel?

Q27–11 Electric rays (genus *Torpedo*) deliver electric shocks to stun their prey and to discourage predators. (In ancient Rome, physicians practiced a primitive form of electroconvulsive therapy by placing electric rays on their patients to cure headaches and gout.) Figure 27–24a shows *Torpedo* as seen from below. The voltage is produced by thin, wafer–like cells called *electrocytes,* each of which acts as a battery with an emf of about 10^{-4} V. Stacks of electrocytes are arranged side-by-side on the underside of *Torpedo* (Fig. 27–24b); in such a stack, the positive face of each electrocyte touches the negative face of the next electrocyte (Fig. 27–24c). What is the advantage of stacking the electrocytes? Of having the stacks side by side?

Q27–12 The emf of a flashlight battery is roughly constant with time, but its internal resistance increases with age and use. What sort of meter should be used to test the freshness of a battery?

FIGURE 27–24 Question Q27–11.

Q27–13 Is it possible to have a circuit in which the potential difference across a battery in the circuit is zero? If so, give an example. If not, explain why not.

Q27–14 For very large resistances it is easy to construct R-C circuits having time constants of several seconds or minutes. How might this fact be used to measure very large resistances, those that are too large to measure by more conventional means?

Q27–15 When a capacitor, battery, and resistor are connected in series, does the resistor affect the maximum charge stored on the capacitor? Why or why not? What purpose does the resistor serve?

Q27–16 A physics student attempted to cook a hot dog by inserting a nail into each end of the hot dog, wrapping a wire around each nail, and inserting the wires into a wall outlet. (*Note:* Don't try this yourself! You could get a potentially fatal shock.) This cooked the hot dog, but it also made the lights in his apartment go out. Explain why.

Q27–17 The greater the diameter of the wire used in household wiring, the greater the maximum current that can safely be carried by the wire. Why is this? Does the maximum permissible current depend on the length of the wire? Does it depend on what the wire is made of? Explain your reasoning.

EXERCISES

SECTION 27–2 **RESISTORS IN SERIES AND IN PARALLEL**

27–1 A 32-Ω resistor and a 20-Ω resistor are connected in parallel, and the combination is connected across a 240-V dc line.

a) What is the resistance of the parallel combination? b) What is the total current through the parallel combination? c) What is the current through each resistor?

27–2 Prove that when two resistors are connected in parallel, the equivalent resistance of the combination is always smaller than that of either resistor.

27–3 Three resistors having resistances of 1.60 Ω, 2.40 Ω, and 4.80 Ω are connected in parallel to a 28.0-V battery that has negligible internal resistance. Find a) the equivalent resistance of the combination; b) the current in each resistor; c) the total current through the battery; d) the voltage across each resistor; e) the power dissipated in each resistor. f) Which resistor dissipates the most power: the one with the greatest resistance or the least resistance? Explain why this should be.

27–4 Now the three resistors of Exercise 27–3 are connected in series to the same battery. Answer the same questions for this situation.

27–5 a) The *power rating* of a resistor is the maximum power the resistor can safely dissipate without too great a rise in temperature. The power rating of a 15-kΩ resistor is 5.0 W. What is the maximum allowable potential difference across the terminals of the resistor? b) A 9.0-kΩ resistor is to be connected across a 120-V potential difference. What power rating is required?

27–6 Compute the equivalent resistance of the network in Fig. 27–25, and find the current in each resistor. The battery has negligible internal resistance.

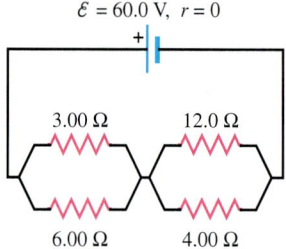

$\mathcal{E} = 60.0$ V, $r = 0$

3.00 Ω 12.0 Ω

6.00 Ω 4.00 Ω

FIGURE 27–25 Exercise 27–6.

27–7 Compute the equivalent resistance of the network in Fig. 27–26, and find the current in each resistor. The battery has negligible internal resistance.

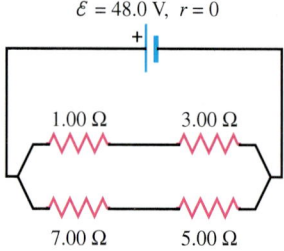

$\mathcal{E} = 48.0$ V, $r = 0$

1.00 Ω 3.00 Ω

7.00 Ω 5.00 Ω

FIGURE 27–26 Exercise 27–7.

27–8 Four resistors and a battery of negligible internal resistance are assembled to make the circuit in Fig. 27–27. Let $\mathcal{E} = 6.00$ V, $R_1 = 3.50$ Ω, $R_2 = 8.20$ Ω, $R_3 = 1.50$ Ω, and $R_4 = 4.50$ Ω. Find a) the equivalent resistance of the network; b) the current in each resistor.

27–9 In the circuit of Fig. 27–27, each resistor represents a light bulb. Let $R_1 = R_2 = R_3 = R_4 = 4.50$ Ω and $\mathcal{E} = 9.00$ V. a) Find

the current in each bulb. b) Find the power dissipated in each bulb. Which bulb or bulbs glows the brightest? c) Bulb R_4 is now removed from the circuit, leaving a break in the wire at its position. Now what is the current in each of the remaining bulbs R_1, R_2, and R_3? d) With bulb R_4 removed, what is the power dissipated in each of the remaining bulbs? e) Which light bulb(s) glow brighter as a result of removing R_4? Which bulb(s) glow less brightly? Discuss why there are different effects on different bulbs.

$\mathcal{E}$ R_1 R_3

R_2 R_4

FIGURE 27–27 Exercises 27–8 and 27–9.

27–10 A Three-Way Light Bulb. A three-way light bulb has three brightness settings (low, medium, and high), but only two filaments. a) A particular three-way light bulb connected across a 120-V line can dissipate 60 W, 120 W, or 180 W. Describe how the two filaments are arranged in the bulb, and calculate the resistance of each filament. b) Suppose the filament with the higher resistance burns out. How much power will the bulb dissipate on each of the three brightness settings? What will be the brightness (low, medium, or high) on each setting? c) Repeat part (b) for the situation where it is the filament with the lower resistance that burns out.

27–11 Light Bulbs in Series and in Parallel. Two light bulbs have resistances of 400 Ω and 800 Ω. If the two light bulbs are connected in series across a 120-V line, find a) the current through each bulb; b) the power dissipated in each bulb, and the total power dissipated in both bulbs. The two light bulbs are now connected in parallel across the 120-V line. Find c) the current through each bulb; d) the power dissipated in each bulb, and the total power dissipated in both bulbs. e) In each situation, which of the two bulbs glows the brightest? In which situation is there a greater total light output from both bulbs combined?

27–12 Light Bulbs in Series. A 60-W, 120-V light bulb and a 200-W, 120-V light bulb are connected in series across a 240-V line. Assume that the resistance of each bulb does not vary with current. (*Note:* This description of a light bulb gives the power it dissipates when connected to the stated potential difference; that is, a 25-W, 120-V light bulb dissipates 25 W when connected to a 120-V line.) a) Find the current through the bulbs. b) Find the power dissipated in each bulb. c) One bulb burns out very quickly. Which one? Why?

SECTION 27–3 KIRCHHOFF'S RULES

27–13 In the circuit shown in Fig. 27–28, find a) the current in resistor R; b) the resistance R; c) the unknown emf $\mathcal{E}$. d) If the circuit is broken at point x, what is the current in resistor R?

27–14 Find the emf's $\mathcal{E}_1$ and $\mathcal{E}_2$ in the circuit of Fig. 27–29, and find the potential difference of point b relative to point a.

27–15 In the circuit shown in Fig. 27–30, find a) the current in the 3.00-Ω resistor; b) the unknown emf's $\mathcal{E}_1$ and $\mathcal{E}_2$; c) the resistance R. Note that three currents are given.

FIGURE 27–28 Exercise 27–13.

FIGURE 27–29 Exercise 27–14.

FIGURE 27–30 Exercise 27–15.

27–16 In the circuit shown in Fig. 27–31, find a) the current in each branch; b) the potential difference V_{ab} of point a relative to point b.

FIGURE 27–31 Exercises 27–16, 27–17, and 27–18.

27–17 The 10.00-V battery in Fig. 27–31 is removed from the circuit and reinserted with the opposite polarity, so that its positive terminal is now next to point a. The rest of the circuit is as shown in the figure. Find a) the current in each branch; b) the potential difference V_{ab} of point a relative to point b.

27–18 The 5.00-V battery in Fig. 27–31 is removed from the circuit and replaced by a 20.00-V battery, with its negative terminal next to point b. The rest of the circuit is as shown in the figure. Find a) the current in each branch; b) the potential difference V_{ab} of point a relative to point b.

27–19 Use the expression $P = I^2 R$ to calculate the total power dissipated in all four resistors in Fig. 27–8a.

27–20 In the circuit shown in Fig. 27–8a (Example 27–3 in Section 27–3), the 12-V battery is removed and reinserted with the opposite polarity, so that its positive terminal is now next to point b. The rest of the circuit is as shown in the figure. Find a) the current in the circuit (magnitude *and* direction); b) the potential difference V_{ab}.

27–21 In the circuit shown in Fig. 27–10 (Example 27–6 in Section 27–3), the 2-Ω resistor is replaced by a 1-Ω resistor, and the center 1-Ω resistor (through which the current is I_3) is replaced by a resistor of unknown resistance R. The rest of the circuit is as shown in the figure. a) Calculate the current in each resistor. Draw a diagram of the circuit and label each resistor with the current through it. b) Calculate the equivalent resistance of the network. c) Calculate the potential difference V_{ab}. d) Your answers in parts (a), (b), and (c) do not depend on the value of R. Explain why.

SECTION 27–4 ELECTRICAL MEASURING INSTRUMENTS

27–22 The resistance of a galvanometer coil is 25.0 Ω, and the current required for full-scale deflection is 500 μA. a) Show in a diagram how to convert the galvanometer to an ammeter reading 20.0 mA full scale, and compute the shunt resistance. b) Show how to convert the galvanometer to a voltmeter reading 500 mV full scale, and compute the series resistance.

27–23 The resistance of the coil of a pivoted-coil galvanometer is 9.36 Ω, and a current of 0.0224 A causes it to deflect full scale. We want to convert this galvanometer to an ammeter reading 20.0 A full scale. The only shunt available has a resistance of 0.0250 Ω. What resistance R must be connected in series with the coil? (See Fig. 27–32.)

FIGURE 27–32 Exercise 27–23.

27–24 A 90.0-V battery has an internal resistance of $r = 8.23\ \Omega$. a) What is the reading of a voltmeter having a resistance of $R_V = 425\ \Omega$ when placed across the terminals of the battery? b) What maximum value may the ratio r/R_V have if the error in the reading of the emf of a battery is not to exceed 4.0%?

27–25 Consider the potentiometer circuit of Fig. 27–15. The resistor between a and b is a uniform wire with length l, with a sliding contact c at a distance x from b. An unknown emf ε_2 is measured by sliding the contact until the galvanometer G reads zero. a) Show that under this condition the unknown emf is given by $\varepsilon_2 = (x/l)\varepsilon_1$. b) Why is the internal resistance of the galvanometer not important? c) Suppose $\varepsilon_1 = 9.15$ V and $l = 1.000$ m. The galvanometer G reads zero when $x = 0.365$ m. What is the emf ε_2?

27–26 Two 150-V voltmeters, one with a resistance of 10.0 kΩ and the other with a resistance of 90.0 kΩ, are connected in series across a 120-V dc line. Find the reading of each voltmeter. (A 150-V voltmeter deflects full scale when the potential difference between its two terminals is 150 V.)

27–27 In the ohmmeter of Fig. 27–14, the coil of the meter has resistance $R_c = 15.0\ \Omega$ and the current required for full-scale deflection is $I_{fs} = 3.60$ mA. The source is a flashlight battery with $\varepsilon = 1.50$ V and negligible internal resistance. The ohmmeter is to show a meter deflection of one-half of full scale when connected to a resistor with $R = 600\ \Omega$. What series resistance R_s is required?

27–28 In the ohmmeter in Fig. 27–33, M is a 2.50-mA meter of resistance 65.0 Ω. (A 2.50-mA meter deflects full scale when the current through it is 2.50 mA.) The battery B has an emf of 1.52 V and negligible internal resistance. R is so chosen that when the terminals a and b are shorted ($R_x = 0$), the meter reads full scale. When a and b are open ($R_x = \infty$), the meter reads zero. a) What is the resistance of the resistor R? b) What current indicates a resistance R_x of 200 Ω? c) What values of R_x correspond to meter deflections of $\frac{1}{4}$, $\frac{1}{2}$, and $\frac{3}{4}$ of full scale if the deflection is proportional to the current through the galvanometer?

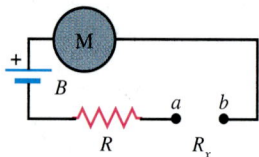

FIGURE 27–33 Exercise 27–28.

SECTION 27–5 **RESISTANCE-CAPACITANCE CIRCUITS**

27–29 Verify that the product RC has dimensions of time.

27–30 A 4.60-μF capacitor that is initially uncharged is connected in series with a 7.50-kΩ resistor and an emf source with $\varepsilon = 125$ V and negligible internal resistance. Just after the circuit is completed, what is a) the voltage drop across the capacitor? b) the voltage drop across the resistor? c) the charge on the capacitor? d) the current through the resistor? e) A long time after the circuit is completed (after many time constants) what are the values of the preceding four quantities?

27–31 A capacitor with capacitance $C = 455$ pF is charged, with charge of magnitude 65.5 nC on each plate. The capacitor is then connected to a voltmeter that has internal resistance 1.28 MΩ. a) What is the current through the voltmeter just after the connection is made? b) What is the time constant of this RC circuit?

27–32 A capacitor is charged to a potential of 12.0 V and is then connected to a voltmeter having an internal resistance of

3.40 MΩ. After a time of 4.00 s the voltmeter reads 3.0 V. What is the capacitance?

27–33 A 12.4-μF capacitor is connected through a 0.895-MΩ resistor to a constant potential difference of 60.0 V. a) Compute the charge on the capacitor at the following times after the connections are made: 0, 5.0 s, 10.0 s, 20.0 s, and 100.0 s. b) Compute the charging currents at the same instants. c) Graph the results of parts (a) and (b) for t between 0 and 20 s.

27–34 In the circuit shown in Fig. 27–34, $C = 5.90\ \mu$F, $\varepsilon = 28.0$ V, and the emf has negligible resistance. Initially the capacitor is uncharged and the switch S is in position 1. The switch is then moved to position 2, so that the capacitor begins to charge. a) What will be the charge on the capacitor a long time after the switch is moved to position 2? b) After the switch has been in position 2 for 3.00 ms, the charge on the capacitor is measured to be 110 μC. What is the value of the resistance R? c) How long after the switch is moved to position 2 will the charge on the capacitor be equal to 99.0% of the final value found in part (a)?

FIGURE 27–34 Exercises 27–34 and 27–35.

27–35 A capacitor with $C = 1.50 \times 10^{-5}$ F is connected as shown in Fig. 27–34 with a resistor with $R = 980\ \Omega$ and an emf source with $\varepsilon = 18.0$ V and negligible internal resistance. Initially the capacitor is uncharged and the switch S is in position 1. The switch is then moved to position 2, so that the capacitor begins to charge. After the switch has been in position 2 for 10.0 ms, the switch is moved back to position 1 so that the capacitor begins to discharge. a) Compute the charge on the capacitor just *before* the switch is thrown from position 2 back to position 1. b) Compute the voltage drops across the resistor and across the capacitor at the instant described in part (a). c) Compute the voltage drops across the resistor and across the capacitor just *after* the switch is thrown from position 2 back to position 1. d) Compute the charge on the capacitor 10.0 ms after the switch is thrown from position 2 back to position 1.

SECTION 27–6 **POWER DISTRIBUTION SYSTEMS: A CASE STUDY IN CIRCUIT ANALYSIS**

27–36 The heating element of an electric dryer is rated at 4.1 kW when connected to a 240-V line. a) What is the current in the heating element? Is 12-gauge wire large enough to supply this current? b) What is the resistance of the dryer's heating element at its operating temperature? c) At 11 cents per kWh, how much does it cost per hour to operate the dryer?

27-37 A 1500-W electric heater is plugged into the outlet of a 120-V circuit that has a 20-A circuit breaker. You plug an electric hair dryer into the same outlet. The hair dryer has power settings of 600 W, 900 W, 1200 W, and 1500 W. You start with the hair dryer on the 600 W setting and increase the power setting until the circuit breaker trips. What power setting caused the breaker to trip?

27-38 How many 90-W, 120-V light bulbs can be connected to a 20-A, 120-V circuit without tripping the circuit breaker? (See the note in Exercise 27–12.)

27-39 The heating element of an electric stove consists of a heater wire embedded within an electrically insulating material, which in turn is inside a metal casing. The heater wire has a resistance of 20 Ω at room temperature (23.0°C) and a temperature coefficient of resistivity $\alpha = 2.8 \times 10^{-3}$ (C°)$^{-1}$. The heating element operates from a 120-V line. a) When the heater element is first turned on, what current does it draw and what electrical power does it dissipate? b) When the heater element has reached an operating temperature of 280°C (536°F), what current does it draw and what electrical power does it dissipate?

PROBLEMS

27-40 a) A resistance R_2 is connected in parallel with a resistance R_1. Derive an expression for the resistance R_3 that must be connected in series with the combination of R_1 and R_2 so that the equivalent resistance is equal to the resistance R_1. Show the arrangement of resistors in a diagram. b) A resistance R_2 is connected in series with a resistance R_1. Derive an expression for the resistance R_3 that must be connected in parallel with the combination of R_1 and R_2 so that the equivalent resistance is equal to R_1. Show the arrangement of resistors in a diagram.

27-41 A 400-Ω, 2.4-W resistor is needed, but only several 400-Ω, 1.2-W resistors are available. (See Exercise 27–5.) a) What two different combinations of the available units give the required resistance and power rating? b) For each of the resistor networks from part (a), what power is dissipated in each resistor when 2.4 W is dissipated by the combination?

27-42 The two identical light bulbs in Example 27–2 (Section 27–2) are connected in parallel to a different source, one with $\varepsilon = 8.0$ V and internal resistance 0.8 Ω. Each light bulb has a resistance $R = 2.0$ Ω (assumed independent of the current through the bulb). a) Find the current through each bulb, the potential difference across each bulb, and the power delivered to each bulb. b) Suppose one of the bulbs burns out, so that its filament breaks and current no longer flows through it. Find the power delivered to the remaining bulb. Does the remaining bulb glow more or less brightly after the other bulb burns out than before?

27-43 Each of the three resistors in Fig. 27–35 has a resistance of 2.4 Ω and can dissipate a maximum of 36 W without becoming excessively heated. What is the maximum power the circuit can dissipate?

FIGURE 27-35 Problem 27–43.

27-44 a) Calculate the equivalent resistance of the circuit of Fig. 27–36 between x and y. b) What is the potential of point a relative to point x if the current in the 8.0-Ω resistor is 2.4 A in the direction from left to right in the figure?

27-45 Calculate the three currents I_1, I_2, and I_3 indicated in the circuit diagram of Fig. 27–37.

27-46 What must be the emf ε in Fig. 27–38 in order for the current through the 7.00-Ω resistor to be 1.80 A? Each emf source has negligible internal resistance.

FIGURE 27-36 Problem 27–44.

FIGURE 27-37 Problem 27–45.

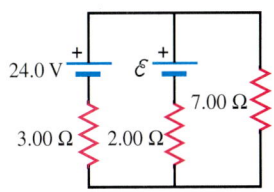

FIGURE 27-38 Problem 27–46.

27-47 Find the current through each of the three resistors of the circuit of Fig. 27–39. The emf sources have negligible internal resistance.

FIGURE 27-39 Problem 27–47.

27–48 a) Find the current through the battery and each resistor in the circuit shown in Fig. 27–40. b) What is the equivalent resistance of the resistor network?

FIGURE 27–40 Problem 27–48.

27–49 a) Find the potential of point a with respect to point b in Fig. 27–41. b) If points a and b are connected by a wire with negligible resistance, find the current in the 12.0-V battery.

FIGURE 27–41 Problem 27–49.

27–50 Three identical resistors are connected in series. When a certain potential difference is applied across the combination, the total power dissipated is 27 W. What power would be dissipated if the three resistors were connected in parallel across the same potential difference?

27–51 Figure 27–42 employs a convention often used in circuit diagrams. The battery (or other power supply) is not shown explicitly. It is understood that the point at the top, labeled "36.0 V," is connected to the positive terminal of a 36.0-V battery having negligible internal resistance, and that the "ground" symbol at the bottom is connected to the negative terminal of the battery. The circuit is completed through the battery, even though it is not shown on the diagram. a) What is the potential difference V_{ab}, the potential of point a relative to point b, when the switch S is

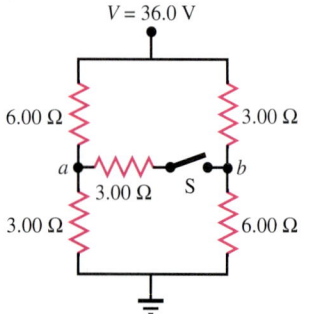

FIGURE 27–42 Problem 27–51.

open? b) What is the current through switch S when it is closed? c) What is the equivalent resistance when switch S is closed?

27–52 (See Problem 27–51.) a) What is the potential of point a with respect to point b in Fig. 27–43 when switch S is open? b) Which point, a or b, is at the higher potential? c) What is the final potential of point b with respect to ground when switch S is closed? d) How much does the charge on each capacitor change when S is closed?

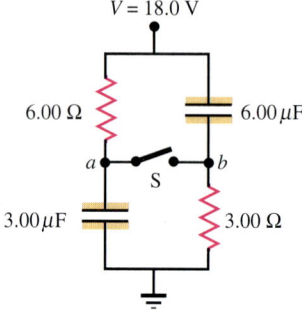

FIGURE 27–43 Problem 27–52.

27–53 (See Problem 27–51.) a) What is the potential of point a with respect to point b in Fig. 27–44 when switch S is open? b) Which point, a or b, is at the higher potential? c) What is the final potential of point b with respect to ground when switch S is closed? d) How much charge flows through switch S when it is closed?

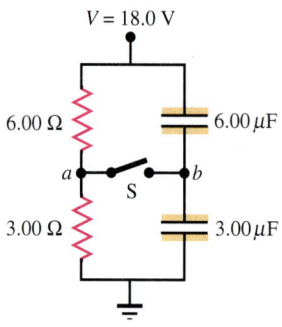

FIGURE 27–44 Problem 27–53.

27–54 A Multirange Ammeter. The resistance of the moving coil of the galvanometer G in Fig. 27–45 is 48.0 Ω, and the galvanometer deflects full scale with a current of 0.0200 A. When the meter is connected to the circuit being measured, one connection is made to the post marked + and the other to the post marked with the desired current range. Find the magnitudes of the resis-

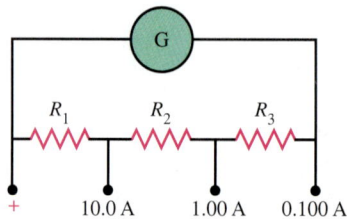

FIGURE 27–45 Exercise 27–54.

tances R_1, R_2, and R_3 required to convert the galvanometer to a multirange ammeter deflecting full scale with currents of 10.0 A, 1.00 A, and 0.100 A.

27-55 A Multirange Voltmeter. Figure 27–46 shows the internal wiring of a "three-scale" voltmeter whose binding posts are marked +, 3.00 V, 15.0 V, and 150 V. When the meter is connected to the circuit being measured, one connection is made to the post marked + and the other to the post marked with the desired voltage range. The resistance of the moving coil, R_G, is 40.0 Ω, and a current of 1.00 mA in the coil causes it to deflect full scale. Find the resistances R_1, R_2, and R_3, and the overall resistance of the meter on each of its ranges.

FIGURE 27–46 Exercise 27–55.

27-56 Point a in Fig. 27–47 is maintained at a constant potential of 400 V above ground. (See Problem 27–51.) a) What is the reading of a voltmeter with the proper range and with resistance 5.00×10^4 Ω when connected between point b and ground? b) What is the reading of a voltmeter with resistance 5.00×10^6 Ω? c) What is the reading of a voltmeter with infinite resistance?

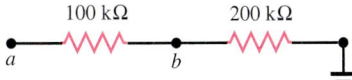

FIGURE 27–47 Problem 27–56.

27-57 A 150-V voltmeter has a resistance of 30,000 Ω. When connected in series with a large resistance R across a 110-V line, the meter reads 68 V. Find the resistance R.

27-58 Let V and I represent, respectively, the readings of the voltmeter and ammeter shown in Fig. 27–13, and let R_V and R_A be their equivalent resistances. Because of the resistances of the meters, the resistance R is not simply equal to V/I. a) When the circuit is connected as in Fig. 27–13a, show that

$$R = \frac{V}{I} - R_A.$$

Explain why the true resistance R is always *less* than V/I. b) When the connections are as in Fig. 27–13b, show that

$$R = \frac{V}{I - (V/R_V)}.$$

Explain why the true resistance R is always *greater* than V/I. c) Show that the power delivered to the resistor in part (a) is $IV - I^2R_A$ and that in part (b) is $IV - (V^2/R_V)$.

27-59 The Wheatstone Bridge. The circuit shown in Fig. 27–48, called a *Wheatstone bridge,* is used to determine the value of an unknown resistor X by comparison with three resistors M, N, and P whose resistance can be varied. For each setting, the resistance of each resistor is precisely known. With

switches K_1 and K_2 closed, these resistors are varied until the current in the galvanometer G is zero; the bridge is then said to be *balanced.* a) Show that under this condition the unknown resistance is given by $X = MP/N$. (This method permits very high precision in comparing resistors.) b) If the galvanometer G shows zero deflection when $M = 850.0$ Ω, $N = 15.00$ Ω, and $P = 33.48$ Ω, what is the unknown resistance X?

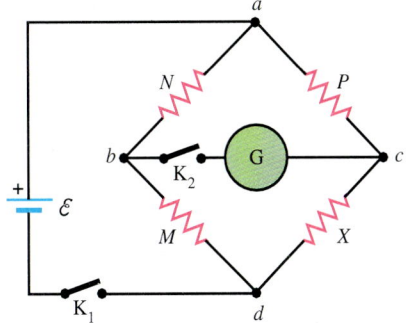

FIGURE 27–48 Problem 27–59.

27-60 A certain galvanometer has a resistance of 65.0 Ω and deflects full scale with a current of 1.50 mA in its coil. This is to be replaced with a second galvanometer that has a resistance of 38.0 Ω and deflects full scale with a current of 3.60 μA in its coil. Devise a circuit incorporating the second galvanometer such that the equivalent resistance of the circuit equals the resistance of the first galvanometer, and the second galvanometer deflects full scale when the current through the circuit equals the full-scale current of the first galvanometer.

27-61 A 224-Ω resistor and a 589-Ω resistor are connected in series across a 90.0-V line. a) What is the voltage across each resistor? b) A voltmeter connected across the 224-Ω resistor reads 23.8 V. Find the voltmeter resistance. c) Find the reading of the same voltmeter if connected across the 589-Ω resistor. d) The readings on this voltmeter are lower than the "true" voltages (that is, without the voltmeter present). Would it be possible to design a voltmeter that gave readings *higher* than the "true" voltages? Explain.

27-62 A 2.36-μF capacitor that is initially uncharged is connected in series with a 4.26-Ω resistor and an emf source with $\mathcal{E} = 120$ V and negligible internal resistance. a) Just after the connection is made, what is i) the rate at which electrical energy is being dissipated in the resistor? ii) the rate at which the electrical energy stored in the capacitor is increasing? iii) the electrical power output of the source? How do the answers to parts (i), (ii), and (iii) compare? b) Answer the same questions as in part (a) at a long time after the connection is made. c) Answer the same questions as in part (a) at the instant when the charge on the capacitor is one-half its final value.

27-63 A 3.40-μF capacitor is initially charged, then connected in series with a 7.25-kΩ resistor and an emf source with $\mathcal{E} = 180$ V and negligible internal resistance. a) A short time later the charge on the capacitor is 815 μC. At this instant, what is the current and what is its direction: toward the positive plate of the capacitor, or toward the negative plate? b) After a long time has elapsed, what will be the charge on the capacitor?

27–64 A 5.88-kΩ resistor is connected to the plates of a charged capacitor that has capacitance $C = 8.55 \times 10^{-10}$ F. The initial current through the resistor, just after the connection is made, is 0.620 A. What magnitude of charge was initially on each plate of the capacitor?

27–65 A capacitor that is initially uncharged is connected in series with a resistor and an emf source with $\mathcal{E} = 110$ V and negligible internal resistance. Just after the circuit is completed the current through the resistor is 6.5×10^{-5} A. The time constant for the circuit is 6.2 s. What are the resistance of the resistor and the capacitance of the capacitor?

27–66 A resistor with $R = 850$ Ω is connected to the plates of a charged capacitor with capacitance $C = 4.62$ μF. Just before the connection is made, the charge on the capacitor is 8.10 mC.
a) What is the energy initially stored in the capacitor? b) What is the electrical power dissipated in the resistor just after the connection is made? c) What is the electrical power dissipated in the resistor at the instant when the energy stored in the capacitor has decreased to half the value calculated in part (a)?

27–67 Strictly speaking, Eq. (27–16) implies that an *infinite* amount of time is required to completely discharge a capacitor. Yet for practical purposes, a capacitor may be considered to be fully discharged after a finite length of time. To be specific, consider a capacitor with capacitance C connected to a resistor R to be fully discharged if its charge q differs from zero by no more than the charge of one electron. a) Calculate the time required to reach this state if $C = 0.920$ μF, $R = 670$ kΩ, and $Q_0 = 7.00$ μC. How many time constants is this? b) For a given Q_0, is the time required to reach this state always the same number of time constants, independent of the values of C and R? Why or why not?

27–68 Two capacitors in series are charged by a 12.0-V battery that has an internal resistance of 1.00 Ω. There is a 5.00-Ω resis-

tor in series between the capacitors (Fig. 27–49). a) What is the time constant of the charging circuit? b) After the switch has been closed for the time determined in part (a), what is the voltage across the 3.00-μF capacitor?

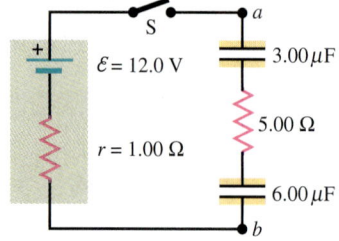

FIGURE 27–49 Problem 27–68.

27–69 The current in a charging capacitor is given by Eq. (27–13). a) The instantaneous power supplied by the battery is $\mathcal{E}i$. Integrate this to find the total energy supplied by the battery. b) The instantaneous power dissipated in the resistor is i^2R. Integrate this to find the total energy dissipated in the resistor. c) Find the final energy stored in the capacitor, and show that this equals the total energy supplied by the battery less the energy dissipated in the resistor, as obtained in parts (a) and (b). d) What fraction of the energy supplied by the battery is stored in the capacitor? How does this fraction depend on R?

27–70 a) Using Eq. (27–17) for the current in a discharging capacitor, derive an expression for the instantaneous power $P = i^2R$ dissipated in the resistor. b) Integrate the expression for P to find the total energy dissipated in the resistor, and show that this is equal to the total energy initially stored in the capacitor.

CHALLENGE PROBLEMS

27–71 According to the theorem of superposition, the response (current) in a circuit is proportional to the stimulus (voltage) that causes it. This is true even if there are multiple sources in a circuit. This theorem can be used to analyze a circuit without resorting to Kirchhoff's rules by considering the currents in the circuit to be the superposition of currents caused by each source independently. In this way the circuit can be analyzed by computing equivalent resistances rather than by using the (sometimes) more cumbersome method of Kirchhoff's rules. Furthermore, by using the superposition theorem it is possible to examine how the modification of a source in one part of the circuit will affect the currents in all parts of the circuit without having to use Kirchhoff's rules to recalculate all of the currents. Consider the circuit shown in Fig. 27–50. If the circuit were redrawn with the 55.0-V and 57.0-V sources replaced by short circuits, the circuit could be analyzed by the method of equivalent resistances without resorting to Kirchhoff's rules, and the current in each branch could be found in a simple manner. Similarly, by redrawing the circuit with the 92.0-V and the 55.0-V sources replaced by short circuits, the circuit could again be analyzed in a simple manner. Finally, by replacing the 92.0-V and the 57.0-V sources with a

short circuit the circuit could once again be analyzed simply. By superimposing the respective currents found in each of the three simplified circuits, the actual current in each branch can be found. a) Using Kirchhoff's rules, find the branch currents in the 140.0-Ω, 210.0-Ω, and 35.0-Ω resistors. b) Using a circuit similar to the circuit of Fig. 27–50, but with the 55.0-V and 57.0-V sources replaced by a short circuit, determine the currents in each resistance. c) Repeat part (b) by replacing the 92.0-V and 55.0-V sources by short circuits, leaving the 57.0-V source intact. d) Repeat part (b) by replacing the

FIGURE 27–50 Challenge Problem 27–71.

92.0-V and 57.0-V sources by short circuits, leaving the 55.0-V source intact. e) Verify the superposition theorem by taking the currents calculated in parts (b), (c), and (d) and comparing them with the currents calculated in part (a). f) If the 57.0-V source is replaced by an 80.0-V source, what will be the new currents in all branches of the circuit? (*Hint:* Using the superposition theorem, recalculate the partial currents calculated in part (c) using the fact that those currents are proportional to the source that is being replaced. Then superpose the new partial currents with those found in parts (b) and (d).)

27–72 A Capacitor Burglar Alarm. The capacitance of a capacitor can be affected by dielectric material that, although not inside the capacitor, is near enough to the capacitor to be polarized by the fringing electric field that exists near a charged capacitor. This effect is usually of the order of picofarads (pF), but it can be used with appropriate electronic circuitry to detect a change in the dielectric material surrounding the capacitor. Such a dielectric material might be the human body, and the effect described above might be used in the design of a burglar alarm. Consider the simplified circuit shown in Fig. 27–51. The voltage source has emf $\mathcal{E} = 1000$ V, and the capacitor has capacitance $C = 10.0$ pF. The electronic circuitry for detecting the current, represented as an ammeter in the diagram, has negligible resistance and is capable of detecting a current that persists at a level of at least $1.00\ \mu$A for at least $200\ \mu$s after the capacitance has changed abruptly from C to C'. The burglar alarm is designed to be activated if the capacitance changes by 10%. a) Determine the charge on the 10.0-pF capacitor when it is fully charged. b) If the capacitor is fully charged before the intruder is detected, assuming that the time taken for the capacitance to change by 10% is small enough to be neglected, derive an equation that expresses the current through the resistor R as a function of the time t since the capacitance has changed. c) Determine the range of values of the resistance R that will meet the design specifications of the burglar alarm. What happens if R is too small? Too large? (*Hint:* You will not be able to solve this part analytically but must use numerical methods. Express R as a logarithmic function of R plus known quantities. Use a trial value of R and calculate from the expression a new value. Continue to do this until the input and output values of R agree to within three significant figures.)

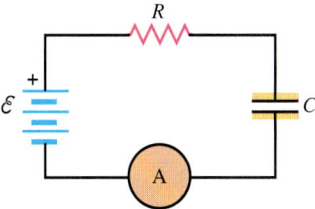

FIGURE 27–51 Challenge Problem 27–72.

27–73 An Infinite Network. As shown in Fig. 27–52, a network of resistors of resistances R_1 and R_2 extends off to infinity to the right. Prove that the total resistance R_T of the infinite network is equal to

$$R_T = R_1 + \sqrt{R_1^{\ 2} + 2R_1R_2}\ .$$

(*Hint:* Since the network is infinite, the resistance of the network to the right of points c and d is also equal to R_T.)

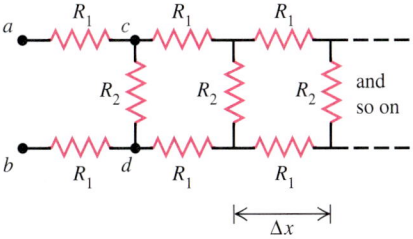

FIGURE 27–52 Challenge Problems 27–73 and 27–75.

27–74 Suppose a resistor R lies along each edge of a cube (12 resistors in all) with connections at the corners. Find the equivalent resistance between two diagonally opposite corners of the cube (points a and b in Fig. 27–53).

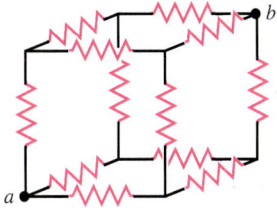

FIGURE 27–53 Challenge Problem 27–74.

27–75 Attenuator Chains and Axons. The infinite network of resistors shown in Fig. 27–52 is known as an *attenuator chain,* since this chain of resistors causes the potential difference between the upper and lower wires to decrease, or attenuate, along the length of the chain. a) Show that if the potential difference between the points a and b in Fig. 27–52 is V_{ab}, then the potential difference between the points c and d is $V_{cd} = V_{ab}/(1 + \beta)$, where $\beta = 2R_1(R_T + R_2)/R_TR_2$ and R_T, the total resistance of the network, is given in Challenge Problem 27–73. (See the hint given in that problem.) b) If the potential difference between the terminals a and b at the left-hand end of the infinite network is V_0, show that the potential difference between the upper and lower wires n segments from the left-hand end is $V_n = V_0/(1 + \beta)^n$. If $R_1 = R_2$, how many segments are needed to decrease the potential difference V_n to less than 1.0% of V_0? c) An infinite attenuator chain provides a model of the propagation of a voltage pulse along a nerve fiber, or axon (see Fig. 26–25 in Section 26–8). Each segment of the network in Fig. 27–52 represents a short segment of the axon of length Δx. The resistors R_1 represent the resistance of the fluid inside and outside the membrane wall of the axon. The resistance of the membrane to current flowing through the wall is represented by R_2. For an axon segment of length $\Delta x = 1.0\ \mu$m, $R_1 = 6.4 \times 10^3\ \Omega$ and $R_2 = 8.0 \times 10^8\ \Omega$ (the membrane wall is a good insulator). Calculate the total resistance R_T and β for an infinitely long axon. (This is a good approximation, since the length of an axon is much greater than its width; the largest axons in the human nervous system are more than 1 m long but only about 10^{-7} m in radius.) d) By what fraction does the potential difference between the inside and outside of the axon decrease over a distance of 2.0 mm?

This attenuation of the potential difference shows that the axon cannot simply be a passive, current-carrying electrical cable; the potential difference must periodically be reinforced along the axon's length by the action potential mechanism shown in Fig. 26–25. e) The action potential mechanism is slow, so a signal propagates along the axon at only about 30 m/s. In situations where faster response is required, axons are covered with a segmented sheath of fatty material called myelin. The segments are about 2 mm long, separated by gaps called the *nodes of Ranvier;* it is only at these nodes that action potentials are generated.

The myelin increases the resistance of a 1.0-μm-long segment of the membrane to $R_2 = 3.3 \times 10^{12}$ Ω. For such a myelinated axon, by what fraction does the potential difference between the inside and outside of the axon decrease over the distance from one node of Ranvier to the next? This smaller attenuation means the pulse is sufficiently intense to generate an action potential at the next node; the propagation speed is increased because the pulse travels as a conventional electrical signal in the myelinated segments.

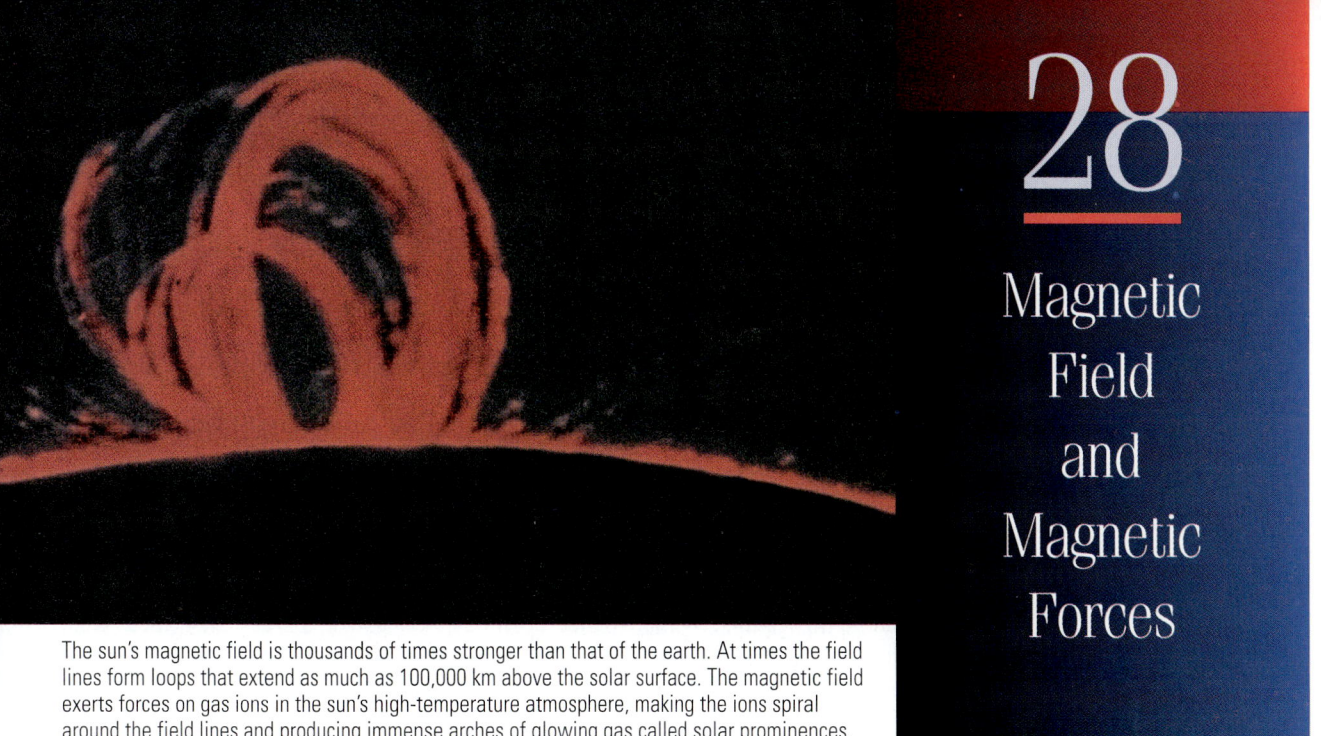

The sun's magnetic field is thousands of times stronger than that of the earth. At times the field lines form loops that extend as much as 100,000 km above the solar surface. The magnetic field exerts forces on gas ions in the sun's high-temperature atmosphere, making the ions spiral around the field lines and producing immense arches of glowing gas called solar prominences.

28

Magnetic Field and Magnetic Forces

28-1 INTRODUCTION

Everybody uses magnetic forces. They are at the heart of electric motors, TV picture tubes, microwave ovens, loud-speakers, computer printers, and disk drives. The most familiar aspects of magnetism are those associated with permanent magnets, which attract unmagnetized iron objects and can also attract or repel other magnets. A compass needle aligning itself with the earth's magnetism is an example of this interaction. But the *fundamental* nature of magnetism is the interaction of moving electric charges. Unlike electric forces, which act on electric charges whether they are moving or not, magnetic forces act only on *moving* charges.

As we did for the electric force, we will describe magnetic forces using the concept of a field. A magnetic field is established by a permanent magnet, by an electric current in a conductor, or by other moving charges. This magnetic field, in turn, exerts forces on moving charges and current-carrying conductors. In this chapter we study the magnetic forces and torques *exerted* on moving charges and currents by magnetic fields. In Chapter 29 we'll see how to calculate the magnetic fields *produced* by particular configurations of moving charges and currents.

28-2 MAGNETISM

Magnetic phenomena were first observed at least 2500 years ago in fragments of magnetized iron ore found near the ancient city of Magnesia (now Manisa, in western Turkey). These fragments were examples of what are now called **permanent magnets;** you probably have several permanent magnets on your refrigerator door at home. Permanent magnets were found to exert forces on each other as well as on pieces of iron that were not magnetized. It was discovered that when an iron rod is brought in contact with a natural magnet, the rod also becomes magnetized. When such a rod is floated on water or suspended by a string from its center, it tends to line itself up in a north-south direction. The needle of an ordinary compass is just such a piece of magnetized iron.

Before the relation of magnetic interactions to moving charges was understood, the interactions of permanent magnets and compass needles were described in terms of *magnetic poles.* If a bar-shaped permanent magnet, or *bar magnet,* is free to rotate, one end points north. This end is called a *north pole* or *N-pole;* the other end is a *south pole* or *S-pole.* Opposite poles attract each other, and like poles repel each other (Fig. 28–1). An object that contains iron but is not itself magnetized (that is, it shows no tendency to point north or south) is attracted by *either* pole of a permanent magnet (Fig. 28–2). This is the attraction that acts between a magnet and the unmagnetized steel door of a refrigerator. By analogy to electric interactions, we describe the interactions in Figs. 28–1 and 28–2 by saying that a bar magnet sets up a *magnetic field* in the space around it and a second body responds to that field. A compass needle tends to align with the magnetic field at the needle's position.

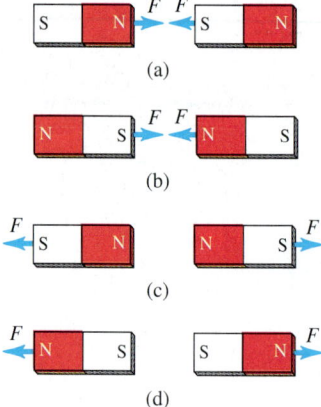

28–1 (a), (b) Two bar magnets attract when opposite poles (N and S, or S and N) are next to each other. (c), (d) When like poles (N and N, or S and S) are next to each other, the bar magnets repel each other.

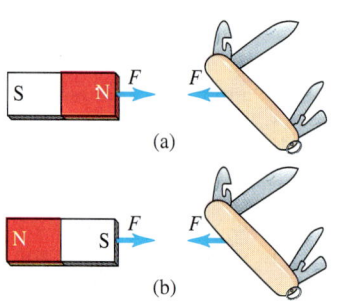

28–2 (a), (b) Either pole of a bar magnet attracts an unmagnetized object that contains iron.

28–3 A compass placed at any point in the earth's magnetic field will point in the direction of the field line at that point. Representing the earth's field as that of a tilted bar magnet is only a crude approximation of its fairly complex configuration. The field, which is thought to be caused by currents in the earth's molten core, changes with time; geologic evidence shows that it reverses direction entirely at irregular intervals of about a half million years.

The earth itself is a magnet. Its north geographical pole is close to a magnetic *south* pole, which is why the north pole of a compass needle points north. The earth's magnetic axis is not quite parallel to its geographic axis (the axis of rotation), so a compass reading deviates somewhat from geographic north. This deviation, which varies with location, is called *magnetic declination* or *magnetic variation*. Also, the magnetic field is not horizontal at most points on the earth's surface; its angle up or down is called *magnetic inclination*. At the magnetic poles the magnetic field is vertical.

Figure 28–3 is a sketch of the earth's magnetic field. The lines, called *magnetic field lines,* show the direction that a compass would point at each location; they are discussed in detail in Section 28–4. The direction of the field at any point can be defined as the direction of the force that the field would exert on a magnetic north pole. In Section 28–3 we'll describe a more fundamental way to define the direction and magnitude of a magnetic field.

The concept of magnetic poles may appear similar to that of electric charge, and north and south poles may seem analogous to positive and negative charge. But the analogy can be misleading. While isolated positive and negative charges exist, there is *no* experimental evidence that a single isolated magnetic pole exists; poles always appear in pairs. If a bar magnet is broken in two, each broken end becomes a pole. The existence of an isolated magnetic pole, or **magnetic monopole,** would have sweeping implications for theoretical physics. Extensive searches for magnetic monopoles have been carried out, but so far without success.

The first evidence of the relationship of magnetism to moving charges was discovered in 1819 by the Danish scientist Hans Christian Oersted. He found that a compass needle was deflected by a current-carrying wire, as shown in Fig. 28–4. Similar investigations were carried out in France by André Ampère. A few years later, Michael Faraday in England and Joseph Henry in the United States discovered that moving a magnet near a conducting loop can cause a current in the loop. We now know that the magnetic forces between two bodies shown in Figs. 28–1 and 28–2 are fundamentally due to interactions between moving electrons in the atoms of the bodies (above and

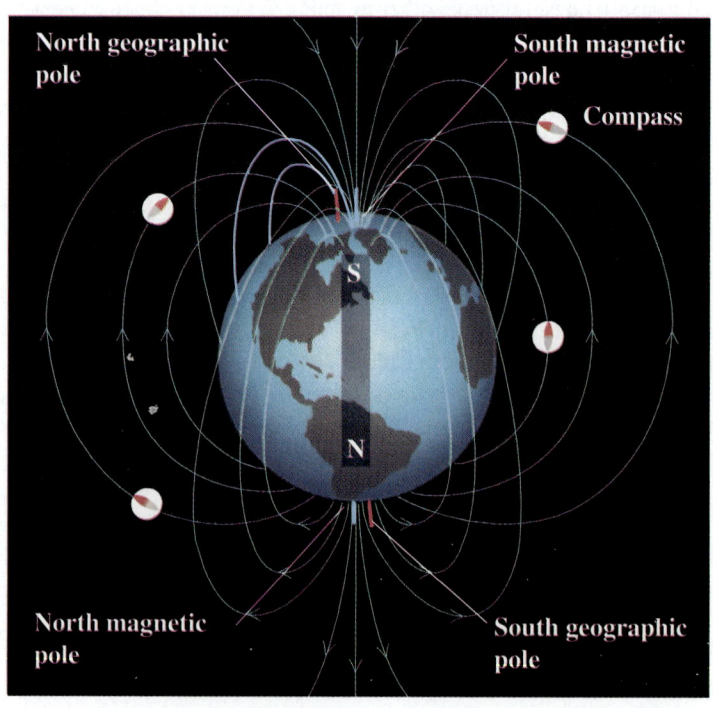

beyond the electric interactions between these charges). Inside a magnetized body such as a permanent magnet, there is a *coordinated* motion of certain of the atomic electrons; in an unmagnetized body these motions are not coordinated. (We'll describe these motions further in Section 28–8, and see how the interactions shown in Figs. 28–1 and 28–2 come about.)

Electric and magnetic interactions prove to be intimately intertwined. Over the next several chapters we will develop the unifying principles of electromagnetism, culminating in the expression of these principles in *Maxwell's equations.* These equations represent the synthesis of electromagnetism, just as Newton's laws of motion are the synthesis of mechanics, and like Newton's laws they represent a towering achievement of the human intellect.

$I = 0$
(a)

28–3 MAGNETIC FIELD

To introduce the concept of magnetic field properly, let's review our formulation of *electric* interactions in Chapter 22, where we introduced the concept of *electric field.* We represented electric interactions in two steps:

1. A distribution of electric charge at rest creates an electric field $\vec{E}$ in the surrounding space.
2. The electric field exerts a force $\vec{F} = q\vec{E}$ on any other charge q that is present in the field.

We can describe magnetic interactions in a similar way:

1. A moving charge or a current creates a **magnetic field** in the surrounding space (in addition to its *electric* field).
2. The magnetic field exerts a force $\vec{F}$ on any other moving charge or current that is present in the field.

In this chapter we'll concentrate on the *second* aspect of the interaction: Given the presence of a magnetic field, what force does it exert on a moving charge or a current? In Chapter 29 we will come back to the problem of how magnetic fields are *created* by moving charges and currents.

Like electric field, magnetic field is a *vector field,* that is, a vector quantity associated with each point in space. We will use the symbol $\vec{B}$ for magnetic field. At any position the direction of $\vec{B}$ is defined as that in which the north pole of a compass needle tends to point. The arrows in Fig. 28–3 suggest the direction of the earth's magnetic field; for any magnet, $\vec{B}$ points out of its north pole and into its south pole.

What are the characteristics of the magnetic force on a moving charge? First, its magnitude is proportional to the magnitude of the charge. If a 1-μC charge and a 2-μC charge move through a given magnetic field with the same velocity, the force on the 2-μC charge is twice as great as that on the 1-μC charge. The magnitude of the force is also proportional to the magnitude, or "strength," of the field; if we double the magnitude of the field (for example, by using two identical bar magnets instead of one) without changing the charge or its velocity, the force doubles.

The magnetic force also depends on the particle's velocity. This is quite different from the electric-field force, which is the same whether the charge is moving or not. A charged particle at rest experiences *no* magnetic force. Furthermore, the magnetic force $\vec{F}$ *does not* have the same direction as the magnetic field $\vec{B}$ but instead is always *perpendicular* to both $\vec{B}$ and the velocity $\vec{v}$. The magnitude F of the force is found to be proportional to the component of $\vec{v}$ perpendicular to the field; when that component is zero (that is, when $\vec{v}$ and $\vec{B}$ are parallel or antiparallel), the force is zero.

(b)   (c)

28–4 In Oersted's experiment, a compass is placed directly over a horizontal wire (here viewed from above). (a) The needle points north when there is no current. (b) The needle swings toward the east when the current flows north. (c) The needle swings west when the current flows south. When the compass is placed directly under the wire, the needle swings are reversed.

$F = 0$

(a)

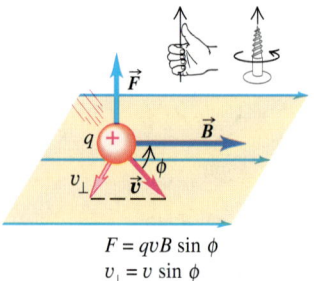

$F = qvB \sin \phi$

$v_\perp = v \sin \phi$

(b)

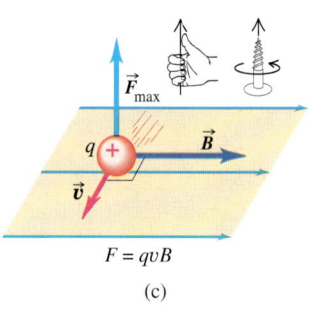

$F = qvB$

(c)

28–5 The magnetic force $\vec{F}$ acting on a positive charge q moving with velocity $\vec{v}$ is perpendicular to both $\vec{v}$ and the magnetic field $\vec{B}$. (a) The magnetic force is zero when $\vec{v}$ is parallel or antiparallel to $\vec{B}$. (b) When $\vec{v}$ is at an angle ϕ to $\vec{B}$, then $F = qvB \sin \phi$. (c) When $\vec{v}$ is perpendicular to $\vec{B}$, then $F = qvB$; for given values of v and B, this orientation gives the largest force.

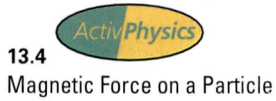

13.4

Magnetic Force on a Particle

Figure 28–5 shows these relationships. The direction of $\vec{F}$ is always perpendicular to the plane containing $\vec{v}$ and $\vec{B}$. Its magnitude is given by

$$F = |q|v_\perp B = |q|vB \sin \phi, \qquad (28\text{–}1)$$

where $|q|$ is the magnitude of the charge and ϕ is the angle measured from the direction of $\vec{v}$ to the direction of $\vec{B}$, as shown in the figure.

This description does not specify the direction of $\vec{F}$ completely; there are always two directions, opposite to each other, that are both perpendicular to the plane of $\vec{v}$ and $\vec{B}$. To complete the description, we use the same right-hand rule that we used to define the vector product in Section 1–11. (It would be a good idea to review that section before you go on.) Draw the vectors $\vec{v}$ and $\vec{B}$ with their tails together, as in Fig. 28–5b. Imagine turning $\vec{v}$ until it points in the direction of $\vec{B}$ (turning through ϕ, the smaller of the two possible angles). Wrap the fingers of your right hand around the line perpendicular to the plane of $\vec{v}$ and $\vec{B}$ so that they curl around with the sense of rotation from $\vec{v}$ to $\vec{B}$. Your thumb then points in the direction of the force $\vec{F}$ on a *positive* charge. (Alternatively, the direction of the force $\vec{F}$ on a positive charge is the direction in which a right-hand-thread screw would advance if turned the same way.)

This discussion shows that the force on a charge q moving with velocity $\vec{v}$ in a magnetic field $\vec{B}$ is given, both in magnitude and in direction, by

$$\vec{F} = q\vec{v} \times \vec{B} \qquad \text{(magnetic force on a moving charged particle).} \qquad (28\text{–}2)$$

This is the first of several vector products we will encounter in our study of magnetic-field relationships. It's important to note that Eq. (28–2) was not deduced theoretically; it is an observation based on experiment.

Equation (28–2) is valid for both positive and negative charges. When q is negative, the direction of the force $\vec{F}$ is opposite to that of $\vec{v} \times \vec{B}$. If two charges with equal magnitude and opposite sign move in the same $\vec{B}$ field with the same velocity (Fig. 28–6), the forces have equal magnitude and opposite direction. Figures 28–5 and 28–6 show several examples of the relationships of the directions of $\vec{F}$, $\vec{v}$, and $\vec{B}$ for both positive and negative charges. Be sure you understand the relationships shown in these figures.

Equation (28–1) gives the magnitude of the magnetic force $\vec{F}$ in Eq. (28–2). We can express this magnitude in a different but equivalent way. Since ϕ is the angle between the directions of vectors $\vec{v}$ and $\vec{B}$, we may interpret $B \sin \phi$ as the component of $\vec{B}$ perpendicular to $\vec{v}$, that is, $B_\perp$. With this notation the force magnitude is

$$F = |q|vB_\perp. \qquad (28\text{–}3)$$

This form is sometimes more convenient, especially in problems involving *currents* rather than individual particles. We will discuss forces on currents later in this chapter.

From Eq. (28–1) the *units* of B must be the same as the units of F/qv. Therefore the SI unit of B is equivalent to $1 \text{ N} \cdot \text{s}/\text{C} \cdot \text{m}$, or, since one ampere is one coulomb per second (1 A = 1 C/s), $1 \text{ N/A} \cdot \text{m}$. This unit is called the **tesla** (abbreviated T), in honor of Nikola Tesla (1857–1943), the prominent Serbian-American scientist and inventor:

$$1 \text{ tesla} = 1 \text{ T} = 1 \text{ N/A} \cdot \text{m}.$$

The cgs unit of B, the **gauss** ($1 \text{ G} = 10^{-4}$ T), is also in common use. Instruments for measuring magnetic field are sometimes called *gaussmeters*.

The magnetic field of the earth is of the order of 10^{-4} T or 1 G. Magnetic fields of the order of 10 T occur in the interior of atoms and are important in the analysis of atomic spectra. The largest steady magnetic field that can be produced at present in the laboratory is about 45 T. Some pulsed-current electromagnets can produce fields of the order of 120 T for short time intervals of the order of a millisecond. The magnetic field at the surface of a neutron star is believed to be of the order of 10^8 T.

To explore an unknown magnetic field, we can measure the magnitude and direction of the force on a *moving* test charge, then use Eq. (28–2) to determine $\vec{B}$. The electron beam in a cathode-ray tube, discussed in Section 24–7, is a convenient device for making such measurements. (An old TV set with the deflection coils disconnected could be used for the same purpose.) The electron gun shoots out a narrow beam of electrons at a known speed. If there is no force to deflect the beam, it strikes the center of the screen.

If a magnetic field is present, in general the electron beam is deflected. But if the beam is parallel or antiparallel to the field, then $\phi = 0$ or π in Eq. (28–1) and $F = 0$; there is no force, and hence no deflection. If we find that the electron beam is not deflected when its direction is parallel to a certain axis as in Fig. 28–7a, the $\vec{B}$ vector must point either up or down along that axis.

If we then turn the tube 90° (Fig. 28–7b), $\phi = \pi/2$ in Eq. (28–1) and the magnetic force is maximum; the beam is deflected in a direction perpendicular to the plane of $\vec{B}$ and $\vec{v}$. The direction and magnitude of the deflection determine the direction and magnitude of $\vec{B}$. We can perform additional experiments in which the angle between $\vec{B}$ and $\vec{v}$ is between zero and 90° to confirm Eq. (28–1) or (28–3) and the accompanying discussion. We note that the electron has a negative charge; the force in Fig. 28–7 is opposite in direction to the force on a positive charge.

When a charged particle moves through a region of space where *both* electric and magnetic fields are present, both fields exert forces on the particle. The total force $\vec{F}$ is the vector sum of the electric and magnetic forces:

$$\vec{F} = q(\vec{E} + \vec{v} \times \vec{B}). \qquad (28\text{–}4)$$

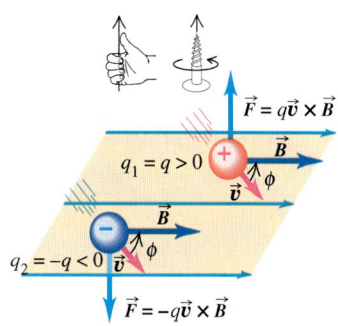

28–6 Two charges of the same magnitude but opposite sign moving with the same velocity in the same magnetic field. The magnetic forces exerted on the charges by the field are equal in magnitude but opposite in direction.

(a)   (b)

28–7 (a) If the electron beam of a cathode-ray tube is undeflected when the beam is parallel to the y-axis, the $\vec{B}$ vector points either up or down along that axis. (b) If the beam is deflected in the negative z-direction when the tube axis is parallel to the x-axis, then the $\vec{B}$ vector points upward. The force $\vec{F}$ on the electrons points along the negative z-axis, opposite to the rule of Fig. 28–5, because q is negative.

Problem–Solving Strategy

MAGNETIC FORCES

1. The biggest difficulty in determining magnetic forces is in relating the directions of the vector quantities. In evaluating $\vec{v} \times \vec{B}$, draw the two vectors $\vec{v}$ and $\vec{B}$ with their tails together so that you can visualize and draw the plane in which they lie. This also helps you to correctly identify the angle ϕ between the two vectors. Then remember that $\vec{F}$ is always perpendicular to this plane. The direction of $\vec{v} \times \vec{B}$ is determined by the right-hand rule; keep referring

to Figs. 28–5 and 28–6 until you're sure you understand this. If q is negative, the force is *opposite* to $\vec{v} \times \vec{B}$.

2. Whenever you can, do the problem in two ways. Do it directly from the geometric definition of the vector product. Then find the components of the vectors in some convenient axis system and calculate the vector product algebraically from the components. Check that the results agree.

EXAMPLE 28–1

A beam of protons ($q = 1.6 \times 10^{-19}$ C) moves at 3.0×10^5 m/s through a uniform magnetic field with magnitude 2.0 T that is directed along the positive z-axis, as in Fig. 28–8. The velocity of each proton lies in the xz-plane at an angle of 30° to the $+z$-axis. Find the force on a proton.

SOLUTION The right-hand rule shows that the direction of the force is along the negative y-axis. The magnitude of the force, from Eq. (28–1), is

$$F = qvB \sin \phi = (1.6 \times 10^{-19} \text{ C})(3.0 \times 10^5 \text{ m/s})(2.0 \text{ T})(\sin 30°)$$
$$= 4.8 \times 10^{-14} \text{ N}.$$

Alternatively, in vector language, with Eq. (28–2),

$$\vec{v} = (3.0 \times 10^5 \text{ m/s})(\sin 30°)\hat{\imath} + (3.0 \times 10^5 \text{ m/s})(\cos 30°)\hat{k},$$
$$\vec{B} = (2.0 \text{ T})\hat{k},$$
$$\vec{F} = q\vec{v} \times \vec{B}$$
$$= (1.6 \times 10^{-19} \text{ C})(3.0 \times 10^5 \text{ m/s})(2.0 \text{ T})$$
$$\times (\sin 30° \, \hat{\imath} + \cos 30° \, \hat{k}) \times \hat{k} = (-4.8 \times 10^{-14} \text{ N})\hat{\jmath}.$$

(Recall that $\hat{\imath} \times \hat{k} = -\hat{\jmath}$ and $\hat{k} \times \hat{k} = 0$.)

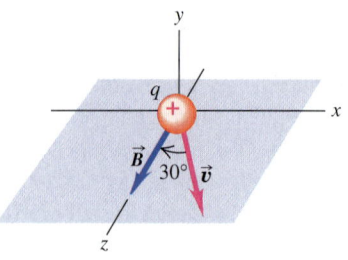

28–8 Directions of $\vec{v}$ and $\vec{B}$ for a proton beam in a magnetic field.

If the beam consists of *electrons* rather than protons, the charge is negative ($q = -1.6 \times 10^{-19}$ C), and the direction of the force is reversed. The force is now directed along the *positive y*-axis, but the magnitude is the same as before, $F = 4.8 \times 10^{-14}$ N.

28–4 MAGNETIC FIELD LINES AND MAGNETIC FLUX

We can represent any magnetic field by **magnetic field lines,** just as we did for the earth's magnetic field in Fig. 28–3. The idea is the same as for the electric field lines we introduced in Section 22–8. We draw the lines so that the line through any point is tangent to the magnetic field vector $\vec{B}$ at that point. Just as with electric field lines, we draw only a few representative lines; otherwise, the lines would fill up all of space. Where adjacent field lines are close together, the field magnitude is large; where these field lines are far apart, the field magnitude is small. Also, because the direction of $\vec{B}$ at each point is unique, field lines never intersect.

CAUTION ▶ Magnetic field lines are sometimes called "magnetic lines of force," but that's not a good name for them; unlike electric field lines, they *do not* point in the direction of the force on a charge. Equation (28–2) shows that the force on a moving charged particle is always perpendicular to the magnetic field, and hence to the magnetic field line that passes through the particle's position. The direction of the force depends on the particle's velocity and the sign of its charge, so just looking at magnetic field lines cannot in itself tell you the direction of the force on an arbitrary moving charged particle. Magnetic field lines *do* have the direction that a compass needle would point at each location; this may help you to visualize them. ◀

Magnetic field lines produced by several common sources of magnetic field are shown in Fig. 28–9. In the gap between the poles of the electromagnet shown in Fig. 28–9c, the field lines are approximately straight, parallel, and equally spaced, showing that the magnetic field in this region is approximately *uniform* (that is, constant in magnitude and direction). Figure 28–10 shows the familiar technique of visualizing magnetic field lines with iron filings.

MAGNETIC FLUX AND GAUSS'S LAW FOR MAGNETISM

We define the **magnetic flux** Φ_B through a surface just as we defined electric flux in connection with Gauss's law in Section 23–3. We can divide any surface into elements of area dA (Fig. 28–11). For each element we determine $B_\perp$, the component of $\vec{B}$ normal to

28-9 Magnetic field lines in a plane through the center of (a) a permanent magnet, (b) a cylindrical coil, (c) an iron-core electromagnet. (d) Magnetic field lines in a plane perpendicular to a long, straight, current-carrying wire. (e) Magnetic field lines in a plane containing the axis of a circular current-carrying loop.

28-10 Magnetic field lines can be visualized by the use of iron filings, which line up tangent to the magnetic field lines like little compass needles.

the surface at the position of that element, as shown. From the figure, $B_\perp = B \cos \phi$, where ϕ is the angle between the direction of $\vec{B}$ and a line perpendicular to the surface. (Be careful not to confuse ϕ with Φ_B.) In general, this component varies from point to point on the surface. We define the magnetic flux $d\Phi_B$ through this area as

$$d\Phi_B = B_\perp \, dA = B \cos \phi \, dA = \vec{B} \cdot d\vec{A}. \qquad (28\text{-}5)$$

The *total* magnetic flux through the surface is the sum of the contributions from the individual area elements:

$$\Phi_B = \int B_\perp \, dA = \int B \cos \phi \, dA = \int \vec{B} \cdot d\vec{A} \qquad \text{(magnetic flux through} \quad (28\text{-}6)$$
$$\text{a surface).}$$

(This equation uses the concepts of vector area and surface integral that we introduced in Section 23-3; you may want to review that discussion.)

Magnetic flux is a *scalar* quantity. In the special case in which $\vec{B}$ is uniform over a plane surface with total area A, $B_\perp$ and ϕ are the same at all points on the surface, and

$$\Phi_B = B_\perp A = BA \cos \phi. \qquad (28\text{-}7)$$

If $\vec{B}$ happens to be perpendicular to the surface, then $\cos \phi = 1$ and Eq. (28-7) reduces to $\Phi_B = BA$. We will use the concept of magnetic flux extensively during our study of electromagnetic induction in Chapter 30.

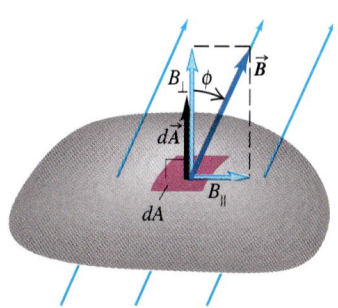

28-11 The magnetic flux through an area element dA is defined to be $d\Phi_B = B_\perp \, dA$.

The SI unit of magnetic flux is equal to the unit of magnetic field (1 T) times the unit of area (1 m^2). This unit is called the **weber** (1 Wb), in honor of the German physicist Wilhelm Weber (1804–1891):

$$1 \text{ Wb} = 1 \text{ T} \cdot \text{m}^2.$$

Also, 1 T = 1 N/A · m, so

$$1 \text{ Wb} = 1 \text{ T} \cdot \text{m}^2 = 1 \text{ N} \cdot \text{m/A}.$$

In Gauss's law the total *electric* flux through a closed surface is proportional to the total electric charge enclosed by the surface. For example, if the closed surface encloses an electric dipole, the total electric flux is zero because the total charge is zero. (You may want to review Section 23–4 on Gauss's law.) By analogy, if there were such a thing as a single magnetic charge (magnetic monopole), the total *magnetic* flux through a closed surface would be proportional to the total magnetic charge enclosed. But we have mentioned that no magnetic monopole has ever been observed, despite intensive searches. We conclude that **the total magnetic flux through a closed surface is always zero.** Symbolically,

$$\oint \vec{B} \cdot d\vec{A} = 0 \qquad \text{(magnetic flux through any closed surface).} \qquad (28\text{–}8)$$

This equation is sometimes called *Gauss's law for magnetism.* You can verify it by examining Fig. 28–9; if you draw a closed surface anywhere in any of the field maps shown in that figure, you will see that every field line that enters the surface also exits from it; the net flux through the surface is zero. It also follows from Eq. (28–8) that magnetic field lines always form closed loops. Unlike electric field lines that begin and end on electric charges, magnetic field lines *never* have end points; such a point would indicate the existence of a monopole.

For Gauss's law, which always deals with *closed* surfaces, the vector area element $d\vec{A}$ in Eq. (28–6) always points *out of* the surface. However, some applications of *magnetic* flux involve an *open* surface with a boundary line; there is then an ambiguity of sign in Eq. (28–6) because of the two possible choices of direction for $d\vec{A}$. In these cases we choose one of the possible sides of the surface to be the "positive" side and use that choice consistently.

If the element of area dA in Eq. (28–5) is at right angles to the field lines, then $B_\perp = B$; calling the area $dA_\perp$, we have

$$B = \frac{d\Phi_B}{dA_\perp}. \qquad (28\text{–}9)$$

That is, the magnitude of magnetic field is equal to *flux per unit area* across an area at right angles to the magnetic field. For this reason, magnetic field $\vec{B}$ is sometimes called **magnetic flux density.**

EXAMPLE 28-2

Figure 28–12a shows a side view of a flat surface with area 3.0 cm^2 in a uniform magnetic field. If the magnetic flux through this area is 0.90 mWb, calculate the magnitude of the magnetic field, and find the direction of the area vector.

SOLUTION Because B and ϕ are the same at all points on the surface, we can use Eq. (28–7): $\Phi_B = BA \cos \phi$. The area A is 3.0×10^{-4} m^2; the direction of $\vec{A}$ is perpendicular to the surface,

so ϕ could be either 60° or 120°. But Φ_B, B, and A are all positive, so cos ϕ must also be positive. This rules out 120°, so $\phi = 60°$, and we find

$$B = \frac{\Phi_B}{A \cos \phi} = \frac{0.90 \times 10^{-3} \text{ Wb}}{(3.0 \times 10^{-4} \text{ m}^2)(\cos 60°)} = 6.0 \text{ T}.$$

The area vector $\vec{A}$ is perpendicular to the area in the direction shown in Fig. 28–12b.

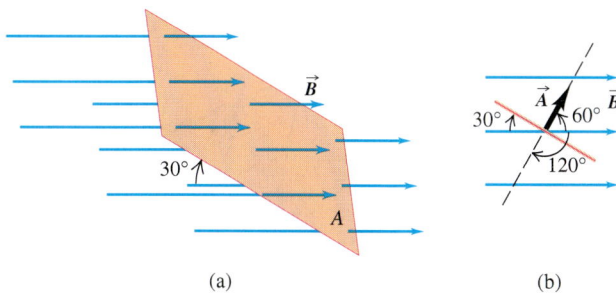

(a)   (b)

28-12 (a) A flat area A in a uniform magnetic field $\vec{B}$. (b) An edge-on view of the area A. The area vector $\vec{A}$ makes an angle $\phi = 60°$ with $\vec{B}$. (If we had chosen $\vec{A}$ to point in the opposite direction, ϕ would have been 120° and the magnetic flux Φ_B would have been negative.) If the area rotates to $\phi = 90°$, Φ_B becomes zero; but if the area rotates so that $\vec{A}$ and $\vec{B}$ are parallel, Φ_B has its maximum value.

28-5 MOTION OF CHARGED PARTICLES IN A MAGNETIC FIELD

When a charged particle moves in a magnetic field, it is acted on by the magnetic force given by Eq. (28–2), and the motion is determined by Newton's laws. Figure 28–13 shows a simple example. A particle with positive charge q is at point O, moving with velocity $\vec{v}$ in a uniform magnetic field $\vec{B}$ directed into the plane of the figure. The vectors $\vec{v}$ and $\vec{B}$ are perpendicular, so the magnetic force $\vec{F} = q\vec{v} \times \vec{B}$ has magnitude $F = qvB$ and a direction as shown in the figure. The force is *always* perpendicular to $\vec{v}$, so it cannot change the *magnitude* of the velocity, only its direction. To put it differently, the magnetic force never has a component parallel to the particle's motion, so the magnetic force can never do *work* on the particle. This is true even if the magnetic field is not uniform. **Motion of a charged particle under the action of a magnetic field alone is always motion with constant speed.**

Using this principle, we see that in the situation shown in Fig. 28–13a the magnitudes of both $\vec{F}$ and $\vec{v}$ are constant. At points such as P and S the directions of force and velocity have changed as shown, but their magnitudes are the same. The particle therefore moves under the influence of a constant-magnitude force that is always at right angles to the velocity of the particle. Comparing these conditions with the discussion of circular motion in Sections 3–5 and 5–5, we see that the particle's path is a *circle,* traced out with constant speed v (Fig. 28–13b). The centripetal acceleration is v^2/R, and the only force acting is the magnetic force, so from Newton's second law,

$$F = |q|vB = m\frac{v^2}{R}, \qquad (28-10)$$

where m is the mass of the particle. Solving Eq. (28–10) for the radius R of the circular

(a)

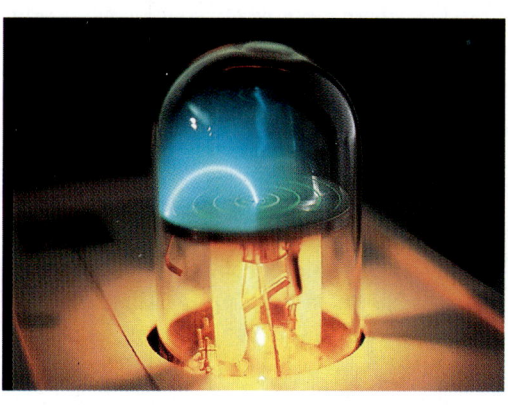

(b)

28-13 (a) The orbit of a charged particle in a uniform magnetic field $\vec{B}$ is a circle when the initial velocity is perpendicular to the field. The crosses represent a uniform magnetic field directed straight into the page. (b) An electron beam (seen as a blue arc) curving in a magnetic field.

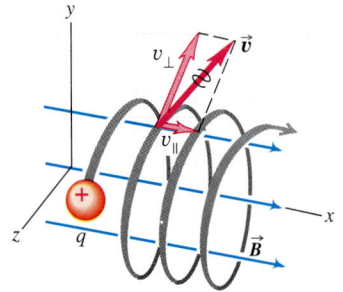

28–14 When a charged particle has velocity components both perpendicular and parallel to a uniform magnetic field, the particle moves in a helical path. The magnetic field does no work on the particle, so its speed and kinetic energy remain constant.

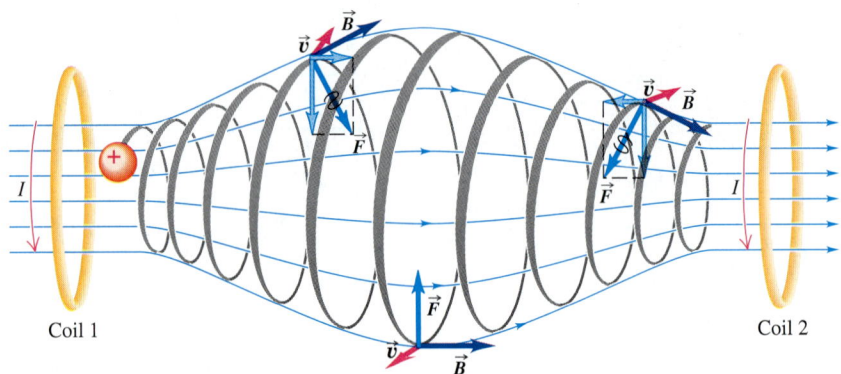

28–15 A magnetic bottle. Particles near either end of the region experience a magnetic force toward the center of the region. This is one way of containing an ionized gas that has a temperature of the order of 10^6 K, which would melt any material container.

path, we find

$$R = \frac{mv}{|q|B} \quad \text{(radius of a circular orbit in a magnetic field).} \quad (28\text{–}11)$$

We can also write this as $R = p/|q|B$, where $p = mv$ is the magnitude of the particle's momentum. If the charge q is negative, the particle moves *clockwise* around the orbit in Fig. 28–13a.

The angular speed ω of the particle can be found from Eq. (9–13), $v = R\omega$. Combining this with Eq. (28–11), we get

$$\omega = \frac{v}{R} = v\frac{|q|B}{mv} = \frac{|q|B}{m}. \quad (28\text{–}12)$$

The number of revolutions per unit time is $f = \omega/2\pi$. This frequency f is independent of the radius R of the path. It is called the **cyclotron frequency;** in a particle accelerator called a *cyclotron,* particles moving in nearly circular paths are given a boost twice each revolution, increasing their energy and their orbital radii but not their angular speed or frequency. Similarly, one type of *magnetron,* a common source of microwave radiation for microwave ovens and radar systems, emits radiation with a frequency equal to the frequency of circular motion of electrons in a vacuum chamber between the poles of a magnet.

If the direction of the initial velocity is *not* perpendicular to the field, the velocity *component* parallel to the field is constant because there is no force parallel to the field. Then the particle moves in a helix (Fig. 28–14). The radius of the helix is given by Eq. (28–11), where v is now the component of velocity perpendicular to the $\vec{B}$ field.

Motion of a charged particle in a non-uniform magnetic field is more complex. Figure 28–15 shows a field produced by two circular coils separated by some distance. Particles near either coil experience a magnetic force toward the center of the region; particles with appropriate speeds spiral repeatedly from one end of the region to the other and back. Because charged particles can be trapped in such a magnetic field, it is called a *magnetic bottle.* This technique is used to confine very hot plasmas with temperatures of the order of 10^6 K. In a similar way the earth's non-uniform magnetic field traps charged particles coming from the sun in doughnut-shaped regions around the earth, as shown in Fig. 28–16. These regions, called the *Van Allen radiation belts,* were discovered in 1958 using data obtained by instruments aboard the Explorer I satellite.

Figure 28–17a is a photograph of the track made in a bubble chamber filled with liquid hydrogen by a high-energy electron moving in a magnetic field perpendicular to the

Charged particles
from sun enter earth's
magnetic field

North
Pole

Protons trapped in
inner radiation belts

South
Pole

Electrons trapped in
outer radiation belts

(a)

28–16 (a) The Van Allen radiation belts around the earth. Near the poles, charged particles can escape from these belts into the atmosphere, producing the aurora borealis ("Northern Lights") and aurora australis ("Southern Lights"). (b) This photo of the aurora, taken from the Space Shuttle, shows a sheet of atmospheric gas atoms set aglow by impacts from solar particles. The shape of the auroral display is controlled by the configuration of the earth's magnetic field, which determines where the particles strike the atmosphere.

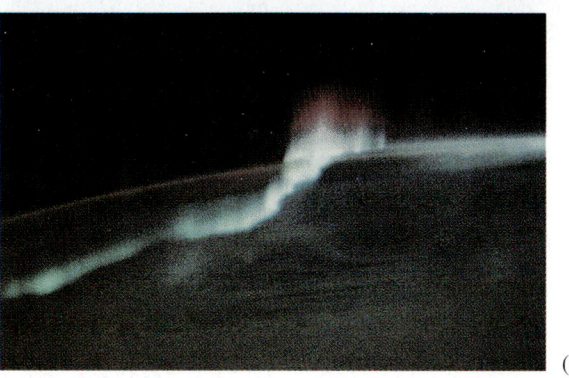

(b)

(a)

(b)

28–17 (a) A track in a liquid-hydrogen bubble chamber made by an electron with initial kinetic energy of about 27 MeV. The magnetic field strength is 1.8 T. As the particle loses energy, the radius of curvature decreases; the maximum radius is about 5 cm. (b) In this bubble chamber image, a high-energy gamma ray (which does not leave a track) comes in from above and collides with an electron in a hydrogen atom. Some of the energy in the collision is transformed into another electron and a positron (positively charged electron). The long, gently curving path is created by the electron from the hydrogen atom as it recoils from the collision, while the spiraling tracks are created by the electron-positron pair; the directions of curvature in the magnetic field show the signs of the charges.

plane of the photo. There are other forces acting in addition to the magnetic force; as the electron plows through the liquid hydrogen, it collides with charged particles, losing energy (and speed). As a result, the radius of curvature decreases as suggested by Eq. (28–11). (The electron's speed is comparable to the speed of light, so Eq. (28–11) isn't directly applicable here.) Figure 28–17b shows tracks due to the production of an electron and a *positron*, or positive electron, also in a bubble chamber. Similar experiments using a cloud chamber provided the first experimental evidence in 1932 for the existence of the positron.

Problem–Solving Strategy

MOTION IN MAGNETIC FIELDS

1. In analyzing the motion of a charged particle in electric and magnetic fields, you will apply Newton's second law of motion, $\Sigma\vec{F} = m\vec{a}$, with $\Sigma\vec{F}$ given by $\Sigma\vec{F} = q(\vec{E} + \vec{v} \times \vec{B})$. Typically, other forces such as gravity can be neglected. Many of the problems are similar to the trajectory and circular-motion problems in Sections 3–4, 3–5, and 5–5; it would be a good idea to review those sections.

2. Often the use of components is the most efficient approach. Choose a coordinate system, and then express all the vector quantities (including $\vec{E}$, $\vec{B}$, $\vec{v}$, $\vec{F}$, and $\vec{a}$) in terms of their components in this system. Then use $\Sigma\vec{F} = m\vec{a}$ in component form: $\Sigma F_x = ma_x$, and so forth. This approach is particularly useful when both electric and magnetic fields are present.

EXAMPLE 28-3

Electron motion in a microwave oven A magnetron in a microwave oven emits electromagnetic waves with frequency $f = 2450$ MHz. What magnetic field strength is required for electrons to move in circular paths with this frequency?

SOLUTION The corresponding angular speed is $\omega = 2\pi f = (2\pi)(2450 \times 10^6 \text{ s}^{-1}) = 1.54 \times 10^{10} \text{ s}^{-1}$. From Eq. (28–12),

$$B = \frac{m\omega}{|q|} = \frac{(9.11 \times 10^{-31} \text{ kg})(1.54 \times 10^{10} \text{ s}^{-1})}{1.60 \times 10^{-19} \text{ C}}$$

$$= 0.0877 \text{ T}.$$

This is a moderate field strength, easily produced with a permanent magnet. Incidentally, 2450-MHz electromagnetic waves are strongly absorbed by water molecules, so they are particularly useful for heating and cooking food.

EXAMPLE 28-4

Helical particle motion In a situation like that shown in Fig. 28–14, the charged particle is a proton ($q = 1.60 \times 10^{-19}$ C, $m = 1.67 \times 10^{-27}$ kg) and the uniform magnetic field is directed along the x-axis with magnitude 0.500 T. Only the magnetic force acts on the proton. At $t = 0$ the proton has velocity components $v_x = 1.50 \times 10^5$ m/s, $v_y = 0$, and $v_z = 2.00 \times 10^5$ m/s. a) At $t = 0$, find the force on the proton and its acceleration. b) Find the radius of the helical path, the angular speed of the proton, and the *pitch* of the helix (the distance traveled along the helix axis per revolution).

SOLUTION a) Since $v_y = 0$, the velocity vector is $\vec{v} = v_x\hat{\imath} + v_z\hat{k}$. Using Eq. (28–2) and recalling that $\hat{\imath} \times \hat{\imath} = 0$ and $\hat{k} \times \hat{\imath} = \hat{\jmath}$, we find

$$\vec{F} = q\vec{v} \times \vec{B} = q(v_x\hat{\imath} + v_z\hat{k}) \times B\hat{\imath} = qv_zB\hat{\jmath}$$

$$= (1.60 \times 10^{-19} \text{ C})(2.00 \times 10^5 \text{ m/s})(0.500 \text{ T})\hat{\jmath}$$

$$= (1.60 \times 10^{-14} \text{ N})\hat{\jmath}.$$

(To check unit consistency, recall from Section 28–3 that

1 T = 1 N/A · m = 1 N · s/C · m.) This may seem like a very weak force, but the resulting acceleration is tremendous because the proton mass is so small:

$$\vec{a} = \frac{\vec{F}}{m} = \frac{1.60 \times 10^{-14} \text{ N}}{1.67 \times 10^{-27} \text{ kg}}\hat{\jmath} = (9.58 \times 10^{12} \text{ m/s}^2)\hat{\jmath}.$$

b) The force is perpendicular to the velocity, so the speed of the proton does not change. The radius of the helical trajectory is given by Eq. (28–11) with v replaced by the component of velocity perpendicular to $\vec{B}$. At $t = 0$ this component is v_z, so

$$R = \frac{mv_z}{|q|B} = \frac{(1.67 \times 10^{-27} \text{ kg})(2.00 \times 10^5 \text{ m/s})}{(1.60 \times 10^{-19} \text{ C})(0.500 \text{ T})}$$

$$= 4.18 \times 10^{-3} \text{ m} = 4.18 \text{ mm}.$$

From Eq. (28–12) the angular speed is

$$\omega = \frac{|q|B}{m} = \frac{(1.60 \times 10^{-19} \text{ C})(0.500 \text{ T})}{1.67 \times 10^{-27} \text{ kg}} = 4.79 \times 10^7 \text{ rad/s}.$$

The time required for one revolution (the period) is $T = 2\pi/\omega = 2\pi/(4.79 \times 10^7 \text{ s}^{-1}) = 1.31 \times 10^{-7}$ s. The pitch is the distance traveled along the *x*-axis during this time, or

$$v_x T = (1.50 \times 10^5 \text{ m/s})(1.31 \times 10^{-7} \text{ s})$$

$$= 0.0197 \text{ m} = 19.7 \text{ mm}.$$

The pitch of the helix is almost five times greater than the radius. This helical trajectory is much more "stretched out" than that shown in Fig. 28–14.

28–6 APPLICATIONS OF MOTION OF CHARGED PARTICLES

This section describes several applications of the principles introduced in this chapter. Study them carefully, watching for applications of the problem-solving strategy outlined in the preceding section.

VELOCITY SELECTOR

In a beam of charged particles produced by a heated cathode or a radioactive material, not all particles move with the same speed. Particles of a specific speed can be selected from the beam using an arrangement of electric and magnetic fields called a *velocity selector*. In Fig. 28–18 a charged particle with mass *m*, charge *q*, and speed *v* enters a region of space where the electric and magnetic fields are perpendicular to the particle's velocity and to each other. The electric field $\vec{E}$ is vertically downward, and the magnetic field $\vec{B}$ is into the plane of the figure. If *q* is positive, the electric force is downward, with magnitude *qE*, and the magnetic force is upward, with magnitude *qvB*. For given field magnitudes *E* and *B*, for a particular value of *v* the electric and magnetic forces will be equal in magnitude; the total force is then zero, and the particle travels in a straight line with constant velocity. For zero total force, $\Sigma F_y = 0$, we need $-qE + qvB = 0$; solving for the speed *v* for which there is no deflection, we find

13.8
Velocity Selector

$$v = \frac{E}{B}. \qquad (28\text{–}13)$$

Only particles with speeds equal to *E/B* can pass through without being deflected by the fields. By adjusting *E* and *B* appropriately, we can select particles having a particular speed for use in other experiments. Because *q* divides out in Eq. (28–13), a velocity selector for positively charged particles also works for electrons or other negatively charged particles.

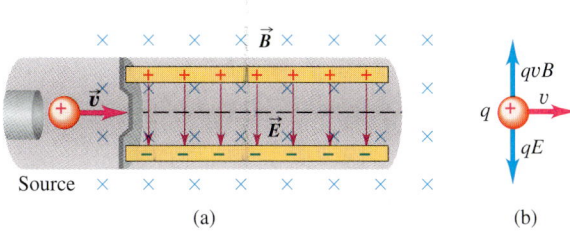

Source

(a)   (b)

28–18 (a) A velocity selector for charged particles uses perpendicular $\vec{E}$ and $\vec{B}$ fields. Only charged particles with *v* = *E/B* move through undeflected. (b) The electric and magnetic forces on a positive charge. The forces are reversed if the charge is negative.

THOMSON'S *e/m* EXPERIMENT

In one of the landmark experiments in physics at the end of the nineteenth century, J. J. Thomson (1856–1940) used the idea just described to measure the ratio of charge to mass for the electron. For this experiment, carried out in 1897 at the Cavendish Laboratory in Cambridge, England, Thomson used the apparatus shown in Fig. 28–19. It is very similar in principle to the cathode-ray tube discussed in Section 24–7; you may want to review that section. In a highly evacuated glass container, electrons from the hot cathode are accelerated and formed into a beam by a potential difference *V* between the two anodes A and A'. The speed *v* of the electrons is determined by the accelerating

28-19 Thomson's apparatus for measuring the ratio e/m for cathode rays. (a) A sketch showing the crossed $\vec{E}$ and $\vec{B}$ fields. (b) A photo of the tube used by Thomson.

potential V, just as in the derivation of Eq. (24–24). The kinetic energy $\frac{1}{2}mv^2$ equals the loss of electric potential energy eV, where e is the magnitude of the electron charge:

$$\frac{1}{2}mv^2 = eV, \qquad \text{or} \qquad v = \sqrt{\frac{2eV}{m}}. \qquad (28\text{–}14)$$

The electrons pass between the plates P and P′ and strike the screen at the end of the tube, which is coated with a material that fluoresces (glows) at the point of impact. The electrons pass straight through when Eq. (28–13) is satisfied; combining this with Eq. (28–14), we get

$$\frac{E}{B} = \sqrt{\frac{2eV}{m}}, \qquad \text{so} \qquad \frac{e}{m} = \frac{E^2}{2VB^2}. \qquad (28\text{–}15)$$

All the quantities on the right side can be measured, so the ratio e/m of charge to mass can be determined. It is *not* possible to measure e or m separately by this method, only their ratio.

The most significant aspect of Thomson's e/m measurements was that he found a *single value* for this quantity. It did not depend on the cathode material, the residual gas in the tube, or anything else about the experiment. This independence showed that the particles in the beam, which we now call electrons, are a common constituent of all matter. Thus Thomson is credited with discovery of the first subatomic particle, the electron. He also found that the *speed* of the electrons in the beam was about one tenth the speed of light, much larger than any previously measured speed of a material particle.

The most precise value of e/m available as of this writing is

$$e/m = 1.75881962(53) \times 10^{11} \text{ C/kg}.$$

In this expression, (53) indicates the likely uncertainty in the last two digits, 62.

Fifteen years after Thomson's experiments, the American physicist Robert Millikan succeeded in measuring the charge of the electron precisely (see Challenge Problem 24–83). This value, together with the value of e/m, enables us to determine the *mass* of the electron. The most precise value available at present is

$$m = 9.1093897(54) \times 10^{-31} \text{ kg}.$$

MASS SPECTROMETERS

Techniques similar to Thomson's e/m experiment can be used to measure masses of ions and thus measure atomic and molecular masses. In 1919, Francis Aston (1877–1945), a student of Thomson's, built the first of a family of instruments called **mass spectrometers.** A variation built by Bainbridge is shown in Fig. 28–20. Positive ions from a source pass through the slits S_1 and S_2, forming a narrow beam. Then the ions pass through a velocity selector with crossed $\vec{E}$ and $\vec{B}$ fields, as we have described, to block all ions except those with speeds v equal to E/B. Finally, the ions pass into a region with a magnetic field $\vec{B}'$ perpendicular to the figure, where they move in circular arcs with radius R determined by Eq. (28–11): $R = mv/qB'$. Ions with different masses strike the photographic plate at different points, and the values of R can be measured. We assume that each ion has lost one electron, so the net charge of each ion is just $+e$. With everything known in this equation except m, we can compute the mass m of the ion.

One of the earliest results from this work was the discovery that neon has two species of atoms, with atomic masses 20 and 22 g/mol. We now call these species **isotopes** of the element. Later experiments have shown that many elements have several isotopes, atoms that are identical in their chemical behavior but different in mass owing to differing numbers of neutrons in their nuclei.

Present-day mass spectrometers can measure masses routinely with a precision of better than one part in 10,000. The photographic plate is replaced by a more sophisticated particle detector that scans across the impact region.

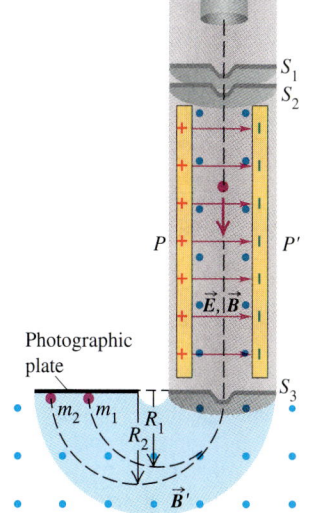

28–20 Bainbridge's mass spectrometer utilizes a velocity selector to produce particles with uniform speed v. In the region of magnetic field B', particles with larger mass travel in paths with larger radius R.

13.7
Mass Spectrometer

EXAMPLE 28–5

An e/m experiment You set out to reproduce Thomson's e/m experiment with an accelerating potential of 150 V and a deflecting electric field of magnitude 6.0×10^6 N/C. a) At what fraction of the speed of light do the electrons move? b) What magnitude of magnetic field will you need? c) With this magnetic field, what will happen to the electron beam if you increase the accelerating potential above 150 V?

SOLUTION a) The electron speed v is related to the accelerating potential by Eq. (28–14):

$$v = \sqrt{2(e/m)V} = \sqrt{2(1.76 \times 10^{11} \text{ C/kg})(150 \text{ V})}$$

$$= 7.27 \times 10^6 \text{ m/s},$$

$$\frac{v}{c} = \frac{7.27 \times 10^6 \text{ m/s}}{3.00 \times 10^8 \text{ m/s}} = 0.024.$$

The electrons are traveling at 2.4% of the speed of light.
b) From Eq. (28–13),

$$B = \frac{E}{v} = \frac{6.00 \times 10^6 \text{ N/C}}{7.27 \times 10^6 \text{ m/s}} = 0.83 \text{ T}.$$

c) Increasing the accelerating potential V increases the electron speed v. In Fig. 28–19 this doesn't change the upward electric force eE, but it increases the downward magnetic force evB. Therefore the electron beam will be bent *downward* and will hit the end of the tube below the undeflected position.

EXAMPLE 28–6

Finding leaks in a vacuum system There is almost no helium in ordinary air, so helium sprayed near a leak in a vacuum system will quickly show up in the output of a vacuum pump connected to such a system. You are designing a leak detector that

uses a mass spectrometer to detect He$^+$ ions (charge $+e$ $= +1.60 \times 10^{-19}$ C, mass 6.65×10^{-27} kg). The ions emerge from the velocity selector with a speed of 1.00×10^5 m/s. They are curved in a semicircular path by a magnetic field B' and are detected at a distance of 10.16 cm from the slit S_3 in Fig. 28–20. Calculate the magnitude of the magnetic field B'.

SOLUTION The radius R of the semicircular path is $\frac{1}{2}(10.16 \times 10^{-2}$ m$) = 5.08 \times 10^{-2}$ m. From Eq. (28–11),

$R = mv/qB'$, we get

$$B' = \frac{mv}{qR} = \frac{(6.65 \times 10^{-27}\ \text{kg})(1.00 \times 10^5\ \text{m/s})}{(1.60 \times 10^{-19}\ \text{C})(5.08 \times 10^{-2}\ \text{m})}$$

$$= 0.0817\ \text{T}.$$

28-7 MAGNETIC FORCE ON A CURRENT-CARRYING CONDUCTOR

What makes an electric motor work? The forces that make it turn are forces that a magnetic field exerts on a conductor carrying a current. The magnetic forces on the moving charges within the conductor are transmitted to the material of the conductor, and the conductor as a whole experiences a force distributed along its length. The moving-coil galvanometer that we described in Section 27–4 also uses magnetic forces on conductors.

We can compute the force on a current-carrying conductor starting with the magnetic force $\vec{F} = q\vec{v} \times \vec{B}$ on a single moving charge. Figure 28–21 shows a straight segment of a conducting wire, with length l and cross-section area A; the current is from bottom to top. The wire is in a uniform magnetic field $\vec{B}$, perpendicular to the plane of the diagram and directed *into* the plane. Let's assume first that the moving charges are positive. Later we'll see what happens when they are negative.

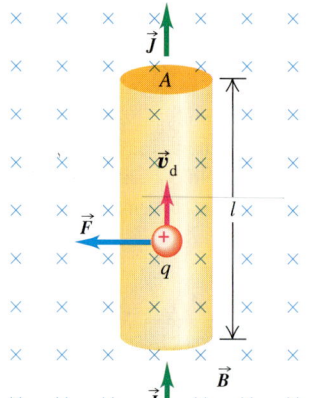

28-21 Force on a moving positive charge in a current-carrying conductor.

The drift velocity $\vec{v}_d$ is upward, perpendicular to $\vec{B}$. The average force on each charge is $\vec{F} = q\vec{v}_d \times \vec{B}$, directed to the left as shown in the figure; since $\vec{v}_d$ and $\vec{B}$ are perpendicular, the magnitude of the force is $F = qv_d B$.

We can derive an expression for the *total* force on all the moving charges in a length l of conductor with cross-section area A, using the same language we used in Eqs. (26–2) and (26–3) of Section 26–2. The number of charges per unit volume is n; a segment of conductor with length l has volume Al and contains a number of charges equal to nAl. The total force $\vec{F}$ on *all* the moving charges in this segment has magnitude

$$F = (nAl)(qv_d B) = (nqv_d A)(lB). \tag{28-16}$$

From Eq. (26–3) the current density is $J = nqv_d$. The product JA is the total current I, so we can rewrite Eq. (28–16) as

$$F = IlB. \tag{28-17}$$

If the $\vec{B}$ field is not perpendicular to the wire but makes an angle ϕ with it, we handle the situation the same way we did in Section 28–3 for a single charge. Only the component of $\vec{B}$ perpendicular to the wire (and to the drift velocities of the charges) exerts a force; this component is $B_\perp = B \sin \phi$. The magnetic force on the wire segment is then

$$F = IlB_\perp = IlB \sin \phi. \tag{28-18}$$

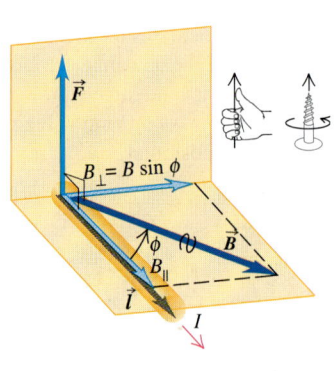

28-22 The magnetic force $\vec{F}$ on a straight wire segment of length $\vec{l}$ that carries a current I (in the direction of $\vec{l}$) is perpendicular both to $\vec{l}$ and to the magnetic field $\vec{B}$.

The force is always perpendicular to both the conductor and the field, with the direction determined by the same right-hand rule we used for a moving positive charge (Fig. 28–22). Hence this force can be expressed as a vector product, just like the force on a single moving charge. We represent the segment of wire with a vector $\vec{l}$ along the wire in the direction of the current; then the force $\vec{F}$ on this segment is

$$\vec{F} = I\vec{l} \times \vec{B} \qquad \text{(magnetic force on a straight wire segment).} \tag{28-19}$$

Figure 28–23 illustrates the directions of $\vec{B}, \vec{l}$, and $\vec{F}$ for several cases.

CAUTION ▶ As we discussed in Section 26–2, the current I is not a vector. The direction of current flow is described by $\vec{l}$, not I. ◀

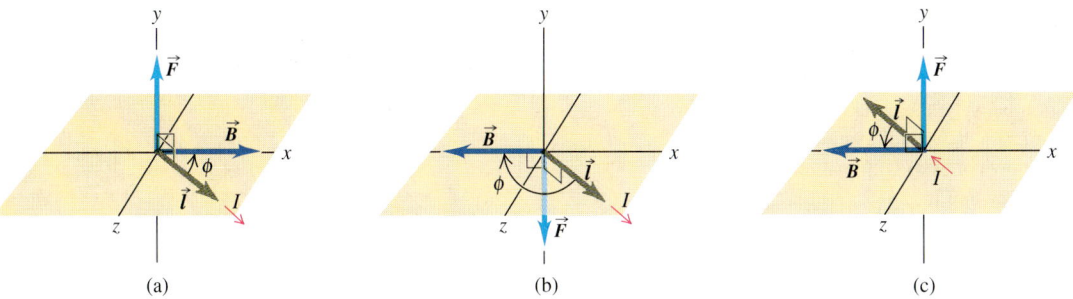

28–23 (a) Magnetic field $\vec{B}$, length $\vec{l}$, and force $\vec{F}$ vectors for a straight wire carrying a current I. (b) Reversing $\vec{B}$ reverses $\vec{F}$. (c) Also reversing the current reverses $\vec{l}$ and returns $\vec{F}$ to the same direction as in (a).

If the conductor is not straight, we can divide it into infinitesimal segments $d\vec{l}$. The force $d\vec{F}$ on each segment is

$$d\vec{F} = I\, d\vec{l} \times \vec{B} \qquad \text{(magnetic force on an infinitesimal wire segment).} \qquad (28–20)$$

13.5
Magnetic Force on a Wire

Then we can integrate this along the wire to find the total force on a conductor of any shape. The integral is a *line integral*, the same mathematical operation we have used to define work (Section 6–4) and electric potential (Section 24–3).

Finally, what happens when the moving charges are negative, such as electrons in a metal? Then in Fig. 28–21 an upward current corresponds to a downward drift velocity. But because q is now negative, the direction of the force $\vec{F}$ is the same as before. Thus Eqs. (28–17) through (28–20) are valid for *both* positive and negative charges and even when *both* signs of charge are present at once. This happens in some semiconductor materials and in ionic solutions.

A common application of the magnetic forces on a current-carrying wire is found in loudspeakers (Fig. 28–24). The radial magnetic field created by the permanent magnet exerts a force on the voice coil that is proportional to the current in the coil; the direction of the force is either to the left or to the right, depending on the direction of the current. The signal from the amplifier causes the current to oscillate in direction and magnitude. The coil and the speaker cone to which it is attached respond by oscillating with an amplitude proportional to the amplitude of the current in the coil. Turning up the volume knob on the amplifier increases the current amplitude and hence the amplitudes of the cone's oscillation and of the sound wave produced by the moving cone.

28–24 (a) Components of a loudspeaker. (b) The permanent magnet creates a magnetic field that exerts forces on the current in the voice coil; for a current I in the direction shown, the force is to the right. If the electric current in the voice coil oscillates, the speaker cone attached to the voice coil oscillates at the same frequency.

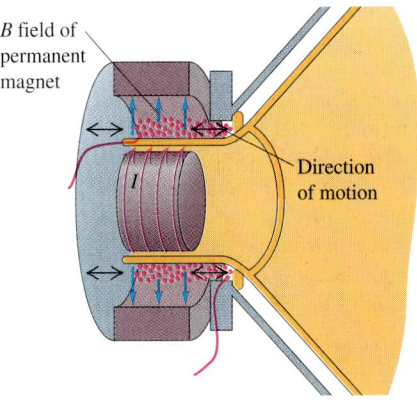

(a)

(b)

EXAMPLE 28-7

A straight horizontal copper rod carries a current of 50.0 A from west to east in a region between the poles of a large electromagnet. In this region there is a horizontal magnetic field toward the northeast (that is, 45° north of east) with magnitude 1.20 T, as shown in an overhead view in Fig. 28–25. a) Find the magnitude and direction of the force on a 1.00-m section of rod. b) While keeping the rod horizontal, how should it be oriented to maximize the magnitude of the force?

SOLUTION a) The angle ϕ between the directions of current and field is 45°. From Eq. (28–18) we obtain

$$F = IlB \sin \phi = (50.0 \text{ A})(1.00 \text{ m})(1.20 \text{ T})(\sin 45°) = 42.4 \text{ N}.$$

The *direction* of the force is perpendicular to the plane of the current and the field, both of which lie in the horizontal plane. Thus the force must be vertical; the right-hand rule shows that it is vertically *upward* (out of the plane of the figure).

Alternatively, we can use a coordinate system with the *x*-axis pointing east, the *y*-axis north, and the *z*-axis up. Then we have

$$\vec{l} = (1.00 \text{ m})\hat{\imath}, \qquad \vec{B} = (1.20 \text{ T})[(\cos 45°)\hat{\imath} + (\sin 45°)\hat{\jmath}],$$

$$\vec{F} = I\vec{l} \times \vec{B}$$

$$= (50 \text{ A})(1.00 \text{ m})\hat{\imath} \times (1.20 \text{ T})[(\cos 45°)\hat{\imath} + (\sin 45°)\hat{\jmath}]$$

$$= (42.4 \text{ N})\hat{k}.$$

If the conductor is in mechanical equilibrium under the action of its weight and the upward magnetic force, its weight is 42.4 N and its mass is

$$m = w/g = (42.4 \text{ N})/(9.8 \text{ m/s}^2) = 4.33 \text{ kg}.$$

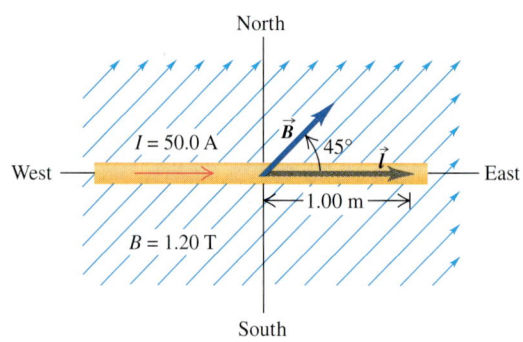

28–25 Overhead view of the copper rod.

b) The magnitude of the force is maximum if $\phi = 90°$ so that $\vec{I}$ and $\vec{B}$ are perpendicular. To have the force still be upward, we rotate the rod clockwise by 45° from its orientation in Fig. 28–25 so that the current runs toward the southeast. Then the magnetic force has magnitude

$$F = IlB = (50.0 \text{ A})(1.00 \text{ m})(1.20 \text{ T}) = 60.0 \text{ N},$$

and the mass of a rod that can be held up against gravity is $m = w/g = (60.0 \text{ N})/(9.8 \text{ m/s}^2) = 6.12 \text{ kg}.$

This is a simple example of magnetic levitation. Magnetic levitation is also used in special high-speed trains. Conventional electromagnetic technology is used to suspend the train over the tracks; the elimination of rolling friction allows the train to achieve speeds in excess of 400 km/h (250 mi/h).

EXAMPLE 28-8

In Fig. 28–26 the magnetic field $\vec{B}$ is uniform and perpendicular to the plane of the figure, pointing out. The conductor has a straight segment with length L perpendicular to the plane of the figure on the right, with the current opposite to $\vec{B}$; followed by a semicircle with radius R, and finally another straight segment with length L parallel to the x-axis, as shown. The conductor carries a current I. Find the total magnetic force on these three segments of wire.

SOLUTION Let's do the easy parts (the straight segments) first. There is *no* force on the segment on the right perpendicular to the plane of the figure because it is antiparallel to $\vec{B}$; $\vec{L} \times \vec{B} = 0$, or $\phi = 180°$ and $\sin \phi = 0$. For the straight segment on the left, $\vec{L}$ points to the left, perpendicular to $\vec{B}$. The force has magnitude $F = ILB$, and its direction is up (the +y-direction in the figure).

The fun part is the semicircle. The figure shows a segment $d\vec{l}$ with length $dl = R \, d\theta$, at angle θ. The direction of $d\vec{l} \times \vec{B}$ is radially outward from the center; make sure you can verify this direction. Because $d\vec{l}$ and $\vec{B}$ are perpendicular, the magnitude dF of the force on the segment $d\vec{l}$ is just $dF = I \, dl \, B$, so we have

$$dF = I(R \, d\theta)B.$$

The components of the force $d\vec{F}$ on segment $d\vec{l}$ are

$$dF_x = IR \, d\theta \, B \cos \theta, \qquad dF_y = IR \, d\theta \, B \sin \theta.$$

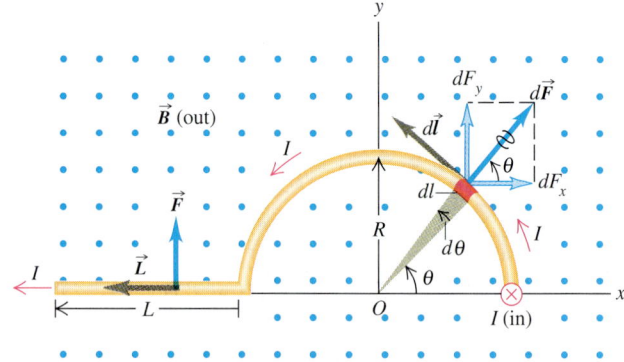

28–26 What is the total magnetic force on the conductor?

To find the components of the total force, we integrate these expressions, letting θ vary from 0 to π to take in the whole semicircle. We find

$$F_x = IRB \int_0^\pi \cos \theta \, d\theta = 0,$$

$$F_y = IRB \int_0^\pi \sin \theta \, d\theta = 2IRB.$$

We could have predicted from symmetry that F_x would be zero. On the right half of the semicircle the x-component of the force is positive (to the right), and on the left half it is negative (to the left); the positive and negative contributions to the integral cancel.

Finally, adding the forces on the straight and semicircular segments, we find the total force:

$$F_x = 0, \qquad F_y = IB(L + 2R),$$

or

$$\vec{F} = IB(L + 2R)\hat{j}.$$

Note that this is the same force that would be exerted if we replaced the semicircle with a straight segment along the x-axis. Do you see why?

28–8 FORCE AND TORQUE ON A CURRENT LOOP

Current-carrying conductors usually form closed loops, so it is worthwhile to use the results of Section 28–7 to find the *total* magnetic force and torque on a conductor in the form of a loop. These results will also help us to understand the behavior of bar magnets described in Section 28–2.

As an example, let's look at a rectangular current loop in a uniform magnetic field. We can represent the loop as a series of straight line segments. We will find that the total *force* on the loop is zero but that there can be a net *torque* acting on the loop, with some interesting properties.

Figure 28–27 shows a rectangular loop of wire with side lengths a and b. A line perpendicular to the plane of the loop (i.e., a *normal* to the plane) makes an angle ϕ with the direction of the magnetic field $\vec{B}$, and the loop carries a current I. The wires leading the current into and out of the loop and the source of emf are omitted to keep the diagram simple.

The force $\vec{F}$ on the right side of the loop (length a) is to the right, in the $+x$-direction as shown. On this side, $\vec{B}$ is perpendicular to the current direction, and the force on this side has magnitude

$$F = IaB. \tag{28–21}$$

28–27 (a) Forces on the sides of a current-carrying loop in a uniform magnetic field. The resultant force is zero; the net torque has magnitude $\tau = IAB \sin \phi$. (b) The torque is maximum when the normal to the loop is perpendicular to $\vec{B}$. (c) When the normal to the loop is parallel to $\vec{B}$, the torque is zero (stable equilibrium). If the normal is antiparallel to $\vec{B}$, the torque is also zero but the equilibrium is unstable.

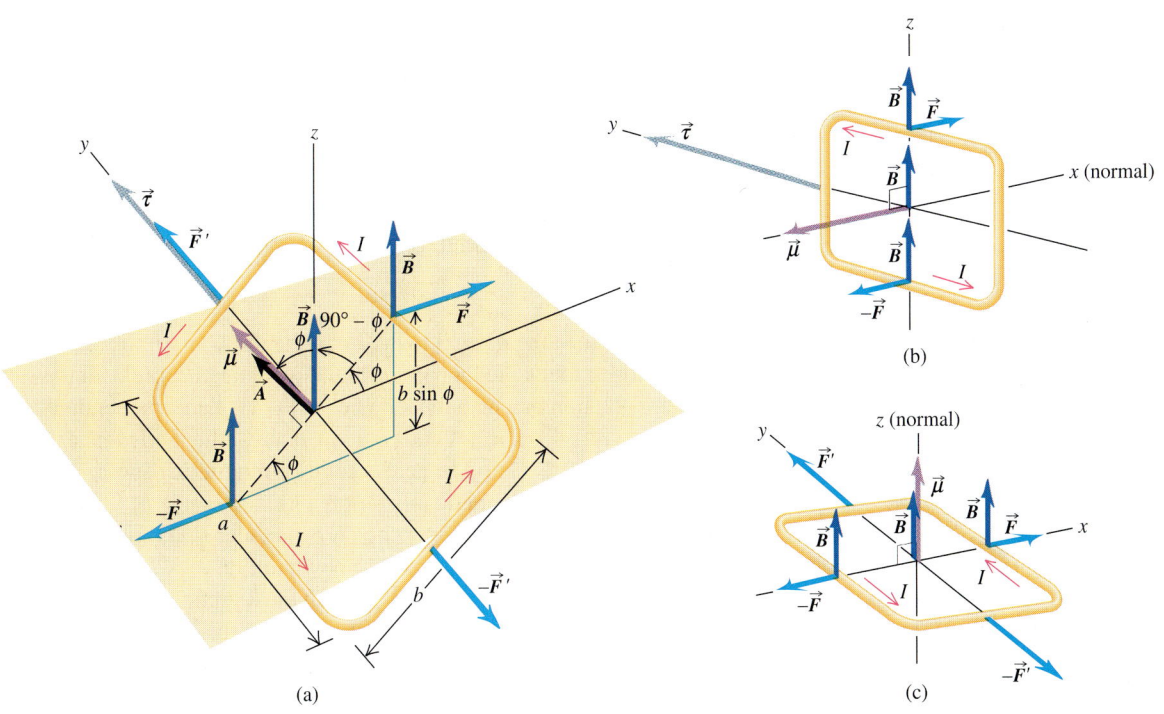

(a)

(b)

(c)

A force $-\vec{F}$ with the same magnitude but opposite direction acts on the opposite side of the loop, as shown in the figure.

The sides with length b make an angle $(90° - \phi)$ with the direction of $\vec{B}$. The forces on these sides are the vectors $\vec{F}'$ and $-\vec{F}'$; their magnitude F' is given by

$$F' = IbB \sin (90° - \phi) = IbB \cos \phi.$$

The lines of action of both forces lie along the y-axis.

13.6

Magnetic Torque on a Loop

The *total* force on the loop is zero because the forces on opposite sides cancel out in pairs. **The net force on a current loop in a uniform magnetic field is zero.** However, the net *torque* is not in general equal to zero. (You may find it helpful at this point to review the discussion of torque in Section 10–2.) The two forces $\vec{F}'$ and $-\vec{F}'$ in Fig. 28–27a lie along the same line and so give rise to zero net torque with respect to any point. The two forces $\vec{F}$ and $-\vec{F}$ lie along different lines, and each gives rise to a torque about the y-axis. According to the right-hand rule for determining the direction of torques, the vector torques due to $\vec{F}$ and $-\vec{F}$ are both in the +y-direction; hence the net vector torque $\vec{\tau}$ is in the +y-direction as well. The moment arm for each of these forces (equal to the perpendicular distance from the rotation axis to the line of action of the force) is $(b/2) \sin \phi$, so the torque due to each force has magnitude $F(b/2) \sin \phi$. If we use Eq. (28–21) for F, the magnitude of the net torque is

$$\tau = 2F(b/2) \sin \phi = (IBa)(b \sin \phi). \tag{28–22}$$

The torque is greatest when $\phi = 90°$, $\vec{B}$ is in the plane of the loop, and the normal to this plane is perpendicular to $\vec{B}$ (Fig. 28–27b). The torque is zero when ϕ is zero or $180°$ and the normal to the loop is parallel or antiparallel to the field (Fig. 28–27c). The value $\phi = 0$ is a stable equilibrium position because the torque is zero there, and when the loop is rotated slightly from this position, the resulting torque tends to rotate it back toward $\phi = 0$. The position $\phi = 180°$ is an *unstable* equilibrium position; if displaced slightly from this position, the loop tends to move farther away from $\phi = 180°$. Figure 28–27 shows rotation about the y-axis, but because the net force on the loop is zero, Eq. (28–22) for the torque is valid for *any* choice of axis (see Problem 28–62).

The area A of the loop is equal to ab, so we can rewrite Eq. (28–22) as

$$\tau = IBA \sin \phi \quad \text{(magnitude of torque on a current loop).} \tag{28–23}$$

The product IA is called the **magnetic dipole moment** or **magnetic moment** of the loop, for which we use the symbol μ (Greek letter "mu"):

$$\mu = IA. \tag{28–24}$$

It is analogous to the electric dipole moment introduced in Section 22–9. In terms of μ, the magnitude of the torque on a current loop is

$$\tau = \mu B \sin \phi, \tag{28–25}$$

where ϕ is the angle between the normal to the loop (the direction of the vector area $\vec{A}$) and $\vec{B}$. The torque tends to rotate the loop in the direction of *decreasing* ϕ, that is, toward its stable equilibrium position in which the loop lies in the xy-plane, perpendicular to the direction of the field $\vec{B}$ (Fig. 28–27c). A current loop, or any other body that experiences a magnetic torque given by Eq. (28–25), is also called a **magnetic dipole.**

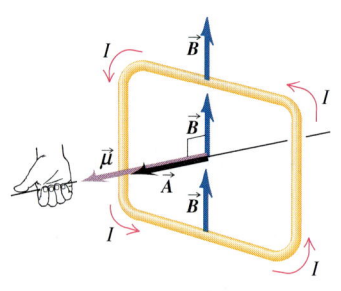

28–28 The right-hand rule determines the direction of the magnetic moment of a current-carrying loop. This is also the direction of the loop's area vector $\vec{A}$; $\vec{\mu} = I\vec{A}$ is a vector equation.

We can also define a vector magnetic moment $\vec{\mu}$ with magnitude IA; this is shown in Fig. 28–27. The direction of $\vec{\mu}$ is defined to be perpendicular to the plane of the loop, with a sense determined by a right-hand rule, as shown in Fig. 28–28. Wrap the fingers of your right hand around the perimeter of the loop in the direction of the current. Then extend your thumb so that it is perpendicular to the plane of the loop; its direction is the direction of $\vec{\mu}$ (and of the vector area $\vec{A}$ of the loop). The torque is greatest when $\vec{\mu}$ and

$\vec{B}$ are perpendicular and is zero when they are parallel or antiparallel. In the stable equilibrium position, $\vec{\mu}$ and $\vec{B}$ are parallel.

Finally, we can express this interaction in terms of the torque vector $\vec{\tau}$, which we used for *electric*-dipole interactions in Section 22–9. From Eq. (28–25) the magnitude of $\vec{\tau}$ is equal to the magnitude of $\vec{\mu} \times \vec{B}$, and reference to Fig. 28–27 shows that the directions are also the same. So we have

$$\vec{\tau} = \vec{\mu} \times \vec{B} \qquad \text{(vector torque on a current loop).} \qquad (28\text{–}26)$$

This result is directly analogous to the result we found in Section 22–9 for the torque exerted by an *electric* field $\vec{E}$ on an *electric* dipole with dipole moment $\vec{p}$: $\vec{\tau} = \vec{p} \times \vec{E}$.

When a magnetic dipole changes orientation in a magnetic field, the field does work on it. In an infinitesimal angular displacement $d\phi$ the work dW is given by $\tau d\phi$, and there is a corresponding change in potential energy. As the above discussion suggests, the potential energy is least when $\vec{\mu}$ and $\vec{B}$ are parallel and greatest when they are antiparallel. To find an expression for the potential energy U as a function of orientation, we can make use of the beautiful symmetry between the electric and magnetic dipole interactions. The torque on an *electric* dipole in an *electric* field is $\vec{\tau} = \vec{p} \times \vec{E}$; we found in Section 22–9 that the corresponding potential energy is $U = -\vec{p} \cdot \vec{E}$. The torque on a *magnetic* dipole in a *magnetic* field is $\vec{\tau} = \vec{\mu} \times \vec{B}$, so we can conclude immediately that the corresponding potential energy is

$$U = -\vec{\mu} \cdot \vec{B} = -\mu B \cos \phi \qquad \text{(potential energy for a magnetic dipole).} \quad (28\text{–}27)$$

With this definition, U is zero when the magnetic dipole moment is perpendicular to the magnetic field.

Although we have derived Eqs. (28–21) through (28–27) for a rectangular current loop, all these relations are valid for a plane loop of any shape at all. Any planar loop may be approximated as closely as we wish by a very large number of rectangular loops, as shown in Fig. 28–29. If these loops all carry equal currents in the same sense, then the forces and torques on the sides of two loops adjacent to each other cancel, and the only forces and torques that do not cancel are due to currents around the boundary. Thus all the above relations are valid for a plane current loop of any shape, with the magnetic moment $\vec{\mu}$ given by $\vec{\mu} = I\vec{A}$.

We can also generalize this whole formulation to a coil consisting of N planar loops close together; the effect is simply to multiply each force, the magnetic moment, the torque, and the potential energy by a factor of N.

An arrangement of particular interest is the **solenoid,** a helical winding of wire, such as a coil wound on a circular cylinder (Fig. 28–30). If the windings are closely spaced, the solenoid can be approximated by a number of circular loops lying in planes at right angles to its long axis. The total torque on a solenoid in a magnetic field is simply the sum of the torques on the individual turns. For a solenoid with N turns in a uniform field B, the magnetic moment is $\mu = NIA$ and

$$\tau = NIAB \sin \phi, \qquad (28\text{–}28)$$

where ϕ is the angle between the axis of the solenoid and the direction of the field. The magnetic moment vector $\vec{\mu}$ is along the solenoid axis. The torque is greatest when the solenoid axis is perpendicular to the magnetic field and zero when they are parallel. The effect of this torque is to tend to rotate the solenoid into a position where its axis is parallel to the field. Solenoids are also useful as *sources* of magnetic field, as we'll discuss in Chapter 29.

The d'Arsonval galvanometer, described in Section 27–4, makes use of a magnetic torque on a coil carrying a current. As Fig. 27–11 shows, the magnetic field is not uniform but is *radial,* so the side thrusts on the coil are always perpendicular to its plane.

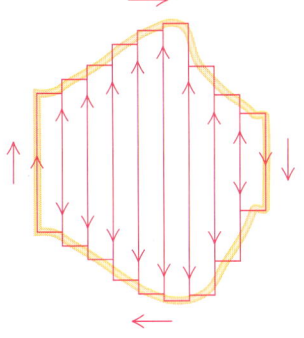

28–29 The collection of rectangles exactly matches the irregular plane loop in the limit as the number of rectangles approaches infinity and the width of each rectangle approaches zero.

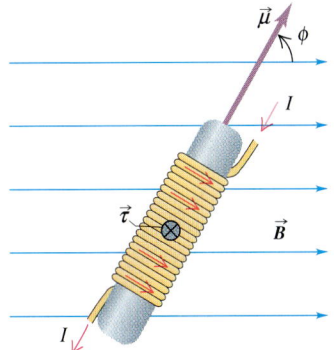

28–30 The torque $\vec{\tau}$ on this solenoid in a uniform magnetic field is clockwise, that is, in the direction of $\vec{\mu} \times \vec{B}$ (straight into the page). It tends to align the magnetic moment and the axis of the solenoid with the magnetic field. An actual solenoid has many more turns, wrapped closely together.

Thus the angle ϕ in Eq. (28–28) is always 90°, and the magnetic torque is directly proportional to the current, no matter what the orientation of the coil. A restoring torque proportional to the angular displacement of the coil is provided by two hairsprings, which also serve as current leads to the coil. When current is supplied to the coil, it rotates along with its attached pointer until the restoring spring torque just balances the magnetic torque. Thus the pointer deflection is proportional to the current. Such instruments can measure currents as small as 10^{-7} A, and modified versions can measure currents as small as 10^{-10} A.

EXAMPLE 28-9

Magnetic torque on a circular coil A circular coil 0.0500 m in radius, with 30 turns of wire, lies in a horizontal plane (Fig. 28–31). It carries a current of 5.00 A in a counterclockwise sense when viewed from above. The coil is in a uniform magnetic field directed toward the right, with magnitude 1.20 T. Find the magnitudes of the magnetic moment and the torque on the coil.

SOLUTION The area of the coil is

$$A = \pi r^2 = \pi(0.0500 \text{ m})^2 = 7.85 \times 10^{-3} \text{ m}^2.$$

The magnetic moment of each turn of the coil is

$$\mu = IA = (5.00 \text{ A})(7.85 \times 10^{-3} \text{ m}^2) = 3.93 \times 10^{-2} \text{ A} \cdot \text{m}^2,$$

and the total magnetic moment of all 30 turns is

$$\mu_{\text{total}} = (30)(3.93 \times 10^{-2} \text{ A} \cdot \text{m}^2) = 1.18 \text{ A} \cdot \text{m}^2.$$

The angle ϕ between the direction of $\vec{B}$ and the direction of $\vec{\mu}$ (which is along the normal to the plane of the coil) is 90°. From Eq. (28–23) the torque on each turn of the coil is

$$\tau = IBA \sin \phi = (5.00 \text{ A})(1.20 \text{ T})(7.85 \times 10^{-3} \text{ m}^2)(\sin 90°)$$

$$= 0.0471 \text{ N} \cdot \text{m},$$

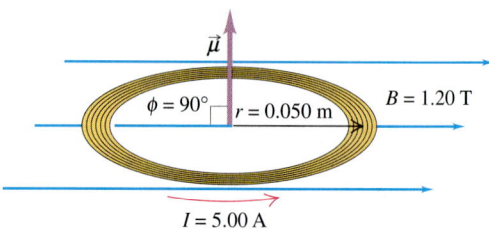

28-31 A circular coil of current-carrying wire in a uniform magnetic field.

and the total torque on the coil is

$$\tau = (30)(0.0471 \text{ N} \cdot \text{m}) = 1.41 \text{ N} \cdot \text{m}.$$

Alternatively, from Eq. (28–25),

$$\tau = \mu_{\text{total}} B \sin \phi = (1.18 \text{ A} \cdot \text{m}^2)(1.20 \text{ T})(\sin 90°)$$

$$= 1.41 \text{ N} \cdot \text{m}.$$

The torque tends to rotate the right side of the coil down and the left side up, into a position where the normal to its plane is parallel to $\vec{B}$.

EXAMPLE 28-10

Potential energy for a coil in a magnetic field If the coil in Example 28–9 rotates from its initial position to a position where its magnetic moment is parallel to $\vec{B}$, what is the change in potential energy?

SOLUTION From Eq. (28–27) the initial potential energy U_1 is

$$U_1 = -\mu_{\text{total}} B \cos \phi_1 = -(1.18 \text{ A} \cdot \text{m}^2)(1.20 \text{ T})(\cos 90°) = 0,$$

and the final potential energy U_2 is

$$U_2 = -\mu_{\text{total}} B \cos \phi_2 = -(1.18 \text{ A} \cdot \text{m}^2)(1.20 \text{ T})(\cos 0°)$$

$$= -1.41 \text{ J}.$$

The change in potential energy is −1.41 J; the potential energy decreases because the rotation is in the direction of the magnetic torque.

EXAMPLE 28-11

What vertical forces applied to the left and right edges of the coil of Fig. 28–31 would be required to maintain it in equilibrium in its initial position?

SOLUTION An upward force of magnitude F at the right side and a downward force of equal magnitude on the left side would

have a total torque of

$$\tau = 2rF = (2)(0.0500 \text{ m})F.$$

This applied torque must be equal in magnitude to the magnetic-field torque magnitude, 1.41 N · m. We find that the required forces have magnitude $F = 14.1$ N.

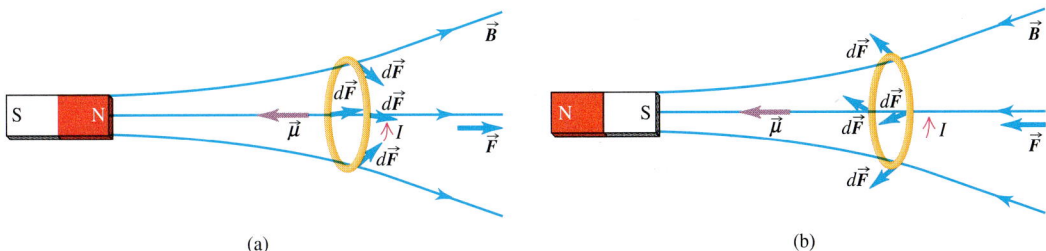

MAGNETIC DIPOLE IN A NON-UNIFORM MAGNETIC FIELD

We have seen that a current loop (that is, a magnetic dipole) experiences zero net force in a uniform magnetic field. Figure 28–32 shows two current loops in the *non-uniform* $\vec{B}$ field of a bar magnet; in both cases the net force on the loop is *not* zero. In Fig. 28–32a the magnetic moment $\vec{\mu}$ is in the direction opposite to the field, and the force $d\vec{F} = I\, d\vec{l} \times \vec{B}$ on a segment of the loop has both a radial component and a component to the right. When these forces are summed to find the net force $\vec{F}$ on the loop, the radial components cancel so that the net force is to the right, away from the magnet. Note that in this case the force is toward the region where the field lines are farther apart and the field magnitude B is less. The polarity of the bar magnet is reversed in Fig. 28–32b, so $\vec{\mu}$ and $\vec{B}$ are parallel; now the net force on the loop is to the left, toward the region of greater field magnitude near the magnet. Later in this section we'll use these observations to explain why bar magnets can pick up unmagnetized iron objects.

MAGNETIC DIPOLES AND MAGNETIC MATERIALS

The behavior of a solenoid in a magnetic field (Fig. 28–30) resembles that of a bar magnet or compass needle; if free to turn, both the solenoid and the magnet orient themselves with their axes parallel to the $\vec{B}$ field. In both cases this is due to the interaction of moving electric charges with a magnetic field; the difference is that in a bar magnet the motion of charge occurs on the microscopic scale of the atom. Think of an electron as being like a spinning ball of charge. In this analogy the circulation of charge around the spin axis is like a current loop, and so the electron has a net magnetic moment. (This analogy, while helpful, is inexact; an electron isn't really a spinning sphere. A full explanation of the origin of an electron's magnetic moment involves quantum mechanics, which is beyond our scope here.) In an iron atom a substantial fraction of the electron magnetic moments align with each other, and the atom has a nonzero magnetic moment. (By contrast, the atoms of most elements have little or no net magnetic moment.) In an unmagnetized piece of iron there is no overall alignment of the magnetic moments of the atoms; their vector sum is zero, and the net magnetic moment is zero (Fig. 28–33a). But in an iron bar magnet the magnetic moments of many of the atoms are parallel, and there is a substantial net magnetic moment $\vec{\mu}$ (Fig. 28–33b). If the magnet is placed in a magnetic field $\vec{B}$, the field exerts a torque given by Eq. (28–26) that tends to align $\vec{\mu}$ with $\vec{B}$ (Fig. 28–33c). A bar magnet tends to align with a $\vec{B}$ field so that a line from the south pole to the north pole of the magnet is in the direction of $\vec{B}$; hence the real significance of a magnet's north and south poles is that they represent the head and tail, respectively, of the magnet's dipole moment $\vec{\mu}$.

The torque experienced by a current loop in a magnetic field also explains how an unmagnetized iron object like that in Fig. 28–33a becomes magnetized. If an unmagnetized iron paper clip is placed next to a powerful magnet, the magnetic moments of the paper clip's atoms tend to align with the $\vec{B}$ field of the magnet. When the paper clip is removed, its atomic dipoles tend to remain aligned, and the paper clip has a net magnetic moment. The paper clip can be demagnetized by being dropped on the floor or heated; the added internal energy jostles and re-randomizes the atomic dipoles.

28–32 (a), (b) Current loops in a non-uniform $\vec{B}$ field. In each case the axis of the bar magnet is perpendicular to the plane of the loop and passes through the center of the loop.

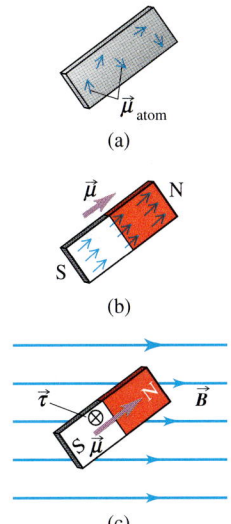

28–33 (a) Randomly oriented atomic magnetic moments in an unmagnetized piece of iron. (Only a few representative atomic moments are shown.) (b) Aligned atomic magnetic moments in a magnetized piece of iron (bar magnet). The net magnetic moment of the bar magnet points from its south pole to its north pole. (c) The torque on a bar magnet in a magnetic field tends to align the magnet's dipole moment with the $\vec{B}$ field.

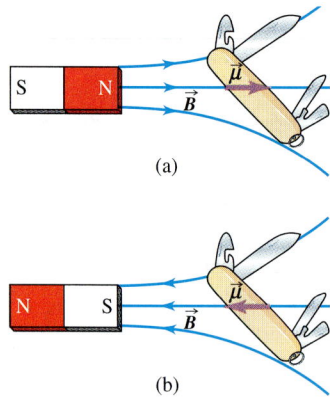

28–34 A bar magnet attracts an unmagnetized iron object in two steps; first, the $\vec{B}$ field of the bar magnet gives rise to a net magnetic moment in the object; second, because the field of the bar magnet is not uniform, this magnetic dipole is attracted toward the magnet. The attraction is the same whether the object is closer to (a) the magnet's north pole or (b) the magnet's south pole.

The magnetic-dipole picture of a bar magnet explains the attractive and repulsive forces between bar magnets shown in Fig. 28–1. The magnetic moment $\vec{\mu}$ of a bar magnet points from its south pole to its north pole, so the current loops in Figs. 28–32a and 28–32b are both equivalent to a magnet with its north pole on the left. Hence the situation in Fig. 28–32a is equivalent to two bar magnets with their north poles next to each other; the resultant force is repulsive, just as in Fig. 28–1c. In Fig. 28–32b we again have the equivalent of two bar magnets end to end, but with the south pole of the left-hand magnet next to the north pole of the right-hand magnet. The resultant force is attractive, as in Fig. 28–1b.

Finally, we can explain how a magnet can attract an unmagnetized iron object (Fig. 28–2). It's a two-step process. First, the atomic magnetic moments of the iron tend to align with the $\vec{B}$ field of the magnet, so the iron acquires a net magnetic dipole moment $\vec{\mu}$ parallel to the field. Second, the non-uniform field of the magnet attracts the magnetic dipole. Figure 28–34a shows an example. The north pole of the magnet is closer to the pocket knife (which contains iron), and the magnetic dipole produced in the knife is equivalent to a loop with a current that circulates in a direction opposite to that shown in Fig. 28–32a. Hence the net magnetic force on the knife is opposite to the force on the loop in Fig. 28–32a, and the knife is attracted toward the magnet. Changing the polarity of the magnet, as in Fig. 28–34b, reverses the directions of both $\vec{B}$ and $\vec{\mu}$. The situation is now equivalent to that shown in Fig. 28–32b; like the loop in that figure, the knife is attracted toward the magnet. Hence a previously unmagnetized object containing iron is attracted to *either* pole of a magnet. By contrast, objects made of brass, aluminum, or wood hardly respond at all to a magnet; the atomic magnetic dipoles of these materials, if present at all, have less tendency to align with an external field.

Our discussion of how magnets and pieces of iron interact has just scratched the surface of a diverse subject known as *magnetic properties of materials*. We'll discuss these properties in more depth in Section 29–9.

*28–9   THE DIRECT-CURRENT MOTOR

No one needs to be reminded of the importance of electric motors in contemporary society. In a motor a magnetic torque acts on a current-carrying conductor, and electric energy is converted to mechanical energy. As an example, let's look at a simple type of direct-current (dc) motor, shown in Fig. 28–35.

The moving part of the motor is the *rotor*, a length of wire formed into an open-ended loop and free to rotate about an axis. The ends of the rotor wires are attached to circular conducting segments that form a *commutator*. In Fig. 28–35a, each of the two commutator segments makes contact with one of the terminals, or *brushes*, of an external circuit that includes a source of emf. This causes a current to flow into the rotor on one side, shown in red, and out of the rotor on the other side, shown in blue. Hence the rotor is a current loop with a magnetic moment $\vec{\mu}$. The rotor lies between opposing poles of a permanent magnet, so there is a magnetic field $\vec{B}$ that exerts a torque $\vec{\tau} = \vec{\mu} \times \vec{B}$ on the rotor. For the rotor orientation shown in Fig. 28–35a the torque causes the rotor to turn counterclockwise, in the direction that will align $\vec{\mu}$ with $\vec{B}$.

In Fig. 28–35b the rotor has rotated by 90° from its orientation in Fig. 28–35a. If the current through the rotor were constant, the rotor would now be in its equilibrium orientation; it would simply oscillate around this orientation. But here's where the commutator comes into play; each brush is now in contact with *both* segments of the commutator. There is no potential difference between the commutators, so at this instant no current flows through the rotor, and the magnetic moment is zero. The rotor contin-

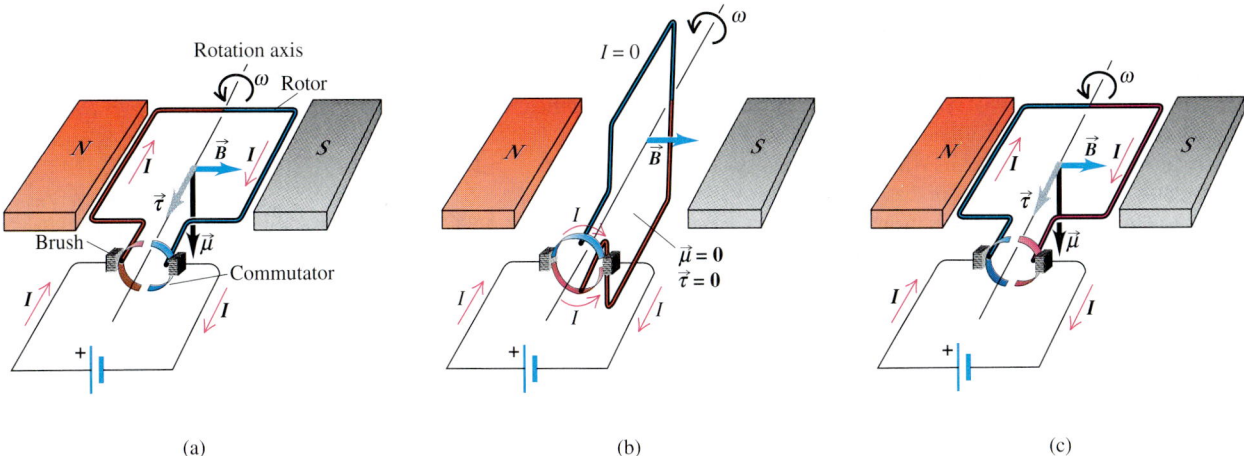

(a)   (b)   (c)

28–35 Schematic diagram of a simple dc motor. The rotor is a wire loop that is free to rotate about an axis; the rotor ends are attached to the two curved conductors that form the commutator. The commutator segments are insulated from one another. (a) The brushes are aligned with the commutator segments, and current flows into one side of the rotor (colored red) and out of the other side (colored blue). The magnetic torque causes the rotor to spin counterclockwise. (b) The rotor has turned by 90°. Each brush is in contact with both commutator segments, so the current bypasses the rotor altogether. (c) As the rotor rotates beyond the position in (b), current again flows through the commutators into the rotor, but now current flows into the blue-colored side and out of the red-colored side. The magnetic torque acts in the same direction as in (a).

ues to rotate counterclockwise because of its inertia, and current again flows through the rotor as in Fig. 28–35c. But now current enters on the *blue*-colored side of the rotor and exits on the *red*-colored side, just the opposite of the situation in Fig. 28–35a. While the direction of the current has reversed with respect to the rotor, the rotor itself has rotated 180° and the magnetic moment $\vec{\mu}$ is in the same direction with respect to the magnetic field. Hence the magnetic torque $\vec{\tau}$ is in the same direction in Fig. 28–35c as in 28–35a. Thanks to the commutator, the current reverses after every 180° of rotation, so the torque is always in the direction to rotate the rotor counterclockwise. When the motor has come "up to speed," the average magnetic torque is just balanced by an opposing torque due to air resistance, friction in the rotor bearings, and friction between the commutator and brushes.

The simple motor shown in Fig. 28–35 has only a single turn of wire in its rotor. In practical motors, the rotor has many turns; this increases the magnetic moment and the torque so that the motor can spin larger loads. The torque can also be increased by using a stronger magnetic field, which is why many motor designs use electromagnets instead of a permanent magnet. Another drawback of the simple design in Fig. 28–35 is that the magnitude of the torque rises and falls as the rotor spins. This can be remedied by having the rotor include several independent coils of wire oriented at different angles.

Because a motor converts electric energy to mechanical energy or work, it requires electric energy input. If the potential difference between its terminals is V_{ab} and the current is I, then the power input is $P = V_{ab}I$. Even if the motor coils have negligible resistance, there must be a potential difference between the terminals if P is to be different from zero. This potential difference results principally from magnetic forces exerted on the currents in the conductors of the rotor as they rotate through the magnetic field. The associated electromotive force $\mathcal{E}$ is called an *induced* emf; it is also called a *back* emf because its sense is opposite to that of the current. In Chapter 30 we will study induced emfs resulting from motion of conductors in magnetic fields.

In a *series* motor the rotor is connected in series with the electromagnet that produces the magnetic field; in a *shunt* motor they are connected in parallel. In a series motor with internal resistance r, V_{ab} is greater than $\mathcal{E}$, and the difference is the potential drop Ir across the internal resistance. That is,

$$V_{ab} = \mathcal{E} + Ir. \qquad (28\text{–}29)$$

Because the magnetic force is proportional to velocity, $\mathcal{E}$ is *not* constant but is proportional to the speed of rotation of the rotor.

EXAMPLE 28-12

A series dc motor A dc motor with its rotor and field coils connected in series has an internal resistance of 2.00 Ω. When
running at full load on a 120-V line, it draws a current of 4.00 A.
a) What is the emf in the rotor? b) What is the power delivered
to the motor? c) What is the rate of dissipation of energy in the
resistance of the motor? d) What is the mechanical power developed? e) What is the efficiency of the motor? f) What happens if
the machine the motor is driving jams and the rotor suddenly
stops turning?

SOLUTION a) From Eq. (28–29), $V_{ab} = \mathcal{E} + Ir$, we have

$$120 \text{ V} = \mathcal{E} + (4.0 \text{ A})(2.0 \,\Omega), \quad \text{and so} \quad \mathcal{E} = 112 \text{ V}.$$

b) The power input P from the source is

$$P_{input} = V_{ab}I = (120 \text{ V})(4.0 \text{ A}) = 480 \text{ W}.$$

c) The power P dissipated in the resistance r is

$$P_{dissipated} = I^2r = (4.0 \text{ A})^2(2.0 \,\Omega) = 32 \text{ W}.$$

d) The mechanical power output is the electric power input
minus the rate of dissipation of energy in the motor's resistance
(assuming that there are no other power losses):

$$P_{output} = P_{input} - P_{dissipated} = 480 \text{ W} - 32 \text{ W} = 448 \text{ W}.$$

e) The efficiency e is the ratio of mechanical power output to
electric power input:

$$e = \frac{P_{output}}{P_{input}} = \frac{448 \text{ W}}{480 \text{ W}} = 0.93 = 93\%.$$

f) With the rotor stalled, the back emf $\mathcal{E}$ (which is proportional to
rotor speed) goes to zero. From Eq. (28–29) the current
becomes

$$I = \frac{V_{ab}}{r} = \frac{120 \text{ V}}{2.0 \,\Omega} = 60 \text{ A},$$

and the power dissipation P in the resistance r becomes

$$P_{dissipated} = I^2r = (60 \text{ A})^2(2 \,\Omega) = 7200 \text{ W}.$$

If this massive overload doesn't blow a fuse or trip a circuit
breaker, the coils will quickly melt. When the motor is first
turned on, there's a momentary surge of current until the motor
picks up speed. This surge causes greater-than-usual voltage
drops ($V = IR$) in the power lines supplying the current. Similar
effects are responsible for the momentary dimming of lights in a
house when an air-conditioner or dishwasher motor starts.

*28-10 THE HALL EFFECT

The reality of the forces acting on the moving charges in a conductor in a magnetic field
is strikingly demonstrated by the **Hall effect,** an effect analogous to the transverse
deflection of an electron beam in a magnetic field in vacuum. (The effect was discovered by the American physicist Edwin Hall in 1879 while he was still a graduate
student.) To describe this effect, let's consider a conductor in the form of a flat strip, as
shown in Fig. 28–36. The current is in the direction of the +x-axis, and there is a uniform magnetic field $\vec{B}$ perpendicular to the plane of the strip, in the +y-direction. The
drift velocity of the moving charges (charge magnitude $|q|$) has magnitude v_d. Figure
28–36a shows the case of negative charges, such as electrons in a metal, and Fig.
28–36b shows positive charges. In both cases the magnetic force is upward, just as the
magnetic force on a conductor is the same whether the moving charges are positive or
negative. In either case a moving charge is driven toward the *upper* edge of the strip by
the magnetic force $F_z = |q|v_d B$.

If the charge carriers are electrons, as in Fig. 28–36a, an excess negative charge
accumulates at the upper edge of the strip, leaving an excess positive charge at its lower
edge. This accumulation continues until the resulting transverse electrostatic field $\vec{E}_e$

28–36 Forces on charge carriers
in a conductor in a magnetic field.
(a) Negative charge carriers (electrons) are pushed toward the top of
the strip, leading to a charge distribution as shown. Point a is at a
higher potential than point b. (b)
With positive charge carriers, the
polarity of the potential difference
is opposite to that of part (a).

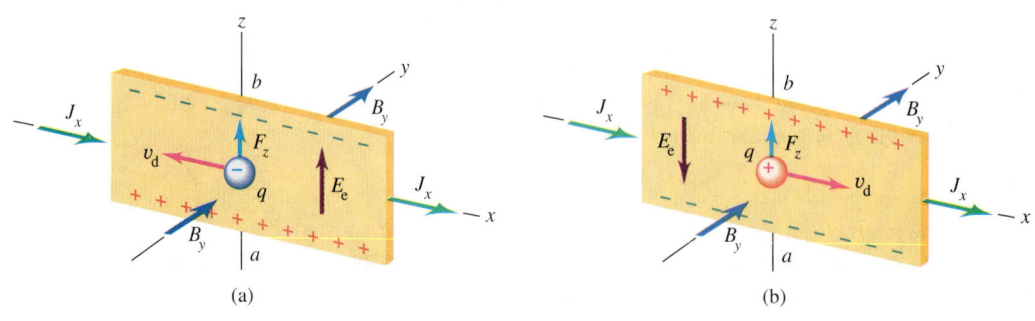

(a)   (b)

becomes large enough to cause a force (magnitude $|q|E_e$) that is equal and opposite to the magnetic force (magnitude $|q|v_dB$). After that, there is no longer any net transverse force to deflect the moving charges. This electric field causes a transverse potential difference between opposite edges of the strip, called the *Hall voltage* or the *Hall emf.* The polarity depends on whether the moving charges are positive or negative. Experiments show that for metals the upper edge of the strip in Fig. 28–36a *does* become negatively charged, showing that the charge carriers in a metal are indeed negative electrons.

However, if the charge carriers are *positive,* as in Fig. 28–36b, then *positive* charge accumulates at the upper edge, and the potential difference is *opposite* to the situation with negative charges. Soon after the discovery of the Hall effect in 1879, it was observed that some materials, particularly some *semiconductors,* show a Hall emf opposite to that of the metals, as if their charge carriers were positively charged. We now know that these materials conduct by a process known as *hole conduction.* Within such a material there are locations, called *holes,* that would normally be occupied by an electron but are actually empty. A missing negative charge is equivalent to a positive charge. When an electron moves in one direction to fill a hole, it leaves another hole behind it. The hole migrates in the direction opposite to that of the electron.

In terms of the coordinate axes in Fig. 28–36b, the electrostatic field $\vec{E}_e$ for the positive-q case is in the $-z$-direction; its z-component E_z is negative. The magnetic field is in the $+y$-direction, and we write it as B_y. The magnetic force (in the $+z$-direction) is qv_dB_y. The current density J_x is in the $+x$-direction. In the steady state, when the forces qE_z and qv_dB_y are equal in magnitude and opposite in direction,

$$qE_z + qv_dB_y = 0, \quad \text{or} \quad E_z = -v_dB_y.$$

This confirms that when q is positive, E_z is negative. The current density J_x is

$$J_x = nqv_d.$$

Eliminating v_d between these equations, we find

$$nq = \frac{-J_xB_y}{E_z}. \qquad (28\text{–}30)$$

Note that this result (as well as the entire derivation) is valid for both positive and negative q. When q is negative, E_z is positive, and conversely.

We can measure J_x, B_y, and E_z, so we can compute the product nq. In both metals and semiconductors, q is equal in magnitude to the electron charge, so the Hall effect permits a direct measurement of n, the concentration of current-carrying charges in the material. The *sign* of the charges is determined by the polarity of the Hall emf, as we have described.

The Hall effect can also be used for a direct measurement of electron drift speed v_d in metals. As we saw in Chapter 26, these speeds are very small, often of the order of 1 mm/s or less. If we move the entire conductor in the opposite direction to the current with a speed equal to the drift speed, then the electrons are at rest with respect to the magnetic field, and the Hall emf disappears. Thus the conductor speed needed to make the Hall emf vanish is equal to the drift speed.

EXAMPLE 28-13

Using the Hall effect You place a slab of copper, 2.0 mm thick and 1.50 cm wide, in a uniform magnetic field with magnitude 0.40 T, as shown in Fig. 28–36a. When you run a 75-A current in the $+x$-direction, you find by careful measurement that the potential at the bottom of the slab is 0.81 μV higher than at the top. From this measurement, determine the concentration of mobile electrons in copper.

SOLUTION We'll use Eq. (28–30) to find n. First we find the current density J_x and the electric field E_z:

$$J_x = \frac{I}{A} = \frac{75 \text{ A}}{(2.0 \times 10^{-3} \text{ m})(1.50 \times 10^{-2} \text{ m})} = 2.5 \times 10^6 \text{ A/m}^2,$$

$$E_z = \frac{V}{d} = \frac{0.81 \times 10^{-6} \text{ V}}{1.5 \times 10^{-2} \text{ m}} = 5.4 \times 10^{-5} \text{ V/m}.$$

Then, from Eq. (28–30),

$$n = \frac{-J_x B_y}{q E_z} = \frac{-(2.5 \times 10^6 \text{ A/m}^2)(0.40 \text{ T})}{(-1.60 \times 10^{-19} \text{ C})(5.4 \times 10^{-5} \text{ V/m})}$$

$$= 11.6 \times 10^{28} \text{ m}^{-3}.$$

The actual value of n for copper is 8.5×10^{28} m^{-3}, which shows that the simple model of the Hall effect in this section, ignoring quantum effects and electron interactions with the ions, must be used with caution. This example also shows that with good conductors, the Hall emf is very small even with large current densities. Hall-effect devices for magnetic-field measurements and other purposes use semiconductor materials, for which moderate current densities give much larger Hall emfs.

SUMMARY

KEY TERMS

permanent magnet, 865

magnetic monopole, 866

magnetic field, 867

tesla, 868

gauss, 868

magnetic field line, 870

magnetic flux, 870

weber, 872

magnetic flux density, 872

cyclotron frequency, 874

mass spectrometer, 879

isotope, 879

magnetic dipole moment, 884

magnetic moment, 884

magnetic dipole, 884

solenoid, 885

Hall effect, 890

- Magnetic interactions are fundamentally interactions between moving charged particles. These interactions are described by the vector magnetic field, denoted by $\vec{B}$. A particle with charge q moving with velocity $\vec{v}$ in a magnetic field $\vec{B}$ experiences a force $\vec{F}$ given by

$$\vec{F} = q\vec{v} \times \vec{B}. \tag{28–2}$$

The SI unit of magnetic field is the tesla (1 T = 1 N/A · m).

- A magnetic field can be represented graphically by magnetic field lines. At each point a magnetic field line is tangent to the direction of $\vec{B}$ at that point. Where field lines are close together, the field magnitude is large.

- Magnetic flux Φ_B through an area is defined as

$$\Phi_B = \int B_\perp \, dA = \int B \cos \phi \, dA = \int \vec{B} \cdot d\vec{A}. \tag{28–6}$$

The SI unit of magnetic flux is the weber (1 Wb = 1 T · m^2).

- The net magnetic flux through any closed surface is zero (Gauss's law for magnetism). As a result, magnetic field lines always form closed loops.

- The magnetic force is always perpendicular to $\vec{v}$; a particle moving under the action of a magnetic field alone moves with constant speed. In a uniform field a particle with initial velocity perpendicular to the field moves in a circle with radius R given by

$$R = \frac{mv}{|q|B}. \tag{28–11}$$

- J. J. Thomson used crossed electric and magnetic fields to measure the charge-to-mass ratio (e/m) for electrons. The electric and magnetic forces exactly cancel when $v = E/B$.

- The force $\vec{F}$ on a straight segment $\vec{l}$ of a conductor carrying current I in a uniform magnetic field $\vec{B}$ is

$$\vec{F} = I\vec{l} \times \vec{B}; \tag{28–19}$$

for an infinitesimal segment $d\vec{l}$ the force is

$$d\vec{F} = I \, d\vec{l} \times \vec{B}. \tag{28–20}$$

- A current loop with area A and current I in a uniform magnetic field $\vec{B}$ experiences no net force but does experience a torque of magnitude

$$\tau = IBA \sin \phi. \tag{28–23}$$

In terms of the magnetic dipole moment $\vec{\mu} = I\vec{A}$ of the loop, the vector torque is

$$\vec{\tau} = \vec{\mu} \times \vec{B}. \tag{28-26}$$

The potential energy U for a magnetic dipole in a magnetic field $\vec{B}$ is

$$U = -\vec{\mu} \cdot \vec{B} = -\mu B \cos \phi. \tag{28-27}$$

The magnetic moment of a loop depends only on the current and the area; it is independent of the shape of the loop.

- In a dc motor a magnetic field exerts a torque on a current in the rotor. Motion of the rotor through the magnetic field causes an induced emf called a back emf. For a series motor, in which the rotor coil is in series with coils that produce the magnetic field, the terminal voltage is the sum of the back emf and the drop Ir across the internal resistance.

- The Hall effect is a potential difference measured perpendicular to the direction of current in a conductor, when the conductor is placed in a magnetic field. The Hall voltage is determined by the requirement that the associated electric force must just balance the magnetic force on a moving charge. Hall-effect measurements can be used to determine the density n of charge carriers and their sign from the relation

$$nq = \frac{-J_x B_y}{E_z}. \tag{28-30}$$

DISCUSSION QUESTIONS

Q28-1 Can a charged particle move through a magnetic field without experiencing any force? If so, how? If not, why not?

Q28-2 At any point in space, the electric field $\vec{E}$ is defined to be in the direction of the electric force on a positively charged particle at that point. Why don't we similarly define the magnetic field $\vec{B}$ to be in the direction of the magnetic force on a moving positively charged particle?

Q28-3 If an electron beam in a cathode-ray tube travels in a straight line, can you be sure that there is no magnetic field present? Why or why not?

Q28-4 The magnetic force on a moving charged particle is always perpendicular to the magnetic field $\vec{B}$. Is the trajectory of a moving charged particle always perpendicular to the magnetic field lines? Explain your reasoning.

Q28-5 A charged particle is fired into a cubical region of space where there is a uniform magnetic field. Outside this region, there is no magnetic field. Is it possible that the particle will remain inside the cubical region? Why or why not?

Q28-6 If the magnetic force does no work on a charged particle, how can it have any effect on the particle's motion? Are there other examples of forces that do no work but have a significant effect on a particle's motion?

Q28-7 A compass in New York points about 15° west of true north; in California, it points about 15° east of true north. Magnetic declination also varies with time. What are some possible explanations for magnetic declination?

Q28-8 Could the electron beam in an oscilloscope tube (cathode-ray tube) be used as a compass? How? What advantages and disadvantages would it have in comparison with a conventional compass?

Q28-9 How might a loop of wire carrying a current be used as a compass? Could such a compass distinguish between north and south? Why or why not?

Q28-10 How could the direction of a magnetic field be determined by making only *qualitative* observations of the magnetic force on a straight wire carrying a current?

Q28-11 Does a magnetic field exert forces on the electrons within atoms? Why or why not? What observable effect might such interactions have on the behavior of the atom?

Q28-12 A loose, floppy loop of wire is carrying current I. The loop of wire is placed on a horizontal table in a uniform magnetic field $\vec{B}$ perpendicular to the plane of the table. This causes the loop of wire to expand into a circular shape while still lying on the table. In a diagram, show all possible orientations of the current I and magnetic field $\vec{B}$ that could cause this to occur. Explain your reasoning.

Q28-13 Equation (28–19) for the force on a current-carrying wire, $\vec{F} = I\vec{l} \times \vec{B}$, was derived using the assumption that the cross-section area of the wire is constant along its length. Is this equation still valid if the cross-section area varies along the wire's length? Why or why not?

Q28-14 A student claims that if lightning strikes a metal flagpole, the force exerted by the earth's magnetic field on the current in the pole can be large enough to bend it. Typical lightning currents are of the order of 10^4 to 10^5 A. Is the student's opinion justified? Explain your reasoning.

Q28-15 A student tried to make an electromagnetic compass by suspending a coil or wire from a thread (with the plane of the coil vertical) and passing a current through it. She expected the coil to align itself perpendicular to the horizontal components of the earth's magnetic field; instead, the coil went into what appeared to be angular simple harmonic motion, oscillating back and forth around the expected direction. What was happening? Was the motion truly simple harmonic?

Q28–16 How could a compass be used for a *quantitative* determination of the magnitude and direction of the magnetic field at a point?

Q28–17 An ordinary loudspeaker such as that shown in Fig. 28–24 should not be placed next to a computer monitor or TV screen. Why not?

Q28–18 A permanent magnet can be used to pick up a string of nails, tacks, or paper clips, even those these are not magnets by themselves. How can this be?

Q28–19 If an emf is produced in a dc motor, would it be possible to use the motor somehow as a generator or source, taking power out of it rather than putting power into it? How might this be done?

Q28–20 When the polarity of the voltage applied to a dc motor is reversed, the direction of motion does *not* reverse. Why not? How *could* the direction of motion be reversed?

Q28–21 In a Hall-effect experiment, is it possible that *no* transverse potential difference will be observed? Under what circumstances might this happen?

Q28–22 Hall-effect voltages are much larger for relatively poor conductors (such as germanium) than for good conductors (such as copper), for comparable currents, fields, and dimensions. Why?

Q28–23 Could an accelerator be built in which *all* the forces on the particles, for steering and for increasing speed, are magnetic forces? Why or why not?

Q28–24 The magnetic force acting on a charged particle can never do work, because at every instant the force is perpendicular to the velocity. The torque exerted by a magnetic field can do work on a current loop when the loop rotates. Explain how these seemingly contradictory statements can be reconciled.

EXERCISES

SECTION 28–3 MAGNETIC FIELD

28–1 A particle with a charge of -1.24×10^{-8} C is moving with instantaneous velocity $\vec{v} = (4.19 \times 10^4 \text{ m/s})\hat{\imath} + (-3.85 \times 10^4 \text{ m/s})\hat{\jmath}$. What is the force exerted on this particle by a magnetic field a) $\vec{B} = (1.40 \text{ T})\hat{\imath}$? b) $\vec{B} = (1.40 \text{ T})\hat{k}$?

28–2 A particle of mass 0.195 g carries a charge of -2.50×10^{-8} C. The particle is given an initial horizontal velocity that is due north and has a magnitude 4.00×10^4 m/s. What are the magnitude and direction of the minimum magnetic field that will keep the particle moving in the earth's gravitational field in the same horizontal, northward direction?

28–3 A particle initially moving south in a vertically downward magnetic field is deflected toward the east. What is the sign of the charge on the particle? Explain your answer with a diagram.

28–4 A particle with a mass of 1.81×10^{-3} kg and a charge of 1.22×10^{-8} C has at a given instant a velocity $\vec{v} = (3.00 \times 10^4 \text{ m/s})\hat{\jmath}$. What are the magnitude and direction of the particle's acceleration produced by a uniform magnetic field $\vec{B} = (1.63 \text{ T})\hat{\imath} + (0.980 \text{ T})\hat{\jmath}$?

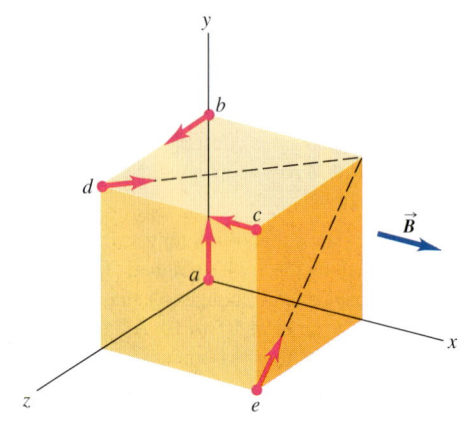

FIGURE 28–37 Exercise 28–5.

28–5 Each of the lettered points at the corners of the cube in Fig. 28–37 represents a positive charge q moving with a velocity of magnitude v in the direction indicated. The region in the figure is in a uniform magnetic field $\vec{B}$, parallel to the x-axis and directed toward the right. Copy the figure, find the magnitude and direction of the force on each charge, and show the force in your diagram.

28–6 An electron moves at 2.50×10^6 m/s through a region in which there is a magnetic field of unspecified direction and magnitude 7.40×10^{-2} T. a) What are the largest and smallest possible magnitudes of the acceleration of the electron due to the magnetic field? b) If the actual acceleration of the electron is one-fourth of the largest magnitude in part (a), what is the angle between the electron velocity and the magnetic field?

28–7 An electron experiences a magnetic force of magnitude 4.60×10^{-15} N when moving at an angle of 60.0° with respect to a magnetic field of magnitude 3.50×10^{-3} T. Find the speed of the electron.

SECTION 28–4 MAGNETIC FIELD LINES AND MAGNETIC FLUX

28–8 The magnetic flux through one face of a cube is +0.120 Wb. a) What must be the total magnetic flux through the other five faces of the cube? b) Why did you not need to know the dimensions of the cube in order to answer part (a)? c) Suppose the magnetic flux is due to a permanent magnet like that shown in Fig. 28–9a. In a sketch, show where the cube in part (a) might be located relative to the magnet.

28–9 A circular area with a radius of 6.50 cm lies in the xy-plane. What is the magnitude of the magnetic flux through this circle due to a uniform magnetic field $B = 0.230$ T a) in the $+z$-direction? b) at an angle of 53.1° from the $+z$-direction? c) in the $+y$-direction?

28–10 The magnetic field $\vec{B}$ in a certain region is 0.128 T, and its direction is that of the $+z$-axis in Fig. 28–38. a) What is the magnetic flux across the surface $abcd$ in the figure? b) What is

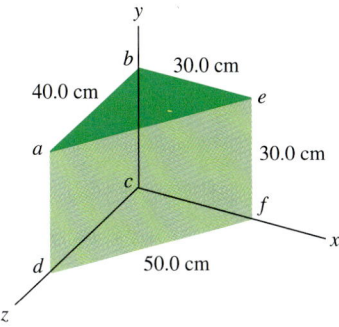

FIGURE 28–38 Exercises 28–10 and 28–11.

the magnetic flux across the surface *befc*? c) What is the magnetic flux across the surface *aefd*? d) What is the net flux through all five surfaces that enclose the shaded volume?

28–11 A physics student claims that she has arranged magnets so that the magnetic field within the shaded volume in Fig. 28–38 is $\vec{B} = (\beta - \gamma y^2)\hat{j}$, where $\beta = 0.300$ T and $\gamma = 2.00$ T/m². a) Find the net flux of $\vec{B}$ through the five surfaces that enclose the shaded volume in Fig. 28–38. b) Is the student's claim plausible? Why or why not?

SECTION 28–5 MOTION OF CHARGED PARTICLES IN A MAGNETIC FIELD

28–12 A particle with charge 6.40×10^{-19} C travels in a circular orbit with radius 4.68 mm due to the force exerted on it by a magnetic field with magnitude 1.65 T and perpendicular to the orbit. a) What is the magnitude of the linear momentum $\vec{p}$ of the particle? b) What is the magnitude of the angular momentum $\vec{L}$ of the particle?

28–13 An electron at point A in Fig. 28–39 has a speed v_0 of 1.41×10^6 m/s. Find a) the magnitude and direction of the magnetic field that will cause the electron to follow the semicircular path from A to B; b) the time required for the electron to move from A to B.

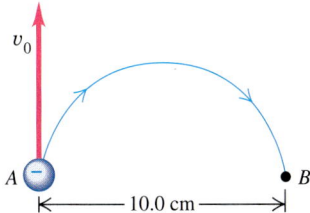

FIGURE 28–39 Exercises 28–13 and 28–14.

28–14 Repeat Exercise 28–13 for the case in which the particle is a proton rather than an electron.

28–15 A deuteron (the nucleus of an isotope of hydrogen) has a mass of 3.34×10^{-27} kg and a charge of $+e$. The deuteron travels in a circular path with a radius of 6.96 mm in a magnetic field with magnitude 2.50 T. a) Find the speed of the deuteron. b) Find the time required for it to make one-half of a revolution. c) Through what potential difference would the deuteron have to be accelerated to acquire this speed?

28–16 Suppose that the electrons in the magnetron of Example 28–3 (Section 28–5) move at 2.80×10^6 m/s in the plane perpendicular to the magnetic field. What are the radii of their orbits?

28–17 A physicist wishes to produce electromagnetic waves of frequency 3.0 THz (1 THz = 1 terahertz = 10^{12} Hz) using a magnetron (see Example 28–3 in Section 28–5). a) What magnetic field would be required? Compare this field with the strongest constant magnetic fields yet produced on earth, about 45 T. b) Would there be any advantage to using protons instead of electrons in the magnetron? Why or why not?

28–18 In the situation shown in Fig. 28–14 the charged particle is a proton ($q = 1.60 \times 10^{-19}$ C, $m = 1.67 \times 10^{-27}$ kg) and the uniform magnetic field of magnitude B is directed along the $+x$-axis. The speed of the proton is v, and its velocity $\vec{v}_0$ at $t = 0$ has zero z-component. What must be the angle of $\vec{v}_0$ with the xz-plane for the pitch of the helix to equal its radius? (See Example 28–4 in Section 28–5.)

28–19 A proton ($q = 1.60 \times 10^{-19}$ C, $m = 1.67 \times 10^{-27}$ kg) moves in a uniform magnetic field $\vec{B} = (0.500$ T$)\hat{\imath}$. At $t = 0$ the proton has velocity components $v_x = 1.50 \times 10^5$ m/s, $v_y = 0$, and $v_z = 2.00 \times 10^5$ m/s (Example 28–4 in Section 28–5). In addition to the magnetic field there is a uniform electric field in the $+x$-direction, $\vec{E} = (+2.00 \times 10^4$ V/m$)\hat{\imath}$. a) Describe the path of the proton. Does the electric field affect the radius of the helix? Explain. b) At $t = T/2$, where T is the period of the circular motion of the proton, what is the x-component of the displacement of the proton from its position at $t = 0$?

28–20 A singly charged ion of ^{7}Li (an isotope of lithium) has a mass of 1.16×10^{-26} kg. It is accelerated through a potential difference of 220 V and then enters a magnetic field with magnitude 0.723 T perpendicular to the path of the ion. What is the radius of the ion's path in the magnetic field?

28–21 An electron in the beam of a TV picture tube is accelerated by a potential difference of 2.00 kV. Then it passes through a region of transverse magnetic field, where it moves in a circular arc with radius 0.180 m. What is the magnitude of the field?

SECTION 28–6 APPLICATION OF MOTION OF CHARGED PARTICLES

28–22 a) What is the speed of a beam of electrons when the simultaneous influence of an electric field of 1.56×10^4 V/m and a magnetic field of 4.62×10^{-3} T, with both fields normal to the beam and to each other, produces no deflection of the electrons? b) In a diagram, show the relative orientation of the vectors $\vec{v}$, $\vec{E}$, and $\vec{B}$. c) When the electric field is removed, what is the radius of the electron orbit? What is the period of the orbit?

28–23 **Determining the Mass of an Isotope.** The electric field between the plates of the velocity selector in a Bainbridge mass spectrometer (Fig. 28–20) is 1.12×10^5 V/m, and the magnetic field in both regions is 0.540 T. A stream of singly charged selenium ions moves in a circular path with a radius of 31.0 cm in the magnetic field. Determine the mass of one selenium ion and the mass number of this selenium isotope. (The mass number is equal to the mass of the isotope in atomic mass units, rounded to the nearest integer. One atomic mass unit = 1 u = 1.66×10^{-27} kg.)

28–24 In the Bainbridge mass spectrometer (Fig. 28–20), the magnetic field magnitude in the velocity selector is 0.650 T, and ions having a speed of 1.82×10^6 m/s pass through undeflected. a) What is the electric field magnitude in the velocity selector? b) If the separation of the plates is 5.20 mm, what is the potential difference between plates P and P'?

SECTION 28–7 MAGNETIC FORCE ON A CURRENT-CARRYING CONDUCTOR

28–25 A horizontal rod of mass m and length L is aligned in a north-south direction. The rod carries a current I in the direction toward the north. When a uniform magnetic field $\vec{B}$ is applied to the entire length of the rod, the magnetic force on the rod keeps it suspended at rest in mid-air. a) Find the minimum magnitude that $\vec{B}$ can have. b) Find the direction of $\vec{B}$ if it has the magnitude found in part (a).

28–26 A straight, vertical wire carries a current of 1.20 A downward in a region between the poles of a large superconducting electromagnet, where the magnetic field has magnitude $B = 0.588$ T and is horizontal. What are the magnitude and direction of the magnetic force on a 1.00-cm section of the wire that is in this uniform magnetic field, if the magnetic field direction is a) east? b) south? c) 30.0° south of west?

28–27 A horizontal rod 0.200 m long is mounted on a balance and carries a current. At the location of the rod a uniform horizontal magnetic field has magnitude 0.067 T and direction perpendicular to the rod. The magnetic force on the rod is measured by the balance and is found to be 0.13 N. What is the current?

28–28 An electromagnet produces a magnetic field of 0.550 T in a cylindrical region of radius 2.50 cm between its poles. A straight wire carrying a current of 10.8 A passes through the center of this region and is perpendicular to both the axis of the cylindrical region and the magnetic field. What magnitude of force is exerted on the wire?

28–29 A wire along the x-axis carries a current of 3.50 A in the negative direction. Calculate the force (expressed in terms of unit vectors) on a 1.00-cm section of the wire exerted by these magnetic fields: a) $\vec{B} = -(0.65 \text{ T})\hat{j}$; b) $\vec{B} = +(0.56 \text{ T})\hat{k}$; c) $\vec{B} = -(0.31 \text{ T})\hat{i}$; d) $\vec{B} = +(0.33 \text{ T})\hat{i} - (0.28 \text{ T})\hat{k}$; e) $\vec{B} = +(0.74 \text{ T})\hat{j} - (0.36 \text{ T})\hat{k}$.

SECTION 28–8 FORCE AND TORQUE ON A CURRENT LOOP

28–30 The plane of a 5.0 cm × 8.0 cm rectangular loop of wire is parallel to a 0.19-T magnetic field. The loop carries a current of 6.2 A. a) What torque acts on the loop? b) What is the magnetic moment of the loop? c) What is the maximum torque that can be obtained with the same total length of wire carrying the same current in this magnetic field?

28–31 A circular coil of wire 8.6 cm in diameter has 15 turns and carries a current of 2.7 A. The coil is in a region where the magnetic field is 0.56 T. a) What orientation of the coil gives the maximum torque on the coil, and what is the maximum torque? b) For what orientation of the coil is the magnitude of the torque 71% that found in part (a)?

28–32 A circular coil with area A and N turns is free to rotate about a diameter that coincides with the x-axis. Current I is cir-

culating in the coil. There is a uniform magnetic field $\vec{B}$ in the positive y-direction. Calculate the magnitude and direction of the torque $\vec{\tau}$ and the value of the potential energy U, as given in Eq. (28–27), when the coil is oriented as shown in parts (a) through (d) of Fig. 28–40.

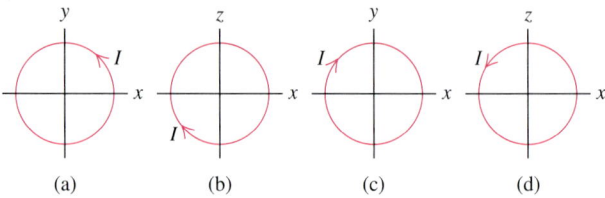

FIGURE 28–40 Exercise 28–32.

28–33 A coil with magnetic moment 1.45 A · m² is oriented initially with its magnetic moment antiparallel to a uniform 0.835-T magnetic field. What is the change in potential energy of the coil when it is rotated 180° so that its magnetic moment is parallel to the field?

*SECTION 28–9 THE DIRECT-CURRENT MOTOR

*28–34 A dc motor with its rotor and field coils connected in series has an internal resistance of 3.2 Ω. When running at full load on a 120-V line, the emf in the rotor is 105 V. a) What is the current drawn by the motor from the line? b) What is the power delivered to the motor? c) What is the mechanical power developed by the motor?

*28–35 In a shunt-wound dc motor with the field coils and rotor connected in parallel (Fig. 28–41), the resistance R_f of the field coils is 106 Ω, and the resistance R_r of the rotor is 5.9 Ω. When a potential difference of 120 V is applied to the brushes and the motor is running at full speed delivering mechanical power, the current supplied to it is 4.82 A. a) What is the current in the field coils? b) What is the current in the rotor? c) What is the induced emf developed by the motor? d) How much mechanical power is developed by this motor?

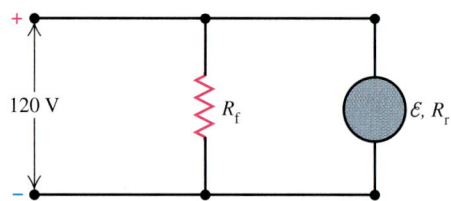

FIGURE 28–41 Exercises 28–35 and 28–36.

*28–36 A shunt-wound dc motor with the field coils and rotor connected in parallel (Fig. 28–41) operates from a 120-V dc power line. The resistance of the field windings, R_f, is 218 Ω. The resistance of the rotor, R_r, is 5.9 Ω. When the motor is running, the rotor develops an emf $\mathcal{E}$. The motor draws a current of 4.82 A from the line. Friction losses amount to 45.0 W. Compute a) the field current; b) the rotor current; c) the emf $\mathcal{E}$; d) the rate of development of thermal energy in the field windings; e) the rate of development of thermal energy in the rotor; f) the power input to the motor; g) the efficiency of the motor.

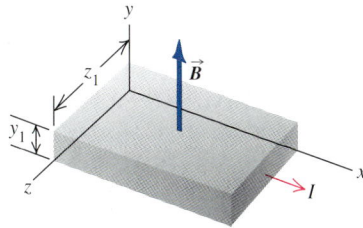

FIGURE 28–42 Exercises 28–37 and 28–38.

*28–37 Figure 28–42 shows a portion of a silver ribbon with $z_1 = 11.8$ mm and $y_1 = 0.23$ mm, carrying a current of 120 A in the $+x$-direction. The ribbon lies in a uniform magnetic field, in the y-direction, with magnitude 0.95 T. Apply the simplified model of the Hall effect presented in Section 28–10. If there are 5.85×10^{28} free electrons per cubic meter, find a) the magnitude of the drift velocity of the electrons in the x-direction; b) the magnitude and direction of the electric field in the z-direction due to the Hall effect; c) the Hall emf.

*28–38 Let Fig. 28–42 represent a strip of an unknown metal of the same dimensions as those of the silver ribbon in Exercise 28–37. When the magnetic field is 2.29 T and the current is 78.0 A, the Hall emf is found to be 131 μV. What does the simplified model of the Hall effect presented in Section 28–10 give for the density of free electrons in the unknown metal?

PROBLEMS

28–39 When a particle of charge $q > 0$ moves with a velocity of $\vec{v}_1$ that is at $45.0°$ from the $+x$-axis in the xy-plane, a uniform magnetic field exerts a force $\vec{F}_1$ along the $-z$-axis (Fig. 28–43). When the same particle moves with a velocity $\vec{v}_2$ with the same magnitude as $\vec{v}_1$ but along the $+z$-axis, a force $\vec{F}_2$ of magnitude F_2 is exerted on it along the $+x$-axis. a) What are the magnitude and direction of the magnetic field? b) What is the magnitude of $\vec{F}_1$ in terms of F_2?

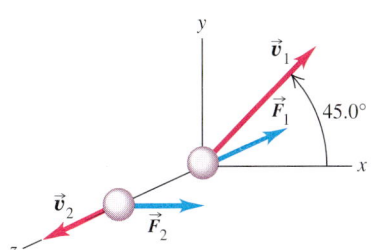

FIGURE 28–43 Problem 28–39.

28–40 The force on a charged particle moving in a magnetic field can be computed as the vector sum of the forces due to each separate component of the magnetic field. As an example, a particle with charge q is moving with speed v in the $-y$-direction. It is moving in a uniform magnetic field $\vec{B} = B_x\hat{i} + B_y\hat{j} + B_z\hat{k}$. a) What are the components of the force $\vec{F}$ exerted on the particle by the magnetic field? b) If $q > 0$, what must be the signs of the components of $\vec{B}$ if the components of $\vec{F}$ are all nonnegative? c) If $q < 0$ and $B_x = B_y = B_z > 0$, find the direction of $\vec{F}$ and find the magnitude of $\vec{F}$ in terms of $|q|$, v, and B_x.

28–41 Crossed $\vec{E}$ and $\vec{B}$ Fields. A particle with initial velocity $\vec{v}_0 = (5.85 \times 10^3 \text{ m/s})\hat{j}$ enters a region of uniform electric and magnetic fields. The magnetic field in the region is $\vec{B} = -(1.35 \text{ T})\hat{k}$. Calculate the magnitude and direction of the electric field in the region if the particle is to pass through undeflected, for a particle of charge a) $+0.640$ nC; b) -0.320 nC. Neglect the weight of the particle.

28–42 In the electron gun of a TV picture tube the electrons (charge $-e$, mass m) are accelerated by a voltage V. After leaving the electron gun, the electron beam travels a distance D to the screen; in this region there is a transverse magnetic field of magnitude B and no electric field. a) Show that the approximate deflection of the beam due to this magnetic field is

$$d = \frac{BD^2}{2}\sqrt{\frac{e}{2mV}} .$$

b) Evaluate this expression for $V = 750$ V, $D = 50$ cm, and $B = 5.0 \times 10^{-5}$ T (comparable to the earth's field). Is this deflection significant?

28–43 The magnetic poles of a small cyclotron produce a magnetic field with magnitude 0.85 T. The poles have a radius of 0.40 m, which is the maximum radius of the orbits of the accelerated particles. a) What is the maximum energy to which protons ($q = 1.60 \times 10^{-19}$ C, $m = 1.67 \times 10^{-27}$ kg) can be accelerated by this cyclotron? Give your answer in electron volts and in joules. b) What is the time for one revolution of a proton orbiting at this maximum radius? c) What would the magnetic field magnitude have to be for the maximum energy to which a proton can be accelerated to be twice that calculated in part (a)? d) For $B = 0.85$ T, what is the maximum energy to which alpha particles ($q = 3.20 \times 10^{-19}$ C, $m = 6.65 \times 10^{-27}$ kg) can be accelerated by this cyclotron? How does this compare to the maximum energy for protons?

28–44 A particle of charge $q > 0$ is moving at speed v in the $+z$-direction through a region of uniform magnetic field $\vec{B}$. The magnetic force on the particle is $\vec{F} = F_0(3\hat{i} + 4\hat{j})$, where F_0 is a positive constant. a) Determine the components B_x, B_y, and B_z, or at least as many of the three components as is possible from the information given. (See Problem 28–40.) b) If it is given in addition that the magnetic field has magnitude $6F_0/qv$, determine as much as you can about the remaining components of $\vec{B}$.

28–45 An electron and a nucleus of ^{7}Li (an isotope of lithium that has three protons and four neutrons) both move in a magnetic field in circular paths with the same tangential speed. Compute the ratio of the number of revolutions the electron makes per second to the number made per second by the ^{7}Li nucleus. The mass of the ^{7}Li nucleus is 1.16×10^{-26} kg.

28–46 A cyclotron is to accelerate protons to an energy of 5.4 MeV. The superconducting electromagnet of the cyclotron produces a 3.5-T magnetic field perpendicular to the proton orbits. a) When the protons have achieved a kinetic energy of 2.7 MeV, what is the radius of their circular orbit and what is their angular speed? b) Repeat part (a) when the protons have achieved their final kinetic energy of 5.4 MeV.

28–47 A particle with negative charge q and mass $m = 2.58 \times 10^{-15}$ kg is traveling through a region containing a uniform magnetic field $\vec{B} = -(0.120 \text{ T})\hat{k}$. At a particular instant of time the velocity of the particle is $\vec{v} = (1.05 \times 10^6 \text{ m/s})(-3\hat{i} + 4\hat{j} + 12\hat{k})$ and the force $\vec{F}$ on the particle has a magnitude of 1.25 N. (See Problem 28–40.) a) Determine the charge q. b) Determine the acceleration $\vec{a}$ of the particle. c) Explain why the path of the particle is a helix, and determine the radius of curvature R of the circular component of the helical path. d) Determine the cyclotron frequency of the particle. e) Although helical motion is not periodic in the full sense of the word, the x- and y-coordinates do vary in a periodic way. If the coordinates of the particle at $t = 0$ are $(x, y, z) = (R, 0, 0)$, determine its coordinates at a time $t = 2T$, where T is the period of the motion in the xy-plane.

28–48 An electron moves in a circular path with radius $r = 4.00$ cm in the space between two concentric cylinders. The inner cylinder is a positively charged wire with radius $a = 1.00$ mm, and the outer cylinder is a negatively charged hollow cylinder with radius $b = 5.00$ cm. The potential difference between the inner and outer cylinders is $V_{ab} = 120$ V, with the wire being at the higher potential. (See Fig. 28–44. Note that the figure is not drawn to scale.) The electric field $\vec{E}$ in the region between the cylinders is radially outward and was shown in Problem 24–57 to have the magnitude $E = V_{ab}/[r \ln (b/a)]$. a) Determine the speed the electron needs to maintain its circular orbit. Neglect both the gravitational and magnetic fields of the earth. b) Now include the effect of the earth's magnetic field. If the axis of symmetry of the cylinders is positioned parallel to the magnetic field of the earth, at what speed must the electron move to maintain the same circular orbit? Assume that the magnetic field of the earth has magnitude 1.30×10^{-4} T and that its direction is out of the plane of the page in Fig. 28–44. c) Redo the calculation of part (b) for the case in which the magnetic field is in the opposite direction to that of part (b).

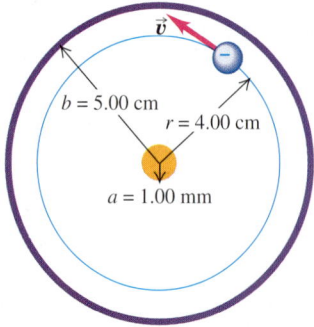

FIGURE 28–44 Problem 28–48.

28–49 Suppose the electric field between the plates P and P' in Fig. 28–20 is 1.88×10^4 V/m and the magnetic field in both

regions is 0.701 T. If the source contains the three isotopes of krypton, ^{82}Kr, ^{84}Kr, and ^{86}Kr, and the ions are singly charged, find the distance between the lines formed by the three isotopes on the photographic plate. Assume the atomic masses of the isotopes (in atomic mass units) are equal to their mass numbers, 82, 84, and 86. (One atomic mass unit = 1 u = 1.66×10^{-27} kg.)

28–50 The force on a charged particle moving in a magnetic field can be computed as the vector sum of the forces due to each separate component of the particle's velocity. (See Problem 28–40.) A particle with charge 9.45×10^{-8} C is moving in a region where there is a uniform magnetic field of 0.450 T in the $+x$–direction. At a particular instant of time the velocity of the particle has components $v_x = -1.68 \times 10^4$ m/s, $v_y = -3.11 \times 10^4$ m/s, and $v_z = 5.85 \times 10^4$ m/s. What are the components of the force on the particle at this time?

28–51 The cube in Fig. 28–45, 75.0 cm on a side, is in a uniform magnetic field of 0.860 T parallel to the x-axis. The wire $abcdef$ carries a current of 6.58 A in the direction indicated. a) Determine the magnitude and direction of the force acting on each of the segments ab, bc, cd, de, and ef. b) What are the magnitude and direction of the total force on the wire?

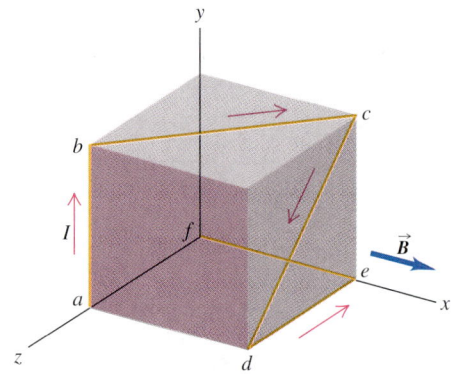

FIGURE 28–45 Problem 28–51.

28–52 An Electromagnetic Rail Gun. A conducting bar of mass m and length L slides over horizontal rails that are connected to a voltage source. The voltage source maintains a constant current I in the rails and bar, and a constant, uniform, vertical magnetic field $\vec{B}$ fills the region between the rails (Fig. 28–46). a) Find the magnitude and direction of the net force on the conducting bar. Ignore friction, air resistance, and electrical resistance. b) If the bar has mass m, find the distance d that the bar must move along the rails from rest to attain speed v.

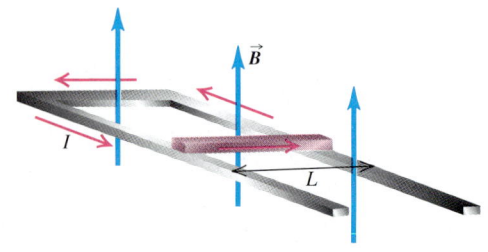

FIGURE 28–46 Problem 28–52.

c) It has been suggested that rail guns based on this principle could accelerate payloads into earth orbit or beyond. Find the distance the bar must travel along the rails if it is to reach the escape speed for the earth (11.2 km/s). Let $B = 0.50$ T, $I = 2.0 \times 10^3$ A, $m = 25$ kg, and $L = 50$ cm.

28-53 A straight piece of conducting wire of mass M and length L is placed on a frictionless incline tilted at an angle θ from the horizontal (Fig. 28–47). There is a uniform, vertical magnetic field $\vec{B}$ at all points (produced by an arrangement of magnets not shown in the figure). To keep the wire from sliding down the incline, a voltage source is attached to the ends of the wire. When just the right amount of current flows through the wire, the wire remains at rest. Determine the magnitude and direction of the current in the wire that will cause the wire to remain at rest. Copy the figure and draw the direction of the current on your copy. In addition, show in a free-body diagram all the forces that act on the wire.

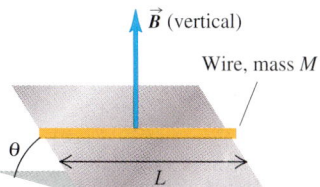

FIGURE 28-47 Problem 28–53.

28-54 As shown in Fig. 28–48, a thin, flexible wire hangs from point P in a region where there is a uniform horizontal magnetic field of magnitude B directed into and perpendicular to the plane of the figure. A weight is attached to the bottom of the wire to provide a uniform tension T throughout the wire. (The weight of the wire itself is negligible). When a current I flows from the top to the bottom of the wire, the wire curves into a circular arc of radius R. a) By considering the forces on a small segment of wire that subtends an angle θ, show that the radius of curvature of the wire is $R = T/IB$. (*Hint:* Recall that if θ is small, $\sin\theta \cong \tan\theta \cong \theta$, where θ is in radians.) b) The wire is now removed. A negatively charged particle of charge $-q$ and mass m is launched from the same point P from which the wire was hung, in the same direction as that in which the wire extended from P. Show that the trajectory of the particle will follow the same circular arc as did the wire if the speed of the particle is $v = qT/mI$. (For this reason, weighted wires are used in designing magnet systems used for steering beams of charged particles. The bending of the current-carrying wire indicates how particle trajectories will bend.)

FIGURE 28-48 Problem 28–54.

28-55 Two positive ions having the same charge q but different masses, m_1 and m_2, are accelerated horizontally from rest through a potential difference V. They then enter a region where there is a uniform magnetic field $\vec{B}$ normal to the plane of the trajectory. a) Show that if the beam entered the magnetic field along the x-axis, the value of the y-coordinate for each ion at any time t is approximately

$$y = Bx^2 \left(\frac{q}{8mV} \right)^{1/2},$$

provided y remains much smaller than x. b) Can this arrangement be used for isotope separation? Why or why not?

28-56 A wire 25.0 cm long lies along the z-axis and carries a current of 9.00 A in the $+z$-direction. The magnetic field is uniform and has components $B_x = -0.242$ T, $B_y = -0.985$ T, and $B_z = -0.336$ T. a) Find the components of the magnetic force on the wire. (As in Problem 28–40, the net magnetic force is the vector sum of the forces due to each component of $\vec{B}$.) b) What is the magnitude of the net magnetic force on the wire?

28-57 The rectangular loop of wire in Fig. 28–49 has a mass of 0.15 g per centimeter of length and is pivoted about side ab as a frictionless axis. The current in the wire is 8.2 A in the direction shown. Find the magnitude and direction of the magnetic field parallel to the y-axis that will cause the loop to swing up until its plane makes an angle of $30.0°$ with the yz-plane.

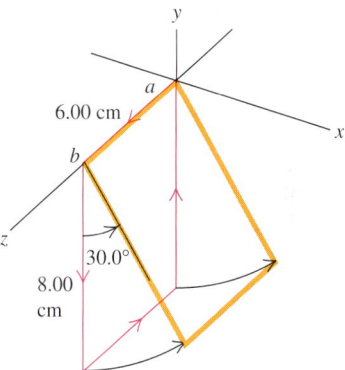

FIGURE 28-49 Problem 28–57.

28-58 The rectangular loop in Fig. 28–50 is pivoted about the y-axis and carries a current of 15.0 A in the direction indicated.

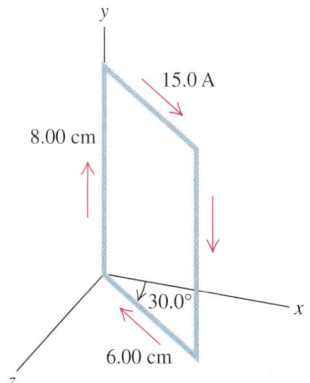

FIGURE 28-50 Problem 28–58.

a) If the loop is in a uniform magnetic field with magnitude 0.48 T in the +x-direction, find the magnitude and direction of the torque required to hold the loop in the position shown. b) Repeat part (a) for the case in which the field is in the −z-direction. c) For each of the above magnetic fields, what torque would be required if the loop were pivoted about an axis through its center, parallel to the y-axis?

28–59 A solenoid with N turns of wire, each of cross-section area A, is connected to a source of emf that maintains a constant current I in the wire. The solenoid is initially placed as shown in Fig. 28–30, with the solenoid axis at an angle ϕ to a uniform magnetic field $\vec{B}$. The solenoid is then released and is free to pivot about an axis perpendicular to the plane of Fig. 28–30. The moment of inertia of the solenoid about this axis is I_S. Determine the period of oscillation of the solenoid. Assume that the angle ϕ is much smaller than is shown in Fig. 28–30. (*Hint:* See Section 13–7.)

28–60 A thin wire of length L is made of an insulating material. The wire is bent to form a circular loop, and a positive charge q is distributed uniformly around the circumference of the loop. The loop is then set into rotation with angular speed ω around an axis perpendicular to the plane of the loop and passing through its center. If the loop is in a region where there is a uniform magnetic field $\vec{B}$ directed parallel to the plane of the loop, calculate the magnitude of the magnetic torque on the loop.

28–61 A Voice Coil. It was shown in Section 28–8 that the net force on a current loop in a *uniform* magnetic field is zero. The magnetic force on the voice coil of a loudspeaker (Fig. 28–24) is nonzero because the magnetic field at the coil is not uniform. A voice coil in a loudspeaker has 50 turns of wire and a diameter of 1.56 cm, and the current in the coil is 0.950 A. Assume that the magnetic field at each point of the coil has a constant magnitude of 0.220 T and is directed at an angle of 60.0° outward from the normal to the plane of the coil (Fig. 28–51). Let the axis of the coil be in the y-direction. The current in the coil is in the direction shown (counterclockwise as viewed from a point above the coil on the y-axis). Calculate the magnitude and direction of the net magnetic force on the coil.

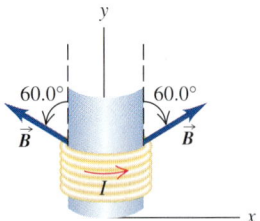

FIGURE 28–51 Problem 28–61.

28–62 Several external forces $\vec{F}_i$ are applied to a rigid body. With respect to an origin O, force $\vec{F}_1$ is applied at the point located at $\vec{r}_1$, force $\vec{F}_2$ at $\vec{r}_2$, and so on. Point P is located at $\vec{r}_P$. If the body is in translational equilibrium, so that $\Sigma \vec{F}_i = 0$, prove that the sum of the torques about point P equals the sum of torques about point O. (This justifies a statement made in Section 28–8: Because there is zero net magnetic force on a cur-

rent loop in a uniform magnetic field, the magnetic torque has the same magnitude and direction for *any* choice of axis.)

28–63 Force on a Current Loop in a Non-Uniform Magnetic Field. It was shown in Section 28–8 that the net force on a current loop in a *uniform* magnetic field is zero. But what if $\vec{B}$ is *not* uniform? Figure 28–52 shows a square loop of wire that lies in the xy-plane. The loop has corners at (0, 0), (0, L), (L, 0), and (L, L), and carries a constant current I in the clockwise direction. The magnetic field has no x-component but has both y- and z-components: $\vec{B} = (B_0 z/L)\hat{\jmath} + (B_0 y/L)\hat{k}$, where B_0 is a positive constant. a) Sketch the magnetic field lines in the yz-plane. b) Find the magnitude and direction of the magnetic force exerted on each of the sides of the loop by integrating Eq. (28–20). c) Find the magnitude and direction of the net magnetic force on the loop.

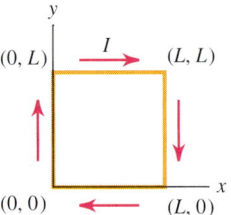

FIGURE 28–52 Problems 28–63 and 28–64.

28–64 Torque on a Current Loop in a Non-Uniform Magnetic Field. In Section 28–8 the expression for the torque on a current loop was derived assuming that the magnetic field $\vec{B}$ was uniform. But what if $\vec{B}$ is *not* uniform? Figure 28–52 shows a square loop of wire that lies in the xy-plane. The loop has corners at (0, 0), (0, L), (L, 0), and (L, L), and carries a constant current I in the clockwise direction. The magnetic field has no z-component but has both x- and y-components: $\vec{B} = (B_0 y/L)\hat{\imath} + (B_0 x/L)\hat{\jmath}$, where B_0 is a positive constant. a) Sketch the magnetic field lines in the xy-plane. b) Find the magnitude and direction of the magnetic force exerted on each of the sides of the loop by integrating Eq. (28–20). (As in Problem 28–40, the net force on each side of the loop is the vector sum of the forces due to each component of $\vec{B}$.) c) If the loop is free to rotate about the x-axis, find the magnitude and direction of the magnetic torque on the loop. d) Repeat part (c) for the case in which the loop is free to rotate about the y-axis. e) Is Eq. (28–26), $\vec{\tau} = \vec{\mu} \times \vec{B}$, an appropriate description of the torque on this loop? Why or why not?

28–65 An insulated wire with mass $m = 5.40 \times 10^{-5}$ kg is bent into the shape of an inverted U such that the horizontal part has a length $l = 15.0$ cm. The bent ends of the wire are partially immersed in two pools of mercury, with 2.5 cm of each end below the mercury surface. The entire structure is in a region containing a uniform 0.00650-T magnetic field directed into the page (Fig. 28–53). An electrical connection from the mercury pools is made through the ends of the wires. The mercury pools are connected to a 1.50-V battery and a switch S. When switch S is closed, the wire jumps 35.0 cm into the air, measured from its initial position. a) Determine the speed v of the wire as it leaves the mercury. b) Assuming that the current I through the wire was constant from the time the switch was closed until the wire left the mercury, determine I. c) Neglecting the resistance of the mercury and the circuit wires, determine the resistance of the moving wire.

FIGURE 28–53 Problem 28–65.

28–66 Quark Model of the Neutron. The neutron is a particle with zero charge. Nonetheless, it has a nonzero magnetic moment with z-component 9.66×10^{-27} A · m^2. This can be explained by the internal structure of the neutron. A substantial body of evidence indicates that a neutron is composed of three fundamental particles called *quarks:* an "up" (*u*) quark, of charge $+2e/3$, and two "down" (*d*) quarks, each of charge $-e/3$. The combination of the three quarks produces a net charge of $2e/3 - e/3 - e/3 = 0$. If the quarks are in motion, they can produce a nonzero magnetic moment. As a very simple model, suppose the *u* quark moves in a counterclockwise circular path and the *d* quarks move in a clockwise path, all of radius r and all with the same speed v (Fig. 28–54). a) Determine the current due to the circulation of the *u* quark. b) Determine the magnitude of the magnetic moment due to the circulating *u* quark. c) Determine the magnitude of the magnetic moment of the three-quark system. (Be careful to use the correct magnetic moment directions.) d) With what speed v must the quarks move if this model is to reproduce the magnetic moment of the neutron? Use $r = 1.20 \times 10^{-15}$ m (the radius of the neutron) for the radius of the orbits.

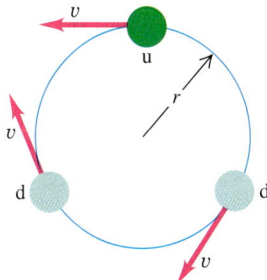

FIGURE 28–54 Problem 28–66.

28–67 A circular loop of wire with area A lies in the xy-plane. As viewed along the z-axis looking in the $-z$-direction toward the origin, a current I is circulating clockwise around the loop. The torque produced by an external magnetic field $\vec{B}$ is given by $\vec{\tau} = D(4\hat{\imath} - 3\hat{\jmath})$, where D is a positive constant, and for this orientation of the loop the magnetic potential energy $U = -\vec{\mu} \cdot \vec{B}$ is

negative. The magnitude of the magnetic field is $B_0 = 13D/IA$. a) Determine the vector magnetic moment of the current loop. b) Determine the components B_x, B_y, and B_z of $\vec{B}$.

28–68 Derivation of Eq. (28–26) for a Circular Current Loop. A wire ring lies in the xy-plane with its center at the origin. The ring carries a counterclockwise current I (Fig. 28–55). A uniform magnetic field $\vec{B}$ is in the $+x$-direction, $\vec{B} = B_x \hat{\imath}$. (The result is easily extended to $\vec{B}$ in an arbitrary direction.) a) In Fig. 28–55, show that the element $d\vec{l} = R\, d\theta\, (-\sin\theta\hat{\imath} + \cos\theta\hat{\jmath})$, and find $d\vec{F} = I\, d\hat{l} \times \vec{B}$. b) Integrate $d\vec{F}$ around the loop to show that the net force is zero. c) From part (a), find $d\vec{\tau} = \vec{r} \times d\vec{F}$, where $\vec{r} = R(\cos\theta\hat{\imath} + \sin\theta\hat{\jmath})$ is the vector from the center of the loop to the element $d\vec{l}$. (Note that $d\vec{l}$ is perpendicular to $\vec{r}$.) d) Integrate $d\vec{\tau}$ over the loop to find the total torque $\vec{\tau}$ on the loop. Show that the result can be written as $\vec{\tau} = \vec{\mu} \times \vec{B}$, where $\mu = IA$. (*Note:* $\int \cos^2 x\, dx = \frac{1}{2}x + \frac{1}{4}\sin 2x$, $\int \sin^2 x\, dx = \frac{1}{2}x - \frac{1}{4}\sin 2x$, and $\int \sin x \cos x\, dx = \frac{1}{2}\sin^2 x$.)

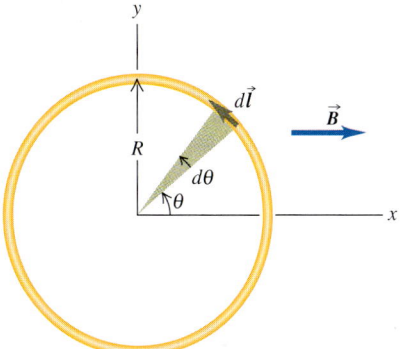

FIGURE 28–55 Problem 28–68.

28–69 Using Gauss's Law for Magnetism. In a certain region of space, the magnetic field $\vec{B}$ is not uniform. The magnetic field has both a z-component and a component that points radially away from or toward the z-axis. The z-component is given by $B_z(z) = \beta z$, where β is a positive constant. The radial component B_r depends only on r, the radial distance from the z-axis. a) Use Gauss's law for magnetism, Eq. (28–8), to find the radial component B_r as a function of r. (*Hint:* Try a cylindrical Gaussian surface of radius r concentric with the z-axis, with one end at $z = 0$ and the other at $z = L$.) b) Sketch the magnetic field lines.

28–70 A circular ring with area 4.45 cm^2 is carrying a current of 12.5 A. The ring is free to rotate about a diameter. The ring, initially at rest, is immersed in a region of uniform magnetic field given by $\vec{B} = (1.15 \times 10^{-2}$ T$)(12\hat{\imath} + 3\hat{\jmath} - 4\hat{k})$. The ring is positioned initially such that its magnetic moment is given by $\vec{\mu}_i = \mu(-0.800\hat{\imath} + 0.600\hat{\jmath})$, where μ is the (positive) magnitude of the magnetic moment. The ring is released and turns through an angle of 90.0°, at which point its magnetic moment is given by $\vec{\mu}_f = -\mu\hat{k}$. a) Determine the decrease in potential energy. b) If the moment of inertia of the ring about a diameter is 8.50×10^{-7} kg · m^2, determine the angular speed of the ring as it passes through the second position.

CHALLENGE PROBLEMS

28–71 A particle with charge 2.15 μC and mass 3.20×10^{-11} kg is initially traveling in the +y-direction with a speed $v_0 = 1.45 \times 10^5$ m/s. It then enters a region containing a uniform magnetic field that is directed into and perpendicular to the page in Fig. 28–56. The magnitude of the field is 0.420 T. The region extends a distance of 25.0 cm along the initial direction of travel; 75.0 cm from the point of entry into the magnetic field region is a wall. The length of the field-free region is thus 50.0 cm. When the charged particle enters the magnetic field, it follows a curved path whose radius of curvature is R. It then leaves the magnetic field after a time t_1, having been deflected a distance Δx_1. The particle then travels in the field-free region and strikes the wall after undergoing a total deflection Δx. a) Determine the radius R of the curved part of the path. b) Determine t_1, the time the particle spends in the magnetic field. c) Determine Δx_1, the horizontal deflection at the point of exit from the field. d) Determine Δx, the total horizontal deflection.

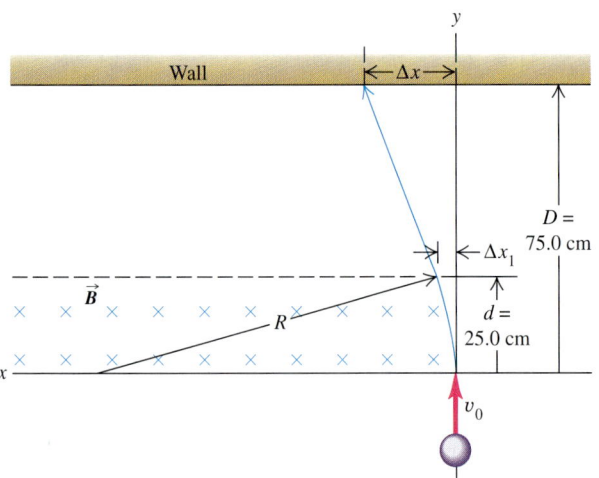

FIGURE 28–56 Challenge Problem 28–71.

28–72 The Electromagnetic Pump. Magnetic forces acting on conducting fluids provide a convenient means of pumping these fluids. For example, this method can be used to pump blood without the damage to the cells that can be caused by a mechanical pump. A horizontal tube with rectangular cross section (height h, width w) is placed at right angles to a uniform magnetic field with magnitude B so that a length l is in the field (Fig. 28–57). The tube is filled with a conducting liquid and an electric current of density J is maintained in the third mutually

perpendicular direction. a) Show that the difference of pressure between a point in the liquid on a vertical plane through ab and a point in the liquid on another vertical plane through cd, under conditions in which the liquid is prevented from flowing, is $\Delta p = JlB$. b) What current density is needed to provide a pressure difference of 1.00 atm between these two points if $B = 2.20$ T and $l = 35.0$ mm?

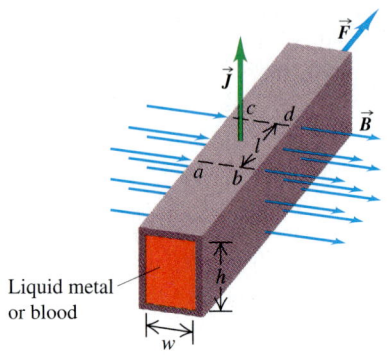

Liquid metal or blood

FIGURE 28–57 Challenge Problem 28–72.

28–73 A Cycloidal Path. A particle with mass m and positive charge q starts from rest at the origin in Fig. 28–58. There is a uniform electric field $\vec{E}$ in the +y-direction and a uniform magnetic field $\vec{B}$ directed out of the page. It is shown in more advanced books that the path is a *cycloid* whose radius of curvature at the top points is twice the y-coordinate at that level. a) Explain why the path has this general shape and why it is repetitive. b) Prove that the speed at any point is equal to $\sqrt{2qEy/m}$. (*Hint:* Use energy conservation.) c) Applying Newton's second law at the top point and taking as given that the radius of curvature here equals $2y$, prove that the speed at this point is $2E/B$.

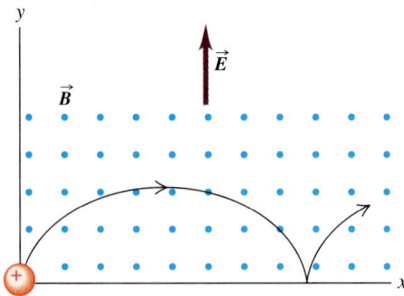

FIGURE 28–58 Challenge Problem 28–73.

The medical diagnostic technique called magnetic resonance imaging (MRI) requires that the patient lie in a strong magnetic field. The first MRI devices used a solenoid to produce the field. Newer devices use two large coils, placed either above and below the patient or at the patient's head and feet. This "open" design provides a much less claustrophobic experience for the patient.

29–1 INTRODUCTION

In Chapter 28 we studied the *forces* exerted on moving charges and on current-carrying conductors in a magnetic field. We didn't worry about how the magnetic field got there; we simply took its existence as a given fact. But how are magnetic fields *created*? We know that both permanent magnets and electric currents in electromagnets create magnetic fields. Now it's time to study these sources of magnetic field in detail.

We've learned that a charge creates an electric field and that an electric field exerts a force on a charge. But a *magnetic* field exerts a force only on a *moving* charge. Is it also true that a charge *creates* a magnetic field only when the charge is moving?

In a word, yes. Our analysis will begin with the magnetic field created by a single moving point charge. We can use this analysis to determine the field created by a small segment of a current-carrying conductor. Once we can do that, we can in principle find the magnetic field produced by *any* shape of conductor.

Then we will introduce Ampere's law, the magnetic analog of Gauss's law in electrostatics. Ampere's law lets us exploit symmetry properties in relating magnetic fields to their sources.

Moving charged particles within atoms respond to magnetic fields and can also act as sources of magnetic field. We'll use these ideas to understand how certain magnetic materials can be used to intensify magnetic fields as well as why some materials such as iron act as permanent magnets. Finally, we will study how a time-varying electric field, which we will describe in terms of a quantity called *displacement current,* can act as a source of magnetic field. This relationship will be one of the key elements in our study of electromagnetic waves in Chapter 33.

29–2 MAGNETIC FIELD OF A MOVING CHARGE

Let's start with the basics, the magnetic field of a single point charge q moving with a constant velocity $\vec{v}$. As we did for electric fields, we call the location of the charge the **source point** and the point P where we want to find the field the **field point.** In Section 22–6 we found that at a field point a distance r from a point charge q, the magnitude of the *electric* field $\vec{E}$ caused by the charge is proportional to the charge magnitude $|q|$ and to $1/r^2$, and the direction of $\vec{E}$ (for positive q) is along the line from source point to field point. The corresponding relationship for the *magnetic* field $\vec{B}$ of a point charge q moving with constant velocity has some similarities and some interesting differences.

Experiments show that the magnitude of $\vec{B}$ is also proportional to $|q|$ and to $1/r^2$. But the *direction* of $\vec{B}$ is *not* along the line from source point to field point. Instead, $\vec{B}$ is perpendicular to the plane containing this line and the particle's velocity vector $\vec{v}$, as shown in Fig. 29–1. Furthermore, the field *magnitude* B is also proportional to

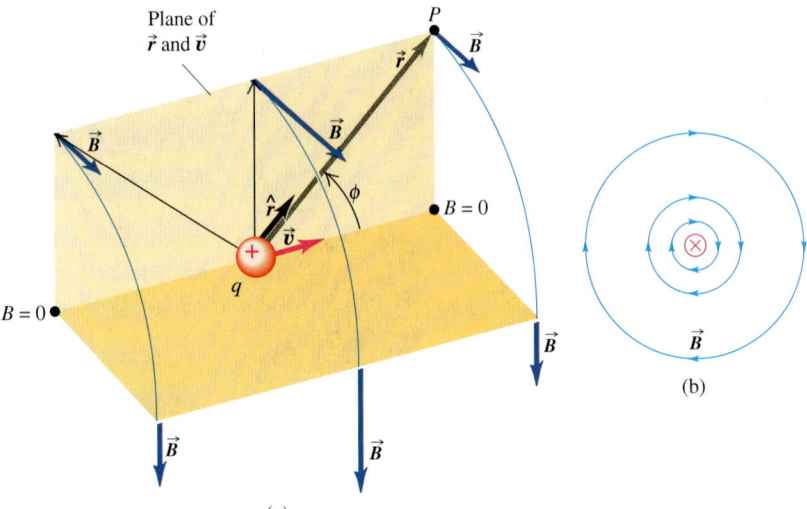

29-1 (a) Magnetic-field vectors due to a moving positive point charge q. At each point, $\vec{B}$ is perpendicular to the plane of $\vec{r}$ and $\vec{v}$, and its magnitude is proportional to the sine of the angle between them. (b) Magnetic field lines in a plane containing a moving positive charge. The charge is moving into the plane of the page.

the particle's speed v and to the sine of the angle ϕ between $\vec{v}$ and $\vec{B}$. Thus the magnetic field magnitude at point P is given by

$$B = \frac{\mu_0}{4\pi}\frac{|q|v\sin\phi}{r^2}, \tag{29-1}$$

where $\mu_0/4\pi$ is a proportionality constant (μ_0 is pronounced "mu-nought" or "mu-sub-zero"). The reason for writing the constant in this particular way will emerge shortly. We did something similar with Coulomb's law in Section 22–5.

We can incorporate both the magnitude and direction of $\vec{B}$ into a single vector equation using the vector product. To avoid having to say "the direction from the source q to the field point P" over and over, we introduce a *unit* vector $\hat{r}$ ("r-hat") that points from the source point to the field point. (We used $\hat{r}$ for the same purpose in Section 22–6). This unit vector is equal to the vector $\vec{r}$ from the source to the field point divided by its magnitude: $\hat{r} = \vec{r}/r$. Then the $\vec{B}$ field of a moving point charge is

$$\vec{B} = \frac{\mu_0}{4\pi}\frac{q\vec{v}\times\hat{r}}{r^2} \quad \text{(magnetic field of a point charge with constant velocity).} \tag{29-2}$$

Figure 29–1 shows the relation of $\hat{r}$ to P and also shows the magnetic field $\vec{B}$ at several points in the vicinity of the charge. At all points along a line through the charge parallel to the velocity $\vec{v}$, the field is zero because $\sin\phi = 0$ at all such points. At any distance r from q, $\vec{B}$ has its greatest magnitude at points lying in the plane perpendicular to $\vec{v}$ because at all such points, $\phi = 90°$ and $\sin\phi = 1$. If the charge q is negative, the directions of $\vec{B}$ are opposite to those shown in Fig. 29–1.

A point charge in motion also produces an *electric* field, with field lines that radiate outward from a positive charge. The *magnetic* field lines are completely different. The above discussion shows that for a point charge moving with velocity $\vec{v}$, the magnetic field lines are *circles* centered on the line of $\vec{v}$ and lying in planes perpendicular to this line. The field line directions for a positive charge are given by the following *right-hand rule,* one of several that we will encounter in this chapter for determining the direction of the magnetic field caused by different sources. Grasp the velocity vector $\vec{v}$ with your right hand so your right thumb points in the direction of $\vec{v}$; your fingers then curl around the line of $\vec{v}$ in the same sense as the magnetic field lines (assuming q is positive).

Figure 29–1a shows parts of a few field lines; Fig. 29–1b shows some field lines in a plane through q, perpendicular to $\vec{v}$, as seen by looking in the direction of $\vec{v}$.

Equations (29–1) and (29–2) describe the $\vec{B}$ field of a point charge moving with *constant* velocity. If the charge *accelerates*, the field can be much more complicated. We won't need these more complicated results for our purposes. (The moving charged particles that make up a current in a wire accelerate at points where the wire bends and the direction of $\vec{v}$ changes. But because the magnitude v_d of the drift velocity in a conductor is typically very small, the acceleration v_d^2/r is also very small, and the effects of acceleration can be ignored.)

As we discussed in Section 28–3, the unit of B is one tesla (1 T):

$$1\,\text{T} = 1\,\text{N}\cdot\text{s/C}\cdot\text{m} = 1\,\text{N/A}\cdot\text{m}.$$

Using this with Eq. (29–1) or (29–2), we find that the units of the constant μ_0 are

$$1\,\text{N}\cdot\text{s}^2/\text{C}^2 = 1\,\text{N/A}^2 = 1\,\text{Wb/A}\cdot\text{m} = 1\,\text{T}\cdot\text{m/A}.$$

In SI units the numerical value of μ_0 is exactly $4\pi \times 10^{-7}$. Thus

$$\mu_0 = 4\pi \times 10^{-7}\,\text{N}\cdot\text{s}^2/\text{C}^2 = 4\pi \times 10^{-7}\,\text{Wb/A}\cdot\text{m} = 4\pi \times 10^{-7}\,\text{T}\cdot\text{m/A}. \quad (29\text{–}3)$$

It may seem incredible that μ_0 has *exactly* this numerical value! In fact this is a *defined* value that arises from the definition of the ampere, as we'll discuss in Section 29–5.

We mentioned in Section 22–5 that the constant $1/4\pi\epsilon_0$ in Coulomb's law is related to the speed of light c:

$$k = \frac{1}{4\pi\epsilon_0} = (10^{-7}\,\text{N}\cdot\text{s}^2/\text{C}^2)c^2.$$

When we study electromagnetic waves in Chapter 33, we will find that their speed of propagation in vacuum, which is equal to the speed of light c, is given by

$$c^2 = \frac{1}{\epsilon_0\mu_0}. \quad (29\text{–}4)$$

If we solve the equation $k = 1/4\pi\epsilon_0$ for ϵ_0, substitute the resulting expression into Eq. (29–4), and solve for μ_0, we indeed get the value of μ_0 stated above. This discussion is a little premature, but it may give you a hint about one of the nice unifying threads that runs through electromagnetic theory.

EXAMPLE 29–1

Forces between two moving protons Two protons move parallel to the x-axis in opposite directions (Fig. 29–2) at the same speed v (small compared to the speed of light c.) At the instant shown, find the electric and magnetic forces on the upper proton, and determine the ratio of their magnitudes.

SOLUTION The electric force is easy; Coulomb's law gives

$$F_E = \frac{1}{4\pi\epsilon_0}\frac{q^2}{r^2}.$$

The forces are repulsive, and the force on the upper proton is vertically upward (in the $+y$-direction). To find the magnetic force, we first find the $\vec{B}$ field caused by the lower proton at the location of the upper one. From the right-hand rule for the cross

product $\vec{v} \times \hat{r}$ in Eq. (29–2), we find that $\vec{B}$ is in the $+z$-direction, as shown in the figure; from Eq. (29–1) the magnitude of $\vec{B}$ is

$$B = \frac{\mu_0}{4\pi}\frac{qv}{r^2},$$

since $\phi = 90°$. Alternatively, from Eq. (29–2),

$$\vec{B} = \frac{\mu_0}{4\pi}\frac{q(v\hat{\imath}) \times \hat{\jmath}}{r^2} = \frac{\mu_0}{4\pi}\frac{qv}{r^2}\hat{k}.$$

The velocity of the upper proton is $-\vec{v}$, and the magnetic force on it is $\vec{F} = q(-\vec{v}) \times \vec{B}$. Combining this with the expressions for $\vec{B}$, we find

$$F_B = \frac{\mu_0}{4\pi}\frac{q^2 v^2}{r^2}$$

or

$$\vec{F}_B = q(-\vec{v}) \times \vec{B} = q(-v\hat{\imath}) \times \frac{\mu_0}{4\pi} \frac{qv}{r^2} \hat{k} = \frac{\mu_0}{4\pi} \frac{q^2v^2}{r^2} \hat{\jmath}.$$

The magnetic interaction in this situation is also repulsive. The ratio of the magnitudes of the two forces is

$$\frac{F_B}{F_E} = \frac{\mu_0 q^2 v^2 / 4\pi r^2}{q^2 / 4\pi \epsilon_0 r^2} = \frac{\mu_0 v^2}{1/\epsilon_0} = \epsilon_0 \mu_0 v^2.$$

Using the relationship $\epsilon_0 \mu_0 = 1/c^2$, Eq. (29–4), we can express our result very simply as

$$\frac{F_B}{F_E} = \frac{v^2}{c^2}.$$

When v is small in comparison to c, the speed of light, the magnetic force is much smaller than the electric force.

Note that it is essential to use the same frame of reference in this entire calculation. We have described the velocities and the fields as they appear to an observer who is stationary in the coordinate system of Fig. 29–2. In a coordinate system that moves with one of the charges, one of the velocities would be zero, so

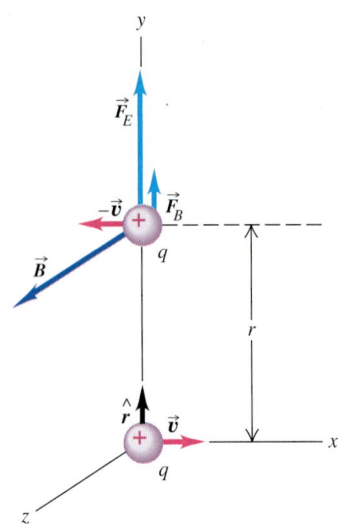

29–2 Electric and magnetic forces on two protons.

there would be *no* magnetic force. The explanation of this apparent paradox provided one of the paths that led to the special theory of relativity.

29–3 MAGNETIC FIELD OF A CURRENT ELEMENT

Just as for the electric field, there is a **principle of superposition of magnetic fields: The total magnetic field caused by several moving charges is the vector sum of the fields caused by the individual charges.** We can use this principle with the results of Section 29–2 to find the magnetic field produced by a current in a conductor.

We begin by calculating the magnetic field caused by a short segment $d\vec{l}$ of a current-carrying conductor, as shown in Fig. 29–3a. The volume of the segment is $A\,dl$, where A is the cross-section area of the conductor. If there are n moving charged particles per unit volume, each of charge q, the total moving charge dQ in the segment is

$$dQ = nqA\,dl.$$

The moving charges in this segment are equivalent to a single charge dQ, traveling with a velocity equal to the *drift* velocity $\vec{v}_d$. (Magnetic fields due to the *random* motions of the charges will, on average, cancel out at every point.) From Eq. (29–1) the magnitude

29–3 (a) Magnetic-field vectors due to a current element $d\vec{l}$. (b) Magnetic field lines in a plane containing the current element $d\vec{l}$. The current is directed into the plane of the page. Compare this figure to Fig. 29–1 for the field of a moving point charge.

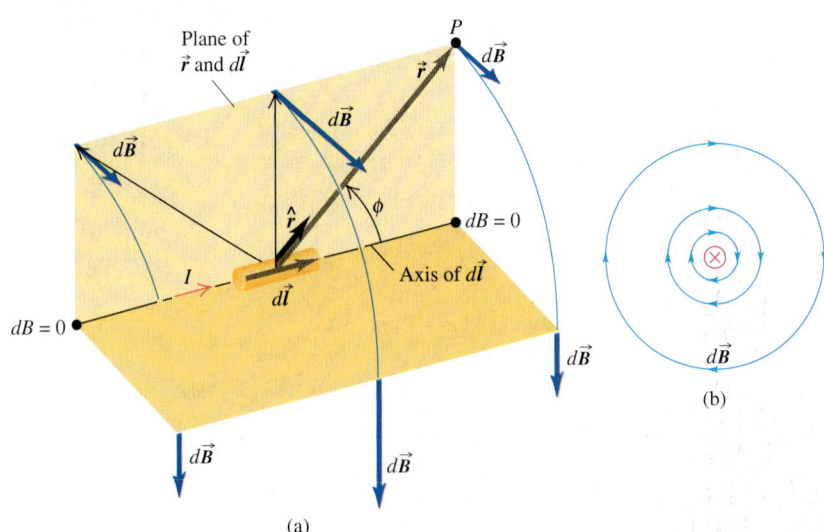

of the resulting field $d\vec{B}$ at any field point P is

$$dB = \frac{\mu_0}{4\pi} \frac{|dQ|v_\text{d} \sin \phi}{r^2} = \frac{\mu_0}{4\pi} \frac{n|q|v_\text{d} A \, dl \sin \phi}{r^2}.$$

But from Eq. (26–2), $n|q|v_\text{d} A$ equals the current I in the element. So

$$dB = \frac{\mu_0}{4\pi} \frac{I \, dl \sin \phi}{r^2}. \tag{29–5}$$

In vector form, using the unit vector $\hat{\boldsymbol{r}}$ as in Section 29–2, we have

$$d\vec{B} = \frac{\mu_0}{4\pi} \frac{I \, d\vec{l} \times \hat{\boldsymbol{r}}}{r^2} \qquad \text{(magnetic field of a current element)}, \tag{29–6}$$

where $d\vec{l}$ is a vector with length dl, in the same direction as the current in the conductor.

Equations (29–5) and (29–6) are called the **law of Biot and Savart** (pronounced "Bee-oh" and "Suh-var"). We can use this law to find the total magnetic field $\vec{B}$ at any point in space due to the current in a complete circuit. To do this, we integrate Eq. (29–6) over all segments $d\vec{l}$ that carry current; symbolically,

$$\vec{B} = \frac{\mu_0}{4\pi} \int \frac{I \, d\vec{l} \times \hat{\boldsymbol{r}}}{r^2}. \tag{29–7}$$

In the following sections we will carry out this vector integration for several examples.

As Fig. 29–3a shows, the field vectors $d\vec{B}$ and the magnetic field lines of a current element are exactly like those set up by a positive charge dQ moving in the direction of the drift velocity $\vec{v}_\text{d}$. The field lines are circles in planes perpendicular to $d\vec{l}$ and centered on the line of $d\vec{l}$. Their directions are given by the same right-hand rule that we introduced for point charges in Section 29–2.

We can't verify Eq. (29–5) or (29–6) directly because we can never experiment with an isolated segment of a current-carrying circuit. What we measure experimentally is the *total* $\vec{B}$ for a complete circuit. But we can still verify these equations indirectly by calculating $\vec{B}$ for various current configurations using Eq. (29–7) and comparing the results with experimental measurements.

If matter is present in the space around a current-carrying conductor, the field at a field point P in its vicinity will have an additional contribution resulting from the *magnetization* of the material. We'll return to this point in Section 29–9. However, unless the material is iron or some other ferromagnetic material, the additional field is small and is usually negligible. Additional complications arise if time-varying electric or magnetic fields are present or if the material is a superconductor; we'll return to these topics later.

Computer cables, or cables for audio-video equipment, create little or no magnetic field. This is because within each cable, closely spaced wires carry current in both directions along the length of the cable. The magnetic fields from these opposing currents cancel each other (see Example 29–4 in the next section).

Problem–Solving Strategy

MAGNETIC FIELD CALCULATIONS

1. Be careful about the directions of vector quantities. The current element $d\vec{l}$ always points in the direction of the current. The unit vector $\hat{\boldsymbol{r}}$ is always directed *from* the current element (the source point) *toward* the point P at which the field is to be determined (the field point).

2. In some situations the $d\vec{B}$'s at point P have the same direction for all the current elements; then the magnitude of the total $\vec{B}$ field is the sum of the magnitudes of the $d\vec{B}$'s. But often the $d\vec{B}$'s have different directions for different current elements. Then you have to set up a coordinate system and represent each $d\vec{B}$ in terms of its components. The integral for the total $\vec{B}$ is then expressed in terms of an integral for each component. Sometimes you can use the symmetry of the situation to prove that one component must vanish. Always be alert for ways to use symmetry to simplify the problem.

3. Look for ways to use the principle of superposition of magnetic fields. Later in this chapter we'll determine the fields produced by certain simple conductor shapes. If you encounter a conductor of a complex shape that can be represented as a combination of these simple shapes, you can use superposition to find the field of the complex shape. Examples include a rectangular loop and a semicircle with straight-line segments on both sides.

EXAMPLE 29-2

Magnetic field of a current segment A copper wire carries a steady current of 125 A to an electroplating tank. Find the magnetic field caused by a 1.0-cm segment of this wire at a point 1.2 m away from it, if the point is a) point P_1, straight out to the side of the segment; b) point P_2, on a line at 30° to the segment, as shown in Fig. 29–4.

SOLUTION a) Although Eqs. (29–5) and (29–6) are strictly to be used with infinitesimal current elements only, we may use them here since the segment's 1.0-cm length is much smaller than the 1.2-m distance to the field point. From the right-hand rule, the direction of $\vec{B}$ at P_1 is *into* the xy-plane of Fig. 29–4. Or, using unit vectors, we note that $d\vec{l} = dl(-\hat{\imath})$. At point P_1, $\hat{r} = \hat{\jmath}$, so in Eq. (29–6),

$$d\vec{l} \times \hat{r} = dl(-\hat{\imath}) \times \hat{\jmath} = dl(-\hat{k}).$$

The negative z-direction is *into* the plane.

To find the magnitude of $\vec{B}$, we use Eq. (29–5). At point P_1,

$$B = \frac{\mu_0}{4\pi} \frac{I\, dl \sin\phi}{r^2} = (10^{-7}\ \text{T·m/A}) \frac{(125\ \text{A})(1.0 \times 10^{-2}\ \text{m})(\sin 90°)}{(1.2\ \text{m})^2}$$

$$= 8.7 \times 10^{-8}\ \text{T}.$$

b) At point P_2 the direction of $\vec{B}$ is again into the xy-plane of the figure. The magnitude is

29–4 Finding the magnetic field at two points due to a 1.0-cm segment of current-carrying wire (not shown to scale).

$$B = (10^{-7}\ \text{T·m/A}) \frac{(125\ \text{A})(1.0 \times 10^{-2}\ \text{m})(\sin 30°)}{(1.2\ \text{m})^2}$$

$$= 4.3 \times 10^{-8}\ \text{T}.$$

Note that these magnetic-field magnitudes are very small; for comparison the magnetic field of the earth is of the order of 10^{-4} T. Note also that the values are not the *total* fields at points P_1 and P_2, but only the contributions from the short segment of conductor described.

29–4 MAGNETIC FIELD OF A STRAIGHT CURRENT-CARRYING CONDUCTOR

An important application of the law of Biot and Savart is finding the magnetic field produced by a straight current-carrying conductor. This result is useful because straight conducting wires are found in essentially all electric and electronic devices. Figure 29–5 shows such a conductor with length $2a$ carrying a current I. We will find $\vec{B}$ at a point a distance x from the conductor on its perpendicular bisector.

We first use the law of Biot and Savart, Eq. (29–5), to find the field $d\vec{B}$ caused by the element of conductor of length $dl = dy$ shown in Fig. 29–5. From the figure, $r = \sqrt{x^2 + y^2}$ and $\sin\phi = \sin(\pi - \phi) = x/\sqrt{x^2 + y^2}$. The right-hand rule for the vector product $d\vec{l} \times \hat{r}$ shows that the *direction* of $d\vec{B}$ is into the plane of the figure, perpendicular to the plane; furthermore, the directions of the $d\vec{B}$'s from *all* elements of the conductor are the same. Thus in integrating Eq. (29–7), we can just add the *magnitudes* of the $d\vec{B}$'s, a significant simplification.

Putting the pieces together, we find that the magnitude of the total $\vec{B}$ field is

$$B = \frac{\mu_0 I}{4\pi} \int_{-a}^{a} \frac{x\, dy}{(x^2 + y^2)^{3/2}}.$$

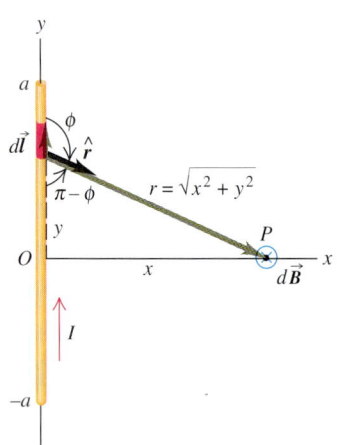

29–5 Magnetic field produced by a straight current-carrying conductor of length $2a$. At point P, $\vec{B}$ is directed into the plane of the page.

We can integrate this by trigonometric substitution or by using an integral table. The final result is

$$B = \frac{\mu_0 I}{4\pi} \frac{2a}{x\sqrt{x^2 + a^2}}. \qquad (29\text{–}8)$$

When the length $2a$ of the conductor is very great in comparison to its distance x from point P, we can consider it to be infinitely long. When a is much larger than x,

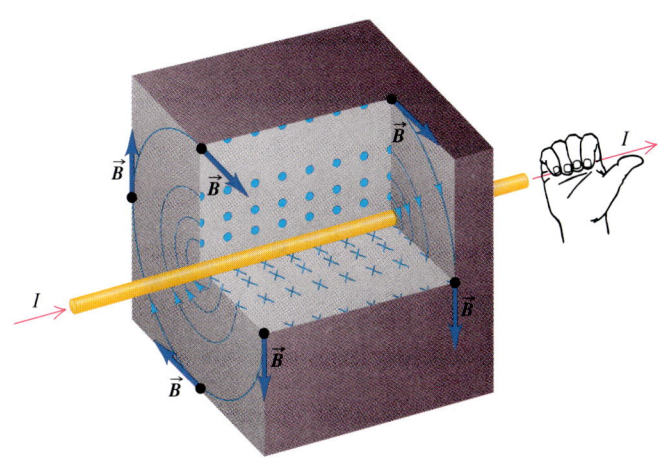

29–6 Magnetic field around a long, straight conductor. The field lines are circles, with directions determined by the right-hand rule.

$\sqrt{x^2 + a^2}$ is approximately equal to a; hence in the limit $a \to \infty$, Eq. (29–8) becomes

$$B = \frac{\mu_0 I}{2\pi x}.$$

The physical situation has axial symmetry about the y-axis. Hence $\vec{B}$ must have the same *magnitude* at all points on a circle centered on the conductor and lying in a plane perpendicular to it, and the *direction* of $\vec{B}$ must be everywhere tangent to such a circle. Thus at all points on a circle of radius r around the conductor, the magnitude B is

$$B = \frac{\mu_0 I}{2\pi r} \qquad \text{(a long, straight, current-carrying conductor).} \qquad (29\text{–}9)$$

Part of the magnetic field around a long, straight, current-carrying conductor is shown in Fig. 29–6.

The geometry of this problem is similar to that of Example 22–11 (Section 22–7), in which we solved the problem of the *electric* field caused by an infinite line of charge. The same integral appears in both problems, and the field magnitudes in both problems are proportional to $1/r$. But the lines of $\vec{B}$ in the magnetic problem have completely different shapes than the lines of $\vec{E}$ in the analogous electrical problem. Electric field lines radiate outward from a positive line charge distribution (inward for negative charges). By contrast, magnetic field lines *encircle* the current that acts as their source. Electric field lines due to charges begin and end at those charges, but magnetic field lines always form closed loops and *never* have end points, irrespective of the shape of the current-carrying conductor that sets up the field. As we discussed in Section 28–4, this is a consequence of Gauss's law for magnetism, which states that the total magnetic flux through *any* closed surface is always zero:

$$\oint \vec{B} \cdot d\vec{A} = 0 \qquad \text{(magnetic flux through any closed surface).} \qquad (29\text{–}10)$$

This implies that there are no isolated magnetic charges or magnetic monopoles. Any magnetic field line that enters a closed surface must also emerge from that surface.

ActivPhysics

13.1
Magnetic Field of a Wire

EXAMPLE 29–3

A long, straight conductor carries a current of 100 A. At what distance from the axis of the conductor is the magnetic field caused by the current equal in magnitude to the earth's magnetic field in Pittsburgh (about 0.5×10^{-4} T)?

SOLUTION We use Eq. (29–9). Everything except r is known, so

we solve for r and insert the appropriate numbers:

$$r = \frac{\mu_0 I}{2\pi B} = \frac{(4\pi \times 10^{-7}\ \text{T} \cdot \text{m/A})(100\ \text{A})}{(2\pi)(0.5 \times 10^{-4}\ \text{T})} = 0.4\ \text{m}.$$

At smaller distances the field becomes stronger; for example, when $r = 0.2$ m, $B = 1.0 \times 10^{-4}$ T, and so on.

EXAMPLE 29-4

Figure 29–7a is an end view of two long, straight, parallel wires perpendicular to the xy-plane, each carrying a current I but in opposite directions. a) Find the magnitude and direction of $\vec{B}$ at points P_1, P_2, and P_3. b) Find the magnitude and direction of $\vec{B}$ at any point on the x-axis to the right of wire 2 in terms of the x-coordinate of the point.

SOLUTION a) We can use Eq. (29–9) to find the magnitude of the fields $\vec{B}_1$ (due to wire 1) and $\vec{B}_2$ (due to wire 2) at any point. We can then use the principle of superposition of magnetic fields to find the *total* magnetic field at that point: $\vec{B}_{total} = \vec{B}_1 + \vec{B}_2$.

Point P_1 is closer to wire 1 (distance $2d$) than to wire 2 (distance $4d$), so at this point the field magnitude B_1 is greater than the magnitude B_2:

$$B_1 = \frac{\mu_0 I}{2\pi(2d)} = \frac{\mu_0 I}{4\pi d}, \qquad B_2 = \frac{\mu_0 I}{2\pi(4d)} = \frac{\mu_0 I}{8\pi d}.$$

The right-hand rule shows that $\vec{B}_1$ is in the negative y-direction and $\vec{B}_2$ is in the positive y-direction. Since B_1 is the larger magnitude, the total field $\vec{B}_{total} = \vec{B}_1 + \vec{B}_2$ is in the negative y-direction, with magnitude

$$B_{total} = B_1 - B_2 = \frac{\mu_0 I}{4\pi d} - \frac{\mu_0 I}{8\pi d} = \frac{\mu_0 I}{8\pi d} \qquad (\text{point } P_1).$$

At point P_2, a distance d from both wires, $\vec{B}_1$ and $\vec{B}_2$ are both in the positive y-direction, and both have the same magnitude:

$$B_1 = B_2 = \frac{\mu_0 I}{2\pi d},$$

so $\vec{B}_{total}$ is also in the positive y-direction and has magnitude

$$B_{total} = B_1 + B_2 = \frac{\mu_0 I}{\pi d} \qquad (\text{point } P_2).$$

Finally, at point P_3 the right-hand rule shows that $\vec{B}_1$ is in the positive y-direction and $\vec{B}_2$ is in the negative y-direction. This point is farther from wire 1 (distance $3d$) than from wire 2 (distance d), so B_1 is less than B_2:

$$B_1 = \frac{\mu_0 I}{2\pi(3d)} = \frac{\mu_0 I}{6\pi d}, \qquad B_2 = \frac{\mu_0 I}{2\pi d}.$$

The total field is in the negative y-direction, the same as $\vec{B}_2$, and has magnitude

$$B_{total} = B_2 - B_1 = \frac{\mu_0 I}{2\pi d} - \frac{\mu_0 I}{6\pi d} = \frac{\mu_0 I}{3\pi d} \qquad (\text{point } P_3).$$

You should be able to use the right-hand rule to verify for yourself the directions of $\vec{B}_1$ and $\vec{B}_2$ for each point.

The fields $\vec{B}_1$, $\vec{B}_2$, and $\vec{B}_{total}$ at each of the three points are shown in Fig. 29–7a. The same technique can be used to find $\vec{B}_{total}$ at any point; for points off the x-axis, caution must be taken in vector addition, since $\vec{B}_1$ and $\vec{B}_2$ need no longer be simply parallel or antiparallel (see Problems 29–46 and 29–47). Figure 29–7b shows some of the magnetic field lines due to this combination of wires.

b) At any point to the right of wire 2 (that is, for $x > d$), $\vec{B}_1$ and $\vec{B}_2$ are in the same directions as at P_3. As x increases, both $\vec{B}_1$ and $\vec{B}_2$ decrease in magnitude, so $\vec{B}_{total}$ must decrease as well. The

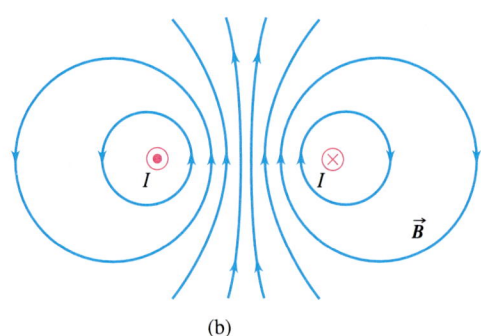

29–7 (a) Two long, straight conductors carrying equal currents in opposite directions. The conductors are seen end-on. (b) Map of the magnetic field produced by the two conductors. The field lines are closest together between the conductors, where the field is strongest.

magnitudes of the fields due to each wire are

$$B_1 = \frac{\mu_0 I}{2\pi(x + d)} \qquad \text{and} \qquad B_2 = \frac{\mu_0 I}{2\pi(x - d)}.$$

At any field point to the right of wire 2, wire 2 is closer than wire 1, and so $B_2 > B_1$. Hence $\vec{B}_{total}$ is in the negative y-direction, the same as $\vec{B}_2$, and has magnitude

$$B_{total} = B_2 - B_1 = \frac{\mu_0 I}{2\pi(x - d)} - \frac{\mu_0 I}{2\pi(x + d)} = \frac{\mu_0 Id}{\pi(x^2 - d^2)},$$

where we combined the two terms using a common denominator.

At points very far from the wires, so that x is much larger than d, the d^2 term in the denominator can be neglected, and

$$B_{total} = \frac{\mu_0 Id}{\pi x^2}.$$

The magnetic-field magnitude for a single wire decreases with distance in proportion to $1/x$, as shown by Eq. (29–9); for two wires carrying opposite currents, $\vec{B}_1$ and $\vec{B}_2$ partially cancel each other, and so the magnitude of $\vec{B}_{total}$ decreases more rapidly, in proportion to $1/x^2$. This effect is used in communication systems such as telephone or computer networks. The wiring is arranged so that a conductor carrying a signal in one direction and the conductor carrying the return signal are side by side, as in Fig. 29–7a, or twisted around each other. As a result, the magnetic field caused *outside* the conductors by these signals is greatly reduced and is less likely to exert unwanted forces on other information-carrying currents.

29–5 FORCE BETWEEN PARALLEL CONDUCTORS

In Example 29–4 (Section 29–4) we showed how to use the principle of superposition of magnetic fields to find the total field due to two long current-carrying conductors. Another important aspect of this configuration is the *interaction force* between the conductors. This force plays a role in many practical situations in which current-carrying wires are close to each other, and it also has fundamental significance in connection with the definition of the ampere. Figure 29–8 shows segments of two long, straight parallel conductors separated by a distance r and carrying currents I and I', respectively, in the same direction. Each conductor lies in the magnetic field set up by the other, so each experiences a force. The diagram shows some of the field lines set up by the current in the lower conductor.

From Eq. (29–9) the lower conductor produces a $\vec{B}$ field that, at the position of the upper conductor, has magnitude

$$B = \frac{\mu_0 I}{2\pi r}.$$

From Eq. (28–19) the force that this field exerts on a length L of the upper conductor is $\vec{F} = I'\vec{L} \times \vec{B}$, where the vector $\vec{L}$ is in the direction of the current I' and has magnitude L. Since $\vec{B}$ is perpendicular to the length of the conductor and hence to $\vec{L}$, the magnitude of this force is

$$F = I'LB = \frac{\mu_0 II'L}{2\pi r},$$

and the force *per unit length F/L* is

$$\frac{F}{L} = \frac{\mu_0 II'}{2\pi r} \quad \text{(two long, parallel, current-carrying conductors)}. \quad (29\text{–}11)$$

Applying the right-hand rule to $\vec{F} = I'\vec{L} \times \vec{B}$ shows that the force on the upper conductor is directed *downward*.

29–8 Parallel conductors carrying currents in the same direction attract each other. The diagrams show the force exerted on the upper conductor by the magnetic field caused by the current in the lower conductor.

(a)   (b)

The current in the upper conductor also sets up a field at the position of the lower one. Two successive applications of the right-hand rule for vector products (one to find the direction of the $\vec{B}$ field due to the upper conductor, as in Section 29–3, and one to find the direction of the force that this field exerts on the lower conductor, as in Section 28–7) show that the force on the lower conductor is *upward*. Thus *two parallel conductors carrying current in the same direction attract each other.* If the direction of either current is reversed, the forces also reverse. *Parallel conductors carrying currents in opposite directions repel each other.*

The attraction or repulsion between two straight, parallel, current-carrying conductors is the basis of the official SI definition of the **ampere:**

> ***One ampere*** **is that unvarying current that, if present in each of two parallel conductors of infinite length and one meter apart in empty space, causes each conductor to experience a force of exactly 2×10^{-7} newtons per meter of length.**

From Eq. (29–11) you can see that this definition of the ampere is what leads us to choose the value of $4\pi \times 10^{-7}$ T · m/A for μ_0. It also forms the basis of the SI definition of the coulomb, which is the amount of charge transferred in one second by a current of one ampere.

This is an *operational definition;* it gives us an actual experimental procedure for measuring current and defining a unit of current. In principle we could use this definition to calibrate an ammeter, using only a meter stick and a spring balance. For high-precision standardization of the ampere, coils of wire are used instead of straight wires, and their separation is only a few centimeters. The complete instrument, which is capable of measuring currents with a high degree of precision, is called a *current balance.*

Mutual forces of attraction exist not only between *wires* carrying currents in the same direction, but also between the longitudinal elements of a single current-carrying conductor. If the conductor is a liquid or an ionized gas (a plasma), these forces result in a constriction of the conductor, as if its surface were acted on by an inward pressure. The constriction of the conductor is called the *pinch effect.* The high temperature produced by the pinch effect in a plasma has been used in one technique to bring about nuclear fusion.

EXAMPLE 29–5

Two straight, parallel, superconducting cables 4.5 mm apart carry equal currents of 15,000 A in opposite directions. Should we worry about the mechanical strength of these wires?

SOLUTION Because the currents are in opposite directions, the two conductors repel each other. From Eq. (29–11) the force per unit length is

$$\frac{F}{L} = \frac{\mu_0 I I'}{2\pi r} = \frac{(4\pi \times 10^{-7} \text{ T} \cdot \text{m/A})(15{,}000 \text{ A})^2}{(2\pi)(4.5 \times 10^{-3} \text{ m})}$$

$$= 1.0 \times 10^4 \text{ N/m}.$$

This is a large force, more than one ton per meter, so mechanical strength of the conductors and insulating materials are certainly a significant consideration. Currents and separations of this magnitude are used in superconducting electromagnets in particle accelerators, and mechanical stress analysis is a crucial part of the design process.

EXAMPLE 29–6

Mechanical stress in a solenoid A *solenoid,* introduced in Section 28–8, is a wire wound into a helical coil. For a solenoid that carries a current I (Fig. 29–9), discuss the directions of the magnetic forces a) between adjacent turns of the coil; b) between opposite sides of the same turn.

SOLUTION a) Adjacent turns carry parallel currents and therefore attract one another, causing a *compressive* stress along the solenoid axis.
b) Opposite sides of the same turn carry current in opposite directions and thus repel each other. Therefore the coil experi-

29-9 A current-carrying solenoid.

ences forces that pull radially outward, creating *tensile* stress in the conductors. These forces are proportional to the square of the current; as we mentioned above, they are an important consideration in electromagnet design. We'll discuss the nature of the *total* magnetic field produced by a solenoid in Section 29–8.

29-6 MAGNETIC FIELD OF A CIRCULAR CURRENT LOOP

If you look inside a doorbell, a transformer, or an electric motor, you will find coils of wire with a large number of turns, spaced so closely that each turn is very nearly a planar circular loop. A current in such a coil is used to establish a magnetic field. So it is worthwhile to derive an expression for the magnetic field produced by a single circular conducting loop carrying a current or by *N* closely spaced circular loops forming a coil. In Section 28–8 we considered the force and torque on such a current loop placed in an external magnetic field produced by other currents; we are now about to find the magnetic field produced by the loop itself.

Figure 29–10 shows a circular conductor with radius *a* that carries a current *I*. The current is led into and out of the loop through two long, straight wires side by side; the currents in these straight wires are in opposite directions, and their magnetic fields very nearly cancel each other (see Example 29–4 in Section 29–4).

We can use the law of Biot and Savart, Eq. (29–5) or (29–6), to find the magnetic field at a point *P* on the axis of the loop, at a distance *x* from the center. As the figure shows, $d\vec{l}$ and $\hat{r}$ are perpendicular, and the direction of the field $d\vec{B}$ caused by this particular element $d\vec{l}$ lies in the *xy*-plane. Since $r^2 = x^2 + a^2$, the magnitude *dB* of the field due to element $d\vec{l}$ is

$$dB = \frac{\mu_0 I}{4\pi} \frac{dl}{(x^2 + a^2)}. \qquad (29\text{--}12)$$

The components of the vector $d\vec{B}$ are

$$dB_x = dB \cos\theta = \frac{\mu_0 I}{4\pi} \frac{dl}{(x^2 + a^2)} \frac{a}{(x^2 + a^2)^{1/2}}, \qquad (29\text{--}13)$$

$$dB_y = dB \sin\theta = \frac{\mu_0 I}{4\pi} \frac{dl}{(x^2 + a^2)} \frac{x}{(x^2 + a^2)^{1/2}}. \qquad (29\text{--}14)$$

The situation has rotational symmetry about the *x*-axis, so there cannot be a component of the total field $\vec{B}$ perpendicular to this axis. For every element $d\vec{l}$ there is a corresponding element on the opposite side of the loop, with opposite direction. These

This electromagnet contains a current-carrying coil with numerous turns of wire. The resulting magnetic field is so strong that the electromagnet can pick up hundreds of steel cans at once.

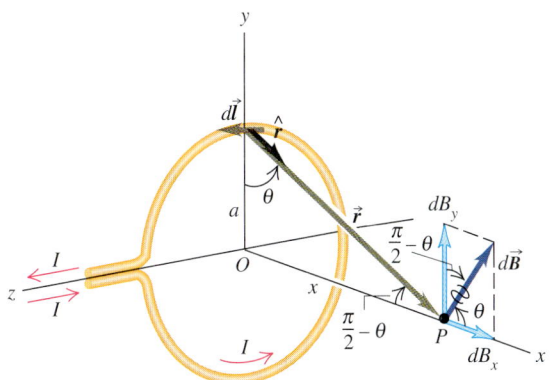

29-10 Magnetic field of a circular loop. The current in the segment $d\vec{l}$ causes the field $d\vec{B}$, which lies in the *xy*-plane. The currents in other $d\vec{l}$'s cause $d\vec{B}$'s with different components perpendicular to the *x*-axis; these add to zero. The *x*-components of the $d\vec{B}$'s combine to give the total $\vec{B}$ field at point *P*.

two elements give equal contributions to the x-component of $d\vec{B}$, given by Eq. (29–13), but *opposite* components perpendicular to the x-axis. Thus all the perpendicular components cancel, and only the x-components survive.

To obtain the x-component of the total field $\vec{B}$, we integrate Eq. (29–13), including all the $d\vec{l}$'s around the loop. Everything in this expression except dl is constant and can be taken outside the integral, and we have

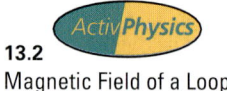

13.2
Magnetic Field of a Loop

$$B_x = \int \frac{\mu_0 I}{4\pi} \frac{a\, dl}{(x^2 + a^2)^{3/2}} = \frac{\mu_0 I a}{4\pi(x^2 + a^2)^{3/2}} \int dl.$$

The integral of dl is just the circumference of the circle, $\int dl = 2\pi a$, and we finally get

$$B_x = \frac{\mu_0 I a^2}{2(x^2 + a^2)^{3/2}} \qquad \text{(on the axis of a circular loop).} \qquad (29\text{–}15)$$

Now suppose that instead of the single loop in Fig. 29–10 we have a coil consisting of N loops, all with the same radius. The loops are closely spaced so that the plane of each loop is essentially the same distance x from the field point P. Each loop contributes equally to the field, and the total field is N times the field of a single loop:

$$B_x = \frac{\mu_0 N I a^2}{2(x^2 + a^2)^{3/2}} \qquad \text{(on the axis of N circular loops).} \qquad (29\text{–}16)$$

The factor N in Eq. (29–16) is the reason why coils of wire, not single loops, are used to produce strong magnetic fields; for a desired field strength, using a single loop might require a current I so great as to exceed the rating of the loop's wire.

Figure 29–11 shows a graph of B_x as a function of x. The maximum value of the field is at $x = 0$, the center of the loop or coil:

$$B_x = \frac{\mu_0 N I}{2a} \qquad \text{(at the center of N circular loops).} \qquad (29\text{–}17)$$

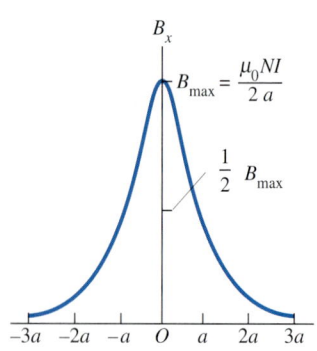

29–11 Graph of the magnetic field along the axis of a circular coil with N turns. When x is much larger than a, the field magnitude decreases approximately as $1/x^3$.

As we go out along the axis, the field decreases in magnitude.

In Section 28–8 we defined the *magnetic dipole moment μ* (or *magnetic moment*) of a current-carrying loop to be equal to IA, where A is the cross-section area of the loop. If there are N loops, the total magnetic moment is NIA. The circular loop in Fig. 29–10 has area $A = \pi a^2$, so the magnetic moment of a single loop is $\mu = I\pi a^2$; for N loops, $\mu = NI\pi a^2$. Substituting these results into Eqs. (29–15) and (29–16), we find that both of these expressions can be written as

$$B_x = \frac{\mu_0 \mu}{2\pi(x^2 + a^2)^{3/2}} \qquad \text{(on the axis of any number of circular loops).} \qquad (29\text{–}18)$$

We described a magnetic dipole in Section 28–8 in terms of its response to a magnetic field produced by currents outside the dipole. But a magnetic dipole is also a *source* of magnetic field; Eq. (29–18) describes the magnetic field *produced* by a magnetic dipole for points along the dipole axis. This field is directly proportional to the magnetic dipole moment μ. Note that the field along the x-axis is in the same direction as the vector magnetic moment $\vec{\mu}$; this is true on both the positive and negative x-axis.

CAUTION ▶ Equations (29–15), (29–16), and (29–18) are valid only on the *axis* of a loop or coil. Don't attempt to apply these equations at other points! ◀

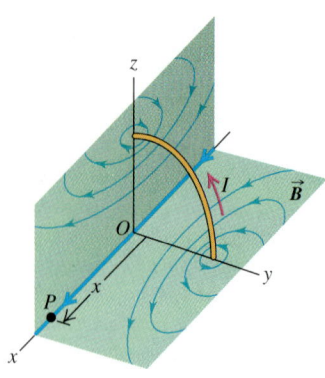

29–12 Magnetic field lines produced by the current in a circular loop. At points on the axis the $\vec{B}$ field has the same direction as the magnetic moment of the loop.

Some of the magnetic field lines surrounding a circular current loop (magnetic dipole), in planes through the axis, are shown in Fig. 29–12. The directions of the field lines are given by the same right-hand rule as for a long, straight conductor. Grab the conductor with your right hand, with your thumb in the direction of the current; your fingers curl around in the same direction as the field lines. The field lines for the circular current loop are closed curves that encircle the conductor; they are *not* circles, however.

EXAMPLE 29–7

A coil consisting of 100 circular loops with radius 0.60 m carries a current of 5.0 A. a) Find the magnetic field at a point along the axis of the coil, 0.80 m from the center. b) Along the axis, at what distance from the center is the field magnitude 1/8 as great as it is at the center?

SOLUTION a) In Eq. (29–16) we have $I = 5.0$ A, $a = 0.60$ m, and $x = 0.80$ m. We find

$$B_x = \frac{(4\pi \times 10^{-7} \text{ T} \cdot \text{m/A})(100)(5.0 \text{ A})(0.60 \text{ m})^2}{2[(0.80 \text{ m})^2 + (0.60 \text{ m})^2]^{3/2}} = 1.1 \times 10^{-4} \text{ T}.$$

Alternatively, we can first find the magnetic moment, then substitute this into Eq. (29–18):

$$\mu = NI\pi a^2 = (100)(5.0 \text{ A})\pi(0.60 \text{ m})^2 = 5.7 \times 10^2 \text{ A} \cdot \text{m}^2,$$

$$B_x = \frac{(4\pi \times 10^{-7} \text{ T} \cdot \text{m/A})(5.7 \times 10^2 \text{ A} \cdot \text{m}^2)}{2\pi[(0.80 \text{ m})^2 + (0.60 \text{ m})^2]^{3/2}} = 1.1 \times 10^{-4} \text{ T}.$$

The magnetic moment μ is relatively large, yet this is a rather small field, comparable in magnitude to the earth's magnetic field. This problem may give you some idea of the difficulty of producing a field of 1 T or more.

b) Considering Eq. (29–16), we want to find a value of x such that

$$\frac{1}{(x^2 + a^2)^{3/2}} = \frac{1}{8} \frac{1}{(0^2 + a^2)^{3/2}}.$$

To solve this for x, we take the reciprocal of the whole thing, then take the 2/3 power of both sides; the result is

$$x = \pm\sqrt{3}\, a = \pm 1.04 \text{ m}.$$

At a distance of about 1.7 radii from the center, the field has dropped off to 1/8 its value at the center.

29–7 AMPERE'S LAW

So far our calculations of the magnetic field due to a current have involved finding the infinitesimal field $d\vec{B}$ due to a current element, then summing all the $d\vec{B}$'s to find the total field. This approach is directly analogous to our *electric*-field calculations in Chapter 22. For the electric-field problem we found that in situations with a highly symmetric charge distribution it was often easier to use Gauss's law to find $\vec{E}$.

There is likewise a law that allows us to more easily find the *magnetic* fields caused by highly symmetric *current* distributions. But the law that allows us to do this, called *Ampere's law,* is rather different in character from Gauss's law.

Gauss's law for electric fields involves the flux of $\vec{E}$ through a closed surface; it states that this flux is equal to the total charge enclosed within the surface, divided by the constant ϵ_0. Thus this law relates electric fields and charge distributions. By contrast, Gauss's law for *magnetic* fields, Eq. (29–10), is *not* a relation between magnetic fields and current distributions; it states that the flux of $\vec{B}$ through *any* closed surface is always zero, whether or not there are currents within the surface. So Gauss's law for $\vec{B}$ can't be used to determine the magnetic field produced by a particular current distribution.

Ampere's law is formulated not in terms of magnetic flux, but rather in terms of the *line integral* of $\vec{B}$ around a closed path, denoted by

$$\oint \vec{B} \cdot d\vec{l}.$$

We used line integrals to define work in Chapter 6 and to calculate electric potential in Chapter 24. To evaluate this integral, we divide the path into infinitesimal segments $d\vec{l}$, calculate the scalar product of $\vec{B} \cdot d\vec{l}$ for each segment, and sum these products. In general, $\vec{B}$ varies from point to point, and we must use the value of $\vec{B}$ at the location of each $d\vec{l}$. An alternative notation is $\oint B_{\parallel} dl$, where $B_{\parallel}$ is the component of $\vec{B}$ parallel to $d\vec{l}$ at each point. The circle on the integral sign indicates that this integral is always computed for a *closed* path, one whose beginning and end points are the same.

To introduce the basic idea of Ampere's law, let's consider again the magnetic field caused by a long, straight conductor carrying a current I. We found in Section 29–4 that the field at a distance r from the conductor has magnitude

$$B = \frac{\mu_0 I}{2\pi r}$$

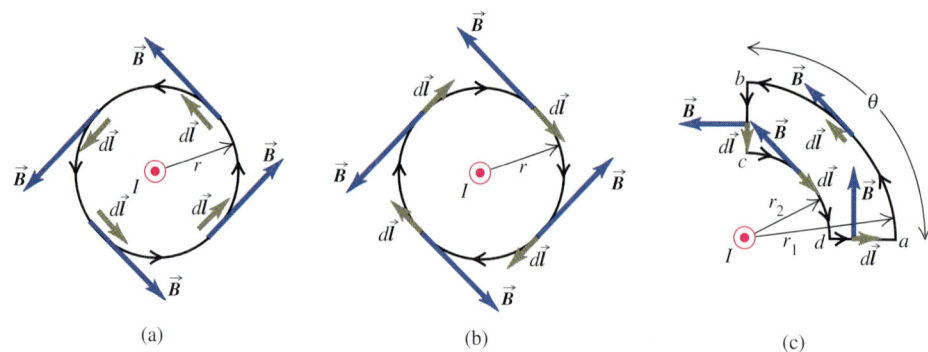

(a)   (b)   (c)

29-13 Some integration paths for the line integral of $\vec{B}$ in the vicinity of a long, straight conductor carrying current I *out* of the plane of the page (as indicated by the circle with a dot). The conductor is seen end-on. (a) The integration path is a circle centered on the conductor; the integration goes around the circle counterclockwise. (b) Same as (a), but now the integration goes around the circle clockwise. (c) An integration path that does not enclose the conductor.

and that the magnetic field lines are circles centered on the conductor. Let's take the line integral of $\vec{B}$ around one such circle with radius r, as in Fig. 29–13a. At every point on the circle, $\vec{B}$ and $d\vec{l}$ are parallel, and so $\vec{B} \cdot d\vec{l} = B\,dl$; since r is constant around the circle, B is constant as well. Alternatively, we can say that $B_{\parallel}$ is constant and equal to B at every point on the circle. Hence we can take B outside of the integral. The remaining integral $\oint dl$ is just the circumference of the circle, so

$$\oint \vec{B} \cdot d\vec{l} = \oint B_{\parallel}\,dl = B\oint dl = \frac{\mu_0 I}{2\pi r}(2\pi r) = \mu_0 I.$$

The line integral is thus independent of the radius of the circle and is equal to μ_0 multiplied by the current passing through the area bounded by the circle.

In Fig. 29–13b the situation is the same, but the integration path now goes around the circle in the opposite direction. Now $\vec{B}$ and $d\vec{l}$ are antiparallel, so $\vec{B} \cdot d\vec{l} = -B\,dl$, and the line integral equals $-\mu_0 I$. We get the same result if the integration path is the same as in Fig. 29–13a, but the direction of the current is reversed. Thus the line integral $\oint \vec{B} \cdot d\vec{l}$ equals μ_0 multiplied by the current passing through the area bounded by the integration path, with a positive or negative sign depending on the direction of the current relative to the direction of integration.

There's a simple rule for the sign of the current; you won't be surprised to learn that it uses your right hand. Curl the fingers of your right hand around the integration path so that they curl in the direction of integration (that is, the direction that you use to evaluate $\oint \vec{B} \cdot d\vec{l}$). Then your right thumb indicates the positive current direction. Currents that pass through in this direction are positive; those in the opposite direction are negative. Using this rule, you should be able to convince yourself that the current is positive in Fig. 29–13a and negative in Fig. 29–13b. Here's another way to say the same thing: Looking at the surface bounded by the integration path, integrate counterclockwise around the path as in Fig. 29–13a. Currents moving toward you through the surface are positive, and those going away from you are negative.

An integration path that does *not* enclose the conductor is used in Fig. 29–13c. Along the circular arc ab of radius r_1, $\vec{B}$ and $d\vec{l}$ are parallel, and $B_{\parallel} = B_1 = \mu_0 I/2\pi r_1$; along the circular arc cd of radius r_2, $\vec{B}$ and $d\vec{l}$ are antiparallel and $B_{\parallel} = -B_2 = -\mu_0 I/2\pi r_2$. The $\vec{B}$ field is perpendicular to $d\vec{l}$ at each point on the straight sections bc and da, so $B_{\parallel} = 0$ and these sections contribute zero to the line integral. The total line integral is then

$$\oint \vec{B} \cdot d\vec{l} = \oint B_{\parallel}\,dl = B_1 \int_a^b dl + (0)\int_b^c dl + (-B_2)\int_c^d dl + (0)\int_d^a dl$$

$$= \frac{\mu_0 I}{2\pi r_1}(r_1\theta) + 0 - \frac{\mu_0 I}{2\pi r_2}(r_2\theta) + 0 = 0.$$

The magnitude of $\vec{B}$ is greater on arc cd than on arc ab, but the arc length is less, so the contributions from the two arcs exactly cancel. Even though there is a magnetic field everywhere along the integration path, the line integral $\oint \vec{B} \cdot d\vec{l}$ is zero if there is no current passing through the area bounded by the path.

We can also derive these results for more general integration paths, such as the one in Fig. 29-14a. At the position of the line element $d\vec{l}$, the angle between $d\vec{l}$ and $\vec{B}$ is ϕ, and

$$\vec{B} \cdot d\vec{l} = B\, dl \cos\phi.$$

From the figure, $dl \cos\phi = r\, d\theta$, where $d\theta$ is the angle subtended by $d\vec{l}$ at the position of the conductor and r is the distance of $d\vec{l}$ from the conductor. Thus

$$\oint \vec{B} \cdot d\vec{l} = \oint \frac{\mu_0 I}{2\pi r}(r\, d\theta) = \frac{\mu_0 I}{2\pi} \oint d\theta.$$

But $\oint d\theta$ is just equal to 2π, the total angle swept out by the radial line from the conductor to $d\vec{l}$ during a complete trip around the path. So we get

$$\oint \vec{B} \cdot d\vec{l} = \mu_0 I. \qquad (29\text{-}19)$$

This result doesn't depend on the shape of the path or on the position of the wire inside it. If the current in the wire is opposite to that shown, the integral has the opposite sign. But if the path doesn't enclose the wire (Fig. 29-14b), then the net change in θ during the trip around the integration path is zero; $\oint d\theta$ is zero instead of 2π, and the line integral is zero.

Equation (29-19) is almost, but not quite, the general statement of Ampere's law. To generalize it even further, suppose *several* long, straight conductors pass through the surface bounded by the integration path. The total magnetic field $\vec{B}$ at any point on the path is the vector sum of the fields produced by the individual conductors. Thus the line integral of the total $\vec{B}$ equals μ_0 times the *algebraic sum* of the currents. In calculating this sum, we use the sign rule for currents described above. If the integration path does not enclose a particular wire, the line integral of the $\vec{B}$ field of that wire is zero, because the angle θ for that wire sweeps through a net change of zero rather than 2π during the integration. Any conductors present that are not enclosed by a particular path may still contribute to the value of $\vec{B}$ at every point, but the *line integrals* of their fields around the path are zero.

Thus we can replace I in Eq. (29-19) with I_{encl}, the algebraic sum of the currents *enclosed* or *linked* by the integration path, with the sum evaluated by using the sign rule just described. Our statement of **Ampere's law** is then

$$\oint \vec{B} \cdot d\vec{l} = \mu_0 I_{\text{encl}} \qquad \text{(Ampere's law)}. \qquad (29\text{-}20)$$

While we have derived Ampere's law only for the special case of the field of several long, straight parallel conductors, Eq. (29-20) is in fact valid for conductors and paths of *any* shape. The general derivation is no different in principle from what we have presented, but the geometry is more complicated.

If $\oint \vec{B} \cdot d\vec{l} = 0$, it *does not* necessarily mean that $\vec{B} = 0$ everywhere along the path, only that the total current through an area bounded by the path is zero. In Figs. 29-13c and 29-14b, the integration paths enclose no current at all; in Fig. 29-15 (page 918) there are positive and negative currents of equal magnitude through the area enclosed by the path. In both cases, $I_{\text{encl}} = 0$ and the line integral is zero.

CAUTION ▶ In Chapter 24 we saw that the line integral of the electrostatic field $\vec{E}$ around any closed path is equal to zero; this is a statement that the electrostatic force

(a)

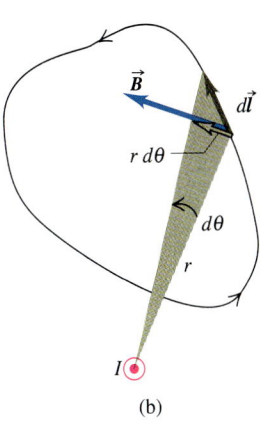

(b)

29-14 (a) A more general integration path for the line integral of $\vec{B}$ around a long, straight conductor carrying current I *out* of the plane of the page. The conductor is seen end-on. (b) A more general integration path that does not enclose the conductor.

29–15 Two long, straight conductors carrying equal currents in opposite directions. The conductors are seen end-on, and the integration path is counterclockwise. The line integral $\oint \vec{B} \cdot d\vec{l}$ gets zero contribution from the upper and lower segments, a positive contribution from the left segment, and a negative contribution from the right segment; the net integral is zero.

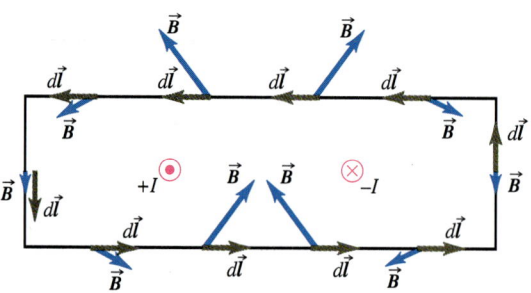

$\vec{F} = q\vec{E}$ on a point charge q is conservative, so this force does zero work on a charge that moves around a closed path that returns to the starting point. You might think that the value of the line integral $\oint \vec{B} \cdot d\vec{l}$ is similarly related to the question of whether the *magnetic* force is conservative. This isn't the case at all. Remember that the magnetic force $\vec{F} = q\vec{v} \times \vec{B}$ on a moving charged particle is always *perpendicular* to $\vec{B}$, so $\oint \vec{B} \cdot d\vec{l}$ is *not* related to the work done by the magnetic force; as stated in Ampere's law, this integral is related only to the total current through a surface bounded by the integration path. In fact, the magnetic force on a moving charged particle is *not* conservative. A conservative force depends only on the position of the body on which the force is exerted, but the magnetic force on a moving charged particle also depends on the *velocity* of the particle. ◀

In the form we have stated it, Ampere's law turns out to be valid only if the currents are steady and if no magnetic materials or time-varying electric fields are present. In Section 29–10 we will see how to generalize Ampere's law for time-varying fields.

29–8 APPLICATIONS OF AMPERE'S LAW

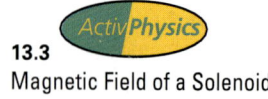

13.3

Magnetic Field of a Solenoid

Ampere's law is useful when we can exploit the symmetry of a situation to evaluate the line integral of $\vec{B}$. Several examples are given below. The following problem-solving strategy is directly analogous to the strategy suggested in Section 23–5 for applications of Gauss's law; we suggest you review that strategy now and compare the two methods.

Problem–Solving Strategy

AMPERE'S LAW

1. The first step is to select the integration path you will use with Ampere's law. If you want to determine $\vec{B}$ at a certain point, then the path must pass through that point.

2. The integration path doesn't have to be any actual physical boundary. Usually, it is a purely geometric curve; it may be in empty space, embedded in a solid body, or some of each.

3. The integration path has to have enough *symmetry* to make evaluation of the integral possible. If the problem itself has cylindrical symmetry, the integration path will usually be a circle coaxial with the cylinder axis.

4. If $\vec{B}$ is tangent to all or some portion of the integration path and has the same magnitude B at every point, then its line integral equals B multiplied by the length of that portion of the path.

5. If $\vec{B}$ is perpendicular to all or some portion of the path, that portion of the path makes no contribution to the integral.

6. The sign of a current enclosed by the integration path is given by a right-hand rule. Curl the fingers of your right hand so that they follow the integration path in the direction that you carry out the integration. Your right thumb then points in the direction of positive current. If $\vec{B}$ is as described in Step 4 and I_{encl} is positive, then the direction of $\vec{B}$ is the same as the direction of the integration path; if instead I_{encl} is negative, $\vec{B}$ is in the direction opposite to that of the integration.

7. In the integral $\oint \vec{B} \cdot d\vec{l}$, $\vec{B}$ is always the *total* magnetic field at each point on the path. This field can be caused partly by currents enclosed by the path and partly by currents outside. If *no* net current is enclosed by the path, the field at points on the path need not be zero, but the integral $\oint \vec{B} \cdot d\vec{l}$ is always zero.

EXAMPLE 29–8

Field of a long, straight, current-carrying conductor In Section 29–7 we derived Ampere's law using Eq. (29–9) for the field of a long, straight, current-carrying conductor. Reverse this process, and use Ampere's law to find the magnitude *and* direction of $\vec{B}$ for this situation.

SOLUTION We exploit the cylindrical symmetry of the situation by taking as our integration path a circle with radius r centered on the conductor and in a plane perpendicular to it, as in Fig. 29–13a (Section 29–7). At each point, $\vec{B}$ is tangent to this circle. Then Ampere's law, Eq. (29–20), becomes

$$\oint \vec{B} \cdot d\vec{l} = \oint B_{\parallel}\, dl = B(2\pi r) = \mu_0 I.$$

Equation (29–9) follows immediately.

Ampere's law determines the direction of $\vec{B}$ as well as its magnitude. Since we go around the integration path in the counterclockwise direction, the positive direction for current is out of the plane of Fig. 29–13a; this is the same as the actual current direction in the figure, so I is positive and the integral $\oint \vec{B} \cdot d\vec{l}$ is also positive. Since the $d\vec{l}$'s run counterclockwise, the direction of $\vec{B}$ must be counterclockwise as well, as shown in Fig. 29–13a.

EXAMPLE 29–9

Field inside a long cylindrical conductor A cylindrical conductor with radius R carries a current I. The current is uniformly distributed over the cross-section area of the conductor. Find the magnetic field as a function of the distance r from the conductor axis for points both inside ($r < R$) and outside ($r > R$) the conductor.

SOLUTION To find the magnetic field *inside* the conductor, we take as our integration path a circle with radius $r < R$ as shown in Fig. 29–16. Because of the cylindrical symmetry, $\vec{B}$ has the same magnitude at every point on this circle and is tangent to it. Thus the magnitude of the line integral is simply $B(2\pi r)$. If we use the right-hand rule for determining the sign of the current, the current through the red area enclosed by the path is positive; hence $\vec{B}$ points in the same direction as the integration path, as shown. To find the current I_{encl} enclosed by the path, note that the current density (current per unit area) is $J = I/\pi R^2$, so $I_{encl} = J(\pi r^2) = Ir^2/R^2$. Finally, Ampere's law gives

$$B(2\pi r) = \mu_0 \frac{Ir^2}{R^2},$$

$$B = \frac{\mu_0 I}{2\pi} \frac{r}{R^2} \qquad \text{(inside the conductor, } r < R). \quad (29\text{–}21)$$

Outside the conductor, we again use a circle for the integration path, but now $r > R$. The right-hand rule gives the direction of $\vec{B}$ as shown in Fig. 29–16. For this path, $I_{encl} = I$, the total current in the conductor. Applying Ampere's law gives the same equation as in Example 29–8, with the same result for B:

$$B = \frac{\mu_0 I}{2\pi r} \qquad \text{(outside the conductor, } r > R). \quad (29\text{–}22)$$

Outside the conductor, the magnetic field is the same as that of a long, straight conductor carrying current I, independent of the radius R over which the current is distributed. Indeed, the magnetic field outside *any* cylindrically symmetric current distribution is the same as if the entire current were concentrated along the axis of the distribution. This is analogous to the results of Examples 23–5 and 23–9 (Section 23–5), in which we found that the *electric* field outside a spherically symmetric *charged* body is the same as though the entire charge were concentrated at the center.

Note that at the surface of the conductor ($r = R$), Eq. (29–21) for $r < R$ and Eq. (29–22) for $r > R$ agree (as they must). Figure 29–17 shows a graph of B as a function of r, both inside and outside the conductor.

29–16 To find the magnetic field at radius $r < R$, we apply Ampere's law to the circle enclosing the red area. The current through the red area is $(r^2/R^2)I$. To find the magnetic field at radius $r > R$, we apply Ampere's law to the circle enclosing the entire conductor.

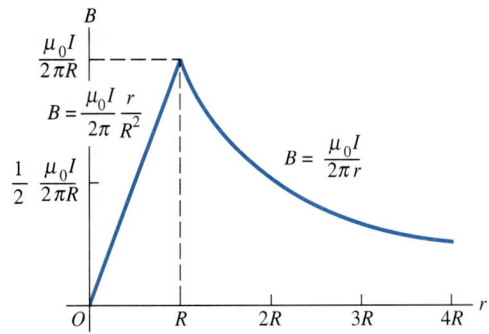

29–17 Magnitude of the magnetic field inside and outside a long, straight cylindrical conductor with radius R carrying a current I.

EXAMPLE 29-10

Field of a solenoid A solenoid, discussed in Example 29–6, consists of a helical winding of wire on a cylinder, usually circular in cross section. There can be hundreds or thousands of closely spaced turns, each of which can be regarded as a circular loop. There may be several layers of windings. For simplicity, Fig. 29–18 shows a solenoid with only a few turns. All turns carry the same current I, and the total $\vec{B}$ field at every point is the vector sum of the fields caused by the individual turns. The figure shows field lines in the xy- and xz-planes. We draw a set of field lines that are uniformly spaced at the center of the solenoid. Exact calculations show that for a long, closely wound solenoid, half of these field lines emerge from the ends and half "leak out" through the windings between the center and the end.

The field lines near the center of the solenoid are approximately parallel, indicating a nearly uniform $\vec{B}$; outside the solenoid, the field lines are spread apart, and the magnetic field is weak. If the solenoid is long in comparison with its cross-section diameter and the coils are tightly wound, the *internal* field near the midpoint of the solenoid's length is very nearly uniform over the cross section and parallel to the axis, and the *external* field near the midpoint is very small.

Use Ampere's law to find the field at or near the center of such a long solenoid. The solenoid has n turns of wire per unit length and carries a current I.

SOLUTION We choose as our integration path the rectangle $abcd$ in Fig. 29–19. Side ab, with length L, is parallel to the axis of the solenoid. Sides bc and da are taken to be very long so that side cd is far from the solenoid; then the field at side cd is negligibly small.

By symmetry the $\vec{B}$ field along side ab is parallel to this side and is constant. In carrying out the Ampere's-law integration, we go along side ab in the same direction as $\vec{B}$. So for this side, $B_\parallel = +B$ and

$$\int_a^b \vec{B} \cdot d\vec{l} = BL.$$

Along sides bc and da, $B_\parallel = 0$ because $\vec{B}$ is perpendicular to these sides; along side cd, $B_\parallel = 0$ because $\vec{B} = \mathbf{0}$. The integral $\oint \vec{B} \cdot d\vec{l}$ around the entire closed path therefore reduces to BL.

The number of turns in length L is nL. Each of these turns passes once through the rectangle $abcd$ and carries a current I, where I is the current in the windings. The total current enclosed by the rectangle is then $I_{encl} = nLI$. From Ampere's law, since the integral $\oint \vec{B} \cdot d\vec{l}$ is positive, I_{encl} must be positive as well; hence the current passing through the surface bounded by the integration path must be in the direction shown in Fig. 29–19. Ampere's law then gives the magnitude B:

$$BL = \mu_0 nLI,$$

$$B = \mu_0 nI \quad \text{(solenoid).} \quad (29\text{–}23)$$

Side ab need not lie on the axis of the solenoid, so this calculation also proves that the field is uniform over the entire cross section at the center of the solenoid's length.

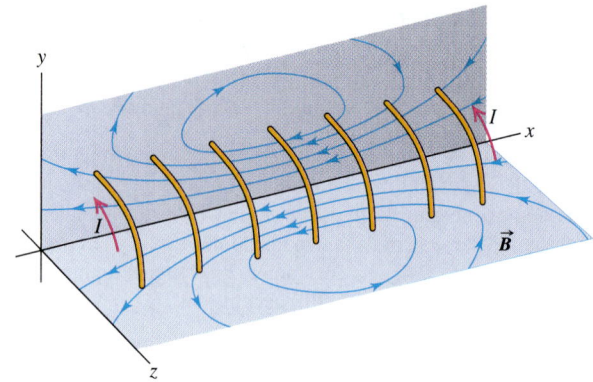

29–18 Magnetic field lines produced by the current in a solenoid. For clarity, only a few turns are shown.

29–19 A section of a long, tightly wound solenoid centered on the x-axis, showing magnetic field lines in the xy- and xz-planes.

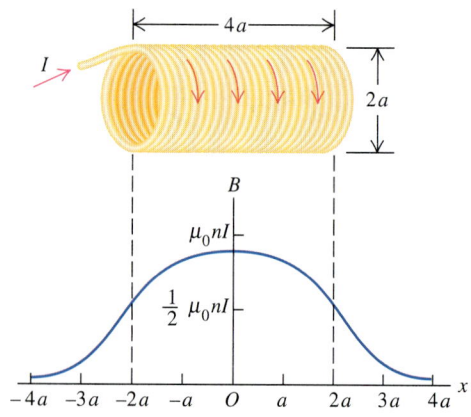

29–20 Magnitude of the magnetic field at points along the axis of a solenoid with length $4a$, equal to four times its radius a. The field magnitude at each end is about half its value at the center. (Compare with Fig. 29–11 for the field of N circular loops.)

Note that the *direction* of $\vec{B}$ inside the solenoid is in the same direction as the solenoid's vector magnetic moment $\vec{\mu}$. This is the same result that we found in Section 29–6 for a single current-carrying loop.

For points along the axis, the field is strongest at the center of the solenoid and drops off near the ends. For a solenoid that is very long in comparison to its diameter, the field at each end is

exactly half as strong as the field at the center. For a short, fat solenoid the relationship is more complicated. Figure 29–20 shows a graph of B as a function of x for points on the axis of a short solenoid.

EXAMPLE 29-11

Field of a toroidal solenoid Figure 29–21a shows a doughnut-shaped **toroidal solenoid,** also called a *toroid,* wound with N turns of wire carrying a current I. In a practical version the turns would be more closely spaced than they are in the figure. Find the magnetic field at all points.

SOLUTION The flow of current around the toroid's circumference produces a magnetic field component perpendicular to the plane of the figure, just as for the current loop discussed in Section 29–6. But if the coils are very tightly wound, we can consider them as circular loops that carry current between the inner and outer radii of the toroidal solenoid; the flow of current around the toroid's circumference is then negligible, and the perpendicular component of $\vec{B}$ is likewise negligible. In this idealized approximation the circular symmetry of the situation tells us that the magnetic field lines must be circles concentric with the axis of the toroid.

To find the field using Ampere's law, we'll use the integration paths shown as black lines in Fig. 29–21b. Consider path 1 first. If the toroidal solenoid produces any field at all in this region, it must be *tangent* to the path at all points, and $\oint \vec{B} \cdot d\vec{l}$ will equal the product of B and the circumference $l = 2\pi r$ of the path. But the total current enclosed by the path is zero, so from Ampere's law the field $\vec{B}$ must be zero everywhere on this path.

Similarly, if the toroidal solenoid produces any field along path 3, it must also be tangent to the path at all points. Each turn of the winding passes *twice* through the area bounded by this path, carrying equal currents in opposite directions. The *net current* I_{encl} enclosed within this area is therefore zero, and hence $\vec{B} = 0$ at all points of the path. Conclusion: *The field of an idealized toroidal solenoid is confined completely to the space enclosed by the windings.* We can think of such an idealized toroidal solenoid as a tightly wound solenoid that has been bent into a circle.

Finally, we consider path 2, a circle with radius r. Again by symmetry we expect the $\vec{B}$ field to be tangent to the path, and $\oint \vec{B} \cdot d\vec{l}$ equals $2\pi r B$. Each turn of the winding passes *once* through the area bounded by path 2. The total current enclosed by the path is $I_{\text{encl}} = NI$, where N is the total number of turns in the winding; I_{encl} is positive for the clockwise direction of integration in Fig. 29–21b, so $\vec{B}$ is in the direction shown. Then, from Ampere's law,

$$2\pi r B = \mu_0 NI,$$

$$B = \frac{\mu_0 NI}{2\pi r} \qquad \text{(toroidal solenoid).} \qquad (29\text{–}24)$$

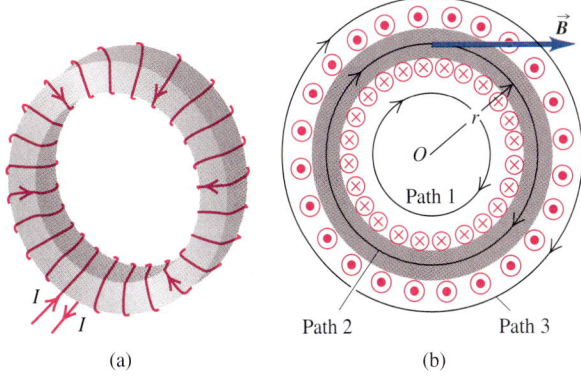

29–21 (a) A toroidal solenoid. For clarity, only a few turns of the winding are shown. (b) Integration paths (black circles) used to compute the magnetic field $\vec{B}$ set up by a current in a toroidal solenoid. The field is very nearly zero at all points except those within the space enclosed by the windings.

The magnetic field is *not* uniform over a cross section of the core, because the radius r is larger at the outer side of the section than at the inner side. However, if the radial thickness of the core is small in comparison to r, the field varies only slightly across a section. In that case, considering that $2\pi r$ is the circumferential length of the toroid and that $N/2\pi r$ is the number of turns per unit length n, the field may be written as

$$B = \mu_0 nI,$$

just as it is at the center of a long, *straight* solenoid.

In a real toroidal solenoid the turns are not precisely circular loops but rather segments of a bent helix. As a result, the field outside is not strictly zero. To estimate its magnitude, we imagine Fig. 29–21a as being roughly equivalent, for points outside the torus, to a circular loop with a single turn and radius r. Then we can use Eq. (29–17) to show that the field at the *center* of the torus is smaller than the field inside by approximately a factor of N/π.

The equations we have derived for the field in a closely wound straight or toroidal solenoid are strictly correct only for windings in *vacuum*. For most practical purposes, however, they can be used for windings in air or on a core of any non-magnetic, nonsuperconducting material. In the next section we will show how they are modified if the core is a magnetic material.

*29–9 MAGNETIC MATERIALS

In discussing how currents cause magnetic fields, we have assumed that the conductors are surrounded by vacuum. But the coils in transformers, motors, generators, and electromagnets nearly always have iron cores to increase the magnetic field and confine it to desired regions. Permanent magnets, magnetic recording tapes, and computer disks depend directly on the magnetic properties of materials; when you store information on a computer disk, you are actually setting up an array of microscopic permanent magnets on the disk. So it is worthwhile to examine some aspects of the magnetic properties of materials. After describing the atomic origins of magnetic properties, we will discuss three broad classes of magnetic behavior that occur in materials; these are called *paramagnetism, diamagnetism,* and *ferromagnetism.*

THE BOHR MAGNETON

As we discussed briefly in Section 28–8, the atoms that make up all matter contain moving electrons, and these electrons form microscopic current loops that produce magnetic fields of their own. In many materials these currents are randomly oriented and cause no net magnetic field. But in some materials an external field (a field produced by currents outside the material) can cause these loops to become oriented preferentially with the field, so their magnetic fields *add* to the external field. We then say that the material is *magnetized.*

Let's look at how these microscopic currents come about. Figure 29–22 shows a primitive model of an electron in an atom. We picture the electron (mass m, charge $-e$) as moving in a circular orbit with radius r and speed v. This moving charge is equivalent to a current loop. In Section 28–8 we found that a current loop with area A and current I has a magnetic dipole moment μ given by $\mu = IA$; for the orbiting electron the area of the loop is $A = \pi r^2$. To find the current associated with the electron, we note that the orbital period T (the time that it takes the electron to make one complete orbit) is the orbit circumference divided by the electron speed: $T = 2\pi r/v$. The equivalent current I is the total charge passing any point on the orbit per unit time, which is just the magnitude e of the electron charge divided by the orbital period T:

$$I = \frac{e}{T} = \frac{ev}{2\pi r}.$$

The magnetic moment $\mu = IA$ is then

$$\mu = \frac{ev}{2\pi r}(\pi r^2) = \frac{evr}{2}. \tag{29–25}$$

It is useful to express μ in terms of the *angular momentum L* of the electron. For a particle moving in a circular path, the magnitude of angular momentum equals the magnitude of momentum mv multiplied by the radius r, that is, $L = mvr$ (Section 10–6). Comparing this with Eq. (29–25), we can write

$$\mu = \frac{e}{2m}L. \tag{29–26}$$

Equation (29–26) is useful in this discussion because atomic angular momentum is *quantized*; its component in a particular direction is always an integer multiple of $h/2\pi$, where h is a fundamental physical constant called *Planck's constant.* The numerical value of h is

$$h = 6.626 \times 10^{-34} \text{ J·s.}$$

The quantity $h/2\pi$ thus represents a fundamental unit of angular momentum in atomic

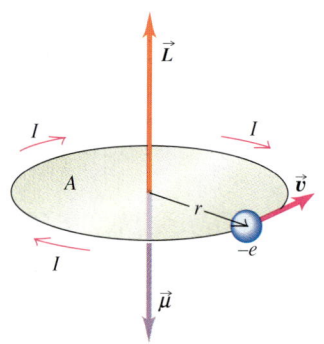

29–22 An electron moving with speed v in a circular orbit of radius r has an angular momentum $\vec{L}$ and an oppositely directed orbital magnetic dipole moment $\vec{\mu}$. It also has a spin angular momentum and an oppositely directed spin magnetic dipole moment.

systems, just as e is a fundamental unit of charge. Associated with the quantization of $\vec{L}$ is a fundamental uncertainty in the *direction* of $\vec{L}$ and therefore of $\vec{\mu}$. In the following discussion, when we speak of the magnitude of a magnetic moment, a more precise statement would be "maximum component in a given direction." Thus to say that a magnetic moment $\vec{\mu}$ is aligned with a magnetic field $\vec{B}$ really means that $\vec{\mu}$ has its maximum possible component in the direction of $\vec{B}$; such components are always quantized.

Equation (29–26) shows that associated with the fundamental unit of angular momentum is a corresponding fundamental unit of magnetic moment. If $L = h/2\pi$, then

$$\mu = \frac{e}{2m}\left(\frac{h}{2\pi}\right) = \frac{eh}{4\pi m}. \qquad (29\text{–}27)$$

This quantity is called the **Bohr magneton,** denoted by μ_B. Its numerical value is

$$\mu_B = 9.274 \times 10^{-24} \ \text{A} \cdot \text{m}^2 = 9.274 \times 10^{-24} \ \text{J/T}.$$

We invite you to verify that these two sets of units are consistent. The second set is useful when we compute the potential energy $U = -\vec{\mu} \cdot \vec{B}$ for a magnetic moment in a magnetic field.

Electrons also have an intrinsic angular momentum, called *spin,* that is not related to orbital motion but that can be pictured in a classical model as spinning on an axis. This angular momentum also has an associated magnetic moment, and its magnitude turns out to be almost exactly one Bohr magneton. (Effects having to do with quantization of the electromagnetic field cause the spin magnetic moment to be about $1.001 \ \mu_B$.)

PARAMAGNETISM

In an atom, most of the various orbital and spin magnetic moments of the electrons add up to zero. However, in some cases the atom has a net magnetic moment that is of the order of μ_B. When such a material is placed in a magnetic field, the field exerts a torque on each magnetic moment, as given by Eq. (28–26): $\vec{\tau} = \vec{\mu} \times \vec{B}$. These torques tend to align the magnetic moments with the field, the position of minimum potential energy, as we discussed in Section 28–8. In this position, the directions of the current loops are such as to *add* to the externally applied magnetic field.

We saw in Section 29–6 that the $\vec{B}$ field produced by a current loop is proportional to the loop's magnetic dipole moment. In the same way, the additional $\vec{B}$ field produced by microscopic electron current loops is proportional to the total magnetic moment $\vec{\mu}_{\text{total}}$ per unit volume V in the material. We call this vector quantity the **magnetization** of the material, denoted by $\vec{M}$:

$$\vec{M} = \frac{\vec{\mu}_{\text{total}}}{V}. \qquad (29\text{–}28)$$

The additional magnetic field due to magnetization of the material turns out to be equal simply to $\mu_0\vec{M}$, where μ_0 is the same constant that appears in the law of Biot and Savart and Ampere's law. When such a material completely surrounds a current-carrying conductor, the total magnetic field $\vec{B}$ in the material is

$$\vec{B} = \vec{B}_0 + \mu_0\vec{M}, \qquad (29\text{–}29)$$

where $\vec{B}_0$ is the field caused by the current in the conductor.

A material showing the behavior just described is said to be **paramagnetic.** The result is that the magnetic field at any point in such a material is greater by a dimensionless factor K_m, called the **relative permeability** of the material, than it would be if the material were replaced by vacuum. The value of K_m is different for different materials; for common paramagnetic solids and liquids at room temperature, K_m is typically 1.00001 to 1.003.

All of the equations in this chapter that relate magnetic fields to their sources can be adapted to the situation in which the current-carrying conductor is embedded in a paramagnetic material. All that need be done is to replace μ_0 by $K_m \mu_0$. This product is usually denoted as μ and is called the **permeability** of the material:

$$\mu = K_m \mu_0. \qquad (29\text{--}30)$$

CAUTION ▶ Equation (29–30) involves some really dangerous notation because we have also used μ for magnetic dipole moment. It's customary to use μ for both quantities, but beware; from now on, every time you see a μ, make sure you know whether it is permeability or magnetic moment. You can usually tell from the context. ◀

The amount by which the relative permeability differs from unity is called the **magnetic susceptibility,** denoted by χ_m:

$$\chi_m = K_m - 1. \qquad (29\text{--}31)$$

Both K_m and χ_m are dimensionless quantities. Values of magnetic susceptibility for several materials are given in Table 29–1. For example, for aluminum, $\chi_m = 2.2 \times 10^{-5}$ and $K_m = 1.000022$. The first group of materials in the table are paramagnetic; we'll discuss the second group of materials, which are called *diamagnetic,* very shortly.

The tendency of atomic magnetic moments to align themselves parallel to the magnetic field (where the potential energy is minimum) is opposed by random thermal motion, which tends to randomize their orientations. For this reason, paramagnetic susceptibility always decreases with increasing temperature. In many cases it is inversely proportional to the absolute temperature T, and the magnetization M can be expressed as

$$M = C \frac{B}{T}. \qquad (29\text{--}32)$$

This relation is called *Curie's law,* after its discoverer, Pierre Curie (1859–1906). The quantity C is a constant, different for different materials, called the *Curie constant.*

As we described in Section 28–8, a body with atomic magnetic dipoles is attracted to the poles of a magnet. In most paramagnetic substances this attraction is very weak due to thermal randomization of the atomic magnetic moments. That's why a magnet can't be used to pick up objects made of aluminum (a paramagnetic substance). But at very low temperatures the thermal effects are reduced, the magnetization increases in accordance with Curie's law, and the attractive forces are greater.

TABLE 29–1

MAGNETIC SUSCEPTIBILITIES OF PARAMAGNETIC AND DIAMAGNETIC MATERIALS AT $T = 20°C$

MATERIAL	$\chi_m = K_m - 1 \ (\times 10^{-5})$
Paramagnetic	
Iron ammonium alum	66
Uranium	40
Platinum	26
Aluminum	2.2
Sodium	0.72
Oxygen gas	0.19
Diamagnetic	
Bismuth	−16.6
Mercury	−2.9
Silver	−2.6
Carbon (diamond)	−2.1
Lead	−1.8
Sodium chloride	−1.4
Copper	−1.0

EXAMPLE 29-12

Show that the units in Eq. (29–29) are consistent.

SOLUTION Magnetization $\vec{M}$ is magnetic moment per unit volume. The units of magnetic moment are current times area $(A \cdot m^2)$, so the units of magnetization are $(A \cdot m^2)/m^3 = A/m$.

From Section 29–2 the units of the constant μ_0 are T · m/A. So the units of $\mu_0\vec{M}$ are the same as the units of $\vec{B}$:

$$(T \cdot m/A)(A/m) = T.$$

EXAMPLE 29-13

Magnetic dipoles in a paramagnetic material Nitric oxide (NO) is a paramagnetic compound. Its molecules have a magnetic moment with a maximum component in any direction of about one Bohr magneton each. In a magnetic field with magnitude $B = 1.5$ T, compare the interaction energy of the magnetic moments with the field to the average translational kinetic energy of the molecules at a temperature of 300 K.

SOLUTION In Section 28–7 we derived an expression for the potential energy for a magnetic moment in a $\vec{B}$ field: $U = -\vec{\mu} \cdot \vec{B} = -(\mu \cos \phi)B$. (This symbol μ denotes a magnetic moment.) In our case the maximum value of the component $\mu \cos \phi$ is about μ_B, so

$$|U|_{max} \approx \mu_B B = (9.27 \times 10^{-24} \text{ J/T})(1.5 \text{ T})$$
$$= 1.4 \times 10^{-23} \text{ J} = 8.7 \times 10^{-5} \text{ eV}.$$

The average translational kinetic energy K is given by the equipartition principle (Section 16–5):

$$K = \frac{3}{2}kT = \frac{3}{2}(1.38 \times 10^{-23} \text{ J/K})(300 \text{ K})$$
$$= 6.2 \times 10^{-21} \text{ J} = 0.039 \text{ eV}.$$

At a temperature of 300 K the magnetic interaction energy is much *smaller* than the random kinetic energy, so we expect only a slight degree of alignment. This is why paramagnetic susceptibilities at ordinary temperature are usually very small.

DIAMAGNETISM

In some materials the total magnetic moment of all the atomic current loops is zero when no magnetic field is present. But even these materials have magnetic effects because an external field alters electron motions within the atoms, causing additional current loops and induced magnetic dipoles comparable to the induced *electric* dipoles we studied in Section 25–6. In this case the additional field caused by these current loops is always *opposite* in direction to that of the external field. (This behavior is explained by Faraday's law of induction, which we will study in Chapter 30. An induced current always tends to cancel the field change that caused it.)

(a) No field

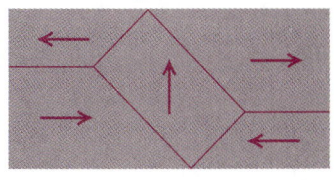
(b) Weak field $\vec{B}$

Such materials are said to be **diamagnetic.** They always have negative susceptibility, as shown in Table 29–1, and relative permeability K_m slightly *less* than unity, typically of the order of 0.99990 to 0.99999 for solids and liquids. Diamagnetic susceptibilities are very nearly temperature-independent.

FERROMAGNETISM

There is a third class of materials, called **ferromagnetic** materials, which includes iron, nickel, cobalt, and many alloys containing these elements. In these materials, strong interactions between atomic magnetic moments cause them to line up parallel to each other in regions called **magnetic domains,** even when no external field is present. Figure 29–23 shows an example of magnetic domain structure. Within each domain, nearly all of the atomic magnetic moments are parallel.

When there is no externally applied field, the domain magnetizations are randomly oriented. But when a field $\vec{B}_0$ (caused by external currents) is present, the domains tend to orient themselves parallel to the field. The domain boundaries also shift; the domains that are magnetized in the field direction grow, and those that are magnetized in other directions shrink. Because the total magnetic moment of a domain may be many

(c) Stronger field $\vec{B}$

29–23 In this drawing adapted from a magnified photo, the arrows show the directions of magnetization in the domains of a single crystal of nickel. Domains that are magnetized in the direction of an applied magnetic field grow larger.

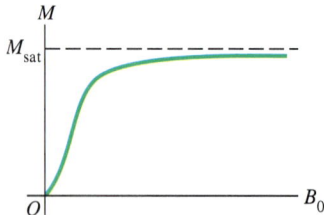

29-24 A magnetization curve for a ferromagnetic material. The magnetization M approaches its saturation value M_{sat} as the magnetic field B_0 (caused by external currents) becomes large.

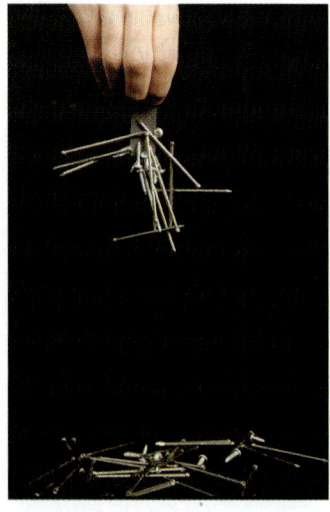

This permanent magnet can pick up screws and nails made of steel (a ferromagnetic material). Nails made of aluminum (which is paramagnetic) would also be attracted to the magnet, but with so little force that the magnet could not pick them up.

thousands of Bohr magnetons, the torques that tend to align the domains with an external field are much stronger than occur with paramagnetic materials. The relative permeability K_m is *much* larger than unity, typically of the order of 1,000 to 100,000.

As the external field is increased, a point is eventually reached at which nearly *all* the magnetic moments in the ferromagnetic material are aligned parallel to the external field. This condition is called *saturation magnetization;* after it is reached, further increase in the external field causes no increase in magnetization or in the additional field caused by the magnetization.

Figure 29-24 shows a "magnetization curve," a graph of magnetization M as a function of external magnetic field B_0, for soft iron. An alternative description of this behavior is that K_m is not constant but decreases as B_0 increases. (Paramagnetic materials also show saturation at sufficiently strong fields. But the magnetic fields required are so large that departures from a linear relationship between M and B_0 in these materials can be observed only at very low temperatures, 1 K or so.)

For many ferromagnetic materials the relation of magnetization to external magnetic field is different when the external field is increasing from when it is decreasing. Figure 29-25a shows this relation for such a material. When the material is magnetized to saturation and then the external field is reduced to zero, some magnetization remains. This behavior is characteristic of permanent magnets, which retain most of their saturation magnetization when the magnetizing field is removed. To reduce the magnetization to zero requires a magnetic field in the reverse direction.

This behavior is called **hysteresis,** and the curves in Fig. 29-25 are called *hysteresis loops.* Magnetizing and demagnetizing a material that has hysteresis involves the dissipation of energy, and the temperature of the material increases during such a process.

Ferromagnetic materials are widely used in electromagnets, transformer cores, and motors and generators, in which it is desirable to have as large a magnetic field as possible for a given current. Because hysteresis dissipates energy, materials that are used in these applications should usually have as narrow a hysteresis loop as possible. Soft iron is often used; it has high permeability without appreciable hysteresis. For permanent magnets a broad hysteresis loop is usually desirable, with large zero-field magnetization and large reverse field needed to demagnetize. Many kinds of steel and many alloys, such as Alnico, are commonly used for permanent magnets. The remaining magnetic field in such a material, after it has been magnetized to near saturation, is typically

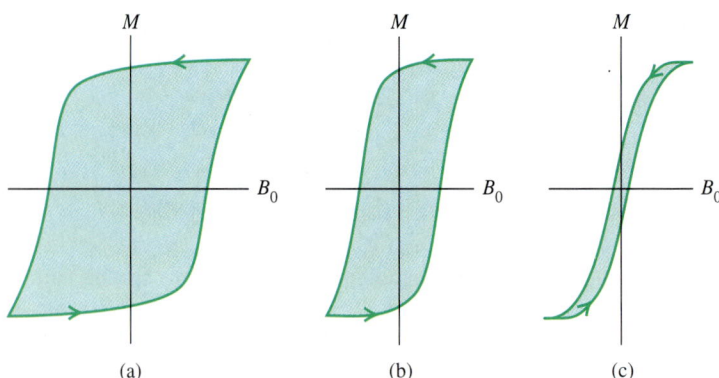

29-25 Hysteresis loops. The materials of both (a) and (b) remain strongly magnetized when B_0 is reduced to zero. Since (a) is also hard to demagnetize, it would be good for permanent magnets. Since (b) magnetizes and demagnetizes more easily, it could be used as a computer memory material. The material of (c) would be useful for transformers and other alternating-current devices in which zero hysteresis would be optimal.

of the order of 1 T, corresponding to a remaining magnetization $M = B/\mu_0$ of about 800,000 A/m.

EXAMPLE 29-14

A permanent magnet is made of a ferromagnetic material with a magnetization M of about 8×10^5 A/m. The magnet is in the shape of a cube of side 2 cm. a) Find the magnetic dipole moment of the magnet. b) Estimate the magnetic field due to the magnet at a point 10 cm from the magnet along its axis.

SOLUTION a) Magnetization equals magnetic dipole moment per unit volume, so the total magnetic moment is

$$\mu_{total} = MV = (8 \times 10^5 \text{ A/m})(2 \times 10^{-2} \text{ m})^3 = 6 \text{ A} \cdot \text{m}^2.$$

b) We found in Section 29-6 that the magnetic field on the axis of a current loop with magnetic moment μ_{total} is given by Eq. (29-18),

$$B = \frac{\mu_0 \mu_{total}}{2\pi(x^2 + a^2)^{3/2}},$$

where x is the distance from the loop and a is its radius. We can use this same expression here, except that a refers to the size of

the permanent magnet. Strictly speaking, there are complications because our magnet does not have the same geometry as a circular current loop. But because $x = 10$ cm is fairly large in comparison to the 2-cm size of the magnet, the term a^2 is negligible in comparison to x^2 and can be ignored. So

$$B \approx \frac{\mu_0 \mu_{total}}{2\pi x^3} = \frac{(4\pi \times 10^{-7} \text{ T} \cdot \text{m/A})(6 \text{ A} \cdot \text{m}^2)}{2\pi(0.1 \text{ m})^3}$$

$$= 1 \times 10^{-3} \text{ T} = 10 \text{ G},$$

which is about ten times stronger than the magnetic field of the earth. Such a magnet can easily deflect a compass needle.

Note that we used μ_0, not the permeability μ of the magnetic material, in calculating B. The reason is that we are calculating B at a point *outside* the magnetic material. You would substitute permeability μ for μ_0 only if you were calculating B *inside* a material with relative permeability K_m, for which $\mu = K_m \mu_0$.

29-10 DISPLACEMENT CURRENT

Ampere's law, in the form stated in Section 29-7, is incomplete. As we learn how it has to be modified for the most general situations, we will uncover some profound and fundamental aspects of the behavior of electric and magnetic fields.

To introduce the problem, let's consider the process of charging a capacitor (Fig. 29-26). Conducting wires lead current i_C into one plate and out of the other; the charge Q increases, and the electric field $\vec{E}$ between the plates increases. The notation i_C indicates *conduction* current to distinguish it from another kind of current we are about to encounter, called *displacement* current i_D. We use lowercase i's and v's to denote instantaneous values of currents and potential differences, respectively, that may vary with time.

Let's apply Ampere's law to the circular path shown. The integral $\oint \vec{B} \cdot d\vec{l}$ around this path equals $\mu_0 I_{encl}$. For the plane circular area bounded by the circle, I_{encl} is just the current i_C in the left conductor. But the surface that bulges out to the right is bounded by the same circle, and the current through that surface is zero. So $\oint \vec{B} \cdot d\vec{l}$ is equal to $\mu_0 i_C$, and at the same time it is equal to zero! This is a clear contradiction.

But something else is happening on the bulged-out surface. As the capacitor charges, the electric field $\vec{E}$ and the electric *flux* Φ_E through the surface are increasing. We can

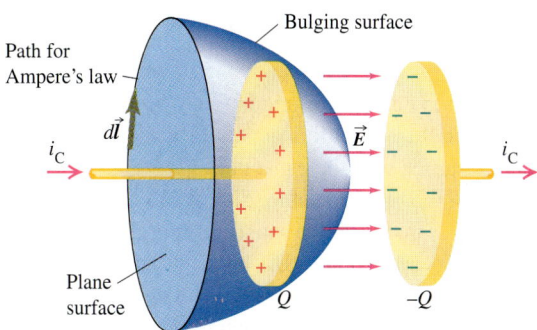

29-26 Parallel-plate capacitor being charged. The conduction current through the plane surface is i_C, but there is no conduction current through the surface that bulges out to pass between the plates. The two surfaces have a common boundary, so this difference in I_{encl} leads to an apparent contradiction in applying Ampere's law.

determine their rates of change in terms of the charge and current. The instantaneous charge is $q = Cv$, where C is the capacitance and v is the instantaneous potential difference. For a parallel-plate capacitor, $C = \epsilon_0 A/d$, where A is the plate area and d is the spacing. The potential difference v between plates is $v = Ed$, where E is the electric field magnitude between plates. (We neglect fringing and assume that $\vec{E}$ is uniform in the region between the plates.) If this region is filled with a material with permittivity ϵ, we replace ϵ_0 by ϵ everywhere; we'll use ϵ in the following discussion.

Substituting these expressions for C and v into $q = Cv$, we can express the capacitor charge q as

$$q = Cv = \frac{\epsilon A}{d}(Ed) = \epsilon EA = \epsilon\Phi_E, \tag{29-33}$$

where $\Phi_E = EA$ is the electric flux through the surface.

As the capacitor charges, the rate of change of q is the conduction current, $i_C = dq/dt$. Taking the derivative of Eq. (29–33) with respect to time, we get

$$i_C = \frac{dq}{dt} = \epsilon\frac{d\Phi_E}{dt}. \tag{29-34}$$

Now, stretching our imagination a little, we invent a fictitious current or pseudocurrent i_D in the region between the plates, defined as

$$i_D = \epsilon\frac{d\Phi_E}{dt} \qquad \text{(displacement current)}. \tag{29-35}$$

That is, we imagine that the changing flux through the curved surface in Fig. 29–26 is somehow equivalent, in Ampere's law, to a conduction current through that surface. We include this fictitious current, along with the real conduction current i_C, in Ampere's law:

$$\oint \vec{B} \cdot d\vec{l} = \mu_0(i_C + i_D)_{\text{encl}} \qquad \text{(generalized Ampere's law)}. \tag{29-36}$$

Ampere's law in this form is obeyed no matter which surface we use in Fig. 29–26. For the flat surface, i_D is zero; for the curved surface, i_C is zero; and i_C for the flat surface equals i_D for the curved surface.

The fictitious current i_D was invented in 1865 by the Scottish physicist James Clerk Maxwell (1831–1879), who called it **displacement current.** There is a corresponding *displacement current density* $j_D = i_D/A$; using $\Phi_E = EA$ and dividing Eq. (29–35) by A, we find

$$j_D = \epsilon\frac{dE}{dt}. \tag{29-37}$$

We have pulled the concept out of thin air, as Maxwell did, but we see that it enables us to save Ampere's law in situations such as that in Fig. 29–26.

Another benefit of displacement current is that it lets us generalize Kirchhoff's junction rule, discussed in Section 27–3. Considering the left plate of the capacitor plate, we have conduction current into it but none out of it. But when we include the displacement current, we have conduction current coming in one side and an equal displacement current coming out the other side. With this generalized meaning of the term *current,* we can speak of current going *through* the capacitor.

You might well ask at this point whether displacement current has any real physical significance or whether it is just a ruse to satisfy Ampere's law and Kirchhoff's junction rule. Here's a fundamental experiment that helps to answer that question. We take a plane circular area between the capacitor plates, as shown in Fig. 29–27. If displacement current really plays the role in Ampere's law that we have claimed, then there ought to

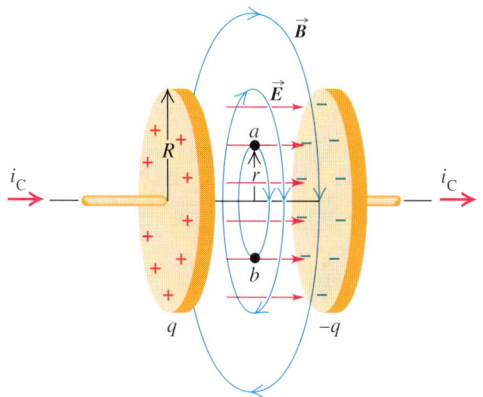

29–27 A capacitor being charged by a current i_C has a displacement current equal to i_C between the plates, with displacement-current density $j_D = \epsilon\, dE/dt$. This can be regarded as the source of the magnetic field between the plates.

be a magnetic field in the region between the plates while the capacitor is charging. We can use our generalized Ampere's law, including displacement current, to predict what this field should be.

To be specific, let's picture round capacitor plates with radius R. To find the magnetic field at a point in the region between the plates at a distance r from the axis, we apply Ampere's law to a circle of radius r passing through the point, with $r < R$. This circle passes through points a and b in Fig. 29–27. The total current enclosed by the circle is j_D times its area, or $(i_D/\pi R^2)(\pi r^2)$. The integral $\oint \vec{B} \cdot d\vec{l}$ in Ampere's law is just B times the circumference $2\pi r$ of the circle, and because $i_D = i_C$ for the charging capacitor, Ampere's law becomes

$$\oint \vec{B} \cdot d\vec{l} = 2\pi r B = \mu_0 \frac{r^2}{R^2} i_C,$$

or

$$B = \frac{\mu_0}{2\pi} \frac{r}{R^2} i_C. \qquad (29\text{–}38)$$

This result predicts that in the region between the plates $\vec{B}$ is zero at the axis and increases linearly with distance from the axis. A similar calculation shows that *outside* the region between the plates (that is, for $r > R$), $\vec{B}$ is the same as though the wire were continuous and the plates not present at all.

When we *measure* the magnetic field in this region, we find that it really is there and that it behaves just as Eq. (29–38) predicts. This confirms directly the role of displacement current as a source of magnetic field. It is now established beyond reasonable doubt that displacement current, far from being just an artifice, is a fundamental fact of nature. Maxwell's discovery was the bold step of an extraordinary genius. Indeed, as we will see later, displacement current was the missing link in the chain of electromagnetic theory that led Maxwell and others to the understanding of electromagnetic waves.

The calculation leading to Eq. (29–38) bears a strong resemblance to the one in Example 29–9 (Section 29–8), in which we used Ampere's law to find the magnetic field inside a long cylindrical conductor. There is no actual charge in motion to constitute a conduction current, but there is a displacement current. Thus it should not be too surprising that Eq. (29–38) is in fact the same result as in Example 29–9.

Two final comments: First, the generalized form of Ampere's law, Eq. (29–36), remains valid in a magnetic material, provided that magnetization is proportional to external field and we replace μ_0 by μ. Second, Ampere's law is valid even in empty space where there is no conduction current at all. This fact has profound implications; it means, among other things, that when $\vec{E}$ and $\vec{B}$ fields vary with time, they are interrelated. In particular, a changing electric field in a particular region of space induces a *magnetic* field in neighboring regions, even when no conduction current and no matter are present. Conversely, we will learn in Chapter 30 that a changing *magnetic* field acts as a source of *electric* field. These relationships, first given in their complete form by Maxwell in 1865, provide the key to a theoretical understanding of electromagnetic radiation and of light as a particular example of this radiation. We return to this topic in Chapter 33.

SUMMARY

- The magnetic field $\vec{B}$ created by a charge q moving with velocity $\vec{v}$ is

$$\vec{B} = \frac{\mu_0}{4\pi} \frac{q\vec{v} \times \hat{r}}{r^2}, \quad (29\text{--}2)$$

where r is the distance from the source point (the location of q) to the field point P and $\hat{r}$ is a unit vector in that direction. The principle of superposition of magnetic fields: The total $\vec{B}$ produced by several moving charges is the vector sum of the fields produced by the individual charges.

- The law of Biot and Savart: The magnetic field $d\vec{B}$ created by an element $d\vec{l}$ of a conductor carrying current I is

$$d\vec{B} = \frac{\mu_0}{4\pi} \frac{I\,d\vec{l} \times \hat{r}}{r^2}. \quad (29\text{--}6)$$

The field created by a finite current-carrying conductor is the integral of this expression over the length of the conductor.

- The magnetic field $\vec{B}$ at a distance r from a long, straight conductor carrying a current I has magnitude

$$B = \frac{\mu_0 I}{2\pi r}. \quad (29\text{--}9)$$

The magnetic field lines are circles coaxial with the wire, with directions given by the right-hand rule.

- The interaction force, per unit length, between two long parallel conductors with currents I and I' has magnitude

$$\frac{F}{L} = \frac{\mu_0 II'}{2\pi r}. \quad (29\text{--}11)$$

The definition of the ampere is based on this relation. Currents in parallel wires attract if they are in the same direction and repel if they are in opposite directions.

- The magnetic field produced by a circular conducting loop with radius a, carrying current I, at a distance x from its center, along its axis, has magnitude

$$B_x = \frac{\mu_0 I a^2}{2(x^2 + a^2)^{3/2}} \quad \text{(circular loop)}. \quad (29\text{--}15)$$

For N loops, this expression is multiplied by N. At the center of the loops, where $x = 0$,

$$B_x = \frac{\mu_0 NI}{2a} \quad \text{(center of } N \text{ circular loops)}. \quad (29\text{--}17)$$

- Ampere's law states that the line integral of $\vec{B}$ around any closed path equals μ_0

times the net current through the area enclosed by the path:

$$\oint \vec{B} \cdot d\vec{l} = \mu_0 I_{\text{encl}}. \qquad (29\text{-}20)$$

The positive sense of current is determined by a right-hand rule.

■ The following table lists magnetic fields caused by several current distributions. In each case the conductor is carrying current I.

CURRENT DISTRIBUTION	POINT IN MAGNETIC FIELD	MAGNETIC FIELD MAGNITUDE
Long, straight conductor	Distance r from conductor	$B = \dfrac{\mu_0 I}{2\pi r}$
Circular loop of radius a	On axis of loop	$B = \dfrac{\mu_0 I a^2}{2(x^2 + a^2)^{3/2}}$
	At center of loop	$B = \dfrac{\mu_0 I}{2a}$ (for N loops, multiply these expressions by N)
Long cylindrical conductor of radius R	Inside conductor, $r < R$	$B = \dfrac{\mu_0 I}{2\pi}\dfrac{r}{R^2}$
	Outside conductor, $r > R$	$B = \dfrac{\mu_0 I}{2\pi r}$
Long, closely wound solenoid with n turns per unit length, near its midpoint	Inside solenoid, near center	$B = \mu_0 n I$
	Outside solenoid	$B \approx 0$
Tightly wound toroidal solenoid (toroid) with N turns	Within the space enclosed by the windings, distance r from symmetry axis	$B = \dfrac{\mu_0 N I}{2\pi r}$
	Outside the space enclosed by the windings	$B \approx 0$

■ When magnetic materials are present, the magnetization of the material causes an additional contribution to $\vec{B}$. For paramagnetic and diamagnetic materials, μ_0 is replaced in magnetic-field expressions by $\mu = K_m \mu_0$, where μ is the permeability of the material and K_m is its relative permeability. The magnetic susceptibility χ_m is defined as $\chi_m = K_m - 1$. Magnetic susceptibilities at room temperature for paramagnetic materials are small positive quantities; those for diamagnetic materials are small negative quantities. For ferromagnetic materials, K_m is much larger than unity and is not constant. Some ferromagnetic materials are permanent magnets, retaining their magnetization even after the external magnetic field is removed.

■ Displacement current i_D acts as a source of magnetic field in exactly the same way as conduction current. It is defined as

$$i_D = \epsilon \frac{d\Phi_E}{dt}. \qquad (29\text{-}35)$$

Ampere's law including displacement current is

$$\oint \vec{B} \cdot d\vec{l} = \mu_0 (i_C + i_D)_{\text{encl}}. \qquad (29\text{-}36)$$

Displacement current plays an essential role in the analysis of electromagnetic waves.

DISCUSSION QUESTIONS

Q29–1 A topic of current interest in physics research is the search (thus far unsuccessful) for an isolated magnetic pole, or magnetic *monopole*. If such an entity were found, how could it be recognized? What would its properties be?

Q29–2 Streams of charged particles emitted from the sun during periods of solar activity create a disturbance in the earth's magnetic field. How does this happen?

Q29–3 The text discussed the magnetic field of an infinitely long, straight conductor carrying a current. Of course, there is no such thing as an infinitely long *anything*. How do you decide whether a particular wire is long enough to be considered infinite?

Q29–4 Two parallel conductors carrying current in the same direction attract each other. If they are permitted to move toward each other, the forces of attraction do work. Where does the energy come from? Does this contradict the assertion in Chapter 28 that magnetic forces on moving charges do no work? Explain.

Q29–5 Pairs of conductors carrying current into or out of the power-supply components of electronic equipment are sometimes twisted together to reduce magnetic field effects. Why does this help?

Q29–6 Suppose you have three long, parallel wires, arranged so that in cross section they are at the corners of an equilateral triangle. Is there any way to arrange the currents so that all three wires attract each other? So that all three wires repel each other? Explain.

Q29–7 In deriving the force on one of the long current-carrying conductors in Section 29–5, why did we use the magnetic field due to only one of the conductors? That is, why didn't we use the *total* magnetic field due to *both* conductors?

Q29–8 Considering the magnetic field of a circular loop of wire, would you expect the field to be greatest at the center, or is it greater at some points in the plane of the loop but off-center? Explain.

Q29–9 Two concentric coplanar circular loops of wire, of different diameter, carry currents in the same direction. Describe the nature of the forces exerted on the inner loop and on the outer loop.

Q29–10 A current was sent through a helical coil spring. The spring contracted, as though it had been compressed. Why?

Q29–11 What are the relative advantages and disadvantages of Ampere's law and the law of Biot and Savart for practical calculations of magnetic fields?

Q29–12 Magnetic field lines never have a beginning or an end. Use this to explain why it is reasonable for the field of a toroidal solenoid to be confined entirely to its interior, while a straight solenoid *must* have some field outside.

*Q29–13 Why should the permeability of a paramagnetic material be expected to decrease with increasing temperature?

*Q29–14 If a magnet is suspended over a container of liquid air, it attracts droplets to its poles. The droplets contain only liquid oxygen; even though nitrogen is the primary constituent of air, it is not attracted to the magnet. Explain what this tells you about the magnetic susceptibilities of oxygen and nitrogen, and explain why a magnet in ordinary, room-temperature air doesn't attract molecules of oxygen *gas* to its poles.

*Q29–15 What features of atomic structure determine whether an element is diamagnetic or paramagnetic? Explain.

*Q29–16 The magnetic susceptibility of paramagnetic materials is quite strongly temperature-dependent, but that of diamagnetic materials is nearly independent of temperature. Why the difference?

*Q29–17 A cylinder of iron is placed so that it is free to rotate around its axis. Initially the cylinder is at rest, and a magnetic field is applied to the cylinder so that it is magnetized in a direction parallel to its axis. If the direction of the external field is suddenly reversed, the direction of magnetization will also reverse and the cylinder will begin rotating around its axis. (This is called the *Einstein–de Haas Effect*.) Explain why the cylinder begins to rotate.

*Q29–18 The discussion of magnetic forces on current loops in Section 28–8 commented that no net force is exerted on a complete loop in a uniform magnetic field, only a torque. Yet magnetized materials, which contain atomic current loops, certainly *do* experience net forces in magnetic fields. How is this discrepancy resolved?

Q29–19 Can one have a displacement current as well as a conduction current within a conductor? Explain.

EXERCISES

SECTION 29–2 MAGNETIC FIELD OF A MOVING CHARGE

29–1 A $+6.00$-μC point charge is moving at a constant 8.00×10^6 m/s in the $+y$-direction relative to a reference frame. At the instant when the point charge is at the origin of this reference frame, what is the magnetic field vector $\vec{B}$ it produces at the following points: a) $x = 0.500$ m, $y = 0$, $z = 0$; b) $x = 0$, $y = -0.500$ m, $z = 0$; c) $x = 0$, $y = 0$, $z = +0.500$ m; d) $x = 0$, $y = -0.500$ m, $z = +0.500$ m?

29–2 Positive point charges $q = +8.00$ μC and $q' = +3.00$ μC are moving relative to an observer at point P, as shown in Fig. 29–28. The distance d is 0.120 m. When the two charges are at

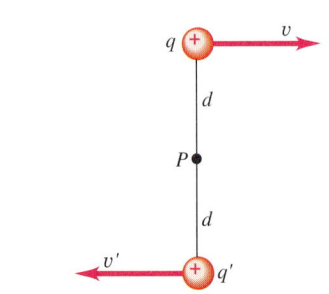

FIGURE 29–28 Exercises 29–2 and 29–4.

the locations shown in the figure, what are the magnitude and direction of the net magnetic field they produce at point P? Take $v = 4.50 \times 10^6$ m/s and $v' = 9.00 \times 10^6$ m/s.

29–3 A pair of point charges, $q = +4.00\ \mu$C and $q' = -1.50\ \mu$C, are moving in a reference frame as shown in Fig. 29–29. At this instant, what are the magnitude and direction of the magnetic field produced at the origin? Take $v = 2.00 \times 10^5$ m/s and $v' = 8.00 \times 10^5$ m/s .

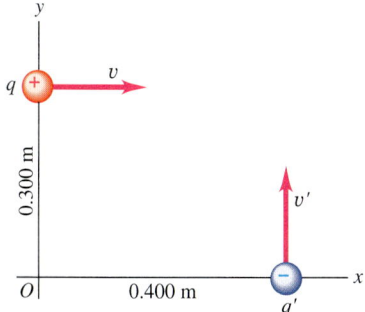

FIGURE 29–29 Exercise 29–3 and Problem 29–40.

29–4 Figure 29–28 shows two point charges q and q' moving relative to an observer at point P. Suppose that the lower charge is actually *negative*, with $q' = -q$. a) Find the magnetic field (magnitude and direction) produced by the two charges at point P if i) $v' = v/2$; ii) $v' = v$; iii) $v' = 2v$. b) Find the direction of the magnetic force that q exerts on q', and find the direction of the magnetic force that q' exerts on q. c) If $v = v' = 3.00 \times 10^5$ m/s, what is the ratio of the magnitude of the magnetic force acting on each charge to that of the Coulomb force acting on each charge?

SECTION 29–3 MAGNETIC FIELD OF A CURRENT ELEMENT

29–5 a) In part (b) of Example 29–2 (Section 29–3), what is the unit vector $\hat{r}$ (expressed in terms of $\hat{i}$ and $\hat{j}$) that points from $d\vec{l}$ to point P_2? b) Compute the magnitude and direction of $d\vec{l} \times \hat{r}$. c) Use the result of part (b) and Eq. (29–6) to find the magnitude and direction of the magnetic field produced at point P_2 by the 1.0-cm segment of the wire.

29–6 A long, straight wire, carrying a current of 200 A, runs through a cubical wooden box, entering and leaving through holes in the centers of opposite faces (Fig. 29–30). The length of each side of the box is 20.0 cm. Consider an element dl of

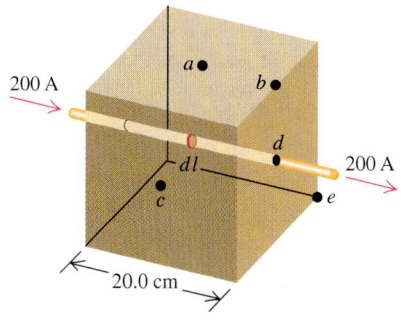

FIGURE 29–30 Exercise 29–6.

the wire 0.100 cm long at the center of the box. Compute the magnitude dB of the magnetic field produced by this element at the points a, b, c, d, and e in Fig. 29–30. Points a, c, and d are at the centers of the faces of the cube; point b is at the midpoint of one edge; and point e is at a corner. Copy the figure and show the directions and relative magnitudes of the field vectors. (*Note:* Assume that dl is small in comparison to the distances from the current element to the points where the magnetic field is to be calculated.)

29–7 A long, straight wire lies along the z-axis and carries a 4.00-A current in the $+z$-direction. Find the magnetic field (magnitude and direction) produced at the following points by a 0.500-mm segment of the wire centered at the origin: a) $x = 2.00$ m, $y = 0$, $z = 0$; b) $x = 0$, $y = 2.00$ m, $z = 0$; c) $x = 2.00$ m, $y = 2.00$ m, $z = 0$; d) $x = 0$, $y = 0$, $z = 2.00$ m.

SECTION 29–4 MAGNETIC FIELD OF A STRAIGHT CURRENT-CARRYING CONDUCTOR

29–8 In Figure 29–7a (Example 29–4 of Section 29–4), what are the magnitude and direction of the net field produced by the wires on the x-axis at a) $x = d/2$? b) $x = -d/2$?

29–9 Two long, straight wires, one above the other, are separated by a distance $2a$ and are parallel to the x-axis. Let the $+y$-axis be in the plane of the wires in the direction from lower wire to the upper wire. Each wire carries current I in the $+x$-direction. What are the magnitude and direction of the net magnetic field of the two wires at a point in the plane of the wires a) midway between them? b) at a distance a above the upper wire? c) at a distance a below the lower wire?

29–10 A long, straight wire lies along the y-axis and carries a current $I = 8.00$ A in the $-y$-direction (Fig. 29–31). In addition to the magnetic field due to the current in the wire, a uniform magnetic field $\vec{B}_0$ with magnitude 1.50×10^{-6} T is in the $+x$-direction. What is the total field (magnitude and direction) at the following points in the xz-plane? a) $x = 0$, $z = 1.00$ m; b) $x = 1.00$ m, $z = 0$; c) $x = 0$, $z = -0.25$ m?

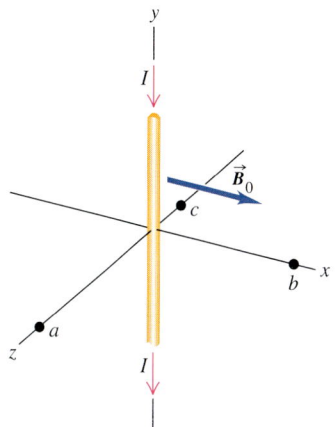

FIGURE 29–31 Exercise 29–10.

29–11 Provide the details of the derivation of Eq. (29–8) from the equation that precedes it.

29–12 You want to produce a magnetic field with a magnitude of 5.50×10^{-4} T at a distance of 0.040 m from a long, straight wire. a) What current is required to produce this field? b) With the current found in part (a), what is the magnitude of the field at a distance of 0.080 m from the wire and at 0.160 m?

29–13 Effect of Transmission Lines. Two hikers are reading a compass under an overhead transmission line that is 5.50 m above the ground and carries a current of 800 A in a horizontal direction from north to south. a) Find the magnitude and direction of the magnetic field at a point on the ground directly under the conductor. b) One hiker suggests they walk on another 50 m to avoid inaccurate compass readings caused by the current. Considering that the magnitude of the earth's field is of the order of 0.5×10^{-4} T, is the current really a problem?

SECTION 29–5 FORCE BETWEEN PARALLEL CONDUCTORS

29–14 Three parallel wires each carry current I in the directions shown in Fig. 29–32. If the separation between adjacent wires is d, calculate the magnitude and direction of the net magnetic force per unit length on each wire.

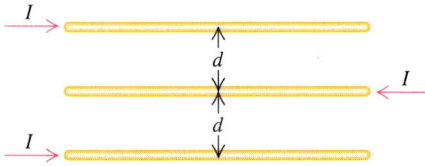

FIGURE 29–32 Exercise 29–14.

29–15 A long, horizontal wire AB rests on the surface of a table and carries a current I. Horizontal wire CD is vertically above wire AB, and is free to slide up and down on the two vertical metal guides C and D (Fig. 29–33). Wire CD is connected through the sliding contacts to another wire that also carries a current I, opposite in direction to the current in wire AB. The mass per unit length of the wire CD is λ. To what equilibrium height h will the wire CD rise, assuming that the magnetic force on it is due wholly to the current in the wire AB?

FIGURE 29–33 Exercise 29–15.

29–16 Two long, parallel wires are separated by a distance of 0.400 m (Fig. 29–34). The currents I_1 and I_2 have the directions shown. a) Calculate the magnitude of the force exerted by each

FIGURE 29–34 Exercise 29–16.

wire on a 1.20-m length of the other. Is the force attractive or repulsive? b) Each current is doubled, so that I_1 becomes 10.0 A and I_2 becomes 4.00 A. Now what is the magnitude of the force that each wire exerts on a 1.20-m length of the other?

29–17 Two long, parallel wires are separated by a distance of 2.50 cm. The force per unit length that each wire exerts on the other is 4.00×10^{-5} N/m, and the wires repel each other. The current in one wire is 0.600 A. a) What is the current in the second wire? b) Are the two currents in the same or in opposite directions?

SECTION 29–6 MAGNETIC FIELD OF A CIRCULAR CURRENT LOOP

29–18 Calculate the magnitude and direction of the magnetic field at point P due to the current in the semicircular section of wire shown in Fig. 29–35. (*Hint:* Does the current in the long, straight section of the wire produce any field at P?)

FIGURE 29–35 Exercise 29–18.

29–19 Calculate the magnitude of the magnetic field at point P of Fig. 29–36 in terms of R, I_1, and I_2. What does your expression give when $I_1 = I_2$?

FIGURE 29–36 Exercise 29–19.

29–20 A closely wound circular coil with a diameter of 4.00 cm has 600 turns and carries a current 0.500 A. What is the magnitude of the magnetic field a) at the center of the coil? b) at a point on the axis of the coil 8.00 cm from its center?

29–21 A closely wound coil has a radius of 6.00 cm and carries a current of 2.50 A. How many turns must it have if, at a point on the coil axis 6.00 cm from the center of the coil, the magnetic field is 6.39×10^{-4} T?

SECTION 29–7 AMPERE'S LAW

29–22 A closed curve encircles several conductors. The line integral $\oint \vec{B} \cdot d\vec{l}$ around this curve is 3.83×10^{-4} T $\cdot$ m. a) What is the net current in the conductors? b) If you were to integrate around the curve in the opposite direction, what would be the value of the line integral? Explain.

29–23 Figure 29–37 shows, in cross section, several conductors that carry currents through the plane of the figure. The currents have the magnitudes $I_1 = 4.0$ A, $I_2 = 6.0$ A, and $I_3 = 2.0$ A, and the directions shown. Four paths, labeled a through d, are shown. What is the line integral $\oint \vec{B} \cdot d\vec{l}$ for each path? Each integral involves going around the path in the counterclockwise direction. Explain your answers.

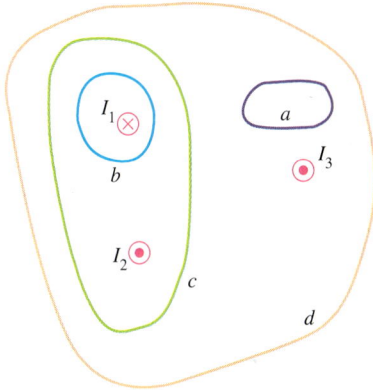

FIGURE 29-37 Exercise 29-23.

SECTION 29-8 APPLICATIONS OF AMPERE'S LAW

29-24 Coaxial Cable. A solid conductor with radius a is supported by insulating disks on the axis of a conducting tube with inner radius b and outer radius c (Fig. 29-38). The central conductor and tube carry equal currents I in opposite directions. The currents are distributed uniformly over the cross sections of each conductor. Derive an expression for the magnitude of the magnetic field a) at points outside the central solid conductor but inside the tube ($a < r < b$); b) at points outside the tube ($r > c$).

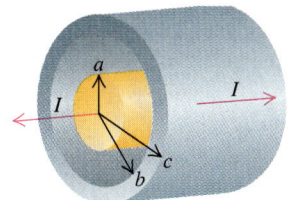

FIGURE 29-38 Exercises 29-24 and 29-25, and Problem 29-58.

29-25 Repeat Exercise 29-24 for the case in which the current in the central solid conductor is I_1, that in the tube is I_2, and these currents are in the same direction rather than in opposite directions.

29-26 A 15.0-cm-long solenoid with radius 2.50 cm is closely wound with 600 turns of wire. The current in the windings is 8.00 A. Compute the magnetic field at a point near the center of the solenoid.

29-27 A solenoid is designed to produce a magnetic field of 0.0270 T at its center. It has radius 1.40 cm and length 40.0 cm, and the wire can carry a maximum current of 12.0 A. a) What minimum number of turns per unit length must the solenoid have? b) What total length of wire is required?

29-28 A toroidal solenoid (Fig. 29-21) has inner radius $r_1 = 15.0$ cm and outer radius $r_2 = 18.0$ cm. The solenoid has 250 turns and carries a current of 8.50 A. What is the magnitude of the magnetic field at the following distances from the center of the torus: a) 12.0 cm; b) 16.0 cm; c) 20.0 cm?

29-29 A wooden ring whose mean diameter is 14.0 cm is wound with a closely spaced toroidal winding of 600 turns. Compute the magnitude of the magnetic field at the center of the cross section of the windings when the current in the windings is 0.650 A.

*SECTION 29-9 MAGNETIC MATERIALS

*29-30 A toroidal solenoid having 400 turns of wire and a mean radius of 6.0 cm carries a current of 0.25 A. The relative permeability of the core is 80. a) What is the magnetic field in the core? b) What part of the magnetic field is due to atomic currents?

*29-31 A toroidal solenoid with 500 turns is wound on a ring having a mean radius of 2.90 cm. Find the current in the winding that is required to set up a magnetic field of 0.350 T in the ring a) if the ring is made of annealed iron ($K_m = 1400$); b) if the ring is made of silicon steel ($K_m = 5200$).

*29-32 The current in the windings of a toroidal solenoid is 2.400 A. There are 500 turns and the mean radius is 25.00 cm. The toroidal solenoid is filled with a magnetic material. The magnetic field inside the windings is found to be 1.940 T. Calculate a) the relative permeability; b) the magnetic susceptibility of the material that fills the toroid.

*29-33 A long solenoid with 60 turns of wire per centimeter carries a current of 0.15 A. The wire that makes up the solenoid is wrapped around a solid core of silicon steel ($K_m = 5200$). (The wire of the solenoid is jacketed with an insulator so that none of the current flows into the core.) a) For a point inside the core, find the magnitudes of i) the magnetic field $\vec{B}_0$ due to the solenoid current, ii) the magnetization $\vec{M}$, and iii) the total magnetic field $\vec{B}$. b) In a sketch of the solenoid and core, show the directions of the vectors $\vec{B}$, $\vec{B}_0$, and $\vec{M}$ inside the core.

*29-34 Show that the units $A \cdot m^2$ and J/T for the Bohr magneton are equivalent.

*29-35 Curie's Law. Experimental measurements of the magnetic susceptibility of iron ammonium alum are given below. Graph $1/\chi_m$ against Kelvin temperature. Does the material obey Curie's law? If so, what is the Curie constant?

T (°C)	χ_m
−258.15	129×10^{-4}
−173	19.4×10^{-4}
−73	9.7×10^{-4}
27	6.5×10^{-4}

SECTION 29-10 DISPLACEMENT CURRENT

29-36 Displacement Current in a Dielectric. Suppose that the parallel plates in Fig. 29-27 have an area of 3.00 cm^2 and are separated by a 2.50-mm-thick sheet of dielectric that completely fills the volume between the plates. The dielectric has dielectric constant 4.70. (Neglect fringing effects.) At a certain instant, the potential difference between the plates is 120 V and the conduction current i_C equals 6.00 mA. At this instant, what is a) the charge q on each plate; b) the rate of change of charge on the plates; c) the displacement current in the dielectric?

29-37 A parallel-plate air-filled capacitor is being charged as in Fig. 29-27. The circular plates have radius 4.00 cm, and at a particular instant the conduction current in the wires is 0.280 A. a) What is the displacement current density j_D in the air space between the plates? b) What is the rate at which the electric field between the plates is changing? c) What is the induced magnetic field between the plates at a distance of 2.00 cm from the axis? d) At 1.00 cm from the axis?

29–38 A long, straight copper wire with a circular cross-section area of 2.1 mm^2 carries a current of 16 A. The resistivity of the material is $2.0 \times 10^{-8}\ \Omega \cdot \text{m}$. a) What is the uniform electric field in the material? b) If the current is changing at the rate of 4000 A/s, at what rate is the electric field in the material changing? c) What is the displacement-current density in the material in part (b)? (*Hint:* Since K for copper is very close to 1, use $\epsilon = \epsilon_0$.) d) If the current is changing as in part (b), what is the magnitude of the magnetic field 6.0 cm from the center of the wire? Note

that both the conduction current and the displacement current should be included in the calculation of B. Is the contribution from the displacement current significant?

29–39 In Fig. 29–27 the capacitor plates have area 5.00 cm^2 and separation 2.00 mm. The plates are in vacuum. The charging current i_C has a *constant* value of 1.80 mA. At $t = 0$ the charge on the plates is zero. a) Calculate the charge on the plates, the electric field between the plates, and the potential difference between the plates when $t = 0.500\ \mu s$. b) Calculate dE/dt, the time rate of change of the electric field between the plates. Does dE/dt vary in time? c) Calculate the displacement current density j_D between the plates, and from this the total displacement current i_D. How do i_C and i_D compare?

PROBLEMS

29–40 A pair of point charges, $q = +8.00\ \mu\text{C}$ and $q' = -5.00\ \mu\text{C}$, are moving in a reference frame as shown in Fig. 29–29 with speeds $v = 9.00 \times 10^4$ m/s and $v' = 6.50 \times 10^4$ m/s. When the point charges are at the locations shown in the figure, what magnetic force (magnitude and direction) does q' exert on q?

29–41 A long, straight wire carries a current of 2.50 A. An electron is traveling in the vicinity of the wire. At the instant when the electron is 4.50 cm from the wire and traveling with a speed of 6.00×10^4 m/s parallel to the wire in the direction opposite to the current, what are the magnitude and direction of the force that the magnetic field of the current exerts on the electron?

29–42 A negative point charge $q = -7.20$ mC is moving in a reference frame. When the point charge is at the origin, the magnetic field it produces at the point $x = 25.0$ cm, $y = 0$, $z = 0$ is $\vec{B} = (6.00\ \mu\text{T})\hat{\jmath}$, and its speed is 800 m/s. a) What are the x-, y-, and z-components of the velocity $\vec{v}_0$ of the charge? b) At this same instant, what is the magnitude of the magnetic field that the charge produces at the point $x = 0$, $y = 25.0$ cm, $z = 0$?

29–43 A neophyte magnet designer tells you that he can produce a magnetic field $\vec{B}$ in a vacuum that points everywhere in the x-direction and that increases in magnitude with increasing x. That is, $\vec{B} = B_0(x/a)\hat{\imath}$, where B_0 and a are constants with units of teslas and meters, respectively. Use Gauss's law for magnetic fields to show that this claim is *impossible*. (*Hint:* Use a Gaussian surface in the shape of a rectangular box, with edges parallel to the x-, y-, and z-axes.)

29–44 Three long, parallel wires are located as shown in Fig. 29–39. Wire 2 is at the origin, wire 1 is on the y-axis at $y = 3.00$ cm, and wire 3 is on the x-axis at $x = 4.00$ cm. The currents are $I_1 = 1.00$ A, $I_2 = 2.00$ A, and $I_3 = 4.00$ A. a) Copy Fig. 29–39 and show the directions of the magnetic fields of wires 2 and 3 in the xy-plane at the location of wire 1. b) Find the x- and y-components of the net magnetic field in the xy-plane at the location of wire 1 due to the currents in wires 2 and 3. c) Find the magnitude and direction of the net force exerted on a 1.00-cm section of wire 1 by the other two wires.

29–45 Two long, straight, parallel wires are 1.00 m apart (Fig. 29–40). The upper wire carries a current I_1 of 6.00 A into the plane of the paper. a) What must be the magnitude and direction

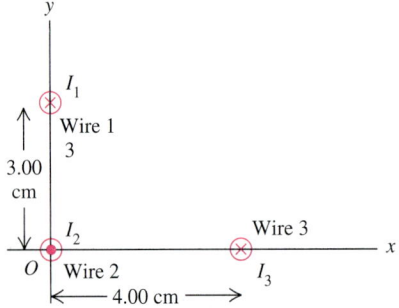

FIGURE 29–39 Problem 29–44.

of the current I_2 for the net field at point P to be zero? b) What are then the magnitude and direction of the net field at Q? c) What is then the magnitude of the net field at S?

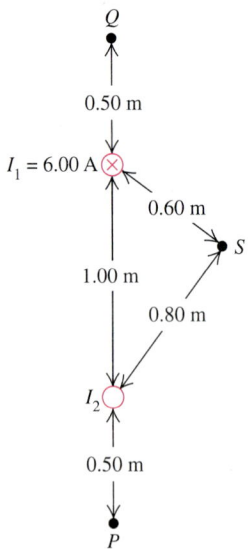

FIGURE 29–40 Problem 29–45.

29–46 Figure 29–41 is an end view of two long, parallel wires perpendicular to the xy-plane, each carrying a current I, but in opposite directions. a) Copy the diagram, and draw vectors to show the $\vec{B}$ field of each wire and the net $\vec{B}$ field at point P. b) Derive the expression for the magnitude of $\vec{B}$ at any point on the x-axis in terms of the x-coordinate of the point. What is the direction of $\vec{B}$? c) Graph the magnitude of $\vec{B}$ at points on the x-axis. d) At what value of x is the magnitude of $\vec{B}$ a maximum? e) What is the magnitude of $\vec{B}$ when $x \gg a$?

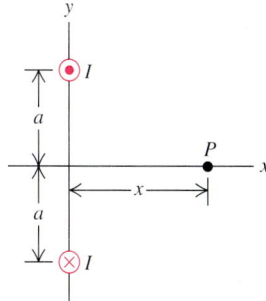

FIGURE 29–41 Problems 29–46, 29–47, and 29–48.

29–47 Repeat Problem 29–46, but with the current in both wires in Fig. 29–41 directed into the plane of the figure.

29–48 Refer to the situation in Problem 29–46. Suppose a third long, straight wire, parallel to the other two, passes through point P (Fig. 29–41), and that each wire carries a current $I = 6.00$ A. Let $a = 40.0$ cm and $x = 60.0$ cm. Find the magnitude and direction of the force per unit length on the third wire, a) if the current in it is directed into the plane of the figure; b) if the current in it is directed out of the plane of the figure.

29–49 Two long, parallel wires hang by 4.00-cm-long cords from a common axis (Fig. 29–42). The wires have a mass per unit length of 0.0125 kg/m and carry the same current in opposite directions. What is the current in each wire if the cords hang at an angle of 6.00° with the vertical?

FIGURE 29–42 Problem 29–49.

29–50 The long, straight wire AB in Fig. 29–43 carries a current of 14.0 A. The rectangular loop whose long edges are parallel to the wire carries a current of 5.00 A. Find the magnitude and direction of the net force exerted on the loop by the magnetic field of the wire.

29–51 A circular wire loop of radius a has N turns and carries a current I. A second loop with N' turns of radius a' carries current I' and is located on the axis of the first loop, a distance x from the center of the first loop. The second loop is tipped so that its axis is at an angle θ from the axis of the first loop. The distance x is large compared to both a and a'. a) Find the magnitude of

FIGURE 29–43 Problem 29–50.

the torque exerted on the second loop by the first loop. b) Find the potential energy for the second loop due to this interaction. c) What simplifications result from having x much larger than a? From having x much larger than a'?

29–52 The wire semicircles in Fig. 29–44 have radii a and b. Calculate the net magnetic field (magnitude and direction) that the current in the wires produces at point P.

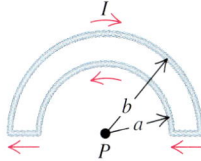

FIGURE 29–44 Problem 29–52.

29–53 Helmholtz Coils. Figure 29–45 is a sectional view of two circular coils with radius a, each wound with N turns of wire carrying a current I, circulating in the same direction in both coils. The coils are separated by a distance a equal to their radii. In this configuration the coils are called Helmholtz coils; they produce a very uniform magnetic field in the region between them. a) Derive the expression for the magnitude B of the magnetic field at a point on the axis a distance x to the right of point P, which is midway between the coils. b) Graph B versus x for $x = 0$ to $x = a/2$. Compare this graph to one for the magnetic field due to the right-hand coil alone. c) From part (a), obtain an expression for the magnitude of the magnetic field at point P. d) Calculate the magnitude of the magnetic field at P if $N = 300$ turns, $I = 6.00$ A, and $a = 8.00$ cm. e) Calculate dB/dx and d^2B/dx^2 at P ($x = 0$). Discuss how your results show that the field is very uniform in the vicinity of P.

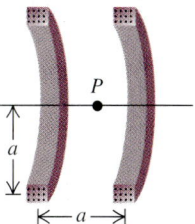

FIGURE 29–45 Problem 29–53.

29–54 Calculate the magnitude and direction of the magnetic field produced at point P in Fig. 29–46 by the current I in the rectangular wire loop. (Point P is at the center of the rectangle.) (*Hint:* The gap on the left-hand side where the wires enter and leave the rectangle is so small that this side of the rectangle can be taken to be a continuous wire with length b.)

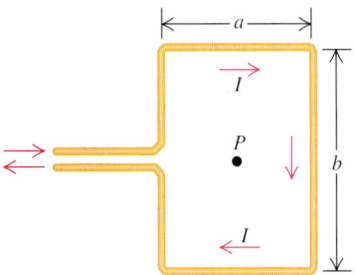

FIGURE 29–46 Problem 29–54.

29–55 The wire in Fig. 29–47 carries current I in the direction shown. The wire consists of a very long, straight section, a quarter-circle with radius R, and another long, straight section. What are the magnitude and direction of the net magnetic field at the center of curvature of the quarter-circle section (point P)?

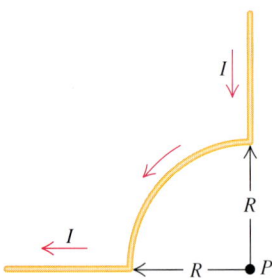

FIGURE 29–47 Problem 29–55.

29–56 The wire in Fig. 29–48 is infinitely long and carries a current I. Calculate the magnitude and direction of the magnetic field that this current produces at point P.

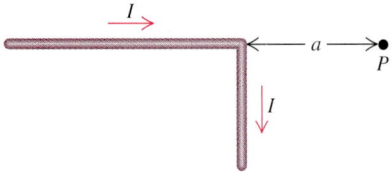

FIGURE 29–48 Problem 29–56.

29–57 A long, straight wire with circular cross section of radius R carries a current I. Assume that the current density is not constant across the cross section of the wire but rather varies as $J = \alpha r$, where α is a constant. a) By the requirement that J integrated over the cross section of the wire gives the total current I, calculate the constant α in terms of I and R. b) Use Ampere's law to calculate the magnetic field $B(r)$ for i) $r \le R$; ii) $r \ge R$. Express your answers in terms of I.

29–58 a) For the coaxial cable of Exercise 29–24, derive an expression for the magnitude of the magnetic field at points inside the central solid conductor ($r < a$). Compare your result

when $r = a$ to the results of part (a) of Exercise 29–24 at that same point. b) For this coaxial cable derive an expression for the field within the tube ($b < r < c$). Compare your result when $r = b$ to part (a) of Exercise 29–24 at that same point. Compare your result when $r = c$ to part (b) of Exercise 29–24 at that same point.

29–59 The electric field of an infinite line of positive charge is directed radially outward from the wire and can be calculated using Gauss's law for the electric field (see Example 23–6 in Section 23–5). Use Gauss's law for magnetism to show that the *magnetic* field of a straight, infinitely long current-carrying *conductor* cannot have a radial component.

29–60 A conductor is made in the form of a hollow cylinder with inner and outer radii a and b, respectively. It carries a current I, uniformly distributed over its cross section. Derive expressions for the magnitude of the magnetic field in the regions a) $r < a$; b) $a < r < b$; c) $r > b$.

29–61 A long, straight solid cylinder, oriented with its axis in the z-direction, carries a current whose current density is $\vec{J}$. The current density, although symmetrical about the cylinder axis, is not constant but varies according to the relation

$$\vec{J} = \frac{2I_0}{\pi a^2}\left[1 - \left(\frac{r}{a}\right)^2\right]\hat{k} \qquad \text{for} \qquad r \le a,$$
$$= 0 \qquad \text{for} \qquad r \ge a,$$

where a is the radius of the cylinder, r is the radial distance from the cylinder axis, and I_0 is a constant having units of amperes. a) Show that I_0 is the total current passing through the entire cross section of the wire. b) Using Ampere's law, derive an expression for the magnitude of the magnetic field $\vec{B}$ in the region $r \ge a$. c) Obtain an expression for the current I contained in a circular cross section of radius $r \le a$ and centered at the cylinder axis. d) Using Ampere's law, derive an expression for the magnitude of the magnetic field $\vec{B}$ in the region $r \le a$. How do your results in parts (b) and (d) compare for $r = a$?

29–62 A long, straight, solid cylinder, oriented with its axis in the z-direction, carries a current whose current density is $\vec{J}$. The current density, although symmetrical about the cylinder axis, is not constant and varies according to the relation

$$\vec{J} = \left(\frac{b}{r}\right)e^{(r-a)/\delta}\hat{k} \qquad \text{for} \qquad r \le a,$$
$$= 0 \qquad \text{for} \qquad r \ge a,$$

where the radius of the cylinder is $a = 5.00$ cm, r is the radial distance from the cylinder axis, b is a constant equal to 600 A · m^{-1}, and δ is a constant equal to 2.50 cm. a) Let I_0 be the total current passing through the entire cross section of the wire. Obtain an expression for I_0 in terms of b, δ, and a. Evaluate your expression to obtain a numerical value for I_0. b) Using Ampere's law, derive an expression for the magnetic field $\vec{B}$ in the region $r \ge a$. Express your answer in terms of I_0 rather than b. c) Obtain an expression for the current I contained in a circular cross section of radius $r \le a$ and centered at the cylinder axis. Express your answer in terms of I_0 rather than b. d) Using Ampere's law, derive an expression for the magnetic field $\vec{B}$ in the region $r \le a$. e) Evaluate the magnitude of the magnetic field at $r = \delta$, $r = a$, and $r = 2a$.

29–63 Integrate B_x as given in Eq. (29–15) from $-\infty$ to $+\infty$. That is, calculate $\int_{-\infty}^{+\infty} B_x\, dx$. Explain the significance of your result.

29–64 In a region of space where there are no conduction or displacement currents, it is impossible to have a uniform magnetic field that abruptly drops to zero. To prove this statement, use the method of contradiction: Assume that such a case *is* possible, then show that your assumption contradicts a law of nature. a) In the bottom half of a piece of paper, draw evenly spaced horizontal lines representing a uniform magnetic field to your right. Use dashed lines to draw a rectangle *abcda* with horizontal side *ab* in the magnetic field region and horizontal side *cd* in the top half of your paper where $B = 0$. b) Show that integration around your rectangle contradicts Ampere's law.

29–65 An Infinite Current Sheet. Long, straight conductors with square cross section and each carrying current I are laid side by side to form an infinite current sheet (Fig. 29–49). The conductors lie in the *xy*-plane, are parallel to the *y*-axis and carry current in the +*y*-direction. There are *n* conductors per unit length measured along the *x*-axis. a) What are the magnitude and direction of the magnetic field a distance a below the current sheet? b) What are the magnitude and direction of the magnetic field a distance a above the current sheet?

FIGURE 29–49 Problem 29–65.

29–66 Long, straight conductors with square cross section, each carrying current I, are laid side by side to form an infinite current sheet with current directed out of the plane of the page (Fig. 29–50). A second infinite current sheet is a distance d below the first and is parallel to it. The second sheet carries current into the plane of the page. Each sheet has *n* conductors per unit length. (Refer to Problem 29–65.) Calculate the magnitude and direction of the net magnetic field at a) point P (above the

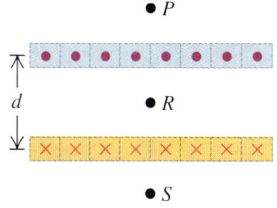

FIGURE 29–50 Problem 29–66.

upper sheet); b) point R (midway between the two sheets); c) point S (below the lower sheet).

*29–67 A piece of iron has magnetization $M = 6.50 \times 10^4$ A/m. Find the average magnetic dipole moment *per atom* in this piece of iron. Express your answer both in A $\cdot$ m^2 and in Bohr magnetons. The density of iron is given in Table 14–1 and the atomic mass of iron (in grams per mole) is given in Appendix D. The chemical symbol for iron is Fe.

*29–68 a) In Section 28–8 we discussed how a magnetic dipole, such as a current loop or a magnetized object, can be attracted or repelled by a permanent magnet (see Figs. 28–32 and 28–34). Use this to explain why *either* pole of a magnet *attracts* both paramagnetic materials and (initially unmagnetized) ferromagnetic materials, but *repels* diamagnetic materials. b) The force that a magnet exerts on an object is directly proportional to the object's magnetic moment. A particular magnet is just strong enough to pick up a cube of annealed iron ($K_m = 1400$) 2.00 cm on a side so that the iron sticks to one of the magnet's poles; that is, the magnet exerts an upward force on the iron cube equal to the cube's weight. If you tried to use this magnet to pick up a 2.00-cm cube of aluminum instead, what would be the upward force on the cube? How does this compare to the weight of the cube? Could the magnet pick up the cube? (*Hint:* You will need to use information from Tables 14–1 and 29–1.) c) If you tried to use the magnet to pick up a 2.00-cm cube of silver, what would be the magnitude and direction of the force on the cube? How does this magnitude compare to the weight of the cube? Would the effects of the magnetic force be noticeable?

29–69 A capacitor has two parallel plates with area A separated by a distance d. The space between plates is filled with a material having dielectric constant K. The material is not a perfect insulator but has resistivity ρ. The capacitor is initially charged with charge of magnitude Q_0 on each plate, which gradually discharges by conduction through the dielectric. a) Calculate the conduction current density $j_C(t)$ in the dielectric. b) Show that at any instant the displacement current density in the dielectric is equal in magnitude to the conduction current density but opposite in direction, so the *total* current density is zero at every instant.

29–70 A rod of pure silicon (resistivity $\rho = 2300\ \Omega \cdot$ m) is carrying a current. The electric field varies sinusoidally with time according to $E = E_0 \sin \omega t$, where $E_0 = 0.450$ V/m, $\omega = 2\pi f$, and the frequency $f = 120$ Hz. a) Find the magnitude of the maximum conduction current density in the wire. b) Assuming $\epsilon = \epsilon_0$, find the maximum displacement current density in the wire, and compare with the result of part (a). c) At what frequency f would the maximum conduction and displacement densities become equal, if $\epsilon = \epsilon_0$ (which is not actually the case)? d) At the frequency determined in part (c), what is the relative *phase* of the conduction and displacement currents?

CHALLENGE PROBLEMS

29–71 Two long, straight conducting wires with linear mass density λ are suspended from cords so that they are each hori-

zontal, parallel to each other, and a distance d apart. The back ends of the wires are connected to one another by a slack low-resistance connecting wire. A charged capacitor (capacitance C) is now added to the system; the positive plate of the capacitor (initial charge $+Q_0$) is connected to the front end of one of the wires, and the negative plate of the capacitor (initial charge $-Q_0$) is connected to the front end of the other wire (Fig. 29–51). Both these connections are also made by slack low-resistance wires. When the connection is made, the wires are pushed aside by the repulsive force between the wires, and each wire has an initial horizontal velocity of magnitude v_0. Assume that the time for the capacitor to discharge is negligible compared to the time it takes for any appreciable displacement in the position of the wires to occur. a) Show that the initial speed v_0 of either wire is given by

$$v_0 = \frac{\mu_0 Q_0^2}{4\pi\lambda RCd},$$

where R is the total resistance of the circuit. b) Determine v_0 numerically if $\lambda = 4.50 \times 10^{-3}$ kg/m, $d = 3.00$ cm, $C = 2.50$ μF, and $R = 0.0480$ Ω, and if the capacitor was initially charged by connecting it to a 3.00-kV source. c) To what height h will each wire rise as a result of the circuit connection?

29–72 A wide, long insulating belt has a uniform positive charge per unit area σ on its upper surface. Rollers at each end move the belt to the right at a constant speed v. Calculate the magnitude and direction of the magnetic field produced by the moving belt at a point just above its surface. (*Hint:* At points near the surface and far from its edges or ends, the moving belt can be considered to be an infinite current sheet like that in Problem 29–65.)

29–73 A Charged Dielectric Disk. A thin disk of dielectric material with radius a has a total charge $+Q$ distributed uniformly over its surface. It rotates n times per second about an axis perpendicular to the surface of the disk and passing through its center. Find the magnetic field at the center of the disk. (*Hint:* Divide the disk into concentric rings of infinitesimal width.)

29–74 A wire in the shape of a semicircle with radius a is oriented in the yz-plane with its center of curvature at the origin (Fig. 29–52). If the current in the wire is I, calculate the magnetic field components produced at point P, a distance x out along the x-axis. (*Note:* Do not forget the contribution from the straight wire at the bottom of the semicircle that runs from $z = -a$ to $z = +a$. You may use the fact that the fields of the two antiparallel currents at $z > a$ cancel, but you must explain *why* they cancel.)

FIGURE 29–51 Challenge Problem 29–71.

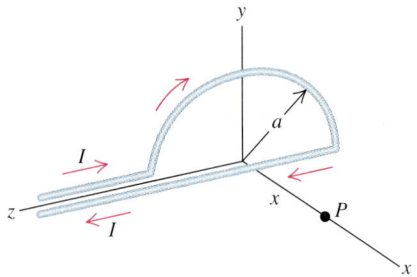

FIGURE 29–52 Challenge Problem 29–74.

A credit card's number, expiration date, and cardholder name are coded into a magnetized pattern in a stripe on the back. When the card is "swiped" through a card reader, the moving stripe bathes the reader's circuitry in a varying magnetic field that induces currents in the circuits. These currents transmit the information in the stripe to the cardholder's bank.

30–1 INTRODUCTION

Almost every modern device or machine, from a computer to a washing machine to a power drill, has electric circuits at its heart. We learned in Chapter 26 that an electromotive force (emf) is required for a current to flow in a circuit; in Chapters 26 and 27 we almost always took the source of emf to be a battery. But for the vast majority of electric devices that are used in industry and in the home (including any device that you plug into a wall socket), the source of emf is *not* a battery but an electrical generating station. Such a station produces electric energy by converting other forms of energy: gravitational potential energy at a hydroelectric plant, chemical energy in a coal- or oil-fired plant, nuclear energy at a nuclear plant. But how is this energy conversion done? In other words, what is the physics behind the production of almost all of our electric energy needs?

The answer is a phenomenon known as *electromagnetic induction:* If the magnetic flux through a circuit changes, an emf and a current are induced in the circuit. In a power-generating station, magnets move relative to coils of wire to produce a changing magnetic flux in the coils and hence an emf. Other key components of electric power systems, such as transformers, also depend on magnetically induced emfs. When electromagnetic induction was discovered in the 1830s, it was a mere laboratory curiosity; today, thanks to its central role in the generation of electric power, it is fundamentally responsible for the nature of our technological society.

The central principle of electromagnetic induction, and the keystone of this chapter, is *Faraday's law.* This law relates induced emf to changing magnetic flux in any loop, including a closed circuit. We also discuss Lenz's law, which helps us to predict the directions of induced emfs and currents. This chapter provides the principles we need to understand electrical energy-conversion devices such as motors, generators, and transformers. We will also present a neat package of formulas, called *Maxwell's equations,* that describe the behavior of electric and magnetic fields in *any* situation. These equations pave the way for the analysis of electromagnetic waves in Chapter 33.

30–2 INDUCTION EXPERIMENTS

During the 1830s, several pioneering experiments with magnetically induced emf were carried out in England by Michael Faraday and in the United States by Joseph Henry (1797–1878), later the first director of the Smithsonian Institution. Figure 30–1 shows several examples. In Fig. 30–1a, a coil of wire is connected to a galvanometer. When the nearby magnet is stationary, the meter shows no current. This isn't surprising; there is no source of emf in the circuit. But when we *move* the magnet, either toward or away from the coil, the meter shows current in the circuit, but *only* while the magnet is moving. If we keep the magnet stationary and move the coil, we again detect a current during the motion. We call this an **induced current,** and the corresponding emf required to cause this current is called an **induced emf.**

In Fig. 30–1b we replace the magnet with a second coil connected to a battery. When the second coil is stationary, there is no current in the first coil. However, when we move

(a)   (b)   (c)

30–1 (a) A magnet moving toward a coil of wire connected to a galvanometer induces a current in the coil. When the magnet and coil move apart, the induced current is in the opposite direction. If the magnet is held stationary, there is *no* induced current. (b) A second coil carrying a constant current moves toward the coil connected to the galvanometer, inducing a current in it. When the coils move apart, the induced current is in the opposite direction. There is no induced current if the coils are held stationary. (c) When the switch is opened or closed, a changing current in the inside coil induces a current through the galvanometer attached to the other coil.

the second coil toward or away from the first or move the first toward or away from the second, there is current in the first coil, but again *only* while one coil is moving relative to the other.

Finally, using the two-coil setup in Fig. 30–1c, we keep both coils stationary and vary the current in the second coil, either by opening and closing the switch or by changing the resistance of the second coil with the switch closed (perhaps by changing the second coil's temperature). We find that as we open or close the switch, there is a momentary current pulse in the first circuit. When we vary the resistance (and thus the current) in the second coil, there is an induced current in the first circuit, but only while the current in the second circuit is changing.

To explore further the common elements in these observations, let's consider a more detailed series of experiments with the situation shown in Fig. 30–2. We connect a coil of wire to a galvanometer, then place the coil between the poles of an electromagnet whose magnetic field we can vary. Here's what we observe:

1. When there is no current in the electromagnet, so that $\vec{B} = 0$, the galvanometer shows no current.
2. When the electromagnet is turned on, there is a momentary current through the meter as $\vec{B}$ increases.
3. When $\vec{B}$ levels off at a steady value, the current drops to zero, no matter how large $\vec{B}$ is.
4. With the coil in a horizontal plane, we squeeze it so as to decrease the cross-section area of the coil. The meter detects current only *during* the deformation, not before or after. When we increase the area to return the coil to its original shape, there is current in the opposite direction, but only while the area of the coil is changing.
5. If we rotate the coil a few degrees about a horizontal axis, the meter detects current during the rotation, in the same direction as when we decreased the area. When we rotate the coil back, there is a current in the opposite direction during this rotation.
6. If we jerk the coil out of the magnetic field, there is a current during the motion, in the same direction as when we decreased the area.
7. If we decrease the number of turns in the coil by unwinding one or more turns, there is a current during the unwinding, in the same direction as when we decreased the area. If we wind more turns onto the coil, there is a current in the opposite direction during the winding.

30–2 When the $\vec{B}$ field is constant and the shape, location, and orientation of the coil do not change, the induced current is zero. A current is induced when any of these factors change.

8. When the magnet is turned off, there is a momentary current in the direction opposite to the current when it was turned on.

9. The faster we carry out any of these changes, the greater the current.

10. If all these experiments are repeated with a coil that has the same shape but different material and different resistance, the current in each case is inversely proportional to the total circuit resistance. This shows that the induced emfs that are causing the current do not depend on the material of the coil but only on its shape and the magnetic field.

The common element in all these experiments is changing *magnetic flux* Φ_B through the coil connected to the galvanometer. In each case the flux changes either because the magnetic field changes with time or because the coil is moving through a non-uniform magnetic field. Check back through the list to verify this statement. Faraday's law of induction, the subject of the next section, states that in all of these situations the induced emf is proportional to the *rate of change* of magnetic flux Φ_B through the coil. The *direction* of the induced emf depends on whether the flux is increasing or decreasing.

Magnetically induced emf's, like the emf's discussed in Section 26–5, are always the result of the action of *non-electrostatic* forces. When these forces are the result of additional electric fields induced by changing magnetic fields, we will have to distinguish carefully between electric fields produced by charges (according to Coulomb's law) and those produced by changing magnetic fields. We'll denote these by $\vec{E}_c$ (where c stands for Coulomb or conservative) and $\vec{E}_n$ (where n stands for non-Coulomb or non-conservative), respectively. We'll return to this distinction later in this chapter and the next.

13.9
Electromagnetic Induction

30-3 FARADAY'S LAW

The common element in all induction effects is changing magnetic flux through a circuit. Before stating the simple physical law that summarizes all of the kinds of experiments described in Section 30–2, let's first review the concept of magnetic flux Φ_B (which we introduced in Section 28–4). For an infinitesimal area element $d\vec{A}$ in a magnetic field $\vec{B}$ (Fig. 30–3), the magnetic flux $d\Phi_B$ through the area is

$$d\Phi_B = \vec{B} \cdot d\vec{A} = B_\perp \, dA = B \, dA \cos \phi,$$

where $B_\perp$ is the component of $\vec{B}$ perpendicular to the surface of the area element and ϕ is the angle between $\vec{B}$ and $d\vec{A}$. (As in Chapter 29, be careful to distinguish between two quantities named "phi," ϕ and Φ_B.) The total magnetic flux Φ_B through a finite area is the integral of this expression over the area:

$$\Phi_B = \int \vec{B} \cdot d\vec{A} = \int B \, dA \cos \phi. \tag{30-1}$$

If $\vec{B}$ is uniform over a flat area $\vec{A}$, then

$$\Phi_B = \vec{B} \cdot \vec{A} = BA \cos \phi. \tag{30-2}$$

In Eqs. (30–1) and (30–2) we have to be careful to define the direction of the vector area $d\vec{A}$ or $\vec{A}$ unambiguously. There are always two directions perpendicular to any given area, and the sign of the magnetic flux through the area depends on which one we choose to be positive.

Faraday's law of induction states:

The induced emf in a closed loop equals the negative of the time rate of change of magnetic flux through the loop.

In symbols, Faraday's law is

$$\varepsilon = -\frac{d\Phi_B}{dt} \qquad \text{(Faraday's law of induction)}. \tag{30-3}$$

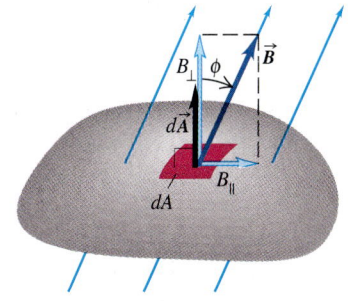

30-3 The magnetic flux through an area element dA is defined to be $d\Phi_B = B_\perp \, dA$.

To understand the negative sign, we have to introduce a sign convention for the induced emf $\mathcal{E}$. But first let's look at a simple example of this law in action.

EXAMPLE 30–1

Emf and current induced in a loop In Fig. 30–4 the magnetic field between the poles of the electromagnet is uniform at any time, but its magnitude is increasing at the rate of 0.020 T/s. The area of the conducting loop in the field is 120 cm^2, and the total circuit resistance, including the meter, is 5.0 Ω. a) Find the induced emf and the induced current in the circuit. b) If the loop is replaced by one made of an insulator, what effect does this have on the induced emf and induced current?

SOLUTION a) The vectors $\vec{A}$ and $\vec{B}$ are parallel and $\vec{B}$ is uniform, so $\Phi_B = \vec{B} \cdot \vec{A} = BA$. The area $A = 0.012$ m^2 is constant, so the rate of change of magnetic flux is

$$\frac{d\Phi_B}{dt} = \frac{d(BA)}{dt} = \frac{dB}{dt}A = (0.020 \text{ T/s})(0.012 \text{ m}^2)$$

$$= 2.4 \times 10^{-4} \text{ V} = 0.24 \text{ mV}.$$

This, apart from a sign that we haven't discussed yet, is the induced emf $\mathcal{E}$.

It's worthwhile to verify unit consistency in this calculation. There are many ways to do this; one is to note that because of the magnetic force relation $\vec{F} = q\vec{v} \times \vec{B}$, 1 T = (1 N)/(1 C · m/s). The units of magnetic flux can then be expressed as (1 T) (1 m^2) = 1 N · s · m/C, and the rate of change of magnetic flux as 1 N · m/C = 1 J/C = 1 V. Thus the unit of $d\Phi_B/dt$ is the volt, as required by Eq. (30–3). Also recall that the unit of magnetic flux is the weber (Wb): 1 T · m^2 = 1 Wb, so 1 V = 1 Wb/s.

Finally, the induced current I in the circuit is

$$I = \frac{\mathcal{E}}{R} = \frac{2.4 \times 10^{-4} \text{ V}}{5.0 \text{ Ω}} = 4.8 \times 10^{-5} \text{ A} = 0.048 \text{ mA}.$$

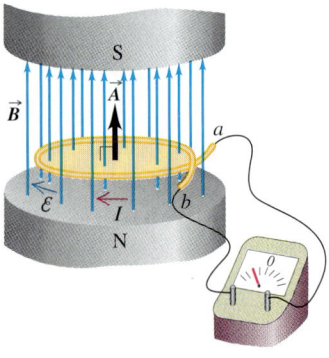

30–4 A stationary conducting loop in an increasing magnetic field.

We haven't found the *direction* of the current yet. We'll see how to do this shortly.

b) Faraday's law, Eq. (30–3), does not involve the resistance of the circuit in any way. So the induced emf does *not* change if the resistance of the loop is made very high. But the current will be decreased, as given by the equation $I = \mathcal{E}/R$. If the loop is made of a perfect insulator with infinite resistance, the induced current will be zero even though an emf is present. This situation is analogous to an isolated battery whose terminals aren't connected to anything: There is an emf present, but no current flows.

DIRECTION OF INDUCED EMF

We can find the direction of an induced emf or current by using Eq. (30–3) together with some simple sign rules. Here's the procedure:

1. Define a positive direction for the vector area $\vec{A}$.
2. From the directions of $\vec{A}$ and the magnetic field $\vec{B}$, determine the sign of the magnetic flux Φ_B and its rate of change $d\Phi_B/dt$. Figure 30–5 shows several examples.
3. Determine the sign of the induced emf or current. If the flux is increasing, so $d\Phi_B/dt$ is positive, then the induced emf or current is negative; if the flux is decreasing, $d\Phi_B/dt$ is negative and the induced emf or current is positive.
4. Finally, determine the direction of the induced emf or current using your right hand. Curl the fingers of your right hand around the $\vec{A}$ vector, with your right thumb in the direction of $\vec{A}$. If the induced emf or current in the circuit is *positive,* it is in the same direction as your curled fingers; if the induced emf or current is *negative,* it is in the opposite direction.

In Example 30–1, in which $\vec{A}$ is upward, a positive $\mathcal{E}$ would be directed counterclockwise around the loop, as seen from above. Both $\vec{A}$ and $\vec{B}$ are upward in this example, so Φ_B is positive; the magnitude B is increasing, so $d\Phi_B/dt$ is positive. Hence

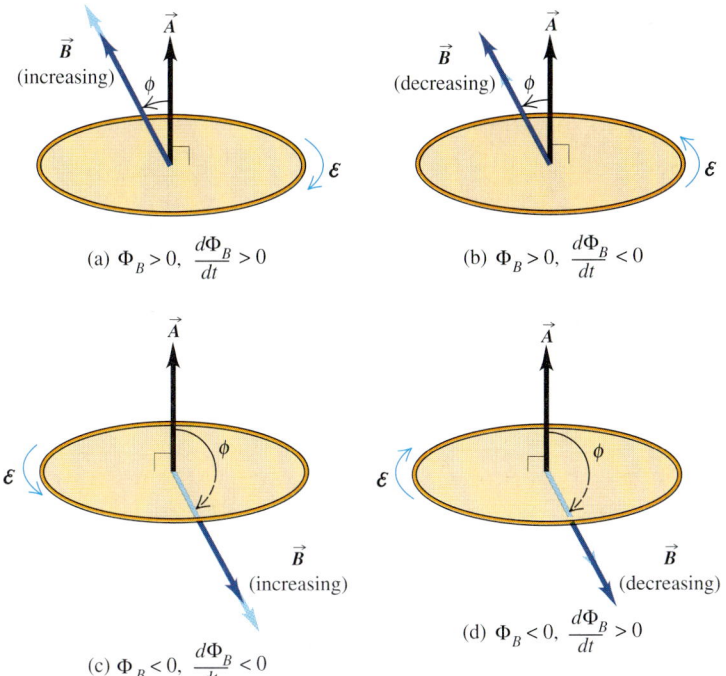

(a) $\Phi_B > 0$, $\dfrac{d\Phi_B}{dt} > 0$

(b) $\Phi_B > 0$, $\dfrac{d\Phi_B}{dt} < 0$

(c) $\Phi_B < 0$, $\dfrac{d\Phi_B}{dt} < 0$

(d) $\Phi_B < 0$, $\dfrac{d\Phi_B}{dt} > 0$

30–5 The magnetic flux is becoming (a) more positive, (b) less positive, (c) more negative, and (d) less negative. Therefore Φ_B is increasing in (a) and (d) and decreasing in (b) and (c). In (a) and (d) the emfs are negative (clockwise as seen from above the vector $\vec{A}$), and in (b) and (c) the emfs are positive (counterclockwise as seen from above the vector $\vec{A}$).

by Eq. (30–3), $\mathcal{E}$ in Example 30–1 is *negative*. Its actual direction is thus *clockwise* around the loop, as shown in Fig. 30–4.

If the loop in Fig. 30–4 is a conductor, an induced current results from this emf; this current is also clockwise. This induced current produces an additional magnetic field through the loop, and the right-hand rule described in Section 29–6 shows that this field is *opposite* in direction to the increasing field produced by the electromagnet. This is an example of a general rule called *Lenz's law,* which says that any induction effect tends to oppose the change that caused it; in this case the change is the increase in the flux of the electromagnet's field through the loop. (We'll study this law in detail in the next section.)

We invite you to check out the signs of the induced emfs and currents for the list of experiments cited at the end of Section 30–2. For example, when the loop in Fig. 30–2 is in a constant field and we tilt it or squeeze it to *decrease* the flux through it, the induced emf and current are counterclockwise, as seen from above.

CAUTION ▶ Since magnetic flux plays a central role in Faraday's law, it's tempting to think that *flux* is the cause of induced emf and that an induced emf will appear in a circuit whenever there is a magnetic field in the region bordered by the circuit. But Eq. (30–3) shows that only a *change* in flux through a circuit, not flux itself, can induce an emf in a circuit. If the flux through a circuit has a constant value, whether positive, negative, or zero, there is no induced emf. ◀

If we have a coil with N identical turns, and if the flux varies at the same rate through each turn, the *total* rate of change through all the turns is N times as large as for a single turn. If Φ_B is the flux through each turn, the total emf in a coil with N turns is

$$\mathcal{E} = -N\frac{d\Phi_B}{dt}. \tag{30-4}$$

As we discussed in Section 30–1, induced emfs play an essential role in the generation of electric power for commercial use. Several of the following examples explore different methods of producing emfs by the motion of a conductor relative to a magnetic field, giving rise to a changing flux through a circuit.

Problem–Solving Strategy

FARADAY'S LAW

1. To calculate the rate of change of magnetic flux, you first have to understand what is making the flux change. Is the loop or coil moving? Is it changing orientation? Is the magnetic field changing? Remember that it's not the flux itself that counts, but its *rate of change.*

2. Choose a direction for the area vector $\vec{A}$ or $d\vec{A}$, and then use it consistently. Remember the sign rules for the positive directions of magnetic flux and emf, and use them

consistently when you implement Eq. (30–3) or (30–4). If your conductor has N turns in a coil, don't forget to multiply by N.

3. Use Faraday's law to obtain the induced emf. Use the sign rules to determine the direction of the induced emf and induced current. If the circuit resistance is known, you can then calculate the magnitude of the current.

Magnitude and direction of an induced emf A coil of wire containing 500 circular loops with radius 4.00 cm is placed between the poles of a large electromagnet, where the magnetic field is uniform and at an angle of 60° with the plane of the coil (Fig. 30–6). The field decreases at a rate of 0.200 T/s. What are the magnitude and direction of the induced emf?

SOLUTION We choose the direction for $\vec{A}$ shown in Fig. 30–6. Then the angle between $\vec{A}$ and $\vec{B}$ is $\phi = 30°$ (*not* 60°). The flux Φ_B at any time is given by $\Phi_B = BA \cos \phi$, and the rate of change of flux is $d\Phi_B/dt = (dB/dt)A \cos \phi$. In our problem, $dB/dt = -0.200$ T/s and $A = \pi(0.0400 \text{ m})^2 = 0.00503 \text{ m}^2$, so

$$\frac{d\Phi_B}{dt} = \frac{dB}{dt} A \cos 30° = (-0.200 \text{ T/s})(0.00503 \text{ m}^2)(0.866)$$

$$= -8.71 \times 10^{-4} \text{ T} \cdot \text{m}^2/\text{s} = -8.71 \times 10^{-4} \text{ Wb/s}.$$

From Eq. (30–4) the induced emf is

$$\mathcal{E} = -N \frac{d\Phi_B}{dt} = -(500)(-8.71 \times 10^{-4} \text{ Wb/s}) = 0.435 \text{ V}.$$

When we look in along the coil axis from the left, in the direction of the area vector $\vec{A}$ (30° above the magnetic field $\vec{B}$), the positive direction for $\mathcal{E}$ is clockwise, according to our right-hand rule. The emf in this example is in fact positive and thus is clockwise. If the ends of the wire are connected together, the direction of current in the coil is clockwise. A clockwise current

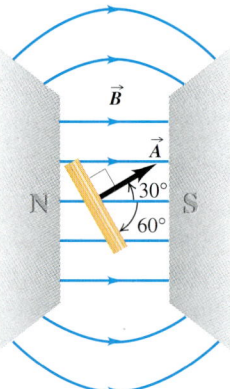

30–6 The magnitude of $\vec{B}$ is decreasing. With the direction of $\vec{A}$ shown, the flux through the coil is decreasing, and $\mathcal{E}$ is positive. The corresponds to clockwise emf and current as you look in from the left along the direction of $\vec{A}$. This additional $\vec{B}$ field caused by the induced current tends to compensate for the decrease in flux.

gives added magnetic flux through the coil in the same direction as the flux from the electromagnet, and therefore tends to oppose the *decrease* in total flux.

The search coil One practical way to measure magnetic field strength uses a small, closely wound coil with N turns called a *search coil*. The coil, of area A, is initially held so that its area vector $\vec{A}$ is aligned with a magnetic field with magnitude B. The coil is then either quickly rotated a quarter-turn about a diameter or quickly pulled out of the field. Explain how this device can be used to measure the value of B.

SOLUTION Initially, the flux through the coil is $\Phi_B = NBA$; when the coil is rotated or pulled from the field, the flux decreases

rapidly from NBA to zero. While the flux is decreasing, there is a momentary induced emf, and a momentary induced current occurs in an external circuit connected to the coil. The rate of change of flux through the coil is proportional to the current, or rate of flow of charge, so it is easy to show that the *total* flux change is proportional to the total charge that flows around the circuit. We can build an instrument that measures this total charge, and from this we can compute B. We leave the details as a problem. Strictly speaking, this method gives only the *average* field over the area of the coil. But if the area is small, this is very nearly equal to the field at the center of the coil.

EXAMPLE 30-4

Generator I: A simple alternator Figure 30–7 shows a simple version of an *alternator,* a device that generates an emf. A rectangular loop is made to rotate with constant angular velocity ω about the axis shown. The magnetic field $\vec{B}$ is uniform and constant. At time $t = 0$, $\phi = 0$. Determine the induced emf.

SOLUTION The flux Φ_B through the loop equals its area A multiplied by $B_\perp = B \cos \phi$, the component of $\vec{B}$ perpendicular to the area, where the angle ϕ equals ωt:

$$\Phi_B = BA \cos \phi = BA \cos \omega t.$$

Then, by Faraday's law,

$$\mathcal{E} = -\frac{d\Phi_B}{dt} = \omega BA \sin \omega t.$$

The induced emf $\mathcal{E}$ varies sinusoidally with time (Fig. 30–7b). When the plane of the loop is perpendicular to $\vec{B}$ ($\phi = 0$ or $180°$), Φ_B reaches its maximum and minimum values. At these times, its instantaneous rate of change is zero and $\mathcal{E}$ is zero. Also, $\mathcal{E}$ is greatest in absolute value when the plane of the loop is parallel to $\vec{B}$ ($\phi = 90°$ or $270°$) and Φ_B is changing most rapidly. Finally, we note that the induced emf does not depend on the *shape* of the loop, but only on its area. Because $\mathcal{E}$ is directly proportional

to ω and B, some tachometers use the emf in a rotating coil to measure rotational speed, and other devices use an emf of this kind to measure magnetic field.

We can use the alternator as a source of emf in an external circuit by use of two *slip rings* S, which rotate with the loop, as shown in Fig. 30–7a. The rings slide against stationary contacts called *brushes,* which are connected to the output terminals a and b. Since the emf varies sinusoidally, the current that results in the circuit is an *alternating* current that also varies sinusoidally in magnitude and direction. An alternator is also called an *alternating-current* (ac) *generator* for this reason. The amplitude of the emf can be increased by increasing the rotation speed, the field magnitude, or the loop area or by using N loops instead of one, as in Eq. (30–4).

Alternators are used in automobiles to generate the currents in the ignition, the lights, and the entertainment system. The arrangement is a little different than in this example; rather than having a rotating loop in a magnetic field, the loop stays fixed and an electromagnet rotates. (The rotation is provided by a mechanical connection between the alternator and the engine.) But the result is the same; the flux through the loop varies sinusoidally, producing a sinusoidally varying emf.

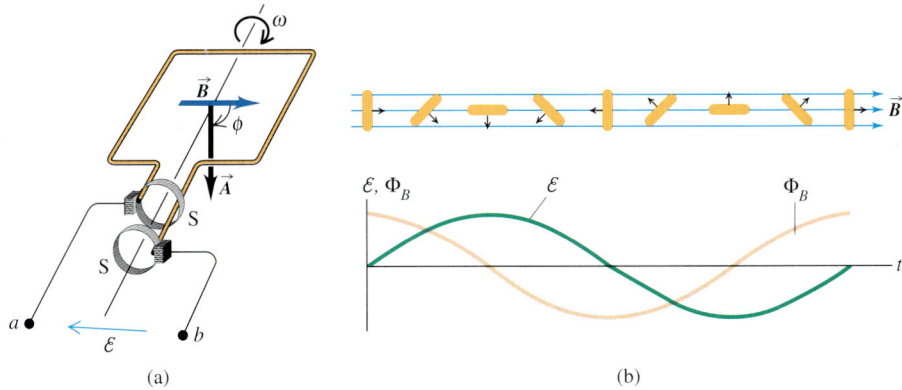

(a)   (b)

30–7 (a) Schematic diagram of a simple alternator, using a conducting loop rotating in a magnetic field. Connections from each end of the loop to the external circuit are made by means of that end's slip ring S. The system is shown at the time when the angle $\phi = \omega t = 90°$. (b) Graph of the flux through the loop and the resulting emf at terminals ab, along with corresponding positions of the loop during one complete rotation. The emf is the negative slope of the flux graph: $\mathcal{E} = -d\Phi_B/dt$.

EXAMPLE 30-5

Generator II: A dc generator and back emf in a motor The alternator in Example 30–4 produces a sinusoidally varying emf and hence an alternating current. We can use a similar scheme to make a *direct-current* (dc) *generator* that produces an emf that always has the same sign. A prototype dc generator is shown in Fig. 30–8a. The arrangement of split rings is called a *commutator;* it reverses the connections to the external circuit at angular positions where the emf reverses. The resulting emf is shown in

Fig. 30–8b. Commercial dc generators have a large number of coils and commutator segments; this arrangement smooths out the bumps in the emf, so the terminal voltage is not only one-directional but also practically constant. This brush-and-commutator arrangement is the same as that in the direct-current motor we discussed in Section 28–9. The motor's *back emf* is just the emf induced by the changing magnetic flux through its rotating coil. Consider a motor with a square coil 10.0 cm on a

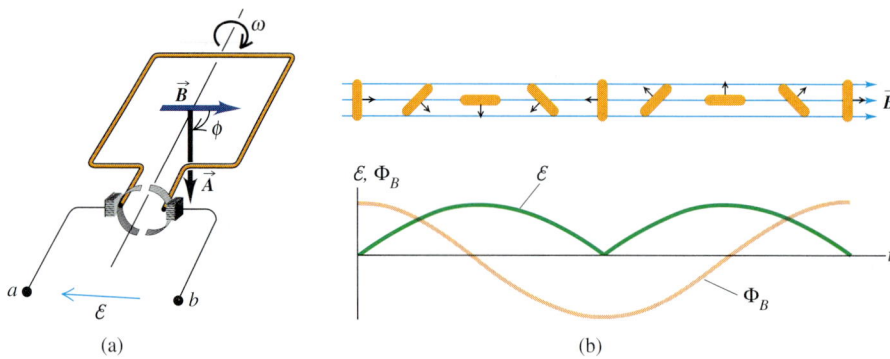

30–8 (a) Schematic diagram of a dc generator, using a split-ring commutator. The ring halves are attached to the loop and rotate with it. (b) Graph of the resulting induced emf at terminals *ab*.

side, with 500 turns of wire. If the magnetic field has magnitude 0.200 T, at what rotation speed will the *average* back emf of the motor be 112 V?

SOLUTION Comparing Figs. 30–7b and 30–8b shows that the back emf of the motor is just the absolute value of the emf found for an alternator in Example 30–4, multiplied by the number of coils *N* as in Eq. (30–4):

$$|\mathcal{E}| = N\omega BA |\sin \omega t|.$$

To find the *average* back emf, we replace $|\sin \omega t|$ by its average value. The average value of the sine function is found by integrating $\sin \omega t$ over half a cycle, from $t = 0$ to $t = T/2 = \pi/\omega$, then dividing by the elapsed time π/ω. During this half cycle, the sine function is positive, so $|\sin \omega t| = \sin \omega t$, and we find

$$\left(|\sin \omega t|\right)_{\text{av}} = \frac{\int_0^{\pi/\omega} \sin \omega t \, dt}{\pi/\omega} = \frac{2}{\pi},$$

or about 0.64. The average back emf is then

$$\mathcal{E}_{\text{av}} = \frac{2 N \omega BA}{\pi}.$$

The back emf is proportional to the rotation speed ω, as was stated without proof in Section 28–9. Solving for ω, we obtain

$$\omega = \frac{\pi \mathcal{E}_{\text{av}}}{2 NBA} = \frac{\pi(112 \text{ V})}{2(500)(0.200 \text{ T})(0.100 \text{ m})^2} = 176 \text{ rad/s}.$$

We used the relationships $1 \text{ V} = 1 \text{ Wb/s} = 1 \text{ T} \cdot \text{m}^2/\text{s}$ from Example 30–1. We were able to add "radians" to the units of the answer because it is a dimensionless quantity, as we discussed in Chapter 9. The rotation speed can also be written as

$$\omega = 176 \text{ rad/s} \frac{1 \text{ rev}}{2\pi \text{ rad}} \frac{60 \text{ s}}{1 \text{ min}} = 1680 \text{ rev/min}.$$

The slower the rotation speed, the less the back emf and the greater the possibility of burning out the motor, as described in Example 28–12 (Section 28–9).

EXAMPLE 30–6

Generator III: The slidewire generator Figure 30–9 shows a U-shaped conductor in a uniform magnetic field $\vec{B}$ perpendicular to the plane of the figure, directed *into* the page. We lay a metal rod with length *L* across the two arms of the conductor, forming a circuit, and move the rod to the right with constant velocity $\vec{v}$. This induces an emf and a current, which is why this device is called a *slidewire generator.* Find the magnitude and direction of the resulting induced emf.

SOLUTION The magnetic flux through the circuit is changing because the area is increasing. In a time *dt* the sliding rod moves a distance $v \, dt$ and the area increases by $dA = Lv \, dt$. Take the positive direction for area to be into the plane, parallel to $\vec{B}$. Then the magnetic flux through the circuit is positive, and in time *dt* it increases by an amount

$$d\Phi_B = B \, dA = BLv \, dt.$$

The induced emf is

$$\mathcal{E} = -\frac{d\Phi_B}{dt} = -BLv.$$

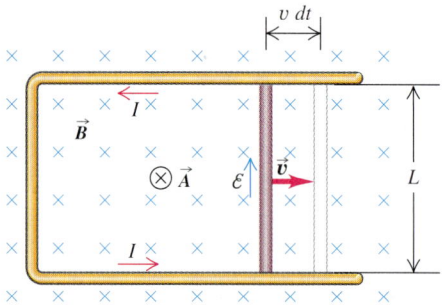

30–9 A slidewire generator. The magnetic field $\vec{B}$ and the vector area $\vec{A}$ are both directed into the figure. The increase in magnetic flux (caused by an increase in area) induces the emf and current shown.

To interpret the negative sign, recall our sign rule. Point your right thumb in the direction of the vector area $\vec{A}$; your fingers then curl in the direction of the positive sense of emf. In our case this is the clockwise direction around the circuit in Fig. 30–9. The negative sign means that the actual emf is opposite to this, or counterclockwise, as shown in the figure. The induced current is also counterclockwise. The emf is constant if the velocity $\vec{v}$ is constant, in which case the slidewire generator acts as a *direct-current* generator.

EXAMPLE 30-7

Work and power in the slidewire generator In the slidewire generator of Example 30–6, energy is dissipated in the circuit owing to its resistance. Let the resistance of the circuit (made up of the moving slidewire and the U-shaped conductor that connects the ends of the slidewire) at a given point in the slidewire's motion be R. Show that the rate at which energy is dissipated in the circuit is exactly equal to the rate at which work must be done to move the rod through the magnetic field.

SOLUTION The current I in the circuit equals the absolute value of the induced emf $\mathcal{E}$ divided by the resistance R, and the rate at which energy is dissipated in the rod is $P_{\text{dissipated}} = I^2 R$. Using the results of Example 30–6, we have

$$I = \frac{BLv}{R} \quad \text{and} \quad P_{\text{dissipated}} = \left(\frac{BLv}{R}\right)^2 R = \frac{B^2 L^2 v^2}{R}.$$

Because the moving rod carries a current in the presence of a magnetic field $\vec{B}$, there is a force on the rod given by $\vec{F} = I\vec{L} \times \vec{B}$; the vector $\vec{L}$ points along the rod in the direction of the current. Figure 30–10 shows that this force is opposite to the velocity of the rod, and so to maintain the motion, a force of equal magnitude must be applied in the direction of $\vec{v}$. Since $\vec{L}$ and $\vec{B}$ are perpendicular, this magnitude is

$$F = ILB = \frac{BLv}{R}LB = \frac{B^2 L^2 v}{R}.$$

The rate at which work is done by this applied force is $P_{\text{applied}} = Fv$, or

$$P_{\text{applied}} = \frac{B^2 L^2 v^2}{R}.$$

This is just equal to the rate at which energy is dissipated in the resistance.

CAUTION ▶ You might think that by reversing the direction of $\vec{B}$ or of $\vec{v}$, it might be possible to have the magnetic force $\vec{F} = I\vec{L} \times \vec{B}$ be in the *same* direction as $\vec{v}$. This would be a pretty neat trick. Once the rod was moving, the changing magnetic flux would induce an emf and a current, and the magnetic force on the rod would make it move even faster, increasing the emf and current; this would go on until the rod was moving at tremendous speed and producing electric power at a prodigious rate. If this seems too good to be true, not to mention a violation of energy conservation, that's because it is. Reversing $\vec{B}$ will also reverse the sign of the induced emf and current and hence the direction of $\vec{L}$, so the magnetic force still opposes the motion of the rod; a similar result holds true if we reverse $\vec{v}$. This behavior is part of Lenz's law, to be discussed in Section 30–4. ◀

30–10 The magnetic force $\vec{F} = I\vec{L} \times \vec{B}$ due to the induced current is to the left, opposite to $\vec{v}$.

Example 30–7 shows that the slidewire generator doesn't produce electric energy out of nowhere; the energy is supplied by whatever body exerts the force that keeps the rod moving. All that the generator does is to *convert* that energy into a different form. The equality between the rate at which *mechanical* energy is supplied to a generator and the rate at which *electric* energy is generated holds for all types of generators. This is true in particular for the alternator described in Example 30–4; we leave the details as a problem. (We are neglecting the effects of friction in the bearings of an alternator or between the rod and the U-shaped conductor of a slidewire generator. If these are included, the conservation of energy demands that the energy lost to friction is not available for conversion to electric energy. In real generators the friction is kept to a minimum to keep the energy conversion process as efficient as possible.)

In Chapter 28 we stated that the magnetic force on moving charges can never do work. But you might think that the magnetic force $\vec{F} = I\vec{L} \times \vec{B}$ in Example 30–7 *is* doing (negative) work on the current-carrying rod as it moves, contradicting our earlier statement. In fact, the work done by the magnetic force is actually zero. The moving charges that make up the current in the rod in Fig. 30–10 have a vertical component of velocity, causing a horizontal component of force on these charges. As a result, there is a

This airplane's ignition system uses Faraday's law. The engine's rotating crankshaft spins a magnet, inducing an emf in an adjacent coil and generating the spark that ignites fuel in the engine cylinders. This keeps the engine running even if the airplane's other electrical systems fail.

horizontal displacement of charge within the rod, the left side acquiring a net positive charge and the right side a net negative charge. The result is a horizontal component of electric field, perpendicular to the length of the rod (analogous to the Hall effect, described in Section 28–10). It is this field, in the direction of motion of the rod, that does work on the mobile charges in the rod and hence indirectly on the atoms making up the rod.

30-4 LENZ'S LAW

Lenz's law is a convenient alternative method for determining the direction of an induced current or emf. Lenz's law is not an independent principle; it can be derived from Faraday's law. It always gives the same results as the sign rules we introduced in connection with Faraday's law, but it is often easier to use. Lenz's law also helps us gain intuitive understanding of various induction effects and of the role of energy conservation. H. F. E. Lenz (1804–1865) was a German scientist who duplicated independently many of the discoveries of Faraday and Henry. **Lenz's law** states:

> **The direction of any magnetic induction effect is such as to oppose the cause of the effect.**

The "cause" may be changing flux through a stationary circuit due to a varying magnetic field, changing flux due to motion of the conductors that make up the circuit, or any combination. If the flux in a stationary circuit changes, as in Examples 30–1 and 30–2, the induced current sets up a magnetic field of its own. Within the area bounded by the circuit, this field is *opposite* to the original field if the original field is *increasing* but is in the *same* direction as the original field if the latter is *decreasing.* That is, the induced current opposes the *change in flux* through the circuit (*not* the flux itself).

If the flux change is due to motion of the conductors, as in Examples 30–3 through 30–7, the direction of the induced current in the moving conductor is such that the direction of the magnetic-field force on the conductor is opposite in direction to its motion. Thus the motion of the conductor, which caused the induced current, is opposed. We saw this explicitly for the slidewire generator in Example 30–7. In all these cases the induced current tries to preserve the *status quo* by opposing motion or a change of flux.

Lenz's law is also directly related to energy conservation. If the induced current in Example 30–7 were in the direction opposite to that given by Lenz's law, the magnetic force on the rod would accelerate it to ever-increasing speed with no external energy source, even though electrical energy is being dissipated in the circuit. This would be a clear violation of energy conservation and doesn't happen in nature.

EXAMPLE 30-8

The slidewire generator, revisited In Fig. 30–9, the induced current in the loop causes an additional magnetic field in the area bounded by the loop. The direction of the induced current is counterclockwise. From the discussion of Section 29–3, we see that the direction of the additional magnetic field caused by this current is *out of* the plane of the figure. Its direction is opposite that of the original magnetic field, so it tends to cancel the effect of that field. This is consistent with the prediction of Lenz's law.

EXAMPLE 30-9

In Fig. 30–11 there is a uniform magnetic field $\vec{B}$ through the coil. The magnitude of the field is increasing, and the resulting induced emf causes an induced current. Use Lenz's law to determine the direction of the induced current.

SOLUTION This situation is the same as in Example 30–1 (Section 30–3). By Lenz's law the induced current must produce a magnetic field $\vec{B}_{induced}$ inside the coil that is downward, opposing the change in flux. Using the right-hand rule described in Section 29–6 for the direction of the magnetic field produced by a circular loop, $\vec{B}_{induced}$ will be in the desired direction if the induced current flows as shown in Fig. 30–11.

Figure 30–12 shows several applications of Lenz's law to the similar situation of a magnet moving near a conducting loop. In each of the four cases shown, the induced current produces a magnetic field of its own, in a direction that opposes the change in flux through the loop due to the magnet's motion.

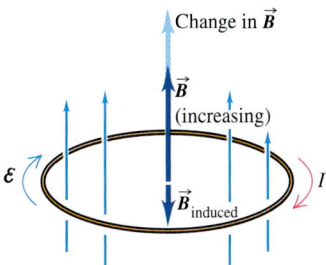

30–11 The induced current due to the change in $\vec{B}$ is clockwise, as seen from above the loop. The added field $\vec{B}_{induced}$ that it causes is downward, opposing the change in the upward field $\vec{B}$.

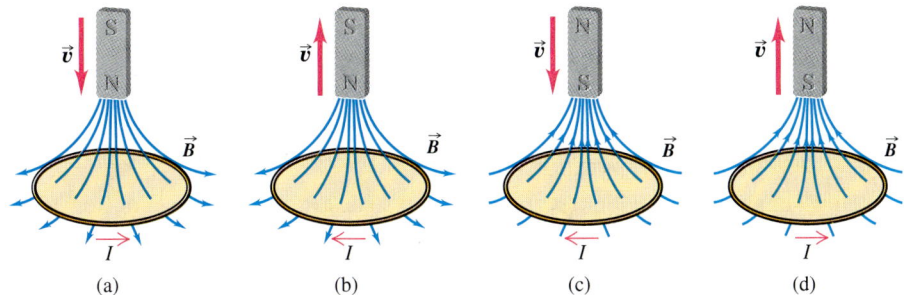

(a)   (b)   (c)   (d)

30–12 Directions of induced currents as a bar magnet moves along the axis of a conducting loop. If the bar magnet is stationary, there is no induced current.

LENZ'S LAW AND THE RESPONSE TO FLUX CHANGES

Since an induced current always opposes any change in magnetic flux through a circuit, how is it possible for the flux to change at all? The answer is that Lenz's law gives only the *direction* of an induced current; the *magnitude* of the current depends on the resistance of the circuit. The greater the circuit resistance, the less the induced current that appears to oppose any change in flux and the easier it is for a flux change to take effect. If the loop in Fig. 30–12 were made out of wood (an insulator), there would be almost no induced current in response to changes in the flux through the loop.

Conversely, the less the circuit resistance, the greater the induced current and the more difficult it is to change the flux through the circuit. If the loop in Fig. 30–12 is a good conductor, an induced current flows as long as the magnet moves relative to the loop. Once the magnet and loop are no longer in relative motion, the induced current very quickly decreases to zero because of the nonzero resistance in the loop.

The extreme case occurs when the resistance of the circuit is *zero*. Then the induced current in Fig. 30–12 will continue to flow even after the induced emf has disappeared, that is, even after the magnet has stopped moving relative to the loop. Thanks to this *persistent current,* it turns out that the flux through the loop is exactly the same as it was before the magnet started to move, so the flux through a loop of zero resistance *never* changes. Exotic materials called *superconductors* do indeed have zero resistance; we discuss these further in Section 30–9.

30–5 MOTIONAL ELECTROMOTIVE FORCE

In situations in which a conductor moves in a magnetic field, as in the generators discussed in Examples 30–4 through 30–7, we can gain additional insight into the origin

(a)

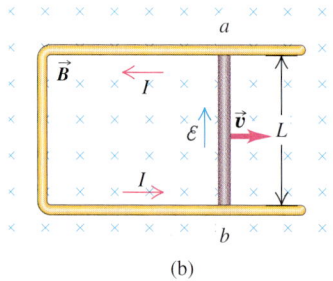

(b)

30–13 A conducting rod moving in a uniform magnetic field. (a) The rod, the velocity, and the field are mutually perpendicular. (b) Direction of induced current in the circuit.

13.10
Motional EMF

of the induced emf by considering the magnetic forces on mobile charges in the conductor. Figure 30–13a shows the same moving rod that we discussed in Example 30–6, separated for the moment from the U-shaped conductor. The magnetic field $\vec{B}$ is uniform and directed into the page, and we move the rod to the right at a constant velocity $\vec{v}$. A charged particle q in the rod then experiences a magnetic force $\vec{F} = q\vec{v} \times \vec{B}$ with magnitude $F = |q|vB$. We'll assume in the following discussion that q is positive; in that case the direction of this force is *upward* along the rod, from b toward a.

This magnetic force causes the free charges in the rod to move, creating an excess of positive charge at the upper end a and negative charge at the lower end b. This in turn creates an electric field $\vec{E}$ within the rod, in the direction from a toward b (opposite to the magnetic force). Charge continues to accumulate at the ends of the rod until $\vec{E}$ becomes large enough for the downward electric force (with magnitude qE) to cancel exactly the upward magnetic force (with magnitude qvB). Then $qE = qvB$, and the charges are in equilibrium.

The magnitude of the potential difference $V_{ab} = V_a - V_b$ is equal to the electric field magnitude E multiplied by the length L of the rod. From the above discussion, $E = vB$, so

$$V_{ab} = EL = vBL, \qquad (30\text{–}5)$$

with point a at higher potential than point b.

Now suppose the moving rod slides along a stationary U-shaped conductor, forming a complete circuit (Fig. 30–13b). No *magnetic* force acts on the charges in the stationary U-shaped conductors, but the charge that was near points a and b redistributes itself along the statonary conductors, creating an *electric* field within them. This field establishes a current in the direction shown. The moving rod has become a source of electromotive force; within it, charge moves from lower to higher potential, and in the remainder of the circuit, charge moves from higher to lower potential. We call this emf a **motional electromotive force,** denoted by $\mathcal{E}$. From the above discussion, the magnitude of this emf is

$$\mathcal{E} = vBL \qquad \begin{array}{l}\text{(motional emf; length and velocity}\\ \text{perpendicular to uniform } \vec{B}\text{)},\end{array} \qquad (30\text{–}6)$$

corresponding to a force per unit charge of magnitude vB acting for a distance L along the moving rod. If the total circuit resistance of the U-shaped conductor and the sliding rod is R, the induced current I in the circuit is given by $vBL = IR$. This is the same result we obtained in Section 30–3 using Faraday's law, and indeed motional emf is a particular case of Faraday's law, one of the several examples described in Section 30–2.

The emf associated with the moving rod in Fig. 30–13 is analogous to that of a battery with its positive terminal at a and its negative terminal at b, although the origins of the two emfs are quite different. In each case a non-electrostatic force acts on the charges in the device, in the direction from b to a, and the emf is the work per unit charge done by this force when a charge moves from b to a in the device. When the device is connected to an external circuit, the direction of current is from b to a in the device and from a to b in the external circuit. While we have discussed motional emf in terms of a closed circuit like that in Fig. 30–13b, a motional emf is also present in the isolated moving rod in Fig. 30–13a, in the same way that a battery has an emf even when it's not part of a circuit.

The direction of the induced emf in Fig. 30–13 can be deduced by using Lenz's law, even if (as in Fig. 30–13a) the conductor does not form a complete circuit. In this case we can mentally complete the circuit between the ends of the conductor and use Lenz's law to determine the direction of the current. From this we can deduce the polarity of the ends of the open-circuit conductor. The direction from the $-$ end to the $+$ end within the conductor is the direction the current would have if the circuit were complete.

If we express v in meters per second, B in teslas, and L in meters, $\mathcal{E}$ is in volts; we invite you to verify this. (Recall that $1\ V = 1\ J/C$.)

We can generalize the concept of motional emf for a conductor with *any* shape, moving in any magnetic field, uniform or not (assuming that the magnetic field at each point does not vary with time). For an element $d\vec{l}$ of conductor, the contribution $d\mathcal{E}$ to the emf is the magnitude dl multiplied by the component of $\vec{v} \times \vec{B}$ (the magnetic force per unit charge) parallel to $d\vec{l}$; that is,

$$d\mathcal{E} = \left(\vec{v} \times \vec{B}\right) \cdot d\vec{l}.$$

For any closed conducting loop, the total emf is

$$\mathcal{E} = \oint \left(\vec{v} \times \vec{B}\right) \cdot d\vec{l}. \tag{30-7}$$

This expression looks very different from our original statement of Faraday's law, Eq. (30–3), which stated that $\mathcal{E} = -d\Phi_B/dt$. In fact, though, the two statements are equivalent. It can be shown that the rate of change of magnetic flux through a moving conducting loop is always given by the negative of the expression in Eq. (30–7). Thus this equation gives us an alternative formulation of Faraday's law. This alternative is often more convenient than the original one in problems with *moving* conductors. But when we have *stationary* conductors in changing magnetic fields, Eq. (30–7) *cannot* be used; in this case, $\mathcal{E} = -d\Phi_B/dt$ is the only correct way to express Faraday's law.

EXAMPLE 30–10

Suppose the length L in Fig. 30–13b is 0.10 m, the velocity v is 2.5 m/s, the total resistance of the loop is 0.030 Ω, and B is 0.60 T. Find $\mathcal{E}$, the induced current, and the force acting on the rod.

SOLUTION From Eq. (30–6) the emf $\mathcal{E}$ is

$$\mathcal{E} = vBL = (2.5\ \text{m/s})(0.60\ \text{T})(0.10\ \text{m}) = 0.15\ \text{V}.$$

The current I in the loop is

$$I = \frac{\mathcal{E}}{R} = \frac{0.15\ \text{V}}{0.030\ \Omega} = 5.0\ \text{A}.$$

Because of this current, a magnetic force $\vec{F}$ acts on the rod in the direction *opposite* to its motion; this can be seen by applying the right-hand rule for vector products to the formula $\vec{F} = I\vec{L} \times \vec{B}$ or by using Lenz's law. From Eq. (28–17) this force has magnitude

$$F = ILB = (5.0\ \text{A})(0.10\ \text{m})(0.60\ \text{T}) = 0.30\ \text{N}.$$

EXAMPLE 30–11

The Faraday disk dynamo A conducting disk with radius R, shown in Fig. 30–14, lies in the xy-plane and rotates with constant angular velocity ω about the z-axis. The disk is in a uniform, constant $\vec{B}$ field parallel to the z-axis. Find the induced emf between the center and the rim of the disk.

SOLUTION The current flow in the disk is mostly along the radial direction, along the negative y-axis in the figure. You may be tempted to use Eq. (30–5) and simply multiply vB by the length of the moving conducting path, which is just the radius R. This wouldn't be right, because segments of this radial line that are at different distances from the center move at different speeds. Instead, let's consider the emf $d\mathcal{E}$ due to a small segment of the moving conductor, with an infinitesimal radial length dr at radius r. The speed v of dr is $v = \omega r$, and the motional emf $d\mathcal{E}$ induced in dr is

$$d\mathcal{E} = vB\ dr = \omega Br\ dr.$$

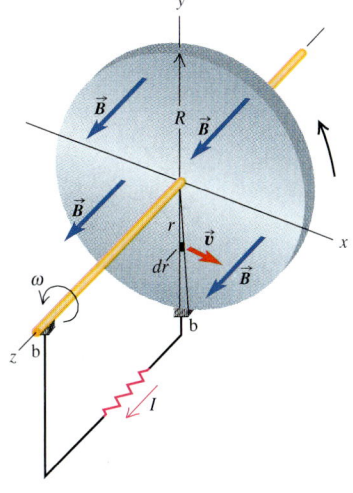

30–14 A rotating conducting disk with radius R. The emf is induced along radial lines of the disk and is connected to an external circuit through the two sliding contacts labeled b. The speed of a small radial segment of length dr at radius r is $v = \omega r$, and the emf induced by its motion is $d\mathcal{E} = vB\ dr = \omega Br\ dr$.

The total emf between center and rim is the sum of all such contributions:

$$\mathcal{E} = \int_0^R \omega B r \, dr = \frac{1}{2} \omega B R^2 .$$

We can use this device as a source of emf in a circuit by completing the circuit through stationary brushes (b in the figure) that contact the disk and its conducting shaft as shown. The emf in such a disk was studied by Faraday; the device is called *a Faraday disk dynamo* or a *homopolar generator.* Unlike the alternator in Example 30–4, the Faraday disk dynamo is a direct-current generator; it produces an emf that is constant in time. Can you use Lenz's law to show that for the direction of rotation in Fig. 30–14, the current in the external circuit must be in the direction shown?

30–6 INDUCED ELECTRIC FIELDS

When a conductor moves in a magnetic field, we can understand the induced emf on the basis of magnetic forces on charges in the conductor, as described in Section 30–5. But an induced emf also occurs when there is a changing flux through a stationary conductor. What is it that pushes the charges around the circuit in this type of situation?

As an example, let's consider the situation shown in Fig. 30–15. A long, thin solenoid with cross-section area A and n turns per unit length is encircled at its center by a circular conducting loop. The galvanometer G measures the current in the loop. A current I in the winding of the solenoid sets up a magnetic field $\vec{B}$ along the solenoid axis, as shown, with magnitude B as calculated in Example 29–10 (Section 29–8): $B = \mu_0 n I$, where n is the number of turns per unit length. If we neglect the small field outside the solenoid, then the magnetic flux Φ_B through the loop is

$$\Phi_B = BA = \mu_0 n I A.$$

When the solenoid current I changes with time, the magnetic flux Φ_B also changes, and according to Faraday's law the induced emf in the loop is given by

$$\mathcal{E} = -\frac{d\Phi_B}{dt} = -\mu_0 n A \frac{dI}{dt} . \tag{30–8}$$

If the total resistance of the loop is R, the induced current in the loop, which we may call I', is $I' = \mathcal{E}/R$.

But what *force* makes the charges move around the loop? It can't be a magnetic force because the conductor isn't moving in a magnetic field and in fact isn't even *in* a magnetic field. We are forced to conclude that there has to be an **induced electric field** in the conductor *caused by the changing magnetic flux.* This may be a little jarring; we are accustomed to thinking about electric field as being caused by electric charges, and now we are saying that a changing magnetic field somehow acts as a source of electric field. Furthermore, it's a strange sort of electric field. When a charge q goes once around the loop, the total work done on it by the electric field must be equal to q times the emf $\mathcal{E}$. That is, the electric field in the loop *is not conservative,* as we used the term in Chapter

30–15 (a) The windings of a long solenoid carry a current I that is increasing at a rate dI/dt. The magnetic flux in the solenoid is increasing at a rate $d\Phi_B/dt$, and this changing flux passes through a wire loop. An emf $\mathcal{E} = -d\Phi_B/dt$ is induced in the loop, inducing a current I' that is measured by the galvanometer G. (b) Cross-section view.

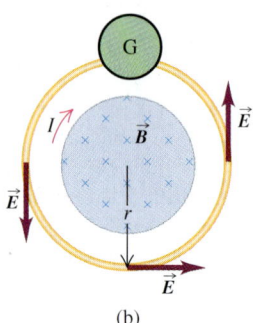

(a)   (b)

24, because the line integral of $\vec{E}$ around a closed path is not zero. Indeed, this line integral, representing the work done by the induced $\vec{E}$ field per unit charge, is equal to the induced emf ε:

$$\oint \vec{E} \cdot d\vec{l} = \varepsilon. \qquad (30\text{–}9)$$

From Faraday's law the emf ε is also the negative of the rate of change of magnetic flux through the loop. Thus for this case we can restate Faraday's law as

$$\oint \vec{E} \cdot d\vec{l} = -\frac{d\Phi_B}{dt} \qquad \text{(stationary integration path).} \qquad (30\text{–}10)$$

Note that Faraday's law is always true in the form $\varepsilon = -d\Phi_B/dt$; the form given in Eq. (30–10) is valid *only* if the path around which we integrate is *stationary*.

As an example of a situation to which Eq. (30–10) can be applied, consider the stationary circular loop in Fig. 30–15b, which we take to have radius r. Because of cylindrical symmetry, the electric field $\vec{E}$ has the same magnitude at every point on the circle and is tangent to it at each point. (Symmetry would also permit the field to be *radial,* but then Gauss's law would require the presence of a net charge inside the circle, and there is none.) The line integral in Eq. (30–10) becomes simply the magnitude E times the circumference $2\pi r$ of the loop, $\oint \vec{E} \cdot d\vec{l} = 2\pi r E$, and Eq. (30–10) gives

$$E = \frac{1}{2\pi r} \left| \frac{d\Phi_B}{dt} \right|. \qquad (30\text{–}11)$$

The directions of $\vec{E}$ at several points on the loop are shown in the figure. We know that $\vec{E}$ has to have the direction shown when $\vec{B}$ in the solenoid is increasing, because $\oint \vec{E} \cdot d\vec{l}$ has to be negative when $d\Phi_B/dt$ is positive. The same approach can be used to find the induced electric field *inside* the solenoid when the solenoid $\vec{B}$ field is changing; we leave the details to you (see Exercise 30–22).

Now let's summarize what we've learned. Faraday's law, Eq. (30–3), is valid for two rather different situations. In one, an emf is induced by magnetic forces on charges when a conductor moves through a magnetic field. In the other, a time-varying magnetic field induces an electric field in a stationary conductor and hence induces an emf; in fact, the $\vec{E}$ field is induced even when no conductor is present. This $\vec{E}$ field differs from an electro*static* field in an important way. It is *nonconservative;* the line integral $\oint \vec{E} \cdot d\vec{l}$ around a closed path is not zero, and when a charge moves around a closed path, the field does a nonzero amount of work on it. It follows that for such a field the concept of *potential* has no meaning. We call such a field a **non-electrostatic field.** In contrast, an electrostatic field is *always* conservative, as we discussed in Section 24–2, and always has an associated potential function. Despite this difference, the fundamental effect of *any* electric field is to exert a force $\vec{F} = q\vec{E}$ on a charge q. This relation is valid whether $\vec{E}$ is a conservative field produced by a charge distribution or a nonconservative field caused by changing magnetic flux.

So a changing magnetic field acts as a source of electric field of a sort that we *cannot* produce with any static charge distribution. This may seem strange, but it's the way nature behaves. This development is analogous to our generalized Ampere's law with displacement current, which shows that a changing *electric* field acts as a source of *magnetic* field. We'll explore this symmetry between the two fields in greater detail in Section 30–8 and in our study of electromagnetic waves in Chapter 33.

If any doubt remains in your mind about the reality of magnetically induced electric fields, consider a few of the many practical applications. In the playback head of a tape deck, currents are induced in a stationary coil as the variously magnetized regions of the tape move past it. Computer disk drives operate on the same principle. Pickups in electric guitars use currents induced in stationary pickup coils by the vibration of nearby ferromagnetic strings. Alternators in most cars use rotating magnets to induce currents in stationary coils. The list goes on and on; whether we realize it or not, magnetically induced electric fields play an important role in everyday life.

EXAMPLE 30-12

Suppose the long solenoid in Fig. 30–15a is wound with 500 turns per meter and the current in its windings is increasing at the rate of 100 A/s. The cross-section area of the solenoid is $4.0 \text{ cm}^2 = 4.0 \times 10^{-4} \text{ m}^2$. a) Find the magnitude of the induced emf in the wire loop outside the solenoid. b) Find the magnitude of the induced electric field within the loop if its radius is 2.0 cm.

SOLUTION a) We can use Eq. (30–8). The induced emf is

$$\mathcal{E} = -\frac{d\Phi_B}{dt} = -\mu_0 nA \frac{dI}{dt}$$

$$= -(4\pi \times 10^{-7} \text{ Wb/A} \cdot \text{m})(500 \text{ turns/m})(4.0 \times 10^{-4} \text{ m}^2)(100 \text{ A/s})$$

$$= -25 \times 10^{-6} \text{ Wb/s} = -25 \times 10^{-6} \text{ V} = -25 \text{ } \mu\text{V}.$$

In Fig. 30–15b the flux *into* the plane of the figure is increasing; according to our right-hand rule, the negative sign of $\mathcal{E}$ shows that the emf is in the counterclockwise direction. We invite you to verify that Lenz's law also predicts this direction.
b) By symmetry the line integral $\oint \vec{E} \cdot d\vec{l}$ has absolute value $2\pi rE$ (disregarding the direction in which we integrate around the loop). This is equal to the absolute value of the emf, so

$$E = \frac{|\mathcal{E}|}{2\pi r} = \frac{25 \times 10^{-6} \text{ V}}{2\pi(2.0 \times 10^{-2} \text{ m})} = 2.0 \times 10^{-4} \text{ V/m}.$$

*30-7 EDDY CURRENTS

In the examples of induction effects that we have studied, the induced currents have been confined to well-defined paths in conductors and other components forming a circuit. However, many pieces of electrical equipment contain masses of metal moving in magnetic fields or located in changing magnetic fields. In situations like these we can have induced currents that circulate throughout the volume of a material. Because their flow patterns resemble swirling eddies in a river, we call these **eddy currents.**

As an example, consider a metallic disk rotating in a magnetic field perpendicular to the plane of the disk but confined to a limited portion of the disk's area, as shown in Fig. 30–16a. Sector Ob is moving across the field and has an emf induced in it. Sectors Oa and Oc are not in the field, but they provide return conducting paths for charges displaced along Ob to return from b to O. The result is a circulation of eddy currents in the disk, somewhat as sketched in Fig. 30–16b.

We can use Lenz's law to decide on the direction of the induced current in the neighborhood of sector Ob. This current must experience a magnetic force $\vec{F} = I\vec{L} \times \vec{B}$ that *opposes* the rotation of the disk, and so this force must be to the right in Fig. 30–16b. Since $\vec{B}$ is directed into the plane of the disk, the current and hence $\vec{L}$ have downward components. The return currents lie outside the field, so they do not experience magnetic forces. The interaction between the eddy currents and the field causes a braking action on the disk. Such effects can be used to stop the rotation of a circular saw quickly when the power is turned off. Some sensitive balances use this effect to damp out vibrations. Eddy current braking is used on some electrically-powered rapid transit vehicles. Electromagnets mounted in the cars induce eddy currents in the rails; the resulting magnetic fields cause braking forces on the electromagnets and thus on the cars.

Eddy currents have many other practical uses. The shiny metal disk in the electric power company's meter outside your house rotates as a result of eddy currents. These currents are induced in the disk by magnetic fields caused by sinusoidally varying currents in a coil. In induction furnaces, eddy currents are used to heat materials in completely sealed containers for processes in which it is essential to avoid the slightest contamination of the materials. The metal detectors used at airport security checkpoints

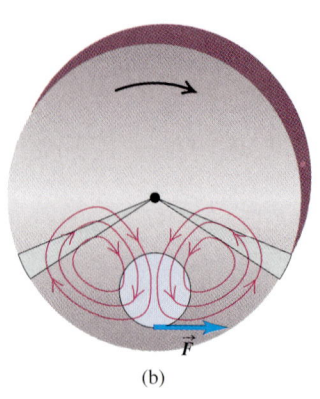

30-16 (a) A metallic disk rotating through a perpendicular magnetic field $\vec{B}$. (b) Eddy currents formed by the induced emf.

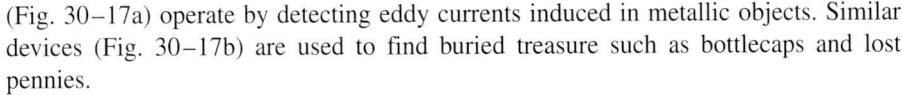

30-17 (a) A metal detector at an airport security checkpoint generates an alternating magnetic field $\vec{B}_0$. This induces eddy currents in a conducting object carried through the detector. The eddy currents in turn produce an alternating magnetic field $\vec{B}'$, and this field induces a current in the detector's receiver coil. (b) Portable metal detectors work on the same principle.

(Fig. 30–17a) operate by detecting eddy currents induced in metallic objects. Similar devices (Fig. 30–17b) are used to find buried treasure such as bottlecaps and lost pennies.

Eddy currents also have undesirable effects. In an alternating-current transformer, coils wrapped around an iron core carry a sinusoidally varying current. The resulting eddy currents in the core waste energy through I^2R heating and themselves set up an unwanted opposing emf in the coils. To minimize these effects, the core is designed so that the paths for eddy currents are as narrow as possible. We'll describe how this is done when we discuss transformers in detail in Section 32–7.

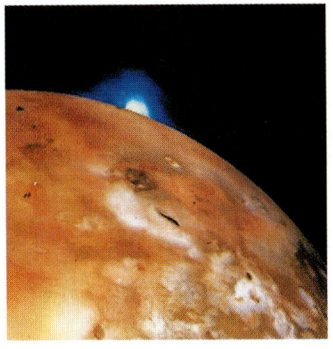

Jupiter's moon Io is constantly moving through Jupiter's strong magnetic field, setting up eddy currents that dissipate 10^{12} W of power within Io. This helps to keep Io's interior hot and so helps cause volcanic eruptions like this one. (Gravitational effects from Jupiter cause even more heating.)

30-8 MAXWELL'S EQUATIONS

We are now in a position to wrap up in a wonderfully neat package all the relationships between electric and magnetic fields and their sources that we have studied in the past several chapters. The package consists of four equations, called **Maxwell's equations.** You may remember Maxwell as the discoverer of the concept of displacement current, which we studied in Section 29–10. Maxwell didn't discover all of these equations single-handedly, but he put them together and recognized their significance, particularly in predicting the existence of electromagnetic waves.

For now we'll state Maxwell's equations in their simplest form, for the case in which we have charges and currents in otherwise empty space. In Chapter 33 we'll discuss how to modify these equations if a dielectric or a magnetic material is present.

Two of Maxwell's equations involve an integral of $\vec{E}$ or $\vec{B}$ over a closed surface. The first is simply Gauss's law for electric fields, Eq. (23–8), which states that the surface integral of $E_\perp$ over any closed surface equals $1/\epsilon_0$ times the total charge Q_{encl} enclosed within the surface:

$$\oint \vec{E} \cdot d\vec{A} = \frac{Q_{\text{encl}}}{\epsilon_0} \qquad \text{(Gauss's law for } \vec{E}\text{)}. \qquad (30\text{--}12)$$

The second is the analogous relation for *magnetic* fields, Eq. (28–8), which states that the surface integral of $B_\perp$ over any closed surface is always zero:

$$\oint \vec{B} \cdot d\vec{A} = 0 \qquad \text{(Gauss's law for } \vec{B}\text{)}. \qquad (30\text{--}13)$$

This statement means, among other things, that there are no magnetic monopoles (single magnetic charges) to act as sources of magnetic field.

The third equation is Ampere's law including displacement current, which we discussed in Section 29–10. This states that both conduction current i_C and displacement current $\epsilon_0 d\Phi_E/dt$, where Φ_E is electric flux, act as sources of magnetic field:

$$\oint \vec{B} \cdot d\vec{l} = \mu_0 \left(i_C + \epsilon_0 \frac{d\Phi_E}{dt} \right)_{\text{encl}} \qquad \text{(Ampere's law)}. \qquad (30\text{–}14)$$

The fourth and final equation is Faraday's law, studied in this chapter. It states that a changing magnetic field or magnetic flux induces an electric field:

$$\oint \vec{E} \cdot d\vec{l} = -\frac{d\Phi_B}{dt} \qquad \text{(Faraday's law)}. \qquad (30\text{–}15)$$

If there is a changing magnetic flux, the line integral in Eq. (30–15) is not zero, which shows that the $\vec{E}$ field produced by a changing magnetic flux is not conservative. Recall that this line integral must be carried out over a *stationary* closed path.

It's worthwhile to look more carefully at the electric field $\vec{E}$ and its role in Maxwell's equations. In general, the total $\vec{E}$ field at a point in space can be the superposition of an electrostatic field $\vec{E}_c$ caused by a distribution of charges at rest and a magnetically induced, non-electrostatic field $\vec{E}_n$. (The subscript c stands for Coulomb or conservative; the subscript n stands for non-Coulomb, non-electrostatic, or non-conservative.) That is,

$$\vec{E} = \vec{E}_c + \vec{E}_n.$$

The electrostatic part $\vec{E}_c$ is *always* conservative, so $\oint \vec{E}_c \cdot d\vec{l} = 0$. This conservative part of the field does not contribute to the integral in Faraday's law, so we can take $\vec{E}$ in Eq. (30–15) to be the *total* electric field $\vec{E}$, including both the part $\vec{E}_c$ due to charges and the magnetically induced part $\vec{E}_n$. Similarly, the nonconservative part $\vec{E}_n$ of the $\vec{E}$ field does not contribute to the integral in Gauss's law, because this part of the field is not caused by static charges. Hence $\oint \vec{E}_n \cdot d\vec{A}$ is always zero. We conclude that in all the Maxwell equations, $\vec{E}$ is the *total* electric field; these equations don't distinguish between conservative and nonconservative fields.

There is a remarkable symmetry in Maxwell's four equations. In empty space where there is no charge, the first two equations (Eqs. (30–12) and (30–13)) are identical in form, one containing $\vec{E}$ and the other containing $\vec{B}$. When we compare the second two equations, Eq. (30–14) says that a changing electric flux creates a magnetic field, and Eq. (30–15) says that a changing magnetic flux creates an electric field. In empty space, where there is no conduction current, $i_C = 0$ and the two equations have the same form, apart from a numerical constant and a negative sign, with the roles of $\vec{E}$ and $\vec{B}$ exchanged in the two equations.

We can rewrite Eqs. (30–14) and (30–15) in a different but equivalent form by introducing the definitions of electric and magnetic flux, $\Phi_E = \int \vec{E} \cdot d\vec{A}$ and $\Phi_B = \int \vec{B} \cdot d\vec{A}$, respectively. In empty space, where there is no charge or conduction current, $i_C = 0$ and $Q_{\text{encl}} = 0$, and we have

$$\oint \vec{B} \cdot d\vec{l} = \epsilon_0 \mu_0 \frac{d}{dt} \int \vec{E} \cdot d\vec{A}, \qquad (30\text{–}16)$$

$$\oint \vec{E} \cdot d\vec{l} = -\frac{d}{dt} \int \vec{B} \cdot d\vec{A}. \qquad (30\text{–}17)$$

Again we notice the symmetry between the roles of $\vec{E}$ and $\vec{B}$ in these expressions.

The most remarkable feature of these equations is that a time-varying field of *either* kind induces a field of the other kind in neighboring regions of space. Maxwell recognized that these relationships predict the existence of electromagnetic disturbances consisting of time-varying electric and magnetic fields that travel or *propagate* from one region of space to another, even if no matter is present in the intervening space. Such disturbances, called *electromagnetic waves,* provide the physical basis for light, radio and television waves, infrared, ultraviolet, x rays, and the rest of the electromagnetic spectrum. We will return to this vitally important topic in Chapter 33.

Although it may not be obvious, *all* the basic relations between fields and their sources are contained in Maxwell's equations. We can derive Coulomb's law from Gauss's law, we can derive the law of Biot and Savart from Ampere's law, and so on. When we add the equation that defines the $\vec{E}$ and $\vec{B}$ fields in terms of the forces that they exert on a charge q, namely,

$$\vec{F} = q(\vec{E} + \vec{v} \times \vec{B}), \qquad (30\text{–}18)$$

we have *all* the fundamental relations of electromagnetism!

Finally, we note that Maxwell's equations would have even greater symmetry between the $\vec{E}$ and $\vec{B}$ fields if single magnetic charges (magnetic monopoles) existed. The right side of Eq. (30–13) would contain the total *magnetic* charge enclosed by the surface, and the right side of Eq. (30–15) would include a magnetic monopole current term. Perhaps you can begin to see why some physicists wish that magnetic monopoles existed; they would help to perfect the mathematical poetry of Maxwell's equations.

The discovery that electromagnetism can be wrapped up so neatly and elegantly is a very satisfying one. In conciseness and generality, Maxwell's equations are in the same league with Newton's laws of motion and the laws of thermodynamics. Indeed, a major goal of science is learning how to express very broad and general relationships in a concise and compact form. Maxwell's synthesis of electromagnetism stands as a towering intellectual achievement, comparable to the Newtonian synthesis we described at the end of Section 12–6 and to the development of relativity and quantum mechanics in the twentieth century.

*30–9 SUPERCONDUCTIVITY

A Case Study in Magnetic Properties

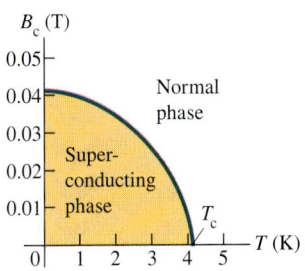

30–18 Phase diagram for pure mercury, showing the critical magnetic field B_c and its dependence on temperature. Superconductivity is impossible above the critical temperature T_c. The curves for other superconducting materials are similar but with different numerical values.

The most familiar property of a superconductor is the sudden disappearance of all electrical resistance when the material is cooled below a temperature called the *critical temperature,* denoted by T_c. We discussed this behavior and the circumstances of its discovery in Section 26–3. But superconductivity is far more than just the absence of measurable resistance. Superconductors also have extraordinary *magnetic* properties. We'll explore some of these properties in this section.

The first hint of unusual magnetic properties was the discovery that for any superconducting material the critical temperature T_c changes when the material is placed in an externally produced magnetic field $\vec{B}_0$. Figure 30–18 shows this dependence for mercury, the first element in which superconductivity was observed. As the external field magnitude B_0 increases, the superconducting transition occurs at lower and lower temperature. When B_0 is greater than 0.0412 T, *no* superconducting transition occurs. The minimum magnitude of magnetic field that is needed to eliminate superconductivity at a temperature below T_c is called the *critical field,* denoted by B_c.

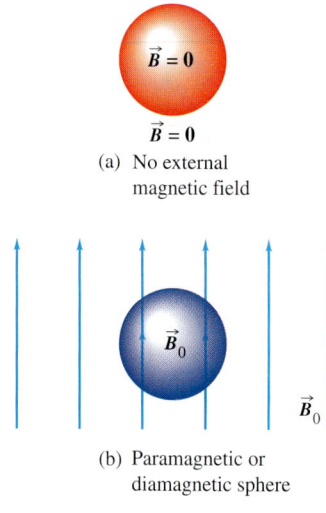

(a) No external
magnetic field

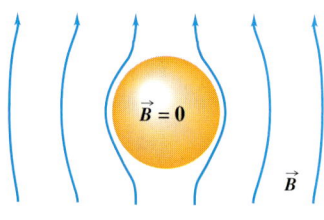

(b) Paramagnetic or
diamagnetic sphere

(c) Superconducting
sphere

30–19 (a) With no external field, $\vec{B}$ is zero everywhere. (b) When an external field $\vec{B}_0$ is applied to a normal paramagnetic substance, the field inside the material is very nearly equal to $\vec{B}_0$. (c) When an external field is applied to a super-conducting substance, the field inside the material is zero.

Figure 30–18 is a *phase diagram;* it shows which phase (normal or superconducting) occurs for any given temperature and magnetic field. It is analogous to the *p-T* phase diagrams that we studied in Section 16–7, which show the phase of a substance (solid, liquid, or gas) that exists for any given temperature and pressure.

Another important aspect of superconductivity is the complete absence of magnetic field inside a material in the superconducting phase. When a homogeneous sphere of a normal diamagnetic or paramagnetic material (Fig. 30–19a) is placed in a magnetic field (Fig. 30–19b), the field inside the material is nearly the same as the field outside. (We choose the spherical shape because the field inside the material then turns out to be uniform, avoiding complications associated with the shape of the specimen.) The relative permeability K_m of such materials (Section 29–9) usually differs from unity by less than 0.1%. But when a superconductor is placed in a magnetic field, we find that there is *no* field inside the material, provided that the external field magnitude is less than B_c. The field lines are distorted as shown in Fig. 30–19c. To confirm the absence of field inside the material, we can wrap a coil of wire around the specimen and then, with the material in the superconducting phase, turn on the external field. We find that no emf is induced in the coil; this shows that there is no flux change through the coil and therefore no $\vec{B}$ field in the specimen when the external field is turned on.

The absence of magnetic field inside a superconductor can be understood on the basis of eddy currents induced in the material. Lenz's law shows that these currents always oppose the change that caused them; in this case the cause is increasing magnetic flux. In a superconductor, which offers no resistance at all to current, the induced eddy currents oppose the flux change so successfully that there is *no* net change in flux.

Our discussion of magnetic materials in Section 29–9 shows that when a material is placed in an external magnetic field $\vec{B}_0$, the magnetic field $\vec{B}$ within the material differs from $\vec{B}_0$ because of the magnetization $\vec{M}$ (magnetic moment per unit volume) in the material. Specifically, for certain simple geometries,

$$\vec{B} = \vec{B}_0 + \mu_0\vec{M}. \tag{30–19}$$

In a paramagnetic material the magnetization $\vec{M}$ results from preferential orientation of permanent magnetic dipoles associated with orbital and spin motion of electrons in the material. The dipoles tend to align themselves with the field direction; $\vec{B}_0$ and $\vec{M}$ have the same direction, and the magnetization adds to the external field. In a diamagnetic material the magnetization results from *induced* magnetic dipoles; $\vec{M}$ is opposite in direction to $\vec{B}_0$, and the magnetization subtracts from the external field. This behavior is in accordance with Lenz's law.

If the magnetization $\vec{M}$ is proportional to the external field $\vec{B}_0$, it is also proportional to the field $\vec{B}$ in the material. This relation can be expressed in terms of the magnetic susceptibility χ_m (defined in Section 29–9) as

$$\mu_0\vec{M} = \frac{\chi_m}{1 + \chi_m}\,\vec{B}. \tag{30–20}$$

Paramagnetic materials have positive susceptibilities; diamagnetic materials have negative susceptibilities. For most paramagnetic materials, χ_m is very small, of the order of 10^{-3} or less. Diamagnetic susceptibilities are smaller still, usually of the order of -10^{-4} or less. In both paramagnetic and diamagnetic materials, $\mu_0\vec{M}$ is usually much smaller in magnitude than $\vec{B}$.

In a superconductor, on the other hand, the induced currents and associated magnetization are large enough to reduce the magnetic field $\vec{B}$ in the material to zero. Rewriting Eq. (30–20) as $\vec{B} = (1 + \chi_m)(\mu_0/\chi_m)\vec{M}$, we see that this situation corresponds to a diamagnetic susceptibility $\chi_m = -1$ and a relative permeability $K_m = 1 + \chi_m = 0$. A superconductor is sometimes called a "perfect diamagnetic material."

Here's another aspect of the magnetic behavior of superconductors. We place a homogeneous sphere of a superconducting material in a uniform applied magnetic field $\vec{B}_0$ at a temperature T greater than T_c. The material is then in the normal phase, not the superconducting phase. The field is as shown in Fig. 30–20a. Now we lower the temperature until the superconducting transition occurs. (We assume that the magnitude of $\vec{B}_0$ is not large enough to prevent the phase transition.) What happens to the field?

Measurements of the field outside the sphere show that the field lines become distorted as in Fig. 30–20b. There is no longer any field inside the material, except possibly in a very thin surface layer a hundred or so atoms thick. If a coil is wrapped around the sphere, the emf induced in the coil shows that during the superconducting transition the magnetic flux through the coil decreases from its initial value to zero; this is consistent with the absence of field inside the material. Finally, if the field is now turned off while the material is still in its superconducting phase, no emf is induced in the coil, and measurements show no field outside the sphere.

We conclude that during a superconducting transition in the presence of the field $\vec{B}_0$, all of the magnetic flux is expelled from the bulk of the sphere, and the magnetic flux Φ_B through the coil becomes zero. This expulsion of magnetic flux is called the *Meissner effect*. As shown in Fig. 30–20b, this expulsion crowds the magnetic field lines closer together to the side of the sphere, increasing $\vec{B}$ there.

The diamagnetic nature of a superconductor has some interesting *mechanical* consequences. A paramagnetic or ferromagnetic material is attracted by a permanent magnet because the magnetic dipoles in the material align with the non-uniform magnetic field of the permanent magnet. (We discussed this in Section 28–8.) For a diamagnetic material the magnetization is in the opposite sense, and a diamagnetic material is *repelled* by a permanent magnet. By Newton's third law the magnet is also repelled by the diamagnetic material. Figure 30–21 (see page 962) shows the repulsion between a specimen of a high-temperature superconductor and a magnet; the magnet is supported ("levitated") by this repulsive magnetic force.

The behavior we have described is characteristic of what are called *type-I superconductors*. There is another class of superconducting materials called *type-II superconductors*. When such a material in the superconducting phase is placed in a magnetic field, the bulk of the material remains superconducting, but thin filaments of material, running parallel to the field, may return to the normal phase. Currents circulate around the boundaries of these filaments, and there *is* magnetic flux inside them. Type-II superconductors are used for electromagnets because they usually have much larger values of B_c than do type-I materials, permitting much larger magnetic fields without destroying the superconducting state. Type-II superconductors have *two* critical magnetic fields; the first, B_{c1}, is the field at which magnetic flux begins to enter the material, forming the filaments just described, and the second, B_{c2}, is the field at which the material becomes normal.

Many important and exciting applications of superconductors are under development. Superconducting electromagnets have been used in research laboratories for several years. Their advantages compared to conventional electromagnets include greater efficiency, compactness, and greater field magnitudes. Once a current is established in the coil of a superconducting electromagnet, no additional power input is required because there is no resistive energy loss. The coils can also be made more compact because there is no need to provide channels for the circulation of cooling fluids. Superconducting magnets routinely attain steady fields of the order of 10 T, much larger than the maximum fields that are available with ordinary electromagnets.

Superconductors are attractive for long-distance electric power transmission and for energy-conversion devices, including generators, motors, and transformers. Very sensitive measurements of magnetic fields can be made with superconducting quantum

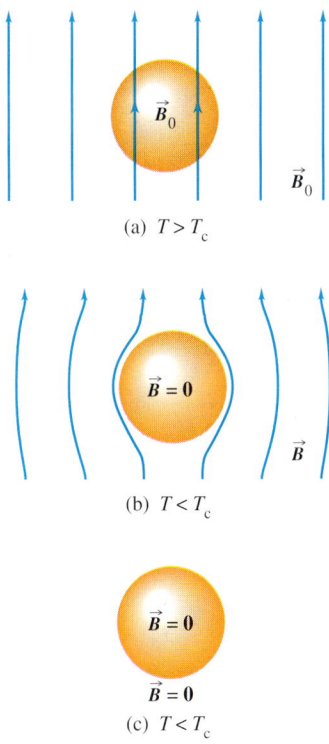

(a) $T > T_c$

(b) $T < T_c$

(c) $T < T_c$

30–20 (a) When a superconducting material that is above the critical temperature is placed in an external magnetic field $\vec{B}_0$, the field inside the material is very nearly equal to $\vec{B}_0$. (b) As the material is cooled through the critical temperature T_c and becomes superconducting, the magnetic flux is expelled from the material, and the field inside it becomes zero. (c) When the external field is turned off with the material in the superconducting phase, the field is zero everywhere; there is no change of magnetic flux in the material.

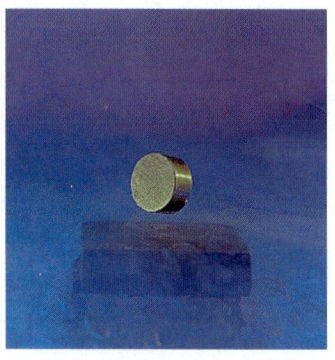

interference devices (SQUIDs), which can detect changes in magnetic flux of less than 10^{-14} Wb; these devices have applications in medicine, geology, and other fields. The number of potential uses for superconductors has increased greatly since the discovery in 1987 of high-temperature superconductors. As discussed in Section 26–3, these materials have critical temperatures that are above the temperature of liquid nitrogen (about 77 K) and so are comparatively easy to attain. Development of practical applications of superconductor science promises to be an exciting chapter in contemporary technology.

30–21 A superconductor (the black slab) exerts a repulsive force on a magnet (the metallic cylinder), supporting the magnet in midair.

SUMMARY

KEY TERMS

induced current, 941

induced emf, 941

Faraday's law of induction, 943

Lenz's law, 950

motional electromotive force, 952

induced electric field, 954

non-electrostatic field, 955

eddy currents, 956

Maxwell's equations, 957

- Faraday's law states that the induced emf in a closed loop equals the negative of the time rate of change of magnetic flux through the loop:

$$\mathcal{E} = -\frac{d\Phi_B}{dt}. \qquad (30\text{–}3)$$

This relation is valid whether the flux change is caused by a changing magnetic field, motion of the loop, or both.

- Lenz's law states that an induced current or emf always tends to oppose or cancel out the change that caused it. Lenz's law can be derived from Faraday's law and is often easier to use.

- If a conductor with length L moves with speed v in a uniform magnetic field with magnitude B, and if the length and velocity are both perpendicular to the field, the induced emf is

$$\mathcal{E} = vBL. \qquad (30\text{–}6)$$

More generally, when part or all of a closed conducting loop moves in a magnetic field $\vec{B}$, the induced emf is

$$\mathcal{E} = \oint \left(\vec{v} \times \vec{B} \right) \cdot d\vec{l}. \qquad (30\text{–}7)$$

- When an emf is induced by a changing magnetic flux through a stationary closed path, there is an induced electric field $\vec{E}$ of non-electrostatic origin such that

$$\oint \vec{E} \cdot d\vec{l} = -\frac{d\Phi_B}{dt}. \qquad (30\text{–}10)$$

This field is nonconservative and cannot be associated with a potential.

- When a bulk piece of conducting material, such as a metal, is in a changing magnetic field or moves through a field, currents called eddy currents are induced in the volume of the material.

- The relationships between electric and magnetic fields and their sources can be stated compactly in four equations called Maxwell's equations:

$$\oint \vec{E} \cdot d\vec{A} = \frac{Q_{encl}}{\epsilon_0}, \qquad (30\text{--}12)$$

$$\oint \vec{B} \cdot d\vec{A} = 0, \qquad (30\text{--}13)$$

$$\oint \vec{B} \cdot d\vec{l} = \mu_0 \left(i_C + \epsilon_0 \frac{d\Phi_E}{dt} \right)_{encl}, \qquad (30\text{--}14)$$

$$\oint \vec{E} \cdot d\vec{l} = -\frac{d\Phi_B}{dt}. \qquad (30\text{--}15)$$

The first equation is Gauss's law of electrostatics; the second is Gauss's law for magnetic fields, stating the nonexistence of single magnetic charges. The third is Ampere's law, generalized by Maxwell to include displacement current; and the fourth is Faraday's law. Together they form a complete basis for the relation of $\vec{E}$ and $\vec{B}$ fields to their sources.

DISCUSSION QUESTIONS

Q30–1 A sheet of copper is placed between the poles of an electromagnet with the magnetic field perpendicular to the sheet. When it is pulled out, a considerable force is required, and the force required increases with speed. Explain.

Q30–2 In Fig. 30–7, if the angular speed ω of the loop is doubled, then the frequency with which the induced current changes direction doubles, and the maximum emf also doubles. Why? Does the torque required to turn the loop change? Explain.

Q30–3 Two circular loops lie side by side in the same plane. One is connected to a source that supplies an increasing current; the other is a simple closed ring. Is the induced current in the ring in the same direction as that in the loop connected to the source, or opposite? What if the current in the first loop is decreasing? Explain.

Q30–4 A farmer claimed that the high-voltage transmission lines running parallel to his fence induced dangerously large voltages on the fence. Is this within the realm of possibility? Explain. (The lines carry alternating current that changes direction 120 times each second.)

Q30–5 A long, straight conductor passes through the center of a metal ring, perpendicular to its plane. If the current in the conductor increases, is a current induced in the ring? Explain.

Q30–6 A student asserted that if a permanent magnet is dropped down a vertical copper pipe, it eventually reaches a terminal velocity even if there is no air resistance. Why should this be? Or should it?

Q30–7 When a conductor moves through a magnetic field, the magnetic forces on the charges in the conductor cause an emf. But if this phenomenon is viewed in a frame of reference moving with the conductor, there is no motion yet there is still an emf. How is this paradox resolved?

Q30–8 Consider the moving rod in Fig. 30–13. If the charge q of charge carriers within the rod were negative rather than posi-

tive, how would this affect the directions of the electric and magnetic forces on these charge carriers? Would this affect the sign of the potential difference V_{ab} or the sign of the motional emf? Explain.

Q30–9 A car stops for a red light at an intersection. As it comes to a halt, it moves over a coil embedded in the surface of the road. As a result, the light soon changes from red to green. Explain how the coil is able to signal the light to change.

Q30–10 In the situation shown in Fig. 30–15, would it be appropriate to ask how much *energy* an electron gains during a complete trip around the wire loop with current I'? Would it be appropriate to ask what *potential difference* the electron moves through during such a complete trip? Explain your answers.

Q30–11 A fellow student, who read this chapter in haste, writes Faraday's law (Eq. (30–15)) with $+d\Phi_B/dt$ on the right-hand side rather than $-d\Phi_B/dt$. If this were truly the form of Faraday's law, what would be the direction of the magnetic force on the sliding wire in Example 30–7 (Section 30–3)? Would energy be conserved? How does this argument tell you that the right-hand side of the equation must be $-d\Phi_B/dt$ rather than $+d\Phi_B/dt$?

Q30–12 If magnetic monopoles existed, the right-hand side of Eq. (30–15) would include a term proportional to the current of magnetic monopoles. Suppose a steady monopole current is moving around a circular loop. Sketch the *electric* field lines that such a current would produce.

Q30–13 If magnetic monopoles existed, the right-hand side of Eq. (30–13) would be proportional to the total enclosed *magnetic* charge, and the right-hand side of Eq. (30–15) would include a term proportional to the current of magnetic monopoles. Suppose an infinite line of magnetic monopoles on the x-axis were moving in the positive x-direction. Sketch the electric and magnetic field lines that this line of monopoles would produce.

EXERCISES

SECTION 30–3 FARADAY'S LAW

30–1 A closely wound rectangular coil of 80 turns has dimensions of 25.0 cm × 40.0 cm. The plane of the coil is rotated in 0.0600 s from a position where it makes an angle of 45.0° with a magnetic field of 1.10 T to a position perpendicular to the field. What is the average emf induced in the coil?

30–2 In a physics laboratory experiment, a coil with 200 turns enclosing an area of 12 cm^2 is rotated in 0.040 s from a position where its plane is perpendicular to the earth's magnetic field to one where its plane is parallel to the field. What average emf is induced if the earth's magnetic field is 6.0×10^{-5} T?

30–3 Search Coils and Credit Cards. a) Derive the equation relating the total charge Q that flows through a search coil (Example 30–3 in Section 30–3) to the magnetic field magnitude B. The search coil has N turns, each with area A, and the flux through the coil is decreased from its initial maximum value to zero in a time Δt. The resistance of the coil is R, and the total charge is $Q = I\Delta t$, where I is the average current induced by the change in flux. b) In a credit card reader, the magnetic stripe on the back of a credit card is rapidly "swiped" past a coil within the reader. Explain, using the same ideas that underlie the operation of a search coil, how the reader can decode the information stored in the pattern of magnetization on the stripe. c) Is it necessary that the credit card be "swiped" through the reader at exactly the right speed? Why or why not?

30–4 The cross-section area of a closely wound search coil (Exercise 30–3) having 90 turns is 2.20 cm^2, and its resistance is 6.80 Ω. The coil is connected through leads of negligible resistance to a charge-measuring instrument having an internal resistance of 12.0 Ω. Find the quantity of charge displaced when the coil is pulled quickly out of a region where $B = 2.05$ T to a point where the magnetic field is zero. The plane of the coil, when in the field, makes an angle of 90° with the magnetic field.

30–5 A closely wound search coil (Exercise 30–3) has an area of 3.20 cm^2, 120 turns, and a resistance of 60.0 Ω. It is connected to a charge-measuring instrument whose resistance is 45.0 Ω. When the coil is rotated quickly from a position parallel to a uniform magnetic field to one perpendicular to the field, the instrument indicates a charge of 3.56×10^{-5} C. What is the magnitude of the field?

30–6 A coil 4.00 cm in radius, containing 500 turns, is placed in a uniform magnetic field that varies with time according to $B = (0.0120 \text{ T/s})t + (3.00 \times 10^{-5} \text{ T/s}^4)t^4$. The coil is connected to a 600-Ω resistor, and its plane is perpendicular to the magnetic field. The resistance of the coil can be neglected. a) Find the magnitude of the induced emf in the coil as a function of time. b) What is the current in the resistor at time $t = 5.00$ s?

30–7 A closely wound coil with N turns of cross-section area A lies in the xy-plane. The coil is placed in a uniform but time-dependent magnetic field $\vec{B} = B_z(t)\hat{k}$. Here $B_z(t)$ equals zero for $t \le 0$, equals $B_0(1 - \cos(2\pi t/T))$ for $0 < t < T$, and equals zero for $t \ge T$; B_0 is a positive constant. a) Find the induced emf in the coil as a function of time. b) At what time(s) t, if any, in the range $0 < t < T$ is the induced emf equal to zero? c) At what

time(s) t in the range $0 < t < T$ is the absolute value of the induced emf a maximum, and what is that maximum value? d) As seen by an observer on the positive z-axis looking back toward the coil, at what times is the induced emf in the clockwise direction? At what times is it counterclockwise?

30–8 A motor with a brush-and-commutator arrangement as described in Example 30–5 (Section 30–3) has a circular coil with radius 2.5 cm and 150 turns of wire. The magnetic field has magnitude 0.060 T and the coil rotates at 440 rev/min. a) What is the maximum emf induced in the coil? b) What is the average back emf?

30–9 If the rotation speed of the motor in Example 30–5 (Section 30–3) is increased to 3360 rev/min, what is the average back emf of the motor?

30–10 The armature of a small generator consists of a flat, square coil with 120 turns and sides with a length of 1.60 cm. The coil rotates in a magnetic field of 0.0750 T. What is the angular speed of the coil if the maximum emf produced is 24.0 mV?

SECTION 30–4 LENZ'S LAW

30–11 Redraw Figure 30–11 in Example 30–9 (Section 30–4) for the case in which the upward magnetic field is *decreasing*.

30–12 A circular loop of wire is in a region of spatially uniform magnetic field, as shown in Fig. 30–22. The magnetic field is directed into the plane of the figure. Determine the direction (clockwise or counterclockwise) of the induced current in the loop when a) B is increasing; b) B is decreasing; c) B is constant with value B_0. Explain your reasoning.

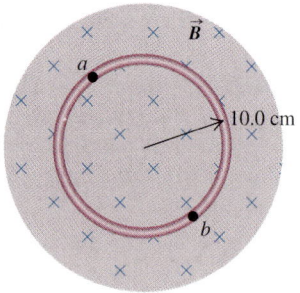

FIGURE 30–22 Exercises 30–12 and 30–23.

30–13 Using Lenz's law, determine the direction of the current in resistor ab of Fig. 30–23 when a) switch S is opened after

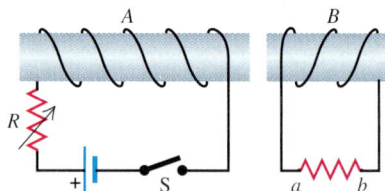

FIGURE 30–23 Exercise 30–13.

having been closed for several minutes; b) coil B is brought closer to coil A with the switch closed; c) the resistance of R is decreased while the switch remains closed.

30–14 A cardboard tube is wrapped with two windings of insulated wire wound in opposite directions, as in Fig. 30–24. Terminals a and b of winding A may be connected to a battery through a reversing switch. State whether the induced current in the resistor R is from left to right or from right to left in the following circumstances: a) the current in winding A is from a to b and is increasing; b) the current in winding A is from b to a and is decreasing; c) the current in winding A is from b to a and is increasing.

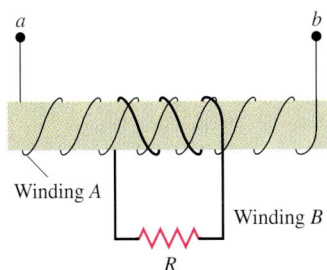

FIGURE 30–24 Exercise 30–14.

SECTION 30–5 MOTIONAL ELECTROMOTIVE FORCE

30–15 For Eq. (30–6), show that if v is in meters per second, B in teslas, and L in meters, then the units of the right-hand side of the equation are joules per coulomb or volts (the correct SI units for $\mathcal{E}$).

30–16 In Fig. 30–25 a conducting rod with length $L = 30.0$ cm moves in a magnetic field $\vec{B}$ of magnitude 0.450 T directed into the plane of the figure. The rod moves with velocity $v = 5.00$ m/s in the direction shown. a) What is the motional emf induced in the rod? b) What is the potential difference between the ends of the rod? c) Which point, a or b, is at higher potential? d) When the charges in the rod are in equilibrium, what is the magnitude and direction of the electric field within the rod? e) When the charges in the rod are in equilibrium, which point, a or b, has an excess of positive charge?

FIGURE 30–25 Exercise 30–16.

30–17 In Fig. 30–13b a rod with length $L = 85.0$ cm moves in a magnetic field with magnitude $B = 0.850$ T. The emf induced in the moving rod is 0.620 V. a) What is the speed of the rod? b) If the total circuit resistance is 0.750 Ω, what is the induced current? c) What force (magnitude and direction) does the field exert on the rod as a result of this current?

30–18 The conducting rod ab in Fig. 30–26 makes contact with metal rails ca and db. The apparatus is in a uniform magnetic field 0.800 T, perpendicular to the plane of the figure. a) Find the magnitude of the emf induced in the rod when it is moving toward the right with a speed 7.50 m/s. b) In what direction does the current flow in the rod? c) If the resistance of the circuit abdc is 1.50 Ω (assumed to be constant), find the force (magnitude and direction) required to keep the rod moving to the right with a constant speed of 7.50 m/s. Neglect friction. d) Compare the rate at which mechanical work is done by the force (Fv) with the rate at which thermal energy is developed in the circuit (I^2R).

FIGURE 30–26 Exercise 30–18.

30–19 A square loop of wire with resistance R is moved at constant speed v across a uniform magnetic field confined to a square region whose sides are twice the length of those of the square loop (Fig. 30–27). a) Graph the external force F needed to move the loop at constant speed as a function of the coordinate x from $x = -2L$ to $x = +2L$. (The coordinate x is measured from the center of the magnetic field region to the center of the loop. It is negative when the center of the loop is to the left of the center of the magnetic field region. Take positive force to be to the right.) b) Graph the induced current in the loop as a function of x. Take counterclockwise currents to be positive.

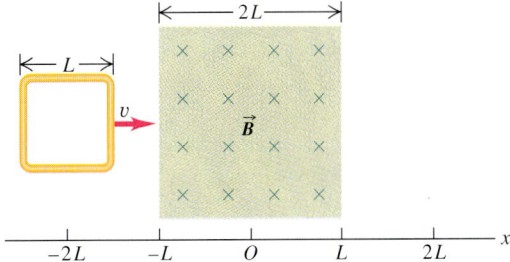

FIGURE 30–27 Exercise 30–19.

30–20 In Fig. 30–13a a 12.0-cm-long rod moves with constant speed 4.50 m/s in the direction shown. The induced emf is 0.450 V. a) What is the magnitude of the magnetic field? b) Which point is at higher potential, a or b?

SECTION 30–6 INDUCED ELECTRIC FIELDS

30–21 A long, thin solenoid has 900 turns per meter and radius 2.50 cm. The current in the solenoid is increasing at a uniform rate of 60.0 A/s. What is the magnitude of the induced electric field at a point near the center of the solenoid and a) 0.500 cm from the axis of the solenoid? b) 1.00 cm from the axis of the solenoid?

30–22 The magnetic field within a long, straight solenoid with a circular cross section and radius R is increasing at a rate of dB/dt. a) What is the rate of change of flux through a circle with radius r_1 inside the solenoid, normal to the axis of the solenoid, and with center on the solenoid axis? b) Find the magnitude of the induced electric field inside the solenoid, at a distance r_1 from its axis. Show the direction of this field in a diagram. c) What is the magnitude of the induced electric field *outside* the solenoid, at a distance r_2 from the axis? d) Graph the magnitude of the induced electric field as a function of the distance r from the axis from $r = 0$ to $r = 2R$. e) What is the magnitude of the induced emf in a circular turn of radius $R/2$ that has its center on the solenoid axis? f) What is the magnitude of the induced emf if the radius in part (e) is R? g) What is the induced emf if the radius in part (e) is $2R$?

30–23 The magnetic field $\vec{B}$ at all points within the colored circle of Fig. 30–22 has an initial magnitude of 0.750 T. (The circle could represent approximately the space inside a long, thin solenoid.) The magnetic field is directed into the plane of the diagram and is decreasing at the rate of −0.0350 T/s. a) What is the shape of the field lines of the induced electric field in Fig. 30–22, within the colored circle? b) What are the magnitude and direction of this field at any point on the circular conducting ring with radius 0.100 m? c) What is the current in the ring if its resistance is 4.00 Ω? d) What is the emf between points a and b on the ring? e) If the ring is cut at some point and the ends are separated slightly, what will be the emf between the ends?

30–24 A long, thin solenoid has 400 turns per meter and radius 1.10 cm. The current in the solenoid is increasing at a uniform rate di/dt. The induced electric field at a point near the center of the solenoid and 3.50 cm from its axis is 8.00×10^{-6} V/m. Calculate di/dt.

30–25 A $+6.50\text{-}\mu\text{C}$ charge moves in a circular path centered on the center of the solenoid in Exercise 30–24. The circular path is 3.50 cm in radius and lies in a plane perpendicular to the axis of the solenoid. a) If the charge moves around the circular path in the direction of the induced electric field, how much work is done on the charge during one complete trip around the path? b) Explain how your answer to part (a) confirms that the electric field is nonconservative. c) How much work is done on the charge during one complete trip around the path if the path radius is 7.00 cm?

30–26 A long, straight solenoid with a cross-section area of 8.00 cm^2 is wound with 90 turns of wire per centimeter, and the windings carry a current of 0.350 A. A second winding of 12 turns encircles the solenoid at its center. The current in the solenoid is turned off such that the magnetic field of the solenoid becomes zero in 0.0400 s. What is the average induced emf in the second winding?

*SECTION **30–9** SUPERCONDUCTIVITY: A CASE STUDY IN MAGNETIC PROPERTIES

****30–27** A long, straight wire made of a type-I superconductor carries a constant current I along its length. Show that the current cannot be uniformly spread over the wire's cross section but instead must all be at the surface.

****30–28** A type-II superconductor in an external field between B_{c1} and B_{c2} has regions that contain magnetic flux and have resistance, and also has superconducting regions. What is the resistance of a long, thin cylinder of such material?

****30–29** At temperatures near absolute zero, B_c approaches 0.142 T for vanadium, a type-I superconductor. The normal phase of vanadium has a magnetic susceptibility close to zero. Consider a long, thin vanadium cylinder with its axis parallel to an external magnetic field $\vec{B}_0$ in the $+x$-direction. At points far from the ends of the cylinder, by symmetry, all the magnetic vectors are parallel to the x-axis. At temperatures near absolute zero, what are the resultant magnetic field $\vec{B}$ and the magnetization $\vec{M}$ inside and outside the cylinder (far from the ends) for a) $\vec{B}_0 = (0.130 \text{ T})\hat{\imath}$? b) $\vec{B}_0 = (0.260 \text{ T})\hat{\imath}$?

****30–30** The compound SiV_3 is a type-II superconductor. At temperatures near absolute zero the two critical fields are $B_{c1} = 55.0$ mT and $B_{c2} = 15.0$ T. The normal phase of SiV_3 has a magnetic susceptibility close to zero. A long, thin SiV_3 cylinder has its axis parallel to an external magnetic field $\vec{B}_0$ in the $+x$-direction. At points far from the ends of the cylinder, by symmetry all the magnetic vectors are parallel to the x-axis. At a temperature near absolute zero the external magnetic field is slowly increased from zero. What are the resultant magnetic field $\vec{B}$ and the magnetization $\vec{M}$ inside the cylinder at points far from its ends a) just before the magnetic flux begins to penetrate the material; b) just after the material becomes completely normal?

PROBLEMS

30–31 A circular coil of wire has a radius of 0.500 m, 20 turns, and a total resistance of 1.57 Ω. The coil lies in the xy-plane. The coil is in a uniform magnetic field $\vec{B}$ that is in the $-z$-direction, which is directed away from you as you view the coil. The magnitude B of the field depends on time as follows: increases at a constant rate from 0 at $t = 0$ to 0.800 T at $t = 0.500$ s; is constant at 0.800 T from $t = 0.500$ s to $t = 1.00$ s; decreases at a constant rate from 0.800 T at $t = 1.00$ s to 0 at $t = 2.00$ s. a) Graph B versus t for t from 0 to 2.00 s. b) Graph the current I induced in the coil versus t for t from 0 to 2.00 s. Let counter-

clockwise currents be positive and clockwise currents be negative. c) What is the maximum induced electric field magnitude in the coil during the 0- to 2.00-s time interval?

30–32 The rectangular loop in Example 30–4 (Section 30–3) is connected to an external circuit, forming a complete current path of resistance R. a) Show that the induced current is $i = (\omega BA/R) \sin \omega t$. b) Calculate the rate at which electrical energy is dissipated due to the resistance of the current path. c) Show that the magnetic dipole moment of the loop has magnitude $\mu = (\omega BA^2/R) \sin \omega t$. d) Show that the external torque

required to maintain a constant ω is $\tau = (\omega B^2 A^2/R) \sin^2 \omega t$.
e) Calculate the rate at which the external torque does work. How do the answers to parts (b) and (e) compare? Explain.

30–33 A circular wire loop of radius a and resistance R initially has a magnetic flux through it due to an external magnetic field. The external field then decreases to zero. A current is induced in the loop while the external field is changing; however, this current does not stop at the instant that the external field stops changing. The reason is that the current itself generates a magnetic field, which gives rise to a flux through the loop. If the current changes, the flux through the loop changes as well, and an induced emf appears in the loop to oppose the change. a) The magnetic field at the center of the loop of radius a produced by a current i in the loop is given by $B = \mu_0 i/2a$. If we use the crude approximation that the field has this same value at all points within the loop, what is the flux of this field through the loop? b) By using Faraday's law, Eq. (30–3), and the relationship $\mathcal{E} = iR$, show that after the external field has stopped changing, the current in the loop obeys the differential equation

$$\frac{di}{dt} = -\left(\frac{2R}{\pi \mu_0 a}\right) i.$$

c) If the current has the value i_0 at $t = 0$, the instant that the external field stops changing, solve the equation in part (b) to find i as a function of time for $t > 0$. (*Hint:* In Section 27–5 we encountered a similar differential equation, Eq. (27–15), for the quantity q. This equation for i may be solved in the same way.) d) If the loop has radius $a = 50$ cm and resistance $R = 0.10$ Ω, how long after the external field stops changing will the current be equal to $0.010i_0$ (that is, $\frac{1}{100}$ of its initial value)? e) In solving the examples in this chapter, we ignored the effects described in this problem. Explain why this is a good approximation.

30–34 Suppose the loop in Fig. 30–28 is a) rotated about the y-axis; b) rotated about the x-axis; c) rotated about an edge parallel to the z-axis. What is the maximum induced emf in each case if $A = 600$ cm^2, $\omega = 35.0$ rad/s, and $B = 0.450$ T?

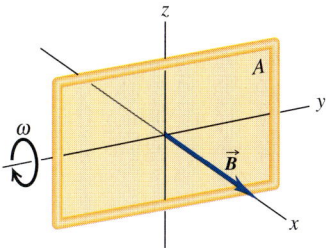

FIGURE 30–28 Problem 30–34.

30–35 A flexible circular loop 6.50 cm in diameter lies in a magnetic field with magnitude 0.950 T, directed into the plane of the page in Fig. 30–29. The loop is pulled at the points indicated by the arrows, forming a loop of zero area in 0.250 s. a) Find the average induced emf in the circuit. b) What is the direction of the current in R, from a to b or from b to a? Explain your reasoning.

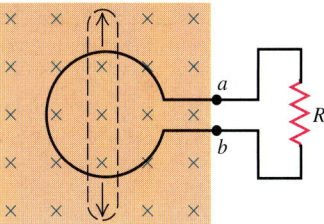

FIGURE 30–29 Problem 30–35.

30–36 The current in the long, straight wire AB of Fig. 30–30 is upward and is increasing steadily at a rate di/dt. a) At an instant when the current is i, what are the magnitude and direction of the field $\vec{B}$ at a distance r to the right of the wire? b) What is the flux $d\Phi_B$ through the narrow shaded strip? c) What is the total flux through the loop? d) What is the induced emf in the loop? e) Evaluate the numerical value of the induced emf if $a = 12.0$ cm, $b = 36.0$ cm, $L = 24.0$ cm, and $di/dt = 9.60$ A/s.

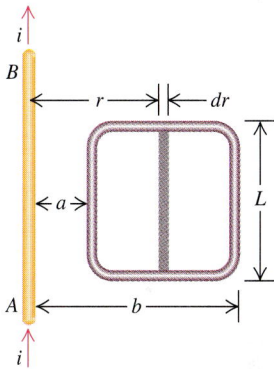

FIGURE 30–30 Problem 30–36.

30–37 A conducting rod with length L, mass m, and resistance R moves without friction on metal rails as shown in Fig. 30–9. A uniform magnetic field $\vec{B}$ is directed into the plane of the figure. The rod starts from rest and is acted on by a constant force $\vec{F}$ directed to the right. The rails are infinitely long and have negligible resistance. a) Graph the speed of the rod as a function of time. b) Find an expression for the terminal speed (the speed when the acceleration of the rod is zero).

30–38 A circular conducting ring with radius $r_0 = 0.0420$ m lies in the xy-plane in a region of uniform magnetic field $\vec{B} = B_0[1 - 3(t/t_0)^2 + 2(t/t_0)^3]\hat{k}$. In this expression, $t_0 = 0.0100$ s and is constant, t is time, $\hat{k}$ is the unit vector in the $+z$-direction, and $B_0 = 0.0800$ T and is constant. At points a and b (see Fig. 30–31) there is a small gap in the ring with wires leading to an external circuit of resistance $R = 12.0$ Ω. There is no magnetic field at the location of the external circuit. a) Derive an expression, as a function of time, for the total magnetic flux Φ_B through the ring. b) Determine the emf induced in the ring at time $t = 5.00 \times 10^{-3}$ s. What is the polarity of the emf?

c) Because of the internal resistance of the ring, the current through R at the time given in part (b) is only 3.00 mA. Determine the internal resistance of the ring. d) Determine the emf in the ring at a time $t = 1.21 \times 10^{-2}$ s. What is the polarity of the emf? e) Determine the time at which the current through R reverses its direction.

FIGURE 30–31 Problem 30–38.

30–39 The long, straight wire in Fig. 30–32a carries constant current I. A metal bar with length L is moving at constant velocity $\vec{v}$, as shown in the figure. Point a is a distance d from the wire. a) Calculate the emf induced in the bar. b) Which point, a or b, is at higher potential? c) If the bar is replaced by a rectangular wire loop of resistance R (Fig. 30–32b) what is the magnitude of the current induced in the loop?

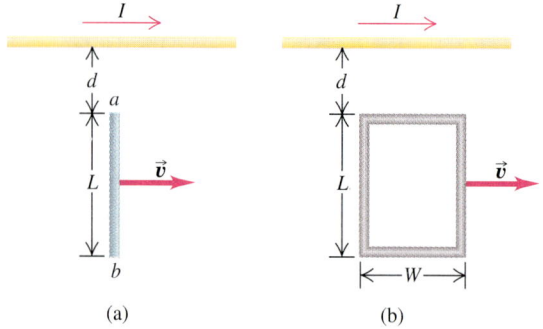

(a)   (b)

FIGURE 30–32 Problem 30–39.

30–40 The cube in Fig. 30–33, 50.0 cm on a side, is in a uniform magnetic field of 0.120 T, directed along the positive y-axis. Wires A, C, and D move in the directions indicated, each with a speed of 0.350 m/s. (Wire A moves parallel to the xy-plane, C moves at an angle of 45.0° below the xy-plane, and D moves parallel to the xz-plane.) What is the potential difference between the ends of each wire?

30–41 A slender rod 0.240 m long rotates with an angular speed of 8.80 rad/s about an axis through one end and perpendicular to the rod. The plane of rotation of the rod is perpendicular to a uniform magnetic field with a magnitude of 0.650 T. a) What is the induced emf in the rod? b) What is the potential difference between its ends? c) Suppose instead the rod

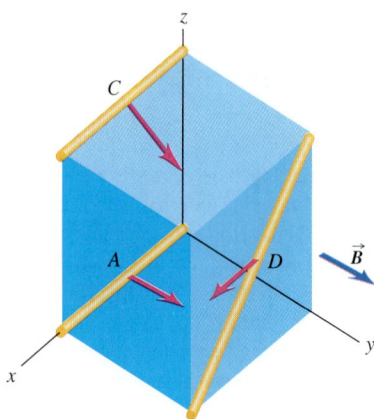

FIGURE 30–33 Problem 30–40.

rotates at 8.80 rad/s about an axis through its center and perpendicular to the rod. In this case, what is the potential difference between the ends of the rod? Between the center of the rod and one end?

30–42 Bar ab moves without friction on conducting rails as shown in Fig. 30–34. A uniform magnetic field with magnitude $B = 0.95$ T is directed into the plane of the page. You want to make an exercise machine out of this apparatus, in which the person exercises by pushing the bar back and forth with a root-mean-square speed of 3.5 m/s. What should be the circuit resistance R if the person moving the bar is to do work at an average rate of 250 W?

FIGURE 30–34 Problem 30–42.

30–43 A rectangular loop with width a and a slide wire with mass m are as shown in Fig. 30–35. A uniform magnetic field $\vec{B}$ is directed perpendicular to the plane of the loop into the plane of the figure. The slide wire is given an initial speed of v_0 and then released. There is no friction between the slide wire and the loop, and the resistance of the loop is negligible in comparison to the resistance R of the slide wire. a) Obtain an expression for F, the magnitude of the force exerted on the wire while it is

FIGURE 30–35 Problem 30–43.

moving at speed v. b) Show that the distance x that the wire moves before coming to rest is $x = mv_0 R / a^2 B^2$.

30–44 A 25.0-cm-long metal rod lies in the xy-plane and makes an angle of 36.9° with the positive x-axis and an angle of 53.1° with the positive y-axis. The rod is moving in the $+x$-direction with a speed of 4.20 m/s. The rod is in a uniform magnetic field $\vec{B} = (0.120 \text{ T})\hat{\imath} - (0.220 \text{ T})\hat{\jmath} - (0.0900 \text{ T})\hat{k}.$ a) What is the magnitude of the emf induced in the rod? b) Indicate in a sketch which end of the rod is at higher potential.

30–45 The magnetic field $\vec{B}$ at all points within a circular region of radius R is uniform in space and directed into the plane of the page in Fig. 30–36. (The region could be a cross section inside the windings of a long, straight solenoid.) If the magnetic field is increasing at a rate dB/dt, what are the magnitude and direction of the force on a stationary positive point charge q located at points a, b, and c? (Point a is a distance r above the center of the region, point b is a distance r to the right of the center, and point c is at the center of the region.)

30–46 It is impossible to have a uniform electric field that abruptly drops to zero in a region of space in which the magnetic

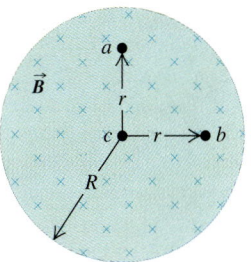

FIGURE 30–36 Problem 30–45.

field is constant. To prove this statement, use the method of contradiction: Assume that such a case *is* possible, then show that your assumption contradicts a law of nature. a) In the bottom half of a piece of paper, draw evenly spaced horizontal lines representing a uniform electric field to your right. Use dashed lines to draw a rectangle $abcda$ with horizontal side ab in the electric field region and horizontal side cd in the top half of your paper where $E = 0$. b) Show that integration around your rectangle contradicts Faraday's law, Eq. (30–15).

CHALLENGE PROBLEMS

30–47 A square, conducting wire loop of side L, total mass m, and total resistance R initially lies in the horizontal xy-plane, with corners at $(x, y, z) = (0, 0, 0)$, $(0, L, 0)$, $(L, 0, 0)$, and $(L, L, 0)$. There is a uniform upward magnetic field $\vec{B} = B\hat{k}$ in the space within and around the loop. The side of the loop that extends from $(0, 0, 0)$ to $(L, 0, 0)$ is held in place on the x-axis; the rest of the loop is free to pivot around this axis. When the loop is released, it begins to rotate due to the gravitational torque. a) Find the *net* torque (magnitude and direction) that acts on the loop when it has rotated through an angle ϕ from its original orientation and is rotating downward at an angular speed ω. b) Find the angular acceleration of the loop at the instant described in part (a). c) Compared to the case with zero magnetic field, does it take the loop a longer or shorter time to rotate through 90°? Explain. d) Is mechanical energy conserved as the loop rotates downward? Explain.

30–48 A square conducting loop, 20.0 cm on a side, is placed in the same magnetic field as in Exercise 30–23. (See Fig. 30–37;

the center of the square loop is at the center of the magnetic field region.) a) Copy Fig. 30–37, and draw vectors to show the directions and relative magnitudes of the induced electric field $\vec{E}$ at points a, b, and c. b) Prove that the component of $\vec{E}$ along the loop has the same value at every point of the loop and is equal to that of the ring of Fig. 30–22 (Exercise 30–23). c) What current is induced in the loop if its resistance is 1.90 Ω? d) What is the potential difference between points a and b?

30–49 A uniform square conducting loop, 20.0 cm on a side, is placed in the same magnetic field as in Exercise 30–23, with side ac along a diameter and with point b at the center of the field (Fig. 30–38). a) Copy Fig. 30–38, and draw vectors to show the direction and relative magnitude of the induced electric field $\vec{E}$ at the lettered points. b) What is the induced emf in side ac? c) What is the induced emf in the loop? d) What is the current in the loop if its resistance is 1.90 Ω? e) What is the potential difference between points a and c? Which is at higher potential?

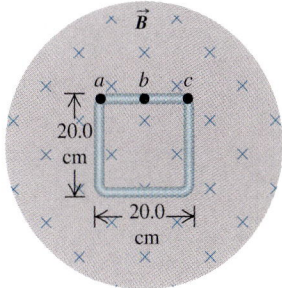

FIGURE 30–37 Challenge Problem 30–48.

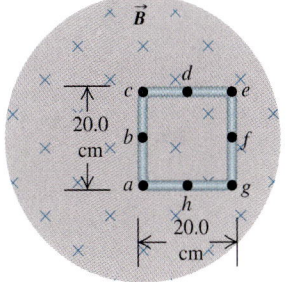

FIGURE 30–38 Challenge Problem 30–49.

30–50 A metal bar with length L, mass m, and resistance R is placed on frictionless metal rails that are inclined at an angle ϕ above the horizontal. The rails have negligible resistance. A uniform magnetic field of magnitude B is directed downward in Fig. 30–39. The bar is released from rest and slides down the rails. a) Is the direction of the current induced in the bar from a to b or from b to a? b) What is the terminal speed of the bar? c) What is the induced current in the bar when the terminal speed has been reached? d) After the terminal speed has been reached, at what rate is electrical energy being converted to thermal energy in the resistance of the bar? e) After the terminal speed has been reached, at what rate is work being done on the bar by gravity? Compare your answer to that in part (d).

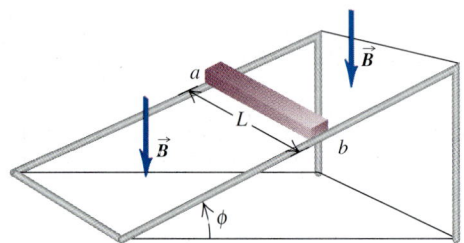

FIGURE 30–39 Challenge Problem 30–50.

*30–51 Consider a uniform metallic disk rotating through a perpendicular magnetic field $\vec{B}$, as shown in Fig. 30–16a. The disk has mass m, radius R, and thickness t, is made of a material with resistivity ρ, and is rotating clockwise in Fig. 30–16a with angular speed ω. The magnetic field is directed into the plane of the disk. Suppose that the region to which the magnetic field is confined is not circular, as shown in Fig. 30–16a, but is a small square with sides of length L ($L \ll R$) centered a distance d from the point O (the center of the disk). The sides of this square are horizontal and vertical in Fig. 30–16a. a) Show that the current induced within the square is approximately equal to $I = \omega dBLt/\rho$. In which direction does this current flow? (*Hint:* Assume that the resistance to the current is confined to the region of the square. The current also encounters resistance as it flows outside the region to which the magnetic field is confined, as shown in Fig. 30–16b; however, this resistance is relatively small since the current can flow through such a wide area. Recall Eq. (26–10) for resistance, given in Section 26–4.) b) Show that the induced current gives rise to a torque of approximate magnitude $\tau = \omega d^2 B^2 L^2 t/\rho$ that opposes the rotation of the disk (that is, a counterclockwise torque). c) What would be the magnitudes and directions of the induced current and torque if the direction of $\vec{B}$ were still into the plane of the disk but the disk rotated counterclockwise? What if the direction of $\vec{B}$ were out of the plane and the disk rotated counterclockwise?

This electric car is being recharged by a 220-V line—but there is no physical contact between the recharging cable and the car's wires. Instead, mutual inductance is used: The alternating current in the recharging cable creates a time-varying magnetic field, which induces a current in the car's circuitry. This induced current recharges the car's batteries.

31

Inductance

31–1 Introduction

How can a 12-volt car battery provide the thousands of volts needed to produce sparks across the gaps of the spark plugs in the engine? If you turn off a vacuum cleaner by yanking its plug out of the wall socket, what's the source of the high voltage that causes breakdown in the air so that sparks fly in the socket? The answers to these questions and many others concerned with varying currents in circuits involve the *induction* effects that we studied in Chapter 30.

A changing current in a coil induces an emf in an adjacent coil. The coupling between the coils is described by their *mutual inductance.* A changing current in a coil also induces an emf in that same coil. Such a coil is called an *inductor;* the relationship of current to emf is described by the *inductance* (also called *self-inductance*) of the coil. If a coil is initially carrying a current, energy is released when the current decreases; this principle is used in automotive ignition systems. We'll find that this released energy was stored in the magnetic field caused by the current that was initially in the coil, and we'll look at some of the practical applications of magnetic-field energy.

We'll also take a first look at what happens when an inductor is part of a circuit. In Chapter 32 we'll go on to study how inductors behave in alternating-current circuits; in that chapter we'll learn why inductors play an essential role in modern electronics, including communication systems, power supplies, fluorescent lights, and many other devices.

31–2 Mutual Inductance

In Section 29–5 we considered the magnetic interaction between two wires carrying *steady* currents; the current in one wire causes a magnetic field, which exerts a force on the current in the second wire. But an additional interaction arises between two circuits when there is a *changing* current in one of the circuits. Consider two neighboring coils of wire, as in Fig. 31–1. A current flowing in coil 1 produces a magnetic field $\vec{B}$ and hence a magnetic flux through coil 2. If the current in coil 1 changes, the flux through coil 2 changes as well; according to Faraday's law, this induces an emf in coil 2. In this way, a change in the current in one circuit can induce a current in a second circuit.

Let's analyze the situation shown in Fig. 31–1 in more detail. We will use lowercase letters to represent quantities that vary with time; for example, a time-varying current is i, often with a subscript to identify the circuit. In Fig. 31–1 a current i_1 in coil 1 sets up a magnetic field (as indicated by the blue lines), and some of these field lines pass through coil 2. We denote the magnetic flux through *each* turn of coil 2, caused by the current i_1 in coil 1, as Φ_{B2}. (If the flux is different through different turns of the coil, then Φ_{B2} denotes the *average* flux.) The magnetic field is proportional to i_1, so Φ_{B2} is also proportional to i_1. When i_1 changes, Φ_{B2} changes; this changing flux induces an emf $\mathcal{E}_2$ in coil 2, given by

$$\mathcal{E}_2 = -N_2 \frac{d\Phi_{B2}}{dt}. \qquad (31\text{–}1)$$

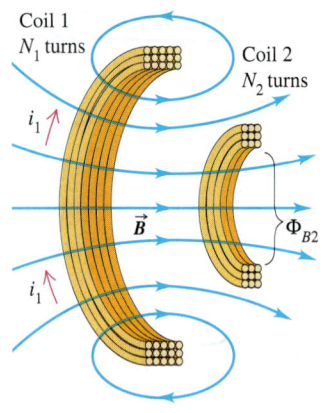

Coil 1
N_1 turns

Coil 2
N_2 turns

i_1

$\vec{B}$

Φ_{B2}

i_1

31–1 A current i_1 in coil 1 gives rise to a magnetic flux through coil 2. If i_1 changes, an emf is induced in coil 2; this is described in terms of mutual inductance.

Gears are often prepared for industrial use by first wrapping a coil of wire around each tooth. Alternating current in the coil induces eddy currents in the tooth that cause localized heating. The gear is then immersed in water or oil, and the rapid cooling hardens the gear teeth.

We could represent the proportionality of Φ_{B2} and i_1 in the form $\Phi_{B2} = (\text{constant})i_1$, but instead it is more convenient to include the number of turns N_2 in the relation. Introducing a proportionality constant M_{21}, called the **mutual inductance** of the two coils, we write

$$N_2\Phi_{B2} = M_{21}i_1, \tag{31–2}$$

where Φ_{B2} is the flux through a *single* turn of coil 2. From this,

$$N_2\frac{d\Phi_{B2}}{dt} = M_{21}\frac{di_1}{dt},$$

and we can rewrite Eq. (31–1) as

$$\mathcal{E}_2 = -M_{21}\frac{di_1}{dt}. \tag{31–3}$$

That is, a change in the current i_1 in coil 1 induces an emf in coil 2 that is directly proportional to the rate of change of i_1.

We may also write the definition of mutual inductance, Eq. (31–2), as

$$M_{21} = \frac{N_2\Phi_{B2}}{i_1}.$$

If the coils are in vacuum, the flux Φ_{B2} through each turn of coil 2 is directly proportional to the current i_1. Then the mutual inductance M_{21} is a constant that depends only on the geometry of the two coils (the size, shape, number of turns, and orientation of each coil and the separation between the coils). If a magnetic material is present, M_{21} also depends on the magnetic properties of the material. If the material has nonlinear magnetic properties, that is, if the relative permeability K_m (defined in Section 29–9) is not constant and magnetization is not proportional to magnetic field, then Φ_{B2} is no longer directly proportional to i_1. In that case the mutual inductance also depends on the value of i_1. In this discussion we will assume that any magnetic material present has constant K_m so that flux *is* directly proportional to current and M_{21} depends on geometry only.

We can repeat our discussion for the opposite case in which a changing current i_2 in coil 2 causes a changing flux Φ_{B1} and an emf $\mathcal{E}_1$ in coil 1. We might expect that the corresponding constant M_{12} would be different from M_{21} because in general the two coils are not identical and the flux through them is not the same. It turns out, however, that M_{12} is *always* equal to M_{21}, even when the two coils are not symmetric. We call this common value simply the mutual inductance, denoted by the symbol M without subscripts; it characterizes completely the induced-emf interaction of two coils. Then we can write

$$\mathcal{E}_2 = -M\frac{di_1}{dt} \quad \text{and} \quad \mathcal{E}_1 = -M\frac{di_2}{dt} \quad \text{(mutually induced emf's)}, \tag{31–4}$$

where the mutual inductance M is

$$M = \frac{N_2\Phi_{B2}}{i_1} = \frac{N_1\Phi_{B1}}{i_2} \quad \text{(mutual inductance)}. \tag{31–5}$$

The negative signs in Eq. (31–4) are a reflection of Lenz's law. The first equation says that a change in current in coil 1 causes a change in flux through coil 2, inducing an emf in coil 2 that opposes the flux change; in the second equation the roles of the two coils are interchanged.

The SI unit of mutual inductance is called the **henry** (1 H), in honor of the American physicist Joseph Henry (1797–1878), one of the discoverers of electromagnetic induction. From Eq. (31–5), one henry is equal to *one weber per ampere*. Other equivalent units, obtained by using Eq. (31–4), are *one volt-second per ampere*, *one ohm-second*, or *one joule per ampere squared*:

$$1 \text{ H} = 1 \text{ Wb/A} = 1 \text{ V} \cdot \text{s/A} = 1 \, \Omega \cdot \text{s} = 1 \text{ J/A}^2.$$

Mutual inductance can be a nuisance in electric circuits, since variations in current in one circuit can induce unwanted emf's in other, nearby circuits. To minimize these effects, multiple-circuit systems must be designed so that M is as small as possible; for example, two coils would be placed far apart or with their planes perpendicular.

Happily, mutual inductance also has many useful applications. A *transformer,* used in alternating-current circuits to raise or lower voltages, is fundamentally no different from the two coils shown in Fig. 31–1. A time-varying alternating current in one coil of the transformer produces an alternating emf in the other coil; the value of M, which depends on the geometry of the coils, determines the amplitude of the induced emf in the second coil and hence the amplitude of the output voltage. (We'll describe transformers in more detail in Chapter 32 after we've discussed alternating current in greater depth.)

EXAMPLE 31–1

The Tesla coil In one form of Tesla coil (a high-voltage generator that you may have seen in a science museum), a long solenoid with length l and cross-section area A is closely wound with N_1 turns of wire. A coil with N_2 turns surrounds it at its center (Fig. 31–2). Find the mutual inductance.

SOLUTION A current i_1 in the solenoid sets up a magnetic field $\vec{B}_1$ at its center; from Example 29–10 (Section 29–8) the magnitude of $\vec{B}_1$ is

$$B_1 = \mu_0 n_1 i_1 = \frac{\mu_0 N_1 i_1}{l}.$$

The flux through a cross section of the solenoid equals $B_1 A$. Since a very long solenoid produces no magnetic field outside of its coil, this is also equal to the flux Φ_{B2} through each turn of the outer, surrounding coil, no matter what the cross-section area of the outer coil. From Eq. (31–5) the mutual inductance M is

$$M = \frac{N_2 \Phi_{B2}}{i_1} = \frac{N_2 B_1 A}{i_1} = \frac{N_2}{i_1} \frac{\mu_0 N_1 i_1}{l} A = \frac{\mu_0 A N_1 N_2}{l}.$$

The mutual inductance of any two coils is always proportional to the product $N_1 N_2$ of their numbers of turns. Notice that the mutual inductance M depends only on the geometry of the two coils, not on the current.

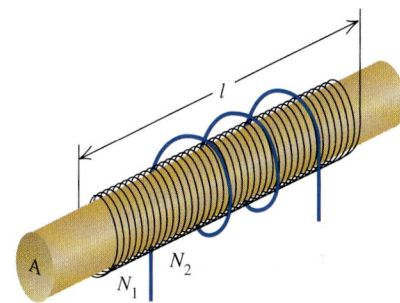

31–2 One form of Tesla coil, a long solenoid with cross-section area A and N_1 turns surrounded at its center by a coil with N_2 turns.

Here's a numerical example to give you an idea of magnitudes. Suppose $l = 0.50$ m, $A = 10$ cm$^2 = 1.0 \times 10^{-3}$ m^2, $N_1 = 1000$ turns, and $N_2 = 10$ turns. Then

$$M = \frac{(4\pi \times 10^{-7} \text{ Wb/A} \cdot \text{m})(1.0 \times 10^{-3} \text{ m}^2)(1000)(10)}{0.50 \text{ m}}$$

$$= 25 \times 10^{-6} \text{ Wb/A} = 25 \times 10^{-6} \text{ H} = 25 \, \mu\text{H}.$$

EXAMPLE 31–2

In Example 31–1, suppose the current i_2 in the outer, surrounding coil is given by $i_2 = (2.0 \times 10^6 \text{ A/s})t$ (currents in wires can indeed increase this rapidly for brief periods). a) At time $t = 3.0 \, \mu$s, what average magnetic flux through each turn of the solenoid is caused by the current in the outer, surrounding coil? b) What is the induced emf in the solenoid?

SOLUTION a) At time $t = 3.0 \, \mu$s $= 3.0 \times 10^{-6}$ s, the current in the outer coil (coil 2) is $i_2 = (2.0 \times 10^6 \text{ A/s})(3.0 \times 10^{-6} \text{ s}) = 6.0$ A. To find the average flux through each turn of the solenoid (coil 1), we solve Eq. (31–5) for Φ_{B1}:

$$\Phi_{B1} = \frac{M i_2}{N_1} = \frac{(25 \times 10^{-6} \text{ H})(6.0 \text{ A})}{1000} = 1.5 \times 10^{-7} \text{ Wb}.$$

Note that this is an *average* value; the flux can vary considerably between the center and the ends of the solenoid.

b) The induced emf ε_1 is given by Eq. (31–4):

$$\varepsilon_1 = -M\frac{di_2}{dt} = -(25 \times 10^{-6}\ \text{H})\frac{d}{dt}((2.0 \times 10^6\ \text{A/s})\,t)$$

$$= -(25 \times 10^{-6}\ \text{H})(2.0 \times 10^6\ \text{A/s}) = -50\ \text{V}.$$

This is a substantial induced emf in response to a very rapid rate of change of current. In an operating Tesla coil, there is a high-frequency alternating current rather than a continuously increasing current as in this example; both di_2/dt and ε_1 alternate as well, with amplitudes than can be thousands of times larger than in this example.

31–3 SELF-INDUCTANCE AND INDUCTORS

In our discussion of mutual inductance we considered two separate, independent circuits; a current in one circuit creates a magnetic field, and this field gives rise to a flux through the second circuit. If the current in the first circuit changes, the flux through the second circuit changes and an emf is induced in the second circuit.

An important related effect occurs even if we consider only a *single* isolated circuit. When a current is present in a circuit, it sets up a magnetic field that causes a magnetic flux through the *same* circuit; this flux changes when the current changes. Thus any circuit that carries a varying current will have an emf induced in it by the variation in *its own* magnetic field. Such an emf is called a **self-induced emf.** By Lenz's law a self-induced emf always opposes the change in the current that caused the emf and so tends to make it more difficult for variations in current to occur. For this reason, self-induced emfs can be of great importance whenever there is a varying current.

Self-induced emf's can occur in *any* circuit, since there will always be some magnetic flux through the closed loop of a current-carrying circuit. But the effect is greatly enhanced if the circuit includes a coil with N turns of wire (Fig. 31–3). As a result of the current i, there is an average magnetic flux Φ_B through each turn of the coil. In analogy to Eq. (31–5) we define the **self-inductance** L of the circuit as follows:

31–3 The current in the circuit causes a magnetic field in the coil and hence a flux through the coil. When the current in the circuit changes, the flux changes also, and a self-induced emf appears in the circuit.

$$L = \frac{N\Phi_B}{i} \qquad \text{(self-inductance).} \qquad (31\text{–}6)$$

When there is no danger of confusion with mutual inductance, the self-inductance is also called simply the **inductance.** Comparing Eqs. (31–5) and (31–6), we see that the units of self-inductance are the same as those of mutual inductance; the SI unit of self-inductance is one henry.

If the current i in the circuit changes, so does the flux Φ_B; from rearranging Eq. (31–6) and taking the derivative with respect to time, the rates of change are related by

$$N\frac{d\Phi_B}{dt} = L\frac{di}{dt}.$$

From Faraday's law for a coil with N turns, Eq. (30–4), the self-induced emf is $\varepsilon = -N\,d\Phi_B/dt$, so it follows that

$$\varepsilon = -L\frac{di}{dt} \qquad \text{(self-induced emf).} \qquad (31\text{–}7)$$

The minus sign in Eq. (31–7) is a reflection of Lenz's law; it says that the self-induced emf in a circuit opposes any change in the current in that circuit. (Later in this section we'll explore the significance of this minus sign in greater depth.) Equation (31–7) also states that **the self-inductance of a circuit is the magnitude of the self-induced emf per unit rate of change of current.**

A circuit device that is designed to have a particular inductance is called an **inductor,** or a *choke.* The usual circuit symbol for an inductor is

Like resistors and capacitors, inductors are among the indispensable circuit elements of modern electronics. Their purpose is to oppose any variations in the current through the circuit. An inductor in a direct-current circuit helps to maintain a steady current despite any fluctuations in the applied emf; in an alternating-current circuit, an inductor tends to suppress variations of the current that are more rapid than desired. In this chapter and the next we will explore the behavior and applications of inductors in circuits in more detail.

To understand the behavior of circuits containing inductors, we need to develop a general principle, analogous to Kirchhoff's loop rule, discussed in Section 27–3. To apply that rule, we go around a conducting loop, measuring potential differences across successive circuit elements as we go. The algebraic sum of these differences around any closed loop must be zero because the electric field produced by charges distributed around the circuit is *conservative*. In Section 30–8 we denoted such a conservative field as $\vec{E}_c$.

When an inductor is included in the circuit, the situation changes. The magnetically induced electric field within the coils of the inductor is *not* conservative; we'll denote it by $\vec{E}_n$. We need to think very carefully about the roles of the various fields. Let's assume we are dealing with an inductor whose coils have negligible resistance. Then a negligibly small electric field is required to make charge move through the coils, so the *total* electric field $\vec{E}_c + \vec{E}_n$ within the coils must be zero, even though neither field is individually zero. Because $\vec{E}_c$ is nonzero, we know there have to be accumulations of charge on the terminals of the inductor and the surfaces of its conductors, to produce this field.

Consider the circuit shown in Fig. 31–4; the box contains some combination of batteries and variable resistors that enables us to control the current i in the circuit. According to Faraday's law, Eq. (30–10), the line integral of $\vec{E}_n$ around the circuit is the negative of the rate of change of flux through the circuit, which in turn is given by Eq. (31–7). Combining these two relations, we get

$$\oint \vec{E}_n \cdot d\vec{l} = -L\frac{di}{dt},$$

where we integrate clockwise around the loop (the direction of the assumed current). But $\vec{E}_n$ is different from zero only within the inductor. Therefore the integral of $\vec{E}_n$ around the whole loop can be replaced by its integral only from a to b through the inductor; that is,

$$\int_a^b \vec{E}_n \cdot d\vec{l} = -L\frac{di}{dt}.$$

Next, because $\vec{E}_c + \vec{E}_n = 0$ at each point within the inductor coils, we can rewrite this as

$$\int_a^b \vec{E}_c \cdot d\vec{l} = L\frac{di}{dt}.$$

But this integral is just the potential V_{ab} of point a with respect to point b, so we finally obtain

$$V_{ab} = V_a - V_b = L\frac{di}{dt}. \qquad (31\text{–}8)$$

We conclude that there is a genuine potential difference between the terminals of the inductor, associated with conservative, electrostatic forces, despite the fact that the electric field associated with the magnetic induction effect is nonconservative. Thus we are justified in using Eq. (31–8) in Kirchhoff's loop rule for circuit analysis.

Note that the self-induced emf does not oppose the current i itself; rather, it opposes any *change* (di/dt) in the current. Thus the circuit behavior of an inductor is quite different from that of a resistor. Figure 31–5 compares the behavior of a resistor and an inductor and summarizes the sign relations.

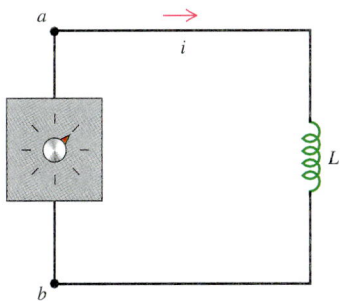

31–4 A circuit containing an energy source and an inductor. The source is variable, so the current i and its rate of change di/dt can be varied.

$V_{ab} = iR$ $V_{ab} = \dfrac{di}{dt}L$

(a)   (b)

31–5 (a) When a current i flows from a to b through a resistor, the potential always drops from a to b when i is positive and $V_{ab} = V_a - V_b = iR$. (b) When current flows from a to b through an inductor, the potential drops from a to b when di/dt is positive (increasing current) and increases from a to b when di/dt is negative (decreasing current). In each case, $V_{ab} = V_a - V_b = L\, di/dt$. When i is constant, $V_{ab} = 0$.

The self-inductance of a circuit depends on its size, shape, and number of turns. For N turns close together, it is always proportional to N^2. It also depends on the magnetic properties of the material enclosed by the circuit. In the following examples we will assume that the circuit encloses only vacuum (or air, which from the standpoint of magnetism is essentially a vacuum). But if the flux is concentrated in a region containing a magnetic material with permeability μ, then in the expression for B we must replace μ_0 (the permeability of vacuum) by $\mu = K_{\mathrm{m}}\mu_0$, as discussed in Section 29–9. If the material is diamagnetic or paramagnetic, this replacement makes very little difference, since K_{m} is very close to 1. If the material is *ferromagnetic,* however, the difference is of crucial importance. A solenoid wound on a soft iron core having $K_{\mathrm{m}} = 5000$ can have an inductance approximately 5000 times as great as that of the same solenoid with an air core. Iron-core and ferrite-core inductors are very widely used in a variety of electronic and electric-power applications.

An added complication is that with ferromagnetic materials the magnetization is in general not a linear function of magnetizing current; in particular, this is true as saturation is approached. As a result, the inductance is not constant but can depend on current in a fairly complicated way. In our discussion we will ignore this complication and assume always that the inductance is constant. This is a reasonable assumption even for a ferromagnetic material if the magnetization remains well below the saturation level.

EXAMPLE 31–3

An air-core toroidal solenoid with cross-section area A and mean radius r is closely wound with N turns of wire (Fig. 31–6). Determine its self-inductance L. Assume that B is uniform across a cross section (that is, neglect the variation of B with distance from the toroid axis).

SOLUTION From Eq. (31–6), which defines inductance, $L = N\Phi_B/i$. From Example 29–11 (Section 29–8) the field magnitude at a distance r from the toroid axis is $B = \mu_0 Ni/2\pi r$. If we assume that the field has this magnitude over the entire cross-section area A, then the magnetic flux through the cross section is

$$\Phi_B = BA = \frac{\mu_0 NiA}{2\pi r}.$$

The flux Φ_B is the same through each turn, and the self-inductance L is

$$L = \frac{N\Phi_B}{i} = \frac{\mu_0 N^2 A}{2\pi r} \qquad \begin{array}{l}\text{(self-inductance of a} \\ \text{toroidal solenoid).}\end{array} \quad (31\text{–}8)$$

Suppose $N = 200$ turns, $A = 5.0 \text{ cm}^2 = 5.0 \times 10^{-4} \text{ m}^2$, and

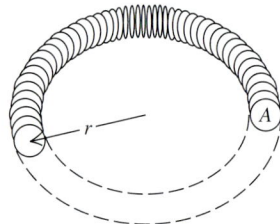

31–6 Determining the self-inductance of a closely wound toroidal solenoid.

$r = 0.10$ m; then

$$L = \frac{(4\pi \times 10^{-7} \text{ Wb/A} \cdot \text{m})(200)^2 (5.0 \times 10^{-4} \text{ m}^2)}{2\pi(0.10 \text{ m})}$$

$$= 40 \times 10^{-6} \text{ H} = 40 \ \mu\text{H}.$$

Later in this chapter we will use the expression $L = \mu_0 N^2 A/2\pi r$ for the inductance of a toroidal solenoid to help develop an expression for the energy stored in a magnetic field.

EXAMPLE 31–4

If the current in the toroidal solenoid in Example 31–3 increases uniformly from zero to 6.0 A in 3.0 μs, find the magnitude and direction of the self-induced emf.

SOLUTION The rate of change of the solenoid current is $di/dt = (6.0 \text{ A})/(3.0 \times 10^{-6} \text{ s}) = 2.0 \times 10^6 \text{ A/s}$. From Eq. (31–7),

$$|\mathcal{E}| = L\left|\frac{di}{dt}\right| = (40 \times 10^{-6} \text{ H})(2.0 \times 10^6 \text{ A/s}) = 80 \text{ V}.$$

The current is increasing, so according to Lenz's law the direction of the emf is opposite to that of the current. This corresponds to the situation in Fig. 31–4b; the emf is in the direction from b to a, like a battery with a as the (+) terminal and b the (−) terminal, tending to oppose the current increase from the external circuit.

31-4 MAGNETIC-FIELD ENERGY

Establishing a current in an inductor requires an input of energy, and an inductor carrying a current has energy stored in it. Let's see how this comes about. In Fig. 31–4, an increasing current i in the inductor causes an emf $\mathcal{E}$ between its terminals, and a corresponding potential difference V_{ab} between the terminals of the source, with point a at higher potential than point b. Thus the source must be adding energy to the inductor, and the instantaneous power P (rate of transfer of energy into the inductor) is $P = V_{ab}i$.

We can calculate the total energy input U needed to establish a final current I in an inductor with inductance L if the initial current is zero. We assume that the inductor has zero resistance, so no energy is dissipated within the inductor. Let the current at some instant be i, and let its rate of change be di/dt; the current is increasing, so $di/dt > 0$. The voltage between the terminals a and b of the inductor at this instant is $V_{ab} = L\, di/dt$, and the rate P at which energy is being delivered to the inductor (equal to the instantaneous power supplied by the external source) is

$$P = V_{ab}i = Li\frac{di}{dt}.$$

The energy dU supplied to the inductor during an infinitesimal time interval dt is $dU = P\, dt$, so

$$dU = Li\, di.$$

The total energy U supplied while the current increases from zero to a final value I is

$$U = L\int_0^I i\, di = \frac{1}{2}LI^2 \qquad \text{(energy stored in an inductor).} \qquad (31\text{–}9)$$

After the current has reached its final steady value I, $di/dt = 0$ and no more energy is input to the inductor. We can think of the energy U as being analogous to a *kinetic energy* associated with the current. When there is no current, this energy is zero; and when the current is I, the energy is $\frac{1}{2}LI^2$.

When the current decreases from I to zero, as in Fig. 31–4c, the inductor acts as a source that supplies a total amount of energy $\frac{1}{2}LI^2$ to the external circuit. If we interrupt the circuit suddenly by opening a switch or yanking a plug from a wall socket, the current decreases very rapidly, the induced emf is very large, and the energy may be dissipated in an arc across the switch contacts. This large emf is the electrical analog of the large force exerted by a car running into a brick wall and stopping very suddenly. The high voltage that is developed in one winding of a car engine's ignition coil results from the sudden flux change caused by interruption of the current in another winding.

CAUTION ▶ It's important not to confuse the behavior of resistors and inductors where energy is concerned. Energy flows into a resistor whenever a current passes through it, whether the current is steady or varying; this energy is dissipated in the form of heat. By contrast, energy flows into an ideal, zero-resistance inductor only when the current in the inductor *increases*. This energy is not dissipated; it is stored in the inductor and released when the current *decreases*. When a steady current flows through an inductor, there is no energy flow in or out. ◀

The energy in an inductor is actually stored in the magnetic field within the coil, just as the energy of a capacitor is stored in the electric field between its plates. We can develop relations for magnetic-field energy analogous to those we obtained for electric-field energy in Section 25–4 (Eqs. (25–9) and (25–11)). We will concentrate on one simple case, the ideal toroidal solenoid. This system has the advantage that its magnetic field is confined completely to a finite region of space within its core. As in Example 31–3, we assume that the cross-section area A is small enough that we can pretend that the magnetic field is uniform over the area. The volume V enclosed by the toroidal

Automotive spark plugs ignite the fuel-air mixture in the cylinders of the engine (see Fig. 18–2 in Section 18–4). The energy for each spark comes from a current-carrying inductor called the ignition coil.

solenoid is approximately equal to the circumference $2\pi r$ multiplied by the area A: $V = 2\pi rA$. From Eq. (31–8) in Example 31–3 the self-inductance of the toroidal solenoid with vacuum within its coils is

$$L = \frac{\mu_0 N^2 A}{2\pi r}.$$

From Eq. (31–9) the energy U stored in the toroidal solenoid when the current is I is

$$U = \frac{1}{2} L I^2 = \frac{1}{2} \frac{\mu_0 N^2 A}{2\pi r} I^2.$$

The magnetic field and therefore this energy are localized in the volume $V = 2\pi rA$ enclosed by the windings. The energy *per unit volume*, or *magnetic energy density*, is $u = U/V$:

$$u = \frac{U}{2\pi rA} = \frac{1}{2} \mu_0 \frac{N^2 I^2}{(2\pi r)^2}.$$

We can express this in terms of the magnitude B of the magnetic field inside the toroidal solenoid. From Eq. (29–24) in Example 29–11 (Section 29–8) this is

$$B = \frac{\mu_0 NI}{2\pi r},$$

and so

$$\frac{N^2 I^2}{(2\pi r)^2} = \frac{B^2}{\mu_0^2}.$$

When we substitute this into the above equation for u, we finally find the expression for **magnetic energy density** in a vacuum:

$$u = \frac{B^2}{2\mu_0} \qquad \text{(magnetic energy density in a vacuum).} \qquad (31\text{–}10)$$

This is the magnetic analog of the energy per unit volume in an *electric* field in a vacuum, $u = \frac{1}{2}\epsilon_0 E^2$, which we derived in Section 25–4.

When the material inside the toroid is not vacuum but a material with (constant) magnetic permeability $\mu = K_m\mu_0$, we replace μ_0 by μ in Eq. (31–10). The energy per unit volume in the magnetic field is then

$$u = \frac{B^2}{2\mu} \qquad \text{(magnetic energy density in a material).} \qquad (31\text{–}11)$$

Although we have derived Eq. (31–11) only for one special situation, it turns out to be the correct expression for the energy per unit volume associated with *any* magnetic-field configuration in a material with constant permeability. For vacuum, Eq. (31–11) reduces to Eq. (31–10). We will use the expressions for electric-field and magnetic-field energy in Chapter 33 when we study the energy associated with electromagnetic waves.

EXAMPLE 31–5

Storing energy in an inductor The electric-power industry would like to find efficient ways to store surplus energy generated during low-demand hours to help meet customer requirements during high-demand hours. Perhaps a large inductor can be used. What inductance would be needed to store 1.00 kWh of energy in a coil carrying a 200-A current?

SOLUTION We have $I = 200$ A and $U = 1.00$ kWh $= (1.00 \times 10^3 \text{ W})(3600 \text{ s}) = 3.60 \times 10^6$ J. Solving Eq. (31–9) for L, we find

$$L = \frac{2U}{I^2} = \frac{2(3.60 \times 10^6 \text{ J})}{(200 \text{ A})^2} = 180 \text{ H}.$$

This is more than a *million* times greater than the self-inductance of the toroidal solenoid of Example 31–3 (Section 31–3).

Conventional wires that are to carry 200 A would have to be of large diameter to keep the resistance low and avoid unacceptable energy losses due to I^2R heating. As a result, a 180-H inductor using conventional wire would be very large (room-size). A superconducting inductor could be much smaller, since the resistance of a superconductor is zero and much thinner wires could be used; one drawback is that the wires would have to be kept at low temperature to remain superconducting, and energy would have to be used to maintain this low temperature. As a result, this scheme is impractical with present technology.

EXAMPLE 31–6

In a proton accelerator used in elementary particle physics experiments, the trajectories of protons are controlled by bending magnets that produce a magnetic field of 6.6 T. What is the energy density in this field in the vacuum between the poles of such a magnet?

SOLUTION In a vacuum, $\mu = \mu_0$, and from Eq. (31–10) we have

$$u = \frac{B^2}{2\mu_0} = \frac{(6.6 \text{ T})^2}{2(4\pi \times 10^{-7} \text{ T} \cdot \text{m/A})} = 1.73 \times 10^7 \text{ J/m}^3.$$

As an interesting comparison, the heat of combustion of natural gas, expressed on an energy per unit volume basis, is about $3.8 \times 10^7 \text{ J/m}^3$.

31–5 THE *R-L* CIRCUIT

Let's look at some examples of the circuit behavior of an inductor. One thing is clear already; an inductor in a circuit makes it difficult for rapid changes in current to occur, thanks to the effects of self-induced emf. Equation (31–7) shows that the greater the rate of change of current di/dt, the greater the self-induced emf and the greater the potential difference between the inductor terminals. This equation, together with Kirchhoff's rules (Section 27–3), gives us the principles we need to analyze circuits containing inductors.

14.1
The *RL* Circuit

Problem–Solving Strategy

INDUCTORS IN CIRCUITS

1. When an inductor is used as a *circuit* device, all the voltages, currents, and capacitor charges are in general functions of time, not constants as they have been in most of our previous circuit analysis. But Kirchhoff's rules, which we studied in Section 27–3, are still valid. When the voltages and currents vary with time, Kirchhoff's rules hold at each instant of time.

2. As in all circuit analysis, getting the signs right is sometimes more challenging than understanding the principles.

We suggest that you review the strategy in Section 27–3. In addition, remember Lenz's law and the sign rule described in Section 31–3 in conjunction with Eq. (31–7) and Figs. 31–4 and 31–5. In Kirchhoff's loop rule, when we go through an inductor in the *same* direction as the assumed current, we encounter a voltage *drop* equal to $L \, di/dt$, so the corresponding term in the loop equation is $-L \, di/dt$.

CURRENT GROWTH IN AN *R-L* CIRCUIT

We can learn several basic things about inductor behavior by analyzing the circuit of Fig. 31–7. A circuit that includes both a resistor and an inductor, and possibly a source of emf, is called an **R-L circuit.** The inductor helps to prevent rapid changes in current, which can be useful if a steady current is required but the external source has a fluctuating emf. The resistor R may be a separate circuit element, or it may be the resistance of the inductor windings; every real-life inductor has some resistance unless it is made of superconducting wire. By closing switch S_1, we can connect the *R-L* combination to a source with constant emf $\mathcal{E}$. (We assume that the source has zero internal resistance, so the terminal voltage equals the emf.) Suppose both switches are open to begin with, and then at some initial time $t = 0$ we close switch S_1. The current cannot change suddenly from zero to some final value, since di/dt and the induced emf in the inductor would both be infinite. Instead, the current begins to grow at a rate that depends only on the value of L in the circuit.

Let i be the current at some time t after switch S_1 is closed, and let di/dt be its rate of change at that time. The potential difference v_{ab} across the resistor at that time is

$$v_{ab} = iR,$$

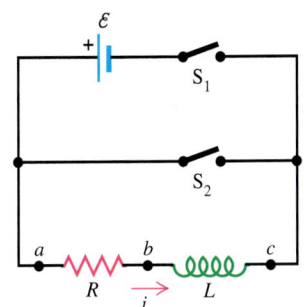

31–7 An *R-L* circuit. Closing switch S_1 connects the *R-L* combination in series with a source of emf $\mathcal{E}$; closing switch S_2 while opening switch S_1 disconnects the combination from the source.

and the potential difference v_{bc} across the inductor is

$$v_{bc} = L\frac{di}{dt}.$$

Note that if the current is in the direction shown in Fig. 31–7 and is increasing, then both v_{ab} and v_{bc} are positive; a is at a higher potential than b, which in turn is at a higher potential than c. (Compare to Fig. 31–5.) We apply Kirchhoff's loop rule, starting at the negative terminal and proceeding counterclockwise around the loop:

$$\mathcal{E} - iR - L\frac{di}{dt} = 0. \tag{31–12}$$

Solving this for di/dt, we find that the rate of increase of current is

$$\frac{di}{dt} = \frac{\mathcal{E} - iR}{L} = \frac{\mathcal{E}}{L} - \frac{R}{L}i. \tag{31–13}$$

At the instant that switch S_1 is first closed, $i = 0$ and the potential drop across R is zero. The initial rate of change of current is

$$\left(\frac{di}{dt}\right)_{\text{initial}} = \frac{\mathcal{E}}{L}.$$

The greater the inductance L, the more slowly the current increases.

As the current increases, the term $(R/L)i$ in Eq. (31–13) also increases, and the *rate* of increase of current becomes smaller and smaller. When the current reaches its final steady-state value I, its rate of increase is zero. Then Eq. (31–13) becomes

$$\left(\frac{di}{dt}\right)_{\text{final}} = 0 = \frac{\mathcal{E}}{L} - \frac{R}{L}I,$$

and

$$I = \frac{\mathcal{E}}{R}.$$

That is, the *final* current I does not depend on the inductance L; it is the same as it would be if the resistance R alone were connected to the source with emf $\mathcal{E}$.

The behavior of the current as a function of time is shown by the graph of Fig. 31–8. To derive the equation for this curve (that is, an expression for current as a function of time), we proceed just as we did for the charging capacitor in Section 27–5. First we rearrange Eq. (31–13) to the form

$$\frac{di}{i - (\mathcal{E}/R)} = -\frac{R}{L}dt.$$

This separates the variables, with i on the left side and t on the right. Then we integrate both sides, renaming the integration variables i' and t' so that we can use i and t as the upper limits. (The lower limit for each integral is zero, corresponding to zero current at the initial time $t = 0$.) We get

$$\int_0^i \frac{di'}{i' - (\mathcal{E}/R)} = -\int_0^t \frac{R}{L}dt',$$

$$\ln\left(\frac{i - (\mathcal{E}/R)}{-\mathcal{E}/R}\right) = -\frac{R}{L}t.$$

Now we take exponentials of both sides and solve for i. We leave the details for you to work out; the final result is

$$i = \frac{\mathcal{E}}{R}\left(1 - e^{-(R/L)t}\right) \qquad \text{(current in an } R\text{-}L \text{ circuit with emf).} \tag{31–14}$$

This is the equation of the curve in Fig. 31–8. Taking the derivative of Eq. (31–14), we find

$$\frac{di}{dt} = \frac{\mathcal{E}}{L}e^{-(R/L)t}. \tag{31–15}$$

At time $t = 0$, $i = 0$ and $di/dt = \mathcal{E}/L$. As $t \to \infty$, $i \to \mathcal{E}/R$ and $di/dt \to 0$, as we predicted.

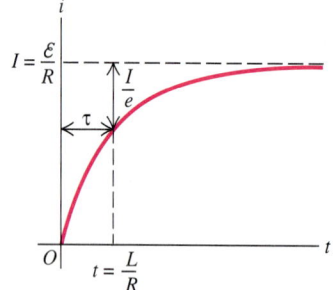

31–8 Graph of i versus t for growth of current in an R-L circuit with an emf in series. The final current is $I = \mathcal{E}/R$; after one time constant the current is $1 - 1/e$ of this value.

As Fig. 31–8 shows, the instantaneous current i first rises rapidly, then increases more slowly and approaches the final value $I = \mathcal{E}/R$ asymptotically. At a time equal to L/R the current has risen to $(1 - 1/e)$, or about 63%, of its final value. The quantity L/R is therefore a measure of how quickly the current builds toward its final value; this quantity is called the **time constant** for the circuit, and denoted by τ:

$$\tau = \frac{L}{R} \qquad \text{(time constant for an } R\text{-}L \text{ circuit).} \qquad (31\text{--}16)$$

In a time equal to 2τ, the current reaches 86% of its final value; in 5τ, 99.3%; and in 10τ, 99.995%. (Compare the discussion in Section 27–5 of charging a capacitor of capacitance C that was in series with a resistor of resistance R; the time constant for that situation was the product RC.)

The graphs of i versus t have the same general shape for all values of L. For a given value of R, the time constant τ is greater for greater values of L. When L is small, the current rises rapidly to its final value; when L is large, it rises more slowly. For example, if $R = 100\ \Omega$ and $L = 10$ H,

$$\tau = \frac{L}{R} = \frac{10\ \text{H}}{100\ \Omega} = 0.10\ \text{s},$$

and the current increases to about 63% of its final value in 0.10 s. (Recall that 1 H $= 1\ \Omega \cdot \text{s}$.) But if $L = 0.010$ H, $\tau = 1.0 \times 10^{-4}$ s $= 0.10$ ms, and the rise is much more rapid.

Energy considerations offer us additional insight into the behavior of an *R-L* circuit. The instantaneous rate at which the source delivers energy to the circuit is $P = \mathcal{E}i$. The instantaneous rate at which energy is dissipated in the resistor is i^2R, and the rate at which energy is stored in the inductor is $iv_{bc} = Li\ di/dt$ (or, equivalently, $(d/dt)(\frac{1}{2}Li^2) = Li\ di/dt$). When we multiply Eq. (31–12) by i and rearrange, we find

$$\mathcal{E}i = i^2R + Li\frac{di}{dt}. \qquad (31\text{--}17)$$

Of the power $\mathcal{E}i$ supplied by the source, part (i^2R) is dissipated in the resistor and part $(Li\ di/dt)$ goes to store energy in the inductor. This discussion is completely analogous to our power analysis for a charging capacitor, given at the end of Section 27–5.

EXAMPLE 31-7

A sensitive electronic device of resistance 175 Ω is to be connected to a source of emf by a switch. The device is designed to operate with a current of 36 mA, but to avoid damage to the device, the current can rise to no more than 4.9 mA in the first 58 μs after the switch is closed. To protect the device, it is connected in series with an inductor as in Fig. 31–7; the switch in question is S$_1$. a) What emf must the source have? Assume negligible internal resistance. b) What inductance is required? c) What is the time constant?

SOLUTION a) The final current I, which does not depend on the inductance L, is given by $I = \mathcal{E}/R$. Using $I = 36$ mA $= 0.036$ A and $R = 175\ \Omega$, we have

$$\mathcal{E} = IR = (0.036\ \text{A})(175\ \Omega) = 6.3\ \text{V}.$$

b) The current as a function of time is given by Eq. (31–14); we want $i = 4.9$ mA at $t = 58\ \mu$s. To find the required inductance, we solve Eq. (31–14) for L. First we multiply through by $(-R/\mathcal{E})$

and then add 1 to both sides to obtain

$$1 - \frac{iR}{\mathcal{E}} = e^{-(R/L)t}.$$

Then we take natural logs of both sides, solve for L, and insert the numbers:

$$L = \frac{-Rt}{\ln(1 - iR/\mathcal{E})}$$

$$= \frac{-(175\ \Omega)(58 \times 10^{-6}\ \text{s})}{\ln[1 - (4.9 \times 10^{-3}\ \text{A})(175\ \Omega)/(6.3\ \text{V})]} = 69\ \text{mH}.$$

c) The time constant τ is given by Eq. (31–16):

$$\tau = \frac{L}{R} = \frac{69 \times 10^{-3}\ \text{H}}{175\ \Omega} = 3.9 \times 10^{-4}\ \text{s} = 390\ \mu\text{s}.$$

We note that 58 μs is much less than the time constant. In 58 μs the current builds up only from zero to 4.9 mA, a small fraction of its final value of 36 mA; after 390 μs the current equals $(1 - 1/e)$ of its final value, or about $(0.63)(36\ \text{mA}) = 23$ mA.

CURRENT DECAY IN AN *R-L* CIRCUIT

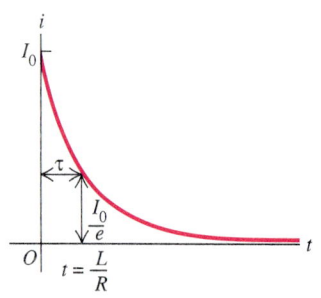

31–9 Graph of i versus t for decay of current in an R-L circuit.

Now suppose switch S_1 in the circuit of Fig. 31–7 has been closed for a while and that the current has reached the value I_0. Resetting our stopwatch to redefine the initial time, we close switch S_2 at time $t = 0$, bypassing the battery. (At the same time we should open S_1 to save the battery from ruin.) The current through R and L does not instantaneously go to zero but decays smoothly, as shown in Fig. 31–9. The Kirchhoff's-rule loop equation is obtained from Eq. (31–12) by simply omitting the $\mathcal{E}$ term. We challenge you to retrace the steps in the above analysis and show that the current i varies with time according to

$$i = I_0 e^{-(R/L)t}, \tag{31–18}$$

where I_0 is the initial current at time $t = 0$. The time constant, $\tau = L/R$, is the time for current to decrease to $1/e$, or about 37%, of its original value. In time 2τ it has dropped to 13.5%, in time 5τ to 0.67%, and in 10τ to 0.0045%.

The energy that is needed to maintain the current during this decay is provided by the energy stored in the magnetic field of the inductor. The detailed energy analysis is simpler this time. In place of Eq. (31–17) we have

$$0 = i^2 R + Li\frac{di}{dt}. \tag{31–19}$$

In this case, $Li\,di/dt$ is negative; Eq. (31–19) shows that the energy stored in the inductor *decreases* at a rate equal to the rate of dissipation of energy i^2R in the resistor.

This entire discussion should look familiar; the situation is very similar to that of a charging and discharging capacitor, analyzed in Section 27–5. It would be a good idea to compare that section with our discussion of the R-L circuit.

EXAMPLE 31–8

Energy in an *R-L* circuit When the current in an R-L circuit is decaying, what fraction of the original energy stored in the inductor has been dissipated after 2.3 time constants?

SOLUTION From Eq. (31–18) the current i at any time t is

$$i = I_0 e^{-(R/L)t}.$$

The energy U in the inductor at *any* time is obtained by substi-

tuting this expression into $U = \frac{1}{2}Li^2$. We obtain

$$U = \frac{1}{2}LI_0^2 e^{-2(R/L)t} = U_0 e^{-2(R/L)t},$$

where $U_0 = \frac{1}{2}LI_0^2$ is the energy at the initial time $t = 0$. When $t = 2.3\tau = 2.3L/R$, we have

$$U = U_0 e^{-2(2.3)} = U_0 e^{-4.6} = 0.010\ U_0.$$

That is, only 0.010 or 1.0% of the energy initially stored in the inductor remains, so 99.0% has been dissipated in the resistor.

31–6 THE *L-C* CIRCUIT

A circuit containing an inductor and a capacitor shows an entirely new mode of behavior, characterized by *oscillating* current and charge. This is in sharp contrast to the *exponential* approach to a steady-state situation that we have seen with both R-C and R-L circuits. In the *L-C circuit* in Fig. 31–10 we charge the capacitor to a potential difference V_m and initial charge $Q = CV_m$ on its left-hand plate, as shown in Fig. 31–10a, and then close the switch. What happens?

The capacitor begins to discharge through the inductor. Because of the induced emf in the inductor, the current cannot change instantaneously; it starts at zero and eventually builds up to a maximum value I_m. During this buildup the capacitor is discharging. At each instant the capacitor potential equals the induced emf, so as the capacitor dis-

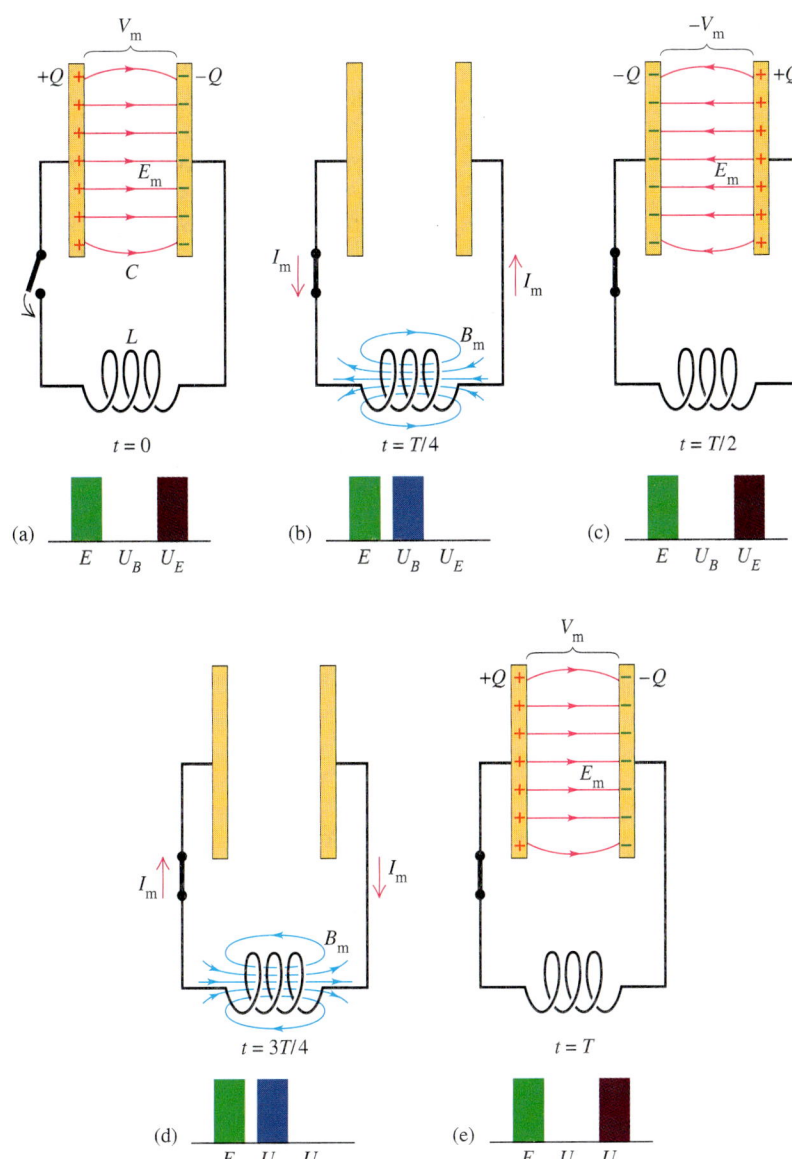

31–10 Energy transfer between the electric field of the capacitor and the magnetic field of the inductor in an oscillating *L-C* circuit. The switch is closed at time $t = 0$. As in simple harmonic motion, the total energy E remains constant; compare Fig. 13–10 in Section 13–4.

charges, the *rate of change* of current decreases. When the capacitor potential becomes zero, the induced emf is also zero, and the current has leveled off at its maximum value I_m. Figure 31–10b shows this situation; the capacitor has completely discharged. The potential difference between its terminals (and those of the inductor) has decreased to zero, and the current has reached its maximum value I_m.

During the discharge of the capacitor the increasing current in the inductor has established a magnetic field in the space around it, and the energy that was initially stored in the capacitor's electric field is now stored in the inductor's magnetic field.

Although the capacitor is completely discharged in Fig. 31–10b, the current persists (it cannot change instantaneously), and the capacitor begins to charge with polarity opposite to that in the initial state. As the current decreases, the magnetic field also decreases, inducing an emf in the inductor in the *same* direction as the current; this slows down the decrease of the current. Eventually, the current and the magnetic field reach zero, and the capacitor has been charged in the sense *opposite* to its initial polarity (Fig. 31–10c), with potential difference $-V_m$ and charge $-Q$ on its left-hand plate.

ActivPhysics

14.2
AC Circuits: The *RLC* Oscillator
(Questions 1–6)

The process now repeats in the reverse direction; a little later, the capacitor has again discharged, and there is a current in the inductor in the opposite direction (Fig. 31–10d). Still later, the capacitor charge returns to its original value (Fig. 31–10e), and the whole process repeats. If there are no energy losses, the charges on the capacitor continue to oscillate back and forth indefinitely. This process is called an **electrical oscillation.**

From an energy standpoint the oscillations of an electrical circuit transfer energy from the capacitor's electric field to the inductor's magnetic field and back. The *total* energy associated with the circuit is constant. This is analogous to the transfer of energy in an oscillating mechanical system from potential energy to kinetic energy and back, with constant total energy. As we will see, this analogy goes much further.

To study the flow of charge in detail, we proceed just as we did for the *R-L* circuit. Figure 31–11 shows our definitions of q and i.

CAUTION ▶ After examining Fig. 31–10, the positive direction for current in Fig. 31–11 may seem backward to you. In fact we've chosen this direction to simplify the relationship between current and capacitor charge. We define the current at each instant to be $i = dq/dt$, the rate of change of the charge on the left-hand capacitor plate. Hence if the capacitor is initially charged and begins to discharge as in Figs. 31–10a and 31–10b, then $dq/dt < 0$ and the initial current i is negative; the direction of the current is then opposite to the (positive) direction shown in Fig. 31–11. ◀

We apply Kirchhoff's loop rule to the circuit in Fig. 31–11. Starting at the lower right corner of the circuit and adding voltages as we go clockwise around the loop, we obtain

$$-L\frac{di}{dt} - \frac{q}{C} = 0.$$

Since $i = dq/dt$, it follows that $di/dt = d^2q/dt^2$. We substitute this expression into the above equation and divide by $-L$ to obtain

$$\frac{d^2q}{dt^2} + \frac{1}{LC}q = 0 \qquad (L\text{-}C \text{ circuit}). \tag{31–20}$$

Equation (31–20) has exactly the same form as the equation we derived for simple harmonic motion in Section 13–3, Eq. (13–4). This equation is $d^2x/dt^2 = -(k/m)x$, or

$$\frac{d^2x}{dt^2} + \frac{k}{m}x = 0.$$

(We suggest that you review Section 13–3 before going on with this discussion.) In the *L-C* circuit the capacitor charge q plays the role of the displacement x, and the current $i = dq/dt$ is analogous to the particle's velocity $v = dx/dt$. The inductance L is analogous to the mass m, and the reciprocal of the capacitance, $1/C$, is analogous to the force constant k.

Pursuing this analogy, we recall that the angular frequency $\omega = 2\pi f$ of the harmonic oscillator is equal to $(k/m)^{1/2}$, and the position is given as a function of time by Eq. (13–13),

$$x = A\cos(\omega t + \phi),$$

where the amplitude A and the phase angle ϕ depend on the initial conditions. In the analogous electrical situation the capacitor charge q is given by

$$q = Q\cos(\omega t + \phi), \tag{31–21}$$

and the angular frequency ω of oscillation is given by

$$\omega = \sqrt{\frac{1}{LC}} \qquad \text{(angular frequency of oscillation in an } L\text{-}C \text{ circuit).} \tag{31–22}$$

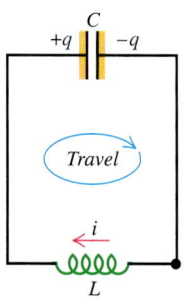

31–11 Applying Kirchhoff's loop rule to the *L-C* circuit. The direction of travel around the loop in the loop equation is shown. Just after the circuit is completed and the capacitor first begins to discharge, the current is negative (opposite to the direction shown).

We invite you to verify that Eq. (31–21) satisfies the loop equation, Eq. (31–20), when ω has the value given by Eq. (31–22). In doing this, you will find that the instantaneous current $i = dq/dt$ is given by

$$i = -\omega Q \sin(\omega t + \phi). \qquad (31\text{–}23)$$

Thus the charge and current in an *L-C* circuit oscillate sinusoidally with time, with an angular frequency determined by the values of L and C. The ordinary frequency f, the number of cycles per second, is equal to $\omega/2\pi$ as always. The constants Q and ϕ in Eqs. (31–21) and (31–23) are determined by the initial conditions. If at time $t = 0$ the left-hand capacitor plate in Fig. 31–11 has its maximum charge Q and the current i is zero, then $\phi = 0$. If $q = 0$ at time $t = 0$, then $\phi = \pm\pi/2$ rad.

We can also analyze the *L-C* circuit using an energy approach. The analogy to simple harmonic motion is equally useful here. In the mechanical problem a body with mass m is attached to a spring with force constant k. Suppose we displace the body a distance A from its equilibrium position and release it from rest at time $t = 0$. The kinetic energy of the system at any later time is $\frac{1}{2}mv^2$, and its elastic potential energy is $\frac{1}{2}kx^2$. Because the system is conservative, the sum of these energies equals the initial energy of the system, $\frac{1}{2}kA^2$. We find the velocity v at any position x just as we did in Section 13–4, Eq. (13–22):

$$v = \pm\sqrt{\frac{k}{m}}\sqrt{A^2 - x^2}. \qquad (31\text{–}24)$$

The *L-C* circuit is also a conservative system. Again let Q be the maximum capacitor charge. The magnetic-field energy $\frac{1}{2}Li^2$ in the inductor at any time corresponds to the kinetic energy $\frac{1}{2}mv^2$ of the oscillating body, and the electric-field energy $q^2/2C$ in the capacitor corresponds to the elastic potential energy $\frac{1}{2}kx^2$ of the spring. The sum of these energies equals the total energy $Q^2/2C$ of the system:

$$\frac{1}{2}Li^2 + \frac{q^2}{2C} = \frac{Q^2}{2C}. \qquad (31\text{–}25)$$

An inductor plays the same role in an *L-C* circuit that a spring like this does when connected to a mass: It provides a restoring force that always tries to return the system to equilibrium.

The total energy in the *L-C* circuit is *constant;* it oscillates between the magnetic and the electric forms, just as the constant total mechanical energy in simple harmonic motion is constant and oscillates between the kinetic and potential forms.

Solving Eq. (31–25) for i, we find that when the charge on the capacitor is q, the current i is

$$i = \pm\sqrt{\frac{1}{LC}}\sqrt{Q^2 - q^2}. \qquad (31\text{–}26)$$

You can verify this equation by substituting q from Eq. (31–21) and i from Eq. (31–23). Comparing Eqs. (31–24) and (31–26), we see that current $i = dq/dt$ and charge q are related in the same way as are velocity $v = dx/dt$ and position x in the mechanical problem.

The analogies between simple harmonic motion and *L-C* circuit oscillations are summarized in Table 31–1. The striking parallel shown there between mechanical and electrical oscillations is one of many such examples in physics. This parallel is so close that we can solve complicated mechanical and acoustical problems by setting up analogous electrical circuits and measuring the currents and voltages that correspond to the mechanical and acoustical quantities to be determined. This is the basic principle of many analog computers. This analogy can be extended to *damped* oscillations, which we consider in the next section; in Chapter 32 we will extend the analogy further to include *forced* electrical oscillations, which occur in all alternating-current circuits.

TABLE 31-1

OSCILLATION OF A MASS-SPRING SYSTEM COMPARED WITH ELECTRICAL OSCILLATION IN AN *L-C* CIRCUIT

MASS-SPRING SYSTEM	INDUCTOR-CAPACITOR CIRCUIT
Kinetic energy $= \frac{1}{2}mv^2$	Magnetic energy $= \frac{1}{2}Li^2$
Potential energy $= \frac{1}{2}kx^2$	Electric energy $= q^2/2C$
$\frac{1}{2}mv^2 + \frac{1}{2}kx^2 = \frac{1}{2}kA^2$	$\frac{1}{2}Li^2 + q^2/2C = Q^2/2C$
$v = \pm\sqrt{k/m}\,\sqrt{A^2 - x^2}$	$i = \pm\sqrt{1/LC}\,\sqrt{Q^2 - q^2}$
$v = dx/dt$	$i = dq/dt$
$\omega = \sqrt{\dfrac{k}{m}}$	$\omega = \sqrt{\dfrac{1}{LC}}$
$x = A\cos(\omega t + \phi)$	$q = Q\cos(\omega t + \phi)$

EXAMPLE 31-9

An oscillating circuit A 300-V dc power supply is used to charge a 25-μF capacitor. After the capacitor is fully charged, it is disconnected from the power supply and connected across a 10-mH inductor. The resistance in the circuit is negligible.
a) Find the frequency and period of oscillation of the circuit.
b) Find the capacitor charge and the circuit current 1.2 ms after the inductor and capacitor are connected.

SOLUTION a) The natural *angular* frequency is

$$\omega = \sqrt{\frac{1}{LC}} = \sqrt{\frac{1}{(10 \times 10^{-3}\ \text{H})(25 \times 10^{-6}\ \text{F})}} = 2.0 \times 10^3\ \text{rad/s}.$$

The frequency f is $1/2\pi$ times this:

$$f = \frac{\omega}{2\pi} = \frac{2.0 \times 10^3\ \text{rad/s}}{2\pi\ \text{rad/cycle}} = 320\ \text{Hz}.$$

The period is the reciprocal of the frequency:

$$T = \frac{1}{f} = \frac{1}{320\ \text{Hz}} = 3.1 \times 10^{-3}\ \text{s} = 3.1\ \text{ms}.$$

b) Since the period of the oscillation is $T = 3.1$ ms, $t = 1.2$ ms equals $0.38T$; this corresponds to a situation intermediate

between Fig. 31-10b ($t = T/4$) and Fig. 31-10c ($t = T/2$). Comparing those figures to Fig. 31-11, we expect the capacitor charge q to be negative (that is, there will be negative charge on the left-hand plate of the capacitor) and the current i to be negative as well (that is, current will be traveling in a counterclockwise direction).

To find the value of q, we use Eq. (31-21). The charge is maximum at $t = 0$, so $\phi = 0$ and $Q = C\mathcal{E} = (25 \times 10^{-6}\ \text{F})(300\ \text{V}) = 7.5 \times 10^{-3}$ C. The charge q at any time is

$$q = (7.5 \times 10^{-3}\ \text{C})\cos\omega t.$$

At time $t = 1.2 \times 10^{-3}$ s,

$$\omega t = (2.0 \times 10^3\ \text{rad/s})(1.2 \times 10^{-3}\ \text{s}) = 2.4\ \text{rad},$$
$$q = (7.5 \times 10^{-3}\ \text{C})\cos(2.4\ \text{rad}) = -5.5 \times 10^{-3}\ \text{C}.$$

The current i at any time is

$$i = -\omega Q \sin\omega t.$$

At time $t = 1.2 \times 10^{-3}$ s,

$$i = -(2.0 \times 10^3\ \text{rad/s})(7.5 \times 10^{-3}\ \text{C})\sin(2.4\ \text{rad}) = -10\ \text{A}.$$

Note that the signs of q and i are both negative, as we predicted.

EXAMPLE 31-10

Energy in an oscillating circuit Consider again the *L-C* circuit of Example 31-9. a) Find the magnetic energy and electric energy at $t = 0$. b) Find the magnetic energy and electric energy at $t = 1.2$ ms.

SOLUTION a) At $t = 0$ there is no current and $q = Q$. Hence there is no magnetic energy, and all the energy in the circuit is in the form of electric energy in the capacitor:

$$U_B = \frac{1}{2}Li^2 = 0, \qquad U_E = \frac{Q^2}{2C} = \frac{(7.5 \times 10^{-3}\ \text{C})^2}{2(25 \times 10^{-6}\ \text{F})} = 1.1\ \text{J}.$$

b) As we mentioned in Example 31-9, $t = 1.2$ ms corresponds to a situation intermediate between Fig. 31-10b ($t = T/4$) and Fig. 31-10c ($t = T/2$). So we expect the energy to be part magnetic

and part electric at this time. From Example 31-9, $i = -10$ A and $q = -5.5 \times 10^{-3}$ C, so

$$U_B = \frac{1}{2}Li^2 = \frac{1}{2}(10 \times 10^{-3}\ \text{H})(-10\ \text{A})^2 = 0.5\ \text{J},$$

$$U_E = \frac{q^2}{2C} = \frac{(-5.5 \times 10^{-3}\ \text{C})^2}{2(25 \times 10^{-6}\ \text{F})} = 0.6\ \text{J}.$$

The magnetic and electric energies are the same at $t = 3T/8 = 0.375T$; the time we are considering here is slightly later, and U_B is slightly less than U_E. We emphasize that at *all* times, the *total* energy $E = U_B + U_E$ has the same value, 1.1 J. An *L-C* circuit without resistance is a conservative system; no energy is dissipated.

31-7 THE *L-R-C* SERIES CIRCUIT

In our discussion of the *L-C* circuit we assumed that there was no *resistance* in the circuit. This is an idealization, of course; every real inductor has resistance in its windings, and there may also be resistance in the connecting wires. Because of resistance, the electromagnetic energy in the circuit is dissipated and converted to other forms such as internal energy of the circuit materials. Resistance in an electric circuit is analogous to friction in a mechanical system.

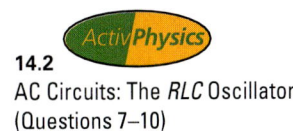

14.2

AC Circuits: The *RLC* Oscillator (Questions 7–10)

Suppose an inductor with inductance *L* and a resistor of resistance *R* are connected in series across the terminals of a charged capacitor, forming a **series *L-R-C* circuit.** As before, the capacitor starts to discharge as soon as the circuit is completed. But because of i^2R losses in the resistor, the magnetic-field energy acquired by the inductor when the capacitor is completely discharged is *less* than the original electric-field energy of the capacitor. In the same way the energy of the capacitor when the magnetic field has decreased to zero is still smaller, and so on.

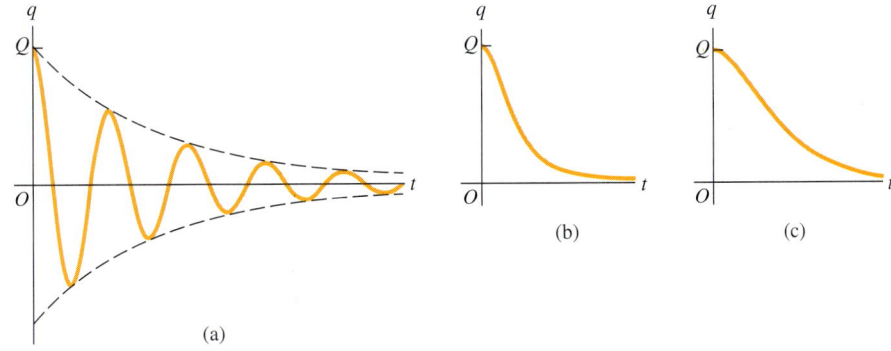

31-12 Graphs of capacitor charge as a function of time in an *L-R-C* series circuit with initial charge *Q*.
(a) underdamped (small *R*);
(b) critically damped (larger *R*);
(c) overdamped (very large *R*).

If the resistance *R* is relatively small, the circuit still oscillates, but with **damped harmonic motion** (Fig. 31–12a), and we say that the circuit is **underdamped.** If we increase *R*, the oscillations die out more rapidly. When *R* reaches a certain value, the circuit no longer oscillates; it is **critically damped** (Fig. 31–12b). For still larger values of *R* the circuit is **overdamped** (Fig. 31–12c), and the capacitor charge approaches zero even more slowly. We used these same terms to describe the behavior of the analogous mechanical system, the damped harmonic oscillator, in Section 13–8.

To analyze *L-R-C* circuit behavior in detail, we consider the circuit shown in Fig. 31–13. It is like the *L-C* circuit of Fig. 31–11 except for the added resistor *R*; we also show the source that charges the capacitor initially. The labeling of the positive senses of *q* and *i* are the same as for the *L-C* circuit.

First we close the switch in the upward position, connecting the capacitor to a source of emf $\mathcal{E}$ for a long enough time to ensure that the capacitor acquires its final charge $Q = C\mathcal{E}$ and any initial oscillations have died out. Then at time $t = 0$ we flip the switch to the downward position, removing the source from the circuit and placing the capacitor in series with the resistor and inductor. Note that the initial current is negative, opposite in direction to the direction of *i* shown in the figure.

To find how *q* and *i* vary with time, we apply Kirchhoff's loop rule. Starting at point *a* and going around the loop in the direction *abcda*, we obtain the equation

$$-iR - L\frac{di}{dt} - \frac{q}{C} = 0.$$

Replacing *i* with *dq/dt* and rearranging, we get

$$\frac{d^2q}{dt^2} + \frac{R}{L}\frac{dq}{dt} + \frac{1}{LC}q = 0. \qquad (31\text{--}27)$$

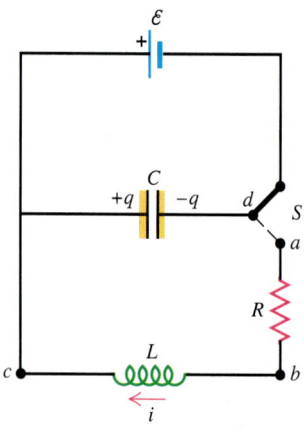

31-13 An L-R-C series circuit.

Note that when $R = 0$, this reduces to Eq. (31–20) for an L-C circuit.

There are general methods for obtaining solutions of Eq. (31–27). The form of the solution is different for the underdamped (small R) and overdamped (large R) cases. When R^2 is less than $4L/C$, the solution has the form

$$q = Ae^{-(R/2L)t} \cos\left(\sqrt{\frac{1}{LC} - \frac{R^2}{4L^2}} \; t + \phi\right), \qquad (31\text{–}28)$$

where A and ϕ are constants. We invite you to take the first and second derivatives of this function and show by direct substitution that it does satisfy Eq. (31–27).

This solution corresponds to the *underdamped* behavior shown in Fig. 31–12a; the function represents a sinusoidal oscillation with an exponentially decaying amplitude. (Note that the exponential factor $e^{-(R/2L)t}$ is *not* the same as the factor $e^{-(R/L)t}$ that we encountered in describing the R-L circuit in Section 31–5.) When $R = 0$, Eq. (31–28) reduces to Eq. (31–21) for the oscillations in an L-C circuit. If R is not zero, the angular frequency of the oscillation is *less* than $1/(LC)^{1/2}$ because of the term containing R. The angular frequency ω' of the damped oscillations is given by

$$\omega' = \sqrt{\frac{1}{LC} - \frac{R^2}{4L^2}} \qquad \text{(underdamped } L\text{-}R\text{-}C \text{ series circuit).} \qquad (31\text{–}29)$$

When $R = 0$, this reduces to Eq. (31–22), $\omega = (1/LC)^{1/2}$. As R increases, ω' becomes smaller and smaller. When $R^2 = 4L/C$, the quantity under the radical becomes zero; the system no longer oscillates, and the case of *critical damping* (Fig. 31–12b) has been reached. For still larger values of R the system behaves as in Fig. 31–12c. In this case the circuit is *overdamped,* and q is given as a function of time by the sum of two decreasing exponential functions.

In the *underdamped* case the phase constant ϕ in the cosine function of Eq. (31–28) provides for the possibility of both an initial charge and an initial current at time $t = 0$, analogous to an underdamped harmonic oscillator given both an initial displacement and an initial velocity (see Exercise 31–34).

We emphasize once more that the behavior of the L-R-C circuit is completely analogous to that of the damped harmonic oscillator studied in Section 13–8. We invite you to verify, for example, that if you start with Eq. (13–41) and substitute L for m, $1/C$ for k, and R for the damping constant b, the result is Eq. (31–27). Similarly, the cross-over point between underdamping and overdamping occurs at $b^2 = 4km$ for the mechanical system and at $R^2 = 4L/C$ for the electrical one. Can you find still other aspects of this analogy?

Other interesting aspects of this circuit's behavior emerge when we include a sinusoidally varying source of emf in the circuit. This is analogous to the *forced oscillations* that we discussed in Section 13–8, and there are analogous *resonance* effects. Such a circuit is called an *alternating-current (ac) circuit;* the analysis of ac circuits is the principal topic of the next chapter.

EXAMPLE 31-11

What resistance R is required (in terms of L and C) to give an L-R-C circuit a frequency that is one half the undamped frequency?

SOLUTION We want the angular frequency ω' given by Eq. (31–29) to be half the undamped angular frequency given by Eq. (31–22):

$$\sqrt{\frac{1}{LC} - \frac{R^2}{4L^2}} = \frac{1}{2}\sqrt{\frac{1}{LC}}.$$

When we square both sides and solve for R, we get

$$R = \sqrt{\frac{3L}{C}}.$$

For example, adding 35 Ω to the circuit of Example 31–9 would reduce the frequency from 320 Hz to 160 Hz.

SUMMARY

- When a changing current i_1 in one circuit causes a changing magnetic flux in a second circuit, an emf $\mathcal{E}_2$ is induced in the second circuit; likewise, a changing current i_2 in the second circuit induces an emf $\mathcal{E}_1$ in the first circuit:

$$\mathcal{E}_2 = -M\frac{di_1}{dt} \quad \text{and} \quad \mathcal{E}_1 = -M\frac{di_2}{dt}. \quad (31\text{--}4)$$

The constant M, called the mutual inductance, depends on the geometry of the two coils and on the material between them. If the circuits are coils of wire with N_1 and N_2 turns, respectively, the mutual inductance can be expressed in terms of the average flux Φ_{B2} through each turn of coil 2 that is caused by the current i_1 in coil 1 or in terms of the average flux Φ_{B1} through each turn of coil 1 that is caused by the current i_2 in coil 2:

$$M = \frac{N_2\Phi_{B2}}{i_1} = \frac{N_1\Phi_{B1}}{i_2}. \quad (31\text{--}5)$$

The SI unit of mutual inductance is the henry, abbreviated H. Equivalent units are

$$1\text{ H} = 1\text{ Wb/A} = 1\text{ V}\cdot\text{s/A} = 1\,\Omega\cdot\text{s}.$$

- A changing current i in any circuit induces an emf $\mathcal{E}$ in that same circuit, called a self-induced emf:

$$\mathcal{E} = -L\frac{di}{dt}. \quad (31\text{--}7)$$

The constant L, called the inductance or self-inductance, depends on the geometry of the circuit and the material surrounding it. The inductance of a coil of N turns is related to the average flux Φ_B through each turn caused by the current i in the coil:

$$L = \frac{N\Phi_B}{i}. \quad (31\text{--}6)$$

A circuit device, usually including a coil of wire, that is intended to have a substantial inductance is called an inductor.

- An inductor with inductance L carrying current I has energy

$$U = \frac{1}{2}LI^2. \quad (31\text{--}9)$$

This energy is associated with the magnetic field of the inductor. If the field is in vacuum, the magnetic energy density u (energy per unit volume) is

$$u = \frac{B^2}{2\mu_0}; \quad (31\text{--}10)$$

in a material with magnetic permeability μ, the magnetic energy density is

$$u = \frac{B^2}{2\mu}. \quad (31\text{--}11)$$

- In an R-L circuit, containing a resistor R, an inductor L, and a source of emf, the growth and decay of current are exponential, with a characteristic time τ called the time constant:

$$\tau = \frac{L}{R}. \quad (31\text{--}16)$$

This is the time required for the current to approach within a fraction $1/e$ of its final value.

■ An *L-C* circuit, which contains inductance *L* and capacitance *C*, undergoes electrical oscillations with angular frequency ω:

$$\omega = \sqrt{\frac{1}{LC}}. \tag{31–22}$$

Such a circuit is analogous to a mechanical harmonic oscillator, with inductance *L* analogous to mass *m*, the reciprocal of capacitance $1/C$ to force constant *k*, charge *q* to displacement *x*, and current *i* to velocity *v*.

■ An *L-R-C* series circuit, which contains inductance, resistance, and capacitance, undergoes damped oscillations for sufficiently small resistance. The frequency ω' of damped oscillations is

$$\omega' = \sqrt{\frac{1}{LC} - \frac{R^2}{4L^2}}. \tag{31–29}$$

As *R* increases, the damping increases; if *R* is greater than a certain value, the behavior becomes overdamped and no longer oscillates. The cross-over between underdamping and overdamping occurs when $R^2 = 4L/C$; when this condition is satisfied, the oscillations are critically damped. There is a direct analogy between every aspect of the behavior of the *L-R-C* circuit and the mechanical damped harmonic oscillator. This analogy and similar ones are widely used in analog computers.

DISCUSSION QUESTIONS

Q31–1 In Fig. 31–1, if coil 2 is turned 90° so that its axis is vertical, does the mutual inductance increase or decrease? Explain.

Q31–2 The tightly wound toroidal solenoid is one of the few configurations for which it is easy to calculate self-inductance. What features of the toroidal solenoid give it this simplicity?

Q31–3 Two identical closely wound circular coils, each having self-inductance *L*, are placed next to each other, so they are coaxial and almost touching. If they are connected in series, what is the self-inductance of the combination? What if they are connected in parallel? Can they be connected so that the total inductance is zero? Explain.

Q31–4 Two closely wound circular coils have the same number of turns, but one has twice the radius of the other. How are the self-inductances of the two coils related? Explain your reasoning.

Q31–5 You are to make a resistor by winding a wire around a cylindrical form. To make the inductance as small as possible, it is proposed that you wind half the wire in one direction and the other half in the opposite direction. Would this achieve the desired result? Why or why not?

Q31–6 If two inductors are separated enough that practically no magnetic field from either passes through the coils of the other, show that the equivalent inductance of two inductors in series or parallel is obtained by the same rules as used for combining resistance.

Q31–7 For the same magnetic field strength *B*, is the energy density larger in vacuum or in a magnetic material? Explain. Does Eq. (31–11) imply that for a long solenoid in which the current is *I* the energy stored is proportional to $1/\mu$? And does

this mean that for the same current less energy is stored when the solenoid is filled with a ferromagnetic material rather than with air? Explain.

Q31–8 In Section 31–6, Kirchhoff's loop rule is applied to an *L-C* circuit where the capacitor is initially fully charged and the equation $-L\,di/dt - q/C = 0$ is derived. But as the capacitor starts to discharge, the current increases from zero. The equation says $L\,di/dt = -q/C$, so it says $L\,di/dt$ is negative. Explain how $L\,di/dt$ can be negative when the current is increasing.

Q31–9 In Section 31–6 the relation $i = dq/dt$ is used in deriving Eq. (31–20). But a flow of current corresponds to a decrease in the charge on the capacitor. Explain, therefore, why this is the correct equation to use in the derivation, rather than $i = -dq/dt$.

Q31–10 In the *R-L* circuit of Fig. 31–7, when the switch S_1 is closed, the potential v_{ab} changes suddenly and discontinuously, but the current does not. Explain why the voltage can change suddenly but the current can't.

Q31–11 In the *R-L* circuit of Fig. 31–7, is the current in the resistor always the same as that in the inductor? How do you know?

Q31–12 Suppose there is a steady current in an inductor. If you attempt to reduce the current to zero instantaneously by quickly opening a switch, an arc can appear at the switch contacts. Why? Is it physically possible to stop the current instantaneously? Explain.

Q31–13 In an *R-L-C* circuit, what criteria could be used to decide whether the system is overdamped or underdamped? For example, could we compare the maximum energy stored during one cycle to the energy dissipated during one cycle? Explain.

EXERCISES

SECTION 31–2 MUTUAL INDUCTANCE

31–1 Two coils have mutual inductance $M = 3.25 \times 10^{-4}$ H. The current i_1 in the first coil increases at a uniform rate of 830 A/s. a) What is the induced emf in the second coil? Is it constant? b) Suppose that the current described is in the second coil rather than the first. What is the magnitude of the induced emf in the first coil?

31–2 A toroidal solenoid has a mean radius r and a cross-section area A and is wound uniformly with N_1 turns. A second toroidal solenoid with N_2 turns is wound uniformly around the first. The two coils are wound in the same direction. What is their mutual inductance? (Neglect the variation of the magnetic field across the cross section of the toroid.)

31–3 Two toroidal solenoids are wound around the same form so that the magnetic field of one passes through the turns of the other. Solenoid 1 has 700 turns and solenoid 2 has 400 turns. When the current in solenoid 1 is 6.52 A, the average flux through each turn of solenoid 2 is 0.0320 Wb. a) What is the mutual inductance of the pair of solenoids? b) When the current in solenoid 2 is 2.54 A, what is the average flux through each turn of solenoid 1?

31–4 Two coils are wound around the same cylindrical form, like the coils in Example 31–1. When the current in the first coil is decreasing at a rate of -0.242 A/s, the induced emf in the second coil has magnitude 1.65×10^{-3} V. a) What is the mutual inductance of the pair of coils? b) If the second coil has 25 turns, what is the flux through each turn when the current in the first coil equals 1.20 A? c) If the current in the second coil increases at a rate of 0.360 A/s, what is the magnitude of the induced emf in the first coil?

31–5 From Eq. (31–5), 1 H = 1 Wb/A and from Eq. (31–4), 1 H = 1 $\Omega \cdot$ s. Show that these two definitions are equivalent.

SECTION 31–3 SELF-INDUCTANCE AND INDUCTORS

31–6 When the current in a toroidal solenoid is changing at a rate of 0.0260 A/s, the magnitude of the induced emf is 12.6 mV. When the current equals 1.40 A, the average flux through each turn of the solenoid is 0.00285 Wb. How many turns does the solenoid have?

31–7 The inductor in Fig. 31–14 has inductance 0.260 H and carries a current in the direction shown that is decreasing at a uniform rate, $di/dt = -0.0180$ A/s. a) Find the self-induced emf. b) Which end of the inductor, a or b, is at a higher potential?

FIGURE 31–14 Exercise 31–7.

31–8 A toroidal solenoid has a cross-section area of 0.480 cm^2, a mean radius of 12.0 cm, and 1800 turns. The space inside the windings is filled with a ferromagnetic material that has a relative permeability 500. a) Calculate the inductance of the solenoid. (Neglect the variation of the magnetic field across the cross section of the toroid.) b) If the material is removed from inside the solenoid and replaced by air, what is the inductance of the solenoid?

31–9 Inductance of a Solenoid. A long, straight solenoid has N turns, uniform cross-section area A, and length l. Derive an approximate expression for the inductance of the solenoid. Assume that the magnetic field is uniform inside the solenoid and zero outside. (Your answer is approximate because B is actually smaller at the ends than at the center. For this reason, your answer is actually an upper limit on the inductance.)

31–10 Show that the two expressions for inductance, $N\Phi_B/i$ and $-\varepsilon/(di/dt)$, have the same units.

31–11 At the instant when the current in an inductor is increasing at a rate of 0.0640 A/s, the magnitude of the self-induced emf is 0.0160 V. a) What is the inductance of the inductor? b) If the inductor is a solenoid with 400 turns, what is the average magnetic flux through each turn when the current is 0.720 A?

SECTION 31–4 MAGNETIC-FIELD ENERGY

31–12 An inductor used in a dc power supply has an inductance of 12.0 H and a resistance of 180 Ω. It carries a current of 0.300 A. a) What is the energy stored in the magnetic field? b) At what rate is thermal energy developed in the inductor? c) Does your answer to part (b) mean that the magnetic-field energy is decreasing with time? Explain.

31–13 An air-filled toroidal solenoid has a mean radius of 15.0 cm and a cross-section area of 5.00 cm^2. When the current is 12.0 A, the energy stored is 0.390 J. How many turns does the winding have?

31–14 It has been proposed to use large inductors as energy storage devices. a) How much electrical energy is converted to light and thermal energy by a 200-W light bulb in one day? b) If the amount of energy calculated in part (a) is stored in an inductor in which the current is 80.0 A, what is the inductance?

31–15 Starting from Eq. (31–9), derive in detail Eq. (31–11) for the energy density in a toroidal solenoid filled with a magnetic material.

31–16 A magnetic field with magnitude $B = 0.560$ T is uniform across a volume of 0.0290 m^3. Calculate the total magnetic energy in the volume if a) the volume is free space; b) the volume is filled with material with relative permeability 450.

31–17 It is proposed to store 1.00 kWh = 3.60×10^6 J of electrical energy in a uniform magnetic field with magnitude 0.600 T. a) What volume (in vacuum) must the magnetic field occupy to store this amount of energy? b) If instead this amount of energy is to be stored in a volume (in vacuum) equivalent to a cube 40.0 cm on a side, what magnetic field is required?

31–18 An air-filled toroidal solenoid has 600 turns and a mean radius of 6.90 cm. Each winding has a cross-section area of 0.0350 cm^2. Assume that the magnetic field is uniform over the cross section of the windings. a) When the current in the solenoid is 2.50 A, what is the magnetic field within the solenoid?

b) Use Eq. (31–10) to calculate the energy density in the solenoid. c) What is the total volume enclosed by the windings? d) Use the volume from part (c) and the energy density from part (b) to calculate the total energy stored in the solenoid. e) What is the inductance of the solenoid? f) Use Eq. (31–9) to calculate the energy stored in the solenoid. Compare your answer to what you obtained in part (d).

SECTION 31–5 THE *R-L* CIRCUIT

31–19 An inductor with an inductance of 2.50 H and a resistance of 8.00 Ω is connected to the terminals of a battery with an emf of 6.00 V and negligible internal resistance. Find a) the initial rate of increase of current in the circuit; b) the rate of increase of current at the instant when the current is 0.500 A; c) the current 0.250 s after the circuit is closed; d) the final steady-state current.

31–20 In Fig. 31–7, switch S_1 is closed while S_2 is kept open. The inductance is $L = 0.115$ H and the resistance is $R = 120$ Ω. a) When the current has reached its final value, the energy stored in the inductor is 0.260 J. What is the emf ε of the battery? b) After the current has reached its final value, S_1 is opened and S_2 is closed. How much time does it take for the energy stored in the inductor to decrease to 0.130 J, half the original value?

31–21 In Fig. 31–7, suppose that $\varepsilon = 60.0$ V, $R = 240$ Ω, and $L = 0.160$ H. With switch S_2 open, switch S_1 is left closed until a constant current is established. Then S_2 is closed and S_1 opened, taking the battery out of the circuit. a) What is the initial current in the resistor, just after S_2 is closed and S_1 is opened? b) What is the current in the resistor at $t = 4.00 \times 10^{-4}$ s? c) What is the potential difference between points b and c at $t = 4.00 \times 10^{-4}$ s? Which point is at a higher potential? d) How long does it take the current to decrease to half its initial value?

31–22 In Fig. 31–7, suppose that $\varepsilon = 60.0$ V, $R = 240$ Ω, and $L = 0.160$ H. Initially, there is no current in the circuit. Switch S_2 is left open, and switch S_1 is closed. a) Just after S_1 is closed, what are the potential differences v_{ab} and v_{bc}? b) A long time (many time constants) after S_1 is closed, what are v_{ab} and v_{bc}? c) What are v_{ab} and v_{bc} at an intermediate time when $i = 0.150$ A?

31–23 Refer to Exercise 31–19. a) What is the power input to the inductor from the battery as a function of time if the circuit is completed at $t = 0$? b) What is the rate of dissipation of energy in the resistance of the inductor as a function of time? c) What is the rate at which the energy of the magnetic field in the inductor is increasing, as a function of time? d) Compare the results of (a), (b), and (c).

31–24 Write an equation corresponding to Eq. (31–13) for the current in Fig. 31–7 just after the switch S_2 is closed and S_1 is opened, if the initial current is I_0. Use integration methods to verify Eq. (31–18).

31–25 Show that L/R has units of time.

SECTION 31–6 THE *L-C* CIRCUIT

31–26 A Radio Tuning Circuit. The minimum capacitance of a variable capacitor in a radio is 4.18 pF. a) What is the inductance of a coil connected to this capacitor if the oscillation

frequency of the *L-C* circuit is 1600×10^3 Hz, corresponding to one end of the AM radio broadcast band, when the capacitor is set to its minimum capacitance? b) The frequency at the other end of the broadcast band is 540×10^3 Hz. What is the maximum capacitance of the capacitor if the oscillation frequency is adjustable over the range of the broadcast band?

31–27 *L-C* Oscillations. A capacitor with capacitance 6.00×10^{-5} F is charged by connecting it to a 12.0-V battery. The capacitor is disconnected from the battery and connected across an inductor with $L = 1.50$ H. a) What are the angular frequency ω of the electrical oscillations and the period of these oscillations (the time for one oscillation)? b) What is the initial charge on the capacitor? c) How much energy is initially stored in the capacitor? d) What is the charge on the capacitor 0.0230 s after the connection to the inductor is made? Interpret the sign of your answer. e) At the time given in part (d), what is the current in the inductor? Interpret the sign of your answer. f) At the time given in part (d), how much electrical energy is stored in the capacitor and how much is stored in the inductor?

31–28 In an *L-C* circuit, $L = 85.0$ mH and $C = 3.20$ μF. During the oscillations the maximum current in the inductor is 0.850 mA. a) What is the maximum charge on the capacitor? b) What is the magnitude of the charge on the capacitor at an instant when the current in the inductor has magnitude 0.500 mA?

31–29 In an *L-C* circuit, $L = 0.640$ H and $C = 3.60$ μF. a) At the instant when the current in the inductor is changing at a rate of 2.80 A/s, what is the magnitude of the charge on the capacitor? b) When the charge on the capacitor is 8.50×10^{-6} C, what is the magnitude of the induced emf in the inductor?

31–30 An *L-C* circuit has a 0.400-H inductor and a 0.250-nF capacitor. During the current oscillations, the maximum current in the inductor is 1.5 A. a) What is the maximum energy stored in the capacitor at any time during the current oscillations? b) How many times per second does the capacitor contain the amount of energy found in part (a)?

31–31 Show that $\sqrt{L/C}$ has units of time.

31–32 Show that the differential equation of Eq. (31–20) is satisfied by the function $q = Q \cos(\omega t + \phi)$, with ω given by $1/\sqrt{L/C}$.

31–33 a) Using Eqs. (31–21) and (31–23) for an *L-C* circuit, write expressions for the energy stored in the capacitor as a function of time and for the energy stored in the inductor as a function of time. b) Using Eq. (31–22) and the trigonometric identity $\sin^2 x + \cos^2 x = 1$, show that the total energy in the *L-C* circuit is constant and equal to $Q^2/2C$.

SECTION 31–7 THE *L-R-C* SERIES CIRCUIT

31–34 a) Take first and second derivatives with respect to time of q given in Eq. (31–28), and show that it is a solution of Eq. (31–27). b) At $t = 0$ the switch in Fig. 31–13 is thrown so that it connects points d and a; at this time, $q = Q$ and $i = dq/dt = 0$. Show that the constants ϕ and A in Eq. (31–28) are given by

$$\tan \phi = -\frac{R}{2L\sqrt{(1/LC) - (R^2/4L^2)}} \quad \text{and} \quad A = \frac{Q}{\cos \phi}.$$

31-35 a) In Eq. (13-41), substitute q for x, L for m, $1/C$ for k, and R for the damping constant b. Show that the result is Eq. (31-27). b) Make these same substitutions in Eq. (13-43) and show that Eq. (31-29) results. c) Make these same substitutions in Eq. (13-42) and show that Eq. (31-28) results.

31-36 Show that the quantity $\sqrt{L/C}$ has units of resistance (ohms).

31-37 An L-R-C circuit has $L = 0.285$ H, $C = 4.60 \times 10^{-4}$ F, and an angular frequency $\omega' = 1/\sqrt{6LC}$. What is the resistance R in the circuit?

31-38 An L-R-C circuit has $L = 0.450$ H, $C = 2.50 \times 10^{-5}$ F, and resistance R. a) What is the angular frequency of the circuit when $R = 0$? b) What value must R have to give a 5.0% decrease in angular frequency compared to the value calculated in part (a)?

PROBLEMS

31-39 A Differentiating Circuit. The current in a resistance-less inductor is caused to vary with time as in the graph of Fig. 31-15. a) Sketch the pattern that would be observed on the screen of an oscilloscope connected to the terminals of the inductor. (The oscilloscope spot sweeps horizontally across the screen at a constant speed, and its vertical deflection is proportional to the potential difference between the inductor terminals.) b) Explain why the inductor can be described as a "differentiating circuit."

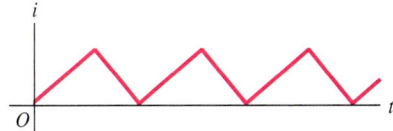

FIGURE 31-15 Problem 31-39.

31-40 A 0.250-H inductor carries a time-varying current given by the expression $i = (124 \text{ mA}) \cos [(240\pi/\text{s})t]$. a) Find an expression for the induced emf as a function of time. Graph the current and induced emf as functions of time for $t = 0$ to $t = \frac{1}{60}$ s. b) What is the maximum emf? What is the current when the induced emf is a maximum? c) What is the maximum current? What is the induced emf when the current is a maximum?

31-41 We previously neglected the variation of the magnetic field across the cross section of a toroidal solenoid. Let's now examine the validity of that approximation. A certain toroidal solenoid has a rectangular cross section (Fig. 31-16). It has N uniformly spaced turns, with air inside. The magnetic field at a point inside the toroid is given by the equation derived in Example 29-11 (Section 29-8). *Do not* assume the field to be uniform over the cross section. a) Show that the magnetic flux through a cross section of the toroid is

$$\Phi_B = \frac{\mu_0 Nih}{2\pi} \ln\left(\frac{b}{a}\right).$$

b) Show that the inductance of the toroidal solenoid is given by

$$L = \frac{\mu_0 N^2 h}{2\pi} \ln\left(\frac{b}{a}\right).$$

c) The fraction b/a may be written as

$$\frac{b}{a} = \frac{a + b - a}{a} = 1 + \frac{b - a}{a}.$$

Use the power series expansion $\ln(1 + z) = z + z^2/2 + \cdots$, valid for $|z| < 1$, to show that when $b - a$ is much less than a, the inductance is approximately equal to

$$L = \frac{\mu_0 N^2 h(b - a)}{2\pi a}.$$

Compare this result with the result given in Eq. (31-8), obtained in Example 31-3 (Section 31-3).

FIGURE 31-16 Problem 31-41.

31-42 A solenoid has length l_1, radius r_1, and number of turns N_1. A second, smaller solenoid with length l_2, radius r_2 ($r_2 < r_1$), and number of turns N_2 is placed at the center of the first solenoid, such that their axes coincide. Assume that the magnetic field of the first solenoid at the location of the second is uniform and has a magnitude given by the equation derived in Example 29-10 (Section 29-8). a) What is the mutual inductance of the pair of solenoids? b) If the current in the large solenoid is increasing at the rate di_1/dt, what is the magnitude of the emf induced in the small solenoid? c) If the current in the smaller solenoid is increasing at the rate di_2/dt, what is the magnitude of the emf induced in the larger solenoid?

31-43 The current in a coil of wire is initially zero but increases at a constant rate; after 12.0 s it is 48.0 A. The changing current induces an emf of magnitude 30.0 V in the coil. a) Determine the inductance of the coil. b) Determine the total magnetic flux through the coil when the current is 48.0 A. c) If the resistance of the coil is 60.0 Ω, determine the ratio of the rate at which energy is being stored in the magnetic field to the rate at which electrical energy is being dissipated by the resistance at the instant when the current is 48.0 A.

31-44 A coil has 400 turns and self-inductance 3.50 mH. The current in the coil varies with time according to $i = (680 \text{ mA}) \cos [\pi t/(0.0250 \text{ s})]$. a) What is the maximum emf induced in the coil? b) What is the maximum average flux through each turn of the coil? c) At $t = 0.0180$ s, what is the magnitude of the induced emf?

31–45 You are given two inductors, one of self-inductance L_1 and the other of self-inductance L_2. a) You connect the two inductors in series and arrange them so that their mutual inductance is negligible. Show that the equivalent inductance of the combination is $L_{eq} = L_1 + L_2$. b) You now connect the two inductors in parallel, again arranging them so that their mutual inductance is negligible. Show that the equivalent inductance of the combination is $L_{eq} = (1/L_1 + 1/L_2)^{-1}$. (*Hint:* For either a series or a parallel combination, the potential difference across the combination is $L_{eq}(di/dt)$, where i is the current through the combination. For a parallel combination, i is the sum of the currents through the two inductors.)

31–46 A Coaxial Cable. A small solid conductor with radius a is supported by insulating, nonmagnetic disks on the axis of a thin-walled tube with inner radius b. The inner and outer conductors carry equal currents i in opposite directions. a) Use Ampere's law to find the magnetic field at any point in the volume between the conductors. b) Write the expression for the flux $d\Phi_B$ through a narrow strip of length l parallel to the axis, of width dr, at a distance r from the axis of the cable and lying in a plane containing the axis. c) Integrate your expression from part (b) over the volume between the two conductors to find the total flux produced by a current i in the central conductor. d) Show that the inductance of a length l of the cable is

$$L = l\,\frac{\mu_0}{2\pi}\ln\left(\frac{b}{a}\right).$$

e) Use Eq. (31–9) to calculate the energy stored in the magnetic field for a length l of the cable.

31–47 Consider the coaxial cable of Problem 31–46. The conductors carry equal currents i in opposite directions. a) Use Ampere's law to find the magnetic field at any point in the volume between the conductors. b) Use the energy density for a magnetic field, Eq. (31–10), to calculate the energy stored in a thin cylindrical shell between the two conductors. Let the cylindrical shell have inner radius r, outer radius $r + dr$, and length l. c) Integrate your result in part (b) over the volume between the two conductors to find the total energy stored in the magnetic field for a length l of the cable. d) Use your result in part (c) and Eq. (31–9) to calculate the inductance L of a length l of the cable. Compare your result to L calculated in part (d) of Problem 31–46.

31–48 Consider the toroidal solenoid of Exercise 31–2. a) Derive an expression for the inductance L_1 when only the first coil is used and an expression for L_2 when only the second coil is used. b) Show that $M^2 = L_1 L_2$.

31–49 Uniform electric and magnetic fields $\vec{E}$ and $\vec{B}$ occupy the same region of free space. If $E = 650$ V/m, what is B if the energy densities in the electric and magnetic fields are equal?

31–50 An inductor is connected to the terminals of a battery that has an emf of 12.0 V and negligible internal resistance. The current is 4.86 mA at 0.725 ms after the connection is completed. After a long time the current is 6.45 mA. What is a) the resistance R of the inductor? b) the inductance L of the inductor?

31–51 Continuation of Exercises 31–19 and 31–23. a) How much energy is stored in the magnetic field of the inductor one time constant after the battery has been connected? Compute

this both by integrating the expression in Exercise 31–23(c) and by using Eq. (31–9), and compare the results. b) Integrate the expression obtained in Exercise 31–23(a) to find the *total* energy supplied by the battery during the time interval considered in part (a). c) Integrate the expression obtained in Exercise 31–23(b) to find the *total* energy dissipated in the resistance of the inductor during the same time period. d) Compare the results obtained in (a), (b), and (c).

31–52 Continuation of Exercise 31–21. a) What is the total energy initially stored in the inductor? b) At $t = 4.00 \times 10^{-4}$ s, at what rate is the energy stored in the inductor decreasing? c) At $t = 4.00 \times 10^{-4}$ s, at what rate is electrical energy being converted into thermal energy in the resistor? d) Obtain an expression for the rate at which electrical energy is being converted into thermal energy in the resistor as a function of time. Integrate this expression from $t = 0$ to $t = \infty$ to obtain the total electrical energy dissipated in the resistor. Compare your result to that of part (a).

31–53 The equation preceding Eq. (31–27) may be converted into an energy relation. Multiply both sides of this equation by $-i = -dq/dt$. The first term then becomes $i^2 R$. Show that the second can be written $d(\frac{1}{2}Li^2)/dt$, and that the third can be written $d(q^2/2C)/dt$. What does the resulting equation say about energy conservation in the circuit?

31–54 At $t = 0$ in an *L-C* circuit, the capacitor has its maximum charge. a) Show that the electric and magnetic energies are equal at $t = 3T/8$, where T is the period of the oscillations. b) What is the next time after $t = 3T/8$ when the electric and magnetic energies again are equal?

31–55 You are asked to design an *L-C* circuit in which the stored energy is 1.60×10^{-4} J and the angular frequency is $\omega = 6.40 \times 10^4$ rad/s. The maximum voltage across the capacitor is to be 25.0 V. What values of C and L are required?

31–56 An *L-C* circuit consists of a 60.0-mH inductor and a 250-μF capacitor. The initial charge on the capacitor is 6.00 μC, and the initial current in the inductor is zero. a) What is the maximum voltage across the capacitor? b) What is the maximum current in the inductor? c) What is the maximum energy stored in the inductor? d) When the current in the inductor has half its maximum value, what is the charge on the capacitor and what is the energy stored in the inductor?

31–57 In the circuit shown in Fig. 31–17, $\mathcal{E} = 60.0$ V, $R_1 = 40.0$ Ω, $R_2 = 25.0$ Ω, and $L = 0.300$ H. Switch S is closed at $t = 0$. Just after the switch is closed, a) what is the potential difference v_{ab} across the resistor R_1? b) which point a or b is at a higher potential? c) what is the potential difference v_{cd} across the inductor L? d) which point, c or d, is at a higher potential? The switch is left closed a long time and then opened. Just after the switch is opened, e) what is the potential difference v_{ab} across the resistor R_1? f) which point, a or b, is at a higher potential? g) what is the potential difference v_{cd} across the inductor L? h) which point, c or d, is at a higher potential?

31–58 In the circuit shown in Fig. 31–17, $\mathcal{E} = 60.0$ V, $R_1 = 40.0$ Ω, $R_2 = 25.0$ Ω, and $L = 0.300$ H. a) Switch S is closed. At some time t afterward the current in the inductor is increasing at a rate of $di/dt = 50.0$ A/s. At this instant, what is the current i_1 through R_1 and the current i_2 through R_2? b) After the switch has been closed a long time, it is opened again. Just after it is opened, what is the current through R_1?

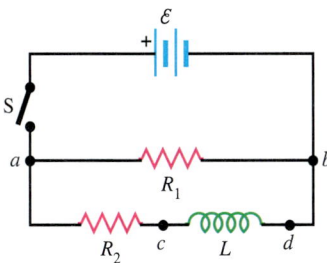

FIGURE 31–17 Problems 31–57, 31–58, and 31–61.

31–59 Consider the circuit of Fig. 31–18. Let $\mathcal{E} = 36.0$ V, $R_0 = 50.0\ \Omega$, $R = 150\ \Omega$, and $L = 4.00$ H. a) Switch S_1 is closed, and S_2 is left open. Just after S_1 is closed, what are the current i_0 through R_0 and the potential differences v_{ac} and v_{cb}? b) After S_1 has been closed a long time (S_2 is still open) so that the current has reached its final steady value, what are i_0, v_{ac}, and v_{cb}? c) Find the expressions for i_0, v_{ac}, and v_{cb} as functions of the time t since S_1 was closed. Your results should agree with part (a) when $t = 0$ and with part (b) when $t \to \infty$. Graph i_0, v_{ac}, and v_{cb} versus time.

FIGURE 31–18 Problems 31–59 and 31–60.

31–60 After the current in the circuit of Fig. 31–18 has reached its final steady value with switch S_1 closed and S_2 open, the switch S_2 is closed, thus short-circuiting the inductor. (Switch S_1 remains closed. See Problem 31–59 for numerical values of the circuit elements.) a) Just after S_2 is closed, what are v_{ac} and v_{cb}, and what are the currents through R_0, R, and S_2? b) A long time after S_2 is closed, what are v_{ac} and v_{cb}, and what are the currents through R_0, R, and S_2? c) Derive expressions for the currents through R_0, R, and S_2 as functions of the time t that has elapsed since S_2 was closed. Your results should agree with part (a) when $t = 0$ and with part (b) when $t \to \infty$. Graph these three currents versus time.

31–61 Demonstrating Inductance. A common demonstration of inductance employs a circuit such as the one shown in Fig. 31–17. Switch S is closed, and the light bulb (represented by resistance R_1) just barely glows. After a period of time, switch S is opened, and the bulb lights up brightly for a short period of time. To understand this effect, think of an inductor as a device that imparts an "inertia" to the current, preventing a discontinuous change in the current through it. a) Derive, as explicit functions of time, expressions for i_1 (the current through the light bulb) and i_2 (the current through the inductor) after the switch S is closed. b) After a long period of time, the currents i_1 and i_2 reach their steady-state values. Obtain expressions for these steady-state currents. c) Switch S is now opened. Obtain an expression for the current through the inductor and light bulb as an explicit function of time. d) You have been asked to design a demonstration apparatus using the circuit shown in Fig. 31–17 with a 22.0-H inductor and a 40.0-W light bulb. You are to connect a resistor in series with the inductor, and R_2 represents the sum of that resistance plus the internal resistance of the inductor. When switch S is opened, a transient current is to be set up that starts at 0.600 A and is not to fall below 0.150 A until after 0.0800 s. For simplicity, assume that the resistance of the light bulb is constant and equals the resistance the bulb must have to dissipate 40.0 W at 120 V. Determine R_2 and $\mathcal{E}$ for the given design considerations. e) With the numerical values determined in part (d), what is the current through the light bulb just before the switch is opened? Does this result confirm the qualitative description of what is observed in the demonstration?

CHALLENGE PROBLEMS

31–62 Two coils are wrapped around each other as in Fig. 31–2. The current travels in the same sense around each coil. One coil has self-inductance L_1, and the other coil has self-inductance L_2. The mutual inductance of the two coils is M. a) Show that if the two coils are connected in series, the equivalent inductance of the combination is $L_{eq} = L_1 + L_2 + 2M$. b) Show that if the two coils are connected in parallel, the equivalent inductance of the combination is

$$L_{eq} = \frac{L_1 L_2 - M^2}{L_1 + L_2 - 2M}.$$

(*Hint:* See the hint for Problem 31–45.)

31–63 Consider the circuit shown in Fig. 31–19. Switch S is closed at time $t = 0$, causing a current i_1 through the inductive branch and a current i_2 through the capacitive branch. The initial charge on the capacitor is zero, and the charge at time t is q_2. a) Derive expressions for i_1, i_2, and q_2 as functions of time. Express your answers in terms of $\mathcal{E}$, L, C, R_1, R_2, and t. For the

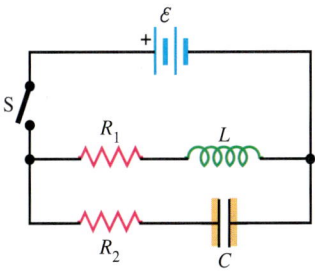

FIGURE 31–19 Challenge Problem 31–63.

remainder of the problem let the circuit elements have the following values: $\mathcal{E} = 48$ V, $L = 8.0$ H, $C = 20$ μF, $R_1 = 25$ Ω, and $R_2 = 5000$ Ω. b) What is the initial current through the inductive branch? What is the initial current through the capacitive branch? c) What are the currents through the inductive and capacitive branches a long time after the switch has been closed? How long is a "long time"? Explain. d) At what time t_1 (accurate to two significant figures) will the currents i_1 and i_2 be equal? (*Hint:* You might consider using series expansions for the exponentials.) e) For the conditions given in part (d), determine i_1. f) The total current through the battery is $i = i_1 + i_2$. At what time t_2 (accurate to two significant figures) will i equal one-half of its final value? (*Hint:* The numerical work is greatly simplified if one makes suitable approximations. A sketch of i_1 and i_2 versus t may help you decide what approximations are valid.)

31–64 Consider the circuit shown in Fig. 31–20. The circuit elements are as follows: $\mathcal{E} = 32.0$ V, $L = 0.640$ H, $C = 2.00$ μF, and $R = 400$ Ω. At time $t = 0$, switch S is closed. The current through the inductor is i_1, the current through the capacitor branch is i_2, and the charge on the capacitor is q_2. a) Using Kirchhoff's rules, verify the circuit equations

$$R(i_1 + i_2) + L\left(\frac{di_1}{dt}\right) = \mathcal{E},$$

$$R(i_1 + i_2) + \frac{q_2}{C} = \mathcal{E}.$$

b) What are the initial values of i_1, i_2, and q_2? c) Show by direct substitution that the following solutions for i_1 and q_2 satisfy the circuit equations from part (a). Also, show that they satisfy the initial conditions

$$i_1 = \left(\frac{\mathcal{E}}{R}\right)[1 - e^{-\beta t}(2\omega RC)^{-1}\sin(\omega t) + \cos(\omega t)],$$

$$q_2 = \left(\frac{\mathcal{E}}{\omega R}\right)e^{-\beta t}\sin(\omega t),$$

where $\beta = (2RC)^{-1}$ and $\omega = [(LC)^{-1} - (2RC)^{-2}]^{1/2}$. d) Determine the time t_1 at which i_2 first becomes zero.

31–65 **A Volume Gauge.** A tank containing a liquid has turns of wire wrapped around it, causing it to act as an inductor. The liquid content of the tank can be measured by using its inductance to determine the height of the liquid in the tank. The inductance of the tank changes from a value of L_0 corresponding to a relative permeability of 1 when the tank is empty to a value of L_f corresponding to a relative permeability of K_m (the relative permeability of the liquid) when the tank is full. The appropriate electronic circuitry can determine the inductance to five significant figures and thus the effective relative permeability of the combined air and liquid within the rectangular cavity of the tank. The four sides of the tank each have width W and height D (Fig. 31–21). The height of the liquid in the tank is d. Neglect any fringing effects and assume that the relative permeability of the material of which the tank is made can be neglected.
a) Derive an expression for d as a function of L, the inductance corresponding to a certain fluid height, L_0, L_f, and D. b) What is the inductance (to five significant figures) for a tank one-quarter full, one-half full, three-quarters full, and completely full if the tank contains liquid oxygen? Take $L_0 = 0.63000$ H. The magnetic susceptibility of liquid oxygen is $\chi_m = 1.52 \times 10^{-3}$. c) Repeat part (b) for mercury. The magnetic susceptibility of mercury is given in Table 29–1. d) For which material is this volume gauge more practical?

FIGURE 31–20 Challenge Problem 31–64.

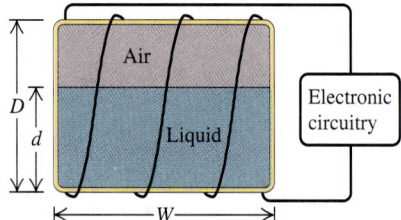

FIGURE 31–21 Challenge Problem 31–65.

Alternating Current

To tune the radio in this classic car, you turn a knob that adjusts a variable capacitance. This in turn adjusts the radio's resonant frequency, which depends on the capacitance as well as the radio's inductance and resistance. When the resonant frequency matches the frequency of the desired radio station, that station's signals come in loud and clear.

32–1 INTRODUCTION

During the 1880s in the United States there was a heated and acrimonious debate between two inventors over the best method of electric-power distribution. Thomas Edison favored direct current (dc), that is, steady current that does not vary with time. George Westinghouse favored **alternating current (ac),** with sinusoidally varying voltages and currents. He argued that transformers (which we will study in this chapter) can be used to step the voltage up and down with ac but not with dc; low voltages are safer for consumer use, but high voltages and correspondingly low currents are best for long-distance power transmission to minimize i^2R losses in the cables.

Eventually, Westinghouse prevailed, and most present-day household and industrial power-distribution systems operate with alternating current. Any appliance that you plug into a wall outlet uses ac, and many battery-powered devices such as radios and cordless telephones make use of the dc supplied by the battery to create or amplify alternating currents. Circuits in modern communication equipment, including pagers and television, also make extensive use of ac.

In this chapter we will learn how resistors, inductors, and capacitors behave in circuits with sinusoidally varying voltages and currents. Many of the principles that we found useful in Chapters 26, 29, and 31 are applicable, along with several new concepts related to the circuit behavior of inductors and capacitors. A key concept in this discussion is *resonance,* which we studied in Chapter 13 for mechanical systems.

32–2 PHASORS AND ALTERNATING CURRENTS

To supply an alternating current to a circuit, a source of alternating emf or voltage is required. An example of such a source is a coil of wire rotating with constant angular velocity in a magnetic field, which we discussed in Example 30–4 (Section 30–3). This develops a sinusoidal alternating emf and is the prototype of the commercial alternating-current generator or *alternator.*

We will use the term **ac source** for any device that supplies a sinusoidally varying voltage (potential difference) v or current i. The usual circuit-diagram symbol for an ac source is

A sinusoidal voltage might be described by a function such as

$$v = V \cos \omega t. \qquad (32\text{–}1)$$

In this expression, v (lowercase) is the *instantaneous* potential difference; V (uppercase) is the maximum potential difference, which we call the **voltage amplitude;** and ω is the *angular frequency,* equal to 2π times the frequency f.

In the United States and Canada, commercial electric-power distribution systems always use a frequency of $f = 60$ Hz, corresponding to $\omega = (2\pi \text{ rad})(60 \text{ s}^{-1}) = 377$ rad/s; in much of the rest of the world, $f = 50$ Hz ($\omega = 314$ rad/s) is used. Similarly, a sinusoidal current might be described as

$$i = I \cos \omega t, \tag{32-2}$$

where i (lowercase) is the instantaneous current and I (uppercase) is the maximum current or **current amplitude.**

To represent sinusoidally varying voltages and currents, we will use rotating vector diagrams similar to those we used in the study of simple harmonic motion in Section 13–3 (see Fig. 13–3). In these diagrams the instantaneous value of a quantity that varies sinusoidally with time is represented by the *projection* onto a horizontal axis of a vector with a length equal to the amplitude of the quantity. The vector rotates counterclockwise with constant angular speed ω. These rotating vectors are called **phasors,** and diagrams containing them are called **phasor diagrams.** Figure 32–1 shows a phasor diagram for the sinusoidal current described by Eq. (32–2). The projection of the phasor onto the horizontal axis at time t is $I \cos \omega t$; this is why we chose to use the cosine function rather than the sine in Eq. (32–2).

A phasor is not a real physical quantity with a direction in space, such as velocity, momentum, or electric field. Rather, it is a *geometric* entity that helps us to describe and analyze physical quantities that vary sinusoidally with time. In Section 13–3 we used a single phasor to represent the position of a point mass undergoing simple harmonic motion. In this chapter we will use phasors to *add* sinusoidal voltages and currents. Combining sinusoidal quantities with phase differences then becomes a matter of vector addition. We will find a similar use for phasors in Chapters 37 and 38 in our study of interference effects with light.

How do we measure a sinusoidally varying current? In Section 27–4 we used a d'Arsonval galvanometer to measure steady currents. But if we pass a *sinusoidal* current through a d'Arsonval meter, the torque on the moving coil varies sinusoidally, with one direction half the time and the opposite direction the other half. The needle may wiggle a little if the frequency is low enough, but its average deflection is zero. Hence a d'Arsonval meter by itself isn't very useful for measuring alternating currents.

To get a measurable one-way current through the meter, we can use *diodes,* which we described in Section 26–4. A diode (or rectifier) is a device that conducts better in one direction than in the other; an ideal diode has zero resistance for one direction of current and infinite resistance for the other. One possible arrangement is shown in Fig. 32–2a. The current through the galvanometer G is always upward, regardless of the direction of the current from the ac source (i.e., which part of the cycle the source is in). The current through G is as shown by the graph in Fig. 32–2b. It pulsates but always has the same direction, and the average meter deflection is *not* zero. This arrangement of diodes is called a *full-wave rectifier circuit.*

The **rectified average current** I_{rav} is defined so that during any whole number of cycles, the total charge that flows is the same as though the current were constant with a value equal to I_{rav}. The notation I_{rav} and the name *rectified average* current emphasize that this is *not* the average of the original sinusoidal current. In Fig. 32–2b the total charge that flows in time t corresponds to the area under the curve of i versus t (recall that $i = dq/dt$, so q is the integral of t); this area must equal the rectangular area with height I_{rav}. We see that I_{rav} is less than the maximum current I; the two are related by

$$I_{rav} = \frac{2}{\pi} I = 0.637I \qquad \text{(rectified average value of a sinusoidal current).} \tag{32-3}$$

(The factor of $2/\pi$ is the average value of $|\cos \omega t|$ or of $|\sin \omega t|$; see Example 30–5 in Section 30–3.) The galvanometer deflection is proportional to I_{rav}. The galvanometer scale can be calibrated to read I, I_{rav}, or, most commonly, I_{rms} (discussed below).

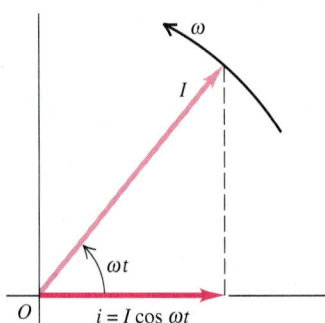

32–1 A phasor diagram. The projection of the rotating vector (phasor) onto the horizontal axis represents the instantaneous current.

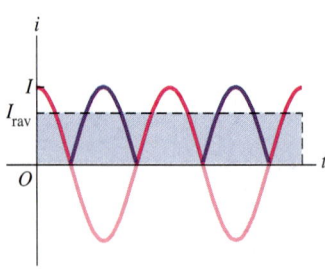

32–2 (a) A full-wave rectifier circuit. The arrow-bar symbol for a diode indicates the directions in which current can and cannot pass. (b) Graph of a full-wave rectified current and its average value, the rectified average current I_{rav}.

A more useful way to describe a quantity that can be either positive or negative is the *root-mean-square (rms) value.* We used rms values in Section 16–4 in connection with the speeds of molecules in a gas. We *square* the instantaneous current i, take the *average* (mean) value of i^2, and finally take the *square root* of that average. This procedure defines the **root-mean-square current,** denoted as I_{rms}. Even when i is negative, i^2 is always positive, so I_{rms} is never zero (unless i is zero at every instant).

Here's how we obtain I_{rms}. If the instantaneous current is given by $i = I \cos \omega t$, then

$$i^2 = I^2 \cos^2 \omega t.$$

Using a double-angle formula from trigonometry,

$$\cos^2 A = \frac{1}{2}(1 + \cos 2A),$$

we find

$$i^2 = I^2 \frac{1}{2}(1 + \cos 2\omega t) = \frac{1}{2} I^2 + \frac{1}{2} I^2 \cos 2\omega t.$$

The average of $\cos 2\omega t$ is zero because it is positive half the time and negative half the time. Thus the average of i^2 is simply $I^2/2$. The square root of this is I_{rms}:

$$I_{rms} = \frac{I}{\sqrt{2}} \qquad \text{(root-mean-square value of a sinusoidal current).} \qquad (32\text{–}4)$$

In the same way, the root-mean-square value of a sinusoidal voltage with amplitude (maximum value) V is

$$V_{rms} = \frac{V}{\sqrt{2}} \qquad \text{(root-mean-square value of a sinusoidal voltage).} \qquad (32\text{–}5)$$

We can convert a rectifying ammeter into a voltmeter by adding a series resistor, just as for the dc case discussed in Section 27–4. Meters used for ac voltage and current measurements are nearly always calibrated to read rms values, not maximum or rectified average. Voltages and currents in power distribution systems are always described in terms of their rms values. The usual household power supply, "120-volt ac," has an rms voltage of 120 V. The voltage amplitude is

$$V = \sqrt{2}\, V_{rms} = \sqrt{2}\, (120\ V) = 170\ V.$$

This wall socket delivers a root-mean-square voltage of 120 V. Sixty times per second, the instantaneous voltage across its terminals varies from $(\sqrt{2})(120\ V) = 170\ V$ to $-170\ V$ and back again.

<hr>

EXAMPLE 32–1

Current in a personal computer The plate on the back of a personal computer says that it draws 2.7 A from a 120-V, 60-Hz line. For this computer, what are a) the average current, b) the average of the square of the current, and c) the current amplitude?

SOLUTION a) The average of *any* sinusoidal alternating current, over any whole number of cycles, is zero.

b) The current given, 2.7 A, is the rms value, which is the *square root* of the *mean* (average) of the *square* of the current. That is,

$$I_{rms} = \sqrt{(i^2)_{av}}, \qquad \text{or} \qquad (i^2)_{av} = (I_{rms})^2 = (2.7\ A)^2 = 7.3\ A^2.$$

c) From Eq. (32–4) the current amplitude I is

$$I = \sqrt{2} I_{rms} = \sqrt{2}(2.7\ A) = 3.8\ A.$$

<hr>

32–3 RESISTANCE AND REACTANCE

In this section we will derive voltage-current relations for individual circuit elements carrying a sinusoidal current. We'll consider resistors, inductors, and capacitors.

(a)

(b)

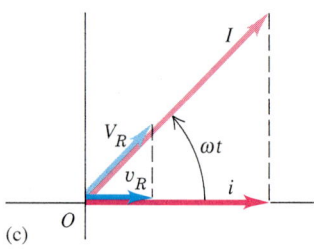

(c)

32–3 (a) Resistance R connected across an ac source. (b) Graphs of instantaneous voltage (blue) and current (red). (c) Phasor diagram. As the phasors rotate counterclockwise, current and voltage remain in phase.

14.3
AC Circuits: The Driven Oscillator
(Questions 1–5)

RESISTOR IN AN AC CIRCUIT

First let's consider a resistor with resistance R through which there is a sinusoidal current given by Eq. (32–2): $i = I \cos \omega t$. The positive direction of current is counterclockwise around the circuit, as in Fig. 32–3a. The current amplitude (maximum current) is I. From Ohm's law the instantaneous potential v_R of point a with respect to point b (that is, the instantaneous voltage across the resistor) is

$$v_R = iR = (IR) \cos \omega t. \qquad (32–6)$$

The maximum voltage V_R, the *voltage amplitude,* is the coefficient of the cosine function:

$$V_R = IR \qquad \text{(amplitude of voltage across a resistor, ac circuit).} \qquad (32–7)$$

Hence we can also write

$$v_R = V_R \cos \omega t. \qquad (32–8)$$

The current i and voltage v_R are both proportional to $\cos \omega t$, so the current is *in phase* with the voltage. Equation (32–7) shows that the current and voltage amplitudes are related in the same way as in a dc circuit.

Figure 32–3b shows graphs of i and v_R as functions of time. The vertical scales for current and voltage are different, so the relative heights of the two curves are not significant. The corresponding phasor diagram is given in Fig. 32–3c. Because i and v_R are *in phase* and have the same frequency, the current and voltage phasors rotate together; they are parallel at each instant. Their projections on the horizontal axis represent the instantaneous current and voltage, respectively.

INDUCTOR IN AN AC CIRCUIT

Next, we replace the resistor in Fig. 32–3 with a pure inductor with self-inductance L and zero resistance (Fig. 32–4a). Again we assume that the current is $i = I \cos \omega t$, with the positive direction of current taken as counterclockwise around the circuit.

Although there is no resistance, there is a potential difference v_L between the inductor terminals a and b because the current varies with time, giving rise to a self-induced emf. The induced emf in the direction of i is given by Eq. (31–7), $\varepsilon = -L \, di/dt$; however, the voltage v_L is *not* simply equal to ε. To see why, notice that if the current in the inductor is in the positive (counterclockwise) direction from a to b and is increasing, then di/dt is positive and the induced emf is directed to the left to oppose the increase in current; hence point a is at higher potential than is point b. Thus the potential of point a with respect to point b is positive and is given by $v_L = +L \, di/dt$, the *negative* of the induced emf. (You should convince yourself that this expression gives the correct sign of v_L in *all* cases, including i counterclockwise and decreasing, i clockwise and increasing, and i clockwise and decreasing; you may also want to review Section 31–3.) So we have

$$v_L = L \frac{di}{dt} = L \frac{d}{dt} (I \cos \omega t) = -I \omega L \sin \omega t. \qquad (32–9)$$

The voltage v_L across the inductor is at any instant proportional to the *rate of change* of the current. The points of maximum voltage on the graph correspond to maximum steepness of the current curve, and the points of zero voltage are the points where the current curve instantaneously levels off at its maximum and minimum values (Fig. 32–4b). The voltage and current are "out of step" or *out of phase* by a quarter-cycle. Since the voltage peaks occur a quarter-cycle earlier than the current peaks, we say that the voltage *leads* the current by 90°. The phasor diagram in Fig. 32–4c also shows this relationship; the voltage phasor is ahead of the current phasor by 90°.

We can also obtain this phase relationship by rewriting Eq. (32–9) using the identity $\cos(A + 90°) = -\sin A$:

$$v_L = I\omega L \cos(\omega t + 90°). \qquad (32\text{–}10)$$

This result shows that the voltage can be viewed as a cosine function with a "head start" of 90° relative to the current.

As we have done in Eq. (32–10), we will usually describe the phase of the *voltage* relative to the *current*, not the reverse. Thus if the current i in a circuit is

$$i = I \cos \omega t$$

and the voltage v of one point with respect to another is

$$v = V \cos(\omega t + \phi),$$

we call ϕ the **phase angle;** it gives the phase of the *voltage* relative to the *current*. For a pure resistor, $\phi = 0$, and for a pure inductor, $\phi = 90°$.

From Eq. (32–9) or (32–10) the amplitude V_L of the inductor voltage is

$$V_L = I\omega L. \qquad (32\text{–}11)$$

We define the **inductive reactance** X_L of an inductor as

$$X_L = \omega L \qquad \text{(inductive reactance).} \qquad (32\text{–}12)$$

Using X_L, we can write Eq. (32–11) in a form similar to Eq. (32–7) for a resistor ($V_R = IR$):

$$V_L = IX_L \qquad \text{(amplitude of voltage across an inductor, ac circuit).} \qquad (32\text{–}13)$$

Because X_L is the ratio of a voltage and a current, its SI unit is one ohm, the same as for resistance.

The inductive reactance X_L is really a description of the self-induced emf that opposes any change in the current through the inductor. From Eq. (32–13), for a given current amplitude I the voltage $v_L = +L\,di/dt$ across the inductor and the self-induced emf $\mathcal{E} = -L\,di/dt$ both have an amplitude V_L that is directly proportional to X_L. According to Eq. (32–12), the inductive reactance and self-induced emf increase with more rapid variation in current (that is, increasing angular frequency ω) and increasing inductance L.

If an oscillating voltage of a given amplitude V_L is applied across the inductor terminals, the resulting current will have a smaller amplitude I for larger values of X_L. Since X_L is proportional to frequency, a high-frequency voltage applied to the inductor gives only a small current, while a lower-frequency voltage of the same amplitude gives rise to a larger current. Inductors are used in some circuit applications, such as power supplies and radio-interference filters, to block high frequencies while permitting lower frequencies or dc to pass through. A circuit device that uses an inductor for this purpose is called a *low-pass filter* (see Problem 32–40).

(a)

(b)

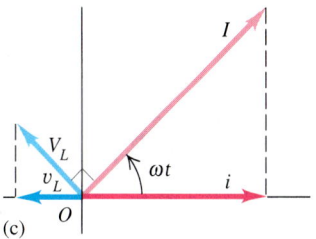

(c)

32–4 (a) Inductance L connected across an ac source. (b) Graphs of instantaneous voltage (blue) and current (red). (c) Phasor diagram. As the phasors rotate counter-clockwise, voltage *leads* current by 90°.

EXAMPLE 32–2

An inductor in an ac circuit Suppose you want the current amplitude in a pure inductor in a radio receiver to be 250 μA when the voltage amplitude is 3.60 V at a frequency of 1.60 MHz (corresponding to the upper end of the AM broadcast band). a) What inductive reactance is needed? What inductance? b) If the voltage amplitude is kept constant, what will be the current amplitude through this inductor at 16.0 MHz? At 160 kHz?

SOLUTION a) From Eq. (32–13),

$$X_L = \frac{V_L}{I} = \frac{3.60 \text{ V}}{250 \times 10^{-6} \text{ A}} = 1.44 \times 10^4 \ \Omega = 14.4 \text{ k}\Omega.$$

From Eq. (32–12), with $\omega = 2\pi f$, we find

$$L = \frac{X_L}{2\pi f} = \frac{1.44 \times 10^4 \ \Omega}{2\pi(1.60 \times 10^6 \text{ Hz})} = 1.43 \times 10^{-3} \text{ H} = 1.43 \text{ mH}.$$

b) Combining Eqs. (32–12) and (32–13), we find that the current amplitude is $I = V_L/X_L = V_L/\omega L = V_L/2\pi fL$. Thus the current amplitude is inversely proportional to the frequency f. Since $I = 250\ \mu$A at $f = 1.60$ MHz, the current amplitude at 16.0 MHz (ten times the original frequency) will be one tenth as great, or

25.0 μA; at 160 kHz = 0.160 MHz (one tenth of the original frequency) the current amplitude is ten times as great, or 2500 μA = 2.50 mA. In general, the lower the frequency of an oscillating voltage applied across an inductor, the greater the amplitude of the oscillating current that results.

CAPACITOR IN AN AC CIRCUIT

Finally, we connect a capacitor with capacitance C to the source, as in Fig. 32–5a, producing a current $i = I \cos \omega t$ through the capacitor. Again, the positive direction of current is counterclockwise around the circuit.

CAUTION ▶ You may object that charge can't really move through the capacitor because its two plates are insulated from each other. True enough, but as the capacitor charges and discharges, there is at each instant a current i into one plate, an equal current out of the other plate, and an equal *displacement* current between the plates just as though the charge were being conducted through the capacitor. (You may want to review the discussion of displacement current in Section 29–10.) Thus we often speak about alternating current *through* a capacitor. ◀

To find the instantaneous voltage v_C across the capacitor, that is, the potential of point a with respect to point b, we first let q denote the charge on the left-hand plate of the capacitor in Fig. 32–5a (so $-q$ is the charge on the right-hand plate). The current i is related to q by $i = dq/dt$; with this definition, positive current corresponds to an increasing charge on the left-hand capacitor plate. Then

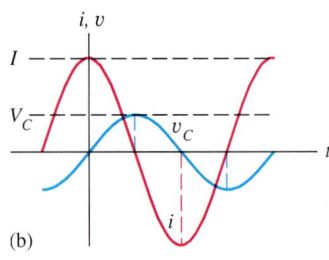

$$i = \frac{dq}{dt} = I \cos \omega t.$$

Integrating this, we get

$$q = \frac{I}{\omega} \sin \omega t. \tag{32–14}$$

Also, from Eq. (25–1) the charge q equals the voltage v_C multiplied by the capacitance, $q = Cv_C$. Using this in Eq. (32–14), we find

$$v_C = \frac{I}{\omega C} \sin \omega t. \tag{32–15}$$

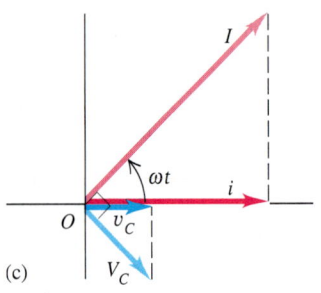

32–5 (a) Capacitor C connected across an ac source. (b) Graphs of instantaneous voltage (blue) and current (red). (c) Phasor diagram. As the phasors rotate counterclockwise, voltage *lags* current by 90°.

The instantaneous current i is equal to the rate of change dq/dt of the capacitor charge q; since $q = Cv_C$, i is also proportional to the rate of change of voltage. (Compare to an inductor, for which the situation is reversed and v_L is proportional to the rate of change of i.) Figure 32–5b shows v_C and i as functions of t. Because $i = dq/dt = C\ dv_C/dt$, the current has its greatest magnitude when the v_C curve is rising or falling most steeply and is zero when the v_C curve instantaneously levels off at its maximum and minimum values.

The capacitor voltage and current are out of phase by a quarter-cycle. The peaks of voltage occur a quarter-cycle *after* the corresponding current peaks, and we say that the voltage *lags* the current by 90°. The phasor diagram in Fig. 32–5c shows this relationship; the voltage phasor is behind the current phasor by a quarter-cycle, or 90°.

We can also derive this phase difference by rewriting Eq. (32–15), using the identity $\cos(A - 90°) = \sin A$:

$$v_C = \frac{I}{\omega C} \cos(\omega t - 90°). \tag{32–16}$$

This corresponds to a phase angle $\phi = -90°$. This cosine function has a "late start" of 90° compared with the current $i = I \cos \omega t$.

Equations (32–15) and (32–16) show that the *maximum* voltage V_C (the voltage amplitude) is

$$V_C = \frac{I}{\omega C}. \tag{32-17}$$

To put this expression in a form similar to Eq. (32–7) for a resistor, $V_R = IR$, we define a quantity X_C, called the **capacitive reactance** of the capacitor, as

$$X_C = \frac{1}{\omega C} \quad \text{(capacitive reactance)}. \tag{32-18}$$

Then

$$V_C = IX_C \quad \text{(amplitude of voltage across a capacitor, ac circuit).} \tag{32-19}$$

The SI unit of X_C is one ohm, the same as for resistance and inductive reactance, because X_C is the ratio of a voltage and a current.

The capacitive reactance of a capacitor is inversely proportional both to the capacitance C and to the angular frequency ω; the greater the capacitance and the higher the frequency, the *smaller* is the capacitive reactance X_C. Capacitors tend to pass high-frequency current and to block low-frequency currents and dc, just the opposite of inductors. A device that preferentially passes signals of high frequency is called a *high-pass filter* (see Problem 32–39).

EXAMPLE 32-3

A resistor and a capacitor in an ac circuit A 200-Ω resistor is connected in series with a 5.0-μF capacitor. The voltage across the resistor is $v_R = (1.20 \text{ V}) \cos (2500 \text{ rad/s})t$. a) Derive an expression for the circuit current. b) Determine the capacitive reactance of the capacitor. c) Derive an expression for the voltage v_C across the capacitor.

SOLUTION a) Circuit elements in series have the same current. Using $v_R = iR$, we find that the current i in the resistor, and hence the current through the circuit as a whole, is

$$i = \frac{v_R}{R} = \frac{(1.20 \text{ V}) \cos (2500 \text{ rad/s})t}{200 \ \Omega}$$

$$= (6.0 \times 10^{-3} \text{ A}) \cos (2500 \text{ rad/s})t.$$

b) To find the capacitive reactance X_C, we use Eq. (32–18), with $\omega = 2500$ rad/s:

$$X_C = \frac{1}{\omega C} = \frac{1}{(2500 \text{ rad/s})(5.0 \times 10^{-6} \text{ F})} = 80 \ \Omega.$$

c) From Eq. (32–19) the amplitude V_C of the voltage across the capacitor is

$$V_C = IX_C = (6.0 \times 10^{-3} \text{ A})(80 \ \Omega) = 0.48 \text{ V}.$$

Although the *current* is the same in the resistor and capacitor, their *voltage* amplitudes differ. The 80-Ω reactance of the capacitor is 40% of the resistor's 200-Ω resistance, so the value of V_C is 40% of V_R. The instantaneous capacitor voltage v_C is given by Eq. (32–16):

$$v_C = V_C \cos(\omega t - 90°) = (0.48 \text{ V}) \cos[(2500 \text{ rad/s})t - \pi/2 \text{ rad}].$$

We converted the 90° to $\pi/2$ rad so that all the angular quantities have the same units. In ac circuit analysis, phase angles are often given in degrees, so be careful to convert to radians when necessary.

Table 32–1 summarizes the relations of voltage and current amplitudes for the three circuit elements we have discussed.

TABLE 32–1

CIRCUIT ELEMENTS WITH ALTERNATING CURRENT

CIRCUIT ELEMENT	AMPLITUDE RELATION	CIRCUIT QUANTITY	PHASE OF v
Resistor	$V_R = IR$	R	In phase with i
Inductor	$V_L = IX_L$	$X_L = \omega L$	Leads i by 90°
Capacitor	$V_C = IX_C$	$X_C = 1/\omega C$	Lags i by 90°

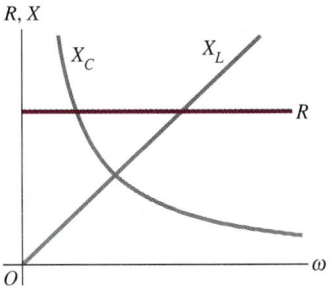

32–6 Graphs of R, X_L, and X_C as functions of angular frequency ω.

CAUTION ▶ Remember that the relations $V_L = IX_L = I\omega L$ for an inductor and $V_C = IX_C = I/\omega C$ for a capacitor express the proportionality between the *amplitudes* of the voltage and current. The *instantaneous* values of voltage v and current i in these devices are *not* proportional! This is because there is a 90° phase difference between the voltage and current in both an inductor (Fig. 32–4b) and a capacitor (Fig. 32–5b). Instantaneous voltage and current *are* proportional in a resistor because the phase difference between v_R and i in a resistor is zero (Fig. 32–3b). ◀

Figure 32–6 shows how the resistance of a resistor and the reactances of an inductor and a capacitor vary with angular frequency ω. Resistance R is independent of frequency, while the reactances X_L and X_C are not. If $\omega = 0$, corresponding to a dc circuit, there is *no* current through a capacitor because $X_C \rightarrow \infty$, and there is no inductive effect because $X_L = 0$. In the limit $\omega \rightarrow \infty$, X_L also approaches infinity, and the current through an inductor becomes vanishingly small; recall that the self-induced emf opposes rapid changes in current. In this same limit, X_C and the voltage across a capacitor both approach zero; the current changes direction so rapidly that no charge can build up on either plate.

Figure 32–7 shows an application of the above discussion to a loudspeaker system. Low-frequency sounds are produced by the *woofer,* which is a speaker with large diameter; the *tweeter,* a speaker of smaller diameter, produces high-frequency sounds. In order to route signals of different frequency to the appropriate speaker, the woofer and tweeter are connected in parallel across the amplifier output. The capacitor in the tweeter branch blocks the low-frequency components of sound but passes the higher frequencies; the inductor in the woofer branch does the opposite.

32–4 THE *L-R-C* SERIES CIRCUIT

Many ac circuits used in practical electronic systems involve resistance, inductive reactance, and capacitive reactance. A simple example is a series circuit containing a resistor,

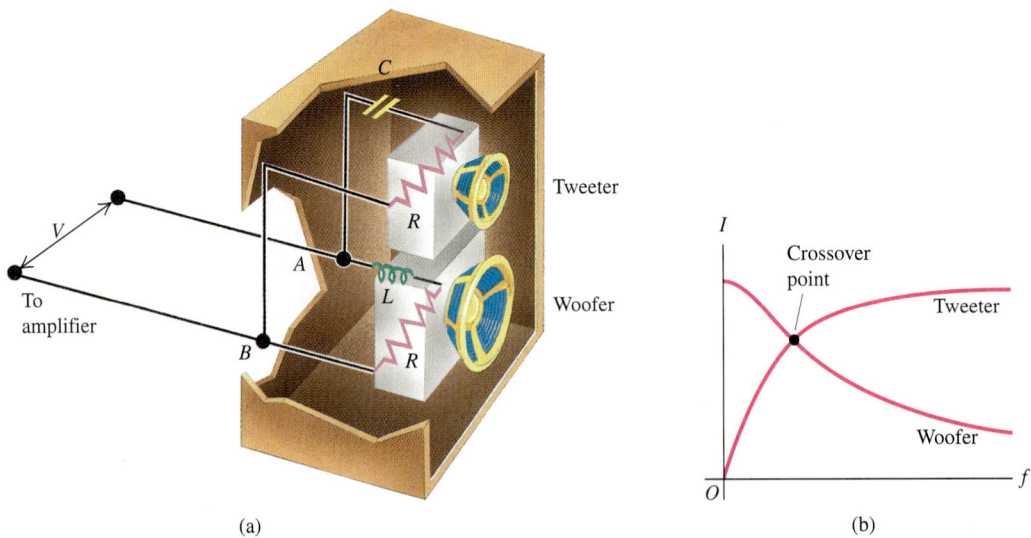

(a)

(b)

32–7 (a) A crossover network in a loudspeaker system. The two speakers are connected in parallel to the amplifier; the inductance and capacitance feed the lower frequencies predominantly to the woofer, the higher frequencies to the tweeter. (b) Graphs of current amplitude in the tweeter and woofer as functions of frequency for a given amplifier voltage amplitude. The point where the two curves cross is called the crossover point.

an inductor, a capacitor, and an ac source, as shown in Fig. 32–8a. (In Section 31–7 we considered the behavior of the current in an *L-R-C* series circuit *without* a source.)

To analyze this and similar circuits, we will use a phasor diagram that includes the voltage and current phasors for each of the components. In this circuit, because of Kirchhoff's loop rule, the instantaneous *total* voltage v_{ad} across all three components is equal to the source voltage at that instant. We will show that the phasor representing this total voltage is the *vector sum* of the phasors for the individual voltages. Complete phasor diagrams for this circuit are shown in Figs. 32–8b and 32–8c. These may appear complex, but we'll explain them one step at a time.

Let's assume that the source supplies a current i given by $i = I \cos \omega t$. Because the circuit elements are connected in series, the current at any instant is the same at every point in the circuit. Thus a *single phasor I*, with length proportional to the current amplitude, represents the current in *all* circuit elements.

As in Section 32–3, we use the symbols v_R, v_L, and v_C for the instantaneous voltages across R, L, and C, and the symbols V_R, V_L, and V_C for the maximum voltages. We denote the instantaneous and maximum *source* voltages by v and V. Then, in Fig. 32–8a, $v = v_{ad}$, $v_R = v_{ab}$, $v_L = v_{bc}$, and $v_C = v_{cd}$.

We have shown that the potential difference between the terminals of a resistor is *in phase* with the current in the resistor and that its maximum value V_R is given by Eq. (32–7):

$$V_R = IR.$$

The phasor V_R in Fig. 32–8b, in phase with the current phasor I, represents the voltage across the resistor. Its projection onto the horizontal axis at any instant gives the instantaneous potential difference v_R.

The voltage across an inductor *leads* the current by 90°. Its voltage amplitude is given by Eq. (32–13):

$$V_L = IX_L.$$

The phasor V_L in Fig. 32–8b represents the voltage across the inductor, and its projection onto the horizontal axis at any instant equals v_L.

The voltage across a capacitor *lags* the current by 90°. Its voltage amplitude is given by Eq. (32–19):

$$V_C = IX_C.$$

The phasor V_C in Fig. 32–8b represents the voltage across the capacitor, and its projection onto the horizontal axis at any instant equals v_C.

The instantaneous potential difference v between terminals a and d is equal at every instant to the (algebraic) sum of the potential differences v_R, v_L, and v_C. That is, it equals the sum of the *projections* of the phasors V_R, V_L, and V_C. But the sum of the projections of these phasors is equal to the *projection* of their *vector sum*. So the vector sum V must be the phasor that represents the source voltage v and the instantaneous total voltage v_{ad} across the series of elements.

To form this vector sum, we first subtract the phasor V_C from the phasor V_L. (These two phasors always lie along the same line, with opposite directions.) This gives the phasor $V_L - V_C$. This is always at right angles to the phasor V_R, so from the Pythagorean theorem the magnitude of the phasor V is

$$V = \sqrt{V_R{}^2 + (V_L - V_C)^2} = \sqrt{(IR)^2 + (IX_L - IX_C)^2},$$

or

$$V = I\sqrt{R^2 + (X_L - X_C)^2}. \tag{32–20}$$

(a)

(b)

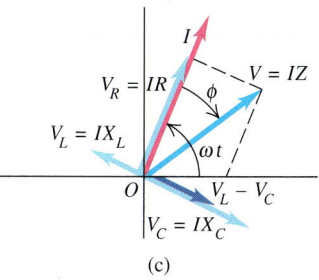

(c)

32-8 (a) An *L-R-C* series circuit with an ac source. (b) Phasor diagram for the case $X_L > X_C$. (c) Phasor diagram for the case $X_C > X_L$.

14.3
AC Circuits: The Driven Oscillator
(Questions 6, 7, and 10)

We define the **impedance** Z of an ac circuit as the ratio of the voltage amplitude across the circuit to the current amplitude in the circuit. From Eq. (32–20) the impedance of the L-R-C series circuit is

$$Z = \sqrt{R^2 + (X_L - X_C)^2},\tag{32–21}$$

so we can rewrite Eq. (32–20) as

$$V = IZ \qquad \text{(amplitude of voltage across an ac circuit).}\tag{32–22}$$

While Eq. (32–21) is valid only for an L-R-C series circuit, we can use Eq. (32–22) to define the impedance of *any* network of resistors, inductors, and capacitors as the ratio of the amplitude of the voltage across the network to the current amplitude. The SI unit of impedance is one ohm.

Equation (32–22) has a form similar to $V = IR$, with impedance Z in an ac circuit playing the role of resistance R in a dc circuit. Note, however, that impedance is actually a function of R, L, and C, as well as of the angular frequency ω. We can see this by substituting Eq. (32–12) for X_L and Eq. (32–18) for X_C into Eq. (32–21), giving the following complete expression for Z for a series circuit:

$$\begin{aligned} Z &= \sqrt{R^2 + (X_L - X_C)^2} \\ &= \sqrt{R^2 + [\omega L - (1/\omega C)]^2} \quad \text{(impedance of an } L\text{-}R\text{-}C \text{ series circuit).}\end{aligned}\tag{32–23}$$

Hence for a given amplitude V of the source voltage applied to the circuit, the amplitude $I = V/Z$ of the resulting current will be different at different frequencies. We'll explore this frequency dependence in detail in Section 32–6.

In the phasor diagram shown in Fig. 32–8b, the angle ϕ between the voltage and current phasors is the phase angle of the source voltage v with respect to the current i; that is, it is the angle by which the source voltage leads the current. From the diagram,

$$\tan \phi = \frac{V_L - V_C}{V_R} = \frac{I(X_L - X_C)}{IR} = \frac{X_L - X_C}{R},$$

$$\tan \phi = \frac{\omega L - 1/\omega C}{R} \quad \text{(phase angle of an } L\text{-}R\text{-}C \text{ series circuit).}\tag{32–24}$$

If the current is $i = I \cos \omega t$, then the source voltage v is

$$v = V \cos(\omega t + \phi).$$

Figure 32–8b shows the behavior of a circuit in which $X_L > X_C$. Figure 32–8c shows the behavior when $X_L < X_C$; the voltage phasor V lies on the opposite side of the current phasor I and the voltage *lags* the current. In this case, $X_L - X_C$ is *negative*, $\tan \phi$ is negative, and ϕ is a negative angle between 0 and $-90°$. Since X_L and X_C depend on frequency, the phase angle ϕ depends on frequency as well. We'll examine the consequences of this in Section 32–6.

All of the expressions that we've developed for an L-R-C series circuit are still valid if one of the circuit elements is missing. If the resistor is missing, we set $R = 0$; if the inductor is missing, we set $L = 0$. But if the capacitor is missing, we set $C = \infty$, corresponding to the absence of any potential difference ($v_C = q/C = 0$) or any capacitive reactance ($X_C = 1/\omega C = 0$).

In this entire discussion we have described magnitudes of voltages and currents in terms of their *maximum* values, the voltage and current *amplitudes*. But we remarked at the end of Section 32–2 that these quantities are usually described in terms of rms values, not amplitudes. For any sinusoidally varying quantity the rms value is always $1/\sqrt{2}$ times the amplitude. All the relations between voltage and current that we have derived in this and the preceding sections are still valid if we use rms quantities throughout

This gas-filled glass sphere has an alternating voltage between its surface and the electrode at its center. The glowing streamers show the resulting alternating current that passes through the gas. When you touch the outside of the sphere, your fingertips and the inner surface of the sphere act as the plates of a capacitor, and the sphere and your body together form an L-R-C series circuit. The current (which is low enough to be harmless) is drawn to your fingers because the path through your body has a low impedance.

instead of amplitudes. For example, if we divide Eq. (32–22) by $\sqrt{2}$, we get

$$\frac{V}{\sqrt{2}} = \frac{I}{\sqrt{2}} Z,$$

which we can rewrite as

$$V_{\text{rms}} = I_{\text{rms}}Z. \qquad (32\text{–}25)$$

We can translate Eqs. (32–7), (32–13), and (32–19) in exactly the same way.

We have considered only ac circuits in which an inductor, a resistor, and a capacitor are in series. You can do a similar analysis for a *parallel L-R-C* circuit; see Problem 32–44.

Finally, we remark that in this section we have been describing the *steady-state* condition of a circuit, the state that exists after the circuit has been connected to the source for a long time. When the source is first connected, there may be additional voltages and currents, called *transients,* whose nature depends on the time in the cycle when the circuit is initially completed. A detailed analysis of transients is beyond our scope. They always die out after a sufficiently long time, and they do not affect the steady-state behavior of the circuit. But they can cause dangerous and damaging surges in power lines, which is why delicate electronic systems such as computers are often provided with power-line surge protectors.

Problem–Solving Strategy

ALTERNATING-CURRENT CIRCUITS

1. In ac circuit problems it is nearly always easiest to work with angular frequency ω. If you are given the ordinary frequency f, expressed in Hz, don't forget to convert it using $\omega = 2\pi f$.

2. Keep in mind a few basic facts about phase relationships. For a resistor, voltage and current are always *in phase,* and the two corresponding phasors in a phasor diagram always have the same direction. For an inductor the voltage always *leads* the current by 90° (that is, $\phi = +90°$), and the voltage phasor is always turned 90° counterclockwise from the current phasor. For a capacitor the voltage always *lags* the current by 90° (that is, $\phi = -90°$), and the voltage phasor is always turned 90° clockwise from the current phasor.

3. Remember that with ac circuits, all the voltages and currents are sinusoidal functions of time instead of being constant, but Kirchhoff's rules hold nonetheless at each instant. Thus in a series circuit the instantaneous current is the same in all circuit elements; in a parallel circuit the instantaneous potential difference is the same across all circuit elements.

4. Inductive reactance, capacitive reactance, and impedance are analogous to resistance; each represents the ratio of voltage amplitude V to current amplitude I in a circuit element or combination of elements. But keep in mind that phase relations play an essential role; the effects of resistance and reactance have to be combined by *vector* addition of the corresponding voltage phasors, as in Figs. 32–8b and 32–8c. When you have several circuit elements in series, for example, you can't just *add* all the numerical values of resistance and reactance to get the impedance; that would ignore the phase relations.

EXAMPLE 32–4

An *L-R-C* series circuit In the series circuit of Fig. 32–8a, suppose $R = 300\ \Omega$, $L = 60$ mH, $C = 0.50\ \mu$F, $V = 50$ V, and $\omega = 10{,}000$ rad/s. Find the reactances X_L and X_C, the impedance Z, the current amplitude I, the phase angle ϕ, and the voltage amplitude across each circuit element.

SOLUTION From Eqs. (32–12) and (32–18),

$$X_L = \omega L = (10{,}000\ \text{rad/s})(60\ \text{mH}) = 600\ \Omega,$$

$$X_C = \frac{1}{\omega C} = \frac{1}{(10{,}000\ \text{rad/s})(0.50 \times 10^{-6}\ \text{F})} = 200\ \Omega.$$

The impedance Z of the circuit is

$$Z = \sqrt{R^2 + (X_L - X_C)^2} = \sqrt{(300\ \Omega)^2 + (600\ \Omega - 200\ \Omega)^2}$$
$$= 500\ \Omega.$$

With source voltage amplitude $V = 50$ V the current amplitude is

$$I = \frac{V}{Z} = \frac{50\ \text{V}}{500\ \Omega} = 0.10\ \text{A}.$$

The phase angle ϕ is

$$\phi = \arctan \frac{X_L - X_C}{R} = \arctan \frac{400\ \Omega}{300\ \Omega} = 53°.$$

Because ϕ is positive, we know that the voltage *leads* the current by 53°; this is like the situation shown in Fig. 32–8b. From Eq. (32–7) the voltage amplitude V_R across the resistor is

$$V_R = IR = (0.10 \text{ A})(300 \ \Omega) = 30 \text{ V}.$$

From Eq. (32–13) the voltage amplitude V_L across the inductor is

$$V_L = IX_L = (0.10 \text{ A})(600 \ \Omega) = 60 \text{ V}.$$

From Eq. (32–19) the voltage amplitude V_C across the capacitor is

$$V_C = IX_C = (0.10 \text{ A})(200 \ \Omega) = 20 \text{ V}.$$

Note that the source voltage amplitude $V = 50$ V is *not* equal to the sum of the voltage amplitudes across the separate circuit elements. (That is, 50 V $\neq$ 30 V + 60 V + 20 V.) Make sure you understand why not!

EXAMPLE 32–5

Instantaneous current and voltages in an *L-R-C* series circuit
For the *L-R-C* series circuit described in Example 32–4, describe the time dependence of the instantaneous current and each instantaneous voltage.

SOLUTION The current and all of the voltages oscillate with the same angular frequency, $\omega = 10,000$ rad/s, and hence with the same period, $2\pi/\omega = 2\pi/(10,000 \text{ rad/s}) = 6.3 \times 10^{-4}$ s = 0.63 ms. We'll describe the current by Eq. (32–2),

$$i = I \cos \omega t = (0.10 \text{ A}) \cos(10,000 \text{ rad/s})t;$$

this choice simply means that we choose $t = 0$ to be an instant when the current is maximum. The resistor voltage is *in phase* with the current, so

$$v_R = V_R \cos \omega t = (30 \text{ V}) \cos(10,000 \text{ rad/s})t.$$

The inductor voltage *leads* the current by 90°, so

$$v_L = V_L \cos(\omega t + 90°) = -V_L \sin \omega t$$
$$= -(60 \text{ V}) \sin(10,000 \text{ rad/s})t.$$

The capacitor voltage *lags* the current by 90°, so

$$v_C = V_C \cos(\omega t - 90°) = V_C \sin \omega t = (20 \text{ V}) \sin(10,000 \text{ rad/s})t.$$

Finally, the source voltage (equal to the voltage across the entire combination of resistor, inductor, and capacitor) *leads* the current by $\phi = 53°$, so

$$v = V \cos(\omega t + \phi)$$

$$= (50 \text{ V}) \cos\left[(10,000 \text{ rad/s})t + \left(\frac{2\pi \text{ rad}}{360°} \right)(53°) \right]$$

$$= (50 \text{ V}) \cos[(10,000 \text{ rad/s})t + 0.93 \text{ rad}].$$

These functions are graphed in Fig. 32–9. The inductor voltage has a larger amplitude than the capacitor voltage, because $X_L > X_C$. While the source voltage amplitude V is not equal to the sum of the individual voltage amplitudes V_R, V_L, and V_C, the *instantaneous* source voltage v is always equal to the sum of the instantaneous voltages v_R, v_L, and v_C. You should verify this by measuring the values of the voltages shown in the graph at different values of the time t.

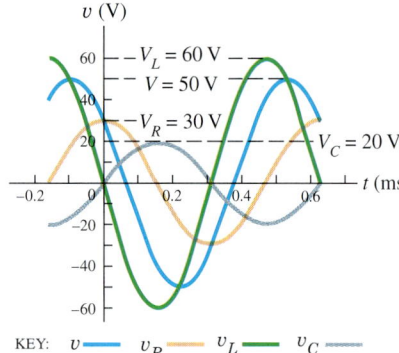

32–9 Graphs of the source voltage v, resistor voltage v_R, inductor voltage v_L, and capacitor voltage v_C as functions of time for the situation of Example 32–4. The current, which is not shown, is in phase with the resistor voltage.

32–5 POWER IN ALTERNATING-CURRENT CIRCUITS

Alternating currents play a central role in systems for distributing, converting, and using electrical energy, so it's important to look at power relationships in ac circuits. For an ac circuit with instantaneous current i and current amplitude I, we'll consider an element of that circuit across which the instantaneous potential difference is v with voltage amplitude V. The instantaneous power p delivered to this circuit element is

$$p = vi.$$

Let's first see what this means for individual circuit elements. We'll assume in each case that $i = I \cos \omega t$.

Suppose first that the circuit element is a *pure resistance R,* as in Fig. 32–3; then $v = v_R$ and i are *in phase.* We obtain the graph representing p by multiplying the heights

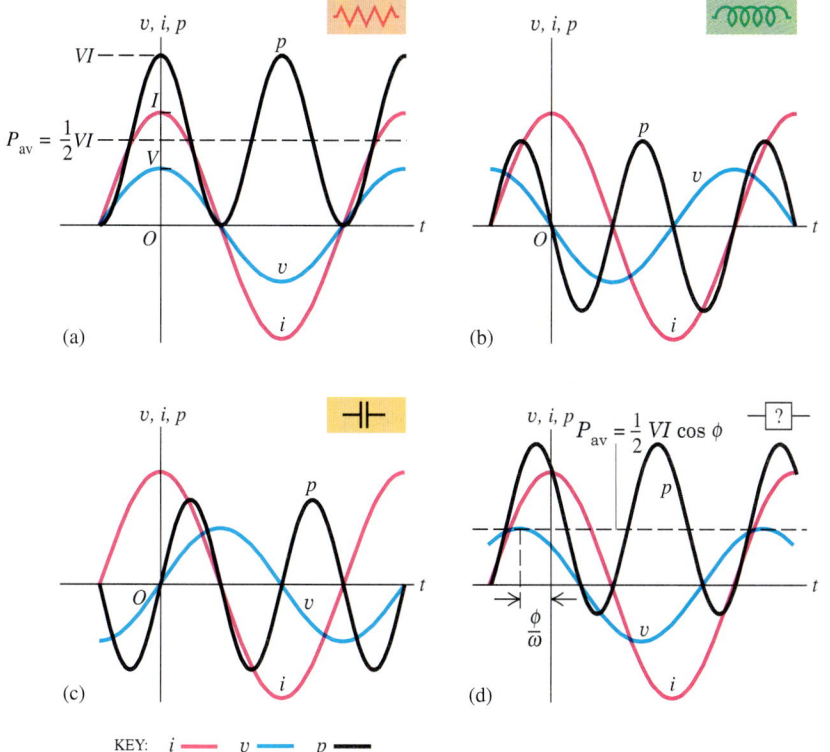

32-10 Graphs of voltage (blue), current (red), and power (black) as functions of time for various circuits. (a) Instantaneous power input to a pure resistor. The average power is $\frac{1}{2}VI = V_{rms}I_{rms}$. (b) Instantaneous power input to a pure inductor. The average power is zero. (c) Instantaneous power input to a pure capacitor. The average power is zero. (d) Instantaneous power input to an arbitrary ac circuit. The average power is $\frac{1}{2}VI\cos\phi = V_{rms}I_{rms}\cos\phi$.

of the graphs of v and i in Fig. 32–3b at each instant. This graph is shown by the black curve in Fig. 32–10a. The product vi is always positive because v and i are always either both positive or both negative. Hence energy is supplied *to* the resistor at every instant for both directions of i, although the power is not constant.

The power curve for a pure resistance is symmetrical about a value equal to one half its maximum value VI, so the *average power* P_{av} is

$$P_{av} = \frac{1}{2}VI \qquad \text{(for a pure resistance).} \qquad (32\text{–}26)$$

An equivalent expression is

$$P_{av} = \frac{V}{\sqrt{2}}\frac{I}{\sqrt{2}} = V_{rms}I_{rms} \qquad \text{(for a pure resistance).} \qquad (32\text{–}27)$$

Also, $V_{rms} = I_{rms}R$, so we can express P_{av} by any of the equivalent forms

$$P_{av} = I_{rms}{}^2 R = \frac{V_{rms}{}^2}{R} = V_{rms}I_{rms} \qquad \text{(for a pure resistance).} \qquad (32\text{–}28)$$

Note that the expressions in Eq. (32–28) have the same form as the corresponding relations for a dc circuit, Eq. (26–18). Also note that they are valid only for pure resistors, not for more complicated combinations of circuit elements.

Next we connect the source to a pure inductor L, as in Fig. 32–4. The voltage $v = v_L$ leads the current i by 90°. When we multiply the curves of v and i, the product vi is *negative* during the half of the cycle when v and i have *opposite* signs. The power curve, shown in Fig. 32–10b, is symmetrical about the horizontal axis; it is positive half the time and negative the other half, and the average power is zero. When p is positive, energy is being supplied to set up the magnetic field in the inductor; when p is negative, the field is collapsing and the inductor is returning energy to the source. The net energy transfer over one cycle is zero.

Finally, we connect the source to a pure capacitor C, as in Fig. 32–5. The voltage $v = v_C$ lags the current i by 90°. Figure 32–10c shows the power curve; the average power is again zero. Energy is supplied to charge the capacitor and is returned to the source when the capacitor discharges. The net energy transfer over one cycle is again zero.

In *any* ac circuit, with any combination of resistors, capacitors, and inductors, the voltage v across the entire circuit has some phase angle ϕ with respect to the current i. Then the instantaneous power p is given by

$$p = vi = [V \cos(\omega t + \phi)][I \cos \omega t]. \qquad (32\text{--}29)$$

The instantaneous power curve has the form shown in Fig. 32–10d. The area between the positive loops and the horizontal axis is greater than that between the negative loops and the horizontal axis, and the average power is positive.

We can derive from Eq. (32–29) an expression for the *average* power P_{av} by using the identity for the cosine of the sum of two angles:

$$p = [V(\cos \omega t \cos \phi - \sin \omega t \sin \phi)][I \cos \omega t]$$

$$= VI \cos \phi \cos^2 \omega t - VI \sin \phi \cos \omega t \sin \omega t.$$

From the discussion in Section 32–2 that led to Eq. (32–4), we see that the average value of $\cos^2 \omega t$ (over one cycle) is 1/2. The average value of $\cos \omega t \sin \omega t$ is zero because this product is equal to $\frac{1}{2} \sin 2\omega t$, whose average over a cycle is zero. So the average power P_{av} is

$$P_{av} = \frac{1}{2} VI \cos \phi = V_{rms} I_{rms} \cos \phi \qquad \begin{matrix}\text{(average power into a} \\ \text{general ac circuit).}\end{matrix} \qquad (32\text{--}30)$$

When v and i are in phase, so $\phi = 0$, the average power equals $\frac{1}{2} VI = V_{rms} I_{rms}$; when v and i are 90° out of phase, the average power is zero. In the general case, when v has a phase angle ϕ with respect to i, the average power equals $\frac{1}{2} I$ multiplied by $V \cos \phi$, the component of the voltage phasor that is *in phase* with the current phasor. Figure 32–11 shows the general relationship of the current and voltage phasors. For the *L-R-C* series circuit, Figs. 32–8b and 32–8c show that $V \cos \phi$ equals the voltage amplitude V_R for the resistor; hence Eq. (32–30) is the average power dissipated in the resistor. On average there is no energy flow into or out of the inductor or capacitor, so none of P_{av} goes into either of these circuit elements.

The factor $\cos \phi$ is called the **power factor** of the circuit. For a pure resistance, $\phi = 0$, $\cos \phi = 1$, and $P_{av} = V_{rms} I_{rms}$. For a pure inductor or capacitor, $\phi = \pm 90°$, $\cos \phi = 0$, and $P_{av} = 0$. For an *L-R-C* series circuit the power factor is equal to R/Z; we leave the proof of this statement as a problem.

A low power factor (large angle ϕ of lag or lead) is usually undesirable in power circuits. The reason is that for a given potential difference, a large current is needed to supply a given amount of power. This results in large $i^2 R$ losses in the transmission lines. Your electric power company may charge a higher rate to a client with a low power factor. Many types of ac machinery draw a *lagging* current; that is, the current drawn by the machinery lags the applied voltage; hence the voltage leads the current, so $\phi > 0$ and $\cos \phi < 1$. The power factor can be corrected toward the ideal value of 1 by connecting a capacitor in parallel with the load. The current drawn by the capacitor *leads* the voltage (that is, the voltage across the capacitor lags the current), which compensates for the lagging current in the other branch of the circuit. The capacitor itself absorbs no net power from the line.

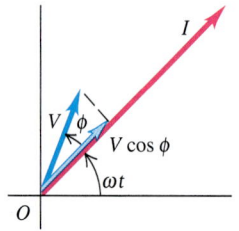

32–11 The average power is half the product of I and the component of V in phase with I.

EXAMPLE 32–6

Power in a hair dryer An electric hair dryer (Fig. 32–12) is rated at 1500 W at 120 V. The rated power of this hair dryer, or of any other ac device, is the *average* power drawn by the device, and the rated voltage is the *rms* voltage. Calculate a) the resistance, b) the rms current, and c) the maximum instantaneous power. Assume that the hair dryer is a pure resistance.

SOLUTION a) Since the hair dryer is assumed to be a pure resistance, we solve Eq. (32–28) for R and substitute the given values:

$$R = \frac{V_{rms}^2}{P_{av}} = \frac{(120 \text{ V})^2}{1500 \text{ W}} = 9.6 \ \Omega.$$

b) From Eq. (32–27),

$$I_{rms} = \frac{P_{av}}{V_{rms}} = \frac{1500 \text{ W}}{120 \text{ V}} = 12.5 \text{ A}.$$

Alternatively, from Eq. (32–7),

$$I_{rms} = \frac{V_{rms}}{R} = \frac{120 \text{ V}}{9.6 \ \Omega} = 12.5 \text{ A}.$$

c) The maximum instantaneous power for a pure resistance is VI; from Eq. (32–26),

$$VI = 2P_{av} = 2(1500 \text{ W}) = 3000 \text{ W}.$$

Some manufacturers of stereo amplifiers state power outputs in terms of the peak value rather than the lower average value, to mislead the unwary consumer.

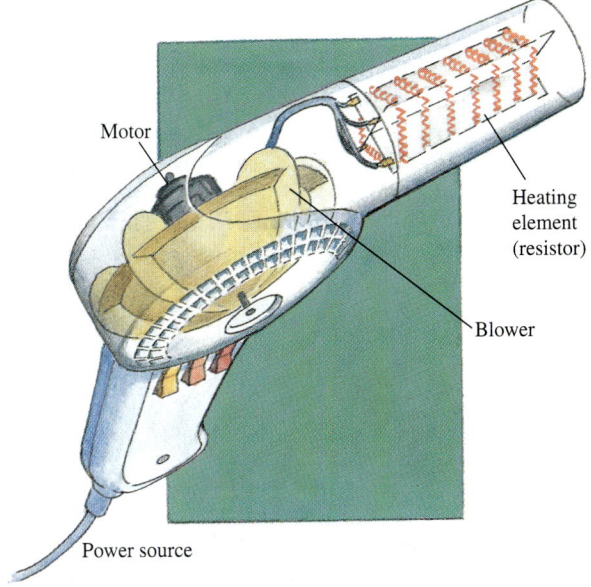

32–12 The average power delivered to this hair dryer is equal to V_{rms}^2/R.

EXAMPLE 32–7

Power in an *L-R-C* series circuit For the *L-R-C* series circuit of Example 32–4, a) calculate the power factor; b) calculate the average power delivered to the entire circuit and to each circuit element.

SOLUTION a) The power factor is $\cos \phi = \cos 53° = 0.60$.
b) From Eq. (32–30) the average power delivered to the

circuit is

$$P_{av} = \frac{1}{2} VI \cos \phi = \frac{1}{2} (50 \text{ V})(0.10 \text{ A})(0.60) = 1.5 \text{ W}.$$

All of this power is dissipated in the resistor; the average power delivered to a pure inductor or capacitor is always zero.

32–6 RESONANCE IN ALTERNATING-CURRENT CIRCUITS

Much of the practical importance of *L-R-C* series circuits arises from the way in which such circuits respond to sources of different angular frequency ω. For example, one type of tuning circuit used in radio receivers is simply an *L-R-C* series circuit. A radio signal of any given frequency will produce a current of the same frequency in the receiver circuit, but the amplitude of the current will be *greatest* if the signal frequency equals the particular frequency to which the receiver circuit is "tuned." This effect is called *resonance*. The circuit is designed so that signals at other than the tuned frequency produce currents that are too small to make an audible sound come out of the radio's speakers.

To see how an *L-R-C* series circuit can be used in this way, suppose we connect an ac source with constant voltage amplitude V but adjustable angular frequency ω across an *L-R-C* series circuit. The current that appears in the circuit has the same

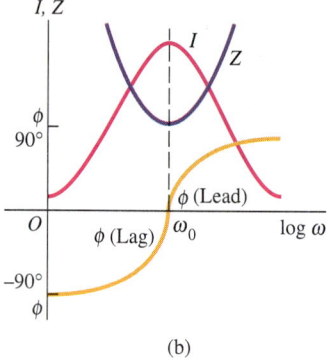

32–13 (a) Reactance, resistance, and impedance as functions of angular frequency (logarithmic angular frequency scale). (b) Impedance, current amplitude, and phase angle as functions of angular frequency (logarithmic angular frequency scale).

14.3

AC Circuits: The Driven Oscillator (Questions 8, 9, and 11)

angular frequency as the source and a current amplitude $I = V/Z$, where Z is the impedance of the L-R-C series circuit. This impedance depends on the frequency, as Eq. (32–23) shows. Figure 32–13a shows graphs of R, X_L, X_C, and Z as functions of ω. We have used a logarithmic angular frequency scale so that we can cover a wide range of frequencies. As the frequency increases, X_L increases and X_C decreases; hence there is always one frequency at which X_L and X_C are equal and $X_L - X_C$ is zero. At this frequency the impedance $Z = \sqrt{R^2 + (X_L - X_C)^2}$ has its *smallest* value, equal simply to the resistance R.

As we vary the angular frequency ω of the source, the current amplitude $I = V/Z$ varies as shown in Fig. 32–13b; the *maximum* value of I occurs at the frequency at which the impedance Z is *minimum*. This peaking of the current amplitude at a certain frequency is called **resonance**. The angular frequency ω_0 at which the resonance peak occurs is called the **resonance angular frequency**. This is the angular frequency at which the inductive and capacitive reactances are equal, so at resonance,

$$X_L = X_C, \quad \omega_0 L = \frac{1}{\omega_0 C}, \quad \omega_0 = \frac{1}{\sqrt{LC}} \quad (L\text{-}R\text{-}C \text{ series circuit at resonance}).$$

$$(32\text{–}31)$$

Note that this is equal to the natural angular frequency of oscillation of an L-C circuit, which we derived in Section 31–6, Eq. (31–22). The **resonance frequency** f_0 is $\omega_0/2\pi$. This is the frequency at which the greatest current appears in the circuit for a given source voltage amplitude; in other words, f_0 is the frequency to which the circuit is "tuned."

It's instructive to look at what happens to the *voltages* in an L-R-C series circuit at resonance. The current at any instant is the same in L and C. The voltage across an inductor always *leads* the current by 90°, or 1/4 cycle, and the voltage across a capacitor always *lags* the current by 90°. Therefore the instantaneous voltages across L and C always differ in phase by 180°, or 1/2 cycle; they have opposite signs at each instant. At the resonance frequency, and *only* at the resonance frequency, $X_L = X_C$ and the voltage amplitudes $V_L = IX_L$ and $V_C = IX_C$ are *equal;* then the instantaneous voltages across L and C add to zero at each instant, and the *total* voltage v_{bd} across the L-C combination in Fig. 32–8a is exactly zero. The voltage across the resistor is then equal to the source voltage. So at the resonance frequency the circuit behaves as if the inductor and capacitor weren't there at all!

The *phase* of the voltage relative to the current is given by Eq. (32–24). At frequencies below resonance, X_C is greater than X_L; the capacitive reactance dominates, the voltage *lags* the current, and the phase angle ϕ is between zero and −90°. Above resonance the inductive reactance dominates; the voltage *leads* the current, and the phase angle is between zero and +90°. This variation of ϕ with angular frequency is shown in Fig. 32–13b.

If we can vary the inductance L or the capacitance C of a circuit, we can also vary the resonance frequency. This is exactly how a radio or television receiving set is "tuned" to receive a particular station. In the early days of radio this was accomplished by use of capacitors with movable metal plates whose overlap could be varied to change C. A more modern approach is to vary L by using a coil with a ferrite core that slides in or out.

In a series L-R-C circuit the impedance reaches its minimum value and the current its maximum value at the resonance frequency. The red curve in Fig. 32–14 is a graph of current as a function of frequency for such a circuit, with source voltage amplitude $V = 100$ V, $L = 2.0$ H, $C = 0.50$ μF, and $R = 500$ Ω. This curve is called a *response curve* or a *resonance curve*. The resonance angular frequency is $\omega_0 = (LC)^{-1/2} = 1000$ rad/s. As we expect, the curve has a peak at this angular frequency.

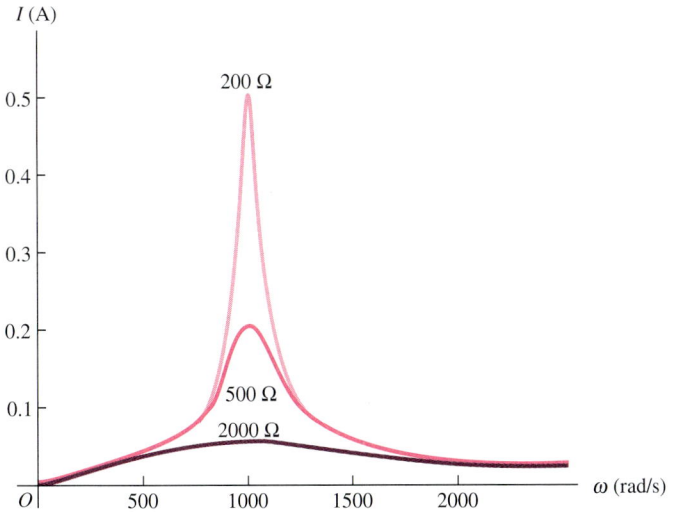

32-14 Graph of current amplitude I as a function of angular frequency ω for an L-R-C series circuit with $V = 100$ V, $L = 2.0$ H, $C = 0.50$ μF, and $R = 500$ Ω (the red curve). The other curves show the relationship for different values of the resistance, $R = 2000$ Ω (dark red curve) and $R = 200$ Ω (light red curve).

The resonance frequency is determined by L and C; what happens when we change R? Figure 32–14 also shows graphs of I as a function of ω for $R = 200$ Ω and for $R = 2000$ Ω. The curves are similar for frequencies far away from resonance, where the impedance is dominated by X_L or X_C. But near resonance, where X_L and X_C nearly cancel each other, the curve is higher and more sharply peaked for small values of R and broader and flatter for large values of R. At resonance, $Z = R$ and $I = V/R$, so the maximum height of the curve is inversely proportional to R.

The shape of the response curve is important in the design of radio and television receiving circuits. The sharply peaked curve is what makes it possible to discriminate between two stations broadcasting on adjacent frequency bands. But if the peak is *too* sharp, some of the information in the received signal is lost, such as the high-frequency sounds in music. The shape of the resonance curve is also related to the overdamped and underdamped oscillations that we described in Section 31–7. A sharply peaked resonance curve corresponds to a small value of R and a lightly damped oscillating system; a broad, flat curve goes with a large value of R and a heavily damped system.

In this section we have discussed resonance in an L-R-C *series* circuit. Resonance can also occur in an ac circuit in which the inductor, resistor, and capacitor are connected in *parallel*. We leave the details to you (see Problem 32–45).

Resonance phenomena occur not just in ac circuits, but in all areas of physics. We discussed examples of resonance in *mechanical* systems in Sections 13–9 and 20–7. The amplitude of a mechanical oscillation peaks when the driving-force frequency is close to a natural frequency of the system; this is analogous to the peaking of the current in an L-R-C series circuit. We suggest that you review the sections on mechanical resonance now, looking for the analogies. Other important examples of resonance occur in atomic and nuclear physics and in the study of fundamental particles (high-energy physics).

EXAMPLE 32-8

Tuning a radio The series circuit in Fig. 32–15 is similar to arrangements that are sometimes used in radio tuning circuits. This circuit is connected to the terminals of an ac source with a constant rms terminal voltage of 1.0 V and a variable frequency.

Find a) the resonance frequency; b) the inductive reactance, the capacitive reactance, and the impedance at the resonance frequency; c) the rms current at resonance; and d) the rms voltage across each circuit element at resonance.

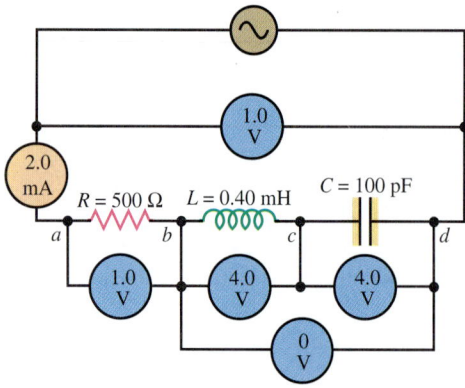

32-15 A radio tuning circuit at resonance.

SOLUTION a) The resonance angular frequency is

$$\omega_0 = \frac{1}{\sqrt{LC}} = \frac{1}{\sqrt{(0.40 \times 10^{-3} \ \mathrm{H})(100 \times 10^{-12} \ \mathrm{F})}}$$

$$= 5.0 \times 10^6 \ \mathrm{rad/s}.$$

The corresponding frequency $f_0 = \omega_0/2\pi$ is

$$f_0 = 8.0 \times 10^5 \ \mathrm{Hz} = 800 \ \mathrm{kHz}.$$

This corresponds to the lower part of the AM radio band.

b) At this frequency,

$$X_L = \omega L = (5.0 \times 10^6 \ \mathrm{rad/s})(0.40 \times 10^{-3} \ \mathrm{H}) = 2000 \ \Omega,$$

$$X_C = \frac{1}{\omega C} = \frac{1}{(5.0 \times 10^6 \ \mathrm{rad/s})(100 \times 10^{-12} \ \mathrm{F})} = 2000 \ \Omega.$$

Since $X_L = X_C$ and $X_L - X_C = 0$, Eq. (32–23) shows that the impedance Z at resonance is equal to the resistance:
$Z = R = 500 \ \Omega$.

c) From Eq. (32–25) the rms current at resonance is

$$I_{\mathrm{rms}} = \frac{V_{\mathrm{rms}}}{Z} = \frac{V_{\mathrm{rms}}}{R} = \frac{1.0 \ \mathrm{V}}{500 \ \Omega} = 0.0020 \ \mathrm{A} = 2.0 \ \mathrm{mA}.$$

d) The rms potential difference across the resistor is

$$V_{R\text{-}\mathrm{rms}} = I_{\mathrm{rms}} R = (0.0020 \ \mathrm{A})(500 \ \Omega) = 1.0 \ \mathrm{V}.$$

The rms potential differences across the inductor and capacitor are, respectively,

$$V_{L\text{-}\mathrm{rms}} = I_{\mathrm{rms}} X_L = (0.0020 \ \mathrm{A})(2000 \ \Omega) = 4.0 \ \mathrm{V},$$

$$V_{C\text{-}\mathrm{rms}} = I_{\mathrm{rms}} X_C = (0.0020 \ \mathrm{A})(2000 \ \Omega) = 4.0 \ \mathrm{V}.$$

The potential differences across the inductor and the capacitor have equal rms values and amplitudes, but are 180° out of phase and so add to zero at each instant. Note also that at resonance, $V_{R\text{-}\mathrm{rms}}$ is equal to the source voltage V_{rms}, while in this example, $V_{L\text{-}\mathrm{rms}}$ and $V_{C\text{-}\mathrm{rms}}$ are both considerably *larger* than V_{rms}.

32-7 TRANSFORMERS

One of the great advantages of ac over dc for electric-power distribution is that it is much easier to step voltage levels up and down with ac than with dc. For long-distance power transmission it is desirable to use as high a voltage and as small a current as possible; this reduces i^2R losses in the transmission lines, and smaller wires can be used, saving on material costs. Present-day transmission lines routinely operate at rms voltages of the order of 500 kV. On the other hand, safety considerations and insulation requirements dictate relatively low voltages in generating equipment and in household and industrial power distribution. The standard voltage for household wiring is 120 V in the United States and Canada and 240 V in many other countries. The necessary voltage conversion is accomplished by the use of **transformers.**

Figure 32–16 shows an idealized transformer. The key components of the transformer are two coils or *windings,* electrically insulated from each other but wound on the same core. The core is typically made of a material, such as iron, with a very large

32-16 Schematic diagram of an idealized step-up transformer. The primary is connected to an ac source with voltage V_1; the secondary is connected to a device with resistance R.

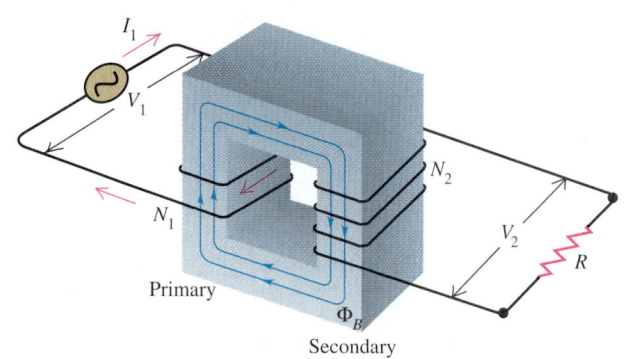

relative permeability K_m. This keeps the magnetic field lines due to a current in one winding almost completely within the core. Hence almost all of these field lines pass through the other winding, maximizing the *mutual inductance* of the two windings (see Section 31–2). The winding to which power is supplied is called the **primary;** the winding from which power is delivered is called the **secondary.** The circuit symbol for a transformer with an iron core, such as those used in power distribution systems, is

Here's how a transformer works. The ac source causes an alternating current in the primary, which sets up an alternating flux in the core; this induces an emf in each winding, in accordance with Faraday's law. The induced emf in the secondary gives rise to an alternating current in the secondary, and this delivers energy to the device to which the secondary is connected. All currents and emf's have the same frequency as the ac source.

Let's see how the voltage across the secondary can be made larger or smaller in amplitude than the voltage across the primary. We neglect the resistance of the windings and assume that all the magnetic field lines are confined to the iron core, so at any instant the magnetic flux Φ_B is the same in each turn of the primary and secondary windings. The primary winding has N_1 turns, and the secondary winding has N_2 turns. When the magnetic flux changes because of changing currents in the two coils, the resulting induced emf's are

$$\mathcal{E}_1 = -N_1 \frac{d\Phi_B}{dt} \quad \text{and} \quad \mathcal{E}_2 = -N_2 \frac{d\Phi_B}{dt}. \tag{32–32}$$

The flux *per turn* Φ_B is the same in both the primary and the secondary, so Eqs. (32–32) show that the induced emf *per turn* is the same in each. The ratio of the secondary emf $\mathcal{E}_2$ to the primary emf $\mathcal{E}_1$ is therefore equal at any instant to the ratio of secondary to primary turns:

$$\frac{\mathcal{E}_2}{\mathcal{E}_1} = \frac{N_2}{N_1}. \tag{32–33}$$

Since $\mathcal{E}_1$ and $\mathcal{E}_2$ both oscillate with the same frequency as the ac source, Eq. (32–33) also gives the ratio of the amplitudes or of the rms values of the induced emf's. If the windings have zero resistance, the induced emf's $\mathcal{E}_1$ and $\mathcal{E}_2$ are equal to the terminal voltages across the primary and the secondary, respectively; hence

$$\frac{V_2}{V_1} = \frac{N_2}{N_1} \quad \text{(terminal voltages of transformer primary and secondary),} \tag{32–34}$$

where V_1 and V_2 are either the amplitudes or the rms values of the terminal voltages. By choosing the appropriate turns ratio N_2/N_1, we may obtain any desired secondary voltage from a given primary voltage. If $N_2 > N_1$, as in Fig. 32–16, then $V_2 > V_1$, and we have a *step-up* transformer; if $N_2 < N_1$, then $V_2 < V_1$, and we have a *step-down* transformer. At a power generating station, step-up transformers are used; the primary is connected to the power source, and the secondary is connected to the transmission lines, giving the desired high voltage for transmission. Near the consumer, step-down transformers lower the voltage to a value suitable for use in home or industry.

If the secondary circuit is completed by a resistance R, then the amplitude or rms value of the current in the secondary circuit is $I_2 = V_2/R$. From energy considerations the power delivered to the primary equals that taken out of the secondary (since there is no resistance in the windings), so

$$V_1 I_1 = V_2 I_2 \quad \text{(currents in transformer primary and secondary).} \tag{32–35}$$

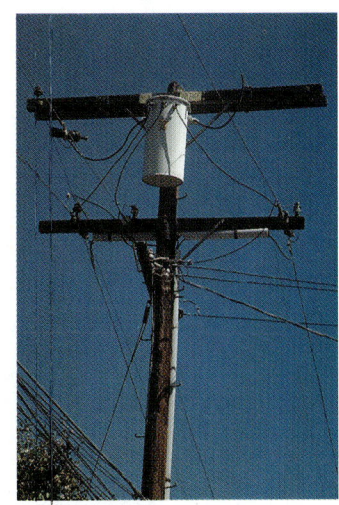

The cylindrical can near the top of this power pole is a step-down transformer. It converts the high-voltage ac in the power lines to low-voltage (120 V) ac, which is then distributed to the surrounding homes and businesses.

We can combine Eqs. (32–34) and (32–35) and the relation $I_2 = V_2/R$ to eliminate V_2 and I_2; we obtain

$$\frac{V_1}{I_1} = \frac{R}{(N_2/N_1)^2}. \qquad (32\text{–}36)$$

This shows that when the secondary circuit is completed through a resistance R, the result is the same as if the *source* had been connected directly to a resistance equal to R divided by the square of the turns ratio, $(N_2/N_1)^2$. In other words, the transformer "transforms" not only voltages and currents, but resistances as well. More generally, we can regard a transformer as "transforming" the *impedance* of the network to which the secondary circuit is completed.

Equation (32–36) has many practical consequences. The power supplied by a source to a resistor depends the resistances of both the resistor and the source. It can be shown that the power transfer is greatest when the two resistances are *equal*. The same principle applies in both dc and ac circuits. When a high-impedance ac source must be connected to a low-impedance circuit, such as an audio amplifier connected to a loudspeaker, the source impedance can be *matched* to that of the circuit by use of a transformer with an appropriate turns ratio N_2/N_1.

Real transformers always have some energy losses. The windings have some resistance, leading to i^2R losses, although superconducting transformers may appear on the horizon in the next few years. There are also energy losses through hysteresis in the core (Section 29–9). Hysteresis losses are minimized by the use of soft iron with a narrow hysteresis loop.

Another important mechanism for energy loss in a transformer core involves eddy currents (Section 30–7). Consider a section AA through an iron transformer core (Fig. 32–17a). Since iron is a conductor, any such section can be pictured as several conducting circuits, one within the other (Fig. 32–17b). The flux through each of these circuits is continually changing, so eddy currents circulate in the entire volume of the core, with lines of flow that form planes perpendicular to the flux. These eddy currents are very undesirable; they waste energy through i^2R heating and themselves set up an opposing flux.

The effects of eddy currents can be minimized by the use of a *laminated* core, that is, one built up of thin sheets or laminae. The large electrical surface resistance of each lamina, due either to a natural coating of oxide or to an insulating varnish, effectively confines the eddy currents to individual laminae (Fig. 32–17c). The possible eddy-current paths are narrower, the induced emf in each path is smaller, and the eddy currents are greatly reduced. The alternating magnetic field exerts forces on the current-carrying lamina that cause them to vibrate back and forth; this vibration causes

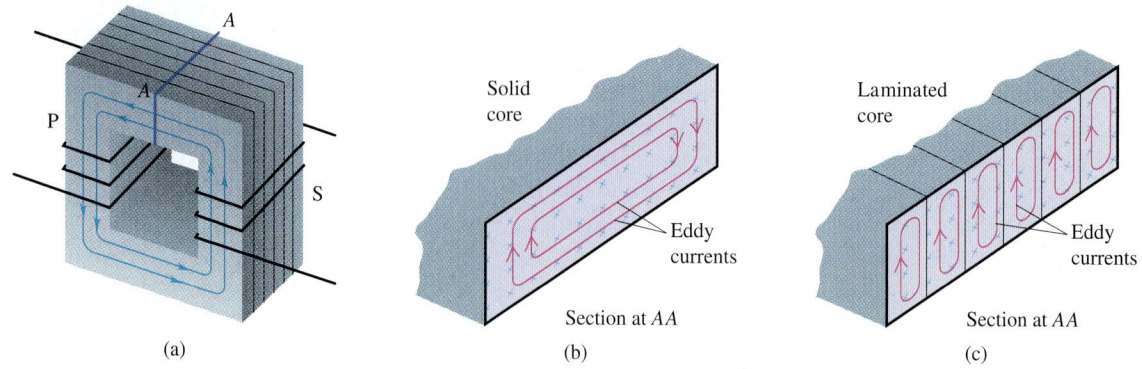

32–17 (a) Primary and secondary windings in a transformer. (b) Eddy currents in the iron core, shown in the cross section at AA. (c) Reduction of eddy currents by use of a laminated core.

32-18 Large transformers at power stations are immersed in tanks of oil for insulation and cooling. The high-voltage coil, low-voltage coil, and connecting core are close together to minimize electrical losses and the cost of materials. For clarity the connections between the bushings and coils are not shown.

the characteristic "hum" of an operating transformer. This type of design is used in small transformers for home appliances as well as in the large transformers used at power stations (Fig. 32–18).

Thanks to the use of soft iron cores and lamination, transformer efficiencies are usually well over 90%; in large installations they may reach 99%.

EXAMPLE 32-9

"Wake up and smell the (transformer)!" A friend brings back from Europe a device that she claims to be the world's greatest coffee maker. Unfortunately, it was designed to operate from a 240-V line to obtain the 960 W of power that it needs. a) What can she do to operate it at 120 V? b) What current will the coffee maker draw from the 120-V line? c) What is the resistance of the coffee maker? (The voltages are rms values.)

SOLUTION a) To get $V_2 = 240$ V with $V_1 = 120$ V, our friend needs a step-up transformer with a turns ratio of $N_2/N_1 = V_2/V_1 = (240 \text{ V})/(120 \text{ V}) = 2$. That is, the secondary coil (connected to the coffee maker) should have twice as many turns as the primary coil (connected to the 120-V line).
b) The rms current I_1 in the 120-V primary is found by using $P_{av} = V_1 I_1$, where P_{av} is the average power drawn by the coffee

maker and hence the power supplied by the 120-V line. (We're assuming that there are no energy losses in the transformer.) Hence $I_1 = P_{av}/V_1 = (960 \text{ W})/(120 \text{ V}) = 8.0$ A. The secondary current is then $I_2 = P_{av}/V_2 = (960 \text{ W})/(240 \text{ V}) = 4.0$ A.
c) We have $V_1 = 120$ V, $I_1 = 8.0$ A, and $N_2/N_1 = 2$, so

$$\frac{V_1}{I_1} = \frac{120 \text{ V}}{8.0 \text{ A}} = 15 \ \Omega.$$

From Eq. (32–36),

$$R = 2^2 (15 \ \Omega) = 60 \ \Omega.$$

As a check, $V_2/R = (240 \text{ V})/(60 \ \Omega) = 4.0$ A $= I_2$, the same value obtained previously. You can also check this result for R by using the expression $P_{av} = V_2^2/R$ for the power drawn by the coffee maker.

SUMMARY

- An alternator or ac source produces an emf that varies sinusoidally with time. A sinusoidal voltage or current can be represented by a phasor, a vector that rotates counterclockwise with constant angular speed ω equal to the angular frequency of the sinusoidal quantity. Its projection on the horizontal axis at any instant represents the instantaneous value of the quantity.

- For a sinusoidal current the rectified average and rms (root-mean-square) currents

KEY TERMS

alternating current (ac), 997
ac source, 997
voltage amplitude, 997
current amplitude, 998
phasor, 998

are related to the current amplitude I by

$$I_{\text{rav}} = \frac{2}{\pi} I = 0.637I, \tag{32-3}$$

$$I_{\text{rms}} = \frac{I}{\sqrt{2}}. \tag{32-4}$$

In the same way, the rms value of a sinusoidal voltage is related to the voltage amplitude V_{rms} by

$$V_{\text{rms}} = \frac{V}{\sqrt{2}}. \tag{32-5}$$

- If the current i in an ac circuit is

$$i = I \cos \omega t,$$

and the voltage v between two points is

$$v = V \cos(\omega t + \phi),$$

then ϕ is called the phase angle of the voltage relative to the current.

- The voltage across a resistor R is in phase with the current, and the voltage and current amplitudes are related by

$$V_R = IR. \tag{32-7}$$

The voltage across an inductor L leads the current by 90°; the voltage and current amplitudes are related by

$$V_L = IX_L, \tag{32-13}$$

where $X_L = \omega L$ is the inductive reactance of the inductor. The voltage across a capacitor C lags the current by 90°; the voltage and current amplitudes are related by

$$V_C = IX_C, \tag{32-19}$$

where $X_C = 1/\omega C$ is the capacitive reactance of the capacitor.

- In an ac circuit the voltage and current amplitudes are related by

$$V = IZ, \tag{32-22}$$

where Z is the impedance of the circuit. In an L-R-C series circuit,

$$Z = \sqrt{R^2 + (X_L - X_C)^2} = \sqrt{R^2 + [\omega L - (1/\omega C)]^2}, \tag{32-23}$$

and the phase angle ϕ of the voltage relative to the current is

$$\tan \phi = \frac{\omega L - 1/\omega C}{R}. \tag{32-24}$$

- The average power input P_{av} to an ac circuit is

$$P_{\text{av}} = \frac{1}{2} VI \cos \phi = V_{\text{rms}} I_{\text{rms}} \cos \phi, \tag{32-30}$$

where ϕ is the phase angle of voltage with respect to current. The quantity $\cos \phi$ is called the power factor.

- In an L-R-C series circuit the current becomes maximum (for a given voltage amplitude) and the impedance becomes minimum at an angular frequency $\omega_0 = 1/(LC)^{1/2}$ called the resonance angular frequency. This phenomenon is called resonance. At resonance the voltage and current are in phase, and the impedance Z is equal to the resistance R.

■ A transformer is used to transform the voltage and current levels in an ac circuit. In an ideal transformer with no energy losses, if the primary winding has N_1 turns and the secondary N_2, the amplitudes (or rms values) of the two voltages are related by

$$\frac{V_2}{V_1} = \frac{N_2}{N_1}. \tag{32-34}$$

The amplitudes (or rms values) of the primary and secondary voltages and currents are related by

$$V_1 I_1 = V_2 I_2. \tag{32-35}$$

DISCUSSION QUESTIONS

Q32–1 Are the equations for the rectified average and rms values of current, Eqs. (32–3) and (32–4), correct when the variation with time is not sinusoidal? Explain.

Q32–2 Some electric-power systems formerly used 25-Hz alternating current instead of the 60 Hz that is now standard. The lights flickered noticeably. Why is this not a problem with 60 Hz?

Q32–3 Household electric power in most of western Europe is supplied at 240 V, rather than the 120 V that is standard in the United States and Canada. What are the advantages and disadvantages of each system?

Q32–4 The current in an ac power line changes direction 120 times per second, and its average value is zero. Explain how it is possible for power to be transmitted in such a system.

Q32–5 Equation (32–14) was derived by using the relation $i = dq/dt$ between the current and the charge on the capacitor. In Fig. 32–5a the positive counterclockwise current increases the charge on the capacitor. When the charge on the left plate is positive but decreasing in time, is $i = dq/dt$ still correct or should it be $i = -dq/dt$? Is $i = dq/dt$ still correct when the right-hand plate has positive charge that is increasing or decreasing in magnitude? Explain.

Q32–6 Fluorescent lights often use an inductor, called a ballast, to limit the current through the tubes. Why is it better to use an inductor rather than a resistor for this purpose?

Q32–7 Equation (32–9) says that $v_{ab} = L\, di/dt$ (Fig. 32–4a). Using Faraday's law, explain why point a is at higher potential than point b when i is in the direction shown in Fig. 32–4a and is increasing in magnitude. What about when i is counterclockwise and decreasing in magnitude, is $v_{ab} = L\, di/dt$ still correct, or should it be $v_{ab} = -L\, di/dt$? Is $v_{ab} = L\, di/dt$ still correct when i is clockwise and increasing or decreasing in magnitude? Explain.

Q32–8 In a series L-R-C circuit, can the instantaneous voltage across the capacitor exceed the source voltage at that same instant? Can this be true for the instantaneous voltage across the inductor? Across the resistor? Explain.

Q32–9 In a series L-R-C circuit, what are the phase angle ϕ and power factor $\cos\phi$ when the resistance is much smaller than the inductive or capacitive reactance and the circuit is operated far from resonance? Explain.

Q32–10 When a series L–R–C circuit is connected across a 120-V ac line, the voltage rating of the capacitor may be exceeded even if it is rated at 200 or 400 V. How can this be?

Q32–11 In Example 32–6 (Section 32–5), a hair dryer was treated as a pure resistance. But because there are coils in the heating element and in the motor that drives the blower fan, a hair dryer also has inductance. Qualitatively, does including an inductance increase or decrease the values of R, I_{rms}, and P?

Q32–12 Electric-power cords, such as lamp cords, always have two conductors that carry equal currents in opposite directions. How might one determine the direction of power transmission in the cord by making measurements only at the midpoints along the lengths of the wires?

Q32–13 Explain, using the ideas of *mechanical* resonance, why the small tweeter in a loudspeaker is preferred for generating high-frequency sounds, while the larger woofer is optimal for producing low-frequency sounds.

Q32–14 Can a transformer be used with dc? Explain. What happens if a transformer designed for 120-V ac is connected to a 120-V dc line?

Q32–15 Some electrical appliances operate equally well on ac or dc, and others work only on ac or only on dc. Give examples of each, and explain the differences.

EXERCISES

SECTION 32–2 PHASORS AND ALTERNATING CURRENTS

32–1 The voltage across the terminals of an ac power supply varies with time according to Eq. (32–1). The voltage amplitude is $V = 45.0$ V. What is a) the root-mean-square potential difference V_{rms}? b) the average potential difference V_{av} between the two terminals of the power supply?

32–2 A sinusoidal current $i = I\cos\omega t$ has an rms value $I_{rms} = 2.10$ A. a) What is the current amplitude? b) The current is passed through a full-wave rectifier circuit. What is the rectified average current? c) Which is larger, I_{rms} or I_{rav}? Explain, using graphs of i^2 and of the rectified current.

SECTION 32–3 RESISTANCE AND REACTANCE

32–3 A 5.00-H inductor with negligible resistance is connected across an ac source whose voltage amplitude is kept constant at 60.0 V but whose frequency can be varied. Find the current amplitude when the angular frequency is a) 100 rad/s; b) 1000 rad/s; c) 10,000 rad/s. d) Show the results of parts (a) through (c) in a plot of log I versus log ω.

32–4 A 2.20-μF capacitor is connected across the ac source of Exercise 32–3. Find the current amplitude when the angular frequency is a) 100 rad/s; b) 1000 rad/s; c) 10,000 rad/s. d) Show the results of parts (a) through (c) in a plot of log I versus log ω.

32–5 a) What is the reactance of a 3.00-H inductor at a frequency of 80.0 Hz? b) What is the inductance of an inductor whose reactance is 120 Ω at 80.0 Hz? c) What is the reactance of a 4.00-μF capacitor at a frequency of 80.0 Hz? d) What is the capacitance of a capacitor whose reactance is 120 Ω at 80.0 Hz?

32–6 a) Compute the reactance of a 0.450-H inductor at frequencies of 60.0 Hz and 600 Hz. b) Compute the reactance of a 2.50-μF capacitor at the same frequencies. c) At what frequency is the reactance of a 0.450-H inductor equal to that of a 2.50-μF capacitor?

32–7 Kitchen Capacitance. The wiring for a refrigerator contains a starter capacitor. A voltage of amplitude 170 V and frequency 60.0 Hz applied across the capacitor is to produce a current amplitude of 0.850 A through the capacitor. What capacitance C is required?

32–8 A Radio Inductor. You want the current amplitude through a 0.450-mH inductor (part of the circuitry for a radio receiver) to be 2.60 mA when a sinusoidal voltage with amplitude 12.0 V is applied across the inductor. What frequency is required?

32–9 A 150-Ω resistor is connected in series with a 0.250-H inductor. The voltage across the resistor is $v_R =$ (3.80 V)cos [(720 rad/s)t]. a) Derive an expression for the circuit current. b) Determine the inductive reactance of the inductor. c) Derive an expression for the voltage v_L across the inductor.

32–10 A 250-Ω resistor is connected in series with a 4.80-μF capacitor. The voltage across the capacitor is $v_C = $ (7.60 V)sin[(120 rad/s)t]. a) Determine the capacitive reactance of the capacitor. b) Derive an expression for the voltage v_R across the resistor.

SECTION 32–4 THE L-R-C SERIES CIRCUIT

32–11 Define the reactance X of an L-R-C circuit to be $X = X_L - X_C$. a) Show that $X = 0$ when the angular frequency ω of the current is equal to the resonance angular frequency ω_0. b) What is the sign of X when $\omega > \omega_0$? c) What is the sign of X when $\omega < \omega_0$? d) Graph X versus ω.

32–12 You have a 200-Ω resistor, a 0.400-H inductor, and a 6.00-μF capacitor. Suppose you take the resistor and inductor and make a series circuit with a voltage source that has voltage amplitude 30.0 V and an angular frequency of 250 rad/s. a) What is the impedance of the circuit? b) What is the current amplitude? c) What are the voltage amplitudes across the resistor and across the inductor? d) What is the phase angle ϕ of the source voltage with respect to the current? Does the source voltage lag or lead the current? e) Construct the phasor diagram.

32–13 Repeat Exercise 32–12 with the circuit consisting of only the resistor and the capacitor in series. For part (c), calculate the voltage amplitudes across the resistor and across the capacitor.

32–14 Repeat Exercise 32–12 with the circuit consisting of only the capacitor and the inductor in series. For part (c), calculate the voltage amplitudes across the capacitor and across the inductor.

32–15 a) For the R-L circuit of Exercise 32–12, graph v, v_R, and v_L versus t for $t = 0$ to $t = 50.0$ ms. The current is given by $i = I \cos \omega t$, so $v = V \cos(\omega t + \phi)$. b) What are v, v_R, and v_L at $t = 20.0$ ms? Compare $v_R + v_L$ to v at this instant. c) Repeat part (b) for $t = 40.0$ ms.

32–16 a) For the R-C circuit of Exercise 32–13, graph v, v_R, and v_C versus t for $t = 0$ to $t = 50.0$ ms. The current is given by $i = I \cos \omega t$, so $v = V \cos(\omega t + \phi)$. b) What are v, v_R, and v_C at $t = 20.0$ ms? Compare $v_R + v_C$ to v at this instant. c) Repeat part (b) for $t = 40.0$ ms.

32–17 The resistor, inductor, and capacitor, and voltage source described in Exercise 32–12 are connected to form an L-R-C series circuit. a) What is the impedance of the circuit? b) What is the current amplitude? c) What is the phase angle of the source voltage with respect to the current? Does the source voltage lag or lead the current? d) What are the voltage amplitudes across the resistor, inductor, and capacitor? e) Explain how it is possible for the voltage amplitude across the capacitor to be greater than the voltage amplitude across the source.

32–18 a) For the L-R-C circuit of Exercise 32–17, graph v, v_R, v_L, and v_C versus t for $t = 0$ to $t = 50.0$ ms. The current is given by $i = I \cos \omega t$, so $v = V \cos(\omega t + \phi)$. b) What are v, v_R, v_L, and v_C at $t = 20.0$ ms? Compare $v_R + v_L + v_C$ to v at this instant. c) Repeat part (b) for $t = 40.0$ ms.

32–19 The resistor, inductor, and capacitor described in Exercise 32–12 are connected with an ac source to form a series L-R-C circuit. Compute the impedance of the circuit and draw the phasor diagram a) at an angular frequency of 400 rad/s; b) at an angular frequency of 800 rad/s. Compute, in each case, the phase angle of the source voltage with respect to the current, and state whether the source voltage lags or leads the current.

32–20 The resistor, inductor, and capacitor described in Exercise 32–12 are connected with an ac source of voltage amplitude 30.0 V to form a series L-R-C circuit. a) Compute the impedance of the circuit at angular frequencies of 1000, 600, and 200 rad/s. b) Describe how the current amplitude varies as the frequency of the source is slowly reduced from 1000 rad/s to 200 rad/s. c) What is the phase angle of the source voltage with respect to the current when $\omega = 1000$ rad/s? d) Construct the phasor diagram when $\omega = 1000$ rad/s. e) Repeat parts (c) and (d) for $\omega = 200$ rad/s.

SECTION 32–5 POWER IN ALTERNATING-CURRENT CIRCUITS

32–21 a) Show that for an L-R-C series circuit the power factor is equal to R/Z. (*Hint:* Use the phasor diagram; see Fig. 32–8b.) b) Show that for any ac circuit, not just one containing pure resistance only, the average power delivered by the voltage source is given by $P = I_{rms}^2 R$.

32–22 An L-R-C series circuit is connected to a 120-Hz ac source that has $V_{rms} = 80.0$ V. The circuit has a resistance of 75.0 Ω and an impedance at this frequency of 105 Ω. What average power is delivered to the circuit by the source?

32–23 An L-R-C series circuit with $L = 0.120$ H, $R = 240\ \Omega$, and $C = 7.30\ \mu$F carries an rms current of 0.450 A with a frequency of 400 Hz. a) What are the phase angle and power factor for this circuit? b) What is the impedance of the circuit? c) What is the rms voltage of the source? d) What average power is delivered by the source? e) What is the average rate at which electrical energy is converted to thermal energy in the resistor? f) What is the average rate at which electrical energy is dissipated (converted to other forms) in the capacitor? g) In the inductor?

32–24 A large 360-Ω, 5.20-H electromagnet coil is connected across the terminals of a source that has voltage amplitude 240 V and frequency 60.0 Hz. a) What is the power factor? b) What is the average power delivered by the source?

SECTION 32–6 RESONANCE IN ALTERNATING-CURRENT CIRCUITS

32–25 In an L-R-C series circuit, $R = 150\ \Omega$, $L = 0.750$ H, and $C = 0.0180\ \mu$F. The source has voltage amplitude $V = 150$ V and a frequency equal to the resonance frequency of the circuit. a) What is the power factor? b) What is the average power delivered by the source? c) The capacitor is replaced by one with $C = 0.0360\ \mu$F and the source frequency is adjusted to the new resonance value. What is then the average power delivered by the source?

32–26 In an L-R-C series circuit, $R = 400\ \Omega$, $L = 0.350$ H, and $C = 0.0120\ \mu$F. a) What is the resonance angular frequency of the circuit? b) The capacitor can withstand a peak voltage of 550 V. If the voltage source operates at the resonance frequency, what maximum voltage amplitude can it have if the maximum capacitor voltage is not exceeded?

32–27 Consider an L-R-C series circuit composed of the inductor, resistor, and capacitor described in Exercise 32–12. The circuit is connected to an ac source with voltage amplitude 30.0 V. a) At what frequency (in Hz) is the circuit in resonance? b) Sketch the phasor diagram at the resonance frequency. c) What is the reading of each voltmeter in Fig. 32–19 when the source frequency equals the resonance frequency? The voltmeters are calibrated to read rms voltages. d) What is the resonance frequency if the resistance is reduced to 100 Ω? e) What is then the rms current at resonance?

32–28 In an L-R-C series circuit, $L = 0.280$ H and $C = 4.00\ \mu$F. The voltage amplitude of the source is 120 V. a) What is the resonance angular frequency of the circuit? b) When the source operates at the resonance angular frequency, the current amplitude in the circuit is 1.70 A. What is the resistance R of the resistor? c) At the resonance angular frequency, what are the peak voltages across the inductor, the capacitor, and the resistor?

SECTION 32–7 TRANSFORMERS

32–29 A Step-Down Transformer. A transformer connected to a 120-V (rms) ac line is to supply 12.0 V (rms) to a portable electronic device. The total equivalent resistance of the system is 5.00 Ω. a) What should be the ratio of primary to secondary turns of the transformer? b) What rms current must the secondary supply? c) What average power is delivered to the load? d) What resistance connected directly across the 120-V line would draw the same power as the transformer? Show that this is equal to 5.00 Ω times the square of the ratio of primary to secondary turns.

32–30 A Step-Up Transformer. A transformer connected to a 120-V (rms) ac line is to supply 13,000 V (rms) for a neon sign. To reduce shock hazard, a fuse is to be inserted in the primary circuit; the fuse is to blow when the rms current in the secondary circuit exceeds 8.50 mA. a) What is the ratio of secondary to primary turns of the transformer? b) What power must be supplied to the transformer when the secondary current is 8.50 mA? c) What current rating should the fuse in the primary circuit have?

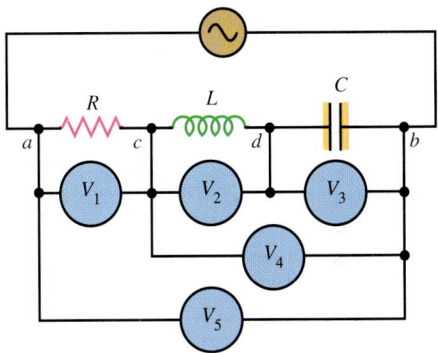

FIGURE 32–19 Exercise 32–27 and Problem 32–34.

32–31 The internal resistance of an ac source is 12.8 kΩ. a) What should be the ratio of primary to secondary turns of a transformer to match the source to a load with a resistance of 8.00 Ω? ("Matching" means that the effective load resistance equals the internal resistance of the source. See Exercise 32–29.) b) If the voltage amplitude of the source is 60.0 V, what is the voltage amplitude in the secondary circuit under open circuit conditions?

PROBLEMS

32–32 Figure 32–7a shows the crossover network in a loudspeaker system. One branch consists of a capacitor C and a resistor R in series (the tweeter). This branch is in parallel with a second branch (the woofer) that consists of an inductor L and a resistor R in series. The same source voltage with angular frequency ω is applied across each parallel branch. a) What is the impedance of the tweeter branch? b) What is the impedance of the woofer branch? c) Explain why the currents in the two branches are equal when the impedances of the branches are equal. d) Derive an expression for the frequency f that corresponds to the crossover point in Fig. 32–7b.

32–33 A coil has a resistance of 48.0 Ω. At a frequency of 80.0 Hz the voltage across the coil leads the current in it by 52.3°. Determine the inductance of the coil.

32–34 Five infinite-impedance voltmeters, calibrated to read rms values, are connected as shown in Fig 32–19. Let $R = 200$ Ω, $L = 0.400$ H, $C = 6.00$ μF, and $V = 30.0$ V. What is the reading of each voltmeter if a) $\omega = 200$ rad/s; b) $\omega = 1000$ rad/s?

32–35 A sinusoidal current is given by $i = I \cos \omega t$. The full-wave rectified current is shown in Fig. 32–2b. a) Let t_1 and t_2 be the two smallest positive times at which the rectified current is zero. Express t_1 and t_2 in terms of ω. b) Find the area under the rectified i versus t curve between t_1 and t_2 by computing the integral $\int_{t_1}^{t_2} i\, dt$. Since $dq = i\, dt$, this area equals the charge that flows during the t_1 to t_2 time interval. c) Set the result in part (b) equal to $I_{rav}(t_2 - t_1)$ and calculate I_{rav} in terms of the current amplitude I. Compare your answer to Eq. (32–3).

32–36 A large electromagnet coil is connected to a 120-Hz ac source. The coil has resistance 400 Ω, and at this source frequency the coil has inductive reactance 250 Ω. a) What is the inductance of the coil? b) What must be the rms voltage of the source if the coil is to consume an average electrical power of 800 W?

32–37 A series circuit has an impedance of 60.0 Ω and a power factor of 0.720 at 50.0 Hz. The source voltage lags the current. a) What circuit element, an inductor or a capacitor, should be placed in series with the circuit to raise its power factor? b) What size element will raise the power factor to unity?

32–38 At a frequency ω_1 the reactance of a certain capacitor equals that of a certain inductor. a) If the frequency is changed to $\omega_2 = 2\omega_1$, what is the ratio of the reactance of the inductor to that of the capacitor? Which reactance is larger? b) If the frequency is changed to $\omega_3 = \omega_1/3$, what is the ratio of the reactance of the inductor to that of the capacitor? Which reactance is larger? c) If the capacitor and inductor are placed in series with a resistor of resistance R to form a series L-R-C circuit, what will be the resonance angular frequency of the circuit?

32-39 A High-Pass Filter. One application of L-R-C series circuits is to high-pass or low-pass filters, which filter out either the low- or high-frequency components of a signal. A high-pass filter is shown in Fig. 32–20, where the output voltage is taken across the L-R combination. (The L-R combination represents an inductive coil that also has resistance due to the large length of wire in the coil.) Derive an expression for V_{out}/V_s, the ratio of the output and source voltage amplitudes, as a function of the angular frequency ω of the source. Show that when ω is small, this ratio is proportional to ω and thus is small, and show that the ratio approaches unity in the limit of large frequency.

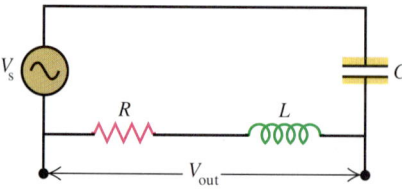

FIGURE 32–20 Problem 32–39.

32–40 A Low-Pass Filter. Figure 32–21 shows a low-pass filter (see Problem 32–39); the output voltage is taken across the

capacitor in an L-R-C series circuit. Derive an expression for V_{out}/V_s, the ratio of the output and source voltage amplitudes, as a function of the angular frequency ω of the source. Show that when ω is large this ratio is proportional to ω^{-2} and thus is very small, and show that the ratio approaches unity in the limit of small frequency.

FIGURE 32–21 Problem 32–40.

32–41 An L-R-C series circuit is connected to an ac source of constant voltage amplitude V and variable angular frequency ω. a) Show that the current amplitude, as a function of ω, is

$$I = \frac{V}{\sqrt{R^2 + (\omega L - 1/\omega C)^2}}.$$

b) Show that the average power dissipated in the resistor is

$$P = \frac{V^2 R/2}{R^2 + (\omega L - 1/\omega C)^2}.$$

c) Show that I and P are *both* maximum when $\omega = 1/\sqrt{LC}$, that is, when the source frequency equals the resonance frequency of the circuit. d) Graph P as a function of ω for $V = 100$ V, $R = 200$ Ω, $L = 2.0$ H, and $C = 0.50$ μF. Compare to the light red curve in Fig. 32–14. Discuss the behavior of I and P in the limits $\omega = 0$ and $\omega \to \infty$.

32–42 An L-R-C series circuit is connected to an ac source of constant voltage amplitude V and variable angular frequency ω. Using the results of Problem 32–41, find an expression for a) the amplitude V_L of the voltage across the inductor as a function of ω; b) the amplitude V_C of the voltage across the capacitor as a function of ω. c) Graph V_L and V_C as functions of ω for $V = 100$ V, $R = 200$ Ω, $L = 2.0$ H, and $C = 0.50$ μF. d) Discuss the behavior of V_L and V_C in the limits $\omega = 0$ and $\omega \to \infty$. For what value of ω is $V_L = V_C$? What is the significance of this value of ω?

32–43 An L-R-C series circuit is connected to an ac source of constant voltage amplitude V and variable angular frequency ω. a) Show that the time-averaged energy stored in the inductor is $U_B = \frac{1}{4}LI^2$ and the time-averaged energy stored in the capacitor is $U_E = \frac{1}{4}CV^2$. b) Use the results of Problems 32–41 and 32–42 to find expressions for U_B and U_E as functions of ω. c) Graph U_B and U_E as functions of ω for $V = 100$ V, $R = 200$ Ω, $L = 2.0$ H, and $C = 0.50$ μF. d) Discuss the behavior of U_B and U_E in the limits $\omega = 0$ and $\omega \to \infty$. For what value of ω is $U_B = U_E$? What is the significance of this value of ω?

32–44 The L-R-C Parallel Circuit. A resistor, inductor, and capacitor are connected in parallel to an ac source with voltage amplitude V and angular frequency ω. Let the source voltage be given by $v = V \cos \omega t$. a) Show that the instantaneous voltages v_R, v_L, and v_C at any instant are each equal to v and that $i = i_R + i_L + i_C$, where i is the current through the source and

i_R, i_L, and i_C are the currents through the resistor, the inductor, and the capacitor, respectively. b) What are the phases of i_R, i_L, and i_C with respect to v? Use current phasors to represent i, i_R, i_L, and i_C. On a phasor diagram, show the phases of these four currents with respect to v. c) Use the phasor diagram of part (b) to show that the current amplitude I for the current i through the source is given by $I = \sqrt{I_R^2 + (I_C - I_L)^2}$. d) Show that the result of part (c) can be written as $I = V/Z$, with $1/Z = \sqrt{1/R^2 + (\omega C - 1/\omega L)^2}$.

32–45 Parallel Resonance. The impedance of an *L-R-C* parallel circuit was derived in Problem 32–44. a) Show that at the resonance angular frequency $\omega_0 = 1/\sqrt{LC}$, $I_C = I_L$ and I is a *minimum*. b) Since I is a minimum at resonance, is it correct to say that the power delivered to the resistor is also a minimum at $\omega = \omega_0$? Explain. c) At resonance, what is the phase angle of the source current with respect to the source voltage? How does this compare to the phase angle for an *L-R-C series* circuit at resonance? d) Draw the circuit diagram for an *L-R-C* parallel circuit. Arrange the circuit elements in your diagram so that the resistor is closest to the ac source. Justify the following statement: When the angular frequency of the source is $\omega = \omega_0$, there is *no* current flowing between i) the part of the circuit that includes the source and the resistor and ii) the part that includes the inductor and capacitor, so you could cut the wires connecting these two parts of the circuit without affecting the currents. e) Is the statement in part (d) still valid if we consider that any real inductor or capacitor also has some resistance of its own? Explain.

32–46 A 400-Ω resistor and a 6.00-μF capacitor are connected in parallel to an ac generator that supplies an rms voltage of 220 V at an angular frequency of 360 rad/s. Use the results of Problem 32–44. Note that since there is no inductor in the circuit, the $1/\omega L$ term is not present in the expression for Z. Find a) the current amplitude in the resistor; b) the current amplitude in the capacitor; c) the phase angle of the source current with respect to the source voltage; d) the amplitude of the current through the generator. e) Does the source current lag or lead the source voltage?

32–47 An *L-R-C* parallel circuit is connected to an ac source of constant voltage amplitude V and variable angular frequency ω. a) Using the results of Problem 32–44, find expressions for the amplitudes I_R, I_L, and I_C of the currents through the resistor, inductor, and capacitor as functions of ω. b) Graph I_R, I_L, and I_C as functions of ω for $V = 100$ V, $R = 200$ Ω, $L = 2.0$ H, and $C = 0.50$ μF. c) Discuss the behavior of I_L and I_C in the limits $\omega = 0$ and $\omega \rightarrow \infty$. Explain why I_L and I_C behave as they do in these limits. d) Calculate the resonance frequency (in Hz) of the circuit, and sketch the phasor diagram at the resonance frequency. e) At the resonance frequency, what is the current amplitude through the source? f) At the resonance frequency, what is the current amplitude through the resistor, through the inductor, and through the capacitor?

32–48 An *L-R-C* series circuit draws 220 W from a 120-V (rms), 50.0-Hz ac line. The power factor is 0.560, and the source voltage leads the current. a) What is the net resistance R of the circuit? b) Find the capacitance of the series capacitor that will result in a power factor of unity when it is added to the original circuit. c) What power will then be drawn from the supply line?

32–49 In an *L-R-C* series circuit the magnitude of the phase angle is 54.0°, with the source voltage lagging the current. The reactance of the capacitor is 350 Ω, and the resistor resistance is 180 Ω. The average power delivered by the source is 140 W. Find a) the reactance of the inductor; b) the rms current; c) the rms voltage of the source.

32–50 An *L-R-C* series circuit has $R = 500$ Ω, $L = 2.00$ H, $C = 0.500$ μF, and $V = 100$ V. a) For $\omega = 800$ rad/s, calculate V_R, V_L, V_C, and ϕ. Using a single set of axes, graph v, v_R, v_L, and v_C as functions of time. Include two cycles of v on your graph. b) Repeat part (a) for $\omega = 1000$ rad/s. c) Repeat part (a) for $\omega = 1250$ rad/s.

32–51 In an *L-R-C* series circuit, the source has a voltage amplitude of 120 V, $R = 80.0$ Ω, and the reactance of the capacitor is 480 Ω. The voltage amplitude across the capacitor is 360 V. a) What is the current amplitude in the circuit? b) What is the impedance? c) What two values can the reactance of the inductor have? d) For which of the two values found in part (c) is the angular frequency less than the resonance angular frequency? Explain.

32–52 Designing an FM Radio Receiver. You enjoy listening to KONG-FM, which broadcasts at 94.1 MHz. You detest listening to KRUD-FM, which broadcasts at 94.0 MHz. You live the same distance from both stations and both transmitters are equally powerful, so both radio signals produce the same 1.0-V source voltage as measured at your house. Your goal is to design an *L-R-C* radio circuit with the following properties: i) It gives the maximum power response to the signal from KONG-FM; ii) the average power delivered to the resistor in response to KRUD-FM is 1.00% of the average power in response to KONG-FM. This limits the power received from the unwanted station, making it inaudible. You are required to use an inductor with $L = 1.00$ μH. Find the capacitance C and resistance R that satisfy the design requirements.

32–53 The current in a certain circuit varies with time as shown in Fig. 32–22. Find the average current and the rms current in terms of I_0.

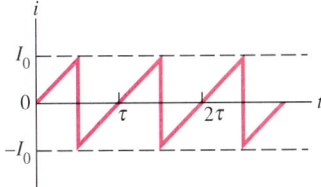

FIGURE 32–22 Problem 32–53.

32–54 The Resonance Width. Consider an *L-R-C* series circuit with a 1.80-H inductor, a 0.900-μF capacitor, and a 300-Ω resistor. The source has terminal rms voltage $V_{rms} = 60.0$ V and variable angular frequency ω. a) What is the resonance angular frequency ω_0 of the circuit? b) What is the rms current through the circuit at resonance, I_{rms-0}? c) For what two values of the angular frequency, ω_1 and ω_2, is the rms current half the resonance value? d) The quantity $|\omega_1 - \omega_2|$ defines the *resonance width*. Calculate I_{rms-0} and the resonance width for $R = 300$ Ω,

30.0 Ω, and 3.00 Ω. Describe how your results compare to the discussion in Section 32–6.

32–55 The Resonance Width II. An L-R-C series circuit has inductance L, resistance R, and capacitance C. The ac source has voltage amplitude V and variable angular frequency ω.
a) When ω equals the resonance angular frequency $\omega_0 = 1/\sqrt{LC}$, what is the amplitude I_0 of the current through the circuit?
b) Suppose the angular frequency is only slightly different than ω_0. Then we can write $\omega = \omega_0 + \Delta\omega$, where $|\Delta\omega|$ is much less than ω_0. Show that the impedance is then approximately equal to $\sqrt{R^2 + 4L^2(\Delta\omega)^2}$. (*Hint:* Use the binomial expansion $(1 + z)^n = 1 + nz + n(n - 1)z^2/2 + \dots$, valid for the case $|z| < 1$.)
c) Using the approximate expression for the impedance in part (b), find the two values of the angular frequency $\omega = \omega_0 + \Delta\omega$ for which the current amplitude is half the resonance value. Explain why these r sults are valid only if R is much less than $\sqrt{4L/3C}$.
d) The difference between the two angular frequencies found in part (c) is the *resonance width*. Find an expression for the resonance width valid in the case $R \ll \sqrt{4L/3C}$. Use your result and the result of part (a) to explain how the shape of the curve in Fig. 32–14 changes as R is decreased. e) Determine I_0, the resonance angular frequency, and the resonance width for an L-R-C series circuit with $L = 2.50$ H, $C = 0.400$ μF, and $V = 120$ V if i) $R = 15.0$ Ω; ii) $R = 1.50$ Ω.

32–56 A resistance R, capacitance C, and inductance L are connected in series to a voltage source with amplitude V and variable angular frequency ω. If $\omega = \omega_0$, the resonance angular frequency, find a) the maximum current in the resistor; b) the maximum voltage across the capacitor; c) the maximum voltage across the inductor; d) the maximum energy stored in the capacitor; and e) the maximum energy stored in the inductor. Give your answers in terms of R, C, L, and V.

32–57 Repeat Problem 32–56 for the case $\omega = \omega_0/2$.

32–58 Repeat Problem 32–56 for the case $\omega = 2\omega_0$.

32–59 In an L-R-C series circuit the current is given by $i = I \cos \omega t$. The voltage amplitudes for the resistor, inductor, and capacitor are V_R, V_L, and V_C. a) Show that the instantaneous power into the resistor is $p_R = V_R I \cos^2 \omega t = \frac{1}{2}V_R I(1 + \cos 2\omega t)$. What does this expression give for the average power into the resistor? b) Show that the instantaneous power into the inductor is $p_L = -V_L I \sin \omega t \cos \omega t = -\frac{1}{2}V_L I \sin 2\omega t$. What does this expression give for the average power into the inductor? c) Show that the instantaneous power into the capacitor is $p_C = V_C I \sin \omega t \cos \omega t = \frac{1}{2}V_C I \sin 2\omega t$. What does this expression give for the average power into the capacitor? d) The instantaneous power delivered by the source is shown in Section 32–5 to be $p = VI \cos \omega t (\cos \phi \cos \omega t - \sin \phi \sin \omega t)$. Show that $p_R + p_L + p_C$ equals p at each instant of time.

CHALLENGE PROBLEMS

32–60 a) At what angular frequency is the voltage amplitude across the *resistor* in an L-R-C series circuit at maximum value? b) At what angular frequency is the voltage amplitude across the *inductor* at maximum value? c) At what angular frequency is the voltage amplitude across the *capacitor* at maximum value? (You may want to refer to the results of Problem 32–42.)

32–61 Complex Numbers in a Circuit. The voltage across a circuit element in an ac circuit is not necessarily in phase with the current through that circuit element. Therefore the voltage amplitudes across the circuit elements in a branch in an ac circuit do not add algebraically. A method that is commonly employed to simplify the analysis of an ac circuit driven by a sinusoidal source is to represent the impedance Z as a *complex* number. The resistance R is taken to be the real part of the impedance, and the reactance $X = X_L - X_C$ is taken to be the imaginary part. Thus, for a branch containing a resistor, inductor, and capacitor in series, the complex impedance is $Z_{cpx} = R + iX$, where $i^2 = -1$. If the voltage amplitude across the branch is V_{cpx}, we define a *complex* current amplitude by $I_{cpx} = V_{cpx}/Z_{cpx}$. The *actual* current amplitude is the absolute value of the complex current amplitude, that is, $I = (I^*_{cpx}I_{cpx})^{1/2}$. The phase angle φ of the current with respect to the source voltage is given by $\tan \phi = \mathrm{Im}(I_{cpx})/\mathrm{Re}(I_{cpx})$. The voltage amplitudes $V_{R\text{-}cpx}$, $V_{L\text{-}cpx}$, and $V_{C\text{-}cpx}$ across the resistance, inductance, and capacitance, respectively, are found by multiplying I_{cpx} by R, iX_L, or $-iX_C$, respectively. Using the complex representation for the voltage amplitudes, the voltage across a branch is just the algebraic sum of the voltages across each circuit element; $V_{cpx} = V_{R\text{-}cpx} + V_{L\text{-}cpx} + V_{C\text{-}cpx}$. The actual value of any current amplitude or voltage amplitude is the absolute value of the cor-

responding complex quantity. Consider the series L-R-C circuit shown in Fig. 32–23. The values of the circuit elements, the source voltage amplitude, and the source angular frequency are as shown. Use the phasor diagram techniques presented in Section 32–2 to solve for a) the current amplitude; b) the phase angle φ of the current with respect to the source voltage. (Note that this angle is the negative of the phase angle defined in Fig. 32–8.) Now analyze the same circuit using the complex-number approach. c) Determine the complex impedance of the circuit, Z_{cpx}. Take the absolute value to obtain Z, the actual impedance of the circuit. d) Take the voltage amplitude of the source, V_{cpx}, to be real, and find the complex current amplitude I_{cpx}. Find the actual current amplitude by taking the absolute value of I_{cpx}. e) Find the phase angle φ of the current with respect to the source voltage by using the real and imaginary parts of I_{cpx}, as explained above. f) Find the complex representations of the voltages across the resistance, the inductance, and the capacitance. g) Adding the answers found in part (f), verify that the sum of these complex numbers is real and equal to 200 V, the voltage of the source.

$L = 0.500$ H

$V = 200$ V
$\omega = 1000$ rad/s

$C = 1.25$ μF

$R = 400$ Ω

FIGURE 32–23 Challenge Problem 32–61.

The cellular phone has revolutionized personal communications for millions of people around the world. Like any mobile radio, a cellular phone transmits and receives electromagnetic waves. Typical cellular phones transmit at frequencies between 824.04 and 848.97 megahertz (MHz) and receive at frequencies between 869.04 and 893.07 MHz.

33

Electromagnetic Waves

33-1 INTRODUCTION

What is light? This question has been asked by humans for countless centuries, but there was no answer until electricity and magnetism were unified into the single discipline of *electromagnetism,* as described by Maxwell's equations. These equations show that a time-varying magnetic field acts as a source of electric field and that a time-varying electric field acts as a source of magnetic field. These $\vec{E}$ and $\vec{B}$ fields can sustain each other, forming an *electromagnetic wave* that propagates through space. Visible light emitted by the glowing filament of a light bulb is one example of an electromagnetic wave; other kinds of electromagnetic waves are produced by sources such as TV and radio stations, microwave oscillators for ovens and radar, x-ray machines, and radioactive nuclei.

In this chapter we'll use Maxwell's equations as the theoretical basis for understanding electromagnetic waves. We'll find that these waves carry both energy and momentum. In sinusoidal electromagnetic waves, the $\vec{E}$ and $\vec{B}$ fields are sinusoidal functions of time and position, with a definite frequency and wavelength. The various types of electromagnetic waves—visible light, radio, x rays, and others—differ only in their frequency and wavelength. Our study of optics in the following chapters will be based in part on the electromagnetic nature of light.

Unlike waves on a string or sound waves in a fluid, electromagnetic waves do not require a material medium; the light that you see coming from the stars at night has traveled without difficulty across tens or hundreds of light years of (nearly) empty space. Nonetheless, electromagnetic waves and mechanical waves have much in common and are described in much the same language. Before reading further in this chapter, you should review the properties of mechanical waves as discussed in Chapters 19 and 20.

33-2 MAXWELL'S EQUATIONS AND ELECTROMAGNETIC WAVES

In the last several chapters we studied various aspects of electric and magnetic fields. We learned that when the fields don't vary with time, such as an electric field produced by charges at rest or the magnetic field of a steady current, we can analyze the electric and magnetic fields independently without considering interactions between them. But when the fields vary with time, they are no longer independent. Faraday's law (Section 30–3) tells us that a time-varying magnetic field acts as a source of electric field, as shown by induced emf's in inductors and transformers. Ampere's law, including the displacement current discovered by Maxwell (Section 29–10) shows that a time-varying electric field acts as a source of magnetic field. This mutual interaction between the two fields is summarized neatly in Maxwell's equations, presented in Section 30–8.

Thus when *either* an electric or a magnetic field is changing with time, a field of the other kind is induced in adjacent regions of space. We are led (as Maxwell was) to consider the possibility of an electromagnetic disturbance, consisting of time-varying electric and magnetic fields, that can propagate through space from one region to another, even when there is no matter in the intervening region. Such a disturbance, if it exists, will have the properties of a *wave,* and an appropriate term is **electromagnetic wave.**

Such waves do exist; radio and television transmission, light, x rays, and many other kinds of radiation are examples of electromagnetic waves. Our goal in this chapter is to see how the existence of such waves is related to the principles of electromagnetism that we have studied thus far and to examine the properties of these waves.

As so often happens in the development of science, the theoretical understanding of electromagnetic waves originally took a considerably more devious path than the one just outlined. In the early days of electromagnetic theory (the early nineteenth century), two different units of electric charge were used, one for electrostatics and the other for magnetic phenomena involving currents. In the system of units used at that time, these two units of charge had different physical dimensions. Their *ratio* had units of velocity, and measurements showed that the ratio had a numerical value that was precisely equal to the speed of light, 3.00×10^8 m/s. At the time, physicists regarded this as an extraordinary coincidence and had no idea how to explain it.

In searching to understand this result, Maxwell proved in 1865 that an electromagnetic disturbance should propagate in free space with a speed equal to that of light and hence that light waves were likely to be electromagnetic in nature. At the same time, he discovered that the basic principles of electromagnetism can be expressed in terms of the four equations that we now call **Maxwell's equations,** which we discussed in Section 30–8. These four equations are (1) Gauss's law for electric fields; (2) Gauss's law for magnetic fields, showing the absence of magnetic monopoles; (3) Ampere's law, including displacement current; and (4) Faraday's law. Maxwell's equations, in the integral form used in this text, are

James Clerk Maxwell (1831–1879) was the first person to truly understand the fundamental nature of light. He also made major contributions to thermodynamics, optics, astronomy, and color photography. Albert Einstein described Maxwell's accomplishments as "the most profound and the most fruitful that physics has experienced since the time of Newton."

$$\oint \vec{E} \cdot d\vec{A} = \frac{Q_{\text{encl}}}{\epsilon_0} \qquad \text{(Gauss's law),} \qquad (23–8)$$

$$\oint \vec{B} \cdot d\vec{A} = 0 \qquad \text{(Gauss's law for magnetism),} \qquad (28–8)$$

$$\oint \vec{B} \cdot d\vec{l} = \mu_0 \left(i_C + \epsilon_0 \frac{d\Phi_E}{dt} \right)_{\text{encl}} \qquad \text{(Ampere's law),} \qquad (29–36)$$

$$\oint \vec{E} \cdot d\vec{l} = -\frac{d\Phi_B}{dt} \qquad \text{(Faraday's law).} \qquad (30–15)$$

These equations apply to electric and magnetic fields *in vacuum.* If a material is present, the permittivity ϵ_0 and permeability μ_0 of free space are replaced by the permittivity ϵ and permeability μ of the material. If the values of ϵ and μ are different at different points in the regions of integration, then ϵ and μ have to be transferred to the left sides of Eqs. (23–8) and (29–36), respectively, and placed inside the integrals. The ϵ in Eq. (29–36) also has to be included in the integral that gives $d\Phi_E/dt$.

According to Maxwell's equations, a point charge at rest produces a static $\vec{E}$ field but no $\vec{B}$ field; a point charge moving with a constant velocity (Section 29–2) produces both $\vec{E}$ and $\vec{B}$ fields. Maxwell's equations can also be used to show that in order for a point charge to produce electromagnetic waves, the charge must *accelerate.* In fact, it's a general result of Maxwell's equations that *every* accelerated charge radiates electromagnetic energy. This is the reason for the shielding required around high-energy particle accelerators and high-voltage power supplies in TV sets.

One way in which a point charge can be made to emit electromagnetic waves is by making it oscillate in simple harmonic motion, so that it has an acceleration at almost

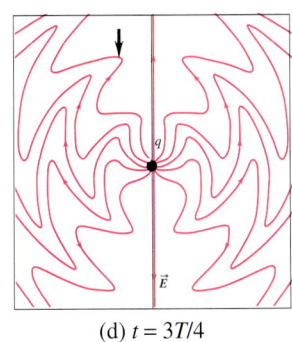

(a) $t = 0$ (b) $t = T/4$ (c) $t = T/2$ (d) $t = 3T/4$

33–1 Electric field lines of a point charge oscillating in simple harmonic motion. The field lines are viewed at four instants during an oscillation period T in a plane containing the point charge's trajectory. At $t = 0$ the point charge is moving upward through its equilibrium position. The arrow in each part shows one "kink" in the lines of $\vec{E}$, which propagates outward from the point charge. For clarity the magnetic field lines are not shown; these are circles that lie in planes perpendicular to these figures and that are concentric with the axis of oscillation.

every instant (the exception is when the charge is passing through its equilibrium position). Figure 33–1 shows some of the electric field lines produced by such an oscillating point charge. Field lines are *not* material objects, but you may nonetheless find it helpful to think of them as behaving somewhat like strings that extend from the point charge off to infinity. Oscillating the charge up and down makes waves that propagate outward from the charge along these "strings." Note that the charge does not emit waves equally in all directions; the waves are strongest at 90° to the axis of motion of the charge, while there are *no* waves along this axis. This is just what the "string" picture would lead you to conclude. There is also a *magnetic* disturbance that spreads outwards from the charge; this is not shown in Fig. 33–1. Because the electric and magnetic disturbances spread or radiate away from the source, the name **electromagnetic radiation** is used interchangeably with the phrase "electromagnetic waves."

Electromagnetic waves with macroscopic wavelengths were first produced in the laboratory in 1887 by the German physicist Heinrich Hertz. As a source of waves, he used charges oscillating in *L-C* circuits of the sort discussed in Section 31–6; he detected the resulting electromagnetic waves with other circuits tuned to the same frequency. Hertz also produced electromagnetic *standing waves* and measured the distance between adjacent nodes (one half-wavelength) to determine the wavelength. Knowing the resonant frequency of his circuits, he then found the speed of the waves from the wavelength-frequency relation $v = \lambda f$. He established that their speed was the same as that of light; this verified Maxwell's theoretical prediction directly. The SI unit of frequency is named in honor of Hertz: One hertz (1 Hz) equals one cycle per second.

The possible use of electromagnetic waves for long-distance communication does not seem to have occurred to Hertz. It remained for the enthusiasm and energy of Marconi and others to make radio communication a familiar household experience. In a radio transmitter, electric charges are made to oscillate along the length of the conducting antenna, producing oscillating field disturbances like those shown in Fig. 33–1. Since many charges oscillate together in the antenna, the disturbances are much stronger than those of a single oscillating charge and can be detected at a much greater distance. In a radio receiver the antenna is also a conductor; the fields of the wave emanating from a distant transmitter exert forces on free charges within the receiver antenna, producing an oscillating current that is detected and amplified by the receiver circuitry.

For most of the remainder of this chapter our concern will be with electromagnetic waves themselves, not with the rather complex problem of how they are produced. In Section 33–9 we'll describe in some detail the character of the electromagnetic waves produced by one important type of transmitting antenna.

33–3 PLANE ELECTROMAGNETIC WAVES AND THE SPEED OF LIGHT

We are now ready to develop the basic ideas of electromagnetic waves and their relation to the principles of electromagnetism as summarized in Maxwell's equations. Our procedure will be to postulate a simple field configuration that has wavelike behavior. We'll assume an electric field $\vec{E}$ that has only a y-component and a magnetic field $\vec{B}$ with only a z-component, and we'll assume that both fields move together in the $+x$-direction with a speed c that is initially unknown. (As we go along, it will become clear why we choose $\vec{E}$ and $\vec{B}$ to be perpendicular to the direction of propagation as well as to each other.) Then we will test whether these fields are physically possible by asking whether they are consistent with Maxwell's equations, particularly Ampere's law and Faraday's law. We'll find that the answer is yes, provided that c has a particular value. We'll also show that the *wave equation,* which we encountered during our study of mechanical waves in Chapter 19, can be derived from Maxwell's equations.

A SIMPLE PLANE ELECTROMAGNETIC WAVE

Using an xyz-coordinate system (Fig. 33–2), we imagine that all space is divided into two regions by a plane perpendicular to the x-axis (parallel to the yz-plane). At every point to the left of this plane there are a uniform electric field $\vec{E}$ in the $+y$-direction and a uniform magnetic field $\vec{B}$ in the $+z$-direction, as shown. Furthermore, we suppose that the boundary plane, which we call the *wave front,* moves to the right in the $+x$-direction with a constant speed c, as yet unknown. Thus the $\vec{E}$ and $\vec{B}$ fields travel to the right into previously field-free regions with a definite speed. The situation, in short, describes a rudimentary electromagnetic wave. A wave such as this, in which at any instant the fields are uniform over any plane perpendicular to the direction of propagation, is called a **plane wave.** In the case shown in Fig. 33–2, the fields are zero for planes to the right of the wave front and have the same values on all planes to the left of the wave front; later we will consider more complex plane waves.

We won't concern ourselves with the problem of actually *producing* such a field configuration. Instead, we simply ask whether it is consistent with the laws of electromagnetism, that is, with Maxwell's equations. We'll consider each of these four equations in turn.

Let us first verify that our wave satisfies Maxwell's first and second equations, that is, Gauss's laws for electric and magnetic fields. To do this, we take as our Gaussian surface a rectangular box with sides parallel to the xy, xz, and yz coordinate planes (Fig. 33–3). The box encloses no electric charge, and you should be able to show that the total electric flux and magnetic flux through the box are both zero; this is true even if part of the box is in the region where $E = B = 0$. This would *not* be the case if $\vec{E}$ or $\vec{B}$ had an x-component, parallel to the direction of propagation. We leave the proof as a problem (see Problem 33–29). Thus in order to satisfy Maxwell's first and second equations, the electric and magnetic fields must be perpendicular to the direction of propagation; that is, the wave must be **transverse.**

The next of Maxwell's equations to be considered is Faraday's law:

$$\oint \vec{E} \cdot d\vec{l} = -\frac{d\Phi_B}{dt}. \tag{33–1}$$

To test whether our wave satisfies Faraday's law, we apply this law to a rectangle $efgh$ that is parallel to the xy-plane (Fig. 33–4a). As shown in Fig. 33–4b, a cross section in the xy-plane, this rectangle has height a and width Δx. At the time shown, the wave front has progressed partway through the rectangle, and $\vec{E}$ is zero along the side ef. In applying Faraday's law we take the vector area $d\vec{A}$ of rectangle $efgh$ to be in the $+z$-direction. With this choice the right-hand rule requires that we integrate $\vec{E} \cdot d\vec{l}$ *counterclockwise*

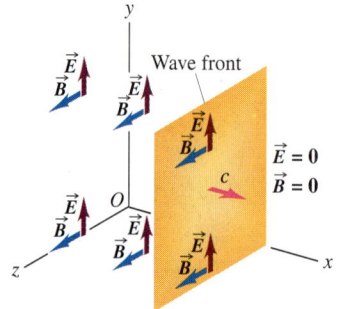

33–2 An electromagnetic wave front. The plane representing the wave front moves to the right with speed c. The $\vec{E}$ and $\vec{B}$ fields are uniform over the region behind the wave front but are zero everywhere in front of it.

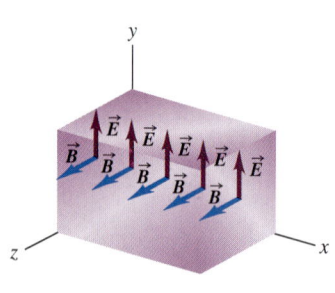

33–3 Gaussian surface for a plane electromagnetic wave. The total electric flux and the total magnetic flux through the surface are both zero.

around the rectangle. At every point on side *ef*, $\vec{E}$ is zero. At every point on sides *fg* and *he*, $\vec{E}$ is either zero or perpendicular to $d\vec{l}$. Only side *gh* contributes to the integral. On this side, $\vec{E}$ and $d\vec{l}$ are opposite, and we obtain

$$\oint \vec{E} \cdot d\vec{l} = -Ea. \qquad (33\text{–}2)$$

Hence the left-hand side of Eq. (33–1) is nonzero.

In order to satisfy Faraday's law, Eq. (33–1), there must be a component of $\vec{B}$ in the *z*-direction (perpendicular to $\vec{E}$) so that there can be a nonzero magnetic flux Φ_B through the rectangle *efgh* and a nonzero derivative $d\Phi_B/dt$. Indeed, in our wave, $\vec{B}$ has *only* a *z*-component. We have assumed that this component is in the *positive z*-direction; let's see whether this assumption is consistent with Faraday's law. During a time interval *dt* the wave front moves a distance *c dt* to the right in Fig. 33–4b, sweeping out an area *ac dt* of the rectangle *efgh*. During this interval the magnetic flux Φ_B through the rectangle *efgh* increases by $d\Phi_B = B(ac\,dt)$, so the rate of change of magnetic flux is

$$\frac{d\Phi_B}{dt} = Bac. \qquad (33\text{–}3)$$

Now we substitute Eqs. (33–2) and (33–3) into Faraday's law, Eq. (33–1); we get

$$-Ea = -Bac,$$

$$E = cB \qquad \text{(electromagnetic wave in vacuum)}. \qquad (33\text{–}4)$$

This shows that our wave is consistent with Faraday's law only if the wave speed *c* and the magnitudes of the perpendicular vectors $\vec{E}$ and $\vec{B}$ are related as in Eq. (33–4). Note that if we had assumed that $\vec{B}$ was in the *negative z*-direction, there would have been an additional minus sign in Eq. (33–4); since *E*, *c*, and *B* are all positive magnitudes, no solution would then have been possible. Furthermore, any component of $\vec{B}$ in the *y*-direction (parallel to $\vec{E}$) would not contribute to the changing magnetic flux Φ_B through the rectangle *efgh* (which is parallel to the *xy* plane) and so would not be part of the wave.

Finally, we carry out a similar calculation using Ampere's law, the remaining member of Maxwell's equations. There is no conduction current ($i_C = 0$), so Ampere's law is

$$\oint \vec{B} \cdot d\vec{l} = \mu_0 \epsilon_0 \frac{d\Phi_E}{dt}. \qquad (33\text{–}5)$$

To check whether our wave is consistent with Ampere's law, we move our rectangle so that it lies in the *xz*-plane, as shown in Fig. 33–5, and we again look at the situation at a time when the wave front has traveled partway through the rectangle. We take the vector area $d\vec{A}$ in the +*y*-direction, and so the right-hand rule requires that we integrate $\vec{B} \cdot d\vec{l}$ counterclockwise around the rectangle. The $\vec{B}$ field is zero at every point along side *ef*, and at each point on sides *fg* and *he* it is either zero or perpendicular to $d\vec{l}$. Only side *gh*, where $\vec{B}$ and $d\vec{l}$ are parallel, contributes to the integral, and we find

$$\oint \vec{B} \cdot d\vec{l} = Ba. \qquad (33\text{–}6)$$

Hence the left-hand side of Ampere's law, Eq. (33–5), is nonzero; the right-hand side must be nonzero as well. Thus $\vec{E}$ must have a *y*-component (perpendicular to $\vec{B}$) so that the electric flux Φ_E through the rectangle and the time derivative $d\Phi_E/dt$ can be nonzero. We come to the same conclusion that we inferred from Faraday's law: In an electromagnetic wave, $\vec{E}$ and $\vec{B}$ must be mutually perpendicular.

In a time interval *dt* the electric flux Φ_E through the rectangle increases by $d\Phi_E = E(ac\,dt)$. Since we chose $d\vec{A}$ to be in the +*y*-direction, this flux change is positive;

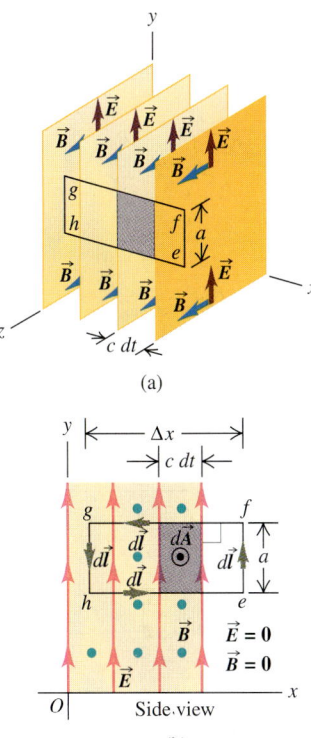

(a)

(b)

33–4 Applying Faraday's law to a plane wave. In time *dt* the wave front moves to the right a distance *c dt*. The magnetic flux through the rectangle in the *xy*-plane increases by an amount $d\Phi_B$ equal to the flux through the shaded rectangle with area *ac dt*; that is, $d\Phi_B = Bac\,dt$.

(a)

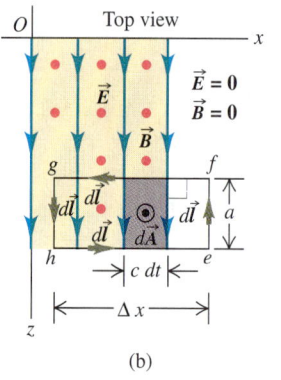

(b)

33–5 Applying Ampere's law to a plane wave. In time dt the electric flux through the rectangle in the xz-plane increases by an amount equal to E times the area $ac\,dt$ of the shaded rectangle; that is, $d\Phi_E = Eac\,dt$. Thus $d\Phi_E/dt = Eac$.

the rate of change of electric field is

$$\frac{d\Phi_E}{dt} = Eac. \qquad (33\text{--}7)$$

Substituting Eqs. (33–6) and (33–7) into Ampere's law, Eq. (33–5), we find

$$Ba = \epsilon_0\mu_0 Eac,$$

$$B = \epsilon_0\mu_0 cE \qquad \text{(electromagnetic wave in vacuum).} \qquad (33\text{--}8)$$

Thus our assumed wave obeys Ampere's law only if B, c, and E are related as in Eq. (33–8).

Our electromagnetic wave must obey *both* Ampere's law and Faraday's law, so Eqs. (33–4) and (33–8) must both be satisfied. This can happen only if $\epsilon_0\mu_0 c = 1/c$, or

$$c = \frac{1}{\sqrt{\epsilon_0\mu_0}} \qquad \text{(speed of electromagnetic waves in vacuum).} \qquad (33\text{--}9)$$

Inserting the numerical values of these quantities, we find

$$c = \frac{1}{\sqrt{(8.85 \times 10^{-12}\ \text{C}^2/\text{N}\cdot\text{m}^2)(4\pi \times 10^{-7}\ \text{N/A}^2)}}$$

$$= 3.00 \times 10^8\ \text{m/s}.$$

Our assumed wave is consistent with all of Maxwell's equations, provided that the wave front moves with the speed given above, which you should recognize as the speed of light! Note that the *exact* value of c is defined to be 299,792,458 m/s; for our purposes, $c = 3.00 \times 10^8$ m/s is sufficiently accurate.

We chose a simple wave for our study in order to avoid mathematical complications, but this special case illustrates several important features of *all* electromagnetic waves:

1. The wave is *transverse;* both $\vec{E}$ and $\vec{B}$ are perpendicular to the direction of propagation of the wave. The electric and magnetic fields are also perpendicular to each other. The direction of propagation is the direction of the vector product $\vec{E} \times \vec{B}$.
2. There is a definite ratio between the magnitudes of $\vec{E}$ and $\vec{B}$: $E = cB$.
3. The wave travels in vacuum with a definite and unchanging speed.
4. Unlike mechanical waves, which need the oscillating particles of a medium such as water or air to transmit a wave, electromagnetic waves require no medium. What's "waving" in an electromagnetic wave are the electric and magnetic fields.

We can generalize this discussion to a more realistic situation. Suppose we have several wave fronts in the form of parallel planes perpendicular to the x-axis, all of which are moving to the right with speed c. Suppose that the $\vec{E}$ and $\vec{B}$ fields are the same at all points within a single region between two planes, but that the fields differ from region to region. The overall wave is a plane wave, but one in which the fields vary in steps along the x-axis. Such a wave could be constructed by superposing several of the simple step waves we have just discussed (shown in Fig. 33–2). This is possible because the $\vec{E}$ and $\vec{B}$ fields obey the superposition principle in waves just as in static situations: When two waves are superposed, the total $\vec{E}$ field at each point is the vector sum of the $\vec{E}$ fields of the individual waves, and similarly for the total $\vec{B}$ field.

We can extend the above development to show that a wave with fields that vary in steps is also consistent with Ampere's and Faraday's laws, provided that the wave fronts all move with the speed c given by Eq. (33–9). In the limit that we make the individual steps infinitesimally small, we have a wave in which the $\vec{E}$ and $\vec{B}$ fields at any instant vary *continuously* along the x-axis. The entire field pattern moves to the right with speed c. In Section 33–4 we will consider waves in which $\vec{E}$ and $\vec{B}$ are *sinusoidal* functions of x and t. Because at each point the magnitudes of $\vec{E}$ and $\vec{B}$ are related by $E = cB$, the periodic variations of the two fields in any periodic traveling wave must be *in phase*.

Electromagnetic waves have the property of **polarization.** In the above discussion the choice of the y-direction for $\vec{E}$ was arbitrary. We could just as well have specified the z-axis for $\vec{E}$; then $\vec{B}$ would have been in the −y-direction. A wave in which $\vec{E}$ is always parallel to a certain axis is said to be **linearly polarized** along that axis. More generally, *any* wave traveling in the x-direction can be represented as a superposition of waves linearly polarized in the y- and z-directions. We will study polarization in greater detail, with special emphasis on polarization of light, in Chapter 34.

*DERIVATION OF THE WAVE EQUATION

Here is an alternative derivation of Eq. (33–9) for the speed of electromagnetic waves. It is more mathematical than our other treatment, but it includes a derivation of the wave equation for electromagnetic waves. This part of the section can be omitted without loss of continuity in the chapter.

During our discussion of mechanical waves in Section 19–4, we showed that a function $y(x, t)$ that represents the displacement of any point in a mechanical wave traveling along the x-axis must satisfy a differential equation, Eq. (19–12):

$$\frac{\partial^2 y(x, t)}{\partial x^2} = \frac{1}{v^2} \frac{\partial^2 y(x, t)}{\partial t^2}. \tag{33–10}$$

This equation is called the **wave equation,** and v is the speed of propagation of the wave.

To derive the corresponding equation for an electromagnetic wave, we again consider a plane wave. That is, we assume that at each instant, E_y and B_z are uniform over any plane perpendicular to the x-axis, the direction of propagation. But now we let E_y and B_z vary continuously as we go along the x-axis; then each is a function of x and t. We consider the values of E_y and B_z on two planes perpendicular to the x-axis, one at x and one at $x + \Delta x$.

Following the same procedure as previously, we apply Faraday's law to a rectangle lying parallel to the xy-plane, as in Fig. 33–6. This figure is similar to Fig. 33–4. Let the left end gh of the rectangle be at position x, and let the right end ef be at position $(x + \Delta x)$. At time t, the values of E_y on these two sides are $E_y(x, t)$ and $E_y(x + \Delta x, t)$, respectively. When we apply Faraday's law to this rectangle, we find that instead of $\oint \vec{E} \cdot d\vec{l} = -Ea$ as before, we have

$$\oint \vec{E} \cdot d\vec{l} = -E_y(x, t)a + E_y(x + \Delta x, t)a$$

$$= a[E_y(x + \Delta x, t) - E_y(x, t)]. \tag{33–11}$$

To find the magnetic flux Φ_B through this rectangle, we assume that Δx is small enough that B_z is nearly uniform over the rectangle. In that case, $\Phi_B = B_z(x, t)A = B_z(x, t)a\,\Delta x$, and

$$\frac{d\Phi_B}{dt} = \frac{\partial B_z(x, t)}{\partial t}a\,\Delta x.$$

We use partial-derivative notation because B_z is a function of both x and t. When we substitute this expression and Eq. (33–11) into Faraday's law, Eq. (33–1), we get

$$a[E_y(x + \Delta x, t) - E_y(x, t)] = -\frac{\partial B_z}{\partial t}a\,\Delta x,$$

$$\frac{E_y(x + \Delta x, t) - E_y(x, t)}{\Delta x} = -\frac{\partial B_z}{\partial t}.$$

Finally, imagine shrinking the rectangle down to a sliver so that Δx approaches zero.

(a)

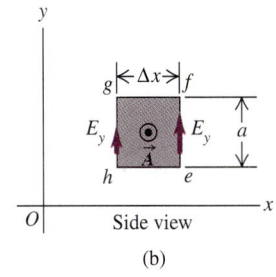

(b)

33–6 Faraday's law applied to a rectangle with height a and width Δx parallel to the xy-plane.

(a)

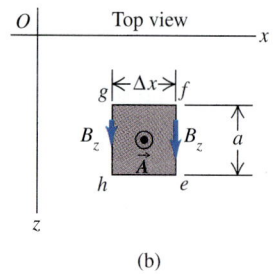

(b)

33–7 Ampere's law applied to a rectangle with height a and width Δx parallel to the xz-plane.

When we take the limit of this equation as $\Delta x \to 0$, we get

$$\frac{\partial E_y(x,t)}{\partial x} = -\frac{\partial B_z(x,t)}{\partial t}. \qquad (33\text{–}12)$$

This equation shows that if there is a time-varying component B_z of magnetic field, there must also a component E_y of electric field that varies with x, and conversely. We put this relation on the shelf for now; we'll return to it soon.

Next we apply Ampere's law to the rectangle shown in Fig. 33–7. The line integral $\oint \vec{B} \cdot d\vec{l}$ becomes

$$\oint \vec{B} \cdot d\vec{l} = -B_z(x+\Delta x, t)a + B_z(x,t)a. \qquad (33\text{–}13)$$

Again assuming that the rectangle is narrow, we approximate the electric flux Φ_E through it as $\Phi_E = E_y(x,t)\,A = E_y(x,t)a\,\Delta x$. The rate of change of Φ_E, which we need for Ampere's law, is then

$$\frac{d\Phi_E}{dt} = \frac{\partial E_y(x,t)}{\partial t}a\,\Delta x.$$

Now substitute this expression and Eq. (33–13) into Ampere's law, Eq. (33–5):

$$-B_z(x+\Delta x, t)a + B_z(x,t)a = \epsilon_0\mu_0 \frac{\partial E_y(x,t)}{\partial t}a\,\Delta x.$$

Again we divide both sides by $a\,\Delta x$ and take the limit as $\Delta x \to 0$. We find

$$-\frac{\partial B_z(x,t)}{\partial x} = \epsilon_0\mu_0 \frac{\partial E_y(x,t)}{\partial t}. \qquad (33\text{–}14)$$

Now comes the final step. We take the partial derivatives with respect to x of both sides of Eq. (33–12), and we take the partial derivatives with respect to t of both sides of Eq. (33–14). The results are

$$-\frac{\partial^2 E_y(x,t)}{\partial x^2} = \frac{\partial^2 B_z(x,t)}{\partial x \partial t},$$

$$-\frac{\partial^2 B_z(x,t)}{\partial x \partial t} = \epsilon_0\mu_0 \frac{\partial^2 E_y(x,t)}{\partial t^2}.$$

Combining these two equations to eliminate B_z, we finally find

$$\frac{\partial^2 E_y(x,t)}{\partial x^2} = \epsilon_0\mu_0 \frac{\partial^2 E_y(x,t)}{\partial t^2} \qquad \begin{array}{c}\text{(electromagnetic wave equation} \\ \text{in vacuum).}\end{array} \qquad (33\text{–}15)$$

This expression has the same form as the general wave equation, Eq. (33–10). Because the electric field E_y must satisfy this equation, it behaves as a wave with a pattern that travels through space with a definite speed. Furthermore, comparison of Eqs. (33–15) and (33–10) shows that the wave speed v is given by

$$\frac{1}{v^2} = \epsilon_0\mu_0, \qquad \text{or} \qquad v = \frac{1}{\sqrt{\epsilon_0\mu_0}}.$$

This agrees with Eq. (33–9) for the speed c of electromagnetic waves.

We can show that B_z also must satisfy the same wave equation as E_y, Eq. (33–15). To prove this, we take the partial derivative of Eq. (33–12) with respect to t and the partial derivative of Eq. (33–14) with respect to x and combine the results. We leave this derivation as a problem (see Problem 33–31).

33–4 SINUSOIDAL ELECTROMAGNETIC WAVES

Sinusoidal electromagnetic waves are directly analogous to sinusoidal transverse mechanical waves on a stretched string, which we studied in Section 19–4. In a sinusoidal electromagnetic wave, $\vec{E}$ and $\vec{B}$ at any point in space are sinusoidal functions of time, and at any instant of time the *spatial* variation of the fields is also sinusoidal.

Some sinusoidal electromagnetic waves are *plane waves;* they share with the waves described in Section 33–3 the property that at any instant the fields are uniform over any plane perpendicular to the direction of propagation. The entire pattern travels in the direction of propagation with speed c. The directions of $\vec{E}$ and $\vec{B}$ are perpendicular to the direction of propagation (and to each other), so the wave is *transverse*. Electromagnetic waves produced by an oscillating point charge, shown in Fig. 33–1, are an example of sinusoidal waves that are *not* plane waves. But if we restrict our observations to a relatively small region of space at a sufficiently great distance from the source, even these waves are well approximated by plane waves; in the same way, the curved surface of the (nearly) spherical earth appears flat to us because of our small size relative to the earth's radius. In this section we'll restrict our discussion to plane waves.

The frequency f, the wavelength λ, and the speed of propagation c of any periodic wave are related by the usual wavelength-frequency relation $c = \lambda f$. If the frequency f is the power-line frequency of 60 Hz, the wavelength is

$$\lambda = \frac{c}{f} = \frac{3 \times 10^8 \text{ m/s}}{60 \text{ Hz}} = 5 \times 10^6 \text{ m} = 5000 \text{ km},$$

which is of the order of the earth's radius! For a wave with this frequency, even a distance of many miles includes only a small fraction of a wavelength. But if the frequency is 10^8 Hz (100 MHz), typical of commercial FM radio broadcasts, the wavelength is

$$\lambda = \frac{3 \times 10^8 \text{ m/s}}{10^8 \text{ Hz}} = 3 \text{ m},$$

and a moderate distance can include many complete waves.

Figure 33–8 shows a sinusoidal electromagnetic wave traveling in the $+x$-direction. The $\vec{E}$ and $\vec{B}$ vectors are shown for only a few points on the positive x-axis. Imagine a plane perpendicular to the x-axis at a particular point, at a particular time; the fields have the same values at all points in that plane. The values are different on different planes. In planes where $\vec{E}$ is in the $+y$-direction, $\vec{B}$ is in the $+z$-direction; where $\vec{E}$ is in the $-y$-direction, $\vec{B}$ is in the $-z$-direction. Note that in all planes the vector product $\vec{E} \times \vec{B}$ is in the direction in which the wave is propagating (the $+x$-direction).

We can describe electromagnetic waves by means of *wave functions,* just as we did in Section 19–4 for waves on a string. One form of the wave function for a transverse wave traveling in the $+x$-direction along a stretched string is Eq. (19–7):

$$y(x, t) = A \sin(\omega t - kx),$$

where $y(x, t)$ is the transverse displacement from its equilibrium position at time t of a point with coordinate x on the string. The quantity A is the maximum displacement, or *amplitude,* of the wave; ω is its *angular frequency,* equal to 2π times the frequency f; and k is the *wave number,* equal to $2\pi/\lambda$, where λ is the wavelength.

Let $E(x, t)$ and $B(x, t)$ represent the instantaneous values of the y-component of $\vec{E}$ and the z-component of $\vec{B}$, respectively, in Fig. 33–8, and let E_{max} and B_{max} represent the maximum values, or *amplitudes,* of these fields. The wave functions for the wave are then

$$E(x, t) = E_{max} \sin(\omega t - kx), \qquad B(x, t) = B_{max} \sin(\omega t - kx)$$

(sinusoidal electromagnetic plane wave, propagating in $+x$-direction). (33–16)

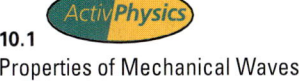

10.1
Properties of Mechanical Waves

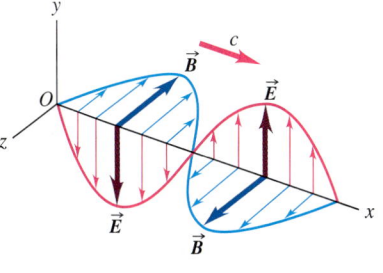

33–8 Representation of the electric and magnetic fields as functions of x for a linearly polarized sinusoidal plane electromagnetic wave. One wavelength of the wave is shown at time $t = 0$. The wave is traveling in the positive x-direction; this is the same as the direction of $\vec{E} \times \vec{B}$. The fields are shown only for points along the x-axis.

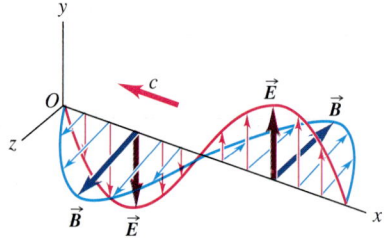

33–9 Representation at $t = 0$ of one wavelength of a linearly polarized sinusoidal plane electromagnetic wave traveling in the negative x-direction. The fields are shown only for points along the x-axis.

We can also write the wave functions in vector form:

$$\vec{E}(x,t) = E_{max}\hat{\jmath}\,\sin(\omega t - kx),$$
$$\vec{B}(x,t) = B_{max}\hat{k}\,\sin(\omega t - kx). \tag{33–17}$$

CAUTION▶ Note the two different k's, the unit vector $\hat{k}$ in the z-direction and the wave number k. Don't get these confused!◀

The sine curves in Fig. 33–8 represent instantaneous values of the electric and magnetic fields as functions of x at time $t = 0$, that is, $\vec{E}(x, t = 0)$ and $\vec{B}(x, t = 0)$. As time goes by, the wave travels to the right with speed c. Equations (33–16) and (33–17) show that at any point the sinusoidal oscillations of $\vec{E}$ and $\vec{B}$ are *in phase*. From Eq. (33–4) the amplitudes must be related by

$$E_{max} = cB_{max} \qquad \text{(electromagnetic wave in vacuum).} \tag{33–18}$$

These amplitude and phase relations are also required for $E(x, t)$ and $B(x, t)$ to satisfy Eqs. (33–12) and (33–14), which came from Faraday's law and Ampere's law respectively. Can you verify this statement? (See Problem 33–30.)

Figure 33–9 shows the electric and magnetic fields of a wave traveling in the *negative x-direction*. At points where $\vec{E}$ is in the positive y-direction, $\vec{B}$ is in the *negative z-direction*; where $\vec{E}$ is in the negative y-direction, $\vec{B}$ is in the *positive z-direction*. We note that the direction of propagation is the direction of $\vec{E} \times \vec{B}$, as mentioned in Section 33–3 in the list of characteristics of electromagnetic waves. The wave functions for this wave are

$$E(x,t) = -E_{max}\sin(\omega t + kx), \qquad B(x,t) = B_{max}\sin(\omega t + kx)$$

(sinusoidal electromagnetic plane wave, propagating in $-x$-direction). (33–19)

As with the wave traveling in the $+x$-direction, at any point that the sinusoidal oscillations of the $\vec{E}$ and $\vec{B}$ fields are *in phase*, and the vector product $\vec{E} \times \vec{B}$ points in the direction of propagation.

The sinusoidal waves shown in Figs. 33–8 and 33–9 are both linearly polarized in the y-direction; the $\vec{E}$ field is always parallel to the y-axis. Example 33–1 concerns a wave that is linearly polarized in the z-direction.

Problem–Solving Strategy

ELECTROMAGNETIC WAVES

1. For problems involving electromagnetic waves, the most important advice we can give is to concentrate on basic relationships, such as the relation of $\vec{E}$ to $\vec{B}$ (both magnitude and direction), how the wave speed is determined, the transverse nature of the waves, and so on. Keep these in mind when working through the mathematical details.

2. For sinusoidal waves, you need to use the language of sinusoidal waves from Chapters 19 and 20. Don't hesitate

to go back and review that material, including the problem-solving strategies suggested in those chapters.

3. Keep in mind the basic relationships for periodic waves, $v = \lambda f$ and $\omega = vk$. For electromagnetic waves in vacuum, $v = c$. Be careful to distinguish between ordinary frequency f, usually expressed in hertz, and angular frequency $\omega = 2\pi f$, expressed in rad/s. Also remember that the wave number is $k = 2\pi/\lambda$.

EXAMPLE 33-1

Fields of a laser beam A carbon dioxide laser emits a sinusoidal electromagnetic wave that travels in vacuum in the negative x-direction (Fig. 33–10). The wavelength is 10.6 μm, and the

$\vec{E}$ field is parallel to the z-axis, with maximum magnitude of 1.5 MV/m. Write vector equations for $\vec{E}$ and $\vec{B}$ as functions of time and position.

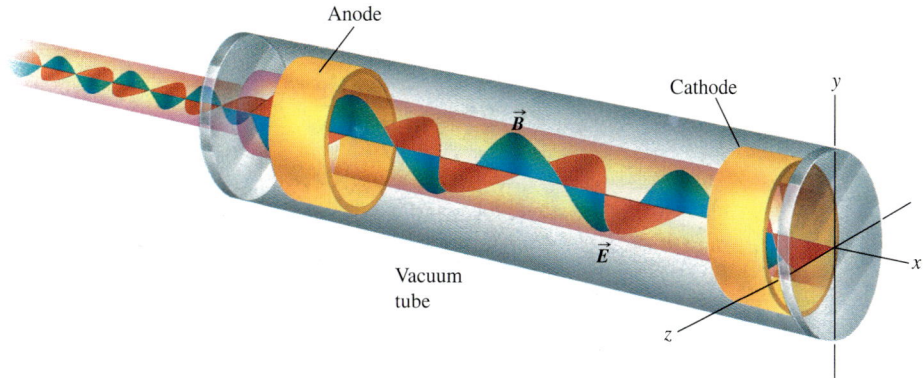

33–10 A sinusoidal electromagnetic wave is emitted from a carbon dioxide laser. The red and blue-green curves represent the electric and magnetic fields, respectively, for points on the *x*-axis at a given instant.

SOLUTION Equations (33–19) describe a wave traveling in the negative *x*-direction with $\vec{E}$ along the *y*-axis; that is, a wave that is linearly polarized along the *y*-axis. By contrast, the wave in this example is linearly polarized along the *z*-axis. At points where $\vec{E}$ is in the positive *z*-direction, $\vec{B}$ must be in the positive *y*-direction in order for the vector product $\vec{E} \times \vec{B}$ to be in the negative *x*-direction (the direction of propagation). A possible pair of wave functions that satisfies these requirements is

$$\vec{E}(x,t) = E_{max}\hat{k}\sin(\omega t + kx), \qquad \vec{B}(x,t) = B_{max}\hat{j}\sin(\omega t + kx).$$

Faraday's law requires $E_{max} = cB_{max}$, so

$$B_{max} = \frac{E_{max}}{c} = \frac{1.5 \times 10^6 \ \text{V/m}}{3.0 \times 10^8 \ \text{m/s}} = 5.0 \times 10^{-3} \ \text{T}.$$

To check unit consistency, note that 1 V = 1 Wb/s and 1 Wb/m^2 = 1 T.

We have $\lambda = 10.6 \times 10^{-6}$ m, so the wave number and angular frequency are

$$k = 2\pi/\lambda = (2\pi \ \text{rad})/(10.6 \times 10^{-6} \ \text{m}) = 5.93 \times 10^5 \ \text{rad/m},$$
$$\omega = ck = (3.00 \times 10^8 \ \text{m/s})(5.93 \times 10^5 \ \text{rad/m})$$
$$= 1.78 \times 10^{14} \ \text{rad/s}.$$

Substituting these values into the above wave functions, we get

$$\vec{E}(x,t) =$$
$$(1.5 \times 10^6 \ \text{V/m})\hat{k}\sin[(1.78 \times 10^{14} \ \text{rad/s})t + (5.93 \times 10^5 \ \text{rad/m})x],$$
$$\vec{B}(x,t) =$$
$$(5.0 \times 10^{-3} \ \text{T})\hat{j}\sin[(1.78 \times 10^{14} \ \text{rad/s})t + (5.93 \times 10^5 \ \text{rad/m})x].$$

With these equations we can find the fields in the laser beam at any particular position and time by substituting specific values of *x* and *t*.

Careful inspection of Fig. 33–10 shows that the $\vec{E}$ and $\vec{B}$ fields *inside* the laser are *not* in phase. The electromagnetic wave inside the laser is an example of a *standing wave*. We'll discuss standing electromagnetic waves in Section 33–7.

33–5 ENERGY AND MOMENTUM IN ELECTROMAGNETIC WAVES

It is a familiar fact that energy is associated with electromagnetic waves; think of the energy in the sun's radiation or in the radiation inside a microwave oven. To derive detailed relationships for the energy in an electromagnetic wave, we begin with the expressions derived in Sections 25–4 and 31–4 for the **energy densities** in electric and magnetic fields; we suggest that you review those derivations now. Specifically, Eqs. (25–11) and (31–9) show that the total energy density *u* in a region of empty space where $\vec{E}$ and $\vec{B}$ fields are present is given by

$$u = \frac{1}{2}\epsilon_0 E^2 + \frac{1}{2\mu_0}B^2, \tag{33–20}$$

where ϵ_0 and μ_0 are the permittivity and permeability, respectively, of free space. For the simple electromagnetic waves we've studied so far, $\vec{E}$ and $\vec{B}$ are related by

$$B = \frac{E}{c} = \sqrt{\epsilon_0\mu_0}\,E. \tag{33–21}$$

Combining Eqs. (33–20) and (33–21), we can also express the energy density u in a simple electromagnetic wave in vacuum as

$$u = \frac{1}{2}\epsilon_0 E^2 + \frac{1}{2\mu_0}\left(\sqrt{\epsilon_0\mu_0}\,E\right)^2 = \epsilon_0 E^2. \tag{33–22}$$

This shows that in a vacuum, the energy density associated with the $\vec{E}$ field in our simple wave is equal to the energy density of the $\vec{B}$ field. In general, the electric-field magnitude E is a function of position and time, as for the sinusoidal wave described by Eqs. (33–16); thus the energy density u of an electromagnetic wave, given by Eq. (33–22), also depends in general on position and time.

ELECTROMAGNETIC ENERGY FLOW AND THE POYNTING VECTOR

Electromagnetic waves such as those we have described are *traveling* waves that transport energy from one region to another. For instance, in the wave described in Section 33–3 the $\vec{E}$ and $\vec{B}$ fields advance with time into regions where originally no fields were present and carry the energy density u with them as they advance. We can describe this energy transfer in terms of energy transferred *per unit time per unit cross-section area,* or *power per unit area,* for an area perpendicular to the direction of wave travel.

To see how the energy flow is related to the fields, consider a stationary plane, perpendicular to the x-axis, that coincides with the wave front at a certain time. In a time dt after this, the wave front moves a distance $dx = c\,dt$ to the right of the plane. Considering an area A on this stationary plane (Fig. 33–11), we note that the energy in the space to the right of this area must have passed through the area to reach the new location. The volume dV of the relevant region is the base area A times the length $c\,dt$, and the energy dU in this region is the energy density u times this volume:

$$dU = u\,dV = (\epsilon_0 E^2)(Ac\,dt).$$

This energy passes through the area A in time dt. The energy flow per unit time per unit area, which we will call S, is

$$S = \frac{1}{A}\frac{dU}{dt} = \epsilon_0 c E^2 \qquad \text{(in a vacuum).} \tag{33–23}$$

Using Eqs. (33–15) and (33–21), we can derive the alternate forms

$$S = \frac{\epsilon_0}{\sqrt{\epsilon_0\mu_0}}E^2 = \sqrt{\frac{\epsilon_0}{\mu_0}}E^2 = \frac{EB}{\mu_0} \qquad \text{(in a vacuum).} \tag{33–24}$$

The derivation of Eq. (33–24) from Eq. (33–23) is left as a problem (see Exercise 33–15). The units of S are energy per unit time per unit area, or power per unit area. The SI unit of S is 1 J/s · m^2 or 1 W/m^2.

We can define a *vector* quantity that describes both the magnitude and direction of the energy flow rate:

$$\vec{S} = \frac{1}{\mu_0}\vec{E}\times\vec{B} \qquad \text{(Poynting vector in a vacuum).} \tag{33–25}$$

The vector $\vec{S}$ is called the **Poynting vector;** it was introduced by the British physicist John Poynting (1852–1914). Its direction is in the direction of propagation of the wave. Since $\vec{E}$ are $\vec{B}$ are perpendicular, the magnitude of $\vec{S}$ is $S = EB/\mu_0$; from Eqs. (33–23) and (33–24) this is the energy flow per unit area and per unit time through a cross-section area perpendicular to the propagation direction. The total energy flow per unit time

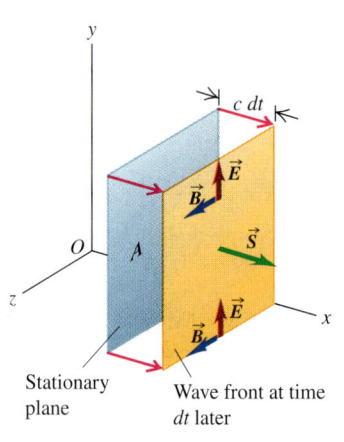

33–11 Wave front at a time dt after it passes through the stationary plane with area A. The volume between the plane and the wave front contains an amount of electromagnetic energy $uAc\,dt$.

(power, P) out of any closed surface is the integral of $\vec{S}$ over the surface:

$$P = \oint \vec{S} \cdot d\vec{A}.$$

For the sinusoidal waves studied in Section 33–4 as well as for other more complex waves, the electric and magnetic fields at any point vary with time, so the Poynting vector at any point is also a function of time. Because the frequencies of typical electromagnetic waves are very high, the time variation of the Poynting vector is so rapid that it's most appropriate to look at its *average* value. The average value of the magnitude of $\vec{S}$ at a point is called the **intensity** of the radiation at that point. The SI unit of intensity is the same as for S, 1 W/m^2 (watt per square meter).

Let's work out the intensity of the sinusoidal wave described by Eqs. (33–16). We first substitute these expressions into Eq. (33–24):

$$S(x,t) = \frac{E(x,t)B(x,t)}{\mu_0} = \frac{E_{max}B_{max}}{\mu_0}\sin^2(\omega t - kx)$$

$$= \frac{E_{max}B_{max}}{2\mu_0}[1 - \cos 2(\omega t - kx)].$$

The time-average value of $\cos 2(\omega t - kx)$ is zero because at any point, it is positive during half a cycle and negative during the other half. So the average value S_{av} of the Poynting vector magnitude over a full cycle is

$$S_{av} = \frac{E_{max}B_{max}}{2\mu_0}.$$

That is, the average value of S for a sinusoidal wave (the intensity I of the wave) is 1/2 the maximum value. By using the relations $E_{max} = B_{max}c$ and $\epsilon_0\mu_0 = 1/c^2$, we can express the intensity in several equivalent forms:

$$I = S_{av} = \frac{E_{max}B_{max}}{2\mu_0} = \frac{E_{max}^2}{2\mu_0 c} = \frac{1}{2}\sqrt{\frac{\epsilon_0}{\mu_0}}E_{max}^2 = \frac{1}{2}\epsilon_0 cE_{max}^2 \qquad (33\text{–}26)$$

(intensity of a sinusoidal wave in a vacuum).

We invite you to verify that these expressions are all equivalent.

For a wave traveling in the $-x$-direction, represented by Eqs. (33–19), the Poynting vector is in the $-x$-direction at every point, but its magnitude is the same as for a wave traveling in the $+x$-direction. Verifying these statements is left to you (see Exercise 33–12).

CAUTION ▶ At any point x, the magnitude of the Poynting vector varies with time. Hence the *instantaneous* rate at which electromagnetic energy in a sinusoidal plane wave arrives at a surface is not constant. This may seem to contradict everyday experience; the light from the sun, a light bulb, or the laser in a grocery-store scanner appears steady and unvarying in strength. In fact the Poynting vector from these sources *does* vary in time, but the variation isn't noticeable because the oscillation frequency is so high (around 5×10^{14} Hz for visible light). All that you sense is the *average* rate at which energy reaches your eye, which is why we commonly use intensity (the average value of S) to describe the strength of electromagnetic radiation. ◀

EXAMPLE 33-2

For the nonsinusoidal wave described in Section 33–3, suppose that $E = 100$ V/m = 100 N/C. Find the value of B, the energy density, and the rate of energy flow per unit area S.

SOLUTION From Eq. (33–4),

$$B = \frac{E}{c} = \frac{100 \text{ V/m}}{3.00 \times 10^8 \text{ m/s}} = 3.33 \times 10^{-7} \text{ T}.$$

From Eq. (33–22),

$$u = \epsilon_0 E^2 = (8.85 \times 10^{-12} \text{ C}^2/\text{N} \cdot \text{m}^2)(100 \text{ N/C})^2$$
$$= 8.85 \times 10^{-8} \text{ N/m}^2 = 8.85 \times 10^{-8} \text{ J/m}^3.$$

The magnitude of the Poynting vector is

$$S = \frac{EB}{\mu_0} = \frac{(100 \text{ V/m})(3.33 \times 10^{-7} \text{ T})}{4\pi \times 10^{-7} \text{ T} \cdot \text{m/A}}$$
$$= 26.5 \text{ V} \cdot \text{A/m}^2 = 26.5 \text{ W/m}^2.$$

Alternatively,

$$S = \epsilon_0 c E^2 = (8.85 \times 10^{-12} \text{ C}^2/\text{N} \cdot \text{m}^2)(3.00 \times 10^8 \text{ m/s})(100 \text{ N/C})^2$$
$$= 26.5 \text{ W/m}^2.$$

As we described in Section 33–3, in this wave the $\vec{E}$ and $\vec{B}$ fields are zero at points in front of the wave front and have constant values at all points behind the wave front; hence u and S also have constant values at all points behind the wave front.

EXAMPLE 33-3

A radio station on the surface of the earth radiates a sinusoidal wave with an average total power of 50 kW (Fig. 33–12). Assuming that the transmitter radiates equally in all directions above the ground (which is unlikely in real situations), find the amplitudes E_{max} and B_{max} detected by a satellite at a distance of 100 km from the antenna.

SOLUTION The first step is to find the magnitude S_{av} of the average Poynting vector. Then we can find E_{max} from Eqs. (33–26) and B_{max} from Eq. (33–4). We imagine a hemisphere centered on the antenna, with radius 100 km = 1.00×10^5 m and area

$$A = 2\pi R^2 = 2\pi (1.00 \times 10^5 \text{ m})^2 = 6.28 \times 10^{10} \text{ m}^2.$$

All the radiated power passes through this surface, so the average power per unit area (that is, the intensity) is

$$I = \frac{P}{A} = \frac{P}{2\pi R^2} = \frac{5.00 \times 10^4 \text{ W}}{6.28 \times 10^{10} \text{ m}^2} = 7.96 \times 10^{-7} \text{ W/m}^2.$$

From Eqs. (33–26), $I = S_{av} = E_{max}^2/2\mu_0 c$, so

$$E_{max} = \sqrt{2\mu_0 c S_{av}}$$
$$= \sqrt{2(4\pi \times 10^{-7} \text{ T} \cdot \text{m/A})(3.00 \times 10^8 \text{ m/s})(7.96 \times 10^{-7} \text{ W/m}^2)}$$
$$= 2.45 \times 10^{-2} \text{ V/m}.$$

From Eq. (33–4),

$$B_{max} = \frac{E_{max}}{c} = 8.17 \times 10^{-11} \text{ T}.$$

33-12 A radio station radiates waves into the hemisphere shown.

Note that the magnitude of E_{max} is comparable to fields commonly seen in the laboratory, but B_{max} is extremely small in comparison to $\vec{B}$ fields we saw in previous chapters. For this reason, most detectors of electromagnetic radiation respond to the effect of the electric field, not the magnetic field. Loop radio antennas are an exception.

ELECTROMAGNETIC MOMENTUM FLOW AND RADIATION PRESSURE

By using the observation that energy is required to establish electric and magnetic fields, we have shown that electromagnetic waves transport energy. It can also be shown that electromagnetic waves carry *momentum p,* with a corresponding momentum density (momentum dp per volume dV) of magnitude

$$\frac{dp}{dV} = \frac{EB}{\mu_0 c^2} = \frac{S}{c^2}. \tag{33–27}$$

This momentum is a property of the field; it is not associated with the mass of a moving particle in the usual sense.

There is also a corresponding momentum flow rate. The volume dV occupied by an electromagnetic wave (speed c) that passes through an area A in time dt is $dV = Ac\, dt$. When we substitute this into Eq. (33–27) and rearrange, we find that the momentum

flow rate per unit area is

$$\frac{1}{A}\frac{dp}{dt} = \frac{S}{c} = \frac{EB}{\mu_0 c} \qquad \text{(flow rate of electromagnetic momentum).} \qquad (33\text{--}28)$$

This is the momentum transferred per unit surface area per unit time. We obtain the *average* rate of momentum transfer per unit area by replacing S in Eq. (33–28) by $S_{av} = I$.

This momentum is responsible for the phenomenon of **radiation pressure.** When an electromagnetic wave is completely absorbed by a surface, the wave's momentum is also transferred to the surface. For simplicity we'll consider a surface perpendicular to the propagation direction. Using the ideas developed in Section 8–2, we see that the rate dp/dt at which momentum is transferred to the absorbing surface equals the *force* on the surface. The average force per unit area due to the wave, or *radiation pressure* p_{rad}, is the average value of dp/dt divided by the absorbing area A. (We use the subscript "rad" to distinguish pressure from momentum, for which the symbol p is also used.) From Eq. (33–28) the radiation pressure is

$$p_{rad} = \frac{S_{av}}{c} = \frac{I}{c} \qquad \text{(radiation pressure, wave totally absorbed).} \qquad (33\text{--}29)$$

If the wave is totally reflected, the momentum change is twice as great, and the pressure is

$$p_{rad} = \frac{2S_{av}}{c} = \frac{2I}{c} \qquad \text{(radiation pressure, wave totally reflected).} \qquad (33\text{--}30)$$

For example, the value of I (or S_{av}) for direct sunlight, before it passes through the earth's atmosphere, is approximately 1.4 kW/m². From Eq. (33–29) the corresponding average pressure on a completely absorbing surface is

$$p_{rad} = \frac{I}{c} = \frac{1.4 \times 10^3 \text{ W/m}^2}{3.0 \times 10^8 \text{ m/s}} = 4.7 \times 10^{-6} \text{ Pa.}$$

From Eq. (33–30) the average pressure on a totally *reflecting* surface is twice this, $2I/c$ or 9.4×10^{-6} Pa. These are very small pressures, of the order of 10^{-10} atm, but they can be measured with sensitive instruments.

The radiation pressure of sunlight is much larger *inside* the sun than at the earth (see Problem 33–37). Inside stars that are much more massive and luminous than the sun, radiation pressure is so great that it substantially augments the gas pressure within the star and so helps to prevent the star from collapsing under its own gravity.

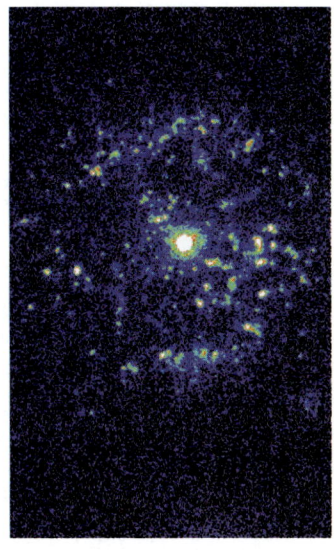

From time to time the unusual star at the center of this image undergoes a dramatic increase in brightness. When this happens, radiation pressure at the star's surface becomes so strong that part of its outer layers are ejected into space. Some of this ejected material appears as glowing blobs surrounding the star.

EXAMPLE 33-4

Power and pressure from sunlight An earth-orbiting satellite has solar-energy-collecting panels with a total area of 4.0 m² (Fig. 33–13). If the sun's radiation is perpendicular to the panels and is completely absorbed, find the average solar power absorbed and the average force associated with radiation pressure.

SOLUTION From the above discussion the intensity I (power per unit area) is 1.4×10^3 W/m². Although the light from the sun is not a simple sinusoidal wave, we can still use the relationship that the average power P is the intensity I times the area A:

$$P = IA = (1.4 \times 10^3 \text{ W/m}^2)(4.0 \text{ m}^2) = 5.6 \times 10^3 \text{ W} = 5.6 \text{ kW.}$$

This is a substantial amount of power, part of which can be used

33–13 Solar-energy-collecting panels on a satellite.

to power the equipment aboard the satellite. The rest goes into heating the panels, either directly or because of inefficiencies in the photocells contained in the panels (see Section 18–11).

We calculated the radiating pressure of sunlight on an absorbing surface in the discussion above: $p_{rad} = 4.7 \times 10^{-6}$ Pa $= 4.7 \times 10^{-6}$ N/m^2. Pressure is force per unit area, so the average force F is the pressure p_{rad} times the area A:

$$F = p_{rad}A = (4.7 \times 10^{-6} \text{ N/m}^2)(4.0 \text{ m}^2) = 1.9 \times 10^{-5} \text{ N}.$$

This is comparable to the weight (on the earth) of a single grain of salt. Over time, however, this small force can have a noticeable effect on the orbit of a satellite like that in Fig. 33–13, and so radiation pressure must be taken into account.

*33–6 ELECTROMAGNETIC WAVES IN MATTER

So far, our discussion of electromagnetic waves has been restricted to waves in *vacuum.* But electromagnetic waves can also travel in *matter;* think of light traveling through air, water, or glass. We can extend our analysis fairly directly to electromagnetic waves in nonconducting materials, that is, *dielectrics.*

In a dielectric the wave speed is not the same as in vacuum, and we denote it by v instead of c. Faraday's law is unaltered, but in Eq. (33–4), derived from Faraday's law, the speed c is replaced by v. In Ampere's law the displacement current is given not by $\epsilon_0 \, d\Phi_E/dt$, where Φ_E is the flux of $\vec{E}$ through a surface, but by $\epsilon \, d\Phi_E/dt = K\epsilon_0 \, d\Phi_E/dt$, where K is the dielectric constant and ϵ is the permittivity of the dielectric. (We introduced these quantities in Section 25–5.) Also, the constant μ_0 in Ampere's law must be replaced by $\mu = K_m\mu_0$, where K_m is the relative permeability of the dielectric and μ is its permeability (see Section 29–9). Hence Eqs. (33–4) and (33–8) are replaced by

$$E = vB \quad \text{and} \quad B = \epsilon\mu vE. \tag{33–31}$$

Following the same procedure as in Section 33–3, we find that the wave speed v is

$$v = \frac{1}{\sqrt{\epsilon\mu}} = \frac{1}{\sqrt{KK_m}} \frac{1}{\sqrt{\epsilon_0\mu_0}} = \frac{c}{\sqrt{KK_m}} \qquad \text{(speed of electromagnetic waves in a dielectric).} \tag{33–32}$$

For most dielectrics the relative permeability K_m is very nearly equal to unity (except for insulating ferromagnetic materials). When $K_m \cong 1$,

$$v = \frac{1}{\sqrt{K}} \frac{1}{\sqrt{\epsilon_0\mu_0}} = \frac{c}{\sqrt{K}}.$$

Because K is always greater than unity, the speed v of electromagnetic waves in a dielectric is always *less* than the speed c in vacuum by a factor of $1/\sqrt{K}$. The ratio of the speed c in vacuum to the speed v in a material is known in optics as the **index of refraction n** of the material. When $K_m \cong 1$,

$$\frac{c}{v} = n = \sqrt{KK_m} \cong \sqrt{K}. \tag{33–33}$$

Usually, we can't use the values of K in Table 25–1 in this equation because those values are measured using *constant* electric fields. When the fields oscillate rapidly, there is usually not time for the re-orientating of electric dipoles that occurs with steady fields. Values of K with rapidly varying fields are usually much *smaller* than the values in the table. For example, K for water is 80.4 for steady fields but only about 1.8 in the frequency range of visible light. Thus the dielectric "constant" K is actually a function of frequency, called the *dielectric function* in more advanced treatments.

When a dielectric is present, we must also modify the expressions for the energy density and the Poynting vector. The energy density is then

$$u = \frac{1}{2}\epsilon E^2 + \frac{1}{2\mu}B^2 = \epsilon E^2. \tag{33–34}$$

The energy densities in the $\vec{E}$ and $\vec{B}$ fields are still equal (see Exercise 33–21). The Poynting vector becomes

$$\vec{S} = \frac{1}{\mu}\vec{E} \times \vec{B} \qquad \text{(Poynting vector in a dielectric)}, \qquad (33\text{–}35)$$

and its magnitude S is (using Eqs. (33–31))

$$S = \frac{EB}{\mu} = \sqrt{\frac{\epsilon}{\mu}}E^2. \qquad (33\text{–}36)$$

The intensity I of a sinusoidal wave in a dielectric is obtained by simply replacing ϵ_0 by ϵ and μ_0 by μ in Eqs. (33–26):

$$I = \frac{E_{max}B_{max}}{2\mu} = \frac{E_{max}^2}{2\mu v} = \frac{1}{2}\sqrt{\frac{\epsilon}{\mu}}E_{max}^2 = \frac{1}{2}\epsilon v E_{max}^2$$

$$\text{(sinusoidal wave in a dielectric)}. \qquad (33\text{–}37)$$

This discussion has been confined to electromagnetic waves in nonconducting dielectrics. In a *conducting* material, electromagnetic waves cannot propagate any appreciable distance because the $\vec{E}$ and $\vec{B}$ fields lead to currents that provide a mechanism for dissipating and reflecting the energy of the wave. For an ideal conductor with zero resistivity, $\vec{E}$ must be zero everywhere inside the material. When an electromagnetic wave strikes such a material, the wave is totally reflected. Real conductors with finite resistivity permit some penetration of the wave into the material, with partial reflection. A polished metal surface such as silver is usually an excellent reflector of electromagnetic waves. As we will see in the next section, such reflectors play an important role in producing *standing* electromagnetic waves.

EXAMPLE 33-5

A wave in a magnetic material An electromagnetic wave with a frequency of 100 MHz (in the FM radio broadcast band) travels in an insulating ferrite (ferromagnetic) material with the properties $K = 10$ and $K_m = 1000$ at this frequency. The intensity of the wave is 2.00×10^{-7} W/m^2. a) What is the speed of propagation? b) What is the wavelength of the wave? c) What are the amplitudes of the electric and magnetic fields in the material?

SOLUTION a) From Eqs. (33–32),

$$v = \frac{c}{\sqrt{KK_m}} = \frac{3.00 \times 10^8 \text{ m/s}}{\sqrt{(10)(1000)}} = 3.00 \times 10^6 \text{ m/s}.$$

This is much slower than the speed in vacuum.
b) The wavelength is

$$\lambda = \frac{v}{f} = \frac{3.00 \times 10^6 \text{ m/s}}{100 \times 10^6 \text{ s}^{-1}} = 0.0300 \text{ m} = 3.00 \text{ cm}.$$

The wavelength in a vacuum for this frequency is

$$\lambda_{vacuum} = \frac{c}{f} = \frac{3.00 \times 10^8 \text{ m/s}}{100 \times 10^6 \text{ s}^{-1}} = 3.00 \text{ m}.$$

c) From Eqs. (33–37) the amplitude of the electric field in the material is

$$E_{max} = \sqrt{\frac{2I}{\epsilon v}} = \sqrt{\frac{2I}{K\epsilon_0 v}}$$

$$= \sqrt{\frac{2(2.00 \times 10^{-7} \text{ W/m}^2)}{(10)(8.85 \times 10^{-12} \text{ C}^2/\text{N} \cdot \text{m}^2)(3.00 \times 10^6 \text{ m/s})}}$$

$$= 3.88 \times 10^{-2} \text{ V/m}.$$

From Eqs. (33–31), $E_{max} = vB_{max}$, so the magnetic-field magnitude is

$$B_{max} = \frac{E_{max}}{v} = \frac{3.88 \times 10^{-2} \text{ V/m}}{3.00 \times 10^6 \text{ m/s}} = 1.29 \times 10^{-8} \text{ T}.$$

By comparison the values of E_{max} and B_{max} for a wave of this intensity in vacuum are, from Eqs. (33–26) and (33–4),

$$E_{max\text{-}vacuum} = \sqrt{\frac{2I}{\epsilon_0 c}} = 1.23 \times 10^{-2} \text{ V/m},$$

$$B_{max\text{-}vacuum} = \frac{E_{max\text{-}vacuum}}{c} = 4.09 \times 10^{-11} \text{ T}.$$

Compared to a wave of the same intensity in vacuum, the wave in the ferrite has a slightly greater electric field and a *much* greater magnetic field.

33-7 STANDING ELECTROMAGNETIC WAVES

Electromagnetic waves can be *reflected;* the surface of a conductor (like a polished sheet of metal) or of a dielectric (such as a sheet of glass) can serve as a reflector. The superposition principle holds for electromagnetic waves just as for electric and magnetic fields. The superposition of an incident wave and a reflected wave forms a **standing wave.** The situation is analogous to standing waves on a stretched string, discussed in Section 20–3; we suggest that you review that discussion.

Suppose a sheet of a perfect conductor (zero resistivity) is placed in the yz-plane of Fig. 33–14 and a linearly polarized electromagnetic wave, traveling in the negative x-direction, strikes it. As we discussed in Section 24–5, $\vec{E}$ cannot have a component parallel to the surface of a perfect conductor. Therefore in the present situation, $\vec{E}$ must be zero everywhere in the yz-plane. The electric field of the *incident* electromagnetic wave is *not* zero at all times in the yz-plane. But this incident wave induces oscillating currents on the surface of the conductor, and these currents give rise to an additional electric field. The *net* electric field, which is the vector sum of this field and the incident $\vec{E}$, *is* zero everywhere inside and on the surface of the conductor.

The currents induced on the surface of the conductor also produced a *reflected* wave that travels out from the plane in the +x-direction. Suppose the incident wave is described by the wave functions of Eqs. (33–19) (a sinusoidal wave traveling in the −x-direction) and the reflected wave Eqs. (33–16) (a sinusoidal wave traveling in the +x-direction). The superposition principle states that the total $\vec{E}$ field at any point is the vector sum of the $\vec{E}$ fields of the incident and reflected waves, and similarly for the $\vec{B}$ field. Therefore the wave functions for the superposition of the two waves are

$$E(x,t) = E_{\text{max}}[-\sin(\omega t + kx) + \sin(\omega t - kx)],$$

$$B(x,t) = B_{\text{max}}[\sin(\omega t + kx) + \sin(\omega t - kx)].$$

We can expand and simplify these expressions, using the identities

$$\sin(A \pm B) = \sin A \cos B \pm \cos A \sin B;$$

the results are

$$E(x,t) = -2E_{\text{max}} \sin kx \cos \omega t, \tag{33-38}$$

$$B(x,t) = 2B_{\text{max}} \cos kx \sin \omega t. \tag{33-39}$$

Equation (33–38) is analogous to Eq. (20–1) for a stretched string. We see that at $x = 0$ the electric field $E(x = 0, t)$ is *always* zero; this is required by the nature of the ideal conductor, which plays the same role as a fixed point at the end of a string. Furthermore, $E(x, t)$ is zero at *all* times at points in those planes perpendicular to the x-axis for which

33-14 Representation of the electric and magnetic fields of a linearly polarized electromagnetic standing wave when $\omega t = \pi/4$ rad. In any plane perpendicular to the x-axis, E is maximum where B is zero, and conversely. As time elapses, the pattern does *not* move along the x-axis; instead, at every point the $\vec{E}$ and $\vec{B}$ vectors simply oscillate.

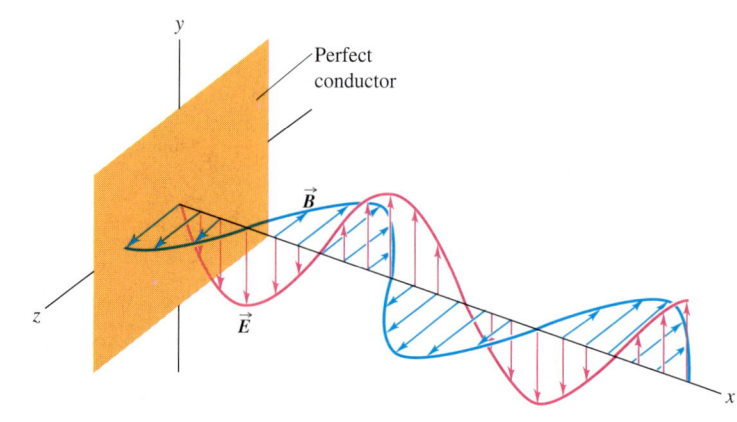

$\sin kx = 0$; that is, $kx = 0, \pi, 2\pi, \ldots$. Since $k = 2\pi/\lambda$, the positions of these planes are

$$x = 0, \frac{\lambda}{2}, \lambda, \frac{3\lambda}{2}, \cdots. \qquad \text{(nodal planes of } \vec{E}) \qquad (33\text{-}40)$$

These planes are called the **nodal planes** of the $\vec{E}$ field; they are the equivalent of the nodes, or nodal points, of a standing wave on a string. Midway between any two adjacent nodal planes is a plane on which $\sin kx = \pm 1$; on each such plane, the magnitude of $E(x, t)$ equals the maximum possible value of $2E_{max}$ twice per oscillation cycle. These are the **antinodal planes** of $\vec{E}$, corresponding to the antinodes of waves on a string.

The total magnetic field is zero at all times at points in planes on which $\cos kx = 0$. This occurs where

$$x = \frac{\lambda}{4}, \frac{3\lambda}{4}, \frac{5\lambda}{4}, \cdots. \qquad \text{(nodal planes of } \vec{B}) \qquad (33\text{-}41)$$

These are the nodal planes of the $\vec{B}$ field; there is an antinodal plane of $\vec{B}$ midway between any two adjacent nodal planes.

Figure 33–14 shows a standing-wave pattern at one instant of time. The magnetic field is *not* zero at the conducting surface ($x = 0$), and there is no reason it should be. The surface currents that must be present to make $\vec{E}$ exactly zero at the surface cause magnetic fields at the surface. The nodal planes of each field are separated by one half-wavelength. The nodal planes of one field are midway between those of the other; hence the nodes of $\vec{E}$ coincide with antinodes of $\vec{B}$, and conversely. Compare this situation to the distinction between pressure nodes and displacement nodes in Section 20–4.

The total electric field is a *cosine* function of t, and the total magnetic field is a *sine* function of t. The sinusoidal variations of the two fields are therefore 90° out of phase at each point. At times when $\cos \omega t = 0$, the electric field is zero *everywhere,* and the magnetic field is maximum. When $\sin \omega t = 0$, the magnetic field is zero everywhere, and the electric field is maximum. This is in contrast to a wave traveling in one direction, as described by Eqs. (33–16) or (33–19) separately, in which the sinusoidal variations of $\vec{E}$ and $\vec{B}$ at any particular point are *in phase.* It is interesting to check that Eqs. (33–38) and (33–39) satisfy the wave equation, Eq. (33–15). They also satisfy Eqs. (33–12) and (33–14) (the equivalents of Faraday's and Ampere's laws); we leave the proofs of these statements as problems.

Pursuing the stretched-string analogy, we may now insert a second conducting plane, parallel to the first and a distance L from it, along the $+x$-axis. This is analogous to a stretched string held at the points $x = 0$ and $x = L$. Both conducting planes must be nodal planes for $\vec{E}$; a standing wave can exist only when the second plane is placed at one of the positions where $E(x, t) = 0$. That is, for a standing wave to exist, L must be an integer multiple of $\lambda/2$. The wavelengths that satisfy this condition are

$$\lambda_n = \frac{2L}{n} \qquad (n = 1, 2, 3, \cdots). \qquad (33\text{-}42)$$

The corresponding frequencies are

$$f_n = \frac{c}{\lambda_n} = n\frac{c}{2L} \qquad (n = 1, 2, 3, \cdots). \qquad (33\text{-}43)$$

Thus there is a set of *normal modes,* each with a characteristic frequency, wave shape, and node pattern. By measuring the node positions, we can measure the wavelength. If the frequency is known, the wave speed can be determined. This technique was first used by Hertz in the 1880s in his pioneering investigations of electromagnetic waves. A laser such as that shown in Fig. 33–10 (Example 33–1) has two mirrors; a standing wave is

A microwave oven sets up a standing electromagnetic wave with $\lambda = 12.2$ cm, a wavelength that is strongly absorbed by the water in food. Because the wave has nodes spaced $\lambda/2 = 6.1$ cm apart, the food must be rotated while cooking. Otherwise, the portion that lies at a node—where the wave amplitude is zero—will remain cold.

set up in the cavity between the mirrors. One of the mirrors has a small, partially transmitting aperture that allows waves to escape from this end of the laser.

Conducting surfaces are not the only reflectors of electromagnetic waves. Reflections also occur at an interface between two insulating materials with different dielectric or magnetic properties. The mechanical analog is a junction of two strings with equal tension but different linear mass density. In general, a wave incident on such a boundary surface is partly transmitted into the second material and partly reflected back into the first. For example, light is transmitted through a glass window, but its surfaces also reflect light.

EXAMPLE 33-6

Prove that the intensity of the standing wave discussed in this section is zero.

SOLUTION The intensity of the wave is the average value S_{av} of the Poynting vector. Let's first find the instantaneous value of S. Using the wave functions of Eqs. (33–38) and (33–39) with Eq. (33–24) for S, we find

$$S(x, t) = \frac{E(x, t)B(x, t)}{\mu_0}$$

$$= \frac{(-2E_{max} \sin kx \cos \omega t)(2B_{max} \cos kx \sin \omega t)}{\mu_0}.$$

Using the identity $\sin 2A = 2 \sin A \cos A$, we can rewrite this as

$$S(x, t) = -\frac{E_{max} B_{max} \sin 2kx \sin 2\omega t}{\mu_0}.$$

The average value of a sine function over any whole number of cycles is zero. Thus *the time average of S at any point is zero;* $I = S_{av} = 0$. This is just what we should expect. We formed our standing wave by superposing two waves with the same frequency and amplitude, traveling in opposite directions. All the energy transferred by one wave is completely cancelled by an equal amount transferred in the opposite direction by the other wave. When we use waves to transmit power, it is important to avoid reflections that give rise to standing waves.

EXAMPLE 33-7

Electromagnetic standing waves are set up in a cavity with two parallel, highly conducting walls separated by 1.50 cm. a) Calculate the longest wavelength and lowest frequency of electromagnetic standing waves between the walls. b) For this longest-wavelength standing wave, where in the cavity does $\vec{E}$ have maximum magnitude? Where is $\vec{E}$ zero? Where does $\vec{B}$ have maximum magnitude? Where is $\vec{B}$ zero?

SOLUTION a) The longest possible wavelength corresponds to $n = 1$ in Eq. (33–42):

$$\lambda_1 = 2L = 2(1.50 \text{ cm}) = 3.00 \text{ cm}.$$

The corresponding frequency is

$$f = \frac{c}{\lambda} = \frac{3.00 \times 10^8 \text{ m/s}}{3.00 \times 10^{-2} \text{ m}} = 1.00 \times 10^{10} \text{ Hz} = 10 \text{ GHz}.$$

Alternatively, using Eq. (33–43) with $n = 1$, we have

$$f = \frac{c}{2L} = \frac{3.00 \times 10^8 \text{ m/s}}{2(1.50 \times 10^{-2} \text{ m})} = 1.00 \times 10^{10} \text{ Hz} = 10 \text{ GHz}.$$

b) With $n = 1$ there is a single half-wavelength between the walls. The electric field has nodal planes ($\vec{E} = 0$) at the walls and an antinodal plane (where the maximum magnitude of $\vec{E}$ occurs) midway between them. The magnetic field has *antinodal* planes at the walls and a nodal plane midway between them.

Such standing waves are used to produce an oscillating $\vec{E}$ field of definite frequency, which in turn is used to probe the behavior of a small sample of material placed inside the cavity. To subject the sample to the strongest possible field, the sample should be placed near the center of the cavity, at the antinode of $\vec{E}$.

33-8 THE ELECTROMAGNETIC SPECTRUM

Electromagnetic waves cover an extremely broad spectrum of wavelength and frequency. This **electromagnetic spectrum** encompasses radio and TV transmission, visible light, infrared and ultraviolet radiation, x rays, and gamma rays. Electromagnetic waves have been detected with frequencies from at least 1 to 10^{24} Hz; the most commonly encountered portion of the spectrum is shown in Fig. 33–15, which gives approximate wavelength and frequency ranges for the various segments. Despite vast differences in their uses and means of production, these are all electromagnetic waves with the general characteristics described in preceding sections, including the propagation speed (in vacuum) $c = 299,792,458$ m/s. Electromagnetic waves may differ in frequency f and wavelength λ, but the relation $c = \lambda f$ in vacuum holds for each.

Wavelengths in m

| 10 | 1 | 10^{-1} | 10^{-2} | 10^{-3} | 10^{-4} | 10^{-5} | 10^{-6} | 10^{-7} | 10^{-8} | 10^{-9} | 10^{-10} | 10^{-11} | 10^{-12} | 10^{-13} |

Radio, TV → ← Infrared → ← X rays →
← Microwave → ← Ultraviolet → ← Gamma rays →

| 10^8 | 10^9 | 10^{10} | 10^{11} | 10^{12} | 10^{13} | 10^{14} | 10^{15} | 10^{16} | 10^{17} | 10^{18} | 10^{19} | 10^{20} | 10^{21} | 10^{22} |

Visible light

Frequencies in Hz

| 700 nm | 650 | 600 | 550 | 500 | 450 | 400 nm |

RED   ORANGE   YELLOW   GREEN   BLUE   VIOLET

We can detect only a very small segment of this spectrum directly through our sense of sight. We call this range **visible light.** Its wavelengths range from about 400 to 700 nm (400 to 700 $\times 10^{-9}$ m), with corresponding frequencies from about 750 to 430 THz (7.5 to 4.3×10^{14} Hz). Different parts of the visible spectrum evoke in humans the sensations of different colors. Wavelengths for colors in the visible spectrum are given (very approximately) in Table 33-1.

Ordinary white light includes all visible wavelengths. However, by using special sources or filters, we can select a narrow band of wavelengths within a range of a few nm. Such light is approximately *monochromatic* (single-color) light. Absolutely monochromatic light with only a single wavelength is an unattainable idealization. When we use the expression "monochromatic light with $\lambda = 550$ nm" with reference to a laboratory experiment, we really mean a small band of wavelengths *around* 550 nm. Light from a *laser* is much more nearly monochromatic than is light obtainable in any other way.

*33–9 RADIATION FROM AN ANTENNA

Our discussion of electromagnetic waves in this chapter has centered mostly on *plane waves,* which propagate in a single direction. In any plane perpendicular to the direction of propagation of the wave, the $\vec{E}$ and $\vec{B}$ fields are uniform at any instant of time. Though easy to describe, plane waves are by no means the simplest to produce experimentally. Any charge or current distribution that oscillates sinusoidally with time, such as the oscillating point charge in Fig. 33–1, produces sinusoidal electromagnetic waves, but in general there is no reason to expect them to be plane waves.

A device that uses an oscillating distribution to produce electromagnetic radiation is called an *antenna.* A simple example of an antenna is an **oscillating electric dipole,** a pair of electric charges that vary sinusoidally with time such that at any instant the two charges have equal magnitude but opposite sign. (You may want to review the definition of an electric dipole in Section 22–9.) One charge could be equal to $Q \sin \omega t$ and the other to $-Q \sin \omega t$. An oscillating dipole antenna can be constructed in various ways, depending on frequency. One technique that works well for radio frequencies is to connect two straight conductors to the terminals of an ac source, as shown in Fig. 33–16.

The radiation pattern from an oscillating electric dipole is fairly complex, but at points far away from the dipole (compared to its dimensions and the wavelength of the radiation) it becomes fairly simple. We'll confine our description to this far region. A key feature of the radiation in the far region is that it is *not* a plane wave, but a wave that

33–15 The electromagnetic spectrum. The frequencies and wavelengths found in nature extend over such a wide range that we have to use a logarithmic scale to show all important bands. The boundaries between bands are somewhat arbitrary.

TABLE 33–1

Wavelengths of Visible light

400 to 440 nm	Violet
440 to 480 nm	Blue
480 to 560 nm	Green
560 to 590 nm	Yellow
590 to 630 nm	Orange
630 to 700 nm	Red

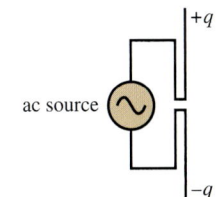

33–16 An oscillating electric dipole antenna. Each terminal of an ac source is connected to a straight conductor; the two conductors together comprise the antenna. As the voltage across the source oscillates, the charges on the two conductors also oscillate. The charges are always equal in magnitude and opposite in sign.

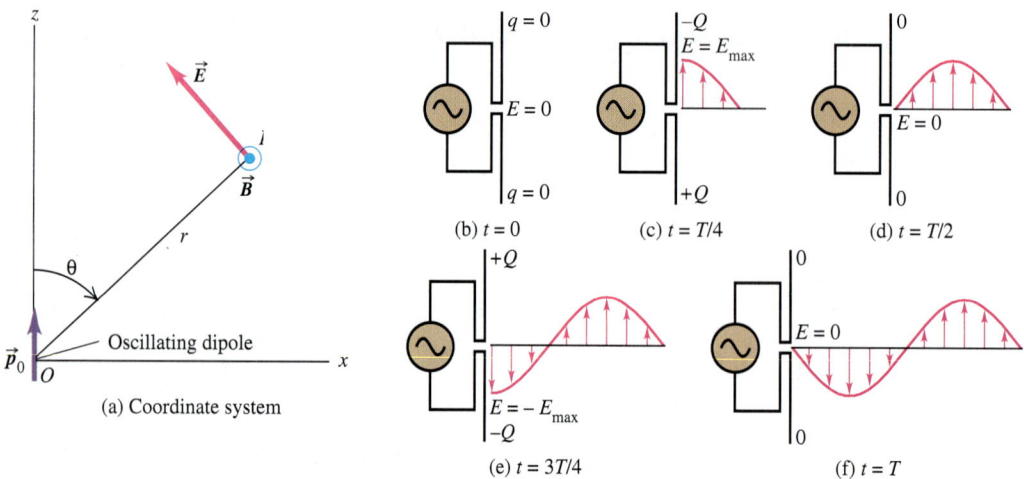

33–17 (a) An oscillating electric dipole oriented along the z-axis. The electric and magnetic fields at a point P are given by Eqs. (33–44). (b)–(f) One cycle in the production of an electromagnetic wave by an oscillating electric dipole antenna. The red curve and arrows depict the $\vec{E}$ field at points on the x-axis (where $\theta = \pi/2$); the magnetic field is not shown. The figure is not to scale.

travels out radially in all directions from the source. The wave fronts are not planes; in the far region they are expanding concentric spheres centered at the source. Figure 33–17 shows an oscillating electric dipole aligned with the z-axis, with maximum dipole moment p_0. The $\vec{E}$ and $\vec{B}$ fields at a point described by the spherical coordinates (r, θ, ϕ) have the directions shown in Fig. 33–17a during half the cycle and the opposite direction during the other half. Their magnitudes in the far region are

$$E(r, \theta, \phi, t) = \frac{p_0 k^2}{4\pi\epsilon_0} \frac{\sin\theta}{r} \sin(\omega t - kr),$$

$$B(r, \theta, \phi, t) = \frac{p_0 k^2}{4\pi\epsilon_0 c} \frac{\sin\theta}{r} \sin(\omega t - kr). \tag{33–44}$$

One distinctive feature of the oscillating-dipole fields given by Eqs. (33–44) is that their magnitudes are proportional to $1/r$. This is in contrast to the $\vec{E}$ field of a *stationary* point charge or the $\vec{B}$ field of a point charge moving with *constant* velocity, both of which are proportional of $1/r^2$. In fact, the complete expressions for the $\vec{E}$ and $\vec{B}$ fields of an oscillating dipole also include terms that are proportional to $1/r^2$; we haven't included these in Eqs. (33–44), since our interest is in the behavior of the fields in the far region, and the $1/r^2$ terms become negligible at greater distances r from the dipole.

Figure 33–18 shows a cross section of the radiation pattern at one instant. At each point, $\vec{E}$ is in the plane of the section and $\vec{B}$ is perpendicular to that plane. The electric field lines form closed loops, as is characteristic of *induced* electric fields; in the far region the electric field is induced by the variation of $\vec{B}$, and the magnetic field is induced by the variation of $\vec{E}$, forming a self-sustaining wave. The field magnitudes are greatest in the directions perpendicular to the dipole, where $\theta = \pi/2$; there is *no* radiation along the axis of the dipole, where $\theta = 0$ or π. We saw a similar result for a single oscillating electric charge in Fig. 33–1 (Section 33–2). The key difference is that Fig. 33–18 shows only the fields that are proportional to $1/r$ in the far region of an oscillating electric dipole, while Fig. 33–1 shows the field in the *near* region of a single oscillating charge; in this near region the $1/r^2$ terms must also be included.

At points very far from the oscillating dipole, $\vec{E}$ and $\vec{B}$ are perpendicular to each other, and the direction of the Poynting vector $\vec{S} = (\vec{E} \times \vec{B})/\mu_0$ is radially outward from the source. Because each field magnitude is proportional to $1/r$, the intensity I (the aver-

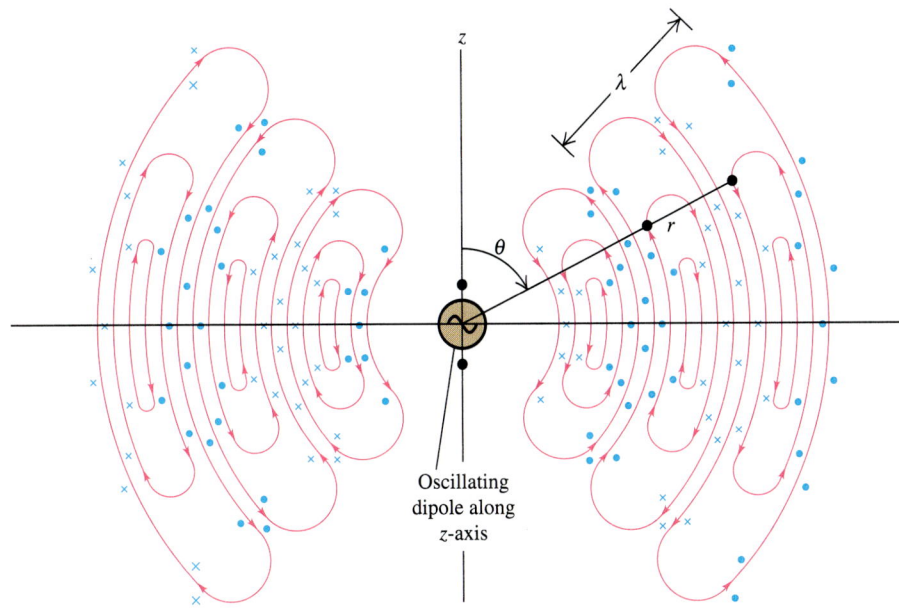

33-18 Representation of the electric field (red lines) and the magnetic field (blue dots and crosses) in a plane containing an oscillating electric dipole. During one period the loop of $\vec{E}$ shown closest to the source moves out and expands to become the loop shown farthest from the source. You can use Eq. (33–25) and the right-hand rule to find the direction of the Poynting vector $\vec{S}$ at each point within the pattern. No energy is radiated along the axis of the dipole.

Oscillating dipole along z-axis

age value of the magnitude of $\vec{S}$) is proportional to $1/r^2$. The net average power radiated by the oscillating dipole through a spherical surface of radius r centered on the dipole is the integral of the intensity over this surface. Since the area of this surface is proportional to r^2, the net average power is proportional to $(1/r^2)(r^2) = 1$; that is, the power radiated by the dipole in all directions is independent of r. This means that the radiated energy does not "get lost" as it spreads outward but continues on to arbitrarily great distances from the source. The intensity is also proportional to $\sin^2 \theta$, which vanishes on the axis of the dipole ($\theta = 0$ or $\theta = \pi$). No energy is radiated along the dipole axis.

Oscillating *magnetic* dipoles also act as radiation sources; an example is a circular loop antenna that uses a sinusoidal current. At sufficiently high frequencies a magnetic dipole antenna is more efficient at radiating energy than is an electric dipole antenna of the same overall size.

SUMMARY

- Maxwell's equations predict the existence of electromagnetic waves that propagate in vacuum with a speed equal to the speed of light. In a plane wave, $\vec{E}$ and $\vec{B}$ are uniform over any plane perpendicular to the propagation direction. Faraday's law requires that

$$E = cB. \qquad (33-4)$$

Ampere's law requires that

$$B = \epsilon_0 \mu_0 cE, \qquad (33-8)$$

where c is the propagation speed. For both of these requirements to be satisfied,

$$c = \frac{1}{\sqrt{\epsilon_0 \mu_0}}. \qquad (33-9)$$

- Electromagnetic waves are transverse; the $\vec{E}$ and $\vec{B}$ fields are perpendicular to the direction of propagation and to each other. The direction of propagation is the direction of $\vec{E} \times \vec{B}$.

KEY TERMS

electromagnetic wave, 1026
Maxwell's equations, 1026
electromagnetic radiation, 1027
plane wave, 1028
transverse wave, 1028
polarization, 1031
linearly polarized, 1031
wave equation, 1031
energy density, 1035
Poynting vector, 1036
intensity, 1037
radiation pressure, 1039
index of refraction, 1040
standing wave, 1042
nodal plane, 1043

- For a sinusoidal plane electromagnetic wave traveling in vacuum in the $+x$-direction,

$$E(x, t) = E_{max} \sin(\omega t - kx), \qquad B(x, t) = B_{max} \sin(\omega t - kx), \qquad (33-16)$$

$$E_{max} = c B_{max}. \qquad (33-18)$$

- The energy flow rate (power per unit area) in an electromagnetic wave in a vacuum is given by the Poynting vector $\vec{S}$:

$$\vec{S} = \frac{1}{\mu_0} \vec{E} \times \vec{B}. \qquad (33-25)$$

The time-average value of the magnitude EB/μ_0 of the Poynting vector is called the intensity I of the wave. For a sinusoidal wave in a vacuum,

$$I = S_{av} = \frac{E_{max} B_{max}}{2\mu_0} = \frac{E_{max}^2}{2\mu_0 c} = \frac{1}{2}\sqrt{\frac{\epsilon_0}{\mu_0}} E_{max}^2 = \frac{1}{2}\epsilon_0 c E_{max}^2. \qquad (33-26)$$

Electromagnetic waves also carry momentum. The rate of transfer of momentum per unit cross-section area is

$$\frac{1}{A}\frac{dp}{dt} = \frac{S}{c} = \frac{EB}{\mu_0 c}. \qquad (33-28)$$

When an electromagnetic wave strikes a surface, it exerts a radiation pressure p_{rad}. If the surface is perpendicular to the wave propagation direction and is totally absorbing, $p_{rad} = I/c$; if the surface is a perfect reflector, $p_{rad} = 2I/c$.

- When an electromagnetic wave travels through a dielectric, the wave speed v is

$$v = \frac{1}{\sqrt{\epsilon\mu}} = \frac{1}{\sqrt{KK_m}}\frac{1}{\sqrt{\epsilon_0\mu_0}} = \frac{c}{\sqrt{KK_m}}. \qquad (33-32)$$

In a dielectric the Poynting vector $\vec{S}$ and the intensity I of a sinusoidal wave are

$$\vec{S} = \frac{1}{\mu} \vec{E} \times \vec{B}, \qquad (33-35)$$

$$I = \frac{E_{max} B_{max}}{2\mu} = \frac{E_{max}^2}{2\mu v} = \frac{1}{2}\sqrt{\frac{\epsilon}{\mu}} E_{max}^2 = \frac{1}{2}\epsilon v E_{max}^2. \qquad (33-37)$$

- If a perfect reflecting surface is placed at $x = 0$, the incident and reflected waves form a standing wave. Nodal planes for $\vec{E}$ occur at $kx = 0, \pi, 2\pi, \ldots$, and nodal planes for $\vec{B}$ occur at $kx = \pi/2, 3\pi/2, 5\pi/2, \ldots$. At each point, the sinusoidal variations of $\vec{E}$ and $\vec{B}$ with time are $90°$ out of phase.

- The electromagnetic spectrum covers a range of frequencies from at least 1 to 10^{24} Hz and a correspondingly broad range of wavelengths. Visible light is a very small part of this spectrum, with wavelengths of 400 to 700 nm.

- An oscillating electric dipole produces a wave that radiates out in all directions. At points that are very far from the source, $\vec{E}$ and $\vec{B}$ are perpendicular to each other and to the radial direction, so this wave is transverse. The intensity depends on direction; it is zero along the dipole axis and greatest in the plane that perpendicularly bisects the dipole.

DISCUSSION QUESTIONS

Q33–1 By measuring the electric and magnetic fields at a point in space where there is an electromagnetic wave, can you determine the direction from which the wave came? Explain.

Q33–2 According to Ampere's law, is it possible to have both a conduction current and a displacement current at the same time? Is it possible for the effects of the two kinds of current to cancel each other exactly so that *no* magnetic field is produced? Explain.

Q33–3 Give several examples of electromagnetic waves that are encountered in everyday life. How are they all alike? How do they differ?

Q33–4 Sometimes neon signs located near a powerful radio station are seen to glow faintly at night, even though they are not turned on. What is happening?

Q33–5 Is polarization a property of all electromagnetic waves, or is it unique to visible light? Can sound waves be polarized? What fundamental distinction in wave properties is involved? Explain.

Q33–6 Suppose that a positive point charge q is initially at rest on the x-axis, in the path of the electromagnetic plane wave described in Section 33–3. Will the charge move after the wave front reaches it? If not, why not? If the charge does move, describe its motion qualitatively. (Remember that $\vec{E}$ and $\vec{B}$ have the same value at all points behind the wave front.)

Q33–7 The light beam from a searchlight may have an electric-field magnitude of 1000 V/m, corresponding to a potential difference of 1500 V between the head and feet of a 1.5-m-tall person on whom the light is shone. Does this cause the person to feel a strong electric shock? Why or why not?

Q33–8 The magnetic-field amplitude of the electromagnetic wave from the laser described in Example 33–1 (Section 33–4)

is about 100 times greater than the earth's magnetic field. If you illuminate a compass with the light from this laser, would you expect the compass to deflect? Why or why not?

Q33–9 In Eq. (33–16), the angular frequency ω appears instead of the ordinary frequency f. Explain why.

Q33–10 We are surrounded by electromagnetic waves emitted by many radio stations. How is a radio receiver able to select a single station among all these waves? What happens inside the receiver when it is tuned to change stations?

Q33–11 Most automobiles have vertical antennas for receiving radio broadcasts. Explain what this tells you about the direction of polarization of $\vec{E}$ in the radio waves used in broadcasting.

Q33–12 If a light beam carries momentum, should a person holding a flashlight feel a recoil analogous to the recoil of a rifle when it is fired? Why is this recoil not actually observed?

Q33–13 Does an electromagnetic *standing* wave have energy? Does it have momentum? Are your answers to these questions the same as for a *traveling* wave? Why or why not?

Q33–14 When driving on the upper level of the Bay Bridge westbound from Oakland to San Francisco, you can easily pick up a number of radio stations on your car radio. But when driving eastbound on the lower level of the bridge, which has steel girders on either side to support the upper level, the radio reception is much worse. Why is there a difference?

Q33–15 When an airplane flies directly over a transmitter on the ground that uses a vertical electric dipole antenna, a radio receiver in the airplane detects no signal from the transmitter. Why is this?

EXERCISES

SECTION 33–3 PLANE ELECTROMAGNETIC WAVES AND THE SPEED OF LIGHT

33–1 a) How much time does it take light to travel from the moon to the earth, a distance of 384,000 km? b) Light from the star Sirius takes 8.61 years to reach the earth. What is the distance to Sirius in kilometers?

33–2 TV Ghosting. In a TV picture, ghost images are formed when the signal from the transmitter travels to the receiver both directly and indirectly after reflection from a building or other large metallic mass. In a 25-inch set, the ghost is about 1.0 cm to the right of the principal image if the reflected signal arrives 0.60 μs after the principal signal. In this case, what is the difference in path lengths for the two signals?

SECTION 33–4 SINUSOIDAL ELECTROMAGNETIC WAVES

33–3 A sinusoidal electromagnetic wave of frequency 6.10×10^{14} Hz travels in vacuum in the $+z$-direction. The $\vec{B}$ field is parallel to the y-axis and has amplitude 5.80×10^{-4} T. Write the vector equations for $\vec{E}(z, t)$ and $\vec{B}(z, t)$.

33–4 An electromagnetic wave of wavelength 435 nm is traveling in vacuum in the $-z$-direction. The electric field has amplitude 2.70×10^{-3} V/m and is parallel to the x-axis. What is a) the frequency? b) the magnetic-field amplitude? c) Write the vector equations for $\vec{E}(z, t)$ and $\vec{B}((z, t))$.

33–5 An electromagnetic wave has an electric field given by $\vec{E}(y, t) = -(3.10 \times 10^5 \text{ V/m})\hat{k} \sin[(2.65 \times 10^{12} \text{ rad/s})t - ky]$. a) In what direction is the wave traveling? b) What is the wavelength of the wave? c) Write the vector equation for $\vec{B}(y, t)$.

33–6 An electromagnetic wave has a magnetic field given by $\vec{B}(x, t) = (8.25 \times 10^{-9} \text{ T})\hat{j} \sin[\omega t + (1.38 \times 10^4 \text{ rad/m})x]$. a) In what direction is the wave traveling? b) What is the frequency f of the wave? c) Write the vector equation for $\vec{E}(x, t)$.

33–7 Radio station WCCO in Minneapolis broadcasts at a frequency of 830 kHz. At a point some distance from the transmitter, the magnetic-field amplitude of the electromagnetic wave from WCCO is 4.82×10^{-11} T. Calculate a) the wavelength; b) the wave number; c) the angular frequency; d) the electric-field amplitude.

33–8 The electric-field amplitude near a certain radio transmitter is 3.85×10^{-3} V/m. What is the amplitude of $\vec{B}$? How does this compare in magnitude with the earth's field?

SECTION 33–5 ENERGY AND MOMENTUM IN ELECTROMAGNETIC WAVES

33–9 At a distance of 25.0 km from a radio station antenna, the electric-field amplitude is 0.0900 V/m. a) What is the magnetic-field amplitude at this point? b) Assuming that the antenna radiates equally in all directions above ground, what is the total power output of the station? c) At what distance from the antenna is $E_{max} = 0.0300$ V/m, one-third the above value?

33–10 A sinusoidal electromagnetic wave emitted by a cellular phone has a wavelength of 35.4 cm and an electric-field amplitude of 5.40×10^{-2} V/m at a distance of 250 m from the antenna. Calculate a) the frequency of the wave; b) the magnetic-field amplitude; c) the intensity of the wave.

33–11 A monochromatic light source with power output 60.0 W radiates light of wavelength 700 nm uniformly in all directions. Calculate E_{max} and B_{max} for the 700-nm light at a distance of 5.00 m from the source.

33–12 For the electromagnetic wave represented by Eq. (33–19), show that the Poynting vector a) is in the same direction as the propagation of the wave; b) has average magnitude given by Eqs. (33–26).

33–13 If the intensity of direct sunlight at a point on the earth's surface is 0.78 kW/m^2, find a) the average momentum density (momentum per unit volume) in the sunlight; b) the average momentum flow rate in the sunlight.

33–14 In the 25-ft Space Simulator facility at NASA's Jet Propulsion Laboratory, a bank of overhead arc lamps can produce light of intensity 2500 W/m^2 at the floor of the facility. (This simulates the intensity of sunlight near the planet Venus.) Find the average radiation pressure (in pascals and in atmospheres) on a) a totally absorbing section of the floor; b) a totally reflecting section of the floor. c) Find the average momentum density (momentum per unit volume) in the light at the floor.

33–15 Verify that all the expressions in Eqs. (33–24) are equivalent to Eq. (33–23).

33–16 Consider each of the electric- and magnetic-field orientations given below. In each case, what is the direction of propagation of the wave? a) $\vec{E} = E\hat{\imath}, \vec{B} = -B\hat{\jmath}$; b) $\vec{E} = E\hat{\jmath}, \vec{B} = B\hat{\imath}$; c) $\vec{E} = -E\hat{k}, \vec{B} = -B\hat{\imath}$; d) $\vec{E} = E\hat{\imath}, \vec{B} = -B\hat{k}$.

*SECTION 33–6 ELECTROMAGNETIC WAVES IN MATTER

*33–17 An electromagnetic wave propagates in a dielectric material. At the frequency of the light, the dielectric constant of the material is 1.74 and the relative permeability is 1.23. If the magnetic-field amplitude is 3.80×10^{-9} T, what is the electric-field amplitude?

*33–18 An electromagnetic wave with frequency 65.0 Hz travels in an insulating magnetic material that has dielectric constant 3.64 and relative permeability 5.18 at this frequency. The electric field has amplitude 7.20×10^{-3} V/m. a) What is the speed of propagation of the wave? b) What is the wavelength of the wave? c) What is the amplitude of the magnetic field? d) What is the intensity of the wave?

*33–19 An electromagnetic wave with frequency 5.70×10^{14} Hz propagates with a speed of 2.17×10^8 m/s in a certain piece of glass. Find a) the wavelength of the wave in the glass; b) the wavelength of a wave of the same frequency propagating in air; c) the index of refraction n of the glass for an electromagnetic wave with this frequency; d) the dielectric constant for glass at this frequency, assuming that the relative permeability is unity.

*33–20 An electromagnetic wave with frequency 38.0 MHz, wavelength 6.15 m, and intensity 2.95×10^{-6} W/m^2 travels in an insulating material that has K_m very nearly equal to unity. a) What is the speed of propagation of the wave? b) What is the dielectric constant of the insulating material at this frequency? c) What are the amplitudes of the electric and magnetic fields in the material?

*33–21 Use Eqs. (33–31) and (33–32) to prove that the electric and magnetic energy densities are equal in a dielectric, as stated in Eq. (33–34).

SECTION 33–7 STANDING ELECTROMAGNETIC WAVES

33–22 An electromagnetic standing wave in air has frequency 75.0 MHz. a) What is the distance between nodal planes of the $\vec{E}$ field? b) What is the distance between a nodal plane of $\vec{E}$ and the closest nodal plane of $\vec{B}$?

33–23 An electromagnetic standing wave in a certain material has frequency 1.20×10^{10} Hz and speed of propagation 2.10×10^8 m/s. a) What is the distance between a nodal plane of $\vec{B}$ and the closest antinodal plane of $\vec{B}$? b) What is the distance between an antinodal plane of $\vec{E}$ and the closest antinodal plane of $\vec{B}$? c) What is the distance between a nodal plane of $\vec{E}$ and the closest nodal plane of $\vec{B}$?

33–24 An electromagnetic standing wave in air of frequency 750 MHz is set up between two conducting planes 80.0 cm apart. At what positions between the planes could a point charge be placed at rest so that it would *remain* at rest? Explain.

33–25 A standing electromagnetic wave in a certain material has frequency 2.20×10^{10} Hz. The nodal planes of $\vec{B}$ are 3.55 mm apart. Find a) the wavelength of the wave in this material; b) the distance between adjacent nodal planes of the $\vec{E}$ field; c) the speed of propagation of the wave.

*33–26 Show that the standing wave electric and magnetic fields for standing waves given by Eqs. (33–38) and (33–39) a) satisfy the wave equation, Eq. (33–15); b) satisfy Eqs. (33–12) and (33–14).

SECTION 33–8 THE ELECTROMAGNETIC SPECTRUM

33–27 For waves propagating in air, what is the wavelength in meters and nanometers of a) gamma rays of frequency 6.50×10^{21} Hz; b) visible light of frequency 5.75×10^{14} Hz?

33–28 For an electromagnetic wave propagating in air, determine the frequency of a wave with a wavelength of a) 5.0 km; b) 5.0 m; c) 5.0 μm; d) 5.0 nm.

PROBLEMS

33–29 Consider a plane electromagnetic wave such as that shown in Fig. 33–2, but in which $\vec{E}$ and $\vec{B}$ also have components in the x-direction (along the direction of wave propagation). Use Gauss's law for electric and magnetic fields to show that the components E_x and B_x must both be equal to zero so that the fields $\vec{E}$ and $\vec{B}$ are both transverse. (*Hint:* Use a Gaussian surface like that shown in Fig. 33–3. Of the two faces parallel to the yz-plane, choose one to be to the left of the wave front and the other to the right of the wave front.)

*33–30 Consider a sinusoidal electromagnetic wave with fields $\vec{E} = E_{max}\hat{j} \sin(\omega t - kx)$ and $\vec{B} = B_{max}\hat{k} \sin(\omega t - kx + \phi)$, with $-\pi \le \phi \le \pi$. Show that if $\vec{E}$ and $\vec{B}$ are to satisfy Eqs. (33–12) and (33–14), then $E_{max} = cB_{max}$ and $\phi = 0$. (The result $\phi = 0$ means the $\vec{E}$ and $\vec{B}$ fields oscillate in phase.)

*33–31 Show that the *magnetic* field $B_z(x, t)$ in a plane electromagnetic wave propagating in the $+x$-direction must satisfy Eq. (33–15). (*Hint:* Take the partial derivative of Eq. (33–12) with respect to t and the partial derivative of Eq. (33–14) with respect to x. Then combine the results.)

33–32 For a sinusoidal electromagnetic wave in vacuum, such as that described by Eqs. (33–16), show that the *average* energy density in the electric field is the same as that in the magnetic field.

33–33 A small mirror with area 5.00 cm² faces a monochromatic light source that is 3.20 m away. At the mirror the electric-field amplitude of the light from the source is 0.0280 V/m. a) How much energy is incident on the mirror in 1.00 s? b) What is the average radiation pressure exerted by the light on the mirror? c) What is the total radiant power output of the source if it is assumed to radiate uniformly in all directions?

33–34 A plane sinusoidal electromagnetic wave in air has a wavelength of 3.84 cm and an $\vec{E}$-field amplitude of 1.35 V/m. a) What is the frequency? b) What is the $\vec{B}$-field amplitude? c) What is the intensity? d) What average force does this radiation exert on a totally absorbing surface with area 0.240 m² perpendicular to the direction of propagation?

33–35 A small helium-neon laser emits red visible light with a power of 3.20 mW in a beam that has a diameter of 2.50 mm. a) What are the amplitudes of the electric and magnetic fields of the light? b) What are the average energy densities associated with the electric field and with the magnetic field? c) What is the total energy contained in a 1.00-m length of the beam?

33–36 The plane of a flat surface is perpendicular to the propagation direction of an electromagnetic wave of intensity I. The surface absorbs a fraction w of the incident intensity, where $0 \le w \le 1$, and reflects the rest. a) Show that the radiation pressure on the surface equals $(2 - w)I/c$. b) Show that this expression gives the correct results for a surface that is i) totally absorbing; ii) totally reflective. c) For an incident intensity of 1.40 kW/m², what is the radiation pressure for 90% absorption? For 90% reflection?

33–37 The sun emits energy in the form of electromagnetic waves at a rate of 3.9×10^{26} W. This energy is produced by nuclear reactions deep in the sun's interior. a) Find the intensity of electromagnetic radiation and the radiation pressure on an absorbing object at the surface of the sun (radius $r = R =$

6.96×10^5 km) and at $r = R/2$, in the sun's interior. Ignore any scattering of the waves as they move radially outward from the center of the sun. Compare to the values given in Section 33–5 for sunlight just before it enters the earth's atmosphere. b) The gas pressure at the sun's surface is about 1.0×10^4 Pa; at $r = R/2$, the gas pressure is calculated from solar models to be about 4.7×10^{13} Pa. Comparing with your results in part (a), would you expect that radiation pressure is an important factor in determining the structure of the sun? Why or why not?

33–38 For a sinusoidal electromagnetic plane wave propagating in the $+x$-direction, the Poynting vector is $\vec{S}(x, t) = S(x, t)\hat{i}$, where $S(x, t) = (E_{max}B_{max}/2\mu_0)[1 - \cos 2(\omega t - kx)]$. a) Is $S(x, t)$ ever negative, corresponding to energy flowing in the direction *opposite* the direction of wave propagation? Explain why or why not. b) Figure 33–8 shows $\vec{E}$ and $\vec{B}$ for a plane electromagnetic wave at $t = 0$. Draw a copy of this figure, and on the copy draw vectors showing $\vec{S}$ for $t = 0$ at the following values of x: 0, $\lambda/8$, $\lambda/4$, $3\lambda/8$, $\lambda/2$, $5\lambda/8$, $3\lambda/4$, $7\lambda/8$, and λ.

33–39 A very long solenoid with n turns per unit length and radius a carries a current i that is increasing at a constant rate of di/dt. a) Calculate the magnetic field and induced electric field at a point inside the solenoid at a distance r from the solenoid axis. b) Compute the magnitude and direction of the Poynting vector $\vec{S}$ at this point. Show that $\vec{S}$ is directed inward towards the solenoid axis. c) Find the magnetic energy stored in a length l of the solenoid, and the rate at which that energy is increasing due to the increase in the current. d) Consider a cylindrical surface of radius a and length l that coincides with the coils of the solenoid. Integrate $\vec{S}$ over this surface to find the total rate at which electromagnetic energy is flowing into the solenoid through the solenoid walls. e) Compare the rate of change of magnetic-field energy from part (c) with the result of part (d). Discuss why the energy stored in a current-carrying solenoid can be thought of a entering through the cylindrical walls of the solenoid.

33–40 Consider the standing wave given by Eqs. (33–38) and (33–39). Let $\vec{E}$ be parallel to the y-axis and $\vec{B}$ be parallel to the z-axis; the equations then give the components of $\vec{E}$ and $\vec{B}$ along each axis. a) Plot the energy density as a function of x, $0 < x < \pi/k$, for the times $t = 0$, $\pi/4\omega$, $\pi/2\omega$, $3\pi/4\omega$, and π/ω. b) Find the direction of $\vec{S}$ in the regions $0 < x < \pi/2k$ and $\pi/2k < x < \pi/k$ at the times $t = \pi/4\omega$, $3\pi/4\omega$. c) Use your results in part (b) to explain the plots obtained in part (a).

33–41 A cylindrical conductor with circular cross section has a radius a and a resistivity ρ and carries a constant current I. a) What are the magnitude and direction of the electric field vector $\vec{E}$ at a point just inside the wire at a distance a from the axis? b) What are the magnitude and direction of the magnetic field vector $\vec{B}$ at the same point? c) What are the magnitude and direction of the Poynting vector $\vec{S}$ at the same point? (The direction of $\vec{S}$ is the direction in which electromagnetic energy flows into or out of the conductor.) d) Use the result in part (c) to find the rate of flow of energy into the volume occupied by a length l of the conductor. (*Hint:* Integrate $\vec{S}$ over the surface of this volume.) Compare your result to the rate of generation of thermal energy in the same volume. Discuss why the energy dissipated in a current-carrying conductor due to its resistance can be thought of as entering through the cylindrical sides of the conductor.

33–42 A capacitor consists of two circular plates of radius r separated by a distance l. Neglecting fringing, show that while the capacitor is being charged, the rate at which energy flows into the space between the plates is equal to the rate at which the electrostatic energy stored in the capacitor increases. (*Hint:* Integrate $\vec{S}$ over the surface of the empty cylindrical volume between the plates.)

33–43 A circular loop of wire can be used as a radio antenna. If a 18.0-cm-diameter antenna is located 2.50 km from a 95.0-MHz source with a total power of 55.0 kW, what is the maximum emf induced in the loop? (Assume that the plane of the antenna loop is perpendicular to the direction of the radiation's magnetic field and that the source radiates uniformly in all directions.)

33–44 It has been proposed to place solar-power-collecting satellites in earth orbit. The power they collect would be beamed down to earth as microwave radiation. For a microwave beam with a cross-section area of 36.0 m^2 and a total power of 2.80 kW at the earth's surface, what is the amplitude of the electric field of the beam at the earth's surface?

33–45 A space-walking astronaut has run out of fuel for her jet-pack and is floating 16.0 m from her spacecraft with zero relative velocity. The astronaut and all her equipment have a total mass of 150 kg. If she uses her 120-W flashlight as a light rocket, how long will it take her to reach the spacecraft?

33–46 The 19th-century inventor Nikola Tesla proposed to transmit electric power via sinusoidal electromagnetic waves. Suppose power is to be transmitted in a beam of cross-section area 100 m^2. What electric- and magnetic-field amplitudes are required to transmit an amount of power comparable to that

handled by modern transmission lines (that carry voltages and currents of the order of 500 kV and 1000 A)?

33–47 Interplanetary space contains many small particles referred to as *interplanetary dust*. Radiation pressure from the sun sets a lower limit on the size of such dust particles. To see the origin of this limit, consider a spherical dust particle of radius R and mass density ρ. a) Write an expression for the gravitational force exerted on this particle by the sun (mass M) when the particle is a distance r from the sun. b) Let L represent the luminosity of the sun, equal to the rate at which it emits energy in electromagnetic radiation. Find the force exerted on the (totally absorbing) particle due to solar radiation pressure. The relevant area is the cross-section area of the particle, *not* the total surface area of the particle. As part of your answer, explain why this is so. c) The mass density of a typical interplanetary dust particle is about 3000 kg/m^3. Find the particle radius R such that the gravitational and radiation forces acting on the particle are equal in magnitude. The luminosity of the sun is 3.9×10^{26} W. Does your answer depend on the distance of the particle from the sun? Why or why not? d) Explain why dust particles with a radius less than that found in part (c) are unlikely to be found in the solar system. (*Hint:* Construct the ratio of the two force expressions found in parts (a) and (b).)

33–48 NASA is giving serious consideration to the concept of *solar sailing*. A solar sailcraft uses a large, low-mass sail and the energy and momentum of sunlight for propulsion. a) Should the sail be absorbing or reflective? Why? b) The total power output of the sun is 3.9×10^{26} W. How large a sail is necessary to propel a 10,000-kg spacecraft against the gravitational force of the sun? Express your result in square kilometers. c) Explain why your answer to part (b) is independent of the distance from the sun.

CHALLENGE PROBLEMS

33–49 Electromagnetic radiation is emitted by accelerating charges. The rate at which energy is emitted from an accelerating charge that has charge q and acceleration a is given by

$$\frac{dE}{dt} = \frac{q^2 a^2}{6\pi\epsilon_0 c^3},$$

where c is the speed of light. a) Verify that this equation is dimensionally correct. b) If a proton with a kinetic energy of 6.0 MeV is traveling in a particle accelerator in a circular orbit of radius 0.750 m, what fraction of its energy does it radiate per second? c) Consider an electron orbiting with the same speed and radius. What fraction of its energy does it radiate per second?

33–50 The Classical Hydrogen Atom. The electron in a hydrogen atom can be considered to be in a circular orbit with a radius of 0.0529 nm and a kinetic energy of 13.6 eV If the electron behaved classically, how much energy would it radiate per second? (See Challenge Problem 33–49.) What does this tell you about the use of classical physics in describing the atom?

*33–51 Electromagnetic waves propagate much differently in *conductors* than they do in dielectrics or in vacuum. If the resistivity of the conductor is sufficiently low (that is, if it is a sufficiently good conductor), the oscillating electric field of the wave gives rise to an oscillating conduction current that is much larger than the displacement current. In this case, the wave equa-

tion for an electric field $\vec{E}(x, t) = Ey(x, t)\hat{j}$ propagating in the $+x$-direction within a conductor is

$$\frac{\partial^2 E_y(x, t)}{\partial x^2} = \frac{\mu}{\rho} \frac{\partial E_y(x, t)}{\partial t},$$

where μ is the permeability of the conductor and ρ is its resistivity. a) A solution to this wave equation is

$$E_y(x, t) = E_{max} e^{-k_C x} \sin(\omega t - k_C x),$$

where $k_C = \sqrt{\omega\mu/2\rho}$. Verify this by substituting $E_y(x, t)$ into the above wave equation. b) The exponential term shows that the electric field decreases in amplitude as it propagates. Explain why this happens. (*Hint:* The field does work to move charges within the conductor. The current of these moving charges causes i^2R heating within the conductor, raising its temperature. Where does the energy to do this come from?). c) Show that the electric-field amplitude decreases by a factor of $1/e$ in a distance $1/k_C = \sqrt{2\rho/\omega\mu}$, and calculate this distance for a radio wave with frequency $f = 1.0$ MHz in copper (resistivity 1.72×10^{-8} $\Omega \cdot$ m, permeability $\mu = \mu_0$). Since this distance is so short, electromagnetic waves of this frequency can hardly propagate at all into copper. Instead, they are reflected at the surface of the metal. This is why radio waves cannot penetrate through copper or other metals, and why radio reception is poor inside a metal structure.

34

The Nature and Propagation of Light

When white light passes through a glass prism, a rainbow of colors emerges. Isaac Newton was the first to show that prisms do not add color to white light, but rather cause light of different colors to bend through different angles. This phenomenon, called dispersion, is a characteristic of glass and other transparent materials.

34–1 INTRODUCTION

Blue lakes, ochre deserts, green forests, and multicolored rainbows can be enjoyed by anyone who has eyes with which to see them. But by studying the branch of physics called **optics,** which deals with the behavior of light and other electromagnetic waves, we can reach a deeper appreciation of the visible world. A knowledge of the properties of light allows us to understand the blue color of the sky and the design of optical devices such as telescopes, microscopes, cameras, eyeglasses, and the human eye. The same basic principles of optics also lie at the heart of modern developments such as the laser, optical fibers, holograms, optical computers, and new techniques in medical imaging.

The importance of optics to physics, and to science and engineering in general, is so great that we will devote the next five chapters to its study. In this chapter we begin with a study of the laws of reflection and refraction and the concepts of dispersion, polarization, and scattering of light. Along the way we compare the various possible descriptions of light in terms of particles, rays, or waves, and we introduce Huygens' principle, an important link that connects the ray and wave viewpoints. In Chapters 35 and 36 we'll use the ray description of light to understand how mirrors and lenses work, and we'll see how mirrors and lenses are used in optical instruments such as cameras, microscopes, and telescopes. We'll explore the wave characteristics of light further in Chapters 37 and 38.

34–2 THE NATURE OF LIGHT

Until the time of Isaac Newton (1642–1727), most scientists thought that light consisted of streams of particles (called *corpuscles*) emitted by light sources. Galileo and others tried (unsuccessfully) to measure the speed of light. Around 1665, evidence of *wave* properties of light began to be discovered. By the early nineteenth century, evidence that light is a wave had grown very persuasive.

In 1873, James Clerk Maxwell predicted the existence of electromagnetic waves and calculated their speed of propagation, as we learned in Chapter 33. This development, along with the experimental work of Heinrich Hertz starting in 1887, showed conclusively that light is indeed an electromagnetic wave.

The wave picture of light is not the whole story, however. Several effects associated with emission and absorption of light reveal a particle aspect, in that the energy carried by light waves is packaged in discrete bundles called *photons* or *quanta*. These apparently contradictory wave and particle properties have been reconciled since 1930 with the development of quantum electrodynamics, a comprehensive theory that includes *both* wave and particle properties. The *propagation* of light is best described by a wave model, but understanding emission and absorption requires a particle approach.

The fundamental sources of all electromagnetic radiation are electric charges in accelerated motion. All bodies emit

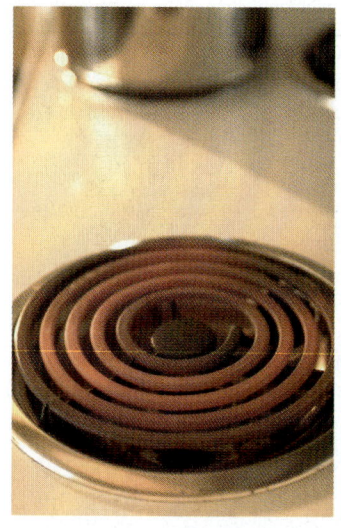

An electric heating element emits primarily infrared radiation. But if its temperature is high enough, it also emits a discernible amount of visible light.

electromagnetic radiation as a result of thermal motion of their molecules; this radiation, called *thermal radiation,* is a mixture of different wavelengths. At sufficiently high temperatures, all matter emits enough visible light to be self-luminous; a very hot body appears "red hot" or "white hot." Thus hot matter in any form is a light source. Familiar examples are a candle flame, hot coals in a campfire, the coils in an electric room heater, and an incandescent lamp filament (which usually operates at a temperature of about 3000°C).

Light is also produced during electrical discharges through ionized gases. The bluish light of mercury-arc lamps, the orange-yellow of sodium-vapor lamps, and the various colors of "neon" signs are familiar. A variation of the mercury-arc lamp is the *fluorescent* lamp. This light source uses a material called a *phosphor* to convert the ultraviolet radiation from a mercury arc into visible light. This conversion makes fluorescent lamps more efficient than incandescent lamps in converting electrical energy into light.

A light source that has attained prominence in the last forty years is the *laser.* In most light sources, light is emitted independently by different atoms within the source; in a laser, by contrast, atoms are induced to emit light in a cooperative, coherent fashion. The result is a very narrow beam of radiation that can be enormously intense and that is much more nearly *monochromatic,* or single-frequency, than light from any other source. Lasers are used by physicians for microsurgery, in CD players and computers to scan the information encoded on a compact disc or CD-ROM, in industry to cut through steel and to fuse high-melting-point materials, and in many other applications.

No matter what its source, electromagnetic radiation travels in vacuum at the speed of light. The first demonstration that the speed of light is not infinite was made in 1676 by the Danish astronomer Ole Rømer, from observations of the motion of one of Jupiter's satellites. The first successful *terrestrial* measurement of the speed of light was made by the French scientist Armand Fizeau in 1849, using a reflected light beam interrupted by a notched rotating disk. More refined versions of this experiment were carried out later in the nineteenth century by Jean Foucault in France and by Albert A. Michelson in the United States.

From analysis of all measurements up to 1983, the most probable value for the speed of light at that time was

$$c = 2.99792458 \times 10^8 \text{ m/s.}$$

As we explained in Section 1–4, the definition of the second based on the cesium clock is precise to within one part in 10 trillion (10^{13}). Up to 1983 the definition of the meter was much less precise, about four parts in a billion (10^9). Any attempt to measure the speed of light with greater precision foundered on this limitation. For this reason, in November 1983 the General Conference of Weights and Measures redefined the meter by *defining* the speed of light in vacuum to be precisely 299,792,458 m/s. One meter is now defined to be the distance traveled by light in a time of 1/299,792,458 s, with the second defined by the cesium clock.

WAVES, WAVE FRONTS, AND RAYS

We often use the concept of a **wave front** to describe wave propagation. We introduced this concept in Section 33–3 to describe the leading edge of a wave. More generally, we define a wave front as *the locus of all adjacent points at which the phase of vibration of a physical quantity associated with the wave is the same.* That is, at any instant, all points on a wave front are at the same part of the cycle of their variation.

When we drop a pebble into a calm pool, the expanding circles formed by the wave crests, as well as the circles formed by the wave troughs between them, are wave fronts. Similarly, when sound waves spread out in still air from a pointlike source, or when electromagnetic radiation spreads out from a pointlike emitter, any spherical surface that is concentric with the source is a wave front, as shown in Fig. 34–1. In diagrams of wave motion we usually draw only parts of a few wave fronts, often choosing consecutive wave fronts that have the same phase and thus are one wavelength apart, such as crests of water waves. Similarly, a diagram for sound waves might show only the "pressure crests," the surfaces over which the pressure is maximum, and a diagram for electromagnetic waves might show only the "crests" on which the electric or magnetic field is maximum.

We will often use diagrams that show the shapes of the wave fronts or their cross sections in some reference plane. For example, when electromagnetic waves are radiated by a small light source, we can represent the wave fronts as spherical surfaces concentric with the source or, as in Fig. 34–2a, by the circular intersections of these surfaces with the plane of the diagram. Far away from the source, where the radii of the spheres have become very large, a section of a spherical surface can be considered as a plane, and we have a *plane* wave like those discussed in Sections 33–3 and 33–4 (Fig. 34–2b).

To describe the directions in which light propagates, it's often convenient to represent a light wave by **rays** rather than by wave fronts. Rays were used to describe light long before its wave nature was firmly established. In a particle theory of light, rays are the paths of the particles. From the wave viewpoint *a ray is an imaginary line along the direction of travel of the wave*. In Fig. 34–2a the rays are the radii of the spherical wave fronts, and in Fig. 34–2b they are straight lines perpendicular to the wave fronts. When waves travel in a homogeneous isotropic material (a material with the same properties in all regions and in all directions), the rays are always straight lines normal to the wave fronts. At a boundary surface between two materials, such as the surface of a glass plate in air, the wave speed and the direction of a ray may change, but the ray segments in the air and in the glass are straight lines.

The next several chapters will give you many opportunities to see the interplay of the ray, wave, and particle descriptions of light. The branch of optics for which the ray description is adequate is called **geometric optics;** the branch dealing specifically with wave behavior is called **physical optics.** This chapter and the two following ones are concerned mostly with geometric optics. In Chapters 37 and 38 we will study wave phenomena and physical optics.

34–3 REFLECTION AND REFRACTION

In this section we'll use the *ray* model of light to explore two of the most important aspects of light propagation: **reflection** and **refraction.** When a light wave strikes a smooth interface separating two transparent materials (such as air and glass or water and glass), the wave is in general partly *reflected* and partly *refracted* (transmitted) into the second material, as shown in Fig. 34–3a (page 1056). For example, when you look into a restaurant window from the street, you see a reflection of the street scene, but a person inside the restaurant can look out through the window at the same scene as light reaches him by refraction.

The segments of plane waves shown in Fig. 34–3a can be represented by bundles of rays forming *beams* of light (Fig. 34–3b). For simplicity we often draw only one ray in each beam (Fig. 34–3c). Representing these waves in terms of rays is the basis of geometric optics. We begin our study with the behavior of an individual ray.

We describe the directions of the incident, reflected, and refracted (transmitted) rays at a smooth interface between two optical materials in terms of the angles they make with the *normal* to the surface at the point of incidence, as shown in Fig. 34–3c. If the interface is rough, both the transmitted light and the reflected light are scattered in various directions, and there is no single angle of transmission or reflection. Reflection at a definite angle from a very smooth surface is called **specular reflection** (from the Latin

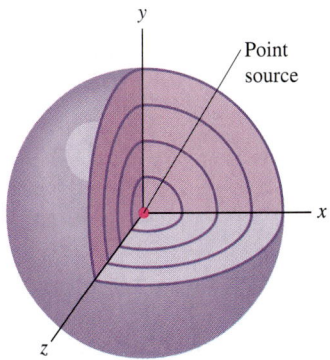

34–1 Spherical wave fronts spread out uniformly in all directions from a point source in a motionless medium, such as still air, that is homogeneous and isotropic (that is, has the same properties in all regions and in all directions). Electromagnetic waves in a vacuum also spread out as shown here.

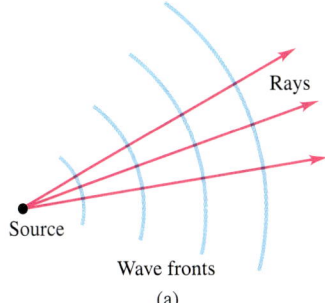

Rays

Source

Wave fronts

(a)

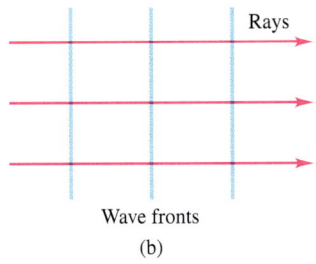

Rays

Wave fronts

(b)

34–2 Wave fronts and rays. (a) When the wave fronts are spherical, the rays radiate out from the center of the spheres. (b) When the wave fronts are planes, the rays are parallel.

34–3 (a) A plane wave is in part reflected and in part refracted at the boundary between two media (in this case, air and glass). The light that reaches the inside of the coffee shop is refracted twice, once when it enters the glass and once when it exits the glass. (b) The waves in the outside air and in glass in (a) are represented by rays. (c) For simplicity, only one example of an incident ray, a reflected ray, and a refracted ray is drawn. For the case shown here, material b has a greater index of refraction that material a ($n_b > n_a$), and the angle θ_b is smaller than θ_a.

word for "mirror"); scattered reflection from a rough surface is called **diffuse reflection.** This distinction is shown in Fig. 34–4. Both kinds of reflection can occur with either transparent materials or *opaque* materials that do not transmit light. (In Section 33–6 we described why metals are opaque.) The vast majority of objects in your environment (including clothing, plants, other people, and this book) are visible to you because they reflect light in a diffuse manner from their surfaces. Our primary concern, however, will be with specular reflection from a very smooth surface such as highly polished glass, plastic, or metal. Unless stated otherwise, when referring to "reflection" we will always mean *specular* reflection.

The **index of refraction** of an optical material (also called the *refractive index*), denoted by n, plays a central role in geometric optics. It is the ratio of the speed of light c in vacuum to the speed v in the material:

$$n = \frac{c}{v} \quad \text{(index of refraction)}. \tag{34–1}$$

Light always travels *more slowly* in a material than in vacuum, so the value of n in anything other other than a vacuum is always greater than unity. For vacuum, $n = 1$. Since n is a ratio of two speeds, it is a pure number without units. (The relation of the value of n to the electric and magnetic properties of a material is described in Section 33–6.)

CAUTION ▶ Keep in mind that the wave speed v is *inversely* proportional to the index of refraction n. The greater the index of refraction in a material, the *slower* the wave speed in that material. Failure to remember this point can lead to hopeless confusion! ◀

LAWS OF REFLECTION AND REFRACTION

Experimental studies of the directions of the incident, reflected, and refracted rays at a smooth interface between two optical materials lead to the following conclusions:

1. **The incident, reflected, and refracted rays and the normal to the surface all lie in the same plane.** The plane of the three rays is perpendicular to the plane of the boundary surface between the two materials. We always draw ray diagrams so that the incident, reflected, and refracted rays are in the plane of the diagram.

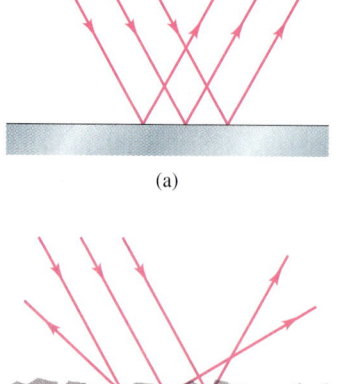

34–4 (a) Specular reflection. (b) Diffuse reflection.

2. **The angle of reflection θ_r is equal to the angle of incidence θ_a for all wavelengths and for any pair of materials.** That is, in Fig. 34–3c,

$$\theta_r = \theta_a \qquad \text{(law of reflection)}. \qquad (34\text{–}2)$$

This relation, together with the observation that the incident and reflected rays and the normal all lie in the same plane, is called the **law of reflection.**

3. For monochromatic light and for a given pair of materials, a and b, on opposite sides of the interface, **the ratio of the sines of the angles θ_a and θ_b, where both angles are measured from the normal to the surface, is equal to the inverse ratio of the two indexes of refraction:**

$$\frac{\sin \theta_a}{\sin \theta_b} = \frac{n_b}{n_a}, \qquad (34\text{–}3)$$

or

$$n_a \sin \theta_a = n_b \sin \theta_b \qquad \text{(law of refraction)}. \qquad (34\text{–}4)$$

This experimental result, together with the observation that the incident and refracted rays and the normal all lie in the same plane, is called the **law of refraction** or **Snell's law,** after the Dutch scientist Willebrord Snell (1591–1626). There is some doubt that Snell actually discovered it. The discovery that $n = c/v$ came much later.

Equations (34–3) and (34–4) show that when a ray passes from one material (a) into another material (b) having a larger index of refraction ($n_b > n_a$) and hence a slower wave speed, the angle θ_b with the normal is *smaller* in the second material than the angle θ_a in the first; hence the ray is bent *toward* the normal (Fig. 34–3c). When the second material has a *smaller* index of refraction than the first material ($n_b < n_a$) and hence a faster wave speed, the ray is bent *away from* the normal (Fig. 34–5). This explains why a partially submerged ruler or drinking straw appears bent; light rays coming from below the surface change in direction at the air-water interface, so the rays appear to be coming from a position above their actual point of origin (Fig. 34–6).

An important special case is refraction that occurs at an interface between vacuum, for which the index of refraction is unity by definition, and a material. When a ray passes from vacuum into a material (b), so that $n_a = 1$ and $n_b > 1$, the ray is always bent *toward* the normal. When a ray passes from a material into vacuum, so that $n_a > 1$ and $n_b = 1$, the ray is always bent *away from* the normal.

No matter what the materials on either side of the interface, the transmitted ray is not bent at all in the special case of *normal* incidence, in which the incident ray is perpendicular to the interface so that $\theta_a = 0$ and $\sin \theta_a = 0$. From Eq. (34–4) this means that θ_b is also equal to zero, so the transmitted ray is also perpendicular to the interface.

The laws of reflection and refraction apply regardless of which side of the interface the incident ray comes from. If a ray of light approaches the interface in Fig. 34–3c or Fig. 34–5 from the right rather than from the left, there are again reflected and refracted

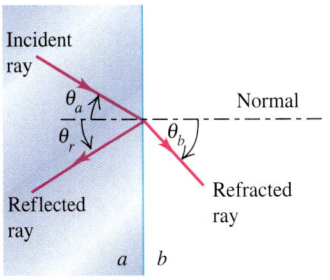

34–5 Reflection and refraction in the case in which material b has a smaller index of refraction than material a, so $n_b < n_a$. The refracted angle θ_b is greater than the incident angle θ_a. Compare to Fig. 34–3c, which shows the situation for $n_b > n_a$.

ActivPhysics

15.1
Reflection and Refraction
15.3
Refraction Applications

34–6 (a) This ruler is actually straight, but it appears to bend at the surface of the water. (b) Light rays from any submerged object bend away from the normal when they emerge into the air. As seen by an observer above the surface of the water, the object appears to be much closer to the surface than it actually is.

(a)

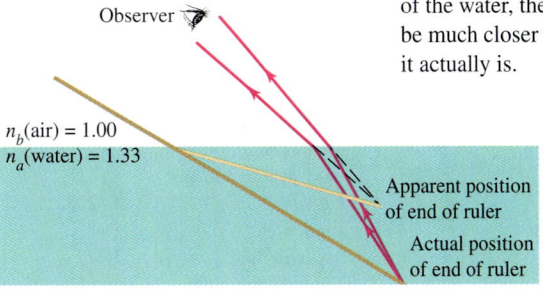

Observer

n_b(air) = 1.00
n_a(water) = 1.33

Apparent position
of end of ruler

Actual position
of end of ruler

(b)

rays; these two rays, the incident ray, and the normal to the surface again lie in the same plane. Furthermore, the path of a refracted ray is *reversible;* it follows the same path when going from *b* to *a* as when going from *a* to *b*. (You can verify this using Eq. (34–4).) Since reflected and incident rays make the same angle with the normal, the path of a reflected ray is also reversible. That's why when you see someone's eyes in a mirror, they can also see you.

The *intensities* of the reflected and refracted rays depend on the angle of incidence, the two indexes of refraction, and the polarization (that is, the direction of the electric field vector) of the incident ray. The fraction reflected is smallest at normal incidence ($\theta_a = 0°$), where it is about 4% for an air-glass interface. This fraction increases with increasing angle of incidence to 100% at grazing incidence, when $\theta_a = 90°$.

Although we have described the laws of reflection and refraction as experimental results, they can also be derived from a wave model using Maxwell's equations. This analysis also enables us to predict the amplitude, intensity, phase, and polarization states of the reflected and refracted waves. This analysis is beyond our scope, however.

The index of refraction depends not only on the substance but also on the wavelength of the light. The dependence on wavelength is called *dispersion;* we will consider it in Section 34–5. Indexes of refraction for several solids and liquids are given in Table 34–1 for a particular wavelength of yellow light.

The index of refraction of air at standard temperature and pressure is about 1.0003, and we will usually take it to be exactly unity. The index of refraction of a gas increases as its density increases. Most glasses used in optical instruments have indexes of refraction between about 1.5 and 2.0. A few substances have larger indexes; two examples are diamond, with 2.417, and rutile (a crystalline form of titanium dioxide), with 2.62.

INDEX OF REFRACTION AND THE WAVE ASPECTS OF LIGHT

We have discussed how the direction of a light ray changes when it passes from one material to another material with a different index of refraction. It's also important to see what happens to the *wave* characteristics of the light when this happens.

First, the frequency *f* of the wave does not change when passing from one material to another. That is, the number of wave cycles arriving per unit time must equal the number leaving per unit time; this is a statement that the boundary surface cannot create or destroy waves.

Second, the wavelength λ of the wave *is* different in general in different materials.

The index of refraction of air is slightly greater than 1, so light rays from the setting sun bend downward slightly when they enter our atmosphere and pass through it to our eyes. The effect is stronger for light coming from the lower limb of the sun (the part closest to the horizon). As a result, the setting sun appears flattened vertically.

TABLE 34–1

INDEX OF REFRACTION FOR YELLOW SODIUM LIGHT ($\lambda_0 = 589$ nm)

SUBSTANCE	INDEX OF REFRACTION, n	SUBSTANCE	INDEX OF REFRACTION, n
Solids		Liquids at 20°C	
Ice (H_2O)	1.309	Methanol (CH_3OH)	1.329
Fluorite (CaF_2)	1.434	Water (H_2O)	1.333
Polystyrene	1.49	Ethanol (C_2H_5OH)	1.36
Rock salt (NaCl)	1.544	Carbon tetrachloride (CCl_4)	1.460
Quartz (SiO_2)	1.544	Turpentine	1.472
Zircon ($ZrO_2 \cdot SiO_2$)	1.923	Glycerine	1.473
Diamond (C)	2.417	Benzene	1.501
Fabulite ($SrTiO_3$)	2.409	Carbon disulfide (CS_2)	1.628
Rutile (TiO_2)	2.62		
Glasses (typical values)			
Crown	1.52		
Light flint	1.58		
Medium flint	1.62		
Dense flint	1.66		
Lanthanum flint	1.80		

This is because in any material, $v = \lambda f$; since f is the same in any material as in vacuum and v is always less than the wave speed c in vacuum, λ is also correspondingly reduced. Thus the wavelength λ of light in a material is *less than* the wavelength λ_0 of the same light in vacuum. From the above discussion, $f = c/\lambda_0 = v/\lambda$. Combining this with Eq. (34-1), $n = c/v$, we find

$$\lambda = \frac{\lambda_0}{n} \qquad \text{(wavelength of light in a material).} \qquad (34\text{-}5)$$

When a wave passes from one material into a second material with larger index of refraction, so that $n_b > n_a$, the wave speed decreases. The wavelength $\lambda_b = \lambda_0/n_b$ in the second material is then shorter than the wavelength $\lambda_a = \lambda_0/n_a$ in the first material. If instead the second material has a smaller index of refraction than the first material, so that $n_b < n_a$, then the wave speed increases. Then the wavelength λ_b in the second material is longer than the wavelength λ_a in the first material. This makes intuitive sense; the waves get "squeezed" (the wavelength gets shorter) if the wave speed decreases and get "stretched" (the wavelength gets longer) if the wave speed increases.

Problem-Solving Strategy

REFLECTION AND REFRACTION

1. In geometric optics problems involving rays and angles, *always* start by drawing a large, neat diagram. Label all known angles and indexes of refraction.

2. Remember to always measure the angles of incidence, reflection, and refraction from the *normal* to the surface where the reflection and refraction occur, *never* from the surface itself.

3. You will often have to use some simple geometry or trigonometry in working out angular relations. The sum of the interior angles in a triangle is 180°, an angle and its complement add to 90°, and so on. Ask yourself "What information am I given?" "What do I need to know to find this angle?" or "What other angles or other quantities can I compute using the information given in the problem?"

EXAMPLE 34-1

Reflection and refraction In Fig. 34-7, material a is water and material b is a glass with index of refraction 1.52. If the incident ray makes an angle of 60° with the normal, find the directions of the reflected and refracted rays.

SOLUTION According to Eq. (34-2), the angle the reflected ray makes with the normal is the same as that of the incident ray, so $\theta_r = \theta_a = 60.0°$.

To find the direction of the refracted ray, we use Snell's law, Eq. (34-4), with $n_a = 1.33$, $n_b = 1.52$, and $\theta_a = 60.0°$. We find

$$n_a \sin \theta_a = n_b \sin \theta_b,$$

$$\sin \theta_b = \frac{n_a}{n_b} \sin \theta_a = \frac{1.33}{1.52} \sin 60.0° = 0.758,$$

$$\theta_b = 49.3°.$$

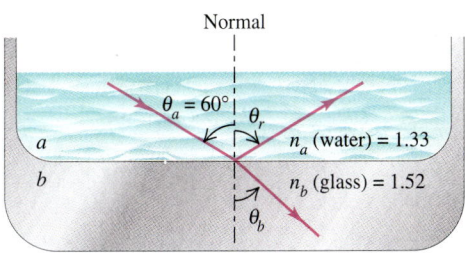

34-7 Reflection and refraction of light passing from water to glass.

The second material has a larger refractive index than the first; the refracted ray is bent toward the normal as the wave slows down upon entering the second material.

EXAMPLE 34-2

Index of refraction in the eye The wavelength of the red light from a helium-neon laser is 633 nm in air but 474 nm in the aqueous humor inside your eyeball. Calculate the index of refraction of the aqueous humor and the speed and frequency of the light in this substance.

SOLUTION The index of refraction of air is very close to unity,

so we assume that the wavelengths in air and vacuum are the same. Then the wavelength λ in the material is given by Eq. (34-5):

$$\lambda = \frac{\lambda_0}{n}, \qquad n = \frac{\lambda_0}{\lambda} = \frac{633 \text{ nm}}{474 \text{ nm}} = 1.34,$$

which is about the same index of refraction as for water. Then

$n = c/v$ gives

$$v = \frac{c}{n} = \frac{3.00 \times 10^8 \text{ m/s}}{1.34} = 2.25 \times 10^8 \text{ m/s}.$$

Finally, from $v = \lambda f$,

$$f = \frac{v}{\lambda} = \frac{2.25 \times 10^8 \text{ m/s}}{474 \times 10^{-9} \text{ m}} = 4.74 \times 10^{14} \text{ Hz}.$$

Note that while the speed and wavelength have different values in air and in the aqueous humor, the *frequency* in air, f_0, is the same as the frequency f in the aqueous humor:

$$f_0 = \frac{c}{\lambda_0} = \frac{3.00 \times 10^8 \text{ m/s}}{633 \times 10^{-9} \text{ m}} = 4.74 \times 10^{14} \text{ Hz}.$$

EXAMPLE 34-3

A twice-reflected ray Two mirrors are perpendicular to each other. A ray traveling in a plane perpendicular to both mirrors is reflected from one mirror, then the other, as shown in Fig. 34–8. What is the ray's final direction relative to its original direction?

SOLUTION For mirror 1 the angle of incidence is θ_1, and this equals the angle of reflection. The sum of interior angles in the triangle shown in the figure is 180°, so we see that the angles of incidence and reflection for mirror 2 are both $90° - \theta_1$. The total change in direction of the ray after both reflections is therefore $2(90° - \theta_1) + 2\theta_1 = 180°$. That is, the ray's final direction is opposite to its original direction.

An alternative viewpoint is that specular reflection reverses the sign of the component of light velocity perpendicular to the surface but leaves the other components unchanged. We invite you to verify this in detail. You should also be able to use this result to show that when a ray of light is successively reflected by three mirrors forming a corner of a cube (a "corner reflector"), its final direction is again opposite to its original direction. This principle is widely used in tail-light lenses and highway signs to improve their night-time visibility. Apollo astronauts placed arrays of corner reflectors on the moon. By use of laser beams reflected from these arrays, the earth-moon distance has been measured to within 0.15 m.

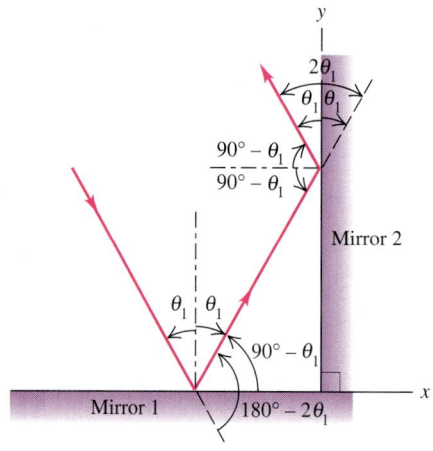

34–8 A ray moving in the *xy*-plane. The first reflection changes the sign of the *y*-component of its velocity, and the second reflection changes the sign of the *x*-component. For a different ray with a *z*-component of velocity, a third mirror (perpendicular to the two shown) could be used to change the sign of that component.

34–4 TOTAL INTERNAL REFLECTION

15.2
Total Internal Reflection

We have described how light is partially reflected and partially transmitted at an interface between two materials with different indexes of refraction. Under certain circumstances, however, *all* of the light can be reflected back from the interface, with none of it being transmitted, even though the second material is transparent. Figure 34–9a shows how this can occur. Several rays are shown radiating from a point source P in material a with index of refraction n_a. The rays strike the surface of a second material b with index n_b, where $n_a > n_b$. (For instance, materials a and b could be water and air, respectively.) From Snell's law of refraction,

$$\sin \theta_b = \frac{n_a}{n_b} \sin \theta_a.$$

Because n_a/n_b is greater than unity, $\sin \theta_b$ is larger than $\sin \theta_a$; the ray is bent *away from* the normal. Thus there must be some value of θ_a *less than* 90° for which Snell's law gives $\sin \theta_b = 1$ and $\theta_b = 90°$. This is shown by ray 3 in the diagram, which emerges just grazing the surface at an angle of refraction of 90°. Compare the diagram in Fig. 34–9a to the photograph of light rays in Fig. 34–9b.

(a)

(b)

34-9 (a) Total internal reflection. The angle of incidence for which the angle of refraction is 90° is called the critical angle; this is the case for ray 3. The reflected portions of rays 1, 2, and 3 are omitted for clarity. (b) Rays of laser light enter the water in the fishbowl from above; they are reflected at the bottom by mirrors tilted at slightly different angles, and one ray undergoes total internal reflection at the air-water interface.

The angle of incidence for which the refracted ray emerges tangent to the surface is called the **critical angle,** denoted by θ_{crit}. (A more detailed analysis using Maxwell's equations shows that as the incident angle approaches the critical angle, the transmitted intensity approaches zero.) If the angle of incidence is *greater than* the critical angle, the sine of the angle of refraction, as computed by Snell's law, would have to be greater than unity, which is impossible. Beyond the critical angle, the ray *cannot* pass into the upper material; it is trapped in the lower material and is completely reflected at the boundary surface. This situation, called **total internal reflection,** occurs only when a ray is incident on the interface with a second material whose index of refraction is *smaller* than that of the material in which the ray is traveling.

We can find the critical angle for two given materials by setting $\theta_b = 90°$ ($\sin \theta_b = 1$) in Snell's law. We then have

$$\sin \theta_{crit} = \frac{n_b}{n_a} \qquad \text{(critical angle for total internal reflection).} \qquad (34\text{-}6)$$

Total internal reflection will occur if the angle of incidence θ_a is greater than or equal to θ_{crit}.

As an example, for a glass-air surface with $n = 1.52$ for the glass,

$$\sin \theta_{crit} = \frac{1}{1.52} = 0.658, \qquad \theta_{crit} = 41.1°.$$

Light propagating within this glass will be *totally reflected* if it strikes the glass-air surface at an angle of 41.1° or greater. Because the critical angle is slightly less than 45°, it is possible to use a prism with angles of 45°–45°–90° as a totally reflecting surface. As reflectors, totally reflecting prisms have some advantages over metallic surfaces such as ordinary coated-glass mirrors. While no metallic surface reflects 100% of the light incident on it, light can be *totally* reflected by a prism. The reflecting properties of a prism have the additional advantages of being permanent and unaffected by tarnishing.

A 45°–45°–90° prism, used as in Fig. 34–10a (page 1062), is called a *Porro* prism. Light enters and leaves at right angles to the hypotenuse and is totally reflected at each of the shorter faces. The total change of direction of the rays is 180°. Binoculars often use combinations of two Porro prisms, as in Fig. 34–10b.

The brilliance of diamond is due in large measure to its very high index of refraction ($n = 2.417$) and correspondingly high critical angle. Light entering a cut diamond is totally internally reflected from facets on its back surface, then emerges from the front surface. "Imitation diamond" gems, such as cubic zirconia, are made from less expensive crystalline materials with comparable indexes of refraction.

When a beam of light enters at one end of a transparent rod (Fig. 34–11), the light can be totally reflected internally if the index of refraction of the rod is greater than that

(a)   (b)

34–10 (a) Total internal reflection in a Porro prism. (b) A combination of two Porro prisms in binoculars.

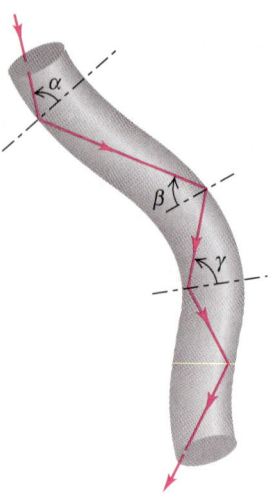

34–11 A transparent rod with refractive index greater than that of the surrounding material. A light ray is "trapped" by internal reflections within the rod, provided that the angles shown exceed the critical angle.

of the surrounding material. The light is "trapped" within the rod even if the rod is curved, provided that the curvature is not too great. Such a rod is sometimes called a *light pipe.* A bundle of fine glass or plastic fibers behaves in the same way and has the advantage of being flexible. A bundle may consist of thousands of individual fibers, each of the order of 0.002 to 0.01 mm in diameter. If the fibers are assembled in the bundle so that the relative positions of the ends are the same (or mirror images) at both ends, the bundle can transmit an image, as shown in Fig. 34–12.

Fiber-optic devices have found a wide range of medical applications in instruments called *endoscopes,* which can be inserted directly into the bronchial tubes, the bladder, the colon, and so on for direct visual examination. A bundle of fibers can be enclosed in a hypodermic needle for study of tissues and blood vessels far beneath the skin.

Fiber optics also have applications in communication systems, in which they are used to transmit a modulated laser beam. The rate at which information can be transmitted by a wave (light, radio, or whatever) is proportional to the frequency. To see

34–12 Image transmission by a bundle of optical fibers.

qualitatively why this is so, consider modulating (modifying) the wave by chopping off some of the wave crests. Suppose each crest represents a binary digit, with a chopped-off crest representing a zero and an unmodified crest representing a one. The number of binary digits we can transmit per unit time is thus proportional to the frequency of the wave. Infrared and visible-light waves have much higher frequency than do radio waves, so a modulated laser beam can transmit an enormous amount of information through a single fiber-optic cable. For example, the Carnegie-Mellon University computer system, which includes several thousand networked personal computers and work stations, is linked partly by fiber-optic cables. Many telephone systems are connected by fiber optics.

Another advantage of fiber-optic cables is that they are electrical insulators. They are immune to electrical interference from lightning and other sources, and they don't allow unwanted currents between source and receiver. They are secure and difficult to "bug," but they are also difficult to splice and tap into.

EXAMPLE 34–4

A leaky periscope A periscope for a submarine uses two totally reflecting 45°–45°–90° prisms with total internal reflection on the sides adjacent to the 45° angles. It springs a leak, and the bottom prism is covered with water. Explain why the periscope no longer works.

SOLUTION The critical angle for water ($n_b = 1.33$) on glass ($n_a = 1.52$) is

$$\theta_{crit} = \arcsin \frac{1.33}{1.52} = 61.0°.$$

The 45° angle of incidence for a totally reflecting prism is *less than* the 61° critical angle, so total internal reflection does not occur at the glass-water boundary. Most of the light is transmitted into the water, and very little is reflected back into the prism.

---

*34–5 DISPERSION

Ordinary white light is a superposition of waves with wavelengths extending throughout the visible spectrum. The speed of light *in vacuum* is the same for all wavelengths, but the speed in a material substance is different for different wavelengths. Therefore the index of refraction of a material depends on wavelength. The dependence of wave speed and index of refraction on wavelength is called **dispersion.**

Figure 34–13 shows the variation of index of refraction n with wavelength for a few common optical materials. Note that the horizontal axis of this figure refers to the wavelength of the light *in vacuum*, λ_0; the wavelength in the material is given by Eq. (34–5), $\lambda = \lambda_0/n$. In most materials the value of n *decreases* with increasing wavelength and decreasing frequency, and thus n *increases* with decreasing wavelength and increasing frequency. In such a material, light of longer wavelength has greater speed than light of shorter wavelength.

Figure 34–14 shows a ray of white light incident on a prism. The deviation (change of direction) produced by the prism increases with increasing index of refraction and frequency and decreasing wavelength. Violet light is deviated most, and red is deviated least; other colors are in intermediate positions. When it comes out of the prism, the light is spread out into a fan-shaped beam, as shown. The light is said to be *dispersed* into a spectrum. The amount of dispersion depends on the *difference* between the refractive indexes for violet light and for red light. From Fig. 34–13 we can see that for a substance such as fluorite, the difference between the indexes for red and violet is small, and the dispersion will also be small. A better choice of material for a prism whose purpose is

Index of refraction (n)

Silicate flint glass

Borate flint glass

Quartz

Silicate crown glass

Fused quartz

Fluorite

Wavelength in vacuum (nm)

34–13 Variation of index of refraction n with wavelength for different transparent materials. The horizontal axis shows the wavelength λ_0 of the light *in vacuum;* the wavelength in the material is equal to $\lambda = \lambda_0/n$.

34–14 Dispersion of light by a prism. The band of colors is called a spectrum.

to produce a spectrum would be silicate flint glass, for which there is a larger difference in the value of n between red and violet.

As we mentioned in Section 34–4, the brilliance of diamond is due in part to its unusually large refractive index; another important factor is its large dispersion, which causes white light entering a diamond to emerge as a multicolored spectrum. Crystals of rutile and of strontium titanate, which can be produced synthetically, have about eight times the dispersion of diamond.

When you experience the beauty of a rainbow, you are seeing the combined effects of dispersion, refraction, and reflection (Fig. 34–15). Sunlight comes from behind you, is refracted into a water droplet, is (partially) reflected from the back surface of the droplet, and is refracted again upon exiting the droplet (Fig. 34–15a). Dispersion causes different colors to be refracted at different angles. When you see a second, slightly larger rainbow with its colors reversed, you are seeing the results of dispersion and *two* reflections from the back surface of the droplet (Fig. 34–15b). Both of these rainbows (called the *primary* and *secondary* bows) can be seen in Fig. 34–15c.

34-6 POLARIZATION

Polarization is a characteristic of all transverse waves. This chapter is about light, but to introduce some basic polarization concepts, let's go back to the transverse waves on a string that we studied in Chapter 19. For a string that in equilibrium lies along the x-axis,

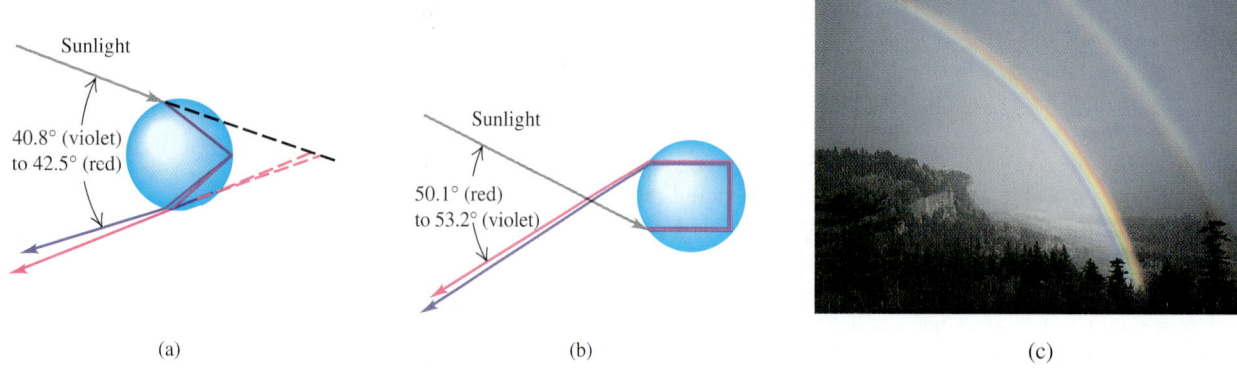

34–15 Rainbows are formed by refraction, reflection, and dispersion in water drops. (a) The outside of the primary bow is red. (b) The inside of the fainter secondary bow is red. (c) A photograph showing the primary and secondary bows.

the displacements may be along the *y*-direction, as in Fig. 34–16a. In this case the string always lies in the *xy*-plane. But the displacements might instead be along the *z*-axis, as in Fig. 34–16b; then the string always lies in the *xz*-plane.

When a wave has only *y*-displacements, we say that it is **linearly polarized** in the *y*-direction; a wave with only *z*-displacements is linearly polarized in the *z*-direction. For mechanical waves we can build a **polarizing filter,** or **polarizer,** that permits only waves with a certain polarization direction to pass. In Fig. 34–16c the string can slide vertically in the slot without friction, but no horizontal motion is possible. This filter passes waves that are polarized in the *y*-direction but blocks those that are polarized in the *z*-direction.

This same language can be applied to electromagnetic waves, which also have polarization. As we learned in Chapter 33, an electromagnetic wave is a *transverse* wave; the fluctuating electric and magnetic fields are perpendicular to each other and to the direction of propagation. We always define the direction of polarization of an electromagnetic wave to be the direction of the *electric*-field vector $\vec{E}$, not the magnetic field, because many common electromagnetic-wave detectors respond to the electric forces on electrons in materials, not the magnetic forces. Thus the electromagnetic wave described by Eq. (33–17),

$$\vec{E}(x,t) = E_{max}\hat{\jmath}\sin(\omega t - kx),$$

$$\vec{B}(x,t) = B_{max}\hat{k}\sin(\omega t - kx),$$

is said to be polarized in the *y*-direction because the electric field has only a *y*-component.

CAUTION▶ It's unfortunate that the same word "polarization" that is used to describe the direction of $\vec{E}$ in an electromagnetic wave is also used to describe the shifting of electric charge within a body, such as in response to a nearby charged body; we described this latter kind of polarization in Section 22–4 (see Fig. 22–5). You should remember that while these two concepts have the same name, they are not the same. ◀

POLARIZING FILTERS

Waves emitted by a radio transmitter are usually linearly polarized. The vertical rod antennas that are used for cellular telephones in automobiles emit waves that, in a horizontal plane around the antenna, are polarized in the vertical direction (parallel to the antenna). Rooftop TV antennas have horizontal elements in the United States and vertical elements in Great Britain because the transmitted waves have different polarizations.

The situation is different for visible light. Light from ordinary sources, such as incandescent light bulbs and fluorescent light fixtures, is *not* polarized. The "antennas" that radiate light waves are the molecules that make up the sources. The waves emitted by any one molecule may be linearly polarized, like those from a radio antenna. But any actual light source contains a tremendous number of molecules with random orientations, so the emitted light is a random mixture of waves linearly polarized in all possible transverse directions. Such light is called **unpolarized light** or **natural light.** To create polarized light from unpolarized natural light requires a filter that is analogous to the slot for mechanical waves in Fig. 34–16c.

Polarizing filters for electromagnetic waves have different details of construction, depending on the wavelength. For microwaves with a wavelength of a few centimeters, a good polarizer is an array of closely spaced, parallel conducting wires that are insulated from each other. (Think of a barbecue grill with the outer metal ring replaced by an insulating one.) Electrons are free to move along the length of the conducting wires and will do so in response to a wave whose $\vec{E}$ field is parallel to the wires. The resulting currents in the wires dissipate energy by I^2R heating; the dissipated energy comes from the wave, so whatever wave passes through the grid is greatly reduced in amplitude. Waves with $\vec{E}$ oriented perpendicular to the wires pass through almost unaffected, since

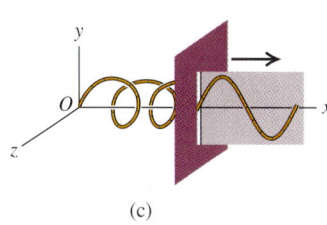

34–16 (a) Transverse wave on a string, polarized in the *y*-direction. (b) Wave polarized in the *z*-direction. (c) A barrier with a frictionless vertical slot passes components that are polarized in the *y*-direction but blocks those that are polarized in the *z*-direction, acting as a polarizing filter.

16.9
Physical Optics: Polarization

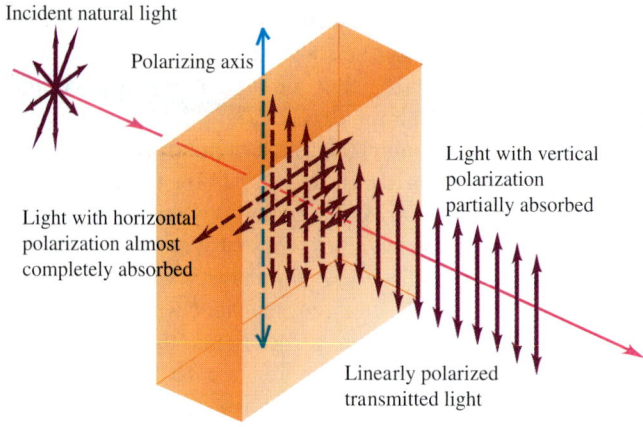

34–17 A Polaroid filter is illuminated by unpolarized natural light (shown by $\vec{E}$ vectors that point in all directions perpendicular to the direction of propagation). The components of $\vec{E}$ that are perpendicular to the polarizing axis are absorbed by the filter; the transmitted light is linearly polarized (shown by $\vec{E}$ vectors along the polarization direction only).

electrons cannot move through the air between the wires. Hence a wave that passes through such a filter will be predominantly polarized in the direction perpendicular to the wires.

The most common polarizing filter for visible light is a material known by the trade name Polaroid, widely used for sunglasses and polarizing filters for camera lenses. Developed originally by the American scientist Edwin H. Land, this material incorporates substances that have **dichroism,** a selective absorption in which one of the polarized components is absorbed much more strongly than the other (Fig. 34–17). A Polaroid filter transmits 80% or more of the intensity of a wave that is polarized parallel to a certain axis in the material, called the **polarizing axis,** but only 1% or less for waves that are polarized perpendicular to this axis. In one type of Polaroid filter, long-chain molecules within the filter are oriented with their axis perpendicular to the polarizing axis; these molecules preferentially absorb light that is polarized along their length, much like the conducting wires in a polarizing filter for microwaves.

An *ideal* polarizing filter (polarizer) passes 100% of the incident light that is polarized in the direction of the filter's polarizing axis but completely blocks all light that is polarized perpendicular to this axis. Such a device is an unattainable idealization, but the concept is useful in clarifying the basic ideas. In the following discussion we will assume that all polarizing filters are ideal. In Fig. 34–18, unpolarized light is incident on a flat polarizing filter. The polarizing axis is represented by the blue line. The $\vec{E}$ vector of the incident wave can be represented in terms of components parallel and

34–18 Unpolarized natural light is incident on the polarizing filter. The intensity of the transmitted linearly polarized light, measured by the photocell, is the same for all orientations of the polarizing filter. For an ideal polarizing filter, the transmitted intensity is half the incident intensity.

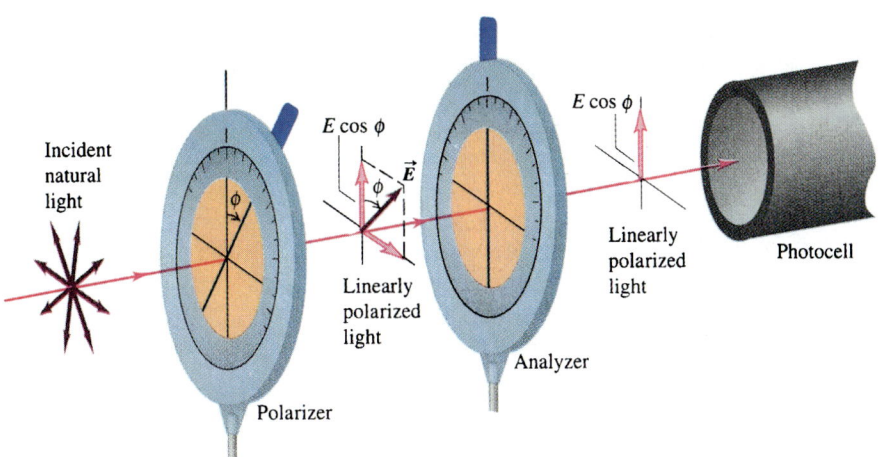

34–19 The ideal analyzer transmits only the component that is parallel to its transmission direction, or polarizing axis.

perpendicular to the polarizer axis; only the component of $\vec{E}$ parallel to the polarizing axis is transmitted. Hence the light emerging from the polarizer is linearly polarized parallel to the polarizing axis.

When unpolarized light is incident on an ideal polarizer as in Fig. 34–18, the intensity of the transmitted light is *exactly half* that of the incident unpolarized light, no matter how the polarizing axis is oriented. Here's why: We can resolve the $\vec{E}$ field of the incident wave into a component parallel to the polarizing axis and a component perpendicular to it. Because the incident light is a random mixture of all states of polarization, these two components are, on average, equal. The ideal polarizer transmits only the component that is parallel to the polarizing axis, so half the incident intensity is transmitted.

What happens when the linearly polarized light emerging from a polarizer passes through a second polarizer, as in Fig. 34–19? Consider the general case in which the polarizing axis of the second polarizer, or *analyzer,* makes an angle ϕ with the polarizing axis of the first polarizer. We can resolve the linearly polarized light that is transmitted by the first polarizer into two components, as shown in Fig. 34–19, one parallel and the other perpendicular to the axis of the analyzer. Only the parallel component, with amplitude $E \cos \phi$, is transmitted by the analyzer. The transmitted intensity is greatest when $\phi = 0$, and it is zero when polarizer and analyzer are *crossed* so that $\phi = 90°$. To determine the direction of polarization of the light transmitted by the first polarizer, rotate the analyzer until the photocell in Fig. 34–19 measures zero intensity; the polarization axis of the first polarizer is then perpendicular to that of the analyzer.

To find the transmitted intensity at intermediate values of the angle ϕ, we recall from our energy discussion in Chapter 33 that the intensity of an electromagnetic wave is proportional to the *square* of the amplitude of the wave (see Eq. (33–26)). The ratio of transmitted to incident *amplitude* is $\cos \phi$, so the ratio of transmitted to incident *intensity* is $\cos^2 \phi$. Thus the intensity of the light transmitted through the analyzer is

$$I = I_{max} \cos^2 \phi \qquad \text{(Malus's law, polarized light passing through an analyzer),} \qquad (34\text{–}7)$$

where I_{max} is the maximum intensity of light transmitted (at $\phi = 0$) and I is the amount transmitted at angle ϕ. This relation, discovered experimentally by Etienne Louis Malus in 1809, is called **Malus's law.** Malus's law applies *only* if the incident light passing through the analyzer is already linearly polarized.

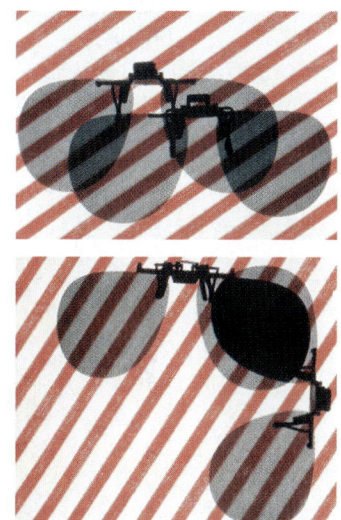

These photos show the view through Polaroid sunglasses whose polarizing axes are aligned ($\phi = 0$, top) and whose axes are perpendicular ($\phi = 90°$, bottom). The transmitted intensity is greatest when the axes are aligned; it is zero when the axes are perpendicular.

Problem–Solving Strategy

1. Remember that in all electromagnetic waves, including light waves, the direction of the $\vec{E}$ field is the direction of polarization and is perpendicular to the propagation direction. When working with polarizing filters, you are really dealing with components of $\vec{E}$ that are parallel and perpendicular to the polarizing axis. Everything you know about components of vectors is applicable here.

2. The intensity (average power per unit area) of a wave is proportional to the *square* of its amplitude. If you find

that two waves differ in amplitude by a certain ratio, their intensities differ by the square of that ratio.

3. Unpolarized light is a random mixture of all possible polarization states, so on average it has equal components in any two perpendicular directions. Partially linearly polarized light is a superposition of linearly polarized and unpolarized light.

EXAMPLE 34–5

Two polarizers in combination In Fig. 34–19 the incident unpolarized light has intensity I_0. Find the intensities transmitted by the first and second polarizers if the angle between the axes of the two filters is 30°.

SOLUTION As we explained above, the intensity of the linearly polarized light transmitted by the first filter is $I_0/2$. According to

Eq. (34–7) with $\phi = 30°$, the second filter reduces the intensity by a factor of $\cos^2 30° = \frac{3}{4}$. Thus the intensity transmitted by the second polarizer is

$$\left(\frac{I_0}{2}\right)\left(\frac{3}{4}\right) = \frac{3}{8} I_0.$$

POLARIZATION BY REFLECTION

Unpolarized light can be polarized, either partially or totally, by *reflection*. In Fig. 34–20, unpolarized natural light is incident on a reflecting surface between two transparent optical materials; the plane containing the incident and reflected rays and the normal to the surface is called the **plane of incidence.** For most angles of incidence, waves for which the electric-field vector $\vec{E}$ is perpendicular to the plane of incidence (that is, parallel to the reflecting surface) are reflected more strongly than those for which $\vec{E}$ lies in this plane. In this case the reflected light is *partially polarized* in the direction perpendicular to the plane of incidence.

34–20 When light is incident on a reflecting surface at the polarizing angle, the reflected light is linearly polarized.

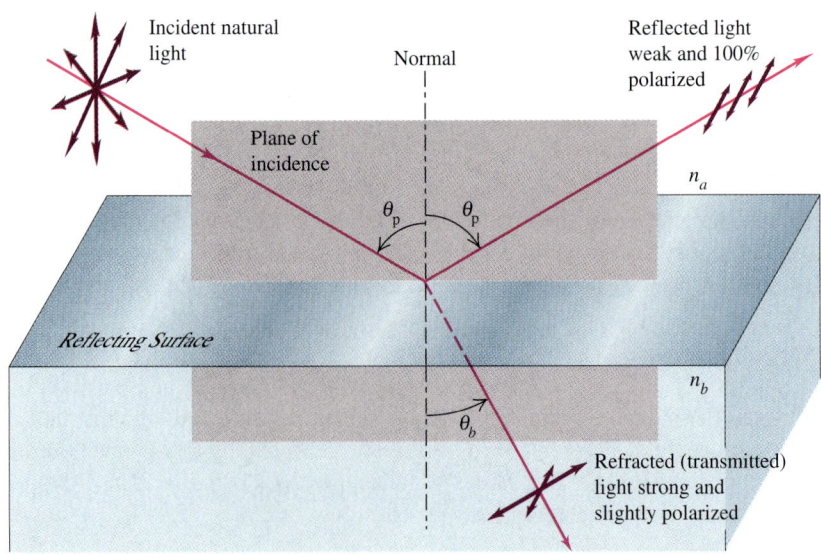

But at one particular angle of incidence, called the **polarizing angle** θ_p, the light for which $\vec{E}$ lies in the plane of incidence is *not reflected at all* but is completely refracted. At this same angle of incidence the light for which $\vec{E}$ is perpendicular to the plane of incidence is partially reflected and partially refracted. The *reflected* light is therefore *completely* polarized perpendicular to the plane of incidence, as shown in Fig. 34–20. The *refracted* light is *partially* polarized parallel to this plane; the refracted light is a mixture of the component parallel to the plane of incidence, all of which is refracted, and the remainder of the perpendicular component.

In 1812 the British scientist Sir David Brewster discovered that when the angle of incidence is equal to the polarizing angle θ_p, the reflected ray and the refracted ray are perpendicular to each other (Fig. 34–21). In this case the angle of refraction θ_b becomes the complement of θ_p, so $\theta_b = 90° - \theta_p$. From the law of refraction,

$$n_a \sin \theta_p = n_b \sin \theta_b,$$

so we find

$$n_a \sin \theta_p = n_b \sin(90° - \theta_p) = n_b \cos \theta_p,$$

$$\tan \theta_p = \frac{n_b}{n_a} \qquad \text{(Brewster's law for the polarizing angle)}. \qquad (34–8)$$

This relation is known as **Brewster's law.** Although discovered experimentally, it can also be *derived* from a wave model using Maxwell's equations.

Polarization by reflection is the reason polarizing filters are widely used in sunglasses. When sunlight is reflected from a horizontal surface, the plane of incidence is vertical, and the reflected light contains a preponderance of light that is polarized in the horizontal direction. When the reflection occurs at a smooth asphalt road surface or the surface of a lake, it causes unwanted glare. Vision can be improved by eliminating this glare. The manufacturer makes the polarizing axis of the lens material vertical, so very little of the horizontally polarized light reflected from the road is transmitted to the eyes. The glasses also reduce the overall intensity of the transmitted light to somewhat less than 50% of the intensity of the unpolarized incident light.

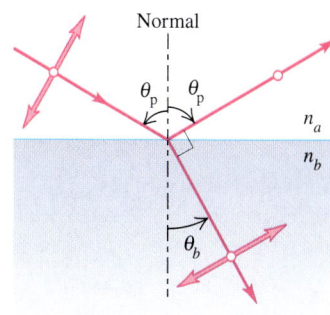

34–21 When light is incident at the polarizing angle, the reflected and refracted rays are perpendicular to each other. The open circles represent a component of $\vec{E}$ that is perpendicular to the plane of the figure (the plane of incidence) and parallel to the surface between the two materials.

EXAMPLE 34-6

Reflection from a swimming pool's surface Sunlight reflects off the smooth surface of an unoccupied swimming pool. a) At what angle of reflection is the light completely polarized? b) What is the corresponding angle of refraction for the light that is transmitted (refracted) into the water? c) At night an underwater floodlight is turned on in the pool. Repeat parts (a) and (b) for rays from the floodlight that strike the smooth surface from below.

34–22 Light striking the water-air interface at the polarizing angle (a) from the air side; (b) from the water side.

SOLUTION a) We're looking for the polarizing angle for light that is first in air, then in water, so $n_a = 1.00$ (air) and $n_b = 1.33$ (water). From Eq. (34–8),

$$\theta_p = \arctan \frac{n_b}{n_a} = \arctan \frac{1.33}{1.00} = 53.1°.$$

The angles are shown in Fig. 34–22a.

b) The incident light is at the polarizing angle, so the reflected and refracted rays are perpendicular; hence

$$\theta_p + \theta_b = 90°,$$

$$\theta_b = 90° - 53.1° = 36.9°.$$

c) Now the light is *first* in the water, then in the air, so $n_a = 1.33$ and $n_b = 1.00$. Again using Eq. (34–8), we have

$$\theta_p = \arctan \frac{1.00}{1.33} = 36.9°,$$

$$\theta_b = 90° - 36.9° = 53.1°.$$

The angles are shown in Fig. 34–22b. We see that the two polarizing angles for this interface add to 90°. This is *not* an accident; do you see why?

CIRCULAR AND ELLIPTICAL POLARIZATION

Light and other electromagnetic radiation can also have *circular* or *elliptical* polarization. To introduce these concepts, let's return once more to mechanical waves on a stretched string. In Fig. 34–16, suppose the two linearly polarized waves in parts (a) and (b) are in phase and have equal amplitude. When they are superposed, each point in the string has simultaneous y- and z-displacements of equal magnitude. A little thought shows that the resultant wave lies in a plane oriented at 45° to the y- and z-axes (i.e., in a plane making a 45° angle with the xy- and xz-planes). The amplitude of the resultant wave is larger by a factor of $\sqrt{2}$ than that of either component wave, and the resultant wave is linearly polarized.

But now suppose the two equal-amplitude waves differ in phase by a quarter-cycle. Then the resultant motion of each point corresponds to a superposition of two simple harmonic motions at right angles, with a quarter-cycle phase difference. The y-displacement at a point is greatest at times when the z-displacement is zero, and vice versa. The motion of the string as a whole then no longer takes place in a single plane. It can be shown that each point on the rope moves in a *circle* in a plane parallel to the yz-plane. Successive points on the rope have successive phase differences, and the overall motion of the string has the appearance of a rotating helix. This is shown to the left of the polarizing filter in Fig. 34–16c. This particular superposition of two linearly polarized waves is called **circular polarization.** By convention, the wave is said to be *right circularly polarized* when the sense of motion of a particle of the string, to an observer looking *backward* along the direction of propagation, is *clockwise;* the wave is said to be *left circularly polarized* if the sense of motion is the reverse.

Figure 34–23 shows the analogous situation for an electromagnetic wave. Two sinusoidal waves of equal amplitude, polarized in the y- and z-directions and with a quarter-cycle phase difference, are superposed. The result is a wave in which the $\vec{E}$ vector at each point has a constant magnitude but *rotates* around the direction of propagation. If the wave in Fig. 34–23 is propagating toward you, it is called a right circularly polarized electromagnetic wave.

If the phase difference between the two component waves is something other than a quarter-cycle, or if the two component waves have different amplitudes, then each point on the string traces out not a circle but an *ellipse*. The resulting wave is said to be **elliptically polarized.**

For electromagnetic waves with radio frequencies, circular or elliptical polarization can be produced by using two antennas at right angles, fed from the same transmitter but with a phase-shifting network that introduces the appropriate phase difference. For light, the phase shift can be introduced by use of a material that exhibits **birefringence,** that is, has different indexes of refraction for different directions of polarization. A common example is calcite ($CaCO_3$). When a calcite crystal is oriented appropriately in a beam of unpolarized light, its refractive index (for a wavelength in vacuum of 589 nm) is 1.658 for one direction of polarization and 1.486 for the perpendicular direction. When two

34–23 Circular polarization. The y-component of $\vec{E}$ lags the z-component by a quarter-cycle. This phase difference results in right circular polarization if the wave is moving toward you (in the positive x-direction).

waves with equal amplitude and with perpendicular directions of polarization enter such a material, they travel with different speeds. If they are in phase when they enter the material, then in general they are no longer in phase when they emerge. If the crystal is just thick enough to introduce a quarter-cycle phase difference, then the crystal converts linearly polarized light to circularly polarized light. Such a crystal is called a **quarter-wave plate.** Such a plate also converts circularly polarized light to linearly polarized light. Can you prove this? (See Problem 34–40.)

PHOTOELASTICITY

Some optical materials that are not normally birefringent become so when they are subjected to mechanical stress. This is the basis of the science of **photoelasticity.** Stresses in girders, boiler plates, gear teeth, and cathedral pillars can be analyzed by constructing a transparent model of the object, usually of a plastic material, subjecting it to stress, and examining it between a polarizer and an analyzer in the crossed position. Very complicated stress distributions can be studied by these optical methods. Figure 34–24 is a photograph of a photoelastic model under stress.

34–24 Photoelastic stress analysis of a model of a cross section of a Gothic cathedral. The masonry construction that was used for this kind of building had great strength in compression but very little in tension. Inadequate buttressing and high winds sometimes caused tensile stresses in normally compressed structural elements, leading to some spectacular collapses.

*34–7 SCATTERING OF LIGHT

The sky is blue. Sunsets are red. Skylight is partially polarized; that's why the sky looks darker from some angles than from others when it is viewed through Polaroid sunglasses. It turns out that one phenomenon is responsible for all of these effects.

When you look at the daytime sky, the light that you see is sunlight that has been absorbed and then re-radiated in a variety of directions. This process is called **scattering.** (If the earth had no atmosphere, the sky would appear as black in the daytime as it does at night, just as it does to an astronaut in space or on the moon; you would see the sun's light only if you looked directly at it, and the stars would be visible in the daytime.) Figure 34–25 shows some of the details of the scattering process. Sunlight, which is unpolarized, comes from the left along the *x*-axis and passes over an observer looking vertically upward along the *y*-axis. (We are viewing the situation from the side.) Consider the molecules of the earth's atmosphere located at point *O*. The electric field in the beam of sunlight sets the electric charges in these molecules into vibration. Since light is a transverse wave, the direction of the electric field in any component of the sunlight lies in the *yz*-plane, and the motion of the charges takes place in this plane. There is no field, and hence no motion of charges, in the direction of the *x*-axis.

An incident light wave with its $\vec{E}$ field at an angle θ with the *z*-axis sets the electric charges in the molecules vibrating along the line of $\vec{E}$, as indicated by the double-ended arrow through point *O*. We can resolve this vibration into two components, one along the *y*-axis and the other along the *z*-axis. Each component in the incident light produces the equivalent of two molecular "antennas," oscillating with the same frequency as the incident light and lying along the *y*- and *z*-axes.

We mentioned in Chapter 33 that an oscillating charge does not radiate in the direction of its oscillation. (See Fig. 33–1 in Section 33–2; you may also want to review the discussion of electric dipole antennas in Section 33–9.) Thus the "antenna" along the *y*-axis does not send any light to the observer directly below it, although it does emit light in other directions. Therefore the only light reaching this observer comes from the other molecular "antenna," corresponding to the oscillation of charge along the *z*-axis. This light is linearly polarized, with its electric field along the *z*-axis (parallel to the "antenna"). The red vectors on the *y*-axis below point *O* in Fig. 34–25 show the direction of polarization of the light reaching the observer.

34–25 The sunbathing observer in the west sees sunlight that has been scattered by 90°. This scattered light is linearly polarized and contains predominantly light from the blue end of the spectrum. The initial white light loses this blue light as it travels through the atmosphere, and the transmitted light seen by the observer in the east contains predominantly light from the red end of the spectrum.

As the original beam of sunlight passes though the atmosphere, its intensity decreases as its energy goes into the scattered light. Detailed analysis of the scattering process shows that the intensity of the light scattered from air molecules increases in proportion to the fourth power of the frequency (inversely to the fourth power of the wavelength). Thus the intensity ratio for the two ends of the visible spectrum is $(700\ \text{nm}/400\ \text{nm})^4 = 9.4$. Roughly speaking, scattered light contains nine times as much blue light as red, and that's why the sky is blue.

Clouds contain a high concentration of water droplets or ice crystals, which also scatter light. Because of this high concentration, light passing through the cloud has many more opportunities for scattering than does light passing through a clear sky. Thus light of *all* wavelengths is eventually scattered out of the cloud, so the cloud looks white. Milk looks white for the same reason; the scattering is due to fat globules in the milk. If you dilute milk by mixing it with enough water, the concentration of fat globules will be so low that only blue light will be substantially scattered; the dilute solution will look blue, not white. (Nonfat milk, which also has a very low concentration of globules, looks somewhat bluish for this same reason.)

Near sunset, when sunlight has to travel a long distance through the earth's atmosphere, a substantial fraction of the blue light is removed by scattering. White light minus blue light looks yellow or red. This explains the yellow or red hue that we so often see from the setting sun (and that is seen by the observer at the far right of Fig. 34–25).

Because skylight is partially polarized, polarizers are useful in photography. The sky in a photograph can be darkened by orienting the polarizer axis to be perpendicular to the predominant direction of polarization of the scattered light. The most strongly polarized light comes from parts of the sky that are 90° away from the sun; for example, from directly overhead when the sun is on the horizon at sunrise or sunset.

Clouds are white because they efficiently scatter sunlight of all wavelengths.

34–8 HUYGENS' PRINCIPLE

The laws of reflection and refraction of light rays that we introduced in Section 34–3 were discovered experimentally long before the wave nature of light was firmly established. However, we can *derive* these laws from wave considerations and show that they are consistent with the wave nature of light. The same kind of analysis that we use here will be of central importance in Chapters 37 and 38 in our discussion of physical optics.

We begin with a principle called **Huygens' principle.** This principle, stated originally by the Dutch scientist Christiaan Huygens in 1678, is a geometrical method for finding, from the known shape of a wave front at some instant, the shape of the wave front at some later time. Huygens assumed that **every point of a wave front may be considered the source of secondary wavelets that spread out in all directions with a speed equal to the speed of propagation of the wave.** The new wave front at a later time is then found by constructing a surface *tangent* to the secondary wavelets or, as it is called, the *envelope* of the wavelets. All the results that we obtain from Huygens' principle can also be obtained from Maxwell's equations. Thus it is not an independent principle, but it is often very convenient for calculations with wave phenomena.

Huygens' principle is shown in Fig. 34–26. The original wave front AA' is traveling outward from a source, as indicated by the small arrows. We want to find the shape of the wave front after a time interval t. Let v be the speed of propagation of the wave; then in time t it travels a distance vt. We construct several circles (traces of spherical wavelets) with radius $r = vt$, centered at points along AA'. The trace of the envelope of these wavelets, which is the new wave front, is the curve BB'. We are assuming that the speed v is the same at all points and in all directions.

To derive the law of reflection from Huygens' principle, we consider a plane wave approaching a plane reflecting surface. In Fig. 34–27a the lines AA', OB', and NC'

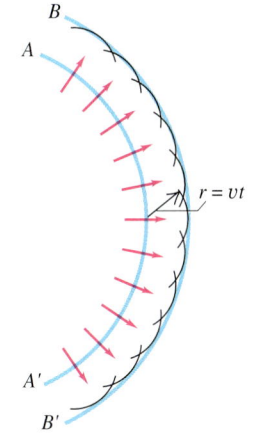

34–26 Applying Huygens' principle to wave front AA' to construct a new wave front BB'.

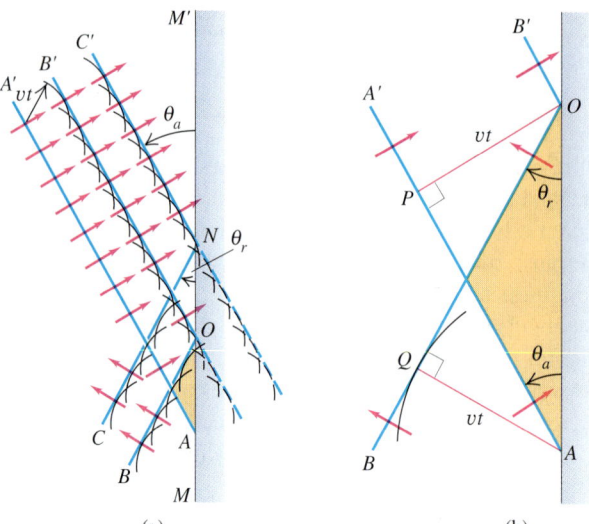

34–27 (a) Successive positions of a plane wave AA' as it is reflected from a plane surface. (b) Magnified portion of part (a).

(a)   (b)

represent successive positions of a wave front approaching the surface MM'. Point A on the wave front AA' has just arrived at the reflecting surface. We can use Huygens' principle to find the position of the wave front after a time interval t. With points on AA' as centers, we draw several secondary wavelets with radius vt. The wavelets that originate near the upper end of AA' spread out unhindered, and their envelope gives the portion OB' of the new wave front. If the reflecting surface were not there, the wavelets originating near the lower end of AA' would similarly reach the positions shown by the broken circular arcs. Instead, these wavelets strike the reflecting surface.

The effect of the reflecting surface is to *change the direction* of travel of those wavelets that strike it, so the part of a wavelet that would have penetrated the surface actually lies to the left of it, as shown by the full lines. The first such wavelet is centered at point A; the envelope of all such reflected wavelets is the portion OB of the wave front. The trace of the entire wave front at this instant is the bent line BOB'. A similar construction gives the line CNC' for the wave front after another interval t.

From plane geometry the angle θ_a between the incident *wave front* and the *surface* is the same as that between the incident *ray* and the *normal* to the surface and is therefore the angle of incidence. Similarly, θ_r is the angle of reflection. To find the relation between these angles, we consider Fig. 34–27b. From O we draw $OP = vt$, perpendicular to AA'. Now OB, by construction, is tangent to a circle of radius vt with center at A. If we draw AQ from A to the point of tangency, the triangles APO and OQA are congruent because they are right triangles with the side AO in common and with $AQ = OP = vt$. The angle θ_a therefore equals the angle θ_r, and we have the law of reflection.

We can derive the law of *refraction* by a similar procedure. In Fig. 34–28a we consider a wave front, represented by line AA', for which point A has just arrived at the boundary surface SS' between two transparent materials a and b, with indexes of refraction n_a and n_b and wave speeds v_a and v_b. (The *reflected* waves are not shown in the figure; they proceed exactly as in Fig. 34–27.) We can apply Huygens' principle to find the position of the refracted wave fronts after a time t.

With points on AA' as centers, we draw several secondary wavelets. Those originating near the upper end of AA' travel with speed v_a and, after a time interval t, are spherical surfaces of radius $v_a t$. The wavelet originating at point A, however, is traveling in the second material b with speed v_b and at time t is a spherical surface of radius $v_b t$. The envelope of the wavelets from the original wave front is the plane whose trace is the bent line BOB'. A similar construction leads to the trace CPC' after a second interval t.

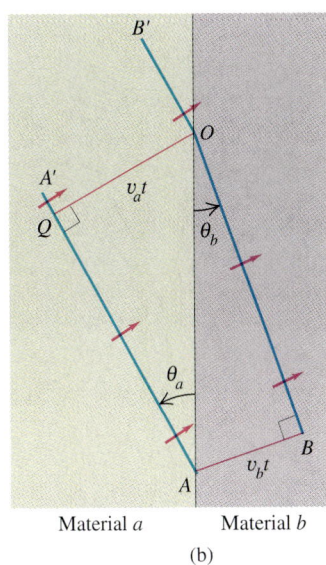

34-28 (a) Successive positions of a plane wave front AA' as it is refracted at a plane surface. (b) Magnified portion of part (a). The case $v_b < v_a$ is shown.

The angles θ_a and θ_b between the surface and the incident and refracted wave fronts are the angle of incidence and the angle of refraction, respectively. To find the relation between these angles, refer to Fig. 34–28b. Draw $OQ = v_a t$, perpendicular to AQ, and draw $AB = v_b t$, perpendicular to BO. From the right triangle AOQ,

$$\sin \theta_a = \frac{v_a t}{AO},$$

and from the right triangle AOB,

$$\sin \theta_b = \frac{v_b t}{AO}.$$

Combining these, we find

$$\frac{\sin \theta_a}{\sin \theta_b} = \frac{v_a}{v_b}. \tag{34–9}$$

We have defined the index of refraction n of a material as the ratio of the speed of light c in vacuum to its speed v in the material: $n_a = c/v_a$ and $n_b = c/v_b$. Thus

$$\frac{n_b}{n_a} = \frac{c/v_b}{c/v_a} = \frac{v_a}{v_b},$$

and we can rewrite Eq. (34–9) as

$$\frac{\sin \theta_a}{\sin \theta_b} = \frac{n_b}{n_a},$$

or

$$n_a \sin \theta_a = n_b \sin \theta_b,$$

which we recognize as Snell's law, Eq. (34–4). So we have derived Snell's law from a wave theory. Alternatively, we may choose to regard Snell's law as an experimental result that defines the index of refraction of a material; in that case this analysis helps to confirm the relationship $v = c/n$ for the speed of light in a material.

Mirages offer an interesting example of Huygens' principle in action. When the surface of pavement or desert sand is heated intensely by the sun, a hot, less dense,

34–29 A mirage is observed because wavelets near the hot surface have slightly greater radii vt, gradually tilting the wave fronts and bending the rays.

smaller-n layer of air forms near the surface. The speed of light is slightly greater in the hotter air near the ground, the Huygens wavelets have slightly larger radii, the wave fronts tilt slightly, and rays that were headed toward the surface with a large angle of incidence (near 90°) can be bent up as shown in Fig. 34–29. Light farther from the ground is bent less and travels nearly in a straight line. The observer sees the object in its natural position, with an inverted image below it, as though seen in a horizontal reflecting surface. Even when the turbulence of the heated air prevents a clear inverted image from being formed, the mind of the thirsty traveler can interpret the apparent reflecting surface as a sheet of water.

It is important to keep in mind that Maxwell's equations are the fundamental relations for electromagnetic wave propagation. But it is a remarkable fact that Huygens' principle anticipated Maxwell's analysis by two centuries. Maxwell provided the theoretical underpinning for Huygens' principle. Every point in an electromagnetic wave, with its time-varying electric and magnetic fields, acts as a source of the continuing wave, as predicted by Ampere's and Faraday's laws.

SUMMARY

KEY TERMS

optics, 1053
wave front, 1054
ray, 1055
geometric optics, 1055
physical optics, 1055

- Light is an electromagnetic wave. When emitted or absorbed, it also shows particle properties. It is emitted by accelerated electric charges that have been given excess energy by heat or electrical discharge. The speed of light is a fundamental physical constant.

- A wave front is a surface of constant phase; wave fronts move with a speed equal to the propagation speed of the wave. A ray is a line along the direction of propagation, perpendicular to the wave fronts. Representation of light by rays is the basis of geometric optics.

■ When light is transmitted from one material to another, the frequency of the light is unchanged, but the wavelength and wave speed can change. The index of refraction n of a material is the ratio of the speed of light in vacuum c to the speed v in the material: $n = c/v$. If λ_0 is the wavelength in a vacuum, the same wave has wavelength $\lambda = \lambda_0/n$ in a medium with index of refraction n.

■ At a smooth interface between two optical materials, the incident, reflected, and refracted rays and the normal to the interface all lie in a single plane called the plane of incidence. The law of reflection states that the angles of incidence and reflection are equal. The law of refraction is

$$n_a \sin \theta_a = n_b \sin \theta_b. \qquad (34\text{--}4)$$

Angles of incidence, reflection, and refraction are always measured from the normal to the surface.

■ When a ray travels in a material of greater index of refraction n_a toward a material of smaller index n_b, total internal reflection occurs at the interface when the angle of incidence exceeds a critical angle θ_{crit} given by

$$\sin \theta_{\text{crit}} = \frac{n_b}{n_a}. \qquad (34\text{--}6)$$

■ The variation of index of refraction n with wavelength λ is called dispersion. Usually, n decreases with increasing λ.

■ The direction of polarization of a linearly polarized electromagnetic wave is the direction of the $\vec{E}$ field. A polarizing filter passes waves that are linearly polarized along its polarizing axis and blocks waves that are polarized perpendicularly to that axis. When polarized light of intensity I_{max} is incident on a polarizing filter used as an analyzer, the intensity I of the light transmitted through the analyzer is given by Malus's law:

$$I = I_{\text{max}} \cos^2 \phi, \qquad (34\text{--}7)$$

where ϕ is the angle between the polarization direction of the incident light and the polarizing axis of the analyzer.

■ When unpolarized light strikes an interface between two materials, Brewster's law states that the reflected light is completely polarized perpendicular to the plane of incidence (parallel to the interface) if the angle of incidence θ_p is given by

$$\tan \theta_p = \frac{n_b}{n_a}. \qquad (34\text{--}8)$$

■ When two linearly polarized waves with a phase difference are superposed, the result is circularly or elliptically polarized light. In this case the $\vec{E}$ vector is not confined to a plane containing the direction of propagation but describes circles or ellipses in planes that are perpendicular to the propagation direction.

■ A birefringent material has different indexes of refraction for two perpendicular directions of polarization. Materials that become birefringent under mechanical stress form the basis of photoelastic stress analysis. A dichroic material has preferential absorption for one polarization direction.

■ Light is scattered by air molecules. The scattered light is partially polarized.

■ Huygens' principle states that if the position of a wave front at one instant is known, the position of the front at a later time can be constructed by imagining the front as a source of secondary wavelets. Huygens' principle can be used to derive the laws of reflection and refraction.

DISCUSSION QUESTIONS

Q34–1 Light requires about 8 minutes to travel from the sun to the earth. Is it delayed appreciably by the earth's atmosphere? Explain.

Q34–2 Sunlight or starlight passing through the earth's atmosphere is always bent toward the vertical. Why? Does this mean that a star is not really where it appears to be? Explain.

Q34–3 The sun or moon usually appears flattened just before it sets. Is this related to refraction in the earth's atmosphere (mentioned in Question 34–2)? Explain.

Q34–4 A student claimed that, because of atmospheric refraction (see Question 34–2), the sun can be seen after it has set and that the day is therefore longer than it would be if the earth had no atmosphere. First, what does she mean by saying that the sun can be seen after it has set? Second, comment on the validity of her conclusion.

Q34–5 When hot air rises from a radiator or heating duct, objects behind it appear to shimmer or waver. What causes this?

Q34–6 Sometimes when looking at a window, you see two reflected images slightly displaced from each other. What causes this?

Q34–7 A ray of light in air strikes a glass surface. Is there a range of angles for which total reflection occurs? Explain.

Q34–8 When light is incident on an interface between two materials, the angle of the refracted ray depends on the wavelength, but the angle of the reflected ray does not. Why should this be?

Q34–9 A salesperson at a bargain counter claims that a certain pair of sunglasses has Polaroid filters; you suspect that they are just tinted plastic. How could you find out for sure?

Q34–10 Does it make sense to talk about the polarization of a *longitudinal* wave, such as a sound wave? Why or why not?

Q34–11 How can you determine the direction of the polarizing axis of a single polarizer?

Q34–12 It has been proposed that automobile windshields and headlights should have polarizing filters to reduce the glare of oncoming lights during night driving. Would this work? How should the polarizing axes be arranged? What advantages would this scheme have? What disadvantages?

Q34–13 For three-dimensional movies, two images are projected on the screen, and the viewers wear special glasses to sort them out. How could this work?

Q34–14 When a sheet of plastic food wrap is placed between two crossed polarizers, no light is transmitted. When the sheet is stretched in one direction, some light passes through the crossed polarizers. What is happening?

Q34–15 If you sit on the beach and look at the ocean through Polaroid sunglasses, the glasses help to reduce the glare from

sunlight reflecting off the water. But if you lie on your side on the beach, there is little reduction in the glare. Explain why there is a difference.

Q34–16 When unpolarized light is incident on two crossed polarizers, no light is transmitted. A student asserted that if a third polarizer is inserted between the other two, some transmission will occur. Does this make sense? How can adding a third filter *increase* transmission?

Q34–17 In Fig. 34–25, since the light that is scattered out of the incident beam is polarized, why is the transmitted beam not also partially polarized?

Q34–18 You are sunbathing in the late afternoon when the sun is relatively low in the western sky. You are lying flat on your back, looking straight up through Polaroid sunglasses. To minimize the amount of sky light reaching your eyes, how should you lie: with your feet pointing north, east, south, west, or in some other direction? Explain your reasoning.

Q34–19 Light scattered from blue sky is strongly polarized because of the nature of the scattering process described in Section 34–7. But light scattered from white clouds is usually *not* polarized. Why not?

Q34–20 Atmospheric haze is due to water droplets or smoke particles ("smog"). Such haze reduces visibility by scattering light, so that the light from distant objects becomes randomized and images become indistinct. Explain why visibility through haze can be improved by wearing red-tinted sunglasses, which filter out blue light.

Q34–21 The explanation given in Section 34–7 for the color of the setting sun should apply equally well to the *rising* sun, since sunlight travels the same distance through the atmosphere to reach your eyes at either sunrise or sunset. Typically, however, sunsets are redder than sunrises. Why? (*Hint:* Particles of all kinds in the atmosphere contribute to scattering.)

Q34–22 Huygens' principle also applies to sound waves. During the day, the temperature of the atmosphere decreases with increasing altitude above the ground. But at night, when the ground cools, there is a layer of air just above the surface in which the temperature *increases* with altitude. Use this to explain why sound waves from distant sources can be heard more clearly at night than in the daytime. (*Hint:* The speed of sound increases with increasing temperature. Use the ideas displayed in Fig. 34–29 for light.)

Q34–23 Television transmission usually uses plane-polarized waves. It has been proposed to use circularly polarized waves to improve reception. Why?

Q34–24 Can water waves be reflected and refracted? Give examples. Does Huygens' principle apply to water waves? Explain.

EXERCISES

SECTION 34–3 REFLECTION AND REFRACTION

34–1 A beam of light has a wavelength of 650 nm in vacuum. a) What is the speed of this light in a liquid whose index of refraction at this wavelength is 1.47? b) What is the wavelength of these waves in the liquid?

34–2 Light with a frequency of 5.80×10^{14} Hz travels in a block of glass that has an index of refraction of 1.52. What is the wavelength of the light a) in the glass; b) in vacuum?

34–3 A light beam travels at 1.94×10^8 m/s in quartz. The wavelength of the light in quartz is 355 nm. a) What is the index of refraction of quartz at this wavelength? b) If this same light travels through air, what is its wavelength there?

34–4 Light of a certain frequency has a wavelength of 438 nm in water. What is the wavelength of this light in benzene?

34–5 A parallel beam of light in air makes an angle of 47.5° with the surface of a glass plate having a refractive index of 1.66. a) What is the angle between the reflected part of the beam and the surface of the glass? b) What is the angle between the refracted beam and the surface of the glass?

34–6 A horizontal, parallel-sided plate of glass having a refractive index of 1.52 is in contact with the surface of water in a tank. A ray coming from above in air makes an angle of incidence of 35.0° with the normal to the top surface of the glass. a) What angle does the ray refracted into the water make with the normal to the surface? b) What is the dependence of this angle on the refractive index of the glass?

34–7 A parallel beam of light is incident on a prism, as shown in Fig. 34–30. Part of the light is reflected from one face and part from another. Show that the angle θ between the two reflected beams is twice the angle A between the two reflecting surfaces.

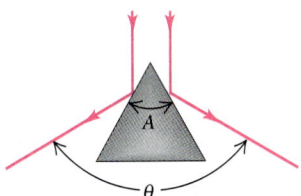

FIGURE 34–30 Exercise 34–7.

34–8 Prove that a ray of light reflected from a plane mirror rotates through an angle of 2θ when the mirror rotates through an angle θ about an axis perpendicular to the plane of incidence.

34–9 A ray of light is incident on a plane surface separating two sheets of glass with refractive indexes 1.70 and 1.58. The angle of incidence is 62.0°, and the ray originates in the glass with $n = 1.70$. Compute the angle of refraction.

34–10 In Example 34–1 (Section 34–3), the water-glass interface is horizontal. If instead this interface were tilted 15.0° above the horizontal, with the right side higher than the left side, what would be the angle from the vertical of the ray in the glass? (The ray in the water still makes an angle of 60.0° with the vertical.)

SECTION 34–4 TOTAL INTERNAL REFLECTION

34–11 The critical angle for total internal reflection at a liquid-air interface is 42.5°. a) If a ray of light traveling in the liquid has an angle of incidence at the interface of 35.0°, what angle does the refracted ray in the air make with the normal? b) If a ray of light traveling in air has an angle of incidence at the interface of 35.0°, what angle does the refracted ray in the liquid make with the normal?

34–12 A point source of light is 85.0 cm below the surface of a body of water. Find the diameter of the largest circle at the surface through which light can emerge from the water.

34–13 The speed of a sound wave is 344 m/s in air and 1320 m/s in water. a) Which medium has the higher "index of refraction" for sound? b) What is the critical angle for a sound wave incident on the surface between air and water? c) For total internal reflection to occur, must the sound wave be traveling in the air or in the water?

34–14 A ray of light in diamond (index of refraction 2.42) is incident on an interface with air. What is the *largest* angle the ray can make with the normal and not be totally reflected back into the diamond?

SECTION 34–6 POLARIZATION

34–15 A parallel beam of unpolarized light in air is incident at an angle of 54.5° (with respect to the normal) on a plane glass surface. The reflected beam is completely linearly polarized. a) What is the refractive index of the glass? b) What is the angle of refraction of the transmitted beam?

34–16 Light traveling in water strikes a glass plate at an angle of incidence of 53.0°; part of the beam is reflected and part is refracted. If the reflected and refracted portions make an angle of 90.0° with each other, what is the index of refraction of the glass?

34–17 Unpolarized light traveling in a liquid with refractive index n is incident on the surface of the liquid, above which there is air. If the light is incident on the surface at an angle of 31.2° with respect to the normal, the light reflected back into the liquid is completely polarized. a) What is the refractive index n of the liquid? b) What angle does the refracted light traveling in air make with the normal to the surface?

34–18 a) At what angle above the horizontal is the sun if sunlight reflected from the surface of a calm body of water is completely polarized? b) What is the plane of the $\vec{E}$ vector in the reflected light?

34–19 Unpolarized light with intensity I_0 is incident on a polarizing filter. The emerging light strikes a second polarizing filter whose axis is at 41.0° to that of the first. Determine a) the intensity of the beam after it has passed through the second polarizer; b) its state of polarization.

34–20 A polarizer and an analyzer are oriented so that the maximum amount of light is transmitted. To what fraction of its maximum value is the intensity of the transmitted light reduced when the analyzer is rotated through a) 22.5°; b) 45.0°; c) 67.5°?

34–21 Three Polarizing Filters. Three polarizing filters are stacked with the polarizing axes of the second and third at 45.0° and 90.0°, respectively, with that of the first. a) If unpolarized light of intensity I_0 is incident on the stack, find the intensity and state of polarization of light emerging from each filter. b) If the second filter is removed, what is the intensity of the light emerging from each remaining filter?

*SECTION **34–7 SCATTERING OF LIGHT**

34–22 A beam of light, after passing through the Polaroid disk P_1 in Fig. 34–31, traverses a cell containing a scattering medium. The cell is observed at right angles through another Polaroid disk P_2. Initially, the disks are oriented until the brightness of the scattered light from the cell as seen by the observer is a maximum. a) Disk P_2 is now rotated through 90°. Does the cell as viewed by the observer appear bright or dark? Explain. b) Disk P_1 is now rotated through 90°. Is the cell bright or dark?

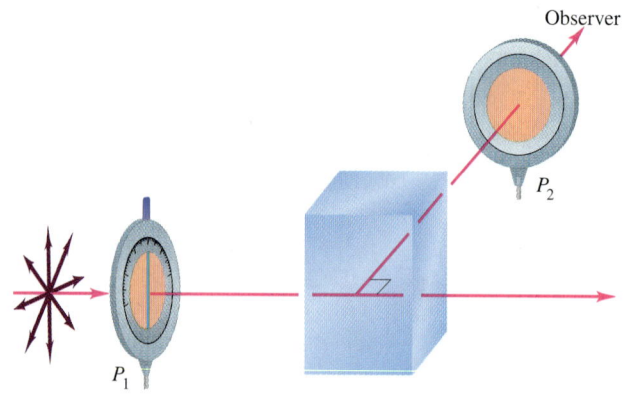

FIGURE 34–31 Exercise 34–22.

Explain. c) Disk P_2 is then restored to its original position. Is the cell bright or dark? Explain.

PROBLEMS

34–23 The Corner Reflector. An inside corner of a cube is lined with mirrors to make a corner reflector (Example 34–3 in Section 34–3). A ray of light is reflected successively from each of three mutually perpendicular mirrors; show that its final direction is always exactly opposite to its initial direction.

34–24 Sonogram of the Heart. Physicians use high-frequency ($f = 1–5$ MHz) sound waves, called ultrasound, to image internal organs. The speed of these ultrasound waves is 1480 m/s in muscle and 344 m/s in air. a) At what angle from the normal does an ultrasound beam enter the heart if it leaves the lungs at an angle of 9.73° from the normal to the heart wall? (Assume that the speed of sound in the lungs is 344 m/s.) b) What is the critical angle for sound waves in air incident on muscle?

34–25 A ray of light is incident in air on a block of a transparent solid whose index of refraction is n. If $n = 1.38$, what is the *largest* angle of incidence θ_a for which total internal reflection will occur at the vertical face (point A in Fig. 34–32)?

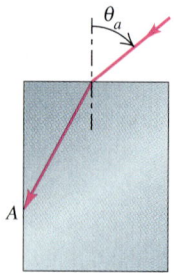

FIGURE 34–32 Problem 34–25.

34–26 A thin beam of light in air is incident on the surface of a lanthanum flint glass plate having a refractive index of 1.80. What is the angle of incidence θ_a with this plate for which the angle of refraction is $\theta_a/2$? Both angles are measured relative to the normal.

34–27 Use the law of reflection to show that specular reflection reverses the sign of the component of light velocity perpendicular to the reflecting surface, but leaves the other components unchanged.

34–28 A glass plate 2.50 mm thick, with an index of refraction of 1.40, is placed between a point source of light with wavelength 540 nm (in vacuum) and a screen. The distance from source to screen is 1.80 cm. How many wavelengths are there between the source and the screen?

34–29 Old photographic plates were made of glass with a light-sensitive emulsion on the front surface. This emulsion was somewhat transparent. When a bright point source is focused on the front of the plate, the developed photograph will show a halo around the image of the spot. If the glass plate is 3.10 mm thick and the halos have an inner radius of 5.34 mm, what is the index of refraction of the glass? (*Hint:* Light from the spot on the front surface is totally reflected at the back surface of the plate and comes back to the front surface.)

34–30 After a long day of driving you take a late-night swim in a motel swimming pool. When you go to your room, you

FIGURE 34–33 Problem 34–30.

realize that you have lost your room key in the pool. You borrow a powerful flashlight and walk around the pool, shining the light into it. The light shines on the key, which is lying on the bottom of the pool, when the flashlight is held 1.2 m above the water surface and is directed at the surface a horizontal distance of 1.5 m from the edge (Fig. 34–33). If the water here is 4.0 m deep, how far is the key from the edge of the pool?

34–31 You sight along the rim of a glass with vertical sides so that the top rim is lined up with the opposite edge of the bottom (Fig. 34–34a). The glass is a thin-walled hollow cylinder 16.0 cm high with a top and bottom diameter of 8.0 cm. While you keep your eye in the same position, a friend fills the glass with a transparent liquid, and you then see a dime that is lying at the center of the bottom of the glass (Fig. 34–34b). What is the index of refraction of the liquid?

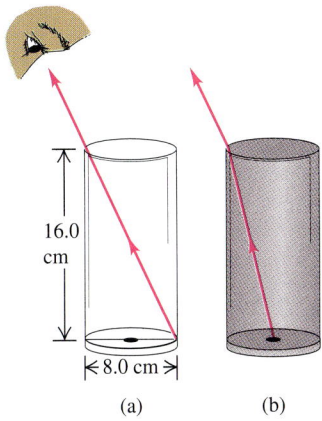

FIGURE 34–34 Problem 34–31.

34–32 A beaker with a mirrored bottom is filled with a liquid whose index of refraction is 1.63. A light beam strikes the top surface of the liquid at an angle of 42.5° from the normal. At what angle from the normal will the beam exit from the liquid after traveling down through the liquid, reflecting from the mirrored bottom, and returning to the surface?

34–33 A thin layer of ice ($n = 1.309$) floats on the surface of water ($n = 1.333$) in a bucket. A ray of light from the bottom of the bucket travels upward through the water. a) What is the largest angle with respect to the normal that the ray can make at the ice-water interface and still pass out into the air above the ice? b) What is this angle after the ice melts?

34–34 A 45°–45°–90° prism is immersed in water. A ray of light is incident normally on one of its shorter faces. What is the minimum index of refraction that the prism must have if this ray is to be totally reflected within the glass at the long face of the prism?

34–35 The prism of Fig. 34–35 has a refractive index of 1.66, and the angles A are 25.0°. Two light rays m and n are parallel as they enter the prism. What is the angle between them after they emerge?

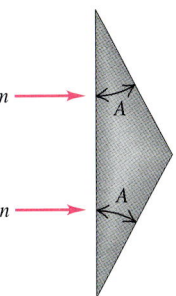

FIGURE 34–35 Problem 34–35.

34–36 Light is incident normally on the short face of a 30°–60°–90° prism (Fig. 34–36). A drop of liquid is placed on the hypotenuse of the prism. If the index of the prism is 1.62, find the maximum index that the liquid may have if the light is to be totally reflected.

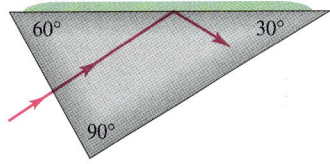

FIGURE 34–36 Problem 34–36.

34–37 A ray of light traveling in a block of glass ($n = 1.52$) is incident on the top surface at an angle of 57.2° with respect to the normal. If a layer of oil is placed on the top surface of the glass, the ray is totally reflected. What is the maximum possible index of refraction of the oil?

34–38 A thin beam of white light is directed at a flat sheet of silicate flint glass at an angle of 20.0° to the surface of the sheet. Due to dispersion in the glass, the beam is spread out as shown into a spectrum (Fig. 34–37). The refractive index of silicate flint glass versus wavelength is graphed in Fig. 34–13. a) The rays a and b in Fig. 34–37 correspond to the extremes of the visible spectrum. Which corresponds to red, and which to violet? Explain your reasoning. b) For what thickness d of the glass sheet will the spectrum be 1.0 mm wide, as shown? (See Problem 34–44.)

FIGURE 34–37 Problem 34–38.

34–39 When the sun is either rising or setting and appears to be just on the horizon, it is in fact *below* the horizon. The explanation for this seeming paradox is that light from the sun bends slightly when entering the earth's atmosphere, as shown in Fig. 34–38. Since our perception is based on the idea that light travels in straight lines, we perceive the light to be coming from an apparent position that is an angle δ above the sun's true position. a) Make the simplifying assumptions that the atmosphere has uniform density, and hence uniform index of refraction n, and extends to a height h above the earth's surface, at which point it abruptly stops. Show that the angle δ is given by

$$\delta = \arcsin\left(\frac{nR}{R+h}\right) - \arcsin\left(\frac{R}{R+h}\right),$$

where $R = 6378$ km is the radius of the earth. b) Calculate δ using $n = 1.0003$ and $h = 20$ km. How does this compare to the angular radius of the sun, which is about one quarter of a degree? (In actuality a light ray from the sun bends gradually, not abruptly, since the density and refractive index of the atmosphere change gradually with altitude.)

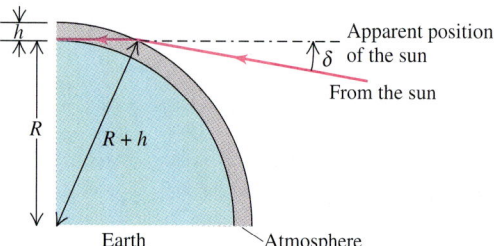

FIGURE 34–38 Problem 34–39.

34–40 A quarter-wave plate converts linearly polarized light to circularly polarized light. Prove that a quarter-wave plate also converts circularly polarized light to linearly polarized light.

34–41 Three polarizing filters are stacked, with the polarizing axes of the second and third at angles θ and 90°, respectively, with that of the first. Unpolarized light with intensity I_0 is incident on the stack. a) Derive an expression for the intensity of light transmitted through the stack as a function of I_0 and θ. b) For what value of θ does the maximum transmission occur?

34–42 Fermat's Principle of Least Time. A ray of light traveling with speed c leaves point 1 of Fig. 34–39 and is reflected to point 2. The ray strikes the reflecting surface a horizontal distance x from point 1. a) Show that the time t required for the light to travel from 1 to 2 is

$$t = \frac{\sqrt{y_1{}^2 + x^2} + \sqrt{y_2{}^2 + (l-x)^2}}{c}.$$

b) Take the derivative of t with respect to x. Set the derivative equal to zero to show that this time reaches its *minimum* value when $\theta_1 = \theta_2$, which is the law of reflection and corresponds to the actual path taken by the light. This is an example of Fermat's *principle of least time,* which states that among all possible paths between two points, the one actually taken by a ray of light is that

for which the time of travel is a *minimum*. (In fact, there are some cases in which the time is a maximum rather than a minimum.)

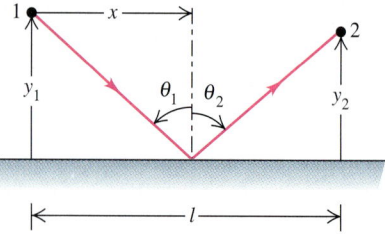

FIGURE 34–39 Problem 34–42.

34–43 A ray of light goes from point A in a medium in which the speed of light is v_1 to point B in a medium in which the speed is v_2 (Fig. 34–40). The ray strikes the interface a horizontal distance x to the right of point A. a) Show that the time required for the light to go from A to B is

$$t = \frac{\sqrt{h_1{}^2 + x^2}}{v_1} + \frac{\sqrt{h_2{}^2 + (l-x)^2}}{v_2}.$$

b) Take the derivative of t with respect to x. Set this derivative equal to zero to show that this time reaches its *minimum* value when $n_1 \sin\theta_1 = n_2 \sin\theta_2$. This is Snell's law, and corresponds to the actual path taken by the light. This is another example of Fermat's principle of least time (see Problem 34–42).

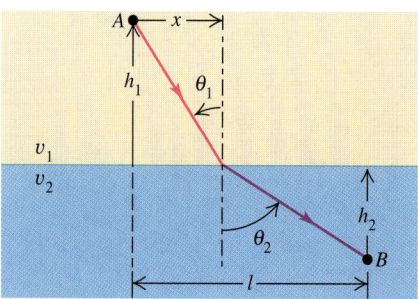

FIGURE 34–40 Problem 34–43.

34–44 Light is incident in air at an angle θ_a (Fig. 34–41) on the upper surface of a transparent plate, the surfaces of the plate being plane and parallel to each other. a) Prove that $\theta_a = \theta_a{}'$. b) Show that this is true for any number of different parallel plates. c) Prove that the lateral displacement d of the emergent beam is given by the relation

$$d = t\,\frac{\sin(\theta_a - \theta_b{}')}{\cos\theta_b{}'},$$

where t is the thickness of the plate. d) A ray of light is incident at an angle of 66.0° on one surface of a glass plate 2.40 cm thick with an index of refraction 1.80. The medium on either side of the plate is air. Find the lateral displacement between the incident and emergent rays.

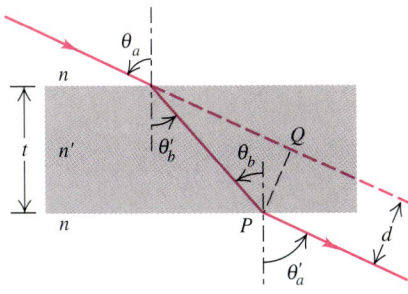

FIGURE 34–41 Problem 34–44.

34–45 Angle of Deviation. The incident angle θ_a in Fig. 34–42 is chosen so that the light passes symmetrically through the prism, which has refractive index n and apex angle A. a) Show that the angle of deviation δ (the angle between the initial and final directions of the ray) is given by

$$\sin \frac{A + \delta}{2} = n \sin \frac{A}{2}.$$

(When the light passes through symmetrically, as shown, the angle of deviation is a minimum.) b) Use the result of part (a) to find the angle of deviation for a ray of light passing symmetrically through a prism having three equal angles ($A = 60.0°$) and $n = 1.52$. c) A certain glass has a refractive index of 1.61 for red light (700 nm) and 1.66 for violet light (400 nm). If both colors pass through symmetrically, as described in part (a), and if $A = 60.0°$, find the difference between the angles of deviation for the two colors.

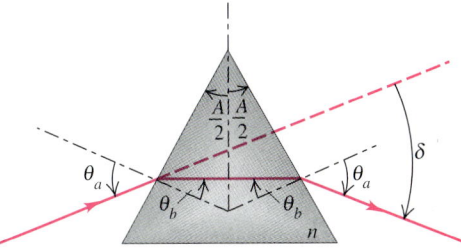

FIGURE 34–42 Problem 34–45.

34–46 The refractive index of a certain glass is 1.66. For what incident angle is light reflected from the surface of this glass completely polarized if the glass is immersed in a) air? b) water?

34–47 Optical Activity. Many biologically important molecules are optically active. When plane-polarized light traverses a solution of these compounds, its plane of polarization is rotated. Some compounds rotate the polarization clockwise; others rotate the polarization counterclockwise. The amount of rotation depends on the amount of material in the light path. The following data give the amount of rotation through two amino acids over a path length of 100 cm. From this data, find the relationship between the concentration C (in grams per 100 mL) and the rotation of the polarization (in degrees) of each amino acid.

ROTATION		CONCENTRATION
l-leucine	*d*-glutamic acid	(g/100 mL)
−0.11°	0.124°	1.0
−0.22°	0.248°	2.0
−0.55°	0.620°	5.0
−1.10°	1.24°	10.0
−2.20°	2.48°	20.0
−5.50°	6.20°	50.0
−11.0°	12.4°	100.0

34–48 A certain birefringent material has indexes of refraction n_1 and n_2 for the two perpendicular components of linearly polarized light passing through it. The corresponding wavelengths are $\lambda_1 = \lambda_0/n_1$ and λ_0/n_2, where λ_0 is the wavelength in vacuum. a) If the crystal is to function as a quarter-wave plate, the number of wavelengths of each component within the material must differ by $\frac{1}{4}$. Show that the minimum thickness for a quarter-wave plate is

$$d = \frac{\lambda_0}{4(n_1 - n_2)}.$$

b) Find the minimum thickness of a quarter-wave plate made of siderite ($FeO \cdot CO_2$) if the indexes of refraction are $n_1 = 1.875$ and $n_2 = 1.635$ and the wavelength in vacuum is $\lambda_0 = 589$ nm.

34–49 A beam of light traveling horizontally is made of an unpolarized component with intensity I_0 and a polarized component with intensity I_p. The plane of polarization of the polarized component is oriented at an angle of θ with respect to the vertical. The following data gives the intensity measured through a polarizer with an orientation of ϕ with respect to the vertical:

ϕ (°)	I_{total} (W/m^2)	ϕ (°)	I_{total} (W/m^2)
0	18.4	100	8.6
10	21.4	110	6.3
20	23.7	120	5.2
30	24.8	130	5.2
40	24.8	140	6.3
50	23.7	150	8.6
60	21.4	160	11.6
70	18.4	170	15.0
80	15.0	180	18.4
90	11.6		

a) What is the orientation of the polarized component? (That is, what is the angle θ?) b) What are the values of I_0 and I_p?

34–50 It is desired to rotate the polarization direction of linearly polarized light by 90°. This is to be done by using N ideal polarizers. The intensity of the original linearly polarized light is I_0. a) Describe how the N filters should best be arranged to accomplish the desired rotation of the polarization direction, and show that the intensity of the light transmitted through the last filter is $I = I_0 \cos^{2N}(\pi/2N)$. b) If $N \gg 1$, use the information given in Appendix B to show that $I \cong I_0(1 - \pi^2/4N)$. Hence show that if N is very large, the final intensity is essentially the same as the initial intensity.

CHALLENGE PROBLEMS

34–51 Consider two vibrations of equal amplitude and frequency but differing in phase, one along the x-axis,

$$x = a \sin(\omega t - \alpha),$$

and the other along the y-axis,

$$y = a \sin(\omega t - \beta).$$

These can be written as follows:

$$\frac{x}{a} = \sin \omega t \cos \alpha - \cos \omega t \sin \alpha, \qquad (1)$$

$$\frac{y}{a} = \sin \omega t \cos \beta - \cos \omega t \sin \beta. \qquad (2)$$

a) Multiply Eq. (1) by sin β and Eq. (2) by sin α, and then subtract the resulting equations. b) Multiply Eq. (1) by cos β and Eq. (2) by cos α, and then subtract the resulting equations. c) Square and add the results of parts (a) and (b). d) Derive the equation $x^2 + y^2 - 2xy \cos \delta = a^2 \sin^2 \delta$, where $\delta = \alpha - \beta$. e) Use the above result to justify each of the diagrams in Fig. 34–43. In the figure, the angle given is the phase difference between two simple harmonic motions of the same frequency and amplitude, one horizontal (along the x-axis) and the other vertical (along the y-axis). The figure thus shows the resultant motion from the superposition of the two perpendicular harmonic motions.

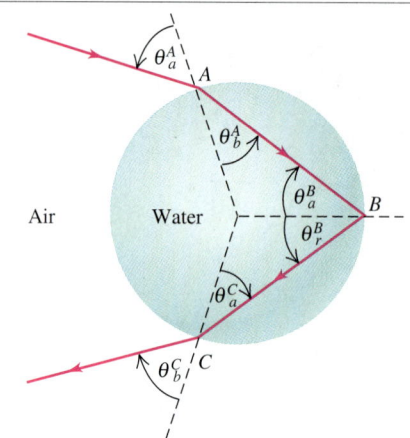

FIGURE 34–44 Challenge Problem 34–52.

this occurs. Show that $\cos^2\theta_1 = \frac{1}{3}(n^2 - 1)$. (*Hint:* You may find the derivative formula $d (\arcsin u(x))/dx = (1 - u^2)^{-1/2}(du/dx)$ helpful.) e) The index of refraction in water is 1.342 for violet light and 1.330 for red light. Use the results of parts (c) and (d) to find θ_1 and Δ for violet and red light. Do your results agree

0	$\dfrac{\pi}{4}$	$\dfrac{\pi}{2}$	$\dfrac{3\pi}{4}$	π	$\dfrac{5\pi}{4}$	$\dfrac{3\pi}{2}$	$\dfrac{7\pi}{4}$	2π

FIGURE 34–43 Challenge Problem 34–51.

34–52 A rainbow is produced by the reflection of sunlight by spherical drops of water in the air. Figure 34–44 shows a ray that refracts into a drop at A, is reflected from the back surface of the drop at B, and refracts back into the air at C. The angles of incidence and refraction, θ_a and θ_b, are shown at points A and C and the angles of incidence and reflection, θ_a and θ_r, are shown at B. a) Show that $\theta_a^B = \theta_b^A$, $\theta_a^C = \theta_b^A$, and $\theta_b^C = \theta_a^A$. b) Show that the angle in radians between the ray before it enters the drop at A and after it exits at C (the total angular deflection of the ray) is $\Delta = 2\theta_a^A - 4\theta_b^A + \pi$. (*Hint:* Find the angular deflections that occur at A, B, and C, and add to get Δ.) c) Use Snell's law to write Δ in terms of θ_a^A and n, the refractive index of the water in the drop. d) A rainbow will form when the angular deflection Δ is *stationary* in the incident angle θ_a^A, that is, when $d\Delta/d\theta_a^A = 0$. If this condition is satisfied, all the rays with incident angles close to θ_a^A will be sent back in the same direction, producing a bright zone in the sky. Let θ_1 be the value of θ_a^A for which

with the angles shown in Fig. 34–15a? When you view the rainbow, which color, red or violet, is higher above the horizon?

34–53 A *secondary rainbow* is formed when the incident light undergoes two internal reflections in a spherical drop of water as shown in Fig. 34–15b. (See Challenge Problem 34–52.) a) In terms of the incident angle θ_a^A and the refractive index n of the drop, what is the angular deflection Δ of the ray? That is, what is the angle between the ray before it enters the drop and after it exits? b) What is the incident angle θ_2 for which the derivative of Δ with respect to the incident angle θ_a^A is zero? c) The indexes of refraction for red and violet light in water are given in part (e) of Challenge Problem 34–52. Use the results of parts (a) and (b) to find θ_2 and Δ for violet and red light. Do your results agree with the angles shown in Fig. 34–15b? When you view a secondary rainbow, is red or violet highest above the horizon?

35

Geometric Optics

Light rays bend upon entering or leaving this glass of water. As a result, the glass acts as a cylindrical lens that forms an image of the cat. Lenses of all kinds use this principle—refraction by a curved surface—to form images. Images can also be formed by the reflection of light from suitably curved mirrors.

35–1 INTRODUCTION

Your reflection in the bathroom mirror, the view of the moon through a telescope, the patterns seen in a kaleidoscope; all of these are examples of *images.* In each case the object that you're looking at appears to be in a different place than its actual position; your reflection is on the other side of the mirror, the moon appears to be much closer when seen through a telescope, and objects seen in a kaleidoscope seem to be in many places at the same time. In each case, light rays that come from a point on an object are deflected by reflection or refraction (or a combination of the two), so they converge toward or appear to diverge from a point called an *image point.* Our goal in this chapter is to see how this is done and to explore the different kinds of images that can be made with simple optical devices.

To understand images and image formation, all we need are the ray model of light, the laws of reflection and refraction, and some simple geometry and trigonometry. The key role played by geometry in our analysis is the reason for the name *geometric optics* that is given to the study of how light rays form images. We'll begin our analysis with one of the simplest of image-forming optical devices, a plane mirror. We'll go on to study how images are formed by curved mirrors, by refracting surfaces, and by thin lenses. Our results will lay the foundation for understanding many familiar optical instruments, including camera lenses, magnifiers, the human eye, microscopes, and telescopes. We will study these instruments in Chapter 36.

35–2 REFLECTION AND REFRACTION AT A PLANE SURFACE

Before discussing what is meant by an image, we first need the concept of **object** as it is used in optics. By an *object* we mean anything from which light rays radiate. This light could be emitted by the object itself if it is *self-luminous,* like the glowing filament of a light bulb. Alternatively, the light could be emitted by another source (such as a lamp or the sun) and then reflected from the object; an example is the light you see coming from the pages of this book. Figure 35–1 shows light rays radiating in all directions from an object at a point *P*. For an observer to see this object directly, there must be no obstruction between the object and the observer's eyes. Note that light rays from the object reach the observer's left and right eyes at different angles; these differences are processed by the observer's brain to infer the *distance* from the observer to the object.

The object in Fig. 35–1 is a **point object** that has no physical extent. Real objects with length, width, and height are called **extended objects.** To start with, we'll consider only an idealized point object, since we can always think of an extended object as being made up of a very large number of point objects.

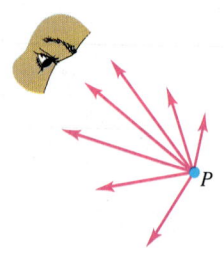

35–1 Light rays radiate from a point object in all directions.

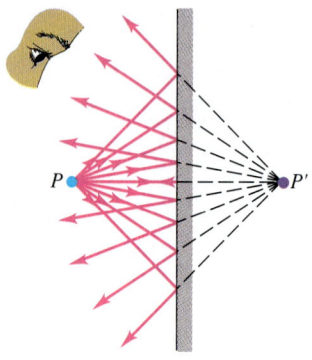

Plane mirror

35–2 The rays entering the eye after reflection from a plane mirror look as though they had come from point P', the image point for object point P.

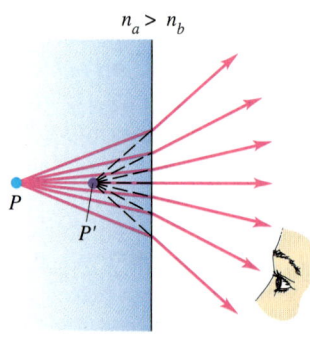

35–3 The rays entering the eye after refraction at the plane interface look as though they had come from point P', the image point for object P. When $n_a > n_b$, as shown here, the image point P' is closer to the surface than the object point P. The angles of incidence have been exaggerated for clarity.

Suppose some of the rays from the object strike a smooth, plane reflecting surface (Fig. 35–2). This could be the surface of a material with a different index of refraction, which reflects part of the incident light, or a polished metal surface that reflects almost 100% of the light that strikes it. We will always draw the reflecting surface as a black line with a shaded area behind it, as in Fig. 35–2. Bathroom mirrors have a thin sheet of glass that lies in front of and protects the reflecting surface; we'll ignore the effects of this thin sheet (but see Problem 35–51).

According to the law of reflection, all rays striking the surface are reflected at an angle from the normal equal to the angle of incidence. Since the surface is plane, the normal is in the same direction at all points on the surface, and we have *specular* reflection. After the rays are reflected, their directions are the same as though they had come from point P'. We call point P an *object point* and point P' the corresponding *image point,* and we say that the reflecting surface forms an **image** of point P. An observer who can see only the rays reflected from the surface, and who doesn't know that he's seeing a reflection, *thinks* that the rays originate from the image point P'. The image point is therefore a convenient way to describe the directions of the various reflected rays, just as the object point P describes the directions of the rays arriving at the surface *before* reflection.

If the surface in Fig. 35–2 were *not* smooth, the reflection would be *diffuse,* and rays reflected from different parts of the surface would go in uncorrelated directions. In this case there would not be a definite image point P' from which all reflected rays seem to emanate. You can't see your reflection in the surface of a tarnished piece of metal, because its surface is rough; polishing the metal smooths the surface so that specular reflection occurs and a reflected image becomes visible.

An image is also formed by a plane *refracting* surface, as shown in Fig. 35–3. Rays coming from point P are refracted at the interface between two optical materials. When the angles of incidence are small, the final directions of the rays after refraction are the same as though they had come from point P', as shown, and again we call P' an *image point.* In Section 34–3 we described how this effect makes underwater objects appear closer to the surface than they really are (see Fig. 34–6).

In both Figs. 35–2 and 35–3 the rays do not actually pass through the image point P'. Indeed, if the mirror in Fig. 35–2 is opaque, there is no light at all on its right side. If the outgoing rays don't actually pass through the image point, we call the image a **virtual image.** Later we will see cases in which the outgoing rays really *do* pass through an image point, and we will call the resulting image a **real image.** The images that are formed on a projection screen, on the photographic film in a camera, and on the retina of your eye are real images.

IMAGE FORMATION BY A PLANE MIRROR

Let's concentrate for now on images produced by *reflection;* we'll return to refraction later in the chapter. To find the precise location of the virtual image P' that a plane mirror forms of an object at P, we use the construction shown in Fig. 35–4. The figure shows two rays diverging from an object point P at a distance s to the left of a plane mirror. We call s the **object distance.** The ray PV is incident normally on the mirror (that is, it is perpendicular to the mirror surface), and it returns along its original path.

The ray PB makes an angle θ with PV. It strikes the mirror at an angle of incidence θ and is reflected at an equal angle with the normal. When we extend the two reflected rays backward, they intersect at point P', at a distance s' behind the mirror. We call s' the **image distance.** The line between P and P' is perpendicular to the mirror. The two triangles are congruent, so P and P' are at equal distances from the mirror, and s and s' have equal magnitudes. The image point P' is located exactly opposite the object point P as far *behind* the mirror as the object point is from the front of the mirror.

We can repeat the construction of Fig. 35–4 for each ray diverging from P. The directions of *all* the outgoing reflected rays are the same as though they had originated at point P', confirming that P' is the *image* of P. No matter where the observer is located, she will always see the image at the point P'.

SIGN RULES

Before we go further, let's introduce some general sign rules. These may seem unnecessarily complicated for the simple case of an image formed by a plane mirror, but we want to state the rules in a form that will be applicable to *all* the situations we will encounter later. These will include image formation by a plane or spherical reflecting or refracting surface, or by a pair of refracting surfaces forming a lens. Here are the rules:

1. **Sign rule for the object distance:** When the object is on the same side of the reflecting or refracting surface as the incoming light, the object distance s is positive; otherwise, it is negative.
2. **Sign rule for the image distance:** When the image is on the same side of the reflecting or refracting surface as the outgoing light, the image distance s' is positive; otherwise, it is negative.
3. **Sign rule for the radius of curvature of a spherical surface:** When the center of curvature C is on the same side as the outgoing light, the radius of curvature is positive; otherwise, it is negative.

For a mirror the incoming and outgoing sides are always the same; for example, in Figs. 35–2 and 35–4 they are both the left side. For the refracting surface in Fig. 35–3 the incoming and outgoing sides are on the left and right sides, respectively, of the interface between the two materials.

In Fig. 35–4 the object distance s is *positive* because the object point P is on the incoming side (the left side) of the reflecting surface. The image distance s' is *negative* because the image point P' is *not* on the outgoing side (the left side) of the surface. The object and image distances s and s' are related simply by

$$s = -s' \quad \text{(plane mirror).} \quad (35\text{–}1)$$

For a plane reflecting or refracting surface, the radius of curvature is infinite and not a particularly interesting or useful quantity; in these cases we really don't need the third sign rule. But this rule will be of great importance when we study image formation by *curved* reflecting and refracting surfaces later in the chapter.

IMAGE OF AN EXTENDED OBJECT–PLANE MIRROR

Next we consider an *extended* object with finite size. For simplicity we often consider an object that has only one dimension, like a slender arrow, oriented parallel to the reflecting surface; an example is the arrow PQ in Fig. 35–5. The distance from the head to the tail of an arrow oriented in this way is called its *height;* in Fig. 35–5 the height is y. The image formed by such an extended object is an extended image; to each point on the object, there corresponds a point on the image. Two of the rays from Q are shown; *all* the rays from Q appear to diverge from its image point Q' after reflection. The image of the arrow is the line $P'Q'$, with height y'. Other points of the object PQ have image points between P' and Q'. The triangles PQV and $P'Q'V$ are congruent, so the object PQ and image $P'Q'$ have the same size and orientation, and $y = y'$.

The ratio of image height to object height, y'/y, in *any* image-forming situation is called the **lateral magnification** m; that is,

$$m = \frac{y'}{y} \quad \text{(lateral magnification).} \quad (35\text{–}2)$$

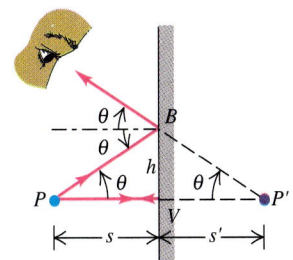

35–4 After reflection at a plane surface, all rays originally diverging from the object point P appear to diverge from the image point P', although they do not actually pass through P'; hence P' is a *virtual image* of P. The image is as far behind the mirror as the object is in front of it.

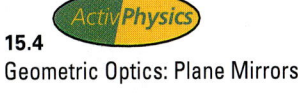

15.4
Geometric Optics: Plane Mirrors

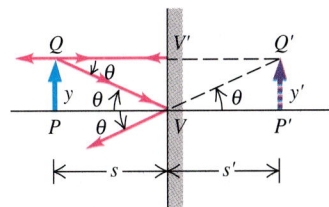

35–5 Construction for determining the height of an image formed by reflection at a plane reflecting surface.

35–6 The image $S'P'Q'R'$ formed by a plane mirror is virtual, erect, and reversed and is the same size as the object $SPQR$.

Thus for a plane mirror the lateral magnification m is unity. When you look at yourself in a plane mirror, your image is the same size as the real you.

In Fig. 35–5 the image arrow points in the *same* direction as the object arrow; we say that the image is **erect.** In this case, y and y' have the same sign, and the lateral magnification m is positive. The image formed by a plane mirror is always erect, so y and y' have both the same magnitude and the same sign; from Eq. (35–2) the lateral magnification of a plane mirror is always $m = +1$. Later we will encounter situations in which the image is **inverted,** that is, the image arrow points in the direction *opposite* to that of the object arrow. For an inverted image, y and y' have *opposite* signs, and the lateral magnification m is *negative.*

The object in Fig. 35–5 has only one dimension, its height y'. Figure 35–6 shows a *three*-dimensional virtual image formed by a plane mirror of a *three*-dimensional object. They are related in the same way as a right hand and a left hand. At this point, you may be asking, "Why does a plane mirror reverse images left and right but not top and bottom?" This question is quite misleading! As Fig. 35–6 shows, the up-down image $P'Q'$ and the left-right image $P'S'$ are parallel to their objects and are not reversed at all! Only the front-back image $P'R'$ is reversed relative to PR. Hence it's most correct to say that a plane mirror reverses *back to front.* To verify this object-image relationship, point your thumbs along PR and $P'R'$, your forefingers along PQ and $P'Q'$, and your middle fingers along PS and $P'S'$. When an object and its image are related in this way, the image is said to be **reversed;** this means that *only* the front-back dimension is reversed.

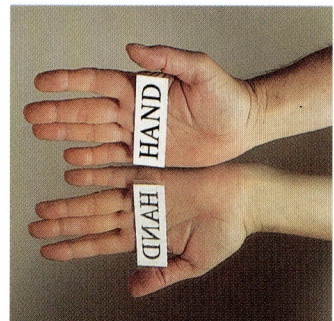

35–7 The image formed by a plane mirror is reversed; the image of a right hand is a left hand, and so on. Are images of the letters H and A reversed?

The reversed image of a three-dimensional object formed by a plane mirror is the same *size* as the object in all its dimensions. When the transverse dimensions of object and image are in the same direction, the image is erect. Thus a plane mirror always forms an erect but reversed image. Figure 35–7 illustrates this point.

An important property of all images formed by reflecting or refracting surfaces is that an *image* formed by one surface or optical device can serve as the *object* for a second surface or device. Figure 35–8 shows a simple example. Mirror 1 forms an image P_1' of the object point P, and mirror 2 forms another image P_2', each in the way we have just discussed. But in addition, the image P_1' formed by mirror 1 serves an object for mirror 2, which then forms an image of this object at point P_3' as shown. Similarly, mirror 1 uses the image P_2' formed by mirror 2 as an object and forms an image of it. We leave it to you to show that this image point is also at P_3'. The idea that an image formed by one device can act as the object for a second device is of great importance in geometric optics. We will use it later in this chapter to locate the image formed by two successive curved-surface refractions in a lens; in Chapter 36 this idea will help us to understand image formation by combinations of lenses, as in a microscope or a refracting telescope.

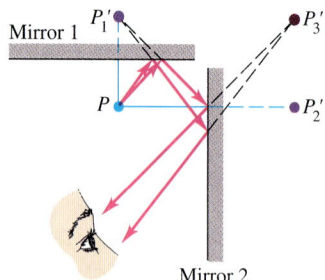

35–8 Images P_1' and P_2' are formed by a single reflection of each ray. Image P_3', located by treating either of the other images as an object, is formed by a double reflection of each ray.

35–3 REFLECTION AT A SPHERICAL SURFACE

A plane mirror produces an image that is the same size as the object. But there are many applications for mirrors in which the image and object must be of different sizes. A magnifying mirror used when applying makeup gives an image that is *larger* than the object, and surveillance mirrors (used in stores to help spot shoplifters) give an image that is *smaller* than the object. There are also applications of mirrors in which a *real* image is desired, so light rays do indeed pass through the image point P'; an example is a reflecting telescope, in which photographic film or an electronic detector is placed at the image point to record the image of a distant star. A plane mirror by itself cannot perform any of these tasks. Instead, *curved* mirrors are used.

We'll consider the special (and easily analyzed) case of image formation by a *spherical* mirror. Figure 35–9a shows a spherical mirror with radius of curvature R, with its concave side facing the incident light. The **center of curvature** of the surface (the center of the sphere of which the surface is a part) is at C, and the **vertex** of the mirror (the center of the mirror surface) is at V. The line CV is called the **optic axis.** Point P is an object point that lies on the optic axis; for the moment, we assume that the distance from P to V is greater than R.

Ray PV, passing through C, strikes the mirror normally and is reflected back on itself. Ray PB, at an angle α with the axis, strikes the mirror at B, where the angles of incidence and reflection are θ. The reflected ray intersects the axis at point P'. We will show shortly that *all* rays from P intersect the axis at the *same* point P', as in Fig. 35–9b, provided that the angle α is small. Point P' is therefore the *image* of object point P. Unlike the reflected rays in Fig. 35–1, the reflected rays in Fig. 35–9b actually do intersect at point P', then diverge from P' *as if* they had originated at this point. Thus P' is a *real* image.

To see the usefulness of having a real image, suppose that the mirror is in a darkened room in which the only source of light is a self-luminous object at P. If you place a small piece of photographic film at P', all the rays of light coming from point P that reflect off the mirror will strike the same point P' on the film; when developed, the film will show a single bright spot, representing a sharply focused image of the object at point P. This principle is at the heart of most astronomical telescopes, which use large concave mirrors to make photographs of celestial objects. With a *plane* mirror like that in Fig. 35–2, placing a piece of film at the image point P' would be a waste of time; the light rays never actually pass through the image point, and the image can't be recorded on film. Real images are *essential* for photography.

Let's now find the location of the real image point P' in Fig. 35–9a and prove the assertion that all rays from P intersect at P' (provided that their angle with the optic axis is small). The object distance, measured from the vertex V, is s; the image distance, also measured from V, is s'; and the radius of curvature of the mirror is R. The signs of s, s', and R are determined by the sign rules given in Section 35–2. The object point P is on the same side as the incident light, so according to the first sign rule, s is positive. The image point P' is on the same side as the reflected light, so according to the second sign rule, the image distance s' is also positive. The center of curvature C is on the same side as the reflected light, so according to the third sign rule, R, too, is positive; R is always positive when reflection occurs at the *concave* side of a surface.

We now use the following theorem from plane geometry: An exterior angle of a triangle equals the sum of the two opposite interior angles. Applying this theorem to triangles PBC and $P'BC$ in Fig. 35–9a, we have

$$\phi = \alpha + \theta, \qquad \beta = \phi + \theta.$$

Eliminating θ between these equations gives

$$\alpha + \beta = 2\phi. \qquad (35\text{–}3)$$

We may now compute the image distance s'. Let h represent the height of point B above the optic axis, and let δ represent the short distance from V to the foot of this vertical line. We now write expressions for the tangents of α, β, and ϕ, remembering that s,

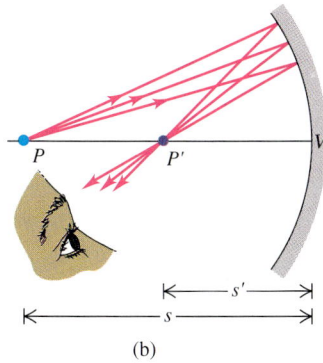

35–9 (a) Construction for finding the position of the image P' formed by a concave spherical mirror of a point object P on the mirror's optic axis. (b) If the angle α is small, *all* rays from P intersect at P'. The eye sees some of the outgoing rays and perceives them as having come from P'.

s', and R are all positive quantities:

$$\tan \alpha = \frac{h}{s - \delta}, \qquad \tan \beta = \frac{h}{s' - \delta}, \qquad \tan \phi = \frac{h}{R - \delta}.$$

These trigonometric equations cannot be solved as simply as the corresponding algebraic equations for a plane mirror. However, *if the angle α is small*, the angles β and ϕ are also small. The tangent of an angle that is much less than one radian is nearly equal to the angle itself (measured in radians), so we can replace $\tan \alpha$ by α, and so on, in the equations above. Also, if α is small, we can neglect the distance δ compared with s', s, and R. So for small angles we have the following approximate relations:

$$\alpha = \frac{h}{s}, \qquad \beta = \frac{h}{s'}, \qquad \phi = \frac{h}{R}.$$

Substituting these into Eq. (35–3) and dividing by h, we obtain a general relation among s, s', and R:

$$\frac{1}{s} + \frac{1}{s'} = \frac{2}{R} \qquad \text{(object-image relation, spherical mirror)}. \qquad (35–4)$$

This equation does not contain the angle α. Hence *all* rays from P that make sufficiently small angles with the axis intersect at P' after they are reflected; this verifies our earlier assertion. Such rays, nearly parallel to the axis and close to it, are called **paraxial rays.** (The term **paraxial approximation** is often used for the approximations we have just described.) Since all such reflected light rays converge on the image point, a concave mirror is also called a *converging mirror.*

Be sure you understand that Eq. (35–4), as well as many similar relations that we will derive later in this chapter and the next, is only *approximately* correct. It results from a calculation containing approximations, and it is valid only for paraxial rays. If we increase the angle α that a ray makes with the optic axis, the point P' where the ray intersects the optic axis moves somewhat closer to the vertex than for a paraxial ray. As a result, a spherical mirror, unlike a plane mirror, does not form a precise point image of a point object; the image is "smeared out." This property of a spherical mirror is called **spherical aberration.** The initially disappointing results from the Hubble Space Telescope when it was first placed in orbit in 1990 resulted in part from errors in the corrections for spherical aberration in its primary mirror (Fig. 35–10a). The performance of the telescope has improved dramatically since the installation of corrective optics in 1993 (Fig. 35–10b).

If the radius of curvature becomes infinite ($R = \infty$), the mirror becomes *plane,* and Eq. (35–4) reduces to Eq. (35–1) for a plane reflecting surface.

35–10 (a) An image of a star 170,000 light years from the earth, made with the original Hubble Space Telescope (HST) after it was placed in orbit in 1990. The 2.4-m-diameter concave mirror was too shallow by about 2 μm (about 1/50 the width of a human hair), leading to spherical aberration of the star's image. (b) An HST image of the same star made after corrective optics were installed in 1993. The effects of spherical aberration have been almost completely eliminated.

(a)

(b)

FOCAL POINT AND FOCAL LENGTH

When the object point P is very far from the spherical mirror ($s = \infty$), the incoming rays are parallel. (The star shown in Fig. 35–10 is an example of such a distant object.) From Eq. (35–4) the image distance s' in this case is given by

$$\frac{1}{\infty} + \frac{1}{s'} = \frac{2}{R}, \qquad s' = \frac{R}{2}.$$

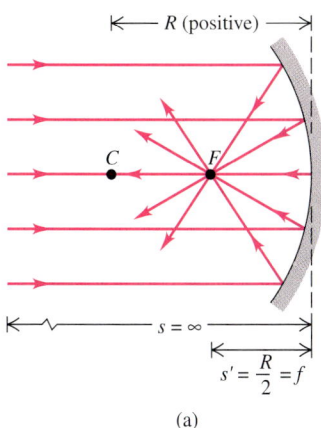

The situation is shown in Fig. 35–11a. The beam of incident parallel rays converges, after reflection from the mirror, to a point F at a distance $R/2$ from the vertex of the mirror. The point F at which the incident parallel rays converge is called the **focal point;** we say that these rays are brought to a focus. The distance from the vertex to the focal point, denoted by f, is called the **focal length.** We see that f is related to the radius of curvature R by

$$f = \frac{R}{2} \qquad \text{(focal length of a spherical mirror).} \tag{35–5}$$

The opposite situation is shown in Fig. 35–11b. Now the *object* is placed at the focal point F, so the object distance is $s = f = R/2$. The image distance s' is again given by Eq. (35–4):

$$\frac{2}{R} + \frac{1}{s'} = \frac{2}{R}, \qquad \frac{1}{s'} = 0, \qquad s' = \infty.$$

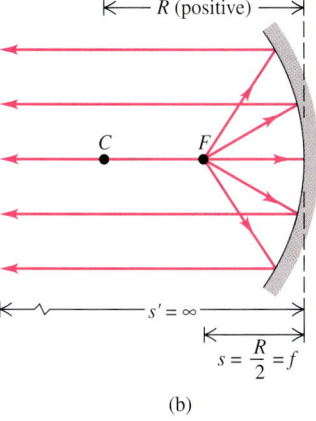

With the object at the focal point, the reflected rays in Fig. 35–11b are parallel to the optic axis; they meet only at a point infinitely far from the mirror, so the image is at infinity.

Thus the focal point F of a spherical mirror has the properties that (1) any incoming ray parallel to the optic axis is reflected through the focal point and (2) any incoming ray that passes through the focal point is reflected parallel to the optic axis. For spherical mirrors these statements are true only for paraxial rays. For parabolic mirrors these statements are *exactly* true; this is why parabolic mirrors are preferred for astronomical telescopes. Spherical or parabolic mirrors are used in flashlights and headlights to form the light from the bulb into a parallel beam. Some solar-power plants use an array of plane mirrors to simulate an approximately spherical concave mirror; light from the sun is collected by the mirrors and directed to the focal point, where a steam boiler is placed. (The concepts of focal point and focal length also apply to lenses, as we'll see in Section 35–6.)

We will usually express the relationship between object and image distances for a mirror, Eq. (35–4), in terms of the focal length f:

$$\frac{1}{s} + \frac{1}{s'} = \frac{1}{f} \qquad \text{(object-image relation, spherical mirror).} \tag{35–6}$$

35–11 (a) Incident rays parallel to the axis converge to the focal point F of a concave mirror. (b) Rays diverging from the focal point F of a concave mirror are parallel to the axis after reflection. The angles are exaggerated for clarity.

IMAGE OF AN EXTENDED OBJECT—SPHERICAL MIRROR

Now suppose we have an object with *finite* size, represented by the arrow PQ in Fig. 35–12, perpendicular to the optic axis CV. The image of P formed by paraxial rays is at P'. The object distance for point Q is very nearly equal to that for point P, so the image $P'Q'$ is nearly straight and perpendicular to the axis. Note that the object and image arrows have different sizes, y and y', respectively, and that they have opposite orientation. In Eq. (35–2) we defined the *lateral magnification m* as the ratio of image size y' to object size y:

$$m = \frac{y'}{y}.$$

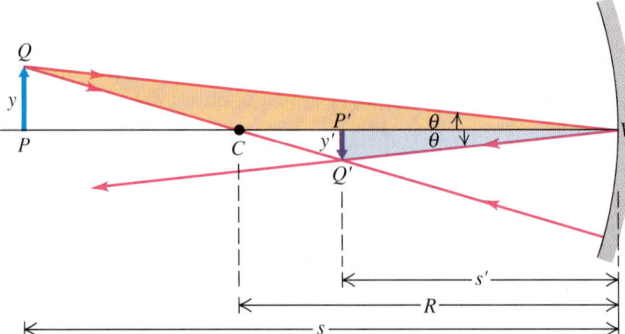

35–12 Construction for determining the position, orientation, and height of an image formed by a concave spherical mirror.

Because triangles PVQ and $P'VQ'$ in Fig. 35–12 are *similar,* we also have the relation $y/s = -y'/s'$. The negative sign is needed because object and image are on opposite sides of the optic axis; if y is positive, y' is negative. Therefore

$$m = \frac{y'}{y} = -\frac{s'}{s} \quad \text{(lateral magnification, spherical mirror).} \tag{35–7}$$

If m is positive, the image is erect in comparison to the object; if m is negative, the image is *inverted* relative to the object, as in Fig. 35–12. For a *plane* mirror, $s = -s'$, so $y' = y$ and $m = +1$; since m is positive, the image is erect, and since $|m| = 1$, the image is the same size as the object.

CAUTION ▶ Although the ratio of image size to object size is called the *magnification,* the image formed by a mirror or lens may be larger than, smaller than, or the same size as the object. If it is smaller, then the magnification is less than unity in absolute value: $|m| < 1$. The image formed by an astronomical telescope mirror or a camera lens is usually *much* smaller than the object. ◀

For three-dimensional objects, the ratio of image-to-object distances measured *along* the optic axis is different from the ratio of *lateral* distances (the lateral magnification). In particular, if m is a small fraction, the three-dimensional image of a three-dimensional object is reduced *along* the axis much more than it is reduced laterally. Figure 35–13 shows this effect. Note that the image formed by a spherical mirror, like that of a plane mirror, is always reversed along the optic axis.

In our discussion of concave mirrors we have so far considered only objects that lie *outside* or at the focal point, so that the object distance s is greater than or equal to the (positive) focal length f. In this case the image point is on the same side of the mirror as the outgoing rays, and the image is real and inverted. If an object is placed *inside* the focal point of a concave mirror, so that $s < f$, the resulting image is *virtual* (that is, the image point is on the opposite side of the mirror from the object), *erect,* and *larger* than the object. Mirrors used when applying makeup (referred to at the beginning of this section) are concave mirrors; in use, the distance from the face to the mirror is less than the

35–13 Schematic diagram of an object and its real, inverted, reduced image formed by a concave mirror. The face of the object closest to the mirror has the greatest magnitude of lateral magnification.

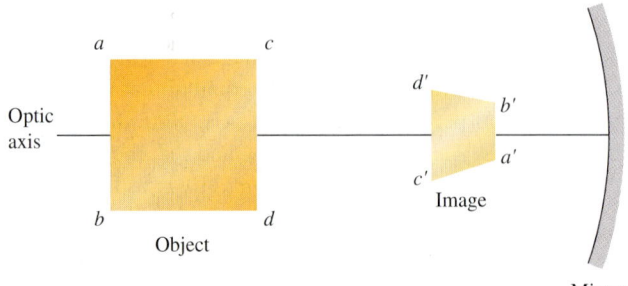

focal length, and an enlarged, erect image is seen. You can prove these statements about concave mirrors by applying Eqs. (35–6) and (35–7) (see Exercise 35–9). We'll also be able to verify these results in the next section, after we've learned some graphical methods for relating the positions and sizes of the object and the image.

EXAMPLE 35–1

Image formation by a concave mirror I A concave mirror forms an image, on a wall 3.00 m from the mirror, of the filament of a headlight lamp 10.0 cm in front of the mirror (Fig. 35–14). a) What are the radius of curvature and focal length of the mirror? b) What is the height of the image if the height of the object is 5.00 mm?

SOLUTION a) Both object distance and image distance are positive; we have $s = 10.0$ cm and $s' = 300$ cm. From Eq. (35–4),

$$\frac{1}{10.0 \text{ cm}} + \frac{1}{300 \text{ cm}} = \frac{2}{R},$$

$$R = \frac{2}{0.100 \text{ cm}^{-1} + 3.33 \times 10^{-3} \text{ cm}^{-1}} = 19.4 \text{ cm}.$$

The focal length of the mirror is $f = R/2 = 9.7$ cm. In a headlight lamp the filament is usually placed close to the focal point, producing a beam of nearly parallel rays.
b) From Eq. (35–7) the lateral magnification is

$$m = \frac{y'}{y} = -\frac{s'}{s} = -\frac{300 \text{ cm}}{10.0 \text{ cm}} = -30.0.$$

35–14 The concave mirror forms a real, enlarged, inverted image of the lamp filament.

Because m is negative, the image is inverted. The height of the image is 30.0 times the height of the object, or $(30.0)(5.00 \text{ mm}) = 150 \text{ mm}$.

EXAMPLE 35–2

Image formation by a concave mirror II In Example 35–1, suppose that the left half of the mirror's reflecting surface is covered with nonreflective soot. What effect will this have on the image of the filament?

SOLUTION It would be natural to guess that the image would now show only half of the filament. But in fact the image will still show the *entire* filament. The explanation can be seen by examining Fig. 35–9b. Light rays coming from any object point P are reflected from *all* parts of the mirror and converge on the corresponding image point P'. If part of the mirror surface is

made nonreflective or is removed altogether, the light rays from the remaining reflective surface still form an image of every part of the object.

The only effect of reducing the reflecting area is that the image becomes dimmer because less light energy reaches the image point. In our example the reflective area of the mirror is reduced by one half, and the image will be one half as bright. *Increasing* the reflective area makes the image brighter; to make reasonably bright images of distant stars, astronomical telescopes use mirrors that are up to several meters in diameter.

CONVEX MIRRORS

In Fig. 35–15a the *convex* side of a spherical mirror faces the incident light. The center of curvature is on the side opposite to the outgoing rays; according to the third sign rule in Section 35–2, R is negative. Ray PB is reflected, with the angles of incidence and reflection both equal to θ. The reflected ray, projected backward, intersects the axis at P'. As with a concave mirror, *all* rays from P that are reflected by the mirror diverge from the same point P', provided that the angle α is small. Therefore P' is the image of P. The object distance s is positive, the image distance s' is negative, and the radius of curvature R is *negative* for a *convex* mirror.

Figure 35–15b shows two ways diverging from the head of the arrow PQ and the virtual image $P'Q'$ of this arrow. The same procedure that we used for a concave mirror can

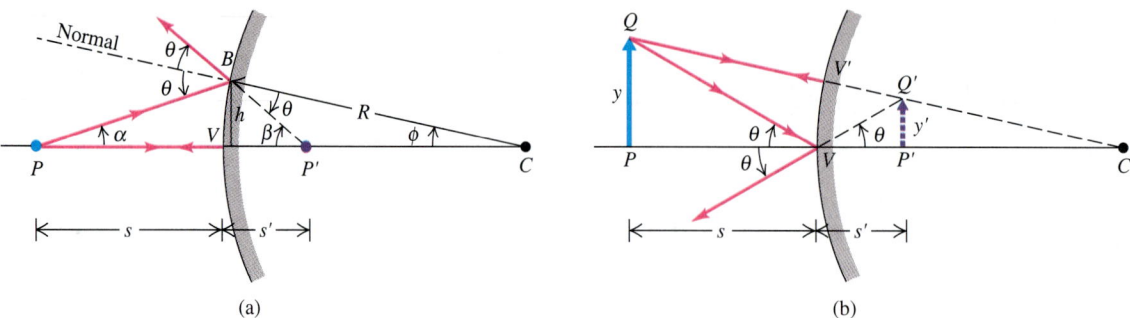

35–15 Constructions for finding (a) the position and (b) the magnification of the image formed by a convex mirror.

be used to show that for a convex mirror,

$$\frac{1}{s} + \frac{1}{s'} = \frac{2}{R},$$

and the lateral magnification is

$$m = \frac{y'}{y} = -\frac{s'}{s}.$$

These expressions are exactly the same as Eqs. (35–4) and (35–7) for a concave mirror; we leave the proof as a problem. Thus when we use our sign rules consistently, Eqs. (35–4) and (35–7) are valid for both concave and convex mirrors.

When R is negative (convex mirror), incoming rays that are parallel to the optic axis are not reflected through the focal point F. Instead, they diverge as though they had come from the point F at a distance f *behind* the mirror, as shown in Fig. 35–16a. In this case, f is the focal length, and F is called a *virtual focal point*. The corresponding image distance s' is negative, so both f and R are negative, and Eq. (35–5), $f = R/2$, holds for convex as well as concave mirrors. In Fig. 35–16b the incoming rays are converging as though they would meet at the virtual focal point F, and they are reflected parallel to the optic axis.

In summary, Eqs. (35–4) through (35–7), the basic relationships for image formation by a spherical mirror, are valid for both concave and convex mirrors, provided that we use the sign rules consistently.

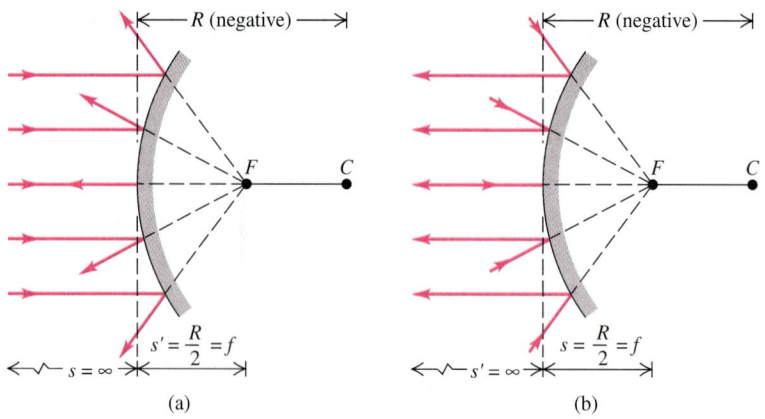

35–16 (a) Incident rays parallel to the axis diverge as if from the virtual focal point F of a convex mirror. (b) Rays aimed at the virtual focal point F of a concave mirror are parallel to the axis after reflection. The angles are exaggerated for clarity.

EXAMPLE 35–3

Santa's image problem Santa checks himself for soot, using his reflection in a shiny silvered Christmas tree ornament 0.750 m away (Fig. 35–17a). The diameter of the ornament is 7.20 cm. Standard reference works state that he is a "right jolly old elf," so we estimate his height to be 1.6 m. Where and how tall is the image of Santa formed by the ornament? Is it erect or inverted?

SOLUTION The surface of the ornament closest to Santa acts as a convex mirror with radius $R = -(7.20 \text{ cm})/2 = -3.60$ cm and focal length $f = R/2 = -1.80$ cm. The object distance is $s = 0.750$ m = 75.0 cm. From Eq. (35–6),

$$\frac{1}{s'} = \frac{1}{f} - \frac{1}{s} = \frac{1}{-1.80 \text{ cm}} - \frac{1}{75.0 \text{ cm}},$$

$$s' = -1.76 \text{ cm}.$$

Because s' is negative, the image is behind the mirror, that is, on the side opposite to the outgoing light (Fig. 35–17b), and it is virtual. The image is about halfway between the front surface of the ornament and its center.

The lateral magnification m is given by Eq. (35–7):

$$m = \frac{y'}{y} = -\frac{s'}{s} = -\frac{-1.76 \text{ cm}}{75.0 \text{ cm}} = 2.34 \times 10^{-2}.$$

Because m is positive, the image is erect. It is only about 0.0234 as tall as Santa himself:

$$y' = my = (0.0234)(1.6 \text{ m}) = 3.8 \times 10^{-2} \text{ m} = 3.8 \text{ cm}.$$

(a)   (b)

35–17 (a) The ornament forms a virtual, reduced, erect image of Santa. (b) Two of the rays forming the image. The angles are exaggerated for clarity.

When the object distance s is positive, a convex mirror *always* forms an erect, virtual, diminished, reversed image. For this reason, convex mirrors are used for shoplifting surveillance in stores, at blind intersections, and as "wide-angle" rear-view mirrors for cars and trucks (including those that bear the legend "objects in mirror are closer than they appear").

35–4   GRAPHICAL METHODS FOR MIRRORS

In the preceding section we used Eqs. (35–6) and (35–7) to find the position and size of the image formed by a mirror. We can also determine the properties of the image by a simple *graphical* method. This method consists of finding the point of intersection of a few particular rays that diverge from a point of the object (such as point Q in Fig. 35–18) and are reflected by the mirror. Then (neglecting aberrations) *all* rays from this object point that strike the mirror will intersect at the same point. For this construction we always choose an object point that is *not* on the optic axis. Four rays that we can usually draw easily are shown in Fig. 35–18. These are called **principal rays.**

(a)

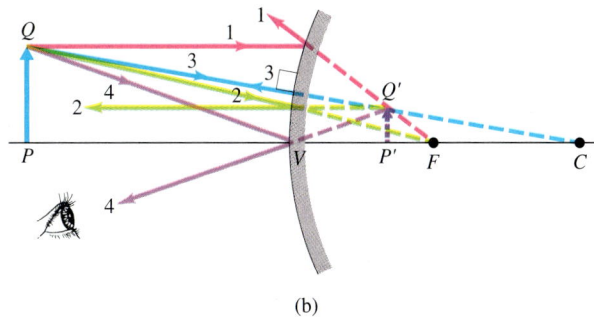

(b)

35–18 Principal-ray diagrams showing the graphical method of locating an image formed by a mirror. (a) A concave mirror; (b) a convex mirror. The colors of the rays are for identification only; they do not refer to specific colors of light.

1. *A ray parallel to the axis,* after reflection, passes through the focal point *F* of a concave mirror or appears to come from the (virtual) focal point of a convex mirror.
2. *A ray through (or proceeding toward) the focal point F* is reflected parallel to the axis.
3. *A ray along the radius* through or away from the center of curvature *C* intersects the surface normally and is reflected back along its original path.
4. *A ray to the vertex V* is reflected forming equal angles with the optic axis.

Once we have found the position of the image point by means of the intersection of any two of these principal rays (1, 2, 3, 4), we can draw the path of any other ray from the object point to the same image point.

CAUTION ▶ Although we've emphasized the principal rays, in fact *any* ray from the object that strikes the mirror will pass through the image point (for a real image) or appear to originate from the image point (for a virtual image). Usually, you only need to draw the principal rays, because these are all you need to locate the image. ◀

Problem-Solving Strategy

IMAGE FORMATION BY MIRRORS

1. The principal-ray diagram is to geometric optics what the free-body diagram is to mechanics. In any problem involving image formation by a mirror, *always* draw a principal-ray diagram first if you have enough information. (The same advice should be followed in dealing with lenses in the following sections.) It is usually best to orient your diagrams consistently with the incoming rays traveling from left to right. Don't draw a lot of other rays at random; stick with the principal rays, the ones you know something about. Use a ruler and measure distances carefully! A freehand sketch will *not* give good results.

2. If your principal rays don't converge at a real image point, you may have to extend them straight backward to

locate a virtual image point, as in Fig. 35–18b. We recommend drawing the extensions with dashed lines. Another useful aid is to color-code the different principal rays, as is done in Fig. 35–18.

3. Pay careful attention to signs on object and image distances, radii of curvature, and object and image heights. A negative sign on any of these quantities *always* has significance; use the equations and the sign rules carefully and consistently, and they will tell you the truth! Make certain you understand that the *same* sign rules work for all four cases in this chapter: reflection and refraction from plane and spherical surfaces.

EXAMPLE 35-4

Concave mirror, different object distances A concave mirror has a radius of curvature with absolute value 20 cm. Find graphically the image of an object in the form of an arrow perpendicular to the axis of the mirror at each of the following object distances: a) 30 cm, b) 20 cm, c) 10 cm, and d) 5 cm. Check the construction by *computing* the size and magnification of each image.

SOLUTION The graphical constructions are shown in the four parts of Fig. 35–19. Study each of these diagrams carefully, comparing each numbered ray with the description above. Several points are worth noting. First, in (b) the object and image distances are equal. Ray 3 cannot be drawn in this case because a ray from *Q* through the center of curvature *C* does not strike the mirror. Ray 2 cannot be drawn in (c) because a ray from *Q* toward *F* also does not strike the mirror. In this case the outgoing rays are parallel, corresponding to an infinite image distance. In (d) the outgoing rays have no real intersection point; they must be extended backward to find the point from which they appear to diverge, that is, from the *virtual image point Q'*.

The case shown in (d) illustrates the general observation that an object placed inside the focal point of a concave mirror produces a virtual image.

Measurements of the figures, with appropriate scaling, give the following approximate image distances: a) 15 cm; b) 20 cm; c) ∞ or −∞ (because the outgoing rays are parallel and do not converge at any finite distance); d) −10 cm. To *compute* these distances, we first note that $f = R/2 = 10$ cm; then we use Eq. (35–6):

a) $$\frac{1}{30 \text{ cm}} + \frac{1}{s'} = \frac{1}{10 \text{ cm}}, \qquad s' = 15 \text{ cm};$$

b) $$\frac{1}{20 \text{ cm}} + \frac{1}{s'} = \frac{1}{10 \text{ cm}}, \qquad s' = 20 \text{ cm};$$

c) $$\frac{1}{10 \text{ cm}} + \frac{1}{s'} = \frac{1}{10 \text{ cm}}, \qquad s' = \infty \ (\text{or} -\infty);$$

d) $$\frac{1}{5 \text{ cm}} + \frac{1}{s'} = \frac{1}{10 \text{ cm}}, \qquad s' = -10 \text{ cm}.$$

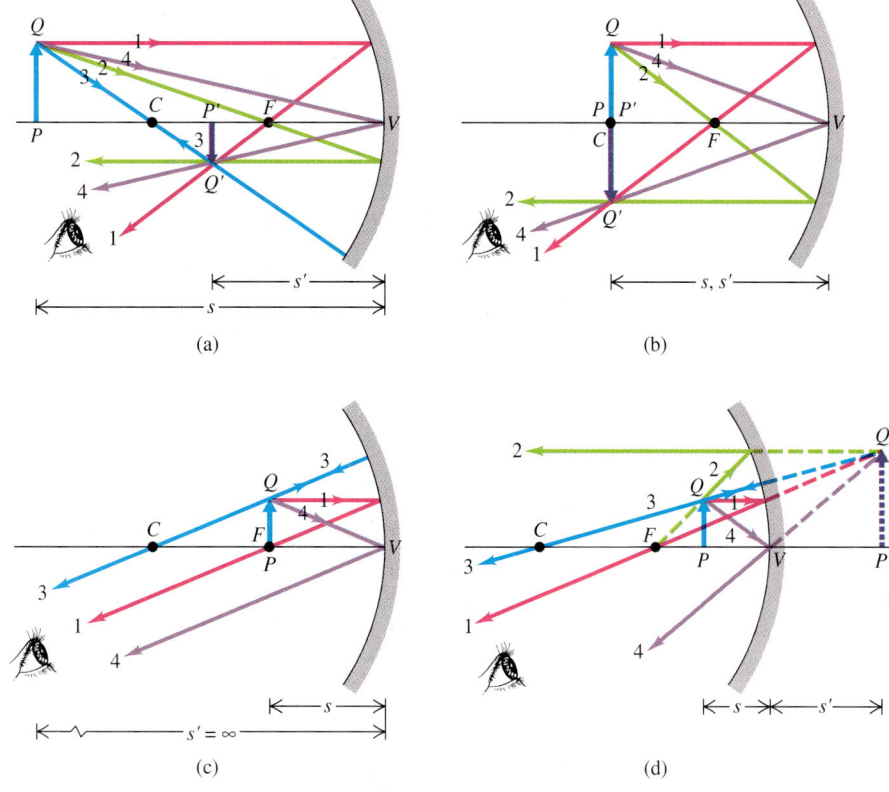

35–19 Image of an object at various distances from a concave mirror, determined using principal rays.

In (a) and (b) the image is real; in (d) it is virtual. In (c) the image is formed at infinity.

The lateral magnifications measured from the figures are approximately a) $-\frac{1}{2}$; b) -1; c) ∞ or $-\infty$ (because the image distance is infinite); d) $+2$. *Computing* the magnifications from Eq. (35–7), we find:

a)
$$m = \frac{-15\ \text{cm}}{30\ \text{cm}} = -\frac{1}{2};$$

b)
$$m = \frac{-20\ \text{cm}}{20\ \text{cm}} = -1;$$

c)
$$m = -\frac{\infty\ \text{cm}}{10\ \text{cm}} = -\infty\ (\text{or}\ +\infty);$$

d)
$$m = -\frac{-10\ \text{cm}}{5\ \text{cm}} = +2.$$

In (a) and (b) the image is inverted; in (d) it is erect.

35–5 REFRACTION AT A SPHERICAL SURFACE

As we mentioned in Section 35–2, images can be formed by refraction as well as by reflection. To begin with, let's consider refraction at a spherical surface, that is, at a spherical interface between two optical materials with different indexes of refraction. This analysis is directly applicable to some real optical systems, such as the human eye. It also provides a stepping-stone for the analysis of lenses, which usually have *two* spherical (or nearly spherical) surfaces.

In Fig. 35–20 a spherical surface with radius R forms an interface between two materials with different indexes of refraction n_a and n_b. The surface forms an image P' of an object point P; we want to find how the object and image distances (s and s') are related. We will use the same sign rules that we used for spherical mirrors. The center of curvature C is on the outgoing side of the surface, so R is positive. Ray PV strikes the vertex V and is perpendicular to the surface (that is, to the plane that is tangent to the surface at the point of incidence V). It passes into the second material without deviation.

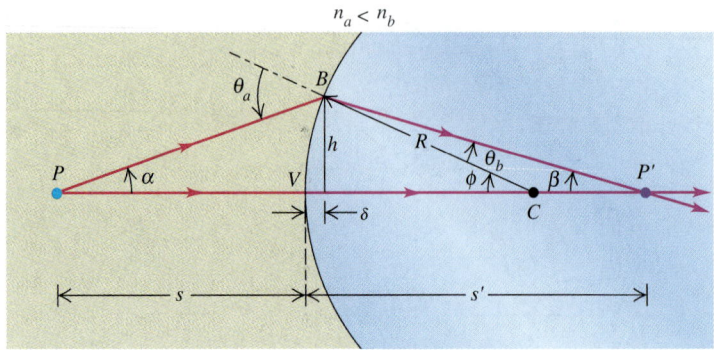

35–20 Construction for finding the position of the image point P' of a point object P formed by refraction at a spherical surface. The materials to the left and right of the interface have refractive lenses n_a and n_b, respectively. In the case shown here, $n_a < n_b$.

Ray PB, making an angle α with the axis, is incident at an angle θ_a with the normal and is refracted at an angle θ_b. These rays intersect at P', a distance s' to the right of the vertex. The figure is drawn for the case $n_a < n_b$. The object and image distances are both positive.

We are going to prove that if the angle α is small, *all* rays from P intersect at the same point P', so P' is the *real image* of P. We use much the same approach as we did for spherical mirrors in Section 35–3. We again use the theorem that an exterior angle of a triangle equals the sum of the two opposite interior angles; applying this to the triangles PBC and $P'BC$ gives

$$\theta_a = \alpha + \phi, \qquad \phi = \beta + \theta_b. \tag{35-8}$$

From the law of refraction,

$$n_a \sin\theta_a = n_b \sin\theta_b.$$

Also, the tangents of α, β, and ϕ are

$$\tan\alpha = \frac{h}{s+\delta}, \qquad \tan\beta = \frac{h}{s'-\delta}, \qquad \tan\phi = \frac{h}{R-\delta}. \tag{35-9}$$

For paraxial rays, θ_a and θ_b are both small in comparison to a radian, and we may approximate both the sine and tangent of either of these angles by the angle itself (measured in radians). The law of refraction then gives

$$n_a\theta_a = n_b\theta_b.$$

Combining this with the first of Eqs. (35–8), we obtain

$$\theta_b = \frac{n_a}{n_b}(\alpha + \phi).$$

When we substitute this into the second of Eqs. (35–8), we get

$$n_a\alpha + n_b\beta = (n_b - n_a)\phi. \tag{35-10}$$

Now we use the approximations $\tan\alpha = \alpha$, and so on, in Eqs. (35–9) and also neglect the small distance δ; those equations then become

$$\alpha = \frac{h}{s}, \qquad \beta = \frac{h}{s'}, \qquad \phi = \frac{h}{R}.$$

Finally, we substitute these into Eq. (35–10) and divide out the common factor h. We obtain

$$\frac{n_a}{s} + \frac{n_b}{s'} = \frac{n_b - n_a}{R} \qquad \text{(object-image relation, spherical refracting surface).} \tag{35-11}$$

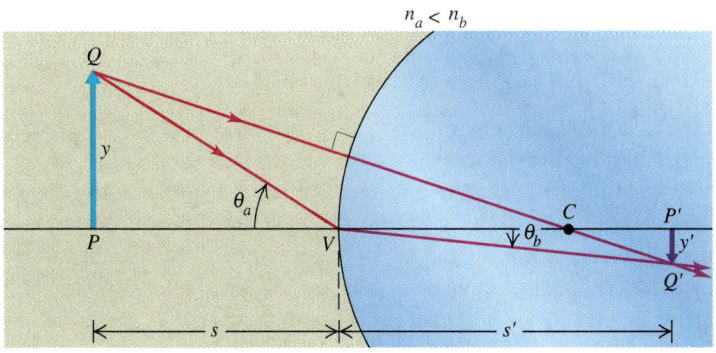

35-21 Construction for determining the height of an image formed by refraction at a spherical surface. In the case shown here, $n_a < n_b$.

This equation does not contain the angle α, so the image distance is the same for *all* paraxial rays emanating from P; this proves our assertion that P' is the image of P.

To obtain the lateral magnification m for this situation, we use the construction in Fig. 35–21. We draw two rays from point Q, one through the center of curvature C and the other incident at the vertex V. From the triangles PQV and $P'Q'V$,

$$\tan \theta_a = \frac{y}{s}, \qquad \tan \theta_b = \frac{-y'}{s'},$$

and from the law of refraction,

$$n_a \sin \theta_a = n_b \sin \theta_b.$$

For small angles,

$$\tan \theta_a = \sin \theta_a, \qquad \tan \theta_b = \sin \theta_b,$$

so finally

$$\frac{n_a y}{s} = -\frac{n_b y'}{s'},$$

or

$$m = \frac{y'}{y} = -\frac{n_a s'}{n_b s} \qquad \text{(lateral magnification, spherical refracting surface).} \quad (35\text{–}12)$$

Light rays refract as they enter the curved sides of this fish bowl, giving the fish a distorted view of the outside world.

Equations (35–11) and (35–12) can be applied to both convex and concave refracting surfaces, provided that you use the sign rules consistently. It doesn't matter whether n_b is greater or less than n_a. To verify these statements, we suggest that you construct diagrams like Figs. 35–20 and 35–21 for the following three cases: (i) $R > 0$ and $n_a > n_b$, (ii) $R < 0$ and $n_a < n_b$, and (iii) $R < 0$ and $n_a > n_b$. Then in each case, use your diagram to again derive Eqs. (35–11) and (35–12).

Here's a final note on the sign rule for the radius of curvature R of a surface. For the convex reflecting surface in Fig. 35–15, we considered R negative, but the convex *refracting* surface in Fig. 35–20 has a *positive* value of R. This may seem inconsistent, but it isn't. The rule is that R is positive if the center of curvature C is on the outgoing side of the surface and negative if C is on the other side. For the convex reflecting surface in Fig. 35–15, R is negative because point C is to the right of the surface but outgoing rays are to the left. For the convex refracting surface in Fig. 35–20, R is positive because both C and the outgoing rays are to the right of the surface.

An important special case of a spherical refracting surface is a *plane* surface between two optical materials. This corresponds to setting $R = \infty$ in Eq. (35–11). In this case,

$$\frac{n_a}{s} + \frac{n_b}{s'} = 0 \qquad \text{(plane refracting surface).} \quad (35\text{–}13)$$

To find the lateral magnification m for this case, we combine this equation with the general relation, Eq. (35–12), obtaining the simple result

$$m = 1.$$

That is, the image formed by a *plane* refracting surface always has the same lateral size as the object, and it is always erect.

An example of image formation by a plane refracting surface is the appearance of a partly submerged drinking straw or canoe paddle. When viewed from some angles, the object appears to have a sharp bend at the water surface because the submerged part appears to be only about three quarters of its actual distance below the surface. (We commented on the appearance of a submerged object in Section 34–3; see Fig. 34–6.)

EXAMPLE 35-5

Image formation by refraction I A cylindrical glass rod in air (Fig. 35–22) has index of refraction 1.52. One end is ground to a hemispherical surface with radius $R = 2.00$ cm. a) Find the image distance of a small object on the axis of the rod, 8.00 cm to the left of the vertex. b) Find the lateral magnification.

SOLUTION We are given $n_a = 1.00$, $n_b = 1.52$, $R = +2.00$ cm, and $s = +8.00$ cm. a) From Eq. (35–11),

$$\frac{1.00}{8.00 \text{ cm}} + \frac{1.52}{s'} = \frac{1.52 - 1.00}{+2.00 \text{ cm}},$$

$$s' = +11.3 \text{ cm}.$$

Because s' is positive, this means that the image is formed 11.3 cm to the *right* of the vertex.
b) From Eq. (35–12),

$$m = -\frac{n_a s'}{n_b s} = -\frac{(1.00)(11.3 \text{ cm})}{(1.52)(8.00 \text{ cm})} = -0.929.$$

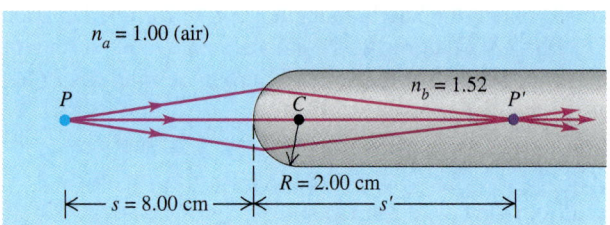

35–22 The rod in air forms a real image.

The image is somewhat smaller than the object, and it is inverted. If the object is an arrow 1.000 mm high, pointing upward, the image is an arrow 0.929 mm high, pointing downward.

EXAMPLE 35-6

Image formation by refraction II The glass rod in Example 35–5 is immersed in water (index of refraction $n = 1.33$), as shown in Fig. 35–23. The other quantities have the same values as before. Find the image distance and lateral magnification.

SOLUTION From Eq. (35–11),

$$\frac{1.33}{8.00 \text{ cm}} + \frac{1.52}{s'} = \frac{1.52 - 1.33}{+2.00 \text{ cm}},$$

$$s' = -21.3 \text{ cm}.$$

The negative value of s' means that after the rays are refracted

by the surface, they are not converging but *appear* to diverge from a point 21.3 cm to the *left* of the vertex. We saw a similar case in the reflection of light from a convex mirror; we call the point a *virtual image*. In this example the surface forms a virtual image 21.3 cm to the left of the vertex.

The magnification in this case is

$$m = -\frac{(1.33)(-21.3 \text{ cm})}{(1.52)(8.00 \text{ cm})} = +2.33.$$

In this case the image is erect (because m is positive) and 2.33 times as large as the object.

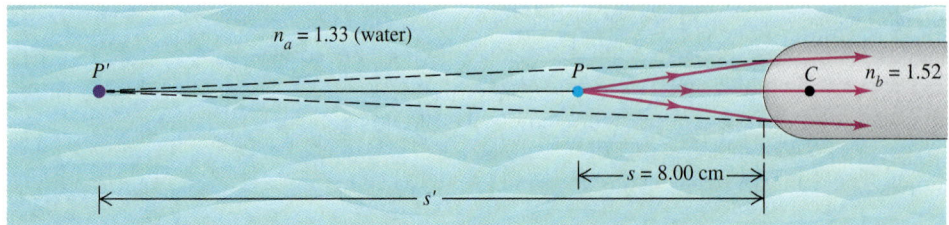

35–23 When immersed in water, the rod forms a virtual image.

EXAMPLE 35-7

Apparent depth of a swimming pool Swimming pool owners know that the pool always looks shallower than it really is and that it is important to identify the deep parts conspicuously so that people who can't swim won't jump into water that's over their heads. If a nonswimmer looks straight down into water that is actually 2.00 m (about 6 ft, 7 in.) deep, how deep does it appear to be?

SOLUTION The situation is shown in Fig. 35-24. Suppose there is an arrow PQ painted on the bottom of the pool at a depth $s = 2.00$ m. The refracting surface forms an image $P'Q'$ of this arrow; because $m = 1$ for a plane refracting surface, the image is the same length as the object arrow, but the *depth* s' of the image is different than the depth s of the object. To find s', we use Eq. (35-13):

$$\frac{n_a}{s} + \frac{n_b}{s'} = \frac{1.33}{2.00 \text{ m}} + \frac{1.00}{s'} = 0,$$

$$s' = -1.50 \text{ m.}$$

The apparent depth is only about three quarters of the actual depth, or about 4 ft, 11 in. A 6-ft nonswimmer who didn't allow for this effect would be in trouble. The negative sign shows that the image is virtual and on the incoming side of the refracting surface.

35-24 Arrow $P'Q'$ is the virtual image of the underwater arrow PQ. The angles of the ray with the vertical are exaggerated for clarity.

35-6 THIN LENSES

The most familiar and widely used optical device (after the plane mirror) is the *lens*. A lens is an optical system with two refracting surfaces. The simplest lens has two *spherical* surfaces close enough together that we can neglect the distance between them (the thickness of the lens); we call this a **thin lens.** If you wear eyeglasses or contact lenses while reading, you are viewing these words through a pair of thin lenses. We can analyze thin lenses in detail using the results of Section 35-5 for refraction by a single spherical surface. However, we postpone this analysis until later in the section so that we can first discuss the properties of thin lenses.

PROPERTIES OF A LENS

A lens of the type shown in Fig. 35-25 has the property that when a beam of rays parallel to the axis passes through the lens, the rays converge to a point F_2 (Fig. 35-25a) and form a real image at that point. Such a lens is called a **converging lens.** Similarly, rays passing through point F_1 emerge from the lens as a beam of parallel rays (Fig. 35-25b). The points F_1 and F_2 are called the first and second *focal points*, and the distance f (measured from the center of the lens) is called the *focal length*. Note the similarities between the two focal points of a converging lens and the single focal point of a concave mirror (Fig. 35-11). As for a concave mirror, the focal length of a converging lens is defined to be a *positive* quantity, and such a lens is also called a *positive lens.*

The central horizontal line in Fig. 35-25 is called the *optic axis,* as with spherical mirrors. The centers of curvature of the two spherical surfaces lie on and define the optic axis. The two focal lengths in Fig. 35-25, both labeled f, *are always equal* for a thin lens, even when the two sides have different curvature. We will derive this somewhat surprising result later in the section, when we derive the relationship of f to the index of refraction of the lens and the radii of curvature of its surfaces.

(a)

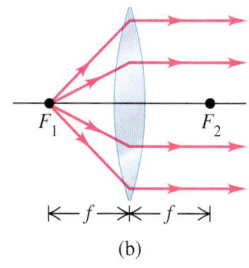

(b)

35-25 First and second focal points of a converging thin lens. The numerical value of f is positive.

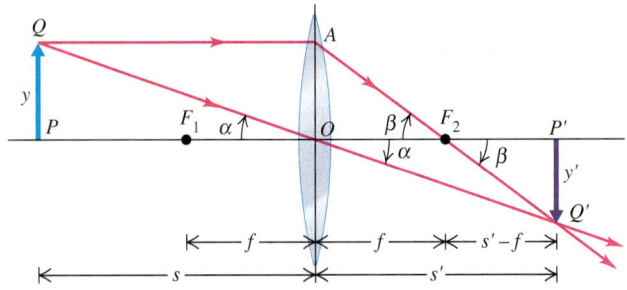

35–26 Construction used to find image position for a thin lens. To emphasize that the thickness of the lens is assumed to be very small, the ray QAQ' is shown as bent at the midplane of the lens rather than at the two surfaces and ray QOQ' is shown as a straight line.

Like a concave mirror, a converging lens can form an image of an extended object. Figure 35–26 shows how to find the position and lateral magnification of an image made by a thin converging lens. Using the same notation and sign rules as before, we let s and s' be the object and image distances, respectively, and let y and y' be the object and image heights. Ray QA, parallel to the optic axis before refraction, passes through the second focal point F_2 after refraction. Ray QOQ' passes undeflected straight through the center of the lens because at the center the two surfaces are parallel and (we have assumed) very close together. There is refraction where the ray enters and leaves the material but no net change in direction.

The two angles labeled α in Fig. 35–26 are equal. Therefore the two right triangles PQO and $P'Q'O$ are *similar,* and ratios of corresponding sides are equal. Thus

$$\frac{y}{s} = -\frac{y'}{s'}, \qquad \text{or} \qquad \frac{y'}{y} = -\frac{s'}{s}. \tag{35–14}$$

(The reason for the negative sign is that the image is below the optic axis and y' is negative.) Also, the two angles labeled β are equal, and the two right triangles OAF_2 and $P'Q'F_2$ are similar, so

$$\frac{y}{f} = -\frac{y'}{s'-f},$$

or

$$\frac{y'}{y} = -\frac{s'-f}{f}. \tag{35–15}$$

We now equate Eqs. (35–14) and (35–15), divide by s', and rearrange to obtain

$$\frac{1}{s} + \frac{1}{s'} = \frac{1}{f} \qquad \text{(object-image relation, thin lens).} \tag{35–16}$$

This analysis also gives the lateral magnification $m = y'/y$ for the lens; from Eq. (35–14),

$$m = -\frac{s'}{s} \qquad \text{(lateral magnification, thin lens).} \tag{35–17}$$

The negative sign tells us that when s and s' are both positive, as in Fig. 35–26, the image is *inverted,* and y and y' have opposite signs.

Equations (35–16) and (35–17) are the basic equations for thin lenses. It is pleasing to note that they are exactly the same as the corresponding equations for spherical mirrors, Eqs. (35–6) and (35–7). As we will see, the same sign rules that we used for spherical mirrors are also applicable to lenses. In particular, consider a lens with a positive focal length (a converging lens). When an object is outside the first focal point F_1 of this lens (that is, when $s > f$), the image distance s' is positive (that is, the image is on the same side as the outgoing rays); this image is real and inverted, as in Fig. 35–26. An object placed inside the first focal point of a converging lens, so that $s < f$, produces an

ActivPhysics

15.9
Thin Lens Ray Diagrams

15.10
Converging Thin Lenses

15.11
Diverging Thin Lenses

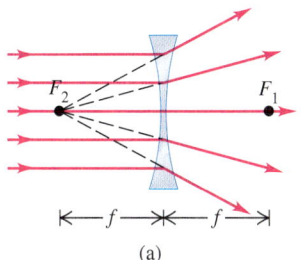

35-27 The image of a three-dimensional object is *not* reversed by a lens.

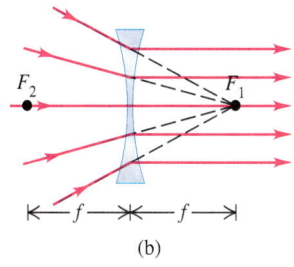

35-28 Second and first focal points of a diverging thin lens. The numerical value of f is negative.

image with a negative value of s'; this image is located on the same side of the lens as the object and is virtual, erect, and larger than the object. You can verify these statements algebraically using Eqs. (35–16) and (35–17); we'll also verify them in the next section, using graphical methods analogous to those introduced for mirrors in Section 35–4.

Figure 35–27 shows how a lens forms a three-dimensional image of a three-dimensional object. Point R is nearer the lens than point P. From Eq. (35–16), image point R' is farther from the lens than is image point P', and the image $P'R'$ points in the same direction as the object PR. Arrows $P'S'$ and $P'Q'$ are reversed relative to PS and PQ.

Let's compare Fig. 35–27 with Fig. 35–6, which shows the image formed by a plane *mirror*. We note that the image formed by the lens is inverted, but it is *not* reversed front to back along the optic axis. That is, if the object is a left hand, its image is also a left hand. You can verify this by pointing your left thumb along PR, your left forefinger along PQ, and your left middle finger along PS. Then rotate your hand 180°, using your thumb as an axis; this brings the fingers into coincidence with $P'Q'$ and $P'S'$. In other words, an *inverted* image is equivalent to an image that has been rotated by 180° about the lens axis.

So far we have been discussing *converging* lenses. Figure 35–28 shows a **diverging lens**; the beam of parallel rays incident on this lens *diverges* after refraction. The focal length of a diverging lens is a negative quantity, and the lens is also called a *negative lens*. The focal points of a negative lens are reversed, relative to those of a positive lens. The second focal point, F_2, of a negative lens is the point from which rays that are originally parallel to the axis *appear to diverge* after refraction, as in Fig. 35–28a. Incident rays converging toward the first focal point F_1, as in Fig. 35–28b, emerge from the lens parallel to its axis.

Equations (35–16) and (35–17) apply to *both* positive and negative lenses. Various types of lenses, both converging and diverging, are shown in Fig. 35–29. Here's an important observation: *Any lens that is thicker at its center than at its edges is a converging lens with positive f; and any lens that is thicker at its edges than at its center is a diverging lens with negative f* (provided that the lens has a greater index of refraction than the surrounding material). We can prove this using the *lensmaker's equation,* which it is our next task to derive.

THE LENSMAKER'S EQUATION

Now we proceed to derive Eq. (35–16) in more detail and at the same time derive the *lensmaker's equation,* which is a relationship between the focal length f, the index of refraction n of the lens, and the radii of curvature R_1 and R_2 of the lens surfaces. We use the principle that an image formed by one reflecting or refracting surface can serve as the object for a second reflecting or refracting surface.

We begin with the somewhat more general problem of two spherical interfaces separating three materials with indexes of refraction n_a, n_b, and n_c, as shown in Fig. 35–30. The object and image distances for the first surface are s_1 and s_1', and those for the

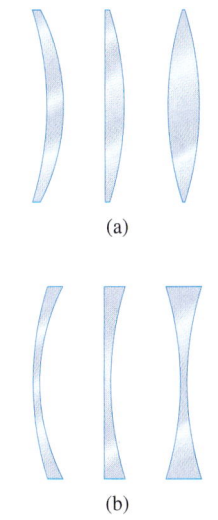

35-29 (a) Meniscus, plano-convex, and double-convex converging lenses. (b) Meniscus, plano-concave, and double-concave diverging lenses. Note that if these lenses are surrounded by a material with an index of refraction *greater* than that of the lens, the lenses in (a) become diverging and those in (b) become converging.

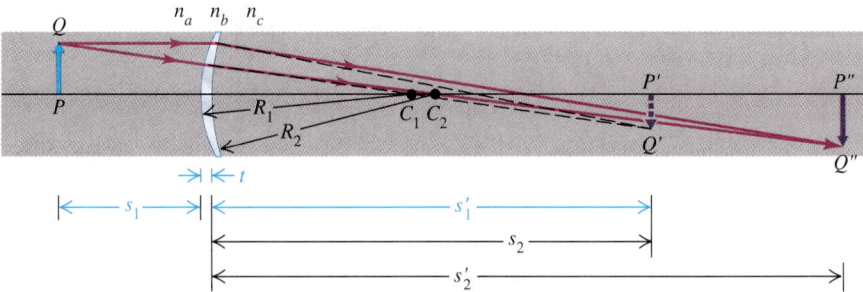

35–30 The image formed by the first surface of a lens serves as the object for the second surface. The distances s_1' and s_2 are taken to be equal; this is a good approximation if the lens thickness t is small.

second surface are s_2 and s_2'. We assume that the lens is thin, so that the distance t between the two surfaces is small in comparison with the object and image distances and can therefore be neglected. Then s_2 and s_1' have the same magnitude but opposite sign. For example, if the first image is on the outgoing side of the first surface, s_1' is positive. But when viewed as an object for the second surface, the first image is *not* on the incoming side of that surface. So we can say that $s_2 = -s_1'$.

We need to use the single-surface equation, Eq. (35–11), twice, once for each surface. The two resulting equations are

$$\frac{n_a}{s_1} + \frac{n_b}{s_1'} = \frac{n_b - n_a}{R_1},$$

$$\frac{n_b}{s_2} + \frac{n_c}{s_2'} = \frac{n_c - n_b}{R_2}.$$

The focal lengths of eyeglass lenses are chosen to compensate for deficiencies in the wearer's vision. The focal length is built into each lens by giving the front and back surfaces the appropriate radii of curvature, R_1 and R_2.

Ordinarily, the first and third materials are air or vacuum, so we set $n_a = n_c = 1$. The second index n_b is that of the lens, which we can call simply n. Substituting these values and the relation $s_2 = -s_1'$, we get

$$\frac{1}{s_1} + \frac{n}{s_1'} = \frac{n - 1}{R_1},$$

$$-\frac{n}{s_1'} + \frac{1}{s_2'} = \frac{1 - n}{R_2}.$$

To get a relation between the initial object position s_1 and the final image position s_2', we add these two equations. This eliminates the term n/s_1', and we obtain

$$\frac{1}{s_1} + \frac{1}{s_2'} = (n - 1)\left(\frac{1}{R_1} - \frac{1}{R_2}\right).$$

Finally thinking of the lens as a single unit, we call the object distance simply s instead of s_1, and we call the final image distance s' instead of s_2'. Making these substitutions, we have

$$\frac{1}{s} + \frac{1}{s'} = (n - 1)\left(\frac{1}{R_1} - \frac{1}{R_2}\right). \qquad (35\text{--}18)$$

Now we compare this with the other thin-lens equation, Eq. (35–16). We see that the object and image distances s and s' appear in exactly the same places in both equations and that the focal length f is given by

$$\frac{1}{f} = (n - 1)\left(\frac{1}{R_1} - \frac{1}{R_2}\right) \qquad \text{(lensmaker's equation for a thin lens).} \qquad (35\text{--}19)$$

This is the **lensmaker's equation.** In the process of rederiving the relationship between object distance, image distance, and focal length for a thin lens, we have also derived an

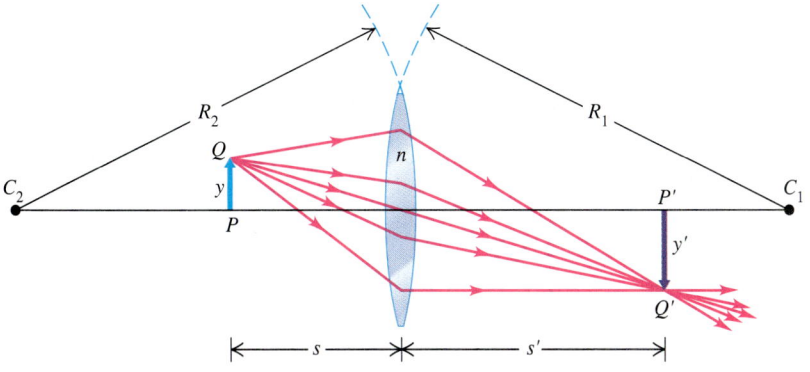

35-31 A thin lens. The radius of curvature of the first surface is R_1, and that of the second surface is R_2. In this case, R_1 is positive (because the center of curvature C_1 is on the same side as the outgoing light) and R_2 is negative (because C_2 is on the side opposite the outgoing light). The focal length f is positive, and the lens is a converging lens. Here, s and s' are also positive, and m is therefore negative.

expression for the focal length f of a lens in terms of its index of refraction n and the radii of curvature R_1 and R_2 of its surfaces. This can be used to show that all the lenses in Fig. 35–29a are converging lenses with positive focal lengths and that all the lenses in Fig. 35–29b are diverging lenses with negative focal lengths (see Exercise 35–32).

We use all our previous sign rules from Section 35–2 with Eqs. (35–18) and (35–19). For example, in Fig. 35–31, s, s', and R_1 are positive, but R_2 is negative.

It is not hard to generalize Eq. (35–19) to the situation in which the lens is immersed in a material with an index of refraction greater than unity. We invite you to work out the lensmaker's equation for this more general situation.

We stress that the paraxial approximation is indeed an approximation! Rays that are at sufficiently large angles to the optic axis of a spherical lens will not be brought to the same focus as paraxial rays; this is the same problem of spherical aberration that plagues spherical *mirrors* (Section 35–3). To avoid this and other limitations of thin spherical lenses, lenses of more complicated shape are used in precision optical instruments.

EXAMPLE 35-8

Determining the focal length of a lens a) Suppose the absolute values of the radii of curvature of the lens surfaces in Fig. 35–31 are both equal to 10 cm and the index of refraction is $n = 1.52$. What is the focal length f of the lens? b) Suppose the lens in Fig. 35–28 also has $n = 1.52$, and the absolute values of the radii of curvature of its lens surfaces are also both equal to 10 cm. What is the focal length of this lens?

SOLUTION a) The center of curvature of the first surface is on the outgoing side, so R_1 is positive: $R_1 = +10$ cm. The center of curvature of the second surface is *not* on the outgoing side, so R_2 is negative: $R_2 = -10$ cm. From the lensmaker's equation, Eq. (35–19),

$$\frac{1}{f} = (1.52 - 1)\left(\frac{1}{+10 \text{ cm}} - \frac{1}{-10 \text{ cm}} \right),$$

$$f = 9.6 \text{ cm}.$$

Since f is positive, this is a converging lens (as we would expect, since the lens is thicker at its center than at its edges).
b) The first surface has its center of curvature on the *incoming* side, so R_1 is negative; for the second surface, the center of curvature is on the outgoing side, and R_2 is positive. Hence $R_1 = -10$ cm and $R_2 = +10$ cm. Again using the lensmaker's equation,

$$\frac{1}{f} = (1.52 - 1)\left(\frac{1}{-10 \text{ cm}} - \frac{1}{+10 \text{ cm}} \right),$$

$$f = -9.6 \text{ cm}.$$

Since f is negative, this is a diverging lens (as expected for a lens that is thicker at its edges than at its center).

35-7 GRAPHICAL METHODS FOR LENSES

We can determine the position and size of an image formed by a thin lens by using a graphical method very similar to the one we used in Section 35–4 for spherical mirrors. Again we draw a few special rays called *principal rays* that diverge from a point of the object that is *not* on the optic axis. The intersection of these rays, after they pass through

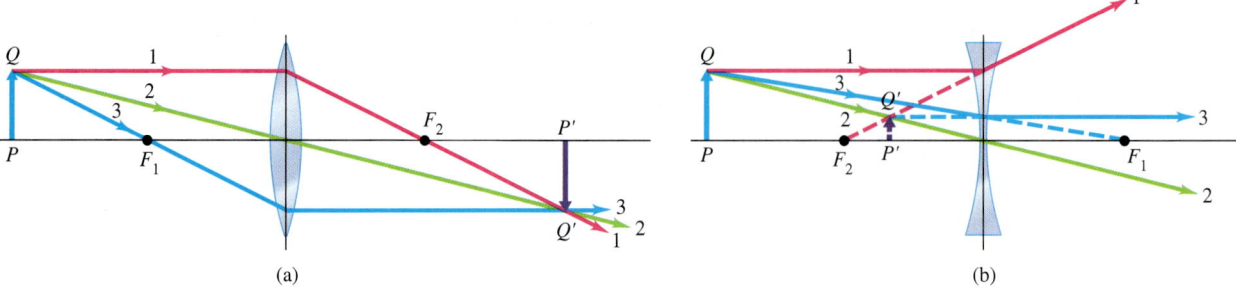

(a)   (b)

35–32 Principal-ray diagrams showing the graphical method of locating an image formed by a thin lens. (a) A converging lens; (b) a diverging lens. The colors of the rays are for identification only; they do not refer to specific colors of light.

the lens, determines the position and size of the image. In using this graphical method, we will consider the entire deviation of a ray as occurring at the midplane of the lens, as shown in Fig. 35–32. This is consistent with the assumption that the distance between the lens surfaces is negligible.

The three principal rays whose paths are usually easy to trace for lenses are shown in Fig. 35–32:

1. *A ray parallel to the axis* emerges from the lens in a direction that passes through the second focal point F_2 of a converging lens, or appears to come from the second focal point of a diverging lens.
2. *A ray through the center of the lens* is not appreciably deviated; at the center of the lens the two surfaces are parallel, so this ray emerges at essentially the same angle at which it enters and along essentially the same line.
3. *A ray through (or proceeding toward) the first focal point F_1* emerges parallel to the axis.

When the image is real, the position of the image point is determined by the intersection of any two rays 1, 2, and 3 (Fig. 35–32a). When the image is virtual, we extend the diverging outgoing rays backward to their intersection point to find the image point (Fig. 35–32b).

CAUTION▶ Keep in mind that *any* ray from the object that strikes the lens will pass through the image point (for a real image) or appear to originate from the image point (for a virtual image). (We made a similar comment about image formation by mirrors in Section 35–4.) We've emphasized the principal rays because they're the only ones you need to draw to locate the image.◀

Figure 35–33 shows several principal-ray diagrams for a converging lens for several object distances. We suggest you study each of these diagrams very carefully, comparing each numbered ray with the above description.

Several points are worth noting in Fig. 35–33. Parts (a), (b), and (c) of this figure help to explain what happens in focusing a camera. For a photograph to be in sharp focus, the image made by the camera's lens must be located at the same position as the film. As the object is brought closer, the distance from the lens to the real image increases, so the film is moved farther behind the lens (or, equivalently, the lens is moved farther in front of the film). In Fig. 35–33d the object is at the focal point; ray 3 cannot be drawn because it does not pass through the lens. The emerging parallel light rays appear to be coming from infinity. In Fig. 35–33e the object distance is less than the focal length. The outgoing rays are divergent, and the image is *virtual;* its position is located by extending the outgoing rays backward. In this case the image distance s' is negative. Note also that the image is erect and larger than the object. (A converging lens used in this way is called a *magnifying glass* or *magnifier.* We'll discuss this in more detail in Chapter 36.) Figure 35–33f corresponds to a *virtual object.* The incoming rays do not diverge from a real object but are *converging* as though they would meet at the tip of the virtual object O on the right side; the object distance s is negative in this case. The

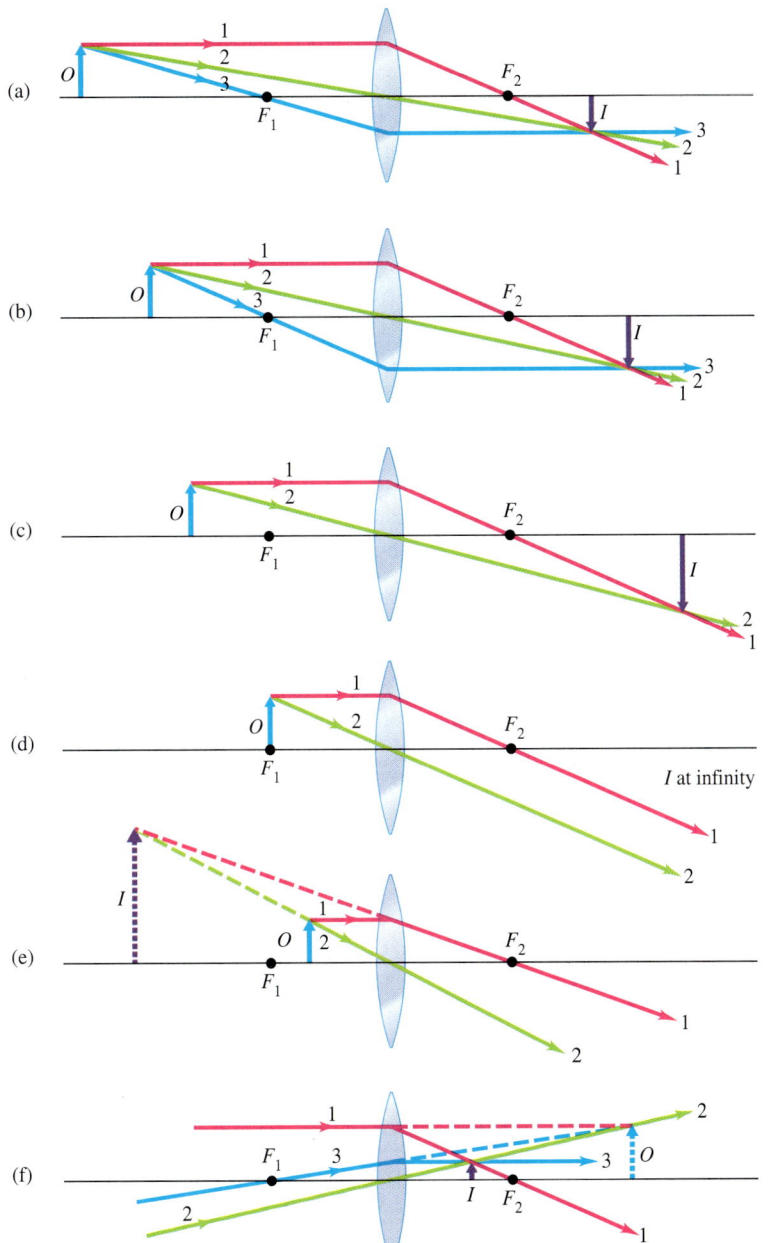

35–33 Formation of images by a thin converging lens for various object distances. The principal rays are numbered.

image is real, since the image distance s' is positive and is located between the lens and the second focal point. This situation can arise if the rays that strike the lens in Fig. 33–33f emerge from another converging lens (not shown) to the left of the figure. The last example in this section, Example 35–12, involves a virtual object.

Problem–Solving Strategy

IMAGE FORMATION BY A THIN LENS

1. The strategy outlined in Section 35–4 is equally applicable here, and we suggest that you review it now. Always begin with a principal-ray diagram if you have enough information. Orient your diagrams consistently so that light travels from left to right. For a lens there are only three principal rays, compared to four for a mirror. Don't just sketch these diagrams; draw the rays with a ruler, and measure the distances carefully. Draw the rays so that they bend at the midplane of the lens, as shown in Fig. 35–32. Be sure to draw *all three* principal rays whenever possible. The intersection of any two determines the image, but if the third doesn't pass through the same intersection point, you know that you've made a mistake. Redundancy can be useful in spotting errors.

2. If the outgoing principal rays don't converge at a real image point, the image is virtual. Then you have to extend the outgoing rays backward to find the virtual image point, which lies on the *incoming* side of the lens.

3. The same sign rules that we have used for mirrors and single refracting surfaces are also applicable for thin lenses. Be extremely careful to get your signs right and to interpret the signs of results correctly.

4. Always determine the image position and size *both* graphically and by calculating. This gives an extremely useful consistency check.

5. The *image* from one lens or mirror may serve as the *object* for another. In that case, be careful in finding the object and image *distances* for this intermediate image; be sure you include the distance between the two elements (lenses and/or mirrors) correctly.

EXAMPLE 35–9

Image location and magnification with a converging lens   A converging lens has a focal length of 20 cm. Find graphically the image location for an object at each of the following distances from the lens: a) 50 cm; b) 20 cm; c) 15 cm; d) –40 cm. Determine the magnification in each case. Check your results by calculating the image position and magnification from Eqs. (35–16) and (35–17).

SOLUTION   The appropriate principal-ray diagrams are shown in Figs. 35–33a, 35–33d, 35–33e, and 35–33f. The approximate image distances, from measurements of these diagrams, are 35 cm, $-\infty$, -40 cm, and 15 cm, and the approximate magnifications are $-\frac{2}{3}$, $+\infty$, and $+3$, and $+\frac{1}{3}$, respectively.

Calculating the image positions from Eq. (35–16), we find

a) $$\frac{1}{50 \text{ cm}} + \frac{1}{s'} = \frac{1}{20 \text{ cm}}, \qquad s' = 33.3 \text{ cm};$$

b) $$\frac{1}{20 \text{ cm}} + \frac{1}{s'} = \frac{1}{20 \text{ cm}}, \qquad s' = \pm\infty;$$

c) $$\frac{1}{15 \text{ cm}} + \frac{1}{s'} = \frac{1}{20 \text{ cm}}, \qquad s' = -60 \text{ cm};$$

d) $$\frac{1}{-40 \text{ cm}} + \frac{1}{s'} = \frac{1}{20 \text{ cm}}, \qquad s' = 13.3 \text{ cm}.$$

The graphical results are fairly close to these except for part (c); the accuracy of the diagram in Fig. 35–33e is limited because the rays extended backward have nearly the same direction. Note that s' is positive for the real images in parts (a) and (d) and negative for the virtual image in part (c).

From Eq. (35–17) the magnifications are

a) $$m = -\frac{33.3 \text{ cm}}{50 \text{ cm}} = -\frac{2}{3},$$

b) $$m = -\frac{\pm\infty \text{ cm}}{20 \text{ cm}} = \mp\infty,$$

c) $$m = -\frac{-60 \text{ cm}}{15 \text{ cm}} = +4,$$

d) $$m = -\frac{13.3 \text{ cm}}{-40 \text{ cm}} = +\frac{1}{3}.$$

The image is inverted in part (a) and erect in parts (c) and (d).

EXAMPLE 35–10

Image formation by a diverging lens   You are given a thin diverging lens. You find that a beam of parallel rays spreads out after passing through the lens, as though all the rays came from a point 20.0 cm from the center of the lens. You want to use this lens to form an erect virtual image that is $\frac{1}{3}$ the height of the object. a) Where should the object be placed? b) Draw a principal-ray diagram.

SOLUTION   a) The observation with incoming parallel rays shows that the focal length is $f = -20.0$ cm. We want the lateral magnification to be $m = +\frac{1}{3}$ (positive because the image is to be erect). From Eq. (35–17), $m = +\frac{1}{3} = -s'/s$, so $s' = -s/3$. From

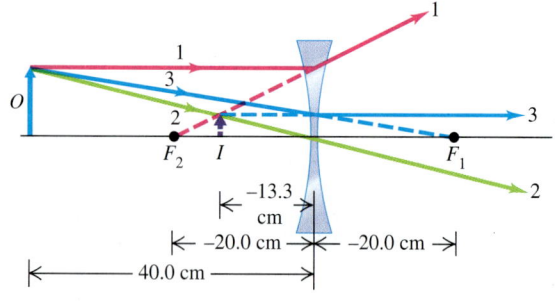

35–34 Principal-ray diagram for an image formed by a thin diverging lens.

Eq. (35–16),

$$\frac{1}{s} + \frac{1}{-s/3} = \frac{1}{-20.0 \text{ cm}},$$

$$s = 40.0 \text{ cm},$$

$$s' = -\frac{s}{3} = -\frac{40.0 \text{ cm}}{3} = -13.3 \text{ cm}.$$

The image distance is negative, so the object and image are on the same side of the lens.

b) Figure 35–34 is a principal-ray diagram for this problem, with the rays numbered in the same way as in Fig. 35–33b.

EXAMPLE 35–11

An image of an image An object 8.0 cm high is placed 12.0 cm to the left of a converging lens of focal length 8.0 cm. A second converging lens of focal length 6.0 cm is placed 36.0 cm to the right of the first lens. Both lenses have the same optic axis. Find the position, size, and orientation of the image produced by the two lenses in combination. (Combinations of converging lenses are used in telescopes and microscopes, to be discussed in Chapter 36.)

SOLUTION The situation is shown in Fig. 35–35. The object O lies outside the first focal point F_1 of the first lens, so this lens produces a real image I. The light rays that strike the second lens diverge from this real image just as if I was a material object. Hence the *image* made by the *first* lens acts as an *object* for the *second* lens.

In Fig. 35–35 we have drawn principal rays 1, 2, and 3 from the head of the object arrow O to find the position of the first image I and principal rays $1'$, $2'$, and $3'$ from the head of the first image to find the position of the second image I' made by the second lens (even though rays $2'$ and $3'$ don't actually exist in this case). Note that the image is inverted *twice*, once by each lens, so the second image I' has the same orientation as the original object.

To *calculate* the position and size of the second image I', we must first find the position and size of the first image I. Applying Eq. (35–16), $1/s + 1/s' = 1/f$, to the first lens gives

$$\frac{1}{12.0 \text{ cm}} + \frac{1}{s_1'} = \frac{1}{8.0 \text{ cm}}, \qquad s_1' = +24.0 \text{ cm}.$$

The first image I is 24.0 cm to the right of the first lens. The magnification is $m_1 = -(24.0 \text{ cm})/(12.0 \text{ cm}) = -2.00$, so the height of the first image is $(-2.0)(8.0 \text{ cm}) = -16.0 \text{ cm}.$

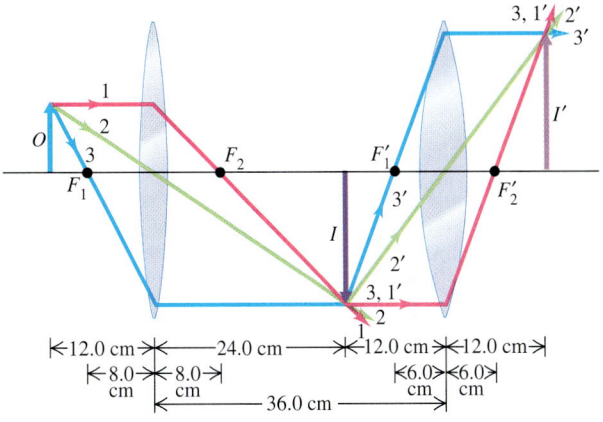

35–35 Principal-ray diagram for a combination of two converging lenses. The first lens makes a real image of the object. This real image acts as an object for the second lens.

The first image is $36.0 \text{ cm} - 24.0 \text{ cm} = 12.0 \text{ cm}$ to the left of the second lens, so the object distance for the second lens is $+12.0 \text{ cm}$. Using Eq. (35–16) for the second lens gives the position of the second and final image:

$$\frac{1}{12.0 \text{ cm}} + \frac{1}{s_2'} = \frac{1}{6.0 \text{ cm}}, \qquad s_2' = +12.0 \text{ cm}.$$

The final image is 12.0 cm to the right of the second lens and 48.0 cm to the right of the first lens.

The magnification produced by the second converging lens is $m_2 = -(12.0 \text{ cm})/(12.0 \text{ cm}) = -1.0$; hence the final image is just as large as the first image but has the opposite orientation. This is also shown in the principal-ray diagram.

EXAMPLE 35–12

An image of an image II: A virtual object In the situation described in Example 35–11, the second lens is moved so that it is only 12.0 cm to the right of the first lens. For this new situation, find the position, size, and orientation of the image produced by the two lenses in combination.

SOLUTION Figure 35–36 shows the new configuration of lenses. As in Example 35–11, the image I of the first lens acts as an object for the second lens, which produces the final image I'. The first image is still 24.0 cm to the right of the first lens, but this is now 12.0 cm to the *right* of the second lens. Thus the rays emerging from the first lens never actually converge at the position of the first image, because the second lens gets in the way.

The situation for the second lens is like that shown in Fig. 35–33f; the light rays that strike the second lens are not diverging from an object point but are converging as though they would meet at I. The image of the first lens acts as a *virtual object* for the second lens.

The principal rays 1, 2, and 3, used for finding the image I formed by the first lens, are shown in Fig. 35–36. Ray 3 enters the second lens parallel to the axis and, after emerging from this lens, passes through its second focal point F_2'; hence we also call this ray $1'$. We need one or two additional principal rays in order to graphically locate the position of the final image. These are found by projecting *backward* from the tip of the first image arrow I. Ray $2'$ passes through the center of the second lens and

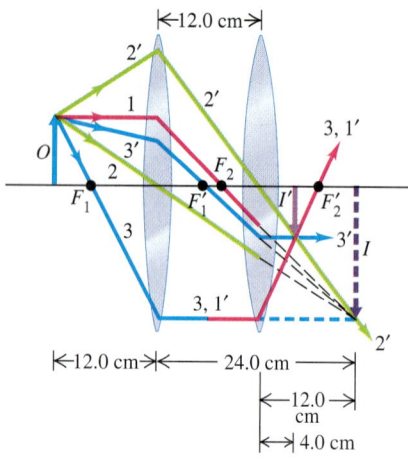

35–36 The second lens in Fig. 35–35 has been moved closer to the first lens. The image produced by the first lens now acts as a virtual object for the second lens.

is undeflected; ray 3′ passes through the first focal point $F_1′$ of the second lens and emerges parallel to the axis. Rays 1′, 2′, and 3′ intersect at the tip of the final image $I′$.

From Example 35–11 the image I is 24.0 cm to the right of the first lens and has height −16.0 cm. Hence the object distance for the second lens is 12.0 cm − 24.0 cm = −12.0 cm, and the image distance $s_2′$ for this lens is given by

$$\frac{1}{-12.0 \text{ cm}} + \frac{1}{s_2′} = \frac{1}{6.0 \text{ cm}}, \qquad s_2′ = 4.0 \text{ cm}.$$

The final image is 4.0 cm to the right of the second lens.

The magnification produced by the second converging lens is $m′ = -(4.0 \text{ cm})/(-12.0 \text{ cm}) = +0.33$, so the height of the final image is $(+0.33)(-16.0 \text{ cm}) = -5.33 \text{ cm}$. The image is inverted in comparison to the original object.

SUMMARY

- When rays diverge from an object point P and are reflected or refracted, the directions of the outgoing rays are the same as though they had diverged from a point $P′$ called the image point. If they actually converge at $P′$ and diverge again beyond it, $P′$ is a real image of P; if they only appear to have diverged from $P′$, it is a virtual image. Images can be either erect or inverted. A virtual image that is formed by a plane or spherical mirror is always reversed; the virtual image of a right hand is a left hand.

- The lateral magnification m in any reflecting or refracting situation is defined as the ratio of image height $y′$ to object height y:

$$m = \frac{y′}{y}. \qquad (35\text{–}2)$$

When m is positive, the image is erect; when m is negative, the image is inverted.

- Object-image relations derived in this chapter are valid only for rays that are close to and nearly parallel to the optic axis; these are called paraxial rays. Nonparaxial rays do not converge precisely to an image point. This effect is called spherical aberration.

- The focal point of a mirror is the point to which rays parallel to the axis converge after reflection from a concave mirror or the point from which they appear to diverge after reflection from a convex mirror. Rays diverging from the focal point of a concave mirror are parallel to the optic axis after reflection, and rays converging toward the focal point of a convex mirror are parallel to the optic axis after reflection. The distance from the focal point to the vertex is called the focal length, denoted as f. The focal points of a lens are defined similarly.

- The formulas for object distance s and image distance $s′$ for plane and spherical mirrors and single refracting surfaces are summarized in the following table. The equation for a plane surface can be obtained from the corresponding equation for a spherical surface by setting $R = \infty$.

	PLANE MIRROR	SPHERICAL MIRROR	PLANE REFRACTING SURFACE	SPHERICAL REFRACTING SURFACE
Object and image distances	$\dfrac{1}{s} + \dfrac{1}{s'} = 0$	$\dfrac{1}{s} + \dfrac{1}{s'} = \dfrac{2}{R} = \dfrac{1}{f}$	$\dfrac{n_a}{s} + \dfrac{n_b}{s'} = 0$	$\dfrac{n_a}{s} + \dfrac{n_b}{s'} = \dfrac{n_b - n_a}{R}$
Lateral magnification	$m = -\dfrac{s'}{s} = 1$	$m = -\dfrac{s'}{s}$	$m = -\dfrac{n_a s'}{n_b s} = 1$	$m = -\dfrac{n_a s'}{n_b s}$

The object-image relation for a thin lens is the same as that for a spherical mirror:

$$\frac{1}{s} + \frac{1}{s'} = \frac{1}{f}. \tag{35–16}$$

The lensmaker's equation is

$$\frac{1}{f} = (n - 1)\left(\frac{1}{R_1} - \frac{1}{R_2} \right). \tag{35–19}$$

■ The following sign rules are used with all plane and spherical reflecting and refracting surfaces:

s is positive when the object is on the incoming side of the surface (a real object) and negative otherwise.

s' is positive when the image is on the outgoing side of the surface (a real image) and negative otherwise.

R is positive when the center of curvature is on the outgoing side of the surface and negative otherwise.

m is positive when the image is erect and negative when it is inverted.

DISCUSSION QUESTIONS

Q35–1 Can you see a real image by looking backward along the direction from which the rays come? Can you see a virtual image? Explain. Can you tell by looking whether an image is real or virtual? If not, how *can* the two be distinguished?

Q35–2 For the situation shown in Fig. 35–3, is the image distance s' positive or negative? Is the image real or virtual? Explain your answers.

Q35–3 The laws of optics also apply to electromagnetic waves invisible to the eye. A satellite TV dish is used to detect radio waves coming from orbiting satellites. Why is a curved reflecting surface (a "dish") used? The dish is always concave, never convex; why? The actual radio receiver is placed on an arm and suspended in front of the dish. How far in front of the dish should it be placed?

Q35–4 Explain why the focal length of a *plane* mirror is infinite, and explain what it means for the focal point to be at infinity.

Q35–5 If a spherical mirror is immersed in water, does its focal length change? Explain.

Q35–6 For what range of object positions does a concave spherical mirror form a real image? What about a convex spherical mirror?

Q35–7 When a room has mirrors on two opposite walls, an infinite series of reflections can be seen. Discuss this phenomenon in terms of images. Why do the distant images appear fainter?

Q35–8 For a spherical mirror, if $s = f$, then $s' = \infty$, and the lateral magnification m is infinite. Does this make sense? If so, what does it mean?

Q35–9 A concave mirror (sometimes surrounded by lights) is often used as an aid for applying cosmetics to the face. Why is such a mirror always concave rather than convex? What considerations determine its radius of curvature?

Q35–10 A student claims that she can start a fire on a sunny day using just the sun's rays and a concave mirror. How is this done? Is the concept of image relevant? Can she do the same thing with a convex mirror? Explain.

Q35–11 A person looks at his reflection in the concave side of a shiny spoon. Is it right side up or inverted? Does it matter how far his face is from the spoon? What if he looks in the convex side? (Try this yourself!)

Q35–12 "See yourself as others see you." Can you do this with an ordinary plane mirror? If not, how *can* you do it?

Q35–13 In Example 35–4 (Section 35–4), there appears to be an ambiguity for the case $s = 10$ cm, as to whether s' is $+\infty$ or $-\infty$ and as to whether the image is erect or inverted. How is this resolved? Or is it?

Q35–14 Suppose that in the situation of Example 35–7 of Section 35–5 (Fig. 35–24) a vertical arrow 2.00 m tall is painted on the side of the pool beneath the water line. According to the calculations in the example, this arrow would appear to the person in Fig. 35–24 to be 1.50 m long. But the discussion following Eq. (35–13) states that the magnification for a plane refracting surface is $m = +1$, which suggests that the arrow would appear to the person to be 2.00 m long. How can you resolve this apparent contradiction?

Q35–15 When observing fish in an aquarium filled with water, one can see clearly only when looking nearly perpendicularly to the glass wall; objects viewed at an oblique angle always appear blurred. Why? Do the fish have the same problem when looking at you? Explain.

Q35–16 There have been reports of round fishbowls starting fires by focusing the sun's rays coming in a window. Is this possible? Explain.

Q35–17 Sometimes a wine glass filled with white wine forms an image of an overhead light on a white tablecloth. Would the same image be formed with an empty glass? With a glass of gin? With a glass of gasoline? Explain.

Q35–18 How could you very quickly make an approximate measurement of the focal length of a converging lens? Could the same method be applied if you wished to use a diverging lens? Explain.

Q35–19 The focal length of a simple lens depends on the color (wavelength) of light passing through it. Why? Is it possible for a lens to have a positive focal length for some colors and negative for others? Explain.

Q35–20 When a converging lens is immersed in water, does its focal length increase or decrease in comparison with the value in air? Explain.

Q35–21 A spherical air bubble in water can function as a lens. Is it a converging or diverging lens? How is its focal length related to its radius?

Q35–22 Can an image formed by one reflecting or refracting surface serve as an object for a second reflection or refraction? Does it matter whether the first image is real or virtual? Explain.

Q35–23 If a piece of photographic film is placed at the location of a real image, the film will record the image. Can this be done with a virtual image? How might one record a virtual image?

Q35–24 How could you make a lens for sound waves?

Q35–25 According to the discussion in Section 34–3, light rays are reversible. Are the formulas in the table in this chapter's summary still valid if object and image are interchanged? What does reversibility imply with respect to the *forms* of the various formulas?

EXERCISES

SECTION 35–2 **REFLECTION AND REFRACTION AT A PLANE SURFACE**

35–1 In Fig. 35–8, mirror 1 uses the image P_2' formed by mirror 2 as an object and forms an image of it. Show that this image is at point P_3' in the figure.

35–2 The image of a tree just covers the length of a plane mirror 4.00 cm tall when the mirror is held 35.0 cm from the eye. The tree is 28.0 m from the mirror. What is its height?

35–3 A candle 4.85 cm tall is 39.2 cm to the left of a plane mirror. Where is the image formed by the mirror, and what is the height of this image?

SECTION 35–3 **REFLECTION AT A SPHERICAL SURFACE**
SECTION 35–4 **GRAPHICAL METHODS FOR MIRRORS**

35–4 A concave mirror has a radius of curvature of 34.0 cm. a) What is its focal length? b) If the mirror is immersed in water (refractive index 1.33), what is its focal length?

35–5 An object 0.600 cm tall is placed 16.5 cm to the left of the vertex of a concave spherical mirror having a radius of curvature of 22.0 cm. a) Draw a principal-ray diagram showing formation of the image. b) Determine the position, size, orientation, and nature (real or virtual) of the image.

35–6 Repeat Exercise 35–5 for the case in which the mirror is convex.

35–7 The diameter of Mars is 6794 km, and its minimum distance from the earth is 5.58×10^7 km. When Mars is at this distance, find the diameter of the image of Mars formed by a spherical concave telescope mirror with a focal length of 1.75 m.

35–8 An object 24.0 cm from the center of a silvered spherical glass Christmas tree ornament 6.00 cm in diameter. What are the position and magnification of its image?

35–9 a) Show that Eq. (35–6) can be written as $s' = sf/(s - f)$ and hence that the lateral magnification, given by Eq. (35–7), can be expressed as $m = f/(f - s)$. b) Consider a concave mirror, which has $f > 0$, and let $s > f$ (that is, let the object lie outside the focal point). Show that the image distance s' is *positive*, and explain why this means that the image is *real*. Also show that the image is *inverted*. c) Show that if the object distance s is greater than $2f$, the image is smaller than the object, and that if s is between f and $2f$, the image is larger than the object. d) For a concave mirror, show that if $0 < s < f$ (that is, if the object lies inside the focal point), the image distance s' is *negative*. Explain why this means that the image is *virtual*. Show also that the image is *erect* and larger than the object.

35–10 Consider a convex mirror, which has $f < 0$, so that $f = -|f|$. Use the results of part (a) of Exercise 35–9 to show that for *any* positive object distance s, the image distance s' is negative and the image is virtual, erect, and smaller than the object.

35-11 a) Using the formulas for s' and m obtained in part (a) of Exercise 35-9, graph s' as a function of s for the case $f > 0$ (a concave mirror). b) For what values of s is s' positive, so that the image is real? c) For what values of s is s' negative, so that the image is virtual? d) Where is the image if the object is just outside the focal point (s slightly greater than f)? e) If the object is just inside the focal point (s slightly less than f)? f) Where is the image if the object is at infinity? g) Where is the image if the object is next to the mirror ($s = 0$)? h) Graph m as a function of s for the case of a concave mirror. i) For what values of s is the image erect? j) For what values of s is the image inverted? k) For what values of s is the image larger than the object? l) For what values of s is the image smaller than the object? m) What happens to the size of the image when the object is placed at the focal point?

35-12 Using the formulas for s' and m obtained in part (a) of Exercise 35-9, graph s' as a function of s, and graph m as a function of s, for the case $f < 0$ (a convex mirror), so that $f = -|f|$. a) For what values of s is s' positive? b) For what values of s is s' negative? c) Where is the image if the object is at infinity? d) Where is the image if the object is next to the mirror ($s = 0$)? For what values of s is the image e) erect; f) inverted; g) larger than the object; h) smaller than the object?

35-13 An object 9.00 mm tall is placed 12.0 cm to the left of the vertex of a convex spherical mirror whose radius of curvature has a magnitude of 20.0 cm. a) Draw a principal-ray diagram showing formation of the image. b) Determine the position, size, orientation, and nature (real or virtual) of the image.

35-14 A spherical concave shaving mirror has a radius of curvature of 32.0 cm. a) What is the magnification of a person's face when it is 12.0 cm to the left of the vertex of the mirror? b) Where is the image? Is the image real or virtual? c) Draw a principal-ray diagram showing formation of the image.

SECTION 35-5 REFRACTION AT A SPHERICAL SURFACE

35-15 A speck of dirt is embedded 3.50 cm below the surface of a sheet of ice ($n = 1.309$). What is its apparent depth when viewed at normal incidence?

35-16 A tank whose bottom is a mirror is filled with water to a depth of 20.0 cm. A small fish floats motionless 7.0 cm under the surface of the water. a) What is the apparent depth of the fish when viewed at normal incidence? b) What is the apparent depth of the image of the fish when viewed at normal incidence?

35-17 Equations (35-11) and (35-12) were derived in the text for the case in which R is positive and $n_a < n_b$. (See Figs. 35-20 and 35-21.) a) Carry through the derivation of these two equations for the case $R > 0$ and $n_a > n_b$. b) Carry through the derivation for $R < 0$ and $n_a < n_b$.

35-18 The left end of a long glass rod 6.00 cm in diameter has a convex hemispherical surface 3.00 cm in radius. The refractive index of the glass is 1.60. Determine the position of the image if an object is placed in air on the axis of the rod at the following distances to the left of the vertex of the curved end: a) infinitely far, b) 12.0 cm, c) 2.00 cm.

35-19 The rod of Exercise 35-18 is immersed in a liquid. An object 90.0 cm from the vertex of the left end of the rod and on

its axis is imaged at a point 1.60 m inside the rod. What is the refractive index of the liquid?

35-20 The left end of a long glass rod 8.00 cm in diameter, with an index of refraction 1.60, is ground and polished to a convex hemispherical surface with a radius of 4.00 cm. An object in the form of an arrow 1.50 mm tall, at right angles to the axis of the rod, is located on the axis 24.0 cm to the left of the vertex of the convex surface. Find the position and height of the image of the arrow formed by paraxial rays incident on the convex surface. Is the image erect or inverted?

35-21 Repeat Exercise 35-20 for the case in which the end of the rod is ground to a *concave* hemispherical surface with radius 4.00 cm.

35-22 A Spherical Fish Bowl. A small tropical fish is at the center of a water-filled spherical fish bowl 28.0 cm in diameter. a) Find the apparent position and magnification of the fish to an observer outside the bowl. The effect of the thin walls of the bowl may be neglected. b) A friend advised the owner of the bowl to keep it out of direct sunlight to avoid blinding the fish, which might swim into the focal point of the parallel rays from the sun. Is the focal point actually within the bowl?

SECTION 35-6 THIN LENSES
SECTION 35-7 GRAPHICAL METHODS FOR LENSES

35-23 A converging lens has a focal length of 14.0 cm. For an object to the left of the lens, at distances of 18.0 cm and 7.00 cm, determine a) the image position; b) the magnification; c) whether the image is real or virtual; d) whether the image is erect or inverted. Draw a principal-ray diagram in each case.

35-24 A lens forms an image of an object. The object is 16.0 cm from the lens. The image is 12.0 cm from the lens on the same side as the object. a) What is the focal length of the lens? Is the lens converging or diverging? b) If the object is 8.50 mm tall, how tall is the image? Is it erect or inverted? c) Draw a principal-ray diagram.

35-25 A converging lens with a focal length of 7.00 cm forms an image of a 4.00-mm-tall real object that is to the left of the lens. The image is 1.30 cm tall and erect. Where are the object and image located? Is the image real or virtual?

35-26 A converging lens with a focal length of 90.0 cm forms an image of a 3.20-cm-tall real object that is to the left of the lens. The image is 4.50 cm tall and inverted. Where are the object and image located in relation to the lens? Is the image real or virtual?

35-27 A diverging meniscus lens (see Fig. 35-29b) with a refractive index of 1.48 has spherical surfaces whose radii are 5.00 cm and 3.50 cm. What is the position of the image if an object is placed 18.0 cm to the left of the lens?

35-28 A photographic slide is to the left of a lens. The lens projects an image of the slide onto a wall 6.00 m to the right of the slide. The image is 80.0 times the size of the slide. a) How far is the slide from the lens? b) Is the image erect or inverted? c) What is the focal length of the lens? Is the lens converging or diverging?

35-29 Two thin lenses with a focal length of magnitude 12.0 cm, the first converging and the second diverging, are placed

9.00 cm apart. An object 2.50 mm tall is placed 20.0 cm to the left of the first (converging) lens. a) How far from this first lens is the final image formed? b) Is the final image real or virtual? c) What is the height of the final image? Is the final image erect or inverted?

35–30 The radii of curvature of the surfaces of a thin converging meniscus lens are $R_1 = +12.0$ cm and $R_2 = +28.0$ cm. The index of refraction is 1.60. a) Compute the position and size of the image of an object in the form of an arrow 5.00 mm tall, perpendicular to the lens axis, 45.0 cm to the left of the lens. b) A second converging lens with the same focal length is placed 3.15 m to the right of the first. Find the position and size of the final image. Is the final image erect or inverted with respect to the original object? c) Repeat part (b) except with the second lens 45.0 cm to the right of the first.

35–31 Sketch the various possible thin lenses that can be obtained by combining two surfaces whose radii of curvature are 4.00 cm and 8.00 cm in absolute magnitude. Which are converging and which are diverging? Find the focal length of each if the surfaces are made of glass with index of refraction 1.60.

35–32 Six lenses in air are shown in Fig. 35–29. Each lens is made of a material with index of refraction $n > 1$. Considering each lens individually, imagine that light enters the lens from the left. Show that the three lenses in Fig. 35–29a have *positive* focal lengths and hence are *converging* lenses. In addition, show that the three lenses in Fig. 35–29b have *negative* focal lengths and hence are *diverging* lenses.

35–33 Exercises 35–9 through 35–12 deal with spherical mirrors. a) Show that the equations for s' and m derived in part (a) of Exercise 35–9 also apply to a thin lens. b) A concave mirror is used in Exercises 35–9 and 35–11. Repeat these exercises for a converging lens. Are there any differences in the results when the mirror is replaced by a lens? Explain. c) A convex mirror is used in Exercises 35–10 and 35–12. Repeat these exercises for a diverging lens. Are there any differences in the results when the mirror is replaced by a lens? Explain.

35–34 A converging lens with a focal length of 12.0 cm forms a virtual image 8.00 mm tall, 17.0 cm to the right of the lens. Determine the position and size of the object. Is the image erect or inverted? Are the object and image on the same side or opposite sides of the lens? Draw a principal-ray diagram for this situation.

35–35 Repeat Exercise 35–34 for the case in which the lens is diverging, with a focal length of –48.0 cm.

35–36 An object is 16.0 cm to the left of a lens. The lens forms an image 36.0 cm to the right of the lens. a) What is the focal length of the lens? Is the lens converging or diverging? b) If the object is 8.00 mm tall, how tall is the image? Is it erect or inverted? c) Draw a principal-ray diagram.

PROBLEMS

35–37 A luminous object is 4.00 m from a wall. You are to use a concave mirror to project an image of the object on the wall, with the image 2.25 times the size of the object. How far should the mirror be from the wall? What should its radius of curvature be?

35–38 An object is placed between two plane mirrors arranged at right angles to each other at a distance d_1 from the surface of one mirror and a distance d_2 from the other. a) How many images are formed? Show the location of the images in a diagram. b) Draw the paths of rays from the object to the eye of an observer.

35–39 If you run away from a plane mirror at 2.40 m/s, at what speed does your image move away from you?

35–40 What is the size of the smallest vertical plane mirror in which a woman of height h can see her full-length image?

35–41 A concave mirror is to form an image of the filament of a headlight lamp on a screen 8.00 m from the mirror. The filament is 6.00 mm tall, and the image is to be 36.0 cm tall. a) How far in front of the vertex of the mirror should the filament be placed? b) What should be the radius of curvature of the mirror?

35–42 A mirror on the passenger side of your car is convex and has a radius of curvature with magnitude 18.0 cm. a) Another car is seen in this side mirror and is 13.0 m behind the mirror. If this car is 1.5 m tall, what is the height of the image? b) The mirror has a warning attached that objects viewed in it are closer than they appear. Why is this so?

35–43 If the light incident from the left onto a convex mirror does not diverge from an object point but instead converges toward a point at a (negative) distance s to the right of the mirror, this point is called a *virtual object.* a) For a convex mirror having a radius of curvature of 24.0 cm, for what range of virtual-object positions is a real image formed? b) What is the orientation of this real image? c) Draw a principal-ray diagram showing formation of such an image.

35–44 Equations (35–6) and (35–7) were derived in the text for the case of a concave mirror. Carry out a similar derivation for a convex mirror, and show that the same equations result if you use the sign convention established in the text.

35–45 Suppose the lamp filament in Example 35–1 (Section 35–3) is moved to a position 8.0 cm in front of the mirror. a) Where is the image located now? Is it real or virtual? b) What is the height of the image? Is it erect or inverted? c) In Example 35–1, the filament is 10.0 cm in front of the mirror, and an image of the filament is formed on a wall 3.00 m from the mirror. If the filament is 8.0 cm from the mirror, can a wall be placed so that an image is formed on it? If so, where should the wall be placed? If not, why not?

35–46 Where must you place an object in front of a concave mirror with radius R so that the image is erect and $2\frac{1}{2}$ times the size of the object? Where is the image?

35–47 **A Three-Dimensional Image.** The *longitudinal* magnification is defined as $m' = ds'/ds$. It relates the longitudinal dimension of a small object to the longitudinal dimension of its image. a) Show that for a spherical mirror, $m' = -m^2$. What is the significance of the fact that m' is *always* negative? b) A wire frame in the form of a small cube 8.00 mm on a side is placed with its center on the axis of a concave mirror with radius of curvature 30.0 cm. The sides of the cube are all either parallel or perpendicular to the axis. The cube face toward the mirror is 50.0 cm to the left of the mirror vertex. Find i) the location of the image of this face and of the opposite face of the cube; ii) the lateral and longitudinal magnifications; iii) the shape and dimensions of each of the six faces of the image.

35–48 Refer to Problem 35–47. Show that the longitudinal magnification m' for refraction at a spherical surface is given by

$$m' = -\frac{n_b}{n_a} m^2.$$

35–49 As you sit in your parked car, you glance in the convex side mirror (radius of curvature 125 cm) and notice a jogger approaching. If the jogger is running at a speed of 2.50 m/s, how fast is her image moving when she is a) 10.0 m away; b) 2.0 m away?

35–50 A microscope is focused on the upper surface of a glass plate. A second plate is then placed over the first. To focus on the bottom surface of the second plate, the microscope must be raised 0.780 mm. To focus on the upper surface, it must be raised another 2.50 mm. Find the index of refraction of the second plate.

35–51 The primary reflecting surface of a bathroom mirror lies behind a thin sheet of glass. This sheet protects the reflecting surface and keeps it from tarnishing. Because some light reflects from the front surface of the glass, and because light refracts upon entering or leaving the glass, the formation of images is somewhat more complex than is shown in Fig. 35–2. Consider a bathroom mirror with a glass sheet of thickness d and index of refraction n. Your face is located a distance h from the front surface of the glass sheet. a) Find the position of the image of your face that you see due to reflection from the front surface of the glass. b) Find the position of the image of your face that you see due to reflection from the mirror surface behind the glass. Take into account the effects of refraction at the glass-air interface. c) How far apart are the images in parts (a) and (b)?

35–52 A Glass Rod. Both ends of a glass rod with an index of refraction 1.60 are ground and polished to convex hemispherical surfaces. The radius of curvature at the left end is 6.00 cm, and the radius of curvature at the right end is 12.0 cm. The length of the rod between vertexes is 40.0 cm. The object for the surface at the left end is an arrow that lies 23.0 cm to the left of the vertex of this surface. The arrow is 1.50 mm tall and at right angles to the axis. a) What constitutes the object for the surface at the right end of the rod? b) What is the object distance for this surface? c) Is the object for this surface real or virtual? (*Hint:* See Example 35–12 (Section 35–7).) d) What is the position of the final image? e) Is the final image real or virtual? Is it erect or inverted with respect to the original object? f) What is the height of the final image?

35–53 The rod in Problem 35–52 is shortened to a distance of 25.0 cm between its vertexes; the curvatures of its ends remain the same. As in Problem 35–52, the object for the surface at the left end is an arrow that lies 23.0 cm to the left of the vertex of this surface. The arrow is 1.50 mm tall and at right angles to the axis. a) What is the object distance for the surface at the right end of the rod? b) Is the object for this surface real or virtual? c) What is the position of the final image? d) Is the final image

real or virtual? Is it erect or inverted with respect to the original object? e) What is the height of the final image?

35–54 A layer of benzene ($n = 1.50$) 2.60 cm deep floats on water ($n = 1.33$) that is 6.50 cm deep. What is the apparent distance from the upper benzene surface to the bottom of the water layer when it is viewed at normal incidence?

35–55 What should be the index of refraction of a transparent sphere in order for paraxial rays from an infinitely distant object to be brought to a focus at the vertex of the surface opposite the point of incidence?

35–56 A transparent rod 50.0 cm long and with a refractive index of 1.60 is cut flat at the right end and rounded to a hemispherical surface with a 15.0-cm radius at the left end. An object is placed on the axis of the rod 12.0 cm to the left of the vertex of the hemispherical end. a) What is the position of the final image? b) What is its magnification?

35–57 A narrow beam of parallel rays enters a solid glass sphere in a radial direction. At what point outside the sphere are these rays brought to a focus? The radius of the sphere is 4.00 cm, and its index of refraction is 1.80.

35–58 A transparent rod 30.0 cm long is cut flat at one end and rounded to a hemispherical surface of radius 10.0 cm at the other end. A small object is embedded within the rod along its axis and halfway between its ends, 15.0 cm from the flat end and 15.0 cm from the vertex of the curved end. When viewed from the flat end of the rod, the apparent depth of the object is 9.50 cm from the flat end. What is its apparent depth when viewed from the curved end?

35–59 A solid glass hemisphere of radius 12.0 cm and index of refraction $n = 1.50$ is placed with its flat face downward on a table. A parallel beam of light with a circular cross section 3.80 mm in diameter travels straight down and enters the hemisphere at the center of its curved surface. a) What is the diameter of the circle of light formed on the table? b) How does your result depend on the radius of the hemisphere?

35–60 A thick-walled wine goblet sitting on a table can be considered to be a hollow glass sphere with an outer radius of 4.00 cm and an inner radius of 3.40 cm. The index of refraction of the goblet glass is 1.50. a) A beam of parallel light rays enters the side of the empty goblet along a horizontal radius. Where, if anywhere, will an image be formed? b) The goblet is filled with white wine ($n = 1.37$). Where is the image formed?

35–61 A thin-walled glass sphere with radius R is filled with water. An object is placed a distance $3R$ from the surface of the sphere. Determine the position of the final image. The effect of the glass wall may be neglected. Use $n = \frac{4}{3}$ for water.

35–62 For refraction at a spherical surface, the first focal length f is defined as the value of s corresponding to $s' = \infty$, as shown in Fig. 35–37a (page 1116). The second focal length f' is defined as the value of s' when $s = \infty$, as shown in Fig. 35–37b. a) Prove that $n_a/n_b = f/f'$. b) Prove that the general relation between object and image distance is

$$\frac{f}{s} + \frac{f'}{s'} = 1.$$

(a)

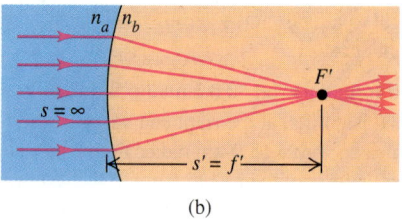

(b)

FIGURE 35–37 Problem 35–62.

35–63 An object to the left of a lens is imaged by the lens on a screen 30.0 cm to the right of the lens. When the lens is moved 4.00 cm to the right, the screen must be moved 4.00 cm to the left to refocus the image. Determine the focal length of the lens.

35–64 A glass rod with a refractive index of 1.55 is ground and polished at both ends to hemispherical surfaces with radii of 6.00 cm. When an object is placed on the axis of the rod, 25.0 cm to the left of the left-hand end, the final image is formed 65.0 cm to the right of the right-hand end. What is the length of the rod measured between the vertexes of the two hemispherical surfaces?

35–65 Three thin lenses, each with a focal length of 40.0 cm, are aligned on a common axis; adjacent lenses are separated by 52.0 cm. Find the position of the image of a small object on the axis, 80.0 cm to the left of the first lens.

35–66 An object is placed 18.0 cm from a screen. a) At what two points between object and screen may a converging lens with a 3.00-cm focal length be placed to obtain an image on the screen? b) What is the magnification of the image for each position of the lens?

35–67 A convex mirror and a concave mirror are placed on the same optic axis, separated by a distance $L = 0.600$ m. The radius of curvature of each mirror has a magnitude of 0.360 m. A light source is located a distance x from the concave mirror, as shown in Fig. 35–38. a) What distance x will result in the rays from the source returning to the source after reflecting first from the convex mirror and then from the concave mirror? b) Repeat part (a), but now let the rays reflect first from the concave mirror and then from the convex one.

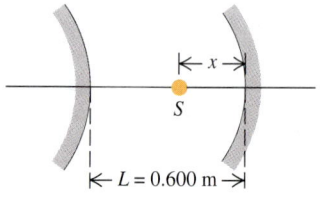

FIGURE 35–38 Problem 35–67.

35–68 In Fig. 35–39 the candle is at the center of curvature of the concave mirror, whose focal length is 10.0 cm. The converging lens has a focal length of 32.0 cm and is 85.0 cm to the right of the candle. The candle is viewed looking through the lens from the right. The lens forms two images of the candle. The first is formed by light passing directly through the lens. The second image is formed from the light that goes from the candle to the mirror, is reflected, and then passes through the lens. a) For each of these two images, draw a principal-ray diagram that locates the image. b) For each image, answer the following questions: i) Where is the image? ii) Is the image real or virtual? iii) Is the image erect or inverted with respect to the original object?

FIGURE 35–39 Problem 35–68.

35–69 An eyepiece consists of two converging thin lenses, each with a focal length 12.0 cm, separated by a distance of 4.00 cm. Where are the first and second focal points of the eyepiece?

35–70 Two Lenses in Contact. a) Prove that when two thin lenses with focal lengths f_1 and f_2 are placed *in contact,* the focal length f of the combination is given by the relation

$$\frac{1}{f} = \frac{1}{f_1} + \frac{1}{f_2}.$$

b) A converging meniscus lens (Fig. 35–29a) has index of refraction of 1.55 and radii of curvature for its surfaces of 4.50 cm and 9.00 cm. The concave surface is placed upward and filled with carbon tetrachloride (CCl_4), which has $n = 1.46$. What is the focal length of the CCl_4-glass combination?

35–71 Rays from a lens are converging toward a point image P located to the right of the lens. What thickness t of glass with an index of refraction 1.60 must be interposed between the lens and P for the image to be formed at P', located 0.30 cm to the right of P? The locations of the piece of glass and of points P and P' are shown in Fig. 35–40.

FIGURE 35–40 Problem 35–71.

35–72 A Lens in a Liquid. A lens obeys Snell's law, bending light rays at each surface an amount determined by the index of refraction of the lens and the index of the medium in which the lens is located. a) Equation (35–19) assumes that the lens is surrounded by air. Consider instead a thin lens immersed in a liquid with refractive index n_{liq}. Prove that the focal length f' is then given by Eq. (35–19) with n replaced by n/n_{liq}. b) A thin lens with index n has focal length f in vacuum. Use the result of part (a) to show that when this lens is immersed in a liquid of index n_{liq}, it will have a new focal length given by

$$f' = \left[\frac{n_{liq}(n-1)}{n - n_{liq}}\right]f.$$

35–73 When an object is placed at the proper distance to the left of a converging lens, the image is focused on a screen 30.0 cm to the right of the lens. A diverging lens is now placed 15.0 cm to the right of the converging lens, and it is found that the screen must be moved 19.2 cm farther to the right to obtain a sharp image. What is the focal length of the diverging lens?

35–74 A convex spherical mirror with a focal length of magnitude 24.0 cm is placed 20.0 cm to the left of a plane mirror. An object 0.250 cm tall is placed midway between the surface of the plane mirror and the vertex of the spherical mirror. The spherical mirror forms multiple images of the object. Where are the two images of the object formed by the spherical mirror that are closest to the spherical mirror, and how tall is each image?

35–75 A glass plate 3.50 cm thick, with an index of refraction 1.55 and plane parallel faces, is held with its faces horizontal and its lower face 6.00 cm above a printed page. Find the position of the image of the page formed by rays making a small angle with the normal to the plate.

35–76 A symmetric double-convex thin lens made of glass with an index of refraction 1.52 has a focal length in air of 40.0 cm. The lens is sealed into an opening in the left-hand end of a tank filled with water. At the right-hand end of the tank, opposite the lens, is a plane mirror 90.0 cm from the lens. The index of refraction of the water is $\frac{4}{3}$. a) Find the position of the image formed by the lens-water-mirror system of a small object outside the tank on the lens axis and 70.0 cm to the left of the lens. b) Is the image real or virtual? c) Is it erect or inverted? d) If the object has a height of 4.00 mm, what is the height of the image?

CHALLENGE PROBLEMS

35–77 An Object at an Angle. A 16.0-cm-long pencil is placed at a 45.0° angle, with its center 15.0 cm above the optic axis and 45.0 cm from a lens with a 20.0-cm focal length as shown in Fig. 35–41. (Note that the figure is not drawn to scale.) Assume that the diameter of the lens is large enough for the paraxial approximation to be valid. a) Where is the image of the pencil? (Give the location of the images of the points A, B, and C on the object, which are located at the eraser, point, and center of the pencil, respectively.) b) What is the length of the image (that is, the distance between the images of points A and B)? c) Show the orientation of the image in a sketch.

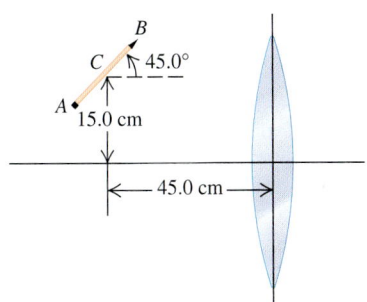

FIGURE 35–41 Challenge Problem 35–77.

35–78 *Spherical aberration* is a blurring of the image formed by a spherical mirror. It occurs because parallel rays striking the mirror far from the optic axis are focused at a different point than are rays near the axis. This problem is usually minimized by using only the center of a spherical mirror. a) Show that for a spherical concave mirror, the focus moves toward the mirror as the parallel rays move toward the outer edge of the mirror. (*Hint:* Derive an analytic expression for the distance from the vertex to the focus of the ray for a particular parallel ray. This expression should be in terms of i) the radius of curvature R of the mirror and ii) the angle θ between the incident ray and a line connecting the center of curvature of the mirror with the point where the ray strikes the mirror.) b) What value of θ produces a 2% change in the location of the focus, compared to the location for θ very close to zero?

35–79 a) For a lens with focal length f, find the smallest distance possible between the object and its real image. b) Graph the distance between the object and real image as a function of the distance of the object from the lens. Does your graph agree with the result you found in part (a)?

35–80 Two mirrors are placed together as shown in Fig. 35–42. a) Show that a point source in front of these mirrors and its two images lie on a circle. b) Find the center of the circle. c) In a diagram, show where an observer should stand so as to be able to see both images.

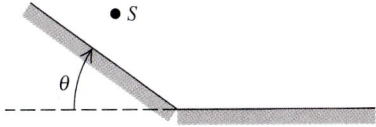

FIGURE 35–42 Challenge Problem 35–80.

36 Optical Instruments

Some of the largest optical instruments in the world are found at the summit of Mauna Kea, an extinct 4300-m (14,000-ft) volcano in Hawaii, where the view of the heavens is exceptionally clear. Unlike an earlier generation of telescopes that used lenses, these telescopes use curved mirrors to bring the light from distant objects to a focus.

36-1 INTRODUCTION

In the previous chapter we learned the basics about how images are formed by mirrors and lenses. Now it's time to apply these ideas to some common optical devices, and to ask how such devices work. How does a camera resemble the human eye? What are the significant differences? What does a photographer or a projector operator have to do to "focus" the picture? How is it that a particular combination of two lenses makes a microscope, but a different combination makes a telescope? We will be able to answer all of these questions and more by applying the basic principles of mirror and lens behavior that we studied in Chapter 35.

The concept of *image*, which was so central to understanding simple optical devices in Chapter 35, plays an equally important role in the analysis of optical instruments. We continue to base our analysis on the *ray* model of light, so the content of this chapter comes under the general heading *geometric optics*.

36-2 CAMERAS AND PROJECTORS

Cameras and projectors are among the most common of optical devices. They use similar optics to perform complementary tasks. Cameras make a small image of an object and record that image on film. This film can then be used as the object by a projector, which produces a large image on a screen.

CAMERAS

The basic elements of a **camera** are a converging lens, a light-tight box ("camera" is a Latin word meaning a room or enclosure), a light-sensitive film to record an image, and a shutter to let the light from the lens strike the film for a prescribed length of time (Fig. 36–1a). The lens forms an inverted real image on the film of the object being photographed. High-quality camera lenses have several elements, permitting partial correction of various *aberrations,* including the dependence of index of refraction on wavelength and the limitations imposed by the paraxial approximation. (We will discuss lens aberrations in Section 36–7.) A classic lens design is the Zeiss "Tessar" design, shown in Fig. 36–1b.

When the camera is in proper *focus,* the position of the film coincides with the position of the real image formed by the lens. The resulting photograph will then be as sharp as possible. With a converging lens, the image distance increases as the object distance decreases (see Figs. 35–33a, 35–33b, and 35–33c and the discussion in Section 35–7). Hence in "focusing" the camera, the lens is moved closer to the film for a distant object and farther from the film for a nearby object. Often this is done by turning the lens in a threaded mount.

The choice of the focal length f for a camera lens depends on the film size and the desired angle of view. Figure 36–2 shows three photographs taken on 35-mm film with the same camera at the same position, but with

36–1 Key elements of a single-lens reflex camera, a typical design used by many amateur and professional photographers. (a) The lens forms a real, inverted image of the object in the plane of the film. The image is brought into focus by moving the lens forward or backward relative to the film. (b) Cross-section view of the compound lens and the pentaprism for viewing through the lens. Because of successive reflections at the surfaces of the hinged mirror and the pentaprism, an erect image is seen through the eyepiece. The hinged mirror swings up out of the way before the shutter opens.

36–2 Effect of using different lenses with different focal lengths. (a), (b), (c) Three photographs taken with the same camera from the same position in the Boston Public Garden using lenses with $f = 28$ mm, 105 mm, and 300 mm. Increasing the focal length increases the image size proportionately. (d) The larger image size with a larger value of f corresponds to a smaller angle of view. The angles shown here are for a camera that uses 35-mm film (image area 24 mm × 36 mm) and refer to the angle of view along the diagonal dimension of the film.

lenses of different focal lengths. A lens of long focal length, called a *telephoto* lens, gives a small angle of view and a large image of a distant object (such as the statue in Fig. 36–2c); a lens of short focal length gives a small image and a wide angle of view (as in Fig. 36–2a) and is called a *wide-angle* lens. To understand this behavior, recall that the focal length is the distance from the lens to the image when the object is infinitely far away. In general, for *any* object distance, using a lens of longer focal length gives a greater image distance. This also increases the height of the image; as was discussed in Section 35–6, the ratio of the image height y' to the object height y (the *lateral magnification*) is equal in absolute value to the ratio of image distance s' to the object distance s (Eq. (35–17)):

$$m = \frac{y'}{y} = -\frac{s'}{s}.$$

With a lens of short focal length, the ratio s'/s is small, and a distant object gives only a small image. When a lens with a long focal length is used, the image of this same object may entirely cover the area of the film. Hence the longer the focal length, the narrower the angle of view (Fig. 36–2d).

The angle of view may be increased by simply increasing the size of the film. When attached to a camera that uses 35-mm film, for which the image area is 24 mm × 36 mm, a lens with $f = 50$ mm gives a 45° angle of view; a lens with this angle of view is called a "normal" lens. When attached to a portrait camera that uses 60 mm × 70 mm film, a lens of this same focal length acts as a wide-angle lens with a 63° angle of view.

In order for the film to record the image properly, the total light energy per unit area reaching the film (the "exposure") must fall within certain limits. This is controlled by the *shutter* and the *lens aperture.* The shutter controls the time interval during which light enters the lens. This is usually adjustable in steps corresponding to factors of about two, often from 1 s to $\frac{1}{1000}$ s.

The intensity of light reaching the film is proportional to the area viewed by the camera lens and to the effective area of the lens. The size of the area that the lens "sees" is proportional to the square of the angle of view of the lens, and so is roughly proportional to $1/f^2$. The effective area of the lens is controlled by means of an adjustable lens aperture, or *diaphragm,* a nearly circular hole with variable diameter D; hence the effective area is proportional to D^2. Putting these factors together, we see that the intensity of light reaching the film with a particular lens is proportional to D^2/f^2. The light-gathering capability of a lens is commonly expressed by photographers in terms of the ratio f/D, called the ***f*-number** of the lens:

$$f\text{-number} = \frac{\text{Focal length}}{\text{Aperture diameter}} = \frac{f}{D}. \qquad (36\text{–}1)$$

For example, a lens with a focal length $f = 50$ mm and an aperture diameter $D = 25$ mm is said to have an *f*-number of 2, or "an aperture of *f*/2." The light intensity reaching the film is *inversely* proportional to the square of the *f*-number.

For a lens with a variable-diameter aperture, increasing the diameter by a factor of $\sqrt{2}$ changes the *f*-number by $1/\sqrt{2}$ and increases the intensity at the film by a factor of 2. Adjustable apertures usually have scales labeled with successive numbers (often called *f-stops*) related by factors of $\sqrt{2}$, such as

$$f/2, \quad f/2.8, \quad f/4, \quad f/5.6, \quad f/8, \quad f/11, \quad f/16,$$

and so on. The larger numbers represent smaller apertures and exposures, and each step corresponds to a factor of two in intensity (Fig. 36–3). The actual *exposure* (total amount of light reaching the film) is proportional both to the aperture area and the time of expo-

f-stops

←*D*→

Adjustable diaphragm   *f*/4 aperture

f/8 aperture

36–3 In a camera lens, larger *f*-numbers mean smaller aperture diameters.

(a)

(b)

(c)

36-4 (a), (b) Principle of the zoom lens, using two elements with variable spacing. The effective focal length depends on the distance between the lens elements. (c) A typical zoom lens for a 35-mm camera, containing twelve elements arranged in four groups.

sure. Thus $f/4$ and $\frac{1}{500}$ s, $f/5.6$ and $\frac{1}{250}$ s, and $f/8$ and $\frac{1}{125}$ s all correspond to the same exposure.

Many photographers use a *zoom lens*, which is not a single lens but a complex collection of several lens elements that give a continuously variable focal length, often over a range as great as 10 to 1. Figure 36–4a shows a simple system with variable focal length, and Fig. 36–4b shows a typical zoom lens for a 35-mm camera. Zoom lenses give a range of image sizes of a given object. It is an enormously complex problem in optical design to keep the image in focus and maintain a constant *f*-number while the focal length changes. When you vary the focal length of a typical zoom lens, two groups of elements move within the lens and a diaphragm opens and closes.

The optical system for a television camera is the same in principle as for an ordinary camera. The film is replaced by an electronic system that, in the format used in North America, scans the image with a series of 525 parallel lines. The image brightness at points along these lines is translated into electrical impulses that can be broadcast, using electromagnetic waves with frequencies of the order of 100 to 400 MHz, or stored on videotape. The entire picture is scanned 30 times each second, so 30×525 or 15,750 lines are scanned each second. Some TV receivers emit a faint high-pitched sound at this scanning frequency (two octaves above the highest B on the piano).

EXAMPLE 36-1

Photographic exposures A common telephoto lens for a 35-mm camera has a focal length of 200 mm and a range of *f*-stops from $f/5.6$ to $f/45$. a) What is the corresponding range of aperture diameters? b) What is the corresponding range of intensity of the image on the film?

SOLUTION a) From Eq. (36–1) the range of diameters is from

$$D = \frac{f}{f\text{-number}} = \frac{200 \text{ mm}}{5.6} = 36 \text{ mm}$$

to

$$D = \frac{200 \text{ mm}}{45} = 4.4 \text{ mm}.$$

b) Because the intensity is proportional to the square of the diameter, the ratio of the intensity at $f/5.6$ to the intensity at $f/45$ is

$$\left(\frac{36 \text{ mm}}{4.4 \text{ mm}}\right)^2 = \left(\frac{45}{5.6}\right)^2 = 65 \quad (\text{about } 2^6).$$

If the correct exposure time at $f/5.6$ is $\frac{1}{1000}$ s, then at $f/45$ it is $(65)(\frac{1}{1000} \text{ s}) = \frac{1}{15}$ s.

PROJECTORS

A **projector** for viewing slides or motion pictures operates very much like a camera in reverse. The essential elements are shown in Fig. 36–5. Light from the source (an incandescent lamp bulb or, in large motion-picture projectors, a carbon-arc lamp) shines through the film, and the projection lens forms a real, enlarged, inverted image of the film on the projection screen. Additional lenses called *condenser* lenses are placed between the lamp and the film. Their function is to direct the light from the source so that most of it enters the projection lens after passing through the film. A concave mirror behind the lamp also helps to direct the light. The condenser lenses must be large

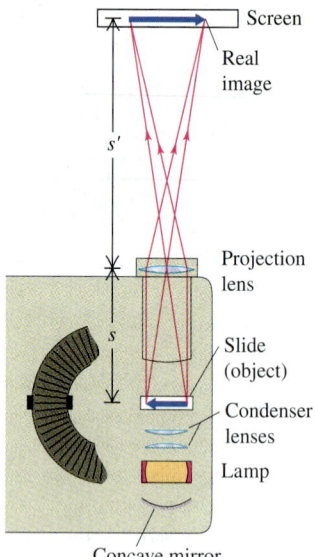

36–5 A slide projector. The concave mirror and the condenser lenses gather and direct the light from the lamp so that it will enter the projection lens after passing through the slide. A fan is usually needed to cool the lamp; this is not shown.

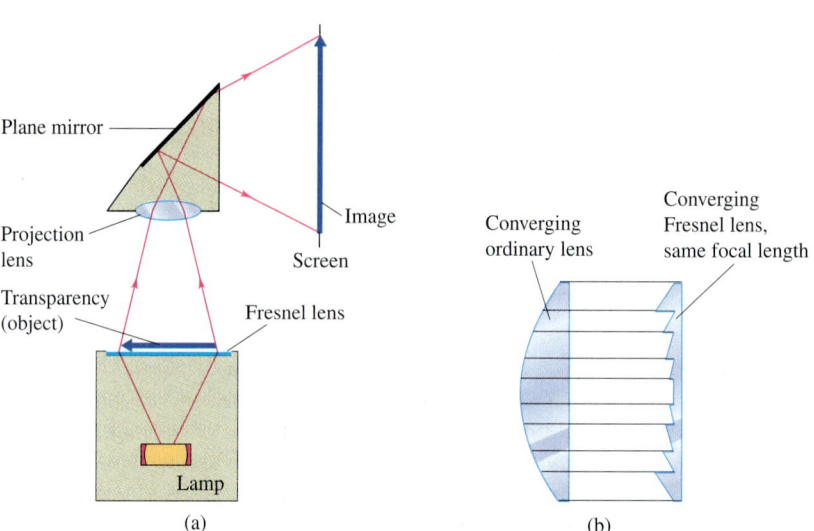

(a)   (b)

36–6 (a) An overhead projector. The light from the lamp passes through a converging Fresnel lens, then through the projection lens before hitting the angled plane mirror. (b) Cross sections of a converging ordinary lens and a converging Fresnel lens.

enough to cover the entire area of the film. The image on the screen is always real and inverted; this is why slides have to be put into a projector upside-down. The position and size of the image projected on the screen are determined by the position and focal length of the projection lens.

Overhead projectors used in classrooms use a similar scheme to produce an image on a projection screen, with two notable differences (Fig. 36–6a). After the light leaves the projection lens, an angled plane mirror reflects and reverses the image so that it appears right-side up on a vertical projection screen. Also, the light from the lamp is directed toward the projection lens by the clear piece of plastic on top of which the transparency is placed. This clear plastic is an example of a *Fresnel lens*. All of the lenses that we've discussed work because of the bending of light rays at the surfaces; no refraction occurs in the interior of the lens. If we could eliminate some of the material between the surfaces, we could greatly reduce the weight of a large lens. This is what a Fresnel lens (Fig. 36–6b) does. Each circular step copies the contour of the corresponding ring of an ordinary lens; you can see these steps by looking at any overhead projector. Fresnel lenses don't usually have high optical quality, but they are very light and inexpensive for their size. They are also used in traffic lights, solar-cell light collectors, flat pocket magnifiers, lighthouse lamps, and many other places.

EXAMPLE 36–2

A slide projector On an ordinary 35-mm color slide, the picture area is 24 mm × 36 mm. What focal-length projection lens would be needed to project an image 1.2 m × 1.8 m on a screen 5.0 m from the lens?

SOLUTION We need a lateral magnification (apart from sign) of (1.2 m)/(24 mm) = 50. From Eq. (35–17) the ratio s'/s must also be 50. (The image is real, so s' is positive.) We are given $s' = 5.0$ m, so $s = (5.0$ m)/50 = 0.10 m. Then from Eq. (35–16),

$$\frac{1}{f} = \frac{1}{s} + \frac{1}{s'} = \frac{1}{0.10 \text{ m}} + \frac{1}{5.0 \text{ m}},$$

$$f = 0.098 \text{ m} = 98 \text{ mm}.$$

A popular focal length for home slide projectors is 100 mm; such lenses are readily available and would be the appropriate choice in this situation. Many projectors are equipped with zoom lenses in order to give a range of image sizes and to be able to cope with different distances from projector to screen.

36–3   THE EYE

15.14
The Eye
15.15
Eyeglasses

The optical behavior of the eye is similar to that of a camera. The essential parts of the human eye, considered as an optical system, are shown in Fig. 36–7. The eye is nearly spherical in shape and about 2.5 cm in diameter. The front portion is somewhat more sharply curved and is covered by a tough, transparent membrane called the *cornea.* The region behind the cornea contains a liquid called the *aqueous humor.* Next comes the *crystalline lens,* a capsule containing a fibrous jelly, hard at the center and progressively softer at the outer portions. The crystalline lens is held in place by ligaments that attach it to the ciliary muscle, which encircles it. Behind the lens, the eye is filled with a thin watery jelly called the *vitreous humor.* The indexes of refraction of both the aqueous humor and the vitreous humor are about 1.336, nearly equal to that of water. The crystalline lens, while not homogeneous, has an average index of 1.437. This is not very different from the indexes of the aqueous and vitreous humors; most of the refraction of light entering the eye occurs at the outer surface of the cornea.

Refraction at the cornea and the surfaces of the lens produces a *real image* of the object being viewed. This image is formed on the light-sensitive *retina,* lining the rear inner surface of the eye. The retina plays the same role as the film in a camera. The *rods* and *cones* in the retina act like an array of miniature photocells; they sense the image and transmit it via the *optic nerve* to the brain. Vision is most acute in a small central region called the *fovea centralis,* about 0.25 mm in diameter.

In front of the lens is the *iris.* It contains an aperture with variable diameter called the *pupil,* which opens and closes to adapt to changing light intensity. The receptors of the retina also have intensity adaptation mechanisms.

For an object to be seen sharply, the image must be formed exactly at the location of the retina. The eye adjusts to different object distances s by changing the focal length f of its lens; the lens-to-retina distance, corresponding to s', does not change. (Contrast this with focusing a camera, in which the focal length is fixed and the lens-to-film distance is changed). For the normal eye, an object at infinity is sharply focused when the ciliary muscle is relaxed. To permit sharp imaging on the retina of closer objects, the tension in the ciliary muscle surrounding the lens increases, the ciliary muscle contracts, the lens bulges, and the radii of curvature of its surfaces decrease; this decreases the focal length. This process is called *accommodation.*

The extremes of the range over which distinct vision is possible are known as the *far point* and the *near point* of the eye. The far point of a normal eye is at infinity. The position of the near point depends on the amount by which the ciliary muscle can increase the curvature of the crystalline lens. The range of accommodation gradually diminishes with age because the crystalline lens grows throughout a person's life (it is about 50% larger at age 60 than at age 20) and the ciliary muscles are less able to distort a larger

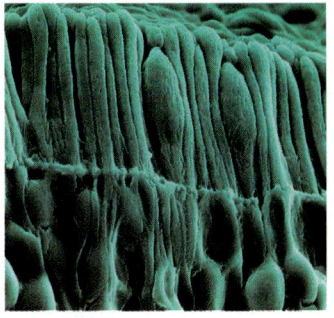

This magnified image shows rods and cones in the human eye. The rods are more sensitive to light than the cones, but only the cones are sensitive to differences in color. A typical human eye contains about 1.3×10^8 rods and about 7×10^6 cones.

36–7 The eye. The ciliary muscle contracts to change the focal length of the crystalline lens in order to image close objects sharply.

(a) Normal eye

(b) Myopic eye

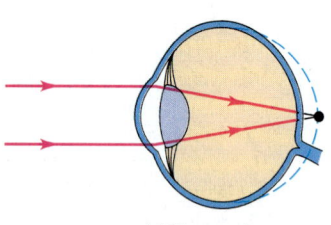

(c) Hyperopic eye

36–8 Refractive errors for a myopic (nearsighted) eye and a hyperopic (farsighted) eye viewing a very distant object. The dashed blue curve indicates the correct position of the retina.

36–9 Vertical lines are imaged in front of the retina by this astigmatic eye.

36–10 (a) An uncorrected hyperopic (farsighted) eye. (b) A positive (converging) lens gives the extra convergence needed for a hyperopic eye to focus the image on the retina. The virtual image formed by the converging lens acts as an object at or past the near point.

TABLE 36–1
RECEDING OF NEAR POINT WITH AGE

AGE (years)	NEAR POINT (cm)
10	7
20	10
30	14
40	22
50	40
60	200

lens. For this reason, the near point gradually recedes as one grows older. This recession of the near point is called *presbyopia.* Table 36–1 shows the approximate position of the near point for an average person at various ages. For example, an average person 50 years of age cannot focus on an object that is closer than about 40 cm.

Several common defects of vision result from incorrect distance relations in the eye. A normal eye forms an image on the retina of an object at infinity when the eye is relaxed (Fig. 36–8a). In the *myopic* (nearsighted) eye, the eyeball is too long from front to back in comparison with the radius of curvature of the cornea (or the cornea is too sharply curved), and rays from an object at infinity are focused in front of the retina (Fig. 36–8b). The most distant object for which an image can be formed on the retina is then nearer than infinity. In the *hyperopic* (farsighted) eye, the eyeball is too short or the cornea is not curved enough, and the image of an infinitely distant object is behind the retina (Fig. 36–8c). The myopic eye produces *too much* convergence in a parallel bundle of rays for an image to be formed on the retina; the hyperopic eye, *not enough* convergence.

Astigmatism refers to a defect in which the surface of the cornea is not spherical but is more sharply curved in one plane than in another. As a result, horizontal lines may be imaged in a different plane from vertical lines (Fig. 36–9). Astigmatism may make it impossible, for example, to focus clearly on the horizontal and vertical bars of a window at the same time.

All of these defects can be corrected by the use of corrective lenses (eyeglasses or contact lenses). The near point of either a presbyopic or a hyperopic eye is *farther* from the eye than normal. To see clearly an object at normal reading distance (often assumed to be 25 cm), we need a lens that forms a virtual image of the object at or beyond the near point. This can be accomplished by a converging (positive) lens, as shown in Fig. 36–10. In effect the lens moves the object farther away from the eye to a point where a sharp retinal image can be formed. Similarly, correcting the myopic eye involves using

(a)

(b)

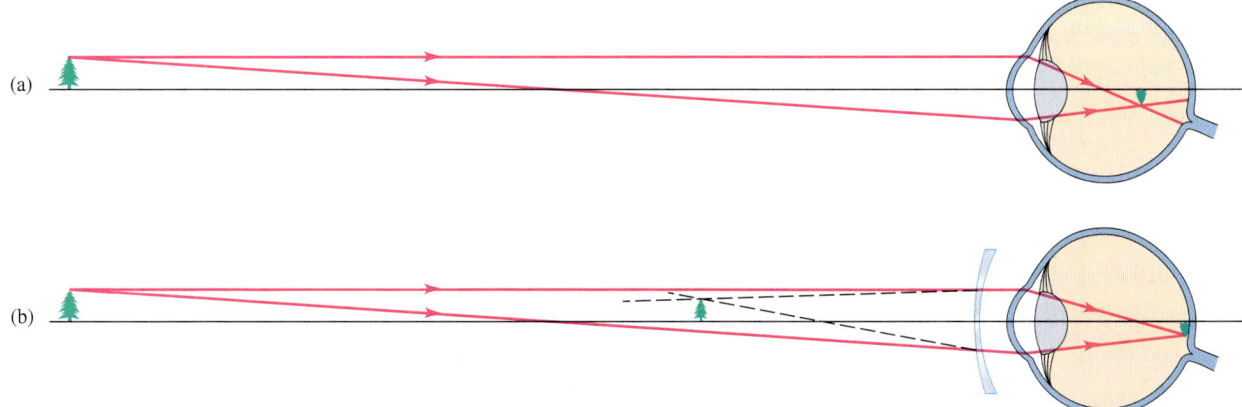

36–11 (a) An uncorrected myopic (nearsighted) eye. (b) A negative (diverging) lens spreads the rays farther apart to compensate for the excessive convergence of the myopic eye. The virtual image formed by the diverging lens acts as an object at or inside the far point.

a diverging (negative) lens to move the image closer to the eye than the actual object, as shown in Fig. 36–11.

Astigmatism is corrected by use of a lens with a *cylindrical* surface. For example, suppose the curvature of the cornea in a horizontal plane is correct to focus rays from infinity on the retina but the curvature in the vertical plane is too great to form a sharp retinal image. When a cylindrical lens with its axis horizontal is placed before the eye, the rays in a horizontal plane are unaffected, but the additional divergence of the rays in a vertical plane causes these to be sharply imaged on the retina, as shown in Fig. 36–12.

Lenses for vision correction are usually described in terms of the **power,** defined as the reciprocal of the focal length expressed in meters. The unit of power is the **diopter.** Thus a lens with $f = 0.50$ m has a power of 2.0 diopters, $f = -0.25$ m corresponds to -4.0 diopters, and so on. The numbers on a prescription for glasses are usually powers expressed in diopters. When the correction involves both astigmatism and myopia or hyperopia, there are three numbers: one for the spherical power, one for the cylindrical power, and an angle to describe the orientation of the cylinder axis.

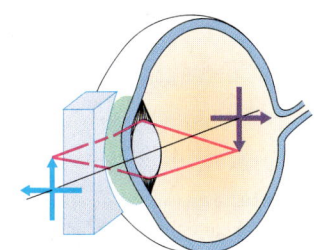

36–12 The diverging cylindrical lens with horizontal axis separates rays farther apart vertically (but not horizontally) so that the vertical line is also imaged on the retina of this astigmatic eye.

EXAMPLE 36–3

Correcting for farsightedness The near point of a certain hyperopic eye is 100 cm in front of the eye. To see clearly an object that is 25 cm in front of the eye, what contact lens is required?

SOLUTION We want the lens to form a virtual image of the object at a location corresponding to the near point of the eye, 100 cm from it. That is, when $s = 25$ cm, we want s' to be

-100 cm. From the basic thin-lens equation, Eq. (35–16),

$$\frac{1}{f} = \frac{1}{s} + \frac{1}{s'} = \frac{1}{+25 \text{ cm}} + \frac{1}{-100 \text{ cm}},$$

$$f = +33 \text{ cm}.$$

We need a converging lens with focal length $f = 33$ cm. The corresponding power is $1/(0.33$ m$)$, or $+3.0$ diopters.

EXAMPLE 36–4

Correcting for nearsightedness The far point of a certain myopic eye is 50 cm in front of the eye. To see clearly an object at infinity, what eyeglass lens is required? Assume that the lens is worn 2 cm in front of the eye.

SOLUTION The far point of a myopic eye is nearer than infinity. To see clearly objects beyond the far point, we need a lens that forms a virtual image of such objects no farther from the eye than the far point. Assume that the virtual image of the object at infinity is formed at the far point, 50 cm in front of the eye and 48 cm in front of the eyeglass lens. Then when $s = \infty$, we want s' to be -48 cm. From Eq. (35–16),

$$\frac{1}{f} = \frac{1}{s} + \frac{1}{s'} = \frac{1}{\infty} + \frac{1}{-48 \text{ cm}},$$

$$f = -48 \text{ cm}.$$

We need a *diverging* lens with focal length -48 cm $= -0.48$ m. The power is -2.1 diopter. Can you see that if a *contact* lens were used instead, we would need $f = -50$ cm?

36–4 THE MAGNIFIER

The apparent size of an object is determined by the size of its image on the retina. If the eye is unaided, this size depends upon the *angle* θ subtended by the object at the eye, called its **angular size** (Fig. 36–13a).

To look closely at a small object, such as an insect or a crystal, you bring it close to your eye, making the subtended angle and the retinal image as large as possible. But your eye cannot focus sharply on objects that are closer than the near point, so the angular size of an object is greatest (that is, it subtends the largest possible viewing angle) when it is placed at the near point. In the following discussion we will assume an average viewer for whom the near point is 25 cm from the eye.

A converging lens can be used to form a virtual image that is larger and farther from the eye than the object itself, as shown in Fig. 36–13b. Then the object can be moved closer to the eye, and the angular size of the image may be substantially larger than the angular size of the object at 25 cm without the lens. A lens used in this way is called a **magnifier,** otherwise known as a *magnifying glass* or a *simple magnifier.* The virtual image is most comfortable to view when it is placed at infinity, so that the ciliary muscle of the eye is relaxed; in the following discussion we assume that this is done.

In Fig. 36–13a the object is at the near point, where it subtends an angle θ at the eye. In Fig. 36–13b a magnifier in front of the eye forms an image at infinity, and the angle subtended at the magnifier is θ'. The usefulness of the magnifier is given by the ratio of the angle θ' (with the magnifier) to the angle θ (without the magnifier). This ratio is called the **angular magnification** M:

$$M = \frac{\theta'}{\theta} \qquad \text{(angular magnification).} \qquad (36\text{–}2)$$

36–13 (a) The subtended angle θ (angular size) is largest when the object is at the near point. (b) The magnifier gives a virtual image at infinity. This virtual image appears to the eye to be a real object subtending a larger angle θ' at the eye.

CAUTION ▶ Don't confuse the angular magnification M with the *lateral* magnification m. Angular magnification is the ratio of the *angular* size of an image to the angular size of the corresponding object; lateral magnification refers to the ratio of the *height* of an image to the height of the corresponding object. For the situation shown in Fig. 36–13b, the angular magnification is about $3\times$, since the insect subtends an angle about three times larger than that in Fig. 36–13a; hence the insect will look about three times

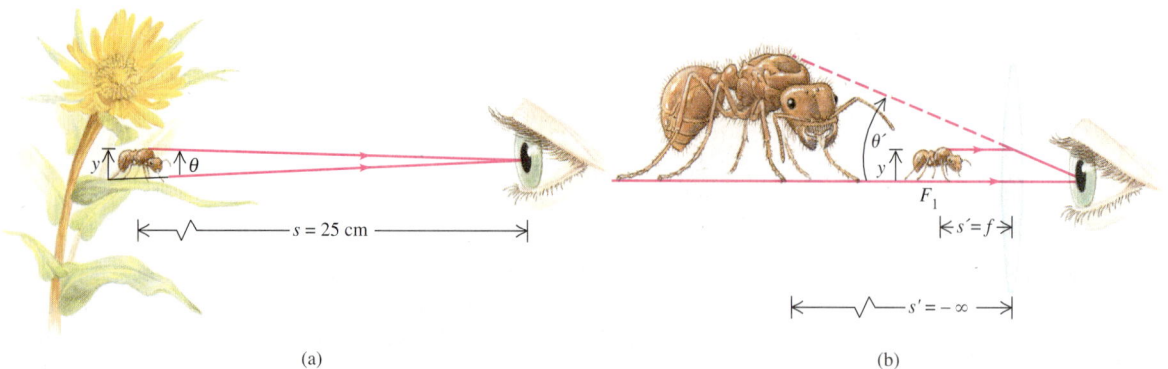

(a)   (b)

larger to the eye. The *lateral* magnification $m = -s'/s$ in Fig. 36–13b is *infinite* because the virtual image is at infinity, but that doesn't mean that the insect looks infinitely large through the magnifier! (That's why we didn't attempt to draw an infinitely large ant in Fig. 36–13.) When dealing with a magnifier, M is useful but m is not. ◄

To find the value of M, we first assume that the angles are small enough that each angle (in radians) is equal to its sine and its tangent. Using Fig. 36–13a and drawing the ray in Fig. 36–13b that passes undeviated through the center of the lens, we find that θ and θ' (in radians) are

$$\theta = \frac{y}{25\ \text{cm}}, \qquad \theta' = \frac{y}{f}.$$

Combining these expressions with Eq. (36–2), we find

$$M = \frac{\theta'}{\theta} = \frac{y/f}{y/25\ \text{cm}} = \frac{25\ \text{cm}}{f} \qquad \text{(angular magnification for a simple magnifier).} \qquad (36\text{–}3)$$

It may seem that we can make the angular magnification as large as we like by decreasing the focal length f. In fact, the aberrations of a simple double-convex lens (to be discussed in Section 36–7) set a limit to M of about 3× to 4×. If these aberrations are corrected, the angular magnification may be made as great as 20× . When greater magnification than this is needed, we usually use a compound microscope, discussed in the next section.

EXAMPLE 36–5

You have two plastic lenses, one double-convex and the other double-concave, each with a focal length with absolute value 10.0 cm. a) Which lens can you use as a simple magnifier? b) What is the angular magnification?

SOLUTION a) The virtual image formation shown in Fig. 36–13 requires a converging lens (positive focal length), so the double-

convex lens is the one to use (see Fig. 35–29 in Section 35–6). b) From Eq. (36–3) the angular magnification M with this double-convex lens is

$$M = \frac{25\ \text{cm}}{10\ \text{cm}} = 2.5, \quad \text{or} \quad 2.5\times.$$

36–5 THE MICROSCOPE

When we need greater magnification than we can get with a simple magnifier, the instrument that we usually use is the **microscope,** sometimes called a *compound microscope.* The essential elements of a microscope are shown in Fig. 36–14a. To analyze this system, we use the principle that an image formed by one optical element such as a lens or mirror can serve as the object for a second element. We used this principle in Section 35–6 when we derived the thin-lens equation by repeated application of the single-surface refraction equation; we used this principle again in Examples 35–11 and 35–12 (Section 35–7), in which the image formed by a lens was used as the object of a second lens.

The object O to be viewed is placed just beyond the first focal point F_1 of the **objective,** a converging lens that forms a real and enlarged image I (Fig. 36–14b). In a properly designed instrument this image lies just inside the first focal point F_1' of a second converging lens called the **eyepiece** or *ocular.* (The reason why the image should lie just *inside* F_1' is left for you to discover; see Problem 36–36.) The eyepiece acts as a simple magnifier, as discussed in Section 36–4, and forms a final virtual image I' of I. The position of I' may be anywhere between the near and far points of the eye. Both the objective and the eyepiece of an actual microscope are highly corrected compound lenses with several optical elements, but for simplicity we show them here as simple thin lenses.

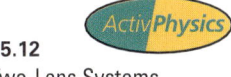

15.12
Two-Lens Systems

36–14 (a) Elements of a microscope. (b) The object O is placed just outside the first focal point of the objective (the distance s_1 has been exaggerated for clarity). The objective forms a real, inverted image inside the first focal point F_1' of the eyepiece. The eyepiece uses that image as an object and forms an enlarged virtual image that remains inverted. Typically, optical microscopes can resolve details as small as 200 nm, about half of the shortest visible wavelength of light.

(a)

(b)

As for a simple magnifier, what matters when viewing through a microscope is the *angular* magnification M. The overall angular magnification of the compound microscope is the product of two factors. The first factor is the *lateral* magnification m_1 of the objective, which determines the linear size of the real image I; the second factor is the *angular* magnification M_2 of the eyepiece, which relates the angular size of the virtual image seen through the eyepiece to the angular size that the real image I would have if you viewed it *without* the eyepiece. The first of these factors is given by

$$m_1 = -\frac{s_1'}{s_1},$$

where s_1 and s_1' are the object and image distances, respectively, for the objective lens. Ordinarily, the object is very close to the focal point, and the resulting image distance s_1' is very great in comparison to the focal length f_1 of the objective lens. Thus s_1 is approximately equal to f_1, and we can write $m_1 = -s_1'/f_1$.

The real image I is close to the focal point F_1' of the eyepiece, so to find the eyepiece angular magnification, we can use Eq. (36–3): $M_2 = (25\ \text{cm})/f_2$, where f_2 is the focal length of the eyepiece (considered as a simple lens). The overall angular magnification M of the compound microscope (apart from a negative sign, which is customarily ignored) is the product of the two magnifications:

$$M = m_1 M_2 = \frac{(25\ \text{cm})s_1'}{f_1 f_2} \qquad \text{(angular magnification for a microscope),} \quad (36\text{–}4)$$

where s_1', f_1, and f_2 are measured in centimeters. The final image is inverted with respect to the object. Microscope manufacturers usually specify the values of m_1 and M_2 for microscope components rather than the focal lengths of the objective and eyepiece.

Equation (36–4) shows that the angular magnification of a microscope can be increased by using an objective of shorter focal length f_1, thereby increasing m_1 and the size of the real image I. Most optical microscopes have a rotating "turret" with three or more objectives of different focal lengths so that the same object can be viewed at different magnifications. The eyepiece should also have a short focal length f_2 to help to maximize the value of M.

36-6 TELESCOPES

The optical system of a **telescope** is similar to that of a compound microscope. In both instruments the image formed by an objective is viewed through an eyepiece. The key difference is that the telescope is used to view large objects at large distances and the microscope is used to view small objects close at hand. Another difference is that many telescopes use a curved mirror, not a lens, as an objective.

An *astronomical telescope* is shown in Fig. 36–15. Because this telescope uses a lens as an objective, it is called a *refracting telescope* or *refractor*. The objective lens forms a real, reduced image I of the object. This image is the object for the eyepiece lens, which forms an enlarged virtual image of I. Objects that are viewed with a telescope are usually so far away from the instrument that the first image I is formed very nearly at the second focal point of the objective lens. If the final image I' formed by the eyepiece is at infinity (for most comfortable viewing by a normal eye), the first image must also be at the first focal point of the eyepiece. The distance between objective and eyepiece, which is the length of the telescope, is therefore the *sum* of the focal lengths of objective and eyepiece, $f_1 + f_2$.

The angular magnification M of a telescope is defined as the ratio of the angle subtended at the eye by the final image I' to the angle subtended at the (unaided) eye by the object. We can express this ratio in terms of the focal lengths of objective and eyepiece. In Fig. 36–15 the ray passing through F_1, the first focal point of the objective, and through F_2', the second focal point of the eyepiece, is shown as a darker color. The object

36–15 Optical system of an astronomical refracting telescope. The objective forms a real, inverted image of the distant object (in this case, the Big Dipper) at its second focal point; this point is also the first focal point of the eyepiece. The eyepiece uses that image as an object to form a magnified virtual image at infinity that remains inverted.

This is the largest refracting telescope ever constructed. The lens is a meter in diameter, and the tube of the telescope is 19.5 m in length. Built in 1897, it is at the University of Chicago's Yerkes Observatory in Wisconsin.

15.3
The Telescope and Angular Magnification

(not shown) subtends an angle θ at the objective and would subtend essentially the same angle at the unaided eye. Also, since the observer's eye is placed just to the right of the focal point F_2', the angle subtended at the eye by the final image is very nearly equal to the angle θ'. Because bd is parallel to the optic axis, the distances ab and cd are equal to each other and also to the height y' of the real image I. Because the angles θ and θ' are small, they may be approximated by their tangents. From the right triangles F_1ab and $F_2'cd$,

$$\theta = \frac{-y'}{f_1}, \qquad \theta' = \frac{y'}{f_2},$$

and the angular magnification M is

$$M = \frac{\theta'}{\theta} = -\frac{y'/f_2}{y'/f_1} = -\frac{f_1}{f_2} \qquad \text{(angular magnification for a telescope). (36–5)}$$

The angular magnification M of a telescope is equal to the ratio of the focal length of the objective to that of the eyepiece. The negative sign shows that the final image is inverted. Equation (36–5) shows that to achieve good angular magnification, a *telescope* should have a *long* objective focal length f_1. By contrast, we found in Section 36–5 that a *microscope* should have a *short* objective focal length. However, a telescope objective with a long focal length should also have a large diameter D so that the f-number f_1/D will not be too large; as described in Section 36–2, a large f-number means a dim, low-intensity image. Telescopes typically do not have interchangeable objectives; instead, the magnification is varied by using different eyepieces with different focal lengths f_2. Just as for a microscope, smaller values of f_2 give larger angular magnifications.

An inverted image is no particular disadvantage for astronomical observations. When we use a telescope or binoculars to view objects on the earth, though, we want the image to be right-side up. Inversion of the image is accomplished in *prism binoculars* by a pair of 45°–45°–90° totally reflecting prisms called *Porro prisms* (introduced in Section 34–4; see Fig. 34–10). These are inserted between objective and eyepiece, as shown in Fig. 36–16. The image is inverted by the four internal reflections from the 45° faces of the prisms. The prisms also have the effect of folding the optical path and making the instrument shorter and more compact than it would otherwise be. Binoculars are usually described by two numbers separated by a multiplication sign, such as 7 × 50. The first number is the angular magnification M, and the second is the diameter of the objective lenses (in millimeters). The diameter helps to determine the light-gathering capacity of the objective lenses and thus the brightness of the image.

In the *reflecting telescope* (Fig. 36–17) the objective lens is replaced by a concave mirror. In large telescopes this scheme has many advantages, both theoretical and practical. Mirrors are inherently free of chromatic aberrations (dependence of focal length on wavelength), and spherical aberrations (associated with the paraxial approximation) are easier to correct than with a lens. The reflecting surface is sometimes parabolic rather

36–16 Inversion of an image in prism binoculars.

Eyepiece

Objective

than spherical. The material of the mirror need not be transparent, and it can be made more rigid than a lens, which has to be supported only at its edges.

The largest reflecting telescopes in the world, the Keck telescopes atop Mauna Kea in Hawaii, each have a mirror of overall diameter 10 m made up of 36 separate hexagonal reflecting elements. Lenses that are larger than 1 m in diameter are usually not practical.

Because the image is formed in a region traversed by incoming rays, this image can be observed directly with an eyepiece only by blocking off part of the incoming beam (Fig. 36–17a); this is practical only for the very largest telescopes. Alternative schemes use a second mirror to reflect the image out the side or through a hole in the primary mirror, as shown in Figs. 36–17b and 36–17c. When a telescope is used for photography, the eyepiece is removed, and either photographic film or an electronic detector is placed at the position of the real image formed by the objective. (Some long-focal-length "lenses" for photography are actually reflecting telescopes used in this way.) Most telescopes used for astronomical research are never used with an eyepiece.

*36–7 LENS ABERRATIONS

An **aberration** is any failure of a mirror or lens to behave precisely according to the simple formulas we have derived. Aberrations can be classified as **chromatic aberrations,** which involve wavelength-dependent imaging behavior, or **monochromatic aberrations,** which occur even with monochromatic (single-wavelength) light. Lens aberrations are not caused by faulty construction of the lens, such as irregularities in its surfaces, but are inevitable consequences of the laws of refraction at spherical surfaces.

Monochromatic aberrations are all related to the *paraxial approximation.* Our derivations of equations for object and image distances, focal lengths, and magnification have all been based on this approximation. We have assumed that all rays are *paraxial,* that is, that they are very close to the optic axis and make very small angles with it. This condition is never obeyed exactly.

For any lens that has an aperture of finite size, the cone of rays that forms the image of any point has a finite size. In general, nonparaxial rays that proceed from a given

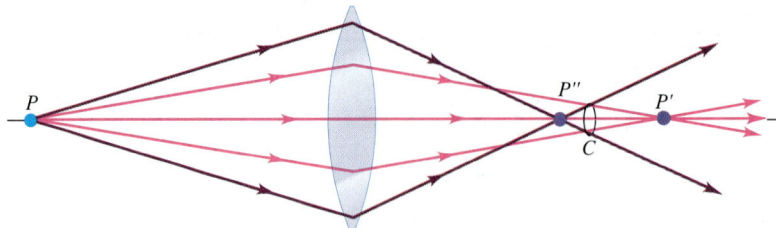

36–18 Spherical aberration for a lens. The circle of least confusion is shown by *C*.

object point *do not* all intersect at precisely the same point after they are refracted by a lens. For this reason, the image that is formed by these rays is never perfectly sharp. *Spherical aberration* is the failure of rays from a point object on the optic axis to converge to a point image. Instead, the rays converge within a circle of minimum radius, called *the circle of least confusion,* and then diverge again, as shown in Fig. 36–18. The corresponding effect for points off the optic axis produces images that are comet-shaped figures rather than circles; this is called *coma.* Note that decreasing the size of the lens aperture (Fig. 36–1a) cuts off the larger-angle rays, thus decreasing spherical aberration.

Spherical aberration is also present in spherical mirrors, as we discussed briefly in Section 35–3. Mirrors that are used in astronomical reflecting telescopes are often parabolic rather than spherical; this shape completely eliminates spherical aberration for points on the axis. Parabolic shapes are much more difficult to fabricate precisely than are spherical shapes. The disappointing results from the Hubble Space Telescope when it was first placed in orbit in 1990 were associated with spherical aberration, the result of errors in measurement during the shaping process (see Fig. 35–10).

Astigmatism is the imaging of a point off the axis as two perpendicular *lines* in different planes. In this aberration the rays from a point object converge at a certain distance from the lens to a line called the *primary image,* which is perpendicular to the plane defined by the optic axis and the object point. At a somewhat different distance from the lens, they converge to a second line, called the *secondary image,* which is *parallel* to this plane. This effect is shown in Fig. 36–19. The circle of least confusion (greatest convergence) appears between these two positions.

The location of the circle of least confusion depends on the object point's *transverse* distance from the axis as well as its *longitudinal* distance from the lens. As a result, object points lying in a plane are in general imaged not in a plane but in some curved surface. This effect is called *curvature of field.*

Finally, the image of a straight line that does not pass through the optic axis may be curved. As a result, the image of a square with the axis through its center may resemble

36–19 Astigmatism of a lens for a point below the optic axis. The lens forms two images of the point, in planes perpendicular to each other.

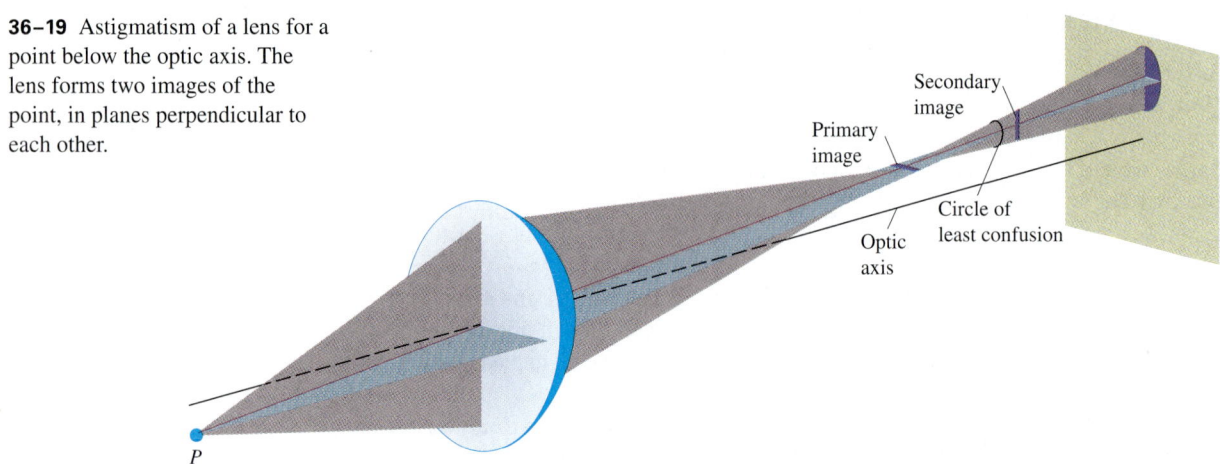

a barrel (sides bent outward) or a pincushion (sides bent inward). This effect, called *distortion,* is not related to lack of sharpness of the image but results from a change in lateral magnification with distance from the axis.

Chromatic aberrations are a result of *dispersion,* the variation of index of refraction with wavelength that we discussed in Section 34–5. Dispersion causes the focal length of a lens to be somewhat different for different wavelengths, so different wavelengths are imaged at different points. The magnification of a lens also varies with wavelength; this effect is responsible for the rainbow-fringed images seen with inexpensive binoculars or telescopes. Mirrors are inherently free of chromatic aberrations, which is one of the reasons for their usefulness in large astronomical telescopes.

EXAMPLE 36-6

Chromatic aberration A glass plano-convex lens has its flat side toward the object. The other side has a radius of curvature of 30.0 cm. The index of refraction of the glass for violet light (wavelength 400 nm) is 1.537, and that for red light (700 nm) is 1.517. The color purple is a mixture of red and violet. If a purple object is placed 80.0 cm from this lens, where are the red and violet images formed?

SOLUTION We use the thin-lens equation in the form given by Eq. (35–18):

$$\frac{1}{s} + \frac{1}{s'} = (n-1)\left(\frac{1}{R_1} - \frac{1}{R_2}\right).$$

In this case, using the sign rules described in Section 35–2, we have $R_1 = \infty$ and $R_2 = -30.0$ cm. For violet light ($n = 1.537$),

$$\frac{1}{80.0 \text{ cm}} + \frac{1}{s'} = (1.537 - 1)\left(\frac{1}{\infty} - \frac{1}{-30.0 \text{ cm}}\right),$$

$$s' = 185 \text{ cm}.$$

For red light ($n = 1.517$) we find $s' = 211$ cm. The violet light is refracted more than the red, and its image is formed closer to the lens. We see that a rather small variation in index of refraction causes a substantial displacement of the image.

SUMMARY

- A camera forms a real, inverted, usually reduced image (on film) of the object being photographed. The amount of light striking the film is controlled by the shutter speed and the aperture. The projector is essentially a camera in reverse; a lens forms a real, inverted, enlarged image (on a screen) of the slide or motion-picture film.

- The intensity of light that reaches the film through a camera lens, or the screen through a projector lens, is inversely proportional to the square of the *f*-number of the lens:

$$f\text{-number} = \frac{\text{Focal length}}{\text{Aperture diameter}} = \frac{f}{D}. \tag{36-1}$$

- In the eye, a real image is formed on the retina and is transmitted to the optic nerve. Adjustment for various object distances is made by the ciliary muscle, which squeezes the lens, making it bulge and decreasing its focal length. A nearsighted eye is too long for its cornea and lens; a farsighted eye is too short. The power of a corrective lens, in diopters, is the reciprocal of the focal length in meters.

- The simple magnifier creates a virtual image whose angular size θ' is larger than the angular size θ of the object itself at a distance of 25 cm, the nominal closest distance for comfortable viewing. The angular magnification produced by a simple magnifier is the ratio of the angular size of the virtual image to that of the object at this distance:

$$M = \frac{\theta'}{\theta}. \tag{36-2}$$

For a simple magnifier of focal length f, $M = (25 \text{ cm})/f$.

KEY TERMS

camera, 1118
f-number, 1120
projector, 1121
power, 1125
diopter, 1125
angular size, 1126
magnifier, 1126
angular magnification, 1126
microscope, 1127
objective, 1127
eyepiece, 1127
telescope, 1129
aberration, 1131
chromatic aberration, 1131
monochromatic aberration, 1131

- In the compound microscope the objective lens forms a first image in the barrel of the instrument, and the eyepiece forms a final virtual image, often at infinity, of the first image. The telescope operates on the same principle, but the object is far away. In a reflecting telescope the objective lens is replaced by a concave mirror, which eliminates chromatic aberrations.

- Lens aberrations describe the failure of a lens to form a perfectly sharp image of an object. Monochromatic aberrations occur because of limitations of the paraxial approximation; chromatic aberrations result from the dependence of index of refraction on wavelength.

DISCUSSION QUESTIONS

Q36–1 When a slide projector is turned on without a slide in it, and the focus adjustment is moved far enough in one direction, a gigantic image of the light bulb filament can be seen on the screen. Explain how this happens.

Q36–2 The human eye is often compared to a camera. In what ways is it similar to a camera? In what ways is it different?

Q36–3 How does a person judge distance? Can a person with vision in only one eye judge distance? What is meant by "binocular vision"?

Q36–4 While lost in the mountains, a person who is very farsighted in one eye and barely farsighted in the other made a crude emergency telescope from the two lenses of his eyeglasses. How did he do this?

Q36–5 Nearsightedness can be corrected by surgical reshaping of the cornea (a process called *radial keratotomy*). Why is it the cornea, not the lens, that is reshaped?

Q36–6 You can't see clearly underwater with the naked eye, but you *can* if you wear a face mask or goggles (with air between your eyes and the mask or goggles). Why is there a difference? Could you instead wear eyeglasses (with water between your eyes and the eyeglasses) in order to see underwater? If so, should the lenses be converging or diverging? Explain.

Q36–7 You are marooned on a desert island and want to use your eyeglasses to start a fire. Can this be done if you are near-sighted? If you are farsighted? Explain.

Q36–8 In using a magnifying glass, is the magnification greater when the glass is close to the object or when it is close to the eye? Explain.

Q36–9 As was discussed in the text, some binoculars use prisms to invert the final image. Why are prisms better than ordinary mirrors for this purpose?

EXERCISES

SECTION 36–2 CAMERAS AND PROJECTORS

36–1 Choosing a Camera Lens. The picture size on ordinary 35-mm camera film is 24 mm × 36 mm. Focal lengths of lenses available for 35-mm cameras typically include 28, 35, 50 (the "normal" lens), 85, 100, 135, 200, and 300 mm, among others. Which of these lenses should be used to photograph the following objects, assuming that the object is to fill most of the picture area? a) A building 240 m tall and 160 m wide at a distance of 600 m. b) A mobile home 9.6 m in length at a distance of 40.0 m.

36–2 When a camera is focused, the lens is moved away from or toward the film. If you take a picture of your friend, who is standing 3.90 m from the lens, using a camera with a lens with a 85-mm focal length, how far from the film is the lens? Will the whole image of your friend, who is 175 cm tall, fit on film that is 24 × 36 mm?

36–3 A camera lens has a focal length of 200 mm. How far from the lens should the subject for the photo be if the lens is 20.4 cm from the film?

36–4 A photographer takes a photograph of a Boeing 747 air-liner (length 70.7 m) when it is flying directly overhead at an altitude of 9.50 km. The lens has a focal length of 5.00 m. How long is the image of the airliner on the film?

36–5 Figure 36–2 shows photographs of the same scene taken with the same camera with lenses of different focal length. If the object is 200 m from the lens, what is the magnitude of the lateral magnification for a lens of focal length a) 28 mm? b) 105 mm? c) 300 mm?

36–6 Consider the simple model of a zoom lens shown in Fig. 36–4a. The converging lens has focal length $f_1 = 12$ cm and the diverging lens has focal length $f_2 = -12$ cm. The lenses are separated by 4 cm as shown in Fig. 36–4a. a) For a distant object, where is the image of the converging lens? b) The image of the converging lens serves as the object for the diverging lens. What is the object distance for the diverging lens? c) Where is the final image? Compare your answer to Fig. 36–4a. d) Repeat parts (a), (b), and (c) for the situation of Fig. 36–4b, in which the lenses are separated by 8 cm.

36–7 The focal length of an $f/4$ camera lens is 300 mm. a) What is the aperture diameter of the lens? b) If the correct exposure of a certain scene is $\frac{1}{250}$ s at $f/4$, what is the correct exposure at $f/8$?

36–8 Camera A has a lens with an aperture diameter of 8.00 mm. It photographs an object using the correct exposure time of $\frac{1}{30}$ s. What exposure time should be used with camera B in photographing the same object with the same film if this camera has a lens with an aperture diameter of 23.1 mm?

36–9 In Example 36–2 (Section 36–2) for a projector lens with a 98.0-mm focal length it was found that the slide needs to be placed 10.0 cm in front of the lens to focus the image on a screen 5.00 m from the lens. Now assume instead that a lens of 200-mm focal length is used. a) If the screen remains 5.00 m from the lens, how far from the lens should the slide be? b) Could the distance between the slide and lens be kept at 10.0 cm and the screen moved to achieve focus? If so, how far and in what direction would it have to be moved?

36–10 The dimensions of the picture on a 35-mm color slide are 24 mm × 36 mm. An image of the slide is projected on a screen 9.00 m from the projector lens. The focal length of the projector lens is 150 mm. a) How far is the slide from the lens? b) What are the dimensions of the image on the screen?

SECTION 36–3 THE EYE

36–11 a) Where is the near point of an eye for which a contact lens with a power of +2.75 diopters is prescribed? b) Where is the far point of an eye for which a contact lens with a power of −1.30 diopters is prescribed for distant vision?

36–12 Curvature of the Cornea. In a simplified model of the human eye, the aqueous and vitreous humors and the lens all have a refractive index of 1.40, and all the refraction occurs at the cornea, whose vertex is 2.60 cm from the retina. What should be the radius of curvature of the cornea such that the image of an object 40.0 cm from the cornea's vertex is focused on the retina?

36–13 Determine the power of the corrective contact lenses required by a) a hyperopic eye whose near point is at 60.0 cm; b) a myopic eye whose far point is at 60.0 cm.

SECTION 36–4 THE MAGNIFIER

36–14 A thin lens with a focal length of 6.00 cm is used as a simple magnifier. a) What angular magnification is obtainable with the lens if the object is at the focal point? b) When an object is examined through the lens, how close may it be brought to the lens? Assume that the image viewed by the eye is at the near point, 25.0 cm from the eye, and that the lens is very close to the eye.

36–15 The focal length of a simple magnifier is 8.00 cm. Assume the magnifier to be a thin lens placed very close to the eye. a) How far in front of the magnifier should an object be placed if the image is formed at the observer's near point, 25.0 cm in front of her eye? b) If the object is 1.00 mm high, what is the height of its image formed by the magnifier?

36–16 You want to view an insect 2.00 mm in length through a magnifier. If the insect is to be at the focal point of the magnifier, what focal length will give the image of the insect an angular size of 0.025 radian?

36–17 You are examining a flea with a magnifying lens that has focal length 4.00 cm. If the image of the flea is 6.50 times the size of the flea, how far is the flea from the lens? Where, relative to the lens, is the image?

SECTION 36–5 THE MICROSCOPE

36–18 Resolution of a Microscope. The image formed by a microscope objective with a focal length of 5.00 mm is 160 mm from its second focal point. The eyepiece has a focal length of 26.0 mm. a) What is the angular magnification of the microscope? b) The unaided eye can distinguish two points at its near point as separate if they are about 0.10 mm apart. What is the minimum separation that can be resolved with this microscope?

36–19 The focal length of the eyepiece of a certain microscope is 18.0 mm. The focal length of the objective is 8.00 mm. The distance between objective and eyepiece is 19.7 cm. The final image formed by the eyepiece is at infinity. Treat all lenses as thin. a) What is the distance from the objective to the object being viewed? b) What is the magnitude of the linear magnification produced by the objective? c) What is the overall angular magnification of the microscope?

36–20 A certain microscope is provided with objectives that have focal lengths of 16 mm, 4 mm, and 1.9 mm and with eyepieces that have angular magnifications of 5× and 10×. Each objective forms an image 120 mm beyond its second focal point. Determine a) the largest overall angular magnification obtainable; b) the least overall angular magnification obtainable.

SECTION 36–6 TELESCOPES

36–21 A telescope is constructed from two lenses with focal lengths of 95.0 cm and 15.0 cm, the 95.0-cm lens being used as the objective. Both the object being viewed and the final image are at infinity. a) Find the angular magnification for the telescope. b) Find the height of the image formed by the objective of a building 60.0 m tall, 3.00 km away. c) What is the angular size of the final image as viewed by an eye very close to the eyepiece?

36–22 The eyepiece of a refracting telescope (Fig. 36–15) has a focal length of 9.00 cm. The distance between objective and eyepiece is 1.80 m, and the final image is at infinity. What is the angular magnification of the telescope?

36–23 The Yerkes refracting telescope of the University of Chicago has an objective 1.02 m in diameter with an f-number of 19.0. (This is the largest-diameter refracting telescope in the world.) What is its focal length?

36–24 In catalogs of telescopes for amateur astronomers, the telescope diameter and f-number are usually specified, but the focal length is not. Determine the focal length of each of the following telescopes: a) $D = 15$ cm, $f/8$; b) $D = 20$ cm, $f/6$; c) $D = 40$ cm, $f/4.5$. d) Which telescope gives the narrowest angle of view? The widest angle of view? Which gives the brightest image? The dimmest? Explain your answers.

36–25 A reflecting telescope (Fig. 36–20a; see page 1136) is to be made by using a spherical mirror with a radius of curvature of 1.30 m and an eyepiece with a focal length of 1.10 cm. The final image is at infinity. a) What should be the distance between the eyepiece and the mirror vertex if the object is taken to be at infinity? b) What will be the angular magnification?

36–26 A reflecting telescope uses an optical system similar to that shown in Fig. 36–20b (Cassegrain system). The image of a distant galaxy is focused on the detector through a hole in the large (primary) mirror. The primary mirror has a focal length of 1.3 m, the distance between the vertexes of the two mirrors is

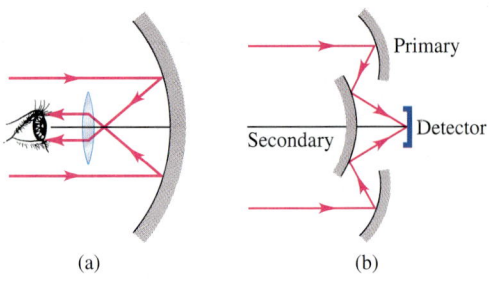

FIGURE 36–20 Exercises 36–25 and 36–26, and Problem 36–40.

75 cm, and the distance from the vertex of the primary mirror to the detector is 12 cm. Should the smaller mirror (the secondary) be concave or convex? What should its radius of curvature be?

36–27 As viewed from the earth, Jupiter subtends an angle of approximately 0.014°. What is the diameter of the image of Jupiter produced by the objective of the Lick Observatory refracting telescope of focal length 18 m?

*SECTION 36–7 LENS ABERRATIONS

*36–28 A diverging double-concave lens (Fig. 35–29b) is ground from the glass of Example 36–6 (Section 36–7). The two radii of curvature each have magnitude 12.0 cm. The purple object of Example 36–6 is placed 25.0 cm to the left of the lens. Where are the red and violet images formed, and what is the distance between these two images?

PROBLEMS

36–29 Your camera has a lens with a 35.0-mm focal length and film that is 36.0 mm wide. In taking a picture of an airplane 22.7 m in length you find that the image of the airplane fills only three quarters of the width of the film. a) How far are you from the airplane? b) How close should you stand if you want to fill the width of the film with the airplane's image?

36–30 Resolution of a Camera. The *resolution* of a camera lens can be defined as the maximum number of lines per millimeter in the image that can barely be distinguished as separate lines. A certain lens has a focal length of 35.0 mm and resolution of 120 lines/mm. What is the minimum separation of two lines in an object 25.0 m away if they are to be visible in the image as separate lines?

36–31 Angle of View. The angle of view for a camera and lens is defined as the angle subtended by the widest object for which the width of the image equals the diagonal dimension d of the film. a) For a distant object for which the image distance is approximately equal to the focal length of the camera lens, show that the angle of view θ is given by $\theta = 2\arctan(d/2f)$. b) The film for a 35-mm camera is 24 mm by 36 mm. Use the result of part (a) to calculate the angle of view for focal lengths of 28 mm, 105 mm, and 300 mm. Compare your results to the angles of view given in Fig. 36–2d.

36–32 a) Show that when two thin lenses are placed in contact, the *power* of the combination in diopters, as defined in Section 36–3, is the sum of the powers of the separate lenses. Is this relation valid even when one lens has positive power and the other negative? b) Two thin converging lenses with 12.0-cm and 24.0-cm focal lengths are in contact. What is the power of the combination?

36–33 You place an object at the focal point of a converging lens. Prove that the angular magnification of the lens is one quarter of the value of its power in diopters.

36–34 A camera with a 90-mm focal length lens is focused on an object 1.30 m from the lens. To refocus on an object 6.50 m from the lens, by how much must the distance between the lens and the film be changed? To refocus on the more distant object, is the lens moved toward or away from the film?

36–35 The derivation of the expression for angular magnification, Eq. (36–2), assumed a near point of 25 cm. In fact, the near point changes with age as shown in Table 36–1. In order to achieve an angular magnification of 2.0×, what focal length should be used by a person of a) age 10? b) age 30? c) age 60? d) If the lens that gives $M = 2.0$ for a 10-year-old is used by a 60-year-old, what angular magnification will the older viewer obtain? e) Does your answer in (d) mean that older viewers are able to see more highly magnified images than younger viewers? Explain.

36–36 In deriving Eq. (36–2) for the angular magnification of a magnifier, we assumed that the object is placed at the focal point of the magnifier so that the virtual image is formed at infinity. Suppose instead that the object is placed so that the virtual image appears at an average viewer's near point of 25 cm, the closest point at which the viewer can bring an object into focus. a) Where should the object be placed to achieve this? Give your answer in terms of the magnifier focal length f. b) What angle θ' will an object of height y subtend at the position found in part (a)? c) Find the angular magnification M with the object at the position found in part (a). The angle θ is the same as in Fig. 36–13a, since it refers to viewing the object *without* the magnifier. d) For the double-convex lens described in Example 36–5 (Section 36–4), what is the value of M with the object at the position found in part (a)? How many times greater is M in this case than in the case where the image is formed at infinity? e) In the description of a compound microscope in Section 36–5, it is stated that in a properly designed instrument, the real image formed by the objective lies *just inside* the first focal point F_1' of the eyepiece. What advantages are gained by having the image formed by the objective be just inside F_1', as opposed to precisely at F_1'? What happens if the image formed by the objective is *just outside* F_1'?

36–37 In one form of cataract surgery the person's natural lens, which has become cloudy, is replaced by an artificial lens. The refracting properties of the replacement lens can be chosen so that the person's eye focuses on distant objects. But there is no accommodation, and glasses or contact lenses are needed for close vision. What is the power, in diopters, of the corrective contact lenses that will enable a person who has had such surgery to focus on the page of a book at a distance of 24 cm?

36–38 A Nearsighted Eye. A certain very nearsighted person cannot focus on anything farther than 36.0 cm from the eye. Consider the simplified model of the eye described in Exercise 36–12. If the radius of curvature of the cornea is 0.75 cm when the eye is focusing on an object 36.0 cm from the cornea vertex and the indexes of refraction are as described in Exercise 36–12, what is the distance from the cornea vertex to the retina? What does this tell you about the shape of the nearsighted eye?

36–39 Focal Length of a Zoom Lens. Figure 36–21 shows a simple version of a zoom lens. The converging lens has focal length f_1, and the diverging lens has focal length $f_2 = -|f_2|$. The two lenses are separated by a variable distance d that is always less than f_1. Also, the magnitude of the focal length of the diverging lens satisfies the inequality $|f_2| > (f_1 - d)$. To determine the effective focal length of the combination lens, consider a bundle of parallel rays of radius r_0 entering the converging lens. a) Show that the radius of the ray bundle decreases to $r_0' = r_0(f_1 - d)/f_1$ at the point that it enters the diverging lens. b) Show that the final image I' is formed a distance $s_2' = |f_2|(f_1 - d)/(|f_2| - f_1 + d)$ to the right of the diverging lens. c) If the rays that emerge from the diverging lens and reach the final image point are extended backward to the left of the diverging lens, they will eventually expand to the original radius r_0 at some point Q. The distance from the final image I' to the point Q is the *effective focal length f* of the lens combination; if the combination were replaced by a single lens of focal length

f placed at Q, parallel rays would still be brought to a focus at I'. Show that the effective focal length is given by $f = f_1 |f_2|/(|f_2| - f_1 + d)$. d) If $f_1 = 12.0$ cm, $f_2 = -18.0$ cm, and the separation d is adjustable between zero and 4.0 cm, find the maximum and minimum focal lengths of the combination. What value of d gives $f = 30.0$ cm?

36–40 A certain reflecting telescope, constructed as in Fig. 36–20a, has a spherical mirror with a radius of curvature of 96.0 cm and an eyepiece with a focal length of 1.20 cm. If the angular magnification has a magnitude of 36 and the object is at infinity, find the position of the eyepiece and the position and nature (real or virtual) of the final image. (*Note:* |M| is *not* equal to $|f_1/f_2|$, so the image formed by the eyepiece is *not* at infinity.)

36–41 A microscope with an objective of focal length 8.00 mm and an eyepiece of focal length 7.50 cm is used to project an image on a screen 2.00 m from the eyepiece. Let the image distance of the objective be 18.0 cm. a) What is the lateral magnification of the image? b) What is the distance between the objective and the eyepiece?

36–42 The Galilean Telescope. Figure 36–22 is a diagram of a *Galilean telescope*, or *opera glass*, with both the object and its final image at infinity. The image I serves as a virtual object for the eyepiece. The final image is virtual and erect. a) Prove that the angular magnification is $M = -f_1/f_2$. b) A Galilean telescope is to be constructed with the same objective lens as in Exercise 36–21. What focal length should the eyepiece have if this telescope is to have the same magnitude of angular magnification as the one in Exercise 36–21? c) Compare the lengths of the telescopes.

FIGURE 36–21 Problem 36–39.

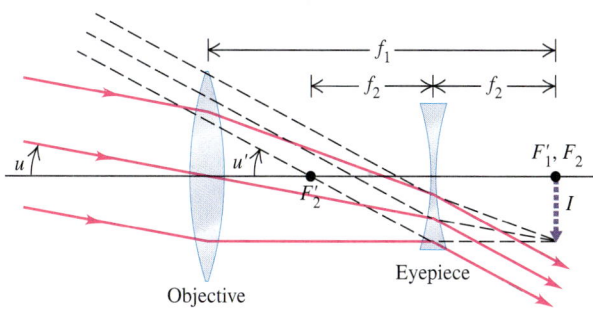

FIGURE 36–22 Problem 36–42.

CHALLENGE PROBLEM

36–43 People with normal vision cannot focus their eyes underwater if they aren't wearing a face mask or goggles and there is water in contact with their eyes (see Question Q36–6). a) Why not? b) With the simplified model of the eye described in Exercise 36–12, what corrective lens (specified by focal length as measured in air) would be needed to enable a person under-

water to focus an infinitely distant object? (Be careful—the focal length of a lens underwater is *not* the same as in air! See Problem 35–72. Assume that the corrective lens has a refractive index of 1.62 and that the lens is used in eyeglasses, not goggles, so there is water on both sides of the lens. Assume that the eyeglasses are 2.00 cm in front of the eye.)

37

Interference

The Very Large Array is a set of 27 radio telescopes in New Mexico. Because the telescopes are at slightly different distances from the object being observed, there are phase differences between the signals received by different telescopes. These differences create an interference pattern that can be decoded to determine the object's structure.

37-1 INTRODUCTION

An ugly black oil spot on the pavement can become a thing of beauty after a rain, when the oil reflects a rainbow of colors. Multicolored reflections can also be seen from the surfaces of soap bubbles and compact discs. These familiar sights give us a hint that there are aspects of light that we haven't yet explored.

In our discussion of lenses, mirrors, and optical instruments we used the model of *geometric optics,* in which we represent light as *rays,* straight lines that are bent at a reflecting or refracting surface. But many aspects of the behavior of light *can't* be understood on the basis of rays. We have already learned that light is fundamentally a *wave,* and in some situations we have to consider its wave properties explicitly. If two or more light waves of the same frequency overlap at a point, the total effect depends on the *phases* of the waves as well as their amplitudes. The resulting patterns of light are a result of the *wave* nature of light and cannot be understood on the basis of rays. Optical effects that depend on the wave nature of light are grouped under the heading **physical optics.**

In this chapter we'll look at *interference* phenomena that occur when two waves combine. The colors seen in oil films and soap bubbles are a result of interference between light reflected from the front and back surfaces of a thin film of oil or soap solution. Effects that occur when *many* sources of waves are present are called *diffraction* phenomena; we'll study these in Chapter 38. In that chapter we'll see that diffraction effects occur whenever a wave passes through an aperture or around an obstacle. They are important in practical applications of physical optics such as diffraction gratings, x-ray diffraction, and holography.

While our primary concern is with light, interference and diffraction can occur with waves of *any* kind. As we go along, we'll point out applications to other types of waves such as sound and water waves.

37-2 INTERFERENCE AND COHERENT SOURCES

As we discussed in Chapter 20, the term **interference** refers to any situation in which two or more waves overlap in space. When this occurs, the total wave at any point at any instant of time is governed by the **principle of superposition,** which we introduced in Section 20–2 in the context of waves on a string. This principle also applies to electromagnetic waves and is the most important principle in all of physical optics, so make sure you understand it well. The principle of superposition states: **When two or more waves overlap, the resultant displacement at any point and at any instant may be found by adding the instantaneous displacements that would be produced at the point by the individual waves if each were present alone.** (In some special physical situations, such as electromagnetic waves propagating in a crystal, this principle may not apply. A discussion of these is beyond our scope.)

We use the term *displacement* in a general sense. With waves on the surface of a liquid, we mean the actual displacement of the surface above or below its normal level. With sound waves, the term refers to the excess or deficiency of pressure. For electromagnetic waves, we usually mean a specific component of electric or magnetic field.

We have already discussed one important case of interference, in which two identical waves propagating in opposite directions combine to produce a *standing wave*. We saw this in Chapter 20 for transverse waves on a string and for longitudinal waves in a fluid filling a pipe; we described the same phenomenon for electromagnetic waves in Section 33–7. In all of these cases the waves propagated along only a single axis: along a string, along the length of a fluid-filled pipe, or along the propagation direction of an electromagnetic plane wave. But light waves can (and do) travel in two or three dimensions, as can any kind of wave that propagates in a two- or three-dimensional medium. In this section we'll see what happens when we combine waves that spread out in two or three dimensions from a pair of identical wave sources.

Interference effects are most easily seen when we combine *sinusoidal* waves with a single frequency f and wavelength λ. Figure 37–1 shows a "snapshot" or "freeze-frame" of a *single* source S_1 of sinusoidal waves and some of the wave fronts produced by this source. The figure shows only the wave fronts corresponding to wave *crests*, so the spacing between successive wave fronts is one wavelength. The material surrounding S_1 is uniform, so the wave speed is the same in all directions, and there is no refraction (and hence no bending of the wave fronts). If the waves are two-dimensional, like waves on the surface of a liquid, the circles in Fig. 37–1 represent circular wave fronts; if the waves propagate in three dimensions, the circles represent spherical wave fronts spreading away from S_1.

In optics, sinusoidal waves are characteristic of **monochromatic light** (light of a single color). While it's fairly easy to make water waves or sound waves of a single frequency, common sources of light *do not* emit monochromatic (single-frequency) light. For example, incandescent light bulbs and flames emit a continuous distribution of wavelengths. However, there are several ways to produce *approximately* monochromatic light. For example, some filters block all but a very narrow range of wavelengths. Gas-discharge lamps, such as the mercury-vapor lamp, emit light with a discrete set of colors, each having a narrow band of wavelengths. The bright green line in the spectrum of a mercury-vapor lamp has a wavelength of about 546.1 nm, with a spread of the order of ±0.001 nm. By far the most nearly monochromatic source that is available at present is the *laser*. The familiar helium-neon laser, which is inexpensive and readily available, emits red light at 632.8 nm with a wavelength range of the order of ±0.000001 nm, or about one part in 10^9. As we analyze interference and diffraction effects in this chapter and the next, we will assume that we are working with monochromatic waves (unless we explicitly state otherwise).

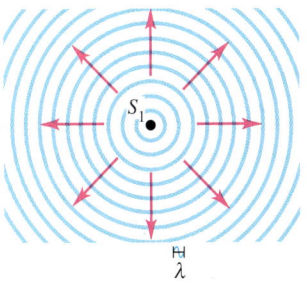

37–1 A "snapshot" of sinusoidal waves of frequency f and wavelength λ spreading out from source S_1 in all directions. The concentric circles are wave fronts representing crests of the wave and are separated by one wavelength. As time passes, the wave fronts spread away from S_1 at the wave speed, equal to the product of f and λ.

CONSTRUCTIVE AND DESTRUCTIVE INTERFERENCE

Two identical sources of monochromatic waves, S_1 and S_2, are shown in Fig. 37–2a. The two sources produce waves of the same amplitude and the same wavelength λ. In addition, the two sources are permanently *in phase;* they vibrate in unison. They might be two synchronized agitators in a ripple tank, two loudspeakers driven by the same amplifier, two radio antennas powered by the same transmitter, or two small holes or slits in an opaque screen, illuminated by the same monochromatic light source. We will see that if there were not a constant phase relationship between the two sources, the phenomena we are about to discuss would not occur. Two monochromatic sources of the same frequency and with any definite, constant phase relation (not necessarily in phase) are said to be **coherent.** We also use the term *coherent waves* (or, for light waves, *coherent light*) to refer to the waves emitted by two such sources.

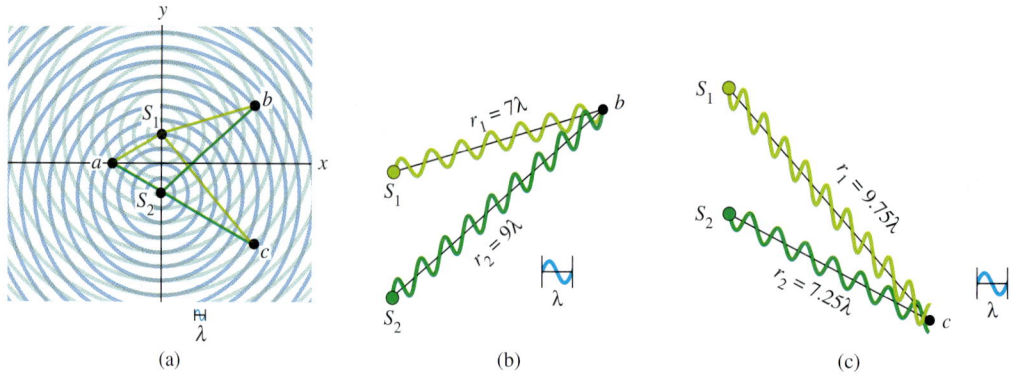

37–2 (a) A "snapshot" of sinusoidal waves spreading out from two coherent sources S_1 and S_2. The sources are in phase. In this example the distance between sources is 4λ. The light green and dark green lines indicate the distances from S_1 and S_2, respectively, to the points a, b, and c. (b) Constructive interference occurs at points where the path difference (the difference in distance from the two sources to the point) is a whole number of wavelengths, $m\lambda$. (c) Destructive interference occurs at points where the path difference is a half-integral number of wavelengths, $\left(m + \frac{1}{2}\right)\lambda$.

If the waves emitted by the two sources are *transverse,* like electromagnetic waves, then we will also assume that the wave disturbances produced by both sources have the same *polarization* (that is, they lie along the same line). For example, the sources S_1 and S_2 in Fig. 37–2a could be two radio antennas in the form of long rods oriented parallel to the z-axis (perpendicular to the plane of the figure); then at any point in the xy-plane the waves produced by both antennas have $\vec{E}$ fields with only a z-component. (We discussed such *dipole antennas* and the waves that they emit in Section 33–9.) Then we need only a single scalar function to describe each wave; this makes the analysis much easier.

We position the two sources of equal amplitude, equal wavelength, and (if the waves are transverse) the same polarization along the y-axis in Fig. 37–2a, equidistant from the origin. Consider a point a on the x-axis. From symmetry the two distances from S_1 to a and from S_2 to a are *equal;* waves from the two sources thus require equal times to travel to a. Hence waves that leave S_1 and S_2 in phase arrive at a in phase. The two waves add, and the total amplitude at a is *twice* the amplitude of each individual wave. This is true for *any* point on the x-axis.

Similarly, the distance from S_2 to point b is exactly two wavelengths *greater* than the distance from S_1 to b. A wave crest from S_1 arrives at b exactly two cycles earlier than a crest emitted at the same time from S_2, and again the two waves arrive in phase. As at point a, the total amplitude is the sum of the amplitudes of the waves from S_1 and S_2.

In general, when waves from two or more sources arrive at a point *in phase,* the amplitude of the resultant wave is the *sum* of the amplitudes of the individual waves; the individual waves reinforce each other. This is called **constructive interference** (Fig. 37–2b). Let the distance from S_1 to any point P be r_1, and let the distance from S_2 to P be r_2. For constructive interference to occur at P, the path difference $r_2 - r_1$ for the two sources must be an integral multiple of the wavelength λ:

$$r_2 - r_1 = m\lambda \quad (m = 0, \pm 1, \pm 2, \pm 3, \ldots) \quad \text{(constructive interference,} \quad (37\text{–}1)$$
$$\text{sources in phase).}$$

In Fig. 37–2a, points a and b satisfy Eq. (37–1) with $m = 0$ and $m = +2$, respectively.

Something different occurs at point c in Fig. 37–2a. At this point, the path difference $r_2 - r_1 = -2.5\lambda$, which is a *half-integral* number of wavelengths. Waves from the two

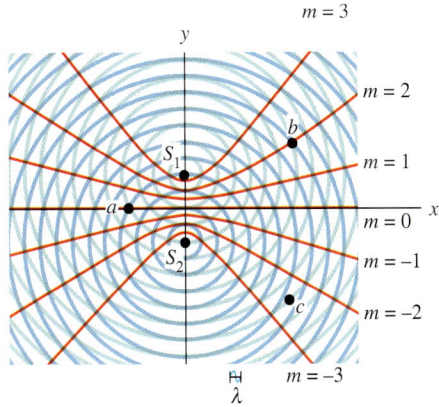

37–3 The same as Fig. 37–2a, but with red antinodal curves superimposed (curves of maximum amplitude). All points on each curve satisfy Eq. (37–1) with the value of m shown. The nodal curves, which lie between each adjacent pair of antinodal curves, are not shown.

sources arrive at point c exactly a half-cycle out of phase. A crest of one wave arrives at the same time as a crest in the opposite direction (a "trough") of the other wave (Fig. 37–2c). The resultant amplitude is the *difference* between the two individual amplitudes. If the individual amplitudes are equal, then the total amplitude is *zero!* This cancellation or partial cancellation of the individual waves is called **destructive interference.** The condition for destructive interference in the situation shown in Fig. 37–2a is

$$r_2 - r_1 = \left(m + \frac{1}{2}\right)\lambda \quad (m = 0, \pm 1, \pm 2, \pm 3, \dots) \quad \text{(destructive interference,} \quad (37\text{–}2)\text{)} \\ \text{sources in phase).}$$

The path difference at point c in Fig. 37–2a satisfies Eq. (37–2) with $m = -3$.

Figure 37–3 shows the same situation as in Fig. 37–2a, but with red curves that denote all points on which *constructive* interference occurs. On each curve, the path difference $r_2 - r_1$ is equal to an integer m times the wavelength, as in Eq. (37–1). These curves are called **antinodal curves.** They are directly analogous to *antinodes* in the standing-wave patterns described in Chapter 20 and Section 33–7. In a standing wave formed by interference between waves propagating in opposite directions, the antinodes are points at which the amplitude is maximum; likewise, the wave amplitude in the situation of Fig. 37–3 is maximum along the antinodal curves. Not shown in Fig. 37–3 are the **nodal curves,** which are the curves denoting points on which *destructive* interference occurs in accordance with Eq. (37–2); these are analogous to the *nodes* in a standing wave pattern. A nodal curve lies between each two adjacent antinodal curves in Fig. 37–3; one such curve, corresponding to $r_2 - r_1 = -2.5\lambda$, passes through point c.

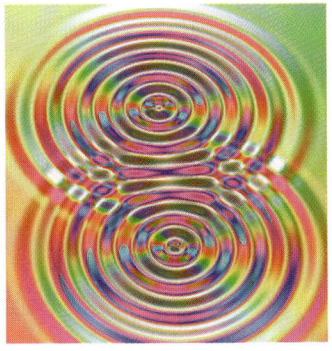

The concepts of constructive interference and destructive interference apply to these water waves as well as to light waves or sound waves.

In some cases, such as two loudspeakers or two radio-transmitter antennas, the interference pattern is three-dimensional. Think of rotating the color curves of Fig. 37–3 around the y-axis; then maximum constructive interference occurs at all points on the resulting surfaces of revolution.

CAUTION ▶ The interference patterns in Figs. 37–2a and 37–3 are *not* standing waves, though they have some similarities to the standing-wave patterns described in Chapter 20 and Section 33–7. In a standing wave, the interference is between two waves propagating in opposite directions; a stationary pattern of antinodes and nodes appears, and there is *no* net energy flow in either direction (the energy in the wave is left "standing"). In the situations shown in Figs. 37–2a and 37–3, there is likewise a stationary pattern of antinodal and nodal curves, but there is a net flow of energy *outward* from the two sources. From the energy standpoint, all that interference does is to "channel" the energy flow so that it is greatest along the antinodal curves and least along the nodal curves. ◀

For Eqs. (37–1) and (37–2) to hold, the two sources must have the same wavelength and must *always* be in phase. These conditions are rather easy to satisfy for sound waves (see Example 20–7 in Section 20–6). But with *light* waves there is no practical way to achieve a constant phase relationship (coherence) with two independent sources. This is because of the way light is emitted. In ordinary light sources, atoms gain excess energy by thermal agitation or by impact with accelerated electrons. An atom that is "excited" in such a way begins to radiate energy and continues until it has lost all the energy it can, typically in a time of the order of 10^{-8} s. The many atoms in a source ordinarily radiate in an unsynchronized and random phase relationship, and the light that is emitted from *two* such sources has no definite phase relation.

However, the light from a single source can be split so that parts of it emerge from two or more regions of space, forming two or more *secondary sources*. Then any random phase change in the source affects these secondary sources equally and does not change their *relative* phase.

The distinguishing feature of light from a *laser* is that the emission of light from many atoms is synchronized in frequency and phase. As a result, the random phase changes mentioned above occur much less frequently. Definite phase relations are preserved over correspondingly much greater lengths in the beam, and laser light is much more coherent than ordinary light.

37–3 TWO-SOURCE INTERFERENCE OF LIGHT

The interference pattern produced by two coherent sources of *water* waves of the same wavelength can be readily seen in a ripple tank with a shallow layer of water. This pattern is not directly visible when the interference is between *light* waves, since light traveling in a uniform medium cannot be seen. (A shaft of afternoon sunlight in a room is made visible by scattering from airborne dust particles.)

One of the earliest quantitative experiments to reveal the interference of light from two sources was performed in 1800 by the English scientist Thomas Young. We will refer back to this experiment several times in this and later chapters, so it's important to understand it in detail. Young's apparatus is shown in perspective in Fig. 37–4a. A light source (not shown) emits monochromatic light; however, this light is not suitable for use in an interference experiment because emissions from different parts of an ordinary source are not synchronized. To remedy this, the light is directed at a screen with a narrow slit S_0, 1 μm or so wide. The light emerging from the slit originated from only a small region of the light source; thus slit S_0 behaves more nearly like the idealized source shown in Fig. 37–1. (In modern versions of the experiment, a laser is used as a source of coherent light, and the slit S_0 isn't needed.) The light from slit S_0 falls on a screen with two other narrow slits S_1 and S_2, each 1 μm or so wide and a few tens or hundreds of micrometers apart. Cylindrical wave fronts spread out from slit S_0 and reach slits S_1 and S_2 *in phase* because they travel equal distances from S_0. The waves *emerging* from slits S_1 and S_2 are therefore also always in phase, so S_1 and S_2 are *coherent* sources. The interference of waves from S_1 and S_2 produces a pattern in space like that to the right of the sources in Figs. 37–2a and 37–3.

To visualize the interference pattern, a screen is placed so that the light from S_1 and S_2 falls on it (Fig. 37–4b). The screen will be most brightly illuminated at points P, where the light waves from the slits interfere constructively, and will be darkest at points where the interference is destructive.

To simplify the analysis of Young's experiment, we assume that the distance R from the slits to the screen is so large in comparison to the distance d between the slits that the

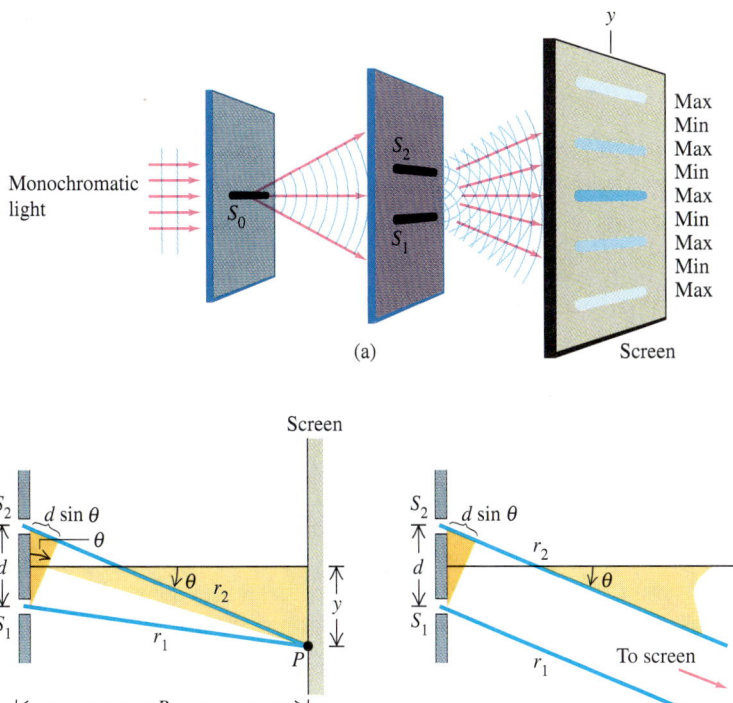

(a)

(b)   (c)

37–4 (a) Young's experiment to show interference of light passing through two slits. The regions on the screen labeled "Max" are where constructive interference of the light from slits S_1 and S_2 occurs, so that the resultant light wave has maximum intensity. (b) Geometrical analysis of Young's experiment. The slits S_1 and S_2 are horizontal and are seen from the side in cross section. (c) Approximate geometry when the distance R to the screen is much greater than the distance d between the slits. In this figure r_2 is larger than r_1, the angle θ is positive, and the illuminated point on the screen is to one side of the screen's center. For points on the other side of the screen's center, r_2 is less than r_1 and θ is negative.

lines from S_1 and S_2 to P are very nearly parallel, as in Fig. 37–4c. This is usually the case for experiments with light; the slit separation is typically a few millimeters, while the screen may be a meter or more away. The difference in path length is then given by

$$r_2 - r_1 = d \sin \theta, \qquad (37\text{–}3)$$

where θ is the angle between a line from slits to screen (shown in blue in Fig. 37–4c) and the normal to the plane of the slits (shown as a thin black line).

We found in Section 37–2 that constructive interference (reinforcement) occurs at points where the path difference $d \sin \theta$ is an integral number of wavelengths, $m\lambda$, where $m = 0, \pm 1, \pm 2, \pm 3, \dots$. So the bright regions on the screen occur at angles θ for which

$$d \sin \theta = m\lambda \qquad (m = 0, \pm 1, \pm 2, \dots) \qquad \text{(constructive interference,} \quad (37\text{–}4)$$
$$\text{two slits).}$$

Similarly, destructive interference (cancellation) occurs, forming dark regions on the screen, at points for which the path difference is a half-integral number of wavelengths, $\left(m + \tfrac{1}{2}\right)\lambda$:

$$d \sin \theta = \left(m + \frac{1}{2}\right)\lambda \qquad (m = 0, \pm 1, \pm 2, \dots) \qquad \text{(destructive interference,} \quad (37\text{–}5)$$
$$\text{two slits).}$$

Thus the pattern on the screen of Figs. 37–4a and 37–4b is a succession of bright and dark bands, or **interference fringes,** parallel to the slits S_1 and S_2. A photograph of such a pattern is shown in Fig. 37–5. The center of the pattern is a bright band corresponding to $m = 0$ in Eq. (37–4); this point on the screen is equidistant from the two slits.

We can derive an expression for the positions of the centers of the bright bands on the screen. In Fig. 37–4b, y is measured from the center of the pattern, corresponding to the distance from the center of Fig. 37–5. Let y_m be the distance from the center of the

m (max)		$m + 1/2$ (min)
5 →		← 11/2
4 →		← 9/2
3 →		← 7/2
2 →		← 5/2
1 →		← 3/2
0 →		← 1/2
−1 →		← −1/2
−2 →		← −3/2
−3 →		← −5/2
−4 →		← −7/2
−5 →		← −9/2
		← −11/2

37–5 Photograph of interference fringes produced on a screen in Young's double-slit experiment.

pattern ($\theta = 0$) to the center of the mth bright band. Let θ_m be the corresponding value of θ; then

$$y_m = R \tan \theta_m.$$

In experiments such as this, the distances y_m are often much smaller than the distance R from the slits to the screen. Hence θ_m is very small, $\tan \theta_m$ is very nearly equal to $\sin \theta_m$, and

$$y_m = R \sin \theta_m.$$

Combining this with Eq. (37–4), we find that *for small angles only,*

$$y_m = R \frac{m\lambda}{d} \quad \text{(constructive interference in Young's experiment).} \quad (37\text{–}6)$$

We can measure R and d, as well as the positions y_m of the bright fringes, so this experiment provides a direct measurement of the wavelength λ. Young's experiment was in fact the first direct measurement of wavelengths of light.

CAUTION▶ While Eqs. (37–4) and (37–5) are valid at any angle, Eq. (37–6) is valid only for *small* angles. It can be used *only* if the distance R from slits to screen in much greater than the slit separation d and if R is much greater than the distance y_m from the center of the interference pattern to the mth bright fringe. **◀**

The distance between adjacent bright bands in the pattern is *inversely* proportional to the distance d between the slits. The closer together are the slits, the more the pattern spreads out. When the slits are far apart, the bands in the pattern are closer together.

While we have described the experiment that Young performed with visible light, the results given in Eqs. (37–4) and (37–5) are valid for *any* type of wave, provided that the resultant wave from two coherent sources is detected at a point that is far away in comparison to the separation d.

EXAMPLE 37–1

Two-slit interference In a two-slit interference experiment, the slits are 0.20 mm apart, and the screen is at a distance of 1.0 m. The third bright fringe (not counting the central bright fringe straight ahead from the slits) is found to be displaced 7.5 mm from the central fringe. Find the wavelength of the light used.

SOLUTION The third bright fringe corresponds to $m = 3$ in Eqs. (37–4) and (37–6). We may use the latter equation, since

$R = 1.0$ m is much greater than either $d = 0.20$ mm or $y_3 = 7.5$ mm. Solving Eq. (37–6) for λ, we find

$$\lambda = \frac{y_m d}{mR} = \frac{(7.5 \times 10^{-3} \text{ m})(0.20 \times 10^{-3} \text{ m})}{(3)(1.0 \text{ m})}$$

$$= 500 \times 10^{-9} \text{ m} = 500 \text{ nm}.$$

This bright fringe could also correspond to $m = -3$; can you show that this gives the same result for λ?

EXAMPLE 37–2

Broadcast pattern of a radio station A radio station operating at a frequency of 1500 kHz $= 1.5 \times 10^6$ Hz (near the top end of the AM broadcast band) has two identical vertical dipole antennas spaced 400 m apart, oscillating in phase. At distances much greater than 400 m, in what directions is the intensity greatest in the resulting radiation pattern? (This is not just a hypothetical problem. It is often desirable to beam most of the radiated energy from a radio transmitter in particular directions rather than uniformly in all directions. Pairs or rows of antennas are

often used to produce the desired radiation pattern.)

SOLUTION The two antennas, seen from above in Fig. 37–6, correspond to sources S_1 and S_2 in Fig. 37–4. The wavelength is $\lambda = c/f = 200$ m. Since the resultant wave is detected at distances much greater than $d = 400$ m, we may use Eq. (37–4) to give the directions of the intensity *maxima*, the values of θ for which the path difference is zero or an integer number of wavelengths.

(Note that we can't use Eq. (37–6) because the angles θ_m aren't necessarily small.) Inserting the numerical values, with $m = 0, \pm 1, \pm 2$, we find

$$\sin\theta = \frac{m\lambda}{d} = \frac{m(200\text{ m})}{400\text{ m}} = \frac{m}{2}, \qquad \theta = 0, \pm 30°, \pm 90°.$$

In this example, values of m greater than 2 or less than -2 give values of $\sin\theta$ greater than 1 or less than -1, which is impossible. There is *no* direction for which the path difference is three or more wavelengths. Thus values of m of ± 3 and beyond have no physical meaning in this example.

The angles for *minimum* intensity (destructive interference) are given by Eq. (37–5), with $m = -2, -1, 0, 1$:

$$\sin\theta = \frac{(m + \frac{1}{2})\lambda}{d} = \frac{m + \frac{1}{2}}{2}, \qquad \theta = \pm 14.5°, \pm 48.6°.$$

Other values of m have no physical significance in this example.

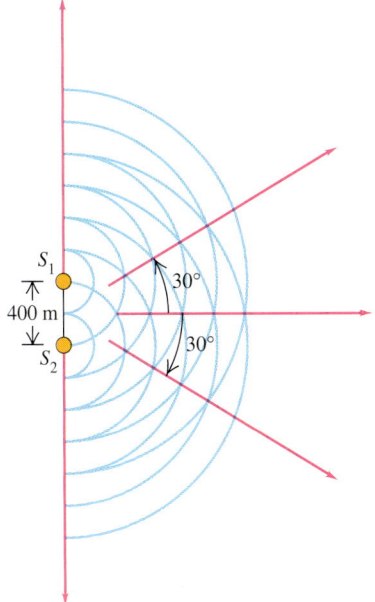

37–6 Two radio antennas broadcasting in phase. The red arrows indicate the directions of maximum intensity. The waves that are emitted to the left of the two sources are not shown.

37–4 INTENSITY IN INTERFERENCE PATTERNS

In Section 37–3 we found the positions of maximum and minimum intensity in a two-source interference pattern. Let's now see how to find the intensity at *any* point in the pattern. To do this, we have to combine the two sinusoidally varying fields (from the two sources) at a point P in the radiation pattern, taking proper account of the phase difference of the two waves at point P, which results from the path difference. The intensity is then proportional to the square of the resultant electric-field amplitude, as we learned in Chapter 33.

To calculate the intensity, we will assume that the two sinusoidal functions (corresponding to waves from the two sources) have equal amplitude E and that the $\vec{E}$ fields lie along the same line (have the same polarization). This assumes that the sources are identical and neglects the slight amplitude difference caused by the unequal path lengths (the amplitude decreases with increasing distance from the source). From Eq. (33–26), each source by itself would give an intensity $\frac{1}{2}\epsilon_0 cE^2$ at point P. If the two sources are in phase, then the waves that arrive at P differ in phase by an amount proportional to the difference in their path lengths, $(r_2 - r_1)$. If the phase angle between these arriving waves is ϕ, then we can use the following expressions for the two electric fields superposed at P:

$$E_1(t) = E\cos(\omega t + \phi),$$

$$E_2(t) = E\cos\omega t.$$

Here is our program. The superposition of the two fields at P is a sinusoidal function with some amplitude E_P that depends on E and the phase difference ϕ. First we'll work on finding the amplitude E_P if E and ϕ are known. Then we'll find the intensity I of the resultant wave, which is proportional to E_P^2. Finally, we'll relate the phase difference ϕ to the path difference, which is determined by the geometry of the situation.

AMPLITUDE IN TWO-SOURCE INTERFERENCE

To add the two sinusoidal functions with a phase difference, we use the same *phasor* representation that we used for simple harmonic motion (Section 13–3) and for voltages

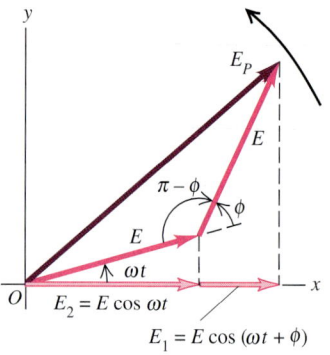

37–7 Phasor diagram for the superposition at a point P of two waves of waves of equal amplitude E with a phase difference ϕ. The resultant wave has amplitude E_P, equal to the magnitude of the phasor shown in dark red.

and currents in ac circuits (Section 32–2). We suggest that you review these sections so that phasors are fresh in your mind. Each sinusoidal function is represented by a rotating vector (phasor) whose projection on the horizontal axis at any instant represents the instantaneous value of the sinusoidal function.

In Fig. 37–7, E_1 is the horizontal component of the phasor representing the wave from source S_1, and E_2 is the horizontal component of the phasor for the wave from S_2. As shown in the diagram, both phasors have the same magnitude E, but E_1 is *ahead* of E_2 in phase by an angle ϕ. Both phasors rotate counterclockwise with constant angular velocity ω, and the sum of the projections on the horizontal axis at any time gives the instantaneous value of the total E field at point P. Thus the amplitude E_P of the resultant sinusoidal wave at P is the magnitude of the dark red phasor in the diagram (labeled E_P); this is the *vector sum* of the other two phasors. To find E_P, we use the law of cosines and the trigonometric identity $\cos(\pi - \phi) = -\cos\phi$:

$$E_P{}^2 = E^2 + E^2 - 2E^2 \cos(\pi - \phi)$$

$$= E^2 + E^2 + 2E^2 \cos\phi.$$

Then, using the identity $1 + \cos\phi = 2\cos^2(\phi/2)$, we obtain

$$E_P{}^2 = 2E^2(1 + \cos\phi) = 4E^2 \cos^2\left(\frac{\phi}{2}\right),$$

$$E_P = 2E\left|\cos\frac{\phi}{2}\right| \qquad \text{(amplitude in two-source interference).} \qquad (37\text{–}7)$$

You can also obtain this result algebraically without using phasors (see Problem 37–40).

When the two waves are in phase, $\phi = 0$ and $E_P = 2E$. When they are exactly a half-cycle out of phase, $\phi = \pi$ rad $= 180°$, $\cos(\phi/2) = \cos(\pi/2) = 0$, and $E_P = 0$. Thus the superposition of two sinusoidal waves with the same frequency and amplitude but with a phase difference yields a sinusoidal wave with the same frequency and an amplitude between zero and twice the individual amplitudes, depending on the phase difference.

INTENSITY IN TWO-SOURCE INTERFERENCE

To obtain the intensity I at point P, we recall from Section 33–5 that I is equal to the average magnitude of the Poynting vector, S_{av}. For a sinusoidal wave with electric-field amplitude E_P, this is given by Eq. (33–26) with E_{max} replaced by E_P. Thus we can express the intensity in any of the following equivalent forms:

$$I = S_{av} = \frac{E_P{}^2}{2\mu_0 c} = \frac{1}{2}\sqrt{\frac{\epsilon_0}{\mu_0}}E_P{}^2 = \frac{1}{2}\epsilon_0 c E_P{}^2. \qquad (37\text{–}8)$$

The essential content of these expressions is that I is proportional to $E_P{}^2$. When we substitute Eq. (37–7) into the last expression in Eq. (37–8), we get

$$I = \frac{1}{2}\epsilon_0 c E_P{}^2 = 2\epsilon_0 c E^2 \cos^2\frac{\phi}{2}. \qquad (37\text{–}9)$$

In particular, the *maximum* intensity I_0, which occurs at points where the phase difference is zero ($\phi = 0$), is

$$I_0 = 2\epsilon_0 c E^2.$$

Note that the maximum intensity I_0 is *four times* (not twice) as great as the intensity $\frac{1}{2}\epsilon_0 c E^2$ from each individual source.

Substituting the expression for I_0 into Eq. (37–9), we can express the intensity I at any point very simply in terms of the maximum intensity I_0:

$$I = I_0 \cos^2\frac{\phi}{2} \qquad \text{(intensity in two-source interference).} \qquad (37\text{–}10)$$

Note that if we average this over all possible phase differences, the result is $I_0/2 = \epsilon_0 cE^2$ (the average of $\cos^2(\phi/2)$ is $\frac{1}{2}$). This is just twice the intensity from each individual source, as we should expect. The total energy output from the two sources isn't changed by the interference effects, but the energy is redistributed (as we mentioned in Section 37–2). For some phase angles the intensity is four times as great as for an individual source, but for others it is zero. It averages out.

PHASE DIFFERENCE AND PATH DIFFERENCE

Now we have to find how the phase difference ϕ between the two fields at point P is related to the geometry of the situation. We know that ϕ is proportional to the difference in path length from the two sources to point P. When the path difference is one wavelength, the phase difference is one cycle, and $\phi = 2\pi$ rad $= 360°$. When the path difference is $\lambda/2$, $\phi = \pi$ rad $= 180°$, and so on. That is, the ratio of the phase difference ϕ to 2π is equal to the ratio of the path difference $r_2 - r_1$ to λ:

$$\frac{\phi}{2\pi} = \frac{r_2 - r_1}{\lambda}.$$

Thus a path difference $(r_2 - r_1)$ causes a phase difference given by

$$\phi = \frac{2\pi}{\lambda}(r_2 - r_1) = k(r_2 - r_1) \qquad \text{(phase difference related to path difference)}, \qquad (37\text{–}11)$$

where $k = 2\pi/\lambda$ is the *wave number* introduced in Section 19–4.

If the material in the space between the sources and P is anything other than vacuum, we must use the wavelength *in the material* in Eq. (37–11). If the material has index of refraction n, then

$$\lambda = \frac{\lambda_0}{n} \qquad \text{and} \qquad k = nk_0, \qquad (37\text{–}12)$$

where λ_0 and k_0 are the wavelength and wave number, respectively, in vacuum.

Finally, if the point P is far away from the sources in comparison to their separation d, the path difference is given by Eq. (37–3):

$$r_2 - r_1 = d \sin \theta.$$

Combining this with Eq. (37–11), we find

$$\phi = k(r_2 - r_1) = kd \sin \theta = \frac{2\pi d}{\lambda} \sin \theta. \qquad (37\text{–}13)$$

When we substitute this into Eq. (37–10), we find

$$I = I_0 \cos^2 \left(\frac{1}{2} kd \sin \theta \right) = I_0 \cos^2 \left(\frac{\pi d}{\lambda} \sin \theta \right) \qquad \text{(intensity far from two sources).} \qquad (37\text{–}14)$$

The directions of *maximum* intensity occur when the cosine has the values ± 1, that is, when

$$\frac{\pi d}{\lambda} \sin \theta = m\pi \qquad (m = 0, \pm 1, \pm 2, \ldots),$$

or

$$d \sin \theta = m\lambda,$$

in agreement with Eq. (37–4). We leave it as an exercise to show that Eq. (37–5) for the zero-intensity directions can also be derived from Eq. (37–14).

37–8 Intensity distribution in the interference pattern from two identical slits. The three scales show the distance y of a point in the pattern from the center ($y = 0$), the phase difference ϕ between the two waves at each point in the pattern, and the path difference $d \sin \phi$ expressed as a multiple of the wavelength λ. The maxima occur at points where ϕ is an integral multiple of 2π and $d \sin \theta$ is an integral multiple of λ.

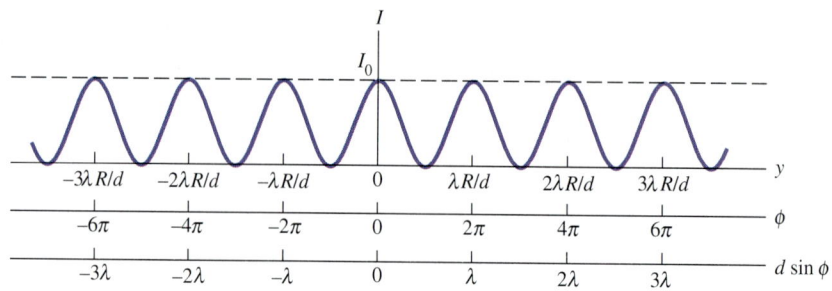

As we noted in Section 37–3, in experiments with light we visualize the interference pattern due to two slits by using a screen placed at a distance R from the slits. We can describe positions on the screen with the coordinate y; the positions of the bright fringes are given by Eq. (37–6), where ordinarily $y \ll R$. In that case, $\sin \theta$ is approximately equal to y/R, and we obtain the following expressions for the intensity at *any* point on the screen as a function of y:

$$I = I_0 \cos^2 \left(\frac{kdy}{2R} \right) = I_0 \cos^2 \left(\frac{\pi dy}{\lambda R} \right) \qquad \text{(intensity in two-slit interference)}.$$

$$(37\text{–}15)$$

Figure 37–8 shows a graph of Eq. (37–15); we can compare this with the photographically recorded pattern of Fig. 37–5. The peaks in Fig. 37–8 all have the same intensity, while those in Fig. 37–5 fade off as we go away from the center. We'll explore the reasons for this variation in peak intensity in Chapter 38.

EXAMPLE 37–3

A directional transmitting antenna array Suppose the two identical radio antennas in Fig. 37–6 are moved to be only 10 m apart and the frequency of the radiated waves is increased to $f = 60$ MHz. The intensity at a distance of 700 m in the +x-direction (corresponding to $\theta = 0$ in Fig. 37–4) is $I_0 = 0.020$ W/m². a) What is the intensity in the direction $\theta = 4.0°$? b) In what direction near $\theta = 0$ is the intensity $I_0/2$? c) In what directions is the intensity zero? (This is an example of using interference to beam radiated energy in certain directions, as mentioned in Example 37–2 in Section 37–3).

SOLUTION Because the distance between the antennas is much less than 700 m, the amplitudes of the two waves are very nearly equal; then we can use Eq. (37–14). First we find the wavelength, using the relation $c = \lambda f$:

$$\lambda = \frac{c}{f} = \frac{3.0 \times 10^8 \text{ m/s}}{60 \times 10^6 \text{ s}^{-1}} = 5.0 \text{ m}.$$

The spacing between sources is $d = 10$ m $= 2\lambda$, and Eq. (37–14) becomes

$$I = I_0 \cos^2 \left(\frac{\pi d}{\lambda} \sin \theta \right)$$

$$= (0.020 \text{ W/m}^2) \cos^2 \left[\frac{\pi(10 \text{ m})}{(5.0 \text{ m})} \sin \theta \right]$$

$$= (0.020 \text{ W/m}^2) \cos^2 ((2\pi \text{ rad}) \sin \theta).$$

a) When $\theta = 4.0°$,

$$I = (0.020 \text{ W/m}^2) \cos^2 ((2\pi \text{ rad}) \sin 4.0°)$$

$$= 0.016 \text{ W/m}^2.$$

This is about 82% of the intensity at $\theta = 0$.

b) The intensity I equals $I_0/2$ when the cosine has the value $\pm 1/\sqrt{2}$. This occurs when $2\pi \sin \theta = \pm \pi/4$ rad, $\sin \theta = \pm \frac{1}{8}$, and $\theta = \pm 7.2°$. (This condition is also satisfied when $\sin \theta = \pm \frac{3}{8}, \pm \frac{5}{8}$, and $\pm \frac{7}{8}$; can you verify this? These aren't the values the problem asked for. Why not?)

c) The intensity is zero when $\cos(2\pi \sin \theta) = 0$. This occurs when $2\pi \sin \theta = \pm \pi/2, \pm 3\pi/2, \pm 5\pi/2, \ldots$, or

$$\sin \theta = \pm \frac{1}{4}, \pm \frac{3}{4}, \pm \frac{5}{4}, \ldots.$$

Values of $\sin \theta$ greater than 1 have no meaning, and we find

$$\theta = \pm 14.5°, \quad \pm 48.6°.$$

37–9 Photograph of a thin film of oil illuminated by white light.

37–5 INTERFERENCE IN THIN FILMS

You often see bright bands of color when light reflects from a soap bubble or from a thin layer of oil floating on water (Fig. 37–9). These are the results of interference. Light waves are reflected from opposite surfaces of such thin films, and constructive interference between the two reflected waves (with different path lengths) occurs in different places for different wavelengths. The situation is shown schematically in Fig. 37–10. Light shining on the upper surface of a thin film with thickness t is partly reflected at the upper surface (path abc). Light *transmitted* through the upper surface is partly reflected at the lower surface (path $abdef$). The two reflected waves come together at point P on the retina of the eye. Depending on the phase relationship, they may interfere constructively or destructively. Different colors have different wavelengths, so the interference may be constructive for some colors and destructive for others. That's why we see colored rings or fringes. The complex shapes of the colored rings in Fig. 37–9 result from variations in the thickness of the oil film.

Here's an example involving *monochromatic* light reflected from two nearly parallel surfaces at nearly normal incidence. Figure 37–11 shows two plates of glass separated by a thin wedge, or film, of air. We want to consider interference between the two light waves reflected from the surfaces adjacent to the air wedge, as shown. (Reflections also occur at the top surface of the upper plate and the bottom surface of the lower plate; to keep our discussion simple, we won't include these.) The situation is the same as in Fig. 37–10 except that the film (wedge) thickness is not uniform. The path difference between the two waves is just twice the thickness t of the air wedge at each point. At points where $2t$ is an integer number of wavelengths, we expect to see constructive interference and a bright area; where it is a half-integer number of wavelengths, we expect to see destructive interference and a dark area. Along the line where the plates are in contact, there is practically *no* path difference, and we expect a bright area.

When we carry out the experiment, the bright and dark fringes appear, but they are interchanged! Along the line where the plates are in contact, we find a *dark* fringe, not a bright one. This suggests that one or the other of the reflected waves has undergone a half-cycle phase shift during its reflection. In that case the two waves that are reflected at the line of contact are a half-cycle out of phase even though they have the same path length.

In fact, this phase shift can be predicted from Maxwell's equations and the electromagnetic nature of light. The details of the derivation are beyond our scope, but here is the result. Suppose a light wave with electric-field amplitude E_i is traveling in an optical material with index of refraction n_a. It strikes, at normal incidence, an interface with another optical material with index n_b. The amplitude E_r of the wave reflected from the

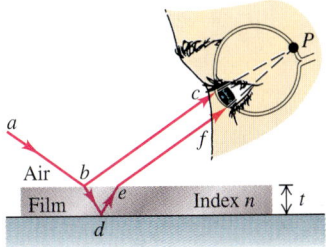

37–10 Interference between rays reflected from the upper and lower surfaces of a thin film. The path differences and phase shifts at the surfaces cause some wavelengths of light to interfere constructively and others to interfere destructively. The angles have been exaggerated for clarity.

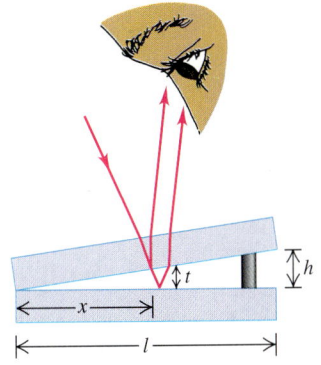

37–11 Interference between two light waves reflected from the two sides of an air wedge separating two glass plates. The path difference is $2t$. The thickness of the air wedge has been exaggerated; the distances h and t are much less than l. The angles have also been exaggerated for clarity; for the case analyzed in the text, the light strikes the upper plate at normal incidence.

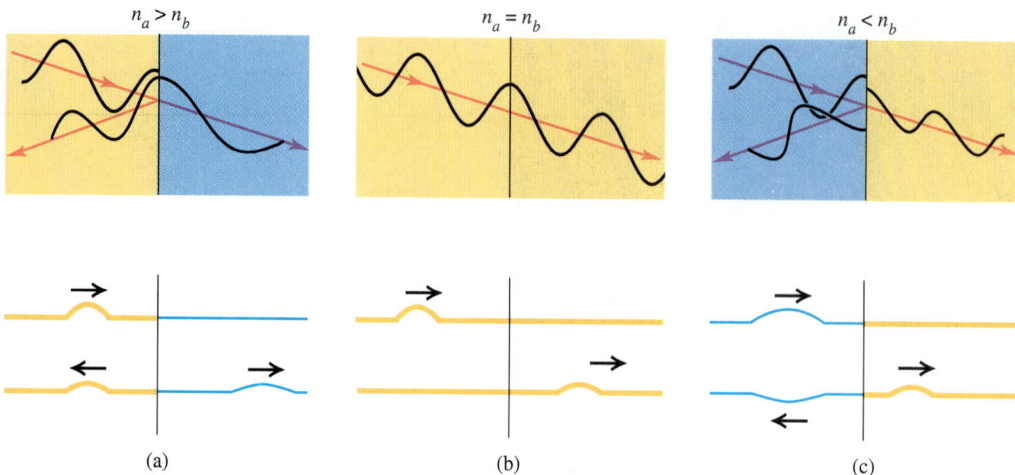

37–12 Upper figures: Electromagnetic waves striking an interface between optical materials at normal incidence (shown as a small angle for clarity). Lower figures: Mechanical wave pulses on ropes. (a) There is no phase change upon reflection when $n_a > n_b$, and the transmitted wave moves faster than the incident wave. (b) There is no reflection when there is no difference in n and no change in wave speed. (c) There is a half-cycle phase change upon reflection when $n_a < n_b$ and the transmitted wave moves more slowly than the incident wave.

interface is proportional to the amplitude E_i of the incident wave and is given by

$$E_r = \frac{n_a - n_b}{n_a + n_b} E_i \qquad \text{(normal incidence).} \qquad (37-16)$$

This result shows that the incident and reflected amplitudes have the same sign when n_a is larger than n_b and opposite sign when n_b is larger than n_a. We can distinguish three cases, as shown in Fig. 37–12:

Figure 37–12a: When $n_a > n_b$, light travels more slowly in the first medium than in the second. In this case, E_r and E_i have the same sign, and the phase shift of the reflected wave relative to the incident wave is zero. This is analogous to reflection of a transverse mechanical wave on a heavy rope at a point where it is tied to a lighter rope or a ring that can move vertically without friction.

Figure 37–12b: When $n_a = n_b$, the amplitude E_r of the reflected wave is zero. The incident light wave can't "see" the interface, and there *is no* reflected wave.

Figure 37–12c: When $n_a < n_b$, light travels more slowly in the second material than in the first. In this case, E_r and E_i have opposite signs, and the phase shift of the reflected wave relative to the incident wave is π rad (180° or a half-cycle). This is analogous to reflection (with inversion) of a transverse mechanical wave on a light rope at a point where it is tied to a heavier rope or a rigid support.

Let's compare with the situation of Fig. 37–11. For the wave reflected from the upper surface of the air wedge, n_a (glass) is greater than n_b, so this wave has zero phase shift. For the wave reflected from the lower surface, n_a (air) is less than n_b (glass), so this wave has a half-cycle phase shift. Waves that are reflected from the line of contact have no path difference to give additional phase shifts, and they interfere destructively; this is what we observe. We invite you to use the above principle to show that for normal incidence, the wave reflected at point b in Fig. 37–10 is shifted by a half-cycle, while the wave reflected at d is not (if there is air below the film).

We can summarize this discussion mathematically. If the film has thickness t, the light is at normal incidence and has wavelength λ in the film; if neither or both of the reflected waves from the two surfaces have a half-cycle reflection phase shift, the condition for constructive interference is

$$2t = m\lambda \qquad (m = 0, 1, 2, \ldots) \qquad \text{(constructive reflection from thin film, no relative phase shift).} \qquad (37-17)$$

However, when *one* of the two waves has a half-cycle reflection phase shift, this equation is the condition for *destructive* interference.

Similarly, if neither wave or both have a half-cycle phase shift, the condition for *destructive* interference in the reflected waves is

$$2t = \left(m + \frac{1}{2}\right)\lambda \quad (m = 0, 1, 2, \ldots) \quad \begin{array}{l}\text{(destructive reflection from thin} \\ \text{film, no relative phase shift).}\end{array} \quad (37\text{--}18)$$

However, if one wave has a half-cycle phase shift, this is the condition for *constructive* interference.

Problem–Solving Strategy

THIN FILMS

1. Be sure you understand the rule for phase changes during reflection. There is a half-cycle phase shift when $n_b > n_a$ and no phase shift when $n_b < n_a$.

2. Be careful to distinguish between transmission and reflection. *Minimum* intensity in the *reflected* wave corresponds to *maximum transmitted* intensity, and vice versa.

3. If the film consists of anything other than vacuum, be sure to use the wavelength of light *in the film* in your calculations. If the film is anything except vacuum, this is smaller than the wavelength in vacuum by a factor of n. (But for air, $n = 1.000$ to four-figure accuracy.)

EXAMPLE 37-4

Suppose the two glass plates in Fig. 37–11 are two microscope slides 10 cm long. At one end they are in contact; at the other end they are separated by a piece of paper 0.020 mm thick. What is the spacing of the interference fringes seen by reflection? Is the fringe at the line of contact bright or dark? Assume monochromatic light with a wavelength in air of $\lambda = \lambda_0 = 500$ nm.

SOLUTION We'll consider only interference between the light reflected from the upper and lower surfaces of the air wedge in Fig. 37–11. The wave reflected from the lower surface of the air wedge has a half-cycle phase shift; the wave from the upper surface has none. Thus the fringe at the line of contact is dark. The condition for *destructive* interference (a dark fringe) is Eq. (37–17):

$$2t = m\lambda_0 \quad (m = 0, 1, 2, \ldots).$$

From similar triangles in Fig. 37–11 the thickness t of the air wedge at each point is proportional to the distance x from the line of contact:

$$\frac{t}{x} = \frac{h}{l}.$$

Combining this with Eq. (37–17), we find

$$\frac{2xh}{l} = m\lambda_0,$$

$$x = m\frac{l\lambda_0}{2h} = m\frac{(0.10 \text{ m})(500 \times 10^{-9} \text{ m})}{(2)(0.020 \times 10^{-3} \text{ m})} = m(1.25 \text{ mm}).$$

Successive dark fringes, corresponding to successive integer values of m, are spaced 1.25 mm apart.

This result shows that the fringe spacing is proportional to the wavelength of the light used; the fringes would be farther apart with red light (larger λ_0) than with blue light (smaller λ_0). If we use white light, the reflected light at any point is a mixture of wavelengths for which constructive interference occurs; the wavelengths that interfere destructively are weak or absent in the reflected light.

We've ignored interference between the light reflected from the upper and lower surfaces of the *glass* plate on top of the air wedge. This is acceptable if the glass plate is very much thicker than a wavelength of light. Example 37–7 shows one reason why interference effects are usually unimportant with a *thick* film.

EXAMPLE 37-5

In Example 37–4, suppose the glass plates have $n = 1.52$ and the space between plates contains water ($n = 1.33$) instead of air. What happens now?

SOLUTION The phase changes are the same as in Example 37–4, and we again find a dark fringe at the line of contact. But

the wavelength in water is

$$\lambda = \frac{\lambda_0}{n} = \frac{500 \text{ nm}}{1.33} = 376 \text{ nm}.$$

The fringe spacing is reduced by a factor 1.33 and is equal to 0.94 mm.

EXAMPLE 37-6

Suppose the upper of the two plates in Example 37–4 is a plastic material with $n = 1.40$, the wedge is filled with a silicone grease having $n = 1.50$, and the bottom plate is a dense flint glass with $n = 1.60$. What happens now?

SOLUTION In this case there are half-cycle phase shifts at *both* surfaces of the wedge of grease, and the line of contact is at a *bright* fringe, not a dark one. The fringe spacing is again determined by the wavelength *in the film* (that is, in the silicone grease), $\lambda = (500 \text{ nm})/1.50 = 333$ nm; we invite you to show that the fringe spacing is 0.83 mm.

EXAMPLE 37-7

Thick films versus thin films We have emphasized interference effects in rather *thin* films. Suppose the film in Fig. 37–10 is a layer of water ($n = 1.33$) 1.00 cm thick on top of a glass surface with $n = 1.52$. What will you see if this rather *thick* film is illuminated from above?

SOLUTION As in Example 37–6, there are half-cycle phase shifts at both the top and bottom surfaces of the film. Hence the condition for constructive interference in the reflected waves is given by Eq. (37–17); the wavelengths in *air* that undergo constructive interference are given by

$$\lambda_0 = n\lambda = \frac{2nt}{m},$$

where n is the index of refraction of water.

Because $t = 1.00$ cm is large in comparison to the wavelengths of visible light, the values of m in this equation will be large. For instance, using $m = 40,000$ in the first equation tells us that light will be strongly reflected if it has a wavelength in air of

$$\lambda = \frac{2nt}{m} = \frac{2(1.33)(1.00 \times 10^{-2} \text{ m})}{40,000} = 6.67 \times 10^{-7} \text{ m} = 667 \text{ nm},$$

corresponding to orange-red light. You might be tempted to conclude that this thick film would appear strongly orange-red; however, there will also be strong reflection of a slightly longer wavelength corresponding to $m = 39,999$ and of a slightly shorter wavelength corresponding to $m = 40,001$. These wavelengths differ from each other by only one part in 40,000, or $(667 \text{ nm})/(40,000) = 0.017$ nm. This is a much smaller wavelength difference than the eye can detect. Hence no one wavelength appears to be emphasized more than any other in the light that is reflected from a *thick* film. If you shine white light on this thick film, the reflected light appears white. For *thin* films that are no more than a few wavelengths in thickness, the integer m in Eqs. (37–17) and (37–18) will be small, the difference between adjacent strongly reflected wavelengths will be substantial, and the preferential reflection of certain wavelengths (as in the thin oil film in Fig. 37–9) will be apparent to the eye.

NEWTON'S RINGS

Figure 37–13a shows the convex surface of a lens in contact with a plane glass plate. A thin film of air is formed between the two surfaces. When you view the setup with monochromatic light, you see circular interference fringes (Fig. 37–13b). These were studied by Newton and are called **Newton's rings.** When you view the setup by reflected light, the center of the pattern is black. Can you see why this should be expected?

We can use interference fringes to compare the surfaces of two optical parts by placing the two in contact and observing the interference fringes. Figure 37–14 is a photograph made during the grinding of a telescope objective lens. The lower, larger-diameter, thicker disk is the correctly shaped master, and the smaller, upper disk is the lens under test. The "contour lines" are Newton's interference fringes; each one indi-

37–13 (a) Air film between a convex lens and a plane surface. The thickness of the film t increases from zero as we move out from the center, giving a series of alternating dark and bright rings for monochromatic light. (b) A photograph of Newton's rings. What do you suppose causes the fainter sets of rings in the photograph?

(a)

(b)

cates an additional distance between the specimen and the master of one half wave-length. At 10 lines from the center spot the distance between the two surfaces is 5 wavelengths, or about 0.003 mm. This isn't very good; high-quality lenses are routinely ground with a precision of less than one wavelength. The surface of the primary mirror of the Hubble Space Telescope was ground to a precision of better than $\frac{1}{50}$ wavelength. Unfortunately, it was ground to incorrect specifications, creating one of the most precise errors in the history of optical technology.

NONREFLECTIVE AND REFLECTIVE COATINGS

Nonreflective coatings for lens surfaces make use of thin-film interference. A thin layer or film of hard transparent material with an index of refraction smaller than that of the glass is deposited on the lens surface, as in Fig. 37–15. Light is reflected from both sur-faces of the layer. In both reflections the light is reflected from a medium of greater index than that in which it is traveling, so the same phase change occurs in both reflec-tions. If the film thickness is a quarter (one fourth) of the wavelength *in the film* (assuming normal incidence), the total path difference is a half-wavelength. Light reflected from the first surface is then a half-cycle out of phase with light reflected from the second, and there is destructive interference.

The thickness of the nonreflective coating can be a quarter-wavelength for only one particular wavelength. This is usually chosen in the central yellow-green portion of the spectrum ($\lambda = 550$ nm), where the eye is most sensitive. Then there is somewhat more reflection at both longer (red) and shorter (blue) wavelengths, and the reflected light has a purple hue. The overall reflection from a lens or prism surface can be reduced in this way from 4–5% to less than 1%. This treatment is particularly important in eliminating stray reflected light in highly corrected photographic lenses with many individual pieces of glass and many air-glass surfaces. It also increases the net amount of light that is *transmitted* through the lens, since light that is not reflected will be transmitted. The same principle is used to minimize reflection from silicon photovoltaic solar cells ($n = 3.5$) by use of a thin surface layer of silicon monoxide (SiO, $n = 1.45$); this helps to increase the amount of light that actually reaches the solar cells.

If a quarter-wavelength thickness a material with an index of refraction *greater* than that of glass is deposited on glass, then the reflectivity is *increased,* and the deposited material is called a **reflective coating.** In this case there is a half-cycle phase shift at the air-film interface but none at the film-glass interface, and reflections from the two sides of the film interfere constructively. For example, a coating with refractive index 2.5 causes 38% of the incident energy to be reflected, compared with 4% or so with no coat-ing. By use of multiple-layer coatings, it is possible to achieve nearly 100% transmission or reflection for particular wavelengths. Some practical applications of these coatings are for color separation in color television cameras and for infrared "heat reflectors" in motion-picture projectors, solar cells, and astronauts' visors. Reflective coatings occur in nature on the scales of herring and other silvery fish; this gives these fish their char-acteristic shiny appearance (see Problem 37–48).

37–14 The surface of a telescope objective under inspection during manufacture.

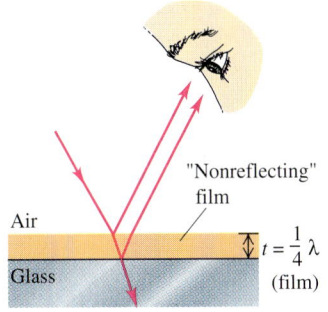

37–15 A nonreflective coating has an index of refraction that is intermediate between those of glass and air. For normal inci-dence, destructive interference results when the film thickness is one fourth of the wavelength in the film. The angles have been exaggerated for clarity.

EXAMPLE 37-8

A commonly used lens coating material is magnesium fluoride, MgF_2, with $n = 1.38$. What thickness should a nonreflective coating have for 550-nm light if it is applied to glass with $n = 1.52$?

SOLUTION With this coating, the wavelength λ of yellow-green

light in the coating is

$$\lambda = \frac{\lambda_0}{n} = \frac{550 \text{ nm}}{1.38} = 400 \text{ nm}.$$

A nonreflective film of MgF_2 should have a thickness of one quarter-wavelength, or 100 nm.

37–6 THE MICHELSON INTERFEROMETER

An important experimental device that uses interference is the **Michelson interferometer.** A century ago, it helped to provide one of the key experimental underpinnings of the theory of relativity. More recently, Michelson interferometers have been used to make precise measurements of wavelengths and of very small distances, such as the minute changes in thickness of an axon when a nerve impulse propagates along its length (Section 26–8). Like the Young two-slit experiment, a Michelson interferometer takes monochromatic light from a single source and divides it into two waves that follow different paths. In Young's experiment, this is done by sending part of the light through one slit and part through another; in a Michelson interferometer a device called a *beam splitter* is used. Interference occurs in both experiments when the two light waves are recombined.

The principal components of a Michelson interferometer are shown schematically in Fig. 37–16. A ray of light from a monochromatic source A strikes the beam splitter C, which is a glass plate with a thin coating of silver on its right side. Part of the light (ray 1) passes through the silvered surface and the compensator plate D and is reflected from mirror M_1. It then returns through D and is reflected from the silvered surface of C to the observer. The remainder of the light (ray 2) is reflected from the silvered surface at point P to the mirror M_2 and back through C to the observer's eye. The purpose of the compensator plate D is to ensure that rays 1 and 2 pass through the same thickness of glass; plate D is cut from the same piece of glass as plate C, so their thicknesses are identical to within a fraction of a wavelength.

The whole apparatus in Fig. 37–16 is mounted on a very rigid frame, and the position of mirror M_2 can be adjusted with a fine, very accurate micrometer screw. If the distances L_1 and L_2 are exactly equal and the mirrors M_1 and M_2 are exactly at right angles, the virtual image of M_1 formed by reflection at the silvered surface of plate C coincides with mirror M_2. If L_1 and L_2 are *not* exactly equal, the image of M_1 is displaced slightly from M_2; and if the mirrors are not exactly perpendicular, the image of M_1 makes a slight angle with M_2. Then the mirror M_2 and the virtual image of M_1 play the same roles as the two surfaces of a wedge-shaped thin film (Section 37–5), and light reflected from these surfaces forms the same sort of interference fringes.

Suppose the angle between mirror M_2 and the virtual image of M_1 is just great

37–16 A schematic Michelson interferometer. The observer sees an interference pattern that results from the difference in path lengths for rays 1 and 2.

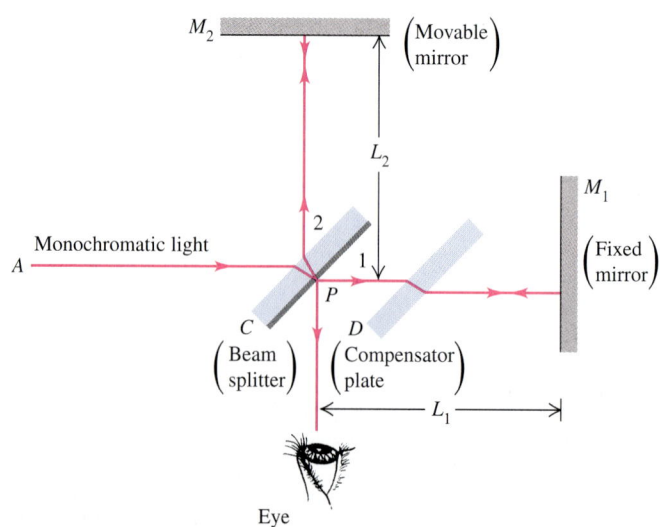

enough that five or six vertical fringes are present in the field of view. If we now move the mirror M_2 slowly either backward or forward a distance $\lambda/2$, the difference in path length between rays 1 and 2 changes by λ, and each fringe moves to the left or right a distance equal to the fringe spacing. If we observe the fringe positions through a telescope with a crosshair eyepiece and m fringes cross the crosshairs when we move the mirror a distance y, then

$$y = m\frac{\lambda}{2}, \quad \text{or} \quad \lambda = \frac{2y}{m}. \tag{37–19}$$

If m is several thousand, the distance y is large enough that it can be measured with good accuracy, and we can obtain an accurate value for the wavelength λ. Alternatively, if the wavelength is known, a distance y can be measured by simply counting fringes when M_2 is moved by this distance. In this way, distances that are comparable to a wavelength of light can be measured with relative ease.

The original application of the Michelson interferometer was to the historic **Michelson-Morley experiment.** Before the electromagnetic theory of light and Einstein's special theory of relativity became established, most physicists believed that the propagation of light waves occurred in a medium called the **ether,** which was believed to permeate all space. In 1887 the American scientists Albert Michelson and Edward Morley used the Michelson interferometer in an attempt to detect the motion of the earth through the ether. Suppose the interferometer in Fig. 37–16 is moving from left to right relative to the ether. According to the ether theory, this would lead to changes in the speed of light in the portions of the path shown as horizontal lines in the figure. There would be fringe shifts relative to the positions that the fringes would have if the instrument were at rest in the ether. Then when the entire instrument was rotated 90°, the other portions of the paths would be similarly affected, giving a fringe shift in the opposite direction.

Michelson and Morley expected that the motion of the earth through the ether would cause a fringe shift of about four tenths of a fringe when the instrument was rotated. The shift that was actually observed was less than a hundredth of a fringe and, within the limits of experimental uncertainty, appeared to be exactly zero. Despite its orbital motion around the sun, the earth appeared to be *at rest* relative to the ether. This negative result baffled physicists until Einstein developed the special theory of relativity in 1905. Einstein postulated that the speed of a light wave in vacuum has the same magnitude c relative to *all* inertial reference frames, no matter what their velocity may be relative to each other. The presumed ether then plays no role, and the concept of an ether has been abandoned.

The theory of relativity is a well-established cornerstone of modern physics, and we will study it in detail in Chapter 39. In retrospect, the Michelson-Morley experiment gives strong experimental support to the special theory of relativity, and it is often called the most significant "negative-result" experiment ever performed.

EXAMPLE 37–9

A Michelson interferometer is used with light of wavelength 605.78 nm. If the observer views the interference pattern through a telescope with a crosshair eyepiece, how many fringes pass the crosshairs when mirror M_2 moves exactly one centimeter?

SOLUTION From Eq. (37–19) the number of fringes is

$$m = \frac{2y}{\lambda} = \frac{2(1.0000 \times 10^{-2}\ \text{m})}{605.78 \times 10^{-9}\ \text{m}} = 33{,}015.$$

That sounds like a lot of counting, but it is easily automated by using a photocell and an electronic counter in place of a human observer.

*37-7 THE PHOTON

A Case Study in Quantum Physics

The interference effects discussed in this chapter, and similar effects to be discussed in the next chapter, show clearly the *wave* nature of light. But many other phenomena show a different aspect of the nature of light, in which it seems to behave as a stream of *particles*. For example, when a photograph of an interference pattern, such as Fig. 37–5, is made by using extremely weak monochromatic light and electronic image amplification, the pattern does not build up uniformly. Instead, first a light spot appears at one point, then another spot appears at another point, and so on. As the pattern emerges, the regions of maximum intensity have the greatest number of spots, the regions of zero intensity have none, and so forth.

This behavior suggests that the energy in a light wave is not a continuous stream, but rather is *quantized* into a succession of little bundles of energy. These bundles are called *quanta* or **photons.** We don't notice this effect in ordinary photography because the total number of spots is extremely large.

The concept of quantization of energy was introduced by the German physicist Max Planck in 1900. He used this concept as a computational technique in a calculation aimed at predicting the energy distribution among various wavelengths in the spectrum of radiation from hot objects (blackbody radiation, mentioned in Section 15–8). Einstein recognized in 1905 that the quantization of energy was far more than a computational trick and that instead it was a fundamental aspect of the nature of light. He used this concept to analyze the photoelectric effect, a process in which electrons are liberated from a conducting surface when light strikes it. Einstein assumed that an electron at the surface absorbs one photon and thus gains enough energy to escape from the surface.

Einstein assumed that the energy E of an individual photon was proportional to the frequency f of the light, with a proportionality constant h that is now called **Planck's constant:**

$$E = hf = \frac{hc}{\lambda} \quad \text{(energy of a photon)}, \tag{37-20}$$

where c is the speed of light and $\lambda = c/f$ is the wavelength of the radiation in vacuum. Detailed measurements of the spectra of blackbody radiation and of the photoelectric effect confirmed the correctness of the photon concept and also enabled physicists to determine the numerical value of Planck's constant. To four significant figures,

$$h = 6.626 \times 10^{-34} \text{ J} \cdot \text{s}.$$

Thus light has a dual personality that includes aspects of *both* waves and particles. Light can form interference patterns that are similar to those of sound and water waves. But if the light in the pattern is examined closely enough, it is found to be made of particle-like photons. This duality is present not only in visible light but in the entire spectrum of electromagnetic radiation. In some experiments and with some frequency ranges, one aspect or the other may predominate, but fundamentally, both are always present. The following example shows why the particle aspects of *visible* light are normally not evident. Example 37–11 shows why these same particle aspects are dominant for gamma rays, which have much higher frequency than visible light.

EXAMPLE 37-10

The familiar red light emitted by a helium-neon laser (used in grocery checkout scanners and many other applications) has a wavelength of 632.8 nm. If the power output is 1.00 mW, how many photons of this light does the laser emit each second?

SOLUTION First we use Eq. (37–20) to find the energy of each photon:

$$E = \frac{hc}{\lambda} = \frac{(6.626 \times 10^{-34} \text{ J} \cdot \text{s})(3.00 \times 10^{8} \text{ m/s})}{632.8 \times 10^{-9} \text{ m}}$$

$$= 3.14 \times 10^{-19} \text{ J}.$$

A power output of 1.00 mW means that the laser is emitting

1.00×10^{-3} J of energy each second; the number of photons per second is

$$\frac{1.00 \times 10^{-3} \text{ J/s}}{3.14 \times 10^{-19} \text{ J/photon}} = 3.18 \times 10^{15} \text{ photons/s}.$$

Even this very low-power source of light emits an *enormous* number of photons each second, far too many to distinguish individually. Quantization is ordinarily not evident for visible light because each photon has only a tiny amount of energy.

EXAMPLE 37–11

A gamma-ray photon emitted during the decay of a radioactive cobalt-60 nucleus has an energy of 2.135×10^{-13} J. What are the frequency and wavelength of this electromagnetic radiation?

SOLUTION From Eq. (37–20) we have

$$f = \frac{E}{h} = \frac{2.135 \times 10^{-13} \text{ J}}{6.626 \times 10^{-34} \text{ J} \cdot \text{s}} = 3.222 \times 10^{20} \text{ Hz}.$$

This is roughly a million times larger than frequencies of visible light. The wavelength λ is

$$\lambda = \frac{c}{f} = \frac{3.00 \times 10^{8} \text{ m/s}}{3.222 \times 10^{20} \text{ Hz}} = 9.31 \times 10^{-13} \text{ m}.$$

Particle aspects are much more evident for gamma rays than for visible light because an individual gamma-ray photon has roughly a million times more energy than a visible-light photon. With this much energy an individual gamma-ray photon is relatively easy to detect with any of several types of particle detectors, such as a Geiger counter or a solid-state detector.

Conversely, the *wave* properties of visible light are easy to see in interference patterns, but interference effects with gamma rays with wavelength of 9.31×10^{-13} m are very difficult to detect. As an example, consider the two-slit interference experiment described in Example 37–1 (Section 37–3), for which the distance between slits is $d = 0.20$ mm and the distance from slits to screen is $R = 1.0$ m. With visible light of wavelength $\lambda = 500$ nm, the third bright fringe is 7.5 mm from the center of the interference pattern. If we instead use gamma rays with $\lambda = 9.31 \times 10^{-13}$ m, the distance from the center of the pattern to the third bright fringe is given by Eq. (37–6) with $m = 3$:

$$y_3 = R \frac{m\lambda}{d} = (1.0 \text{ m}) \frac{3(9.31 \times 10^{-13} \text{ m})}{0.20 \times 10^{-3} \text{ m}} = 1.4 \times 10^{-8} \text{ m}.$$

This is too small to measure easily, which shows that interference effects for these gamma rays are hard to detect. Put another way, to get the same fringe spacing for gamma rays as was found with visible light in Example 37–1, you would need a slit spacing of only 0.37 nm!

MOMENTUM OF PHOTONS AND PARTICLES

Any material particle that has kinetic energy has momentum as well, and photons also have momentum. Einstein showed, as part of his special theory of relativity (which we will study in Chapter 39), that a photon with energy E has momentum $\vec{p}$ with magnitude p given by

$$E = pc. \tag{37–21}$$

Using the relation $c = \lambda f$ and rearranging, we find

$$p = \frac{E}{c} = \frac{hf}{c} = \frac{h}{\lambda}. \tag{37–22}$$

The *direction* of $\vec{p}$ is the direction in which the electromagnetic wave is traveling.

In Section 33–5 we discussed the phenomenon of *radiation pressure*, which is associated with the momentum carried by electromagnetic waves. Equation (37–22) shows us that this momentum is quantized. Just as the pressure that a gas exerts on a wall of its container results from the momentum change of gas molecules during impact, radiation pressure results from the momentum change of photons when they strike a surface and are absorbed or reflected.

We've seen that electromagnetic waves have particle-like properties, with the magnitude of the photon momentum given by $p = h/\lambda$. Nature is full of beautiful and sometimes surprising symmetries. In 1924 the French physicist Louis de Broglie suggested that particles as well as electromagnetic waves might have the dual wave-particle nature that we have described. If $p = h/\lambda$ holds for particles such as electrons as well as for photons, then a moving particle has a wavelength. Let's consider an electron (mass $m = 9.11 \times 10^{-31}$ kg) in the electron beam of a cathode-ray tube (described in Section 24–7). An electron accelerated through a potential difference of 100 V has a speed $v = 5.93 \times 10^6$ m/s. The momentum of this electron is

$$p = mv = (9.11 \times 10^{-31} \text{ kg})(5.93 \times 10^6 \text{ m/s})$$
$$= 5.40 \times 10^{-24} \text{ kg} \cdot \text{m/s}.$$

According to de Broglie's hypothesis, the wavelength λ is

$$\lambda = \frac{h}{p} = \frac{6.626 \times 10^{-34} \text{ J} \cdot \text{s}}{5.40 \times 10^{-24} \text{ kg} \cdot \text{m/s}}$$
$$= 1.23 \times 10^{-10} \text{ m} = 0.123 \text{ nm}.$$

This value is comparable to the spacing of atoms in a crystal. In 1927, the American physicists Clinton Davisson and Lester Germer observed interference effects in the scattering of an electron beam from the surface of a crystal of nickel. They inferred correctly that their results demonstrated the wave properties of electrons. Measurements of the interference pattern enabled them to determine the wavelengths of the electrons and to confirm de Broglie's hypothesis directly.

Electron beams are now used routinely in electron microscopes. The small wavelengths make it possible to see details that are far smaller than the wavelengths of visible light and hence beyond the range of optical microscopes.

The number of photons in the universe is not a conserved quantity; photons are created and destroyed in many processes. This lack of permanence may seem to distinguish photons from other particles such as electrons that we think of as having a permanent existence. But electrons (and all other particles) can also be created and destroyed, within the limitations imposed by various conservation laws including energy, momentum, angular momentum, electric charge, and others. Thus, ultimately, there is no fundamental distinction between photons and the other particles found in nature. Each type of particle has its individual characteristics and modes of interaction, but all are described with the same general language.

SUMMARY

KEY TERMS

physical optics, 1138

interference, 1138

principle of superposition, 1138

monochromatic light, 1139

coherent, 1139

constructive interference, 1140

destructive interference, 1141

antinodal curves, 1141

nodal curves, 1141

interference fringes, 1143

Newton's rings, 1152

nonreflective coating, 1153

reflective coating, 1153

■ Monochromatic light is light with a single frequency. Coherence is a definite, unchanging phase relationship between two waves or wave sources. The overlap of waves from two coherent sources of monochromatic light forms an interference pattern. The principle of superposition states that the total wave disturbance at any point is the sum of the disturbances from the separate waves.

■ When two sources are in phase, constructive interference occurs at points where the difference in path length from the two sources is zero or an integer number of wavelengths; destructive interference occurs at points where the path difference is a half-integer number of wavelengths. If the sources are both very far from a point P, the line from the sources to a point P makes an angle θ with the line perpendicular to the line of the sources. If the distance between sources is d, the condition for constructive interference at P is

$$d \sin \theta = m\lambda \qquad (m = 0, \pm 1, \pm 2, \ldots). \qquad (37\text{–}4)$$

The condition for destructive interference is

$$d \sin \theta = \left(m + \frac{1}{2}\right)\lambda \qquad (m = 0, \pm 1, \pm 2, \ldots). \qquad (37\text{--}5)$$

When θ is very small, the position y_m of the mth bright fringe on a screen located a distance R from the sources is given by

$$y_m = R\frac{m\lambda}{d}. \qquad (37\text{--}6)$$

- When two sinusoidal waves with equal amplitude E and phase difference ϕ are superimposed, the resultant amplitude E_P is

$$E_P = 2E\left|\cos\frac{\phi}{2}\right|, \qquad (37\text{--}7)$$

and the intensity I is

$$I = I_0 \cos^2\frac{\phi}{2}. \qquad (37\text{--}10)$$

When two sources emit waves in phase, the phase difference of the waves arriving at point P is related to the difference in path length $(r_2 - r_1)$ by

$$\phi = \frac{2\pi}{\lambda}(r_2 - r_1) = k(r_2 - r_1). \qquad (37\text{--}11)$$

- When light is reflected from both sides of a thin film of thickness t and no phase shift occurs at either surface, constructive interference between the reflected waves occurs when

$$2t = m\lambda \qquad (m = 0, 1, 2, \ldots). \qquad (37\text{--}17)$$

If a half-cycle phase shift occurs at one surface, this is the condition for destructive interference. A half-cycle phase shift occurs during reflection whenever the index of refraction of the second material is greater than that of the first.

- The Michelson interferometer uses a monochromatic source and can be used for high-precision measurements of wavelengths. Its original purpose was to detect motion of the earth relative to a hypothetical ether, the supposed medium for electromagnetic waves. The ether has never been detected, and the concept has been abandoned; the speed of light is the same relative to all observers. This is part of the foundation of the special theory of relativity.

- The energy in an electromagnetic wave is carried in bundles called photons. The energy E of one photon is proportional to the frequency f:

$$E = hf = \frac{hc}{\lambda}, \qquad (37\text{--}20)$$

where h is a fundamental constant called Planck's constant.

DISCUSSION QUESTIONS

Q37–1 A two-slit interference experiment is set up, and the fringes are displayed on a screen. Then the whole apparatus is immersed in the nearest swimming pool. How does the fringe pattern change?

Q37–2 Could an experiment similar to Young's two-slit experiment be performed with sound? How might this be carried out? Does it matter that sound waves are longitudinal and electromagnetic waves are transverse? Explain.

Q37–3 You place a bright flashlight inside a shoe box and cut very narrow slits a few micrometers apart in one end. When the light from the two slits falls on a screen in a dark room, will interference fringes be observed? What if the flashlight is replaced by a laser? Explain your answers.

Q37–4 Would the headlights of a distant car form a two-source interference pattern? If so, how might it be observed? If not, why not?

Q37–5 The two sources S_1 and S_2 in Fig. 37–3 emit waves of the same wavelength λ and are in phase with other. Suppose S_1 is a weaker source, so that the waves emitted by S_1 have half the amplitude of the waves emitted by S_2. How would this affect the positions of the antinodal lines and nodal lines? Would there be total reinforcement at points on the antinodal curves? Would there be total cancellation at points on the nodal curves? Explain your answers.

Q37–6 Could the Young two-slit interference experiment be performed with gamma rays? If not, why not? If so, discuss differences in the experimental design compared to the experiment with visible light.

Q37–7 Coherent red light illuminates two narrow slits that are 25 cm apart. Will a two-slit interference pattern be observed when the light from the slits falls on a screen? Explain.

Q37–8 Coherent light of wavelength λ falls on two narrow slits separated by a distance d. If d is less than some minimum value, no dark fringes are observed. Explain. In terms of λ, what is this minimum value of d?

Q37–9 A fellow student who values memorizing equations above understanding them combines Eqs. (37–4) and (37–13) to "prove" that ϕ can *only* equal $2\pi m$. How would you explain to this student that ϕ can have values other than $2\pi m$?

Q37–10 If the monochromatic light in Fig. 37–4a were replaced by white light, would a two-slit interference pattern be seen on the screen? Explain.

Q37–11 In using the superposition principle to calculate intensities in interference patterns, could you add the intensities of the waves instead of their amplitudes? Explain.

Q37–12 A glass windowpane with a thin film of water on it reflects less than when it is perfectly dry. Why?

Q37–13 A *very* thin soap film ($n = 1.33$), whose thickness is much less than a wavelength of visible light, looks black; it appears to reflect no light at all. Why? By contrast, an equally thin layer of soapy water ($n = 1.33$) on glass ($n = 1.50$) appears quite shiny. Why is there a difference?

Q37–14 Interference can occur in thin films. Why is it important that the films be *thin*? Why don't you get these effects with a relatively *thick* film? Where should you put the dividing line between "thin" and "thick"? Explain your reasoning.

Q37–15 If we shine white light on an air wedge like that shown in Fig. 37–11, the colors that are weak in the light *reflected* from any point along the wedge are strong in the light *transmitted* through the wedge. Explain why this should be so.

Q37–16 Monochromatic light is directed at normal incidence on a thin film. There is destructive interference for the reflected light, so the intensity of the reflected light is very low. What happened to the energy of the incident light?

Q37–17 When a thin oil film spreads out on a puddle of water, the thinnest part of the film looks dark in the resulting interference pattern. What does this tell you about the relative magnitudes of the refractive indices of oil and water?

EXERCISES

SECTION **37–2 INTERFERENCE AND COHERENT SOURCES**

37–1 Consider Fig. 37–3, which could represent interference between water waves in a ripple tank. Pick at least three points on the antinodal curve labeled "$m = 3$," and make measurements from the figure to show that Eq. (37–1) is indeed satisfied. Explain what measurements you made and how you measured the wavelength λ.

37–2 Figure 37–3 depicts the *antinodal curves* in the wave pattern produced by two identical coherent sources emitting waves of wavelength λ. The sources are separated by a distance 4λ. Each curve is the set of points on which the path difference $r_2 - r_1$ is equal to a given integer multiple of λ. a) Why are no antinodal curves shown in Fig. 37–3 for $m > 3$ or $m < -3$? b) On a copy of Fig. 37–3, sketch all possible *nodal curves*. On each nodal curve, the path difference is equal to $(m + \frac{1}{2})\lambda$, where m is an integer. Label each curve in your drawing by its value of m. c) For the antinodal curves, what is the maximum possible value of m? The minimum possible value? Explain how you determined these answers.

37–3 Figure 37–3 shows the wave pattern produced by two identical coherent sources emitting waves of wavelength λ and separated by a distance $d = 4\lambda$. a) Explain why the positive y-axis above S_1 constitutes an antinodal curve with $m = +4$ and why the negative y-axis below S_2 constitutes an antinodal curve with $m = -4$. b) Draw the wave pattern produced when the separation between the sources is reduced to 3λ. On your drawing, sketch all antinodal curves, that is, the curves on which $r_2 - r_1 = m\lambda$. Label each curve by its value of m. c) In general, what determines the maximum (most positive) and minimum (most negative) values of the integer m that labels the antinodal lines? d) Suppose the separation between the sources is increased to $7\frac{1}{2}\lambda$. How many antinodal curves will there be? To what values of m do they correspond? Explain your reasoning. (You should not have to make a drawing to answer these questions.)

37–4 Radio Interference. Two radio antennas A and B radiate in phase. Antenna B is 120 m to the right of antenna A. Consider point Q along the extension of the line connecting the antennas, a horizontal distance of 40 m to the right of antenna B. The frequency, and hence the wavelength, of the emitted waves can be varied. a) What is the longest wavelength for which there will be destructive interference at point Q? b) What is the longest wavelength for which there will be constructive interference at point Q?

37–5 A radio transmitting station operating at a frequency of 120 MHz has two identical antennas that radiate in phase. Antenna B is 9.00 m to the right of antenna A. Consider point P between the antennas and along the line connecting them, a horizontal distance x to the right of antenna A. For what values of x will constructive interference occur at point P?

37–6 Two light sources can be adjusted to emit monochromatic light of any visible wavelength. The two sources are coherent, 2.04 μm apart, and in line with an observer, so that one source is 2.04 μm farther from the observer than the other. a) For what visible wavelengths (400 to 700 nm) will the observer see the brightest light, owing to constructive interference? b) How would your answers to part (a) be affected if the two sources were not in line with the observer, but were still arranged so that one source is 2.04 μm farther away from the observer than the other?

37–7 In part (a) of Exercise 37–6, for what visible wavelengths will there be destructive interference at the location of the observer?

SECTION 37–3 TWO-SOURCE INTERFERENCE OF LIGHT

37–8 An FM radio station has a frequency of 107.9 MHz and uses two identical antennas mounted at the same elevation, 12.0 m apart. The antennas radiate in phase. The resulting radiation pattern has a maximum intensity along a horizontal line perpendicular to the line joining the antennas and midway between them. Assume that the intensity is observed at distances from the antennas that are much larger than 12.0 m. a) At what other angles (measured from the line of maximum intensity) is the intensity maximum? b) At what angles is it zero?

37–9 Coherent light from a sodium-vapor lamp is passed through a filter that blocks everything except for light of a single wavelength. It then falls on two slits separated by 0.460 mm. In the resulting interference pattern on a screen 2.20 m away, adjacent bright fringes are separated by 2.82 mm. What is the wavelength?

37–10 Young's experiment is performed with light from excited helium atoms ($\lambda = 502$ nm). Fringes are measured carefully on a screen 1.20 m away from the double slit, and the center of the twentieth fringe (not counting the central bright fringe) is found to be 10.6 mm from the center of the central bright fringe. What is the separation of the two slits?

37–11 Two slits spaced 0.450 mm apart are placed 75.0 cm from a screen. What is the distance between the second and third dark lines of the interference pattern on the screen when the slits are illuminated with coherent light with a wavelength of 500 nm?

SECTION 37–4 INTENSITY IN INTERFERENCE PATTERNS

37–12 Consider two antennas separated by 9.00 m, as described in Exercise 37–5. The two antennas radiate in phase at 15.0 MHz, and the resulting interference pattern is observed at distances from the antennas that are much greater than 9.00 m. a) In what directions is the intensity maximum, considering points in a horizontal plane? b) Calling the maximum intensity in part (a) I_0, determine in terms of I_0 the intensity in directions

making angles of 30°, 45°, 60°, and 90° to the direction of maximum intensity.

37–13 Two slits spaced 0.260 mm apart are placed 0.700 m from a screen and illuminated by coherent light that has a wavelength of 660 nm. The intensity at the center of the central maximum ($\theta = 0°$) is I_0. a) What is the distance on the screen from the center of the central maximum to the first minimum? b) What is the distance on the screen from the center of the central maximum to the point where the intensity has fallen to $I_0/2$?

37–14 Consider two antennas separated by 9.00 m that radiate in phase at 120 MHz, as described in Exercise 37–5. A receiver placed 150 m from both antennas measures an intensity I_0. The receiver is moved so that it is 1.8 m closer to one antenna than to the other. a) What is the phase difference ϕ between the two radio waves produced by this path difference? b) In terms of I_0, what is the intensity measured by the receiver at its new position?

37–15 Two slits spaced 0.130 mm apart are placed 90.0 cm from a screen and illuminated by coherent light of wavelength 550 nm. The intensity at the center of the central maximum ($\theta = 0°$) is 4.00×10^{-6} W/m^2. a) What is the distance on the screen from the center of the central maximum to the first minimum? b) What is the intensity at a point midway between the center of the central maximum and the first *minimum*?

37–16 Show that Eq. (37–14) gives zero-intensity directions that agree with Eq. (37–5).

SECTION 37–5 INTERFERENCE IN THIN FILMS

37–17 Two rectangular pieces of plane glass are laid one upon the other on a table. A thin strip of paper is placed between them at one edge so that a very thin wedge of air is formed. The plates are illuminated at normal incidence by 546-nm light from a mercury-vapor lamp. Interference fringes are formed, with 15.0 fringes per centimeter. Find the angle of the wedge.

37–18 A plate of glass 9.00 cm long is placed in contact with a second plate and is held at a small angle with it by a metal strip 0.0800 mm thick placed under one end. The space between the plates is filled with air. The glass is illuminated from above with light having a wavelength in air of 656 nm. How many interference fringes are observed per centimeter in the reflected light?

37–19 What is the thinnest film of a coating with $n = 1.42$ on glass ($n = 1.52$) for which destructive interference of the red component (650 nm) of an incident white light beam in air can take place by reflection?

37–20 A thin film of polystyrene is used as a nonreflecting coating for fabulite. (See Table 34–1.) What is the minimum thickness of the film required? Assume that the wavelength of the light in air is 480 nm.

37–21 Suppose the top plate in Example 37–4 (Section 37–5) is plastic with $n = 1.40$, the wedge is filled with silicone grease having $n = 1.50$, and the bottom plate is glass with $n = 1.60$. Calculate the spacing between the dark fringes.

37–22 Light with wavelength 480.0 nm in air is incident perpendicularly from air on a film 6.412 μm thick and with refractive index 1.400. Part of the light is reflected from the first surface of the film, and part enters the film and is reflected back at the second surface, where the film is again in contact with air. a) How many waves are contained along the path of this second part of the light in its round-trip through the film? b) What is the phase difference between these two parts of the light as they leave the film?

37–23 The human ear is especially sensitive to sounds at frequencies around 3500 Hz. Show that this can be understood by regarding the ear's auditory canal, which extends about 2.5 cm from the outside ear to the eardrum, as a "nonreflecting coating" for sound. (This same phenomenon can also be understood in terms of standing waves; see Problem 20–38. See Fig. 21–3 for a diagram of the human ear.)

37–24 What is the thinnest soap film (excluding the case of zero thickness) that appears black when illuminated with light with a wavelength of 480 nm? The index of refraction of the film is 1.33, and there is air on both sides of the film.

37–25 The walls of a soap bubble have about the same index of refraction as that of plain water, $n = 1.33$. There is air both inside and outside the bubble. a) What wavelength (in air) of visible light is most strongly reflected from a point on a soap bubble where its wall is 290 nm thick? To what color does this correspond? (See Fig. 33–15 and Table 33–1.) b) Repeat part (a) for a wall thickness of 340 nm.

37–26 A plastic film with index of refraction 1.85 is put on the surface of a car window to increase the reflectivity and this to keep the interior of the car cooler. The window glass has index of refraction 1.52. a) What minimum thickness is required if light of wavelength 550 nm in air reflected from the two sides of the film is to interfere constructively? b) It is found to be difficult to manufacture and install coatings as thin as calculated in part (a). What is the next greatest thickness for which there will also be constructive interference?

SECTION 37–6 THE MICHELSON INTERFEROMETER

37–27 How far must the mirror M_2 (Fig. 37–16) of the Michelson interferometer be moved so that 1800 fringes of He-Ne laser light ($\lambda = 633$ nm) move across a line in the field of view?

37–28 Jan first uses a Michelson interferometer with the 606-nm light from a krypton-86 lamp. He displaces the movable mirror away from him, counting 818 fringes moving across a line in his field of view. Then Linda replaces the krypton lamp with filtered 502-nm light from a helium lamp and displaces the movable mirror toward her. She also counts 818 fringes, but they move across the line in her field of view opposite to the direction they moved for Jan. Assume that both Jan and Linda counted to 818 correctly. a) What distance did each person move the mirror? b) What is the resultant displacement of the mirror?

*SECTION 37–7 THE PHOTON: A CASE STUDY IN QUANTUM PHYSICS

*37–29 Reaction Force on a Laser. a) The photons leaving a laser exert a reaction force of magnitude F on the laser. Show that $F = P/c$, where P is the power output of the laser. b) An extraordinarily powerful laser used in controlled fusion experiments emits pulses of light with a peak power of 2.8×10^{13} W. What is the magnitude of the reaction force at this peak power?

*37–30 How is the photon flux Φ (the number of photons per perpendicular area per time) related to the intensity I and frequency f of a beam of monoenergetic photons?

*37–31 A cesium-137 nucleus has a mass of 2.27×10^{-25} kg. It recoils with a kinetic energy of 2.75×10^{-19} J after emitting a gamma ray photon. Find the magnitude of the momentum of the gamma ray, as well as its energy, frequency, and wavelength.

*37–32 A particle with mass m and speed v has momentum with magnitude mv. a) In terms of these quantities, what is the de Broglie wavelength of the particle? b) What is the energy of a photon with the same wavelength as the particle? c) A proton is traveling with speed 3.00×10^5 m/s. What is the energy of the photon that has the same wavelength as the proton? Express your answer in eV.

*37–33 What is the range of photon energies and photon momenta for visible light, for which the wavelength range is 400–700 nm?

*37–34 An AM radio station broadcasts at 690 kHz with an output power of 50,000 W. How many photons does it emit during a 30-s commercial?

PROBLEMS

37–35 If the entire apparatus of Exercise 37–11 (slits, screen, and space in between) is immersed in water, what is then the distance between the second and third dark lines?

37–36 Coherent light of wavelength 450 nm falls on a double slit. On a screen 1.80 m away the distance between dark fringes is 4.20 mm. What is the separation of the slits?

37–37 Two identical audio speakers connected to the same amplifier produce monochromatic sound waves with a frequency that can be varied between 300 and 600 Hz. The speed of sound is 340 m/s. You find that where you are standing, you hear minimum-intensity sound. a) Explain why you hear minimum-intensity sound. b) If one of the speakers is moved 39.8 cm

toward you, the sound you hear has maximum intensity. What is the frequency of the sound? c) How much closer to you from the position in part (b) must the speaker be moved to the next position where you hear maximum intensity?

37–38 Two radio antennas radiating in phase are located at points A and B, 200 m apart (Fig. 37–17). The radio waves have a frequency of 5.80 MHz. A radio receiver is moved out from point B along a line perpendicular to the line connecting A and B (line BC in Fig. 37–17). At what distances from B will there be *destructive* interference? (*Note:* The distance of the receiver from the sources is not large in comparison to the separation of the sources, so Eq. (37–5) does not apply.)

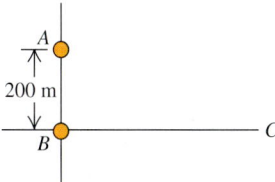

FIGURE 37–17 Problem 37–38.

37–39 a) In Fig. 37–3, suppose source S_2 is *not* in phase with S_1, but instead is *out* of phase by one half cycle. In this situation, Eq. (37–1) is the condition for *destructive* interference, and Eq. (37–2) is the condition for *constructive* interference. Explain why this is so. b) Suppose S_2 *leads* S_1 by a phase angle ϕ; that is, if the displacement of source S_1 is given by $x_1(t) = A \sin \omega t$, then the displacement of source S_2 is $x_2(t) = A \sin (\omega t + \phi)$. (In the situation of part (a), $\phi = \pi$.) Find expressions for the values of the path difference $r_2 - r_1$ that corresponds to constructive interference and to destructive interference.

37–40 The electric fields received at point P from two identical, coherent wave sources are $E_1(t) = E \cos (\omega t + \phi)$ and $E_2(t) = E \cos \omega t$. a) Use one of the trigonometric identities in Appendix B to show that the resultant wave is $E_P(t) = 2E \cos (\phi/2) \cos (\omega t + \phi/2)$. b) Show that the amplitude of this resultant wave is given by Eq. (37–7). c) Use the result of part (a) to show that at an interference maximum, the amplitude of the resultant wave is in phase with the original waves $E_1(t)$ and $E_2(t)$. d) Use the result of part (a) to show that near an interference minimum, the resultant wave is approximately one quarter cycle out of phase with either of the original waves. e) Show that the *instantaneous* Poynting vector at point P has magnitude $S = 4\epsilon_0 cE^2 \cos^2 (\phi/2) \cos^2 (\omega t + \phi/2)$, and that the *time-averaged* Poynting vector is given by Eq. (37–9).

37–41 Let the two sources S_1 and S_2 shown in Fig. 37–3 be located at $y = d$ and $y = -d$, respectively. a) Rewrite Eq. (37–1) in terms of the x- and y-coordinates of a point P in Fig. 37–3 at which constructive interference occurs. b) Your expression in part (a) is the equation for the antinodal curves shown in Fig. 37–3. Show that these curves are hyperbolas. (*Hint:* You may want to review the definition of a hyperbola in analytic geometry.) c) Repeat part (a) for Eq. (37–2), which describes points at which *destructive* interference occurs, and show that the *nodal* curves (not shown in Fig. 37–3) are also hyperbolas.

37–42 Consider a two-slit interference experiment in which the two slits are of different widths. As measured on a distant screen, the amplitude of the wave from the first slit is E, while the amplitude of the wave from the second slit is $2E$. a) Show that the intensity at any point in the interference pattern is

$$I = I_0 \left(\frac{5}{9} + \frac{4}{9} \cos \phi \right),$$

where ϕ is the phase difference between the two waves as measured at a particular point on the screen and I_0 is the maximum intensity in the pattern. b) Graph I versus ϕ (like Fig. 37–8). What is the minimum value of the intensity, and for what values of ϕ does it occur?

37–43 Two very narrow slits are spaced $1.80 \ \mu m$ apart and are placed 35.0 cm from a screen. What is the distance between the first and second dark lines of the interference pattern when the slits are illuminated with coherent light with $\lambda = 550$ nm? (*Hint:* The angle θ in Eq. (37–5) is *not* small.)

37–44 Red light of wavelength 700 nm is passed through a two-slit apparatus. At the same time, monochromatic visible light of another wavelength passes through the same apparatus. As a result, most of the pattern that appears on the screen is a mixture of two colors; however, the center of the third bright fringe ($m = 3$) of the red light appears pure red, with none of the other color. What is the wavelength of the second type of visible light? Do you need to know the slit spacing to answer this question? Why or why not?

37–45 Consider a two-slit interference pattern, for which the intensity distribution is given by Eq. (37–14). Let θ_m be the angular position of the mth bright fringe, where the intensity is I_0. Assume that θ_m is small, so that $\sin \theta_m \cong \theta_m$. Let θ_m^+ and θ_m^- be the two angles on either side of θ_m for which $I = \frac{1}{2} I_0$. The quantity $\Delta \theta_m = |\theta_m^+ - \theta_m^-|$ is the half-width of the mth fringe. Calculate $\Delta \theta_m$. How does $\Delta \theta_m$ depend on m?

37–46 White light reflects at normal incidence from the top and bottom surface of a glass plate ($n = 1.52$). There is air above and below the plate. Constructive interference is observed for light whose wavelength in air is 477.0 nm. What is the thickness of the plate if the next longer wavelength for which there is constructive interference is 540.6 nm?

37–47 A source S of monochromatic light and a detector D are both located in air a distance h above a horizontal plane sheet of glass, and are separated by a horizontal distance x. Waves reaching D directly from S interfere with waves that reflect off the glass. The distance x is small compared to h so that the reflection is at close to normal incidence. a) Show that the condition for constructive interference is $\sqrt{x^2 + 4h^2} - x = (m + \frac{1}{2})\lambda$, and the condition for destructive interference is $\sqrt{x^2 + 4h^2} - x = m\lambda$. (*Hint:* Take into account the phase change on reflection.) b) Let $h = 24$ cm and $x = 14$ cm. What is the longest wavelength for which there will be constructive interference?

37–48 Reflective Coatings and Herring. Herring and related fish have a brilliant silvery appearance that camouflages them while swimming in a sunlit ocean. The silveriness is due to *platelets* attached to the surfaces of these fish. Each platelet is made up of several alternating layers of crystalline guanine ($n = 1.80$) and of cytoplasm ($n = 1.333$, the same as water), with a guanine layer on the outside in contact with the surrounding water (Fig. 37–18). In one typical platelet, the guanine layers are 74 nm thick and the cytoplasm layers are 100 nm thick. a) For light striking the platelet surface at normal incidence, for what vacuum wavelengths of visible light will all of the reflections R_1, R_2, R_3, R_4, and R_5 shown in Fig. 37–18 be approximately in phase? If white light is shone on this platelet, what color will be most strongly reflected? (See Fig. 33–15.) The surface of a herring has very many platelets side by side with layers of different thickness, so that *all* visible wavelengths are reflected. b) Explain why such a "stack" of layers is more reflective than a single layer of guanine with cytoplasm underneath. (A stack of five guanine layers separated by cytoplasm layers reflects over

80% of incident light at the wavelength for which it is "tuned.") c) The color that is most strongly reflected from a platelet depends on the angle at which it is viewed. Explain why this should be so. (You can see these changes in color by examining a herring from different angles. Most of the platelets on these fish are oriented in the same way, so that they are vertical when the fish is swimming.)

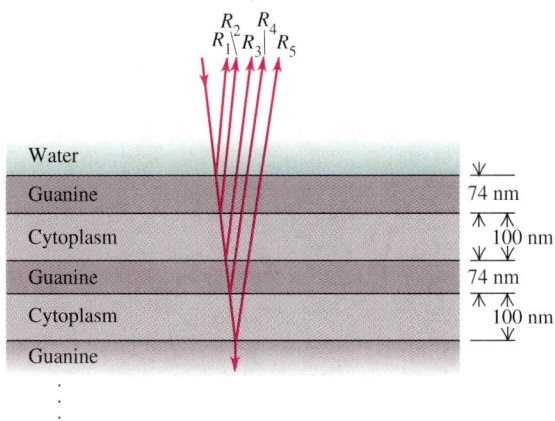

FIGURE 37–18 Problem 37–48.

37–49 A glass plate ($n = 1.53$) that is 0.485 μm thick and surrounded by air is illuminated by a beam of white light normal to the plate. a) What wavelengths (in air) within the limits of the visible spectrum ($\lambda = 400$ to 700 nm) are intensified in the reflected beam? b) What wavelengths within the visible spectrum are intensified in the transmitted light?

37–50 An oil tanker spills a large amount of oil ($n = 1.45$) into the sea ($n = 1.33$). a) If you look down onto the oil spill from overhead, what predominant wavelength of light do you see at a

point where the oil is 380 nm thick? What color is the light? (*Hint*: See Table 33–1.) b) In the water under the slick, what visible wavelength (as measured in air) is predominant in the transmitted light at the same place in the slick as in part (a)?

37–51 The radius of curvature of the convex surface of a plano-convex lens is 95.2 cm. The lens is placed convex side down on a perfectly flat glass plate that is illuminated from above with red light having a wavelength of 580 nm. Find the diameter of the second bright ring in the interference pattern.

37–52 Newton's rings can be seen when a plano-convex lens is placed on a flat glass surface (Problem 37–51). For a particular lens with an index of refraction of $n = 1.50$ and a glass plate with an index of $n = 1.80$, the diameter of the third bright ring is 0.850 mm. If water ($n = 1.33$) now fills the space between the lens and the plate, what is the new diameter of this ring?

37–53 In a Young's two-slit experiment a piece of glass with an index of refraction n and a thickness L is placed in front of the upper slit. a) Describe qualitatively what happens to the interference pattern. b) Derive an expression for the intensity I of the light at points on a screen as a function of n, L, and θ. Here θ is the usual angle measured from the center of the two slits. That is, determine the equation analogous to Eq. (37–14) but that also involves L and n for the glass plate. c) From your result in part (b) derive an expression for the values of θ that locate the maxima in the interference pattern. (That is, derive an equation analogous to Eq. (37–4).)

37–54 An initially evacuated tube is placed in one arm of a Michelson interferometer. As a gas is slowly added to the tube, 48 fringes are seen to cross the telescope crosshairs. If the length of the tube is 5.00 cm and the light source is a mercury-vapor lamp (546 nm), what is the index of refraction of the final quantity of gas within the tube? (*Note*: For gases it is convenient to give the value of $n - 1$ rather than n itself, since n is very close to unity.)

CHALLENGE PROBLEMS

37–55 The index of refraction of a glass rod is 1.48 at $T = 20.0°C$ and varies linearly with temperature, with a coefficient of $2.50 \times 10^{-5}/C°$. The coefficient of linear expansion of the glass is $5.00 \times 10^{-6}/C°$. At 20.0°C the length of the rod is 3.00 cm. A Michelson interferometer has this glass rod in one arm, and the rod is being heated so that its temperature increases at a rate of 5.00 C°/min. The light source has wavelength $\lambda = 589$ nm, and the rod initially is at $T = 20.0°C$. How many fringes cross the field of view each minute?

37–56 Figure 37–19 shows an interferometer known as *Fresnel's biprism*. The magnitude of the prism angle A is extremely small. a) If S_0 is a very narrow source slit, show that the separation of the two virtual coherent sources S_1 and S_2 is given by $d = 2aA(n - 1)$, where n is the index of refraction of the material of the prism. b) Calculate the spacing of the fringes of green light with wavelength 500 nm on a screen 2.00 m from the biprism. Take $a = 0.200$ m, $A = 3.50$ mrad, and $n = 1.50$.

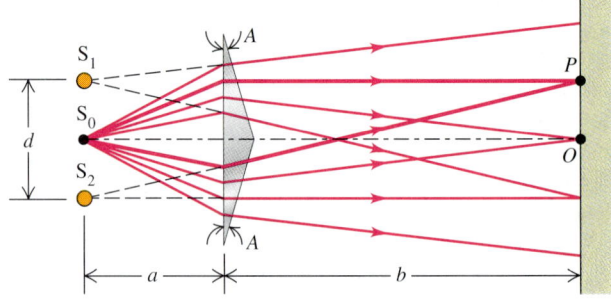

FIGURE 37–19 Challenge Problem 37–56.

Diffraction—the bending of waves around an obstacle or through an aperture—helps you hear an instructor who faces the blackboard. The sound waves used in speech have a long wavelength (about 1 m) and can easily bend around the instructor's head. By contrast, light waves have very short wavelengths and undergo very little diffraction. Hence you can't *see* around his head!

38

Diffraction

38-1 INTRODUCTION

Everyone is used to the idea that sound bends around corners. If sound didn't behave this way, you couldn't hear a police siren that's out of sight around a corner or the speech of a person whose back is turned to you. What may surprise you (and certainly surprised many scientists of the early nineteenth century) is that *light* can bend around corners as well. When light from a point source falls on a straight edge and casts a shadow, the edge of the shadow is never perfectly sharp. Some light appears in the area that we expect to be in the shadow, and we find alternating bright and dark fringes in the illuminated area. In general, light emerging from apertures doesn't behave precisely according to the predictions of the straight-line ray model of geometric optics.

The reason for these effects is that light, like sound, has wave characteristics. In Chapter 37 we studied the interference patterns that can arise when two light waves are combined. In this chapter we'll investigate interference effects due to combining *many* light waves. Such effects are referred to as *diffraction*. We'll find that the behavior of waves after they pass through an aperture is an example of diffraction; each infinitesimal part of the aperture acts as a source of waves, and the resulting pattern of light and dark is a result of interference among the waves emanating from these sources.

Light emerging from arrays of apertures also forms interesting light-and-dark patterns, and we'll see that measuring these enables a precise determination of the

wavelength of the light. We'll explore similar effects with x rays that are used to study the atomic structure of solids and liquids. Finally, we'll look at the physics of a *hologram*, which is nothing more than a special kind of interference pattern recorded on photographic film and reproduced. When properly illuminated, it forms a three-dimensional image of the original object.

38-2 FRESNEL AND FRAUNHOFER DIFFRACTION

According to geometric optics, when an opaque object is placed between a point light source and a screen, as in Fig. 38–1 (page 1166), the shadow of the object forms a perfectly sharp line. No light at all strikes the screen at points within the shadow, and the area outside the shadow is illuminated nearly uniformly. But as we saw in Chapter 37, the *wave* nature of light causes effects that can't be understood with the simple model of geometric optics. An important class of such effects occurs when light strikes a barrier that has an aperture or an edge. The interference patterns formed in such a situation are grouped under the heading **diffraction.**

An example of diffraction is shown in Fig. 38–2. The photograph in Fig. 38–2a (on the right) was made by placing a razor blade halfway between a pinhole, illuminated by monochromatic light, and a photographic film. The film recorded the shadow cast by the blade. Figure 38–2b (on

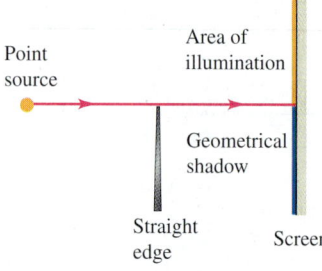

38–1 Geometric optics predicts that a straight edge should give a shadow with a sharp boundary and a region of relatively uniform illumination above it.

the left) is an enlargement of a region near the shadow of the left edge of the blade. The position of the *geometrical* shadow line is indicated by arrows. The area outside the geometric shadow is bordered by alternating bright and dark bands. There is some light in the shadow region, although this is not very visible in the photograph. The first bright band in Fig. 38–2b, just to the left of the geometrical shadow, is considerably brighter than in the region of uniform illumination to the extreme left. This simple experiment gives us some idea of the richness and complexity of what might seem to be a simple idea, the casting of a shadow by an opaque object.

We don't often observe diffraction patterns such as Fig. 38–2 in everyday life because most ordinary light sources are not monochromatic and are not point sources. If we use a white frosted light bulb instead of a point source in Fig. 38–1, each wavelength of the light from every point of the bulb forms its own diffraction pattern, but the patterns overlap to such an extent that we can't see any individual pattern.

Figure 38–3 shows a diffraction pattern formed by a steel ball about 3 mm in diameter. Note the rings in the pattern, both outside and inside the geometrical shadow area, and the *bright* spot at the very *center* of the shadow. The existence of this spot was predicted in 1818, on the basis of a wave theory of light, by the French mathematician Siméon-Denis Poisson during an extended debate in the French Academy of Sciences concerning the nature of light. Ironically, Poisson was *not* a believer in the wave theory of light, and he published this *apparently* absurd prediction as a death blow to the wave theory. But the prize committee of the Academy arranged for an experimental test, and soon the bright spot was actually observed. (It had in fact been seen as early as 1723, but those earlier experiments had gone unnoticed.)

Diffraction patterns can be analyzed by use of Huygens' principle (Section 34–8). Let's review that principle briefly. Every point of a wave front can be considered the source of secondary wavelets that spread out in all directions with a speed equal to the speed of propagation of the wave. The position of the wave front at any later time is the *envelope* of the secondary waves at that time. To find the resultant displacement at any point, we combine all the individual displacements produced by these secondary waves, using the superposition principle and taking into account their amplitudes and relative phases.

In Fig. 38–1, both the point source and the screen are relatively close to the obstacle forming the diffraction pattern. This situation is described as *near-field diffraction* or **Fresnel diffraction,** pronounced "Freh-nell" (after the French scientist Augustin Jean Fresnel, 1788–1827). If the source, obstacle, and screen are far enough away that all lines from the source to the obstacle can be considered parallel and all lines from the

38–2 (a) Actual shadow of a razor blade illuminated by monochromatic light from a point source. (b) Enlarged shadow of the straight edge. The arrows show the position of the *geometric* shadow.

(a)

(b)

obstacle to a point in the pattern can be considered parallel, the phenomenon is called *far-field diffraction* or **Fraunhofer diffraction** (after the German physicist Joseph von Fraunhofer, 1787–1826). We will restrict the following discussion to Fraunhofer diffraction, which is usually simpler to analyze in detail than Fresnel diffraction.

Diffraction is sometimes described as "the bending of light around an obstacle." But the process that causes diffraction is present in the propagation of *every* wave. When part of the wave is cut off by some obstacle, we observe diffraction effects that result from interference of the remaining parts of the wave fronts. Optical instruments typically use only a limited portion of a wave; for example, a telescope uses only the part of a wave that is admitted by its objective lens or mirror. Thus diffraction plays a role in nearly all optical phenomena.

Finally, we emphasize that there is no fundamental distinction between *interference* and *diffraction.* In Chapter 37 we used the term *interference* for effects involving waves from a small number of sources, usually two. *Diffraction* usually involves a *continuous* distribution of Huygens' wavelets across the area of an aperture, or a very large number of sources or apertures. But both categories of phenomena are governed by the same basic physics of superposition and Huygens' principle.

38–3 Fresnel diffraction pattern formed by a steel ball 3 mm in diameter. The Poisson bright spot is seen at the center of the shadow area.

38-3 DIFFRACTION FROM A SINGLE SLIT

In this section we'll discuss the diffraction pattern formed by plane-wave (parallel-ray) monochromatic light when it emerges from a long, narrow slit, as shown in Fig. 38–4. We call the narrow dimension the *width,* even though in this figure it is a vertical dimension.

According to geometric optics, the transmitted beam should have the same cross section as the slit, as in Fig. 38–4a. What is *actually* observed is the pattern shown in Fig. 38–4b. The beam spreads out vertically after passing through the slit. The diffraction pattern consists of a central bright band, which may be much broader than the width of the slit, bordered by alternating dark and bright bands with rapidly decreasing intensity. About 85% of the power in the transmitted beam is in the central bright band, whose width is found to be *inversely* proportional to the width of the slit. In general, the smaller the width of the slit, the broader the entire diffraction pattern. (The *horizontal* spreading of the beam in Fig. 38–4b is negligible because the horizontal dimension of the slit is relatively large.) You can easily observe a similar diffraction pattern by looking at a point source, such as a distant street light, through a narrow slit formed between your two thumbs held in front of your eye; the retina of your eye corresponds to the screen.

16.6
Single-Slit Diffraction

38–4 (a) Geometric "shadow" of a horizontal slit. (b) Diffraction pattern of a horizontal slit. The slit width has been greatly exaggerated.

(a)

(b)

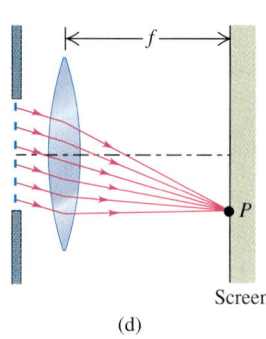

(a)   (b)   (c)   (d)

38–5 Diffraction by a single rectangular slit. The long sides of the slit are perpendicular to the figure. (a) Plane waves are incident on the slit; each strip of the slit acts as a source of cylindrical secondary wavelets. (b) Fresnel (near-field) diffraction occurs when the rays from each strip to point P are not parallel. (c) Fraunhofer (far-field) diffraction occurs when those rays can be considered parallel, as when the screen is very far from the slit. (d) Using a converging cylindrical lens to obtain a Fraunhofer diffraction pattern on a nearby screen.

Figure 38–5 shows a side view of the same setup; the long sides of the slit are perpendicular to the figure, and plane waves are incident on the slit from the left. According to Huygens' principle, each element of area of the slit opening can be considered as a source of secondary waves. In particular, imagine dividing the slit into several narrow strips of equal width, parallel to the long edges and perpendicular to the page. Two such strips are shown in Fig. 38–5a. Cylindrical secondary wavelets, shown in cross section, spread out from each strip.

In Fig. 38–5b a screen is placed to the right of the slit. We can calculate the resultant intensity at a point P on the screen by adding the contributions from the individual wavelets, taking proper account of their various phases and amplitudes. It's easiest to do this calculation if we assume that the screen is far enough away that all the rays from various parts of the slit to a particular point P on the screen are parallel, as in Fig. 38–5c. An equivalent situation is Fig. 38–5d, in which the rays to the lens are parallel and the lens forms a reduced image of the same pattern that would be formed on an infinitely distant screen without the lens. We might expect that the various light paths through the lens would introduce additional phase shifts, but in fact it can be shown that all the paths have *equal* phase shifts, so this is not a problem.

The situation of Fig. 38–5b is Fresnel diffraction; those in Figs. 38–5c and 38–5d, where the outgoing rays are considered parallel, are Fraunhofer diffraction. We can derive quite simply the most important characteristics of the Fraunhofer diffraction pattern from a single slit. First consider two narrow strips, one just below the top edge of the drawing of the slit and one at its center, shown in end view in Fig. 38–6. The difference in path length to point P is $(a/2) \sin \theta$, where a is the slit width and θ is the angle between the perpendicular to the slit and a line from the center of the slit to P. Suppose this path difference happens to be equal to $\lambda/2$; then light from these two strips arrives at point P with a half-cycle phase difference, and cancellation occurs.

38–6 Side view of a horizontal slit. (a) When the distance x to the screen is much greater than the slit width a, the rays from a distance $a/2$ apart may be considered parallel. (b) Enlarged view of half of the slit. The ray from the middle of the slit travels a distance $(a/2) \sin \theta$ farther to point P than does the ray from the top edge of the slit.

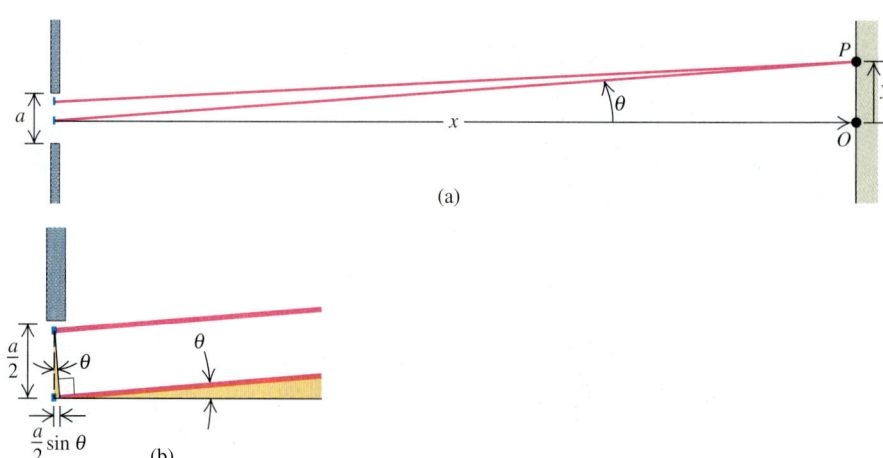

Similarly, light from two strips immediately *below* the two in the figure also arrives at P a half-cycle out of phase. In fact, the light from *every* strip in the top half of the slit cancels out the light from a corresponding strip in the bottom half. The result is complete cancellation at P for the combined light from the entire slit, giving a dark fringe in the interference pattern. That is, a dark fringe occurs whenever

$$\frac{a}{2}\sin\theta = \pm\frac{\lambda}{2}, \quad \text{or} \quad \sin\theta = \pm\frac{\lambda}{a}. \tag{38-1}$$

The plus-or-minus ($\pm$) sign in Eq. (38–1) says that there are symmetrical dark fringes above and below point O in Fig. 38–6a. The upper fringe ($\theta > 0$) occurs at a point P where light from the bottom half of the slit travels $\lambda/2$ farther to P than does light from the top half; the lower fringe ($\theta < 0$) occurs where light from the *top* half travels $\lambda/2$ farther than light from the *bottom* half.

We may also divide the screen into quarters, sixths, and so on, and use the above argument to show that a dark fringe occurs whenever $\sin\theta = \pm 2\lambda/a$, $\pm 3\lambda/a$, and so on. Thus the condition for a *dark* fringe is

$$\sin\theta = \frac{m\lambda}{a} \quad (m = \pm 1, \pm 2, \pm 3, \ldots) \quad \begin{array}{l}\text{(dark fringes in single-slit} \\ \text{diffraction).}\end{array} \tag{38-2}$$

For example, if the slit width is equal to ten wavelengths ($a = 10\lambda$), dark fringes occur at $\sin\theta = \pm\frac{1}{10}, \pm\frac{2}{10}, \pm\frac{3}{10}, \ldots$. Between the dark fringes are bright fringes. We also note that $\sin\theta = 0$ corresponds to a *bright* band; in this case, light from the entire slit arrives at P in phase. Thus it would be wrong to put $m = 0$ in Eq. (38–2). The central bright fringe is wider than the other bright fringes, as Fig. 38–4 shows. In the small-angle approximation that we will use below, it is exactly *twice* as wide.

With light, the wavelength λ is of the order of 500 nm $= 5 \times 10^{-7}$ m. This is often much smaller than the slit width a; a typical slit width is 10^{-2} cm $= 10^{-4}$ m. Therefore the values of θ in Eq. (38–2) are often so small that the approximation $\sin\theta \approx \theta$ (where θ is in radians) is a very good one. In that case we can rewrite this equation as

$$\theta = \frac{m\lambda}{a}, \quad (m = \pm 1, \pm 2, \pm 3, \ldots) \quad \text{(for small angles } \theta\text{),}$$

where θ is in *radians*. Also, if the distance from slit to screen is x, as in Fig. 38–6a, and the vertical distance of the mth dark band from the center of the pattern is y_m, then $\tan\theta = y_m/x$. For small θ we may also approximate $\tan\theta$ by θ (in radians), and we then find

$$y_m = x\frac{m\lambda}{a} \quad \text{(for } y_m \ll x\text{).} \tag{38-3}$$

CAUTION ▶ This equation has the same form as the equation for the two-slit pattern, Eq. (37–6), except that in Eq. (38–3) we use x rather than R for the distance to the screen. But Eq. (38–3) gives the positions of the *dark* fringes in a *single-slit* pattern rather than the *bright* fringes in a *double-slit* pattern. Also, $m = 0$ is *not* a dark fringe. Be careful! ◀

EXAMPLE 38–1

You pass 633-nm laser light through a narrow slit and observe the diffraction pattern on a screen 6.0 m away. You find that the distance on the screen between the centers of the first minima outside the central bright fringe is 32 mm. How wide is the slit?

SOLUTION In this situation the distances between points on the

screen are much smaller than the distance from the slit to the screen, so the angle θ is very small. Hence we can use the approximate relation of Eq. (38–3). Figure 38–6 shows the various distances. The distance y_1 from the central maximum to the first minimum on either side is half the distance between the two first minima, so $y_1 = (32 \text{ mm})/2$. Solving Eq. (38–3) for the slit width a and substituting $m = 1$, we find

$$a = \frac{x\lambda}{y_1} = \frac{(6.0 \text{ m})(633 \times 10^{-9} \text{ m})}{(32 \times 10^{-3} \text{ m})/2}$$

$$= 2.4 \times 10^{-4} \text{ m} = 0.24 \text{ mm}.$$

Can you show that the distance between the *second* minima on the two sides is 2(32 mm), and so on?

38–4 Intensity in the Single-Slit Pattern

(a) $\theta = 0$

(b)

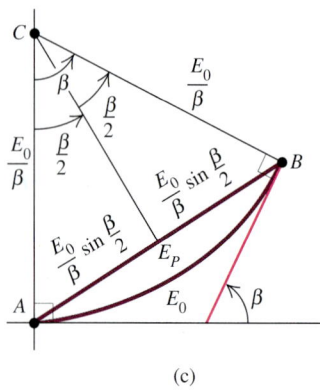

(c)

38–7 Phasor diagrams used to find the amplitude of the $\vec{E}$ field in single-slit diffraction. Each phasor represents the $\vec{E}$ field from a single strip within the slit. (a) Diagram when all phasors are in phase ($\theta = 0$, $\beta = 0$). (b) Diagram when each phasor differs in phase slightly from the preceding one. The total phase difference between the first and last phasors is β. (c) Limit reached by the phasor diagram when the slit is subdivided into infinitely many strips.

We can derive an expression for the intensity distribution for the single-slit diffraction pattern by the same phasor-addition method that we used in Section 37–4 to obtain Eqs. (37–10) and (37–14) for the two-slit interference pattern. We again imagine a plane wave front at the slit subdivided into a large number of strips. We superpose the contributions of the Huygens wavelets from all the strips at a point P on a distant screen at an angle θ from the normal to the slit plane. To do this, we use a phasor to represent the sinusoidally varying $\vec{E}$ field from each individual strip. The magnitude of the vector sum of the phasors at each point P is the amplitude E_P of the total $\vec{E}$ field at that point. The intensity at P is proportional to E_P^2.

At the point O shown in Figure 38–6a, corresponding to the center of the pattern where $\theta = 0$, there are negligible path differences for $x \gg a$; the phasors are all essentially *in phase* (that is, have the same direction). In Fig. 38–7a we draw the phasors at time $t = 0$ and denote the resultant amplitude at O by E_0. In this illustration we have divided the slit into 14 strips.

Now consider wavelets arriving from different strips at point P in Fig. 38–6a, at an angle θ from point O. Because of the differences in path length, there are now phase differences between wavelets coming from adjacent strips; the corresponding phasor diagram is shown in Fig. 38–7b. The vector sum of the phasors is now part of the perimeter of a many-sided polygon, and E_P, the amplitude of the resultant electric field at P, is the *chord*. The angle β is the total phase difference between the wave from the top strip of Fig. 38–6a and the wave from the bottom strip; that is, β is the phase of the wave at P from the top strip with respect to the wave received at P from the bottom strip.

We may imagine dividing the slit into narrower and narrower strips. In the limit that there is an infinite number of infinitesimally narrow strips, the curved trail of phasors becomes an *arc of a circle* (Fig. 38–7c), with arc length equal to the length E_0 in Fig. 38–7a. The center C of this arc is found by constructing perpendiculars at A and B. From the relationship between arc length, radius, and angle, the radius of the arc is E_0/β; the amplitude E_P of the resultant electric field at P is equal to the chord AB, which is $2(E_0/\beta) \sin (\beta/2)$. (Note that β *must* be in radians!) We then have

$$E_P = E_0 \frac{\sin(\beta/2)}{\beta/2} \qquad \text{(amplitude in single-slit diffraction)}. \qquad (38–4)$$

The intensity at each point on the screen is proportional to the square of the amplitude given by Eq. (38–4). If I_0 is the intensity in the straight-ahead direction where $\theta = 0$ and $\beta = 0$, then the intensity I at any point is

$$I = I_0 \left[\frac{\sin(\beta/2)}{\beta/2} \right]^2 \qquad \text{(intensity in single-slit diffraction)}. \qquad (38–5)$$

We can express the phase difference β in terms of geometric quantities, as we did for the two-slit pattern. From Eq. (37–11) the phase difference is $2\pi/\lambda$ times the path difference. Figure 38–6 shows that the path difference between the ray from the top of the slit and the ray from the middle of the slit is $(a/2) \sin \theta$. The path difference between the rays from the top of the slit and the bottom of the slit is twice this, so

$$\beta = \frac{2\pi}{\lambda} a \sin \theta, \qquad (38–6)$$

(a)   (b)

38–8 (a) Intensity distribution for a single slit. (b) Photograph of the Fraunhofer diffraction pattern of a single horizontal slit.

and Eq. (38–5) becomes

$$I = I_0\left\{\frac{\sin[\pi a(\sin\theta)/\lambda]}{\pi a(\sin\theta)/\lambda}\right\}^2 \qquad \text{(intensity in single-slit diffraction)}. \qquad (38\text{–}7)$$

This equation expresses the intensity directly in terms of the angle θ. In many calculations it is easier first to calculate the phase angle β, using Eq. (38–6), and then to use Eq. (38–5).

Equation (38–7) is plotted in Fig. 38–8a, and a photograph of an actual diffraction pattern is shown in Fig. 38–8b. Note that the central intensity peak is much larger than any of the others and that the peak intensities decrease rapidly as we go away from the center of the pattern.

The dark fringes in the pattern are the places where $I = 0$. These occur at points for which the numerator of Eq. (38–5) is zero so that β is a multiple of 2π. From Eq. (38–6) this corresponds to

$$\frac{a\sin\theta}{\lambda} = m \qquad (m = \pm1, \pm2, \ldots),$$

$$\sin\theta = \frac{m\lambda}{a} \qquad (m = \pm1, \pm2, \ldots). \qquad (38\text{–}8)$$

This agrees with our previous result, Eq. (38–2). Note again that $\beta = 0$ (corresponding to $\theta = 0$) is *not* a minimum. Equation (38–5) is indeterminate at $\beta = 0$, but we can evaluate the limit as $\beta \to 0$ using L'Hôpital's rule. We find that at $\beta = 0$, $I = I_0$, as we should expect.

INTENSITY MAXIMA IN THE SINGLE-SLIT PATTERN

We can also use Eq. (38–5) to calculate the positions of the peaks, or *intensity maxima*, and the intensities at these peaks. This is not quite as simple as it may appear. We might expect the peaks to occur where the sine function reaches the value ±1, namely, where $\beta = \pm\pi, \pm3\pi, \pm5\pi$, or in general,

$$\beta \approx \pm(2m + 1)\pi \qquad (m = 0, 1, 2, \ldots). \qquad (38\text{–}9)$$

This is *approximately* correct, but because of the factor $(\beta/2)^2$ in the denominator of Eq. (38–5), the maxima don't occur precisely at these points. When we take the derivative of Eq. (38–5) with respect to β and set it equal to zero to try to find the maxima and

These water waves passing through a small aperture behave exactly like light waves in single-slit diffraction. Most of the intensity is within the central maximum of the diffraction pattern. That is, most of the diffracted wave remains within an angle θ from the perpendicular to the aperture, where $\sin\theta = \lambda/a$ (the first diffraction minimum).

minima, we get a transcendental equation that has to be solved numerically. In fact there is *no* maximum near $\beta = \pm\pi$. The first maxima on either side of the central maximum, near $\beta = \pm 3\pi$, actually occur at $\pm 2.860\pi$. The second side maxima, near $\beta = \pm 5\pi$, are actually at $\pm 4.918\pi$, and so on. The error in Eq. (38–9) vanishes in the limit of large m, that is, for intensity maxima far from the center of the pattern.

To find the intensities at the side maxima, we substitute these values of β back into Eq. (38–5). Using the approximate expression in Eq. (38–9), we get

$$I_m \approx \frac{I_0}{\left(m + \dfrac{1}{2}\right)^2 \pi^2},$$ (38–10)

where I_m is the intensity of the mth side maximum and I_0 is the intensity of the central maximum. Equation (38–10) gives the series of intensities

$$0.0450 I_0, \qquad 0.0162 I_0, \qquad 0.0083 I_0,$$

and so on. As we have pointed out, this equation is only approximately correct. The actual intensities of the side maxima turn out to be

$$0.0472 I_0, \qquad 0.0165 I_0, \qquad 0.0083 I_0, \qquad \dots$$

Note that the intensities of the side maxima decrease very rapidly, as Fig. 38–8 also shows. Even the first side maxima have less than 5% of the intensity of the central maximum.

WIDTH OF THE SINGLE-SLIT PATTERN

For small angles the angular spread of the diffraction pattern is inversely proportional to the slit width a or, more precisely, to the ratio of a to the wavelength λ. Figure 38–9 shows graphs of intensity I as a function of the angle θ for various values of the ratio a/λ.

With light waves, the wavelength λ is often much smaller than the slit width a, and the values of θ in Eqs. (38–6) and (38–7) are so small that the approximation $\sin\theta = \theta$ is very good. With this approximation the position θ_1 of the first minimum beside the central maximum, corresponding to $\beta/2 = \pi$, is, from Eq. (38–7),

$$\theta_1 = \frac{\lambda}{a}.$$ (38–11)

This characterizes the width (angular spread) of the central maximum, and we see that it is *inversely* proportional to the slit width a. When the small-angle approximation is valid, the central maximum is exactly twice as wide as each side maximum. When a is of the order of a centimeter or more, θ_1 is so small that we can consider practically all the light to be concentrated at the geometrical focus. But when a is less than λ, the central maximum spreads over 180°, and the fringe pattern is not seen at all.

It's important to keep in mind that diffraction occurs for *all* kinds of waves, not just light. Sound waves undergo diffraction when they pass through a slit or aperture such as an ordinary doorway. The sound waves used in speech have wavelengths of about a meter or greater, and a typical doorway is less than 1 m wide; in this situation, a is less than λ, and the central intensity maximum extends over 180°. This is why the sounds coming through an open doorway can easily be heard by an eavesdropper hiding out of sight around the corner. By contrast, there is essentially no diffraction of visible light through such a doorway because the width a is very much greater than the wavelength λ (of order 5×10^{-7} m). You can *hear* around corners because typical sound waves have relatively long wavelengths; you cannot *see* around corners because the wavelength of visible light is very short.

(a) $a = \lambda$

(b) $a = 5\lambda$

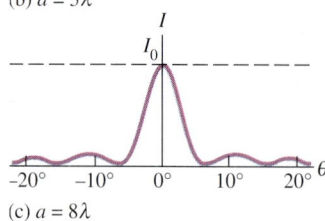

(c) $a = 8\lambda$

38–9 (a) When the slit width a is less than or equal to the wavelength λ, the central maximum of the diffraction pattern is spread out from +90° to −90°. (b), (c) The angular width of the central maximum decreases when the ratio a/λ is increased, either by increasing the slit width or by decreasing the wavelength.

EXAMPLE 38-2

a) In a single-slit diffraction pattern, what is the intensity at a point where the total phase difference between wavelets from the top and bottom of the slit is 66 rad? b) If this point is 7.0° away from the central maximum, how many wavelengths wide is the slit?

SOLUTION a) We are given $\beta = 66$ rad, so $\beta/2 = 33$ rad. Then Eq. (38–5) becomes

$$I = I_0 \left[\frac{\sin(33 \text{ rad})}{33 \text{ rad}} \right]^2 = (9.2 \times 10^{-4}) I_0.$$

This is the intensity of the *tenth* side maximum; it is very much less than the intensity I_0 of the central maximum. The actual location of this maximum is at $\beta = 65.91$ rad $= 20.98\pi$, approximately midway between the minima at $\beta = 20\pi$ and $\beta = 22\pi$.

b) We solve Eq. (38–6) for a:

$$a = \frac{\beta\lambda}{2\pi \sin\theta} = \frac{(66 \text{ rad})\lambda}{(2\pi \text{ rad}) \sin 7.0°} = 86\lambda.$$

For example, for 550-nm light, the slit width a is $(86)(550 \text{ nm}) = 4.7 \times 10^{-5}$ m $= 0.047$ mm, or roughly $\frac{1}{20}$ mm.

EXAMPLE 38-3

In the experiment described in Example 38–1 (Section 38–3), what is the intensity at a point on the screen 3.0 mm from the center of the pattern? The intensity at the center of the pattern is I_0.

SOLUTION From Example 38–1 the first intensity minimum is (32 mm)/2 from the center of the pattern, so the point in question here lies within the central maximum. To find the intensity, we must first find the angle θ for our point. Referring to

Fig. 38–6a, we have $y = 3.0$ mm and $x = 6.0$ m, so $\tan\theta = y/x = (3.0 \times 10^{-3} \text{ m})/(6.0 \text{ m}) = 5.0 \times 10^{-4}$; since this is so small, the values of $\tan\theta$, $\sin\theta$, and θ (in radians) are all nearly the same. Then, using Eq. (38–7), we have

$$\frac{\pi a \sin\theta}{\lambda} = \frac{\pi(2.4 \times 10^{-4} \text{ m})(5.0 \times 10^{-4})}{6.33 \times 10^{-7} \text{ m}} = 0.60,$$

$$I = I_0 \left(\frac{\sin 0.60}{0.60} \right)^2 = 0.89 I_0.$$

38–5 MULTIPLE SLITS

In Sections 37–3 and 37–4 we analyzed interference from two point sources or from two very narrow slits; in this analysis we ignored effects due to the finite (that is, nonzero) slit width. In Sections 38–3 and 38–4 we considered the diffraction effects that occur when light passes through a single slit of finite width. Additional interesting effects occur when we have two slits with finite width or when there are several very narrow slits.

TWO SLITS OF FINITE WIDTH

Let's take another look at the two-slit pattern in the more realistic case in which the slits have finite width. If the slits are narrow in comparison to the wavelength, we can assume that light from each slit spreads out uniformly in all directions to the right of the slit. We used this assumption in Section 37–3 to calculate the interference pattern described by Eq. (37–10) or (37–15), consisting of a series of equally spaced, equally intense maxima. However, when the slits have finite width, the peaks in the two-slit interference pattern are modulated by the single-slit diffraction pattern characteristic of the width of each slit.

Figure 38–10a shows the intensity in a single-slit diffraction pattern with slit width a. The *diffraction minima* are labeled by the integer $m_d = \pm 1, \pm 2, \ldots$ ("d" for "diffraction"). Figure 38–10b shows the pattern formed by two very narrow slits with distance d between slits, where d is four times as great as the single-slit width a in Fig. 38–10a; that is, $d = 4a$. The *interference maxima* are labeled by the integer $m_i = 0, \pm 1, \pm 2, \ldots$ ("i" for "interference"). We note that the spacing between adjacent minima in the single-slit pattern is four times as great as in the two-slit pattern. Now suppose we widen each of the narrow slits to the same width a as that of the single slit in Fig. 38–10a. Figure 38–10c (page 1174) shows the pattern from two slits with width a, separated by a distance (between centers) $d = 4a$. The effect of the finite width of the slits is to superimpose

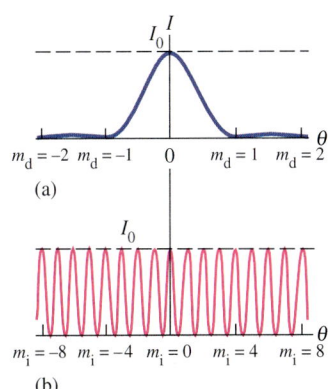

38–10 (a) Single-slit diffraction pattern for a slit of width a. (b) Double-slit interference pattern for narrow slits whose separation d is four times the width of the slit in (a). *(Continued on next page.)*

(c)

(d)

38–10 (continued) (c) Calculated double-slit pattern for $d = 4a$. The intensity as a function of θ is shown in red; the purple curve shows the outer limits or "envelope" of the intensity function. (d) Actual photograph of the pattern calculated in (c). For $d = 4a$, every fourth interference maximum at the sides is missing.

the two patterns, that is, to multiply the two intensities at each point. The two-slit peaks are in the same positions as before, but their intensities are modulated by the single-slit pattern. The expression for the intensity shown in Fig. 38–10c is proportional to the product of the two-slit and single-slit expressions, Eqs. (37–10) and (38–5):

$$I = I_0 \cos^2 \frac{\phi}{2} \left[\frac{\sin(\beta/2)}{\beta/2} \right]^2 \qquad \text{(two slits of finite width),} \qquad (38\text{–}12)$$

where, as before,

$$\phi = \frac{2\pi d}{\lambda} \sin \theta, \qquad \beta = \frac{2\pi a}{\lambda} \sin \theta.$$

Note that in Fig. 38–10c, every fourth interference maximum at the sides is *missing* because these interference maxima ($m_i = \pm 4, \pm 8, \ldots$) coincide with diffraction minima ($m_d = \pm 1, \pm 2, \ldots$). This can also be seen in Fig. 38–10d, which is a photograph of an actual pattern with $d = 4a$. You should be able to convince yourself that there will be "missing" maxima whenever d is an integer multiple of a.

Figures 38–10c and 38–10d show that as you move away from the central bright maximum of the two-slit pattern, the intensity of the maxima decreases. This is a result of the single-slit modulating pattern shown in Fig. 38–10a; mathematically, the decrease in intensity arises from the factor $(\beta/2)^2$ in the denominator of Eq. (38–12). This decrease in intensity can also be seen in Fig. 37–5 (Section 37–3). The narrower the slits, the broader the single-slit pattern (as in Fig. 38–9) and the slower the decrease in intensity from one interference maximum to the next.

Shall we call the pattern in Fig. 38–10d *interference* or *diffraction?* It's really both, since it results from superposition of waves coming from various parts of the two apertures. There is no truly fundamental distinction between interference and diffraction.

SEVERAL SLITS

Next let's consider patterns produced by *several* very narrow slits. As we will see, systems of narrow slits are of tremendous practical importance in *spectroscopy,* the determination of the particular wavelengths of light coming from a source. Assume that each slit is narrow in comparison to the wavelength, so its diffraction pattern spreads out nearly uniformly. Figure 38–11 shows an array of eight narrow slits, with distance d between adjacent slits. Constructive interference occurs for rays at angle θ to the normal that arrive at point P with a path difference between adjacent slits equal to an integer

38–11 In multiple-slit diffraction, rays from every slit arrive in phase, giving a sharp maximum if the path difference between adjacent slits is a whole number of wavelengths. Here a lens is used to give a Fraunhofer pattern on a nearby screen, as in Fig. 38–5d.

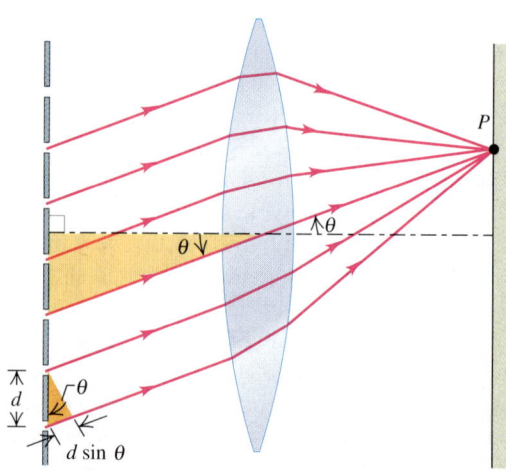

number of wavelengths,

$$d \sin \theta = m\lambda \qquad (m = 0, \pm 1, \pm 2, \ldots).$$

This means that reinforcement occurs when the phase difference ϕ at P for light from adjacent slits is an integer multiple of 2π. That is, the maxima in the pattern occur at the *same* positions as for *two* slits with the same spacing. To this extent the pattern resembles the two-slit pattern.

But what happens *between* the maxima? In the two-slit pattern, there is exactly one intensity minimum located midway between each pair of maxima, corresponding to angles for which the phase difference between waves from the two sources is π, 3π, 5π, and so on. In the eight-slit pattern these are also minima because the light from adjacent slits cancels out in pairs, corresponding to the phasor diagram in Fig. 38–12a. But these are not the only minima in the eight-slit pattern. For example, when the phase difference ϕ from adjacent sources is $\pi/4$, the phasor diagram is as shown in Fig. 38–12b; the total (resultant) phasor is zero, and the intensity is zero. When $\phi = \pi/2$, we get the phasor diagram of Fig. 38–12c, and again both the total phasor and the intensity are zero. More generally, the intensity with eight slits is zero whenever ϕ is an integer multiple of $\pi/4$, *except* when ϕ is a multiple of 2π. Thus there are seven minima for every maximum.

Detailed calculation shows that the eight-slit pattern is as shown in Fig. 38–13b. The large maxima, called *principal maxima*, are in the same positions as for the two-slit pattern of Fig. 38–13a but are much narrower. If the phase difference ϕ between adjacent slits is slightly different from a multiple of 2π, the waves from slits 1 and 2 will be only a little out of phase; however, the phase difference between slits 1 and 3 will be greater, that between slits 1 and 4 will be greater still, and so on. This leads to a partial cancellation for angles that are only slightly different from the angle for a maximum, giving the narrow maxima in Fig. 38–13b. The maxima are even narrower with sixteen slits (Fig. 38–13c).

We invite you to show that when there are N slits, there are $(N-1)$ minima between each pair of principal maxima and a minimum occurs whenever ϕ is an integral multiple of $2\pi/N$ (except when ϕ is an integral multiple of 2π, which gives a principal maximum). There are small *secondary* intensity maxima between the minima; these become smaller in comparison to the principal maxima as N increases. The greater the value of N, the narrower the principal maxima become. From an energy standpoint the total power in the entire pattern is proportional to N. The height of each principal maximum is proportional to N^2, so from energy conservation the width of each principal

$\phi = \pi = 180°$

(a)

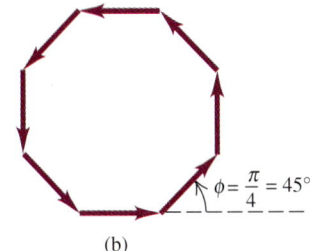

$\phi = \dfrac{\pi}{4} = 45°$

(b)

$\phi = \dfrac{\pi}{2} = 90°$

(c)

38–12 Phasor diagrams for light passing through eight narrow slits. Intensity maxima occur when the phase difference $\phi = 0$, 2π, 4π, $\ldots$. Between the maxima at $\phi = 0$ and $\phi = 2\pi$ are seven minima, corresponding to $\phi = \pi/4$, $\pi/2$, $3\pi/4$, π, $5\pi/4$, $3\pi/2$, and $7\pi/4$. Phasor diagrams are shown for (a) $\phi = \pi$, (b) $\phi = \pi/4$, (c) $\phi = \pi/2$. Can you draw phasor diagrams for the other minima?

(a) $N = 2$

(b) $N = 8$

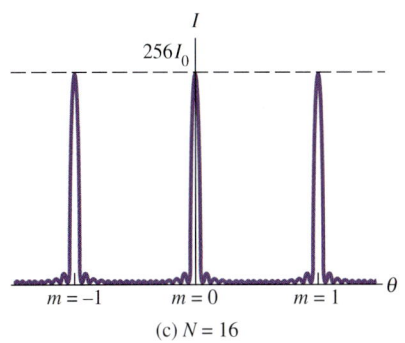

(c) $N = 16$

38–13 Interference patterns for N equally spaced very narrow slits. (a) Two slits. One minimum occurs between adjacent maxima. (b) Eight slits. Seven minima occur between adjacent pairs of the narrower large maxima. (c) Sixteen slits. The vertical scales are different for each graph; I_0 is the maximum intensity for a single slit, and the maximum intensity for N slits is $N^2 I_0$. The width of each peak is proportional to $1/N$.

maximum must be proportional to $1/N$. As we will see in the next section, the narrowness of the principal maxima in a multiple-slit pattern is of great practical importance in physics and astronomy.

38-6 THE DIFFRACTION GRATING

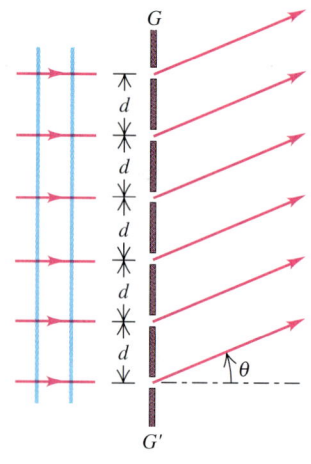

38-14 A portion of a transmission diffraction grating.

We have just seen that increasing the number of slits in an interference experiment (while keeping the spacing of adjacent slits constant) gives interference patterns in which the maxima are in the same positions, but progressively sharper and narrower, than with two slits. Because these maxima are so sharp, their angular position, and hence the wavelength, can be measured to very high precision. An array of a large number of parallel slits, all with the same width a and spaced equal distances d between centers, is called a **diffraction grating.** The first one was constructed by Fraunhofer using fine wires. Gratings can be made by using a diamond point to scratch many equally spaced grooves on a glass or metal surface, or by photographic reduction of a pattern of black and white stripes on paper. For a grating, what we have been calling *slits* are often called *rulings* or *lines.*

In Fig. 38–14, GG' is a cross section of a *transmission grating;* the slits are perpendicular to the plane of the page, and an interference pattern is formed by the light that is transmitted through the slits. The diagram shows only six slits; an actual grating may contain several thousand. The spacing d between centers of adjacent slits is called the *grating spacing.* A plane monochromatic wave is incident normally on the grating from the left side. We assume far-field (Fraunhofer) conditions; that is, the pattern is formed on a screen that is far enough away that all rays emerging from the grating and going to a particular point on the screen can be considered to be parallel.

We found in Section 38–5 that the principal intensity maxima with multiple slits occur in the same directions as for the two-slit pattern. These are the directions for which the path difference for adjacent slits is an integer number of wavelengths. So the positions of the maxima are once again given by

$$d \sin \theta = m\lambda \qquad (m = 0, \pm 1, \pm 2, \pm 3, \ldots) \qquad \text{(intensity maxima,} \quad \text{(38–13)} \atop \text{multiple slits).}$$

Intensity patterns for two, eight, and sixteen slits are displayed in Fig. 38–13 (Section 38–5), showing the progressive increase in sharpness of the maxima as the number of slits increases.

When a grating containing hundreds or thousands of slits is illuminated by a beam of parallel rays of monochromatic light, the pattern is a series of very sharp lines at angles determined by Eq. (38–13). The $m = \pm 1$ lines are called the *first-order lines,* the $m = \pm 2$ lines the *second-order lines,* and so on. If the grating is illuminated by white light with a continuous distribution of wavelengths, each value of m corresponds to a continuous spectrum in the pattern. The angle for each wavelength is determined by Eq. (38–13); for a given value of m, long wavelengths (the red end of the spectrum) lie at larger angles (that is, are deviated more from the straight-ahead direction) than do the shorter wavelengths at the violet end of the spectrum.

As Eq. (38–13) shows, the sines of the deviation angles of the maxima are proportional to the ratio λ/d. For substantial deviation to occur, the grating spacing d should be of the same order of magnitude as the wavelength λ. Gratings for use with visible light (λ from 400 to 700 nm) usually have about 1000 slits per millimeter; the value of d is the *reciprocal* of the number of slits per unit length, so d is of the order of $\frac{1}{1000}$ mm = 1000 nm.

In a *reflection grating,* the array of equally spaced slits shown in Fig. 38–14 is replaced by an array of equally spaced ridges or grooves on a reflective screen. The

16.4
The Grating: Introduction and Questions

16.5
The Grating: Problems

reflected light has maximum intensity at angles where the phase difference between light waves reflected from adjacent ridges or grooves is an integral multiple of 2π. If light of wavelength λ is incident normally on a reflection grating with a spacing d between adjacent ridges or grooves, the *reflected* angles at which intensity maxima occur are given by Eq. (38–13). The iridescent colors of certain butterflies arise from microscopic ridges on the butterfly's wings that form a reflection grating. When the wings are viewed from different angles, corresponding to varying θ in Eq. (38–13), the wavelength and color that are predominantly reflected to the viewer's eye vary as well.

The rainbow-colored reflections that you see from the surface of a compact disc or CD-ROM are a reflection-grating effect. The "grooves" are tiny pits 0.1 μm deep in the surface of the disc, with a uniform radial spacing of $d = 1.60\ \mu\text{m} = 1600$ nm. Information is coded on the CD by varying the *length* of the pits; the reflection-grating aspect of the disc is merely an aesthetic side benefit.

EXAMPLE 38-4

Width of a grating spectrum The wavelengths of the visible spectrum are approximately 400 nm (violet) to 700 nm (red). Find the angular width of the first-order visible spectrum produced by a plane grating with 600 slits per millimeter when white light falls normally on the grating.

SOLUTION The first-order spectrum corresponds to $m = 1$. The grating spacing d is

$$d = \frac{1}{600 \text{ slits/mm}} = 1.67 \times 10^{-6} \text{ m.}$$

From Eq. (38–13), with $m = 1$, the angular deviation θ_v of the violet light (400 nm or 400×10^{-9} m) is

$$\sin \theta_v = \frac{400 \times 10^{-9} \text{ m}}{1.67 \times 10^{-6} \text{ m}} = 0.240,$$

$$\theta_v = 13.9°.$$

The angular deviation θ_r of the red light (700 nm) is

$$\sin \theta_r = \frac{700 \times 10^{-9} \text{ m}}{1.67 \times 10^{-6} \text{ m}} = 0.419,$$

$$\theta_r = 24.8°.$$

So the angular width of the first-order visible spectrum is

$$24.8° - 13.9° = 10.9°.$$

EXAMPLE 38-5

In the situation of Example 38–4, show that the violet end of the third-order spectrum overlaps the red end of the second-order spectrum.

SOLUTION From Eq. (38–13), with a grating spacing of d the angular deviation of the third-order violet end ($m = 3$) is

$$\sin \theta_v = \frac{(3)(400 \times 10^{-9} \text{ m})}{d} = \frac{1.20 \times 10^{-6} \text{ m}}{d}.$$

The deviation of the second-order red ($m = 2$) is

$$\sin \theta_r = \frac{(2)(700 \times 10^{-9} \text{ m})}{d} = \frac{1.40 \times 10^{-6} \text{ m}}{d}.$$

This shows that no matter what the value of the grating spacing d, the largest angle (at the red end) for the second-order spectrum is always greater than the smallest angle (at the violet end) for the third-order spectrum, so the second and third orders *always* overlap.

GRATING SPECTROMETERS

Diffraction gratings are widely used to measure the spectrum of light emitted by a source, a process called *spectroscopy* or *spectrometry*. Light incident on a grating of known spacing is dispersed into a spectrum. The angles of deviation of the maxima are then measured, and Eq. (38–13) is used to compute the wavelength. By using a grating with many slits, very sharp maxima are produced, and the angle of deviation (and hence the wavelength) can be measured very precisely. When a gas is heated, it emits light of wavelengths that are characteristic of the particular atoms and molecules that constitute the gas; measuring these wavelengths with a diffraction grating allows the chemical composition of the gas to be determined. One of the many applications of this technique is to astronomy, where it is used to make chemical assays of distant stars and gas clouds.

38–15 A diffraction-grating spectrometer. The lenses between source and grating form a beam of parallel rays incident on the grating. The beam, when normal to the grating, is diffracted into various orders in directions that satisfy the equation $d \sin \theta = m\lambda$ ($m = 0$, $\pm 1, \pm 2, \ldots$). The diffracted beam is observed with a telescope with crosshairs, permitting precise measurements of the angle θ.

A typical setup for spectroscopy is shown in Fig. 38–15. A transmission grating is used in the figure, but in other setups, reflection gratings are used. A prism can also be used to disperse the various wavelengths through different angles, because the index of refraction always varies with wavelength. But there is no simple relationship that describes this variation, so a spectrometer using a prism has to be calibrated with known wavelengths that are determined in some other way. Another difference is that a prism deviates red light the least and violet the most, while a grating does the opposite.

RESOLUTION OF A GRATING SPECTROMETER

In spectroscopy it is often important to distinguish slightly differing wavelengths. The minimum wavelength difference $\Delta\lambda$ that can be distinguished by a spectrometer is described by the **chromatic resolving power** R, defined as

$$R = \frac{\lambda}{\Delta\lambda} \qquad \text{(chromatic resolving power).} \qquad (38\text{–}14)$$

A spectrometer that can barely distinguish the two lines with wavelengths 589.00 nm and 589.59 nm in the spectrum of sodium (called the *sodium doublet*) has a chromatic resolving power of 589/0.59 or 1000. (These wavelengths are commonly seen in cooking on a gas range; if soup boils over onto the flame, the yellow light of the sodium doublet is emitted from the flame by sodium ions from dissolved table salt.)

We can derive an expression for the resolving power of a diffraction grating used as a spectrometer. Two different wavelengths give diffraction maxima at slightly different angles. As a reasonable (though arbitrary) criterion, let's assume that we can distinguish them as two separate peaks if the maximum of one coincides with the first minimum of the other.

From our discussion in Section 38–5 the mth-order maximum occurs when the phase difference ϕ for adjacent slits is $\phi = 2\pi m$. The first minimum beside that maximum occurs when $\phi = 2\pi m + 2\pi/N$, where N is the number of slits. Now ϕ is also given by $\phi = (2\pi d \sin \theta)/\lambda$, so the angular interval $d\theta$ corresponding to a small increment $d\phi$ in the phase shift can be obtained from the differential of this equation:

$$d\phi = \frac{2\pi d \cos \theta \, d\theta}{\lambda}.$$

When $d\phi = 2\pi/N$, this corresponds to the angular interval $d\theta$ between a maximum and the first adjacent minimum. Thus $d\theta$ is given by

$$\frac{2\pi}{N} = \frac{2\pi d \cos \theta \, d\theta}{\lambda}, \qquad \text{or} \qquad d \cos \theta \, d\theta = \frac{\lambda}{N}.$$

CAUTION ▶ Don't confuse the spacing d with the differential "d" in the angular interval $d\theta$ or in the phase shift increment $d\phi$! ◀

Now we need to find the angular spacing $d\theta$ between maxima for two slightly different wavelengths. This is easy; we have $d \sin \theta = m\lambda$, so the differential of this equation gives

$$d \cos \theta \, d\theta = m \, d\lambda.$$

According to our criterion, the limit or resolution is reached when these two angular spacings are equal. Equating the two expressions for the quantity $(d \cos \theta \, d\theta)$, we find

$$\frac{\lambda}{N} = m \, d\lambda \quad \text{and} \quad \frac{\lambda}{d\lambda} = Nm.$$

If $\Delta\lambda$ is small, we can replace $d\lambda$ by $\Delta\lambda$, and the resolving power R is given simply by

$$R = \frac{\lambda}{\Delta\lambda} = Nm. \tag{38–15}$$

The greater the number of slits N, the better the resolution; also, the higher the order m of the diffraction-pattern maximum that we use, the better the resolution.

EXAMPLE 38-6

What minimum number of slits would be required in a grating to resolve the sodium doublet in the first order? In the fourth order?

SOLUTION As we found above, the required resolving power is $R = 1000$. In the first order ($m = 1$) we need 1000 slits, but in the fourth order we need only 250. These numbers are only approximate because of the arbitrary nature of our criterion for resolution and because real gratings always have slight imperfections in the shapes and spacings of the slits.

38-7 X-RAY DIFFRACTION

X rays were discovered by Wilhelm Röntgen (1845–1923) in 1895, and early experiments suggested that they were electromagnetic waves with wavelengths of the order of 10^{-10} m. At about the same time, the idea began to emerge that in a crystalline solid the atoms are arranged in a regular repeating pattern, with spacing between adjacent atoms also of the order of 10^{-10} m. Putting these two ideas together, Max von Laue (1879–1960) proposed in 1912 that a crystal might serve as a kind of three-dimensional diffraction grating for x rays. That is, a beam of x rays might be scattered (that is, absorbed and re-emitted) by the individual atoms in a crystal, and the scattered waves might interfere just like waves from a diffraction grating.

The first **x-ray diffraction** experiments were performed in 1912 by Friederich, Knipping, and von Laue, using the experimental setup sketched in Fig. 38–16a (page 1180). The scattered x rays *did* form an interference pattern, which they recorded on photographic film. Figure 38–16b is a photograph of such a pattern. These experiments verified that x rays *are* waves, or at least have wavelike properties, and also that the atoms in a crystal *are* arranged in a regular pattern (Fig. 38–17). Since that time, x-ray diffraction has proved to be an invaluable research tool, both for measuring x-ray wavelengths and for the study of crystal structure.

To introduce the basic ideas, we consider first a two-dimensional scattering situation, as shown in Fig. 38–18a, in which a plane wave is incident on a rectangular array of scattering centers. The situation might be a ripple tank with an array of small posts, 3-cm microwaves striking an array of small conducting spheres, or x rays incident on an array of atoms. In the case of electromagnetic waves, the wave induces an oscillating

(a)   (b)

38–16 (a) In an x-ray diffraction experiment, most x rays pass straight through the crystal. But some x rays are scattered and form an interference pattern that exposes the film in a pattern related to the atomic arrangement in the crystal. (b) Diffraction pattern (or *Laue pattern*) formed by directing a beam of x rays at a thin section of quartz crystal.

38–17 Model of the arrangement of ions in a crystal of NaCl (table salt). Black spheres are Na; red spheres are Cl. The spacing of adjacent atoms is 0.282 nm. The electron clouds of the atoms actually overlap slightly, but the atoms are represented as solid spheres for clarity.

electric dipole moment in each scatterer. These dipoles act like little antennas, emitting scattered waves. The resulting interference pattern is the superposition of all these scattered waves. The situation is different from that with a diffraction grating, in which the waves from all the slits are emitted *in phase* (for a plane wave at normal incidence). Here the scattered waves are *not* all in phase because their distances from the *source* are different. To compute the interference pattern, we have to consider the *total* path differences for the scattered waves, including the distances from source to scatterer and from scatterer to observer.

As Fig. 38–18b shows, the path length from source to observer is the same for all the scatterers in a single row if the two angles θ_a and θ_r are equal. Scattered radiation from *adjacent* rows is *also* in phase if the path difference for adjacent rows is an integer number of wavelengths. Figure 38–18c shows that this path difference is $2d \sin \theta$, where

38–18 (a) Scattering of waves from a rectangular array. (b) Interference of waves scattered from adjacent atoms in a row is constructive when $a \cos \theta_a = a \cos \theta_r$, that is, when the angle of incidence θ_a equals the angle of reflection θ_r. Both angles are measured from the *surface* of the crystal, not from its normal. (c) Interference from adjacent rows is also constructive when the path difference $2d \sin \theta$ equals an integral number of wavelengths, as in Eq. (38–16).

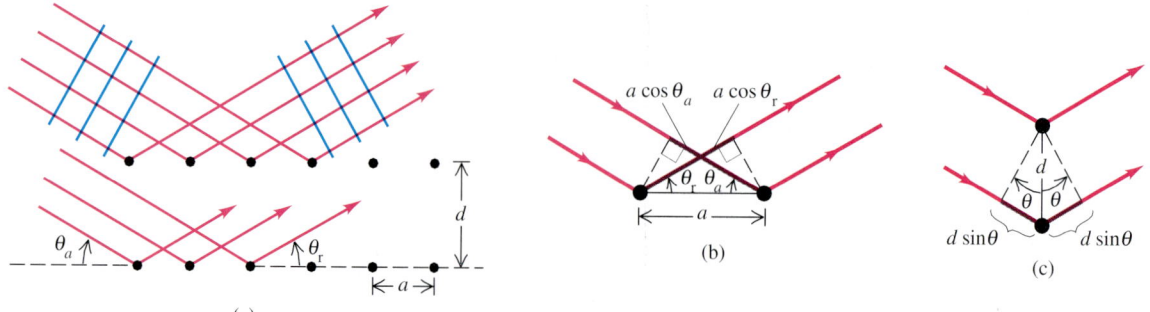

(a)

(b)

(c)

θ is the common value of θ_a and θ_r. Therefore the conditions for radiation from the *entire array* to reach the observer in phase are (1) the angle of incidence must equal the angle of scattering and (2) the path difference for adjacent rows must equal $m\lambda$, where m is an integer. We can express the second condition as

$$2d\sin\theta = m\lambda \quad (m = 1, 2, 3, \ldots) \qquad \text{(Bragg condition for constructive interference from an array).} \qquad (38\text{-}16)$$

CAUTION ▶ In this equation the angle θ is measured with respect to the *surface* of the crystal, rather than with respect to the *normal* to the plane of an array of slits or a grating. Also, note that the path difference in Eq. (38–16) is $2d\sin\theta$, not $d\sin\theta$ as in Eq. (38–13) for a diffraction grating. ◀

In directions for which Eq. (38–16) is satisfied, we see a strong maximum in the interference pattern. We can describe this interference in terms of *reflections* of the wave from the horizontal rows of scatterers in Fig. 38–18a. Strong reflection (constructive interference) occurs at angles such that the incident and scattered angles are equal and Eq. (38–16) is satisfied.

We can extend this discussion to a three-dimensional array by considering *planes* of scatterers instead of *rows*. Figure 38–19 shows two different sets of parallel planes that pass through all the scatterers. Waves from all the scatterers in a given plane interfere constructively if the angles of incidence and scattering are equal. There is also constructive interference between planes when Eq. (38–16) is satisfied, where d is now the distance between adjacent planes. Because there are many different sets of parallel planes, there are also many values of d and many sets of angles that give constructive interference for the whole crystal lattice. This phenomenon is called **Bragg reflection,** and Eq. (38–16) is called the **Bragg condition,** in honor of Sir William Bragg and his son Laurence Bragg, two pioneers in x-ray analysis.

CAUTION ▶ While we are using the term *reflection,* remember that we are dealing with an *interference* effect. In fact, the reflections from various planes are closely analogous to interference effects in thin films (Section 37–5). ◀

As Fig. 38–16b shows, in x-ray diffraction there is nearly complete cancellation in all but certain very specific directions in which constructive interference occurs and forms bright spots. Such a pattern is usually called an x-ray *diffraction* pattern, although *interference* pattern might be more appropriate.

We can determine the wavelength of x rays by examining the diffraction pattern for a crystal of known structure and known spacing between atoms, just as we determined wavelengths of visible light by measuring patterns from slits or gratings. (The spacing between atoms in simple crystals of known structure, such as sodium chloride, can be found from the density of the crystal and Avogadro's number.) Then, once we know the x-ray wavelength, we can use x-ray diffraction to explore the structure and determine the spacing between atoms in crystals with unknown structure.

X-ray diffraction is by far the most important experimental tool in the investigation of crystal structure of solids. X-ray diffraction also plays an important role in studies of the structures of liquids and of organic molecules. It has been one of the chief experimental techniques in working out the double-helix structure of DNA and subsequent advances in molecular genetics.

(a)

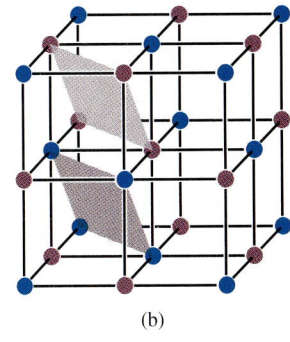
(b)

38–19 A cubic crystal and two different families of crystal planes. The spacing of the planes in (a) is $d = a/\sqrt{2}$; that of the planes in (b) is $a/\sqrt{3}$. There are also three sets of planes parallel to the cube faces, with spacing a.

EXAMPLE 38–7

You direct a beam of x rays with wavelength 0.154 nm at certain planes of a silicon crystal. As you increase the angle of incidence from zero, you find the first strong interference maximum from these planes when the beam makes an angle of 34.5° with the planes. a) How far apart are the planes? b) Will you find other interference maxima from these planes at larger angles?

SOLUTION a) To find the plane spacing d, we solve the Bragg equation, Eq. (38–16), for d and set $m = 1$:

$$d = \frac{m\lambda}{2 \sin \theta} = \frac{(1)(0.154 \text{ nm})}{2 \sin 34.5°} = 0.136 \text{ nm}.$$

This is the distance between adjacent planes.

b) To calculate other angles, we solve Eq. (38–16) for $\sin \theta$:

$$\sin \theta = \frac{m\lambda}{2d} = m \frac{0.154 \text{ nm}}{2(0.136 \text{ nm})} = m(0.566).$$

Values of m of 2 or greater give values of $\sin \theta$ greater than unity, which is impossible. Hence there are no other angles for interference maxima for this particular set of crystal planes.

38–8 CIRCULAR APERTURES AND RESOLVING POWER

We have studied in detail the diffraction patterns formed by long, thin slits or arrays of slits. But an aperture of *any* shape forms a diffraction pattern. The diffraction pattern formed by a *circular* aperture is of special interest because of its role in limiting how well an optical instrument can resolve fine details. In principle, we could compute the intensity at any point P in the diffraction pattern by dividing the area of the aperture into small elements, finding the resulting wave amplitude and phase at P, and then integrating over the aperture area to find the resultant amplitude and intensity at P. In practice, the integration cannot be carried out in terms of elementary functions. We will simply *describe* the pattern and quote a few relevant numbers.

The diffraction pattern formed by a circular aperture consists of a central bright spot surrounded by a series of bright and dark rings, as shown in Fig. 38–20. We can describe the pattern in terms of the angle θ, representing the angular radius of each ring. If the aperture diameter is D and the wavelength is λ, the angular radius θ_1 of the first *dark* ring is given by

$$\sin \theta_1 = 1.22 \frac{\lambda}{D} \qquad \text{(diffraction by a circular aperture).} \qquad (38–17)$$

The angular radii of the next two dark rings are given by

$$\sin \theta_2 = 2.23 \frac{\lambda}{D}, \qquad \sin \theta_3 = 3.24 \frac{\lambda}{D}. \qquad (38–18)$$

Between these are bright rings with angular radii given by

$$\sin \theta = 1.63 \frac{\lambda}{D}, \qquad 2.68 \frac{\lambda}{D}, \qquad 3.70 \frac{\lambda}{D}, \qquad (38–19)$$

and so on. The central bright spot is called the **Airy disk,** in honor of Sir George Airy (1801–1892), Astronomer Royal of England, who first derived the expression for the

38–20 Diffraction pattern formed by a circular aperture of diameter D. The pattern consists of a central bright spot and alternating dark and bright rings. The angular radius θ_2 of the second dark ring is shown. (This diagram is not drawn to scale.)

Airy disk

intensity in the pattern. The angular radius of the Airy disk is that of the first dark ring, given by Eq. (38–17).

The intensities in the bright rings drop off very quickly with increasing angle. When D is much larger than the wavelength λ, the usual case for optical instruments, the peak intensity in the first ring is only 1.7% of the value at the center of the Airy disk, and the peak intensity of the second ring is only 0.4%. Most (85%) of the light energy falls within the Airy disk. Figure 38–21 shows a diffraction pattern from a circular aperture 1.0 mm in diameter.

Diffraction has far-reaching implications for image formation by lenses and mirrors. In our study of optical instruments in Chapter 36 we assumed that a lens with focal length f focuses a parallel beam (plane wave) to a *point* at a distance f from the lens. This assumption ignored diffraction effects. We now see that what we get is not a point but the diffraction pattern just described. If we have two point objects, their images are not two points but two diffraction patterns. When the objects are close together, their diffraction patterns overlap; if they are close enough, their patterns overlap almost completely and cannot be distinguished. The effect is shown in Fig. 38–22, which shows the patterns for four (very small) "point" sources of light. In Fig. 38–22a the images of sources 1 and 2 are well separated, but the images of sources 3 and 4 on the right have merged. In Fig. 38–22b, with a larger aperture diameter and resulting smaller Airy disks, images 3 and 4 are barely resolved. In Fig. 38–22c, with a still larger aperture, they are well resolved.

A widely used criterion for resolution of two point objects, proposed by the English physicist Lord Rayleigh (1842–1919) and called **Rayleigh's criterion,** is that the objects are just barely resolved (that is, distinguishable) if the center of one diffraction pattern coincides with the first minimum of the other. In that case the angular separation of the image centers is given by Eq. (38–17). The angular separation of the *objects* is the

38-21 Diffraction pattern formed by a circular aperture 1.0 mm in diameter. The Airy disk is over-exposed so that other rings may be seen.

(a)

(b)

(c)

38-22 Diffraction patterns of four "point" sources. From left to right in each photograph, the sources are numbered 1, 2, 3, and 4. The photographs were made with a circular aperture in front of the lens. (a) Here the aperture is so small that the patterns of sources 3 and 4 overlap and are barely resolved by Rayleigh's criterion. Increasing the size of the aperture decreases the size of the diffraction patterns, as shown in (b) and (c).

same as that of the *images* made by a telescope, microscope, or other optical device. So two point objects are barely resolved, according to Rayleigh's criterion, when their angular separation is given by Eq. (38–17).

The minimum separation of two objects that can just be resolved by an optical instrument is called the **limit of resolution** of the instrument. The smaller the limit of resolution, the greater the *resolution*, or **resolving power**, of the instrument. Diffraction sets the ultimate limits on resolution of lenses. *Geometric* optics may make it seem that we can make images as large as we like. Eventually, though, we always reach a point at which the image becomes larger but does not gain in detail. The images in Fig. 38–22 would not become sharper with further enlargement.

CAUTION ▶ Be careful not to confuse the resolving power of an optical instrument with the *chromatic* resolving power of a grating (described in Section 38–6). Resolving power refers to the ability to distinguish the images of objects that appear close to each other, either when looking through an optical instrument or in a photograph made with the instrument. Chromatic resolving power describes how well different wavelengths can be distinguished in a spectrum formed by a diffraction grating. **◀**

Rayleigh's criterion combined with Eq. (38–17) shows that resolution (resolving power) improves with larger diameter; it also improves with shorter wavelengths. Ultraviolet microscopes have higher resolution than visible-light microscopes. In electron microscopes the resolution is limited by the wavelengths associated with the wavelike behavior of electrons (Section 37–7). These wavelengths can be made 100,000 times smaller than wavelengths of visible light, with a corresponding gain in resolution. One reason for building very large telescopes is to increase the aperture diameter and thus minimize diffraction effects.

Diffraction is an important consideration for satellite "dishes," parabolic reflectors designed to receive satellite transmission. Satellite dishes have to be able to pick up transmissions from two satellites that are only a few degrees apart, transmitting at the same frequency; the need to resolve two such transmissions determines the minimum diameter of the dish. As higher frequencies are used, the needed diameter decreases. For example, when two satellites 5.0° apart broadcast 7.5-cm microwaves, the minimum dish diameter to resolve them (by Rayleigh's criterion) is about 1.0 m.

The effective diameter of a telescope can be increased in some cases by using arrays of smaller telescopes. The Very Long Baseline Array (VLBA), a group of ground-based radio telescopes, has a maximum separation of about 8000 km and can resolve radio signals to 10^{-8} rad. This astonishing resolution is comparable, in the optical realm, to seeing a human hair at a distance of 10 km. NASA has proposed placing a number of infrared telescopes in space to form an array 6000 km across. The superb resolution of this array would allow us to see continents on planets many light years away.

EXAMPLE 38-8

Resolving power of a camera lens A camera lens with focal length $f = 50$ mm and maximum aperture $f/2$ forms an image of an object 9.0 m away. a) If the resolution is limited by diffraction, what is the minimum distance between two points on the object that are barely resolved, and what is the corresponding distance between image points? b) How does the situation change if the lens is "stopped down" to $f/16$? Assume that $\lambda = 500$ nm in both cases.

SOLUTION a) From Eq. (36–1) the f-number is the focal length divided by the aperture diameter D; hence $D = (50$ mm$)/2 = 25$ mm $= 25 \times 10^{-3}$ m. From Eq. (38–17) the angular separation

θ of two object points that are barely resolved is given by

$$\theta \approx \sin\theta = 1.22\frac{\lambda}{D} = 1.22\frac{500 \times 10^{-9}\ \text{m}}{25 \times 10^{-3}\ \text{m}}$$

$$= 2.4 \times 10^{-5}\ \text{rad}.$$

Let y be the separation of the object points, and let y' be the separation of the corresponding image points. We know from our thin-lens analysis in Section 35–6 that, apart from sign, $y/s = y'/s'$. Thus the angular separations of the object points and the corresponding image points are both equal to θ. Because the object distance s is much greater than the focal length

$f = 50$ mm, the image distance s' is approximately equal to f. Thus

$$\frac{y}{9.0 \text{ m}} = 2.4 \times 10^{-5}, \qquad y = 2.2 \times 10^{-4} \text{ m} = 0.22 \text{ mm};$$

$$\frac{y'}{50 \text{ mm}} = 2.4 \times 10^{-5}, \qquad y' = 1.2 \times 10^{-3} \text{ mm}$$

$$= 0.0012 \text{ mm} \approx \frac{1}{800} \text{ mm}.$$

b) The aperture diameter is now (50 mm)/16, or one eighth as large as before. The angular separation between barely-resolved points is eight times as great, and the values of y and y' are also eight times as great as before:

$$y = 1.8 \text{ mm}, \qquad y' = 0.0096 \text{ mm} = \frac{1}{100} \text{ mm}.$$

Only the best camera lenses can approach this resolving power.

Many photographers always use the smallest possible aperture for maximum sharpness, since lens aberrations cause light rays that are far from the optic axis to converge to a different image point than do rays near the axis (Section 36–7). Photographers should be aware that diffraction effects become more significant at small apertures. One cause of fuzzy images has to be balanced against another.

38–9 HOLOGRAPHY

Holography is a technique for recording and reproducing an image of an object through the use of interference effects. Unlike the two-dimensional images recorded by an ordinary photograph or television system, a holographic image is truly three-dimensional. Such an image can be viewed from different directions to reveal different sides and from various distances to reveal changing perspective. If you had never seen a hologram, you wouldn't believe it was possible!

The basic procedure for making a hologram is shown in Fig. 38–23a. We illuminate the object to be holographed with monochromatic light, and we place a photographic film so that it is struck by scattered light from the object and also by direct light from the source. In practice, the light source must be a laser, for reasons we will discuss later. Interference between the direct and scattered light leads to the formation and recording of a complex interference pattern on the film.

To form the images, we simply project light through the developed film, as shown in Fig. 38–23b. Two images are formed, a virtual image on the side of the film nearer the source and a real image on the opposite side.

A complete analysis of holography is beyond our scope, but we can gain some insight into the process by looking at how a single point is holographed and imaged. Consider the interference pattern that is formed on a sheet of photographic negative film

38–23 (a) A hologram is the record on film of the interference pattern formed with light from the coherent source and light scattered from the object. (b) Images are formed when light is projected through the hologram. The observer sees the virtual image formed behind the hologram.

(a)   (b)

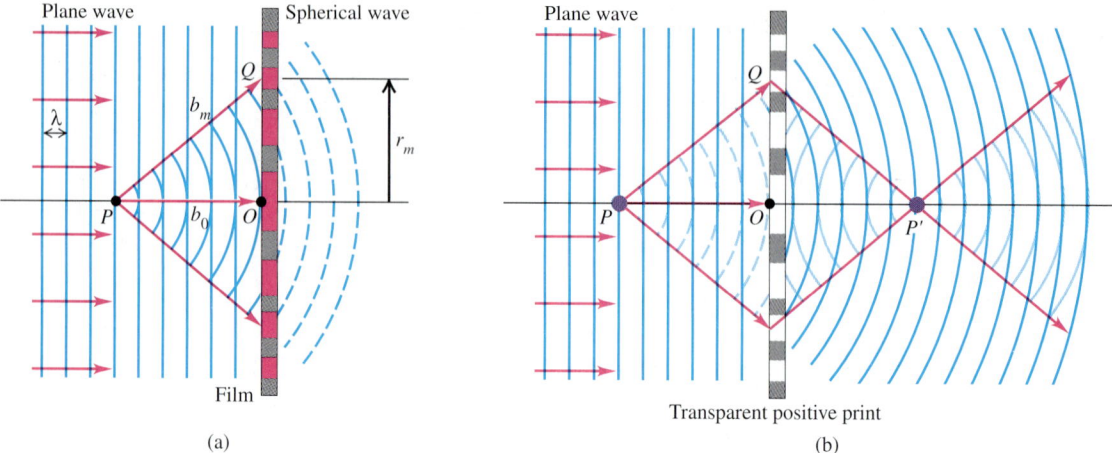

(a)   (b)

38–24 (a) Constructive interference of the plane and spherical waves occurs in the plane of the film at every point Q for which the distance b_m from P is greater than the distance b_0 from P to O by an integral number of wavelengths $m\lambda$. For the point Q shown, $m = 2$. (b) When a plane wave strikes a transparent positive print of the developed film, the diffracted wave consists of a wave converging to P' and then diverging again and a diverging wave that appears to originate at P. These waves form the real and virtual images, respectively.

by the superposition of an incident plane wave and a spherical wave, as shown in Fig. 38–24a. The spherical wave originates at a point source P at a distance b_0 from the film; P may in fact be a small object that scatters part of the incident plane wave. We assume that the two waves are monochromatic and coherent and that the phase relation is such that constructive interference occurs at point O on the diagram. Then constructive interference will *also* occur at any point Q on the film that is farther from P than O is by an integer number of wavelengths. That is, if $b_m - b_0 = m\lambda$, where m is an integer, then constructive interference occurs. The points where this condition is satisfied form circles on the film centered at O, with radii r_m given by

$$b_m - b_0 = \sqrt{b_0{}^2 + r_m{}^2} - b_0 = m\lambda \qquad (m = 1,\ 2,\ 3, \ldots). \qquad (38\text{--}20)$$

Solving this for $r_m{}^2$, we find

$$r_m{}^2 = \lambda(2mb_0 + m^2\lambda).$$

Ordinarily, b_0 is very much larger than λ, so we neglect the second term in parentheses and obtain

$$r_m = \sqrt{2m\lambda b_0} \qquad (m = 1,\ 2,\ 3, \ldots). \qquad (38\text{--}21)$$

The interference pattern consists of a series of concentric bright circular fringes with radii given by Eq. (38–21). Between these bright fringes are dark fringes.

Now we develop the film and make a transparent positive print, so the bright-fringe areas have the greatest transparency on the film. Then we illuminate it with monochromatic plane-wave light of the same wavelength λ that we used initially. In Fig. 38–24b, consider a point P' at a distance b_0 along the axis from the film. The centers of successive bright fringes differ in their distances from P' by an integer number of wavelengths, and therefore a strong *maximum* in the diffracted wave occurs at P'. That is, light converges to P' and then diverges from it on the opposite side. Therefore P' is a *real image* of point P.

This is not the entire diffracted wave, however. The interference of the wavelets that spread out from all the transparent areas forms a second spherical wave that is diverging rather than converging. When this wave is traced back behind the film in Fig. 38–24b, it appears to be spreading out from point P. Thus the total diffracted wave from the hologram is a superposition of a spherical wave converging to form a real image at P' and a spherical wave that diverges as though it had come from the virtual image point P.

Because of the principle of superposition for waves, what is true for the imaging of a single point is also true for the imaging of any number of points. The film records the superposed interference pattern from the various points, and when light is projected

(a)

(b)

through the film, the various image points are reproduced simultaneously. Thus the images of an extended object can be recorded and reproduced just as for a single point object. Figure 38–25 shows photographs of a holographic image from two different angles, showing the changing perspective in this three-dimensional image.

In making a hologram, we have to overcome several practical problems. First, the light used must be *coherent* over distances that are large in comparison to the dimensions of the object and its distance from the film. Ordinary light sources *do not* satisfy this requirement, for reasons that we discussed in Section 37–2. Therefore laser light is essential for making a hologram. Second, extreme mechanical stability is needed. If any relative motion of source, object, or film occurs during exposure, even by as much as a quarter of a wavelength, the interference pattern on the film is blurred enough to prevent satisfactory image formation. These obstacles are not insurmountable, however, and holography has become important in research, entertainment, and a wide variety of technological applications.

38–25 Two views of a hologram. (a) From this angle you can look through the magnifying glass and see the row of connector tabs at the edge of the printed circuit board. (b) The angle of view has changed, so when you look through the glass, you see the manufacturer's name.

SUMMARY

- Diffraction occurs when light passes through an aperture or around an edge. When the source and observer are so far away from the obstructing surface that the outgoing rays can be considered parallel, it is called Fraunhofer diffraction. When the source or observer is relatively close to the obstructing surface, it is Fresnel diffraction.

- For a single narrow slit with width a the condition for destructive interference (a dark fringe) at a point P at angle θ is

$$\sin \theta = \frac{m\lambda}{a} \qquad (m = \pm 1, \pm 2, \pm 3, \ldots). \tag{38–2}$$

The intensity I at any angle θ is

$$I = I_0 \left\{ \frac{\sin[\pi a (\sin \theta)/\lambda]}{\pi a (\sin \theta)/\lambda} \right\}^2. \tag{38–7}$$

- A diffraction grating consists of a large number of thin parallel slits, spaced a distance d apart. The condition for maximum intensity in the interference pattern is

$$d \sin \theta = m\lambda \qquad (m = 0, \pm 1, \pm 2, \pm 3, \ldots). \tag{38–13}$$

This is the same condition as for the two-source pattern, but the maxima for the grating are very sharp and narrow.

- A crystal serves as a three-dimensional diffraction grating for x rays with wavelengths of the same order of magnitude as the spacing between atoms in the crystal. For a set of crystal planes spaced a distance d apart, constructive interference occurs when the angles of incidence and scattering (measured from the crystal planes) are equal and when

$$2d \sin \theta = m\lambda \qquad (m = 1, 2, 3, \ldots). \qquad (38\text{--}16)$$

This is called the Bragg condition.

- The diffraction pattern from a circular aperture of diameter D consists of a central bright spot, called the Airy disk, and a series of concentric dark and bright rings. The angular radius θ_1 of the first dark ring, equal to the angular size of the Airy disk, is given by

$$\sin \theta_1 = 1.22 \frac{\lambda}{D}. \qquad (38\text{--}17)$$

Diffraction sets the ultimate limit on resolution (image sharpness) of optical instruments. According to Rayleigh's criterion, two point objects are just barely resolved when their angular separation θ is given by Eq. (38–17).

- A hologram is a photographic record of an interference pattern formed by light scattered from an object and direct light coming from the source. A hologram forms a true three-dimensional image of the object.

DISCUSSION QUESTIONS

Q38-1 The sound waves used predominantly in human speech have wavelengths in the range from 1 to 3 meters. Using the ideas of diffraction, explain how it is possible to hear a person's voice even when they are facing away from you.

Q38-2 In a diffraction experiment with waves of wavelength λ, there will be *no* intensity minima (that is, no dark fringes) if the slit width is small enough. What is the maximum slit width for which this occurs? Explain your answer.

Q38-3 A rainbow ordinarily shows a range of colors (see Section 34–5). But if the water droplets that form the rainbow are small enough, the rainbow will appear white. Explain why, using diffraction ideas. How small do you think the raindrops would have to be for this to occur?

Q38-4 Some loudspeaker horns for outdoor concerts (at which the entire audience is seated on the ground) are wider vertically than horizontally. Use diffraction ideas to explain why this is more efficient at spreading the sound uniformly over the audience than either a square speaker horn or a horn that is wider horizontally than vertically. Would this still be the case if the audience was seated at different elevations, as in an amphitheater? Why or why not?

Q38-5 Figure 32–7 (Section 32–3) shows a loudspeaker system. Low-frequency sounds are produced by the *woofer*, which is a speaker with large diameter; the *tweeter*, a speaker of smaller diameter, produces high-frequency sounds. Use diffraction ideas to explain why the tweeter is more effective for distributing high-frequency sounds uniformly over a room than is the woofer.

Q38-6 Information is stored on an audio compact disc, CD-ROM, or DVD disc in a series of pits on the disc. These pits are scanned by a laser beam. An important limitation on the amount of information that can be stored on such a disc is the width of the laser beam. Explain why this should be, and explain how using a shorter-wavelength laser allows more information to be stored on a disc of the same size.

Q38-7 With which kind of light can the Hubble Space Telescope see finer detail in a distant astronomical object: red, blue, or ultraviolet? Explain your answer.

Q38-8 A typical telescope used by amateur astronomers has a mirror 20 cm in diameter. With such a telescope (and a filter to cut the intensity of sunlight to a safe level for viewing), fine details can be seen on the surface of the sun. Explain why a *radio* telescope would have to be *much* larger to "see" comparable details on the sun.

Q38-9 Could x-ray diffraction effects with crystals be observed by using visible light instead of x rays? Why or why not?

Q38-10 Why is a diffraction grating better than a two-slit setup for measuring wavelengths of light?

Q38-11 One sometimes sees rows of evenly spaced radio antenna towers. A student remarked that these act like diffraction gratings. What did she mean? Why would one want them to act like a diffraction grating?

Q38-12 If a hologram is made using 600-nm light and then viewed with 500-nm light, how will the images look compared to those observed when viewed with 600-nm light? Explain.

Q38–13 A hologram is made using 600-nm light and is then viewed by using white light from an incandescent bulb. What will be seen? Explain.

Q38–14 Ordinary photographic film reverses black and white, in the sense that the most brightly illuminated areas become blackest upon development (hence the term *negative*). Suppose a hologram negative is viewed directly, without making a positive transparency. How will the resulting images differ from those obtained with the positive? Explain.

EXERCISES

SECTION 38–3 DIFFRACTION FROM A SINGLE SLIT

38–1 Monochromatic light from a distant source is incident on a slit 0.750 mm wide. On a screen 2.00 m away, the distance from the central maximum of the diffraction pattern to the first minimum is measured to be 1.35 mm. Calculate the wavelength of the light.

38–2 Parallel rays of green mercury light with a wavelength of 546 nm pass through a slit covering a lens with a focal length of 60.0 cm. In the focal plane of the lens the distance from the central maximum to the first minimum is 10.2 mm. What is the width of the slit?

38–3 In the situation described in Example 38–1 (Section 38–3), what is the angle θ of the first minimum (dark fringe)?

38–4 Light of wavelength 633 nm from a distant source is incident on a slit 0.750 mm wide, and the resulting diffraction pattern is observed on a screen 3.50 m away. What is the distance between the two dark fringes on either side of the central bright fringe?

38–5 Diffraction occurs for all types of waves, including sound waves. High-frequency sound from a distant source with wavelength 9.00 cm passes through a narrow slit 12.0 cm wide. A microphone is placed 40.0 cm directly in front of the center of the slit, corresponding to point O in Fig. 38–6a. The microphone is then moved in a direction perpendicular to the line from the center of the slit to point O. At what distances from O will the intensity detected by the microphone be zero?

38–6 Monochromatic electromagnetic radiation with wavelength λ from a distant source passes through a slit. The diffraction pattern is observed on a screen 2.50 m from the slit. If the width of the central maximum is 6.00 mm, what is the slit width a if the wavelength is a) 500 nm (visible light)? b) 50.0 μm (infrared radiation)? c) 0.500 nm (x rays)?

38–7 Red light of wavelength 633 nm from a helium-neon laser passes through a slit 0.350 mm wide. The diffraction pattern is observed on a screen 3.00 m away. Define the width of a bright fringe as the distance between the minima on either side. a) What is the width of the central bright fringe? b) What is the width of the first bright fringe on either side of the central one?

SECTION 38–4 INTENSITY IN THE SINGLE-SLIT PATTERN

38–8 Monochromatic light of wavelength $\lambda = 620$ nm from a distant source passes through a slit 0.450 mm wide. The diffraction pattern is observed on a screen 3.00 m from the slit. In terms of the intensity I_0 at the peak of the central maximum, what is the intensity of the light at the screen the following distances from the center of the central maximum: a) 1.00 mm; b) 3.00 mm; c) 5.00 mm?

38–9 A slit 0.240 mm wide is illuminated by parallel light rays of wavelength 540 nm. The diffraction pattern is observed on a screen that is 3.00 m from the slit. The intensity at the center of the central maximum ($\theta = 0°$) is 6.00×10^{-6} W/m². a) What is the distance on the screen from the center of the central maximum to the first minimum? b) What is the intensity at a point on the screen midway between the center of the central maximum and the first minimum?

38–10 A diffraction pattern is formed by passing parallel rays of 600-nm light through a slit 0.150 mm wide. What is the phase angle β (the phase difference between wavelets from the top and bottom of the slit) at a) the center of the central maximum; b) the second minimum out from the central maximum; c) 7.0° from the central maximum?

38–11 In a single-slit diffraction pattern produced by electromagnetic radiation from a distant source passing through a slit 0.320 mm wide, the phase angle β (the phase difference between wavelets from the top and bottom of the slit) is $\pi/2$ rad at an angle of 0.24° from the central maximum. What is the wavelength of the electromagnetic radiation?

38–12 Consider a single-slit diffraction experiment in which the amplitude of the wave at point O in Fig. 38–6a is E_0. For each of the following cases, draw a phasor diagram like that in Fig. 38–7b and determine *graphically* the amplitude of the wave at the point in question. (*Hint:* Use Eq. (38–6) to determine the value of β for each case.) Compute the intensity and compare to Eq. (38–5). a) $\sin \theta = \lambda/2a$; b) $\sin \theta = \lambda/a$; c) $\sin \theta = 3\lambda/2a$.

38–13 A single-slit diffraction pattern is formed by monochromatic electromagnetic radiation from a distant source passing through a slit 0.105 mm wide. At the point in the pattern 3.25° from the center of the central maximum, the total phase difference between wavelets from the top and bottom of the slit is 56.0 rad. a) What is the wavelength of the radiation? b) What is the intensity at this point, if the intensity at the center of the central maximum is I_0?

SECTION 38–5 MULTIPLE SLITS

38–14 **Diffraction in an Interference Pattern.** Consider the interference pattern produced by two parallel slits with width a and separation d. Let $d = 4a$. a) Ignoring diffraction effects due to the slit width, at what angles θ from the central maximum will the next five maxima in the two-slit interference pattern occur? (Your answer will be in terms of d and the wavelength λ of the light.) b) Now include the effects of diffraction. If the intensity at $\theta = 0$ is I_0, what is the intensity at each of the angles calculated in part (a)? Compare your results to Fig. 38–10c.

38–15 **Number of Fringes in a Diffraction Maximum.** In Fig. 38–10c the central diffraction maximum contains exactly seven interference fringes, and in this case $d/a = 4$. a) What must the ratio d/a be if the central maximum contains exactly five fringes? b) In the case considered in part (a), how many fringes are contained within the first diffraction maximum on one side of the central maximum?

38–16 An interference pattern is produced by eight parallel and equally spaced narrow slits. There is an interference minimum when the phase difference ϕ between light from adjacent slits is $\pi/4$. The phasor diagram is given in Fig. 38–12b. For which pairs of slits is there totally destructive interference?

38–17 An interference pattern is produced by light of wavelength 580 nm from a distant source incident on two identical parallel slits separated by a distance (between centers) of 0.530 mm. a) If the slits are very narrow, what would be the angular positions of the first-order and second-order two-slit interference maxima? b) Let the slits have width 0.320 mm. In terms of the intensity I_0 at the center of the central maximum, what is the intensity at each of the angular positions in part (a)?

38–18 An interference pattern is produced by light of wavelength 620 nm from a distant source incident on two parallel slits of width 0.280 mm and separated by a distance (between centers) of $d = 0.800$ mm. In terms of the intensity I_0 at the center of the central maximum, what is the intensity at the following angular positions θ in the pattern: a) 0.125 mrad; a) 0.250 mrad; a) 0.300 mrad? (1 mrad = 10^{-3} rad)

38–19 An interference pattern is produced by four parallel and equally spaced narrow slits. By drawing appropriate phasor diagrams, show that there is an interference minimum when the phase difference ϕ from adjacent slits is i) $\pi/2$; ii) π; iii) $3\pi/2$. In each case, for which pairs of slits is there totally destructive interference?

38–20 An interference pattern is produced by two identical parallel slits of width a and separation (between centers) $d = 3a$. Which interference maxima m_i will be missing in the pattern?

38–21 An interference pattern is produced by monochromatic light of wavelength 630 nm from a distant source incident on two identical parallel slits separated by a distance (between centers) of 0.840 mm. a) It is found that the $n = 3$ interference maximum is missing from the pattern. What is the width of each slit? b) Which other interference maxima should also be missing from the pattern? c) If the wavelength is decreased from 630 nm to 420 nm, which interference maxima will be missing from the pattern?

SECTION 38–6 THE DIFFRACTION GRATING

38–22 Identifying Isotopes by Spectra. Different isotopes of the same element emit light at slightly different wavelengths. A wavelength in the emission spectrum of a hydrogen atom is 656.45 nm; for deuterium the corresponding wavelength is 656.27 nm. a) What minimum number of slits is required to resolve these two wavelengths in second order? b) If the grating has 500.00 slits/mm, find the angles and angular separation of these two wavelengths in the second order.

38–23 Measuring Wavelengths with a CD. A laser beam of wavelength $\lambda = 632.8$ nm shines at normal incidence on the reflective side of a compact disc. The tracks of tiny pits in which information is coded onto the CD are 1.60 μm apart. For what

angles of reflection (measured from the normal) will the intensity of light be maximum?

38–24 a) What is the wavelength of light that is deviated in the first order through an angle of 13.5° by a transmission grating having 5000 slits/cm? b) What is the second-order deviation of this wavelength? Assume normal incidence.

38–25 Plane monochromatic waves with wavelength 520 nm are incident normally on a plane transmission grating having 350 slits/mm. Find the angles of deviation in the first, second, and third orders.

38–26 The wavelength range of the visible spectrum is approximately 400 nm to 700 nm. White light falls at normal incidence on a diffraction grating that has 350 slits/mm. Find the angular width of the visible spectrum in a) the first order; b) the third order. (*Note:* An advantage of working in higher orders is the greater angular spread and better resolution. A disadvantage is the overlapping of different orders, as shown in Example 38–5 in Section 38–6.)

38–27 A plane transmission grating is ruled with 4000 slits/cm. Assume normal incidence. The α and β lines emitted by atomic hydrogen have wavelengths 656 nm and 486 nm, respectively. Compute the angular separation in degrees between these lines in a) the first-order spectrum; b) the second-order spectrum.

38–28 The light from an iron arc includes many different wavelengths. Two of these are at $\lambda = 587.9782$ nm and $\lambda = 587.8002$ nm. You wish to resolve these spectral lines in first order using a grating 1.20 cm in length. What minimum number of slits per centimeter must the grating have?

SECTION 38–7 X-RAY DIFFRACTION

38–29 X rays of wavelength 0.0850 nm are scattered from the atoms of a crystal. The second-order maximum in the Bragg reflection occurs when the angle θ in Fig. 38–18 is 21.5°. What is the spacing between adjacent atomic planes in the crystal?

38–30 Monochromatic x rays are incident on a crystal for which the spacing of the atomic planes is 0.440 nm. The first-order maximum in the Bragg reflection occurs when the incident and reflected x rays make an angle of 39.4° with the crystal planes. What is the wavelength of the x rays?

SECTION 38–8 CIRCULAR APERTURES AND RESOLVING POWER

38–31 Due to blurring caused by atmospheric distortion, the best resolution that can be obtained by a normal earth-based visible-light telescope is about 0.3 arcsecond (there are 60 arcminutes in a degree and 60 arcseconds in an arcminute). a) Using Rayleigh's criterion, calculate the diameter of an earth-based telescope that gives this resolution with 550-nm light. b) Increasing the telescope diameter beyond the value found in part (a) will increase the light-gathering power of the telescope, allowing more distant and dimmer astronomical objects to be studied, but will not improve the resolution. In what ways are the Keck telescopes (each of 10-m diameter) atop Mauna Kea in Hawaii superior to the Hale Telescope (5-m diameter) on Palomar Mountain in California? In what ways are they *not* superior? Explain.

38–32 If you can read the bottom row of your doctor's eye chart, your eye has a resolving power of one arcminute, equal to $\frac{1}{60}$ degree. If this resolving power is diffraction-limited, to what effective diameter of your eye's optical system does this correspond? Use Rayleigh's criterion and assume $\lambda = 550$ nm.

38–33 Two satellites at an altitude of 1200 km are separated by 28 km. If they broadcast 3.6-cm microwaves, what minimum receiving dish diameter is needed to resolve (by Rayleigh's criterion) the two transmissions?

38–34 The VLBA can resolve (by Rayleigh's criterion) signals from sources separated by 1.0×10^{-8} rad. If the effective diameter of the receiver is 8000 km, what is the wavelength of the signal?

38–35 Monochromatic light with wavelength 620 nm passes through a circular aperture with diameter 7.4 μm. The resulting diffraction pattern is observed on a screen that is 4.5 m from the aperture. What is the diameter of the Airy disk on the screen?

38–36 Figure 38–26 shows two point sources of light, a and b, that are 25.0 m from lens L and 8.00 mm apart. The two sources produce images at c that are just resolved according to Rayleigh's criterion. The focal length of the lens is 180 mm.

What is the diameter of the diffraction circles at c? (Note that the figure is not to scale.)

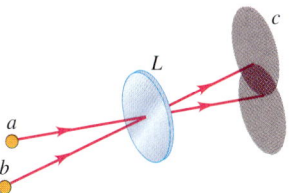

FIGURE 38–26 Exercise 38–36.

38–37 Observing Jupiter. You are asked to design a space telescope for earth orbit. When Jupiter is 5.93×10^8 km away (its closest approach to the earth), the telescope is to resolve by Rayleigh's criterion features on Jupiter that are 250 km apart. What minimum diameter mirror is required? Assume a wavelength of 500 nm.

38–38 A converging lens 7.20 cm in diameter has a focal length of 300 mm. If the resolution is diffraction-limited, how far away can an object be if points on it 4.00 mm apart are to be resolved (according to Rayleigh's criterion)? Use $\lambda = 550$ nm.

PROBLEMS

38–39 Consider a single-slit diffraction pattern. The center of the central maximum, where the intensity is I_0, is located at $\theta = 0$. a) Let θ_+ and θ_- be the two angles on either side of $\theta = 0$ for which $I = \frac{1}{2}I_0$. $\Delta\theta = |\theta_+ - \theta_-|$ is called the *full width at half maximum*, or *FWHM*, of the central diffraction maximum. Solve for $\Delta\theta$ when the ratio between the slit width a and wavelength λ is i) $a/\lambda = 2$, ii) $a/\lambda = 5$, iii) $a/\lambda = 10$. (*Hint:* Your equation for θ_+ or θ_- cannot be solved analytically. You must use trial and error or solve it graphically.) b) The width of the central maximum can alternatively be defined as $2\theta_0$, where θ_0 is the angle that locates the minimum on one side of the central maximum. Calculate $2\theta_0$ for each case considered in part (a), and compare to $\Delta\theta$.

38–40 Suppose the entire apparatus (slits, screen, and space in between) in Exercise 38–4 is immersed in water ($n = 1.33$). Then what is the distance between the two dark fringes?

38–41 A slit 0.360 mm wide is illuminated by parallel rays of light that has a wavelength of 540 nm. The diffraction pattern is observed on a screen that is 1.20 m from the slit. The intensity at the center of the central maximum ($\theta = 0°$) is I_0. a) What is the distance on the screen from the center of the central maximum to the first minimum? b) What is the distance on the screen from the center of the central maximum to the point where the intensity has fallen to $I_0/2$? (See Problem 38–39a for a hint about how to solve for the phase angle β.)

38–42 The intensity of light in the Fraunhofer diffraction pattern of a single slit is

$$I = I_0\left(\frac{\sin\gamma}{\gamma}\right)^2,$$

where

$$\gamma = \frac{\pi a \sin\theta}{\lambda}.$$

a) Show that the equation for the values of γ at which I is a maximum is $\tan\gamma = \gamma$. b) Determine the three smallest positive values of γ that are solutions of this equation. (*Hint:* You can use a trial-and-error procedure. Guess a value of γ and adjust your guess to bring $\tan\gamma$ closer to γ. A graphical solution of the equation is very helpful in locating the solutions approximately, to get good initial guesses.)

38–43 Angular Width of a Principal Maximum. Consider N evenly spaced narrow slits. Use the small-angle approximation $\sin\theta = \theta$ (for θ in radians) to prove the following: For an intensity maximum that occurs at an angle θ, the intensity minima immediately adjacent to this maximum are at angles $\theta + \lambda/Nd$ and $\theta - \lambda/Nd$, so that the angular width of the principal maximum is $2\lambda/Nd$. This is proportional to $1/N$, as we concluded in Section 38–5 on the basis of energy conservation.

38–44 Calculating the Intensity for Three Slits. An opaque sheet has three narrow, parallel slits width a distance d between adjacent slits. Plane waves of light with wavelength λ are incident on the sheet, and the light that passes through the slits produces a Fraunhofer pattern on a distant screen. a) Let E_0 be the electric field amplitude of the wave reaching the screen from each individual slit, and let ϕ be the phase difference between waves from adjacent slits arriving at a point P on the screen. Draw a phasor diagram for the *total* wave at P, and show that the amplitude of this total wave is $E_P = E_0(1 + 2\cos\phi)$. (*Hint:* The three phasors of length E_0 and the total phasor of length E_P form a trapezoid.) b) Calculate the intensity measured on the screen as a function of d, λ, and the angle θ (the angle between the straight-ahead direction and a line extending from the center slit to point P.) Graph the intensity as a function of ϕ. c) Show that the intensity function calculated in part (b) has the following properties: i) The intensity at the principal maximum is nine times greater than the intensity from a single slit. ii) The locations of the principal maxima are given by $d \sin \theta = m\lambda$, where $m = 0, \pm1, \pm2, \ldots$, corresponding to values of ϕ that are integral multiples of 2π, just as in the two-slit pattern. iii) There are $(3 - 1) = 2$ minima between each pair of principal maxima. iv) The minima are located at $d \sin \theta = \pm\lambda/3, \pm2\lambda/3, \pm4\lambda/3, \pm5\lambda/3, \ldots$, corresponding to values of ϕ that are integral multiples of $2\pi/3$ but *not* integral multiples of 2π. v) Between each pair of principal maxima there is a *secondary* maximum at which the intensity is only one-ninth the value at a principal maximum. Show that these conclusions are in agreement with the general statements about N slits made in Section 38–5. d) Draw the phasor diagram for the following cases: i) $\phi = 2\pi/3$; ii) $\phi = \pi$; iii) $\phi = 4\pi/3$; iv) $\phi = 2\pi$. State which of these cases are intensity maxima, which are minima, and which are neither.

38–45 Phasor Diagram for Eight Slits. An interference pattern is produced by eight equally spaced narrow slits. Figure 38–12 shows phasor diagrams for the cases in which the phase difference ϕ between light from adjacent slits is $\phi = \pi$, $\phi = \pi/4$, and $\phi = \pi/2$. Each of these cases gives an intensity minimum. The caption for Fig. 38–12 also claims that a minimum occurs for $\phi = 3\pi/4$, $\phi = 5\pi/4$, $\phi = 3\pi/2$, and $\phi = 7\pi/4$. a) Draw the phasor diagram for each of these four cases, and explain why each diagram proves that there is in fact a minimum. (Note: You may find it helpful to use a different colored pencil for each slit!) b) For each of the four cases $\phi = 3\pi/4$, $\phi = 5\pi/4$, $\phi = 3\pi/2$, and $\phi = 7\pi/4$, for which pairs of slits is there totally destructive interference?

38–46 An interference pattern is produced by six equally spaced narrow slits. a) Draw the phasor diagram for the case in which the phase difference between light from adjacent slits is zero. In this case, if the electric field amplitude due to each individual slit is E, what is the resultant amplitude due to all six slits? If the intensity of light through each slit is I, what is the resultant intensity? b) Repeat part (a) for the case in which $\phi = 2\pi$. c) For what values of ϕ between 0 and 2π is there an interference minimum? For each ϕ draw the phasor diagram.

38–47 At the end of Section 38–5, the following statements were made about an array of N slits. Explain, using phasor diagrams, why each statement is true. a) A minimum occurs whenever ϕ is an integral multiple of $2\pi/N$, except when ϕ is an integral multiple of 2π (which gives a principal maximum). b) There are $(N - 1)$ minima between each pair of principal maxima.

38–48 In Eq. (38–12), consider the case in which $d = a$. In a sketch, show that in this case the two slits reduce to a single slit with width $2a$. Then show that Eq. (38–12) reduces to Eq. (38–5) with slit width $2a$.

38–49 What is the longest wavelength that can be observed in the third order for a transmission grating having 6500 slits/cm? Assume normal incidence.

38–50 a) Figure 38–14 shows plane waves of light incident *normally* on a diffraction grating. If instead the light strikes the grating at an angle of incidence θ' (measured from the normal), show that the condition for an intensity maximum is *not* Eq. (38–13), but rather

$$d(\sin \theta + \sin \theta') = m\lambda \qquad (m = 0, \pm1, \pm2, \pm3, \ldots).$$

b) For the grating described in Example 38–4 (Section 38–6), with 600 slits per millimeter, find the angles of the maxima cor-responding to $m = 0$, 1, and –1 with red light ($\lambda = 650$ nm) for the cases $\theta' = 0$ (normal incidence) and $\theta' = 20.0°$.

38–51 A diffraction grating has 650 slits/mm. What is the highest order that contains the entire visible spectrum? (The wavelength range of the visible spectrum is approximately 400 nm to 700 nm.)

38–52 X-Ray Diffraction of Salt. X rays with a wavelength of 0.125 nm are scattered from a cubic array (of a sodium chloride crystal), for which the spacing of adjacent atoms is $a = 0.282$ nm. a) If diffraction from planes parallel to a cube face is considered, at what angles θ of the incoming beam relative to the crystal planes will maxima be observed? b) Repeat part (a) for diffraction produced by the planes shown in Fig. 38–19b, which are separated by $a/\sqrt{2}$.

38–53 Phased-Array Radar. In one common type of radar installation, a rotating antenna sweeps a radio beam around the sky. But in a *phased-array* radar system, the antennas remain stationary and the beam is swept electronically. To see how this is done, consider an array of N antennas that are arranged along the horizontal x-axis at $x = 0, \pm d, \pm2d, \ldots, \pm(N-1)d/2$. (The number N is odd.) Each antenna emits radiation uniformly in all directions in the horizontal xy-plane. The antennas all emit radiation coherently, with the same amplitude E_0 and the same wavelength λ. The relative phase δ of the emission from adjacent antennas can be varied, however. If the antenna at $x = 0$ emits a signal that is given by $E_0 \cos \omega t$ as measured at a point next to the antenna, the antenna at $x = d$ emits a signal given by $E_0 \cos(\omega t + \delta)$ as measured at a point next to that antenna. The corresponding quantity for the antenna at $x = -d$ is $E_0 \cos(\omega t - \delta)$; for the antennas at $x = \pm2d$, it is $E_0 \cos(\omega t \pm 2\delta)$; and so on. a) If $\delta = 0$, the interference pattern at a distance from the antennas large compared to d has a principal maximum at $\theta = 0$ (that is, in the $+y$-direction, perpendicular to the line of the antennas). Show that if $d < \lambda$, this is the *only* principal interference maximum in the angular range $-90° < \theta < 90°$. Hence this principal maximum describes a beam emitted in the direction $\theta = 0$. As described in Section 38–5, if N is large, the beam will have a large intensity and be quite narrow. b) If $\delta \neq 0$, show that the principal intensity maximum described in part (a) is located at

$$\theta = \arcsin\left(\frac{\delta\lambda}{2\pi d}\right),$$

where δ is measured in radians. Thus by varying δ from positive to negative values and back again, which can easily be done electronically, the beam can be made to sweep back and forth around $\theta = 0$. c) A weather radar unit to be installed on an airplane emits radio waves at 8800 MHz. The unit uses 15 antennas in an array 28.0 cm long (from the antenna at one end of the array to the antenna at the other end). What must be the maximum and minimum values of δ (that is, the most positive and most negative values) if the radar beam is to sweep 45° to the left or right of the airplane's direction of flight? Give your answer in radians.

38–54 Resolution of Radio and Optical Telescopes. Diffraction occurs for circular antennas and for circular mirrors, just as for circular apertures. What is the minimum angular separation θ of two sources of electromagnetic radiation that can be resolved (according to Rayleigh's criterion) by a) the 8.3-m-diameter Subaru reflecting telescope atop Mauna Kea, Hawaii, for a wavelength of 500 nm; b) the 305-m-diameter radio telescope at Arecibo, Puerto Rico, for a wavelength of 21.0 cm? (This shows that radio telescopes must be much larger than optical ones to achieve the same diffraction-limited resolution.)

38-55 A telescope is used to observe two distant point sources 2.50 m apart with light of wavelength 600 nm. The objective of the telescope is covered with a slit of width 0.350 mm. What is the maximum distance in meters at which the two sources may be distinguished if the resolution is diffraction limited and Rayleigh's criterion is used?

38-56 *Quasars,* an abbreviation for *quasi-stellar radio sources,* are distant objects that look like stars through a telescope but that emit far more electromagnetic radiation than an entire normal galaxy of stars. An example is the bright object below and to the left of center in Fig. 38–27; the other elongated objects in this image are normal galaxies. The leading model for the structure of a quasar is a galaxy with a supermassive black hole at its center. In this model, the radiation is emitted by interstellar gas and dust within the galaxy as this material falls toward the black hole. The radiation is thought to emanate from a region just a few light years in diameter. (The diffuse glow surrounding the bright quasar in Fig. 38–27 is thought to be this quasar's host galaxy.) To investigate this model of quasars and to study other exotic astronomical objects, the Russian Space Agency plans to place a radio telescope in an orbit that extends to 77,000 km from the earth. When the signals from this telescope are combined with signals from the ground-based telescopes of the VLBA, the resolution will be that of a single radio telescope 77,000 km in diameter. What is the size of the smallest detail that this arrangement could resolve in the quasar 3C 405, which is 7.2×10^8 light years from earth, using radio waves at a frequency of 1665 MHz? (*Hint:* Use Rayleigh's criterion.) Give your answer in light years and in kilometers.

38-57 An astronaut in orbit can just resolve two point sources on earth that are 75.0 m apart. Assume that the resolution is diffraction-limited and use Rayleigh's criterion. What is the astronaut's altitude above the earth? Treat her eye as a circular aperture with a diameter of 4.00 mm (the diameter of her pupil), and take the wavelength of the light to be 500 nm.

38-58 Observing Planets Beyond Our Solar System. NASA is considering a project called *Planet Imager* that would give astronomers the ability to see details on planets orbiting other stars. Using the same principle as the Very Long Baseline Array (see Section 38–8), *Planet Imager* will use an array of infrared telescopes spread over thousands of kilometers of space. (Visible light would give even better resolution. Unfortunately, at visible wavelengths stars are so bright that a planet would be lost in the glare. This is less of a problem at infrared wavelengths.) a) If *Planet Imager* has an effective diameter of 6000 km and

FIGURE 38–27 Exercise 38–56.

observes infrared radiation at a wavelength of 10 μm, what is the greatest distance at which it would be able to observe details as small as 250 km across (about the size of the greater Los Angeles area) on a planet? Give your answer in light years (see Appendix E). (*Hint:* Use Rayleigh's criterion.) b) For comparison, consider the resolution of a single infrared telescope in space with a diameter of 1.0 m and that observes 10-μm radiation. What is the size of the smallest details that such a telescope could resolve at the distance of the nearest star to the sun, Proxima Centauri, which is 4.22 light years distant? How does this compare to the diameter of the earth (1.27×10^4 km)? To the average distance from the earth to the sun (1.50×10^8 km)? Would a single telescope of this kind be able to detect the presence of a planet like the earth, in an orbit the size of the earth's orbit, around *any* other star? Explain. c) Suppose *Planet Imager* is used to observe a planet orbiting the star 70 Virginis, which lies 59 light years from our solar system. A planet (though not an earthlike one) has in fact been detected orbiting this star, not by imaging it directly but by observing the slight "wobble" of the star as both it and the planet orbit their common center of mass. What is the size of the smallest details that *Planet Imager* could hope to resolve on the planet of 70 Virginis? How does this compare to the diameter of the planet, assumed to be comparable to that of Jupiter (1.38×10^5 km)? (Although the planet of 70 Virginis is thought to be at least 6.6 times more massive than Jupiter, its radius is probably not too different from that of Jupiter. The reason is that such large planets are thought to be composed primarily of gases, not rocky material, and hence can be greatly compressed by the mutual gravitational attraction of different parts of the planet.)

CHALLENGE PROBLEMS

38-59 It is possible to calculate the intensity in the single-slit Fraunhofer diffraction pattern *without* using the phasor method of Section 38–4. Let y' represent the position of a point within the slit of width a in Fig. 38–6a, with $y' = 0$ at the center of the slit so that the slit extends from $y' = -a/2$ to $y' = a/2$. We imagine dividing the slit up into infinitesimal strips of width dy', each of which acts as a source of secondary wavelets. a) The amplitude of the total wave at the point O on the distant screen in Fig. 38–6a is E_0. Explain why the amplitude of the wavelet from each infinitesimal strip within the slit is $E_0(dy'/a)$, so that the electric field of the wavelet a distance x from the infinitesimal strip is $dE = E_0(dy'/a) \sin(\omega t - kx)$. b) Explain why the wavelet from each strip as detected at point P in Fig. 38–6a can be expressed as

$$dE = E_0 \frac{dy'}{a} \sin[\omega t - k(D - y' \sin\theta)],$$

where D is the distance from the center of the slit to point P and $k = 2\pi/\lambda$. c) By integrating the contributions dE from all parts of the slit, show that the total wave detected at point P is

$$E = E_0 \sin(\omega t - kD) \frac{\sin[ka(\sin\theta)/2]}{ka(\sin\theta)/2}$$

$$= E_0 \sin(\omega t - kD) \frac{\sin[\pi a(\sin\theta)/\lambda]}{\pi a(\sin\theta)/\lambda}.$$

(The trigonometric identities in Appendix B will be useful.) Show that at $\theta = 0$, corresponding to point O in Fig. 38–6a, the wave is $E = E_0 \sin(\omega t - kD)$ and has amplitude E_0, as stated in part (a). d) Use the result of part (c) to show that if the intensity at point O is I_0, then the intensity at a point P is given by Eq. (38–7).

38–60 Intensity Pattern of N Slits. a) Consider an arrangement of N slits with a distance d between adjacent slits. The slits emit coherently and in phase at wavelength λ. Show that at a time t, the electric field at a distant point P is

$$E_P(t) = E_0 \cos(\omega t - kR) + E_0 \cos(\omega t - kR + \phi)$$
$$+ E_0 \cos(\omega t - kR + 2\phi) + \cdots$$
$$+ E_0 \cos(\omega t - kR + (N - 1)\phi),$$

where E_0 is the amplitude at P of the electric field due to an individual slit, $\phi = (2\pi d \sin\theta)/\lambda$, θ is the angle of the rays reaching P (as measured from the perpendicular bisector of the slit arrangement), and R is the distance from P to the most distant slit. In this problem, assume that R is much larger than d. b) To carry out the sum in part (a), it is convenient to use the complex-number relationship

$$e^{iz} = \cos z + i \sin z,$$

where $i = \sqrt{-1}$. In this expression, $\cos z$ is the *real part* of the complex number e^{iz}, and $\sin z$ is its *imaginary part*. Show that

the electric field $E_P(t)$ is equal to the real part of the complex quantity

$$\sum_{n=0}^{N-1} E_0 e^{i(\omega t - kR + n\phi)}.$$

c) Using the properties of the exponential function that $e^A e^B = e^{(A+B)}$ and $(e^A)^n = e^{nA}$, show that the sum in part (b) can be written as

$$E_0 \left(\frac{e^{iN\phi} - 1}{e^{i\phi} - 1} \right) e^{i(\omega t - kR)} = E_0 \left(\frac{e^{iN\phi/2} - e^{-iN\phi/2}}{e^{i\phi/2} - e^{-i\phi/2}} \right) e^{i[\omega t - kR + (N-1)\phi/2]}.$$

Then, using the relationship $e^{iz} = \cos z + i \sin z$, show that the (real) electric field at point P is

$$E_P(t) = \left[E_0 \frac{\sin(N\phi/2)}{\sin(\phi/2)} \right] \cos[\omega t - kR + (N-1)\phi/2].$$

The quantity in the first square brackets in this expression is the amplitude of the electric field at P. d) Use the result for the electric field amplitude in part (c) to show that the intensity at an angle θ is

$$I = I_0 \left[\frac{\sin(N\phi/2)}{\sin(\phi/2)} \right]^2,$$

where I_0 is the maximum intensity for an individual slit. e) Check the result in part (d) for the case $N = 2$. It will help to recall that $\sin 2A = 2 \sin A \cos A$. Explain why your result differs from Eq. (37–10), the expression for the intensity in two-source interference, by a factor of 4. (*Hint*: Is I_0 defined in the same way in both expressions?)

38–61 Intensity Pattern of N Slits, continued. Part (d) of the previous problem gives an expression for the intensity in the interference pattern of N identical slits. Use this result to verify the following statements. a) The maximum intensity in the pattern is $N^2 I_0$. b) The principal maximum at the center of the pattern extends from $\phi = -2\pi/N$ to $\phi = 2\pi/N$, so its width is inversely proportional to $1/N$. c) A minimum occurs whenever ϕ is an integral multiple of $2\pi/N$, except when ϕ is an integral multiple of 2π (which gives a principal maximum). d) There are $(N - 1)$ minima between each pair of principal maxima. e) Halfway between two principal maxima, the intensity can be no greater than I_0; that is, it can be no greater than $1/N^2$ times the intensity at a principal maximum.

This fuzzy blob of light, called a quasar, lies some 2×10^9 light years from earth. Emerging from the lower right of the quasar is a bright jet (shown by the box) roughly 200,000 light years in length. This jet is made of material that has been ejected from the quasar at speeds comparable to the speed of light. To describe such motions, the special theory of relativity is needed.

39

Relativity

39–1 INTRODUCTION

When the year 1905 began, Albert Einstein was an unknown 25-year-old clerk in the Swiss patent office. By the end of that amazing year he had published three papers of extraordinary importance. One was an analysis of Brownian motion; a second (for which he was awarded the Nobel Prize) was on the photoelectric effect. In the third, Einstein introduced his **special theory of relativity,** proposing drastic revisions in the Newtonian concepts of space and time.

The special theory of relativity has made wide-ranging changes in our understanding of nature, but Einstein based it on just two simple postulates. One states that the laws of physics are the same in all inertial frames of reference; the other states that the speed of light in vacuum is the same in all inertial frames. These innocent-sounding propositions have far-reaching implications. Here are three: (1) Events that are simultaneous for one observer may not be simultaneous for another. (2) When two observers moving relative to each other measure a time interval or a length, they may not get the same results. (3) For the conservation principles for momentum and energy to be valid in all inertial systems, Newton's second law and the equations for momentum and kinetic energy have to be revised.

Relativity has important consequences in *all* areas of physics, including thermodynamics, electromagnetism, optics, atomic and nuclear physics, and high-energy physics. Although many of the results derived in this chapter may run counter to your intuition, the theory is in solid agreement with experimental observations.

39–2 INVARIANCE OF PHYSICAL LAWS

EINSTEIN'S FIRST POSTULATE

Einstein's first postulate, called the **principle of relativity,** states: **The laws of physics are the same in every inertial frame of reference.** If the laws differed, that difference could distinguish one inertial frame from the others or make one frame somehow more "correct" than another. Here are two examples. Suppose you watch two children playing catch with a ball while the three of you are aboard a train moving with constant velocity. Your observations of the motion *of the ball,* no matter how carefully done, can't tell you how fast (or whether) the train is moving. This is because the laws of mechanics (Newton's laws) are the same in every inertial system.

Another example is the electromotive force (emf) induced in a coil of wire by a nearby moving permanent magnet. In the frame of reference in which the *coil* is stationary (Fig. 39–1a), the moving magnet causes a change of magnetic flux through the coil, and this induces an emf. In a different frame of reference in which the *magnet* is stationary (Fig. 39–1b), the motion of the coil through a magnetic field induces the emf. According to the principle of relativity, both of these points of view have equal validity, and both must predict the same induced emf. As we saw in Chapter 30, Faraday's law of induction can be applied to either description, and it does indeed satisfy this requirement.

(a)

(b)

39–1 The same result from two different reference frames. (a) In the frame of reference of the coil, the moving magnet increases the magnetic flux through the stationary coil, inducing an emf. (b) In the reference frame of the magnet, the moving coil increases the magnetic flux to induce the emf.

If the moving-magnet and moving-coil situations *did not* give the same results, we could use this experiment to distinguish one inertial frame from another. This would contradict the principle of relativity.

Equally significant is the prediction of the speed of electromagnetic radiation, derived from Maxwell's equations (Section 33–3). According to this analysis, light and all other electromagnetic waves travel in vacuum with a constant speed, now defined to equal exactly 299,792,458 m/s. (We often use the approximate value $c = 3.00 \times 10^8$ m/s, which is within one part in 1000 of the exact value.) As we will see, the speed of light in vacuum plays a central role in the theory of relativity.

EINSTEIN'S SECOND POSTULATE

During the nineteenth century, most physicists believed that light traveled through a hypothetical medium called the *ether,* just as sound waves travel through air. If so, the speed of light measured by observers would depend on their motion relative to the ether and would therefore be different in different directions. The Michelson-Morley experiment, described in Section 37–6, was an effort to detect motion of the earth relative to the ether. Einstein's conceptual leap was to recognize that if Maxwell's equations are valid in all inertial frames, then the speed of light in vacuum should also be the same in all frames and in all directions. In fact, Michelson and Morley detected *no* ether motion across the earth, and the ether concept has been discarded. Although Einstein may not have known about this negative result, it supported his bold hypothesis of the constancy of the speed of light in vacuum.

Thus Einstein was led to his second postulate: **The speed of light in vacuum is the same in all inertial frames of reference and is independent of the motion of the source.** Let's think about what this means. Suppose two observers measure the speed of light in vacuum. One is at rest with respect to the light source, and the other is moving away from it. Both are in inertial frames of reference. According to the principle of relativity, the two observers must obtain the same result, despite the fact that one is moving with respect to the other.

If this seems too easy, consider the following situation. A spacecraft moving past the earth at 1000 m/s fires a missile straight ahead with a speed of 2000 m/s (relative to the spacecraft). (Fig. 39–2). What is the missile's speed relative to the earth? Simple, you say; this is an elementary problem in relative velocity (Section 3–6). The correct answer, according to Newtonian mechanics, is 3000 m/s. But now suppose the spacecraft turns

39–2 A spaceship (S') moves with speed $v_{S'/E} = 1000$ m/s relative to the earth (E). It fires a missile (M) with speed $v_{M/S'} = 2000$ m/s relative to the spaceship. (a) Newtonian mechanics tells us that the missile moves at a speed of 3000 m/s relative to the earth. (b) Newtonian mechanics tells us that the light beam emitted by the spaceship moves at a speed greater than c relative to the earth; this contradicts Einstein's second postulate.

on a searchlight, pointing in the same direction in which the missile was fired. An observer on the spacecraft measures the speed of light emitted by the searchlight and obtains the value c. According to our previous discussion, the motion of the light after it has left the source cannot depend on the motion of the source. So the observer on earth who measures the speed of this same light must also obtain the value c, *not* $c + 1000$ m/s. This result contradicts our elementary notion of relative velocities, and it may not appear to agree with common sense. But "common sense" is intuition based on everyday experience, and this does not usually include measurements of the speed of light.

Einstein's second postulate immediately implies the following result: *It is impossible for an inertial observer to travel at* c, *the speed of light in vacuum.* We can prove this by showing that travel at c implies a logical contradiction. Suppose that the spacecraft S' in Fig. 39–2b is moving at the speed of light relative to an observer on the earth, so that $v_{S'/E} = c$. If the spacecraft now turns on a headlight, the second postulate now asserts that the earth observer E measures the headlight beam to be also moving at c. Thus this observer measures that the headlight beam and the spacecraft move together and are always at the same point in space. But Einstein's second postulate also asserts that the headlight beam moves at a speed c relative to the spacecraft, so they *cannot* be at the same point in space. This contradictory result can be avoided only if it is impossible for an inertial observer, such as a passenger on the spacecraft, to move at c. As we go through our discussion of relativity, you may find yourself asking the question Einstein asked himself as a 16-year-old student, "What would I see if I were traveling at the speed of light?" Einstein realized only years later that his question's basic flaw was that he could *not* travel at c.

DEFINING INERTIAL FRAMES OF REFERENCE

Let's restate this argument symbolically, using two inertial frames of reference, labeled S for the observer on earth and S' for the moving spacecraft, as shown in Fig. 39–3. To keep things as simple as possible, we have omitted the z-axes. The x-axes of the two frames lie along the same line, but the origin O' of frame S' moves relative to the origin O of frame S with constant velocity u along the common x-x'-axis. We on earth set our clocks so that the two origins coincide at time $t = 0$, so their separation at a later time t is ut.

CAUTION ▶ Many of the equations derived in this chapter are true *only* if you define your inertial reference frames as stated in the preceding paragraph. For instance, the positive x-direction must be the direction in which the origin O' moves relative to the origin O. In Fig. 39–3 this direction is to the right; if instead O' moves to the left relative to O, you must define the positive x-direction to be to the left. ◀

Now think about how we describe the motion of a particle P. This might be an exploratory vehicle launched from the spacecraft or a pulse of light from a laser. We can describe the *position* of this particle by using the earth coordinates (x, y, z) in S or the spacecraft coordinates (x', y', z') in S'. Figure 39–3 shows that these are simply

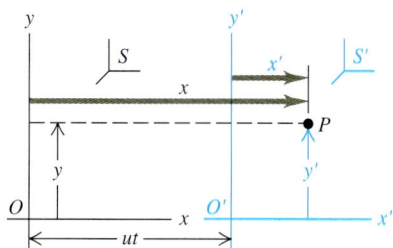

39–3 The position of particle P can be described by the coordinates x and y in frame of reference S or by x' and y' in S'. Frame S' moves relative to S with constant velocity u along the common x-x'-axis. The two origins O and O' coincide at time $t = 0 = t'$.

related by

$$x = x' + ut, \qquad y = y', \qquad z = z' \qquad \text{(Galilean coordinate} \qquad (39\text{-}1)$$
$$\text{transformation).}$$

These equations, based on the familiar Newtonian notions of space and time, are called the **Galilean coordinate transformation.**

If particle P moves in the x-direction, its instantaneous velocity component v as measured by an observer stationary in S is given by $v = dx/dt$. Its velocity component v' measured by an observer at rest in S' is $v' = dx'/dt$. We can derive a relation between v and v' by taking the derivative with respect to t of the first of Eqs. (39–1):

$$\frac{dx}{dt} = \frac{dx'}{dt} + u.$$

Now dx/dt is the velocity v measured in S, and dx'/dt is the velocity v' measured in S', so we get the *Galilean velocity transformation* for one-dimensional motion:

$$v = v' + u \qquad \text{(Galilean velocity transformation).} \qquad (39\text{-}2)$$

Although the notation differs, this result agrees with our discussion of relative velocities in Section 3–6.

Now here's the fundamental problem. Applied to the speed of light in vacuum, Eq. (39–2) says that $c = c' + u$. Einstein's second postulate, supported subsequently by a wealth of experimental evidence, says that $c = c'$. This is a genuine inconsistency, not an illusion, and it demands resolution. If we accept this postulate, we are forced to conclude that Eqs. (39–1) and (39–2) *cannot* be precisely correct, despite our convincing derivation. These equations have to be modified to bring them into harmony with this principle.

The resolution involves some very fundamental modifications in our kinematic concepts. The first idea to be changed is the seemingly obvious assumption that the observers in frames S and S' use the same *time scale,* formally stated as $t = t'$. Alas, we are about to show that this everyday assumption cannot be correct; the two observers *must* have different time scales. (See also Fig. 39–4.) We must define the velocity v' in frame S' as $v' = dx'/dt'$, not as dx'/dt; the two quantities are not the same. The difficulty lies in the concept of *simultaneity,* which is our next topic. A careful analysis of simultaneity will help us develop the appropriate modifications of our notions about space and time.

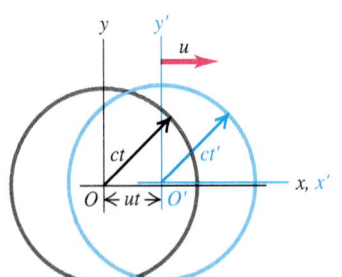

39–4 According to Einstein's second postulate, observers in both frames of reference must measure a spherical wave front expanding with speed c. For this to be true, $x - ct$ must equal $x' - ct'$, in disagreement with the Galilean coordinate transformation and $t = t'$.

39-3 RELATIVITY OF SIMULTANEITY

Measuring times and time intervals involves the concept of **simultaneity.** In a given frame of reference, an **event** is an occurrence that has a definite position and time. When you say that you awoke at seven o'clock, you mean that two events (your awakening and your clock showing 7:00) occurred *simultaneously.* The fundamental problem in measuring time intervals is that, in general, two events that are simultaneous in one frame of reference are *not* simultaneous in a second frame that is moving relative to the first, even if both are inertial frames.

This may seem to be contrary to common sense. To illustrate the point, here is a version of one of Einstein's *thought experiments*—mental experiments that follow concepts to their logical conclusions. Imagine a train moving with a speed comparable to c, with uniform velocity (Fig. 39–5). Two lightning bolts strike a passenger car, one near each end. Each bolt leaves a mark on the car and one on the ground at the instant the bolt hits. The points on the ground are labeled A and B in the figure, and the corresponding points on the car are A' and B'. Stanley is stationary on the ground at O, midway between A and B. Mavis is moving with the train at O' in the middle of the passenger car, midway

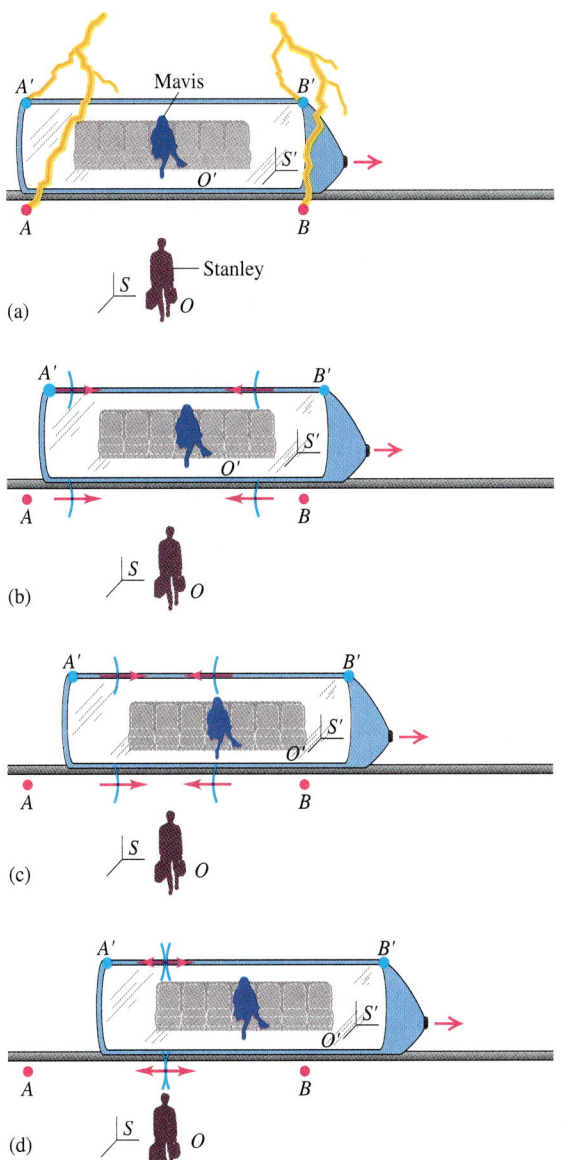

39–5 (a) Lightning bolts strike the passenger car and ground at each end. (b) Mavis moves toward the wave from the front and away from the wave from the rear. (c) She sees the light from the front and concludes that the bolt from the front struck first. (d) Stationary Stanley sees the light from the two bolts arrive at the same time. He concludes that the lightning bolts struck simultaneously. The light from the rear bolt has not yet reached Mavis.

between A' and B'. Both Stanley and Mavis see both light flashes emitted from the points where the lightning strikes.

Suppose the two wave fronts from the lightning strikes reach Stanley at O simultaneously. He knows that he is the same distance from B and A, so Stanley concludes that the two bolts struck B and A simultaneously. Mavis agrees that the two wave fronts reached Stanley at the same time, but she disagrees that the flashes were emitted simultaneously.

Stanley and Mavis agree that the two wave fronts do not reach Mavis at the same time. Mavis at O' is moving to the right with the train, so she runs into the wave front from B' *before* the wave front from A' catches up to her. However, because she is in the middle of the passenger car equidistant from A' and B', her observation is that both wave fronts took the same time to reach her because both moved the same distance at the same speed c. (Recall that the speed of each wave front with respect to *either* observer is c.) Thus she concludes that the lightning bolt at B' struck *before* the one at A'. Stanley at O measures the two events to be simultaneous, but Mavis at O' does not! *Whether or not two events at different x-axis locations are simultaneous depends on the state of motion of the observer.*

You may want to argue that in this example the lightning bolts really *are* simultaneous and that if Mavis at O' could communicate with the distant points without the time delay caused by the finite speed of light, she would realize this. But that would be erroneous; the finite speed of information transmission is not the real issue. If O' is midway between A' and B', then in her frame of reference the time for a signal to travel from A' to O' is the same as that from B' to O'. Two signals arrive simultaneously at O' only if they were emitted simultaneously at A' and B'. In this example they *do not* arrive simultaneously at O', and so Mavis must conclude that the events at A' and B' were *not* simultaneous.

Furthermore, there is no basis for saying that Stanley is right and Mavis is wrong or vice versa. According to the principle of relativity, no inertial frame of reference is more correct than any other in the formulation of physical laws. Each observer is correct *in his or her own frame of reference.* In other words, simultaneity is not an absolute concept. Whether two events are simultaneous depends on the frame of reference. As we mentioned at the beginning of this section, simultaneity plays an essential role in measuring time intervals. It follows that *the time interval between two events may be different in different frames of reference.* So our next task is to learn how to compare time intervals in different frames of reference.

39–4 RELATIVITY OF TIME INTERVALS

17.1
Relativity of Time

To derive a quantitative relation between time intervals in different coordinate systems, let's consider another thought experiment. As before, a frame of reference S' moves along the common x-x′ axis with constant speed u relative to a frame S. As discussed in Section 39–2, u must be less than the speed of light c. Mavis, who is riding along with frame S', measures the time interval between two events that occur at the *same* point in space. Event 1 is when a flash of light from a light source leaves O'. Event 2 is when the flash returns to O', having been reflected from a mirror a distance d away, as shown in Fig. 39–6a. We label the time interval Δt_0, using the subscript zero as a reminder that the apparatus is at rest, with zero velocity, in frame S'. The flash of light moves a total distance $2d$, so the time interval is

39–6 (a) A light pulse emitted from a source at O' and reflected back along the same line, as observed by Mavis in S'. (b) Path of the same light pulse, as observed by Stanley in S. The positions of O' at the times of departure and return of the pulse are shown. The speed of the pulse is the same in S as in S', but the path is longer in S.

$$\Delta t_0 = \frac{2d}{c}. \qquad (39–3)$$

The round-trip time measured by Stanley in frame S is a different interval Δt; in his frame of reference the two events occur at *different* points in space. During the time Δt, the source moves relative to S a distance $u \Delta t$ (Fig. 39–6b). In S' the round-trip distance

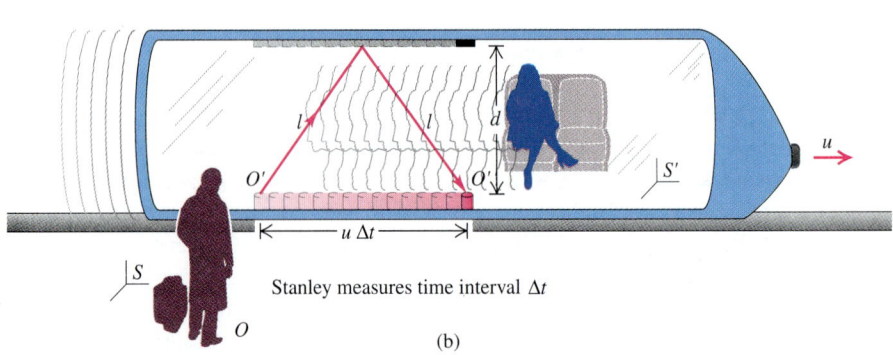

(a)   (b)

is $2d$ perpendicular to the relative velocity, but the round-trip distance in S is the longer distance $2l$, where

$$l = \sqrt{d^2 + \left(\frac{u\,\Delta t}{2}\right)^2}.$$

In writing this expression, we have assumed that both observers measure the same distance d. We will justify this assumption in the next section. The speed of light is the same for both observers, so the round-trip time measured in S is

$$\Delta t = \frac{2l}{c} = \frac{2}{c}\sqrt{d^2 + \left(\frac{u\,\Delta t}{2}\right)^2}. \tag{39-4}$$

We would like to have a relation between Δt and Δt_0 that is independent of d. To get this, we solve Eq. (39-3) for d and substitute the result into Eq. (39-4), obtaining

$$\Delta t = \frac{2}{c}\sqrt{\left(\frac{c\,\Delta t_0}{2}\right)^2 + \left(\frac{u\,\Delta t}{2}\right)^2}. \tag{39-5}$$

Now we square this and solve for Δt; the result is

$$\Delta t = \frac{\Delta t_0}{\sqrt{1 - u^2/c^2}}.$$

We may generalize this important result. In a particular frame of reference, suppose that two events occur at the same point in space. The time interval between these events, as measured by an observer at rest in this same frame (which we call the *rest frame* of this observer), is Δt_0. Then an observer in a second frame moving with constant speed u relative to the rest frame will measure the time interval to be Δt, where

$$\Delta t = \frac{\Delta t_0}{\sqrt{1 - u^2/c^2}} \qquad \text{(time dilation).} \tag{39-6}$$

We recall that no inertial observer can travel at $u = c$ and note that $\sqrt{1 - u^2/c^2}$ is imaginary for $u > c$. Thus Eq. (39-6) gives sensible results only when $u < c$. The denominator of Eq. (39-6) is always smaller than 1, so Δt is always *larger* than Δt_0. Thus we call this effect **time dilation.**

Think of an old-fashioned pendulum clock that has one second between ticks, as measured by Mavis in the clock's rest frame; this is Δt_0. If the clock's rest frame is moving relative to Stanley, he measures a time between ticks Δt that is longer than one second. In brief, *observers measure any clock to run slow if it moves relative to them.* Note that this conclusion is a direct result of the fact that the speed of light in vacuum is the same in both frames of reference.

PROPER TIME

There is only one frame of reference in which a clock is at rest, and there are infinitely many in which it is moving. Therefore the time interval measured between two events (such as two ticks of the clock) that occur at the same point in a particular frame is a more fundamental quantity than the interval between events at different points. We use the term **proper time** to describe the time interval Δt_0 between two events that occur *at the same point.*

It is important to note that the time interval Δt in Eq. (39-6) involves events that occur *at different space points* in the frame of reference S. Note also that any differences between Δt and the proper time Δt_0, are *not* caused by differences in transit times from those space points to an observer at rest in S. We assume that our observer is able to

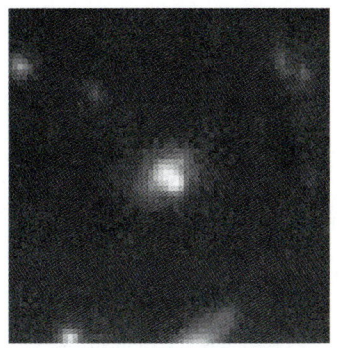

This image shows an exploding star called a *supernova* within a galaxy billions of light years away. The brightness of a typical supernova decays at a certain rate. But supernovae that are moving away from us at a substantial fraction of the speed of light, like the one shown here, decay noticeably more slowly. The slowdown is in quantitative agreement with Eq. (39-6) and the principle that moving clocks—in this case, the decaying light from the supernova—run slow.

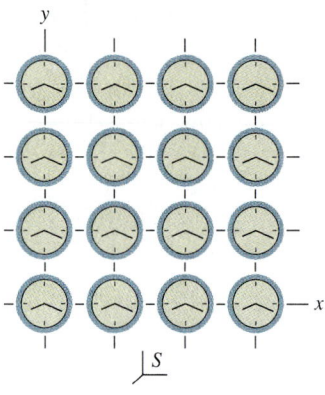

39–7 A frame of reference pictured as a coordinate system with a set of synchronized clocks. For the third dimension, picture identical planes of these clocks that are all parallel to the page, in front of and behind it, connected by grid lines perpendicular to the page.

correct for transit times just as an astronomer who's observing the sun understands that an event observed now on earth actually occurred 500 s ago on the sun's surface. Alternatively, we can use *two* observers, one stationary at the location of the first event and the other at the second, each with his or her own clock. We can synchronize these two clocks without difficulty, as long as they are at rest in the same frame of reference. For example, we could send a light pulse simultaneously to the two clocks from a point midway between them. When the pulses arrive, the observers set their clocks to a pre-arranged time. (But note that clocks that are synchronized in one frame of reference *are not* in general synchronized in any other frame.)

In thought experiments, it's often helpful to imagine many observers with synchronized clocks at rest at various points in a particular frame of reference. We can picture a frame of reference as a coordinate grid with lots of synchronized clocks distributed around it, as suggested by Fig. 39–7. Only when a clock is moving relative to a given frame of reference do we have to watch for ambiguities of synchronization or simultaneity.

Problem–Solving Strategy

TIME DILATION

1. First, make sure you understand the difference between the proper time Δt_0 and the dilated time Δt. The proper time Δt_0 in a particular reference frame is the time interval between two events that occur at the *same point* in that reference frame. Then the longer time Δt is the time interval between the same two events as measured in a second reference frame that has a speed u relative to the first frame. The two events will be at different points in this second frame.

2. Check that your answers make sense; Δt is never smaller than Δt_0, and u is never larger than c.

EXAMPLE 39-1

Time dilation at 0.990c You are on earth as a spaceship flies past at a speed of 0.990c (about 2.97×10^8 m/s) relative to the earth. A high-intensity signal light (perhaps a pulsed laser) on the ship blinks on and off; each pulse lasts 2.20×10^{-6} s as measured on the spaceship. What do you measure as the duration of each light pulse?

SOLUTION Let S be the earth's frame of reference, and let S' be that of the spaceship. The time between laser pulses measured by an observer on the spaceship (in which the laser is at rest) is 2.20×10^{-6} s. This is a proper time in S', referring to two events, the starting and stopping of the pulse, that occur at the same point relative to S'. In the notation of Eq. (39–6), $\Delta t_0 = 2.20 \times 10^{-6}$ s. The corresponding interval Δt that you

measure on earth (S) is given by Eq. (39–6):

$$\Delta t = \frac{\Delta t_0}{\sqrt{1 - u^2/c^2}} = \frac{2.20 \times 10^{-6} \text{ s}}{\sqrt{1 - (0.990)^2}} = 15.6 \times 10^{-6} \text{ s}.$$

Thus the time dilation in S is about a factor of seven.

An experiment with similarities to the spaceship example provided the first direct experimental confirmation of Eq. (39–6). *Muons* are unstable subatomic particles that were first observed in cosmic rays. They decay with a mean lifetime of 2.20×10^{-6} s as measured in a frame of reference in which they are at rest. But in cosmic-ray showers they move very fast. The mean lifetime of muons with a speed of 0.990c is measured to be 15.6×10^{-6} s. Note that the numbers are identical to those in the spaceship example; the duration of the light pulse is replaced by the mean lifetime of a muon.

EXAMPLE 39-2

Time dilation at jetliner speeds An airplane flies from San Francisco to New York (about 4800 km, or 4.80×10^6 m) at a steady speed of 300 m/s (about 670 mi/h). How much time does

the trip take, as measured by an observer on the ground? By an observer in the plane?

SOLUTION San Francisco and New York are two different

points, so the time interval measured by the ground observers corresponds to Δt in Eq. (39–6). To find it, we simply divide the distance by the speed:

$$\Delta t = \frac{4.80 \times 10^6 \text{ m}}{300 \text{ m/s}} = 1.60 \times 10^4 \text{ s} \qquad \text{(about } 4\tfrac{1}{2} \text{ hours).}$$

In the airplane's frame, San Francisco and New York passing under the plane occur at the *same* point (the position of the plane). The time interval in the airplane is a proper time, corresponding to Δt_0 in Eq. (39–6). We have

$$\frac{u^2}{c^2} = \frac{(300 \text{ m/s})^2}{(3.00 \times 10^8 \text{ m/s})^2} = 1.00 \times 10^{-12},$$

and from Eq. (39–6),

$$\Delta t_0 = (1.60 \times 10^4 \text{ s})\sqrt{1 - 1.00 \times 10^{-12}}.$$

The radical can't be evaluated with adequate precision with an ordinary calculator. But we can approximate it using the bino-

mial theorem (Appendix B):

$$(1 - 1.00 \times 10^{-12})^{1/2} = 1 - \left(\frac{1}{2}\right)(1.00 \times 10^{-12}) + \cdots.$$

The remaining terms are of the order of 10^{-24} or smaller and can be discarded. The approximate result for Δt_0 is

$$\Delta t_0 = (1.60 \times 10^4 \text{ s})(1 - 0.50 \times 10^{-12}).$$

The proper time Δt_0, measured in the airplane, is very slightly less (by less than one part in 10^{12}) than the time measured on the ground.

We don't notice such effects in everyday life. But we mentioned in Section 1–4 that present-day atomic clocks can attain a precision of about one part in 10^{13}. A cesium clock traveling a long distance in a Boeing 747 has been used to measure this effect and thereby verify Eq. (39–6) even at speeds much less than c.

EXAMPLE 39-3

Just when is it proper? Mavis boards a spaceship, then zips past Stanley on earth at a relative speed of 0.600c. At the instant she passes, both start timers. a) At the instant when Stanley measures that Mavis has traveled 9.00×10^7 m past him, what does Mavis's timer read? b) At the instant when Mavis reads 0.400 s on her timer, what does Stanley read on his?

SOLUTION a) Relative to Stanley, Mavis moves at $0.600c = 0.600(3.00 \times 10^8 \text{ m/s}) = 1.80 \times 10^8 \text{ m/s}$ and travels the 9.00×10^7 m in a time of $(9.00 \times 10^7 \text{ m})/(1.80 \times 10^8 \text{ m/s}) = 0.500$ s. The two events, Mavis passing the earth and Mavis reaching $x = 9.00 \times 10^7$ m, are not at the same point in Stanley's frame of reference, so 0.500 s equals Δt. Mavis's timer is at rest relative to her, so her timer reads an elapsed time of

$$\Delta t_0 = \Delta t\sqrt{1 - u^2/c^2} = 0.500 \text{ s}\sqrt{1 - (0.600)^2} = 0.400 \text{ s}.$$

b) It is tempting—but wrong—to answer that Stanley's timer reads 0.500 s. We are now considering a *different* pair of events,

the starting and the reading of Stanley's timer, that both occur on earth at the same point. Mavis measures that 0.400 s elapses between these two events, but they occur at different places in her reference frame, so 0.400 s $= \Delta t$. (In her frame, Stanley passes her at time zero and is a distance behind her of $(1.80 \times 10^8 \text{ m/s})(0.400 \text{ s}) = 7.20 \times 10^7$ m at time 0.400 s.) The time on Stanley's timer is now the proper time:

$$\Delta t_0 = \Delta t\sqrt{1 - u^2/c^2} = 0.400 \text{ s}\sqrt{1 - (0.600)^2} = 0.320 \text{ s}.$$

If the difference between 0.500 s and 0.320 s still troubles you, consider the following: Stanley, taking proper account of the transit time of a signal from $x = 9.00 \times 10^7$ m, says that Mavis passed that point and his timer read 0.500 s at the same instant. But Mavis says that those two events occurred at different positions and were *not* simultaneous—she passed the point at the instant his timer read 0.320 s. This example shows the relativity of simultaneity.

The quantity $1/(1 - u^2/c^2)^{1/2}$ appears so often in relativity that it is given its own symbol γ ("gamma"):

$$\gamma = \frac{1}{\sqrt{1 - u^2/c^2}}. \qquad (39\text{–}7)$$

For example, we can then express the time dilation equation, Eq. (39–6), simply as

$$\Delta t = \gamma \Delta t_0 \qquad \text{(time dilation).} \qquad (39\text{–}8)$$

To simplify equations, u/c is sometimes given the symbol β; then $\gamma = 1/(1 - \beta^2)^{1/2}$.

When the relative speed u of two frames of reference is very small in comparison to c, u^2/c^2 is much smaller than 1 and γ is very nearly *equal* to 1. In that limit, Eqs. (39–6) and (39–8) approach the Newtonian relation $\Delta t = \Delta t_0$, corresponding to the same time scale in all frames of reference. Figure 39–8 shows a graph of γ as a function of u.

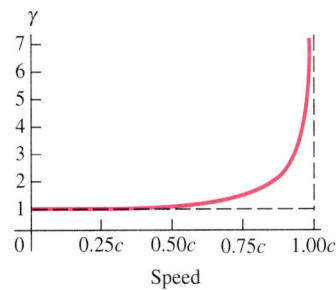

39–8 Because u is always less than or equal to c, γ is always greater than or equal to 1. It approaches infinity as u approaches c.

THE TWIN PARADOX

Equations (39–6) and (39–8) for time dilation suggest an apparent paradox called the **twin paradox.** Consider identical twin astronauts named Eartha and Astrid. Eartha remains on earth while her twin Astrid takes off on a high-speed trip through the galaxy. Because of time dilation, Eartha sees Astrid's heartbeat and all other life processes proceeding more slowly than her own. Thus to Eartha, Astrid ages more slowly; when Astrid returns to earth she is younger (has aged less) than Eartha.

Now here is the paradox: All inertial frames are equivalent. Can't Astrid make exactly the same arguments to conclude that Eartha is in fact the younger? Then each twin measures the other to be younger when they're back together, and that's a paradox.

To resolve the paradox, we recognize that the twins are *not* identical in all respects. While Eartha remains in an approximately inertial frame at all times, Astrid must accelerate with respect to that inertial frame during parts of her trip in order to leave, turn around, and return to earth. Eartha's reference frame is always approximately inertial; Astrid's is often far from inertial. Thus there is a real physical difference between the circumstances of the two twins. Careful analysis shows that Eartha is correct; when Astrid returns, she *is* younger than Eartha.

39–5 RELATIVITY OF LENGTH

17.2
Relativity of Length

LENGTHS PARALLEL TO THE RELATIVE MOTION

Not only does the time interval between two events depend on the observer's frame of reference, the *distance* between two points may also depend on the observer's frame of reference. The concept of simultaneity is involved. Suppose you want to measure the length of a moving car. One way is to have two assistants make marks on the pavement at the positions of the front and rear bumpers. Then you measure the distance between the marks. But your assistants have to make their marks *at the same time*. If one marks the position of the front bumper at one time and the other marks the position of the rear bumper half a second later, you won't get the car's true length. Since we've learned that simultaneity isn't an absolute concept, we have to proceed with caution.

To develop a relation between lengths that are measured parallel to the direction of motion in various coordinate systems, we consider another thought experiment. We attach a light source to one end of a ruler and a mirror to the other end. The ruler is at rest in reference frame S', and its length in this frame is l_0 (Fig. 39–9a). Then the time Δt_0

39–9 (a) A light pulse is emitted from a source at one end of a ruler, reflected from a mirror at the opposite end, and returned to the source position. (b) Motion of the light pulse as seen by Stanley in S. The distance traveled from source to mirror is greater, by the amount $u \, \Delta t_1$, than the length l measured in S, as shown.

required for a light pulse to make the round trip from source to mirror and back is

$$\Delta t_0 = \frac{2l_0}{c}. \tag{39–9}$$

This is a proper time interval because departure and return occur at the same point in S'.

In reference frame S the ruler is moving to the right with speed u during this travel of the light pulse (Fig. 39–9b). The length of the ruler in S is l, and the time of travel from source to mirror, as measured in S, is Δt_1. During this interval the ruler, with source and mirror attached, moves a distance $u\,\Delta t_1$. The total length of path d from source to mirror is not l, but rather

$$d = l + u\,\Delta t_1. \tag{39–10}$$

The light pulse travels with speed c, so it is also true that

$$d = c\,\Delta t_1. \tag{39–11}$$

Combining Eqs. (39–10) and (39–11) to eliminate d, we find

$$c\,\Delta t_1 = l + u\,\Delta t_1,$$

or

$$\Delta t_1 = \frac{l}{c - u}. \tag{39–12}$$

(Dividing the distance l by $c - u$ does *not* mean that light travels with speed $c - u$, but rather that the distance the pulse travels in S is greater than l.)

In the same way we can show that the time Δt_2 for the return trip from mirror to source is

$$\Delta t_2 = \frac{l}{c + u}. \tag{39–13}$$

The *total* time $\Delta t = \Delta t_1 + \Delta t_2$ for the round trip, as measured in S, is

$$\Delta t = \frac{l}{c - u} + \frac{l}{c + u} = \frac{2l}{c(1 - u^2/c^2)}. \tag{39–14}$$

We also know that Δt and Δt_0 are related by Eq. (39–6), because Δt_0 is a proper time in S'. Thus Eq. (39–9) for the round-trip time in the rest frame S' of the ruler becomes

$$\Delta t \sqrt{1 - \frac{u^2}{c^2}} = \frac{2l_0}{c}. \tag{39–15}$$

Finally, combining Eqs. (39–14) and (39–15) to eliminate Δt and simplifying, we obtain

$$l = l_0 \sqrt{1 - \frac{u^2}{c^2}} \qquad \text{(length contraction)}. \tag{39–16}$$

Thus the length l measured in S, in which the ruler is moving, is *shorter* than the length l_0 measured in its rest frame S'.

CAUTION ▶ This is *not* an optical illusion! The ruler really is shorter in reference frame S than it is in S'. ◀

A length measured in the frame in which the body is at rest (the rest frame of the body) is called a **proper length**; thus l_0 is a proper length in S', and the length measured in any other frame moving relative to S' is *less than* l_0. This effect is called **length contraction**. In terms of the quantity $\gamma = 1/(1 - u^2/c^2)^{1/2}$ defined in Eq. (39–7), we can also

express Eq. (39–16) as

$$l = \frac{l_0}{\gamma} \qquad \text{(length contraction).} \qquad (39\text{–}17)$$

When u is very small in comparison to c, γ approaches 1. Thus in the limit of small speeds we approach the Newtonian relation $l = l_0$. This and the corresponding result for time dilation show that Eqs. (39–1), the Galilean coordinate transformation, are usually sufficiently accurate for relative speeds much smaller than c.

LENGTHS PERPENDICULAR TO THE RELATIVE MOTION

We have derived Eq. (39–16) for lengths measured in the direction *parallel* to the relative motion of the two frames of reference. Lengths that are measured *perpendicular* to the direction of motion are *not* contracted. To prove this, consider two identical meter sticks. One stick is at rest in frame S and lies along the positive y-axis with one end at O, the origin of S. The other is at rest in frame S' and lies along the positive y'-axis with one end at O', the origin of S'. Frame S' moves in the positive x-direction relative to frame S. Observers Stanley and Mavis, at rest in S and S' respectively, station themselves at the 50-cm mark of their sticks. At the instant the two origins coincide, the two sticks lie along the same line. At this instant, Mavis makes a mark on Stanley's stick at the point that coincides with her own 50-cm mark, and Stanley does the same to Mavis's stick.

Suppose for the sake of argument that Stanley sees Mavis's stick as longer than his own. Then the mark Stanley makes on her stick is *below* its center. In that case, Mavis will think Stanley's stick has become shorter, since half of its length coincides with *less* than half her stick's length. So Mavis sees moving sticks getting shorter and Stanley sees them getting longer. But this implies an asymmetry between the two frames that contradicts the basic postulate of relativity that tells us all inertial frames are equivalent. We conclude that consistency with the postulates of relativity requires that both observers see the rulers as having the same length, even though to each observer one of them is stationary and the other is moving (see Fig. 39–10). So *there is no length contraction perpendicular to the direction of relative motion of the coordinate systems.* We used this result in our derivation of Eq. (39–6) in assuming that the distance d is the same in both frames of reference.

For example, suppose a moving rod of length l_0 makes an angle θ_0 with the direction of relative motion (the x-axis) as measured in its rest frame. Its length component in that frame parallel to the motion, $l_0 \cos \theta_0$, is contracted to $(l_0 \cos \theta_0)/\gamma$. However, its length component perpendicular to the motion, $l_0 \sin \theta_0$, remains the same.

39–10 The meter sticks are perpendicular to the relative velocity, so for any value of u, both Stanley and Mavis measure either meter stick to have a length of one meter.

Problem-Solving Strategy

LENGTH CONTRACTION

1. Make sure you understand the difference between the proper length l_0 and the contracted length l. The proper length l_0 in a particular reference frame is a length measured on an object at rest in that reference frame. Then the length l is the length between the same two points on the object as measured in a second reference frame that has a speed u relative to the first frame.

2. In Eqs. (39–16) and (39–17), both l and l_0 are lengths measured parallel to the direction of motion. Any length that is perpendicular to the relative motion is unchanged.

3. Check that your answers make sense; l is never larger than l_0, and u is never larger than c.

EXAMPLE 39-4

How long is the spaceship? A crew member on the spaceship of Example 39–1 (Section 39–4) measures its length, obtaining the value 400 m. What length do observers measure on earth?

SOLUTION The 400-m length of the spaceship is the proper length l_0 because it is measured in the frame in which the spaceship is at rest. We want to find the length l measured by observers on earth. From Eq. (39–16),

$$l = l_0\sqrt{1 - \frac{u^2}{c^2}} = (400\text{ m})\sqrt{1 - (0.990)^2}$$

$$= 56.4\text{ m.}$$

To find this quantity, two observers with synchronized clocks could measure the positions of the two ends of the spaceship simultaneously in the earth's reference frame, as shown in Fig. 39–11. (These two measurements will *not* appear simultaneous to an observer in the spaceship.)

39–11 The two observers on the earth (frame S) must measure x_2 and x_1 simultaneously to correctly obtain the length in their frame of reference using $l = x_2 - x_1$.

EXAMPLE 39-5

How far apart are the observers? The two observers mentioned in Example 39–4 are 56.4 m apart on the earth. How far apart does the spaceship crew measure them to be?

SOLUTION The answer is *not* 400 m! Think of a 56.4-m-long telephone cable stretching from one observer to the other. This cable is at rest in the earth frame, and so 56.4 m is the proper length l_0. Relative to the crew members, the earth zooms by at $u = 0.990c$, so Eq. (39–16) gives

$$l = l_0\sqrt{1 - \frac{u^2}{c^2}} = (56.4\text{ m})\sqrt{1 - (0.990)^2}$$

$$= 7.96\text{ m.}$$

This answer does *not* say that the crew measures their spaceship to be both 400 m long and 7.96 m long. The observers on earth measure the spaceship to have a contracted length of 56.4 m because they are 56.4 m apart when they measure the positions of the ends at what they measure to be the same instant. (Viewed from the spaceship frame, the observers do not measure those positions simultaneously.) The crew then measure the 56.4-m proper length to be contracted to 7.96 m. Similarly to Example 39–3, two different measurements are being made.

EXAMPLE 39-6

Moving with a muon As was stated in Example 39–1, a muon has, on average, a proper lifetime of 2.20×10^{-6} s and a dilated lifetime of 15.6×10^{-6} s in a frame in which its speed is $0.990c$. Multiplying constant speed by time to find distance gives $0.990(3.00 \times 10^8 \text{ m/s})(2.20 \times 10^{-6} \text{ s}) = 653$ m and $0.990(3.00 \times 10^8 \text{ m/s})(15.6 \times 10^{-6} \text{ s}) = 4630$ m. Interpret these two distances.

SOLUTION If an average muon moves at $0.990c$ past observers, they will measure it to be created at one point, then to decay 15.6×10^{-6} s later at another point 4630 m away. For example, this muon could be created level with the top of a mountain, then move straight down to decay at its base 4630 m below.

However, an observer moving with an average muon will say that it traveled only 653 m because it existed for only

2.20×10^{-6} s. To show that this answer is completely consistent, consider the mountain. The 4630-m distance is its height, a proper length in the direction of motion. Relative to the observer traveling with this muon, the mountain moves up at $0.990c$ with the 4630-m length contracted to

$$l = l_0 \sqrt{1 - \frac{u^2}{c^2}} = (4630 \text{ m})\sqrt{1 - (0.990)^2}$$
$$= 653 \text{ m}.$$

Thus we see that length contraction is consistent with time dilation.

Finally, let's think a little about the visual appearance of a moving three-dimensional body. If we could see the positions of all points of the body simultaneously, it would appear to shrink only in the direction of motion. But we *don't* see all the points simultaneously; light from points farther from us takes longer to reach us than does light from points near to us, so we see the farther points at the positions they had at earlier times.

Suppose we have a rectangular rod with its faces parallel to the coordinate planes. When we look end-on at the center of the closest face of such a rod at rest, we see only that face. (See the center rod in computer-generated Fig. 39–12a.) But when that rod is moving past us toward the right at relativistic speed, we may also see its left side because of the earlier-time effect just described. That is, we can see some points that we couldn't see when the rod was at rest because the rod moves out of the way of the light rays from those points to us. Conversely, some light that can get to us when the rod is at rest is blocked by the moving rod. Because of all this, the rods in Figs. 39–12b and 39–12c appear rotated and distorted.

39–6 THE LORENTZ TRANSFORMATIONS

In Section 39–2 we discussed the Galilean coordinate transformation equations, Eqs. (39–1). They relate the coordinates (x, y, z) of a point in frame of reference S to the coordinates (x', y', z') of the point in a second frame S'. The second frame moves with constant speed u relative to S in the positive direction along the common x-x' axis. This transformation also assumes that the time scale is the same in the two frames of reference, as expressed by the additional relation $t = t'$. This Galilean transformation, as we have seen, is valid only in the limit when u approaches zero. We are now ready to derive more general transformations that are consistent with the principle of relativity. The more general relations are called the **Lorentz transformations.**

Our first question is this: When an event occurs at point (x, y, z) at time t, as observed in a frame of reference S, what are the coordinates (x', y', z') and time t' of the event as observed in a second frame S' moving relative to S with constant speed u in the $+x$-direction?

39–12 Computer simulation of the appearance of an array of 25 rods with square cross section. The center rod is viewed end-on. (a) The array at rest; (b) the array moving to the right at $0.2c$; (c) the array moving to the right at $0.9c$. The simulation ignores color changes in the array caused by the Doppler effect.

(a)   (b)   (c)

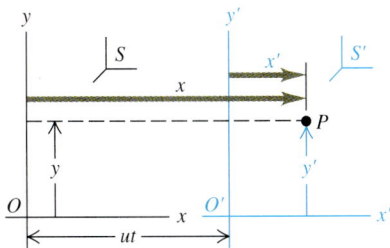

39–13 As measured in reference frame S, x' is contracted to x'/γ, so $x = ut + x'/\gamma$ and $x' = \gamma(x - ut)$.

To derive the coordinate transformation, we refer to Fig. 39–13, which is the same as Fig. 39–3. As before, we assume that the origins coincide at the initial time $t = 0 = t'$. Then in S the distance from O to O' at time t is still ut. The coordinate x' is a *proper length* in S', so in S it is contracted by the factor $1/\gamma = (1 - u^2/c^2)^{1/2}$, as in Eq. (39–17). Thus the distance x from O to P, as seen in S, is not simply $x = ut + x'$, as in the Galilean coordinate transformation, but

$$x = ut + x'\sqrt{1 - \frac{u^2}{c^2}}. \tag{39–18}$$

Solving this equation for x', we obtain

$$x' = \frac{x - ut}{\sqrt{1 - u^2/c^2}}. \tag{39–19}$$

Equation (39–19) is part of the Lorentz coordinate transformation; another part is the equation giving t' in terms of x and t. To obtain this, we note that the principle of relativity requires that the *form* of the transformation from S to S' be identical to that from S' to S. The only difference is a change in the sign of the relative velocity component u. Thus from Eq. (39–18) it must be true that

$$x' = -ut' + x\sqrt{1 - \frac{u^2}{c^2}}. \tag{39–20}$$

We now equate Eqs. (39–19) and (39–20) to eliminate x'. This gives us an equation for t' in terms of x and t. We leave the algebraic details for you to work out; the result is

$$t' = \frac{t - ux/c^2}{\sqrt{1 - u^2/c^2}}. \tag{39–21}$$

As we discussed previously, lengths perpendicular to the direction of relative motion are not affected by the motion, so $y' = y$ and $z' = z$.

Collecting all these transformation equations, we have

$$x' = \frac{x - ut}{\sqrt{1 - u^2/c^2}} = \gamma(x - ut),$$

$$y' = y,$$
$$z' = z,$$
(Lorentz coordinate transformation) $\qquad$ (39–22)

$$t' = \frac{t - ux/c^2}{\sqrt{1 - u^2/c^2}} = \gamma(t - ux/c^2).$$

These equations are the *Lorentz coordinate transformation*, the relativistic generalization of the Galilean coordinate transformation, Eqs. (39–1) and $t = t'$. For values of u that approach zero, the radicals in the denominators and γ approach 1, and the ux/c^2 term approaches zero. In this limit, Eqs. (39–22) become identical to Eqs. (39–1) along with

$t = t'$. In general, though, both the coordinates and time of an event in one frame depend on its coordinates and time in another frame. *Space and time have become intertwined; we can no longer say that length and time have absolute meanings independent of the frame of reference.*

We can use Eqs. (39–22) to derive the relativistic generalization of the Galilean velocity transformation, Eq. (39–2). We consider only one-dimensional motion along the x-axis and use the term "velocity" as being short for the "x-component of the velocity." Suppose that in a time dt a particle moves a distance dx, as measured in frame S. We obtain the corresponding distance dx' and time dt' in S' by taking differentials of Eqs. (39–22):

$$dx' = \gamma(dx - u\,dt),$$

$$dt' = \gamma(dt - u\,dx/c^2).$$

We divide the first equation by the second and then divide the numerator and denominator of the result by dt to obtain

$$\frac{dx'}{dt'} = \frac{\dfrac{dx}{dt} - u}{1 - \dfrac{u}{c^2}\dfrac{dx}{dt}}.$$

Now dx/dt is the velocity v in S, and dx'/dt' is the velocity v' in S', so we finally obtain the relativistic generalization

$$v' = \frac{v - u}{1 - uv/c^2} \qquad \text{(Lorentz velocity transformation).} \qquad (39\text{–}23)$$

When u and v are much smaller than c, the denominator in Eq. (39–23) approaches 1, and we approach the nonrelativistic result $v' = v - u$. The opposite extreme is the case $v = c$; then we find

$$v' = \frac{c - u}{1 - uc/c^2} = \frac{c(1 - u/c)}{1 - u/c} = c.$$

This says that anything moving with speed $v = c$ measured in S also has speed $v' = c$ measured in S', despite the relative motion of the two frames. So Eq. (39–23) is consistent with Einstein's postulate that the speed of light in vacuum is the same in all inertial frames of reference.

The principle of relativity tells us there is no fundamental distinction between the two frames S and S'. Thus the expression for v in terms of v' must have the same form as Eq. (39–23), with v changed to v' and vice versa and the sign of u reversed. Carrying out these operations with Eq. (39–23), we find

$$v = \frac{v' + u}{1 + uv'/c^2} \qquad \text{(Lorentz velocity transformation).} \qquad (39\text{–}24)$$

This can also be obtained algebraically by solving Eq. (39–23) for v. Both Eqs. (39–23) and (39–24) are *Lorentz velocity transformations* for one-dimensional motion.

When u is less than c, the Lorentz velocity transformations show us that a body moving with a speed less than c in one frame of reference always has a speed less than c in *every other* frame of reference. This is one reason for concluding that no material body may travel with a speed equal to or greater than that of light in vacuum, relative to *any* inertial frame of reference. The relativistic generalizations of energy and momentum, which we will explore later, give further support to this hypothesis.

Problem-Solving Strategy

LORENTZ TRANSFORMATIONS

1. The Lorentz coordinate transformation equations tell you how to relate measurements made in different inertial frames of reference, S and S'. Remember that S' moves at a constant velocity u in the $+x$-direction relative to S. When you use these equations, it helps to make a list of coordinates and times of events in the two frames, such as x_1, x_1', t_1, t_1', and so on. List and label carefully what you know and don't know. Do you know the coordinates in one frame? The time of an event in one frame? What are your knowns and unknowns?

2. In velocity-transformation problems, if you have two observers measuring the motion of a body, decide which you want to call S and which S', identify the velocities v and v' clearly, and make sure you know the velocity u of S' relative to S. Use either form of the velocity transformation equation, Eq. (39–23) or (39–24), whichever is more convenient.

3. Don't be discouraged if some of your results don't seem to make sense or if they disagree with your intuition. It takes time to develop reliable intuition about relativity; keep trying. But if your result is a speed greater than c, something is wrong.

EXAMPLE 39-7

Was it received before it was sent? Winning an interstellar race, Mavis pilots her spaceship across a finish line in space at a speed of $0.600c$ relative to that line. A "hooray" message is sent from the back of her ship (event 2) at the instant (in her frame of reference) that the front of her ship crosses the line (event 1). She measures the length of her ship to be 300 m. Stanley is at the finish line and is at rest relative to it. When and where does he measure events 1 and 2 to occur?

SOLUTION Our derivation of the Lorentz transformation assumes that the origins of frames S and S' coincide at $t = 0 = t'$. Thus for simplicity we fix the origin of S at the finish line and the origin of S' at the front of the spaceship so that Stanley and Mavis measure event 1 to be at $x = 0 = x'$ and $t = 0 = t'$.

Mavis in S' measures her spaceship to be 300 m long, so she has the "hooray" sent from 300 m behind her spaceship's front at the instant she measures the front to cross the finish line. That is, she measures event 2 at $x' = -300$ m and $t' = 0$.

Now we want to find x and t, the corresponding values of event 2 that Stanley measures in S as he properly takes into account the transit time. The hard way would be to solve the first and last of the Lorentz coordinate transformation Eqs. (39–22) to find new equations for x and t in terms of x' and t'. The easier way is to use the principle of relativity in the same way that we did in obtaining Eq. (39–24) from Eq. (39–23). Relative to S', Mavis is at rest while Stanley in S moves in the $-x'$-direction at

the opposite relative velocity. Thus we modify the first and last of Eqs. (39–22) by removing the primes from x' and t', adding primes to x and t, and changing all u to $-u$, getting

$$x = \gamma(x' + ut') \quad \text{and} \quad t = \gamma(t' + ux'/c^2)$$

From Eq. (39–7), $\gamma = 1.25$ for $u = 0.600c = 1.80 \times 10^8$ m/s. We also substitute $x' = -300$ m, $t' = 0$, $c = 3.00 \times 10^8$ m/s, and $u = 1.80 \times 10^8$ m/s in the equations for x and t to find $x = -375$ m at $t = -7.50 \times 10^{-7}$ s $= -0.750$ μs for event 2.

Mavis says that the events are simultaneous, but Stanley disagrees. In fact, he says that the "hooray" was sent *before* Mavis crossed the finish line. This does not mean that the effect preceded the cause. The fastest Mavis can send a signal the length of her ship is 300 m/3.00×10^8 m/s = 1.00 μs. She cannot send a signal from the front at the instant it crosses the finish line that would cause a "hooray" to be broadcast from the back at the same instant. She would have to send that signal from the front at least 1.00 μs before then, so she had to slightly anticipate her success.

Relativity *is* consistent: In his frame, Stanley measures Mavis's ship to be $l_0/\gamma = 300$ m/1.25 = 240 m long with its back at $x = -375$ m at $t = -0.750$ μs $= -7.50 \times 10^{-7}$ s when the "hooray" is sent. At that instant he thus measures the front of her 240-m-long ship to be a distance of $(375 - 240)$m = 135 m from the finish line. However, since $(1.80 \times 10^8$ m/s$)(7.50 \times 10^{-7}$ s$) = 135$ m, the front does cross the line at $t = 0$.

EXAMPLE 39-8

Relative velocities A spaceship moving away from the earth with speed $0.90c$ fires a robot space probe in the same direction as its motion, with speed $0.70c$ relative to the spaceship. What is the probe's speed relative to the earth?

SOLUTION Let the earth's frame of reference be S, and let the spaceship's frame of reference be S' (Fig. 39–14). Then

$u = 0.90c$ and $v' = 0.70c$. The nonrelativistic Galilean velocity addition formula would give a speed relative to the earth of $1.60c$; but this value is greater than the speed of light and must be incorrect. The correct relativistic result, from Eq. (39–24), is

$$v = \frac{v' + u}{1 + uv'/c^2} = \frac{0.70c + 0.90c}{1 + (0.90c)(0.70c)/c^2} = 0.98c.$$

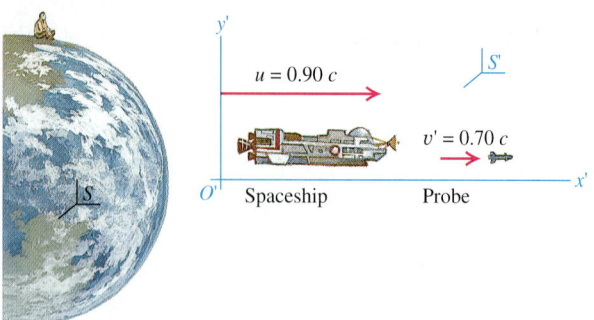

39–14 The spaceship and the robot space probe.

EXAMPLE 39–9

A scoutship from the earth tries to catch up with the missile-firing spaceship of Example 39–8 by traveling at $0.95c$ relative to the earth. What is its speed relative to the spaceship?

SOLUTION Again we let the earth's frame of reference be S and the spaceship's frame of reference be S'. Again we have $u = 0.90c$, but now $v = 0.95c$. According to nonrelativistic veloc-

ity addition, the scoutship's velocity relative to the spaceship would be $0.05c$. We get the correct result from Eq. (39–23):

$$v' = \frac{v - u}{1 - uv/c^2} = \frac{0.95c - 0.90c}{1 - (0.90c)(0.95c)/c^2} = 0.34c.$$

The relativistically correct value of the relative velocity is nearly seven times as large as the incorrect Newtonian value.

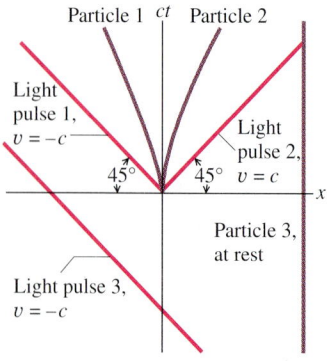

39–15 A spacetime diagram showing worldlines of three light pulses and three particles. Particles 1 and 2 leave $x = 0$ at $t = 0$, accelerating from rest in opposite directions.

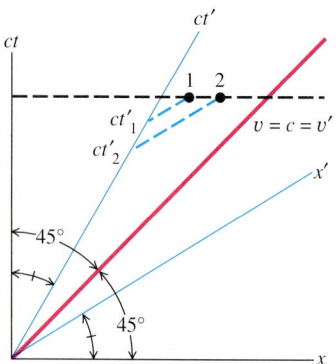

39–16 The ct'- and x'-axes drawn on our ct-x spacetime diagram. Notice the two sets of equal angles.

*39–7 SPACETIME DIAGRAMS

We found as early as Chapter 2 that graphical representations can make kinematic concepts less abstract and also give useful information. For example, not only does a v-t graph for one-dimensional motion show the velocity at any instant, but its slopes give accelerations, and areas under it give displacements. For relativity the transformations $y' = y$ and $z' = z$ are easy to understand, so we'll just consider the $\pm x$ directions.

The usual convention in relativity is to graph ct on the vertical axis and x on the horizontal axis. Such a graph provides us with a **spacetime diagram.** We use ct rather than t so that both scales can have the same unit and the same scale. The path of a particle forms a line, called its **worldline**, as the particle moves in one-dimensional motion. At any point, the slope of the worldline is $d(ct)/dx = (c\,dt)/(v\,dt) = c/v$. Thus a light pulse with $v = \pm c$ has a slope of ± 1 on a spacetime diagram, giving angles of $45°$ with the $\pm x$-axes. Since material particles have speeds less than c, all worldlines for material particles are steeper than those $45°$ angles. That is, nothing known has a worldline with a slope between -1 and 1. The worldline of a particle at rest is vertical and so has infinite slope. Figure 39–15 shows six worldlines, three of light pulses and three of particles. Can you show that these six worldlines agree with the statements made in this paragraph about their slopes?

How does the S' reference frame appear on our ct-x spacetime diagram? Recall that we *always* set $x' = 0$ at $x = 0$ when $t' = 0 = t$ and let S' move at a speed u in the $+x$-direction. But $x' = 0$ all along the ct'-axis, so $x' = 0$ and the ct'-axis have a worldline of slope c/u on our ct-x spacetime diagram. For example, if $u = 0.600c$, the ct'-axis is at an angle of $\arctan(1/0.600) = 59.0°$ from the x-axis or $90.0° - 59.0° = 31.0°$ from the ct-axis.

Surprisingly enough, the x'-axis is not drawn perpendicular to the ct'-axis on our ct-x spacetime diagram. Since $ct' = 0$ (so $t' = 0$) all along the x'-axis, the last of Eqs. (39–22) gives $(t - ux/c^2) = 0$ or $ct = (u/c)x$ for the x'-axis. Thus the x'-axis is drawn with a slope of u/c on our ct-x spacetime diagram. For $u = 0.600c$ the x'-axis is at an angle of $\arctan(0.600) = 31.0°$ from the x-axis. That is, the x'-axis makes the same angle with the x-axis as the ct'-axis makes with the ct-axis. We see in Fig. 39–16 that the worldline of a light pulse leaving $x' = 0 = x$ at $t' = 0 = t$ with a velocity $+c$ bisects the angle between either set of axes.

EXAMPLE 39-10

Simultaneity on a spacetime diagram Stanley measures events 1 and 2 to occur simultaneously in S at positions x_1 and x_2, where $x_2 > x_1$. Use our spacetime diagram to show that Mavis, who moves in the positive x-direction relative to Stanley, measures event 2 to occur before event 1.

SOLUTION Events that are simultaneous in S have the same time t, so in Fig. 39–16 we draw a dashed line parallel to the x-axis (constant t). We put a dot on that line for event 1, and farther

from the ct-axis we put another dot for event 2 (because $x_2 > x_1$). How do we read a value of a point on a graph? We draw a line through that point parallel to one axis and measure where it intercepts the other axis. Thus to measure the times of the two events in Mavis's S' frame, in Fig. 39–16 we draw dashed lines parallel to the x'-axis that intercept the ct'-axis at ct_1' and ct_2'. We see that $ct_2' < ct_1'$, so $t_2' < t_1'$. The events are not simultaneous in S', and Mavis measures event 2 to occur before event 1.

On our spacetime diagram, the scale for the S' axes is not the same as the scale for the S axes. For example, consider the dashed line $x' = 1$ in Fig. 39–17, which must be drawn parallel to the ct'-axis. (We have left off the unit for generality; it could be $x' = 1$ meter, $x' = 1$ light year, or whatever is convenient.) This dashed line intercepts the x-axis at $ct = 0$. Substituting $t = 0$ in the Lorentz transformation $x' = \gamma(x - ut)$ gives $x = 1/\gamma$ for the $x' = 1$ line. In Fig. 39–17, $u = 0.60c$ and this intercept is at $x = 0.80$. We can see that the symmetry of our spacetime diagram gives us the same scaling ratio for the ct'- and ct-axes. To summarize, in comparison to the ct- and x-axes, the ct'- and x'-axes are rotated through an angle arctan u/c toward the common $v = c = v'$ line at 45° and are stretched in scale so that the $x' = 1$ line intercepts the x-axis at $x = 1/\gamma$.

Let's finish this discussion with a simple example of length contraction. Mavis, at rest in frame S', holds a meter stick with its left end at $x' = 0$ and its right end at $x' = 1$ m. Thus in Fig. 39–17 the units are meters. At any time t' measured in frame S', the left end is at $x' = 0$ (on the ct'-axis) and the right end is on the $x' = 1$ dashed line. In frame S the positions of both ends of the meter stick are measured at the same time t, then subtracted to find the length. For instance, at $t = 0$ (on the x-axis) we see from Fig. 39–17 that the left end of her stick is at $x = 0$ and the right end is at $x = (1 \text{ m})/\gamma$. Thus in S the meter stick has a contracted length of $(1 \text{ m})/\gamma$, in accordance with Eq. (39–17).

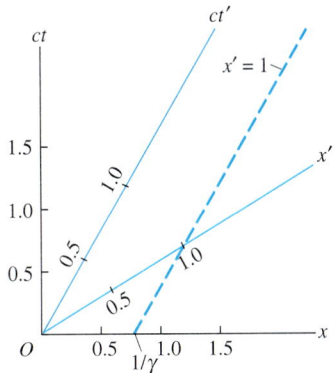

39–17 A spacetime diagram for $u = 0.600c$. The dashed line $x' = 1$ intercepts the x-axis at $x = 1/\gamma = 0.800$. The scale of the S'-axes is greater than that of the S-axes.

*39–8 THE DOPPLER EFFECT FOR ELECTROMAGNETIC WAVES

An additional important consequence of relativistic kinematics is the Doppler effect for electromagnetic waves. In our previous discussion of the Doppler effect (Section 21–5) we quoted without proof the formula, Eq. (21–18), for the frequency shift that results from motion of a source of electromagnetic waves relative to an observer. We can now derive that result.

Here's a statement of the problem. A source of light is moving with constant speed u toward Stanley, who is stationary in an inertial frame (Fig. 39–18). As measured in its rest frame, the source emits light waves with frequency f_0 and period $T_0 = 1/f_0$. What is the frequency f of these waves as received by Stanley?

Let T be the time interval between emission of successive wave crests as observed in Stanley's reference frame. Note that this is *not* the interval between the *arrival* of successive crests at his position, because the crests are emitted at different points in

39–18 The Doppler effect for light. A moving light source emits a wave crest, then travels a distance uT toward an observer and emits the next crest. In Stanley's reference frame S, the second crest is a distance λ behind the first, and the crests arrive a time λ/c apart, where $\lambda = (c - u)T$ and $T = \gamma T_0$.

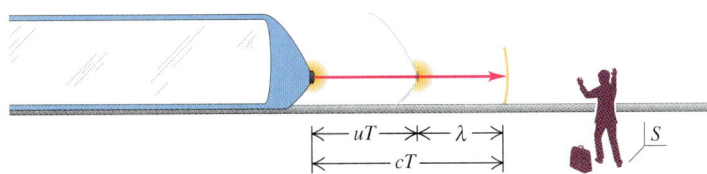

Stanley's frame. In measuring only the frequency f he receives, he does not take into account the difference in transit times for successive crests. Therefore the frequency he receives is not $1/T$. What is the equation for f ?

During a time T the crests ahead of the source move a distance cT, and the source moves a shorter distance uT in the same direction. The distance λ between successive crests—that is, the wavelength—is thus $\lambda = (c - u)T$, as measured in Stanley's frame. The frequency that he measures is c/λ. Therefore

$$f = \frac{c}{(c - u)T}. \tag{39–25}$$

So far we have followed a pattern similar to that for the Doppler effect for sound from a moving source (Section 21–5). In that discussion our next step was to equate T to the time T_0 between emissions of successive wave crests by the source. However, it is not relativistically correct to equate T to T_0. The time T_0 is measured in the rest frame of the source, so it is a proper time. From Eq. (39–6), T_0 and T are related by

$$T = \frac{T_0}{\sqrt{1 - u^2/c^2}} = \frac{cT_0}{\sqrt{c^2 - u^2}}.$$

or, since $T_0 = 1/f_0$,

$$\frac{1}{T} = \frac{\sqrt{c^2 - u^2}}{cT_0} = \frac{\sqrt{c^2 - u^2}}{c} f_0.$$

Remember, $1/T$ is not equal to f. We must substitute this expression for $1/T$ into Eq. (39–25) to find f:

$$f = \frac{c}{c - u} \frac{\sqrt{c^2 - u^2}}{c} f_0.$$

Using $c^2 - u^2 = (c - u)(c + u)$ gives

$$f = \sqrt{\frac{c + u}{c - u}} f_0 \qquad \text{(Doppler effect, electromagnetic waves).} \tag{39–26}$$

This shows that when the source moves *toward* the observer, the observed frequency f is *greater* than the emitted frequency f_0. The difference $f - f_0 = \Delta f$ is called the Doppler frequency shift. When u/c is much smaller than 1, the fractional shift $\Delta f/f$ is approximately equal to u/c:

$$\frac{\Delta f}{f} = \frac{u}{c}.$$

This relationship was found in Problem 21–31, with some changes in notation.

When the source moves *away from* the observer, we change the sign of u in Eq. (39–26) to get

$$f = \sqrt{\frac{c - u}{c + u}} f_0. \tag{39–27}$$

This agrees with Eq. (21–18), which we quoted previously, with minor notation changes.

With light, unlike sound, there is no distinction between motion of source and motion of observer; only the *relative* velocity of the two is significant. The last three paragraphs of Section 21–5 discuss several practical applications of the Doppler effect with light and other electromagnetic radiation; we suggest you review those paragraphs now.

Modern weather radar uses the Doppler effect to measure the velocity of falling rain. By seeing how rapidly the rain swirls within a storm, meteorologists can make better predictions about the severity of storms and thus provide more accurate weather warnings.

EXAMPLE 39-11

We are observing the radiation from a beam of sodium atoms moving directly toward us. The frequency in the rest frame of the atoms is $f_0 = 5.08 \times 10^{14}$ Hz, but the frequency we observe is $f = 10.16 \times 10^{14}$ Hz. At what speed are the atoms moving toward us?

SOLUTION We are given $f/f_0 = 2.00$, and this is related to the relative velocity u by Eq. (39–26). The simplest procedure is to solve this equation for u. That takes a little algebra; we'll leave it as an exercise for you to show that the result is

$$u = \frac{(f/f_0)^2 - 1}{(f/f_0)^2 + 1} c.$$

In our case we find

$$u = \frac{(2.00)^2 - 1}{(2.00)^2 + 1} c = 0.600c.$$

When the source moves toward us at $0.600c$, the frequency that we measure is double the value measured in the rest frame of the source. We leave it as a further exercise to show that when the source moves *away from* us at $0.600c$, the frequency that we measure is *half* the value measured in the rest frame of the source.

39-9 RELATIVISTIC MOMENTUM

Newton's laws of motion have the same form in all inertial frames of reference. When we use transformations to change from one inertial frame to another, the laws should be *invariant* (unchanging). But we have just learned that the principle of relativity forces us to replace the Galilean transformations with the more general Lorentz transformations. As we will see, this requires corresponding generalizations in the laws of motion and the definitions of momentum and energy.

The principle of conservation of momentum states that *when two bodies interact, the total momentum is constant,* provided that the net external force acting on the bodies in an inertial reference frame is zero (for example, if they form an isolated system, interacting only with each other). If conservation of momentum is a valid physical law, it must be valid in *all* inertial frames of reference. Now, here's the problem: Suppose we look at a collision in one inertial coordinate system S and find that momentum is conserved. Then we use the Lorentz transformation to obtain the velocities in a second inertial system S'. We find that if we use the Newtonian definition of momentum ($\vec{p} = m\vec{v}$), momentum is *not* conserved in the second system! If we are convinced that the principle of relativity and the Lorentz transformation are correct, the only way to save momentum conservation is to generalize the *definition* of momentum.

We won't derive the correct relativistic generalization of momentum, but here is the result. Suppose we measure the mass of a particle to be m when it is at rest relative to us: we often call m the **rest mass.** We will use the term *material particle* for a particle that has a nonzero rest mass. When such a particle has a velocity $\vec{v}$, its **relativistic momentum** $\vec{p}$ is

$$\vec{p} = \frac{m\vec{v}}{\sqrt{1 - v^2/c^2}} \quad \text{(relativistic momentum)}. \tag{39–28}$$

When the particle's speed v is much less than c, this is approximately equal to the Newtonian expression $\vec{p} = m\vec{v}$, but in general the momentum is greater in magnitude than mv (Fig. 39–19). In fact, as v approaches c, the momentum approaches infinity.

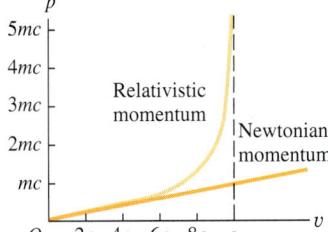

39-19 Graph of the magnitude of the momentum of a particle of rest mass m as a function of speed v.

NEWTON'S SECOND LAW

What about the relativistic generalization of Newton's second law? In Newtonian mechanics the most general form of the second law is

$$\vec{F} = \frac{d\vec{p}}{dt}. \tag{39–29}$$

That is, the net force $\vec{F}$ on a particle equals the time rate of change of its momentum.

Experiments show that this result is still valid in relativistic mechanics, provided that we use the relativistic momentum given by Eq. (39–28). That is, the relativistically correct generalization of Newton's second law is

$$\vec{F} = \frac{d}{dt} \frac{m\vec{v}}{\sqrt{1 - v^2/c^2}}. \tag{39–30}$$

Because momentum is no longer directly proportional to velocity, rate of change of momentum is no longer directly proportional to acceleration. As a result, *constant force does not cause constant acceleration.* For example, when the net force and the velocity are both along the *x*-axis, Eq. (39–30) gives

$$F = \frac{m}{(1 - v^2/c^2)^{3/2}} a \qquad (\vec{F} \text{ and } \vec{v} \text{ along the same line}), \tag{39–31}$$

where *a* is the acceleration, also along the *x*-axis. Solving Eq. (39–31) for the acceleration *a* gives

$$a = \frac{F}{m}\left(1 - \frac{v^2}{c^2}\right)^{3/2}.$$

We see that as a particle's speed increases, the acceleration caused by a given force continuously *decreases.* As the speed approaches *c*, the acceleration approaches zero, no matter how great a force is applied. Thus it is impossible to accelerate a particle with nonzero rest mass to a speed equal to or greater than *c*. The speed of light in vacuum is sometimes called "the ultimate speed."

Equation (39–28) for relativistic momentum is sometimes interpreted to mean that a rapidly moving particle undergoes an increase in mass. If the mass at zero velocity (the rest mass) is denoted by *m*, then the "relativistic mass" m_{rel} is given by

$$m_{\text{rel}} = \frac{m}{\sqrt{1 - v^2/c^2}}.$$

Indeed, when we consider the motion of a system of particles (such as rapidly moving ideal gas molecules in a stationary container), the total rest mass of the system is the sum of the relativistic masses of the particles, not the sum of their rest masses. Also with relativistic mass, the famous equation $E = mc^2$ can be applied to all types of energy, not just most types.

However, if blindly applied, the concept of relativistic mass has its pitfalls. As Eq. (39–30) shows, the relativistic generalization of Newton's second law is *not* $\vec{F} = m_{\text{rel}}\vec{a}$, and we will show in Section 39–9 that the relativistic kinetic energy of a particle is *not* $K = \frac{1}{2}m_{\text{rel}}v^2$. The use of relativistic mass has its supporters and detractors, some quite strong in their opinions. We will mostly deal with individual particles, so we will sidestep the controversy and use Eq. (39–28) as the generalized definition of momentum with *m* as a constant for each particle, independent of its state of motion.

We will use the abbreviation

$$\gamma = 1/\sqrt{1 - v^2/c^2}.$$

When we used this abbreviation in Section 39–4 with *u* in place of *v*, *u* was the relative speed of two coordinate systems. Here *v* is the speed of a particle in a particular coordinate system, that is, the speed of the particle's *rest frame* with respect to that system. In terms of γ, Eq. (39–28) becomes

$$\vec{p} = \gamma m\vec{v} \qquad (\text{relativistic momentum}), \tag{39–32}$$

and Eq. (39–31)

$$F = \gamma^3 ma \qquad (\vec{F} \text{ and } \vec{v} \text{ along the same line}). \tag{39–33}$$

In linear accelerators (used in medicine as well as nuclear and elementary-particle physics) the net force $\vec{F}$ and the velocity $\vec{v}$ of the accelerated particle are along the same straight line. But for much of the path in most circular accelerators the particle moves in uniform circular motion at constant speed v. Then the net force and velocity are perpendicular, so the force can do no work on the particle and the kinetic energy and speed remain constant. Thus the denominator in Eq. (39–30) is constant, and we obtain

$$F = \frac{m}{(1 - v^2/c^2)^{1/2}} a = \gamma m a \qquad (\vec{F} \text{ and } \vec{v} \text{ perpendicular}). \qquad (39\text{–}34)$$

Recall from Section 3–5 that if the particle moves in a circle, the net force and acceleration are directed inward along the radius r, and $a = v^2/r$.

What about the general case in which $\vec{F}$ and $\vec{v}$ are neither along the same line nor perpendicular? Then we can resolve the net force $\vec{F}$ at any instant into components parallel to and perpendicular to $\vec{v}$. The resulting acceleration will have corresponding components obtained from Eqs. (39–33) and (39–34). Because of the different γ^3 and γ factors, the acceleration components will not be proportional to the net force components. That is, *unless the net force on a relativistic particle is either along the same line as the particle's velocity or perpendicular to it, the net force and acceleration vectors are not parallel.*

EXAMPLE 39–12

Relativistic dynamics of an electron An electron (rest mass 9.11×10^{-31} kg, charge -1.60×10^{-19} C) is moving opposite to an electric field of magnitude $E = 5.00 \times 10^5$ N/C. All other forces are negligible in comparison to the electric field force. a) Find the magnitudes of momentum and of acceleration at the instants when $v = 0.010c$, $0.90c$, and $0.99c$. b) Find the corresponding accelerations if a net force of the same magnitude is perpendicular to the velocity.

SOLUTION a) We can use Eqs. (39–32) and (39–33). First we calculate the values of γ for the three speeds given; we find $\gamma = 1.00$, 2.29, 7.09. The values of momentum p are

$$p_1 = \gamma_1 m v_1 = (1.00)(9.11 \times 10^{-31} \text{ kg})(0.010)(3.00 \times 10^8 \text{ m/s})$$
$$= 2.7 \times 10^{-24} \text{ kg} \cdot \text{m/s at } v_1 = 0.010\,c,$$

$$p_2 = (2.29)(9.11 \times 10^{-31} \text{ kg})(0.90)(3.00 \times 10^8 \text{ m/s})$$
$$= 5.6 \times 10^{-22} \text{ kg} \cdot \text{m/s at } v_2 = 0.90\,c,$$

$$p_3 = (7.09)(9.11 \times 10^{-31} \text{ kg})(0.99)(3.00 \times 10^8 \text{ m/s})$$
$$= 1.9 \times 10^{-21} \text{ kg} \cdot \text{m/s at } v_3 = 0.99\,c.$$

At higher speeds, the relativistic values of momentum differ more and more from the nonrelativistic values computed from $p = mv$. The momentum at $0.99c$ is over three times as great as at $0.90c$ because of the increase in the factor γ.

The magnitude of the force on the electron is

$$F = |q|E = (1.60 \times 10^{-19} \text{ C})(5.00 \times 10^5 \text{ N/C}) = 8.00 \times 10^{-14} \text{ N}.$$

From Eq. (39–33), $a = F/\gamma^3 m$. When $v = 0.010c$ and $\gamma = 1.00$,

$$a_1 = \frac{8.00 \times 10^{-14} \text{ N}}{(1.00)^3(9.11 \times 10^{-31} \text{ kg})} = 8.8 \times 10^{16} \text{ m/s}^2.$$

The other accelerations are smaller by factors of γ^3:

$$a_2 = 7.3 \times 10^{15} \text{ m/s}^2, \qquad a_3 = 2.5 \times 10^{14} \text{ m/s}^2.$$

These last two accelerations are only 8.3% and 0.28%, respectively, of the values predicted by nonrelativistic mechanics.

We note again that as v approaches c the acceleration drops off very quickly. At the Stanford Linear Accelerator Center a path length of 3 km is needed to give electrons the speed that according to Newtonian physics they could acquire in 1.5 cm. b) From Eq. (39–34), $a = F/\gamma m$ if $\vec{F}$ and $\vec{v}$ are perpendicular. When $v = 0.010c$ and $\gamma = 1.00$,

$$a_1 = \frac{8.00 \times 10^{-14} \text{ N}}{(1.00)(9.11 \times 10^{-31} \text{ kg})} = 8.8 \times 10^{16} \text{ m/s}^2.$$

The other accelerations are smaller by a factor of γ.

$$a_2 = 3.8 \times 10^{16} \text{ m/s}^2, \qquad a_3 = 1.2 \times 10^{16} \text{ m/s}^2.$$

These accelerations are larger than the corresponding ones in part (a) by factors of γ^2.

39–10 RELATIVISTIC WORK AND ENERGY

When we developed the relationship between work and kinetic energy in Chapter 6, we used Newton's laws of motion. When we generalize these laws according to the principle of relativity, we need a corresponding generalization of the equation for kinetic energy.

We use the work-energy principle, beginning with the definition of work. When the net force and displacement are in the same direction, the work done by that force is $W = \int F \, dx$. We substitute the expression for F from Eq. (39–31), the applicable relativistic version of Newton's second law. In moving a particle of rest mass m from point x_1 to point x_2,

$$W = \int_{x_1}^{x_2} F \, dx = \int_{x_1}^{x_2} \frac{ma \, dx}{(1 - v^2/c^2)^{3/2}}. \tag{39–35}$$

To derive the generalized expression for kinetic energy K as a function of speed v, we would like to convert this to an integral on v. But first we remember that the kinetic energy of a particle equals the net work done on it in moving it from rest to the speed v: $K = W$. Thus we let the speeds be zero at point x_1 and v at point x_2. So as not to confuse the variable of integration with the final speed, we change v to v_x in Eq. (39–35). That is, v_x is the varying x-component of the velocity of the particle as the net force accelerates it from rest to a speed v. We also realize that dx and dv_x are the infinitesimal changes in x and v_x, respectively, in the time interval dt. Because $v_x = dx/dt$ and $a = dv_x/dt$, we can rewrite $a \, dx$ in Eq. (39–35) as

$$a \, dx = \frac{dv_x}{dt} \, dx = dx \frac{dv_x}{dt} = \frac{dx}{dt} \, dv_x = v_x \, dv_x.$$

Making these substitutions gives us

$$K = W = \int_0^v \frac{mv_x \, dv_x}{(1 - v_x^2/c^2)^{3/2}}. \tag{39–36}$$

We can evaluate this integral by a simple change of variable; the final result is

$$K = \frac{mc^2}{\sqrt{1 - v^2/c^2}} - mc^2 = (\gamma - 1)mc^2 \qquad \text{(relativistic kinetic energy)}. \tag{39–37}$$

If Eq. (39–37) is correct, it must also approach the Newtonian expression $K = \frac{1}{2}mv^2$ when v is much smaller than c (see Fig. 39–20). To verify this, we expand the radical, using the binomial theorem in the form

$$(1 + x)^n = 1 + nx + n(n - 1)x^2/2 + \cdots.$$

In our case, $n = -1/2$ and $x = -v^2/c^2$, and we get

$$\gamma = \left(1 - \frac{v^2}{c^2}\right)^{-1/2} = 1 + \frac{1}{2}\frac{v^2}{c^2} + \frac{3}{8}\frac{v^4}{c^4} + \cdots.$$

Combining this with $K = (\gamma - 1) \, mc^2$, we find

$$K = \left(1 + \frac{1}{2}\frac{v^2}{c^2} + \frac{3}{8}\frac{v^4}{c^4} + \cdots - 1\right)mc^2$$

$$= \frac{1}{2}mv^2 + \frac{3}{8}\frac{mv^4}{c^2} + \cdots. \tag{39–38}$$

When v is much smaller than c, all the terms in the series except the first are negligibly small, and we obtain the Newtonian expression $\frac{1}{2}mv^2$.

REST ENERGY

Equation (39–37) for the kinetic energy, the energy due to the motion of the particle, includes an energy term $mc^2/\sqrt{1 - v^2/c^2}$ that depends on the motion and a second energy term mc^2 that is independent of the motion. It seems that the kinetic energy of a particle is the difference between some **total energy** E and an energy mc^2 that it has even when

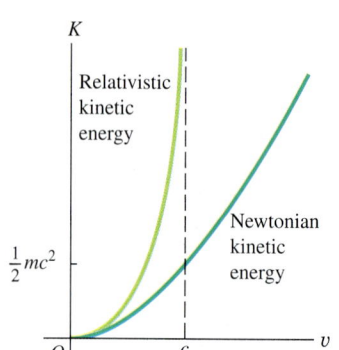

39–20 The relativistic kinetic energy at any speed is greater than the Newtonian equation predicts. For a material particle, as v approaches c, the relativistic kinetic energy and the work needed to provide it approach infinity. Thus v is always less than c for material particles.

it is at rest. Thus we can rewrite Eq. (39–37) as

$$E = K + mc^2 = \frac{mc^2}{\sqrt{1 - v^2/c^2}} = \gamma mc^2 \qquad \text{(total energy of a particle)}. \quad (39\text{–}39)$$

For a particle at rest ($K = 0$), we see that $E = mc^2$. The energy mc^2 associated with rest mass m rather than motion is called the **rest energy** of the particle.

There is in fact direct experimental evidence that rest energy really does exist. The simplest example is the decay of a neutral *pion*. This is an unstable subatomic particle of rest mass m_π; when it decays, it disappears and electromagnetic radiation appears. If a neutral pion has no kinetic energy before its decay, the total energy of the radiation after its decay is found to equal exactly $m_\pi c^2$. In many other fundamental particle transformations the sum of the rest masses of the particles changes. In every case there is a corresponding energy change, consistent with the assumption of a rest energy mc^2 associated with a rest mass m.

Historically, the principles of conservation of mass and of energy developed quite independently. The theory of relativity now shows that they are actually two special cases of a single broader conservation principle, the *principle of conservation of mass and energy*. In some physical phenomena, neither the sum of the rest masses of the particles nor the total energy other than rest energy is separately conserved, but there is a more general conservation principle: In an isolated system, when the sum of the rest masses changes, there is always a change in $1/c^2$ times the total energy other than the rest energy. This change is equal in magnitude but opposite in sign to the change in the sum of the rest masses.

This more general mass-energy conservation law is the fundamental principle involved in the generation of power through nuclear reactions. When a uranium nucleus undergoes fission in a nuclear reactor, the sum of the rest masses of the resulting fragments is *less than* the rest mass of the parent nucleus. An amount of energy is released that equals the mass decrease multiplied by c^2. Most of this energy can be used to produce steam to operate turbines for electric power generators.

We can also relate the total energy E of a particle (kinetic energy plus rest energy) directly to its momentum by combining Eq. (39–28) for relativistic momentum and Eq. (39–39) for total energy to eliminate the particle's velocity. The simplest procedure is to rewrite these equations in the following forms:

$$\left(\frac{E}{mc^2}\right)^2 = \frac{1}{1 - v^2/c^2} \qquad \text{and} \qquad \left(\frac{p}{mc}\right)^2 = \frac{v^2/c^2}{1 - v^2/c^2}.$$

A nuclear power plant like this one converts part of the rest energy of nuclei into thermal energy.

Subtracting the second of these from the first and rearranging, we find

$$E^2 = (mc^2)^2 + (pc)^2 \qquad \text{(total energy, rest energy, and momentum)}. \quad (39\text{–}40)$$

Again we see that for a particle at rest ($p = 0$), $E = mc^2$.

Equation (39–40) also suggests that a particle may have energy and momentum even when it has no rest mass. In such a case, $m = 0$ and

$$E = pc \qquad \text{(zero rest mass)}. \qquad (39\text{–}41)$$

In fact, zero rest mass particles do exist. Such particles always travel at the speed of light in vacuum. One example is the *photon*, the quantum of electromagnetic radiation discussed in Section 37–7. Photons are emitted and absorbed during changes of state of an atomic or nuclear system when the energy and momentum of the system change.

EXAMPLE 39–13

Energetic electrons a) Find the rest energy of an electron ($m = 9.109 \times 10^{-31}$ kg, $q = -e = -1.602 \times 10^{-19}$ C) in joules and in electron volts. b) Find the speed of an electron that has been accelerated by an electric field, from rest, through a potential increase of 20 kV (typical of TV picture tubes) or of 5.0 MV (a high-voltage x-ray machine).

SOLUTION a) The rest energy is

$$mc^2 = (9.109 \times 10^{-31} \text{ kg})(2.998 \times 10^8 \text{ m/s})^2$$

$$= 8.187 \times 10^{-14} \text{ J}.$$

From the definition of the electron volt in Section 24–3, $1 \text{ eV} = 1.602 \times 10^{-19}$ J. Using this, we find

$$mc^2 = (8.187 \times 10^{-14} \text{ J}) \frac{1 \text{ eV}}{1.602 \times 10^{-19} \text{ J}}$$

$$= 5.11 \times 10^5 \text{ eV} = 0.511 \text{ MeV}.$$

b) In calculations such as this, it is often convenient to work with the quantity γ defined from the modified Eq. (39–7):

$$\gamma = \frac{1}{\sqrt{1 - v^2/c^2}}.$$

Solving this for v, we get

$$v = c\sqrt{1 - (1/\gamma)^2}.$$

The total energy E of the accelerated electron is the sum of its rest energy mc^2 and the kinetic energy eV_{ba} that it gains from the work done on it by the electric field in moving from point a to point b:

$$E = \gamma mc^2 = mc^2 + eV_{ba},$$

or

$$\gamma = 1 + \frac{eV_{ba}}{mc^2}.$$

An electron accelerated through a potential increase of $V_{ba} = 20$ kV gains an amount of energy 20 keV, so for this electron we have

$$\gamma = 1 + \frac{20 \times 10^3 \text{ eV}}{0.511 \times 10^6 \text{ eV}}$$

$$= 1.039$$

and

$$v = c\sqrt{1 - (1/1.039)^2} = 0.272c$$

$$= 8.15 \times 10^7 \text{ m/s}.$$

The added energy is less than 4% of the rest energy, and the final speed is about a quarter of the speed of light in vacuum.

Repeating the calculation for $V_{ba} = 5.0$ MV, we find $eV_{ba}/mc^2 = 9.78$, $\gamma = 10.78$, and $v = 0.996c$. The kinetic energy (5.0 MeV) is much larger than the rest energy (0.511 MeV), and the speed is very close to c. At the Stanford Linear Accelerator Center, electrons attain a kinetic energy of 50 GeV. This energy corresponds approximately to $\gamma = 100,000$ and $v = 0.99999999995c$. Such a particle is said to be in the *extreme relativistic range*.

CAUTION ▶ The electron that accelerated from rest through a potential increase of 5.0 MV had a kinetic energy of 5.0 MeV. By convention we call such an electron a "5.0-MeV electron." A 5.0-MeV electron has a rest energy of 0.511 MeV (as do all electrons), a kinetic energy of 5.0 MeV, and a total energy of 5.5 MeV. Be careful not to mix these energies up. ◀

EXAMPLE 39-14

Two protons (each with $M = 1.67 \times 10^{-27}$ kg) are initially moving with equal speeds in opposite directions. They continue to exist after a head-on collision that also produces a neutral pion ($m = 2.40 \times 10^{-28}$ kg). If the protons and the pion are at rest after the collision, find the initial speed of the protons. Energy is conserved in the collision.

SOLUTION Energy is conserved, so we equate the total energy of the two protons before the collision to the rest energy of the two protons and the pion after the collision:

$$2(\gamma Mc^2) = 2(Mc^2) + mc^2,$$

$$\gamma = 1 + \frac{m}{2M} = 1 + \frac{2.40 \times 10^{-28} \text{ kg}}{2(1.67 \times 10^{-27} \text{ kg})} = 1.072,$$

so

$$v = c\sqrt{1 - (1/\gamma)^2} = 0.360c.$$

The initial kinetic energy of each proton is $(\gamma - 1)Mc^2 = 0.072Mc^2$. The rest energy of a proton is 938 MeV, so the kinetic energy is $(0.072)(938 \text{ MeV}) = 67.5$ MeV. (These are "67.5-MeV protons.") We invite you to verify that the rest energy of the pion is twice this, or 135 MeV. All the kinetic energy "lost" in this completely inelastic collision is transformed to the rest energy of the pion.

39-11 NEWTONIAN MECHANICS AND RELATIVITY

The sweeping changes required by the principle of relativity go to the very roots of Newtonian mechanics, including the concepts of length and time, the equations of motion, and the conservation principles. Thus it may appear that we have destroyed the foundations on which Newtonian mechanics is built. In one sense this is true, yet the Newtonian formulation is still accurate whenever speeds are small in comparison with the speed of light in vacuum. In such cases, time dilation, length contraction, and the modifications of the laws of motion are so small that they are unobservable. In fact, every one of the principles of Newtonian mechanics survives as a special case of the more general relativistic formulation.

The laws of Newtonian mechanics are not *wrong;* they are *incomplete.* They are a limiting case of relativistic mechanics. They are *approximately* correct when all speeds are small in comparison to c, and they become exactly correct in the limit when all speeds approach zero. Thus relativity does not completely destroy the laws of Newtonian mechanics but *generalizes* them. Newton's laws rest on a very solid base of experimental evidence, and it would be very strange to advance a new theory that is inconsistent with this evidence. This is a common pattern in the development of physical theory. Whenever a new theory is in partial conflict with an older, established theory, the new must yield the same predictions as the old in areas in which the old theory is supported by experimental evidence. Every new physical theory must pass this test, called the **correspondence principle.**

At this point we may ask whether the special theory of relativity gives the final word on mechanics or whether *further* generalizations are possible or necessary. For example, inertial frames have occupied a privileged position in our discussion. Can the principle of relativity be extended to noninertial frames as well?

Here's an example that illustrates some implications of this question. A student decides to go over Niagara Falls while enclosed in a large wooden box. During her free fall she can float through the air inside the box. She doesn't fall to the floor because both she and the box are in free fall with a downward acceleration of 9.8 m/s^2. But an alternative interpretation, from her point of view, is that she doesn't fall to the floor because her gravitational interaction with the earth has suddenly been turned off. As long as she remains in the box and it remains in free fall, she cannot tell whether she is indeed in free fall or whether the gravitational interaction has vanished.

A similar problem appears in a space station in orbit around the earth. Objects in the space station *seem* to be weightless, but without looking outside the station there is no way to determine whether gravity has been turned off or whether the station and all its contents are accelerating toward the center of the earth. Figure 39–21 makes a similar point for a spaceship that is not in free fall but may be accelerating relative to an inertial frame or be at rest on a planet's surface.

These considerations form the basis of Einstein's **general theory of relativity.** If we cannot distinguish experimentally between a uniform gravitational field at a particular location and a uniformly accelerated reference frame, then there cannot be any real distinction between the two. Pursuing this concept, we may try to represent *any*

39–21 (a) An astronaut drops his watch in the spaceship. (b) In gravity-free space the floor accelerates upward at $a = g$ and hits the watch. (c) On the planet's surface the watch accelerates downward at $a = g$ and hits the floor. Without information from outside the spaceship, the astronaut cannot distinguish situation (b) from situation (c).

(a) (b) (c)

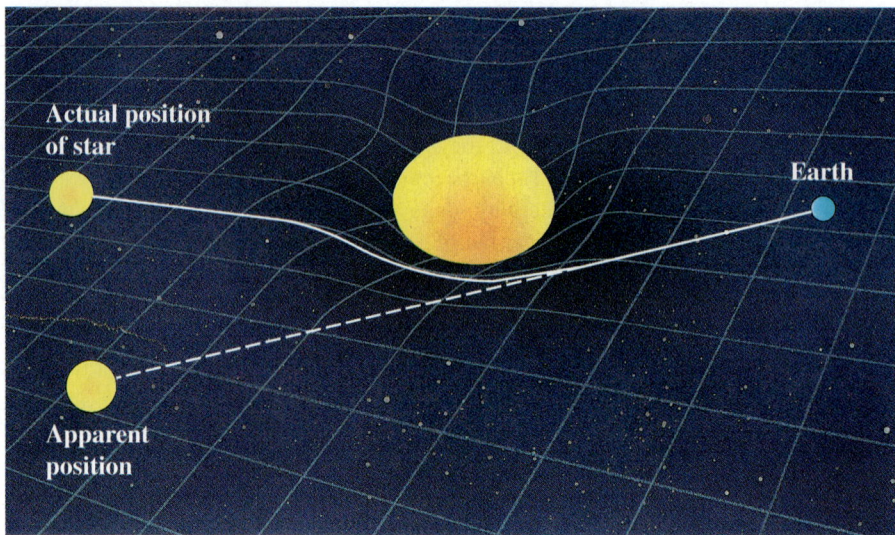

39–22 A two-dimensional representation of a curved space. We imagine the space (a plane) as being distorted as shown by a massive object (the sun). Light from a distant star (solid line) follows the distorted surface on its way to the earth. The dashed line shows the direction from which the light *appears* to be coming. The effect is greatly exaggerated; for the sun the maximum deviation is only 0.00048°.

gravitational field in terms of special characteristics of the coordinate system. This turns out to require even more sweeping revisions of our space-time concepts than did the special theory of relativity. In the general theory of relativity the geometric properties of space are in general non-Euclidean (Fig. 39–22).

The general theory of relativity has passed several experimental tests, including three proposed by Einstein. One test has to do with understanding the rotation of the axes of the planet Mercury's elliptical orbit, called the *precession of the perihelion.* (The perihelion is the point of closest approach to the sun.) Another test concerns the apparent bending of light rays from distant stars when they pass near the sun, and the third test is the *gravitational red shift,* the increase in wavelength of light proceeding outward from a massive source. Some details of the general theory are more difficult to test, but this theory has played a central role in cosmological investigations of the structure of the universe, the formation and evolution of stars, black holes, and related matters. In recent years the theory has also been confirmed by several purely *terrestrial* experiments.

SUMMARY

- All the fundamental laws of physics have the same form in all inertial frames of reference. The speed of light in vacuum is the same in all inertial frames and is independent of the motion of the source. Simultaneity is not an absolute concept; two events that are simultaneous in one frame are not necessarily simultaneous in a second frame moving relative to the first.

- If Δt_0 is the time interval between two events that occur at the same space point in a particular frame of reference, Δt_0 is called a proper time interval. If this frame moves with constant velocity u relative to a second frame, the time interval Δt between the events, as observed in the second frame, is

$$\Delta t = \frac{\Delta t_0}{\sqrt{1 - u^2/c^2}} = \gamma \Delta t_0, \qquad \text{(39–6), (39–8)}$$

where

$$\gamma = \frac{1}{\sqrt{1 - u^2/c^2}}. \qquad \text{(39–7)}$$

This effect is called time dilation.

- If l_0 is the distance between two points that are at rest in a particular frame of reference, l_0 is called a proper length. If this frame moves with constant speed u relative to a second frame and the distances are measured parallel to the motion, the distance l between the points as measured in the second frame is

$$l = l_0 \sqrt{1 - \frac{u^2}{c^2}} = \frac{l_0}{\gamma}. \qquad (39\text{–}16),\ (39\text{–}17)$$

This effect is called length contraction.

- The Lorentz coordinate transformation relates the coordinates and time of an event in an inertial coordinate system S to the coordinates and time of the same event as observed in a second inertial system S' moving at speed u relative to the first:

$$x' = \frac{x - ut}{\sqrt{1 - u^2/c^2}} = \gamma(x - ut),$$

$$y' = y,$$

$$z' = z, \qquad (39\text{–}22)$$

$$t' = \frac{t - ux/c^2}{\sqrt{1 - u^2/c^2}} = \gamma(t - ux/c^2).$$

For one-dimensional motion a particle's velocities v in S and v' in S' are related by

$$v' = \frac{v - u}{1 - uv/c^2}, \qquad (39\text{–}23)$$

$$v = \frac{v' + u}{1 + uv'/c^2}. \qquad (39\text{–}24)$$

- The Doppler effect is the frequency shift from a source due to the relative motion of source and observer. For a source emitting electromagnetic waves of frequency f_0 as it moves toward the observer with speed u, the received frequency f is

$$f = \sqrt{\frac{c + u}{c - u}}\, f_0. \qquad (39\text{–}26)$$

- For momentum conservation in collisions to hold in all inertial frames of reference, the definition of momentum must be generalized. For a particle of rest mass m moving with velocity $\vec{v}$, the momentum $\vec{p}$ is defined as

$$\vec{p} = \frac{m\vec{v}}{\sqrt{1 - v^2/c^2}} = \gamma m\vec{v}. \qquad (39\text{–}28),\ (39\text{–}32)$$

- Applying the work-energy relation gives the relativistic equation for kinetic energy K. For a particle of rest mass m moving with speed v,

$$K = \frac{mc^2}{\sqrt{1 - v^2/c^2}} - mc^2 = (\gamma - 1)mc^2. \qquad (39\text{–}37)$$

This form suggests assigning a rest energy mc^2 to a particle, so the total energy E, kinetic energy plus rest energy, is

$$E = K + mc^2 = \frac{mc^2}{\sqrt{1 - v^2/c^2}} = \gamma mc^2. \qquad (39\text{–}39)$$

The total energy E and magnitude of momentum p for a particle with rest mass m are related by

$$E^2 = (mc^2)^2 + (pc)^2. \qquad (39\text{–}40)$$

■ The special theory of relativity is a generalization of Newtonian mechanics. All the principles of Newtonian mechanics are present as limiting cases in which all the speeds are small in comparison to c. Further generalization to include noninertial frames of reference and their relation to gravitational fields leads to the general theory of relativity.

DISCUSSION QUESTIONS

Q39–1 What are the fundamental distinctions between an inertial and a noninertial frame of reference?

Q39–2 If simultaneity is not an absolute concept, does that mean that we must discard the concept of causality? If event A is to *cause* event B, A must occur first. Is it possible that in some frames, A appears to be the cause of B, and in others, B appears to be the cause of A? Explain.

Q39–3 A new book purports to refute the special theory of relativity. The author begins with a premise that might be paraphrased as follows: "Either two events occur at the same time, or they don't. That's just common sense." In the light of our discussion of the relative nature of simultaneity, how would you respond to this?

Q39–4 What do you think would be different in everyday life if the speed of light were 10 m/s instead of 3.00×10^8 m/s?

Q39–5 The average life span in the United States is about 70 years. Does this mean that it is impossible for an average person to travel a distance greater than 70 light years away from the earth? (A light year is the distance light travels in a year.) Explain.

Q39–6 A physicist claims that it is impossible to define what is meant by a *rigid body* in a relativistically correct way. Why should this be?

Q39–7 Two events occur at the same space point in a particular inertial frame of reference and are simultaneous in that frame. Is it possible that they may not be simultaneous in a different inertial frame? Explain.

Q39–8 As Mavis flies past Stanley at $0.80c$, she holds a meter stick that is aligned parallel to her flight path. She says that her meter stick is 1.00 m in length, but Stanley says it is $(1.00 \text{ m})\sqrt{1-(0.80)^2} = 0.60$ m in length. Which observer is correct?

Q39–9 The theory of relativity sets an upper limit on the speed that a particle can have. Are there also limits on the energy and momentum of a particle? Explain.

Q39–10 When a monochromatic light source moves toward an observer, its wavelength appears to be shorter than the value measured when the source is at rest. Does this contradict the hypothesis that the speed of light is the same for all observers? Explain.

Q39–11 The speed of light relative to still water is 2.25×10^8 m/s. If the water is moving past us, the speed of light we measure depends on the speed of the water. Do these facts violate Einstein's second postulate? Explain.

Q39–12 A student asserts that a material particle must always have a speed less than that of light, and a massless particle must always move at exactly the speed of light. Is she correct? If so, how do massless particles such as photons and neutrinos acquire this speed? Can't they start from rest and accelerate? Explain.

Q39–13 In principle, does a hot gas have more mass than the same gas when it is cold? Explain. In practice, would this be a measurable effect? Explain.

Q39–14 Why do you think the development of Newtonian mechanics preceded the more refined relativistic mechanics by so many years?

EXERCISES

SECTION 39–3 **RELATIVITY OF SIMULTANEITY**

39–1 Suppose the two lightning bolts in Fig. 39–5a are simultaneous to an observer on the train. Show that they are *not* simultaneous to an observer on the ground. Which lightning strike does the ground observer measure to come first?

SECTION 39–4 **RELATIVITY OF TIME INTERVALS**

39–2 The positive muon (μ^+), an unstable particle, lives on average 2.20×10^{-6} s (measured in its own frame of reference) before decaying. a) If such a particle is moving with respect to the laboratory with a speed of $0.900c$, what average lifetime is measured in the laboratory? b) What average distance, measured in the laboratory, does the particle move before decaying?

39–3 Everyday Time Dilation. Two atomic clocks are carefully synchronized. One remains in New York, and the other is loaded on an airliner that travels at an average speed of 250 m/s

and then returns to New York. When the plane returns, the elapsed time on the clock that stayed behind is 4.00 h. By how much will the reading of the two clocks differ, and which clock will show the shorter elapsed time? (*Hint:* Since $u \ll c$, you can simplify $\sqrt{1-u^2/c^2}$ by a binomial expansion.)

39–4 A spaceship flies past Mars with a speed of $0.978c$ relative to the surface of the planet. When the spaceship is directly overhead at an altitude of 1200 km, a very bright signal light on the Martian surface blinks on and then off. An observer on Mars measures that the signal light was on for 82.4μs. What is the duration of the light pulse measured by the pilot of the spaceship?

39–5 The negative pion (π^-) is an unstable particle with an average lifetime of 2.60×10^{-8} s (measured in the rest frame of the pion). a) If the pion is made to travel at very high speed relative to a laboratory, its average lifetime is measured in the laboratory to be 4.20×10^{-7} s. Calculate the speed of the pion expressed as a fraction of c. b) What distance, measured in the laboratory, does the pion travel during its average lifetime?

39–6 As you pilot your space utility vehicle at a constant speed toward the moon, a race pilot flies past you in her spaceracer at a constant speed of $0.800c$ relative to you. At the instant the spaceracer passes you, both of you start timers at zero. a) At the instant when you measure that the spaceracer has traveled 1.20×10^8 m past you, what does the race pilot read on her timer? b) When the race pilot reads the value calculated in part (a) on her timer, what does she measure to be your distance from her? c) At the instant when the race pilot reads the value calculated in part (a) on her timer, what do you read on yours?

39–7 A spacecraft flies away from the earth with a speed of 4.80×10^6 m/s relative to the earth and then returns at the same speed. The spacecraft carries an atomic clock that has been carefully synchronized with an identical clock that remains at rest on earth. The spacecraft returns to its starting point 365 days (1 year) later, as measured by the clock that remained on earth. What is the difference in the elapsed times on the two clocks, measured in hours? Which clock, the one in the spacecraft or the one on earth, shows the smallest elapsed time?

39–8 An alien spacecraft is flying overhead at a great distance as you stand in your backyard. You see its searchlight blink on for 0.190 s. The first officer on the spacecraft measures that the searchlight is on for 12.0 ms. a) Which of these two measured times is the proper time? b) What is the speed of the spacecraft relative to the earth expressed as a fraction of the speed of light c?

SECTION 39–5 RELATIVITY OF LENGTH

39–9 A muon is created 55.0 km above the surface of the earth (as measured in the earth's frame). The average lifetime of a muon, measured in its own rest frame, is 2.20×10^{-6} s; the muon we are considering has this lifetime. In the frame of the muon the earth is moving toward the muon with a speed of $0.9860c$. a) In the muon's frame, what is its initial height above the surface of the earth? b) In the muon's frame, how much closer does the earth get during the lifetime of the muon? What fraction is this of the muon's original height as measured in the muon's frame? c) In the earth's frame, what is the lifetime of the muon? In the earth's frame, how far does the muon travel during its lifetime? What fraction is this of the muon's original height in the earth's frame?

39–10 An unstable particle is created in the upper atmosphere from a cosmic ray and travels straight down toward the surface of the earth with a speed of $0.99540c$ relative to the earth. A scientist at rest on the earth's surface measures that the particle is created at an altitude of 45.0 km. a) As measured by the scientist, how much time does it take the particle to travel the 45.0 km to the surface of the earth? b) Use the length contraction formula to calculate the distance from where the particle is created to the surface of the earth as measured in the particle's frame. c) In the particle's frame, how much time does it take the particle to travel from where it is created to the surface of the earth? Calculate this time both by the time dilation formula and also from the distance calculated in part (b). Do the two results agree?

39–11 As measured by an observer on earth, a spacecraft runway on earth has a length of 3600 m. a) What is the length of the runway as measured by a pilot of a spacecraft flying past at a speed of 4.00×10^7 m/s relative to the earth? b) An observer on earth measures the time interval from when the spacecraft is directly over one end of the runway until it is directly over the other end. What result does she get? c) The pilot of the spacecraft measures the time it takes him to travel from one end of the runway to the other end. What value does he get?

39–12 A meter stick moves past you at great speed. Its motion relative to you is parallel to its long axis. If you measure the length of the moving meter stick to be 1.00 ft (1 ft = 0.3048 m), for example by comparing it to a one-foot ruler that is at rest relative to you, at what speed is the meter stick moving relative to you?

39–13 A spacecraft of the Trade Federation flies past the planet Coruscant at a speed of $0.600c$. A scientist on Coruscant measures the length of the moving spacecraft to be 74.0 m. The spacecraft later lands on Coruscant, and the same scientist measures the length of the now stationary spacecraft. What value does she get?

SECTION 39–6 THE LORENTZ TRANSFORMATIONS

39–14 Solve Eqs. (39–22) to obtain x and t in terms of x' and t', and show that the resulting transformation has the same form as the original one except for a change of sign for u.

39–15 An observer in frame S' is moving to the right at speed $u = 0.600c$ away from a stationary observer in frame S. The observer in S' measures the speed v' of a particle moving to the right away from her. What speed v does the observer in S measure for the particle if a) $v' = 0.400c$? b) $v' = 0.900c$? c) $v' = 0.990c$?

39–16 Space pilot Mavis zips past Stanley at a constant speed relative to him of $0.800c$. Mavis and Stanley start timers at zero when the front of Mavis's ship is directly above Stanley. When Mavis reads 5.00 s on her timer, she turns on a bright light under the front of her spaceship. a) Use the Lorentz coordinate transformation derived in Exercise 39–14 and Example 39–7 (Section 39–6) to calculate x and t as measured by Stanley for the event of the light turning on. b) Use the time dilation formula, Eq. (39–6), to calculate the time interval between the two events (the front of the spaceship passing overhead and the light turning on) as measured by Stanley. Compare to the value of t you calculated in part (a). c) Multiply the time interval by Mavis's speed, both as measured by Stanley, to calculate the distance she has traveled as measured by him when the light turns on. Compare to the value of x you calculated in part (a).

39–17 Show in detail the derivation of Eq. (39–21) from Eqs. (39–19) and (39–20).

39–18 Show the details of the derivation of Eq. (39–24) from (39–23).

39–19 A pursuit spacecraft from the planet Tatooine is attempting to catch up with a Trade Federation cruiser. As measured by an observer on Tatooine, the cruiser is traveling away from the planet with a speed of $0.600c$. The pursuit ship is traveling at a speed of $0.800c$ relative to Tatooine, in the same direction as the cruiser. What is the speed of the cruiser relative to the pursuit ship?

39–20 Two particles in a high-energy accelerator experiment are approaching each other head-on, each with a speed $0.9520c$ as measured in the laboratory. What is the magnitude of the velocity of one particle relative to the other?

39–21 Two particles are created in a high-energy accelerator and move off in opposite directions. The speed of one particle, as measured in the laboratory, is $0.650c$, and the speed of each particle relative to the other is $0.950c$. What is the speed of the second particle, as measured in the laboratory?

39–22 An enemy spaceship is moving toward your starfighter with a speed, as measured in your frame, of $0.400c$. The enemy ship fires a missile toward you at a speed of $0.700c$ relative to the enemy ship (Fig. 39–23). a) What is the speed of the missile relative to you? Express your answer in terms of the speed of light. b) If you measure that the enemy ship is 8.00×10^6 km away from you when the missile is fired, how much time, measured in your frame, will it take the missile to reach you?

Enemy   Starfighter

FIGURE 39–23 Exercise 39–22.

39–23 An imperial spaceship moving at high speed relative to the planet Arrakis fires a rocket toward the planet with a speed of $0.920c$ relative to the spaceship. An observer on Arrakis measures that the rocket is approaching with a speed of $0.360c$. What is the speed of the spaceship relative to Arrakis? Is the spaceship moving toward or away from Arrakis?

*SECTION **39–7** SPACETIME DIAGRAMS

*__39–24__ Mavis moves at constant speed u relative to Stanley. Let Mavis's frame be S' and Stanley's be S; the two frames coincide at $t = t' = 0$. Mavis measures events 1 and 2 as occurring simultaneously in S' at positions x_1' and x_2', where $x_2' > x_1'$. Draw a spacetime diagram and use it to decide whether Stanley measures the two events as occurring simultaneously in S. If they do not occur simultaneously in his frame, which event occurs first? Explain.

*__39–25__ Mavis moves at constant speed u relative to Stanley. Let Mavis's frame be S' and Stanley's be S; the two frames coincide at $t = t' = 0$. Stanley measures events 1 and 2 as occurring at the same position x in S and at different times t_1 and t_2. According to Stanley, $t_2 > t_1$, so event 1 occurs before event 2. a) On a spacetime diagram, show the times t_1' and t_2' that Mavis measures for these two events. Do Mavis and Stanley agree about which event occurs first? Explain, using your diagram. b) On the spacetime diagram, show the positions x_1' and x_2' that Mavis measures for the positions of these two events. According to Mavis, do the two events occur at the same position? Explain.

*SECTION **39–8** THE DOPPLER EFFECT FOR ELECTROMAGNETIC WAVES

*__39–26__ In terms of c, what relative velocity u between a source of electromagnetic waves and an observer produces a) a 2.0%

decrease in frequency? b) an increase by a factor of four of the frequency of the observed light?

*__39–27__ a) How fast must you be approaching a red traffic light ($\lambda = 675$ nm) for it to appear yellow ($\lambda = 575$ nm)? Express your answer in terms of the speed of light. b) If you used this as an excuse for not getting a ticket for running a red light, how much of a fine would you get for speeding? Assume that the fine is $1.00 for each kilometer per hour that your speed exceeds the posted limit of 90 km/h.

*__39–28__ Show that when the source of electromagnetic waves moves away from us at $0.600c$, the frequency we measure is half the value measured in the rest frame of the source.

SECTION **39–9** RELATIVISTIC MOMENTUM

39–29 a) A particle with mass m moves along a straight line under the action of a force F directed along the same line. Evaluate the derivative in Eq. (39–30) to show that the acceleration $a = dv/dt$ of the particle is given by $a = (F/m)(1 - v^2/c^2)^{3/2}$. b) Evaluate the derivative in Eq. (39–30) to find the expression for the magnitude of the acceleration in terms of F, m, and v/c if the force is perpendicular to the velocity.

39–30 *Relativistic Baseball.* Calculate the magnitude of the force required to give a 0.145-kg baseball an acceleration $a = 1.00$ m/s^2 in the direction of the baseball's initial velocity when this velocity has a magnitude of a) 10.0 m/s; b) $0.900c$; c) $0.990c$. d) Repeat parts (a), (b), and (c) if the force and acceleration are perpendicular to the velocity.

39–31 a) At what speed is the momentum of a particle twice as great as the result obtained from the nonrelativistic expression mv? Express your answer in terms of the speed of light. b) A force is applied to a particle along its direction of motion. At what speed is the magnitude of force required to produce a given acceleration twice as great as the force required to produce the same acceleration when the particle is at rest? Express your answer in terms of the speed of light.

39–32 a) At what speed does the momentum of a particle differ by 1.0% from the value obtained using the nonrelativistic expression mv? b) Is the correct relativistic value greater or less than that obtained from the nonrelativistic expression?

SECTION **39–10** RELATIVISTIC WORK AND ENERGY

39–33 What is the speed of a particle whose kinetic energy is equal to a) its rest energy; b) five times its rest energy?

39–34 In *proton-antiproton annihilation* a proton and an antiproton (a negatively charged proton) collide and disappear, producing electromagnetic radiation. If each particle has a mass of 1.67×10^{-27} kg and they are at rest just before the annihilation, find the total energy of the radiation. Give your answers in joules and in electron volts.

39–35 What is the speed of a particle if its kinetic energy is 2.0% larger than $\frac{1}{2}mv^2$? (*Hint:* Use Eq. (39–38).)

39–36 a) How much work must be done on a particle with mass m to accelerate it a) from rest to a speed of $0.090c$? b) from a speed $0.900c$ to a speed $0.990c$? (Express the answers in terms of mc^2.) c) How do your answers in parts (a) and (b) compare?

39–37 A particle has rest mass 6.64×10^{-27} kg and momentum 2.10×10^{-18} kg · m/s. a) What is the total energy (kinetic plus rest energy) of the particle? b) What is the kinetic energy of the particle? c) What is the ratio of the kinetic energy to the rest energy of the particle?

39–38 Starting from Eq. (39–40), show that in the classical limit ($pc \ll mc^2$) the energy approaches the classical kinetic energy $\frac{1}{2}mv^2$ plus the rest mass energy mc^2.

39–39 Compute the kinetic energy of a proton (mass 1.67×10^{-27} kg) using both the nonrelativistic and relativistic expressions, and compute the ratio of the two results (relativistic divided by nonrelativistic) for speeds of a) 8.00×10^7 m/s; b) 2.85×10^8 m/s.

39–40 A ψ ("psi") particle has mass 5.52×10^{-27} kg. Compute the rest energy of the ψ in MeV.

39–41 An 0.420-MeV electron in an x-ray tube strikes an anode. When it arrives at the anode, what is a) its kinetic energy in eV? b) its total energy in eV? c) its speed? d) What is the speed of the electron, calculated classically?

39–42 Energy of Fusion. In a hypothetical nuclear-fusion reactor, two deuterium nuclei combine or "fuse" to form one helium nucleus. The mass of a deuterium nucleus, expressed in atomic mass units (u), is 2.0136 u; that of a helium nucleus is 4.0015 u (1 u = $1.6605402 \times 10^{-27}$ kg). a) How much energy is released when 1.0 kg of deuterium undergoes fusion? b) The annual consumption of electrical energy in the United States is of the order of 1.0×10^{19} J. How much deuterium must react to produce this much energy?

PROBLEMS

39–43 After being produced in a collision between elementary particles, a positive pion (π^+) must travel down a 1.20-km-long tube to reach an experimental area. A π^+ particle has an average lifetime (measured in its rest frame) of 2.60×10^{-8} s; the π^+ we are considering has this lifetime. a) How fast must the π^+ travel if it is not to decay before it reaches the end of the tube? (Since u will be very close to c, write $u = (1 - \Delta)c$ and give your answer in terms of Δ rather than u.) b) The π^+ has a rest energy of 139.6 MeV. What is the total energy of the π^+ at the speed calculated in part (a)?

39–44 A cube of metal with sides of length a sits at rest in a frame S with one edge parallel to the x-axis. Therefore, in S the cube has volume a^3. Frame S' moves along the x-axis with a speed u. As measured by an observer in frame S', what is the volume of the metal cube?

39–45 The starships of the Solar Federation are marked with the symbol of the Federation, a circle, while starships of the Denebian Empire are marked with the Empire's symbol, an ellipse whose major axis is 1.40 times its minor axis ($a = 1.40b$ in Fig. 39–24). How fast relative to an observer does a Federation ship have to travel for its markings to be confused with those of an Empire ship?

 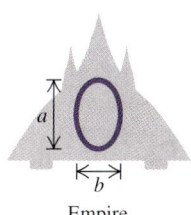

Federation   Empire

FIGURE 39–24 Exercise 39–45.

39–46 A space probe is sent to the vicinity of the star Capella, which is 42.2 light years from earth. (A light year is the distance light travels in a year.) The probe travels with a speed of $0.9910c$. An astronaut recruit on board is 19 years old when the probe leaves earth. What is her biological age when the probe reaches Capella?

39–47 A particle is said to be *extremely relativistic* when its kinetic energy is much larger than its rest energy. a) What is the speed of a particle (expressed as a fraction of c) such that the total energy is ten times the rest energy? b) What is the percentage difference between the left and right sides of Eq. (39–40) if $(mc^2)^2$ is neglected for a particle with the speed calculated in part (a)?

39–48 A nuclear bomb containing 8.00 kg of plutonium explodes. The sum of the rest masses of the products of the explosion is less than the original rest mass by one part in 10^4. a) How much energy is released in the explosion? b) If the explosion takes place in 4.00 μs, what is the average power developed by the bomb? c) What mass of water could the released energy lift to a height of 1.00 km?

39–49 By what minimum amount does the mass of 4.00 kg of ice increase when the ice melts at 0°C to form water at 0°C?

39–50 A photon with energy E is emitted by an atom with mass m, which recoils in the opposite direction. a) Assuming that the motion of the atom can be treated nonrelativistically, compute the recoil speed of the atom. b) From the result of part (a), show that the recoil speed is much smaller than c whenever E is much smaller than the rest energy mc^2 of the atom.

39–51 Čerenkov Radiation. The Russian physicist P. A. Čerenkov discovered that a charged particle traveling in a solid with a speed exceeding the speed of light in that material radiates electromagnetic radiation. (This is analogous to the sonic boom produced by an aircraft moving faster than the speed of sound in air; see Section 21–6. Čerenkov shared the 1958 Nobel Prize for this discovery.) What is the minimum kinetic energy (in electron volts) that an electron must have while traveling inside a slab of crown glass ($n = 1.52$) in order to create this Čerenkov radiation?

39–52 Albert in Wonderland. Einstein and Lorentz, being avid tennis players, play a fast-paced game on a court where they stand 20.0 m from each other. Being very skilled players, they play without a net. The tennis ball has a mass of 0.0580 kg. Neglect gravity and assume that the ball travels parallel to the ground as it travels between the two players. Unless specified otherwise, all measurements are made by the two men. a) Lorentz serves the ball at 80.0 m/s. What is the ball's kinetic energy? b) Einstein slams a return at 1.80×10^8 m/s. What is the ball's kinetic energy? c) During Einstein's return of the ball in part (a), a white rabbit runs beside the court in the direction from Einstein to Lorentz. The rabbit has a speed of 2.20×10^8 m/s relative to the two men. What is the speed of the rabbit relative to the ball? d) What does the rabbit measure as the distance from Einstein to Lorentz? e) How much time does it take for the rabbit to run 20.0 m, according to the players? f) The white rabbit carries a pocket watch. He uses this watch to measure the time (as he sees it) for the distance from Einstein to Lorentz to pass by under him. What time does he measure?

39–53 Frame S' has an x-component of velocity u relative to frame S and at $t = t' = 0$ the two frames coincide (Fig. 39–3). A light pulse is emitted at the origin of S' at time $t' = 0$. Its distance x' from the origin after a time t' is given by $x'^2 = c^2 t'^2$. Use the Lorentz coordinate transformation to transform this equation to an equation in x and t, and show that the result is $x^2 = c^2 t^2$; that is, the motion appears exactly the same in the frame of reference S as it does in S'. (See Fig. 39–4.)

39–54 An astronaut, her rocket pack, and her spacesuit have a combined rest mass of 90.0 kg and a speed of 3.00×10^4 m/s. a) What is the difference between her correct relativistic kinetic energy and her Newtonian kinetic energy, $\frac{1}{2}mv^2$? Note that you cannot simply calculate both values and subtract, as the precision of most calculators is insufficient. Instead, use Eq. (39–38). b) What fraction of the Newtonian kinetic energy is this difference?

39–55 A particle with mass m accelerated from rest by a constant force F will, according to Newtonian mechanics, continue to accelerate without bound. That is, as $t \to \infty$, $v \to \infty$. Show that according to relativistic mechanics, the particle's speed approaches c as $t \to \infty$. (*Note:* A useful integral is $\int (1 - x^2)^{-3/2}\, dx = x/\sqrt{1 - x^2}$.)

39–56 Two events are observed in a frame of reference S to occur at the same space point, the second occurring 1.80 s after the first. In a second frame S' moving relative to S, the second event is observed to occur 2.35 s after the first. What is the difference between the positions of the two events as measured in S'?

39–57 Two events observed in a frame of reference S have positions and times given by (x_1, t_1) and (x_2, t_2), respectively. a) Frame S' moves along the x-axis just fast enough that the two events occur at the same position in S'. Show that in S', the time interval $\Delta t'$ between the two events is given by

$$\Delta t' = \sqrt{(\Delta t)^2 - \left(\frac{\Delta x}{c}\right)^2},$$

where $\Delta x = x_2 - x_1$ and $\Delta t = t_2 - t_1$. Hence show that if $\Delta x > c\, \Delta t$, there is *no* frame S' in which the two events occur at

the same point. The interval $\Delta t'$ is sometimes called the *proper time interval* for the events. Is this term appropriate? b) Show that if $\Delta x > c\, \Delta t$, there is a different frame of reference S' in which the two events occur *simultaneously.* Find the distance between the two events in S'; express your answer in terms of Δx, Δt, and c. This distance is sometimes called a *proper length.* Is this term appropriate? c) Two events are observed in a frame of reference S' to occur simultaneously at points separated by a distance of 2.50 m. In a second frame S moving relative to S' along the line joining the two points in S', the two events appear to be separated by 5.00 m. What is the time interval between the events as measured in S? (*Hint:* Apply the result obtained in part (b).)

39–58 Tachyons. Tachyons are hypothetical particles that can travel faster than light. Consider two observers, Sebulba (at rest in frame S, which uses coordinates x, y, z, and t) and Watto (at rest in frame S', which uses coordinates x', y', z', and t'). Watto moves relative to Sebulba with speed $v = 0.600c$. At the instant they pass each other, they both set their clocks to zero. a) At time $t_1 = 100$ s in Sebulba's frame, Sebulba transmits a tachyon message to Watto. At this instant, how far away is Watto from Sebulba (as measured in Sebulba's frame)? b) The speed of the tachyons relative to Sebulba is $u = 4.00c$. What is the time t_2 (in Sebulba's frame) at which Watto receives the message? c) Watto is displeased with the message that Sebulba sends him, and with zero response time sends Sebulba a very unpleasant tachyon message in return. The tachyons have the same speed as those in part (b), so that their speed relative to Watto is $v' = 4.00c$. What is the velocity (magnitude and direction) *relative to Sebulba* of Watto's return message? d) What is the time t_3 (in Sebulba's frame) at which Sebulba receives Watto's message? e) The message that Watto sent overloaded the circuits of Sebulba's transmitter, converting it into a pile of useless metal and making it impossible for Selbulba to send tachyon signals. How do t_1 (the time when Sebulba sent the original signal) and t_3 (the time at which Sebulba's transmitter was destroyed) compare? Discuss your result and the possible existence of tachyons.

*39–59 One of the wavelengths of light emitted by hydrogen atoms under normal laboratory conditions is $\lambda = 656.3$ nm, in the red portion of the electromagnetic spectrum. In the light emitted from a distant galaxy this same spectral line is observed to be Doppler-shifted to $\lambda = 953.4$ nm, in the infrared portion of the spectrum. How fast are the emitting atoms moving relative to the earth? Are they approaching the earth or receding from it?

*39–60 Measuring Speed by Radar. A baseball coach uses a radar device to measure the speed of an approaching pitched baseball. This device sends out electromagnetic waves with frequency f_0 and then measures the shift in frequency Δf of the waves reflected from the moving baseball. If the fractional frequency shift produced by a baseball is $\Delta f/f_0 = 2.86 \times 10^{-7}$, what is the baseball's speed in km/h?

39–61 A particle of mass m moves at speed v. Construct a right triangle in which one of the angles is α, where $\sin \alpha = v/c$. If the length of the side of the triangle adjacent to α equals the rest energy mc^2, show that a) the length of the hypotenuse equals the total energy; b) the length of the side opposite α equals c times the relativistic momentum. c) Describe a simple graphical procedure for finding the kinetic energy K.

*39–62 A spaceship moving at constant speed u relative to us broadcasts a radio signal at constant frequency f_0. As the spaceship approaches us, we receive a higher frequency f; after it has passed, we receive a lower frequency. a) As the spaceship passes by, so it is instantaneously neither moving toward or away from us, show that the frequency we receive is not f_0, and derive an expression for the frequency we do receive. Is the frequency we receive larger or smaller than f_0? (*Hint:* In this case, successive wave crests move the same distance to the observer and so they have the same transit time. Thus f equals $1/T$. Use the time dilation formula to relate the periods in the stationary and moving frames.) b) A spaceship emits electromagnetic waves of frequency $f_0 = 345$ MHz as measured in a frame moving with the ship. The spaceship is moving at a constant speed $0.758c$ relative to us. What frequency f do we receive when the spaceship is approaching us? When it is moving away? In each case what is the shift in frequency, $f - f_0$? c) Use the result of part (a) to calculate the frequency f and frequency shift $f - f_0$ we receive at the instant that the ship passes by us. How does the shift in frequency calculated here compare to the shifts calculated in part (b)?

*39–63 The Pole and Barn Paradox. Suppose a *very* fast runner ($v = 0.600c$) holding a long horizontal pole runs through a barn open at both ends. The length of the pole (in its rest frame) is 6.00 m and the length of the barn (in *its* rest frame) is 5.00 m. In the barn's reference frame, the pole will undergo length contraction and can all fit inside the barn at the same time. But in the runner's reference frame, the *barn* will undergo length contraction and the entire pole can *never* be entirely within the barn at any time! Explain the resolution of this paradox. (A spacetime diagram may be useful.)

39–64 The French physicist Armand Fizeau was the first to measure the speed of light accurately. He also found experimentally that the speed relative to the lab frame of light traveling in a tank of water that is itself moving at a speed V relative to the lab frame is

$$v = (c/n) + kV,8$$

where $n = 1.333$ is the index of refraction of water. Fizeau called k the dragging coefficient and obtained an experimental value of $k = 0.44$. What value of k do you calculate from relativistic transformations?

CHALLENGE PROBLEMS

39–65 Lorentz Transformation for Acceleration. Using a method analogous to the one in the text to find the Lorentz transformation formula for velocity, we can find the Lorentz transformation for *acceleration*. Let frame S' have a constant x-component of velocity u relative to frame S. An object moves relative to frame S along the x-axis with instantaneous velocity v and instantaneous acceleration a. a) Show that its instantaneous acceleration in frame S' is

$$a' = a\left(1 - \frac{u^2}{c^2}\right)^{3/2}\left(1 - \frac{uv}{c^2}\right)^{-3}.$$

(*Hint:* Express the acceleration in S' as $a' = dv'/dt'$. Then use Eq. (39–22) to express dt' in terms of dt and dx, and use Eq. (39–23) to express dv' in terms of u and dv. The velocity of the object in S is $v = dx/dt$.) b) Show that the acceleration in frame S can be expressed as

$$a = a'\left(1 - \frac{u^2}{c^2}\right)^{3/2}\left(1 + \frac{uv'}{c^2}\right)^{-3},$$

where $v' = dx'/dt'$ is the velocity of the object in frame S'.

39–66 A Realistic Version of the Twin Paradox. A rocket ship leaves the earth on January 1, 2100. Stella, one of a pair of twins born in the year 2075, pilots the rocket (reference frame S'); the other twin, Terra, stays on the earth (reference frame S). The rocket ship has an acceleration of constant magnitude g in its own reference frame (this makes the pilot feel at home, since it simulates earth gravity). The path of the rocket ship is a straight line. a) Using the results of Challenge Problem 39–65, show that in Terra's earth frame S, the rocket's acceleration is

$$\frac{du}{dt} = g\left(1 - \frac{u^2}{c^2}\right)^{3/2},$$

where u is the rocket's instantaneous velocity in frame S.

b) Write the result of part (a) in the form $dt = f(u)du$, where $f(u)$ is a function of u, and integrate both sides. (*Hint:* Use the integral given in Problem 39–55.) Show that in Terra's frame, the time when Stella attains a velocity v_1 is

$$t_1 = \frac{v_1}{g\sqrt{1 - v_1^2/c^2}}.$$

c) Use the time dilation formula to relate dt and dt' (infinitesimal time intervals measured in frames S and S', respectively). Combine this result with the result of (a) and integrate as in (b) to show the following: When Stella attains a velocity v_1 relative to Terra, the time t_1' that has elapsed in frame S' is

$$t_1' = \frac{c}{g}\,\text{arctanh}\left(\frac{v_1}{c}\right).$$

Here arctanh is the inverse hyperbolic tangent. (*Hint:* Use the integral given in Challenge Problem 5–112.) d) Combine the results of parts (b) and (c) to find t_1 in terms of t_1', g, and c alone. e) Stella accelerates in a straight-line path for 5 years (by her clock), slows down at the same rate for 5 years, turns around, accelerates for 5 years, slows down for 5 years, and lands back on earth. According to Stella's clock, the date is January 1, 2120. What is the date according to Terra's clock?

*39–67 Determining the Masses of Stars. Many of the stars in the sky are actually *binary stars,* in which two stars orbit about their common center of mass. If the orbital speeds of the stars are high enough, the motion of the stars can be detected by the Doppler shifts of the light that they emit. Stars for which this is the case are called *spectroscopic binary stars.* Figure 39–25 shows the simplest case of a spectroscopic binary star: two identical stars, each of mass m, orbiting their center of mass in a circle of radius R. The plane of the stars' orbits is edge-on to the line of sight of an observer on the earth. a) The light produced by heated hydrogen gas in a laboratory on earth has a frequency of 4.568110×10^{14} Hz. In the light received from the stars by a telescope on earth, hydrogen light is observed to vary in frequency between 4.567710×10^{14} Hz and 4.568910×10^{14} Hz. Determine whether the binary star system as a whole is moving toward or away from the earth, the speed of this motion, and the orbital speeds of the stars. (*Hint:* The speeds involved are much less than c, so you may use the approximate result $\Delta f/f = u/c$ given in Section 39–8.) b) The light from each star in the binary system varies from its maximum frequency to its minimum frequency and back again in 11.0 days. Determine the orbital radius R and the mass m of each star. Give your answer for m in kilograms and as a multiple of the mass of the sun, 1.99×10^{30} kg. Compare the value of R to the distance from the earth to the sun, 1.50×10^{11} m. (This technique is actually used in astronomy to determine the masses of stars. In practice, the problem is more complicated because the two stars in a binary system are usually not identical, the orbits are usually not circular, and the plane of the orbits is usually tilted with respect to the line of sight from earth.)

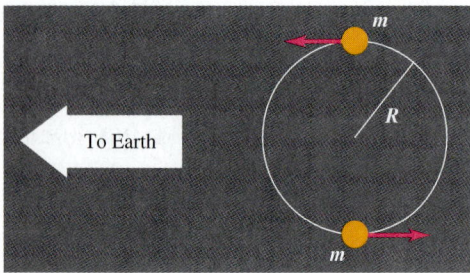

FIGURE 39–25 Challenge Problem 39–67.

39–68 Relativity and the Wave Equation. a) Consider the Galilean transformation along the x-direction: $x' = x - vt$ and $t' = t$. In frame S the wave equation for electromagnetic waves in a vacuum is

$$\frac{\partial^2 E(x,t)}{\partial x^2} - \frac{1}{c^2}\frac{\partial^2 E(x,t)}{\partial t^2} = 0.$$

where E represents the electric field in the wave. Show that by using the Galilean transformation the wave equation in frame S'

is found to be

$$\left(1 - \frac{v^2}{c^2}\right)\frac{\partial^2 E(x',t')}{\partial x'^2} + \frac{2v}{c^2}\frac{\partial^2 E(x',t')}{\partial x' \partial t'} - \frac{1}{c^2}\frac{\partial^2 E(x',t')}{\partial t'^2} = 0.$$

This has a different form than the wave equation in S. Hence the Galilean transformation *violates* the first relativity postulate that all physical laws have the same form in all inertial reference frames. (*Hint:* Express the derivatives $\partial/\partial x$ and $\partial/\partial t$ in terms of $\partial/\partial x'$ and $\partial/\partial t'$ by use of the chain rule.) b) Repeat the analysis of part (a), but use the Lorentz coordinate transformations, Eqs. (39–22), and show that in frame S' the wave equation has the same form as in frame S:

$$\frac{\partial^2 E(x',t')}{\partial x'^2} - \frac{1}{c^2}\frac{\partial^2 E(x',t')}{\partial t'^2} = 0.$$

Explain why this shows that the speed of light in vacuum is c in both frames S and S'.

39–69 Kaon Production. In high-energy physics, new particles can be created by collisions of fast–moving projectile particles with stationary particles. Some of the kinetic energy of the incident particle is used to create the mass of the new particle. A proton-proton collision can result in the creation of a negative kaon (K^-) and a positive kaon (K^+):

$$p + p \rightarrow p + p + K^- + K^+.$$

a) Calculate the minimum kinetic energy of the incident proton that will allow this reaction to occur if the second (target) proton is initially at rest. The rest energy of each kaon is 493.7 MeV, and the rest energy of each proton is 938.3 MeV. (*Hint:* It is useful to work in the frame in which the total momentum is zero is useful here. See Problem 8–86, but note that here the Lorentz transformation must be used to relate the velocities in the laboratory frame to those in the zero-total-momentum frame.) b) How does this calculated minimum kinetic energy compare with the total rest mass energy of the created kaons? c) Suppose that instead the two protons are both in motion with velocities of equal magnitude and opposite direction. Find the minimum combined kinetic energy of the two protons that will allow the reaction to occur. How does this calculated minimum kinetic energy compare with the total rest mass energy of the created kaons? (This example shows that when colliding beams of particles are used instead of a stationary target, the energy requirements for producing new particles are reduced substantially.)

Photons, Electrons, and Atoms

Though this glowing gas cloud is 5000 light years away, we can nonetheless tell that it contains mostly hydrogen. The key is the particular red color of the gas, which is emitted by heated hydrogen and by no other element. In their quest to understand the spectrum of hydrogen, 20th-century physicists revolutionized our picture of what light is and what atoms are.

40-1 INTRODUCTION

In Chapter 33 we saw how Maxwell, Hertz, and others established firmly that light is an electromagnetic wave. Phenomena such as interference, diffraction, and polarization, discussed in Chapters 37 and 38, further demonstrate this *wave nature* of light.

But many phenomena, particularly those concerned with emission and absorption of electromagnetic radiation, show a completely different aspect of light. We find that the energy of an electromagnetic wave is *quantized;* it is emitted and absorbed in particle-like packages of definite energy, called *photons* or *quanta.* The energy of a single photon is proportional to the frequency of the radiation.

The internal energy of atoms is also quantized. For a given kind of individual atom the energy can't have just any value; only discrete values called *energy levels* are possible.

The basic ideas of photons and energy levels take us a long way toward understanding a wide variety of otherwise puzzling observations. Among these are the unique sets of wavelengths emitted and absorbed by gaseous elements, the emission of electrons from a surface by light, the operation of lasers, and the production and scattering of x rays. Our studies of photons and energy levels will take us to the threshold of *quantum mechanics,* which involves some radical changes in our views of the nature of electromagnetic radiation and of matter itself.

40-2 EMISSION AND ABSORPTION OF LIGHT

How is light produced? In Chapter 33 we discussed how Heinrich Hertz produced electromagnetic waves using oscillations in a resonant *L-C* circuit similar to those we studied in Chapter 31. He used frequencies of the order of 10^8 Hz, but visible light has frequencies of the order of 10^{15} Hz, far higher than any frequency that can be attained with conventional electronic circuits. At the end of the nineteenth century, some physicists speculated that waves in this frequency range might be produced by oscillating electric charges within individual atoms. However, their speculations were unable to give explanations for some key experimental data. Among the great challenges facing physicists in 1900 were how to explain line spectra, the photoelectric effect, and the production of x rays. We'll describe each of these in turn.

LINE SPECTRA

We can use a prism or a diffraction grating to separate the various wavelengths in a beam of light into a spectrum. If the light source is a hot solid (such as a light-bulb filament) or liquid, the spectrum is *continuous;* light of all wavelengths is present (Fig. 40–1a). But if the source is a gas carrying an electric discharge (as in a neon sign) or a volatile salt heated in a flame (as when table salt is thrown into a campfire), only a few colors appear, in the form of isolated sharp parallel lines (Fig. 40–1b). (Each "line" is an

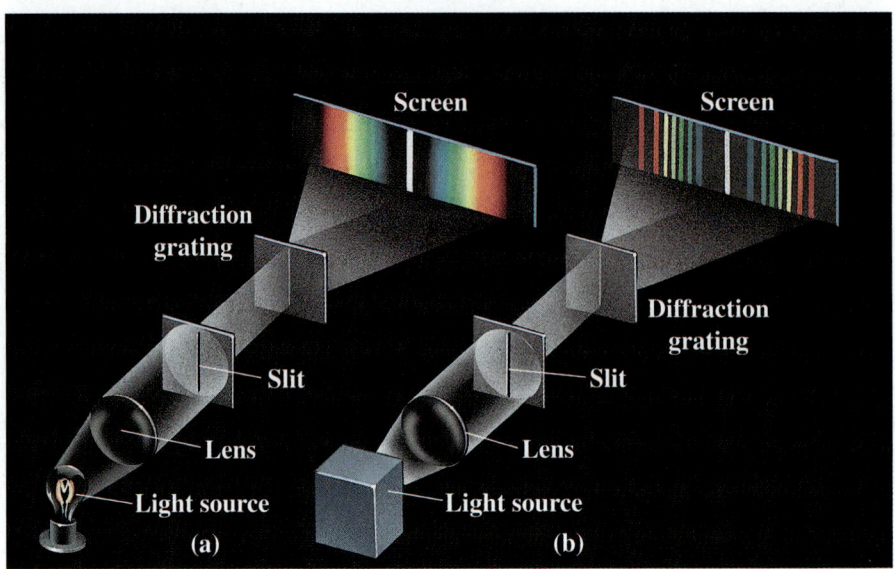

40–1 (a) Continuous spectrum emitted by a glowing light-bulb filament. (b) Line spectrum emitted by a mercury vapor lamp. The slit and diffraction grating are shown.

image of the spectrograph slit, deviated through an angle that depends on the wavelength of the light forming that image; see Section 38–6.) A spectrum of this sort is called a **line spectrum.** Each line corresponds to a definite wavelength and frequency.

It was discovered early in the nineteenth century that each element in its gaseous state has a unique set of wavelengths in its line spectrum. The spectrum of hydrogen always contains a certain set of wavelengths; sodium produces a different set, iron still another, and so on. Scientists find the use of spectra to identify elements and compounds to be an invaluable tool. For instance, astronomers have detected the spectra from more than 100 different molecules in interstellar space, including some that are not found naturally on earth. The characteristic spectrum of an atom was presumably related to its internal structure, but attempts to understand this relation solely on the basis of *classical* mechanics and electrodynamics—the physics summarized in Newton's three laws and Maxwell's four equations—were not successful.

17.3
Photoelectric Effect

PHOTOELECTRIC EFFECT

There were also mysteries associated with *absorption* of light. In 1887, during his electromagnetic-wave experiments, Hertz discovered the *photoelectric effect.* When light struck a metal surface, some electrons near the surface absorbed enough energy to overcome the attraction of the positive ions in the metal and escape into the surrounding space. Detailed investigation of this effect revealed some puzzling features that couldn't be understood on the basis of classical optics. We will discuss these in the next section.

X RAYS

Other unsolved problems in the emission and absorption of radiation centered on the production and scattering of *x rays,* discovered in 1895. These rays were produced in high-voltage electric discharge tubes, but no one understood how or why they were produced or what determined their wavelengths (which are much shorter those of visible light). Even worse, when x rays collided with matter, the scattered rays sometimes had longer wavelengths than the original ray. This is analogous to a beam of blue light striking a mirror and reflecting back as red!

PHOTONS AND ENERGY LEVELS

All these phenomena (and several others) pointed forcefully to the conclusion that classical optics, successful though it was in explaining lenses, mirrors, interference, and polarization, had its limitations. We now understand that all these phenomena result from the *quantum* nature of radiation. Electromagnetic radiation, along with its *wave* nature, has properties resembling those of *particles*. In particular, the energy in an electromagnetic wave is always emitted and absorbed in packages called *photons* or *quanta,* with energy proportional to the frequency of the radiation.

The two common threads that are woven through this chapter are the quantization of electromagnetic radiation and the existence of discrete energy levels in atoms. In the remainder of this chapter we will show how these two concepts contribute to understanding the phenomena mentioned above. We are not yet ready for a comprehensive theory of atomic structure; that will come in Chapters 42 and 43. But we'll look at the Bohr model of the hydrogen atom, one attempt to predict atomic energy levels on the basis of atomic structure.

40–3 THE PHOTOELECTRIC EFFECT

The **photoelectric effect** is the emission of electrons when light strikes a surface. The liberated electrons absorb energy from the incident radiation and are thus able to overcome the attraction of positive charges. This attraction causes a potential-energy barrier that normally confines the electrons inside the material. Think of this barrier as being like a rounded curb separating a flat street from a raised sidewalk. The curb will keep a slowly moving soccer ball in the street. But if the ball is kicked hard enough, it can roll up onto the sidewalk, with the work done against the gravitational attraction (the gain in gravitational potential energy) equal to its loss in kinetic energy.

The photoelectric effect was first observed in 1887 by Hertz, quite by accident. He noticed that a spark would jump more readily between two electrically charged spheres when their surfaces were illuminated by the light from another spark. Light shining on the surfaces somehow facilitated the escape of what we now know to be electrons. This idea in itself was not revolutionary. The existence of the surface potential-energy barrier was already known. In 1883, Thomas Edison had discovered *thermionic emission,* in which the escape energy is supplied by heating the material to a very high temperature, liberating electrons by a process analogous to boiling a liquid. The *minimum* amount of energy an individual electron has to gain to escape from a particular surface is called the **work function** for that surface, denoted by ϕ. However, the surfaces that Hertz used were not at the high temperatures needed for thermionic emission.

The photoelectric effect was investigated in detail by the German physicists Wilhelm Hallwachs and Philipp Lenard during the years 1886–1900; their results were quite unexpected. We will describe their work in terms of a more modern phototube (Fig. 40–2). Two conducting electrodes, the anode and the cathode, are enclosed in an evacuated glass tube. The battery or other source of potential difference creates an electric field in the direction from anode to cathode in Fig. 40–2a. Light (indicated by the green arrows) falling on the surface of the cathode causes a current in the external circuit; the current is measured by the galvanometer (G). Hallwachs and Lenard studied how this *photocurrent* varies with voltage and with the frequency and intensity of the light.

After the discovery of the electron in 1897, it became clear that the light causes electrons to be emitted from the cathode. Because of their negative charge $-e$, the emitted *photoelectrons* are then pushed toward the anode by the electric field. A high vacuum with residual pressure of 0.01 Pa (10^{-7} atm) or less is needed to minimize collisions of the electrons with gas molecules.

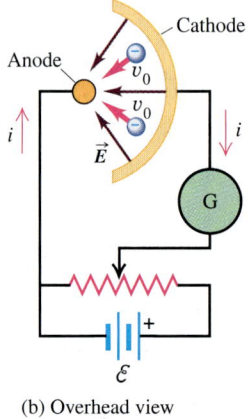

40–2 Schematic diagram of a phototube circuit. (a) Electrons emitted from the cathode are pushed toward the anode by the electric-field force. (b) Even when the direction of the field is reversed, some electrons still reach the anode. The stopping potential V_0 is the minimum absolute value of the reverse potential difference that gives zero current.

(a)

(b) Overhead view with $\vec{E}$ field reversed

Hallwachs and Lenard found that when monochromatic light fell on the cathode, *no* photoelectrons at all were emitted unless the frequency of the light was greater than some minimum value called the **threshold frequency.** This minimum frequency depends on the material of the cathode. For most metals the threshold frequency is in the ultraviolet (corresponding to wavelengths λ between 200 and 300 nm), but for potassium and cesium oxides it is in the visible spectrum (λ between 400 and 700 nm).

When the frequency f is *greater* than the threshold frequency, some electrons are emitted from the cathode with substantial initial speeds. This can be shown by reversing the polarity of the battery (Fig. 40–2b) so that the electric-field force on the electrons is back toward the cathode. If the magnitude of the field is not too great, the highest-energy emitted electrons still reach the anode and there is still a current. We can determine the *maximum* kinetic energy of the emitted electrons by making the potential of the anode relative to the cathode, V_{AC}, just negative enough so that the current stops. This occurs for $V_{AC} = -V_0$, where V_0 is called the **stopping potential.** As an electron moves from the cathode to the anode, the potential decreases by V_0 and negative work $-eV_0$ is done on the (negatively charged) electron; the most energetic electron leaves the cathode with kinetic energy $K_{max} = \frac{1}{2}mv_{max}^2$ and has zero kinetic energy at the anode. Using the work-energy theorem, we have

$$W_{tot} = -eV_0 = \Delta K = 0 - K_{max},$$

$$K_{max} = \frac{1}{2}mv_{max}^2 = eV_0 \qquad \text{(maximum kinetic energy of photoelectrons).} \quad (40\text{–}1)$$

Hence by measuring the stopping potential V_0, we can determine the maximum kinetic energy with which electrons leave the cathode. (We are ignoring any effects due to differences in the materials of the cathode and anode.)

Figure 40–3 shows graphs of photocurrent as a function of potential difference V_{AC} for light of constant frequency and two different intensities. When V_{AC} is sufficiently large and positive, the curves level off, showing that *all* the emitted electrons are being collected by the anode. The reverse potential difference $-V_0$ needed to reduce the current to zero is shown.

If the intensity of light is increased while its frequency is kept the same, the current levels off at a higher value, showing that more electrons are being emitted per time. But the stopping potential V_0 is found to be the same.

Figure 40–4 shows current as a function of potential difference for two different frequencies, with the same intensity in each case. We see that when the frequency of the incident monochromatic light is increased, the stopping potential V_0 increases. In fact, V_0 turns out to be a linear function of the frequency f.

These results are hard to understand on the basis of classical physics. When the intensity (average energy per unit area per unit time) increases, electrons should be able to gain more energy, increasing the stopping potential V_0. But V_0 was found *not* to depend on intensity. Also, classical physics offers no explanation for the threshold frequency. In Section 33–5 we found that the intensity of an electromagnetic wave such as light does not depend on frequency, so an electron should be able to acquire its needed escape energy from light of any frequency. Thus there should not be a threshold frequency f_0. Finally, we would expect it to take a while for an electron to collect enough energy from extremely faint light. But experiment also shows that electrons are emitted as soon as *any* light with $f \geq f_0$ hits the surface.

The correct analysis of the photoelectric effect was developed by Albert Einstein in 1905. Building on an assumption made five years earlier by Max Planck (Section 40–9), Einstein postulated that a beam of light consists of small packages of energy called **photons** or *quanta*. The energy E of a photon is equal to a constant h times its frequency f. From $f = c/\lambda$ for electromagnetic waves in vacuum we have

$$E = hf = \frac{hc}{\lambda} \quad \text{(energy of a photon)}, \quad (40\text{–}2)$$

where h is a universal constant called **Planck's constant.** The numerical value of this constant, to the accuracy known at present, is

$$h = 6.6260755(40) \times 10^{-34} \text{ J·s}.$$

A photon arriving at the surface is absorbed by an electron. This energy transfer is an all-or-nothing process, in contrast to the continuous transfer of energy in the classical theory; the electron gets all the photon's energy or none at all. If this energy is greater than the work function ϕ, the electron may escape from the surface. Greater intensity at a particular frequency means a proportionally greater number of photons per second absorbed, thus a proportionally greater number of electrons emitted per second and the proportionally greater current seen in Fig. 40–3.

Recall that ϕ is the *minimum* energy needed to remove an electron from the surface. Thus Einstein applied conservation of energy to find that the *maximum* kinetic energy $K_{max} = \frac{1}{2}mv_{max}^2$ for an emitted electron is the energy hf gained from a photon minus the work function ϕ:

$$K_{max} = \frac{1}{2}mv_{max}^2 = hf - \phi. \quad (40\text{–}3)$$

Substituting $K_{max} = eV_0$ from Eq. (40–1), we find

$$eV_0 = hf - \phi \quad \text{(photoelectric effect)}. \quad (40\text{–}4)$$

We can measure the stopping potential V_0 for each of several values of frequency f

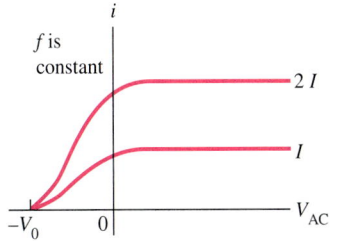

40–3 Photocurrent i as a function of the potential V_{AC} of the anode with respect to the cathode for a constant light frequency f. The stopping potential V_0 is independent of the light intensity I, but the photocurrent for large positive V_{AC} is directly proportional to the intensity.

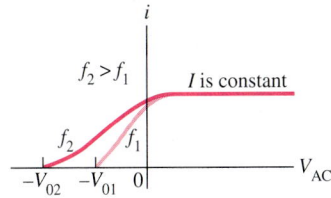

40–4 Photocurrent i as a function of the potential V_{AC} of an anode with respect to a cathode for two different light frequencies f_1 and f_2 with the same intensity. The stopping potential V_0 (and therefore the maximum kinetic energy of the photoelectrons) increases linearly with frequency.

for a given cathode material. A graph of V_0 as a function of f turns out to be a straight line, verifying Eq. (40–4), and from such a graph we can determine both the work function ϕ for the material and the value of the quantity h/e. (We do this graphical analysis in Example 40–3.) After the electron charge $-e$ was measured by Robert Millikan in 1909, Planck's constant h could also be determined from these measurements.

Electron energies and work functions are usually expressed in electron volts (eV), defined in Section 24–3. To four significant figures,

$$1 \text{ eV} = 1.602 \times 10^{-19} \text{ J}.$$

To this accuracy, Planck's constant is

$$h = 6.626 \times 10^{-34} \text{ J} \cdot \text{s} = 4.136 \times 10^{-15} \text{ eV} \cdot \text{s}.$$

Table 40–1 lists a few typical work functions of elements. These values are approximate because they are very sensitive to surface impurities. The greater the work function, the higher the minimum frequency needed to emit photoelectrons.

TABLE 40–1
WORK FUNCTIONS OF SEVERAL ELEMENTS

ELEMENT	WORK FUNCTION (eV)	ELEMENT	WORK FUNCTION (eV)
Aluminum	4.3	Nickel	5.1
Carbon	5.0	Silicon	4.8
Copper	4.7	Silver	4.3
Gold	5.1	Sodium	2.7

A night vision scope makes use of the photoelectric effect. Photons entering the scope strike a plate, producing photoelectrons. These electrons pass through a thin disk in which there are millions of tiny channels. The current through each channel is amplified electronically, then directed toward a screen that glows when hit by electrons. The image formed on the screen (shown below), which is a combination of these millions of glowing spots, is thousands of times brighter than the naked-eye view.

We have discussed photons mostly in the context of light. However, the quantization concept applies to *all* regions of the electromagnetic spectrum, including radio waves, x rays, and so on. A photon of any electromagnetic radiation with frequency f and wavelength λ has energy E given by Eq. (40–2). Furthermore, according to the special theory of relativity, every particle that has energy must also have momentum, even if it has no rest mass. Photons have zero rest mass. As we saw in Eq. (39–41), a photon with energy E has momentum with magnitude p given by $E = pc$. Thus the wavelength λ of a photon and the magnitude of its momentum p are related simply by

$$p = \frac{E}{c} = \frac{hf}{c} = \frac{h}{\lambda} \qquad \text{(momentum of a photon)}. \qquad (40\text{–}5)$$

The direction of the photon's momentum is simply the direction in which the electromagnetic wave is moving.

Problem–Solving Strategy

PHOTONS

1. The electron volt is an important unit. We used it in Chapter 39 and will use it even more in this and the next three chapters. Recall that one electron volt (1 eV) is the amount of kinetic energy gained by an electron in moving freely through an increase of potential of one volt. Thus $1 \text{ eV} = 1.602 \times 10^{-19} \text{ J}.$

2. Remember that photons are associated with electromagnetic waves. Their frequency f and wavelength λ are related by $f = c/\lambda$. The energy E of a photon can be expressed as hf or hc/λ, whichever is more convenient for the problem at hand. Be careful with units; use c in meters/second, λ in meters, and f in hertz or 1/seconds. With E in electron volts, use $h = 4.136 \times 10^{-15} \text{ eV} \cdot \text{s}$, and with E in joules, use $h = 6.626 \times 10^{-34} \text{ J} \cdot \text{s}$.

3. The magnitudes are in such unfamiliar ranges that common sense may not help if your calculation is wrong by a factor of 10^{19}. It helps to remember that a visible-light photon with $\lambda = 600 \text{ nm}$ and $f = 5 \times 10^{14} \text{ Hz}$ has an energy E of about 2 eV, or about $3 \times 10^{-19} \text{ J}$.

EXAMPLE 40–1

Photon energy and wavelength Silicon films become better electrical conductors when illuminated by photons with energies of 1.14 eV or greater. (This behavior is called *photoconductivity.*) What is the corresponding wavelength?

SOLUTION Using Eq. (40–2) as $E = hc/\lambda$, we find

$$\lambda = \frac{hc}{E} = \frac{(4.136 \times 10^{-15} \text{ eV} \cdot \text{s})(3.00 \times 10^{8} \text{ m/s})}{1.14 \text{ eV}}$$

$$= 1.09 \times 10^{-6} \text{ m} = 1090 \text{ nm}.$$

This is in the infrared region of the spectrum. The *minimum* energy of 1.14 eV corresponds to the *maximum* wavelength that causes photoconductivity in silicon. Thus photons of visible light, which have higher energy and shorter wavelength, also cause this photoconductivity.

EXAMPLE 40–2

A photoelectric experiment While conducting a photoelectric effect experiment with light of a certain frequency, you find that a reverse potential difference of 1.25 V is required to reduce the current to zero. Find a) the maximum kinetic energy; b) the maximum speed of the emitted photoelectrons.

SOLUTION a) From Eq. (40–1),

$$K_{max} = eV_0 = (1.60 \times 10^{-19} \text{ C})(1.25 \text{ V}) = 2.00 \times 10^{-19} \text{ J}.$$

To check unit consistency, recall that 1 V = 1 J/C. In terms of electron volts,

$$K_{max} = eV_0 = e(1.25 \text{ V}) = 1.25 \text{ eV},$$

since the electron volt (eV) is the magnitude of the electron charge e times one volt (1 V).

b) From $K_{max} = \frac{1}{2}mv_{max}^2$ we get

$$v_{max} = \sqrt{\frac{2K_{max}}{m}} = \sqrt{\frac{2(2.00 \times 10^{-19} \text{ J})}{9.11 \times 10^{-31} \text{ kg}}}$$

$$= 6.63 \times 10^{5} \text{ m/s}.$$

This speed is about 1/500 the speed of light c, so we are justified in using the nonrelativistic expression for kinetic energy. An equivalent justification is that the electron's 1.25-eV kinetic energy is much smaller than its rest energy $mc^2 = 0.511$ MeV.

EXAMPLE 40–3

Determining ϕ and h experimentally For a certain cathode material in a photoelectric-effect experiment, you measure a stopping potential of 1.0 V for light of wavelength 600 nm, 2.0 V for 400 nm, and 3.0 V for 300 nm. Determine the work function for this material and the value of Planck's constant.

SOLUTION According to Eq. (40–4), a graph of V_0 as a function of f should be a straight line. We rewrite that equation as

$$V_0 = \frac{h}{e} f - \frac{\phi}{e}.$$

In this form we see that the slope of the line is h/e and the intercept on the vertical axis (corresponding to $f = 0$) is at $-\phi/e$. The frequencies, obtained from $f = c/\lambda$ and $c = 3.00 \times 10^{8}$ m/s, are 0.50×10^{15} Hz, 0.75×10^{15} Hz, and 1.0×10^{15} Hz, respectively. The graph is shown in Fig. 40–5. From it we find

$$-\frac{\phi}{e} = \text{vertical intercept} = -1.0 \text{ V},$$

$$\phi = 1.0 \text{ eV} = 1.6 \times 10^{-19} \text{ J},$$

and

$$\text{Slope} = \frac{\Delta V_0}{\Delta f} = \frac{3.0 \text{ V} - (-1.0 \text{ V})}{1.00 \times 10^{15} \text{ s}^{-1} - 0} = 4.0 \times 10^{-15} \text{ J} \cdot \text{s/C},$$

$$h = \text{slope} \times e = (4.0 \times 10^{-15} \text{ J} \cdot \text{s/C})(1.60 \times 10^{-19} \text{ C})$$

$$= 6.4 \times 10^{-34} \text{ J} \cdot \text{s}.$$

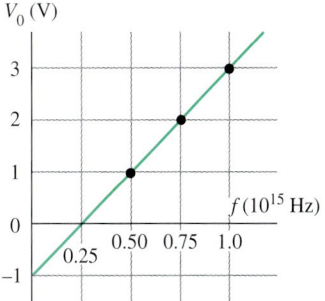

40–5 Stopping potential as a function of frequency. For a different cathode material having a different work function, the line would be displaced up or down but would have the same slope, equal to h/e within experimental error.

This experimental value differs by about 3% from the accepted value. The small value of ϕ tells us that the cathode surface is not composed solely of one of the elements in Table 40–1.

EXAMPLE 40-4

FM radio photons The radio station WQED in Pittsburgh broadcasts at 89.3 MHz with a radiated power of 43.0 kW. a) What is the magnitude of the momentum of each photon? b) How many photons does it emit each second?

SOLUTION a) From Eq. (40–2) the energy of each photon emitted is

$$E = hf = (6.626 \times 10^{-34} \text{ J} \cdot \text{s})(89.3 \times 10^6 \text{ /s})$$
$$= 5.92 \times 10^{-26} \text{ J}.$$

Thus each photon has a momentum of magnitude

$$p = \frac{E}{c} = \frac{5.92 \times 10^{-26} \text{ J}}{3.00 \times 10^8 \text{ m/s}} = 1.97 \times 10^{-34} \text{ kg} \cdot \text{m/s}.$$

This very small value is about the magnitude of the momentum an electron would have if it crawled along at a rate of one meter every hour. b) The station sends out 43.0×10^3 joules each second. The rate at which photons are emitted is therefore

$$\frac{43.0 \times 10^3 \text{ J/s}}{5.92 \times 10^{-26} \text{ J/photon}} = 7.26 \times 10^{29} \text{ photons/s}.$$

With this huge number of photons leaving the station each second, the discreteness of the tiny individual packages of energy isn't noticed; the radiated energy appears to be a continuous flow.

40-4 ATOMIC LINE SPECTRA AND ENERGY LEVELS

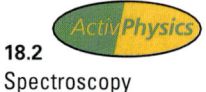

18.2
Spectroscopy

The origin of *line spectra,* described in Section 40–2, can be understood in general terms on the basis of two central ideas. One is the photon concept; the other is the concept of *energy levels* of atoms. These two ideas were combined by the Danish physicist Niels Bohr in 1913.

Bohr's hypothesis represented a bold breakaway from nineteenth century ideas. His reasoning went like this. The line spectrum of an element results from the emission of photons with specific energies from the atoms of that element. During the emission of a photon, the internal energy of the atom changes by an amount equal to the energy of the photon. Therefore, said Bohr, each atom must be able to exist only with certain specific values of internal energy. Each atom has a set of possible **energy levels.** An atom can have an amount of internal energy equal to any one of these levels, but it *cannot* have an energy *intermediate* between two levels. All isolated atoms of a given element have the same set of energy levels, but atoms of different elements have different sets. In electric discharge tubes atoms are raised, or *excited,* to higher energy levels mainly through inelastic collisions with electrons.

According to Bohr, an atom can make a *transition* from one energy level to a lower level by emitting a photon with energy equal to the energy *difference* between the initial and final levels (Fig. 40–6a). If E_i is the initial energy of the atom before such a transition, E_f is its final energy after the transition, and the photon's energy is $hf = hc/\lambda$, then conservation of energy gives

(a)

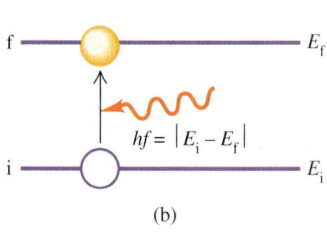

(b)

40-6 (a) The atom drops from an initial level i to a lower-energy final level f by emitting a photon with energy equal to $E_i - E_f$. (b) The atom is raised from initial level i to a higher-energy final level f by absorbing a photon with energy equal to $E_f - E_i$.

$$hf = \frac{hc}{\lambda} = E_i - E_f \qquad \text{(energy of emitted photon).} \qquad (40–6)$$

For example, when a krypton atom emits a photon of orange light with wavelength $\lambda = 606$ nm, the corresponding photon energy is

$$E = \frac{hc}{\lambda} = \frac{(6.63 \times 10^{-34} \text{ J} \cdot \text{s})(3.00 \times 10^8 \text{ m/s})}{606 \times 10^{-9} \text{ m}}$$
$$= 3.28 \times 10^{-19} \text{ J} = 2.05 \text{ eV}.$$

This photon is emitted during a transition like that in Fig. 40–6a between two levels of the atom that differ in energy by 2.05 eV.

THE HYDROGEN SPECTRUM

By 1913 the spectrum of hydrogen, the simplest and least massive atom, had been studied intensively. In an electric discharge tube, atomic hydrogen emits the series of lines

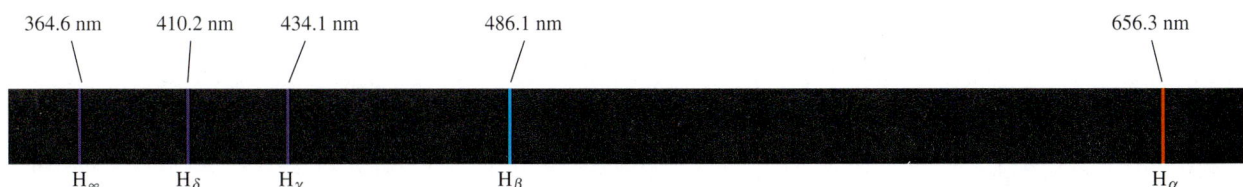

364.6 nm   410.2 nm   434.1 nm   486.1 nm   656.3 nm

H$_\infty$ H$_\delta$ H$_\gamma$ H$_\beta$ H$_\alpha$

40–7 The Balmer series of spectrum lines for atomic hydrogen. The H$_\alpha$, H$_\beta$, H$_\gamma$, and H$_\delta$ lines are in the visible portion of the spectrum; all other Balmer-series lines are in the ultraviolet.

shown in Fig. 40–7. The visible line with longest wavelength, or lowest frequency, is in the red and is called H$_\alpha$; the next line, in the blue-green, is called H$_\beta$; and so on. In 1885 the Swiss teacher Johann Balmer (1825–1898) found (by trial and error) a formula that gives the wavelengths of these lines, which are now called the *Balmer series*. We may write Balmer's formula as

$$\frac{1}{\lambda} = R\left(\frac{1}{2^2} - \frac{1}{n^2}\right), \tag{40–7}$$

where λ is the wavelength, R is a constant called the **Rydberg constant** (chosen to make Eq. (40–7) fit the measured wavelengths), and n may have the integer values 3, 4, 5, $\cdots$. If λ is in meters, the numerical value of R is

$$R = 1.097 \times 10^7 \text{ m}^{-1}.$$

Letting $n = 3$ in Eq. (40–7), we obtain the wavelength of the H$_\alpha$ line:

$$\frac{1}{\lambda} = (1.097 \times 10^7 \text{ m}^{-1})\left(\frac{1}{4} - \frac{1}{9}\right), \quad \text{or} \quad \lambda = 656.3 \text{ nm}.$$

For $n = 4$ we obtain the wavelength of the H$_\beta$ line, and so on. For $n = \infty$ we obtain the *smallest* wavelength in the series, $\lambda = 364.6$ nm.

Balmer's formula has a very direct relation to Bohr's hypothesis about energy levels. Using the relation $E = hc/\lambda$, we can find the *photon energies* corresponding to the wavelengths of the Balmer series. Multiplying Eq. (40–7) by hc, we find

$$E = \frac{hc}{\lambda} = hcR\left(\frac{1}{2^2} - \frac{1}{n^2}\right) = \frac{hcR}{2^2} - \frac{hcR}{n^2}. \tag{40–8}$$

Equations (40–6) and (40–8) for the photon's energy agree most simply if we identify $-hcR/n^2$ as the initial energy E_i of the atom and $-hcR/2^2$ as its final energy E_f, in a transition in which a photon with energy $E_i - E_f$ is emitted. The energies of the levels are negative because we choose the potential energy to be zero when the electron and nucleus are infinitely far apart. The Balmer (and other) series suggest that the hydrogen atom has a series of energy levels, which we may call E_n, given by

$$E_n = -\frac{hcR}{n^2}, \quad n = 1, 2, 3, 4, \cdots \quad \text{(energy levels of the hydrogen atom).} \tag{40–9}$$

ActivPhysics
18.1
The Bohr Model

Each wavelength in the Balmer series corresponds to a transition from a level with n equal to 3 or greater to the level with $n = 2$.

The numerical value of the product hcR is

$$hcR = (6.626 \times 10^{-34} \text{ J} \cdot \text{s})(2.998 \times 10^8 \text{ m/s})(1.097 \times 10^7 \text{ m}^{-1})$$

$$= 2.179 \times 10^{-18} \text{ J} = 13.60 \text{ eV}.$$

Thus the magnitudes of the energy levels given by Eq. (40–9) are −13.60 eV, −3.40 eV, −1.51 eV, −0.85 eV, $\cdots$.

Other spectral series for hydrogen have been discovered. These are known, after their discoverers, as the Lyman, Paschen, Brackett, and Pfund series. Their wavelengths can be represented by formulas similar to Balmer's formula:

Lyman series

$$\frac{1}{\lambda} = R\left(\frac{1}{1^2} - \frac{1}{n^2}\right), \qquad n = 2, \ 3, \ 4, \cdots;$$

Paschen series

$$\frac{1}{\lambda} = R\left(\frac{1}{3^2} - \frac{1}{n^2}\right), \qquad n = 4, \ 5, \ 6, \cdots;$$

Brackett series

$$\frac{1}{\lambda} = R\left(\frac{1}{4^2} - \frac{1}{n^2}\right), \qquad n = 5, \ 6, \ 7, \cdots;$$

Pfund series

$$\frac{1}{\lambda} = R\left(\frac{1}{5^2} - \frac{1}{n^2}\right), \qquad n = 6, \ 7, \ 8, \cdots.$$

The Lyman series is in the ultraviolet, and the Paschen, Brackett, and Pfund series are in the infrared. We see that the Balmer series fits into the scheme between the Lyman and Paschen series.

40–8 (a) "Permitted" orbits of an electron in the Bohr model of a hydrogen atom (not to scale). The transitions that are responsible for some of the lines of the various series are indicated by arrows. (b) Energy-level diagram, showing some transitions corresponding to the various series.

Thus *all* the spectral series of hydrogen can be understood on the basis of Bohr's picture of transitions from one energy level (and corresponding electron orbit) to another, with the energy levels given by Eq. (40–9) with $n = 1, 2, 3, \cdots$. For the Lyman series the final level is always $n = 1$; for the Paschen series it is $n = 3$; and so on. The relation of the various spectral series to the energy levels and to electron orbits is shown in Fig. 40–8. Taken together, these spectral series give very strong support to Bohr's picture of energy levels in atoms.

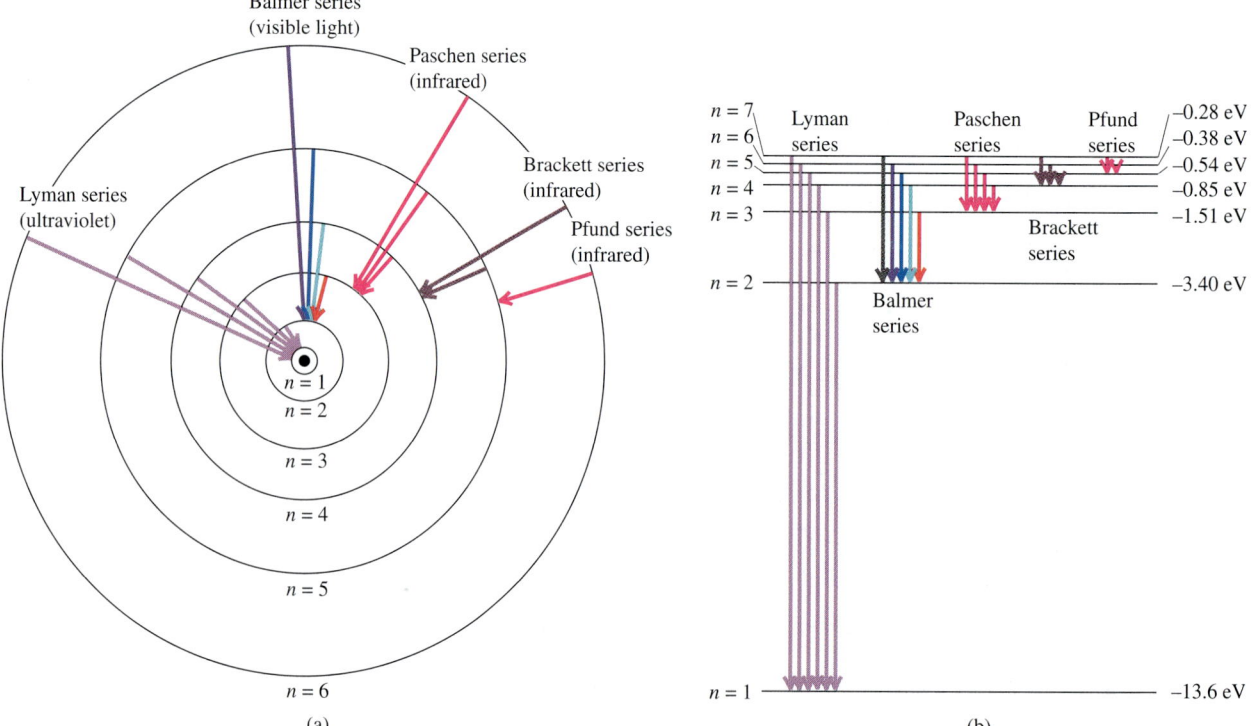

(a)   (b)

CAUTION ▶ The lines of a spectrum, such as the hydrogen spectrum shown in Fig. 40–7, are *not* all produced by a single atom. The sample of hydrogen gas that produced the spectrum in Fig. 40–7 contained a large number of atoms; these were excited in an electric discharge tube to various energy levels. The spectrum of the gas shows the light emitted from all the different transitions that occurred in different atoms of the sample. ◀

We haven't yet discussed any way to *predict* what the energy levels for a particular atom should be. Neither have we shown how to derive Eq. (40–9) from fundamental theory or to relate the Rydberg constant R to other fundamental constants. We'll return to these problems later.

THE FRANCK-HERTZ EXPERIMENT

In 1914, James Franck and Gustav Hertz found even more direct experimental evidence for the existence of atomic energy levels. Franck and Hertz studied the motion of electrons through mercury vapor under the action of an electric field. They found that when the electron kinetic energy was 4.9 eV or greater, the vapor emitted ultraviolet light of wavelength 0.25 μm. Suppose mercury atoms have an energy level 4.9 eV above the lowest energy level. An atom can be raised to this level by collision with an electron; it later decays back to the lowest energy level by emitting a photon. According to Eq. (40–2), the wavelength of the photon should be

$$\lambda = \frac{hc}{E} = \frac{(4.136 \times 10^{-15} \text{ eV} \cdot \text{s})(3.00 \times 10^8 \text{ m/s})}{4.9 \text{ eV}} = 2.5 \times 10^{-7} \text{ m} = 0.25 \ \mu\text{m}.$$

This is equal to the measured wavelength, confirming the existence of this energy level of the mercury atom. Similar experiments with other atoms yield the same kind of evidence for atomic energy levels.

ENERGY LEVELS

Only a few atoms and ions (such as hydrogen, singly ionized helium, doubly ionized lithium) have spectra whose wavelengths can be represented by a simple formula such as Balmer's. But it is *always* possible to analyze the more complicated spectra of other elements in terms of transitions among various energy levels and to deduce the numerical values of these levels from the measured spectrum wavelengths.

Every atom has a lowest energy level that includes the *minimum* internal energy state that the atom can have. This is called the *ground-state level,* or **ground level,** and all higher levels are called **excited levels.** A photon corresponding to a particular spectrum line is emitted when an atom makes a transition from a state in an excited level to a state in a lower excited level or the ground level.

Some energy levels for sodium are shown in Fig. 40–9 with energies relative to the ground level. You may have noticed the yellow-orange light emitted by sodium vapor street lights. Sodium atoms emit this characteristic yellow-orange light with wavelengths 589.0 and 589.6 nm when they make transitions from the two closely spaced levels labeled *lowest excited levels* to the ground level.

A sodium atom in the ground level can also *absorb* a photon with wavelength 589.0 or 589.6 nm. To demonstrate this process, we pass a beam of light from a sodium-vapor lamp through a bulb containing sodium vapor. The atoms in the vapor absorb the 589.0-nm or 589.6-nm photons from the beam, reaching the lowest excited levels; after a short time they return to the ground level, emitting photons in all directions and causing the sodium vapor to glow with the characteristic yellow light. The average time spent in an excited level is called the *lifetime* of the level; for the lowest excited levels of the sodium atom, the lifetime is about 1.6×10^{-8} s.

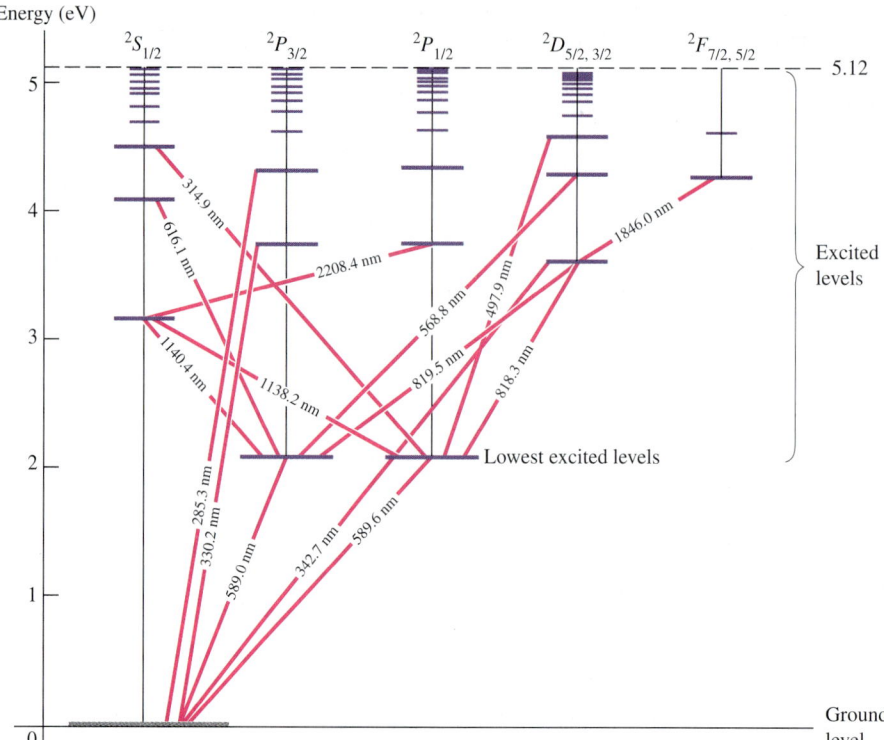

40–9 Energy levels of the sodium atom relative to the ground level. Numbers on the lines between levels are wavelengths. The column labels, such as $^2S_{1/2}$, refer to some quantum states of the valence electron (to be discussed in Chapter 43).

More generally, a photon *emitted* when an atom makes a transition from an excited level to a lower level can also be *absorbed* by a similar atom that is initially in the lower level (Fig. 40–6b, page 1236). If we pass white (continuous-spectrum) light through a gas and look at the *transmitted* light with a spectrometer, we find a series of dark lines corresponding to the wavelengths that have been absorbed, as shown in Fig. 40–10. This is called an **absorption spectrum.**

A related phenomenon is *fluorescence.* An atom absorbs a photon (often in the ultra-violet region) to reach an excited level and then drops back to the ground level in steps, emitting two or more photons with smaller energy and longer wavelength. For example, the electric discharge in a fluorescent tube causes the mercury vapor in the tube to emit ultraviolet radiation. This radiation is absorbed by the atoms of the coating on the inside of the tube. The coating atoms then re-emit light in the longer-wavelength visible portion of the spectrum. Fluorescent lamps are more efficient than incandescent lamps in converting electrical energy to visible light because they do not waste as much energy producing (invisible) infrared photons.

The Bohr hypothesis established the relation of wavelengths to energy levels, but it provided no general principles for *predicting* the energy levels of a particular atom. Bohr provided a partial analysis for the hydrogen atom; we will discuss this in Section 40–6.

40–10 The absorption spectrum of the sun. The darker lines are absorption lines from the relatively cool atmosphere surrounding the sun's glowing surface. The particular dark lines seen in the spectrum tell astronomers what kinds of atoms are present in the solar atmosphere.

A more general understanding of atomic structure and energy levels rests on the concepts of *quantum mechanics,* which we will introduce in Chapters 41 and 42. Quantum mechanics provides all the principles needed to calculate energy levels from fundamental theory. Unfortunately, for many-electron atoms the calculations are so complex that they can be carried out only approximately.

EXAMPLE 40-5

Emission and absorption spectra A hypothetical atom has three energy levels: the ground level and 1.00 eV and 3.00 eV above the ground level. a) Find the frequencies and wavelengths of the spectrum lines that this atom can emit when excited. b) What wavelengths can be absorbed by this atom if it is initially in its ground level?

SOLUTION a) Figure 40–11a shows an energy-level diagram. The possible photon energies, corresponding to the transitions shown, are 1.00 eV, 2.00 eV, and 3.00 eV. For 1.00 eV we have, from Eq. (40–2),

$$f = \frac{E}{h} = \frac{1.00 \text{ eV}}{4.136 \times 10^{-15} \text{ eV} \cdot \text{s}} = 2.42 \times 10^{14} \text{ Hz}.$$

For 2.00 eV and 3.00 eV, $f = 4.84 \times 10^{14}$ Hz and 7.25×10^{14} Hz, respectively. We can find the wavelengths using $\lambda = c/f$. For 1.00 eV,

$$\lambda = \frac{c}{f} = \frac{3.00 \times 10^8 \text{ m/s}}{2.42 \times 10^{14} \text{ Hz}} = 1.24 \times 10^{-6} \text{ m} = 1240 \text{ nm},$$

in the infrared region of the spectrum. For 2.00 eV and 3.00 eV the wavelengths are 620 nm (red) and 414 nm (violet), respectively (Fig. 40–11b).
b) From the atom's ground level, only a 1.00-eV or 3.00-eV photon can be absorbed; a 2.00-eV photon cannot be because there is no energy level 2.00 eV above the ground level. As we have calculated, the wavelengths of 1.00-eV and 3.00-eV photons are 1240 nm and 414 nm, respectively. Passing light from a hot solid through a cool gas of these atoms would result in a continuous spectrum with dark absorption lines at 1240 nm and 414 nm.

40–11 (a) Energy-level diagram, showing the possible transitions for emission from excited levels and for absorption from the ground level. (b) Emission spectrum of this hypothetical atom. The absorption spectrum from the ground level would not contain the 620-nm line, corresponding to a 2.00-eV photon.

If a gas of these atoms were at a high enough temperature, collisions would excite a number of atoms into the 1.00-eV energy level. Such excited atoms *can* absorb 2.00-eV photons, and an absorption line at 620 nm would appear in the spectrum. Thus the observed spectrum of a given substance depends on its energy levels and its temperature.

40-5 THE NUCLEAR ATOM

Before we can make further progress in relating the energy levels of an atom to its internal structure, we need to have a better idea of what the inside of an atom is like. We know that atoms are much smaller than the wavelengths of visible light, so there is no hope of actually *seeing* an atom using that light. But we can still describe how the mass and electric charge are distributed throughout the volume of the atom.

Here's where things stood in 1910. J. J. Thomson had discovered the electron and measured its charge-to-mass ratio (e/m) in 1897; and by 1909, Millikan had completed his first measurements of the electron charge $-e$. These and other experiments showed that almost all the mass of an atom had to be associated with the *positive* charge, not with the electrons. It was also known that the overall size of atoms is of the order of 10^{-10} m and that all atoms except hydrogen contain more than one electron. What was *not* known was how the mass and charge were distributed within the atom. Thomson had proposed a model in which the atom consisted of a sphere of positive charge, of the order of

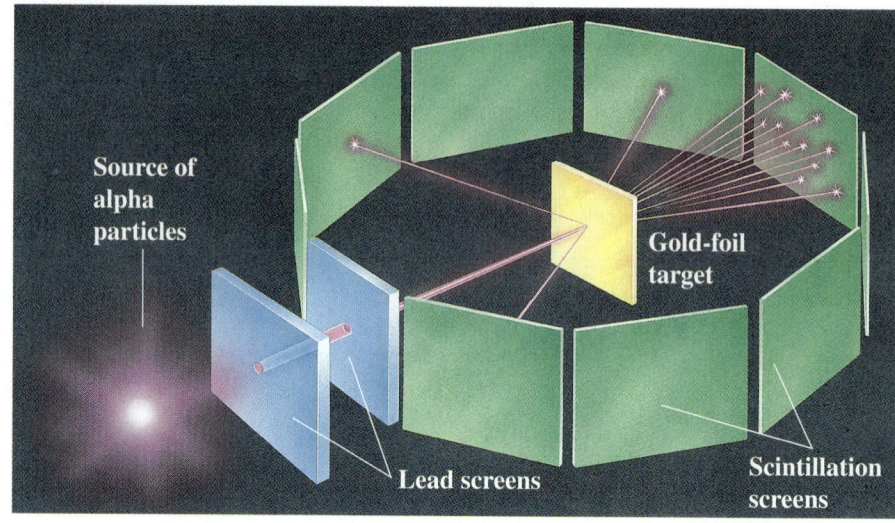

40–12 The scattering of alpha particles by a thin metal foil. The source of alpha particles is a radioactive element such as radium. The two lead screens with small holes form a narrow beam of alpha particles, which are scattered by the gold foil. The directions of the scattered particles are determined from the flashes on the scintillation screens.

ActivPhysics

19.1
Particle Scattering

Born in New Zealand, Ernest Rutherford (1871–1937) spent his professional life in England and Canada. Before carrying out the experiments that established the existence of atomic nuclei, he shared (with Frederick Soddy) the 1908 Nobel Prize in chemistry for showing that radioactivity results from the disintegration of atoms.

10^{-10} m in diameter, with the electrons embedded in it like raisins in a more or less spherical muffin.

The first experiments designed to probe the interior structure of the atom were the **Rutherford scattering experiments,** carried out in 1910–1911 by Ernest Rutherford and two of his students, Hans Geiger and Ernest Marsden, at the University of Manchester in England. These experiments consisted of projecting charged particles at thin foils of the elements under study and observing the deflections of the particles. The particle accelerators that are now in common use in laboratories had not yet been invented, and Rutherford's projectiles were *alpha particles* emitted from naturally radioactive elements. Alpha particles are identical with the nuclei of most helium atoms: two protons and two neutrons bound together. They are ejected from unstable nuclei with speeds of the order of 10^7 m/s, and they can travel several centimeters through air or 0.1 mm or so through solid matter before they are brought to rest by collisions.

Rutherford's experimental setup is shown schematically in Fig. 40–12. A radioactive substance at the left emits alpha particles. Thick lead screens stop all particles except those in a narrow beam defined by small holes. The beam then passes through a target consisting of a thin gold, silver, or copper foil and strikes screens coated with zinc sulfide, similar in principle to the screen of a TV picture tube. A momentary flash, or *scintillation,* can be seen on the screen whenever it is struck by an alpha particle. Rutherford and his students counted the numbers of particles deflected through various angles.

Think of the atoms of the target material as being packed together like marbles in a box. An alpha particle can pass through a thin sheet of metal foil, so the alpha particle must be able actually to pass through the interiors of atoms. The *total* electric charge of the atom is zero, so outside the atom there is little force on the alpha particle. Within the atom there are electrical forces caused by the electrons and by the positive charge. But the mass of an alpha particle is about 7300 times that of an electron. Momentum considerations show that the alpha particle can be scattered only a very small amount by its interaction with the much lighter electrons. It's like throwing a pebble through a swarm of mosquitoes; the mosquitoes don't deflect the pebble very much. Only interactions with the *positive* charge, which is tied to almost all of the mass of the atom, can deflect the alpha particle appreciably.

In the Thomson model, the positive charge and the negative electrons are distributed through the whole atom. Hence the electric field inside the atom should be quite small, and the electric force on an alpha particle that enters the atom should be quite weak. The maximum deflection to be expected is then only a few degrees (Fig. 40–13a).

The results were very different from this and were totally unexpected. Some alpha particles were scattered by nearly 180°, that is, almost straight backward (Fig. 40–13b). Rutherford later wrote:

> It was quite the most incredible event that ever happened to me in my life. It was almost as incredible as if you had fired a 15-inch shell at a piece of tissue paper and it came back and hit you.

Back to the drawing board! Suppose the positive charge, instead of being distributed through a sphere with atomic dimensions (of the order of 10^{-10} m), is all concentrated in a much *smaller* space. Then it would act like a point charge down to much smaller distances. The maximum electric field repelling the alpha particle would be much larger, and the amazing large-angle scattering would be possible. Rutherford called this concentration of positive charge the **nucleus.** He again computed the numbers of particles expected to be scattered through various angles. Within the accuracy of his experiments, the computed and measured results agreed, down to distances of the order of 10^{-14} m. His experiments therefore established that the atom does have a nucleus, a very small, very dense structure, no larger than 10^{-14} m in diameter. The nucleus occupies only about 10^{-12} of the total volume of the atom or less, but it contains *all* the positive charge and at least 99.95% of the total mass of the atom.

Figure 40–14 shows a computer simulation of the scattering of 5.0-MeV alpha particles from a gold nucleus of radius 7.0×10^{-15} m (the actual value) and from a nucleus with a hypothetical radius ten times this great. In the second case there is *no* large-angle scattering. The presence of large-angle scattering in Rutherford's experiments thus attested to the small size of the nucleus.

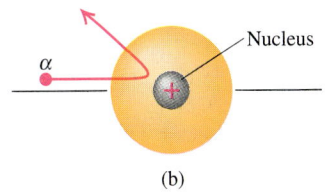

40–13 (a) Thomson's model of the atom: An alpha particle is scattered through only a small angle. (b) Rutherford's model of the atom: An alpha particle can be scattered through a large angle by the dense, positively charged nucleus (not drawn to scale).

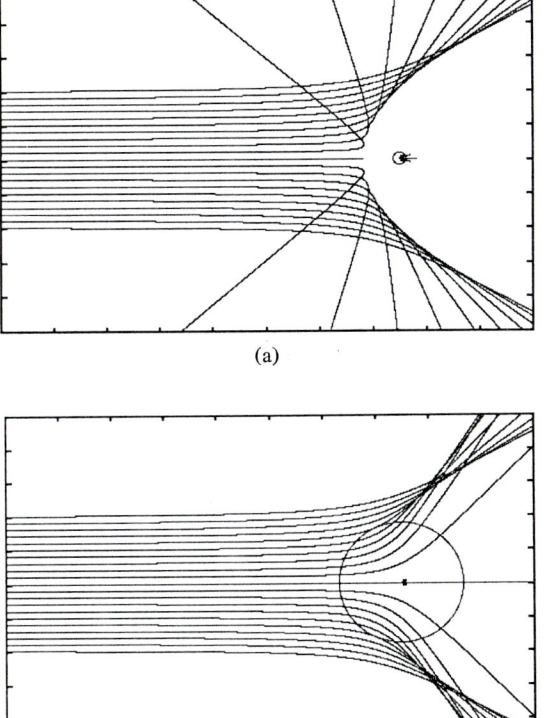

(a)

(b)

40–14 Computer simulation of scattering of 5.0-MeV alpha particles from a gold nucleus. Each curve shows a possible alpha-particle trajectory. (a) A gold nucleus with radius 7.0×10^{-15} m (its actual size); (b) a nucleus with ten times the radius of that in (a) shows *no* large-angle scattering.

EXAMPLE 40–6

A Rutherford experiment An alpha particle is aimed directly at a gold nucleus. An alpha particle has two protons and a charge of $2e = 2(1.60 \times 10^{-19}$ C), while a gold nucleus has 79 protons and a charge of $79e = 79(1.60 \times 10^{-19}$ C) . What minimum initial kinetic energy must the alpha particle have in order to approach within 5.0×10^{-14} m of the center of the gold nucleus? Assume that the gold nucleus, which has about 50 times the rest mass of an alpha particle, remains at rest. Thus at the point of closest approach, all of the alpha particle's initial kinetic energy is transformed to electric potential energy.

SOLUTION We first find the electric potential energy of the system when the alpha particle is 5.0×10^{-14} m from the center of the gold nucleus. From Eq. (24–9),

$$U = \frac{1}{4\pi\epsilon_0} \frac{qq_0}{r}$$

$$= (9.0 \times 10^9 \text{ N} \cdot \text{m}^2/\text{C}^2) \frac{(2)(79)(1.60 \times 10^{-19} \text{ C})^2}{5.0 \times 10^{-14} \text{ m}}$$

$$= 7.3 \times 10^{-13} \text{ J} = 4.6 \times 10^6 \text{ eV} = 4.6 \text{ MeV}.$$

If the alpha particle is to approach closer than 5.0×10^{-14} m from the center of the gold nucleus before stopping, it must have more than 4.6 MeV of kinetic energy when it is far away from the nucleus. In fact, alpha particles emitted from naturally occurring radioactive elements typically have energies in the range 4 to 6 MeV. For example, the common isotope of radium, ^{226}Ra, emits an alpha particle with energy 4.78 MeV.

40–6 THE BOHR MODEL

At the same time (1913) that Bohr established the relationship between spectrum wavelengths and energy levels, he also proposed a model of the hydrogen atom. He developed his ideas while working in Rutherford's laboratory. Using this model, now known as the **Bohr model,** he was able to *calculate* the energy levels of hydrogen and obtain agreement with values determined from spectra.

Rutherford's discovery of the atomic nucleus raised a serious question. What kept the negatively charged electrons at relatively large distances ($\sim 10^{-10}$ m) away from the very small ($\sim 10^{-14}$ m), positively charged nucleus despite their electrostatic attraction? Rutherford suggested that perhaps the electrons *revolve* in orbits about the nucleus, just as the planets revolve around the sun.

But according to classical electromagnetic theory, any accelerating electric charge (either oscillating or revolving) radiates electromagnetic waves. An example is the oscillating electric dipole in Section 33–9. The energy of an orbiting electron should therefore decrease continuously, its orbit should become smaller and smaller, and it should spiral rapidly into the nucleus (Fig. 40–15). Even worse, according to classical theory the *frequency* of the electromagnetic waves emitted should equal the frequency of revolution. As the electrons radiated energy, their angular speeds would change continuously, and they would emit a *continuous* spectrum (a mixture of all frequencies), not the *line* spectrum actually observed.

40–15 Classical physics predicts that the orbiting electron should continuously radiate electromagnetic waves and spiral into the nucleus. The electron's angular speed varies with its radius, so classical physics predicts that the frequency of the emitted radiation should change continuously.

18.1
The Bohr Model

STABLE ORBITS

To solve this problem, Bohr made a revolutionary proposal. He postulated that an electron in an atom can move around the nucleus in certain circular *stable orbits, without emitting radiation,* contrary to the predictions of classical electromagnetic theory. According to Bohr, there is a definite energy associated with each stable orbit, and an atom radiates energy only when it makes a transition from one of these orbits to another. The energy is radiated in the form of a photon with energy and frequency given by Eq. (40–6), $hf = E_i - E_f$.

As a result of a rather complicated argument that related the angular frequency of the light emitted to the angular speeds of the electron in highly excited energy levels, Bohr found that the magnitude of the electron's angular momentum is *quantized,* that this magnitude for the electron must be an integral multiple of $h/2\pi$. (Note that because $1 \text{ J} = 1 \text{ kg} \cdot \text{m}^2/\text{s}^2$, the SI units of Planck's constant h, J $\cdot$ s, are the same as the SI units of angular momentum, usually written as kg $\cdot$ m^2/s.) From Section 10–6, Eq. (10–28), the magnitude of the angular momentum is $L = mvr$ for a particle with mass m moving with speed v in a circle of radius r (thus with $\phi = 90°$). So Bohr's argument led to

$$mvr = n\frac{h}{2\pi},$$

where $n = 1, 2, 3, \cdots$. Each value of n corresponds to a permitted value of the orbit radius, which we denote from now on by r_n, and a corresponding speed v_n. The value of n for each orbit is called the **principal quantum number** for the orbit. With this notation the above equation becomes

$$mv_n r_n = n\frac{h}{2\pi} \qquad \text{(quantization of angular momentum.)} \qquad (40\text{–}10)$$

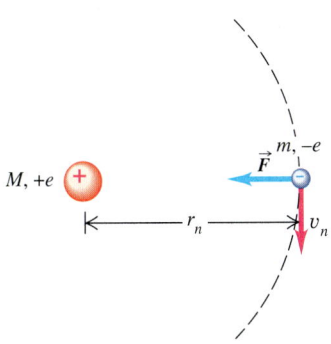

40–16 Bohr model of the hydrogen atom. The proton is assumed to be stationary; the electron revolves in a circle of radius r_n with speed v_n. The electrostatic attraction provides the necessary centripetal acceleration.

Now let's consider a model of the hydrogen atom (Fig. 40–16) that is Newtonian in spirit but incorporates this quantization assumption. This atom consists of a single electron with mass m and charge $-e$, revolving around a single proton with charge $+e$. The proton is nearly 2000 times as massive as the electron, so we have been assuming that the proton does not move. We learned in Section 5–5 that when a particle with mass m moves with speed v_n in a circular orbit with radius r_n, its radially inward acceleration is v_n^2/r_n. According to Newton's second law, a radially inward net force with magnitude $F = mv_n^2/r_n$ is needed to cause this acceleration. We discussed in Section 12–5 how the gravitational attraction provided that force for satellite orbits. In hydrogen the force F is provided by the electrical attraction between the positive proton and the negative electron. From Coulomb's law (Eq. 22–2),

$$F = \frac{1}{4\pi\epsilon_0}\frac{e^2}{r_n^2},$$

so Newton's second law states that

$$\frac{1}{4\pi\epsilon_0}\frac{e^2}{r_n^2} = \frac{mv_n^2}{r_n}. \qquad (40\text{–}11)$$

When we solve Eqs. (40–10) and (40–11) simultaneously for r_n and v_n, we get

$$r_n = \epsilon_0\frac{n^2 h^2}{\pi m e^2} \qquad \text{(orbit radii in the Bohr model),} \qquad (40\text{–}12)$$

$$v_n = \frac{1}{\epsilon_0}\frac{e^2}{2nh} \qquad \text{(orbital speeds in the Bohr model).} \qquad (40\text{–}13)$$

Equation (40–12) shows that the orbit radius r_n is proportional to n^2; the smallest orbit radius corresponds to $n = 1$. We'll denote this minimum radius, called the *Bohr radius*, as a_0:

$$a_0 = \epsilon_0\frac{h^2}{\pi m e^2}. \qquad (40\text{–}14)$$

Then Eq. (40–12) can be written as

$$r_n = n^2 a_0. \qquad (40\text{–}15)$$

The permitted orbits have radii a_0, $4a_0$, $9a_0$, and so on.

The numerical values of the quantities on the right side of Eq. (40–14) are given in Appendix F. Using these values, we find that the radius a_0 of the smallest Bohr orbit is

$$a_0 = \frac{(8.854 \times 10^{-12}\ \text{C}^2/\text{N}\cdot\text{m}^2)(6.626 \times 10^{-34}\ \text{J}\cdot\text{s})^2}{(3.142)(9.109 \times 10^{-31}\ \text{kg})(1.602 \times 10^{-19}\ \text{C})^2}$$

$$= 5.29 \times 10^{-11}\ \text{m}.$$

This result, giving an atomic diameter of about 10^{-10} m = 0.1 nm, is consistent with atomic dimensions estimated by other methods.

We can use Eq. (40–13) to find the orbital *speed* of the electron. We leave this calculation as an exercise; the result is that for the $n = 1$ state, $v_1 = 2.19 \times 10^6$ m/s. This, the greatest possible speed of the electron in the hydrogen atom, is less than 1% of the speed of light, showing that relativistic considerations aren't significant.

ENERGY LEVELS

We can use Eqs. (40–13) and (40–12) to find the kinetic and potential energies K_n and U_n when the electron is in the orbit with quantum number n:

$$K_n = \frac{1}{2}mv_n^2 = \frac{1}{\epsilon_0^2}\frac{me^4}{8n^2h^2},$$

$$U_n = -\frac{1}{4\pi\epsilon_0}\frac{e^2}{r_n} = -\frac{1}{\epsilon_0^2}\frac{me^4}{4n^2h^2}.$$

The total energy E_n is the sum of the kinetic and potential energies:

$$E_n = K_n + U_n = -\frac{1}{\epsilon_0^2}\frac{me^4}{8n^2h^2}. \tag{40–16}$$

The potential energy has a negative sign because we have taken the potential energy to be zero when the electron is infinitely far from the nucleus. We are interested only in energy *differences,* so the reference position doesn't matter.

The orbits and energy levels are depicted in Fig. 40–8 (page 1238). The possible energy levels of the atom are labeled by values of the quantum number n. For each value of n there are corresponding values of orbit radius r_n, speed v_n, angular momentum $L_n = nh/2\pi$, and total energy E_n. The energy of the atom is least when $n = 1$ and E_n has its most negative value. This is the *ground level* of the atom; it is the level with the smallest orbit, with radius a_0. For $n = 2, 3, \cdots$, the absolute value of E_n is smaller and the energy is progressively larger (less negative). The orbit radius increases as n^2, as shown by Eqs. (40–12) and (40–15), while the speed decreases as $1/n$, as shown by Eq. (40–13).

Comparing the expression for E_n in Eq. (40–16) with Eq. (40–9) (deduced from the measured spectrum of hydrogen), we see that they agree only if the coefficients are equal:

$$hcR = \frac{1}{\epsilon_0^2}\frac{me^4}{8h^2}, \qquad \text{or} \qquad R = \frac{me^4}{8\epsilon_0^2h^3c}. \tag{40–17}$$

This equation therefore shows us how to *calculate* the value of the Rydberg constant from the fundamental physical constants m, c, e, h, and ϵ_0, all of which can be determined quite independently of the Bohr theory. When we substitute the numerical values of these quantities, we obtain the value $R = 1.097 \times 10^7$ m^{-1}. To four significant figures, this is the value determined from wavelength measurements. This agreement provides very strong and direct confirmation of Bohr's theory. We invite you to substitute numerical values into Eq. (40–17) and compute the value of R to confirm these statements.

The *ionization energy* of the hydrogen atom is the energy required to remove the electron completely. Ionization corresponds to a transition from the ground level ($n = 1$) to an infinitely large orbit radius ($n = \infty$) and thus equals $-E_1$. Substituting the constants from Appendix F into Eq. (40–16) gives an ionization energy of 13.606 eV. The ionization energy can also be measured directly; the result is 13.60 eV. These two values agree within 0.1%.

EXAMPLE 40–7

Find the kinetic, potential, and total energies of the hydrogen atom in the first excited level, and find the wavelength of the photon emitted in the transition from the first excited level to the ground level.

SOLUTION We first note that the constant that appears in Eq. (40–16) should equal hcR, which appears in Eq. (40–9) and experimentally equals 13.60 eV:

$$\frac{me^4}{8\epsilon_0^2 h^2} = hcR = 13.60 \text{ eV}.$$

Using this expression, we can rewrite Eq. (40–16) and the two preceding equations as

$$K_n = \frac{13.60 \text{ eV}}{n^2}, \qquad U_n = \frac{-27.20 \text{ eV}}{n^2}, \qquad E_n = \frac{-13.60 \text{ eV}}{n^2}.$$

The first excited level is the $n = 2$ level, which has $K_2 = 3.40$ eV, $U_2 = -6.80$ eV, and $E_2 = -3.40$ eV.

The ground level is the $n = 1$ level, for which the energy is $E_1 = -13.60$ eV. The energy of the emitted photon is $E_2 - E_1 = -3.40$ eV $-(-13.60$ eV$) = 10.20$ eV. The photon energy equals hc/λ, so we find

$$\lambda = \frac{hc}{E_2 - E_1} = \frac{(4.136 \times 10^{-15} \text{ eV} \cdot \text{s})(3.00 \times 10^8 \text{ m/s})}{10.2 \text{ eV}}$$

$$= 1.22 \times 10^{-7} \text{ m} = 122 \text{ nm}.$$

This is the wavelength of the "Lyman alpha" line, the longest-wavelength line in the Lyman series of ultraviolet lines in the hydrogen spectrum.

REDUCED MASS

The values of the Rydberg constant R and the ionization energy of hydrogen predicted by Bohr's analysis are within 0.1% of the measured values. The agreement would be even better if we had not assumed that the nucleus (a proton) remains at rest. Rather, the proton and electron both revolve in circular orbits about their common *center of mass* (Section 8–6), as shown in Fig. 40–17. It turns out that we can take this motion into account very simply by using in Bohr's equations not the electron rest mass m but a quantity called the **reduced mass** m_r of the system. For a system composed of two bodies of masses m_1 and m_2, the reduced mass is defined as

$$m_r = \frac{m_1 m_2}{m_1 + m_2}. \qquad (40\text{–}18)$$

For ordinary hydrogen we let m_1 equal m and m_2 equal the proton mass, $m_p = 1836.2m$. Thus the proton-electron system of ordinary hydrogen has a reduced mass of

$$m_r = \frac{m(1836.2m)}{m + 1836.2m} = 0.99946m.$$

When this value is used instead of the electron mass m in the Bohr equations, the predicted values agree very well with the measured values.

In an atom of deuterium, also called *heavy hydrogen,* the nucleus is not a single proton but a proton and a neutron bound together to form a composite body called the *deuteron.* The reduced mass of the deuterium atom turns out to be $0.99973m$. Equations (40–6) and (40–16) (with m replaced by m_r) show that all energies and frequencies are proportional to m_r, while all wavelengths are inversely proportional to m_r. Thus the spectrum wavelengths of deuterium should be those of hydrogen divided by $(0.99973m)/(0.99946m) = 1.00027$. This is a small effect but well within the precision of modern spectrometers. This small wavelength shift led the American scientist Harold Urey to the discovery of deuterium in 1932, an achievement that earned him the 1934 Nobel Prize in chemistry.

The concept of reduced mass has other applications. A *positron* has the same rest mass as an electron but a charge $+e$. A positronium "atom" (Fig. 40–18) consists of an electron and a positron, each with mass m, in orbit around their common center of mass. This structure lasts only about 10^{-6} s before the two particles annihilate one another and disappear, but this is enough time to study the positronium spectrum. The reduced mass

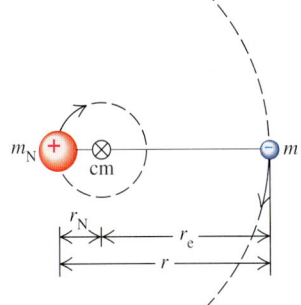

40–17 The nucleus and the electron revolve about their common center of mass. The distance r_N has been exaggerated for clarity; it actually equals $r_e/1836.2$ for ordinary hydrogen.

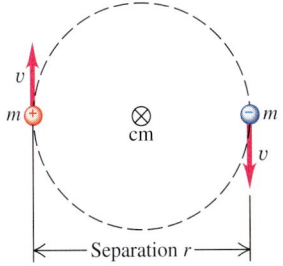

40–18 Applying the Bohr model to positronium. The electron and the positron revolve about their common center of mass, which is located midway between them because they have equal mass. Their separation is twice either orbit radius.

is $m/2$, so the energy levels and photon frequencies have exactly half the values for the simple Bohr model with infinite proton mass. The existence of positronium atoms was confirmed by observation of the corresponding spectrum lines.

HYDROGENLIKE ATOMS

We can extend the Bohr model to other one-electron atoms, such as singly ionized helium (He^+), doubly ionized lithium (Li^{2+}), and so on. Such atoms are called *hydrogen-like* atoms. In such atoms, the nuclear charge is not e but Ze, where Z is the *atomic number,* equal to the number of protons in the nucleus. The effect in the previous analysis is to replace e^2 everywhere by Ze^2. In particular, the orbit radii r_n given by Eq. (40–12) become smaller by a factor of Z, and the energy levels E_n given by Eq. (40–16) are multiplied by Z^2. We invite you to verify these statements. The reduced-mass correction in these cases is even less than 0.1% because the nuclei are more massive than the single proton of ordinary hydrogen. Figure 40–19 compares the energy levels for H and for He^+, which has $Z = 2$. Can you see why H and He^+ have many spectrum lines that have almost the same wavelengths?

Atoms of the *alkali* metals have one electron outside a core consisting of the nucleus and the inner electrons, with net core charge $+e$. These atoms are approximately hydrogenlike, especially in excited levels. Even trapped electrons and electron vacancies in semiconducting solids act somewhat like hydrogen atoms.

Although the Bohr model predicted the energy levels of the hydrogen atom correctly, it raised as many questions as it answered. It combined elements of classical physics with new postulates that were inconsistent with classical ideas. The model provided no insight into what happens *during* a transition from one orbit to another; the angular speeds of the electron motion were not in general the angular frequencies of the emitted radiation, a result that is contrary to classical electrodynamics. Attempts to extend the model to atoms with two or more electrons were not successful. An electron moving in one of Bohr's circular orbits forms a current loop and should produce a magnetic moment. However, a hydrogen atom in its ground level has *no* magnetic moment due to orbital motion. In Chapters 41 and 42 we will find that an even more radical departure

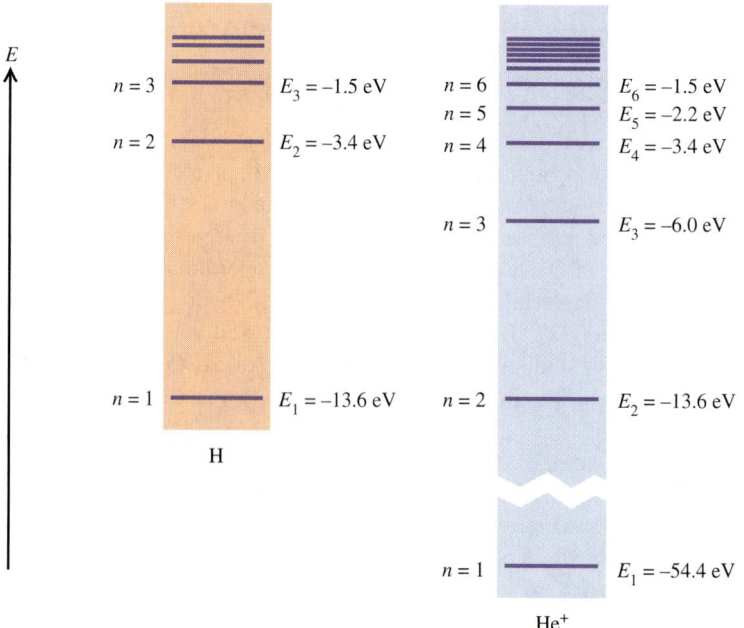

40–19 Energy levels of H and He^+. Because of the additional factor Z^2 in the energy expression, Eq. (40–16), the energy of the He^+ ion with a given n is almost exactly four times that of the H atom with the same n. There are small differences (of the order of 0.05%) because the reduced masses are slightly different.

from classical concepts was needed before the understanding of atomic structure could progress further.

40–7   THE LASER

The **laser** is a light source that produces a beam of highly coherent and very nearly monochromatic light as a result of cooperative emission from many atoms. The name *laser* is an acronym for "light amplification by stimulated emission of radiation." We can understand the principles of laser operation on the basis of photons and atomic energy levels. During the discussion we'll also introduce two new concepts: *stimulated emission* and *population inversion.*

If an atom has an excited level an energy E above the ground level, the atom in its ground level can absorb a photon with frequency f given by $E = hf$. This process is shown schematically in Fig. 40–20a, which shows a gas in a transparent container. Three atoms A each absorb a photon, reaching an excited level and being denoted as A*. Some time later, the excited atoms return to the ground level by each emitting a photon with the same frequency as the one originally absorbed. This process is called *spontaneous emission;* the direction and phase of the emitted photons are random (Fig. 40–20b).

In **stimulated emission** (Fig. 40–20c), each incident photon encounters a previously excited atom. A kind of resonance effect induces each atom to emit a second photon with the same frequency, direction, phase, and polarization as the incident photon, which is not changed by the process. For each atom there is one photon before a stimulated emission and two photons after; thus the name *light amplification*. Because the two photons have the same phase, they emerge together as *coherent* radiation. The laser makes use of stimulated emission to produce a beam consisting of a large number of such coherent photons.

To discuss stimulated emission from atoms in excited levels, we need to know something about how many atoms are in each of the various energy levels. First, we need to make the distinction between the terms *energy level* and *state*. A system may have more than one way to attain a given energy level; each different way is a different **state.** For instance, there are two ways of putting an ideal unstretched spring in a given energy level. Remembering that $U = \frac{1}{2}kx^2$, we could compress the spring by $x = -b$ or we could stretch it by $x = +b$ to get the same $U = \frac{1}{2}kb^2$. The Bohr model had only one state in each energy level, but we will find in Section 43–5 that the hydrogen atom actually has two states in its −13.60-eV ground level, eight states in its −3.40-eV first excited level, and so on.

18.4
The Laser

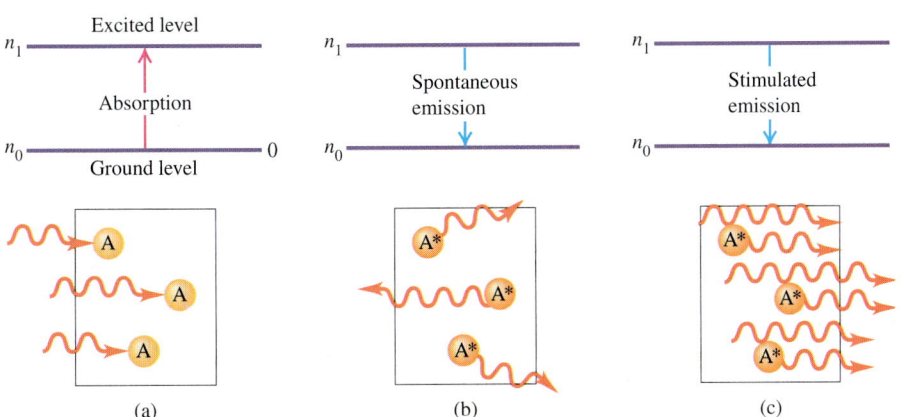

40–20 Three interaction processes between an atom and electromagnetic waves.

The Maxwell-Boltzmann distribution function (Section 16–6) determines the number of atoms in a given state in a gas. The function tells us that when the gas is in thermal equilibrium at absolute temperature T, the number n_i of atoms in a state with energy E_i equals $Ae^{-E_i/kT}$, where k is Boltzmann's constant and A is another constant determined by the total number of atoms in the gas. (In Section 16–6, E was the kinetic energy $\frac{1}{2}mv^2$ of a gas molecule; here we're talking about the internal energy of an atom.) Because of the negative exponent, fewer atoms are in higher-energy states, as we should expect. If E_g is a ground-state energy and E_{ex} is the energy of an excited state, then the ratio of numbers of atoms in the two states is

$$\frac{n_{ex}}{n_g} = \frac{Ae^{-E_{ex}/kT}}{Ae^{-E_g/kT}} = e^{-(E_{ex}-E_g)/kT}. \tag{40–19}$$

For example, suppose $E_{ex} - E_g = 2.0$ eV $= 3.2 \times 10^{-19}$ J, the energy of a 620-nm visible-light photon. At $T = 3000$ K (the temperature of the filament in an incandescent light bulb),

$$\frac{E_{ex} - E_g}{kT} = \frac{3.2 \times 10^{-19} \text{ J}}{(1.38 \times 10^{-23} \text{ J/K})(3000 \text{ K})} = 7.73$$

and

$$e^{-(E_{ex}-E_g)/kT} = e^{-7.73} = 0.00044.$$

That is, the fraction of atoms in a state 2.0 eV above a ground state is extremely small, even at this high temperature. The point is that at any reasonable temperature there aren't enough atoms in excited states for any appreciable amount of stimulated emission from these states to occur. Rather, absorption is much more probable.

We could try to enhance the number of atoms in excited states by sending through the container a beam of radiation with frequency $f = E/h$ corresponding to the energy difference $E = E_{ex} - E_g$. Some of the atoms absorb photons of energy E and are raised to the excited state, and the population ratio n_{ex}/n_g momentarily increases. But because n_g is originally so much larger than n_{ex}, an enormously intense beam of light would be required to momentarily increase n_{ex} to a value comparable to n_g. The rate at which energy is *absorbed* from the beam by the n_g ground-state atoms far outweighs the rate at which energy is added to the beam by stimulated emission from the relatively rare (n_{ex}) excited atoms.

We need to create a non-equilibrium situation in which the number of atoms in a higher-energy state is greater than the number in a lower-energy state. Such a situation is called a **population inversion.** Then the rate of energy radiation by stimulated emission can *exceed* the rate of absorption, and the system acts as a net *source* of radiation with photon energy E. Furthermore, because the photons are the result of stimulated emission, they all have the same frequency, phase, polarization, and direction of propagation. The resulting radiation is therefore very much more coherent than light from ordinary sources, in which the emissions of individual atoms are not coordinated. This coherent emission is exactly what happens in a laser.

The necessary population inversion can be achieved in a variety of ways. A familiar example is the helium-neon laser, a common and inexpensive laser that is available in many undergraduate laboratories. A mixture of helium and neon, typically at a pressure of the order of 10^2 Pa (10^{-3} atm), is sealed in a glass enclosure provided with two electrodes. When a sufficiently high voltage is applied, an electric discharge occurs. Collisions between ionized atoms and electrons carrying the discharge current excite atoms to various energy states.

Figure 40–21a shows an energy-level diagram for the system. The labels for the various energy levels, such as $1s$, $3p$, and $5s$, refer to states of the electrons in the levels. (We'll discuss this notation in Chapter 43; we don't need detailed knowledge of the

40–21 (a) Energy-level diagram for a helium-neon laser. (b) Operation of a laser. The light emitted by electrons returning to lower-energy states is reflected between mirrors, so it continues to simulate emission of more coherent light. One mirror is partially transmitting and allows the high-intensity light beam to leave the laser cavity.

nature of the states to understand laser action.) Because of restrictions imposed by conservation of angular momentum, a helium atom with an electron excited to a $2s$ state cannot return to a ground ($1s$) state by emitting a 20.61-eV photon, as you might expect it to do. Such a state, in which single-photon emission is impossible, has an unusually long lifetime and is called a **metastable state.** Helium atoms "pile up" in the metastable $2s$ states, creating a population inversion relative to the ground states.

However, excited helium atoms *can* lose energy by energy-exchange collisions with neon atoms that are initially in a ground state. A helium atom, excited to a $2s$ state 20.61 eV above its ground states and possessing a little additional kinetic energy, can collide with a neon atom in a ground state, exciting the *neon* atom to a $5s$ excited state at 20.66 eV and dropping the *helium* atom back to a ground state. Thus we have the necessary mechanism for a population inversion in neon, with more neon atoms per $5s$ state than per $3p$ state.

Stimulated emissions during transitions from a $5s$ to a $3p$ state then result in the emission of highly coherent light at 632.8 nm, as shown in the diagram. In practice the beam is sent back and forth through the gas many times by a pair of parallel mirrors (Fig. 40–21b), so as to stimulate emission from as many excited atoms as possible. One of the mirrors is partially transparent, so a portion of the beam emerges as an external beam. The net result of all these processes is a beam of light that (1) can be quite intense, (2) has very parallel rays, (3) is highly monochromatic, and (4) is spatially *coherent* at all points within a given cross section.

Other types of lasers use different processes to achieve the necessary population inversion. In a semiconductor laser the inversion is obtained by driving electrons and holes to a *p-n* junction (to be discussed in Section 44–8) with a steady electric field. In one type of *chemical* laser a chemical reaction forms molecules in metastable excited states. In a carbon dioxide gas dynamic laser the population inversion results from rapid expansion of the gas. There are even x-ray lasers, with the population inversion caused by small nuclear explosions. A *maser* (acronym for "microwave amplification by stimulated emission of radiation") operates on the basis of population inversions in molecules, using closely spaced energy levels. The corresponding emitted radiation is in the microwave range. Maser action even occurs naturally in interstellar molecular clouds.

In recent years, lasers have found a wide variety of practical applications. A high-intensity laser beam makes a convenient drill. For example, such a laser beam can drill a very small hole in a diamond for use as a die in drawing very small-diameter wire. Because a laser beam has very parallel rays, it can travel long distances without appreciable spreading. Surveyors often use lasers, especially in situations requiring great precision such as drilling a long tunnel from both ends.

Lasers are widely used in medicine. A laser with a very narrow intense beam can be used in the treatment of a detached retina; a short burst of radiation damages a small area of the retina, and the resulting scar tissue "welds" the retina back to the membrane from which it has become detached. Laser beams are used in surgery; blood vessels cut by the beam tend to seal themselves off, making it easier to control bleeding. Lasers are also used for selective destruction of tissue, as in the removal of tumors.

40–8 X-Ray Production and Scattering

The production and scattering of x rays provide additional examples of the quantum nature of electromagnetic radiation. X rays are produced when rapidly moving electrons that have been accelerated through a potential difference of the order of 10^3 to 10^6 V strike a metal target. They were first produced in 1895 by Wilhelm Röntgen (1845–1923), using an apparatus similar in principle to the setup shown in Fig. 40–22a. Electrons are "boiled off" from the heated cathode by thermionic emission and are accelerated toward the anode (the target) by a large potential difference V_{AC}. The bulb is evacuated (residual pressure 10^{-7} atm or less), so the electrons can travel from the cathode to the anode without colliding with air molecules. When V_{AC} is a few thousand volts or more, a very penetrating radiation is emitted from the anode surface. A more modern x-ray device is shown in Fig. 40–22b.

X-RAY PHOTONS

Because they are emitted by accelerated charges, it is clear that x rays are electromagnetic waves. Like light, x rays are governed by quantum relations in their interaction with matter. Thus we can talk about x-ray photons or quanta, and the energy of an x-ray photon is related to its frequency and wavelength in the same way as for photons of light, $E = hf = hc/\lambda$. Typical x-ray wavelengths are 0.001 to 1 nm (10^{-12} to 10^{-9} m). X-ray wave-

40–22 (a) Apparatus used to produce x rays, similar to Röntgen's apparatus. Electrons are emitted thermionically from the heated cathode and are accelerated toward the anode; when they strike it, x rays are produced. (b) Cross section of an x-ray device. The anode is liquid-cooled to prevent overheating.

Heated cathode   Anode

Power supply for heater

X-ray beam

Accelerating voltage V_{AC}

(a)

Heated cathode

Electron beam

Evacuated tube

Tungsten anode (target)

Oil bath absorbs heat

Lead casing

Emitted x-rays   Window

(b)

lengths can be measured quite precisely by crystal diffraction techniques, which we discussed in Section 38–7.

X-ray emission is the inverse of the photoelectric effect. In photoelectric emission there is a transformation of the energy of a photon into the kinetic energy of an electron; in x-ray production there is a transformation of the kinetic energy of an electron into the energy of a photon. The energy relationships are similar. In x-ray production we usually neglect the work function of the target and the initial kinetic energy of the "boiled-off" electrons because they are very small in comparison to the other energies.

Two distinct processes are involved in x-ray emission. In the first process, some electrons are slowed down or stopped by the target, and part or all of their kinetic energy is converted directly to a continuous spectrum of photons, including x rays. This process is called *bremsstrahlung* (German for "braking radiation"). Classical physics is completely unable to explain why the x rays that are emitted in a bremsstrahlung process have a maximum frequency f_{max} and a corresponding minimum wavelength λ_{min}, much less predict their values. With quantum concepts it is easy. An electron has charge $-e$ and gains kinetic energy eV_{AC} when accelerated through a potential increase V_{AC}. The most energetic photon (highest frequency and shortest wavelength) is produced when all the electron's kinetic energy goes to produce one photon; that is,

$$eV_{AC} = hf_{max} = \frac{hc}{\lambda_{min}} \quad \text{(bremsstrahlung limits).} \qquad (40\text{–}20)$$

Note that the maximum frequency and minimum wavelength in the bremsstrahlung process do *not* depend on the target material.

The second process gives peaks in the x-ray spectrum at characteristic frequencies and wavelengths that *do* depend on the target material. Other electrons, if they have enough kinetic energy, can transfer that energy partly or completely to individual atoms within the target. These atoms are left in excited levels; when they decay back to their ground levels, they may emit x-ray photons. Since each element has a unique set of atomic energy levels, each also has a characteristic x-ray spectrum. The energy levels associated with x rays are rather different in character from those associated with visible spectra. They involve vacancies in the inner electron configurations of complex atoms. The energy differences between these levels can be hundreds or thousands of electron volts, rather than a few electron volts as is typical for optical spectra. We will return to energy levels and spectra associated with x rays in Section 43–6.

EXAMPLE 40-8

An x-ray wavelength Electrons in an x-ray tube are accelerated by a potential difference of 10.0 kV. If an electron produces one photon on impact with the target, what is the minimum wavelength of the resulting x rays? Answer using both SI units and electron volts.

SOLUTION We solve Eq. (40–20) for the minimum wavelength and use SI values:

$$\lambda_{min} = \frac{hc}{eV_{AC}} = \frac{(6.626 \times 10^{-34} \text{ J} \cdot \text{s})(3.00 \times 10^8 \text{ m/s})}{(1.602 \times 10^{-19} \text{ C})(10.0 \times 10^3 \text{ V})}$$

$$= 1.24 \times 10^{-10} \text{ m} = 0.124 \text{ nm.}$$

Using electron volts, we have

$$\lambda_{min} = \frac{hc}{eV_{AC}} = \frac{(4.136 \times 10^{-15} \text{ eV} \cdot \text{s})(3.00 \times 10^8 \text{ m/s})}{e(10.0 \times 10^3 \text{ V})}$$

$$= 1.24 \times 10^{-10} \text{ m} = 0.124 \text{ nm.}$$

The wavelength is smaller than that of typical visible-light photons by a factor of about 5000. The photon energy and frequency are greater by the same factor. Note that the "e" in the unit eV is canceled by the "e" for the magnitude of the electron charge, because the electron volt (eV) is the magnitude of the electron charge e times one volt (1 V).

COMPTON SCATTERING

A phenomenon called **Compton scattering,** first explained in 1923 by the American physicist A. H. Compton, provides additional direct confirmation of the quantum nature of x rays. When x rays strike matter, some of the radiation is *scattered,* just as visible

40–23 A Compton-effect experiment.

17.4
Compton Scattering

(a)

(b)

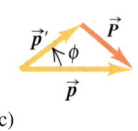

(c)

40–24 Schematic diagram of Compton scattering showing (a) an incident photon with wavelength λ and momentum $\vec{p}$ approaching an electron at rest; (b) a scattered photon with longer wavelength λ' and momentum $\vec{p}'$ and a recoiling electron with momentum $\vec{P}$. The direction of the scattered photon makes an angle ϕ with that of the incident photon, so the angle between $\vec{p}$ and $\vec{p}'$ is also ϕ. (c) Vector diagram for the conservation of momentum relation $\vec{p} = \vec{p}' + \vec{P}$.

light falling on a rough surface undergoes diffuse reflection. Compton and others discovered that some of the scattered radiation has smaller frequency (longer wavelength) than the incident radiation and that the change in wavelength depends on the angle through which the radiation is scattered. Specifically, if the scattered radiation emerges at an angle ϕ with respect to the incident direction (Fig. 40–23) and if λ and λ' are the wavelengths of the incident and scattered radiation, respectively, we find that

$$\lambda' - \lambda = \frac{h}{mc}(1 - \cos\phi) \quad \text{(Compton scattering)}, \quad (40\text{–}21)$$

where m is the electron rest mass. The quantity h/mc that appears in Eq. (40–21) has units of length. Its numerical value is

$$\frac{h}{mc} = \frac{6.626 \times 10^{-34} \text{ J} \cdot \text{s}}{(9.109 \times 10^{-31} \text{ kg})(2.998 \times 10^8 \text{ m/s})} = 2.426 \times 10^{-12} \text{ m}.$$

We cannot explain Compton scattering using classical electromagnetic theory, which predicts that the scattered wave has the same wavelength as the incident wave. However the quantum theory provides a beautifully clear explanation. We imagine the scattering process as a collision of two *particles,* the incident photon and an electron that is initially at rest, as in Fig. 40–24a. The incident photon disappears, giving part of its energy and momentum to the electron, which recoils as a result of this impact. The remainder goes to a new scattered photon that therefore has less energy, smaller frequency, and longer wavelength than the incident one (Fig. 40–24b).

We can derive Eq. (40–21) from the principles of conservation of energy and momentum. We outline the derivation below; we invite you to fill in the details (see Exercise 40–33). The electron recoil energy may be in the relativistic range, so we have to use the relativistic energy-momentum relations, Eqs. (39–40) and (39–41). The incident photon has momentum $\vec{p}$, with magnitude p, and energy pc. The scattered photon has momentum $\vec{p}'$, with magnitude p', and energy $p'c$. The electron is initially at rest, so its initial momentum is zero and its initial energy is its rest energy mc^2. The final electron momentum $\vec{P}$ has magnitude P, and the final electron energy E is given by $E^2 = (mc^2)^2 + (Pc)^2$. Then energy conservation gives us the relation

$$pc + mc^2 = p'c + E,$$

or

$$(pc - p'c + mc^2)^2 = E^2 = (mc^2)^2 + (Pc)^2. \quad (40\text{–}22)$$

We may eliminate the electron momentum $\vec{P}$ from this equation by using momentum conservation:

$$\vec{p} = \vec{p}' + \vec{P},$$

or

$$\vec{P} = \vec{p} - \vec{p}'. \quad (40\text{–}23)$$

By taking the scalar product of each side of Eq. (40–23) with itself or by using the law of cosines with the vector diagram in Fig. 40–24c, we find

$$P^2 = p^2 + p'^2 - 2pp' \cos \phi. \qquad (40\text{--}24)$$

We now substitute this expression for P^2 into Eq. (40–22) and multiply out the left side. We divide out a common factor c^2; several terms cancel, and when the resulting equation is divided through by (pp'), the result is

$$\frac{mc}{p'} - \frac{mc}{p} = 1 - \cos \phi. \qquad (40\text{--}25)$$

Finally, we substitute $p' = h/\lambda'$ and $p = h/\lambda$, then multiply by h/mc to obtain Eq. (40–21).

When the wavelengths of x rays scattered at a certain angle are measured, the curve of intensity per unit wavelength as a function of wavelength has two peaks (Fig. 40–25). The longer-wavelength peak represents Compton scattering. The shorter-wavelength peak, labeled λ_0, is at the wavelength of the incident x rays and corresponds to x-ray scattering from tightly bound electrons. In such scattering processes the entire atom must recoil, so the m in Eq. (40–21) is the mass of the entire atom rather than of a single electron. The resulting wavelength shifts are negligible.

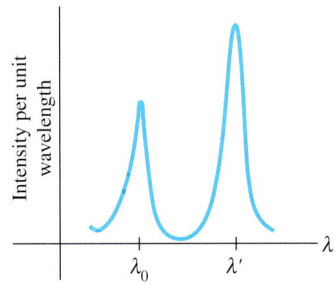

40–25 Intensity as a function of wavelength for photons scattered at an angle of 135° in the Compton effect.

EXAMPLE 40-9

Compton scattering You use the x-ray photons in Example 40–8 (with $\lambda = 0.124$ nm) in a Compton scattering experiment. At what angle is the wavelength of the scattered x rays 1.0% longer than that of the incident x rays? At what angle is it 0.050% greater?

SOLUTION In Eq. (40–21) we want $\Delta\lambda = \lambda' - \lambda$ to be 1.0% of 0.124 nm. That is, $\Delta\lambda = 0.00124$ nm $= 1.24 \times 10^{-12}$ m. Using the value $h/mc = 2.426 \times 10^{-12}$ m, we find

$$\Delta\lambda = \frac{h}{mc}(1 - \cos \phi),$$

$$\cos \phi = 1 - \frac{\Delta\lambda}{(h/mc)} = 1 - \frac{1.24 \times 10^{-12} \text{ m}}{2.426 \times 10^{-12} \text{ m}} = 0.4889,$$

$$\phi = 60.7°.$$

For $\Delta\lambda$ to be 0.050% of 0.124 nm, or 6.2×10^{-14} m,

$$\cos \phi = 1 - \frac{6.2 \times 10^{-14} \text{ m}}{2.426 \times 10^{-12} \text{ m}} = 0.9744,$$

$$\phi = 13.0°.$$

Smaller angles give smaller wavelength shifts. Thus in a grazing collision the photon energy loss and the electron recoil energy are smaller than when the scattering angle is larger.

X rays have many practical applications in medicine and industry. Because they can penetrate several centimeters of solid matter, they can be used to visualize the interiors of materials that are opaque to ordinary light, such as broken bones or defects in structural steel. The object to be visualized is placed between an x-ray source and a large sheet of photographic film; the darkening of the film is proportional to the radiation exposure. A crack or air bubble allows greater transmission and shows as a dark area. Bones appear lighter than the surrounding soft tissue because they contain greater proportions of elements with higher atomic number (and greater absorption); in the soft tissue the light elements carbon, hydrogen, and oxygen predominate.

Recently, several vastly improved x-ray techniques have been developed. One widely used system is *computerized axial tomography;* the corresponding instrument is called a CAT scanner. The x-ray source produces a thin, fan-shaped beam that is detected on the opposite side of the subject by an array of several hundred detectors in a line. Each detector measures absorption along a thin line through the subject. The entire apparatus is rotated around the subject in the plane of the beam during a few seconds. The changing photon-counting rates of the detectors are recorded digitally; a computer processes this information and reconstructs a picture of absorption over an entire cross section of the subject. Differences as small as 1% can be detected with CAT scans, and tumors and other anomalies that are much too small to be seen with older x-ray techniques can be detected.

X rays cause damage to living tissues. As x-ray photons are absorbed in tissues, their energy breaks molecular bonds and creates highly reactive free radicals (such as neutral H and OH), which in turn can disturb the molecular structure of proteins and especially genetic material. Young and rapidly growing cells are particularly susceptible; thus x rays are useful for selective destruction of cancer cells. Conversely, however, a cell may be damaged by radiation but survive, continue dividing, and produce generations of defective cells; thus x rays can *cause* cancer.

Even when the organism itself shows no apparent damage, excessive radiation exposure can cause changes in the organism's reproductive system that will affect its offspring. The use of x rays in medical diagnosis has been an area of great concern for many years; a careful assessment of the balance between risks and benefits of radiation exposure is essential in each individual case.

40-9 CONTINUOUS SPECTRA

Line spectra are emitted by matter in the gaseous state, in which the atoms are so far apart that interactions between them are negligible and each atom behaves as an isolated system. Hot matter in condensed states (solid or liquid) nearly always emits radiation with a *continuous* distribution of wavelengths rather than a line spectrum. An ideal surface that absorbs all wavelengths of electromagnetic radiation incident upon it is also the best possible emitter of electromagnetic radiation at any wavelength. Such an ideal surface is called a *blackbody,* and the continuous-spectrum radiation that it emits is called **blackbody radiation.** By 1900 this radiation had been studied extensively, and several characteristics had been established.

First, the total intensity I (the average rate of radiation of energy per unit surface area or average power per area) emitted from the surface of an ideal radiator is proportional to the fourth power of the absolute temperature. We studied this relationship in Section 15–8 during our study of heat transfer mechanisms. This total intensity I emitted at absolute temperature T is given by the **Stefan-Boltzmann law:**

$$I = \sigma T^4 \qquad \text{(Stefan-Boltzmann law for a blackbody)}, \qquad (40\text{–}26)$$

where σ is a fundamental physical constant called the *Stefan-Boltzmann constant.* In SI units,

$$\sigma = 5.67051(19) \times 10^{-8} \frac{\text{W}}{\text{m}^2 \cdot \text{K}^4}.$$

Second, the intensity is not uniformly distributed over all wavelengths. Its distribution can be measured and described by the intensity per wavelength interval $I(\lambda)$, called the *spectral emittance.* Thus $I(\lambda)\,d\lambda$ is the intensity corresponding to wavelengths in the interval from λ to $\lambda + d\lambda$. The *total* intensity I, given by Eq. (40–26), is the *integral* of the distribution function $I(\lambda)$ over all wavelengths, which equals the area under the $I(\lambda)$ versus λ curve:

$$I = \int_0^\infty I(\lambda)\,d\lambda. \qquad (40\text{–}27)$$

CAUTION ▶ Although we use the symbol $I(\lambda)$ for spectral emittance, keep in mind that spectral emittance is *not* the same thing as intensity I. Intensity is power per unit area, with units W/m^2; spectral emittance is power per unit area per unit wavelength interval, with units W/m^3. ◀

This close-up view of the sun's surface shows two dark sunspots. Their temperature is about 4000 K, while the surrounding solar material is at $T = 5800$ K. From the Stefan-Boltzmann law, the intensity from a given area of sunspot is only (4000 K/5800 K)^4 = 0.23 as great as the intensity from the same area of the surrounding material—which is why sunspots appear dark.

Measured spectral emittances $I(\lambda)$ for three different temperatures are shown in Fig. 40–26. Each has a peak wavelength λ_m at which the emitted intensity per wavelength interval is largest. Experiment shows that λ_m is inversely proportional to T, so their product is constant. This result is called the **Wien displacement law.** The experimental value of the constant is 2.90×10^{-3} m · K:

$$\lambda_m T = 2.90 \times 10^{-3} \text{ m} \cdot \text{K} \qquad \text{(Wien displacement law).} \qquad (40\text{–}28)$$

As the temperature rises, the peak of $I(\lambda)$ becomes higher and shifts to shorter wavelengths. A body that glows yellow is hotter and brighter than one that glows red; yellow light has shorter wavelengths than red light. Finally, experiments show that the *shape* of the distribution function is the same for all temperatures; we can make a curve for one temperature fit any other temperature by simply changing the scales on the graph.

During the last decade of the nineteenth century, many attempts were made to *derive* these empirical results from basic principles. In one attempt, Rayleigh considered light enclosed in a rectangular box with perfectly reflecting sides. Such a box has a series of possible *normal modes* for electromagnetic waves, as discussed in Section 33–7. It seemed reasonable to assume that the distribution of energy among the various modes was given by the equipartition principle (Section 16–5), which had been successfully used in the analysis of heat capacities. A small hole in the box would behave as an ideal blackbody radiator.

Including both the electric- and magnetic-field energies, Rayleigh assumed that the total energy of each normal mode was equal to kT. Then by computing the *number* of normal modes corresponding to a wavelength interval $d\lambda$, Rayleigh could predict the distribution of wavelengths in the radiation within the box. Finally, he could compute the intensity distribution $I(\lambda)$ of the radiation emerging from a small hole in the box. His result was quite simple:

$$I(\lambda) = \frac{2\pi c k T}{\lambda^4}. \qquad (40\text{–}29)$$

At large wavelengths this formula agrees quite well with the experimental results shown in Fig. 40–26, but there is serious disagreement at small wavelengths. The experimental curve falls toward zero at small λ, but Rayleigh's curve goes the opposite direction, approaching infinity as $1/\lambda^4$, a result that was called in Rayleigh's time the "ultraviolet catastrophe." Even worse, the integral of Eq. (40–29) over all λ is infinite, indicating an infinitely large *total* radiated intensity. Clearly, something is wrong.

Finally, in 1900, Planck succeeded in deriving a function, now called the **Planck radiation law,** that agreed very well with experimental intensity distribution curves. In his derivation he made what seemed at the time to be a crazy assumption. Planck assumed that electromagnetic oscillators (electrons) in the walls of Rayleigh's box vibrating at a frequency f could have only certain values of energy equal to nhf, where $n = 0, 1, 2, 3, \cdots$ and h turned out to be the constant that now bears Planck's name. These oscillators were in equilibrium with the electromagnetic waves in the box. His assumption gave quantized energy levels and was in sharp contrast to Rayleigh's point of view, which was that each normal mode could have any amount of energy.

Planck was not comfortable with this *quantum hypothesis;* he regarded it as a calculational trick rather than a fundamental principle. In a letter to a friend, he called it "an act of desperation" into which he was forced because "a theoretical explanation had to be found at any cost, whatever the price." But five years later, Einstein identified the energy change hf between levels as the energy of a photon to explain the photoelectric effect (Section 40–3), and other evidence quickly mounted. By 1915 there was little doubt about the validity of the quantum concept and the existence of photons. By discussing atomic spectra *before* continuous spectra, we have departed from the historical

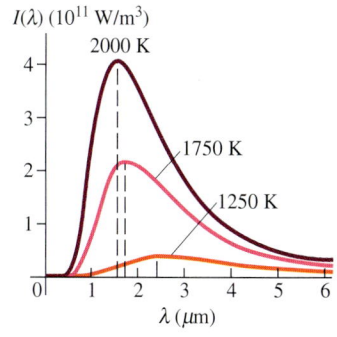

40–26 Spectral emittance $I(\lambda)$ for radiation from a blackbody. The average power per unit surface area between λ and $\lambda + d\lambda$ is $I(\lambda)\,d\lambda$. The total area under the curve at any temperature represents the total radiated average power per unit surface area and is proportional to T^4. Curves are plotted for three different temperatures, with darker color corresponding to higher temperature. The dashed vertical lines show the value of λ_m in Eq. (40–28) for each temperature. As the temperature increases, the peak grows larger and shifts to shorter wavelengths.

order of things. The credit for inventing the concept of quantization of energy levels goes to Planck, even though he didn't believe it at first.

We won't go into the details of Planck's derivation of the intensity distribution. Here is his result:

$$I(\lambda) = \frac{2\pi hc^2}{\lambda^5(e^{hc/\lambda kT} - 1)} \qquad \text{(Planck radiation law)}, \qquad (40\text{--}30)$$

where h is Planck's constant, c is the speed of light, k is Boltzmann's constant, T is the *absolute* temperature, and λ is the wavelength. This function turns out to agree well with experimental intensity curves such as those in Fig. 40–26.

The Planck radiation law also contains the Wien displacement law and the Stefan-Boltzmann law as consequences. To derive the Wien law, we take the derivative of Eq. (40–30) and set it equal to zero to find the value of λ at which $I(\lambda)$ is maximum. We leave the details as a problem (see Exercise 40–39); the result is

$$\lambda_{\text{m}} = \frac{hc}{4.965kT}. \qquad (40\text{--}31)$$

To obtain this result, you have to solve the equation

$$5 - x = 5e^{-x}. \qquad (40\text{--}32)$$

The root of this equation, found by trial and error or more sophisticated means, is 4.965 to four significant figures. We invite you to evaluate the constant $hc/4.965k$ and show that it agrees with the experimental value of 2.90×10^{-3} m · K given in Eq. (40–28).

We can obtain the Stefan-Boltzmann law for a blackbody by integrating Eq. (40–30) over all λ to find the *total* radiated intensity (see Problem 40–65). This is not a simple integral; the result is

$$I = \int_0^\infty I(\lambda)\,d\lambda = \frac{2\pi^5 k^4}{15c^2 h^3}T^4 = \sigma T^4, \qquad (40\text{--}33)$$

in agreement with Eq. (40–26). This result also shows that the constant σ in that law can be expressed as a combination of other fundamental constants:

$$\sigma = \frac{2\pi^5 k^4}{15c^2 h^3}. \qquad (40\text{--}34)$$

We invite you to substitute the values of k, c, and h from Appendix F and verify that you obtain the Stefan-Boltzmann constant $\sigma = 5.6705 \times 10^{-8}$ W/m² · K⁴.

The general form of Eq. (40–31) is what we should expect from kinetic theory. If photon energies are typically of the order of kT, as suggested by the equipartition theorem, then for a typical photon we would expect

$$E \approx kT \approx \frac{hc}{\lambda}, \qquad \text{and} \qquad \lambda \approx \frac{hc}{kT}. \qquad (40\text{--}35)$$

Indeed, a photon with wavelength given by Eq. (40–31) has an energy $E = 4.965kT$.

The Planck radiation law, Eq. (40–30), looks so different from the unsuccessful Rayleigh expression, Eq. (40–29), that it may seem unlikely that they would agree at large values of λ. But when λ is large, the exponent in the denominator of Eq. (40–30) is very small. We can then use the approximation $e^x \approx 1 + x$ (for $x \ll 1$). We invite you to verify that when this is done, the result approaches Eq. (40–29), showing that the two expressions *do* agree in the limit of very large λ. We also note that the Rayleigh expression does not contain h. At very long wavelengths and correspondingly very small photon energies, quantum effects become unimportant.

EXAMPLE 40–10

Light from the sun The surface of the sun has a temperature of approximately 5800 K. To a good approximation we may treat it as a blackbody. a) What is the peak-intensity wavelength λ_m? b) What is the total radiated power per unit area?

SOLUTION a) We use the Wien displacement law, Eq. (40–28):

$$\lambda_m = \frac{2.90 \times 10^{-3} \text{ m} \cdot \text{K}}{T} = \frac{2.90 \times 10^{-3} \text{ m} \cdot \text{K}}{5800 \text{ K}}$$

$$= 0.500 \times 10^{-6} \text{ m} = 500 \text{ nm}.$$

This wavelength is near the middle of the visible spectrum, a not surprising result; our eyes evolved to take maximum advantage of natural light.

b) We get the total intensity from the Stefan-Boltzmann law, Eq. (40–26):

$$I = \sigma T^4 = (5.67 \times 10^{-8} \text{ W/m}^2 \cdot \text{K}^4)(5800 \text{ K})^4$$

$$= 6.42 \times 10^7 \text{ W/m}^2 = 64.2 \text{ MW/m}^2.$$

This enormous value is the intensity at the *surface* of the sun. When the radiated power reaches the earth, the intensity drops to about 1.4 kW/m² because the power spreads out over the much larger area of a sphere with a radius equal to that of the earth's orbit.

EXAMPLE 40–11

Find the intensity of light emitted from the surface of the sun in the wavelength range 600.0 to 605.0 nm.

SOLUTION For an exact result we should integrate Eq. (40–30) between the limits 600.0 and 605.0 nm, finding the area under the $I(\lambda)$ curve between these limits. This integral can't be evaluated in terms of familiar functions, so we *approximate* the area by the height of the curve at the median wavelength $\lambda = 602.5$ nm multiplied by the width of the interval ($\Delta\lambda = 5.0$ nm). First, we evaluate $hc/\lambda kT$ at $\lambda = 602.5$ nm $= 6.025 \times 10^{-7}$ m; then we obtain the height of the $I(\lambda)$ curve using Eq. (40–30):

$$\frac{hc}{\lambda kT} = \frac{(6.626 \times 10^{-34} \text{ J} \cdot \text{s})(2.998 \times 10^8 \text{ m/s})}{(6.025 \times 10^{-7} \text{ m})(1.381 \times 10^{-23} \text{ J/K})(5800 \text{ K})} = 4.116,$$

$$I(\lambda) = \frac{2\pi(6.626 \times 10^{-34} \text{ J} \cdot \text{s})(2.998 \times 10^8 \text{ m/s})^2}{(6.025 \times 10^{-7} \text{ m})^5 (e^{4.116} - 1)}$$

$$= 7.81 \times 10^{13} \text{ W/m}^3.$$

The intensity in the 5.0-nm range from 600.0 to 605.0 nm is then approximately

$$I(\lambda)\Delta\lambda = (7.81 \times 10^{13} \text{ W/m}^3)(5.0 \times 10^{-9} \text{ m})$$

$$= 3.9 \times 10^5 \text{ W/m}^2 = 0.39 \text{ MW/m}^2.$$

We see from this result and Example 40–10 that about 0.6% of the total intensity from the sun is in the range 600 to 605 nm.

40–10 WAVE-PARTICLE DUALITY

We have studied many examples of the behavior of light and other electromagnetic radiation. Some, including the interference and diffraction effects described in Chapters 37 and 38, demonstrate conclusively the *wave* nature of light. Others, the subject of the present chapter, point with equal force to the *particle* nature of light. At first glance these two aspects seem to be in direct conflict. How can light be a wave and a particle at the same time?

We can find the answer to this apparent wave-particle conflict in the **principle of complementarity,** first stated by Bohr in 1928. The wave descriptions and the particle descriptions are complementary. That is, we need both descriptions to complete our model of nature, but we will never need to use both descriptions at the same time to describe a single part of an occurrence.

Let's start by considering again the diffraction pattern for a single slit, which we analyzed in Sections 38–3 and 38–4. Instead of recording the pattern on photographic film, we use a detector called a *photomultiplier* that can actually detect individual photons. Using the setup shown in Fig. 40–27, we place the photomultiplier at various positions

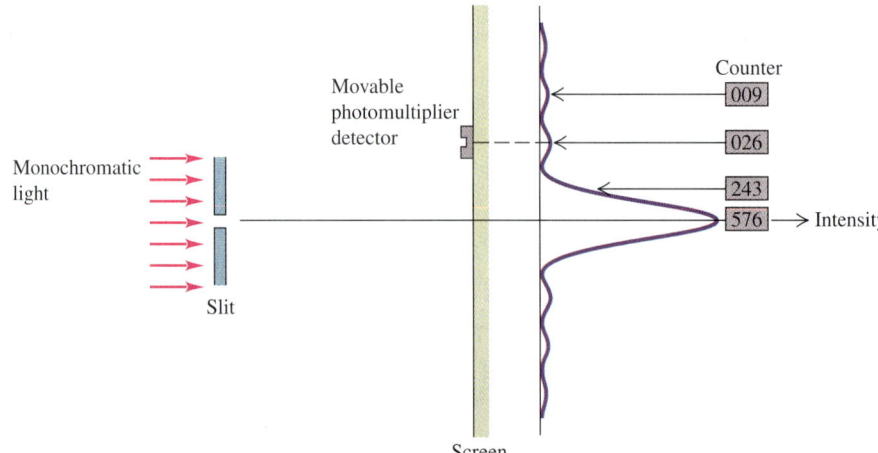

40–27 Single-slit diffraction pattern observed with a movable photomultiplier. The curve shows the intensity distribution predicted by the wave picture, and the photon distribution is shown by the numbers of photons counted at various positions.

These images record the positions where individual photons in a two-slit interference experiment strike the screen. The top image shows the pattern after 21 photons have reached the screen; the middle image, after 1000 photons; and the bottom image, after 10,000 photons. As more photons reach the screen, a recognizable interference pattern appears.

for equal time intervals, count photons at each position, and plot out the intensity distribution.

We find that, on average, the distribution of photons agrees with our predictions from Section 38–4. At points corresponding to the maxima of the pattern, we count many photons; at minimum points, we count almost none, and so on. The graph of the counts at various points gives us the same diffraction pattern that we predicted with Eq. (38–7).

But suppose we now reduce the intensity to the point at which only a few photons per second pass through the slit. With only a few photons we can't expect to get the smooth diffraction curve that we found with very large numbers. In fact, there is no way to predict exactly where any individual photon will go. To reconcile the wave and particle aspects of this pattern, we have to regard the pattern as a *statistical* distribution that tells us how many photons, on average, go various places, or the *probability* for an individual photon to land in each of several places. But we can't predict exactly where an individual photon will go.

Now let's take a brief look at a quantum interpretation of a *two-slit* optical interference pattern, which we studied in Section 37–3. We can again trace out the pattern using a photomultiplier and a counter. We reduce the light intensity to a level of a few photons per second. Again we can't predict exactly where an individual photon will go; the interference pattern is a *statistical distribution.*

How does the principle of complementarity apply to these diffraction and interference experiments? The wave description, not the particle description, explains the single- and double-slit patterns. But the particle description, not the wave description, explains why the photomultiplier measures the pattern to be built up by discrete packages of energy. The two descriptions complete our understanding of the results.

In these experiments, light goes through a slit or slits, and photons are detected at various positions on a screen. Suppose we ask the following question: "We just detected a photon at a certain position; when the photon went through the slit, how did it know which way to go?" The problem with the question is that it is asked in terms of a *particle* description. The wave nature of light, not the particle nature, determines the distribution of photons. Asking that question is trying to force a particle description (the photon being somehow directed which way to go) on a wave phenomenon (the formation of the pattern).

We know from experiments that sending electromagnetic waves through slits gives interference patterns and that these patterns are built up by individual photons. But if we didn't know the experimental results, how would we predict when to apply the wave description and when to apply the particle description? We need a theory that includes both these descriptions and predicts as well as explains both types of behavior. Such a comprehensive theory has been developed only in about the past 50 years, in the branch of physics called *quantum electrodynamics* (QED). In this theory the concept of energy levels of an atomic system is extended to electromagnetic fields. Just as an atom exists only in certain definite energy states, so the electromagnetic field has certain well-defined energy states, corresponding to the presence of various numbers of photons with assorted energies, momenta, and polarizations. QED came to full flower 50 years after Planck's quantum hypothesis of 1900 gave quantum mechanics its conceptual birth.

In the following chapters we will find that particles such as electrons also have a dual wave-particle personality. One of the great achievements of quantum mechanics has been to reconcile these apparently incompatible aspects of behavior of photons, electrons, and other constituents of matter.

SUMMARY

- The energy in an electromagnetic wave is carried in packages called photons or quanta. The energy E of one photon, for a wave with frequency f and wavelength λ, is

$$E = hf = \frac{hc}{\lambda}. \tag{40-2}$$

- In the photoelectric effect, a surface can absorb a photon and eject an electron if the photon energy hf is greater than or equal to the work function ϕ. The stopping potential V_0 for the emitted electrons is related to these quantities by

$$eV_0 = hf - \phi. \tag{40-4}$$

- When an atom makes a transition from an energy level E_i to a level E_f by emitting a photon, the frequency f and wavelength λ of the photon are given by

$$hf = \frac{hc}{\lambda} = E_i - E_f. \tag{40-6}$$

The energy levels of the hydrogen atom are

$$E_n = -\frac{hcR}{n^2} = -\frac{13.60 \text{ eV}}{n^2}, \qquad n = 1, 2, 3, 4, \cdots. \tag{40-9}$$

where R is a constant called the Rydberg constant. All the observed spectral series of hydrogen can be understood in terms of these levels.

- The Rutherford scattering experiment shows that at the center of an atom is a dense nucleus, much smaller than the overall size of the atom but containing all of the positive charge and almost all of the mass.

- In the Bohr model of the hydrogen atom, the permitted values of angular momentum are

$$mv_n r_n = n\frac{h}{2\pi}, \tag{40-10}$$

where $n = 1, 2, 3, \cdots$, and n is called the principal quantum number for the level. The corresponding orbital radii r_n and speeds v_n are

$$r_n = \epsilon_0 \frac{n^2 h^2}{\pi m e^2} = n^2 a_0 = n^2(5.29 \times 10^{-11} \text{ m}), \tag{40-12}$$

$$v_n = \frac{1}{\epsilon_0} \frac{e^2}{2nh} = \frac{2.19 \times 10^6 \text{ m/s}}{n}. \tag{40-13}$$

KEY TERMS

line spectrum, 1232
photoelectric effect, 1233
work function, 1233
threshold frequency, 1234
stopping potential, 1234
photon, 1235
Planck's constant, 1235
energy level, 1238
Rydberg constant, 1239
ground level, 1241
excited level, 1241
absorption spectrum, 1242
Rutherford scattering experiments, 1244
nucleus, 1245
Bohr model, 1246
principal quantum number, 1247
reduced mass, 1249
laser, 1251
stimulated emission, 1251
state, 1251
population inversion, 1252
metastable state, 1253
Compton scattering, 1255
blackbody radiation, 1258
Stefan-Boltzmann law, 1258
Wien displacement law, 1259
Planck radiation law, 1259
principle of complementarity, 1261

- The laser operates on the principle of stimulated emission, in which many photons with identical wavelength and phase are emitted. Laser operation requires the creation of a non-equilibrium condition called a population inversion, in which the number of atoms in a higher-energy state is greater than the number in a lower-energy state.

- X rays can be produced by electron impact on a target. For electrons accelerated through a potential increase V_{AC}, the maximum frequency and minimum wavelength are given by

$$eV_{AC} = hf_{max} = \frac{hc}{\lambda_{min}}. \tag{40–20}$$

- Compton scattering is scattering of x-ray photons by electrons. For free electrons the wavelengths λ and λ' of incident and scattered photons are related to the scattering angle ϕ by

$$\lambda' - \lambda = \frac{h}{mc}(1 - \cos \phi), \tag{40–21}$$

where $h/mc = 2.426 \times 10^{-12}$ m.

- The Stefan-Boltzmann law states that the total radiated intensity I from a blackbody surface is related to the absolute temperature T by

$$I = \sigma T^4, \tag{40–26}$$

where $\sigma = 5.67 \times 10^{-8}$ W/m$^2 \cdot$ K^4 is called the Stefan-Boltzmann constant. In blackbody radiation the wavelength λ_m at which the spectral emittance is maximum is inversely proportional to the absolute temperature T, such that

$$\lambda_m T = 2.90 \times 10^{-3} \text{ m} \cdot \text{K}. \tag{40–28}$$

The Planck radiation law gives the intensity per wavelength interval $I(\lambda)$ in blackbody radiation:

$$I(\lambda) = \frac{2\pi hc^2}{\lambda^5(e^{hc/\lambda kT} - 1)}. \tag{40–30}$$

- Electromagnetic radiation behaves as both waves and particles, and a comprehensive theory must include both these aspects of its behavior.

DISCUSSION QUESTIONS

Q40–1 In what ways do photons resemble other particles such as electrons? In what ways do they differ? Do photons have mass? Do they have electric charge? Can they be accelerated? What mechanical properties do they have?

Q40–2 There is a certain probability that a single electron may simultaneously absorb *two* identical photons from a high-intensity laser. How would such an occurrence affect the threshold frequency and the equations of Section 40–3? Explain.

Q40–3 Would you expect effects due to the photon nature of light to be generally more important at the low-frequency end of the electromagnetic spectrum (radio waves) or at the high-frequency end (x rays and gamma rays)? Why?

Q40–4 Most black-and-white photographic film (with the exception of some special-purpose films) is less sensitive to red light than blue light, and has almost no sensitivity to infrared. How can these properties be understood on the basis of photons?

Q40–5 Human skin is relatively insensitive to visible light, but ultraviolet radiation can cause severe burns. Does this have anything to do with photon energies? Explain.

Q40–6 Explain why Fig. 40–3 shows that most photoelectrons have kinetic energies less than $hf - \phi$, and also explain how these smaller kinetic energies occur.

Q40–7 Figure 40–4 shows that in a photoelectric-effect experiment, the photocurrent i for large positive values of V_{AC} has the same value no matter what the light frequency f (provided that f is greater than the threshold frequency f_0). Explain why.

Q40–8 On a graph of K_{max} versus f for the photoelectric effect, what is the physical significance of a) the slope? b) the intercept with the horizontal (f) axis? c) the intercept with the vertical (K_{max}) axis?

Q40–9 The materials called *phosphors* that coat the inside of a fluorescent lamp convert ultraviolet radiation (from the mercury-vapor discharge inside the tube) into visible light. Could one also make a phosphor that converts visible light to ultraviolet? Explain.

Q40–10 The total internal energy (kinetic plus potential) of the hydrogen atom is negative. What significance does this have?

Q40–11 Galaxies tend to be strong emitters of Lyman-α photons (from the $n = 2$ to $n = 1$ transition in atomic hydrogen). But the intergalactic medium—the very thin gas between the galaxies—tends to *absorb* Lyman-α photons. What can you infer from these observations about the temperature in these two environments? Explain.

Q40–12 A doubly ionized lithium atom (Li^{++}) is one that has had two of its three electrons removed. The energy levels of the remaining single-electron ion are closely related to those of the hydrogen atom. The nuclear charge for lithium is $+3e$ instead of just $+e$. How are the energy levels related to those of hydrogen? How is the *radius* of the ion in the ground level related to that of the hydrogen atom? Explain.

Q40–13 The emission of a photon by an isolated atom is a recoil process in which momentum is conserved. Thus Eq. (40–6) should include a recoil kinetic energy K_r for the atom. Why is this energy negligible in that equation?

Q40–14 How might the energy levels of an atom be measured directly, that is, without recourse to analysis of spectra?

Q40–15 Consider the line spectrum emitted from a gas discharge tube such as a neon sign or a sodium-vapor or mercury-vapor lamp. It is found that when the pressure of the vapor is increased, the spectrum lines spread out over a larger range of wavelengths, that is, are less monochromatic. Why?

Q40–16 Elements in the gaseous state emit line spectra with well-defined wavelengths. But hot solid bodies always emit a continuous spectrum, that is, a continuous smear of wavelengths. Can you account for this difference?

Q40–17 As a body is heated to a very high temperature and becomes self-luminous, the apparent color of the emitted radiation shifts from red to yellow and finally to blue as the temperature increases. Why does the color shift? What other changes in the character of the radiation occur?

Q40–18 Can Compton scattering occur with protons as well as electrons? For example, suppose a beam of x rays is directed at a target of liquid hydrogen. (Recall that the nucleus of hydrogen consists of a single proton.) Compared to Compton scattering with electrons, what similarities and differences would you expect? Explain.

Q40–19 Why must engineers and scientists shield against x-ray production in high-voltage equipment?

EXERCISES

SECTION 40–3 THE PHOTOELECTRIC EFFECT

40–1 A photon of green light has a wavelength of 520 nm. Find the photon's frequency, magnitude of momentum, and energy. Express the energy both in joules and electron volts.

40–2 A laser used to weld detached retinas emits light with a wavelength of 652 nm in pulses that are 20.0 ms in duration. The average power during each pulse is 0.600 W. a) How much energy is in each pulse in joules? In electron volts? b) What is the energy of one photon in joules? In electron volts? c) How many photons are in each pulse?

40–3 An excited nucleus emits a gamma-ray photon with an energy of 2.45 MeV. a) What is the photon frequency? b) What is the photon wavelength? c) How does the wavelength compare with a typical nuclear diameter of 10^{-14} m?

40–4 The predominant wavelength emitted by an ultraviolet lamp is 248 nm. If the total power emitted at this wavelength is 12.0 W, how many photons are emitted per second?

40–5 A clean nickel surface is exposed to light of wavelength 235 nm. What is the maximum speed of the photoelectrons emitted from this surface? Use Table 40–1.

40–6 The photoelectric threshold wavelength of a tungsten surface is 272 nm. Calculate the maximum kinetic energy of the electrons ejected from this tungsten surface by ultraviolet radiation of frequency 1.45×10^{15} Hz. Express the answer in electron volts.

40–7 When ultraviolet light with a wavelength of 254 nm falls upon a clean copper surface, the stopping potential necessary to stop emission of photoelectrons is 0.181 V. a) What is the photo-electric threshold wavelength for this copper surface? b) What is the work function for this surface, and how does your calculated value compare with that given in Table 40–1?

40–8 The photoelectric work function of potassium is 2.3 eV. If light having a wavelength of 250 nm falls on potassium, find a) the stopping potential in volts; b) the kinetic energy in electron volts of the most energetic electrons ejected; c) the speeds of these electrons.

40–9 A photon has momentum of magnitude 8.24×10^{-28} kg · m/s. a) What is the energy of this photon? Give your answer in joules and in electron volts. b) What is the wavelength of this photon? In what region of the electro-magnetic spectrum does it lie?

40–10 a) In the photoelectric effect, what is the relation between the threshold frequency f_0 and the work function ϕ? b) The threshold wavelength for producing photoelectrons from a metal surface is 372 nm. What is the work function for this surface, in eV?

SECTION 40–4 ATOMIC LINE SPECTRA AND ENERGY LEVELS

40–11 a) An atom initially in an energy level with $E = -6.52$ eV absorbs a photon that has wavelength 860 nm. What is the internal energy of the atom after it absorbs the photon? b) An atom initially in an energy level with $E = -2.68$ eV emits a photon that has wavelength 420 nm. What is the internal energy of the atom after it emits the photon?

40–12 The energy-level scheme for the hypothetical one-electron element Searsium is shown in Fig. 40–28. The potential energy is taken to be zero for an electron at an infinite distance from the nucleus. a) How much energy (in electron volts) does it take to ionize an electron from the ground level? b) An 18-eV photon is absorbed by a Searsium atom in its ground level. When the atom returns to its ground level, what possible energies can the emitted photons have? c) What will happen if a photon with an energy of 8 eV strikes a Searsium atom in its ground level? Why? d) Photons emitted in the Searsium transitions $n = 3 \rightarrow n = 2$ and $n = 3 \rightarrow n = 1$ will eject photoelectrons from an unknown metal, but the photon emitted from the transition $n = 4 \rightarrow n = 3$ will not. What are the limits (maximum and minimum possible values) of the work function of the metal?

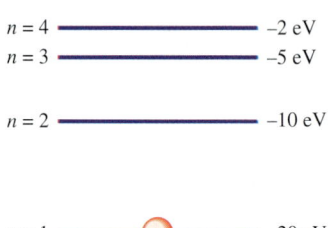

FIGURE 40–28 Exercise 40–12.

40–13 Use Balmer's formula to calculate a) the wavelength, b) the frequency, and c) the photon energy for the H_γ line of the Balmer series for hydrogen.

40–14 Find the longest and shortest wavelengths in the Lyman and Paschen series for hydrogen. In what region of the electromagnetic spectrum does each series lie?

40–15 Using the data shown in Fig. 40–9, determine the *difference* in energy (in electron volts) between the two lowest excited levels of the sodium atom.

SECTION 40–5 THE NUCLEAR ATOM

40–16 A 4.78-MeV alpha particle from a radium ^{226}Ra decay makes a head-on collision with a uranium nucleus. A uranium nucleus has 92 protons. a) What is the distance of closest approach of the alpha particle to the center of the nucleus? Assume that the uranium nucleus remains at rest and that the distance of closest approach is much larger than the radius of the uranium nucleus. b) What is the force on the alpha particle at the instant when it is at the distance of closest approach?

40–17 A beam of alpha particles is incident on a target of lead. A particular alpha particle comes in "head-on" to a particular lead nucleus and stops 6.50×10^{-14} m away from the center of the nucleus. (This point is well outside the nucleus). Assume that the lead nucleus, which has 82 protons, remains at rest. The mass of the alpha particle is 6.64×10^{-27} kg. a) Calculate the electrostatic potential energy at the instant that the alpha particle stops. Express your result in joules and in MeV. b) What initial kinetic energy (in joules and in MeV) did the alpha particle have? c) What was the initial speed of the alpha particle?

SECTION 40–6 THE BOHR MODEL

40–18 A hydrogen atom initially in the ground level absorbs a photon, which excites it to the $n = 4$ level. Determine the wavelength and frequency of the photon.

40–19 A triply ionized beryllium ion, Be^{3+} (a beryllium atom with three electrons removed), behaves very much like a hydrogen atom except that the nuclear charge is four times as great. a) What is the ground-level energy of Be^{3+}? How does this compare to the ground-level energy of the hydrogen atom? b) What is the ionization energy of Be^{3+}? How does this compare to the ionization energy of the hydrogen atom? c) For the hydrogen atom the wavelength of the photon emitted in the $n = 2$ to $n = 1$ transition is 122 nm (Example 40–7 of Section 40–6). What is the wavelength of the photon emitted when a Be^{3+} ion undergoes this transition? d) For a given value of n, how does the radius of an orbit in Be^{3+} compare to that for hydrogen?

40–20 a) What is the angular momentum L of the electron in a hydrogen atom, with respect to an origin at the nucleus, when the atom is in its lowest energy level? b) Repeat part (a) for the ground level of He^+. Compare to the answer in part (a).

40–21 a) Using the Bohr model, calculate the speed of the electron in a hydrogen atom in the $n = 1$, 2, and 3 levels. b) Calculate the orbital period in each of these levels. c) The average lifetime of the first excited level of a hydrogen atom is 1.0×10^{-8} s. In the Bohr model, how many orbits does an electron in the $n = 2$ level complete before returning to the ground level?

40–22 According to the Bohr model (with the nucleus assumed to be at rest), the Rydberg constant R is equal to $me^4/8\epsilon_0^2 h^3 c$. a) Calculate R in m^{-1} and compare with the experimental value. b) Calculate the energy (in electron volts) of a photon whose wavelength equals R^{-1}. (This quantity is known as the *Rydberg energy*.) c) To account for the motion of the nucleus, replace the mass of the electron m with the reduced mass m_r for ordinary hydrogen and repeat parts (a) and (b).

SECTION 40–7 THE LASER

40–23 A large number of neon atoms are in thermal equilibrium. What is the ratio of the number of atoms in a $5s$ state to the number in a $3p$ state at a) 300 K; b) 600 K; c) 1200 K? The energies of these states are given in Fig. 40–21a. d) At any of these temperatures, the rate at which a neon gas will spontaneously emit 632.8-nm radiation is quite low. Explain why.

40–24 Figure 40–9 shows the energy levels of the sodium atom. The two lowest excited levels are shown in columns labeled $^2P_{3/2}$ and $^2P_{1/2}$. Find the ratio of the number of atoms in a $^2P_{3/2}$ state to the number in a $^2P_{1/2}$ state for a sodium gas in thermal equilibrium at 500 K. In which state are more atoms found?

40–25 How many photons per second are emitted by a 7.50-mW CO_2 laser that has a wavelength of 10.6 μm?

40–26 Using the information in Fig. 40–21a, compute the energy difference for the $5s$ to $3p$ transition in neon. Express your result in electron volts and in joules. Compute the wavelength of a photon having this energy, and compare your result with the observed wavelength of the laser light.

SECTION 40–8 X-RAY PRODUCTION AND SCATTERING

40–27 Protons are accelerated from rest by a potential difference of 4.00 kV and strike a metal target. If a proton produces one photon on impact, what is the minimum wavelength of the resulting x rays? How does your answer compare to the minimum wavelength if 4.00-keV electrons are used instead? Why do x-ray tubes use electrons rather than protons to produce x rays?

40–28 a) What is the minimum potential difference between the filament and the target of an x-ray tube if the tube is to produce x rays with a wavelength of 0.150 nm? b) What is the shortest wavelength produced in an x-ray tube operated at 30.0 kV?

40–29 The cathode-ray tubes that generated the picture in early color television sets were sources of x rays. If the acceleration voltage in a TV tube is 15.0 kV, what are the shortest wavelength x rays produced by the television? (Modern TV sets contain shielding to stop these x rays.)

40–30 A beam of x rays with wavelength 0.0500 nm are Compton-scattered by the electrons in a sample. At what angle from the incident beam should you look to find x rays with a wavelength of a) 0.0542 nm; b) 0.0521 nm; c) 0.0500 nm?

40–31 X rays with initial wavelength 0.0665 nm undergo Compton scattering. What is the largest wavelength found in the scattered x rays? At what scattering angle is this wavelength observed?

40–32 X rays are produced in a tube operating at 18.0 kV. After emerging from the tube, x rays with the minimum wavelength produced strike a target and are Compton-scattered through an angle of 45.0°. a) What is the original x-ray wavelength? b) What is the wavelength of the scattered x rays? c) What is the energy of the scattered x rays (in electron volts)?

40–33 Complete the derivation of the Compton-scattering formula, Eq. (40–21), following the outline given in Eqs. (40–22) through (40–25).

SECTION 40–9 CONTINUOUS SPECTRA

40–34 The shortest visible wavelength is about 400 nm. What is the temperature of an ideal radiator whose spectral emittance peaks at this wavelength?

40–35 Radiation has been detected from space that is characteristic of an ideal radiator at $T = 2.728$ K. (This radiation is a relic of the "Big Bang" at the beginning of the universe.) For this temperature, at what wavelength does the Planck distribution peak? In what part of the electromagnetic spectrum is this wavelength?

40–36 Determine λ_m, the wavelength at the peak of the Planck distribution, and the corresponding frequency f, at the following Kelvin temperatures: a) 3.00 K; b) 300 K; c) 3000 K.

40–37 Show that for large values of λ the Planck distribution, Eq. (40–30), agrees with the Rayleigh distribution, Eq. (40–29).

40–38 Light is emitted from an ideal blackbody that has temperature T. Use the approximate method of Example 40–11 (Section 40–9) to calculate the percent of the total spectral emittance that is in the wavelength range 500.0 nm to 501.0 nm for a) $T = 300$ K; b) $T = 2000$ K; c) $T = 6000$ K. d) How do the answers to parts (a), (b), and (c) compare? What does this tell you about the change in the intensity distribution with temperature?

40–39 a) Show that the maximum in the Planck distribution, Eq. (40–30), occurs at a wavelength λ_m given by $\lambda_m = hc/4.965kT$ (Eq. 40–31). As discussed in the text, 4.965 is the root of Eq. (40–32). b) Evaluate the constants in the expression derived in part (a) to show that $\lambda_m T$ has the numerical value given in the Wien displacement law, Eq. (40–28).

40–40 An ideal radiator radiates with a total intensity of $I = 6.94$ MW/m^2. At what wavelength does the spectral emittance $I(\lambda)$ peak?

PROBLEMS

40–41 Exposing Photographic Film. The light-sensitive compound on most photographic films is silver bromide, AgBr. A film is "exposed" when the light energy absorbed dissociates this molecule into its atoms. (The actual process is more complex, but the quantitative result does not differ greatly.) The energy of dissociation of AgBr is 1.00×10^5 J/mol. For a photon that is just able to dissociate a molecule of silver bromide, find a) the photon energy in electron volts, b) the wavelength of the photon, and c) the frequency of the photon. d) What is the energy in electron volts of a photon having a frequency of 100 MHz? e) Light from a firefly can expose photographic film, but the radiation from an FM station broadcasting 50,000 W at 100 MHz cannot. Explain why this is so.

40–42 An atom of mass m emits a photon of wavelength λ. a) What is the recoil speed of the atom? b) What is the kinetic energy K of the recoiling atom? c) Find the ratio K/E, where E is the energy of the emitted photon. If this ratio is much less than unity, the recoil of the atom can be neglected in the emission process. Is the recoil of the atom more important for small or large atomic masses? For long or short wavelengths?

d) Calculate K (in electron volts) and K/E for a hydrogen atom (mass 1.67×10^{-27} kg) that emits an ultraviolet photon of energy 10.2 eV. Is recoil an important consideration in this emission process?

40–43 The directions of spontaneous emission of photons from a source of radiation are random. According to the wave theory, the intensity of radiation from a point source varies inversely as the square of the distance from the source. Show that the number of photons from a point source passing out through a unit area is also given by an inverse-square law.

40–44 The photoelectric work functions for particular samples of certain metals are as follows: cesium, 2.1 eV; copper, 4.7 eV; potassium, 2.3 eV; and zinc, 4.3 eV. a) What is the threshold wavelength for each metal surface? b) Which of these metals could *not* emit photoelectrons when irradiated with visible light (400–700 nm)?

40–45 When a certain photoelectric surface is illuminated with light of different wavelengths, the following stopping potentials are observed:

Wavelength (nm)	Stopping potential (V)
366	1.48
405	1.15
436	0.93
492	0.62
546	0.36
579	0.24

Plot the stopping potential on the vertical axis against the frequency of the light on the horizontal axis. Determine a) the threshold frequency; b) the threshold wavelength; c) the photoelectric work function of the material (in electron volts); d) the value of Planck's constant h (assuming that the value of e is known).

40–46 a) If the average frequency emitted by a 200-W light bulb is 5.00×10^{14} Hz, and 10.0% of the input power is emitted as visible light, approximately how many visible-light photons are emitted per second? b) At what distance would this correspond to 1.00×10^{11} visible-light photons per square centimeter per second if the light is emitted uniformly in all directions?

40–47 a) The wavelength of light incident on a metal surface is reduced from λ_1 to λ_2. (Both λ_1 and λ_2 are less than the threshold wavelength for the surface.) When the wavelength is reduced in this way, what is the change in the stopping potential for photoelectrons emitted from the surface? b) Evaluate the change in stopping potential for $\lambda_1 = 295$ nm and $\lambda_2 = 265$ nm.

40–48 Consider a hydrogenlike atom with nuclear charge Ze. a) For what value of Z (rounded to the nearest integer value) is the Bohr speed of the electron in the ground level equal to 10.0% of the speed of light? b) For what value of Z (rounded to the nearest integer value) is the ionization energy of the ground level equal to 1.0% of the rest energy of the electron?

40–49 The negative muon has a charge equal to that of an electron but a mass that is 207 times as great. Consider a hydrogenlike atom consisting of a proton and a muon. a) What is the reduced mass of the atom? b) What is the ground-level energy (in electron volts)? c) What is the wavelength of the radiation emitted in the transition from the $n = 2$ level to the $n = 1$ level?

40–50 An unknown element has a spectrum for absorption from its ground level with lines at 2.0, 5.0, and 9.0 eV. Its ionization energy is 10.0 eV. a) Draw an energy-level diagram for this element. b) If a 9.0-eV photon is absorbed, what energies can the subsequently emitted photons have?

40–51 A sample of hydrogen atoms is irradiated with light with a wavelength of 85.5 nm, and electrons are observed leaving the gas. a) If each hydrogen atom were initially in its ground level, what would be the maximum kinetic energy in electron volts of these photoelectrons? b) A few electrons are detected with energies as much as 10.2 eV greater than the maximum kinetic energy calculated in part (b). How can this be?

40–52 Bohr Orbits of a Satellite. A 20.0-kg satellite circles the earth once every 2.00 h in an orbit having a radius of 8060 km. a) Assuming that Bohr's angular-momentum result ($L = nh/2\pi$) applies to satellites just as it does to an electron in the hydrogen atom, find the quantum number n of the orbit of the satellite. b) Show from Bohr's angular momentum result and Newton's law of gravitation that the radius of an earth-satellite orbit is directly proportional to the square of the quantum number, $r = kn^2$, where k is the constant of proportionality. c) Using the result from part (b), find the distance between the orbit of the satellite in this problem and its next "allowed" orbit. (Calculate a numerical value.) d) Comment on the possibility of observing the separation of the two adjacent orbits. e) Do quantized and classical orbits correspond for this satellite? Which is the "correct" method for calculating the orbits?

40–53 A nonrelativistic particle of mass m is held in circular orbit around the origin by an attractive force $F(r) = -Dr$, where D is a positive constant. Use the Bohr-model idea that only certain values of the angular momentum are allowed to answer the following questions. a) What are the allowed values of the orbital radius? b) What are the allowed energies? (Let the potential energy be zero when $r = 0$.) c) If the particle is excited, it can decay back down to the ground level by emitting one or more photons. What are the possible photon energies? d) Describe a possible physical situation to which this problem corresponds.

40–54 The mass of a deuteron is 3.34359×10^{-27} kg, the mass of a proton is 1.672623×10^{-27} kg, and the mass of an electron is 9.10939×10^{-31} kg. a) Calculate the reduced mass of a deuterium atom and express your result in terms of the electron mass m. b) Consider the $n = 2$ to $n = 1$ transition in atomic hydrogen. Calculate the difference in the wavelength of light emitted in this transition by a hydrogen atom and by a deuterium atom.

40–55 a) What is the least amount of energy in electron volts that must be given to a hydrogen atom initially in its ground level so that it can emit the H_α line in the Balmer series? b) How many different possibilities of spectral-line emissions are there for this atom when the electron starts in the $n = 3$ level and eventually ends up in the ground level? Calculate the wavelength of the emitted photon in each case.

40–56 A large number of hydrogen atoms are in thermal equilibrium. Let n_2/n_1 be the ratio of the number of atoms in an $n = 2$ excited state to the number of atoms in an $n = 1$ ground state. At what temperature is n_2/n_1 equal to a) 10^{-12}; b) 10^{-8}; c) 10^{-4}? d) Like the sun, other stars have continuous spectra with dark absorption lines (see Fig. 40–10). The absorption takes place in the star's atmosphere, which in all stars is composed primarily of hydrogen. Explain why the Balmer absorption lines are relatively weak in stars with low atmospheric temperatures such as the sun (atmosphere temperature = 5800 K) but strong in stars with higher atmospheric temperatures.

40–57 When a photon is emitted by an atom, the atom must recoil to conserve momentum. This means that the photon and the recoiling atom share the transition energy. a) For an atom of mass m, calculate the correction $\Delta\lambda$ due to recoil to the wavelength of an emitted photon. Let λ be the wavelength of the photon if recoil is not taken into consideration. (*Hint:* The correction is very small, as Problem 40–42 suggests, so $|\Delta\lambda|/\lambda \ll 1$. Use this fact to obtain an approximate but very accurate expression for $\Delta\lambda$.) b) Evaluate the correction for a hydrogen atom in which an electron in the nth level returns to the ground level. How does the answer depend on n?

40–58 A photon with a wavelength of 0.1800 nm is Compton-scattered through an angle of 180°. a) What is the wavelength of the scattered photon? b) How much energy is given to the electron? c) What is the recoil speed of the electron? Is it necessary to use the relativistic kinetic-energy relationship?

40–59 a) Calculate the maximum increase in photon wavelength that can occur during Compton scattering. b) What is the energy (in electron volts) of the smallest-energy x-ray photon for which Compton scattering could result in doubling the original wavelength?

40–60 An x-ray tube is operating at voltage V and current I. a) If only a fraction p of the electric power supplied is converted into x rays, at what rate is energy being delivered to the target? b) If the target has mass m and specific heat capacity c (in J/kg · K), at what average rate would its temperature rise if there were no thermal losses? c) Evaluate your results from parts (a) and (b) for an x-ray tube operating at 18.0 kV and 60.0 mA that converts 1.0% of the electric power into x rays. Assume that the 0.250-kg target is made of lead ($c = 130$ J/kg · K). d) What must be the physical properties of a practical target material? What would be some suitable target elements?

40–61 A photon with wavelength 0.1100 nm collides with a free electron that is initially at rest. After the collision the wavelength is 0.1132 nm. a) What is the kinetic energy of the electron after the collision? What is its speed? b) If the electron is suddenly stopped (for example, in a solid target), all of its kinetic energy is used to create a photon. What is the wavelength of this photon?

40–62 a) Derive an expression for the total shift in photon wavelength after two successive Compton scatterings from electrons at rest. The photon is scattered by an angle θ_1 in the first scattering and by θ_2 in the second. b) In general, is the total shift in wavelength produced by two successive scatterings of an angle $\theta/2$ the same as by a single scattering of θ? If not, are there any specific values of θ, other than $\theta = 0°$, for which the total shifts are the same? c) Use the result of part (a) to calculate the total wavelength shift produced by two successive Compton scatterings of 30.0° each. Express your answer in terms of h/mc. d) What is the wavelength shift produced by a single Compton scattering of 60.0°? Compare to the answer in part (c).

40–63 Nuclear fusion reactions at the center of the sun produce gamma-ray photons with energies of order 1 MeV (10^6 eV). By contrast, what we see emanating from the sun's surface are visible-light photons with wavelengths of order 500 nm. A simple model that explains this difference in wavelength is that a photon undergoes Compton scattering many times—in fact,

about 10^{26} times, as suggested by models of the solar interior—as it travels from the center of the sun to its surface. a) Estimate the increase in wavelength of a photon in an average Compton-scattering event. b) Find the angle in degrees through which the photon is scattered in the scattering event described in part (a). (*Hint:* A useful approximation is $\cos\phi \approx 1 - \phi^2/2$, which is valid for $\phi \ll 1$. Note that ϕ is in radians in this expression.) c) It is estimated that a photon takes about 10^6 years to travel from the core to the surface of the sun. Find the average distance that light can travel within the interior of the sun without being scattered. (This distance is roughly equivalent to how far you could see if you were inside the sun and could survive the extreme temperatures there. As your answer shows, the interior of the sun is *very* opaque.)

40–64 An x-ray photon is scattered from a free electron (mass m) at rest. The wavelength of the scattered photon is λ', and the final speed of the struck electron is v. a) What was the initial wavelength λ of the photon? Express your answer in terms of λ', v, and m. (*Hint:* Use the relativistic expression for the electron kinetic energy.) b) Through what angle ϕ is the photon scattered? Express your answer in terms of λ, λ', and m. c) Evaluate your results in parts (a) and (b) for a wavelength of 5.10×10^{-3} nm for the scattered photon and a final electron speed of 1.80×10^8 m/s. Give ϕ in degrees.

40–65 a) Write the Planck distribution law in terms of the frequency f rather than the wavelength λ, to obtain $I(f)$. b) Show that

$$\int_0^\infty I(\lambda)\, d\lambda = \frac{2\pi^5 k^4}{15c^2h^3}\, T^4,$$

where $I(\lambda)$ is the Planck distribution formula of Eq. (40–30). (*Hint:* Change the integration variable from λ to f.) You will need to use the following tabulated integral:

$$\int_0^\infty \frac{x^3}{e^{\alpha x}-1}\, dx = \frac{1}{240}\left(\frac{2\pi}{\alpha}\right)^4.$$

c) The result of (b) is I, and has the form of the Stefan-Boltzmann law, $I = \sigma T^4$ (Eq. 40–26). Evaluate the constants in (b) to show that σ has the value given in Section 40–9.

40–66 An Ideal Blackbody. A large cavity with a very small hole and maintained at a temperature T is a good approximation to an ideal radiator or blackbody. Radiation can pass into or out of the cavity only through the hole. The cavity is a perfect absorber, since any radiation incident on the hole becomes trapped inside the cavity. Such a cavity at 200°C has a hole with area 4.00 mm². How long does it take for the cavity to radiate 100 J of energy through the hole?

CHALLENGE PROBLEMS

40–67 a) Show that in the Bohr model, the frequency of revolution of an electron in its circular orbit around a stationary hydrogen nucleus is $f = me^4/4\epsilon_0^2 n^3h^3$. b) In classical physics, the frequency of revolution of the electron is equal to the frequency of the radiation that it emits. Show that when n is very large, the frequency of revolution does indeed equal the radiated

frequency calculated from Eq. (40–6) for a transition from $n_1 = n + 1$ to $n_2 = n$. (This illustrates Bohr's *correspondence principle,* which is often used as a check on quantum calculations. When n is small, quantum physics gives results that are very different from those of classical physics. When n is large, the differences are not significant, and the two methods then "correspond." In fact, when Bohr first tackled the hydrogen atom problem, he sought to determine f as a function of n such that it would correspond to classical results for large n.)

40–68 Consider a beam of monochromatic light with intensity I incident on a perfectly absorbing surface oriented perpendicular to the beam. Use the photon concept to show that the radiation pressure exerted by the light on the surface is given by I/c.

40–69 Consider Compton scattering of a photon by a *moving* electron. Before the collision the photon has wavelength λ and is moving in the $+x$-direction, and the electron is moving in the $-x$-direction with total energy E (including its rest energy mc^2). The photon and electron collide head-on. After the collision, both are moving in the $-x$-direction (that is, the photon has been scattered by 180°). a) Derive an expression for the wavelength λ' of the scattered photon. Show that if $E \gg mc^2$, where m is the rest mass of the electron, your result reduces to

$$\lambda' = \frac{hc}{E}\left(1 + \frac{m^2c^4\lambda}{4hcE}\right).$$

b) A beam of infrared radiation from a CO_2 laser ($\lambda = 10.6\ \mu m$) collides head-on with a beam of electrons, each of total energy $E = 10.0$ GeV (1 GeV $= 10^9$ eV). Calculate the wavelength λ' of the scattered photons, assuming a 180° scattering angle. c) What kind of scattered photons are these (infrared, microwave, ultraviolet, etc.)? Can you think of an application of this effect?

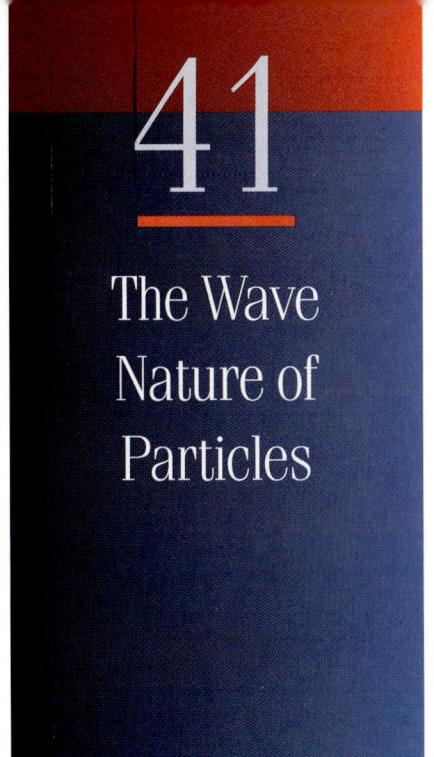

41 The Wave Nature of Particles

Both of these images appear to show the diffraction pattern formed when light shines on a sharp edge (see Fig. 38–2). Indeed, the image on the top was made in just this way. But the image on the bottom was made by a beam of electrons directed at a sharp edge. This experiment and others show that electrons and other particles undergo diffraction—that is, they behave like waves.

41–1 INTRODUCTION

In preceding chapters we've seen one aspect of nature's wave-particle duality: Light and other electromagnetic radiation sometimes act like waves and sometimes like particles. Interference and diffraction demonstrate wave behavior, while emission and absorption of photons demonstrate the particle behavior.

A complete theory should also be able to *predict,* on theoretical grounds, the energy levels of any particular atom. The 1913 Bohr model of the hydrogen atom was a step in this direction. But it combined classical principles with new ideas that were inconsistent with classical theory, and it raised as many questions as it answered. More drastic departures from classical concepts were needed.

A successful drastic departure is *quantum mechanics,* a theory that began to emerge in the 1920s. Besides waves that sometimes act like particles, quantum mechanics extends the concept of wave-particle duality to include particles that sometimes show *wavelike* behavior. In these situations a particle is modeled as an inherently spread-out entity that can't be described as a point with a perfectly definite position and velocity.

Quantum mechanics is the key to understanding atoms and molecules, including their structure, spectra, chemical behavior, and many other properties. It has the happy effect of restoring unity and symmetry to our description of both particles and radiation. We will learn about quantum mechanics and use its results throughout the remainder of this book.

41–2 DE BROGLIE WAVES

A major advance in the understanding of atomic structure began in 1924, about ten years after the Bohr model, with a proposition made by a French physicist and nobleman, Prince Louis de Broglie (pronounced "de broy"). His reasoning, freely paraphrased, went like this: Nature loves symmetry. Light is dualistic in nature, behaving in some situations like waves and in others like particles. If nature is symmetric, this duality should also hold for matter. Electrons and protons, which we usually think of as *particles,* may in some situations behave like *waves.*

If a particle acts like a wave, it should have a wavelength and a frequency. De Broglie postulated that a free particle with rest mass m, moving with nonrelativistic speed v, should have a wavelength λ related to its momentum $p = mv$ in exactly the same way as for a photon, as expressed by Eq. (40–5): $\lambda = h/p$. The **de Broglie wavelength** of a particle is then

$$\lambda = \frac{h}{p} = \frac{h}{mv} \quad \begin{array}{l}\text{(de Broglie wavelength} \\ \text{of a particle),}\end{array} \quad (41\text{–}1)$$

where h is Planck's constant. If the particle's speed is an appreciable fraction of the speed of light c, we use Eq. (39–28) to replace mv in Eq. (41–1) with $\gamma mv = mv/\sqrt{1 - v^2/c^2}$. The frequency f, according to de Broglie, is also related to the particle's energy E in the same way as for a photon, namely,

$$E = hf.$$

Thus the relations of wavelength to momentum and of frequency to energy, in de Broglie's hypothesis, are exactly the same for particles as for photons.

CAUTION ▶ The relationship $E = hf$ must be carefully applied to nonzero-rest-mass particles such as electrons and protons. Unlike a photon, they do *not* travel at speed c, so neither of the equations $f = c/\lambda$ nor $E = pc$ apply to them! ◀

To appreciate the enormous significance of de Broglie's proposal, we have to realize that at the time there was no direct experimental evidence that particles have wave characteristics. It is one thing to suggest a new hypothesis to explain experimental observations; it is quite another to propose such a radical departure from established concepts on theoretical grounds alone. But it was clear that a radical idea was needed. The dual nature of electromagnetic radiation had led to adoption of the photon concept, also a radical idea. The relatively complete lack of success in understanding atomic structure indicated that a similar revolution was needed in the mechanics of particles.

De Broglie's hypothesis was the beginning of that revolution. Within a few years after 1924 it was developed by Heisenberg, Schrödinger, Dirac, Born, and many others into a detailed theory called **quantum mechanics.** This development was well underway even before direct experimental evidence for the wave properties of particles was found.

Quantum mechanics involves sweeping revisions of our fundamental concepts of the description of matter. A particle is not a geometric point but an entity that is spread out in space. The spatial distribution of a particle is defined by a function called a **wave function,** which is closely analogous to the wave functions we used for mechanical waves in Chapter 19 and for electromagnetic waves in Chapter 33. The wave function for a *free* particle with definite energy has a recurring wave pattern with definite wavelength and frequency. The wave and particle aspects are not inconsistent; the *principle of complementarity,* which we discussed in Section 40–9, tells us that we need both the particle model and the wave model for a complete description of nature.

In the Bohr model, we pictured the energy levels of the hydrogen atom in terms of definite electron orbits, as shown in Fig. 40–8. This is an oversimplification and should not be taken literally. But the most important idea in Bohr's theory was the existence of discrete energy levels and their relation to the frequencies of emitted photons. The new quantum mechanics still assigns only certain allowed energy states to an atom, but with a more general description of the electron motion in terms of wave functions. In the hydrogen atom the energy levels predicted by quantum mechanics turn out to be the same as those given by Bohr's theory. In more complicated atoms, for which the Bohr theory does not work, the quantum-mechanical picture is in excellent agreement with observation.

The de Broglie wave hypothesis has an interesting relation to the Bohr model. We can use Eq. (41–1) to obtain the Bohr quantum condition that the angular momentum $L = mvr$ must be an integer multiple of Planck's constant h. The method is analogous to determining the normal-mode frequencies of standing waves. We discussed this problem in Sections 20–4, 20–5, and 33–7; the central idea was to satisfy the **boundary conditions** for the waves. For example, for standing waves on a string that is fixed at both ends, the ends are always nodes, and there will be additional nodes along the string for all but the fundamental mode. For the boundary conditions to be satisfied, the total length of the string must equal some *integral* number of half-wavelengths.

A standing wave on a string transmits no energy, and electrons in Bohr's orbits radiate no energy. So think of an electron as a standing wave fitted around a circle in one of the Bohr orbits. For the wave to "come out even" and join onto itself smoothly, the

Louis-Victor de Broglie, the 7th Duke de Broglie (1892–1987), broke with family tradition by choosing a career in physics rather than as a diplomat. His revolutionary proposal that particles have wave characteristics—for which de Broglie won the 1929 Nobel Prize in physics—was published in his doctoral thesis.

18.3
Standing Electron Waves

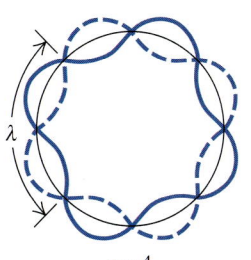

$n = 2$ $n = 3$ $n = 4$

41–1 Diagrams showing the idea of fitting a standing wave around a circular orbit. For the wave to join onto itself smoothly, the circumference of the orbit must be an integral number n of wavelengths. Examples are shown for $n = 2$, 3, and 4.

circumference of this circle must include some *whole number* of wavelengths, as suggested by Fig. 41–1. For an orbit with radius r and circumference $2\pi r$, we must have $2\pi r = n\lambda$, where $n = 1, 2, 3, \ldots$. According to the de Broglie relation, Eq. (41–1), the wavelength λ of a particle with rest mass m, moving with nonrelativistic speed v, is $\lambda = h/mv$. Combining $2\pi r = n\lambda$ and $\lambda = h/mv$, we find $2\pi r = nh/mv$, or

$$mvr = n\frac{h}{2\pi}. \qquad (41–2)$$

We recognize Eq. (41–2) as being identical to Eq. (40–10), Bohr's result that the magnitude of the angular momentum $L = mvr$ must equal an integer n times $h/2\pi$. Thus a wave-mechanical picture leads naturally to the quantization of the electron's angular momentum.

To be sure, the idea of fitting a standing wave around a circular orbit is a rather vague notion. But the agreement of Eq. (41–2) with Bohr's result is much too remarkable to be a coincidence. It strongly suggests that the wave properties of electrons do indeed have something to do with atomic structure.

Later we will learn how wave functions for specific systems are determined by solution of a wave equation called the Schrödinger equation. Boundary conditions play a central role in finding solutions of this equation and thus in determining possible energy levels, values of angular momentum, and other properties.

Problem–Solving Strategy

QUANTUM MECHANICS

1. In atomic physics the orders of magnitude of physical quantities are so unfamiliar that often common sense isn't much help in judging the reasonableness of a result. In working out problems, be very careful to handle powers of ten properly. A gross error may not be obvious. To check your results, it helps to remind yourself of some typical orders of magnitude:

   Size of an atom: 10^{-10} m;
   Mass of an atom: 10^{-26} kg;
   Mass of an electron: 10^{-30} kg;
   Energy magnitude of an atomic state: 1 to 10 eV (10^{-18} J) for outer electrons (but some interaction energies are much smaller);
   Speed of an electron in the Bohr atom: 10^{6} m/s;
   Electron charge magnitude: 10^{-19} C;
   kT at room temperature: 1/40 eV.

   You may want to add items to this list. These approximate values will also help you in Chapter 45, in which we will deal with magnitudes that are characteristic of *nuclear* rather than atomic structure; these are often different by factors of 10^{4} to 10^{6}.

2. As in Chapter 40, energies may be expressed in either joules or electron volts. Be sure you use consistent units. Lengths, such as wavelengths, are always in meters if you use the other quantities consistently in SI units, such as $h = 6.626 \times 10^{-34}$ J · s. If you want nanometers or something else, don't forget to convert. In some problems it's useful to express h in eV · s: $h = 4.136 \times 10^{-15}$ eV · s.

3. Nonrelativistic kinetic energy can be expressed either as $K = \frac{1}{2}mv^2$ or (because $p = mv$) as $K = p^2/2m$. The latter form is often useful in calculations involving the de Broglie wavelength.

4. Aside from these calculational details, the main challenges of this chapter are conceptual, not computational. Try to keep an open mind when you encounter new and sometimes jarring ideas. Photons and electrons both *do* have wavelike and particlelike properties. It takes time to develop an understanding of the concepts and results of quantum mechanics.

EXAMPLE 41-1

Energy of a thermal neutron   Find the speed and kinetic energy of a neutron ($m = 1.675 \times 10^{-27}$ kg) that has a de Broglie wavelength $\lambda = 0.200$ nm, approximately the atomic spacing in many crystals. Compare the energy with the average kinetic energy of a gas molecule at room temperature ($T = 20°C = 293$ K).

SOLUTION   From Eq. (41–1),

$$v = \frac{h}{\lambda m} = \frac{6.626 \times 10^{-34} \text{ J} \cdot \text{s}}{(0.200 \times 10^{-9} \text{ m})(1.675 \times 10^{-27} \text{ kg})}$$

$$= 1.98 \times 10^{3} \text{ m/s}.$$

Because this speed is much less than the speed of light, we are justified in using the nonrelativistic form of Eq. (41–1). Thus the kinetic energy is

$$K = \frac{1}{2} mv^{2} = \frac{1}{2}(1.675 \times 10^{-27} \text{ kg})(1.98 \times 10^{3} \text{ m/s})^{2}$$

$$= 3.28 \times 10^{-21} \text{ J} = 0.0204 \text{ eV}.$$

The average translational kinetic energy of a molecule of an ideal gas is given by Eq. (16–16):

$$\frac{1}{2} m(v^{2})_{\text{av}} = \frac{3}{2} kT = \frac{3}{2}(1.38 \times 10^{-23} \text{ J/K})(293 \text{ K})$$

$$= 6.07 \times 10^{-21} \text{ J} = 0.0379 \text{ eV}.$$

The two energies are comparable in magnitude. In fact, a neutron with kinetic energy in this range is called a *thermal neutron*. Diffraction of thermal neutrons, which we'll discuss in the next section, is used to study crystal and molecular structure in the same way as x-ray diffraction. Neutron diffraction has proved to be especially useful in the study of large organic molecules.

41–3 ELECTRON DIFFRACTION

De Broglie's wave hypothesis, radical though it seemed, almost immediately received experimental confirmation. The first direct evidence involved a diffraction experiment with electrons that was analogous to the x-ray diffraction experiments that we described in Section 38–7. In those experiments, atoms in a crystal act as a three-dimensional diffraction grating for x rays. An x-ray beam is strongly reflected when it strikes a crystal at an angle that gives constructive interference among the waves scattered from the various atoms in the crystal. These interference effects demonstrate the *wave* nature of x rays.

In 1927, Clinton Davisson and Lester Germer, working at the Bell Telephone Laboratories, were studying the surface of a piece of nickel by directing a beam of *electrons* at the surface and observing how many electrons bounced off at various angles. Figure 41–2a shows an experimental setup like theirs. The specimen was polycrystalline; like many ordinary metals, it consisted of many microscopic crystals bonded together with random orientations. The experimenters expected that even the smoothest surface attainable would still look rough to an electron and that the electron beam would be diffusely reflected, with a smooth distribution of intensity as a function of the angle θ.

During the experiment an accident occurred that permitted air to enter the vacuum chamber, and an oxide film formed on the metal surface. To remove this film, Davisson and Germer baked the specimen in a high-temperature oven, almost hot enough to melt it. Unknown to them, this had the effect of *annealing* the specimen, creating large single-crystal regions with crystal planes that were continuous over the width of the electron beam.

When the observations were repeated, the results were quite different. Strong maxima in the intensity of the reflected electron beam occurred at specific angles (Fig. 41–2b), in contrast to the smooth variation of intensity with angle that Davisson and Germer had observed before the accident. The angular positions of the maxima depended on the accelerating voltage V_{ba} used to produce the electron beam. Davisson and Germer were familiar with de Broglie's hypothesis, and they noticed the similarity of this behavior to x-ray diffraction. This was not the effect they had been looking for,

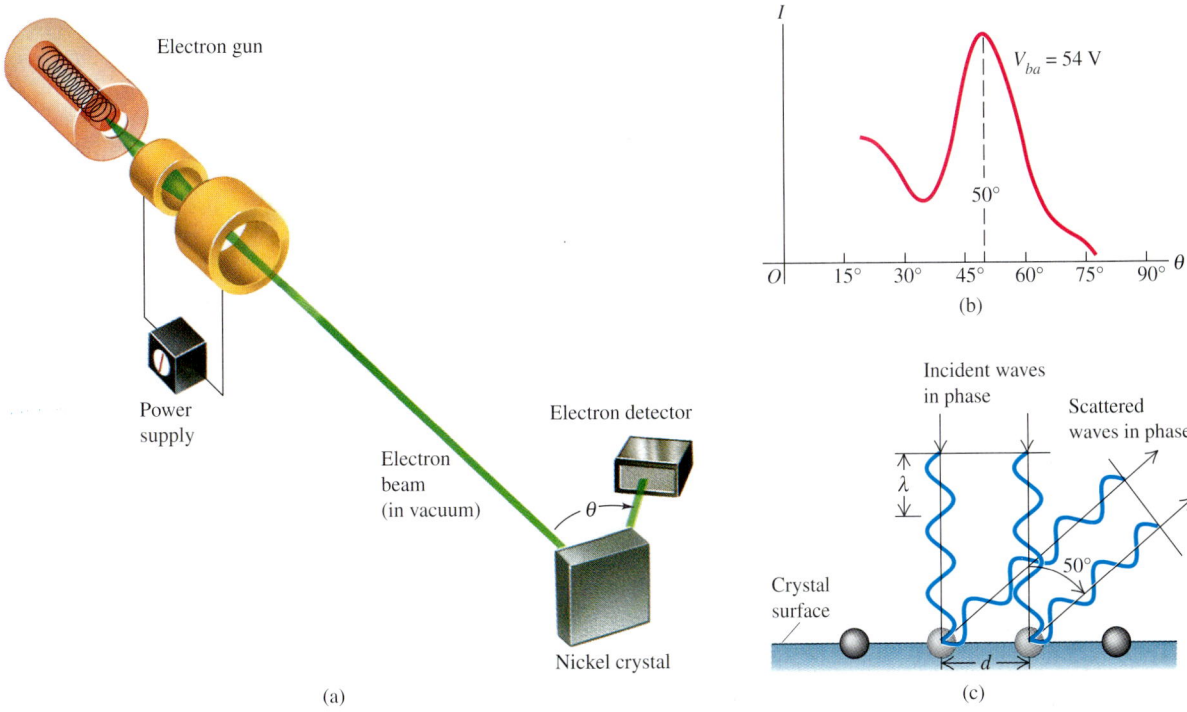

Electron gun

Power supply

Electron beam (in vacuum)

θ

Electron detector

Crystal surface

Nickel crystal

(a)

I

$V_{ba} = 54$ V

$50°$

O $15°$ $30°$ $45°$ $60°$ $75°$ $90°$ θ

(b)

Incident waves in phase

Scattered waves in phase

λ

$50°$

d

(c)

41–2 (a) An apparatus similar to that used by Davisson and Germer. Electrons emitted from the heated filament are accelerated by the electrodes in the electron gun and directed at the crystal, and the electrons in the scattered beam are observed by a detector. The orientation of the detector, described by the angle θ, can be varied. (b) A graph of intensity of the scattered beam as a function of the angle θ. The sharp peak at $\theta = 50°$ results from constructive interference between electron waves scattered from various atoms in the surface layer of the crystal. (c) Constructive interference of waves scattered from two adjacent atoms for the case in which the extra distance $d \sin 50°$ for the top wave equals 1λ.

but they immediately recognized that the electron beam was being *diffracted*. They had discovered a very direct experimental confirmation of the wave hypothesis.

Davisson and Germer could determine the speeds of the electrons from the accelerating voltage, so they could compute the de Broglie wavelength from Eq. (41–1). The electrons were scattered primarily by the planes of atoms at the surface of the crystal. Atoms in a surface plane are arranged in rows, with a distance d that can be measured by x-ray diffraction techniques. These rows act like a reflecting diffraction grating; the angles at which strong reflection occurs are the same as for a grating with center-to-center distance d between its slits. From Eq. (38–13) the angles of maximum reflection are given by

$$d \sin \theta = m\lambda \qquad (m = 1, 2, 3, \cdots), \qquad (41–3)$$

where θ is the angle shown in Fig. 41–2a. The angles predicted by this equation, using the de Broglie wavelength, were found to agree with the observed values (Fig. 41–2b). Thus the accidental discovery of **electron diffraction** was the first direct evidence confirming de Broglie's hypothesis.

The de Broglie wavelength of a nonrelativistic particle is $\lambda = h/p = h/mv$. We can also express λ in terms of the particle's kinetic energy. For example, consider an electron freely accelerated from rest at point a to point b through a potential increase $V_b - V_a = V_{ba}$. The work done on the electron eV_{ba} equals its kinetic energy K. Using $K = p^2/2m$, we have

$$eV_{ba} = \frac{p^2}{2m}, \qquad p = \sqrt{2meV_{ba}},$$

and the de Broglie wavelength of the electron is

$$\lambda = \frac{h}{p} = \frac{h}{\sqrt{2meV_{ba}}} \qquad \text{(de Broglie wavelength of an electron).} \qquad (41–4)$$

EXAMPLE 41-2

In a particular electron-diffraction experiment using an accelerating voltage of 54 V, an intensity maximum occurs when the angle θ in Fig. 41–2a is 50° (see Fig. 41–2b). The initial kinetic energy of the electrons is negligible. The rows of atoms have been found by x-ray diffraction to have a distance $d = 2.15 \times 10^{-10}$ m = 0.215 nm. Find the electron wavelength a) from Eq. (41–3) (assuming that $m = 1$); b) from the de Broglie formula. Compare your results.

SOLUTION a) From Eq. (41–3), with $m = 1$,

$$\lambda = d \sin \theta = (2.15 \times 10^{-10} \text{ m}) \sin 50°$$
$$= 1.65 \times 10^{-10} \text{ m}.$$

b) From Eq. (41–4) the electron wavelength is

$$\lambda = \frac{6.626 \times 10^{-34} \text{ J} \cdot \text{s}}{\sqrt{2(9.109 \times 10^{-31} \text{ kg})(1.602 \times 10^{-19} \text{ C})(54 \text{ V})}}$$
$$= 1.67 \times 10^{-10} \text{ m}.$$

The two numbers agree within the accuracy of the experimental results. Note that this electron wavelength is less than the spacing between the atoms.

17.5

Electron Interference

In 1928, just a year after the Davisson-Germer discovery, the English physicist G. P. Thomson carried out electron diffraction experiments using a thin polycrystalline metallic foil as a target. Debye and Sherrer had used a similar technique several years earlier to study x-ray diffraction from polycrystalline specimens. Because of the random orientations of the individual microscopic crystals in his foil, the diffraction pattern consisted of intensity maxima forming rings around the direction of the incident beam. Thomson's results again confirmed the de Broglie relationship. Figure 41–3 shows both x-ray and electron diffraction patterns for a polycrystalline aluminum foil. (It is interesting to note that G. P. Thomson was the son of J. J. Thomson, who 31 years earlier had performed the definitive experiment to establish the *particle* nature of electrons.)

Additional experiments were soon carried out in many laboratories. In Germany, Estermann and Stern demonstrated diffraction of alpha particles. More recently, diffraction experiments have been performed with various ions and low-energy neutrons (see Example 41–1). Thus the wave nature of particles, so strange in 1924, became firmly established in the years that followed.

41–3 X-ray and electron diffraction. The upper half of the photo shows the diffraction pattern for 71-pm x rays passing through aluminum foil. The lower half, with a different scale, shows the diffraction pattern for 600-eV electrons from aluminum.

41–4 PROBABILITY AND UNCERTAINTY

The discovery of the dual wave-particle nature of matter has forced us to reevaluate the kinematic language we use to describe the position and motion of a particle. In classical Newtonian mechanics we think of a particle as a point. We can describe its location and state of motion at any instant with three spatial coordinates and three components of velocity. But in general such a specific description is not possible. When we look on a small enough scale, there are fundamental limitations on the precision with which we can determine the position and velocity of a particle. Many aspects of a particle's behavior can be stated only in terms of *probabilities*.

SINGLE-SLIT DIFFRACTION

To try to get some insight into the nature of the problem, let's review the optical single-slit diffraction experiment described in Section 38–3. Suppose the wavelength λ is much less than the slit width a. Then most (85%) of the light in the diffraction pattern is concentrated in the central maximum, bounded on either side by the first intensity minimum. We use θ_1 to denote the angle between the central maximum and the first minimum. Using Eq. (38–2) with $m = 1$, we find that θ_1 is given by $\sin \theta_1 = \lambda/a$. Since we assume $\lambda \ll a$, it follows that θ_1 is very small, $\sin \theta_1$ is very nearly equal to θ_1 (in radians), and

$$\theta_1 = \frac{\lambda}{a}. \qquad (41\text{–}5)$$

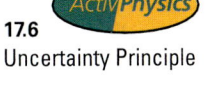

17.6
Uncertainty Principle

Now we perform the same experiment again, but using a beam of *electrons* instead of a beam of monochromatic light (Fig. 41–4). We have to do the experiment in vacuum (10^{-7} atm or less) so that the electrons don't bounce off air molecules. We can produce the electron beam with a setup that is similar in principle to the electron gun in a cathode-ray tube. This produces a narrow beam of electrons that all have very nearly the same direction and speed and therefore also the same de Broglie wavelength.

The result of this experiment, recorded on photographic film or by use of more sophisticated detectors, is a diffraction pattern identical to the one shown in Fig. 38–8. This pattern gives us additional direct evidence of the *wave* nature of electrons. About 85% of the electrons strike the film within the central maximum; the remainder strike the film within the subsidiary maxima on both sides.

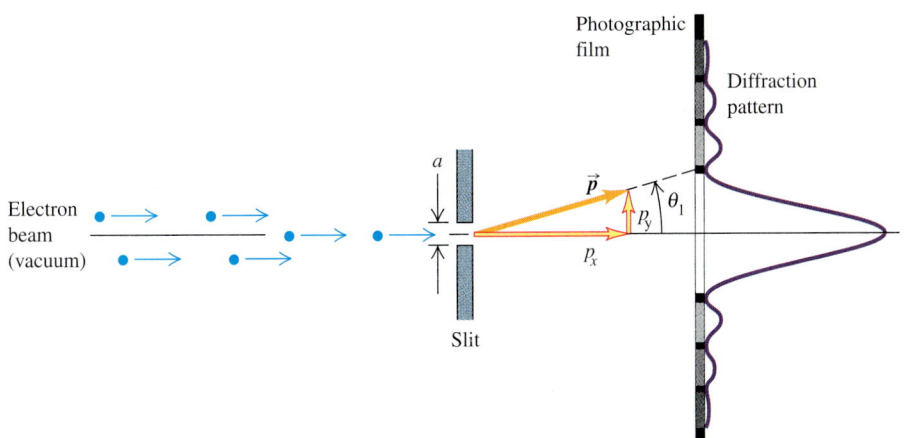

41–4 An electron diffraction experiment. The graph at the right shows the degree of exposure of the film, which in any region is proportional to the number of electrons striking that region. The components of momentum of an electron striking the outer edge of the central maximum, at angle θ_1, are shown.

If we believe that electrons are waves, the wave behavior in this experiment isn't surprising. But if we try to interpret it in terms of *particles,* we run into very serious problems. First, the electrons don't all follow the same path, even though they all have the same initial state of motion. In fact, we can't predict the exact trajectory of any individual electron from knowledge of its initial state. The best we can do is to say that *most* of the electrons go to a certain region, *fewer* go to other regions, and so on. That is, we can describe only the *probability* that an individual electron will strike each of various areas on the film. This fundamental indeterminacy has no counterpart in Newtonian mechanics, in which the motion of a particle can always be well predicted if we know the initial position and motion with sufficient accuracy.

Second, there are fundamental *uncertainties* in both the position and the momentum of an individual particle, and these two uncertainties are related inseparably. To clarify this point, let's go back to Fig. 41–4. An electron that strikes the film at the outer edge of the central maximum, at angle θ_1, must have a component of momentum p_y in the y-direction, as well as a component p_x in the x-direction, despite the fact that initially the beam was directed along the x-axis. From the geometry of the situation the two components are related by $p_y/p_x = \tan\theta_1$. Since θ_1 is small, we may use the approximation $\tan\theta_1 = \theta_1$, and

$$p_y = p_x\theta_1. \tag{41–6}$$

Substituting Eq. (41–5), $\theta_1 = \lambda/a$, we have

$$p_y = p_x\frac{\lambda}{a}. \tag{41–7}$$

For the 85% of the electrons that strike the film within the central maximum (that is, at angles between $-\lambda/a$ and $+\lambda/a$), we see that the y-component of momentum is spread out over a range from $-p_x\lambda/a$ to $+p_x\lambda/a$. Now let's consider *all* the electrons that pass through the slit and strike the film. Again, they may hit above or below the center of the pattern, so their component p_y may be positive or negative. However the symmetry of the diffraction pattern shows us the average value $(p_y)_{av} = 0$. There will be an *uncertainty* Δp_y in the y-component of momentum at least as great as $p_x\lambda/a$. That is,

$$\Delta p_y \ge p_x\frac{\lambda}{a}. \tag{41–8}$$

The narrower the slit width a, the broader is the diffraction pattern and the greater is the uncertainty in the y-component of momentum p_y.

The electron wavelength λ is related to the momentum $p_x = mv_x$ by the de Broglie relation, Eq. (41–1), which we can rewrite as $\lambda = h/p_x$. Using this relation in Eq. (41–8) and simplifying, we find

$$\Delta p_y \ge p_x\frac{h}{p_x a} = \frac{h}{a},$$

$$\Delta p_y a \ge h. \tag{41–9}$$

What does this result mean? The slit width a represents an uncertainty in the y-component of the *position* of an electron as it passes through the slit. We don't know exactly *where* in the slit each particle passes through. So both the y-position and the y-component of momentum have uncertainties, and the two uncertainties are related by Eq. (41–9). We can reduce the *momentum* uncertainty Δp_y only by reducing the width of the diffraction pattern. To do this, we have to increase the slit width a, which increases the *position* uncertainty. Conversely, when we *decrease* the position uncertainty by narrowing the slit, the diffraction pattern broadens and the corresponding momentum uncertainty *increases*.

You may protest that it doesn't seem to be consistent with common sense for a particle not to have a definite position and momentum. We reply that what we call *common sense* is based on familiarity gained through experience. Our usual experience includes very little contact with the microscopic behavior of particles. Sometimes we have to accept conclusions that violate our intuition when we are dealing with areas that are far removed from everyday experience.

THE UNCERTAINTY PRINCIPLE

In more general discussions of uncertainty relations, the uncertainty of a quantity is usually described in terms of the statistical concept of *standard deviation,* which is a measure of the spread or dispersion of a set of numbers around their average value. Suppose we now begin to describe uncertainties in this way (neither Δp_y nor a in Eq. (41–9) is a standard deviation). If a coordinate x has an uncertainty Δx and if the corresponding momentum component p_x has an uncertainty Δp_x, then those standard-deviation uncertainties are found to be related in general by the inequality

$$\Delta x\, \Delta p_x \geq \frac{h}{2\pi} \quad \begin{array}{l}\text{(Heisenberg uncertainty principle for} \\ \text{position and momentum).}\end{array} \quad (41\text{–}10)$$

Equation (41–10) is one form of the **Heisenberg uncertainty principle.** It states that, in general, neither the position nor the momentum of a particle can be determined with arbitrarily great precision, as classical physics would predict. Instead, the uncertainties in the two quantities play complementary roles, as we have described. Figure 41–5 shows the relationship between the two uncertainties.

It is tempting to suppose that we could get greater precision by using more sophisticated detectors of position and momentum. This turns out not to be possible. To detect a particle, the detector must *interact* with it, and this interaction unavoidably changes the state of motion of the particle, introducing uncertainty about its original state. For example, if we were to bounce shorter-wavelength photons off a particle to better locate its position, the larger photon momentum h/λ would make the particle recoil more, giving us greater uncertainty in its momentum. A more detailed analysis of such hypothetical experiments shows that the uncertainties we have described are fundamental and intrinsic. They *cannot* be circumvented *even in principle* by any experimental technique, no matter how sophisticated.

There is nothing special about the x-axis. In a three-dimensional situation with coordinates (x, y, z) there is an uncertainty relation for each coordinate and its corresponding momentum component: $\Delta x\, \Delta p_x \geq h/2\pi$, $\Delta y\, \Delta p_y \geq h/2\pi$, and $\Delta z\, \Delta p_z \geq h/2\pi$. However, the uncertainty in one coordinate is *not* related to the uncertainty in a different component of momentum. For example, Δx is not related directly to Δp_y.

For a particle moving along a radius, we can replace x in Eq. (41–10) with r, giving $\Delta r\, \Delta p_r \geq h/2\pi$. In the Bohr model, an electron moves in a circle of exact radius r, giving $\Delta r = 0$ and $\Delta p_r = 0$. Thus the Bohr model violates the Heisenberg uncertainty principle. We'll give a more correct description of atomic structure in Chapter 43; happily, it turns out that the energy-level predictions of the Bohr model *are* correct.

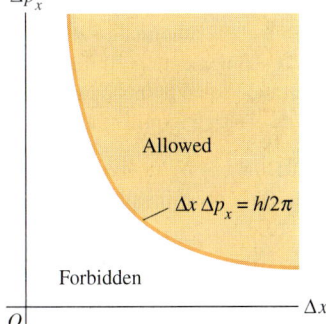

41–5 The Heisenberg uncertainty principle for position and momentum components. The allowed region has values such that the product $\Delta x\, \Delta p_x$ is greater than or equal to $h/2\pi$. In the forbidden region, $\Delta x\, \Delta p_x$ is less than $h/2\pi$.

UNCERTAINTY IN ENERGY

There is also an uncertainty principle for *energy.* It turns out that the energy of a system also has inherent uncertainty. The uncertainty ΔE depends on the *time interval* Δt dur-

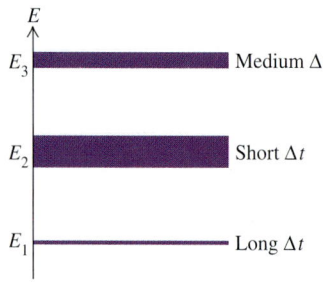

41-6 The longer the lifetime of an energy state, the smaller is its spread in energy.

ing which the system remains in the given state. The relation is

$$\Delta E \Delta t \geq \frac{h}{2\pi} \qquad \text{(Heisenberg uncertainty principle for} \atop \text{energy and time interval).}} \qquad (41-11)$$

A system that remains in a metastable state for a very long time (large Δt) can have a very well-defined energy (small ΔE), but if it remains in a state for only a short time (small Δt), the uncertainty in energy must be correspondingly greater (large ΔE). Figure 41-6 illustrates this idea.

EXAMPLE 41-3

An electron is confined within a region with $\Delta x = 1.0 \times 10^{-10}$ m. a) Estimate the minimum uncertainty in the x-component of the electron's momentum. b) If the electron has momentum with magnitude equal to the uncertainty found in part (a), what is its kinetic energy? Express the result in joules and in electron volts.

SOLUTION a) We're given information about x-components. In the uncertainty principle we have $\Delta x = 1.0 \times 10^{-10}$ m. From Eq. (41-10),

$$(\Delta p_x)_{min} = \frac{h}{2\pi \Delta x} = \frac{6.63 \times 10^{-34} \text{ J} \cdot \text{s}}{2\pi (1.0 \times 10^{-10} \text{ m})} = 1.1 \times 10^{-24} \text{ kg} \cdot \text{m/s}.$$

b) An electron with this magnitude of momentum has kinetic energy

$$K = \frac{p_x^{\,2}}{2m} = \frac{(1.1 \times 10^{-24} \text{ kg} \cdot \text{m/s})^2}{2(9.11 \times 10^{-31} \text{ kg})}$$

$$= 6.1 \times 10^{-19} \text{ J} = 3.8 \text{ eV}.$$

The region is roughly the same width as an atom, and the energy is of the same order of magnitude as typical electron energies in atoms. This is a very rough calculation, but it is reassuring that the energy has a reasonable order of magnitude. If our result had differed from this by a factor of 10^6, we would have had reason to be alarmed.

In a type of radioactive decay called *beta-minus decay*, an electron is emitted from a nucleus. Although it seems reasonable to assume that the electron was confined within the nucleus before the decay, that would require the position of the electron to be known to within an uncertainty of $\Delta x = 10^{-14}$ m or so. But this would give the confined electron a value of Δp_x that is 10^4 times greater than for the electron in this example and an expected kinetic energy so large that we would need to use relativistic equations to calculate it. This high energy is one reason to believe that there are no electrons confined in nuclei. In beta-minus decay the electron is actually *produced* within the nucleus when a neutron changes into a proton.

EXAMPLE 41-4

A sodium atom is in a state in one of the lowest excited levels shown in Fig. 40-9. It remains in that state for an average time of 1.6×10^{-8} s before it makes a transition back to a ground state, emitting a photon with wavelength 589.0 nm and energy 2.105 eV. What is the uncertainty in energy of that excited state? What is the wavelength spread of the corresponding spectrum line?

SOLUTION From Eq. (41-11),

$$\Delta E = \frac{h}{2\pi \Delta t} = \frac{6.626 \times 10^{-34} \text{ J} \cdot \text{s}}{2\pi (1.6 \times 10^{-8} \text{ s})}$$

$$= 6.6 \times 10^{-27} \text{ J} = 4.1 \times 10^{-8} \text{ eV}.$$

The atom remains an indefinitely long time in the ground state, so there is *no* fundamental uncertainty there. The fractional

uncertainty of the photon energy is

$$\frac{4.1 \times 10^{-8} \text{ eV}}{2.105 \text{ eV}} = 1.95 \times 10^{-8}.$$

The corresponding spread in wavelength, or "width," of the spectrum line is approximately

$$\Delta \lambda = (1.95 \times 10^{-8})(589.0 \text{ nm}) = 0.000011 \text{ nm}.$$

The irreducible uncertainty $\Delta \lambda$ in Example 41-4 is called the *natural line width* of this particular spectrum line. Though very small, it is within the limits of resolution of present-day spectrometers. Ordinarily, the natural line width is much smaller than the line width from other causes such as the Doppler effect and collisions among the rapidly moving atoms.

(a)

After 28 electrons   After 1000 electrons   After 10,000 electrons

(b)

41–7 (a) Formation of an interference pattern for electrons incident on two slits, (b) after 28 electrons, after 1000 electrons, and after 10,000 electrons.

TWO-SLIT INTERFERENCE

Now let's take a brief look at a quantum interpretation of a *two-slit* optical interference pattern. We studied these patterns in detail for light in Sections 37–3 and 37–4, and in Section 40–10 we discussed their interpretation in terms of the probability that photons strike various regions of the screen where the pattern is formed.

In view of our discussion of the wave properties of electrons, it is natural to ask what happens when we do a two-slit interference experiment with electrons. The answer is: exactly the same thing as we saw in Section 40–10 with photons! We can again use photographic film (Fig. 41–7) or particle counters to trace out the interference pattern, as we did with photons. The principle of complementarity, introduced in Section 40–10, again tells us that we cannot simultaneously attempt to apply the wave model and the particle model to describe a single part of this experiment. Thus we *cannot* predict exactly where in the pattern (a wave phenomenon) any individual electron (a particle) will land. We can't even ask which slit an individual electron passed through in building up the two-slit interference pattern. If we do determine the slits that the electrons pass through by scattering photons off them, the electrons recoil, and the two-slit interference pattern is not built up.

41–5 THE ELECTRON MICROSCOPE

The **electron microscope** offers an important and interesting example of the interplay of wave and particle properties of electrons. An electron beam can be used to form an image of an object in much the same way as a light beam. A ray of light can be bent by reflection or refraction, and an electron trajectory can be bent by an electric or magnetic field. Rays of light diverging from a point on an object can be brought to convergence by a converging lens or concave mirror, and electrons diverging from a small region can be brought to convergence by electric and/or magnetic fields.

The analogy between light rays and electrons goes deeper. The *ray* model of geometric optics is an approximate representation of the more general *wave* model. Geometric optics (ray optics) is valid whenever interference and diffraction effects can be neglected. Similarly, the model of an electron as a point particle following a line trajectory is an approximate description of the actual behavior of the electron; this model is useful when we can neglect effects associated with the wave nature of electrons.

How is an electron microscope superior to an optical microscope? The *resolution* of an optical microscope is limited by diffraction effects, as we discussed in Section 38–8. Using wavelengths around 500 nm, an optical microscope can't resolve objects smaller than a few hundred nanometers, no matter how carefully its lenses are made. The resolution of an electron microscope is similarly limited by the wavelengths of the electrons, but these wavelengths may be many thousands of times *smaller* than wavelengths of visible light. As a result, the useful magnification of an electron microscope can be thousands of times as great as that of an optical microscope.

Note that the ability of the electron microscope to form a magnified image *does not* depend on the wave properties of electrons. Within the limitations of the Heisenberg uncertainty principle, we can compute the electron trajectories by treating them as classical charged particles under the action of electric- and magnetic-field forces (in analogy to ray optics). Only when we talk about *resolution* do the wave properties become important.

EXAMPLE 41–5

The nonrelativistic electron beam in an electron microscope is formed by a setup similar to the electron gun in a cathode-ray tube (Section 24–8). What accelerating voltage is needed to produce electrons with wavelength 10 pm = 0.010 nm (roughly 50,000 times smaller than typical visible-light wavelengths)? The initial kinetic energy of the electrons is negligible.

SOLUTION The accelerating voltage is the quantity V_{ba} in Eq. (41–4),

$$\lambda = \frac{h}{\sqrt{2meV_{ba}}}.$$

Solving for V_{ba} and inserting the appropriate numbers, we find

$$V_{ba} = \frac{h^2}{2me\lambda^2}$$

$$= \frac{(6.626 \times 10^{-34} \text{ J} \cdot \text{s})^2}{2(9.109 \times 10^{-31} \text{ kg})(1.602 \times 10^{-19} \text{ C})(10 \times 10^{-12} \text{ m})^2}$$

$$= 1.5 \times 10^4 \text{ V} = 15,000 \text{ V}.$$

This is approximately equal to the accelerating voltage for the electron beam in a TV picture tube. This example shows, incidentally, that the sharpness of a TV picture is *not* limited by electron diffraction effects. Also note that this 15-kV voltage will increase the kinetic energy of the electrons from a relatively small value to 15 keV. Since electrons have a rest energy of 0.511 MeV = 511 keV, we can accurately describe these 15-keV electrons as nonrelativistic.

THE TRANSMISSION ELECTRON MICROSCOPE

Except within their electron guns, most practical electron microscopes use magnetic fields rather than electric fields as "lenses" for focusing the beam. A common setup for *transmission electron microscopes* includes three such lenses in a compound-microscope arrangement, as shown in Fig. 41–8. Electrons are emitted from a hot cathode and accelerated by a potential difference, typically 10 to 100 kV. The electrons pass through a condensing lens and are formed into a parallel beam before passing through the specimen or object to be viewed. The specimen to be viewed is very thin, typically 10 to 100 nm, so the electrons are not slowed appreciably as they pass through. The objective lens then forms an intermediate image of this object, and the projection lens produces a final real image of that image. The objective and projection lenses play the roles of the objective and eyepiece lenses, respectively, of a compound optical microscope. The final image is recorded on photographic film or projected onto a fluorescent screen for viewing or photographing. The entire apparatus, including the specimen, must be enclosed in a vacuum container, just as with the cathode-ray tube; otherwise, electrons would scatter off air molecules and muddle the image.

We might think that when the electron wavelength is 0.01 nm (as in Example 41–5), the resolution would also be about 0.01 nm. In fact, it is seldom better than 0.5 nm, for

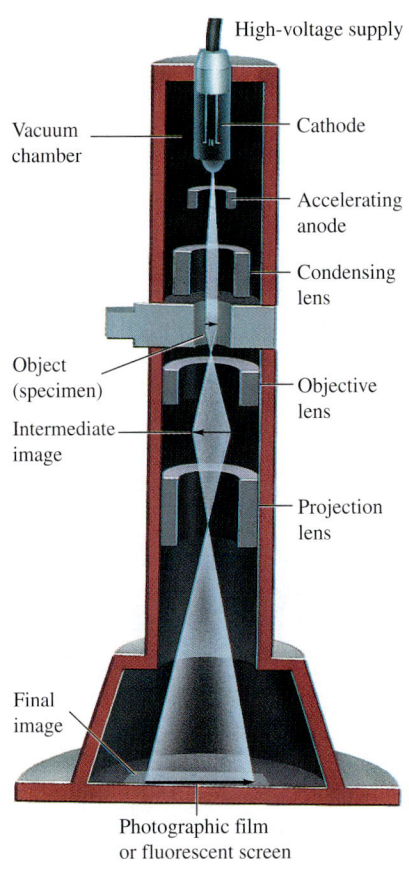

High-voltage supply

Vacuum chamber

Cathode

Accelerating anode

Condensing lens

Object (specimen)

Objective lens

Intermediate image

Projection lens

Final image

Photographic film or fluorescent screen

(a)

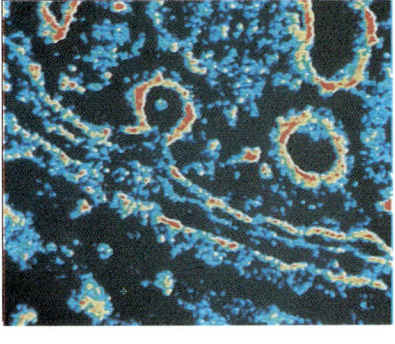

(b)

41–8 (a) Schematic diagram of a transmission electron microscope. The magnetic lenses, consisting of coils of wire carrying currents, are shown in cross section. The condensing lens forms a parallel beam of electrons that strikes the object. The objective lens forms an intermediate image that serves as the object for the final real image formed by the projection lens. The final image is projected onto photographic film, a fluorescent screen, or the screen of a video camera. The magnification of each lens may be of the order of 100×, and the overall magnification may be of the order of 10,000×. The angles of the electron paths with the optic axis are greatly exaggerated; in actual instruments these angles are usually less than 0.01 rad (0.5°). The entire apparatus is enclosed in a vacuum chamber, not shown in the diagram. (b) An electron micrograph of a cell membrane. The image has been digitally processed so that red indicates high concentration and blue indicates low concentration.

several reasons. Large-aperture magnetic lenses have aberrations analogous to those of optical lenses, as we discussed in Section 36–7. The focal length of a magnetic lens depends on the current in the coil, which must be controlled precisely. The focal length also depends on the electron speed, which is never exactly the same for all electrons in the beam. This effect is analogous to chromatic aberration in an optical system (Section 36–7).

THE SCANNING ELECTRON MICROSCOPE

An important variation is the *scanning electron microscope* (Fig. 41–9a). The electron beam is focused to a very fine line and is swept across the specimen, just as the electron beam in a TV picture tube traces out the picture. As the beam scans the specimen, electrons are knocked off and are collected by a collecting anode that is kept at a potential a few hundred volts positive with respect to the specimen. The current in the collecting anode is amplified and used to modulate the electron beam in a cathode-ray tube, which is swept in synchronization with the microscope beam. Thus the cathode-ray tube traces out a greatly magnified image of the specimen. This scheme has several advantages. The specimen can be thick because the beam does not need to pass through it. Also, the knock-off electron production depends on the *angle* at which the beam strikes the surface. Thus scanning electron micrographs have an appearance that is much more three dimensional than conventional visible-light micrographs. The resolution is typically of the order of 10 nm, still much finer than the best optical microscopes. Figure 41–9b shows a photograph made with a scanning electron microscope.

High-voltage supply

Cathode

Vacuum chamber

Accelerating anode

Condensing lens

Beam deflector

Scan generator

Objective lens

Amplifier

Object (specimen)

Image

Cathode-ray tube

(a)

(b)

41–9 (a) Schematic diagram of a scanning electron microscope. The beam deflector uses the signals from the scan generator to scan the beam across the specimen, and these signals simultaneously scan the electron beam in the video display. The signal received by the detector is used to modulate the video display beam, creating light and dark areas on the screen. (b) A scanning electron micrograph of a sponge spicule.

41–6 WAVE FUNCTIONS

We have now seen persuasive evidence that, on an atomic or subatomic scale, a particle such as an electron can't be described simply as a point that has three position coordinates and three velocity components. In some situations, such a particle behaves like a wave, and we have spoken a few times about using a wave function to describe the state of a particle. Let's now describe more specifically the quantum-mechanical language that we use to replace the classical scheme of coordinates and velocity components.

Our new scheme for describing the state of a particle has a lot in common with the language of classical wave motion. In Chapter 19 we described transverse waves on a string by specifying the position of each point in the string at each instant of time by means of a *wave function* (Section 19–4). If y represents the displacement from equilibrium of a point on the string, then the function $y(x, t)$ represents that displacement at any distance x from the origin and at any time t. Once we know the wave function for a particular wave motion, we know everything there is to know about the motion. We can find the position and velocity of any point on the string at any time, and so on. We worked out specific forms for these functions for *sinusoidal* waves, in which each particle undergoes simple harmonic motion.

We followed a similar pattern for sound waves in Chapter 21. The wave function $p(x, t)$ for a wave traveling along the x-direction represented the pressure variation at any point x at any time t. We used this language once more in Section 33–4, in which we used two wave functions to describe the electric and magnetic fields of *electromagnetic* waves at any point in space at any time.

Thus it is natural to use a wave function as the central element of our new quantum-mechanical language. The symbol that is usually used for this wave function is Ψ or ψ, the Greek letters "psi" (pronounced "sigh"). In general, we will use Ψ for a function of the space coordinates and time, and will use ψ for a function of the space coordinates only, *not* of time. Just as we can use the wave function $y(x, t)$ for mechanical waves on a string to provide a complete description of the motion of the particles of the string, we can use the quantum-mechanical wave function $\Psi(x, y, z, t)$ for a particle to give us information about that particle.

CAUTION ▶ Keep in mind that a quantum-mechanical wave function is unlike any wave you've yet encountered. Mechanical waves require a medium to travel through; the stretched string is the medium for transverse waves on a string, and air is the medium for sound waves. But the wave function for a particle is *not* a mechanical wave that needs some material medium in order to propagate. The wave function describes the particle, but we can't define the function itself in terms of anything material. We can only describe how it is related to physically observable effects. ◀

INTERPRETATION OF THE WAVE FUNCTION

The wave function can give us the probability distribution of a particle in space, just as the wave functions for an electromagnetic wave can give us the distribution of the electric and magnetic fields. When we worked out interference and diffraction patterns in Chapters 37 and 38, we found that the intensity I of the radiation at any point in a pattern is proportional to the square of the electric-field magnitude, that is, to E^2. In the photon interpretation of interference and diffraction (Section 40–10), the intensity at each point is proportional to the number of photons striking around that point or, alternatively, to the *probability* that any individual photon will strike around the point. Thus the square of the electric-field magnitude at each point is proportional to the probability of finding a photon around that point.

In much the same way, the square of the wave function of a particle at each point represents the probability of finding the particle around that point. More precisely, we should say the square of the *absolute value* of the wave function, $|\Psi|^2$. This is necessary because, as we'll see later, Ψ isn't necessarily a real quantity. It may be a *complex* quantity with real and imaginary parts. (The imaginary part of the function is a real function multiplied by the imaginary number $i = \sqrt{-1}$.) For a particle moving in three dimensions, the quantity $|\Psi|^2\, dV$ is the probability that the particle will be found within a volume dV around the point at which $|\Psi|^2$ is evaluated. The particle is most likely to be found in regions where $|\Psi|^2$ is large, and so on. We have already used this interpretation in our discussion of electron diffraction experiments. For a particle with charge, such as the electron, $|\Psi|^2$ is also proportional to the *charge density* at any point in space. This discussion assumes that Ψ is *normalized.* For a normalized wave function the integral of $|\Psi|^2\, dV$ over all space equals exactly 1; that is, the probability is exactly 1, or 100%, that the particle is *somewhere* in the universe.

The wave function Ψ can give us information about a particle when it is in a particular state. In the mathematical theory of quantum mechanics there are definite procedures for determining the average position of the particle, its average velocity, and dynamic quantities such as momentum, energy, and angular momentum. The required techniques are beyond the scope of this discussion, but they are well established and well supported by experimental results.

For a moving free particle the wave function Ψ is always a function of both the space coordinates and time. The value of $|\Psi|^2$ at a particular point also varies with time, corresponding to the fact that the place a moving particle is most likely to be found changes with time. But in some cases, such as an electron in an atom in a definite energy state, the value of $|\Psi|^2$ at each point is *constant,* independent of time. In such cases the

In 1926, the German physicist Max Born (1882–1970) devised the interpretation that $|\Psi|^2\, dV$ is the probability of finding a particle within a given volume dV. He also coined the term "quantum mechanics" (in the original German, *Quantenmechanik*). For his contributions, Born shared (with Walther Bothe) the 1954 Nobel Prize in physics.

probability distribution for the particle doesn't change with time. This is always true with states that have a definite energy, and we call such a state a **stationary state.** We will show later that we can describe a stationary state using a time-independent wave function ψ, and that ψ can be found by solving a differential equation that doesn't involve time explicitly. This is often a lot easier than solving a more general equation that involves both the space coordinates and time.

Problem–Solving Strategy

FINDING THE SQUARE OF THE ABSOLUTE VALUE OF A COMPLEX QUANTITY

1. The first step is to find the *complex conjugate* of the quantity, represented with an asterisk. To do so, you simply replace all i with $-i$. For example, if the complex quantity is $z = x + iy$, where x and y are real, then the complex conjugate of z is $z^* = x + (-i)y = x - iy$.

2. The second and final step is to multiply the complex conjugate by the original quantity. In our example,

$$|z|^2 = z^*z = (x - iy)(x + iy) = x^2 - (iy)^2 = x^2 + y^2.$$

The last step occurs because $i^2 = -1$ by definition.

EXAMPLE 41–6

Show that $\Psi = \psi e^{-i\omega t}$ is a wave function of a stationary state.

SOLUTION If Ψ is the wave function of a stationary state, then the value of $|\Psi|^2$ at each point must be constant, independent of time. To find $|\Psi|^2$, we first take the complex conjugate of $\Psi = \psi e^{-i\omega t}$, which is $\Psi^* = \psi^* e^{+i\omega t}$. Then

$$|\Psi|^2 = \Psi^*\Psi = (\psi^* e^{+i\omega t})(\psi e^{-i\omega t}) = \psi^*\psi e^0 = |\psi|^2.$$

Recall that ψ is *not* a function of time, so $|\psi|^2$ is also independent of time. We have just shown that $|\Psi|^2 = |\psi|^2$, so $|\Psi|^2$ is independent of time and $\Psi = \psi e^{-i\omega t}$ is a wave function of a stationary state.

We have now reached a stage in our discussion comparable to the end of Chapter 3. At that point, we had learned the kinematic language for *describing* the motion of a particle but had not yet studied the *dynamic* principles (Newton's laws) that determine what motions are possible. We need a general dynamic principle, comparable to Newton's laws or Maxwell's equations, that will enable us to determine the wave function, or the possible wave functions, for specific physical situations. The needed principle is the *Schrödinger equation,* developed by Erwin Schrödinger in 1925. Physically possible states of a system are represented by wave functions that are solutions of this equation, just as physically possible waves on a string have wave functions that are solutions of the corresponding wave equation, Eq. (19–18). We'll study the Schrödinger equation in the next chapter. Among other things, we will learn that the solutions of this equation for any particular system yield a set of allowed *energy levels.* This discovery is of the utmost importance. Before the development of the Schrödinger equation, there was no way to predict energy levels from any fundamental theory except for the Bohr model for hydrogen, which had very limited success.

WAVE PACKETS

Finally, let's think once more about how an electron can be a particle and a wave at the same time. Armed with the idea of a wave function, we can be a little more specific about how to reconcile these seemingly incompatible aspects of particle behavior. First we note that a particle with a definite wavelength λ also has a definite momentum p, as shown by the de Broglie relation $\lambda = h/p$. For such a state, there is *no* uncertainty in momentum: $\Delta p = 0$. The uncertainty principle, Eq. (41–11), says that $\Delta x \, \Delta p_x \geq h/2\pi$. If Δp is zero, Δx must be infinite. If we know the particle's momentum precisely, we have no idea at all *where* the particle is. Such a state is represented by a sinusoidal wave function with no beginning and no end.

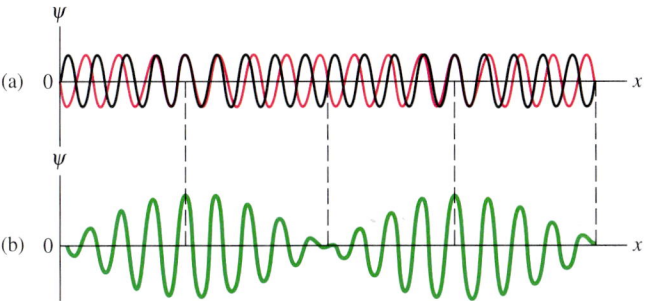

41–10 (a) Two sinusoidal waves with slightly different wavelengths, shown at one instant of time. (b) The superposition of these waves has a wavelength equal to the average of the two wavelengths but with varying amplitude, giving it a lumpy character not possessed by either individual wave.

We can superpose two or more sinusoidal functions to make a wave function that is more localized in space. To keep things simple, we'll imagine doing this only in one dimension (x) and at one instant of time. Our wave functions are then functions only of the spatial coordinate x, so we denote them as ψ. In our discussion of beats in Section 21–4, we superposed two sinusoidal waves with slightly different frequencies (Fig. 21–6). The result was a wave that had a lumpy character that the individual waves do not possess. Imagine doing the same thing with two particle waves. Superposing two waves with slightly different wavelengths (Fig. 41–10a) gives the wave shown in Fig. 41–10b. A particle represented by this function is more likely to be found in some regions than in others, but the particle's momentum no longer has a definite value because we began with two different wavelengths.

It's not hard to imagine superposing two additional sinusoidal waves with different wavelengths and amplitudes so as to reinforce alternate lumps in Fig. 41–10b and cancel out the in-between ones. Finally, if we superpose waves with a very large number of different wavelengths, we can construct a wave with only one lump (Fig. 41–11). Then, finally, we have something that begins to look like both a particle and a wave. It is a particle in the sense that it is localized in space; if we look from a distance, it may look like a point. But it also has a periodic structure that is characteristic of a wave.

Such a wave pulse is called a **wave packet.** We can represent such a superposition by an expression such as

$$\psi(x) = \int_0^\infty A(\lambda) \cos \frac{2\pi x}{\lambda} \, d\lambda, \qquad (41\text{–}12)$$

where $\cos (2\pi x/\lambda)$ is a sinusoidal wave with wavelength λ. The integral represents a superposition in which we add a very large number of such waves with different values of λ, each with an amplitude $A(\lambda)$ that depends on λ.

It turns out that there is a very important relation between the two functions $\psi(x)$ and $A(\lambda)$. It is shown qualitatively in Fig. 41–12. If the function $A(\lambda)$ is sharply peaked, as in Fig. 41–12a, we are superposing only a narrow range of wavelengths. The resulting wave pulse $\psi(x)$ is then relatively broad (Fig. 41–12b). But if we use a wider range of values of λ, so that the function $A(\lambda)$ is broader (Fig. 41–12c), then the wave pulse is more narrowly localized (Fig. 41–12d).

What we are seeing is the uncertainty principle in action. A narrow range of λ means a narrow range of p_x and thus a small Δp_x; the result is a relatively large Δx. A broad range of λ corresponds to a large Δp_x, and the resulting Δx is smaller. Thus we see that the uncertainty principle $\Delta x \, \Delta p_x \geq h/2\pi$ is an inevitable consequence of the de Broglie relation and the properties of integrals such as Eq. (41–12). These integrals are called *Fourier integrals;* they are a generalization of the concept of Fourier series, which we mentioned in Section 20–4. In both cases we are representing a complex wave form as a superposition of sinusoidal functions. With Fourier series we use a sequence of frequencies (or values of $1/\lambda$) that are integer multiples of some basic value, while with Fourier integrals we superpose functions with a *continuous* distribution of values of λ.

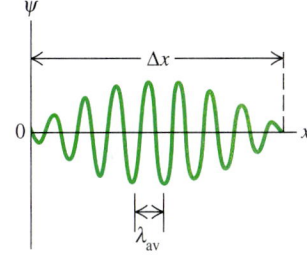

41–11 Superposing a large number of sinusoidal waves with differing wavelengths and appropriate amplitudes can produce a wave pulse that has a periodic nature with an average wavelength λ_{av} and yet is localized in a region of space Δx. Is this a particle or a wave?

17.7
Wave Packets

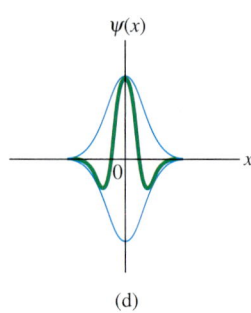

(a)   (b)   (c)   (d)

41-12 Two different functions: $A(\lambda)$ and the corresponding functions $\psi(x)$. The width of $\psi(x)$ is inversely proportional to that of $A(\lambda)$.

KEY TERMS

de Broglie wavelength, 1271

quantum mechanics, 1272

wave function, 1272

boundary condition, 1272

electron diffraction, 1275

Heisenberg uncertainty principle, 1279

electron microscope, 1281

stationary state, 1286

wave packet, 1287

SUMMARY

- Electrons and other particles have wave properties. The de Broglie wavelength of a particle is inversely proportional to the magnitude of the particle's momentum:

$$\lambda = \frac{h}{p} = \frac{h}{mv}. \tag{41-1}$$

- The state of a particle is described not by its coordinates and velocity components but by a wave function, in general a function of the space coordinates and time.

- A crystalline solid serves as a three-dimensional diffraction grating for both x rays and particles; diffraction of an electron beam from the surface of a metallic crystal provided the first direct confirmation of the wave nature of particles and of the correctness of Eq. (41–1). Diffraction experiments have also been carried out with neutrons, ions, and other particles. The wavelength of a nonrelativistic electron that has been accelerated from rest through a potential difference V_{ba} is

$$\lambda = \frac{h}{p} = \frac{h}{\sqrt{2meV_{ba}}}. \tag{41-4}$$

- It is impossible to make exact determinations of a coordinate of a particle and of the corresponding momentum component at the same time. The precision of such measurements is limited by the Heisenberg uncertainty principle:

$$\Delta x \, \Delta p_x \geq \frac{h}{2\pi} \tag{41-10}$$

for the x-components with corresponding relations for the y- and z-components. The uncertainty ΔE in energy of a state that is occupied for a time Δt is given by

$$\Delta E \, \Delta t \geq \frac{h}{2\pi}. \tag{41-11}$$

- The electron microscope uses an electron beam to produce greatly enlarged images of objects. The overall image formation does not depend on wave properties, but the resolution in the image is ultimately limited by the electron wavelengths, which can be thousands of times smaller than the wavelengths for visible light.

- We can use the wave function $\Psi(x, y, z, t)$ for a particle to give us information about that particle. A wave function can be localized in a certain region of space and still have wave properties, giving it both particle and wave aspects.

DISCUSSION QUESTIONS

Q41-1 If a proton and an electron have the same speed, which has the longer de Broglie wavelength? Explain.

Q41-2 If a proton and an electron have the same kinetic energy, which has the longer de Broglie wavelength? Explain.

Q41-3 Does a photon have a de Broglie wavelength? If so, how is it related to the wavelength of the associated electromagnetic wave? Explain.

Q41-4 You have been asked to design a magnet system to steer a beam of 54-V electrons like those described in Example 41–2 (Section 41–3). The goal is to be able to direct the electron beam to a specific target location with an accuracy of ±1.0 mm. In your design, do you need to take the wave nature of electrons into account? Explain.

Q41–5 The upper half of the electron diffraction pattern shown in Fig. 41–4 is the mirror image of the lower half of the pattern. Would it be correct to say that the upper half of the pattern is caused by electrons that pass through the upper half of the slit? Explain.

Q41–6 A particular electron in the experimental setup shown in Fig. 41–4 lands on the photographic film a distance x above the center of the pattern. Given the value of x, is it possible to calculate the precise trajectory that the electron followed? Explain.

Q41–7 Does the uncertainty principle have anything to do with marksmanship? That is, is the accuracy with which a bullet can be aimed at a target limited by the uncertainty principle? Explain.

Q41–8 What happens to the uncertainty in the y-component of the momentum of a particle as Δy approaches zero? As Δz approaches zero? Explain.

Q41–9 Suppose a two-slit interference experiment is carried out using an electron beam. Would the same interference pattern result if one slit at a time is uncovered instead of both at once? If not, why not? Does not each electron go through one slit or the other? Or does every electron go through both slits? Discuss the latter possibility in light of the principle of complementarity.

Q41–10 Equation (41–11) states that the energy of a system can have uncertainty. Does this mean that the principle of conservation of energy is no longer valid? Explain.

Q41–11 Laser light results from transitions from long-lived metastable states. Why is it more monochromatic than ordinary light?

Q41–12 Could an electron-diffraction experiment be carried out using three or four slits? Using a grating with many slits? What sort of results would you expect with a grating? Would the uncertainty principle be violated? Explain.

Q41–13 As the lower half of Fig. 41–3 shows, the diffraction

pattern made by electrons that pass through an aluminum foil is a series of concentric rings. But if the aluminum foil is replaced by a single crystal of aluminum, only certain points on these rings appear in the pattern. Explain.

Q41–14 Why can an electron microscope have greater magnification than an ordinary microscope?

Q41–15 Sometimes the incorrect statement is made that "All molecular motion ceases at absolute zero." If all molecular motion were to cease, what would the uncertainty in the momentum be for a molecule? What then would the uncertainty be in the molecule's position? Explain why there must still be some molecular motion even at absolute zero.

Q41–16 If quantum mechanics replaces the language of Newtonian mechanics, why do we not have to use wave functions to describe the motion of macroscopic bodies such as baseballs and cars?

Q41–17 A student remarks that the relation of ray optics to the more general wave picture is analogous to the relation of Newtonian mechanics, with well-defined particle trajectories, to quantum mechanics. Comment on this remark.

Q41–18 When you check the air pressure in a tire, a little air always escapes; the process of making the measurement changes the quantity being measured. Think of other examples of measurements that change or disturb the quantity being measured.

Q41–19 If a particle is in a stationary state, does that mean that the particle is not moving? If a particle moves in empty space with constant momentum $\vec{p}$ and hence constant energy $E = p^2/2m$, is it in a stationary state? Explain your answers.

Q41–20 Some lasers emit light in pulses that are only 10^{-12} s in duration. The length of such a pulse is $(3 \times 10^8 \text{ m/s})(10^{-12} \text{ s}) = 3 \times 10^{-4}$ m = 0.3 mm. Can pulsed laser light be as monochromatic as the light from a laser that emits a steady, continuous beam? Explain.

EXERCISES

SECTION 41–2 DE BROGLIE WAVES

41–1 An electron has a de Broglie wavelength of 2.80×10^{-10} m. Determine a) the magnitude of its momentum; b) its kinetic energy (in joules and in electron volts).

41–2 Wavelength of an Alpha Particle. An alpha particle ($m = 6.64 \times 10^{-27}$ kg) emitted in the radioactive decay of uranium-238 has an energy of 4.20 MeV. What is its de Broglie wavelength?

41–3 a) An electron moves with a speed of 4.70×10^6 m/s. What is its de Broglie wavelength? b) A proton moves with the same speed. Determine its de Broglie wavelength.

41–4 For crystal diffraction experiments (discussed in Section 41–3), wavelengths on the order of 0.20 nm are often appropriate. Find the energy in electron volts for a particle with this wavelength if the particle is a) a photon; b) an electron; c) an alpha particle ($m = 6.64 \times 10^{-27}$ kg).

41–5 In the Bohr model of the hydrogen atom, what is the de Broglie wavelength for the electron when it is in a) the $n = 1$

level? b) the $n = 4$ level? In each case, compare the de Broglie wavelength to the circumference $2\pi r_n$ of the orbit.

41–6 a) A nonrelativistic free particle of mass m has kinetic energy K. Derive an expression for the de Broglie wavelength of the particle in terms of m and K. b) What is the de Broglie wavelength of an 800-eV electron?

41–7 Wavelength of a Bullet. Calculate the de Broglie wavelength of a 5.00-g bullet that is moving at 340 m/s. Will the bullet exhibit wavelike properties?

41–8 What is the de Broglie wavelength for an electron with speed a) $v = 0.480c$; b) $v = 0.960c$? (*Hint:* Use the correct relativistic expression for linear momentum if necessary.)

SECTION 41–3 ELECTRON DIFFRACTION

41–9 A beam of neutrons that all have the same energy scatters from the atoms in the surface plane of a crystal that have a spacing of 0.0910 nm. The $m = 1$ intensity maximum occurs when the angle θ in Fig. 41–2a is 28.6°. What is the kinetic energy in eV of each neutron in the beam?

41–10 A beam of 188-eV electrons is directed at normal incidence onto a crystal surface as shown in Fig. 41–2c. The $m = 2$ intensity maximum occurs at an angle $\theta = 60.6°$. a) What is the spacing between adjacent atoms on the surface? b) At what other angle or angles is there an intensity maximum? c) For what electron energy (in eV) would the $m = 1$ intensity maximum occur at $\theta = 60.6°$? For this energy, is there an $m = 2$ intensity maximum? Explain.

41–11 A beam of 840-eV alpha particles ($m = 6.64 \times 10^{-27}$ kg) scatters from the atoms in the surface plane of a crystal that have a spacing of 0.0834 nm. At what angle θ in Fig. 41–2a does the $m = 1$ intensity maximum occur?

41–12 A CD-ROM is used instead of a crystal in an electron-diffraction experiment like that shown in Fig. 41–2. The surface of the CD-ROM has tracks of tiny pits with a uniform spacing of 1.60 μm. a) If the speed of the electrons is 1.26×10^4 m/s, at which values of θ will the $m = 1$ and $m = 2$ intensity maxima appear? b) The scattered electrons in these maxima strike at normal incidence a piece of photographic film that is 50.0 cm from the CD-ROM. What is the spacing on the film between these maxima?

SECTION 41–4 PROBABILITY AND UNCERTAINTY

41–13 By extremely careful measurement, you determine the x-coordinate of a car's center of mass with an uncertainty of only 1.00 μm. The car has a mass of 1200 kg. a) What is the minimum uncertainty in the x-component of the velocity of the car's center of mass as prescribed by the Heisenberg uncertainty principle? b) Does the uncertainty principle impose a practical limit on our ability to make simultaneous measurements of the positions and velocities of ordinary objects like cars, books, and people? Explain.

41–14 a) The uncertainty in the y-component of a proton's position is 2.0×10^{-12} m. What is the minimum uncertainty in a simultaneous measurement of the y-component of the proton's velocity? b) The uncertainty in the z-component of an electron's velocity is 0.250 m/s. What is the minimum uncertainty in a simultaneous measurement of the z-coordinate of the electron?

41–15 A scientist has devised a new method of isolating individual particles. He claims that this method enables him to detect simultaneously the position of a particle along an axis with a standard deviation of 0.12 nm and its momentum component along this axis with a standard deviation of 3.0×10^{-25} kg · m/s. Use the Heisenberg uncertainty principle to evaluate the validity of this claim.

41–16 a) The x-coordinate of an electron is measured with an uncertainty of 0.20 mm. What is the x-component of the electron's velocity, v_x, if the minimum percentage uncertainty in a simultaneous measurement of v_x is 1.0%? b) Repeat part (a) for a proton.

41–17 The spacing between adjacent atoms on the surface of a nickel crystal (Fig. 41–2c) is 0.215 nm. If the line of atoms shown in Fig. 41–2c lies along the x-direction, we can regard the uncertainty in the x-coordinate of each atom as approximately half the spacing. The mass of a single nickel atom is 9.75×10^{-26} kg. a) Estimate the minimum uncertainty in the x-component of momentum of a nickel atom in the crystal. b) If the magnitude of the momentum of a nickel atom is equal to the uncertainty found in part (a), what is its kinetic energy? Express the result in joules and in electron volts. c) If every atom in a 1.00-kg nickel crystal had the kinetic energy found in part (b), what would be the combined kinetic energy (in joules) of all the atoms? (The motion of each atom is in a random direction, so the center of mass of the crystal has zero momentum on average.) d) If all of this kinetic energy could be converted to gravitational potential energy, how high would the nickel crystal rise? e) Use the Heisenberg uncertainty principle to explain why the energy conversion described in part (d) is impossible.

41–18 Particle Lifetime. The unstable W^+ particle has a rest energy of 80.41 GeV (1 GeV = 10^9 eV) and an uncertainty in rest energy of 2.06 GeV. Estimate the lifetime of the W^+ particle.

41–19 The ψ ("psi") particle has a rest energy of 3097 MeV (1 MeV = 10^6 eV). The ψ is unstable with a lifetime of 7.6×10^{-21} s. Estimate the uncertainty in rest energy of the ψ particle. Express your answer in MeV and as a fraction of the rest energy of the particle.

41–20 An atom in a metastable state has a lifetime of 5.2 ms. What is the uncertainty in energy of the metastable state?

41–21 An unstable particle produced in a high-energy collision has a mass that is 4.50 times that of the proton and an uncertainty in mass that is 14.5% of the particle's mass. Using the relationship $E = mc^2$ between rest mass and energy, estimate the lifetime of the particle.

SECTION 41–5 THE ELECTRON MICROSCOPE

41–22 a) In an electron microscope, what accelerating voltage is needed to produce electrons with wavelength 0.0600 nm? b) If protons are used instead of electrons, what accelerating voltage is needed to produce protons with wavelength 0.0600 nm? (*Hint:* In each case the initial kinetic energy is negligible.)

41–23 a) What is the de Broglie wavelength of an electron that has been accelerated from rest through a potential increase of 800 V? b) What is the de Broglie wavelength of a proton accelerated from rest through a potential decrease of 800 V?

SECTION 41–6 WAVE FUNCTIONS

41–24 Compute $|\Psi|^2$ for $\Psi = \psi \sin \omega t$, where ψ is time independent and ω is a real constant. Is this a wave function for a stationary state? Why or why not?

41–25 Consider a wave function given by $\psi(x) = A \sin kx$, where $k = 2\pi/\lambda$ and A is a real constant. a) For what values of x is there the highest probability of finding the particle described by this wave function? Explain. b) For what values of x is the probability *zero*? Explain.

41–26 A particle is described by a wave function $\psi(x) = Ae^{-\alpha x^2}$, where A and α are real, positive constants. If the value of α is increased, what effect does this have on a) the particle's uncertainty in position? b) the particle's uncertainty in momentum? Explain your answers.

41–27 Consider the complex-valued function $f(x, y) = (x - iy)/(x + iy)$. Calculate $|f|^2$.

41–28 Particle A is described by the wave function $\psi(x, y, z)$. Particle B is described by the wave function $\psi(x, y, z)e^{i\phi}$, where ϕ

is a real constant. How does the probability of finding particle A within a volume dV around a certain point in space compare with the probability of finding particle B within this same volume?

41–29 At points within the region $0 \leq x \leq L$, $0 \leq y \leq L$, $0 \leq z \leq L$, a particle is described by the normalized wave function $\psi(x, y, z)$ $= (2/L)^{3/2} \sin(\pi x/L) \sin(\pi y/L) \sin(\pi z/L)$. Outside this region, $\psi(x, y, z) = 0$. Calculate the probability that the particle will be found within a cube of side $0.01L$ a) centered on the point $(L/4, L/4, L/4)$; b) centered on the point $(L/2, L/2, L/2)$. (*Hint:* Assume that $\psi(x, y, z)$ is uniform over the volume of such a small cube.)

PROBLEMS

41–30 A beam of 40-eV electrons traveling in the $+x$-direction passes through a slit that is parallel to the y-axis and 5.0 μm wide. The diffraction pattern is recorded on a screen 2.5 m from the slit. a) What is the de Broglie wavelength of the electrons? b) How much time does it take the electrons to travel from the slit to the screen? c) Use the width of the central diffraction pattern to calculate the uncertainty in the y-component of momentum of an electron just after it has passed through the slit. d) Use the result of part (c) and the Heisenberg uncertainty principle (Eq. 41–10 for y) to estimate the minimum uncertainty in the y-coordinate of an electron just after it has passed through the slit. Compare your result to the width of the slit.

41–31 What is the de Broglie wavelength of a red blood cell, with a mass of 1.00×10^{-11} g, that is moving with a speed of 0.400 cm/s? Do we need to be concerned with the wave nature of the blood cells when we describe the flow of blood in the body?

41–32 High-speed electrons are used to probe the interior structure of the atomic nucleus. For such electrons the expression $\lambda = h/p$ still holds, but we must use the relativistic expression for momentum, $p = mv/\sqrt{1 - v^2/c^2}$. a) Show that the speed of an electron that has de Broglie wavelength λ is

$$v = \frac{c}{\sqrt{1 + \left(\dfrac{mc\lambda}{h}\right)^2}}.$$

b) The quantity h/mc equals 2.426×10^{-12} m. (As we saw in Section 40–8, this same quantity appears in Eq. (40–21), the expression for Compton scattering of photons by electrons.) If λ is small compared to h/mc, the denominator in the expression found in part (a) is close to unity and the speed v is very close to c. In this case it is convenient to write $v = (1 - \Delta)c$ and express the speed of the electron in terms of Δ rather than v. Find an expression for Δ valid when $\lambda \ll h/mc$. (*Hint:* Use the binomial expansion $(1 + z)^n = 1 + nz + n(n - 1)z^2/2 + \cdots$, valid for the case $|z| < 1$.) c) How fast must an electron move for its de Broglie wavelength to be 1.00×10^{-15} m, comparable to the size of a proton? Express your answer in the form $v = (1 - \Delta)c$, and state the value of Δ.

41–33 a) What is the de Broglie wavelength of an electron accelerated from rest through a potential increase of 125 V? b) What is the de Broglie wavelength of an alpha particle $(q = +2e$, $m = 6.64 \times 10^{-27}$ kg) accelerated from rest through a potential drop of 125 V?

41–34 For relativistic particles the de Broglie relation $\lambda = h/p$ still holds, but the magnitude of momentum p is related to the total energy E by $E^2 = (pc)^2 + (mc^2)^2$ (Eq. 39–40). The kinetic energy is $K = E - mc^2$. a) Show that the de Broglie wavelength of a particle of kinetic energy K and rest mass m is

$$\lambda = \frac{hc}{\sqrt{K(K + 2mc^2)}}.$$

b) Find approximate expressions for λ as a function of K in the special cases i) $K \ll mc^2$ (nonrelativistic limit) and ii) $K \gg mc^2$ (extreme relativistic limit). c) Calculate the de Broglie wavelength for a proton with kinetic energy 7.00 GeV. d) Repeat part (c) for an electron with kinetic energy 25.0 MeV.

41–35 a) A particle of mass m has kinetic energy equal to three times its rest energy. What is the de Broglie wavelength of this particle? (*Hint:* You must use the relativistic expression for kinetic energy. See Problem 41–34.) b) Determine the numerical value of the kinetic energy (in MeV) and the wavelength (in meters) if the particle in part (a) is i) an electron; ii) a proton.

41–36 Suppose that the uncertainty of position of an electron is equal to the radius of the $n = 1$ Bohr orbit for hydrogen. Calculate the simultaneous minimum uncertainty of the corresponding momentum component, and compare this with the magnitude of the momentum of the electron in the $n = 1$ Bohr orbit. Discuss your results.

41–37 Suppose that the uncertainty in position of a particle in a particular direction is 40% of its de Broglie wavelength. Show that in this case the simultaneous minimum uncertainty in its momentum component along the same direction is 40% of its momentum.

41–38 Proton Energy in a Nucleus. The radii of atomic nuclei are of the order of 5.0×10^{-15} m. a) Estimate the minimum uncertainty in the momentum of a proton if it is confined within a nucleus. b) Take this uncertainty in momentum to be an estimate of the magnitude of the momentum. Use the relativistic relationship between energy and momentum, Eq. (39–40), to obtain an estimate of the kinetic energy of a proton confined within a nucleus. c) For a proton to remain bound within a nucleus, what must be the magnitude of the (negative) potential energy for a proton within the nucleus? Give your answer in eV and in MeV. Compare to the potential energy for an electron in a hydrogen atom, which has a magnitude of a few tens of eV. (This shows why the interaction that binds the nucleus together is called the "strong nuclear force.")

41–39 Electron Energy in a Nucleus. The radii of atomic nuclei are of the order of 5.0×10^{-15} m. a) Estimate the minimum uncertainty in the momentum of an electron if it is confined within a nucleus. b) Take this uncertainty in momentum to be an estimate of the magnitude of the momentum. Use the relativistic relationship between energy and momentum, Eq. (39–40), to obtain an estimate of the kinetic energy of an electron confined within a nucleus. c) Compare the energy calculated in part (b) to the magnitude of the Coulomb potential energy of a proton and an electron separated by 5.0×10^{-15} m. On the basis of your result, could there be electrons within the nucleus? (*Note:* It is interesting to compare this result to that of Problem 41–38.)

41–40 In a TV picture tube the accelerating voltage is 15.0 kV, and the electron beam passes through an aperture 0.50 mm in diameter to a screen 0.300 m away. a) What is the uncertainty in position of the point where the electrons strike the screen? b) Does this uncertainty affect the clarity of the picture significantly? (Use nonrelativistic expressions for the motion of the electrons. This is fairly accurate and is certainly adequate for obtaining an estimate of uncertainty effects.)

41–41 The neutral pion (π^0) is an unstable particle produced in high-energy particle collisions. Its mass is about 264 times that of the electron, and it exists for an average lifetime of 8.4×10^{-17} s before decaying into two gamma-ray photons. Using the relationship $E = mc^2$ between rest mass and energy, find the uncertainty in the mass of the particle and express it as a fraction of the mass.

41–42 a) The average kinetic energy of a molecule in a gas at absolute temperature T is $\frac{3}{2}kT$. Find the de Broglie wavelength of a molecule of mass m with this (nonrelativistic) kinetic energy. b) A volume V of the gas contains N molecules. Explain why the average distance between atoms is approximately $(V/N)^{1/3}$. c) The classical description of an ideal gas is based on the premise that molecules can be regarded as particles. This premise will break down if the average de Broglie wavelength of a molecule is greater than the average distance between molecules, since in this case the wave nature of the molecules cannot be neglected. If a gas at temperature T is comprised of N molecules, each of mass m, find an approximate expression for the minimum volume at which it becomes necessary to use a wave description of the molecules. d) Evaluate the volume in part (c) for a gas consisting of 1.00 mol of oxygen molecules (mass 2.66×10^{-26} kg) at room temperature (293 K). How does this compare to the volume that such a gas actually occupies, as found using the ideal-gas equation? Is the wave nature of oxygen molecules important at room temperature and ordinary pressures? e) The conduction electrons within a metal can be regarded as a gas. Consider the metal silver, which has density 1.05×10^4 kg/m^3. A single silver atom has mass 1.79×10^{-25} kg, and there is about one conduction electron per atom. Evaluate the volume in part (c) for a gas made up of all the conduction electrons in 1.00 kg of silver at 293 K. How does this compare with the actual volume of 1.00 kg of silver? In describing the behavior of conduction electrons in silver, is it necessary to account for their wave nature? (In Chapter 44 we will study the behavior of electrons in solids in greater depth.)

41–43 Doorway Diffraction. If your wavelength were 1.0 m, you would undergo considerable diffraction in moving through a doorway. a) What must be your speed for you to have this wavelength? (Assume that your mass is 60.0 kg.) b) At the speed calculated in part (a), how many years would it take you to move 0.80 m (one step)? Will you notice diffraction effects as you walk through doorways?

41–44 Atomic Spectra Uncertainties. A certain atom has an energy level 2.58 eV above the ground level. Once excited to this level, the atom remains in this level for 1.64×10^{-7} s (on average) before emitting a photon and returning to the ground level. a) What is the energy of the photon (in electron volts)? What is its wavelength (in nanometers)? b) What is the smallest possible uncertainty in energy of the photon? Give your answer in electron volts. c) Show that $|\Delta E/E| = |\Delta\lambda/\lambda|$ if $|\Delta\lambda/\lambda| \ll 1$. Use this to calculate the magnitude of the smallest possible uncertainty in the wavelength of the photon. Give your answer in nanometers.

41–45 In another universe the value of Planck's constant is 6.63×10^{-22} J · s. Assume that the physical laws and all other physical constants are the same as in our universe. In the other universe, an atom is in an excited state 4.50 eV above the ground state. The lifetime of this excited state (the average time the electron stays in this state) is 2.24×10^{-3} s. What is the minimum uncertainty (in eV) in the energy of the photon emitted when the atom makes the transition from this excited state to the ground state?

41–46 For x rays with wavelength 0.0300 nm, the $m = 1$ intensity maximum for a crystal occurs when the angle θ in Fig. 38–20 is 35.8°. At what angle θ does the $m = 1$ maximum occur when a beam of 4.50-keV electrons is used instead? Assume that the electrons also scatter from the atoms in the surface plane of this same crystal.

41–47 Electron diffraction can also take place when there is interference between electron waves that scatter from atoms on the surface of a crystal and waves that scatter from atoms in the next plane below the surface, a distance d from the surface (see Fig. 38–18c). a) Find an equation for the angles θ at which there is an intensity maximum for electron waves of wavelength λ. b) The spacing between crystal planes in a certain metal is 0.091 nm. If 71.0-eV electrons are used, find the angle at which there is an intensity maximum due to interference between scattered waves from adjacent crystal planes. The angle is measured as in Fig. 38–18c. c) The actual angle of the intensity maximum is slightly different from your result in part (b). The reason is the work function ϕ of the metal (see Section 40–3), which changes the electron potential energy by $-e\phi$ when it moves from vacuum into the metal. If the effect of the work function is taken into account, is the angle of the intensity maximum greater than or less than the value found in part (b)? Explain.

41–48 Zero-Point Energy. Consider a particle of mass m moving in a potential $U = \frac{1}{2}kx^2$, as in a mass-spring system. The total energy of the particle is $E = p^2/2m + \frac{1}{2}kx^2$. Assume that p and x are approximately related by the Heisenberg uncertainty principle, $px \approx h$. a) Calculate the minimum possible value of the energy E, and the value of x that gives this minimum E. This lowest possible energy, which is not zero, is called the *zero-point energy*. b) For the x calculated in part (a), what is the ratio of the kinetic to the potential energy of the particle?

41–49 A particle of mass m moves in a potential $U(x) = A|x|$, where A is a positive constant. In a simplified picture, quarks (the constituents of protons, neutrons, and other particles, as will be described in Chapter 46) have a potential energy of interaction of approximately this form, where x represents the separation between a pair of quarks. Because $U(x) \to \infty$ as $x \to \infty$, it's not possible to separate quarks from each other (a phenomenon called *quark confinement*). a) Classically, what is the force acting on this particle as a function of x? b) Using the uncertainty principle as in Problem 41–48, determine approximately the zero-point energy of the particle.

41–50 Example 41–6 (Section 41–6) showed that an example of a wave function of a stationary state is $\Psi = \psi e^{-i\omega t}$, where ψ is time independent and ω is a real (not complex) constant. Consider the wave function $\Psi = \psi_1 e^{-i\omega_1 t} + \psi_2 e^{-i\omega_2 t}$, where ψ_1 and ψ_2 are different time-independent functions and ω_1 and ω_2 are different real constants. Assume that ψ_1 and ψ_2 are real-valued functions, so that $\psi_1{}^* = \psi_1$ and $\psi_2{}^* = \psi_2$. Is this Ψ a wave function for a stationary state? Why or why not?

41–51 Imagine another universe in which the value of Planck's constant is 0.0663 J · s, but in which the physical laws and all other physical constants are the same as in our universe. In this universe, two physics students are playing catch. They are 12 m apart, and one throws a 0.25-kg ball directly toward the other with a speed of 6.0 m/s. a) What is the uncertainty in the ball's horizontal momentum, in a direction perpendicular to that in which it is being thrown, if the student throwing the ball knows that it is located within a cube with volume 125 cm³ at the time she throws it? b) By what horizontal distance could the ball miss the second student?

41–52 A particle is described by the normalized wave function $\psi(x, y, z) = A x e^{-\alpha x^2} e^{-\beta y^2} e^{-\gamma z^2}$, where A, α, β, and γ are all real, positive constants. The probability that the particle will be found in the infinitesimal volume $dx\, dy\, dz$ centered at the point (x_0, y_0, z_0) is $|\psi(x_0, y_0, z_0)|^2\, dx\, dy\, dz$. a) At what value of x_0 is the particle most likely to be found? b) Are there values of x_0 for which the probability of the particle being found is zero? If so, at what x_0?

41–53 A particle is described by the normalized wave function $\psi(x, y, z) = A e^{-\alpha(x^2 + y^2 + z^2)}$, where A and α are real, positive constants. a) Determine the probability of finding the particle at a distance between r and $r + dr$ from the origin. (*Hint:* See Problem 41–52. Consider a spherical shell centered on the origin with inner radius r and thickness dr.) b) For what value of r does the probability in part (a) have its maximum value? Is this the

same value of r for which $|\psi(x, y, z)|^2$ is a maximum? Explain any differences.

41–54 Consider the wave packet defined in Eq. (41–12). It is often convenient to use the wave number $k = 2\pi/\lambda$ rather than the wavelength, so the equation becomes

$$\psi(x) = \int_0^\infty B(k) \cos kx\, dk.$$

Let $B(k) = e^{-\alpha^2 k^2}$. a) The function $B(k)$ has its maximum value at $k = 0$. Let k_h be the value of k at which $B(k)$ has fallen to half its maximum value, and define the width of $B(k)$ as $w_k = k_h$. In terms of α, what is w_k? b) Use integral tables to evaluate the integral that gives $\psi(x)$. For what value of x is $\psi(x)$ maximum? c) Define the width of $\psi(x)$ as $w_x = x_h$, where x_h is the positive value of x where $\psi(x)$ has fallen to half its maximum value. Calculate w_x in terms of α. d) The momentum p is equal to $hk/2\pi$, so the width of B in momentum is $w_p = hw_k/2\pi$. Calculate the product $w_p w_x$ and compare to the Heisenberg uncertainty principle.

41–55 a) Using the integral in Problem 41–54, determine the wave function $\psi(x)$ for a function $B(k)$ given by

$$B(k) = \begin{cases} 0, & k < 0 \\ 1/k_0, & 0 \le k \le k_0 \\ 0, & k > k_0. \end{cases}$$

This represents an equal combination of all wave numbers between 0 and k_0. Thus $\psi(x)$ represents a particle with average wave number $k_0/2$, with a total spread or uncertainty in wave number of k_0. We will call this spread the *width* w_k of $B(k)$, so $w_k = k_0$. b) Graph $B(k)$ versus k and $\psi(x)$ versus x for the case $k_0 = 2\pi/L$, where L is a length. Locate the point where $\psi(x)$ has its maximum value and label this point on your graph. Locate the two points closest to this maximum (one on each side of it) where $\psi(x) = 0$, and define the distance along the x-axis between these two points as w_x, the width of $\psi(x)$. Indicate the distance w_x on your graph. What is the value of w_x if $k_0 = 2\pi/L$? c) Repeat part (b) for the case $k_0 = \pi/L$. d) The momentum p is equal to $hk/2\pi$, so the width of B in momentum is $w_p = hw_k/2\pi$. Calculate the product $w_p w_x$ for each of the cases $k_0 = 2\pi/L$ and $k_0 = \pi/L$. Discuss your results in light of the Heisenberg uncertainty principle.

CHALLENGE PROBLEMS

41–56 The wave nature of particles results in the quantum mechanical situation that a particle confined in a box can only assume wavelengths that result in standing waves in the box, with nodes at the box walls. a) Show that an electron confined in a one-dimensional box of length L will have energy levels given by

$$E_n = \frac{n^2 h^2}{8mL^2}.$$

(*Hint:* Recall that the relation between the de Broglie wavelength and the speed of a nonrelativistic particle is $mv = h/\lambda$. The energy of the particle is $\frac{1}{2}mv^2$.) b) If a hydrogen atom is modeled as a one-dimensional box with length equal to the Bohr radius, what is the energy (in electron volts) of the lowest energy level of the electron?

41–57 You have entered a contest in which the contestants drop a marble with a mass of 20.0 g from the roof of a building onto a small target 25.0 m below. From uncertainty considerations, what is the typical distance by which you will miss the target, given that you aim with the highest possible precision? (*Hint:* The uncertainty Δx_f in the x-coordinate of the marble when it reaches the ground comes in part from the uncertainty Δx_i in the x-coordinate initially and in part from the initial uncertainty in v_x. The latter gives rise to an uncertainty Δv_x in the horizontal motion of the marble as it falls. The values of Δx_i and Δv_x are related by the uncertainty principle. A small Δx_i gives rise to a large Δv_x, and vice versa. Find the value of Δx_i that gives the smallest total uncertainty in x at the ground. Ignore any effects of air resistance.)

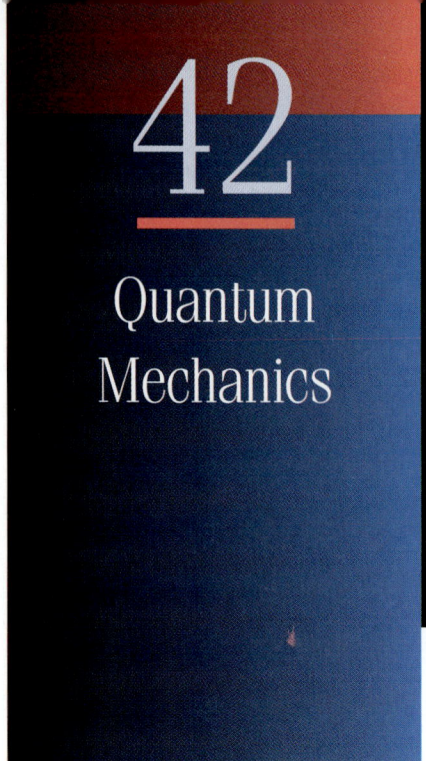

42

Quantum Mechanics

This STM (scanning tunneling microscope) image of a copper surface is graphic evidence of the wave behavior of electrons. The "elevation" at each point in the image indicates the density of electrons at that point. Forty-eight iron atoms (shown as yellow peaks) have been placed in a circle on the surface. The electrons within the circle form a clear standing-wave pattern.

42–1 INTRODUCTION

In the preceding chapter we found that particles sometimes behave like waves and that they can be described by wave functions. Now we're ready for a systematic analysis of particles in bound states (such as electrons in atoms), including finding their possible wave functions and energy levels.

Our analysis involves finding solutions of a fundamental equation called the *Schrödinger equation.* The wave functions for any system must be solutions of the Schrödinger equation for that system. Each solution corresponds to a definite energy, so solving the Schrödinger equation automatically gives the possible energy levels for a system. We'll discuss several simple one-dimensional applications of the Schrödinger equation.

Besides energies, solving the Schrödinger equation also gives us the probabilities of finding a particle in various regions. One surprising result of solving the Schrödinger equation is that there is a nonzero probability that microscopic particles will pass through thin barriers, even though such a process is forbidden by Newtonian mechanics.

Finally, we'll generalize the Schrödinger equation to three dimensions. This will pave the way for describing the wave functions for the hydrogen atom in Chapter 43. The hydrogen-atom wave functions in turn form the foundation for our analysis of more complex atoms, of the periodic table of the elements, of x-ray energy levels, and of other properties of atoms.

42–2 PARTICLE IN A BOX

In this chapter we want to learn how to find wave functions and energy levels for various systems. As often happens, the simplest problems may not correspond exactly to any situation found in nature, but they often serve as enlightening first models.

Our first example fits that description, serving as a first approximation to the behavior of an electron that is free to move within a long, straight molecule or along a very thin wire. Our system consists of a particle confined between two rigid walls separated by a distance L (Fig. 42–1). We'll make it a one-dimensional problem, with the particle moving always along the x-axis and the walls located at $x = 0$ and $x = L$. We'll assume that the energy E of the particle and the magnitude of its momentum p are constant. The potential energy corresponding to the rigid walls is infinite, and the particle cannot escape. The situation is often described succinctly as a **"particle in a box."**

We begin with some assumptions. Because the particle is confined to the region $0 \le x \le L$, we expect the particle's wave function to be zero outside that region. Also, it seems physically reasonable that the wave function should be a *continuous* function of x. If it is, then it must be zero at the region's boundaries, $x = 0$ and $x = L$. These two conditions are *boundary conditions* for the problem. They should look familiar, since these are the same conditions that we used to find normal modes of a vibrating string in Section 20–4 (Fig. 42–2) and the electric field in a standing electromagnetic wave in Section 33–7; you may want to review those discussions.

The energy and the magnitude of momentum are constant for our particle. Thus we consider a one-dimensional wave function ψ that does not depend on time. Through the de Broglie relation $\lambda = h/p$, a definite wavelength λ corresponds to a definite magnitude of momentum p, so it's reasonable to assume that ψ is a *sinusoidal* function of x. If so, a possible form is

$$\psi = A \sin kx, \tag{42-1}$$

where k is the *wave number* introduced in Section 19–4, Eq. (19–5):

$$k = \frac{2\pi}{\lambda}. \tag{42-2}$$

(We'll discuss the significance of the constant A later.)

Equation (42–1) satisfies the boundary condition that ψ should be zero at $x = 0$. To make ψ equal to zero at $x = L$, we choose values of k such that $kL = n\pi$ ($n = 1, 2, 3, \ldots$). The possible values of k and λ are therefore

$$k = \frac{n\pi}{L} \quad \text{and} \quad \lambda = \frac{2\pi}{k} = \frac{2L}{n} \quad (n = 1, 2, 3, \ldots). \tag{42-3}$$

Now we are only two short steps from determining the possible energy levels. From λ we can find the momentum p using the de Broglie relation $p = h/\lambda$; the energy E is all kinetic and thus equals $p^2/2m$. For each value of n there are corresponding values of p, λ, and E; let's call them p_n, λ_n, and E_n. Putting the pieces together, we get

$$p_n = \frac{h}{\lambda_n} = \frac{nh}{2L} \tag{42-4}$$

and

$$E_n = \frac{p_n^2}{2m} = \frac{n^2 h^2}{8mL^2} \quad (n = 1, 2, 3, \ldots) \quad \begin{array}{l}\text{(energy levels, particle}\\ \text{in a box).}\end{array} \tag{42-5}$$

These are the possible energy levels for a particle in a box. Each energy level has its own value of the quantum number n and a corresponding wave function, which we denote by ψ_n. When we replace k in Eq. (42–1) by $n\pi/L$ from Eq. (42–3), we find

$$\psi_n = A \sin \frac{n\pi x}{L} \quad (n = 1, 2, 3, \ldots). \tag{42-6}$$

Figure 42–3a (page 1296) shows graphs of the wave functions for $n = 1, 2, 3, 4$, and 5, and Fig. 42–3b shows the energy-level diagram for this system. The levels of successively higher energies, proportional to n^2, are spaced farther and farther apart. There is an infinite number of levels because the walls are perfectly rigid; even a particle of infinitely great kinetic energy is confined within the box.

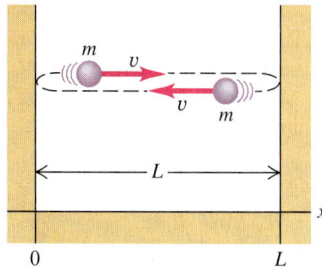

42–1 The Newtonian view of a particle in a box. A particle with mass m moves along a straight line at constant speed, bouncing between two rigid walls that are a distance L apart.

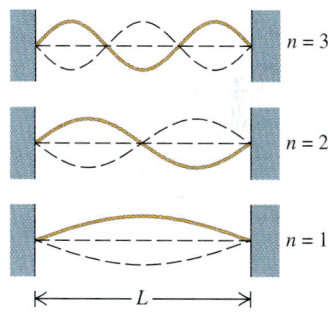

42–2 Normal modes of vibration for a string with length L, held at both ends. Each end is a node, and there may be $n - 1$ additional nodes. Thus the length is an integer number of half-wavelengths; $L = n\lambda_n/2$.

20.2
Particle in a Box

EXAMPLE 42-1

Electron in an atom-size box Find the lowest energy level for a particle in a box if the particle is an electron in a box 5.0×10^{-10} m across, or a little bigger than an atom.

SOLUTION Placing $n = 1$ in Eq. (42–5) to find the smallest E

(that is, the ground level), we get

$$E_1 = \frac{h^2}{8mL^2} = \frac{(6.626 \times 10^{-34} \text{ J} \cdot \text{s})^2}{8(9.109 \times 10^{-31} \text{ kg})(5.0 \times 10^{-10} \text{ m})^2}$$

$$= 2.4 \times 10^{-19} \text{ J} = 1.5 \text{ eV}.$$

A particle trapped in a box is rather different from an electron bound in an atom, but it is reassuring that this energy is of the same order of magnitude as actual atomic energy levels.

You should be able to show that replacing the electron with a proton or a neutron ($m = 1.67 \times 10^{-27}$ kg) in a box the width of a medium-sized nucleus ($L = 1.1 \times 10^{-14}$ m) gives $E_1 = 1.7$ MeV. This shows us that the energies for particles in the nucleus are about a million times larger than the energies of electrons in atoms, giving us a clue as to why each nuclear fission and fusion reaction provides so much more energy than each chemical reaction.

If you repeat this calculation for a billiard ball ($m = 0.2$ kg) bouncing back and forth between the cushions of an frictionless, perfectly elastic billiard table ($L = 1.5$ m), you'll find the separation between energy levels to be as small as 4×10^{-67} J (Exercise 42–1). This negligible value shows that quantum effects won't have much effect on your billiard game.

PROBABILITY AND NORMALIZATION

Several additional points need to be discussed. First is the *probability* interpretation of the wave function ψ, which we discussed in Section 41–6. In our one-dimensional situation the quantity $|\psi|^2 \, dx$ (with ψ evaluated at a particular value of x) is proportional to the probability that the particle will be found within a small interval dx about x. For a particle in a box,

$$|\psi|^2 \, dx = A^2 \sin^2 \frac{n\pi x}{L} \, dx.$$

Both ψ and $|\psi|^2$ are plotted in Fig. 42–4 for $n = 1, 2$, and 3. We note that not all positions are equally likely. This is in contrast to the situation in classical mechanics, in which all positions between $x = 0$ and $x = L$ are equally likely. We see in Fig. 42–4b that $|\psi|^2 = 0$ at some points, so the probability is zero of finding the particle exactly at

42–3 (a) Wave functions for a particle in a box, with $n = 1, 2, 3,$ 4, and 5. The horizontal dashed lines represent $\psi = 0$ for each of the five levels. (b) Energy-level diagram for a particle in a box. Each energy is $n^2 E_1$, where E_1 is the energy of the ground level; n is the quantum number for each level.

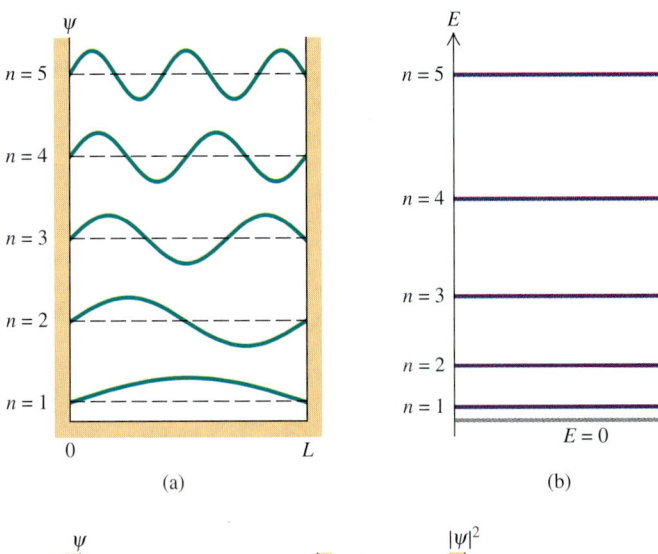

42–4 Graphs of (a) ψ and (b) $|\psi|^2$ for the first three wave functions ($n = 1, 2, 3$) for a particle in a box. The horizontal dashed lines represent $\psi = 0$ and $|\psi|^2 = 0$ for each of the three levels. The value of $|\psi|^2 \, dx$ at each point is proportional to the probability of finding the particle in a small interval dx about the point.

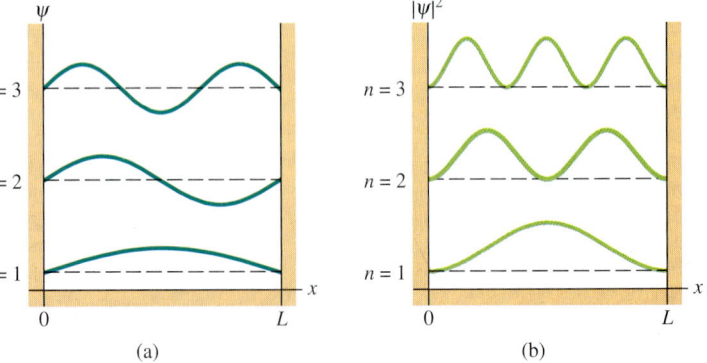

these points. Don't let that bother you; the uncertainty principle has already shown us that we can't measure position exactly. The particle is localized only to be somewhere between $x = 0$ and $x = L$, not to be exactly at some particular value of x.

CAUTION ▶ Note that the energy of a particle in a box cannot be zero. Equation (42–5) shows us that $E = 0$ requires $n = 0$, and $n = 0$ in Eq. (42–6) gives a zero wave function and thus no probability of finding the particle with that energy. ◀

We know that the particle *must* be somewhere in the universe, that is, between $x = -\infty$ and $x = +\infty$. So the *sum* of the probabilities for all the dx's everywhere (the *total* probability of finding the particle) must equal 1. That is,

$$\int_{-\infty}^{\infty} |\psi|^2 \, dx = 1 \qquad \text{(normalization condition)}. \tag{42–7}$$

A wave function is said to be *normalized* if it has a constant such as A in Eq. (42–6) that is calculated to make the total probability equal 1 in Eq. (42–7). The process of determining the constant is called **normalization,** as was mentioned in Section 41–6. Why bother with normalization? Because for a normalized wave function, $|\psi|^2 \, dx$ is not merely proportional to, but *equals,* the probability of finding the particle in dx.

Now let's normalize the wave function ψ given by Eq. (42–6) for the particle in a box. Since ψ is zero except between $x = 0$ and $x = L$, Eq. (42–7) becomes

$$\int_0^L A^2 \sin^2 \frac{n\pi x}{L} \, dx = 1. \tag{42–8}$$

You can evaluate this integral using the trigonometric identity $\sin^2 \theta = \frac{1}{2}(1 - \cos 2\theta)$; the result is $A^2 L/2$. Thus our probability interpretation of the wave function demands that $A^2 L/2 = 1$, or $A = (2/L)^{1/2}$; the constant A is *not* arbitrary. (This is in contrast to the classical vibrating string problem, in which A represents an amplitude that depends on initial conditions.) Thus the normalized wave functions for a particle in a box are

$$\psi_n = \sqrt{\frac{2}{L}} \sin \frac{n\pi x}{L} \qquad (n = 1, \ 2, \ 3, \ \ldots) \qquad \text{(particle in a box)}. \tag{42–9}$$

Next, let's check whether our results for the particle in a box are consistent with the uncertainty principle. We can write the position of the particle as $x = L/2 \pm L/2$, so we can estimate the uncertainty in position as $\Delta x \approx L/2$. From Eq. (42–4) the magnitude of momentum in state n is $p_n = nh/2L$. A reasonable estimate of the momentum uncertainty is the difference in momentum of two levels that differ by 1 in their n values, that is, $\Delta p_x \approx h/2L$. Then the product $\Delta x \, \Delta p_x$ is

$$\Delta x \, \Delta p_x \approx \frac{h}{4}.$$

This is consistent with the uncertainty principle, Eq. (41–10), $\Delta x \, \Delta p_x \geq h/2\pi$, in which the uncertainties were more precisely defined as standard deviations.

TIME DEPENDENCE

Finally, we ask whether the wave functions for this problem have any *time* dependence. We recall that the wave functions for standing waves on a string, Eq. (20–1), contained a time factor $\cos \omega t$. Does the wave functions Ψ have a similar time dependence here? The answer is that there *is* a type of sinusoidal time dependence, but it can't have the form $\sin \omega t$ or $\cos \omega t$. If it did, all the normalization integrals that are equivalent to Eq. (42–7) would also depend on time. If such a time-dependent function Ψ were normalized at one instant, it wouldn't be normalized at all other times.

It turns out that the correct expression for the time dependence is

$$e^{-i\omega t} = \cos \omega t - i \sin \omega t,$$

where ω is determined by the de Broglie energy-frequency relation $E = hf$, so $\omega = 2\pi f = 2\pi E/h$. The exponential function is a complex quantity that equals the sum of a real function and a pure imaginary function, $\cos \omega t$ and $-i \sin \omega t$, respectively. This relation is called *Euler's formula;* if you haven't encountered it yet in your study of calculus, you probably will do so soon.

As we saw in Example 41–6, the *absolute value* of $e^{-i\omega t}$ is unity. If we multiply each ψ by this factor, it doesn't change the value of $|\Psi|^2$. So in calculations with states that have definite energy, we are justified in omitting the time factor. We pointed out in Section 41–6 that a state with definite energy is a *stationary state* with the property that $|\Psi|^2$ is independent of time at each point.

Using $\Psi(x, t)$ and $\psi(x)$ for Ψ and ψ to emphasize whether or not they depend on time, we can write

$$\Psi(x, t) = \psi(x)e^{-i(2\pi E/h)t} \qquad \text{(time-dependent wave function, stationary state).} \qquad (42\text{--}10)$$

We'll be concerned mainly with stationary states in our applications of quantum mechanics.

EXAMPLE 42–2

Find the expression for the normalized time-dependent wave functions for a particle in a box.

SOLUTION From Eq. (42–5) the term $2\pi E/h$ in the exponent in Eq. (42–10) equals $\pi n^2 h/4mL^2$. Also, the time-independent wave functions of Eq. (42–9) remain normalized when multiplied by

the quantity $e^{-i(2\pi E/h)t}$, which has an absolute value of 1. Thus our answer is

$$\Psi_n = \sqrt{\frac{2}{L}} \sin \frac{n\pi x}{L} e^{-i(\pi n^2 h/4mL^2)t}.$$

You should be able to show that these are wave functions of stationary states.

42–3 The Schrödinger Equation

The Schrödinger equation, developed by the Austrian physicist Erwin Schrödinger (1887–1961), is the basic relationship for determining wave functions and energy levels. We will apply it to several systems, including the particle in a box (which we've already discussed), the harmonic oscillator, and the hydrogen atom. We won't pretend that we can *derive* this equation from known principles. We can't; it is a new principle. But we can show how it is related to the de Broglie equations, and we can make it seem plausible. The ultimate test of the Schrödinger equation is to compare its predictions with experimental observations. As we will see, it passes such tests with flying colors.

We'll begin with a one-dimensional problem, a particle moving freely along the x-axis between two walls like a particle in a box. From Eqs. (42–1) and (42–2) the wave functions for a particle in a box have the general form

$$\psi = A \sin kx = A \sin \frac{2\pi x}{\lambda}. \qquad (42\text{--}11)$$

By using the de Broglie relation, $p = h/\lambda$, the energy E of the corresponding level can be expressed as

$$E = \frac{p^2}{2m} = \frac{h^2}{2\lambda^2 m} = \frac{h^2 k^2}{8\pi^2 m}. \qquad (42\text{--}12)$$

Notice that the second derivative of Eq. (42–11) is $-k^2 A \sin kx$, that is, the original function multiplied by $-k^2$. So taking the second derivative of ψ and then multiplying by the factor $(-h^2/8\pi^2 m)$ have the same effect as multiplying ψ by the factor $(-h^2/8\pi^2 m)(-k^2)$, which equals E. That is,

$$-\frac{h^2}{8\pi^2 m}\frac{d^2\psi}{dx^2} = E\psi. \qquad (42\text{–}13)$$

We invite you to verify that Eq. (42–11) satisfies this equation, no matter what the value of k, if E is given by Eq. (42–12).

Equation (42–13) is the simplest form of the **Schrödinger equation.** We obtained it in a somewhat contrived way, but we can see that it is consistent with what we know about wave properties of particles and about the wave functions for a particle in a box.

We can avoid writing a lot of factors of 2π in later discussions by using the abbreviation $\hbar$ (pronounced "h-bar") for Planck's constant divided by 2π:

$$\hbar = \frac{h}{2\pi} = 1.055 \times 10^{-34}\ \text{J}\cdot\text{s} = 6.582 \times 10^{-16}\ \text{eV}\cdot\text{s}. \qquad (42\text{–}14)$$

We can re-express the de Broglie relations $p = h/\lambda$ and $E = hf$ in terms of $\hbar$, the wave number k, and the angular frequency ω:

$$p = \hbar k, \qquad E = \hbar\omega. \qquad (42\text{–}15)$$

In terms of $\hbar$, Eq. (42–13) is

$$-\frac{\hbar^2}{2m}\frac{d^2\psi}{dx^2} = E\psi \qquad \text{(Schrödinger equation, particle in a box).} \qquad (42\text{–}16)$$

So far, we've been talking about a *free* particle, a particle that moves along the x-axis with no force acting on it (except for the forces that act at the instant a particle in a box strikes a wall). Now suppose the particle is acted on by a force with an x-component F_x that depends on x (but not on time). We'll assume the force is conservative so that there is a corresponding potential energy U. In guessing how to generalize the Schrödinger equation for this more general problem, we start with the classical energy relation for motion under a conservative force: $K + U = E$, or

$$\frac{p^2}{2m} + U = E.$$

Comparing this with Eq. (42–16), we see that a reasonable guess for the generalized one-dimensional Schrödinger equation would be

$$-\frac{\hbar^2}{2m}\frac{d^2\psi}{dx^2} + U\psi = E\psi \qquad \text{(one-dimensional Schrödinger equation).} \qquad (42\text{–}17)$$

In this equation, ψ and U are, in general, functions of x, while E is a constant for a given energy level. How do we know that this equation is correct? Because it works. Predictions made using this equation agree with experimental results and thus confirm that this is indeed the correct way to include conservative, time-independent interactions in the Schrödinger equation. We'll apply this equation to several problems in the following sections.

In working with this more general form of the Schrödinger equation, we will insist that the wave functions be *normalized,* that is, $\int_{-\infty}^{\infty}|\psi|^2\,dx = 1$. Note that if ψ is a solution of the Schrödinger equation, $C\psi$ is also a solution, where C is any constant. We can always choose the value of C to satisfy the normalization requirement.

Again we state that the function ψ must be *continuous.* Its first derivative, $d\psi/dx$, must also be continuous except where the potential energy becomes infinite (as at the walls of the box). These requirements ensure that the wave function is a mathematically well-behaved solution to the Schrödinger equation. They are analogous to the requirements that the wave functions for a vibrating string, and their derivatives, should be

Erwin Schrödinger (1887–1961) developed the equation that bears his name in 1926, an accomplishment for which he shared (with the British physicist P. A. M. Dirac) the 1933 Nobel Prize in physics. In later life he did research into the philosophy and history of science.

continuous at points where the linear mass density or tension in the string changes. An exception occurs at the end points of the string, where only the function itself, not its derivative, needs to be continuous.

APPLYING THE SCHRÖDINGER EQUATION TO THE PARTICLE IN A BOX

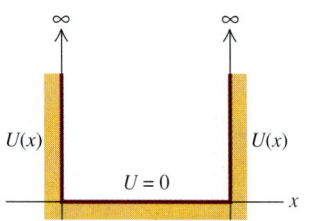

42–5 The potential-energy function for a particle in a box. The potential energy U is zero in the interval $0 \le x \le L$ and is infinite everywhere outside this interval.

We can interpret the particle in a box in terms of Eq. (42–17) by representing the interaction with the walls in terms of the potential-energy function shown in Fig. 42–5. The potential energy U is zero for $0 \le x \le L$ (in the box) but infinitely great everywhere else. Then the wave function must be zero everywhere except in the box, where Eq. (42–17) reduces to Eq. (42–16). We see that the continuity requirement then demands that ψ be zero at $x = 0$ and $x = L$, as we assumed in our initial discussion of this problem.

To obtain the wave functions and energies for the particle in a box directly from Eq. (42–16), we note that ψ must satisfy this equation and the conditions $\psi = 0$ at both $x = 0$ and $x = L$. We rewrite Eq. (42–16) in the form

$$\frac{d^2\psi}{dx^2} = -\frac{2mE}{\hbar^2}\psi.$$

The solutions of this equation can have the forms $A \sin kx$ and $B \cos kx$, where $k = \sqrt{2mE}/\hbar$ and A and B are constants. The condition $\psi = 0$ at $x = 0$ is satisfied only by $\sin kx$, so B must be zero. To satisfy $\psi = 0$ at $x = L$, we must set $kL = n\pi$. So both boundary conditions are satisfied only when

$$\frac{\sqrt{2mE}}{\hbar} = \frac{n\pi}{L}.$$

Solving this relation for E, we find

$$E = \frac{n^2\pi^2\hbar^2}{2mL^2} \qquad \text{(energy levels, particle in a box).} \qquad (42\text{–}18)$$

Substituting $\hbar = h/2\pi$ in Eq. (42–18), we see that this energy expression agrees with our previous result, Eq. (42–5). The values of k in the wave function are the same as we found previously, and the normalization is carried out just as before.

EXAMPLE 42–3

A nonsinusoidal wave function? a) Show that $\psi = Ax + B$, where A and B are constants, is a solution to the Schrödinger equation for an $E = 0$ energy level of a particle in a box. b) Show, however, that the probability of finding a particle with this wave function is zero.

SOLUTION a) The Schrödinger equation for a particle in a box is Eq. (42–16). Differentiating $\psi = Ax + B$ twice with respect to x gives $d^2\psi/dx^2 = 0$ for the left side of that equation. Also $E = 0$ gives zero for the right side. Since $0 = 0$, $\psi = Ax + B$ is a solution to this Schrödinger equation for $E = 0$.

b) The wave function ψ equals zero outside the box ($x < 0$ and

$x > L$). In order that $\psi = Ax + B$ may be continuous at $x = 0$, it must be true that $\psi = 0$ at $x = 0$, so $A(0) + B = 0$, or $B = 0$. Similarly, in order that ψ may be continuous at $x = L$, it must be true that $\psi = 0$ at $x = L$, so $A(L) + 0 = 0$, or $A = 0$. With both A and B equal to zero, $\psi = Ax + B = 0$. Thus the wave function equals zero inside the box as well as outside the box, and the probability of finding the particle anywhere with this wave function is zero.

The moral of this story is that a wave function for a particle must satisfy *both* the Schrödinger equation *and* the boundary conditions, and be nonzero in some region of space.

There is also a version of the Schrödinger equation that includes time dependence. However, we don't need that version to calculate the energy levels and wave functions of stationary states. If we do need a time-dependent wave function for a stationary state, we simply use Eq. (42–10), as was done in Example 42–2. We do need a time-dependent Schrödinger equation to study the details of *transitions* between states. In more advanced courses you may study the quantum mechanics of time-dependent phenomena such as photon emission and absorption and lifetimes of states.

42–4 POTENTIAL WELLS

A **potential well** is a potential-energy function $U(x)$ that has a minimum. We introduced this term in Section 7–7, and we also used it in our discussion of periodic motion in Chapter 13. In Newtonian mechanics a particle trapped in a potential well can vibrate back and forth with periodic motion. Our first application of the Schrödinger equation, the particle in a box, involved a rudimentary potential well with a function $U(x)$ that is zero within a certain interval and infinite everywhere else. This function corresponds to a few situations found in nature, as mentioned in Section 42–2, but the correspondence is only approximate.

A potential well that serves as a better approximation to several actual physical situations is a well with straight sides but *finite* height. Figure 42–6 shows a potential-energy function that is zero in the interval $0 \le x \le L$ and has the value U_0 outside this interval. This function is often called a **square-well potential.** It could serve as a simple model of an electron within a metallic sheet with thickness L, moving perpendicular to the surfaces of the sheet. The electron can move freely inside the metal but has to climb a potential-energy barrier with height U_0 to escape from either surface of the metal. The energy U_0 is related to the *work function* that we discussed in Section 40–3 in connection with the photoelectric effect. In a three-dimensional application a spherical version of a potential well can be used to approximately describe the motions of protons and neutrons within a nucleus.

In Newtonian mechanics the particle is trapped (localized) in a well if the total energy E is less than U_0. In quantum mechanics, such a trapped state is often called a **bound state.** All states are bound for an infinitely deep well, but if E is greater than U_0 for a finite well, the particle is *not* bound.

For a finite square well, let's consider bound-state solutions of the Schrödinger equation, corresponding to $E < U_0$. The easiest approach is to consider separately the regions where $U = 0$ and where $U = U_0$. Where $U = 0$, the Schrödinger equation reduces to Eq. (42–16). Rearranging this equation, we find

$$\frac{d^2\psi}{dx^2} = -\frac{2mE}{\hbar^2}\,\psi. \qquad (42\text{–}19)$$

The solutions of this equation are sinusoidal, with the forms

$$\psi = A\sin\frac{\sqrt{2mE}}{\hbar}\,x + B\cos\frac{\sqrt{2mE}}{\hbar}\,x \qquad \text{(inside the well),} \qquad (42\text{–}20)$$

where A and B are constants. So far, this looks a lot like the particle-in-a-box analysis in Section 42–3.

In the regions $x < 0$ and $x > L$ we use Eq. (42–17) with $U = U_0$. Rearranging, we get

$$\frac{d^2\psi}{dx^2} = \frac{2m(U_0 - E)}{\hbar^2}\,\psi. \qquad (42\text{–}21)$$

The quantity $U_0 - E$ is positive, so the solutions of this equation are exponential. Using κ (the Greek letter "kappa") as positive in the abbreviation $\kappa = [2m(U_0 - E)]^{1/2}/\hbar$, we can write the solutions as

$$\psi = Ce^{\kappa x} + De^{-\kappa x} \qquad \text{(outside the well),} \qquad (42\text{–}22)$$

where C and D are constants with different values in the $x < 0$ and $x > L$ regions.

We see that the bound-state wave functions for this system are sinusoidal inside the well and exponential outside. We have to use the positive exponent in the $x < 0$ region and the negative exponent in the $x > L$ region. That is, $D = 0$ for $x < 0$ and $C = 0$ for $x > L$. Without this choice of constants, ψ would approach infinity as $|x|$ approached

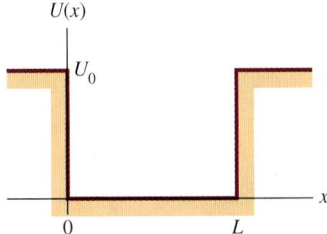

42–6 A finite potential well, or square-well potential. The potential-energy function $U(x)$ is zero in the interval $0 \le x \le L$ and has the constant value U_0 everywhere outside this interval.

20.1
Potential Energy Diagrams

infinity, and the normalization condition, Eq. (42–7), could not be satisfied. We also have to match the wave functions so that they satisfy the boundary conditions that we mentioned in Section 42–3: ψ and $d\psi/dx$ must be continuous at the boundary points ($x = 0$ and $x = L$). If the wave function ψ or the slope $d\psi/dx$ were to change discontinuously at a point, the second derivative $d^2\psi/dx^2$ would be *infinite* at that point. But that would violate the Schrödinger equation, which says that at every point, $d^2\psi/dx^2$ is proportional to $U - E$. In our situation, $U - E$ is finite everywhere, so $d^2\psi/dx^2$ must also be finite everywhere.

COMPARING FINITE AND INFINITE SQUARE WELLS

Matching the sinusoidal and exponential functions at the boundary points so that they join smoothly is possible only for certain specific values of the total energy E, so this requirement determines the possible energy levels of the finite square well. There is no simple formula for the energy levels as there was for the infinitely deep well. Finding the levels is a fairly complex mathematical problem that requires solving a transcendental equation by numerical approximation; we won't go into the details. Figure 42–7 shows the general shape of a possible wave function. The most striking features of this wave function are the "exponential tails" that extend outside the well into regions that are forbidden by Newtonian mechanics (because in those regions the particle would have negative kinetic energy). We see that there is some probability for finding the particle *outside* the potential well, despite the fact that in classical mechanics this is impossible. This penetration into classically forbidden regions is a quantum effect that has no classical analog for particles. We will discuss an amazing result of this effect in Section 42–5.

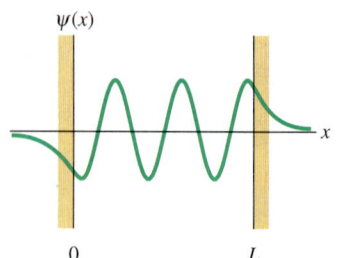

$\psi(x)$

0   L

42–7 A possible wave function for a particle in a finite potential well. The function is sinusoidal inside the well ($0 \leq x \leq L$) and exponential outside. It approaches zero asymptotically at large $|x|$. The functions must join smoothly at $x = 0$ and $x = L$; the wave function and its derivative must be continuous.

EXAMPLE 42-4

Well versus box Show that Eq. (42–22), $\psi = Ce^{\kappa x} + De^{-\kappa x}$, is consistent with the corresponding particle-in-a-box wave function.

SOLUTION Equation (42–22) gives the possible wave functions outside the finite potential well. Since the wave function outside the box (an infinite potential well) equals zero, we need to show

that $\psi = Ce^{\kappa x} + De^{-\kappa x}$ approaches zero as the depth of the potential well (U_0) approaches infinity. As U_0 approaches infinity, $\kappa = [2m(U_0 - E)]^{1/2}/\hbar$ also approaches infinity. When κ approaches infinity, the wave functions $\psi = Ce^{\kappa x}$ for $x < 0$ and $\psi = De^{-\kappa x}$ for $x > L$ approach zero as required.

We previously *assumed* $\psi = 0$ outside the box. This analysis validates our assumption.

Let's continue our comparison of the finite-depth potential well with the infinitely deep well. First, because the wave functions for the finite well aren't zero at $x = 0$ and $x = L$, the wavelength of the sinusoidal part of each wave function is *longer* than it would be with an infinite well. From $p = h/\lambda$ this increase in λ corresponds to reduced magnitude of momentum and therefore reduced energy. Thus each energy level, including the ground level, is *lower* for a finite well than for an infinitely deep well with the same width.

Second, a well with finite depth U_0 has only a *finite* number of bound states and corresponding energy levels, compared to the *infinite* number for an infinitely deep well. How many levels there are depends on the magnitude of U_0 in comparison with the ground-level energy for the infinite well, which we call E_∞. From Eq. (42–18),

$$E_\infty = \frac{\pi^2 \hbar^2}{2mL^2}. \qquad (42\text{–}23)$$

When U_0 is much *larger* than E_∞ (a very deep well), there are many bound states, and the energies of the lowest few are nearly the same as the energies for the infinitely deep well. When U_0 is only a few times as large as E_∞, there are only a few bound states. (But there

42–8 (a) Wave functions for the three bound states for a particle in a finite potential well with depth U_0, for the case $U_0 = 6E_\infty$, where E_∞ is the energy of the ground level for a particle in an infinitely deep well with the same width. The horizontal violet line for each wave function corresponds to $\psi = 0$. The vertical placement of the violet lines indicates the energy of each bound state. (b) Energy-level diagram for this system. The energies are expressed both as multiples of E_∞ and as fractions of U_0. All energies greater than U_0 are possible; states with $E > U_0$ form a continuum.

is always at least one bound state, no matter how shallow the well.) As with the infinitely deep well, there is *no* state with $E = 0$; such a state would violate the uncertainty principle.

Figure 42–8 shows the particular case in which $U_0 = 6E_\infty$; for this case there are three bound states. The energy levels are expressed both as fractions of the well depth U_0 and as multiples of E_∞. Note that if the well were infinitely deep, the lowest three levels, as given by Eq. (42–18), would be E_∞, $4E_\infty$, and $9E_\infty$. The wave functions for the three bound states are also shown.

It turns out that when U_0 is less than E_∞, there is only one bound state. In the limit when U_0 is *much smaller* than E_∞ (a very shallow or very narrow well), the energy of this single state is approximately $E = 0.68U_0$.

Figure 42–9 shows graphs of the probability distributions, that is, the values of $|\psi|^2$, for the wave functions shown in Fig. 42–8a. As with the infinite well, not all positions are equally likely. We have already commented on the possibility of finding the particle outside the well, in the classically forbidden regions.

There are also states for which E is *greater than* U_0. In this case the particle is not bound but is free to move through all values of x. Thus *any* energy E greater than U_0 is possible. These free-particle states thus form a *continuum* rather than a discrete set of states with definite energy levels. The free-particle wave functions are sinusoidal both inside and outside the well. The wavelength is shorter inside the well than outside, corresponding to greater kinetic energy in the well than outside.

20.3
Potential Wells

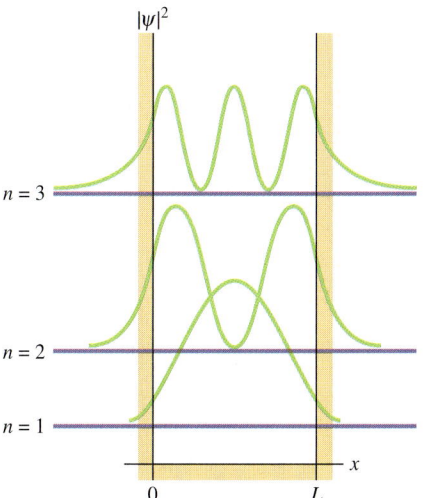

42–9 Probability distributions (values of $|\psi|^2$) for the wave functions shown in Fig. 42–8 for a particle in a square-well potential. The horizontal violet line for each wave function corresponds to $|\psi|^2 = 0$.

The square-well potential described in this section has a number of practical applications. We mentioned earlier the example of an electron inside a metallic sheet. A three-dimensional version, in which U is zero inside a spherical region with radius R and has the value U_0 outside, provides a simple model to represent the interaction of a neutron with a nucleus in neutron-scattering experiments. In this context the model is called the *crystal-ball model* of the nucleus, because neutrons interacting with such a potential are scattered in a way that's analogous to scattering of light by a crystal ball.

Problem–Solving Strategy

THE SCHRÖDINGER EQUATION

1. Carefully distinguish between $h = 6.626 \times 10^{-34}$ J · s and $\hbar = h/2\pi = 1.055 \times 10^{-34}$ J · s. Usually, $\hbar$ is more convenient in the Schrödinger equation, but in working with photon energies or wavelengths you may need h. Don't get the two confused.

2. When you have energies expressed in electron volts, it's often convenient to use the values $h = 4.136 \times 10^{-15}$ eV · s and $\hbar = 6.582 \times 10^{-16}$ eV · s.

3. Think about the magnitudes of the various quantities. Atomic dimensions are typically tenths of a nanometer, and typical energy levels for the outer electrons in atoms are a few electron volts. Visible-light photons have energies near 2 eV, corresponding to wavelengths near 600 nm and frequencies near 5×10^{14} Hz. If your calculations give you numbers that are wildly different from these, recheck your work. Finally, always carry units through your calculations to check unit consistency.

EXAMPLE 42–5

Electron in a square well An electron is trapped in a square well with a width of 0.50 nm (comparable to a few atomic diameters). a) Find the ground-level energy if the well is infinitely deep. b) If the actual well depth is six times the ground-level energy found in part (a), find the energy levels. c) If the atom makes a transition from a state with energy E_2 to a state with energy E_1 by emitting a photon, find the wavelength of the photon. In what region of the electromagnetic spectrum does the photon lie? d) If the electron is initially in its ground level and absorbs a photon, what is the minimum energy the photon must have to liberate the electron from the well? In what region of the spectrum does the photon lie?

SOLUTION a) The ground-level energy E_∞ for an infinitely deep well is given by Eq. (42–23):

$$E_\infty = \frac{\pi^2 \hbar^2}{2mL^2} = \frac{\pi^2 (1.055 \times 10^{-34} \text{ J} \cdot \text{s})^2}{2(9.11 \times 10^{-31} \text{ kg})(0.50 \times 10^{-9} \text{ m})^2}$$

$$= 2.4 \times 10^{-19} \text{ J} = 1.5 \text{ eV}.$$

b) We are given $U_0 = 6E_\infty$, so $U_0 = 6(1.5 \text{ eV}) = 9.0$ eV. This is the same ratio as in the example shown in Fig. 42–8b, so we can simply read off the energy levels. Our answers are

$$E_1 = 0.104 U_0 = 0.104(9.0 \text{ eV}) = 0.94 \text{ eV},$$

$$E_2 = 0.405 U_0 = 0.405(9.0 \text{ eV}) = 3.6 \text{ eV},$$

$$E_3 = 0.848 U_0 = 0.848(9.0 \text{ eV}) = 7.6 \text{ eV}.$$

Alternatively,

$$E_1 = 0.625 E_\infty = 0.625(1.5 \text{ eV}) = 0.94 \text{ eV},$$

$$E_2 = 2.43 E_\infty = 0.243(1.5 \text{ eV}) = 3.6 \text{ eV},$$

$$E_3 = 5.09 E_\infty = 5.09(1.5 \text{ eV}) = 7.6 \text{ eV}.$$

(Note that if the well had been infinitely deep, the energies would have been

$$E_1 = E_\infty = 1.5 \text{ eV}, \qquad E_2 = 4E_\infty = 6.0 \text{ eV},$$

$$\text{and} \qquad E_3 = 9E_\infty = 13.5 \text{ eV}.$$

As we mentioned earlier, the finite depth of the well *lowers* the energy levels compared to the values for an infinitely deep well.) c) The photon energy is

$$E_2 - E_1 = 3.6 \text{ eV} - 0.94 \text{ eV} = 2.7 \text{ eV}.$$

We determine the wavelength from $E = hf = hc/\lambda$:

$$\lambda = \frac{hc}{E} = \frac{(4.136 \times 10^{-15} \text{ eV} \cdot \text{s})(3.00 \times 10^8 \text{ m/s})}{2.7 \text{ eV}} = 460 \text{ nm}.$$

This photon is in the blue region of the visible spectrum.
d) We see from Fig. 42–8b that the minimum energy needed to lift the electron out of the well from its $n = 1$ ground level is the well depth ($U_0 = 9.0$ eV) minus the electron's initial energy ($E_1 = 0.94$ eV), or 8.1 eV. Since 8.1 eV is three times the 2.7-eV photon energy of part (c), the corresponding photon wavelength is one third of 460 nm, or 150 nm, which is in the ultraviolet region of the spectrum.

42–5 POTENTIAL BARRIERS AND TUNNELING

A **potential barrier** is the opposite of a potential well; it is a potential-energy function with a *maximum*. Figure 42–10 shows an example. In Newtonian mechanics, if the total energy is E_1, a particle that is initially to the left of the barrier must remain to the left of the point $x = a$. If it were to move to the right of this point, the potential energy U would be greater than the total energy E. Because $K = E - U$, the kinetic energy would be negative, and this is impossible since a negative value of $K = \frac{1}{2}mv^2$ would require negative mass or imaginary speed.

If the total energy is greater than E_2, the particle can get past the barrier. A roller-coaster car can surmount a hill if it has enough kinetic energy at the bottom. If it doesn't, it stops partway up and then rolls back down.

Quantum mechanics supplies us with a peculiar and very interesting phenomenon in connection with potential barriers. A particle that encounters such a barrier is not necessarily turned back; there is some probability for it to emerge on the other side, even when it doesn't have enough kinetic energy to surmount the barrier according to Newtonian mechanics. This penetration of a barrier is called **tunneling.** It's a natural name; if you dig a tunnel through a hill, you don't have to go over the top. However, in quantum-mechanical tunneling, the particle does not actually bore a hole through the barrier and loses no energy in the tunneling process.

To understand how tunneling can occur, let's look at the potential-energy function $U(x)$ shown in Fig. 42–11. It's like Fig. 42–6 turned upside-down; the potential energy is zero everywhere except in the range $0 \le x \le L$, where it has the value U_0. This might represent a simple model for an electron and two slabs of metal separated by an air gap of thickness L. We let the potential energy be zero within either slab but equal to U_0 in the gap between them.

Let's consider solutions of the Schrödinger equation for this potential-energy function for the case in which E is less than U_0. We can use our results from Section 42–4. In the regions $x < 0$ and $x > L$ the solution is sinusoidal and given by Eq. (42–20). Within the barrier ($0 \le x \le L$) the solution is exponential, as in Eq. (42–22). Just as with the finite potential well, the functions have to join smoothly at the boundary points $x = 0$ and $x = L$. That is, the function and its derivative have to be continuous at these points.

These requirements lead to a wave function like the one shown in Fig. 42–12. The function is *not* zero inside the barrier (the region forbidden by Newtonian mechanics). Even more remarkable, a particle that is initially to the *left* of the barrier has some probability of being found to the *right* of the barrier. How great a probability? That depends on the width L of the barrier and the particle's energy E (all kinetic) in comparison with the barrier height U_0. The probability that the particle gets through the barrier, called the *transmission coefficient T,* is proportional to the square of the ratio of the amplitudes of sinusoidal wave functions on the two sides of the barrier. These amplitudes are determined by matching wave functions and their derivatives at the boundary points, a fairly involved mathematical problem. When T is much smaller than unity, it is given

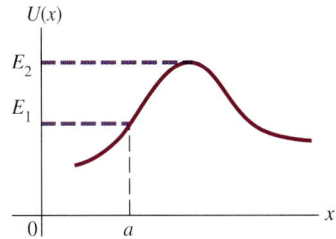

42–10 A potential-energy barrier. According to Newtonian mechanics, if the total energy is E_1, a particle that is on the left side of the barrier can go no farther than $x = a$. If the total energy is greater than E_2, the particle can pass the barrier.

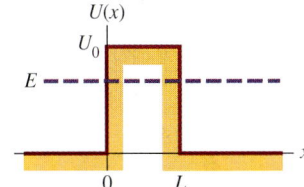

42–11 A potential-energy barrier with width L and height U_0. According to Newtonian mechanics, if the total energy E is less than U_0, a particle cannot pass over this barrier but is confined to the side where it starts.

20.4
Potential Barriers

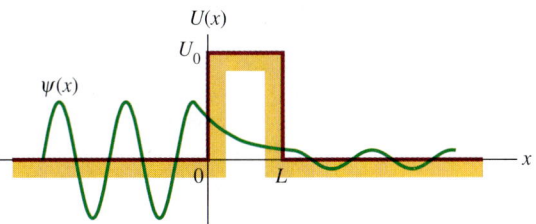

42–12 A possible wave function for a particle tunneling through the potential-energy barrier shown in Fig. 42–11. The function is exponential within the barrier ($0 \le x \le L$) and sinusoidal outside the barrier. The functions must join smoothly at $x = 0$ and $x = L$; the function and its derivative (slope) must be continuous.

approximately by

$$T = Ge^{-2\kappa L}, \quad \text{where} \quad G = 16\frac{E}{U_0}\left(1 - \frac{E}{U_0}\right) \quad \text{and} \quad \kappa = \frac{\sqrt{2m(U_0 - E)}}{\hbar} \quad (42\text{-}24)$$

(probability of tunneling).

The probability decreases rapidly with increasing barrier width L. It also depends critically on the energy difference $U_0 - E$, representing the additional kinetic energy the particle would need to be able to climb over the barrier in a Newtonian analysis.

EXAMPLE 42-6

A 2.0-eV electron encounters a barrier with height 5.0 eV. What is the probability that it will tunnel through the barrier if the barrier width is a) 1.00 nm; b) 0.50 nm (roughly ten and five atomic diameters, respectively)?

SOLUTION First we evaluate G and κ in Eq. (42-24), using $E = K = 2.0$ eV:

$$G = 16\frac{2.0 \text{ eV}}{5.0 \text{ eV}}\left(1 - \frac{2.0 \text{ eV}}{5.0 \text{ eV}}\right) = 3.8,$$

$$U_0 - E = 5.0 \text{ eV} - 2.0 \text{ eV} = 3.0 \text{ eV} = 4.8 \times 10^{-19} \text{ J},$$

$$\kappa = \frac{\sqrt{2(9.11 \times 10^{-31} \text{ kg})(4.8 \times 10^{-19} \text{ J})}}{1.055 \times 10^{-34} \text{ J} \cdot \text{s}} = 8.9 \times 10^9 \text{ m}^{-1}.$$

(Note that the eV energy units canceled in calculating G, but we had to convert eV to J = kg $\cdot$ m^2/s^2 to find κ in m^{-1}.)
a) When $L = 1.00$ nm $= 1.00 \times 10^{-9}$ m, $2\kappa L = 2(8.9 \times 10^9 \text{ m}^{-1})(1.00 \times 10^{-9} \text{ m}) = 17.8$, and

$$T = Ge^{-2\kappa L} = 3.8e^{-17.8} = 7.1 \times 10^{-8}.$$

b) When $L = 0.50$ nm, half of 1.00 nm, $2\kappa L$ is half of 17.8, or 8.9, and

$$T = 3.8e^{-8.9} = 5.2 \times 10^{-4}.$$

Halving the width of this barrier increases the tunneling probability by a factor of nearly ten thousand.

APPLICATIONS OF TUNNELING

Tunneling is significant in many areas of physics, including some with considerable practical importance. For example, when you twist two copper wires together or close the contacts of a switch, current passes from one conductor to the other despite a thin layer of nonconducting copper oxide between them. The electrons tunnel through this thin insulating layer. The *tunnel diode* is a semiconductor device in which electrons tunnel through a potential barrier. The current can be switched on and off very quickly (within a few picoseconds) by varying the height of the barrier. The *Josephson junction* consists of two superconductors separated by an oxide layer a few atoms (1 to 2 nm) thick. Electron pairs in the superconductors can tunnel through the barrier layer, giving such a device unusual circuit properties. Josephson junctions are useful for establishing precise voltage standards and measuring tiny magnetic fields.

The scanning tunneling electron microscope uses electron tunneling to create images of surfaces down to the scale of individual atoms. An extremely sharp conducting needle is brought very close to the surface, within 1 nm or so (Fig. 42–13a). When

42–13 (a) Schematic diagram of the probe of a scanning tunneling electron microscope. The probe, an extremely sharp conducting needle, is brought within about 1 nm of the surface. As it is scanned across the surface, it is moved perpendicular to the surface to maintain constant tunneling current. Its positions are recorded and used to construct the image. (b) A color-enhanced image showing iodine atoms absorbed on the surface of a platinum crystal. The yellow spot indicates a missing atom.

Electron cloud · Probe
L
Specimen
(a)

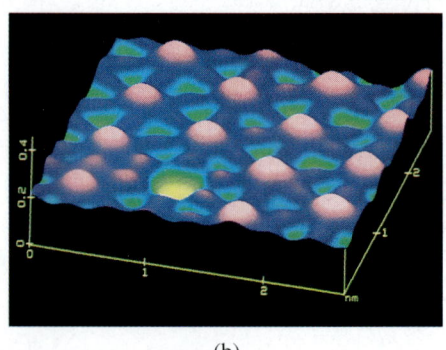

(b)

the needle is at a positive potential with respect to the surface, electrons can tunnel through the surface potential-energy barrier and reach the needle. As Example 42–6 shows, the tunneling probability and hence the tunneling current depend strongly on the width L of the barrier (the distance between the surface and the needle tip). In one mode of operation the needle is scanned across the surface and at the same time is moved perpendicular to the surface to maintain a constant tunneling current. The needle motion is recorded, and after many parallel scans, an image of the surface can be reconstructed. Extremely precise control of needle motion, including isolation from vibration, is essential. Figure 42–13b shows a scanning tunneling microscope image of iodine atoms on the surface of a platinum crystal.

In the realm of nuclear physics a nuclear fusion reaction can occur when two nuclei tunnel through the barrier caused by their electrical repulsion and approach each other closely enough for the attractive nuclear force to cause them to fuse. Fusion reactions occur in the cores of stars, including the sun; without tunneling, the sun wouldn't shine. The emission of alpha particles from unstable nuclei also involves tunneling. An alpha particle at the surface of a nucleus encounters a potential barrier that results from the combined effect of the attractive nuclear force and the electrical repulsion of the remaining part of the nucleus (Fig. 42–14). The alpha particle tunnels through this barrier. Because the tunneling probability depends so critically on the barrier height and width, the lifetimes of alpha-emitting nuclei vary over an extremely wide range. We'll return to alpha decay in Chapter 45.

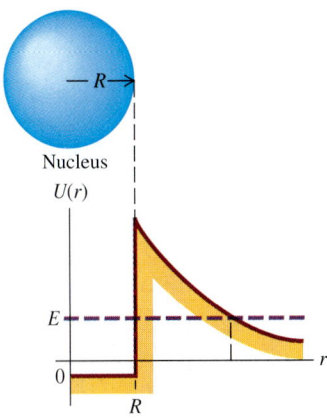

42–14 Approximate potential-energy function for an alpha particle in a nucleus. The nucleus is represented as a sphere with radius R. When the alpha particle is inside the nucleus ($r < R$), the potential energy resembles a square well. When the alpha particle is outside, the potential energy is a $1/r$ function associated with the electrical repulsion. If the energy E inside the well is greater than zero, an alpha particle can tunnel through the barrier and escape from the nucleus.

42–6 THE HARMONIC OSCILLATOR

In Newtonian mechanics a **harmonic oscillator** is a particle with mass m acted on by a conservative force component $F_x = -k'x$. (In this discussion we will use k' for the *force constant* to minimize confusion with the wave number $k = 2\pi/\lambda$.) This force component is proportional to the particle's displacement x from its equilibrium position, $x = 0$. The corresponding potential-energy function is $U = \frac{1}{2}k'x^2$ (Fig. 42–15). When the particle is displaced from equilibrium, it undergoes sinusoidal motion with angular frequency $\omega = (k'/m)^{1/2}$. We studied this system in detail in Chapter 13; you may want to review that discussion.

A quantum-mechanical analysis of the harmonic oscillator using the Schrödinger equation is an interesting and useful study. The solutions provide insight into vibrations of molecules, the quantum theory of heat capacities, atomic vibrations in crystalline solids, and many other situations.

Before we get into the details, let's make an enlightened guess about the energy levels. The energy E of a *photon* is related to its angular frequency ω by $E = \hbar\omega$. The harmonic oscillator has a characteristic angular frequency ω, at least in Newtonian mechanics. So a reasonable guess would be that in the quantum-mechanical analysis the energy levels of a harmonic oscillator would be multiples of the quantity

$$\hbar\omega = \hbar\sqrt{\frac{k'}{m}}.$$

These are the energy levels that Planck assumed in deriving his radiation law (Section 40–9). It was a good assumption; the energy levels are in fact half-integer ($\frac{1}{2}, \frac{3}{2}, \frac{5}{2}, \ldots$) multiples of $\hbar\omega$.

For the harmonic oscillator we write $\frac{1}{2}k'x^2$ in place of U in the one-dimensional Schrödinger equation, Eq. (42–17):

$$-\frac{\hbar^2}{2m}\frac{d^2\psi}{dx^2} + \frac{1}{2}k'x^2\psi = E\psi \qquad \text{(Schrödinger equation for the harmonic oscillator).} \qquad (42\text{–}25)$$

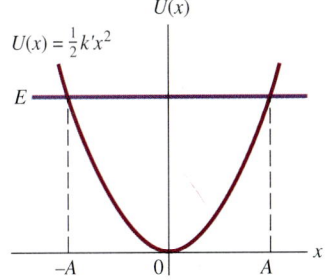

42–15 Potential energy function for the harmonic oscillator, $U = \frac{1}{2}k'x^2$. The curve is a parabola. In Newtonian mechanics the amplitude A is related to the total energy E by $E = \frac{1}{2}k'A^2$, and the particle is restricted to the range from $x = -A$ to $x = A$.

The solutions of this equation are wave functions for the physically possible states of the system.

WAVE FUNCTIONS, BOUNDARY CONDITIONS, AND ENERGY LEVELS

In the discussion of square-well potentials in Section 42–4 we found that the energy levels are determined by boundary conditions at the walls of the well. However, the harmonic-oscillator potential has no walls as such; what, then, are the appropriate boundary conditions? Classically, $|x|$ cannot be greater than the amplitude A, which is the maximum displacement from equilibrium. Quantum mechanics does allow some penetration into classically forbidden regions, but the probability decreases as that penetration increases. Thus the wave functions must approach zero as $|x|$ grows large.

Satisfying the requirement that $\psi \rightarrow 0$ as $|x| \rightarrow \infty$ is not as trivial as it may seem. Suppose we compute solutions of Eq. (42–25) *numerically*. We start at some point x, choosing values of ψ and $d\psi/dx$ at that point. Using the Schrödinger equation to evaluate $d^2\psi/dx^2$ at the point, we can compute ψ and $d\psi/dx$ at a neighboring point $x + \Delta x$, and so on. By using a computer to iterate this process many times with sufficiently small steps Δx, we can compute the wave function with great precision. There is no guarantee, however, that a function obtained in this way will approach zero at large x. To see what kinds of trouble we can get into, let's rewrite Eq. (42–25) in the form

$$\frac{d^2\psi}{dx^2} = \frac{2m}{\hbar^2}\left(\frac{1}{2}k'x^2 - E\right)\psi. \tag{42–26}$$

In this form, the equation shows that when x is large enough (either positive or negative) to make the quantity $(\frac{1}{2}k'x^2 - E)$ positive, the function ψ and its second derivative $d^2\psi/dx^2$ have the same sign.

The second derivative of ψ is the rate of change of the *slope* of ψ. Consider a point with $x > A$ for which $\frac{1}{2}k'x^2 - E > 0$. If ψ is positive, so is $d^2\psi/dx^2$ and the function is *concave upward*. Figure 42–16 shows four possible kinds of behavior beginning at a point $x > A$. If the slope is initially positive, the function curves upward more and more steeply (curve *a*) and goes to infinity. If the slope is initially negative at the point, there are three possibilities. If the slope changes too quickly (curve *b*), the curve bends up and again goes to infinity. If the slope doesn't change quickly enough, the curve bends down and crosses the x-axis. After it has crossed, ψ and $d^2\psi/dx^2$ are both negative (curve *c*) and the concave-downward curve heads for *negative* infinity. Between these infinities is the chance that the curve may bend just enough to glide in asymptotically to the x-axis (curve *d*). In this case, ψ, $d\psi/dx$, and $d^2\psi/dx^2$ all approach zero at large x; in this possibility lies the only hope of satisfying the boundary condition that $\psi \rightarrow 0$ as $|x| \rightarrow \infty$, and it occurs only for certain very special values of the constant E.

This qualitative discussion offers some insight into how the boundary conditions for this problem determine the possible energy levels. Equation (42–25) can also be solved

42–16 Possible behaviors of harmonic-oscillator wave functions in the region $\frac{1}{2}k'x^2 > E$. In this region, ψ and $d^2\psi/dx^2$ have the same sign. The curve is concave upward when $d^2\psi/dx^2$ is positive and concave downward when $d^2\psi/dx^2$ is negative. Only curve (d), which approaches the x-axis asymptotically for large x, is an acceptable wave function for this system.

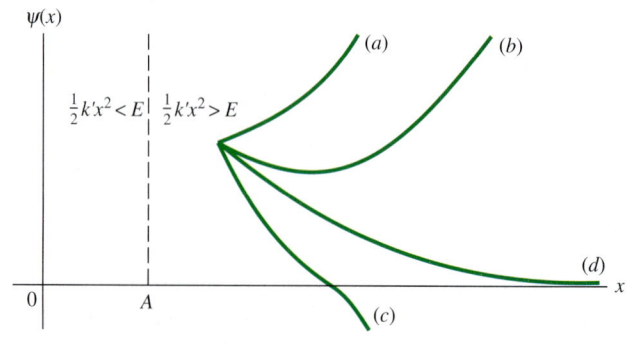

exactly. The solutions, although not encountered in elementary calculus courses, are well known to mathematicians; they are called *Hermite functions*. Each one is an exponential function multiplied by a polynomial in x. The state with lowest energy (the ground level) has the wave function

$$\psi = Ce^{-\sqrt{mk'}x^2/2\hbar}. \tag{42–27}$$

The constant C is chosen to normalize the function, that is, to make $\int_{-\infty}^{\infty}|\psi|^2 dx = 1$. You can find C using, from integral tables,

$$\int_{-\infty}^{\infty} e^{-a^2x^2} dx = \frac{\sqrt{\pi}}{a}.$$

The corresponding energy, which we'll call E_0, is the ground-state energy; it is

$$E_0 = \frac{1}{2}\hbar\omega = \frac{1}{2}\hbar\sqrt{\frac{k'}{m}}. \tag{42–28}$$

You may not believe that Eq. (42–27) really *is* a solution of Eq. (42–25) (the Schrödinger equation for the harmonic oscillator) with the energy given by Eq. (42–28). We invite you to calculate the second derivative of Eq. (42–27), substitute it into Eq. (42–25), and verify that it really is a solution if Eq. (42–28) gives E_0 (Exercise 42–24). It's a little messy, but the result is satisfying and worth the effort.

Further analysis of the Schrödinger equation for the harmonic oscillator shows that it can be written in the form

$$\frac{d^2\psi}{dx^2} = \frac{2m}{\hbar^2}\left(\frac{1}{2}k'x^2 - (n + \frac{1}{2})\hbar\sqrt{\frac{k'}{m}}\right)\psi.$$

Comparing this complicated expression to Eq. (42–26) gives us a pleasant surprise: The energy levels, which we'll call E_n, are given by the simple formula

$$E_n = (n + \frac{1}{2})\hbar\sqrt{\frac{k'}{m}} = (n + \frac{1}{2})\hbar\omega \quad (n = 0, 1, 2, \ldots) \quad \text{(energy levels,} \atop \text{harmonic oscillator),} \tag{42–29}$$

where n is the quantum number identifying each state and energy level.

Equation (42–29) confirms our guess that the energy levels are multiples of $\hbar\omega$. Adjacent energy levels are separated by a constant interval of $\hbar\omega = hf$, as Planck assumed in 1900. There are infinitely many levels; this shouldn't be surprising, because we are dealing with an infinitely deep potential well. As $|x|$ increases, $U = \frac{1}{2}k'x^2$ increases without bound.

Figure 42–17 shows the lowest six energy levels and the potential-energy function $U(x)$. For each level n, the value of $|x|$ at which the horizontal line representing the total energy E_n intersects $U(x)$ gives the amplitude A_n of the corresponding Newtonian oscillator.

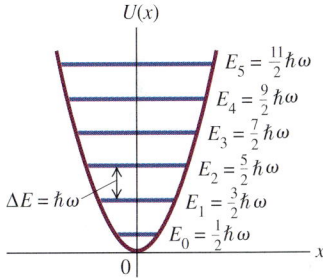

42–17 Energy levels for the harmonic oscillator. The levels are equally spaced, with spacing $\Delta E = \hbar\omega$ between adjacent levels. The energy of the ground level is $E_0 = \frac{1}{2}\hbar\omega$.

EXAMPLE 42–7

Vibration in a sodium crystal A sodium atom of mass 3.82×10^{-26} kg vibrates with simple harmonic motion in a crystal. The potential energy increases by 0.0075 eV when the atom is displaced 0.014 nm from its equilibrium position. a) Find the angular frequency, according to Newtonian mechanics. b) Find the spacing of adjacent energy levels in electron volts. c) If an

atom emits a photon during a transition from one vibrational level to the next lower level, what is the wavelength of the emitted photon? In what region of the electromagnetic spectrum does it lie?

SOLUTION a) The angular frequency is $\omega = (k'/m)^{1/2}$. We don't know k', but we can find it from the fact that $U = 0.0075$ eV $= 1.2 \times 10^{-21}$ J when $x = 0.014 \times 10^{-9}$ m. We have

$$U = \frac{1}{2} k'x^2,$$

$$k' = \frac{2U}{x^2} = \frac{2(1.2 \times 10^{-21} \text{ J})}{(0.014 \times 10^{-9} \text{ m})^2} = 12.2 \text{ N/m}.$$

(Surprisingly, force constants for interatomic forces aren't very different in order of magnitude from those of household springs.)

The angular frequency is

$$\omega = \sqrt{\frac{k'}{m}} = \sqrt{\frac{12.2 \text{ N/m}}{3.82 \times 10^{-26} \text{ kg}}} = 1.79 \times 10^{13} \text{ rad/s}.$$

b) From Eq. (42–29) and Fig. 42–17 the spacing between adjacent energy levels is

$$\hbar\omega = (1.054 \times 10^{-34} \text{ J} \cdot \text{s})(1.79 \times 10^{13} \text{ s}^{-1})$$

$$= 1.88 \times 10^{-21} \text{ J} = 0.0118 \text{ eV}.$$

c) From $E = hc/\lambda$ the corresponding wavelength is

$$\lambda = \frac{hc}{E} = \frac{(4.136 \times 10^{-15} \text{ eV} \cdot \text{s})(3.00 \times 10^8 \text{ m/s})}{0.0118 \text{ eV}}$$

$$= 1.05 \times 10^{-4} \text{ m} = 105 \ \mu\text{m}.$$

This photon is in the infrared region of the spectrum.

QUANTUM AND NEWTONIAN OSCILLATORS

Figure 42–18 shows the first four harmonic-oscillator wave functions. Each graph also shows the amplitude A of a Newtonian harmonic oscillator with the same energy, that is, the value of A determined from

$$\frac{1}{2} k'A^2 = \left(n + \frac{1}{2}\right)\hbar\omega. \qquad (42–30)$$

In each case there is some penetration of the wave function into the regions $|x| > A$ that are forbidden by Newtonian mechanics. This is similar to the effect that we noted with the particle in a finite square well.

CAUTION▶ Be very careful with your algebraic symbols. This Newtonian amplitude A (in units of m) is *not* the same as the quantum-mechanical normalizing constant A (in $\text{m}^{-1/2}$) that is used in several wave functions in this chapter. Also the angular frequency ω (in rad/s) of the harmonic oscillator is *not* the same as the angular frequency ω (also in rad/s) obtained from the de Broglie relation $E = \hbar\omega$ (which is why we expressed Eq. (42–10) in terms of $2\pi E/h = E/\hbar$ rather than in terms of ω). Finally, we have distinguished the force constant k' (in N/m or kg/s^2) from the wave number k (in rad/m). Neither of these is the same as κ (in m^{-1}) used in Eqs. (42–22) and (42–24) or the kinetic energy K (in J or eV). ◀

42–18 The first four wave functions for the harmonic oscillator. The amplitude A of a Newtonian oscillator with the same total energy is shown for each. Each wave function penetrates somewhat into the classically forbidden regions $|x| > A$. The total number of finite maxima and minima for each function is $1 + n$, one more than the quantum number.

Figure 42–19 shows the probability distributions $|\psi|^2$ for these same states. Each graph also shows the probability distributions determined from Newtonian analysis, in which the probability of finding the particle near a randomly chosen point is inversely proportional to its speed at that point. If we average out the wiggles in the quantum-mechanical probability curves, the results for $n > 0$ resemble the Newtonian predictions. This agreement improves with increasing n; Fig. 42–20 shows the classical and quantum-mechanical probability functions for $n = 10$.

In the Newtonian analysis of the harmonic oscillator the minimum energy is zero, with the particle at rest at its equilibrium position. This is not possible in quantum

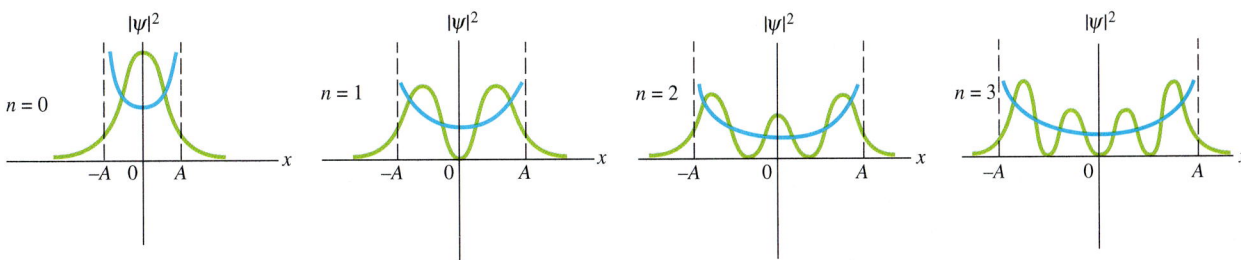

42–19 Probability distributions $|\psi|^2$ for the harmonic-oscillator wave functions shown in Fig. 42–18. The amplitude A of the Newtonian motion with the same energy is shown for each. The blue lines show the corresponding probability distributions for the Newtonian motion. As n increases, the averaged-out wave functions resemble the Newtonian curves more and more.

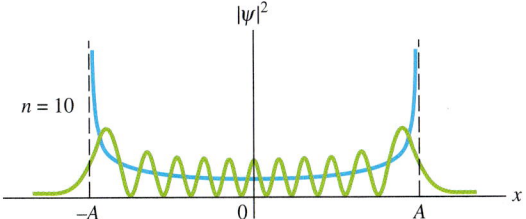

42–20 Newtonian and quantum-mechanical probability distributions for a harmonic oscillator for the state $n = 10$. The Newtonian amplitude A is also shown.

mechanics; no solution of the Schrödinger equation has $E = 0$ and satisfies the boundary conditions. Furthermore, if there were such a state, it would violate the uncertainty principle because there would be no uncertainty in either position or momentum. Indeed, the energy must be at least $\frac{1}{2}\hbar\omega$ for the system to conform to the uncertainty principle. To see qualitatively why this is so, consider a Newtonian oscillator with total energy $\frac{1}{2}\hbar\omega$. We can find the amplitude A and the maximum velocity just as we did in Section 13–4. The appropriate relations are Eqs. (13–21) for the total energy and (13–23) for the maximum speed. Setting $E = \frac{1}{2}\hbar\omega$, we find

$$E = \frac{1}{2}k'A^2 = \frac{1}{2}\hbar\omega = \frac{1}{2}\hbar\sqrt{\frac{k'}{m}}, \qquad A = \sqrt{\frac{\hbar}{\sqrt{k'm}}},$$

$$p_{max} = mv_{max} = m\sqrt{\frac{k'}{m}}A = m\sqrt{\frac{k'}{m}}\sqrt{\frac{\hbar}{\sqrt{k'm}}} = \sqrt{\hbar\sqrt{k'm}}.$$

Finally, if we assume that A represents the uncertainty Δx in position and p_{max} is the corresponding uncertainty Δp_x in momentum, then the product of the two uncertainties is

$$\Delta x\,\Delta p_x = \sqrt{\frac{\hbar}{\sqrt{k'm}}}\sqrt{\hbar\sqrt{k'm}} = \hbar.$$

So the product equals the minimum value allowed by Eq. (41–10), $\Delta x\,\Delta p_x \geq h/2\pi$, and thus satisfies the uncertainty principle. If the energy had been less than $\frac{1}{2}\hbar\omega$, the product $\Delta x\,\Delta p_x$ would have been less than $\hbar$, and the uncertainty principle would have been violated.

Even when a potential-energy function isn't precisely parabolic in shape, we may be able to approximate it by the harmonic-oscillator potential for sufficiently small displacements from equilibrium. Figure 42–21 shows a typical potential-energy function for an interatomic force in a molecule. At large separations it levels off, corresponding to the absence of force at great distances. But it is approximately parabolic near the minimum point (the equilibrium position of the atoms). Near equilibrium the molecular vibration is approximately simple harmonic with energy levels given by Eq. (42–29), as we assumed in Example 42–7.

ActivPhysics

20.1.6
Potential Energy
Diagrams, Question 6

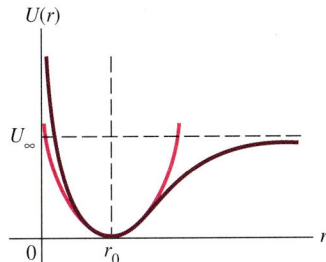

42–21 A potential-energy function describing the interaction of two atoms in a diatomic molecule. At each point, the force F is $F = -dU/dr$. The equilibrium position is at $r = r_0$. When r is greater than r_0, the force is attractive; when r is less than r_0, it is repulsive. When r is near r_0, the curve is approximately parabolic (as shown by the bright red curve), and the motion is approximately simple harmonic. The potential energy needed to dissociate the molecule is U_∞.

42–7 THREE-DIMENSIONAL PROBLEMS

We have discussed the Schrödinger equation and its applications only for *one-dimensional* problems, the analog of a Newtonian particle moving along a straight line. The straight-line model is adequate for some applications, but to understand atomic structure, we need a three-dimensional generalization.

It's not difficult to guess what the three-dimensional Schrödinger equation should look like. First, the wave function ψ is a function of all three space coordinates (x, y, z). In general, the potential-energy function also depends on all three coordinates and can be written as $U(x, y, z)$. Next, recall that we obtained the term containing $d^2\psi/dx^2$ in the one-dimensional equation, Eq. (42–17), by a line of reasoning based on the relation of kinetic energy K to momentum p: $K = p^2/2m$. If the particle's momentum has three components (p_x, p_y, p_z), then the corresponding relation in three dimensions is

$$K = \frac{p_x^{\,2}}{2m} + \frac{p_y^{\,2}}{2m} + \frac{p_z^{\,2}}{2m}. \qquad (42\text{–}31)$$

These observations, taken together, suggest that the correct generalization of the Schrödinger equation to three dimensions is

$$-\frac{\hbar^2}{2m}\left(\frac{\partial^2\psi}{\partial x^2} + \frac{\partial^2\psi}{\partial y^2} + \frac{\partial^2\psi}{\partial z^2}\right) + U\psi = E\psi \qquad \begin{array}{l}\text{(three-dimensional}\\ \text{Schrödinger equation).}\end{array} \qquad (42\text{–}32)$$

In this equation it is understood that ψ is, and U may be, a function of x, y, and z (hence the partial-derivative notation), while E is constant for a state of particular energy.

We won't pretend that we have *derived* Eq. (42–32). Like the one-dimensional version, this equation has to be tested by comparison of its predictions with experimental results. As we will see in later chapters, Eq. (42–32) passes this test with flying colors, so we are confident that it *is* the correct equation.

In many practical problems, in atomic structure and elsewhere, the potential-energy function is *spherically symmetric;* it depends only on the distance $r = (x^2 + y^2 + z^2)^{1/2}$ from the origin of coordinates. To take advantage of this symmetry, we use *spherical coordinates* (r, θ, ϕ) (Fig. 42–22) instead of Cartesian coordinates (x, y, z). Then a spherically symmetric potential-energy function is a function only of r, not of θ or ϕ, so $U = U(r)$. This fact turns out to simplify greatly the problem of finding solutions of the Schrödinger equation, even though the derivatives in Eq. (42–32) are considerably more complex when expressed in terms of spherical coordinates. Be careful; many math texts exchange the angles θ and ϕ shown in Fig. 42–22.

For the hydrogen atom the potential-energy function $U(r)$ is the familiar Coulomb's-law function:

$$U(r) = -\frac{1}{4\pi\epsilon_0}\frac{e^2}{r}. \qquad (42\text{–}33)$$

We will find that for *all* spherically symmetric potential-energy functions $U(r)$, each possible wave function can be expressed as a product of three functions: one a function only of r, one only of θ, and one only of ϕ. Furthermore, the functions of θ and ϕ are *the same* for *every* spherically symmetric potential-energy function. This result is directly related to the problem of finding the possible values of *angular momentum* for the various states. We'll discuss these matters more in the next chapter.

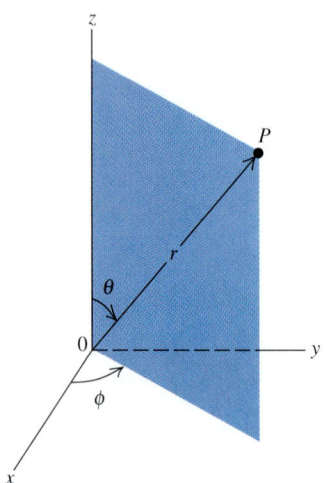

42–22 The position of a point P in space can be described by the rectangular coordinates (x, y, z) or by the spherical coordinates (r, θ, ϕ). Spherical coordinates are particularly useful in quantum mechanics in problems in which the potential energy depends only on r, the distance from the origin.

SUMMARY

- The energy levels for a particle of mass m in a box (an infinitely deep square potential well) with width L are

$$E_n = \frac{p_n^2}{2m} = \frac{n^2 h^2}{8mL^2} \qquad (n = 1, 2, 3, \ldots). \qquad (42\text{–}5)$$

The corresponding normalized wave functions of the particle are

$$\psi_n = \sqrt{\frac{2}{L}} \sin \frac{n\pi x}{L} \qquad (n = 1, 2, 3, \ldots). \qquad (42\text{–}9)$$

- The Schrödinger equation for a particle moving in a one-dimensional system with potential energy U is

$$-\frac{\hbar^2}{2m}\frac{d^2\psi}{dx^2} + U\psi = E\psi. \qquad (42\text{–}17)$$

- The wave function ψ and its derivative $d\psi/dx$ must be continuous everywhere, except when U has an infinite discontinuity. Wave functions are usually normalized so that the total probability for finding the particle somewhere is unity:

$$\int_{-\infty}^{\infty} |\psi|^2 dx = 1. \qquad (42\text{–}7)$$

- In a potential well with finite depth U_0 the energy levels are lower than those for an infinitely deep well with the same width, and the number of energy levels corresponding to bound states is finite. The levels are obtained by matching wave functions at well walls to satisfy boundary conditions.

- There is a certain probability that a particle will penetrate a potential energy barrier although its initial kinetic energy is less than the barrier height; this process is called tunneling.

- The solutions of the Schrödinger equation for the harmonic oscillator, for which $U = \frac{1}{2}k'x^2$, give a set of energy levels

$$E_n = (n + \tfrac{1}{2})\hbar\omega = (n + \tfrac{1}{2})\hbar\sqrt{\frac{k'}{m}}, \qquad (42\text{–}29)$$

where $n = 0, 1, 2, 3, \ldots$ is the quantum number for each state.

- The Schrödinger equation in three dimensions is

$$-\frac{\hbar^2}{2m}\left(\frac{\partial^2\psi}{\partial x^2} + \frac{\partial^2\psi}{\partial y^2} + \frac{\partial^2\psi}{\partial z^2}\right) + U\psi = E\psi. \qquad (42\text{–}32)$$

DISCUSSION QUESTIONS

Q42-1 For the particle in a box, we chose $k = n\pi/L$ with $n = 1, 2, 3, \ldots$ to fit the boundary condition that $\psi = 0$ at $x = L$. However, $n = 0$ and $n = -1, -2, -3, \ldots$ also satisfy that boundary condition. Why didn't we also choose those values of n?

Q42-12 If ψ is normalized, what is the physical significance of the area under a graph of $|\psi|^2$ versus x between x_1 and x_2? What is the total area under the graph of $|\psi|^2$ when all x are included? Explain.

Q42-3 For a particle in a box, what would the probability distribution function $|\psi|^2$ look like if the particle behaved as a classical (Newtonian) particle? Do the actual probability distributions approach this classical form when n is very large? Explain.

Q42-4 In Chapter 20 we represented a standing wave as a superposition of two waves traveling in opposite directions. Can the wave functions for a particle in a box also be thought of as a combination of two traveling waves? Why or why not? What physical interpretation does this representation have? Explain.

Q42-5 A particle in a box is in the ground level. What is the probability of finding the particle in the right half of the box?

(Refer to Fig. 42–4, but don't evaluate an integral.) Is the answer the same if the particle is in an excited level? Explain.

Q42–6 The wave functions for a particle in a box (Fig. 42–4a) are zero at certain points. Does this mean that the particle can't move past one of these points? Explain.

Q42–7 For a particle confined to an infinite square well, is it correct to say that each state of definite energy is also a state of definite wavelength? Is it also a state of definite momentum? Explain. (*Hint:* Remember that momentum is a vector.)

Q42–8 For a particle in a finite potential well, is it correct to say that each bound state of definite energy is also a state of definite wavelength? Is it a state of definite momentum? Explain.

Q42–9 Example 42–2 (Section 42–2) gives the time-dependent wave functions for a particle in a box. How can a wave function that has both a real and an imaginary part describe the position (a real number) of a particle?

Q42–10 In Fig. 42–4b, the probability function is zero at the points $x = 0$ and $x = L$, the "walls" of the box. Does this mean that the particle never strikes the walls? Explain.

Q42–11 In Fig. 42–7, the second derivative $d^2\psi/dx^2$ for x slightly less than zero has the opposite sign to the second derivative for x slightly greater than zero. Use the properties of the Schrödinger equation, Eq. (42–17), to explain why this should be.

Q42–12 Compare the wave functions for the first three energy levels for a particle in a box of width L (Fig. 42–4a) to the corresponding wave functions for a finite potential well of the same width (Fig. 42–8a). How does the wavelength in the interval $0 \le x \le L$ for the $n = 1$ level of the particle in a box compare to the corresponding wavelength for the $n = 1$ level of the finite potential well? Use this to explain why E_1 is less than E_∞ in the situation depicted in Fig. 42–8b.

Q42–13 It is stated in Section 42–4 that a finite potential well always has at least one bound level, no matter how shallow the well. Does this mean that as $U_0 \to 0$, $E_1 \to 0$? Does this violate the Heisenberg uncertainty principle? Explain.

Q42–14 Figure 42–8a shows that the greater the energy of a bound state for a finite potential well, the more the wave function extends outside the well (into the intervals $x < 0$ and $x > L$). Explain why this happens.

Q42–15 In classical (Newtonian) mechanics, the total energy E of a particle can never be less than the potential energy U because the kinetic energy K cannot be negative. Yet in barrier tunneling (Section 42–5) a particle passes through regions where E is less than U. Is this a contradiction? Explain.

Q42–16 In Example 42–6, halving L increased T by a factor of nearly 10,000. Keeping L constant, would halving the barrier height U_0 have the same effect? The initial kinetic energy of the particle is kept fixed at 2.0 eV. Explain.

Q42–17 Qualitatively, how would you expect the probability for a particle to tunnel through a potential barrier to depend on the height of the barrier? Explain.

Q42–18 The wave function shown in Fig. 42–12 is nonzero for both $x < 0$ and $x > L$. Does this mean that the particle splits into two parts when it strikes the barrier, with one part tunneling through the barrier and the other part bouncing off the barrier? Explain.

Q42–19 The probability distributions for the harmonic oscillator wave functions (Figs. 42–19 and 42–20) begin to resemble the classical (Newtonian) probability distribution when the quantum number n becomes large. Would the distributions become the same as in the classical case in the limit of very large n? Explain.

Q42–20 For a particle in a harmonic oscillator potential, what is the probability of finding the particle at any $x \ge 0$ if the particle is in the ground level? (Refer to Fig. 42–19, but do not evaluate an integral.) Is the answer the same if the particle is in an excited level? Explain.

Q42–21 Compare the allowed energy levels for the hydrogen atom, the particle in a box, and the harmonic oscillator. What are the values of the quantum number n for the ground level and the second excited level of each system?

Q42–22 In the hydrogen atom the potential-energy function depends only on distance r from the nucleus, not on direction. That is, it is spherically symmetric. Would you expect all the corresponding wave functions for the electron in the hydrogen atom to be spherically symmetric? Explain.

Q42–23 The hydrogen atom, particle in a box, and harmonic oscillator are all bound systems with discrete energy levels. Why are the bound-state energies E_n negative for the hydrogen atom but positive for the other two systems?

EXERCISES

SECTION 42–2 PARTICLE IN A BOX

42–1 Ground-Level Billiards. a) Find the lowest energy level for a particle in a box if the particle is a billiard ball ($m = 0.20$ kg) and the box has a width of 1.5 m, the size of a billiard table. (Assume that the billiard ball slides without friction rather than rolls. That is, ignore the *rotational* kinetic energy.) b) Since the energy in part (a) is all kinetic, what speed does this correspond to? How much time would it take at this speed for the ball to move from one side of the table to the other? c) What is the difference in energy between the $n = 2$ and $n = 1$ levels? d) Are quantum-mechanical effects important for the game of billiards?

42–2 A proton is in a box of width L. What must be the width of the box for the ground-level energy to be 5.0 MeV, a typical value for the energy with which the particles in a nucleus are bound? Compare your result to the size of a nucleus, which is on the order of 10^{-14} m.

42–3 It takes 3.0 eV of energy to excite an electron in a box from the ground level to the first excited level. What is the width L of the box?

42–4 Recall that $|\psi|^2\, dx$ is the probability of finding the particle that has normalized wave function $\psi(x)$ in the interval x to $x + dx$. Consider a particle in a box with rigid walls at $x = 0$ and

$x = L$. Let the particle be in the ground level and use ψ_n as given in Eq. (42–9). a) For what values of x, if any, in the range from 0 to L is the probability of finding the particle zero? b) For what values of x is the probability largest? c) In parts (a) and (b) are your answers consistent with Fig. 42–8? Explain.

42–5 Repeat Exercise 42–4 for the particle in the first excited level.

42–6 a) Find the excitation energy from the ground level to the third excited level for an electron confined to a box that has a width of 0.125 nm. b) The electron makes a transition from the $n = 1$ to $n = 4$ level by absorbing a photon. Calculate the wavelength of this photon.

42–7 An electron is in a box of width 3.0×10^{-10} m. What are the de Broglie wavelength and the magnitude of the momentum of the electron if it is in a) the $n = 1$ level; b) the $n = 2$ level; c) the $n = 3$ level? In each case how does the wavelength compare to the width of the box?

SECTION 42–3 THE SCHRÖDINGER EQUATION

42–8 Let ψ be a solution of Eq. (42–17) with energy E. Show that the function $\psi' = A\psi$ is also a solution with the same E, for any values of the constant A. (This result shows that normalizing a wave function (Section 42–2) does not change the energy associated with the wave function.)

42–9 **Linear Combinations of Wave Functions.** Let ψ_1 and ψ_2 be two solutions of Eq. (42–17) with the same energy E. Show that $\psi = B\psi_1 + C\psi_2$ is also a solution with energy E, for any values of the constants B and C.

42–10 a) Show that $\psi = A \sin kx$ is a solution to Eq. (42–13) if $k = \sqrt{2mE}/\hbar$. b) Explain why this is an acceptable wave function for a box with rigid walls at $x = 0$ and $x = L$ only if k is an integer multiple of π/L.

42–11 a) Repeat Exercise 42–10 for $\psi = A \cos kx$. b) Explain why this cannot be an acceptable wave function for a particle in a box with rigid walls at $x = 0$ and $x = L$ no matter what the value of k.

42–12 Let ψ_1 and ψ_2 be two solutions of Eq. (42–17) with energies E_1 and E_2, respectively, where $E_1 \neq E_2$. Is $\psi = A\psi_1 + B\psi_2$, where A and B are nonzero constants, a solution to Eq. (42–17)? Explain your answer.

SECTION 42–4 POTENTIAL WELLS

42–13 a) Show that $\psi = A \sin kx$, where k is a constant, is *not* a solution of Eq. (42–17) for $U = U_0$ and $E < U_0$. b) Is this ψ a solution for $E > U_0$?

42–14 An electron is bound in a square well of width 1.50 nm and depth $U_0 = 6E_\infty$. If the electron is initially in the ground level and absorbs a photon, what maximum wavelength can the photon have and still liberate the electron from the well?

42–15 A proton is bound in a square well of width 4.0 fm = 4.0×10^{-15} m. The depth of the well is six times the ground level energy E_∞ of the corresponding infinite well. If the proton makes a transition from the level with energy E_1 to the level with energy E_3 by absorbing a photon, find the wavelength of the photon.

42–16 An electron is bound in a square well with a depth equal to six times the ground-level energy E_∞ of an infinite well of the same width. When the electron makes a transition from the level with energy E_2 to the level with energy E_1, it emits a photon of wavelength 455 nm. Determine the width of the well.

42–17 Calculate $d^2\psi/dx^2$ for the wave function of Eq. (42–20), and show that the function is a solution of Eq. (42–19).

42–18 Calculate $d^2\psi/dx^2$ for the wave function of Eq. (42–22), and show that for any values of C and D it is a solution of Eq. (42–17) for $x \leq 0$ and $x \geq L$ when $U = U_0$ and $E < U_0$.

SECTION 42–5 POTENTIAL BARRIERS AND TUNNELING

42–19 An electron with initial kinetic energy 6.0 eV encounters a barrier with height 11.0 eV. What is the probability of tunneling if the width of the barrier is a) 0.80 nm; b) 0.40 nm?

42–20 An electron with initial kinetic energy 5.0 eV encounters a barrier with height U_0 and width 0.60 nm. What is the transmission coefficient if a) $U_0 = 7.0$ eV; b) $U_0 = 9.0$ eV; c) $U_0 = 13.0$ eV?

42–21 a) An electron with initial kinetic energy 32 eV encounters a square barrier with height 41 eV and width 0.25 nm. What is the probability that the electron will tunnel through the barrier? b) A proton with the same kinetic energy encounters the same barrier. What is the probability that the proton will tunnel through the barrier?

42–22 **Alpha Decay.** In a simple model for a radioactive nucleus, an alpha particle ($m = 6.64 \times 10^{-27}$ kg) is trapped by a square barrier that has width 2.0 fm and height 30.0 MeV. a) What is the tunneling probability when the alpha particle encounters the barrier if its kinetic energy is 1.0 MeV below the top of the barrier (Fig. 42–23)? b) What is the tunneling probability if the energy of the alpha particle is 10.0 MeV below the top of the barrier?

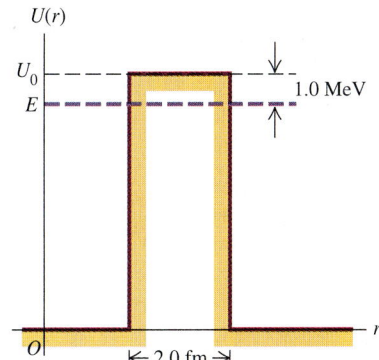

FIGURE 42–23 Exercise 42–22.

SECTION 42–6 THE HARMONIC OSCILLATOR

42–23 A wooden block with mass 0.250 kg is oscillating on the end of a spring that has force constant 110 N/m. Calculate the ground-level energy and the energy separation between adjacent levels. Express your results in joules and in electron volts. Are quantum effects important?

42–24 Show that $\psi(x)$ given by Eq. (42–27) is a solution to Eq. (42–25) with energy $E_0 = \hbar\omega/2$.

42–25 An atom with mass 9.4×10^{-26} kg vibrates with simple harmonic motion in a crystal lattice. If the atom absorbs a photon with wavelength 525 μm when it makes a transition from the ground level to the first excited level, what is the force constant?

42–26 For the ground-level harmonic oscillator wave function $\psi(x)$ given in Eq. (42–27), $|\psi|^2$ has a maximum at $x = 0$.

a) Compute the ratio of $|\psi|^2$ at $x = +A$ to $|\psi|^2$ at $x = 0$, where A is given by Eq. (42–30) with $n = 0$ for the ground level.
b) Compute the ratio of $|\psi|^2$ at $x = +2A$ to $|\psi|^2$ at $x = 0$. In each case is your result consistent with what is shown in Fig. 42–19?

42–27 In Section 42–6 it is shown that for the ground level of a harmonic oscillator, $\Delta x \, \Delta p_x = \hbar$. Do a similar analysis for an excited level that has quantum number n. How does the uncertainty product $\Delta x \, \Delta p_x$ depend on n?

PROBLEMS

42–28 A particle is in the ground level of a box that extends from $x = 0$ to $x = L$. a) What is the probability of finding the particle in the region between 0 and $L/4$? Calculate this by integrating $|\psi(x)|^2 \, dx$, where ψ is normalized, from $x = 0$ to $x = L/4$. b) What is the probability of finding the particle in the region $x = L/4$ to $x = L/2$? c) How do the results of parts (a) and (b) compare? Explain. d) Add the probabilities calculated in parts (a) and (b). e) Are your results in parts (a), (b) and (d) consistent with Fig. 42–4b? Explain.

42–29 What is the probability of finding a particle in a box of length L in the region between $x = L/4$ and $x = 3L/4$ when the particle is in a) the ground level; b) the first excited level? (*Hint:* Integrate $|\psi(x)|^2 \, dx$, where ψ is normalized, between $L/4$ and $3L/4$.) c) Are your results in parts (a) and (b) consistent with Fig. 42–4b? Explain.

42–30 Consider a particle in a box with rigid walls at $x = 0$ and $x = L$. Let the particle be in the ground level. Calculate the probability $|\psi|^2 \, dx$ that the particle will be found in the interval x to $x + dx$ for a) $x = L/4$; b) $x = L/2$; c) $x = 3L/4$.

42–31 Repeat Problem 42–30 for a particle in the first excited level.

42–32 Let ΔE_n be the energy difference between the adjacent energy levels E_n and E_{n+1} for a particle in a box. The ratio $R_n = \Delta E_n / E_n$ compares the energy of a level to the energy separation of the next higher energy level. a) For what value of n is R_n largest, and what is this largest R_n? b) What does R_n approach as n becomes very large? How does this result compare to the classical value for this quantity?

42–33 **Photon in a Dye Laser.** An electron in a long organic molecule used in a dye laser behaves approximately like a particle in a box with width 4.18 nm. What is the wavelength of the photon emitted when the electron undergoes a transition a) from the first excited level to the ground level? b) from the second excited level to the first excited level?

42–34 Consider a particle in a box that extends from $x = 0$ to $x = L$. The particle is in its ground level. a) What is $d\psi/dx$ for $x > L$? b) Calculate $d\psi/dx$ for $0 < x < L$ and from this result obtain $d\psi/dx$ as $x \rightarrow L$. c) Is $d\psi/dx$ continuous at $x = L$? Use Eq. (42–17) to show that $d\psi/dx$ need not be continuous at points where U becomes infinite. (*Hint:* Solve Eq. (42–17) for $d^2\psi/dx^2$.)

42–35 It is stated in Section 42–3 that $d\psi/dx$ need not be continuous at points where U becomes infinite (see Problem 42–34). For a particle in a box with rigid walls at $x = 0$ and $x = L$, what is $d\psi/dx$ for x slightly greater than zero if a) the particle is in the $n = 1$ level? b) the particle is in the $n = 2$ level? What is $d\psi/dx$ for x slightly less than L if c) the particle is in the $n = 1$ level? d) the particle is in the $n = 2$ level? e) For which of these levels does the slope of the wave function near the walls have the greater magnitude? Do your results agree with Fig. 42–4a?

42–36 A particle is confined within a box with perfectly rigid walls at $x = 0$ and $x = L$. Although the magnitude of the instantaneous force exerted on the particle by the walls is infinite and the time over which it acts is zero, the impulse (which involves a product of force and time) is both finite and quantized. Show that the impulse exerted by the wall at $x = 0$ is $(nh/L)\hat{\imath}$ and that the impulse exerted by the wall at $x = L$ is $-(nh/L)\hat{\imath}$. (*Hint:* You may wish to review Section 8–2.)

42–37 A fellow student proposes that a possible wave function for a free particle of mass m (one for which the potential-energy function $U(x)$ is zero) is

$$\psi(x) = \begin{cases} e^{+\kappa x}, & x < 0, \\ e^{-\kappa x}, & x \geq 0, \end{cases}$$

where κ is a positive constant. a) Graph this proposed wave function. b) Show that the proposed wave function satisfies the Schrödinger equation for $x < 0$ if the energy is $E = -\hbar^2\kappa^2/2m$, that is, if the energy of the particle is *negative*. c) Show that the proposed wave function also satisfies the Schrödinger equation for $x \geq 0$ with the same energy as in part (b). d) Explain why the proposed wave function is nonetheless *not* an acceptable solution of the Schrödinger equation for a free particle. (*Hint:* What is the behavior of the function at $x = 0$?) It is in fact impossible for a free particle (one for which $U(x) = 0$) to have an energy less than zero.

42–38 **The Time-Dependent Schrödinger Equation.** Equation (42–17) is the time-independent Schrödinger equation in one dimension. The time-dependent Schrödinger equation is

$$-\frac{\hbar^2}{2m}\frac{\partial^2\Psi(x,t)}{\partial x^2} + U(x)\Psi(x,t) = i\hbar\frac{\partial\Psi(x,t)}{\partial t}.$$

a) If $\psi(x)$ is a solution to Eq. (42–17) with energy E, show that the time-dependent function $\Psi(x, t) = \psi(x)e^{-i\omega t}$ is a solution to the time-dependent Schrödinger equation if ω is chosen appropriately. What is the value of ω that makes Ψ a solution? b) Show that $|\Psi(x, t)|^2 = |\psi(x)|^2$ for the time-dependent function $\Psi(x, t)$ described in part (a). (This says that the probability of finding the particle in any given region along the x-axis is independent of time, provided it is in a state with definite energy E. For this reason, a quantum-mechanical state of definite energy is called a *stationary state*.)

42–39 Time-Dependent Wave Function for a Free Particle. One example of a time-dependent wave function is that for a free particle (one for which $U(x) = 0$ for all x) of energy E and x-component of momentum p. From the de Broglie relations (Section 41–2), such a particle has associated with it a frequency $f = E/h$ and a wavelength $\lambda = h/p$. A reasonable first guess for the time-dependent wave function for such a particle is $\Psi(x, t) = A \sin(\omega t - kx)$, where A is a constant, $\omega = 2\pi f$ is the angular frequency, and $k = 2\pi/\lambda$ is the wave number. This is the same function we used to describe a mechanical wave (see Eq. (19–7)) or an electromagnetic wave propagating in the x-direction (see Eq. (33–16)). a) Show that $\omega = E/\hbar$, $k = p/\hbar$, and $\omega = \hbar k^2/2m$. (*Hint:* The energy is purely kinetic, so $E = p^2/2m$.) b) To check this guess for the time-dependent wave function, substitute $\Psi(x, t) = A \sin(\omega t - kx)$ into the time-dependent Schrödinger equation (Problem 42–38) with $U(x) = 0$ (so the particle is free). Show that this guess for $\Psi(x, t)$ does *not* satisfy this equation, and so is not a suitable wave function for a free particle. c) Use the procedure described in part (b) to show that a second guess, $\Psi(x, t) = A \cos(\omega t - kx)$, is also not a suitable wave function for a free particle. d) Consider a combination of the functions proposed in parts (b) and (c):

$$\Psi(x, t) = A \cos(\omega t - kx) + B \sin(\omega t - kx).$$

By using the procedure described in part (b), show that this wave function is a solution to the time-dependent Schrödinger equation with $U(x) = 0$, but only if $B = -iA$. (*Hint:* To satisfy the time-dependent Schrödinger equation for all x and t, the coefficients of $\cos(\omega t - kx)$ on both sides of the equation must be equal. The same is true for the coefficients of $\sin(\omega t - kx)$ on both sides of the equation.) This is an example of the general result that time-dependent wave functions always have both a real part and an imaginary part.

42–40 Figure 42–8a shows several bound-state wave functions for a particle in a finite potential well. a) By inspection of the wave functions, verify that ψ and its second derivative $d^2\psi/dx^2$ have opposite signs at each point inside the well ($0 \leq x \leq L$), but have the same sign at each point outside the well ($x < 0$ and $x > L$). b) Use the time-independent Schrödinger equation, Eq. (42–17), to explain the observations in part (a) about ψ and $d^2\psi/dx^2$.

42–41 a) For the finite potential well of Fig. 42–6, what relations among the constants A and B of Eq. (42–20) and C and D of Eq. (42–22) are obtained by applying the boundary condition that ψ be continuous at $x = 0$ and at $x = L$? b) What relations among A, B, C, and D are obtained by applying the boundary condition that $d\psi/dx$ be continuous at $x = 0$ and at $x = L$?

42–42 An electron with initial kinetic energy 5.5 eV encounters a square potential barrier with height 10.0 eV. What is the width of the barrier if the electron has a 0.10% probability of tunneling through the barrier?

42–43 A particle of mass m and total energy E tunnels through a square barrier of height U_0 and width L. When the transmis-

sion coefficient is *not* much smaller than unity, it is given by

$$T = \left[1 + \frac{(U_0 \sinh \kappa L)^2}{4E(U_0 - E)}\right]^{-1},$$

where $\sinh \kappa L = (e^{\kappa L} - e^{-\kappa L})/2$ is the hyperbolic sine of κL. a) Show that if $\kappa L \gg 1$, this expression for T approaches Eq. (42–24). b) Explain why the restriction $\kappa L \gg 1$ in part (a) implies either that the barrier is relatively wide or that the energy E is relatively small compared to U_0. c) Show that as the particle's incident kinetic energy E approaches the barrier height U_0, T approaches $[1 + (kL/2)^2]^{-1}$, where $k = \sqrt{2mE}/\hbar$ is the wave number of the incident particle. (*Hint:* If $|z| \ll 1$, $\sinh z \approx z$.)

42–44 A particle of mass m and total energy E tunnels through a square barrier of height U_0 and width L. Show that the wave number k of the tunneling particle is imaginary when the particle is within the barrier. That is, show that $k = i\kappa$, where κ is as given in Eq. (42–24).

42–45 For small amplitudes of oscillation the motion of a pendulum is simple harmonic. For a pendulum with a period of 0.500 s, find the ground-level energy and the energy difference between adjacent energy levels. Express your results in joules and in electron volts. Are these values detectable?

42–46 A harmonic oscillator consists of a 0.020-kg mass on a spring. Its frequency is 1.50 Hz, and the mass has a speed of 0.360 m/s as it passes the equilibrium position. a) What is the value of the quantum number n for its energy level? b) What is the difference in energy between the levels E_n and E_{n+1}? Is this difference detectable?

42–47 a) Show that $\psi(x) = Cxe^{-m\omega x^2/2\hbar}$, where C is a normalization constant, is a solution to Eq. (42–25) with energy $E_1 = \frac{3}{2}\hbar\omega$. b) Sketch $\psi(x)$ as a function of x. Which of the graphs in Fig. 42–18 does your sketch resemble most closely?

42–48 The wave function for the first excited level of a harmonic oscillator is given by $\psi(x) = Cxe^{-m\omega x^2/2\hbar}$, where C is a normalization constant (Problem 42–47). a) At what values of x is $|\psi|^2$ a maximum? b) At what values of x is $|\psi|^2$ zero? Compare your results to the corresponding graph in Fig. (42–19).

42–49 A Three-Dimensional Isotropic Harmonic Oscillator. An isotropic harmonic oscillator has the potential energy function $U(x, y, z) = \frac{1}{2}k'(x^2 + y^2 + z^2)$. (*Isotropic* means that the force constant k' is the same in all three coordinate directions.) a) Show that for this potential, a solution to Eq. (42–25) is given by $\psi = \psi_{n_x}(x)\psi_{n_y}(y)\psi_{n_z}(z)$. In this expression, $\psi_{n_x}(x)$ is a solution to the one-dimensional harmonic oscillator Schrödinger equation, Eq. (42–25), with energy $E_{n_x} = (n_x + \frac{1}{2})\hbar\omega$. The functions $\psi_{n_y}(y)$ and $\psi_{n_z}(z)$ are analogous one-dimensional wave functions for oscillations in the y- and z-directions. Find the energy associated with this ψ. b) From your results in part (a) what are the ground level and first excited-level energies of the three-dimensional isotropic oscillator? c) Show that there is only one state (one set of quantum numbers n_x, n_y, and n_z) for the ground level but three states for the first excited level.

42–50 Three-Dimensional Anisotropic Harmonic Oscillator. An oscillator has the potential energy function

$U(x, y, z) = \frac{1}{2}k_1'(x^2 + y^2) + \frac{1}{2}k_2'z^2$, where $k_1' > k_2'$. This oscillator is called *anisotropic* because the force constant is not the same in all three coordinate directions. a) Find a general expression for the energy levels of the oscillator. (See Problem 42–49). b) From your results in part (a), what are the ground level and

first excited-level energies of this oscillator? c) How many states (different sets of quantum numbers n_x, n_y, and n_z) are there for the ground level and for the first excited level? Compare to part (c) of Problem 42–49.

CHALLENGE PROBLEMS

42–51 Section 42–2 considered a box with walls at $x = 0$ and $x = L$. Consider now a box with width L but centered at $x = 0$, so that it extends from $x = -L/2$ to $x = +L/2$ (Fig. 42–24). Note that this box is symmetric about $x = 0$. a) Consider possible wave functions of the form $\psi(x) = A \sin kx$. Apply the boundary conditions at the wall to obtain the allowed energy levels. b) Another set of possible wave functions are functions of the form $\psi(x) = A \cos kx$. Apply the boundary conditions at the wall to obtain the allowed energy levels. c) Compare the energies obtained in parts (a) and (b) to the set of energies given in Eq. (42–5). d) An odd function f satisfies the condition $f(x) = -f(-x)$, and an even function g satisfies $g(x) = g(-x)$. Of the wave functions from parts (a) and (b), which are even and which are odd?

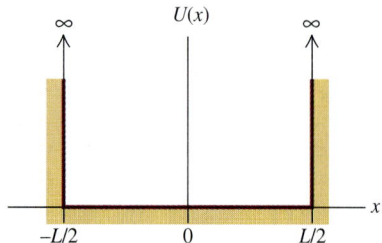

FIGURE 42–24 Challenge Problem 42–51.

42–52 Consider a potential well defined as follows: $U(x) = \infty$ for $x < 0$, $U(x) = 0$ for $0 < x < L$, and $U(x) = U_0 > 0$ for $x > L$ (Fig. 42–25). Consider a particle with mass m and kinetic energy $E < U_0$ that is trapped in the well. a) The boundary condition at the infinite wall ($x = 0$) is $\psi(0) = 0$. What must be the form of the function $\psi(x)$ for $0 < x < L$ in order to satisfy both the Schrödinger equation and this boundary condition? b) The wave function must remain finite as $x \to \infty$. What must be the form of the function $\psi(x)$ for $x > L$ in order to satisfy both the Schrödinger equation and this boundary condition at infinity? c) Impose the boundary conditions that ψ and $d\psi/dx$ are continuous at $x = L$. Show that the energies of the allowed levels are obtained from solutions of the equation $k \cot kL = -\kappa$, where $k = \sqrt{2mE}/\hbar$ and $\kappa = \sqrt{2m(U_0 - E)}/\hbar$.

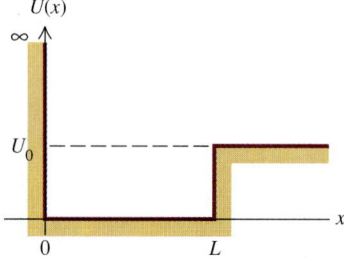

FIGURE 42–25 Challenge Problem 42–52.

42–53 The WKB Approximation. It can be a challenge to solve the Schrödinger equation for the bound-state energy levels of an arbitrary potential well. An alternative approach that can yield good approximate results for the energy levels is the *WKB approximation* (named for the physicists Gregor Wentzel, Hendrik Kramers, and Léon Brillouin, who pioneered its application to quantum mechanics). The WKB approximation begins from three physical statements: i) According to de Broglie, the magnitude of momentum p of a quantum-mechanical particle is $p = h/\lambda$. ii) The magnitude of momentum is related to the kinetic energy K by the relation $K = p^2/2m$. iii) If there are no nonconservative forces, then in Newtonian mechanics the energy E for a particle is constant and equal at each point to the sum of the kinetic and potential energies at that point: $E = K + U(x)$, where x is the coordinate. a) Combine these three relations to show that the wavelength of the particle at a coordinate x can be written as

$$\lambda(x) = \frac{h}{\sqrt{2m(E - U(x))}}$$

Thus we envision a quantum-mechanical particle in a potential well $U(x)$ as being like a free particle, but with a wavelength $\lambda(x)$ that is a function of position. b) When the particle moves into a region of increasing potential energy, what happens to its wavelength? c) At a point where $E = U(x)$, Newtonian mechanics says that the particle has zero kinetic energy and must be instantaneously at rest. Such a point is called a *classical turning point*, since this is where a Newtonian particle must stop its motion and reverse direction. As an example, an object oscillating in simple harmonic motion with amplitude A moves back and forth between the points $x = -A$ and $x = +A$; each of these is a classical turning point, since there the potential energy $\frac{1}{2}k'x^2$ equals the total energy $\frac{1}{2}k'A^2$. In the WKB expression for $\lambda(x)$, what is the wavelength at a classical turning point? d) For a particle in a box of length L, the walls of the box are classical turning points (see Fig. 42–1). Furthermore, the number of wavelengths that fit within the box must be a half-integer (see Fig. 42–3a), so that $L = (n/2)\lambda$ and hence $L/\lambda = n/2$, where $n = 1, 2, 3, \ldots$. (Note that this is a restatement of Eq. (42–3).) The WKB scheme for finding the allowed bound-state energy levels of an *arbitrary* potential well is an extension of these observations. It demands that for an allowed energy E, there must be a half-integer number of wavelengths between the classical turning points for that energy. Since the wavelength in the WKB approximation is not a constant but depends on x, the number of wavelengths between the classical turning points a and b for a given value of the energy is the integral of $1/\lambda(x)$ between those points:

$$\int_a^b \frac{dx}{\lambda(x)} = \frac{n}{2}, \quad n = 1, 2, 3, \ldots .$$

Using the expression for $\lambda(x)$ you found in part (a), show that the WKB *condition for an allowed bound-state energy* can be written as

$$\int_a^b \sqrt{2m(E - U(x))} \; dx = \frac{nh}{2}, \; n = 1, 2, 3, \dots .$$

e) As a check on the expression in part (d), apply it to a particle in a box with walls at $x = 0$ and $x = L$. Evaluate the integral and show that the allowed energy levels according to the WKB approximation are the same as those given by Eq. (42–5). (*Hint:* Since the walls of the box are infinitely high, the points $x = 0$ and $x = L$ are classical turning points for *any* energy E. Inside the box, the potential energy is zero.) f) For the finite square well shown in Fig. 42–6, show that the WKB expression given in part (d) predicts the *same* bound-state energies as for an infinite square well of the same width. (*Hint:* Assume $E < U_0$. Then the classical turning points are at $x = 0$ and $x = L$.) This shows that the WKB approximation does a poor job when the potential energy function changes discontinuously, as for a finite potential well. In the next two problems we consider situations in which the potential energy function changes gradually and the WKB approximation is much more useful.

42–54 The WKB approximation (Challenge Problem 42–53) can be used to calculate the energy levels for a harmonic oscillator. In this approximation, the energy levels are the solutions to the equation

$$\int_a^b \sqrt{2m(E - U(x))} \; dx = \frac{nh}{2}, \; n = 1, 2, 3, \dots .$$

Here E is the energy, $U(x)$ is the potential energy function, and $x = a$ and $x = b$ are the classical turning points (the points at which E is equal to the potential energy, so the Newtonian kinetic energy would be zero). a) Determine the classical turning points for a harmonic oscillator with energy E and force constant k'. b) Carry out the integral in the WKB approximation and show that the energy levels in this approximation are $E_n = \hbar\omega$, where $\omega = \sqrt{k'/m}$ and $n = 1, 2, 3, \dots$. (*Hint:* Recall that $\hbar = h/2\pi$. A useful standard integral is

$$\int \sqrt{A^2 - x^2} \; dx = \frac{1}{2}\left[x\sqrt{A^2 - x^2} + A^2 \arcsin\left(\frac{x}{|A|}\right) \right],$$

where arcsin denotes the inverse sine function. Note that the integrand is even, so the integral from $-x$ to x is equal to twice

the integral from 0 to x.) c) How do the approximate energy levels found in part (b) compare with the true energy levels given by Eq. (42–29)? Does the WKB approximation give an underestimate or an overestimate of the energy levels?

42–55 Protons, neutrons, and many other particles are made of more fundamental particles called *quarks* and *antiquarks* (the antimatter equivalent of quarks). A quark and an antiquark can form a bound state with a variety of different energy levels, each of which corresponds to a different particle observed in the laboratory. As an example, the ψ particle is low-energy bound state of a so-called charm quark and its antiquark, with a rest energy of 3097 MeV; the ψ(2S) particle is an excited state of this same quark-antiquark combination, with a rest energy of 3686 MeV. A simplified representation of the potential energy of interaction between a quark and an antiquark is $U(x) = A|x|$, where A is a positive constant and x represents the distance between the quark and the antiquark. You can use the WKB approximation (Challenge Problem 42–53) to determine the bound-state energy levels for this potential energy function. In the WKB approximation, the energy levels are the solutions to the equation

$$\int_a^b \sqrt{2m(E - U(x))} \; dx = \frac{nh}{2}, \; n = 1, 2, 3, \dots .$$

Here E is the energy, $U(x)$ is the potential energy function, and $x = a$ and $x = b$ are the classical turning points (the points at which E is equal to the potential energy, so the Newtonian kinetic energy would be zero). a) Determine the classical turning points for the potential $U(x) = A|x|$ and for an energy E. b) Carry out the above integral and show that the allowed energy levels in the WKB approximation are given by

$$E_n = \frac{1}{2m}\left(\frac{3mAh}{4} \right)^{2/3} n^{2/3}, \;\; n = 1, 2, 3, \dots .$$

(*Hint:* The integrand is even, so the integral from $-x$ to x is equal to twice the integral from 0 to x.) c) Does the difference in energy between successive levels increase, decrease, or remain the same as n increases? How does this compare to the behavior of the energy levels for the harmonic oscillator? For the particle in a box? Can you suggest a simple rule that relates the difference in energy between successive levels to the shape of the potential energy function?

43

Atomic Structure

Why is seawater so salty? That is, why does salt dissolve so readily in water? The answer lies in the electronic structure of the sodium and chlorine atoms that make up salt molecules. Sodium can easily lose an electron to form an Na^+ ion, just as chlorine can easily form a Cl^- ion by gaining an electron. These ions are attracted to the polar ends of water molecules, holding the ions in solution.

43–1 INTRODUCTION

Some physicists claim that all of chemistry is contained in the Schrödinger equation. This is somewhat of an exaggeration, but this equation can teach us a great deal about the chemical behavior of elements and the nature of chemical bonds. We can gain insight into the periodic table of the elements and the microscopic basis of magnetism.

We can learn a great deal about the structure and properties of *all* atoms from the solutions to the Schrödinger equation for the hydrogen atom. These solutions have quantized values of angular momentum; we don't need to make a separate statement about quantization as we did with the Bohr model. We label the states with a set of quantum numbers, which we'll use later with many-electron atoms as well. We'll see that the electron also has an intrinsic "spin" angular momentum in addition to the orbital angular momentum associated with its motion.

We'll introduce the exclusion principle, a kind of microscopic zoning ordinance that is the key to understanding many-electron atoms. This principle says that no two electrons in an atom can have the same quantum-mechanical state. Finally, we'll use the principles of this chapter to explain the characteristic x-ray spectra of atoms.

43–2 THE HYDROGEN ATOM

Let's continue the discussion of the hydrogen atom that we began in Chapter 40. In the Bohr model, electrons moved in circular orbits like Newtonian particles, but with quantized values of angular momentum. While this model gave the correct energy levels of the hydrogen atom, as deduced from spectra, it had many conceptual difficulties. It mixed classical physics with new and seemingly contradictory concepts. It provided no insight into the process by which photons are emitted and absorbed. It could not be generalized to atoms with more than one electron. It predicted the wrong magnetic properties for the hydrogen atom. And perhaps most important, its picture of the electron as a localized point particle was inconsistent with the more general view we developed in Chapters 41 and 42.

Now let's apply the Schrödinger equation to the hydrogen atom. As we discussed in Section 40–6, we include the motion of the nucleus by simply replacing the electron mass m with the reduced mass m_r. We discussed the three-dimensional version of the Schrödinger equation in Section 42–7. The hydrogen-atom problem is best formulated in spherical coordinates (r, θ, ϕ), shown in Fig. 42–22; the potential energy is then simply

$$U = -\frac{1}{4\pi\epsilon_0}\frac{e^2}{r}. \qquad (43\text{–}1)$$

The Schrödinger equation with this potential-energy function can be solved exactly; the solutions are combinations of familiar functions. Without going into a lot of detail, we can describe the most important features of the procedure and the results.

First, the solutions are obtained by a method called *separation of variables,* in which we express the wave function $\psi(r, \theta, \phi)$ as a product of three functions, each one a function of only one of the three coordinates:

$$\psi(r, \theta, \phi) = R(r)\,\Theta(\theta)\,\Phi(\phi). \tag{43-2}$$

That is, the function $R(r)$ depends *only* on r, $\Theta(\theta)$ depends *only* on θ, and $\Phi(\phi)$ depends *only* on ϕ. When we substitute Eq. (43–2) into the Schrödinger equation, we get three separate equations, each containing only one of the coordinates. This is an enormous simplification; it reduces the problem of solving a fairly complex *partial* differential equation with three independent variables to the much simpler problem of solving three separate *ordinary* differential equations with one independent variable each.

The physically acceptable solutions of these three equations are determined by *boundary conditions.* The radial function $R(r)$ must approach zero at large r, because we are describing *bound states* of the electron that are localized near the nucleus. This is analogous to the requirement that the harmonic-oscillator wave functions (Section 42–6) must approach zero at large x. The angular functions $\Theta(\theta)$ and $\Phi(\phi)$ must be *periodic.* For example, (r, θ, ϕ) and $(r, \theta, \phi + 2\pi)$ describe the same point, so $\Phi(\phi + 2\pi)$ must equal $\Phi(\phi)$. Also, the angular functions must be *finite* for all relevant values of the angles. For example, there are solutions of the Θ equation that become infinite at $\theta = 0$ and $\theta = \pi$; these are unacceptable, since $\psi(r, \theta, \phi)$ must be normalizable.

The radial functions $R(r)$ turn out to be an exponential function $e^{-\alpha r}$ (where α is positive) multiplied by a polynomial in r. The functions $\Theta(\theta)$ are polynomials containing various powers of $\sin\theta$ and $\cos\theta$, and the functions $\Phi(\phi)$ are simply proportional to $e^{im_l\phi}$, where $i = \sqrt{-1}$ and m_l is an integer that may be positive, zero, or negative.

In the process of finding solutions that satisfy the boundary conditions, we also find the corresponding energy levels. Their energies, which we denote by E_n ($n = 1, 2, 3, \ldots$), turn out to be *identical* to those from the Bohr model, as given by Eq. (40–16), with the electron rest mass m replaced by the reduced mass m_r. Rewriting that equation using $\hbar$, we have

$$E_n = -\frac{1}{(4\pi\epsilon_0)^2}\frac{m_r e^4}{2n^2\hbar^2} = -\frac{13.60\text{ eV}}{n^2} \qquad \text{(energy levels of hydrogen).} \tag{43-3}$$

As in Section 40–6, we call n the **principal quantum number** for the level of energy E_n.

The result that Eq. (43–3) can be obtained from the Schrödinger equation is of critical importance. The Schrödinger analysis is quite different from the Bohr model, both formally and conceptually, yet both yield the same energy-level scheme, a scheme that agrees with the energies determined from spectra. This result is a very significant confirmation of the validity of the Schrödinger approach. As we will see, the Schrödinger analysis can explain many more aspects of the hydrogen atom than can the Bohr model.

QUANTIZATION OF ORBITAL ANGULAR MOMENTUM

The solutions that satisfy the boundary conditions mentioned above also have quantized values of *orbital angular momentum.* That is, only certain discrete values of the magnitude and components of orbital angular momentum are permitted. In discussing the Bohr model in Section 40–6, we mentioned that quantization of angular momentum was a result with little fundamental justification. With the Schrödinger equation it appears automatically!

The possible values of the magnitude L of orbital angular momentum $\vec{L}$ are determined by the requirement that the $\Theta(\theta)$ function must be finite at $\theta = 0$ and $\theta = \pi$. In a level with energy E_n and principal quantum number n, the possible values of L are

$$L = \sqrt{l(l + 1)}\hbar \qquad (l = 0,\ 1,\ 2,\ \dots, n - 1) \qquad \text{(magnitude of orbital} \atop \text{angular momentum).}} \quad (43\text{--}4)$$

The *orbital angular-momentum quantum number* l is called the **orbital quantum number** for short. In the Bohr model, each energy level corresponded to a single value of angular momentum. Equation (43–4) shows that in fact there are n different possible values of L for the nth energy level.

An interesting feature of Eq. (43–4) is that the orbital angular momentum is *zero* for $l = 0$ states. This result disagrees with the Bohr model, in which the electron always moved in a circle of definite radius and L was never zero. The $l = 0$ wave functions ψ depend only on r; the $\Theta(\theta)$ and $\Phi(\phi)$ functions for these states are constants. Thus the wave functions for $l = 0$ states are spherically symmetric; there is nothing in their probability distribution $|\psi|^2$ to favor one direction over any other, and there is no orbital angular momentum.

The permitted values of the *component* of $\vec{L}$ in a given direction, say the z-component L_z, are determined by the requirement that the $\Phi(\phi)$ function must equal $\Phi(\phi + 2\pi)$. The possible values of L_z are

$$L_z = m_l\hbar \qquad (m_l = 0,\ \pm 1,\ \pm 2, \dots,\ \pm l) \qquad \text{(components of orbital} \atop \text{angular momentum).}} \quad (43\text{--}5)$$

We see that m_l can be zero or a positive or negative integer up to, but no larger in magnitude than, l. That is, $|m_l| \le l$. For example, if $l = 1$, m_l can equal 1, 0, or -1. For reasons that will emerge later, we call m_l the *orbital magnetic quantum number* or **magnetic quantum number** for short.

The component L_z can never be quite as large as L (unless both are zero). For example, when $l = 2$, the largest possible value of m_l is also 2; then Eqs. (43–4) and (43–5) give

$$L = \sqrt{2(2 + 1)}\hbar = \sqrt{6}\hbar = 2.45\hbar,$$

$$L_z = 2\hbar.$$

Figure 43–1 shows the situation. The minimum value of the angle θ_L between the vector $\vec{L}$ and the z-axis is given by

$$\theta_L = \arccos\frac{L_z}{L} = \arccos\frac{2}{2.45} = 35.3°.$$

43–1 (a) When $l = 2$, the magnitude of $\vec{L}$ is $\sqrt{2(2 + 1)}\ \hbar = 2.45\hbar$, but the direction of $\vec{L}$ is not definite. In this semiclassical vector picture, $\vec{L}$ makes an angle of $35.3°$ with the z-axis when the z-component has its maximum value of $2\hbar$. (b) Cones of the possible directions for $\vec{L}$.

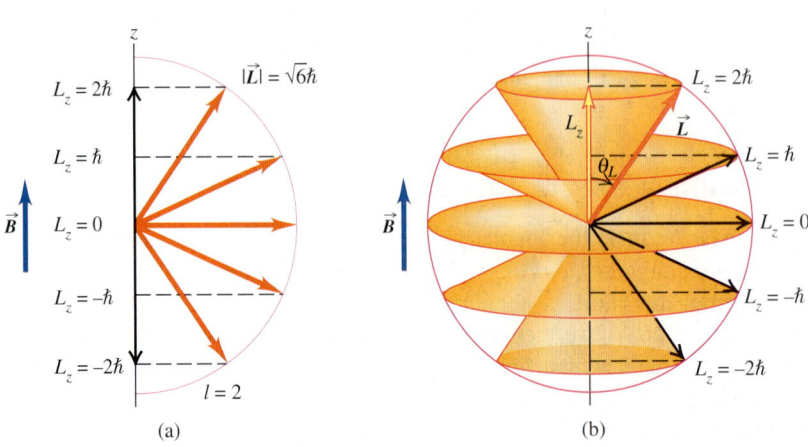

That $|L_z|$ is always less than L is also required by the uncertainty principle. Suppose we could know the precise *direction* of the orbital angular momentum vector. Then we could let that be the direction of the z-axis, and L_z would equal L. This corresponds to a particle moving in the xy-plane only, in which case the z-component of the linear momentum $\vec{p}$ would be zero with no uncertainty Δp_z. Then the uncertainty principle $\Delta z\, \Delta p_z \geq \hbar$ requires infinite uncertainty Δz in the coordinate z. This is impossible for a localized state; we conclude that we can't know the direction of $\vec{L}$ precisely. Thus, as we've already stated, the component of $\vec{L}$ in a given direction can never be quite as large as its magnitude L. Also, if we can't know the direction of $\vec{L}$ precisely, we can't determine the components L_x and L_y precisely. Thus we show *cones* of possible directions for $\vec{L}$ in Fig. 43–1b.

You may wonder why we have singled out the z-axis for special attention. There's no fundamental reason for this; the atom certainly doesn't care what coordinate system we use. The point is that we can't determine all three components of orbital angular momentum with certainty, so we arbitrarily pick one as the component we want to measure. When we discuss interactions of the atom with a magnetic field, we will consistently orient the field along the z-axis as in Fig. 43–1.

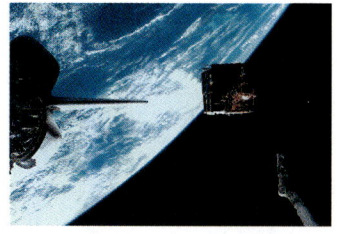

The energy for an orbiting satellite deployed from the space shuttle depends on the average distance between the satellite and the center of the earth. It does *not* depend on whether the orbit is circular (with a large orbital angular momentum) or elliptical (in which case the orbital angular momentum is smaller). In the same way, the energy of a hydrogen atom does not depend on the orbital angular momentum.

QUANTUM NUMBER NOTATION

The wave functions for the hydrogen atom are determined by the values of three quantum numbers n, l, and m_l. The energy E_n is determined by the principal quantum number n according to Eq. (43–3). The magnitude of orbital angular momentum is determined by the orbital quantum number l, as in Eq. (43–4). The component of orbital angular momentum in a specified axis direction (customarily the z-axis) is determined by the magnetic quantum number m_l, as in Eq. (43–5). For each energy level E_n given by Eq. (43–3), there is more than one distinct state having the same energy but different quantum numbers. The existence of more than one distinct state with the same energy is called **degeneracy;** it has no counterpart in the Bohr model.

States with various values of the orbital quantum number l are often labeled with letters, according to the following scheme:

$l = 0$: s states,

$l = 1$: p states,

$l = 2$: d states,

$l = 3$: f states,

$l = 4$: g states,

and so on alphabetically. This seemingly irrational choice of the letters s, p, d, and f originated in the early days of spectroscopy and has no fundamental significance. In an important form of *spectroscopic notation* that we'll use often, a state with $n = 2$ and $l = 1$ is called a $2p$ state; a state with $n = 4$ and $l = 0$ is a $4s$ state, and so on. Only s states ($l = 0$) are spherically symmetric.

Here's another bit of notation. The radial extent of the wave functions increases with the principal quantum number n, and we can speak of a region of space associated with a particular value of n as a **shell.** Especially in discussions of many-electron atoms, these shells are given letters:

$n = 1$: K shell,

$n = 2$: L shell,

$n = 3$: M shell,

$n = 4$: N shell,

and so on alphabetically. For each n, different values of l correspond to different *sub-shells.* For example, the L shell ($n = 2$) contains the $2s$ and $2p$ subshells.

TABLE 43–1

QUANTUM STATES OF THE HYDROGEN ATOM

n	l	m_l	SPECTROSCOPIC NOTATION	SHELL
1	0	0	$1s$	K
2	0	0	$2s$	L
2	1	$-1, 0, 1$	$2p$	
3	0	0	$3s$	M
3	1	$-1, 0, 1$	$3p$	
3	2	$-2, -1, 0, 1, 2$	$3d$	
4	0	0	$4s$	N

and so on.

Table 43–1 shows some of the possible combinations of the quantum numbers n, l, and m_l for hydrogen-atom wave functions. The spectroscopic notation and the shell notation for each are also shown.

Problem–Solving Strategy

ATOMIC STRUCTURE

1. Be sure you know the possible values of the quantum numbers n, l, and m_l for states of the hydrogen atom. They are all integers: n is always greater than zero, l can be zero or positive up to $n - 1$, and m_l can range from $-l$ to l in steps of one. Be able to count the number of (n, l, m_l) states in each shell and subshell. You should know how to derive, not simply memorize, Table 43–1.

2. Familiarizing yourself with some numerical magnitudes is useful. For example, the electric potential energy of a proton and an electron that are 0.10 nm apart (typical of atomic dimensions) is about –15 eV. Wavelengths of visi-

ble light are around 500 nm, and their frequencies are around 5×10^{14} Hz. Keeping numbers such as these in mind helps you know what kinds of magnitudes to expect in atomic physics.

3. As in the last several chapters, you'll need to use both electron volts and joules. The conversion $1 \text{ eV} = 1.602 \times 10^{-19}$ J is useful, as is Planck's constant in eV, $h = 4.136 \times 10^{-15}$ eV $\cdot$ s. Nanometers are convenient for atomic and molecular dimensions, but don't forget to convert to meters in calculations.

EXAMPLE 43-1

Counting hydrogen states How many distinct (n, l, m_l) states of the hydrogen atom with $n = 3$ are there? Find the energy of these states.

SOLUTION When $n = 3$, l can be 0, 1, or 2. When $l = 0$, m_l can be only 0 (1 state). When $l = 1$, m_l can be -1, 0, or 1 (3 states). When $l = 2$, m_l can be -2, -1, 0, 1, or 2 (5 states). The total number of (n, l, m_l) states with $n = 3$ is therefore $1 + 3 + 5 = 9$. (In Section 43–4 we'll find that the total number of $n = 3$ states is in

fact twice this, or 18, because of electron spin.)

The energies of these states are all the same because the energy depends only on n. According to Eq. (43–3), with $n = 3$,

$$E_3 = \frac{-13.60 \text{ eV}}{3^2} = -1.51 \text{ eV}.$$

It's useful to remember that the ground level of hydrogen has $n = 1$ and $E_1 = -13.6$ eV.

EXAMPLE 43-2

Angular momentum in an excited level of hydrogen Consider the $n = 4$ states of hydrogen. a) What is the maximum magnitude L of the orbital angular momentum? b) What is the maximum value of L_z? c) What is the minimum angle between $\vec{L}$ and the z-axis? Give your answers to parts (a) and (b) in terms of $\hbar$.

SOLUTION a) When $n = 4$, the maximum value of the orbital angular-momentum quantum number l is $(n - 1) = (4 - 1) = 3$;

from Eq. (43–4),

$$L = \sqrt{3(3 + 1)}\hbar = \sqrt{12}\hbar = 3.464\hbar.$$

b) For $l = 3$ the maximum value of the magnetic quantum number m_l is 3; from Eq. (43–5),

$$L_z = 3\hbar.$$

c) The *minimum* allowed angle between $\vec{L}$ and the z-axis corresponds to the *maximum* allowed values of L_z and m_l (Fig. 43–1b shows an $l = 2$ example). For the state with $l = 3$ and $m_l = 3$,

$$\theta_{min} = \arccos \frac{(L_z)_{max}}{L} = \arccos \frac{3\hbar}{3.464\hbar} = 30.0°.$$

We invite you to verify that the angle is greater than 30.0° for all states with smaller values of l.

ELECTRON PROBABILITY DISTRIBUTIONS

Rather than picturing the electron as a point particle moving in a precise circle, the Schrödinger equation gives a *probability distribution* surrounding the nucleus. Because the hydrogen-atom probability distributions are three-dimensional, they are harder to visualize than the two-dimensional circular orbits of the Bohr model. It's helpful to look at the *radial probability distribution* $P(r)$, that is, the probability per radial length for the electron to be found at various distances from the proton. From Section 42–6 the probability for finding the electron in a small volume element dV is $|\psi|^2\, dV$. (We assume that ψ is normalized, that is, that the integral of $|\psi|^2\, dV$ over all space equals unity so that there is 100% probability of finding the electron somewhere in the universe.) Let's take as our volume element a thin spherical shell with inner radius r and outer radius $r + dr$. The volume dV of this shell is approximately its area $4\pi r^2$ multiplied by its thickness dr:

$$dV = 4\pi r^2 dr. \qquad (43\text{–}6)$$

We denote by $P(r)\, dr$ the probability of finding the particle within the radial range dr; then

$$P(r)\, dr = |\psi|^2 dV = |\psi|^2 4\pi r^2 dr \qquad \begin{array}{l}\text{(probability that the electron} \\ \text{is between } r \text{ and } r + dr).\end{array} \qquad (43\text{–}7)$$

For wave functions that depend on θ and ϕ as well as r, we use the value of $|\psi|^2$ averaged over all angles in Eq. (43–7).

Figure 43–2 shows graphs of $P(r)$ for several hydrogen-atom wave functions. The r scales are labeled in multiples of a, the smallest distance between the electron and the nucleus in the Bohr model,

$$a = \frac{\epsilon_0 h^2}{\pi m_r e^2} = \frac{4\pi\epsilon_0 \hbar^2}{m_r e^2} = 5.29 \times 10^{-11} \text{ m} \qquad \text{(smallest } r \text{, Bohr model).} \qquad (43\text{–}8)$$

Just as for a particle in a box (Section 42–2), there are some positions where the probability is zero. But again, the uncertainty principle tells us not to worry; we can't localize the electron exactly anyway. Note that for the states having the largest possible l for each n (such as $1s$, $2p$, $3d$ and $4f$ states), $P(r)$ has a single maximum at n^2a. For these states, the electron is most likely to be found at the distance from the nucleus that is predicted by the Bohr model, $r = n^2a$.

  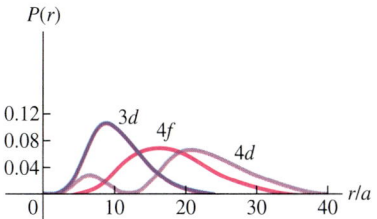

43–2 Radial probability distribution functions $P(r)$ for several hydrogen-atom wave functions, plotted as functions of the ratio r/a, where a is the Bohr separation. For each function, the number of maxima is $(n - l)$. The curves for which $l = n - 1$ have only one maximum, located at n^2a.

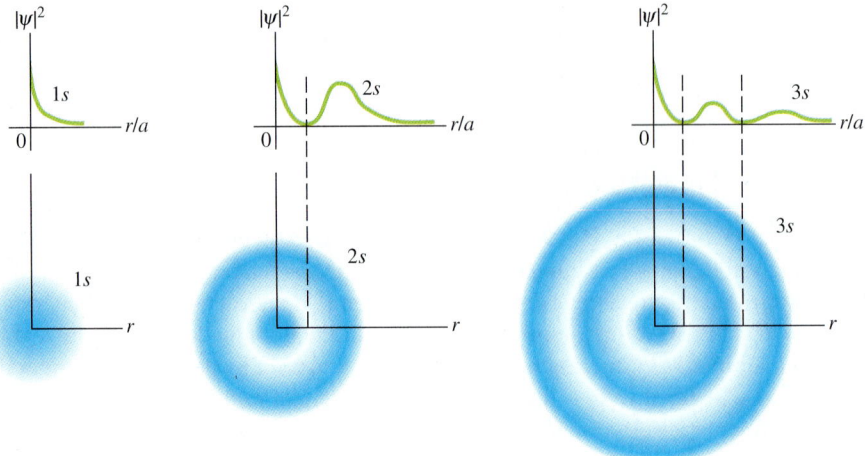

43–3 Probability distribution $|\psi|^2$ for the spherically symmetric $1s$, $2s$, and $3s$ wave functions for the hydrogen atom.

Figure 43–3 shows the spherically symmetric probability distributions $|\psi|^2$ of the lowest three s subshells of hydrogen. Figure 43–4 shows sketches of cross sections of $|\psi|^2$ containing the z-axis for a few hydrogen-atom wave functions. Recalling that $\Phi(\phi)$ is proportional to $e^{im_l\phi}$, can you show why $|\psi|^2 = \psi^*\psi$ does not depend on ϕ?

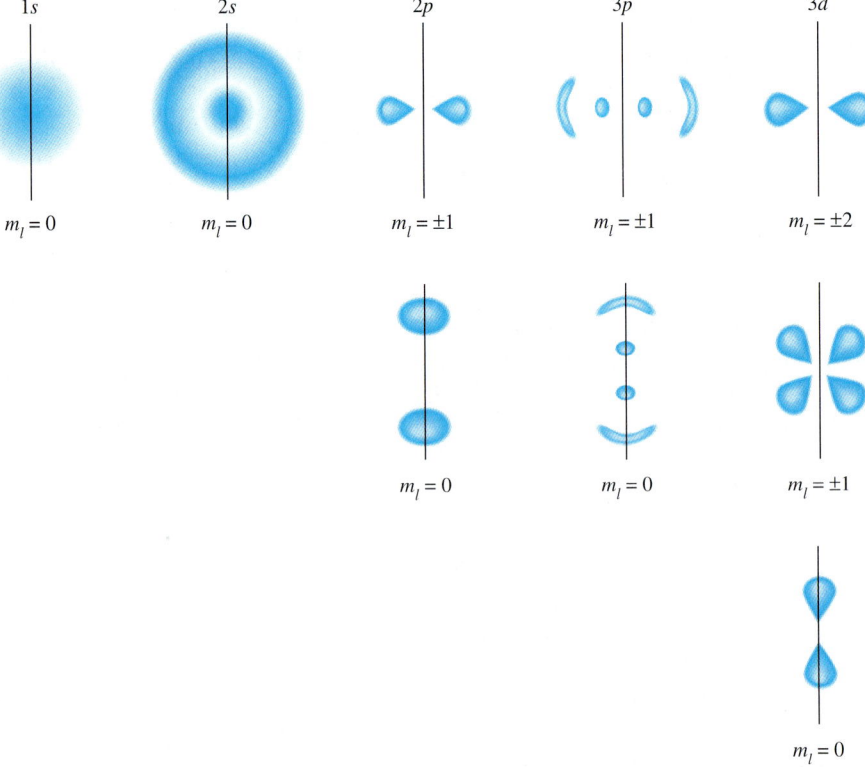

43–4 Sketches of cross sections of three-dimensional probability distributions for a few quantum states of the hydrogen atom. They are not to the same scale. The vertical black line is the z-axis; mentally rotate the sketches about this axis to obtain the three-dimensional representation of $|\psi|^2$. For example, the $2p$, $m_l = \pm 1$ probability distribution looks like a fuzzy doughnut.

EXAMPLE 43-3

The ground-state wave function for hydrogen (a $1s$ state) is

$$\psi_{1s}(r) = \frac{1}{\sqrt{\pi a^3}} \, e^{-r/a}.$$

a) Verify that this function is normalized. b) What is the probability that the electron will be found at a distance less than a from the nucleus?

SOLUTION a) The normalization condition is $\int_0^\infty |\psi|^2 \, dV = 1$. Using the volume element $dV = 4\pi r^2 \, dr$ given by Eq. (43–6), we find

$$\int_0^\infty |\psi_{1s}|^2 \, dV = \int_0^\infty \frac{1}{\pi a^3} \, e^{-2r/a} (4\pi r^2 \, dr) = \frac{4}{a^3} \int_0^\infty r^2 e^{-2r/a} \, dr.$$

The following indefinite integral can be found in a table of integrals or by integrating by parts:

$$\int r^2 e^{-2r/a} \, dr = \left(-\frac{ar^2}{2} - \frac{a^2 r}{2} - \frac{a^3}{4} \right) e^{-2r/a}.$$

Evaluating this between the limits $r = 0$ and $r = \infty$ is simple; it is zero at $r = \infty$ because of the exponential factor, and at $r = 0$ only

the last term in the parentheses survives. Thus the value of the integral is $a^3/4$. Putting it all together, we find

$$\int_0^\infty |\psi_{1s}|^2 \, dV = \frac{4}{a^3} \int_0^\infty r^2 e^{-2r/a} \, dr = \frac{4}{a^3} \frac{a^3}{4} = 1.$$

The wave function *is* normalized.

b) To find the probability P that the electron is found at a distance less than a, we carry out the same integration but with the limits 0 and a. We'll leave the details as an exercise. From the upper limit we get $-5e^{-2}a^3/4$; the final result is

$$P = \int_0^a |\psi_{1s}|^2 4\pi r^2 \, dr = \frac{4}{a^3} \left(-\frac{5a^3 e^{-2}}{4} + \frac{a^3}{4} \right)$$

$$= 1 - 5e^{-2} = 0.323.$$

Thus in a ground state we expect to find the electron at a distance from the nucleus less than a about $\frac{1}{3}$ of the time and at a greater distance about $\frac{2}{3}$ of the time. It's hard to tell, but in Fig. 43–2, about $\frac{2}{3}$ of the area under the $1s$ curve is at distances greater than a (that is, $r/a > 1$).

Two generalizations that we discussed with the Bohr model in Section 40–6 are equally valid in the Schrödinger analysis. First, if the "atom" is not composed of a single proton and a single electron, the reduced mass m_r of the system will give changes in Eqs. (43–3) and (43–8) that are substantial with some exotic systems. Examples include *muonium,* made up of a proton and a muon, and *positronium,* composed of a positron and an electron. Second, our analysis is applicable to single-electron ions, such as He^+, Li^{2+}, and so on. For such ions we replace e^2 by Ze^2 in Eqs. (43–3) and (43–8), where Z is the number of protons (the **atomic number**). This replacement will far overshadow the small change in the reduced mass m_r in these equations.

The Schrödinger analysis of the hydrogen atom is a lot more complex, both conceptually and mathematically, than the Newtonian analysis of planetary motion or the semiclassical Bohr model. It deals with probabilities rather than certainties, and it predicts discrete rather than continuous values of energy and angular momentum. But this analysis enables us to understand phenomena for which classical mechanics and electromagnetism are inadequate. Added complexity is the price we pay for expanded understanding.

43-3 THE ZEEMAN EFFECT

The **Zeeman effect** is the splitting of atomic energy levels and the associated spectrum lines when the atoms are placed in a magnetic field (Fig. 43–5). This effect confirms experimentally the quantization of angular momentum. The discussion in this section, which assumes that the only angular momentum is the *orbital* angular momentum of a single electron, also shows why we call m_l the magnetic quantum number.

Atoms contain charges in motion, so it should not be surprising that magnetic forces cause changes in that motion and in the energy levels. As early as the middle of the nineteenth century, physicists speculated that the sources of visible light might be vibrating electric charge on an atomic scale. In 1862, Michael Faraday placed light sources in a magnetic field in an attempt to observe changes in spectrum lines. His spectroscopic techniques were not refined enough to observe any effect. But in 1896 the Dutch

43–5 Magnetic effects on the spectrum of light from the sun. (a) The slit of a spectrograph is positioned along the black line crossing a portion of a sunspot. (b) The Zeeman effect in the sunspot splits the middle single spectral line into three lines. The splitting indicates that this (typical) sunspot has a maximum magnetic field of over 0.4 T, a thousand times greater than the earth's field.

(a)   (b)

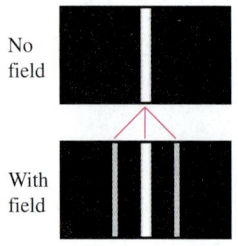

43–6 The normal Zeeman effect. When the source of radiation is placed in a magnetic field, the interactions of orbital magnetic moments with the field split individual spectral lines into sets of three lines.

physicist Pieter Zeeman, using improved instruments, found that in the presence of a magnetic field, some spectrum lines were split into groups of closely spaced lines (Fig. 43–6). This effect now bears his name.

Let's begin our analysis of the Zeeman effect by reviewing the concept of *magnetic dipole moment* or *magnetic moment,* introduced in Section 28–8. A plane current loop with vector area $\vec{A}$ carrying current I has a magnetic moment $\vec{\mu}$ given by

$$\vec{\mu} = I\vec{A}. \tag{43–9}$$

When a magnetic dipole of moment $\vec{\mu}$ is placed in a magnetic field $\vec{B}$, the field exerts a torque $\vec{\tau} = \vec{\mu} \times \vec{B}$ on the dipole. The potential energy U associated with this interaction is given by Eq. (28–27):

$$U = -\vec{\mu} \cdot \vec{B}. \tag{43–10}$$

Now let's use Eqs. (43–9) and (43–10) and the Bohr model to look at the interaction of a hydrogen atom with a magnetic field. The orbiting electron (speed v) is equivalent to a current loop with radius r and area πr^2. The average current I is the average charge per unit time that passes a point of the orbit. This is equal to the charge magnitude e divided by the time T for one revolution, given by $T = 2\pi r/v$. Thus $I = ev/2\pi r$, and from Eq. (43–9) the magnitude μ of the magnetic moment is

$$\mu = IA = \frac{ev}{2\pi r}\pi r^2 = \frac{evr}{2}. \tag{43–11}$$

We can also express this in terms of the magnitude L of the orbital angular momentum. From Eq. (10–28), $L = mvr$ for a particle in a circular orbit, so Eq. (43–11) becomes

$$\mu = \frac{e}{2m}L. \tag{43–12}$$

The ratio of the magnitude of $\vec{\mu}$ to the magnitude of $\vec{L}$ is $\mu/L = e/2m$ and is called the *gyromagnetic ratio.*

In the Bohr model, $L = nh/2\pi = n\hbar$, where $n = 1, 2, \ldots$. For an $n = 1$ state (a ground state), Eq. (43–12) becomes $\mu = (e/2m)\hbar$. This quantity is a natural unit for magnetic moment; it is called one **Bohr magneton,** denoted by μ_B:

$$\mu_B = \frac{e\hbar}{2m} \qquad \text{(definition of the Bohr magneton).} \tag{43–13}$$

Evaluating Eq. (43–13) gives

$$\mu_B = 5.788 \times 10^{-5} \text{ eV/T} = 9.274 \times 10^{-24} \text{ J/T or A} \cdot \text{m}^2.$$

Note that the units J/T and A · m² are equivalent. We defined this quantity previously in (optional) Section 29–9.

EXAMPLE 43–4

Find the interaction potential energy predicted by the Bohr model when a hydrogen atom in its ground state is placed in a magnetic field with magnitude 2.00 T, if the field is perpendicular to the plane of the orbit and parallel to the magnetic dipole moment.

SOLUTION In the Bohr model, $\mu = \mu_B$ in the ground state. From Eq. (43–10) the interaction energy U when $\vec{\mu}$ and $\vec{B}$ are parallel is

$$U = -\vec{\mu} \cdot \vec{B} = -\mu_B B \cos 0 = -(5.788 \times 10^{-5} \text{ eV/T})(2.00 \text{ T})(1)$$
$$= -1.16 \times 10^{-4} \text{ eV} = -1.85 \times 10^{-23} \text{ J}.$$

The magnetic interaction shifts the energy level of the state by this amount. When $\vec{\mu}$ and $\vec{B}$ are antiparallel, the energy is $+1.16 \times 10^{-4}$ eV. These energies are *smaller* than the energy levels of the atom by a factor of the order of 10^4, even though 2.00 T would be a strong field in most laboratories.

For a hydrogen atom in a ground state, or any other *s* state, both *l* and *L* are zero. Thus orbital motion in *s* states gives μ and U equal to zero also. As the preceding example shows, the orbital magnetic effect predicted by the Bohr model is *wrong*. Hence we need to describe the states by Schrödinger wave functions. It turns out that in the Schrödinger formulation, electrons have the same ratio of μ to L (gyromagnetic ratio) as in the Bohr model, namely, $e/2m$. Suppose the magnetic field $\vec{B}$ is directed along the $+z$-axis. From Eq. (43–10) the interaction energy U of the atom's magnetic moment with the field is

$$U = -\mu_z B, \tag{43–14}$$

where μ_z is the *z*-component of the vector $\vec{\mu}$.

Now we use Eq. (43–12) to find μ_z, recalling that *e* is the *magnitude* of the electron charge and that the actual charge is $-e$. Because the electron charge is negative, the orbital angular momentum and magnetic moment vectors are opposite. We find

$$\mu_z = -\frac{e}{2m} L_z. \tag{43–15}$$

For the Schrödinger wave functions, $L_z = m_l \hbar$, with $m_l = 0, \pm 1, \pm 2, \dots, \pm l$, so

$$\mu_z = -\frac{e}{2m} L_z = -m_l \frac{e\hbar}{2m}. \tag{43–16}$$

CAUTION ▶ Be careful not to confuse the electron mass *m* with the magnetic quantum number m_l. ◀

Finally, we can express the interaction energy, Eq. (43–14), as

$$U = -\mu_z B = m_l \frac{e\hbar}{2m} B \qquad (m_l = 0, \pm 1, \pm 2, \dots, \pm l) \qquad \text{(orbital magnetic} \tag{43–17}$$
$$\text{interaction energy).}$$

In terms of the Bohr magneton $\mu_B = e\hbar/2m$,

$$U = m_l \mu_B B \qquad \text{(orbital magnetic interaction energy).} \tag{43–18}$$

The effect of the magnetic field is to shift the energy of each orbital state by an amount U. The interaction energy U depends on the value of m_l because m_l determines the orientation of the orbital magnetic moment relative to the magnetic field. This dependence is the reason m_l is called the magnetic quantum number.

Because the values of m_l range from $-l$ to $+l$ in steps of one, an energy level with a particular value of the orbital quantum number *l* contains $(2l + 1)$ different orbital states. Without a magnetic field these states all have the same energy; that is, they are

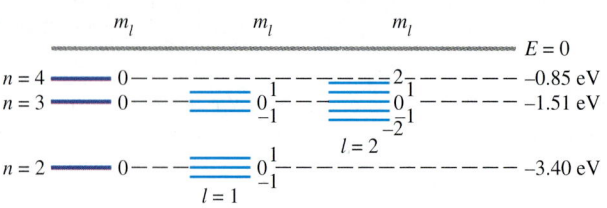

43–7 Energy-level diagram for hydrogen, showing the splitting of energy levels resulting from the interaction of the magnetic moment of the electron's orbital motion with an external magnetic field. Values of m_l are shown adjacent to the various levels. Relative magnitudes of splittings are exaggerated for clarity. The $n = 4$ splittings are not shown; can you draw them in?

degenerate. The magnetic field removes this degeneracy. In the presence of a magnetic field they are split into $2l + 1$ distinct energy levels; adjacent levels differ in energy by $(e\hbar/2m)B = \mu_B B$. The effect on the energy levels of hydrogen is shown in Fig. 43–7. Spectrum lines corresponding to transitions from one set of levels to another set are correspondingly split and appear as a series of three closely spaced spectrum lines replacing a single line.

EXAMPLE 43–5

An atom in a state with $l = 1$ emits a photon with wavelength 600.000 nm as it decays to a state with $l = 0$. If the atom is placed in a magnetic field with magnitude $B = 2.00$ T, determine the shifts in the energy levels and in the wavelength resulting from the interaction of the magnetic field and the atom's orbital magnetic moment.

SOLUTION The energy of a 600-nm photon is

$$E = \frac{hc}{\lambda} = \frac{(4.14 \times 10^{-15} \text{ eV} \cdot \text{s})(3.00 \times 10^8 \text{ m/s})}{600 \times 10^{-9} \text{ m}} = 2.07 \text{ eV}.$$

The $l = 0$ state is not split by the field. For $l = 1$, the splitting of levels is given by Eq. (43–18):

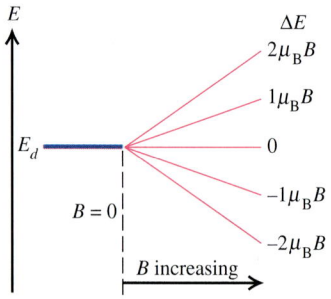

43–8 Splitting of the energy levels of a d state caused by an applied magnetic field, assuming only an orbital magnetic moment.

$$U = m_l \mu_B B = m_l (5.788 \times 10^{-5} \text{ eV/T})(2.00 \text{ T})$$
$$= m_l (1.16 \times 10^{-4} \text{ eV}) = m_l (1.85 \times 10^{-23} \text{ J}).$$

When $l = 1$, the possible values of m_l are -1, 0, and $+1$, and the three corresponding levels are separated by equal intervals of 1.16×10^{-4} eV. This is a small fraction of the 2.07-eV photon energy:

$$(1.16 \times 10^{-4} \text{ eV})/(2.07 \text{ eV}) = 5.60 \times 10^{-5}.$$

The corresponding *wavelength* shifts are approximately $(5.60 \times 10^{-5})(600 \text{ nm}) = 0.034$ nm. The original 600.000-nm line is split into a triplet with wavelengths 599.966, 600.000 and 600.034 nm. This splitting is well within the limit of resolution of modern spectrometers.

SELECTION RULES

Figure 43–8 shows what happens to a set of d states ($l = 2$) as the magnetic field increases. The five states, $m_l = -2, -1, 0, 1,$ and 2, are degenerate (have the same energy) with zero field, but the increasing field spreads the states out and removes their degeneracy. Figure 43–9 shows the splittings of the $3d$ and $2p$ states. Equal energy differences $(e\hbar/2m)B = \mu_B B$ separate adjacent levels. In the absence of a magnetic field, a transition from a $3d$ to a $2p$ state would yield a single spectrum line with photon energy $E_i - E_f$. With the levels split as shown, it might seem that there are five possible photon energies.

In fact, there are only three possibilities. Not all combinations of initial and final levels are possible, because of a restriction associated with conservation of angular momentum. The photon ordinarily carries off one unit ($\hbar$) of angular momentum, which leads to the requirements that in a transition l must change by 1 and m_l must change by

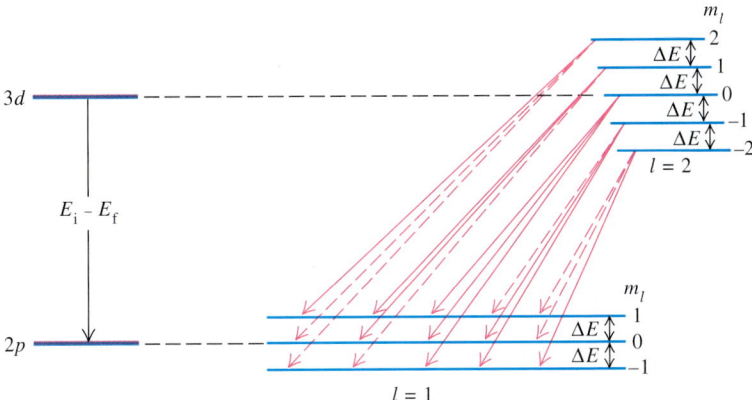

43-9 The cause of the normal Zeeman effect. The magnetic field splits the levels, but selection rules allow transitions with only three different energy changes, giving three different photon frequencies and wavelengths. Solid lines show allowed transitions; dashed lines show forbidden transitions.

0 or ± 1. These requirements are called **selection rules.** Transitions that obey these rules are called *allowed transitions;* those that don't are *forbidden transitions.* The allowed transitions are shown by solid arrows in Fig. 43–9. We invite you to count the possible transition energies to convince yourself that the nine solid arrows give only three possible energies; the zero-field value $E_i - E_f$, and that value plus or minus $\Delta E = (e\hbar/2m)B = \mu_B B$. The corresponding spectrum lines are shown.

What we have described is called the *normal* Zeeman effect. It is based entirely on the orbital angular momentum of the electron. In fact, there's nothing particularly *normal* about it. It leaves out a very important consideration: the *spin* angular momentum, the subject of the next section.

43-4 ELECTRON SPIN

Despite the success of the Schrödinger equation in predicting the energy levels of the hydrogen atom, several experimental observations indicate that it doesn't tell the whole story of the behavior of electrons in atoms. First, spectroscopists have found magnetic-field splitting into other than the three lines we've explained, sometimes unequally spaced. Before this effect was understood, it was called the *anomalous* Zeeman effect to distinguish it from the "normal" effect discussed in the preceding section. Both kinds of splittings are shown in Fig. 43–10.

Second, some energy levels show splittings that resemble the Zeeman effect even when there is *no* external magnetic field. For example, when the lines in the hydrogen spectrum are examined with a high-resolution spectrometer, some lines are found to

43-10 Illustrations of the normal and anomalous Zeeman effects for two elements. The brackets under each illustration show the "normal" splitting predicted by neglecting the effect of electron spin.

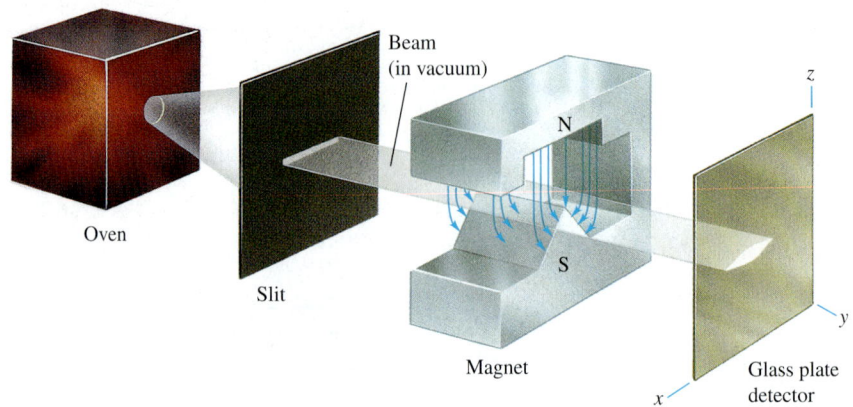

43–11 The Stern-Gerlach experiment. A beam of atoms is directed parallel to the *y*-axis. The specially shaped magnet poles produce a strongly non-uniform magnetic field that exerts a net force on the atomic magnetic dipoles, deflecting the atoms upward or downward according to the orientation of the magnetic moment.

consist of sets of closely spaced lines called *multiplets*. Similarly, the orange-yellow line of sodium, corresponding to the transition $4p \rightarrow 3s$ of the outer electron, is found to be a doublet ($\lambda = 589.0, 589.6$ nm), suggesting that the $4p$ level might in fact be two closely spaced levels. The Schrödinger equation in its original form didn't predict any of this.

Similar anomalies appeared in 1922 in atomic-beam experiments performed in Germany by Otto Stern and Walter Gerlach. When they passed a beam of neutral atoms through a nonuniform magnetic field (Fig. 43–11), atoms were deflected according to the orientation of their magnetic moments with respect to the field. These experiments demonstrated the quantization of angular momentum in a very direct way. If there were only orbital angular momentum, the deflections would split the beam into an odd number ($2l + 1$) of different components. However, some atomic beams were split into an *even* number of components. If we use a different symbol j for an angular momentum quantum number, setting $2j + 1$ equal to an even number gives $j = \frac{1}{2}, \frac{3}{2}, \frac{5}{2}, \ldots$, suggesting a half-integer angular momentum. This can't be understood on the basis of the Bohr model and similar pictures of atomic structure.

In 1925, two graduate students in the Netherlands, Samuel Goudsmidt and George Uhlenbeck, proposed that the electron might have some additional motion. Using a semiclassical model, they suggested that the electron might behave like a spinning sphere of charge instead of a particle. If so, it would have additional *spin* angular momentum and magnetic moment. If these were quantized in much the same way as *orbital* angular momentum and magnetic moment, they might help to explain the observed energy-level anomalies.

To introduce the concept of **electron spin,** let's start with an analogy. The earth travels in a nearly circular orbit around the sun, and at the same time it *rotates* on its axis. Each motion has its associated angular momentum, which we call the *orbital* and *spin* angular momentum, respectively. The total angular momentum of the earth is the vector sum of the two. If we were to model the earth as a single point, it would have no moment of inertia about its spin axis and thus no spin angular momentum. But when our model includes the finite size of the earth, spin angular momentum becomes possible.

In the Bohr model, suppose the electron is not just a point charge, but a small spinning sphere moving in orbit. Then the electron has not only orbital angular momentum but also spin angular momentum associated with the rotation of its mass about its axis. The sphere carries an electric charge, so the spinning motion leads to current loops and to a magnetic moment, as we discussed in Section 28–8. In a magnetic field, the *spin* magnetic moment has an interaction energy in addition to that of the *orbital* magnetic moment (the normal Zeeman-effect interaction discussed in Section 43–3). We should see additional Zeeman shifts due to the spin magnetic moment.

As we mentioned above, such shifts *are* indeed observed in precise spectroscopic analysis. This and a variety of other experimental evidence have shown conclusively that the electron *does* have spin angular momentum and a spin magnetic moment that do not depend on its orbital motion but are intrinsic to the electron itself.

SPIN QUANTUM NUMBERS

Like orbital angular momentum, the spin angular momentum of an electron (denoted by $\vec{S}$) is found to be *quantized.* Suppose we have some apparatus that measures a particular component of $\vec{S}$, say the z-component S_z. We find that the only possible values are

$$S_z = \pm\frac{1}{2}\hbar \qquad \text{(components of spin angular momentum).} \qquad (43\text{--}19)$$

This relation is reminiscent of the expression $L_z = m_l\hbar$ for the z-component of orbital angular momentum, except that $|S_z|$ is *one half* of $\hbar$ instead of an *integer* multiple. Equation (43–19) also suggests that the magnitude S of the spin angular momentum is given by an expression analogous to Eq. (43–4) with the orbital quantum number l replaced by the **spin quantum number** $s = \frac{1}{2}$:

$$S = \sqrt{\frac{1}{2}\left(\frac{1}{2}+1\right)}\hbar = \sqrt{\frac{3}{4}}\hbar \qquad \text{(magnitude of spin angular momentum).} \quad (43\text{--}20)$$

The electron is often called a "spin-$\frac{1}{2}$ particle."

The Bohr model is an oversimplified picture of electron behavior. In quantum mechanics, in which the Bohr orbits are superseded by probability distributions $|\psi|^2$, we can't really *picture* electron spin. If we visualize a probability distribution as a cloud surrounding the nucleus, then we can imagine many tiny spin arrows distributed throughout the cloud, either all with components in the $+z$-direction or all with components in the $-z$-direction. But don't take this picture too seriously.

To label completely the state of the electron in a hydrogen atom, we now need a fourth quantum number m_s, to specify the electron spin orientation. For an electron we give m_s the values $+\frac{1}{2}$ or $-\frac{1}{2}$ to agree with Eq. (43–19):

$$S_z = m_s\hbar \quad \left(m_s = \pm\frac{1}{2}\right) \qquad \text{(allowed values of } m_s \text{ and } S_z \text{ for an electron).} \qquad (43\text{--}21)$$

The spin angular momentum vector $\vec{S}$ can have only two orientations in space relative to the z-axis: "*spin up*" with a z-component of $+\frac{1}{2}\hbar$ and "*spin down*" with a z-component of $-\frac{1}{2}\hbar$.

The z-component of the associated spin magnetic moment (μ_z) turns out to be related to S_z by

$$\mu_z = -(2.00232)\frac{e}{2m}S_z, \qquad (43\text{--}22)$$

where $-e$ and m are (as usual) the charge and mass of the electron. When the atom is placed in a magnetic field, the interaction energy $-\vec{\mu} \cdot \vec{B}$ of the spin magnetic dipole moment with the field causes further splittings in energy levels and in the corresponding spectrum lines.

Equation (43–22) shows that the gyromagnetic ratio $|\mu_z/S_z| = (2.00232)e/2m$ for electron spin is approximately *twice* as great as the value $e/2m$ for *orbital* angular momentum and magnetic dipole moment. This result has no classical analog. But in 1928, Paul Dirac developed a relativistic generalization of the Schrödinger equation for electrons. His equation gave a spin gyromagnetic ratio of exactly $2(e/2m)$. It took another two decades to develop the area of physics called *quantum electrodynamics,* abbreviated QED, that predicts the value we've given to "only" six significant figures as 2.00232. In fact, QED now predicts a value that agrees with a recent measurement of 2.002319304386(20), making QED the most precise theory in all science.

EXAMPLE 43-6

Spin magnetic interaction energy Calculate the interaction energy for an electron in an $l = 0$ state (with no orbital magnetic moment) in a magnetic field with magnitude 2.00 T.

SOLUTION The interaction energy is $U = -\vec{\mu} \cdot \vec{B}$, where $\vec{\mu}$ is the electron's spin magnetic moment. Taking the z-axis as the direction of $\vec{B}$, this energy is $-\mu_z B$. From Eqs. (43–13), (43–19), and (43–22),

$$U = -(2.00232)\left(\frac{e}{2m}\right)\left(\pm\frac{1}{2}\hbar\right)B$$

$$= \mp\frac{1}{2}(2.00232)\left(\frac{e\hbar}{2m}\right)B = \mp(1.00116)\mu_B B$$

$$= \mp(1.00116)(9.274 \times 10^{-24}\ \text{J/T})(2.00\ \text{T})$$

$$= \mp 1.86 \times 10^{-23}\ \text{J} = \mp 1.16 \times 10^{-5}\ \text{eV}.$$

With $(1.00116)\,\mu_B$ evaluated as 5.795×10^{-5} eV/T, the effect on an energy level is shown in Fig. 43–12. Note that the z-component of the electron's spin magnetic dipole moment is

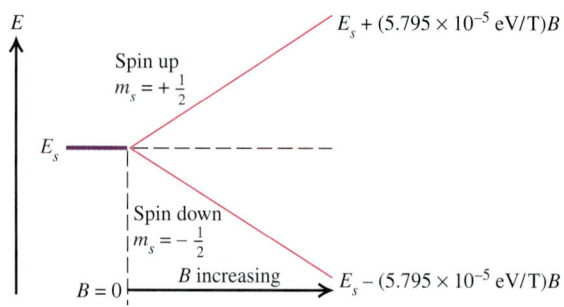

43–12 An $l = 0$ level of a single electron is split by interaction of the spin magnetic moment with a magnetic field.

approximately one Bohr magneton even though the z-component of angular momentum is only $\pm\frac{1}{2}\hbar$. The reason is that the gyromagnetic ratio is approximately twice as great for spin angular momentum as for orbital angular momentum.

SPIN-ORBIT COUPLING

As has already been mentioned, the spin magnetic dipole moment also gives splitting of energy levels even when there is *no* external field. One cause involves the orbital motion of the electron. In the Bohr model, observers moving with the electron would see the positively charged nucleus revolving around them (just as to earthbound observers the sun seems to be orbiting the earth). This apparent motion of charge causes a magnetic field at the location of the electron, as measured in the electron's moving frame of reference. The resulting interaction with the spin magnetic moment causes a twofold splitting of this level, corresponding to the two possible orientations of electron spin.

Discussions based on the Bohr model can't be taken too seriously, but a similar result can be derived from the Schrödinger equation. The interaction energy U can be expressed in terms of the scalar product of the angular-momentum vectors $\vec{L}$ and $\vec{S}$. This effect is called **spin-orbit coupling**; it is responsible for the small energy difference between the two closely spaced lowest excited levels of sodium shown in Fig. 40–9 and for the corresponding doublet (589.0, 589.6 nm) in the spectrum of sodium.

EXAMPLE 43-7

An effective magnetic field Calculate the effective magnetic field experienced by the electron in the $3p$ levels of the sodium atom.

SOLUTION The two lines in the sodium doublet result from transitions from the two $3p$ levels, which are split by spin-orbit coupling, to the $3s$ level, which is *not* split because it has $L = 0$. The energies of the two photons are

$$E = \frac{hc}{\lambda} = \frac{(4.136 \times 10^{-15}\ \text{eV}\cdot\text{s})(2.998 \times 10^8\ \text{m/s})}{589.0 \times 10^{-9}\ \text{m}} = 2.1052\ \text{eV}$$

and

$$E = \frac{hc}{\lambda} = \frac{(4.136 \times 10^{-15}\ \text{eV}\cdot\text{s})(2.998 \times 10^8\ \text{m/s})}{589.6 \times 10^{-9}\ \text{m}} = 2.1031\ \text{eV}.$$

The energy difference is

$$2.1052\ \text{eV} - 2.1031\ \text{eV} = 0.0021\ \text{eV} = 3.4 \times 10^{-22}\ \text{J}.$$

This is the energy difference between the two $3p$ levels. The spin-orbit interaction raises one level and lowers the other, each by 1.7×10^{-22} J, half this difference. From Example 43–6,

$$B = \left|\frac{U}{(1.00116)\mu_B}\right| = \frac{1.7 \times 10^{-22}\ \text{J}}{9.28 \times 10^{-24}\ \text{J/T}} = 18\ \text{T}.$$

This is a very strong field; to produce a steady field of this magnitude in the laboratory requires state-of-the-art electromagnets.

The orbital and spin angular momenta ($\vec{L}$ and $\vec{S}$, respectively) can combine in various ways. We define the total angular momentum $\vec{J}$ as

$$\vec{J} = \vec{L} + \vec{S}. \qquad (43\text{–}23)$$

The possible values of the magnitude J are given in terms of a quantum number j by

$$J = \sqrt{j(j + 1)}\hbar. \qquad (43\text{–}24)$$

We can then have states in which $j = |l \pm \frac{1}{2}|$. The $l + \frac{1}{2}$ states correspond to the case in which the vectors $\vec{L}$ and $\vec{S}$ have parallel z-components; while for the $l - \frac{1}{2}$ states, $\vec{L}$ and $\vec{S}$ have antiparallel z-components. For example, when $l = 1$, j can be $\frac{1}{2}$ or $\frac{3}{2}$. In another spectroscopic notation these p states are labeled $^2P_{1/2}$ and $^2P_{3/2}$, respectively. The superscript is the number of possible spin orientations, the letter P (now capitalized) indicates states with $l = 1$, and the subscript is the value of j. We used this scheme to label the energy levels of the sodium atom in Fig. 40–9.

The various line splittings resulting from magnetic interactions are collectively called *fine structure*. There are also additional, much smaller splittings associated with the fact that the *nucleus* of the atom has a magnetic dipole moment that interacts with the orbital and/or spin magnetic dipole moments of the electrons. These effects are called *hyperfine structure*. For example, the ground level of hydrogen is split into two states, separated by only 5.9×10^{-6} eV. The photon that is emitted in the transitions between these states gives 21-cm radiation, used by radio astronomers to locate interstellar clouds of hydrogen gas (which are too cold to emit visible light).

43-5 MANY-ELECTRON ATOMS AND THE EXCLUSION PRINCIPLE

This false-color image shows 21-cm radio emission from a large cloud of interstellar hydrogen about 7000 light years from the earth. The dark areas within the cloud are regions where hydrogen is at such a low temperature—just a few kelvins—that it cannot emit even low-energy 21-cm photons, but absorbs them instead.

So far, our analysis of atomic structure has concentrated on the hydrogen atom. That's natural; neutral hydrogen, with only one electron, is the simplest atom. If we can't understand hydrogen, we certainly can't understand anything more complex. But now let's move to many-electron atoms.

In general, an atom in its normal (electrically neutral) state has Z electrons and Z protons. Recall from Section 43–2 that we call Z the *atomic number*. The total electric charge of such an atom is exactly zero because the neutron has no charge while the proton and electron charges have the same magnitude but opposite sign.

We can apply the Schrödinger equation to this general atom. However, the complexity of the analysis increases very rapidly with increasing Z. Each of the Z electrons interacts not only with the nucleus but also with every other electron. The wave functions and the potential energy are functions of $3Z$ coordinates, and the equation contains second derivatives with respect to all of them. The mathematical problem of finding solutions of such equations is so complex that it has not been solved exactly even for the neutral helium atom, which has only two electrons.

Fortunately, various approximation schemes are available. The simplest approximation is to ignore all interactions between electrons and consider each electron as moving under the action only of the nucleus (considered to be a point charge). In this approximation the wave function for each electron is a function like those for the hydrogen atom, specified by four quantum numbers (n, l, m_l, m_s); the nuclear charge is Ze instead of e. This requires replacement of every factor of e^2 in the wave functions and the energy levels by Ze^2. In particular, the energy levels are given by Eq. (43–3) with e^4 replaced by Z^2e^4:

$$E_n = -\frac{1}{(4\pi\epsilon_0)^2}\frac{m_r Z^2 e^4}{2n^2\hbar^2} = -\frac{Z^2}{n^2}(13.6\text{ eV}). \qquad (43\text{–}25)$$

This approximation is fairly drastic; when there are many electrons, their interactions with each other are as important as the interaction of each with the nucleus. So this model isn't very useful for quantitative predictions.

THE CENTRAL-FIELD APPROXIMATION

A less drastic and more useful approximation is to think of all the electrons together as making up a charge cloud that is, on average, *spherically symmetric.* We can then think of each individual electron as moving in the total electric field due to the nucleus and this averaged-out cloud of all the other electrons. There is a corresponding spherically symmetric potential-energy function $U(r)$. This picture is called the **central-field approximation;** it provides a useful starting point for the understanding of atomic structure. If you are disappointed that we have to make approximations at such an early stage in our discussion, keep in mind that we are dealing with problems that initially defied all attempts at analysis, with or without approximations.

In the central-field approximation we can again deal with one-electron wave functions. The Schrödinger equation differs from the equation for hydrogen only in that the $1/r$ potential-energy function is replaced by a different function $U(r)$. But it turns out that $U(r)$ does not enter the differential equations for $\Theta(\theta)$ and $\Phi(\phi)$, so those angular functions are exactly the same as for hydrogen, and the orbital angular-momentum *states* are also the same as before. The quantum numbers l, m_l, and m_s have the same meaning as before, and the magnitude and z-component of the orbital angular momentum are again given by Eqs. (43–4) and (43–5).

The radial wave functions and probabilities are different than for hydrogen because of the change in $U(r)$, so the energy levels are no longer given by Eq. (43–3). We can still label a state using the four quantum numbers (n, l, m_l, m_s). In general, the energy of a state now depends on both n and l, rather than just on n as with hydrogen. The restrictions on values of the quantum numbers are the same as before:

$$n \geq 1,\ 0 \leq l \leq n-1,\ |m_l| \leq l,\ m_s = \pm\frac{1}{2} \qquad \text{(allowed values of quantum numbers).} \qquad (43\text{–}26)$$

THE EXCLUSION PRINCIPLE

To understand the structure of many-electron atoms, we need an additional principle, the *exclusion principle.* To see why this principle is needed, let's consider the lowest-energy state or *ground state* of a many-electron atom. In the one-electron states of the central-field model, there is a lowest-energy state (corresponding to an $n = 1$ state of hydrogen). We might expect that in the ground state of a complex atom, *all* the electrons should be in this lowest state. If so, then we should see only gradual changes in physical and chemical properties when we look at the behavior of atoms with increasing numbers of electrons (Z).

Such gradual changes are *not* what is observed. Instead, properties of elements vary widely from one to the next with each element having its own distinct personality. For example, the elements fluorine, neon, and sodium have 9, 10, and 11 electrons, respectively, per atom. Fluorine ($Z = 9$) is a *halogen;* it tends strongly to form compounds in which each fluorine atom acquires an extra electron. Sodium ($Z = 11$) is an *alkali metal;* it forms compounds in which each sodium atom *loses* an electron. Neon ($Z = 10$) is an *noble gas,* forming no compounds at all. Such observations show that in the ground state of a complex atom the electrons *cannot* all be in the lowest-energy states. But why not?

The key to this puzzle, discovered by the Austrian physicist Wolfgang Pauli in 1925, is called the **exclusion principle.** This principle states that **no two electrons can occupy the same quantum-mechanical state** in a given system. That is, **no two electrons in an atom can have the same values of all four quantum numbers** (n, l, m_l, m_s). Each

TABLE 43–2

QUANTUM STATES OF ELECTRONS IN THE FIRST FOUR SHELLS

n	l	m_l	SPECTROSCOPIC NOTATION	NUMBER OF STATES		SHELL
1	0	0	$1s$	2		K
2	0	0	$2s$	2	8	L
2	1	$-1, 0, 1$	$2p$	6		
3	0	0	$3s$	2	18	M
3	1	$-1, 0, 1$	$3p$	6		
3	2	$-2, -1, 0, 1, 2$	$3d$	10		
4	0	0	$4s$	2	32	N
4	1	$-1, 0, 1$	$4p$	6		
4	2	$-2, -1, 0, 1, 2$	$4d$	10		
4	3	$-3, -2, -1, 0, 1, 2, 3$	$4f$	14		

quantum state corresponds to a certain distribution of the electron "cloud" in space. Therefore the principle also says, in effect, that no more than two electrons with opposite values of the quantum number m_s can occupy the same region of space. We shouldn't take this last statement too seriously because the electron probability functions don't have sharp, definite boundaries. But the exclusion principle limits the amount that electron wave functions can overlap. Think of it as the quantum-mechanical analog of a university rule that allows only one student per desk.

CAUTION ▶ Don't confuse the exclusion principle with the electric repulsion between electrons. While both effects tend to keep electrons within an atom separated from each other, they are very different in character. Two electrons can always be pushed closer together by adding energy to combat electric repulsion; in contrast, *nothing* can overcome the exclusion principle and force two electrons into the same quantum-mechanical state. ◀

Table 43–2 lists some of the sets of quantum numbers for electron states in an atom. It's similar to Table 43–1 (Section 43–1), but we've added the number of states in each subshell and shell. Because of the exclusion principle, the "number of states" is the same as the *maximum* number of electrons that can be found in those states. For each state, m_s can be either $+\frac{1}{2}$ or $-\frac{1}{2}$.

As with the hydrogen wave functions, different states correspond to different spatial distributions; electrons with larger values of n are concentrated at larger distances from the nucleus. Figure 43–2 (page 1325) shows this effect. When an atom has more than two electrons, they can't all huddle down in the low-energy $n = 1$ states nearest to the nucleus because there are only two of these states; the exclusion principle forbids multiple occupancy of a state. Some electrons are forced into states farther away, with higher energies. Each value of n corresponds roughly to a region of space around the nucleus in the form of a spherical *shell.* Hence we speak of the K shell as the region that is occupied by the electrons in the $n = 1$ states, the L shell as the region of the $n = 2$ states, and so on. States with the same n but different l form *subshells,* such as the $3p$ subshell.

The key to understanding the periodic table of the elements was the discovery by Wolfgang Pauli (1900–1958) of the exclusion principle. Pauli received the 1945 Nobel Prize in physics for his accomplishment. This photo shows Pauli (on the left) and Niels Bohr watching a toy top spinning on the floor.

THE PERIODIC TABLE

We can use the exclusion principle to derive the most important features of the structure and chemical behavior of multielectron atoms, including the periodic table of the elements. Let's imagine constructing a neutral atom by starting with a bare nucleus with Z protons and adding Z electrons, one by one. To obtain the ground state of the atom as a whole, we fill the lowest-energy electron states (those closest to the nucleus, with the smallest values of n and l) first, and we use successively higher states until all the electrons are in place. The chemical properties of an atom are determined principally by

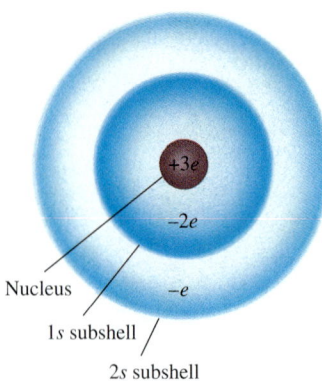

43–13 Schematic representation of charge distribution in a lithium atom. The nucleus has a charge of $3e$; the two $1s$ electrons are closer to the nucleus than is the $2s$ electron, which moves in a field roughly equal to that of a point charge of $3e - 2e = e$.

interactions involving the outermost, or *valence,* electrons, so we particularly want to learn how these electrons are arranged.

Let's look at the ground-state electron configurations for the first few atoms (in order of increasing Z). For hydrogen the ground state is $1s$; the single electron is in a state $n = 1$, $l = 0$, $m_l = 0$, and $m_s = \pm\frac{1}{2}$. In the helium atom ($Z = 2$), *both* electrons are in $1s$ states, with opposite spins; one has $m_s = -\frac{1}{2}$ and the other has $m_s = +\frac{1}{2}$. We denote the helium ground state as $1s^2$. (The superscript 2 is not an exponent; the notation $1s^2$ tells us that there are two electrons in the $1s$ subshell. Also, the superscript 1 is understood, as in $2s$.) For helium the K shell is completely filled, and all others are empty. Helium is a noble gas; it has no tendency to gain or lose an electron, and it forms no compounds.

Lithium ($Z = 3$) has three electrons. In its ground state, two are in $1s$ states and one is in a $2s$ state, so we denote the lithium ground state as $1s^2 2s$. On average, the $2s$ electron is considerably farther from the nucleus than are the $1s$ electrons, as shown schematically in Fig. 43–13. According to Gauss's law, the *net* charge Q_{encl} attracting the $2s$ electron is nearer to $+e$ than to the value $+3e$ it would have without the two $1s$ electrons present. As a result, the $2s$ electron is loosely bound; only 5.4 eV is required to remove it, compared with the 30.6 eV given by Eq. (43–25) with $Z = 3$ and $n = 2$. In chemical behavior, lithium is an *alkali metal*. It forms ionic compounds in which each lithium atom loses an electron and has a valence of $+1$.

Next is beryllium ($Z = 4$); its ground-state configuration is $1s^2 2s^2$, with its two valence electrons filling the s subshell of the L shell. Beryllium is the first of the *alkaline-earth* elements, forming ionic compounds in which the valence of the atoms is $+2$.

Table 43–3 shows the ground-state electron configurations of the first 30 elements. The L shell can hold eight electrons. At $Z = 10$, both the K and L shells are filled, and there are no electrons in the M shell. We expect this to be a particularly stable configuration, with little tendency to gain or lose electrons. This element is neon, a noble gas with no known compounds. The next element after neon is sodium ($Z = 11$), with filled K and L shells and one electron in the M shell. Its "noble-gas-plus-one-electron" structure resembles that of lithium; both are alkali metals. The element *before* neon is fluorine, with $Z = 9$. It has a vacancy in the L shell and has an affinity for an extra electron to fill the shell. Fluorine forms ionic compounds in which it has a valence of -1. This behavior is characteristic of the *halogens* (fluorine, chlorine, bromine, iodine, and astatine), all of which have "noble-gas-minus-one" configurations.

Proceeding down the list, we can understand the regularities in chemical behavior displayed by the **periodic table of the elements** (Appendix D) on the basis of electron configurations. The similarity of elements in each *group* (vertical column) of the periodic table is the result of similarity in outer electron configuration. All the noble gases (helium, neon, argon, krypton, xenon, and radon) have filled-shell or filled-shell plus filled p subshell configurations. All the alkali metals (lithium, sodium, potassium, rubidium, cesium, and francium) have "noble-gas-plus-one" configurations. All the alkaline-earth metals (beryllium, magnesium, calcium, strontium, barium, and radium) have "noble-gas-plus-two" configurations, and all the halogens (fluorine, chlorine, bromine, iodine, and astatine) have "noble-gas-minus-one" structures.

A slight complication occurs with the M and N shells because the $3d$ and $4s$ subshell levels ($n = 3$, $l = 2$, and $n = 4$, $l = 0$, respectively) have similar energies. (We'll discuss in the next subsection why this happens.) Argon ($Z = 18$) has all the $1s$, $2s$, $2p$, $3s$, and $3p$ subshells filled, but in potassium ($Z = 19$) the additional electron goes into a $4s$ energy state rather than a $3d$ state (because the $4s$ state has slightly lower energy).

The next several elements have one or two electrons in the $4s$ subshell and increasing numbers in the $3d$ subshell. These elements are all metals with rather similar chemical and physical properties; they form the first *transition series,* starting with

TABLE 43–3

GROUND-STATE ELECTRON CONFIGURATIONS

ELEMENT	SYMBOL	ATOMIC NUMBER (Z)	ELECTRON CONFIGURATION
Hydrogen	H	1	$1s$
Helium	He	2	$1s^2$
Lithium	Li	3	$1s^2 2s$
Beryllium	Be	4	$1s^2 2s^2$
Boron	B	5	$1s^2 2s^2 2p$
Carbon	C	6	$1s^2 2s^2 2p^2$
Nitrogen	N	7	$1s^2 2s^2 2p^3$
Oxygen	O	8	$1s^2 2s^2 2p^4$
Fluorine	F	9	$1s^2 2s^2 2p^5$
Neon	Ne	10	$1s^2 2s^2 2p^6$
Sodium	Na	11	$1s^2 2s^2 2p^6 3s$
Magnesium	Mg	12	$1s^2 2s^2 2p^6 3s^2$
Aluminum	Al	13	$1s^2 2s^2 2p^6 3s^2 3p$
Silicon	Si	14	$1s^2 2s^2 2p^6 3s^2 3p^2$
Phosphorus	P	15	$1s^2 2s^2 2p^6 3s^2 3p^3$
Sulfur	S	16	$1s^2 2s^2 2p^6 3s^2 3p^4$
Chlorine	Cl	17	$1s^2 2s^2 2p^6 3s^2 3p^5$
Argon	Ar	18	$1s^2 2s^2 2p^6 3s^2 3p^6$
Potassium	K	19	$1s^2 2s^2 2p^6 3s^2 3p^6 4s$
Calcium	Ca	20	$1s^2 2s^2 2p^6 3s^2 3p^6 4s^2$
Scandium	Sc	21	$1s^2 2s^2 2p^6 3s^2 3p^6 4s^2 3d$
Titanium	Ti	22	$1s^2 2s^2 2p^6 3s^2 3p^6 4s^2 3d^2$
Vanadium	V	23	$1s^2 2s^2 2p^6 3s^2 3p^6 4s^2 3d^3$
Chromium	Cr	24	$1s^2 2s^2 2p^6 3s^2 3p^6 4s 3d^5$
Manganese	Mn	25	$1s^2 2s^2 2p^6 3s^2 3p^6 4s^2 3d^5$
Iron	Fe	26	$1s^2 2s^2 2p^6 3s^2 3p^6 4s^2 3d^6$
Cobalt	Co	27	$1s^2 2s^2 2p^6 3s^2 3p^6 4s^2 3d^7$
Nickel	Ni	28	$1s^2 2s^2 2p^6 3s^2 3p^6 4s^2 3d^8$
Copper	Cu	29	$1s^2 2s^2 2p^6 3s^2 3p^6 4s 3d^{10}$
Zinc	Zn	30	$1s^2 2s^2 2p^6 3s^2 3p^6 4s^2 3d^{10}$

scandium ($Z = 21$) and ending with zinc ($Z = 30$), for which all the $3d$ and $4s$ subshells are filled.

Something similar happens with $Z = 57$ through $Z = 71$, which have one or two electrons in the $6s$ subshell but only partially filled $4f$ and $5d$ subshells. These are the *rare-earth* elements; they all have very similar physical and chemical properties. Yet another such series, called the *actinide* series, starts with $Z = 91$.

SCREENING

We have mentioned that in the central-field picture, the energy levels depend on l as well as n. Let's take sodium ($Z = 11$) as an example. If ten of its electrons fill its K and L shells, the energies of some of the states for the remaining electron are found experimentally to be

$3s$ states:	-5.138 eV,
$3p$ states:	-3.035 eV,
$3d$ states:	-1.521 eV,
$4s$ states:	-1.947 eV.

The $3s$ states are the lowest (most negative); one is the ground state for the eleventh electron in sodium. The energy of the $3d$ states is quite close to the energy of the $n = 3$ state in hydrogen. The surprise is that the 4s state energy is 0.426 eV *below* the 3d state, even though the 4s state has larger n.

We can understand these results quite well using Gauss's law and the radial probability distribution. For any spherically symmetric charge distribution the electric-field magnitude at a distance r from the center is $(1/4\pi\epsilon_0)Q_{encl}/r^2$, where Q_{encl} is the total charge enclosed within a sphere with radius r. Mentally remove the outer (valence) electron atom from a sodium atom. What you have left is a spherically symmetric collection of 10 electrons (filling the K and L shells) and 11 protons, so $Q_{encl} = -10e + 11e = +e$. If the eleventh electron is completely outside this collection of charges, it is attracted by an effective charge of $+e$, not $+11e$.

This effect is called **screening**; the 10 electrons *screen* 10 of the 11 protons, leaving an effective net charge of $+e$. In general, an electron that spends all its time completely outside a positive charge $Z_{eff}e$ has energy levels given by the hydrogen expression with e^2 replaced by $Z_{eff}e^2$. From Eq. (43–25) this is

$$E_n = -\frac{Z_{eff}^2}{n^2}(13.6\text{ eV}) \qquad \text{(energy levels with screening).} \qquad (43\text{–}27)$$

If the eleventh electron in the sodium atom is completely outside the remaining charge distribution, then $Z_{eff} = 1$.

CAUTION ▶ Equations (43–3), (43–25), and (43–27) all give values of E_n in terms of $(13.6\text{ eV})/n^2$, but they don't apply in general to the same atoms. Equation (43–3) is *only* for hydrogen, Eq. (43–25) is only for the case in which there is no interaction with any other electron (and is thus accurate only when the atom has just one electron), and Eq. (43–27) is useful when one electron is screened from the nucleus by other electrons. ◀

Now let's use the radial probability functions shown in Fig. 43–2 to explain why the energy of a sodium 3d state is approximately the same as the $n = 3$ value of hydrogen, −1.51 eV. The distribution for the 3d state (for which l has the maximum value $n − 1$) has one peak, and its most probable radius is *outside* the positions of the electrons with $n = 1$ or 2. (Those electrons also are pulled closer to the nucleus than in hydrogen because they are less effectively screened from the positive charge $11e$ of the nucleus.) Thus in sodium a 3d electron spends most of its time well outside the $n = 1$ and $n = 2$ states (the K and L shells). The 10 electrons in these shells screen almost all but one of the 11 protons, leaving a net charge of about $Z_{eff}e = (1)e$. Then, from Eq. (43–27), the corresponding energy is approximately $−(1)^2(13.6\text{ eV})/3^2 = −1.51\text{ eV}$. This approximation is very close to the experimental value of −1.521 eV.

Looking again at Fig. 43–2, we see that the radial probability density for the 3p state (for which $l = n − 2$) has two peaks and that for the 3s state ($l = n − 3$) has three peaks. For sodium the first small peak in the 3p distribution gives a 3p electron a higher probability (compared to the 3d state) of being *inside* the charge distributions for the electrons in the $n = 2$ states. That is, a 3p electron is less completely screened from the nucleus than is a 3d electron because it spends some of its time within the filled K and L shells. Thus for the 3p electrons, Z_{eff} is greater than unity. From Eq. (43–27) the 3p energy is lower (more negative) than the 3d energy of −1.521 eV. The actual value is −3.035 eV. A 3s electron spends even more time within the inner electron shells than a 3p electron does, giving an even larger Z_{eff} and an even more negative energy.

EXAMPLE 43-8

Determining Z_{eff} experimentally   The measured energy of a 3s state of sodium is −5.138 eV. Calculate the value of Z_{eff}.

SOLUTION   We use Eq. (43–27). Solving for Z_{eff}, we have

$$Z_{eff}^2 = -\frac{n^2 E_n}{13.6\text{ eV}} = -\frac{3^2(-5.138\text{ eV})}{13.6\text{ eV}} = 3.40,$$

$$Z_{eff} = 1.84.$$

The effective charge attracting a $3s$ electron is $1.84e$. Sodium's 11 protons are screened by an average of $11 − 1.84 = 9.16$ electrons instead of 10 electrons because of the penetration of the inner (K and L) shells by the $3s$ electron.

Each alkali metal (lithium, sodium, potassium, rubidium, and cesium) has one more electron than the corresponding noble gas (helium, neon, argon, krypton, and xenon). This extra electron is mostly outside the other electrons in the filled shells and subshells. Therefore all the alkali metals behave similarly to sodium.

Energies for a valence electron The valence electron in potassium has a $4s$ ground state. Calculate the approximate energy of the state having the smallest Z_{eff}, and discuss the relative energies of the $4s$, $4p$, $4d$, and $4f$ states.

SOLUTION A $4f$ state has $n = 4$ and $l = 3 = 4 − 1$. Thus it is the state of largest orbital angular momentum for $n = 4$, and thus the state in which the electron spends the most time outside the electron charge clouds of the inner filled shells and subshells. This makes Z_{eff} for a $4f$ state close to unity. Equation (43–27) then gives

$$E_4 = -\frac{Z_{eff}^2}{n^2}(13.6 \text{ eV}) = -\frac{1^2}{4^2}(13.6 \text{ eV}) = -0.85 \text{ eV}.$$

This approximation agrees with the measured energy to the precision given.

An electron in a $4d$ state spends a bit more time within the inner shells, and its energy is therefore a bit more negative (measured to be $−0.94$ eV). For the same reason, a $4p$ state has an even lower energy (measured to be $−2.73$ eV), and a $4s$ state has the lowest energy (measured to be $−4.339$ eV).

We can extend this analysis to the singly ionized alkaline earths: Be^+, Mg^+, Ca^+, Sr^+, and Ba^+. For any allowed value of n, the highest-l state ($l = n − 1$) of the one remaining outer electron sees an effective charge of almost $+2e$, so for these states, $Z_{eff} = 2$. A $3d$ state for Mg^+, for example, has an energy of about $−2^2(13.6 \text{ eV})/3^2 = −6.0$ eV.

43–6 X-RAY SPECTRA

X-ray spectra provide yet another example of the richness and power of the Schrödinger equation and of the model of atomic structure that we derived from it in the preceding section. In Section 40–8 we discussed x-ray production on the basis of the photon concept. With the development of x-ray diffraction techniques (Section 38–7) by von Laue, Bragg, and others, beginning in 1912, it became possible to measure x-ray wavelengths quite precisely (to within 0.1% or less).

Detailed studies of x-ray spectra showed a continuous spectrum of wavelengths (Fig. 43–14), with *minimum* wavelength (corresponding to *maximum* frequency and photon energy) determined by the accelerating voltage V_{AC} in the x-ray tube, according to the relation derived in Section 40–8 for *bremsstrahlung* processes:

$$\lambda_{min} = \frac{hc}{eV_{AC}}.$$

(43–28)

This continuous-spectrum radiation is nearly independent of the target material in the x-ray tube.

Depending on the accelerating voltage and the target element, we may find sharp peaks superimposed on this continuous spectrum, as in Fig. 43–15. These peaks are at different wavelengths for different elements; they form what is called a *characteristic x-ray spectrum* for each target element. In 1913, the British scientist H. G. J. Moseley studied these spectra in detail using x-ray diffraction techniques. He found that the most intense short-wavelength line in the characteristic x-ray spectrum from a particular target element, called the K_α line, varied smoothly with that element's atomic number Z (Fig. 43–16). This is in sharp contrast to optical spectra, in which elements with adjacent Z values have spectra that often bear no resemblance to each other.

Moseley found that the relationship could be expressed in terms of x-ray frequencies f by a simple formula called *Moseley's law:*

$$f = (2.48 \times 10^{15} \text{ Hz})(Z − 1)^2 \qquad \text{(Moseley's law).}$$

(43–29)

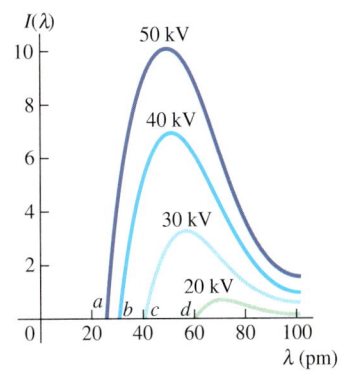

43–14 Graphs of x-ray intensity per unit wavelength as a function of wavelength, showing a continuous x-ray spectrum, for various accelerating voltages with a tungsten target. The minimum wavelength for each voltage is shown by the points a, b, c, and d. Compare these values of λ_{min} with the predictions of Eq. (43–28).

Moseley's discovery of this formula is comparable to Balmer's discovery of the empirical formula, Eq. (40–8), for the wavelengths in the spectrum of hydrogen.

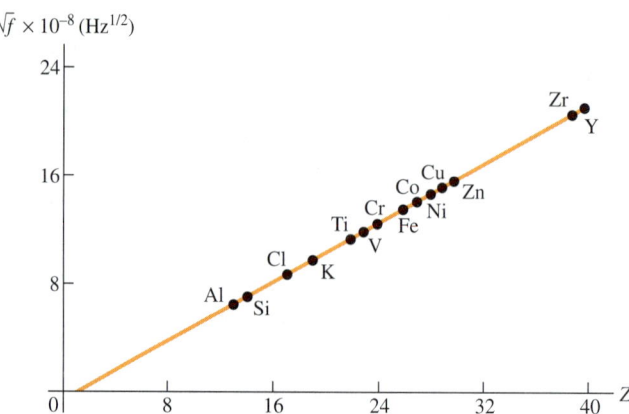

43–15 Graph of intensity per unit wavelength as a function of wavelength for x rays produced with an accelerating voltage of 35 kV and a molybdenum target. The curve is a smooth function similar to Fig. 43–14 with the addition of two sharp spikes corresponding to part of the characteristic x-ray spectrum for this element.

43–16 The square root of Moseley's measured frequencies of the K_α line for 14 elements. The graph of $\sqrt{f}$ versus Z is a straight line with an intercept at $Z = 1$, confirming Moseley's law, Eq. (43–29).

But Moseley went far beyond this empirical relationship; he showed how characteristic x-ray spectra could be understood on the basis of energy levels of atoms in the target. His analysis was based on the Bohr model, published in the same year. We will recast it somewhat, using the ideas of atomic structure discussed in Section 43–5. First recall that the *outer* electrons of an atom are responsible for optical spectra. Their excited states are usually only a few electron volts above their ground state. In transitions from excited states to the ground state, they usually emit photons in or near the visible region.

Characteristic x rays, by contrast, are emitted in transitions involving the *inner* shells of a complex atom. We mentioned these briefly in Section 40–8. In an x-ray tube the electrons may strike the target with enough energy to knock electrons out of the inner shells of the target atoms. These inner electrons are much closer to the nucleus than are the electrons in the outer shells; they are much more tightly bound, and hundreds or thousands of electron volts may be required to remove them.

Suppose one electron is knocked out of the K shell. This process leaves a vacancy, which we'll call a *hole*. (One electron remains in the K shell.) The hole can then be filled by an electron falling in from one of the outer shells, such as the $L, M, N, \ldots$ shell. This transition is accompanied by a decrease in the energy of the atom (because *less* energy would be needed to remove an electron from an $L, M, N, \ldots$ shell), and an x-ray photon is emitted with energy equal to this decrease. Each state has definite energy, so the emitted x rays have definite wavelengths; the emitted spectrum is a *line spectrum*.

We can estimate the energy and frequency of K_α x-ray photons using the concept of *screening*. A K_α x-ray photon is emitted when an electron in the L shell ($n = 2$) drops down to fill a hole in the K shell ($n = 1$). As the electron drops down, it is attracted by the Z protons in the nucleus screened by the one remaining electron in the K shell. We therefore approximate the energy by Eq. (43–27), with $Z_{eff} = Z - 1$, $n_i = 2$, and $n_f = 1$. The energy before the transition is

$$E_i \approx -(Z-1)^2 (13.6 \text{ eV})/2^2 = -(Z-1)^2 (3.4 \text{ eV}),$$

and the energy after the transition is

$$E_f \approx -(Z-1)^2 (13.6 \text{ eV})/1^2 = -(Z-1)^2 (13.6 \text{ eV}).$$

The energy of the K_α x-ray photon is $E_{K\alpha} = E_i - E_f \approx (Z-1)^2 (-3.4 \text{ eV} + 13.6 \text{ eV})$. That is,

$$E_{K\alpha} \approx (Z-1)^2 (10.2 \text{ eV}). \tag{43–30}$$

43–17 Wavelengths of the K_α, K_β, and K_γ lines of tungsten, molybdenum, and copper.

The frequency of any photon is its energy divided by Planck's constant,

$$f = \frac{E}{h} \approx \frac{(Z-1)^2 10.2 \text{ eV}}{4.136 \times 10^{-15} \text{ eV} \cdot \text{s}} = (2.47 \times 10^{15} \text{ Hz})(Z-1)^2.$$

This relation agrees almost exactly with Moseley's experimental law, Eq. (43–29). Indeed, considering the approximations we have made, the agreement is better than we have a right to expect. But our calculation does show how Moseley's law can be understood on the bases of screening and transitions between energy levels.

The hole in the K shell may also be filled by an electron falling from the M or N shell, assuming that these are occupied. If so, the x-ray spectrum of a large group of atoms of a single element shows a series, named the K series, of three lines, called the K_α, K_β, and K_γ lines. These three lines result from transitions in which the K-shell hole is filled by an L, M, or N electron, respectively. Figure 43–17 shows the K series for tungsten ($Z = 74$), molybdenum ($Z = 42$), and copper ($Z = 29$).

There are other series of x-ray lines, called the L, M, and N series that are produced after the ejection of electrons from the L, M, and N shells rather than the K shell. Electrons in these outer shells are farther away from the nucleus and are not held as tightly as are those in the K shell. Their removal requires less energy, and the x-ray photons that are emitted when these vacancies are filled have lower energy than those in the K series.

EXAMPLE 43-10

Chemical analysis by x-ray emission You measure the K_α wavelength for an unknown element, obtaining the value 0.0709 nm. What is the element?

SOLUTION The corresponding frequency is

$$f = \frac{c}{\lambda} = \frac{3.00 \times 10^8 \text{ m/s}}{0.0709 \times 10^{-9} \text{ m}} = 4.23 \times 10^{18} \text{ Hz}.$$

From Moseley's law, Eq. (43–29),

$$Z = 1 + \sqrt{\frac{f}{2.48 \times 10^{15} \text{ Hz}}} = 1 + \sqrt{\frac{4.23 \times 10^{18} \text{ Hz}}{2.48 \times 10^{15} \text{ Hz}}} = 42.3.$$

We know that Z has to be an integer; we conclude that $Z = 42$, corresponding to the element molybdenum.

We can also observe x-ray *absorption* spectra. Unlike optical spectra, the absorption wavelengths are usually not the same as those for emission, especially in many-electron atoms, and do not give simple line spectra. For example, the K_α emission line results from a transition from the L shell to a hole in the K shell. The reverse transition doesn't occur in atoms with $Z \geq 10$ because in the atom's ground state, there is no vacancy in the L shell. To be absorbed, a photon must have enough energy to move an electron to an empty state. Since empty states are only a few electron volts in energy below the free-electron continuum, the minimum absorption energies in many-electron atoms are about the same as the minimum energies that are needed to remove an electron from its shell. Experimentally, if we gradually increase the accelerating voltage and hence the maximum photon energy, we observe sudden increases in absorption when we reach these minimum energies. These sudden jumps of absorption are called *absorption edges* (Fig. 43–18).

Characteristic x-ray spectra provide a very useful analytical tool. Satellite-borne x-ray spectrometers are used to study x-ray emission lines from highly excited atoms in

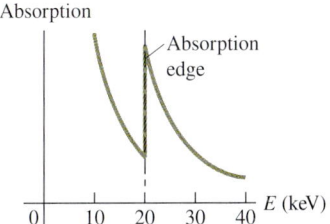

43–18 X-ray absorption by molybdenum. A beam of x rays is passed through a slab of molybdenum, and the absorption is measured for various energies. A sharp increase in absorption occurs at the K absorption edge, corresponding to excitation of an electron that is originally in the K shell 20 keV below the free-electron continuum.

distant astronomical sources. X-ray spectra are also used in air-pollution monitoring and in studies of the abundance of various elements in rocks.

SUMMARY

■ The Schrödinger equation for the hydrogen atom gives certain allowed energy levels

$$E_n = -\frac{1}{(4\pi\epsilon_0)^2}\frac{m_r e^4}{2n^2\hbar^2} = -\frac{13.60 \text{ eV}}{n^2}. \tag{43-3}$$

If the nucleus has charge Ze, there is also a factor Z^2 in the numerator.

■ The possible magnitudes of orbital angular momentum are

$$L = \sqrt{l(l+1)}\hbar \qquad (l = 0, 1, 2, \ldots, n-1). \tag{43-4}$$

The possible values of the z-component of orbital angular momentum are

$$L_z = m_l\hbar \qquad (m_l = 0, \pm1, \pm2, \ldots, \pm l). \tag{43-5}$$

■ The probability that the electron is between r and $r + dr$ from the nucleus is

$$P(r)\,dr = |\psi|^2\,dV = |\psi|^2\,4\pi r^2\,dr. \tag{43-7}$$

Atomic distances are often measured in units of the smallest distance between the electron and the nucleus in the Bohr model:

$$a = \frac{\epsilon_0 h^2}{\pi m_r e^2} = \frac{4\pi\epsilon_0\hbar^2}{m_r e^2} = 0.529 \times 10^{-10} \text{ m}. \tag{43-8}$$

■ The magnetic moment of the ground state in the Bohr model of hydrogen is called one Bohr magneton μ_B:

$$\mu_B = \frac{e\hbar}{2m}. \tag{43-13}$$

■ The interaction energy of an electron with magnetic quantum number m_l in a magnetic field B along the $+z$ direction is

$$U = -\mu_z B = m_l\frac{e\hbar}{2m}B \qquad (m_l = 0, \pm1, \pm2, \ldots, \pm l) \tag{43-17}$$

or

$$U = m_l\mu_B B. \tag{43-18}$$

■ The spin angular momentum of an electron has magnitude

$$S = \sqrt{\frac{1}{2}\left(\frac{1}{2}+1\right)}\hbar = \sqrt{\frac{3}{4}}\hbar \tag{43-20}$$

with z-component $S_z = m_s\hbar$, where $m_s = \pm\frac{1}{2}$:

$$S_z = \pm\frac{1}{2}\hbar. \tag{43-19}$$

■ In atoms the quantum numbers n, l, m_l, m_s have certain allowed values:

$$n \geq 1, \qquad 0 \leq l \leq n-1, \qquad |m_l| \leq l, \qquad m_s = \pm\frac{1}{2}. \tag{43-26}$$

■ In the central-field approximation, each electron moves in the electric field of the nucleus and of the averaged-out, spherically symmetric charge distribution of all the remaining electrons. The quantum numbers are the same as for hydrogen-atom states, but the energy levels depend on both n and l because of screening, the partial cancellation of the field of the nucleus by the inner electrons.

- If the effective charge attracting an electron is $Z_{eff}e$, the energies of the levels are approximately

$$E_n = -\frac{Z_{eff}^2}{n^2} (13.6 \text{ eV}). \qquad (43-27)$$

The value of Z_{eff} for a particular electron depends on n and l.

- Moseley's law states that the frequency of a K_α x ray from a target with atomic number Z is

$$f = (2.48 \times 10^{15} \text{ Hz})(Z - 1)^2. \qquad (43-29)$$

Characteristic x-ray spectra result from transitions to a hole in an inner energy level.

DISCUSSION QUESTIONS

Q43–1 Do gravitational forces play a significant role in atomic structure? Explain.

Q43–2 What are the most significant differences between the Bohr model of the hydrogen atom and the Schrödinger analysis? What are the similarities?

Q43–3 Why is the analysis of the helium atom much more complex than that of the hydrogen atom, either in a Bohr type of model or using the Schrödinger equation?

Q43–4 The Stern-Gerlach experiment is always performed with beams of *neutral* atoms. Wouldn't it be easier to form beams using *ionized* atoms? Why won't this work?

Q43–5 In the Stern-Gerlach experiment, why is it essential for the magnetic field to be *inhomogeneous* (that is, nonuniform)?

Q43–6 In the ground state of the helium atom one electron must have "spin down" and the other "spin up." Why?

Q43–7 An electron in a hydrogen atom is in an s level and the atom is in a magnetic field $\vec{B} \times B\hat{k}$. Explain why the "spin up" state ($m_s = +\frac{1}{2}$) has a higher energy than the "spin down" state ($m_s = -\frac{1}{2}$).

Q43–8 Example 43–4 uses the Bohr model to calculate a magnetic interaction energy of 1.16×10^{-5} eV for $B = 2.00$ T. Examples 43–5 and 43–7 also give 1.16×10^{-5} eV for the same magnetic field. Does that mean that the Bohr model is correct after all? Explain.

Q43–9 The central-field approximation is more accurate for alkali metals than for transition metals such as iron, nickel, or copper. Why?

Q43–10 Table 43–3 shows that for the ground state of the potassium atom, the outermost electron is in a $4s$ state. What does this tell you about the relative energies of the $3d$ and $4s$ levels for this atom? Explain.

Q43–11 A student asserted that any filled shell must have zero total angular momentum and hence must be spherically symmetric.

Do you believe this to be true? What about a filled *subshell*? Explain.

Q43–12 Why do the transition elements ($Z = 21$ to 30) all have similar chemical properties?

Q43–13 Use Table 43–3 to help determine the ground-state electron configuration of the neutral gallium atom (Ga), as well as the ions Ga$^+$ and Ga$^-$. Gallium has an atomic number of 31.

Q43–14 On the basis of the Pauli exclusion principle, the structure of the periodic table of the elements shows that there must be a fourth quantum number in addition to n, l, and m_l. Explain.

Q43–15 The ionization energies of the alkali metals (that is, the smallest energy required to remove one outer electron when the atom is in its ground state) are about 4 or 5 eV, while those of the noble gases are in the range from 11 to 25 eV. Why the difference?

Q43–16 The energy required to remove the $3s$ electron from a sodium atom in its ground state is about 5 eV. Would you expect the energy required to remove an additional electron to be about the same, or more, or less? Why?

Q43–17 The nucleus of a gold atom contains 79 protons. How does the energy required to remove a $1s$ electron completely from a gold atom compare with the energy required to remove the electron from the ground level in a hydrogen atom? In what region of the electromagnetic spectrum would a photon with this energy for each of these two atoms lie?

Q43–18 What is the basic distinction between x-ray energy levels and ordinary energy levels?

Q43–19 An atom in its ground level absorbs a photon with energy equal to the K absorption edge. Does absorbing this photon ionize this atom? Explain.

Q43–20 Can a hydrogen atom emit x rays? If so, how? If not, why not?

EXERCISES

SECTION 43–2 THE HYDROGEN ATOM

43–1 The orbital angular momentum of an electron has a magnitude of 4.716×10^{-34} kg · m²/s. What is the angular-momentum quantum number l for this electron?

43–2 Consider states with angular-momentum quantum number $l = 2$. a) In units of $\hbar$, what is the largest possible value of L_z? b) In units of $\hbar$, what is the value of L? Which is larger, L or the maximum possible L_z? c) For each allowed value of L_z, what angle does the vector $\vec{L}$ make with the $+z$-axis? How does the minimum angle for $l = 2$ compare to the minimum angle for $l = 3$ calculated in Example 43–2 (Section 43–2)?

43–3 Calculate, in units of $\hbar$, the magnitude of the maximum orbital angular momentum for an electron in a hydrogen atom for states with a principal quantum number of 2, 20, and 200. Compare each with the value of $n\hbar$ postulated in the Bohr model. What trend do you see?

43–4 a) Make a chart showing all the possible sets of quantum numbers l and m_l for the states of the electron in the hydrogen atom when $n = 5$. How many combinations are there? b) What are the energies of these states?

43–5 Step 2 of the Problem-Solving Strategy in Section 43–2 claims that the electric potential energy of a proton and an electron 0.10 nm apart has magnitude 14.4 eV. Verify this claim.

43–6 a) What is the probability that an electron in the $1s$ state of a hydrogen atom will be found at a distance less than $a/2$ from the nucleus? b) Use the results of part (a) and of Example 43–3 (Section 43–2) to calculate the probability that the electron will be found at distances between $a/2$ and a from the nucleus.

43–7 a) For the wave function $\psi(r, \theta, \phi) = R(r)\Theta(\theta)\Phi(\phi)$ with $\Phi(\phi) = Ae^{im_l\phi}$, show that $|\psi|^2$ is independent of ϕ. b) What value must A have if $\Phi(\phi)$ is to satisfy the normalization condition $\int_0^{2\pi} |\Phi(\phi)|^2 \, d\phi = 1$?

43–8 For ordinary hydrogen, the reduced mass of the electron and proton is $m_r = 0.99946m$, where m is the electron mass (Section 40–6). For each of the following cases, find the numerical coefficient of $-1/n^2$ in Eq. (43–3) and the energy of the photon emitted in a transition from the $n = 2$ level to the $n = 1$ level. a) A hydrogen atom in which the nucleus is taken to be infinitely massive, so $m_r = m$. b) Positronium (Section 40–6), for which $m_r = m/2$ exactly. c) Muonium, an atom composed of a muon and a proton (see Problem 40–49), for which $m_r = 185.8m$.

43–9 Find the numerical value of a in Eq. (43–8) for a) a hydrogen atom in which the nucleus is taken to be infinitely massive, so $m_r = m$; b) positronium (Section 40–6), for which $m_r = m/2$ exactly; c) muonium, an atom composed of a muon and a proton (see Problem 40–49), for which $m_r = 185.8m$. In part (a), a becomes a_0, the Bohr radius.

43–10 Show that $\Phi(\phi) = e^{im_l\phi} = \Phi(\phi + 2\pi)$ (that is, show that $\Phi(\phi)$ is periodic with period 2π) if and only if m_l is restricted to the values $0, \pm1, \pm2, \ldots$. (*Hint:* Euler's formula states that $e^{i\phi} = \cos\theta + i\sin\theta$.)

43–11 In Example 43–3 (Section 43–2), fill in the missing details that show that $P = 1 - 5e^{-2}$.

SECTION **43–3 THE ZEEMAN EFFECT**

43–12 A hydrogen atom is in a d state. In the absence of an external magnetic field the states with different m_l have (approximately) the same energy. Consider the interaction of the magnetic field with the atom's orbital magnetic dipole moment. a) Calculate the splitting in electron volts of the m_l levels when the atom is put in a 0.400-T magnetic field that is in the $+z$-direction. b) Which m_l level will have the lowest energy? c) Draw an energy-level diagram that shows the d levels with and without the external magnetic field.

43–13 A hydrogen atom in the $5g$ state is placed in a magnetic field of 0.600 T that is in the z-direction. a) Into how many levels is this state split by the interaction of the atom's orbital magnetic dipole moment with the magnetic field? b) What is the energy separation between adjacent levels? c) What is the energy separation between the level of lowest energy and the level of highest energy?

43–14 a) In the Bohr model, what is the magnitude of the magnetic moment of a hydrogen atom in the $n = 3$ state? b) Use the result of part (a) to calculate in the Bohr model the magnetic interaction energy when a hydrogen atom in the $n = 3$ state is placed in a magnetic field with magnitude 1.20 T if the magnetic moment of the atom and the magnetic field are antiparallel.

43–15 A hydrogen atom in a $3p$ state is placed in a uniform external magnetic field $\vec{B}$. Consider the interaction of the magnetic field with the atom's orbital magnetic dipole moment. a) What field magnitude B is required to split the $3p$ state into multiple levels with an energy difference of 2.71×10^{-5} eV between adjacent levels? b) How many levels will there be?

SECTION **43–4 ELECTRON SPIN**

43–16 A hydrogen atom in the $n = 1$, $m_s = -\frac{1}{2}$ state is placed in a magnetic field with a magnitude of 0.480 T in the $+z$-direction. a) Find the magnetic interaction energy (in electron volts) of the electron with the field. b) Is there any orbital magnetic dipole moment interaction for this state? Explain. Can there be an orbital magnetic dipole moment interaction for $n \neq 1$?

43–17 Calculate the energy difference between the $m_s = \frac{1}{2}$ ("spin up") and $m_s = -\frac{1}{2}$ ("spin down") levels of a hydrogen atom in the $1s$ state when it is placed in a 1.45-T magnetic field in the *negative* z-direction. Which level, $m_s = \frac{1}{2}$ or $m_s = -\frac{1}{2}$, has the lower energy?

43–18 List the different possible combinations of l and j for a hydrogen atom in the $n = 3$ level.

43–19 A hydrogen atom in a particular orbital angular momentum state is found to have j quantum numbers $\frac{7}{2}$ and $\frac{9}{2}$. What is the letter that labels the value of l for the state?

43–20 The hyperfine interaction in a hydrogen atom between the magnetic dipole moment of the proton and the spin magnetic dipole moment of the electron splits the ground level into two levels separated by 5.9×10^{-6} eV. a) Calculate the wavelength and frequency of the photon emitted when the atom makes a transition between these states and compare your answer to the value given at the end of Section 43–4. In what part of the electromagnetic spectrum does this lie? Such photons are emitted by cold hydrogen clouds in interstellar space; by detecting these photons, astronomers can learn about the number and density of such clouds. b) Calculate the effective magnetic field experienced by the electron in these states (see Fig. 43–12). Compare your result to the effective magnetic field due to the spin-orbit coupling calculated in Example 43–7.

43–21 Classical Electron Spin. a) If you treat an electron as a classical spherical object with a radius of 1.0×10^{-17} m, what angular speed is necessary to produce a spin angular momentum of magnitude $\sqrt{\frac{3}{4}}\hbar$? b) Use $v = r\omega$ and the result of part (a) to calculate the speed v of a point at the electron's equator. What does your result suggest about the validity of this model?

SECTION 43–5 MANY-ELECTRON ATOMS AND THE EXCLUSION PRINCIPLE

43–22 For magnesium the first ionization potential is 7.6 eV. The second ionization potential (additional energy required to remove a second electron) is almost twice this, 15 eV, and the third ionization potential is much larger, about 80 eV. How can these numbers be understood?

43–23 The 5s electron in rubidium (Rb) sees an effective charge of 2.771e. Calculate the ionization energy of this electron.

43–24 For germanium (Ge, $Z = 32$), make a list of the number of electrons in each subshell ($1s$, $2s$, $2p$, ...). Use the allowed values of the quantum numbers along with the exclusion principle; do *not* refer to Table 43–3.

43–25 Make a list of the four quantum numbers n, l, m_l, and m_s for each of the ten electrons in the ground state of the neon atom. Do *not* refer to Tables 43–2 or 43–3.

43–26 The energies of the $4s$, $4p$, and $4d$ levels of potassium are given in Example 43–9 (Section 43–5). Calculate Z_{eff} for each state. What trend does your results show? How can you explain this trend?

43–27 a) The doubly charged ion N^{2+} is formed by removing two electrons from a nitrogen atom. What is the ground-state electron configuration for the N^{2+} ion? b) Estimate the energy of the least strongly bound level in the L shell of N^{2+}. c) The doubly charged ion P^{2+} is formed by removing two electrons from a phosphorus atom. What is the ground-state electron configuration for the P^{2+} ion? d) Estimate the energy of the least strongly bound level in the M shell of P^{2+}.

43–28 a) The energy of the $2s$ state of lithium is −5.391 eV. Calculate the value of Z_{eff} for this state. b) The energy of the $4s$ state of potassium is −4.339 eV. Calculate the value of Z_{eff} for this state. c) Compare Z_{eff} for the $2s$ state of lithium, the $3s$ state of sodium (Example 43–8 in Section 43–5), and the $4s$ state of potassium. What trend do you see? How can you explain this trend?

43–29 Estimate the energy of the highest-l state for a) the L shell of Be^+; b) the N shell of Ca^+.

SECTION 43–6 X-RAY SPECTRA

43–30 A K_α x ray emitted from a sample has an energy of 7.46 keV. Of which element is the sample made?

43–31 Calculate the frequency, energy (in keV), and wavelength of the K_α x ray for the elements a) calcium (Ca, $Z = 20$); b) cobalt (Co, $Z = 27$); c) cadmium (Cd, $Z = 48$).

PROBLEMS

43–32 For a hydrogen atom the probability $P(r)$ of finding the electron within a spherical shell with inner radius r and outer radius $r + dr$ is given by Eq. (43–7). For a hydrogen atom in the $1s$ ground state, at what value of r does $P(r)$ have its maximum value? How does your result compare to the distance between the electron and the nucleus for the $n = 1$ state in the Bohr model (Eq. 43–8)?

43–33 Consider a hydrogen atom in the $1s$ state. a) For what value of r is the potential energy $U(r)$ equal to the total energy E? Express your answer in terms of a. This value of r is called the *classical turning point*, since this is where a Newtonian particle would stop its motion and reverse direction. b) For r greater than the classical turning point, $U(r) > E$. Classically, the particle cannot be in this region since the kinetic energy cannot be negative. Calculate the probability of the electron being found in this classically forbidden region.

43–34 Rydberg Atoms. Rydberg atoms are atoms whose outermost electron is in an excited state with a *very* large principal quantum number. Rydberg atoms have been produced in the laboratory and detected in interstellar space. a) Why do all neutral Rydberg atoms with the same n value have essentially the same ionization energy, independent of the total number of electrons in the atom? b) What is the ionization energy for a Rydberg atom with a principal quantum number of 350? What is the radius in the Bohr model of the Rydberg electron's orbit? c) Repeat part (b) for $n = 650$.

43–35 The wave function for a hydrogen atom in the $2s$ state is

$$\psi_{2s}(r) = \frac{1}{\sqrt{32\pi a^3}}\left(2 - \frac{r}{a}\right)e^{-r/2a}.$$

a) Verify that this function is normalized. b) In the Bohr model, the distance between the electron and the nucleus in the $n = 2$ state is exactly $4a$. Calculate the probability that an electron in the $2s$ state will be found at a distance less than $4a$ from the nucleus.

43–36 The normalized wave function for a hydrogen atom in the $2s$ state is given in Problem 43–35. a) For a hydrogen atom in the $2s$ state, at what value of r is $P(r)$ maximum? How does your result compare to $4a$, the distance between the electron and the nucleus in the $n = 2$ state of the Bohr model? b) At what value of r (other than $r = 0$ or $r = \infty$) is $P(r)$ equal to zero, so that the probability of finding the electron at that separation from the nucleus is zero? Compare your result to Fig. 43–3.

43–37 a) For an excited state of hydrogen, show that the smallest angle that the orbital angular momentum vector $\vec{L}$ can have with the z-axis is

$$(\theta_L)_{min} = \arccos\left(\frac{n-1}{\sqrt{n(n-1)}}\right).$$

b) What is the corresponding expression for $(\theta_L)_{max}$, the largest possible angle between $\vec{L}$ and the z-axis?

43–38 a) If the value of L_z is known, we cannot know either L_x or L_y precisely. But we *can* know the value of the quantity $\sqrt{L_x^2 + L_y^2}$. Write an expression for this quantity in terms of l, m_l, and $\hbar$. b) What is the meaning of $\sqrt{L_x^2 + L_y^2}$? c) For a state of nonzero orbital angular momentum, find the maximum and minimum values of $\sqrt{L_x^2 + L_y^2}$. Explain your results.

43–39 Consider the transition from a $3d$ to a $2p$ state of hydrogen in an external magnetic field. Assume that the effects of electron spin can be neglected (which is not actually the case) so that the magnetic field interacts only with the orbital angular momentum. Identify each allowed transition by the m_l values of the initial and final states. For each of these allowed transitions, determine the shift of the transition energy from the zero-field value and show that there are three different transition energies.

43–40 A hydrogen atom makes a transition from a $n = 3$ state to a $n = 2$ state (the Balmer H_α line) while in a magnetic field in the $+z$-direction and with magnitude 1.40 T. a) If the magnetic quantum number is $m_l = 2$ in the initial ($n = 3$) state and $m_l = 1$ in the final ($n = 2$) state, by how much is each energy level shifted from the zero-field value? b) By how much is the wavelength of the H_α line shifted from the zero-field value? Is the wavelength increased or decreased? Disregard the effect of electron spin. (*Hint:* Use the result of Problem 41–44(c).)

43–41 A large number of hydrogen atoms in $1s$ states are placed in an external magnetic field that is in the $+z$-direction. Assume that the atoms are in thermal equilibrium at room temperature, $T = 300$ K. According to the Boltzmann distribution (Section 40–7), what is the ratio of the number of atoms in the $m_s = \frac{1}{2}$ state to the number in the $m_s = -\frac{1}{2}$ when the magnetic field magnitude is a) 5.00×10^{-5} T (approximately the earth's field); b) 0.500 T; c) 5.00 T?

43–42 For an ion with nuclear charge Z and a single electron, the electric potential energy is $-Ze^2/4\pi\epsilon_0 r$ and the expression for the energies of the states and for the normalized wave functions are obtained from those for hydrogen by replacing e^2 by Ze^2. Consider the N^{6+} ion, with seven protons and one electron. a) What is the ground-state energy in electron volts? b) What is the ionization energy, the energy required to remove the electron from the N^{6+} ion if it is initially in the ground state? c) What is the distance a (given for hydrogen by Eq. (43–8)) for this ion?

d) What is the wavelength of the photon emitted when the N^{6+} ion makes a transition from the $n = 2$ state to the $n = 1$ ground state?

43–43 A hydrogen atom in an $n = 2$, $l = 1$, $m_l = -1$ state emits a photon when it decays to an $n = 1$, $l = 0$, $m_l = 0$ ground state. a) In the absence of an external magnetic field, what is the wavelength of this photon? b) If the atom is in a magnetic field in the $+z$-direction and with a magnitude of 2.20 T, what is the shift in the wavelength of the photon from the zero-field value? Does the magnetic field increase or decrease the wavelength? Disregard the effect of electron spin. (*Hint:* Use the result of Problem 41–44(c).)

43–44 A lithium atom has three electrons, and the $^2S_{1/2}$ ground-state electron configuration is $1s^2 2s$. The $1s^2 2p$ excited state is split into two closely spaced levels, $^2P_{3/2}$ and $^2P_{1/2}$, by the spin-orbit interaction (Example 43–7 in Section 43–4). A photon with wavelength 67.09608 μm is emitted in the $^2P_{3/2} \to {}^2S_{1/2}$ transition, and a photon with wavelength 67.09761 μm is emitted in the $^2P_{1/2} \to {}^2S_{1/2}$ transition. Calculate the effective magnetic field seen by the electron in the $1s^2 2p$ state of the lithium atom. How does your result compare to that for the $3p$ level of sodium found in Example 43–8?

43–45 Estimate the minimum and maximum wavelengths of the characteristic x rays emitted by a) vanadium ($Z = 23$); b) rhenium ($Z = 45$). Discuss any approximations that you make.

43–46 Electron Spin Resonance. Electrons in the lower of two spin states in a magnetic field can absorb a photon of the right frequency and move to the higher state. a) Find the magnetic field magnitude B required for this transition in a hydrogen atom with $n = 1$ and $l = 0$ to be induced by microwaves with wavelength λ. b) Calculate the value of B for a wavelength of 3.50 cm.

43–47 a) Show that the total number of atomic states (including different spin states) in a shell of principal quantum number n is $2n^2$. (*Hint:* The sum of the first N integers $1 + 2 + 3 + \cdots + N$ is given by $N(N + 1)/2$.) b) Which shell has 50 states?

CHALLENGE PROBLEMS

43–48 Consider a simple model of the helium atom in which two electrons, each of mass m, move around the nucleus (charge $+2e$) in the same circular orbit. Each electron has orbital angular momentum $\hbar$ (that is, the orbit is the smallest-radius Bohr orbit), and the two electrons are always on opposite sides of the nucleus. Ignore the effects of spin. a) Determine the radius of the orbit and the orbital speed of each electron. (*Hint:* Follow the procedure used in Section 40–6 to derive Eqs. (40–12) and (40–13). Each electron experiences an attractive force from the nucleus and a repulsive force from the other electron.) b) What is the total kinetic energy of the electrons? c) What is the potential energy of the system (the nucleus and the two electrons)? d) In this model, how much energy is required to remove both electrons to infinity? How does this compare to the experimental value of 79.0 eV?

43–49 Each of $2N$ electrons (mass m) is free to move along the x-axis. The potential-energy function for each electron is $U(x) = \frac{1}{2} k'x^2$, where k' is a positive constant. The electric and magnetic interactions between electrons can be ignored. Use the exclusion principle to show that the minimum energy of the system of $2N$ electrons is $\hbar N^2 \sqrt{k'/m}$. (*Hint:* See Section 42–6 and the hint given in Problem 43–47.)

43–50 Repeat the calculation of Problem 43–33 for a one-electron ion with nuclear charge Z. (See Problem 43–42.) How does the probability of the electron being found in the classically forbidden region depend on Z?

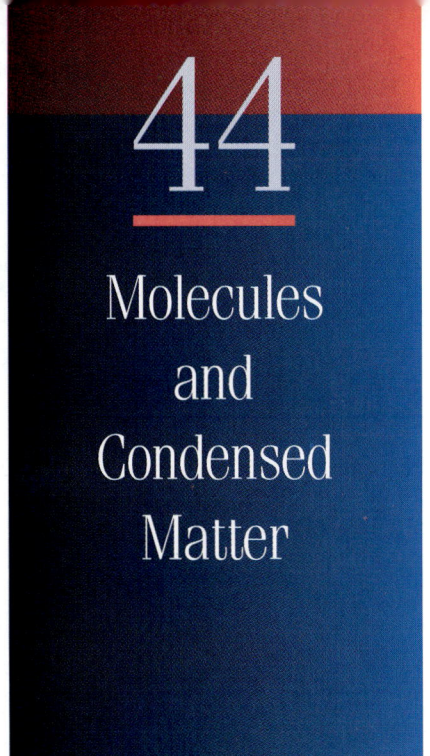

ozone abundance

Ozone molecules (O_3) absorb ultraviolet radiation, but ordinary oxygen molecules (O_2) do not. Ultraviolet radiation from the sun is extremely harmful to living beings, so there is great alarm over the loss of ozone in our atmosphere due to man-made chemicals. These images show how the ozone abundance over Antarctica decreased between 1979 (left) and 1998 (right).

44-1 INTRODUCTION

Can you *design* a material with any particular set of physical properties you happen to need at the moment? To do that, you have to understand how the microscopic structure of matter determines its bulk behavior; transparent or opaque, conductor or insulator, hard or soft, gas, liquid, or solid. This understanding is to be found in the nature of interatomic interactions.

In Chapter 43 we discussed the structure and properties of isolated atoms. But such atoms are the exception; usually, we find atoms combined to form molecules or more extended structures we call condensed matter (liquid or solid). It's the attractive forces between atoms, called molecular bonds, that causes them to combine. In this chapter we'll study several kinds of bonds as well as the energy levels and spectra associated with diatomic molecules. Then we'll apply the same principles to the study of condensed matter, in which various types of bonding occur. We'll explore the concept of energy bands and see how it helps understanding the properties of solids. Then we'll look more closely at the properties of a special class of solids called semiconductors. Devices using semiconductors are found in every radio, TV, pocket calculator, and computer used today; they have revolutionized the entire field of electronics during the past half-century.

44-2 TYPES OF MOLECULAR BONDS

We can use our discussion of atomic structure in Chapter 43 as a basis for exploring the nature of *molecular bonds,* the interactions that hold atoms together to form stable structures such as molecules and solids.

IONIC BONDS

The **ionic bond,** also called the *electrovalent* or *heteropolar* bond, is an interaction between oppositely charged *ionized* atoms. The most familiar example is sodium chloride (NaCl), in which the sodium atom gives its one $3s$ electron to the chlorine atom, filling the vacancy in the $3p$ subshell of chlorine.

Let's look at the energy balance in this transaction. Removing the $3s$ electron from a neutral sodium atom requires 5.138 eV of energy; this is called the *ionization energy* or *ionization potential* of sodium. The neutral chlorine atom can attract an extra electron into the vacancy in the $3p$ subshell, where it is incompletely screened by the other electrons and therefore is attracted to the nucleus. This state has 3.613 eV lower energy than a neutral chlorine atom and a distant free electron; 3.613 eV is the magnitude of the *electron affinity* of chlorine. Thus creating the well-separated Na^+ and Cl^- ions requires a net investment of only 5.138 eV − 3.613 eV = 1.525 eV. When the two oppositely charged ions are brought together by their mutual attraction, the magnitude of their negative potential energy is determined by how closely they can approach

44–1 At large values of the separation r between two oppositely charged ions, (a) the attractive interaction force $F(r)$ between two oppositely charged ions is proportional to $1/r^2$, and (b) the potential energy $U(r)$ is proportional to l/r, as for point charges. As r decreases, the charge clouds overlap, and the force becomes less attractive. The force becomes repulsive when r is less than the equilibrium separation r_0.

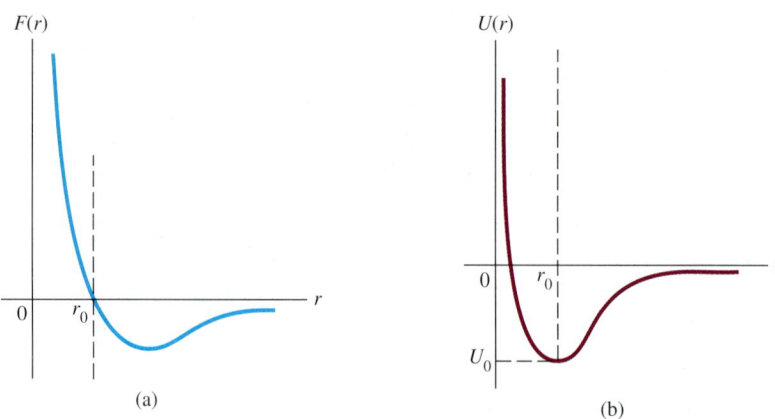

(a)   (b)

each other. This in turn is limited by the exclusion principle, which forbids extensive overlap of the electron clouds of the two ions. As the distance decreases, the exclusion principle distorts the charge clouds, so the ions no longer interact like point charges and the interaction eventually becomes repulsive (Fig. 44–1).

The minimum electrical potential energy for NaCl turns out to be -5.7 eV at a separation of 0.24 nm. The net energy released in creating the ions and letting them come together to the equilibrium separation of 0.24 nm is 5.7 eV $-$ 1.525 eV = 4.2 eV. Thus, if the kinetic energy of the ions is neglected, 4.2 eV is the *binding energy* of the NaCl molecule, the energy that is needed to dissociate the molecule into separate neutral atoms.

EXAMPLE 44–1

Electric potential energy of the NaCl molecule Find the electric potential energy of Na$^+$ and Cl$^-$ ions separated by 0.24 nm if they can be treated as point charges.

SOLUTION Using Eq. (24–9), Section 24–2, the electric potential energy of two point charges at this separation distance is

$$U = -\frac{1}{4\pi\epsilon_0}\frac{e^2}{r_0} = -(9.0 \times 10^9 \text{ N} \cdot \text{m}^2/\text{C}^2)\frac{(1.6 \times 10^{-19}\text{ C})^2}{0.24 \times 10^{-9}\text{ m}}$$

$$= -9.6 \times 10^{-19} \text{ J} = -6.0 \text{ eV}.$$

This agrees fairly well with the observed value of -5.7 eV. The difference is because at the equilibrium separation the ions don't behave exactly like point charges; at this separation the electron clouds of the two ions are overlapping.

COVALENT BONDS

Ionic bonds are interactions between charge distributions that are nearly spherically symmetric; hence they are not highly directional. They can involve more than one electron per atom. The alkaline-earth elements form ionic compounds in which an atom loses *two* electrons; an example is $Mg^{2+}(Cl^-)_2$. Loss of more than two electrons is relatively rare; instead, a different kind of bond comes into operation.

The **covalent bond** is characterized by a more egalitarian participation of the two atoms than occurs with the ionic bond. The simplest covalent bond is found in the hydrogen molecule, a structure containing two protons and two electrons. This bond is shown schematically in Fig. 44–2. As the separate atoms (Fig. 44–2a) come together, the electron wave functions are distorted and become more concentrated in the region between the two protons (Fig. 44–2b). The net attraction of the electrons for each proton more than balances the repulsion of the two protons and of the two electrons.

The attractive interaction is then supplied by a *pair* of electrons, one contributed by each atom, with charge clouds that are concentrated primarily in the region between the two atoms. This bond is also called a *homopolar, shared-electron,* or *electron-pair bond.* The energy of the covalent bond in the hydrogen molecule H_2 is -4.48 eV.

As we saw in the preceding chapter, the exclusion principle permits two electrons to occupy the same region of space (that is, to be in the same spatial quantum state) only when they have opposite spins. When the spins are parallel, the exclusion principle forbids the molecular state that would be most favorable from energy considerations (with both electrons in the region between atoms). Opposite spins are an essential requirement for a covalent bond, and no more than two electrons can participate in such a bond.

However, an atom with several electrons in its outermost shell can form several covalent bonds. The bonding of carbon and hydrogen atoms, of central importance in organic chemistry, is an example. In the *methane* molecule (CH_4) the carbon atom is at the center of a regular tetrahedron, with a hydrogen atom at each corner. The carbon atom has four electrons in its L shell, and each of these four electrons forms a covalent bond with one of the four hydrogen atoms, as shown in Fig. 44–3. Similar patterns occur in more complex organic molecules.

Because of the role played by the exclusion principle, covalent bonds are highly directional. In the methane molecule the wave function for each of carbon's four valence electrons is a combination of the $2s$ and $2p$ wave functions called a *hybrid wave function.* The probability distribution for each one has a lobe protruding toward a corner of a tetrahedron. This symmetric arrangement minimizes the overlap of wave functions for the electron pairs, minimizing their repulsive potential energy.

Ionic and covalent bonds represent two extremes in molecular bonding, but there is no sharp division between the two types. Often there is a *partial* transfer of one or more electrons from one atom to another. As a result, many molecules that have dissimilar atoms have electric dipole moments, that is, a preponderance of positive charge at one end and of negative charge at the other. Such molecules are called *polar* molecules. Water molecules have large electric dipole moments; these are responsible for the exceptionally large dielectric constant of liquid water.

VAN DER WAALS BONDS

Ionic and covalent bonds, with typical bond energies of 1 to 5 eV, are called *strong bonds.* There are also two types of weaker bonds. One of these, the **van der Waals bond,** is an interaction between the electric dipole moments of atoms or molecules; typical energies are 0.1 eV or less. The bonding of water molecules in the liquid and solid state results partly from dipole-dipole interactions.

No atom has a permanent electric dipole moment, nor do many molecules. However, fluctuating charge distributions can lead to fluctuating dipole moments; these in turn can induce dipole moments in neighboring structures. Overall, the resulting dipole-dipole interaction is attractive, giving a weak bonding of atoms or molecules. The interaction potential energy drops off very quickly with distance r between molecules, usually as $1/r^6$. The liquefaction and solidification of the inert gases and of molecules such as H_2, O_2, and N_2 is due to induced-dipole van der Waals interactions. Not much thermal-agitation energy is needed to break these weak bonds, so such substances usually exist in the liquid and solid states only at very low temperatures.

HYDROGEN BONDS

In another type of weak bond, the **hydrogen bond,** a proton (H^+ ion) gets between two atoms, polarizing them and attracting them by means of the induced dipoles. This bond is unique to hydrogen-containing compounds because only hydrogen has a singly ionized state with no remaining electron cloud; the hydrogen ion is a bare proton, much smaller than any other singly ionized atom. The bond energy is usually less than 0.5 eV. The hydrogen bond plays an essential role in many organic molecules, including the

(a)

(b)

44–2 Covalent bond in a hydrogen molecule. (a) Two separate hydrogen atoms. (b) The covalent bond; the charge clouds for the two electrons with opposite spins are concentrated in the region between the nuclei (protons).

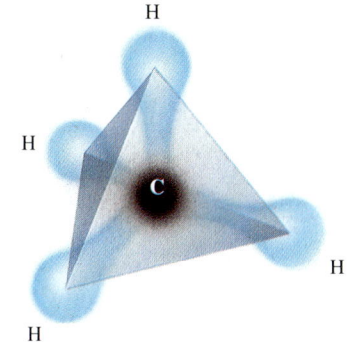

44–3 Schematic diagram of the methane (CH_4) molecule. The carbon atom is at the center of a regular tetrahedron and forms four covalent bonds with the hydrogen atoms at the corners. Each covalent bond includes two electrons with opposite spins, forming a charge cloud that is concentrated between the carbon atom and a hydrogen atom.

cross-linking of polymer chains such as polyethylene and cross-link bonding between the two strands of the double-helix DNA molecule. Hydrogen bonding also plays a role in the structure of ice.

All these bond types hold the atoms together in *solids* as well as in molecules. Indeed, a solid is in many respects a giant molecule. Still another type of bonding, the *metallic bond,* comes into play in the structure of metallic solids. We'll return to this subject in Section 44–4.

44–3 MOLECULAR SPECTRA

Molecules have energy levels that are associated with rotational motion of a molecule as a whole and with vibrational motion of the atoms relative to each other. Just as transitions between energy levels in atoms lead to atomic spectra, transitions between rotational and vibrational levels in molecules lead to *molecular spectra.*

ROTATIONAL ENERGY LEVELS

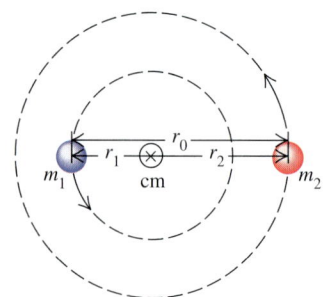

In this discussion we'll concentrate mostly on *diatomic* molecules, to keep things as simple as possible. In Fig. 44–4 we picture a diatomic molecule as a rigid dumbbell (two point masses m_1 and m_2 separated by a constant distance r_0) that can *rotate* about axes through its center of mass, perpendicular to the line joining them. What are the energy levels associated with this motion?

We showed in Section 10–6 that when a rigid body rotates with angular speed ω about a perpendicular axis through its center of mass, the magnitude L of its angular momentum is given by Eq. (10–31), $L = I\omega$, where I is its moment of inertia about that symmetry axis. Its kinetic energy is given by Eq. (9–17), $K = \frac{1}{2}I\omega^2$. Combining these two equations, we find $K = L^2/2I$. There is no potential energy U, so the kinetic energy K is equal to the total mechanical energy E:

$$E = \frac{L^2}{2I}. \qquad (44\text{–}1)$$

44–4 Model of a diatomic molecule as two point masses m_1 and m_2 connected by a light rigid rod with length r_0. The distances of the masses from the center of mass are r_1 and r_2, where $r_1 + r_2 = r_0$.

Zero potential energy U means no dependence of U on θ or ϕ. But the potential-energy function U in the hydrogen atom also has no dependence on θ or ϕ. Thus the angular solutions to the Schrödinger equation for rigid-body rotation are the same as for the hydrogen atom, and the angular momentum is quantized in the same way. As in Eq. (43–4),

$$L^2 = l(l + 1)\hbar^2 \qquad (l = 0, 1, 2, \ldots). \qquad (44\text{–}2)$$

Combining Eqs. (44–1) and (44–2), we obtain the *rotational energy levels:*

$$E = l(l + 1)\frac{\hbar^2}{2I} \qquad (l = 0, 1, 2, \ldots) \qquad \begin{array}{l}\text{(rotational energy levels,} \\ \text{diatomic molecule).}\end{array} \qquad (44\text{–}3)$$

Figure 44–5 is an energy-level diagram showing these rotational levels. The ground level has zero quantum number l and zero energy E, corresponding to zero angular momentum (no rotation). The spacing of adjacent levels increases with increasing l. The energy difference between levels $l - 1$ and l is $l\hbar^2/I$. Can you prove this?

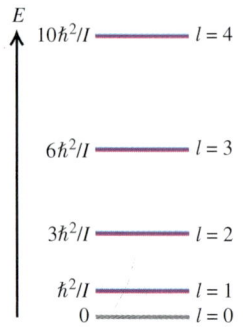

We can express the moment of inertia I in Eqs. (44–1) and (44–3) in terms of the *reduced mass* m_r of the molecule:

$$m_r = \frac{m_1 m_2}{m_1 + m_2}. \qquad (44\text{–}4)$$

44–5 The ground level and first four excited rotational energy levels for a diatomic molecule. The levels are not equally spaced.

We introduced this quantity in Section 40–6 to accommodate the finite nuclear mass of the hydrogen atom. In Fig. 44–4 the distances r_1 and r_2 are the distances from the center of mass to the nuclei of the atoms. By definition of the center of mass, $m_1 r_1 = m_2 r_2$, and the figure also shows that $r_0 = r_1 + r_2$. Solving these equations for r_1 and r_2, we find

$$r_1 = \frac{m_2}{m_1 + m_2} r_0, \qquad r_2 = \frac{m_1}{m_1 + m_2} r_0. \tag{44–5}$$

The moment of inertia is $I = m_1 r_1^2 + m_2 r_2^2$; substituting Eq. (44–5), we find

$$I = m_1 \frac{m_2^2}{(m_1 + m_2)^2} r_0^2 + m_2 \frac{m_1^2}{(m_1 + m_2)^2} r_0^2 = \frac{m_1 m_2}{m_1 + m_2} r_0^2,$$

or

$$I = m_r r_0^2 \qquad \text{(moment of inertia of a diatomic molecule).} \tag{44–6}$$

The reduced mass enables us to reduce this two-body problem to an equivalent one-body problem (a particle of mass m_r moving around a circle with radius r_0), just as we did with the hydrogen atom. Indeed, the only difference between this problem and the hydrogen atom is the difference in the radial forces. To conserve angular momentum, the allowed transitions are determined by the same selection rule as for the hydrogen atom: in allowed transitions, l must change by exactly one unit.

EXAMPLE 44–2

Rotational spectrum of carbon monoxide The two nuclei in the carbon monoxide (CO) molecule are 0.1128 nm apart. The mass of the most common carbon atom is exactly 12 u, or 1.993×10^{-26} kg. The mass of the most common oxygen atom is 15.995 u = 2.656×10^{-26} kg. a) Find the energies of the lowest three rotational energy levels. Express your results in electron volts. b) Find the wavelength of the photon emitted in the transition from the $l = 2$ to $l = 1$ state.

SOLUTION a) The reduced mass, from Eq. (44–4), is $m_r = 1.139 \times 10^{-26}$ kg. From Eq. (44–6),

$$I = m_r r_0^2 = (1.139 \times 10^{-26} \text{ kg})(0.1128 \times 10^{-9} \text{ m})^2$$

$$= 1.449 \times 10^{-46} \text{ kg} \cdot \text{m}^2.$$

The rotational levels are given by Eq. (44–3):

$$E_l = l(l + 1)\frac{\hbar^2}{2I} = l(l + 1)\frac{(1.0546 \times 10^{-34} \text{ J} \cdot \text{s})^2}{2(1.449 \times 10^{-46} \text{ kg} \cdot \text{m}^2)}$$

$$= l(l + 1)(3.838 \times 10^{-23} \text{ J}) = l(l + 1)0.2395 \text{ meV}.$$

Substituting $l = 0, 1, 2$, we find

$$E_0 = 0, \qquad E_1 = 0.479 \text{ meV}, \qquad E_2 = 1.437 \text{ meV}.$$

b) The photon energy is

$$E = E_2 - E_1 = 0.958 \text{ meV}.$$

The photon wavelength is

$$\lambda = \frac{hc}{E} = \frac{(4.136 \times 10^{-15} \text{ eV} \cdot \text{s})(3.00 \times 10^8 \text{ m/s})}{0.958 \times 10^{-3} \text{ eV}}$$

$$= 1.29 \times 10^{-3} \text{ m} = 1.29 \text{ mm}.$$

This photon is in the microwave region of the spectrum. Try working this example backward to show how we can determine the equilibrium separation r_0 (also called the *bond length*) by measuring wavelengths.

As Example 44–2 shows, the two lowest rotational energy levels in carbon monoxide are less than $\frac{1}{2}$ meV apart. This separation is much *smaller* than the atomic energy level differences (a few electron volts) that are typically associated with optical spectra. Photon energies for transitions among rotational levels in other molecules are also small, falling in the microwave and far infrared region of the spectrum.

Rotational levels of water molecules play an important role in the operation of microwave ovens. The magnetron oscillator in such an oven generates microwaves with a frequency of 2450 MHz; the corresponding photon energy is 1.01×10^{-5} eV. The water molecule has an $l = 1$ rotational level 1.01×10^{-5} eV above the nonrotating ($l = 0$) ground level, so the microwaves are strongly absorbed. Energy transfer from the rotating molecules to their surroundings then heats up the substance.

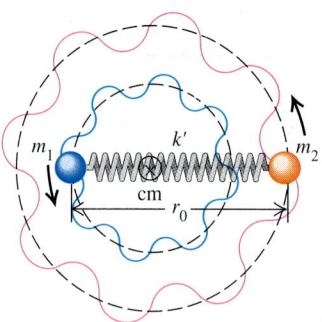

44–6 Model of a diatomic molecule as two point masses m_1 and m_2 connected by a spring with force constant k'.

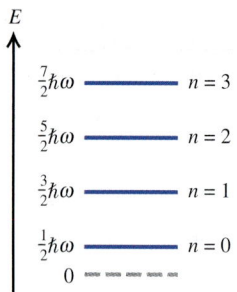

44–7 The ground level and first three excited levels for vibration in the one-dimensional harmonic oscillator approximation. The levels are equally spaced, with spacing $\Delta E = \hbar\omega$.

VIBRATIONAL ENERGY LEVELS

Molecules are never completely rigid. In a more realistic model of a diatomic molecule we represent the connection between atoms not as a rigid rod but as a *spring* (Fig. 44–6). Then in addition to rotating, the atoms of the molecule can also *vibrate* about their equilibrium positions along the line joining them. For small oscillations the restoring force can be taken as proportional to the displacement from the equilibrium separation r_0 (like a spring that obeys Hooke's law with a force constant k'), and the system is a harmonic oscillator. We discussed the quantum-mechanical harmonic oscillator in Section 42–6. The energy levels are given by Eq. (42–29), with the mass m replaced with the reduced mass m_r:

$$E_n = \left(n + \frac{1}{2}\right)\hbar\omega = \left(n + \frac{1}{2}\right)\hbar\sqrt{\frac{k'}{m_r}} \qquad (n = 0, 1, 2, \ldots) \qquad (44\text{--}7)$$

(vibrational energy levels of a diatomic molecule).

This represents a series of levels equally spaced in energy, with an energy separation of

$$\Delta E = \hbar\omega = \hbar\sqrt{\frac{k'}{m_r}}. \qquad (44\text{--}8)$$

Figure 44–7 is an energy-level diagram showing these vibrational levels.

   CAUTION ▶ We're again using k' for the force constant, this time to minimize confusion with Boltzmann's constant k, the gas constant per molecule (introduced in Section 16–4). Besides the quantities k and k', we also use the absolute temperature unit 1 K = 1 kelvin. ◀

EXAMPLE 44–3

Force constant of carbon monoxide For the carbon monoxide molecule of Example 44–2, the spacing of vibrational energy levels is found to be $\Delta E = 0.2690$ eV. Find the force constant k' for the interatomic force.

This corresponds to a fairly loose spring; to stretch a macroscopic spring with this force constant by 1.0 cm would require a pull of 19 N (about 4 lb). Force constants for diatomic molecules are typically about 100 to 2000 N/m.

SOLUTION Solving Eq. (44–8) for k', we find

$$k' = m_r\left(\frac{\Delta E}{\hbar}\right)^2 = (1.139 \times 10^{-26}\ \text{kg})\left(\frac{0.2690\ \text{eV}}{6.582 \times 10^{-16}\ \text{eV}\cdot\text{s}}\right)^2$$

$$= 1902\ \text{N/m}.$$

44–8 Energy-level diagram for vibrational and rotational energy levels of a diatomic molecule. For each vibrational level (n) there is a series of more closely spaced rotational levels (l). Several transitions corresponding to a single band in a band spectrum are shown. These transitions obey the selection rule $\Delta l = \pm 1$.

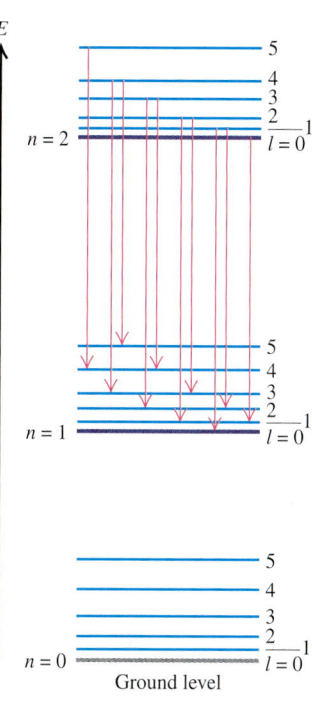

Visible-light photons have energies between 1.77 eV and 3.10 eV. The 0.269-eV energy difference in Example 44–3 corresponds to a photon in the infrared region of the spectrum, though closer to the visible region than the photon in the *rotational* transition in Example 44–2. Vibrational energy differences, while usually much smaller than those that produce atomic spectra, are usually much *larger* than the rotational energy differences. When we include both rotational and vibrational energies, the energy levels for our diatomic molecule are

$$E = l(l + 1)\frac{\hbar^2}{2I} + \left(n + \frac{1}{2}\right)\hbar\sqrt{\frac{k'}{m_r}}. \qquad (44\text{–}9)$$

The energy-level diagram is shown in Fig. 44–8. For each value of n there are many values of l, forming a series of closely spaced levels. Besides the $\Delta l = \pm 1$ selection rule for the rotational levels, when the vibrational level changes there is also a $\Delta n = 1$ selection rule for photon absorption and a $\Delta n = -1$ selection rule for photon emission.

EXAMPLE 44–4

Vibration-rotation spectrum of carbon monoxide Again consider the CO molecule of Examples 44–2 and 44–3. Find the photon wavelength that is emitted when there is a change in its vibrational energy, and its rotational energy is (a) initially zero and (b) finally zero.

SOLUTION From Example 44–3 the energy available from the change in *vibrational* levels is $\Delta E = 0.2690$ eV (corresponding to $\Delta n = -1$). From Example 44–2 the energy difference between the zero energy *rotational* level and the first excited *rotational* level is 0.479 meV = 0.000479 eV (corresponding to $\Delta l = \pm 1$).
a) To increase the rotational energy from zero to 0.000479 eV, we must spend 0.000479 eV of the 0.2690 eV available, leaving

$E = 0.2685$ eV for the photon. The photon's wavelength is then

$$\lambda = \frac{hc}{E} = \frac{(4.136 \times 10^{-16} \text{ eV} \cdot \text{s})(2.998 \times 10^8 \text{ m/s})}{0.2685 \text{ eV}}$$
$$= 4.618 \times 10^{-6} \text{ m} = 4.618 \ \mu\text{m}.$$

b) Now we gain 0.2690 eV from the decrease in vibrational energy and 0.000479 eV from the decrease in the rotational energy for a total of $E = 0.2695$ eV for the photon. The photon's wavelength is then

$$\lambda = \frac{hc}{E} = \frac{(4.136 \times 10^{-16} \text{ eV} \cdot \text{s})(2.998 \times 10^8 \text{ m/s})}{0.2695 \text{ eV}}$$
$$= 4.601 \times 10^{-6} \text{ m} = 4.601 \ \mu\text{m}.$$

Transitions between states with various pairs of n values give different series of spectrum lines, and the resulting spectrum has a series of *bands*. Each band corresponds to a particular vibrational transition, and each individual line in a band represents a particular rotational transition, with the selection rule $\Delta l = \pm 1$. A typical *band spectrum* is shown in Fig. 44–9. All molecules can also have excited states of the *electrons* in addition to the rotational and vibrational states that we have described. In general, these lie at higher energies than the rotational and vibrational states, and there is no simple rule relating them. When there is a transition between electronic states, the $\Delta n = \pm 1$ selection rule for the vibrational levels no longer holds.

44–9 Typical molecular band spectrum.

We can apply these same principles to more complex molecules. A molecule with three or more atoms has several different kinds or *modes* of vibratory motion. Each mode has its own set of energy levels, related to its frequency by Eq. (44–7). In nearly all cases the associated radiation lies in the infrared region of the electromagnetic spectrum. Infrared spectroscopy has proved to be an extremely valuable analytical tool. It provides information about the strength, rigidity, and length of molecular bonds and the structure of complex molecules. Also, because every molecule (like every atom) has its characteristic spectrum, infrared spectroscopy can be used to identify unknown compounds.

44–4 STRUCTURE OF SOLIDS

The term *condensed matter* includes both solids and liquids. In both states, the interactions between atoms or molecules are strong enough to give the material a definite volume that changes relatively little with applied stress. In condensed matter, adjacent atoms attract one another until their outer electron charge clouds begin to overlap significantly. Thus the distances between adjacent atoms in condensed matter are about the same as the diameters of the atoms themselves, typically 0.1 to 0.5 nm. Also, when we speak of the distances between atoms, we mean the center-to-center, that is, nucleus-to-nucleus distances.

Ordinarily, we think of a liquid as a material that can flow and of a solid as a material with a definite shape. However, if you heat a horizontal glass rod in the flame of a burner, you'll find that the rod begins to sag (flow) more and more easily as its temperature rises. Glass has no definite transition from solid to liquid, and no definite melting point. On this basis, we can consider glass at room temperature as being an extremely viscous liquid. Tar and butter show similar behavior.

What is the microscopic difference between materials like glass or butter and solids like ice or copper, which do have definite melting points? Ice and copper are examples of *crystalline solids,* solids in which the atoms have *long-range order,* a recurring pattern of atomic positions that extends over many atoms. This pattern is called the *crystal structure.* In contrast, glass at room temperature is an example of an *amorphous* solid, one that has no long-range order, but only *short-range order* (correlations between neighboring atoms or molecules). Liquids also have only short-range order. The boundaries between crystalline solid, amorphous solid, and liquid may be sometimes blurred. Some solids, crystalline when perfect, can form with so many imperfections in their structure that they have almost no long-range order. Conversely, some liquids have a fairly high degree of long-range order; a familiar example is the type of *liquid crystal* illustrated in Fig. 44–10.

44–10 (a) The simplest *liquid-crystal display* (LCD) consists of a thin layer of an organic compound whose cylindrical molecules tend to line up parallel to each other. The layer is confined between two parallel glass plates between two polarizers. Fine scratches on the surface of the glass plates tend to align the molecules at each surface, leading to the twisted array shown. The array rotates the plane of polarization of the light that has passed through the first polarizer, allowing it to also pass through the second, crossed polarizer. (b) Each glass plate also has a pattern of thin electrodes. A potential difference between the plates can give an electric field that orients the molecules so that they no longer rotate the plane of polarization of the light. Then no light passes through, and a bright area becomes dark. Adding color filters makes a full-color picture possible.

(a)

(b)

(c)

(d)

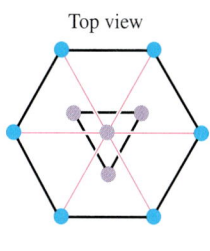
Top view

Nearly everything we know about crystal structure has been learned from diffraction experiments, initially with x rays, later with electrons and neutrons. A typical distance between atoms is of the order of 0.1 nm. We leave it as an exercise for you to show that 12.4-keV x rays, 150-eV electrons, and 0.0818-eV neutrons all have wavelengths of $\lambda = 0.1$ nm.

44–11 Portions of some common types of crystal lattices. (a) Simple cubic (sc); (b) face-centered cubic (fcc); (c) body-centered cubic (bcc); (d) hexagonal close-packed (hcp) in two views.

CRYSTAL LATTICES

A *crystal lattice* is a repeating pattern of mathematical points that extends throughout space. There are 14 general types of such patterns; Fig. 44–11 shows small portions of a few common examples. The *simple cubic lattice* (sc) has a lattice point at each corner of a cubic array (Fig. 44–11a). The *face-centered cubic lattice* (fcc) is like the simple cubic but with an additional lattice point at the center of each cube face (Fig. 44–11b). The *body-centered cubic lattice* (bcc) is like the simple cubic but with an additional point at the center of each cube (Fig. 44–11c). The *hexagonal close-packed lattice* has layers of lattice points in hexagonal patterns, each hexagon made up of six equilateral triangles (Fig. 44–11d).

CAUTION ▶ Realize that Fig. 44–11 shows just enough lattice points to easily visualize the pattern; the lattice, a mathematical abstraction, extends throughout space. Thus the lattice points shown repeat endlessly in all directions. **◀**

In a crystal structure, a single atom or a group of atoms is associated with *each* lattice point. The group may contain the same or different kinds of atoms. This atom or group of atoms is called a *basis.* Thus a complete description of a crystal structure includes both the lattice and the basis. We initially consider *perfect crystals,* or *ideal single crystals,* in which the crystal structure extends uninterrupted throughout space.

The bcc and fcc structures are two common simple crystal structures. The alkali metals have a bcc structure, that is, a bcc lattice with a basis of one atom at each lattice point. Each atom in a bcc structure has eight nearest neighbors (Fig. 44–12a). The elements Al, Ca, Cu, Ag, and Au have an fcc structure, that is, an fcc lattice with a basis of one atom at each lattice point. Each atom in an fcc structure has 12 nearest neighbors (Fig. 44–12b).

Figure 44–13 shows a representation of the structure of sodium chloride (NaCl, ordinary salt). It may look like a simple cubic structure, but it isn't. The sodium and chloride ions each form an fcc structure, so we can think roughly of the sodium chloride structure as being composed of two interpenetrating fcc structures. More correctly, the sodium chloride crystal structure of Fig. 44–13 has a fcc lattice with one chloride ion at each lattice point and one sodium ion half a cube length above it. That is, its basis consists of one chloride and one sodium ion.

Another example is the *diamond structure;* it's called that because it is the crystal structure of carbon in the diamond form. It's also the crystal structure of silicon, germanium, and gray tin (all four are group IV elements in the periodic table). The diamond lattice is fcc; the basis consists of one atom at each lattice point and a second *identical* atom displaced a quarter of a cube length in each of the three cube-edge directions.

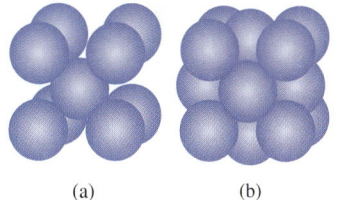
(a)   (b)

44–12 (a) The bcc *structure* is composed of a bcc *lattice* with a basis of one atom for each lattice point. (b) The fcc *structure* is composed of an fcc *lattice* with a basis of one atom for each lattice point. These structures repeat precisely to make up perfect crystals.

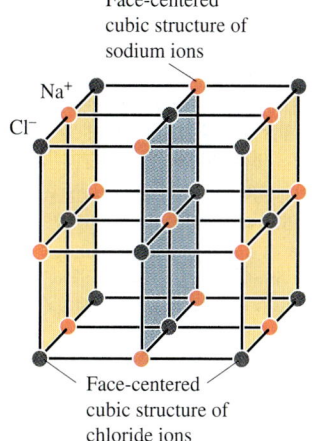
Face-centered cubic structure of sodium ions

Na⁺

Cl⁻

Face-centered cubic structure of chloride ions

44–13 Representation of part of the sodium chloride crystal structure. The distances between ions are exaggerated.

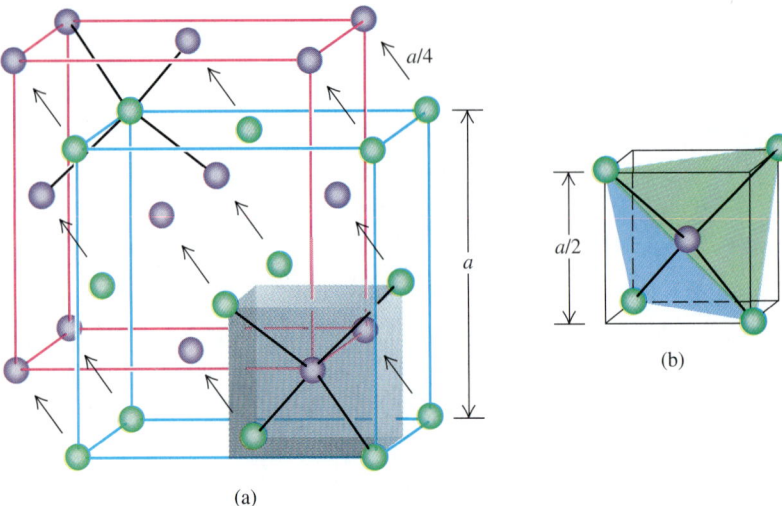

(a)

(b)

44–14 (a) The diamond structure, shown as two interpenetrating face-centered cubic structures with distances between atoms exaggerated. Relative to the corresponding green atom, each purple atom is shifted up, back, and to the left by a distance $a/4$. (b) The front lower right eighth of (a), showing four green atoms at the corners of a tetrahedron inscribed in a cube with side length $a/2$, with a purple atom at the center. In the diamond structure, both purple and green spheres represent *identical* atoms, for example, both silicon. In the cubic zinc sulfide structure, the purple spheres represent one type of atom and the green spheres represent a *different* type, for example, purple for gallium and green for arsenic.

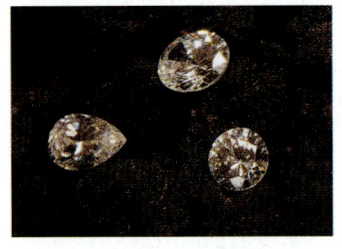

The crystal structure of diamond is similar to that of many other, less highly coveted materials. What makes diamond unique are its beautiful optical properties. A large index of refraction makes diamond sparkle due to total internal reflection, and a large amount of dispersion turns white light into a rainbow of colors.

Figure 44–14a will help you visualize this. Fig. 44–14b shows the bottom right front eighth of the basic cube; the four atoms at alternate corners of this cube are at the corners of a regular tetrahedron, and there is an additional atom at the center. Thus each atom in the diamond structure is at the center of a regular tetrahedron with four nearest-neighbor atoms at the corners.

The cubic zinc sulfide structure is similar to the diamond structure. Its lattice is again fcc, but its basis consists of two *different* atoms, one of zinc and one of sulfur. Each zinc atom is at the center of a regular tetrahedron with four sulfur atoms at the corners, and vice versa. Gallium arsenide and other III-V compounds also have this structure.

BONDING IN SOLIDS

The forces that are responsible for the regular arrangement of atoms in a crystal are the same as those involved in molecular bonds, plus one additional type. Not surprisingly, *ionic* and *covalent* molecular bonds are found in ionic and covalent crystals, respectively. The most familiar *ionic crystals* are the alkali halides, such as ordinary salt (NaCl). The positive sodium ions and the negative chloride ions occupy alternate positions in a cubic arrangement (Fig. 44–13). The attractive forces are the familiar Coulomb's-law forces between charged particles. These forces have no preferred direction, and the arrangement in which the material crystallizes is partly determined by the relative size of the two ions. It's not difficult to prove that such a structure is *stable* in the sense that it has lower total energy than the separated ions. Example 44–5 gives a hint as to how the proof can be carried out. The negative potential energies of pairs of opposite charges are greater in absolute value than the positive energies of pairs of like charges because the pairs of unlike charges are closer together, on average.

EXAMPLE 44-5

Potential energy of an ionic crystal Consider a fictitious one-dimensional ionic crystal consisting of a very large number of alternating positive and negative ions with charges e and $-e$, with equal spacing a along a line. Prove that the total interaction potential energy is negative.

SOLUTION Let's pick an ion somewhere in the middle of the string and add up the potential energies of its interactions with all the ions to one side of it. We get the series

$$U = -\frac{e^2}{4\pi\epsilon_0}\frac{1}{a} + \frac{e^2}{4\pi\epsilon_0}\frac{1}{2a} - \frac{e^2}{4\pi\epsilon_0}\frac{1}{3a} + \cdots$$
$$= -\frac{e^2}{4\pi\epsilon_0 a}\left(1 - \frac{1}{2} + \frac{1}{3} - \frac{1}{4} + \cdots\right).$$

You may notice the resemblance of the series in parentheses to

the Taylor series for $\ln(1 + x)$:

$$\ln(1 + x) = x - \frac{x^2}{2} + \frac{x^3}{3} - \frac{x^4}{4} + \cdots.$$

When $x = 1$, we have the series in the parentheses, and

$$U = -\frac{e^2}{4\pi\epsilon_0 a}\ln 2.$$

This is certainly a negative quantity. The atoms on the other side of the ion that we're considering make an equal contribution to the potential energy. And if we include the potential energies of all pairs of atoms, the sum is certainly negative. So this structure has lower energy than the zero electric potential energy that is obtained when all the ions are infinitely separated from each other.

Carbon, silicon, germanium, and tin in the diamond structure are simple examples of *covalent crystals.* These elements are in Group IV of the periodic table, meaning that each atom has four electrons in its outermost shell. Each atom forms a covalent bond with each of four adjacent atoms at the corners of a tetrahedron (Fig. 44–14b). These bonds are strongly directional because of the asymmetrical electron distributions dictated by the exclusion principle, and the result is the tetrahedral diamond structure.

A third crystal type, less directly related to the chemical bond than are ionic or covalent crystals, is the **metallic crystal.** In this structure, one or more of the outermost electrons in each atom become detached from the parent atom (leaving a positive ion) and are free to move through the crystal. These electrons are not localized near the individual ions. The corresponding electron wave functions extend over many atoms.

Thus we can picture a metallic crystal as an array of positive ions immersed in a sea of freed electrons whose attraction for the positive ions holds the crystal together (Fig. 44–15). These electrons also give metals their high electrical and thermal conductivities. This sea of electrons has many of the properties of a gas, and indeed we speak of the *electron-gas model* of metallic solids. The simplest version of this model is the *free-electron model,* which ignores interactions with the ions completely (except at the surface). We'll return to this model in Section 44–6.

In a metallic crystal the freed electrons are not localized but are shared among *many* atoms. This gives a bonding that is neither localized nor strongly directional. The crystal structure is determined primarily by considerations of *close packing,* that is, the maximum number of atoms that can fit into a given volume. The two most common metallic crystal lattices, the face-centered cubic and the hexagonal close-packed, are shown in Figs. 44–11b and 44–11d. In structures composed of these lattices with a basis of one atom, each atom has 12 nearest neighbors.

As we mentioned in Section 44–2, van der Waals interactions and hydrogen bonding also play a role in the structure of some solids. In polyethylene and similar polymers, covalent bonding of atoms forms long-chain molecules, and hydrogen bonding forms cross-links between adjacent chains. In solid water, both van der Waals forces and hydrogen bonds are significant, and together they determine the crystal structures of ice. Many other examples might be cited.

Our discussion has centered on perfect crystals, or ideal single crystals. Real crystals show a variety of departures from this idealized structure. Materials are often *polycrystalline,* composed of many small single crystals bonded together at *grain boundaries.* Within a single crystal, *interstitial* atoms may occur in places where they do

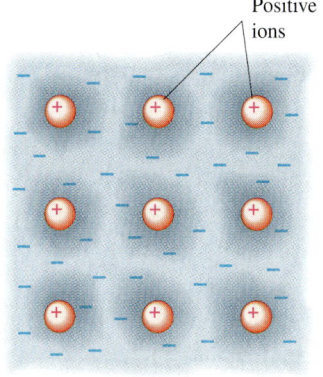

Positive ions

44–15 Representation of a metallic solid. One or more electrons are detached from each atom and are free to wander around the crystal, forming the "electron gas." The wave functions for these electrons extend over many atoms. The positive ions vibrate around fixed locations in the crystal.

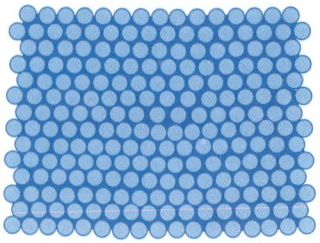

44-16 An edge dislocation in two dimensions. The irregularity is seen most easily by viewing the figure from various directions at a grazing angle with the page. In three dimensions an edge dislocation would look like an extra plane of atoms slipped partway into the crystal.

not belong, and there may be *vacancies,* positions that should be occupied by an atom but are not. A point defect of particular interest in semiconductors, which we will discuss in Section 44–7, is the *substitutional impurity,* a foreign atom replacing a regular atom (for example, arsenic in a silicon crystal).

There are several basic types of extended defects called *dislocations.* One type is the *edge dislocation,* shown schematically in Fig. 44–16, in which one plane of atoms slips relative to another. The mechanical properties of metallic crystals are influenced strongly by the presence of dislocations. The ductility and malleability of some metals depend on the presence of dislocations that can move through the crystal during plastic deformations. Solid-state physicists often point out that the biggest extended defect of all, present in *all* real crystals, is the surface of the material with its dangling bonds and abrupt change in potential energy.

44-5 ENERGY BANDS

The **energy band** concept is a great help in understanding several properties of solids. To introduce the idea, suppose we have a large number N of identical atoms, far enough apart that their interactions are negligible. Every atom has the same energy-level diagram. We can draw an energy-level diagram for the *entire system.* It looks just like the diagram for a single atom, but the exclusion principle, applied to the entire system, permits each state to be occupied by N electrons instead of just one.

Now we begin to push the atoms uniformly closer together. Because of the electrical interactions and the exclusion principle, the wave functions begin to distort, especially those of the outer, or *valence,* electrons. The corresponding energies also shift, some upward and some downward, by varying amounts, as the valence electron wave functions become less localized and extend over more and more atoms. Thus the valence states that formerly gave the *system* a state with a sharp energy level that could accommodate N electrons now give a *band* containing N closely spaced levels (Fig. 44–17). Ordinarily, N is very large, somewhere near the order of Avogadro's number (10^{24}), so we can accurately treat the levels as forming a *continuous* distribution of energies within a band. Between adjacent energy bands are gaps or forbidden regions where there are *no* allowed energy levels. The inner electrons in an atom are affected much less by nearby atoms than are the valence electrons, and their energy levels remain relatively sharp.

44-17 Origin of energy bands in a solid. (a) As the atoms move together, the energy levels spread into bands. The vertical line at r_0 shows the actual atomic spacing in the crystal. (b) Symbolic representation of energy bands.

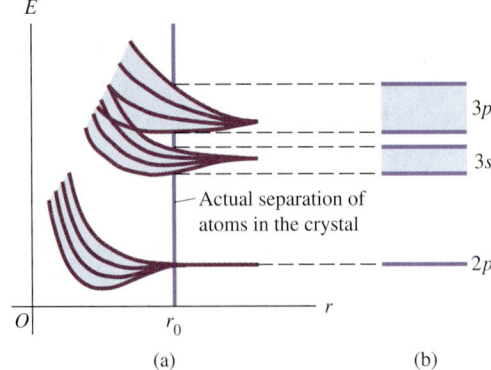

INSULATORS AND SEMICONDUCTORS

The nature of the energy bands determines whether the material is an electrical conductor or an insulator. In insulators and semiconductors at absolute zero temperature the valence electrons completely fill the highest occupied band, called the **valence band.**

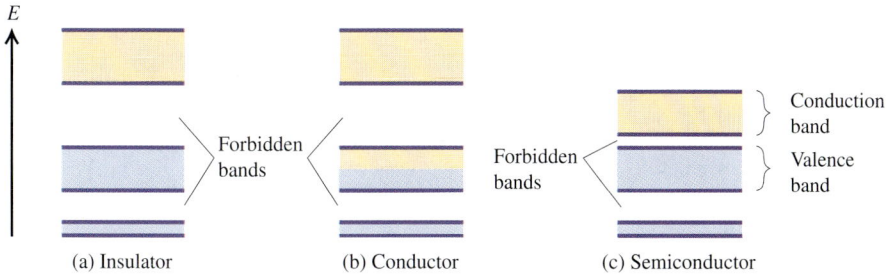

E

(a) Insulator (b) Conductor (c) Semiconductor

Forbidden bands

Forbidden bands

Conduction band

Valence band

44–18 Three types of energy-band structure. (a) An insulator at absolute zero. A completely full valence band is separated by a gap of several electron volts from a completely empty conduction band, and electrons in the valence band cannot move. At finite temperatures, extremely few electrons can reach the upper band. (b) A conductor at any temperature. There is a partially filled valence band, and electrons in this band are free to move when an electric field is applied. (c) A semiconductor at absolute zero. A completely filled valence band is separated by a smaller gap of 1 eV or so from an empty conduction band. At finite temperatures, substantial numbers of electrons can reach the conduction band, where they are free to move.

The next higher band, called the **conduction band,** is completely empty at absolute zero.

The energy gap separating the valence band and the conduction band may be of the order of 1 to 5 eV. This situation is shown in Figs. 44–18a and 44–18c. The electrons in a full valence band are not free to move in response to an applied electric field. To move, an electron would have to go to a different quantum state with slightly different energy, but it can't do that because all the neighboring states are already occupied. The only way such an electron can move is to jump into the conduction band. This would require an additional energy of a few electron volts or so, and that much energy is not ordinarily available. The situation is like a completely filled floor below an empty floor in a parking garage; none of the cars can move because there is no place to go. If a car could jump up to the empty floor, it could move!

However, at any temperature above absolute zero there is some probability that an electron can gain enough energy from thermal motion to jump to the conduction band. Once in the conduction band, an electron is free to move in response to an applied electric field because there are plenty of nearby empty states available to permit it to gain or lose energy in small increments. Furthermore, as the temperature increases, the population in the conduction band and the resulting conductivity increase very rapidly. For example, near room temperature the number of conduction electrons doubles for just a 10 C° rise for a pure semiconductor with an energy gap of 1 eV.

A distinguishing characteristic of all conductors, including metals, is that the valence band is only partly filled. The metal sodium is an example. An analysis of the energy-level diagram of Fig. 40–9 shows that for an isolated sodium atom, the six $3p$ lowest excited states for the valence electron are about 2.1 eV above the two $3s$ ground states. But in solid sodium the $3s$ and $3p$ *bands* spread out enough that they actually overlap, forming a single band (Fig. 44–17 with a smaller r_0). Because each sodium atom has only one valence electron but eight $3s$ and $3p$ states, that single band is only $\frac{1}{8}$ filled. The situation is similar to the one shown in Fig. 44–18b. Electrons in states near the top of the filled portion of the band have many adjacent unoccupied states available, and they can easily gain or lose small amounts of energy in response to an applied electric field. Therefore these electrons are mobile and can contribute to electrical and thermal conductivity. Metallic crystals always have partly filled bands. In the *ionic* NaCl crystal, on the other hand, there is no overlapping of bands. The Na^+ and Cl^- ions both have noble gas electron configurations corresponding to completely filled bands, and solid sodium chloride is an insulator.

As we saw in Section 25–5, an insulating material that is subjected to a large enough electric field becomes a conductor; this is called *dielectric breakdown.* If the electric field in an insulator is so large that there is a potential difference of a few volts over a distance comparable to atomic sizes (that is, a field of the order of 10^{10} V/m), then the field can do enough work on a valence electron to boost it over the forbidden region and into the conduction band. Dielectric breakdown occurs in real insulators for fields that are much *less* than 10^{10} V/m because of imperfections that provide some energy states in the forbidden region.

The concept of energy bands is very useful in understanding the properties of semiconductors, which we will study in a later section.

The concept of energy bands was first developed by the Swiss-American physicist Felix Bloch (1905–1983) in his doctoral thesis. Our modern understanding of electrical conductivity stems from that landmark work. However, it was Bloch's work in nuclear physics that brought him (along with Edward Purcell) the 1952 Nobel Prize in physics.

EXAMPLE 44–6

Photoconductivity in germanium Even at room temperature, pure germanium has an almost completely filled valence band separated by a gap of 0.67 eV from an almost completely empty conduction band. It is a poor electrical conductor, but its conductivity increases substantially when it is irradiated with electromagnetic waves of certain maximum wavelength. Why? What maximum wavelength is appropriate?

SOLUTION An electron at the top of the valence band can absorb a photon with energy of 0.67 eV (no less) and move to the bot-

tom of the conduction band, where it is a mobile charge. Thus the maximum wavelength is

$$\lambda_{max} = \frac{hc}{E_{min}} = \frac{(4.136 \times 10^{-15} \text{ eV} \cdot \text{s})(3.00 \times 10^8 \text{ m/s})}{0.67 \text{ eV}}$$

$$= 1.9 \times 10^{-6} \text{ m} = 1.9 \text{ }\mu\text{m} = 1900 \text{ nm}.$$

This wavelength is in the infrared region of the spectrum. Semiconductor crystals are widely used as photocells, with variations to be discussed in Section 44–8.

44–6 FREE-ELECTRON MODEL OF METALS

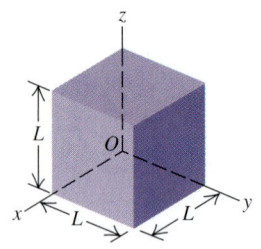

44–19 A cubical box with rigid walls and side length L. This is the three-dimensional version of the infinite square well discussed in Section 42–2. The energy levels for a particle in this box are given by Eq. (44–11).

Studying the energy states of electrons in metals can give us a lot of insight into their electrical and magnetic properties, the electron contributions to heat capacities, and other behavior. As we discussed in Section 44–4, one of the distinguishing features of a metal is that one or more valence electrons are detached from their home atom and can move freely within the metal, with wave functions that extend over many atoms.

The **free-electron model** assumes that these electrons are completely free inside the material, that they don't interact at all with the ions or with each other, but that there are infinite potential-energy barriers at the surfaces. The wave functions and energy levels are then the three-dimensional versions of those for the particle in a box that we analyzed in Sections 42–2 and 42–3 in one-dimension. Suppose the box is a cube with side length L (Fig. 44–19). Then the possible wave functions, analogous to Eq. (42–6), are

$$\psi(x, y, z) = A \sin \frac{n_x \pi x}{L} \sin \frac{n_y \pi y}{L} \sin \frac{n_z \pi z}{L}, \quad (44\text{–}10)$$

where (n_x, n_y, n_z) is a set of three positive-integer quantum numbers that identify the state. We invite you to verify that these functions are zero at the surfaces of the cube, satisfying the boundary conditions. You can also substitute Eq. (44–10) into the three-dimensional Schrödinger equation, Eq. (42–32), to show that the energies of the states are

$$E = \frac{(n_x^2 + n_y^2 + n_z^2)\pi^2 \hbar^2}{2mL^2}. \quad (44\text{–}11)$$

This equation is the three-dimensional analog of Eq. (42–18) for the energy levels of a particle in a box.

DENSITY OF STATES

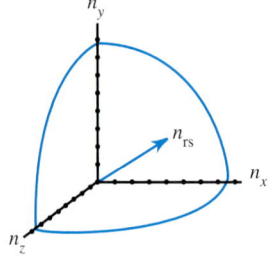

44–20 The allowed values of n_x, n_y, and n_z are positive integers for the electron states in the free-electron gas model. Including spin, there are two states for each unit volume in n space.

Later we'll need to know the *number dn* of quantum states that have energies in a given range dE. The number of states per unit energy range dn/dE is called the **density of states,** denoted by $g(E)$. We'll begin by working out an expression for $g(E)$. Think of a three-dimensional space with coordinates (n_x, n_y, n_z) (Fig. 44–20). The radius n_{rs} of a sphere centered at the origin in that space is given by $n_{rs}^2 = n_x^2 + n_y^2 + n_z^2$. Each point with integer coordinates in that space represents one spatial quantum state. Thus each point corresponds to one unit of volume in the space, and the total number of points with integer coordinates inside a sphere equals the volume of the sphere, $\frac{4}{3}\pi n_{rs}^3$. Because all our n's are positive, we must take only one *octant* of the sphere, with $\frac{1}{8}$ the total volume, or $(\frac{1}{8})(\frac{4}{3}\pi n_{rs}^3) = \frac{1}{6}\pi n_{rs}^3$. The particles are electrons, so each point corresponds to *two states* with opposite spin components ($m_s = \pm\frac{1}{2}$), and the total number n of electron states corresponding to points inside the octant is twice $\frac{1}{6}\pi n_{\text{rs}}^3$, or

$$n = \frac{\pi n_{\text{rs}}^3}{3}. \tag{44–12}$$

The energy E of states at the surface of the sphere can be expressed in terms of n_{rs}. Equation (44–11) becomes

$$E = \frac{n_{\text{rs}}^2 \pi^2 \hbar^2}{2mL^2}. \tag{44–13}$$

We can combine Eqs. (44–12) and (44–13) to get a relation between E and n that doesn't contain n_{rs}. We'll leave the details as an exercise (Exercise 44–20); the result is

$$n = \frac{(2m)^{3/2} V E^{3/2}}{3\pi^2 \hbar^3}, \tag{44–14}$$

where $V = L^3$ is the volume of the box. Equation (44–14) gives the total number of states with energies of E or less.

To get the number of states dn in an energy interval dE, we take differentials of both sides of Eq. (44–14). We get

$$dn = \frac{(2m)^{3/2} V E^{1/2}}{2\pi^2 \hbar^3} dE. \tag{44–15}$$

The density of states $g(E)$ is equal to dn/dE, so from Eq. (44–15) we get

$$g(E) = \frac{(2m)^{3/2} V}{2\pi^2 \hbar^3} E^{1/2} \qquad \text{(density of states, free-electron model)}. \tag{44–16}$$

FERMI-DIRAC DISTRIBUTION

Now we need to know how the electrons are distributed among the various quantum states at any given temperature. The *Maxwell-Boltzmann distribution* states that the average number of particles in a state of energy E is proportional to $e^{-E/kT}$ (page 1250). However, it wouldn't be right to use the Maxwell-Boltzmann distribution, for two very important reasons. The first reason is the *exclusion principle*. At absolute zero the Maxwell-Boltzmann function predicts that *all* the electrons would go into the two ground states of the system, with $n_x = n_y = n_z = 1$ and $m_s = \pm\frac{1}{2}$. But the exclusion principle allows only one electron in each state. At absolute zero the electrons can fill up the lowest *available* states, but there's not enough room for *all* of them to go into the lowest states. Thus a reasonable guess as to the shape of the distribution would be Fig. 44–21. At absolute zero temperature the states are filled up to some value E_{F0}, and all states above this value are empty.

The second reason we can't use the Maxwell-Boltzmann distribution is more subtle. That distribution assumes that we are dealing with *distinguishable* particles. It might seem that we could put a tag on each electron and know which is which. But overlapping electrons in a system such as a metal are *indistinguishable*. Suppose we have two electrons; a state in which the first is in an energy level E_1 and the second is in level E_2 is not distinguishable from a state in which the two electrons are reversed, because we can't tell which electron is which.

The statistical distribution function that emerges from the exclusion principle and the indistinguishability requirement is called (after its inventors) the **Fermi-Dirac distribution.** Because of the exclusion principle, the probability that a particular state with energy E is occupied by an electron is the same as $f(E)$, the fraction of states with that energy that are occupied:

$$f(E) = \frac{1}{e^{(E-E_{\text{F}})/kT} + 1} \qquad \text{(Fermi-Dirac distribution).} \tag{44–17}$$

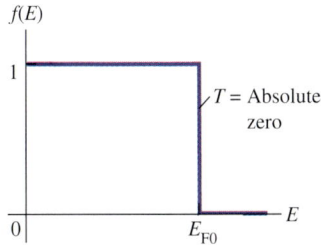

44–21 A guess as to the probability distribution for occupation of free-electron energy states at absolute zero. All the states are occupied (probability one) up to the value E_{F0}, and all those above E_{F0} are empty (occupation probability zero). As the temperature increases, we might expect more and more electrons to be in states with $E > E_{\text{F0}}$.

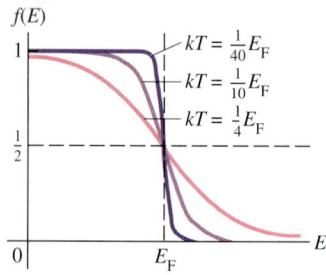

$f(E)$

$kT = \frac{1}{40}E_F$
$kT = \frac{1}{10}E_F$
$kT = \frac{1}{4}E_F$

E

E_F

44–22 Graphs of the Fermi-Dirac distribution function for various values of kT, assuming that the Fermi energy E_F is independent of the temperature T. As T increases, more and more of the electrons are excited to states above E_F. For solid metals at ordinary temperatures the graph looks much more like Fig. 44–21. For example, copper *melts* at $kT \approx \frac{1}{60}E_F$.

The energy E_F is called the **Fermi energy** or the *Fermi level;* we'll discuss its significance below. We use E_{F0} for its value at absolute zero ($T = 0$) and E_F for other temperatures. We can accurately let $E_F = E_{F0}$ for metals because the Fermi energy does not change much with temperature for solid conductors. However, it is not safe to assume that $E_F = E_{F0}$ for semiconductors, in which the Fermi energy usually does change with temperature.

Figure 44–22 shows graphs of Eq. (44–17) for several temperatures. The trend of this function as kT approaches zero confirms our guess. When $E = E_F$, the exponent is zero and $f(E_F) = \frac{1}{2}$. That is, the probability is $\frac{1}{2}$ that a state at the Fermi energy contains an electron. Alternatively, at $E = E_F$, half the states are filled (and half are empty).

For $E < E_F$ the exponent is negative, and $f(E) > \frac{1}{2}$. For $E > E_F$ the exponent is positive, and $f(E) < \frac{1}{2}$. The shape depends on the ratio E_F/kT. At $T << E_F/k$ this ratio is very large. Then for $E < E_F$ the curve very quickly approaches 1, and for $E > F_F$ it quickly approaches zero. When T is larger, the changes are more gradual. When T is zero, all the states up to the Fermi level E_{F0} are filled, and all states above that level are empty.

EXAMPLE 44-7

Two probabilities in the free-electron model For free electrons in a solid, at what energy is the probability that a particular state is occupied equal to a) 0.01; b) 0.99?

SOLUTION We solve Eq. (44–17) for E, obtaining

$$E = E_F + kT \ln\left(\frac{1}{f(E)} - 1\right).$$

a) When $f(E) = 0.01$,

$$E = E_F + kT \ln\left(\frac{1}{0.01} - 1\right) = E_F + 4.6kT.$$

A state $4.6kT$ above the Fermi level is occupied only 1% of the time.

b) When $f(E) = 0.99$,

$$E = E_F + kT \ln\left(\frac{1}{0.99} - 1\right) = E_F - 4.6kT.$$

A state $4.6kT$ below the Fermi level is occupied 99% of the time. At very low temperatures, $4.6kT$ becomes very small. Then levels even slightly below E_F are nearly always full, and levels even slightly above E_F are nearly always empty. In general, if the probability is P that a state with an energy ΔE *above* E_F is occupied, then the probability is $1 - P$ that a state ΔE *below* E_F is occupied. We leave the proof as a problem (Problem 44–42).

Equation (44–17) gives the probability that any specific state with energy E is occupied at a temperature T. To get the actual number of electrons in any energy range dE, we have to multiply this probability by the number dn of states in that range $g(E)\, dE$. Thus the number dN of electrons with energies in the range dE is

$$dN = g(E)f(E)\, dE = \frac{(2m)^{3/2} VE^{1/2}}{2\pi^2\hbar^3} \frac{1}{e^{(E-E_F)/kT} + 1}\, dE. \qquad (44\text{–}18)$$

The Fermi energy E_F is determined by the total number N of electrons; at any temperature the electron states are filled up to a point at which all electrons are accommodated. At absolute zero there is a simple relation between E_{F0} and N. All states below E_{F0} are filled; in Eq. (44–14) we set n equal to the total number of electrons N and E to the Fermi energy at absolute zero E_{F0}:

$$N = \frac{(2m)^{3/2} VE_{F0}^{3/2}}{3\pi^2\hbar^3}. \qquad (44\text{–}19)$$

Solving for E_{F0}, we get

$$E_{F0} = \frac{3^{2/3}\pi^{4/3}\hbar^2}{2m}\left(\frac{N}{V}\right)^{2/3}. \qquad (44\text{–}20)$$

The quantity N/V is the number of free electrons per unit volume. It is called the *electron concentration* and is usually denoted by n.

CAUTION ▶ Don't confuse the electron concentration n with any quantum number n. Furthermore, the number of states n is not in general the same as the total number of electrons N. ◀

If we replace N/V with n, Eq. (44–20) becomes

$$E_{F0} = \frac{3^{2/3}\pi^{4/3}\hbar^2 n^{2/3}}{2m}.$$ (44–21)

EXAMPLE 44–8

The Fermi energy in copper At low temperatures, copper has a free-electron concentration of 8.45×10^{28} m^{-3}. Using the free-electron model, find the Fermi energy for solid copper.

SOLUTION Because copper is a metal, we can accurately replace E_{F0} with E_F in Eq. (44–21):

$$E_F = \frac{3^{2/3}\pi^{4/3}(1.055 \times 10^{-34}\text{ J·s})^2(8.45 \times 10^{28}\text{ m}^{-3})^{2/3}}{2(9.11 \times 10^{-31}\text{ kg})}$$

$$= 1.126 \times 10^{-18}\text{ J} = 7.03\text{ eV}.$$

This energy is much *larger* than kT at ordinary temperatures, so it is a good approximation to take almost all the states below E_F as completely full and almost all those above E_F as completely empty.

We can also use Eq. (44–16) to find $g(E)$ if E and V are known. We invite you to show that if $E = 7$ eV and $V = 1$ cm^3, $g(E)$ is about 2×10^{22} states/eV. This huge number shows why we were justified in treating n and E as a continuous variables in our density-of-states derivation.

We can calculate the *average* free-electron energy in a metal at absolute zero using the same ideas that were used to find E_{F0}. From Eq. (44–18) the number dN of electrons with energies in the range dE is $g(E)f(E)\,dE$. The energy of these electrons is $E\,dN = Eg(E)f(E)\,dE$. At absolute zero we substitute $f(E) = 1$ from $E = 0$ to $E = E_{F0}$ and $f(E) = 0$ for all other energies. Therefore the total energy E_{tot} of all the N electrons is

$$E_{tot} = \int_0^{E_{F0}} Eg(E)(1)\,dE + \int_{E_{F0}}^{\infty} Eg(E)(0)\,dE = \int_0^{E_{F0}} Eg(E)\,dE.$$

The simplest way to evaluate this expression is to compare Eqs. (44–16) and (44–20), noting that

$$g(E) = \frac{3NE^{1/2}}{2E_{F0}^{3/2}}.$$

Substituting this expression into the integral and using $E_{av} = E_{tot}/N$, we get

$$E_{av} = \frac{3}{2E_{F0}^{3/2}} \int_0^{E_{F0}} E^{3/2}\,dE = \frac{3}{5}E_{F0}.$$ (44–22)

That is, at absolute zero the average free-electron energy equals $\frac{3}{5}$ of the corresponding Fermi energy.

EXAMPLE 44–9

Free electrons in copper a) Find the average energy of the free electrons in copper at absolute zero (Example 44–8). b) If the equipartition principle were valid for this system, what would be the temperature of the electrons? c) What is the speed of an electron with a kinetic energy equal to the Fermi energy?

SOLUTION a) The Fermi energy in copper is 1.126×10^{-18} J $=$ 7.03 eV. According to Eq. (44–22), the average energy is $\frac{3}{5}$ of this, or 6.76×10^{-19} J $= 4.22$ eV.

b) If the equipartition principle were applicable (which it isn't), the average energy would equal $\frac{3}{2}kT$. In that case,

$$T = \frac{2E_{av}}{3k} = \frac{2(6.76 \times 10^{-19}\text{ J})}{3(1.381 \times 10^{-23}\text{ J/K})} = 3.26 \times 10^4\text{ K}.$$

Copper *vaporizes* at 2868 K, so it couldn't be a solid at 3.26×10^4 K.

c) We use $E_F = \frac{1}{2}mv_F^2$:

$$v_F = \sqrt{\frac{2E_F}{m}} = \sqrt{\frac{2(1.126 \times 10^{-18} \text{ J})}{9.11 \times 10^{-31} \text{ kg}}} = 1.57 \times 10^6 \text{ m/s}.$$

The quantity v_F is called the *Fermi speed.* We invite to you show, for comparison, that electrons with the average kinetic energy

$\frac{3}{2}kT$ (as incorrectly predicted by the equipartition principle and the Maxwell speed distribution) even at room temperature would have an rms speed of 1.17×10^5 m/s, less than $\frac{1}{10}$ of the above result.

Example 44–9 shows that the average energies and Fermi speeds of free electrons in metals are generally determined almost entirely by the exclusion principle; even at ordinary temperature the behavior is very far from what the equipartition principle predicts. We could make a similar analysis to determine the electronic contributions to heat capacities in a solid. If there is one conduction electron per atom, the equipartition principle would predict an additional molar heat capacity at constant volume of $3R/2$ from electron kinetic energies. But when kT is much smaller than E_F, the usual situation in metals, then only the few electrons near the Fermi level can find empty states and change energies appreciably when the temperature changes. The number of such electrons is proportional to kT/E_F, so we should expect the electron heat capacity at constant volume to be proportional to the product $(kT/E_F)(3R/2) = (3kT/2E_F)R$. A more detailed analysis shows that in fact the electron contribution to the molar heat capacity at constant volume of a metal is

$$C_V = \frac{\pi^2 kT}{2E_F} R \qquad \text{(contribution of the free electrons),}$$

not far from our prediction. We invite you to verify that if $T = 290$ K ($kT = 1/40$ eV) and if $E_F = 7.0$ eV, then $C_V = 0.018R$, which is only 1.2% of the $3R/2$ prediction of the equipartition principle.

Fermi energies for metals typically fall in the range from 1.6 to 14 eV, and typical Fermi speeds are from 0.8 to 2.2×10^6 m/s.

44-7 SEMICONDUCTORS

A **semiconductor** has an electrical resistivity that is intermediate between those of good conductors and of good insulators. The tremendous importance of semiconductors in present-day electronics stems in part from the fact that their electrical properties are very sensitive to very small concentrations of impurities. We'll discuss the basic concepts using the semiconductor elements silicon (Si) and germanium (Ge) as examples.

Silicon and germanium are in Group IV of the periodic table. Each has four electrons in the outermost electron subshells ($3s^2 3p^2$ for Si, $4s^2 4p^2$ for Ge). Both crystallize in the diamond structure (Section 44–4) with covalent bonding; each atom lies at the center of a regular tetrahedron, forming a covalent bond with each of four nearest neighbors at the corners of the tetrahedron. Because all the valence electrons are involved in the bonding, at absolute zero the band structure (Section 44–5) has a completely filled valence band separated by an energy gap from an empty conduction band (Fig. 44–18c). This distribution makes these materials insulators at very low temperatures; their electrons have no nearby states available into which they can move in response to an applied electric field.

However, the energy gap E_g between the valence and conduction bands is small in comparison to the gap of 5 eV or more for many insulators; room temperature values are 1.42 eV for gallium arsenide, 1.12 eV for silicon, and only 0.67 eV for germanium. Thus even at room temperature a substantial number of electrons can gain enough energy to jump the gap to the conduction band, where they are dissociated from their parent atoms and are free to move about the crystal. The number of these electrons increases rapidly with temperature.

EXAMPLE 44–10

Jumping the gap Consider a material with the band structure described above, with its Fermi energy in the middle of the gap (Fig. 44–23). Find the probability that a state at the bottom of the conduction band is occupied at a temperature of 300 K if the band gap is a) 0.200 eV; b) 1.00 eV; c) 5.00 eV. Repeat the calculations for a temperature of 310 K.

SOLUTION We use the Fermi-Dirac distribution function, Eq. (44–17) with $E - E_F = E_g/2$. a) When $E_g = 0.200$ eV,

$$\frac{E - E_F}{kT} = \frac{0.100 \text{ eV}}{(8.617 \times 10^{-5} \text{ eV/K})(300 \text{ K})} = 3.87,$$

$$f(E) = \frac{1}{e^{3.87} + 1} = 0.0205.$$

With $E_g = 0.200$ eV and $T = 310$ K, the exponent is 3.74 and $f(E) = 0.0231$, a 13% increase for a temperature rise of only 10 K.

b) When $E_g = 1.00$ eV, both exponents are five times as large as before, 19.3 and 18.7; the values of $f(E)$ are 4.0×10^{-9} and 7.4×10^{-9}. In this case the probability nearly doubles with a temperature rise of 10 K.

c) When $E_g = 5.00$ eV, the exponents are 96.7 and 93.6, the values of $f(E)$ are 1.0×10^{-42} and 2.3×10^{-41}. The probability increases by a factor of 23 for a 10-K temperature rise, but it is still extremely small. Pure diamond, with a 5.47-eV band gap, has essentially no electrons in the conduction band and is an excellent insulator.

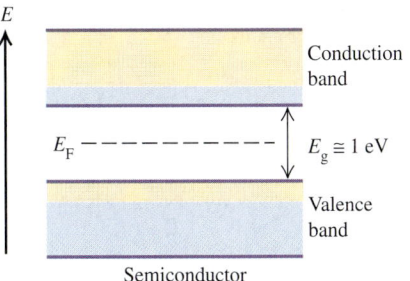

44–23 Band structure of a semiconductor. At absolute zero a completely filled valence band is separated by a narrow energy gap E_g of 1 eV or so from a completely empty conduction band. At ordinary temperatures, many electrons are excited to the conduction band.

This example illustrates two important points. First, the probability of finding an electron in a state at the bottom of the conduction band is extremely sensitive to the width of the band gap. When the gap is 0.20 eV the chance is about 2%, but when it is 1.00 eV, the chance is a few in a thousand million, and for a band gap of 5.0 eV it is essentially zero. Second, for any given band gap the probability is very temperature dependent, more so for large gaps than for small.

In principle, we could continue the calculation in Example 44–10 to find the actual density $n = N/V$ of electrons in the conduction band at any temperature. To do this, we would have to evaluate the integral $\int g(E)f(E) \, dE$ from the bottom of the conduction band to its top. First we would need to know the density of states function $g(E)$. It wouldn't be correct to use Eq. (44–16) because the energy-level structure and the density of states for real solids are more complex than those for the simple free-electron model. However, there are theoretical methods for predicting what $g(E)$ should be near the bottom of the conduction band, and such calculations have been carried out. Once we know n, we can *begin* to determine the resistivity of the material (and its temperature dependence) using the analysis of Section 26–3, which you may want to review. But next we'll see that the electrons in the conduction band don't tell the whole story about conduction in semiconductors.

HOLES

When an electron is removed from a covalent bond, it leaves a vacancy behind. An electron from a neighboring atom can move into this vacancy, leaving the neighbor with the vacancy. In this way the vacancy, called a **hole,** can travel through the material and serve as an additional current carrier. It's like describing the motion of a bubble in a liquid. In a pure, or *intrinsic,* semiconductor, valence-band holes and conduction-band electrons are always present in equal numbers. When an electric field is applied, they move in opposite directions (Fig. 44–24). Thus a hole in the valence band behaves like a positively charged particle, even though the moving charges in that band are electrons. The conductivity that we just described for a pure semiconductor is called *intrinsic conductivity.* Another kind of conductivity, to be discussed in the next subsection, is due to impurities.

44-24 Motion of electrons in the conduction band and of holes in the valence band of a semiconductor under the action of an applied electric field $\vec{E}$.

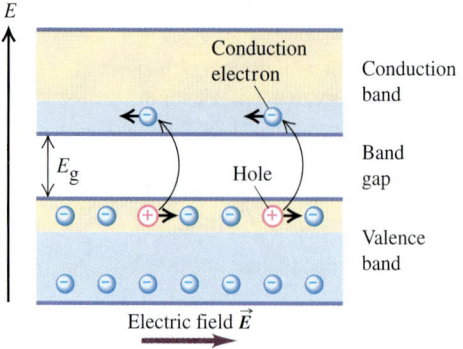

The parking-garage analogy that we mentioned in Section 44–5 helps to picture conduction in an intrinsic semiconductor. The valence band at absolute zero is like a completely filled floor. No cars (electrons) can move because there is nowhere for them to go. But if one car is moved to the vacant floor above, it can move freely, just as electrons can move freely in the conduction band. Also, the empty space that it leaves permits cars to move on the nearly filled floor, thereby moving the empty space just as holes move in the normally filled valence band.

IMPURITIES

Suppose we mix into melted germanium ($Z = 32$) a small amount of arsenic ($Z = 33$), the next element after germanium in the periodic table. This deliberate addition of impurity elements is called *doping*. Arsenic is in Group V; it has *five* valence electrons. When one of these electrons is removed, the remaining electron structure is essentially identical to that of germanium. The only difference is that it is smaller; the arsenic nucleus has a charge of $+33e$ rather than $+32e$, and it pulls the electrons in a little more. An arsenic atom can comfortably take the place of a germanium atom as a substitutional impurity. Four of its five valence electrons form the necessary nearest-neighbor covalent bonds.

The fifth valence electron is very loosely bound (Fig. 44–25a); it doesn't participate in the covalent bonds, and it is screened from the nuclear charge of $+33e$ by the 32 elec-

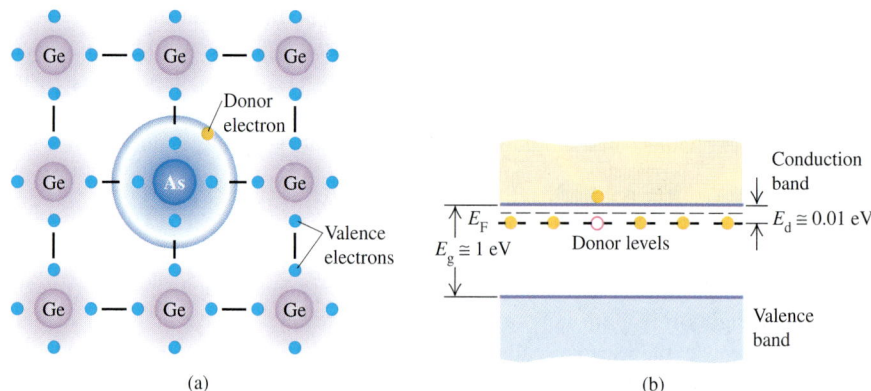

44-25 (a) A donor (*n*-type) impurity atom has a fifth valence electron that does not participate in the covalent bonding and is very loosely bound. In fact, if its probability density were drawn to scale, the maximum would be well off this page. (b) Energy-band diagram for an *n*-type semiconductor at a low temperature. The donor levels lie just below the conduction band. One donor electron has been excited into the conduction band. The Fermi level is between the donor levels and the bottom of the conduction band; as the temperature increases, it will move closer to the center of the energy gap.

trons, leaving a net effective charge of about $+e$. We might guess that the binding energy would be of the same order of magnitude as the energy of the $n = 4$ level in hydrogen, that is, $(\frac{1}{4})^2(13.6 \text{ eV}) = 0.85 \text{ eV}$. In fact, it is much smaller than this, only about 0.01 eV, because the electron probability distribution actually extends over many atomic diameters and the polarization of intervening atoms provides additional screening.

The energy level of this fifth electron corresponds in the band picture to an isolated energy level lying in the gap, about 0.01 eV below the bottom of the conduction band (Fig. 44–25b). This level is called a *donor level*, and the impurity atom that is responsible for it is simply called a *donor*. All Group V elements, including N, P, As, Sb, and Bi, can serve as donors. At room temperature, kT is about 0.025 eV. This is substantially greater than 0.01 eV, so at ordinary temperatures, most electrons can gain enough energy to jump from donor levels into the conduction band, where they are free to wander through the material. The remaining ionized donor stays at its site in the structure and does not participate in conduction.

Example 44–10 shows that at ordinary temperatures and with a band gap of 1.0 eV, only a very small fraction (of the order of 10^{-9}) of the states at the bottom of the conduction band in a pure semiconductor contain electrons to participate in intrinsic conductivity. Thus we expect the conductivity of such a semiconductor to be about 10^9 smaller than that of good metallic conductors, and measurements bear out this prediction. However, a concentration of donors as small as one part in 10^8 can increase the conductivity so drastically that conduction due to impurities becomes by far the dominant mechanism. In this case the conductivity is due almost entirely to *negative* charge (electron) motion. We call the material an **n-type semiconductor,** with *n*-type impurities.

Adding atoms of an element in Group III (B, Al, Ga, In, Tl), with only *three* valence electrons, has an analogous effect. An example is gallium ($Z = 31$); as a substitutional impurity in germanium, the gallium atom would like to form four covalent bonds, but it has only three outer electrons. It can, however, steal an electron from a neighboring germanium atom to complete the required four covalent bonds (Fig. 44–26a). The resulting atom has the same electron configuration as Ge but is somewhat larger because gallium's nuclear charge is smaller, $+31e$ instead of $+32e$.

(a) (b)

44–26 (a) An acceptor (*p*-type) impurity atom has only three valence electrons. It can borrow an electron from a neighboring atom to complete four covalent bonds. The resulting hole is then free to move about the crystal. (b) Energy-band diagram for a *p*-type semiconductor at a low temperature. The acceptor levels lie just above the valence band. One level has accepted an electron from the valence band, leaving a hole behind. The Fermi level is between the top of the valence band and the acceptor levels; as the temperature increases, it will move closer to the center of the energy gap.

This theft leaves the neighboring atom with a *hole,* or missing electron. The hole acts as a positive charge that can move through the crystal just as with intrinsic conductivity. The stolen electron is bound to the gallium atom in a level called an *acceptor level* about 0.01 eV above the top of the valence band (Fig. 44–26b). The gallium atom, called an *acceptor,* thus accepts an electron to complete its desire for four covalent bonds. This extra electron gives the previously neutral gallium atom a net charge of $-e$. The resulting gallium ion is *not* free to move. In a semiconductor that is doped with acceptors, we consider the conductivity to be almost entirely due to *positive* charge (hole) motion. We call the material a ***p*-type semiconductor,** with *p*-type impurities. Some semiconductors are doped with *both n*- and *p*-type impurities. Such materials are called *compensated* semiconductors.

CAUTION ▶ Saying that a material is a *p*-type semiconductor does *not* mean that the material has a positive charge; ordinarily, it would be neutral. Rather, it means that its *majority carriers* of current are positive holes (and therefore its *minority carriers* are negative electrons). The same idea holds for an *n*-type semiconductor; ordinarily, it will *not* have a negative charge, but its majority carriers are negative electrons. ◀

We can verify the assertion that the current in *n* and *p* semiconductors really *is* carried by electrons and holes, respectively, by using the Hall effect (optional Section 28–10). The sign of the Hall emf is opposite in the two cases. Hall-effect devices constructed from semiconductor materials are used in probes to measure magnetic fields and the currents that cause those fields.

44–8 SEMICONDUCTOR DEVICES

Semiconductor devices play an indispensable role in contemporary electronics. In the early days of radio and television, transmitting and receiving equipment relied on vacuum tubes, but these have been almost completely replaced in the last three decades by solid-state devices, including transistors, diodes, integrated circuits, and other semiconductor devices. The only surviving vacuum tubes in radio and TV equipment are the picture tube in most TV receivers, imaging devices in studio TV cameras, and high-power transmitting equipment.

A thin slab of semiconductor can serve as a *photocell.* When the material is irradiated with an electromagnetic wave whose photons have at least as much energy as the band gap between the valence and conduction bands, an electron in the valence band can absorb a photon and jump to the conduction band, where it (and the hole it left behind) contribute to the conductivity. The conductivity therefore increases with wave intensity. Detectors for charged particles operate on the same principle. An external circuit applies a voltage across a semiconductor. An energetic charged particle passing through the semiconductor collides inelastically with valence electrons, exciting them from the valence to the conduction band and creating pairs of holes and conduction electrons. The conductivity increases momentarily, causing a pulse of current in the external circuit. Solid-state detectors are widely used in nuclear and high-energy physics research; without them, some of the most recent discoveries could not have been made.

THE *p-n* JUNCTION

In many semiconductor devices the essential principle is the fact that the conductivity of the material is controlled by impurity concentrations, which can be varied within wide limits from one region of a device to another. An example is the ***p-n* junction** at the boundary between one region of a semiconductor with *p*-type impurities and another region containing *n*-type impurities. One way of fabricating a *p-n* junction is to deposit some *n*-type material on the *very* clean surface of some *p*-type material. (We can't just

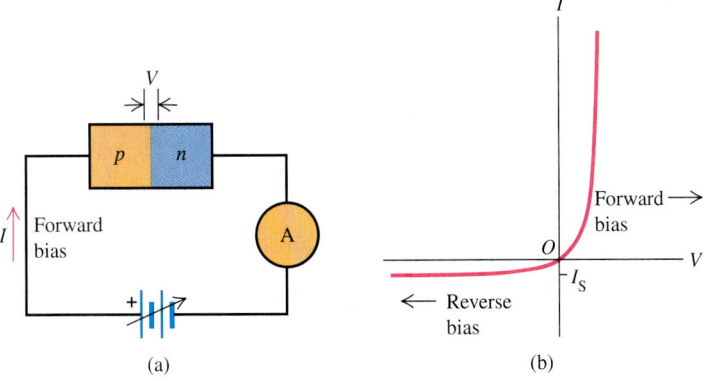

44-27 (a) A semiconductor *p-n* junction in a circuit. (b) Graph showing the asymmetric current-voltage relationship. The curve is described by Eq. (44-23).

stick *p*- and *n*-type pieces together and expect the junction to work properly, because of the impossibility of matching their surfaces at the atomic level.)

When a *p-n* junction is connected to an external circuit, as in Fig. 44-27a, and the potential difference $V_p - V_n = V$ across the junction is varied, the current *I* varies as shown in Fig. 44-27b. In striking contrast to the symmetric behavior of resistors that obey Ohm's law and give a straight line on an *I-V* graph, a *p-n* junction conducts much more readily in the direction from *p* to *n* than the reverse. Such a (mostly) one-way device is called a **diode rectifier.** Later we'll discuss a simple model of *p-n* junction behavior that predicts a current-voltage relation in the form

$$I = I_S(e^{eV/kT} - 1) \qquad \text{(current through a } p\text{-}n \text{ junction).} \qquad (44\text{-}23)$$

In the exponent, *e* is the quantum of charge, *k* is Boltzmann's constant, and *T* is absolute temperature.

CAUTION ▶ In $e^{eV/kT}$ the base of the exponent also uses the symbol *e*, standing for the base of the natural logarithms, 2.71828 ... . This *e* is quite different from *e* = 1.602×10^{-19} C in the exponent. ◀

Equation (44-23) is valid for both positive and negative values of *V*; note that *V* and *I* always have the same sign. As *V* becomes very negative, *I* approaches the value $-I_S$. The magnitude I_S (always positive) is called the *saturation current.*

EXAMPLE 44-11

Is a *p-n* junction diode always a one-way device? At a temperature of 290 K, a certain *p-n* junction diode has a saturation current $I_S = 0.500$ mA. Find the current at this temperature when the voltage is 1.00 mV, −1.00 mV, 100 mV, and −100 mV.

SOLUTION At *T* = 290 K, *kT* = 0.0250 eV = 25.0 meV. When *V* = 1.00 mV, *eV/kT* = *e*(1.00 mV)/(25.0 meV) = 0.0400. From Eq. (44-23) the current is

$$I = (0.500 \text{ mA})(e^{0.0400} - 1) = 0.0204 \text{ mA}.$$

When *V* = −1.00 mV,

$$I = (0.500 \text{ mA})(e^{-0.0400} - 1) = -0.0196 \text{ mA}.$$

The values of *I* for the other two voltages are obtained in the same way; when *V* = 100 mV, *I* = 26.8 mA; and when *V* = −100 mV, *I* = −0.491 mA. We summarize the data in the following table, also calculating the resistance *R* = *V/I*.

V (mV)	I (mA)	R (Ω)
+1.00	+0.0204	49.0
−1.00	−0.0196	51.0
+100	+26.8	3.73
−100	−0.491	204

Note that at $|V| = 1.00$ mV the current has nearly the same magnitude for both directions. That is, when $|V| \ll kT/e$ (near the origin of Fig. 44-27b), the curve approaches a straight line, and this junction diode acts more like a 50.0-Ω resistor than like a rectifier. However, as the voltage increases, the directional asymmetry becomes more and more pronounced. At $|V| = 0.100$ V the negative current is nearly equal to the saturation value and has a magnitude that is less than 2% of the positive current.

CURRENTS THROUGH A *p-n* JUNCTION

We can understand the behavior of a *p-n* junction diode qualitatively on the basis of the mechanisms for conductivity in the two regions. Suppose, as in Fig. 44–27a, you connect the positive terminal of the battery to the *p* region and the negative terminal to the *n* region. Then the *p* region is at higher potential than the *n*, corresponding to positive V in Eq. (44–23), and the resulting electric field is in the direction *p* to *n*. This is called the *forward* direction, and the positive potential difference is called *forward bias*. Holes, plentiful in the *p* region, flow easily across the junction into the *n* region, and free electrons, plentiful in the *n* region, easily flow into the *p* region; these movements of charge constitute a *forward* current. Connecting the battery with the opposite polarity gives *reverse bias,* and the field tends to push electrons from *p* to *n* and holes from *n* to *p*. But there are very few free electrons in the *p* region and very few holes in the *n* region. As a result, the current in the *reverse* direction is much smaller than that with the same potential difference in the forward direction.

Suppose you have a box with a barrier separating the left and right sides: You fill the left side with PF_3 gas and the right side with N_2 gas. What happens if the barrier leaks? PF_3 diffuses to the right, and N_2 diffuses to the left. A similar diffusion occurs across a *p-n* junction. First consider the equilibrium situation with no applied voltage (Fig. 44–28). The many holes in the *p* region act like a hole gas that diffuses across the junction into the *n* region. Once there, the holes recombine with some of the many free electrons. Similarly, electrons diffuse from the *n* region to the *p*-region and fall into some of the many holes there. The hole and electron diffusion currents lead to a net positive charge in the *n* region and a net negative charge in the *p* region, causing an electric field in the direction from *n* to *p* at the junction. The potential energy associated with this field raises the electron energy levels in the *p* region relative to the same levels in the *n* region.

There are four currents across the junction, as shown. The diffusion processes lead to *recombination currents* of holes and electrons, labeled i_{pr} and i_{nr} in Fig. 44–28. At the same time, electron-hole pairs are generated in the junction region by thermal excitation. The electric field described above sweeps these electrons and holes out of the junction; electrons are swept opposite the field to the *n* side, and holes are swept in the same direction as the field to the *p* side. The corresponding currents, called *generation currents,* are labeled i_{pg} and i_{ng}. At equilibrium the magnitudes of the generation and recombination currents are equal:

$$\left| i_{pg} \right| = \left| i_{pr} \right| \qquad \text{and} \qquad \left| i_{ng} \right| = \left| i_{nr} \right|. \tag{44–24}$$

At thermal equilibrium the Fermi energy is the same at each point across the junction.

44–28 A *p-n* junction in equilibrium, with no externally applied field or potential difference. The generation and recombination currents exactly balance. The Fermi energy E_F is the same on both sides of the junction. There is an excess of positive charge on the *n* side and of negative charge on the *p* side, resulting in an electric field $\vec{E}$ in the direction shown. (The energy E increases when the electric potential decreases because electrons have a negative charge.)

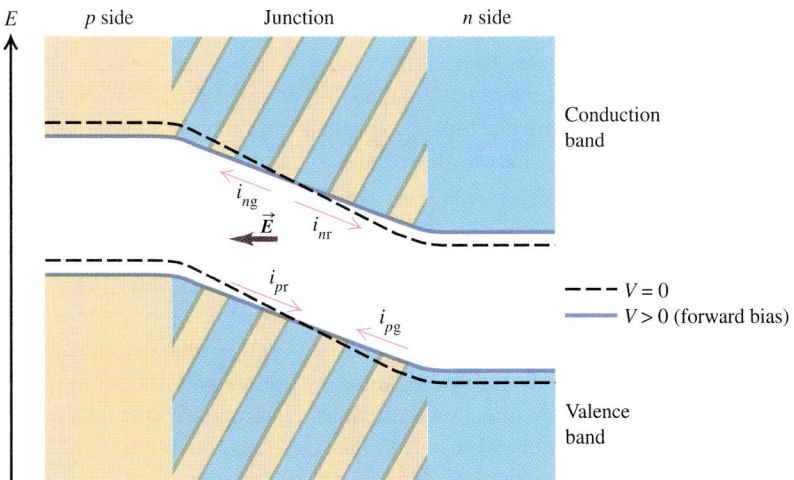

E

p side   Junction n side

Conduction band

i_{ng}
$\vec{E}$ i_{nr}

i_{pr}

i_{pg}

--- $V = 0$
— $V > 0$ (forward bias)

Valence band

44-29 A p-n junction under forward-bias conditions. The electric field in the junction region, and therefore the potential difference between p and n regions, is reduced. The recombination currents increase but the generation currents are nearly constant, causing a net current from left to right.

Now we apply a forward bias, a positive potential difference V across the junction. A forward bias *decreases* the electric field in the junction region. It also decreases the difference between the energy levels on the p and n sides (Fig. 44–29) by an amount $\Delta E = -eV$. It becomes easier for the electrons in the n region to climb the potential-energy hill and diffuse into the p region and for the holes in the p region to diffuse into the n region. This effect increases both recombination currents by the Maxwell-Boltzmann factor $e^{-\Delta E/kT} = e^{eV/kT}$. (We don't have to use the Fermi-Dirac distribution because most of the available states for the diffusing electrons and holes are empty, so the exclusion principle has little effect.) The generation currents don't change appreciably, so the net hole current is

$$
\begin{aligned}
i_{p\text{tot}} &= i_{pr} - \left| i_{pg} \right| \\
&= \left| i_{pg} \right| e^{eV/kT} - \left| i_{pg} \right| \\
&= \left| i_{pg} \right| (e^{eV/kT} - 1).
\end{aligned}
\tag{44-25}
$$

The net electron current $i_{n\text{tot}}$ is given by a similar expression, so the total current $I = i_{p\text{tot}} + i_{n\text{tot}}$ is

$$
I = I_S(e^{eV/kT} - 1),
\tag{44-26}
$$

in agreement with Eq. (44–23). This entire discussion can be repeated for reverse bias (negative V and I) with the same result. Therefore Eq. (44–23) is valid for both positive and negative values.

Several effects make the behavior of practical p-n junction diodes more complex than this simple analysis predicts. One effect, *avalanche breakdown,* occurs under large reverse bias. The electric field in the junction is so great that the carriers can gain enough energy between collisions to create electron-hole pairs during inelastic collisions. The electrons and holes then gain energy and collide to form more pairs, and so on. (A similar effect occurs in dielectric breakdown in insulators, discussed in Section 44–5.)

A second type of breakdown begins when the reverse bias becomes large enough that the top of the valence band in the p region is just higher in energy than the bottom of the conduction band in the n region (Fig. 44–30). If the junction region is thin enough, the probability becomes large that electrons can *tunnel* from the valence band of the p region to the conduction band of the n region. This process is called *Zener breakdown*. It occurs in Zener diodes, which are widely used for voltage regulation and protection against voltage surges.

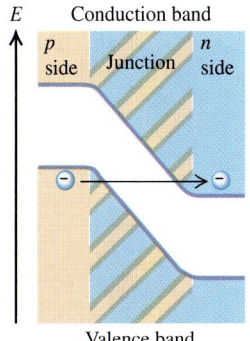

E Conduction band

p side   Junction n side

Valence band

44-30 Under reverse-bias conditions the potential-energy difference between the p and n sides of a junction is greater than at equilibrium. If this difference is great enough, the bottom of the conduction band on the n side may actually be below the top of the valence band on the p side. If the junction is thin enough, electrons can easily tunnel through from the valence band to the conduction band; this process is called Zener breakdown.

SEMICONDUCTOR DEVICES AND LIGHT

A *light-emitting diode (LED)* is, as the name implies, a *p-n* junction diode that emits light. When the junction is forward biased, many holes are pushed from their *p* region to the junction region, and many electrons are pushed from their *n* region to the junction region. In the junction region the electrons fall into holes (recombine). In recombining, the electron can emit a photon with energy approximately equal to the band gap. This energy (and therefore the photon wavelength and the color of the light) can be varied by using materials with different band gaps. Light-emitting diodes are widely used for digital displays in clocks, electronic equipment, automobile instrument panels, and many other applications.

The reverse process is called the *photovoltaic effect.* Here the material absorbs photons, and electron-hole pairs are created. Pairs that are created in the *p-n* junction, or close enough to migrate to it without recombining, are separated by the electric field we described above that sweeps the electrons to the *n* side and the holes to the *p* side. We can connect this device to an external circuit, where it becomes a source of emf and power. Such a device is often called a *solar cell,* although sunlight isn't required. *Any* light with photon energies greater than the band gap will do. You might have a calculator powered by such cells. Production of low-cost photovoltaic cells for large-scale solar energy conversion is a very active field of research. The same basic physics is used in charge-coupled device (CCD) image detectors, used in digital cameras and camcorders and for astronomical research.

TRANSISTORS

A *bipolar junction transistor* includes two *p-n* junctions in a "sandwich" configuration, which may be either *p-n-p* or *n-p-n*. Such a *p-n-p* transistor is shown in Fig. 44–31. The three regions are called the emitter, base, and collector, as shown. When there is no current in the left loop of the circuit, there is only a very small current through the resistor *R* because the voltage across the base-collector junction is in the reverse direction. But when a forward bias is applied between emitter and base, as shown, most of the holes traveling from emitter to base travel *through* the base (which is typically both narrow and lightly doped) to the second junction, where they come under the influence of the collector-to-base potential difference and flow on through the collector to give an increased current to the resistor.

In this way the current in the collector circuit is *controlled* by the current in the emitter circuit. Furthermore, V_c may be considerably larger than V_e, so the *power* dissipated in *R* may be much larger than the power supplied to the emitter circuit by the battery V_e. Thus the device functions as a *power amplifier.* If the potential drop across *R* is greater than V_e, it may also be a voltage amplifier.

In this configuration the *base* is the common element between the "input" and "output" sides of the circuit. Another widely used arrangement is the *common-emitter* circuit, shown in Fig. 44–32. In this circuit the current in the collector side of the circuit is much larger than that in the base side, and the result is current amplification.

The *field-effect transistor* (Fig. 44–33) is an important type. In one variation a slab of *p*-type silicon is made with two *n*-type regions on the top, called the *source* and the *drain;* a metallic conductor is fastened to each. A third electrode called the *gate is* separated from the slab, source, and drain by an insulating layer of SiO_2. When there is no charge on the gate and a potential difference of either polarity is applied between the source and the drain, there is very little current because one of the *p-n* junctions is reverse biased.

Now we place a positive charge on the gate. With dimensions of the order of 10^{-6} m, it takes little charge to provide a substantial electric field. Thus there is very little current into or out of the gate. There aren't many free electrons in the *p*-type material, but

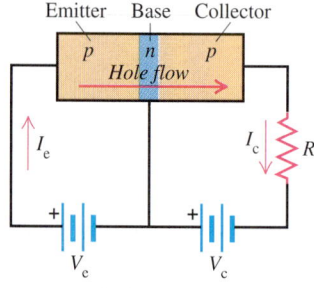

44–31 Schematic diagram of a *p-n-p* transistor and circuit. When $V_e = 0$, the current in the collector circuit is very small. When a potential V_e is applied between emitter and base, holes travel from emitter to base, as shown; when V_c is sufficiently large, most of the holes continue into the collector. The collector current I_c is controlled by the emitter current I_e.

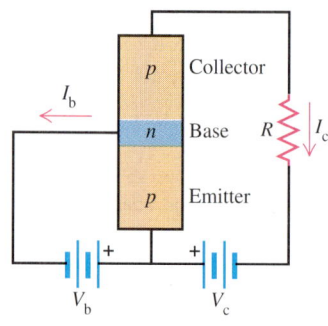

44–32 A common-emitter circuit. When $V_b = 0$, I_c is very small, and most of the voltage V_c appears across the base-collector junction. As V_b increases, the base-collector potential decreases, and more holes can diffuse into the collector; thus I_c increases. Ordinarily, I_c is much larger than I_b.

44–33 A field-effect transistor. The current from source to drain is controlled by the potential difference between the source and the drain and by the charge on the gate; no current flows through the gate.

there are some, and the effect of the field is to attract them toward the positive gate. The resulting greatly enhanced concentration of electrons near the gate (and between the two junctions) permits current to flow between the source and the drain. The current is very sensitive to the gate charge and potential, and the device functions as an amplifier. The device just described is called an *enhancement-type MOSFET* (metal-oxide-semiconductor field-effect transistor).

INTEGRATED CIRCUITS

A further refinement in semiconductor technology is the *integrated circuit.* By successively depositing layers of material and etching patterns to define current paths, we can combine the functions of several MOSFETs, capacitors, and resistors on a single square of semiconductor material that may be only a few millimeters on a side. An elaboration of this idea leads to *large-scale integrated circuits* and *very-large-scale integration* (VLSI). The resulting integrated circuit chips are the heart of all pocket calculators and present-day computers, large and small. An example is shown in Fig. 44–34.

The first semiconductor devices were invented in 1947. Since then, they have completely revolutionized the electronics industry through miniaturization, reliability, speed, energy usage, and cost. They have found applications in communications, computer systems, control systems, and many other areas. In transforming these areas, they have changed, and continue to change, human civilization itself.

44–34 Large-scale integrated circuits. A silicon chip that is less than 1 cm on a side (about the size of a ladybug) can contain millions of transistors, providing more computing power for much less money than early room-sized vacuum-tube computers did.

44–9 SUPERCONDUCTIVITY

Superconductivity is the complete disappearance of all electrical resistance at low temperatures. We described this property at the end of Section 26–3 and the magnetic properties of type I and type II superconductors in Section 30–9. In this section we'll relate superconductivity to the structure and energy-band model of a solid.

Although superconductivity was discovered in 1911, it was not well understood on a theoretical basis until 1957. In that year, the American physicists John Bardeen, Leon Cooper, and Robert Schrieffer published the theory of superconductivity, now called the BCS theory, that was to earn them the Nobel Prize in physics in 1972. (It was Bardeen's second Nobel Prize; he shared his first for his work on the development of the transistor.) The key to the BCS theory is an interaction between *pairs* of conduction electrons, called *Cooper pairs,* caused by an interaction with the positive ions of the crystal. Here's a rough qualitative picture of what happens. A free electron exerts attractive forces on nearby positive ions, pulling them slightly closer together. The resulting slight concentration of positive charge then exerts an attractive force on another free electron with momentum opposite to the first. At ordinary temperatures this electron-pair interaction is very small in comparison to energies of thermal motion, but at very low temperatures it becomes significant.

Bound together this way, the pairs of electrons cannot *individually* gain or lose very small amounts of energy, as they would ordinarily be able to do in a partly filled conduction band. Their pairing gives an energy gap in the allowed electron quantum levels, and at low temperatures there is not enough collision energy to jump this gap. Therefore the electrons can move freely through the crystal without any energy exchange through collisions, that is, with zero resistance.

Researchers in the field have not yet reached a consensus on whether or not some modification of the BCS theory can explain the properties of the high-T_C superconductors that have been discovered since 1986. There *is* evidence for pairing, but possibly of holes rather than electrons. Furthermore, the original pairing mechanism of the BCS theory seems too weak to explain the high transition temperatures and critical fields of these new superconductors.

SUMMARY

- The principal types of molecular bonds are ionic, covalent, van der Waals, and hydrogen bonds.

- In a diatomic molecule the rotational energy levels are

$$E = l(l+1)\frac{\hbar^2}{2I} \qquad (l = 0, 1, 2, \ldots). \tag{44-3}$$

The moment of inertia of a diatomic molecule is

$$I = m_r r_0{}^2, \tag{44-6}$$

where m_r is the reduced mass,

$$m_r = \frac{m_1 m_2}{m_1 + m_2}. \tag{44-4}$$

The vibrational energy levels are

$$E_n = \left(n + \frac{1}{2}\right)\hbar\omega = \left(n + \frac{1}{2}\right)\hbar\sqrt{\frac{k'}{m_r}} \qquad (n = 0, 1, 2, \ldots). \tag{44-7}$$

- Interatomic bonds in solids are of the same types as in molecules plus one additional type, the metallic bond. Associating the basis with each lattice point gives the crystal structure.

- When atoms are bound together in condensed matter, their outer energy levels spread out into bands. At absolute zero, insulators and semiconductors have a completely filled valence band separated by an energy gap from an empty conduction band. Conductors, including metals, have partially filled conduction bands.

- In the free-electron model of the behavior of conductors, the electrons are treated as completely free particles inside the material. In this model, the density of states is

$$g(E) = \frac{(2m)^{3/2} V}{2\pi^2 \hbar^3} E^{1/2}. \tag{44-16}$$

The distribution of electrons in the various energy states is determined by the Fermi-Dirac distribution:

$$f(E) = \frac{1}{e^{(E - E_F)/kT} + 1}, \tag{44-17}$$

where E_F is the Fermi energy. This distribution takes into account the exclusion principle and the indistinguishability of electrons.

- A semiconductor has an energy gap of about 1 eV between its valence and conduction bands. Its electrical properties may be drastically changed by the addition of small concentrations of donor impurities, giving an n-type semiconductor, or acceptor impurities, giving an p-type semiconductor.

- Many semiconductor devices, including diodes, transistors, and integrated circuits use one or more p-n junctions. The voltage-current relation for an ideal p-n junction diode is

$$I = I_S(e^{eV/kT} - 1). \qquad (44\text{--}23)$$

DISCUSSION QUESTIONS

Q44–1 Ionic bonds result from the electrical attraction of oppositely charged particles. Are other types of molecular bonds also electrical in nature, or is some other interaction involved? Explain.

Q44–2 In ionic bonds, an electron is transferred from one atom to another and thus no longer "belongs" to the atom from which it came. Are there similar transfers of ownership of electrons with other types of molecular bonds? Explain.

Q44–3 Van der Waals bonds occur in many molecules, but hydrogen bonds occur only with materials containing hydrogen. Why is this type of bond unique to hydrogen?

Q44–4 The bonding of gallium arsenide (GaAs) is said to be 31% ionic and 69% covalent. Explain.

Q44–5 The H_2^+ molecule consists of two hydrogen nuclei and a single electron. What kind of molecular bond do you think holds this molecule together? Explain.

Q44–6 The moment of inertia for an axis through the center of mass of a diatomic molecule calculated from the wavelength emitted in an $l = 19 \rightarrow l = 18$ transition is different from the moment of inertia calculated from the wavelength of the photon emitted in an $l = 1 \rightarrow l = 0$ transition. Explain this difference. Which transition corresponds to the larger moment of inertia?

Q44–7 Analysis of the photon absorption spectrum of a diatomic molecule shows that the vibrational energy levels for small values of n are very nearly equally spaced but the levels for large n are not equally spaced. Discuss the reason for this observation. Would you expect the adjacent levels to move closer together or farther apart as n increases? Explain.

Q44–8 Discuss the differences between the rotational and vibrational energy levels of the deuterium ("heavy hydrogen") molecule D_2 and those of the ordinary hydrogen molecule H_2. A deuterium atom has 2.00 the mass of an ordinary hydrogen atom.

Q44–9 Various organic molecules have been discovered in interstellar space. Why were these discoveries made with radio telescopes rather than optical telescopes?

Q44–10 The air you are breathing contains primarily nitrogen (N_2) and oxygen (O_2). Many of these molecules are in excited rotational energy levels ($l = 1, 2, 3, \ldots$), but almost all of them are in the vibrational ground level ($n = 0$). Explain this difference between the rotational and vibrational behavior of the molecules.

Q44–11 In what ways do atoms in a diatomic molecule behave as though they were held together by a spring? In what ways is this a poor description of the interaction between the atoms?

Q44–12 What factors determine whether a material is a conductor of electricity or an insulator? Explain.

Q44–13 Ionic crystals are often transparent, while metallic crystals are always opaque. Why?

Q44–14 Example 44–5 (Section 44–4) shows why the attractive ionic bonding energy for each ion must be summed over the entire crystal. Why, then, is the repulsive energy for each ion just summed over its nearest neighbors?

Q44–15 The structure of a given perfect crystal may not have as high a degree of symmetry as the lattice. Explain.

Q44–16 Speeds of molecules in a gas vary with temperature, while speeds of electrons in the conduction band of a metal are nearly independent of temperature. Why are these behaviors so different?

Q44–17 Use the band model to explain how it is possible for some materials to undergo a semiconductor-to-metal transition as the temperature or pressure varies.

Q44–18 An isolated zinc atom has a ground-state electron configuration of filled $1s$, $2s$, $2p$, $3s$, $3p$, and $4s$ subshells. How can zinc be a conductor if its valence subshell is full?

Q44–19 Why are materials that are good thermal conductors also good electrical conductors? What kinds of problems does this pose for the design of appliances such as clothes irons and electric heaters? Are there materials that do not follow this general rule?

Q44–20 What is the essential characteristic for an element to serve as a donor impurity in a semiconductor such as Si or Ge? For it to serve as an acceptor impurity? Explain.

Q44–21 There are several methods for removing electrons from the surface of a semiconductor. Can holes be removed from the surface? Explain.

Q44–22 A student asserts that silicon and germanium become good insulators at very low temperatures and good conductors at very high temperatures. Do you agree? Explain your reasoning.

Q44–23 The electric conductivities of most metals decrease gradually with increasing temperature, but the intrinsic conductivity of semiconductors always *increases* rapidly with increasing temperature. What causes the difference?

Q44–24 How could you make compensated silicon that has twice as many acceptors as donors?

Q44–25 For electronic devices such as amplifiers, what are some advantages of transistors compared to vacuum tubes? What are some disadvantages? Are there any situations in which vacuum tubes *cannot* be replaced by solid-state devices? Explain your reasoning.

Q44–26 Why does tunneling limit the miniaturization of MOSFETs?

Q44–27 The saturation current I_S for a *p-n* junction (Eq. (44–23)) depends strongly on temperature. Explain why.

EXERCISES

SECTION 44–2 TYPES OF MOLECULAR BONDS

44–1 We know from Chapter 16 that the average kinetic energy of an ideal gas atom or molecule at Kelvin temperature T is $\frac{3}{2}kT$. For what value of T does this energy correspond to a) the bond energy of the van der Waals bond in He_2 (7.9×10^{-4} eV); b) the bond energy of the covalent bond in H_2 (4.48 eV)? c) The kinetic energy in a collision between molecules can go into dissociating one or both molecules, provided the kinetic energy is greater than the bond energy. At room temperature (300 K), is it likely that He_2 molecules will remain intact after a collision? What about H_2 molecules? Explain.

44–2 An Ionic Bond. a) Calculate the electric potential energy for a K^+ ion and a Br^- ion separated by a distance of 0.29 nm, the equilibrium separation in the KBr molecule. Treat the ions as point charges. b) The ionization energy of the potassium atom is 4.3 eV. Atomic bromine has an electron affinity of 3.5 eV. Use these data and the results of part (a) to estimate the binding energy of the KBr molecule. Do you expect the actual binding energy to be larger or smaller than your estimate? Explain your reasoning.

SECTION 44–3 MOLECULAR SPECTRA

44–3 A lithium atom has mass 1.17×10^{-26} kg, and a hydrogen atom has mass 1.67×10^{-27} kg. The equilibrium separation between the two nuclei in the LiH molecule is 0.159 nm. a) What is the difference in energy between the $l = 3$ and $l = 4$ rotational levels? b) What is the wavelength of the photon emitted in a transition from the $l = 4$ to $l = 3$ level?

44–4 Two atoms of cesium (Cs) can form a Cs_2 molecule. The equilibrium distance between the nuclei in a Cs_2 molecule is 0.447 nm. Calculate the moment of inertia about an axis through the center of mass of the two nuclei and perpendicular to the line joining them. The mass of a cesium atom is 2.21×10^{-25} kg.

44–5 a) Show that the energy difference between rotational levels with angular momentum quantum numbers l and $l - 1$ is $l\hbar^2/I$. b) In terms of l, $\hbar$, and I, what is the frequency of the photon emitted in the pure rotation transition $l \rightarrow l - 1$?

44–6 If a sodium chloride (NaCl) molecule could undergo an $n \rightarrow n - 1$ vibrational transition with no change in rotational quantum number, a photon with wavelength 20.0 μm would be emitted. The mass of a sodium atom is 3.82×10^{-26} kg, and the mass of a chlorine atom is 5.81×10^{-26} kg. Calculate the force constant k' for the interatomic force in NaCl.

44–7 The rotational energy levels of the CO molecule are calculated in Example 44–2 and the vibrational level energy

differences are given in Example 44–3 (Section 44–3). The vibration-rotation energies are given by Eq. (44–9). Calculate the wavelength of the photon absorbed by CO in each of the following vibration-rotation transitions: a) $n = 0$, $l = 1 \rightarrow n = 1$, $l = 2$; b) $n = 0$, $l = 2 \rightarrow n = 1$, $l = 1$; c) $n = 0$, $l = 3 \rightarrow n = 1$, $l = 2$.

44–8 The force constant for the O_2 molecule is 1180 N/m. The mass of an oxygen atom is 2.66×10^{-26} kg. What is the energy separation in electron volts between adjacent vibrational levels of the molecule?

44–9 The vibration frequency for the molecule HF is 1.24×10^{14} Hz. The mass of a hydrogen atom is 1.67×10^{-27} kg, and the mass of a fluorine atom is 3.15×10^{-26} kg. a) What is the force constant k' for the interatomic force? b) What is the spacing between adjacent vibrational energy levels in joules and in electron volts? c) What is the wavelength of a photon of energy equal to the energy difference between two adjacent vibrational levels? In what region of the spectrum does it lie?

SECTION 44–4 STRUCTURE OF SOLIDS

44–10 Calculate the wavelength of a) a 6.20-keV x ray; b) a 37.6-eV electron; c) a 0.0205-eV neutron.

44–11 Density of NaCl. The spacing of adjacent atoms in a crystal of sodium chloride is 0.282 nm. The mass of a sodium atom is 3.82×10^{-26} kg, and the mass of a chlorine atom is 5.89×10^{-26} kg. Calculate the density of sodium chloride.

44–12 Potassium bromide, KBr, has a density of 2.75×10^3 kg/m^3 and the same crystal structure as NaCl. The mass of a potassium atom is 6.49×10^{-26} kg, and the mass of a bromine atom is 1.33×10^{-25} kg. Calculate the density of sodium chloride. a) Calculate the average spacing between adjacent atoms in a KBr crystal. b) How does the value calculated in part (a) compare with the spacing in NaCl (Exercise 44–11)? Is the relation between the two values qualitatively what you would expect? Explain.

SECTION 44–5 ENERGY BANDS

44–13 The gap between valence and conduction bands in silicon is 1.12 eV. A nickel nucleus in an excited state emits a gamma-ray photon with wavelength 9.31×10^{-4} nm. How many electrons can be excited from the top of the valence band to the bottom of the conduction band by the absorption of this gamma ray?

44–14 The gap between valence and conduction bands in diamond is 5.47 eV. a) What is the maximum wavelength of a photon that can excite an electron from the top of the valence band into the conduction band? In what region of the electromagnetic spectrum does this photon lie? b) Explain why pure diamond is transparent and colorless. c) Most gem diamonds have a yellow color. Explain how impurities in the diamond can cause this color.

44-15 Consider a photocell that uses pure silicon rather than germanium (Example 44–9, Section 44–6). a) If the gap between valence and conduction bands in silicon is 1.12 eV, what maximum wavelength can be detected by a silicon photocell? In what region of the electromagnetic spectrum does this photon lie? b) Explain why pure silicon is opaque.

SECTION 44–6 FREE-ELECTRON MODEL OF METALS

44-16 Calculate v_{rms} for free electrons with average kinetic energy $\frac{3}{2}kT$ at a temperature of 300 K. How does your result compare to the speed of an electron with a kinetic energy equal to the Fermi energy of copper, calculated in Example 44–9 (Section 44–6). Why is there such a difference between these speeds?

44-17 a) Show that the wave function ψ given in Eq. (44–10) is a solution of the three-dimensional Schrödinger equation (Eq. (42–32)) with the energy as given by Eq. (44–11). b) What are the energies of the ground level and the lowest two excited levels? What is the degeneracy of each of these levels? (Include the factor of 2 in the degeneracy that is due to the two possible spin states.)

44-18 What is the value of the constant A in Eq. (44–10) that makes $\psi(x, y, z)$ normalized?

44-19 Calculate the density of states $g(E)$ for the free-electron model of a metal if $E = 5.0$ eV and $V = 1.0$ cm^3. Express your answer in units of states per electron volt.

44-20 Supply the details in the derivation of Eq. (44–14) from Eqs. (44–13) and (44–12).

44-21 Calculate n_{rs} from Eq. (44–13) for $E = 7.0$ eV and $L = 1.0$ cm.

44-22 The Fermi energy of sodium is 3.23 eV. a) Find the average energy E_{av} of the electrons at absolute zero. b) What is the speed of an electron that has energy E_{av}? c) At what Kelvin temperature T is kT equal to E_F? (This is called the *Fermi temperature* for the metal. It is approximately the temperature at which molecules in a classical ideal gas would have the same kinetic energy as the fastest-moving electron in the metal.)

44-23 Silver has a Fermi energy of 5.48 eV. Calculate the electron contribution to the molar heat capacity at constant volume of silver, C_V, at 300 K. Express your result a) as a multiple of R; b) as a fraction of the actual value for silver, $C_V = 25.3$ J/mol · K. c) Is the value of C_V due principally to the electrons? If not, what is it due to? (*Hint:* See Section 16–5.)

SECTION 44–7 SEMICONDUCTORS

44-24 Pure germanium has a band gap of 0.67 eV. The Fermi energy is in the middle of the gap. a) For temperatures of 250 K, 300 K, and 350 K, calculate the probability $f(E)$ that a state at the bottom of the conduction band is occupied. b) For each temperature in part (a), calculate the probability that a state at the top of the valence band is empty.

44-25 Germanium has a band gap of 0.67 eV. Doping with arsenic adds donor levels in the gap 0.01 eV below the bottom of the conduction band. At a temperature of 300 K, the probability is 4.4×10^{-4} that an electron state is occupied at the bottom of the conduction band. Where is the Fermi level relative to the conduction band in this case?

SECTION 44–8 SEMICONDUCTOR DEVICES

44-26 A *p-n* junction has a saturation current of 3.60 mA. a) At a temperature of 300 K, what voltage is needed to produce a positive current of 40.0 mA? b) For a voltage equal to the negative of the value calculated in part (a), what is the negative current?

44-27 a) A forward-bias voltage of 15.0 mV produces a positive current of 9.25 mA through a *p-n* junction at 300 K. What does the positive current become if the forward-bias voltage is reduced to 10.0 mV? b) For reverse-bias voltages of −15.0 mV and −10.0 mV, what is the reverse-bias negative current?

PROBLEMS

44-28 When a diatomic molecule undergoes a transition from the $l = 2$ to the $l = 1$ rotational state, a photon with wavelength 63.8 μm is emitted. What is the moment of inertia of the molecule for an axis through its center of mass and perpendicular to the line connecting the nuclei?

44-29 a) The equilibrium separation of the two nuclei in a NaCl molecule is 0.24 nm. If the molecule is modeled as charges $+e$ and $-e$ separated by 0.24 nm, what is the electric dipole moment of the molecule (see Section 22–9)? b) The measured electric dipole moment of an NaCl molecule is 3.0×10^{-29} C · m. If this dipole moment arises from point charges $+q$ and $-q$ separated by 0.24 nm, what is q? c) A definition of the *fractional ionic character* of the bond is q/e. If sodium atom had charge $+e$ and the chlorine atom has charge $-e$, the fractional ionic character would be equal to 1. What is the actual fractional ionic character for the bond in NaCl? d) The equilibrium distance between nuclei in the hydrogen iodide (HI) molecule is 0.16 nm, and the measured electric dipole moment of the molecule is 1.5×10^{-30} C · m. What is the fractional ionic character for the bond in HI? How does your answer compare to that for NaCl calculated in part (c)? Discuss reasons for the difference in these results.

44-30 The binding energy of a potassium chloride molecule (KCl) is 4.43 eV. The ionization energy of a potassium atom is 4.3 eV, and the electron affinity of chlorine is 3.6 eV. Use these data to estimate the equilibrium separation between the two atoms in the KCl molecule. Explain why your result is only an estimate and not a precise value.

44-31 a) For the sodium chloride molecule (NaCl) discussed at the beginning of Section 44–2, what is the maximum separation of the ions for stability if they may be regarded as point charges? That is, what is the largest separation for which the energy of an

Na$^+$ ion and a Cl$^-$ ion, calculated in this model, is lower than the energy of the two separate atoms Na and Cl? b) Calculate this distance for the potassium bromide molecule, described in Exercise 44–2.

44–32 The rotational spectrum of HCl contains the following wavelengths (among others): 60.4 μm, 69.0 μm, 80.4 μm, 96.4 μm, and 120.4 μm. Use this spectrum to find the moment of inertia of the HCl molecule about an axis through the center of mass and perpendicular to the line joining the two nuclei.

44–33 a) Use the result of Problem 44–32 to calculate the equilibrium separation of the atoms in an HCl molecule. The mass of a chlorine atom is 5.81×10^{-26} kg, and the mass of a hydrogen atom is 1.67×10^{-27} kg. b) The value of l changes by ± 1 in rotational transitions. What is the value of l for the upper level of the transition that gives rise to each of the wavelengths listed in Problem 44–32? c) What is the longest-wavelength line in the rotational spectrum of HCl? d) Calculate the wavelengths of the emitted light for the corresponding transitions in the deuterium chloride (DCl) molecule. In this molecule the hydrogen atom in HCl is replaced by an atom of deuterium, an isotope of hydrogen with a mass of 3.34×10^{-27} kg. Assume that the equilibrium separation between the atoms is the same as for HCl.

44–34 When a NaF molecule makes a transition from the $l = 3$ to $l = 2$ rotational level with no change in vibrational quantum number or electronic state, a photon with wavelength 3.83 mm is emitted. A sodium atom has mass 3.82×10^{-26} kg, and a fluorine atom has mass 3.15×10^{-26} kg. Calculate the equilibrium separation between the nuclei in a NaF molecule. How does your answer compare with the value for NaCl given in Section 44.2? Is this result reasonable? Explain.

44–35 Consider a gas of diatomic molecules (moment of inertia I) at an absolute temperature T. If E_g is a ground-state energy and E_{ex} is the energy of an excited state, then the Boltzmann distribution (Section 40–7) predicts that the ratio of the numbers of molecules in the two states is

$$\frac{n_{ex}}{n_g} = e^{-(E_{ex} - E_g)/kT}.$$

a) Explain why the ratio of the number of molecules in the lth rotational energy *level* to the number of molecules in the ground ($l = 0$) rotational level is

$$\frac{n_l}{n_0} = (2l + 1)e^{-(l(l+1)\hbar^2)/2IkT}.$$

(*Hint:* For each value of l, how many states are there with different values of m_l?) b) Determine the ratio n_l/n_0 for a gas of CO molecules at 300 K for the cases i) $l = 1$; ii) $l = 2$; iii) $l = 10$; iv) $l = 20$; v) $l = 50$. The moment of inertia of the CO molecule is given in Example 44–2 (Section 44–3). c) Your results in part (b) show that as l is increased, the ratio n_l/n_0 first increases, then decreases. Explain why.

44–36 Our galaxy contains many *molecular clouds*, regions many light years in extent in which the density is high enough and the temperature low enough for atoms to form into mole-

cules. Most of the molecules are H$_2$, but a small fraction of the molecules are carbon monoxide (CO). Such a molecular cloud in the constellation Orion is shown in Fig. 44–35. The left-hand image was made with an ordinary visible-light telescope; the right-hand image shows the molecular cloud in Orion as imaged with a radio telescope tuned to a wavelength emitted by CO in a rotational transition. The different colors in the radio image indicate regions of the cloud that are moving either toward us (blue) or away from us (red) relative to the motion of the cloud as a whole, as determined by the Doppler shift of the radiation. (Since a molecular cloud has about 10,000 hydrogen molecules for each CO molecule, it might seem more reasonable to tune a radio telescope to emissions from H$_2$ than to emissions from CO. Unfortunately, it turns out that the H$_2$ molecules in molecular clouds do not radiate in either the radio or visible portions of the electromagnetic spectrum.) a) Using the data in Example 44–2 (Section 44–3), calculate the energy and wavelength of the photon emitted by a CO molecule in an $l = 1 \rightarrow l = 0$ rotational transition. b) As a rule, molecules in a gas at temperature T will be found in a certain excited rotational energy level provided the energy of that level is no greater than kT (see Problem 44–35). Use this rule to explain why astronomers can detect radiation from CO in molecular clouds even though the typical temperature of a molecular cloud is a very low 20 K.

FIGURE 44–35 Problem 44–36.

44–37 Spectral Lines from Isotopes. The equilibrium separation for NaCl is 0.2361 nm. The mass of a sodium atom is 3.8176×10^{-26} kg. Chlorine has two stable isotopes, ^{35}Cl and ^{37}Cl, that have different masses but identical chemical properties. The atomic mass of ^{35}Cl is 5.8068×10^{-26} kg, and the atomic mass of ^{37}Cl is 6.1384×10^{-26} kg. a) Calculate the wavelength of the photon emitted in the $l = 2 \rightarrow l = 1$ and $l = 1 \rightarrow l = 0$ transitions for Na^{35}Cl. b) Repeat part (a) for Na^{37}Cl. What are the differences in the wavelengths for the two isotopes?

44–38 When an OH molecule undergoes a transition from the $n = 0$ to the $n = 1$ vibrational level, its internal vibrational energy increases by 0.463 eV. Calculate the frequency of vibration and the force constant for the interatomic force. (The mass of an oxygen atom is 2.66×10^{-26} kg, and the mass of a hydrogen atom is 1.67×10^{-27} kg.)

44–39 The force constant for the internuclear force in a hydrogen molecule (H$_2$) is $k' = 576$ N/m. A hydrogen atom has mass 1.67×10^{-27} kg. Calculate the zero-point vibrational energy for H$_2$ (that is, the vibrational energy the molecule has in the $n = 0$ ground vibrational level). How does this energy compare in magnitude with the H$_2$ bond energy of -4.48 eV?

44–40 Suppose the hydrogen atom in HF (see Exercise 44–9) is replaced by an atom of deuterium, an isotope of hydrogen with a mass of 3.34×10^{-27} kg. The force constant is determined by the electron configuration, so it is the same as for the normal HF molecule. a) What is the vibrational frequency of this molecule? b) What wavelength of light corresponds to the energy difference between the $n = 1$ and $n = 0$ levels? In what region of the spectrum does this wavelength lie?

44–41 The hydrogen iodide (HI) molecule has equilibrium separation 0.160 nm and vibrational frequency 6.93×10^{13} Hz. The mass of a hydrogen atom is 1.67×10^{-27} kg, and the mass of an iodine atom is 2.11×10^{-25} kg. a) Calculate the moment of inertia of HI about a perpendicular axis through its center of mass. b) Calculate the wavelength of the photon emitted in each of the following vibration-rotation transitions: i) $n = 1, l = 1 \rightarrow n = 0$, $l = 0$; ii) $n = 1, l = 2 \rightarrow n = 0, l = 1$; iii) $n = 2, l = 2 \rightarrow n = 1, l = 3$.

44–42 Prove the following statement: For free electrons in a solid, if a state that is at an energy ΔE above E_F has probability P of being occupied, then the probability is $1 - P$ that a state at an energy ΔE below E_F is occupied.

44–43 Compute the Fermi energy of potassium by making the simple approximation that each atom contributes one free electron. The density of potassium is 851 kg/m^3, and the mass of a single potassium atom is 6.49×10^{-26} kg.

44–44 The one-dimensional calculation of Example 44–5 (Section 44–4) can be extended to three dimensions. For the three-dimensional fcc NaCl lattice, the result for the potential energy of a pair of Na$^+$ and Cl$^-$ ions due to the electrostatic interaction with all of the ions in the crystal is $U = -\alpha e^2/4\pi\epsilon_0 r$, where $\alpha = 1.75$ is the *Madelung constant*. Another contribution to the potential energy is a repulsive interaction at small ionic separation r due to overlap of the electron clouds. This contribution can be represented by A/r^8, where A is a positive constant. so the expression for the total potential energy is

$$U_{tot} = -\frac{\alpha e^2}{4\pi\epsilon_0 r} + \frac{A}{r^8}.$$

a) Let r_0 be the value of the ionic separation r for which U_{tot} is a minimum. Use this definition to find an equation that relates r_0 and A, and use this to write U_{tot} in terms of r_0. For NaCl, $r_0 = 0.281$ nm. Obtain a numerical value (in electron volts) of U_{tot} for NaCl. b) The quantity $-U_{tot}$ is the energy required to remove a Na$^+$ ion and a Cl$^-$ ion from the crystal. To form a pair of neutral atoms from this pair of ions involves a release of 5.14 eV (the ionization energy of Na) and the expenditure of 3.61 eV (the electron affinity of Cl). Use the result of part (a) to calculate the energy required to remove a pair of neutral Na and Cl atoms from the crystal. The experimental value for this quantity is 6.39 eV; how well does your calculation agree?

44–45 Consider a system of N free electrons within a volume V. Even at absolute zero, such a system exerts a pressure p on its surroundings due to the motion of the electrons. To calculate this pressure, imagine that the volume increases by a small amount dV. The electrons will do an amount of work $p\,dV$ on their surroundings, which means that the total energy E_{tot} of the electrons will change by an amount $dE_{tot} = -p\,dV$. Hence $p = -dE_{tot}/dV$. a) Show that the pressure of the electrons at absolute zero is

$$p = \frac{3^{2/3}\pi^{4/3}\hbar^2}{5m}\left(\frac{N}{V}\right)^{5/3}.$$

b) Evaluate this pressure for copper, which has a free-electron concentration of 8.45×10^{28} m^{-3}. Express your result in pascals and in atmospheres. c) The pressure you found in part (b) is extremely high. Why, then, don't the electrons in a piece of copper simply explode out of the metal?

44–46 When the pressure p on a material increases by an amount Δp, the volume of the material will change from V to $V + \Delta V$, where ΔV is negative. The *bulk modulus B* of the material is defined to be the ratio of the pressure change Δp to the absolute value $|\Delta V/V|$ of the fractional volume change. The greater the bulk modulus, the greater the pressure increase required for a given fractional volume change, and the more incompressible the material (Section 11–6). Since $\Delta V < 0$, the bulk modulus can be written as $B = -\Delta p/(\Delta V/V_0)$. In the limit that the pressure and volume changes are very small, this becomes

$$B = -V\frac{dp}{dV}.$$

a) Use the result of Problem 44–45 to show that the bulk modulus for a system of N free electrons in a volume V at low temperatures is $B = \frac{5}{3}p$. (*Hint:* The quantity p in the expression $B = -V(dp/dV)$ is the *external* pressure on the system. Can you explain why this is equal to the *internal* pressure of the system itself, as found in Problem 44–45?) b) Evaluate the bulk modulus for the electrons in copper, which has a free-electron concentration of 8.45×10^{28} m^{-3}. Express your result in pascals. c) The actual bulk modulus of copper is 1.4×10^{11} Pa. Based on your result in part (b), what fraction of this is due to the free electrons in copper? (This result shows that the free electrons in a metal play a major role in making the metal resistant to compression.) What do you think is responsible for the remaining fraction of the bulk modulus?

44–47 In the discussion of free electrons in Section 44–6, we assumed that we could ignore the effects of relativity. This is not a safe assumption if the Fermi energy is more than about $\frac{1}{100} mc^2$ (that is, more than about 1% of the rest energy of an electron). a) Assume that the Fermi energy at absolute zero, as given by Eq. (44–20), is equal to $\frac{1}{100} mc^2$. Show that the electron concentration is

$$\frac{N}{V} = \frac{2^{3/2} m^3 c^3}{3000 \pi^3 \hbar^3},$$

and determine the numerical value of N/V. b) Is it a good approximation to ignore relativistic effects for electrons in a metal such as copper, for which the electron concentration is

8.45×10^{28} m^{-3}? Explain. c) A *white dwarf star* is what is left behind by a star like the sun after it has ceased to produce energy by nuclear reactions. (Our own sun will become a white dwarf in another 6×10^9 years or so.) A typical white dwarf has a mass of 2×10^{30} kg (comparable to the sun) and a radius of 6000 km (comparable to that of the earth). The gravitational attraction of different parts of the white dwarf for each other tends to compress the star; what prevents it from compressing is the pressure of free electrons within the star (see Problem 44–45). Estimate the electron concentration within a typical white dwarf using the following assumptions: i) the white dwarf is made of carbon, which has a mass per atom of 1.99×10^{-26} kg; ii) all six of the electrons from each carbon atom are able to move freely throughout the star. d) Is it a good approximation to ignore relativistic effects in the structure of a white dwarf star? Explain.

CHALLENGE PROBLEMS

44–48 a) Consider the hydrogen molecule (H_2) to be a simple harmonic oscillator with an equilibrium spacing of 0.074 nm, and estimate the vibrational energy-level spacing for H_2. The mass of a hydrogen atom is 1.67×10^{-27} kg. (*Hint:* Estimate the force constant by equating the change in Coulomb repulsion of the protons, when the atoms move slightly closer together than r_0, to the "spring" force. That is, assume that the chemical binding force remains approximately constant as r is decreased slightly from r_0.) b) Use the results of part (a) to calculate the vibrational energy level spacing for the deuterium molecule, D_2. Assume that the spring constant is the same for H_2 and D_2. The mass of a deuterium atom is 3.34×10^{-27} kg.

44–49 Van der Waals bonds arise from the interaction between two permanent or induced electric dipole moments in a pair of atoms or molecules. a) Consider two identical dipoles, each consisting of charges $+q$ and $-q$ separated by a distance d and oriented as shown in Fig. 44–36a. Calculate the electric potential energy, expressed in terms of the electric dipole moment $p = qd$, for the

situation where $r \gg d$. Is the interaction attractive or repulsive, and how does this potential energy vary with r, the separation between the centers of the two dipoles? b) Repeat part (a) for the orientation of the dipoles shown in Fig. 44–36b. The dipole interaction is more complicated when we have to average over the relative orientations of the two dipoles due to thermal motion or when the dipoles are induced rather than permanent.

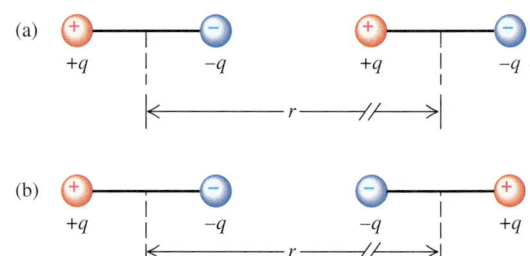

FIGURE 44–36 Challenge Problem 44–49.

45

Nuclear Physics

Using radioactive dating techniques, scientists have learned that this domesticated squash seed—one of several found in a cave in Mexico—is 8,000–10,000 years old. This discovery shows that agriculture began in the Americas at least as early as it did in Europe and Asia. This is just one example of how nuclear physics has become an essential tool for archaeology.

45–1 INTRODUCTION

During this century, applications of nuclear physics have had enormous effects on humankind, some beneficial, some catastrophic. Many people have strong opinions about applications such as bombs and reactors. Ideally, those opinions should be based on understanding, not on prejudice or emotion, and we hope this chapter will help you to reach that ideal.

Every atom contains at its center an extremely dense, positively charged *nucleus,* which is much smaller than the overall size of the atom but contains most of its total mass. We will look at several important general properties of nuclei and of the nuclear force that holds them together. The stability or instability of a particular nucleus is determined by the competition between the attractive nuclear force among the protons and neutrons and the repulsive electrical interactions among the protons. Unstable nuclei *decay,* transforming themselves spontaneously into other structures, by a variety of decay processes. Structure-altering nuclear reactions can also be induced by impact on a nucleus of a particle or another nucleus. Two classes of reactions of special interest are *fission* and *fusion.* We could not survive without the 3.90×10^{26}-watt output of one nearby fusion reactor, our sun.

45–2 PROPERTIES OF NUCLEI

As we described in Section 40–5, Rutherford found that the nucleus is tens of thousands of times smaller in radius than the atom itself. Since Rutherford's initial experiments, many additional scattering experiments have been performed, using high-energy protons, electrons, and neutrons as well as alpha particles (helium-4 nuclei). These experiments show that we can model a nucleus as a sphere with a radius R that depends on the total number of *nucleons* (neutrons and protons) in the nucleus. This number is called the **nucleon number,** A. The radii of most nuclei are represented quite well by the equation

$$R = R_0 A^{1/3} \quad \text{(radius of a nucleus)}, \quad (45-1)$$

where R_0 is an experimentally determined constant:

$$R_0 = 1.2 \times 10^{-15} \text{ m} = 1.2 \text{ fm}.$$

The nucleon number A in Eq. (45–1) is also called the **mass number** because it is the nearest whole number to the mass of the nucleus measured in unified atomic mass units (u). (The proton mass and the neutron mass are both approximately 1 u.) The best current conversion factor is

$$1 \text{ u} = 1.6605402(10) \times 10^{-27} \text{ kg}.$$

In Section 45–3 we'll discuss the masses of nuclei in more detail. Note that when we speak of the masses of nuclei and particles, we mean their rest masses.

NUCLEAR DENSITY

The volume V of a sphere is equal to $4\pi R^3/3$, so Eq. (45–1) shows that the *volume* of a nucleus is proportional to A. Dividing A (the approximate mass in u) by the volume gives us the approximate density and cancels out A. Thus *all nuclei have approximately the same density.* This fact is of crucial importance in understanding nuclear structure.

EXAMPLE 45-1

The most common kind of iron nucleus has a mass number of 56. Find the radius, approximate mass, and approximate density of the nucleus.

SOLUTION From Eq. (45–1) the radius is

$$R = R_0 A^{1/3} = (1.2 \times 10^{-15} \text{ m})(56)^{1/3} = 4.6 \times 10^{-15} \text{ m} = 4.6 \text{ fm.}$$

The mass number A is the approximate mass in u, so the mass is approximately

$$m \approx (56)(1.66 \times 10^{-27} \text{ kg}) = 9.3 \times 10^{-26} \text{ kg.}$$

The volume is

$$V = \frac{4}{3}\pi R^3 = \frac{4}{3}\pi (4.6 \times 10^{-15} \text{ m})^3 = 4.1 \times 10^{-43} \text{ m}^3,$$

and the density ρ is approximately

$$\rho = \frac{m}{V} \approx \frac{9.3 \times 10^{-26} \text{ kg}}{4.1 \times 10^{-43} \text{ m}^3} = 2.3 \times 10^{17} \text{ kg/m}^3.$$

The density of solid iron is about 7000 kg/m^3, so we see that the nucleus is over 10^{13} times as dense as the bulk material. Densities of this magnitude are also found in *neutron stars,* which are similar to gigantic nuclei made almost entirely of neutrons. A 1-cm cube of material with this density would have a mass of 2.3×10^{11} kg, or 230 million metric tons!

NUCLIDES AND ISOTOPES

The basic building blocks of the nucleus are the proton and the neutron. In a neutral atom, the nucleus is surrounded by one electron for every proton in the nucleus. These particles were introduced in Section 22–3; we recount the discovery of the neutron in Section 46–2. The masses of these particles are

$$m_p = 1.007276 \text{ u} = 1.672623 \times 10^{-27} \text{ kg,}$$

$$m_n = 1.008665 \text{ u} = 1.674929 \times 10^{-27} \text{ kg,}$$

$$m_e = 0.000548580 \text{ u} = 9.10939 \times 10^{-31} \text{ kg.}$$

The number of protons in a nucleus is the **atomic number** Z. The number of neutrons is the **neutron number** N. The nucleon number or mass number A is the sum of the number of protons Z and the number of neutrons N,

$$A = Z + N. \tag{45–2}$$

A single nuclear species having specific values of both Z and N is called a **nuclide.** Table 45–1 lists values of A, Z, and N for a few nuclides. The electron structure of an atom, which is responsible for its chemical properties, is determined by the charge Ze of the nucleus. The table shows some nuclides that have the same Z but different N. These nuclides are called **isotopes** of that element; they have different masses because they have different numbers of neutrons in their nuclei. A familiar example is chlorine (Cl, $Z = 17$). About 76% of chlorine nuclei have $N = 18$; the other 24% have $N = 20$. Different isotopes of an element usually have slightly different physical properties such as melting and boiling temperatures and diffusion rates. The two common isotopes of uranium with $A = 235$ and 238 are usually separated industrially by taking advantage of the different diffusion rates of gaseous uranium hexafluoride (UF_6) containing the two isotopes.

Table 45–1 also shows the usual notation for individual nuclides, the symbol of the element, with a pre-subscript equal to Z and a pre-superscript equal to the mass number A. The general format for an element El is ^A_ZEl. The isotopes of chlorine mentioned

TABLE 45–1

COMPOSITIONS OF SOME COMMON NUCLIDES

NUCLEUS	MASS NUMBER (TOTAL NUMBER OF NUCLEONS), A	ATOMIC NUMBER (NUMBER OF PROTONS), Z	NEUTRON NUMBER, $N = A - Z$
$^{1}_{1}\text{H}$	1	1	0
$^{2}_{1}\text{D}$	2	1	1
$^{4}_{2}\text{He}$	4	2	2
$^{6}_{3}\text{Li}$	6	3	3
$^{7}_{3}\text{Li}$	7	3	4
$^{9}_{4}\text{Be}$	9	4	5
$^{10}_{5}\text{B}$	10	5	5
$^{11}_{5}\text{B}$	11	5	6
$^{12}_{6}\text{C}$	12	6	6
$^{13}_{6}\text{C}$	13	6	7
$^{14}_{7}\text{N}$	14	7	7
$^{16}_{8}\text{O}$	16	8	8
$^{23}_{11}\text{Na}$	23	11	12
$^{65}_{29}\text{Cu}$	65	29	36
$^{200}_{80}\text{Hg}$	200	80	120
$^{235}_{92}\text{U}$	235	92	143
$^{238}_{92}\text{U}$	238	92	146

TABLE 45–2

NEUTRAL ATOMIC MASSES FOR SOME LIGHT NUCLIDES

ELEMENT AND ISOTOPE	ATOMIC NUMBER Z	NEUTRON NUMBER N	ATOMIC MASS (u)	MASS NUMBER A
Hydrogen ($^{1}_{1}\text{H}$)	1	0	1.007825	1
Deuterium ($^{2}_{1}\text{H}$)	1	1	2.014102	2
Tritium ($^{3}_{1}\text{H}$)	1	2	3.016049	3
Helium ($^{3}_{2}\text{He}$)	2	1	3.016029	3
Helium ($^{4}_{2}\text{He}$)	2	2	4.002603	4
Lithium ($^{6}_{3}\text{Li}$)	3	3	6.015122	6
Lithium ($^{7}_{3}\text{Li}$)	3	4	7.016004	7
Beryllium ($^{9}_{4}\text{Be}$)	4	5	9.012182	9
Boron ($^{10}_{5}\text{B}$)	5	5	10.012937	10
Boron ($^{11}_{5}\text{B}$)	5	6	11.009305	11
Carbon ($^{12}_{6}\text{C}$)	6	6	12.000000	12
Carbon ($^{13}_{6}\text{C}$)	6	7	13.003355	13
Nitrogen ($^{14}_{7}\text{N}$)	7	7	14.003074	14
Nitrogen ($^{15}_{7}\text{N}$)	7	8	15.000109	15
Oxygen ($^{16}_{8}\text{O}$)	8	8	15.994915	16
Oxygen ($^{17}_{8}\text{O}$)	8	9	16.999132	17
Oxygen ($^{18}_{8}\text{O}$)	8	10	17.999160	18

Source: A. H. Wapstra and G. Audi, *Nuclear Physics* **A595,** 4 (1995).

above, with $A = 35$ and 37, are written $^{35}_{17}\text{Cl}$ and $^{37}_{17}\text{Cl}$ and pronounced "chlorine-35" and "chlorine-37" respectively. This notation is redundant because the name of the element determines the atomic number Z, so the pre-subscript Z is sometimes omitted, as in ^{35}Cl.

The masses of some common atoms, including their electrons, are shown in Table 45–2. This table gives masses of *neutral* atoms (with Z electrons) rather than masses of *bare* nuclei, because it is much more difficult to measure masses of bare nuclei with high precision. The mass of a neutral carbon-12 atom is exactly 12 u; that's how the unified atomic mass unit is defined. The masses of other atoms are *approximately* equal to A atomic mass units, as we stated earlier. You may notice that the atomic masses are *less* than sum of the masses of their parts (the Z protons, the Z electrons, and the N neutrons). We'll explain this very important mass difference in the next section.

NUCLEAR SPINS AND MAGNETIC MOMENTS

Like electrons, protons and neutrons are also spin-$\frac{1}{2}$ particles with spin angular momentum given by the same equations as in Section 43–4. The magnitude of the spin angular momentum $\vec{S}$ is

$$S = \sqrt{\frac{1}{2}\left(\frac{1}{2} + 1\right)}\hbar = \sqrt{\frac{3}{4}}\hbar, \tag{45–3}$$

and the z-component is

$$S_z = \pm\frac{1}{2}\hbar. \tag{45–4}$$

In addition to the spin angular momentum of the nucleons, there may be *orbital* angular momentum associated with their motions within the nucleus. The orbital angular momentum of the nucleons is quantized in the same way as that of electrons in atoms.

The *total* angular momentum $\vec{J}$ of the nucleus has magnitude

$$J = \sqrt{j(j + 1)}\hbar \tag{45–5}$$

and z-component

$$J_z = m_j\hbar \qquad (m_j = -j, -j + 1, \ldots, j - 1, j). \tag{45–6}$$

When the total number of nucleons A is *even*, j is an integer; when it is *odd*, j is a half-integer. All nuclides for which both Z and N are even have $J = 0$, which suggests that pairing of particles with opposite spin components may be an important consideration in nuclear structure. The total nuclear angular momentum quantum number j is usually called the *nuclear spin,* even though in general it refers to a combination of the orbital and spin angular momenta of the nucleons that make up the nucleus.

Associated with nuclear angular momentum is a *magnetic moment.* When we discussed *electron* magnetic moments in Section 43–3, we introduced the Bohr magneton $\mu_\text{B} = e\hbar/2m_\text{e}$ as a natural unit of magnetic moment. We found that the magnitude of the z-component of the electron-spin magnetic moment is almost exactly equal to μ_B. That is, $\left|\mu_{sz}\right|_\text{electron} \approx \mu_\text{B}$. In discussing *nuclear* magnetic moments, we can define an analogous quantity, the **nuclear magneton** μ_n:

$$\mu_\text{n} = \frac{e\hbar}{2m_\text{p}} = 5.05079 \times 10^{-27} \text{ J/T} = 3.15245 \times 10^{-8} \text{ eV/T} \tag{45–7}$$
$$\text{(nuclear magneton)},$$

where m_p is the proton mass. Because the proton mass m_p is 1836 times larger than the electron mass m_e, the nuclear magneton μ_n is 1836 times smaller than the Bohr magneton μ_B.

We might expect the magnitude of the z-component of the spin magnetic moment of the proton to be approximately μ_n. Instead, it turns out to be

$$\left|\mu_{sz}\right|_{\text{proton}} = 2.7928\mu_n.$$

Even more surprising, the neutron, which has no charge, has a corresponding magnitude of

$$\left|\mu_{sz}\right|_{\text{neutron}} = 1.9130\mu_n.$$

The proton has a positive charge; as expected, its spin magnetic moment $\vec{\mu}$ is parallel to its spin angular momentum $\vec{S}$. However, $\vec{\mu}$ and $\vec{S}$ are opposite for a neutron, as would be expected for a negative charge distribution. These *anomalous* magnetic moments arise because the proton and neutron aren't really fundamental particles, but are made of simpler particles called *quarks*. We'll discuss quarks in some detail in the next chapter.

The magnetic moment of an entire nucleus is typically a few nuclear magnetons. When a nucleus is placed in an external magnetic field $\vec{B}$, there is an interaction energy $U = -\vec{\mu} \cdot \vec{B} = -\mu_z B$ just as with atomic magnetic moments. The components of the magnetic moment in the direction of the field μ_z are quantized, so a series of energy levels results from this interaction.

EXAMPLE 45–2

Proton spin flips Protons are placed in a magnetic field in the z-direction with magnitude 2.30 T. a) What is the energy difference between a state with the z-component of proton spin angular momentum parallel to the field and one with the component antiparallel to the field? b) A proton can make a transition from one of these states to the other by emitting or absorbing a photon with energy equal to the energy difference of the two states. Find the frequency and wavelength of such a photon.

SOLUTION a) When the z-component of $\vec{S}$ (and $\vec{\mu}$) is parallel to the field, the interaction energy is

$$U = -\left|\mu_z\right|B = -(2.7928)(3.152 \times 10^{-8}\ \text{eV/T})(2.30\ \text{T})$$

$$= -2.025 \times 10^{-7}\ \text{eV}.$$

When the components are antiparallel to the field, the energy is $+2.025 \times 10^{-7}$ eV, and the energy *difference* between the two states is

$$\Delta E = 2(2.025 \times 10^{-7}\ \text{eV}) = 4.05 \times 10^{-7}\ \text{eV}.$$

b) The corresponding photon frequency and wavelength are

$$f = \frac{\Delta E}{h} = \frac{4.05 \times 10^{-7}\ \text{eV}}{4.136 \times 10^{-15}\ \text{eV}\cdot\text{s}} = 9.79 \times 10^{7}\ \text{Hz} = 97.9\ \text{MHz},$$

$$\lambda = \frac{c}{f} = \frac{3.00 \times 10^{8}\ \text{m/s}}{9.79 \times 10^{7}\ \text{s}^{-1}} = 3.06\ \text{m}.$$

This frequency is in the middle of the FM radio band. When a hydrogen specimen is placed in a 2.30-T magnetic field and then irradiated with radiation of this frequency, the proton *spin flips* can be detected by the absorption of energy from the radiation.

Spin-flip experiments of the sort referred to in Example 45–2 are called *nuclear magnetic resonance* (NMR). They have been carried out with many different nuclides. Frequencies and magnetic fields can be measured very precisely, so this technique permits precise measurements of nuclear magnetic moments. An elaboration of this basic idea leads to *magnetic-resonance imaging* (MRI), a noninvasive imaging technique that discriminates among various body tissues on the basis of the differing environments of protons in the tissues. The principles of MRI are shown in Fig. 45–1.

The magnetic moment of a nucleus is also the *source* of a magnetic field. In an atom the interaction of an electron's magnetic moment with the field of the nucleus's magnetic moment causes additional splittings in atomic energy levels and spectra. We called this effect *hyperfine structure* in Section 43–4. Measurements of the hyperfine structure may be used to directly determine the nuclear spin.

45–3 NUCLEAR BINDING AND NUCLEAR STRUCTURE

Because energy must be added to a nucleus to separate it into its individual protons and neutrons, the total rest energy E_0 of the separated nucleons is greater than the rest energy of the nucleus. The energy that must be added to separate the nucleons is called the

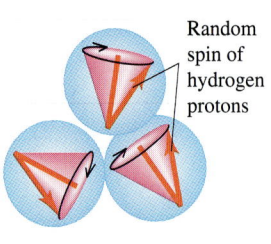

Random spin of hydrogen protons

Hydrogen atoms (mostly in water)

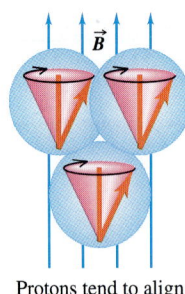

Protons tend to align with uniform B-field

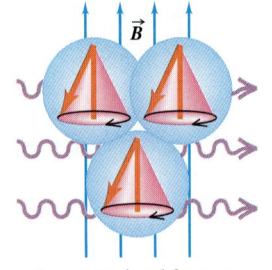

Resonant signal from an electromagnetic wave causes protons to flip

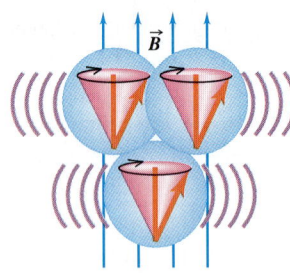

Protons emit signal as they realign with field

(a)

(b)

Main coil supplies uniform B-field

x Coil varies B-field from left to right

z Coil varies B-field from head to toe

y Coil varies B-field from top to bottom

Transceiver sends and receives signals that create image

(c)

45–1 Magnetic-resonance imaging (MRI). (a) Protons, the nuclei of hydrogen atoms in the tissue under study, normally have random spin orientations. In the presence of a strong magnetic field, they become aligned with a component parallel to the field. A brief radio signal flips the spins; as their components reorient parallel to the field, they emit signals that are picked up by sensitive detectors. The differing magnetic environment in various regions permits reconstruction of an image showing the types of tissue present. (b) A color-enhanced MRI image showing a cross section through a patient's head. (c) An electromagnet used for MRI imaging.

binding energy E_B; it is the magnitude of the energy by which the nucleons are bound together. Thus the rest energy of the nucleus is $E_0 - E_B$. Using the equivalence of rest mass and energy (Section 39–10), we see that the total mass of the nucleons is always greater than the mass of the nucleus by an amount E_B/c^2 called the *mass defect*. The binding energy for a nucleus containing Z protons and N neutrons is defined as

$$E_B = (ZM_H + Nm_n - {}_Z^A M)c^2 \qquad \text{(nuclear binding energy)}, \qquad (45-8)$$

where ${}_Z^A M$ is the mass of the *neutral* atom containing the nucleus, the quantity in the parentheses is the mass defect, and $c^2 = 931.5$ MeV/u. Note that Eq. (45–8) does not include Zm_p, the mass of Z protons. Rather, it contains ZM_H, the mass of Z protons and Z electrons combined as Z neutral ${}_1^1 H$ atoms, to balance the Z electrons included in ${}_Z^A M$, the mass of the neutral atom.

The simplest nucleus is that of hydrogen, a single proton. Next comes the nucleus of ${}_1^2 H$, the isotope of hydrogen with mass number 2, usually called *deuterium*. Its nucleus consists of a proton and a neutron bound together to form a particle called the *deuteron.* By using values from Table 45–2 in Eq. (45–8), the binding energy of the deuteron is

$$E_B = (1.007825\ u + 1.008665\ u - 2.014102\ u)(931.5\ \text{MeV/u}) = 2.224\ \text{MeV.}$$

This much energy would be required to pull the deuteron apart into a proton and a neutron. An important measure of how tightly a nucleus is bound is the *binding energy per nucleon, E_B/A.* At (2.224 MeV)(2 nucleons) = 1.112 MeV per nucleon, ^2_1H has the smallest binding energy per nucleon of all nuclides.

Problem–Solving Strategy

NUCLEAR PROPERTIES

1. Familiarity with numerical magnitudes is helpful. The scale of things within nuclei is very different from that within atoms. Protons and neutrons are about 1840 times as massive as electrons. The radius of a nucleus is of the order of 10^{-15} m; the repulsive electric potential energy of two protons at this distance is of the order of 10^{-13} J, or 1 MeV. Thus typical nuclear interaction energies are of the order of a few MeV, rather than a few eV as with atoms. The typical binding energy per nucleon is roughly 1% of the rest energy of a nucleon. For comparison, the ionization energy of the hydrogen atom is only 0.003% of the electron's rest energy.

2. Angular momentum is of the same order of magnitude in both atoms and nuclei because it is determined by the value of Planck's constant h. But magnetic moments of nuclei are about a thousand times *smaller* than those of electrons in atoms because the nucleons are so much more massive than electrons.

3. When doing energy calculations involving the binding energy and binding energy per nucleon, note that mass tables nearly always list the masses of *neutral* atoms, including their full complements of electrons. To compensate for this, use the mass of a ^1_1H atom, rather than the mass of a bare proton. The binding energies of the electrons in the neutral atoms are much smaller and tend to cancel in the subtraction, so we won't worry about them. Binding energy calculations often involve subtracting two nearly equal quantities. To get enough precision in the difference, you often have to carry seven to nine significant figures, if that many are available. If not, you may have to be content with an approximate result.

4. Nuclear masses are usually measured in atomic mass units (u). To convert from a mass defect in u to a binding energy in MeV, use $c^2 = 931.5$ MeV/u.

EXAMPLE 45-3

The most strongly bound nuclide Because it has the highest binding energy per nucleon of all nuclides, $^{62}_{28}Ni$ may be described as the most strongly bound. Its neutral atomic mass is 61.928349 u. Find its mass defect, its total binding energy, and its binding energy per nucleon.

SOLUTION We use $Z = 28$, $M_H = 1.007825$ u, $N = A - Z = 62 - 38 = 34$, $m_n = 1.008665$ u, and $^A_ZM = 61.928349$ u in the parentheses of Eq. (45–8) to find a mass defect of 0.585361 u. Then

$$E_B = (0.585361\ u)(931.5\ \text{MeV/u})$$
$$= 545.3\ \text{MeV.}$$

It would require a minimum of 545.3 MeV to pull a $^{62}_{28}Ni$ nucleus completely apart into 62 separate nucleons. The binding energy *per nucleon* is 1/62 of this, or 8.795 MeV per nucleon.

Nearly all stable nuclides, from the lightest to the most massive, have binding energies in the range of 7 to 9 MeV per nucleon. Figure 45–2 is a graph of binding energy per nucleon as a function of the mass number A. Note the spike at $A = 4$, showing the unusually large binding energy per nucleon of the ^4_2He nucleus (alpha particle) relative to its neighbors. To explain this curve, we must consider the interactions among the nucleons.

THE NUCLEAR FORCE

The force that binds protons and neutrons together in the nucleus, despite the electrical repulsion of the protons, is an example of the *strong interaction* that we mentioned in

19.2
Nuclear Binding Energy

45–2 Approximate binding energy per nucleon as a function of mass number A (the total number of nucleons) for stable nuclides. The curve reaches a peak of about 8.8 MeV/nucleon at $A = 62$, corresponding to the element nickel. The spike at $A = 4$ shows the unusual stability of the $^4_2$He structure.

Section 5–6. In the context of nuclear structure, this interaction is called the *nuclear force.* Here are some of its characteristics. First, it does not depend on charge; neutrons as well as protons are bound, and the binding is the same for both. Second, it has short range, of the order of nuclear dimensions, that is, 10^{-15} m. (Otherwise, the nucleus would grow by pulling in additional protons and neutrons.) But within its range, the nuclear force is much stronger than electrical forces; otherwise, the nucleus could never be stable. It would be nice if we could write a simple equation like Newton's law of gravitation or Coulomb's law for this force, but physicists have yet to fully determine its dependence on the separation r. Third, the nearly constant density of nuclear matter and the nearly constant binding energy per nucleon of larger nuclides show that a particular nucleon cannot interact simultaneously with *all* the other nucleons in a nucleus, but only with those few in its immediate vicinity. This is different from electrical forces; *every* proton in the nucleus repels every other one. This limited number of interactions is called *saturation;* it is analogous to covalent bonding in molecules and solids. Finally, the nuclear force favors binding of *pairs* of protons or neutrons with opposite spins and of *pairs of pairs,* that is, a pair of protons and a pair of neutrons, each pair having opposite spins. Hence the alpha particle (two protons and two neutrons) is an exceptionally stable nucleus for its mass number. We'll see other evidence for pairing effects in nuclei in the next subsection. (In Section 44–9 we described an analogous pairing that binds opposite-spin electrons in Cooper pairs in the BCS theory of superconductivity.)

The analysis of nuclear structure is more complex than the analysis of many-electron atoms. Two different kinds of interactions are involved (electrical and nuclear), and the nuclear force is not yet completely understood. Even so, we can gain some insight into nuclear structure by the use of simple models. We'll discuss briefly two rather different but successful models, the *liquid-drop model* and the *shell model.*

THE LIQUID-DROP MODEL

The **liquid-drop model,** first proposed in 1935 by George Gamow, is suggested by the observation that all nuclei have nearly the same density. The individual nucleons are analogous to molecules of a liquid, held together by short-range interactions and surface-tension effects. We can use this simple picture to derive a formula for the estimated total binding energy of a nucleus. We'll include five contributions:

1. We've remarked that nuclear forces show *saturation;* an individual nucleon interacts only with a few of its nearest neighbors. This effect gives a binding energy term that is proportional to the number of nucleons. We write this term as C_1A, where C_1 is an experimentally determined constant.

2. The nucleons on the surface of the nucleus are less tightly bound than those in the interior because they have no neighbors outside the surface. This decrease in the binding energy gives a *negative* energy term proportional to the surface area $4\pi R^2$. Because R is proportional to $A^{1/3}$, this term is proportional to $A^{2/3}$; we write it as $-C_2A^{2/3}$, where C_2 is another constant.

3. Every one of the Z protons repels every one of the $(Z - 1)$ other protons. The total repulsive electric potential energy is proportional to $Z(Z - 1)$ and inversely proportional to the radius R and thus to $A^{1/3}$. This energy term is negative because the nucleons are less tightly bound than they would be without the electrical repulsion. We write this correction as $-C_3Z(Z - 1)/A^{1/3}$.

4. To be in a stable, low-energy state, the nucleus must have a balance between the energies associated with the neutrons and with the protons. This means that N is close to Z for small A and N is greater than Z (but not too much greater) for larger A. We need a negative energy term corresponding to the difference $|N - Z|$. The best agreement with observed binding energies is obtained if this term is proportional to $(N - Z)^2/A$. If we use $N = A - Z$ to express this energy in terms of A and Z, this correction is $-C_4(A - 2Z)^2/A$.

5. Finally, the nuclear force favors *pairing* of protons and of neutrons. This energy term is positive (more binding) if both Z and N are even, negative (less binding) if both Z and N are odd, and zero otherwise. The best fit to the data occurs with the form $\pm C_5A^{-4/3}$ for this term.

The total estimated binding energy E_B is the sum of these five terms:

$$E_B = C_1A - C_2A^{2/3} - C_3\frac{Z(Z - 1)}{A^{1/3}} - C_4\frac{(A - 2Z)^2}{A} \pm C_5A^{-4/3} \qquad (45\text{-}9)$$

(nuclear binding energy).

The constants C_1, C_2, C_3, C_4, and C_5, chosen to make this formula best fit the observed binding energies of nuclides, are

$$C_1 = 15.75 \text{ MeV},$$

$$C_2 = 17.80 \text{ MeV},$$

$$C_3 = 0.7100 \text{ MeV},$$

$$C_4 = 23.69 \text{ MeV},$$

$$C_5 = 39 \text{ MeV}.$$

The constant C_1 is the binding energy per nucleon due to the saturated nuclear force. This energy is almost 16 MeV per nucleon, about double the *total* binding energy per nucleon in most nuclides.

If we estimate the binding energy E_B using Eq. (45–9), we can solve Eq. (45–8) to use it to estimate the mass of any neutral atom:

$$^A_Z M = ZM_H + Nm_n - E_B/c^2 \qquad \text{(semi-empirical mass formula).} \qquad (45\text{-}10)$$

Equation (45–10) is called the *semi-empirical mass formula*. The name is apt; it is *empirical* in the sense that the C's have to be determined empirically (experimentally), yet it does have a sound theoretical basis.

EXAMPLE 45-4

Estimating the binding energy and mass Consider the nuclide $^{62}_{28}\text{Ni}$ of Example 45–3. a) Calculate the five terms in the binding energy and the total estimated binding energy. b) Find its neutral atomic mass using the semi-empirical mass formula.

SOLUTION a) We have $Z = 28$, $A = 62$, and $N = 34$. We substitute the given numbers into Eq. (45–9). The individual terms are

1. $C_1 A = (15.75 \text{ MeV})(62) = 976.5 \text{ MeV}$;

2. $-C_2 A^{2/3} = -(17.80 \text{ MeV})(62)^{2/3} = -278.8 \text{ MeV}$;

3. $-C_3 \dfrac{Z(Z-1)}{A^{1/3}} = -(0.7100 \text{ MeV})\dfrac{(28)(27)}{(62)^{1/3}} = -135.6 \text{ MeV}$;

4. $-C_4 \dfrac{(A-2Z)^2}{A} = -(23.69 \text{ MeV})\dfrac{(62-56)^2}{62} = -13.8 \text{ MeV}$;

5. $+C_5 A^{-4/3} = (39 \text{ MeV})(62)^{-4/3} = 0.2 \text{ MeV}$.

For this nuclide the pairing correction is positive because Z and N are both even and is small in comparison to the other terms. The total estimated binding energy is the sum of these five terms, or 548.5 MeV. This is about 0.6% larger than the value of 545.3 MeV determined in Example 45–3.

b) Now we use the $E_B = 548.5$ MeV value in Eq. (45–10) to find

$$M = 28(1.007825 \text{ u}) + 34(1.008665 \text{ u}) - \frac{548.5 \text{ MeV}}{931.5 \text{ MeV/u}}$$

$$= 61.925 \text{ u}.$$

This calculated mass is only about 0.005% smaller than the measured value of 61.928349 u.

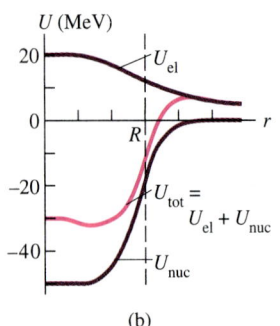

45–3 Approximate potential-energy functions for nucleons in a nucleus. The approximate nuclear radius is R. (a) The potential energy due to the nuclear force. This is the same for protons and neutrons and is the *total* potential energy for neutrons. (b) The total potential energy U_{tot} for a proton is the sum of the nuclear (U_{nuc}) and electric (U_{el}) potential energies.

The liquid-drop model and the mass formula derived from it are quite successful in correlating nuclear masses, and we will see later that they are a great help in understanding decay processes of unstable nuclides. Some other aspects of nuclei, such as angular momentum and excited states, are better approached with different models.

THE SHELL MODEL

The **shell model** of nuclear structure is analogous to the central-field approximation in atomic physics (Section 43–5). We picture each nucleon as moving in a potential that represents the averaged-out effect of all the other nucleons. This may not seem to be a very promising approach; the nuclear force is very strong, very short-range, and therefore strongly distance-dependent. However, in some respects, this model turns out to work fairly well.

The potential-energy function for the nuclear force is the same for protons as for neutrons. A reasonable assumption as to the shape of this function is shown in Fig. 45–3a. This function is a three-dimensional version of the square well we discussed in Section 42–4. The corners are somewhat rounded because the nucleus doesn't have a sharply defined surface. For protons there is an additional potential energy associated with electrical repulsion. We consider each proton to interact with a sphere of uniform charge density, with radius R and total charge $(Z-1)e$. Figure 45–3b shows the nuclear, electric, and total potential energies for a proton as functions of the distance r from the center of the nucleus.

In principle, we could solve the Schrödinger equation for a proton or neutron moving in such a potential. For any spherically symmetric potential energy, the angular-momentum states are the same as for the electrons in the central-field approximation in atomic physics. In particular, we can use the concept of *filled shells and subshells* and their relation to stability. In atomic structure we found that the values $Z = 2, 10, 18, 36, 54,$ and 86 (the atomic numbers of the noble gases) correspond to particularly stable electron arrangements.

A comparable effect occurs in nuclear structure. The numbers are different because the potential-energy function is different and the nuclear spin-orbit interaction is much stronger and of opposite sign than in atoms, so the subshells fill up in a different order from that for electrons in an atom. It is found that when the number of neutrons *or* the number of protons is 2, 8, 20, 28, 50, 82, or 126, the resulting structure is unusually stable, that is, has an unusually great binding energy. (Nuclides with $Z = 126$ have not been observed in nature.) These numbers are called *magic numbers*. Nuclides in which Z is a magic number tend to have an above-average number of stable isotopes. There are several nuclides for which both Z and N are magic, including

$$_2^4\text{He}, \quad _8^{16}\text{O}, \quad _{20}^{40}\text{Ca}, \quad _{20}^{48}\text{Ca}, \quad \text{and} \quad _{82}^{208}\text{Pb}.$$

All these nuclides have substantially larger binding energy per nucleon than do nuclides with neighboring values of N or Z. They also all have zero nuclear spin. The magic numbers correspond to filled-shell or -subshell configurations of nucleon energy levels with a relatively large jump in energy to the next allowed level.

45–4 NUCLEAR STABILITY AND RADIOACTIVITY

Among about 2500 known nuclides, fewer than 300 are stable. The others are unstable structures that decay to form other nuclides by emitting particles and electromagnetic radiation, a process called **radioactivity.** The time scale of these decay processes ranges from a small fraction of a microsecond to billions of years. The *stable* nuclides are shown by dots on the graph in Fig. 45–4, where the neutron number N and proton number (or atomic number) Z for each nuclide are plotted. Such a chart is called a *Segrè chart,* after its inventor, the Italian-American physicist Emilio Segrè (1905–1989).

45–4 Segrè chart, showing neutron number and proton number for stable nuclides. In stable nuclides the number of neutrons exceeds the number of protons by an amount that increases with atomic number Z.

19.2
Nuclear Binding Energy

Each blue line perpendicular to the line $N = Z$ represents a specific value of the mass number $A = Z + N$. Most lines of constant A pass through only one or two stable nuclides; that is, there is usually a very narrow range of stability for a given mass number. The lines at $A = 20$, $A = 40$, $A = 60$, and $A = 80$ are examples. In four cases these lines pass through *three* stable nuclides, namely, at $A = 96$, 124, 130, and 136.

Our four stable nuclides have both odd Z and odd N:

$$^2_1\text{H}, \quad ^6_3\text{Li}, \quad ^{10}_5\text{B}, \quad ^{14}_7\text{N}.$$

These are called *odd-odd nuclides*. The absence of other odd-odd nuclides shows the influence of pairing. Also, there is *no* stable nuclide with $A = 5$ or $A = 8$. The doubly magic ^4_2He nucleus, with a pair of protons and a pair of neutrons, has no interest in accepting a fifth particle into its structure, and collections of eight nucleons decay to smaller nuclides, with a ^8_4Be nucleus immediately splitting into two ^4_2He nuclei.

The points on the Segrè chart representing stable nuclides define a rather narrow stability region. For low mass numbers, the numbers of protons and neutrons are approximately equal, $N \approx Z$. The ratio N/Z increases gradually with A, up to about 1.6 at large mass numbers, because of the increasing influence of the electrical repulsion of the protons. Points to the right of the stability region represent nuclides that have too many protons relative to neutrons to be stable. In these cases, repulsion wins, and the nucleus comes apart. To the left are nuclides with too many neutrons relative to protons. In these cases the energy associated with the neutrons is out of balance with that associated with the protons, and the nuclides decay in a process that converts neutrons to protons. The graph also shows that no nuclide with $A > 209$ or $Z > 83$ is stable. A nucleus is unstable if it is too big. We also note that there is no stable nuclide with $Z = 43$ (technetium) or 61 (promethium). Figure 45–5, a three-dimensional version of the Segrè chart, shows the "valley of stability" for light nuclides (up to $Z = 22$).

ALPHA DECAY

Nearly 90% of the 2500 known nuclides are *radioactive;* they are not stable but decay into other nuclides. When unstable nuclides decay into different nuclides, they usually emit alpha (α) or beta (β) particles. An **alpha particle** is a ^4He nucleus, two protons and two neutrons bound together, with total spin zero. Alpha emission occurs principally with nuclei that are too large to be stable. When a nucleus emits an alpha particle, its N

45–5 A three-dimensional Segrè chart for light nuclides, up to $Z = 22$ (titanium). The quantity plotted on the third axis is $(M - A)c^2$, where M is the nuclide mass expressed in u. This quantity is related to the binding energy by a different constant for each nuclide. The lower points on the raised surface represent especially stable nuclides.

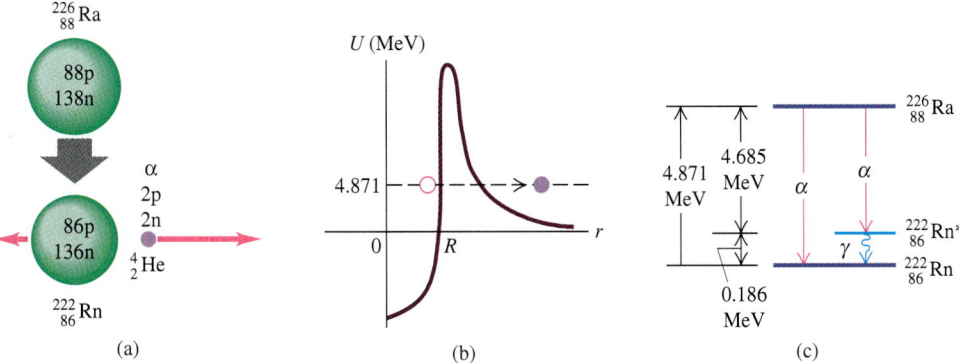

(a)   (b)   (c)

45–6 (a) The nuclide $^{226}_{88}$Ra decays by alpha emission to $^{222}_{86}$Rn. (b) Potential-energy curve for α particle and $^{222}_{86}$Rn nucleus. The particle tunnels through the potential-energy barrier. (c) Energy-level diagram for the system, showing the excited level $^{222}_{86}$Rn* at an energy 0.186 MeV above the ground state. The system can decay from this level to the ground level $^{222}_{86}$Rn by emission of a γ photon with energy 0.186 MeV.

and Z values each decrease by two and A decreases by four, moving it closer to stable territory on the Segrè chart.

A familiar example of an alpha emitter is radium, $^{226}_{88}$Ra (Fig. 45–6a). The speed of the emitted alpha particle, determined from the curvature of its path in a transverse magnetic field, is about 1.52×10^7 m/s. This speed, although large, is only 5% of the speed of light, so we can use the nonrelativistic kinetic-energy expression $K = \frac{1}{2}mv^2$:

$$K = \frac{1}{2}(6.64 \times 10^{-27} \text{ kg})(1.52 \times 10^7 \text{ m/s})^2 = 7.7 \times 10^{-13} \text{ J} = 4.8 \text{ MeV}.$$

As we described in Section 8–8, alpha particles are always emitted with definite kinetic energies, determined by conservation of momentum and energy. Because of their charge and mass, alpha particles can travel only several centimeters in air, or a few tenths or hundredths of a millimeter through solids, before they are brought to rest by collisions.

Some nuclei can spontaneously decay by emission of α particles because energy is released in their alpha decay. You can use conservation of mass-energy to show that *alpha decay is possible whenever the mass of the original neutral atom is greater than the sum of the masses of the final neutral atom and the neutral helium-4 atom.* In alpha decay, the α particle tunnels through a potential-energy barrier, as shown in Fig. 45–6b. You may want to review the discussion of tunneling in Section 42–5.

EXAMPLE 45–5

Alpha decay of radium You are given the following neutral atomic masses:

$$^{226}_{88}\text{Ra: } 226.025403 \text{ u,}$$

$$^{222}_{86}\text{Rn: } 222.017571 \text{ u.}$$

Show that alpha emission is energetically possible and that the calculated kinetic energy of the emitted α particle agrees with the experimentally measured value of 4.78 MeV.

SOLUTION Alpha emission is possible if the $^{226}_{88}$Ra mass is greater than the sum of the $^{222}_{86}$Rn mass and the $^4_2$He mass. We use Table 45–2 to find the third mass we need:

$$^4_2\text{He: } 4.002603 \text{ u.}$$

Then the difference in mass between the original nucleus and the decay products is

$$226.025403 \text{ u} - (222.017571 \text{ u} + 4.002603 \text{ u}) = +0.005229 \text{ u.}$$

Since this is positive, alpha decay is energetically possible.

The energy equivalent of 0.005229 u is

$$E = (0.005229 \text{ u})(931.5 \text{ Mev/u}) = 4.871 \text{ MeV.}$$

Thus we expect the decay products to emerge with total kinetic energy 4.871 MeV. Momentum is also conserved; if the parent nucleus is at rest, the daughter and the α particle have momenta of equal magnitude p but opposite direction. Kinetic energy is $K = p^2/2m$, so since p is the same for the two particles, the kinetic energy divides inversely as their masses. The α particle gets 222/(222 + 226) of the total, or 4.78 MeV, equal to the observed α-particle energy.

BETA DECAY

There are three different simple types of *beta decay: beta-minus, beta-plus,* and *electron capture.* A **beta-minus particle** (β^-) is an electron. It's not obvious how a nucleus can emit an electron if there aren't any electrons in the nucleus. Emission of a β^- involves *transformation* of a neutron into a proton, an electron, and a third particle called an *antineutrino.* In fact, if you freed a neutron from a nucleus, it would decay into a proton, an electron, and an antineutrino in an average time of about 15 minutes.

Beta particles can be identified and their speeds can be measured with techniques that are similar to the Thomson experiments we described in Section 28–6. The speeds of beta particles range up to 0.9995 of the speed of light, so their motion is highly relativistic. They are emitted with a continuous spectrum of energies. This would not be possible if the only two particles were the β^- and the recoiling nucleus, since energy and momentum conservation would then require a definite speed for the β^-. (We discussed this matter in Section 8–8 also; you may want to review that discussion.) Thus there must be a third particle involved. From conservation of charge, it must be neutral, and from conservation of angular momentum, it must be a spin-$\frac{1}{2}$ particle.

This third particle is an antineutrino, the *antiparticle* of a **neutrino.** The symbol for a neutrino is ν (the Greek letter "nu"). Both the neutrino and the antineutrino have zero charge and zero (or very small) mass and therefore produce very little observable effect when passing through matter. Both evaded detection until 1953, when Frederick Reines and Clyde Cowan succeeded in observing the antineutrino directly. We now know that there are at least three varieties of neutrinos, each with its corresponding antineutrino; one is associated with beta decay and the other two are associated with the decay of two unstable particles, the muon and the tau particle. We'll discuss these particles in more detail in Section 46–5. The antineutrino that is emitted in β^- decay is denoted as $\overline{\nu}_e$. The basic process of β^- decay is

$$\text{n} \rightarrow \text{p} + \beta^- + \overline{\nu}_e. \tag{45–11}$$

Beta-minus decay usually occurs with nuclides for which the neutron-to-proton ratio N/Z is too large for stability. In β^- decay, N decreases by one, Z increases by one, and A doesn't change. You can use conservation of mass-energy to show that *beta-minus decay can occur whenever the neutral atomic mass of the original atom is larger than that of the final atom.*

EXAMPLE 45–6

Why cobalt-60 is a beta-minus emitter The nuclide $^{60}_{27}\text{Co}$, an odd-odd unstable nucleus, is used in medical applications of radiation. Show that it is unstable relative to β^- decay. The following masses are given:

$^{60}_{27}\text{Co}$:   59.933822 u,

$^{60}_{28}\text{Ni}$:   59.930791 u.

SOLUTION The original nuclide is $^{60}_{27}\text{Co}$. In β^- decay, Z increases by one from 27 to 28 and A remains at 60, so the final nuclide is $^{60}_{28}\text{Ni}$. Its mass is less than that of $^{60}_{27}\text{Co}$ by 0.003031 u, so β^- decay *can* occur.

We have noted that β^- decay occurs with nuclides that have too large a neutron-to-proton ratio N/Z. Nuclides for which N/Z is too *small* for stability can emit a *positron,* the electron's antiparticle, which is identical to the electron but with positive charge. (We mentioned the positron in connection with positronium in Section 40–6 and will discuss it in more detail in Section 46–2.) The basic process, called *beta-plus decay* (β^+), is

$$\text{p} \rightarrow \text{n} + \beta^+ + \nu_e, \tag{45–12}$$

where β^+ is a positron and ν_e is the electron neutrino. *Beta-plus decay can occur whenever the neutral atomic mass of the original atom is at least two electron masses larger than that of the final atom;* you can show this using conservation of mass-energy.

The third type of beta decay is *electron capture.* There are a few nuclides for which β^+ emission is not energetically possible but in which an orbital electron (usually in the K shell) can combine with a proton in the nucleus to form a neutron and a neutrino. The neutron remains in the nucleus and the neutrino is emitted. The basic process is

$$p + \beta^- \rightarrow n + \nu_e. \tag{45-13}$$

You can use conservation of mass-energy to show that *electron capture can occur whenever the neutral atomic mass of the original atom is larger than that of the final atom.* In all types of beta decay, A remains constant. However, in beta-plus decay and electron capture, N increases by one and Z decreases by one as the neutron-proton ratio increases toward a more stable value. The reaction of Eq. (45–13) also helps to explain the formation of a neutron star, mentioned in Example 45–1.

CAUTION ▶ The beta-decay reactions given by Eqs. (45–11), (45–12), and (45–13) occur *within* a nucleus. Although the decay of a neutron outside the nucleus proceeds through the reaction of Eq. (45–11), the reaction of Eq. (45–12) is forbidden by conservation of mass-energy for a proton outside the nucleus. The reaction of Eq. (45–13) can occur outside the nucleus only with the addition of some extra energy, as in a collision. ◀

EXAMPLE 45-7

Why cobalt-57 is not a beta-plus emitter The nuclide $^{57}_{27}\text{Co}$, an odd-even unstable nucleus, is often used as a source of radiation in a nuclear process called the *Mössbauer effect.* Show that this nuclide is stable relative to β^+ decay but can decay by electron capture. The following masses are given:

$^{57}_{27}\text{Co}$: 56.936296 u,

$^{57}_{26}\text{Fe}$: 56.935399 u.

SOLUTION The original nuclide is $^{57}_{27}\text{Co}$. In β^+ decay and electron capture, Z decreases by one from 27 to 26, and A remains at 57. Thus the final nuclide is $^{57}_{26}\text{Fe}$. Its mass is less than that of $^{57}_{27}\text{Co}$ by 0.000897 u, a value smaller than 0.001097 u (two electron masses), so β^+ decay *cannot* occur. However, the mass of the original atom is greater than the mass of the final atom, so electron capture *can* occur. In Section 45–5 we'll see how to relate the probability that electron capture will occur to the *half-life* of this nuclide.

GAMMA DECAY

The energy of internal motion of a nucleus is quantized. A typical nucleus has a set of allowed energy levels, including a *ground state* (state of lowest energy) and several *excited states.* Because of the great strength of nuclear interactions, excitation energies of nuclei are typically of the order of 1 MeV, compared with a few eV for atomic energy levels. In ordinary physical and chemical transformations the nucleus always remains in its ground state. When a nucleus is placed in an excited state, either by bombardment with high-energy particles or by a radioactive transformation, it can decay to the ground state by emission of one or more photons called **gamma rays** or *gamma-ray photons,* with typical energies of 10 keV to 5 MeV. This process is called *gamma (γ) decay.* For example, alpha particles emitted from ^{226}Ra have two possible kinetic energies, either 4.784 MeV or 4.602 MeV. Including the recoil energy of the resulting ^{222}Rn nucleus, these correspond to a total released energy of 4.871 MeV or 4.685 MeV, respectively. When an alpha particle with the smaller energy is emitted, the ^{222}Rn nucleus is left in an excited state. It then decays to its ground state by emitting a gamma-ray photon with energy

$$(4.871 - 4.685) \text{ MeV} = 0.186 \text{ MeV}.$$

A photon with this energy is observed during this decay (Fig. 45–6c).

CAUTION ▶ In both α and β decay, the Z value of a nucleus changes and the nucleus of one element becomes the nucleus of a different element. In γ decay, the element does *not* change; the nucleus merely goes from an excited state to a less excited state. ◀

NATURAL RADIOACTIVITY

Many radioactive elements occur in nature. For example, you are very slightly radioactive because of unstable nuclides such as carbon-14 and potassium-40 that are present throughout your body. The study of natural radioactivity began in 1896, one year after Röntgen discovered x rays. Henri Becquerel discovered a radiation from uranium salts that seemed similar to x rays. Intensive investigation in the following two decades by Marie and Pierre Curie, Ernest Rutherford, and many others revealed that the emissions consist of positively and negatively charged particles and neutral rays; they were given the names *alpha, beta,* and *gamma* because of their differing penetration characteristics.

The decaying nucleus is usually called the *parent nucleus;* the resulting nucleus is the *daughter nucleus.* When a radioactive nucleus decays, the daughter nucleus may also be unstable. In this case a *series* of successive decays occurs until a stable configuration is reached. Several such series are found in nature. The most abundant radioactive nuclide found on earth is the uranium isotope ^{238}U, which undergoes a series of 14 decays, including eight α emissions and six β^- emissions, terminating at a stable isotope of lead, ^{206}Pb.

Radioactive decay series can be represented on a Segrè chart, as in Fig. 45–7. The neutron number N is plotted vertically, and the atomic number Z is plotted horizontally. In alpha emission, both N and Z decrease by two. In β^- emission, N decreases by one and Z increases by one. The decays can also be represented in equation form; the first two decays in the series are written as

$$^{238}\text{U} \rightarrow {}^{234}\text{Th} + \alpha,$$

$$^{234}\text{Th} \rightarrow {}^{234}\text{Pa} + \beta^- + \bar{\nu}_e,$$

or more briefly as

$$^{238}\text{U} \xrightarrow{\alpha} {}^{234}\text{Th},$$

$$^{234}\text{Th} \xrightarrow{\beta^-} {}^{234}\text{Pa}.$$

In the second process, the beta decay leaves the daughter nucleus ^{234}Pa in an excited state, from which it decays to the ground state by emitting a gamma-ray photon. An excited state is denoted by an asterisk, so we can represent the γ emission as

$$^{234}\text{Pa}^* \rightarrow {}^{234}\text{Pa} + \gamma$$

or

$$^{234}\text{Pa}^* \xrightarrow{\gamma} {}^{234}\text{Pa}.$$

Earthquakes are caused in part by the radioactive decay of ^{238}U. As ^{238}U in the earth's interior decays, it releases energy that helps to keep the interior molten. The earth's solid crust is able to slide over this molten material. When two adjacent sections of crust slide past each other, an earthquake results.

An interesting feature of the ^{238}U decay series is the branching that occurs at ^{214}Bi. This nuclide decays to ^{210}Pb by emission of an α and a β^-, which can occur in either order. We also note that the series includes unstable isotopes of several elements that also have stable isotopes, including thallium (Tl), lead (Pb), and bismuth (Bi). The unstable isotopes of these elements that occur in the ^{238}U series all have too many neutrons to be stable.

Many other decay series are known. Two of these occur in nature, one starting with the uncommon isotope ^{235}U and ending with ^{207}Pb, the other starting with thorium (^{232}Th) and ending with ^{208}Pb.

45-7 Segrè chart showing the uranium ^{238}U decay series, terminating with the stable nuclide ^{206}Pb. The times are half-lives (discussed in the next section), given in years (y), days (d), hours (h), minutes (m), or seconds (s).

45-5 ACTIVITIES AND HALF-LIVES

Suppose you need to dispose of some radioactive waste that contains a certain number of a particular radioactive nuclide. If no more are produced, that number decreases in a simple manner as the nuclei decay. This decrease is a statistical process; there is no way to predict when any individual nucleus will decay. No change in physical or chemical environment, such as chemical reactions or heating or cooling, greatly affects most decay rates. The rate varies over an extremely wide range for different nuclides.

Let $N(t)$ be the (very large) number of radioactive nuclei in a sample at time t, and let $dN(t)$ be the (negative) change in that number during a short time interval dt. (We'll use $N(t)$ to minimize confusion with the neutron number N.) The number of decays during the interval dt is $-dN(t)$. The rate of change of $N(t)$ is the negative quantity $dN(t)/dt$; thus $-dN(t)/dt$ is called the *decay rate* or the **activity** of the specimen. The larger the number of nuclei in the specimen, the more nuclei decay during any time interval. That is, the activity is directly proportional to $N(t)$; it equals a constant λ multiplied by $N(t)$:

$$-\frac{dN(t)}{dt} = \lambda N(t). \tag{45-14}$$

The constant λ is called the **decay constant,** and it has different values for different nuclides. A large value of λ corresponds to rapid decay; a small value corresponds to slower decay. Solving Eq. (45–14) for λ shows us that λ is the ratio of the number of decays per time to the number of remaining radioactive nuclei; λ can then be interpreted as the *probability per time* that any individual nucleus will decay.

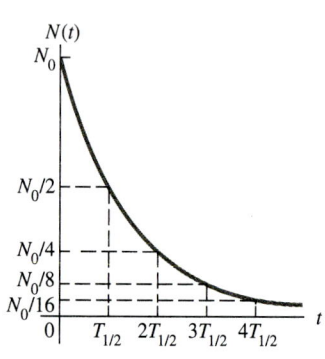

The situation is reminiscent of a discharging capacitor, which we studied in Section 27–5. Equation (45–14) has the same form as the negative of Eq. (27–15), with q and $1/RC$ replaced by $N(t)$ and λ. Then we can make the same substitutions in Eq. (27–16), with the initial number of nuclei $N(0) = N_0$, to find the exponential function:

$$N(t) = N_0 e^{-\lambda t} \qquad \text{(number of remaining nuclei).} \tag{45-15}$$

Figure 45–8 is a graph of this function, showing the number of remaining nuclei $N(t)$ as a function of time.

45–8 The number of nuclei in a sample of a radioactive element as a function of time. The sample's activity has an exponential decay curve with the same shape.

The **half-life** $T_{1/2}$ is the time required for the number of radioactive nuclei to decrease to one-half the original number N_0. Then half of the remaining radioactive nuclei decay during a second interval $T_{1/2}$, and so on. The numbers remaining after successive half-lives are $N_0/2, N_0/4, N_0/8, \ldots$.

To get the relation between the half-life $T_{1/2}$ and the decay constant λ, we set $N(t)/N_0 = 1/2$ and $t = T_{1/2}$ in Eq. (45–15), obtaining

$$\frac{1}{2} = e^{-\lambda T_{1/2}}.$$

We take logarithms of both sides and solve for $T_{1/2}$:

$$T_{1/2} = \frac{\ln 2}{\lambda} = \frac{0.693}{\lambda}. \tag{45-16}$$

The mean lifetime T_{mean}, generally called the *lifetime,* of a nucleus or unstable particle is proportional to the half-life $T_{1/2}$:

19.4
Radioactivity

$$T_{mean} = \frac{1}{\lambda} = \frac{T_{1/2}}{\ln 2} = \frac{T_{1/2}}{0.693} \qquad \begin{array}{l}\text{(lifetime } T_{mean}, \text{ decay constant } \lambda, \\ \text{and half-life } T_{1/2}).\end{array} \tag{45-17}$$

In particle physics the life of an unstable particle is usually described by the lifetime, not the half-life.

Because the activity $-dN(t)/dt$ at any time equals $\lambda N(t)$, Eq. (45–15) tells us that the activity also depends on time as $e^{-\lambda t}$. Thus the graph of activity versus time has the same shape as Fig. 45–8. Also, after successive half-lives, the activity is one half, one fourth, one eighth, and so on of the original activity.

CAUTION▶ It is sometimes implied that any radioactive sample will be safe after a half-life has passed. That's wrong. If your radioactive waste initially has ten times too much activity for safety, it is not safe after one half-life, when it still has five times too much. Even after three half-lives it still has 25% more activity than is safe. The number of radioactive nuclei and the activity approach zero only as t approaches infinity. ◀

A common unit of activity is the **curie,** abbreviated Ci, which is defined to be 3.70×10^{10} decays per second. This is approximately equal to the activity of one gram of radium. The SI unit of activity is the *becquerel,* abbreviated Bq. One becquerel is one decay per second, so

$$1 \text{ Ci} = 3.70 \times 10^{10} \text{ Bq} = 3.70 \times 10^{10} \text{ decays/s.}$$

EXAMPLE 45-8

Activity of ^{57}Co The radioactive isotope ^{57}Co decays by electron capture with a half-life of 272 days. a) Find the decay constant and the lifetime. b) If you have a radiation source containing ^{57}Co, with activity 2.00 μCi, how many radioactive nuclei does it contain? c) What will be the activity of your source after one year?

SOLUTION a) To simplify the units, we convert the half-life to seconds:

$$T_{1/2} = (272 \text{ days})(86,400 \text{ s/day}) = 2.35 \times 10^7 \text{ s}.$$

From Eq. (45–17) the lifetime is

$$T_{mean} = \frac{T_{1/2}}{\ln 2} = \frac{2.35 \times 10^7 \text{ s}}{0.693} = 3.39 \times 10^7 \text{ s}.$$

The decay constant is

$$\lambda = \frac{1}{T_{mean}} = 2.95 \times 10^{-8} \text{ s}^{-1}.$$

b) The activity is $-dN(t)/dt$. This is given as 2.00 μCi, so

$$-\frac{dN(t)}{dt} = 2.00 \text{ } \mu\text{Ci} = (2.00 \times 10^{-6})(3.70 \times 10^{10} \text{ s}^{-1})$$

$$= 7.40 \times 10^4 \text{ decays/s}.$$

From Eq. (45–14) this is equal to $\lambda N(t)$, so we find

$$N(t) = -\frac{dN(t)/dt}{\lambda} = \frac{7.40 \times 10^4 \text{ s}^{-1}}{2.95 \times 10^{-8} \text{ s}^{-1}} = 2.51 \times 10^{12} \text{ nuclei}.$$

This is 4.17×10^{-12} mol or 2.38×10^{-10} g, a far smaller mass than even the most sensitive balance can measure. If you feel we're being too cavalier about the "units" decays and nuclei, you can use decays/(nucleus·s) as the unit for λ.

c) From Eq. (45–15) the number $N(t)$ of nuclei remaining after one year (3.156×10^7 s) is

$$N(t) = N_0 e^{-\lambda t} = N_0 e^{-(2.95 \times 10^{-8} \text{ s}^{-1})(3.156 \times 10^7 \text{ s})}$$

$$= 0.394 N_0.$$

The number of nuclei has decreased to 0.394 of the original number, so the activity has decreased to $(0.394)(2.00 \text{ } \mu\text{Ci}) = 0.788 \text{ } \mu\text{Ci}$.

RADIOACTIVE DATING

An interesting application of radioactivity is the dating of archeological and geological specimens by measuring the concentration of radioactive isotopes. The most familiar example is *carbon dating.* The unstable isotope ^{14}C, produced during nuclear reactions in the atmosphere that result from cosmic-ray bombardment, gives a small proportion of ^{14}C in the CO_2 in the atmosphere. Plants that obtain their carbon from this source contain the same proportion of ^{14}C as the atmosphere. When a plant dies, it stops taking in carbon, and its ^{14}C β^- decays to ^{14}N with a half-life of 5730 years. By measuring the proportion of ^{14}C in the remains, we can determine how long ago the organism died.

One difficulty with radiocarbon dating is that the ^{14}C concentration in the atmosphere changes over long time intervals. Corrections can be made on the basis of other data such as measurements of tree rings that show annual growth cycles. Similar radioactive techniques are used with other isotopes for dating geologic specimens. Some rocks, for example, contain the unstable potassium isotope ^{40}K, a beta emitter that decays to the stable nuclide ^{40}Ar with a half-life of 2.4×10^8 y. The age of the rock can be determined by comparing the concentrations of ^{40}K and ^{40}Ar.

EXAMPLE 45-9

Radiocarbon dating Before 1900 the activity per mass of atmospheric carbon due to the presence of ^{14}C averaged about 0.255 Bq per gram of carbon. a) What fraction of carbon atoms were ^{14}C? b) In analyzing an archeological specimen containing 500 mg of carbon, you observe 174 decays in one hour. What is the age of the specimen, assuming that its activity per mass of carbon when it died was that average value of the air?

SOLUTION a) We'll use Eq. (45–14) to find $N(t)$, and then compare it with N_0. First we find λ from Eq. (45–16):

$$T_{1/2} = 5730 \text{ y} = (5730 \text{ y})(3.156 \times 10^7 \text{ s/y}) = 1.808 \times 10^{11} \text{ s};$$

$$\lambda = \frac{\ln 2}{T_{1/2}} = \frac{0.693}{1.808 \times 10^{11} \text{ s}} = 3.83 \times 10^{-12} \text{ s}^{-1}.$$

Alternatively,

$$\lambda = \frac{0.693}{5730 \text{ y}} = 1.209 \times 10^{-4} \text{ y}^{-1}.$$

Then, from Eq. (45–14),

$$N(t) = \frac{-dN/dt}{\lambda} = \frac{0.255 \text{ s}^{-1}}{3.83 \times 10^{-12} \text{ s}^{-1}} = 6.66 \times 10^{10} \text{ atoms}.$$

The *total* number of C atoms in one gram (1/12.011 mol) is $(1/12.011)(6.022 \times 10^{23}) = 5.01 \times 10^{22}$. The ratio of ^{14}C atoms to all C atoms is

$$\frac{6.66 \times 10^{10}}{5.01 \times 10^{22}} = 1.33 \times 10^{-12}.$$

Only four carbon atoms in every three million million are ^{14}C.
b) Assuming that the activity per gram of carbon in the specimen when it died was 0.255 Bq/g = (0.255 s^{-1} · g^{-1})(3600 s/h) = 918 h^{-1} · g^{-1}, the activity of 500 mg of carbon then was (0.500 g)(918 h^{-1} · g^{-1}) = 459 h^{-1}. The observed activity now, at time t later, is 174 h^{-1}. Since the activity is proportional to the number of radioactive nuclei, the activity ratio 174/459 = 0.379 equals the number ratio $N(t)/N_0$.

Now we solve Eq. (45–15) for t and insert values for $N(t)/N_0$ and λ:

$$t = \frac{\ln (N(t)/N_0)}{-\lambda} = \frac{\ln 0.379}{-1.209 \times 10^{-4} \text{ y}^{-1}} = 8020 \text{ y}.$$

After 8020 y the ^{14}C activity has decreased from 459 to 174 decays per hour. The specimen died and stopped taking CO_2 out of the air about 8000 years ago.

A serious health hazard in some areas is the accumulation in houses of ^{222}Rn, an inert, colorless, odorless radioactive gas. Looking at the ^{238}U decay chain (Fig. 45–7), we see that the half-life of ^{222}Rn is 3.82 days. If so, why not just move out of the house for a while and let it decay away? The answer is that ^{222}Rn is continuously being *produced* by the decay of ^{226}Ra, which is found in minute quantities in the rocks and soil on which houses are built. It's a dynamic equilibrium situation, in which the rate of production equals the rate of decay. The reason why ^{222}Rn is a bigger hazard than the other elements in the ^{238}U decay series is that it's a gas. During its short half-life of 3.82 days it can migrate from the soil into your house. If a ^{222}Rn nucleus decays in your lungs, it emits a damaging α particle and its daughter nucleus ^{218}Po, which is *not* chemically inert and is likely to stay in your lungs until it decays, emits another damaging α particle and so on down the ^{238}U decay series.

How much of a hazard is radon? Although reports indicate values as high as 3500 pCi/L, the average activity per volume in the air inside American homes due to ^{222}Rn is about 1.5 pCi/L (over a thousand decays each second in an average-sized room). *If* your environment has this level of activity, it has been estimated that a lifetime exposure would reduce your life expectancy by about 40 days. For comparison, smoking one pack of cigarettes per day reduces life expectancy by 6 years, and the average emission from all the nuclear power plants in the world reduces life expectancy by anywhere from 0.01 day to 5 days, depending on which estimates you believe. These figures include catastrophies such as the 1986 nuclear reactor disaster at Chernobyl, for which the *local* effect on life expectancy is much greater.

45–6 BIOLOGICAL EFFECTS OF RADIATION

The above discussion of radon introduced the interaction of radiation with living organisms, a topic of vital interest and importance. Under *radiation* we include radioactivity (alpha, beta, gamma, and neutrons) and electromagnetic radiation such as x rays. As these particles pass through matter, they lose energy, breaking molecular bonds and creating ions, hence the term *ionizing radiation*. Charged particles interact directly with the electrons in the material. X rays and γ rays interact by the photoelectric effect, in which an electron absorbs a photon and breaks loose from its site, or by Compton scattering (Section 40–8). Neutrons cause ionization indirectly through collisions with nuclei or absorption by nuclei with subsequent radioactive decay of the resulting nuclei.

These interactions are extremely complex. It is well known that excessive exposure to radiation, including sunlight, x rays, and all the nuclear radiations, can destroy tissues. In mild cases it results in a burn, as with common sunburn. Greater exposure can cause very severe illness or death by a variety of mechanisms, including massive destruction of tissue cells, alterations of genetic material, and destruction of the components in bone marrow that produce red blood cells.

CALCULATING RADIATION DOSES

Radiation dosimetry is the quantitative description of the effect of radiation on living tissue. The *absorbed dose* of radiation is defined as the energy delivered to the tissue per

TABLE 45–3
RELATIVE BIOLOGICAL EFFECTIVENESS (RBE) FOR SEVERAL TYPES OF RADIATION

RADIATION	RBE (Sv/Gy or rem/rad)
X rays and γ rays	1
Electrons	1.0–1.5
Slow neutrons	3–5
Protons	10
α particles	20
Heavy ions	20

unit mass. The SI unit of absorbed dose, the joule per kiligram, is called the *gray* (Gy); 1 Gy = 1 J/kg. Another unit, in more common use at present, is the *rad*, defined as 0.01 J/kg:

$$1 \text{ rad} = 0.01 \text{ J/kg} = 0.01 \text{ Gy}.$$

Absorbed dose by itself is not an adequate measure of biological effect because equal energies of different kinds of radiation cause different extents of biological effect. This variation is described by a numerical factor called the **relative biological effectiveness (RBE)**, also called the *quality factor* (QF), of each specific radiation. X rays with 200 keV of energy are defined to have an RBE of unity, and the effects of other radiations can be compared experimentally. Table 45–3 shows approximate values of RBE for several radiations. All these values depend somewhat on the kind of tissue in which the radiation is absorbed and on the energy of the radiation.

The biological effect is described by the product of the absorbed dose and the RBE of the radiation; this quantity is called the *biologically equivalent dose,* or simply the equivalent dose. The SI unit of equivalent dose for humans is the Sievert (Sv):

$$\text{Equivalent dose (Sv)} = \text{RBE} \times \text{absorbed dose (Gy)}. \qquad (45\text{–}18)$$

A more common unit, corresponding to the rad, is the rem (röntgen equivalent for man):

$$\text{Equivalent dose (rem)} = \text{RBE} \times \text{absorbed dose (rad)}. \qquad (45\text{–}19)$$

Thus the unit of the RBE is 1 Sv/Gy or 1 rem/rad, and 1 rem = 0.01 Sv.

EXAMPLE 45–10

A medical x-ray exam During a diagnostic x-ray examination a 1.2-kg portion of a broken leg receives an equivalent dose of 0.40 mSv. a) What is the equivalent dose in mrem? b) What is the absorbed dose in mrad and mGy? c) If the x-ray energy is 50 keV, how many x-ray photons are absorbed?

SOLUTION a) Since 1 rem = 0.01 Sv, the equivalent dose in mrem is

$$\frac{0.40 \text{ mSv}}{0.01 \text{ Sv/rem}} = 40 \text{ mrem}.$$

b) For x rays, RBE = 1 rem/rad or 1 Sv/Gy, so the absorbed dose is

$$\frac{40 \text{ mrem}}{1 \text{ rem/rad}} = 40 \text{ mrad},$$

$$\frac{0.40 \text{ mSv}}{1 \text{ Sv/Gy}} = 0.40 \text{ mGy} = 4.0 \times 10^{-4} \text{ J/kg}.$$

c) The total energy absorbed is

$$(4.0 \times 10^{-4} \text{ J/kg})(1.2 \text{ kg}) = 4.8 \times 10^{-4} \text{ J} = 3.0 \times 10^{15} \text{ eV}.$$

The number of x-ray photons is

$$\frac{3.0 \times 10^{15} \text{ eV}}{5.0 \times 10^{4} \text{ eV/photon}} = 6.0 \times 10^{10} \text{ photons}.$$

If the ionizing radiation had been a beam of α particles, for which RBE = 20, the absorbed dose needed for an equivalent dose of 0.40 mSv would be 0.020 mGy, corresponding to a total absorbed energy of 2.4×10^{-5} J.

RADIATION HAZARDS

Here are a few numbers for perspective. To convert from Sv to rem, simply multiply by 100. An ordinary chest x-ray exam delivers about 0.20 to 0.40 mSv to about 5 kg of tissue. Radiation exposure from cosmic rays and natural radioactivity in soil, building materials, and so on is of the order of 1.0 mSv per year at sea level and twice that at an elevation of 1500 m (5000 ft). A whole-body dose of up to about 0.20 Sv causes no immediately detectable effect. A short-term whole-body dose of 5 Sv or more usually causes death within a few days or weeks. A localized dose of 100 Sv causes complete destruction of the exposed tissues.

The long-term hazards of radiation exposure in causing various cancers and genetic defects have been widely publicized, and the question of whether there is any "safe" level of radiation exposure has been hotly debated. U.S. government regulations are based on a maximum *yearly* exposure, from all except natural resources, of 2 to 5 mSv. Workers with occupational exposure to radiation are permitted 50 mSv per year. Recent studies suggest that these limits are too high and that even extremely small exposures carry hazards, but it is very difficult to gather reliable statistics on the effects of low dosages. It has become clear that any use of x rays for medical diagnosis should be preceded by a very careful estimation of the relation of risk to possible benefit.

Another sharply debated question is that of radiation hazards from nuclear power plants. The radiation level from these plants is *not* negligible. However, to make a meaningful evaluation of hazards, we must compare these levels with the alternatives, such as coal-powered plants. The health hazards of coal smoke are serious and well documented, and the natural radioactivity in the smoke from a coal-fired power plant is believed to be roughly 100 times as great as that from a properly operating nuclear plant with equal capacity. But the comparison is not this simple; the possibility of a nuclear accident and the very serious problem of safe disposal of radioactive waste from nuclear plants must also be considered. It is clearly impossible to eliminate *all* hazards to health. Our goal should be to try to take a rational approach to the problem of *minimizing* the hazard from all sources. Figure 45–9 shows a recent estimate of the various sources of radiation exposure for the U.S. population. Ionizing radiation is a two-edged sword; it poses very serious health hazards, yet it also provides many benefits to humanity, including the diagnosis and treatments of disease and a wide variety of analytical techniques.

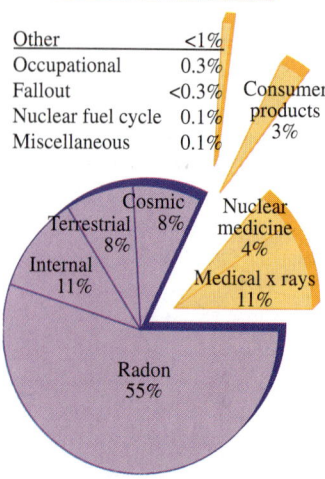

45–9 Contribution of various sources to the total average radiation exposure in the U.S. population, expressed as percentages of the total. (Source: National Council on Radiation Protection and Measurements, Report No. 93, 1987.)

BENEFICIAL USES OF RADIATION

Radiation is widely used in medicine for intentional selective destruction of tissue such as tumors. The hazards are considerable, but if the disease would be fatal without treatment, any hazard may be preferable. Artificially produced isotopes are often used as sources. Such isotopes have several advantages over naturally radioactive isotopes. They may have shorter half-lives and correspondingly greater activity. Isotopes can be chosen that emit the type and energy of radiation desired. Some artificial isotopes have been replaced by photon and electron beams from linear accelerators.

Nuclear medicine is an expanding field of application. Radioactive isotopes have virtually the same electron configurations and resulting chemical behavior as stable isotopes of the same element. But the location and concentration of radioactive isotopes can easily be detected by measurements of the radiation they emit. A familiar example is the use of radioactive iodine for thyroid studies. Nearly all the iodine ingested is either eliminated or stored in the thyroid, and the body's chemical reactions do not discriminate between the unstable isotope ^{131}I and the stable isotope ^{127}I. A minute quantity of ^{131}I is fed or injected into the patient, and the speed with which it becomes concentrated in the thyroid provides a measure of thyroid function. The half-life is 8.02 days, so there are

(a)

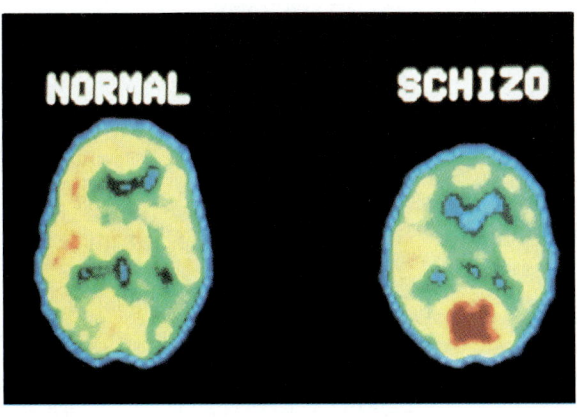

(b)

45–10 (a) In PET (positron-emission tomography) scans, positron-emitting isotopes of common elements such as carbon, nitrogen, and oxygen are administered to the patient. When a positron meets an electron, they annihilate one another, producing two γ photons. These photons are detected by the circular array of detectors, and an image of the cross section is constructed by a computer. (b) PET scans of two brains, one a normal patient's, the other a schizophrenic's, showing differences in glucose usage, a measure of biological activity. Yellow to red colors indicate high values; green to blue colors show low values.

no long-lasting radiation hazards. By use of more sophisticated scanning detectors, one can also obtain a "picture" of the thyroid, which shows enlargement and other abnormalities. This procedure, a type of *autoradiography,* is comparable to photographing the glowing filament of an incandescent light bulb using the light emitted by the filament itself. If this process discovers cancerous thyroid nodules, they can be destroyed by much larger quantities of ^{131}I.

Similar radiographic techniques are used to visualize coronary arteries. A thin tube or *catheter* is threaded through a vein in the arm into the heart, and a radioactive material is injected. Narrowed or blocked arteries can actually be photographed by use of a scanning detector; such a picture is called an *angiogram* or *arteriogram.* A useful isotope for such purposes is technetium ^{99}Tc. This nuclide is formed in an excited state from the β^- decay of molybdenum ^{99}Mo. It decays to its ground state by emitting a γ-ray photon with energy 143 keV. The half-life is 6.01 hours, unusually long for γ emission. (Its ground state is also unstable with a half-life of 2.11×10^5 y; it decays by β^- emission to the stable nuclide ^{99}Ru.) Scanning detectors, often called *gamma cameras,* have been used for studies of the brain, kidneys, and numerous other organs. Figure 45–10 shows another imaging technique.

Radioactive isotopes are used in *tracer* techniques. Tritium, a hydrogen isotope, ^{3}H, is used to tag molecules in complex organic reactions; radioactive tags on pesticide molecules can be used to trace their passage through food chains. In the world of machinery, radioactive iron can be used to study piston-ring wear. Laundry detergent manufacturers have even tested the effectiveness of their products using radioactive dirt.

Many direct effects of radiation are also useful, such as strengthening of polymers by cross-linking, sterilizing surgical tools, dispersion of unwanted static electricity in the air, and intentional ionization of air in smoke detectors. Gamma rays are also being used to sterilize and preserve some food products.

45–7 NUCLEAR REACTIONS

In the preceding sections we studied the decay of unstable nuclei, especially spontaneous emission of an α or β particle, sometimes followed by γ emission. Nothing was done to initiate this decay, and nothing could be done to control it. This section examines some *nuclear reactions,* rearrangements of nuclear components that result from a bombardment by a particle rather than a spontaneous natural process. Rutherford suggested in 1919 that a massive particle with sufficient kinetic energy might be able to penetrate a nucleus. The result would be either a new nucleus with greater atomic number and mass number or a decay of the original nucleus. Rutherford bombarded nitrogen (^{14}N) with α particles and obtained an oxygen (^{17}O) nucleus and a proton:

$$^{4}_{2}\text{He} + {}^{14}_{7}\text{N} \rightarrow {}^{17}_{8}\text{O} + {}^{1}_{1}\text{H}. \qquad (45\text{–}20)$$

Rutherford used alpha particles from naturally radioactive sources. In Section 46–3 we'll describe some of the particle accelerators that are used nowadays to initiate nuclear reactions.

Nuclear reactions are subject to several *conservation laws*. The classical conservation principles for charge, momentum, angular momentum, and energy (including rest energies) are obeyed in all nuclear reactions. An additional conservation law, not anticipated by classical physics, is conservation of the total number of nucleons. The numbers of protons and neutrons need not be conserved separately; we have seen that in β decay, neutrons and protons change into one another. We'll study the basis of the conservation of nucleon number in Chapter 46.

When two nuclei interact, charge conservation requires that the sum of the initial atomic numbers must equal the sum of the final atomic numbers. Because of conservation of nucleon number, the sum of the initial mass numbers must also equal the sum of the final mass numbers. In general, these are *not* elastic collisions, and, correspondingly, the total initial mass does *not* equal the total final mass.

REACTION ENERGY

The difference between the masses before and after the reaction corresponds to the **reaction energy**, according to the mass-energy relation $E = mc^2$. If initial particles A and B interact to produce final particles C and D, the reaction energy Q is defined as

$$Q = (M_A + M_B - M_C - M_D)c^2 \qquad \text{(reaction energy).} \qquad (45\text{–}21)$$

To balance the electrons, we use the neutral atomic masses in Eq. (45–21). That is, we use the mass of $^{1}_{1}$H for a proton, $^{2}_{1}$H for a deuteron, $^{4}_{2}$He for an α particle, and so on. When Q is positive, the total mass decreases and the total kinetic energy increases. Such a reaction is called an *exoergic reaction*. When Q is negative, the mass increases and the kinetic energy decreases, and the reaction is called an *endoergic reaction*. The terms *exothermal* and *endothermal*, borrowed from chemistry, are also used. In an endoergic reaction the reaction cannot occur at all unless the initial kinetic energy in the center of mass reference frame is at least as great as $|Q|$. That is, there is a **threshold energy,** the minimum kinetic energy to make an endoergic reaction go.

EXAMPLE 45–11

An exoergic reaction When lithium (^{7}Li) is bombarded by a proton, two alpha particles (^{4}He) are produced. Find the reaction energy.

SOLUTION The reaction can be written

$$^{1}_{1}\text{H} + {}^{7}_{3}\text{Li} \rightarrow {}^{4}_{2}\text{He} + {}^{4}_{2}\text{He}.$$

Here are the initial and final masses (from Table 45–2):

A:	$^{1}_{1}$H	1.007825 u	C:	$^{4}_{2}$He	4.002603 u
B:	$^{7}_{3}$Li	7.016004 u	D:	$^{4}_{2}$He	4.002603 u
		8.023829 u			8.005206 u

We see that

$$M_A + M_B - M_C - M_D = 0.018623 \text{ u.}$$

Then Eq. (45–21) gives a reaction energy of

$$Q = (0.018623 \text{ u})(931.5 \text{ MeV/u}) = 17.35 \text{ MeV.}$$

This is an exoergic reaction; the final total kinetic energy of the two separating alpha particles is 17.35 MeV *greater* than the initial total kinetic energy of the proton and the lithium nucleus.

EXAMPLE 45–12

An endoergic reaction Calculate the reaction energy for the nuclear reaction represented by Eq. (45–20).

SOLUTION The masses of the various particles are

A:	$^4_2$He	4.002603 u	C:	$^{17}_8$O 16.999132 u
B:	$^{14}_7$N	14.003074 u	D:	$^1_1$H 1.007825 u
		18.005677 u		18.006957 u

We see that the mass increases by 0.001280 u, and the corresponding reaction energy is

$$Q = (-0.001280 \text{ u})(931.5 \text{ MeV/u}) = -1.192 \text{ MeV}.$$

In the center-of-mass system, that is, in a head-on collision with zero total momentum, the minimum total initial kinetic energy for this reaction to occur is 1.192 MeV.

Ordinarily, the endoergic reaction of Example 45–12 would be produced by bombarding stationary ^{14}N nuclei with alpha particles from an accelerator. In this case an alpha's kinetic energy must be *greater than* 1.191 MeV. If all the alpha's kinetic energy went solely to increasing the rest energy, the final kinetic energy would be zero, and momentum would not be conserved. When a particle with mass m and kinetic energy K collides with a stationary particle with mass M, the total kinetic energy K_{cm} in the center-of-mass coordinate system (the energy available to cause reactions) is

$$K_{cm} = \frac{M}{M + m} K. \qquad (45\text{–}22)$$

This expression assumes that the kinetic energies of the particles and nuclei are much less than their rest energies. We leave the derivation of Eq. (45–22) as a problem. In the present example, $K_{cm} = (14.00/18.01)K$, so K must be at least $(18.01/14.00) \times (1.191 \text{ MeV}) = 1.532$ MeV.

For a charged particle such as a proton or an α particle to penetrate the nucleus of another atom and cause a reaction, it must usually have enough initial kinetic energy to overcome the potential-energy barrier caused by the repulsive electrostatic forces. In the reaction of Example 45–11, if we treat the proton and the ^{7}Li nucleus as spherically symmetric charges with radii given by Eq. (45–1), their centers will be 3.5×10^{-15} m apart when they touch. The repulsive potential energy of the proton (charge $+e$) and the ^{7}Li nucleus (charge $+3e$) at this separation r is

$$U = \frac{1}{4\pi\epsilon_0} \frac{(e)(3e)}{r} = (9.0 \times 10^9 \text{ N} \cdot \text{m}^2/\text{C}^2) \frac{(3)(1.6 \times 10^{-19} \text{ C})^2}{3.5 \times 10^{-15} \text{ m}}$$

$$= 2.0 \times 10^{-13} \text{ J} = 1.2 \text{ MeV}.$$

Even though the reaction is exoergic, the proton must have a minimum kinetic energy of about 1.2 MeV for the reaction to occur; unless the proton *tunnels* through the barrier (see Section 42–5).

NEUTRON ABSORPTION

Absorption of *neutrons* by nuclei forms an important class of nuclear reactions. Heavy nuclei bombarded by neutrons in a nuclear reaction can undergo a series of neutron absorptions alternating with beta decays, in which the mass number A increases by as much as 25. Some of the *transuranic elements,* elements having Z larger than 92, are produced in this way. These elements have not been found in nature. Many transuranic elements, having Z possibly as high as 118, have been identified.

The analytical technique of *neutron activation analysis* uses similar reactions. When bombarded by neutrons, many stable nuclides absorb a neutron to become unstable and then undergo β^- decay. The energies of the β^- and γ emissions depend on the unstable nuclide and provide a means of identifying it and the original stable nuclide. Quantities of elements that are far too small for conventional chemical analysis can be detected in this way.

45-8 NUCLEAR FISSION

Nuclear fission is a decay process in which an unstable nucleus splits into two fragments of comparable mass. Fission was discovered in 1938 through the experiments of Otto Hahn and Fritz Strassman. Pursuing earlier work by Fermi, they bombarded uranium ($Z = 92$) with neutrons. The resulting radiation did not coincide with that of any known radioactive nuclide. Urged on by Lise Meitner, they used meticulous chemical analysis to reach the astonishing but inescapable conclusion that they had found a radioactive isotope of barium ($Z = 56$). Later, radioactive krypton ($Z = 36$) was also found. Meitner and Otto Frisch correctly interpreted these results as showing that uranium nuclei were splitting into two massive fragments called *fission fragments*. Two or three free neutrons usually appear along with the fission fragments and, very occasionally, a light nuclide such as ^{3}H.

Both the common isotope (99.3%) ^{238}U and the uncommon isotope (0.7%) ^{235}U (as well as several other nuclides) can be easily split by neutron bombardment: ^{235}U by slow neutrons but ^{238}U only by neutrons with a minimum of about 1 MeV of energy. Fission resulting from neutron absorption is called *induced fission*. Some nuclides can also undergo *spontaneous fission* without initial neutron absorption, but this is quite rare. When ^{235}U absorbs a neutron, the resulting nuclide ^{236}U* is in a highly excited state and splits into two fragments almost instantaneously. Strictly speaking, it is ^{236}U*, not ^{235}U, that undergoes fission, but it's usual to speak of the fission of ^{235}U.

Over 100 different nuclides, representing more than 20 different elements, have been found among the fission products. Figure 45-11 shows the distribution of mass numbers for fission fragments from the fission of ^{235}U. Most of the fragments have mass numbers from 90 to 100 and from 135 to 145; fission into two fragments with nearly equal mass is unlikely.

We invite you to check the following two typical fission reactions for conservation of nucleon number and charge:

$$^{235}_{92}\text{U} + {}^1_0\text{n} \rightarrow {}^{236}_{92}\text{U}^* \rightarrow {}^{144}_{56}\text{Ba} + {}^{89}_{36}\text{Kr} + 3{}^1_0\text{n},$$

$$^{235}_{92}\text{U} + {}^1_0\text{n} \rightarrow {}^{236}_{92}\text{U}^* \rightarrow {}^{140}_{54}\text{Xe} + {}^{94}_{38}\text{Sr} + 2{}^1_0\text{n}.$$

The total kinetic energy of the fission fragments is enormous, about 200 MeV (compared to typical α and β energies of a few MeV). The reason for this is that nuclides at the high end of the mass spectrum (near $A = 240$) are less tightly bound than those nearer

45-11 Mass distribution of fission fragments from the fission of ^{236}U* resulting from neutron absorption by ^{235}U. The vertical scale is logarithmic.

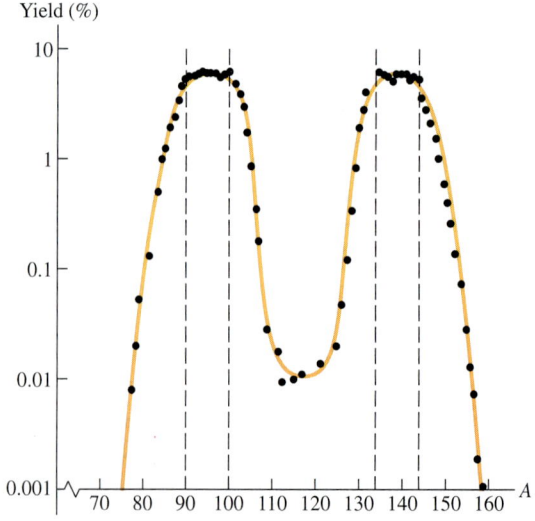

the middle (*A* = 90 to 145). Referring to Fig. 45–2, we see that the average binding energy per nucleon is about 7.6 MeV at *A* = 240 but about 8.5 MeV at *A* = 120. Therefore a rough estimate of the expected *increase* in binding energy during fission is about 8.5 MeV − 7.6 MeV = 0.9 MeV per nucleon, or a total of (235)(0.9 MeV) ≈ 200 MeV.

CAUTION ▶ It may seem to be a violation of conservation of energy to have an increase in both the binding energy and the kinetic energy during a fission reaction. But relative to the total rest energy E_0 of the separated nucleons, the rest energy of the nucleus is E_0 *minus* E_B. Thus an *increase* in binding energy corresponds to a *decrease* in rest energy as rest energy is converted to the kinetic energy of the fission fragments. ◀

Fission fragments always have too many neutrons to be stable. We noted in Section 45–4 that the neutron/proton ratio (*N/Z*) for stable nuclides is about 1 for light nuclides but almost 1.6 for the heaviest nuclides because of the increasing influence of the electrical repulsion of the protons. The *N/Z* value for stable nuclides is about 1.3 at *A* = 100 and 1.4 at *A* = 150. The fragments have about the same *N/Z* as ^{235}U, about 1.55. They usually respond to this surplus of neutrons by undergoing a series of β^- decays (each of which increases *Z* by one and decreases *N* by one) until a stable value of *N/Z* is reached. A typical example is

$$^{140}_{54}\text{Xe} \xrightarrow{\beta^-} {}^{140}_{55}\text{Cs} \xrightarrow{\beta^-} {}^{140}_{56}\text{Ba} \xrightarrow{\beta^-} {}^{140}_{57}\text{La} \xrightarrow{\beta^-} {}^{140}_{58}\text{Ce}.$$

The nuclide ^{140}Ce is stable. This series of β^- decays produces, on average, about 15 MeV of additional kinetic energy. The neutron excess of fission fragments also explains why two or three free neutrons are released during the fission.

Fission appears to set an upper limit on the production of transuranic nuclei, mentioned in Section 45–7, that are relatively stable. There are theoretical reasons to expect that nuclei near *Z* = 114, *N* = 184 or 196, might be stable with respect to spontaneous fission. In the shell model (Section 45–3), these numbers correspond to filled shells and subshells in the nuclear energy-level structure. Such *superheavy nuclei* would still be unstable with respect to alpha emission, but they might live long enough to be identified. As of this writing, there is evidence that nuclei with *Z* = 114 have been produced in the laboratory. Whether they exist in nature is still an open question.

LIQUID-DROP MODEL

We can understand fission qualitatively on the basis of the liquid-drop model of the nucleus (Section 45–3). The process is shown in Fig. 45–12 in terms of an electrically charged liquid drop. These sketches shouldn't be taken too literally, but they may help to develop your intuition about fission. A ^{235}U nucleus absorbs a neutron (Fig. 45–12a), becoming a ^{236}U* nucleus with excess energy (Fig. 45–12b). This excess energy causes violent oscillations (Fig. 45–12c), during which a neck between two lobes develops. The electrical repulsion of these two lobes stretches the neck farther (Fig. 45–12d), and finally two smaller drops are formed (Fig. 45–12e) that move rapidly apart.

This qualitative picture has been developed into a more quantitative theory to explain why some nuclei undergo fission and others don't. Figure 45–13 shows a hypothetical

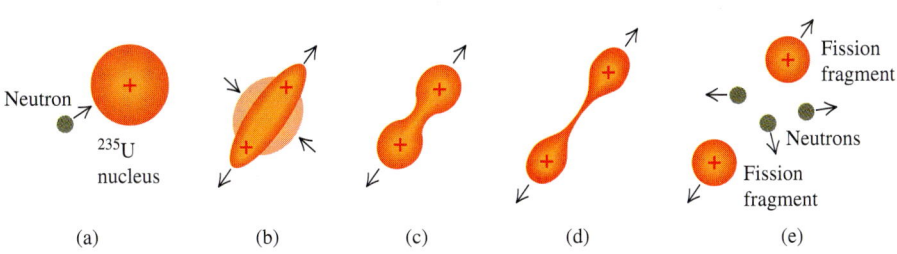

(a)   (b)   (c)   (d)   (e)

45–12 (a) A ^{235}U nucleus absorbs a neutron. (b) The resulting ^{236}U* nucleus is in a highly excited state and oscillates strongly. (c) A neck develops, and electrical repulsion pushes the two lobes apart. (d) The two lobes separate, forming fission fragments. (e) Neutrons are emitted from the fission fragments at the time of fission or occasionally a few seconds later.

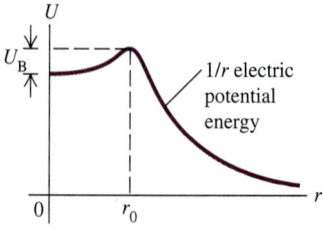

45–13 Hypothetical potential energy function for two fission fragments in a fissionable nucleus. At distances r beyond the range of the nuclear force, the potential energy varies approximately as $1/r$. Fission occurs if there is an excitation energy greater than U_B or an appreciable probability for tunneling through this potential-energy barrier.

potential-energy function for two possible fission fragments. If neutron absorption results in an excitation energy greater than the energy barrier height U_B, fission occurs immediately. Even when there isn't quite enough energy to surmount the barrier, fission can take place by quantum-mechanical *tunneling,* discussed in Section 42–5. In principle, many stable heavy nuclei can fission by tunneling. But the probability depends very critically on the height and width of the barrier. For most nuclei this process is so unlikely that it is never observed.

CHAIN REACTIONS

Fission of a uranium nucleus, triggered by neutron bombardment, releases other neutrons that can trigger more fissions, suggesting the possibility of a **chain reaction** (Fig. 45–14). The chain reaction may be made to proceed slowly and in a controlled manner in a nuclear reactor or explosively in a bomb. The energy release in a nuclear chain reaction is enormous, far greater than that in any chemical reaction. (In a sense, *fire* is a chemical chain reaction.) For example, when uranium is "burned" to uranium dioxide in the chemical reaction

$$U + O_2 \rightarrow UO_2,$$

the heat of combustion is about 4500 J/g. Expressed as energy per atom, this is about

45–14 Schematic diagram of a nuclear fission chain reaction.

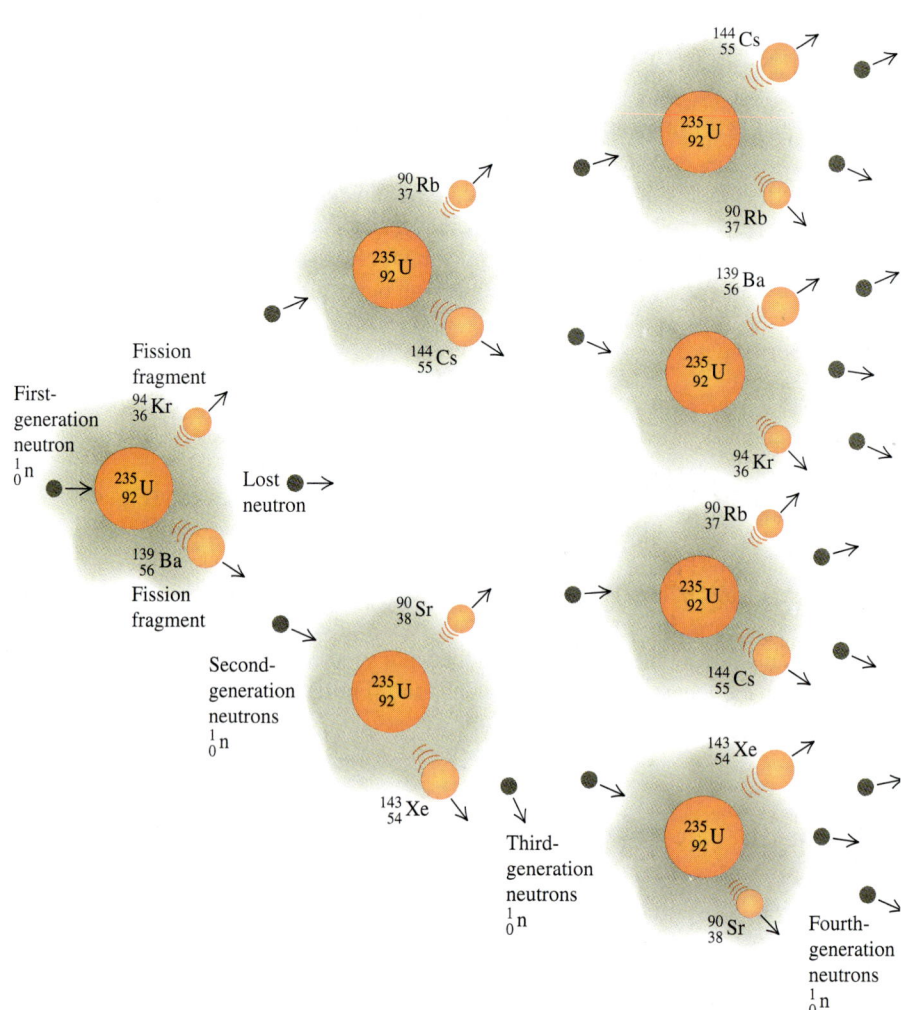

11 eV per atom. By contrast, fission liberates about 200 MeV per atom, nearly 20 million times as much energy.

NUCLEAR REACTORS

A *nuclear reactor* is a system in which a controlled nuclear chain reaction is used to liberate energy. In a nuclear power plant, this energy is used to generate steam, which operates a turbine and turns an electrical generator.

On average, each fission of a ^{235}U nucleus produces about 2.5 free neutrons, so 40% of the neutrons are needed to sustain a chain reaction. A ^{235}U nucleus is much more likely to absorb a low-energy neutron (less than 1 eV) than one of the higher-energy neutrons (1 MeV or so) that are liberated during fission. In a nuclear reactor the higher-energy neutrons are slowed down by collisions with nuclei in the surrounding material, called the *moderator,* so they are much more likely to cause further fissions. In nuclear power plants, the moderator is often water, occasionally graphite. The *rate* of the reaction is controlled by inserting or withdrawing *control rods* made of elements (such as boron or cadmium) whose nuclei *absorb* neutrons without undergoing any additional reaction. The isotope ^{238}U can also absorb neutrons, leading to ^{239}U*, but not with high enough probability for it to sustain a chain reaction by itself. Thus uranium that is used in reactors is often "enriched" by increasing the proportion of ^{235}U above the natural value of 0.7%, typically to 3% or so, by isotope-separation processing.

The most familiar application of nuclear reactors is for the generation of electric power. As was noted above, the fission energy appears as kinetic energy of the fission fragments, and its immediate result is to increase the internal energy of the fuel elements and the surrounding moderator. This increase in internal energy is transferred as heat to generate steam to drive turbines, which spin the electrical generators. Figure 45–15 is a schematic diagram of a nuclear power plant. The energetic fission fragments heat the water surrounding the reactor core. The steam generator is a heat exchanger that takes heat from this highly radioactive water and generates nonradioactive steam to run the turbines.

45–15 Schematic diagram of a nuclear power plant.

A typical nuclear plant has an electric-generating capacity of 1000 MW (or 10^9 W). The turbines are heat engines and are subject to the efficiency limitations imposed by the second law of thermodynamics, discussed in Chapter 18. In modern nuclear plants the overall efficiency is about one third, so 3000 MW of thermal power from the fission reaction are needed to generate 1000 MW of electrical power.

EXAMPLE 45-13

Uranium consumption in a nuclear reactor What mass of ^{235}U has to undergo fission each day to provide 3000 MW of thermal power?

SOLUTION Each second, we need 3000 MJ or 3000×10^6 J. Each fission provides 200 MeV, which is

$$(200 \text{ MeV})(1.6 \times 10^{-13} \text{ J/MeV}) = 3.2 \times 10^{-11} \text{ J}.$$

The number of fissions needed each second is

$$\frac{3000 \times 10^6 \text{ J}}{3.2 \times 10^{-11} \text{ J}} = 9.4 \times 10^{19}.$$

Each ^{235}U atom has a mass of $(235 \text{ u})(1.66 \times 10^{-27} \text{ kg/u}) = 3.9 \times 10^{-25}$ kg, so the mass of ^{235}U needed each second is

$$(9.4 \times 10^{19})(3.9 \times 10^{-25} \text{ kg}) = 3.7 \times 10^{-5} \text{ kg} = 37 \text{ } \mu\text{g}.$$

In one day (86,400 s), the total consumption of ^{235}U is

$$(3.7 \times 10^{-5} \text{ kg/s})(86,400 \text{ s}) = 3.2 \text{ kg}.$$

For comparison, note that the 1000-MW coal-fired power plant that we described in Section 18–9 burns 10,600 tons (about ten million kg) of coal per day!

Nuclear fission reactors have many other practical uses. Among these are the production of artificial radioactive isotopes for medical and other research, production of high-intensity neutron beams for research in nuclear structure, and production of fissionable nuclides such as ^{239}Pu from the common isotope ^{238}U. The last is the function of *breeder reactors,* which can produce more fuel than they use.

We mentioned above that about 15 MeV of the energy released after fission of a ^{235}U nucleus comes from the β^- decays of the fission fragments. This fact poses a serious problem with respect to control and safety of reactors. Even after the chain reaction has been completely stopped by insertion of control rods into the core, heat continues to be evolved by the β^- decays, which cannot be stopped. For a 3000-MW reactor this heat power is initially very large, about 200 MW. In the event of total loss of cooling water, this power is more than enough to cause a catastrophic meltdown of the reactor core and possible penetration of the containment vessel. The difficulty in achieving a "cold shutdown" following an accident at the Three Mile Island nuclear power plant in Pennsylvania in March 1979 was a result of the continued evolution of heat due to β^- decays.

The catastrophe of April 26, 1986, at Chernobyl reactor No. 4 in Ukraine resulted from a combination of an inherently unstable design and several human errors committed during a test of the emergency core cooling system. Too many control rods were withdrawn to compensate for a decrease in power caused by a buildup of neutron absorbers such as ^{135}Xe. The power level rose from 1% of normal to 100 times normal in 4 seconds; a steam explosion ruptured pipes in the core cooling system and blew the heavy concrete cover off the reactor. The graphite moderator caught fire and burned for several days, and there was a meltdown of the core. The total activity of the radioactive material released into the atmosphere has been estimated as about 10^8 Ci.

45-9 NUCLEAR FUSION

19.3
Fusion

In a **nuclear fusion** reaction, two or more small light nuclei come together, or *fuse,* to form a larger nucleus. Fusion reactions release energy for the same reason as fission reactions; the binding energy per nucleon after the reaction is greater than before. Referring to Fig. 45–2, we see that the binding energy per nucleon increases with A up to about $A = 60$, so fusion of nearly any two light nuclei to make a nucleus with A less than 60 is likely to be an exoergic reaction. In comparison to fission, we are moving

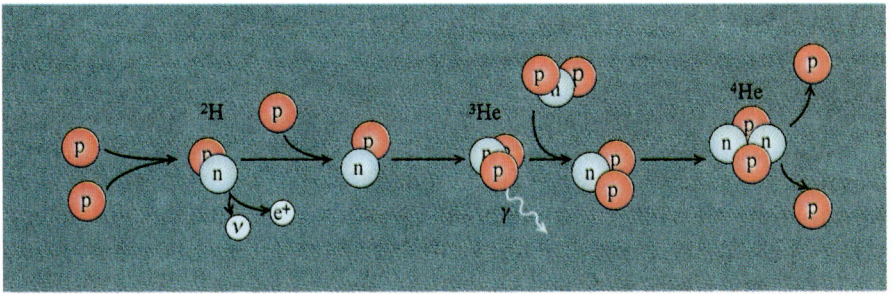

45–16 The proton-proton chain.

toward the peak of this curve from the opposite side. Another way to express the energy relations is that the total mass of the products is less than that of the initial particles.

Here are three examples of energy-liberating fusion reactions, written in terms of the neutral atoms:

$$^1_1H + ^1_1H \rightarrow ^2_1H + \beta^+ + \nu_e,$$

$$^2_1H + ^1_1H \rightarrow ^3_2He + \gamma,$$

$$^3_2He + ^3_2He \rightarrow ^4_2He + ^1_1H + ^1_1H.$$

In the first reaction, two protons combine to form a deuteron (^2H), with the emission of a positron (β^+) and an electron neutrino. In the second, a proton and a deuteron combine to form the nucleus of the light isotope of helium, ^3He, with the emission of a gamma ray. Now double the first two reactions to provide the two ^3He nuclei that fuse in the third reaction to form an alpha particle (^4He) and two protons. Together the reactions make up the process called the *proton-proton chain* (Fig. 45–16).

The net effect of the chain is the conversion of four protons into one α particle, two positrons, two electron neutrinos, and two γ's. We can calculate the energy release from this part of the process: The mass of an α particle plus two positrons is the mass of neutral ^4He, the neutrinos have zero (or negligible) mass, and the gammas have zero mass.

Mass of four protons	4.029106 u
Mass of ^4He	4.002603 u
Mass difference and energy release	0.026503 u and 24.69 MeV.

The two positrons that are produced during the first step of the proton-proton chain collide with two electrons; mutual annihilation of the four particles takes place, and their rest energy is converted into 4(0.511 MeV) = 2.044 MeV of gamma radiation. Thus the total energy release is (24.69 + 2.044) MeV = 26.73 MeV. The proton-proton chain is believed to take place in the interior of the sun and other stars. Each gram of the sun's mass contains about 4.5×10^{23} protons. If all of these protons were fused into helium, the energy released would be about 130,000 kWh. If the sun were to continue to radiate at its present rate, it would take about 75×10^9 years to exhaust its supply of protons. We hope that it causes you no anxiety that current models suggest that the sun won't be able to fuse *all* its protons and will last only about one fifteenth as long, or 5×10^9 years.

The energy released as starlight comes from fusion reactions deep within a star's interior. When a star is first formed and for most of its life, it converts the hydrogen in its core into helium. As a star ages, the core temperature can become high enough for additional fusion reactions that convert helium into carbon, oxygen, and other elements.

EXAMPLE 45–14

A fusion reaction Two deuterons fuse to form a *triton* (a nucleus of tritium, or ^3H) and a proton. How much energy is liberated?

SOLUTION This is a nuclear reaction of the type discussed in Section 45–7. Adding one electron to each particle gives four neutral atoms; we find their masses in Table 45–2 and substitute into Eq. (45–21):

$Q = [2(2.014102 \text{ u}) - 3.016049 \text{ u} - 1.007825 \text{ u}](931.5 \text{ MeV/u})$

$\quad = 4.03 \text{ MeV}.$

Thus 4.03 MeV is released in the reaction; the triton and proton together have 4.03 MeV more kinetic energy than the two deuterons had together.

ACHIEVING FUSION

For two nuclei to undergo fusion, they must come together to within the range of the nuclear force, typically of the order of 2×10^{-15} m. To do this, they must overcome the electrical repulsion of their positive charges. For two protons at this distance, the corresponding potential energy is about 1.2×10^{-13} J or 0.7 MeV; this represents the total initial *kinetic* energy that the fusion nuclei must have, for example, 0.6×10^{-13} J each in a head-on collision.

Atoms have this much energy only at extremely high temperatures. The discussion of Section 16–4 showed that the average translational kinetic energy of a gas molecule at temperature T is $\frac{3}{2}kT$, where k is Boltzmann's constant. The temperature at which this is equal to $E = 0.6 \times 10^{-13}$ J is determined by the relation

$$E = \frac{3}{2} kT,$$

$$T = \frac{2E}{3k} = \frac{2(0.6 \times 10^{-13} \text{ J})}{3(1.38 \times 10^{-23} \text{ J/K})} = 3 \times 10^9 \text{ K.}$$

Fusion reactions are possible at lower temperatures because the Maxwell-Boltzmann distribution function (Section 16–6) gives a small fraction of protons with kinetic energies much higher than the average value. The proton-proton reaction occurs at "only" 1.5×10^7 K in the sun, making it an extremely low-probability process; but that's why the sun is expected to last so long. At these temperatures the fusion reactions are called *thermonuclear* reactions.

Intensive efforts are underway in many laboratories to achieve controlled fusion reactions, which potentially represent an enormous new resource of energy. At the temperatures mentioned, light atoms are fully ionized, and the resulting state of matter is called a *plasma*. In one kind of experiment using *magnetic confinement*, a plasma is heated to extremely high temperature by an electrical discharge, while being contained by appropriately shaped magnetic fields. In another, using *inertial confinement*, pellets of the material to be fused are heated by a high-intensity laser beam. Some of the reactions being studied are

$$^2_1\text{H} + {}^2_1\text{H} \rightarrow {}^3_1\text{H} + {}^1_1\text{H} + 4.0 \text{ MeV,} \tag{1}$$

$$^3_1\text{H} + {}^2_1\text{H} \rightarrow {}^4_2\text{He} + {}^1_0\text{n} + 17.6 \text{ MeV,} \tag{2}$$

$$^2_1\text{H} + {}^2_1\text{H} \rightarrow {}^3_2\text{He} + {}^1_0\text{n} + 3.3 \text{ MeV,} \tag{3}$$

$$^3_2\text{He} + {}^2_1\text{H} \rightarrow {}^4_2\text{He} + {}^1_1\text{H} + 18.3 \text{ MeV.} \tag{4}$$

We described the first reaction in Example 45–14; two deuterons fuse to form a triton and a proton. In the second, a triton combines with another deuteron to form an alpha particle and a neutron. The result of both of these reactions together is the conversion of three deuterons into an alpha particle, a proton, and a neutron, with the liberation of 21.6 MeV of energy. Reactions (3) and (4) together achieve the same conversion. In a plasma that contains deuterons, the two pairs of reactions occur with roughly equal probability. As yet, no one has succeeded in producing these reactions under controlled conditions in such a way as to yield a net surplus of usable energy.

Methods of achieving fusion that don't require high temperatures are also being studied; these are called *cold fusion*. One scheme that does work uses an unusual hydrogen molecule ion. The usual H_2^+ ion consists of two protons bound by one shared electron; the nuclear spacing is about 0.1 nm. If the protons are replaced by a deuteron (^2H) and a triton (^3H) and the electron by a *muon,* which is 208 times as massive as the electron, the spacing is made smaller by a factor of 208. The probability then becomes appreciable for the two nuclei to tunnel through the thin repulsive potential-energy barrier and fuse in reaction (2) above. The prospect of making this process, called *muon-catalyzed fusion,* into a practical energy source is still distant.

SUMMARY

- Nuclei are composed of nucleons (protons and neutrons). All nuclei have about the same density. The radius of a nucleus with mass number A is, approximately,

$$R = R_0 A^{1/3}, \qquad (45\text{–}1)$$

where $R_0 = 1.2 \times 10^{-15}$ m. A single nuclear species is called a nuclide. Isotopes are nuclides of the same element that have different numbers of neutrons. Nuclear masses are measured in atomic mass units. Nucleons have angular momentum and a magnetic moment.

- The mass of a nucleus is always less than the mass of the protons and neutrons in it. The mass difference multiplied by c^2 gives the binding energy:

$$E_{\mathrm{B}} = (ZM_{\mathrm{H}} + Nm_{\mathrm{n}} - {}_Z^A M)c^2. \qquad (45\text{–}8)$$

- The nuclear force is short-range and favors pairs of particles. A nucleus is unstable if A or Z is too large or if the ratio N/Z is wrong. Two widely used models of the nucleus are the liquid-drop model and the shell model; the latter is analogous to the central-field approximation in atomic structure.

- Unstable nuclides usually emit an alpha or beta particle in changing to another nuclide, sometimes followed by a gamma-ray photon. The rate of decay of an unstable nucleus is described by the decay constant λ or the half-life $T_{1/2}$. If the number of nuclei at time $t = 0$ is N_0 and no more are produced, the number at time t is

$$N(t) = N_0 e^{-\lambda t}. \qquad (45\text{–}15)$$

The lifetime T_{mean} is related to λ and $T_{1/2}$ by

$$T_{\mathrm{mean}} = \frac{1}{\lambda} = \frac{T_{1/2}}{\ln 2} = \frac{T_{1/2}}{0.693}. \qquad (45\text{–}17)$$

- The biological effect of any radiation depends on the product of the energy absorbed per unit mass and the relative biological effectiveness (RBE), which is different for different radiations.

- In a nuclear reaction, two nuclei or particles collide to produce two new nuclei or particles. Reactions can be exoergic or endoergic. Several conservation laws, including charge, energy, momentum, angular momentum, and nucleon number, are obeyed.

- Nuclear fission is the splitting of a heavy nucleus into two lighter, always unstable, nuclei, with the release of energy. Nuclear fusion is the combining of two light nuclei into a heavier nucleus, with the release of energy.

DISCUSSION QUESTIONS

Q45–1 Neutrons have a magnetic dipole moment and can undergo spin flips by absorbing electromagnetic radiation. Why, then, are protons rather than neutrons used in MRI of body tissues? (See Fig. 45–1.)

Q45–2 How can you be sure that nuclei are not made of protons and electrons, rather than protons and neutrons?

Q45–3 Why aren't the masses of all nuclei integer multiples of the mass of a single nucleon?

Q45–4 Can you tell from the value of the mass number A whether to use a plus value, a minus value, or zero for the fifth term of Eq. (45–9)? Explain.

Q45–5 The binding energy per nucleon for most nuclides doesn't vary much (see Fig. 45–2). Is there similar consistency in the *atomic* energy of atoms, on an "energy per electron" basis? If so, why? If not, why not?

Q45–6 Heavy unstable nuclei usually decay by emitting an α or β particle. Why do they not usually emit a single proton or neutron?

Q45–7 The only two stable nuclides with more protons than neutrons are $^{1}_{1}\text{H}$ and $^{3}_{2}\text{He}$. Why is $Z > N$ so uncommon?

Q45–8 Since lead is a stable element, why doesn't the ^{238}U decay series shown in Fig. 45–7 stop at lead, ^{214}Pb?

Q45–9 In the ^{238}U decay chain shown in Fig. 45–7, some nuclides in the chain are found much more abundantly in nature than others, even though every ^{238}U nucleus goes through every step in the chain before finally becoming ^{206}Pb. Why don't the intermediate nuclides all have the same abundance?

Q45–10 Compared to α particles with the same energy, β particles can much more easily penetrate through matter. Why is this?

Q45–11 Some decays that are not possible from a nucleus in its ground state are possible if the nucleus is in an excited state. Explain why.

Q45–12 If $^{A}_{Z}\text{El}_i$ represents the initial nuclide, what is the decay process or processes if the final nuclide is a) $^{A}_{Z+1}\text{El}_f$? b) $^{A-4}_{Z-2}\text{El}_f$? c) $^{A}_{Z-1}\text{El}_f$?

Q45–13 In a nuclear decay equation, why can we represent an electron as $^{0}_{-1}\beta^{-}$? What are the equivalent representations for a positron, a neutrino, and an antineutrino?

Q45–14 Why is the alpha, beta, or gamma decay of an unstable nucleus unaffected by the *chemical* situation of the atom, such as the nature of the molecule or solid in which it is bound? The chemical situation of the atom can, however, have an effect on the half-life in electron capture. Why is this?

Q45–15 In the process of *internal conversion,* a nucleus decays from an excited state to a ground state by giving the excitation energy directly to an atomic electron rather than emitting a gamma-ray photon. Why can this process also produce x-ray photons?

Q45–16 In Example 45–9 (Section 45–5), the activity of atmospheric carbon *before* 1900 was given. Discuss why this activity may have changed since 1900.

Q45–17 The most common radium isotope found on earth, ^{226}Ra, has a half-life of about 1600 years. If the earth was formed well over 10^9 years ago, why is there any radium left now?

Q45–18 Fission reactions only occur for nuclei with large nucleon numbers, while exoergic fusion reactions only occur for nuclei with small nucleon numbers. Why is this?

Q45–19 In the interiors of some stars, fusion reactions take place in which three $^{4}_{2}\text{He}$ combine to form a $^{12}_{6}\text{C}$ nucleus. These reactions can only take place if the interior temperature of the star is substantially higher than that of the sun. Why does the temperature play a role?

Q45–20 An *antiproton* has the same rest energy as a proton but an opposite charge $-e$. What are the relative directions of $\vec{\mu}$ and $\vec{S}$ for an antiproton?

EXERCISES

SECTION 45–2 PROPERTIES OF NUCLEI

45–1 How many protons and how many neutrons are there in a nucleus of the most common isotope of a) silicon, $^{28}_{14}\text{Si}$; b) rubidium, $^{85}_{37}\text{Rb}$; c) thallium, $^{205}_{81}\text{Tl}$?

45–2 Consider the three nuclei of Exercise 45–1. Estimate a) the radius, b) the surface area, and c) the volume of each nucleus. Determine d) the mass density (in kg/m^3) and e) the nucleon density (in nucleons per cubic meter) for each nucleus. Assume that the mass of each nucleus is A atomic mass units.

45–3 Hydrogen atoms are placed in an external magnetic field. The protons can make transitions between states in which the nuclear spin component is parallel and antiparallel to the field by absorbing or emitting a photon. What magnetic field magnitude is required for this transition to be induced by photons with frequency 22.7 MHz?

45–4 Neutrons are placed in a magnetic field with magnitude 2.30 T. a) What is the energy difference between the states with the nuclear spin angular momentum components parallel and antiparallel to the field? Which state is lower in energy, the one with its spin component parallel to the field or the one with its spin component antiparallel to the field? How do your results compare with the energy states for a proton in the same field (Example 45–2 in Section 45–2)? b) The neutrons can make transitions from one of these states to the other by emitting or absorbing a photon with energy equal to the energy difference of the two states. Find the frequency and wavelength of such a photon.

45–5 Hydrogen atoms are placed in an external 1.65-T magnetic field. a) The *protons* can make transitions between states where the nuclear spin component is parallel and antiparallel to the field by absorbing or emitting a photon. Which state lies lower in energy, the state with the nuclear spin component parallel or antiparallel to the field? What are the frequency and wavelength of the photon? In which region of the electromagnetic spectrum does it lie? b) The *electrons* can make transitions between states where the electron spin component is parallel and antiparallel to the field by absorbing or emitting a photon. Which state lies lower in energy, the state with the electron spin component parallel or antiparallel to the field? What are the frequency and wavelength of the photon? In which region of the electromagnetic spectrum does it lie?

SECTION 45–3 NUCLEAR BINDING AND NUCLEAR STRUCTURE

45–6 a) Show that the ionization energy of the hydrogen atom (13.6 eV) is 0.0027% of an electron's rest energy. b) We found in Example 45–3 (Section 45–3) that the binding energy per nucleon for $^{62}_{28}\text{Ni}$ is 8.795 MeV. What fraction is this of the rest energy of a proton?

45–7 What is the maximum wavelength of a γ ray that could break a deuteron into a proton and a neutron? (This process is called photodisintegration.)

45–8 Calculate the mass defect, the binding energy (in MeV), and the binding energy per nucleon of a) the nitrogen nucleus, $^{14}_{7}$N; b) the helium nucleus, $^{4}_{2}$He. c) How do the results of parts (a) and (b) compare?

45–9 The most common isotope of boron is $^{11}_{5}$B. a) Determine the total binding energy of $^{11}_{5}$B from Table 45–2 in Section 45–2. b) Calculate this binding energy from Eq. (45–9). (Why is the fifth term zero?) Compare to the result you obtained in part (a). What is the percent difference? Compare the accuracy of Eq. (45–9) for $^{11}_{5}$B to its accuracy for $^{62}_{28}$Ni (Example 45–4 in Section 45–3).

45–10 The most common isotope of copper is $^{63}_{29}$Cu. The measured mass of the neutral atom is 62.929601 u. a) From the measured mass, determine the mass defect, and use it to find the total binding energy and the binding energy per nucleon. b) Calculate the binding energy from Eq. (45–9). (Why is the fifth term zero?) Compare to the result you obtained in part (a). What is the percent difference? What do you conclude about the accuracy of Eq. (45–9)?

45–11 What are the six known elements for which Z is a magic number? Discuss what properties these elements have as a consequence of their special values of Z.

45–12 The most common isotope of uranium, $^{238}_{92}$U, has atomic mass 238.050783 u. Calculate a) the mass defect; b) the binding energy (in MeV); c) the binding energy per nucleon.

SECTION 45–4 NUCLEAR STABILITY AND RADIOACTIVITY

45–13 What nuclide is produced in the following radioactive decays? a) α decay of $^{239}_{94}$Pu; b) β^- decay of $^{24}_{11}$Na; c) β^+ decay of $^{15}_{8}$O.

45–14 a) Is the decay $n \rightarrow p + \beta^- + \bar{\nu}_e$ energetically possible? If not, explain why not. If so, calculate the total energy released. b) Is the decay $p \rightarrow n + \beta^+ + \nu_e$ energetically possible? If not, explain why not. If so, calculate the total energy released.

45–15 Calculate the mass defect for the decay of $^{8}_{4}$Be into two alpha particles, $^{8}_{4}$Be $\rightarrow$ $^{4}_{2}$He $+$ $^{4}_{2}$He. The mass of $^{8}_{4}$Be is 8.005305 u.

45–16 What particle (α particle, electron, or positron) is emitted in the following radioactive decays? a) $^{27}_{14}$Si $\rightarrow$ $^{27}_{13}$Al; b) $^{238}_{92}$U $\rightarrow$ $^{234}_{90}$Th; c) $^{74}_{33}$As $\rightarrow$ $^{74}_{34}$Se.

45–17 The atomic mass of ^{14}C is 14.003242 u. Show that the β^- decay of ^{14}C is energetically possible, and calculate the energy released in the decay.

45–18 a) Calculate the energy released by the electron-capture decay of $^{57}_{27}$Co (Example 45–7 in Section 45–4). b) A negligible amount of this energy goes to the resulting $^{57}_{26}$Fe atom as kinetic energy. About 90% of the time, the $^{57}_{26}$Fe nucleus emits two successive gamma-ray photons after the electron-capture process, of energies 0.122 MeV and 0.014 MeV, respectively, in decaying to its ground state. What is the energy of the neutrino emitted in this case?

45–19 Tritium Decay. Tritium, $^{3}_{1}$H, is an unstable isotope of hydrogen produced in nuclear reactors. a) Show that tritium

must be unstable with respect to β^- decay. b) Determine the total kinetic energy of the decay products.

SECTION 45–5 ACTIVITIES AND HALF-LIVES

45–20 An unstable isotope of cobalt, ^{60}Co, has one more neutron in its nucleus than does the stable isotope ^{59}Co and is a β^- emitter with a half-life of 5.27 y. This isotope is used in medicine. A certain radiation source in a hospital contains 0.0360 mg of ^{60}Co. a) What is the decay constant for this isotope? b) How many atoms are in the source? c) How many decays occur per second? d) What is the activity of the source in curies? How does this compare with the activity of an equal mass of radium (^{226}Ra, half-life 1600 y)?

45–21 The radioactive nuclide ^{199}Pt has a half-life of 30.8 minutes. A sample is prepared that has an initial activity of 7.56×10^{11} Bq. a) How many ^{199}Pt nuclei are initially present in the sample? b) How many are present after 30.8 minutes? What is the activity at this time? c) Repeat part (b) for a time 92.4 minutes after the sample is first prepared.

45–22 Radiocarbon Dating. A sample from timbers at an archeological site containing 500 g of carbon provides 3070 decays per minute. What is the age of the sample?

45–23 The unstable isotope ^{40}K is used for dating of rock samples. Its half-life is 1.28×10^9 y. a) How many decays occur per second in a sample containing 1.63×10^{-6} g of ^{40}K? b) What is the activity of the sample in curies?

45–24 If you are of average mass, about 360 million nuclei in your body undergo radioactive decay each day. Express your activity in curies.

45–25 The most common (99.3%) isotope of uranium, ^{238}U, has a half-life of 4.47×10^9 y, decaying to ^{234}Th by α emission. a) What is the decay constant? b) What mass of uranium would be required for an activity of 12.0 μCi? c) How many α particles are emitted per second by 60.0 g of ^{238}U?

SECTION 45–6 BIOLOGICAL EFFECTS OF RADIATION

45–26 In an industrial accident a 65-kg person receives a lethal whole-body equivalent dose of 5.4 Sv from x rays. a) What is the equivalent dose in rem? b) What is the absorbed dose in rad? c) What is the total energy absorbed by the person's body? How does this amount of energy compare to the amount of energy required to raise the temperature of 65 kg of water 0.010°C?

45–27 A 50-kg person accidentally ingests 0.35 Ci of tritium. a) Assume that the tritium spreads uniformly over the body and that each decay leads on the average to the absorption of 5.0 keV of energy from the electrons emitted in the decay. The half-life of tritium is 12.3 y, and the RBE of the electrons is 1.0. Calculate the absorbed dose in rad and the equivalent dose in rem during one week. b) The β^- decay of tritium releases more than 5.0 keV of energy (see Exercise 45–19). Why is the average energy absorbed less than the total energy released in the decay?

45–28 In an experimental radiation therapy, an equivalent dose of 0.900 rem is given by 0.800-MeV protons to a localized area of tissue with mass 0.150 kg. a) What is the absorbed dose in rad? b) How many protons are absorbed by the tissue? c) How many α particles of the same energy of 0.800 MeV are required to deliver the same equivalent dose of 0.900 rem?

45–29 In a diagnostic x-ray procedure, 6.50×10^{10} photons are absorbed by tissue with mass 0.600 kg. The x-ray wavelength is 0.0200 nm. a) What is the total energy absorbed by the tissue? b) What is the equivalent dose in rem?

45–30 An amount of a radioactive source with a very long lifetime and activity 0.72 μCi is ingested into a person's body. The radioactive material lodges in the lungs, where all of the 4.0-MeV α particles emitted are absorbed within a 0.50-kg mass of tissue. Calculate the absorbed dose and the equivalent dose for one year.

SECTION 45–7 NUCLEAR REACTIONS

SECTION 45–8 NUCLEAR FISSION

SECTION 45–9 NUCLEAR FUSION

45–31 Consider the nuclear reaction

$$^{2}_{1}\text{H} + ^{9}_{4}\text{Be} \rightarrow \text{X} + ^{4}_{2}\text{He},$$

where X is a nuclide. a) What are the values of Z and A for the nuclide X? b) How much energy is liberated? c) Estimate the threshold energy for this reaction.

45–32 Energy from Nuclear Fusion. Calculate the energy released in the fusion reaction

$$^{3}_{2}\text{He} + ^{2}_{1}\text{H} \rightarrow ^{4}_{2}\text{He} + ^{1}_{1}\text{H}.$$

45–33 Consider the nuclear reaction

$$^{2}_{1}\text{H} + ^{14}_{7}\text{N} \rightarrow \text{X} + ^{10}_{5}\text{B},$$

where X is a nuclide. a) What are Z and A for the nuclide X? b) Calculate the reaction energy Q (in MeV). c) If the $^{2}_{1}\text{H}$ nucleus is incident on a stationary $^{14}_{7}\text{N}$ nucleus, what minimum kinetic energy must it have for the reaction to occur?

PROBLEMS

45–39 Use conservation of mass-energy to show that the energy released in alpha decay is positive whenever the mass of the original neutral atom is greater than the sum of the masses of the final neutral atom and the neutral ^{4}He atom. (*Hint:* Let the parent nucleus have atomic number Z and nucleon number A. First write the reaction in terms of the nuclei and particles involved, then add Z electron masses to both sides of the reaction and allot them as needed to arrive at neutral atoms.)

45–40 Use conservation of mass-energy to show that the energy released in β^{-} decay is positive whenever the neutral atomic mass of the original atom is larger than that of the final atom. (See the *Hint* in Problem 45–39.)

45–41 Use conservation of mass-energy to show that the energy released in β^{+} decay is positive whenever the neutral atomic mass of the original atom is at least two electron masses larger than that of the final atom. (See the *Hint* in Problem 45–39.)

45–42 Use conservation of mass-energy to show that the energy released in electron capture is positive whenever the neutral atomic mass of the original atom is larger than that of the final atom. (See the *Hint* in Problem 45–39, except add $Z-1$ electron masses to both sides of the reaction.)

45–43 The atomic mass of $^{25}_{12}\text{Mg}$ is 24.985837 u, and the atomic mass of $^{25}_{13}\text{Al}$ is 24.990429 u. a) Which of these nuclei will decay

45–34 Heat of Fission. In the fission of one ^{238}U nucleus, 200 MeV of energy is released. Express this energy in joules per mole, and compare with typical heats of combustion, which are on the order of 1.0×10^{5} J/mol.

45–35 A $^{235}_{92}\text{U}$ nucleus at rest absorbs a low-energy neutron. What is the internal excitation energy of the $^{236}_{92}\text{U}^{*}$ nucleus that is produced? The atomic masses of the neutral atoms in their nuclear ground states are 235.043923 u for $^{235}_{92}\text{U}$ and 236.045562 u for $^{236}_{92}\text{U}$.

45–36 Consider the nuclear reaction

$$^{4}_{2}\text{He} + ^{7}_{3}\text{Li} \rightarrow \text{X} + ^{1}_{0}\text{n},$$

where X is a nuclide. a) What are Z and A for the nuclide X? b) Is energy absorbed or liberated? How much?

45–37 The second reaction in the proton-proton chain (Fig. 45–16) produces a $^{3}_{2}\text{He}$ nucleus. A $^{3}_{2}\text{He}$ nucleus produced in this way can combine with a $^{4}_{2}\text{He}$ nucleus:

$$^{3}_{2}\text{He} + ^{4}_{2}\text{He} \rightarrow ^{7}_{4}\text{Be} + \gamma.$$

Calculate the energy liberated in this process. (This is shared between the energy of the photon and the recoil kinetic energy of the beryllium nucleus.) The mass of a $^{7}_{4}\text{Be}$ atom is 7.016929 u.

45–38 Consider the nuclear reaction

$$^{28}_{14}\text{Si} + \gamma \rightarrow ^{24}_{12}\text{Mg} + \text{X},$$

where X is a nuclide. a) What are Z and A for the nuclide X? b) Ignoring the effects of recoil, what minimum energy must the photon have for this reaction to occur? The mass of a $^{28}_{14}\text{Si}$ atom is 27.976927 u, and the mass of a $^{24}_{12}\text{Mg}$ atom is 23.985042 u.

into the other? b) What type of decay will occur? Explain how you determined this. c) How much energy (in MeV) is released in the decay?

45–44 The polonium isotope $^{210}_{84}\text{Po}$ has atomic mass 209.982857 u. Other atomic masses are $^{206}_{82}\text{Pb}$, 205.974449 u; $^{209}_{83}\text{Bi}$, 208.980383 u; $^{210}_{83}\text{Bi}$, 209.984105 u; $^{209}_{84}\text{Po}$, 208.982416 u; and $^{210}_{85}\text{At}$, 209.987131 u. a) Show that the alpha decay of $^{210}_{84}\text{Po}$ is energetically possible, and find the energy of the emitted α particle. b) Is $^{210}_{84}\text{Po}$ energetically stable with respect to emission of a proton? Why or why not? c) Is $^{210}_{84}\text{Po}$ energetically stable with respect to emission of a neutron? Why or why not? d) Is $^{210}_{84}\text{Po}$ energetically stable with respect to β^{-} decay? Why or why not? e) Is $^{210}_{84}\text{Po}$ energetically stable with respect to β^{+} decay? Why or why not?

45–45 The experimentally determined mass of the neutral ^{24}Na atom is 23.990963 u. Calculate the mass from the semi-empirical mass formula (Eq. 45–10). What is the percent error of the result as compared to the experimental value? What percent error is made if the E_{B} term is neglected entirely?

45–46 Thorium $^{230}_{90}\text{Th}$ decays to radium $^{226}_{88}\text{Ra}$ by α emission. The masses of the neutral atoms are 230.033127 u for $^{230}_{90}\text{Th}$ and 226.025403 u for $^{226}_{88}\text{Ra}$. If the parent thorium nucleus is at rest, what is the kinetic energy of the emitted α particle? (Be sure to account for the recoil of the daughter nucleus.)

45–47 Gold $^{198}_{79}$Au undergoes β^- decay to an excited state of $^{198}_{80}$Hg. If the excited state decays by emission of a γ photon with energy 0.412 MeV, what is the maximum kinetic energy of the electron emitted in the decay? This maximum occurs when the antineutrino has negligible energy. (The recoil energy of the $^{198}_{80}$Hg nucleus can be neglected. The masses of the neutral atoms in their ground states are 197.968225 u for $^{198}_{79}$Au and 197.966752 u for $^{198}_{80}$Hg.)

45–48 Calculate the mass defect for the β^+ decay of $^{11}_{6}$C. Is this decay energetically possible? Why or why not? The atomic mass of $^{11}_{6}$C is 11.011434 u.

45–49 Calculate the mass defect for the β^+ decay of $^{13}_{7}$N. Is this decay energetically possible? Why or why not? The atomic mass of $^{13}_{7}$N is 13.005739 u.

45–50 The results of activity measurements on a radioactive sample are given below. a) Find the half-life. b) How many radioactive nuclei were present in the sample at $t = 0$? c) How many were present after 7.0 h?

Time (h)	Decays/s
0	20,000
0.5	14,800
1.0	11,000
1.5	8,130
2.0	6,020
2.5	4,460
3.0	3,300
4.0	1,810
5.0	1,000
6.0	550
7.0	300

45–51 If $A(t)$ is the activity of a sample at some time t and A_0 is the activity at $t = 0$, show that $A(t) = A_0 e^{-\lambda t}$.

45–52 Show that Eq. (45–15) may be written as $N(t) = N_0 \left(\frac{1}{2}\right)^n$, where $n = t/T_{1/2}$ is the number of half-lives that have elapsed since $t = 0$. (This expression is valid even if n is not a whole number.)

45–53 Measurements indicate that 27.83% of all rubidium atoms currently on earth are the radioactive ^{87}Rb isotope. The rest are the stable ^{85}Rb isotope. The half-life of ^{87}Rb is 4.75×10^{10} y. Assuming that no rubidium atoms have been formed since, what percentage of rubidium atoms were ^{87}Rb when our solar system was formed 4.6×10^9 y ago?

45–54 A 70.0-kg person experiences a whole-body exposure to α radiation with energy 4.77 MeV. A total of 6.25×10^{12} α particles are absorbed. a) What is the absorbed dose in rad? b) What is the equivalent dose in rem? c) If the source is 0.0320 g of ^{226}Ra (half-life 1600 y) somewhere in the body, what is the activity of this source? d) If all the alpha particles produced are absorbed, what time is required for this dose to be delivered?

45–55 A ^{60}Co source with activity 2.6×10^{-4} Ci is imbedded in a tumor that has a mass of 0.500 kg. The source emits γ photons with average energy 1.25 MeV. Half the photons are absorbed in the tumor, and half escape. a) What energy is delivered to the tumor per second? b) What absorbed dose (in rad) is delivered per second? c) What equivalent dose (in rem) is delivered

per second if the RBE for these γ rays is 0.70? d) What exposure time is required for an equivalent dose of 200 rem?

45–56 The nucleus $^{15}_{8}$O has a half-life of 122.2 s; $^{19}_{8}$O has a half-life of 26.9 s. If at some time a sample contains equal amounts of $^{15}_{8}$O and $^{19}_{8}$O, what is the ratio of $^{15}_{8}$O to $^{19}_{8}$O a) after 4.0 minutes? b) after 15.0 minutes?

45–57 A bone fragment found in a cave believed to have been inhabited by early humans contains 0.21 times as much ^{14}C as an equal amount of carbon in the atmosphere when the organism containing the bone died. (See Example 45–9 in Section 45–5.) Find the approximate age of the fragment.

45–58 A patient is given a 4.2 μCi dose of a radioactive isotope with a half-life of 2.0 h. Assuming that the entire dose remains in the body, how much time must elapse before the activity of the radioactive isotope is 8.5 counts/minute, which is about three times the normal background of 3 counts/minute?

45–59 Consider the fusion reaction

$$^2_1\text{H} + {}^2_1\text{H} \rightarrow {}^3_2\text{He} + {}^1_0\text{n}.$$

a) Estimate the barrier energy by calculating the repulsive electrostatic potential energy of the two $^2_1$H nuclei when they touch. b) Compute the energy liberated in this reaction in MeV and in joules. c) Compute the energy liberated *per mole* of deuterium, remembering that the gas is diatomic, and compare with the heat of combustion of hydrogen, about 2.9×10^5 J/mol.

45–60 Separation Energy. a) The proton separation energy is the minimum amount of energy that must be supplied to a nucleus to remove a proton. Calculate the proton separation energy for $^{16}_{8}$O. b) The neutron separation energy is the minimum amount of energy that must be supplied to a nucleus to remove a neutron. Calculate the neutron separation energy for $^{16}_{8}$O. The atomic mass of $^{15}_{8}$O is 15.003065 u. c) Compare the answers to parts (a) and (b). Which takes less energy to remove from $^{16}_{8}$O, a proton or a neutron?

45–61 Natural Body Radioactivity. On average, 0.21% of the mass of the human body is potassium, of which 0.012% is radioactive ^{40}K with a half-life of 1.28×10^9 y. From each decay, an average of 0.50 MeV of β and γ ray radiation is absorbed by the body. Take the RBE of the β radiation to be 1.0. Calculate the absorbed dose in rad and the equivalent dose in rem for 50 y from the potassium in the body.

45–62 In the 1986 disaster at the Chernobyl reactor in the Soviet Union (now Ukraine), about $\frac{1}{8}$ of the ^{137}Cs present in the reactor was released. The isotope ^{137}Cs has a half-life for β decay of 30.07 y and decays with the emission of a total of 1.17 MeV of energy per decay. Of this, 0.51 MeV goes to the emitted electron and the remaining 0.66 MeV to a γ ray. The radioactive ^{137}Cs is absorbed by plants which are eaten by livestock and humans. How many ^{137}Cs atoms would need to be present in each kilogram of body tissue if an equivalent dose for one week is 3.5 Sv? Assume that all of the energy from the decay is deposited in that 1.0 kg of tissue and that the RBE of the electrons is 1.5.

45–63 a) Prove that when a particle with mass m and kinetic energy K collides with a stationary particle with mass M, the total kinetic energy K_{cm} in the center-of-mass coordinate system (the energy available to cause reactions) is

$$K_{cm} = \frac{M}{M + m} K.$$

Assume that the kinetic energies of the particles and nuclei are much less than their rest energies. b) If K_{th} is the minimum, or threshold, kinetic energy to cause an endoergic reaction to occur in the situation of part (a), show that

$$K_{th} = -\frac{M + m}{M} Q.$$

45–64 A $^{186}_{76}$Os nucleus at rest decays by the emission of a 2.76-MeV α particle. Calculate the atomic mass of the daughter nuclide produced by this decay, assuming that it is produced in its ground state. The atomic mass of $^{186}_{76}$Os is 185.953838 u.

45–65 Calculate the energy released in the fission reaction

$$^{235}_{92}\text{U} + {}^{1}_{0}\text{n} \rightarrow {}^{140}_{54}\text{Xe} + {}^{94}_{38}\text{Sr} + 2{}^{1}_{0}\text{n}.$$

Neglect the initial kinetic energy of the absorbed neutron. The atomic masses are $^{235}_{92}$U, 235.043923 u; $^{140}_{54}$Xe, 139.921636 u; and $^{94}_{38}$Sr, 93.915360 u.

CHALLENGE PROBLEMS

45–66 The results of activity measurements on a mixed sample of radioactive elements are shown below. a) How many different nuclides are present in the mixture? b) What are their half-lives? c) How many nuclei of each type are initially present in the sample? d) How many of each type are present at $t = 5.0$ h?

Time (h)	Decays/s
0	7500
0.5	4120
1.0	2570
1.5	1790
2.0	1350
2.5	1070
3.0	872
4.0	596
5.0	414
6.0	288
7.0	201
8.0	140
9.0	98
10.0	68
12.0	33

45–67 In an experiment, the isotope ^{128}I is created by the irradiation of ^{127}I with a beam of neutrons that creates 1.5×10^6 ^{128}I nuclei per second. Initially no ^{128}I nuclei are present. The half-life of ^{128}I is 25 minutes. a) Graph the number of ^{128}I nuclei present as a function of time. b) What is the activity of the sample 1, 10, 25, 50, 75, and 180 minutes after irradiation is begun? c) What is the maximum number of ^{128}I atoms that can be created in the sample after it is irradiated for a long time? (This steady-state situation is called *saturation*.) d) What is the maximum activity that can be produced?

45–68 **Industrial Radioactivity.** Radioisotopes are used in a variety of manufacturing and testing techniques. Wear measurements can be made using the following method. An automobile engine is produced using piston rings with a total mass of 100 g, which includes 9.4 μCi of ^{59}Fe whose half-life is 45 days. The engine is test run for 1000 hours, after which the oil is drained and its activity is measured. If the activity of the engine oil is 84 decays/s, how much mass was worn from the piston rings per hour of operation?

North Galactic Hemisphere   South Galactic Hemisphere

$-100\ \mu K$ $+100\ \mu K$

Where did we come from? This false-color map of microwave radiation from the entire sky provides an answer. This radiation was emitted 300,000 years after the Big Bang, before any stars or galaxies existed. Regions shown in blue were slightly cooler and denser than their surroundings. Galaxies, including our own Milky Way galaxy, are thought to have formed within these regions.

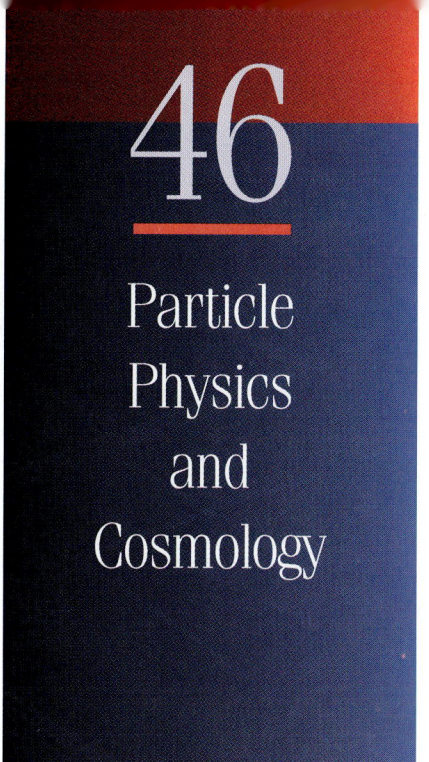

46

Particle Physics and Cosmology

46–1 INTRODUCTION

What is the world made of? What are the most fundamental constituents of matter? Philosophers and scientists have been asking these questions for at least 2500 years. We still don't have anything that could be called a final answer, but we've come a long way. This final chapter is a progress report on where we are now and where we hope to go.

The chapter title, "Particle Physics and Cosmology," may seem strange. Fundamental particles are the *smallest* things in the universe, and cosmology deals with the *biggest* thing there is—the universe itself. Nonetheless, we'll see in Sections 46–7 and 46–8 that there are very close ties between these two areas.

Fundamental particles, we'll find, are not permanent entities; they can be created and destroyed. The development of high-energy particle accelerators and associated detectors has played an essential role in our emerging understanding of particles. We can classify particles and their interactions in several ways in terms of conservation laws and symmetries, some of which are absolute and others of which are obeyed only in certain kinds of interactions. We'll try in this final chapter to convey some idea of the remaining frontiers in this vital and exciting area of fundamental research.

46–2 FUNDAMENTAL PARTICLES— A HISTORY

The idea that the world is made of *fundamental particles* has a long history. In about 400 B.C. the Greek philosophers Democritus and Leucippus suggested that matter is made of indivisible particles that they called *atoms,* a word derived from *a-* (not) and *tomos* (cut or divided). This idea lay dormant until about 1804, when the English scientist John Dalton (1766–1844), often called the father of modern chemistry, discovered that many chemical phenomena could be explained on the basis of atoms of each element as the basic, indivisible building blocks of matter.

THE ELECTRON AND THE PROTON

Toward the end of the nineteenth century it became clear that atoms are *not* indivisible. The existence of characteristic atomic spectra of elements suggested that atoms have internal structure, and J. J. Thomson's discovery of the *electron* in 1897 showed that atoms could be taken apart into charged particles. The hydrogen nucleus was identified as a *proton,* and in 1911 the sizes of nuclei were measured by Rutherford's experiments. Quantum mechanics, including the Schrödinger equation, blossomed during the next 15 years. Scientists were on their way to understanding the principles that underlie atomic structure, although many details remained to be worked out.

THE PHOTON

Einstein explained the photoelectric effect in 1905 by assuming that the energy of electromagnetic waves is quantized; that is, it comes in little bundles called *photons* with energy $E = hf$. Atoms and nuclei can emit (create) and absorb (destroy) photons. Considered as particles, photons have zero charge and zero rest mass. (Note that any discussions of a particle's mass in this chapter will refer to its rest mass.) In particle physics, a photon is denoted by the symbol γ (the Greek letter "gamma").

THE NEUTRON

The discovery of the neutron was an important milestone. In 1930, two German physicists, W. Bothe and H. Becker, observed that when beryllium, boron, or lithium was bombarded by α particles from radioactive polonium, the target material emitted a radiation that had much greater penetrating power than the original α particles. Experiments by James Chadwick in 1932 showed that the emitted particles were electrically neutral, with mass approximately equal to that of the proton. Chadwick christened these particles *neutrons*. A typical reaction of the type studied by Bothe and Becker, using a beryllium target, is

$$\,_2^4\text{He} + \,_4^9\text{Be} \rightarrow \,_6^{12}\text{C} + \,_0^1\text{n}. \tag{46-1}$$

It was difficult to detect neutrons because they have no charge. Hence they produce little ionization when they pass through matter and they are not deflected by electric or magnetic fields. Neutrons almost always interact only with nuclei; they can be slowed down during scattering, and they can penetrate the nucleus. Slow neutrons can be detected by means of a nuclear reaction in which a neutron is absorbed and an α particle is emitted. An example is

$$\,_0^1\text{n} + \,_5^{10}\text{B} \rightarrow \,_3^7\text{Li} + \,_2^4\text{He}. \tag{46-2}$$

The ejected α particle is easy to detect because it is charged.

In Section 45–2 we stated that neutrons, as well as protons and electrons, are spin-$\frac{1}{2}$ particles. That is, the magnitude of the spin angular momentum $\vec{S}$ of a neutron is

$$S = \sqrt{s(s+1)}\hbar = \sqrt{\frac{1}{2}\left(\frac{1}{2}+1\right)}\hbar = \sqrt{\frac{3}{4}}\hbar,$$

where s is the *spin quantum number,* often just called the *spin.*

The neutron was a welcome discovery because it cleared up a mystery about the composition of the nucleus. Before 1930 the mass of a nucleus was thought to be due only to protons, but no one understood why the charge-to-mass ratio was not the same for all nuclides. It soon became clear that all nuclides (except $\,_1^1\text{H}$) contain both protons and neutrons. In fact, the proton, the neutron, and the electron are the building blocks of atoms. One might think that would be the end of the story. On the contrary, it is barely the beginning. These are not the only particles, and they can do more than build atoms.

THE POSITRON

The positive electron, or *positron,* was discovered by the American physicist Carl D. Anderson in 1932, during an investigation of particles bombarding the earth from space. Figure 46–1 shows a historic photograph made with a *cloud chamber,* an instrument that is used to visualize the tracks of charged particles. The chamber contained a supercooled vapor; ions created by the passage of charged particles through the vapor served as nucleation centers, and liquid droplets formed around them, making a visible track.

The cloud chamber in Fig. 46–1 is in a magnetic field directed into the plane of the photograph. The particle has passed through a thin lead plate (which extends from left

46–1 Photograph of the cloud-chamber track made by the first positron ever identified. The photograph was made by Carl D. Anderson in 1932. The lead plate is 6 mm thick.

to right in the figure) that lies within the chamber. The curvature of the track is greater above the plate than below it, showing that the speed was less above the plate than below it. Therefore the particle had to be moving upward; it could not have gained energy passing through the lead. The thickness and curvature of the track suggested that its mass and the magnitude of its charge equaled those of the electron. But the directions of the magnetic field and the velocity in the magnetic force equation $\vec{F} = q\vec{v} \times \vec{B}$ showed that the particle had *positive* charge. Anderson christened this particle the *positron.*

To theorists, the appearance of the positron was a welcome development. In 1928, the English physicist Paul Dirac had developed a relativistic generalization of the Schrödinger equation for the electron. In Section 43–4 we discussed how Dirac's generalization helped to explain the spin magnetic moment of the electron.

One of the puzzling features of the Dirac equation was that for a *free* electron it predicted not only a continuum of energy states greater than its rest energy $m_e c^2$, as should be expected, but also a continuum of *negative* energy states *less than* $-m_e c^2$ (Fig. 46–2a). That posed a problem. What was to prevent an electron from emitting a photon with energy $2m_e c^2$ or greater and hopping from a positive state to a negative state? It wasn't clear what these negative-energy states meant, and there was no obvious way to get rid of them. Dirac's ingenious but somewhat implausible interpretation was that all the negative-energy states that were filled with electrons were for some reason unobservable. The exclusion principle would then forbid a transition to a state that was already occupied.

A vacancy in a negative-energy state would act like a positive charge, just as a hole in the valence band of a semiconductor (Section 44–7) acts like a positive charge. Initially, Dirac tried to argue that such vacancies were protons. But after Anderson's discovery it became clear that the vacancies were observed physically as *positrons.* Furthermore, the Dirac energy-state picture provides a mechanism for the *creation* of positrons. When an electron in a negative-energy state absorbs a photon with energy greater than $2m_e c^2$, it goes to a positive state (Fig. 46–2b), in which it becomes observable. The vacancy that it leaves behind is observed as a positron; the result is the creation of an electron-positron pair. Similarly, when an electron in a positive-energy state falls into a vacancy, both the electron and the vacancy (that is, the positron) disappear, and photons are emitted (Fig. 46–2c). Thus the Dirac theory leads naturally to the conclusion that, like photons, *electrons can be created and destroyed.* While photons can be created and destroyed singly, electrons can be produced or destroyed only in electron-positron pairs or in association with other particles. We'll see why later.

In 1949, the American physicist Richard Feynman showed that a positron could be described mathematically as an electron traveling backward in time. His reformulation of the Dirac theory eliminated difficult calculations involving the infinite sea of negative-energy states and put electrons and positrons on the same footing. But the creation and destruction of electron-positron pairs remain. The Dirac theory is

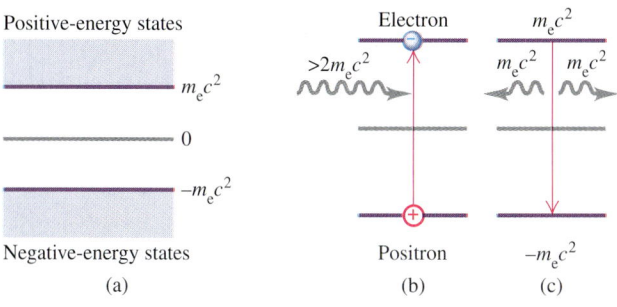

Positive-energy states

$m_e c^2$

0

$-m_e c^2$

Negative-energy states

(a)

Electron

$>2m_e c^2$

Positron

(b)

$m_e c^2$

$m_e c^2$ $m_e c^2$

$-m_e c^2$

(c)

46–2 (a) Energy states for a free electron predicted by the Dirac equation. There is a continuum of states with energies greater than $m_e c^2$ and another continuum of negative-energy states with energies less than $-m_e c^2$. (b) Raising an electron from a negative-energy state to a positive-energy state corresponds to electron-positron pair production. (c) An electron dropping from a positive-energy state to a vacant negative-energy state corresponds to electron-positron pair annihilation.

(a)

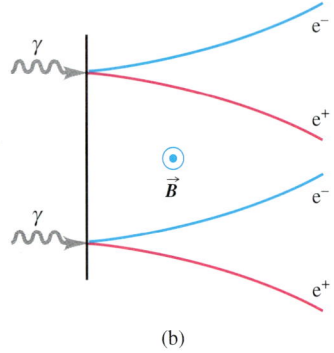

(b)

46–3 (a) Photograph of bubble-chamber tracks of electron-positron pairs that are produced when 300-MeV photons strike a lead sheet. A magnetic field directed out of the photograph made the electrons and positrons curve in opposite directions. (b) Diagram showing the pair-production process for two of the photons.

inherently a *many-particle* theory, and it provides the beginning of a theoretical framework for creation and destruction of all fundamental particles.

Current experiments and theory tell us that the masses of the positron and electron are identical, and that their charges are equal in magnitude but opposite in sign. The positron's spin angular momentum $\vec{S}$ and magnetic moment $\vec{\mu}$ are parallel; they are opposite for the electron. However, $\vec{S}$ and $\vec{\mu}$ have the same magnitude for both particles because they have the same spin. We use the term **antiparticle** for a particle that is related to another particle as the positron is to the electron. Each kind of particle has a corresponding antiparticle. For a few kinds of particles (necessarily all neutral) the particle and antiparticle are identical, and we can say that they are their own antiparticles. The photon is an example; there is no way to distinguish a photon from an antiphoton. We'll use the standard symbols e^- for the electron and e^+ for the positron, and the generic term *electron* will often include both electrons and positrons. Other antiparticles are often denoted by a bar over the particle's symbol; for example, an antiproton is $\bar{p}$. We'll see several other examples of antiparticles later.

Positrons don't occur in ordinary matter. Electron-positron pairs are produced during high-energy collisions of charged particles or γ rays with matter, an example of the process called e^+e^- *pair production* (Fig. 46–3). Electric charge is conserved, and enough energy E must be available to account for the rest energy $2m_ec^2$ of the two particles. The minimum energy for electron-positron pair production is

$$E_{min} = 2m_ec^2 = 2(9.109 \times 10^{-31} \text{ kg})(2.998 \times 10^8 \text{ m/s})^2$$

$$= 1.637 \times 10^{-13} \text{ J} = 1.022 \text{ MeV}.$$

The inverse process, e^+e^- *pair annihilation,* occurs when a positron and an electron collide. Both particles disappear, and two (or occasionally three) photons can appear, with total energy of at least $2m_ec^2 = 1.022$ MeV. Decay into a *single* photon is impossible because such a process could not conserve both energy and momentum. It's easiest to analyze e^+e^- annihilation processes in the frame of reference called the *center-of-momentum system,* in which the total momentum is zero. It is the relativistic generalization of the center-of-mass system that we discussed in Section 8–6.

EXAMPLE 46–1

Pair annihilation An electron and positron are moving in opposite directions with the same speed. They collide head-on, annihilating each other and producing two photons. Find the energies, wavelengths, and frequencies of the two photons if the initial kinetic energies of the e^- and e^+ are a) both negligibly small; b) both 5.000 MeV. The rest energy of an electron is 0.511 MeV, much greater than the initial electrical potential energy of the two particles.

SOLUTION The initial total momentum is zero (that is, we are in the center-of-momentum system). For the total momentum to remain zero, the two photons must have momenta of equal magnitude and opposite direction. From Section 37–7 we have the relations $E = pc$ and $E = hf = hc/\lambda$ for a photon. The relation $E = pc$ tells us that equal momenta p mean equal energies. The relation $E = hf = hc/\lambda$ tells us that the two equal-energy photons also have the same frequency and wavelength. From conservation of energy we know that the initial total energy of the two particles (including their rest energies) must equal the final total energy of the two photons.

a) With negligible kinetic and potential energy the initial total energy is the sum of the rest energies: 0.511 MeV + 0.511 MeV = 1.022 MeV. The final total energy is the sum of the photon energies: $E + E = 2E$. Thus 1.022 MeV = 2E, or $E = 0.511$ MeV. Then

$$\lambda = \frac{hc}{E} = \frac{(4.136 \times 10^{-15} \text{ eV} \cdot \text{s})(3.00 \times 10^8 \text{ m/s})}{0.511 \times 10^6 \text{ eV}}$$

$$= 2.43 \times 10^{-12} \text{ m} = 2.43 \text{ pm},$$

$$f = \frac{E}{h} = \frac{0.511 \times 10^6 \text{ eV}}{4.136 \times 10^{-15} \text{ eV} \cdot \text{s}} = 1.24 \times 10^{20} \text{ Hz}.$$

b) We repeat the same analysis, except that the initial *total* energy is 10.000 MeV larger than 1.022 MeV, or 11.022 MeV. This total gives an energy, wavelength, and frequency of 5.511 MeV, 0.2250 pm, and 1.333×10^{21} Hz, respectively, for each photon.

Implicit in our discussion is the idea that momentum is conserved in e^+e^- pair annihilation. Why do you think this is the case?

46–4 (a) Two skaters exert repulsive forces on each other by tossing a ball back and forth. (b) Two skaters exert attractive forces on each other when one tries to grab the ball out of the other's hands.

Positrons also occur in the decay of some unstable nuclei, in which they are called beta-plus particles (β^+). We discussed β^+ decay in Section 45–4. Isotopes that are β^+ emitters can be used in positron emission tomography (PET), which was mentioned in Section 45–6.

It's often convenient to represent particle masses in terms of the equivalent rest energy using $m = E/c^2$. Then typical mass units are MeV/c^2, for example, 0.511 MeV/c^2 for an electron or positron. We'll use these units frequently in the following discussions.

PARTICLES AS FORCE MEDIATORS

In classical physics we describe the interaction of charged particles in terms of Coulomb's-law forces. In quantum mechanics we can describe this interaction in terms of emission and absorption of photons. Two electrons repeal each other as one emits a photon and the other absorbs it, just as two skaters can push one another apart by tossing a large ball back and forth between them (Fig. 46–4a). For an electron and a proton, in which the charges are opposite and the force is attractive, we imagine the skaters trying to grab the ball away from one other (Fig. 46–4b). The electromagnetic interaction between two charged particles is *mediated* or transmitted by photons.

If charged-particle interactions are mediated by photons, where does the energy to create the photons come from? Recall from our discussion of the uncertainty principle (Section 41–4) that a state that exists for a short time Δt has an uncertainty ΔE in its energy such that

$$\Delta E \, \Delta t \geq \hbar. \tag{46–3}$$

This uncertainty permits the creation of a photon with energy ΔE, provided that it lives no longer than the time Δt given by Eq. (46–3). A photon that can exist for a short time because of this energy uncertainty is called a *virtual photon*. It's as though there were an energy bank; you can borrow energy, provided that you pay it back within the time limit. According to Eq. (46–3), the more you borrow, the sooner you have to pay it back. Later we'll discuss other virtual particles that live for a short time on "borrowed" energy.

MESONS

Is there a particle that mediates the *nuclear* force? In 1935 the nuclear force between two nucleons (neutrons or protons) appeared to be described by a potential energy $U(r)$ with the general form

$$U(r) = -f^2 \frac{e^{-r/r_0}}{r} \qquad \text{(nuclear potential energy)}. \tag{46–4}$$

The constant f characterizes the strength of the interaction, and r_0 describes its range.

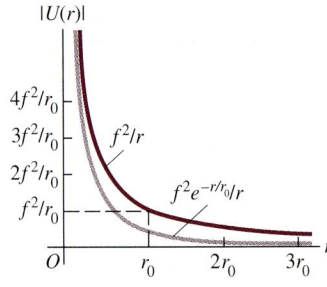

46–5 Graph of the magnitude of the Yukawa potential-energy function for nuclear forces, $|U(r)| = f^2 e^{-r/r_0}/r$. The function approaches zero very rapidly for $r > r_0$. For comparison, the function $U(r) = f^2/r$, proportional to the potential energy for Coulomb's law, is also shown. The two functions are similar at small r, but the Yukawa potential energy drops off much more quickly at large r.

Figure 46–5 shows a graph of the absolute value of this function and compares it with the function f^2/r, which would be analogous to the *electrical* interaction of two protons:

$$U(r) = \frac{1}{4\pi\epsilon_0} \frac{e^2}{r} \quad \text{(electrical potential energy).} \quad (46\text{–}5)$$

In 1935 the Japanese physicist Hideki Yukawa suggested that a hypothetical particle that he called a **meson** might act as a mediator of the nuclear force. He showed that the range of the force was related to the mass of the particle. His argument went like this: The particle must live for a time Δt long enough to travel a distance comparable to the range of the nuclear force. This range was known from the sizes of nuclei and other information to be of the order of $r_0 = 1.5 \times 10^{-15}$ m = 1.5 fm. If we assume that the particle's speed is comparable to c, its lifetime Δt must be of the order of

$$\Delta t = \frac{r_0}{c} = \frac{1.5 \times 10^{-15} \text{ m}}{3.0 \times 10^8 \text{ m/s}} = 5.0 \times 10^{-24} \text{ s.}$$

From Eq. (46–3) the minimum necessary uncertainty ΔE in energy is

$$\Delta E = \frac{\hbar}{\Delta t} = \frac{1.05 \times 10^{-34} \text{ J} \cdot \text{s}}{5.0 \times 10^{-24} \text{ s}} = 2.1 \times 10^{-11} \text{ J} = 130 \text{ MeV.}$$

The mass equivalent Δm of this energy is

$$\Delta m = \frac{\Delta E}{c^2} = \frac{2.1 \times 10^{-11} \text{ J}}{(3.00 \times 10^8 \text{ m/s})^2} = 2.3 \times 10^{-28} \text{ kg,}$$

or 130 MeV/c^2. This is about 250 times the electron mass, and Yukawa postulated that a particle with this mass serves as the messenger for the nuclear force. This was a courageous act; at the time, there was not a shred of experimental evidence that such a particle existed.

A year later, Anderson and Neddermeyer discovered in cosmic radiation two new particles, now called **muons.** The μ^- has charge equal to that of the electron, and its antiparticle the μ^+ has a positive charge with equal magnitude. The two particles have equal mass, about 207 times the electron mass. But it soon became clear that muons were *not* Yukawa's particles because they interacted with nuclei only very weakly.

In 1947 a family of three particles, called π *mesons* or **pions,** were discovered. Their charges are $+e$, $-e$, and zero, and their masses are about 270 times the electron mass. The pions interact strongly with nuclei, and they *are* the particles predicted by Yukawa. Other, heavier mesons, the ω and ρ, evidently also act as shorter-range messengers of the nuclear force. The complexity of this explanation suggests that it has simpler underpinnings; these involve the quarks and gluons that we'll discuss in Section 46–5. Before discussing mesons further, we'll describe some particle accelerators and detectors to see how mesons and other particles are created in a controlled fashion and observed.

46–3 PARTICLE ACCELERATORS AND DETECTORS

Early nuclear physicists used alpha and beta particles from naturally occurring radioactive elements for their experiments, but they were restricted in energy to the few MeV that are available in such random decays. Present-day particle accelerators can produce precisely controlled beams of particles, from electrons and positrons up to heavy ions, with a wide range of energies. These beams have three main uses. First, high-energy particles can collide to produce new particles, just as a collision of an electron and a positron can produce photons. Second, a high-energy particle has a short de Broglie wavelength and so can probe the small-scale interior structure of other particles, just as

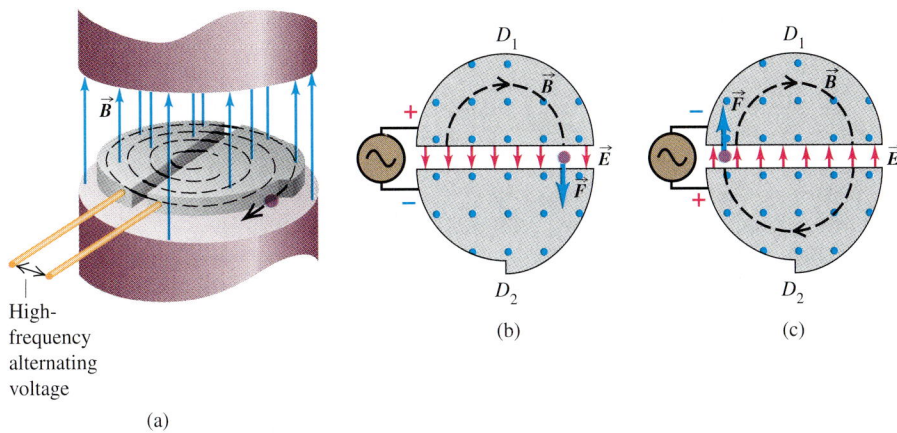

46-6 (a) Schematic diagram of a cyclotron. (b) As the positive particle reaches the gap, it is accelerated by the electric-field force, and the next semicircular orbit has a larger radius. (c) By the time the particle reaches the gap again, the dee voltage has reversed, and the particle is again accelerated.

electron microscopes (Section 41–5) can give better resolution than optical microscopes. Third, they can be used to produce nuclear reactions of scientific or medical use.

LINEAR ACCELERATORS

Particle accelerators use electric and magnetic fields to accelerate and guide beams of charged particles. A *linear accelerator* (linac) accelerates particles in a straight line. The earliest examples were J. J. Thomson's cathode-ray tubes and William Coolidge's x-ray tubes. More sophisticated linacs use a series of electrodes with gaps to give the beam of particles a series of boosts. Most present-day high-energy linear accelerators use a traveling electromagnetic wave; the charged particles "ride" the wave in more or less the way that a surfer rides an incoming ocean wave. In the highest-energy linac in the world today, at the Stanford Linear Accelerator Center (SLAC), electrons and positrons are accelerated to 50 GeV in a tube 3 km long. At this energy their de Broglie wavelengths are 0.025 fm, much smaller than the size of a proton or a neutron.

THE CYCLOTRON

Many accelerators use magnets to deflect the charged particles into circular paths. The first of these was the *cyclotron,* invented in 1931 by E. O. Lawrence and M. Stanley Livingston at the University of California. In the cyclotron, shown schematically in Fig. 46–6a, particles with mass m and charge q move inside a vacuum chamber in a uniform magnetic field $\vec{B}$ that is perpendicular to the plane of their paths. In Section 28–5 we showed that in such a field, a particle with speed v moves in a circular path with radius r given by

$$r = \frac{mv}{|q|B},\tag{46-6}$$

and with angular speed (angular frequency) ω given by

$$\omega = \frac{v}{r} = \frac{|q|B}{m}.\tag{46-7}$$

An alternating potential difference is applied between the two hollow electrodes D_1 and D_2 (called *dees*), creating an electric field in the gap between them. The polarity of the potential difference and electric field is changed precisely twice each revolution (Fig. 46–6b and 46–6c), so that the particles get a push each time they cross the gap. The pushes increase their speed and kinetic energy, boosting them into paths of larger radius. The maximum speed v_{max} and kinetic energy K_{max} are determined by the radius R of the largest possible path. Solving Eq. (46–6) for v, we find $v = |q|Br/m$ and $v_{max} = |q|BR/m$.

Assuming nonrelativistic speeds, we have

$$K_{max} = \frac{1}{2} m v_{max}^2 = \frac{q^2 B^2 R^2}{2m}. \tag{46-8}$$

EXAMPLE 46-2

A proton cyclotron One cyclotron built during the 1930s has a path of maximum radius 0.500 m and a magnetic field of magnitude 1.50 T. If it is used to accelerate protons, a) calculate the frequency of the alternating voltage applied to the dees; b) find the maximum particle energy.

SOLUTION For protons, $q = 1.60 \times 10^{-19}$ C and $m = 1.67 \times 10^{-27}$ kg.

a) The frequency f is the angular frequency ω divided by 2π. Using Eq. (46-7), we have

$$f = \frac{\omega}{2\pi} = \frac{|q|B}{2\pi m} = \frac{(1.60 \times 10^{-19} \text{ C})(1.50 \text{ T})}{2\pi(1.67 \times 10^{-27} \text{ kg})}$$

$$= 2.3 \times 10^7 \text{ Hz} = 23 \text{ MHz}.$$

b) From Eq. (46-8) the maximum kinetic energy is

$$K_{max} = \frac{(1.60 \times 10^{-19} \text{ C})^2 (1.50 \text{ T})^2 (0.50 \text{ m})^2}{2(1.67 \times 10^{-27} \text{ kg})}$$

$$= 4.3 \times 10^{-12} \text{ J} = 2.7 \times 10^7 \text{ eV} = 27 \text{ MeV}.$$

This energy, which is much larger than energies that are available from natural radioactivity, can cause a variety of interesting nuclear reactions.

From Eq. (46-6) or (46-7), the proton speed is $v = 7.2 \times 10^7$ m/s, which is about 25% of the speed of light. At such speeds, relativistic effects are beginning to become important. Since we ignored these effects in our calculation, the above result for f and K_{max} are in error by a few percent; this is why we kept only two significant figures.

The maximum energy that can be attained with a cyclotron is limited by relativistic effects. The relativistic version of Eq. (46-7) is

$$\omega = \frac{|q|B}{m} \sqrt{1 - v^2/c^2}.$$

As the particles speed up, their angular frequency ω *decreases,* and their motion gets out of phase with the alternating dee voltage. In the *synchrocyclotron* the particles are accelerated in bursts. For each burst, the frequency of the alternating voltage is decreased as the particles speed up, maintaining the correct phase relation with the particles' motion.

Another limitation of the cyclotron is the difficulty of building very large electromagnets. The largest synchrocyclotron ever built has a vacuum chamber that is about 8 m in diameter and accelerates protons to energies of about 600 MeV.

THE SYNCHROTRON

To attain higher energies, another type of machine, called the *synchrotron,* is more practical. Particles move in a vacuum chamber in the form of a thin doughnut, called the *accelerating ring.* The particle beam is bent to follow the ring by a series of electromagnets placed around the ring. As the particles speed up, the magnetic field is increased so that the particles retrace the same trajectory over and over. The Tevatron at the Fermi National Accelerator Laboratory (Fermilab) in Batavia, Illinois, is currently the highest-energy accelerator in the world; it can accelerate protons to a maximum energy of 1 TeV (10^{12} eV). The accelerating ring is 2 km in diameter and uses superconducting electromagnets (Fig. 46-7). In each machine cycle, a few seconds long, the Tevatron accelerates approximately 10^{13} protons.

As we pointed out in Section 40-6, accelerated charges radiate electromagnetic energy. Examples include bremsstrahlung and transmissions from radio and TV antennas. In an accelerator in which the particles move in curved paths, this radiation is often called *synchrotron radiation.* High-energy accelerators are typically constructed underground to provide protection from this radiation. From the accelerator standpoint, synchrotron radiation is undesirable since the energy given to an accelerated particle is radiated right back out. It can be minimized by making the accelerator radius r large so

46–7 The Tevatron, the 1-TeV (1000-GeV) accelerator at the Fermi National Accelerator Laboratory, Batavia, Illinois, is the highest-energy accelerator in the world. (top) An aerial view of the main accelerator ring. (bottom) A section of the main tunnel. The original upper ring of conventional magnets (red and blue) and the newer lower ring of superconducting magnets can be seen. The conventional ring is used to inject protons and antiprotons into the superconducting ring. The proton and antiproton beams travel in opposite directions as they are accelerated to 1 TeV, and then they collide at interaction points, making 2 TeV available to create new particles.

that the centripetal acceleration v^2/r is small. On the positive side, synchrotron radiation is used as a source of well-controlled high-frequency electromagnetic waves.

AVAILABLE ENERGY

When a beam of high-energy particles collides with a stationary target, not all the kinetic energy of the incident particles is *available* to form new particle states. Because momentum must be conserved, the particles emerging from the collision must have some net motion and thus some kinetic energy. The discussion following Example 45–12 presented a nonrelativistic example of this principle. The maximum available energy is the kinetic energy in the frame of reference in which the total momentum is zero. We called this the *center-of-momentum system* in Section 46–2. In this system the total kinetic energy after the collision can be zero, so that the maximum amount of the initial kinetic energy becomes available to cause the reaction being studied.

Consider the *laboratory system,* in which a target particle with mass M is initially at rest and is bombarded by a particle with mass m and total energy (including rest energy) E_m. The total available energy E_a in the center-of-momentum system (including rest energies of all the particles) can be shown to be given by

$$E_a{}^2 = 2Mc^2E_m + (Mc^2)^2 + (mc^2)^2 \qquad \text{(available energy).} \qquad (46\text{--}9)$$

When the masses of the target and projectile particles are equal, this can be simplified to

$$E_a{}^2 = 2mc^2(E_m + mc^2) \qquad \text{(available energy, equal masses).} \qquad (46\text{--}10)$$

In the *extreme-relativistic range,* in which the kinetic energy of the bombarding particle is much larger than its rest energy, available energy is a very severe limitation. Let's look again at Eq. (46–10), the case in which beam and target particles have equal masses. When E_m is much greater than mc^2, we can neglect the second term in the parentheses. Then E_a is

$$E_a = \sqrt{2mc^2 E_m} \quad \text{(available energy, equal masses, } E_m \gg mc^2). \quad (46\text{–}11)$$

EXAMPLE 46-3

Threshold energy for pion production A proton (rest energy 938 MeV) with kinetic energy K collides with another proton at rest. Both protons survive the collision, but in addition a neutral pion (π^0, rest energy 135 MeV) is produced. What is the threshold energy (minimum value of K) needed for this process?

SOLUTION The final state includes the two original protons (mass m) and the pion (mass m_π). The threshold energy corresponds to the minimum-energy case in which all three particles are at rest in the center-of-momentum system. The total available energy must be at least their total rest energy,

$$E_a = (2m + m_\pi)c^2.$$

We substitute this expression into Eq. (46–10), simplify, and solve for E_m:

$$4m^2c^4 + 4mm_\pi c^4 + m_\pi^2 c^4 = 2mc^2 E_m + 2(mc^2)^2,$$

$$E_m = mc^2 + m_\pi c^2 \left(2 + \frac{m_\pi}{2m}\right).$$

The first term in the expression for E_m is the rest energy of the bombarding proton, and the remaining terms give its kinetic energy. We see that the kinetic energy must be somewhat greater than twice the rest energy of the pion we want to create. Using $mc^2 = 938$ MeV and $m_\pi c^2 = 135$ MeV in this expression, we find $m_\pi/2m = 0.072$ and

$$E_m = mc^2 + (135 \text{ MeV})(2 + 0.072) = mc^2 + 280 \text{ MeV}.$$

To create a pion with rest energy 135 MeV, a bombarding proton needs a kinetic energy of at least 280 MeV. We suggest that you compare this result with the result of Example 39–14 (Section 39–10), which required only 67.5 MeV of kinetic energy for each proton in a head-on collision. We discuss the energy advantage of such collisions in the next subsection.

EXAMPLE 46-4

Increasing the available energy a) The Fermilab accelerator was originally designed for a proton beam energy of 800 GeV (800×10^9 eV) on a stationary target. Find the available energy in a proton-proton collision. b) If the proton beam energy is increased to 1000 GeV, what is the available energy?

SOLUTION We use Eq. (46–11) because, for the proton, $mc^2 = 938$ MeV $= 0.938$ GeV is much less than either 800 GeV or 1000 GeV.

a) When $E_m = 800$ GeV,

$$E_a = \sqrt{2(0.938 \text{ GeV})(800 \text{ GeV})} = 38.7 \text{ GeV}.$$

b) When $E_m = 1000$ GeV,

$$E_a = \sqrt{2(0.938 \text{ GeV})(1000 \text{ GeV})} = 43.3 \text{ GeV}.$$

With a stationary proton target, increasing the proton beam energy by 200 GeV increases the available energy by only 4.6 GeV!

COLLIDING BEAMS

The limitation illustrated by Example 46–4 is circumvented in *colliding-beam* experiments. In these experiments there is no stationary target; instead, beams of particles moving in opposite directions are tightly focused onto one another so that head-on collisions can occur. Usually the two colliding particles have momenta of equal magnitude and opposite direction, so the total momentum is zero. Hence the laboratory system is also the center-of-momentum system, and the available energy is maximized. If one beam contains particles and the other beam contains antiparticles (for instance, electrons and positrons or protons and antiprotons), the available energy E_a is the *total* energy of the two colliding particles.

One type of particle-antiparticle collider is at the Stanford Linear Accelerator Center (SLAC) (Fig. 46–8). Beams of electrons and positrons are accelerated alternately in the linac, and magnets are used to steer and focus the beams into head-on collisions with total available energy E_a of nearly 100 GeV. In such a linear collider, the colliding beams

(a)   (b)

46–8 (a) The 3-km-long linear accelerator at the Stanford Linear Accelerator Center (SLAC) accelerates alternate bunches of positrons and electrons to 50 GeV; then magnets separate them and bend them around in the arcs (dashed lines) to provide head-on e^+e^- collisions. About 1 GeV of energy is radiated away from the particles as they move around the arcs. (b) Sketch showing the main components. The positrons are produced by pair production when part of the electron beam is diverted about two thirds of the way along the accelerator and is deflected back to the start. The damping rings and focusing magnets help to bunch and focus the beams, which have cross-section radii of about 1 μm. The bending magnets actually extend over all regions where the beam is bent.

can be very intense but have only a single pass at each other. Other laboratories studying electron-positron collisions use *storage rings,* in which bunches of electrons and positrons are kept circulating in opposite directions around a circular tunnel, giving them many opportunities to interact. At present the largest e^+e^- storage ring is the Large Electron-Positron Collider (LEP), situated in an underground tunnel 27 km (about 17 miles) in circumference at the European Laboratory for Particle Physics (CERN) in Geneva, Switzerland. The total available energy E_a is 180 GeV.

The highest-energy colliding beams presently in use are at the Tevatron at Fermilab. A beam of 1-TeV protons collides head-on with a beam of 1-TeV antiprotons to give a total energy of 2 TeV. These colliding beams were used in the discovery of the top quark, to be discussed in Section 46–5. Even higher energies will be possible with the Large Hadron Collider (LHC) at CERN, which will use the same tunnel as LEP but will accelerate colliding beams of protons to 7 TeV each, giving a total of 14 TeV. The LHC is scheduled to go into operation in 2005. (An even larger colliding-beam accelerator, the Superconducting Supercollider (SSC), was intended to collide beams of 20 TeV protons with each other. However, funding for this project was canceled in 1993 by the U.S. Congress.)

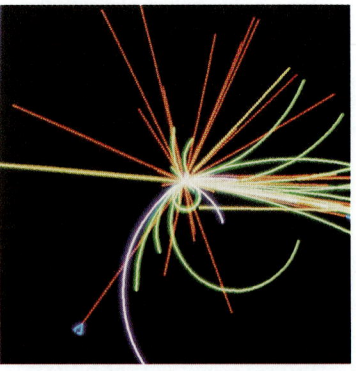

46–9 This computer-generated image shows a typical result of a proton-antiproton collision in the Tevatron recorded by the Collider Detector. Decay products of the collision include pions, kaons, muons, and others, each indicated by a different color.

DETECTORS

Ordinarily, we can't see or feel individual subatomic particles or photons. How, then, do we measure their properties? A wide variety of different devices have been designed. Many detectors use the ionization caused by charged particles as they move through a gas, liquid, or solid. The ions along the particle's path act as nucleation centers for droplets of liquid in the supersaturated vapor of a cloud chamber (Fig. 46–1) or cause small volumes of vapor in the superheated liquid of a bubble chamber (Fig. 46–3a). In a semiconducting solid the ionization can take the form of electron-hole pairs. We discussed their detection in Section 44–8. *Wire chambers* contain arrays of closely spaced wires that detect the ions. The charge collected and time information from each wire are processed using computers to reconstruct the particle trajectories. Examples are shown in Fig. 46–9.

COSMIC RAY EXPERIMENTS

Large numbers of particles called *cosmic rays* continually bombard the earth from sources both within and beyond our galaxy. These particles consist mostly of neutrinos, protons, and heavier nuclei, with energies ranging from less than 1 MeV to more than 10^{20} eV. The earth's atmosphere and magnetic field protect us from much of this radiation. However, this means that cosmic-ray experimentation often must be carried out above all or most of the atmosphere by means of rockets or high-altitude balloons.

In contrast, neutrino detectors are buried below the earth's surface in tunnels or mines or submerged deep in the ocean. This is done to screen out all other types of particles so that only neutrinos, which interact only very weakly with matter, reach the detector. It would take a light-years thickness of lead to absorb a sizable fraction of a beam of neutrinos. Thus neutrino detectors consist of huge amounts of matter: 400,000 liters of C_2Cl_4 (perchloroethylene, a cleaning fluid) in a detector to be discussed in Section 46–6, 60,000 kilograms of gallium in another.

Cosmic rays were important in early particle physics, and their study currently brings us important information about the rest of the universe. Although cosmic rays provide a source of high-energy particles that does not depend on expensive accelerators, most particle physicists use accelerators because the high-energy cosmic-ray particles they want are too few and too random.

46–4 PARTICLES AND INTERACTIONS

This section may seem like an expedition into the wilderness. From the security of 1932, when there were thought to be three permanent, unchanging fundamental particles, we enter the partly mapped territory of present-day particle physics. Particles are *not* permanent; they can be created and destroyed. Many particles are *unstable,* decaying spontaneously into other particles. Virtual particles, existing on borrowed energy allowed by the uncertainty relation, serve as mediators or transmitters of the various interactions.

FOUR FORCES AND THEIR MEDIATING PARTICLES

Particles can be classified in terms of the ways in which they interact. One classification scheme involves four types of forces or interactions, first discussed in Section 5–6. They are, in order of decreasing strength,

1. the strong interaction,
2. the electromagnetic interaction,
3. the weak interaction, and
4. the gravitational interaction.

The *electromagnetic* and *gravitational* interactions are familiar from classical physics. Both are characterized by a $1/r^2$ dependence on distance. In this scheme, the mediating particles for both interactions have mass zero and are stable as ordinary particles. The mediating particle for the electromagnetic interaction is the familiar photon, which has spin 1. That particle for the gravitational force is the spin-2 *graviton,* which has not yet been observed experimentally because the gravitational force is very much weaker than the electromagnetic force. For example, the gravitational attraction of two protons is smaller than their electrical repulsion by a factor of about 10^{36}. The gravitational force is of primary importance in the structure of stars and the large-scale behavior of the universe, but it is not believed to play a significant role in particle interactions at the energies that are currently attainable.

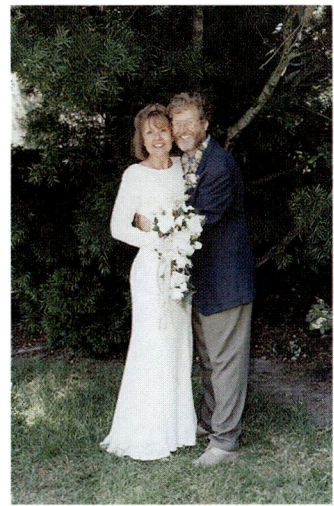

The other two forces are less familiar. One, usually called the *strong interaction,* is responsible for the nuclear force and also for the production of pions and several other particles in high-energy collisions. At the most fundamental level, the mediating particle for the strong interaction is called a *gluon.* However, the force between nucleons is more easily described in terms of mesons as the mediating particles. We'll discuss the spin-1, massless gluon in Section 46–5.

Equation (46–4) is a possible potential-energy function for the nuclear force. The strength of the interaction is described by the constant f^2, which has units of energy times distance. A better basis for comparison with other forces is the dimensionless ratio $f^2/\hbar c$, called the *coupling constant* for the interaction. (We invite you to verify that this ratio is a pure number and so must have the same value in all systems of units.) The observed behavior of nuclear forces suggests that $f^2/\hbar c \approx 1$. The dimensionless coupling constant for *electromagnetic* interactions is

$$\frac{1}{4\pi\epsilon_0} \frac{e^2}{\hbar c} = 7.297 \times 10^{-3} = \frac{1}{137.0}. \tag{46–12}$$

The ties that bind us together originate in the fundamental interactions of nature. The nuclei within our bodies are held together by the strong interaction. The electromagnetic interaction binds nuclei and electrons together to form atoms, binds atoms together to form molecules, and binds molecules together to form us.

Thus the strong interaction is roughly 100 times as strong as the electromagnetic interaction; however, it drops off with distance more quickly than $1/r^2$.

The fourth interaction is called the *weak* interaction. It is responsible for beta decay, such as the conversion of a neutron into a proton, an electron, and an antineutrino. It is also responsible for the decay of many unstable particles (pions into muons, muons into electrons, and so on). Its mediating particles are the short-lived particles W^+, W^-, and Z^0. The existence of these particles was confirmed in 1983 in experiments at CERN, for which Carlo Rubbia and Simon van der Meer were awarded the Nobel Prize in 1984. The $W^\pm$ and Z^0 have spin 1 like the gluon, but they are *not* massless. In fact, they have enormous masses, $80.4~\text{GeV}/c^2$ for the W's and $91.2~\text{GeV}/c^2$ for the Z^0. With such

TABLE 46–1
FOUR FUNDAMENTAL INTERACTIONS

INTERACTION	RELATIVE STRENGTH	RANGE	MEDIATING PARTICLE Name	MEDIATING PARTICLE Mass	MEDIATING PARTICLE Charge	Spin
Strong	1	Short (~ 1 fm)	Gluon	0	0	1
Electromagnetic	$\frac{1}{137}$	Long ($1/r^2$)	Photon	0	0	1
Weak	10^{-9}	Short (~ 0.001 fm)	$W^{\pm}, Z^0$	80.4, 91.2 GeV/c^2	$\pm e, 0$	1
Gravitational	10^{-38}	Long ($1/r^2$)	Graviton	0	0	2

massive mediating particles the weak interaction has a much shorter range than the strong interaction. It also lives up to its name by being weaker than the strong interaction by a factor of about 10^9.

Table 46–1 compares the main features of these four fundamental interactions.

MORE PARTICLES

In Section 46–2 we mentioned the discoveries in cosmic rays of muons in 1937 and of pions in 1947. The electric charges of the muons and the charged pions have the same magnitude e as the electron charge. The positive muon μ^+ is the antiparticle of the negative muon μ^-. Each has spin $\frac{1}{2}$ and a mass of about $207m_e = 106$ MeV/c^2. Muons are unstable; each decays with a lifetime of 2.2×10^{-6} s into an electron of the same sign, a neutrino, and an antineutrino.

There are three kinds of pions, all with spin 0. The π^+ and π^- have masses of $273m_e = 140$ MeV/c^2. They are unstable; each $\pi^{\pm}$ decays with a lifetime of 2.6×10^{-8} s into a muon of the same sign along with a neutrino for the π^+ and an antineutrino for the π^-. The π^0 is somewhat less massive, $264m_e = 135$ MeV/c^2, and it decays with a lifetime of 8.4×10^{-17} s into two photons. The π^+ and π^- are antiparticles of one another, while the π^0 is its own antiparticle. (That is, there is no distinction between particle and antiparticle for the π^0.)

The existence of the *antiproton* $\bar{\text{p}}$ had been suspected ever since the discovery of the positron. The $\bar{\text{p}}$ was found in 1955, when proton-antiproton ($\text{p}\bar{\text{p}}$) pairs were created by use of a beam of 6-GeV protons from the Bevatron at the University of California, Berkeley. The *antineutron* $\bar{\text{n}}$ was found soon afterward. After 1960, as higher-energy accelerators and more sophisticated detectors were developed, a veritable blizzard of new unstable particles were identified. To describe and classify them, we need a small blizzard of new terms.

Initially, particles were classified by mass into three categories: (1) leptons ("light ones" such as electrons); (2) mesons ("intermediate ones" such as pions); and (3) baryons ("heavy ones" such as nucleons and more massive particles). But this scheme has been superseded by a more useful one in which particles are classified in terms of their *interactions*. For instance, *hadrons* (which include mesons and baryons) have strong interactions, and *leptons* do not.

In the following discussion we will also distinguish between **fermions,** which have half-integer spins, and **bosons,** which have zero or integer spins. Fermions obey the exclusion principle, on which the Fermi-Dirac distribution function (Section 44–6) is based. Bosons do not obey the exclusion principle and have a different distribution function, the Bose-Einstein distribution.

TABLE 46–2

THE LEPTONS

PARTICLE NAME	SYMBOL	ANTI-PARTICLE	MASS (MeV/c^2)	L_e	L_μ	L_τ	LIFETIME (s)	PRINCIPAL DECAY MODES
Electron	e^-	e^+	0.511	+1	0	0	Stable	
Electron neutrino	ν_e	$\overline{\nu}_e$	0(?)	+1	0	0	Stable	
Muon	μ^-	μ^+	105.7	0	+1	0	2.20×10^{-6}	$e^- \overline{\nu}_e \nu_\mu$
Muon neutrino	ν_μ	$\overline{\nu}_\mu$	0(?)	0	+1	0	Stable	
Tau	τ^-	τ^+	1777	0	0	+1	3.0×10^{-13}	$\mu^- \overline{\nu}_\mu \nu_\tau$ or $e^- \overline{\nu}_e \nu_\tau$
Tau neutrino	ν_τ	$\overline{\nu}_\tau$	0(?)	0	0	+1	Stable	

LEPTONS

The **leptons,** which do not have strong interactions, include six particles; the electron (e^-) and its neutrino (ν_e), the muon (μ^-) and its neutrino (ν_μ), and the tau particle (τ^-) and its neutrino (ν_τ). Each of the six particles has a distinct antiparticle. All leptons have spin $\frac{1}{2}$ and thus are fermions. The family of leptons is shown in Table 46–2. The taus have mass $3478m_e = 1777$ MeV/c^2. Taus and muons are unstable; a τ^- often decays into a μ^- plus a tau neutrino and a muon antineutrino, and a μ^- decays into a electron plus a muon neutrino and an electron antineutrino. If we consider their opportunities to decay, they have relatively long lifetimes because their decays are mediated by the weak interaction. Despite their zero charge, a neutrino is distinct from an antineutrino; the spin angular momentum of a neutrino has a component that is opposite its linear momentum, while for an antineutrino that component is parallel to its linear momentum. Until recently, all the neutrinos were believed to have zero rest mass; there is speculation and some experimental evidence that they may have small nonzero masses. We'll return to this point later.

Leptons obey a *conservation principle.* Corresponding to the three pairs of leptons are three lepton numbers L_e, L_μ, and L_τ. The electron e^- and the electron neutrino ν_e are assigned $L_e = 1$, and their antiparticles e^+ and $\overline{\nu}_e$ are given $L_e = -1$. Corresponding assignments of L_μ and L_τ are made for the μ and τ particles and their neutrinos. **In all interactions, each lepton number is separately conserved.** For example, in the decay of the μ^-, the lepton numbers are

$$\mu^- \quad \rightarrow \quad e^- \quad + \quad \overline{\nu}_e \quad + \quad \nu_\mu,$$

$$L_\mu = 1 \quad L_e = 1 \quad L_e = -1 \quad L_\mu = 1.$$

These conservation principles have no counterpart in classical physics.

EXAMPLE 46–5

Check conservation of lepton numbers for the following decay schemes:

a) $\mu^+ \rightarrow e^+ + \nu_e + \overline{\nu}_\mu$;

b) $\pi^- \rightarrow \mu^- + \overline{\nu}_\mu$;

c) $\pi^0 \rightarrow \mu^- + e^+ + \nu_e$.

SOLUTION In each of the three decays, there are no τ particles nor any τ neutrinos, so $L_\tau = 0$ both before and after each decay; L_τ is therefore conserved.

a) The initial lepton numbers are $L_e = 0$ and $L_\mu = -1$. The final values are $L_e = -1 + 1 + 0 = 0$ and $L_\mu = 0 + 0 + (-1) = -1$. All lepton numbers are conserved.

b) All lepton numbers are initially zero (the π^- is *not* a lepton). The final values are $L_\mu = 1 + (-1) = 0$ and $L_e = 0$. Again, all lepton numbers are conserved.

c) The initial L's are again all zero; in the final state, $L_e = 0 + (-1) + 1 = 0$ and $L_\mu = 1 + 0 + 0 = 1$. Thus L_μ is *not* conserved; this decay is forbidden by conservation of muon lepton number, and it has never been observed.

TABLE 46–3

SOME HADRONS AND THEIR PROPERTIES

PARTICLE	MASS (MeV/c^2)	CHARGE RATIO Q/e	SPIN	BARYON NUMBER, B	STRANGENESS S	MEAN LIFETIME (s)	TYPICAL DECAY MODES	QUARK CONTENT
Mesons								
π^0	135.0	0	0	0	0	8.4×10^{-17}	$\gamma\,\gamma$	$u\bar{u}, d\bar{d}$
π^+	139.6	+1	0	0	0	2.60×10^{-8}	$\mu^+ \nu_\mu$	$u\bar{d}$
π^-	139.6	−1	0	0	0	2.60×10^{-8}	$\mu^- \bar{\nu}_\mu$	$\bar{u}d$
K^+	493.7	+1	0	0	+1	1.24×10^{-8}	$\mu^+ \nu_\mu$	$u\bar{s}$
K^-	493.7	−1	0	0	−1	1.24×10^{-8}	$\mu^- \bar{\nu}_\mu$	$\bar{u}s$
η^0	547.3	0	0	0	0	$\approx 10^{-18}$	$\gamma\,\gamma$	$u\bar{u}, d\bar{d}, s\bar{s}$
Baryons								
p	938.3	+1	$\frac{1}{2}$	1	0	Stable	–	uud
n	939.6	0	$\frac{1}{2}$	1	0	887	$pe^- \bar{\nu}_e$	udd
Λ^0	1116	0	$\frac{1}{2}$	1	−1	2.63×10^{-10}	$p\pi^-$ or $n\pi^0$	uds
Σ^+	1189	+1	$\frac{1}{2}$	1	−1	7.99×10^{-11}	$p\pi^0$ or $n\pi^+$	uus
Σ^0	1193	0	$\frac{1}{2}$	1	−1	7.4×10^{-20}	$\Lambda^0 \gamma$	uds
Σ^-	1197	−1	$\frac{1}{2}$	1	−1	1.48×10^{-10}	$n\pi^-$	dds
Ξ^0	1315	0	$\frac{1}{2}$	1	−2	2.90×10^{-10}	$\Lambda^0 \pi^0$	uss
Ξ^-	1321	−1	$\frac{1}{2}$	1	−2	1.64×10^{-10}	$\Lambda^0 \pi^-$	dss
Δ^{++}	1231	+2	$\frac{3}{2}$	1	0	$\approx 10^{-23}$	$p\pi^+$	uuu
Ω^-	1672	−1	$\frac{3}{2}$	1	−3	8.2×10^{-11}	$\Lambda^0 K^-$	sss
Λ_c^+	2285	+1	$\frac{1}{2}$	1	0	2.1×10^{-13}	$pK^- \pi^+$	udc

HADRONS

Hadrons, the strongly interacting particles, are a more complex family than leptons. Each hadron has an antiparticle, often denoted with an overbar, as with the antiproton $\bar{p}$. There are two subclasses of hadrons: *mesons* and *baryons*. Table 46–3 shows some of the many hadrons that are currently known. (We'll discuss what is meant by *strangeness* and *quark content* later in this section and in the next one.)

Mesons include the pions that have already been mentioned, K mesons or *kaons, η* mesons, and others that we will discuss later. Mesons have spin 0 or 1 and therefore are all bosons. There are no stable mesons; all can and do decay to less massive particles, obeying all the conservation laws for such decays.

Baryons include the nucleons and several particles called *hyperons,* including the Λ, Σ, Ξ, and Ω. These resemble the nucleons but are more massive. Baryons have half-integer spin, and therefore all are fermions. The only stable baryon is the proton; a free neutron decays to a proton, and the hyperons decay to other hyperons or to nucleons by various processes. Baryons obey the principle of *conservation of baryon number,* analogous to conservation of lepton numbers, again with no counterpart in classical physics. We assign a baryon number $B = 1$ to each baryon (p, n, Λ, Σ, and so on) and $B = -1$ to each antibaryon ($\bar{p}$, $\bar{n}$, $\bar{\Lambda}$, $\bar{\Sigma}$, and so on). **In all interactions, the total baryon number is conserved.** This principle is the reason why the mass number A was conserved in all the nuclear reactions that we studied in Chapter 45.

EXAMPLE 46-6

Which of the following reactions obey the principle of conservation of baryons?

a) $\qquad n + p \rightarrow n + p + p + \bar{p},$

b) $\qquad n + p \rightarrow n + p + \bar{p}.$

SOLUTION In both reactions the initial baryon number is $1 + 1 = 2.$

a) The final baryon number is $1 + 1 + 1 + (-1) = 2.$ Baryon number is conserved and this process can occur (provided that enough energy is available in the n + p collision).

b) The final baryon number for this reation is $1 + 1 + (-1) = 1.$ Baryon number is *not* conserved, and this reaction has never been observed.

EXAMPLE 46-7

Antiproton creation Antiprotons can be produced by bombarding a stationary proton target (liquid hydrogen) with a beam of protons. Find the minimum beam energy that is needed for this reaction to occur.

SOLUTION In this reaction, conservation of charge and of baryon number forbids the creation of an antiproton by itself; it must be created as part of a proton-antiproton pair. The appropriate reaction is

$$p + p \rightarrow p + p + p + \bar{p}.$$

For this reaction to occur, the minimum available energy E_a in Eq. (46–10) is the final rest energy $4mc^2.$ With that substitution,

Eq. (46–10) gives

$$(4mc^2)^2 = 2mc^2(E_m + mc^2),$$
$$E_m = 7mc^2.$$

The energy E_m of the bombarding particle includes its rest energy $mc^2,$ so its minimum *kinetic* energy must be $6mc^2 = 6(938 \text{ MeV}) = 5.63 \text{ GeV}.$

The search for the antiproton was one of the principal reasons for the construction of the Bevatron at the University of California, Berkeley, with beam energy of 6 GeV. The search was successful; in 1959, Emilio Segrè and Owen Chamberlain were awarded the Nobel Prize for its discovery.

STRANGENESS

The K mesons and the Λ and Σ hyperons were discovered during the late 1950s. Because of their unusual behavior they were called *strange particles*. They were produced in high-energy collisions such as $\pi^- + p,$ and a K meson and a hyperon were always produced *together*. The relatively high rate of production of these particles suggested that it was a *strong*-interaction process, but their relatively long lifetimes suggested that their decay was a *weak*-interaction process. The K^0 appeared to have *two* lifetimes, one about 9×10^{-11} s and another nearly 600 times longer. Where the K mesons strongly interacting hadrons or not?

The search for the answer to this questions led physicists to introduce a new quantity called **strangeness.** The hyperons Λ^0 and $\Sigma^{\pm,0}$ were assigned a strangeness quantum number $S = -1,$ and the associated K^0 and K^+ mesons were assigned $S = +1.$ The corresponding antiparticles had opposite strangeness, $S = +1$ for $\bar{\Lambda}^0$ and $\bar{\Sigma}^{\pm,0}$ and $S = -1$ for $\bar{K}^0$ and $K^-.$ Then strangeness was *conserved* in production processes such as

$$p + \pi^- \rightarrow \bar{\Sigma}^- + K^-,$$

$$p + \pi^- \rightarrow \Lambda^0 + K^0.$$

The process

$$p + \pi^- \rightarrow p + K^-$$

does not conserve strangeness, and it doesn't occur.

When strange particles decay individually, strangeness is usually *not* conserved. Typical processes include

$$\Sigma^+ \rightarrow n + \pi^+,$$

$$\Lambda^0 \rightarrow p + \pi^-,$$

$$K^- \rightarrow \pi^+ + \pi^- + \pi^-.$$

In each of these decays, the initial strangeness is 1 or −1, and the final value is zero. All observations of these particles are consistent with the conclusion that *strangeness is conserved in strong interactions but it can change by zero or one unit in weak interactions.* There is no counterpart to the strangeness quantum number in classical physics.

CONSERVATION LAWS

19.5
Particle Physics

The decay of strange particles provides our first example of a conditional conservation law, one that is obeyed in some interactions and not in others. Let's review *all* the conservation laws we know about and see what conclusions we can draw from them.

Several conservation laws are obeyed in *all* interactions. These include the familiar conservation laws; energy, momentum, angular momentum, and electric charge. These are called *absolute conservation laws.* Baryon number and the three lepton numbers are also conserved in all interactions. Strangeness is conserved in strong and electromagnetic interactions but *not* in all weak interactions.

Two other quantities, which are conserved in some but not all interactions, are useful in classifying particles and their interactions. One is *isospin,* a quantity that is used to describe the charge independence of the strong interactions. The other is *parity,* which describes the comparative behavior of two systems that are mirror images of each other. Isospin is conserved in strong interactions, which are charge-independent, but not in electromagnetic or weak interactions. (The electromagnetic interaction is certainly *not* charge-independent.) Parity is conserved in strong and electromagnetic interactions but not in weak ones. The Chinese-American physicists T.D. Lee and C.N. Yang received the Nobel Prize in 1957 for laying the theoretical foundations for nonconservation of parity in weak interactions.

This discussion shows that conservation laws provide another basis for classifying particles and their interactions. Each conservation law is also associated with a *symmetry* property of the system. A familiar example is angular momentum. If a system is in an environment that has spherical symmetry, there can be no torque acting on it because the direction of the torque would violate the symmetry. In such a system, total angular momentum is *conserved.* When a conservation law is violated, the interaction is often described as a *symmetry-breaking interaction.*

46–5 QUARKS AND THE EIGHTFOLD WAY

The leptons form a fairly neat package: three particles and three neutrinos, each with its antiparticle, and a conservation law relating their numbers. Physicists believe that leptons are genuinely fundamental particles. The hadron family, by comparison, is a mess. Table 46–3 contains only a sample of well over 100 hadrons that have been discovered since 1960, and it has become clear that these particles *do not* represent the most fundamental level of the structure of matter.

Our present understanding of the structure of hadrons is based on a proposal made initially in 1964 by the American physicist Murray Gell-Mann and his collaborators. In this proposal, hadrons are not fundamental particles but are composite structures whose constituents are spin-$\frac{1}{2}$ fermions called **quarks.** (The name is found in the line "Three quarks for Muster Mark!" from *Finnegan's Wake* by James Joyce.) Each baryon is composed of three quarks (qqq), each antibaryon of three antiquarks ($\overline{q}\,\overline{q}\,\overline{q}$), and each meson of a quark-antiquark pair ($q\overline{q}$). Table 46–3 of the previous section gives the quark content of many hadrons. No other compositions seem to be necessary. This scheme requires that quarks have electric charges with magnitudes $\frac{1}{3}$ and $\frac{2}{3}$ of the electron charge e, which had previously been thought to be the smallest unit of charge. Each quark also has a fractional value $\frac{1}{3}$ for its baryon number B, and each antiquark has a baryon-number value $-\frac{1}{3}$. In a meson, a quark and antiquark combine with net baryon number 0

TABLE 46–4
PROPERTIES OF THE THREE ORIGINAL QUARKS

SYMBOL	Q/e	SPIN	BARYON NUMBER, B	STRANGE-NESS, S	CHARM, C	BOTTOM-NESS, B'	TOPNESS, T
u	$\frac{2}{3}$	$\frac{1}{2}$	$\frac{1}{3}$	0	0	0	0
d	$-\frac{1}{3}$	$\frac{1}{2}$	$\frac{1}{3}$	0	0	0	0
s	$-\frac{1}{3}$	$\frac{1}{2}$	$\frac{1}{3}$	-1	0	0	0

and can have their spin angular momentum components parallel to form a spin-1 meson or antiparallel to form a spin-0 meson. Similarly, the three quarks in a baryon combine with net baryon number 1 and can form a spin-$\frac{1}{2}$ baryon or a spin-$\frac{3}{2}$ baryon.

THE THREE ORIGINAL QUARKS

The first (1964) quark theory included three types (called *flavors*) of quarks, labeled u (up), d (down), and s (strange). Their principal properties are listed in Table 46–4. The corresponding antiquarks $\bar{u}, \bar{d}$ and $\bar{s}$ have opposite values of Q, B, and S. Protons, neutrons, π and K mesons, and several hyperons can be constructed from these three quarks. For example, the proton quark content is uud. Checking Table 46–4, we see that the values of Q/e add to 1 and that the values of the baryon number B also add to 1, as we should expect. The neutron is udd, with total $Q = 0$ and $B = 1$. The π^+ meson is $u\bar{d}$, with $Q/e = 1$ and $B = 0$, and the K^+ meson is $u\bar{s}$. Checking the values of S for the quark content, we see that the proton, neutron, and π^+ have strangeness 0 and that the K^+ has strangeness 1, in agreement with Table 46–3. The antiproton is $\bar{p} = \bar{u}\bar{u}\bar{d}$, the negative pion is $\pi^- = \bar{u}d$, and so on. The quark content can also be used to explain dynamic properties of hadrons such as their excited states and magnetic moments. Figure 46–10 shows the quark content of two baryons and two mesons.

EXAMPLE 46–8

Given that they contain only $u, d, s, \bar{u}, \bar{d}$, and/or $\bar{s}$, use Table 46–4 to find the quark content of a) Σ^+; b) $\bar{\Lambda}^0$. The Σ^+ and Λ^0 particles are both baryons with strangeness $S = -1$.

SOLUTION Baryons contain three quarks; if $S = -1$, one and only one of the three must be an s quark.
a) The Σ^+ has $Q/e = +1$, so the other two quarks must both be u. Hence the quark content of Σ^+ is uus.

b) First we find the quark content of the Λ^0. For zero total charge, the other two quarks must be u and d, so the quark content of the Λ^0 is uds. The $\bar{\Lambda}^0$ is the antiparticle of the Λ^0, so its quark content is $\bar{u}\,\bar{d}\,\bar{s}$.

Note that while the Λ^0 and $\bar{\Lambda}^0$ are both electrically neutral, and both have the same mass, they are different particles: Λ^0 has $B = 1$ and $S = -1$, while $\bar{\Lambda}^0$ has $B = -1$ and $S = 1$.

What caused physicists to suspect that hadrons were made up of something smaller? The magnetic moment of the neutron (Section 45–2) was one of the first reasons. In Section 28–8 we learned that a magnetic moment results from a circulating current (a motion of electric charge). But the neutron has *no* charge, or, to be more accurate, no *total* charge. It could be made up of smaller particles whose charges add to zero. The quantum motion of these particles within the neutron would then give its surprising

Proton (p)

Neutron (n)

Positive pion (π^+)

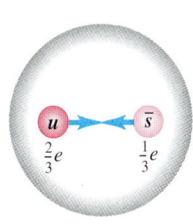
Positive kaon (K$^+$)

46–10 Quark content for four hadrons. The various color combinations that are needed for color neutrality are not shown.

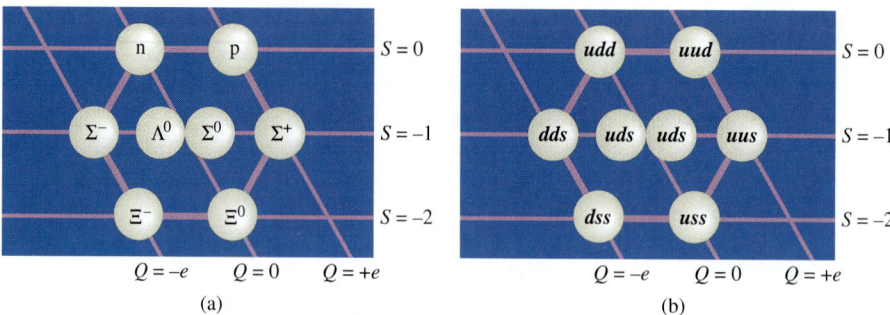

46–11 (a) Plot of S and Q values for spin-$\frac{1}{2}$ baryons showing the symmetry pattern of the eightfold way. (b) Quark content of each spin-$\frac{1}{2}$ baryon. The quark contents of the Σ^0 and Λ^0 are the same (see Table 46–3 in Section 46–4); the Σ^0 is an excited state of the Λ^0 and can decay into it by γ emission.

magnetic moment. To verify this hypothesis by "seeing" inside a neutron, we need a probe with a wavelength that is much less than the neutron's size of about a femtometer. This probe should not be affected by the strong interaction, so that it won't interact with the neutron as a whole but will penetrate into it and interact electromagnetically with these supposed smaller charged particles. A probe with these properties is an electron with energy above 10 GeV. In experiments carried out at SLAC, such electrons were scattered from neutrons and protons to help show that nucleons are indeed made up of fractionally charged, spin-$\frac{1}{2}$, pointlike particles.

THE EIGHTFOLD WAY

Symmetry considerations play a very prominent role in particle theory. Here are two examples. Consider the eight spin-$\frac{1}{2}$ baryons we've mentioned: the familiar p and n, the strange Λ^0, Σ^+, Σ^0, and Σ^-; and the doubly strange Ξ^0 and Ξ^-. For each we plot the value of strangeness S versus the value of charge Q in Fig. 46–11. The result is a hexagonal pattern. A similar plot for the nine spin-0 mesons (six shown in Table 46–3 plus three others not included in that table) is shown in Fig. 46–12; the particles fall in exactly the same hexagonal pattern! In each plot, all the particles have masses that are within about $\pm 200 \text{ MeV}/c^2$ of the median mass value of that plot, with variations due to differences in quark masses and internal potential energies.

The symmetries that lead to these and similar patterns are collectively called the **eightfold way.** They were discovered in 1961 by Murray Gell-Mann and independently by Yu'val Ne'eman. (The name is a slightly irreverent reference to the Noble Eightfold Path, a set of principles for right living in Buddhism.) A similar pattern for the spin-$\frac{3}{2}$ baryons contains *ten* particles, arranged in a triangular pattern like pins in a bowling alley. When this pattern was first discovered, one of the particles was missing. But Gell-Mann gave it a name anyway (Ω^-), predicted the properties it should have, and told experimenters what they should look for. Three years later, the particle was found during an experiment at Brookhaven National Laboratory, a spectacular success for Gell-Mann's theory. The whole series of events is reminiscent of the way in which

46–12 (a) Plot of S and Q values for nine spin-0 mesons, showing the symmetry pattern of the eightfold way. Each particle is on the opposite side of the hexagon from its antiparticle; each of the three particles in the center is its own antiparticle. (b) Quark content of each spin-0 meson. The particles in the center are different mixtures of the three quark-antiquark pairs shown.

Mendeleev used gaps in the periodic table of the elements to predict properties of undiscovered elements and to guide chemists in their search for these elements.

What binds quarks to one another? The attractive interactions among quarks are mediated by massless spin-1 bosons called **gluons** in much the same way that photons mediate the electromagnetic interaction or that pions mediated the nucleon-nucleon force in the old Yukawa theory.

COLOR

Quarks, having spin $\frac{1}{2}$, are fermions and so are subject to the exclusion principle. This would seem to forbid a baryon having two or three quarks with the same flavor and same spin component. To avoid this difficulty, it is assumed that each quark comes in three varieties, which are whimsically called *colors*. Red, green, and blue are the usual choices. The exclusion principle applies separately to each color. A baryon always contains one red, one green, and one blue quark, so the baryon itself has no net color. Each gluon has a color-anticolor combination (for example, blue-antired) that allows it to transmit color when exchanged, and color is conserved during emission and absorption of a gluon by a quark. The gluon-exchange process changes the colors of the quarks in such a way that there is always one quark of each color in every baryon. The color of an individual quark changes continually as gluons are exchanged.

Similar processes occur in mesons such as pions. The quark-antiquark pairs of mesons have canceling color and anticolor (for example, blue and antiblue), so mesons also have no net color. Suppose a pion initially consists of a blue quark and an antiblue antiquark. The blue quark can become a red quark by emitting a blue-antired virtual gluon. The gluon is then absorbed by the antiblue antiquark, converting it to an antired antiquark (Fig. 46–13). Color is conserved in each emission and absorption, but a blue-antiblue pair has become a red-antired pair. Such changes occur continually, so we have to think of a pion as a superposition of three quantum states, blue-antiblue, green-antigreen, and red-antired. On a larger scale, the strong interaction between nucleons was described in Section 46–4 as due to the exchange of virtual mesons. In terms of quarks and gluons, these mediating virtual mesons are quark-antiquark systems bound together by the exchange of gluons.

The theory of strong interactions is known as *quantum chromodynamics* (QCD). No one has yet been able to isolate an individual quark for study. Most QCD theories contain phenomena associated with the binding of quarks that make it impossible to obtain a free quark. An impressive body of experimental evidence supports the correctness of the quark structure of hadrons and the belief that quantum chromodynamics is the key to understanding the strong interactions.

THREE MORE QUARKS

Before the tau particles were discovered, there were four known leptons. This fact, together with some puzzling decay rates, led to the speculation that there might be a fourth quark flavor. This quark is labeled **c** (the *charmed* quark); it has $Q/e = \frac{2}{3}$, $B = \frac{1}{3}$, $S = 0$, and a new quantum number **charm** $C = +1$. This was confirmed in 1974 by the observation at both SLAC and the Brookhaven National Laboratory of a meson, now named ψ, with mass 3097 MeV/c^2. This meson was found to have several decay modes, decaying into e^+e^-, $\mu^+\mu^-$, or into hadrons. The mean lifetime was found to be about 10^{-20} s. These results are consistent with ψ being a spin-1 $c\bar{c}$ system. Almost immediately after this, similar mesons of greater mass were observed and identified as excited states of the $c\bar{c}$ system. A few years later, individual mesons with a nonzero net charm quantum number were also observed. These mesons, D^0 ($c\bar{u}$) and D^+ ($c\bar{d}$), and their excited states are now firmly established. A charmed baryon, Λ_c^+ (**udc**), has also been observed.

(a)

(b)

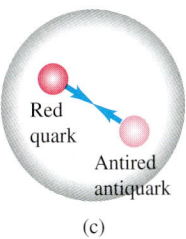

(c)

46–13 (a) A pion containing a blue quark and an antiblue antiquark. (b) The blue quark emits a blue-antired gluon, changing to a red quark. (c) The gluon is absorbed by the antiblue antiquark, which becomes an antired antiquark. The pion now consists of a red-antired quark-antiquark pair. The actual quantum state of the pion is an equal superposition of red-antired, green-antigreen, and blue-antiblue pairs.

TABLE 46–5

PROPERTIES OF QUARKS

SYMBOL	Q/e	SPIN	BARYON NUMBER, B	STRANGE-NESS, S	CHARM, C	BOTTOM-NESS, B'	TOPNESS, T
u	$\frac{2}{3}$	$\frac{1}{2}$	$\frac{1}{3}$	0	0	0	0
d	$-\frac{1}{3}$	$\frac{1}{2}$	$\frac{1}{3}$	0	0	0	0
s	$-\frac{1}{3}$	$\frac{1}{2}$	$\frac{1}{3}$	-1	0	0	0
c	$\frac{2}{3}$	$\frac{1}{2}$	$\frac{1}{3}$	0	$+1$	0	0
b	$-\frac{1}{3}$	$\frac{1}{2}$	$\frac{1}{3}$	0	0	$+1$	0
t	$\frac{2}{3}$	$\frac{1}{2}$	$\frac{1}{3}$	0	0	0	$+1$

In 1977 a meson with mass 9460 MeV/c^2, called upsilon (Υ) was discovered at Brookhaven. Because it had properties similar to ψ, it was conjectured that the meson was really the bound system of a new quark, b (the *bottom* quark), and its antiquark, $\bar{b}$. The bottom quark has the value 1 of a new quantum number B' (not to be confused with baryon number B) called *bottomness*. Excited states of the Υ were soon observed, and the B^+ ($\bar{b}u$) and B^0 ($\bar{b}d$) mesons are now well established.

With five flavors of quarks (u, d, s, c, b) and the six flavors of leptons (e, μ, τ, ν_e, ν_μ, and ν_τ) it was an appealing conjecture that nature is symmetric in its building blocks and that therefore there should be a *sixth* quark. This quark, labeled t (top), would have $Q/e = \frac{2}{3}$, $B = \frac{1}{3}$, and a new quantum number, $T = 1$. In 1995, groups using two different detectors at Fermilab's Tevatron announced the discovery of the top quark. The groups collided 0.9-TeV protons with 0.9-TeV antiprotons, but even with 1.8 TeV of available energy, a top-antitop ($t\bar{t}$) pair was detected in fewer than two of every 10^{11} collisions! Table 46–5 lists some properties of the six quarks. Each has a corresponding antiquark with opposite values of Q, B, S, C, B', and T.

46–6 THE STANDARD MODEL AND BEYOND

The particles and interactions that we've discussed in this chapter provide a reasonably comprehensive picture of the fundamental building blocks of nature. There is enough confidence in the basic correctness of this picture that it is called the **standard model.**

The standard model includes three families of particles: (1) the six leptons, which have no strong interactions; (2) the six quarks, from which all hadrons are made; and (3) the particles that mediate the various interactions. These mediators are gluons for the strong interaction among quarks, photons for the electromagnetic interaction, the $W^\pm$ and Z^0 particles for the weak interaction, and the graviton for the gravitational interaction.

ELECTROWEAK UNIFICATION

Theoretical physicists have long dreamed of combining all the interactions of nature into a single unified theory. As a first step, Einstein spent much of his later life trying to develop a field theory that would unify gravitation and electromagnetism. He was only partly successful.

Between 1961 and 1967, Sheldon Glashow, Abdus Salam, and Steven Weinberg developed a theory that unifies the weak and electromagnetic forces. One outcome of their **electroweak theory** is a prediction of the weak-force mediator particles, the Z^0 and $W^\pm$ bosons, including their masses. The basic idea is that the mass difference between photons (zero mass) and the weak bosons (≈ 100 GeV/c^2) makes the electromagnetic and weak interactions behave quite differently at low energies. At sufficiently high energies (well above 100 GeV), however, the distinction disappears, and the two merge into a single interaction. This prediction was verified experimentally in 1983 by two experimental groups working at the p$\overline{p}$ collider at CERN, as was mentioned in Section 46–4. The weak bosons were found, again with the help provided by the theoretical description, and their observed masses agreed with the predictions of the electroweak theory, a wonderful convergence of theory and experiment. The electroweak theory and quantum chromodynamics form the backbone of the standard model. Glashow, Salam, and Weinberg received the Nobel Prize in 1979.

A remaining difficulty in the electroweak theory is that photons are massless but the weak bosons are very massive. To account for the broken symmetry among these interaction mediators, a particle called the Higgs boson has been proposed. Its mass is expected to be less than 1 TeV/c^2, but to produce it in the laboratory may require a much greater available energy. The search for the Higgs particle (or particles) will be part of the mission of the Large Hadron Collider at CERN.

GRAND UNIFIED THEORIES

Perhaps at sufficiently high energies the strong interaction and the electroweak interaction have a convergence similar to that between the electromagnetic and weak interactions. If so, they can be unified to give a comprehensive theory of strong, weak, and electromagnetic interactions. Such schemes, called **grand unified theories** (GUTs), lean heavily on symmetry considerations, and they are still speculative.

One interesting feature of some grand unified theories is that they predict the decay of the proton (in violation of conservation of baryon number), with an estimated lifetime of more than 10^{28} years. (For comparison the age of the universe is estimated to be of the order of 10^{10} years.) With a lifetime of 10^{28} years, six metric tons of protons would be expected to have only one decay per day, so huge amounts of material must be examined. Some of the neutrino detectors that we mentioned in Section 46–3 originally looked for, and failed to find, evidence of proton decay. Nevertheless, experimental work continues, with current estimates setting the proton lifetime well over 10^{30} years. Some GUTs also predict the existence of magnetic monopoles, which we mentioned in Chapter 28. At present there is no confirmed experimental evidence that magnetic monopoles exist, but the search goes on.

In the standard model, the neutrinos have zero mass. Nonzero values are controversial because experiments to determine neutrino masses are difficult both to perform and to analyze. In most GUTs the neutrinos *must* have nonzero masses. If neutrinos do have mass, transitions called *neutrino oscillations* can occur, in which one type of neutrino (ν_e, ν_μ or ν_τ) changes into another type. In 1998, scientists using the Super-Kamiokande neutrino detector in Japan reported the discovery of oscillations between muon neutrinos and tau neutrinos. This discovery is the first solid evidence for exciting new physics beyond that contained in the standard model.

Neutrino oscillations may clear up a mystery about the flux of neutrinos from the sun. Since the 1960s, physicists have been using sensitive detectors to look for electron neutrinos produced as by-products of nuclear fusion reactions in the sun's core (see Section 45–9). While such *solar neutrinos* are detected, the observed flux is only a fraction of the predicted value. The resolution of this mystery may be that electron neutrinos transform in flight into muon or tau neutrinos. Since the first solar neutrino detectors were only sensitive to electron neutrinos, they could not have detected these transformed neutrinos. At this writing, a new generation of more sophisticated detectors is coming into operation. Data from these detectors may help to verify the reality of neutrino oscillations.

SUPERSYMMETRIC THEORIES AND TOES

The ultimate dream of theorists is to unify all four fundamental interactions, adding gravitation to the strong and electroweak interactions that are included in GUTs. Such a unified theory is whimsically called a Theory of Everything (TOE). One popular

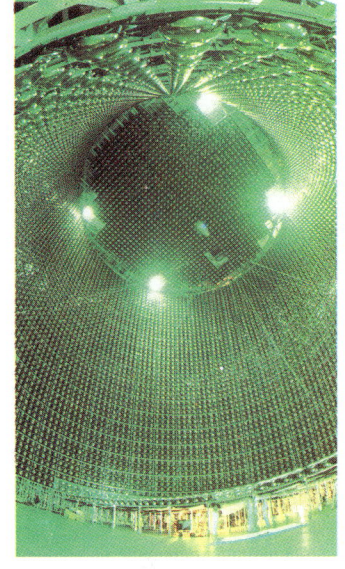

This photo shows the interior of the Super-Kamiokande neutrino detector in Japan. When in operation, the detector is filled with 10^8 kg of water. A neutrino entering the detector can produce a faint flash of light, which is detected by the 13,000 photomultiplier tubes lining the detector walls. Data from this detector were the first to indicate that neutrinos have mass.

ingredient of a TOE is a space-time continuum with more than four dimensions, containing structures called *strings*. Another element is *supersymmetry,* which gives every boson and fermion a "superpartner" of the other spin type. Such concepts lead to the prediction of whole new families of particles, including sleptons, photinos, and squarks bound together by gluinos. None of these new particles have been found, and theorists are still working toward a satisfactory TOE.

46-7 THE EXPANDING UNIVERSE

In the last two sections of this chapter we'll explore briefly the connections between the early history of the universe and the interactions of fundamental particles. It is remarkable that there are such close ties between physics on the smallest scale that we've explored experimentally (the range of the weak interaction, of the order of 10^{-18} m) and physics on the largest scale (the universe itself, of the order of at least 10^{26} m).

Gravitational interactions play an essential role in the large-scale behavior of the universe. One of the great achievements of Newtonian mechanics, including the law of gravitation, was the understanding it brought to the motion of planets in the solar system. Astronomical evidence, such as observations of the motions of binary stars around their common center of mass, shows that gravitational interactions also operate in larger astronomical systems, including stars, galaxies, and nebulae.

Until early in the twentieth century it was usually assumed that the universe was *static;* stars might move relative to each other, but there was not thought to be any overall expansion or contraction. But stars have gravitational attractions. If everything is initially sitting still in the universe, why doesn't gravity just pull it all together into one big clump? Newton himself recognized the seriousness of this troubling question.

Measurements that were begun in 1912 by Vesto Slipher at Lowell Observatory in Arizona, and continued in the 1920s by Edwin Hubble with the help of Milton Humason at Mount Wilson in California, indicated that the universe is *not* static. The motions of galaxies relative to the earth can be measured by observing the shifts in the wavelengths of their spectra. For distant galaxies these shifts are always toward longer wavelength, so they appear to be receding from us and from each other. Astronomers first assumed that these were Doppler shifts and used a relation between the wavelength λ_0 of light measured now from a source receding at speed v and the wavelength λ_S measured in the rest frame of the source when it was emitted. We can derive this relation by inverting Eq. (39–26) for the Doppler effect, making subscript changes, and using $\lambda = c/f$; the result is

$$\lambda_0 = \lambda_S \sqrt{\frac{c + v}{c - v}}. \tag{46–13}$$

Wavelengths from receding sources are always shifted toward longer wavelengths; this increase in λ is called the **redshift.** We can solve Eq. (46–13) for v. We leave the details for a problem (Exercise 46–28); the result is

$$v = \frac{(\lambda_0/\lambda_S)^2 - 1}{(\lambda_0/\lambda_S)^2 + 1} c. \tag{46–14}$$

CAUTION ▶ We want to emphasize that Eqs. (46–13) and (46–14) are from the *special* theory of relativity and are for the Doppler effect. As we'll see, the redshift from *distant* galaxies is caused by an effect that is explained by the *general* theory of relativity and is *not* a Doppler shift. However, as the ratio v/c and the fractional wavelength change $(\lambda_0 - \lambda_S)/\lambda_S$ become small, the general theory's equations approach Eqs. (46–13) and (46–14), and those equations may be used. ◀

EXAMPLE 46-9

Recession speed of a galaxy The spectral lines of various elements are detected in light from a galaxy in the constellation Ursa Major. An ultraviolet line from singly ionized calcium (λ_s = 393 nm) is observed with a wavelength λ = 414 nm, redshifted into the visible portion of the spectrum. At what speed is this galaxy receding from the earth?

SOLUTION First we calculate λ_0/λ_s = (414 nm)/(393 nm) = 1.053. This is only a 5.3% increase, so we can use Eq. (46–14) with reasonable accuracy:

$$v = \frac{(1.053)^2 - 1}{(1.053)^2 + 1} c = 0.0516c = 1.55 \times 10^7 \text{ m/s}.$$

The galaxy is receding from the earth at about one nineteenth the speed of light.

HUBBLE'S LAW

Analysis of redshifts from many distant galaxies led Edwin Hubble to a remarkable conclusion: The speed of recession v of a galaxy is proportional to its distance r from us (Fig. 46–14). This relation is now called **Hubble's law;** its symbolic statement is

$$v = H_0 r, \tag{46–15}$$

where H_0 is an experimental quantity commonly called the *Hubble constant,* since at any given time it is constant over space. Determining H_0 has been a key goal of the Hubble Space Telescope, which can measure distances to galaxies with unprecedented accuracy. The current best value is $2.1 \times 10^{-18}\text{s}^{-1}$, with an uncertainty of 10%.

Astronomical distances are often measured in *light years* (ly). One light year is the distance light travels in one year,

$$1 \text{ ly} = 9.46 \times 10^{15} \text{ m}.$$

The Hubble constant may be expressed without exponents in the mixed units (km/s)/Mly, where 1 Mly = 10^6 ly:

$$H_0 = 20 \frac{\text{km/s}}{\text{Mly}}.$$

46–14 Graph of recession speed as a function of distance for several galaxies. The best-fit straight line illustrates Hubble's law. The slope of the line is the Hubble constant, H_0.

EXAMPLE 46-10

Find the distance of the galaxy in Example 46–9 from the earth, according to Hubble's law.

SOLUTION From Eq. (46–15),

$$r = \frac{v}{H_0} = \frac{1.55 \times 10^7 \text{ m/s}}{20 \times 10^3 \text{(m/s)/Mly}}$$

$$= 7.8 \times 10^2 \text{ Mly} = 6.4 \times 10^{24} \text{ m}.$$

A distance of 780 million light years is truly stupendous, but many galaxies lie much farther away. In contrast, the closest known galaxy to ours is the Sagittarius Dwarf, only 80,000 light years from us—so close that it is held in gravitational orbit around our galaxy. To appreciate the immensity of even this relatively small distance, consider that our farthest-ranging unmanned spacecraft have traveled only about 0.001 ly from our planet.

Another aspect of Hubble's observations was that, *in all directions,* distant galaxies appeared to be receding from us. There is no particular reason to think that our galaxy is at the very center of the universe; if we lived in some other galaxy, every distant galaxy would still seem to be moving away. That is, at any give time, *the universe looks more or less the same, no matter where in the universe we are.* This important idea is called the **cosmological principle.** There are local fluctuations in density, but on average, the universe looks the same from all locations. Thus the Hubble constant is constant in space although not necessarily constant in time, and the laws of physics are the same everywhere.

THE BIG BANG

An appealing hypothesis that is suggested by Hubble's law is that at some time in the past, all the matter in the universe was far more concentrated than it is today. It was then blown apart in an immense explosion called the **Big Bang,** giving all observable matter more or less the velocities that we observe today. When did this happen? According to Hubble's law, matter at a distance r away from us is traveling with speed $v = H_0 r$. The time t needed to travel a distance r is

$$t = \frac{r}{v} = \frac{r}{H_0 r} = \frac{1}{H_0} = 4.8 \times 10^{17} \text{ s} = 1.5 \times 10^{10} \text{ y.}$$

By this hypothesis the Big Bang occurred about 15 billion years ago. It assumes that all speeds are *constant* after the Big Bang, that is, it neglects any slowing down due to gravitational attraction. Whether there is appreciable slowing depends on the average density of matter. We'll return to this point later.

EXPANDING SPACE

The general theory of relativity takes a radically different view of the expansion just described. According to this theory, the increased wavelength is *not* caused by a Doppler shift as the universe expands into a previously empty void. Rather, the increase comes from *the expansion of space itself* and everything in intergalactic space, including the wavelengths of light traveling to us from distant sources. This is not an easy concept to grasp, and if you haven't encountered it before, it may sound like doubletalk.

Here's an analogy that may help to develop some intuition on this point. Imagine we are all bugs crawling around on a horizontal surface. We can't leave the surface, and we can see in any direction along the surface, but not up or down. We are then living in a two-dimensional world; some writers have called such a world *flatland*. If the surface is a plane, we can locate our position with two Cartesian coordinates (x, y). If the plane extends indefinitely in both the x- and y-directions we described our space as having *infinite* extent, or as being *unbounded*. No matter how far we go, we never reach an edge or a boundary.

An alternative habitat for us bugs would be the surface of a sphere with radius R. The space would still seem infinite in the sense that we could crawl forever and never reach an edge or a boundary. Yet in this case the space is *finite* or *bounded*. To describe the location of a point in this space, we could still use two coordinates: latitude and longitude, or the spherical coordinates θ and ϕ shown in Fig. 42–22.

Now suppose the spherical surface is that of a balloon (Fig. 46–15). As we inflate the balloon more and more, increasing the radius R, the coordinates of a point don't change, yet the distance between any two points gets larger and larger. Furthermore, as R increases, the *rate of change* of distance between two points (their recession speed) is proportional to their distance apart. *The recession speed is proportional to the distance,* just as with Hubble's law. For example, the distance from Pittsburgh to Miami is twice as great as the distance from Pittsburgh to Boston. If the earth were to begin to swell, Miami would recede from Pittsburgh twice as fast as Boston would.

We see that although the quantity R isn't one of the two coordinates giving the position of a point on the balloon's surface, it nevertheless plays an essential role in any discussion of distance. It is the radius of curvature of our two-dimensional space, and it is also a varying *scale factor* that changes as this two-dimensional universe expands.

Generalizing this picture to three dimensions isn't so easy. We have to think of our three-dimensional space as being embedded in a space with four or more dimensions, just as we visualized the two-dimensional spherical flatland as being embedded in a three-dimensional Cartesian space. Our real three-space is *not Cartesian;* to describe its characteristics in any small region requires at least one additional parameter, the curvature of space, which is analogous to the radius of the sphere. In a sense, this scale factor, which we'll continue to call R, describes the *size* of the universe, just as the radius of the sphere described the size of our two-dimensional spherical universe. Whether the real universe is *finite* is open to conjecture; we'll return to this question later.

(a)

(b)

46–15 (a) Points (representing positions of astronomical objects) on the surface of a spherical balloon are described by their latitude and longitude coordinates. (b) The radius R of the balloon has increased. The coordinates of the points are the same, but the distance between them has increased. The rate of recession for any two points is proportional to the distance between them.

Any length that is measured in intergalactic space is proportional to R, so the wavelength of light traveling to us from a distant galaxy increases along with every other dimension as the universe expands. That is,

$$\frac{\lambda_0}{\lambda} = \frac{R_0}{R}. \tag{46-16}$$

The zero subscripts refer to the values of the wavelength and scale factor *now*, just as H_0 is the current value of the Hubble constant. The quantities λ and R without subscripts are the values at *any* time, past, present, or future. In the situation described in Example 46-9, we have $\lambda_0 = 414$ nm and $\lambda = \lambda_S = 393$ nm, so Eq. (46-16) gives $R_0/R = 1.053$. That is, the scale factor *now* (R_0) is 5.3% larger than it was 780 million years ago when the light was emitted from that galaxy in Ursa Major. This increase of wavelength with time as the scale factor increases in our expanding universe is called the *cosmological redshift*. The farther away an object is, the longer its light takes to get to us and the greater the change in R and λ. The current largest measured wavelength ratio is about 6, meaning that the volume of space itself is about $6^3 \approx 200$ times larger than it was when the light was emitted. Do *not* attempt to substitute $\lambda_0/\lambda_S = 6$ into Eq. (46-14) to find the recession speed; that equation is accurate only for small cosmological redshifts and $v \ll c$. Currently, the actual value v can only be estimated because it depends on the density of the universe, the value of H_0, and the specific version of the general theory of relativity.

Here's a surprise for you: If the distance from us in Hubble's law is large enough, then the speed of recession will be greater than the speed of light! This does *not* violate the special theory of relativity because the recession speed is *not* caused by the motion of the astronomical object relative to some coordinates in its region of space. Rather, we can have $v > c$ when two sets of coordinates move apart fast enough as space itself expands. In other words, there are objects whose coordinates have been moving away from our coordinates so fast that light from them hasn't had enough time in the entire history of the universe to reach us. What we see is just the *observable* universe; we have no direct evidence about what lies beyond its horizon.

CAUTION ▶ We have spoken of the expansion of space with time, but in an attempt to find the real meaning of the Big Bang, we can also extrapolate backward to the initial time when all of space was concentrated in a very small region, or perhaps a single point. This region (or point) contained the entire universe at that time. It's important to understand that the Big Bang was not an expansion *in* space; it was an expansion *of* space itself. As space expanded, objects in intergalactic space expanded along with it. ◀

CRITICAL DENSITY

We've mentioned that the law of gravitation isn't consistent with a static universe. We need to look at the role of gravity in an *expanding* universe. Gravitational attractions should slow the initial expansion, but by how much? If these attractions are strong enough, the universe should expand more and more slowly, eventually stop, and then begin to contract, perhaps all the way down to what's been called a *Big Crunch*. The universe might then rebound with a *Big Bounce* to start all over again. Some cosmological theories picture the universe this way, as a series of cataclysmic pulsations with a period of perhaps 10^{11} years. On the other hand, if gravitational forces are much weaker, they slow the expansion only a little, and the universe continues to expand forever.

The situation is analogous to the problem of escape velocity of a projectile launched from the earth; we studied this problem in Section 12-4, and you may want to review that discussion. The total energy $E = K + U$ when a projectile with speed v is at a distance r from the center of the earth is

$$E = \frac{1}{2}mv^2 - \frac{Gmm_E}{r}.$$

If E is positive, the projectile has enough kinetic energy to move infinitely far from the earth ($r \rightarrow \infty$) and have some kinetic energy left over. If E is negative, the kinetic energy $K = \frac{1}{2}mv^2$ becomes zero and the projectile stops when $r = -Gmm_E/E$. In that case, no greater value of r is possible, and the projectile can't escape the earth's gravity.

We can carry out a similar analysis for the universe. Whether the universe continues to expand indefinitely depends on the average *density* of matter. If matter is relatively dense, there is a lot of gravitational attraction to slow and eventually stop the expansion and make the universe contract again. If not, the expansion continues indefinitely. We can derive an expression for the *critical density* ρ_c needed to just barely stop the expansion.

Here's a calculation based on Newtonian mechanics; it isn't relativistically correct, but it illustrates the idea. Consider a large sphere with radius R, containing many galaxies (Fig. 46–16), with total mass M. Suppose our own galaxy has mass m and is located at the surface of this sphere. According to the cosmological principle, the average distribution of matter within the sphere is spherically symmetric. The total gravitational force on our galaxy is just the force due to the mass M inside the sphere. The force on our galaxy and potential energy U due to this spherically symmetric distribution are the same as though m and M were both points, so $U = -GmM/R$, just as in Section 12–4. The net force from all the spherically symmetric distribution of mass *outside* the sphere is zero, so we'll ignore it.

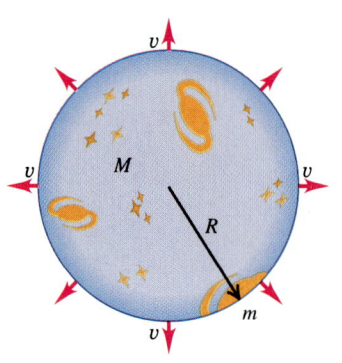

The total energy E (kinetic plus potential) for our galaxy is

$$E = \frac{1}{2}mv^2 - \frac{GmM}{R}. \qquad (46\text{–}17)$$

46–16 The gravitational force on our galaxy (mass m) is the force exerted by the mass M within the sphere with radius R as if it were all concentrated at the center. Because of the spherical symmetry, the mass outside this sphere exerts no net force on our galaxy.

If E is *positive,* our galaxy has enough energy to escape from the gravitational attraction of the mass M inside the sphere; in this case the universe keeps expanding forever. If E is negative, our galaxy cannot escape and the universe is eventually pulled back together. The crossover between these two cases occurs when $E = 0$, so that

$$\frac{1}{2}mv^2 = \frac{GmM}{R}. \qquad (46\text{–}18)$$

The total mass M inside the sphere is the volume $4\pi R^3/3$ times the density ρ_c:

$$M = \frac{4}{3}\pi R^3 \rho_c.$$

We'll assume that the speed v of our galaxy relative to the center of the sphere is given by Hubble's law: $v = H_0 R$. Substituting these expressions for M and v into Eq. (46–18), we get

$$\frac{1}{2}m(H_0 R)^2 = \frac{Gm}{R}\left(\frac{4}{3}\pi R^3 \rho_c\right),$$

or

$$\rho_c = \frac{3H_0{}^2}{8\pi G} \qquad \text{(critical density of the universe).} \qquad (46\text{–}19)$$

This is the *critical density.* If the average density is less than ρ_c, the universe will continue to expand indefinitely; if it is greater, the universe will eventually stop expanding and begin to contract, possibly leading to the Big Crunch and then another Big Bang.

Putting numbers into Eq. (46–19), we find

$$\rho_c = \frac{3(2.1 \times 10^{-18}\text{ s}^{-1})^2}{8\pi(6.67 \times 10^{-11}\text{ kg} \cdot \text{m}^2/\text{s}^2)} = 7.9 \times 10^{-27}\text{ kg/m}^3.$$

The mass of a hydrogen atom is 1.67×10^{-27} kg, so this corresponds to about five hydrogen atoms per cubic meter.

DARK MATTER

Attempts have been made to estimate the actual average density of matter in the universe. We won't attempt to discuss the details; the estimated total of the mass of *luminous matter* (stars and such) and the mass equivalent of radiation energy gives a density of the order of $0.05\rho_c$. Some theorists believe that the universe must be closed and that the average density must be *equal* to ρ_c; then the expansion would approach zero after a long time. But for this to happen, there would have to be a large amount of unseen *dark matter* in the universe. The nature of this missing matter is at present a mystery. Massive neutrinos are one possibility, black holes are another, and WIMPs (weakly interacting massive particles) are a third. The presence of dark matter is also suggested by measurements of the Doppler shift from opposite sides of rotating galaxies. As much as ten times more mass than is visible is needed to provide the gravitational force to hold the galaxies together. Thus it is a fair statement that the nature of the majority of the matter in our universe remains a mystery to science.

46–8 THE BEGINNING OF TIME

What an odd title for the very last section of a book! We will describe in general terms some of the current theories about the very early history of the universe and their relation to fundamental particle interactions. We'll find that an astonishing amount happened in the very first second. A lot of loose ends will be left untied, and many questions will be left unanswered. This is, after all, one of the frontiers of theoretical physics, and continually brings new theories, new discoveries, and new questions.

TEMPERATURES

The early universe was extremely dense and extremely hot, and the average particle energies were extremely large, all many orders of magnitude beyond anything that exists in the present universe. We can compare particle energy E and absolute temperature T using the equipartition principle

$$E = \frac{3}{2}kT, \tag{46–20}$$

where k is Boltzmann's constant, which we'll often express in eV/K:

$$k = 8.617 \times 10^{-5} \text{ eV/K}.$$

Thus we can replace Eq. (46–20) by $E \approx (10^{-4} \text{ eV/K})T = (10^{-13} \text{ GeV/K})T$ when we're discussing orders of magnitude.

EXAMPLE 46–11

Temperature and energy a) What is the average kinetic energy in electron volts of particles at room temperature ($T = 290$ K) and at the surface of the sun ($T = 5800$ K)? b) What approximate temperature corresponds to the ionization energy of the hydrogen atom, to the rest energy of the electron, and to the rest energy of the proton?

SOLUTION a) From Eq. (46–20),

$$E = \frac{3}{2}kT = \frac{3}{2}(8.617 \times 10^{-5} \text{ eV/K})(290 \text{ K})$$

$$= 0.0375 \text{ eV}.$$

The temperature at the sun's surface is larger by a factor of

(5800 K)/(290 K) = 20, so the average kinetic energy there is 20(0.0375 eV) = 0.75 eV.

b) The ionization energy of hydrogen is 13.6 eV. Using the approximation $E \approx (10^{-4} \text{ eV/K})T$, we have

$$T \approx \frac{E}{10^{-4} \text{ eV/K}} = \frac{13.6 \text{ eV}}{10^{-4} \text{ eV/K}}$$

$$\approx 10^5 \text{ K}.$$

The rest energies of the electron and proton are 0.511 MeV and 938 MeV, respectively. Repeating the calculation for these values gives temperatures of 10^{10} K, corresponding to the electron rest energy, and 10^{13} K, corresponding to the proton rest energy.

UNCOUPLING OF INTERACTIONS

The evolution of the universe has been characterized by a continual increase of the scale factor R, which we can think of very roughly as characterizing the *size* of the universe, and by a corresponding decrease in average density. As the total gravitational potential energy increased during expansion, there were corresponding *decreases* in temperature and average particle energy. As this happened, the basic interactions became progressively uncoupled.

To understand the uncouplings, recall that the unification of the electromagnetic and weak interactions occurs at energies that are large enough that the differences in mass among the various spin-1 bosons that mediate the interactions become insignificant by comparison. The electromagnetic interaction is mediated by the massless photon, and the weak interaction is mediated by the weak bosons $W^\pm$ and Z^0 with masses of the order of 100 GeV/c^2. At energies much *less* than 100 GeV the two interactions seem quite different, but at energies much *greater* than 100 GeV they become part of a single interaction.

The grand unified theories (GUTs) provide a similar behavior for the strong interaction. It becomes unified with the electroweak interaction at energies of the order of 10^{14} GeV, but at lower energies the two appear quite distinct. One of the reasons GUTs are still very speculative is that there is no way to do controlled experiments in this energy range, which is larger by a factor of 10^{11} than energies available with any current accelerator.

Finally, at sufficiently high energies and short distances, it is assumed that gravitation becomes unified with the other three interactions. The distance at which this happens is thought to be of the order of 10^{-35} m. This distance, called the *Planck length* l_P, is determined by the speed of light c and the fundamental constants of quantum mechanics and gravitation, h and G, respectively. The Planck length l_P is defined as

$$l_P = \sqrt{\frac{\hbar G}{c^3}} = 1.616 \times 10^{-35} \text{ m.} \tag{46–21}$$

We invite you to verify that this combination of constants does indeed have units of length. The *Planck time* $t_P = l_P/c$ is the time required for light to travel a distance l_P:

$$t_P = \frac{l_P}{c} = \sqrt{\frac{\hbar G}{c^5}} = 0.539 \times 10^{-43} \text{ s.} \tag{46–22}$$

If we mentally go backward in time, we have to stop when we reach $t = 10^{-43}$ s because we have no adequate theory that unifies all four interactions. So as yet we have no way of knowing what happened or how the universe behaved at times earlier than the Planck time or when its size was less than the Planck length. In fact, because time is not an absolute quantity, we can't even say whether there *was* time as we know it before 10^{-43} s.

THE STANDARD MODEL OF THE HISTORY OF THE UNIVERSE

The description that follows is called the *standard model* of the history of the universe. The title may be slightly optimistic, but it does indicate that there are substantial areas of theory that rest on solid experimental foundations and are quite generally accepted. The figure on pages 1452–1453 is a graphical description of this history, with the characteristic sizes, particle energies, and temperatures at various times. Referring to this chart frequently will help you to understand the following discussion.

In this standard model, the temperature of the universe at time $t = 10^{-43}$ s (the Planck time) was about 10^{32} K, and the average energy per particle was approximately

$$E \approx (10^{-13} \text{ GeV/K})(10^{32} \text{ K}) = 10^{19} \text{ GeV.}$$

In a totally unified theory this is about the energy below which gravity begins to behave as a separate interaction. This time therefore marked the transition from any proposed TOE to the GUT period.

During the GUT period, roughly $t = 10^{-43}$ to 10^{-35} s, the strong and electroweak forces were still unified, and the universe consisted of a soup of quarks and leptons transforming into each other so freely that there was no distinction between the two families of particles. Other, much more massive particles may also have been freely created and destroyed. One important characteristic of GUTs is that at sufficiently high energies, baryon number is not conserved. (We mentioned earlier the proposed decay of the proton, which has not yet been observed.) Thus by the end of the GUT period the numbers of quarks and antiquarks may have been unequal. This point has important implications; we'll return to it at the end of the section.

By $t = 10^{-35}$ s the temperature had decreased to about 10^{27} K and the average energy to about 10^{14} GeV. At this energy the strong force separated from the electroweak force (Fig. 46–17), and baryon number and lepton numbers began to be separately conserved. In some models, called *inflationary models,* this separation of the strong force was analogous to a *phase change,* such as boiling of a liquid, with an associated heat of vaporization. Think of it as being similar to boiling a heavy nucleus, pulling the particles apart beyond the short range of the nuclear force. As a result, the inflationary models predict that there was a very rapid expansion. In one model, the scale factor R increased by a factor of 10^{50} in 10^{-32} s.

At $t = 10^{-32}$ s the universe was a mixture of quarks, leptons, and the mediating bosons (gluons, photons, and the weak bosons $W^{\pm}$ and Z^0). It continued to expand and cool from the inflationary period to $t = 10^{-6}$ s, when the temperature was about 10^{13} K and typical energies were about 1 GeV (comparable to the rest energy of a nucleon). At this time the quarks began to bind together to form nucleons and antinucleons. Also there were still enough photons of sufficient energy to produce nucleon-antinucleon pairs to balance the process of nucleon-antinucleon annihilation. However, by about $t = 10^{-2}$ s, most photon energies fell well below the threshold energy for such pair production. There was a slight excess of nucleons over antinucleons; as a result, virtually all of the antinucleons and most of the nucleons annihilated one another. A similar equilibrium occurred later between the production of electron-positron pairs from photons and the annihilation of such pairs. At about $t = 14$ s the average energy dropped to around 1 MeV, below the threshold for e^+e^- pair production. After pair production ceased, virtually all of the

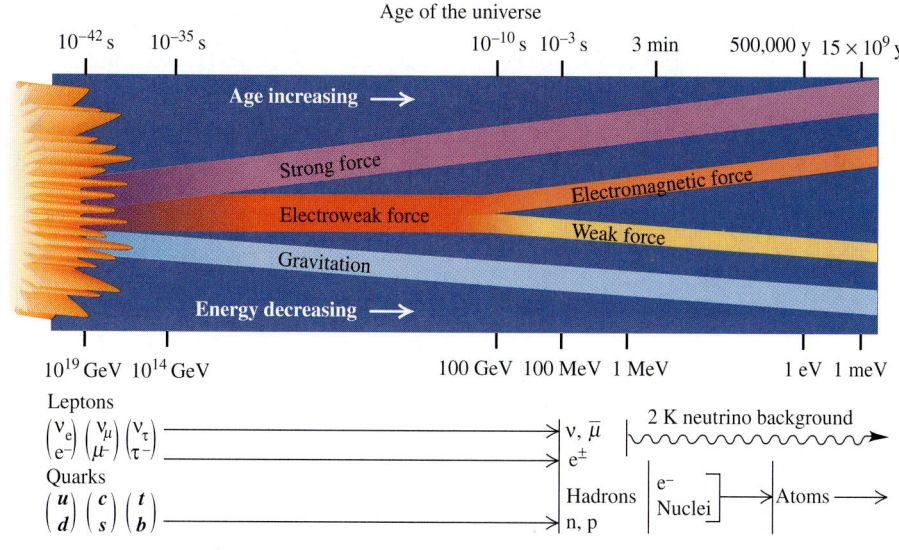

46–17 Schematic diagram showing the times and energies at which the various interactions became uncoupled. The energy scale is backward because the average energy decreased as the age of the universe increased.

AGE OF QUARKS AND GLUONS (GUT Period)
Dense concentration of matter and antimatter; gravity a separate force; more quarks than antiquarks. Inflationary period (10^{-35}s): rapid expansion, strong force separates from electroweak force.

AGE OF NUCLEONS AND ANTINUCLEONS
Quarks bind together to form nucleons and antinucleons; energy too low for nucleon-antinucleon pair production at 10^{-2}s

AGE OF NUCLEOSYNTHESIS
Stable deuterons; matter 74% H, 25% He, 1% heavier nuclei

AGE OF LEPTONS
Leptons distinct from quarks; $W^{\pm}$ and Z^0 bosons mediate weak force (10^{-12}s)

BIG BANG 10^{-43}s 10^{-32}s 10^{-6}s   225 s 10^3s

Neutrino

Quarks

Antineutrino   Antiquarks

Neutron

Proton

Antineutron

Antiproton

γ

γ

γ

γ

n ^{2}H

p

^{4}He
^{3}H
e^-

← TOE →	← GUT →	Electroweak unification	Forces separate	Matter domination

| 10^{-42} s | 10^{-36} s | 10^{-30} s | 10^{-24} s | 10^{-18} s | 10^{-12} s | 10^{-6} s | 1 s | 10^6 s 1 y 10^3 y | 10^6 y | 10^9 y | t |

T

| 10^{30} K | | 10^{25} K | | 10^{20} K | | 10^{15} K | Nuclear binding energy | 10^{10} K | Atomic binding energy | 10^5 K | | Solar system forms | 1 K |

| 10^{18} GeV | 10^{15} GeV | 10^{12} GeV | 10^9 GeV | 10^6 GeV | 1 TeV | 1 GeV | 1 MeV | 1 keV | 1 eV | 1 meV | E |

Size

| 10^{-30} | | 10^{-25} | | 10^{-20} | | 10^{-15} | | 10^{-10} | | 10^{-5} | | 1 |

Nucleosynthesis

Logarithmic scales show characteristic temperature, energy, and size of the universe as functions of time.

A Brief History of the Universe

AGE OF IONS
Expanding, cooling
gas of ionized
H and He

AGE OF ATOMS
Neutral atoms form, pulled
together by gravity; universe becomes
transparent to most light

**AGE OF STARS
AND GALAXIES**
Thermonuclear fusion
begins in stars, forming
heavier nuclei

10^{13} s 10^{15} s

NOW

remaining positrons were annihilated, leaving the universe with many more protons and electrons than the antiparticles of each.

Up until about $t = 1$ s, neutrons and neutrinos could be produced in the endoergic reaction

$$e^- + p \rightarrow n + \nu_e.$$

After this time, most electrons no longer had enough energy for this reaction. The average neutrino energy also decreased, and as the universe expanded, equilibrium reactions that involved *absorption* of neutrinos (which occurred with decreasing probability) became inoperative. At this time, in effect, the flux of neutrinos and antineutrinos throughout the universe uncoupled from the rest of the universe. Because of the extraordinarily low probability for neutrino absorption, most of this flux is still present today, although cooled greatly by expansion. The standard model of the universe predicts a present neutrino temperature of about 2 K, but no one has been able to carry out an experiment to test this prediction.

NUCLEOSYNTHESIS

At about $t = 1$ s, the ratio of protons to neutrons was determined by the Boltzmann distribution factor $e^{-\Delta E/kT}$, where ΔE is the difference between the neutron and proton rest energies, $\Delta E = 1.294$ MeV. At a temperature of about 10^{10} K, this distribution factor gives about 4.5 times as many protons as neutrons. However, as we have discussed, free neutrons (with a lifetime of 887 s) decay spontaneously to protons. This decay caused the proton-neutron ratio to increase until about $t = 225$ s. At this time, the temperature was about 10^9 K, and the average energy was well below 2 MeV.

This energy distribution was critical because the binding energy of the *deuteron* (a neutron and a proton bound together) is 2.22 MeV (see Section 45–3). A neutron bound in a deuteron does not decay spontaneously. As the average energy decreased, a proton and a neutron could combine to form a deuteron, and there were fewer and fewer photons with 2.22 MeV or more of energy to dissociate the deuterons again. Therefore the combining of protons and neutrons into deuterons halted the decay of free neutrons.

The formation of deuterons starting at about $t = 225$ s marked the beginning of the period of formation of nuclei, or *nucleosynthesis*. At this time, there were about seven protons for each neutron. The deuteron (^{2}H) can absorb a neutron and form a triton (^{3}H), or it can absorb a proton and form ^{3}He. Then ^{3}H can absorb a proton, and ^{3}He can absorb a neutron, each yielding ^{4}He (the alpha particle). A few ^{7}Li nuclei may also have been formed by fusion of ^{3}H and ^{4}He nuclei. According to the theory, essentially all the ^{1}H and ^{4}He in the present universe was formed at this time. But then the building of nuclei almost ground to a halt. The reason is that *no* nuclide with mass number $A = 5$ has a half-life greater than 10^{-21} s. Alpha particles simply do not permanently absorb neutrons or protons. The nuclide ^{8}Be that is formed by fusion of two ^{4}He nuclei is unstable, with an extremely short half-life, about 7×10^{-17} s. Note also that at this time, the average energy was still much too large for electrons to be bound to nuclei; there were not yet any atoms.

EXAMPLE 46–12

The relative abundance of hydrogen and helium in the universe
Nearly all of the protons and neutrons in the seven-to-one ratio at $t = 225$ s either formed ^{4}He or remained as ^{1}H. After this time, what was the relative abundance of ^{1}H and ^{4}He by mass?

SOLUTION The ^{4}He nucleus contains two protons and two neutrons. For every two neutrons we have 14 protons. The two neutrons and two of the 14 protons make up one ^{4}He nucleus,

leaving 12 protons (^{1}H nuclei). So at this time where were 12 ^{1}H nuclei for every ^{4}He nucleus. The masses of ^{1}H and ^{4}He are about 1 u and 4 u, respectively, so there were 12 u of ^{1}H for every 4 u of ^{4}He. Therefore the mix, by mass, was 75% ^{1}H and 25% ^{4}He. This result agrees very well with estimates of the present H-He ratio in the universe, an important confirmation of this part of the theory.

Further nucleosynthesis did not occur until very much later, well after the time $t = 2 \times 10^{13}$ s (about 700,000 y). At that time, the temperature was about 3000 K, and the average energy was a few tenths of an electron volt. Because the ionization energies of hydrogen and helium *atoms* are 13.6 eV and 24.5 eV, respectively, almost all the hydrogen and helium was electrically neutral (not ionized). With the electrical repulsions of the nuclei canceled out, gravitational attraction could slowly pull the neutral atoms together to form clouds of gas and eventually stars. Thermonuclear reactions in stars are believed to have produced all of the more massive nuclei. In Section 45–9 we discussed one cycle of thermonuclear reactions in which ^{1}H becomes ^{4}He; this cycle is one of the sources of the energy radiated by stars.

As a star uses up its hydrogen, the inward gravitational pressure exceeds the outward radiation and gas pressure, and the star's core begins to contract. As it does so, the gravitational potential energy decreases, and the kinetic energies of the star's atoms increase. For stars with sufficient mass, there are both enough energy and sufficient density to begin another process, *helium fusion*. First two ^{4}He nuclei fuse to form ^{8}Be. The extremely short lifetime of this unstable nuclide is compensated by the density of the nuclear core and by an usually large probability for absorption of another ^{4}He nucleus with a specific energy, a kind of resonance effect. Thus a reasonable fraction of the ^{8}Be nuclei fuse with ^{4}He to form the stable nuclide ^{12}C. The net result is the fusion of three ^{4}He nuclei to form one ^{12}C, the *triple-alpha process*. Then successive fusions with ^{4}He give ^{16}O, ^{20}Ne, and ^{24}Mg. All these reactions are exoergic. They release energy to heat up the star, and ^{12}C and ^{16}O can fuse to form elements with higher and higher atomic number.

For nuclides that can be created in this manner, the binding energy per nucleon peaks at mass number $A = 56$ with the nuclide ^{56}Fe, so exoergic fusion reactions stop with Fe. But successive neutron captures followed by beta decays can continue the synthesis of more massive nuclei. If the star is massive enough, it may eventually explode as a *supernova* (Fig. 46–18), sending out into space the heavy elements that were produced by the earlier processes. In space, the debris and other interstellar matter can gravitationally bunch together to form new stars and planets. That is how the stuff of our earth was formed.

BACKGROUND RADIATION

In 1965, Arno Penzias and Robert Wilson, working at Bell Telephone Laboratories in New Jersey on satellite communications, turned a microwave antenna skyward and found a background signal that had no apparent preferred direction. (This signal gives

46–18 This Hubble Space Telescope image shows a ring of glowing gas 1.7 ly in diameter surrounding supernova SN 1987A. The light from this supernova first reached the earth in February 1987 after traveling across 170,000 ly of space. The glowing ring, shown here as it appeared three years after the supernova, is made of material ejected from the progenitor star before it exploded into a supernova. The intense radiation from the supernova ionized the gas in the ring, which subsequently emitted visible photons as electrons recombined with the ions.

about 1% of the "hash" you see on a TV screen when you turn to an unused channel.) Further research has shown that the radiation that is received has a frequency spectrum that fits Planck's blackbody radiation law, Eq. (40–30) (see Section 40–9). The wavelength of peak intensity is 1.063 mm (in the microwave region of the spectrum), with a corresponding absolute temperature $T = 2.728$ K. Penzias and Wilson contacted physicists at nearby Princeton University who had begun the design of an antenna to search for radiation that was a remnant from the early evolution of the universe. We mentioned above that neutral atoms began to form at about $t = 700,000$ years when the temperature was 3000 K. With far fewer charged particles present than previously, the universe became transparent at this time to electromagnetic radiation of long wavelength. The 3000-K blackbody radiation therefore survived, cooling to its present 2.728-K temperature as the universe expanded. The *microwave background radiation* is among the most clear-cut experimental confirmations of the Big Bang theory.

EXAMPLE 46–13

Expansion of the universe By approximately what factor has the universe expanded since $t = 700,000$ y?

SOLUTION We can rewrite Equation (40–28), the Wien displacement law, to show that the peak wavelength λ_m in blackbody radiation is inversely proportional to absolute temperature:

$$\lambda_m = \frac{2.90 \times 10^{-3} \text{ m} \cdot \text{K}}{T}.$$

As the universe expands, all intergalactic wavelengths (including λ_m) increase in proportion to the scale factor R. The temperature decreased by a factor of $(3000 \text{ K})/(2.7 \text{ K})$, so λ_m and the scale factor must have *increased* by this factor. Thus between $t = 700,000$ y and the present, the universe expanded by a factor of $(3000 \text{ K})/(2.7 \text{ K})$, or about 1100. Thus any particular intergalactic volume increased by a factor $(1100)^3 = 1.3 \times 10^9$.

MATTER AND ANTIMATTER

One of the most remarkable features of our universe is the asymmetry between matter and antimatter. One might think that the universe should have equal numbers of protons and antiprotons and of electrons and positrons, but this doesn't appear to be the case. There is no evidence for the existence of substantial amounts of antimatter (matter composed of antiprotons, antineutrons, and positrons) anywhere in the universe. Theories of the early universe must explain this imbalance.

We've mentioned that most GUTs include violation of conservation of baryon number at energies at which the strong and electroweak interactions have converged. If particle-antiparticle symmetry is also violated, we have a mechanism for making more quarks than antiquarks, more leptons than antileptons, and eventually more matter than antimatter. One serious problem is that any asymmetry that is created in this way during the GUT era might be wiped out by the electroweak interaction after the end of the GUT era. If so, there must be some mechanism that creates particle-antiparticle asymmetry at a much *later* time. The problem of the matter-antimatter asymmetry is still very much an open one.

We hope that this qualitative discussion has conveyed at least a hint of the close connections between particle theory and cosmology. There are still lots of unanswered questions. Is the universe open or closed? There's not enough visible matter in sight to stop its expansion, but perhaps there is enough dark matter. There are so many electron neutrinos in the universe (about 100 million per cubic meter) that even a 20-eV/c^2 neutrino mass would provide enough additional matter. The exotic particles that are postulated in connection with the unification theories (Section 46–6) may also be relevant for the dark matter problem.

Extrapolating back to the Big Bang, we have no idea what happened before 10^{-43} s because we have no quantum theory of gravity. The electroweak theory is on very firm ground, but there are many versions of grand unification theories from which to choose.

And we are still very far from having a suitable theory that unifies all four interactions in nature. But this search for understanding of our physical world continues to be one of the most exciting adventures of the human mind.

SUMMARY

- Each particle has an antiparticle; some particles are their own antiparticle. Particles can be created and destroyed, some of them (including electrons and positrons) only in pairs or in conjunction with other particles and antiparticles.

- Particles serve as mediators for the fundamental interactions. The photon is the mediator of the electromagnetic interaction. Yukawa proposed the existence of mesons to mediate the nuclear interaction. Mediating particles that can exist only because of the uncertainty principle for energy are called virtual particles.

- Cyclotrons, synchrotrons, and linear accelerators are used to accelerate charged particles to high energies for experiments with particle interactions. Only part of the beam energy is available to cause reactions with targets at rest. This problem is avoided in colliding-beam experiments.

- Four fundamental interactions are found in nature; they are the strong, electromagnetic, weak, and gravitational interactions. Particles can be described in terms of their interactions and of quantities that are conserved in all or some of the interactions.

- Fermions have half-integer spins; bosons have integer spins. Leptons have no strong interactions. Strongly interacting particles are called hadrons. They include mesons, which are also bosons, and baryons, which are always fermions. There are conservation laws for three different lepton numbers and for baryon number.

- Additional quantum numbers, including strangeness, are conserved in some interactions and not in others.

- Hadrons are composed of quarks; mesons of quark-antiquark combinations and baryons of three quarks. There are six flavors of quarks. The interaction between quarks is mediated by gluons. Quarks and gluons have an additional attribute called color.

- Symmetry considerations play an indispensable role in all fundamental-particle theories. The electromagnetic and weak interactions become unified at high energies into the electroweak interaction. In grand unified theories the strong interaction is also unified with these, but at much higher energies.

- According to the Big Bang model, the universe was originally contained in a very small space, possibly a single point. The expansion of space itself is described in terms of temperature, average particle energies, and distances. Whether or not space will continue to expand forever depends on the average density of the universe compared to a value called the critical density.

- The standard model of the universe shows the close connections between particle interactions and the evolution of the universe.

KEY TERMS

antiparticle, 1424
meson, 1426
muon, 1426
pion, 1426
fermion, 1434
boson, 1434
lepton, 1435
hadron, 1436
baryon, 1436
strangeness, 1437
quark, 1438
eightfold way, 1440
gluon, 1441
charm, 1441
standard model, 1442
electroweak theory, 1442
grand unified theory, 1443
redshift, 1444
Hubble's law, 1445
cosmological principle, 1445
Big Bang, 1445

DISCUSSION QUESTIONS

Q46–1 Is it possible that some parts of the universe contain antimatter whose atoms have nuclei made of antiprotons and antineutrons, surrounded by positrons? How could we detect this condition without actually going there? Can we detect these antiatoms by identifying the light they emit as composed of antiphotons? Explain. What problems might arise if we actually *did* go there?

Q46–2 What are the similarities between a neutrino and a photon? What are the differences?

Q46–3 When they were first discovered during the 1930s and 1940s, there was confusion as to the identities of the pions and the muons. In retrospect, what are the similarities? What are the most significant differences?

Q46–4 The gravitational force between two electrons is weaker than the electrical force by the order of 10^{-40}. Yet the gravitational interactions of matter were observed and analyzed long before electrical interactions were understood. Why?

Q46–5 When a π^0 decays to two photons, what happens to the quarks of which it was made?

Q46–6 Why can't an electron decay into two photons? Into two neutrinos?

Q46–7 According to the standard model of the fundamental particles, what are the similarities between baryons and leptons? What are the most important differences?

Q46–8 According to the standard model of the fundamental particles, what are the similarities between quarks and leptons? What are the most important differences?

Q46–9 What are the main advantages of colliding-beam accelerators compared with those using stationary targets? What are the main disadvantages?

Q46–10 Does the weak force act on hadrons, or do they experience only the strong force? Explain.

Q46–11 In what ways is the inflationary period in the evolution of the universe analogous to a phase transition? How is it analogous to the phenomenon of superheating, the heating of a liquid above its normal boiling temperature, as occurs in bubble chambers?

Q46–12 Does the universe have a center? Explain.

Q46–13 Assume that the universe has an edge. Placing yourself at that edge in a thought experiment, explain why this assumption violates the cosmological principle.

Q46–14 Explain why the cosmological principle requires that H_0 must have the same value everywhere in space, but does not require that it be constant in time.

EXERCISES

SECTION 46–2 FUNDAMENTAL PARTICLES—A HISTORY

46–1 A neutral pion at rest decays into two photons. Find the approximate energy, frequency, and wavelength of each photon. In which part of the electromagnetic spectrum does each photon lie? (Use the pion mass given in terms of the electron mass in Section 46–2.)

46–2 Two equal-energy photons collide head-on and annihilate one another, producing a $\mu^+\mu^-$ pair. The muon mass is given in terms of the electron mass in Section 46–2. a) Calculate the maximum wavelength of the photons for this to occur. If the photons have this wavelength, describe the motion of the μ^+ and μ^- immediately after they are produced. b) What happens if the photons have shorter wavelength than this maximum but still annihilate and produce a $\mu^+\mu^-$ pair?

46–3 A positive pion at rest decays into a positive muon and a neutrino. a) Approximately how much energy is released in the decay? (Assume the neutrino has zero rest mass. Use the muon and pion masses given in terms of the electron mass given in Section 46–2.) b) Why can't a positive muon decay into a positive pion?

46–4 a) A proton and an antiproton annihilate, producing two photons. Find the energy, frequency, and wavelength of each photon emitted in the center-of-momentum reference frame a) if the initial kinetic energies of the proton and antiproton are negligible; b) if each particle has an initial kinetic energy of 830 MeV.

46–5 "**Maximum Power, Scotty!**" The starship *Enterprise,* of television and movie fame, is powered by the controlled combination of matter and antimatter. If the entire 400-kg antimatter fuel supply of the *Enterprise* combines with matter, how much energy is released?

SECTION 46–3 PARTICLE ACCELERATORS AND DETECTORS

46–6 In the head-on collision of an electron and a positron of equal energy, the total available energy is 100 GeV. a) What is the total energy of the electron or positron? b) If an electron with

the total energy calculated in part (a) is incident on a stationary positron, what is the available energy?

46–7 Deuterons in a cyclotron travel in a circle with radius 32.0 cm just before emerging from the dees. The frequency of the applied alternating voltage is 9.00 MHz. Find a) the magnetic field; b) the kinetic energy and speed of the deuterons upon emergence.

46–8 The magnetic field in a cyclotron that accelerates protons is 1.30 T. a) How many times per second should the potential across the dees reverse? (This is twice the frequency of the circulating protons.) b) The maximum radius of the cyclotron is 0.250 m. What is the maximum speed of the proton? c) Through what potential difference would the proton have to be accelerated from rest to give it the same speed as calculated in part (b)?

46–9 a) A high-energy beam of alpha particles collides with a stationary helium gas target. What must be the total energy of a beam particle if the available energy in the collision is 16.0 GeV? b) What must be the energy of each beam to produce the same available energy in a colliding-beam experiment?

46–10 a) What is the speed of a proton that has total energy 1000 GeV? b) What is the angular frequency ω of a proton with the speed calculated in part (a) in a magnetic field of 4.00 T? Use both the nonrelativistic Eq. (46–7) and the correct relativistic expression, and compare the results.

46–11 In Example 46–4 (Section 46–3) it was shown that a proton beam with an 800-GeV beam energy gives an available energy of 38.7 GeV for collisions with a stationary proton target. a) You are asked to design an upgrade of the accelerator that will double the available energy in stationary-target collisions. What beam energy is required? b) In a colliding-beam experiment, what total energy of each beam is needed to give an available energy of 2(38.7 GeV) = 77.4 GeV?

46–12 Calculate the threshold kinetic energy for the production of a η^0 particle in the collision of a proton beam with a stationary proton target: $p + p \rightarrow p + p + \eta^0$. The rest energy of the η^0 is 547.3 MeV (Table 46–3).

SECTION 46–4 PARTICLES AND INTERACTIONS

46–13 If a Σ^+ at rest decays into a proton and a π^0, what is the total kinetic energy of the decay products?

46–14 If a muon at rest decays into an electron and two neutrinos, what is the total kinetic energy of the decay products? Assume that the neutrinos have zero rest mass.

46–15 In which of the following decays are the three lepton numbers conserved? In each case, explain your reasoning.

a) $\mu^- \rightarrow e^- + \nu_e + \overline{\nu}_\mu$ c) $\pi^+ \rightarrow e^+ + \gamma$
b) $\tau^- \rightarrow e^- + \overline{\nu}_e + \nu_\tau$ d) $n \rightarrow p + e^- + \overline{\nu}_e$

46–16 In which of the following reactions or decays is baryon number conserved? In each case, explain your reasoning.

a) $n \rightarrow p + e^- + \overline{\nu}_e$ c) $p \rightarrow \pi^+ + \pi^0$
b) $p + n \rightarrow p + \pi^0$ d) $p + p \rightarrow p + p + \pi^0$

46–17 In which of the following reactions or decays is strangeness conserved? In each case, explain your reasoning.

a) $K^+ \rightarrow \mu^+ + \nu_\mu$ c) $K^+ + K^- \rightarrow \pi^0 + \pi^0$
b) $n + K^+ \rightarrow p + \pi^0$ d) $p + K^- \rightarrow \Lambda^0 + \pi^0$

46–18 a) Show that the coupling constant for the electromagnetic interaction, $e^2/4\pi\epsilon_0\hbar c$, is dimensionless and has the numerical value 1/137.0. b) Show that in the Bohr model (Section 40–6), the orbital speed of an electron in the $n = 1$ orbit is equal to c times the coupling constant $e^2/4\pi\epsilon_0\hbar c$.

46–19 Show that the nuclear force coupling constant $f^2/\hbar c$ is dimensionless.

SECTION 46–5 QUARKS AND THE EIGHTFOLD WAY

46–20 Nine of the spin-$\frac{3}{2}$ baryons are four Δ particles, each with mass 1232 MeV/c^2, strangeness 0, and charges $+2e$, $+e$, 0, and $-e$; three Σ^* particles, each with mass 1385 MeV/c^2, strangeness -1, and charges $+e$, 0, and $-e$; and two Ξ^* particles, each with mass 1530 MeV/c^2, strangeness -2, and charges 0 and $-e$. a) Place these particles on a plot of S versus Q. Deduce the Q and S values of the tenth spin-$\frac{3}{2}$ baryon, the Ω^- particle, and place it on your diagram. Also label the particles with their masses. The mass of the Ω^- is 1672 MeV/c^2; is this value consistent with your diagram? b) Deduce the three-quark combinations (of u, d, and s) that comprise each of these ten particles. Redraw the plot of S versus Q from part (a) with each particle labeled by its quark content. What regularities do you see?

46–21 Determine the electric charge, baryon number, strangeness, and charm quantum numbers for the following quark combinations a) uds; b) $c\overline{u}$; c) ddd; d) $d\overline{c}$. Explain your reasoning.

46–22 Given that each particle contains only combinations of $u, d, s, \overline{u}, \overline{d}$, and $\overline{s}$, use the method of Example 46–8

(Section 46–5) to deduce the quark content of a) a particle with charge $+e$, baryon number 0, and strangeness $+1$; b) a particle with charge $+e$, baryon number -1, and strangeness $+1$; c) a particle with charge 0, baryon number $+1$, and strangeness -2.

46–23 The quark content of the neutron is udd. a) What is the quark content of the antineutron? Explain your reasoning. b) Is the neutron its own antiparticle? Why or why not? c) The quark content of the ψ is $c\overline{c}$. Is the ψ its own antiparticle? Explain your reasoning.

46–24 What is the total kinetic energy of the decay products when an upsilon particle at rest decays to $\tau^+ + \tau^-$?

SECTION 46–7 THE EXPANDING UNIVERSE

46–25 A galaxy in the constellation Pisces is 5210 Mly from the earth. a) Use Hubble's law to calculate the speed at which this galaxy is receding from earth. b) What redshifted ratio λ_0/λ_S is expected for light from this galaxy?

46–26 a) According to Hubble's law, what is the distance r from us for galaxies that are receding away from us with a speed c? b) Explain why the distance calculated in part (a) is the size of our observable universe (neglecting any slowing of the expansion of the universe due to gravitational attraction).

46–27 The spectrum of the sodium atom is detected in the light from a distant galaxy. a) If the 590.0-nm line is redshifted to 658.5 nm, at what speed is the galaxy receding from the earth? b) Use Hubble's law to calculate the distance of the galaxy from earth.

46–28 Derive Eq. (46–14) from (46–13).

SECTION 46–8 THE BEGINNING OF TIME

46–29 Calculate the energy released in each of the following reactions: a) $p + {}^2H \rightarrow {}^3He$; b) $n + {}^3He \rightarrow {}^4He$.

46–30 Calculate the energy (in MeV) released in the triple-alpha process $3\,{}^4He \rightarrow {}^{12}C$.

46–31 Calculate the reaction energy Q (in MeV) for the reaction

$$e^- + p \rightarrow n + \nu_e .$$

Is this reaction endoergic or exoergic?

46–32 Calculate the reaction energy Q (in MeV) for the nucleosynthesis reaction

$${}^{12}_{6}C + {}^{4}_{2}He \rightarrow {}^{16}_{8}O.$$

Is this reaction endoergic or exoergic?

46–33 The 2.728 K blackbody radiation has its peak wavelength at 1.062 mm. What was the peak wavelength at $t = 700,000$ y when the temperature was 3000 K?

46–34 a) Show that the expression for the Planck length, $\sqrt{\hbar G/c^3}$, has dimensions of length. b) Evaluate the numerical value of $\sqrt{\hbar G/c^3}$, and verify the value given in Eq. (46–21).

PROBLEMS

46–35 In the LHC, each proton will be accelerated to a kinetic energy of 7.0 TeV. a) In the colliding beams, what is the available energy E_a in a collision? b) In a fixed-target experiment in

which a beam of protons is incident on a stationary proton target, what must be the total energy (in TeV) of the particles in the beam to produce the same available energy as in part (a)?

46–36 A proton and an antiproton collide head-on with equal kinetic energies. In the center of momentum frame, two γ rays with wavelengths of 0.780 fm are produced. Calculate the kinetic energy of the incident proton.

46–37 An electron with kinetic energy K collides with another electron at rest. Both electrons survive the collision, and a neutral pion (π^0) is produced. What is the threshold energy (minimum value of K) needed for this reaction?

46–38 Calculate the threshold kinetic energy for the reaction

$$\pi^- + p \rightarrow \Sigma^0 + K^0$$

if a π^- beam is incident on a stationary proton target. The K^0 has a mass of 497.7 MeV/c^2.

46–39 Calculate the threshold kinetic energy for the reaction

$$K^- + p \rightarrow \Lambda^0 + K^+ + K^-$$

if a K^- beam is incident on a stationary proton target.

46–40 An η^0 meson at rest decays into three π mesons. a) What are the allowed combinations of π^0, π^+, and π^- as decay products? b) Find the total kinetic energy of the π mesons.

46–41 a) What energy is released when a negative pion (π^-) at rest decays to stable products? See Tables 46–2 and 46–3 in Section 46–4. b) Why do the antineutrinos and neutrino carry away most of this energy?

46–42 The ϕ meson has a mass of 1019.4 MeV/c^2 and a measured energy width of 4.4 MeV/c^2. Using the uncertainty principle, Eq. (41–11), estimate the lifetime of the ϕ meson.

46–43 A ϕ meson (Problem 46–42) at rest decays via $\phi \rightarrow K^+ + K^-$. It has strangeness 0. a) Find the kinetic energy of the K^+ meson. (Assume that the two decay products share kinetic energy equally, since their masses are equal.) b) Suggest a reason why the decay $\phi \rightarrow K^+ + K^- + \pi^0$ has not been observed. c) Suggest reasons why the decays $\phi \rightarrow K^+ + \pi^-$ and $\phi \rightarrow K^+ + \mu^-$ have not been observed.

46–44 Each reaction listed below is missing a single particle. Calculate the baryon number, charge, strangeness, and the three lepton numbers (where appropriate) of the missing particle and from this identify the particle. a) $p + p \rightarrow p + \Lambda^0 + ?$ b) $K^- + n \rightarrow \Lambda^0 + ?$ c) $p + \bar{p} \rightarrow n + ?$ d) $\bar{\nu}_\mu + p \rightarrow n + ?$

46–45 Estimate the energy width (energy uncertainty) of the ψ if its mean lifetime is 7.6×10^{-21} s. What fraction is this of its rest energy?

46–46 Proton Decay. Proton decay is a feature of some grand unification theories. One possible decay could be $p^+ \rightarrow e^+ + \pi^0$, which violates both baryon and lepton number conservation, so the proton lifetime is expected to be very long. Suppose the proton half-life were 1.0×10^{18} y. a) Calculate the energy deposited per kilogram of body tissue (in rad) due to the decay of the protons in your body in one year. Model your body as consisting entirely of water. Only the two protons in the hydrogen atoms in each H_2O molecule would decay in the manner shown; do you see why? Assume that the π^0 decays into two γ rays, that the positron annihilates with an electron, and that all the energy produced in the primary decay and these secondary decays remains in your body. b) Calculate the equivalent dose (in rem) assuming a RBE of 1.0 for all the radiation products and compare with the 0.1 rem due to the natural background and the 5.0 rem guideline for industrial workers. Based on your calculation, can the proton lifetime be as short as 1.0×10^{18} y?

46–47 A Ξ^- particle at rest decays into a Λ^0 and a π^-. a) Find the total kinetic energy of the decay products. b) What fraction of the energy is carried off by each particle? (For simplicity, use nonrelativistic momentum and kinetic-energy expressions.)

46–48 Consider the spherical balloon model of a two-dimensional expanding universe (Fig. 46–15 in Section 46–7). The shortest distance between two points on the surface, measured along the surface, is the arc length r, where $r = R\theta$. As the balloon expands, its radius R increases but the angle θ between the two points remains constant. a) Explain why, at any given time, $(dR/dt)/R$ is the same for all points on the balloon. b) Show that $v = dr/dt$ is directly proportional to r at any instant. c) From your answer to part (b), what is the expression for the Hubble constant H_0 in terms of R and dR/dt? d) The expression for H_0 you found in part (c) is constant in space. How would R have to depend on time for H_0 to be constant in time? e) Is your answer to part (d) consistent with the gravitational attraction of matter in the universe?

46–49 Suppose all the conditions are the same as in Problem 46–48, except that $= dr/dt$ is constant for a given θ, rather than H_0 being constant in time. If so, show that the Hubble constant is $H_0 = 1/t$ and hence that the current value is $1/T$, where T is the age of the universe.

CHALLENGE PROBLEMS

46–50 Consider a collision in which a stationary particle with mass M is bombarded by a particle with mass m, speed v_0, and total energy (including rest energy) E_m. a) Use the Lorentz transformation to write the velocities v_m and v_M of particles m and M in terms of the speed v_{cm} of the center of momentum. b) Use the fact that the total momentum in the center-of-momentum frame is zero to obtain an expression for v_{cm} in terms of m, M, and v_0. c) Combine the results of parts (a) and (b) to obtain Eq. (46–9) for the total energy in the center-of-momentum frame.

46–51 A Λ^0 hyperon at rest decays into a neutron and a π^0. a) Find the total kinetic energy of the decay products. b) What fraction of the total kinetic energy is carried off by each particle? (*Hint:* Use the relativistic expressions for momentum and kinetic energy.)

The International System of Units

The Système International d'Unités, abbreviated SI, is the system developed by the General Conference on Weights and Measures and adopted by nearly all the industrial nations of the world. It is based on the mksa (meter-kilogram-second-ampere) system. The following material is adapted from B. N. Taylor, ed., National Bureau of Standards Technol. Spec. Pub. 330 (U.S. Govt. Printing Office, Washington, DC, 1991).

Quantity	Name of unit	Symbol	
	SI base units		
length	meter	m	
mass	kilogram	kg	
time	second	s	
electric current	ampere	A	
thermodynamic temperature	kelvin	K	
amount of substance	mole	mol	
luminous intensity	candela	cd	
	SI derived units		**Equivalent units**
area	square meter	m^2	
volume	cubic meter	m^3	
frequency	hertz	Hz	s^{-1}
mass density (density)	kilogram per cubic meter	kg/m^3	
speed, velocity	meter per second	m/s	
angular velocity	radian per second	rad/s	
acceleration	meter per second squared	m/s^2	
angular acceleration	radian per second squared	rad/s^2	
force	newton	N	$kg \cdot m/s^2$
pressure (mechanical stress)	pascal	Pa	N/m^2
kinematic viscosity	square meter per second	m^2/s	
dynamic viscosity	newton-second per square meter	$N \cdot s/m^2$	
work, energy, quantity of heat	joule	J	$N \cdot m$
power	watt	W	J/s
quantity of electricity	coulomb	C	$A \cdot s$
potential difference, electromotive force	volt	V	J/C, W/A
electric field strength	volt per meter	V/m	N/C
electric resistance	ohm	Ω	V/A
capacitance	farad	F	$A \cdot s/V$
magnetic flux	weber	Wb	$V \cdot s$
inductance	henry	H	$V \cdot s/A$
magnetic flux density	tesla	T	Wb/m^2
magnetic field strength	ampere per meter	A/m	
magnetomotive force	ampere	A	
luminous flux	lumen	lm	$cd \cdot sr$
luminance	candela per square meter	cd/m^2	
illuminance	lux	lx	lm/m^2
wave number	1 per meter	m^{-1}	
entropy	joule per kelvin	J/K	
specific heat capacity	joule per kilogram-kelvin	$J/kg \cdot K$	
thermal conductivity	watt per meter-kelvin	$W/m \cdot K$	

Quantity	Name of unit	Symbol	Equivalent units
radiant intensity	watt per steradian	W/sr	
activity (of a radioactive source)	becquerel	Bq	s^{-1}
radiation dose	gray	Gy	J/kg
radiation dose equivalent	sievert	Sv	J/kg
SI supplementary units			
plane angle	radian	rad	
solid angle	steradian	sr	

DEFINITIONS OF SI UNITS

meter (m)   The *meter* is the length equal to the distance traveled by light, in vacuum, in a time of 1/299,792,458 second.

kilogram (kg)   The *kilogram* is the unit of mass; it is equal to the mass of the international prototype of the kilogram. (The international prototype of the kilogram is a particular cylinder of platinum-iridium alloy that is preserved in a vault at Sèvres, France, by the International Bureau of Weights and Measures.)

second (s)   The *second* is the duration of 9,192,631,770 periods of the radiation corresponding to the transition between the two hyperfine levels of the ground state of the cesium-133 atom.

ampere (A)   The *ampere* is that constant current that, if maintained in two straight parallel conductors of infinite length, of negligible circular cross section, and placed 1 meter apart in vacuum, would produce between these conductors a force equal to 2×10^{-7} newton per meter of length.

kelvin (K)   The *kelvin*, unit of thermodynamic temperature, is the fraction 1/273.16 of the thermodynamic temperature of the triple point of water.

ohm (Ω)   The *ohm* is the electric resistance between two points of a conductor when a constant difference of potential of 1 volt, applied between these two points, produces in this conductor a current of 1 ampere, this conductor not being the source of any electromotive force.

coulomb (C)   The *coulomb* is the quantity of electricity transported in 1 second by a current of 1 ampere.

candela (cd)   The *candela* is the luminous intensity, in a given direction, of a source that emits monochromatic radiation of frequency 540×10^{12} hertz and that has a radiant intensity in that direction of 1/683 watt per steradian.

mole (mol)   The *mole* is the amount of substance of a system that contains as many elementary entities as there are carbon atoms in 0.012 kg of carbon 12. The elementary entities must be specified and may be atoms, molecules, ions, electrons, other particles, or specified groups of such particles.

newton (N)   The *newton* is that force that gives to a mass of 1 kilogram an acceleration of 1 meter per second per second.

joule (J)   The *joule* is the work done when the point of application of a constant force of 1 newton is displaced a distance of 1 meter in the direction of the force.

watt (W)   The *watt* is the power that gives rise to the production of energy at the rate of 1 joule per second.

volt (V)   The *volt* is the difference of electric potential between two points of a conducting wire carrying a constant current of 1 ampere, when the power dissipated between these points is equal to 1 watt.

weber (Wb)   The *weber* is the magnetic flux that, linking a circuit of one turn, produces in it an electromotive force of 1 volt as it is reduced to zero at a uniform rate in 1 second.

lumen (lm)   The *lumen* is the luminous flux emitted in a solid angle of 1 steradian by a uniform point source having an intensity of 1 candela.

farad (F)   The *farad* is the capacitance of a capacitor between the plates of which there appears a difference of potential of 1 volt when it is charged by a quantity of electricity equal to 1 coulomb.

henry (H)   The *henry* is the inductance of a closed circuit in which an electromotive force of 1 volt is produced when the electric current in the circuit varies uniformly at a rate of 1 ampere per second.

radian (rad)   The *radian* is the plane angle between two radii of a circle that cut off on the circumference an arc equal in length to the radius.

steradian (sr)   The *steradian* is the solid angle that, having its vertex in the center of a sphere, cuts off an area of the surface of the sphere equal to that of a square with sides of length equal to the radius of the sphere.

SI Prefixes   The names of multiples and submultiples of SI units may be formed by application of the prefixes listed in Appendix F.

APPENDIX B

Useful Mathematical Relations

ALGEBRA

$$a^{-x} = \frac{1}{a^x} \qquad a^{(x+y)} = a^x a^y \qquad a^{(x-y)} = \frac{a^x}{a^y}$$

Logarithms: If $\log a = x$, then $a = 10^x$. $\log a + \log b = \log (ab)$ $\log a - \log b = \log (a/b)$ $\log (a^n) = n \log a$
If $\ln a = x$, then $a = e^x$. $\ln a + \ln b = \ln (ab)$ $\ln a - \ln b = \ln (a/b)$ $\ln (a^n) = n \ln a$

Quadratic formula: If $ax^2 + bx + c = 0$, $x = \dfrac{-b \pm \sqrt{b^2 - 4ac}}{2a}$.

BINOMIAL THEOREM

$$(a + b)^n = a^n + na^{n-1}b + \frac{n(n-1)a^{n-2}b^2}{2!} + \frac{n(n-1)(n-2)a^{n-3}b^3}{3!} + \cdots$$

TRIGONOMETRY

In the right triangle ABC, $x^2 + y^2 = r^2$.

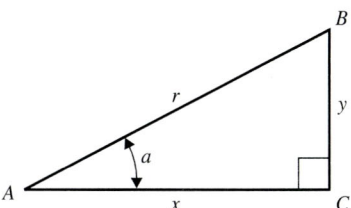

Definitions of the trigonometric functions: $\sin a = y/r$ $\cos a = x/r$ $\tan a = y/x$

Identities: $\sin^2 a + \cos^2 a = 1$ $\tan a = \dfrac{\sin a}{\cos a}$

$\sin 2a = 2 \sin a \cos a$ $\cos 2a = \cos^2 a - \sin^2 a = 2 \cos^2 a - 1$
$= 1 - 2 \sin^2 a$

$\sin \tfrac{1}{2}a = \sqrt{\dfrac{1 - \cos a}{2}}$ $\cos \tfrac{1}{2}a = \sqrt{\dfrac{1 + \cos a}{2}}$

$\sin (-a) = -\sin a$ $\sin (a \pm b) = \sin a \cos b \pm \cos a \sin b$
$\cos (-a) = \cos a$ $\cos (a \pm b) = \cos a \cos b \mp \sin a \sin b$
$\sin (a \pm \pi/2) = \pm \cos a$ $\sin a + \sin b = 2 \sin \tfrac{1}{2}(a + b) \cos \tfrac{1}{2}(a - b)$
$\cos (a \pm \pi/2) = \mp \sin a$ $\cos a + \cos b = 2 \cos \tfrac{1}{2}(a + b) \cos \tfrac{1}{2}(a - b)$

GEOMETRY

Circumference of circle of radius r: $C = 2\pi r$
Area of circle of radius r: $A = \pi r^2$
Volume of sphere of radius r: $V = 4\pi r^3/3$
Surface area of sphere of radius r: $A = 4\pi r^2$
Volume of cylinder of radius r and height h: $V = \pi r^2 h$

CALCULUS

Derivatives:

$$\frac{d}{dx}x^n = nx^{n-1}$$

$$\frac{d}{dx}\sin ax = a\cos ax$$

$$\frac{d}{dx}\cos ax = -a\sin ax$$

$$\frac{d}{dx}e^{ax} = ae^{ax}$$

$$\frac{d}{dx}\ln ax = \frac{1}{x}$$

$$\int \frac{dx}{\sqrt{a^2-x^2}} = \arcsin\frac{x}{a}$$

$$\int \frac{dx}{\sqrt{x^2+a^2}} = \ln\left(x+\sqrt{x^2+a^2}\right)$$

$$\int \frac{dx}{x^2+a^2} = \frac{1}{a}\arctan\frac{x}{a}$$

$$\int \frac{dx}{(x^2+a^2)^{3/2}} = \frac{1}{a^2}\frac{x}{\sqrt{x^2+a^2}}$$

$$\int \frac{x\,dx}{(x^2+a^2)^{3/2}} = -\frac{1}{\sqrt{x^2+a^2}}$$

Power series (convergent for range of x shown):

$$(1+x)^n = 1 + nx + \frac{n(n-1)x^2}{2!} + \frac{n(n-1)(n-2)}{3!}x^3 + \cdots \qquad (|x|<1)$$

Integrals:

$$\int x^n\,dx = \frac{x^{n+1}}{n+1} \quad (n\neq -1)$$

$$\int \frac{dx}{x} = \ln x$$

$$\int \sin ax\,dx = -\frac{1}{a}\cos ax$$

$$\int \cos ax\,dx = \frac{1}{a}\sin ax$$

$$\int e^{ax}dx = \frac{1}{a}e^{ax}$$

$$\sin x = x - \frac{x^3}{3!} + \frac{x^5}{5!} - \frac{x^7}{7!} + \cdots \qquad (\text{all } x)$$

$$\cos x = 1 - \frac{x^2}{2!} + \frac{x^4}{4!} - \frac{x^6}{6!} + \cdots \qquad (\text{all } x)$$

$$\tan x = x + \frac{x^3}{3} + \frac{2x^2}{15} + \frac{17x^7}{315} + \cdots \qquad (|x|<\pi/2)$$

$$e^x = 1 + x + \frac{x^2}{2!} + \frac{x^3}{3!} + \cdots \qquad (\text{all } x)$$

$$\ln(1+x) = x - \frac{x^2}{2} + \frac{x^3}{3} + \frac{x^4}{4} + \cdots \qquad (|x|<1)$$

APPENDIX C

The Greek Alphabet

Name	Capital	Lowercase	Name	Capital	Lowercase
Alpha	A	α	Nu	N	ν
Beta	B	β	Xi	Ξ	ξ
Gamma	Γ	γ	Omicron	O	o
Delta	Δ	δ	Pi	Π	π
Epsilon	E	ϵ	Rho	P	ρ
Zeta	Z	ζ	Sigma	Σ	σ
Eta	H	η	Tau	T	τ
Theta	Θ	θ	Upsilon	Υ	υ
Iota	I	ι	Phi	Φ	ϕ
Kappa	K	κ	Chi	X	χ
Lambda	Λ	λ	Psi	Ψ	ψ
Mu	M	μ	Omega	Ω	ω

Periodic Table of the Elements

Group 1 2 3 4 5 6 7 8 9 10 11 12 13 14 15 16 17 18
Period

Period	1	2	3	4	5	6	7	8	9	10	11	12	13	14	15	16	17	18
1	1 **H** 1.008																	2 **He** 4.003
2	3 **Li** 6.941	4 **Be** 9.012											5 **B** 10.811	6 **C** 12.011	7 **N** 14.007	8 **O** 15.999	9 **F** 18.998	10 **Ne** 20.180
3	11 **Na** 22.990	12 **Mg** 24.305											13 **Al** 26.982	14 **Si** 28.086	15 **P** 30.974	16 **S** 32.066	17 **Cl** 35.453	18 **Ar** 39.948
4	19 **K** 39.098	20 **Ca** 40.078	21 **Sc** 44.956	22 **Ti** 47.867	23 **V** 50.942	24 **Cr** 51.996	25 **Mn** 54.938	26 **Fe** 55.845	27 **Co** 58.933	28 **Ni** 58.693	29 **Cu** 63.546	30 **Zn** 65.39	31 **Ga** 69.723	32 **Ge** 72.61	33 **As** 74.922	34 **Se** 78.96	35 **Br** 79.904	36 **Kr** 83.8
5	37 **Rb** 85.468	38 **Sr** 87.62	39 **Y** 88.906	40 **Zr** 91.224	41 **Nb** 92.906	42 **Mo** 95.94	43 **Tc** (98)	44 **Ru** 101.07	45 **Rh** 102.905	46 **Pd** 106.42	47 **Ag** 107.868	48 **Cd** 112.411	49 **In** 114.818	50 **Sn** 118.71	51 **Sb** 121.76	52 **Te** 127.6	53 **I** 126.904	54 **Xe** 131.29
6	55 **Cs** 132.905	56 **Ba** 137.327	71 **Lu** 174.967	72 **Hf** 178.49	73 **Ta** 180.948	74 **W** 183.84	75 **Re** 186.207	76 **Os** 190.23	77 **Ir** 192.217	78 **Pt** 195.078	79 **Au** 196.967	80 **Hg** 200.59	81 **Tl** 204.38	82 **Pb** 207.2	83 **Bi** 208.98	84 **Po** (210)	85 **At** (210)	86 **Rn** (222)
7	87 **Fr** (223)	88 **Ra** (226)	103 **Lr** (262)	104 **Rf** (261)	105 **Db** (262)	106 **Sg** (266)	107 **Bh** (264)	108 **Hs** (269)	109 **Mt** (268)	110 **Uun** (269)	111 **Uuu** (272)	112 **Uub** (277)	113 **Uut**	114 **Uuq** (285)	115 **Uup**	116 **Uuh** (289)	117 **Uus**	118 **Uuo** (293)

Lanthanoids

57 **La** 138.905	58 **Ce** 140.16	59 **Pr** 140.908	60 **Nd** 144.24	61 **Pm** (145)	62 **Sm** 150.36	63 **Eu** 151.964	64 **Gd** 157.25	65 **Tb** 158.925	66 **Dy** 162.5	67 **Ho** 164.930	68 **Er** 167.26	69 **Tm** 168.934	70 **Yb** 173.04

Actinoids

89 **Ac** (227)	90 **Th** (232)	91 **Pa** (231)	92 **U** (238)	93 **Np** (237)	94 **Pu** (244)	95 **Am** (243)	96 **Cm** (247)	97 **Bk** (247)	98 **Cf** (251)	99 **Es** (252)	100 **Fm** (257)	101 **Md** (258)	102 **No** (259)

For each element the average atomic mass of the mixture of isotopes occurring in nature is shown. For elements having no stable isotope, the approximate atomic mass of the longest-lived isotope is shown in parentheses. For elements that have been predicted but not yet detected, no atomic mass is given. All atomic masses are expressed in atomic mass units (1 u = $1.6605402(10) \times 10^{-27}$ kg), equivalent to grams per mole (g/mol).

Unit Conversion Factors

LENGTH

$1 \text{ m} = 100 \text{ cm} = 1000 \text{ mm} = 10^6 \text{ } \mu\text{m} = 10^9 \text{ nm}$
$1 \text{ km} = 1000 \text{ m} = 0.6214 \text{ mi}$
$1 \text{ m} = 3.281 \text{ ft} = 39.37 \text{ in.}$
$1 \text{ cm} = 0.3937 \text{ in.}$
$1 \text{ in.} = 2.540 \text{ cm}$
$1 \text{ ft} = 30.48 \text{ cm}$
$1 \text{ yd} = 91.44 \text{ cm}$
$1 \text{ mi} = 5280 \text{ ft} = 1.609 \text{ km}$
$1 \text{ Å} = 10^{-10} \text{ m} = 10^{-8} \text{ cm} = 10^{-1} \text{ nm}$
$1 \text{ nautical mile} = 6080 \text{ ft}$
$1 \text{ light year} = 9.461 \times 10^{15} \text{ m}$

AREA

$1 \text{ cm}^2 = 0.155 \text{ in}^2$
$1 \text{ m}^2 = 10^4 \text{ cm}^2 = 10.76 \text{ ft}^2$
$1 \text{ in.}^2 = 6.452 \text{ cm}^2$
$1 \text{ ft} = 144 \text{ in.}^2 = 0.0929 \text{ m}^2$

VOLUME

$1 \text{ liter} = 1000 \text{ cm}^3 = 10^{-3} \text{ m}^3 = 0.03531 \text{ ft}^3 = 61.02 \text{ in.}^3$
$1 \text{ ft}^3 = 0.02832 \text{ m}^3 = 28.32 \text{ liters} = 7.477 \text{ gallons}$
$1 \text{ gallon} = 3.788 \text{ liters}$

TIME

$1 \text{ min} = 60 \text{ s}$
$1 \text{ h} = 3600 \text{ s}$
$1 \text{ d} = 86{,}400 \text{ s}$
$1 \text{ y} = 365.24 \text{ d} = 3.156 \times 10^7 \text{ s}$

ANGLE

$1 \text{ rad} = 57.30° = 180°/\pi$
$1° = 0.01745 \text{ rad} = \pi/180 \text{ rad}$
$1 \text{ revolution} = 360° = 2\pi \text{ rad}$
$1 \text{ rev/min (rpm)} = 0.1047 \text{ rad/s}$

SPEED

$1 \text{ m/s} = 3.281 \text{ ft/s}$
$1 \text{ ft/s} = 0.3048 \text{ m/s}$
$1 \text{ mi/min} = 60 \text{ mi/h} = 88 \text{ ft/s}$
$1 \text{ km/h} = 0.2778 \text{ m/s} = 0.6214 \text{ mi/h}$
$1 \text{ mi/h} = 1.466 \text{ ft/s} = 0.4470 \text{ m/s} = 1.609 \text{ km/h}$
$1 \text{ furlong/fortnight} = 1.662 \times 10^{-4} \text{ m/s}$

ACCELERATION

$1 \text{ m/s}^2 = 100 \text{ cm/s}^2 = 3.281 \text{ ft/s}^2$
$1 \text{ cm/s}^2 = 0.01 \text{ m/s}^2 = 0.03281 \text{ ft/s}^2$
$1 \text{ ft/s}^2 = 0.3048 \text{ m/s}^2 = 30.48 \text{ cm/s}^2$
$1 \text{ mi/h} \cdot \text{s} = 1.467 \text{ ft/s}^2$

MASS

$1 \text{ kg} = 10^3 \text{ g} = 0.0685 \text{ slug}$
$1 \text{ g} = 6.85 \times 10^{-5} \text{ slug}$
$1 \text{ slug} = 14.59 \text{ kg}$
$1 \text{ u} = 1.661 \times 10^{-27} \text{ kg}$
$1 \text{ kg has a weight of } 2.205 \text{ lb when } g = 9.80 \text{ m/s}^2$

FORCE

$1 \text{ N} = 10^5 \text{ dyn} = 0.2248 \text{ lb}$
$1 \text{ lb} = 4.448 \text{ N} = 4.448 \times 10^5 \text{ dyn}$

PRESSURE

$1 \text{ Pa} = 1 \text{ N/m}^2 = 1.450 \times 10^{-4} \text{ lb/in.}^2 = 0.209 \text{ lb/ft}^2$
$1 \text{ bar} = 10^5 \text{ Pa}$
$1 \text{ lb/in.}^2 = 6895 \text{ Pa}$
$1 \text{ lb/ft}^2 = 47.88 \text{ Pa}$
$1 \text{ atm} = 1.013 \times 10^5 \text{ Pa} = 1.013 \text{ bar}$
$\qquad = 14.7 \text{ lb/in.}^2 = 2117 \text{ lb/ft}^2$
$1 \text{ mm Hg} = 1 \text{ torr} = 133.3 \text{ Pa}$

ENERGY

$1 \text{ J} = 10^7 \text{ ergs} = 0.239 \text{ cal}$
$1 \text{ cal} = 4.186 \text{ J (based on 15° calorie)}$
$1 \text{ ft} \cdot \text{lb} = 1.356 \text{ J}$
$1 \text{ Btu} = 1055 \text{ J} = 252 \text{ cal} = 778 \text{ ft} \cdot \text{lb}$
$1 \text{ eV} = 1.602 \times 10^{-19} \text{ J}$
$1 \text{ kWh} = 3.600 \times 10^6 \text{ J}$

MASS–ENERGY EQUIVALENCE

$1 \text{ kg} \leftrightarrow 8.988 \times 10^{16} \text{ J}$
$1 \text{ u} \leftrightarrow 931.5 \text{ MeV}$
$1 \text{ eV} \leftrightarrow 1.074 \times 10^{-9} \text{ u}$

POWER

$1 \text{ W} = 1 \text{ J/s}$
$1 \text{ hp} = 746 \text{ W} = 550 \text{ ft} \cdot \text{lb/s}$
$1 \text{ Btu/h} = 0.293 \text{ W}$

APPENDIX F

Numerical Constants

FUNDAMENTAL PHYSICAL CONSTANTS*

Name	Symbol	Value
Speed of light	c	2.99792458×10^8 m/s
Magnitude of charge of electron	e	$1.60217733(49) \times 10^{-19}$ C
Gravitational constant	G	$6.67259(85) \times 10^{-11}$ N $\cdot$ m^2/kg^2
Planck's constant	h	$6.6260755(40) \times 10^{-34}$ J $\cdot$ s
Boltzmann constant	k	$1.380658(12) \times 10^{-23}$ J/K
Avogadro's number	N_A	$6.0221367(36) \times 10^{23}$ molecules/mol
Gas constant	R	$8.314510(70)$ J/mol $\cdot$ K
Mass of electron	m_e	$9.1093897(54) \times 10^{-31}$ kg
Mass of proton	m_p	$1.6726231(10) \times 10^{-27}$ kg
Mass of neutron	m_n	$1.6749286(10) \times 10^{-27}$ kg
Permeability of free space	μ_0	$4\pi \times 10^{-7}$ Wb/A $\cdot$ m
Permittivity of free space	$\epsilon_0 = 1/\mu_0 c^2$	$8.854187817 \ldots \times 10^{-12}$ C^2/N $\cdot$ m^2
	$1/4\pi\epsilon_0$	$8.987551787 \ldots \times 10^9$ N $\cdot$ m^2/C^2

OTHER USEFUL CONSTANTS*

Name	Symbol	Value
Mechanical equivalent of heat		4.186 J/cal ($15°$ calorie)
Standard atmospheric pressure	1 atm	1.01325×10^5 Pa
Absolute zero	0 K	$-273.15°$C
Electron volt	1 eV	$1.60217733(49) \times 10^{-19}$ J
Atomic mass unit	1 u	$1.6605402(10) \times 10^{-27}$ kg
Electron rest energy	$m_e c^2$	$0.51099906(15)$ MeV
Volume of ideal gas ($0°$C and 1 atm)		$22.41410(19)$ liter/mol
Acceleration due to gravity (standard)	g	9.80665 m/s^2

*Source: E. R. Cohen and B. R. Taylor, *Reviews of Modern Physics* **57**, 1121 (1987). Numbers in parentheses show the uncertainty in the final digits of the main number; for example, the number 1.6454(21) means 1.6454 ± 0.0021. Values shown without uncertainties are exact.

ASTRONOMICAL DATA[†]

Body	Mass (kg)	Radius (m)	Orbit radius (m)	Orbit period
Sun	1.99×10^{30}	6.96×10^{8}	—	—
Moon	7.35×10^{22}	1.74×10^{6}	3.84×10^{8}	27.3 d
Mercury	3.30×10^{23}	2.44×10^{6}	5.79×10^{10}	88.0 d
Venus	4.87×10^{24}	6.05×10^{6}	1.08×10^{11}	224.7 d
Earth	5.97×10^{24}	6.38×10^{6}	1.50×10^{11}	365.3 d
Mars	6.42×10^{23}	3.40×10^{6}	2.28×10^{11}	687.0 d
Jupiter	1.90×10^{27}	6.91×10^{7}	7.78×10^{11}	11.86 y
Saturn	5.69×10^{26}	6.03×10^{7}	1.43×10^{12}	29.42 y
Uranus	8.66×10^{25}	2.56×10^{7}	2.88×10^{12}	83.75 y
Neptune	1.03×10^{26}	2.48×10^{7}	4.50×10^{12}	163.7 y
Pluto	$1.5 \ \times 10^{22}$	1.15×10^{6}	5.92×10^{12}	248.0 y

[†]Source: *The Astronomical Almanac for 1995* (US Government Printing Office, Washington, DC, 1994), p. E88, and P. Kenneth Seidelmann, ed., *Explanatory Supplement to the Astronomical Almanac* (University Science Books, Mill Valley, CA, 1992), pp. 704–706. For each body, "radius" is its radius at its equator and "orbit radius" is its average distance from the sun (for the planets) or from the earth (for the moon).

PREFIXES FOR POWERS OF 10

Power of ten	Prefix	Abbreviation	Pronunciation
10^{-24}	yocto-	y	*yoc*-toe
10^{-21}	zepto-	z	*zep*-toe
10^{-18}	atto-	a	*at*-toe
10^{-15}	femto-	f	*fem*-toe
10^{-12}	pico-	p	*pee*-koe
10^{-9}	nano-	n	*nan*-oe
10^{-6}	micro-	μ	*my*-crow
10^{-3}	milli-	m	*mil*-i
10^{-2}	centi-	c	*cen*-ti
10^{3}	kilo-	k	*kil*-oe
10^{6}	mega-	M	*meg*-a
10^{9}	giga-	G	*jig*-a or *gig*-a
10^{12}	tera-	T	*ter*-a
10^{15}	peta-	P	*pet*-a
10^{18}	exa-	E	*ex*-a
10^{21}	zetta-	Z	*zet*-a
10^{24}	yotta-	Y	*yot*-a

Examples:

1 femtometer = 1 fm = 10^{-15} m

1 picosecond = 1 ps = 10^{-12} s

1 nanocoulomb = 1 nC = 10^{-9} C

1 microkelvin = 1 μK = 10^{-6} K

1 millivolt = 1 mV = 10^{-3} V

1 kilopascal = 1 kPa = 10^{3} Pa

1 megawatt = 1 MW = 10^{6} W

1 gigahertz = 1 GHz = 10^{9} Hz

ANSWERS TO ODD-NUMBERED PROBLEMS

Chapter 1

1–1:   1.61 km
1–3: 3.34×10^3 ns
1–5:   5.36 L
1–7:   a) 2330 km/h b) 648 m/s
1–9:   35.3 mi/gal
1–11:   a) $1.1 \times 10^{-3}\%$ b) no
1–13:   a) 0.1% b) 0.008% c) 0.03%
1–15:   a) 2.8 ± 0.3 cm^3 b) 170 ± 20
1–17:   Ten thousand
1–19: 10^6
1–21: 10^5 (with a full head of hair)
1–23:   about 10^8
1–27: 9×10^8; about $3.50 per person
1–29:   7.8 m, 38° north of east
1–31:   104 m, 43° south of west
1–33: $A_x = 7.22$ m, $A_y = 9.58$ m; $B_x = 11.5$ m, $B_y = -9.64$ m; $C_x = -3.00$ m, $C_y = -5.20$ m
1–35:   a) 11.1 m, 77.6° b) 11.1 m, 77.6° c) 28.5 m, 202.3° d) 28.5 m, 22.3°
1–37:   5.06 km, 20.2° north of west
1–39:   a) 2.48 cm, 18.3° b) 4.10 cm, 83.7° c) 4.10 cm, 263.7°
1–41: $\vec{A} = (7.2 \text{ m})\hat{\imath} + (9.6 \text{ m})\hat{\jmath}$, $\vec{B} = (11.5 \text{ m})\hat{\imath} - (9.6 \text{ m})\hat{\jmath}$, $\vec{C} = (-3.0 \text{ m})\hat{\imath} - (5.2 \text{ m})\hat{\jmath}$
1–43:   a) $A = 5.00$, $B = 5.39$ b) $-1.00\hat{\imath} + 5.00\hat{\jmath}$ c) 5.10, 101.3°
1–45:   a) 14.00 b) 58.7°
1–47:   a) out of the page b) into the page
1–49: $-23.0\hat{k}$; 23.0
1–51:   a) 640 acres/mi^2 b) 43,560 ft^2/acre c) 3.26×10^5 gal/acre-ft
1–53:   a) 7.04×10^{-10} s b) 5.11×10^{12} cycle/h c) 2.1×10^{26} d) 4.6×10^4 s
1–55:   About 1 dentist per 1,000 population
1–57:   b) $A_x = 3.03$ cm, $A_y = 8.10$ cm c) 8.65 cm, 69.5° measured in the sense from the +x-axis toward the +y-axis
1–59:   144 m, 41° south of west
1–61:   b) 1.45 km
1–63:   a) (87, 258) b) 137 pixels, 35° below straight left
1–67:   b) 90°
1–69:   a) $A = 5.39$, $B = 4.36$ b) $-5.00\hat{\imath} + 2.00\hat{\jmath} + 7.00\hat{k}$ c) 8.83; yes
1–73:   a) 54.7° b) 35.3°
1–75:   b) 72.2
1–77:   38.5 yd, 24.6° right of downfield
1–79:   a) 76 ly b) 129°

Chapter 2

2–1:   a) 197 m/s b) 169 m/s
2–3:   1 h 10 min
2–5:   a) 14.0 m/s b) 11.4 m/s
2–7:   a) 12.0 m/s b) i) 0 m/s ii) 15.0 m/s iii) 12.0 m/s c) 13.3 s
2–9:   a) (in m/s^2) 0, 1.0, 2.0, 2.0, 3.0, 1.5, 1.5, 0; no; yes b) 2.5 m/s^2, 1.5 m/s^2, 0
2–11:   a) near 5 s b) 30 s to 40 s c) 0 d) -1.7 m/s^2
2–13:   about 10 m/s^2
2–17:   a) 5.0 m/s b) 1.43 m/s^2
2–19:   1.70 m
2–21:   a) 1.7 m/s^2 b) 12 s c) 240 m
2–25:   a) 0, 6.3 m/s^2, -11.2 m/s^2 b) 100 m, 230 m, 320 m
2–27:   a) 1.80×10^4 m/s b) 0.957 c) 6 h 11 min
2–29:   b) 1 s, 3 s d) 2 s e) 3 s f) 1 s
2–31:   b) $d/4$
2–33:   a) 2.94 m/s b) 0.599 s
2–35:   a) $t = \sqrt{2d/g}$ b) 0.190 s
2–37:   a) 5.56 m/s, downward b) 9.80 m/s^2, downward c) 2.16 s

---

d) 16.1 m/s
2–39:   a) 25.6 m/s b) 31.6 m c) 15.2 m/s
2–41:   a) 249 m/s^2 b) 25.4 c) 101 m d) no
2–43:   a) 7.5 m b) 180 m c) 20 m
2–45:   a) $x(t) = (0.250 \text{ m/s}^3)t^3 - (0.0100 \text{ m/s}^4)t^4$, $v(t) = (0.750 \text{ m/s}^3)t^2 - (0.0400 \text{ m/s}^4)t^3$ b) 39.1 m/s
2–47:   b) 0.627 s, 1.60 s c) negative at 0.627 s, positive at 1.60 s d) 1.11 s e) 2.45 m f) 2.00 s, 0 s
2–49:   a) 82 km/h b) 31 km/h
2–51:   a) 3.5 m/s^2 b) 0 c) 1.5 m/s^2
2–53:   brake pedal
2–55:   4.6 m/s^2
2–57:   a) 6.17 s b) 24.8 m c) $v_{\text{truck}} = 13.0$ m/s, $v_{\text{auto}} = 21.0$ m/s
2–59:   a) 17 m/s b) 1.6 s
2–61:   a) 15.9 s b) 393 m c) 29.5 m/s
2–63:   a) -4.00 m/s b) 12.0 m/s
2–65:   28.6 m
2–67:   a) no b) yes, 14.4 m/s; not physically attainable
2–69:   a) 13.3 m b) 1.65 s
2–71:   a) 7.59 m/s b) 5.14 m c) 1.60 s
2–73:   a) 92.2 m b) 75.1 m/s
2–75:   a) A b) 2.27 s, 5.73 s c) 1.00 s, 4.33 s d) 2.67 s
2–77:   a) 9.55 s, 47.8 m b) 1.62 m/s d) 8.38 m/s e) no f) 3.69 m/s, 21.7 s, 80.0 m
2–79:   a) 8.18 m/s b) i) 0.411 m ii) 1.15 km c) 9.80 m/s d) 4.90 m/s

Chapter 3

3–1:   a) $(v_{\text{av}})_x = 1.4$ m/s, $(v_{\text{av}})_y = -1.3$ m/s b) 1.9 m/s, $-43°$
3–3:   a) 7.1 cm/s, 45° b) 5.0 cm/s, 90°; 7.1 cm/s, 45°; 11 cm/s, 27°
3–5:   b) $(a_{\text{av}})_x = -8.67$ m/s^2, $(a_{\text{av}})_y = -2.33$ m/s^2 c) 8.98 m/s^2, 195°
3–7:   b) $\vec{v} = \alpha\hat{\imath} - 2\beta t\hat{\jmath}, \vec{a} = -2\beta\hat{\jmath}$ c) $v = 5.4$ m/s, $-63°$; $a = 2.4$ m/s^2, $-90°$ d) speeding up and turning right
3–9:   a) 0.600 m b) 0.385 m c) $v_x = 1.10$ m/s, $v_y = -3.43$ m/s; $v = 3.60$ m/s, 72.2° below the horizontal
3–11:   a) 1.08 s b) 6.18 m, 4.51 m; 11.5 m, 5.74 m; 16.8 m, 4.51 m c) 11.7 m/s, +24.8°; 10.6 m/s, 0°; 11.7 m/s, $-24.8°$ d) parallel: -4.11 m/s^2, 0, 4.11 m/s^2; perpendicular: 8.90 m/s^2, 9.80 m/s^2, 8.90 m/s^2
3–13:   a) 1.4 km b) 8.5 km
3–15:   a) 0.682 s, 2.99 s b) 24.0 m/s, 11.3 m/s; 24.0 m/s, -11.3 m/s c) 30.0 m/s, $-36.9°$
3–17:   a) 1.5 m b) -0.89 m/s
3–19:   a) 13.6 m b) 34.6 m/s c) 103 m
3–21:   a) 0.034 m/s^2 = $0.0034g$ b) 1.4 h
3–23:   a) 3.07 s b) 1.68 s
3–25:   a) 3.50 m/s^2, upward b) 3.50 m/s^2, downward c) 12.6 s
3–27:   b) no c) where the car is farthest from the center of the ellipse
3–29:   a) 14 s b) 70 s
3–31:   0.36 m/s, 38° west of south
3–33:   a) 4.7 m/s, 25° south of east b) 190 s c) 380 m
3–35:   a) -7.1 m/s, -42 m/s b) 43 m/s, 9.5° west of south
3–37:   a) $t = 0$ b) $t = 0$ c) $t = 0$, $x = 0$, $y = 15.0$ m; $t = 5.21$ s, $x = 6.25$ m, $y = 1.44$ m d) 6.41 m, 5.21 s
3–39:   a) 1.0 m/s^2 b) 5.4° c) 2.3 m/s d) 31 s
3–41:   22 m/s
3–43:   274 m
3–45:   a) 42.8 m/s b) 42.0 m
3–47: $4\sqrt{2}\,D$

3–49:   c) less than 45°
3–51:   b) 15°, 75°
3–53:   a) 17.8 m/s   b) in river, 28.4 m from the near bank
3–55:   a) 2.23 m   b) 3.84 m   c) 8.65 m/s   d) 3.09 m, 0.62 m
3–57:   c) $(18/25)gT^2$ d) $(3/2)gT^2$
3–61:   b) $v_x = R\omega(1 - \cos \omega t)$, $v_y = R\omega \sin \omega t$, $a_x = R\omega^2 \sin \omega t$,
        $a_y = R\omega^2 \cos \omega t$ c) $t = 0, 2\pi/\omega, 4\pi/\omega, ...; x = 0, 2\pi R, 4\pi R, ...$;
        $y = 0; a = R\omega^2$ in the +y-direction   d) no
3–63:   30 km
3–65:   a) 44.7 km/h, 26.6° west of south   b) 10.5° north of west
3–67:   a) 0.782 s   b) 7.67 m/s   c) 5.17 m/s   d) 1.04 m
3–69:   a) 80 m   b) 1.56×10^{-3} c) overall effect is to reduce the
        radius
3–71:   a) $\left(\dfrac{2v_0^2}{g}\right)[\tan(\theta + \phi) - \tan \theta]\dfrac{\cos^2(\theta + \phi)}{\cos \theta}$ b) $\dfrac{\pi}{4} - \dfrac{\theta}{2}$
3–73: $\Delta t = 0.5$ s: 9.589 m/s², 118.6°; $\Delta t = 0.1$ s: 9.983 m/s², 95.73°;
        $\Delta t = 0.05$ s: 9.996 m/s², 92.86°
3–75:   a) 1.5 km/h   b) 3.5 km/h

Chapter 4

4–1:   a) 0   b) 90°   c) 180°
4–3:   7.1 N horizontal, 7.1 N vertical
4–5:   494 N, 31.7°
4–7:   2.2 m/s²
4–9:   16.0 kg
4–11:   a) 3.13 m, 3.13 m/s   b) 21.9 m, 6.25 m/s
4–13:   a) $\Sigma \vec{F} = \mathbf{0}$
4–15: 2.94×10^3 N
4–17:   a) 4.49 kg   b) 4.49 kg, 8.13 N
4–19:   825 N, blocks
4–23: 7.4×10^{-23} m/s²
4–25:   5.83 m/s²
4–27:   a) 25.0 N   b) 10.0 N
4–29:   1.47 m/s², upward
4–31: $F_x(t) = -6mBt$
4–33:   1840 N, 135°
4–35:   a) 17 N, 90° clockwise from +x-direction   b) 840 N
4–37: 2.36×10^3 N
4–39:   a) 4.4 m   b) i) 2.7×10^4 N   ii) 9.0×10^3 N
4–41:   a) $4m\vec{a}$ b) $3m\vec{a}$ c) $2m\vec{a}$ d) $-m\vec{a}$ e) same
        magnitudes but opposite directions
4–43:   a) 2.93 m/s²   b) 11.1 m/s²
4–45:   b) 5.9×10^2 N   c) 2.2×10^3 N
4–47:   b) 3.53 m/s²   c) 120 N   d) 93.3 N
4–49: $(k_1 t/m)\hat{i} + (k_2 t^4/4m)\hat{j}$
4–51: $\vec{r} = \left(\dfrac{k_1}{2m}t^2 + \dfrac{k_2 k_3}{120m^2}t^5\right)\hat{i} + \left(\dfrac{k_3}{6m}t^3\right)\hat{j}$

        $\vec{v} = \left(\dfrac{k_1}{m}t + \dfrac{k_2 k_3}{24m^2}t^4\right)\hat{i} + \left(\dfrac{k_3}{2m}t^2\right)\hat{j}$

Chapter 5

5–1:   a) 25.0 N   b) 50.0 N
5–3:   a) 2.54×10^3 N   b) 1.01°
5–7:   a) 5.23×10^4 N   b) 3.36×10^4 N
5–9:   523 N
5–11:   a) 84.9 N   b) 60.0 N, 60.0 N
5–13:   a) $w \sin \alpha$ b) $2w \sin \alpha$ c) $w \cos \alpha$ for each block
5–15:   b) 2.96 m/s²   c) 191 N; more than the bricks, less than the
        counterweight
5–17:   b) 2.50 m/s²   c) 1.37 kg   d) $T < mg$

5–19:   a) 1.78 m/s², downward   b) 2.14 m/s², upward   c) yes
5–21:   a) in the direction down the slope   b) no deflection
5–25:   a) 22 N   b) 3.1 m
5–27:   a) 7.06 N   b) 8.14 N   c) 1.17 N, 2.25 N
5–29:   a) pulled down   b) 379 N
5–31:   11 times farther
5–33:   a) $\mu_k(m_A + m_B)g$ b) $\mu_k m_A g$
5–35:   a) 0.218 m/s   b) 11.7 N
5–37:   a) $\mu_k mg/(\cos \theta - \mu_k \sin \theta)$ b) $1/(\tan \theta) = \cot \theta$
5–39:   b) 8.75 N   c) 30.8 N   d) 1.54 m/s²
5–41:   a) 0.44 kg/m   b) 42 m/s
5–43:   26.0 m/s
5–45:   a) 75°   b) no
5–47:   20.7°
5–49:   a) 1.49 rev/min   b) 0.918 rev/min
5–51:   a) 138 km/h   b) 3580 N
5–53:   2.43 m/s
5–55:   b) 10 m
5–57:   43 m/s
5–59:   14 m
5–61: $mg \tan \alpha$
5–63: $F[M + m(1 - x/L)]/(M + m)$
5–65:   a) 12.0 N   b) 15.0 N
5–67:   a) 16.9 N   b) 10.2 N
5–69: $\mu_s/(1 + \mu_s)$
5–71:   a) 66 N, northward   b) 59 N, southward
5–75: 3.09×10^3 N
5–77:   2.52 N
5–79:   a) 12.9 kg   b) 47.2 in left-hand cord, 101 N in right-hand
        cord
5–81: $a_1 = 2m_2 g/(4m_1 + m_2)$, $a_2 = m_2 g/(4m_1 + m_2)$
5–83:   1.46 m above the floor
5–85: g/μ_s
5–87:   b) 0.450
5–89:   a) $(F \cos \alpha)/(m_1 + m_2)$
5–91:   a) 21 m/s   b) 8.5 m/s
5–93:   a) move up   b) remains constant   c) remains constant
        d) stop
5–95:   a) 6.00 m/s²   b) 3.80 m/s²   c) 7.36 m/s   d) 8.18 m/s
        e) 7.78 m, 6.29 m/s, 1.38 m/s²   f) 3.14 s
5–97:   a) 120 N   b) 3.79 m/s
5–99:   b) 0.28   c) no
5–101: a) right   b) 117 m
5–103: a) 81.1°   b) no   c) bead rides at bottom of loop ($\beta = 0$)
5–107: $T_{\max} = 2\pi \sqrt{\left(\dfrac{h \tan \beta}{g}\right)\left(\dfrac{(\sin \beta + \mu_s \cos \beta)}{(\cos \beta - \mu_s \sin \beta)}\right)}$

        $T_{\max} = 2\pi \sqrt{\left(\dfrac{h \tan \beta}{g}\right)\left(\dfrac{(\sin \beta - \mu_s \cos \beta)}{(\cos \beta + \mu_s \sin \beta)}\right)}$
5–109: $(M + m)g \tan \alpha$
5–111: a) $F = \dfrac{\mu_k w}{\cos \theta - \mu_k \sin \theta}$ c) $\theta = \tan^{-1} \mu_k = 14.0°$
5–113: a) $a_3 = g\dfrac{(-4m_1 m_2 + m_2 m_3 + m_3 m_1)}{(4m_1 m_2 + m_2 m_3 + m_3 m_1)}$ b) $a_B = -a_3$

        c) $a_1 = g\dfrac{(4m_1 m_2 - 3m_2 m_3 + m_3 m_1)}{(4m_1 m_2 + m_2 m_3 + m_3 m_1)}$

        d) $a_2 = g\dfrac{(4m_1 m_2 + m_2 m_3 - 3m_3 m_1)}{(4m_1 m_2 + m_2 m_3 + m_3 m_1)}$

        e) $T_A = \frac{1}{2}T_C$ f) $T_C = \dfrac{8gm_1 m_2 m_3}{(4m_1 m_2 + m_2 m_3 + m_3 m_1)}$

        g) $a_1 = a_2 = a_3 = a_B = 0$, $T_C = 2m_2 g$, $T_A = m_2 g$; yes
5–115: $\cos^2 \beta$

Chapter 6

6–1:	a) 3.60 J b) −0.900 J c) 2.70 J
6–3:	300 J
6–5:	a) 99.2 N b) 387 J c) −387 J d) zero; zero e) zero
6–7:	2.6×10^9 J
6–9:	a) 4.3×10^3 J b) 40 km/h
6–11:	74.2 J
6–13:	a) $-(15/16)K_1$ b) no
6–15:	16.8 cm
6–17:	a) −28.4 J b) 15.3 m/s c) no
6–19:	a) 4.96 m/s b) $a = 1.43$ m/s^2, $v = 4.96$ m/s
6–21:	a) $v_0^2/2\mu_k g$ b) 51.3 m
6–23:	a) 48.0 N, 64.0 N b) 0.360 J, 0.640 J
6–25:	a) 2.8 m/s b) 3.5 m/s
6–27:	a) 1.76 b) 0.67 m/s
6–29:	a) 4.0 J b) zero c) −1.0 J d) 3.0 J e) −1.0 J
6–31:	a) 2.83 m/s b) 2.40 m/s
6–33:	a) 5.65 cm b) no; 0.57 J
6–35:	743 W = 0.995 hp
6–37:	0.23
6–39:	28
6–41:	8.1×10^6 N
6–43:	b) 9 c) 108 W
6–45:	b) 11 L c) 0.10 L/km = 0.044 gal/mi
6–47:	a) 358 N b) 47.2 hp c) 4.06 hp d) 2.03%
6–49:	a) 532 J b) −315 J c) zero d) −203 J e) 14.7 J f) 1.21 m/s
6–51:	a) $1/\sin\alpha$ b) $W_{in} = W_{out}$
6–53:	a) 2.59×10^{12} J b) 4800 J
6–55:	2.59 m
6–57:	a) 4.00 N b) 32.0 N c) −15.0 J
6–59:	a) 0.15 N b) 9.4 N c) 0.44 J
6–61:	a) 2.56 m/s b) 5.28 N c) 19.7 J
6–63:	a) −910 J b) 3.17×10^3 J
6–65:	1.0×10^5 N/m
6–67:	1.1 m from where the spring is released
6–69:	a) 1.02×10^4 N/m, 8.16 m
6–71:	a) 0.600 m b) 1.50 m/s
6–73:	0.786
6–75:	a) 1.10×10^5 J b) 1.30×10^5 J c) 3.99 kW
6–77:	3.6 h
6–79:	1.30×10^3 m^3/s
6–81:	a) 1.26×10^5 J b) 1.46 W
6–83:	a) 2.4 MW b) 61 MW c) 6.0 MW
6–85:	a) 513 W b) 355 W c) 52.1 W
6–87:	a) $\frac{1}{6}Mv^2$ b) 6.1 m/s c) 3.9 m/s d) $K_{ball} = 0.40$ J, $K_{spring} = 0.60$ J
6–89:	a) 2.0×10^5 J b) 2.8×10^5 J c) 2.8×10^5 J d) 5 km/h

Chapter 7

7–1:	3.45×10^6 J
7–5:	a) 24.0 m/s b) 24.0 m/s c) part (b)
7–9:	2.8 m/s
7–11:	a) 880 J b) −157 J c) 471 J d) 253 J e) $a = 3.16$ m/s^2, $v = 7.11$ m/s, $\Delta K = 253$ J
7–13:	a) 80.0 J b) 5.00 J
7–15:	a) 6.32 cm b) 12 cm
7–17:	1.72 m
7–19:	a) 0.500 m/s b) 0.500 m
7–21:	a) 23.6 m/s b) 69.2 m/s^2, upward
7–23:	a) -1.2×10^2 J b) $+1.2 \times 10^2$ J c) zero d) conservative
7–25:	a) zero b) −0.10 J c) +0.10 J d) conservative; $U(x) = \frac{1}{3}\alpha x^3$

7–27:	a) −3.6 J b) −3.6 J c) −7.2 J d) nonconservative
7–29:	a) −59 J b) −42 J c) −59 J d) nonconservative
7–31:	$-6C_6/x^7$, attractive
7–33:	$\vec{F} = -(2kx + k'y)\hat{\imath} - (k'x + 2ky)\hat{\jmath}$
7–35:	a) $F(r) = (12a/r^{13}) - (6b/r^7)$ b) $(2a/b)^{1/6}$, stable c) $b^2/4a$ d) $a = 6.68 \times 10^{-138}$ J · m^{12}, $b = 6.41 \times 10^{-78}$ J · m^6
7–37:	a) 229 N b) 823 J
7–39:	0.41
7–41:	15.5 m/s
7–43:	4.4 m/s
7–45:	a) 919 N/m, 39.8 m b) 17.0 m/s^2
7–47:	a) 7.00 m/s b) 2.94 N
7–49:	a) $mg(1 - h/d)$ b) 4.4×10^2 N c) $\sqrt{2gh(1 - y/d)}$
7–51:	48.2°
7–53:	a) 0.392 b) −0.832 J
7–55:	a) $U(x) = (30.0$ N/m$)x^2 + (6.00$ N/m$^2)x^3$ b) 7.85 m/s
7–57:	7.01 m/s
7–59:	a) $m(g + a)^2/2gh$ b) $2gh/(g + a)$
7–61:	119 J
7–63:	a) 3.87 m/s b) 0.10 m
7–65:	a) 3.1 m/s b) 1.4 m c) no
7–67:	a) −50.6 J b) −67.5 J c) nonconservative
7–69:	a) no b) $x_0 = F/k$ d) no e) $3F/k$, $-F/k$ f) $v_{max} = 2F/\sqrt{mk}$ at $x = x_0 = F/k$
7–71:	b) $v(x) = \left[\dfrac{2\alpha}{mx_0^2}\left(\left(\dfrac{x_0}{x}\right) - \left(\dfrac{x_0}{x}\right)^2\right)\right]^{1/2}$ c) $x = 2x_0$, $v = \sqrt{\alpha/2mx_0^2}$ d) zero e) $v(x) = \left[\dfrac{2\alpha}{mx_0^2}\left(\left(\dfrac{x_0}{x}\right) - \left(\dfrac{x_0}{x}\right)^2 - \dfrac{2}{9}\right)\right]^{1/2}$ f) first case: x_0, ∞; second case: $3x_0/2, 3x_0$

Chapter 8

8–1:	a) 1.20×10^5 kg · m/s b) i) 60.0 m/s ii) 26.8 m/s
8–3:	b) baseball, 0.525 c) woman, 0.643
8–5:	0.256 kg · m/s, $-y$-direction
8–7:	562 N, no
8–9:	a) 10.8 m/s to the right b) 0.75 m/s to the left
8–11:	a) $(18.8$ kg · m/s$)\hat{\imath}$ b) $(-3.55 \times 10^{-3}$ kg · m/s$)\hat{\jmath}$ c) $(7.50 \times 10^3$ N$)\hat{\imath}$ d) $(13.0\hat{\imath} - 0.73\hat{\jmath})$ kg · m/s, $(89.3\hat{\imath} - 5.0\hat{\jmath})$ m/s
8–13:	a) $At_2 + (B/3)t_2^3$ b) $(A/m)t_2 + (B/3m)t_2^3$
8–15:	a) 5.68×10^{-2} m/s b) 0.103 m/s
8–17:	a) 4.66 m/s in the direction opposite the defender's original motion b) −6580 J
8–19:	a) 3.60 m/s b) 8.64 J
8–21:	a) $v_B = (m_A/m_B)v_A$
8–23:	525 N
8–25:	a) $v_{A2} = 29.3$ m/s, $v_{B2} = 20.7$ m/s b) 0.196 = 19.6%
8–27:	a) 6.44 m/s, eastbound b) 2.49 m/s c) -2.81×10^5 J, -1.38×10^5 J; part (a)
8–29:	35.3 km/h, 19.3° west of south
8–31:	229 m/s
8–33:	0.150-kg glider: 3.2 m/s to the left; 0.300-kg glider: 0.20 m/s to the left
8–37:	a) $9.00m$ b) 3.00×10^6 m/s
8–39:	7.42×10^8 m from the center of the sun; outside
8–41:	a) 0.30 kg b) $(2.0$ kg · m/s$)\hat{\imath}$ c) $(6.7$ m/s$)\hat{\imath}$
8–43:	a) 1.25 kg b) $(1.5$ m/s$^3)t\hat{\imath}$ c) $(5.6$ N$)\hat{\imath}$
8–45:	a) 80.0 N b) yes
8–47:	75.0 kg
8–49:	a) 0.442 b) 800 m/s c) 530 m/s
8–51:	a) 7.2×10^{-66} b) 0.223

8–53: 3.21×10^{-25} kg

8–55:   a) 6.32×10^{-22} kg • m/s   b) 5.71×10^{-19} J

8–57: $(0.521\hat{i} + 7.81\hat{j})$ m/s

8–59:   15

8–61:   30° east of north, 30° east of south

8–63:   a) 5.00 m/s, east   b) 5.71 m/s, east   c) 3.78 m/s, east

8–65:   a) Nash, 1.68   b) Packard, 0.933   c) F_p is larger, 0.933
      d) F_N is larger, 1.68

8–67:   23.2 cm

8–69:   a) 25.8 m/s, 35.5°   b) no

8–71:   a) 0.222   b) –291 J   c) 0.784 J

8–73:   b) $M = m$ c) zero

8–75:   b) $\frac{1}{2} M v_{cm}^2$

8–77:   a) 3.56 m/s   b) 5.22 m/s   c) 4.67 m/s

8–79:   *A:* 13.6 m/s; *B:* 6.34 m/s, 65.0°

8–81:   a) $(L/2)\cos{(\alpha/2)}$, along axis from apex   b) $(L/3)$, along
      bisector from bottom
      c) $L/\sqrt{8}$ along bisector   d) $L/\sqrt{12}$ from each side

8–83:   0.400 m/s

8–85:   222 m/s, 1.01×10^3 m/s; $v_{Kr} = 1.5 v_{Ba}$

8–87:   a) zero   b) 1   d) 0.87 m   f) 0.089 m

8–89:   a) yes   b) no; kinetic energy decreases by 4.8×10^3 J

8–91:   a) $1.37 v_{ex}$ b) $1.18 v_{ex}$ c) $2.38 v_{ex}$ d) 2.94 km/s

8–93:   a) $v = \begin{cases} (2400 \text{ m/s})\ln\left[\dfrac{1}{1 - t/(120 \text{ s})}\right], & 0 \leq t \leq 90 \text{ s} \\ 3.33 \text{ km/s}, & t \geq 90 \text{ s} \end{cases}$

      b) $(20 \text{ m/s}^2)/[1 - t/(120 \text{ s})]$ for $0 \leq t \leq 90$ s
      c) 6000 N vs. 735 N on earth

8–95:   b) $2L/3$

8–97:   a) $l^2 \lambda g/32$ b) $l^2 \lambda g/32$

Chapter 9

9–1:   a) 0.600 rad $= 34.4°$ b) 6.27 cm   c) 1.05 m

9–3:   a) 42 rad/s²   b) 74 rad/s

9–5:   a) $\omega(t) = (0.400 \text{ rad/s}) + (0.0360 \text{ rad/s}^3)t^2$ b) 0.400 rad/s
      c) $\omega = 1.30$ rad/s, $\omega_{av} = 0.700$ rad/s

9–7:   a) $\alpha(t) = 2b - 6ct$ b) $b/3c$

9–9:   a) –1.25 rev/s², 23.3 rev   b) 2.67 s

9–11:   a) 24.0 s   b) 68.8 rev

9–13:   10.5 rad/s

9–15:   9.00 rev

9–17:   a) 540 rad   b) 12.3 s   c) –8.17 rad/s²

9–19:   a) 47.1 m/s   b) 47.3 m/s

9–21:   a) $a_{rad} = 18.0$ m/s²   b) $v = 3.00$ m/s, $a_{rad} = 18.0$ m/s²

9–23:   a) 0.180 m/s², 0, 0.180 m/s²   b) 0.180 m/s², 0.377 m/s²,
      0.418 m/s²
      c) 0.180 m/s², 0.754 m/s², 0.775 m/s²

9–25:   10.7 cm; no

9–27:   a) 0.831 m/s   b) 109 m/s²

9–29:   a) 2.29   b) 1.51   c) 15.7 m/s, 1.06×10^3 m/s² $= 108g$

9–31: $(M/12 + m/2)L^2$

9–33:   a) 0.0640 kg • m²   b) 0.0320 kg • m²   c) 0.0320 kg • m²

9–35:   0.193 kg • m²

9–37:   b) $K = \pi^2 I \omega^2/1800$

9–39:   0.600 kg • m²

9–41: 7.35×10^4 J

9–43:   a) $K = 2\pi^2 I/T^2$ b) $dK/dt = (-4\pi^2 I/T^3)(dT/dt)$ c) 70 J
      b) –0.56 J/s

9–45:   823 J

9–47:   an axis parallel to a diameter and $\left(2/\sqrt{15}\right)R$ from the center
      of the sphere

9–49: $\frac{1}{3} M(a^2 + b^2)$

9–51: $\frac{1}{3} M(L^2 - 3Lh + 3h^2)$

9–53: $\frac{1}{3} ML^2$

9–57:   a) $(6.40 \text{ rad/s}^2)t - (1.50 \text{ rad/s}^3)t^2$
      b) $6.40 \text{ rad/s}^2 - (3.00 \text{ rad/s}^3)t$
      c) $\omega_{max} = 6.83$ rad/s at $t = 2.13$ s

9–59:   a) 35.0 km/h $= 9.72$ m/s   b) 8.51 J   c) 652 rad/s

9–61:   a) 211 rev/min   b) 800 W

9–63:   a) 1.70 m/s   b) 84.8 rad/s

9–65:   b) 2.00 m/s²   d) 0.208 kg • m²

9–67:   a) 2.14×10^{29} J   b) 2.66×10^{33} J

9–69:   a) –0.784 J   b) 5.42 rad/s   c) 5.42 m/s
      d) particle speed = 4.43 m/s

9–71: $\sqrt{(2gd)(m_B - \mu_k m_A)/(m_A + m_B + I/R^2)}$

9–73: $\sqrt{(g/R)(1 - \cos\beta)}$

9–75:   a) 2.25×10^{-3} kg • m²   b) 3.40 m/s   c) 4.95 m/s

9–77:   a) $(247/512)MR^2$ b) $(383/512)MR^2$

9–79:   b) $\frac{1}{4} M(R_1^2 + R_2^2)$

9–81:   a) $\frac{3}{5} MR^2$ b) larger

9–83:   b) 5.97×10^{24} kg   c) $0.334 MR^2$

9–85:   a) $s = r_0\theta + \beta\theta^2/2$

      b) $\theta = \left(\dfrac{1}{\beta}\right)\left[\sqrt{r_0^2 + 2\beta vt} - r_0\right]$

      c) $\omega = \dfrac{v}{\sqrt{r_0^2 + 2\beta vt}}$, $\alpha = -\dfrac{\beta v^2}{(r_0^2 + 2\beta vt)^{3/2}}$; no
      d) $r_0 = 2.50$ cm, $\beta = 0.247$ μm/rad, 2.13×10^4 rev

Chapter 10

10–1:   a) 40.0 N • m, out of the page   b) 34.6 N • m, out of the page
      c) 20.0 N • m, out of the page   d) 17.3 N • m, into the page
      e) zero   f) zero

10–3   2.50 N • m, counterclockwise

10–5:   c) $(-1.05 \text{ N} \cdot \text{m})\hat{k}$

10–7:   1.2 m/s

10–9:   a) $g(M + 3m)/(1 + 2m/M)$ b) less   c) no effect

10–11:   0.482

10–13:   a) 7.5 N in horizontal part, 18.2 N in hanging part
      b) 0.0160 kg • m²

10–15:   a) 0.882 N   b) 0.553 s   c) 33.9 rad/s

10–17:   a) 1/3   b) 2/7   c) 2/5   d) 5/13

10–19:   11.7 m

10–21:   a) 0.309 rad/s   b) 100 J   c) 6.67 W

10–23:   a) 0.38 N • m   b) 160 rad   c) 59 J   d) 59 J

10–25:   b) 65.6 N

10–27:   a) 358 N • m   b) 1.79×10^3 N   c) 83.8 m/s

10–29:   a) 115 kg • m²/s, into the page   b) 125 kg • m²/s², out of
      the page

10–31: 4.71×10^{-6} kg • m²/s

10–33:   b) 7.00 rad/s   c) 1.03×10^{-2} J   d) 1.03×10^{-2} J

10–35:   0.60 rev

10–37:   a) 1.4 rad/s   b) 1080 J before, 500 J after

10–41:   a) 36.8 min   b) 1.10×10^5 N • m

10–43: 5.4×10^{22} N • m

10–45:   a) 0.955 kg • m²   b) –0.0800 N • m   c) 104 rev

10–47:   b) 4500 W   d) 2600 W

10–49:   0.675 s

10–51:   a) $L/4$ from the end with the clay   b) $(9g/8L) \sin\theta$
      c) $(3g/2L) \sin\theta$

10–53:   1200 N

10–55: a) 1.12 m/s^2 b) 14.0 N
10–57: $a = F/2m, f = F/2$
10–59: a) 0.957 m
10–61: a) $(27R - 17r)/10$ b) $(5R - 3r)/2$
10–63: a) 1.76 N b) 123 rad/s^2 c) 9.80 m/s^2
d) T would be unchanged, α and a would double
10–65: a) 9.34 m/s b) 18.7 m/s c) zero d) 5.60 m
10–67: 3.4 m
10–69: $g/3$
10–71: a) $6v/19L$ b) 3/19
10–73: a) 5.46 rad/s b) 3.17 cm c) 1.01×10^3 m/s
10–75: 3200 J
10–77: 0.30 rad/s, clockwise
10–81: 12.7 rad/s
10–85: a) $a = +\mu_k g$, $\alpha = -2\mu_k g/R$ b) $\omega_0^2 R^2/18\mu_k g$ c) $-M\omega_0^2 R^2/6$
10–87: a) $mv_1^2 r_1^2/r^3$ b) $(mv_1^2/2)[(r_1/r_2)^2 - 1]$ c) same

Chapter 11

11–1: 0.387 m from center of 1.00-kg sphere
11–5: 5450 N
11–7: 1000 N, 1.20 m from end where 400-N force is applied
11–11: a) 1920 N b) 1140 N
11–13: a) $T = 2.60 w$; $F_{pivot} = 3.28 w$ at 37.6° b) $T = 4.10 w$;
$F_{pivot} = 5.38 w$ at 48.8°
11–15: 140 N by each hinge
11–17: $T_{left} = 270$ N, $T_{right} = 304$ N, $\theta = 40°$
11–19: a) 0.800 m b) clockwise c) 0.800 m, clockwise
11–21: 1.4 mm
11–23: 2.0×10^{11} Pa
11–25: a) upper: 3.1×10^{-3}; lower: 2.0×10^{-3} b) upper: 1.6 mm;
lower: 0.98 mm
11–27: 9.1×10^6 N
11–29: a) 3.33×10^6 Pa b) 1.33×10^5 N
11–31: 4.8×10^9 Pa, $2.1 \times 10^{-10} \text{ Pa}^{-1}$
11–33: 1/9
11–35: 3.41×10^7 Pa
11–37: 10.2 m/s^2
11–41: wing force: 7300 N upward; tail force: 600 N downward
11–43: a) 590 N b) neutral
11–47: 120 N to the left, 160 N downward
11–49: a) fore: 37 N; hind: 58 N b) fore: 53 N; hind: 55 N
11–51: a) $\tau = Fl \sin \phi$ b) 12.6 N • m with either method of
calculation
11–53: a) $V = mg + w$, $H = T = (w + mg/4) \cot \theta$ b) 9.50×10^2 N
c) 4.00°
11–55: a) 4.90 m b) 60 N
11–57: a) 300 N upward, 225 N • m counterclockwise
b) $T = 132$ N; $F_{pivot} = 194$ N, 71.2° above horizontal
11–59: a) 60.0 N b) 53.1°
11–61: person above: 590 N; person below: 1370 N; above
11–63: a) 268 N b) 232 N c) 366 N
11–65: a) tips at 27°, slips at 31°; the bale tips before it slips
b) tips at 27°, slips at 22°; the bale slips before it tips
11–67: a) $F_A = 80$ N, $F_B = 870$ N b) 1.92 m
11–69: a) 3.7 kN, 2.0 kN vertically upward
11–71: a) 5.4 mm b) 4.2 mm
11–73: a) 0.70 m from wire A b) 0.45 m from wire B
11–75: a) 1.20 m b) brass: 2.00×10^8 Pa; nickel: 4.00×10^8 Pa
c) brass: 2.22×10^{-3}; nickel: 1.90×10^{-3}
11–77: a) 0.36 mm b) 0.045 mm c) 0.33 mm
11–79: a) $(F \cos^2 \theta)/A$ b) $(F \sin 2\theta)/2A$ c) 0 d) 45°
11–81: a) 600 N b) 13.5 kN
c) slide: $\mu_s w/(\sin \theta - \mu_s \cos \theta)$; tip: $w/[(1/9)\cos \theta + 2 \sin \theta]$; 66°
11–83: the lesser of $h^2/L + L/2$ and L

11–85: $[(A^2 x/F) - k_0 V_0]/V_s$
11–87: a) 0.662 mm b) 2.20×10^{-2} J c) 8.33×10^{-3} J
d) -3.04×10^{-2} J e) 3.04×10^{-2} J

Chapter 12

12–1: 2.18
12–3: F_{12}
12–5: a) 2.59×10^8 m
12–7: 2.4×10^{-3} N
12–9: a) 6.30×10^{20} N, toward sun b) 4.77×10^{20} N, 24.6° toward
earth from sun
c) 2.37×10^{20} N, toward sun
12–11: $2.1 \times 10^{-9} \text{ m/s}^2$, down
12–13: 1.38×10^7 m
12–15: a) 0.37 m/s^2 b) 1700 kg/m^3
12–17: 5.98×10^{24} kg
12–19: 0.83 m/s; yes
12–21: a) 5.02×10^3 m/s b) 6.06×10^4 m/s
12–23: a) 175 min b) 3.71 m/s²
12–25: 7.46×10^3 m/s
12–27: 8.3×10^4 m/s b) 1.3×10^6 s
12–29: b) 4.45×10^{12} m/s, 4.55×10^{12} m b) 248 y
12–33: a) $-GMm/\sqrt{a^2 + x^2}$ c) $GMmx/(a^2 + x^2)^{3/2}$ toward the center
of the ring
e) $-GMm/a$, zero
12–35: a) 53 N b) 52 N
12–37: a) 4.3×10^{37} kg = 2.1×10^7 M_{sun} b) no c) 6.3×10^{10} m; yes
12–39: 1.39×10^{-9}
12–41: a) 2.20×10^{-10} N, 163° from +x-direction toward
+y-direction
b) $x = 0$, $y = 1.39$ m
12–43: a) 3.40 N, 0.61° b) 1.31×10^9 J
12–45: 3.83×10^8 m
12–47: a) 3.58×10^7 m
12–49: 1.8×10^2 m/s
12–51: $0.01R_E = 64$ km
12–53: 6.06×10^3 km/h
12–55: $\sqrt{2Gm_E h/(R_E^2 + hR_E)}$
12–57: a) 13.7 km/s b) 13.3 km/s c) 13.2 km/s
12–59: a) i) 2.8 y ii) 6.1 y b) 4.90×10^8 km c) 4.22×10^8 km
12–61: a) $GM^2/4R^2$ b) $\sqrt{GM/4R}$, $4\pi \sqrt{R^3/GM}$ c) $GM^2/4R$
12–63: 6.8×10^4 m/s
12–65: a) 7.91×10^3 s b) 1.53 c) 8.43×10^3 m/s, 5.51×10^3 m/s
d) 2.41×10^3 m/s, 3.26×10^3 m/s; perigee
12–67: 3.22×10^9 J
12–69: 1.4×10^{14} m
12–71: $GMm/(x^2 + xL)$
12–73: b) 22 min c) 2.5×10^8 s = 2900 d = 7.9 y
12–75: a) 2.1 kN b) no
12–77: perpendicular: $GMm/(a\sqrt{L^2 + a^2})$; parallel: zero

Chapter 13

13–1: a) 4.54×10^{-3} s, 1.38×10^3 rad/s b) 1.14×10^{-3} s,
5.53×10^3 rad/s
13–3: 5.53×10^3 rad/s, 1.14×10^{-3} s
13–5: a) 0.167 s b) 37.7 rad/s c) 8.44×10^{-2} kg
13–7: a) 0.375 s b) 2.66 Hz c) 16.7 rad/s
13–9: a) $x = (3.0 \text{ mm}) \cos [2\pi(440 \text{ Hz})t]$ b) 8.3 m/s, $2.3 \times 10^4 \text{ m/s}^3$
c) jerk = $(6.3 \times 10^7 \text{ m/s}^3) \sin [2\pi(440 \text{ Hz})t]$, $6.3 \times 10^7 \text{ m/s}^3$
13–11: a) 0.383 m b) 1.02 rad = 58.5°
c) $x = (0.383 \text{ m}) \cos [(12.2 \text{ rad/s})t + 1.02 \text{ rad}]$
13–13: 0.353 m

13–17: a) $A/\sqrt{2}$, $\omega A/\sqrt{2}$ b) four times; $\pi/2\omega$ c) 3/4 kinetic, 1/4 potential

13–19: a) 5.13 m/s^2, 0.961 m/s b) −2.57 m/s^2, 0.833 m/s c) 0.137 s d) v, a can be found with energy approach; t cannot

13–21: a) 2.84×10^{-2} J b) 0.014 m c) 0.615 m/s

13–23: a) 5.31×10^3 N/m b) 0.695 s

13–25: 3.97 cm

13–27: a) 2.7×10^{-8} kg $\cdot$ m^2 b) 4.3×10^{-6} N $\cdot$ m/rad

13–29: 5.12×10^{-2} kg $\cdot$ m^2

13–33: 2.60 s

13–35: 2.00 m

13–37: 10.7 m/s^2

13–41: 0.58 s

13–43: a) 0.393 Hz b) 1.73 kg/s

13–45: a) A b) magnitude = $bA/2m$, in $-x$-direction; slope is negative

c) $a(0) = A\left(\dfrac{b^2}{2m^2} - \dfrac{k}{m}\right)$; if $b < \sqrt{2mk}$, $a(0) < 0$;

if $b = \sqrt{2mk}$, $a(0) = 0$; if $b > \sqrt{2mk}$, $a(0) > 0$

13–47: a) kg/s b) i) $5.0F_{max}/k$ ii) $2.5F_{max}/k$

13–49: a) 6.72×10^3 m/s^2 b) 3.02 kN c) 18.3 m/s, 75.6 J d) 17.6 kW e) 12.1 kN, 36.7 m/s, 302 J, 141 kW

13–51: a) all unchanged b) 1/4 as large c) halved d) $1/\sqrt{5}$ as great e) U: unchanged; K: 1/5 as large

13–53: a) 1.68 s b) 0.090 m c) 0.143

13–55: $\mu_s(m + M)g/k$

13–59: a) 0.150 m/s b) 0.112 m/s^2, downward c) 0.700 s d) 4.38 m

13–61: a) 2.6 m/s b) 0.21 m c) 0.49 s

13–63: 0.933 s

13–65: $t_1 = T/4$

13–71: d) $\sqrt{A/\pi^2 mR_0^3}$

13–73: $2\pi\sqrt{3M/2k}$

13–75: 0.88 m

13–77: a) 3.97 m

13–79: c) 0.430 m

13–81: a) spring 1: 0.350 m; spring 2: 0.250 m b) 0.702 s

13–83: a) $\Delta T/T = -\frac{1}{2}(\Delta g/g)$ b) 9.7791 m/s^2

13–85: a) $T/T_0 = \sqrt{(L^2 + 3y^2)/(L^2 + 2yL)}$ b) $y = 2L/3$

13–87: a) $kl_0/(k - m\omega'^2)$

Chapter 14

14–1: $w = 41.8$ N; no

14–3: 7.02×10^3 kg/m^3; yes

14–5: a) 706 Pa b) 3160 Pa

14–7: a) 2.52×10^6 Pa b) 1.78×10^5 N

14–9: a) 1.07×10^5 Pa b) 1.03×10^5 Pa c) 1.03×10^5 Pa d) 5.33×10^3 Pa

14–11: 2.3×10^5 N

14–13: 4.14 m

14–15: 0.562 m^3

14–17: a) $\rho < \rho_{fluid}$ c) submerged: ρ/ρ_{fluid}; above: $(\rho_{fluid} - \rho)/\rho_{fluid}$ d) 32%

14–19: a) 116 Pa b) 921 Pa c) 0.822 kg, 822 kg/m^3

14–21: 6.67 Pa

14–23: 0.1 N, 0.01 kg

14–25: 4.14×10^{-3} N

14–27: a) 17.0 m/s b) 0.317 m

14–29: 28.4 m/s

14–31: 1.47×10^5 Pa

14–33: 500 N downward

14–35: 0.41 cm

14–37: a) 1.88 m/s b) 0

14–39: a) 0.128 m^3/s b) 9.72×10^4 Pa c) 0.275 m^3/s

14–41: 5.96 mm/s

14–43: a) 16 times greater b) 1/2 as great c) twice as great d) twice as great e) 1/2 as great

14–45: a) 19.4 m/s, 0, 14.6 m/s b) 152 d c) in (a), 4; in (b), 1/16

14–47: a) 1.10×10^8 Pa b) 1080 kg/m^3, 5%

14–49: 2.61×10^4 N $\cdot$ m

14–51: $(p - p_0)VR^2/Gmd$

14–53: a) 12,700 kg/m^3, 3140 kg/m^3

14–55: a) 1470 Pa b) 13.9 cm

14–57: 9.8×10^6 kg, yes

14–59: a) 30% b) 70%

14–61: 4.66×10^{-4} m^3, 5.27 kg

14–63: a) 1.10×10^4 m^3 b) 112 kN

14–65: a) 0.107 m b) 2.42 s

14–67: 0.0958 kg

14–69: 33.5 N

14–71: b) 12.2 N c) 11.8 N

14–73: b) 2.52×10^{-4} m^3, 0.124

14–75: risen by 5.57×10^{-4} m

14–77: a) $1 - \rho_B/\rho_L$ b) $(\rho_L - \rho_B)L/(\rho_L - \rho_W)$ c) 4.60 cm

14–79: a) la/g b) $w^2l^2/2g$

14–83: 7.0 N $\cdot$ m

14–85: 13.1 cm

14–87: 133 m/s

14–89: $3h_1$

14–91: a) $r = r_0(1 + 2gy/v_0^2)^{-1/4}$ b) 1.10 m

14–93: a) 0.0130 m/s b) 2.16 m/s

14–95: a) 6.86×10^{-5} m^3/s b) cd: 0.686 m/s; ef: 1.71 m/s; gh: 3.43 m/s c) c and d: 0.576 m; e and f: 0.450 m g and h: 0 d) 0.0264 m e) 0.165 m f) cd: 1.37 m/s ef: 3.43 m/s gh: 6.86 m/s

14–97: a) 80.4 N

14–99: a) $\sqrt{2gh}$ b) $(p_a/\rho g) - h$

Chapter 15

15–1: a) −81.0°F b) 134.1°F c) 88.0°F

15–3: −40°, winter

15–5: a) 104.4°F, yes b) 54°F

15–7: a) 216.5 K b) 325.9 K c) 205.4 K

15–9: 27.23 K

15–11: 444 mm Hg

15–13: a) −282°C b) no, 4.76×10^4 Pa

15–15: 0.39 m

15–17: a) 1.9014 cm b) 1.8964 cm

15–19: 2.3×10^{-5} (C°)$^{-1}$

15–21: 11 L

15–23: 1.7×10^{-5} (C°)$^{-1}$

15–25: a) 1.431 cm^2 b) 1.437 cm^2

15–27: a) 3.2×10^{-5} (C°)$^{-1}$ b) 2.5×10^9 Pa

15–29: a) 38 J b) 4.5×10^4 J

15–31: 0.526 C°

15–33: 3.03×10^5 J

15–35: 5.79×10^5 J

15–37: a) 2.51×10^3 J/kg $\cdot$ K b) overestimate

15–39: 8 min

15–41: 3.64×10^4 J = 8.69×10^3 cal = 34.5 Btu

15–43: 2.39×10^4 Btu/h = 7.01×10^3 W

15–45: 357 m/s
15–47: 3.45 L
15–49: 27.5°C
15–51: 0.0940 kg
15–53: 2.10 kg
15–55: 190 g
15–57: a) 222 K/m b) 10.7 W c) 73.3°C
15–59: a) −5.8°C b) 11 W/m^2
15–61: 7.1×10^2 Btu $= 7.5 \times 10^5$ J
15–63: 105.5°C
15–65: 167 W
15–67: 2.1 cm^2
15–69: 6.0×10^3 W/m^2
15–71: −6°C
15–73: 5.71 K/W
15–75: b) 5.0×10^{-3} cm^3, −23 kg/m^3
15–77: 23.0 cm, 7.0 cm
15–79: b) 1.9×10^8 Pa
15–81: a) 87°C b) −80°C
15–83: 20.2°C
15–85: a) 54.3
15–87: a) 83.6 J b) 1.86 J/mol • K c) 5.60 J/mol • K
15–89: a) 2.7×10^7 K b) 6.89 K c) 19.3 K
15–91: a) 4.7×10^{-3} kg b) yes
15–93: a) 86.1°C b) no ice, 0.130 kg liquid water, no steam
15–95: a) 100°C b) 0.0214 kg steam, 0.219 kg liquid water
15–97: a) 100°C b) no ice, 0.419 kg liquid water, 0.0264 kg steam
c) 0°C, 0.154 kg ice, 0.408 kg liquid water, no steam
15–99: a) 94 W b) 1.3
15–101: 2.9
15–103: c) 6.0×10^5 s (about 170 h) d) 1.5×10^{10} s (about 500 y); no
15–105: 305 min
15–107: 5.82×10^{-3} kg
15–109: a) 69.6°C
15–111: a) $H = 4\pi kab\Delta T/(b-a)$ b) $T(r) = \dfrac{[T_1 b(r-a) + T_2 a(b-r)]}{r(b-a)}$

c) $H = 2\pi kL\Delta T/\ln(b/a)$ d) $T(r) = \dfrac{[T_1 \ln(r/a) + T_2 \ln(b/r)]}{\ln(b/a)}$
15–113: b) uniform 0°C d) 3.14×10^3 C°/m e) 121 W f) 0
g) 1.1×10^{-4} m^2/s
h) −11 C°/s i) 9.1 s j) decrease k) −7.7 C°/s
15–115: 20.42 K

Chapter 16

16–1: a) 56.2 mol b) 6.81×10^6 Pa $= 67.2$ atm
16–3: 0.959 atm
16–5: 503.0°C
16–7: 3.36×10^5 Pa
16–9: 0.159 L
16–11: 1.05 atm
16–13: a) 70.2°C b) yes
16–15: 2.28×10^4 Pa
16–17: a) 8.2×10^{-17} atm b) no
16–19: 55.6 mol, 3.35×10^{26} molecules
16–21: a) 9.00×10^{-5} m^3 b) 3.10×10^{-10} m c) about the same
16–23: b) 1.004
16–25: a) 6.21×10^{-21} J b) 2.34×10^5 m^2/s^2 c) 484 m/s
d) 2.57×10^{-23} kg • m/s e) 1.24×10^{-19} N
f) 1.24×10^{-17} Pa g) 8.15×10^{21} h) 2.45×10^{22}
16–27: 3800°C
16–29: a) 24.9 J/mol • K
16–31: a) 1560 J b) 935 J
16–33: a) 987 K b) 438 K c) 110 K
16–37: 2500 m

16–39: solid and vapor, with no liquid
16–41: a) 6.9×10^{-3} L b) yes
16–43: 0.213 kg
16–45: 0.00123 g
16–47: 1.92 atm
16–49: a) 31 b) 8.41×10^3 N c) 7.80×10^3 N
16–51: a) 26.2 m/s b) 16.1 m/s, 5.44 m/s c) 1.74 m
16–53: 5.0×10^{27}
16–55: a) 303 mol/m^3 c) van der Waals
16–57: b) r_2 c) $r_1 = R_0/2^{1/6}$, $r_2 = R_0$, $r_1/r_2 = 2^{-1/6}$ d) U_0 e) $r \to \infty$
16–59: a) 517 m/s b) 299 m/s
16–61: b) 1.40×10^5 K, 1.01×10^4 K c) 3.18×10^3 K, 2.30×10^2 K
16–63: a) 1.24×10^{-14} kg b) 4.16×10^{11} d) 2.95×10^{-6} m, no
16–65: a) $2R$ b) less
16–67: CO$_2$: 20.79 J/mol • K, 0.270; SO$_2$: 24.94 J/mol • K, 0.205;
H$_2$S, 24.94 J/mol • K, 0.039
16–71: b) 0.0420N c) $(2.94 \times 10^{-21})N$ d) 0.0297N, $(2.08 \times 10^{-21})N$
e) 0.0595N, $(4.15 \times 10^{-21})N$
16–73: 42.6%
16–75: a) 4.5×10^{11} m b) 703 m/s, 6.4×10^8 s (about 20 y), no
c) 1.4×10^{-14} Pa
d) about 650 m/s; evaporate f) 2×10^5 K; no
16–77: d) $T_c = 8a/27Rb$, $(V/n)_c = 3b$ e) $p_c = a/27b^2$ f) 8/3
g) 3.28, 3.44, 4.35

Chapter 17

17–1: b) 1330 J
17–3: b) 0
17–5: b) 0 for first process, -4.00×10^4 J for second process
17–7: 327 J
17–9: a) -1.15×10^5 J b) 2.55×10^5 J, out of gas c) no
17–11: a) 16.4 min b) 139 m/s $= 501$ km/h
17–13: a) absorbs b) 7200 J c) liberates 7200 J
17–15: a) yes b) no c) negative
17–17: a) 948 K b) 900 K
17–19: a) −605 J b) 0 c) yes, 605 J, liberates
17–21: $C_V = 65.47$ J/mol • K, $C_p = 73.78$ J/mol • K
17–23: a) 553 J b) 454 J
17–25: b) 495°C, 2.62 atm
17–27: b) 224 J c) $Q = 0$ d) −224 J
17–29: a) 4.50×10^4 Pa b) 345 J c) 0.818
17–31: a) 45.0 J b) liberates 65.0 J c) $Q_{ad} = 23.0$ J, $Q_{db} = 22.0$ J
17–33: a) $W_{ab} = 0$, $W_{bc} = p_c(V_c - V_a)$, $W_{ad} = p_a(V_c - V_a)$, $W_{dc} = 0$
b) $Q_{ab} = U_b - U_a$, $Q_{bc} = U_c - U_b + p_c(V_c - V_a)$,
$Q_{ad} = U_d - U_a + p_a(V_c - V_a)$, $Q_{dc} = U_c - U_d$
c) $Q_{abc} = U_c - U_a + p_c(V_c - V_a)$, $W_{abc} = p_c(V_d - V_a)$;
$Q_{adc} = U_c - U_a + p_a(V_c - V_a)$, $W_{adc} = p_a(V_d - V_a)$; abc, abc
17–35: −0.226 m^3
17–37: a) 4.32×10^{-4} m^3 b) 648 J c) 7.15×10^5 J d) 7.14×10^5 J
e) no substantial difference
17–39: 3.4×10^5 J/kg
17–41: b) 13.7 C°
17–43: a) 0.173 m b) 206°C c) 7.46×10^4 J
17–45: a) $Q = 300$ J, $\Delta U = 0$ b) $Q = 0$, $\Delta U = -300$ J c) $Q = 750$ J,
$\Delta U = 450$ J
17–47: a) $W = 738$ J, $Q = 2588$ J, $\Delta U = 1850$ J
b) $W = 0$, $Q = -1850$ J, $\Delta U = -1850$ J
c) 0
17–49: a) $W = -187$ J, $Q = -654$ J, $\Delta U = -467$ J
b) $W = 113$ J, $Q = 0$, $\Delta U = -113$ J
c) $W = 0$, $Q = 580$ J, $\Delta U = 580$ J
17–51: b) 3.60×10^3 J c) 0 d) 3.60×10^3 J e) 0.114 m^3
17–53: a) $W = nRT \ln[(V_2 - nb)/(V_1 - nb)] + an^2(1/V_2 - 1/V_1)$
b) i) 2.80×10^3 J ii) 3.11×10^3 J
c) 310 J; larger for ideal gas

Chapter 18

18–1: a) 6500 J   b) 0.34 = 34%

18–3: a) 0.23 = 23%   b) 12,400 J   c) 0.350 g
d) 222 kW = 298 hp

18–5: a) 0.25 = 25%   b) 970 MW

18–7: a) 453.2°C   b) 1.99×10^6 Pa

18–9: a) 1.62×10^4 J   b) 5.02×10^4 J

18–11: a) 767 W   b) 7.27

18–13: a) 215 J   b) 378 K   c) 0.391 = 39.1%

18–15: a) 3.09×10^7 J   b) 2.49×10^6 J

18–17: a) −665 J, refrigerator   b) −190 J, refrigerator
c) 285 J, engine

18–19: b) $e \to 1$, $K = 0$; $e \to 0$, $K \to \infty$

18–21: 47.4 J/K

18–23: −644 J/K

18–25: a) 109 J/K   b) N_2: 72.8 J/K; Ag: 103 J/K; Hg: 86.6 J/K

18–27: 6.74 J/K

18–29: a) no   b) 18.3 J/K   c) 18.3 J/K

18–31: 142 m^2, yes

18–33: a) 15 L   b) 5

18–35: a) 34.1 m^3 b) 8.02 m^3

18–37: a) the product   b) 23%   c) better

18–39: a) $p_1 = 1.00$ atm, $V_1 = 8.62 \times 10^{-3}$ m^3; $p_2 = 2.00$ atm,
$V_2 = 8.62 \times 10^{-3}$ m^3; $p_3 = 1.00$ atm, $V_3 = 1.31 \times 10^{-2}$ m^3
b) $1 \to 2$: $Q = 2183$ J, $W = 0$, $\Delta U = 2183$ J; $2 \to 3$: $Q = 0$,
$W = 1055$ J, $\Delta U = -1055$ J; $3 \to 1$: $Q = -1579$ J, $W = -451$ J,
$\Delta U = -1128$ J   c) 604 J   d) 604 J
e) $e = 27.7\%$, $e_{\text{Carnot}} = 50.0\%$

18–41: 10.5%

18–43: a) 7.36×10^5 J   b) 9.33×10^5 J   c) 1.98×10^5 J   d) 3.73
e) 3.71

18–45: a) $a \to b$: $Q = -nRT_1 \ln r$, $W = -nRT_1 \ln r$, $\Delta U = 0$;
$b \to c$: $Q = nC_V(T_2 - T_1)$, $W = 0$, $\Delta U = nC_V(T_2 - T_1)$;
$c \to d$: $Q = nRT_2 \ln r$, $W = nRT_2 \ln r$, $\Delta U = 0$;
$d \to a$: $Q = -nC_V(T_2 - T_1)$, $W = 0$, $\Delta U = -nC_V(T_2 - T_1)$
c) $1 - T_1/T_2$, same, no

18–47: b) 14 m/s = 50 km/h

18–49: a) 122 J, 78 J   b) 5.10×10^{-4} m^3
c) $p_b = 2.32 \times 10^6$ Pa, $V_b = 4.81 \times 10^{-5}$ m^3, $T_b = 771$ K;
$p_c = 4.00 \times 10^6$ Pa, $V_c = 4.81 \times 10^{-5}$ m^3, $T_c = 1333$ K;
$p_d = 1.47 \times 10^5$ Pa, $V_d = 5.10 \times 10^{-4}$ m^3, $T_d = 518$ K
d) $e_{\text{Carnot}} = 77.5\%$, $e = 61.1\%$

18–51: 10.5 J/K

18–55: a) 150 J/K   b) −120 J/K   c) 30 J/K   d) irreversible

18–57: a) $e = 1 - \dfrac{1}{\gamma}\left[\dfrac{(T_c/T_a)^\gamma r^{\gamma(1-\gamma)} - 1}{(T_c/T_a) - r^{\gamma-1}}\right]$ b) 70.8%

Chapter 19

19–1: a) 1.2 m/s   b) 0.31 m   c) a): unchanged; b): 0.15 m

19–3: a) 0.439 m   b) 5.25×10^6 Hz

19–5: a) 17.2 m, 0.0172 m   b) 74.0 m, 0.0740 m

19–7: a) $f = 25.0$ Hz, $T = 0.0400$ s, $k = 19.6$ rad/m

b) $y(x, t) = (0.0700 \text{ m}) \sin 2\pi\left(\dfrac{t}{0.0400 \text{ s}} + \dfrac{x}{0.320 \text{ m}}\right)$

c) −0.0495 m   d) 0.0050 s
e) 0.0200 s

19–9: c) +x-direction for both
d) $v_y(x, t) = -\omega A \sin(\omega t + kx)$, $a_y(x, t) = -\omega^2 A \cos(\omega t + kx)$

19–11: a) 4.0 mm   b) 0.040 s   c) 0.14 m, 3.6 m/s   d) 0.24 m,
6.0 m/s   e) no

19–13: b) +x-direction

19–15: a) 16.3 m/s   b) 0.136 m   c) both increase by $\sqrt{2}$

19–17: 0.390 s

19–19: a) 1.33×10^{10} Pa   b) 9.47×10^{10} Pa

19–21: 90.8 m

19–23: a) −82°C   b) no

19–25: 0.208 s

19–27: $Y/900$

19–29: a) 16.0 m   b) 3.29×10^{-8} m   c) 4.55×10^{-5} m/s

19–31: a) 20.0 Hz, 126 rad/s, 3.49 rad/m
b) $y(x, t) = (2.50 \text{ mm}) \sin [(126 \text{ rad/s})t - (3.49 \text{ rad/m})x]$
c) $y(t) = (2.50 \text{ mm}) \sin [(126 \text{ rad/s})t]$
d) $y(t) = (2.50 \text{ mm}) \sin [(126 \text{ rad/s})t - 3\pi/2]$
e) 0.314 m/s   f) 0, −0.314 m/s

19–33: a) $(7L/2)\sqrt{\mu_1/F}$ b) no

19–35: a) $4\pi^2 F \Delta x/\lambda^2$

19–37: a) 1, 0; 2, +; 3, −; 4, 0; 5, −; 6, +
b) 1, −; 2, +; 3, −; 4, +; 5, −; 6; 0
c) (a): answers would reverse sign; (b): no change

19–39: c) C/B

19–41: b) k decreases by factor of $2\sqrt{2}$; ω decreases by factor of $\sqrt{2}$

19–43: $\alpha = (v_1^2 - v_2^2)\,\rho/Y\Delta T$

19–45: b) $(1/2)\mu\omega^2 A^2 \cos^2(\omega t - kx)$ e) $(1/2)Fk^2 A^2 \cos^2(\omega t - kx)$

19–47: $\pi/\omega\sqrt{2}$

Chapter 20

20–3: a) $(1.33 \text{ m})n$, $n = 0, 1, 2, 3, \ldots$
b) $(1.33 \text{ m})(n + 1/2)$, $n = 0, 1, 2, 3, \ldots$

20–5: a) $y(x, t) = (0.850 \text{ cm}) \sin [2\pi x/(0.300 \text{ m})] \cos [2\pi t/(0.0750 \text{ s})]$
b) 4.00 m/s   c) 0.688 cm

20–9: a) 96.0 m/s   b) 461 N

20–11: a) 45.0 cm   b) no

20–13: 3.00 m, 16.0 Hz   b) 1.00 m, 48.0 Hz   c) 0.750 m, 64.0 Hz

20–15: a) *fundamental:* displacement node at 0.60 m, pressure
nodes at 0 and 1.20 m; *first overtone:* displacement nodes
at 0.30 m and 0.90 m, pressure nodes at 0, 0.60 m, and 1.20
m; *second overtone:* displacement nodes at 0.20 m, 0.60 m,
and 1.00 m, pressure nodes at 0, 0.40 m, 0.80 m, and 1.20 m
b) *fundamental:* displacement node at 0, pressure node at
1.20 m; *first overtone:* displacement nodes at 0 and 0.80 m,
pressure nodes at 0.40 m and 1.20 m; *second overtone:*
displacement nodes at 0, 0.48 m, and 0.96 m, pressure nodes
at 0.24 m, 0.72 m, and 1.20 m

20–17: 506 Hz, 1520 Hz, 2530 Hz

20–21: a) 172 Hz   b) 86 Hz

20–23: a) 614 Hz   b) 1.23 kHz

20–25: c) yes

20–27: c) $2A$, $2A\omega$, $2A\omega^2$

20–29: a) 20.0 m/s   b) four times per second

20–31: a) stopped   b) $n = 7$, $n = 9$ c) 43.9 cm

20–33: a) $v/2L$, v/L, $3v/2L$

20–35: a) 0, L b) 0, $L/2$, L d) no

20–37: a) 375 m/s   b) 1.39   c) 0.8 cm

20–39: a) 148 N   b) 26%

20–41: a) 0.246 m   b) 56.3°C

20–43: a) 0.026 m, 0.534 m, 1.27 m, 2.71 m, 9.01 m
b) 0.262 m, 0.861 m, 1.84 m, 4.34 m, ∞   c) 86 Hz

20–45: a) 99.4 N   c) −4.25 Hz, falls

Chapter 21

21–1: a) 0.344 m   b) 1.2×10^{-5} m   c) 6.9 m, 50 Hz

21–3: a) 7.78 Pa   b) 77.8 Pa   c) 778 Pa

21–5: a) 1.94 Pa b) 4.58×10^{-3} W/m^2 c) 96.6 dB

21–7: 4.36×10^{-12} W/m^2, 6.40 dB

21–9: 14.0 dB

21–11: a) 20.0

21–13: a) 4 beats/s b) 3.0×10^{-8} m, 0

21–15: 267 Hz

21–17: a) 0.798 m b) 0.922 m c) 431 Hz d) 373 Hz

21–19: a) 302 Hz b) 228 Hz

21–23: a) 0.0114 Pa b) 7.51×10^{-9} m c) 63.0 m

21–25: a) $I = 2\pi^2\sqrt{\rho B}\, f^2 (\Delta R)^2$ b) $P = 8\pi^3\sqrt{\rho B}\, f^2 R^2 (\Delta R)^2$

c) $A = (R/d)\Delta R$, $\ p_{max} = 2\pi\sqrt{\rho B}\,(Rf/d)\Delta R$,

$I = 2\pi^2\sqrt{\rho B}\,(fR/d)^2(\Delta R)^2$

21–27: a) 6.74 cm b) 147 Hz

21–29: b) 2.0 m/s

21–31: c) 56.8 m/s

21–33: a) greater, 109 Hz b) greater, 218 Hz

21–37: a) 180° b) A: 3.98×10^{-6} W/m^2, 66.0 dB;
B: 5.31×10^{-7} W/m^2, 57.2 dB
c) 1.60×10^{-6} W/m^2, 62.0 dB

Chapter 22

22–1: a) 2.00×10^{10} b) 8.58×10^{-13}

22–3: 2×10^{28} electrons, 3×10^{9} C

22–5: -1.74×10^{5} C

22–7: a) 7.42×10^{-7} C on each sphere
b) 3.71×10^{-7} C on one, 1.48×10^{-6} C on the other

22–9: 5.08 m

22–11: charge on the left exerts a 0.172-N force to the right on the right-hand charge, charge on the right exerts a 0.172-N force to the left on the left-hand charge

22–13: 2.58×10^{-6} C, $-y$-direction

22–15: b) $F_x = 0$, $F_y = +2kqQa/(x^2 + a^2)^{3/2}$ c) $2kqQ/a^2$, $+y$-direction

22–17: b) $kq^2(1 + 2\sqrt{2})/2L^2$

22–19: a) -21.9 μC b) 1.02×10^{-7} N/C

22–21: a) 0.113×10^{-4} N/C, upward b) 5.24×10^{-22} N, upward

22–23: a) 364 N/C b) no, 2.73 μm downward

22–25: a) $mg = 8.93 \times 10^{-30}$ N b) 1.63×10^{-19} kg $= 1.79 \times 10^{14}$ m_e c) no

22–27: a) $-\hat{\jmath}$ b) $(\hat{\imath} + \hat{\jmath})/\sqrt{2}$ c) $-0.390\hat{\imath} + 0.921\hat{\jmath}$

22–29: a) i) 574 N/C, $+x$-direction ii) 268 N/C, $-x$-direction
iii) 404 N/C, $-x$-direction
b) i) 9.20×10^{-17} N, $-x$-direction ii) 4.30×10^{-17} N,
$+x$-direction iii) 6.48×10^{-17} N, $+x$-direction

22–31: a) 0 b) $E_x = -2kq(x^2 + a^2)/(x^2 - a^2)^2$ for $x < -a$;
$E_x = -4kqax/(x^2 - a^2)^2$ for $-a < x < +a$;
$E_x = +2kq(x^2 + a^2)/(x^2 - a^2)^2$ for $x > +a$

22–33: a) $E_x = E_y = E = 0$
b) $E_x = +2.66 \times 10^{3}$ N/C, $E_y = 0$, $E = 2.66 \times 10^{3}$ N/C,
$+x$-direction
c) $E_x = +129$ N/C, $E_y = -510$ N/C, $E = 526$ N/C, 284°
clockwise from $+x$-direction
d) $E_x = 0$, $E_y = E = +1.38 \times 10^{3}$ N/C, $+y$-direction

22–35: a) $E_x = -4.79 \times 10^{3}$ N/C, $E_y = 0$, $E = 4.79 \times 10^{3}$ N/C,
$-x$-direction
b) $E_x = +2.13 \times 10^{3}$ N/C, $E_y = 0$, $E = 2.13 \times 10^{3}$ N/C,
$+x$-direction
c) $E_x = -129$ N/C, $E_y = -164$ N/C, $E = 209$ N/C, 128°
clockwise from $+x$-direction
d) $E_x = -1.04 \times 10^{3}$ N/C, $E_y = 0$, $E = 1.04 \times 10^{3}$ N/C,
$-x$-direction

22–41: 226 N/C, toward sheet

22–47: a) 1.4×10^{-11} C $\cdot$ m, from q_1 toward q_2 b) 860 N/C

22–49: a) $+8.0 \times 10^{-24}$ J b) 0.39 K

22–53: a) $(1/2)\sqrt{qQ/\pi^3\varepsilon_0 a^3}$ b) accelerates along y-axis away from origin

22–55: b) 2.80×10^{-6} C b) 39.5°

22–57: a) 2.09×10^{21} N b) 5.90×10^{23} m/s^2 c) no

22–59: a) $6kq^2/L^2$, away from vacant corner
b) $(3kq^2/2L^2)(1 + 2\sqrt{2})$, toward center of square

22–61: a) $(2kq/x^2)[1 - (1 + a^2/x^2)^{-3/2}]$, $-x$-direction b) $3kqa^2/x^4$

22–63: a) 480 C b) 8.3×10^{13} N, repulsive; 8.3×10^{13} m/s^2

22–65: a) $(mv_0^2 \sin^2 \alpha)/2eE$ b) $(mv_0^2 \sin^2 2\alpha)/eE$ c) $h_{max} = 0.418$ m,
$d = 2.89$ m

22–67: $+73.1$ nC

22–69: a) $E_x = \dfrac{kQ}{x\sqrt{x^2 + a^2}}$, $\ E_y = \dfrac{+kQ}{a}\left(\dfrac{1}{x} - \dfrac{1}{\sqrt{x^2 + a^2}}\right)$

b) $F_x = \dfrac{-kqQ}{x\sqrt{x^2 + a^2}}$, $\ F_y = \dfrac{+kqQ}{a}\left(\dfrac{1}{x} - \dfrac{1}{\sqrt{x^2 + a^2}}\right)$

22–71: a) $-(7850$ N/C$)\hat{\imath}$ b) smaller c) 18 cm

22–73: a) $+(0.89$ N/C$)\hat{\imath}$ c) smaller d) i) 1.2% ii) 4.7%

22–75: a) $F = \dfrac{qQ}{2\pi\varepsilon_0 a}\left(\dfrac{1}{y} - \dfrac{1}{\sqrt{a^2 + y^2}}\right)$, $-x$-direction

b) $F = \dfrac{qQ}{4\pi\varepsilon_0 a}\left(\dfrac{1}{x - a} + \dfrac{1}{x + a} - \dfrac{2}{x}\right)$, $+x$-direction

22–77: $E_x = E_y = 2Q/4\pi^2\varepsilon_0 a^2$

22–79: a) 1.43×10^{3} kg/C b) 1.15×10^{10}

22–81: a) 5.65×10^{8} N/C, to the left b) 1.69×10^{9} N/C, to the right
c) 2.28×10^{9} N/C, to the right d) 5.65×10^{8} N/C, to the right

22–83: $\vec{E} = \dfrac{\sigma}{2\varepsilon_0}\left[-\left(\dfrac{x}{|x|}\right)\hat{\imath} + \left(\dfrac{z}{|z|}\right)\hat{k}\right]$

22–85: b) $q_1 < 0$, $q_2 > 0$ c) 0.844 μC d) 56.2 N

22–87: a) $\dfrac{kQ}{L}\left[\dfrac{1}{x + (a/2)} - \dfrac{1}{x + L + (a/2)}\right]$

Chapter 23

23–1: a) 1.75 N $\cdot$ m^2/C b) no c) i) 0 ii) 90°

23–3: a) S_1: $-CL^2$ S_2: $-DL^2$ S_3: CL^2 S_4: DL^2 S_5: $-BL^2$ S_6: BL^2
b) 0

23–5: a) 2.71×10^{5} N $\cdot$ m^2/C b) 2.71×10^{5} N $\cdot$ m^2/C
c) 5.42×10^{5} N $\cdot$ m^2/C

23–7: a) -4.07×10^{5} N $\cdot$ m^2/C b) 6.91 nC c) no

23–9: b) no

23–11: a) 1.81×10^{6} N $\cdot$ m^2/C b) no change

23–13: a) 1.62 m

23–15: 2.04×10^{10}

23–17: between: $E = \sigma/\epsilon_0$; outside: $E = 0$

23–19: a) $\lambda = 2\pi R\sigma$ b) $\sigma R/r\epsilon_0$

23–21: b) $q/24\epsilon_0$

23–23: a) -5.98×10^{-10} C

23–25: a) $0 < r < R$, $E = 0$; $R < r < 2R$, $E = Q/4\pi\epsilon_0 r^2$, radially
outward; $r > 2R$, $E = 2Q/4\pi\epsilon_0 r^2$, radially outward

23–27: a) i) 0 ii) 0 iii) $q/2\pi\epsilon_0 r^2$, radially outward iv) 0
v) $3q/2\pi\epsilon_0 r^2$, radially outward
b) i) 0 ii) $+2q$ iii) $-2q$ iv) $+6q$

23–29: a) i) 0 ii) 0 iii) $q/2\pi\epsilon_0 r^2$, radially outward iv) 0
v) $q/2\pi\epsilon_0 r^2$, radially inward
b) i) 0 ii) $+2q$ iii) $-2q$ iv) $-2q$

23–31: a) $Qq/4\pi\epsilon_0 r^2$, toward the center of the shell b) 0

23–33: a) $\lambda/2\pi\epsilon_0 r$, radially outward b) $\lambda/2\pi\epsilon_0 r$, radially outward
d) inner: $-\lambda$; outer: $+\lambda$

23–35: a) i) $\alpha/2\pi\epsilon_0 r$, radially outward ii) 0 iii) 0 b) i) $-\alpha$ ii) 0

23–39: $d = R/2$

23–41:   a) 1.0×10^{32} m/s^2 b) 4.1×10^{32} m/s^2 c) 2.1×10^{32} m/s^2
     d) 0

23–43:   b) $|x| \leq d$: $\vec{E} = (\rho_0 x^3/3\epsilon_0 d^2)\hat{\imath}$; $|x| \geq d$, $\vec{E} = (\rho_0 d/3\epsilon_0)(x/|x|)\hat{\imath}$

23–45:   c) $E(r) = \dfrac{Q}{\pi\epsilon_0 R^2}\left(\dfrac{r}{R} - \dfrac{3r^2}{4R^2}\right)$

     e) $E_{max} = Q/3\pi\epsilon_0 R^2$ at $r = 2R/3$

23–49:   b) $\rho\vec{b}/3\epsilon_0$

23–51:   a) $-(Q/16\pi\epsilon_0 R^2)\hat{\imath}$ b) $(Q/72\pi\epsilon_0 R^2)\hat{\imath}$ c) 0   d) $(5Q/18\pi\epsilon_0 R^2)\hat{\imath}$

23–53:   a) $\alpha = 8Q/5\pi R^3$;   b) $r \leq R/2$: $E(r) = 8Qr/15\pi\epsilon_0 R^3$;
     $R/2 \leq r \leq R$: $E(r) = (Q/60\pi\epsilon_0 r^2)[64(r/R)^3 - 48(r/R)^4 - 1]$;
     $r \geq R$: $E(r) = Q/4\pi\epsilon_0 r^2$ c) 4/15   e) $2\pi\sqrt{15m\pi\epsilon_0 R^3/8Qe}$
     f) no

Chapter 24

24–1: -0.356 J

24–3:   a) 12.5 m/s   b) 0.323 m

24–5:   a) 0.198 J   b) i) 26.6 m/s   ii) 36.7 m/s   iii) 37.6 m/s

24–7:   a) -3.60×10^{-7} J   b) $x = 0.0743$ m

24–9: $-q/2$

24–11:   a) 0   b) $+7.50 \times 10^{-4}$ J   c) -2.06×10^{-3} J

24–13:   a) 2.50 mm   b) 7.49 mm

24–15:   a) -737 V   b) -704 V   c) $+8.2 \times 10^{-8}$ J

24–17:   b) 0   d) 0

24–19:   b) $V = \dfrac{q}{4\pi\epsilon_0}\left(\dfrac{1}{|x|} - \dfrac{2}{|x-a|}\right)$ d) $x = -a, x = a/3$ e) $-q/4\pi\epsilon_0 x$

24–21: 1.02×10^7 m/s

24–23:   a) b b) 800 V/m   c) -4.8×10^{-5} J

24–25:   a) oscillatory   b) 1.67×10^7 m/s

24–27:   a) 8.00 kV/m   b) 1.92×10^{-5} N   c) 8.64×10^{-7} J
     d) 8.64×10^{-7} J

24–29:   a) 1.58 mm   b) 2.66×10^{-5} C/m^2

24–31:   a) $E_x = -Ay + 2Bx$, $E_y = -Ax + C$, $E_z = 0$
     b) $x = C/A$, $y = 2BC/A^2$, any value of z

24–33:   a) $\vec{E} = -C\vec{k}$ for $0 < z < d$, $\vec{E} = 0$ otherwise
     b) parallel plates at $z = 0$ (charge density $-C\epsilon_0$) and $z = d$
     (charge density $= +C\epsilon_0$)

24–35:   a) 7.63×10^{-10} C

24–39:   a) 1.76×10^{-16} N, downward   b) 1.93×10^{14} m/s^2, downward
     c) 8.24 mm   d) 15.4°   e) 4.12 cm

24–49:   a) -2.15×10^{-5} J   b) -2.83 kV   c) 3.54×10^4 V/m

24–51:   a) 7.85×10^4 V/m$^{4/3}$ b) $\vec{E} = (-1.05 \times 10^5 \text{ V/m}^{4/3})x^{1/3}\hat{\imath}$
     c) $\vec{F} = (3.13 \times 10^{-15} \text{ N})\hat{\imath}$

24–53:   a) $-1.46q^2/\pi\epsilon_0 d$

24–55:   a) -8.62×10^{-18} J   b) 2.87×10^{-11} m

24–57:   a) i) $V = (\lambda/2\pi\epsilon_0)\ln(b/a)$ ii) $V = (\lambda/2\pi\epsilon_0)\ln(b/r)$
     iii) $V = 0$
     d) $(\lambda/2\pi\epsilon_0)\ln(b/a)$

24–59:   a) 9.71×10^4 V/m   b) 3.03×10^{-11} C

24–61: $(\sigma/2\epsilon_0)\left(\sqrt{x^2 + R^2} - x\right)$

24–63:   a) $r \leq R$: $V = (\lambda/4\pi\epsilon_0)[1 - (r/R)^2]$;
     $r \geq R$, $V = -(\lambda/2\pi\epsilon_0)\ln(r/R)$

24–65: $Q/4\pi\epsilon_0\sqrt{x^2 + a^2}$

24–67:   a) 0   b) 0.135 V; S_4 c) 0.225 V; S_6

24–69: $Q/8\pi\epsilon_0 R$

24–71:   0

24–73:   a) $(Q/4\pi\epsilon_0 a)\ln[1 + (a/x)]$

     b) $(Q/4\pi\epsilon_0 a)\ln\left[(a/y) + \sqrt{1 + (a/y)^2}\right]$

     c) in (a), $Q/4\pi\epsilon_0 x$; in (b), $Q/4\pi\epsilon_0 y$

24–75:   a) 1/3   b) 3

24–77:   a) $E = Q_1/4\pi\epsilon_0 R_1^2$, $V = Q_1/4\pi\epsilon_0 R_1$
     b) sphere 1: $Q_1 R_1/(R_1 + R_2)$; sphere 2: $Q_1 R_2/(R_1 + R_2)$
     c) $V = Q_1/4\pi\epsilon_0(R_1 + R_2)$ for either sphere
     d) sphere 1: $E = Q_1/4\pi\epsilon_0 R_1(R_1 + R_2)$;
     sphere 2: $E = Q_1/4\pi\epsilon_0 R_2(R_1 + R_2)$

24–79:   a) 7.6×10^6 m/s   b) 7.3×10^6 m/s   c) 2.3×10^9 K;
     6.4×10^9 K

24–81: 1.01×10^{-12} m, 1.11×10^{-13} m, 2.54×10^{-14} m

24–83:   c) 3 electrons, 0.507 μm

Chapter 25

25–1: $+1.82 \times 10^{-4}$ C on one plate, -1.82×10^{-4} C on the other

25–3:   a) 604 V   b) 90.8 cm^2 c) 1.84×10^6 V/m   d) 16.3 μC/m^2

25–5:   a) 66 pF/m   b) inner: $+64$ pC; outer: -64 pC

25–7:   a) 89.5 pF   b) 6.08×10^3 V/m   c) 4.47×10^3 V/m   d) no

25–9:   a) $Q_1 = Q_2 = 22.4$ μC, $Q_3 = 44.8$ μC, $Q_4 = 67.2$ μC
     b) $V_1 = V_2 = 5.60$ V, $V_3 = 11.2$ V, $V_4 = 16.8$ V   c) 11.2 V

25–11: $Q_1 = 156$ μC, $Q_2 = 260$ μC   b) $V_1 = V_2 = 52$ V

25–15:   57 μF

25–17:   0.0283 J/m^3

25–19:   19.6 J

25–21:   a) $Q^2 x/2\epsilon_0 A$ b) $(Q^2/2\epsilon_0 A)dx$ c) $Q^2/2\epsilon_0 A$

25–23:   a) 1.60 nC   b) 8.0

25–25:   a) 1.64×10^{-4} J/m^3 b) 8.83×10^{-5} J/m^3 c) no

25–27:   a) 6.20×10^{-7} C/m^2 b) 1.28

25–29:   0.0135 m^2

25–31:   a) 2.3×10^{-11} C^2/N $\cdot$ m^2 b) 40 kV   c) $\sigma = 4.6 \times 10^{-4}$ C/m^2,
     $\sigma_i = 2.8 \times 10^{-4}$ C/m^2

25–33:   a) 10.1 V   b) 2.25

25–35:   a) $Q/K\epsilon_0 A$ b) $Qd/K\epsilon_0 A$ c) $K\epsilon_0 A/d$

25–37:   a) 2.4×10^{-11} F   b) 2.9×10^{-10} C   c) 1.3×10^3 V/m
     d) 1.7×10^{-9} J

25–39:   a) 421 J   b) 5.39×10^{-2} F

25–43:   a) 4.00 μF: 2.64×10^{-3} C, 660 V; 6.00 μF: 3.96×10^{-3} C, 660 V
     b) 4.00 μF: 5.28×10^{-4} C, 132 V; 6.00 μF: 7.92×10^{-4} C, 132 V

25–45:   a) 2.5 μF   b) $Q_1 = 5.5 \times 10^{-4}$ C, $V_1 = 66$ V; $Q_2 = 3.7 \times 10^{-4}$ C,
     $V_2 = 88$ V; $Q_3 = 1.8 \times 10^{-4}$ C, $V_3 = 44$ V; $Q_4 = 1.8 \times 10^{-4}$ C,
     $V_4 = 44$ V; $Q_5 = 5.5 \times 10^{-4}$ C, $V_5 = 66$ V

25–47:   a) 76 μC   b) 1.4×10^{-3} J   c) 11 V   d) 1.2×10^{-3} J

25–49:   a) 2.3 μF   b) C_1: 9.7×10^{-4} C; C_2: 6.4×10^{-4} C   c) 47 V

25–51:   c) 710 μF

25–53:   a) $Q^2 r^2/32\pi^2\epsilon_0 R^6$ b) $Q^2/32\pi^2\epsilon_0 r^4$ c) $3Q^2/20\pi\epsilon_0 R$

25–57:   a) 1.0×10^7 V/m   b) 0.052 V, outer   c) 1.4×10^{-15} J

25–59:   b) 14.0 μF   c) 72.0 μF: 505 μC, 7.02 V; 28.0 μF: 259 μC,
     9.24 V; 21.0 μF: 246 μC, 11.7 V; 18.0 μF: 229 μC, 12.7 V;
     27.0 μF: 276 μC, 10.2 V; 6.0 μF: 14.9 μC, 2.49 V

25–61:   a) $(\epsilon_0 L/D)[L + (K - 1)x]$

25–63:   b) 2.38×10^{-9} F

Chapter 26

26–1: 3.9×10^4 C

26–3:   a) 110 min   b) 440 min

26–5:   a) 330 C   b) 41 A

26–7:   0.125 Ω

26–9:   a) 5.16×10^{-3} V/m   b) 2.70×10^{-3} V/m

26–11:   0.205 mm

26–13:   a) 11 A   b) 3.1 V   c) 0.28 Ω

26–15:   a) 1.25 V/m   b) 2.84×10^{-8} $\Omega \cdot$ m

26–17:   a) 99.54 Ω b) 0.0158 Ω

26–19:   a) 4.67×10^{-8} Ω b) 6.72×10^{-4} Ω

26–21:   a) 0   b) 5.0 V   c) 5.0 V

26–23: $\varepsilon = 3.08$ V, $r = 0.067$ Ω, $R = 1.80$ Ω
26–25: a) 1.41 A, from a toward b b) 13.7 V c) -1.0 V
26–27: b) yes c) 3.88 Ω
26–29: a) 0.101 Ω b) 0.221 Ω c) 0.0126 Ω
26–31: 520 W
26–33: a) EJ b) ρJ^2 c) E^2/ρ
26–35: a) 2.6×10^6 J b) 0.063 L c) 5.8×10^3 s
26–37: a) 24 W b) 4 W c) 20 W
26–39: a) 26.7 Ω b) 4.50 A c) 454 W d) greater
26–41: a) 1.2 A b) 1.4×10^4 A c) 1.4×10^7 A
26–43: a) 3.65×10^{-8} $\Omega \cdot$ m b) 172 A c) 2.58×10^{-3} m/s
26–45: a) 2.5 mA b) 2.1×10^{-5} V/m c) 8.6×10^{-5} V/m
        d) 1.8×10^{-4} V
26–47: a) $\rho h/\pi r_1 r_2$
26–51: a) 0.057 Ω b) 3.3×10^{-8} $\Omega \cdot$ m c) 0.86 mm d) 0.0024 Ω
        e) 1.1×10^{-3} $(C°)^{-1}$
26–53: a) 0.20 Ω b) 8.7 V
26–55: 1.4 A
26–57: a) $\varepsilon I/(r + R)$ b) 0.043 Ω
26–59: b) 8 gauge c) 106 W d) \$19
26–61: a) 0.40 A b) 1.6 W c) 4.8 W d) 3.2 W
26–63: a) $|q|/m = a/E$ b) $|q|/m = aL/V_{bc}$ c) c d) 3.5×10^8 m/s^2
26–65: a) 14.4 V b) 2.6×10^6 J c) 4.3×10^5 J d) 0.96 Ω
        e) 1.7×10^6 J f) 4.3×10^5 J
26–67: b) $a = 8.0 \times 10^{-5}$ $\Omega \cdot$ m $\cdot$ K^n, $n = 0.15$ c) 4.3×10^{-5} $\Omega \cdot$ m,
        3.2×10^{-5} $\Omega \cdot$ m
26–69: a) $R = (\rho_0 L/A)(1 - 1/e)$, V_0/R b)$(V_0/L)(1 - 1/e)e^{-x/L}$
        c) $V_0(e^{-x/L} - 1/e)/(1 - 1/e)$

Chapter 27

27–1:   a) 12.3 Ω b) 19.5 A c) 32-Ω resistor: 7.5 A; 20-Ω resistor:
        12.0 A
27–3:   a) 0.80 Ω
        b) 1.60-Ω resistor: 17.5 A; 2.40-Ω resistor, 11.7 A; 5.8 A
        c) 35.0 A d) 28.0 V for each e) 1.60-Ω resistor: 490 W;
        2.40-Ω resistor: 327 W; 4.80-Ω resistor: 163 W
        f) least resistance
27–5:   a) 270 V b) 1.6 W
27–7: $R_{eq} = 3.00$ Ω;
        1.00-Ω resistor: 12.0 A; 3.00-Ω resistor: 12.0 A;
        7.00-Ω resistor: 4.0 A;
        5.00-Ω resistor: 4.0 A
27–9:   a) $I_1 = 1.50$ A, $I_2 = I_3 = I_4 = 0.50$ A b) $P_1 = 10.1$ W,
        $P_2 = P_3 = P_4 = 1.12$ W
        c) $I_1 = 1.33$ A, $I_2 = I_3 = 0.667$ A d) $P_1 = 8.00$ W,
        $P_2 = P_3 = 2.00$ W
        e) R_2 and R_3 are brighter, R_1 is dimmer
27–11: a) 0.100 A for each b) 400-Ω bulb: 4.00 W; 800-Ω bulb:
        8.00 W; total: 12.0 W
        c) 400-Ω bulb: 0.300 A; 800-Ω bulb: 0.150 A
        d) 400-Ω bulb: 36.0 W; 800-Ω bulb: 18.0 W; total: 54.0 W
        e) in series, 800-Ω bulb is brighter; in parallel, 400-Ω bulb is
        brighter and total light output is greater
27–13: a) 2.00 A b) 5.00 Ω c) 42.0 V d) 3.50 A
27–15: a) 8.00 A b) $\varepsilon_1 = 36.0$ V, $\varepsilon_2 = 54.0$ V c) 9.00 Ω
27–17: a) 1.60 A, 1.40 A, 0.20 A b) 10.4 V
27–19: 4.0 W
27–21: a) 6.5 A in each 1-Ω resistor, 0 in R b) 1.0 Ω c) 0
27–23: 12.9 Ω
27–25: c) 3.34 V
27–27: 218 Ω
27–31: a) 0.112 mA b) 0.582 ms

27–33: a) 0, 2.70×10^{-4} C, 4.42×10^{-4} C, 6.21×10^{-4} C,
        7.44×10^{-4} C
        b) 6.70×10^{-5} A, 4.27×10^{-5} A, 2.72×10^{-5} A, 1.11×10^{-5} A,
        8.19×10^{-9} A
27–35: a) 1.33×10^{-4} C b) $V_R = 9.12$ V, $V_C = 8.88$ V
        c) $V_R = V_C = 8.88$ V
        d) 6.75×10^{-5} C
27–37: 900 W
27–39: a) 6.0 A, 720 W b) 3.5 A, 420 W
27–41: a) two in parallel in series with two in parallel;
        two in series in parallel with two in series b) 0.60 W
27–43: 54 W
27–45: $I_1 = 0.848$ A, $I_2 = 2.14$ A, $I_3 = 0.171$ A
27–47: 2.00-Ω resistor: 5.21 A; 4.00-Ω resistor: 1.11 A; 5.00-Ω
        resistor: 6.32 A
27–49: a) 0.222 V b) 0.464 A
27–51: a) -12.0 V b) 1.71 A c) 4.20 Ω
27–53: a) -6.0 V b) b c) $+6.0$ V d) 54 μC
27–55: $R_1 = 2960$ Ω, $R_2 = 12.0$ kΩ, $R_3 = 135$ kΩ;
        $R_{overall} = 3000$ Ω (3.00-V range), 15.0 kΩ (15.0-V range),
        150 kΩ (150-V range)
27–57: 18.5 kΩ
27–59: b) 1897 Ω
27–61: a) 224-Ω resistor: 24.8 V; 589-Ω resistor: 65.2 V b) 3.87 kΩ
        c) 62.6 V d) no
27–63: a) 8.24 mA toward negative plate b) 612 μC
27–65: 1.7 MΩ, $C = 3.7$ μF
27–67: a) 19.4 s $= 31.4\tau$ b) yes
27–69: a) $C\varepsilon^2$ b) $C\varepsilon^2/2$ c) $C\varepsilon^2/2$ d) 1/2
27–71: a) 0.300 A, 0.500 A, 0.200 A b) 0.541 A, 0.077 A, -0.464 A
        c) -0.287 A, 0.192 A, 0.479 A d) 0.046 A, 0.231 A, 0.185 A
        f) 0.184 A, 0.577 A, 0.393 A
27–75: b) at least four c) $R_T = 3.2$ MΩ, $\beta = 4.0 \times 10^{-3}$
        d) 3.4×10^{-4} e) 0.88

Chapter 28

28–1:   a) $(-6.68 \times 10^{-4}$ N$)\hat{k}$ b) $(+6.68 \times 10^{-4}$ N$)\hat{i} + (7.27 \times 10^{-4}$ N$)\hat{j}$
28–3:   positive
28–5: a: $-qvB\hat{k}$; b: $+qvB\hat{j}$; c: $\mathbf{0}$; d: $(-qvB/\sqrt{2})\hat{j}$;
        e: $(-qvB/\sqrt{2})(\hat{j} + \hat{k})$
28–7: 9.47×10^6 m/s
28–9:   a) 3.05×10^{-3} Wb b) 1.83×10^{-3} Wb c) 0
28–11: a) -0.0108 Wb b) no
28–13: a) 1.60×10^{-4} T, into page b) 1.11×10^{-7} s
28–15: a) 8.35×10^5 m/s b) 2.62×10^{-8} s c) 7.26 kV
28–17: a) 107 T b) no
28–19: a) no b) 1.40 cm
28–21: 8.38×10^{-4} T
28–23: 1.29×10^{-25} kg, 78
28–25: a) mg/IL b) due west
28–27: 9.7 A
28–29: a) $(0.023$ N$)\hat{k}$ b) $(0.020$ N$)\hat{j}$ c) 0 d) $(0.0098$ N$)\hat{j}$
        e) $(-0.013$ N$)\hat{j} + (-0.026$ N$)\hat{k}$
28–31: a) the normal to the plane of the coil is perpendicular to $\vec{B}$;
        0.13 N $\cdot$ m
        b) the normal to the plane is 45° from the direction of $\vec{B}$
28–33: -2.42 J
28–35: a) 1.13 A b) 3.69 A c) 98.2 V d) 362 W
28–37: a) 4.7 mm/s b) 4.5×10^{-3} V/m in the $+z$-direction c) 53 μV
28–39: F_2/qv_1 in the $-y$-direction b) $F_2/\sqrt{2}$
28–41: a) 7.90×10^3 V/m in the $+x$-direction
        b) 7.90×10^3 V/m in the $+x$-direction
28–43: a) 8.9×10^{-13} J $= 5.5 \times 10^6$ eV b) 7.7×10^{-8} s c) 1.2 T
        d) same as in (a)

28–45: 4.24×10^3

28–47:   a) -1.98×10^{-6} C   b) $(9.69 \times 10^{13}$ m/s$^2)(4\hat{\imath} + 3\hat{\jmath})$
        c) $R = 5.69$ cm
        d) 9.23×10^7 rad/s   e) $(R, 0, 1.72$ m)

28–49:   1.6 mm

28–51:   a) ab: $(-4.24$ N$)\hat{k}$; bc: $(-4.24$ N$)\hat{\jmath}$; cd: $(+4.24$ N$)(\hat{\jmath} + k)$;
        de: $(-4.24$ N$)\hat{\jmath}$; ef: 0   b) $(-4.24$ N$)\hat{\jmath}$

28–53: $(Mg \tan \theta)/LB$, right to left

28–57:   0.0242 T, $+y$-direction

28–59: $2\pi\sqrt{\dfrac{I_S}{NIAB}}$

28–61:   0.444 N, $-y$-direction

28–63:   b) side $(0, 0)$ to $(0, L)$: $(B_0 IL/2)\hat{\imath}$; side $(0, L)$ to (L, L): $-B_0 IL\hat{\jmath}$;
        side (L, L) to $(L, 0)$: $(-B_0 IL/2)\hat{\imath}$; side $(L, 0)$ to $(0, 0)$: **0**
        c) $-B_0 IL\hat{\jmath}$

28–65:   a) 2.52 m/s   b) 7.60 A   c) 0.197 Ω

28–67:   a) $\vec{\mu} = -IA\hat{k}$ b) $B_x = 3D/IA$, $B_y = 4D/IA$, $B_z = -12D/IA$

28–69: $-\beta r/2$

28–71:   a) 5.14 m   b) 1.72×10^{-6} s   c) 6.09 mm   d) 3.04 cm

Chapter 29

29–1:   a) $(-1.92 \times 10^{-5}$ T$)\hat{k}$ b) 0
        c) $(1.92 \times 10^{-5}$ T$)\hat{\imath}$ d) $(6.79 \times 10^{-6}$ T$)\hat{\imath}$

29–3: 1.64×10^{-6} T, into page

29–5:   a) $-0.866\hat{\imath} + 0.500\hat{\jmath}$ b) $(-5.00 \times 10^{-3}$ m$)\hat{k}$
        c) $(-4.34 \times 10^{-8}$ T$)\hat{k}$

29–7:   a) $(5.00 \times 10^{-11}$ T$)\hat{\jmath}$ b) $(-5.00 \times 10^{-11}$ T$)\hat{\imath}$
        c) $(-1.77 \times 10^{-11}$ T$)(\hat{\imath} - \hat{\jmath})$ d) **0**

29–9:   a) 0   b) $2\mu_0 I/3\pi a$, $+z$-direction   c) $2\mu_0 I/3\pi a$, $-z$-direction

29–13:   a) 2.91×10^{-5} T, east   b) yes

29–15: $\mu_0 I^2/2\pi\lambda g$

29–17:   a) 8.33 A   b) opposite

29–19: $\mu_0 |I_1 - I_2|/4R$, zero

29–21   69

29–23: a: 0 b: -5.0×10^{-6} T $\cdot$ m c: $+2.5 \times 10^{-6}$ T $\cdot$ m
        d: $+5.0 \times 10^{-6}$ T $\cdot$ m

29–25:   a) $\mu_0 I_1/2\pi r$ b) $\mu_0 (I_1 + I_2)/2\pi r$

29–27:   a) 1790 turns/m   b) 63.0 m

29–29: 1.11×10^{-3} T

29–31:   a) 0.0725 A   b) 0.0195 A

29–33:   a) i) 1.1×10^{-3} T   ii) 4.7×10^6 A/m   iii) 5.9 T

29–35:   yes, $C = 1.55 \times 10^5$ K $\cdot$ A/T $\cdot$ m

29–37:   a) 55.7 A/m^2 b) 6.29×10^{12} V/m $\cdot$ s   c) 7.00×10^{-7} T
        d) 3.50×10^{-7} T

29–39:   a) 0.900 nC, 2.03×10^5 V/m, 407 V   b) 4.07×10^{11} V/m $\cdot$ s; no
        c) 3.60 A $\cdot$ m^2, $i_D = 1.80$ mA, $i_D = i_C$

29–41: 1.07×10^{19} N, toward wire

29–45:   a) 2.00 A, out of paper   b) 2.13×10^{-6} T, to the right
        c) 2.06×10^{-6} T

29–47:   b) $\mu_0 Ix/\pi(a^2 + x^2)$, $-y$-direction   c) $x = \pm a$ c) $\mu_0 I/\pi x$

29–49:   23.2 A

29–51:   a) $\mu_0 \pi NN' II' a^2 a'^2 (\sin \theta)/2x^3$ b) $-\mu_0 \pi NN' II' a^2 a'^2 (\cos \theta)/2x^3$

29–53:   a) $(\mu_0 NIa^2/2)[((x + a/2)^2 + a^2)^{-3/2} + ((x - a/2)^2 + a^2)^{-3/2}]$
        b) $(\mu_0 NI/a)(4/5)^{3/2}$ c) 0.0202 T   e) 0, 0

29–55: $\mu_0 I/8R$, out of paper

29–57:   a) $3I/2\pi R^3$ b) i) $\mu_0 Ir^2/2\pi R^3$ ii) $\mu_0 I/2\pi r$

29–61:   b) $\mu_0 I_0/2\pi r$ c) $(I_0 r^2/a^2)(2 - r^2/a^2)$ d) $(\mu_0 I_0/2\pi a^2)(2 - r^2/a^2)$

29–63: $\mu_0 I$

29–65:   a) $\mu_0 nI/2$, $+x$-direction   b) $\mu_0 nI/2$, $-x$-direction

29–67: 7.73×10^{-25} J/T $= 0.0833 \mu_B$

29–69:   a) $(Q_0/AK\rho\epsilon_0) \exp(-t/K\epsilon_0\rho)$

29–71:   b) 0.347 m/s   c) 6.15 mm

29–73: $\mu_0 Qn/a$

Chapter 30

30–1:   43.0 V

30–3:   a) $Q = NBA/R$ c) no

30–5:   0.0973 T

30–7:   a) $\varepsilon = -(2\pi NB_0 A/T) \sin (2\pi t/T)$ for $0 \le t \le T$, $\varepsilon = 0$ for $t \le 0$
        and $t \ge T$
        b) $t = T/2$ c) $t = T/4$ and $t = 3T/4$; $|\varepsilon_{\max}| = 2\pi NB_0 A/T$
        d) clockwise for $0 < t < T/2$, counterclockwise for $T/2 < t < T$

30–9:   224 V

30–13:   a) a to b b) b to a c) b to a

30–17:   a) 0.858 m/s   b) 0.827 A   c) opposite $\vec{v}$ (to the left)

30–21:   a) 1.70×10^{-4} V/m   b) 3.39×10^{-4} V/m

30–23:   a) concentric circles
        b) 1.75×10^{-3} V/m, clockwise
        c) 2.75×10^{-4} A   d) 5.50×10^{-4} V   e) 1.10×10^{-3} V

30–25:   a) 1.14×10^{-11} J   c) 1.14×10^{-11} J

30–29:   a) inside: $\vec{B} = 0$, $\vec{M} = -(1.03 \times 10^5$ A/m$)\hat{\imath}$;
        outside: $\vec{B} = (+0.130$ T$)\hat{\imath}$, $\vec{M} = 0$
        b) $\vec{B} = (+0.130$ T$)\hat{\imath}$, $\vec{M} = 0$ both inside and outside

30–31:   c) 0.400 V/m

30–33:   a) $\mu_0 i\pi a/2$ c) $i = i_0 \exp(-2Rt/\mu_0\pi a)$ d) 45 μs

30–35:   a) 0.0126 V   b) a to b

30–37:   b) $FR/B^2 L^2$

30–39:   a) $(\mu_0 Iv/2\pi) \ln ((L + d)/d)$ b) a c) 0

30–41:   a) 0.165 V   b) 0.165 V   c) zero; 0.0412 V

30–43:   a) $B^2 a^2 v/R$

30–45: a: $(qr/2)(dB/dt)$, to the left; b: $(qr/2)(dB/dt)$, upward; c: 0

30–47:   a) $\tau = (mgL/2) \sin \phi - (\omega B^2 L^4/R) \sin^2 \phi$; in $+x$-direction
        b) $\alpha = (6g/5L) \sin \phi - (12\omega B^2 L^2/5mR) \sin^2 \phi$
        c) longer   d) no

30–49:   b) 0   c) 1.40×10^{-3} V   d) 7.37×10^{-4} A   e) 3.50×10^{-4} V, a

30–51:   a) radially outward   c) in both cases, magnitudes of I and τ
        are unchanged: in first case, I radially inward and τ
        clockwise; in second case, I radially outward and τ clockwise

Chapter 31

31–1:   a) 0.270 V, yes   b) 0.270 V

31–3:   a) 1.96 H   b) 7.12×10^{-3} Wb

31–7:   a) 4.68 mV   b) a

31–9: $\mu_0 N^2 A/l$

31–11:   a) 0.250 H   b) 4.50×10^{-4} Wb

31–13:   2850

31–17:   a) 25.1 m^3 b) 11.9 T

31–19:   a) 2.40 A/s   b) 0.800 A/s   c) 0.413 A   d) 0.750 A

31–21:   a) 0.250 A   b) 0.137 A   c) 32.9 V, c d) 4.62×10^{-4} s

31–23:   a) $(4.50$ W$)[(1 - \exp(-(3.20$ s$^{-1})t)]$
        b) $(4.50$ W$)[(1 - \exp(-(3.20$ s$^{-1})t]^2$
        c) $(4.50$ W$)[\exp(-(3.20$ s$^{-1})t) - \exp(-(6.40$ s$^{-1})t)]$

31–27:   a) 105 rad/s, 59.6 ms   b) 7.20×10^{-4} C   c) 4.32×10^{-3} J
        d) -543μC   e) -49.9 mA   f) 2.45×10^{-3} J in the capacitor,
        1.87×10^{-3} J in the inductor

31–29:   a) 6.45 μC   b) 2.36 V

31–33:   a) $U_C = (Q^2/2C) \cos^2 (\omega t + \phi)$, $U_L = (Q^2/2C) \sin^2 (\omega t + \phi)$

31–37:   45.4 Ω

31–43:   a) 7.50 H   b) 360 Wb   c) 0.0104

31–47:   a) $\mu_0 i/2\pi r$ b) $(\mu_0 i^2 l/4\pi r) dr$ c) $(\mu_0 i^2 l/4\pi) \ln (b/a)$

31–49: 2.17×10^{-6} T

31–51:   a) 0.281 J   b) 0.517 J   c) 0.236 J

31–55: $C = 0.512 \mu$F, $L = 0.477$ mH

31–57:   a) 60.0 V   b) a c) 60.0 V   d) c e) -96.0 V   f) b
        g) -156 V   h) d

31–59:   a) $i_0 = 0$, $v_{ac} = 0$, $v_{cb} = 36.0$ V   b) $i_0 = 0.180$ A, $v_{ac} = 9.00$ V,
        $v_{cb} = 27.0$ V

c) $i_0 = (0.180 \text{ A})[1 - \exp(-(50.0 \text{ s}^{-1})t)]$,
$v_{ac} = (9.00 \text{ V})[1 - \exp(-(50.0 \text{ s}^{-1})t)]$,
$v_{cb} = 27.0 \text{ V} + (9.00 \text{ V})\exp(-(50.0 \text{ s}^{-1})t)$

31–61: a) $i_1 = \mathcal{E}/R_1$, $i_2 = (\mathcal{E}/R_2)[1 - \exp(-R_2 t/L)]$
b) $i_1 = \mathcal{E}/R_1$, $i_2 = \mathcal{E}/R_2$
c) $i = (\mathcal{E}/R_2)[1 - \exp(-(R_1 + R_2)t/L)]$
d) $R_2 = 21.2 \ \Omega$, $\mathcal{E} = 12.7 \text{ V}$ e) 0.0354 A

31–63: a) $i_1 = (\mathcal{E}/R_1)[1 - \exp(-R_1 t/L)]$, $i_2 = (\mathcal{E}/R_2) \exp(-t/R_2 C)$,
$q_2 = C\mathcal{E}[1 - \exp(-t/R_2 C)]$ b) 0, 9.6 mA c) 1.9 A, 0
d) 1.6 ms e) 9.4 mA f) 0.22 s

31–65: a) $d = [(L - L_0)/(L_f - L_0)]D$
b) 0.63024 H, 0.63048 H, 0.63072 H, 0.63096 H
c) 0.63000 H, 0.62999 H, 0.62999 H, 0.62998 H

Chapter 32

32–1: a) 31.8 V b) 0
32–3: a) 0.120 A b) 1.20×10^{-2} A c) 1.20×10^{-3} A
32–5: a) 1.51 kΩ b) 0.239 H c) 497 Ω d) 16.6 μF
32–7: 13.3 μF
32–9: a) $i = (0.0253 \text{ A}) \cos [(720 \text{ rad/s})t]$ b) 180 Ω
c) $v_L = (-4.56 \text{ V}) \sin [(720 \text{ rad/s})t]$
32–11: b) positive c) negative
32–13: a) 696 Ω b) 0.0431 A c) $V_R = 8.62$ V, $V_C = 28.7$ V
d) $-73.3°$
32–15: b) $v = 20.5$ V, $v_R = 7.6$ V, $v_L = 12.9$ V
c) $v = -15.2$ V, $v_R = -22.5$ V, $v_L = 7.3$ V
32–17: a) 601 Ω b) 49.9 mA c) $-70.6°$, lags
d) $V_R = 9.98$ V, $V_L = 4.99$ V, $V_C = 33.3$ V
32–19: a) $Z = 325 \ \Omega$, $\phi = -52.1°$, lags b) $Z = 229 \ \Omega$, $\phi = +29.2°$,
leads
32–23: a) $+45.8°$, 0.697 b) 344 Ω c) 155 V d) 48.6 W
e) 48.6 W f) 0 g) 0
32–25: a) 1.00 b) 75.0 W c) 75.0 W
32–27: a) 103 Hz c) $V_1 = 21.2$ V, $V_2 = 27.4$ V, $V_3 = 27.4$ V, $V_4 = 0$,
$V_5 = 21.2$ V d) 103 Hz e) 0.212 A
32–29: a) 10 b) 2.40 A c) 28.8 W d) 500 Ω
32–31: a) 40 b) 1.50 V
32–33: 0.124 H
32–35: a) $t_1 = \pi/2\omega$, $t_2 = 3\pi/2\omega$ b) $2I\omega$ c) $I_{rav} = 2I/\pi$
32–37: a) inductor b) 0.133 H
32–39: $\sqrt{(R^2 + \omega^2 L^2)/[R^2 + (\omega L - 1/\omega C)^2]}$
32–43: b) $U_B = LV^2/4[R^2 + (\omega L - 1/\omega C)^2]$,
$U_E = V^2/4\omega^2 C[R^2 + (\omega L - 1/\omega C)^2]$
d) $\omega = 0$: $U_B = 0$, $U_E = CV^2/4$; $\omega \to \infty$: both U_B and $U_E \to 0$;
$U_B = U_E$ at $\omega = \omega_0 = 1/\sqrt{LC}$
32–47: a) $I_R = V/R$, $I_L = V/\omega L$, $I_C = \omega CV$
c) $\omega = 0$: $I_L \to \infty$, $I_C = 0$; $\omega \to \infty$, $I_L = 0$, $I_C \to \infty$
d) 159 Hz e) 0.50 A f) $I_R = 0.50$ A, $I_L = I_C = 0.050$ A
32–49: a) 102 Ω b) 0.882 A c) 270 V
32–51: a) 0.750 A b) 160 Ω c) 619 Ω, 341 Ω d) 341 Ω
32–53: $i_{av} = 0$, $i_{rms} = I_0/\sqrt{3}$
32–55: a) $I_0 = V/R$ c) $\omega_1 = \omega_0 + \sqrt{3}R/2L$, $\omega_2 = \omega_0 - \sqrt{3}R/2L$
d) $\omega_1 - \omega_2 = \sqrt{3}R/L$ e) i) $I_0 = 8.00$ A, $\omega_0 = 1000$ rad/s,
$\omega_1 - \omega_2 = 10.4$ rad/s
ii) $I_0 = 80.0$ A, $\omega_0 = 1000$ rad/s, $\omega_1 - \omega_2 = 1.04$ rad/s
32–57: a) $V/\sqrt{R^2 + 9L/4C}$ b) $\left[2V/\sqrt{R^2 + 9L/4C}\right]\sqrt{L/C}$
c) $\left[V/2\sqrt{R^2 + 9L/4C}\right]\sqrt{L/C}$ d) $2LV^2/(R^2 + 9L/4C)$
e) $LV^2/2(R^2 + 9L/4C)$
32–59: a) $V_R I/2$ b) 0 c) 0
32–61: a) 0.400 A b) 36.9° c) $Z_{cpx} = (400 \ \Omega) - i(300 \ \Omega)$, $Z = 500 \ \Omega$

d) $I_{cpx} = (0.320 \text{ A}) + i(0.240 \text{ A})$, $I = 0.400$ A e) 36.9°
f) $V_{Rcpx} = (128 \text{ V}) + i(96 \text{ V})$, $V_{Lcpx} = (-120 \text{ V}) + i(160 \text{ V})$,
$V_{Ccpx} = (192 \text{ V}) - i(256 \text{ V})$

Chapter 33

33–1: a) 1.28 s b) 8.15×10^{13} km
33–3: $\vec{E}(z, t) = (1.74 \times 10^5 \text{ V/m})\hat{\imath} \sin [(3.83 \times 10^{15} \text{ rad/s})t -$
$(1.28 \times 10^7 \text{ rad/m})z]$,
$\vec{B}(z, t) = (5.80 \times 10^{-4} \text{ T})\hat{\jmath} \sin [(3.83 \times 10^{15} \text{ rad/s})t -$
$(1.28 \times 10^7 \text{ rad/m})z]$
33–5: a) $+y$-direction b) 7.11×10^{-4} m
c) $\vec{B}(y, t) = (-1.03 \times 10^{-3} \text{ T})\hat{\imath} \sin [(2.65 \times 10^{12} \text{ rad/s})t -$
$(8.84 \times 10^3 \text{ rad/m})y]$
33–7: a) 361 m b) 0.0174 rad/m c) 5.22×10^6 rad/s
d) 0.0144 V/m
33–9: a) 3.00×10^{-10} T b) 4.22×10^4 W c) 75.0 km
33–11: $E_{max} = 12.0$ V/m, $B_{max} = 4.00 \times 10^{-8}$ T
33–13: a) 8.68×10^{-15} kg/m$^2 \cdot$ s b) 2.60×10^{-6} kg/m $\cdot$ s^2
33–17: 0.779 V/m
33–19: a) 381 nm b) 526 nm c) 1.38 d) 1.91
33–23: a) 4.38 mm b) 4.38 mm c) 4.38 mm
33–25: a) 7.10 mm b) 3.55 mm c) 1.56×10^8 m/s
33–27: a) 4.41×10^{-14} m $= 4.41 \times 10^{-5}$ nm
b) 5.21×10^{-7} m $= 521$ nm
33–33: a) 5.20×10^{-10} J b) 6.94×10^{-15} Pa c) 1.34×10^{-4} W
33–35: a) $E_{max} = 701$ V/m, $B_{max} = 2.34 \times 10^{-6}$ T
b) $u_E = u_B = 1.09 \times 10^{-6}$ J/m^3 c) 1.07×10^{-11} J
33–37: a) $r = R$: $I = 6.4 \times 10^7$ W/m^2, $p_{rad} = 0.21$ Pa;
$r = R/2$: $I = 2.6 \times 10^8$ W/m^2, $p_{rad} = 0.85$ Pa b) no
33–39: a) $B = \mu_0 ni$, $E = (\mu_0 nr/2)(di/dt)$
b) $S = (\mu_0 n^2 ri/2)(di/dt)$, radially inward
c) $U_B = (1/2)\mu_0 n^2 \pi a^2 li^2$, $dU_B/dt = (\mu_0 n^2 \pi a^2 li)(di/dt)$
d) $(\mu_0 n^2 \pi a^2 li)(di/dt)$
33–41: a) $I\rho/\pi a^2$, in direction of current b) $\mu_0 I/2\pi a$,
counterclockwise if current out of page, clockwise if current
into page c) $I^2 \rho/2\pi^2 a^3$, radially inward
d) $I\rho l/\pi a^2 = I^2 R$
33–43: 0.0368 V
33–45: 1.10×10^5 s $= 30.6$ h
33–47: a) $4\pi R^3 \rho \ GM/3r^2$ b) $LR^2/4r^2 c$ c) 1.9×10^{-7} m, independent
of r
33–49: b) 1.4×10^{-11} s^{-1} c) 2.6×10^{-8} s^{-1}
33–51: c) 0.066 mm

Chapter 34

34–1: a) 2.04×10^8 m/s b) 442 nm
34–3: a) 1.55 b) 549 nm
34–5: a) 47.5° b) 66.0°
34–9: 71.8°
34–11: a) 58.1° b) 22.8°
34–13: a) air b) 15.1° c) air
34–15: a) 1.40 b) 35.5°
34–17: a) 1.65 b) 58.8°
34–19: a) $0.285I_0$ b) linearly polarized along axis of second filter
34–21: a) first: $I = I_0/2$, linearly polarized along axis of first filter;
second: $I = 0.25I_0$, linearly polarized along axis of second
filter; third: $I = 0.125I_0$, linearly polarized along axis of third
filter b) first: $I_0/2$; third: zero
34–25: 72.0°
34–29: 1.53
34–31: 1.8
34–33: 48.6°

34–35:   39.1°

34–37:   1.28

34–39:   b) 0.22°

34–41:   a) $I = (1/8)I_0 \sin^2 2\theta$ b) 45°

34–45:   b) 38.9°   c) 5.0°

34–47:   *l*-leucine: $-(0.11° \, 100 \text{ mL/g})C$; *d*-glutamic acid, $-(0.124° \, 100 \text{ mL/g})C$

34–49:   a) 35°   b) 10.1 W/m², 19.9 W/m²

34–53:   a) $\Delta = 2\theta_a^A - 6 \arcsin\left(\dfrac{1}{n} \sin \theta_a^A\right) + 2\pi$

b) $\cos^2 \theta_2 = (n^2 - 1)/8$

c) red: $\theta_2 = 71.9°$, $\Delta = 230.1°$; violet: $\theta_2 = 71.6°$, $\Delta = 233.2°$; violet

Chapter 35

35–3:   39.2 cm to right of mirror; 4.85 cm

35–5:   b) 33.0 cm to left of vertex, 1.20 cm tall, inverted, real

35–7:   0.213 mm

35–11:   b) $s > f$ c) $s < f$ d) s' at ∞ e) s' at $-\infty$ f) $s' = f$ g) $s' = 0$ i) $s < f$ j) $s > f$ k) $0 < s < 2f$ l) $s > 2f$ k) $m \to \infty$

35–13:   b) 5.46 cm to right of vertex, 4.09 mm tall, erect, virtual

35–15:   2.67 cm

35–19:   1.52

35–21:   8.35 cm to left of vertex, 0.326 mm, erect

35–23: $s = 18.0$ cm: a) 63.0 cm to right of lens b) -3.50 c) real d) inverted, $s = 7.00$ cm: a) 14.0 cm to left of lens b) $+2.00$ c) virtual d) erect

35–25:   4.85 cm to left of lens; image: 15.8 cm to left of lens; virtual

35–27:   10.3 cm to left of lens

35–29:   a) 19.0 cm to left b) virtual c) 5.00 mm, erect

35–31: $f = \pm 4.44$ cm, ± 13.3 cm

35–35:   object is 28.8 cm from the lens and has height 1.28 cm; image is erect; same side

35–37:   7.20 m, 4.43 m

35–39:   4.80 m/s

35–41:   a) 13.3 cm b) 26.2 cm

35–43:   a) -12.0 cm $< s < 0$ b) erect

35–45:   a) 46.2 cm from mirror, on opposite side of mirror; virtual b) 2.88 cm, erect c) no

35–47:   b) i) 120.00 cm from mirror, 119.96 cm from mirror ii) $m = -0.600$, $m' = -0.360$ iii) faces perpendicular to axis: squares with side 0.600 mm; faces parallel to axis: rectangles with sides of length 0.360 mm (parallel to axis) and 0.600 mm (perpendicular to axis)

35–49:   a) 8.65 mm/s b) 14.2 cm/s

35–51:   a) distance h behind front surface of glass b) distance $(h + 2d/n)$ behind front surface of glass c) $2d/n$

35–53:   a) -3.3 cm b) virtual c) 1.9 cm to right of vertex at right end of rod d) real, inverted e) 1.05 mm

35–55:   2.00

35–57:   0.50 cm to right of vertex opposite where rays enter

35–59:   a) 2.53 mm

35–61:   outside sphere, $3R$ from vertex opposite the object

35–63:   10.6 cm

35–65:   134 cm to left of object

35–67:   a) 0.24 m b) 0.24 m

35–69:   4.8 cm to left of first lens, 4.8 cm to right of second lens

35–71:   0.80 cm

35–73: -26.7 cm

35–75:   1.24 cm above page

35–77:   a) The distances from the lens and the distance from the axis for each point are *A:* (33.0 cm, 6.10 cm); *B:* (40.7 cm, 21.4 cm); *C:* (36.0 cm, 12.0 cm) b) 17.1 cm

35–79:   a) $4f$

Chapter 36

36–1:   a) 85 mm b) 135 mm

36–3:   10.2 m

36–5:   a) 1.40×10^{-4} b) 5.25×10^{-4} c) 1.50×10^{-3}

36–7:   a) 75.0 mm b) (1/62.5) s

36–9:   a) 20.8 cm b) no

36–11:   a) 80.0 cm b) 76.9 cm

36–13:   a) +2.33 diopters b) -1.67 diopters

36–15:   a) 6.06 cm b) 4.12 nm

36–17:   flea: 3.38 cm from lens; image: 22.0 cm from lens, on same side as flea

36–19:   a) 8.37 mm b) 21.4 c) 297

36–21:   a) -6.33 b) 1.90 cm c) 0.126 rad = 7.22°

36–23:   19.4 m

36–25:   a) 66.1 cm b) -59.1

36–27:   4.4 mm

36–29:   a) 29.5 m b) 22.1 m

36–31:   b) 75°, 23°, 8.2°

36–35:   a) 3.5 cm b) 7 cm c) 100 cm d) 57 e) no

36–37:   4.17 diopters

36–39:   d) $f_{max} = 36.0$ cm, $f_{min} = 21.6$ cm; $d = 1.2$ cm

36–41:   a) 552 b) 25.8 cm

36–43:   b) 1.34 cm

Chapter 37

37–3:   c) the greatest integer that is less than or equal to d/λ d) 15, $-7 \le m \le 7$

37–5:   0.75 m, 2.00 m, 3.25 m, 4.50 m, 5.75 m, 7.00 m, 8.25 m

37–7:   453 nm, 583 nm

37–9:   590 nm

37–11:   0.83 mm

37–13:   a) 0.888 mm b) 0.444 mm

37–15:   a) 1.90 mm b) 2.00×10^{-6} W/m²

37–17:   0.0235°

37–19:   114 nm

37–21:   0.83 mm

37–25:   a) 514 nm; green b) 603 nm; orange

37–27:   0.570 cm

37–29:   b) 93 kN

37–31: 3.53×10^{-22} kg · m/s, 1.06×10^{-13} J, 1.60×10^{20} Hz, 1.88×10^{-12} m

37–33: $\lambda = 400$ nm: $E = 3.10$ eV $= 4.97 \times 10^{-19}$ J, $p = 1.66 \times 10^{-27}$ kg · m/s; $\lambda = 700$ nm: $E = 1.77$ eV $= 2.84 \times 10^{-19}$ J, $p = 9.47 \times 10^{-28}$ kg · m/s

37–35:   0.63 mm

37–37:   b) 427 Hz c) 79.6 cm

37–39:   b) constructive: $r_2 - r_1 = (m + \phi/2\pi)\lambda$, $m = 0, \pm 1, \pm 2, ...$; destructive: $r_2 - r_1 = (m + \frac{1}{2} + \phi/2\pi)\lambda$, $m = 0, \pm 1, \pm 2, ...$

37–41:   a) $\sqrt{x^2 + (y + d)^2} - \sqrt{x^2 + (y - d)^2} = m\lambda$

c) $\sqrt{x^2 + (y + d)^2} - \sqrt{x^2 + (y - d)^2} = \left(m + \frac{1}{2}\right)\lambda$

37–43:   12.6 cm

37–45: $\lambda/2d$, independent of m

37–47:   b) 72 cm

37–49:   a) 424 nm, 594 nm b) 495 nm

37–51: 1.82 mm

37–53: a) pattern moves down the screen
b) $I = I_0 \cos^2[(\pi/\lambda)(d \sin\theta + (n-1)L)]$
c) $d \sin\theta = m\lambda - (n-1)L$

37–55: 14.0

Chapter 38

38–1: 506 nm

38–3: 0.15°

38–5: ±45.4 cm

38–7: a) 10.8 mm   b) 5.4 mm

38–9: a) 6.75 mm   b) 2.43×10^{-6} W/m^2

38–11: 5.4 μm

38–13: a) 668 nm   b) $9.36 \times 10^{-5} I_0$

38–15: a) 3   b) 2

38–17: a) ±0.0627°, ±0.125°   b) $0.249 I_0$, $0.0256 I_0$

38–19: cases (i), (iii): slits 1 and 3 and slits 2 and 4;
case (ii): slits 1 and 2 and slits 3 and 4

38–21: a) 0.280 mm   b) $m = 6, 9, 12, \ldots$ c) $m = 3, 6, 9, 12, \ldots$

38–23: 23.3°, 52.3°

38–25: 10.5°, 21.3°, 33.1°

38–27: a) 4.00°   b) 8.77°

38–29: 0.232 nm

38–31: a) 0.461 m

38–33: 1.9 m

38–35: 92 cm

38–37: 1.45 m

38–39: a) i) 25.6°   ii) 10.2°   iii) 5.1°   b) i) 60.0°   ii) 23.1°
iii) 11.5°

38–41: a) 1.80 mm   b) 0.798 mm

38–45: b) for $3\pi/2$, any two slits separated by one other slit;
for the other cases, any two slits separated by three other slits

38–49: 513 nm

38–51: second order

38–53: c) ±2.6 rad

38–55: 1.46 km

38–57: 492 km

Chapter 39

39–1: Flash at AA'

39–3: 5.00×10^{-9} s, clock on airliner

39–5: a) $0.998c$ b) 126 m

39–7: 1.12 h, clock on spacecraft

39–9: a) 9.17 km   b) 0.651 km, 0.0710
c) 1.32×10^{-5} s, 3.90 km, 0.0710

39–11: a) 3.57 km   b) 9.00×10^{-5} s   c) 8.92×10^{-5} s

39–13: 92.5 m

39–15: a) $0.806c$ b) $0.974c$ c) $0.997c$

39–19: $0.385c$

39–21: $0.784c$

39–23: $0.837c$, away

39–25: a) yes   b) no

39–27: a) 0.159c   b) 1.72×10^8

39–29: b) $a = (F/m)(1 - v^2/c^2)^{1/2}$

39–31: a) $(\sqrt{3}/2)\,c = 0.866c$ b) $\left(\sqrt{1 - \left(\frac{1}{2}\right)^{2/3}}\right) c = 0.608c$

39–33: a) $(\sqrt{3}/2)\,c = 0.866c$ b) $\sqrt{35/36}\,c = 0.986c$

39–35: $0.16c$

39–37: a) 8.68×10^{-10} J   b) 2.71×10^{-10} J   c) 0.453

39–39: a) nonrelativistic: 5.34×10^{-12} J; relativistic:
5.65×10^{-12} J; 1.06
b) nonrelativistic: 6.78×10^{-11} J; relativistic:
3.31×10^{-10} J; 4.88

39–41: a) 4.20×10^5 eV   b) 9.31×10^5 eV   c) 2.51×10^8 m/s
d) 3.84×10^8 m/s

39–43: a) $\Delta = 2.11 \times 10^{-5}$ b) 2.15×10^4 MeV

39–45: $0.700c$

39–47: a) $0.995c$ b) 1.0%

39–49: 1.48×10^{-11} kg

39–51: 1.68×10^5 eV

39–57: b) $\Delta x' = \sqrt{(\Delta x)^2 - (c\Delta t)^2}$ c) 1.44×10^{-8} s

39–59: $0.357c$, receding

39–67: a) toward the earth at 13.1 km/s, 39.4 km/s
b) 5.96×10^9 m, 5.54×10^{29} kg $= 0.278 \times$ (mass of the sun)

39–69: a) 2494 MeV   b) 2.526 times larger   c) 987.4 MeV, the same

Chapter 40

40–1: 5.77×10^{14} Hz, 1.27×10^{-27} kg $\cdot$ m/s, 3.82×10^{-19} J $= 2.38$ eV

40–3: a) 5.92×10^{20} Hz   b) 5.06×10^{-13} m

40–5: 2.49×10^5 m/s

40–7: a) 264 nm   b) 4.70 eV

40–9: a) 2.47×10^{-19} J $= 1.54$ eV   b) 804 nm, infrared

40–11: a) -5.08 eV   b) -5.63 eV

40–13: a) 434.1 nm   b) 6.906×10^{14} Hz
c) 4.576×10^{-19} J $= 2.856$ eV

40–15: 0.002 eV

40–17: a) 5.82×10^{-13} J $= 3.63$ MeV   b) 5.82×10^{-13} J $= 3.63$ MeV
c) 1.32×10^7 m/s

40–19: a) -218 eV, 16 times greater   b) 218 eV, 16 times greater
c) 7.63 nm
d) 1/4 as large

40–21: a) $v_1 = 2.19 \times 10^6$ m/s, $v_2 = 1.09 \times 10^6$ m/s, $v_3 = 7.29 \times 10^5$ m/s
b) $T_1 = 1.52 \times 10^{-16}$ s, $T_2 = 1.22 \times 10^{-15}$ s, $T_3 = 4.10 \times 10^{-15}$ s
c) 8.22×10^6

40–23: a) 1.2×10^{-33} b) 3.4×10^{-17} c) 5.9×10^{-9}

40–25: 4.00×10^{17}

40–27: 0.310 nm, the same

40–29: 0.0827 nm

40–31: 0.0714 nm, 180°

40–35: 1.06 mm, microwave

40–41: a) 1.04 eV   b) 1.20 μm   c) 2.51×10^{14} Hz   d) 4.14×10^{-7} eV

40–45: a) 4.59×10^{14} Hz   b) 653 nm   c) 1.89 eV   d) 6.59×10^{-34} J $\cdot$ s

40–47: a) $hc(\lambda_1 - \lambda_2)/e\lambda_1\lambda_2$ b) 0.476 V

40–49: a) 1.69×10^{-28} kg   b) -2.53 keV   c) 0.653 nm

40–51: a) 0.90 eV

40–53: a) $r_n = (n^2 h^2/4\pi^2 mD)^{1/4}$ b) $E_n = (nh/2\pi)\sqrt{D/m}$
c) integer multiples of $(h/2\pi)\sqrt{D/m}$

40–55: a) 12.09 eV   b) 3 possibilities:
$3 \rightarrow 2$ (656 nm), $3 \rightarrow 1$ (103 nm),
$2 \rightarrow 1$ (122 nm)

40–57: a) $\Delta\lambda = h/2mc$ b) 6.61×10^{-16} m for any n

40–59: a) 4.85×10^{-12} m   b) 255 keV

40–61: a) 5.10×10^{-17} J $= 319$ eV, 1.06×10^7 m/s   b) 3.89 nm

40–63: a) 5×10^{-33} m   b) $(4 \times 10^{-9})°$ c) 0.1 mm

40–69: a) $\lambda' = \dfrac{\lambda(E - Pc) + 2hc}{E + Pc}$, where $Pc = \sqrt{E^2 - m^2 c^4}$
b) 7.04×10^{-15} m   c) gamma rays

Chapter 41

41–1: a) 2.37×10^{-24} kg $\cdot$ m/s   b) 3.07×10^{-18} J $= 19.2$ eV

41–3: a) 1.55×10^{-10} m   b) 8.44×10^{-14} m

41–5: a) 3.32×10^{-10} m   b) 1.33×10^{-9} m

41–7: 3.90×10^{-34} m; no

41–9: 0.431 eV

41–11: 0.340°

41–13: a) 8.79×10^{-32} m/s b) no
41–15: claim is not valid
41–17: a) 9.81×10^{-25} kg $\cdot$ m/s b) 4.94×10^{-24} J $= 3.08 \times 10^{-5}$ eV
        c) 50.6 J d) 5.16 m
41–19: 0.087 MeV $= (2.8 \times 10^{-5})E$
41–21: 1.08×10^{-24} s
41–23: a) 4.34×10^{-11} m b) 1.01×10^{-12} m
41–25: a) $x = (2n + 1)(\lambda/4)$, n an integer b) $x = n\lambda/2$, n an integer
41–27: 1
41–29: a) 1.0×10^{-6} b) 8.0×10^{-6}
41–31: 1.66×10^{-17} m
41–33: a) 1.10×10^{-10} m b) 9.09×10^{-13} m
41–35: a) $\left(1/\sqrt{15}\right)(h/mc)$

        b) i) 1.53 MeV, 6.26×10^{-13} m ii) 2.81×10^{3} MeV, 3.41×10^{-19} m
41–39: a) 2.1×10^{-20} kg $\cdot$ m/s b) 39 MeV c) no
41–41: 1.4×10^{-35} kg $= 5.8 \times 10^{-8} m_{\text{pion}}$
41–43: a) 1.1×10^{-35} m/s b) 2.3×10^{27} y, no
41–45: 0.294 eV
41–47: a) $2d \sin\theta = m\lambda$, $m = 1, 2, 3, \dots$ b) 53.1° c) less
41–49: a) $-A|x|/x$, $x \neq 0$ b) $(3/2)(h^2 A^2/m)^{1/3}$
41–51: a) 0.21 kg $\cdot$ m/s b) 1.7 m
41–53: a) $4\pi A^2 r^2 e^{-2\alpha r^2}\, dr$ b) $1/\sqrt{2\alpha}$, no
41–55: a) $\psi(x) = (\sin k_0 x)/k_0 x$ b) L c) $2L$ d) h, h
41–57: 2.2×10^{-16} m

Chapter 42

42–1: a) 1.2×10^{-67} J b) 1.1×10^{-33} m/s, 1.4×10^{33} s
       c) 3.7×10^{-67} J d) no
42–3: 0.61 nm
42–5: a) $0, L/2, L$ b) $L/4, 3L/4$ c) yes
42–7: a) 6.0×10^{-10} m, 1.1×10^{-24} kg $\cdot$ m/s
       b) 3.0×10^{-10} m, 2.2×10^{-24} kg $\cdot$ m/s
       c) 2.0×10^{-10} m, 3.3×10^{-24} kg $\cdot$ m/s
42–15: 2.2×10^{-14} m
42–19: a) 4.3×10^{-8} b) 4.2×10^{-4}
42–21: a) 1.3×10^{-3} b) 10^{-143}
42–23: 1.11×10^{-33} J $= 6.90 \times 10^{-15}$ eV; 2.21×10^{-33} J $= 1.38 \times 10^{-14}$ eV
42–25: 1.2 N/m
42–27: $\Delta x \Delta p = (2n + 1)\, \hbar$
42–29: a) $(1/2) + (1/\pi)$ b) $1/2$ c) yes
42–31: a) $2dx/L$ b) 0 c) $2dx/L$
42–33: a) 19.2 μm b) 11.5 μm
42–35: a) $\sqrt{2\pi^2/L^3}$ b) $\sqrt{8\pi^2/L^3}$ c) $-\sqrt{2\pi^2/L^3}$ d) $\sqrt{8\pi^2/L^3}$
        e) $n = 2$, yes
42–41: a) $B = C$,
        $A \sin\dfrac{\sqrt{2mE}}{\hbar} L + B \cos\dfrac{\sqrt{2mE}}{\hbar} L = De^{-\kappa L}$

        b) $\dfrac{\sqrt{2mE}}{\hbar} A = \kappa C$,

        $\dfrac{\sqrt{2mE}}{\hbar}\left(A \cos\dfrac{\sqrt{2mE}}{\hbar} L - B \sin\dfrac{\sqrt{2mE}}{\hbar} L\right) = -\kappa De^{-\kappa L}$
42–45: 6.63×10^{-34} J $= 4.14 \times 10^{-15}$ eV, 1.33×10^{-33} J $= 8.27 \times 10^{-15}$ eV, no
42–49: a) $(n_x + n_y + n_z + (3/2))\, \hbar\omega$ b) $(3/2)\, \hbar\omega$, $(5/2)\, \hbar\omega$
42–51: a) $E_n = n^2 h^2/8mL^2$, $n = 2, 4, 6, \dots$ b) $E_n = n^2 h^2/8mL^2$, $n = 1, 3, 5, \dots$
        c) same d) odd in part (a), even in part (b)

42–53: b) increases c) infinite
42–55: a) $-E/A$, $+E/A$ c) decrease

Chapter 43

43–1: $l = 4$
43–3: $1.414\,\hbar$, $19.49\,\hbar$, $199.5\,\hbar$
43–7: b) $1/\sqrt{2\pi}$
43–9: a) 5.29×10^{-11} m b) 1.06×10^{-10} m c) 2.85×10^{-13} m
43–13: a) 9 b) 3.47×10^{-5} eV c) 2.78×10^{-4} eV
43–15: a) 0.468 T b) 3
43–17: 1.68×10^{-4} eV; $m_s = 1/2$
43–19: g
43–21: a) 2.5×10^{30} rad/s b) 2.5×10^{13} m/s, not valid
43–23: 4.18 eV
43–25: see Tables 43–2 and 43–3
43–27: a) $1s^2 2s^2 2p$ b) -30.6 eV c) $1s^2 2s^2 2p^6 3s^2 3p$ d) -13.6 eV
43–29: a) -13.6 eV b) -3.4 eV
43–31: a) 8.95×10^{17} Hz, 3.70 keV, 0.335 nm
        b) 1.68×10^{18} Hz, 6.93 keV, 0.179 nm
        c) 5.48×10^{18} Hz, 22.7 keV, 0.0547 nm
43–33: a) $2a$ b) 0.238
43–35: b) 0.176
43–37: b) $\arccos\left[-\sqrt{1 - (1/n)}\right]$
43–39: $2 \to 1$, $1 \to 0$, $0 \to -1$, $e\,\hbar B/2m$;
        $1 \to 1$, $0 \to 0$, $-1 \to -1$, 0;
        $0 \to 1$, $-1 \to 0$, $-2 \to -1$, $-e\,\hbar B/2m$
43–41: a) $1 - 2.2 \times 10^{-7}$ b) 0.998 c) 0.978
43–43: a) 122 nm b) 1.52×10^{-3} nm; increases
43–45: a) 0.188 nm, 0.250 nm b) 0.0471 nm, 0.0624 nm
43–47: b) O shell

Chapter 44

44–1: a) 6.1 K b) 3.47×10^{4} K
44–3: a) 1.20×10^{-21} J $= 7.52 \times 10^{-3}$ eV b) 0.165 mm
44–5: b) $\hbar l/2\pi I$
44–7: a) 4.593 μm b) 4.626 μm c) 4.634 μm
44–9: a) 963 N/m b) 8.22×10^{-20} J $= 0.513$ eV c) 2.42 μm, infrared
44–11: 2.16×10^{3} kg/m^3
44–13: 1.19×10^{6}
44–15: a) 1.11 μm, infrared
44–17: b) ground: $E = 3\pi^2\,\hbar^2/2mL^2$, 2; first: $E = 6\pi^2\,\hbar^2/2mL^2$, 6;
        second: $E = 9\pi^2\,\hbar^2/2mL^2$, 6
44–19: 1.5×10^{22} states/eV
44–21: 4.3×10^{7}
44–23: a) $0.0233R$ b) 7.65×10^{-3} c) no; the ions
44–25: 0.20 eV below the band
44–27: a) 5.56 mA b) -5.18 mA, 3.77 mA
44–29: a) 3.8×10^{-29} C $\cdot$ m b) 1.3×10^{-19} C c) 0.78 d) 0.059
44–31: a) 0.94 nm b) 1.8 nm
44–33: a) 0.129 nm b) 8, 7, 6, 5, 4 c) 484 μm
        d) 118 μm, 135 μm, 157 μm, 189 μm, 236 μm
44–35: b) i) 2.94 ii) 4.73 iii) 7.58 iv) 0.837 v) 5.5×10^{-9}
44–37: a) 1.147 cm, 2.293 cm b) 1.172 cm, 2.344 cm; 0.025 cm, 0.050 cm
44–39: 4.38×10^{-20} J $= 0.273$ eV
44–41: a) 4.24×10^{-47} kg $\cdot$ m^2 b) i) 4.30 μm ii) 4.28 μm
        iii) 4.40 μm
44–43: 2.03 eV
44–45: b) 3.80×10^{10} Pa $= 3.75 \times 10^{5}$ atm

44–47:   a) 1.66×10^{33} m^{-3} b) yes   c) 7×10^{35} m^{-3} d) no

44–49:   a) $-p^2/2\pi\epsilon_0 r^3$ b) $+p^2/2\pi\epsilon_0 r^3$

Chapter 45

45–1:   a) $Z = 14$, $N = 14$ b) $Z = 37$, $N = 48$ c) $Z = 81$, $N = 124$

45–3:   0.533 T

45–5:   a) parallel, 70.3 MHz, 4.27 m, radio

        b) antiparallel, 46.2 GHz, 6.48 mm, microwave

45–7: 5.575×10^{-13} m

45–9:   a) 76.21 MeV   b) 76.67 MeV   c) 0.6%

45–11:   He, O, Ca, Ni, Sn, Pb

45–13:   a) $^{235}_{92}$U   b) $^{24}_{12}$Mg   c) $^{15}_{7}$N

45–15: -9.9×10^{-5} u

45–17:   156 keV

45–19:   b) 19 keV

45–21:   a) 2.02×10^{15} b) 1.01×10^{15}, 3.78×10^{11} Bq   c) 2.52×10^{14},

        9.45×10^{10} Bq

45–23:   a) 0.421 Bq   b) 1.14×10^{-11} Ci

45–25:   a) 4.91×10^{-18} s^{-1} b) 35.7 g   c) 7.46×10^{5}

45–27:   a) 12.5 rad, 12.5 rem   b) the antineutrinos are not absorbed

45–29:   a) 6.46×10^{-4} J   b) 0.108 rem

45–31:   a) $Z = 3$, $A = 7$ b) 7.151 MeV   c) 1.4 MeV

45–33:   a) $Z = 3$, $A = 6$ b) -10.14 MeV   c) 11.60 MeV

45–35:   6.545 MeV

45–37:   1.586 MeV

45–43:   a) $^{25}_{13}$Al to $^{25}_{12}$Mg   b) β^+ or electron capture   c) 3.255 MeV

        or 4.277 MeV

45–45:   23.9858 u, 0.021%, 0.9%

45–47:   0.960 MeV

45–49: 1.287×10^{-3} u

45–53:   29.2%

45–55:   a) 9.6×10^{-7} J   b) 1.9×10^{-4} rad   c) 1.3×10^{-4} rem

        d) 1.5×10^{6} s = 17 days

45–57: 1.3×10^{4} y

45–59:   a) 0.48 MeV   b) 3.270 MeV = 5.239×10^{-11} J

        c) 3.155×10^{11} J/mol

45–61:   0.82 rad, 0.82 rem

45–65:   185 MeV

45–67:   b) 4.1×10^{4} Bq, 3.6×10^{5} Bq, 7.5×10^{5} Bq, 1.1×10^{6} Bq,

        1.3×10^{6} Bq, 1.5×10^{6} Bq   c) 3.2×10^{9} d) 1.5×10^{6} Bq

Chapter 46

46–1:   69 MeV, 1.7×10^{22} Hz, 1.8×10^{-14} m, gamma ray

46–3:   a) 32 MeV

46–5: 7.2×10^{19} J

46–7:   a) 1.18 T   b) 3.41 MeV, 1.81×10^{7} m/s

46–9:   a) 30.6 GeV   b) 8.0 GeV

46–11:   a) 3.2 TeV   b) 38.7 GeV

46–13:   116 MeV

46–15:   a) no   b) yes   c) no   d) yes

46–17:   a) no   b) no   c) yes   d) yes

46–21:   a) 0, 1, -1, 0   b) 0, 0, 0, 1   c) $-e$, $+1$, 0, 0   d) $-e$, 0, 0, -1

46–23:   a) $\bar{u}d\bar{d}$ b) no   c) yes

46–25:   a) 1.04×10^{8} m/s   b) 1.44

46–27:   a) 3.28×10^{7} m/s   b) 1640 Mly

46–29:   a) 5.494 MeV   b) 20.58 MeV

46–31: -783 keV, endoergic

46–33:   0.966 μm

46–35:   a) 14.0 TeV   b) 1.0×10^{5} TeV

46–37:   18.1 GeV

46–39:   1265 MeV

46–41:   a) 139.1 MeV   b) momentum conservation

46–43:   a) 16.0 MeV

46–45:   87 keV, 2.8×10^{-5}

46–47:   a) 65 MeV   b) 89% by the π^-, 11% by the Λ^0

46–51:   a) 41 MeV   b) 14% by the neutron, 86% by the π^0

PHOTO CREDITS

National Optical Astronomy Observatories; U20-04: John Bacchus/Pearson Education Corporate Digital Archive; U20-05: Martin Bough/Fundamental Photographs.

Chapter 21 Opener: Brian Smith/Liaison Agency, Inc.; 21-15c: Leonard M. Weinstein/NASA/Langley Research Center; U21-01: Raytheon Aircraft Co.; U21-02: Dennis O'Clair/Tony Stone Images; U21-03: Robert A. Hoover/NASA/Dryden Flight Research Center.

Chapter 22 Opener: Definitive Stock; 22-23a: Reproduced from *PSSC Physics*, 2nd ed. (1965), D.C. Heath & Company with Educational Development Center, Inc., Newton, Massachusetts; U22-01: The Cavendish Laboratory, University of Cambridge, Cambridge, England; U22-02: Photo by Matzene Studio, Chicago/Courtesy of the Archives, California Institute of Technology; U22-03: PhotoDisc, Inc.

Chapter 23 Opener: Department of Energy/Photo Researchers, Inc.; 23-25b: Russ Kinne/Comstock; U23-01: Definitive Stock; U23-02: AKG London Ltd.

Chapter 24 Opener: Gene Peach/Tony Stone Images; 24-16: U.S. Geological Survey/U.S. Department of the Interior; U24-01: ©Ellis Herwig/Stock Boston/PNI; U24-02: Addison Wesley Longman, Inc.; U24-03: French Government Tourist Office; U24-04: Fermilab Visual Media Services.

Chapter 25 Opener: Richard Olivier/Corbis; U25-01: Paul Silverman/Fundamental Photographs; U25-02: David Young-Wolff/PhotoEdit; U25-03: George H.H. Huey/Corbis; U25-04: Stanford Linear Accelerator/Science Photo Library/Photo Researchers, Inc.

Chapter 26 Opener: Nick Galante/NASA/Dryden Flight Research Center; U26-01: Fundamental Photographs; U26-02: Definitive Stock; U26-03: Richard Megna/Fundamental Photographs; U26-04: Definitive Stock.

Chapter 27 Opener: Nick Koudis/PhotoDisc, Inc.; 27-04: John P. Surey; U27-01: Mario Nette/SPS/Vandystadt/Photo Researchers, Inc.; U27-02: John P. Surey.

Chapter 28 Opener: National Solar Observatory, Sacramento Peak/National Optical Astronomy Observatories; 28-10a, b: Richard Megna/Fundamental Photographs; 28-13b: Courtesy of Cenco; 28-16b: Space Shuttle photograph by Robert F. Overmyer and Don L. Lind, courtesy of Thomas J. Hallinan, Geophysical Institute, University of Alaska, Fairbanks; 28-17a, b: Ernest Orlando, Lawrence Berkeley National Laboratory; 28-19b: The Cavendish Laboratory, University of Cambridge, Cambridge, England.

Chapter 29 Opener: Hitachi Medical Systems America, Inc.; U29-01: Definitive Stock; U29-02: George Haling/Photo Researchers, Inc.; U29-03: Bill Aron/PhotoEdit.

Chapter 30 Opener: Shaun Egan/Tony Stone Images; 30-24: Ken Gatherum/Boeing Computer Services; U30-01: Cessna Aircraft Company; U30-02: NASA Headquarters.

Chapter 31 Opener: Spencer Grant/PhotoEdit; U31-01: James Harrington/Tony Stone Images; U31-02: Definitive Stock; U31-03: Definitive Stock.

Chapter 32 Opener: John P. Surey/Caroline J. Robillard; U32-01: John P. Surey; U32-02: Lawrence Migdale/Photo Researchers, Inc.; U32-03: John P. Surey.

Chapter 33 Opener: Definitive Stock; U33-01: Corbis; U33-02: M. Shara, B. Williams, and D. Zurek (STScI); R. Gilmozzi (ESO); D. Prialnik (Tel Aviv University); and NASA; U33-03: Tony Freeman/PhotoEdit.

Chapter 34 Opener: Runk/Schoenberger/Grant Heilman Photography, Inc.; 34-06a: Roger A. Freedman; 34-09b: Copyright The Exploratorium, photograph by Susan Schwartzenberg; 34-12: Barry Blanchard/Barry Blanchard; 34-15c: Richard Johnston/FPG International LLC; 34-24: Sepp Seitz/Woodfin Camp & Associates; 34-29U: K. Nomachi/Photo Researchers, Inc.; U34-01: Michael Newman/PhotoEdit; U34-02: Spencer Grant/PhotoEdit; U34-03-01: Diane Hirsch/Fundamental Photographs; U34-03-02: Diane Hirsch/Fundamental Photographs; U34-04: Camerique/H. Armstrong Roberts.

Chapter 35 Opener: Nina Leen/Life Magazine ©Time Inc.; 35-07: Marshall Henrichs; 35-10a, b: NASA Headquarters; U35-01: Definitive Stock; U35-02: Arnold & Brown.

Chapter 36 Opener: Richard J. Wainscoat/Richard J. Wainscoat; 36-02a–c: Marshall Henrichs/Marshall Henrichs; 36-04: THK Photo Products, Inc.; U36-01: Photo Lennart Nilsson/Albert Bonniers Forlag AB; U36-02: Yerkes Observatory.

Chapter 37 Opener: Joel Gordon Photography; 37-05: From Sears, Zemansky, and Young, *College Physics*, 7th ed., ©1991 by Addison-Wesley Publishing Company. Reprinted by permission of Addison Wesley Longman, Reading, Massachusetts; 37-09: Tom Branch/Photo Researchers, Inc.; 37-13b: Bausch & Lomb Incorporated; 37-14: Bausch & Lomb Incorporated.

Chapter 38 Opener: Dane Hornsleth; 38-02a, b: From Sears, Zemansky, and Young, *College Physics*, 7th ed., ©1991 by Addison-Wesley Publishing Company. Reprinted by permission of Addison Wesley Longman, Reading, Massachusetts; 38-03: From Hecht, *Optics*, 2nd ed., ©1987 by Addison-Wesley Publishing Company. Reprinted by permission of Addison Wesley Longman, Reading, Massachusetts; 38-08b, d: From Sears, Zemansky, and Young, *College Physics*, 7th ed., ©1991 by Addison-Wesley Publishing Company. Reprinted by permission of Addison Wesley Longman, Reading, Massachusetts; 38-16b: Estate of Bertram Eugene Warren; 38-21: From Sears, Zemansky, and Young, *College Physics*, 7th ed., ©1991 by Addison-Wesley Publishing Company. Reprinted by permission of Addison Wesley Longman, Reading, Massachusetts; 38-22a–c: From Sears, Zemansky, and Young, *College Physics*, 7th ed., ©1991 by Addison-Wesley Publishing Company. Reprinted by permission of Addison Wesley Longman, Reading, Massachusetts; 38-25a, b: Ronald R. Erickson/Media Interface, Ltd.; 38-27: John Bahcall (Princeton), NASA/Space Telescope Science Institute; U38-01: Richard Megna/Fundamental Photographs.

Chapter 39 Opener: Association of Universities for Research in Astronomy; 39-12a–c: Image originally supplied by Ping-Kang Hsiung, reproduced with permission of his thesis advisor Robert H. Thibadeau, Carnegie-Mellon University; U39-01: Megan Donahue/Space Telescope Science Institute; U39-02: David P. Wojtowicz, Jr., University of Illinois at Champaign-Urbana; U39-03: O. Johnson/Camerique/H. Armstrong Roberts.

Chapter 40 Opener: Anglo-Australian Observatory, Photograph by David Malin; 40-07: Bausch & Lomb Incorporated; 40-10: National Optical Astronomy Observatories; U40-01: ITT Industries Night Vision; U40-02: American Institute of Physics/Emilio Segré Visual Archives; U40-03: W. Livingston/National Solar Observatory; U40-04a–c: From Huggins, *Physics 1*, ©1968 by W.A. Benjamin, Inc. Reprinted by permission of Addison Wesley Longman, Reading, Massachusetts.

Chapter 41 Opener: Reproduced by permission of Springer-Verlag from H. Raether, "*Elektroninterferenzen,*" *Handbuch der Physik*, Vol. 32, 1957. ©1957 by Springer-Verlag GmbH & Co.

KG, Heidelberg, Germany; 41-03: Reproduced from PSSC *Physics*, 2nd ed. (1965), D.C. Heath & Company with Educational Development Center, Inc., Newton, Massachusetts; 41-07b: From Huggins, *Physics 1*, ©1968 by W.A. Benjamin, Inc. Reprinted by permission of Addison Wesley Longman, Reading, Massachusetts; 41-08b: Carl Zeiss, Inc., Thornwood, NY 10594; 41-09b: Carl Zeiss, Inc., Thornwood, NY 10594; U41-01: AIP Meggers Gallery of Nobel Laureates; U41-02: American Institute of Physics/Niels Bohr Library.

Chapter 42   Opener: Courtesy of International Business Machines Corporation. Unauthorized use not permitted.; 42-13b: Digital Instruments; U42-01: UPI/Corbis.

Chapter 43   Opener: Warren Bolster/Tony Stone Images; 43-05a, b: National Optical Astronomy Observatories; U43-01: NASA/Johnson Space Center; U43-02: Steven J. Gibson and A. Russell Taylor, University of Calgary; U43-03: American Institute of Physics/Emilio Segré Visual Archives.

Chapter 44   Opener: NASA/Goddard Space Flight Center; 44-10a, b: R.C. Herman; 44-34: Hewlett-Packard Company; 44-35: John Bally, University of Colorado; U44-01: Definitive Stock; U44-02: Stanford University, courtesy AIP Emilio Segré Visual Archives.

Chapter 45   Opener : Photo by Carl Hensen, courtesy of Bruce D. Smith, Department of Anthropology, National Museum of Natural History, Smithsonian Institution. ©Smithsonian Institution; 45-01b: Science Photo Library/Photo Researchers, Inc.; 45-10b: NIH/Science Source/Photo Researchers, Inc.; U45-01: Lloyd Cluff/Corbis; U45-02: Anglo-Australian Observatory/Royal Observatory Edinburgh, photograph made from UK Schmidt plates by David Malin.

Chapter 46   Opener: NASA/Goddard Space Flight Center; 46-01: Carl D. Anderson/Ernest Orlando, Lawrence Berkeley National Laboratory; 46-03a: Ernest Orlando, Lawrence Berkeley National Laboratory; 46-07U, W: Fermilab Visual Media Services; 46-08a: Stanford Linear Accelerator Center; 46-09: Brookhaven National Laboratory; 46-18: NASA Headquarters; U46-01: Mark R. Lowenstine; U46-02: Institute for Cosmic Ray Research, University of Tokyo.

INDEX

Numerical Constants

FUNDAMENTAL PHYSICAL CONSTANTS*

Name	Symbol	Value
Speed of light	c	2.99792458×10^8 m/s
Magnitude of charge of electron	e	$1.60217733(49) \times 10^{-19}$ C
Gravitational constant	G	$6.67259(85) \times 10^{-11}$ N $\cdot$ m^2/kg^2
Planck's constant	h	$6.6260755(40) \times 10^{-34}$ J $\cdot$ s
Boltzmann constant	k	$1.380658(12) \times 10^{-23}$ J/K
Avogadro's number	N_A	$6.0221367(36) \times 10^{23}$ molecules/mol
Gas constant	R	$8.314510(70)$ J/mol $\cdot$ K
Mass of electron	m_e	$9.1093897(54) \times 10^{-31}$ kg
Mass of proton	m_p	$1.6726231(10) \times 10^{-27}$ kg
Mass of neutron	m_n	$1.6749286(10) \times 10^{-27}$ kg
Permeability of free space	μ_0	$4\pi \times 10^{-7}$ Wb/A $\cdot$ m
Permittivity of free space	$\epsilon_0 = 1/\mu_0 c^2$	$8.854187817 \ldots \times 10^{-12}$ C^2/N $\cdot$ m^2
	$1/4\pi\epsilon_0$	$8.987551787 \ldots \times 10^9$ N $\cdot$ m^2/C^2

OTHER USEFUL CONSTANTS*

Mechanical equivalent of heat		4.186 J/cal (15° calorie)
Standard atmospheric pressure	1 atm	1.01325×10^5 Pa
Absolute zero	0 K	$-273.15°$C
Electron volt	1 eV	$1.60217733(49) \times 10^{-19}$ J
Atomic mass unit	1 u	$1.6605402(10) \times 10^{-27}$ kg
Electron rest energy	$m_e c^2$	0.51099906(15) MeV
Volume of ideal gas (0°C and 1 atm)		22.41410(19) liter/mol
Acceleration due to gravity (standard)	g	9.80665 m/s^2

*Source: E. R. Cohen and B. R. Taylor, *Reviews of Modern Physics* **57**, 1121 (1987). Numbers in parentheses show the uncertainty in the final digits of the main number; for example, the number 1.6454(21) means 1.6454 ± 0.0021. Values shown without uncertainties are exact.